BIOLOGY

THE FOLLOWING BIOLOGISTS CONTRIBUTED WRITTEN MATERIAL, ADVICE, AND UPDATING:

Lynn Carpenter	University of California at Irvine (Ecology)
Sara Fultz	Stanford University (Fungi)
Lee Gass	University of British Columbia (Behavior)
Earl D. Hanson	Wesleyan University (Origin of Life, Monera, Protists)
Richard E. Jones	University of Colorado (Animal Hormones)
John F. McDonald	University of Georgia (Evolution)
Samuel N. Postlethwait	Purdue University (Plant Diversity)
Peter H. Raven	Missouri Botanical Garden (Consultant for Chapter 51)
Erik Trinkaus	University of New Mexico (Human Origins)
Virginia Walbot	Stanford University (Plant Science)
Jeffrey Wine	Stanford University (Nervous System and the Brain)

RANDOM HOUSE CONSULTING EDITORS FOR THE LIFE SCIENCES

Howard A. Schneiderman	Monsanto Company and the University of California at Irvine
John H. Postlethwait	University of Oregon

BIOLOGY

Norman K. Wessells
Stanford University

Janet L. Hopson

RANDOM HOUSE, INC., NEW YORK

To our parents and spouses, whose love and support at different times in our lives made this book possible: Norman W. Wessells, Grace M. Wessells, Catherine P. B. Wessells, David W. Hopson, Ruth L. Hopson, and Michael A. Rogers.

First Edition
987654321
Copyright © 1988 by Random House, Inc.

Library of Congress Cataloging-in-Publication Data

Wessells, Norman K.
    Biology.

    Bibliography: p.
    Includes index.
    1. Biology. I. Hopson, Janet L. II. Title.
QH308.2.W47   1988   574   87-28529
ISBN: 0-394-33732-8

Designer: John Lennard
Production Manager: Laura Lamorte
Composition and Color Separations:
   York Graphic Services, Inc.
Printed and bound at Rand McNally/Versailles, Kentucky

Manufactured in the United States of America

Cover photograph: Broad-billed hummingbird (*Cynanthus latirostris*) by Bob and Clara Calhoun/Bruce Coleman. Computer-generated model of the DNA double helix by Dan McCoy/Rainbow. The acknowledgments and credits for all other photographs and the line illustrations appear on pages A32–A38, which constitute an extension of the copyright page.

Prologue

Biology! At no time in history has the science of life been so visible and so important to human life and the future of our planet. Newspapers, magazines, and television feature biology prominently every day. Biological issues are discussed in Congress, the courts, on Wall Street, at the World Bank, at the United Nations, at summit meetings of heads of state as well as in classrooms, laboratories, hospitals, agricultural centers. We hear about viral diseases. About repairing brain damage. About the ozone layer and skin cancer. Desertification. Genetic engineering and frost-resistant strawberry plants. How memory works. Organ transplants. Drugs that will prevent heart attacks. In vitro fertilization. The destruction of rain forests. Acid rain. Crops that require no insecticides. The extinction of dinosaurs after an asteroid crashed into Earth. Chemicals produced by plants that protect them from their enemies. The fate of whales. The language of wolves. The durability of cockroaches. The origin of man. The future of our planet.

This book is an introduction to the worlds of biology. Its authors are superb guides for the journey.

Norman Wessells has taught and written about biology since he joined Stanford University in 1962. He has made important research discoveries in developmental and cell biology, vertebrate biology, and neurobiology. He is best known for his insightful studies of the mechanisms by which developing tissues and organs interact during mammalian development. His biological subjects have included chick and mouse embryos, hormones, cell locomotion, growth of nerve cells, lung development, the ultrastructure of algae, and the impact of advances in biology on society.

But, preeminently, Norman Wessells is a master teacher, treasured by students for his wonderful teaching. Three special talents have secured his effectiveness as a teacher and writer. First, he is able to explain complex ideas in plain language that can be understood and remembered. Second, he has a terrific sense for selecting durable and important subjects and examples to write about. And, finally, he has a clear view of how new-found knowledge can be put to use by the reader and by the larger society.

Janet Hopson is a marvelously talented science writer whose articles have appeared in the *New York Times Magazine, Smithsonian, Reader's Digest, Psychology Today, Rolling Stone, Cosmopolitan*, and numerous other major publications. Her work has also appeared in a dozen books, including six textbooks, and she has taught and lectured about writing for several years at the University of California, Berkeley, U.C., Santa Cruz, Mills College, and other schools.

Janet Hopson has three special techniques. One is the declarative sentence, which she uses with skill and grace to convey ideas clearly and memorably. The second is the ability to capture and hold the attention of the reader with marvelous examples drawn from an encyclopedic knowledge of biology. The third is the ability to relate the biology she writes about to the world's work, to commerce, archeology, medicine, agriculture, conservation, behavior, and other subjects.

She has contributed both exciting ideas and graceful writing to *Biology*.

The contributing authors, all leaders in their fields, lend a special authority and currency to many of the chapters. All of them have paid attention to biology's pervasive impact on much that transpires in our world today. This view appears throughout the book in insightful comments on how rapidly advancing discoveries in biology impact other fields of knowledge such as sociology, psychology, economics and business, law, ethics, and the humanities, as well as biology's more obvious relatives such as physics, chemistry, mathematics, computer and information science, medicine, agriculture, engineering, and natural resource management.

This book will provide the reader with a powerful background for further studies in biology, medicine, agriculture, and the behavioral and social sciences. Beyond that it imparts the generic skills of problem solving and scientific reasoning. We believe the book will give its readers a distinct advantage as they continue their careers. And beyond that, it will contribute to the way the reader views the world: the world will look different, it will have more texture, more connectedness, a certain inner logic.

As consulting editors, we have contributed to the lively ferment that went into the book. It has been both challenging and exciting and we are delighted with the finished product. We hope you will enjoy it.

Howard A. Schneiderman
St. Louis, Missouri

John H. Postlethwait
Eugene, Oregon

Preface

Each person's daily world is filled with living things—roommates, trees and shrubs of various kinds, butterflies, stray dogs, flowers, mosquitoes. Much of our physical culture, moreover, is based on biological materials: wooden buildings and furniture; wool, cotton, and linen clothes, carpets, curtains, and upholstery; food, beverages, spices, and pharmaceuticals are but a few examples. Absent its living things, the world would be nothing but a ball of rock, water, and gases, and we humans would never have evolved with our sensibilities, our curiosities, and our dependencies upon other organisms.

Biology, the study of living things, no doubt began as soon as consciousness and curiosity first stirred in our ancient ancestors, and as they applied their intelligence to the problems of collecting animals and plants for food. Biology became more formalized as an intellectual endeavor soon after people could record their knowledge in pictures or words, and it continues today as a fundamental part of a good education. Understanding what defines a living thing, how and why it functions as it does, and how each of the millions of kinds of organisms arose on Earth intensifies our perspectives on and pleasures in the surrounding world. Anyone can watch with interest and even inspiration as a bee lands on a fragrant flower. But the experience is far richer for the observer who understands that the shapes of the insect and the blossom have evolved as complements to each other; that the flower is the plant's showy, tasty, fragrant advertisement, attracting animals that will inadvertently assist in the plant's reproduction; and that the bee has elaborate mechanisms for finding the flower and communicating its location to other members of the hive. The study of biology has vast practical application as well, in understanding one's body and personal health, in grappling with the ethical questions that face us as citizens, and in sensing both our place in the web of interdependent living things and our need to help protect the delicate ecological balance that sustains us all.

The Strategy

By chance and good fortune, students today live amid a revolution in the biological sciences. Significant, sometimes startling discoveries come almost weekly as scientists and their students around the world exploit new techniques, theories, and approaches. To find his or her way in the fast-moving field of modern biology, the novice requires a positive strategy. Our strategy to meet this challenge is fourfold. We feel that a beginning student needs a book that has:

Authority—accuracy and selectivity, with a consistency of scientific taste expressed in the facts, concepts, and explanations selected for inclusion

Breadth of coverage—a complete survey of the disciplines that make up modern biology, including biochemistry, genetics, development, physiology, ecology, evolution, behavior, and an overview of the full range of living organisms

Depth—explanations detailed enough to teach the reader not simply what happens in living things, but what the mechanisms are, and why they function as they do

High-quality presentation—writing that the reader can enjoy and appreciate while exploring the science of life

The Organization

Any successful approach to the teaching of biology rests on the ordering of subjects. The organization of *Biology* is based on the levels of organization within a living thing and its environment.

In the first part, From Atoms to Cells, we discuss atoms and molecules, the building blocks of all matter; biological molecules, the stuff of cells and organisms; the flow of energy in living things; the parts of cells, their basic functions, and how energy propels them; and, in the final two chapters, cellular respiration and photosynthesis, the central energy pathways that sustain all living organisms, directly or indirectly. We describe respiration first because it evolved before photosynthesis and because it is universal: all cells break down chemical compounds in a similar way to acquire usable energy for their survival. Photosynthesis is described next because, while it is crucial to virtually all life on Earth, it is carried out by only a subset of living things—the plants, algae, and certain microbes.

The second part, Like Begets Like, covers cellular reproduction, the mechanisms of heredity, and the way genes and chromosomes control the daily functions of living cells. While some books begin their coverage of genetics with the structure and function of the genetic information molecules called DNA, we feel that before students can fully appreciate the shape and role of DNA molecules they need to understand how cells divide, what chromosomes are and do, and, in general, how organisms inherit their parents' traits. We trace the chronological story of how genetics developed as a field because it shows with unparalleled drama how scientific experiments bring about revised views of nature and how new facts and ideas are built upon a foundation of old. Human genetics, a subject of generally high interest, is our last topic in genetics. We place it here so that we can apply all the concepts presented earlier—from cell division to recombinant DNA—to its understanding. Immediately following the coverage of genetics is a block of chapters on development. With this arrangement, the student can see how genes carry out their foremost task—controlling the formation and unfolding of the embryo and young organism. And development is the conceptual bridge between genetics and the remaining book topics, which are all at the level of whole organisms, their systems, or their populations.

The third part, Order in Diversity, starts with the origins of life on this planet and progresses through the five kingdoms of organisms that emerged, describing the fascinating diversity of living things, their evolutionary relationships, and how each major group may have arisen. We think that students need a clear picture of the wide spectrum of organisms and their basic properties before they can learn, in later chapters, how those organisms survive the daily struggle for existence and how they interact with their environments and with each other.

The fourth and fifth parts, Plant Biology and Animal Biology, describe the physiology or day-to-day functioning of the most complex groups, the plants and the animals. We place plants first because they are excellent examples of the inseparability of structure and function in living things; because plants and their ecological roles are so fundamental to all life on Earth; and because the placement provides important perspectives as we study the analogous and more familiar systems in animals in the subsequent part.

The eleven chapters of the final part, Population Biology, provide an in-depth treatment of the sciences of evolution and ecology—from the way populations change over time, to Earth's physical environment, to the way organisms relate to their surroundings and to each other. This part includes a discussion of how animals behave. Some books consider animal behavior as a final subject in their physiological sections. We feel, however, that an animal's full behavioral repertoire is best viewed in an ecological context, since that is where behavior contributes to survival and, indirectly, to evolutionary success. Chapter 50, on human origins and evolution, is also positioned to build on the part's foundation of environment, ecology, and behavior—the powerful influences that shaped our own species' origin and history. Finally, Chapter 51 is an extended essay that explores how human evolution and activity have affected Earth's ecosystems, and how our future actions will continue to influence them—for better or worse—in the coming centuries.

The Approach

To achieve our goals of authoritative coverage, breadth, depth, and high-quality presentation within the organizational framework we have just described, we began by analyzing contemporary biological knowledge, relying, in part, on input from faculty consultants across the country. Next, experts in various subdisciplines (including Wessells) wrote first drafts of the 51 chapters. Then, based on careful reviews by a large panel of biologists at a range of institutions, the authors rewrote all of the chapters. In so doing, we tried to ensure high readability, a uniform style, a consistent level of presentation, and an emphasis on the interrelations between biological topics. Themes that recur in many chapters include evolution, the key place of development and cellular biology, and the importance of adaptation and ecology in every organism's life and every species' history. The explosion of biological knowledge in the late twentieth century—particularly in our ability to manipulate genes and engineer new genetic combinations—comes amid similar explosions in computer science, material science, particle physics, and other areas. So rapid is the present rate of discovery that it is difficult for scientist and student alike to avoid becoming overwhelmed with new information. With the help of our original contributors and many faculty reviewers, our task has been to present the fundamentals of biology—some dating several centuries, but most developed since the mid-1800s—as well as to distinguish from among the very recent findings those we believe to be lasting and significant, not simply new and glamorous. We hope our strategy will be a successful one for the users of this book. And we hope our work will provide a foundation of biological knowledge from which the reader—whether future scientific professional or informed citizen in a nonscientific field—can understand the stream of discoveries sure to come in biology in the decades ahead and participate in the democratic process of regulating and utilizing the fruits of those discoveries.

Acknowledgments

We are indebted to literally hundreds of people for help in undertaking and completing this project. None has done more than our consultants, John Postlethwait of the University of Oregon, and Howard Schneiderman of the University of California, Irvine, and Monsanto Company, who were the primary formulators of the book's outline and organization. In addition, Dr. Postlethwait wrote the chapter on genetic engineering, and contributed heavily to the other chapters in the genetics section, and both he and Dr. Schneiderman provided invaluable advice on matters large and small throughout the project's development. Special thanks go to former Stanford colleague Peter Raven, of Washington University in St. Louis and director of the Missouri Botanical Garden. Dr. Raven provided advice, perspectives, and constructive criticism for the book's final chapter.

We also extend our warmest appreciation to the original contributors of chapters; their names are listed on the title page. Other scientific contributors who helped us update and reshape the material include David Graber, National Park Service, Sequoia National Park; Lewis J. Feldman, University of California, Berkeley; Russ Fernald, University of Oregon; Kent Holsinger, University of Connecticut; Rhoda Love, University of Oregon; Douglas Miller, Stanford University; Joyce Owen, University of Oregon; William B. Sistrom, University of Oregon.

More than one hundred reviewers from various colleges, universities, and institutions provided critical feedback and recommendations for revision. Their input was very important to implementing our goals of authoritativeness and effective presentation, and is much appreciated. A list of those scientific reviewers follows this preface.

Many scientists, photographers, and artists contributed to the book's art program by giving permission to use or modify drawings, and to print or reprint photographs. Their names and the figure references for their work appear in the Credits and Acknowledgments section at the end of the book. In particular, the late Geoffrey V. Goldin made numerous imaginative contributions to the art program, and we are deeply grateful for his work.

A long and complex science book such as this, couched in readable prose and illustrated with some 1,200 photos, drawings, charts, graphs, tables, boxes, and appendixes, demands the tender loving care of a talented team of professionals at all stages of development. Preeminent have been Mary M. Shuford, whose extraordinary organizational skills and saintly patience and concern for detail have had numerous impacts on every page, and Judith Kromm, who devoted great care and skill to the developmental editing of the book. We are also indebted for the editorial, design, and production efforts of Marjorie Anderson, Sally Beckham, Kathy Bendo, Leon Bolognese, Jacqueline Bryan, Mary Louise Byrd, Jane Edsell, Jackie Estrada, Beverly Fraknoi, Cele Gardner, Andrea Lévy, Michael C. Kennedy, Laura Lamorte, John Lennard, Della Mancuso, Roberta Meyer, Irene Pavitt, Elaine Romano, Jennifer Soyke, Phyl Stevens, Suzanne Thibodeau, Peter Veres, Ruth Veres, Lesley Walsh, and Betty White.

Our joint and deepest gratitude goes to Eirik Børve, our collaborating publisher, who has, at every step of the way, placed his intelligence, energy, and exceptional management skills behind our goal of teaching biology in the most effective manner possible. Mr. Børve's encour-

agement and good spirits kept us stimulated throughout the seven-year-long project, and his support for both a high degree of scientific rigor and a student-oriented presentation was unstinting. So much teaching and learning at the college level depends on carefully published textbooks, and we feel that Mr. Børve represents the very best in his field.

Finally, we express warmest appreciation to one another. A collaboration between a professional biologist and a widely published writer is truly a beneficial education for both parties.

Even with the careful contributions of our aforementioned friends and colleagues, errors of facts or interpretation may have found their way into the book. For these, we alone assume responsibility, and stand ready to correct them.

We sincerely hope that the students and professors who use this book will find it a stimulating introduction to the intricate, fascinating, and beautiful world of life on Earth.

Norman K. Wessells
Janet L. Hopson

Academic Reviewers

John Alcock, Cornell University
Betty D. Allamong, Ball State University
Glenn Aumann, University of Houston, Central Campus
David Barrington, University of Vermont
Penelope H. Bauer, Colorado State University
Stanley Bayley, McMaster University
Paul Biersuck, Nassau Community College
Sharon Bradish-Miller, College of DuPage
Osmond P. Breland, University of Texas, Austin
T. E. Cartwright, University of Pittsburgh
Robert C. Cashner, University of New Orleans
Steve Chalgren, Radford University
Douglas T. Cheeseman, De Anza College
James S. Clegg, University of Miami
Mary U. Connell, Appalachian State University
Murray W. Coulter, Texas Tech University
Bradner Coursen, College of William and Mary
Larry Crawshaw, Portland State University
Sidney Crow, Georgia State University
William Crumpton, Iowa State University
J. M. Cubina, New York Institute of Technology
Donald J. Defler, Portland Community College
Anthony Dickinson, Memorial University, St. Johns, Newfoundland
Gary Dolph, Indiana University at Kokomo
Warren D. Dolphin, Iowa State University

Marvin Druger, Syracuse University
Robert C. Eaton, University of Colorado
D. Craig Edwards, University of Massachusetts, Amherst
David W. Eldridge, Baylor University
James R. Estes, University of Oklahoma
Russell D. Fernald, University of Oregon
Michael Filosa, Scarborough College, University of Toronto
Conrad Firling, University of Minnesota at Duluth
Kathleen M. Fisher, University of California at Davis
Arlene Foley, Wright State University
Lawrence D. Friedman, University of Missouri at St. Louis
Larry Fulton, American River College
Arthur W. Galston, Yale University
Florence H. Gardner, University of Texas, Permian Basin
Lawrence G. Gilbert, University of Texas, Austin
Elizabeth Godrick, Boston University
Michael Gold, University of California at San Francisco
Jonathan Goldthwaite, Boston College
Judith Goodenough, University of Massachusetts, Amherst
Corey Goodman, Stanford University
D. Bruce Gray, University of Rhode Island
Margaret Hartman, California State University at Los Angeles
Robert Hehman, University of Cincinnati
Steven R. Heidemann, Michigan State University
Walter Hempfling, University of Rochester
Robert W. Hoshaw, University of Arizona
Dale Hoyt, University of Georgia
June D. Hudis, Suffolk County Community College
Robert J. Huskey, University of Virginia
Alice Jacklet, State University of New York at Albany
Robert J. Jonas, Washington State University
Pia Kallas-Harvey, University of Toronto
Jerry L. Kaster, University of Wisconsin at Milwaukee
Donald Kraft, Bemidji State University
T. C. Lacalli, University of Saskatchewan
Meredith A. Lane, University of Colorado
Joseph D. Laufersweiler, University of Dayton
William H. Leonard, Clemson University
Georgia Lesh-Laurie, Cleveland State University
Joseph S. Levine, Boston College
Joseph LoBue, New York University
Ellis R. Loew, Cornell University
William F. Loomis, University of California at San Diego
Carl E. Ludwig, California State University, Sacramento

John H. Lyford, Jr., Oregon State University
Henry Merchant, George Washington University
Helen C. Miller, Oklahoma State University
William J. Moody, University of Washington
Frank L. Moore, Oregon State University
Randy Moore, Baylor University
Robert E. Moore, Montana State University
Dorothy B. Mooren, University of Wisconsin at Milwaukee
David Nanney, University of Illinois, Urbana
Maimon Nasatir, University of Toledo
Robert Neill, University of Texas, Arlington
Bette Nicotri, University of Washington
Herman Nixon, Jackson State University
Frank Nordlie, University of Florida, Gainesville
J. R. Nursall, University of Alberta
Ralph Ockerse, Purdue University
William D. O'Dell, University of Nebraska, Omaha
Rollin C. Richmond, Indiana University
Ezequiel Rivera, University of Lowell
Gerald G. Robinson, University of South Florida
Martin Roeder, Florida State University, Tallahassee
Thomas B. Roos, Dartmouth University
Ian Ross, University of California at Santa Barbara
Richard Russell, University of Pittsburgh
Roger H. Sawyer, University of South Carolina
Carl A. Scheel, Central Michigan University
John A. Schmitt, Ohio State University
Richard J. Shaw, Utah State University
Peter Shugarman, University of Southern California
Warren Smith, Central State University
David Stetler, Virginia Polytechnic Institute
Daryl Sweeney, University of Illinois, Champaign
Harry D. Thiers, San Francisco State University
John Thomas, Stanford University
Sidney Townsley, University of Hawaii at Manoa
James Turpen, University of Nebraska Medical Center, Omaha
Joseph Vanable, Purdue University
Dan B. Walker, University of California at Los Angeles
Jack Ward, Illinois State University
Larry G. Williams, Kansas State University
David Wilson, Miami University
Kathryn Wilson, Indiana-Purdue University
Leslie Wilson, University of California at Santa Barbara
Thomas Wilson, University of Vermont
G. A. Wistreich, East Los Angeles College
Daniel E. Wivagg, Baylor University
Keith H. Woodwick, California State University, Fresno
John Zimmerman, Kansas State University

Contents in Brief

Full Contents

PART I   FROM ATOMS TO CELLS · 23

  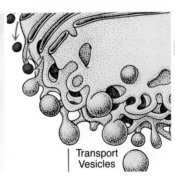 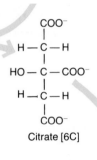

Transport
Vesicles

$$
\begin{array}{c}
COO^- \\
| \\
H - C - H \\
| \\
HO - C - COO^- \\
| \\
H - C - H \\
| \\
COO^-
\end{array}
$$
Citrate [6C]

PART II LIKE BEGETS LIKE   195

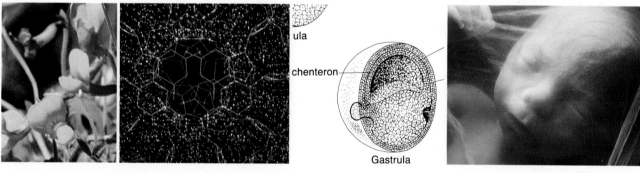

ula

chenteron

Gastrula

PART III   ORDER IN DIVERSITY   432

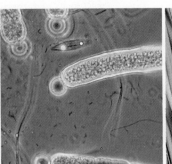
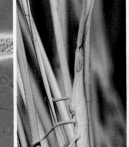

Chapter 19   The Origin and Diversity of
                Life   **434**

PART IV   PLANT BIOLOGY   640

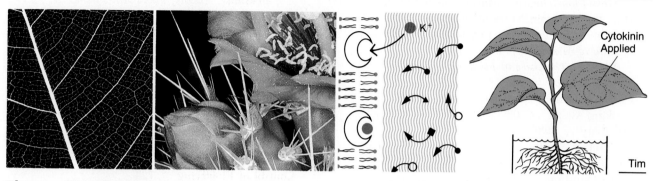

Cytokinin Applied

Tim

PART V   ANIMAL BIOLOGY   732

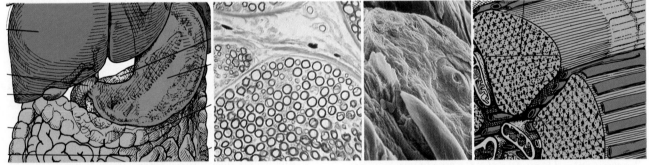

PART VI POPULATION BIOLOGY 1008

BIOLOGY

1

THE STUDY OF LIFE

*Sit down before fact as a little child, be prepared to give up
every preconceived notion, follow humbly wherever and to
whatever abyss nature leads, or you shall learn nothing.*

Thomas H. Huxley, Letter (September 1860)

This lovely orchid grows as a parasite on tropical tree species in Central America. Such
fragrant, brilliantly colored flowers can lure insects of a certain size, shape, and seasonal
activity to the plant. The insect visitors sip nectar or consume nutritious pollen grains,
then wander off to another plant of the same species, inadvertently dusted with pollen
grains that may fertilize eggs in the next set of flowers.

A kingfisher dives headlong into a rushing mountain stream and pulls out a young trout with its beak.

A tropical orchid growing on the trunk of a giant mahogany tree unfolds delicate and fragrant blooms to lure one type of pollen-eating beetle.

The Lake Titicaca frog, found high in the Andes, has very small lungs and skin that is thrown into exaggerated, thin folds. It lives immersed in an ice-cold mountain lake 280 meters deep and may not come to the surface for days or weeks at a time. It breathes through its skin while underwater.

A rich growth of small, brownish-white mushrooms springs up on elk droppings in a Wyoming meadow. The fungi, which are found only on the elk dung, are gathered by a homeward-bound flyfisherman, destined for tomorrow's breakfast with a rainbow trout.

An eighteen-year-old woman develops a condition of premature aging and a year later dies of this disease.

A man with bone cancer experiences a spontaneous remission and lives a normal life for the next decade.

These examples, and hundreds of thousands more like them, are the subjects of biology, the study of life. Diving birds, perfumed orchids, bizarre frogs, and human behavior and diseases represent the more traditional concerns of a science that dates back to Aristotle. In recent decades, biologists have acquired new tools and knowledge that enable them to engineer and explore life in unprecedented ways. Geneticists already can construct bacteria that manufacture human hormones like insulin, which is needed to treat diabetes. Biochemists soon will be able to transfer hereditary information from bean plants to wheat and corn so that these critically needed food crops can generate their own fertilizer from nitrogen in the air. And physiologists are beginning to probe the circuits and functions of the most complex object in the known universe—the human brain—with its millions of interacting nerve cells. Biologists are discovering how that fantastic organ generates memory, emotions, and creative thought; how it grows in an embryo; how in the future we may repair it; and how it developed the capacity for such remarkable functions in our species. As far as we know, we are the only species in the 4-billion-year history of Earth with a conscious awareness of ourselves and of the natural world around us. In relative terms, humans are mere babes in the long story of life, but because of our unique consciousness, we have amassed a great wealth of information about the life process and ourselves. This book is an introduction to that body of knowledge.

Because the life sciences are in the midst of an accelerating scientific revolution, the modern biology student is presented with both a privilege and a challenge. The newcomer will learn facts and concepts not even imagined by the leaders of biology just a generation ago and will be an informed witness—perhaps even a participant—in the important discoveries that are certain to come in the next decades. Yet because the field of biology is moving so fast, the amount to be learned by the beginner is substantial and grows more so each year. How can you avoid being overwhelmed and left behind in the face of these new discoveries? The approach taken in this book is to present those concepts that promise to be enduring, not ephemeral, and so to provide a foundation for understanding biology as it continues to unfold. Despite the fast pace, the challenge to the student is a worthy one, for you join a venerable tradition of philosophers, artists, theologians, and scientists who have been, over the centuries, deeply curious about the process called life.

In this chapter, we will discuss the basic characteristics of living things, how life arose, and how organisms became so diverse on our planet. Then we will introduce the concept of *evolution*, the theory that pervades all the life sciences and satisfactorily explains an immense amount about the biological world. We will consider the *scientific method*—an organized system of common sense—and learn how it allows us to investigate and understand the origins, evolution, and diversity of life, and to distinguish fallacies and unscientific notions from the facts and theories of science. Finally, we will discuss the themes of this text and, in so doing, launch your study of biology in this era of unprecedented progress and promise.

WHAT IS LIFE?

What exactly is "life"? Everyone has an intuitive sense about what living things do. Many living things move—fish swim, birds fly, and plants bend toward light. Most grow taller, wider, and heavier. Most reproduce by means of eggs or seeds or living young. Obviously, steel girders and pieces of wood fail to exhibit such tendencies, and we consider them to be nonliving. But what about dried-out seeds or virus particles—are they living or nonliving? One way to answer this question is to construct a list of characteristics that put some boundaries around this elusive concept we call life.

Living things have a complex organization.

Living things take in and use energy.

Living things grow and develop.

Living things reproduce.

(a)

(b)

Figure 1-1   LIFE'S DIVERSITY.
The fantastic diversity of life is evident in a spectrum of organismal shapes, colors, and textures. (a) The violet-eared hummingbird, a jewel of the Brazilian jungle, has hundreds of tiny soft feathers that reflect light with a metallic glint. (b) This *Lima* clam with its hard, furrowed shell and pliant tentacles, sits in a clump of golden coral just offshore in the Florida Keys.

and structures, whether observed in the diving kingfisher, the radiance of an orchid, or the human brain at work. But to speak of interrelated processes and structures is to say nothing about what they are; hence, to give full meaning to the term "life" requires the hundreds of pages that follow. In a very real sense, it takes an entire book to define life adequately.

Another way to approach the question "What is life?" is to think about the form, color, and texture of common organisms: the iridescent sheen of a hummingbird's feathers as it darts from flower to flower, the softness of rabbit fur, the hardness of clam shells, the roughness of tree bark, the slimy feel of a raw oyster (Figure 1-1). Taken as a group, living things make up a magnificent assemblage—at least 2.5 million species—and countless trillions of individuals. And practically no spot on our planet is devoid of life. Condors soar at 7,500 meters above sea level over the peaks of ancient Andean volcanoes where the air pressure is only one-quarter that at sea level. Salt-loving bacteria thrive in intensely salty inland seas and salt pans—environments that would destroy most other organisms. Other bacteria have been found dividing and growing at temperatures of 200°C (centigrade) in volcanic chimneys miles below the ocean surface. And a type of mite, a relative of the common tick, has been discovered living beneath rocks near the South Pole. Nevertheless, despite this spectacular diversity of form and life style, all living things share the specific set of characteristics listed earlier. It is these characteristics, therefore, that best describe, if not define, the living state.

Living things have a complex organization. Organisms have an intricacy of form—from the levels of atoms and molecules to those of cells, tissues, and entire organ-

Living things show variations based on heredity.

Living things are adapted to their environments and ways of life.

Living things are responsive.

Each of these statements is the result of observation and testing, and probably also makes sense intuitively. Furthermore, this simple list tells us a good deal more than does the historical notion known as *vitalism*, which construes life as a kind of magical drive or vital force that inhabits the bodies of organisms. Some people still accept vitalism, but most contemporary biologists define **life** as *a particular set of processes that result from the organization of matter.* The essential difference between the biologist's definition and the concept of a vital force is that to the biologist, life is not something *different* and *separate from* the organized processes and structures in a living thing. Instead, life is the sum of those processes

Figure 1-2   THE ARCHITECTURE OF LIFE.
The marine diatoms, shown here magnified about 450 times, are microscopic organisms with fragile shells that look like lacy blown glass, and in fact are composed largely of silicon. Each species has its own delicate markings that reflect the precise and intricate arrangement of molecules in the shell.

isms—that is not found in the nonliving world. Quartz crystals and snowflakes are organized assemblages of like atoms, the building blocks of all matter, but even the tiniest living cell contains numerous different kinds of complex substances arranged in special spatial relationships—that arrangement is the key to their coordinated function (Figure 1-2). And, importantly, continuance of the organism's life depends on the maintenance of this complex organization. It has been said that the body of an adult human contains about 85 liters (20 gallons) of water and about $6 worth of minerals. Obviously, one cannot just stir the chemicals into solution and come up with a person, a pine tree, or a toadstool. The characteristics of the organism depend on the way the substances are arranged and organized in space.

Living things take in and use energy. Like elevators, sports cars, and other complex entities, living organisms tend to break down and gradually fall into disrepair unless their organized arrangement of substances is maintained. That continual maintenance depends absolutely on energy. So the organism must take in energy, most of which ultimately derives from the sun's light and heat (Figure 1-3). In plants, solar energy is converted into chemical energy and stored as nutrient molecules. In all organisms, including plants, energy from such nutrients is released and used for maintenance and for various life activities during a series of chemical events called *metabolism*, which we will discuss in detail in later chapters.

Living things grow and develop. One of the most important activities supported by metabolism is *growth*, an

(a)

(b)

Figure 1-4 GROWTH AND DEVELOPMENT.
(a) Sprouting acorns develop into stalwart oak trees (b). The growth and the change in physical form during maturation are easily observed characteristics of most living things.

Figure 1-3 SUNLIGHT: THE ENERGY THAT DRIVES LIFE.
Pine needles, leaves on a maple, or tiny green organisms floating in the sea absorb the vital energy of sunlight to manufacture the nutrients that support all life forms.

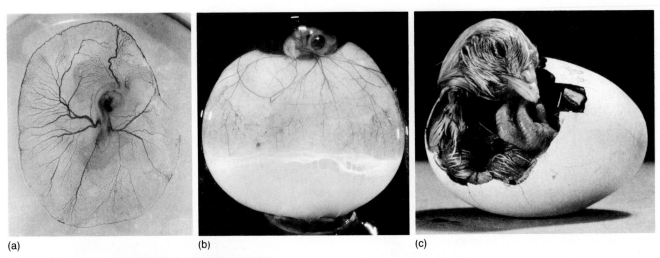

(a)   (b)   (c)

Figure 1-5   LIKE BEGETS LIKE: REPRODUCTION.
Three stages in the development of a bird. The tiny early embryo (a) grows and develops
all the organs found in the hatchling (c). Nutrients in the yellow yolk are carried through
the red blood vessels into the maturing embryo (a, b).

increase in mass, size, or organization (Figure 1-4). In living organisms, growth depends on internal processes and on duplication of many diverse structures; this is quite distinct from the growth of, say, a crystal, during which additional molecules attach to preexisting facets. Organisms also *develop;* they become more complex and take on a series of new forms, such as when a fertilized egg cell develops into a chick inside a shell, and then, after hatching, continues to develop into a hen (Figure 1-5).

Living things reproduce. All types of organisms generate offspring in a process called *reproduction.* This can be a simple matter of the division of the organism in two, or a more elaborate process that includes courtship, mating, fertilization, internal incubation, and live birth. In every case, however, the organism begets a like organism—the redwood tree reproduces a redwood and not a pine; the starfish, a new starfish and not an oyster or a trout, and so on. This reproductive fidelity depends on a set of chemical blueprints contained within each organism—the hereditary or *genetic* information, which specifies what form the new individual will take and, in part, how it will develop and grow.

Living things show variations. A basic feature of reproduction is that it results in *variation;* the offspring almost always differ in various ways from one or both parents (Figure 1-6). One reason is that the hereditary materials are sensitive to agents in the environment, such as cosmic rays or certain noxious chemicals. In addition, shuffling of the hereditary information and slight irregularities in the inheritance process occur normally. As we shall see, this variation is not always beneficial to the individual organism. Nevertheless, it is critical to the

continuation of a species because it gives a lineage of organisms a better chance of surviving over vast spans of time as local and global environments change. In other words, it permits evolutionary change.

Living things are adapted to their environments. Every living thing is organized and functions in such a way that it can exploit and cope with features of its physical surroundings—water, air, heat or cold, light or darkness, predators, competitors, to name some important ones. To survive, an organism must be specifically suited to its environment and way of life. Such *adaptation* is the result of evolutionary change, the accumulation of inher-

Figure 1-6   VARIATION BETWEEN GENERATIONS.
Offspring like these mixed-breed puppies are usually somewhat different from their parents—a fact that underlies the inevitable variations that are a basis for evolution.

(a)

(b)

Figure 1-7 ADAPTATION: COPING WITH THE ENVIRONMENT.
Organisms are physically and functionally suited to their surroundings. (a) Flightless emperor penguins thrive in the frigid Antarctic, where inhabitants of temperate forests such as the Gila woodpecker (b) would quickly perish. This bird, however, is uniquely suited to its own habitat, which is evident in its particular adaptations for tree drilling—strong beak, thick skull, and powerful neck.

ited variations over time. Examples of adaptation are the emperor penguin's ability to swim in icy Antarctic seas and the red-headed woodpecker's strong beak and powerful neck muscles, which allow it to drill into an oak to find insects (Figure 1-7).

Living things are responsive. Flowers bend toward the sun. A baby pulls its hand away from a hot radiator. A trout darts away from a shadow cast on the edge of a stream. A bacterium orients itself in a magnetic field, as does a migrating starling. These are examples of *responsiveness*, the ability to detect and adjust to certain features of the environment. Although movement is a common form of response, some organisms respond to stimuli in other ways, such as by releasing a chemical or changing color (Figure 1-8).

One other very special feature of living organisms is their *history.* Every living thing on Earth today is a descendant of an organism that lived before it. Each is a member of an unbroken lineage stretching backward in time to the era, billions of years ago, when life processes

first became associated with organized sets of matter. Thus a knowledge of evolutionary history is important to our understanding of many characteristics of present-day organisms.

LIFE ON EARTH: A BRIEF HISTORY

The evolutionary history of life forms usually is depicted as a branching tree, as shown in Figure 1-9. At the ends of the branches are the many modern groups. The supporting branches beneath them represent the groups that emerged earlier and gave rise to the organisms at the branch ends. The trunk supporting all the branches represents modern descendants of the earliest and simplest living organisms that arose on the young planet Earth and were the progenitors of all later forms of life.

Early Beliefs About the Origin of Life

The treelike picture of evolutionary history, which is widely accepted by biologists, is based on fossil evidence, precise anatomies, comparisons of genetic blueprints of different organisms, and studies of the probable origin of life, which we will discuss shortly. For many centuries, however, people's day-to-day observations led them to attribute the origin of life to a far different process: **spontaneous generation.** The sudden, seemingly miraculous appearance of molds on bread, worms on aging meat, or toadstools on decaying wood led peo-

Figure 1-8 A CLOUD FOR SELF-DEFENSE.
This boreal squid ejects a cloud of smoky gray pigment and jet propels away when threatened.

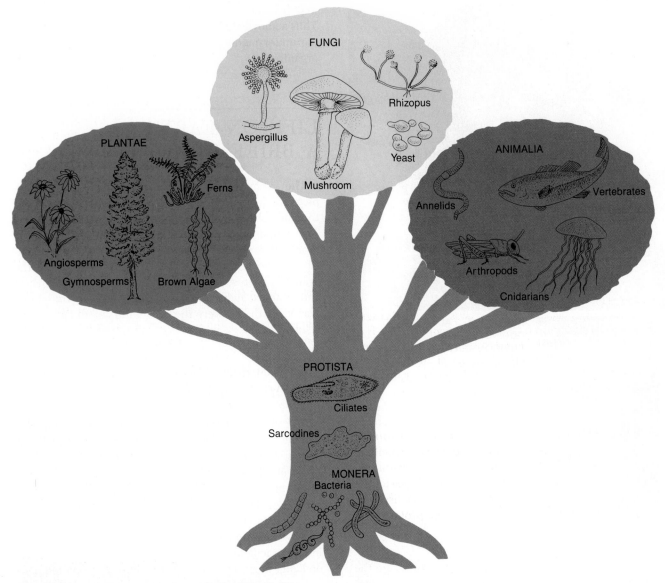

Figure 1-9   ALL LIVING THINGS AS BRANCHES OF ONE TREE.
This tree of life depicts the origins of all life forms from simple cells—the "roots"—and the subsequent evolution of plants, fungi (mushrooms), and animals—the major "branches."

ple to the idea that life arises anew if conditions are conducive.

The belief in spontaneous generation held sway from well before the era of the Greek and Roman scholars until the seventeenth century, when an Italian named Francesco Redi published a careful refutation based on direct tests. He suspected that the maggots in meat actually arise from the eggs of flies that land on the spoiling food. However, in his words, "belief would be in vain without the confirmation of experiment." So to test his

theory, Redi put fresh pieces of fish, veal, eels, and snakes into glass flasks, leaving some open and sealing others carefully. Flies visited the open flasks, and worms soon appeared on the meat; but in the sealed flasks, worms never developed (Figure 1-10). Redi concluded that maggots come only from eggs and generalized further that life forms originated only once and that all living organisms are direct descendants of preexisting individuals.

Despite Redi's well-founded arguments, the idea of spontaneous generation did not die gracefully. Around the time of Redi's experiments, the microscope, the biologist's single most important tool, was invented. Ironically, by allowing scientists to peer into a previously invisible world, the microscope helped to cloud the issue for two more centuries. A microscopist could witness

Figure 1-10.   FRANCESCO REDI.
The myth of spontaneous generation was seriously damaged by Francesco Redi's careful experimentation—but beliefs die hard. Although Redi proved that maggots do not arise from meat in sealed flasks, the discovery of microorganisms fostered the old notion of a life force in the air or water that could animate the inanimate.

strange new organisms in the realm of a water droplet; perhaps *these* minute creatures might arise spontaneously. It was not until the mid-nineteenth century in France that Louis Pasteur finally laid the old idea to rest.

Pasteur observed that the microscopic organisms living in a solution of food molecules (sugars, proteins, and water) can be killed by boiling the broth and that they do not reappear as long as the opening to the flask is heated (to sterilize it) and then sealed tightly. Pasteur showed that the broth in sealed flasks can remain uncontaminated for as long as eighteen months. If the flasks are opened to the air, however, the solutions inside will teem with bacteria within a day or two. To Pasteur, this proved that microbes floating on dust particles in the air must enter the newly opened flasks and begin to multiply. Nevertheless, some of his contemporaries remained unconvinced and countered with the argument that a life-giving force in the air enters the flasks and causes spontaneous generation to occur in the broth.

In a brilliant move that finally silenced his critics, Pasteur designed a special type of flask with a long, downward curving neck that allows air to reach the solution but traps dust particles and microbes in its lower part (Figure 1-11). He showed that a solution boiled in such a flask remains free of organisms despite the free exchange of air. Yet when the neck is removed by the stroke of a file and dust can enter the solution, the growth of organisms becomes apparent within a matter of hours.

As a result of Pasteur's experiments, the belief in spontaneous generation was replaced by the idea that all life comes from preexisting life. However, logic demanded, then and now, that this unbroken chain of life had some beginning, somewhere, at some time. One area of modern biological research is concerned with the origin of life and relies on data from paleontology, geology, astronomy, physics, and other fields. We will describe the latest theories and experimental results in detail in Chapter 19. Here we will simply preview the topic.

Figure 1-11   LOUIS PASTEUR.
Pasteur propelled biology past myths and mysteries with his brilliant disproof of spontaneous generation. He showed that sterilized broth in a swan-necked flask remains free of microbes for days—even years. But as soon as the curving glass neck is broken, the broth becomes clouded with bacterial growth within hours. Pasteur showed that air carries not a life force but life itself, in the form of floating organisms, and that life alone begets life.

Figure 1-12   FROM ORGANIC COMPOUNDS TO LIVING CELLS. Organic chemicals, precursors of biological molecules, existed on our planet long before life arose. About four billion years ago energy from the sun and heat from Earth's core aided the formation of complex chemicals, which are the building blocks of living things. Once living cells formed and proliferated, they slowly and irrevocably changed the face of Earth. This geyser basin with its hot sulfurous water—reminiscent, perhaps, of primal scenes on the early Earth—teems with bacteria, algae, and other life forms.

A Modern View of the Origin of Life

As a result of the exploratory efforts of the American space program, scientists are virtually certain that all the other major planets of our solar system, and probably their moons as well, are devoid of life. Someday, when space communication or travel extends beyond our tiny corner of the universe, humans may contact life on thousands of planets much like Earth circling stars similar to our sun. For now, however, we are faced with the fact that in our solar system, Earth seems to be a uniquely hospitable place for life. The reasons include Earth's size, gravity, atmosphere, and surface temperature. The size of our planet and its distance from the sun dictate that gravitational force will hold an atmosphere near the surface. The atmosphere not only screens out much ultraviolet light, which destroys many chemicals and damages living organisms, but also helps maintain surface temperatures on the Earth within a range of about 0° to 100°C. Within this range, the major constituent of most living organisms—water—is a liquid, not a gas or a solid. At temperatures much below 0°C and above 100°C, most life processes as we know them cannot occur.

The origin of life is intertwined with the history of Earth. Although Earth was once a hot cloud of condensing matter not compatible with life as we know it, evidence suggests that the planet passed through a stage about 4 billion years ago during which energy from the sun and the heat of the Earth's molten core aided the formation of complex chemicals from simple atoms and molecules. Those complex materials became the building blocks of living things (Figure 1-12). Other such chemicals may have been delivered to Earth during the frequent bombardment by meteors that marked the early history of our planet. Biologists believe that aggregations of those building-block chemicals led to the first organized systems with the characteristics of life we discussed earlier. These first *cells*, the fundamental units of all living things, possessed the capacity to produce copies of themselves (Figure 1-13). However, imperfections

Figure 1-13   DESCENDANTS OF EARLY CELLS. Early cells may have resembled modern bacteria in fundamental ways. This bacterial colony with its sausagelike chains of cells is magnified about 2,000 times its normal size with the scanning electron microscope.

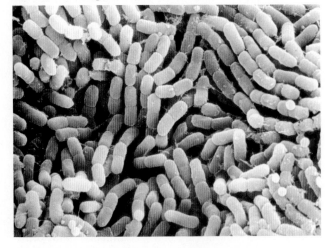

in this hereditary process led to inevitable variations among the generations of cells that followed and set the stage for new kinds of organisms. Eventually, over vast spans of time, the major and minor branches on life's evolutionary tree arose.

This quick overview of life's origin shows that the notion of spontaneous generation was not entirely wrong; life did arise spontaneously from nonliving matter on the early Earth. However, as we shall discuss in Chapter 19, the first organisms that appeared—various kinds of primitive bacteria—acquired new properties over time. For instance, over hundreds of millions of years, certain water-dwelling bacteria released into the atmosphere huge quantities of oxygen, which blocked out deadly ultraviolet light from the sun. This protective covering allowed organisms to live near the surface of the sea or even in moist places on land. Thus life and Earth began a process of **coevolution,** or simultaneous, interrelated change, which continues today. Significantly, the physical attributes of the infant planet that led to life's origin from nonliving matter were *changed permanently* in the process and were no longer conducive to further spontaneous generation. In a sense, then, living things altered the very world that had given rise to them in a way that would prevent the same process from happening again.

Through reproduction was forged an unbroken chain of living organisms, each passing to its progeny the structures and processes of the living state. From the original common stock of single cells arose colonies of cells that exhibited plantlike or animal-like characteristics. Those colonies, over many millions of years, gave rise to an enormous spectrum of organisms that inhabit water, land, and air. But despite the profound differences that exist among, say, a bacterium growing on a cactus, a moss carpeting a rock, a tuna darting through deep ocean waters, and a stork sunning its broad wings, all living organisms continue to share many fundamental chemical traits. Among these are the structure of the hereditary material, *deoxyribonucleic acid* or *DNA* (Figure 1-14), and the fuel that drives life processes, *adenosine triphosphate* or *ATP*. Our mutual descent from the first living cells explains the universal occurrence of such properties.

EVOLUTION: A THEORY THAT CHANGED BIOLOGY AND HUMAN THOUGHT

Most educated people accept the idea that today's creatures are descendants of yesterday's organisms. Most of us have seen fossils in museums; read about the extinction of dinosaurs, apelike humans, and other

Figure 1-14   DNA: LIFE'S HEREDITARY MATERIAL.
This drop of fluid contains DNA that has been extracted from living cells and purified. Modern biotechnology uses such DNA, and is rapidly changing agriculture and medicine and expanding biological knowledge in unprecedented ways.

groups; and know that many species have come and gone as the Earth itself changed. But the idea that one group gives rise to another over time was not always accepted. One of the great intellectual adventures of human history was the explanation of the rise of immensely varied living things, each species adapted to its particular habitat and way of life. Charles Darwin's theory of evolution changed the way people think about their origin and their place in the scheme of nature. It is, without a doubt, the most important unifying concept in biology.

The Intellectual Climate in Darwin's Time

In an ironic way, an enlightened trend that began during the Renaissance of the sixteenth century had a dampening effect on the study of life and its history. For 2,000 years prior to that era, scholars had accepted the writings of Aristotle and other ancient philosophers, as well as certain Church doctrines, to be unfaltering truths about the natural world. It took some of the greatest minds in human history, including Copernicus, Galileo, and Newton, to shake this dominion of dogma and to replace it with theories and laws based on direct observation of nature. Nicolaus Copernicus, and later Galileo Galilei, made calculations that proved that the Earth and other planets circle the sun, and Sir Isaac Newton discovered the universal laws of gravity and motion. These thinkers showed that the forces of nature can be described in precise, quantifiable terms and that matter is governed by impersonal physical laws. However, similar direct observations of living organisms stimulated by this enlightened, scientific approach seemed to reveal a *purposefulness* that was missing from the nonliving world:

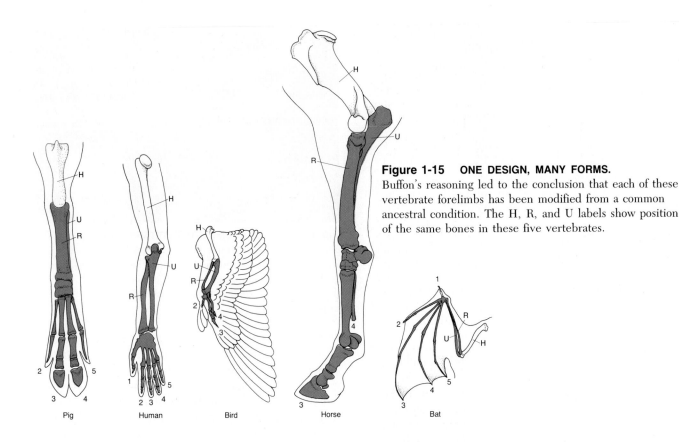

Figure 1-15   ONE DESIGN, MANY FORMS.
Buffon's reasoning led to the conclusion that each of these
vertebrate forelimbs has been modified from a common
ancestral condition. The H, R, and U labels show position
of the same bones in these five vertebrates.

Pig

Human

Bird

Horse

Bat

legs clearly were "designed" for running, wings for flying, and fins for swimming; and organisms lived in just
the environments suited to such structures—mice in
fields, birds in trees, and fish in ponds. Direct observation seemed to show that the living world was imbued
with vital forces and governed by benevolent purpose
and order. And most world religions agreed; for instance, the Old Testament describes how all living things
were created specifically for their place in nature and
their usefulness to humankind.

It was in this intellectual climate that the young Englishman Charles Darwin was to travel widely and then
spend most of his life formulating the theory of evolution
by natural selection, which is the basis of contemporary
evolutionary thought. However, Darwin was not the
first to consider an alternative to the special creation of
living things, to fulfill a particular role in the scheme of
nature. His intellectual forebears included the French
naturalists Georges Louis Leclerc de Buffon and Jean
Baptiste Lamarck, both of whom believed that variations
among living creatures were the consequence of some
mechanism of heredity.

Buffon is reputed to have believed in special creation,
but he puzzled over such unexplainable, apparently
functionless appendages as the two little side toes on a
pig's foot, which never touch the ground. What purpose
could these toes serve when compared with the func-

tional and very similar toe bones in the limbs of most
reptiles, mammals, and birds? Buffon concluded that the
pig's extra toes, as well as other variations in the number
and shape of limbs among vertebrates (animals with
backbones), could be explained only if one assumed that
all such limbs had been inherited from ancestors whose
limbs were made up of fully functional parts. Thus one
basic limb design must have been modified in different
creatures for running, flying, digging, or swimming (Figure 1-15). All vertebrates, Buffon concluded, are descendants of a common ancestor. This was a remarkable
idea for the 1760s, but it did not catch on because Buffon
offered no evidence.

Lamarck believed that simple organisms had given
rise to more complex ones in a natural progression of
species, but he based his systematic evolutionary theory
(the first ever proposed) on two ideas that were discred-
ited later: (1) some natural force operates to purposefully
move organisms up the ladder of complexity toward the
place held by humankind, and (2) an organism can acquire a new characteristic during its lifetime and then
pass it on to offspring. Lamarck's most famous example
was the giraffe's neck, which the French naturalist
claimed can grow longer as an individual animal
stretches for leaves. The acquired trait of the longer
neck, he said, can then be passed to the animal's young
through the mechanisms of heredity (which at that time

were unknown). Although Lamarck's reasons were wrong, his theory challenged the accepted belief that each form of life was immutable in form.

As important as these early evolutionary theories were in shaping young Darwin's ideas about nature, certain geological evidence was even more provocative to him. Briefly, this evidence implied that the Earth is very much older than the 6,000 years that Christian scholars had calculated from the Bible. Earth's history has been marked by constant slow change, which is documented in the successive layers of rock that were deposited as the Earth aged. Embedded in these layers are the remains of all sorts of extinct plants and animals that once inhabited the planet. These fossils sometimes reveal modifications and changes in body organization and structure from one layer of rock to another. And, finally, many of the animals and plants alive in Darwin's day are not represented at all in the fossil record, implying that they did not live during ancient times (Table 1-1).

Darwin carried along this assortment of biological, geological, and theological ideas when he signed on as the naturalist aboard the HMS *Beagle* in 1831 and began a five-year exploratory voyage to South America and other continents (Figure 1-16). But it was his careful observations of the natural world and much reading and reflecting after his return that led him some twenty years later to write his famous treatise, *On the Origin of Species by Means of Natural Selection.*

Table 1-1 THE ORDER OF APPEARANCE OF LIFE FORMS IN THE FOSSIL RECORD

Life Form	Millions of Years Since First Known Appearance (approximate)
Microbial (prokaryotic cells)	2,700
Complex (eukaryotic cells)	1,400
First multicellular animals	670
Shell-bearing animals	540
Vertebrates (fishes)	490
Amphibians	350
Reptiles	310
Mammals	200
Nonhuman primates	60
Earliest apes	25
Australopithecine ancestors	5
Homo sapiens sapiens (modern humans)	0.05 (50,000 years)

Source: The U.S. National Academy of Sciences

Darwin's Theory

Darwin's book really offered two related theories: (1) the **theory of evolution,** which states that all living things have evolved from a common ancestor that diverged into millions of species by means of a gradual process of change and variation; and (2) the **theory of natural selection,** which states that natural events "select" organisms in such a way that the better-adapted individuals tend to survive and reproduce, whereas the less well-adapted ones tend not to contribute to later generations. Darwin proposed that natural selection is the primary underlying mechanism that brings about evolution.

Important evidence cited by Darwin centered on the selective breeding of farm animals. Since the domestication of animals began more than 10,000 years ago, farmers have selected the best milk producer, the best egg layer, the strongest burro or plow horse, the best watchdog. The farmer could slowly "improve" the breed for a desired characteristic by allowing a chosen pair of prize animals—but not others—to mate. This evidence alone proved to Darwin that a given kind of organism is not physically immutable (Figure 1-17). But if a farmer could act as an "artificial selector," causing a lineage of animals or plants to "evolve" in a certain way, then perhaps, Darwin reasoned, there is a "natural farmer" that selects in a nonpurposeful way certain plants and animals—but

Figure 1-16 DARWIN ON THE GALAPAGOS.
Upon reaching the volcanic archipelago lying due west of Ecuador in the Pacific Ocean, Darwin marveled at the unique wildlife and observed it carefully. He clocked the walking speed of the massive Galapagos tortoise, for example, and recorded its habit of eating prickly cactus pads.

**Figure 1-17
ARTIFICIAL SELECTION.**
Generations of farmers
have selected and bred
domestic cattle for
specific traits,
producing within
the same species
a remarkable range of
variation. (a) Herefords
and (b) Longhorns are
just two breeds of many
dozens.

(a)   (b)

not others—for successful breeding. He called this hypothetical process *natural selection.*

Historians believe that Darwin's theory was greatly influenced by the writings of Thomas Malthus, an English economist and clergyman. Malthus deduced that populations of organisms tend to increase in size because organisms produce an excess of offspring—hundreds of seeds from one flower, thousands of eggs from one pair of salmon, and so on. But the world is not overwhelmed with organisms, as one might expect from such overproduction, because the hazards of living wipe out most of the offspring (Figure 1-18). Examples are easy to find: a seed lodges in a soil-less rocky crevice and fails to take root; a juvenile monkey topples from a branch, breaks a leg, and succumbs to infection; a new mushroom is consumed by a grazing moose. Examples like these led Darwin to realize that such chance events could be the "natural farmer" working not just to reduce the population

Figure 1-18   NATURAL POPULATION CONTROL.
Elk populations are held in check partly by coyotes, which
tend to attack the young, the old, and the sick.

size, but also to eliminate those individuals less fit than the remaining population to survive predation and environmental hardships.

Darwin recognized that the key issue was survival to reproductive age and then successful reproduction. Natural selection, he concluded, is the Grim Reaper that lets some individuals achieve reproductive success, while others die.

Darwin's theory provided a mechanism to explain the evolution of diverse life forms. Since all offspring vary slightly from their parents and from one another, some will be better able to survive environmental challenges than others. Thus the survivors, on the average, will live to reproduce and pass along their unique and "selected" genetic endowment, whereas their siblings who are not so well adapted, on the average, will meet untimely fates and fail to pass on their hereditary substance. The line of survivors evolves through history, while other organisms and lineages die out.

Darwin's ideas were disconcerting to his nineteenth-century contemporaries, and some people still reject them. It somehow diminishes their sense of life, and they regard it as demeaning to the human species to accept the idea that we evolved from what they consider to be lower life forms and that the same impartial principles of chance and physical laws that govern the falling apple or the orbiting planet might also determine the shapes, activities, and histories of living things.

But Darwin's theories have helped biologists understand and organize an incredible array of facts and observations during the past century, including the obvious physical adaptations that all organisms show to their environments; thus the purposefulness that stumped sixteenth-century scholars has proved to be easily explainable on scientific grounds. Evolution explains both the unity and the diversity of life—why *all* organisms share many characteristics, while at the same time, great variability arises. And evolution gives us historical reasons for such vestigial structures as the pig's dangling toes.

There is so much accumulated evidence for the theory of evolution by natural selection and it explains so much—from the properties of molecules to the characteristics of entire communities of diverse organisms—that almost all biologists view evolution as a scientific fact. But while the fact of evolution is believed to be beyond dispute, the mechanism (that is, "what causes it") is still under intensive investigation. Many bright young scientists now follow in the footsteps of Charles Darwin to investigate natural selection and other possible mechanisms of evolutionary change.

THE SCIENTIFIC METHOD: ONE APPROACH TO EXTENDING KNOWLEDGE

Charles Darwin, like other great scientists before and after his time, had a restless curiosity about the natural world. He made thousands of individual observations, organized and analyzed his findings, and drew creative conclusions from them about the laws of nature. But a fascination with nature is not solely the province of scientists. Creative geniuses of many kinds have focused their attention on nature and given humankind great and lasting works: Ludwig van Beethoven's *Pastoral* Symphony; Claude Monet's *Water Lilies;* Henry David Thoreau's *Walden;* St. Augustine's *De Naturae Et Gratia (Of Nature and Grace)* (Figure 1-19). There are, nevertheless, important differences between the ways artists, philosophers, and theologians approach nature and the way a scientist studies it. The nonscientist often works through inspiration, sentiment, or formal logic, while the scientist relies primarily on the **scientific method,** a kind of organized common sense that we shall see applied over and over in this book. The scientific method begins with observations of natural phenomena. On the basis of these observations, the researcher poses questions, designs experiments, gathers results, and formulates explanations.

Just as Darwin began by observing the world around him, so all science begins with observations that stir the curiosity of scientists. The pea seedling grows rapidly; the oak tree, slowly. The Lake Titicaca frog has folds of skin; the tree frog, a smooth skin. Such observations lead to a different sort of question—answerable by experiment and measurement—from that which the artist, theologian, or philosopher would pose. Why does the seedling grow fast? What purpose do the frog's skin folds serve? Scientists make observations, take measurements, and establish the basic facts. Then, in mulling over the results, they use **inductive reasoning**—go from the specific observations and facts to a general explanation—and formulate an educated guess, or **hypothesis,** that is a probable answer to the question. The framing of a scientific hypothesis is no less creative an act than is painting a picture or developing an ethical theme. For many scientists, the wonderful excitement that comes at the moment of insight makes the entire process worthwhile. But it is the next step—*testing the hypothesis*—that sets science apart from other disciplines and enables scientists to produce accurate and enduring explanations of natural phenomena.

All scientists must design tests that could *disprove* a hypothesis, in case it is actually an incorrect guess. Suppose a scientist hypothesizes that red light causes plants to develop flowers. If that hypothesis is correct, then red light, and only red light, should stimulate flowering. To test that prediction, the scientist exposes one set of plants to red light, another set to blue light, and other

Figure 1-19 NATURE'S MANY IMAGINATIVE FORMS.

Claude Monet's *Water Lilies* are but one way to view life's beauty; Darwin and Pasteur described other, equally beautiful aspects of life.

sets to green or yellow light; raises one set in the dark; and so on (Figure 1-20). The scientist then watches and records the results. If the only plants to flower are the ones exposed to red light, then the hypothesis is supported. If some of the other plants also flower, then the original hypothesis must be modified. Critical in this process are the **control experiments,** in which the conditions imposed are different from the ones set forth in the hypothesis; in this example, the use of blue and green light, no light at all, and so on, provides the necessary controls. Only if many different approaches fail to disprove a hypothesis will it be considered accurate, and even then, the presumption of accuracy is tentative. A new technique or discovery may force modification of the hypothesis years later. Thus the aim of a scientist is not to *prove* something, but to test it again and again in order to move toward a more and more accurate explanation.

In addition, for any hypothesis to be considered tentatively accurate, the evidence in its favor must be *reproducible;* that is, an independent scientist must be able to repeat the same experiments and get essentially the same results in his or her own lab. It is for this reason that a report in a scientific journal usually has a section called "methods and materials." In it, the scientist is, in effect, giving a recipe that another scientist could follow to carry out an independent test of the hypothesis.

The final stage in the scientific method is the postulation of a **theory,** a general statement that usually is based on a number of tested hypotheses and is designed to explain a range of observations—even some not yet made. In this sense, a good theory is not just inductive, but can be used for **deductive reasoning,** a process of predicting new facts or relations, for which new experiments can be designed and new information collected. For example, examine Table 1-1. As paleontologists have searched the Earth's rocks for fossils of vertebrates, they have found fossils of only ancient fish in the oldest rocks, of amphibians in somewhat younger rocks, of reptiles in still younger rocks, and of mammals in the very youngest rock strata. This has helped to define the evolutionary sequence of backboned animals. If mammalian fossils ever were discovered in the oldest rock layer, the entire theory would need revision. Furthermore, the fossil sequence leads to other predictions. For instance, molecules of the hereditary material (DNA) or proteins should be most dissimilar if fish and mammal are compared, most alike if fish and amphibian are compared, and so on. Indeed, these predictions are supported precisely by molecular analyses. Hence the sequential evolution of vertebrates has still more support.

When a theory such as the theory of evolution is used and tested again and again and corroborated in diverse ways, it is eventually accepted as a **natural law,** or a scientific fact. That is what we mean when we say that

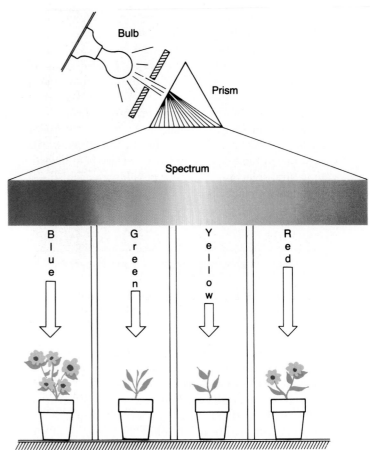

Figure 1-20   TESTING A HYPOTHESIS.
White light can be broken by a prism into its component colors. When experimenters expose plants to four colors of light, only the plants that receive blue and red lights flower. This helps support the original hypothesis that red light is required for plant flowering; but the hypothesis must be modified substantially because even better flowering occurs under blue light. Clearly biologists must do more experiments to explain why two colors of light induce flowering, not just one.

almost all biologists consider evolution to be far more than a theory, in the lay sense of the word. It is as much a fact as is the physicist's theory that matter is built of particles called atoms and the astronomer's theory that the Earth circles the sun.

You can see from this discussion that both the creative part of the scientific method (posing hypotheses) and the steps that follow (testing, recording, communicating to one's peers, theorizing) are interdependent: observations lead via inductive reasoning to hypotheses and theories; theories lead via deductive reasoning to predictions and tests. The real impetus to this scientific process is curiosity about nature. The truly creative scientist always keeps an eye open for the bizarre, the unexpected, the chance observation that may lead to sudden

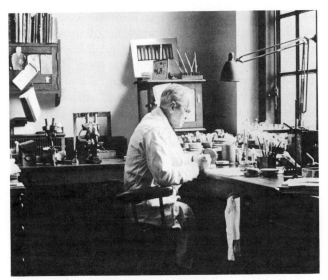

Figure 1-21   ALEXANDER FLEMING.
Sir Alexander Fleming, discoverer of penicillin, did most of his historic work on the penicillin mold at St. Mary's Hospital in London.

new insights. For instance, in 1928, when Sir Alexander Fleming noticed "halos" of killed bacteria around certain molds growing in culture dishes, he sought to learn why. His hypothesis that the molds produced an antibacterial agent resulted in the discovery of penicillin, the antibiotic that has saved millions of human lives (Figure 1-21).

BIOLOGY, SOCIETY, AND YOUR FUTURE

The organized system of common sense we call the scientific method can be applied on many levels. Every day each of us must solve problems and make decisions. The scientific method can lead to wiser choices if we list possible solutions (hypotheses) or courses of action, collect as much information as possible, and eliminate the least satisfactory possibilities. The scientific method also has served as midwife to virtually all the great advances in biology, from the unraveling of such fundamental mysteries as the nature of heredity and the functioning of cells to the exciting applications of such knowledge to modern agriculture, medicine, and environmental science.

Despite the power of this disciplined approach to ordering thought, however, the scientific method has limitations; by itself, it cannot solve many important social problems. As science and technology advance, human society is faced with more and more difficult ethical and moral choices. For example, it is now possible to deter-

mine during pregnancy whether a fetus is grossly deformed and probably would have a short and perhaps painful life, and it is also possible to terminate most such pregnancies without risk to the mother (Figure 1-22). But should the parents choose to abort? And should society allow it? In addition, physicians can prolong—sometimes for years—the lives of people whose conscious brain functions have been permanently destroyed by accident or disease. But should they do so in every case when this procedure is possible? Or should the physicians "pull the plug" after a certain amount of time? Finally, biologists soon will be able to transfer a number of specific, heritable traits from one person to another. An example would be substituting a normal gene for the defective gene (genetic material) that carries the disease sickle cell anemia. But is it ethical to manipulate human genes?

Scientists and physicians can provide factual information and informed counsel about the problems and the procedures just described, but the final ethical decision inevitably involves nonscientific considerations. Decisions such as these must be made by informed individual citizens or perhaps by governmental, professional, or

Figure 1-22   SCIENCE AND MORALITY.
The monitoring of human fetuses, now routine, can reveal deformities, such as those caused by Thalidomide, long before the baby is due. But how should that knowledge be used?

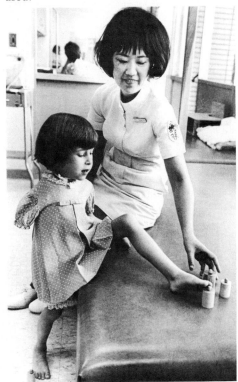

legal groups. The scientific method is not directly applicable to ethical and moral questions, although it does provide a framework for rigorous, common-sense thinking that may aid decision making in these areas of human concern.

The scientific method also is not usually applicable to metaphysical questions or claims of intangible phenomena, since they cannot be tested. For example, the claim that a vital force inhabits and animates all living things lies outside the realm of science because no one can prove or disprove that vital forces exist. Scientists have no more right to assert that artistic or religious beliefs in vital spirits are disproved by science than do the adherents of vitalism have a right to insist that such ideas have a legitimate place in science. Informed and educated citizens have a responsibility to understand the capabilities and limitations of science. Only then can we make decisions on issues that affect us as voters, parents, and members of an increasingly complex technological society.

An understanding of science can have two other important consequences. First, studying life science may lead to an interesting career in medicine, teaching, research, or applied biology. Physicians, dentists, veterinarians, and many farmers have a broad knowledge of biology. Professors and researchers in specific biological fields have this same broad background as well as extensive training in a single area. A trained biologist also may find a career in applied areas such as forest and range management, agricultural economics, genetic engineering of new organisms, science and technical writing, science education, and laboratory technical work (Figure 1-23). The pace of the biological revolution is so rapid that whole new careers may be available within the decade. The student with a firm foundation in biology, chemistry, and computer science, who has the ability to think and write clearly, will be able to exploit those opportunities.

Second, regardless of one's professional path, biology is a liberal art, and its study is an end in itself. It is a source of the intellectual pleasure and perspective that make our lives richer and more enjoyable. The perceptive biology student will forever after view the world in a different way. To know how a tree or a butterfly func-

Figure 1-23   CAREERS IN APPLIED BIOLOGY. Trained biologists work in dozens of areas from agriculture to zoology. The production of wheat and other staple crops for our burgeoning global population will depend increasingly on research and biotechnology.

Figure 1-24   THE STUDY OF BIOLOGY—A WINDOW ON THE WORLD.
The patterns of nature are rich and varied—and more easily appreciated by the informed observer.

tions, to know why migrating birds fly in a V formation, to know why a turkey has light meat and dark meat—such knowledge can help us appreciate the beauty and complexity of the world around us (Figure 1-24).

OUR APPROACH TO BIOLOGY

Our exploration of modern biology will be divided into six parts. Although each part covers a different area of biology, several themes run through the entire book. Understanding these themes is as much a key to appreciating biology and the living state as is learning the facts and concepts in each chapter.

1. *Evolution*, the foundation of modern biological thought.

2. *The central role of development* in translating genetic instructions into a functioning organism. Through developmental biology, we can understand how the structures and life processes on which evolution operates actually arise in the individual and the species.

3. *The integration of the biological and physical worlds.* Living matter obeys the same universal physical laws that govern nonliving matter, and these laws set limits on the characteristics of living

things (Figure 1-25). These physical laws, along with biological laws, explain the great diversity of living things and the ways they live, reproduce, and evolve on Earth.

4. *The coevolution of organisms with one another and with the physical environment.* Plants, animals, and other organisms evolve simultaneously, with a dependence on one another. But in a broader sense, changes in the physical environment bring about changes in organisms, and organisms in turn alter the Earth and its atmosphere.

We shall see throughout our study of biology that the grand scheme of biological evolution is complemented in a few species, most notably our own, by **cultural evolution,** a process involving transfer of information from generation to generation in a nongenetic way. Thus we will discover how the two great languages—the language of heredity laid down in DNA and human language—interact in remarkable ways to accelerate the rate of evolutionary change (Figure 1-26). The application of the scientific method, the invention and use of computers, and the manipulation of hereditary information are all parts of this cultural evolution. They give us the power to destroy the environment, other organisms, and ourselves or to improve the quality of life. It is up to educated citizens, professional biologists, physicians, and others to influence which choices are made and thus to contribute to human cultural evolution in a positive way.

Figure 1-25   PHYSICAL LAWS AND THE PROPERTIES OF WATER.
Universal physical laws govern living and nonliving matter alike. The properties of water are the same whether it takes the form of snow in a Canadian meadow, steam issuing from a bison's nostrils, or hot fluid in an animal's coursing blood.

```
GAGTTTTATCGCTTCCATGACGCAGAAGTTAACACTTTCGGATATTTCTGATGAGTCGAA
AAATTATCTTGATAAAGCAGGAATTACTACTGCTTGTTTACGAATTAAATCGAAGTGGAC
TGCTGGCGGAAAATGAGAAAATTCGACCTATCCTTGCGCAGCTCGAGAAGCTCTTACTTT
GCGACCTTTCGCCATCAACTAACGATTCTGTCAAAAACTGACGCGTTGGATGAGGAGAAG
TGGCTTAATATGCTTGGCACGTTCGTCAAGGACTGGTTTAGATATGAGTCACATTTTGTT
CATGGTAGAGATTCTCTTGTTGACATTTTAAAAGAGCGTGGATTACTATCTGAGTCCGAT
GCTGTTCAACCACTAATAGGTAAGAAATCATGAGTCAAGTTACTGAACAATCCGTACGTT
TCCAGACCGCTTTGGCCTCTATTAAGCTCATTCAGGCTTCTGCCGTTTTGGATTAACCG
AAGATGATTTCGATTTTCTGACGAGTAACAAAGTTTGGATTGCTACTGACCGCTCTCGTG
CTCGTCGCTGCGTTGAGGCTTGCGTTTATGGTACGCTGGACTTTGTGGGATACCCTCGCT
TTCCTGCTCCTGTTGAGTTTATTGCTGCCGTCATTGCTTATTGTTATTATTGCTTATCTTGCTCAACA
TTCAAACGGCCTGTCTCATCATGGAAGGCGCTGAATTTACGGAAAACATTATTAATGGCG
TCGAGCGTCCGGTTAAAGCCGCTGAATTGTTCGCGTTTACCTTGCGTGTACGCGCAGGAA
ACACTGACGTTCTTACTGACGCAGAAGAAAACGTGCGTCAAAAATTACGTGCGGAAGGAG
TGATGTAATGTCTAAAGGTAAAAAACGTTCTGGCGCTCGCCCTGGTCGTCCGCAGCCGTT
GCGAGGTACTAAAGGCAAGCGTAAAGGCGCTCGTCTTTGGTATGTAGGTGGTCAACAATT
TTAATTGCAGGGGCTTCGGCCCCTTACTTGAGGATAAATTATGTCTAATATTCAAACTGG
CGCCGAGCGTATGCCGCATGACCTTTCCCATCTTGGCTTCCTTGCTGGTCAGATTGGTCG
TCTTATTACCATTTCAACTACTCCGGTTATCGCTGGCGACTCCTTCGAGATGGACGCCGT
TGGCGCTCTCCGTCTTTCTCCATTGCGTCGTGGCCTTGCTATTGACTCTACTGTAGACAT
TTTTACTTTTTATGTCCCTCATCGTCACGTTTATGGTGAACAGTGGATTAAGTTCATGAA
GGATGGTGTTAATGCCACTCCTCTCCCGACTGTTAACACTACTGGTTATATTGACCATGC
CGCTTTTCTTGGCACGATTAACCCTGATACCAATAAAATCCCTAAGCATTTGTTTCAGGG
TTATTGAATATCTATAACAACTATTTTAAAGCGCCGTGGATGCCTGACCGTACCGAGGC
TAACCCTAATGAGCTTAATCAAGATGATGCTCGTTATGGTTTCCGTTGCTGCCATCTCAA
AAACATTTGGACTGCTCCGCTTCCTCCTGAGACTGAGCTTTCTCGCCAAATGACGACTTC
TACCACATCTATTGACATTATGGGTCTGCAAGCTGCTTATGCTAATTTGCATACTGACCA
AGAACGTGATTACTTCATGCAGCGTTACCATGATGTTATTTCTTCATTTGGAGGTAAAAC
CTCTTATGACGCTGACAACCGTCCTTTACTTGTCATGCGCTCTAATCTCTGGGCATCTGG
CTATGATGTTGATGGAACTGACCAAACGTCGTTAGGCCAGTTTTCTGGTCGTGTTCAACA
GACCTATAAACATTCTGTGCCGACTTTCTTTGTTCCTGAGCATGGCACTATGTTTACTCT
TGCGCTTGTTCGTTTTCCGCCTACTGCGACTAAAGAGATTCAGTACCTTAACGCTAAAGG
TGCTTTGACTTATACCGATATTGCTGGCGACCCTGTTTTGTATGGCAACTTGCCGCCGCG
TGAAAATTCTATGAAGGATGTTTCCGTTCTGGTGATTCGTCTGAAGAAGTTTAAGATTGC
TGAGGGTCAGTGGTATCGTTATGCGCCTTCGTATGTTTCTCCTGCTTATCACCTTCTTGA
AGGCTTCCCATTCATTCAGGAACCGCCTTCTGGTGATTTGCAAGAACGCGTACTTATTCG
CCACCATGATTATGACCAGTGTTTCCAGTCCGTTCAGTTGTTGCAGTGGAATGTCAAGGT
TAAATTTAATGTGACCGTTTATCGCAATCTGCCGACCACTCGCGATTCAATCATGACTTC
GTGATAAAAGATTGAGTGTGAGGTTATAACGCCGAAGCGGTAAAAATTTAATTTTTGCC
GCTGAGGGGTTGACCAAGCGAAGCGCGGTAGGTTTTCTGCTTAGGAGTTTAATCATGTTT
CAGACTTTTATTTCTCGCCATAATTCAAACTTTTTTTCTGATAAGCTGGTTCTCACTTCT
GTTACTCCAGCTTCTTCGGCACCTGTTTTACAGACACCTAAAGCTACATCGTCAACGTTA
TATTTTGATAGTTTGACGGTTAATGCTGGTAATGGTGGTTTTCTTCATTGCATTCAGATG
GATACATCTGTCAACGCCGCTAATCAGGTTGTTTCTGTTGGTGCTGATATTGCTTTTGAT
GCCGACCCTAAATTTTTTGCCTGTTTGGTTCGCTTTGAGTCTTCTTCGGTTCCGACTACC
CTCCCGACTGCCTATGATGTTTATCCTTTGAATGGTCGCCATGATGGTGGTTATTATACC
GTCAAGGACTGTGTGACTATTGACGTCCTTCCCCGTACGCCGGGCAATAACGTTTATGTT
GGTTTCATGGTTGGTCAACTTTACCGCTACTAAATGCCGCGGATTGGTTTCGCTGAAT
CAGGTTATTAAAGAGATTATTTGTCTCCAGCCACTTAAGTGAGGTGATTTATGTTTGGTG
CTATTGCTGGCGGTATTGCTTCTGCTCTTGCTGGTGGCGCCATGTCTAAATTGTTTGGAG
GCGGTCAAAAAGCCGCCTCCGGTGGCATTCAAGGTGATGTGCTTGCTACCGATAACAATA
CTGTAGGCATGGGTGATGCTGGTATTAAATCTGCCATTCAAGGCTCTAATGTTCCTAACC
CTGATGAGGCCGCCCCTAGTTTTGTTTCTGGTGCTATGGCTAAAGCTGGTAAAGGACTTC
TTGAAGGTACGTTGCAGGCTGGCACTTCTGCCGTTTCTGATAAGTTGCTTGATTTGGTTG
GACTTGGTGGCAAGTCTGCCGCTGATAAAGGAAAGGATACTCGTGATTATCTTGCTGCTG
CATTTCCTGAGCTTAATGCTTGGGAGCGTGCTGGTGCTGATGCTTCGTTATCACAAGGA
TTGACGCCGGATTTGAGAATCAAAAAGAGCTTACTAAAATGCAACTGGACAATCAGAAAG
AGATTGCCGAGATGCAAAATGAGACTCAAAAAGAGATTGCTGGCATTCAGTCGGCGACTT
CACGCCAGAATACGAAAGACCAGGTATATGCACAAAATGAGATGCTTGCTTATCAACACA
AGGAGTCTACTGCTCGCGTTGCGTCTATTATGGAAAACACCAATCTTTCCAAGCAACAGC
AGGTTTCCGAGATTATGCGCCAAATGCTTACTCAAGCTCAAACGGCTGGTCAGTATTTTA
CCAATGACCAAATCAAAGAAATGACTCGCAAGGTTAGTGCTGAGGTTGACTTAGTTCATC
AGCAAACGCAGAATCAGCGGTATGGCTCTTCTCATATTGGCGCTACTGCAAAGGATATTT
CTAATGTCGTCACTGATGCTGCTTCTGGTGTGGTTGATATTTTTCATGGTATTGATAAAG
CTGTTGCCGATACGTTGGAACAATTTCTGGAAGACGGTAGAGTTGATGGTTGGTTCATC
ATTTGTCTAGGAAATAACCGTCAGGATTGACACCCTCCCAATTGTATGTTTTCATGCCTC
CAAATCTTGGAGGCTTTTTTATGGTTCGTTCTTATTACCCTTCTGAATGTCACGCTGATT
ATTTTGACTTTGAGCGTATCGAGGCTCTTAAACCTGCTATTGAGGCTTGTGGCATTTCATC
CTCTTTCTCAATCCCCAATGCTTGGCTTCCATAAGCAGATGGATAACCGCATCAAGCTCT
TGGAAGAGATTCTGTCTTTTCGTATGCAGGGCGTTGAGTTCGATAATGGTGATATGTATG
TTGACGGCCATAAGGCTGCTTCTGACGTTCGTGATGAGTTTGTATCTGTTACTGAGAAGT
TAATGGATGAATTGGCACAATGCTACAATGTGCTCCCCCAACTTGATATTAATAACACTA
TAGACCACCGCCCCGAAGGGGACGAAAAATGGTTTTTAGAGAACGAGAAGACGGTTCAGC
AGTTTTGCCGCAAGCTGGCTGCTGCTGAACGCCCTCTTAAGGATATTCGCGATGAGTATAATT
ACCCCAAAAAGAAAGGTATTAAGGATGAGTGTTCAAGATTGCTGGAGGCCTCCACTATGA
AATCGCGTAGAGGCTTTGCTATTCAGCGTTTGATGAATGCAATGCGACAGGCTCATGCTG
ATGGTTGGTTTATCGTTTTTGACACTCTCACGTTGGCTGACGACCGATTAGAGGCGTTTT
ATGATAATCCCAATGCTTTGCGTGACTATTTTCGTGATATTGGTCGTATGGTTCTTGCTG
CCGAGGGTCGCAAGGCTAATGATTCACACGCCGACTGCTATCAGTATTTTTGTGTGCCTG
AGTATGGTACAGCTAATGGCCGTCTCATTTCCATGCGGTGCACTTTATGCGGACAACTTC
CTACAGGTAGCGTTGACCCTAATTTTGGTCGTCGGGTACGCAATCGCCGCCAGTTAAATA
GCTTGCAAAATACGTGGCCTTATGGTTACAGTATGCCCATCGCAGTTCGCTACACGCAGG
ACGCTTTTTCACGTTCTGGTTGGTTGTGGCCTGTTGATGCTAAAGGTGAGCGCTTAAAG
CTACCAGTTATATGGCTGTTGGTTTCTATGTGGCTAAATACGTTAACAAAAAGTCAGATA
TGGACCTTGCTGCTAAAGGTCTAGGAGCTAAAGAATGGAACAACTCACTAAAAACCAAGC
TGTCGCTACTTCCCAAGAAGCTGTTCAGAATCAGAATGAGCCGCAACTTCGGGATGAAAA
TGCTCACAATGACAAATCTGTCTCACGGAGTGCTTAATCCAACTTACCAAGCTGGGTTACG
ACGCGACGCCGTTCAACCAGATATTGAAGCAGAACGCAAAAAGAGAGATGAGATTGAGGC
TGGGAAAAGTTACTGTAGCCGACGTTTTGGCGGCGCAACCTGTGACGACAAATGCTGCTCA
AATTTATGCGCGCTTCGATAAAAATGATTGGCGTATCCAACCTGCA
```

(a)

To be, or not to be: that is the question:
Whether 'tis nobler in the mind to suffer
The slings and arrows of outrageous fortune,
Or to take arms against a sea of troubles,
And by opposing end them. To die, to sleep—
No more—and by a sleep to say we end
The heartache, and the thousand natural shocks
That flesh is heir to! 'Tis a consummation
Devoutly to be wished. To die, to sleep—
To sleep—perchance to dream: ay, there's the
    rub,
For in that sleep of death what dreams may
    come
When we have shuffled off this mortal coil,
Must give us pause. There's the respect
That makes calamity of so long life:
For who would bear the whips and scorns of
    time,
Th' oppressor's wrong, the proud man's
    contumely,
The pangs of despised love, the law's delay,
The insolence of office, and the spurns
That patient merit of th' unworthy takes,
When he himself might his quietus make
With a bare bodkin? Who would fardels bear,
To grunt and sweat under a weary life,
But that the dread of something after death,
The undiscovered country, from whose bourn
No traveler returns, puzzles the will,
And makes us rather bear those ills we have,
Than fly to others that we know not of?
Thus conscience does make cowards of us all,
And thus the native hue of resolution
Is sicklied o'er with the pale cast of thought,
And enterprises of great pitch and moment,
With this regard their currents turn awry,
And lose the name of action.—Soft you now,
The fair Ophelia!—Nymph, in thy orisons
Be all my sins remembered.

(b)

Figure 1-26 GENETIC LANGUAGE AND HUMAN LANGUAGE: TWO MODES OF COMMUNICATION.
(a) The block of letters represents all the genetic information in the DNA of a virus that infects bacteria. That information is contained in the order of four kinds of molecules called nucleotide bases, whose names are abbreviated here with the four letters A,T,G, and C. (b) The most famous passage from Shakespeare's *Hamlet* exemplifies human language at its finest. Just as in genetic language, the information in human language is contained in the order of letters within words, and words within sentences. Due to the precise way molecules join, however, genetic language is unambiguous, while human language has richness and subtlety of meaning.

SUMMARY

1. Living things have a complex organization, take in and use energy, grow and develop, reproduce, show variations based on heredity, are adapted to their environments and ways of life, and are responsive.

2. *Life* is a particular set of processes resulting from the organization of matter.

3. Any particular organism is a product of the unique biological history that links that creature to its full lineage of ancestors.

4. Redi and Pasteur proved that *spontaneous generation* of live organisms is not possible in the modern world. Nevertheless, the unique set of physical and chemical conditions that prevailed on the early Earth was conducive to the spontaneous generation of the first living cells on our planet.

5. The Earth and its living inhabitants have undergone a lengthy *coevolution*. Living organisms have radically altered the lands, waters, and atmosphere, and the changing Earth has affected how and where organisms may live.

6. The descent of all contemporary living microbes, fungi, plants, and animals from the first cells on Earth explains why all organisms share many chemical constituents.

7. Darwin's *theory of evolution* states that all living things have evolved from a common ancestor over the millions of years of Earth's history.

8. Darwin's *theory of natural selection* states that natural events "select" organisms so that better-adapted ones tend to survive to reproduce, while less well-adapted ones tend not to contribute to subsequent generations.

9. *Cultural evolution* involves the transmission of information between generations by nongenetic means.

10. The *scientific method* begins with observations. They stimulate curiosity that leads to *inductive reasoning* and the development of a *hypothesis* to explain the facts and observations. Tests of a hypothesis include *control experiments*, which could support or disprove the hypothesis.

11. *Theories* are general statements designed to explain sets of related hypotheses that have withstood experimental verification. Theories lead to *deductive reasoning*, which involves the making of predictions about natural phenomena. If those predictions are borne out by experimental testing, the theory is strengthened further. If the predictions are not supported, the theory must be modified to account for the new observations.

12. Theories that withstand numerous, diverse tests may be called *natural laws*. The theory of evolution falls into this category.

13. The scientific method is useful in daily decision making and complements moral and ethical processes that are involved in the resolution of social problems.

KEY TERMS

coevolution
control experiment
cultural evolution
deductive reasoning
hypothesis
inductive reasoning
life
natural law
scientific method
spontaneous generation
theory
theory of evolution
theory of natural selection

SUGGESTED READINGS

DARWIN, C. *On the Origin of Species: A Facsimile of the First Edition*. Cambridge, Mass.: Harvard University Press, 1975.

This marvelous book makes fascinating reading and is a cornerstone of the literature of Western culture.

GARDNER, E. J. *History of Biology*. 3d ed. Minneapolis: Burgess, 1972.

This is one of the best histories of the science and the scientists.

HARRE, R. *Great Scientific Experiments*. Oxford: Oxford University Press, 1983.

Twenty experiments, including those of Aristotle, Newton, and Pasteur; the flow of sap in plants; the process of digestion; and much more—an ideal way to learn how science is conducted.

PITTENDRIGH, C. S. In *Life*, edited by G. G. Simpson, C. S. Pittendrigh, and L. H. Tiffany. New York: Harcourt Brace Jovanovich, 1957.

Chapter 2, on the scientific method, has been used as the text for whole courses on the subject— marvelously clear and simple.

Part ONE

FROM ATOMS TO CELLS

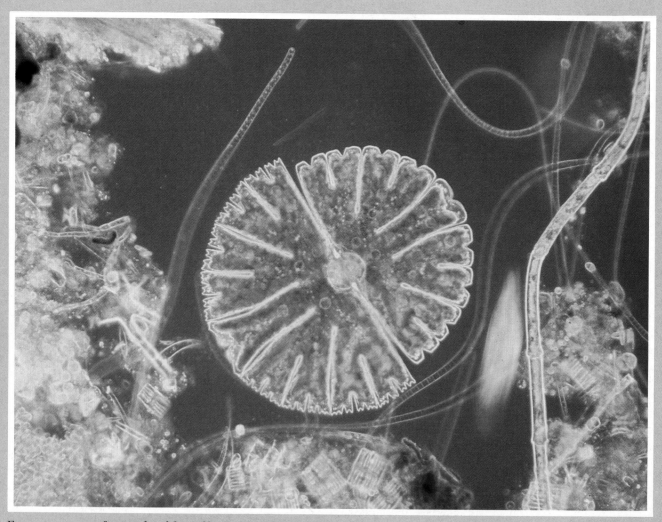

From atoms come form, color, life itself—in this case, a desmid, a free-floating inhabitant of the sea surface.

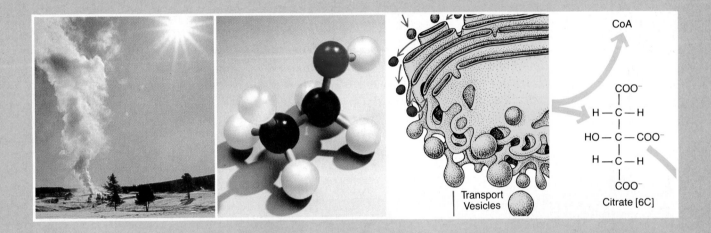

Transport
Vesicles

CoA

COO⁻
|
H — C — H
|
HO — C — COO⁻
|
H — C — H
|
COO⁻

Citrate [6C]

A living thing is an assemblage of biological materials, some tough, some exceedingly delicate, and most precisely structured and highly ordered. An organism is also a living test tube in which these ordered materials undergo a fantastically complicated set of chemical reactions, with old substances breaking down and new ones assembling, moving, and changing. The materials that make up a living thing are inseparable from these constant, life-sustaining chemical activities; and underlying both are atoms and energy. Atoms are the building blocks of all matter—biological and nonbiological—and energy powers atomic interactions, including the complex ones that bring about metabolism, growth, reproduction, and the other fundamental properties of all living things. To understand these properties, we must begin with the chemistry of life.

Within any organism, the basic compartments—the units that actually live—are the cells. We will study the biological molecules that make up cell parts, the roles each part plays in the cellular economy, how the parts function individually and in concert, and how energy and materials move into and out of the cell. We will see how cells can retrieve and use the energy stored in nutrient molecules to maintain their structural integrity and fuel their life processes. And we will discuss the way plants and other photosynthetic organisms capture and convert the energy in sunlight and store it in the nutrients all living things rely on, directly or indirectly.

It will be evident, again and again, that life obeys the laws of the physical universe. Atoms behave in the same ways and energy flows in the same manner, whether part of an organism or a speeding meteorite. In living things, however, matter is organized in increasingly complex arrays: atoms combine in molecules; molecules assemble into cell parts; these parts form living units; cells may be joined in interdependent "colonies"—a sponge, an oak, a spider monkey; and those "colonies" may be parts of still higher-order populations—reefs, forests, troops. With each broader level of organization, new properties emerge; yet each new level and property can be better understood because we have studied the chemistry of life.

2

ATOMS, MOLECULES, AND LIFE

Biology has been fortunate in discovering within the span of one hundred years two great and seminal ideas. One was Darwin's and Wallace's theory of evolution by natural selection. The other was the discovery by our own contemporaries of how to express the cycles of life in a chemical form that links them with nature as a whole.

Jacob Bronowski, *The Ascent of Man* (1974)

Ice crystals in the Olympic Forest: Assemblies of molecules of life's precious solvent, water.

Perhaps the most striking characteristic of life on our planet is its diversity. Twenty-two centuries ago, Aristotle began cataloging animals and plants that inhabited the areas around the Mediterranean. Since then, biologists have identified more than 250,000 species of flowering plants; 20,000 types of mosses; almost 50,000 kinds of fish, amphibians, reptiles, birds, and mammals; and nearly 1 million species of insects. In total, there are at least 1.5 million species on Earth, and some biologists estimate that 2 or 3 million more remain to be identified in tropical zones.

What accounts for this grand diversity? The natural philosophers of early science believed that all living things were divine creations, made up of unique matter and sparked by a vital force. That belief changed after eighteenth-century chemists proved that all matter, living and nonliving, is composed of particles called atoms. The discovery of atoms had a profound and permanent effect on the study of biology, as well as on that of chemistry and physics. Over time, biologists discovered that every organism contains the same types of atoms arranged in different ways. Thus the diversity of life is a consequence of the myriad ways that the basic units of matter combine and interact.

Chemical interactions underlie all life activities—harvesting energy from the sun, breaking down food particles, moving, growing, reproducing, and so on—so it is appropriate to begin our study of biology with atoms. In this chapter, we will discuss the structure of atoms and the ways in which these building blocks of all matter are assembled into more complex structures. We will see how they interact in what are called *chemical reactions*. And finally we will consider the structure and properties of water, a chemically unique substance without which life as we know it could not exist.

ELEMENTS AND ATOMS: BUILDING BLOCKS OF ALL MATTER

Working in the late 1700s and early 1800s, the French chemist Antoine Lavoisier, the English chemist John Dalton, and others set forth the two basic principles of chemistry.

1. All matter, living and nonliving, is made up of **elements,** substances that cannot be decomposed by chemical processes into simpler substances. Chemists have identified ninety-two chemical elements in nature and have created thirteen more in the laboratory. Some examples of elements are hydrogen (symbolized H), oxygen (O), sulfur (S), gold (Au), iron (Fe), and carbon (C).

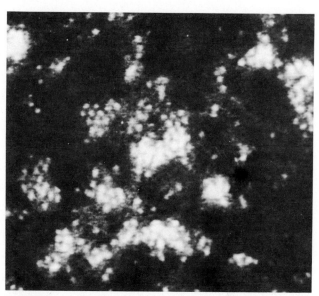

Figure 2-1   ATOMS: THE FUNDAMENTAL UNITS OF ELEMENTS.
This electron micrograph of uranium atoms is magnified about 7 million times; each small dot is an atom.

2. Each element is composed of identical particles called **atoms,** which are the smallest units of matter which still display the characteristic properties of the element. Atoms are so minute that they have been seen on only rare occasions with an electron microscope (Figure 2-1). All the atoms in a given sample of an element, such as a brick of pure gold, are identical to one another but different from all the atoms in samples of other elements, such as a lump of carbon or an ingot of iron. The properties of each element, such as weight, density, color, and so on, are based on the structure of its individual atoms, which we will discuss shortly.

The Elements of Life

A natural question arose from the pioneering work of Lavoisier and Dalton: are living and nonliving things made up of the same groups of elements? Are we truly "dust from dust," or is our chemical make-up different from that of rocks, planets, and stars? Living things share a special subset of the ninety-two naturally occurring elements, but if we compare the subset of the atoms of life with atoms in the Earth's crust, we find them in very different proportions. The most abundant elements in the crust are oxygen, silicon (Si), and aluminum (Al), while living organisms are made up predominantly of hydrogen, oxygen, and carbon (Figure 2-2). While fully 98 percent of the atoms in the Earth's crust are the elements oxygen, silicon, aluminum, iron, calcium (Ca),

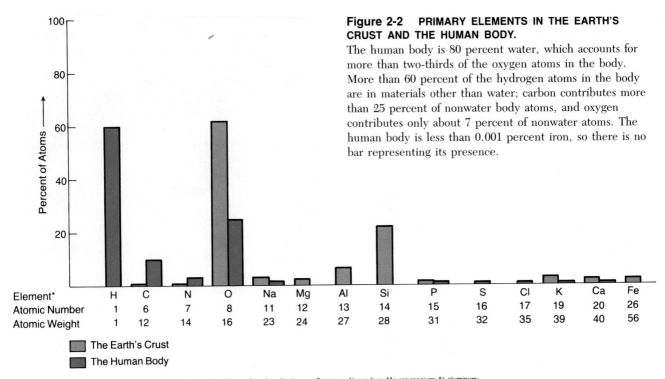

Figure 2-2   PRIMARY ELEMENTS IN THE EARTH'S CRUST AND THE HUMAN BODY.
The human body is 80 percent water, which accounts for more than two-thirds of the oxygen atoms in the body. More than 60 percent of the hydrogen atoms in the body are in materials other than water; carbon contributes more than 25 percent of nonwater body atoms, and oxygen contributes only about 7 percent of nonwater atoms. The human body is less than 0.001 percent iron, so there is no bar representing its presence.

Element*	H	C	N	O	Na	Mg	Al	Si	P	S	Cl	K	Ca	Fe
Atomic Number	1	6	7	8	11	12	13	14	15	16	17	19	20	26
Atomic Weight	1	12	14	16	23	24	27	28	31	32	35	39	40	56

The Earth's Crust
The Human Body

*The symbols for the elements are as follows: H: hydrogen; C: carbon; N: nitrogen; O: oxygen; Na: sodium; Mg: magnesium; Al: aluminum; Si: silicon; P: phosphorus; S: sulfur; Cl: chlorine; K: potassium; Ca: calcium; Fe: iron.

sodium (Na), potassium (K), and magnesium (Mg), 99 percent of the atoms in a typical organism are the markedly different subset carbon, hydrogen, nitrogen (N), and oxygen, with sodium, calcium, phosphorus (P), and sulfur making up most of the remaining 1 percent. Still others are present in trace amounts (Figure 2-3). (A handy acronym for the six basic elements in living things is SPONCH.)

Biologists are not certain why the chemical subsets of living and nonliving things are so different, but the structure and properties of the atoms of their constituent elements provide important clues.

Let us now look at the way atoms are structured in order to understand how their architecture determines the physical properties of elements and, in turn, the properties of live organisms.

Atomic Structure

Atoms are extremely small: about 3 million atoms sitting side by side probably would cover the period at the end of this sentence. The physicist Gerald Feinberg once calculated that there are more atoms in the human body than there are stars in the known universe. Although minuscule in size, each atom is made up of three types of subatomic particles: protons, neutrons, and

Figure 2-3   IRON IS A COMMON TRACE MINERAL.
Iron gives the red color to the blood of many animals, and it sometimes accumulates in cells. Here, a large crystal of iron-rich ferritin (magnified about 10,000 times) is found in a cell of an insect's gut. To be useful biologically, iron must dissolve from this crystal and become associated with biological molecules.

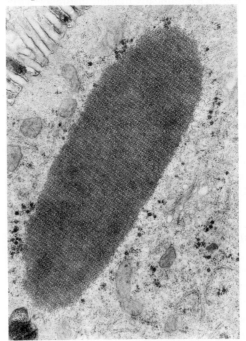

electrons. **Protons** have a positive (+) charge; **neutrons** have no electrical charge (they are neutral); and **electrons** have a negative (−) charge. Since these subatomic particles are only parts of atoms, none of them displays properties of elements. The protons and neutrons are clustered in a small dense body at the center of the atom called the *nucleus* (the diameter of an atom is about 100,000 times larger than the nucleus). The outer limits of the atom are defined by the paths of its electrons, which are continuously moving about the nucleus.

Electrons, protons, and neutrons are themselves made up of a dozen or more smaller subatomic particles. Nuclear physicists study these particles and the forces that hold together these elementary building blocks of matter.

Protons and Neutrons: The Nucleus

Atoms of each kind of element—carbon, oxygen, and so on—are composed of different numbers of protons, neutrons, and electrons. The atoms of each element have a unique number of protons in their nuclei: carbon has six, for example, and oxygen has eight. This unique number of protons is called the **atomic number** of the element. Since each proton carries a positive charge and neutrons have no electrical charge, the atom's nucleus has an overall positive charge equal in magnitude to the number of protons it contains; thus the nucleus of a carbon atom has a charge of +6, and the nucleus of an oxygen atom has a charge of +8. Protons and neutrons have about the same mass, and under normal conditions, each atom of an element usually has approximately the same number of protons and neutrons in its nucleus. The sum of an atom's neutrons and protons is called its *atomic mass*. Thus carbon, with six protons and six neutrons, has an atomic number of 6 and an atomic mass of 12, and hydrogen, with one proton and no neutrons, has the atomic number 1 and an atomic mass of 1. We will use the term **atomic weight** synonymously with that of "atomic mass," recognizing that weight is due to the attraction of gravity and so would be less on the moon, say, than on Earth.

The chemical behavior of atoms primarily involves the interaction of electrically charged particles and is not affected by the uncharged neutrons. However, neutrons do impart a property to atoms that scientists have found particularly useful. Whereas the number of protons in atoms of a particular element always remains the same, the number of neutrons can vary. Most natural samples of elements are, in fact, mixtures of atoms that contain identical numbers of protons but different numbers of neutrons. These atoms have the same atomic number but different atomic weights and are called **isotopes** of the element. One isotope that you may know is carbon

14, in which the carbon atom contains eight neutrons, giving it an atomic weight of 14 instead of 12 (written as ^{14}C). Some isotopes, including ^{14}C, are *radioactive;* they emit energy that can be detected.

Modern medicine and biology would be in a vastly more primitive state were it not for the use of isotopes to chemically mark and trace biological molecules. For instance, radioactively tagged building blocks of proteins are built into newly synthesized proteins at one site in cells; those tagged proteins then can be traced through other cellular compartments prior to being secreted (see Box). Isotopes also permit dating of certain biological and geological events. Thus isotopes that decay very slowly, such as isotopes of chlorine or potassium, can be measured in layers of rock containing fossil plants or animals, and these rocks and fossils can be dated.

Electrons, Atomic Orbitals, and Energy Levels

Now that we have described the nucleus of the atom, let us turn to the electrons, the third type of subatomic particle. In an atom, the number of electrons equals the number of protons in the nucleus. Thus an electrically neutral carbon atom has six electrons, and an oxygen atom has eight electrons. The negatively charged electrons are very much smaller than are protons and neutrons and add little to the mass of an atom (Table 2-1). In addition, electrons are in constant motion. Because each electron's negative charge is equal to the positive charge of each proton, an isolated atom of any element is said to be electrically balanced, or neutral.

Table 2-1   **PROPERTIES OF SUBATOMIC PARTICLES**

Particle	Mass (dalton)*	Weight (gram)	Electrical Charge
Proton	1.00728	1.673×10^{-24}	+1
Neutron	1.00867	1.675×10^{-24}	0
Electron	0.000486	9.1095×10^{-28}	−1

*One dalton equals about 1.65×10^{-24} g, or one-twelfth the mass of an atom of carbon 12.

Electrons are so small that if we could collect and weigh them, just 1.0 gram (g) of electrons (about the weight of three aspirin tablets) would contain 10^{26} electrons (10^{26} is a convenient notation for multiplying $10 \times 10 \times 10 \ldots$, twenty-six times). This is a very large number; the radius of the known universe is 10^{25} meters (m). The diameter of an electron is vanishingly small—about 10^{-12}, or 0.000 000 000 001, centimeter (cm). One German chemist uses the example that if an artisan had started threading electrons onto a necklace in 3000 B.C., adding one electron every second for eight hours a day,

ATOMS IN MEDICINE

In 1895, the German physicist Wilhelm Röntgen started a medical revolution with his discovery of x-rays. He found that this energetic form of radiation could pass through an opaque object, such as his own hand, expose a photographic plate, and thereby reveal dense structures inside the object, such as Röntgen's bones (see Figure A). Doctors suddenly had a window on the human body never before possible, and within a few weeks of Röntgen's great discovery, x-rays were being used to diagnose broken bones.

Today, another medical revolution is under way, based on a host of new and sophisticated imaging techniques. They are more complicated than simple x-rays and have curious sounding names like CT, NMR, and PET. But like x-rays, these techniques are based on the physical properties of atoms, and promise to make the human body practically transparent to the diagnostician.

Computerized tomography, or CT, is an imaging technique that combines x-rays and computers. With CT, a physician can make high-resolution images based on a series of x-rays through cross sections, or "slices," of any part of the living body. CT can reveal tumors, blood clots, or other abnormalities in soft tissues as well as in bones. CT has been useful in the early detection of cancer. However, exposure to x-rays carries its own health risks.

Nuclear-magnetic-resonance imaging, or NMR, uses a harmless process to create pictures of the interior of the body. The technique is based on the fact that protons in the nuclei of certain atoms spin like tops and create small magnetic fields. Placing a sample of a chemical such as hydrogen near a magnet can cause all the axes of the individual spinning "tops" to align along the lines of magnetic force (Figure B-a). The introduction of a second magnetic field with lines of force perpendicular to the first causes the spinning protons to flip over (Figure B-b); they flip back again when the second magnet is removed. As they return to their original position, the protons give off a faint signal that can be detected and can reveal the presence and concentration of the chemical (Figure B-c). NMR devices have been built that are large enough to surround the human body with ring-shaped magnets (Figure B-d). The patient is exposed to only a magnetic field from these magnets and to a second field produced by pulses of harmless low-energy radio waves. The signals created by the flipping of protons in hydrogen atoms enable doctors to watch blood (which is mainly H_2O) flow through vessels in tissues

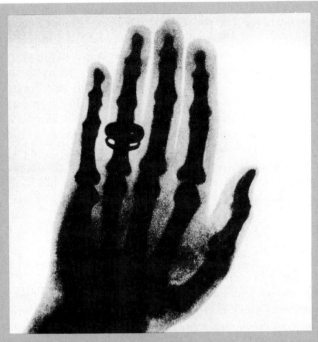

Figure A
Wilhelm Röntgen, the discoverer of a form of radioactivity called x-radiation, took the world's first x-ray of his own hand in 1895. Energetic x-radiation can pass through opaque objects, such as skin and muscles, and expose a photographic plate.

and organs. In this way, they can diagnose blood clots, tumors, clogged vessels, cancers in deeply embedded tissues, and damage to nerves.

The next generation of NMR imagers will routinely detect the flipping of protons in phosphorus as well as in hydrogen. Phosophorus is present in molecules that supply chemical energy for the contraction of muscles and most other life processes. Thus doctors will be able to study subtle changes in the muscles, stomach, brain, and other organs without cutting, irradiating, or otherwise harming the tissues.

The technique called positron-emission tomography, or PET, promises a still clearer window into the body and its biological processes. Positrons are positively charged particles equal in mass to electrons; they are given off by isotopes of certain elements. When positrons collide with electrons, they release a form of energy that can be detected, thus pinpointing the location of the collision. Radioactive isotopes that give off positrons can be injected into a patient and then such collisions detected with an array of scanners that surround the patient's body. In this way, the types and rates of important chemical reactions can be

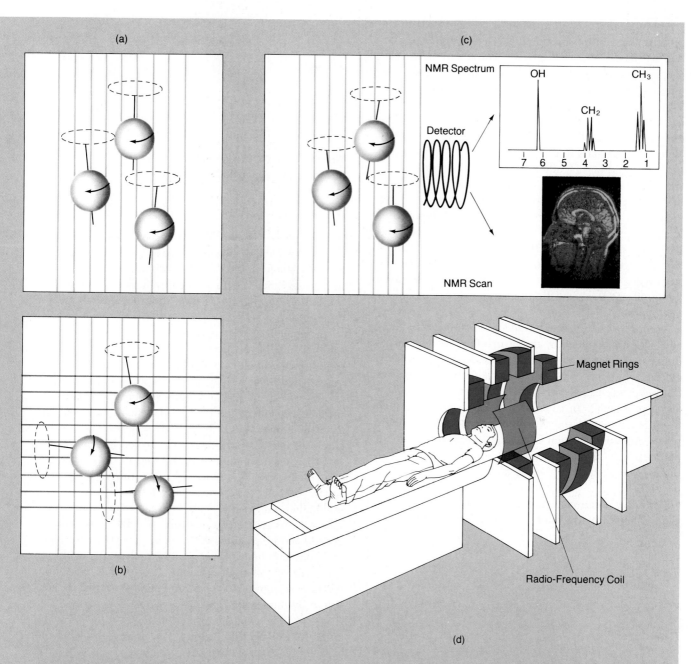

(a)

(c)

NMR Spectrum

OH   CH₃

Detector

CH₂

7 6 5 4 3 2 1

NMR Scan

(b)

Magnet Rings

Radio-Frequency Coil

(d)

studied *as they occur* in the body. Physicians hope that PET will enable them to diagnose diseases of the brain and heart in their earliest stages and to probe the underlying causes of cancer and aging.

These new techniques are an indication of the long way that medical technology has come since Wilhelm Röntgen first saw an image of the bones in his hand. The techniques undoubtedly will become common tools of the physician in the decades to come and make dramatic contributions to the understanding of basic biological processes.

Figure B
NMR imaging reveals the soft and hard tissues of the human head to help detect diseases and abnormal growths. (a) Protons align in a magnetic field. (b) Protons flip when a second magnetic field is superimposed. (c) When the second field is removed, protons flip back and give off a faint signal that can be detected and turned into images of the body. (d) The powerful NMR imaging magnet surrounds the patient. The smaller radio-frequency coil generates the signal that flips protons so that they can be detected.

the length of the necklace today would be only 0.2 millimeter (mm) (less than 0.01 inch)!

Although the mass of an atom is concentrated in the protons and neutrons in the nucleus, the atom's properties are based on the electrons. These tiny negatively charged particles occupy a *volume* as they zip about the nucleus. This volume is much larger than the nucleus alone: if a hydrogen atom could be magnified to the size of a bushel basket, the nucleus in its center would be like an extremely small grain of sand, yet it would contain almost all the mass of the entire atom. If the hydrogen nucleus could be magnified to the size of an orange, its single electron would be zipping around it in volume having a radius of one-third of a mile.

We have said that electrons move around the nucleus of the atom. But why don't they fly off into space? And what path do they follow? It is the electrical attraction of the positively charged protons in the nucleus and the negatively charged electrons that prevents them from flying off. At the same time, electrons are repelled by one another.

According to the accepted model of atomic structure, electrons are confined in **atomic orbitals,** specific three-dimensional zones around the nucleus. The path of an electron cannot be precisely defined; in other words, we can never say exactly where an electron will be or how fast it is moving. In contrast, the Earth moves around the sun at a regular speed in an orbit confined to a single plane. An atomic orbital is thus best defined as a cloudlike region in which there is a 90 percent probability of finding the electron in motion around the nucleus (Figure 2-4).

Only two electrons can occupy an orbital at any one time. Therefore, the more electrons an atom has, the

Figure 2-4   ATOMIC ORBITAL.
An atomic orbital is a three-dimensional region surrounding an atom's nucleus in which there is a 90 percent probability an orbiting electron will be found. The nucleus of this hydrogen atom, in the center of the cloudlike orbital, has a diameter of about one ten-thousandth the diameter of the entire atom. The probability of finding an electron is greatest in the heavily shaded region immediately surrounding the nucleus and falls off as the distance from the nucleus increases.

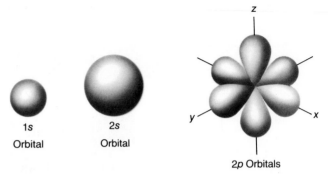

Figure 2-5   ATOMIC ORBITALS.
Here are orbitals for the first and second energy levels.

more orbitals it possesses. Atoms with more than two electrons have a series of orbitals at increasing distances from the nucleus. Each atomic orbital and the electrons in it are associated with a specific amount of energy, and the farther an electron is from the nucleus, the greater its energy.

An orbital that contains two electrons—the maximum number possible—is said to be filled. The orbital that is closest to the nucleus is the first to fill. This orbital, the one at the lowest energy level, is spherical and is called the 1s orbital. At the next higher energy level are four orbitals, capable of holding a total of eight electrons. The 2s orbital, like the 1s orbital, is spherical. The other three second-level orbitals, each shaped like an inflated ball that has been squeezed and tied in the center, are termed 2p orbitals (Figure 2-5). The third energy level has the capacity to accommodate eighteen electrons, and still higher energy levels can hold increasing numbers of electrons in greater numbers of orbitals. This complex system of orbitals allows electrons to keep their distance from one another and yet to stay as close as possible to the positively charged nucleus.

The distribution of electrons among orbitals is governed by four basic rules.

1. An electron will occupy the lowest available energy level.

2. Lower energy level orbitals are filled completely before higher energy level orbitals are occupied.

3. In any one energy level, a simpler orbital will be filled before an orbital of more complex shape is occupied.

4. Orbitals of similar shape at the same energy level must have one electron each before any of them can be filled.

Let us see how these rules apply to the atoms of specific elements, starting with the simplest element, hy-

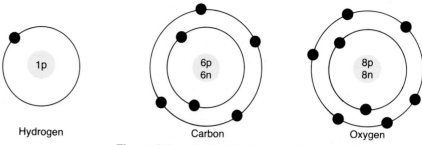

Figure 2-6   BOHR MODELS OF SOME ATOMS.
Bohr models show clearly the energy levels and the number of subatomic particles in each atom, but they depict an overly simplified picture of the locations of the electrons. Electrons are not found in one definite location a specific distance from the nucleus, but travel rapidly in a three-dimensional space around the nucleus in a random fashion.

drogen, and working our way up through elements with more and more electrons. These rules and their application stem from the original work of the Danish physicist Niels Bohr, who explained how electrons could exist in energy levels around hydrogen nuclei. His work led to the use of so-called *Bohr models* to depict the distribution of electrons in orbitals (Figure 2-6).

Hydrogen (H), atomic number 1, has a nucleus consisting of only one proton. It is the only atom with no neutrons. Hydrogen's single electron occupies the 1s orbital—the orbital closest to the nucleus and of simplest (spherical) shape. The next larger atom is helium (He), atomic number 2, with two protons and two electrons. The second electron completes the 1s orbital. Helium is an extremely stable element; it does not react with other elements. All elements whose outermost energy level is completely filled with electrons are essentially like helium in that they tend to be chemically *inert* and quite unreactive with other elements. This is one way in which the distribution of electrons in the outermost energy level affects the chemical behavior of atoms.

Lithium (Li), atomic number 3, has three electrons. The first two occupy the 1s orbital, completing the first energy level, and the third occupies the simplest orbital of the second energy level—the spherical 2s orbital (Figure 2-7). In beryllium (Be), atomic number 4, the fourth electron completes the 2s orbital. (Remember that according to rule 3, the orbital having the simplest shape must be filled before the more complex orbitals can be

occupied, and the 2s is spherical, whereas the 2p orbitals are shaped like squashed balls.) In boron (B), atomic number 5, the fifth electron is in one of the 2p orbitals, and in carbon (C), the sixth electron occupies the second 2p orbital. In nitrogen (N), atomic number 7, the seventh electron occupies the third 2p orbital. Only when each of the three 2p orbitals has one electron does doubling up begin. In oxygen (O), the eighth electron fills the first 2p orbital; in fluorine (F), the ninth fills the second 2p; and in neon (Ne), the tenth fills the third 2p orbital. Like helium, neon is inert, since its outermost energy level (level 2) is completely filled.

If we continued our orbit-filling exercise, by moving

Figure 2-7   THE FILLING OF ELECTRON ORBITALS FOR THE FIRST EIGHTEEN ELEMENTS.
The orbitals are shown as small circles, the electrons are depicted as dots, and filled orbitals (those with two electrons) are yellow. With the exception of hydrogen, atoms in the same column have identical electronic configurations in their outermost orbitals and therefore have similar chemical properties.

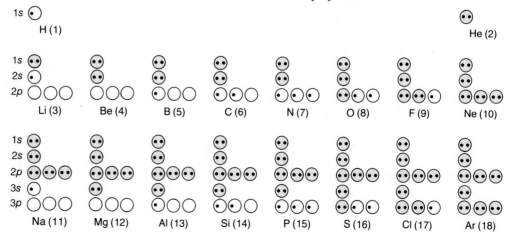

on to the third energy level, then the fourth, and so on, we would see several patterns emerge. For example, the so-called *inert gases*—helium, neon, and argon (Ar)—have filled outer energy levels. The larger atoms of krypton (Kr), xenon (Xe), and radon (Rn) also have filled outer energy levels and are relatively inert chemically; consequently, these gases also are referred to as inert gases. Lithium, sodium (Na), and potassium (K) all have one electron in the outermost orbital and have similar properties; they appear as whitish metals and are highly reactive. The arrangement of the outermost electrons also accounts for the reactivity of these and other elements, for, as we shall see, the interaction of these outermost electrons brings about the formation of diverse substances. These substances, in turn, give shape, size, color, and function to the myriad forms of living things.

MOLECULES AND COMPOUNDS: AGGREGATES OF ATOMS

Moving along the continuum of size and complexity from protons, neutrons, electrons, and atoms, we shall now study groups of atoms. These groups form the building blocks of all living things, as well as those of most nonliving matter.

In nature, atoms link up with other atoms in various ways. Two or more atoms bound together comprise a **molecule.** Identical atoms may be bound to each other—such as oxygen to oxygen, nitrogen to nitrogen (the O_2 and N_2 in the air we breathe), or carbon to carbon, as in the extremely hard crystal we call a diamond. But dissimilar atoms also can combine, as carbon and oxygen do to form the poisonous gas carbon monoxide (CO). Chemical **compounds** contain atoms of more than one element and can be decomposed into these elements. An example is CO_2, carbon dioxide. Molecules and compounds display properties not found in the constituent elements; thus, the gases hydrogen and oxygen form the compound water (H_2O), with its unique characteristics.

How do individual atoms, with their dense nuclei and orbiting electron "clouds," interact to form molecules? What holds them together? And what determines the chemical properties of molecules and compounds? The answers lie in the behavior of orbiting electrons.

Chemical Bonds: The Glue That Holds Molecules Together

Consider hydrogen gas, a trace component of the air around you. Hydrogen usually exists in the form of H_2

molecules. A hydrogen molecule is more stable than two separate hydrogen atoms; energy therefore has to be expended in order to break a hydrogen molecule into its component atoms.

Why are H_2 molecules more stable than individual H atoms? Recall from the last section that the most stable arrangement of electrons is found in the inert gases, in which the outermost orbital is filled. Two hydrogen atoms, each of which has one electron, can *share* the electrons so that each effectively has two electrons in the 1s orbital, thereby completing it and establishing the most stable arrangement. As the atoms approach each other, each nucleus begins to attract the electron held by the other nucleus (Figure 2-8). Eventually, the electron clouds overlap and fuse into one **molecular orbital.** Like an atomic orbital, a molecular orbital is most stable when filled by a pair of electrons. This shared orbital acts as a *chemical bond* between the two atoms and resembles a strong spring in its properties: it can be compressed or stretched to a certain extent without breaking. An atom can form as many bonds as there are unpaired electrons in its outermost orbital. The unpaired electrons involved

Figure 2-8   THE FORMATION OF NONPOLAR COVALENT BONDS.

When the electrons of two hydrogen atoms fuse into one molecular orbital, they form H_2, molecular hydrogen.

in bond formation are called *valence electrons*, and the orbital in which they are found is often referred to as the *valence shell*.

The bond between two atoms of hydrogen is called a *covalent bond*. There are two other types of chemical bonds, *ionic* and *polar*, which differ from covalent bonds in the way electrons are distributed around the nuclei of paired atoms. As we shall now see, the kind of bond that holds together a molecule or compound has a direct bearing on the substance's properties.

Covalent Bonds

Because the bond between two H atoms involves shared electrons, it is called a **covalent bond** (the prefix *co-* indicates a shared condition). Hydrogen can also share its electron—that is, form a covalent bond—with an atom of another element. One biologically important element to which hydrogen can bond is carbon. As we have noted, carbon has six electrons: two completely fill the first energy level, and the remaining four fill orbitals at the second energy level, which can accommodate a total of eight electrons; these four at the second energy level are carbon's valence electrons. Thus carbon is "looking" for four additional electrons to fill its $2p$ orbitals and give it maximum stability. The unfilled orbitals of four hydrogen atoms can fuse with the unfilled orbitals of one carbon atom to form four molecular orbitals with a shared electron pair in each, so that all five atoms (one carbon plus four hydrogens) achieve a stable energy state in which their outermost orbitals are filled. Because of the directionality of carbon's $2p$ orbitals, the four molecular orbitals of CH_4, methane, point to the corners of a regular tetrahedron (a three-dimensional structure with four triangular sides) with the carbon atom at the center, as shown in Figure 2-9. Methane is the primary component of marsh gas.

The kinds of bonds formed in CH_4 are *single bonds*, meaning that only one pair of electrons is shared between two atoms (C and H). But two atoms can share two or three pairs of valence electrons, forming *double* or

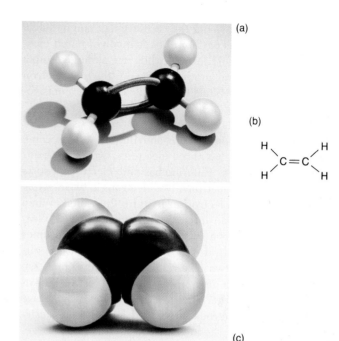

(a)

(b)

(c)

Figure 2-10   THE STRUCTURE OF ETHYLENE.
C_2H_4 can be represented in different ways. (a) The ball-and-stick structure depicts the length and angles of the carbon-hydrogen bonds. (b) In this structural model, each electron pair is represented by a line. The double bond consists of four electrons. (c) Space-filling models depict the three-dimensional space occupied by molecules, including the electron orbitals, but show little about the bonds.

triple bonds. Carbon atoms often form double bonds. For example, in ethylene (C_2H_4), an energetically stable compound, two carbon atoms share two pairs of electrons. Two hydrogen atoms also form covalent bonds with each carbon (Figure 2-10). Ethylene is a naturally occurring molecule that plays a major role in fruit ripening.

In single bonds, the two connected atoms can rotate around the bond much as wheels rotate on an axle. In double bonds, the connected atoms cannot rotate; in ethylene, for example, the two carbon atoms are fixed in space relative to each other. In compounds with triple bonds, such as N≡N (N_2, or nitrogen gas), three electron pairs are shared, the bonds are stronger than a double bond, and the atoms are even more rigidly fixed in space. In Chapter 3, we will see how this structural rigidity affects the behavior of biologically important molecules.

Ionic Bonds

Covalent and ionic bonds form for the same reason—the resulting molecules are more stable units of matter

Figure 2-9   THE STRUCTURE OF METHANE.
CH_4 is held together by four covalent bonds. The orbital structure shows the overlap of atomic orbitals.

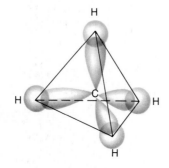

than the free atoms—but they achieve that end in different ways. Whereas covalent bonds involve shared electrons, **ionic bonds** are formed when one atom gives up a valence electron and another atom adds the free electron to its outermost orbital, thereby holding the atoms together in an energetically stable unit.

To understand how ionic bonds form, let us consider what happens when lithium and fluorine atoms come together. Lithium has three electrons: two 1s electrons and one 2s electron. Fluorine has nine electrons: two 1s electrons, two 2s electrons, two electrons in two of its 2p orbitals, and one electron in its third 2p orbital. When lithium and fluorine interact, lithium donates an electron to fluorine. As a result, the lithium atom becomes positively charged; it has three protons but is surrounded by only two electrons. At the same time, the fluorine atom, with nine protons and ten electrons, takes on a negative charge. Such charged atoms, which have lost or gained electrons, are called **ions** and are designated by a + or a − to indicate the nature of the charge. In our example, the ions are Li^+ and F^-. (Ions of oxygen and magnesium are designated O^{2-} and Mg^{2+}, respectively, to indicate that they have two valence electrons.)

The strong attraction between positive and negative ions gives rise to an ionic bond, and in the case of lithium and fluorine, to the ionic compound lithium fluoride (LiF). Without its lone 2s electron, lithium fulfills the requirements for a stable atom (like helium), and fluorine now has a full complement of eight electrons at its third energy level and thus is also stable (Figure 2-11).

Ionic bonds hold together many common compounds, such as NaCl sodium chloride (table salt). When ionic compounds are dissolved in a solvent (for example, when salt is dissolved in water for cooking spaghetti), they may *dissociate*, or break down, and free ions (Na^+ and Cl^- or Li^+ and F^-, for example) then would be found in the solution. If the solution is dried by evaporating the water, the ions reassociate to form crystals in which the ions are once again bonded together. Ions play an important role in many biological processes, as we shall see throughout this book.

Polar Bonds

Covalent bonds and ionic bonds are like opposite sides of a coin when it comes to sharing electrons: covalently bonded atoms share electrons equally, and ionically bonded atoms do not share electrons at all. Many bonds fall between these two extremes, however, and are characterized by a partial transfer of electrons. That is, electrons are shared, but they tend to spend more time orbiting one nucleus than orbiting the other (Figure 2-12). Such bonds are called **polar bonds**. In a polar bond, the electrical charge from the cloud of moving electrons is asymmetrical; one atom is slightly negatively charged

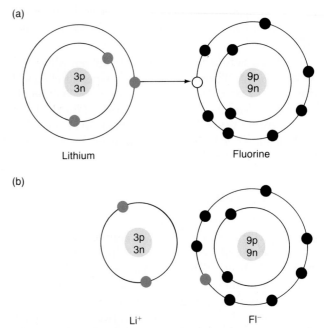

Figure 2-11   BOHR MODELS OF AN IONIC BOND.
Lithium donates an electron to fluorine; the latter takes on a net negative charge while the lithium has a positive charge.

and one is slightly positively charged, so the electrons are more likely to be found near the nucleus of the negative atom than near the nucleus of the positive atom.

The ability to attract electrons from other atoms in a molecule is called **electronegativity.** If a bond forms between two atoms with differing degrees of electronegativity, the shared electron will be located most of the time near the nucleus of the more electronegative atom. Such a bond is said to be *polarized*. In a polar molecule such as H_2O, for example, the electrons spend more time near the oxygen atom than near the two hydrogens; this leaves the H atoms somewhat positively charged and the O atom somewhat negatively charged. This is unlike

Figure 2-12   THE POLAR BONDS OF A WATER MOLECULE.
The electrons spend more time orbiting the oxygen nucleus than orbiting the two hydrogen nuclei, leaving the oxygen more negatively charged and the hydrogen more positively charged.

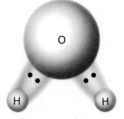

the situation in covalent bonds such as C–H, in which the electrons are distributed symmetrically and are equally likely to reside near the H or the C nucleus.

You can see that there is no sharp distinction between covalent and ionic bonds, but rather a gradient between them. An ionic bond can actually be thought of as an extreme form of polar bonding in which there is no sharing of electrons. The gradient of polarity in chemical bonds has a particularly interesting effect on the behavior of water molecules, as we will see a bit later.

Bond Strength

Now that we have some understanding of the various kinds of bonds, let us consider just how firmly those bonds hold atoms together. Some chemical bonds are strong, while others are weak. The strength of a chemical bond is its **bond energy,** that is, the amount of energy needed to break it. Chemists measure bond strength in **kilocalories** (kcal): 1 kilocalorie is the amount of heat energy required to raise the temperature of 1 kilogram (kg) of water by 1°C. Bond energy is expressed in kilocalories per **mole** (kcal/mole): a mole of a substance is the number of grams equal to the sum of the atomic weights of the constituent elements. For example, 1 mole of H_2O = 2 H (atomic weight 1) + 1 O (atomic weight 16) = 18. This quantity is the **gram molecular weight.** Thus 18 grams of H_2O = 1 mole of H_2O.

One mole of any substance contains 6.023×10^{23} molecules—an enormous quantity known as *Avogadro's number.* Therefore, both 1 mole of water, which weighs 18 g, and 1 mole of O_2 molecules, which weighs 32 g (2×16), contain 6.023×10^{23} molecules. To understand the enormity of this number, and thus the minute size of atoms and molecules, consider that if each H_2O molecule in a mole of water were the size of a pea, Avogadro's number would flood the Earth with a sea of peas 600 feet deep!

Let us now return to the concept of bond energy. Table 2-2 lists the bond energies of various molecules we have discussed. It takes a substantial amount of energy to break a stable bond. For example, the bond energy in a mole of hydrogen molecules (104 kilocalories in the

6.023×10^{23} H–H molecules comprising 1 mole) equals the energy needed to run a motor that could lift a 2-ton Cadillac 60 feet in the air or to keep a 100-watt light bulb glowing for an hour. The bond energies in molecules with more than two atoms are more complicated; for instance, it takes 120 kcal/mole to break the first O–H bond in H_2O, and 102 kcal/mole to break the second O–H bond. Both of these bonds in water, as well as all covalent and ionic bonds, are classified as **strong bonds.**

Weak bonds, which are easily broken by chance collisions or require very few kilocalories per mole to break, include two biologically important types: *hydrogen bonds,* which have a bond energy of about 4 kcal/mole, and *van der Waals forces,* which are very weak attractions that arise when any two atoms chance to come close together. We shall encounter hydrogen bonds a bit later when we discuss the chemistry of water, and van der Waals forces will come up in our discussions of the chemistry of living things in Chapters 3 and 4.

Chemical Formulas and Equations

Chemists have created various shorthand systems for representing the composition and formation of molecules and compounds. Among them are molecular formulas, structural formulas, and chemical equations, all of which we will find useful in studying biology.

A **molecular formula** is an abbreviated way of showing both how many atoms of each type are present in a molecule and whether any of the atoms occur in certain common groups. As we have seen, water is represented by the molecular formula H_2O, meaning that two atoms of hydrogen are bonded to one atom of oxygen. In a molecule of carbon dioxide, CO_2, there is one carbon atom and two oxygen atoms. A molecule of the sugar glucose, $C_6H_{12}O_6$, has twelve atoms of hydrogen and six atoms each of carbon and oxygen. The substance called urea is represented as $(NH_2)_2CO$, meaning that two NH_2 groups are present along with one atom each of C and O.

Another way of representing molecules is the **structural formula,** which shows the approximate arrangement of the constituent atoms in space and the number of bonds between them. Figure 2-13 shows the structural formulas of water, carbon dioxide, glucose, urea, and several other substances.

To show how molecules and ions interact with each other to form new substances in a process called a **chemical reaction,** chemists use molecular formulas to write chemical equations. For example, the following equation represents the reaction between sodium hydroxide (NaOH) and hydrochloric acid (HCl) to yield sodium chloride (NaCl) and water:

$$NaOH + HCl \rightarrow NaCl + H_2O$$

Table 2-2   **SOME BOND ENERGIES**

Bond	Example	Energy (kcal/mole)
H–H	Hydrogen molecule	104
C–H	Methane	99
C=C	Ethylene	125*
C–O	Carbon monoxide	84

*Each bond in ethylene has a bond energy of 125 kcal/mole.

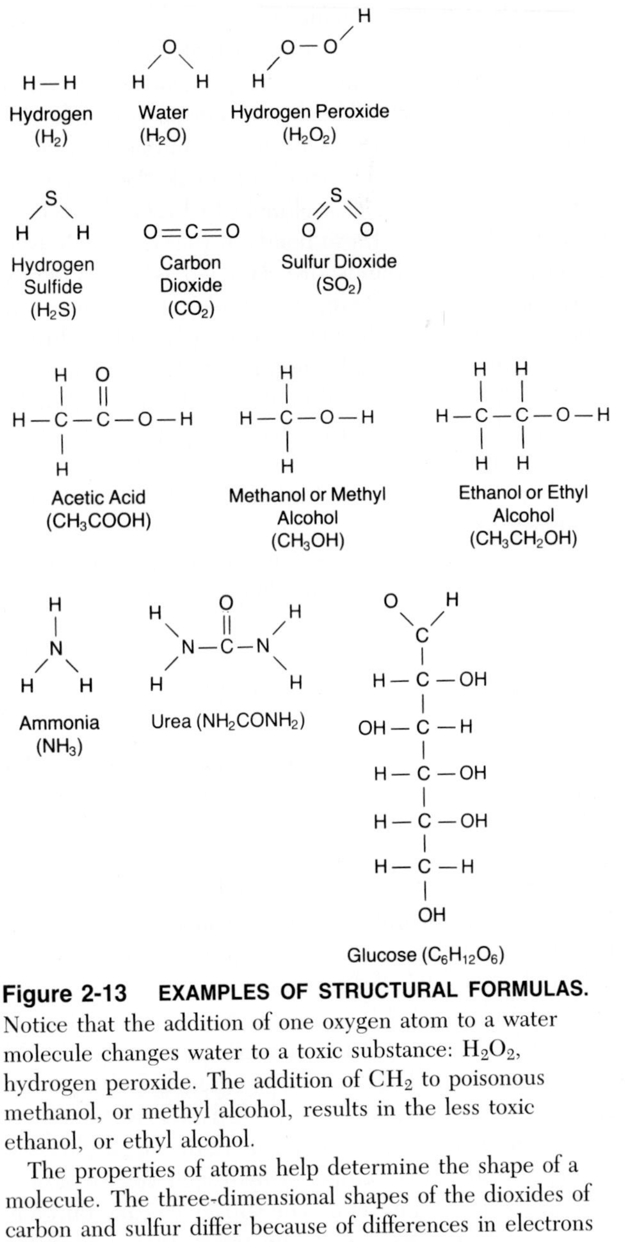

Figure 2-13   EXAMPLES OF STRUCTURAL FORMULAS.
Notice that the addition of one oxygen atom to a water
molecule changes water to a toxic substance: H_2O_2,
hydrogen peroxide. The addition of CH_2 to poisonous
methanol, or methyl alcohol, results in the less toxic
ethanol, or ethyl alcohol.

The properties of atoms help determine the shape of a
molecule. The three-dimensional shapes of the dioxides of
carbon and sulfur differ because of differences in electrons
in carbon versus sulfur. Glucose also is found in a ring
structure (as shown in Figure 2-20).

The substances on the left side of the arrow are called
reactants, and those on the right, which are formed as a
result of the reaction, are called **products.** (As we will
see in Chapter 4, however, this and all biologically relevant reactions are reversible to some degree.) In all
chemical equations, the total number of each kind of
atom must be equal on both sides of the arrow. Thus, in
the above example, there are one Na, one O, one Cl,
and two H atoms on each side, although they become
distributed differently among the compounds as a result
of the reaction process.

Chemical reactions underlie all life processes—from
the harvesting of energy from the sun to growth and

reproduction—and we will study them further in Chapters 4, 7, and 8. Now we will turn to the chemistry of
water, a vital ingredient in most life-sustaining processes.

WATER: LIFE'S PRECIOUS NECTAR

The naturalist Loren Eiseley once wrote that people,
being composed of more than 80 percent H_2O, are a
"way that water has of going about beyond the reach of
rivers." The same could be said for all organisms,
whether bacteria, mushrooms, oak trees, or rainbow
trout. Any substance that is present in such abundance
must have profound consequences for the chemistry of
an organism. In fact, life processes depend absolutely on
water, and most of the molecules and ions in a living
thing are dissolved in this liquid.

Water is one of the most remarkable compounds in
the universe (Figure 2-14). It has several unique physical
and chemical properties, all based on its structure. Let
us first review the properties of water, which are probably familiar to most readers. Then we shall discuss its
structure and its unusual chemical behavior.

Physical Properties of Water

In principle, all compounds can exist as solids, liquids,
and gases, but the temperatures and pressures needed
to bring about a change from one state to another can be
extreme. For example, hydrogen sulfide, H_2S, is a gas
above $-61°C$, a liquid between $-61°C$ and $-86°C$, and a
solid below $-86°C$. Water, however, has the remarkable
property of occurring in all three states—gas, liquid, and
solid—within the range of normal environmental temperatures found on Earth. Water has an unusually high
freezing point ($0°C$) and a high boiling point ($100°C$) in
comparison with similar small molecules, such as CO_2.
Because temperatures on the Earth's surface commonly
fall between water's melting and boiling points ($0°C$ and
$100°C$), water exists as a liquid most of the time. In its
liquid state, it covers more than half of the Earth's surface.

The thermal properties of water have important biological implications. Two particularly significant characteristics are a high **heat of vaporization,** the amount of
heat needed to turn a given amount of liquid water into
water vapor (gas), and a high **specific heat,** the amount of
heat needed to raise the temperature of 1 gram of water
by $1°C$. (Water's heat of vaporization is 580 calories per
gram, and its specific heat is 1 calorie per gram per degree centigrade.) A high heat of vaporization means that

Figure 2-14   LIFE'S PRINCIPAL INGREDIENT: WATER.
Living things are from 50 to 99 percent water, and the
properties of the H_2O molecule are reflected in the
structure and behavior of living matter. This vista from
Yellowstone National Park shows water in its three physical
states—as a liquid in the sparkling river, as a solid in the
ground cover of snow, and as vapor rising from the geyser
that soon condenses back into tiny liquid water droplets.

in order for water to reach the gaseous state, it must
absorb a great deal of heat from the surroundings. For
many plants and animals, this property is the basis of a
natural cooling system: water evaporating from leaves,
skin, or lungs uses up heat from the organism in the
process of changing from liquid to gas. That is the reason
mammals have evolved sweat glands: when the body is
overheated, the glands pour watery "sweat" onto the
skin; as the water evaporates, large amounts of body heat
are used up and the body is cooled.

Due to its high specific heat, water is slow to undergo
changes in temperature. Much heat must be added or
removed before the temperature of water changes
much. This property provides special insurance for living creatures. The temperature of sea water surrounding
a kelp or a barnacle and of the water within organisms
tends to change more slowly than does that of surrounding air or soil so that the living cells are buffered somewhat against temperature fluctuations. This kind of protection is important because many biochemical reactions

will take place only within a narrow range of temperatures.

Water has several other properties that make it important to living things. For example, water molecules exhibit strong cohesion, adhesion, tensile strength, and
capillarity. **Cohesion** is the tendency of like molecules to
cling to one another (such as water to water). **Adhesion** is
the tendency of *unlike* molecules to cling together (such
as water to the molecules of silicon dioxide on the walls
of a drinking glass). **Tensile strength** is related to cohesion and is a measure of the resistance of molecules to
being pulled apart. And **capillarity** is the tendency of
molecules to move upward in a narrow space against the
tug of gravity.

Let us see how these properties apply to some easily
observed phenomena and to some common organisms,
such as ferns, trees, and tomato plants. When you pour
water into a flowerpot full of soil, you first observe the
results of adhesion. The water sinks in and "wets" the
soil, rather than remaining on top, because of water's
inherent "stickiness," its tendency to adhere to the dissimilar molecules in the soil. Furthermore, you observe
that much of the water remains in the soil rather than
running straight through. This is explained by adhesion,
cohesion, and capillarity. The molecules of water can
resist the downward tug of gravity by first adhering to
the surface of soil particles and then moving into tiny air
spaces between them via capillarity. The water molecules enter the spaces and pull others along by means of
cohesion. Thus much of the water stays in these tiny
spaces rather than flowing out the bottom of the pot.

Adhesion, cohesion, and capillarity also explain the
upward movement of water in a thin glass tube. Water
molecules on the periphery of the cylinder of liquid adhere to the silicon dioxide molecules of glass and move
upward, pulling along by means of cohesion the water
molecules inside the column of water. If you place in
water one end of a set of glass tubes with different diameters, the fluid rises highest in the narrowest tube and
lowest in the widest tube because in the wider tubes, a
higher percentage of molecules is inside the water column, away from the glass. Thus there is lower capillarity
and relatively less adhesive force in wider tubes than in
narrower tubes. Moreover, the interior water molecules
tend to pull downward on the peripheral molecules
creeping up the glass.

In our detailed study of plant anatomy later in the
book, we will see that plants have exceedingly thin
transport vessels through which water travels from the
tips of the roots through the stems to the highest leaves.
These narrow vessels function well partly because of the
features of capillarity we just discussed. Water molecules are able to reach the highest leaves partly because
they adhere to the sugar molecules that are assembled
into the walls of the vessels and partly because the water
molecules cohere to one another in unbroken chains.

The chains of water molecules are drawn upward by means of capillarity, link by link, as water evaporates from the leaves. The great tensile strength of water supports the weight of the unbroken chains of molecules, even in 300-foot-tall redwood trees. Adhesion, cohesion, capillarity, and tensile strength are part of the reason that plants can grow higher than a few feet.

No doubt you have observed yet another of water's inherent physical properties: drips from a faucet always form spherical droplets before they fall. This occurs because of **surface tension,** which is defined as the tendency of a liquid to minimize the surface area. The strong cohesion among water molecules causes them to be more attracted to one another than to air molecules. Because of these unequal attractive forces, the surface of a drop of liquid water acts like an elastic skin to confine the molecules in an area that has the least surface area—a sphere. The "skin" of the droplet resists changes in surface area that would allow the water molecules to assume the shape of a cube or anything other than a sphere. Many living organisms, including the water strider pictured in Figure 2-15, are able to exploit the surface tension of water by nimbly walking on the water surface of ponds and streams without breaking through. If a water strider weighed more or had sharp, pointed legs, the surface film would rupture or be punctured and the insect would sink.

Water has yet another familiar, but unique, physical property: its solid form, ice, floats. The solid state of almost every substance is denser than the liquid form; for example, a cube of frozen CO_2 (also called "dry ice") will sink to the bottom of a beaker of liquid CO_2. Ice, however, is less dense than liquid water and floats in it. The biological significance of this property is that if solid water sank, lakes and seas would freeze from the bottom up, and many bodies of water would remain frozen nearly year round. Instead, the ice layer floats, acting as insulation to retard freezing at lower depths. This insulating effect allows many aquatic plants and animals to survive in the chilly water beneath the ice until the surface begins to melt.

Molecular Structure of Water

What makes water such a special substance? What accounts for its high boiling and melting points, its high heat of vaporization and high specific heat, its strong cohesion and surface tension, its adhesive qualities, and its tendency to float when solid? These properties derive from the structure of the water molecule itself, as well as from the kinds of bonds that water molecules tend to form with one another.

In an H_2O molecule, the oxygen atom shares the two single electrons in the two outermost ($2p$) unfilled orbitals with the single electrons from two hydrogen atoms. Two covalent bonds are formed, as shown in Figure 2-16. The two pairs of electrons that are not involved in the covalent bonds are repelled by the electron clouds of the covalent bonds and by each other. Thus while the two hydrogen atoms occupy two corners of a regular tetrahedron, with the oxygen atom at the center, the so-called lone pairs of electrons keep to the other two corners of the tetrahedron. We see, then, that although a water molecule is electrically neutral, it has two charged sides: one positive (the side with the two hydrogen atoms) and the other negative (the side with the lone pairs of electrons). As we saw earlier, a molecule with this type of asymmetrical distribution of charges is called a polar molecule.

Figure 2-15   THE WATER STRIDER. Surface tension allows the water strider to walk on water. Cohesion between H_2O molecules creates an elastic "skin" that supports the weight of small insects like water striders and boatmen with their oarlike movements.

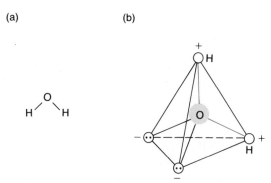

(a)   **(b)**

Figure 2-16   BASIC STRUCTURE OF A WATER MOLECULE.
(a) Simple structure, showing bond angles. (b) Tetrahedral representation, showing oxygen's unbonded electron pairs.

Hydrogen Bonding

Because of its polarity, one water molecule can interact with four other water molecules: two toward the negative side and two toward the positive side. The molecules are in as close a configuration as possible without the electron clouds of the five molecules forming covalent or ionic bonds (Figure 2-17). The attraction between the oxygen atom of one molecule and a hydrogen atom of another results in the formation of a weak bond called a **hydrogen bond**. The name "hydrogen bond" comes from the fact that, in a sense, one hydrogen atom is shared by two molecules. A hydrogen bond is almost twice as long as the O–H covalent bond in a single H_2O molecule and is ten to twenty times weaker than a covalent bond. The energy needed to break a hydrogen bond is 4 to 5 kcal/mole, which is quite close to the average energy of motion of water molecules in liquid water. Thus in liquid water, only a fraction of the possible hydrogen bonds exist at any given time, since hydrogen bonds that form in one instant are torn apart in the next. We will see that hydrogen bonds can form between many biologically

vital molecules and that these bonds play critical roles in controlling the shape and function of such molecules.

The nature of hydrogen bonds explains the special properties of water, including the ability of ice to float. In ice, water molecules are immobilized in a regular, latticelike arrangement in which all possible hydrogen bonds are formed and stable (Figure 2-18). The hydrogen bonds actually hold the water molecules farther apart from one another than in liquid water; thus a piece of ice consisting of 1 mole of water occupies a larger volume than does the same amount of liquid. Consequently, ice is less dense than water. This fact explains why water pipes crack in winter and soft drink bottles explode in the freezer—water expands as it freezes. When the lattice of ice once again melts, the lattice structure collapses like a Tinkertoy tower from which the connecting rods are removed. The same number of water molecules can suddenly fit into a smaller volume, and the density of the liquid gradually increases until a peak is reached at about 4°C. At high temperatures, water molecules possess higher energy. Consequently, they can move farther away from one another, and the volume expands.

Hydrogen bonding also accounts for water's high specific heat and high heat of vaporization. In a substance that lacks hydrogen bonds, all the heat energy supplied increases the motion of the molecules and thereby increases its temperature. In water, much of the energy

Figure 2-18   THE LATTICELIKE STRUCTURE OF ICE.
Hydrogen bonds are shown by dashed lines. Hydrogen is represented by the larger circles, oxygen by the smaller circles.

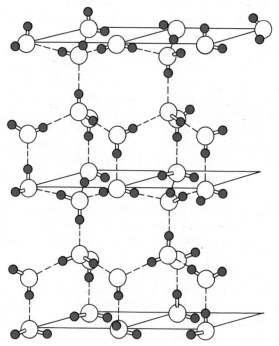

Figure 2-17   HYDROGEN BONDING AMONG WATER MOLECULES.
One water molecule can form hydrogen bonds (dotted lines) with four other water molecules in a tetrahedron.

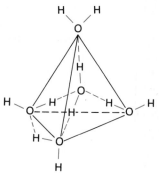

supplied is used up in stretching or breaking hydrogen bonds rather than in increasing molecular motion. Thus water is slow to increase in temperature, as its high specific heat indicates. The high heat of vaporization is explained similarly: before water can turn to vapor, all hydrogen bonds must be broken and the individual molecules must be set in rapid motion so they can break free of the fluid and "fly away" as gas molecules. This requires more energy than it takes to vaporize most other molecules in liquid form.

Finally, hydrogen bonding accounts for water's properties of cohesion, adhesion, tensile strength, and surface tension: hydrogen bonds are the "glue" that holds together groups of water molecules, makes them resistant to penetration, and helps them adhere to other kinds of molecules. Not all substances form hydrogen bonds with water, however, and that is why water "beads" rather than wets a newly waxed car.

Water as a Solvent

On Earth, water is the most widespread **solvent,** or substance capable of forming a homogeneous mixture with molecules of another substance. Indeed, more substances will dissolve in water than in any other known solvent. Substances that dissolve in solvents are called **solutes.** As a solute dissolves in water, the individual molecules disperse through the water and become surrounded by clusters of water molecules. The amount of solute in a solution is less than the amount of solvent. The amount of solute in a solution—expressed as grams per milliliter (ml) or moles per liter (l)—is referred to as the *concentration* of the solution.

The simplest example of this process is seen in ionic compounds such as LiF and NaCl. When added to water, they tend to dissociate into ions. The resultant positive ions, called *cations* (Li$^+$, Na$^+$), are attracted to the negative side of water molecules, while the negative ions, called *anions* (F$^-$, Cl$^-$), are attracted to the positive side of water molecules. In solution, the cations become surrounded by tight clusters of water molecules whose negatively charged oxygen atoms are oriented toward the cation (Figure 2-19). The anions are surrounded tightly by water molecules with their positively charged hydrogen atoms pointed inward. In each case, the surrounded ion is said to be *hydrated,* and the groupings formed are called *hydration spheres.* Complex molecules such as proteins (which we will describe later) also can dissolve in water if they have sufficient polar groups or ionized groups on the surface to form hydrogen bonds to water molecules.

The physical properties of solutions differ from those of pure water. In solutions, water molecules may form an orderly lattice around solute molecules or form hydrogen bonds with molecules or ions. Such water is not free

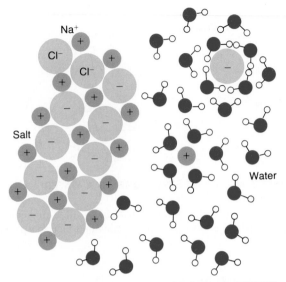

Figure 2-19   DISSOLUTION OF TABLE SALT (SODIUM CHLORIDE, NaCl) AND FORMATION OF HYDRATION SPHERES AROUND Na$^+$ AND Cl$^-$ IONS.
The solid is shown on the left; the water solution on the right. Hydrogen bonds among water molecules have been omitted for clarity.

to participate in boiling, freezing, or other processes and is said to be *bound.*

A good example of this phenomenon occurs when salt is added to water; the boiling point of the solution increases, and the freezing point decreases. When salt is spread on an icy pavement and begins to dissolve, it breaks up the water lattice and melts the ice even though the air temperature may be lower than 0°C. Antifreeze works in a car by preventing the formation of the large numbers of hydrogen bonds necessary to stabilize radiator water into an ice lattice. Many plants, insects, and other organisms have antifreeze substances in their body fluids that act in just this way. Much of the water inside living cells is bound water in an orderly, immobilized array around various organic and inorganic solutes. Although there can be a great abundance of water inside and immediately outside living cells, the bound state of much of that water means that its properties are somewhat different from those we studied for pure water.

Compounds that dissolve readily in water (such as NaCl and LiF) are called **hydrophilic** (water-loving) compounds. Polar molecules such as proteins, sugars, and many other large biological molecules are also hydrophilic; they form hydrogen bounds with water molecules. The dissolving of sugar is illustrated in Figure 2-20.

Compounds with nonpolar, covalent bonds (that is, covalent bonds in which the bonding electrons are shared equally by both nuclei), such as oils, waxes, and

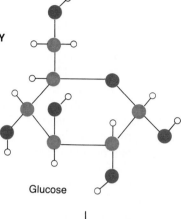

**Figure 2-20
HYDROPHILIC PROPERTY
OF SUGARS.**
Sugars dissolve in water
by forming weak
hydrogen bonds
with water molecules.

Glucose

In Water

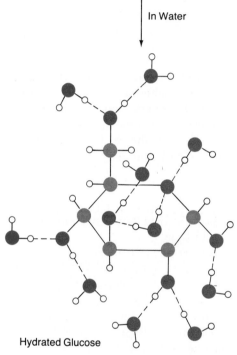

Hydrated Glucose

some plastics, tend to be insoluble in water; therefore, they are called **hydrophobic** (water-fearing) compounds. Being nonpolar, hydrophobic compounds form neither hydrogen bonds nor purely electrical attractions with water molecules. When a drop of oil is put on the surface of water, the two fluids form a boundary, or interface, because of their mutual insolubility. Cohesion tends to keep oil molecules with oil molecules, and water with water, and little, if any, adhesion holds the drops together. In fact, depending on the amount of oil present, it will either stay together as a droplet or spread in a layer across the water surface, just as oil slicks form on the sea following an oil spill. The utilitarian side of this property was well known in the nineteenth century when cloth treated with oil was commonly used to make rain slickers (Figure 2-21). The same principle is a basic feature of life; a hydrophobic layer of lipids (types of fats) covers the surface of *every* living cell on Earth—in fact, the large amount of water in a terrestrial organism's body could not be retained if it were not for hydrophobic compounds in the outer walls of every cell, of organs, and of the body itself (Figure 2-22). In a sense, an organism's structural integrity is based in part on insolubility: if its "walls" could dissolve in water, a land organism would become a puddle, while an aquatic organism would disperse into its watery surroundings. Furthermore, many of the important chemical processes that sustain life go on at the interfaces between polar and nonpolar substances. The principles of solubility are just one example of the precept that Lavoisier and Dalton discovered two centuries ago: living systems obey the same chemical and physical laws that govern nonliving matter.

Figure 2-21   OIL SLICKERS: MAN-MADE HYDROPHOBIC PROTECTIVE LAYERS.
People facing a storm-tossed sea need extra waterproofing. Winslow Homer's fishermen (*The Herring Net*, 1885) wear additional hydrophobic layers—oil slickers made of cloth impregnated with whale grease.

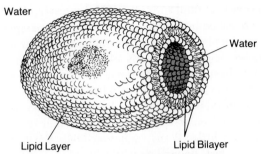

Figure 2-22   HYDROPHOBIC LIPIDS IN CELLS.
Lipids are not soluble in water and form a water-impermeable membrane, or coating, at the surface of every living cell. Lipid layers also surround various specialized regions within cells.

Dissociation of Water: Acids and Bases

Water has still another remarkable property of interest to biologists—a slight tendency to "fall apart." We have seen that ionic compounds such as NaCl have a tendency to dissociate in the presence of water. Curiously enough, water molecules themselves dissociate to a small extent: thus about 2 out of every 10^9 H_2O molecules separate into hydrogen ions (H^+) and *hydroxyl* ions (OH^-). The H^+ ions by and large associate with individual water molecules, thus forming *hydronium* ions, H_3O^+. For all practical purposes, we can consider H^+ and H_3O^+ as equivalent in this discussion. Like Na^+ and Cl^- in water, H^+ and OH^- ions become hydrated. Once dissociated, these ions can again combine, or associate, to re-form water molecules. Thus two reactions proceed at the same time:

$$H_2O \rightarrow H^+ + OH^-$$

$$H^+ + OH^- \rightarrow H_2O$$

The number of hydrogen and hydroxyl ions in a sample of H_2O is equal and constant: in 1 liter of pure water, there is 10^{-7} mole of H^+ and an equal amount, 10^{-7} mole, of OH^-.

Adding a solute to water may change the concentration of H^+. Any substance that gives up, or donates, H^+ ions in solution, and thereby increases the H^+ content of the solution, is called an **acid**. Compounds that decrease the amount of H^+ in solution are called **bases**–bases take up, or accept, H^+ from the solution (for instance, added $-OH^-$, a base, would bind H^+ to form water).

The concentration of H^+ in acid and base solutions is expressed in terms of **pH,** a scale running from 0 to 14. The pH of a solution is the negative logarithm of the molar concentration of H^+: pH $= -\log[H^+]$ (where the brackets [ ] mean the concentration of the substance bracketed). Thus pure water, with an H^+ concentration of 10^{-7} mole per liter, has a pH of 7. More H^+ means a *lower* pH. Hence acidic solutions have pHs less than 7. And because bases decrease H^+ concentration, they have pHs greater than 7. Water, being neither basic nor acidic, is said to be *neutral*.

Since the pH scale is a logarithmic scale, each unit represents a tenfold change in the concentration of H^+ ions. A solution with a pH of 6, for example, is ten times more acidic than pure water, at a pH of 7, and it is ten times less acidic than sour milk, at a pH of 5.

Acids and bases are extremely important in biology because the chemical reactions that go on within living cells and tissues are sensitive to H^+ concentrations (see Figure 2-23). (Who hasn't heard about "acid stomach" and ways to "neutralize stomach acid"?) Similarly, the structural integrity of important components of living cells can be altered as pH changes. Some living tissue, such as the fruit of lemons and limes, is extremely acidic, as is the fluid in the stomachs of most animals. Other fluids, such as the sea water in which many organisms live, are basic, or *alkaline*. However, the fluids of most

Figure 2-23   THE pH SCALE AND THE pH's OF COMMON FLUIDS AND SUBSTANCES.

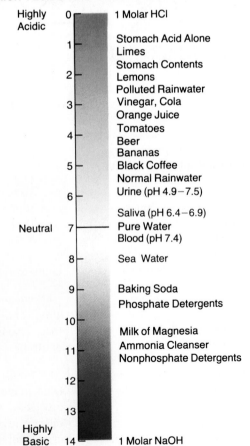

living cells and organisms have an almost neutral pH, between about 6.5 and 7.5, despite the organism's consumption of acidic or basic substances and its exposure to hydrogen and hydroxyl ions in the environment.

What accounts for this absence of fluctuation in pH? Chemical substances called **buffers** are immensely important to cells and organisms because they help them to resist changes in pH when acids or bases are produced or added (as from food). Buffers are substances that bind H^+ when concentrations of H^+ are high and release H^+ when concentrations of H^+ are low. By these processes, they tend to maintain a constant pH. Suppose we have a solution that has a pH of 6.3 and that contains a buffer. If we add more H^+ ions to the solution, the buffer tends to absorb them, thus keeping the pH constant at about 6.3. If another test solution containing a buffer had a pH of 8 and a base, OH^-, was added, the buffer would release H^+, thereby neutralizing the added OH^- and maintaining the pH near 8. In the human body, red blood cells contain the bicarbonate ion HCO_3^-, a buffer that acts as an H^+ acceptor, helping to neutralize acidity and prevent a major shift away from the normal blood pH of about 7.2. We will describe this buffering system in greater detail in Chapter 31.

ATOMS TO ORGANISMS: A CONTINUUM OF ORGANIZATION

We began this chapter with a discussion of life's grand diversity and the foundation of that diversity—Earth's ninety-two naturally occurring chemical elements. These elements join in a virtually unlimited number of combinations to form substances with a spectrum of properties: gaseous, liquid, solid, inert, reactive, sticky, dry, wet, acidic, basic, soluble, insoluble, and so on. These various properties are based on the architecture of the individual atoms and on the configuration of their electrons.

Throughout this book, we will emphasize life's diversity and structural complexity. But keep one thing in mind while you study the chapters that follow, whether the focus is on the birds and flowering plants of North America, the locomotory apparatus of the land snail, the labyrinthine construction of the human brain, or the minute structures inside a living cell: *all matter of significance to life is part of a continuum of organization.* This continuum begins with subatomic particles—protons, neutrons, and electrons—and progresses to atoms, with their cloudlike orbitals; atoms are then built into molecules; small molecules are assembled into giant "macromolecules"; assemblages of macromolecules build parts of cells called organelles and membranes; those cell parts, when added together, form cells; cells put together yield tissues and organs; and, finally, assemblages of tissues and organs form individual organisms of all shapes and sizes. This spectrum of increasing complexity is fascinating in itself, but it is the "base of the pyramid," atoms with their unique chemical properties, that underlie it all. Keeping in mind that atoms obey the same physical and chemical laws, whether they are part of a mineral crystal or a living cell, let us move along the scale of organization now and consider in Chapter 3 how small molecules combine to make the large molecules found in living things.

SUMMARY

1. The universe is composed of ninety-two naturally occurring chemical *elements*, fundamental substances that cannot be broken down by chemical means. *Atoms* are the smallest units into which matter can be broken without changing its properties.

2. Each atom consists of a nucleus, which contains *protons* and *neutrons*, and of negatively charged *electrons*, which are distributed in cloudlike orbitals around the nucleus. The number of positively charged protons determines an element's *atomic number;* the sum of protons and neutrons is an element's *atomic weight*, or atomic mass. Although the number of neutrons usually is the same as that of protons, this figure varies in *isotopes* of an element.

3. The number of electrons in an atom is equal to the number of protons, making the atom electrically neutral.

4. *Atomic orbitals*, in which electrons move, surround the nucleus at distances associated with certain energy levels. The higher the energy level of an electron, the farther from the nucleus it is likely to be. Each orbital can hold a maximum of two electrons.

5. The distribution of electrons in orbitals is governed by a few basic rules: the simplest orbital of the lowest energy level is filled first; one electron must be present in each orbital of the same shape at one energy level before any of them can be filled; and so on.

6. The inert gases are the most stable elements because their outermost energy level is occupied by the maximum number of electrons.

7. *Molecules* are precisely ordered arrangements of atoms. *Compounds* are molecules made up of several types of atoms.

8. The atoms in molecules are held together by chemical bonds that arise when unpaired valence electrons in the outermost atomic orbitals fuse into a common *molecular orbital*. In *covalent bonds*, electrons are shared by atoms; such sharing helps to stabilize the atoms by increasing the number of electrons in their valence shell. In *ionic bonds*, electrons are transferred from one atom to another. Atoms that lose or gain electrons prior to the formation of an ionic bond are called *ions*. Ions carry a net positive charge if they have lost an electron or a net negative charge if they have gained an electron. The attraction between ions having opposite charges results in ionic-bond formation.

9. *Electronegativity* is a measure of the ability of atoms to attract other atoms in the same molecule. In molecules made up of atoms with differing electronegativities, the bond is said to be *polar*, and the shared electrons spend more time near the more electronegative nucleus.

10. The strength of chemical bonds is expressed as *bond energy* and is measured in *kilocalories per mole* (a mole is equal to the *gram molecular weight* of a substance). Covalent and ionic bonds are *strong bonds* that require a substantial amount of energy to break. *Weak bonds* include hydrogen bonds and van der Waals forces.

11. *Molecular formulas* are used to represent the components of molecules and can be used to describe *chemical reactions* that involve *reactants* and *products*. *Structural formulas* show the approximate arrangement in space of the constituents of a molecule.

12. Water is the most abundant compound in living things. It has a high melting point, boiling point, *specific heat*, and *heat of vaporization*. Water molecules also show strong *cohesion*, *tensile strength*, *adhesion*, *capillarity*, and *surface tension*. Furthermore, water is less dense in its solid state than in its liquid state. All these properties can be explained by the structure of the water molecule, which is characterized by a high degree of *hydrogen bonding*.

13. Water is an important *solvent*: it is capable of dissolving many kinds of *solutes*. Water dissolves ionic bonds and forms hydration spheres around ions. Ionic compounds are examples of *hydrophilic* substances, which are readily hydrated. *Hydrophobic* substances, such as compounds with nonpolar bonds, are insoluble in water.

14. Water can dissociate into H^+ and OH^- ions. These ions are equal in concentration in pure water, which has a neutral *pH* of 7. An *acid* is a substance that donates H^+ ions when dissolved in water; acidic solutions have pHs below 7. A *base* is a substance that accepts H^+ ions; basic solutions have pHs above 7. A *buffer* binds or releases H^+ ions depending on the H^+ concentration in its vicinity. Buffers function to keep the pH of solutions relatively constant.

KEY TERMS

acid

adhesion

atom

atomic number

atomic orbital

atomic weight

base

bond energy

buffer

capillarity

chemical reaction

cohesion

compound

covalent bond

electron

electronegativity

element

gram molecular weight

heat of vaporization

hydrogen bond

hydrophilic

hydrophobic

ion

ionic bond

isotope

kilocalorie

mole

molecular formula

molecular orbital

molecule

neutron

pH

polar bond

product

proton

reactant

solute

solvent

specific heat

strong bond

structural formula

surface tension

tensile strength

weak bond

QUESTIONS

1. Draw atoms of two different elements and label the components.

2. Draw a molecule composed of two different elements and describe how it is held together.

3. Draw several molecules of liquid water, showing the bonds that form between the molecules.

4. Can you think of a way to break some of the hydrogen bonds that link water molecules together?

5. What force is strong enough to break the covalent bonds in water and to separate each water molecule into hydrogen and oxygen?

6. When water boils a gas is formed. How is this gas like liquid water? How is it different? Does the gas consist of water molecules or does it contain separate molecules of hydrogen and oxygen?

7. Water forms a film on a clean drinking glass, but beads up on a greasy glass. It also forms beads on a plastic "glass." Which is more hydrophobic: glass or plastic?

8. What is an acid? A base? A buffer?

ESSAY QUESTION

1. Discuss the properties of water that make it a vital component of living organisms. How do hydrogen bonds contribute to these properties?

SUGGESTED READINGS

HENDERSON, L. S. *The Fitness of the Environment.* Boston: Beacon Press, 1958.

This is a classic in biology writing. The chapter on water is worth careful reading.

RODELLA, T. D., et al. *Through the Molecular Maze.* Los Altos, Calif.: W. Kaufmann Press, 1975.

This is a helpful guide to chemistry for beginning life-science students.

WHITE, E. H. *Chemical Background for Biological Sciences.* 2d ed. Englewood Cliffs, N.J.: Prentice-Hall, 1972.

Both inorganic and organic chemistry are treated here in an exceptionally lucid way.

3

THE MOLECULES OF LIVING THINGS

This sameness of composition (encountered in all living beings from bacteria to man) is one of the most striking illustrations of the fact that the prodigious diversity of macroscopic structures of living beings rests in fact on a profound and no less remarkable unity of microscopic makeup.

Jacques Monod, *Chance and Necessity* (1972)

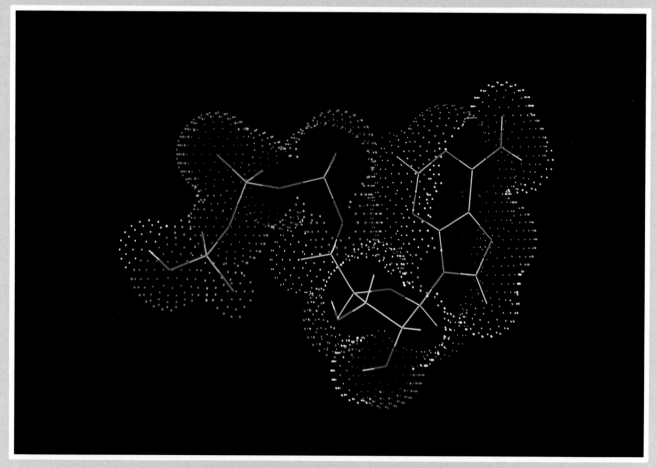

Fuel for the fire of life: A computer's view of ATP.

Just as a novel is made up of words and words are made up of individual letters, the phenomenon we call life is written in a language of molecules and atoms. Regardless of whether an organism is a plant or animal, whether it is microscopic or enormous, whether it walks on land, swims in the ocean, or grows on a rock, it contains the same basic kinds of biological molecules. The shapes and behaviors of these molecules account for the physical characteristics and activities of living organisms in much the same way that the structure of atoms accounts for the properties of molecules and compounds.

In Chapter 2, we saw that atoms are the fundamental units of all matter and that atoms come together to form molecules. In this chapter, we shall discuss the four major classes of large molecules that are unique to living things. In subsequent chapters, we will see how these diverse molecules are assembled into cell parts and how these parts compose whole cells. Still later, we will consider how cells make up organisms and how groups of organisms constitute populations. It is here, however, at the level of biological molecules, that we first encounter the transition zone between chemistry and biology, between the nonliving and the living.

THE FUNDAMENTAL COMPONENTS OF BIOLOGICAL MOLECULES

Most of the compounds that make up living things are of four main types: carbohydrates, lipids, proteins, and nucleic acids. Many of the molecules in these four classes are **macromolecules**—extremely large molecules with molecular weights of about 10,000 daltons or more. We shall describe each of these groups in detail in this chapter, but we first need to examine the components common to all biological molecules and learn how these components are arranged.

Carbon: The Indispensable Element

All four types of biological molecules contain carbon. In fact, life on Earth cannot be separated from carbon and its unique chemistry. Any compound that contains carbon is called an **organic compound**; one without carbon is an *inorganic compound*. (A few simple molecules that contain carbon but no hydrogen, such as CO_2, are considered inorganic.) All biological molecules are organic, but not all organic molecules are manufactured by living things; some plastics are organic compounds not made by living cells.

The unique structure of the carbon atom ultimately accounts for the great diversity of molecules in living things and, in turn, for the diversity of life forms on our planet. As we saw in Chapter 2, the carbon atom is able to form as many as four covalent bonds with other atoms because of the four vacancies in its $2p$ orbitals. Carbon can bond to other carbon atoms, producing long, straight chains, branched chains, rings, and a variety of other shapes, as shown in Figure 3-1. Carbon atoms can also form covalent bonds with oxygen, nitrogen, hydrogen, sulfur, and occasionally phosphorus to produce a great variety of organic molecules. In all, carbon atoms can form the chemical skeletons for more than 1 million compounds—ten times more than the other ninety-one elements combined. The structure and bonding properties of carbon are indeed special. As we shall see, carbon

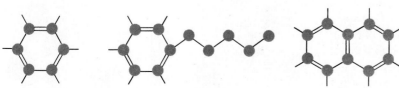

Unbranched Chain with Single Bonds   **Unbranched Chain with Single and Double Bonds**   **Branched Chain**   **Ring with Single Bonds**

Ring with Single and Double Bonds   **Ring with Side Chain**   **Fused Rings**

Figure 3-1   SOME OF THE DIVERSE PATTERNS IN WHICH CARBON ATOMS CAN BE COVALENTLY LINKED.

Figure 3-2   SHAPES OF A FIVE-CARBON CHAIN.
A chain of five carbons can assume many possible shapes, or conformations, as it rotates around the carbon-carbon bonds.

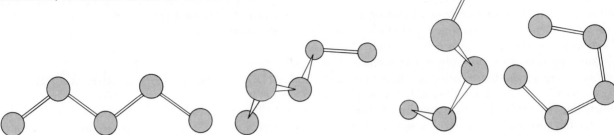

atoms are the backbone in carbohydrates, lipids, proteins, and nucleic acids.

The structure and function of biological molecules are influenced by the kinds of bonds that carbon atoms form. Although molecular formulas and models on paper can lead us to believe that molecules are stable and rigid, the atoms in molecules are constantly jiggling, bending back and forth as chemical bonds vibrate rapidly. Atoms rotate around the axis of a bond like spokes on a bicycle wheel whizzing around the hub. In small molecules, the effect of this rotation is not very great, but in a large organic molecule with a long chain of singly bonded carbon atoms, each of which is free to rotate relative to its neighbors, the molecule can sweep out a large volume as it assumes many shapes, or conformations. Figure 3-2 shows some of the possible conformations of a five-carbon chain; when dissolved in liquid, molecules in a five-carbon chain do indeed rapidly assume all these and other shapes. When carbon atoms are linked by double bonds, they are not free to rotate and change conformation. Consequently, all other atoms directly attached to double-bonded carbons lie in a single plane. Carbon rings are typically flat if double bonds are present. Carbon rings occur in many important biological molecules, including those that carry hereditary information and those in plants that trap the energy from sunlight. The flatness of the structures radically affects the way they interact with other molecules, as we shall see.

Another significant characteristic of carbon is the tetrahedral structure that results when it forms four single covalent bonds with four other atoms. We saw in Chapter 2, for example, that in the methane (CH_4) molecule, four hydrogen atoms are covalently bonded to carbon and a tetrahedron forms, with the carbon nucleus at the center and the hydrogen nuclei at the corners. The tetrahedral shape has some interesting features. Perhaps the most significant is that a molecule composed of one carbon atom with four different atoms attached to it can be arranged in two ways that are mirror images of each other. These mirror image arrangements of atoms are called **stereoisomers** (Figure 3-3). All stereoisomers of a compound share the same chemical properties. Two

compounds with exactly the same formula but different arrangements of atoms are called **isomers.** Isomers are not mirror images and often have very different chemical properties, as we will see shortly.

If stereoisomers of a molecule have the same chemical properties, why should we be concerned with them? There are many important biological stereoisomers, which can exist in either a right-handed, D-form (from *dextro*, meaning "right") or a left-handed, L-form (from *levo*, meaning "left"), and these forms can have different effects in biological systems. For example, a substance called L-adrenalin fits precisely in the pocket-shaped portions of certain molecules in the walls of blood vessels and causes them to constrict. D-adrenalin, however, does not fit so well in the receptors and is some twelve times weaker than L-adrenalin in causing vessels to constrict. Although right-handed and left-handed molecules have the same chemical properties, only the L-form is found in the molecules that make up proteins. This is true for every type of organism yet studied, from the simplest bacterium to the most complex plant or animal. No one knows *why* proteins are built from only left-handed molecules, but this form must have been selected very early in the evolution of life. If you were to pass through the "looking glass" and suddenly become

Figure 3-3   STEREOISOMERS.
These two imaginary molecules are stereoisomers. Their chemical composition is the same; but since their atoms are arranged differently, their structure is different. They are mirror images of each other, like a person's right and left hands.

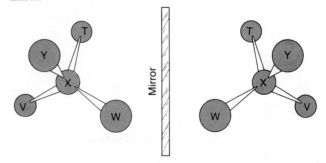

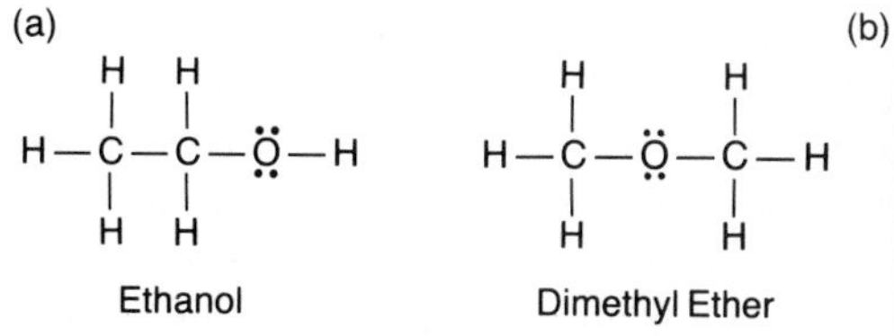

(a)

H   H   H   H
| | | |
H—C—C—Ö—H   H—C—Ö—C—H
| | | |
H   H   H   H

Ethanol   Dimethyl Ether

Figure 3-4   STRUCTURAL ISOMERS ETHANOL (ETHYL ALCOHOL) AND DIMETHYL ETHER.
(a) The structural formulas show the pattern of covalent bonds. (b) The ball-and-stick models show the three-dimensional arrangement of the atoms in space, including the angles at which the molecule is bent.

your own mirror image, right down to the level of molecules, you could still breathe air and drink water, but you might not be able to digest ordinary food, which contains left-handed molecules!

Compounds that have the same molecular formula but different properties are called **structural isomers.** Consider, for example, the isomers ethanol and dimethyl ether: both have the formula C_2H_6O. Ethanol (also called ethyl alcohol) is the colorless liquid that gives an intoxicating effect to wines, beers, and liquors. It is produced in yeast cells and some bacteria. Dimethyl ether, which is nonbiological in origin, is also colorless, but its boiling point is so low ($-23°C$) that it could boil and turn to vapor in a kitchen freezer. Dimethyl ether sometimes is used as an industrial refrigerant, but it never would be used in a cocktail. The arrangements of atoms in ethanol and dimethyl ether, shown in Figure 3-4, are quite different and account for the dissimilar properties of the two isomers. Notice that the oxygen atom in ethanol is bonded to one carbon atom and one hydrogen atom, while the oxygen in dimethyl ether is bonded to two carbon atoms. The position of the adjacent oxygen and hydrogen in ethanol gives this molecule special properties similar to those of water molecules: weak hydrogen bonds can form between ethanol molecules. Because of these hydrogen bonds, ethanol has a higher melting point, boiling point, and heat of vaporization than does the dimethyl ether isomer.

These compounds, and isomers in general, illustrate the shortcomings of molecular formulas and two-dimensional diagrams. We cannot truly understand molecules and appreciate how they function or fit together in living organisms unless we can visualize them in three dimensions.

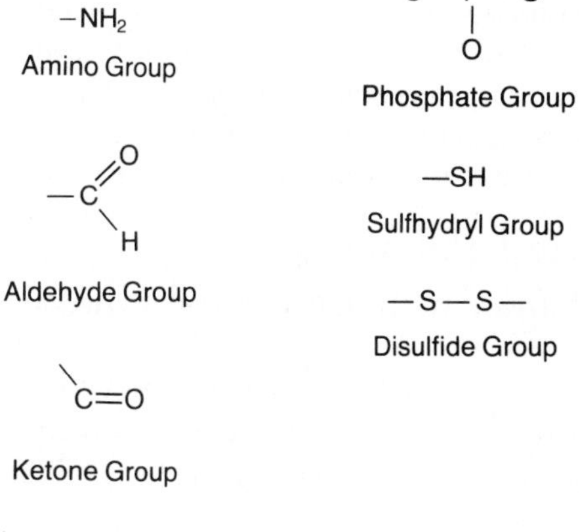

Figure 3-5   FUNCTIONAL GROUPS.
Such groups are responsible for the characteristic properties of compounds that contain the groups.

Functional Groups: The Key to Chemical Reactivity

In a sense, the real key to the chemical behavior of most biologically important molecules is the presence of specially arranged clusters of atoms, such as the –OH we just encountered in ethanol. These clusters are called **functional groups.** The specific structure of each functional group imparts a similar chemical behavior to all molecules to which it is attached. Figure 3-5 shows the structures of nine biologically important functional groups. Let us look briefly at each.

The first functional group we will consider is the *hydroxyl group,* –OH, which figured in our discussions of the dissociation of water (Chapter 2) and the structure of ethanol. Ethanol and other carbon-containing compounds that have a hydroxyl group are classified as *alco-*

–OH
Hydroxyl Group

Carboxyl Group

–NH₂
Amino Group

Aldehyde Group

C=O
Ketone Group

Methyl Group

Phosphate Group

—SH
Sulfhydryl Group

—S—S—
Disulfide Group

hols. Because of the hydroxyl group, the molecules of all alcohols are polar (at least in the –OH region) and tend to be soluble in water, and they form weak hydrogen bonds. Thus we can predict some aspects of a compound's chemical behavior simply by knowing that it has one or more –OH functional groups.

The next group shown in Figure 3-5 is one of the most common in biological compounds, the *carboxyl group,* –COOH. The carboxyl group usually does not exist as –COOH because it donates H^+ ions in solution and therefore exists as $-COO^-$. Like other substances that give up H^+ in solution, molecules containing the carboxyl group are acids, specifically called *carboxylic acids.* Acetic acid, CH_3COOH, is a common carboxylic acid. Dissolved in water, it is the substance we call vinegar.

When groups such as –COOH occur in a molecule along with other functional groups, such as –OH, the characteristics of both are exhibited. For example, lactic acid, the substance that builds up in muscles and causes cramping during sustained periods of exercise, has both hydroxyl and carboxyl groups and displays properties of both an alcohol and a carboxylic acid.

The third functional group on our list is the *amino group,* $-NH_2$. It is an important constituent of proteins, and like the carboxyl group, it is one of the most common functional groups in living organisms. The amino group may act as a base, accepting hydrogen ions to become ammonium ion, $-NH^{3+}$. Urea, a waste product of metabolism, has two $-NH_2$ groups and is perhaps the most common amino compound.

The next two functional groups—*aldehyde,* –CHO, and *ketone,* –C=O—characterize the class of biological compounds called sugars, which are the fundamental constituents of carbohydrates. Both groups are polar and thus render the compounds soluble in water. In contrast, the *methyl group,* $-CH_3$, which is an important component of lipids and oils, is hydrophobic and insoluble in water. The *phosphate group,* $-PO_4^{3-}$, is another important part of certain lipids as well as a component of nucleic acids and proteins. Another common functional group is the *sulfhydryl group,* –SH, which is a key determinant of protein structure. Two sulfhydryl groups on different parts of one large molecule, or even on separate "chains" of a molecule, may be joined together into a *disulfide group,* –S–S–. Disulfides stabilize the shape of many proteins.

Monomers and Polymers: Molecular Links in a Biological Chain

The combination of single and double bonds with the presence of functional groups yields great diversity in the structure and function of organic molecules. Even greater diversity is made possible by the linkage of many molecules to form very large *macromolecules.*

We have seen that all biological molecules contain the element carbon and that various functional groups may be attached to carbon atoms. Macromolecules share another feature: they are all **polymers** (meaning "many parts"). Polymers are long chains, much like strings of beads, made up of simpler units called **monomers** ("single parts") linked in a specific sequence by covalent bonds. The order in which monomers are joined is important, for it helps determine the function of biological polymers.

Polymers are formed by means of **condensation reactions:** as two monomers join and a covalent bond forms between them, one monomer loses an –OH group and the other loses a hydrogen atom. The hydroxyl group and the hydrogen atom combine to produce a water molecule as a by-product of the condensation reaction. The opposite occurs when polymers are broken down into monomers: a hydrogen is added to one monomer and an –OH group is added to the adjoining monomer as they split apart. Because parts of water molecules are added during this process, it is called **hydrolysis,** which means "splitting with water" (Figure 3-6). Condensation takes place in the cells of leaves, for example, as individual sugar molecules combine to form the more complex storage sugars called *carbohydrates.* Hydrolysis takes place in the stomach of an animal that later eats those leaves; the individual sugar molecules are removed one by one from the carbohydrate during the process of digestion. The process of eliminating or adding water molecules

Figure 3-6   CONDENSATION.
Condensation is a process by which monomers are linked into polymers. A water molecule is produced for each bond formed between monomers. The resulting polymer can be split into its constituent monomers by means of hydrolysis.

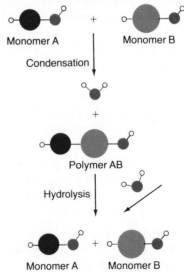

Monomer A   +   Monomer B

Condensation

+

Polymer AB

Hydrolysis

Monomer A   +   Monomer B

has been called a chemical zipper that "zips" molecules together or "unzips" them. This zipper operates during the formation and splitting of all biological polymers.

Biological molecules, then, share some important features—a carbon "backbone," a chainlike polymer structure, and functional groups that impart characteristic properties as well as the ways in which the molecules are formed and broken apart. Despite their similarities, however, the four types of biological molecules—carbohydrates, lipids, proteins, and nucleic acids—differ profoundly in several ways. Keeping in mind that the key feature of these substances is the *specific order* of the constituent monomers, let us see what those differences are and how each type of molecule functions in living organisms.

CARBOHYDRATES: SOURCES OF STORED ENERGY

In terms of sheer numbers, the most abundant carbon compounds in living organisms are **carbohydrates,** which include the sugars and starches. The reason they are the most abundant is that plants are largely constructed of carbohydrates. Immense offshore kelp beds, grassy plains and farmland, and Earth's vast tracts of forest can be thought of as huge living cities built of carbohydrates. Since the sun's energy is trapped by plant leaves and stored in carbohydrates, these macromolecules ultimately provide all living things with the energy needed to sustain life. We will explore this role of carbohydrates in Chapter 8. Here, we will focus mainly on the structure of these molecules.

A carbohydrate is composed of C, H, and O in the ratio of 1:2:1 (CH_2O). This formula gives the group its name, "hydrate of carbon." Carbohydrates consist of a carbon backbone with various functional groups attached. The basic carbohydrate subunits are sugar molecules called **monosaccharides** ("single sugars"); they function as monomers that can be joined together to form the more complex **disaccharides** ("two sugars") and **polysaccharides** ("many sugars").

Monosaccharides: Simple Sugars

Monosaccharides have a distinctive structure, which can be seen in Figure 3-7. The carbon backbone can be made of three, four, five, or six carbon atoms; one of the carbons forms a double bond with an oxygen atom, while each of the other carbons in the chain is bonded to one hydrogen atom and one hydroxyl group. The carbon to oxygen double bond may occur at the head of the chain in an aldehyde, or it may occur in the middle of the chain as a ketone group. The hydroxyl groups make the molecule polar and thus readily soluble in water. (Why? Recall hydrogen bonding in water from Chapter 2.)

The simplest monosaccharide is the three-carbon compound glyceraldehyde, depicted in Figure 3-7 along with its structural isomer, dihydroxyacetone. As structural isomers, they have the same chemical formula, $C_3H_6O_3$, but different configurations and properties. Several simple three-, four-, and five-carbon monosaccharides are involved in metabolism and photosynthesis, which we shall discuss in Chapters 7 and 8, respectively. The five- and six-carbon monosaccharides are of the greatest importance biologically because they form the units in long carbohydrate molecules and also participate in metabolic reactions.

The universal cellular fuel burned by plants and animals for energy and by animals for body heat is a six-carbon monosaccharide called **glucose,** or grape sugar. Glucose can easily be degraded to yield its stored energy or easily built into large storage polymers. It can exist as a straight-chain compound with an aldehyde group, as shown in Figure 3-7, but it occurs more commonly in the ring configuration illustrated in Figure 3-8. Another six-carbon monosaccharide, *fructose* (Figures 3-7 and 3-8), is a structural isomer of glucose. Fructose, or fruit sugar, is common in sweet fruits and vegetables. Fructose also indirectly enables reproduction to occur in animals; the recognition between sperm and egg depends on the presence of fructose in molecules on the surface of the egg. Two other important six-carbon sugars, *galactose*

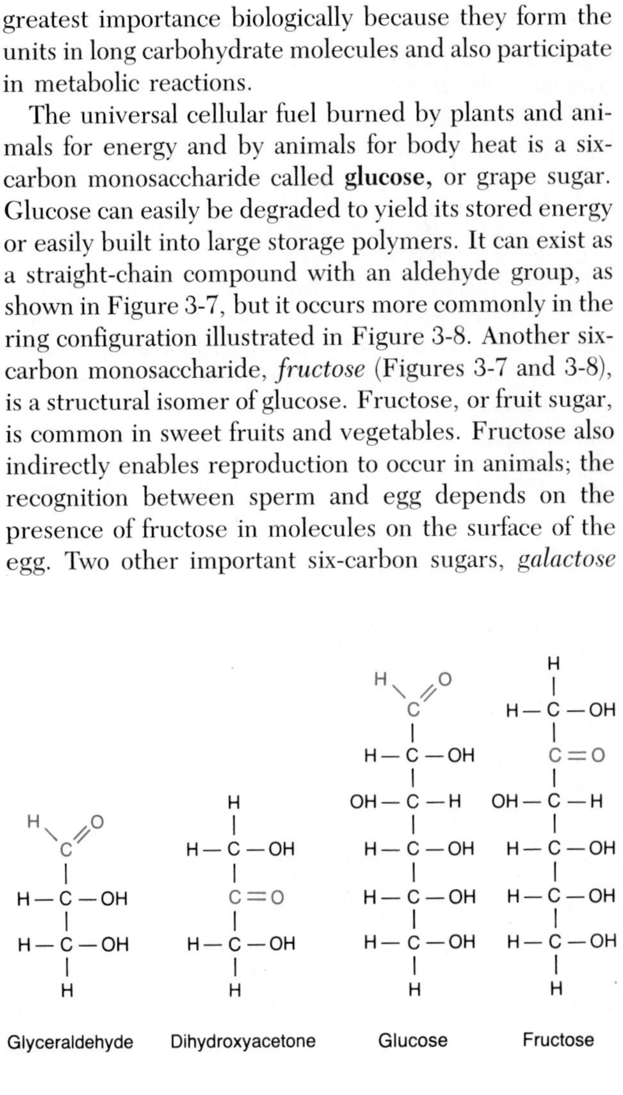

Figure 3-7 STRUCTURAL FORMULAS OF FOUR SIMPLE SUGARS, OR MONOSACCHARIDES: GLYCERALDEHYDE AND DIHYDROXYACETONE (THREE CARBONS), AND GLUCOSE AND FRUCTOSE (SIX CARBONS).
Note that glyceraldehyde and glucose contain an aldehyde group, whereas dihydroxyacetone and fructose contain a ketone group.

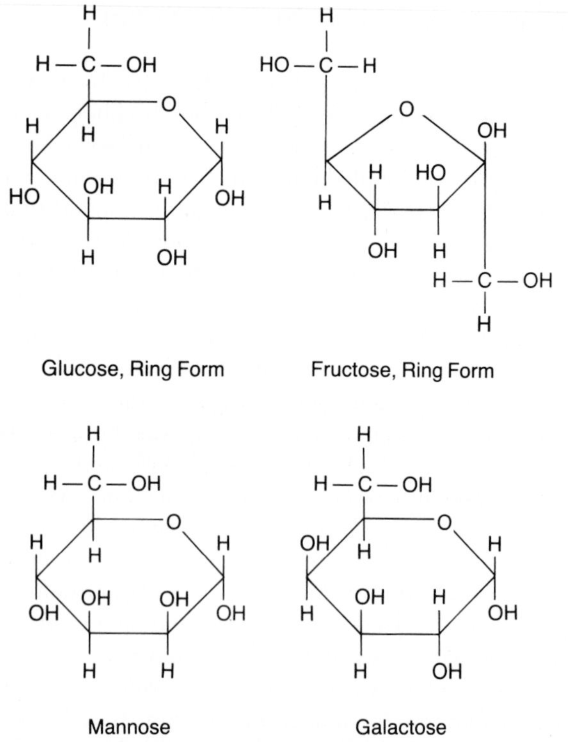

Glucose, Ring Form   Fructose, Ring Form

Mannose   Galactose

Figure 3-8   STRUCTURAL FORMULAS OF FOUR MONOSACCHARIDES SHOWN IN THE FORM OF RINGS IN SOLUTION.
Each point on the ring where the bonds converge represents a carbon atom. (By convention, the carbon atoms in the rings are not indicated in simple line drawings.) Glucose and fructose are structural isomers of each other. Note that in addition to the reversed position of an –OH group, the fructose molecule has a ring with four carbons, whereas glucose has a ring with five carbons. Galactose and mannose are monosaccharides that differ from glucose only in the placement of the –OH groups and are therefore stereoisomers.

and *mannose*, are also depicted in Figure 3-8. Galactose is abundant in milk, and mannose is found in certain sugars and polysaccharides.

The monosaccharides, then, are a direct source of energy for living organisms. They also serve as building blocks for carbohydrate polymers, as we shall see, and they are the raw material from which many other macromolecules are manufactured; for example, there are monosaccharide units in nucleic acids.

Disaccharides: Sugars Built of Two Monosaccharides

The potential for real diversity in the shapes and properties of carbohydrates comes from the linking of monosaccharide monomers into larger molecules. Two monosaccharides joined together form a disaccharide. For example, common table sugar, *sucrose,* is made up of a glucose molecule linked to a fructose molecule and has the formula $C_{12}H_{22}O_{12}$. Sucrose occurs abundantly in honey and in the saps of sugar cane, maple trees, and sugar beets. Sucrose is the product of a condensation reaction in which a molecule of water is formed from two –OH groups as the two sugars join. In the process, the two hydrogen atoms and one oxygen atom form water, and the remaining oxygen atom is shared in a bond that links the two monosaccharides (Figure 3-9). This covalent C–O–C link is called a **glycosidic bond.**

Other important disaccharides include *lactose,* which is made up of galactose and glucose units and is the predominant sugar in milk, and *maltose,* which is composed of two glucose molecules and is the substance that makes malted barley sweet. Lactose and maltose are also shown in Figure 3-9.

Disaccharides are a common form in which sugars are transported in plants and animals. More important, they are sources of sugars to be built into larger polymers or burned during cellular reactions to yield the energy needed for life processes.

Polysaccharides: Storage Depots and Structural Scaffolds

Living organisms form long-chain carbohydrates called polysaccharides by linking large numbers of monosaccharides into polymers by means of glycosidic bonds. Without this natural linkage process, life on Earth would be very different. The protective walls and coatings of bacteria and of all the cells in higher organisms are built in good part from sugar polymers. And inside cells, polysaccharides serve as an efficient way to store large quantities of nutrients that can be used later to fuel cellular processes. At a different level, human culture has evolved with a continuing dependence on polysaccharides, not only for food, but also for the wood we use for building and burning, for paper, and for many kinds of fibers used in making cotton, linen, and other fabrics.

The most important polysaccharides are starch, glycogen, and cellulose. Other biologically significant polysaccharides include *chitin,* a major component of the shells of insects and of crustaceans such as lobsters and crabs, and *glycosaminoglycans,* which are found on the surface of animal cells. We will discuss chitin and glycosaminoglycans in later chapters.

Starch

The major nutrient reserve of plants is **starch,** which is actually a mixture of two polysaccharides, *amylose* and

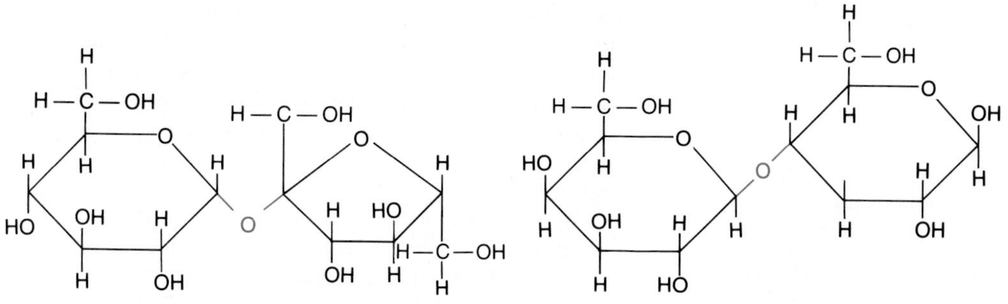

(Glucose+Fructose)
Sucrose

(Galactose+Glucose)
Lactose

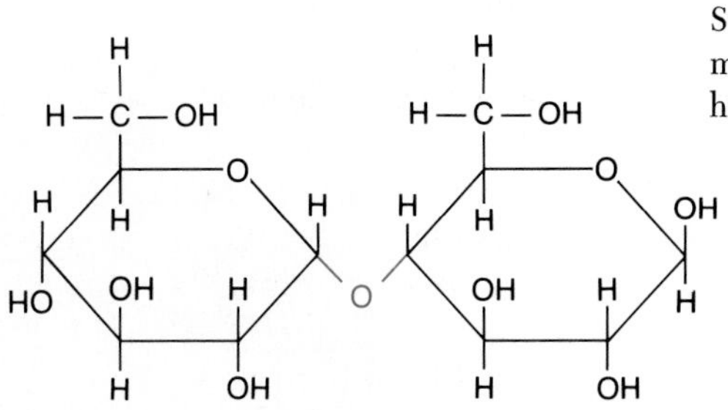

Figure 3-9   THREE COMMON DISACCHARIDES.
Sucrose is ordinary table sugar and lactose is the sugar in milk. Maltose is found in malt. Glycosidic bonds are highlighted in color.

(Glucose + Glucose)
Maltose

amylopectin. The proportions of these polysaccharides in starches vary from one kind of plant to another. Amylose occurs as long, straight polymer chains composed of units of glucose (the "beads" on that chain) joined by glycosidic bonds, as shown in Figure 3-10. An amylose molecule usually is composed of at least 1,000 units of glucose, but this number can vary. Amylopectin is also made up of glucose units bonded together, as in amylose, except that there are many branches joined to the string by cross linkages, as shown in Figure 3-10. This results in separate chains of glucose units being bound together. An amylopectin polymer usually contains about 20,000 glucose monomers.

Starch inside seeds is the nutrient reserve that supports early growth of tiny plantlets. Starch also makes grains (the seeds of wheat, corn, rice, and so on) a rich source of nutrition for humans and other animals. Within the seed are *enzymes* (a class of protein that speeds up chemical reactions) that can hydrolyze the glycosidic bonds in starch, releasing the disaccharide maltose and, in turn, glucose monomers for use by the cells of the growing plant. Thus, as barley seeds begin to grow in warm soil, the disaccharide maltose accumulates as enzymes break down starch. Humans use malted barley seeds to flavor beer. Similar enzymes in human saliva also break starch into smaller saccharides and glucose units; thus a cracker begins to taste sweet if you hold it in your mouth for a few minutes.

Glycogen

We said earlier that in animal cells, glucose is the chemical currency that provides energy for life processes and serves as the basic building block from which other monomers (of lipids, proteins, and nucleic acids) are built. Animals must store this raw material to avoid the necessity of nonstop eating. However, it must be stored in a more stable, less reactive form than free glucose molecules, huge numbers of which would create many problems for living cells. That storage form is a polysaccharide called **glycogen.** Glycogen is synthesized only in animals and is made up of several thousand glucose units. It often is stored in the cells of the liver, lungs, and other organs, where it serves as a reserve of energy-rich sugar molecules. Glycogen chains have many branches, each of which may be only about ten to twelve glucose units long, and the branched chains may be held together by cross linkages like the ones in amylopectin (Figure 3-11a). The way glycogen molecules are built up and broken down in animals will be described in Chapter 37.

Cellulose

Cellulose is the fibrous structural material of plant cells and wood. Like starch, cellulose is composed of long chains of glucose units connected by glycosidic

(a)

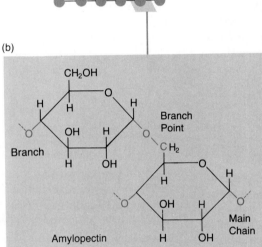

Amylose

(b)

Branch

Branch Point

Amylopectin

Main Chain

Figure 3-10   AMYLOSE AND AMYLOPECTIN.

Starch, one major nutrient reserve of plants, is a mixture of two kinds of polysaccharide chains: amylose and amylopectin. (a) Amylose consists of straight chains of glucose units, each linked to its neighbor by a glycosidic bond, as shown in the enlarged view. (b) Amylopectin is also a polymer of glucose. It differs from amylose in having occasional branches. A branch point is shown in the enlarged view. (c) Starch molecules cluster into granules or grains. An elongated granule can be seen clearly across the bottom of this portion of a plant cell under high magnification (about 10,000 times). A smaller starch granule is visible above it on the right.

(c)

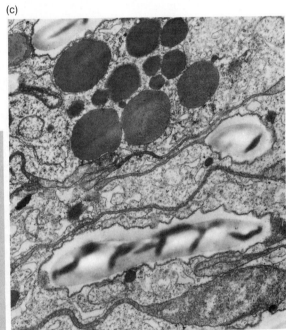

(a)

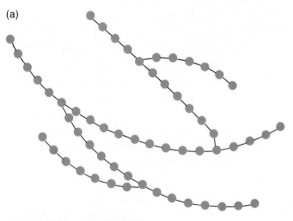

Figure 3-11   GLYCOGEN.

Glycogen is the nutrient reserve of animals in the same way that starch is the nutrient reserve of plants. (a) Made up of long, branched chains of glucose units, glycogen is held together by cross linkages like those found in amylopectin (Figure 3-10). (b) Dark glycogen granules dot this amphibian liver cell, shown magnified about 13,000 times.

(b)

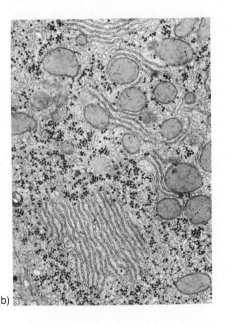

bonds, yet the two polysaccharides have very different properties: we eat starch in potatoes and bread, but we build houses from cellulosic wood. These physical differences are due entirely to the orientation in space of a single bond, as shown in Figure 3-12. In starch and glycogen, the orientation of the bond allows the glucose chains to twist into compact spirals. In cellulose, the orientation of the bond prevents twisting, so that the chains are straight. This molecular rigidity explains the fibrous quality of cellulose and, in turn, the strength of leaves and stems and the hardness of wood. Reactive –OH molecules projecting from both sides of the cellulose molecule allow it to link up with adjacent molecules, thereby forming stable, tough fibers (Figure 3-12).

Many organisms have the enzymes necessary to break the glycosidic bonds in starch and glycogen, but not those for breaking the bonds in cellulose. That is why a cracker will break down in your mouth, but wood or cotton never will, no matter how long it remains there. A few organisms, such as termites and cows, have microorganisms in their guts that produce the necessary enzymes for hydrolyzing cellulose; thus these animals are able to feed on wood or on grass and leaves. Despite what some health-food stores may claim, humans cannot digest cellulose; since we lack the appropriate microorganisms, cellulose passes through the gut without being degraded. Cellulose does, however, serve as roughage, or fiber, that aids in moving food through the digestive tract.

At the beginning of this chapter, we said that life is written in a language of biological molecules and that the meaning of these biological words lies in the order of the letters, the units of the polymers. Let us consider what these polymers say: they are, in a sense, long-winded repetitions of the same few letters. For example, if glucose is the letter *A*, then starch, glycogen, and cellulose all say the same thing: "AAAAAAAAA." Although there is much additional information in the three-dimensional shapes of these polymers (straight, branched, and spiraling chains) and in the structural properties that are based

on these shapes, the range and complexity of information that can be communicated by the repetition of a monomer are limited.

Now let us consider a second type of biological molecule, the lipids.

LIPIDS: ENERGY, INTERFACES, AND SIGNALS

Ice cream and cheeseburgers are fatty because of lipids. Humans, despite their sparse body hair, can stay warm because of deposits of lipid insulation in the skin. And every cell in the body is able to retain its watery contents because water-repellent lipid molecules surround it with a raincoatlike membrane. In fact, **lipids** share one particularly important attribute: they are insoluble in water. Let us look at the three main types of lipids: fats and oils, phospholipids, and steroids.

Fats and Oils: Energy Storehouses of Cells

Fats and oils are familiar substances that serve as nutrient reserves in animals and plants and are essential elements of the diet for most animals. Corn oil and olive oil—both yellowish liquids at room temperature—are typical lipids extracted from plant tissue. Lard and butter—whitish solids at room temperature—are typical fats prepared from animals.

Fats and oils are made up of two basic units: glycerol and fatty acids. **Glycerol** (Figure 3-13b) is a three-carbon alcohol with three hydroxyl groups attached. Because –OH groups readily form hydrogen bonds with water

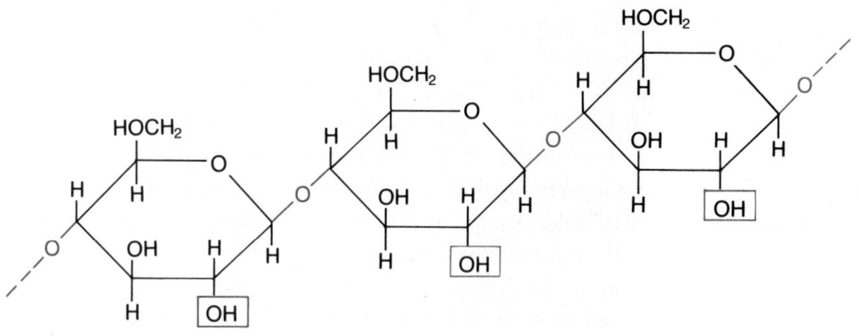

Figure 3-12 CELLULOSE.
Like amylose, the structure of the plant fiber cellulose consists of straight chains of glucose units. Cellulose owes its rigidity to the orientation of the bonds that join the carbon atoms to the oxygen atom that forms the bridge (shown in color). The –OH groups link with adjacent fibers, helping to form the stable, tough fibers.

molecules, glycerol is highly water soluble. **Fatty acids** (Figure 3-13a) are long chains of carbon atoms attached to a carboxyl group (–COOH), which gives fatty acids their acidic properties. Because three fatty acids usually are joined to one glycerol molecule, fats and oils are called **triglycerides** (Figure 3-13c).

Although both glycerol and the –COOH group are soluble in water, the entire oil molecule is insoluble because the hydroxyl and carboxyl groups of the two kinds of units react with each other and thus are no longer available to form hydrogen bonds with water. The bonds that link the glycerol to the fatty acids in a triglyceride are **ester bonds,** which form between any alcohol and any carboxylic acid. As with the formation of glycosidic bonds between monosaccharides, the formation of ester bonds results from a condensation reaction in which one molecule of H_2O is eliminated as each ester bond forms.

An important feature of the fatty-acid molecules in lipids is the number of double bonds (C=C) in their carbon chains. If there are no double bonds, then each carbon in the chain is linked to a maximum number of hydrogen atoms. Such chains are said to be *saturated* with hydrogen atoms. In contrast, if some adjacent carbon atoms are linked by double covalent bonds, then the fatty acid is *unsaturated* because double-bonded carbons are not bound to the maximum number of hydrogens.

The degree of saturation (the proportion of double versus single bonds in the chain) determines important properties of the lipid. One is the melting point: the greater the saturation, the higher the melting point. Farmers have long known that *neat's-foot oil,* a highly unsaturated lipid extracted from the hooves of cattle, has a low melting point. When applied to shoes, neat's-foot oil keeps them soft and pliable in cold weather. It is no wonder that unsaturated fats are present in the feet of cattle, reindeer, and penguins, all of which can stand for days on ice or snow. Oils from these animals' warm interiors, however, tend to be saturated. Therefore, these oils have a high melting point; they are quite stiff and inflexible when chilled; and they are not good lubricants.

Fats and oils are energy-rich compounds because they contain so many C–H covalent bonds. All organic compounds, including the lipids, can be burned, or *oxidized*—most only by fire, but others by means of stepwise breakdown processes inside living cells. During the oxidation process, C–H and C–C bonds are broken and replaced by C–O bonds, the end product being CO_2. Fats and oils provide more energy than do sugars and proteins because they have fewer C–O bonds to begin with; thus almost all the bonds in the molecule become oxidized, and energy is released as each C–H or C–C bond is broken. Due to this great capacity for storing energy, fats and oils serve as long-term nutrient reserves in plants and animals, while carbohydrates provide glucose and immediate fuels for most cellular processes.

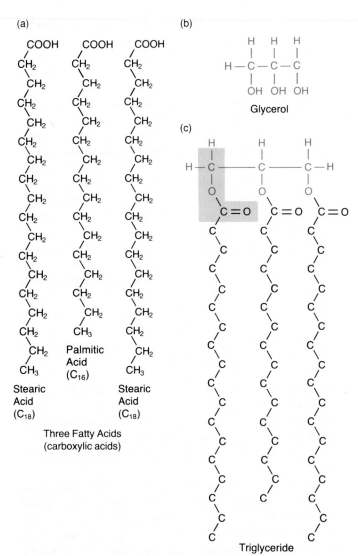

Figure 3-13.   THE STRUCTURE OF AN OIL.
Three fatty acids (a), connected by ester bonds (outlined in yellow) to a glycerol molecule (b), form a triglyceride (c).

Fats usually are not broken down into monomers and used as fuel unless carbohydrate levels are low. In financial terms, we could say that fats and oils are an organism's savings account, whereas carbohydrates are its checking account. Layers of fat or reservoirs of oil or other lipids often accumulate in organisms before long periods in which renewal of carbohydrate stores would be difficult or impossible—hibernation, migration, seed and plant dormancy, and embryonic development.

A variation on oils is the *waxes,* molecules made up of a long-chain alcohol linked to the carboxyl group of a fatty acid. Large numbers of wax molecules packed together form a waterproof outer layer on the leaves of plants, the bills and feathers of some birds, and other structures in living organisms. Waxes are also lightweight and provide buoyancy in certain deep-water fishes and other animals.

structural integrity. No cell can continue to exist if its lipid "skin" is punctured badly or damaged.

Phospholipids: The Ambivalent Lipids

Whereas fats and oils are made up entirely of carbon, hydrogen, and oxygen atoms, **phospholipids** contain two additional kinds of atoms: nitrogen and phosphorus. The main compound in phospholipids is *glycerol phosphoric acid* (Figure 3-14a). This is essentially glycerol with a phosphate group where one fatty acid would normally be found in an oil molecule (compare Figures 3-13 and 3-14). Phospholipid molecules exemplify the significance of functional groups: the charged phosphate group ($-PO_4^{3-}$) constitutes a water-soluble (hydrophilic) region on a water-insoluble (hydrophobic) molecule. This local property is the basis for the structure and function of cell membranes.

As in fats and oils, the two fatty acids in phospholipids are joined to the glycerol with ester bonds. In many phospholipids, the phosphate group attached to the glycerol is also bound to a nitrogen-containing functional group (choline, a common one, is shown in Figure 3-14b). The overall shape of a phospholipid molecule is something like a head with two long, thin tails streaming from the nape of the neck; the glycerol phosphate forms the hydrophilic head, while the fatty-acid chains form the hydrophobic tails. The behavior of phospholipids is rooted in this shape. Since the tails are hydrophobic and the head is hydrophilic, the molecules have an ambivalent approach to water: if phospholipids are introduced into the zone between a layer of oil and a layer of water, the hydrophobic tails will orient themselves to extend into the oil, while the hydrophilic heads will face the water. Thus the phospholipids form an interface, or separating layer, between the oil and the water (Figure 3-15). Soaps and detergents also have water-hating and water-loving regions that act to chemically separate oils and dirt from clothes and dishes. The fence-straddling behavior of phospholipids accounts for the structure of cell membranes, which are double-layered phospholipid barriers surrounding living cells. Without such waterproof layers, as we said in Chapter 2, the fluid contents of cells would escape, and the organism would lose its

Steroids: Regulatory Molecules

Steroids, a third type of lipid, are much less abundant in living cells than are fats, oils, and phospholipids, but they are no less important. Unlike the other lipids, ste-

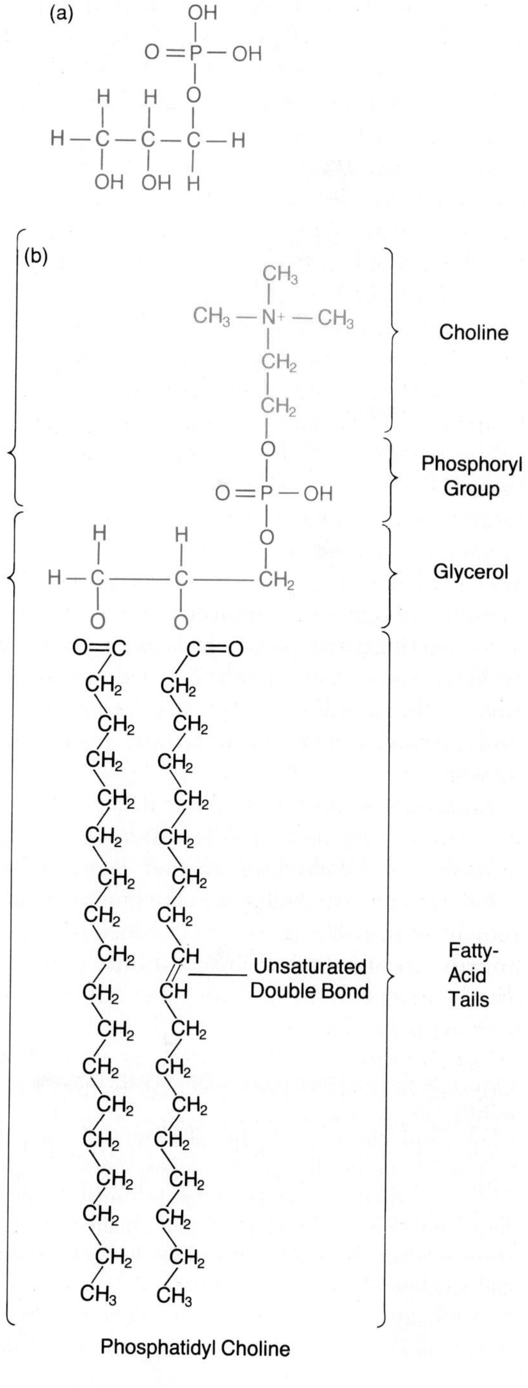

Figure 3-14   PHOSPHOLIPIDS.
(a) Structural formula of glycerol phosphoric acid. (b) The structure of a phospholipid. Phospholipids are the major structural component of membranes and are responsible for the retention of water in our bodies. Choline can be replaced by other molecules, such as ethanolamine, and other fatty acids may be substituted for those shown.

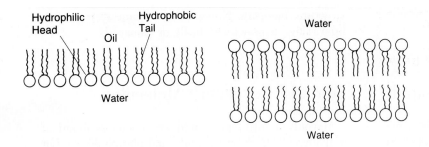

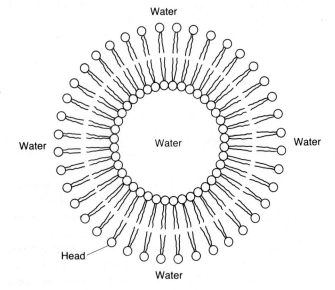

Figure 3-15   BEHAVIOR OF PHOSPHOLIPIDS IN WATER.
The hydrophilic heads face the water, and the hydrophobic fatty-acid tails are turned away from the water. Two layers of phospholipids may have their fatty-acid tails facing toward each other and away from water. This may occur in lipid droplets, as shown below, in which a volume of water is enclosed inside the lipid sphere. This arrangement is reminiscent of the composition of a living cell, with its lipid membrane "raincoat."

Deposits of cholesterol and other substances can build up inside blood vessels and slow down or block the flow of blood, thus increasing blood pressure and the risk of heart attack and stroke. Many nutritionists believe that eating saturated lipids such as butter, cheese, cream, and animal fats increases the formation and deposition of cholesterol in some people. However, the precise causes and mechanisms of this abnormal build-up are yet to be discovered.

The diverse lipids—whether solid or fluid, whether charged or nonpolar, whether saturated or unsaturated—play many vital roles in life's economy. Equally versatile, however, are the proteins—the structural components and agents that speed the chemical reactions of every living cell.

roids are not made up of glycerols and fatty acids. Rather, they are interconnected rings of carbon atoms with functional groups attached, as illustrated in Figure 3-16. Like most other lipids, however, steroids are nonpolar and hydrophobic; thus they are insoluble in water and can dissolve in oils or in lipid membranes themselves.

Some steroids act as vitamins (Chapter 34), and others, such as estrogen and testosterone, are hormones (Chapter 37). *Cholesterol,* another common steroid, has important beneficial effects on the fluidity of many cellular membranes. But this steroid, which is regularly manufactured in the body, may contribute to heart disease.

PROTEINS: THE BASIS OF LIFE'S DIVERSITY

When the water is removed from almost any living thing (except a woody plant), about 50 percent of the remaining matter is protein (Table 3-1). **Proteins** are a class of diverse macromolecules that determine many characteristics of cells and, in turn, of whole organisms. They may be classified according to the functions they perform. *Structural proteins,* for example, help to form bones, muscles, shells, leaves, roots, and even the gossamer cell "skeleton" that provides shape and allows cell

Figure 3-16   STRUCTURES OF SOME COMMON STEROIDS.
Cholesterol affects the fluidity of membranes and is believed to be involved in the deposition of fatty materials within blood vessels. Testosterone is a male hormone that is responsible for the development and maintenance of secondary sex characteristics and for the maturation and function of accessory sex organs.

Cholesterol

Testosterone

Table 3-1   **MOLECULAR COMPOSITION OF LIVING MATERIAL**

Component	Dry Weight (percent)
Carbohydrate	15
Lipid	10
Protein	50
Ribonucleic acid	15
Deoxyribonucleic acid	3
Small molecules and inorganic compounds	7

movement. Other examples are *protein hormones*, which serve as chemical messengers; *antibodies*, which fight infections; and *transport proteins*, which act as carriers of other substances (the hemoglobin protein in blood, for example, transports oxygen). One of the most important kinds of proteins is the **enzymes,** which speed up, or catalyze, chemical reactions. Because chemical reactions underlie every biological activity, enzymes are essential agents of the life process itself. In reality, the actions of an organism and its cells are inseparable from its particular constellation of enzymes, each of which has a specific function.

All proteins are polymers constructed of subunits called *amino acids*. There are twenty types of amino acids in proteins. Thus the biological language expressed in proteins is not the same droning repetition of "AAAAAAAAA," as in starch and glycogen, but a huge vocabulary of complex "words" based on an alphabet (the twenty amino acids) that is almost as large as our own; like the twenty-six letters in the Roman alphabet, the twenty amino acids are capable of a virtually infinite number of combinations.

As with words whose meanings are determined by the arrangement of letters, the meaning of a protein rests in the exact order of its amino acids. The order gives the molecule its special characteristics as a structural component in a cell, as an enzyme, as a carrier, or whatever. Within those categories, the order of amino acids gives the protein its unique properties as, for instance, a carrier of oxygen rather than of potassium. We can see from the huge number of protein "words" that both individuality and diversity in nature stem largely from this class of biological molecules. For example, the unique spectrum of proteins in one species of oak tree accounts for its distinguishing traits. It is easy to understand why the British chemist J. A. V. Butler regarded proteins as the substances most typical of life.

Amino Acids: The Building Blocks of Proteins

Proteins are built from monomers called **amino acids.** In each amino acid there are one central carbon atom, called the *alpha (α) carbon,* bound to a hydrogen atom,

an amino group (–NH_2), a carboxyl group (–COOH), and a side chain, represented by R.

In water, the carboxyl group may dissociate, giving up a hydrogen ion. Meanwhile the amino group, a base, can accept a hydrogen ion. Thus a molecule forms that has one positive group and one negative group.

These events have important consequences because such charged groups keep these essential monomers from passing freely through lipid cell membranes.

Each of the twenty amino acids commonly found in proteins has a distinctive R group. As Figure 3-17 shows, R can be as simple as a single hydrogen atom, as in glycine, or as complex as the double ring structure in tryptophan. Some amino acid side chains are hydrophobic; some are hydrophilic; and the others are ambivalent. Note in Figure 3-17 that seven amino acids have nonpolar, uncharged side chains and therefore are hydrophobic. Nine other amino acids are hydrophilic either because they have charged side chains or because they have properties that allow hydrogen bonding to other molecules. The remaining four amino acids are ambivalent in that they are partially soluble in water. The properties of a protein are determined in good measure by the R groups of its constituent amino acids. Thus, for example, mostly hydrophobic side chains will make a protein quite insoluble.

Some proteins may carry various molecules or groups of atoms. Without these so-called *prosthetic groups,* which differ radically from the R groups of the amino acids, the protein could not function properly. A prosthetic group can be a single atom, such as of copper or zinc, or a complex organic compound, such as the iron-containing heme group found in the blood protein hemoglobin. Many essential biological functions occur at the site of a prosthetic group, as in the binding of oxygen to the heme group in hemoglobin.

Earlier in the chapter, we said that stereoisomers always have a central carbon atom attached to four different atoms or groups of atoms. Since this is the general structure of the amino acids, each has two stereoisomers, the L-form and the D-form. The remarkably consistent occurrence of L-amino acids in all life forms suggests there may have been a powerful selection of the L-form billions of years ago when life began on Earth. We will consider this selection in Chapter 19.

(a) Hydrophobic Amino Acids

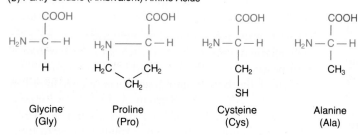

Valine (Val) Leucine (Leu) Isoleucine (Ile) Methionine (Met) Phenylalanine (Phe) Tyrosine (Tyr) Tryptophan (Trp)

(b) Partly Soluble (Ambivalent) Amino Acids

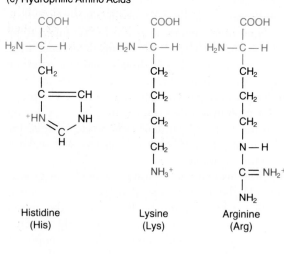

Glycine (Gly)   Proline (Pro)   Cysteine (Cys)   Alanine (Ala)

Figure 3-17   AMINO ACIDS.

Amino acids are the building blocks of proteins. This figure gives the structural formulas and conventional three-letter abbreviations of the twenty amino acids.

An amino acid consists of an amino group, a carboxyl group, a hydrogen atom, and a unique side chain R group bonded to a central atom, which is known as the α-carbon. The common backbone of each amino acid is shown in the figure in blue. In proline, the nitrogen atom and the α-carbon atom of the backbone are part of a ring of atoms.

(a) The hydrophobic amino acids. (b) The ambivalent amino acids. (c) The hydrophilic amino acids. Glutamic acid and aspartic acid each have a carboxyl group on their side chain (highlighted in color). In the cellular environment, the carboxyl group usually is negatively charged because the proton is dissociated. Histidine, lysine, and arginine have basic groups on their side chain (highlighted). These basic groups usually are positively charged because they accept protons. The rest of the hydrophilic amino acids can form hydrogen bonds.

(c) Hydrophilic Amino Acids

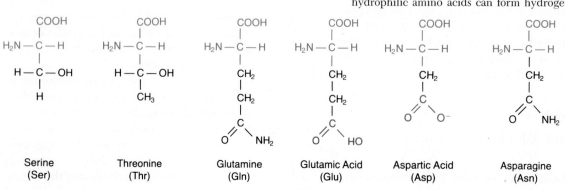

Histidine (His)   Lysine (Lys)   Arginine (Arg)

Serine (Ser)   Threonine (Thr)   Glutamine (Gln)   Glutamic Acid (Glu)   Aspartic Acid (Asp)   Asparagine (Asn)

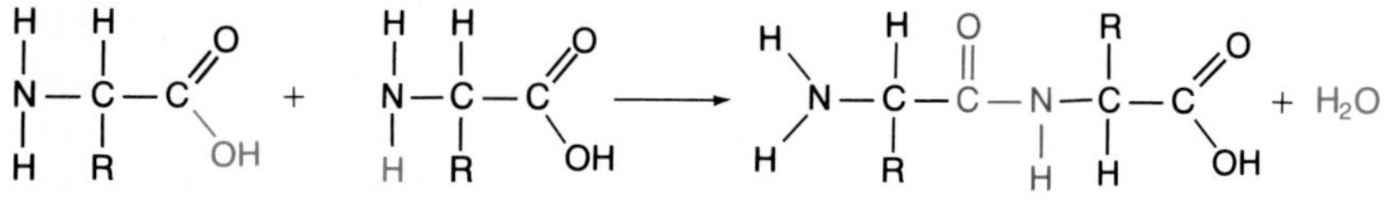

Figure 3-18   PEPTIDE BONDS.
The peptide bond, shown in red on the right-hand side of the reaction, is formed by a condensation reaction. The two bonded amino acids make up a dipeptide.

Polypeptides: Amino-Acid Chains

Just as polysaccharides are made up of monomers strung together like strings of beads, proteins are chains of amino acids linked covalently by **peptide bonds.** Such bonds are the result of a condensation reaction between the carboxyl group of one amino acid and the amino group of another, as Figure 3-18 shows.

Two joined amino-acid units, or residues, are called a *dipeptide*. As Figure 3-18 shows, a carboxyl group protrudes from one end of a dipeptide and an amino group protrudes from the other end. Thus a third amino acid can join at either end, by means of another condensation reaction, to form a *tripeptide,* which also has a carboxyl terminus and an amino terminus. Condensation reactions can be repeated again and again so that long chains of amino-acid residues form, all connected by peptide bonds. These polymer chains are called **polypeptides.** A polypeptide chain typically contains from fifty to hundreds of amino-acid residues. No matter how long or short, the chain will have an amino terminus and a carboxyl terminus.

A protein molecule can consist of one polypeptide chain, two such chains, or several chains bound to one another in various ways. The formation of peptide bonds is the key to protein manufacture in living things, and as such unlocks many secrets of reproduction, heredity, and ongoing life processes.

Many proteins include a second kind of covalent bond, called a *disulfide bond* (–S–S–). As stated earlier, a disulfide bond results from the linking of two sulfhydryl (–SH)

groups, which are found only in cysteine residues (Figure 3-19a). Disulfide bonds can form between two cysteine residues on a polypeptide chain, causing a kink or loop in the chain, or they can join two polypeptide chains into one molecule; both possibilities are shown in Figure 3-19b.

Theoretically, a polypeptide could be made up solely of one type of amino acid, as starch is made up solely of glucose monomers, or it could contain two or more amino acids in various orders. There are no chemical restrictions on the number of times an amino acid can appear in a protein or on where it can be located along a chain, and there is nothing in the chemistry of the amino acids themselves that restricts the lengths of chains. Yet there are only twenty amino acids found in a seemingly endless variety of proteins. How can just twenty amino acids account for the millions of different kinds of proteins that biologists have observed in the living world? The answer lies in the order of the amino acids in polypeptide chains. Think, for a minute, about the size of the *Oxford English Dictionary,* a 15,486-page tome filled with English words composed of twenty-six letters of the

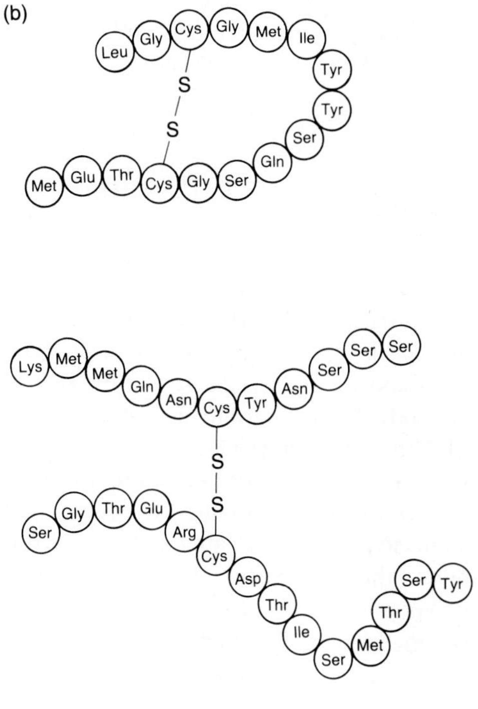

(b)

Figure 3-19   DISULFIDE BONDS.
(a) The formation of a disulfide bond. (b) The disulfide bond between two cysteine residues can link two parts of one polypeptide chain, as shown at the top, or it can link two polypeptide chains.

(a)

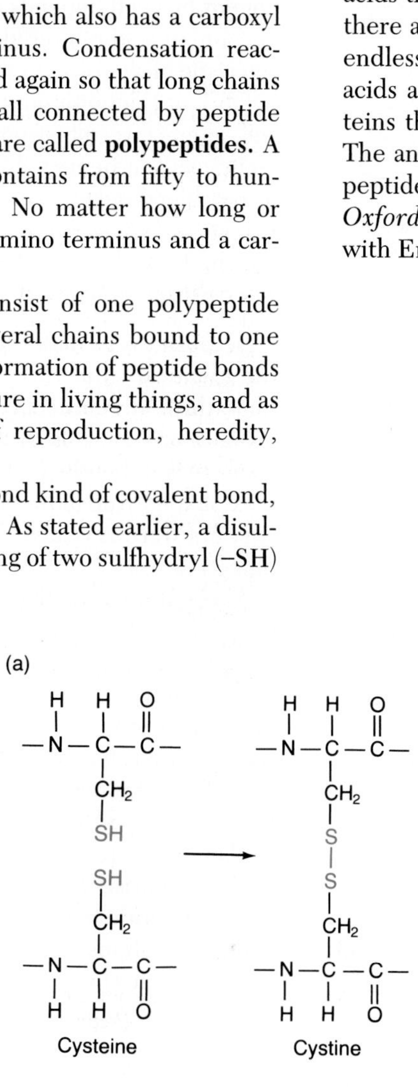

Roman alphabet. Now consider the number of different polypeptides that could be formed by linking fifty amino-acid molecules in a chain. If we think of the polypeptide as a row of fifty little boxes, each of which can be filled by one of the twenty amino acids, we can begin to calculate the number of possible polypeptide chains. The first box can be filled in twenty ways, as can the second: $20 \times 20 = 20^2$, or 400, combinations that we could use to fill just the first two boxes. Thus there are 400 possible dipeptides. Filling the third box would result in 20^3, or 8,000, possible tripeptides. Multiplied by all fifty boxes, the number of possible amino-acid sequences in a polypeptide only fifty subunits long is 20^{50}, which can also be expressed as 10^{65}. In nature, a polypeptide with fifty amino acids would be a fairly short chain. Proteins usually contain between 100 and 10,000 amino acids and have molecular weights of 10,000 to 1 million daltons. The presence of so many amino-acid residues means that the range of protein types, each with its individual properties based on the order of amino acids, is, for all practical purposes, infinite.

In nature, however, not every conceivable polypeptide is formed. There are millions of kinds of proteins (10^7 or 10^8) in the Earth's living organisms, but nowhere near 10^{65}. Many of the potential amino-acid sequences do not occur in real polypeptides, just as most potential combinations of the twenty-six letters do not form recognizable words in English or in any other language. The complex human body, for example, with its myriad tissues and organs, is estimated to contain only 30,000 to 50,000 types of proteins.

Natural proteins are never built of a random order of amino acids. The precise sequence of amino acids, whether common or rare, is dictated by an organism's hereditary material. We will explore this central principle of biology in great detail in later chapters.

The Structure of Proteins

One of the great discoveries in the history of biology was made by Frederick Sanger of Cambridge University in the early 1950s. He and his colleagues studied the protein *insulin*, a hormone that is produced in the pancreas and helps to regulate the breakdown of sugar by the body. Sanger's group chemically fragmented insulin, which is a relatively small protein, into pieces of varying sizes and then determined the exact amino-acid sequence for each fragment. By comparing how the fragment sequences overlapped, they were able to determine the sequence of amino acids for the full polypeptide chain. In 1958, Sanger was awarded the Nobel Prize for this important research.

Sanger provided a new way to study proteins, and his work led to the realization that the order of amino acids is the key to the structure and function of proteins and that a specific sequence characterizes each type of protein. Before Sanger's discovery, biochemists had concentrated on measuring the relative quantities of the different amino acids in proteins. This would be equivalent to analyzing the meaning of the word "banana" by counting the number of *a*'s, *b*'s, and *n*'s rather than noting their order. Within a few years, biologists discovered that sequencing is fundamental not only to protein structure, but also to the hereditary information present in all cells and, furthermore, that the sequence of the subunits in these informational molecules controls the amino-acid sequence in proteins.

Primary Structure

The linear sequence of amino acids in a protein is referred to as its *primary structure*. Whether a protein functions as an enzyme, a hormone, or a structural component of a cell depends on its primary structure. If just one amino acid is altered in a protein, an essential characteristic of the protein might change. For instance, the painful and debilitating human disease known as sickle cell anemia results from the substitution of a single amino acid in hemoglobin, the oxygen-carrying protein in red blood cells.

Secondary Structure

Because every molecule of a given kind of protein has the identical amino-acid sequence, they all assume the same three-dimensional shape. But why? How does the order of amino acids govern the way a long polypeptide chain twists, bends, and folds into a characteristic complex shape? And how does shape affect function? The answers were revealed through some brilliant scientific detective work.

At about the time that Sanger was studying amino-acid sequences, the chemists Linus Pauling and Robert Corey were studying the spatial relationships of amino acids in polypeptide chains. They knew that the subunits of some proteins tend to occur in regularly repeating patterns (like the bends in a coiled spring). These patterns form the *secondary structure* of the protein. Pauling and Corey's experiments were designed to probe and explain these repeating configurations.

The scientists found that atoms are *not* free to rotate around peptide bonds. This means that the peptide bond and the six atoms on either side of it are in a single plane, even though atoms and side chains attached to the α-carbon may rotate. A polypeptide chain thus is like a string of playing cards, with each card free to rotate relative to its neighbors but unable to bend and twist itself.

Pauling and Corey also examined the formation of hydrogen bonds that can form between –NH and –C=O groups along the main drain. The researchers had the flash of insight that marks great science: they reasoned that a polypeptide chain should assume a regular conformation that allows the maximum number of hydrogen bonds to form. They speculated that the inability of atoms to rotate about peptide bonds limits the ways a chain can fold to just a few basic shapes—the repeating patterns that form the secondary structure. Working with precise scale models of atoms, peptide bonds, and

amino acids, they fit the components together in various ways to work out the possible and probable secondary foldings. As a result of this model building, they predicted the two most likely secondary folding patterns: the α-helix and the β-pleated sheet.

In the **α-helix,** shown in Figure 3-20, the long chain of amino-acid residues is wrapped like a spiral staircase into a *helix.* The chain is held in position by hydrogen bonds joining the N–H group of one peptide bond with the C–O group of a peptide bond four subunits down the chain. The spiral α-helix is a more stable configuration than other conformations that do not allow such hydrogen bonds to form. Hence, regions of polypeptide chains

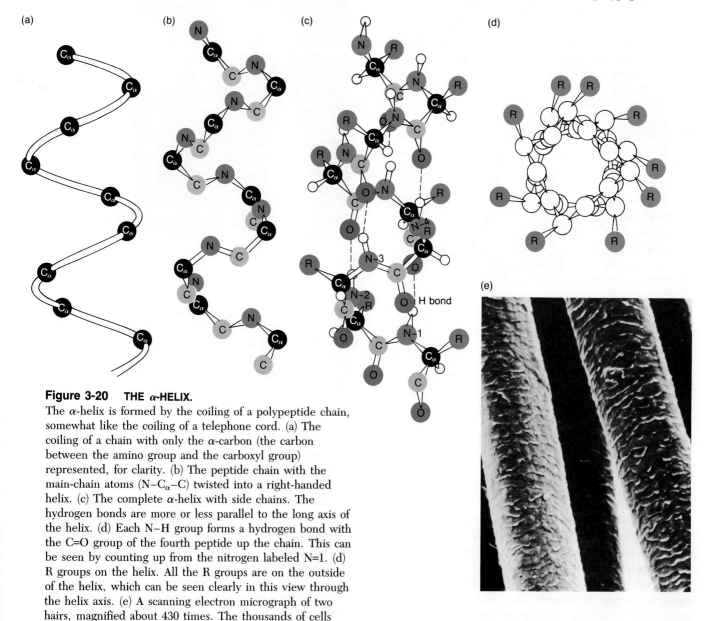

Figure 3-20 THE α-HELIX.
The α-helix is formed by the coiling of a polypeptide chain, somewhat like the coiling of a telephone cord. (a) The coiling of a chain with only the α-carbon (the carbon between the amino group and the carboxyl group) represented, for clarity. (b) The peptide chain with the main-chain atoms (N–Cα–C) twisted into a right-handed helix. (c) The complete α-helix with side chains. The hydrogen bonds are more or less parallel to the long axis of the helix. (d) Each N–H group forms a hydrogen bond with the C=O group of the fourth peptide up the chain. This can be seen by counting up from the nitrogen labeled N=1. (d) R groups on the helix. All the R groups are on the outside of the helix, which can be seen clearly in this view through the helix axis. (e) A scanning electron micrograph of two hairs, magnified about 430 times. The thousands of cells seen here are filled with keratin protein in the α-helical configuration.

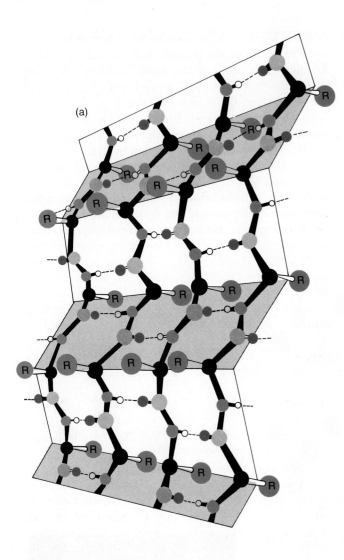

(a)

Figure 3-21   β-PLEATED SHEETS.

A silk scarf is composed of fibers arranged in β-pleated sheets. (a) The β-pleated sheet is an almost fully stretched out polypeptide chain that forms an accordionlike sheet of connected molecules. Hydrogen bonds are formed between adjacent parallel chains (dashed lines). The R groups point alternately up and down and therefore do not touch one another. (b) This silk fiber (magnified about 40,000 times) consists of proteins that are mostly β-pleated sheets.

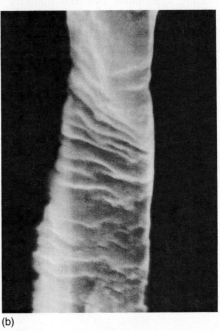

(b)

capable of forming α-helices do so spontaneously, even as the polypeptide chain elongates during its synthesis.

In the configuration called the **β-pleated sheet,** two or more polypeptides lying side by side become cross linked by hydrogen bonds and form an accordionlike sheet of connected molecules (Figure 3-21). Both the α-helix and the β-pleated sheet may repeat at intervals in a protein and contribute to its secondary structure, as shown in Figure 3-22 for the enzyme lysozyme.

A few important structural proteins occur almost entirely in one of these two secondary forms. Keratin, for example, is an α-helical protein found in mammalian hair, fingernails, horns, and claws, in porcupine quills, and in birds' feathers, among other sites. Fibroin, a protein in the silk strands spun by silkworms and spiders, is made up largely of β-sheets. These structures help explain the observable physical properties of hair and silk. When damp and warm, hair—sheep's wool in particular—can be stretched to almost twice its original length

as the α-helices stretch and numerous hydrogen bonds break. Silk, however, does not stretch when damp and warm because in the β-sheet structure, the molecules are already almost fully extended. Stretching silk would require breaking covalent bonds, a much harder task than breaking weak hydrogen bonds.

Tertiary Structure

Proteins have a higher order of organization than primary and secondary structure. An example is *myoglobin,* an oxygen-storing protein of muscle cells. Myoglobin contains eight α-helices folded into the complex shape shown in Figure 3-23. Myoglobin has eight stretches of amino acids with the α-helix structure connected by short segments of peptide chains. Almost 75 percent of the amino acids in myoglobin are stabilized in these helices by their own hydrogen bonds. The stretches of myoglobin with no regular α-helix structure are extremely important: at each such region, there is a bend in the polypeptide chain. It is because of these bends that the helical sections fold into the characteristic shape shown

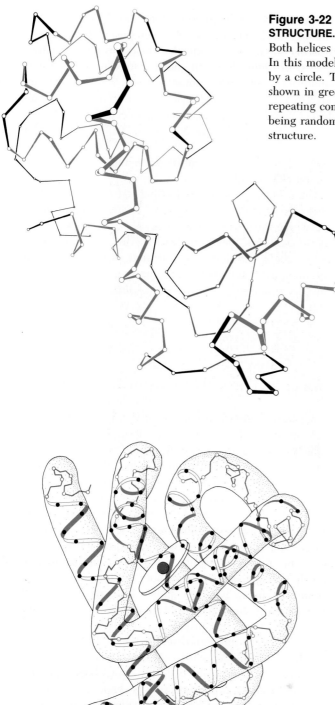

Figure 3-22   HELICES AND SHEETS IN A PROTEIN'S SECONDARY STRUCTURE.
Both helices and sheets may contribute to the secondary structure of a protein. In this model of the enzyme lysozyme, each amino-acid residue is represented by a circle. The helix regions are shown in blue, and the sheet regions are shown in green. The areas shown in black are not composed of regularly repeating configurations such as the helix and the sheet and are referred to as being random. Each lysozyme molecule has exactly the same random areas of structure.

in Figure 3-23. This shape is called the *tertiary structure.*

The three-dimensional folding of the tertiary structure is, of course, imposed on the primary structure (the precise amino-acid sequence) and the secondary structure (the spiraling or pleating of parts of the chain). Virtually all water-soluble proteins like myoglobin form a compact, *globular* shape as a result of this tertiary folding. These include blood proteins, enzymes, and antibodies. Several detailed studies show that a globular protein is a roughly spherical molecule with an irregular, crinkly surface studded with various chemical groups. Those chemical groups are often hydrophilic side chains with functional groups such as $-COOH$ or $-NH_2$. The internal core of the globular molecule is relatively dry; it is filled predominantly with portions of polypeptides that generally have hydrophobic side chains.

Water-soluble globular proteins are not the only type with tertiary structure. The most abundant protein in the animal kingdom is *collagen,* the insoluble structural protein that makes up the fibrous component of skin, tendons, ligaments, cartilage, and bone. Collagen gives tendons the tensile strength of steel wires, makes the corneal covering of the eye extremely tough even though it is clear as glass, and gives leather its toughness. This protein is made up of molecules shaped like long, thin cigars that are some 200 times as long as they are wide. Huge numbers of these long, thin molecules are aligned along the axis of each collagen fiber (Figure 3-24). In turn, the fibers are arranged in parallel bundles. Just as globular proteins are soluble because of their exposed hydrophilic side chains, collagen in a fiber is insoluble because of the bonding of the side chains between adjacent collagen molecules.

Quaternary Structure

Some proteins have yet another level of organization, called quaternary structure. These molecules are made up of two, three, or more polypeptides, each folded into secondary and tertiary shapes and then intertwined in a complex multichain unit like pieces of a three-dimensional puzzle. The *quaternary structure* is the arrangement of these separate polypeptides in a three-dimen-

Figure 3-23   α-HELICES IN PROTEIN'S TERTIARY STRUCTURE.
The protein myoglobin demonstrates the tertiary structure of protein molecules. Eight α-helices are connected by short stretches that do not have a regular structure; it is at these points that the molecule folds into its characteristic shape. Only seven of the α-helices can be seen in this view. Note the heme prosthetic group (red), which carries oxygen in this muscle protein.

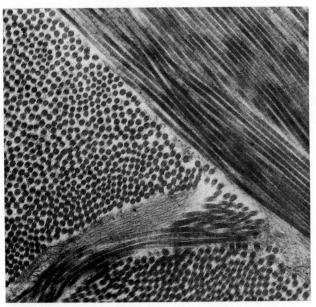

Figure 3-24   COLLAGEN: MOST ABUNDANT PROTEIN IN THE ANIMAL KINGDOM.
The structural protein collagen gives strength to skin, tendons, ligaments, cartilage, and bone. Thousands of collagen molecules are aligned into fibers, which are arrayed in parallel bundles. Here, collagen fibers (magnified about 30,000 times) from a jawless fish called a lamprey can be seen in cross-section (dots) and in longitudinal bundles.

Figure 3-25   THE QUATERNARY STRUCTURE OF HEMOGLOBIN.
The quaternary structure of hemoglobin results when four polypeptides (α_1, α_2, β_1, β_2), each twisted into its characteristic tertiary shape, intertwine and join by means of weak bonds. The heme prosthetic groups shown in red carry oxygen in this blood protein.

Heme Group

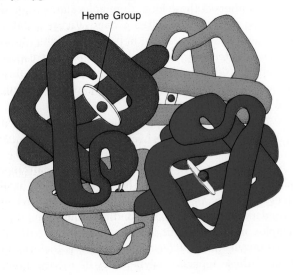

sional shape held together by weak bonds. The hemoglobin molecule, which is composed of four heme-bearing polypeptide chains, has the quaternary structure shown in Figure 3-25. In addition to forming such true quaternary structures, some proteins form various aggregates. One such aggregate is seen in a virus that infects tobacco plants. The protein molecules in that virus contain 2,130 polypeptides, each with a molecular weight of more than 17,000 daltons! When assembled, these chains form the entire protective coat of the virus particle.

Quaternary structure is the highest order of protein structure. Figure 3-26 traces the levels of organization on which quaternary structure depends back to the amino-acid residues. The precise linear sequence of amino acids, the side chains present, and the particular foldings and interactions combine to yield the marvelously complex structure we see as a final, functional protein.

The Causes of Proteins' Specific Three-Dimensional Shapes

Once biologists had elucidated the four levels of protein structure, they began to question why these particular shapes exist as they do. Are the characteristic shapes molded by some mechanism in cells, or do proteins assume these shapes automatically? And once assembled, why do proteins stay kinked, folded, and conjoined in complex shapes instead of relaxing and intertwining like spaghetti in a pot of boiling water?

Biologists investigating these questions found, first of all, that proteins assume their shapes automatically, given the appropriate conditions in the solution surrounding them. This was shown by experiments with ribonuclease, an enzyme that breaks down a kind of nucleic acid. This small protein has a polypeptide chain of 124 amino acids and four disulfide bonds, like the ones in insulin. If ribonuclease is treated with certain chemicals, it *denatures;* that is, the disulfide bonds break, and the enzyme loses its properties and its shape (Figure 3-27). The three-dimensional structure collapses during denaturation, and the chains of ribonuclease go limp as a pile of cooked spaghetti. If the denaturing chemicals are removed, the ribonuclease molecules regain their native (original) shape and activity. This simple experiment is significant because it shows that a protein's amino-acid sequence alone is enough to specify its three-dimensional structure. In other words, the linear sequence leads to *self-assembly* of the native state.

A second important finding is that the final shape of a protein is held and stabilized by chemical bonds and

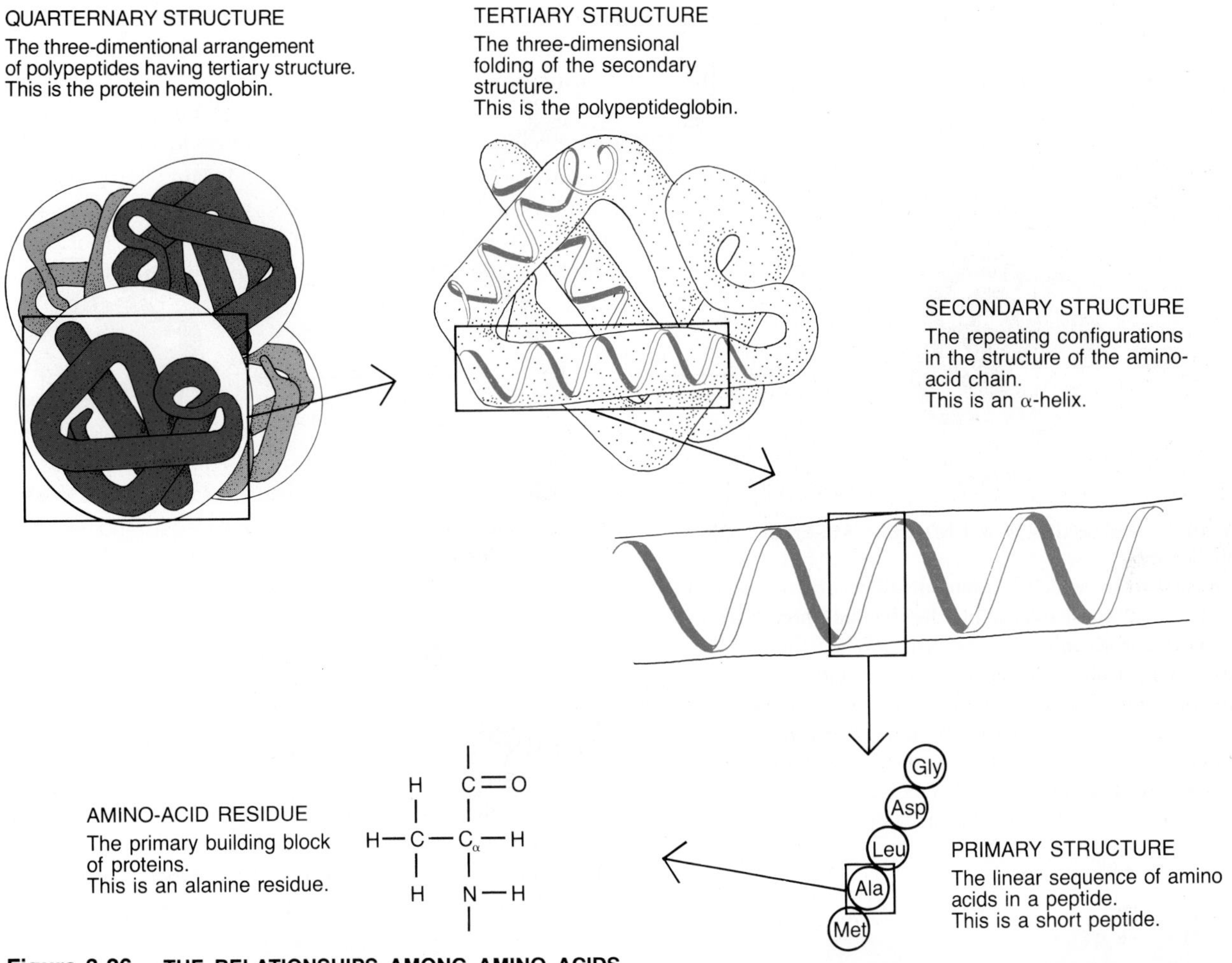

QUARTERNARY STRUCTURE
The three-dimentional arrangement
of polypeptides having tertiary structure.
This is the protein hemoglobin.

TERTIARY STRUCTURE
The three-dimensional
folding of the secondary
structure.
This is the polypeptideglobin.

SECONDARY STRUCTURE
The repeating configurations
in the structure of the amino-
acid chain.
This is an α-helix.

AMINO-ACID RESIDUE
The primary building block
of proteins.
This is an alanine residue.

PRIMARY STRUCTURE
The linear sequence of amino
acids in a peptide.
This is a short peptide.

Figure 3-26   THE RELATIONSHIPS AMONG AMINO ACIDS AND THE PRIMARY, SECONDARY, TERTIARY, AND QUATERNARY STRUCTURES OF A PROTEIN.

forces. Covalent bonds between individual amino acids can contribute to folding or stabilizing shape—the disulfide bonds found between cysteine residues on a chain are an example—but most of the bonds that contribute to higher-order shapes in protein molecules are weak bonds of four types: hydrogen bonds, ionic bonds, van der Waals forces, and hydrophobic interactions.

We already have talked about the role that hydrogen bonds play in the formation of secondary structure, including α-helices and β-pleated sheets. Ionic bonds form between charged groups, such as the acidic or basic side chains of many hydrophilic amino acids. For example, the carboxyl group on the side chain of glutamic acid has a negative charge, while the amino group on the side chain of lysine has a positive charge. If glutamic acid and lysine are brought close together as a polypeptide chain folds, an ionic bond can form between them, stabilizing the loop or fold in the protein's shape.

If two parts of a molecule fit together closely—for example, if an outfolding in one part of a chain closely fits an infolding in another part—the electron clouds of the atoms in each part interact, and weak attractive forces, called *van der Waals forces*, are generated. Van der Waals forces can cause the parts of the molecule to pack more tightly, helping to hold the protein in its tertiary structure (Figure 3-28).

Another very important weak force that helps to hold a protein's shape once it has been attained is *hydrophobic interaction*, the tendency of hydrophobic side chains to tuck themselves inside the dry interiors of globular protein molecules. This "fear of water" is so great that as the hydrophobic amino acids move inward toward the dry center of the protein molecule, they pull on the main chain groups to which they are attached. As a result, many hydrophilic side chains are pulled free of the hydrogen bonds they formed with water. As the hydro-

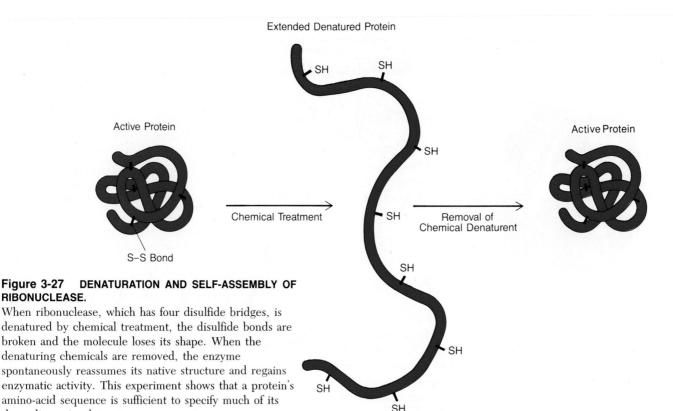

Figure 3-27   DENATURATION AND SELF-ASSEMBLY OF RIBONUCLEASE.
When ribonuclease, which has four disulfide bridges, is denatured by chemical treatment, the disulfide bonds are broken and the molecule loses its shape. When the denaturing chemicals are removed, the enzyme spontaneously reassumes its native structure and regains enzymatic activity. This experiment shows that a protein's amino-acid sequence is sufficient to specify much of its three-dimensional structure.

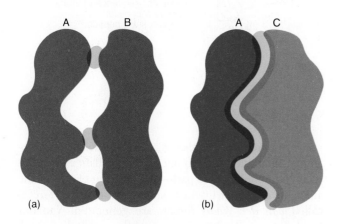

Figure 3-28   VAN DER WAALS FORCES.
Van der Waals forces are effective over only very short distances; therefore, the force can hold together two molecules or parts of molecules only if they are complementary in shape. (a) Molecules A and B do not have complementary shapes; surface contact is minimal; and van der Waals forces are very weak. (b) Molecules A and C have complementary shapes; surface contact is extensive; and van der Waals forces are effective. Note that van der Waals is a force of attraction rather than a bond. The yellow highlight denotes the area in which the attractive forces are effective.

philic side chains assume positions in the interior of the folded protein, they can reestablish hydrogen bonds, but this time with each other rather than with water. Thus, the main force responsible for the stability of globular proteins is hydrophobic interaction. The precise shape of such a stabilized molecule will result from the formation of a maximal number of weak bonds.

Despite their seeming stability, proteins usually are on the brink of falling apart because of these various types of weak bonds. Proteins are stable only within a narrow range of temperatures and degrees of acidity (pH). Outside that range, they change shape or denature completely. Egg white, for example, is a slimy, transparent protein solution that, on heating, denatures into a rubbery, opaque mass. Although egg white denatures permanently, many other proteins can regain their native properties if subjected to somewhat gentler treatments.

If protein molecules were held in shape by strong covalent bonds, they would be enormously stable. However, the proteins also would be almost entirely unable to function as enzymes or to carry out other tasks essential to life. Thus reversible and rapid changes in shape are a hallmark of proteins and a necessity for life processes. Indeed, the structural instability due to weak bonds is actually a major advantage and a precondition for their functioning within living things.

Another advantage of weak bonds is that they allow self-assembly. A polypeptide can literally try out many configurations before chancing into the native form, which then persists because the maximum number of weak bonds form. Without weak bonds, easily made and broken, this trial-and-error testing could not take place. The alternative—using covalent bonds—would be extraordinarily complicated, would have extremely high energy costs, and would not allow the flexing and changing so critical to life.

The Importance of Protein Shape

The key to understanding life is the realization that the molecules and macromolecules of living organisms are assembled in precise ways. They are then built into cell components that carry out specific tasks and, in turn, into complete cells, the smallest living units. Cells can function independently as whole organisms (such as a bacterium or an amoeba), or millions of cells may function together as multicellular organisms. At each level of assembly, the underlying chemistry of molecules remains the same. Thus the correct secondary, tertiary, and quaternary folding and assembly of a given kind of protein may be critical to the function of one part of a cell: only if that function is normal can the cell be normal, and only if the cell is normal can the organism be normal.

We see, then, that the precise ordering of amino acids in protein chains has dramatic, unexpected, but essential consequences for all the higher levels of organization, including the final outcome, life itself. We will return to the significance of proteins' three-dimensional structure at the end of this chapter, when we discuss the importance of structure in the function of all types of macromolecules.

NUCLEIC ACIDS: THE CODE OF LIFE

We come, now, to the fourth class of biological molecules, **nucleic acids,** the information-bearing "code of life." Like proteins, nucleic acids have a specific linear sequence of subunits, a language of chemical "letters."

These letters spell out instructions both for characteristics passed on to offspring and for translating that hereditary message into proteins that will be built into new cell parts, cells, and organisms. Nucleic acids are polymer chains, as are the complex carbohydrates and proteins, made up of building blocks called **nucleotides.** Each nucleotide consists of a nitrogen-containing base, a five-carbon sugar, and a phosphate group, as shown in Figures 3-29 and 3-30. There are two types of nucleic acids: *deoxyribonucleic acid* (DNA) and *ribonucleic acid* (RNA). The sugar molecule in the nucleotides that make up DNA is a five-carbon monosaccharide called *deoxyribose;* a similar sugar, called *ribose*, is found in RNA. These sugars become bonded to the nitrogen-containing bases through a condensation reaction, creating a *nucleoside.* The addition of a phosphate group to the sugar by means of another condensation reaction yields a nucleotide. There are four nitrogen-containing bases commonly found in nucleotides. In DNA, they are *adenine, guanine, cytosine*, and *thymine* (Figure 3-29). RNA also contains adenine, guanine, and cytosine; but instead of

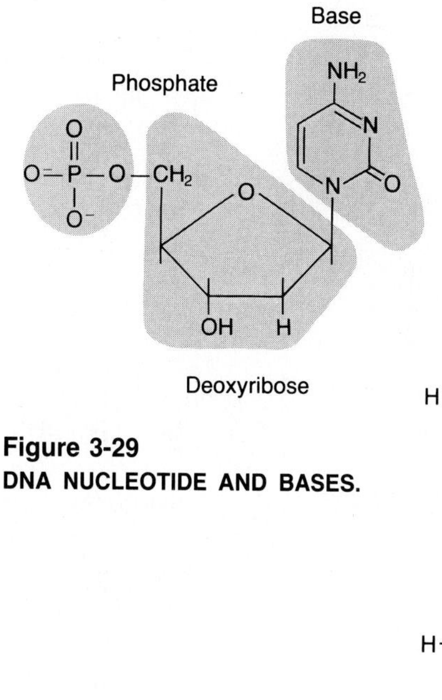

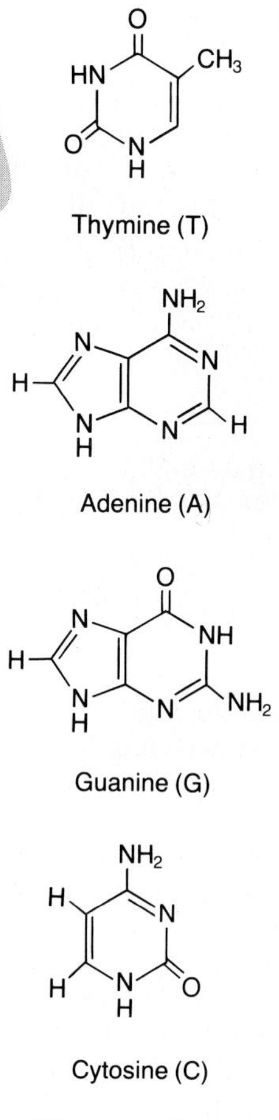

**Figure 3-29
DNA NUCLEOTIDE AND BASES.**

thymine, its fourth base is *uracil* (Figure 3-30). These are the "letters" that make up the nucleic-acid alphabet, and although there are fewer than the twenty amino acids in proteins, they are equally capable of encoding diversity, as we shall see.

Condensation reactions like those that produce proteins, carbohydrates, and some lipids join nucleotides together to form DNA and RNA. During these reactions, the nucleotides attach to each other in such a way that the sugar of one nucleotide is always bonded to the phosphate of the next nucleotide in the chain (Figure 3-31). Therefore, just as polypeptides have amino and carboxyl terminal ends, nucleic acids have a hydroxyl group at one end and a phosphate group at the other.

As with proteins, the range of possible nucleic acids is infinite; that is, both the length of the chain and the sequence of the four nucleotides are theoretically without limit. But just as with proteins, nucleic-acid "words" in nature have neither random size nor random sequence; the order of the letters, the nucleotides, in each nucleic-acid molecule is quite precise and is determined by preexisting nucleic acid of the same type in a preexisting cell. This statement is simple, but it shows that the process of heredity is unique to living organisms. As we shall see many times, life is a process that only can be handed down directly from the living, and it is based on the information contained in the precise order of nucleotides in nucleic acids.

In its native state, DNA consists of two chains of nucleotides twisted around each other in a double-helix conformation and held together by hydrogen bonds. RNA molecules, in contrast, are made up of single chains that may fold into complex shapes or remain stretched out as long threads. The sequence of nucleotides in DNA ultimately determines the structures of every protein, lipid, and carbohydrate molecule in every living thing. A great intellectual adventure led biologists to discover the meaning of the nucleic-acid language. This adventure will be described in Chapter 13, along with a more detailed description of nucleic-acid structure and the process by which DNA and RNA function as the "code of life."

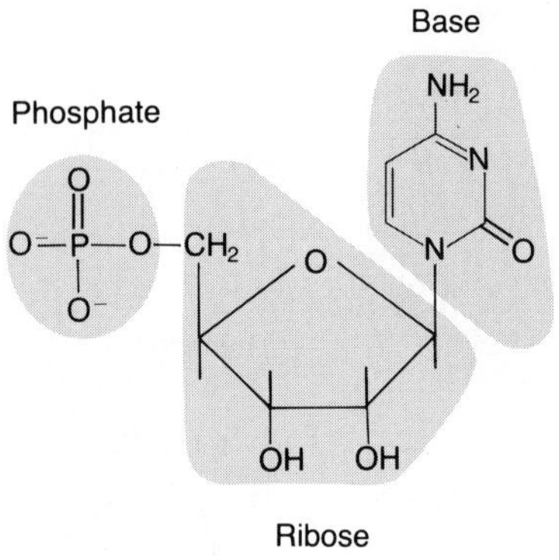

Figure 3-30
RNA NUCLEOTIDE AND BASES.

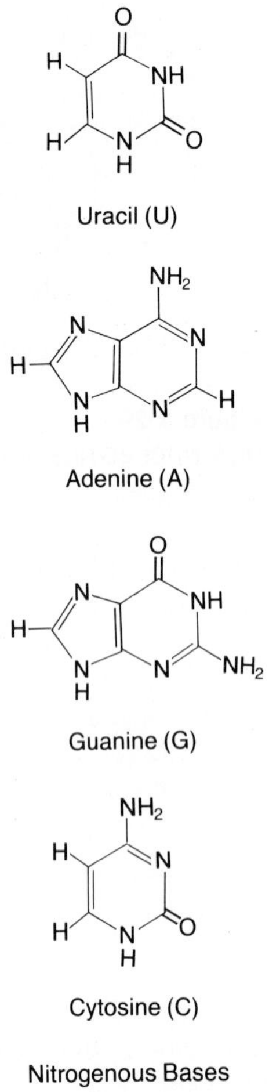

THE SPECIFICITY AND COMPLEMENTARITY OF MACROMOLECULES

Throughout this chapter, we have talked about the importance of molecule shape in biology. We have said that a protein's shape determines its properties and that molecular structure is therefore critical to the functioning and survival of cells. But how exactly does the three-dimensional shape of proteins and other molecules affect their function? To answer that question, we must understand two concepts: specificity and complementarity.

Organisms are not random collections of atoms or molecules that behave in random ways. Each individual is made up of specific substances; it manufactures or consumes food in specific ways; it has specific strategies of reproduction; and so on. Underlying each of these major life processes are hundreds to millions of biological molecules acting in distinctly characteristic ways. This is what is meant by *specificity*. This combined molecular activity forms the basis of highly ordered phenomena, such as physical traits and behaviors.

An organism's specific structures and reactions are based on the recognition of one molecule by another. This recognition comes about because of *complementarity*, the tendency of groups of atoms in molecules with

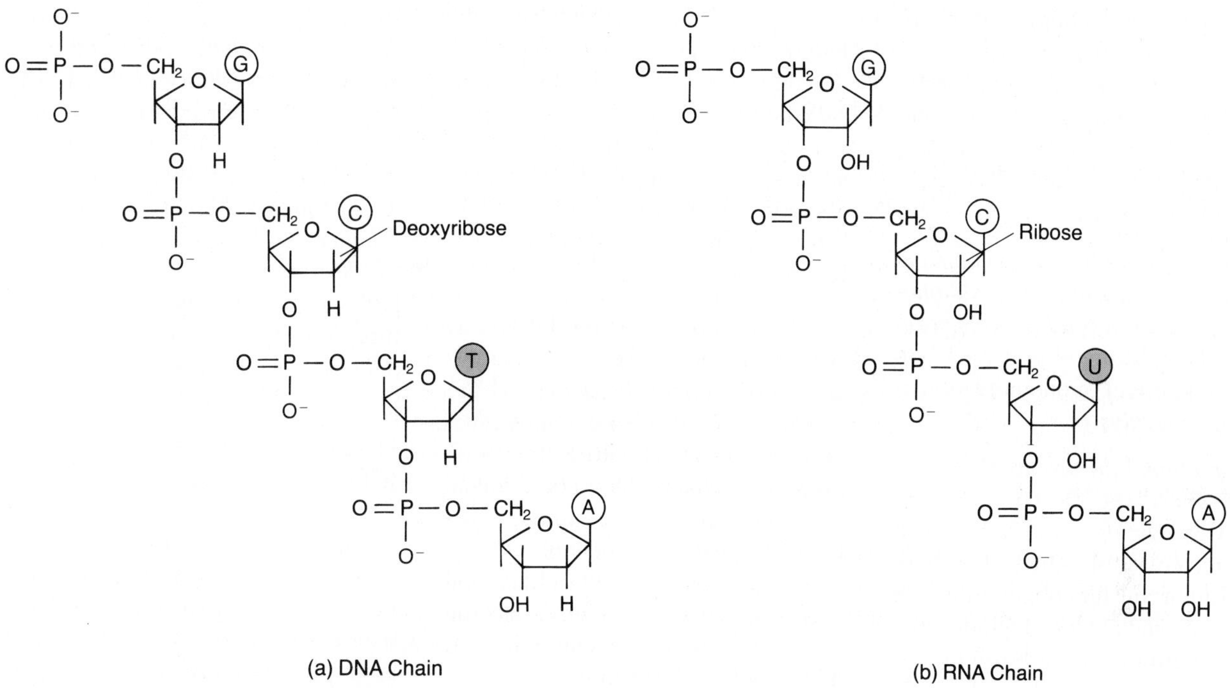

(a) DNA Chain

(b) RNA Chain

Figure 3-31   DNA AND RNA CHAINS.
(a) DNA is the fundamental building block of genetic material. (b) RNA also is involved in genetic processes. Both DNA and RNA are formed by condensation reactions: first the nitrogen-containing base combines with a five-carbon sugar to make a nucleoside; then the phosphate group bonds to the nucleoside by means of another condensation reaction to form a nucleotide. The nucleotides join with one another into chains also by means of condensation reactions. In reality, a DNA molecule is composed of two such chains wrapped about each other.

three-dimensional conformations to form bonds by "fitting" one another the way a key fits into a lock. When the key fits, a chemical reaction can take place, or a structure such as a fiber can begin to take shape. The complementary fit of proteins with specific shapes allows molecules to form the fibers of the tissues that make skin and tendons tough and resilient. The complementary fit of enzymes to other molecules inside a living cell allows the splitting of polysaccharides into monosaccharides to provide energy. And as we shall see later in this book, the complementarity of molecular shapes also explains how genetic material operates, why a cell can give rise to an identical daughter cell, and why children have some of the physical traits of both parents.

LOOKING AHEAD

In this chapter, we have seen that just as great novels are made up of separate words, life is written in a molecular alphabet. The "letters" that make up lipids, proteins, carbohydrates, and nucleic acids dictate not only the shapes, properties, and functions of these molecules, but also the assembly of "words" into articulate "phrases" and profound "ideas" expressed in flesh, blood, wood, hair, shells, feathers, and so on. But just as a novel must have scenes and action, life's macromolecular characters must meet and interact. This is the subject of Chapter 4.

SUMMARY

1. The *macromolecules* of life (carbohydrates, lipids, proteins, and nucleic acids) are made up of carbon, and are *organic compounds*. Carbon's unique properties allow it to form as many as four bonds with other atoms. Thus strings or rings of carbon atoms often form the skeletons of macromolecules.

2. *Isomers* are compounds with identical chemical formulas but different arrangements of atoms. *Stereoisomers* are mirror-image molecules with the same properties, while *structural isomers* have quite different properties.

3. Macromolecules are *polymers* formed by the linking of many *monomers* by means of *condensation reactions*. The splitting of polymers into their component monomers occurs through *hydrolysis*.

4. *Functional groups* are groups of atoms whose specific structure imparts a specific chemical behavior to the molecules of which they are part. Important functional groups in biological molecules include the hydroxyl group, the carboxyl group, the amino group, the aldehyde group, the ketone group, the methyl group, the phosphate group, the sulfhydryl group, and the disulfide group.

5. *Carbohydrates* are macromolecules that consist solely of carbon, hydrogen, and oxygen. Individual carbohydrate monomers, called *monosaccharides*, can be combined into very large carbohydrate polymers. Monosaccharides are three-, four-, five-, or six-carbon sugars with one C–O bond as part of an aldehyde or a ketone group. Monosaccharides, such as *glucose*, play important roles in the transfer of energy within cells.

6. *Disaccharides*, such as sucrose, are composed of two monosaccharides linked by *glycosidic bonds*. *Polysaccharides* are long polymers of monosaccharides linked by glycosidic bonds. The most important polysaccharides are *starch*, which is a storage material in plants; *glycogen*, a storage substance in animals; and *cellulose*, the fibrous structural material of plants.

7. *Lipids* are water-insoluble carbon compounds. Important lipids are *fats, oils, phospholipids,* and *steroids.* Fats and oils contain a *glycerol* molecule attached to three *fatty-acid* chains by *ester bonds;* thus they are *triglycerides.* Phospholipids are composed of a phosphate group and two fatty-acid chains attached to glycerol, making them hydrophilic at one end and hydrophobic at the other. Steroids have interconnected ring structures.

8. *Proteins* act as structural material, *enzymes*, chemical messengers, antibodies, and transport molecules. The monomers that make up proteins are the twenty amino acids.

9. *Amino acids* have a central carbon atom bonded to four functional groups: a carboxyl group; an amino group; a hydrogen atom; and an R side chain, which determines the particular amino acid's properties. Amino acids join by means of *peptide bonds* to form *polypeptides.* The enormous variety of proteins found in living things arises from the seemingly infinite ways in which the twenty amino acids can be arranged in long chains.

10. Every polypeptide has a unique amino-acid sequence, dictated by an organism's hereditary material. This sequence is referred to as the protein's primary structure. A protein's three-dimensional shape is based on its secondary structure (such as α-helices and β-pleated sheets), its tertiary structure (precise folding patterns), and its quaternary structure (intertwining of several polypeptide chains into one functional unit).

11. Proteins have the property of self-assembly; that is, they assume their native shape automatically in the appropriate conditions. The final shape of a protein is held and stabilized by weak hydrogen and ionic bonds, by van der Waals forces, by hydrophobic interaction, and sometimes by strong disulfide bonds. A protein's shape determines its specific behavior and function in a cell.

12. *Nucleic acids*, such as DNA and RNA, are polymers of *nucleotides*, which consist of a nitrogen-containing base, a five-carbon sugar, and a phosphate group. DNA and RNA differ in three respects: each contains a different sugar; one of their four bases is different; and DNA forms a double-helix shape, while RNA usually stays in single chains. The sequence of nucleotides in nucleic acids, which is inherited, determines the structures of proteins.

KEY TERMS

α-helix

amino acid

β-pleated sheet

carbohydrate

cellulose

condensation reaction

disaccharide

enzyme

ester bond

fatty acid

functional group

glucose

glycerol

glycogen

glycosidic bond

hydrolysis

isomer

lipid

macromolecule

monomer

monosaccharide

nucleic acid

nucleotide

organic compound

peptide bond

phospholipid

polymer

polypeptide

polysaccharide

protein

starch

stereoisomer

steroid

structural isomer

triglyceride

QUESTIONS

1. Give an example of two isomers of a compound that have different properties. Explain how they can contain exactly the same atoms but have different properties.

2. Many polymers are formed by the linking of subunits in a condensation reaction. What small molecule is produced in this reaction?

3. What subunits make up cellulose? Starch? Glycogen? How do the structures of these three polymers differ? What function does each polymer serve?

4. Can you think of an example of a saturated fat? An unsaturated fat? What is meant by the term "polyunsaturated"? Which type is more likely to be liquid at room temperature?

5. Give examples of some lipids, and explain their functions in living organisms.

6. What is meant by the primary structure of a protein? What determines the order of amino acids in a protein?

7. When a protein is denatured, what kinds of bonds are broken? When a protein is broken down into its component amino acids, what kinds of bonds are broken?

8. Give an example of a globular protein. Is it soluble or insoluble in water? Give an example of a fibrous protein. Give an example of a protein composed of several polypeptide chains.

9. How is the sugar in nucleic acids different from the sugar in starch?

10. Suppose you have a model-building kit that contains many nucleotides of four different kinds. How many different dinucleotides can you build? How many trinucleotides?

ESSAY QUESTIONS

1. Polysaccharides usually contain only a single kind of subunit, whereas polypeptides contain twenty different subunits. Furthermore, the subunits of different polypeptides are arranged in different order. Explain why polypeptides have such complex structure and are so diverse. Why does each cell need so many different polypeptides?

2. The change of a single amino acid in a protein (containing hundreds of amino acids) can alter the structure and function of the protein. How can a single change have this effect? Is it possible to change an amino acid without affecting the function of the protein? Explain why or why not.

SUGGESTED READINGS

CALVIN, M., and W. A. PRYOR. *Organic Chemistry of Life: Readings from "Scientific American."* San Francisco: Freeman, 1973.

This collection of articles covers most of the important molecules of living things in an easy-to-understand fashion.

DICKERSON, R. E., and I. GEIS. *The Structure and Action of Proteins.* New York: Harper & Row, 1970.

The excellent illustrations and clear text make this a fine book from which to learn about proteins.

STRYER, L. *Biochemistry.* 2d ed. San Francisco: Freeman, 1981.

A superbly illustrated, relatively simple text about the molecules and chemical processes of life.

4
CHEMICAL REACTIONS, ENZYMES, AND METABOLISM

Laws of Thermodynamics:
1. You cannot win.
2. You cannot break even.
3. You cannot get out of the game.

Anonymous

Sunlight, leaves, and browsing giraffe: Energy flows through all living things.

Life involves continuous change at every level. In the realm of molecules within living cells, simple sugars are built into starch, which in turn is digested. Proteins are synthesized, only to be broken down later to individual amino acids. Fats are synthesized and stockpiled. The information in nucleic acids is translated into proteins, which in turn are determinants of cellular structure and activity. At the level of whole organisms, a tree sprouts new leaves, which grow; the trunk widens; the bark deepens; and the branches stretch farther outward each year. A fish moves about in search of food, eats and digests, grows larger, reproduces, and ages. Except in a few rare cases, the organism that is no longer changing is no longer living.

Underlying every change in the living world are **chemical reactions,** transformations of sets of molecules into other kinds of molecules. We encountered two biologically significant types of reactions in Chapter 3: condensation reactions, which link simple sugars or amino acids into complex polysaccharides and proteins; and hydrolysis reactions, which split the basic units apart. During these reactions, molecules are transformed—complex chains are woven from simpler units or are dismantled piece by piece. But why do such reactions take place? What happens as they occur? And under what conditions can they occur? We shall see that underlying all chemical reactions and all change within living things is energy. For most reactions to take place within cells, energy must be expended. This is true whether a cell builds starch from sugar, or a plant builds new leaves with that starch.

In this chapter, we shall discuss chemical reactions and their **energetics.** How much energy is required to bring about a reaction? What form does the energy take? We also will examine the speed of chemical reactions and the process by which enzymes, the agents of biological change, hasten them. In fact, most reactions would take place too slowly to sustain life if it were not for enzymes. Thus our careful consideration of the energetics of chemical reactions will build a framework for understanding the activities of enzymes. Finally, we shall discuss metabolism, the web of simultaneous chemical reactions that go on every second of life. Appropriately, the word "metabolism" comes from the Greek word for "change": without metabolism, there is no life. So energetics, enzymes, and metabolism will provide us with the background to understand the living cell, the subject of Chapters 5 and 6, and the processes of cellular respiration and photosynthesis, which we will discuss in Chapters 7 and 8.

THE ENERGETICS OF CHEMICAL REACTIONS

As we saw in Chapter 2, molecules react in a specific way during a chemical reaction. *Reactants* interact with each other and are converted into *products* by means of the making and breaking of chemical bonds:

$$A + B \rightarrow C + D$$

reactants   products

A classic example of a chemical reaction occurs when a cell first ingests and then digests a protein. One of the last steps in that process involves splitting a dipeptide into two amino acids. During the reaction, as Figure 4-1 shows, two bonds are broken and two new bonds form. What makes this hydrolysis reaction—or, for that matter, any other type of reaction—take place? The answer is energy.

Energy Transformations and the Laws of Nature

Chemical reactions are always accompanied by energy transformations—that is, the change of energy from one form to another. Energy, which is the capacity to do work, occurs in many forms. *Light energy, heat energy,* and *electrical energy* are familiar from everyday life. **Kinetic energy** is the energy of motion, such as the energy generated by rushing water, a rolling rock, or moving molecules (Figure 4-2a). **Potential energy** is stored energy—the capacity to do work later. Water in a tank on top of a building, or a rock poised at the top of a hill, has potential energy (Figure 4-2b). Potential energy sometimes is called energy of position. **Chemical energy** is the energy stored in atoms and molecules and their bonds, so it is a kind of potential energy. For convenience, we sometimes shall refer to such energy as being stored in chemical bonds. In actuality, however, chemical energy is not localized at any specific site, but instead

Figure 4-1   HYDROLYSIS OF A DIPEPTIDE.
A hydrolysis reaction cleaves both the dipeptide and the water molecule. The H and –OH from water are joined to the separated amino acids.

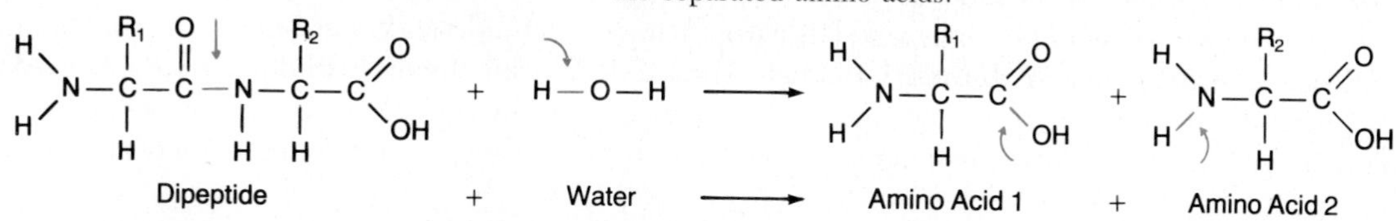

Figure 4-2   KINETIC AND POTENTIAL ENERGY.
(a) A thundering waterfall has kinetic energy. Each day, millions of liters of water tumble over spectacular Victoria Falls on the border between Zambia and Zimbabwe. This moving wall of water has tremendous energy to accomplish work such as turning the wheels of a power generator or simply wearing away the rocks below. (b) Stand beneath this rock, look up, and you will experience potential energy first hand. This giant boulder, poised on a pinnacle, has the capacity to accomplish work, such as crushing cars, houses, and people. Potential energy can be stored for long periods; this rock has been in place for centuries.

is a property of the atoms and the bonds that link them together.

The First Law of Thermodynamics

The different forms of energy share an important relationship that is described by the **first law of thermodynamics,** which also is called the law of conservation of energy. (Thermodynamics is the theory of the relations between heat and other forms of energy.) The first law of thermodynamics states that energy can change from one form to another form, but it can never be created or destroyed. Energy transformations take place every instant: light energy strikes plant cells and is converted, in part, to chemical energy, which is stored in the bonds of carbohydrate molecules. Chemical energy is released as heat energy when wood is burned in a fireplace or when sugar is "burned" in a cell. Heat energy is turned into kinetic energy in a steam engine. Potential energy is converted to kinetic energy when the

spigot on a water tank is opened and the water rushes out. In each case, energy changes form, but it never is created anew, nor is it destroyed—it is *conserved*.

No energy transformation in the universe is 100 percent efficient. Whenever energy changes from one form to another, some portion is converted to heat energy and is lost; it is no longer available to do useful work. Heat energy can cause the rapid, random movement of molecules and thus represents the most disordered form of energy. When a rock rolls down a hill, for example, and energy of position (potential energy) is converted to energy of motion (kinetic energy), some energy is lost because friction between the rock and the hill transforms kinetic energy into heat energy. This heat dissipates into the environment, is lost to the rock, and can never be recaptured. Similarly, when proteins are digested in an animal's stomach and dipeptides are split into separate amino acids, some of the energy in the dipeptide and water molecules is released and lost as heat. This heat may temporarily warm the animal, but it eventually dissipates into the surrounding environment.

The Second Law of Thermodynamics

The fundamental fact that heat is a by-product of *every* energy conversion is reflected in another universal principle: the **second law of thermodynamics.** This law states that the total energy in a system decreases inevitably as conversions take place and heat dissipates. A result is a tendency toward increasing randomness, or *disorder*, in the universe. In other words, because of heat loss during the energy conversion associated with all chemical reactions, the random motion of all the atoms in the universe is continuously increasing. Therefore, the universe itself is becoming increasingly disordered. The amount of disorder in a system is known as **entropy.** Because the unusable energy in the universe is continually increasing, the universe is becoming more and more disordered. Consider a simple illustration. When gas is burned as a blue flame in a stove, the chemical energy stored in gas bonds is converted to light and heat. Some of the heat energy can be used to boil water while some is dissipated and is no longer available to do work. This unusable energy represents increased entropy.

If the universe is heading toward disorder, how can we explain the complex structures of the molecules, cells, and organs that make up living things and the precise activities of their component parts? They are clearly the epitome of order, rather than disorder. For example, as we saw in Chapter 3, a protein is much more than a disorganized pile of amino acids, and each of these subunits is more ordered than the same atoms free of one another. So a cell—let alone a tree or a fish—is an incredible oasis of order in a desert of entropy, the disorder that surrounds the cell in the universe.

The explanation for the presence of order in the midst of disorder lies in the constant change we spoke of earlier. To stay alive and counteract the tendency of complex molecules to degrade, every cell must carry out chemical reactions continuously. Moreover, additional chemical energy must be expended to rebuild the molecules that have degraded. All the reactions that maintain the cell's structure and function give off heat. Some of that heat may be used by the cell or the organism, and some dissipates into the surroundings, increasing the kinetic energy of the atoms and molecules that make up the cell's or organism's environment, as the second law specifies. Thus the cell unavoidably contributes to the disorder of the universe as it labors to keep its own integrity. The price of this process is dear: sufficient energy in the form of nutrient molecules must be put into the cell both to drive the reactions that maintain the cell and to satisfy the second law of thermodynamics. Ultimately, the source of this energy is the sun, for it is the solar energy that is captured and stored by plants and other photosynthetic organisms for use in their own cells that moves up the food chain to plant-consuming and carnivorous animals. We will see in Chapter 7 how the reactions of metabolism tap the stored energy in nutrients, and we will learn in Chapter 8 how the sun's energy is captured and stored in those nutrients. Here the point is that at each level of consumption, energy is lost as heat, and entropy increases.

Eventually, certain of the essential reactions that support life stop, normal chemical changes come to a halt, and the cell or organism dies. Entropy wins out, as the order of the living thing fades, and disorder ensues. The animal or plant disintegrates back to "dust"—to the molecules and atoms of which it is formed. So the next time you gaze at a graceful oak, at a flitting butterfly, or at your face in the mirror, remember that there is a high price being paid every second for the maintenance of biological order.

FREE-ENERGY CHANGES

Earlier we said that when an energy conversion such as a chemical reaction takes place, some work is accomplished (the forming of new bonds), but because of the universal inefficiency of all energy transfers, some heat is lost as well. A system of equations allows scientists to determine the maximum amount of **free energy,** the energy available—or "free"—to do work as a result of a chemical reaction. The concept of free energy was developed by the Yale physicist Josiah Willard Gibbs, and in Gibbs's honor, it is symbolized G. The change in free energy, symbolized ΔG (delta G), equals the total usable energy in the products *minus* the total usable energy in the reactants.

But how is total energy or free energy measured? A molecule with its chemical bonds between atoms can be thought of as a box crammed with coiled springs. Each coiled spring has potential energy and can pop open (convert to kinetic energy) if conditions are right. Suppose you could open that molecular box and measure the energy of each coiled spring by breaking the chemical bonds in a molecule one by one and measuring their energy. If you summed the amounts of these bond energies released, you would have a number representing the total energy of that molecule. If you repeated this measuring exercise for reactants and products of various chemical reactions, you would find that in virtually *all* reactions, the total energy of reactant molecules (say, oxygen and glucose) differs from the total energy of products (say, carbon dioxide plus water, which result when glucose is "burned" metabolically).

When the total bond energy of the products of a reaction is less than the total energies of the reactants, heat is given off and entropy increases. Therefore, less total energy and less usable energy (free energy) remain in the products after the reaction. This type of reaction is said to be **exergonic** (meaning "energy out"). Because there is a decrease, or negative change, in free energy during an exergonic reaction, the ΔG of such a reaction is expressed as a negative number. For example, during the burning of 1 mole of glucose molecules to CO_2 and H_2O, ΔG, = -686 kcal/mole (Figure 4-3a).

In an **endergonic** ("energy in") reaction, the products have more total energy and more free energy than the reactants. This means there must be an input of energy, or the reaction cannot occur. Because there is an in-

Figure 4-3   EXERGONIC AND ENDERGONIC REACTIONS.
(a) If a person consumes a quick-energy drink while, say, playing tennis, the glucose in the drink may be broken down in the player's cells to CO_2 and H_2O. These products have less free, or usable, energy than did the glucose. Some of the energy in the glucose will drive the player's muscles, while some will be released as heat. In general, exergonic reactions are occurring. (b) Glucose taken in by the tennis player while at rest can be stored in the form of glucose–6–phosphate. This product has more free energy than did the glucose. Energy derived from other foods was added to drive the reaction. Therefore, such reactions are endergonic.

crease, or positive change, in free energy during an endergonic reaction, ΔG is a positive number. For example, when energy is provided, 1 mole of glucose molecules will react with phosphoric acid to form water and glucose-6-phosphate, a form of sugar that has more energy in its bonds and atoms than does glucose. For this reaction, $\Delta G = +3.3$ kcal/mole (Figure 4-3b).

The concept of free-energy change helps us understand why energy-producing reactions and energy-consuming reactions very frequently occur simultaneously within living cells (Chapters 7 and 8). In fact, the more we learn about cells and their chemical economy, the more evident it will become that free-energy changes underlie all life processes.

Dependence of Equilibrium Between Reactants and Products on Free Energy

Some chemical reactions will take place spontaneously, and some will not. Those of you who cook may have observed that vinegar (an acid) mixed with baking soda (a base) will react instantly, forming a salt and CO_2 and bubbling vigorously in the process. Why is this reaction spontaneous while many others are not? The answer, quite simply, is that the reaction between vinegar and baking soda is exergonic (ΔG is negative).

Let us consider what happens during an exergonic process. Does such a reaction proceed to completion so there are no reactant molecules left, or does it spontaneously stop at some point? In our vinegar and baking soda example, the reactants will bubble vigorously for a while, then less vigorously, and finally stop.

Chemists discovered long ago that as a reaction takes place, the combined free energy of the reactants drops until it equals the combined free energy of the products. At this point, $\Delta G = 0$, and the reaction is said to be at **equilibrium:** no further *net* conversion of reactants to products takes place. This does not mean that the combined concentrations of two reactants, A and B, will be equal to the concentrations of the products, C and D. (Concentrations are expressed as moles per liter.) If the reaction is strongly exergonic, virtually all the reactants will be converted to products before the equilibrium point is reached. Another way to say this is that products (C and D) must accumulate until their concentrations are high; that is, until their combined free energy equals that of the remaining reactants. In this situation, the concentrations of A and B will be extremely small at equilibrium, while the concentrations of C and D will be very large. This can be expressed as $A + B \rightarrow C + D$. In reality, the reverse reaction does occur at equilibrium, so a more accurate representation is $A + B \leftrightarrows C + D$. (Note: An arrow pointing to the left represents a reverse reaction; the relative lengths of the arrows reflect the proportions of reactants and products at equilibrium.)

If a reaction is only mildly exergonic, the free energy of the reactants might equal that of the products when the concentrations of the reactants, A and B, are about equal to those of the products, C and D. This is expressed as $A + B \leftrightarrows C + D$. In this case, at equilibrium the forward reaction, $A + B \rightarrow C + D$, occurs simultaneously and at about the same rate as the reverse reaction, $C + D \rightarrow A + B$. In fact, these equations make clear that in the mildly exergonic forward reaction, the ΔG is of the same magnitude as the ΔG in the mildly endergonic reverse reaction; in other words, about the same amount of energy is freed in the exergonic reaction as must be put into the reverse endergonic reaction. Finally, at equilibrium in a strongly endergonic reaction, the concentration of reactants remains substantially higher than that of products, a situation that can be represented as $A + B \leftrightarrows C + D$.

Chemists use the relative lengths of the arrows as an indirect indication of which set of molecules has a lower combined free energy and has to accumulate to a greater degree before equilibrium is reached; the long arrow points toward the side of the equation where the free energy is lower.

Another way to understand chemical equilibria is to think of energy as a third product of an exergonic reaction. That tells us that the reverse, or endergonic, reaction is less likely to occur and will require input of the "third product"—energy. Another feature of chemical equilibrium is the implication of a static condition. But a reaction at equilibrium is actually a dynamic condition in which both forward and reverse reactions occur, but at equal rates. Hence, no further net change in concentration of reactants or products occurs.

These considerations bring us to an important principle: in biological systems, the energy freed during one reaction frequently is utilized to drive another reaction. In other words, an exergonic reaction yields energy to bring about an endergonic one. Such pairs of reactions are said to be *coupled*. Coupling is the basis for many of the interrelated reactions in living cells. For example, energy from the breakdown of nutrients is stored in reaction products and used to fuel energy-requiring reactions, such as the building of proteins or fats. Coupling makes highly improbable (endergonic) reactions likely so that the macromolecules of life can be synthesized and maintained. For this reason and because the products of a given reaction generally become participants in other reactions, true chemical equilibrium for a given reaction within a cell is rarely achieved. This is especially so for the sequential reactions of metabolism, which we will discuss later in this chapter. It is important for our study of biology to understand chemical reactions and their equilibria so that we can predict whether reactions will occur spontaneously.

RATES OF CHEMICAL REACTIONS

So far in our discussion of energy flow during chemical reactions, we have talked about the energies of reactants and products as though all the molecules of a certain substance have identical energy states. This is an oversimplification. In a population of identical molecules, there is a distribution of energy such that some molecules vibrate and jostle about rapidly, others move more slowly, and still others have energy levels somewhere in between (Figure 4-4). The free-energy value of a compound really represents the *average* free energy of the molecules in a given population.

For purposes of discussion, we also implied that free-energy levels are the sole determinants of whether a re-

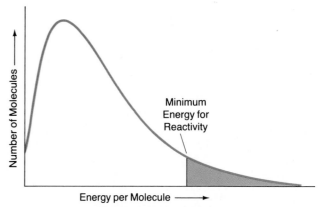

Figure 4-4 DISTRIBUTION OF ENERGY IN A POPULATION OF MOLECULES.
At the temperatures found in organisms, similar molecules have different energies and vibrate at different speeds. Only the molecules in the shaded area have enough energy to react with one another.

action will take place. It is true that unless the entropy (disorder) of the system increases, a reaction will not "go" spontaneously. But it is two very different things to say that a reaction is *energetically possible* and to say that it indeed takes place at a measurable rate. Over years, a large bone resting in a museum case will spontaneously break down into its constituent molecules. But the rate is so slow that even after a few centuries, the bone will look virtually unchanged. Under what we consider normal environmental temperatures and sea-level atmospheric pressure, most reactions take place extremely slowly, if at all. The reason is that there is an energy barrier that must be overcome, even if the free energies are favorable to the reaction. Another way of saying this is that many molecules just do not have enough energy to react reasonably rapidly. Consider two molecules, MN and OP. In order to react, they must collide. Colliding molecules have kinetic energy and can undergo reactions more easily than can molecules that are not adjacent. Some of the energy of the collision between molecules MN and OP goes into breaking the bonds of the atoms so that the entirely different molecules MO and NP can form. Just what occurs when MN and OP react to form MO and NP? For a fleeting instant as MN and OP collide, the bonds are so distorted that it is difficult to determine whether M is bonded to N or O or whether N is bonded to M or P. This intermediate state is called the *transition state*. When reacting molecules are in the transition state, they may be thought of as being at the top of a hill between the reactants and the products; the hill itself represents the energy barrier to the reaction (Figure 4-5).

Significantly, the energy (both kinetic and potential) in most colliding covalent molecules is not high enough

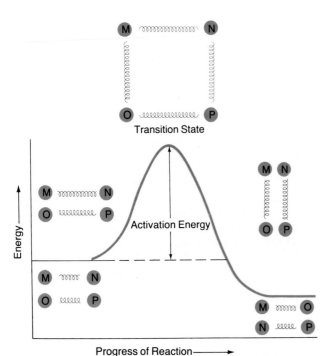

Figure 4-5   STAGES IN THE REACTION BETWEEN MOLECULES.

Bonds between atoms are rearranged during the transition state. There are several stages in the reaction between molecules. (1) molecules MN and OP collide; (2) the kinetic energy of the collision begins to distort the bonds (stretch the "springs") between atoms; (3) in the transition state, it is difficult to determine which atoms are bonded to each other (four stretched springs); (4) after the reaction, the new bonds between the atoms of the products MO and NP momentarily remain distorted; (5) at the lowest energy level, the bonds between the atoms of the new product molecules, MO and NP, are stable.

for them to reach the transition state. The natural repulsion between the electrons of the colliding molecules cannot be overcome in a low-energy collision, and the molecules simply bounce off one another. Also, the orientation of the colliding molecules often chances to be wrong so that the sides bumping together are not the most suitable for bond formation. Only molecules that collide with sufficient impact can make it to the transition state and go over the energy barrier. They must have a certain minimum amount of energy, called **activation energy,** to do so. The activation energy of a molecule is the amount needed to break its bonds. Most molecules that bump into one another as they are jostling about in their environment do not attain this energy level, and thus collisions among them are unproductive—no reactions take place. Productive collisions—those that do lead to reactions—occur among *activated molecules* having the necessary activation energy.

Two questions arise from this consideration of colliding molecules: Why is the number of productive collisions important, and what factors influence whether a collision will be productive or unproductive? The number of productive collisions in a system is important because *it determines the rate of a chemical reaction*, and that rate is critical to survival of living organisms. In a young rabbit, for example, carbohydrate molecules from the grass it nibbles must be broken down quickly enough to provide energy to move its muscles so the animal can outrun a hungry fox. For the most part, the rates of reactions in living organisms are extremely fast and must be so if the organism is to carry out the continuous changes associated with life.

If most reactions are slow because the number of productive molecular collisions is comparatively small, what factors allow organisms to carry out chemical reactions at a fast enough rate to sustain themselves? There are three influences on the rate of a reaction: temperature, concentration of the reactants, and catalysts. We will discuss each in turn.

Temperature and Reaction Rates

Heat can supply the energy needed to activate molecules—to push them over the energy hill—so that a collision between them is productive. If a population of molecules is heated, the molecules move about more quickly, and more high-energy molecules chance to collide (Figure 4-6a). The reason is that heat energy dramatically increases both the frequency with which molecules collide and the force of the collisions. As a result, many more collisions attain activation-energy levels, and the reaction accelerates. Even though the speed of some moving molecules goes up only about 3 percent with each 10°C rise in temperature, the rate of a chemical reaction goes up 100 to 300 percent; clearly, temperature affects the rate of a reaction far more strongly than it affects the speed of molecules.

Temperature changes affect every reaction differently because of differences in the activation energies of different kinds of molecules. One reaction might undergo a fourfold increase in reaction rate in response to an increase in temperature, while the same temperature increase speeds up the rate of a second reaction by sixteenfold (Figure 4-6b).

The normal internal temperatures of most cells and organisms are too low to promote very rapid chemical reactions. If elevated temperature was the sole strategy available to speed reactions in living things, problems would arise because the very high temperatures required probably would denature essential macromolecules. Fortunately, living cells can use changes in concentration and enzyme catalysts to speed up vital reactions in the ice-cold depths of the sea, high on a mountain, or in other cold places on Earth.

(a)

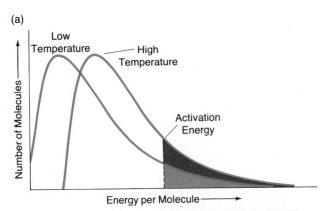

Figure 4-6   REACTIONS SPEEDED BY TEMPERATURE.
A rise in temperature speeds reactions. (a) Only a fraction of molecules in a given population has sufficient kinetic energy to react (blue area). At a higher temperature, the kinetic energy of all the molecules is higher, and thus a much larger fraction of molecules attains activation energy. (b) Rises in temperature increase reaction rates, but each set of reacting molecules has its own characteristic response. An increase of 20°C causes the rate of reaction 1 to increase to four times the original rate. The same temperature change causes the rate of reaction 2 to increase sixteenfold.

(b)

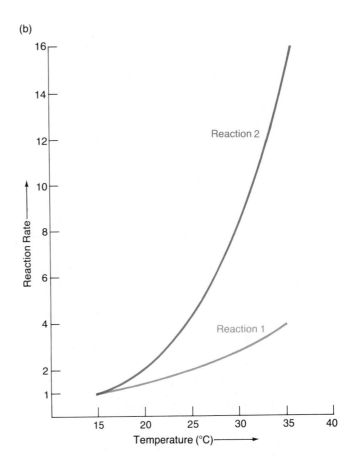

Concentration and Reaction Rates

It makes sense that the more molecules there are in a given volume, the more collisions will occur among them—including ones of sufficiently high energy to bring about a reaction. In fact, increasing the number of molecules in a given volume does increase the chances that activated molecules will have productive collisions, even if the temperature is held constant (Figure 4-7). Increasing concentrations also affect the rates of the reverse reactions. As A and B are converted to C and D, the concentrations of C and D increase; the "crowding" of C and D can be high enough to increase the frequency of the reverse reaction (C + D → A + B), if the free-energy levels of the reactants and the products are about equal. Nevertheless, within living cells, concentration alone, like temperature alone is usually not sufficient to bring about reaction rates fast enough to sustain life. The most important influence on rates of reactions in organisms rests with catalysts.

Catalysts and Reaction Rates

A **catalyst** is a molecule that increases the rate of a reaction without being used up during that reaction. It takes part but emerges unscathed. Thus a small amount of catalyst can be used again and again to speed up a

reaction. A typical industrial catalyst is the metal platinum. A finely ground powder of platinum provides a huge surface area on which individual platinum atoms are exposed. The platinum atoms have loosely held electrons that can be shared with atoms of hydrogen, oxygen, carbon, or other elements, all of which thereby become reactive and form new bonds with nearby atoms other than those of platinum. If a mixture of H_2 and O_2 gases is exposed to a surface covered with platinum pow-

Figure 4-7   THE EFFECT OF HIGHER CONCENTRATIONS ON REACTION RATES.
Higher molecular concentrations lead to more reactions. Like dancers on a dance floor, the more molecules in a given area, the more likely they are to collide and subsequently react with one another.

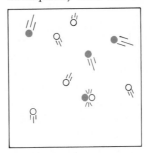
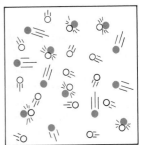

der, there is an instantaneous reaction—indeed, an explosion! The highly exergonic reaction between the H_2 and O_2 catalyzed by the platinum atoms yields H_2O and releases so much energy so quickly that an explosion results.

The effect of catalysts on reaction rates is quite different from the effects of temperature and concentration. If you picture the energy hill that stands between reactants and products, then both temperature and concentration push reactants up and over the energy hill by increasing both the energy of and the frequency of collisions among molecules, which then exceed the activation-energy level represented by the peak of the hill. However, catalysts actually lower the hill itself; *they reduce the activation energy necessary for the reaction to proceed.* This reduction has a dramatic effect on the rate of a reaction. For example, if the activation energy is reduced from 10 kcal/mole to 9 kcal/mole, many more collisions that would have been unproductive now have sufficient energy to overcome the energy barrier, and thus to react. This 1 kcal/mole reduction yields a fivefold increase in the reaction rate.

In living things, catalysts play a much greater role in speeding up reactions than do temperature and concentration. Biological catalysts are called *enzymes.* Most enzymes are globular protein molecules. Biological catalysts do not function in exactly the same way as inorganic ones such as platinum, which, as we saw, lowers activation energy by sharing electrons with other atoms. Nevertheless, as we shall see in the next section, some enzymes can speed up biological reactions a millionfold or more by lowering the activation-energy barrier for those conversions (Figure 4-8). However, certain kinds of polyribonucleotides (RNA) function enzymatically to carry out essential steps in nucleic-acid processing.

Figure 4-8   THE KEY TO THE CHEMISTRY OF LIFE: ENZYME CATALYSTS.
Enzymes lower the activation-energy barrier, allowing reactions to proceed at a much faster rate than they would otherwise.

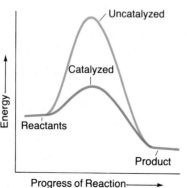

Enzymes are essential for life primarily because the bonds that hold together most biological molecules are very stable, so that high activation energies must be achieved for their rupture. Would it be more efficient for organisms to employ relatively unstable molecules that could react at low activation energies? No; their bonds would break spontaneously, leading to molecular chaos in the cell. Because most bonds of biological molecules are stable, amino acids, lipids, sugars, nucleic-acid bases, and the macromolecules made of these building blocks are relatively stable. Most important, chemical reactions involving such substances will not occur in an uncontrolled way in the cell.

The price of this stability is the cost to the cell of producing hundreds of kinds of enzymes to ensure biologically reasonable reaction rates. And this price is a high one; the manufacture and maintenance of enzymes come at the expense of a good deal of the cell's available energy. But there is no alternative. Life as we know it could not exist if covalent bonds did not hold biological molecules together stably, or if enzymes did not exist to speed up the reactions that underlie all biological change at the levels of molecules, of cells, and of whole organisms.

ENZYMES AND HOW THEY WORK

"Life," according to biochemist Ernest Borek, "may be defined as a system of integrated, cooperating enzyme reactions." Every life process we shall examine in this book begins and ends with reactions that are governed by the principles of energy flow and facilitated by enzyme catalysts. The ability of enzymes to speed up reactions can be quite remarkable. For example, when the Egyptian pharaoh Tutankhamen was interred in his gold sarcophagus in an underground tomb, slaves laid out a sumptuous ritual breakfast for the pharaoh's passage into the afterlife; thirty-three centuries later, when the tomb was opened, the breakfast remained—completely dried out but still recognizable. In a cool, dry room, oxidation reactions in some types of food can be immeasurably slow. In this case, they were not complete after more than 3,000 years. Had Tutankhamen been alive to eat that breakfast when it was prepared, the digestive enzymes in his mouth, stomach, and intestines would have hydrolyzed the food molecules by lunchtime.

Digestive enzymes were the first enzymes to be studied. William Beaumont, an army surgeon, treated a French-Canadian soldier in 1822 for a severe gunshot wound to the abdomen. The man survived and the

Figure 4-9   THE FIRST "WINDOW" ON DIGESTION.
William Beaumont collects stomach fluids from Alexis St.
Martin and observes digestive juices at work.

wound healed, but it left a gaping hole in his upper abdomen (Figure 4-9). Through this "window," Beaumont was able to obtain samples of stomach fluids and to observe the secretion of digestive juices containing enzymes and their activity on foods. Seventy years later, researchers showed that enzymes from yeast cells could break down sugars even after the yeast cells had been disrupted. They coined the word "enzyme" during this experiment; to these researchers, an enzyme was simply a substance *en zyme*—from the Greek for "leavened," or "in yeast."

Since the turn of the century, biologists have learned a great deal more about the roles of enzymes in living cells and about enzyme structure and function. In addition to speeding up chemical reactions by lowering the activation-energy barrier, as do platinum and other inorganic catalysts, enzymes have two unique characteristics of vital importance to living things. First, *enzymes are specific.* A given enzyme can act on only one type of compound or pair of reacting compounds, which is called its **substrate;** and it usually can catalyze only one type of reaction, such as condensation or hydrolysis. Second, *enzymes can be controlled* by the presence or absence of critical compounds. Let us examine in detail how the structure and function of these frequently studied molecules enable them to catalyze the processes of life.

Enzyme Structure

Most enzymes are globular proteins, and the substrates on which they act are often much smaller molecules than the enzymes themselves. Each protein enzyme has a unique three-dimensional shape arising from its primary, secondary, tertiary, and (sometimes) quaternary structure (Chapter 3). On the surface of each enzyme molecule is one small area (or sometimes a few areas) called the **active site.** The key to enzyme specificity is the shape of that site. The conformation of the active site complements the shape of the substrate(s) the enzyme acts on in much the same way that the keyhole of a lock fits around a key. It is this reciprocal matching of the three-dimensional shapes of active sites and substrates that accounts for enzyme specificity. (The shape and active sites of RNA enzymes are not yet known; though the specificity of the catalyzed reactions implies rough similarities to protein enzymes.)

The active site is often a deep groove or pocket shape on the enzyme's surface. It is made up of amino acids from several parts of the enzyme molecule, brought into close proximity by the folding of the polypeptide chain. Thus the active site is created by the enzyme's tertiary (and sometimes quaternary) structure. The active site may also contain a *prosthetic group* (Chapter 3) that is essential to the enzyme's activity. Prosthetic groups may be atoms of zinc or magnesium, or they may be ring-shaped organic compounds that include metals. For example, catalase (the names of most enzymes end in -ase), an enzyme that cleaves the potentially toxic chemical hydrogen peroxide, H_2O_2, is composed of a protein plus an iron-containing porphyrin molecule identical to the heme group in hemoglobin.

Enzymes that lack permanently bound prosthetic groups depend instead on **cofactors,** special substrates that bind temporarily to a site on the enzyme and take part in the reaction and the formation of products. Most cofactors are *coenzymes,* small molecules associated with and essential to the activity of some enzymes. Coenzymes are always changed in some way; thus their original form must be restored by other reactions before they can function again. Vitamins—that is, compounds that cannot be manufactured in an organism and so are essential trace parts of its diet—are often converted to coenzymes within cells. We will describe the critical role of coenzymes in the basic energy metabolism of all cells in Chapter 7 and discuss the role of vitamins in Chapter 34.

Because activity of protein enzymes depends on the precise folding of the polypeptide chain, factors that affect the protein's three-dimensional shape can also affect, or even destroy, enzyme activity. For example, excessive concentrations of Na^+ and Cl^- ions can break the ionic bonds that help maintain an enzyme's three-dimensional structure, thereby disrupting the integrity of the active site and its ability to bind a substrate. In particular, pH (H^+ concentration) can be favorable or unfavorable to enzyme function. Most enzymes have a *pH optimum,* or level of H^+ concentration at which they function best. Pepsin, for example, a digestive enzyme found in the human stomach, functions best at an acidic

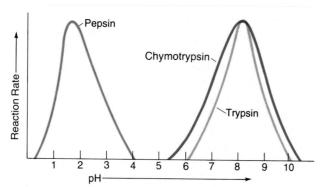

Figure 4-10   OPTIMUM pH FOR THREE DIGESTIVE ENZYMES.
Many enzymes work best over a narrow range of pH. These peaks illustrate the optimum pH for three common digestive enzymes and show why pepsin works best in acid stomach juice and trypsin and chymotrypsin, in alkaline intestinal fluid.

pH near 2, while chymotrypsin and trypsin, digestive enzymes found in the small intestine, function most efficiently in an alkaline environment at pH 8 (Figure 4-10). Cells and organisms have abundant mechanisms to help keep salt and H^+ concentrations within narrow ranges in order to protect the structure and function of enzymes.

Enzyme Function

All catalysts lower the activation-energy barrier between reacting molecules. But how do large globular enzymatic proteins with grooves called active sites achieve this effect? Enzymes function as catalysts by (1) forming complexes with the reacting molecules; (2) increasing the local concentration of the molecules; (3) orienting the molecules correctly so that the reaction can take place most efficiently; and (4) distorting the shape of the molecules slightly, thereby raising their free energies and helping them reach the transition state. Let us look at each part of the catalysis process in more detail.

Just as other molecules must collide before they can react, reactants must collide with enzymes before a reaction can be catalyzed. However, enzymes and reactants do more than just bump into each other; they form a complex held together by weak bonds: the **enzyme–substrate complex (ES)**. Formation of an enzyme–substrate complex is the essential first step in enzyme catalysis. Originally described in 1913 by Leonor Michaelis and Maud Menten, this process can be summarized as follows:

Enzyme + Substrate ⇌
    Enzyme–Substrate Complex → Products + Enzyme

Once the ES has formed, catalysis can take place and yield a product plus the intact enzyme.

To explore the Michaelis–Menten model, let us go back to our basic reaction, A + B ⇆ C + D (Figure 4-11). In order for the ES to form, a collision must take place in such a way that two substrates (A and B) contact the active site. That site holds the substrates close together and in just the right orientation so as to lower the activation-energy barrier and facilitate the reaction between them.

The number of weak bonds formed between substrate and enzyme determines the stability of the binding between a substrate molecule and an active site. Because weak bonds can form only when two atoms are very close together, the active site and the substrate must fit together well if such bonds are to form. The number of close contacts depends on the precise arrangement of amino acids in the active site. Both the main chain and

Figure 4-11   FORMATION OF AN ES COMPLEX DURING CATALYSIS.
The substrates (reactants A and B) bind to the enzyme at the active site, which has a shape reciprocal to those of the substrates. The substrates are converted to products C and D, which then leave the active site. The enzyme's active site is now empty and ready to repeat the process. For a typical enzyme, this cycle of events takes about one-thousandth of a second.

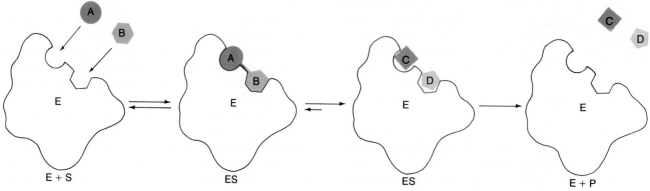

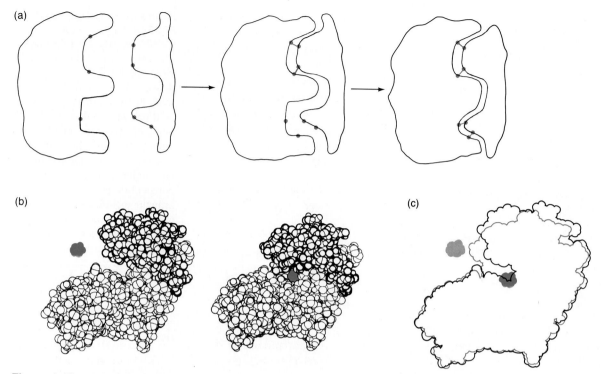

Figure 4-12   INDUCED FIT.
Enzymes change shape to accommodate the substrate.
(a) One portion of the active site of the enzyme is
complementary to part of the substrate. After these
complementary regions bind to each other, the shape of the
enzyme changes so that the remaining portion of the active
site is able to bind to the rest of the substrate. The dots
represent weak-bond forming groups. (b) A space-filling
model of the enzyme hexokinase both before binding its
substrate, glucose (shown in blue), and after changing shape
to improve the fit between the active site and the substrate.
(c) A line drawing of the outline of the hexokinase model
shows the molecule both before it binds the substrate
(black) and after its change in shape, called induced fit
(blue).

the side chains, or R groups, on the amino acids that
protrude into the active site serve as the locations of
weak bonding to the substrate(s). The active site sometimes changes shape as the substrate binds to it, allowing
more weak bonds to form between enzyme and substrate. This change in shape improves the fit between
active site and substrate, thereby increasing the binding
affinity, or strength with which the substrate is bound to
the enzyme. Such an adjustment in enzyme shape is
called **induced fit** (Figure 4-12).

Once the ES has formed, catalysis can take place. Several factors affect the reaction. First, the binding of two
substrates to an enzyme brings them in close proximity
to each other. In effect, then, enzymes greatly increase
the effective *concentration* of the substrates. For example, two substrates present in solution at concentrations

of $10^{-4}\,M$ may within the active sites of enzymes have
effective concentrations of $4\,M$, or 40,000 times greater
than in the solution. This is because the volume around
and in the active site is so minute that the presence of
substrate(s) in it raises the local concentration to a very
high level. Because concentration affects the rate of a
reaction, this simple mechanism of bringing molecules
together greatly speeds reaction rate.

A second function of the enzyme is to orient the substrate molecules so that the reaction may proceed readily. The position of the molecules is important because
bonds can break and re-form only at specific regions of
the substrates. For the reaction to occur, these regions
must lie very close together. It is statistically improbable
that the correct orientation will occur during a random
collision of the reactants in solution. That is one reason
why uncatalyzed reactions are so slow.

The formation of the enzyme–substrate complex has a
third important effect: it may distort the three-dimensional shapes of the substrates. The weak bonds between
enzyme and substrates strain the substrates; this strain
on their geometry alters the distribution of their orbiting
electrons (Figure 4-13), thus reducing the free energies
of the bound substrates. Therefore, the amount of energy needed to reach the transition state—the peak of
the activation-energy hill—is less. Hence, more molecules can cross the barrier separating reactants from
products, and the rate of reaction is greatly increased.

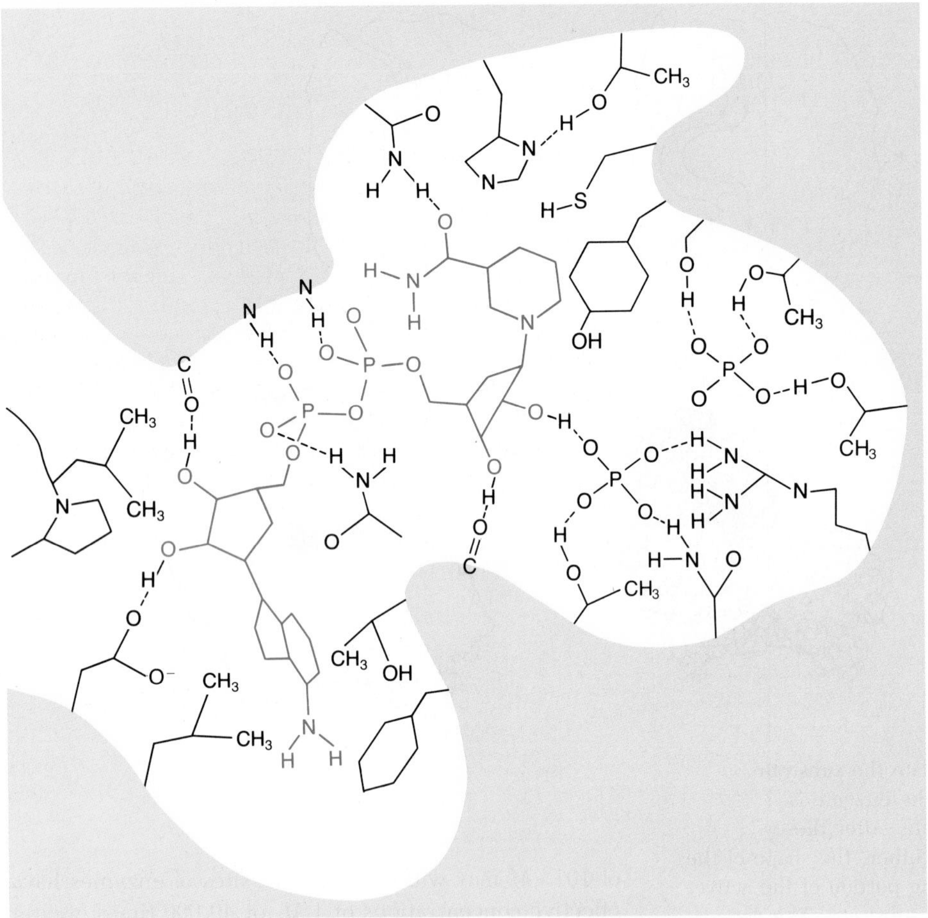

Figure 4-13   STABILIZING PROPERTIES OF WEAK BONDS IN THE ES COMPLEX. The active site of the enzyme glyceraldehyde phosphate dehydrogenase shows the many weak bonds (dashed lines) that hold one of its substrates— here, nicotinamide adenine dinucleotide—to the enzyme. The active site is formed by very different regions of the polypeptide chain that are brought together by the folding of the protein into its tertiary structure. Van der Waals forces are also present between the enzyme and the substrate. The large number of hydrogen bonds, in addition to van der Waals forces, distorts the substrate, which becomes reactive.

After the substrates have reacted, the product of the catalyzed reaction is released, and the enzyme can reassume its initial conformation (if it has changed shape) and thus be ready to catalyze another reaction. This entire process can take place thousands of times a second. For example, the enzyme in red blood cells, carbonic anhydrase, can catalyze the reaction $CO_2 + H_2O \rightarrow H_2CO_3$ more than *600,000 times a second!* The same number of reactions would take many minutes without a catalyst. Carbonic anhydrase is exceptionally fast; most enzymes catalyze between 1 and 10,000 reactions a second.

Now that we have outlined the basic steps in catalysis, let us consider a real example: *lysozyme*, the enzyme that helps mammals guard against nose and eye infections and protects the embryo in the eggs of birds and reptiles by hydrolyzing an essential polysaccharide component of bacterial cell walls. The hydrolysis reaction of lysozyme (shown in Figure 4-14) so weakens the encompassing cell wall that the bacterium falls apart, or lyses (hence the name "lysozyme"; Figure 4-15).

Lysozyme is a roughly spherical protein made up of a polypeptide chain of 125 amino acids (Figure 4-16). The active site is a cleft long enough to accommodate six sugar units of the bacterial polysaccharide. The sugars

are held in the active site by hydrogen bonds. Each of the six subsites within the active site corresponds to one of the six sugars, which we will call A, B, C, D, E, and F. Only the covalent bond linking sugars D and E is hydrolyzed during the reaction. Therefore, the amino-acid groups necessary to catalyze this particular reaction must be near subsites D and E on the enzyme molecule. Measurements show that sugar D is bound less tightly to the enzyme than are the other sugars. Sugar D, which has a ring shape formed by five carbon atoms and one oxygen atom, normally assumes a chairlike shape (Figure 4-17b). In order to fit into the active site of lysozyme, this ring must bend into a new configuration; the distortion occurs because weak bonds form between the active site and the sugar molecule. The bent shape, called a "sofa" configuration (Figure 4-17c), represents the transition state; when that state is reached, the activation-energy barrier has been surmounted. The energy needed to distort the sugar ring from chair to sofa makes up a large part of the activation-energy cost of the reaction. Without a catalyst, the activation energy probably would not be achieved. By causing the substrate to bend to fit the active site, lysozyme reduces the activation energy of the reaction so that hydrolysis can proceed.

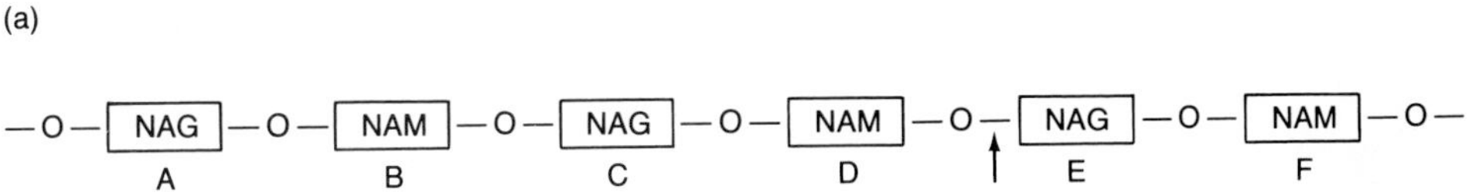

(a)

—O—[NAG]—O—[NAM]—O—[NAG]—O—[NAM]—O—[NAG]—O—[NAM]—O—
   A   B   C   D   ↑ E   F

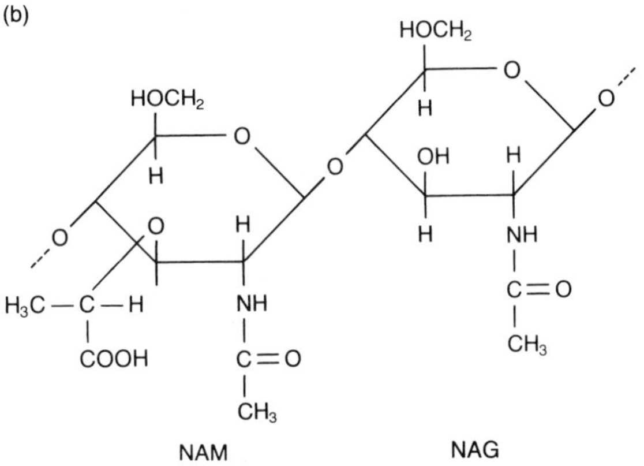

(b)

Figure 4-14   DISMANTLING A SUGAR CHAIN.
Lysozyme hydrolyzes a polysaccharide chain built of two derivatives of glucose. (a) A portion of the polysaccharide that fits the lysozyme active site. The six sugar units accommodated in the active site of the enzyme are labeled A through F. Lysozyme hydrolyzes only the bond between sugar units D and E (indicated by the arrow). (b) Two units of the polysaccharide: on the left, N–acetylmuramic acid (NAM); on the right, N–acetylglucosamine (NAG). N–acetyl refers to the –NHCOCH₃ group attached to a specific carbon atom of each sugar unit. (c) The products of lysozyme's action. Although lysozyme splits only the bond between the D and E units, repeated hydrolysis of the polysaccharide chain by other mechanisms can break the chain into small pieces.

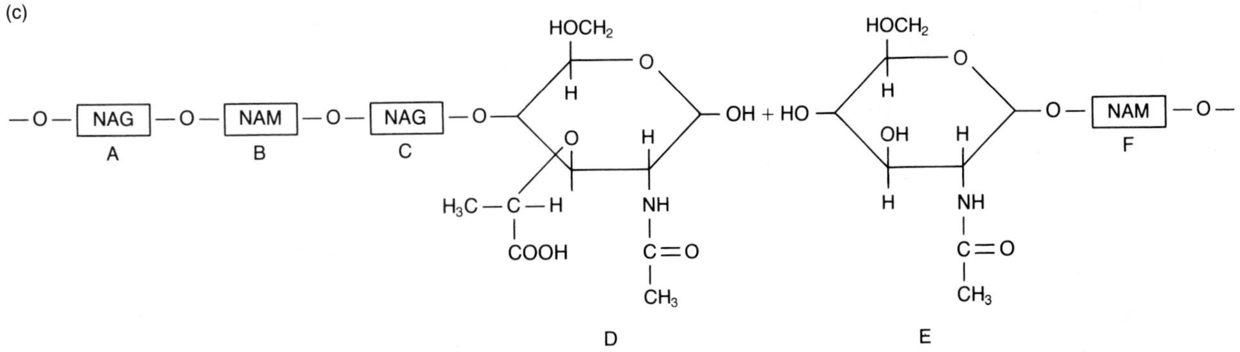

(c)

Figure 4-15   LYSOZYME: "INVISIBLE SHIELD" FOR THE DEVELOPING YOUNG.
As this scanning electron micrograph (magnified about 35 times) shows, an eggshell is porous enough to provide adequate gas exchange, but it also occasionally allows invaders to enter.

Enzymes and Reaction Rates

Although an enzyme can effectively lower the activation energy of a reaction, it cannot change an endergonic reaction into an exergonic one. In other words, enzymes speed up only those reactions that could take place between activated molecules in a productive collision. They cannot bring about reactions that are energetically unfavorable. This emphasizes that the differences between free energies of reactants and products, not the activation energy, govern the equilibrium attained in reactions.

(a)

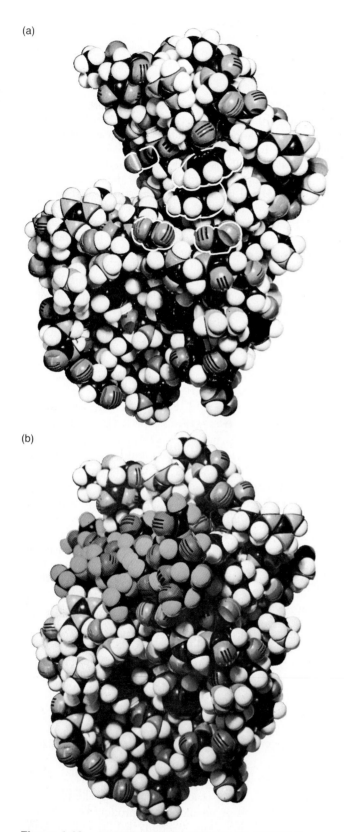

(b)

Figure 4-16   SUBSTRATES BOUND TO AN ENZYME'S ACTIVE SITE.

These space-filling models show the T4 bacteriophage lysozyme (a) without substrate and (b) bound with a polysaccharide substrate.

(a)

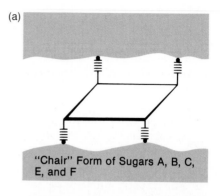

"Chair" Form of Sugars A, B, C, E, and F

(b)

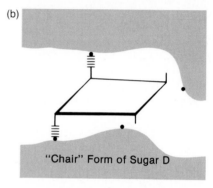

"Chair" Form of Sugar D

(c)

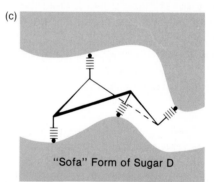

"Sofa" Form of Sugar D

Figure 4-17   THE KEY TO ENZYME CATALYSIS: ADDING ENERGY BY BENDING THE SUBSTRATE.

For simplicity, the "chair" form is shown as a rectangle with four weak bonds holding it to the lysozyme active site, and the "sofa" form is shown as a twisted rectangle. (a) The chair form as it is bound to the enzyme in sugars A, B, C, E, and F. (b) Sugar D in chair configuration does not fit precisely into the active site of the enzyme; hence, only two weak bonds form. (c) Sugar D twists into the sofa configuration as four weak bonds form. The energy required to change the sugar's configuration makes up a large part of the activation energy required for the hydrolysis of the sugar. By causing bending of sugar D of the substrate, the enzyme lysozyme lowers the activation energy of the reaction, which then proceeds.

Among reactions that *are* energetically favorable, the speed with which enzymes catalyze the reaction can vary, depending on the factors we surveyed earlier. These factors are enzyme and substrate concentrations and temperature.

The reaction rate is directly proportional to the amount of enzyme present. For example, if we have enough of the enzyme sucrase to catalyze the hydrolysis of 2.1×10^{-3} mole of sucrose in ten minutes, then twice that amount of sucrase will break down twice as much sucrose in the same amount of time (if the temperature is held constant).

As with uncatalyzed reactions, temperature affects the rate of enzyme-mediated reactions. Because an enzyme reduces but does not eliminate the activation-energy barrier, the additional energy provided by elevated temperature can further speed up a reaction (Figure 4-18). Putting dough in a warm place causes it to rise faster than it would in a cool place; the added heat energy speeds up the enzyme-catalyzed reactions in yeast cells that produce the CO_2 gas bubbles that make the dough expand. Many enzymes have an optimal temperature at which they can bind substrate, catalyze a reaction, and release the products most rapidly. One reason for this optimum is that at high and low temperatures, the weak chemical bonds so central to enzyme shape and function may be altered. At extremes of temperature in particular, the enzyme may be denatured, or permanently altered, to a nonfunctional shape.

Figure 4-18   EFFECT OF TEMPERATURE ON ENZYME-CATALYZED REACTIONS.

Elevating temperature has a profound effect on enzyme-catalyzed reactions, but reactions can be accelerated only to a point. The reaction rate increases with temperature to an optimum point for the enzyme's speed (red line). This behavior is the result of two opposing effects. The rate of the catalyzed reaction increases with temperature, as does any chemical reaction (green). However, the amount of active enzyme decreases with temperature because the enzyme is denatured by heat (blue).

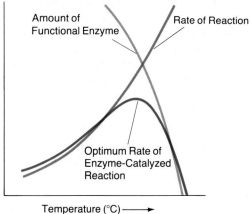

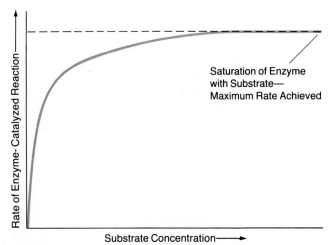

Figure 4-19   ENZYME SATURATION.
Reaction rates are limited by enzyme saturation. An enzyme is "saturated" when the active sites of all the enzyme molecules are occupied most of the time. At the saturation point, the reaction will not speed up, no matter how much additional substrate is added to a solution.

Finally, the rate of an enzyme-mediated reaction will increase as the concentration of the substrate increases, but only up to a point. If, for example, we put a given amount of enzyme in a solution and begin adding substrate, the reaction will go faster and faster in direct proportion to the amount of substrate added. Eventually, however, the reaction rate will gradually stop increasing. In fact, if we add enough substrate, the rate will no longer increase at all, but will remain constant. Why? At some point, so much substrate will be present that the active site in every one of the enzyme molecules will be occupied most of the time. As soon as an individual reaction is completed and products leave a site, new substrate binds to it. At this point the enzyme is said to be *saturated* (Figure 4-19). No matter how much more substrate is added, the reaction cannot go faster. But if we add more of the enzyme, the rate of reaction again rises. The phenomenon of **enzyme saturation** sets enzyme-mediated reactions apart from noncatalyzed reactions, whose rates continue to increase with increasing substrate concentration.

The speed of ES complex formation and the reaction speed are key factors in enzyme saturation. In the case of lysozyme, an individual ES complex exists for only about 10^{-5} second; within that time, either the complex dissociates or the reaction occurs. Most of the time, the reaction occurs. However, if the ES does dissociate, the enzyme is free to quickly form a new ES complex with another substrate molecule.

The speed with which the ES complex forms, the reaction occurs, and the products are released is critical to life. To appreciate how important it is, we have only to

LUCIFERASE: A LUMINOUS ENZYME

The greenish-yellow glow of a firefly on a summer night is produced by one of the most interesting enzymes ever discovered. It is called luciferase, and it catalyzes the breakdown of a protein called luciferin. During this reaction, most of the energy is released as light rather than heat, causing the glow we can see in the dark. Colonies of bacteria that contain a similar light-releasing enzyme inhabit four species of "flashlight fishes"—deep-sea fishes with glowing pockets directly beneath their eyes (see Figure A). Each fish provides housing and nourishment for the millions of bacteria in its eye pockets; in return, the luminous bacteria collectively create a beacon that helps the fish find prey and mate in the inky reaches of the deep ocean.

Luciferase acts on luciferin only if energy is present in the form of adenosine triphosphate (ATP) molecules. This universal cellular fuel, which we will discuss in later chapters, provides the energy needed to drive the endergonic reaction

$$\text{Luciferin} \xrightarrow[\text{Luciferase}]{\text{ATP}} \text{Light} + \text{Heat}$$

Researchers have been using this glowing enzyme to detect how much ATP is present in certain systems. For example, luciferase and luciferin can be added to blood stored in blood banks. If the red blood cells are stored too long, they begin to degenerate

Figure A
This flashlight fish has a pocket of luminous bacteria that lights its path through the darkness of the deep sea.

and leak ATP molecules into the surrounding fluid. These molecules will in turn drive the breakdown of luciferin by luciferase, and the blood will glow in the dark. This same assay is used to detect the presence of bacteria in urine, so that infections may be treated, and to study sources and the levels of chemical energy in tumor cells, so that a rapidly growing malignancy can be spotted. Luciferase and similar enzymes are truly illuminating tools for biologists, as well as headlights for fish and taillights for fireflies.

---

consider that the biologist David Kirk has calculated that the enzyme urease breaks down in one second the same amount of urea that would hydrolyze spontaneously in 3 million years. The speeds with which nutrient molecules are broken down, cellular "fuel" is synthesized and used to drive muscles or organs, and wastes are rendered harmless are just as great. No bacterium, fungus, plant, or animal could maintain itself and win the battle against entropy if its enzymes did not work so astonishingly fast to lower the activation energies of countless reactions.

METABOLIC PATHWAYS

Growth, food manufacture, digestion, movement, and other life-sustaining activities are the consequence of myriad chemical reactions, all going on at once. These reactions take place in orderly, interrelated patterns that are controlled by enzymes. Enzymes themselves are

controlled by a variety of mechanisms, as we shall soon see. The combination of simultaneous, interrelated chemical reactions taking place at any given time in a cell is referred to as cellular **metabolism.** Metabolism is like a tapestry woven of numerous reaction "threads" that provides all cells and organisms with the energy and biological molecules they need to sustain life.

We have seen that some reactions require energy, while others release it. Most energy-requiring reactions in cells are involved in the synthesis of necessary biological molecules—amino acids, nucleic acids, fats, proteins, and carbohydrates. Together, these biosynthetic reactions are called **anabolism** (meaning "to build up"). Conversely, most energy-yielding reactions in cells break down molecules to obtain building blocks, release energy, or digest waste products. Together, these degradative reactions are called **catabolism** ("to tear down"). In cells, all energy-requiring anabolic reactions are coupled to energy-releasing catabolic ones.

Although we have been discussing enzymes as though each of the reactions they mediate occurs in isolation, many enzymes in cells carry out specific individual steps in a chain of reactions. As a result, various compounds are progressively built up or broken down. These series of reactions are called **metabolic pathways.** "Pathway" is a good term for these sets of reactions because they have a starting material (path's beginning), an end product (path's end), and a series of reactions in between (steps along the path).

At each step of a metabolic pathway, an enzyme catalyzes a reaction that changes the starting material a little bit more—adds a phosphate group, removes an –OH group, and so on—until the final product is reached. A general pathway can be diagrammed as follows:

$$\begin{array}{ccccccc} & \text{Enzyme}_a & & \text{Enzyme}_b & & \text{Enzyme}_c & & \text{Enzyme}_d \\ & \Downarrow & & \Downarrow & & \Downarrow & & \Downarrow \\ A & \longrightarrow & B & \longrightarrow & C & \longrightarrow & D & \longrightarrow & P \end{array}$$

This pathway has a starting material (A); three intermediates (B, C, and D); four enzymes (a, b, c, and d); and a product (P). For each separate reaction in the pathway, and for the pathway as a whole, the free-energy change must be negative ($\Delta G < 0$). For such exergonic reactions, energy does not have to be added to drive the series of reactions. In other cases, individual reactions may be endergonic ($\Delta G > 0$) and require energy input. These energy-requiring steps in a pathway are coupled in some manner to exergonic reactions, which can release and transfer energy to steps that could not proceed without additional energy input.

As an example of just a short segment of a metabolic pathway, let us look at the steps by which the amino acid valine is made (Figure 4-20). A molecule of the first substrate, pyruvate, is joined to a two-carbon chain in an enzyme-catalyzed reaction. Then three more reactions occur, each with its own enzyme catalyst. The end product, valine, can serve as a building block for proteins. This simple metabolic pathway is costly because energy equivalent to that contained in three molecules of the biological fuel adenosine triphosphate (ATP) is consumed each time that a pyruvate molecule is transformed into one molecule of valine. This simple series is typical of many reaction pathways in cells.

Metabolic pathways are interconnected; the product or intermediate of one pathway may be the starting material for another. The metabolism of a cell can be thought of as a seamless interwoven chemical "fabric." We unravel this fabric when we examine individual pathways in isolation, but a list of isolated pathways no more indicates the complex metabolism of a cell than a pile of threads suggests a dress or a shirt. For full efficiency, the sets of enzymes that are involved in the same pathways are segregated in distinctive compartments in cells, separated from enzymes that catalyze other reactions. This association allows the enzymes to work much more efficiently than if they floated freely and bumped into substrates at random. Furthermore, metabolic pathways probably work by direct transfer of metabolites from one enzyme to the next. Instead of dissociation of a metabolic product from its enzyme and random diffusion to the next enzyme's active site, enzyme-to-enzyme complexes form and pass on the metabolite directly. As a result, large quantities of each intermediate in a pathway do not

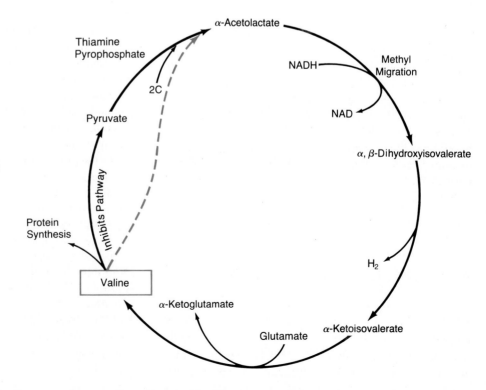

Figure 4-20   THE VALINE PATHWAY FROM PYRUVATE.
Pyruvate is a raw material for the manufacture of several amino acids. Each reaction in each metabolic pathway is catalyzed by a different enzyme. Note that a vitamin (thiamine), a hydrogen donor (NADH), and another amino acid (glutamate) are essential cofactors in the synthesis of valine. Valine itself acts as an inhibitor of the first reaction in this pathway.

have to build up. These concepts will be fundamental to our understanding of respiration and photosynthesis in Chapters 7 and 8.

Control of Enzymes and Metabolic Pathways

As we mentioned earlier, the controllability of enzymes is a feature that sets them apart from other catalysts. Enzymes and metabolic pathways must be regulated if a cell is to function as an integrated whole and to have available the appropriate amounts of energy and specific molecules it requires. Furthermore, individual cells in multicellular organisms must be controlled so that their enzymatic machinery will work to benefit the whole organism.

Cellular metabolism is subject to both external and internal controls. In multicellular organisms, hormones produced in one part of the body can regulate enzyme activities within cells elsewhere in the body. These external agents, in plants and animals, are discussed in Chapters 30 and 37. Here we shall be primarily concerned with internal control mechanisms of individual cells. Let us look at an example of one such internal control circuit that operates on two metabolic pathways: the main pathway by which amino acids are linked into proteins (which is described in detail in Chapter 13) and the pathway that produces molecules of the amino acid valine, which feeds into the growing polypeptide chain:

$$A \xrightarrow{\text{Enzyme}_a} B \xrightarrow{\text{Enzyme}_b} C \xrightarrow{\text{Enzyme}_c} D \xrightarrow{\text{Enzyme}_d} \text{Valine} \nearrow \text{Protein}$$

Other
Amino
Acids

It is advantageous for cells to produce amino acids at the same rate as the amino acids are used up during protein synthesis. Failure to control this rate would result in the build-up of unneeded amino acids or, if there is a deficiency of the necessary amino acid, in the slow-down of protein synthesis. An undesirable outcome can be avoided by a system of regulation called **negative feedback.** With negative feedback, a build-up of the amino acid reduces the rate at which it is synthesized, while a depletion of the amino acid increases the rate. Negative feedback is the basic means of internal control within cells. It is "feedback" because the concentration of the product "feeds back" to control the rate of synthesis. It is "negative" because an increase in product causes a decrease in production and vice versa.

In our example, if the rate of protein synthesis increases, the available valine will be used up more quickly; its concentration will fall; and the rate of valine production will increase. Thus the amino acid can con-

tinue to be incorporated into the protein chain. Conversely, if the rate of protein synthesis slows down, less valine will be required; its concentration will build up; and the rate of valine production will decrease.

Negative feedback usually operates on the first enzyme in the pathway to inhibit formation of the first ES complex. Limiting amino-acid synthesis at this point prevents the wasteful or dangerous build-up of intermediates that would occur if other enzymes in the pathway were the primary control targets. It is also most economical, because it prevents any waste of energy or of substrates along the pathway.

How can a small molecule like an amino acid control an enzyme's activity? Experiments show that the amino acid does not prevent the enzyme's functioning by binding to the active site and thereby blocking the substrates. Rather, the amino acid binds to the enzyme at a *different* site, one specific for that amino acid. When the amino acid occupies that separate *allosteric* binding site, it causes the enzyme to change to a shape that is not compatible with active-site functioning. But this is a reversible shape change; when the amino acid leaves the allosteric site, the enzyme reverts to its native shape, which is compatible with normal functioning. As a reaction pathway functions, the concentration of the final product will increase if the product molecules are not being used. In this case, the excess product molecules will bind to allosteric sites on the first enzyme in the pathway, thereby inhibiting the enzyme and causing the pathway to slow and to come to a halt.

Enzymes whose activities are controlled in this way are called **allosteric enzymes** (meaning "other shape") (Figure 4-21). It should not be surprising to learn that if binding to a separate site can inhibit enzyme function, it also can activate an enzyme. Some allosteric enzymes are inactive prior to the binding of the control substance, or activator. Binding at the allosteric site activates the enzyme, causing it to change shape so that its active site is available to bind substrate optimally. Allosteric activation and inhibition are part of a broader category of control called **noncompetitive inhibition.**

In another type of enzyme control, called competitive inhibition, the active site itself is occupied by a compound other than the normal substrate, thereby preventing binding of that substrate (Figure 4-22). However, an increase in the concentration of the normal substrate can enable it to compete successfully for the active sites.

Methanol poisoning is an example of competitive inhibition. Methanol (CH_3OH), or "wood alcohol," can bind to the active site of alcohol dehydrogenase, whose normal substrate is ethanol (CH_3CH_2OH). When methanol binds to the active site, it is converted to formaldehyde (H_2CO), a highly reactive compound that can denature proteins and damage delicate tissues in the body, such as

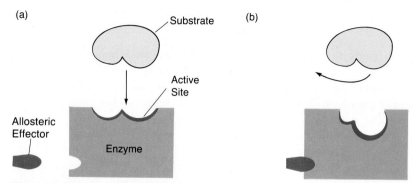

Figure 4-21   ALLOSTERIC ENZYMES AND NONCOMPETITIVE INHIBITION.

The enzyme has an active site, to which it binds its substrate, and another binding site for an allosteric effector, such as an amino acid. (a) When the effector is not bound to the enzyme, the active site and the substrate have complementary shapes, so binding and catalysis take place. (b) But the binding of the effector to its site induces a shape change in the enzyme, distorting the active site so that it binds the substrate less tightly or not at all. This process is noncompetitive inhibition.

The effect of binding an allosteric effector is not always inhibitory, as shown here. In some allosteric enzymes, an allosteric effector may induce a shape change in the active site that enhances the binding of the substrate.

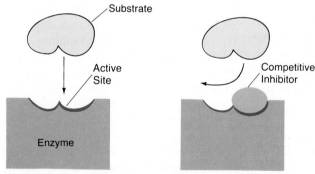

Figure 4-22   COMPETITIVE INHIBITION: A BLOCKAGE OF THE ACTIVE SITE.

A competitive inhibitor is shaped somewhat like the enzyme's normal substrate, with which it competes for binding in the active site. Cells do not regulate metabolism by means of competitive inhibition, but the mechanism is applied extensively in the design of drugs and medical treatments.

the visual receptors in the eye. Thus methanol poisoning can lead to blindness. A person who has accidentally swallowed methanol usually is treated by being given large doses of ethanol, which competes with methanol for the active site and thus reduces the breakdown of methanol to formaldehyde. Although many cases of competitive inhibition have been investigated, it is likely that cells do not use it as a means of regulating metabolism, but more commonly employ the "fine tuning" of noncompetitive inhibition.

Negative feedback usually takes the form of noncompetitive inhibition. But do not forget that enzyme activity can also depend on other factors, such as the presence or absence of such activators as metal ions, prosthetic groups, or coenzymes, and such aspects of the environment as temperature and pH. Although these other factors are not used for control of enzyme activity, they are part of the essential background, which in combination with competitive and noncompetitive inhibitions govern the enzymes so central to the life processes in cells.

LOOKING AHEAD

In this chapter, we have seen how the continuous changes associated with life involve chemical reactions and depend on a favorable energy flow. But entropy always wins; unless "disorder" is ultimately served and heat is lost to the environment, reactions do not take place. Yet living organisms are the epitome of order, and their internal metabolic reactions are complex, controlled, and interrelated processes. In Chapters 5 and 6, we shall explore the orderly micro-universe within the living cell, the basic unit of life on Earth and the site of life-sustaining energy transformations. Then, once we understand cells and their parts, we shall describe, in Chapters 7 and 8, the intricate series of chemical reactions responsible for the breakdown and synthesis of nutrients in cells, and explain how those series conform to the energy principles that govern both the living and the nonliving universe.

SUMMARY

1. Biological activities are based on *chemical reactions*—transformations of sets of molecules into other kinds of molecules. In a reaction, the formation of products from reactants always involves the making and breaking of chemical bonds.

2. *Energetics* refers to the study of energy changes during chemical reactions. Energy may be *kinetic* (energy of motion) or *potential* (stored energy). Energy in atoms, bonds, and molecules is called *chemical energy*.

3. Molecular transformations are always accompanied by energy changes. According to the *first law of thermodynamics*, energy can change form, but it can never be created or destroyed. According to the *second law of thermodynamics*, during energy transformations, there is a loss in the usable energy of a system as heat dissipates, resulting in an increase in *entropy*, or disorder.

4. The *free energy* (G) of a compound is the usable energy that is freed when the compound is broken down into its constituent elements. Free energy can be determined for reactants and products, and the change in free energy (ΔG) can be calculated for a reaction. If energy is released during a reaction, the reaction is said to be *exergonic* ($\Delta G < 0$). If energy input is required for a reaction, it is said to be *endergonic* ($\Delta G > 0$).

5. When A → B is equal to B → A, the free-energy change is zero. Reactions are at *equilibrium* when the combined free energies of reactants and products are equal. At equilibrium, there is no longer net conversion of reactants to products or products to reactants, though reactions in both directions continue to occur.

6. Most collisions between molecules are unproductive—they do not bring about a reaction. In order for covalent bonds to break and for a reaction to take place, colliding molecules must have a minimum amount of energy, called the *activation energy*. Activation energy can be likened to a hill that reactants must scale in order to become products. When activated molecules reach a transition state, at which new bonds can begin to form, they can be said to be at the top of the energy hill.

7. The rate of a chemical reaction can be affected by temperature, concentration of reactants, concentration of products, and the presence of *catalysts*. Catalysts are molecules that speed up a reaction without being changed during the reaction.

8. In living cells, enzymes serve as catalysts by reducing the activation energy necessary for biochemical reactions to take place. Enzymes are specific: a given enzyme can act on only one kind of compound or pair of compounds, called *substrates*, and it can catalyze only one type of reaction. Most enzymes are proteins, although certain RNAs function as critically important enzymes in nucleic acid processing.

9. Each protein enzyme has an *active site*, a groove or pocket whose shape is reciprocal to the shape of a specific substrate(s). Because the activity of an enzyme depends on the precise folding of its polypeptide chain, factors that affect the weak bonds that maintain the molecule's three-dimensional shape can also alter enzyme activity. Some enzymes have *cofactors*, or coenzymes, substances that bind temporarily to enzymes and take part in the catalyzed reactions.

10. The first step in enzyme function is the formation of an *enzyme–substrate complex (ES)*, in which the amino acids in the active site bond to the reactants. The active site sometimes changes shape in such a way that more weak bonds can form between enzyme and substrate; such a change is called *induced fit*. Once the ES complex has formed, the reaction proceeds, since the binding of substrates to an enzyme increases the effective concentration of the substrates, orients the reacting substrate molecules, or changes the shape of the substrate molecules. Once the reaction is catalyzed, the products are released, and the enzyme is ready to catalyze another reaction.

11. Rates of enzyme-catalyzed reactions are affected by temperature, concentration of enzyme, and concentration of substrates. Enzyme reactions show *saturation*, a rate that cannot further speed up even if more substrates are added.

12. The combination of chemical reactions that take place in a cell is called *metabolism*. Biosynthetic reactions together make up *anabolism*; degradative and energy-yielding reactions make up *catabolism*. Series of reactions, with each step having a particular role in building up or breaking down compounds, are called *metabolic pathways*.

13. Metabolic pathways can be controlled by external agents such as hormones, or they can be controlled internally, by substances present within the cell. Most internal control is by means of *negative feedback*, wherein the build-up of a product inhibits its further production. When levels of the product drop, production resumes. In *allosteric enzymes*, the product binds to a site other than the active site on the enzyme, changing the enzyme's shape and either activating or inhibiting enzyme function. This is a type of *noncompetitive inhibition*. During *competitive inhibition*, a substance binds to the active site, thereby prohibiting the binding of the enzyme's normal substrates.

KEY TERMS

activation energy
active site
allosteric enzyme
anabolism
catabolism
catalyst
chemical energy
chemical reaction
cofactor
competitive inhibition

endergonic reaction
energetics
entropy
enzyme saturation
enzyme–substrate complex (ES)
equilibrium (chemical)
exergonic reaction
first law of thermodynamics
free energy
induced fit

kinetic energy
metabolic pathway
metabolism
negative feedback
noncompetitive inhibition
potential energy
second law of thermodynamics
substrate

QUESTIONS

1. With respect to rocks, dammed up water, and organic compounds such as sugars, distinguish between potential, kinetic, and chemical energies.

2. Explain in terms of the second law of thermodynamics why some energy is lost during every chemical reaction of every living cell. How does that loss relate to entropy?

3. Explain the difference between exergonic and endergonic reactions. How and why are the two types so frequently paired in living cells?

4. What is the activation energy of a molecule? What sorts of things affect the activation energy of molecules so that a reation can take place? Which is most important in living cells?

5. Define "enzyme," indicate the unique properties of enzyme-catalyzed reactions, and explain how an enzyme such as lysozyme works.

6. Explain the equation Metabolism = Anabolism + Catabolism.

7. Describe metabolic pathways and the ways that they can be controlled. What exactly is negative feedback? What would be the consequence of positive feedback on a pathway?

ESSAY QUESTIONS

1. How are the terms "active site," "induced fit," and "ES complex" related?

2. Why is life an "oasis of order in a desert of disorder" in the universe? Does a living cell contradict the second law of thermodynamics? Why or why not?

SUGGESTED READINGS

BECKER, W. M. *Energy and the Living Cell.* Philadelphia: Lippincott, 1977.

This book provides a balanced view of different interpretations of enzymes, energetics, and so on.

BLUM, H. F. *Time's Arrow and Evolution.* New York: Harper & Row, 1962.

This is a classic book about the chemical basis of life. Especially relevant is the coverage of energy and entropy.

CHRISTENSEN, H. H., and R. A. CELLARUS. *Introduction to Bioenergetics: Thermodynamics for the Biologist.* San Francisco: Freeman, 1972.

Also for the biology student, this is a programmed text that makes learning easy.

SRIVASTAVA, D. K., and S. A. BERNHARD. "Metabolite Transfer via Enzyme-Enzyme Complexes." *Science* 234 (1986): 1081–1086.

The exciting hypothesis of metabolite transfer is described here.

STRYER, L. *Biochemistry.* 2d ed. San Francisco: Freeman, 1981.

A relatively nonmathematical and beautifully illustrated treatment of enzymes, energy, and reactions.

5

CELLS: THEIR PROPERTIES, SURFACES, AND INTERCONNECTIONS

The living cell is to biology what the electron and the proton are to physics. Apart from cells and from aggregates of cells there are no biological phenomena.

Alfred North Whitehead
Science and the Modern World (1925)

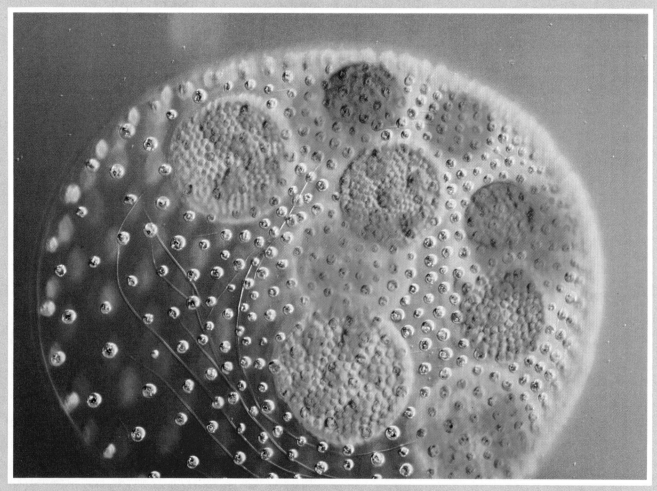

The borderline of multicellularity: Like miniature plants suspended in space, individual cells in the *Volvox* colony have minimal interaction with neighbors. The larger multifaceted green spheres inside are reproductive stages.

In 1665, Robert Hooke, a thirty-year-old physicist and amateur botanist, gazed in fascination through the lens of a primitive microscope. His subject was a thin sheet of tissue sliced from dried cork. The tiny, hollow honeycomb chambers he saw through the microscope reminded him of the small rooms, or cells, that monks inhabited in monasteries. Hooke reported to his colleagues that "these pores, or cells, were not very deep, but consisted of a great many little boxes, separated out of one continued long pore, by certain diaphragms." Ironically, Hooke was seeing the empty spaces left behind in dead cells.

Other microscopists who later examined living plant and animal tissues found that every creature they studied consisted of what Hooke had called cells. Some organisms, such as bacteria, amoebae, and certain algae, are free-living single cells. However, many kinds of living things, from tiny fresh-water colonies of plant cells to massive whales and redwoods, are composed of many cells that function together as a coordinated population.

In this and the next chapter, we move from the submicroscopic realm of atoms, molecules, and their chemical reactions to the basic units of life on Earth—the molecular "municipalities" known as **cells.** We will see that cells are minute entities, consider the reasons for their small size, and discuss how biologists can study such tiny objects. Next, we will examine the major kinds of cells and look closely at the cell surface—a set of structures that enclose and protect the cell and govern the vital commerce of materials into and out of these "municipalities." Finally, we will see how cells are connected to and communicate with one another—the keys to understanding how they make up the smoothly functioning tissues and organs that we recognize as leaves, stems, livers, or kidneys.

In Chapter 6, we will move beyond this basic orientation to look at the myriad structures within cells. Collectively, these internal parts carry on the life processes in the cell; individually, they can account for the cell's unique characteristics and roles. In these two chapters, we will discover how the cell's parts and their coordinated functioning account for the survival of life's most basic units.

HOW CELLS ARE STUDIED

Most cells are extremely small, and so it is not surprising that the discovery of these basic units of life followed closely after the invention of the microscope. Because sight is our primary sense, biologists still rely heavily on microscopes to detect cell structures. But, as we shall

see, many newer devices are also used to weigh, test, and measure cells and to take them apart and learn how they function.

Microscopy

Much of what is known about cells was discovered by biologists peering into microscopes, just as Hooke did more than 300 years ago. The history of cell biology parallels the invention of and improvements in microscopes, just as astronomy grows along with the new and different types of telescopes.

One might assume that the earliest microscopes—called "flea glasses" by their inventors—were crude instruments. One might also expect that these early scientists sliced, stained, and mounted specimens in haphazard ways. However, a recent find in the vaults of the British Museum of Natural History reveals just the opposite. Among the letters, microscopes, and memorabilia of Anton van Leeuwenhoek, the seventeenth-century Dutch "father of the microscope," modern scholars found well-preserved and delicately prepared specimens of algae filaments, cotton seeds, cork, thin slices of elder branches, and the optic nerve of a cow (Figure 5-1). Leeuwenhoek's drawings of these specimens are so accurate that the magnifying power and clarity of his hand-held, postage-stamp-size microscopes cannot be questioned.

Figure 5-1   RIGID WALLS OF PLANT CELLS.
This specimen, magnified about 600 times, reveals the regular compartments Anton van Leeuwenhoek originally saw in a thin slice of cork. The appearance of such plant materials led Robert Hooke to coin the term "cells."

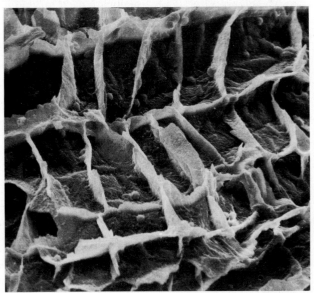

The microscopes used by Hooke, Leeuwenhoek, and other pioneers in cell biology were *light microscopes.* These instruments contain specially ground glass lenses that bend light so that an object being viewed through them is magnified, or enlarged. Although light microscopes are still used extensively in biological research, they are seldom used to magnify objects more than about 1,200 times normal size because of the limitations of resolving power—the ability of the human eye to distinguish adjacent objects. To understand such limitations, try this exercise: draw two dots a small distance apart on a piece of paper. Then draw two dots closer together; then two more, still closer. Continue until the final dots are so close to each other that they appear to have merged. They can no longer be distinguished as separate dots because their proximity has gone below the resolving power of the unaided human eye (a distance of about 0.1 mm). If you were to examine those last two merged dots under a light microscope, they would be easily resolvable; indeed, if magnified sufficiently, they might appear so far apart as to not come into view at the same time. The best light microscope is about 500 times better than the human eye at resolving dots, lines, or other small objects; thus it can be used to reveal extremely small structures. Even so, resolving power is limited by the wavelength of visible light; the finest light microscope is limited to resolving objects spaced more than about 0.2 micrometer (μm) apart. The long slender tail of a sperm cell, for example, is about 1.5 μm in diameter, so it can be seen with the light microscope, but the smaller structures inside the tail cannot.

Such limitations led scientists to design microscopes that use beams of electrons instead of light to irradiate specimens. These microscopes have a resolving power 100,000 times greater than that of the human eye. *Transmission electron microscopes* (TEMs) permit examination of a cell's interior. They usually are used to send a beam of electrons through an ultrathin slice cut from a specimen embedded in plastic. Some electrons are scattered or absorbed by molecules of the specimen. Others pass through the specimen to produce a negative image of the object on a small fluorescent screen.

With *scanning electron microscopes* (SEMs), scientists are able to view the surfaces of cells and organisms. SEMs bombard the surface of a specimen with electrons. The surface atoms become excited and emit electrons, which can be captured and used to form a three-dimensional image of the specimen's surface, with all its holes, ridges, spaces, and textures revealed. Although the resolving power of SEMs is not nearly as great as that of TEMs, the depth of field—the thickness of the specimen that is in perfect focus for optimal viewing—is much greater. Consequently, SEMs have provided our first good three-dimensional views of the surfaces of biological objects. Figure 5-2 compares the images obtained with a light microscope, a TEM, and an SEM.

In order for a watery, translucent cell to be examined through a microscope and distinguished from the background, it must have contrast—distinct areas of light and dark. In light microscopy, contrast is produced in living

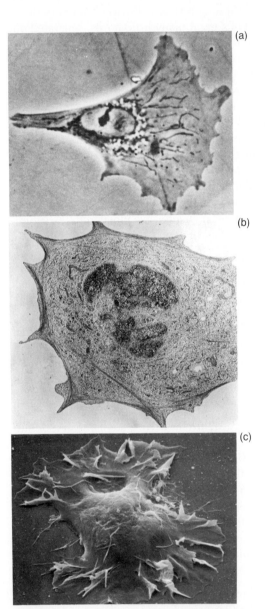

Figure 5-2   SINGLE CELLS FROM A CHICK EMBRYO VIEWED WITH THREE TYPES OF MICROSCOPES.
(a) A living cell viewed through a light microscope, magnified about 775 times. (b) A thin section of a "fixed" cell viewed through a transmission electron microscope, magnified about 3,100 times. (c) A fixed cell, caught in the process of ruffling its edges as it moved, viewed with a scanning electron microscope, magnified about 2,800 times. Each microscopic technique reveals different aspects of the cell to the viewer.

cells by varying the light beam with prisms and other physical devices; dyes are used to produce contrast in dead cells. In electron microscopy, researchers stain specimens by soaking them in solutions of heavy metals, such as lead or uranium salts, so that electrons will be absorbed more readily in denser areas of the object and thus look darker.

The electron gun and beam in a TEM or an SEM must be operated in a high vacuum; specimens in such a vacuum must be prepared in special ways. They commonly are fixed with preservative chemicals and then dried. TEM specimens usually are embedded in a block of plastic and cut with an exceedingly sharp diamond into ultrathin sections (a pile of 1,000 such sections would be no thicker than this page). Alternatively, specimens may be quick-frozen and then literally cracked apart with a sharp blow. When coated with metal, these freeze-fracture preparations give good surface-relief views of cell parts. SEM specimens are fixed, dehydrated, and coated with a thin layer of gold (about 0.2 nanometer [nm] thick), which covers every nook and cranny of the surface and emits the electrons that yield a bright image of the specimen.

Means of Studying Cell Function

Viewing cells through a microscope is a good way to observe structures, but to understand a cell's living functions requires other instruments and techniques. One important method for determining function is *cell fractionation*, in which cells are ground or broken up and analyzed. Figure 5-3 shows how this is done with plant cells, using a centrifuge to separate cellular components from one another. The chemistry of each component, or fraction, may then be investigated. With this process, a given enzyme might be shown to be active in one component (organelle) but not in others; alternatively, most DNA, the genetic material, would be found in the centrifugal fraction that contains cell nuclei.

Another procedure is *radioactive-isotope labeling* to trace ions, molecules, and individual chemical reactions in cells and organisms. For example, a given amino acid with a radioactive isotope of carbon (^{14}C) or hydrogen (^{3}H) in place of its normal C and H atoms can be prepared. Within seconds or a few minutes of entering a cell, the labeled amino acid can be built into a protein and shuttled through various cellular compartments prior to being secreted from the cell or incorporated into a cellular structure. If the cell is fractionated during these processes, the fate of the labeled amino acid can be traced by measuring the radioactivity of the various fractions. Even more precise localization of the labeled, incorporated amino acid is possible using *autoradiogra-*

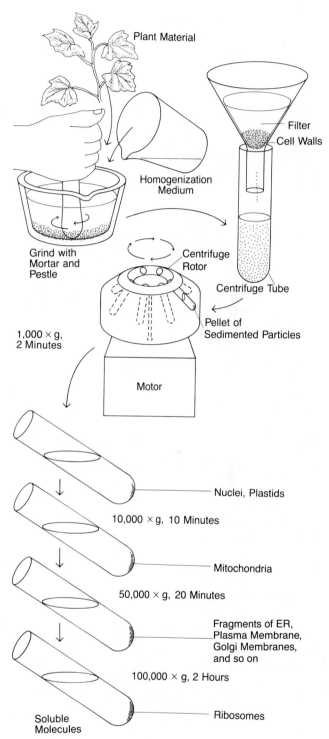

Figure 5-3   CELL-FRACTIONATION TECHNIQUES: SEPARATING CELLULAR COMPONENTS.

In this case, plant leaves are homogenized and then filtered to remove the hard cell walls. Then the material is spun at progressively faster centrifugation speeds until a series of "pellets" forms. The label 1,000 × g signifies centrifugal force 1,000 times stronger than gravity. The first pellet contains the heaviest organelles, such as the nuclei; the last, high-speed run yields the smaller, lighter ribosomes.

phy, in which a thin section of the killed cell is coated with a photographic emulsion; the emulsion is then exposed to radioactive emission from the incorporated amino acid, just as light exposes the emulsion on photographic film.

CHARACTERISTICS OF CELLS

Perhaps the most important outcome of the three centuries of cell research that began with Robert Hooke is the modern **cell theory.** Generations of cell biologists applied the available technology—first microscopes, then cell fractionation, and later radioactive-isotope labeling—to probe the nature of cells. From some of this ground-breaking work, nineteenth-century biologists developed a set of statements that encapsulates the essential characteristics of cells.

The German biologists Lorenza Oken, Matthias Schleiden, Theodor Schwann, and Rudolf Virchow wove loose threads from old ideas and observations about cells into a new synthesis. Their contribution to the modern cell theory states:

1. Cells are the basic units of life on Earth. No organism has ever been found on Earth that shows the attributes of life and yet is not composed of cells.

2. All organisms are constructed of cells. Every living thing on Earth is either a single cell or a population of cells.

3. Except at the origin of life itself, all cells arise from preexisting cells. Cells arise only by division of living cells, never by aggregation of cell parts and cell chemicals ("from life, comes life").

More recent research suggests three additions to the cell theory that apply to higher organisms and thus are not quite as general as the first three:

4. Cells are the functional units of life, in which all of the chemical reactions necessary for the maintenance and the reproduction of life take place.

5. Cells of multicellular organisms sometimes are interconnected so that the resultant populations can function as single units.

6. Cells of multicellular animals must stick to solid surfaces in order to divide, move, assume specialized shapes, and carry out necessary functions.

Like the theories of evolution by natural selection and of the particulate nature of matter, the cell theory is a cornerstone of science: the basic units of matter are

SORTING CELLS WITH A GLOWING TECHNIQUE

Biologists who study cells and cell organelles face some giant obstacles, since the subjects of their interest are usually too small to be seen with the unaided eye or moved about with needles or scalpels. The invention of high-powered microscopes and micromanipulators helped a great deal; but until the mid-1970s, one obstacle stubbornly remained: it often was impossible to sort out a given cell type from a mixture of similar types.

A team of cell biologists at Stanford University solved this problem by inventing a technique called *fluorescence-activated cell sorting.* It employs fluorescent dyes and some of the structures on the cell surface.

Simplifying somewhat, the method works as follows (Figure A): the researcher labels with fluorescein (a dye that glows yellowish-green under certain types of light), a mixture of the many cell types in an organ such as a lymph node. Using a fluorescence microscope, he or she identifies the desired cell type and notes its size and degree of fluorescence, based on the binding of dye to sugar–protein surface markers or to other structures on the cell surface. The labeled cell mixture, suspended in fluid, is then squirted through a nozzle that vibrates rapidly and breaks the fluid into tiny droplets, each just big enough to hold one cell. A stream of these droplets passes through a blue laser light, which excites the yellow-green glow of the dye particles attached to the surface of certain cells. This glow is converted to an electrical signal, which puts a negative electrical charge on droplets containing fluorescent cells; a positive charge is put on droplets with nonfluorescent cells or no cells. Other droplets remain uncharged. The stream of droplets then passes through an electric field. The negatively charged droplets are deflected into one collector tube and the positively charged droplets into another. Uncharged droplets go into a third. Thus one tube will contain only labeled cells of the desired type, alive and well, while the other tubes will hold all the undesired cell types from the original mixture.

Fluorescence-activated cell sorting, which can process 5,000 cells a second, has been immensely valuable to cell biology, making possible some of the most important cell research being conducted today. It has allowed biologists to isolate certain types of cancerous cells and cells of the immune system, capable of mass-producing specific targeted molecules called antibodies that can be used to fight diseases. We will learn more about these and other applications in later chapters.

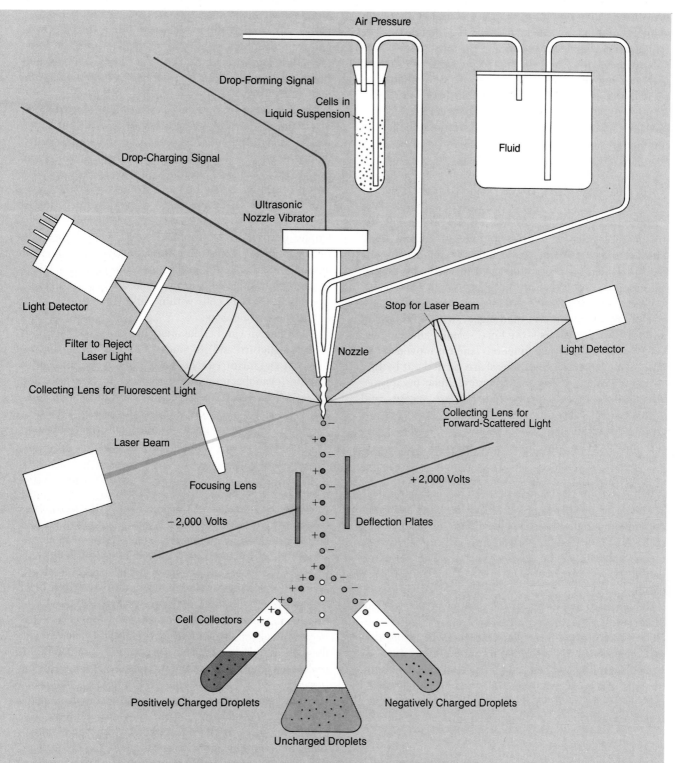

Air Pressure

Drop-Forming Signal

Cells in
Liquid Suspension

Fluid

Drop-Charging Signal

Ultrasonic
Nozzle Vibrator

Light Detector

Stop for Laser Beam

Filter to Reject
Laser Light

Nozzle

Light Detector

Collecting Lens for Fluorescent Light

Collecting Lens for
Forward-Scattered Light

Laser Beam

Focusing Lens

+ 2,000 Volts

− 2,000 Volts

Deflection Plates

Cell Collectors

Positively Charged Droplets

Uncharged Droplets

Negatively Charged Droplets

Figure A

A diagrammatic view of the cell sorter. Tiny droplets
containing as few as a single cell are deflected to the left
or right depending on electrical charges.

atoms; the units of compounds are molecules; the units of life are cells. But cells are not simply building blocks of life, as atoms are of molecules; the cells themselves *are what is alive in organisms*. They comprise the parts of all living things and make possible the life functions of organization, movement, energy use, reproduction, response to stimuli, growth, and so on. Thus we cannot truly understand living things unless we study their simplest living units, their cells. The cell theory marked a major turning point for biology: by focusing on cells, biologists could begin to pose specific questions about how life operates and discover some profound answers.

The Nature and Diversity of Cells

Cells are the most highly ordered assemblages of dissimilar molecules on Earth—perhaps in the universe. This high degree of organization allows them to carry out life functions and resist entropy, the tendency toward increased disorder that we talked about in Chapter 4. Cells can exist as discrete, free-living, single-celled species of bacteria, protozoa, and algae, or as members of multicellular organisms—the fungi, plants, and animals. Within each cell, special functions (such as movement or protein export or photosynthesis) are carried out by tiny structures called *organelles*. In multicellular organisms, large groups of cells arrayed as *tissues, organs,* and *organ systems* carry out particular functions (such as digestion or excretion) in the "economy" of the whole creature and thereby contribute to its survival.

Types of Cells

There are thousands of kinds of cells. Most live as independent single-celled species, but a wide variety make up multicellular organisms. All these diverse types, however, can be classified in two ways: (1) by fundamental elements of structure and (2) by the way they obtain energy.

Structurally speaking, cells are either **prokaryotes** or **eukaryotes.** The word "prokaryote" means "before nucleus," and it describes cells in which DNA is localized in a region but is not bounded by a membrane. In other words, a prokaryotic cell lacks a true, membrane-bound nucleus. All prokaryotes are independent, single-celled organisms, and include the thousands of species of bacteria and cyanobacteria (blue-green algae). We will discuss other distinguishing features of prokaryotes in Chapters 6 and 19.

A eukaryotic ("true nucleus") cell has a membrane-enclosed nucleus containing DNA found in complex structures called chromosomes. In addition, eukaryotic cells contain several other types of membrane-bound internal structures, or **organelles.** Some of these are equivalent in function to unbounded regions within the prokaryotic cell, while others have functions found only in eukaryotes. Finally, eukaryotes have a special network of minute filaments and tubules called the *cytoskeleton*, which gives shape to the cell and allows movement. All single-celled organisms other than bacteria and cyanobacteria are eukaryotic cells, as are the basic units that comprise all multicellular plants, animals, and fungi. Any organism made up of one or more eukaryotic cells is a eukaryote.

The second system of cell classification—by methods of obtaining energy—categorizes cells as either **autotrophs** or **heterotrophs.** Autotrophs ("self-feeders") use light energy or chemical energy to manufacture their own sugars, fats, and proteins. A few types of bacteria are autotrophic, as are the more than 400,000 plant species on Earth. Heterotrophs ("other feeders") derive their energy by taking in foods in the form of whole autotrophs or other heterotrophs, their parts, or their waste products. Many kinds of bacteria and all the millions of animal and fungal species are heterotrophs. The terms "autotroph" and "heterotroph" can apply to individual free-living cells, to cells within multicellular organisms, or to whole organisms.

An organism can be classified simultaneously by both its cell structure and its mode of energy procurement. Thus an organism can be both prokaryotic and autotrophic (cyanobacteria), prokaryotic and heterotrophic (intestinal bacteria), eukaryotic and autotrophic (a spinach plant), or eukaryotic and heterotrophic (a rabbit). We will use these designations again and again throughout the book.

Components of Eukaryotic Cells

While the presence of a true nucleus is the most obvious trait of a eukaryotic cell, a typical plant or animal cell has several kinds of organelles, some common to all eukaryotes and others characteristic of either plants *or* animals. Figures 5-4 and 5-5 show generalized animal and plant cells with organelles labeled. Let us take a brief tour of these cells and identify the cell parts that we will discuss in detail later in this chapter or in the next.

An animal cell is enclosed by the *plasma membrane*, a flexible, double layer that is surrounded by a coating of molecules, the *glycocalyx*, which mediates the cell's interactions with the environment (Figure 5-4). The semifluid interior of the cell, the *cytoplasm*, contains a *cytoskeleton* in which organelles are suspended. The cytoskeleton is a dynamic scaffolding composed of *microfilaments* and *microtubules* that support the cell's shape and often change it by their activity. The suspended organelles include the *mitochondria*, in which high-energy compounds are formed; the *ribosomes*, sites of protein synthesis; the *endoplasmic reticulum*, an assembly line for making certain types of proteins and fats; the *Golgi complex*, a depot where export materials are packaged;

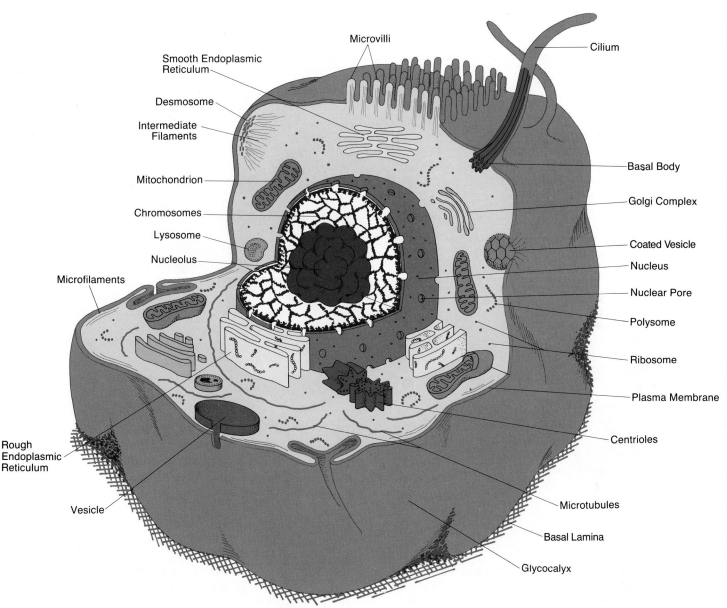

Figure 5-4   COMPONENTS OF A TYPICAL ANIMAL CELL.
The central nucleus contains hereditary material and a nucleolus, a site of ribosome manufacture. The watery cytoplasm has a cytoskeleton composed of microtubules, microfilaments, and other components. Cytoplasmic organelles include mitochondria, the energy powerhouses; rough endoplasmic reticulum, a site of protein synthesis; the Golgi complex, a place for modifying proteins; lysosomes, cellular digestive bodies; and centrioles, organizing bodies for microtubules. The cell surface includes the plasma membrane and attached glycocalyx. The surface may extend outward as cylindrical microvilli or as longer, motile cilia.

the *lysosomes*, which serve as digestion and waste-disposal systems for the cell; and the *nucleus*, the repository of the genetic material contained in chromosomes. The cell surface may possess one or more *cilia*, which act like tiny oars that move the whole cell or sweep materials past it, and it may also have fingerlike projections called *microvilli*, which usually facilitate absorption.

Although Figure 5-4 portrays the cell as static, a living cell is actually quite dynamic. Simultaneously, its cilia may wave about; its microvilli may lengthen and shorten; parts of the cytoskeleton may assemble or disas-

semble; its nucleus may rotate; its mitochondria and other organelles may be moved about. And all the while, at submicroscopic, molecular levels, tens of thousands of chemical reactions are occurring each second.

The components of a generalized plant cell are depicted in Figure 5-5. By comparing the two cell types, we can easily spot the similarities and differences. The plant cell's plasma membrane is surrounded by a protective, supportive *cell wall*. The cytoplasm contains not only the same sorts of organelles found in animal cells, but also sacs called *plastids*, including *chloroplasts*, the

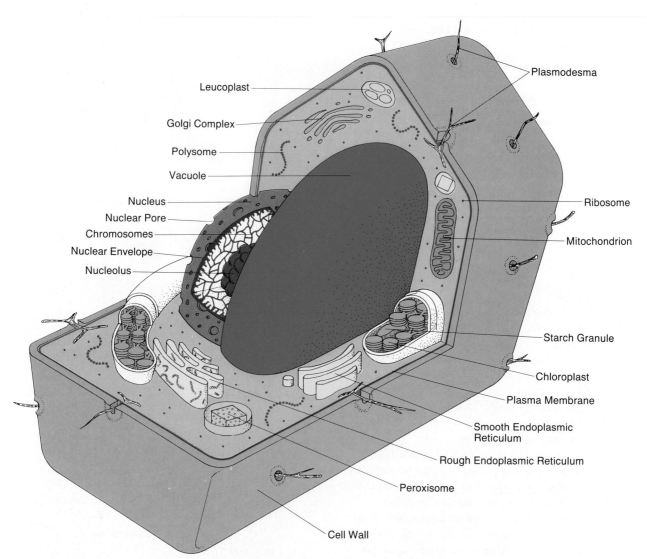

Figure 5-5   COMPONENTS OF A TYPICAL PLANT CELL.
Organelles present in animal cells are also found here. In
addition, chloroplasts, sites of photosynthesis; leucoplasts,
storage depots; and the large central vacuole are present.
Plasmodesmata link plant cells together. Although it appears
static here, the cytoplasm of a living cell streams around the
edge of the central, fluid-filled vacuole.

sites where nutrient molecules are built using the sun's
energy, and *leucoplasts*, where nutrients are stored. The
central portion of the plant cell, the *vacuole*, is a large
drop of fluid enclosed by a membrane and filled with
dissolved inorganic and organic substances. Neighboring
plant cells may be interconnected by gossamer bridges
of cytoplasm called *plasmodesmata*. Like animal cells,
plant cells are extremely dynamic, and through a microscope, it is often possible to view the cytoplasm and nucleus streaming around and around the central vacuolar
cavity.

With this brief tour of animal and plant cells as back-

ground, we can move on to consider some fundamental
facts about cell size and the cell surface.

Limits on Cell Size

One might assume that large organisms are built from
large cells, and small organisms, from small cells. However, measurements show that the difference between a
whale and a mouse or between a redwood and a petunia
lies in the total number of cells, not in the sizes of individual cells. A human, for instance, contains more than
100 trillion (10^{14}) cells, while a rat—possessing the same
set of organs—has about 100 billion (10^{11}) cells. This is
not to suggest that all cells are the same size. The largest
single cell alive today is the ostrich egg yolk, which is
more than 3 cm in diameter, while bacterial cells can be
less than 0.2 μm, or 0.0000002 cm, in length. Within the
human body, there are more than 100 types of cells,
ranging from small white blood cells 3 to 4 μm long to
individual nerve cells almost 1 m long. Most single cells,

Table 5-1 **SIZES OF CELLS, ORGANELLES, AND MOLECULES**

Cell, Organelle, or Molecule	Diameter or Length (μm)
Average plant cell	35.0
Average animal cell	20.0
Bacterium	2.0
Mitochondrion	1.0–8.0
Virus	0.02–0.2
Ribosome	0.025
Microtubule	0.022
Hemoglobin molecule	0.0064
Hydrogen ion	0.0001

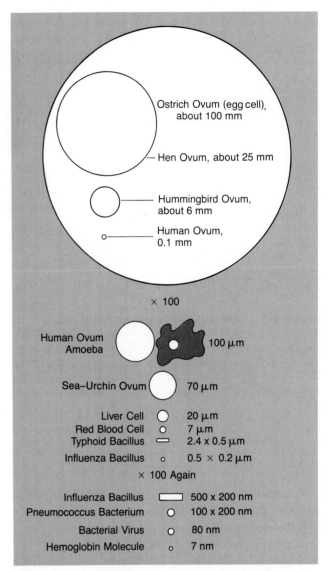

Figure 5-6   THE SIZES OF VARIOUS CELLS AND MOLECULES COMPARED.
A millimeter is $\frac{1}{1000}$ of a meter; a micrometer is $\frac{1}{1,000,000}$ of a meter; a nanometer is $\frac{1}{1,000,000,000}$ of a meter.

however, whether free living or part of a multicellular organism, do not exceed a diameter of about 30 μm. Thus 1,000 typical cells lined up neatly would just about span your thumbnail. Table 5-1 and Figure 5-6 compare some average cell sizes with the sizes of a few common cell organelles and individual molecules.

It may seem puzzling at first that larger size is achieved by increasing cell number rather than cell size. Why do humans not have 100 billion large cells or 100 million gigantic cells instead of 100 trillion small ones? There are several good reasons why cells are both minute and numerous in multicellular organisms—and also why a single-celled blob could never really grow big enough to engulf an automobile or Pittsburgh.

The primary reason cells are rarely larger than 30 μm in diameter is because bigger cells do not have a large enough surface area to exchange sufficient amounts of nutrients, gases, ions, and wastes with the environment. As Figure 5-7 shows, for each doubling in length, the cell's surface area is squared, but its volume is cubed. Conversely, as the cell gets smaller, its volume decreases faster than its surface area. Thus the **surface-to-volume ratio** is greater in small cells. Since the cell surface is the site of interchange between the cell and the external environment, a larger relative surface area allows the cell to more readily absorb oxygen and nutrients for metabolism and to more easily excrete carbon dioxide and other waste products. Another way to visualize these relationships is to imagine a jumbled mass of wet clothes just removed from a washing machine. If they are left in a pile (large volume, small surface area), they will take days to dry; but if they are hung up or spread out, the water will evaporate quickly because of the huge increase in surface area. An advantage of greater surface area and small size in cells is that gases and nutrients have a shorter distance to travel from the edge to the center of the cell, where they are utilized for life processes.

Careful examination reveals that even extraordinarily large cells do not violate the principle of surface-to-volume ratio established for more common small cells. The yolk of a hen's egg is an enormous single cell some 2.5 cm in diameter. However, most of the mass is relatively inert stored nutrient. A thin layer of cytoplasm surrounds the yolk. Similarly, the individual nerve cells that extend from the lower back to the toe of a 7-foot-tall basketball player are exceedingly long, thin columns of cytoplasm. Even though their total cellular volume is great, these nerve cells have a correspondingly huge surface area.

Exactly what is the cell surface, the area of which is so important? As we will see, it is at once a raincoat, a guardian, and a sieve surrounding every living cell.

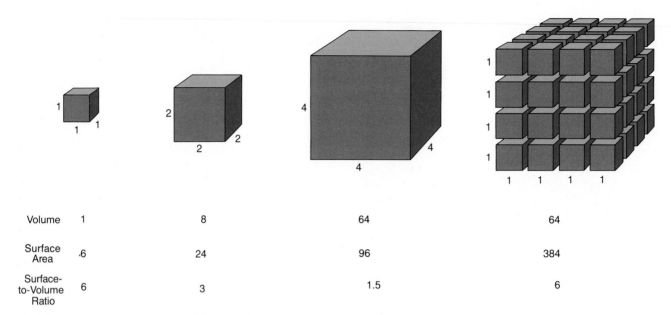

Volume	1	8	64	64
Surface Area	6	24	96	384
Surface-to-Volume Ratio	6	3	1.5	6

Figure 5-7   SURFACE-TO-VOLUME RATIO IN CELLS.
Subdividing a large cube demonstrates the concept of surface-to-volume ratio and shows why cells are small. As the length of each side of this cube doubles from 1 to 2 to 4, the volume increases more rapidly than does the surface area. Hence, the ratio of surface area to volume goes down. If the final large cube (4 × 4 × 4) is cut into sixteen small cubes (1 × 1 × 1), the total surface area becomes four times as large. If these were cells, each of the small cells would be better able to exchange materials with its environment than would the one large cell.

THE CELL SURFACE

It is apt that we begin our study of cell structure at the outside. The cell surface is, of course, the first part of a cell that we encounter, just as we see the brown paper on a package before discovering what is inside. The cell surface is far more than a static wrapper, however; it is a dynamic barrier that governs all traffic into and out of the cell. Just as whole animals need to eat, drink water, and excrete to survive, cells must take in and expel various materials—all through the cell surface. Cells must also recognize and communicate with one another, especially within tissues and organs. The cell surface governs these activities as well. In this section, we will examine the cell surface in detail and look at the many processes by which this dynamic barrier controls the two-way flow of materials.

The Plasma Membrane

Despite many kinds of walls and coatings on the surface of various cell types, there is one invariant component of every cell surface: the **plasma membrane,** or cell membrane. This thin layer of lipid and protein molecules surrounds the watery cell contents, the cytoplasm. Today we know that the plasma membrane is built of two layers of lipid molecules; at normal temperatures, the

lipids behave as a fluid in which proteins of the plasma membrane "float." Let us trace the history of this exciting concept.

In the 1930s British biologists H. Davson and J. F. Danielli considered the evidence then available on the movement of materials into and out of cells and proposed a theory for the structure of the barrier at the cell surface. They hypothesized that cells are covered by a thin, flexible envelope composed of two layers and that this **bilayer** contains phospholipid molecules and proteins. They believed—correctly, it turns out—that the hydrophilic, or water-loving, heads of the phospholipid molecules are oriented outward in these layers and serve as a relatively waterproof boundary for the cell. Their speculations about the protein component of the membrane were incorrect, as we will see shortly. But the hypothetical lipid bilayer was corroborated by an early electron microscopist, who observed and photographed an 8-nm-thick double line at the cell surface (Figure 5-8a). And later physical and chemical tests confirmed that plasma membranes and membranes surrounding many cell organelles consist of a bilayer of phospholipid molecules in which charged hydrophilic heads point outward and uncharged hydrophobic tails point inward (Figure 5-8b).

Additional tests have shown that the rigidity of the membrane depends on the degree of saturation—that is, the proportion of double and single bonds—between adjacent carbon atoms in the fatty-acid chains of the phospholipids. The more double bonds in the chain, and

(a)

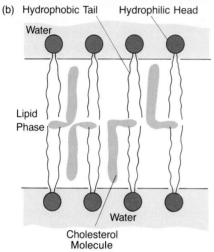

(b) Hydrophobic Tail   Hydrophilic Head

Water

Lipid Phase

Water

Cholesterol Molecule

Figure 5-8   THE PLASMA MEMBRANE: FRAGILE BARRIER ESSENTIAL TO A CELL'S SURVIVAL.
(a) In the plasma membrane of this red blood cell, magnified 195,000 times, the two dark lines are sites where electron-dense stains have bound to the hydrophilic ends of the lipid molecules. The light-colored area between them is the central, hydrophobic portion of the lipid bilayer. (b) The molecular structure of the plasma membrane. Cholesterol molecules reduce membrane fluidity at normal temperatures.

thus the more unsaturated they are, the more flexible and fluid the membrane, and the more easily substances in the membrane can move.

In their lipid-bilayer theory, Davson and Danielli speculated that sheets of proteins coated the phospholipid layers like two slices of bread on each side of a layer of butter. However, later studies showed that membrane proteins are not complete layers, but are individual molecules attached to the inner or outer membrane surface—so-called **extrinsic proteins**—or embedded in it—so-called **intrinsic proteins.** Some extrinsic proteins serve as links to sugar–protein markers, or name tags, on the cell surface, while some intrinsic proteins serve as vehicles that move ions or molecules across the membrane and others moor the membrane to the cell's inner

scaffolding (the cytoskeleton) and to various molecules outside the cell.

While the studies of membrane structure were significant, a far-reaching theory of membrane function proposed in 1972 has revolutionized our understanding of the dynamic barrier at the cell surface. Biologists Jonathan Singer and Garth Nicolson developed the **fluid-mosaic model** (Figure 5-9) from experimental evidence gathered by Harden McConnell, L. D. Frye, M. Edidin, and others. Singer and Nicolson proposed that at normal biological temperatures, the plasma membrane behaves as a thin layer of fluid covering the cell surface. Individual lipid molecules move about fairly freely within the plane of the membrane, thereby contributing to the fluid state. Scattered through the membrane, as in a patchwork or a mosaic, are proteins or groups of proteins, many of which appear to be free to move or diffuse in the lipid plane, like icebergs floating in a lipid sea. Not all proteins are free to diffuse, however, due to special structures associated with the cell surface, which we will discuss later.

In 1970, Frye and Edidin reasoned that if a membrane is fluid, perhaps they could label and watch molecules moving around in it. Using human cells, they labeled certain sugar–protein surface markers with a dye that looks red under ultraviolet light. Next, they labeled surface markers of mouse cells with a green dye. Finally, employing special virus particles that cause cells to fuse, they created giant hybrid cells—part human, part mouse. At first, the red and green markers remained on opposite sides of the fused cells; but within forty minutes, they had intermingled, and the entire cell surface was a mosaic of red and green dots. Frye and Edidin wondered whether the labeled markers had moved laterally in the membrane itself, or whether they had passed into the cytoplasm and then been reinserted into the membrane at a different spot. Such insertion would require the expenditure of energy by burning the fuel ATP, so Frye and Edidin employed a variety of drugs and treatments to block ATP expenditures by the cells. These treatments effectively blocked *other* cellular processes but *not* the movement of the red and green markers. They then chilled the cells so that the membrane lipids—if fluid to begin with—would lose their fluidity. And, indeed, the colored markers stopped moving through the chilled membrane and remained separated. This was strong evidence that membranes are fluid structures and helped lead Singer and Nicolson to their theory.

The thin membranes surrounding various organelles within cells are also composed of lipid bilayers, are fluid in nature, and may have scattered protein "icebergs." However, there are certain chemical differences between the plasma membrane and these other membranes: (1) cholesterol, which reduces membrane fluidity at normal temperatures, is present in the plasma

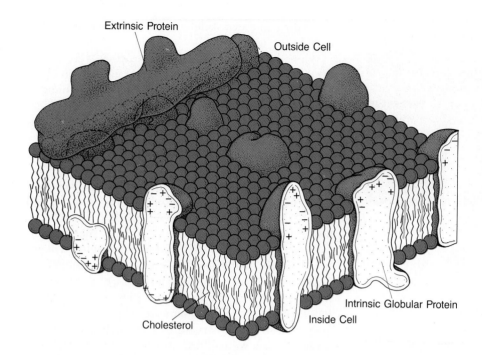

Extrinsic Protein

Outside Cell

Intrinsic Globular Protein

Cholesterol

Inside Cell

Figure 5-9   THE FLUID MOSAIC MODEL.
Globular intrinsic proteins may protrude above or below the lipid bilayer and may move about in the plane of the membrane. Extrinsic proteins are bound to intrinsic ones. If membrane proteins are linked to other molecules (as at sites 1 and 2), their diffusion in the membrane is slowed or prohibited. Notice that the charged regions of the proteins are positioned next to the charged hydrophilic head groups of the lipid molecules, while there are no surface charges on portions of protein in the core of the bilayer, where the uncharged hydrophobic lipid tails reside.

membranes of many eukaryotes, but not in organelle membranes; (2) different hydrocarbon chains compose the tails of the hydrophobic ends of the lipids; and (3) the overall percentages of various other lipids differ. One consequence of these differences is that the membranes surrounding organelles are considerably more fluid than is the plasma membrane. We shall explore the value of this extra fluidity in Chapter 6. For now, it is important to remember that all membranes are built on the same molecular principle: they are lipid barriers that stand between the watery world outside the cell and the watery cytoplasm within.

Movement of Materials into and out of Cells: Role of the Plasma Membrane

Life on Earth evolved in water some 3 billion years ago, and in a very real sense, the cells of multicellular organisms have never left that primal ocean. Inside all plants, animals, and fungi are salty, nutrient-filled liquids that take the form of droplets and layers and sometimes course through vessels. Inside the individual cells are the *intracellular fluids*, which make up the fluid part of the cytoplasm. Outside the cells are layers of clear tissue fluids, vessels full of blood or sap, or reservoirs of other liquids; these *extracellular fluids* bathe the surface of each cell like a shallow, miniature internal ocean. Without moisture, both inside and out, the integrity of the plasma membrane cannot be maintained, and the cell dies.

Dissolved in a large animal's extracellular and intra-

cellular fluids are several grams of salt ions, sugars, proteins, hormones, and other substances. However, some ions and molecules must remain at higher concentrations outside the cell than inside, while others must be more heavily concentrated inside the cell than outside. For example, there are 100 times more potassium ions inside human cells than in the blood outside those cells, but there are 100 times more sodium ions in the blood than in the intracellular fluid. Even slight changes in these concentrations can affect the health of or even kill the cell. Therefore, a major activity at the cell membrane is the maintenance of constant properties of the fluids inside and outside cells by regulating the two-way flow of materials, despite short-term vagaries in an organism's diet, activities, or environment.

Because plasma membranes are composed primarily of lipid molecules, it is difficult for materials that are not lipid soluble to move through them. It is as though the cell were surrounded by an oilskin raincoat. The reason the lipid raincoat is resistant to passage of ions or proteins is that those substances are polar—they bear unequally distributed charges or net electrical charges on their surfaces (Chapter 2). Such charges prevent entry into the charged surface of the lipid bilayer. Nevertheless, certain mechanisms selectively permit some ions and molecules to enter the cell. These mechanisms are the passive processes of simple diffusion and carrier-facilitated diffusion and the energy-requiring process of active transport.

Simple Diffusion

Diffusion is the tendency of a substance to move from a place of high concentration to one of lower concentra-

THE SECRETS OF WINTER WHEAT

The United States has become the "breadbasket" of the world, exporting enormous amounts of wheat each year despite the fact that most of the wheat-growing areas in the West and Midwest are subjected to long, harsh winters and a foreshortened growing season. Farmers in these regions can produce wheat so successfully because most of them grow a hardy grain plant called winter wheat. From seed sown in September, tender green seedlings sprout in October, and the tiny plants staunchly survive the frigid winter months with roots frozen in the soil and leaves buried beneath deep snow. Plant breeders developed winter wheat by selecting hardier and hardier wheat plants over a number of generations. Until recently, however, no one knew what special characteristics allow winter wheat to survive in areas where regular wheat plants die with the first autumn frost. A researcher has discovered the answer, and it lies in the plant's extraordinary cell membranes.

Peter Steponkus of Cornell University compared plants that can survive freezing, so-called *cold-tolerant plants*, and plants that cannot survive freezing, or *cold-sensitive plants*. If a normal cold-sensitive strain of wheat is sown in, say, northern Nebraska, the plants will freeze in November; when the snow melts in spring, they will collapse into limp, lifeless clumps in the field. Winter-wheat plants in the same field will be firm, green, and quite alive as the snows recede.

Steponkus studied the cell membranes of both types of plants and found some very distinct differences. As the cells of a cold-sensitive plant freeze, water from the cytoplasm migrates outward and the cell membrane shrinks, causing the cell to shrivel. As the membrane shrinks, fat droplets appear to be squeezed from it into the cell's interior and to remain suspended within the cytoplasm. When the cell later thaws, water is reabsorbed and the cell expands again; but the fat droplets in the cytoplasm do not rejoin the membrane, and holes appear in the membrane where the droplets were located. The holes weaken the membrane, which eventually splits, and the cell's contents spill out. This explains why the plant wilts and dies in the spring.

When the cells of a cold-tolerant plant, such as winter wheat, freeze, water migrates out and the membranes shrink, just as in the cells of a cold-sensitive plant. However, Steponkus found that the lipids squeezed from the membrane do not move inward and float freely inside the cell's cytoplasm. Instead, those fat droplets are extruded outward and remain "tethered" to the membrane's outer surface in miniature spheres, filaments, and pockets. When the cell thaws, the fats are reabsorbed into the lipid bilayers. Because of this tethering and reabsorption, no holes form in the membrane, which does not split, and the cell remains intact. Thus when the spring thaw comes, a winter-wheat plant can resume active growth, having suffered no damage.

It is amazing to think that this tethering process in delicate cell membranes is indirectly responsible for the success of the United States as a wheat-growing and wheat-exporting nation. Such knowledge is important to plant breeders; with it, they may be able to select and modify other staple food crops to better withstand cold temperatures. This would potentially prolong the plants' growing seasons and expand their geographical ranges. Ultimately, such "super" plants could provide badly needed food for the world's increasingly hungry population and help ensure America's position as a major food provider.

---

tion. A demonstration of how diffusion works is presented in Figure 5-10. The two sides of the beaker are separated by a *permeable membrane*, a porous partition—like a sheet of cellophane punctured with tiny holes—that allows molecules or ions to move from one side to the other. Suppose that both sides contain slightly salty water and that a tablespoon of salt is sprinkled into the one on the right. In water, salt (NaCl) dissociates into two soluble ions, Na^+ and Cl^-. The ions on both sides of the membrane move about at random. Since there are many more ions in the more concentrated solution on the right, they will randomly encounter the tiny holes in the membrane more frequently, on the average, than will the less concentrated ions on the left. Thus more ions will move from right to left (high to low concentration) than from left to right.

In addition, because water molecules are attracted to dissolved polar substances, such as sugar and salt, the mobility of the water molecules is reduced in the immediate vicinity of dissolved ions. That effect lowers the free energy of water, in this case, to move. In the system depicted in Figure 5-10, some water molecules are subjected to this drag effect where the salt concentration is higher, so there is more drag on the right than on the left. Since there are fewer encumbered water molecules on the left, more water molecules are free to encounter pores in the membrane and so to move from left to right. Thus water moves from regions of high free energy to

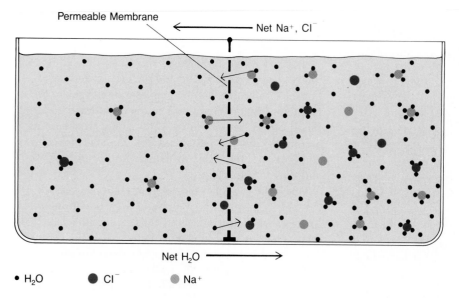

Permeable Membrane

Net Na⁺, Cl⁻

Net H₂O ⟶

- • H₂O   ● Cl⁻   ● Na⁺

Figure 5-10   A SIMPLE DIFFUSION SYSTEM.
The two sides of the beaker contain solutions with high and low salt concentrations, separated by a porous membrane. Free water molecules and sodium and chloride ions move randomly. The more concentrated ions on the right strike pores in the permeable membrane more frequently than do the less concentrated ions on the left. As a result, more sodium and chloride ions move from right to left than the reverse. Water behaves the same way: it moves from a region of high concentration of water molecules (the more dilute solution on the left where water's free energy is higher) to a region of lower concentration (where water's free energy is lower). So, net movements of water and of Na^+ and Cl^- ions occur in opposite directions until concentrations on both sides are equal.

regions of low free energy. This is true not only for a nonliving, experimental system such as a beaker and a cellophane membrane, but also for the plasma membrane of living cells.

The diffusion of ions and water molecules to and from areas of higher and lower concentration will continue until the concentrations on both sides are equal. At that point, the same number of random collisions among water, salt, and the pores of the membrane will occur on each side. Movement of individual ions and water molecules will continue, but at the same rate in both directions. In a living cell, once equal concentrations have been attained, a continuous flux of water and ions takes place through the plasma membrane. For example, 100 times the volume of water in a red blood cell passes back and forth across its membrane every second.

Simple diffusion in living cells can be seen in the movement of lipid-soluble substances across the plasma membrane. Such molecules lack significant surface charge and do not interact strongly with water; ether, a chemical used medically as an anesthetic, is a good ex-

ample. Ether molecules can dissolve freely in the lipids of the plasma membrane and then cross the membrane into the cell. Eventually, the concentrations of ether on the two sides of the membrane become equal, and net transfer of the chemical across the membrane stops. The passage of molecules across the plasma membrane is called *lipid diffusion*. Lipid diffusion accounts for only a small percentage of molecular passages across cellular membranes but involves important biological molecules, including certain steroid hormones and lipid-soluble vitamins (for example, vitamin D, which is absorbed by intestinal cells). Because cell membranes are permeable only to certain substances and not to others, they are said to be *semipermeable*.

What happens to substances that are not lipid soluble? Diffusion of these molecules must occur through temporary and permanent openings in the plasma membrane—gates in the sea wall, if you will. Because the lipids of the plasma membrane are in a dynamic, fluid state, they are continuously moving to some degree. Apparently, as this movement occurs, channels open temporarily to permit polar molecules such as water to enter or leave the cell's interior. In addition, some membrane proteins pack together in such a way that their hydrophilic regions form an actual *pore*, a neatly rimmed hole through the plasma membrane. Because the pore is formed by proteins, a polar molecule passing through the membrane does not interact with the lipids and thus is not impeded by electric charges. Pores are known to be 0.7 to 1 nm in diameter, since water and urea molecules, with diameters somewhat less than 0.4 nm, can pass through, while larger substances cannot.

Carrier-Facilitated Diffusion

Some substances can cross the plasma membrane by combining with so-called carrier proteins, much as a substrate combines with an enzyme (Chapter 4). Carrier proteins are often called *permeases* because they render the cell more permeable. The chemical complex of carrier and substance moves through the lipid bilayer, and the substance is released on the other side. Because the substance is moved into or out of the cell faster and more efficiently than could occur by simple diffusion processes, this process is called **carrier-facilitated diffusion.** No energy is required for this diffusion of carrier and cargo. The direction of carrier-facilitated diffusion is wholly governed by the relative concentrations of the cargo substance inside and outside the cell: movement is along a **concentration gradient,** from areas of higher concentration to areas of lower concentration.

The number and distribution of carrier molecules in a membrane govern the rate of carrier-facilitated diffusion. If enough of a cargo substance is present to form complexes with all the available carrier molecules, the carrier becomes saturated, and the rate of carrier-facilitated diffusion reaches a maximum. (There is no maximum rate for simple diffusion.) Thus there are limits to the speed with which cells can take in or extrude certain substances—limits that can affect metabolic rate, growth, and other activities.

How do carrier proteins function? Discovery of a new class of molecules has provided some insights. Produced by fungi, these so-called *ionophores* act as antibiotics to kill bacteria by drastically upsetting the balance of ions inside those single-celled organisms. Each doughnut-shaped ionophore molecule has a charged hole that can interact with an ion that fits inside. The outer surface of the doughnut is uncharged and so can pass easily through the lipid bilayer, thereby carrying the charged cargo (a single ion) into the cell. The production of ionophores is indeed a deadly defensive strategy for fungi, for these carrier molecules can transport a variety of ions into and out of the hapless bacterium that tries to infect a fungal cell. How the fungal cell protects itself from its own ionophores remains a mystery. Ionophores are the best model of carrier-facilitated diffusion found thus far. Other molecules that move back and forth across membranes must have the same properties—an uncharged exterior and a hole or slot to carry the cargo.

Diffusion has a profound effect on all life, even though it is a passive physical process that depends on the concentration of substances or the availability of carriers. Without the tendency of many substances to move passively into and out of cells, the "cell-state" would quickly become a static kingdom: the interrelated reactions of metabolism would slow to a halt, and life processes would cease. Alternatively, the cell would have to expend precious energy to transport every bit of cargo that passes through its membrane. As we shall see, some molecules must be moved actively, but the life of the cell and its commerce with the outside environment hang in

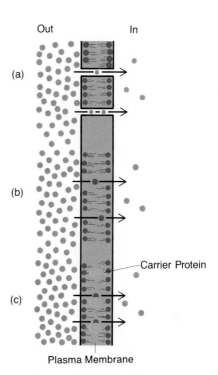

Figure 5-11   THREE WAYS FOR SUBSTANCES TO PASS THROUGH A PLASMA MEMBRANE.
Molecules can move into and out of cells (a) through pores in the plasma membrane; (b) by dissolving and diffusing through the membrane; and (c) by binding to a carrier protein and being ferried across.

a delicate balance between energy-free passive diffusion and energetically costly active transport. Figure 5-11 summarizes the mechanisms of passive transport.

Active Transport

Transporting some ions through the plasma membrane requires energy, particularly when this movement occurs *against* concentration gradients (from areas of low concentration to areas of high concentration). Figure 5-12 shows a simple but reasonable model for **active transport:** a carrier binds, transports, and releases its cargo. The carrier changes its configuration as the cargo is released, and can return to its active shape and state only when the special energy-storing molecule called ATP is hydrolyzed. Fueled by ATP's energy, a phosphate group is added to the carrier as that carrier protein changes shape into a highly strained configuration that can accept a new cargo molecule or ion. Transport, release, and reversion to the inactive shape follow. Another ATP is hydrolyzed, and the cycle begins again. The pump enzymes that catalyze these processes hydrolyze ATP and so are referred to as ATPases. The cycling involved in transport is a costly process that burns much

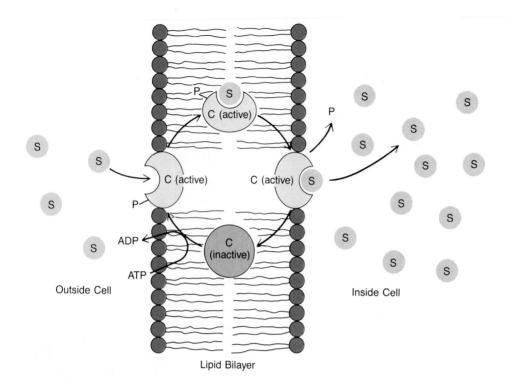

Lipid Bilayer

Figure 5-12   A MODEL FOR ACTIVE TRANSPORT.
When ATP is broken down, adenosine diphosphate (ADP) forms, a phosphate (P) is transferred to a carrier protein, and the protein changes shape into a highly strained configuration. The cargo (S) binds to the carrier protein, which moves the cargo across the plasma membrane. As the cargo and phosphate are released, the carrier reverts to an inactive configuration.

cellular ATP. It is a primary reason that most cells require food and oxygen—they need to replenish ATP supplies for active transport.

Sometimes active-transport carriers work only when a *counter-ion* moves in the opposite direction. Thus every time sodium ions are pumped out of a cell, potassium ions must pass in. If extracellular potassium is not present, the sodium pump halts. The pump cannot function in the absence of the counter-ion—probably because phosphate is released from the carrier and the carrier changes to a shape that will accept a sodium ion only when potassium binds and is transported into the cell. We will learn in the next section why the sodium pump is vital to the prevention of cell swelling.

Another type of active transport uses energy indirectly to allow sugars or amino acids to enter cells. In **co-transport,** both sodium ions and an amino-acid or a sugar molecule bind to a carrier, which transports them together and discharges them inside the cell. Neither the sugar nor the amino acid can be transported alone. However, co-transport causes the level of sodium ions in the cell to rise. This concentration increase is counteracted by the ATPase sodium pump, as Figure 5-13 shows. In this case, energy from ATP is used to keep the salt content low so that more salt and sugar or amino acids will enter the cell. This process may seem bizarre, but it is the common and relatively efficient way that eukaryotic cells, including human ones, obtain their sugar and amino-acid building blocks. Indeed, the active uptake of many nutrients in the human and other animal guts depends on exactly this kind of cotransport and ATPase

pump enzymes. Nevertheless, this movement across the cell surface requires that energy be expended, or transport of many critical substances will stop and the cell will die. Pumping ions and molecules against concentration gradients is one of the most basic features of cell activity, and hence of life itself.

Osmosis and Cell Integrity

Now that we have described how the plasma membrane, the interface between the cell and its environment, regulates the movement of substances into and out of the cell, let us turn to a major result of the transport of ions and molecules—the movement of water that accompanies that process. The diffusion of water through a semipermeable membrane like the cell membrane is called **osmosis.** It is a simple process that is nevertheless central to life on Earth because if it fails, the cell swells and bursts or shrinks and dies. On a larger scale, osmosis is related to thirst and drinking, to the wilting of plants, and to the survival of life forms where water is scarce or salt levels are high. Osmosis is central to life because water is central to life.

Usually, water moves from a dilute solution to a more concentrated one. Thus if the salt concentration inside a cell is high in relation to that outside a cell, water tends to cross the cell membrane and dilute the more concentrated solution inside. At the same time, the excess salt inside tends to move out of the cell. These processes continue until equilibrium is reached. The influx of water causes the cell to swell and exerts pressure on the

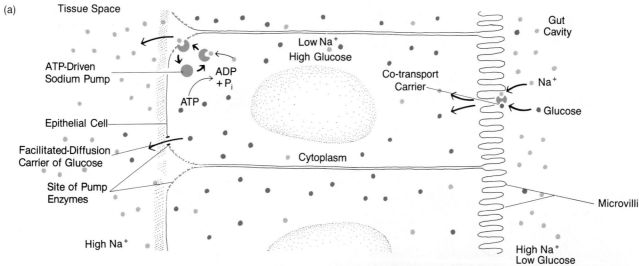

(a)

Tissue Space

ATP-Driven Sodium Pump

Epithelial Cell

Facilitated-Diffusion Carrier of Glucose

Site of Pump Enzymes

High Na⁺

Low Na⁺ High Glucose

ADP + P$_i$

ATP

Cytoplasm

Co-transport Carrier

Gut Cavity

Na⁺

Glucose

Microvilli

High Na⁺ Low Glucose

Figure 5-13   CO-TRANSPORT IN AN INTESTINAL EPITHELIAL CELL.

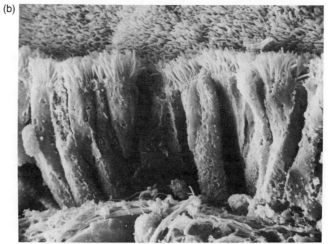

(b)

(a) The ATP-dependent sodium pump at the left end of the cell lowers the concentration of Na⁺ in the cytoplasm. As a result, Na⁺ moves into the cell at the right end. However, Na⁺ can pass in only when a sugar or an amino acid enters at the same time. The co-transport of salt and sugar takes place because the pump at the opposite end of the cell sets up an ion concentration gradient (low inside, compared with outside). For convenience, the pump and the carrier molecules are shown in the cytoplasm. In fact, those molecules probably are associated with the plasma membrane; the probable sites of the sodium pump enzymes are shown. (b) A scanning electron micrograph, magnified about 1,350 times, shows the many microvilli on intestinal epithelial cells; the co-transport carriers are associated with the microvilli. From *Tissues and Organs: A Text-Atlas of Scanning Electron Microscopy* by Richard G. Kessel and Randy H. Kardon. W. H. Freeman and Company. © 1979.

surface. The force exerted outward by this increased internal water is called **osmotic pressure.** In plants and bacteria, the cell membrane is surrounded by a rigid wall, which permits the cell to swell only until the membrane pushes against it. This outward pressure against cell walls is called *turgor;* it is the reason why a well-watered plant is stiff and erect.

Figure 5-14 shows how the concentration of salt inside and outside a cell affects the movement of water and hence the volume of the cell itself. If the salt concentrations (or *tonicity*) inside and outside animal and plant cells are equal, the extracellular fluid is said to be **isotonic** (the same) in relation to the intracellular fluid, and no net water movement occurs. But if the salt concentration outside the cell is less than that inside, the extracellular fluid is considered to be **hypotonic.** Under these circumstances, water tends to rush into the cell, causing it to swell. Although this swelling is limited by the cell wall in bacterial and plant cells, it is less limited

in animal cells, which are bounded by only the cell membrane and a wispy protein–sugar coat. For this reason, an animal cell with a high internal salt concentration can actually swell so much that it bursts. If the salt concentration outside the cell is greater than that inside, the extracellular fluid is said to be **hypertonic** relative to the cell contents. In this situation, water tends to leave the cell, and the cell shrinks.

Many cells contain proteins and other molecules that cannot easily pass out through the semipermeable plasma membrane. These molecules attract water and so exert osmotic pressure. Their presence in the cell leads to a tendency for water to enter the cell continuously throughout its life. What, then, prevents swelling, overhydration, and the dilution of cell components? The answer lies in special membrane enzymes that pump ions out of the cell. For instance, the sodium pump is an enzyme system that continuously pumps Na⁺ outward to lower the total internal ion content and raise the ex-

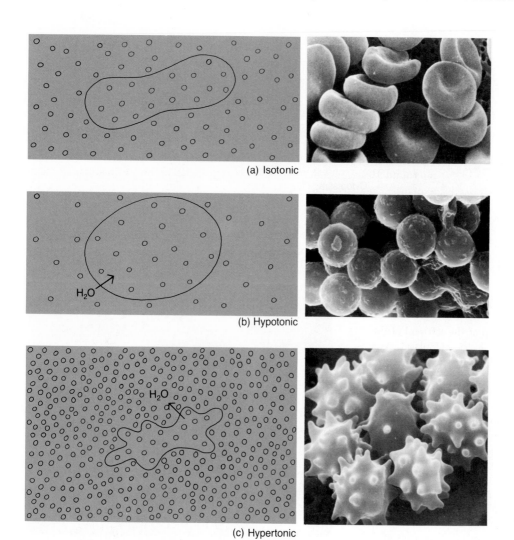

(a) Isotonic

(b) Hypotonic

H_2O

(c) Hypertonic

H_2O

Figure 5-14   EFFECT OF SALT CONCENTRATION ON CELL VOLUME.

The concentration of salt in the fluid surrounding a cell may determine the cell's volume. (a) An isotonic solution, with the same salt concentration as the cytoplasm, has little effect on cell volume or shape. (Cells are magnified about 2,250 times.) (b) A hypotonic (low-salt) solution results in an influx of water, which causes the cell to swell. (Cells are magnified about 1,500 times.) (c) A hypertonic (high-salt) solution causes water to leave the cell and results in a shrunken appearance. (Cells are magnified about 2,730 times.)

ternal ion concentration, thereby reducing the tendency for water to flow in. But the enzyme portion of the pump splits an ATP molecule each time a sodium ion is extruded. The ATP supply must then be replenished by metabolic processes, often ones that utilize oxygen to burn organic food molecules such as glucose. This pumping process is a basic feature of life; whether in a single algal cell floating in a lake or in a tissue cell deep in the human body, work must be done continuously to pump ions out and so preserve the osmotic conditions conducive to cell health and function. If a poison such as cyanide halts ATP production, the sodium pump halts, and the cell swells and dies.

Just as the concentration of ions and other substances inside cells affect cell volume and osmotic pressure, so do the tonicities of the tissue fluids that bathe cells. Consider the potentially drastic consequences if the balance between salt and water in the body's tissues was altered—too much salt in tissue fluid, for example, would tend to

dehydrate cells. Thus it is no wonder that complex animals have kidneys of one sort or another—organs specialized to keep ion levels constant in the blood and, indirectly, in extracellular tissue fluid. As a result of the continuous balancing of ions and water, tissues are not subjected to overwhelming osmotic stresses, and the organism can survive.

In summary, the dynamic plasma membrane, by virtue of its structure, regulates the passage of materials into and out of the cell, and prevents the cytoplasmic fluid from leaking out. The charged exterior and hydrophobic interior of the lipid bilayer allow lipid-soluble substances to diffuse across, but most ions and molecules must pass through pores, be carried passively, or be transported actively across the membrane at the cost to the cell of ATP. Although this dynamic barrier might seem sufficient to support the life of the cell, it is not the only protective structure found at the cell surface. Let us look next at its neighbors.

CELL WALLS AND THE GLYCOCALYX

We have seen how the unique construction of the plasma membrane helps regulate the vital traffic into and out of cells. But biologists hardly ever observe completely uncovered or unprotected plasma membranes in normal living cells. Molecular aggregates surround the plasma membranes of all cells. In bacteria, plants, and some fungi, these aggregates form a rigid wall. In animals, they form a discontinuous, patchy coating on the surface of the cell. Both the walls and patches vitally affect cell function and behavior and together with other surface structures govern much of cell life.

Cell Walls

As Robert Hooke observed more than 300 years ago on a slice of cork, plant cells are surrounded by **cell walls.** So are bacterial cells and some fungal cells. The plasma membrane, acting as a selective sieve, controls the movement of substances into and out of cells, but it is the rigid outer cell wall that actually protects the plant cell. Together, the millions of cell walls in an entire plant lend vertical support, without which plant life as we know it could not exist.

The cell walls of plants and some fungi are made largely of **cellulose,** a high-molecular-weight polysaccharide that is arranged in multiple layers over the cell's plasma membrane. Plant cell walls have three portions: the middle lamella, the primary wall, and the secondary wall (Figure 5-15a). The *middle lamella* is the first layer

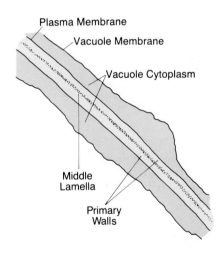

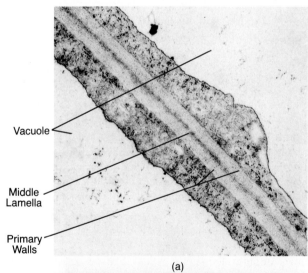

(a)

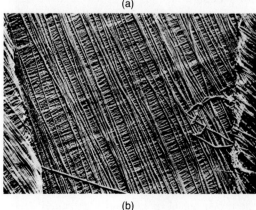

(b)

Figure 5-15   THE PLANT CELL WALL.
(a) A primary wall arising just after the division of two plant cells. The electron micrograph (magnified about 17,000 times) and corresponding drawing show the middle lamella and the primary walls arising between two plant cells. The secondary wall forms later and is much thicker than the primary wall; the secondary wall is also the major source of strength, rigidity, and support for plant tissue. (b) Ultrastructure of plant secondary cell walls (magnified about 9,300 times). Cellulose microfibrils are welded together at almost right angles to one another with lignin and hemicellulose in woody tissue. Huge numbers of such cell walls together can support a tree trunk hundreds of feet tall.

to form when a plant cell divides into two new cells. It is composed largely of *pectin*, a gluelike polysaccharide that helps to hold adjacent cell walls together. *Primary walls* form on each side of the middle lamella as the new plant cells grow and mature. For a while, the primary wall of a plant cell remains flexible and stretchable; this enables the young cell to elongate and grow. The third layer, called the *secondary wall*, forms on the inner side of the primary wall after cell growth and the formation of new cells have stopped, particularly in the outer parts of plant tissue. The secondary wall tends to be very strong because the layers of cellulose within it lie at nearly right angles to one another, like the layers in plywood (Figure 5-15b). The cellulose molecules form tiny ropelike *microfibrils* that are glued together with a hardening substance called *hemicellulose. Lignin*, a complex carbon-containing substance, acts as a further stiffening agent for the secondary wall. Together, the cellulose and lignin in the walls of millions of individual plant cells make up the woody tissue of trees and parts such as pine cones and nuts.

The cell walls that surround bacteria are composed of polysaccharides, lipids, and short chains of amino acids and sugars. The amino acids and sugars, which are called *peptidoglycans*, link the components of the entire wall so that it is quite rigid. In fact, the bacterial cell wall can even be thought of as one giant macromolecule that seamlessly encloses the plasma membrane and bacterial cytoplasm (Figure 5-16). This wall permits bacteria to invade other organisms and to withstand environmental conditions that would kill a typical animal cell. It also imparts to bacterial species their characteristic shapes (spheres, rods, commas, and spirals).

Fungal cell walls are complex combinations of cellulose, **chitin,** and other glucose polymer polysaccharides. Chitin—built from glucose-amine, a nitrogen containing sugar—is also a main ingredient of the external skeletons of lobsters, spiders, houseflies, and their kin.

Glycocalyx

Animal cells lack a continuous wall or rigid coat, and thus are always potentially mobile and plastic. However, on their outer surface, they do have patches of large molecules that act as glue and as molecular name tags. These molecules—complexes of sugar polymers, proteins, and sometimes lipids—are secreted by the cell and are renewed continuously. Collectively, they are referred to as the **glycocalyx,** or "sugar coat" (Figure 5-17). Gluelike glycocalyx molecules promote the adhesion of cells to one another and to external structures, such as the collagen fibers that twist like strong threads through the connective tissue of most organs. The complexly ar-

Figure 5-16   BACTERIAL CELL WALL.
(a) The bacterial Gram-positive cell wall encircles, strengthens, and protects the cell (magnified about 60,000 times).
(b) The main wall components are an interconnected set of sugar polymers. The diagram shows a tiny portion of the wall, composed of peptidoglycans (short chains of amino acids and sugars). The small red spheres represent cross-linking glycine molecules. The entire wall can be considered a single macromolecule.

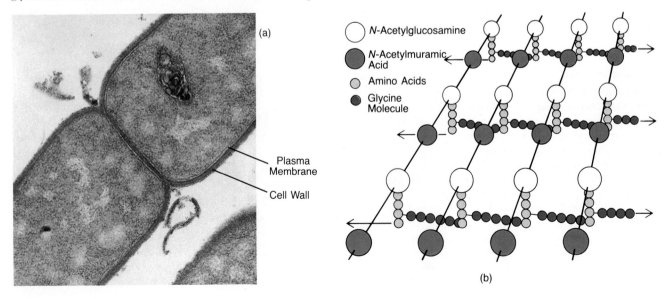

(a)

Plasma
Membrane

Cell Wall

○ *N*-Acetylglucosamine

● *N*-Acetylmuramic Acid

○ Amino Acids

● Glycine Molecule

(b)

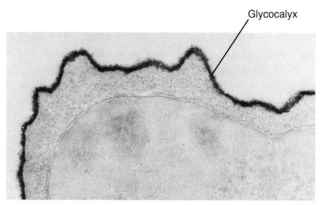

Figure 5-17   THE GLYCOCALYX.
A special stain, ruthenium red, reveals the glycocalyx on a mammalian cell (magnified about 30,400 times). The process obscures the plasma membrane so that it can no longer be seen.

ranged and distinctively shaped groups of sugar molecules of the glycocalyx act as a molecular fingerprint for each cell type and are involved in cell recognition and in the social behavior of cells within both tissues and the organism. The glycocalyx is thus a key component in coordinating cell behavior in animals.

A variation of the glycocalyx is found on one surface of cell populations arranged in sheets called **epithelia.** Every sheet of epithelium is attached to a **basal lamina,** a feltlike layer that consists of highly ordered sugar–

Figure 5-18   BASAL LAMINA.
The basal lamina is a flexible mat of ordered molecules to which epithelial cells attach. The basal lamina seen in this electron micrograph of rat lung tissue (magnified about 14,500 times) is composed of huge sugar polymers and collagen. The numerous large dots visible beneath the basal lamina are cross-sections of collagen fibers that run through connective-tissue spaces and anchor the basal lamina and epithelium in place.

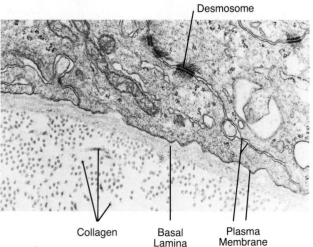

Collagen   Basal   Plasma
                Lamina   Membrane

protein complexes and a special type of collagen fiber. Many types of epithelial cells cannot carry out important synthetic activities, divide, or develop correctly unless their glycocalyx molecules attach to complementary binding molecules of their basal lamina (Figure 5-18).

Most cells of multicellular animals must attach to surfaces such as basal lamina, collagen fibers, or other cells. Contact and adhesion are essential if cells are to assemble an internal scaffolding to form specialized shapes, to divide, and to move about. This need to adhere to a solid surface is called **anchorage dependence.** Thus if animal cells are experimentally placed in a fluid solution, they usually become spherical, cease to divide, and stop moving. Since the cells are truly dependent on anchorage in order to carry out these basic cellular activities, anchorage dependence has a place among the tenets of the cell theory.

In light of the importance of the glycocalyx as a prerequisite for cell division, shape, and movement, it is intriguing that various cancer cells have both altered glycocalyx molecules and reduced anchorage dependence of cell division and locomotion. Some biologists have suggested that these deviations allow certain cancer cells to grow rapidly and invade many body tissues.

LINKAGE AND COMMUNICATION BETWEEN CELLS

We move now from cell membranes, walls, and coatings to specialized structures on those surfaces that hold cells firmly together in tissues, allow cells to communicate with one another and the environment, and prevent fluid leakage in certain tissues. Several types of junctions hold cells together and provide channels for intercellular communication: zonulae adherens, desmosomes, tight junctions, gap junctions, and plasmodesmata. Populations of animal epithelial cells provide an excellent place to study most of these junctions.

Zonulae adherens and desmosomes serve mainly to bind cells together (Figure 5-19). **Zonulae adherens** ("zones of adhesion") are sites of firm physical contact between cells. They are beltlike bands that run around most epithelial cells. In addition to linking adjacent cells, zonulae serve as sites for insertion of important scaffolding filaments of the cytoskeleton. Since zonulae link cells and anchor the internal cytoskeleton, they help to control the shape of cells and tissues.

Desmosomes are analogous to tiny spot welds, rivets, or buttons between cells. These small junctions are made up of unidentified molecules that apparently glue together adjacent plasma membranes. The side of the desmosomes that faces the cytoplasm contains an anchoring material in which certain elements of the cytoskeleton are often embedded. Desmosomes are particularly abundant in tissues subjected to mechanical stresses, such as the outer layers of the human skin and the cervix of the uterus. Diseases that affect the stability of desmosomes can cause epithelial tissue to become fragile or epithelial cells to be shed too easily.

Tight junctions are seals that encompass the lateral surfaces of cells in epithelia and act as barriers to fluid leakage. Tight junctions between adjacent epithelial cells of the bladder, for example, prevent urine from seeping between the cells back into the body-tissue spaces. In the lung, these seals prevent extracellular

fluid of the lung tissue from passing between the epithelial cells and out into the air spaces.

The outer lipid layers of the plasma membranes of adjacent cells actually appear in the electron microscope to touch at tight junctions. The areas of contact are a network of ridges. The traditional view has been that intrinsic proteins embedded in the two membranes' lipid bilayers are bound tightly next to one another at the ridges like the teeth of a zipper. However, more recent observations suggest that the ridges may be membranous tubes that literally link the adjacent plasma membranes with lipid welds. Despite this structural complexity, tight junctions can form between cells in less than five minutes. They are a tidy piece of molecular masonry that is as essential to the integrity of the body as rubber seals are to the functioning of many machines.

The primary communication junction between animal cells is the **gap junction,** a perforated channel that permits easy exchange of small molecules, ions, and electric currents across cell membranes, and thus allows cells to communicate in the electrical, ionic, and molecular language they "speak." Gap junctions occur between many cell types in embryos and adults. These sites of cellular permeability are marked by tightly packed clusters of particles within the membranes, as shown in Figure 5-20. Experiments using dyes and fluorescent molecules of different shapes and sizes have revealed that in insect cells, the channels perforating gap junctions are 2 to 3 nm in diameter, while those between mammalian cells are smaller—1.6 to 2 nm. These channels pass through

Figure 5-19   CELLULAR JUNCTIONS.

Cellular junctions occur in several forms and in a precise order. Zonulae adherens provide anchors for the cytoskeleton. Desmosomes serve as strong "welds" between cells. Tight junctions can take several forms, but all serve as barriers to prevent leakage of substances into the space between cells. Gap junctions are patches where exchange of materials between cells may take place (see Figure 5-20). The position of the various types of junction represented here is typical for many kinds of epithelial cells.

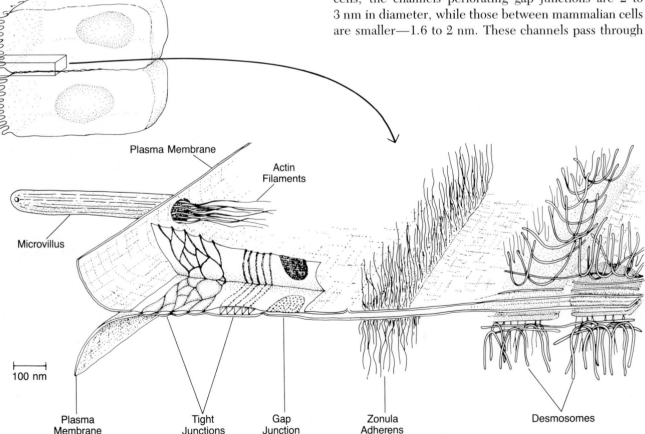

Plasma Membrane

Actin Filaments

Microvillus

100 nm

Plasma Membrane

Tight Junctions

Gap Junction

Zonula Adherens

Desmosomes

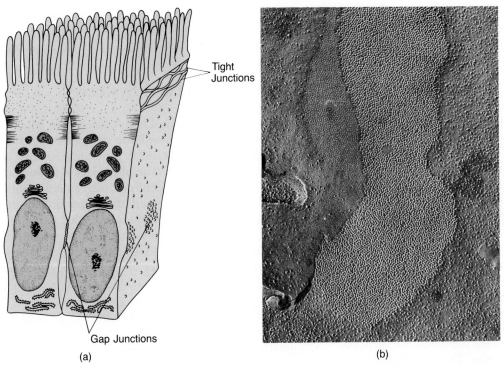

(a)

(b)

Figure 5-20   GAP JUNCTIONS ON THE LATERAL SURFACES OF EPITHELIAL CELLS.
(a) Gap junctions may permit easy exchange of ions and small organic molecules between cells. (b) In this high-magnification electron micrograph (61,500 times) prepared by freeze-fracturing a mouse liver cell, we can see closely packed particles that are believed to surround the minute channels between cells.

the center of protein complexes, called *connexons*, that span both layers of the plasma membrane. The intercellular pipeline formed by gap junctions permits ions, molecules up to the size of polypeptides with fifteen to twenty amino acids, and electric currents to move freely between cells, thus constituting an important form of intercellular communication that may coordinate various cellular activities.

Many kinds of cells in higher plants appear to be able to engage in intercellular exchange more easily than can animal cells. Bridges of cytoplasm called **plasmodesmata** (singular, *plasmodesma*) connect adjacent plant cells (Figure 5-21). These bridges normally arise as a plant cell divides; the two new cells fail to separate completely, leaving threads of cytoplasm, the plasmodesmata, and with them, direct means of exchanging molecules, ions, and so on. Although biologists assume that passage of materials through plasmodesmata provides a direct means of intercellular communication and integration, no one has actually demonstrated the exchange of organelles or molecules across these plant-cell bridges. But if a gardener grafts the stem of a fragile but

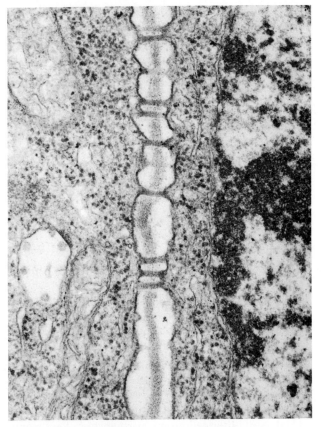

Figure 5-21   PLASMODESMATA—VITAL CHANNELS BETWEEN PLANT CELLS.
In this electron micrograph (magnified 50,350 times), plasmodesmata can be seen interconnecting the cytoplasms of two young root tip cells from a timothy plant.

beautiful rose stock onto a hardy root stock, the graft "takes" only when plasmodesmata bridges become established between graft- and host-tissue cells. Success appears to involve communication between the two tissues via these special channels.

To summarize the ways in which cells are linked, zonulae adherens and desmosomes couple cells mechanically and provide anchorage sites for the internal cytoskeleton. Tight junctions seal the lateral surfaces of cells in epithelial layers to prevent leakage of materials. Gap junctions and plasmodesmata act as communication channels, permitting the passage of certain kinds of molecules and ions and of electric currents between cells. The significance of linkage by gap junctions cannot be overemphasized. In fact, recent studies of animal- and plant-tissue cells linked by gap junctions and plasmodesmata have led to a modification of the traditional cell theory. Cells traditionally have been regarded as units that are independent in structure and function. But that assumption turns out not to be true in some respects for many cells in tissues. When linked by gap junctions, all the cells of a population, not the individual cell, become the unit of response and function. Thus a regulatory molecule acting on one cell may trigger responses in adjacent cells because the "message" is passed through gap junctions. This is indeed a remarkable property and helps explain the coordination of cellular activity that is so essential in tissues and organisms made up of millions of cells.

MULTICELLULAR ORGANIZATION

The junctions between cells we have just studied are found, of course, in multicellular organisms. Whereas the first living organisms on Earth were almost certainly single-celled, evolutionary processes led to the appearance of multicellular colonies and, in turn, to organisms in which millions and sometimes trillions of cells coexist and function in an interrelated way. During this evolution, discrete subpopulations of cells within organisms began to carry out specialized tasks. For instance, the shapes and properties of some cells such as bone cells in an animal's leg and wood cells in a tree trunk allow them to provide physical support. Other groups of cells transport fluids: blood vessels in an animal and vessel cells in a plant. Still other cell groups participate in food manufacture, digestion, perception, movement, or reproduction. This division of labor led one of the founders of the cell theory, Rudolf Virchow, to compare the multicellular organism to a "cell-state," a country or state composed of individual cellular plumbers, delivery persons, managers, and so on.

Tissues, Organs, and Systems

The millions or trillions of individual cells in a multicellular organism are arranged into groups that function collectively as **tissues.** Tissues, in turn, are combined in *organs* and *organ systems*. The physiological division of labor in the "cell-state" depends on such arrangements.

Each tissue type is composed of cells performing the same or closely related functions. Epithelial tissue, for instance, consists of sheets of cells that cover organisms and organs and line cavities (Figure 5-22). Different types of epithelial tissue cover the human body and line the mouth, gut, lungs, heart, and blood vessels. Epithelial tissues also cover leaves and stems of plants.

Epithelial cells can be squarish *(cuboidal)*, rectangular *(columnar)*, or flat *(squamous)*. Despite differences in shape, one feature common to all epithelial cells is that they usually adhere tightly to the similar cells on either side of them, joined by zonulae, desmosomes, and other junctions. Tight junctions between epithelial cells seal off a tissue from adjacent open spaces.

Another common tissue is **connective tissue,** the fibrous component of most animal organs. Related to connective tissue are bone and cartilage, tissues that are built on a fibrous base. Other types of tissue include muscle, nerve, and mesophyll, the light-absorbing green tissue inside leaves.

Organs are body parts composed of several tissues that work in concert to perform specific functions within the organism. For example, the skin is an organ that is made up of epithelial tissue, connective tissue, muscle tissue (such as the tiny muscles that move hairs), and blood tissue. Similarly, a leaf is a light-collecting and carbohydrate-producing organ made up of waterproof epithelia, conducting tissues, sugar-storing cells, and so on. Tissues in plants do not have formal boundaries; nevertheless, specialized groups of cells carry out tasks of support, transport, synthesis, and reproduction analogous to those in animals.

The final level of this hierarchy is the **organ system,** a collection of organs that carry out the various aspects of a complicated activity. In animals, the digestive system, for example, includes the diverse parts of the gut (esophagus, stomach, intestines, and so on) plus the salivary glands, liver, pancreas, and other organs. (The major organ systems of animals are described in detail in Part 5.) In plants, the water-conduction system, for example, includes root hairs, vessels that extend up the root and stem into the leaf, and guard cells on the underside of

Figure 5-22   EPITHELIUM AND CONNECTIVE TISSUE.

Typical animal tissues are epithelium and connective tissue (here magnified about 650 times). The vertically oriented epithelial cells, visible at the top of this specimen of mammalian lung tissue, have many microvilli on their outer surface. The connective tissue below the epithelium has a variety of cell types plus tough fibers that give the tissue strength. Blood vessels are also visible. From *Tissues and Organs: A Text-Atlas of Scanning Electron Microscopy* by Richard G. Kessel and Randy H. Kardon. W. H. Freeman and Company. © 1979.

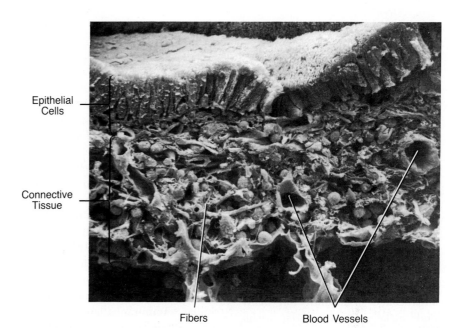

Epithelial Cells

Connective Tissue

Fibers

Blood Vessels

the leaf that regulate evaporation. (Plant tissues, organs, and organ systems are described in Part 4.)

The Price of Multicellularity

The division of labor among cells and tissues in multicellular organisms allows great functional efficiency and survival. However, it also involves one limitation: mortality. An entire single-celled organism participates in reproduction by dividing in half and therefore is in a sense immortal. But only certain cells in the reproductive organs of a multicellular plant or animal contribute the sperm and egg that yield the next generation. The remainder of most multicellular colonies is mortal and dies. In effect, death is a price paid for multicellularity, for large size, and for a range of life styles much greater than single-celled organisms can pursue.

Another price of multicellularity is absolute cellular dependence. Single cells isolated from a plant or an animal cannot live as independent organisms. If a child skins her knee after falling off a bicycle, all the cells that are left on the asphalt will die. None can survive free of its neighbors. Of course, the same blood, skin, and other cell types could be cultured in a laboratory in a nutrient broth, and they or their progeny might survive for weeks, years, or even decades. The culture chamber and the nutrient medium would become the new "organism" in which such cells resided. Eukaryotic cells of multicellular organisms truly are social creatures, dependent for their very survival on the unique internal environment found only within an intact organism. The fluid, nutrients, gases, salt content, pH, and many other aspects of that environment are the keys to cell survival. When we study plant and animal physiology, it will become evident that the multicellular creature daily expends much energy keeping this special inner environment constant and compatible and the resident cells, healthy and long lived: if the environment of the basic units of life, the cells, is adverse and the cells die off, the entire organism also dies.

SUMMARY

1. Most cells are small, and so their discovery and subsequent examination followed the invention and development of the microscope. The light microscope, which has been used by cell biologists for more than 300 years, has been supplemented by the transmission electron microscope (TEM) and the scanning electron microscope (SEM). Microscopes offer the advantages of magnification and greater resolving power. Specimens must be prepared for microscopic viewing using techniques such as dehydrating, fixing, staining, and freeze fracturing. Other methods of studying cells include cell fractionation, radioactive-

isotope labeling, and autoradiography.

2. The modern *cell theory* can be summarized as: (a) cells are the basic structural and functional units of life; (b) all organisms are composed of cells; and (c) all cells arise from pre-existing cells. Two additional points apply to multicellular organisms: (d) populations of cells may behave as functional units; and (e) normal animal cells must adhere to solid substrates in order to divide, move, assume specialized shapes, and carry out certain special functions.

3. Some cells are free-living, single-celled organisms, but many kinds make up multicellular organisms. A *prokaryotic cell* lacks a membrane-bound nucleus, whereas a *eukaryotic cell* has a membrane-bound nucleus containing its hereditary material. Eukaryotic cells also have *organelles* and a cytoskeleton, whereas prokaryotic cells do not.

4. *Autotrophs* can manufacture their own energy-rich organic compounds, while *heterotrophs* must obtain their energy by taking in autotrophs or other heterotrophs.

5. An animal cell is enclosed by the *plasma membrane*, which itself is surrounded by the *glycocalyx*, a coating of molecules that mediates the cell's interaction with its environment. The cytoplasm, or semifluid interior of the cell, contains a cytoskeleton, various organelles, and the nucleus.

6. A plant cell is enclosed by the plasma membrane, which is surrounded by the *cell wall*, which gives rigidity to the cell. The cytoplasm contains many of the same kinds of organelles found in an animal cell. Neighboring cells may be connected to one another by *plasmodesmata*, bridges of cytoplasm.

7. Cell size is limited by *surface-to-volume ratio*. As cells get larger, either much of their contents is rela-

tively inert nutrient, as in a hen's egg, or their surface area increases accordingly, as in a nerve cell.

8. The plasma membrane surrounds the cytoplasm in all cells. It is a lipid *bilayer* in which proteins are suspended. *Intrinsic* proteins have hydrophobic exteriors and are found within the bilayer. *Extrinsic* proteins are attached to either surface of the bilayer. According to the *fluid-mosaic model* of the plasma membrane, lipid molecules move about fairly freely in the plane of the membrane, and many of the proteins may also be free to move, as if floating, in the membrane. Basically similar membranes are associated with various cytoplasmic organelles.

9. Molecules and ions cross the plasma membrane by means of a number of processes. *Diffusion*—the tendency of a substance to move from a region to high concentration to a region of low concentration—accounts for the movement of lipid-soluble substances across the semipermeable cell membrane. Non-lipid-soluble substances enter the cell either through pores or with the help of carriers. Some such *carriers* simply *facilitate diffusion*, while others consume energy (ATP) in order to carry out *active transport* against the *concentration gradient*. *Co-transport* of substances in the same or in opposite directions also occurs.

10. *Osmosis* is the passage of water through semipermeable membranes from a solution of low salt concentration to one of higher salt concentration. Osmosis goes on continuously during the life of every cell and plays an important role in the health and function of cells. When the salt concentrations of the extracellular and the intracellular fluids are the same, they are said to be *isotonic*, and there is no net movement of water. When the salt concentration outside the cell is less than that inside, the

extracellular fluid is *hypotonic:* water rushes into the cell, which swells and can even burst. When the salt concentration outside the cell is greater than that inside, the extracellular fluid is *hypertonic:* water leaves the cell, which shrinks. Water that enters cells exerts a pressure outword—this is *osmotic pressure.*

11. Plant, bacterial, and some fungal cells are encased and protected by *cell walls*, which are composed largely of *cellulose* in plants and of polysaccharides, lipids, amino acids, and sugars in bacteria. Cellulose, *chitin*, and other polysaccharides are found in fungal cell walls. Animal cells possess on their plasma membranes patchily distributed complexes of sugar polymers and proteins—the glycocalyx. This "sugar coat" affects cellular adhesion, locomotion, division, recognition, shape, and other activities. The *basal lamina* layer to which *epithelial cells* attach is a highly ordered variant of the glycocalyx. The cell's need to adhere to a solid substrate in order to perform these functions is called *anchorage dependence.*

12. Junctional complexes link cells in various ways. *Zonulae adherens* and *desmosomes* provide firm structural linkage between some animal cells. *Tight junctions* seal plasma membranes together at ridges and so prevent leakage across epithelial layers. *Gap junctions* are sites of pores that permit passage of ions, small molecules, and electric currents. Plasmodesmata are large bridges between plant cells that may permit exchange of many substances.

13. Cells of multicellular creatures are organized into *tissues*, *organs*, and *organ systems* so that the activities of individual cells can be integrated into the economy of the whole organism. This physiological division of labor occurs at the expense of immortality (in a sense) and of cellular independence.

KEY TERMS

active transport

anchorage dependence

autotroph

basal lamina

bilayer (lipid)

carrier-facilitated diffusion

cell

cell theory

cellulose

cell wall

chitin

concentration gradient

connective tissue

co-transport

desmosome

diffusion

epithelium

eukaryotic cell

extrinsic protein

fluid-mosaic model

gap junction

glycocalyx

heterotroph

hypertonic solution

hypotonic solution

intrinsic protein

isotonic solution

organ

organelle

organ system

osmosis

osmotic pressure

plasma membrane

plasmodesma

prokaryotic cell

surface-to-volume ratio

tight junction

tissue

zonulae adherens

QUESTIONS

1. The theory of evolution is one cornerstone of biology; the cell theory is another. Explain the latter theory and why it is so central to science.

2. Draw typical prokaryotic and eukaryotic cells and label the major parts that distinguish one from the other.

3. Distinguish between the cell wall, the glycocalyx, and the plasma membrane. Which would you prefer to "wear" as a raincoat? Why? Which would best support a tree trunk? Why?

4. How does the surface area to volume ratio help to explain why cells are so small?

5. With labeled diagrams show how various charged and uncharged substances can enter cells. Which processes require energy? Why don't the others?

6. Kidney disease might result in wild fluctuations of the salt concentration in an animal body's extracel-lular fluids. What would be the effects on cells and why?

7. How does tonicity relate to osmosis? What is osmotic pressure and what is an example of its effects?

8. Animal and plant cells are connected by various types of junctions. Prepare a table listing the types, their structures, and their functions.

9. How are cells, tissues, and organs related? What is a heart? The skin surface? Red blood cells?

ESSAY QUESTIONS

1. The fluid-mosaic model explains much about cell function. What would happen if the mean temperature of all life was suddenly lowered so the plasma membranes of cells were no longer fluid? Why?

2. Why are redwood trees and whales not constructed from the same numbers of cells as rosebushes and minnows?

SUGGESTED READINGS

ALBERTS, B. et al., *Molecular Biology of the Cell*. New York: Garland, 1983.

An easy-to-understand, comprehensive treatment of cells in all their aspects.

CUTTER, E. G. *Plant Anatomy: Experiment and Interpretation*. Reading, Mass.: Addison-Wesley, 1969.

This two-part book has major sections on plant cells, tissues, and organs.

FAWCETT, D. *The Cell*. Philadelphia: Saunders, 1981.

A marvelous collection of eukaryotic cells as viewed with electron microscopes.

SINGER, S. J., and G. NICOLSON. "The Fluid-Mosaic Model of the Structure of Cell Membranes." *Science* 175 (1972): 720–731.

A classic paper on cell biology.

THOMAS, L. *The Lives of a Cell: Notes of a Biology Watcher*. New York: Viking Press, 1974.

A set of essays to stimulate thought about cells.

6

INSIDE THE LIVING CELL: STRUCTURE AND FUNCTION OF INTERNAL CELL PARTS

The cell is the basic "module," the common denominator of all the immense variety of living forms. One must not forget, however, that all cells also play specialized roles over the entire range of diversity in biological form and function.

Ariel G. Loewy and Philip Siekevitz,
Cell Structure and Function (1969)

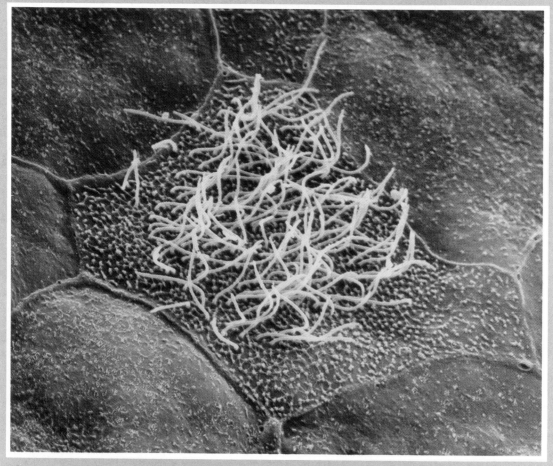

Cilia, microtubule-containing, hairlike structures, beat and propel fluid past these cells on the surface of a salamander embryo (magnified about 160 times).

Every living cell is like a miniature factory, whether that cell is a self-contained, independent organism such as an amoeba, or just one tiny unit in a yellow dandelion petal or a gossamer dragonfly wing. In this living cellular factory, raw materials are turned into products to be stored, used in the cell, or exported. Like all factories, the cell consumes energy and creates wastes. But the outcome of its manufacturing process is not the production of gadgets to sell; it is the ephemeral quality we call life.

We saw in Chapter 5 that cells are incredibly complex organizations. The cell surface is highly intricate and encloses a droplet of cytoplasm. And the cellular factory within the cytoplasm contains far more than a random jumble of proteins, carbohydrates, fats, and nucleic acids. The cell is a model of organization within organization. The tasks of the cellular factory are carried out by a number of special units, the organelles. The organelles, in turn, are complex structures, built from orderly arrays of macromolecules. Just as atoms form small molecular subunits, and subunits form macromolecules, macromolecules may assemble into organelles.

In a sense, cell organelles are the bridge between chemistry and biology, between molecules and life. Although no single organelle is alive, the precise arrangement of structures within the cell and the interrelated, simultaneous accomplishment of specialized functions by these structures account for the life process and, with it, for the cell's survival, reproduction, and function.

In this chapter, we pass inside the cell surface and study the detailed structures of cell parts and the functions they perform. First we will look at the main substance of cells, the cytoplasm. Then we will examine the genetic materials that act as instructions or blueprints. We will move on to consider the organelles, one by one. Finally, we will discuss the structures that are responsible for shape, movement, and division in plant and animal cells—structures that we previewed in Figures 5-4 and 5-5.

CYTOPLASM: THE DYNAMIC, MOBILE FACTORY

Most of the mass of every prokaryotic and eukaryotic cell consists of **cytoplasm,** a semifluid substance bounded on the outside by the plasma membrane. Like moths in a spider web, the organelles are suspended within the cytoplasm. These internal cell parts—shaped like tubes, spheres, and small sacs—carry out much of the business of life. The organelles are supported by a dynamic, three-dimensional framework of filamentous organelles called the *cytoskeleton.* (Later in this chapter, we shall describe the functions of the cytoskeleton and

its component structures.) Dissolved in the cytoplasmic fluid around the cytoskeleton and organelles are nutrients, ions, soluble proteins, enzymes, and other raw materials needed for the work of the cellular factory.

In the nineteenth century, biologists wondered whether the living state derives directly from properties of the translucent cytoplasm or whether the nucleus, with its genetic material, continuously instills eukaryotic cells with life. They wondered what would happen if the nucleus was removed from the cytoplasm: Would that semifluid substance die instantly or continue to exhibit lifelike activity for some period of time? To study this question, they and their twentieth-century successors developed various ways to remove the nucleus from large eukaryotic cells that could be manipulated fairly easily. Using modern experimental techniques, a researcher can suspend single cells in a nutrient fluid and treat them with a drug that weakens the cell surface and cytoskeleton. Then the living cells are centrifuged. Each nucleus literally pops out, surrounded by a thin halo of cytoplasm, and leaves the remainder of the cytoplasm intact within the plasma membrane. The nucleus soon degenerates. In contrast, the enucleated cytoplasm usually survives for many hours as a perfectly viable entity. It can move about normally, absorb and use nutrients and oxygen, synthesize proteins, excrete waste products, and so on. Not until the isolated cytoplasm needs further instructions from the genes that reside in its nucleus does the absence of that organelle become apparent. Only then does the enucleated cytoplasm gradually degenerate and die. These experiments show that most of the properties that we associate with life (Chapter 1) are properties of the cytoplasm. Nevertheless, cellular reproduction depends on heredity, and thus the nucleus is critical to the maintenance of life in eukaryotic cells.

THE NUCLEUS: INFORMATION CENTRAL

The cell **nucleus** serves as the control room for most operations of the eukaryotic cell. The largest cell organelle, the nucleus is often a rounded structure, and it is the first cell part to be recognized under a microscope. In a sense, the nucleus is the paramount organelle because it houses **chromosomes,** long strands of coiled DNA and proteins that contain **genes**—the basic blueprints for all cellular proteins. All cells depend on proteins for metabolism, shape, special functions, cell division, and other processes. Consequently, eukaryotic cells are indirectly dependent on the nucleus for most day-to-day operations.

Genes are never directly used as patterns, or tem-

(b)

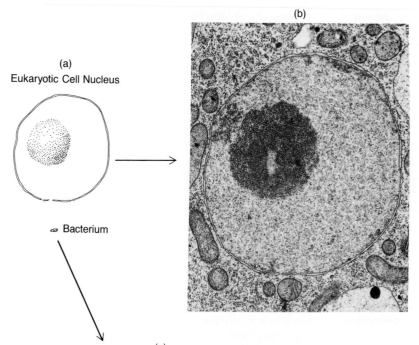

(a)
Eukaryotic Cell Nucleus

Bacterium

(c)

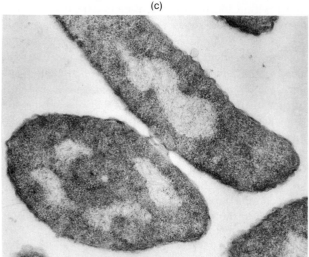

Figure 6-1   NUCLEUS VERSUS NUCLEOID: EUKARYOTIC AND PROKARYOTIC GENETIC CENTERS.
(a) The drawing shows the relative sizes of a eukaryotic cell nucleus and a bacterium. (b) The nucleus of a eukaryotic root tip cell can be seen here magnified about 14,500 times normal size. Note the double-layered nuclear envelope; the perforations in it are called nuclear pores. The large, darkened area in the nucleus is the nucleolus. (c) The light-colored, unbounded, irregular-shaped nucleoid regions are clearly visible in both of these mating prokaryotic bacterial cells, magnified about 40,000 times.

plates, for the production of proteins. In prokaryotes as well as eukaryotes, the genetic plans are copied on an intermediary molecule called *messenger RNA* (mRNA). In eukaryotic cells, mRNA bearing these instructions moves from the nucleus into the cytoplasm, where proteins are synthesized. In the next section, we will see how this information transfer occurs; the process by which mRNAs are formed in the nucleus will be presented in detail in Chapter 13.

It is not clear why chromosomes are segregated in the nucleus of a eukaryotic cell. The reason may have to do with the fact that the quantity of chromosomal material present is quite large; the immense length of a full set of chromosomes might cause them to become tangled or broken were they not segregated in a compartment. More likely, the separation of chromosomes from cyto-

plasm allows more effective regulation of genes and a special processing of mRNA, found only in eukaryotic nuclei. In prokaryotic cells, one long, circular strand of DNA serves as a single chromosome. It is usually concentrated in one dense, unbounded area of the cell called the **nucleoid**. Figure 6-1 compares a eukaryotic nucleus and a prokaryotic nucleoid. The absence of a membrane-bounded nucleus in prokaryotes may help explain why the extremely complex processing of RNA carried out in the eukaryotic nucleus does not occur in prokaryotes.

In addition to chromosomes, the nucleus contains one or two organelles known as **nucleoli** (singular, *nucleolus*, meaning "a small kernel"). Just after a cell divides, one or two new nucleoli appear in association with specific chromosomes of each daughter cell. (We will discuss the fate of the nucleoli during cell division in Chapter 9.) Within each nucleolus are minute fibers and granular components. These granules are precursors of *ribosomes*, organelles located in the cytoplasm that are instrumental as the actual sites of protein synthesis.

The nucleus and its precious contents are isolated from the cytoplasm of eukaryotic cells by the **nuclear envelope**, a flattened, double-layered sac filled with fluid and perforated with pores. The membrane of the nuclear envelope is similar to the plasma membrane and the membranes of other organelles in that it is a lipid bilayer in which proteins float. Small molecules and ions can be transported through the double membrane, but large molecules, such as mRNA and proteins, or ribosomes, can leave or enter the nucleus only through the pores. Nuclear pores are not simply holes; each is composed of a precisely ordered array of globular and filamentous proteins comprising nuclear pore granules,

(a)

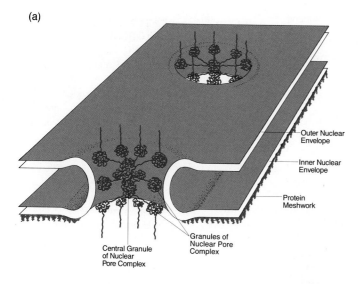

Outer Nuclear Envelope

Inner Nuclear Envelope

Protein Meshwork

Central Granule of Nuclear Pore Complex

Granules of Nuclear Pore Complex

(b)

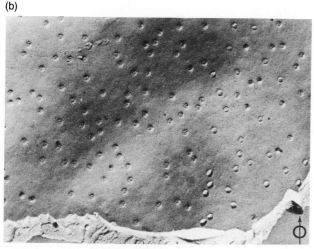

Figure 6-2   NUCLEAR PORES.
Nuclear pores allow materials to pass between the nucleus and cytoplasm. (a) An artist's interpretation of pore structure, revealing how the pore spans the two-layered nuclear envelope. Even structures as large as ribosomal subunits can pass through the pores. Protein granules around the edges and in the center of the pore govern what passes through the pore. (b) In this freeze-fracture preparation of the nuclear envelope from a guinea pig cell, magnified about 14,250 times, the numerous crater-like nuclear pores give the structure an appearance vaguely like a golf ball.

some of which serve to seal off the pore and govern transport of large molecules into and out of the nucleus by an unknown mechanism (Figure 6-2a). Nuclear pores sometimes account for 5 to 30 percent or more of the surface area of the nuclear envelope, making the nucleus look a bit like a golf ball (Figure 6-2b).

Biologists recently discovered that specific chromosomes or specific regions of chromosomes may be associated with a microscopic skeletal meshwork of protein at-

tached to the inner layer of the nuclear envelope (Figure 6-2). This arrangement may impose substantial order on the chromosomes within the nucleus; in other words, they probably do not drift about randomly inside the nuclear sap, but are moored in this meshwork. The attachment of certain parts of chromosomes near the envelope possibly facilitates the passage into the cytoplasm of RNA or RNA molecules complexed to proteins.

ORGANELLES: SPECIALIZED WORK UNITS

Suspended in the cell's semifluid cytoplasm are a variety of organelles, each with a special function in the cell factory. These organelles include ribosomes, the endoplasmic reticulum, the Golgi complex, vacuoles, lysosomes, mitochondria, and plastids (in plant cells), all of which we will discuss here. Although eukaryotic cells vary in shape, size, and function, they all contain most of the various kinds of organelles. The inclusion of "compartments" with specialized tasks—factory departments, if you will—seems to be a winning formula, especially for eukaryotic cells. As we have already seen, compartmentalization of the chromosomes and genes may prevent their tangling or may aid their functioning, but it has other important implications, which we shall come to shortly.

Ribosomes: Protein Synthesis

Ribosomes are perhaps the most numerous organelles within a cell. Anywhere from a few hundred to many hundreds of thousands of ribosomes may be present in the cytoplasm, either floating freely or attached to various membranes. The abundance of ribosomes is related to the importance of their function in the cell: they are the sites at which amino acids are assembled into proteins destined for incorporation into various cell structures or export from the cell. A cell with many ribosomes has the capacity for rapid protein synthesis.

Ribosomes of both prokaryotic and eukaryotic cells have a common structure. A complete ribosome is a submicroscopic particle only 25 nm—about one-millionth of an inch—in diameter. It weighs about the same as 100 to 150 average protein molecules, and it is composed of two globular subunits that differ in size and function. Each subunit contains large structural RNA molecules and many structural proteins. When an mRNA molecule leaves the cell nucleus, a large and a small ribosomal subunit join it at one of its ends, as shown in Figure 6-3a. Once the subunits combine, the single functioning ribosome moves along the length of the mRNA, as though reading the sequence of RNA instructions sent from the

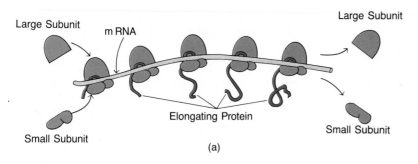

Figure 6-3   HOW RIBOSOMES BUILD PROTEINS.
(a) Two ribosomal subunits—one small, one large—attach to one end of an mRNA template. As each ribosome moves along the mRNA, "reading" the sequence coded in it, the new protein chain elongates. When the protein is released, the ribosomal subunits drop off the mRNA and reenter the pool of free subunits in the cytoplasm. (b) An electron micrograph of a free polysome (a group of ribosomes assembled on an mRNA molecule), magnified about 31,200 times, in the cytoplasm of a chick embryo's nerve cell. This unusually large polysome has about thirty-six ribosomes, indicating a very long mRNA molecule, and a large protein product.

nucleus and translating the instructions into a protein. In accordance with each instruction on the mRNA molecule, a specific amino acid is linked to the end of an elongating protein chain by means of a peptide bond. This lengthening chain appears to protrude through a hole in the large ribosomal subunit. When the ribosome has finished reading the entire mRNA molecule and the protein copy is complete, the ribosomal subunits are released from the mRNA into the cytoplasm, ready to associate with another mRNA molecule and to participate in the next round of protein synthesis. The products of protein synthesis are a bit like steel manufactured either for use within a steel mill itself or for export. The finished proteins may be released directly into the cytoplasm, where they begin to function, or they may be passed to other organelles for transport within the cell or to the surface for export.

A number of ribosomes may move along a single mRNA molecule at the same time. One mRNA molecule plus several ribosomes is called a **polysome** (Figure 6-3b). Most *cellular proteins*—those released into the cytoplasm—are made on polysomes located free in the cytoplasm. Cellular proteins include common enzymes, some structural proteins, some membrane proteins, regulatory proteins for use by the cell, and even nuclear proteins that are parts of chromosomes or of newly manufactured ribosomal subunits. All cells make cellular proteins to replace proteins that undergo chance damage. This is essential to the cell's health and continued functioning. But cells also make two other types of proteins: *exportable proteins*, such as hormones, digestive enzymes, and structural proteins that become part of the cell wall, the glycocalyx, or the tissue spaces; and *membrane proteins*, which are incorporated into the

plasma membrane and the membranes of organelles. Exportable proteins and membrane proteins are not manufactured by the free polysomes, which synthesize cellular proteins. Instead, they are made on membrane-bound polysomes. This compartmentalized arrangement apparently ensures that exportable and membrane proteins do not mix with functioning soluble cellular proteins. This arrangement also allows the exportable and membrane proteins to be modified chemically and facilitates their transport to specific sites in the cell, including sites of export. The special compartment where these processes occur is the endoplasmic reticulum.

Endoplasmic Reticulum: Production and Transport

The cytoplasm once was thought to be a semifluid sap in which organelles floated freely. In the mid-1950s, with the aid of the electron microscope, Rockefeller University researchers Keith Porter and George Palade discovered a lacy array of membranous sacs, tubules, and vesicles within the cytoplasm. They called this array the **endoplasmic reticulum** (meaning "intracellular web or network"), or **ER.** Further studies showed that the ER could be rough or smooth. The *rough endoplasmic reticulum* (RER) is studded with ribosomes, while the *smooth endoplasmic reticulum* (SER) is not. Broad membranous sheets of RER form channels that may be continuous with the outer membrane of the nuclear envelope. The ER is held in place by the cytoskeleton, which gives the cell its shape. It is in the RER and SER that proteins destined for export or for special sites in the cell are made and modified.

Rough Endoplasmic Reticulum

The outer surface (the cytoplasmic side) of the RER is dotted with thousands of polysomes, each of which consists of an mRNA molecule loaded with a variable number of ribosomes. As protein is synthesized on these bound polysomes, the elongating peptide chains pass into channels, or *cisternae*, between the RER membranes (Figure 6-4). These channels act as a transportation network for proteins within the cell. Once inside the RER, the proteins move through the membrane system, to other membranous organelles where they may be modified or packaged for storage, export, or modification.

In most proteins destined for export or for insertion in membranes, the first fifteen to thirty amino acids of the protein chain appear to act as a *signal peptide.* When a protein is synthesized, the signal peptide emerges from the ribosome first and functions as an address label. It binds to a membrane receptor on the RER, thereby binding the mRNA molecule and ribosome to the RER

membrane. This ensures the passage of the elongating protein through the lipid bilayer of the membrane and into the RER cisterna (Figure 6-5). Once released, it can travel within the channel to other places in the cell. Since several ribosomes are associated with each mRNA molecule, the result of signal-peptide binding is the binding of polysomes to the RER membrane. The signal peptides make it possible for just one type of ribosome to manufacture both cellular and exportable proteins. Thus ribosomes involved in the synthesis of an exportable protein or a membrane protein, with its signal peptide, are subsequently released into the cytoplasm. There, by chance, they may associate with an mRNA molecule that carries instructions for the synthesis of a cellular protein; lacking a signal peptide, that mRNA–ribosome complex remains free of the RER membrane. This means that cells do not have to possess two classes of ribosomes: one for reading only mRNAs with instructions for the manufacture of cellular proteins and one for reading only mRNAs with those for exportable and membrane proteins.

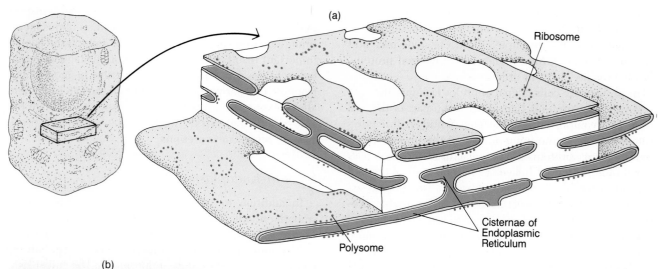

(a)

Ribosome

Cisternae of Endoplasmic Reticulum

Polysome

(b)

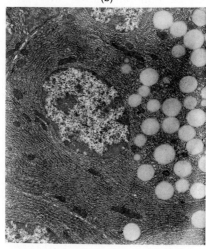

Figure 6-4   ROUGH ENDOPLASMIC RETICULUM: A PROTEIN-BUILDING FACTORY.

(a) This diagram shows an enlarged fraction of the RER. Stacked sacs serve as sites for ribosome attachment and channels for protein transport. The string of ribosomes are individual polysomes; elongating protein chains pass from the polysomes into the cavity of the RER. (b) This electron micrograph shows the RER in a mouse pancreas cell, magnified about 4,000 times. Digestive enzymes are synthesized on these stacked membranes, and are stored in the large, circular storage granules.

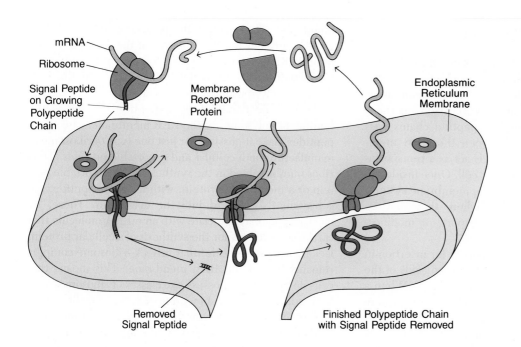

mRNA

Ribosome

Signal Peptide
on Growing
Polypeptide
Chain

Membrane
Receptor
Protein

Endoplasmic
Reticulum
Membrane

Removed
Signal Peptide

Finished Polypeptide Chain
with Signal Peptide Removed

Figure 6-5   THE SIGNAL-PEPTIDE HYPOTHESIS.
The signal-peptide hypothesis explains how certain proteins may enter the RER. The two ribosomal subunits attach to an mRNA molecule. As the ribosome "reads" the instructions for protein synthesis, the signal peptide binds to a recognition site on the RER membrane; then the elongating polypeptide chain "grows" into the RER, ultimately to be released.

Proteins that have just been released into the RER cisternae can be modified by any of some thirty or forty enzymes associated with the endoplasmic reticulum of different cells. For example, these enzymes may add various chemical groups to newly synthesized proteins or may modify them in other ways.

In prokaryotic cells, signal peptides facilitate the correct export of proteins that can become part of the cell wall. To accomplish this task, signal peptides cause polysomes to bind to the inner surface of the plasma membrane. There, proteins are manufactured and then pass out through the lipid bilayer to become part of the cell wall or material on its outer surface. This simple arrangement to allow direct secretion of exportable proteins through the membrane as they are manufactured may have been an evolutionary precursor of the more complex RER system of nucleated, eukaryotic cells.

The RER of eukaryotic cells has another important function besides its role in the synthesis and transport of proteins: it apparently gives rise to the nuclear envelope. As we will see in Chapter 9, when a eukaryotic cell undergoes cell division, the nuclear envelope breaks into small vesicles that cannot be traced. After division, sacs of RER membrane surround the chromosomes of the two daughter cells and give rise to new nuclear envelopes. The rough surface of the RER, studded with ribosomes, forms the outside of the reconstituted nuclear membrane, but ribosomes drop off the inner surface, which becomes smooth.

Let us summarize our discussion of the rough endoplasmic reticulum. Almost every kind of cell makes at least some proteins for export—to the cell wall, the glycocalyx, or a distant place within a multicellular organ-

ism. The RER is a hollow system of membranous channels that serves as the site of manufacture and transport of such proteins. The compartmentalization of this membrane-bounded organelle within the cytoplasm has an important result: it separates the proteins tagged by signal peptides for export or for insertion in membranes from the cellular proteins dissolved in the cytoplasm and allows exportable proteins to be transported through the cell toward the outside world of the organism's tissue spaces, blood, or secretions.

Smooth Endoplasmic Reticulum

In addition to RER, most eukaryotic cells contain smooth endoplasmic reticulum, or SER. As is evident in Figure 6-6, SER consists of a set of flattened sacs that lack ribosomes and, therefore, are smooth. The advantages of compartmentalization within the cytoplasm apply to the SER as well as to the nucleus and the RER. The SER and the enzymes associated with it carry out a variety of tasks, including transportation, synthesis, and the metabolism (chemical modification) of small molecules.

Radioactive tracers reveal that newly synthesized proteins may be transported from the RER to the SER en route to still another transport and packaging system, the Golgi complex. Besides being a cytoplasmic passageway, the SER is abundant in cells that synthesize fats and steroids. In the cells of mammalian ovaries and testes, for instance, the SER serves as a site of production of steroid sex hormones. Finally, the SER is involved in the oxidation of toxic substances. Thus certain enzymes associated with the SER in a number of cell

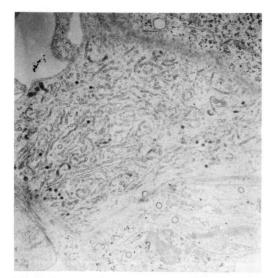

Figure 6-6   SMOOTH ENDOPLASMIC RETICULUM IN A NERVE CELL.
Membranous sacs are present without attached ribosomes (magnified about 5,400 times).

types render other molecules more polar so that they can be excreted from the cell and the body more quickly. For instance, in fish, frogs, reptiles, birds, and mammals, the liver helps to detoxify the blood. If an animal swallows a toxic chemical such as carbon tetrachloride (dry-cleaning fluid) or a barbiturate, the SER in the liver cells becomes very prominent and active, functioning at maximum capacity while the toxic compound is broken down and excreted from the body.

Smooth endoplasmic reticulum also performs specialized functions in certain cell types. In liver cells, the SER contains a large quantity of an enzyme that helps modify glucose so that it can pass through the membrane and into the SER. Once inside the membranous channels of the SER, the sugar can be transported to the cell surface and out to needy cells throughout the body. Another specialized function occurs in skeletal-muscle cells, where a special type of SER triggers muscle-cell contraction in response to nerve impulses. Specialized functions like these enable groups of cells in tissues or organs to carry out unique tasks for the whole organism. Emphasis or use of certain organelles, or enzymes associated with those organelles, can be the key feature in very different cell types—muscle cells, nerve cells, liver cells, and so on.

Golgi Complex: Packaging for Export

What happens to the products of the ER when they leave this system of membranes with internal cavities? Transport vesicles, which are tiny membranous bubbles,

pinch off from the RER or SER and carry exportable molecules to the **Golgi complex,** another membranous organelle that serves as a receiving, packaging, and distributing center for the cell factory. Each Golgi complex is a stack of saucer-shaped, baglike membranes surrounded by vesicles, as shown in Figure 6-7.

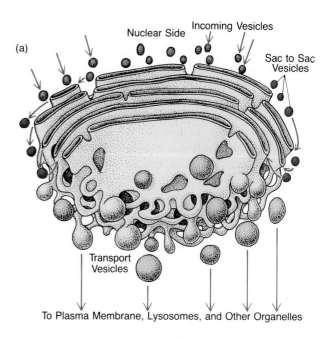

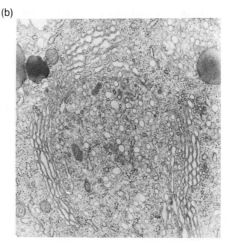

Figure 6-7   GOLGI COMPLEX: PACKAGING CENTER FOR THE CELL.
(a) Small vesicles containing newly synthesized proteins fuse with the edges of the flat Golgi sacs. Other small vesicles carry protein being processed from one sac to others. Finally, transport vesicles carry finished proteins to the plasma membrane, large storage vesicles, lysosomes, or other membranous organelles. (b) In this cell from a green alga plant, magnified about 15,000 times, the Golgi stacks are easily seen and are similar in appearance and function to those in animal cells. Although this organelle looks static, the small transport vesicles are constantly shuttling proteins to and fro.

For years after its discovery by the Nobel Prize winner Camillo Golgi, biologists believed the Golgi complex to be a dynamic organelle to which new flat sacs were continuously being added on the side closest to the nucleus while on the other side mature sacs were breaking off into vesicles that could move toward the cell surface and release exportable molecules. Abundant new evidence does not support this idea, however. Instead, the flat sacs of the Golgi complex seem to be stable structures to which molecules are brought via small transport vesicles from the RER or SER (Figure 6-7). There, the exportable molecules are modified by Golgi enzymes. Sugars, lipids, phosphate groups, or sulfate groups may be added or removed, or the basic structure of the molecule may be altered. In animals, the molecule that undergoes such modification is usually a protein, a fat, or a steroid. In plants, it is a protein or a complex carbohydrate, such as cellulose, destined to be incorporated into the cell wall. New transport vesicles may pinch off from the side of one Golgi sac and carry the contents to another Golgi sac for further modification. Then the cargo can be packaged in vesicles and transported, as Figure 6-8 shows, to any of several destinations, including membranes and large storage granules such as those found in the pancreas, an organ that makes, stores, and secretes digestive enzymes. The prominence and activity of Golgi complexes and storage granules in pancreatic cells are still other examples of special uses of a common organelle that help characterize cell types and, in turn, the function of the entire organ.

Among the proteins that pass through and are modified by the Golgi complex are many of the membrane proteins of the nuclear envelope, the plasma membrane, and the membranes of other organelles. Other enzymatic proteins go to the lysosomes, which we will describe shortly. There must be some remarkable recognition system, involving the equivalent of nametags and addresses on the transport vesicles and on the diverse destinations, for the different kinds of cargo molecules to be delivered to the correct cellular targets.

A major question that still exists about the Golgi complex concerns the origin of the membranes. Do they arise from the ER, or do they arise separately? Some cell biologists believe that (1) the nuclear envelope continuously generates the RER and SER; (2) both types of ER yield Golgi-complex membranes; and (3) the Golgi-derived vesicles contribute to still other organelles and to the plasma membrane itself (Figure 6-8). However, differences in the lipid and protein constituents of some of these membranes suggest that each membrane is a discrete cellular component that is inherited by the cell at "birth" and maintained thereafter. This model, which is favored by recent evidence, requires a distribution system for newly made membrane components and involves the sort of address markers that we have suggested must exist for efficient protein distribution.

Vacuoles: Food and Fluid Storage and Processing

The organelles that we have discussed so far—nucleus, ribosomes, RER, SER, Golgi complex—control and carry out various production tasks for the cell factory, including molecular synthesis, transport, packaging, and export. We turn now to organelles involved in the basic life processes, with functions analogous to eating, drinking, digesting, and excreting. The first organelle we will consider is the **vacuole.** The word "vacuole" is derived from a Latin root meaning "vacant," and, indeed, the several types of membrane-bound bodies classified as vacuoles have so little internal structure that they appear to be empty sacs. However, they are filled with fluids and soluble molecules. They play a variety of roles depending on the type of cell in which they are found.

Vacuoles are critically important to single-celled organisms such as amoebae. Such organisms face problems of continuous water entry because the osmotic pressure is higher in their cytoplasm than in the fresh water they inhabit. These one-celled creatures possess *contractile vacuoles*, membranous sacs into which cytoplasmic water flows by osmosis. The swelling of these vacuoles ultimately triggers the discharge of the contents, at which point the vacuolar and cell membranes fuse, and the contents of the vacuole are expelled. This process is repeated over and over so that the single-celled organism can avoid being drowned by the perpetual influx of water.

The most prominent vacuoles appear in plant cells. Viewed at low magnification, many plant cells seem to contain little more than a huge central vacuole surrounded by a thin halo of cytoplasm, as Figure 6-9 shows. The vacuole serves as a water reservoir and as a storage site for sugars, proteins, and pigments responsible for the bright colors of many fruits and flowers. The fluid inside plant-cell vacuoles (sometimes called cell sap) also contributes to the turgor that keeps the cell stiff.

Many animal cells also use vacuoles for feeding. Inside one-celled eukaryotes, membranous vacuolar sacs fill with food and pinch off from the large feeding organelles, commonly called gullets. This vacuolar feeding is a form of **phagocytosis,** the engulfment of particulate matter by animal cells (Figure 6-10). When the human white blood cells called macrophages accumulate at the site of an infection, they ingest the bacterial invaders by means

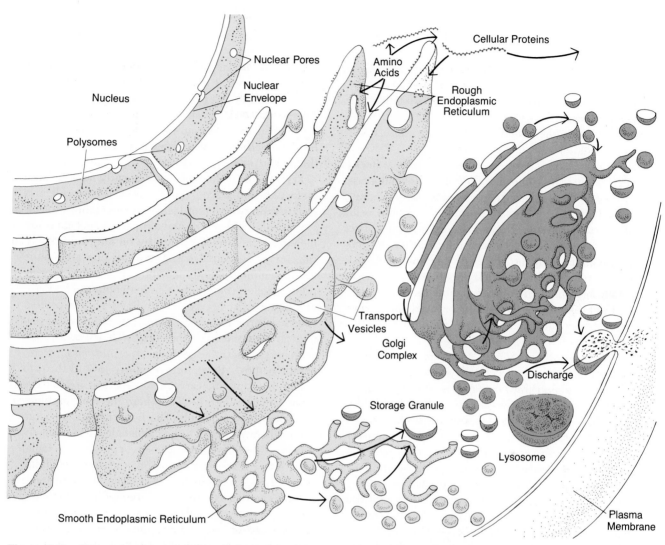

Figure 6-8   THE FLOW OF EXPORTABLE MATERIALS IN THE CYTOPLASMIC MEMBRANE SYSTEMS.
The nuclear envelope may be continuous with the RER (the site of considerable protein synthesis). The RER, in turn, contributes vesicles to the Golgi complex (the packaging site for exportable protein) or is continuous with the SER. Vesicles from the SER and from the Golgi complex fuse with the plasma membrane so that their contents may be expelled from the cell. Other vesicles carry digestive enzymes to the lysosomes.

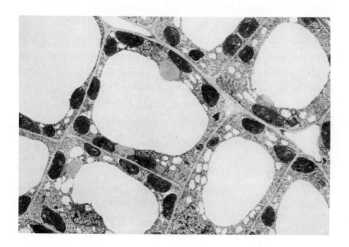

Figure 6-9   CENTRAL VACUOLES IN CELLS OF A TOBACCO LEAF.
The cytoplasm, with its many prominent chloroplasts and other organelles, streams, or circulates, around these large, central vacuoles (magnified about 2,000 times).

TRACING CELLULAR "TRADE ROUTES"

The cell is the center of biological commerce—of life processes—and, as such, is the site of constant export and import of critical materials. Two important pieces of scientific detective work—one classic, the other new—addressed the precise way these trade routes into and out of the cell operate.

In 1964, George Palade and his colleagues at Rockefeller University confirmed the role of the endoplasmic reticulum and the Golgi complex in processing, packaging, and transporting proteins from the cell. Palade was awarded a Nobel Prize for this work. The team used pancreas cells from a guinea pig and tagged the amino acids in the cells with radioactive labels. Palade knew that the labeled amino acids would be quickly incorporated into new proteins, but where would this incorporation take place? And along what routes would the labeled amino acids travel in the cell?

Employing a photographic technique that reveals the locations of the radioactively labeled amino acids as black grains on the photographs, the team found that the amino acids always appeared first over the RER. Timed series of photos showed that black grains appeared next over vesicles near the RER, then over the SER and surrounding vesicles, next over the Golgi complex and large storage granules, and finally outside the cells. Clearly, the amino acids were being incorporated into exportable proteins and secreted from the cells by means of the following route: RER → SER → Golgi complex → secretory granules → cell exterior. Later research revealed that the secretory granules or vesicles travel to the cell membrane and fuse with it; then the contents are liberated outside the cell. This entire sequence, from RER to expulsion, is called exocytosis and is the main trade route for moving materials out of the cell.

What about the importation of materials? Cells can take in liquid by means of *pinocytosis*, or "cell drinking," and small molecules by means of active transport directly across the cell membrane. But how does the cell take up large molecules such as hormones and large proteins that are too big to move through the membrane? Almost twenty years after Palade's classic experiments, Ira Pastan, Mark Willingham, and other researchers, working at the National Cancer Institute, studied this question and developed a model for the import of such materials.

Pastan and Willingham fluorescently labeled large molecules and traced their route into the cell in a sophisticated version of Palade's work, using a television camera, videotape, and a computer. All the large molecules they traced begin by binding to receptor sites on the cell surface. The receptor sites are mobile in the semifluid membrane and can travel to specific regions called *coated pits*—indentations in the outer membrane that are coated or lined with a protein called clathrin (Figure A). These pits form special vesicles around the large incoming molecules. The researchers named these vesicles *receptosomes*, or "receptor bodies." Receptosomes move inward through the cell cytoplasm, carrying their cargo toward the Golgi complex and the ER, where the molecules can be structurally modified. The import route for large materials thus seems to be: molecule → mobile receptor site → coated pit → receptosome → Golgi complex → ER. This is roughly the reverse of the export route, and appears to involve simply a more complex passage through the membrane.

Many details of both processes remain to be explained before we can completely understand the traffic of supplies into and out of cells. Current work centers on such subjects as the molecular name tags on coated pits and other membranous transport vesicles, since the name tags determine where transported materials are delivered in the cell. And newer work seeks to understand how various viruses, some of which cause cancer, enter cells using coated-pit-like mechanisms, but avoid being shuttled to a wrong site where cellular enzymes might destroy the unwelcome invaders.

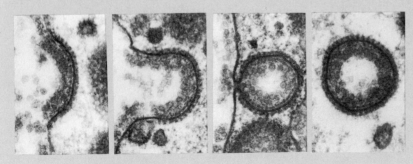

Figure A

Ingestion of extracellular substances by a coated pit (left). Protein particles are taken into a hen's egg cell to form yolk. The coated pit separates from the cell surface (middle) to form a coated vesicle (right). Yolk particles are seen inside and skeletal proteins dot the surface of the coated vesicle (all magnified about 60,000 times).

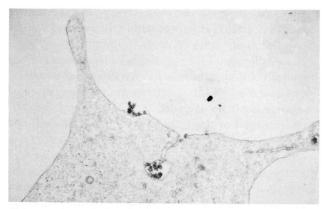

Figure 6-10   PHAGOCYTOSIS.
A cell from a chicken embryo, magnified about 20,500 times, engulfs a cluster of protein molecules. The lower portion of this cell surface indentation later will pinch off as a separate vesicle.

of phagocytosis. The vacuoles containing food then receive digestive enzymes from the Golgi complex, delivered in small vesicles. The food or bacterium is degraded, and its amino acids, lipids, and other components are passed to the cell cytoplasm or expelled as wastes.

Still another function of vacuoles is **pinocytosis**, or

"cell drinking." In multicellular animals, this process can be essential to the functioning of whole organs. One thoroughly studied example of pinocytosis occurs in narrow human blood vessels called capillaries. Figure 6-11 shows how the membranes of cells that line capillaries pinch off vacuoles filled with blood serum, the yellowish fluid that surrounds red and white blood cells. The vacuoles carry the protein-rich serum across the cytoplasm of the cell and expel it from the capillary. The discharge, or "spitting out," of the fluid vacuolar contents is often called *reverse pinocytosis*. By coupling pinocytosis and reverse pinocytosis, capillary cells can transport serum from the blood vessel to the surrounding tissue space without exposing the serum components to the cytoplasm, where it might be chemically changed.

The uptake and discharge of fluids and materials are important tasks in virtually all eukaryotic cells, as we noted in Chapter 5 in our discussion of passive- and active-transport mechanisms at the molecular level. The engulfing and disgorging mechanisms we just described involve whole sections of the plasma membrane, not just pores or carrier molecules, and thus can transport relatively large amounts of material. They are an important adjunct to passive and active transport. Phagocytosis and pinocytosis are examples of **endocytosis** (taking into the cell), while reverse pinocytosis is an example of **exocytosis** (putting out of the cell). Both mechanisms can

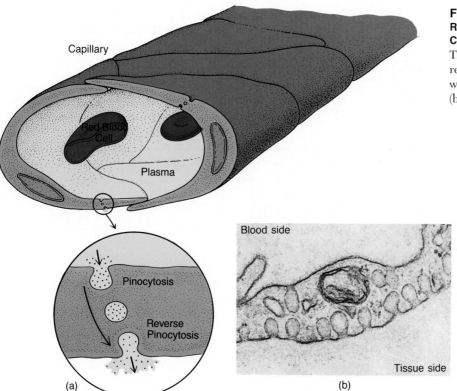

Figure 6-11   PINOCYTOSIS AND REVERSE PINOCYTOSIS IN A HUMAN CAPILLARY.
The drawings (a) depict the events recorded by the electron micrograph, which is magnified about 15,000 times (b).

Capillary

Red Blood Cell

Plasma

Pinocytosis

Reverse Pinocytosis

(a)

Blood side

Tissue side

(b)

be used by vacuoles, vesicles, and other organelles. If these organelles are altered by mutation, disease, or environmental factors (pollutants, narcotics, and so on), the cells—and, in turn, the entire organism—are endangered.

Lysosomes: Digestion and Degradation

Another class of vacuoles plays a role in the digestion of food and of damaged cell parts. When a food morsel is engulfed into a vacuole, an influx of digestive enzymes from the Golgi complex is needed to help break down the ingested substance. Some fifty kinds of digestive or hydrolytic enzymes can be formed in the RER and SER and packaged by the Golgi complex in **lysosomes,** spherical membrane-bound bags. A lysosome fuses with a phagocytic vacuole and incorporates it into its structure, mingling the digestive enzymes it contains with the food morsels from the vacuole. The food contents now in the lysosome are then digested, since the lysosomal enzymes can attack most known biological macromolecules. The rest of the cytoplasm is protected from the enzymes by the lysosomal membrane. (Why the enzymes do not attack *this* membrane from the inside is not known.) Once a lysosome has fused with a vacuole or vesicle, the resultant structure is called a *secondary lysosome* (Figure 6-12). These structures again illustrate

the benefits of compartmentalization; powerful enzymes can digest certain substances within the cell and yet not digest the entire cell's contents in the process.

Besides helping to "feed" the cell, lysosomes play an important custodial role when cell components wear out. Membranes, ribosomes, proteins, and a variety of other components can be degraded by lysosomes and the subunits returned to the cytoplasm for reuse in synthetic processes. Sometimes, within injured or old cells, lysosomes also break open and free their enzymes, literally digesting the cell from the inside out. For this reason, Christian de Duve, the Belgian biologist who discovered lysosomes, called these organelles potential "suicide bags."

Lysosomes cannot degrade all cellular waste products, however. Some cells in multicellular organisms age along with the entire organism, accumulating in the process **lipofuscin granules,** lysosomelike vesicles that contain indigestible lipids and proteins as well as hydrolytic enzymes. Lipofuscin granules sometimes build up to a remarkable degree in body tissues; for instance, they may make up 6 to 10 percent of the total volume of heart-muscle cells in an eighty-year-old human. It is no wonder that such an accumulation can contribute to the loss of function in the aged heart and other organs. A serious disease called silicosis also can result from the inability of lysosomes to digest certain materials. Miners sometimes inhale silica fibers, which are engulfed by lung cells. Lysosomes in the cells then incorporate the

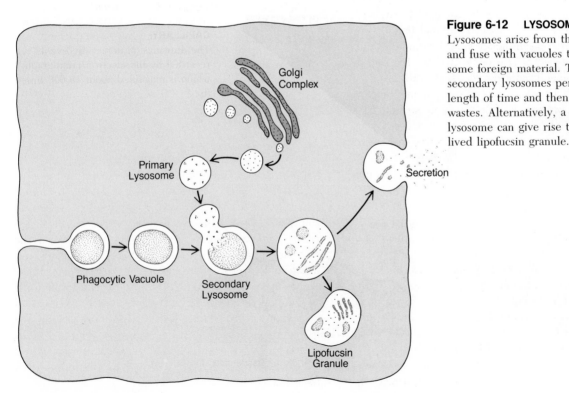

Figure 6-12   LYSOSOME LIFE HISTORY.
Lysosomes arise from the Golgi complex and fuse with vacuoles that have engulfed some foreign material. The resultant secondary lysosomes persist for a varying length of time and then discharge their wastes. Alternatively, a secondary lysosome can give rise to a very long-lived lipofuscin granule.

Golgi Complex

Primary Lysosome

Secretion

Phagocytic Vacuole

Secondary Lysosome

Lipofuscin Granule

fibers but do not break them down. The silica fibers can poke holes in the lysosomal membrane, allowing damaging enzymes to leak into the lung cells, destroy them, and ultimately cause the victim's death.

Eukaryotic (but not prokaryotic) cells often possess additional small membranous vesicles called **microbodies** (also known as *peroxisomes* and *glyoxysomes*). Like lysosomes, microbodies break down cellular waste products. About 1,000 microbodies are present in a typical mammalian liver cell. The microbodies of various animal and plant cells contain a number of different enzymes that, in general, carry out chemical processes called oxidation (Chapter 7). One such enzyme is *catalase*, which splits hydrogen peroxide (H_2O_2) into water and oxygen but oxidizes an organic compound as it does so. Hydrogen peroxide is potentially toxic and arises naturally within cells when hydrogen is cleaved from amino acids or uric acid and then combined with oxygen. Biologists do not yet fully understand microbodies, but their presence in most animal, plant, and fungal cells and their special enzyme constituents suggest that these tiny spheres play a major role in cellular metabolism and detoxification.

Mitochondria and Plastids: Power Generators

Just like any manufacturing concern, the cell factory requires energy for its activities. Two types of organelles transform and store energy in forms usable by the cell: mitochondria and plastids. **mitochondria** (singular, *mitochondrion*) are sites of chemical reactions that harvest the energy from food molecules and generate high-energy compounds—such as ATP—that can be used directly to meet the cell's energy needs. **Plastids** are present in plant cells that use light energy to manufacture energy-rich carbon compounds, such as sugars, from simple inorganic raw materials. In Chapters 7 and 8, we will study in depth the life-sustaining chemical processes that go on within mitochondria and plastids. Here we will consider only their structures and general functions.

Mitochondria

The power plants of all eukaryotic cells—plant, animal, and fungal—are the mitochondria. Mitochondria vary in shape from small spheres to long, sausage-shaped bodies about 1 μm wide and up to 8 μm long. Figure 6-13 shows the unique structure of a mitochondrion: it has a smooth, continuous outer membrane and an inner membrane thrown into folds called *cristae*. The outer mitochondrial membrane is quite permeable, even to molecules that weigh 10,000 daltons (such as polypeptides containing ninety amino acids). In contrast, the

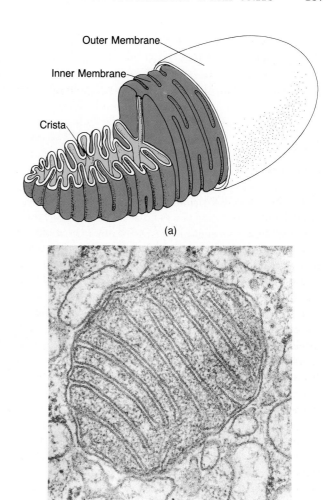

(a)

(b)

Figure 6-13   MITOCHONDRION: CELLULAR POWERHOUSE.

(a) An artist's depiction shows the cristae of the inner membrane in three dimensions. The inner matrix that bathes the cristae contains DNA, ribosomes, and many kinds of enzymes. (b) An electron micrograph reveals the membranous nature of the cristae, magnified about 54,000 times.

inner mitochondrial membrane is relatively impermeable and highly responsive to osmotic pressure. It unfolds and refolds in a variety of ways as water enters and leaves the space between the inner and outer membranes. The cristae provide a large surface area on which many of the enzymes involved in generating ATP molecules are found. The proximity and spatial arrangements of these enzymes are critical to the cell's energy transformations, as we shall see in Chapter 7. The number and size of mitochondria in a cell, as well as the number of cristae in each, depend on the cell's energy requirements. Cells with high energy requirements have large mitochondria with many cristae, whereas cells with low energy needs have few or smaller mitochondria.

ON MITOCHONDRIA

"My mitochondria comprise a very large proportion of me. I cannot do the calculation, but I suppose there is almost as much of them in sheer dry bulk as there is the rest of me. Looked at in this way, I could be taken for a very large, motile colony of respiring bacteria, operating a complex system of nuclei, microtubules, and neurons for the pleasure and sustenance of their families, and running, at the moment, a typewriter. . . .

"It is no good standing on dignity in a situation like this, and better not to try. It is a mystery. There they are, moving about in my cytoplasm, breathing for my own flesh, but strangers. They are much less closely related to me than to each other and to the free-living bacteria out under the hill. They feel like strangers, but the thought comes that the same creatures, precisely the same, are out there in the cells of sea gulls, and whales, and dune grass, and seaweed, and hermit crabs, and further inland in the leaves of the beech in my backyard, and in the family of skunks beneath the back fence, and even in that fly on the window. Through them, I am connected; I have close relatives, once removed, all over the place. This is a new kind of information, for me, and I regret somewhat that I cannot be in closer touch with my mitochondria. If I concentrate, I can imagine that I feel them; they do not quite squirm, but there is, from time to time, a kind of tingle. I cannot help thinking that if only I knew more about them, and how they maintain our synchrony, I would have a new way to explain music to myself."

Source: Lewis Thomas, *Lives of a Cell: Notes of a Biology Watcher* (New York: Viking-Penguin, Inc., 1974).

The cristae protrude into a semifluid **matrix** with a gel-like consistency that may stem from its high protein content. The matrix contains ribosomes constructed of RNAs and proteins that differ from those in the cytoplasm. Mitochondrial ribosomes are like those of prokaryotic bacteria. Every mitochondrion has its own genetic material in the form of DNA that is arranged in a circle or more than one identical circles; the bacterial chromosome also is a circle. In mitochondria of all eukaryotes, the genetic code is probably unique, at least in part. Thus mitochondrial DNA is distinct from that of either eukaryotic or prokaryotic chromosomes.

Mitochondrial DNA codes for various mitochondrial proteins, including some of the subunits of the ATP-generating enzymes. Surprisingly, other subunits of those enzymes are encoded in the cell's nuclear DNA; the latter subunits are made on cytoplasmic polysomes and somehow enter the mitochondria, where they link up to the mitochondrial subunits to form the final enzyme molecules.

Mitochondria are self-replicating bodies; that is, they reproduce independently of the division of the rest of the cell. They can do so because they have their own genetic material. Indeed, it is very likely that mitochondria are descendants of unknown prokaryotes that invaded eukaryotic cells billions of years ago. (We will say more about this interesting hypothesis in Chapter 19.)

Mitochondria are good examples of both compartmentalization and cell specialization. Eukaryotic cells are able to generate the large number of ATP molecules they need only because special sets of energy-harvesting enzymes are arrayed on the membranes of these sausage-shaped organelles. Without such specialized energy-producing compartments, the efficiency of the cell's ATP production might be greatly lower. That cells with high energy requirements contain many large mitochondria is yet another example of how the presence of specific organelles is compatible with the mode of specialization of a cell and, in turn, with the unique function of a tissue or an organ.

Plastids

The surest way to tell a plant cell from an animal cell is to look for plastids; every plant cell has at least one form of plastid in its cytoplasm in addition to the other organelles we have been discussing. Plastids turn plant cells into photosynthetic, carbohydrate-producing factories: they are responsible for capturing light energy to produce sugar and for storing sugar in the form of starch. The presence of plastids in the cells of plants allows them to produce their own food molecules from simple raw materials—CO_2, H_2O, and minerals. The absence of plastids in the cells of animals and fungi has profound consequences: both types of organisms ultimately depend on plants for the energy-containing food molecules they need to sustain life.

There are two types of plastids: those that lack pigments are the *leucoplasts*, and those that contain various pigments are the *chromoplasts*. The colorless leucoplasts serve as storage bins; they accumulate starch, proteins, and oils that can be tapped by the plant as needed. Much

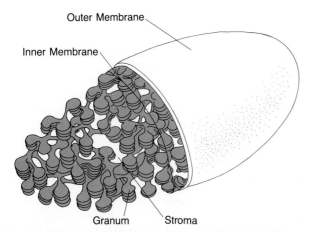

Outer Membrane

Inner Membrane

Granum   Stroma

Figure 6-14   CHLOROPLAST: THE GREEN PLANT'S CARBOHYDRATE FACTORY.
With the outer membrane of the chloroplast partially removed, a large number of stacks of flat sacs can be seen. Each stack, or granum, has a huge surface area so that photons of light are likely to be captured and thus energy can be used to build carbohydrates (Chapter 8). The stroma fluid phase contains soluble enzymes, DNA, and ribosomes.

of the food value of potatoes, carrots, and beets is stored in leucoplasts. Similarly, nutrient-filled leucoplasts form the bulk of cells in seeds and nuts; the nutrients both fuel the development of embryonic plants as they emerge from seeds and provide high-calorie food for animals.

Chromoplasts can contain many pigments, including *carotenoids*—the colored molecules that account for the brilliant reds and yellows of maples, oaks, aspens, and other trees in autumn; the colors of ripening fruit; and the spectrum of pink, yellow, and red hues in flower petals. The most important chromoplasts are the **chloroplasts,** the green chlorophyll-containing organelles in which photosynthesis takes place (Figure 6-14).

Chloroplasts are as large as an animal's red blood cells—some 3 to 8 μm in size. In a typical cell of a higher plant, thirty or more chloroplasts are anchored in the cytoplasm close to the plasma membrane (Figure 5-5). Two lipid-bilayer membranes enclose each chloroplast. Notice in Figure 6-14 that much of the interior of a chloroplast is filled with flattened membranous sacs, arranged much like stacks of coins. Each stack is called a *granum* (plural, *grana*). Enzymes required for the light reactions of photosynthesis (Chapter 8) are situated on the grana membranes. The architectural arrangement of these enzymes is essential to the function of photosynthetic enzymes, since the entire process of photosynthesis immediately stops if chloroplast structure is disrupted. The grana are surrounded by a matrix, called the **stroma.** This matrix contains several components: (1) enzymes used in photosynthesis, (2) small prokaryotelike

ribosomes, and (3) circular strands of DNA. Like mitochondria, chloroplasts are self-replicating organelles that are largely self-reliant but do require some proteins encoded by the cell's nuclear DNA and synthesized in the cytoplasm.

Despite differences in their pigments and enzymes, all plastids are members of a single organelle family. They arise from pigmentless *proplastids* and are to some extent interconvertible: for example, chloroplasts can lose their chlorophyll after prolonged exposure to darkness, while leucoplasts occasionally can be induced to form chlorophyll. Some biologists trace this similarity to the hypothetical ancestors of all plastids: prokaryotic cells that took up residence in some of the first eukaryotic cells on the early Earth (Chapter 9).

THE CYTOSKELETON

We have described most of the major organelles—the intricate, nonliving units within the cell that bridge the mysterious gap between molecules and living cells. Next we turn to the **cytoskeleton,** the dynamic three-dimensional weblike structure that fills the cytoplasm and in which the organelles are suspended. This intracellular scaffolding acts as both muscle and skeleton for the cell. Without it, the cell, its complex surface, and its organelles could not move. Furthermore, the parts of the cell could not remain in proper spatial relationship to one another, and the cell would lose its normal shape. Just as plastids are the hallmark of plant cells, the cytoskeleton is a hallmark of *all* eukaryotic cells—plant, animal, and fungal. The cytoskeleton enables eukaryotic cells to carry out activities impossible for prokaryotic organisms.

The cytoskeleton is actually a convoluted latticework of microscopic filaments and tubules that seems to occupy all available space in the cell. This latticework is assembled and disassembled in the course of various cell activities. The most ubiquitous elements in the cytoskeleton are the **microfilaments,** extremely fine, threadlike rotein fibers only 3 to 6 nm in diameter. A network of these filaments is shown in Figure 6-15. Composed predominantly of a *contractile protein* called **actin,** microfilaments are involved in many types of intracellular movements in plant and animal cells. Regular arrays of microfilaments in skeletal- and heart-muscle cells interact with other, thicker filaments made up of another contractile protein called **myosin.** This interaction brings about the contractions that move an animal's limbs or pump its blood. Myosin molecules are present in other kinds of animal and plant cells as well, but usually not in the form of long, thick filaments. Similarly, a number of other contractile proteins are found in many kinds of cells. Interaction of these proteins with actin may power

(a)

Figure 6-15   THE CYTOSKELETON.
The cytoskeleton is a tangled web of structural and contractile elements. (a) A small portion of an animal cell's cytoskeleton, as viewed with an electron microscope and magnified about 20,000 times, after the enveloping plasma membrane has been removed. The narrowest filaments seen here are microfilaments. (b) A model of the cytoskeleton, showing the three-dimensional arrangement of filaments and microtubules.

(b)

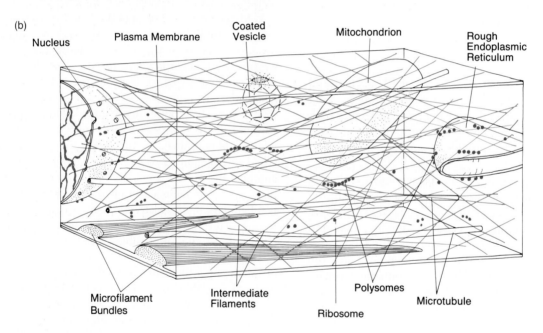

Nucleus   Plasma Membrane   Coated Vesicle   Mitochondrion   Rough Endoplasmic Reticulum

Microfilament Bundles   Intermediate Filaments   Ribosome   Polysomes   Microtubule

various types of cell movements. The inclusion of a large quantity of contractile proteins in muscle cells is another good example of cell specialization. The basic contractile apparatus of the early eukaryotic cell was used for a variety of purposes, such as cell locomotion, propulsion of the stream of moving cytoplasm in plant cells, and so on. Only later in the evolution of animals did the more highly ordered arrays of actin and myosin emerge, allowing the exaggerated contractions that characterize true muscle cells.

Microtubules are long, cylindrical tubes 20 to 25 nm in diameter and thus are considerably wider than microfilaments. Microtubules are composed of subunits of the

globular protein **tubulin,** which are stacked in a spiral to form the long microtubules. The arrangement of microtubules varies according to cell type because they act primarily as a scaffold that helps stabilize the shape of the cell (Figure 6-16). The long, thin cells that make up the lens in the eye of an embryonic mouse contain many microtubules that run parallel to the long axis of the cells and help maintain their shape. If the microtubules are disrupted, the cells assume a short, round configuration. Microtubules are the main component of the *spindle*, the apparatus that moves chromosomes when cells divide (Chapter 9). Finally, microtubules are also arranged in geometric patterns inside the whiplike flagella and

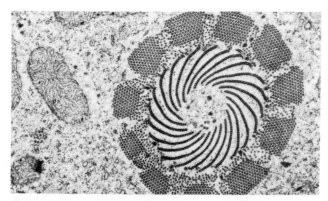

Figure 6-16   MICROTUBULES.
Microtubules are skeletal elements in cells that may be assembled in a variety of patterns. These regularly shaped bundles in the cytopharynx of a protozoan are shown in cross section magnified about 17,000 times. For size comparison, note the portion of a mitochondrion to the left.

hairlike cilia that are used in certain kinds of cell locomotion, as we shall soon see. Evidence suggests that cytoplasmic microtubules lengthen as a result of the rapid addition of tubulin subunits and that this polymerization may exert enough force to move the cell surface or internal organelles. Conversely, the rapid depolymerization of tubulin leads to shortening of microtubules and may cause them to pull on organelles. Cytoplasmic microtubules are highly dynamic structures: so-called catastrophic depolymerization may lead to the disappearance of microtubules 20 μm long in only one minute. In fact, the only stable microtubules may be those in which the ends are capped by special proteins or organelles.

Single microtubules can be isolated from nerve cells, and if supplied with ATP, can glide along the surface of a glass slide. This is due to presence of a **mechanoenzyme** (an enzyme that exerts mechanical forces, as does myosin) attached to the microtubule surface.

Intermediate filaments (about 10 nm in diameter) show up on electron micrographs as numerous wavy lines crisscrossing the interior of most kinds of animal cells. These filaments are made of a family of proteins; those of nerves differ from those of muscle cells or of connective-tissue cells. It seems likely that intermediate filaments impart tensile strength to the cytoplasm, since they are associated with the desmosomal spot welds between epithelial cells described in Chapter 5. Three proteins of the intermediate filament family help form the structural meshwork on the inner surface of the nuclear envelope to which chromosomes may attach.

Although the cytoskeleton is composed of these three types of filaments and tubules, it is not the rigid, permanent structure that the word "skeleton" implies. Bundles of filaments and sets of microtubules do at times serve as stiff, bonelike assemblies, but they are also physically dynamic, capable of dispersing to component molecules and quickly reassembling elsewhere in the cell or reorganizing into a less ordered form. Because the cytoskeleton does not have great intrinsic stability, the shape of an animal cell is changeable, depending on the substrate to which the cell adheres or on the communications it receives from neighboring cells. This is one aspect of anchorage dependence, which we discussed in Chapter 5. In a real sense, the eukaryotic animal cell is shaped by its environment, as sketched in Figure 6-17. In plant cells, microtubules, located just inside the plasma membrane, serve as anchorage sites for plastids. Bundles of microfil-

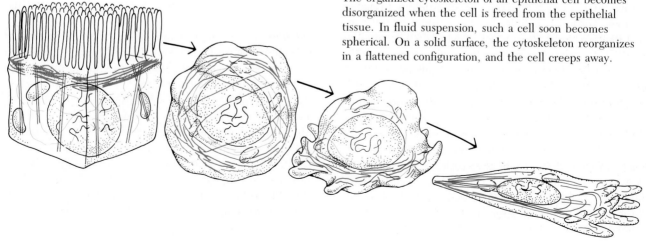

Figure 6-17   THE CYTOSKELETON OF AN EPITHELIAL CELL.
The organized cytoskeleton of an epithelial cell becomes disorganized when the cell is freed from the epithelial tissue. In fluid suspension, such a cell soon becomes spherical. On a solid surface, the cytoskeleton reorganizes in a flattened configuration, and the cell creeps away.

Table 6-1 **COMPONENTS OF PROKARYOTIC, PLANT, AND ANIMAL CELLS**

Component	Prokaryote	Plant Cell	Animal Cell
Cell wall	Present	Present	Absent
Glycocalyx	Absent	Absent	Present
Plasma membrane	Present	Present	Present
Cytoskeleton	Absent	Present	Present
Nucleus	Absent	Present	Present
Chromosomes	Single	Multiple	Multiple
Mitochondria	Absent	Present	Present
Plastids	Absent	Often present	Absent
Ribosomes	Present	Present	Present
Endoplasmic reticulum	Absent	Present	Present
Golgi complex	Absent	Present	Present
Vacuoles	Absent	Present	Present
Lysosomes	Absent	Often absent	Present
Cilia (9 + 2)	Absent	Absent in most	Present in some cells
Flagellum	Often present, unique type	Absent	Present in some cells
Centrioles	Absent	Absent in most	Present

aments, attached to the other side of the plastids, function as tracks along which various organelles move when myosin attached to an organelle pushes on the actin track.

The cytoskeleton is just as important a distinguishing characteristic of eukaryotic cells as is the envelope-bound nucleus or the discrete protein-containing chromosomes (Table 6-1). Without this dynamic network of girders and cables spanning the cell's interior, the larger organelles probably would respond to gravity by sinking toward the bottom of the cell, the cell would not be able to move, and many cellular functions would be impossible. Bacterial prokaryotes lack a cytoskeleton, but because of the minute size of bacteria, diffusion is an adequate means of moving cytoplasmic nutrients, water, and so on; thus the cytoskeleton is not needed for this purpose. Also, bacterial cells have no large organelles, and small ones remain suspended in the cytoplasm as they are stirred about by the random movements of water molecules. In eukaryotes, the activities of the cytoskeleton and the movement of vesicles, vacuoles, and so on may stir the cytoplasm. This motion, together with diffusion, distributes nutrients, ions, and water molecules throughout these larger cells.

CELLULAR MOVEMENTS

Movement is a fundamental feature of life; it is one of the most striking characteristics of living organisms and often is the most obvious sign that an organism is indeed alive. The microtubules and filaments of the cytoskeleton, along with other contractile proteins, enable individual cells and their organelles to move. Bacteria move forward and backward in search of food molecules, propelled by minute "rotors." Sperm swim toward eggs by means of lashing tail movements. An amoeba creeps toward a food source and engulfs it. Within eukaryotic cells, vesicles pinch off from the Golgi complex and shuttle toward the cell membrane, and during cell division, the cell membrane pinches in two as the new cells form.

These same cytoskeletal activities, going on simultaneously within millions of individual cells, enable large organisms to move. Fluids flow quickly in plant cells. Animal muscles contract, moving limbs, eyes, or mouth. The cells of an embryonic animal migrate or divide as the tiny organism grows and takes on the recognizable shape of a rabbit or a tadpole. Most activities of living organisms, whether single-celled or multicellular, would cease were it not for movements of the cytoskeleton and contractile proteins. Let us see how these movements are generated.

Creeping and Gliding Cell Movements

Free-living single-celled amoebae, as well as the great majority of animal cells, will glide and creep along solid surfaces. The capacity to move is such a basic property of cells that even after being immobilized for many months or perhaps years in a human kidney or liver, a cell could activate its locomotor machinery within minutes if freed from constraints. Thus the primitive cellular capacity for movement displayed by free-living amoebae persists even in highly specialized cells of functioning adult tissues.

Creeping and gliding movements take place only at specific times, such as when the cell is migrating in an embryo or a wound. Furthermore, the locomotion is anchorage dependent—it can take place only when there is a solid surface to which the cell can attach. If animal cells are suspended in tissue fluids or culture liquids, they cannot swim; they can begin true locomotion only on contacting and adhering to solid surfaces. Figure 6-18 shows how. At the front of these moving cells, the cell surface protrudes forward in highly dynamic, thin, sheetlike ruffles. A three-dimensional, space-filling network of actin microfilaments is attached to the inner,

(a)

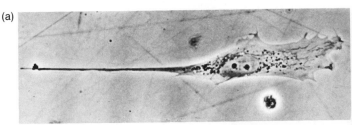

Figure 6-18   CREEPING CELL MOVEMENTS.
(a) A cell from a chick embryo (magnified about 450 times) is stretched lengthwise as it creeps forward on a flat substrate. (b) The edge of a spreading cell (magnified about 10,500 times), such as the epithelial cell drawn in Figure 6-17, ruffles and flutters dynamically, and sends out fingerlike protrusions.

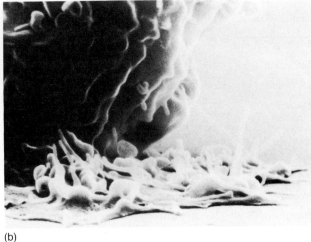

(b)

cytoplasmic side of the ruffling plasma membrane. At the lateral margins of the cell and within the lower cytoplasm, actin microfilaments often make up taut bundles or cables. The ends of the cables may be linked to the less organized space-filling actin network and so indirectly to the inner surface of the plasma membrane; in addition, these cables are somehow stabilized by nearby microtubules. The cell's leading ruffled edge thrusts forward, and with each advance, the lower surface of the protrusion sticks to the substratum. This forward motion applies tension, so that the rear end of the cell is stretched rather like a rubber band. The rear end snaps free, the cytoplasm shifts forward, and the whole process starts again with new protrusions at the front end. Thus the cell moves along somewhat like an inchworm.

If the cell's microtubules are experimentally dispersed by applying a drug, the cell rounds up. The actin cables seem to reorganize into microfilament networks on all sides of the cell, not just at the front. Immediately, the entire surface of the cell protrudes with many active ruffles, making the full periphery a 360-degree "leading edge." In the absence of a single leading edge, the cell is effectively immobilized. Thus both microfilaments and microtubules participate in gliding motion: the former are involved at the leading edge and in snapping the rear forward; the latter immobilize the sides and channel movement in a single direction.

Cells that creep or glide, such as amoebae and animal cells, appear to obey a few rules:

1. They must adhere to a solid surface in order to move.

2. They can be guided by the geometry of the surface on which they move; that is, they can glide or creep along bundles of fibers, much as a squirrel runs along a telephone wire, or they can move along grooves on the surfaces of epithelia or other cell populations. In this way, certain cells in developing embryos migrate from one part of the organism to another, helping to reshape the entire body as it grows and matures.

3. The movement of cells can be turned on or off by internal or external signals. Cell biologists are searching for the chemical signals that activate or halt cell movement.

4. Some cells exhibit **chemotaxis**—the ability to move toward or away from the source of a diffusing chemical. A good example is the tiny soil organism called the slime mold (Chapter 20). Single slime-mold cells creep toward the source of the molecule cyclic-AMP (a derivative of the biological fuel ATP) when that substance diffuses outward from a forming aggregation center. Soon, a large number of cells congregate at the source of cyclic-AMP and begin a new phase of the slime-mold life cycle.

Swimming Cell Movements

While many eukaryotic cells can only creep along solid substrates—even while immersed in tissue fluids—others can swim freely in liquid environments. These include the sperm cells of most multicellular animals and some plants, unicellular organisms such as *Paramecium* (Figure 6-19), spores of some fungi, certain algae, and many types of prokaryotic bacteria. Propulsion in a fluid is accomplished by means of flagella or cilia. **Flagella** (singular, *flagellum*) are fine, whiplike organelles that undulate to move a cell forward or backward. A cell usually has one flagellum or only a few flagella. **Cilia** (singular, *cilium*) are shorter than flagella. One cell may have a

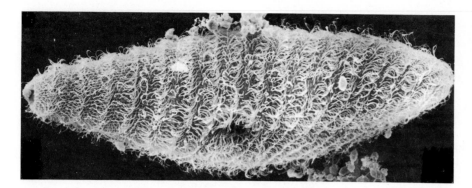

Figure 6-19   *PARAMECIUM.*
This single-celled eukaryotic organism, magnified about 400 times, is propelled by the synchronized beating of its many cilia.

dozen or fewer cilia or as many as a few hundred. Cilia on a single cell usually beat in synchrony with one another.

Aside from length, motion, and number per cell, flagella and cilia of eukaryotes are physically identical. Both protrude from the cell surface and are covered by the plasma membrane. Both have the same internal structure, shown in Figure 6-20: nine pairs of microtubules called *doublets* are arranged in a ring and extend the length of the cilium or flagellum. Two more microtubules run down the center of the doublet ring. Notice that the two central microtubules and the nine outer doublets are interconnected by protein strands arranged much like the spokes of a wagon wheel. A final structural similarity is that every cilium and flagellum grows only from the cell surface at a site where a **basal body** is located. Basal bodies are one of a number of types of microtubule-organizing centers in animal and plant cells. As Figure 6-20 shows, a basal body has nine microtubule triplets (instead of doublets) in a circular arrangement, with no central singlets. Near basal bodies, the tubulin subunits that make up the two central microtubules and the nine doublets of cilia and flagella are assembled.

How do cilia and flagella move and thus enable a cell to swim? Movement is based on tiny side arms that extend from one of the microtubules of each doublet. Each side arm is actually **dynein,** an ATP-cleaving mechano-enzyme. The outer, unattached end of each dynein side arm interacts with the surface of the adjacent microtubular doublet. When dynein binds and splits ATP, energy is released, and the dynein side arms apparently change shape. The change in shape exerts a pushing force against the adjacent microtubular doublet and causes it to slide, just as a person poling a boat pushes against the river bottom with the pole. The movements of the microtubular doublet result in a slight bending of the cilium or flagellum. If thousands of dynein arms are splitting ATP and moving in sequence, the entire cilium or flagellum bends. The central microtubules act like a spring that resists the bending and controls the actual shape of the beating cilia or flagella. Coordination of

dynein arms in a cilium results in undulating or stroking movements that exert force against the fluid and thereby propel the cell forward or backward, just as an eel or a long fish undulates through water. The basic mechanism causing this cellular movement (change in the shape of enzyme side arms when ATP is cleaved) apparently has been adopted at least twice in evolution—once for cilia and flagella and once for muscle. We shall learn in Chapter 39 how myosin in muscle cells also changes shape, assuming a "cocked" position as ATP is split; subsequently, it "fires" and exerts force against nearby actin filaments.

The crucial role of dynein in the swimming movements of cilia and flagella was demonstrated in studies of men of the Maori in Samoa and New Zealand who were unable to impregnate their wives. Careful analysis showed that their sperm had straight tails and thus could not swim and reach eggs for fertilization. Electron-micrographs of sperm tails from those men revealed that the 9 + 2 microtubule arrangement was normal in the sperm flagella but that the dynein arms of the outer doublets were missing. It seems likely that a genetic mutation accounts for the dynein deficiency in some Maori males and in a few other cases reported elsewhere in the world.

While flagella are virtually always involved in cellular swimming, cilia can play a different role: in some tissues, they act to sweep fluids past stationary cells. Thousands of cilia occur on the epithelial cells that line the trachea, the tube that carries air to and from the lungs of many land vertebrates. These cilia beat upward, sweeping mucus, dust particles, and contaminants out of the lungs and throat. Likewise, in the tubes of the human female reproductive tract, cilia beat downward to help propel an egg toward incoming sperm. Interestingly, the sterile Maori males frequently have diseases of the respiratory tract: the dynein deficiency that prevents their sperm from swimming also prevents normal ciliary movements crucial to keeping the lungs healthy and clean. Stationary cells with cilia, though, cannot use their cilia for swimming. If single tracheal epithelial cells are removed

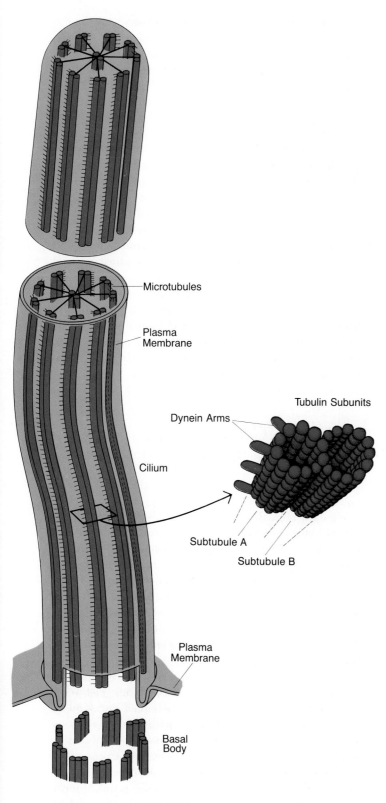

Figure 6-20   INTERNAL STRUCTURE OF CILIA AND FLAGELLA.
The dynein arms push against the adjacent microtubule doublet to cause bending. The two tubules in the center of the cilium or flagellum function as stiff springs.

from an organism and suspended in a fluid medium, they can move only when they settle and begin creeping forward on the substrate, as would most other animal cells without cilia. Clearly, cilia have two distinct roles in different cell types.

Many types of prokaryotic bacteria swim by means of flagella, but these flagella neither have the 9 + 2 internal structure shared by eukaryotic cilia and flagella nor are they covered by the bacterial plasma membrane. Bacterial flagella are true molecular assemblages attached to the cell surface. These cylindrical protein strands do not bend back and forth but rotate in miniature sockets, turning like propellers in a clockwise or counterclockwise direction to move the cell forward or backward. We will learn more about these fascinating protein propellers in Chapter 20.

Internal Cell Movements

All eukaryotic cells, whether mobile or not, need an internal transport system. We have already seen how "eating" and "drinking" vacuoles arise from the cell surface and shuttle inward to join lysosomes for "digestion," and how vesicles pinch off from the Golgi complex and move to different destinations in the cytoplasm. In addition, ribosomes, mRNAs, soluble proteins, and other molecules must be distributed from the nuclear pores or from their sites of manufacture. Nutrients, ions, and building blocks also must be distributed so that an even and constant internal environment can be maintained.

Diffusion plays a role in moving some materials, but cytoskeletal microfilaments and microtubules are responsible for almost all major cytoplasmic movements. For example, in most plant cells, rapidly moving streams of cytoplasm continuously flow around the central fluid-filled vacuole. This movement, called **cytoplasmic streaming,** circulates nutrients, proteins, and other cellular materials. Cytoplasmic streaming occurs because myosin proteins attached to vesicles and other organelles push against microfilaments arrayed through the plant cell, propelling the organelles along the stream.

The movement of organelles inside cells is often important for an organism's survival. In the angelfish, for example, the results of organelle movements in cells are visible to the naked eye, as well as to the eyes of predatory fish or birds. This fish turns from bright red to a less visible pale pink when it is frightened. In a placid fish, nerve impulses trigger pigment granules inside the scale cells to glide outward from a central reservoir along microtubular "railroad tracks" that extend into the long arms of the cells. The dispersed red granules make the cells—and, in turn, the scales of the entire fish—appear red. When a predator is near, however, a different im-

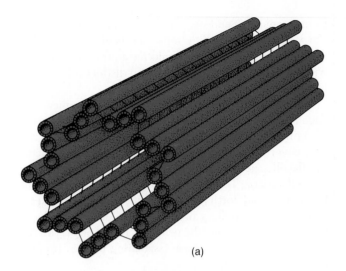

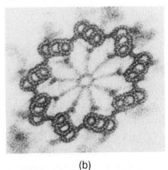

(b)

Figure 6-21   CENTRIOLES.
(a) Centrioles are made up of nine triplets, or groups of three microtubules. (b) This electron micrograph, magnified about 160,000 times, shows a centriole in cross section. Spindle microtubules originate from centrioles. Each time a eukaryotic cell divides, its centrioles are duplicated so that each daughter cell receives a pair. The spindle moves the chromosomes during cell division.

pulse stimulates actin microfilaments to cause the granules to stream back to each cell's central reservoir so that the cell, scale, and fish look pale. Thus the movement of pigment in opposite directions depends on different cytoskeletal components. The means by which pigment granules or other structures move along microtubules has been studied using the single microtubules isolated from nerve cells described earlier. Even as such tubules glide about, particles chancing to alight on the tubules are swept in one direction or the other along microtracks on the tubule surface. This is due to the mechanoenzyme *kinesin*, which cleaves ATP for the energy needed to move the particles or the gliding tubule itself.

Division of animal cells also depends on both actin microfilaments and microtubules. Actin actually helps pinch the dividing cell in two (Chapter 9). Microtubules of the spindle move the chromosomes. These microtubules are assembled from tubulin subunits near organ-

elles called **centrioles,** which occur in pairs near the nuclear envelope. Like the basal bodies of flagella, centrioles are composed of nine microtubular triplets and ill-defined but important substances that seed the assembly of microtubules (Figure 6-21). The role of centrioles can be observed if they are isolated from an animal cell and added to a solution of tubulin subunits: spectacular elongation of microtubules occurs outward in all directions from the centriolar pair. Nevertheless, we still do not know precisely how basal bodies and centrioles function in the assembly of microtubules. The answer may prove valuable in the control of division by cancer cells or in the production of a male contraceptive that would prevent the development of sperm tails.

Most animal cells have centrioles, basal bodies, and spindles (and many can form cilia). In contrast, cells of higher plants usually lack centrioles, although they do have microtubule-organizing centers that facilitate the polymerization of spindle or cytoplasmic microtubules from tubulin. They also lack basal bodies and the cilia that grow from them. Biologists have not yet been able to explain why animal cells have centrioles and basal bodies, while higher plant cells do not.

BRIDGING THE GAP BETWEEN MOLECULES AND LIFE

We have seen that movement is a fundamental property of living cells made possible in prokaryotes and eukaryotes by assemblies of special molecules. The action of the filaments and microtubules that make up the eukaryotic cytoskeleton shuttles nutrients, secretory products, and other materials to and from the cell surface in vesicles and vacuoles. Cytoskeletal action moves the cell surface by means of protrusions, contractions, and adhesions. It causes the bending and lashing of cilia and flagella, and it stirs vital nutrients, ions, and molecules within the cell. The architecture and activities of the cytoskeleton vary among different cell types, just as do the other organelles. The accentuation of one kind of organelle or another is associated with special functions of certain cells and, in turn, of the tissues and organs in which those cells reside. Muscle cells have exaggerated cytoskeletal elements and can contract; pancreatic cells have prominent RER, Golgi complexes, and storage granules and so can secrete digestive enzymes; plant cells have plastids that enable them to carry out photosynthesis. The organized aggregates of macromolecules that form the cytoskeleton and the cytoplasmic organelles are not themselves alive, but their coordinated activity contributes to the ephemeral state we call life. So too, other organelles that synthesize proteins, engulf food particles, or excrete waste molecules bridge the gap between molecules and life. When all are enclosed in the lipid-bilayer plasma membrane—with its permeability, transport capabilities, and fluid properties—and when hereditary material is present, the combination comprises the incredibly organized "factory" we call a cell, imbued with the state we call life.

SUMMARY

1. The *cytoplasm*, a semifluid substance bounded by the plasma membrane, contains organelles and molecules that are responsible for metabolism and many cell functions.

2. The organelles of eukaryotic cells include the nucleus, nucleoli, ribosomes, endoplasmic reticulum, Golgi complex, vacuoles, lysosomes, mitochondria, plastids in plant cells, cilia, flagella, and elements of the cytoskeleton.

3. The *nucleus* of eukaryotic cells is delimited by the double-layered *nuclear envelope*, is the repository of genetic material (DNA) located on *chromosomes*, and is the primary center for the synthesis of RNA. *Nucleoli* are sites of ribosome manufacture in nuclei. Prokaryotes lack nuclei; their hereditary material is found in the *nucleoid*.

4. *Ribosomes*, which are composed of two subunits of RNA and structural proteins, are the sites of protein synthesis. Ribosomes move along the length of an mRNA molecule, reading the sequence of mRNA components and translating them into protein. One mRNA molecule plus a number of attached ribosomes is called a *polysome*.

5. Polysomes that float freely in the cytoplasm are sites of synthesis for many kinds of cellular proteins, which are released into the cytoplasm. But two classes of proteins are made on membrane-bound polysomes: exportable proteins and membrane proteins.

6. The *endoplasmic reticulum*—an array of sacs, tubules, and vesicles—can be rough or smooth. The rough endoplasmic reticulum (RER), whose outer surface is studded with polysomes, serves as a site of protein synthesis and modification, and a transport network for exportable proteins or proteins that must be transported to specific sites in the cell. It seems also to give rise to the nuclear envelope after cells divide. Smooth endoplasmic reticulum (SER) lacks attached ribosomes and is involved in the transport and synthesis of fats and steroids and the detoxification of certain harmful molecules.

7. The *Golgi complex* is the site at which proteins, fats, steroids, and (in plants) complex carbohydrates, made and passed through the RER and SER, are further modified and then packaged for export or for delivery elsewhere in the cytoplasm.

8. *Vacuoles* can store engulfed fluids and nutrients or discharge fluids or wastes. Vacuolar feeding is a form of *phagocytosis*, or engulfment of particles, while vacuolar drinking is termed *pinocytosis*. Taking things into cells is also called *endocytosis*, and expelling things, *exocytosis*.

9. *Lysosomes* contain a variety of digestive enzymes that are used to degrade both materials ingested by the cell and cellular debris. Cellular waste products that cannot be degraded accumulate in *lipofuscin granules*. Eukaryotic cells also contain other enzyme-containing vesicles called *microbodies*, which are detoxifiers.

10. *Mitochondria* are self-replicating organelles found in all eukaryotic cells; they generate ATP, the common high-energy compound of all cells. Each mitochondrion has a smooth, continuous outer membrane and an inner membrane thrown into folds called cristae, which provide a large surface area for the enzymes involved in ATP metabolism. The cristae protrude into a central region, the *matrix*. Mitochondrial functioning is directed by two sets of DNA: the cell's nuclear DNA and mitochondrial DNA, which is unlike eukaryotic or prokaryotic chromosomal DNA.

11. *Plastids* are sites of nutrient generation by photosynthesis in plant cells and of storage of a variety of nutrients and pigments. Plastids with pigments, such as *chloroplasts* (the organelles in which photosynthesis takes place), are termed chromoplasts; plastids without pigments are termed leucoplasts. Both types arise from proplastids, contain DNA, and are self-replicating. Chloroplasts have stacks of membranes called *grana* that reside in the *stroma*, the site of enzymes and DNA.

12. The eukaryotic cell's *cytoskeleton* is composed of *microfilaments*, *microtubules*, and *intermediate filaments;* it is this highly dynamic set of structures that gives the cell its shape. The cytoskeleton is also a scaffolding to which many cell organelles are attached, and it is involved in a variety of cell movements. Microfilaments are composed of the protein *actin*. Microtubules are composed of the globular protein *tubulin*. *Mechanoenzymes* such as myosin, dynein, and kinesin interact

with the cellular filaments and tubules to generate forces that cause movements.

13. Microfilaments and microtubules and their associated proteins are responsible for the creeping and swimming movements of whole cells and for most internal cell movements. Cells may be guided by the shape of the surface on which they creep, or by diffusable chemicals (*chemotaxis*).

14. Swimming cell movements are accomplished by means of *flagella* or *cilia*. Both have the same internal structure: nine microtubular doublets surrounding two microtubules, connected by protein strands in a spokelike arrangement. *Basal bodies* serve as organizing centers for flagella and cilia. Movement of flagella and cilia involves the binding and splitting of ATP by *dynein* side arms on the doublets.

15. Internal cell movements may take the form of *cytoplasmic streaming*, as in plant cells, or of other means of distributing molecules, nutrients, and so on throughout the cell by action of various cytoskeleton components. Cell division also depends on microtubules and microfilaments. In animal cells, microtubules of the chromosomal spindle are assembled near organizing centers called *centrioles*.

KEY TERMS

actin	exocytosis	myosin
basal body	flagellum	nuclear envelope
centriole	gene	nucleoid
chemotaxis	Golgi complex	nucleolus
chloroplast	intermediate filament	nucleus
chromosome	lipofuscin granule	phagocytosis
cilium	lysosome	pinocytosis
cytoplasm	matrix	plastid
cytoplasmic streaming	mechanoenzyme	polysome
cytoskeleton	microbody	ribosome
dynein	microfilament	stroma
endocytosis	microtubule	tubulin
endoplasmic reticulum (ER)	mitochondrion	vacuole

QUESTIONS

1. Draw typical plant and animal cells, including their major organelles. Label the parts.

2. What are the parts and contents of a eukaryotic cell's nucleus?

3. What are the similarities and dissimilarities of the rough endoplasmic reticulum and free polysomes?

4. How are the RER and the Golgi complex related in structure and in function?

5. Various terms describe the taking in or passing out of materials from eukaryotic cells. List and define the terms. Which occur in the wall of one of your blood capillaries?

6. Cells would be in grave difficulty if a mutation eliminated their lysosomes. Why?

7. Draw a mitochondrion and a chloroplast, label their parts, and then identify similarities and dissimilarities between them.

8. The cytoskeleton has structures and functions. What are they? How are mechanoenzymes related to the structural components of the cytoskeleton?

9. How do cells creep, swim, or stream internally? What would adding an ATP poison do to such processes? Why?

ESSAY QUESTIONS

1. Why is contact with a solid surface important to basic cell functions?

2. How are vacuoles and various types of phagocytosis related to the transport of ions and other substances across the plasma membrane described in Chapter 5?

3. Recall the various types of junctions between cells in Chapter 5. What processes would need to go on if cell A synthesizes a small protein and (1) excretes it so it can be picked up by cell B; or (2) creeps to cell B and passes it directly?

SUGGESTED READINGS

ALBERTS, B., et al. *Molecular Biology of the Cell.* New York: Garland, 1983.

The clearest account of cellular organelles from the functional point of view.

Journal of Cell Biology 91 (1981).

This classic issue of the leading scientific journal in cell biology includes reviews of the critical literature on most of the topics in this chapter.

KARP, G. *Cell Biology.* New York: McGraw-Hill, 1979.

A readable, well-illustrated account of organelle structure and function.

7

HARVESTING THE ENERGY STORED IN NUTRIENTS: FERMENTATION AND CELLULAR RESPIRATION

A living cell, like Lewis Carroll's Red Queen, has to run at top speed to stay in the same place. Without a constant input of energy, either from an outside source or from its own storage reservoirs, it will die.

Richard E. Dickerson, *Scientific American* (March 1980)

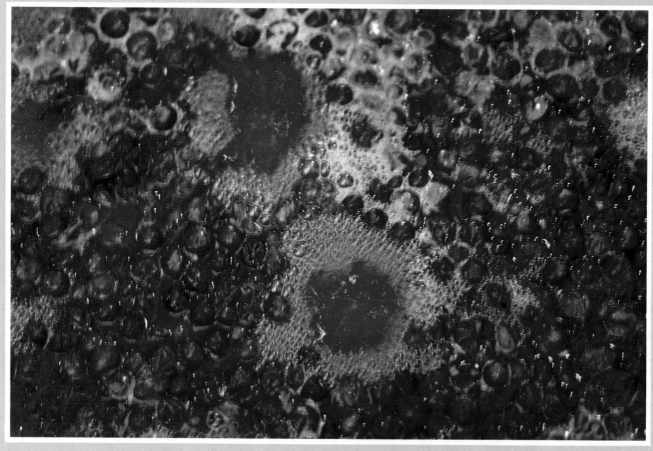

Fermentation: crushed grapes, frothing fluid, and yeast at work in an early stage of wine production.

Every second in the life of a cell, nutrients are being broken down to release energy. Much of the energy is "invested" in a molecular currency called ATP, which can be "spent" to construct and maintain the cell's membranes, mitochondria, ribosomes, and other intricate structures, and to fuel the cell's myriad activities. The simultaneous, integrated actions of enzymes and organelles allow the entire cell to maintain itself, grow, replace worn structures, process wastes, and perhaps reproduce and move. But the total energy costs are large and continuous; indeed, like the Red Queen in *Through the Looking-Glass*, the cell must "run" to stay in one place—if not at full speed, then at least at maintenance speed.

How do cells harness the energy they need to pay these high costs? The answer lies in the general flow of energy we discussed in Chapter 4: energy from the sun is stored in nutrient molecules and then is released by the metabolism of living cells and "spent" for survival. Solar energy is trapped by plants and other autotrophic organisms, and is then converted and stored in the chemical bonds of carbohydrate molecules. This is the process of *photosynthesis*, which supports virtually all life on Earth, either directly or indirectly. Carbohydrates are a temporary repository of transformed solar energy ready to be tapped by the cells of photosynthetic organisms themselves or by the cells of nonphotosynthesizers such as animals, fungi, and microbes, which consume plant matter or one another to stay alive (Figure 7-1).

The subject of this chapter is the way that energy is tapped. We will see that it requires a series of chemical reactions—facilitated by enzymes—during which the bonds of the carbohydrate molecules are broken, and the energy is released. As we saw in Chapter 4, such energy-releasing reactions are *exergonic*, and, significantly, they power the energy-consuming *endergonic* reactions that underlie all the cellular activity described in earlier chapters—movement, maintenance, protein synthesis, ion pumping, and so on. The metabolic breakdown of nutrient molecules through a chain of mostly exergonic reactions is called *catabolism*. Much of the energy freed during catabolism is stored as ATP and other energy intermediates and forms a reservoir that is continuously replenished by this metabolic process and simultaneously depleted by the cell's activities.

Since the flow of energy is from the sun to nutrient molecules to metabolic breakdown of those nutrients, it might seem logical to study photosynthesis first and nutrient breakdown second. However, catabolism is universal, and photosynthesis is not: all living things break down nutrients to harvest life-sustaining energy, whereas only certain kinds of cells can carry out the photosynthetic process of building carbohydrates. Moreover, prokaryotic cells on the ancient Earth broke down

Figure 7-1   RICE: A STORE OF THE SUN'S ENERGY IN CARBOHYDRATE FORM.
Every animal is ultimately dependent on such plant food for the energy needed to manufacture ATP.

nutrients and survived for hundreds of millions of years before photosynthesis evolved. For these reasons, we will study the processes of catabolism and energy storage in this chapter and the more specialized phenomenon of photosynthesis in Chapter 8.

Our first task in this chapter is to understand ATP, the cell's principal energy intermediate and the chemical currency on which the cell's energy economy is largely based. We will go on to discuss oxidation and reduction reactions, which are integral to the metabolic pathways and electron bucket brigades involved in breaking down nutrients and forming ATP. With that as background, we will consider in detail the most fundamental metabolic chain, *glycolysis*, and see how a nutrient such as glucose is broken down step by step. We then consider how the product of glycolysis is further broken down either in the absence of molecular oxygen—a process called *fermentation*—or in the presence of oxygen—the process of *cellular respiration*.

The principle behind all these steps is that the energy in glucose bonds is not liberated in a single step, since the heat of a sudden, highly exergonic reaction might "burn up" the cell. Instead, energy is released from nutrients gradually in the precise sequence of steps we shall discuss. A second fundamental principle of cellular energy processing is that structure and function are strongly linked: the various chains of metabolic reactions take place in specific parts of cells, and the sequence that generates the most ATP occurs on only certain mitochondrial membranes or on the plasma membranes of prokaryotic cells. These high-energy yields are dependent on membrane structure and, in turn, make possible high-level activity that most cells and organisms could not sustain with less efficient harvests of ATP. Let us turn to that critically important energy intermediate.

ATP: THE CELL'S ENERGY CURRENCY

Cells work constantly to maintain a rich supply of the energy-storage molecule **adenosine triphosphate,** or **ATP.** At any one instant, a typical animal or plant cell may have 10 billion molecules of ATP dissolved in its cytoplasm (Figure 7-2). The energy stored in this relatively simple molecule is released when the molecule is cleaved into the related compound *adenosine diphosphate,* or *ADP,* plus inorganic phosphate. Instead of being completely lost as heat, some of the energy released during this reaction is conserved by being used to drive various endergonic reactions (Figure 7-3). Because the energy demands of a cell are very high, ATP is continuously in flux—some being synthesized, some being destroyed—so the ATP "pool" is truly a dynamic one.

The Structure of ATP

The ability of ATP to store and release energy stems from the molecule's structure, which is pictured in Figure 7-4. Notice that ATP, ADP, and *AMP (adenosine monophosphate)* are nucleotides, the building blocks of nucleic acids, which we discussed in Chapter 3. The

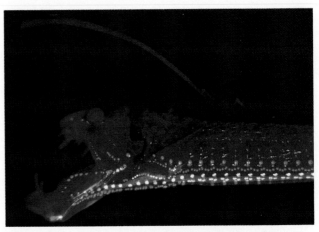

Figure 7-2   LUMINESCENCE IN DEEP-WATER FISH: LIGHT FROM ATP.
The cells of every organism contain millions of ATP molecules, with enormous numbers being manufactured and just as many being consumed every second of every day. Here Sloan's viperfish emits its luminescent communication display, consuming much ATP to do so.

double ringlike portion of all three molecules is the purine base adenine; the five-sided ring is the sugar ribose; and the tail is made up of phosphate groups—three groups in ATP, two in ADP, and one in AMP. Each phosphate group is represented by the symbol Ⓟ.

Figure 7-3   THE "MAKING" AND "SPENDING" OF ATP.
Glycolysis, cellular respiration, fermentation, and photosynthesis are processes that manufacture ATP. Cells use this ATP to fuel vital activities, including the synthesis of molecules, active transport of substances across cell membranes, cell division, secretory activities, and movement.

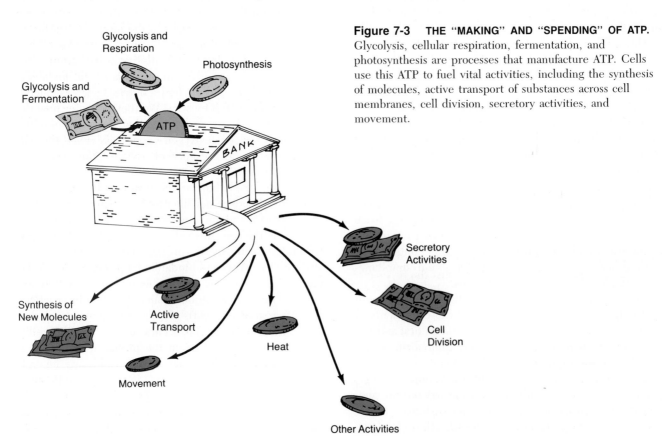

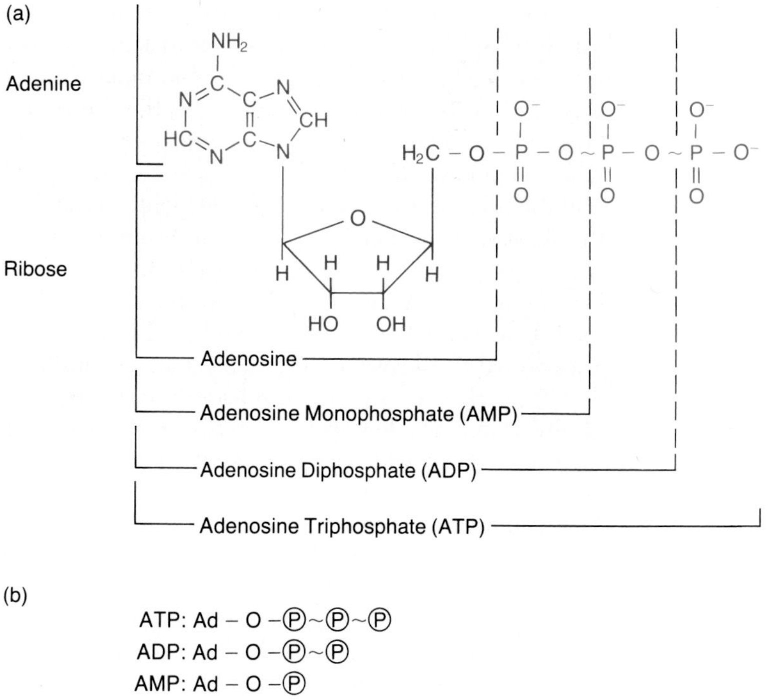

(a)

Adenine

Ribose

Adenosine

Adenosine Monophosphate (AMP)

Adenosine Diphosphate (ADP)

Adenosine Triphosphate (ATP)

(b)

ATP: Ad — O —(P)~(P)~(P)
ADP: Ad — O —(P)~(P)
AMP: Ad — O —(P)

Figure 7-4   THE STRUCTURE OF ATP.
(a) ATP consists of the purine base adenine, the sugar ribose, and three phosphate groups. ADP has two phosphate groups, and AMP has one phosphate group. The high-energy bonds between the phosphate groups are designated by wavy lines. (b) Adenosine is abbreviated "Ad," and each phosphate group is symbolized as (P).

The energy-storage function of ATP is associated with the tail of phosphate groups and with the interconversion of ATP, ADP, and AMP. To add phosphate groups to AMP or ADP requires energy; conversely, as phosphate groups are removed from ATP or ADP and transferred to other compounds, energy is released.

It may seem backward, but we can best understand the synthesis of ATP if we first understand its breakdown. If the appropriate enzyme is present, the terminal phosphate group of an ATP molecule can be transferred to a variety of other compounds. The transfer of any phosphate group is called **phosphorylation.** The transfer of the terminal phosphate group from ATP to a water molecule takes place as follows:

Ad–O–(P)–(P)–(P) + HO–H →
    (ATP)   Ad–O–(P)–(P) + HO–(P) + H$^+$ + heat

                    (ADP)

This reaction yields ADP plus HO–(P), or inorganic phosphate (symbolized by P_i). In fact, just the reverse of this reaction is the most common phosphorylation reaction in cells: an inorganic phosphate group is *added to* the ADP molecule, forming ATP:

HO–(P) + Ad–O–(P)–(P) + energy → ATP + H_2O

This reaction requires about 8 kcal/mole, or the same amount of energy released when ATP is split. How efficient is this synthesis of ATP? The rule of thumb is that about half the energy given off during exergonic reac-

tions can be stored in ATP molecules via phosphorylation.

As we saw in Chapter 3, the cleaving of a molecule in the presence of water is a *hydrolysis* reaction. Therefore, the same reaction can be called either the hydrolysis of ATP or the phosphorylation of water. The key here is that *this reaction releases a large amount of energy*— about 8 kcal for each mole of ATP cleaved. In the hydrolysis of ATP to ADP, the energy is released as heat. However, cells usually are not so wasteful; in many instances, the energy released when ATP is hydrolyzed is stored in other energy-intermediate molecules and used to power subsequent biological processes.

If the ADP formed during the hydrolysis of ATP is itself hydrolyzed and the second phosphate group is cleaved so that AMP remains, an equivalent amount of energy is released—approximately 8 kcal/mole. However, if the remaining AMP is now hydrolyzed and the third phosphate group is transferred to another molecule, much less energy is liberated—only 2 kcal/mole. Since an ATP molecule's first two phosphate groups release so much free energy when cleaved, they are called *high-energy groups,* and the bonds linking them are known as *high-energy phosphate bonds.* These bonds usually are designated by a wavy line:

Ad–O–(P)~(P)~(P)
(ATP)

While the ~ symbol is useful, keep in mind that the energy is not localized in the bond itself; it is a property of the entire molecule and is simply released as the phosphate bond is broken.

ATP and the Harvesting of Energy

The energy released by the cleavage of ATP to ADP drives the vast majority of endergonic biological reactions in cells. As ATP is continuously cleaved to drive life processes, new ATP molecules must be formed. On the average, each of the 20 to 30 trillion cells in the human body cleaves from 1 to 2 billion ATP molecules into ADP every minute—the equivalent of about 90 pounds (40 kg) of ATP a day! No person could possibly eat enough food to allow complete synthesis of this much ATP. So both ADP and AMP molecules are continuously recycled back into ATP.

Now let us address directly the way ATP can be used to drive typical endergonic reactions. Consider the synthesis of sucrose, a reaction in which a significant amount of energy—about 7 kcal/mole of sucrose formed—is required for enzymes to catalyze the synthesis of this disaccharide from the monosaccharides glucose and fructose. In order for such synthetic reactions to proceed spontaneously at biologically reasonable concentrations of all starting materials (recall our discussion of reactant concentrations; Chapter 4), energy is required. It is ATP that supplies this energy. Remember that the appropriate enzymes merely speed the reactions; enzymes cannot bring about energetically unfavorable reactions.

Basically, the synthesis of sucrose involves two steps, as shown in Figure 7-5. First, glucose is phosphorylated; that is, the terminal phosphate group of an ATP molecule is transferred to glucose, yielding ADP and glucose-1-phosphate. During the phosphorylation process, some of the chemical energy of ATP is stored in the new glucose-1-phosphate molecule. Next, if the proper enzyme is present, glucose-1-phosphate and fructose react to form sucrose and inorganic phosphate. In these two reactions, glucose-1-phosphate is a *common intermediate*: it is a product of the first reaction and a reactant in the second. By means of this common intermediate, energy derived from ATP is conserved in a form that can successfully drive the second, endergonic reaction. Common intermediates are always involved in biological reactions driven by ATP energy; they are the means by which the molecular-energy currency of all cells is "spent" to sustain life.

But the process is not 100 percent efficient. Recall that only about half the energy released in exergonic reactions is channeled into the phosphorylation of ADP to ATP. The remaining energy will be wasted as heat. The same is true when ATP is later cleaved, as the phosphate is transferred to glucose or other molecules. Again, heat is lost. This cleavage of ATP to release heat can be biologically useful. In a few special cases, enzymes in living organisms bring about the cleavage of ATP to produce heat. For example, the blood of newborn rabbits and other mammals is warmed by flowing through special regions of fat tissue in which the splitting of ATP and other heat-producing reactions take place. In general, however, heat is not a reason for ATP cleavage, nor is it retained in most organisms or cells. Instead, the inevitable freeing and loss of heat during ATP cleavage is just an inefficiency, an extra "cost" of remaining alive.

OXIDATION-REDUCTION REACTIONS

We have seen that ATP is formed by the phosphorylation of ADP and that this conversion requires energy. But exactly how is energy from nutrients channeled into ATP? The answer is **oxidation-reduction reactions.** In fact, all key metabolic reactions in cells involve **oxidation** (the removal of electrons from an atom or a compound) and **reduction** (the addition of electrons). These reactions always occur simultaneously, so that one partner molecule is oxidized and the other, reduced (Figure 7-6a). The flow of electrons from one molecule to the other can be thought of as a current that carries energy.

Glucose + Ad − O − (P) ~ (P) ~ (P) $\longrightarrow$ Glucose − 1 − (P) + Ad − O − (P) ~ (P)

Glucose − 1 − (P) + Fructose $\longrightarrow$ Sucrose + P_i

Figure 7-5   THE SYNTHESIS OF SUCROSE AND THE "SPENDING" OF AN ATP.
The terminal phosphate group of the ATP molecule first is transferred to glucose. Some of the energy of ATP now is in glucose-1-phosphate, and the rest is in the one remaining high-energy bond of the ADP molecule. Glucose-1-phosphate and fructose then form the disaccharide sucrose, with inorganic phosphate left over. These two steps have a common intermediate, glucose-1-phosphate.

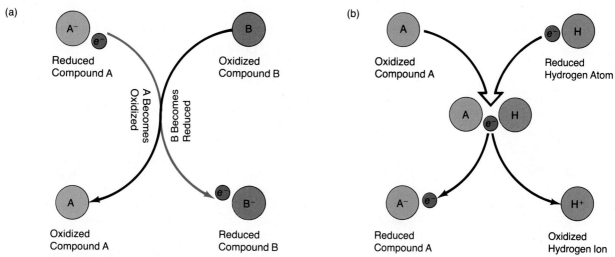

Figure 7-6   OXIDATION-REDUCTION.
Oxidation-reduction reactions involve a flow of electrons. Oxidation is the removal of electrons from an atom or a molecule, and reduction is the addition of electrons. (a) As compound A is oxidized, compound B is reduced. These processes always are paired, and the flow of electrons follows the shaded pathway. (b) The transfer of a hydrogen atom can be the equivalent of the transfer of electrons. The hydrogen atom shares its one electron with compound A. On the dissociation of A and H, a free hydrogen ion (H^+) forms as the electron remains with A. That compound has been reduced, while hydrogen has been oxidized.

The flow moves from an area of higher free energy to one of lower free energy. In biological oxidations and reductions, as well as in other reactions, the change in free energy is expressed in kilocalories.

In many biological oxidation-reduction reactions, the electrons are transferred in the form of hydrogen atoms, which consist of one proton (H^+) and one electron (e^-) (Figure 7-6b). Like the transfer of electrons, the transfer of hydrogen atoms takes place in paired reactions. An oxidation reaction that involves the removal of a hydrogen atom is called *dehydrogenation*, and the concomitant reduction reaction, which involves the addition of a hydrogen atom, is known as *hydrogenation*.

Two particularly important coenzymes serve as electron carriers in many metabolic oxidations and reductions; recall that coenzymes are molecules that must be present for an enzyme to catalyze a reaction (Chapter 4). They are **NAD+** (nicotinamide adenine dinucleotide) and **FAD** (flavin adenine dinucleotide), which transfer electrons to a chainlike series of reactions that provides much of a cell's energy supply of ATP. The structures of NAD^+ and FAD are shown in Figures 7-7 and 7-8. The oxidation and reduction of NAD^+ occurs in its nicotinamide group. The reduction of NAD^+ can be written as follows:

$$NAD^+ + H^+ + 2e^- \rightarrow NADH$$

During the reduction of NAD^+ to NADH, two electrons are added but only one proton is (Figure 7-7b). During

the reduction of FAD to $FADH_2$, two protons (and two electrons) are transferred (Figure 7-8b). Thus NAD^+ participates only in reactions in which one hydrogen atom is transferred, while FAD can participate in exchanges of two hydrogen atoms. As coenzymes, NAD^+, NADH, FAD, and $FADH_2$ are loosely bound to enzymes called *dehydrogenases*, which are dissolved in the cell's cytoplasm.

These coenzymes are important because, as we shall soon see, they carry electrons and hydrogen away from the metabolic chains of glycolysis and the Krebs cycle. Thus they serve as energy intermediates that feed electrons and hydrogen into marvelously organized molecular machinery embedded in mitochondrial membranes. There the hydrogen is used to drive the synthesis of ATP, the "currency" that runs the energy economy of cells.

So far, we have examined some of the molecules that play major roles in the transfer of energy from nutrient molecules to ATP, and from ATP to other substances. Now it is time to explore the pathways themselves. The starting point is glycolysis, the initial breakdown of glucose, a molecule that is synthesized in photosynthetic plants as well as in other cells. Glycolysis is a series of steps that both generates a modest amount of ATP and feeds products into either fermentation or cellular-respiration pathways. As we study glycolysis, the roles of coenzymes, oxidation-reduction reactions, and energy intermediates will become clear.

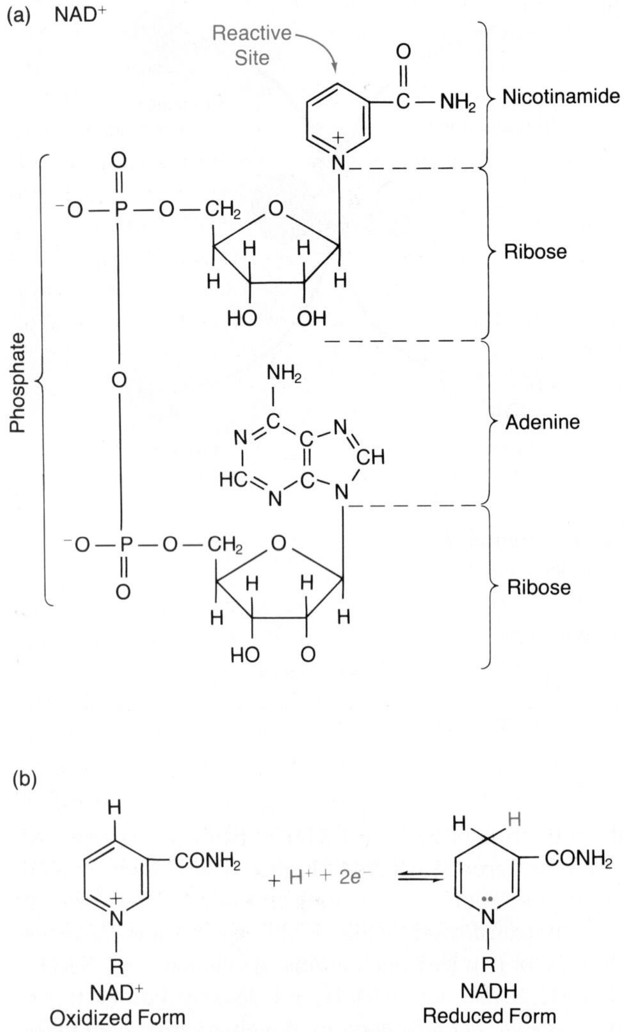

(a)   NAD$^+$

(b)

Figure 7-7   THE ELECTRON CARRIER NAD$^+$.
(a) The oxidized form of NAD$^+$. The reactive site is labeled in blue. (b) The reduction of NAD$^+$ to NADH. Only the nicotinamide group is shown; the rest of the molecule is designated "R." The parts involved in the reaction are shown in blue.

GLYCOLYSIS: THE FIRST PHASE OF ENERGY METABOLISM

Glycolysis is the initial sequence of reactions used by virtually all cells to break six-carbon glucose molecules into two molecules of the three-carbon compound **pyruvate.** Glycolysis is the best understood metabolic pathway. It serves as a model for the stepwise build-up and breakdown of most organic and biological molecules and for the reaction pathways that underlie most biological activities.

Figure 7-9 provides an overview of glycolysis and shows how it fits into the entire process of cellular metabolism. The steps in these pathways at which ATP is generated are indicated. Note that glycolysis and fer-

Figure 7-8   THE ELECTRON CARRIER FAD.
(a) The oxidized form of FAD. (b) The reduction of FAD to FADH$_2$. Only the isoalloxazine group (the reactive site) is shown; the rest is designated "R." The parts involved in the reaction are shown in blue.

(a)
FAD

(b)

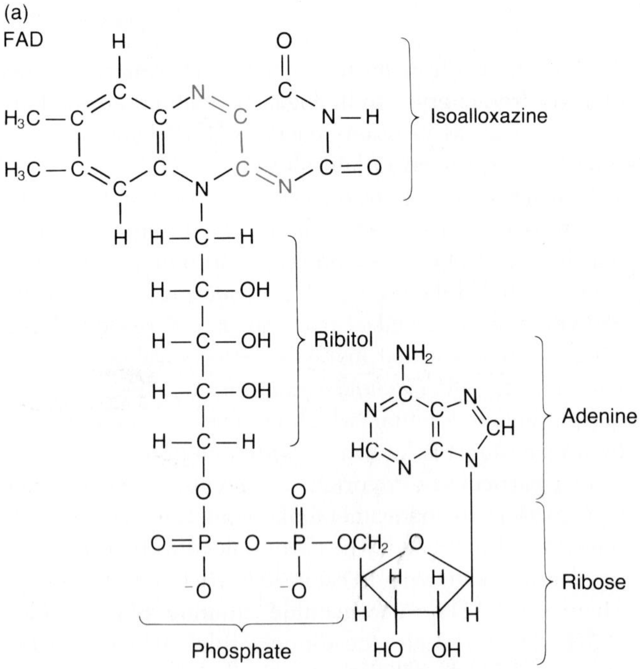

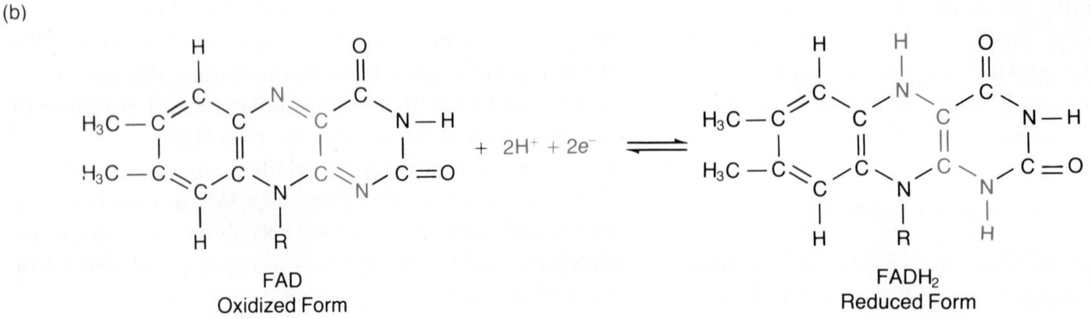

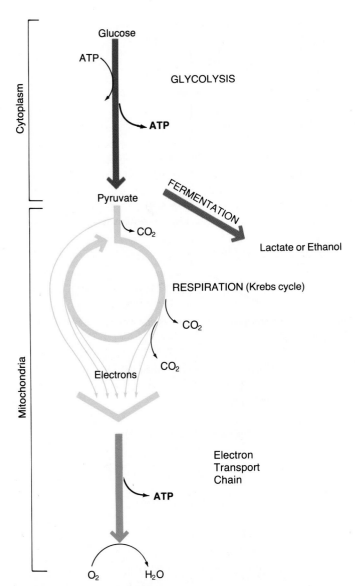

Figure 7-9   AN OVERVIEW OF CELLULAR ENERGY HARVEST: GLYCOLYSIS, FERMENTATION, AND CELLULAR RESPIRATION.
Glycolysis takes place in the cytoplasm; it begins with the breakdown of glucose and ends with pyruvate. The first step in the cellular respiration cycle, or Krebs cycle, is the conversion of pyruvate to a high-energy compound. In the absence of sufficient oxygen in the cell, however, pyruvate will undergo fermentation instead of entering the Krebs cycle.

The Krebs cycle removes electrons from the intermediates in the cycle and passes them to the electron transport chain. Both of these processes take place in the mitochondria. The electron transport chain produces most of the ATP generated from a molecule of glucose.

mentation take place in one part of the cell, and aerobic respiration, in another part. Glycolysis and fermentation occur in the cytoplasm because the enzymes that bring about each reaction are dissolved in the watery cell fluid and are not localized inside membranous organelles. Respiration takes place in the mitochondria in eukaryotic cells and along the inner surface of the plasma membrane in aerobic prokaryotes. The architecture of the mitochondrion, the "powerhouse organelle" in which most ATP is synthesized in eukaryotes, is closely associated with important mechanisms of cellular respiration and provides an excellent example of the relationship between structure and function in biology.

Splitting Glucose: The Steps of Glycolysis

Glycolysis is a prelude and, indeed, a part of both fermentation and respiration. Thus it is the basis for energy metabolism in virtually all living creatures.

During glycolysis, the six-carbon framework of a glucose molecule is broken down to two molecules of pyruvate in a series of nine reaction steps, which are outlined in Figure 7-10. Also illustrated are the structures of all the intermediate compounds formed along the way. You may not be required to memorize every step in the pathway in its entirety, but understanding the basic features of the sequence will make it easier to understand cellular metabolism in general.

Step 1

The initiation of glycolysis requires a high-energy form of glucose. Thus in step 1, a molecule of ATP is cleaved, forming both ADP—when its terminal phosphate group is transferred to the sixth carbon of the glucose molecule—and the *activated* molecule glucose-6-phosphate. Note that a specific enzyme (hexokinase) catalyzes this reaction, just as specific enzymes catalyze *each* step in a metabolic pathway.

Steps 2 and 3

In step 2, the glucose-6-phosphate molecule undergoes a change in structure and is converted to its close relative, fructose-6-phosphate. During Step 3, a molecule of ATP is cleaved, as in step 1, and its terminal phosphate group is transferred to carbon number 1, at the other end of the sugar carbon chain. We now have a molecule of fructose-1,6-diphosphate, which has been generated at the "cost" of two ATPs. This initial energy utilization in steps 1 and 3 of glycolysis is like making an investment—it provides the chemical energy "capital"

Figure 7-10   THE GLYCOLYTIC PATHWAY.
The sequence of the pathway (a) in simplified form and (b) with structural formulas. Note that four ATPs are produced (steps 6 and 9) and two ATPs are used up (steps 1 and 3) for a net gain of two ATPs. Two NADHs are produced (step 5) and are used to make more ATP in the electron transport chain, a sequence of reactions described later in this chapter. In step 4, fructose-1,6-diphosphate is broken into two molecules: dihydroxyacetone phosphate and glyceraldehyde-3-phosphate (PGAL). The dihydroxyacetone phosphate is immediately converted to glyceraldehyde-3-phosphate, resulting in two PGALs. Both PGALs go through the rest of the pathway, resulting in two pyruvate molecules for each glucose molecule broken down.

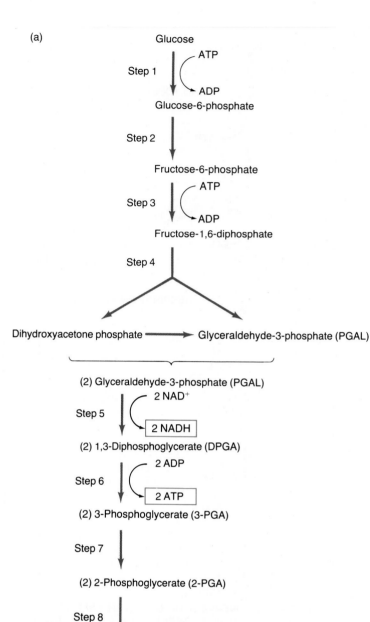

that will ultimately result in a net energy gain by the cell.

Step 4

Fructose-1,6-diphosphate is split into two similar three-carbon molecules during step 4. One is PGAL (glyceraldehyde-3-phosphate), and the other is dihydroxyacetone phosphate, which is converted immediately into another molecule of PGAL. From this point on, each step in the pathway must take place twice—once for each PGAL derived from the original glucose molecule.

Step 5

Steps 5 and 6 are the key steps of glycolysis. During step 5, PGAL is both oxidized and phosphorylated. The aldehyde group $-\text{C} \overset{H}{\underset{O}{\Big<}}$ is oxidized as NAD^+ accepts two electrons in the form of hydrogen atoms; NAD^+ becomes the reduced compound NADH, while the oxidized PGAL reacts with phosphate. A great deal of energy is released during the exergonic oxidation reaction involving PGAL and NAD^+; that energy is trapped immediately as a phosphate group joins PGAL to produce the high-energy molecule 1,3-diphosphoglycerate, or DPGA:

$$\text{PGAL} + \text{NAD}^+ + \text{HO-} \textcircled{P} \rightarrow \text{DPGA} + \text{NADH}$$

Note that the phosphate needed to generate DPGA, shown in red in Figure 7-10b, comes from inorganic phosphate (P_i) dissolved in the cell's cytoplasm, and not from ATP.

Step 6

In step 6, both 3-phosphoglycerate (3-PGA) and two molecules of ATP are formed as one phosphate group

from each DPGA phosphorylates an ADP. This is possible because DPGA has an even higher energy content than does ATP. This last reaction "pays back" the earlier energy investment that set the stage for the exergonic reactions of glycolysis. The overall equation for steps 1 through 6 is:

$$\text{Glucose} + 2\text{ATP} + 2\text{HO-} \textcircled{P} + 2\text{NAD}^+ \rightarrow$$
$$2(3\text{-PGA}) + 2\text{ATP} + 2\text{NADH}$$

(b)

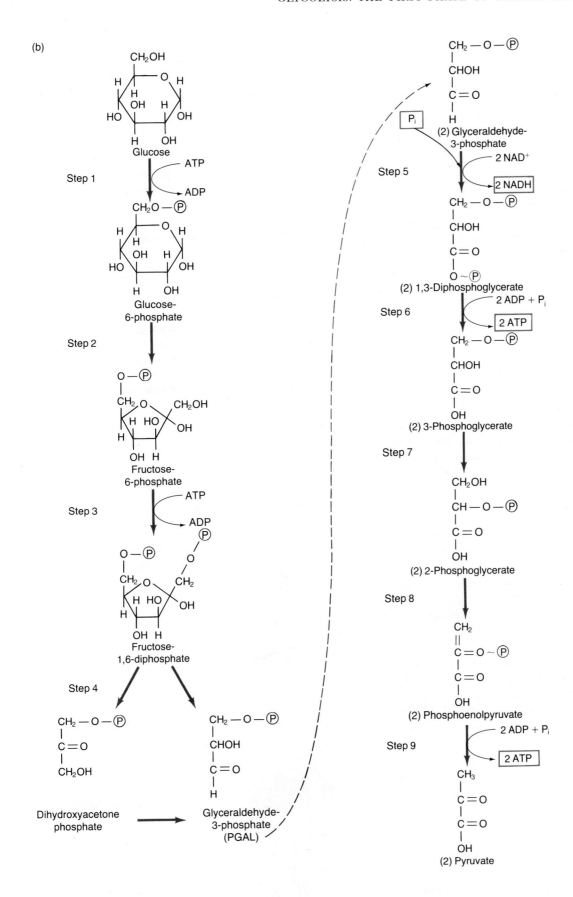

Steps 7 Through 9

In steps 7 and 8, further reactions convert 3-phosphoglycerate to phosphoenolpyruvate (PEP). Like DPGA, PEP is a high-energy phosphate compound, and its energy is used to convert an ADP to ATP. This conversion takes place in step 9:

$$2PEP + 2ADP \rightarrow 2 \text{ pyruvate} + 2ATP$$

Thus for every molecule of glucose that enters the glycolysis pathway, four molecules of ATP are formed. However, since two ATP molecules are used up in steps 1 and 3, the net energy yield from glycolysis is two ATP molecules. In addition, for each molecule of glucose, glycolysis generates two molecules of NADH and two molecules of pyruvate. The pyruvate molecules can be acted on further during the reaction sequences of fermentation or of cellular respiration.

Fermentation Reactions

The biochemical point of **fermentation,** so to speak, is to keep glycolysis—with its generation of ATP—functioning under anaerobic conditions. Two types of cells can carry out fermentation: *anaerobic cells* and *facultatively aerobic cells.* Anaerobic cells include certain types of bacteria that can survive only in the strict absence of molecular oxygen, while facultative aerobes include various other bacteria and yeasts, as well as animal muscle and other cells that can ferment nutrients when oxygen is absent but carry out more efficient processes when oxygen is present.

Before fermentation can occur, glycolysis must generate two net ATPs plus pyruvate. Then the pyruvate is modified to lactate, ethanol, or one of several other organic end products, depending on the fermenting organism (Figure 7-11).

A curious thing about the steps of fermentation is that the organism gains no further ATPs beyond the two already harvested during the initial nine steps of glycolysis, making clear that fermentation is really a set of reactions that regenerates NAD$^+$, the essential cofactor for glycolysis. NAD$^+$ is a limited commodity in cells but is critical for step 5 of glycolysis. During fermentation, pyruvate accepts electrons from NADH so that NAD$^+$ is regenerated.

In certain microbes, pyruvate is converted to lactate. This conversion is responsible for producing yogurt and many cheeses, and is part of the fermentation process used to create pickled foods, soy sauce, sourdough breads, and even chocolate. Another significant fermentation pathway yields ethanol; the wine and beer industries depend on anaerobic yeasts to ferment the carbohydrates in fruits and grains to this alcohol. Although the glycolysis–fermentation pathway yields a net gain of only two ATP molecules for each glucose molecule, potential energy remains stored in the products of fermentation. Hence humans use them as foods and beverages both because they retain considerable caloric content and because they taste good and contain mood-altering alcohol.

A form of fermentation also takes place when a sprinter in a track meet cannot take in enough oxygen to meet the energy needs of hard-working leg-muscle cells. In this situation, the muscle cells become increasingly anaerobic; fermentation speeds up in these facultatively aerobic cells; and lactate begins to build up. Since lactate inhibits muscle function, the runner may end up with

Figure 7-11   THE PROCESS OF FERMENTATION.
The purpose of fermentation is to regenerate NAD$^+$, which is needed to drive certain steps of glycolysis and other metabolic processes. Fermentation can result in the buildup of lactate in the leg muscles of a runner, or it can produce ethanol, the alcohol in beer, wine, and distilled liquors.

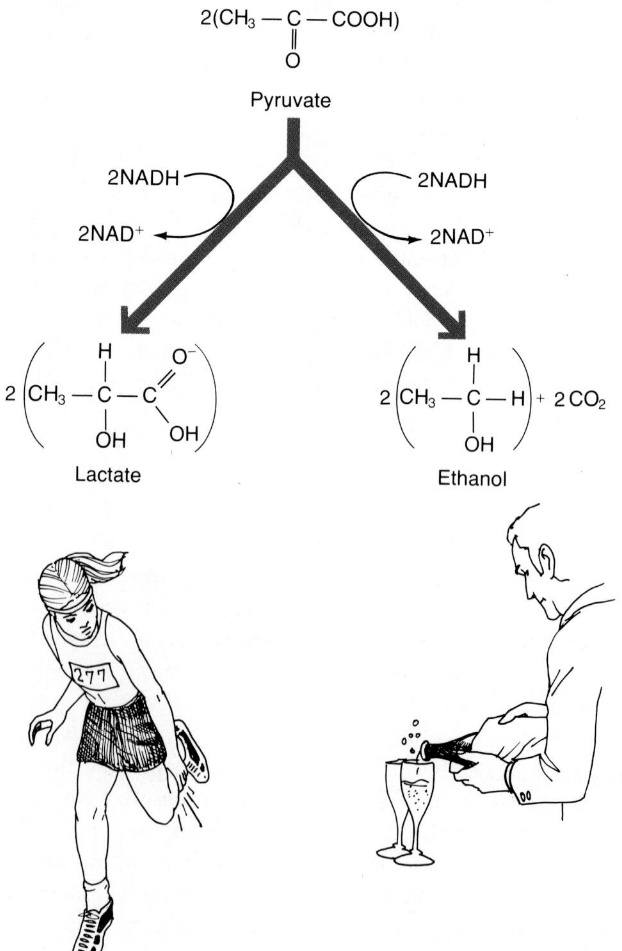

leg cramps. When the race is over and the person's leg muscles are no longer being strenuously exerted, an adequate supply of O_2 again reaches the muscles via the blood, and the products of glycolysis can once more proceed through the reactions of cellular respiration rather than fermentation. Also, the body's lactate can then be used to regenerate glucose, especially in the liver. During so-called aerobic exercise, moderate exertion is maintained at a steady pace; the flow of oxygen to muscle cells is sufficient to keep fermentation at a slow rate so that the level of lactate build-up stays low.

At this point, you may be wondering why so many reactions are biologically necessary when only certain steps lead to the formation of ATP. There are two answers: first, the intervening steps rearrange the molecules so that their atoms are in configurations that allow the specific energy-yielding reactions to occur. Second, the release of energy must be gradual and in small packets, since no biological molecule can absorb the large quantity of energy released by an explosive, one-step burning of glucose. One mole of glucose molecules (about 180 grams) contains 686 kcal of potential energy in the bonds between C, H, and O. This amount of energy could be released in one large step by combustion in a glowing flame or in many small steps by the "cold flame" of cellular metabolism. Clearly, the gradual method is the only practical alternative for a living organism.

The well-nigh universal occurrence of the fermentation and glycolytic pathways in cells has important implications in the quest to explain the origin of life. Biologists regard fermentation as a biochemical antique, a mechanism that almost certainly evolved very early in the history of life on Earth when the atmosphere contained little or no oxygen. The ubiquity of these pathways is strong evidence of the common descent of Earth's diverse creatures from primitive cells in which glycolysis and fermentation were first perfected.

CELLULAR RESPIRATION

The cells and tissues of multicellular plants and animals consume large amounts of ATP, much more than they could produce by means of glycolysis and fermentation. These organisms could not exist without obtaining large amounts of ATP from the glucose and other nutrients they metabolize via **cellular respiration.** Cellular respiration is actually twenty times more efficient than is fermentation in harvesting energy. Put another way, an animal whose cells carried out only fermentation would have to find, eat, digest, and utilize almost twenty times as much food as an animal of similar size whose cells were capable of respiration. This is because organic molecules serve as the final electron acceptors during fer-

mentation, whereas during cellular respiration, oxygen plays this role, and the final metabolic waste products are CO_2 and H_2O. These inorganic compounds contain far less free energy than do organic electron acceptors, and thus most of the chemical energy originally in glucose is released during cellular respiration and trapped in ATP molecules or lost as heat.

In cellular respiration, the pyruvate molecules derived from glycolysis are shunted into a metabolic pathway, the *Krebs cycle;* other products go to the *electron transport chain.* During this aerobic metabolism, oxygen accepts electrons and is reduced to H_2O. The overall equation for the oxidation of glucose is:

$$C_6H_{12}O_6 + 6O_2 \rightarrow 6CO_2 + 6H_2O \text{ (plus 36 ATPs and heat)}$$
(glucose)

During this reaction, thirty-six molecules of ATP are formed for each molecule of glucose consumed. This compares with a net of two ATP molecules produced during glycolysis.

Like glycolysis, cellular respiration involves a number of reactions, each catalyzed by a specific enzyme and organized into a metabolic pathway. The **Krebs cycle**— also known as the *citric-acid cycle* and the *tricarboxylic-acid cycle*—is a series of ten reactions during which the pyruvate from glycolysis is oxidized to CO_2. Both NAD^+ and FAD act as hydrogen and electron acceptors and are reduced to NADH and $FADH_2$. During this phase, three CO_2 molecules are produced for each pyruvate molecule, but only some of the energy is stored as ATP. The rest is in NADH and $FADH_2$ and is harvested during the final phase of respiration as the **electron transport chain** functions. When the reduced compounds NADH and $FADH_2$ are oxidized, their electrons are passed along a chain of oxidation-reduction steps to the final acceptor, O_2. This process releases a great deal of energy, and most of the thirty-six ATPs are formed during this phase.

As mentioned earlier, cellular respiration in eukaryotes takes place in mitochondria. The Krebs cycle occurs in the mitochondrial matrix, the central region of each mitochondrion (Figure 7-12). Most of the enzymes that catalyze these reactions are dissolved in the matrix fluid. In prokaryotes, the Krebs cycle enzymes are suspended in the cytoplasm. Finally, the proteins that bring about the reactions of the electron transport chain are bound to the inner mitochondrial membrane in eukaryotes and to the plasma membrane in prokaryotes. Many of the enzyme molecules extend all the way through the lipid bilayer of the inner mitochondrial membrane. Thus one part of each protein contacts the solution in the mitochondrial matrix, and the other part contacts the solution in the *outer compartment,* the fluid-filled space between the inner and outer mitochondrial membranes. We will

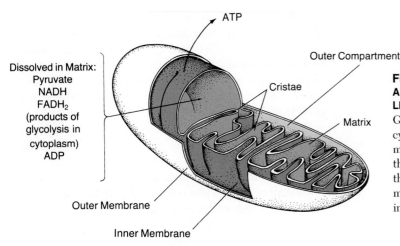

Figure 7-12   MITOCHONDRIAL ARCHITECTURE AND RESPIRATORY PROCESSES: THE CRITICAL LINK BETWEEN STRUCTURE AND FUNCTION.
Glycolysis and fermentation take place in the cell's cytoplasm, while respiration occurs in the mitochondrial matrix, the space within the foldings of the inner mitochondrial membrane. The processes in the electron transport chain are carried out by molecular complexes that are integral parts of the inner membrane itself.

see shortly why this arrangement is essential to the formation of ATP. Pyruvate, oxygen, ADP, and inorganic phosphate continuously diffuse into mitochondria. In turn, the waste products of energy metabolism—CO_2 and H_2O—and the energy-storage product—ATP—diffuse outward into the cytoplasm.

Oxidation of Pyruvate

Molecules of pyruvate generated by glycolysis pass easily through the highly permeable outer membrane of a mitochondrion and into the outer compartment. The molecules then are moved through the inner membrane by facilitated diffusion (which, as we saw in Chapter 5, involves carriers but does not require energy). Once in the mitochondrial matrix, pyruvate is oxidized to CO_2 and an activated form of acetate. In the process, NAD^+ is reduced to NADH, and the molecule of acetate becomes attached temporarily to a molecule of a coenzyme called *coenzyme A (CoA)*. Like ADP and ATP, this new compound, *acetyl CoA,* is a high-energy compound; it conserves a large share of the energy available from the oxidation of pyruvate and, with that energy, drives the next reaction in the pathway, the first step of the Krebs cycle.

The Krebs Cycle

The Krebs cycle, named after its discoverer, the British biochemist Sir Hans Krebs, is shown in general outline in Figure 7-13 and in more detail in Figure 7-14. Working out the cycle was a masterpiece of detective work, for in 1937, the technique of isotope labeling to trace atoms and molecules was not yet practicable. Instead, Krebs used the knowledge of enzymes, enzyme inhibitors, and reaction products to reason that a cycle exists and to outline its main steps. It was not until

Figure 7-13   MAIN EVENTS OF THE KREBS CYCLE: MOLECULAR CONVERSIONS AND A HARVEST OF ENERGY CARRIERS.
This simplified overview of the Krebs cycle shows the number of carbons in each intermediate, as well as the NADH, $FADH_2$, and ATP produced (outlined in red). Note that a two-carbon compound enters the cycle and that two CO_2 molecules are generated each time the cycle "turns." The last four-carbon structure in the pathway joins an incoming two-carbon molecule to begin the cycle again.

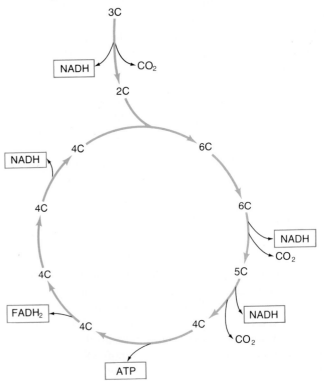

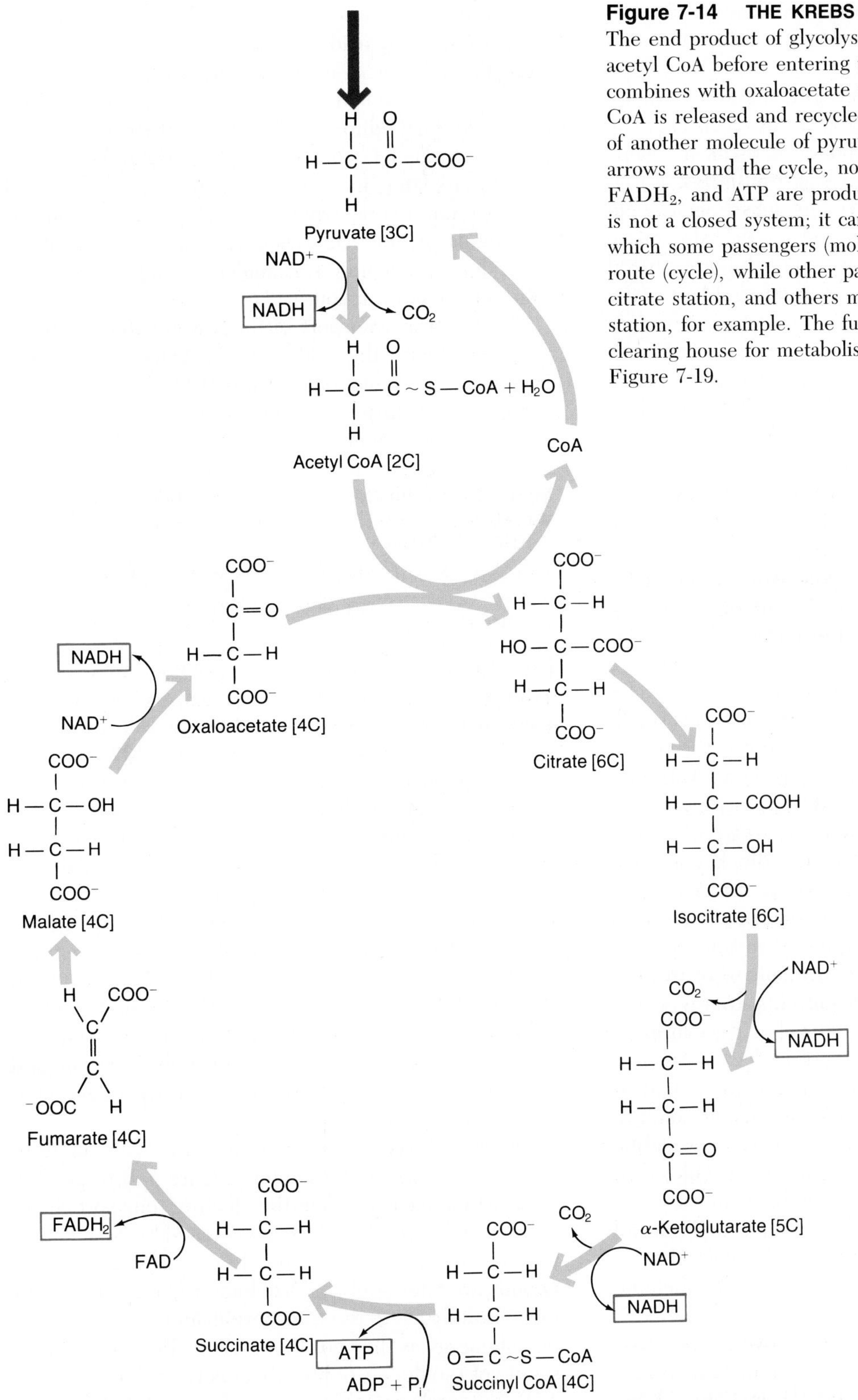

Figure 7-14   THE KREBS CYCLE.
The end product of glycolysis, pyruvate, is oxidized to acetyl CoA before entering the Krebs cycle. Acetyl CoA combines with oxaloacetate to form citrate. In this process, CoA is released and recycled to participate in the oxidation of another molecule of pyruvate. As you follow the yellow arrows around the cycle, note the steps at which NADH, $FADH_2$, and ATP are produced (outlined in red). The cycle is not a closed system; it can be compared to a train in which some passengers (molecules) will ride the entire route (cycle), while other passengers will get on at the citrate station, and others might get off at the succinate station, for example. The function of the Krebs cycle as a clearing house for metabolism is shown in more detail in Figure 7-19.

Pyruvate [3C]

NAD^+

NADH

CO_2

Acetyl CoA [2C]

CoA

Oxaloacetate [4C]

NADH

NAD^+

Citrate [6C]

Malate [4C]

Isocitrate [6C]

CO_2

NAD^+

NADH

Fumarate [4C]

α-Ketoglutarate [5C]

CO_2

NAD^+

NADH

$FADH_2$

FAD

Succinate [4C]

ATP

$ADP + P_i$

Succinyl CoA [4C]

thirty-five years later, when a variety of isotopes became available through the nuclear industry, that the incontrovertible proof of Krebs's elegant reasoning was provided. The Nobel Prize committee waited only twenty years to recognize Krebs's contribution.

The first reaction of the Krebs cycle is actually a mini-

cycle; in it, the two-carbon acetyl group of the acetyl CoA molecule generated from each molecule of pyruvate reacts with a four-carbon compound called oxaloacetate, which is present in mitochondria. A six-carbon compound, citrate, is formed; in the process, one molecule of CO_2 is released, and free aceytl CoA is regenerated and can participate once again in the oxidation of pyruvate. The energy stored in acetyl CoA drives this first reaction of the cycle.

The reactions of the Krebs cycle generate two molecules of CO_2 and remove four pairs of electrons. The CO_2 diffuses out of the mitochondrion as an inorganic end product of respiration. At the sites indicated in Figure 7-14, electrons are transferred by dehydrogenases to NAD^+ acceptors. The enzyme succinic dehydrogenase transfers electrons to FAD. That enzyme is not dissolved in the mitochondrial matrix but is an integral part of the inner mitochondrial membrane. You can see in Figure 7-14 that during the oxidation of one intermediate compound, α-ketoglutarate, to succinate, one molecule of ATP is formed for each molecule of acetyl CoA. This means that two ATPs are formed for each glucose molecule as a result of the reactions of the Krebs cycle. As the reactions continue, succinate is converted to fumarate, malate, and eventually oxaloacetate, completing the cycle.

At this point, all six carbons originally in one molecule of glucose have been oxidized, and some of the energy stored in the glucose has been transferred to four molecules of ATP. Two of these ATPs were formed during glycolysis, and two were formed during the Krebs cycle. The total free-energy change in the two phases is small: −62 kcal per molecule of glucose. Where is the rest of the energy that was once bound up in glucose? Most of it remains stored in NADH and $FADH_2$. As Figure 7-14 shows, the oxidation of pyruvate to acetyl CoA has produced two molecules of NADH, and the Krebs cycle reactions have generated six molecules of NADH and two molecules of $FADH_2$ from one molecule of glucose. In addition, glycolysis has generated two molecules of NADH. During the third, and final, phase of respiration, the electron transport chain, this energy storehouse is unlocked and large amounts of ATP are formed.

The Krebs cycle not only is a critical phase in the oxidation of glucose, but also serves as a "clearing house" for metabolism. Nutrients other than glucose, such as subunits of fats and proteins, also enter the oxidation pathway in the Krebs cycle and can be catabolized to release energy. In addition, intermediates in the cycle can pass out of the mitochondrion into the cytoplasm and serve as precursors for the synthesis of biological compounds. Thus the Krebs cycle is central to the metabolism of most cell types.

The Electron Transport Chain

The third phase of respiration is of special interest because it was the subject of some of the great research accomplishments that led to modern biology. In the 1880s, C. A. MacMunn described a class of substances in animal tissues whose properties changed when they were acted on by oxidizing and reducing agents. These were the substances of the respiratory electron chain. Unfortunately, MacMunn's work was derided by an intolerant journal editor and ignored until 1925, when the iron-containing *cytochrome* molecule was rediscovered. A few years later, in 1930, Otto Warburg did his Nobel Prize-winning work, which showed that cytochrome is the cellular agent that uses oxygen during respiration. Now, let us see how this molecule and the electron transport chain function.

The electron transport chain is a metabolic cascade that begins with the oxidation of NADH and $FADH_2$, ends with the reduction of O_2 to H_2O, and in between harvests ATP during a series of steps in which small amounts of free energy are released gradually. For the first time during cellular respiration, oxygen plays a role, acting as an electron acceptor for both NADH and $FADH_2$.

$$NADH + H^+ + \tfrac{1}{2} O_2 \rightarrow H_2O + NAD^+$$

$$FADH_2 + \tfrac{1}{2} O_2 \rightarrow H_2O + FAD$$

Notice two things about these equations. First, NAD^+ and FAD are regenerated and thus are ready to take part again in the reaction steps of glycolysis and the Krebs cycle. Second, the proton set free when NADH is oxidized joins another proton (released in an earlier reaction) to provide the hydrogen atoms in H_2O.

As NADH and $FADH_2$ are oxidized, the electrons they lose are passed along, like water in a bucket brigade, by a series of electron carriers (Figure 7-15). During each consecutive transfer from one carrier to the next, an electron releases a bit of its energy. At three points along the electron bucket brigade, energy is used to form ATP from ADP, a process called **oxidative phosphorylation.** This "downhill" series of electron transfers gradually lowers the level of energy in the electron, and when most of its energy is spent, the electron is accepted by oxygen.

The **cytochromes**—pigment proteins with an iron-containing *heme* at the center (Figure 7-16)—are the most distinctive carriers in the electron transport chain. As Figure 7-15 shows, electrons donated by NADH pass from one cytochrome compound to the next as the compounds are alternately oxidized and reduced. Electrons move rapidly along the transport chain in this way, giving off energy as they go. Biologists really do not understand yet the details of how that energy is funneled into ATP, but each pair of electrons from a NADH passes through all three sites of ATP generation, and so the oxidation of a molecule of NADH generates three molecules of ATP. When the last carrier in the chain (cytochrome a_3) is reduced, it reacts with free protons and O_2 molecules to form water.

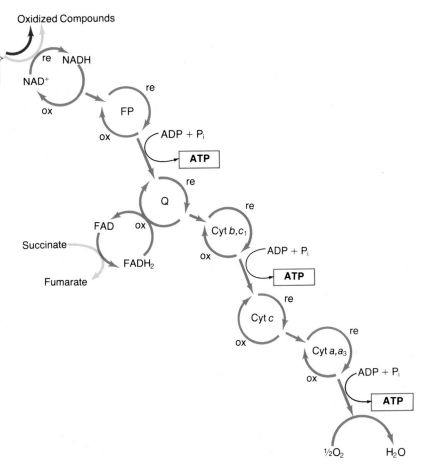

Figure 7-15   THE ELECTRON TRANSPORT CHAIN: AN ENERGY CASCADE.

The oxidation of glyceraldehyde-3-phosphate in glycolysis and pyruvate, isocitrate, α-ketoglutarate, and malate in the Krebs cycle supply the NADH that provides electrons for the electron transport chain. The oxidation of succinate yields $FADH_2$.

As NADH and $FADH_2$ are oxidized, the electrons they lose are passed along by a series of electron carriers (blue circles)—flavoprotein (FP), ubiquinone (Q), and the cytochromes (cytochrome b, cytochrome c_1, cytochrome c, cytochrome a, cytochrome a_3). As the electrons move down the chain, their energy levels decrease until the electrons are finally accepted by oxygen, and H_2O forms.

Oxidative phosphorylation is the process by which ATP is produced by means of this transport of electrons. The oxidation of each NADH molecule in this chain generates three ATPs. An exception is the NADH from the oxidation of pyruvate; that NADH enters the chain at the same site as $FADH_2$. Consequently that NADH only generates two ATPs, just as is the case for $FADH_2$.

The Energy Score for Respiration

We can now tally up the ATPs formed during the complete respiration of 1 mole of glucose. As Table 7-1 shows, the cell realizes a net gain of two ATPs from glycolysis, and two more are produced during the Krebs cycle. The oxidation of NADH and $FADH_2$ in the electron transport chain yields a total of thirty-two ATPs: the eight NADH molecules generated during the Krebs cycle each yield three ATPs, for a total of twenty-four, as they pass down the electron transport chain. The two molecules of NADH formed during glycolysis yield only four ATPs because each "uses up" some of its energy in reactions that transport the NADH electrons across the inner mitochondrial membrane and into the matrix; as a

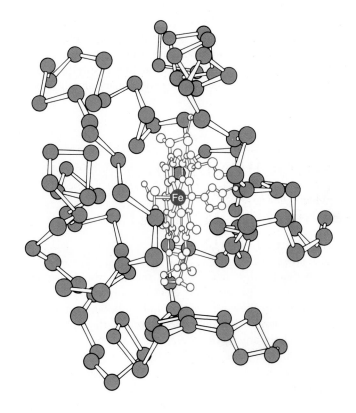

Figure 7-16   THE STRUCTURE OF CYTOCHROME C.
Cytochrome c is one electron carrier in the electron transport chain. Note the heme group (red), an iron-containing prosthetic group, at the center of the molecule.

Table 7-1   **APPROXIMATE MAXIMAL ATP YIELD FROM THE COMPLETE RESPIRATION OF 1 MOLE OF GLUCOSE**

Process	ATP Gain per Glucose
Glycolysis	2 ATP + 2 NADH
Krebs cycle	2 ATP + 8 NADH + 2 FADH$_2$
Electron transport chain from 2 NADH generated from glycolysis	4 ATP
Electron transport chain from 8 NADH generated from Krebs cycle	24 ATP
Electron transport chain from 2 FADH$_2$ generated from Krebs cycle	4 ATP
Total	36 ATP

result, electrons from this NADH enter the electron transport chain after by-passing the highest energy coupling site (they enter where FADH$_2$ electrons pass in). The oxidation of two molecules of FADH$_2$ (produced during the Krebs cycle) yields four more ATPs; FADH$_2$ provides less energy than does NADH because the FADH$_2$ molecules enter the electron transport chain at a lower energy level in the downhill series of transfers.

Theoretically, at least, the respiration of a molecule of glucose in the presence of oxygen has generated thirty-six molecules of ATP to power life processes in the cell. In reality, fewer than that may be made because all the H$^+$ derived from NADH and FADH$_2$ is not available to drive ATP synthesis. Even so, the number of ATP molecules produced per molecule of glucose is quite large compared with the number harvested by the glycolysis pathway alone. This is the reason that respiration and mitochondria have such great evolutionary significance.

MITOCHONDRIAL MEMBRANES AND THE MITCHELL HYPOTHESIS

The molecules that make up the electron transport chain are able to pass along electrons and generate ATP only because they are arrayed and embedded in a particular order in the inner membrane of the mitochondrion. The study of these electron transport molecules and their role in controlling the processes of energy harvest in the cell is one of the exciting frontiers of modern biology.

Although the complete mechanism of electron transport has yet to be clarified, we know that all but two of the energy carriers in the electron transport chain are parts of large multienzyme complexes; ubiquinone and cytochrome c act as connecting carriers to link the complexes together. Each complex is an integral part of the inner mitochondrial membrane, and the passage of an electron pair through each of the three complexes yields enough energy for the oxidative phosphorylation of about three ADPs to ATP. These enzyme complexes are the sites of ATP synthesis mentioned earlier. A simplified version of the arrangement of these enzymes is shown in Figure 7-17.

Elements of precision and chance operate in this electron transport chain. Components of the three multienzyme complexes must assemble precisely for normal function; for instance, the cytochrome oxidase complex is a dimer, each half of which has seven polypeptide chains. When assembled correctly it sits properly in the membrane, accepts electrons, pumps protons, and generates H$_2$O from O$_2$ and protons (Figure 7-18). But chance, too, operates: the three large complexes constantly diffuse about in the membrane, moving distances six times their own diameters every 5 milliseconds. Ubiquinone and cytochrome c move ten times as fast. It is only when these electron carriers chance to collide that pairs of electrons can be transferred from one to another. Despite these odds, electron pairs are received and donated by each multienzyme complex once every 5 to 20 milliseconds. (In our study of photosynthesis in Chapter 8, we will see that the spatial order of molecules in chloroplast membranes is likewise essential for the synthesis of glucose.)

But if the inner membrane of the mitochondrion is the physical site for electron transport, does it also play a role in the actual formation of ATP during respiration? This question has challenged biochemists for decades, and over the years, a number of models have been developed to explain how the energy of the transferred electrons is captured. Today, the most widely accepted of these models is the **chemiosmotic coupling hypothesis,** proposed in 1961 by a British biochemist, Peter Mitchell. The chemiosmotic coupling hypothesis offers an explanation for a puzzling feature of cellular respiration: the formation of ATP by the electron transport chain does not involve phosphorylated compounds. Recall that during glycolysis, large amounts of energy are temporarily stored in phosphorylated compounds such as DPGA and PEP and then transferred to ADP to form ATP. However, despite twenty-five years of biochemical sleuthing, no similar energy-rich compounds are known to be associated with the components of the electron transport chain, where a great deal of ATP is manufactured.

Mitchell proposed that the movement of electrons

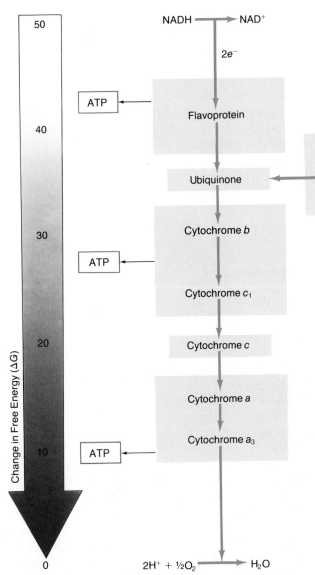

Figure 7-17   CLOSE-UP OF THE ELECTRON CARRIERS IN THE ELECTRON TRANSPORT CHAIN.
The electron carriers are large multienzyme complexes within the mitochondrial membrane. The oxidation of one molecule of NADH yields enough energy for the oxidative phosphorylation of about three ADPs to ATP. Ubiquinone and cytochrome c are connecting carriers. Note that the energy level drops as the electrons progress down the chain.

through the electron transport chain is accompanied by a *proton-pumping* mechanism that sets up an energy gradient across the inner mitochondrial membrane, in other words, a type of current. The energy released from this gradient is conserved by the formation of ATP.

The essential features of this proton-pumping model are shown in Figure 7-19. As electrons move down the electron transport chain, a gradient of hydrogen ions (protons) is created across the inner mitochondrial membrane. These protons are pumped from the inner compartment (matrix) of the mitochondrion to the outer compartment. This mechanism operates to keep the proton concentration in the matrix lower than that in the outer compartment. As we saw in Chapter 5, the movement of a substance against a concentration gradient requires energy. Thus it takes an input of energy to keep up the flow of protons. This energy comes from passage of the electrons down the electron transport chain. The pumping of positively charged protons to the outer compartment of the mitochondrion also leaves the matrix electrically negative in comparison with the outer compartment. This chemical and electrical imbalance represents potential energy: like the proverbial boulder poised at the top of a hill, the protons concentrated in the outer compartment tend to flow "down," back into the more negative matrix, creating a tiny but measurable current. The exergonic process of protons flowing back across the inner membrane releases free energy, just as a rolling rock can release energy. The Mitchell hypothesis suggests that this electrochemical gradient provides a means of conserving some of the free energy released by the steps of electron transport and making it available for ATP synthesis as protons reenter the matrix.

Biologists believe that the protons flow back through the usually impermeable inner membrane through large, complex molecules called **coupling factor.** Somehow, as protons move through these channels, the energy is conserved by the generation of ATP from ADP and P_i. But that is not the only job of the H^+ electrochemical gradient. At other sites on the inner membrane, H^+ passes inward accompanied by pyruvate (en route to the Krebs cycle enzymes), while at still other ports of entry, H^+ and inorganic phosphate (en route to ATP production sites) enter. Thus proton pumping and reentry are central to several mitochondrial functions.

Coupling factor and the reaction of ATP synthesis act as a major control point for ATP synthesis: the protons flow through the channels only as long as enough ADP is

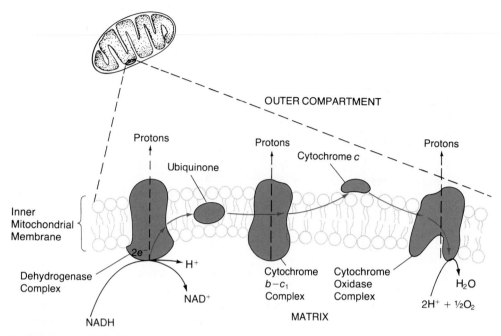

Figure 7-18   STYLIZED VIEW OF THE ELECTRON TRANSPORT CHAIN.
For each NADH oxidized, about nine protons pass from the matrix to the outer compartment
through all three enzyme complexes as a pair of electrons flows from NADH to H_2O.
Although the complexes appear motionless in this drawing, they, ubiquinone, and
cytochrome c all diffuse freely in the membrane; electron transfer only occurs when they
chance to collide.

present and available for conversion to ATP (P_i must also
be present). Therefore, this channeling mechanism automatically adjusts the rate of respiration to the rate at
which ATP is being used in the cell cytoplasm and nucleus. ADP from those sources diffuses in and is supplied
to the coupling-factor channels only as fast as ATP produced in mitochondria is used up in the cytoplasm. The
faster a cell uses ATP, the more ADP will be available
within the mitochondria. The more ADP available, the
more ATP can be made as protons stream inward
through the coupling-factor channels.

In studying the electron transport chain and its relation to proton pumping, we are looking at the real engine that drives most living cells. Chemiosmosis permits
respiration—and, as we shall see in Chapter 8, photosynthesis—to operate and to support the large power
needs of most cells. Without large-scale ATP manufacture by mitochondria, life would be at an incredibly slow
pace and most organisms common today would never
have evolved.

METABOLISM OF FATS
AND PROTEINS

The breakdown of glucose may be the most common
metabolic pathway in living cells, but what happens

when an organism consumes fats or proteins instead of
carbohydrates? The answer is that these nutrients, too,
are broken down and converted to compounds that serve
as intermediates in the Krebs cycle. This is why the
Krebs cycle can be considered a clearing house through
which the building blocks of all organic energy sources—
carbohydrates, proteins, and fats—are ultimately processed in aerobic cells (Figure 7-20).

Proteins are degraded in plant and animal cells by
being hydrolyzed to their component amino acids during
the intracellular digestion process. Then these monomers can be degraded further. The amino group is removed and converted to ammonia, a toxic substance that
is excreted from the organism. The rest of the molecule
is changed to various metabolic intermediates, depending on the chemical structure of the amino acid. Some
fragments are converted to pyruvate; others, to acetate
or acetyl CoA; and still others, to an intermediate in the
Krebs cycle. Regardless of how an amino acid enters the
metabolic pathway, its carbon skeleton is ultimately dismantled and oxidized to CO_2. On average, 1 gram of
protein yields about as much ATP as does 1 gram of glucose.

Fats have a similar but more complicated fate. Recall
from Chapter 3 that fats and oils are built of long-chain
fatty acids and glycerol—they also are called *triglycerides*. The breakdown of fats begins with hydrolysis of
triglycerides to glycerol and free fatty acids. Glycerol

(a)

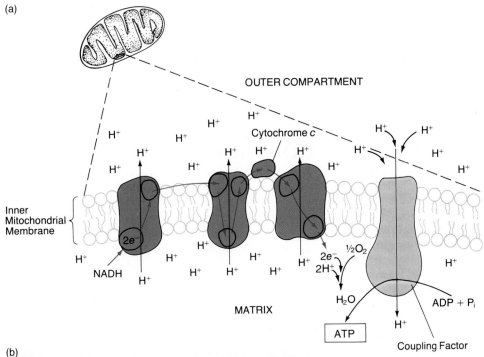

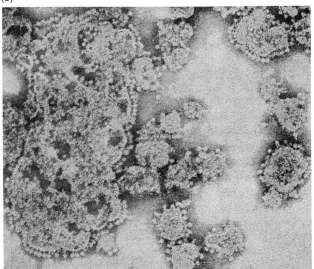

Figure 7-19   MITCHELL'S CHEMIOSMOTIC COUPLING HYPOTHESIS.
(a) In this model, the flow of protons toward the outer compartment of the mitochondrion is coupled to movement of electrons across each transport complex. The return of protons to the matrix via coupling factor is linked to the formation of ATP. (b) Inner mitochondrial membranes isolated in an "inside-out" configuration (magnified about 115,000 times). The lollipop-shaped structures protrude from the original inner surface of the membrane; they are the coupling factor that catalyzes synthesis of ATPs as H^+ ions flow through.

enters the glycolysis pathway at the level of glyceraldehyde-3-phosphate (step 4), while inside the mitochondrion, carbons are removed two by two from fatty acids to form acetyl CoA plus NADH and $FADH_2$. The acetyl CoA is oxidized via the Krebs cycle, and the NADH and $FADH_2$ via the electron transport chain. One gram of fatty-acid molecules provides 2.5 times more ATP than does 1 gram of sugar or protein, which is why many animals and plants store some of their food reserves as fats or oils. It also explains why cutting down on fats is a part of every dieter's scheme for shedding pounds. The richness of fatty acids as an energy source can be explained by the greater number of hydrogen atoms per unit weight compared with other nutrients. For this same reason, petroleum is a more efficient fuel than coal.

The Krebs cycle is important not just in the catabolism of nutrients, but also indirectly in the manufacture of proteins, fats, and carbohydrates for export from the cell or for structural roles inside the cell (Figure 7-21). The building up (*anabolic*) process is called **biosynthesis.** Various intermediates in the Krebs cycle and their immediate precursors, such as pyruvate and acetyl CoA, can serve as the starting compounds in anabolic pathways. For example, α-ketoglutarate and ammonia from the breakdown of amino acids serve as precursors for the biosynthesis of glutamate. Oxaloacetate provides the carbon skeleton for another amino acid, aspartate. Such amino acids can, in turn, form part of a growing chain of protein. Glucose can be converted into fats (again, as would-be dieters know) by being broken down into acetyl CoA, which then is used to synthesize long-chain fatty acids.

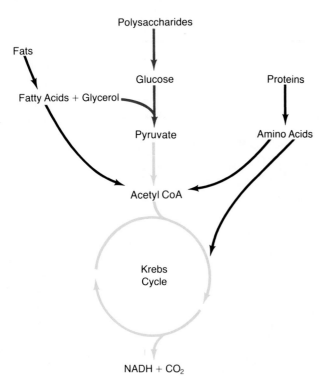

Figure 7-20   THE KREBS CYCLE: A METABOLIC CLEARING HOUSE.
Fats, carbohydrates, and proteins are broken down to compounds that serve as intermediates in the Krebs cycle. Thus the Krebs cycle is considered a clearing house: all metabolites eventually must pass through this sequence of reactions.

The complexity of so many reactions and pathways proceeding simultaneously within the cell is not easy to comprehend. In the next section, however, cellular metabolism may become a bit easier to understand as we consider some of the ways in which metabolic pathways are controlled.

CONTROL OF METABOLISM

Cells must be able to adjust the rates at which molecules are broken down or built up in metabolic pathways, so that energy and other resources will not be depleted and so that the products of the various pathways are available when needed. To accomplish this, cells have available both long-term and short-term regulatory mechanisms. Over the long term, cells can regulate the amount of a particular enzyme produced—a process that involves genetic control mechanisms that we will consider in detail in Chapter 13. In the short run, as we shall

Figure 7-21   THE KREBS CYCLE AND BIOSYNTHETIC PATHWAYS IN CELLS.
Intermediates in the Krebs cycle can be drawn off to participate in a biosynthetic pathway. Some of these same intermediates can be replaced in the cycle by the breakdown of fats, polysaccharides, and proteins (Figure 7-18). Just a few examples are shown here.
    Acetyl CoA is a precursor for the biosynthesis of long-chain fatty acids. Since acetyl CoA is a product of glucose breakdown as well as an intermediary in its synthesis, glucose can be converted to fat in our bodies.
    Pyruvate, α-ketoglutarate, and oxaloacetate are intermediates of the Krebs cycle that serve as precursors for the biosynthesis of amino acids.
    Succinyl CoA is a precursor in the biosynthesis of heme pigments, including cytochrome c, a carrier protein in the electron transport chain.

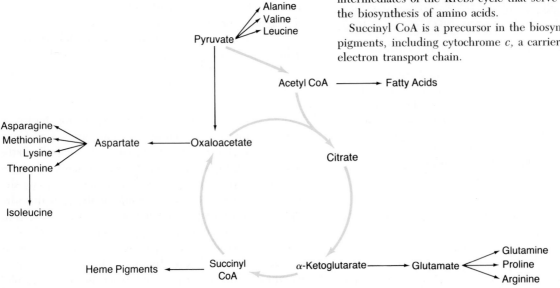

see here, cells can regulate the activity of enzymes that already have been produced.

In Chapter 4, we saw that most cellular controls are based on negative feedback, by which the product of a pathway affects the rate at which the pathway itself operates; the concentration of the product has an inverse effect on the rate. Thus if the pathway is

$$A \xrightarrow{\text{(enzyme 1)}} B \xrightarrow{\text{(enzyme 2)}} C$$

when C builds up in concentration, it acts on enzyme 1 to slow down the enzyme's activity. As a result, the pathway functions less rapidly, and less C is formed. As the concentration of C drops, enzyme 1 is released from inhibition once again, and the production of C speeds up. The effect of negative feedback is to maintain a fairly constant concentration of the product (here, C) regardless of how fast it is being used in the cell.

Cells use two distinct mechanisms to carry out negative-feedback control over metabolic pathways: allosteric enzymes and covalent modification of enzymes.

Allosteric enzymes (described in Chapter 4) have two binding sites. The first, the active site, binds the substrate, and the second binds a molecule (the effector) that changes the activity of the enzyme. A good example of metabolic control by means of allosteric enzymes occurs during glycolysis (Figure 7-22). The activity of one glycolytic enzyme, *phosphofructokinase*, or *PFK*, can be influenced by the build-up of ATP or of citrate. Since the end product of glycolysis, pyruvate, furnishes the carbon skeleton for the Krebs cycle, it is not surprising that citrate, an intermediate product of the Krebs cycle, also regulates PFK. As ATP or citrate accumulates, either can bind to the allosteric site on the PFK enzyme and reduce enzyme activity. This, in turn, slows down the glycolytic chain and, with it, the production of ATP and citrate. As the cell uses up its store of these compounds, molecules of ATP and citrate leave the allosteric site on PFK, and the enzyme's activity once again speeds up.

Covalent modification of enzymes also involves a temporary change in the chemical structure of these biological catalysts. By means of this mechanism, an enzyme is inactivated by the attachment of a chemical group that reduces the enzyme's catalytic function. Phosphate groups can have this effect on certain enzymes that are involved in respiration: when ATP accumulates, the enzymes become phosphorylated, and their activity is inhibited. When the level of ATP falls, another enzyme that is especially efficient in the presence of abundant pyruvate removes the phosphate groups, and the inhibited enzymes are reactivated.

Dozens of anabolic and catabolic pathways function simultaneously in every cell, using up and producing ATP, fats, carbohydrates, proteins, and other molecules

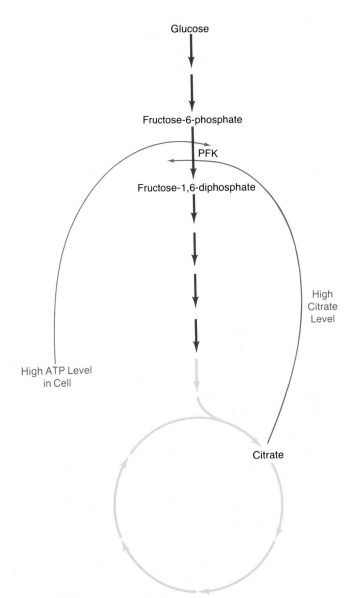

Figure 7-22   ALLOSTERIC ENZYMES AND METABOLIC CONTROL.
Phosphofructokinase (PFK), an enzyme that catalyzes the third step in the glycolytic pathway, can be inhibited by high levels of ATP or citrate in the cell, which signal the enzyme to shut down, thus slowing metabolism. This is an example of negative feedback.

by means of the kinds of multistep processes described in this chapter. The speed and complexity of these simultaneous events are truly amazing. For example, a yeast cell contains about 80 million molecules of ATP, and, on average, each ATP is used up in an energy-requiring reaction and then regenerated every five seconds. When you consider that the human body contains 20 to 30 trillion cells, each metabolizing at a rapid pace, the overall energy economy of an entire organism starts to seem impossibly complicated. Yet it goes on imperceptibly and automatically, every second of a cell's life.

SUMMARY

1. Nutrient molecules are a temporary repository for solar energy trapped by plants and other autotrophs. Metabolic pathways in cells break down nutrients to release this energy in gradual steps, each mediated by a specific enzyme.

2. *ATP (adenosine triphosphate)* is an energy-storage molecule that transfers energy from exergonic reactions, such as nutrient catabolism, to endergonic reactions, such as synthesis of biological molecules. ATP contains three phosphate groups; the related compounds ADP and AMP contain two and one, respectively. The third and second phosphates are linked to the rest of the molecule by high-energy bonds. Energy is released when the bonds between the phosphate groups are broken. Addition of a phosphate group to a compound is termed *phosphorylation.*

3. All key metabolic reactions involve *oxidation*, the removal of electrons from an atom or a compound, coupled to *reduction*, the addition of electrons. The flow of electrons may be viewed as a current that carries energy. Electrons can be transferred in the form of hydrogen atoms in the reciprocal processes of hydrogenation and dehydrogenation. Two important electron carriers in *oxidation-reduction reactions* are the coenzymes *NAD*$^+$ and *FAD*.

4. *Glycolysis* is the initial sequence of reactions used by virtually all cells to metabolize glucose. It involves nine reaction steps during which glucose is broken down to two molecules of *pyruvate*. In the process, a net of two molecules of ATP are formed from ADP and P_i, and two molecules of NAD$^+$ are reduced to NADH.

5. *Fermentation* is a reaction sequence beginning with glycolysis in which organic molecules rather than O_2 function as electron acceptors. Pyruvate is converted to ethanol, lactate, or other end products with no gain of ATPs. The conversions of fermentation are important because they regenerate NAD$^+$, which can then participate once again in critical steps of the glycolysis pathway.

6. During *cellular respiration*, glucose is oxidized completely to CO_2 and H_2O. The pyruvate generated from glycolysis is oxidized—releasing one CO_2 molecule for each pyruvate—to an activated form of acetate, acetyl CoA, which enters the *Krebs cycle*. In a series of reactions, the carbon skeleton of each acetyl group of the acetyl CoA molecule is rearranged, and two CO_2 molecules are produced for each acetyl CoA molecule. NAD$^+$ and FAD act as electron acceptors, becoming reduced to NADH and FADH$_2$. Only two additional ATPs form during the Krebs cycle.

7. NADH and FADH$_2$ are oxidized during the reactions of the *electron transport chain*, a series of electron carrier molecules embedded in the inner membrane of the mitochondrion. Electrons are passed from one of these pigment proteins to the next to the final acceptor, oxygen. During this sequence, thirty-two molecules of ATP are formed. In total, thirty-six ATPs form in an aerobic cell for each molecule of glucose consumed: two during glycolysis, two during the Krebs cycle, and thirty-two during the reactions of the electron transport chain.

8. The current theory for how ATP is formed from ADP and phosphate by *oxidative phosphorylation* is Mitchell's *chemiosmotic coupling hypothesis*. According to the Mitchell hypothesis, the flow of electrons down the electron transport chain in the inner mitochondrial membrane results in the pumping of protons outward through the membrane. Then, as protons pass back into the mitochondrial matrix through channels called *coupling factor*, ADP is phosphorylated to ATP.

9. The Krebs cycle is a metabolic clearing house: as amino acids and fats are broken down, portions of these molecules are converted to intermediates that can enter the Krebs cycle to be broken down. Various Krebs cycle intermediates can also leave the mitochondrion and serve as precursors for the manufacture of biological molecules, a process called *biosynthesis*.

10. Metabolism is controlled so that an appropriate amount of energy is stored in ATP and appropriate raw materials are available to the cell when needed. Allosteric enzymes and covalent modification of enzymes are the two methods used by cells to carry out negative-feedback control over the rate of function of metabolic pathways.

KEY TERMS

adenosine triphosphate (ATP)

biosynthesis

cellular respiration

chemiosmotic coupling hypothesis

coupling factor

cytochrome

electron transport chain

FAD (flavin adenine dinucleotide)

fermentation

glycolysis

Krebs cycle

NAD$^+$ (nicotinamide adenine dinucleotide)

oxidation

oxidation-reduction reaction
oxidative phosphorylation
phosphorylation
pyruvate
reduction

QUESTIONS

1. Explain how the cell's breakdown of glucose is similar to burning fuel in an engine. Explain a few important differences as well.

2. What is the function of each of the following in cell metabolism: ATP? NAD^+? FAD? Krebs cycle? Cytochromes? Oxygen?

3. Which of the following can proceed in the absence of oxygen: Glycolysis? Fermentation? Krebs cycle? Electron transport?

4. What extra metabolic step does fermentation accomplish?

5. Which molecules in a cell act as cyclic electron acceptors? What kinds of molecules act as terminal electron acceptors in fermentation? In respiration?

6. Electron transport system enzymes are embedded in membranes, whereas the enzymes involved in glycolysis are dissolved in the cytoplasm. Which set of enzymes would you expect to be more hydrophobic? Which would be easier to study in a test tube, using standard aqueous salt solutions?

7. Which molecule becomes phosphorylated during oxidative phosphorylation? What is finally oxidized? Which molecule is finally reduced?

8. Which is a more efficient fuel: a highly reduced molecule (like a fatty acid), a highly oxidized one (like carbon dioxide), or one in an intermediate state of oxidation (like a carbohydrate)? Explain why. Which substance is a better substrate for fermentation? Explain why.

9. The ATP, NAD^+, and FAD molecules are essential for cellular energy metabolism. Compare the structures of these molecules. Which contain the sugar ribose? Which contain the base adenine? Which contain phosphate? How do these three molecules differ from the cytochromes?

10. If the respiratory system in a cell is poisoned so that electron transport is uncoupled from ATP production, what happens to the energy produced?

ESSAY QUESTIONS

1. How does Mitchell's chemiosmotic pump work and what relationship does it have to the concept of potential energy?

2. Yeast can ferment sugars and grow in the absence of oxygen. If a yeast cell were to lose its mitochondria, would it still be able to grow? Would it grow as fast as a normal cell?

3. Some bacteria carry out *anaerobic* respiration. They can harness energy by electron transport in the absence of oxygen by using inorganic sulfate or nitrate as the terminal electron acceptors. Discuss how this process differs from fermentation.

SUGGESTED READINGS

ALBERTS, B., et al. *Molecular Biology of the Cell*. New York: Garland, 1983.

This excellent book on cell biology has an easy-to-follow description of energy metabolism and relates structure to function in fine fashion.

HINKLE, P. C., and R. E. MCCARTY. "How Cells Make ATP." *Scientific American*, March 1978, pp. 104–123.

Straightforward coverage of ATP manufacture.

KREBS, H. A. "The History of the Tricarboxylic Acid Cycle." *Perspectives in Biology and Medicine* 14 (1970): 154–170.

The detective story in the words of the Nobel Prize-winning biochemist.

RACKER, E. "From Pasteur to Mitchell, a Hundred Years of Bioenergetics." *Federation Proceedings* 39 (1980): 210–215.

A broad overview of the key steps that led to our understanding of energy metabolism.

8

PHOTOSYNTHESIS: HARNESSING SOLAR ENERGY TO PRODUCE CARBOHYDRATES

The great invention of the plant kingdom, the factor which differentiates it totally from all animals of whatever complexity, apart from a few borderline cases among single-celled organisms, is chlorophyll. Whether it is a one-celled alga or a giant forest tree, a plant will contain this substance.

Anthony Huxley, *Plant and Planet* (1974)

Sunlight, air, the green of chlorophyll in Douglas Firs: The components of photosynthesis, the sustainer of life on Earth.

Every carbon atom in your body and every oxygen molecule you breathe once cycled through the tissues of a plant. The significance of plants, algae, and other photosynthetic organisms to the entire balance of life on Earth cannot be overstated because they convert solar energy to chemical energy in the form of carbohydrates. If plants suddenly stopped producing carbohydrates and liberating oxygen, much of the world's carbon would be oxidized to CO_2 gas within about 300 years, and the oceans and atmosphere would be devoid of free oxygen within about 2,000 years—mere instants in the vast stretch of geologic time. Clearly, as aerobic, nonphotosynthetic organisms, we depend on plants. And plants are dependably productive: they generate more than 150 billion tons of carbohydrates every year. That amount would fill a string of boxcars stretching from the Earth to the moon and back fifty times!

The process of cellular respiration, which we discussed in Chapter 7, is the chemical opposite of the process of photosynthesis: in glycolysis, fermentation, and respiration, carbohydrates are oxidized to release energy; in photosynthesis, the sun's energy is trapped to form carbohydrates. These two sets of chemical processes are reverse sides of the same coin, and both are needed for the transfer of environmental energy to life activities.

During the evolution of life on Earth, fermentation apparently developed first. Then, almost certainly, photosynthesis evolved and released massive amounts of oxygen into the atmosphere. That oxygen provided the essential precondition for the evolution of aerobic respiration, with its electron transport chain, which can produce so much ATP. Once oxygen-dependent cellular respiration had evolved and more ATP had been made available, the diversity of organisms greatly increased to include a wide variety of complex multicellular plants (the so-called higher plants) and animals, and a global cycling of carbon atoms from atmosphere to plant to animal and back to atmosphere developed.

We will describe the global carbon cycle later in this chapter. But first, we will discuss the chemistry of photosynthesis, which, like glycolysis, fermentation, and respiration, proceeds in a series of reaction steps mediated by enzymes. Along the way, we will find out why plants store solar energy in carbohydrates instead of in ATP or other molecules, and how atmospheric oxygen, so crucial to the respiration of animal and plant cells, can interfere with the photosynthetic process in plant cells. Throughout, the beautiful symmetry of metabolism will be evident—the way opposite processes build up and break down nutrients and provide virtually all living things with the energy they need to survive.

AN OVERVIEW OF PHOTOSYNTHESIS

Photosynthesis is a metabolic process that occurs only in cells of green plants, algae, and certain protists and bacteria. During the process, energy from the sun is trapped and used to convert the inorganic raw materials CO_2 and H_2O to carbohydrates and O_2; a pigment called *chlorophyll* is the key to that process. Overall, photosynthesis is a conversion of light energy to chemical energy stored in the form of molecular bonds. The process may be summarized by a deceptively simple equation in which $[CH_2O]$ stands for carbohydrate:

$$CO_2 + H_2O \xrightarrow{\text{light}} [CH_2O] + O_2$$

This equation implies that carbon dioxide is reduced to carbohydrate and that water is oxidized as molecular oxygen forms. The equation also shows that photosynthesis is chemically the reverse of cellular respiration, which can be represented as:

$$[CH_2O] + O_2 \longrightarrow CO_2 + H_2O$$

If these basic equations for photosynthesis and respiration are multiplied by six (because there are six carbon atoms in glucose), the new equations describe the synthesis and breakdown of glucose, $C_6H_{12}O_6$:

photosynthesis: $6CO_2 + 6H_2O \xrightarrow{\text{light}} C_6H_{12}O_6 + 6O_2$

respiration: $C_6H_{12}O_6 + 6O_2 \longrightarrow 6CO_2 + 6H_2O$

Recall from Chapter 7 that the oxidation of 1 mole of glucose molecules to CO_2 is an exergonic process that releases 686 kcal of free energy. It should not be surprising that the reverse process—building glucose molecules from inorganic raw materials—is a highly endergonic process. (The actual amount of energy required is about 2,000 kcal/mole of glucose formed; the extra energy is lost during the series of photosynthetic reactions.)

The Two Basic Reactions of Photosynthesis

Biologists have known the general equation for photosynthesis and the basic starting and ending materials for almost 200 years. However, the first details of photosynthesis were not revealed until the early twentieth century. It was clear from the general equation that CO_2

and H_2O are raw materials, but how those molecules were rearranged during the process was obscure. The earliest assumption was that CO_2 molecules are somehow cleaved, that the O_2 is released to the environment, and that the leftover carbon atoms are then joined to the H_2O molecules to form CH_2O, the basic carbohydrate unit. But was this assumption correct? And did the reaction take place in one step or in several?

In 1929, C. B. van Niel of Stanford University was able to answer these questions by studying photosynthetic bacteria, which are similar to plants in several ways. They need light in order to grow; they contain the blueish pigment, bacterial chlorophyll; and they obtain carbon from CO_2. However, the bacteria he studied use the sulfur compound H_2S instead of H_2O during photosynthesis and release sulfur, not O_2. On the basis of a brilliant set of experiments, van Niel concluded that photosynthesis in these bacteria involves the oxidation of H_2S to sulfur plus hydrogen atoms and the use of the hydrogens to reduce CO_2 to carbohydrate:

$$CO_2 + 2H_2S \xrightarrow{\text{light}} [CH_2O] + 2S + H_2O$$

Note that water is also produced and that its oxygen must come from CO_2, for there is no oxygen in H_2S.

Generalizing from his work on bacteria, van Niel proposed that the photosynthetic processes in bacteria and green plants are fundamentally similar. Thus an oxidation-reduction model for all types of photosynthesis could be written:

$$CO_2 + 2H_2A \xrightarrow{\text{light}} [CH_2O] + 2A + H_2O$$

A can stand for sulfur, as in bacterial photosynthesis, or for oxygen, as in plant photosynthesis. By analogy, then, it was possible for van Niel to write an equation for photosynthesis in plants:

$$6CO_2 + 12H_2O \xrightarrow{\text{light}} C_6H_{12}O_6 + 6O_2 + 6H_2O$$

Comparing this equation with our earlier one, notice that the amount of water is doubled here ($12H_2O$ versus $6H_2O$) and that water also appears among the reaction products. This final equation is a better reflection of the actual step-by-step chemistry of photosynthesis, as we shall see later.

Photosynthesis involves two sets of partial reactions. In the first, water is oxidized; in the second, carbon dioxide is reduced. When water is split, O_2 is released to the environment. H atoms then reduce CO_2 to carbohydrate. The "new" water molecules that arise contain oxygen derived from CO_2 during its reduction. Van Niel's predictions—that all the oxygen given off during photosynthesis comes from water and that there are two basic

steps of photosynthesis—were validated in 1941 by labeling the oxygen with isotopes and tracing its paths directly. Van Niel's original work was a feat of scientific reasoning that still stands as a milestone in biological research.

Chloroplasts: Sites of Photosynthesis

In the half-century since van Niel's seminal experiments, biologists have discovered a great deal about photosynthesis, including the role of light and the fact that the two distinctive reactions of photosynthesis occur at different sites within the cell. Light is clearly necessary for photosynthesis because plants stop growing and eventually die if deprived of light for long periods. But is light used directly during both sets of reactions or during just one of them? Studies have shown that the first stage of photosynthesis, when water is oxidized and O_2 released, is absolutely dependent on direct use of light energy; these reactions are therefore called the **light reactions** (Figure 8-1). The second stage, during which CO_2 is reduced to carbohydrate, consists of reactions that depend on NADPH (another coenzyme electron carrier) and ATP, products of the first stage, but that can proceed in the dark; these are called the **dark reactions.** The designation "dark" does not mean that such reactions occur at night, but merely that the reduction of CO_2 to carbohydrate does not use light directly, whereas the oxidation of H_2O does.

Both light and dark reactions take place in the **chloroplasts,** organelles we first encountered in Chapter 5. One proof of this came in 1939, when a young British biologist, R. Hill, isolated these greenish, membrane-bound organelles from a number of flowering plants, illuminated them, and showed that chloroplasts can produce oxygen and hydrogen atoms even in the absence of CO_2. Many studies before and after Hill's time, employing light microscopes, electron microscopes, and biochemical methods, have revealed the precise structure and function of chloroplasts.

Chloroplasts have a variety of shapes but, in general, may be thought of as resembling a banana. They are somewhat larger than mitochondria, and from twenty to eighty—usually about forty—are present in a typical plant cell. Figure 8-2 shows a chloroplast in cross section. As in mitochondria, two lipid-bilayer membranes surround the chloroplast; internally, a gel-like matrix called the *stroma* contains ribosomes, DNA, and machinery for protein synthesis. The most prominent internal structures in chloroplasts are the stacks of flattened sacs called *grana* (meaning "grains"). These greenish "grains" were identified under the light microscope more than 150 years ago. Each flattened sac in a granum

tion. Most of the enzymes that catalyze the dark reactions are found in the stroma, or matrix, surrounding the stacks of thylakoids.

HOW LIGHT ENERGY REACHES PHOTOSYNTHETIC CELLS

A browsing animal that eats a fresh green leaf from a bush is consuming at least some of the light energy that

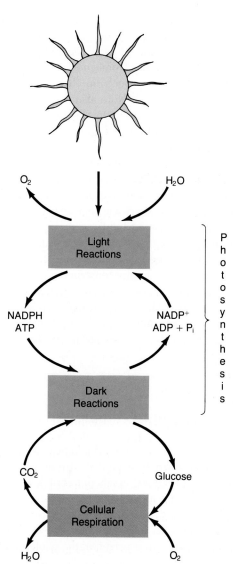

Figure 8-1   AN OVERVIEW OF PHOTOSYNTHESIS.
Light energy drives a series of light reactions, which produce the energy intermediates ATP and NADPH. Those compounds then power the dark reactions, which build carbohydrates such as glucose. Cellular respiration (Chapter 7) ultimately uses O_2 to regenerate the CO_2.

Figure 8-2   THE STRUCTURE OF CHLOROPLASTS: SITES OF PHOTOSYNTHESIS.
In this electron micrograph of a chloroplast (magnified about 15,000 times), note the internal stacks of membranes, the grana. Each granum consists of flattened sacs called thylakoids; adjacent grana are interconnected by the thylakoid membrane. Most of the enzymes and pigments for the light reactions of photosynthesis are embedded in the thylakoid membranes. The stroma, a gel-like matrix, surrounds the grana. The enzymes for the dark reactions of photosynthesis as well as chloroplast DNA and ribosomes and other substances are located in the stroma.

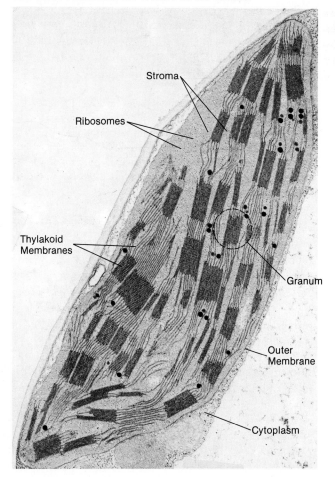

is called a **thylakoid** (from the Greek word for "sack"). The internal space, or *lumen,* of each sac is surrounded by the *thylakoid membrane.* Tubes of thylakoid membrane material connect the thylakoids of one granum with the stacks surrounding it, so that the thylakoid internal spaces are all interconnected.

The enzymes and cofactors that facilitate the light reactions of photosynthesis are embedded in the thylakoid membrane, where they facilitate the movement of protons across the membrane in much the same way as electron transport complexes extrude protons through the inner mitochondrial membrane during cellular respira-

left the sun just eight minutes earlier, traveled through space, and was trapped and transferred almost instantaneously to the bonds of recently synthesized carbohydrate molecules inside the leaf. In Chapter 4, we talked about various kinds of energy—kinetic, potential, mechanical, chemical, and so on. But what exactly is light energy, and how does it interact with the molecules of a living leaf?

The Nature of Light

Visible light is a form of electromagnetic radiation. All such radiation has the properties of both particles and waves. The distance particles, or **photons,** travel during one complete vibration is defined as the *wavelength* of that light. The less energy a photon has, the more slowly it vibrates and the longer the distance of a complete vibration, or wavelength. In other words, the energy of a photon is inversely proportional to its wavelength.

Photons are *packets* of light energy. The intensity, or brightness, of a light beam is a measure of the total number of photons, or energy per unit time.

The full range of light energies and wavelengths is represented by the electromagnetic spectrum, which extends from highly energetic gamma rays, with ex-

tremely short wavelengths of about 10^{-3} nanometer (nm)—smaller than an atom—to low-energy radio waves, with wavelengths of more than a kilometer. Figure 8-3 shows the full electromagnetic spectrum. Notice that the portion of the spectrum that is visible to the human eye is very small. Note also that wavelengths of about 400 nm appear violet to the human eye, while wavelengths of around 700 nm appear red. We perceive light of various wavelengths as having distinct colors because particles of light energy vibrating at different frequencies affect light-receptor cells in our eyes differently. What we see as blue light, for instance, is made up of photons that have shorter wavelengths and more energy than photons of red light.

It is only in the narrow range of the visible part of the spectrum that the energy of photons can be captured by biological molecules to do constructive work. This is because higher energy wavelengths such as gamma rays tend to break apart chemical bonds, while lower energy wavelengths such as radio waves have too little energy to affect bonds or accomplish biologically useful work, such as photosynthesis.

The Absorption of Light by Photosynthetic Pigments

In order for light to drive a biological reaction, photons must be absorbed in photosynthetic cells by compounds called pigments. Each **pigment** has distinctive properties based on its molecular structure. The principal pigment active in photosynthesis is **chlorophyll,** which appears green to the human eye because it absorbs light of most other wavelengths (violet, blue, yel-

Figure 8-3   THE SPECTRUM OF ELECTROMAGNETIC RADIATION.

Electromagnetic radiation is measured in wavelengths and is expressed in nanometers (1 nm = 10^{-7} cm). The shorter the wavelength, the higher the energy per photon. The spectrum of visible light is expanded to show the relationship of color and wavelength. Only this small segment of the electromagnetic spectrum is visible to the human eye.

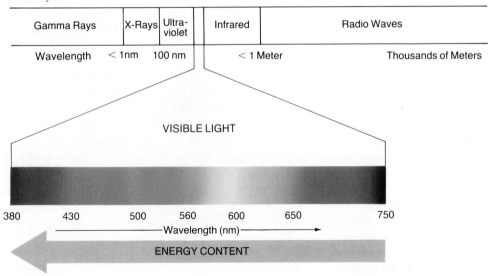

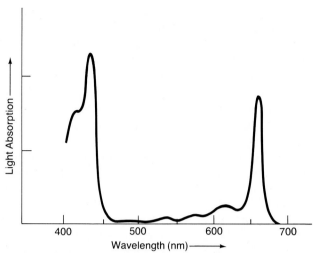

Figure 8-4   ABSORPTION SPECTRUM OF CHLOROPHYLL A.

These curves show the amount of light absorbed at each wavelength when a solution of chlorophyll is illuminated. The chlorophyll absorbs photons more in the blue (400–450 nm) and red (650–700 nm) regions of the spectrum, and less in the green (intermediate) wavelengths. As a result, a solution of chlorophyll appears green to the human eye because the green light is transmitted, while other wavelengths are absorbed.

low, red) but *transmits* light with wavelengths in the green band of the visible spectrum. All pigments work the same way: a black object absorbs all wavelengths and reflects none; a red object absorbs all wavelengths except those in the red region; and so on. Figure 8-4 depicts the **absorption spectrum** of chlorophyll *a*—its ability to absorb the various light wavelengths. Chlorophyll *a* absorbs a large fraction of the photons in both the red (650–700 nm) and blue (400–450) ends of the spectrum, but it does not absorb the green light of intermediate wavelengths.

When a photon of light energy is absorbed, it changes the electron configuration of the absorbing molecule. Recall from Chapter 2 that the nucleus of an atom is surrounded by a cloud of electrons restricted to certain orbitals. Ordinarily, the inner, lowest energy orbitals of an atom in a molecule are occupied by electrons, while the higher energy orbitals are unoccupied. A molecule with all its electrons in their most stable low-energy orbitals is said to be in its **ground state.** A photon of the correct wavelength striking a pigment molecule in the ground state can transfer its energy to an electron and boost it to a higher energy orbital. This happens only if the energy of the photon is precisely equal to the difference in energy between the two orbitals. In other words, the absorption of light is an all-or-none phenomenon; two low-energy photons cannot "add" together to kick an electron to the higher state, nor can a high-energy pho-

ton give one-half or two-thirds of its energy to an electron. Thus whether a photon of a given wavelength is absorbed depends on the precise energy differences between a pigment molecule's ground state and higher energy orbitals.

Curiously, a photon that has been absorbed no longer exists. Its energy has not disappeared, however, but has been captured by the pigment molecule. That molecule, with an electron from one of its atoms in a higher energy orbital, is now in an *excited* state. It takes the incredibly short time of about 10^{-15} second to excite a molecule with a photon of visible light. But a molecule cannot stay excited for very long. It loses its excess energy and returns to the ground state in one of four ways. It may lose energy to neighboring molecules in the form of molecular motion (that is, heat). It may emit energy as a photon of light of slightly lower energy and longer wavelength in a process that is the reverse of absorption: *fluorescence.* It may pass its excitation energy to a neighboring molecule, boosting one of its electrons to a higher orbital so that the neighboring molecule reaches an excited state. Finally, it may make use of its excess energy to drive a chemical reaction. It is the last two mechanisms— transmitting excitation energy to other molecules and transforming it to chemical energy—that are important during the light reactions of photosynthesis.

Types of Photosynthetic Pigments

Plants, algae, and photosynthetic bacteria have a rainbow of pigments in cells, leaves, fronds, and fruits; many of these pigments participate in photosynthesis as *accessory pigments* to chlorophyll. Chlorophyll itself comes in many varieties: the primary photosynthetic pigment in green plants is chlorophyll *a*. Its structure is shown in Figure 8-5. The ring portion of the molecule contains many alternating double and single bonds, which, despite the impression given by the drawing, are not static. The sharing of electrons puts these bonds in constant flux; this dynamism gives all biological pigments, including chlorophyll, the capacity to absorb low-energy photons in the visible spectrum.

Chlorophyll *a* is present in all eukaryotic plants and algae as well as in the prokaryotic cyanobacteria. A similar pigment, chlorophyll *b*, differs only in the presence of an aldehyde group rather than a methyl group at the top of the ring. Chlorophyll *b* is found in higher plants, in green algae, and in one prokaryote. Other classes of chlorophylls, whose absorption spectra differ from those of chlorophyll *a* and chlorophyll *b*, are in the brown, golden-brown, and red algae and in photosynthetic prokaryotes other than cyanobacteria (Chapter 21).

All photosynthetic organisms contain both chlorophyll and one or more **carotenoid pigments** (Figure 8-6). Chemically unrelated to the chlorophylls, carotenoids

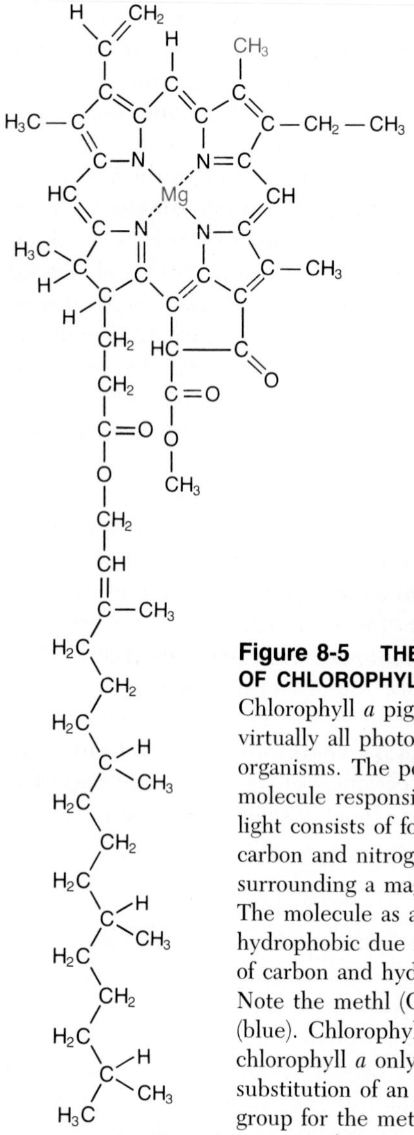

Figure 8-5   THE STRUCTURE OF CHLOROPHYLL A.
Chlorophyll *a* pigment is found in virtually all photosynthetic organisms. The portion of the molecule responsible for absorbing light consists of four rings of carbon and nitrogen atoms surrounding a magnesium atom. The molecule as a whole is hydrophobic due to the long chain of carbon and hydrogen atoms. Note the methl (CH$_3$) group (blue). Chlorophyll *b* differs from chlorophyll *a* only in the substitution of an aldehyde (CHO) group for the methyl group.

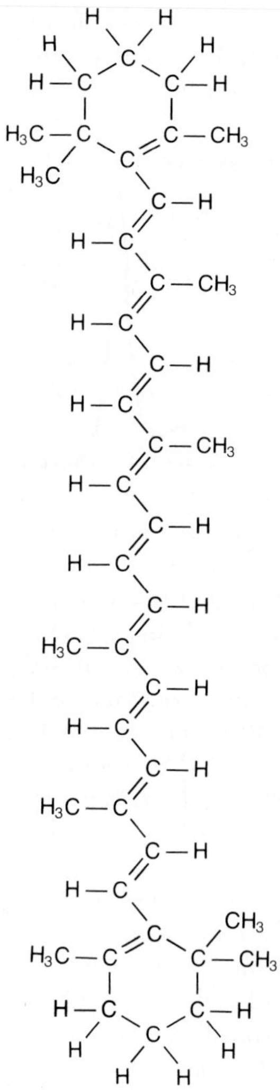

Figure 8-6   THE STRUCTURE OF β-CAROTENE.
β-carotene is a bright yellow pigment found in leaves and stems; red, orange, and yellow vegetables, fruit, and flowers; the skins and skeletons of some animals; and the feathers of some birds. β-carotene absorbs light in the 460 to 550 nm range and transmits the energy of this absorbed light to a photosynthetic reaction center.

absorb photons of wavelengths from 460 to 550 nm; therefore, they appear to our eyes as red, orange, or yellow (the wavelengths of light that carotenoids do not absorb). Unlike the chlorophylls, these accessory pigments are not restricted to photosynthetic organisms or organelles. They are present in many microorganisms and in some animals—pink carotenoids derived from a diet of aquatic animals color the feathers of flamingos. These pigments are also abundant in tomatoes and carrots, and they emblazon most autumn leaves after chlorophyll breaks down. Other pigments—purple and blue—that are neither carotenoids nor chlorophylls also participate in photosynthesis in some organisms.

Experiments have determined the precise wavelengths of light that are active in photosynthesis. This range of wavelengths is called the **action spectrum.** The action spectrum of photosynthesis is very similar to the absorption spectrum of chlorophyll; that correspondence helps establish that chlorophylls are the primary absorbers of light during photosynthesis.

Complexes of Photosynthetic Pigments

If chlorophyll *a* is the main photosynthetic pigment but other chlorophylls, carotenoids, and other pigments also participate in photosynthesis, one might expect the pigments to be associated physically and to function in concert. This, indeed, is so, and is one of the most interesting aspects of photosynthesis.

Hundreds of pigment molecules form compact aggregates on each thylakoid disk; each of these groups, called *antenna complexes*, contains chlorophyll molecules, proteins, and sometimes red, orange, blue, or purple pigments, depending on the organism. The name antenna complexes refers to the molecules' functioning together to gather light energy, just as an antenna gathers radio signals. Because various colored pigments absorb light in different parts of the visible spectrum, light energy of a broader range of wavelengths can be collected than if chlorophyll *a* did the job by itself. This occurs even though only specific wavelengths are directly useful in the photosynthetic event at chlorophyll *a*.

The molecules of an antenna complex are physically associated in such a way that energy from light striking any one of the antenna pigment molecules is funneled to a special chlorophyll *a* molecule, called a **reaction-center chlorophyll,** which directly participates in the events of photosynthesis (Figure 8-7). The reason energy from light is funneled to the reaction-center chlorophyll molecule is that all the other associated pigments in a given antenna complex have higher excitation energies. Thus they are able to pass the energy they absorb from high-energy photons "downhill" to the reaction-center pigment. This pigment molecule can absorb only slightly longer wavelength, lower energy light than does a normal chlorophyll *a* molecule. That may be because it is bound to different proteins than are the antenna chlorophylls.

Some biologists have likened the transfer of energy within an antenna complex to the type of atomic vibration that passes energy from molecule to molecule in a metal antenna. Others have compared the transfer to the movement of electrons in a semiconductor such as the silicon-based material in a computer chip or a solar cell. Regardless of the actual mode of energy transfer within the antenna complex, the passage is amazingly fast: each molecule-to-molecule excitation event occurs in about 10^{-12} second, or 1,000 times faster than it would take for the energy to be released as fluorescence or heat. Overall, the antenna complex operates so that a high proportion of the light energy striking chloroplasts is channeled to reaction-center chlorophylls, which pass it to an electron acceptor. Only then can it be stored as chemical energy in carbohydrate.

Within most photosynthetic organisms, there are two kinds of reaction-center chlorophyll molecules, each associated with its own light-harvesting pigments in the antenna complex. One type most strongly absorbs light with a wavelength of around 700 nm; it is called P700 (for pigment 700). The other type most strongly absorbs light with a wavelength of around 680 nm; it is called P680. Each type of reaction-center molecule is associated in the thylakoid membrane with two other molecules, an electron acceptor and an electron donor. Together, the reaction-center chlorophyll, the acceptor, and the donor are the major constituents in a *photosystem;* P700 is the

Figure 8-7   ANTENNA COMPLEX OF A PHOTOSYNTHETIC ORGANELLE.

An antenna complex is a group of photosynthetic molecules that gather and transmit light energy to a special type of chlorophyll *a* molecule, the reaction-center chlorophyll. The reaction-center chlorophyll uses the energy transferred from the other pigments to participate directly in photosynthesis.

Each circle represents a pigment molecule. After a photon is absorbed by one pigment, the energy from the photon is passed along to adjacent molecules (the yellow circles show the path) until it reaches a reaction-center chlorophyll (red). The energy of a photon sometimes escapes from the complex without exciting a reaction center (upper left).

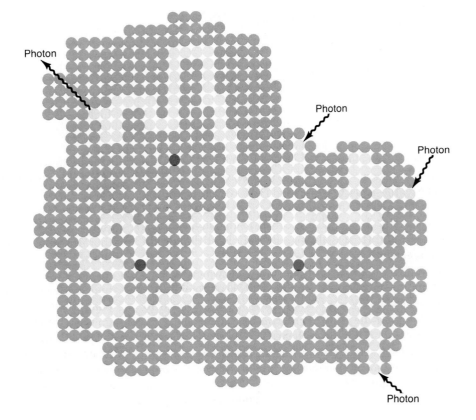

reaction center in **photosystem I,** while P680 is the reaction center in **photosystem II.**

When a P700 or a P680 molecule is excited by energy funneled down the associated antenna complex, one of its electrons is ejected and passed to the acceptor molecule; the nearby electron donor molecule then quickly fills the "hole" in the chlorophyll molecule with one of its own electrons. The physical association of the reaction-center, electron-acceptor, and electron-donor molecules in photosystems I and II is the key to the conversion of light energy to chemical energy during photosynthesis: the passage of an electron to the acceptor constitutes a reduction of the acceptor, while the contribution of an electron by the donor constitutes an oxidation of the donor. Therefore, when photons of light are trapped by antenna pigments, and the energy is passed to the reaction-center chlorophyll in a photosystem, there can be a conversion to chemical energy via oxidation-reduction reactions. Eventually, this energy is stored in carbohydrates—the basic nutrient source for all living things—during the dark reactions. First, however, it is stored in ATP and in another high-energy storage molecule called NADPH (the reduced form of $NADP^+$) during the light reactions.

Now that we have set the physical stage for photosynthesis, we are ready to consider the light and dark reactions in detail.

THE LIGHT REACTIONS: CONVERTING SOLAR ENERGY TO CHEMICAL-BOND ENERGY

The essence of the light reactions is the packaging of light energy in chemical compounds. During the course of this packaging, as we saw in the general equation for photosynthesis, H_2O is split and O_2 is released. As H_2O is cleaved, electrons are donated and eventually flow in an energetically "uphill" direction to the storage compound NADPH. It is the energy from light that drives this overall uphill, endergonic set of reactions. The process requires both photosystems I and II, which are embedded close to each other in the thylakoid membrane and are linked by an electron transport chain similar to that found in the inner membrane of a mitochondrion. Experiments reveal that photosystem II is responsible for the removal of electrons from H_2O and the release of O_2, while photosystem I brings about the reduction of the coenzyme $NADP^+$ to NADPH.

The Zigzag (Z) Scheme of Electron Flow

The light reactions of photosynthesis have been characterized as a zigzag, or Z, pathway, shown in Figure 8-8. In the daytime, when light strikes the pigment molecules associated with photosystem II, energy is funneled down through the antenna complexes, causing electrons to be ejected from P680 reaction centers. Immediately, electrons donated by an H_2O molecule fill the holes. The water molecule is cleaved by a water-splitting enzyme, and an oxygen atom is released during the electron donation—the oxidation of H_2O. Four photons of light are required for each pair of electrons donated by a single H_2O molecule. Thus, two H_2O's must be split, using eight photons, to yield an O_2 molecule.

The electrons ejected from the P680 reaction-center chlorophyll molecule are passed to the electron acceptor *plastoquinone* (PQ), which is reduced. The high-energy electrons deposited in plastoquinone next descend an electron transport chain composed of a series of electron-carrier molecules embedded in the thylakoid membrane between photosystems II and I that become sequentially oxidized and reduced by the descending electrons. Note that during the passage of electrons, two ATP molecules are formed for each two H_2O molecules cleaved and for each four electrons transported.

The final electron acceptor in the electron transport chain is P700, the reaction-center chlorophyll in photosystem I. Packets of light energy that are continually funneled to this molecule from its associated antenna complex cause electrons to be ejected sequentially. An electron passing down the electron transport chain from photosystem II can be accepted by P700 and boosted to a very high energy state—higher in energy than when it started down the chain from plastoquinone, and much higher than when it left P680 a fraction of a second earlier. Four photons from P700 are needed to boost the four electrons from the original two molecules of H_2O to a high enough energy level to be accepted by the electron acceptor for photosystem I.

Thus the streams of electrons continually donated to P680 from cleaved H_2O molecules are boosted one after the other because of the light striking photosystem II; they drop in energy as they flow down the electron transport chain, and they are finally boosted once again to a much higher energy level by light striking photosystem I and P700 (four more photons are needed at this point). The pathway can therefore be pictured as an upward zigzag or Z.

The electron acceptor for photosystem I (the top of the "zag") is an iron-containing protein called *ferredoxin,* which becomes reduced as it accepts high-energy electrons boosted from P700. Reduced ferredoxin is rapidly

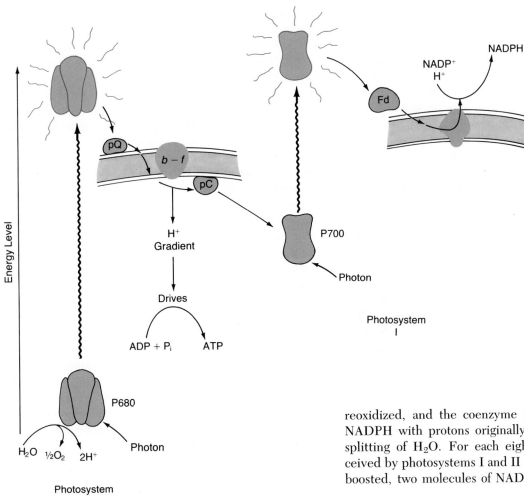

Figure 8-8   THE LIGHT REACTIONS OF PHOTOSYNTHESIS.
In this zigzag (Z) scheme, there is a net flow of electrons from water to NADP+. The process occurs in two series of reactions. In the first, water donates electrons to a photon-activated P680 molecule in photosystem II. High-energy electrons are passed to plastoquinone (pQ) and, from there, down an electron transport chain (b–f) to plastocyanin (pC). As that occurs, H+ is pumped through the thylakoid membrane, just as in the mitochondrial inner membrane. The resulting electrochemical gradient generates ATP. The electrons continue their energy descent to P700 of photosystem I, where another photon has provided the energy to activate the electrons to an even higher energy level. In the second series of reactions, the electrons pass to ferredoxin (Fd) and then are used to generate NADPH, which feeds into the dark reactions of photosynthesis. Because the Z scheme results in ATP production, it is also called *noncyclic photophosphorylation.*

reoxidized, and the coenzyme NADP+ is reduced to NADPH with protons originally generated during the splitting of H2O. For each eight photons of light received by photosystems I and II and each four electrons boosted, two molecules of NADPH are generated:

$$4e^- + 2H^+ + 2NADP^+ + 8 \text{ photons} \longrightarrow 2NADPH$$

These two energy-storing molecules as well as ATP formed during electron transport are used during the dark reactions of photosynthesis to reduce CO_2 to carbohydrate.

Recapping the events of the light reactions, for each eight photons that strike the P680 reaction center in photosystem II, two H2O molecules are cleaved, four electrons are donated, and one O2 molecule is released. The electrons boosted from P680 are accepted first by plastoquinone and then by successive carriers as they descend an electron transport chain, along which two ATP molecules are formed for each four electrons they pass down. The final acceptor in the chain is the P700 reaction center of photosystem I. Four photons of incoming light once again boost the energy of the four electrons, and ferredoxin accepts them and becomes reduced. As ferredoxin is reoxidized, the coenzyme NADP+ is reduced to NADPH; two NADPH molecules are formed for each eight photons of light fed into photosystems II and I. The ATPs and NADPHs then participate in the dark reactions.

Cyclic and Noncyclic Photophosphorylation

The production of ATP from the transport of electrons excited by light energy down an electron transport chain is called **photophosphorylation.** Photophosphorylation in chloroplasts is very similar to oxidative phosphorylation in mitochondria, described in Chapter 7. Free energy released by the electron transport chain is used to generate and maintain a proton gradient across the thylakoid membrane. Electron transport causes protons to be pumped through the membrane into the thylakoid inner space, thereby lowering the pH inside the thylakoid (Figure 8-9); thus this space is equivalent to the outer compartment of the mitochondrion, where protons also accumulate. Because this sets up a proton gradient across the membrane, protons flow back into the chloroplast stroma, passing through coupling-factor proteins embedded in the thylakoid membrane. As protons flow out through the coupling-factor channel, ATP is formed. (Note that coupling factor is oriented the same way relative to proton movement and ATP production in chloroplasts and mitochondria—as protons flow through, down their concentration gradient, ATP is produced.) Therefore, the Mitchell chemiosmotic theory applies to chloroplasts as well as mitochondria.

The zigzag scheme of electron flow through photosystems II and I is called **noncyclic photophosphorylation.** What does "noncyclic" mean? In the zigzag scheme, electrons flow from H_2O to NADPH in a one-way sequence through a series of pigments, proteins, and energy carriers. As we saw, the starting point is P680 and the ending point is ferredoxin; this pathway is therefore linear rather than cyclic. Apparently, most of the ATP produced in chloroplasts of photosynthetic organisms is the result of noncyclic photophosphorylation. (Of course, mitochondria are also present in eukaryotic photosynthetic cells, and they also produce ATP.)

Biologists have discovered that plants derive additional ATP by means of a marvelous trick called **cyclic photophosphorylation.** When the bulk of the $NADP^+$ molecules have been reduced to NADPH, electrons can be cycled from photosystem I back through the electron transport chain between photosystems II and I (Figure 8-10). Light energy striking P700 in photosystem I boosts the electrons, which are shunted back from ferredoxin to an earlier carrier in the electron transport chain (plastoquinone). As the electrons again descend the chain toward P700, ATP is generated. Since photosystem II is not involved, no further NADPH is generated and no O_2 is released. Incoming light can boost electrons through this same cycle again and again with a net production of ATP. Thus *cyclic* photophosphorylation is an

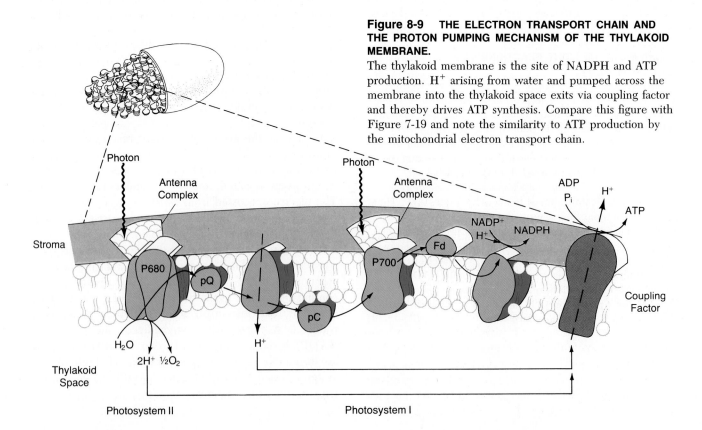

Figure 8-9   THE ELECTRON TRANSPORT CHAIN AND THE PROTON PUMPING MECHANISM OF THE THYLAKOID MEMBRANE.

The thylakoid membrane is the site of NADPH and ATP production. H^+ arising from water and pumped across the membrane into the thylakoid space exits via coupling factor and thereby drives ATP synthesis. Compare this figure with Figure 7-19 and note the similarity to ATP production by the mitochondrial electron transport chain.

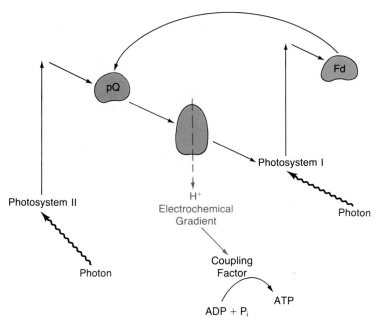

Figure 8-10   CYCLIC PHOTOPHOSPHORYLATION.
High-energy electrons can be reused (cycled) to generate additional ATP. The same molecules seen in Figures 8-8 and 8-9 are present here. But instead of being accepted by ferredoxin, high-energy electrons are shuttled back to plastoquinone: as they pass along the electron transport chain to photosystem I, their energy drives the proton pump and thus ATP synthesis. If photons continue to add energy to P700, this closed loop of electron flow will cycle again and again, producing many ATP molecules. If it does, however, the electrons will not pass from ferredoxin to $NADP^+$, and therefore the NADPH needed for the photosynthetic dark reactions will not be made.

ATP-generating pathway driven only by light energy.

In summary, the two types of photophosphorylation produce a minimum of four moles of ATP for every two moles of NADPH; cyclic photophosphorylation yields only ATP, while noncyclic photophosphorylation yields both ATP and NADPH. This ratio of ATP to NADPH is appropriate to provide the energy-rich compounds required for the synthesis of carbohydrates in the dark reactions. Therefore, we can say that the light reactions—in keeping with the splendid economy of nature—package enough light energy in an entirely appropriate chemical form, ready to be stored in the sugar molecules that ultimately fuel all life activities.

THE DARK REACTIONS: BUILDING CARBOHYDRATES

The dark reactions of photosynthesis, which take place in the chloroplast stroma, use the organic products of the light reactions to store chemical energy in carbohydrates. On the surface, it seems logical that chemical energy would be stored in carbohydrates, since we know that fermentation and cellular respiration act specifically on the glucose subunits of carbohydrates to release chemical energy. However, throughout the book, ATP—not carbohydrates—has been called the "universal energy currency" in all cells. If photosynthetic organisms stored energy directly in the form of ATP, neither they nor the majority of living species would need complex pathways for nutrient breakdown. So why do they not do just that?

There are several reasons. For one thing, ATP is a relatively unstable, reactive compound. As a polar, charged molecule, it is difficult to move across membranes. In contrast, glucose and related compounds can be built into densely packed, inert storage macromolecules, such as the starch of plants and the glycogen of animals. In these inert forms, even a huge number of glucose molecules will not affect the critical osmotic properties of cellular or extracellular fluids. Moreover, carbohydrates are a good starting point for constructing the subunits of proteins, nucleic acids, and lipids. Equally important, carbohydrates are a highly efficient energy reservoir—when "burned," one molecule of glucose, which is smaller than one of ATP, releases energy equivalent to about thirty-six ATPs. In effect, carbohydrates are much more compact and effective storage and transport units than is ATP; it is as though carbohydrates are "one hundred dollar bills," not commonly used in daily life, whereas ATPs are the "coins" used so much in the daily cellular economy.

Earlier we mentioned that the dark reactions of photosynthesis convert six CO_2 molecules to glucose, $C_6H_{12}O_6$. Furthermore, we said that certain products of the light reactions also contribute to the dark reactions; these products are the energy-storing molecules ATP and NADPH, which drive the dark reactions. Straightforward as that sounds, elucidating the pathway by which CO_2 is reduced to carbohydrate in the presence of ATP and NADPH was a major advance in our understanding of photosynthesis. Melvin Calvin and Andrew Benson at the University of California, Berkeley, worked out the pathway in the early 1950s by tracing the fate of carbon dioxide labeled with radioactive carbon (^{14}C). They made the unexpected discovery that a three-carbon molecule, 3-phosphoglycerate (abbreviated PGA, for the acid form, phosphoglyceric acid), is the first *stable* intermediate in the pathway. They also discovered that a short-lived precursor of PGA, an electron acceptor for atmospheric CO_2, is a five-carbon compound called **ribulose bisphosphate (RuBP)**. Subsequent research revealed that **ribulose bisphosphate carboxylase** catalyzes the first reaction in this cycle. This protein is a key enzyme in a pathway so central to life on Earth that it accounts for up to 50 percent of all the protein in chloroplasts. In addition, it works slowly, so that many molecules are needed in each chloroplast. All in all, it probably is the single most abundant protein found in nature.

In recognition of their contributions to our knowledge of photosynthesis, the dark reactions are also called the **Calvin-Benson cycle.** During this series of reactions, atmospheric carbon dioxide is first *fixed*, or incorporated into an organic compound, and then reduced to carbohydrate subunits. The first several steps of the cycle produce the precursors of glucose and other carbohydrates. The last steps regenerate the precise quantity of ribulose bisphosphate needed to keep the cycle going. The main steps of the pathway are outlined in Figure 8-11.

The abundant enzyme ribulose bisphosphate carboxylase catalyzes the reaction of five-carbon RuBP with atmospheric CO_2. An unstable six-carbon intermediate is formed and breaks down immediately to two molecules of 3-phosphoglycerate. In the next step, two ATPs are "spent" to phosphorylate this three-carbon sugar into an activated form, 1,3-diphosphoglycerate (DPGA). We last saw DPGA as an intermediate in glycolysis (Chapter 7). In fact, the next steps of the Calvin-Benson cycle correspond to glycolytic steps in reverse. First, DPGA can accept electrons and hydrogens from the NADPH produced by the light reactions of photosynthesis, and thereby it is reduced to glyceraldehyde-3-phosphate (GLYC-P). After an intermediate step, two of those three-carbon molecules condense to the six-carbon compound fructose diphosphate. That completes the reduction of CO_2 to carbohydrate. Fructose diphosphate can be converted to glucose or built into the disaccharide sucrose or larger polymers such as starch. For this cycle to continue, it is essential to regenerate RuBP. That is done using GLYC-P, additional ATPs, and a series of reactions.

Overall, in order to convert six molecules of CO_2 to one molecule of the six-carbon sugar fructose diphosphate, the cycle must go around six times. This is because only one carbon atom of each six carbons in RuBP plus CO_2 can be used for purposes *other than* regenerating the five-carbon RuBP needed for the cycle to continue. The cost of six turns of the cycle is eighteen molecules of ATP and twelve molecules of NADPH. This can be summarized:

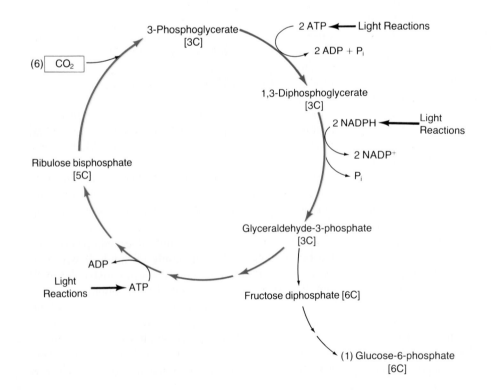

Figure 8-11
THE CALVIN-BENSON CYCLE.
The Calvin-Benson cycle is the set of dark reactions of photosynthesis that builds carbohydrates. The cycle must go around six times in order to convert six molecules of CO_2 to one molecule of the six-carbon sugar fructose diphosphate. For each turn of the cycle, ribulose bisphosphate must be regenerated from the other five carbons. The cycle is driven by energy from the light reactions in the form of NADPH and ATP. The number of carbon atoms in each compound is shown in brackets [ ]; the numbers of CO_2 and glucose molecules are enclosed in parentheses ( ).

$$6CO_2 + 6RuBP + 18ATP + 12NADPH + 6H_2O \rightarrow$$
$$\text{Fructose diphosphate} + 10GLYC\text{-}P + 12P_i + 18ADP + 12NADP^+$$

This is a very large cost indeed. Nevertheless, the Calvin-Benson cycle is much more efficient than are most human-made machines: twelve NADPH and eighteen ATPs represent an expenditure of 750 kcal/mole, but about 680 kcal, or 90 percent, of that energy is stored in the glucose molecule that is formed.

The fact that so much ATP is consumed in running the cycle emphasizes how important the products of the light reactions are to the overall photosynthetic scheme. Note how the dark reactions dovetail with the light reactions: the ATP and NADPH produced by the light reactions fuel the dark reactions, with little waste of chemical energy. And thanks to the organization of chloroplasts, that energy is always in the right place at the right time.

THE KEY ROLE OF CHLOROPLAST ARCHITECTURE IN PHOTOSYNTHESIS

Chloroplasts are an excellent example of the interplay of form and function in a biological system. These organelles have been compared to solar electric cells, which also absorb light energy but convert it to electrical energy instead of chemical energy. Also, chloroplasts act as charge separators, establishing a gradient of positive and negative charges that carries an energy current, just as do the wires in a solar electric cell. However, chloroplasts are far more complex than any silicon semiconductor device. Photosynthetic pigments and proteins are arrayed in the organelle for maximum functional efficiency. In fact, the highly ordered arrangement of these molecules is the key to their ability to trap light and to fix and reduce CO_2.

The light reactions, as we saw, take place in the thylakoid membrane and enclosed thylakoid space; the dark reactions occur in the stroma. The thylakoid membrane is unique to the chloroplasts: the space it encloses has no counterpart in other organelles. The pigments and enzymes for all the steps of the light reactions are integral parts of that unique membrane.

By using detergents to disrupt thylakoid membranes, biologists have isolated molecular complexes associated with both reaction centers P680 and P700. This is strong evidence that the light-harvesting antenna complexes are not simply mental constructs to help us visualize the funneling of light energy, but are actual physical entities as well as functional units.

Experimental evidence also indicates that elements of both photosystems span the thylakoid membrane. For example, the water that acts as an electron donor to P680 is oxidized inside the thylakoid space; but plastoquinone, which accepts excited electrons from P680, is embedded on the outer, stromal side of the thylakoid membrane. Thus the electrons must migrate through the membrane in this first stage of photosynthesis. Similarly, the electron donor for P700 resides on the thylakoid space side of the membrane, while the acceptor of excited electrons from P700, ferredoxin, lies on the outer, stromal side of the membrane. As in mitochondria, it is during these movements of electrons across the membrane that H^+ ions are pumped into the thylakoid space. The final reaction involving ferredoxin thus leaves NADPH in the stroma, where the dark reactions take place. Moreover, as we explained earlier, the proton gradient by which ATP is produced functions through coupling factors that span the thylakoid membrane. This arrangement of molecules ensures that the chemical energy produced by the light reactions ends up in the stroma, where it is needed for carbohydrate production.

Most of the enzymes active in the Calvin-Benson cycle are dissolved in the stroma, which is a solution of proteins and small molecules, and also contains chloroplast DNA and protein-synthetic machinery. It is thought that RuBP carboxylase, the pivotal enzyme that fixes CO_2 to RuBP, may be loosely associated with the exterior of the thylakoid membrane. It is this marvelous ordering of photosynthetic molecules that allows their controlled interaction and so photosynthesis itself.

OXYGEN: AN INHIBITOR OF PHOTOSYNTHESIS

The storage depots of chemical energy we call carbohydrates are the vital products of photosynthesis; they make the solar energy harvested by the light reactions available to cells. The photosynthetic process also yields inorganic waste products, H_2O and O_2. Ironically, even though O_2 is essential for the aerobic respiration of glucose and other nutrients within the plant, too much O_2 can be harmful.

In the presence of excited chlorophyll molecules, oxygen can assume high-energy states that are toxic to the cell. In normal plants, carotenoid pigments protect against this phenomenon, known as *photooxidation*. Mutant plants without carotenoids eventually die from photooxidation.

High levels of O_2 can also cause **photorespiration,** an inefficient form of the dark reactions that fixes O_2 instead of CO_2 and does not produce carbohydrates. When a plant is drying out because of high temperatures, low humidity, and bright sunlight, tiny openings in the leaves called *stomata* (singular, *stoma*) close to prevent the evaporative loss of water vapor. Photosynthesis can continue at a greatly reduced rate, but the closure of these "portholes" prevents the waste product O_2 from escaping and most CO_2 molecules from entering. Thus there is a build-up of O_2 and a depletion of CO_2.

Under these conditions, the important enzyme RuBP carboxylase begins to join O_2 to some molecules of RuBP in the first step of the Calvin-Benson cycle, instead of fixing only CO_2. Thus oxygen and carbon dioxide compete for the active site of the enzyme; the more oxygen present, the more will be joined to RuBP. This competition takes place to some degree even under normal conditions, since air contains 500 times more O_2 than CO_2. However, in a plant with partly or fully closed stomata due to drought, the ratio of O_2 to CO_2 fixed is even higher.

When O_2 is added to RuBP, both *phosphoglycolate* and some 3-phosphoglycerate are formed. The phosphoglycolate molecules can be oxidized to CO_2 in refuse organelles called *peroxisomes* (Chapter 6). However, no ATP is generated; no net carbohydrate is produced; and CO_2, originally fixed at such great expense, is lost. Thus the plant uses precious energy to metabolize phosphoglycolate and suffers a net loss of fixed carbon as a result. Meanwhile, some photosynthesis goes on: the CO_2 molecules in low concentration in the leaf tissue are fixed by RuBP carboxylase, reduced to carbohydrates, and stored for the plant's later energy or structural needs.

REPRIEVE FROM PHOTORESPIRATION: THE C_4 PATHWAY

Because of photorespiration, the majority of plants experience decreased carbohydrate production and thus slowed growth when the weather is hot and dry. These plants are called C_3 **plants** because the first stable intermediates in the Calvin-Benson cycle—whether the productive 3-phosphoglycerate or the nonproductive phosphoglycerate—are three-carbon sugars. Another group, the C_4 **plants,** have a special leaf anatomy and a unique biochemical pathway that begins with a stable four-carbon intermediate. These adaptations allow such C_4 plants as crab grass, corn, and sugar cane to photosynthesize at a faster rate in hot, dry climates than can the C_3 plants (Figure 8-12).

The leaf anatomy of C_3 and C_4 plants is compared in

Figure 8-12 C_3 AND C_4 PLANTS.
(a) Most of the grass in this ordinary lawn is C_3 plants, but clumps of crab grass, a hardy and ubiquitous C_4 plant, are present, too. (b) By midsummer under conditions of minimal watering, most of the C_3 grass is dried and brown because C_3 plants cannot photosynthesize under hot, dry conditions. The C_4 crab grass, however, is still green and growing, since it can carry on photosynthesis despite the blistering weather.

Figure 8-13. Both have a watertight, airtight "skin" called the *cuticle* and *stomata*, through which gases and water vapor enter and exit the leaf. Both also have large *mesophyll cells*, which contain chloroplasts, and a network of hollow veins, which transport water, nutrients, and minerals throughout the plant. In C_3 plants, the mesophyll cells are concentrated beneath the upper leaf surface but also occur in a spongy layer throughout the

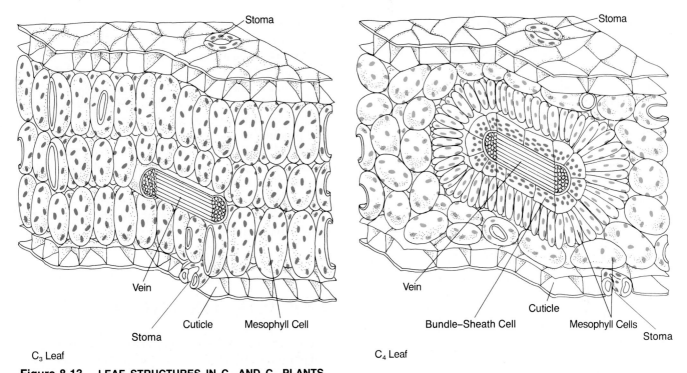

C₃ Leaf   C₄ Leaf

Figure 8-13   LEAF STRUCTURES IN C₃ AND C₄ PLANTS.
Both C₃ and C₄ plants have mesophyll cells, which contain numerous bright-green chloroplasts. However, the C₄ plants also have an inner layer of bundle-sheath cells. The dark reactions of photosynthesis take place only in these bundle-sheath cells, and in C₄ plants, the surrounding mesophyll cells act to "pump" CO_2 inward toward them.

leaf. All photosynthesis in C₃ plants is carried out in these mesophyll cells. In C₄ plants, the veins are surrounded by two layers—an outer layer of mesophyll cells and an inner layer of *bundle-sheath cells*—both of which contain chloroplasts. However, the Calvin-Benson cycle takes place only in the bundle-sheath cells because they contain all the RuBP carboxylase enzyme in the entire leaf. Carbohydrate production therefore occurs only in a layer of cells deep inside the leaf that is insulated from the high concentrations of oxygen in the air. In addition, the mesophyll cells actually carry out a kind of CO_2 "pumping" and thereby concentrate CO_2 and supply its carbon to the bundle-sheath cells in the form of a four-carbon compound. Thus there is little competition by O_2 for the active site of the RuBP carboxylase enzymes located in chloroplasts of the bundle-sheath cells.

The pumping of CO_2 by mesophyll cells begins with a special enzyme called phosphoenolpyruvate carboxylase, which is present in the cytoplasm of the mesophyll cells in C₄ plants. (It is absent in C₃ mesophyll cells.) This enzyme fixes CO_2 molecules to molecules of phosphoenolpyruvate (PEP), a three-carbon compound. (Note that the term "fix" refers to the combining of a gaseous molecule with another compound.) The result is

the four-carbon compound oxaloacetate. (Notice that some of these compounds are also intermediates in the Krebs cycle.) In many C₄ plants, this compound is transported into chloroplasts of the same mesophyll cells, where it is reduced by NADPH to another four-carbon compound, malate. Malate acts as a carbon "delivery person"; it diffuses through plasmodesmata (Chapter 5) into the bundle-sheath cells, where it is cleaved to the three-carbon compound pyruvate. In the process, CO_2 is released and can enter the Calvin-Benson cycle (Figure 8-14).

Once CO_2 is released from the malate "delivery person" and pyruvate is formed, this three-carbon "leftover" is transported back to the mesophyll cells, where it is transformed once again to PEP, ready for a new cycle of CO_2 incorporation and transport. Not surprisingly, the extra CO_2-fixing pathway in C₄ plants requires energy to regenerate PEP. This energy is the price the plant must pay to concentrate CO_2, avoid photorespiration, and carry out carbon fixation at a fast rate despite drought or intense sunlight. However, the gain in photosynthetic efficiency under adverse conditions more than offsets the cost.

The ability of C₄ plants to thrive in arid conditions makes them superior candidates for agriculture, and

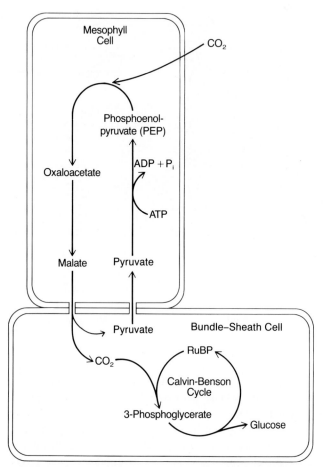

Figure 8-14   THE PATHWAY OF C₄ PHOTOSYNTHESIS: A SPECIAL CO₂ PUMP.

The C_4 pathway involves some molecules that are not involved in C_3 photosynthesis but that are present during glycolysis and the Krebs cycle. The enzyme PEP carboxylase fixes CO_2 molecules to PEP, a three-carbon compound. The result is a four-carbon compound called oxaloacetate, which in turn is reduced to a different four-carbon compound, malate. Malate diffuses into the bundle-sheath cells, where it is converted to the three-carbon molecule pyruvate. In the process, CO_2 is released to take part in the Calvin-Benson cycle. PEP is regenerated with the expenditure of energy from ATP.

they grow faster than C_3 plants. Research is currently under way to transform such important C_3 crop plants as tomatoes, soybeans, and cereals into C_4 plants. However, even with a combination of classical breeding techniques and the latest methods of genetic engineering, this will be a difficult task because the C_4 pathway involves not only special enzymes, but also specialized cellular structure and leaf anatomy.

Although oxygen can have a harmful effect on the energy-storing process of photosynthesis, it is important to remember that oxygen is an integral component of the energy commerce of all aerobic organisms, including autotrophs. Animals, fungi, and most bacteria as well as plants and algae use the atmospheric oxygen derived from photosynthesis to generate large amounts of ATP during cellular respiration. Thus mitochondria in the same cells with chloroplasts are major users of O_2 for ATP production. Large size, high growth rate, and locomotion of multicellular organisms evolved only after photosynthetically produced O_2 made cellular respiration possible. Globally, then, the disadvantages of oxygen for photosynthesis are more than offset by the physiological advantages of oxygen as the driving force of cellular respiration.

THE CARBON CYCLE

The dependence of most life forms on carbohydrates and oxygen is so complete that it is difficult to imagine a time when photosynthetic organisms did not exist. The earliest cells, however, did not contain chlorophyll and did not carry out photosynthesis. As we shall see in Chapter 19, early cells consumed the original organic molecules present in the environment and due to changes in environmental conditions, no more molecules could be formed. Such cells did not give off O_2, and very little O_2 was present in the atmosphere. Ultimately, the organic molecules became depleted, and life probably would have slowly died out had the photosynthetic pathway not evolved.

The emergence of photosynthetic organisms literally changed the face of the planet. By fixing carbon dioxide, they helped reduce high levels of atmospheric CO_2, which can trap heat near the Earth's surface in the so-called *greenhouse effect*. This reduction helped cool the surface, thereby increasing the size of the polar ice caps and lowering the sea levels. As the waters receded and land was exposed, new habitats opened up for organisms. By freeing oxygen in huge quantities, photosynthetic organisms literally revolutionized the atmosphere and everything bathed in it. That free oxygen created the potential for a great diversity of aerobic life forms with their large ATP needs. The simultaneous activities of photosynthesis and respiration established the **carbon cycle**: photosynthetic organisms fix CO_2 into carbohydrate molecules, which are taken in by all organisms, eventually resulting in catabolism and the release of CO_2 back to the atmosphere (Figure 8-15). That replenishment makes enough CO_2 available for continuing rounds of photosynthesis in millions of acres of forests, grasslands, lakes, and sea surfaces: as CO_2 is fixed, O_2 is re-

Figure 8-15   THE GLOBAL CARBON CYCLE.
Photosynthetic organisms fix CO_2 into organic molecules, which are then used as energy compounds by both photosynthetic and nonphotosynthetic organisms. As organic molecules are "burned" in the fires of respiration, O_2 is used and CO_2 is released back into the atmosphere. The burning of fossil fuels derived from dead organisms also uses O_2 and releases CO_2. Decomposers include fungi and various microorganisms.

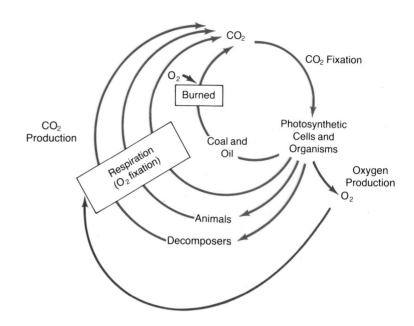

leased; then as O_2 is fixed during aerobic respiration, CO_2 is released back to the atmosphere. Of the 150 billion tons of carbohydrates generated by photosynthesis each year, biologists estimate that 80 or 90 percent is produced by algae. This should not be surprising since 87.5 percent of the Earth is covered by water, but if we look at the vast tracts of forests, agricultural crops, and grasses around us we can begin to appreciate the enormity of the total carbohydrate production.

The continuous cycling of CO_2 maintains the delicate ecological balance on which almost all living organisms depend. That is why ecologists and geologists caution against unrestricted burning of fossil fuels, uncontrolled use of the internal-combustion engine, and large-scale deforestation. All three tamper with the carbon cycle, causing more CO_2 to be released into the atmosphere than photosynthetic organisms are capable of fixing.

It is quite remarkable that reciprocal metabolic events such as photosynthesis and cellular respiration—occurring in a molecular realm far too small to be witnessed directly—could produce such global effects. The carbon cycle demonstrates on the broadest scale the importance of metabolism. But that importance is equally manifest in the simplest activities of living things. Without the building up and breaking down of nutrients, the energy of the sun would not flow through the living world in a usable form, and all life dependent on solar energy would cease to be.

SUMMARY

1. Virtually all contemporary life forms on Earth depend on photosynthetic organisms, directly or indirectly, for the chemical energy and organic building blocks that sustain vital activities.

2. *Photosynthesis* in green plants transforms molecules of water and carbon dioxide into carbohydrate and gaseous oxygen. Sunlight supplies the energy for this endergonic process.

3. Van Niel's study of photosynthetic bacteria provided a general model for all types of photosynthesis. Experimental proof of van Niel's model confirmed that the photosynthetic equation actually summarizes two partial reactions. In the first reaction, water is oxidized to produce ATP, NADPH, and oxygen. In the second, carbon dioxide is reduced, using ATP and NADPH to produce carbohydrate.

4. The reactions of the first stage of photosynthesis use light directly and thus are called *light reactions*. Although the reactions of the second stage depend on the products of the light reactions, they do not require light directly and thus are called *dark reactions*. Both series of reactions take place in the *chloroplast*. The *thylakoid* membranes are sites of the light reactions, and the central stroma of the dark reactions.

5. The essence of the light reactions is the conversion of solar energy to chemical energy. The conversion begins with the interaction of sunlight and the pigment *chlorophyll*, in which *photons* of light are absorbed, causing excitation of electrons. The *absorption spectrum* of chlorophyll or other compounds reveals the degree to which different wavelengths of light are absorbed.

6. More than one kind of pigment absorbs the light energy that powers photosynthesis. Several types of chlorophyll plus *carotenoids* are active in the process. Antenna complexes, composed of many light-harvesting pigment molecules and a *reaction-center chlorophyll* molecule, ensure the high efficiency of light capture and conversion.

7. Reaction-center chlorophylls are the sites where energy from photons energizes electrons, which then can be passed down electron transport chains, or a series of oxidation-reduction reactions that produce ATP and NADPH. In plants, reaction-center chlorophyll P680 is associated with *photosystem II*, and reaction center P700 is associated with *photosystem I*.

8. The electron transport chains provide free energy that is used to generate ATP and NADPH for the dark reactions. The production of ATP from the transport of electrons excited by sunlight is called *photophosphorylation*. Two types of photophosphorylation, *cyclic* and *noncyclic*, produce enough ATP to drive the dark reactions. ATP is produced in the thylakoids of chloroplasts by the same chemiosmotic process that operates in mitochondria.

9. In the production of NADPH, four electrons and two protons derived from two molecules of water are passed to two molecules of $NADP^+$. The energy of each electron is boosted by two photons, one at P680 and one at P700.

10. The dark reactions reduce CO_2 to carbohydrate in what is known as the *Calvin-Benson cycle*. To form one new molecule of the six-carbon sugar fructose diphosphate, six molecules of CO_2 must enter the pathway. Regeneration of *ribulose bisphosphate* (RuBP) completes the cycle each time a CO_2 molecule passes through it. The enzyme *ribulose bisphosphate carboxylase* catalyzes the key CO_2 fixation step.

11. The highly ordered arrangement of enzymes, pigments, and coupling factor within chloroplasts ensures the outcome of photosynthesis. The light reactions, which are carried out in association with the thylakoid membrane, produce chemical energy (ATP, NADPH) in the stroma surrounding the thylakoid stacks, where it is used to fuel the dark reactions.

12. The oxygen produced in photosynthesis can harm chlorophyll-containing cells by forming compounds that are toxic to the cell or by contributing to an inefficient form of the dark reactions known as *photorespiration*. During photorespiration, oxygen rather than carbon dioxide is fixed, and carbohydrates are not produced.

13. C_4 *plants* reduce photorespiration by carrying out photosynthesis only in cells well insulated from high O_2 levels. They also use a novel method of CO_2 fixation, which is not available to C_3 *plants*. The ability of C_4 plants to thrive in conditions of high light intensity and low water availability makes them ideal crop plants.

14. Photosynthesis works in tandem with respiration to foster the *carbon cycle*, an energy cycle that sustains most life on Earth.

KEY TERMS

absorption spectrum
action spectrum
C_3 plant
C_4 plant
Calvin-Benson cycle
carbon cycle
carotenoid pigment
chlorophyll
chloroplast
cyclic photophosphorylation
dark reactions
ground state
light reactions
noncyclic photophosphorylation
photon
photophosphorylation
photorespiration
photosynthesis
photosystem I
photosystem II
pigment
reaction center chlorophyll
ribulose bisphosphate (RuBP)
ribulose bisphosphate carboxylase
thylakoid

QUESTIONS

1. How did the study of photosynthesis in purple sulfur bacteria enable van Niel to understand photosynthesis in green plants? What are the reactants in each case? What are the final products? What is the source of the H atoms that reduce CO_2 to carbohydrates? What is the source of the waste product O_2 or S? What is the source of the energy that drives the reaction?

2. Photosynthesis consists of two sets of reactions: the light reactions

and the dark reactions. What is accomplished by the light reactions? By the dark reactions? Which set of reactions is also called the Calvin-Benson cycle?

3. What is the carbon source for most plants? The energy source? What do most animals use as a carbon source? An energy source?

4. What is an antenna complex, and what is its function? What is a reaction-center chlorophyll?

5. Examine the molecular structures of chlorophyll *a* and *β*-carotene. Are these molecules hydrophobic or hydrophilic? How do you know? Would you expect to find these molecules embedded in membranes or dissolved in the cytoplasm?

6. The light reactions are often pictured as a zigzag pathway lying on its side. In the first ascending vertical line, electrons are ejected by reaction-center chlorophyll P680 of photosystem II, in response to light. What molecule donates electrons to fill the holes in P680? As the electrons descend the diagonal path, some of their energy is captured. In what form? The electrons are then accepted by reaction-center chlorophyll P700 of photosystem I, which boosts them to a much higher energy level. As the electrons start down a second diagonal, their reducing power is captured by what important coenzyme?

7. The Z scheme is called noncyclic photophosphorylation. What is

cyclic photophosphorylation? What parts of the Z are involved?

8. The first step in the Calvin-Benson cycle is the reaction of CO_2 with ribulose bisphosphate, a 5-carbon compound. Which abundant enzyme (possibly the most abundant protein on Earth) catalyzes this reaction? Which molecule contributes the electrons and hydrogen atoms that reduce sugars in the Calvin-Benson cycle? Which molecule was the original donor of the hydrogen atoms (in the light reactions)? In what sense is the Calvin-Benson cycle cyclic?

9. Do C_4 and C_3 plants share the same light reaction enzymes? Do both use the Calvin-Benson cycle? How are they different?

10. Describe our planet's carbon and oxygen cycles. Which compounds contain these atoms as gases, liquids, or solids? What is meant by the "greenhouse effect"? What first gave rise to Earth's oxygen-rich atmosphere?

ESSAY QUESTIONS

1. Why are pigment molecules so critical to photosynthesis? What would it mean if the molecules of the antenna complex appeared orange instead of green to the human eye?

2. What is photophosphorylation? Where on the Z pathway does it take place? In what ways is it similar to oxidative phosphorylation?

3. Would you expect that cells containing chloroplasts would also contain the enzymes for glycolysis, the Krebs cycle, or oxidative phosphorylation? Explain your answer.

SUGGESTED READINGS

BIDWELL, R. G. S. *Plant Physiology.* 2d ed. New York: Macmillan, 1979.

This text contains an in-depth treatment of photosynthesis.

BOGORAD, L. "Chloroplasts." *Journal of Cell Biology* 91 (1981): 256s–270s.

A superb review of the structure of chloroplasts and a guide to much other literature.

BONNER, J., and J. E. VARNER *Plant Biochemistry.* 3d ed. New York: Academic Press, 1976.

Several chapters in this book give an excellent description of the energetics and biochemistry of photosynthesis.

MILLER, K. R. "The Photosynthetic Membrane." *Scientific American,* October 1979, pp. 102–113.

This well-illustrated paper provides a good summary of photosynthesis and references to earlier *Scientific American* papers on the subject.

Part TWO

LIKE BEGETS LIKE

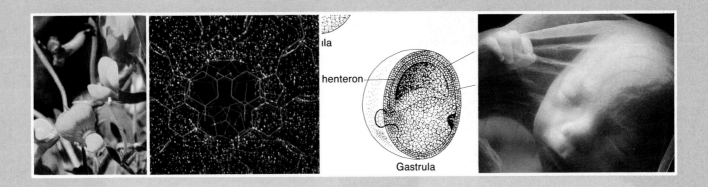

ıla

henteron

Gastrula

An aged oak topples, but in its place an acorn germinates and a new tree grows. New generations, resembling the old but living on as the former die, are central to the continuation of life. At the core of this succession lies reproduction, the generation of offspring. Reproduction depends on inheritance, the passing on of traits from adult to young, and on development, the unfolding of the traits as one cell gives rise to a complex many-celled organism.

All reproduction involves the division of cells—each cell splitting into two daughter cells—and depends on a few basic principles. First, proteins and hereditary molecules (DNA and RNA) contain information stored in the linear order of their building blocks. Second, a molecular copying mechanism transfers that information from the hereditary molecules in a parent cell to those in daughter cells and translates the information into cellular structures and activities. Third, hereditary material is packaged in units (chromosomes); higher-order processes within a dividing cell help distribute one copy of the units to one daughter and an identical copy to another.

The discovery of these principles, all of which we will explore in detail, has led to some of humankind's greatest insights into the workings of nature. They are the basis of understanding the elegant rules of heredity in pea plants, fruit flies, molds, bacteria, domestic animals, and people, and of the genetic engineering techniques that today allow us to alter living things in unprecedented ways. The principles also underlie the process of development, during which a marvelous coordination of events triggers the right genetic information to be activated at the right time so that an animal's heart, appendages, and brain form in the correct positions and operate appropriately for the organism's survival. (Chapter 28 in Part IV will describe the analogous developmental processes in plants.)

The study, in Part I, of atoms and molecules, energy transformations, and the forms and functions of living cells, continues, as we consider cellular reproduction, the chemical basis of heredity, and the supremely complex but beautiful unfolding of genetic information in the developing embryo. An understanding of genetics—how like begets like—prepares the way for all the remaining topics in biology: the variety of life forms on Earth, how they evolved, how their internal systems function, and how they relate to each other and to their physical environment.

9

CELLULAR REPRODUCTION: MITOSIS AND MEIOSIS

The cell cycle is the fundamental unit of temporal organization in the history of a population of well-nourished cells.

John J. Tyson and Wilhelm Sachsenmaier,
Cell Cycle Clocks (1984)

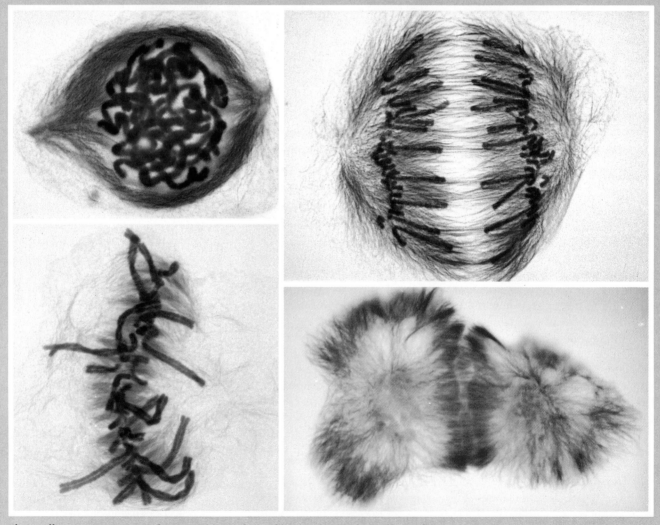

Plant cells in various stages of mitosis: (a) prophase; (b) metaphase; (c) anaphase; (d) telophase (all magnified about 2,700 times).

Each minute of every day, 2 million red blood cells are generated in a healthy person's bone marrow. The new cells are almost identical and have the same unmistakable concave disk shape and reddish color. One might wonder why shape does not vary, so that some red blood cells turn out square or triangular, and why each cell so predictably contains reddish hemoglobin pigment instead of green chlorophyll or yellow carotene. The answer lies in a basic characteristic of life on Earth—*like begets like*—and it extends to all forms of cellular and organismal reproduction, whether the form involves the generation of identical replacement cells within the body or the production of offspring by parents. Just as a single algal cell always gives rise to new algae, snails give rise only to new snails, not to sea urchins, and cardinals produce baby cardinals, not bluebirds, in the spring. This ability to breed true is based on a fundamental biological principle: when cells reproduce, they pass along instructions for constructing new cells that are almost identical to the parental cell. This legacy of information is apportioned equally and faithfully to each new cell and accounts for the striking resemblance from one generation to the next.

In an evolutionary sense, reproduction is the most important property of life; a given individual can live and die without reproducing, but the continuation of a species depends on the replacement of at least some adult members with genetically similar new members that grow, mature, and, in turn, reproduce. Whether an organism is unicellular or multicellular, its growth and reproduction depend on *cell division,* the separation of one cell into two. The precise impetus for cell division is not yet understood. If we observe cells over a period of hours or days, we see that they pass through cell cycles: they grow, prepare for division, and divide; then the two new cells, the *daughters,* repeat the cycle. Growth-stimulator molecules initiate the cycle in some cells, whereas in others, increasing size may be the trigger. Since cell volume grows faster than cell surface area, division into two smaller cells reestablishes a larger relative surface area in the daughters.

Biologists know a great deal about what goes on during cell division, and the remarkable ways in which the chromosomes, nucleus, and cytoskeleton of eukaryotic cells participate to ensure the orderly distribution of genetic material to the new generation. (The division of prokaryotic cells is slightly different from that of eukaryotic cells and will be described in Chapter 13.) In this chapter, we will discuss **mitosis,** the process of cell division in which genetic information is distributed equally to identical daughter cells in eukaryotes. Mitosis is the means of reproduction of most single-celled eukaryotic organisms, as well as the process responsible for the growth of plants, animals, and fungi. Such processes as regeneration and repair of wounds depend on mitosis.

We shall also describe a special type of cell division called *meiosis,* the process that results in the formation of special reproductive cells in sexually reproducing organisms. In meiosis, we shall begin to see how like begets "not quite like" in the living world and thereby helps generate the variability essential for evolution.

THE NUCLEUS AND CHROMOSOMES

Since cellular reproduction distributes genetic material, we begin our discussion at the main cellular repository of genetic information, the nucleus. In Chapter 6, we saw that the most prominent organelle in the vast majority of eukaryotic cells is the *nucleus,* a spherical body with its own envelope dotted by special pores. Within the nucleus lie the *chromosomes,* which contain the blueprints for the cell's proteins and thus, indirectly, for much of its function. This seems like a simple fact, but it took many decades for biologists to establish that the nucleus is indeed the repository of the cell's hereditary materials.

The Role of the Nucleus

The role of the nucleus in cellular reproduction began to unfold in the 1870s, when Oscar Hertwig, a German embryologist, used a microscope to observe sperm fertilizing the eggs of a sea urchin. As a sperm penetrated the egg, Hertwig saw that the sperm nucleus entered the egg and merged with the egg nucleus. From such observations, Hertwig and others concluded that the nucleus alone carries the male parent's genetic contribution to the embryo.

A series of experiments conducted in the 1930s by the German biologist Joachim Haemmerling confirmed and extended Hertwig's conclusions. These experiments were done with two species of *Acetabularia,* a unicellular green alga that lives on the ocean floor in warm, shallow water. Although *Acetabularia* is a single-celled organism, it is quite large (25–50 mm tall), and its architecture is complex. Each cell has a rootlike base, a stalk, and a cap, as shown in Figure 9-1. The shape of the cap differs in the two species that Haemmerling studied: *Acetabularia mediterranea* has an umbrellalike cap, whereas *A. crenulata* has a ragged, petal-like cap.

When Haemmerling cut off the upper stalk and cap of individuals of either species of *Acetabularia,* each cell regenerated the appropriate umbrella- or petal-shaped cap. He wondered whether it was the cytoplasm or the nucleus that controlled the shape of the new cap. To

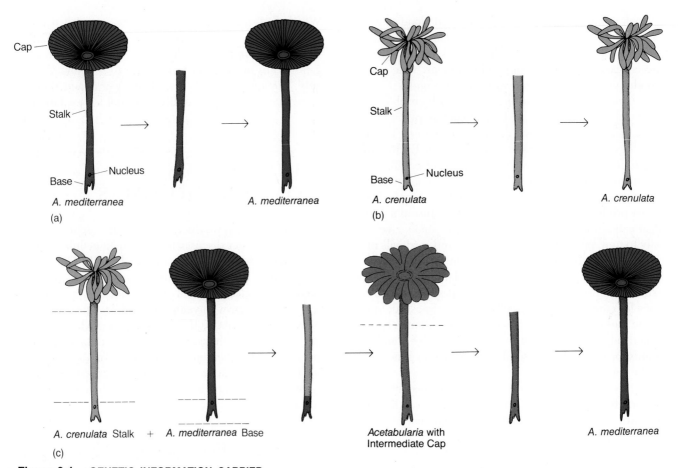

Figure 9-1   GENETIC INFORMATION CARRIED IN THE NUCLEUS.

Haemmerling's experiments with *Acetabularia* revealed the essential role of the nucleus in heredity.

Acetabularia is a single-celled alga with a rootlike base containing the nucleus, a stalk, and a cap. (a) If the upper stalk and cap are cut from *Acetabularia mediterranea*, the cell regenerates the characteristic umbrellalike cap. (b) If the upper stalk and cap are cut from *Acetabularia crenulata*, the cell regenerates the characteristic petal-like cap. (c) If an *A. crenulata* stalk without its cap is transplanted onto the base cut from an *A. mediterranea* cell, the alga regenerates a cap that is intermediate in appearance between the petal-like and umbrellalike caps. If this intermediate cap is then cut off, the algal cell regenerates an umbrellalike cap of the *A. mediterranea* type, and any further amputations of the cap result in the regeneration of only this type. Since the base, which contains the cell nucleus, came from an *A. mediterranea* cell, the experiment shows that the nucleus must contain information governing the cell's architecture.

answer this question, he transplanted a large *A. crenulata* stalk without its cap onto a small *A. mediterranea* base, where the nucleus is located at certain times in the alga's life cycle. He found that the cap that formed was intermediate between the petal and the umbrella shapes. Next, he cut off the new cap to learn what kind would regenerate. The combination algal cell produced an umbrellalike cap, like that on the original *A. mediterranea* base, and further amputations resulted in regeneration of *only* this type of cap.

Clearly, the information from the nucleus present in the base ultimately governed the cell's architecture, including the shape of the cap. The single intermediate cap (half umbrella-, half petal-shaped) was a residual effect of information derived from the original *A. crenulata* nucleus; the stability of that lingering effect was not great, however, and the resident *A. mediterranea* nucleus had fully taken over by the time of the second cap regeneration. From Haemmerling's experiments, and abundant evidence amassed since then, we can conclude that the nucleus is indeed the primary site of the information that determines the structure of the cell.

Chromosomes and DNA

Studies since those of Hertwig and of Haemmerling have also revealed the nature of the genetic material within the nucleus that directs protein production and hence cell structure. The material is *DNA*—deoxyribonucleic acid—an extremely long molecule that consists of two strands of nucleotides wound around each other in a helix. We encountered DNA in Chapter 3, and we will examine the structure of DNA in detail in Chapter 13. Here we shall focus on **chromosomes,** the structures that contain DNA in eukaryotic cell nuclei. The word "chromosome" means "colored body" and was applied by early microscopists to structures that take up basic red or purple dyes in the nucleus of a dividing cell. Modern techniques using fluorescent dyes reveal bands of distinctive size and location that identify each human chromosome unambiguously. Under the light microscope, stained chromosomes may look like solid, flexible rods, as shown in Figure 9-2a. Actually, they are not solid structures at all; rather, each is composed of a long, tightly coiled strand of DNA and associated proteins, as you can see in Figure 9-2b.

How long is the coiled strand of DNA in a given chromosome? If a chromosome were the length of five letters on this page, the single slender DNA molecule it contains would stretch the length of a football field. Clearly, the DNA molecule in a chromosome is very efficiently packaged. But how? Short stretches of the long, continuous DNA molecule wind around clusters of positively charged proteins called **histones,** forming beadlike complexes called **nucleosomes** (Figure 9-3). This arrangement condenses the DNA molecule in much the same way a rubber band shortens when it is twisted tighter and tighter. The nucleosomes are believed to interact with one another to promote still higher orders of coiling and supercoiling (Figure 9-4). The result is the dense chromosome structure we can observe through a microscope in the nucleus of a dividing cell.

In later chapters, we will take a close look at *genes,* short stretches of DNA along the continuous length of that molecule, which determine given physical traits in cells and organisms. For now, just remember that thousands of genes can lie along the single DNA molecule, which coils and wraps around proteins to form each chromosome.

In addition to histones, other chromosomal proteins are present in the cell nucleus. These provide a scaffolding for the chromosome, regulate DNA function, and

Figure 9-2 CHROMOSOMES IN THE NUCLEUS OF A EUKARYOTIC CELL.

(a) Near the time of mitosis (so-called metaphase), chromosomes (magnified about 200 times) appear to be solid bodies within the nucleus. (b) A higher magnification (about 20,000 times) reveals that each body or rod is actually a coiled mass of threads, or strands of DNA and proteins. This chromosome has already been duplicated so that the two identical chromatids lie side by side, attached by the centromere, the narrow region of the chromatids (see pp. 202, 204).

(a)

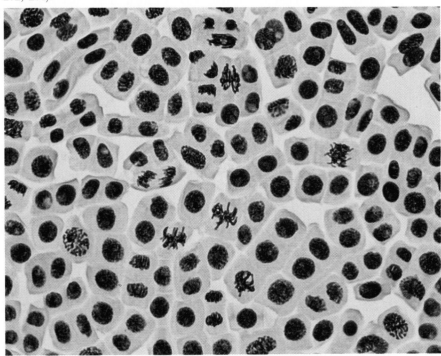

(b)

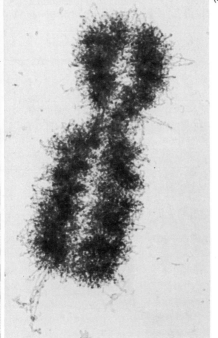

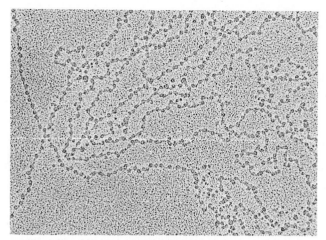

Figure 9-3   NUCLEOSOMES: TINY BEADS IN THE BIOCHEMICAL STRAND OF LIFE—DNA.
Stretches of the long continuous DNA molecule of a fruit fly wind around clusters of histone proteins, forming beadlike complexes, the nucleosomes (magnified about 80,000 times).

may participate in the coiling and supercoiling process. Each long strand of DNA combines with histones and nonhistone proteins to make up a substance called **chromatin.** When a cell is in the resting phase of the cell cycle, between divisions, the chromatin that composes all the chromosomes in the nucleus appears to be un-

wound and tangled like a mass of yarn, and no distinct chromosomes are visible in the nucleus (although, of course, they are really present in the tangled mass). As the time for cell division nears, individual chromosomes separate from the tangled ball and enter the supercoiled, condensed condition, in which they assume flexible rod-like shapes. This tight coiling of chromatin makes chromosomes much more compact during the active phases of cell division, and it permits the orderly allocation of genetic information to the two daughter cells that result from the division of the parental cell.

When an organism's chromosomes are in the coiled, condensed state, a pictorial display of the number, sizes, and shapes of chromosomes can be made. This display is called a **karyotype** (Figure 9-5). From a karyotype, it is immediately apparent that for most cells each chromosome (with one notable exception) is present as two copies; they are called **homologous** ("same shaped") pairs. The exception occurs with the sex-determining chromosomes of some animals. For example, humans have twenty-two pairs of homologous **autosomes,** or non-sex chromosomes, and one pair of **sex chromosomes** in most cells of the body. In human females, the sex chromosomes are homologous. In males, however, the pair is not homologous; there is one X chromosome and one stubby Y chromosome. Therefore, the karyotype of a normal human cell shows twenty-two pairs of autosomes and one pair of sex chromosomes, either XX or XY.

Figure 9-4   A CHROMOSOME UNWOUND: FROM "COLORED BODY" TO DNA.
A chromosome (a) contains highly coiled and supercoiled DNA strands (b). (c) A close-up of a section of the chromosome shows supercoils, each made up of coils of DNA complexed with proteins. (d) The coils of DNA have a "beads-on-a-string" arrangement of nucleosomes. (e) A detailed view of the nucleosomes shows the DNA strands wrapped around clusters of 8 proteins called histones. The DNA double-helix structure is shown in (e).

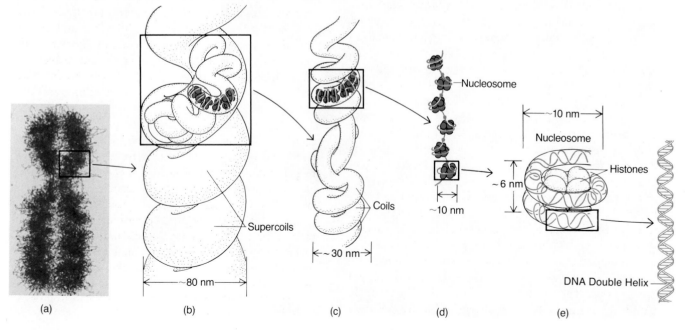

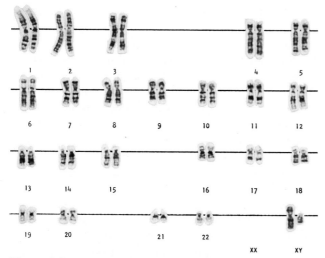

Figure 9-5   KARYOTYPE: AN ORGANIZED REPRESENTATION OF A SET OF CHROMOSOMES.
To produce a karyotype, a geneticist treats mitotic cells with a dye such as Giemsa that has an affinity for DNA, and that fluoresces (glows) when viewed with ultraviolet light. The researcher then photographs the stained chromosomes, cuts the individual chromosomes from the picture, and arranges them roughly in descending order of size. Since Giemsa binds more tightly to some DNA regions than others, each chromosome has brightly fluorescing bands (G-bands). Because the G-banding pattern of each chromosome is unique, each can be identified unambiguously. In this human karyotype, there are twenty-two pairs of autosomes and one pair of sex chromosomes, an X and a Y. The presence of a Y reveals that the chromosome set comes from a male.

Organisms with cells containing two sets of parental chromosomes are called **diploid** (meaning "two"). Organisms with cells containing just one set of parental chromosomes are called **haploid.** Both terms will appear again and again throughout this book. In most plants and animals, the conspicuous adult is diploid. In animals, the eggs, sperms, and sometimes a few associated cells are haploid. In typical eukaryotic, sexually reproducing organisms, diploid and haploid phases alternate during the *life cycle;* the life cycle of a species is defined as the passage from one adult through the reproductive stages to the adult of the next generation. A significant phase of the life cycle of many plants is spent as a haploid organism. For instance, among mosses, the adult you see growing in damp soil or on shaded logs is haploid. We will study life cycles, haploidy, and diploidy of plants in Chapter 23. We should note, however, that in the majority of plants and in virtually all animals, the body cells of the diploid adult contain two sets of chromosomes, with one member of each pair donated by the female parent (in the nucleus of the egg cell) and the other by the male parent (in the nucleus of the sperm cell or by cells in a

plant's pollen grain). The combination, during fertilization, of two haploid chromosome sets results in a diploid organism. Clearly, chromosome pairs are a basic feature of sexual reproduction among higher organisms, and reproduction is a basic characteristic of life. As we shall see, the new set of chromosome pairs that arises during sexual reproduction is the basis for the development of genetically unique individuals and helps account for genetic diversity within species—the key to evolution and long-term survival.

THE CELL CYCLE

The Columbia University biochemist and author Ernest Borek once wrote, "There has been a relentless pressure from time immemorial [within most cells] to surge toward growth and duplication and thus ensure the continuity of life."* In the introduction to this chapter, we described this relentless "surge" as the passage of cells through **cell cycles,** regular sequences during which cells grow, prepare for division, and divide to form two daughter cells, which repeat the sequence. Free-living single-celled eukaryotes, such as amoebas, are essentially immortal because of such cycles. As such cells divide, they distribute hereditary information to their daughter cells, which, in turn, distribute virtually the same information to their progeny, and so on for potentially millions of generations. Within multicellular plants and animals, some cells continue to grow and divide for the life of the organism. These include cells at the tips of roots, which probe ever deeper into the soil, and cells that line the small intestine of the digestive tract, the daughters of which are continually sloughed off during the digestion of food. However, most cells in multicellular organisms either slow the cell cycle greatly or "break out" of it altogether; they stop dividing and remain locked in one part of the sequence until either they die individually of old age or disease or the entire organism dies. These include muscle cells and nerve cells in animals, and cells that form sugar-conducting vessels in plants.

Within the normal cell cycle, the period between cell divisions can be quite long, while the actual division phase takes place quickly. As Figure 9-6 shows, the cycle begins with a period of normal metabolism, called G_1 (gap 1). During this phase, all the components of the cell are synthesized, assembled, and used for normal cell function. Cells in G_1 go about their business in the economy of the whole organism, as discussed in Chapters 5 and 6. But actual increase in cell mass may occur, so that when the time for division comes, there is a full stock of

The Atoms within Us (New York: Columbia University Press, 1961).

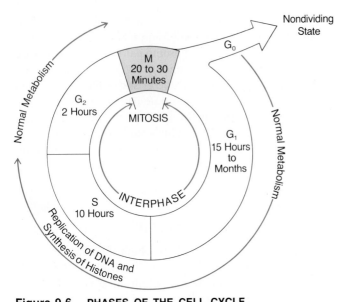

Figure 9-6   PHASES OF THE CELL CYCLE.
The cell cycle includes mitosis (designated M), a period of active division, and interphase, a period of nondivision during which other processes take place. Interphase is usually divided into three phases. During G_1, normal components of the cell are synthesized and metabolized, often resulting in cell growth. Those processes continue during the S phase, but in addition chromosomal DNA is replicated and histones are synthesized. During G_2, normal metabolism and cell growth continue until M is reached. Some cells remain for long periods in a nondividing state (G_0) after completing mitosis.

components with which the daughter cells are constructed. Next comes the **S phase** (synthesis phase), during which normal syntheses continue, and, in addition, DNA is replicated and histones are synthesized. The amount of DNA in the nucleus is doubled, while histones and other proteins that are synthesized in the cytoplasm pass through the pores of the nuclear membrane and combine with the DNA, forming chromatin. This chromatin comprises two identical sets of chromosomes instead of the one diploid set present during the earlier G_1. Each copy of a chromosome is called a **chromatid.** At the end of the S phase, the cell enters G_2 (gap 2), a brief period of cell metabolism and growth during which more proteins and other substances are manufactured and the cytoplasm may increase slightly in volume. During G_1, S, and G_2, the nucleus is said to be in **interphase,** between nuclear divisions. Once a cell enters S, it is normally committed to proceed through S and G_2 and to divide during M.

When the cell enters the **M phase,** the period of mitosis, the chromosomes in the nucleus condense, move about in special ways, and become apportioned equally to the two new nuclei. This phase ends with *cytokinesis*,

the separation of the cytoplasm into the two daughter cells. No measurable growth takes place during the brief M phase, nor can cells carry out complex activities that depend on the cytoskeleton. We will study the details of mitosis and cytokinesis shortly. For now, remember that the characteristic features of the S and M phases are, respectively, DNA replication and chromosome condensation and movement, while the G_1 and G_2 phases are defined by the absence of these activities but, more important, by various synthetic and functional events.

How long do the various phases of the cell cycle last in absolute time? Typical embryonic cells of a rat or a chick may be cultured in dishes to facilitate observation. The results, as indicated in Figure 9-6, normally are: G_1, fifteen hours; S, ten hours; G_2, two hours; and M, twenty to thirty minutes. Of these phases, G_1 varies the most; it may last for days or weeks in some mature functional cells, such as those in the liver or kidney, or it may last for only hours in other cells, such as those in the intestines or bone marrow.

One of the most elusive problems in biology is understanding how the cell cycle is regulated. Most animal and some plant cells spend varying numbers of hours, days, or months in the G_1 phase before the S and M phases are initiated. And in nerve, muscle, and sugar-transport-vessel cells, the functional adult cell remains in a permanent nondividing condition, sometimes called G_0.

What prevents these cell types from dividing? And in cells that do divide, what triggers the end of G_1 and the initiation of S and M? Is there some inherent limitation in the nucleus, or does the cytoplasm have properties that prevent initiation of a cell cycle? Biologists have found that the nuclei of permanently *nondividing* cells retain the ability to participate in the S phase. If nuclei from nerve cells are transplanted into a host cell in the S phase, within minutes, the chromosomes in the transplanted nerve-cell nucleus begin to participate in DNA replication. Such experiments suggest that *the state of the cell's cytoplasm* controls the activity of the nucleus and helps determine its phase in the cell cycle.

What, then, determines the state of the cytoplasm? One factor is the cell's external environment. We saw in Chapter 5, for instance, that normal cells of multicellular animals are *contact dependent*—they can move and divide only if adhering to certain types of solid substrates. Adhesion is essential for the cell's cytoskeleton and surface to carry out the M phase activities, including cytokinesis. Interestingly, many cancer cells have lost this contact dependence and so can divide repeatedly while suspended in fluid. That may be part of the reason for the deadly spread of some tumor cells in the body.

Outside factors may either stimulate or inhibit the cell cycle. For example, in higher plants, hormones called cytokinins stimulate mitotic activity in growth zones, so that the rate of cell production increases substantially, and the root or stem or leaf grows faster. In complex animals, the hormones somatomedin, insulin, and epidermal growth factor can stimulate cells to divide more rapidly. Inhibitors of cell division may include a class of substances called **chalones**. Skin and kidney cells are thought to produce such self-inhibitors that hold mitosis in check in those tissues. When the skin is injured, such as in a cut finger, epidermal cells are lost, and the amount of chalone is apparently decreased locally; as a result, the remaining epidermal cells near the wound begin to divide. When the original tissue has been rebuilt, the new cells produce sufficient chalone to reduce mitotic rates to normal resting levels. Because experiments with chalones have been difficult to replicate, this explanation, although appealing, remains controversial.

Understanding what controls the cell cycle is particularly important in the study of cancer. In some types of malignancy, the normal controls on mitosis seem to be suppressed, so that large numbers of long-lived tumor cells are produced. This is not because cancer cells divide faster than normal cells, but because a higher proportion of tumor cells actively divide than do normal cells of the same type. In addition, daughter cells also divide, cycle after cycle. Intensive research is currently under way to ascertain how normal and cancerous cells differ from each other in structure and function, since such differences may provide a significant clue to the conquest of the dread disease (Figure 9-7).

Now that we have surveyed the complete cell cycle, let us focus more closely on the actual division phase, mitosis.

Figure 9-7   IDENTIFYING CANCEROUS CELLS.
The division process in normal and cancer cells is identical; thus it cannot be determined whether this dividing cell is cancerous or not from this electron micrograph (magnified 1,300 times). Other tests must be used.

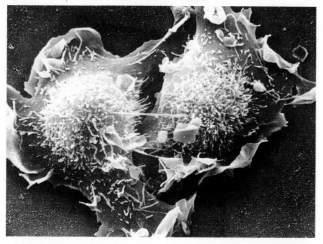

MITOSIS: PARTITIONING THE HEREDITARY MATERIAL

Let us return for a moment to the concept of "like begets like" and consider two interesting facts:

1. Certain algae cells found in rocks more than 1 billion years old look strikingly similar to algae cells alive today.

2. The skin on your right leg is literally a different population of cells from what it was last month or last year, yet those cells are virtually identical to those on your leg skin ten years ago or when you first began to walk.

Cells, whether comprising colonies of algae or part of human appendages, show this amazing fidelity from one generation to the next largely because of the intricate and highly accurate mechanism of mitosis.

We stated in the chapter introduction that chromosomes within the cell nucleus participate in cell division in such a way that genetic information is apportioned equally to each daughter cell. Cells of each species possess a characteristic number of chromosomes: human cells have 46 (22 pairs of autosomes plus 1 pair of sex chromosomes); cotton plants, 52; turkeys, 82; and some ferns, more than 1,000. During normal cell division, the two daughter cells must receive the same number of chromosomes—all cotton cells must get 52 chromosomes; all human cells, 46. It is also critical that each daughter cell receive two copies of each chromosome type: two of chromosome number 1, two of number 2, and so on. The movements of chromosomes during mitosis ensure that each cell gets a complete and identical set of genetic information.

The stages and events of mitosis in eukaryotic organisms were first observed and recorded about the time that European biologists were beginning to probe the role of the nucleus. Walther Fleming, a German cytologist, described the "dance of the chromosomes," a series of spectacular movements during which the chromosomes jostle about in the middle of the nucleus and then, as if participating in an English country reel, split neatly into two groups, which proceed to glide to opposite ends of the cell. The cell then separates into two, leaving each daughter cell with a full complement of chromosomes. Each daughter cell then enters the G_1 phase of the cell cycle.

Although the events of mitosis take place in one continuous sequence, for convenience, biologists divide the mitotic cycle into four phases: *prophase, metaphase,*

(a)
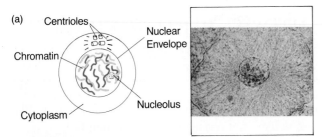

INTERPHASE: Prior to mitosis, the chromosomes are uncoiled and fill the nucleus like tangled yarn. DNA is replicated during the interphase stage.

(b)
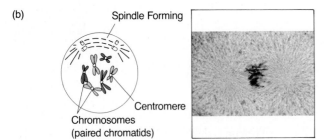

PROPHASE: The chromosomes begin to condense and double, forming two chromatids. The nuclear envelope and nucleolus disappear, and the centrioles begin to separate.

(c)
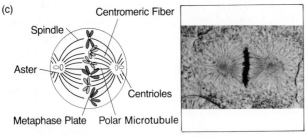

METAPHASE: The centrioles have organized the spindle microtubules, which cause chromosomes to align on the metaphase plate.

(d)
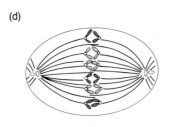
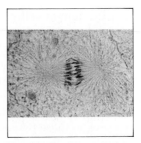

ANAPHASE: The centromeres separate, and each pulls along a single attached chromatid toward a spindle pole.

(e)
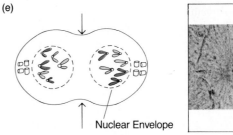

TELOPHASE: The chromatids reach the poles, and two nuclei form. The division of the cytoplasm begins.

(f)
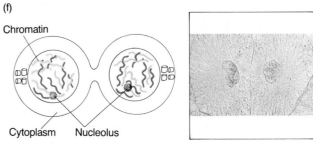

CYTOKINESIS AND RETURN TO INTERPHASE: Chromosomes begin to decondense as division of the cytoplasm occurs. Nuclear envelopes and nucleoli reappear in the two daughter cells as they enter interphase.

Figure 9-8   THE PHASES OF MITOSIS.
After being replicated, the paired chromosomes align on the metaphase plate, then separate and move apart, finally becoming enclosed in nuclei of the daughter cells. For simplicity, the diagrams depict only three pairs of chromosomes. The accompanying photographs show mitosis in a whitefish (magnified about 800 times).

anaphase, and *telophase.* These events are illustrated in Figure 9-8. You should refer to this figure as we describe each of these phases.

Prophase

Prior to mitosis, the genetic material, DNA, must be replicated. During the S phase of the cell cycle, the DNA of each chromosome is copied precisely. Chromosomal proteins are added to yield an exact replica of each original chromosome. The originals and the copies are now called chromatids.

After S and G₂ are over, the first phase of mitosis (M), **prophase** ("before form"), starts; individual chromosomes become recognizable as the tangled yarnlike chromatin begins to supercoil and condense. Each chromosome can be seen under a microscope to consist of two identical parts, the identical chromatids lying side by side. The constricted area at which the two chromatids are attached to each other is called the **centromere,** after the Greek word for "central part." As prophase progresses, the coiling of the chromosomes becomes tighter, the nucleolus disappears, and the nuclear envelope usually breaks down.

Metaphase

As prophase ends and **metaphase** ("middle form") begins, the fully condensed chromosomes become associated with the **spindle.** The spindle is a football-shaped set of microtubules, stretching across the center of the cell from opposite poles. Additional arrays of microtubules fan out in a star-shaped pattern from both poles. The polar complexes are called *asters.* Some spindle microtubules were once thought to extend from one pole to the other. In reality, however, these microtubules extend only just past the equator between the two poles, so that there is a region of overlap at the center of the spindle. In this region, an electron microscope reveals cross bridges of unidentified materials linking the ends of the polar microtubules. Still other bundles of microtubules, called *centromeric fibers,* are attached at one end to the centromeres of the chromosomes and extend to the poles. The chromosomes, attached like marionettes to the delicate yet rigid centromeric fibers, appear to jostle about during early metaphase and eventually become arranged in a plane at a right angle to the spindle fibers in the center of the nuclear region. This plane is called the **metaphase plate** and might be visualized as a disk extending through the planet Earth at the level of the equator.

Anaphase

Anaphase ("again form") begins when the centromeres that link the sister chromatids split, thereby allowing each chromatid to behave as a separate chromosome. One chromatid from each pair is drawn toward each pole. Both sets of chromatids appear to float apart, with floppy arms lagging behind the centromeres like a worm on the end of a fishing line being reeled through a pond. Chromosome separation occurs in minutes, and anaphase ends as the two sets of chromatids have parted and the two spindle poles move farther and farther apart.

Telophase

Telophase ("end form") starts as the spindle fibers disperse and nuclear envelopes begin to form around each

set of chromosomes. The chromosomes apparently become attached to the protein scaffolding near the inner membrane of the nuclear envelope. They uncoil once again into a tangle of chromatin, and nuclei gradually appear. Division of the cytoplasm then takes place. We will discuss details of this division in the section on cytokinesis.

The Nature of the Spindle

Just as a dancing puppet is operated by moving strings, the dance of the chromosomes is choreographed and controlled by the action of the spindle fibers. Chromosomes are incredibly complex structures; but, like other organelles, they are not alive and cannot move by themselves. It is the important task of the spindle to ensure that the paired and the separated chromatids move in the right directions at the proper times. It could be disastrous to the daughter cells if both chromatids of a chromosome moved toward the same pole, or if one chromatid lagged behind the others and was left outside the re-forming nuclear envelope. Textbooks traditionally have spoken little of the spindle. However, the great advances in cell biology during the 1970s revealed that the spindle as well as other elements of the cytoskeleton, nuclear envelope, and plasma membrane all play essential roles during both mitosis and cytokinesis. Just as the other organelles we studied in Chapters 5 and 6 are made up of aggregates of molecules and accomplish critical cellular tasks, so the spindle has a unique architecture that allows it to coordinate the chromosome movements of mitosis (Figure 9-9).

Figure 9-9   MOVING CHROMOSOMES: THE SPINDLE AT WORK.

(a) In each half of the spindle, polar microtubules extend from the pole to the metaphase-plate region, where the two sets overlap. Other microtubules reach from the spindle poles to structures within the centromeres, the kinetochores. These microtubules are called centromeric fibers. (b) Two activities occur as the chromatids move toward the poles in anaphase. The region of overlap of the polar microtubules decreases, apparently moving the poles further apart, and the centromeric fibers shorten and can pull the chromatids toward the poles.

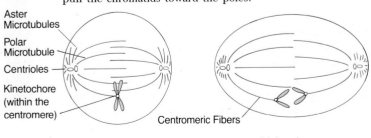

Aster Microtubules
Polar Microtubule
Centrioles
Kinetochore (within the centromere)
Centromeric Fibers

(a) Metaphase   (b) Anaphase

In most cells, the spindle forms in the cytoplasm during prophase. It first appears as parallel sets of fibers, the proteinaceous microtubules, extending from the poles around the nucleus. The slender microtubules form and elongate as subunit molecules of tubulin are added to one end. Each half of the spindle forms, as we have noted, as microtubules extend from each pole to the region of the metaphase plate, where the two halves of the spindle overlap (Figure 9-9a). As the nuclear envelope disperses during prophase, other microtubules, the centromeric fibers, extend outward from the spindle poles to structures on the chromosomes called **kinetochores.** This region within the centromere of each chromatid is the actual attachment point for the centromeric fibers. During metaphase, the chromosomes gradually become aligned on the metaphase plate. Then, during anaphase, after the centromeres themselves divide, the fibers joining the kinetochores and the poles of the spindle begin to shorten, and the chromatids begin to move apart.

During anaphase, two processes move the chromatids toward the poles. The region of overlap of the polar microtubules decreases as the microtubules slide past each other (Figure 9-9b). An energy-dependent process causes the poles to move farther apart and, at the same time, helps to pull the two sets of chromatids apart. Meanwhile, the centromeric fibers connected to the kinetochores shorten, pulling the chromatids closer to the poles. The second process does not require energy. The actual mechanisms by which these movements are generated are not yet known.

The spindle forms differently in animal and plant cells. In animal cells, spindle formation is associated with the *centriole* (Chapter 6), a cytoplasmic organelle similar to the basal bodies of cilia and flagella. During prophase, two pairs of centrioles are moved to opposite sides of the nucleus, marking the positions of the spindle poles. Forces generated by microtubule elongation apparently push the centrioles and asters to their final positions. Once they are in place, the centrioles probably are involved in the formation of the remainder of the spindle, especially as sites where tubulin polymerizes into microtubules.

In most plant and fungal cells, spindle formation proceeds as in animal cells, but the cells lack centrioles. However, they do have regions called *microtubule organizing centers.* These centers may contain at least one kind of molecule also found in centrioles; indeed, their function seems to be analogous to that of the centrioles. It is important in plant, fungal, and animal cells that the organizing centers or centrioles invariably be distributed to both daughter cells; if they are not, the daughter that does not have such sites is unlikely to be able to form a spindle and divide successfully.

CYTOKINESIS: PARTITIONING THE CYTOPLASM

Following the spectacular chromosome movements of mitosis, two nuclei, each with a duplicate set of chromosomes, are formed. However, that is just one aspect of cell division. The other part is **cytokinesis,** the division of the cytoplasm so that two individual cells result. This process is quite different in animal and plant cells, and the reason is straightforward: animal cells, bound by a plasma membrane, can be "pinched" into two cells fairly easily, whereas plant cells, bound by a rigid cell wall, cannot. In plant cells, a new cell wall must be constructed as one cell is partitioned into two.

In animal cells, cytokinesis results from the activity of a contractile ring composed of actin filaments. This beltlike array of filaments—part of the cytoskeleton—appears during telophase. The ring lies midway between the two spindle poles and asters, and actin filaments of the ring are linked to the inner surface of the plasma membrane so that their contractile activity pulls the cell surface inward. This contraction forms the cleavage, or division, furrow that can be seen in Figure 9-10.

The many filaments in the contractile ring can be likened to a purse string pulled by an unseen hand or to a belt cinched tighter and tighter around the waist of the

Figure 9-10   CYTOKINESIS: MOLECULAR PURSE STRINGS CLOSING.
The contractile ring of actin and myosin pinches the animal cell (magnified about 570 times) in two.

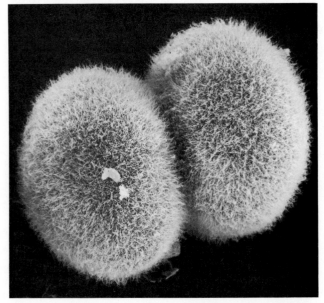

dividing cell. In reality, the "hand" that cinches the belt is probably the interaction of the contractile proteins actin and myosin. As this pinching and tightening proceeds, for a short time the two daughter cells remain connected by a thin neck of cytoplasm. This neck eventually dwindles and disappears, and the cells become fully separated.

Cytokinesis in plants is similar in concept but different in detail. Once a plant cell's chromosomes have moved apart due to the activities of the spindle, the cytoplasm is divided in two by a process that builds a new plasma membrane, called a **cell plate,** across the cell at the equator, as shown in Figure 9-11. The cell plate is composed of membranous sacs that fuse with one another and gradually extend across the cell and outward toward the plasma membrane that lines the rigid cell wall on all

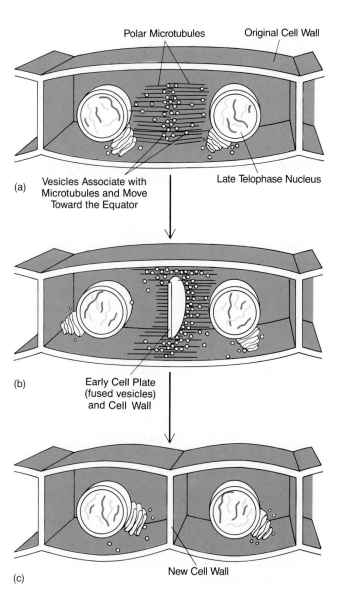

Polar Microtubules   Original Cell Wall

(a) Vesicles Associate with Microtubules and Move Toward the Equator

Late Telophase Nucleus

(b) Early Cell Plate (fused vesicles) and Cell Wall

(c) New Cell Wall

sides. A cell wall begins to appear in the cell plate. The peripheral sacs fuse with the plasma membrane, thereby linking the new cell plate with the sides of the cell. As a result, each daughter cell is fully enclosed within its own intact plasma membrane. Following this, additional cell wall material is deposited in the region of the cell plate; a rigid cell wall forms; and each daughter cell is fully protected in its own tough shell of cellulose and other substances.

Interestingly enough, cytokinesis can occur independently of chromosome duplication and nuclear division in both animals and plants. An animal cell from which the nucleus has been removed can still undergo cytokinesis: the centrioles orchestrate microtubule assembly, and the contractile ring forms and functions. In fact, in some early embryos, the centrioles, spindles, and contractile rings can carry out a lengthy succession of cleavage divisions on a precise and normal timetable even after the nuclei or chromosomes have been removed experimentally. Conversely, chromosomes can be duplicated and nuclear divisions can occur without cytokinesis. The result is multiple nuclei in one cytoplasm, multiple chromosome sets in a nucleus, or chromosomes composed of many DNA copies. This takes place in pollen grains of flowering plants, in many types of insect embryos or cells, and in human liver cells. Clearly, the remarkable cytoskeletal components involved in mitosis and cytokinesis have great autonomy and do not depend on the nucleus or chromosomes to carry out their complex activities. An important but virtually unstudied area of cell biology involves explaining how the cytoskeletal and the nuclear chromosomal activities are coordinated to yield normal mitotic cell division or nuclear divisions without cytokinesis.

MEIOSIS: THE BASIS OF SEXUAL REPRODUCTION

We have been looking at the way cells divide to reproduce themselves, whether the cell is a free-living uni-

Figure 9-11   CYTOKINESIS: PARTITIONING WITH MEMBRANES.
Both a new plasma membrane and cell wall form between the two daughter cells. (a) Vesicles from the Golgi complex move on the polar microtubules to the former sites of the metaphase plate. (b) The vesicles fuse, forming the early cell plate; the cell wall begins to form in the plate. (c) These processes continue until a full plate and cell wall are formed.

cellular alga or part of the multicellular body of a plant or an animal. In each such division, if the parental cell had six, eighteen, or forty-six chromosomes, each normal daughter cell would have six, eighteen, or forty-six chromosomes.

Now let us shift our focus slightly. As you know, the vast majority of multicellular organisms do not simply divide in half to reproduce, as do single cells, but undergo sexual reproduction. In animals, *male gametes* (sperm) fertilize *female gametes* (egg cells). In plants, meiosis yields spores that mature into male or female gametophytes; they, in turn, develop the gametes that fuse at fertilization. What does gamete fusion mean with respect to chromosome number? If the full set of 46 chromosomes in humans were passed along to the offspring by *both* parents, the progeny would have 92 chromosomes per cell, and future generations would have 184, 368, 736, and so on. Such a sequence could end only in genetic confusion and, eventually, the death of the lineage. But that pattern of chromosome accumulation does not occur because of a special kind of cell division called **meiosis,** which takes place only in the reproductive organs that generate sperm, eggs, or spores. In humans and most other animals, meiosis occurs only in certain cells of the ovaries and testes. In flowering plants, pollen grains (male gametophytes) are produced in anthers, and eggs are formed in an ovary at the base of each flower. Finally, as we will see, some plants spend most of their lives as haploid organisms; they may produce short-lived diploid stages prior to undergoing meiosis and producing a new generation of haploid organisms.

Like mitosis, meiosis begins after the S phase in the cell cycle, that is, *after* DNA replication has occurred. Two successive nuclear divisions take place, designated *meiosis I* and *meiosis II.* Each has prophase, metaphase, anaphase, and telophase segments similar to but slightly different from those of mitosis. Spindles function during meiosis just as they do during mitosis. Whereas the result of mitosis is *two* daughter cells, each with the *same* number of chromosomes as the parental cell, the two nuclear divisions of meiosis result in *four* daughter cells, each with *half* the number of chromosomes as the parental cell. Moreover, these four daughter cells are not genetically identical. One reason is that a remarkable process called crossing over often occurs and results in exchanges of genetic material between chromosomes. Crossing over and the reduction in chromosome number are prime characteristics of meiosis and distinguish it from mitosis. A consequence is that the homologous chromosomes, which are distributed to different progeny cells, are not identical. The phases of meiosis are presented in Figure 9-12.

Prophase I

Just as in the mitotic division of cells, DNA replication takes place during the S phase of the cell cycle before meiosis begins. As a result, there are two chromatids for each chromosome at the start of prophase I of meiosis. The sister chromatids of each chromosome are connected at the centromere, as in mitosis. As we saw earlier, each chromosome in the nucleus of a diploid cell is part of a pair in which one chromosome is derived from the female parent and one from the male. These are called homologous pairs, and each carries corresponding genes for corresponding traits. One way to imagine homologous chromosomes is to think of the two shoes of a pair, the "left shoe" from the mother and the "right" from the father. The duplication process during the S phase of the cell cycle results in two identical "left shoes" and two identical "right shoes," each set (for example, the two "lefts") knotted together at its "laces"—the centromere. At the start of prophase I, the homologous chromosomes themselves come together and pair side by side to form a four-chromatid structure known as a *tetrad.* Now the two "right shoes" and the two "left shoes" are joined. The pairing of homologous chromosomes during prophase I is called **synapsis,** which literally means "union."

Synapsis is one of the primary differences between meiosis and mitosis. Synapsis of the homologous chromosomes may begin either at the ends or midway along sister chromatids. Pairing is brought about by a bridging structure of proteins and RNA called the **synaptinemal complex,** which acts like a zipper to bring together in intimate association corresponding regions on the homologous chromosomes. As the synaptinemal complex forms, pairing spreads, zipperlike, often from the ends of the individual chromatids toward the centromeres. We will discuss the importance of synapsis a little later in the chapter.

As in mitotic prophase, the nuclear envelope disappears in both plant and animal cells, and in animal cells, the centrioles separate during prophase I of meiosis. As prophase I comes to an end, the synaptinemal complex dissolves, and the coiling of the unzipped chromatids increases, making the individual chromosomes even more distinct. The complex holding together the tetrads is soon lost, but the pairing of the homologous chromosomes remains. It is at this stage that an "arrest" may occur, the cells remaining in this condition for hours, weeks, years, or decades, depending on the species. At this stage the genetic material of these chromosomes frequently is copied into large quantities of RNA as part of cell maturation. Ultimately, the chromosomes move to the metaphase plate as metaphase I is reached.

(a)

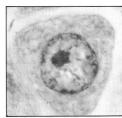

Nucleolus

Centrioles

Cytoplasm

Nuclear
Envelope

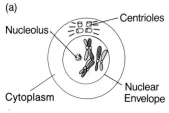

The cell nucleus contains homologous chromosomes: one set from the mother and one set from the father.

(b)

Chromatin

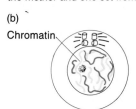

INTERPHASE: DNA is doubled.

(c)

Tetrad

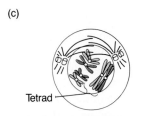

MEIOSIS I, PROPHASE I: Meiosis I begins with prophase I. The chromosomes condense and pair with their homologues. (A phenomenon called crossing over may occur as corresponding chromosome regions are exchanged.) The groups of four chromatids are called tetrads. Cells may remain in a state of prophase arrest for months, years, or decades in animals.

(d)

METAPHASE I: The tetrads move to the metaphase plate in the middle of the cell.

(e)

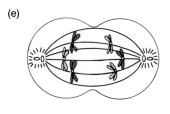

ANAPHASE I: The homologous chromosomes separate without a cleavage of the centromeres. This halves the number of chromosomes in each daughter cell—from six to three in this example.

Figure 9-12   THE PHASES OF MEIOSIS.
After the chromosomes are replicated, two divisions ensue: the first meiotic division reduces the number of chromosomes, whereas the second meiotic division results in the chromatids separating. Four haploid cells result.

(f)

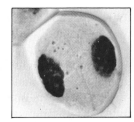

TELOPHASE I: The nuclear envelopes re-form as cytokinesis occurs.

(g)

MEIOTIC INTERPHASE: After cytokinesis, the two daughter cells enter meiotic interphase. No DNA replication occurs during this interphase.

(h)

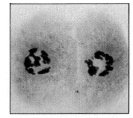

MEIOSIS II, PROPHASE II: The cells enter prophase II with the same number of chromosome sets they had at telophase I.

(i)

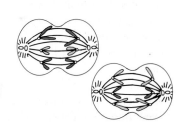

METAPHASE II (left) and ANAPHASE II (right) resemble these same phases of mitosis. The chromosomes align on the metaphase plate, and the centromeres and chromatids separate.

(j)

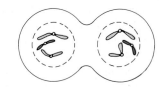

TELOPHASE II: Four haploid nuclei are formed.

(k)

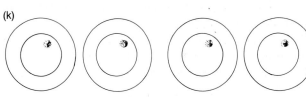

CYTOKINESIS: The haploid nuclei are segregated into four cells that participate in fertilization (details vary among organisms).

Metaphase I and Anaphase I

In metaphase I, the homologous chromosomes remain associated in pairs as they align on the metaphase plate. Recall that during anaphase of mitosis, the centromere of each chromosome splits, allowing the two sister chromatids to move to opposite poles. Pursuing our shoe analogy, this splitting is akin to the laces being untied, so that one from each set of left shoes goes to each pole as does one from each set of right shoes. In contrast, during anaphase I of meiosis, the centromeres do *not* divide. Therefore, the *two* chromatids of each chromosome move to one pole. In other words, the "laces" remain tied, so that the two left shoes go to one pole, and the two right shoes go to the opposite pole. But note that for any particular pair, the two right or two left shoes may go at random to either pole. The important result of this unusual sequence of events is that the chromosome number in each newly forming nucleus is half the chromosome number of the parental cell.

Telophase I

During the next phase of meiosis, telophase I, nuclear envelopes enclose the chromosomes in nuclei. Although the daughter cells contain only half the number of chromosomes that the parental cell contained, each chromosome still consists of paired chromatids because the centromeres did not divide in anaphase I. For instance, during meiosis I in human sperm-producing cells, the original forty-six chromosomes form twenty-three tetrads; on completion of telophase I, twenty-three chromosomes are present in each daughter cell, each composed of two chromatids. In the small desert plant *Haplopappus gracilis*, which has only four chromosomes, there are two paired chromatids in each cell at the close of telophase I.

Meiotic Interphase and Prophase II

Cytokinesis follows telophase I in most species. In the nuclei of the daughter cells, the chromosomes may then disperse during interphase. The chromosomes recondense during prophase II of meiosis in most species. *In no case, however, does further DNA replication occur.* Thus, there is never an S phase before prophase II, since the chromosomes are already duplicated.

Metaphase II and Anaphase II

In all species, as metaphase II begins, the chromosomes again align on a metaphase plate positioned by a new set of spindle microtubules. At the beginning of anaphase II, the centromeres finally divide, and each sister chromatid moves to one of the poles of the spindle. So, in this process, the "laces" are "untied," and one "left shoe" moves to a pole, while its duplicate moves to the other pole. The same occurs in the other daughter cells that arose from meiosis I; all the pairs of shoes separate.

Telophase II

Telophase II follows as nuclear envelopes re-form; the chromosomes expand to the interphase state; and cytokinesis is concluded. Four cells, each with a nucleus containing a haploid complement of chromosomes (twenty-three in humans), have arisen from a single diploid nucleated cell. Each of the four cells has only *one* of the four chromatids present as meiosis started. In male animals, these four cells can mature into sperm, each carrying the haploid (1N) chromosome number; in the male parts of flowers, they can develop into pollen grains. In female animals or female flower parts, the maturation of gametes is more complicated, as we shall see later, but a haploid egg nucleus eventually is produced. It is only when the nuclei from male and female parents unite that the correct diploid (2N) chromosome number is restored in the fertilized egg.

To summarize, the differences between meiosis and mitosis (Figure 9-13) are as follows:

1. In a mitotic cell, chromosomes align independently on the metaphase plate; in a meiotic cell, homologous chromosomes pair and align.

2. In mitosis, the centromeres split, allowing the chromatids to separate; in meiosis, the chromatids remain attached at the centromere during the first division and do not split until the second division.

3. The end products of mitosis are *two* progeny cells, each with a *diploid* set of chromosomes identical to each other and to that of the parental cell; the end products of meiosis are *four* daughter cells, each with a *haploid* set of chromosomes that are different from those of the parental cell and from each other.

Now let us see how such differences arise.

Recombination in Meiosis

Having described the steps of meiosis, we can ask a fundamental question: What is the *significance* of meiosis? Is it just to reduce the chromosome number in ga-

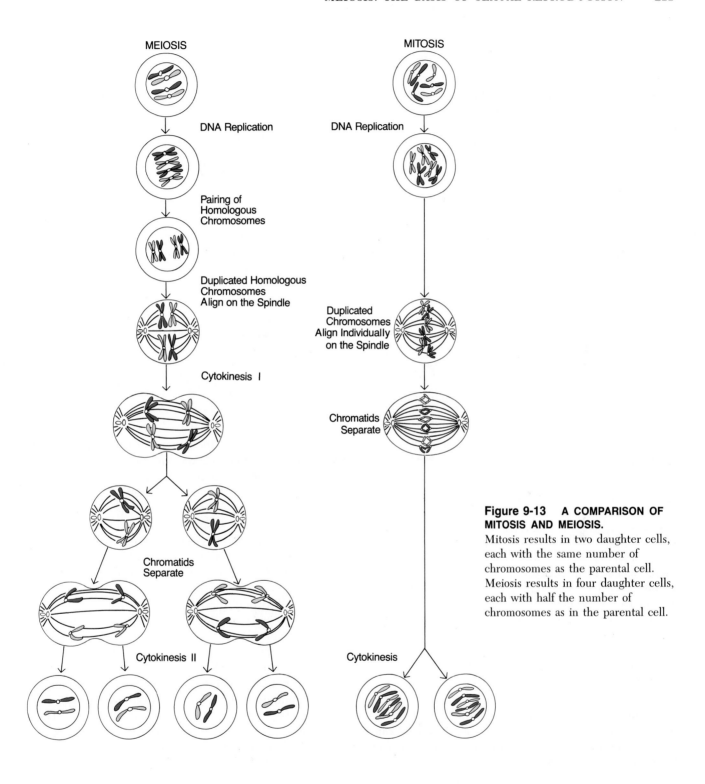

MEIOSIS

DNA Replication

Pairing of
Homologous
Chromosomes

Duplicated Homologous
Chromosomes
Align on the Spindle

Cytokinesis I

Chromatids
Separate

Cytokinesis II

MITOSIS

DNA Replication

Duplicated
Chromosomes
Align Individually
on the Spindle

Chromatids
Separate

Cytokinesis

Figure 9-13   A COMPARISON OF MITOSIS AND MEIOSIS.
Mitosis results in two daughter cells, each with the same number of chromosomes as the parental cell. Meiosis results in four daughter cells, each with half the number of chromosomes as in the parental cell.

metes from diploid to haploid? No, there is much more to meiosis than just the diploid-to-haploid reduction. Meiosis is a primary means of ensuring the hereditary variability among members of a species that lies at the core of evolutionary change in organisms. Without such variability, organisms would be much less capable of adapting to changing environments, and the species

would be less able to survive in the long term. How, then, does meiosis contribute to genetic variability?

One of the most significant results of meiosis is *the random distribution of parental chromosomes*. In a newly fertilized human egg, or zygote, twenty-three chromosomes come from the mother and twenty-three from the father. When the individual who develops from

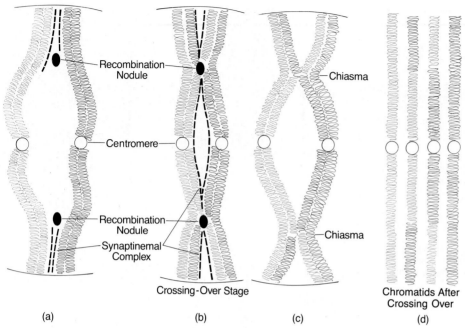

Recombination Nodule

Centromere

Recombination Nodule

Synaptinemal Complex

Chiasma

Chiasma

Crossing-Over Stage

Chromatids After Crossing Over

(a)　　　(b)　　　(c)　　　(d)

Figure 9-14 CHROMOSOMES CROSSING OVER DURING MEIOSIS.
Recombination of genetic information occurs during the crossing over of chromosomes in prophase I. (a, b) In early prophase I, homologous chromosomes begin to pair and become tightly joined by proteins of the synaptinemal complexes. The duplicated chromosomes are attached to the nuclear envelope at both ends and recombination nodules are located at sites where nonsister chromatid exchanges occur. (c) Later in prophase I the synaptinemal complex disperses, as do the recombination nodules; however, the nonsister chromatids remain connected at chiasmata, the sites where crossing over has occurred. The chromatids detach from the nuclear envelope so they can take position on the spindle of metaphase I. (d) The results of the crossing-over process are seen in these chromatids aligned next to each other. In this example, two of the four chromatids have exchanged portions large enough to contain thousands of genes.

that zygote grows into a man or a woman and produces sperm or eggs, the maternal and paternal type chromosomes are distributed to the four daughter cells *at random* during meiosis I: although every gamete receives twenty-three chromosomes, anywhere from none to all twenty-three can be of paternal or of maternal origin. The randomness of distribution of maternal and paternal chromosomes to the daughter cells ensures that most individual gametes will receive different combinations of the original maternal and paternal chromosomes and genes, instead of simply receiving all from one source or all from the other.

In Chapter 10, we will see how such chromosomal distribution helps explain the inheritance of certain traits. If, for instance, the father's sperm contributes the haploid chromosome set Abc, and the mother's egg contributes the haploid set aBC, the child will have the diploid chromosome combination AaBbCc. Years later, when this individual produces eggs or sperm and meiosis separates the pairs randomly, 2^3, or eight, haploid combinations are possible in egg or sperm cells: Abc, ABc, AbC, ABC, abc, aBc, aBC, and abC. If all these possible combinations can exist for just three pairs of homologous chromosomes, imagine the huge number of possible variations with twenty-three homologous pairs: 2^{23} is more than 8 million potential chromosome combinations in the sperm or egg.

Another phenomenon that occurs during meiosis makes the number of possibilities larger still. During synapsis in prophase I, the homologous chromosomes—each composed of two identical chromatids—are "zipped together" by the synaptinemal complex. Then nonsister chromatids (that is, paternal and maternal ones) exchange corresponding pieces of genetic material, as illustrated in Figure 9-14. This process, which is called **crossing over,** involves recombination nodules, large multienzyme aggregates that cut and stitch the maternal and paternal chromatids together. The visible points at which homologous chromosomes break and rejoin are called *chiasmata* (singular, *chiasma*). There is no apparent gain or loss of total genetic material during crossing over, just an equal exchange of corresponding chromosomal regions. This exchange is an important source of genetic variation as we shall see in Chapter 11.

ASEXUAL VERSUS SEXUAL REPRODUCTION

As we have seen, mitosis and meiosis make possible simple cell division and sexual reproduction. Each means of passing along hereditary information has certain advantages in what could be called the gamble of evolution.

Asexual Reproduction

Mitosis in a single-celled creature such as a yeast or an amoeba produces two daughter cells that function as independent and genetically identical organisms. Some multicellular eukaryotes can also reproduce by means of

Figure 9-15   ASEXUAL REPRODUCTION IN THE STRAWBERRY PLANT.
Strawberry runners demonstrate a form of asexual reproduction used by plants. The plants grow new runners, or horizontal stems, which send their own roots down into the ground to form a new plant.

processes based on mitosis. Many plants reproduce by growing new plantlets on their roots or even on their leaves (Figure 9-15). Some types of trees that have toppled in a forest, for example, can send down new roots where branches touch the ground, thereby forming new trees. In animals with simple body organization such as hydra, a relative of the jellyfish, buds grow by means of simple mitotic cell division, pinch off, and subsequently develop into a new hydra through further mitotic division (Figure 9-16). These examples are all forms of *asexual reproduction*—reproduction without sex or meiosis.

(a)

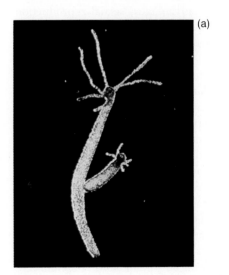

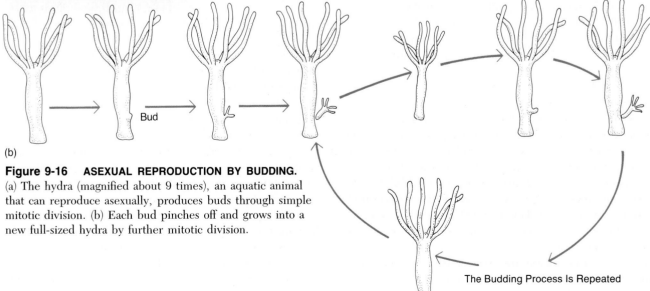

Bud

The Budding Process Is Repeated

(b)

Figure 9-16   ASEXUAL REPRODUCTION BY BUDDING.
(a) The hydra (magnified about 9 times), an aquatic animal that can reproduce asexually, produces buds through simple mitotic division. (b) Each bud pinches off and grows into a new full-sized hydra by further mitotic division.

Thanks to the regularity and precision of mitosis, asexual reproduction can produce many successive generations of organisms, all genetically identical to the parent. These offspring are often called **clones.** (We will study methods of propagating clones—*cloning*—in later chapters.) Asexual reproduction has several advantages: it is often more rapid than sexual reproduction; and it requires less or no specialization of reproductive organs. Most important, asexual reproduction preserves an organism's winning genetic formula. Every such organism gambles, or competes in the world, with the identical "cards" (genes) dealt to it by its parent. As long as environmental conditions remain stable, a genetically successful clone can enlarge rapidly and compete successfully with other organisms in the same environment.

Without the tremendous genetic variability bestowed by meiosis and sexual processes, however, a population of genetically identical organisms stands a greatly increased chance of being wiped out by one disease or one unusual environmental insult—say, a long drought. A line of asexually reproducing organisms can cope with changing conditions only through the relatively rare events of spontaneous mutations (alterations in the genetic material) that prove to be beneficial. However, most mutations are detrimental or even lethal, and herein lies one of the greatest problems of asexual reproduction: all such mutations will be passed on to every offspring along with the normal, unmutated genes. Consequently, the typical asexual organism has only one "good" copy of each hereditary unit (gene); the one on the homologous chromosome will be a mutated form that is inactive or potentially lethal. Nevertheless, asexual reproduction remains common, especially in plants and in some invertebrates (animals without backbones), although in both, it may be coupled with sexual reproduction at certain stages of the life cycle. Finally, many organisms reproduce asexually when environmental conditions remain stable but switch to sexual reproduction when induced to do so by changing conditions.

Sexual Reproduction

Sexual reproduction generates genetically unique progeny that have some traits from one parent, some traits from the other, and, significantly, some totally new traits based on novel combinations of traits from both parents. Further variability can be provided by spontaneous mutations. The sexually reproducing organism in a sense gambles on giving its offspring a better hand; its genetic "cards" are "reshuffled" and "redealt" instead of being passed along in original form.

New combinations of traits can arise much more rapidly in sexually than in asexually reproducing organisms and increase the chances of the species surviving sudden significant environmental changes. In contrast to the way asexual reproduction can retain mutations, copies of deleterious and lethal mutations tend to be eliminated from sexually reproducing populations. (For example, when two parents—each possessing one copy of a lethal mutation—reproduce, the offspring may inherit a lethal copy from each, die as a result, and thus eliminate these copies from the population.) Another advantage of diploid sexual reproduction is that "new" forms of genes can arise and spread within populations. One copy of a gene, say for the protein actin, may be normal, while the copy on the homologous chromosome functions like a "spare." By mechanisms discussed in later chapters, the spare may be duplicated and mutated. Thereby whole "families" of hereditary units may arise, such as the actin, tubulin, and hemoglobin families. This hereditary enrichment process adds more and more new "cards" to the deck that meiosis reshuffles and redeals.

Despite these great advantages, sexual reproduction does have drawbacks. An organism that cannot reproduce asexually—a mammal, for example—can never bequeath its own exact set of genetic material—no matter how successful—to its progeny, the way a prize-winning strawberry plant, for example, can pass along its hereditary complement to a clone. Meiosis bestows on the progeny a reassortment of maternal and paternal chromosomes with areas of crossing over and perhaps a few mutations. Thus the very mixing process that created the successful gene combination in the adult works to dismantle it partially in the offspring.

Cattle ranchers have confronted this problem for decades in trying to breed cattle with desirable sets of traits, such as producing large amounts of tender meat on minimal feed. Yet a rancher has no way to guarantee that the full set of these traits will be passed along from a breeding bull to its thousands of offspring. If the bull could be propagated asexually—that is, by mitosis instead of meiosis—the rancher could have an entire feedlot full of cattle with the desired traits—clones genetically identical to the breeding bull.

Recent technological advances may someday make possible the cloning of bulls and other mammals. In experiments with frogs in the early 1960s, the British biologist John Gurdon produced a clone of adult frogs by implanting the nucleus of a body cell into a frog egg which had its own nucleus destroyed; that egg divided and produced many cells whose nuclei could be transplanted into other eggs, which, in turn, matured into tadpoles and later into genetically identical adult frogs (Figure 9-17). More recent attempts to clone mammals have met technical difficulties arising from the use of adult cells for donor nuclei. It is probably just a matter of time, however, before cloning from the cells of an adult bull or other mammal is possible.

Figure 9-17   A CLONED FROG.
This African clawed frog grew from a fertilized egg in which the original nucleus was destroyed. Then a nucleus from a frog intestinal cell was injected and served as the sole source of nuclear hereditary material used in producing this animal. In turn, nuclei from this frog could have been used to generate dozens of genetically identical clones.

LOOKING AHEAD

In this chapter, we have seen that the resemblances between parents and offspring depend on the behavior of chromosomes and the cycling of cells through periods of growth, synthesis, and division. The process of mitosis allows astonishing fidelity in the passage of genetic information from one cell generation to the next. During meiosis, on the contrary, two successive nuclear divisions follow just one period of chromosome duplication. Meiosis results in a halving of the chromosome number and includes a period of crossing over during which chromosomes usually exchange corresponding regions, thus increasing genetic variability. Despite their inherent differences, mitosis and meiosis both display a similar and spectacular dance of the chromosomes—a dance that makes it possible for the remarkable packages of inheritance, the chromosomes, to be passed to daughter cells in precise ways. In the next chapter, we shall begin to see how biologists unraveled the mysteries of inheritance.

SUMMARY

1. Experiments such as Haemmerling's work with *Acetabularia* proved that the nucleus is the repository of hereditary information. Subsequent research showed that this information is contained in DNA, which is organized in structures called *chromosomes.*

2. Chromosomes are made up of a substance called *chromatin*, which is a combination of DNA, *histones*, and nonhistone proteins. *Nucleosomes* are sites where DNA is wrapped about sets of histone molecules. In dividing cells, the chromatin condenses through extensive coiling and supercoiling, so that individual chromosomes become visible in the light microscope.

3. In most eukaryotic organisms, there are two copies of each chromosome; the two comprise a *homologous* pair. The exception is the *sex chromosomes*, which are either identical (XX), as in female mammals, or dissimilar (XY), as in male mammals. Nonsex chromosomes are termed *autosomes*. Cells and organisms that contain homologous pairs of chromosomes are said to be *diploid*; those that contain only one set of chromosomes are said to be *haploid*. The full set of a cell's or organism's chromosomes is called the *karyotype*; the human karyotype consists of twenty-two pairs of autosomes plus two sex chromosomes.

4. The mitotic *cell cycle* consists of four phases: G_1, a period of normal metabolism; S, the phase of DNA replication plus metabolism; G_2, a brief period of further cell growth; and M, mitosis. The nonmitotic stages (G_1, S, G_2) are referred to collectively as *interphase*. *Mitosis* itself is the nuclear division process. The cell cycle apparently is controlled by cytoplasmic properties and by external stimulators and inhibitors of mitosis.

5. The process that ensures an orderly and accurate distribution of chromosomes during normal cell division is *mitosis*. The events of mitosis can be divided into four phases: *prophase, metaphase, anaphase*, and *telophase*.

6. At the beginning of mitosis, each chromosome consists of two identical *chromatids*. These two chromatids, joined at the *centromere*, become associated with a set of microtubules called the *spindle* and align in the middle of the cell along the *metaphase plate*. In anaphase, the centromeres divide, and the two chromatids are drawn toward opposite poles. With the chromosomes safely separated, two nuclear envelopes form, and the chromosomes begin to uncoil.

7. Two types of spindle microtubules seem to be involved in moving the chromosomes during mitosis. One set, the polar microtubules, extends from the poles and overlaps at the equator. As the region of overlap decreases, the poles are moved farther apart. A second set of microtubules, the centromeric fibers, extends from the *kinetochores* in the

centromere of the chromatids toward the spindle poles. These fibers shorten in order to pull the chromatids toward the poles.

8. In animals, spindle formation is associated with centrioles; in most plants and fungi, it is associated with microtubule organizing centers.

9. Following mitosis, the cell undergoes *cytokinesis*, or division of the cytoplasm. In animal cells, cytokinesis results from the activity of a contractile ring of actin filaments at the cell's equator. In plants, cytokinesis involves the building of a *cell plate* across the cell's equator. Cytokinesis can occur in the absence of a nucleus or chromosomes.

10. *Meiosis* is a special type of cell division that reduces the chromosome number by half. Two successive nuclear divisions take place, meiosis I and meiosis II. In meiotic prophase I, the two chromatids of each homologous chromosome come together in a process called *synapsis* and are held together by the *synaptinemal complex*, a bridge of proteins. The homologous pairs remain together when they align on the metaphase plate, and then in anaphase I, the homologous chromosomes separate to opposite poles. Following telophase I and during meiosis II, the chromosomes align independently on the metaphase plate, and the centromeres finally divide, each sister chromatid of the pair moving to one of the poles. The result is four haploid cells.

11. Meiosis allows for the random distribution of homologous parental chromosomes in offspring. Furthermore, the *crossing over* between nonsister chromatids that occurs during synapsis gives rise to additional combinations of traits in offspring.

12. Meiosis produces haploid cells that may then develop into spores, pollen grains, sperm, or eggs.

13. Asexual reproduction, in which new organisms arise from mitotic processes such as budding and *cloning*, is usually more rapid than sexual reproduction and preserves an organism's "winning genetic formula" in a particular environment. But it results in deleterious mutations accumulating in a species. Sexual reproduction, which involves meiosis and gamete production, assures greater variability in offspring, eliminates many copies of lethal mutations, and ensures hereditary enrichment as new genes and gene families arise. The resultant hereditary variability and richness increase a species's chances of surviving sudden environmental changes and of adapting to new environments.

KEY TERMS

anaphase
autosome
cell cycle
cell plate
centromere
chalone
chromatid
chromatin
chromosome
clone
crossing over
cytokinesis
diploid
G_1 phase
G_2 phase
haploid
histone
homologous chromosome
interphase
karyotype
kinetochore

M phase
meiosis
metaphase
metaphase plate
mitosis
nucleosome
prophase
S phase
sex chromosome
spindle
synapsis
synaptinemal complex
telophase

QUESTIONS AND PROBLEMS

1. What part of the cell contains the hereditary information? Which structures contain this information? Which molecules make up these structures? Does the hereditary information reside in a specific molecule within the structure?

2. The cell cycle consists of four phases: G_1, S, G_2, and M. What do these letters stand for? What occurs during each phase? During which phase are individual chromosomes visible in the light microscope? Which is the shortest phase? Which has the most variable length?

3. What are some substances that can stimulate or inhibit cell division? Describe an experiment which shows that cytoplasmic substances may render certain cells permanently nondividing.

4. What is the outcome of mitosis? Does each daughter cell receive identical chromosomes?

5. What is the outcome of meiosis? Does each haploid cell receive identical chromosomes?

6. Which of the following statements apply to mitosis, which to meiosis, and which to both?

  a. DNA replication occurs before this process starts.

  b. When the chromosomes first become visible they are already doubled.

  c. Homologous chromosomes pair.

  d. Each daughter cell receives an identical complement of chromosomes.

  e. Centromeric fibers attach to kinetochores.

  f. The centromere is the last part of the chromosome to divide.

7. Consider a large population of gamete-producing germ cells in a reproductive organ about to undergo meiosis. Each cell contains a pair of chromosomes that can be represented by a right and a left shoe. After meiosis, each haploid cell will contain only a right or a left shoe. What fraction of the cells will contain a right shoe? A left shoe? If the germ cells contain two pairs of shoes, a brown pair and a black pair, will each haploid cell contain two right or two left shoes, or will they be randomly assorted? What fraction of the haploid cells will contain two right shoes? Two left shoes? A brown right and a black left? A black right and a brown left? How many assortments are possible? Now consider germ cells with three pairs of shoes: black, brown, and white. How many assortments are possible in the resulting haploid gamete cells? What fraction of the haploid cells will contain a black right, brown right, and

white left? If the germ cells contain twenty-three pairs of shoes, how many assortments are possible in the haploid cells? Imagine that when the shoes pair up during meiosis each pair exchanges parts, so that each left shoe will contain pieces of its right partner, and each right shoe will contain pieces of its left partner. In each cell, different parts of each pair are exchanged. Now try to visualize how many different assortments are possible in the haploid cells. Among millions of meiotic products, what are the chances that any two will be identical?

8. Biologists have described two kinds of microtubules that separate the chromosomes during cell division. What are these microtubules, and how do they operate?

9. Describe the function of the centriole or microtubular organizing center during cell division.

10. Describe the process of cytokinesis first in a dividing animal cell, then in a plant cell.

ESSAY QUESTIONS

1. Did you inherit equal numbers of chromosomes from your two parents? From your four grandparents?

2. Is a species more likely to survive in a changing environment if all its members are identical or if they are diverse? What is accomplished by sexual reproduction? What are the advantages of asexual reproduction?

SUGGESTED READINGS

ALBERTS, B., et al. *Molecular Biology of the Cell*. New York: Garland, 1983.

Contains two easy-to-understand chapters on mitosis and meiosis; ideal for students.

JOHN, B., and K. R. LEWIS. *The Meiotic Mechanism*. Oxford: Oxford Biology Readers, 1976.

A well-illustrated, simple summary of this complex process.

JOHN, P. C. L. *The Cell Cycle*. Cambridge: Cambridge University Press, 1981.

A comprehensive review and introduction to this important subject.

PRESCOTT, D. M. *Reproduction of Eukaryotic Cells*. New York: Academic Press, 1976.

A leader in the field presents a broad overview of this topic.

WHITEHOUSE, H. L. *Towards an Understanding of the Mechanism of Heredity*. 3d ed. London: St. Martin's Press, 1973.

A superb description of chromosome movement during meiosis.

ZIMMERMAN, A. M., and A. FORER, eds. *Mitosis/Cytokinesis*. New York: Academic Press, 1981.

A series of review papers covering molecular research, in particular.

10
FOUNDATIONS OF GENETICS

As a consequence of the application of Mendel's principles, that vast medley of seemingly capricious facts which have been recorded as to heredity and variation is rapidly being shaped into an orderly and consistent whole. A new world of intricate order previously undreamt of is disclosed. We are thus endowed with an instrument of peculiar range and precision, and we reach to certainty in problems of physiology which we might have supposed destined to continue for ages inscrutable.

William Bateson, *Mendel's Principles of Heredity* (1909)

The fruit fly *Drosophila:* The solution to a thousand genetic puzzles.

During the 1880s, a German biologist named August Weismann did a seemingly curious thing: three or four times a year for seven years in a row, he selected a few newborn mice from the mouse colony he maintained in his laboratory. Then, holding ice to the base of their tails (as a local anesthetic) he lopped off the tails, one by one, with a sharp knife.

This exercise did not hurt the mice—at least not for long—but it did help the early study of heredity. Weismann had wondered whether a tail-less mouse would sire offspring with normal tails or whether the young might inherit the tail-less trait. During his seven-year experiment, he removed the tails from the mice of twenty-two consecutive generations and found that the tails of the mice in the twenty-third generation looked as normal as had those of the first. The implication was clear: information for constructing a tail does not reside solely in the tail itself and cannot therefore be lost by removing the tail. Instead, such information must lie in another part of the body and somehow be passed from parent to offspring.

Weismann's conclusions may seem obvious today, but in his time, a long, carefully planned experiment to probe the nature of inheritance was quite innovative. People had, of course, been observing the similarities between parents and offspring for many centuries. Domestication and selective breeding of crop plants, pets, and livestock had gone on for more than ten centuries and is, indeed, linked with the beginning of civilization. However, casual observations about heredity failed to produce more than imaginative explanations.

In this chapter, we first will look at some of those early explanations for inheritance of traits and then shall see how, in the late nineteenth century, an extraordinary monk named Gregor Mendel discovered the basic mechanism of heredity. Heredity is a key feature of life as we know it. Cells divide to reproduce, as we saw in Chapter 9, and yet their characteristics do not vary widely or indiscriminately from one generation to the next. The mechanisms of heredity ensure great stability to lineages of organisms, so that they may remain adapted to their environments if the environments remain stable. Yet at the same time, these mechanisms allow some genetic change and the potential to survive long-term environmental fluctuations. Mendel's experiments and the principles he discovered help explain not only this simultaneous genetic stability and flexibility, but also the dance of the chromosomes that we saw in our discussions of mitosis and meiosis. Mendel's laws are fundamental to our understanding of modern biology, for the rules of heredity are in a real sense ground rules for all life on Earth.

EARLY THEORIES OF INHERITANCE

Some of the first recorded discussions of heredity are found in the writings of Aristotle (384–322 B.C.), who criticized a theory that had been proposed earlier by Hippocrates (ca. 460–ca. 360 B.C.): the theory of **pangenesis.** According to this idea, each part of the body produces a characteristic "semen," or "seed," that somehow travels to the reproductive organs. On copulation, these seeds are released, and the combined fluids from the parents directly form the respective body parts of the offspring. In his criticism of pangenesis, Aristotle pointed out that the loss of an arm or a foot by one parent does not result in a similar deformity in the offspring. Moreover, he realized that characteristics not yet present in a young father, such as a heavy beard or baldness, might nevertheless be inherited and show up later in offspring.

Despite Aristotle's objections, the theory of pangenesis persisted in one form or another for more than 2,000 years, until the late nineteenth century. At that time, Charles Darwin modified the idea to some degree, but even the greatest evolutionary thinker of that era accepted the notion of pangenesis in principle. Darwin believed that each organ of the adult body produces *gemmules,* tiny packets of hereditary information. These, he thought, collect in the male semen and blend with the mother's gemmules at the time of fertilization. The gametes (sperm and eggs) from both parents thereby transmit the actual distilled physical body traits to the progeny.

Not until August Weismann began to lop off mouse tails was the theory of pangenesis put to a scientific test. His results directly contradicted the older theories: clearly, tails do not have to be present to pass out tail "semen" or "gemmules." On the basis of his results and of those of several brilliant experiments by other researchers, Weismann proposed the **germ plasm theory.** According to this theory, only hereditary information in the "germ plasm" of the gametes transmits traits to the progeny; other adult body cells (somatic cells) do not make a contribution. The fertilized egg gives rise to new lines of germ cells and somatic cells in the embryo during its development. The somatic cells—muscle, blood, brain, and so on—make up most of the body. The germ cells present in female ovaries or male testes divide and produce eggs or sperm. Hence, only the germ-cell line of each new individual makes a contribution to *its* offspring, as diagrammed in Figure 10-1. In Weismann's theory, the germ plasm of the gametes is perpetuated

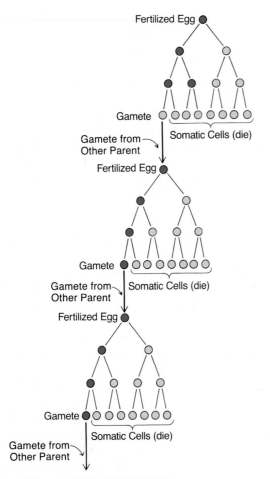

Figure 10-1   AUGUST WEISMANN'S GERM PLASM THEORY.
A sexually reproducing organism, has two types of cells, somatic and germ cells. Germ cells, which become located in the gonads, produce the gametes, eggs or sperm. The germ line is shown in blue. Somatic cells are all the other cells in the body; these make up its various parts, and expire as a group when the organism dies. Only the germ cells contribute directly to the next generation.

generation after generation, whereas somatic cells' sole function is to protect and nourish the germ-cell lineage.

Weismann's theory differed from the theory of pangenesis in that it emphasized the special nature of germ cells, eggs, and sperm. These cells are indeed exceptional in many ways, an especially important one being that eggs and sperm arise by meiosis rather than mitosis. However, Weismann's theory and the older ones shared one important erroneous element: they held that heritable traits of both parents blend, or fuse, thereby losing the distinct characteristics of each.

If the **blending theory** were correct, a simple experiment should prove it. Suppose we mated a purebred black dog and a purebred white dog; the theory predicts

that gray puppies should always result. But they do not: first, most progeny of black and white dogs are not gray; and second, gray dogs that do arise can yield offspring that are black or white. That is, the original color traits may disappear in the first generation, but reappear in the second generation. Another problem with the blending theory is that it is inconsistent with Darwin's theory of natural selection. If blending always occurred, all variation within a population would be blended away over time. All dogs with black and white forebears eventually would be a muddy gray and would be average in height, weight, and intelligence. There would be no superior traits to be favored by natural selection because all traits eventually would average out, and all individuals would be similar. Nevertheless, despite these problems with the blending theory, prior to the twentieth century, no one could explain heredity more satisfactorily, and the belief in the "fluid" nature of heredity persisted. Even today, this ancient view is reflected in our vocabulary, when people refer to human marriage and reproduction as "mixing blood."

Ironically, the view of heredity as fluid and blendable contrasted sharply with the prevalent views in physics and chemistry in the 1880s. Those fields emphasized the particulate nature of matter—the idea that all substances are made up of molecules and atoms. This difference in outlook was probably fostered by *vitalism*, the widespread belief that laws governing living things are essentially different from those governing nonliving matter. At the time that Gregor Mendel made his discoveries, vitalism and the blending theory formed the philosophical backdrop for the study of inheritance. Against this backdrop, his conclusions were all the more revolutionary.

GREGOR MENDEL AND THE BIRTH OF GENETICS

Gregor Johann Mendel (Figure 10-2) was born in 1822 in a small Austrian village that is now part of Czechoslovakia. In school, Mendel was recognized to be a bright student, and he was encouraged by his teachers. His family was so poor, however, that his sister had to renounce part of her dowry so that Mendel could study for entry to the university. Unfortunately, he was still unable to afford a university education. Instead, in 1843, he joined the Augustinian monastery at Brünn (now Brno, Czechoslovakia).

This turned out to be an unparalleled—and, of course, free—educational opportunity for Mendel. Throughout most of the nineteenth century, the Brünn monastery had been a center of enlightenment and scientific

MENDEL'S CLASSIC EXPERIMENTS

Mendel's experiments were carried out with the garden pea, a choice he made deliberately and, in retrospect, wisely. He obtained thirty-four distinct strains of peas from farmers and raised generations of the plants for two years in order to select for his experiments only those strains that bred true. ("Breeding true" means that each offspring is identical to the parent in the trait of interest.)

It was easy for Mendel to select and maintain true-breeding strains because the pea is self-fertilizing. That is, each single pea flower has both male and female reproductive organs, enclosed by the delicate petals. Pollen grains, the sites of sperm nuclei, are formed in *anthers*. The anthers are near the *stigma*, the female structure inside the flower that receives pollen. Because the pea flower does not open fully, pollen usually reaches the stigma within the same flower and does not pass from one flower to another (Figure 10-3). Although self-fertilization is the normal mode, an experimenter can induce artificial cross-fertilization by carefully opening the flower bud and snipping off the anthers before they mature. Pollen from the flower of a second pea plant can then be transferred to the stigma of the first to cause controlled fertilization with pollen from that second plant.

In addition to choosing only true-breeding self-fertilizing plants for his experiments, Mendel focused on the inheritance of several distinct traits, rather than on the entire appearance of each plant in relation to that of the parents. A single pea plant, for example, might be tall, bear red flowers, and produce yellow peas in plump green pods that grow on the sides of each stem. Another plant might be a dwarf, bear white flowers, and produce green peas in shriveled yellow pods that grow at the ends of each stem. Analyzing so many different traits at once would have been a statistical nightmare. So Mendel chose seven characters, or pairs of traits, and studied only one or two of those seven from any given plant and its progeny. The traits he studied are depicted in Figure 10-4. (The significance of the categories "dominant" and "recessive" will soon become apparent.)

Mendel was not the first to study genetics through plant-breeding experiments. Years earlier, Charles Darwin had carried out plant-breeding experiments in England, in an attempt to explain the apparent vagaries of inheritance. However, Darwin did not have Mendel's mathematical background, nor had he planned his experiments as carefully. Mendel's education in mathematics and statistics largely made the difference between failure and success.

Figure 10-2   FATHER OF GENETICS: GREGOR MENDEL (1822–1884).

thought. Much of its reputation was due to the efforts of Abbot Cyrill Napp, who was a proficient plant breeder as well as leader of Church affairs. The abbot often participated in scientific discussions on the nature of heredity, stating at one point that breeding practices and theories were less important than was the question of "what is inherited and how."

In 1851, Napp sent Mendel to the University of Vienna, where he studied physics, mathematics (including the new field of statistics), chemistry, botany, and plant physiology. It was during these studies in Vienna that Mendel encountered theories about the particulate nature of matter. When he returned to the monastery, he began a series of carefully planned experiments that eventually demonstrated the particulate nature of heredity. Abbot Napp had a large greenhouse built for Mendel's work, and as Mendel labored alone for many years, the abbot continued to provide encouragement. In 1866, Mendel presented the results of his experiments on the nature of inheritance to the Brünn Society for Natural History. However, his ideas were neither accepted nor understood in his lifetime. He died in 1884, convinced that it would "not be long before the whole world acknowledges" his discovery.

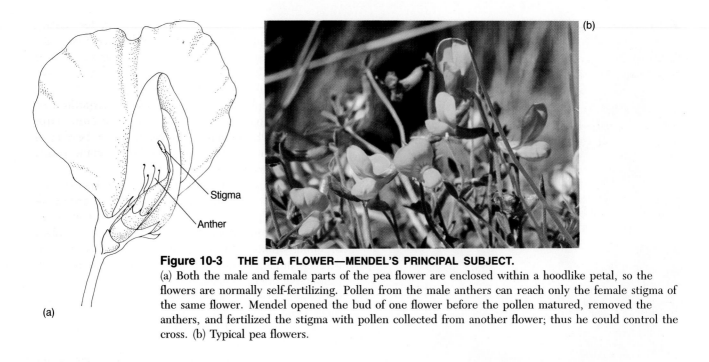

Figure 10-3   THE PEA FLOWER—MENDEL'S PRINCIPAL SUBJECT.
(a) Both the male and female parts of the pea flower are enclosed within a hoodlike petal, so the flowers are normally self-fertilizing. Pollen from the male anthers can reach only the female stigma of the same flower. Mendel opened the bud of one flower before the pollen matured, removed the anthers, and fertilized the stigma with pollen collected from another flower; thus he could control the cross. (b) Typical pea flowers.

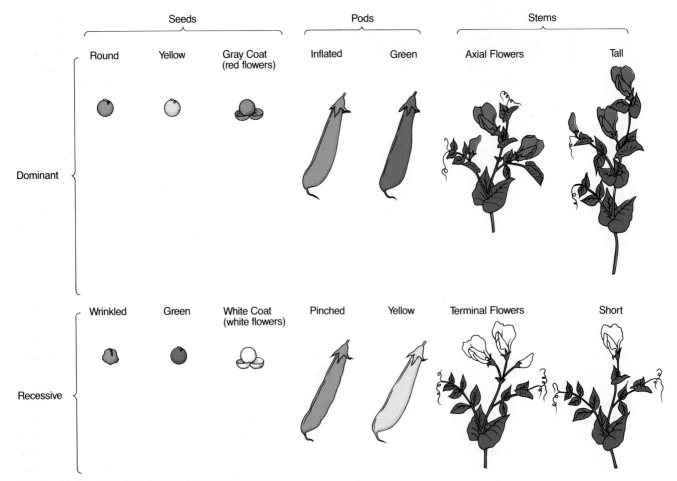

Figure 10-4   THE PEA PLANT'S FAMOUS TRAITS.
Mendel studied seven clearly differentiated traits, or characters, of the pea plant in pure-breeding individuals of each strain. The seven traits and the dominant and recessive forms of each are shown here.

Dominant and Recessive Traits

Mendel's experiments had three unique and important features.

1. He studied characters of pea plants that offered just two clear-cut possibilities: tall or short height, red or white flower color, green or yellow pea seed color.

2. He traced and recorded the type and number of all the progeny produced from each pair of parent pea plants that he crossbred. Thus he could determine ratios of characters appearing in the progeny; for example, white flowers versus red flowers from a given cross.

3. Finally, he followed results of each cross for two generations, not just one—a simple but profound decision that led to the downfall of the blending theory of inheritance.

In a set of experiments on the inheritance of seed color, Mendel crossed a strain of pea plant that produces yellow seeds with a strain that produces green seeds. These strains are called the **P₁, or parental, generation,** and the progeny from such a cross are designated the **F₁, or first filial, generation.** Mendel observed that in the cross between P₁ plants with green seeds and with yellow seeds, the progeny produced *only* yellow seeds (Figure 10-5). No yellowish-green seeds were found—an important result that contradicted the blending theory of inheritance. Somehow, the yellow trait had dominated over the green trait in the F₁ generation.

But Mendel was not content simply to disprove the blending theory. He also wanted to know what would happen in future generations. So he allowed the F₁ generation yellow seeds to sprout, grow into mature plants, self-fertilize, and produce the next generation, which we now designate as the **F₂, or second filial, generation.** In Mendel's experiment, plants producing green seeds "reappeared" along with a great number of yellow-producing plants in the second generation. Mendel's mathematical training now came into play. Among a total of 8,023 F₂ plant seeds, Mendel found that 6,022 were yellow and 2,001 were green. Since three-fourths of the plants produced yellow seeds and one-fourth produced green, this represented an almost exact 3:1 ratio.

Mendel noted two important points from this experiment:

1. Although the green-seed trait had disappeared in F₁, it *reappeared* in F₂.

2. When the green-seed trait reappeared it was *unchanged* from its appearance in the P₁ parent.

He therefore reasoned that information for making green seeds must have been present but invisible in the F₁ plants and that although hidden, it was not altered during its residence there. Mendel inferred that each original P₁ plant contributed information for producing seed color to the F₁ generation, and he hypothesized that the information for yellow seed color donated by one parent was **dominant** over that for green seed color donated by the other parent. The nondominant trait of green seed color he termed **recessive.**

For each of the seven characters he studied, Mendel found one trait to be dominant over the other. For example, the round seed trait was dominant over the wrinkled seed trait, and tall stems were dominant over short stems. Moreover, each recessive trait reappeared in the F₂ progeny in the ratio of three dominant to one recessive.

Genetic Alleles

Mendel deduced that the 3:1 ratio of dominant to recessive traits in the F₂ generation could occur if each individual possesses only two hereditary units that supply information for each character, one unit received from each parent. How did he reach this conclusion?

Mendel knew that a factor causing the dominant yellow trait was present in the F₁ generation, since the seeds were yellow, and yellow seeds also appeared in the next generation. But in addition to the yellow factor, there must have been a second factor present in the F₁ to account for the appearance of green seeds in the F₂ progeny. Thus, at least two factors were present in the F₁: one for the dominant yellow and the other for the recessive green. If the F₁ generation had two factors, it was logical to assume that the P₁ generation had two as well. In each P₁, which you will recall had bred true in color for several generations, both factors must have been the same, either two dominant factors for yellow in the yellow-pea lineage or two recessive factors for green in the green-pea lineage.

The factors that Mendel hypothesized are now called *alleles.* **Alleles** are alternative forms of **genes,** the basic units of heredity. One allele comes from each parent and is responsible for either a dominant or a recessive trait. We can assign the letter *Y* to the allele for yellow seed color and *y* to the allele for green seed color. (In genetic notation, a capital letter is used to represent the dominant trait and the lower case of the same letter is used to represent the recessive trait.) The true-breeding yellow parental strain has the alleles *YY,* and the true-breeding green parental strain, the alleles *yy.* Both true-breeding parents are said to be **homozygous** ("similar pair") for alleles of the seed-color character; a *homozygote* is an organism with two identical alleles for a particular trait.

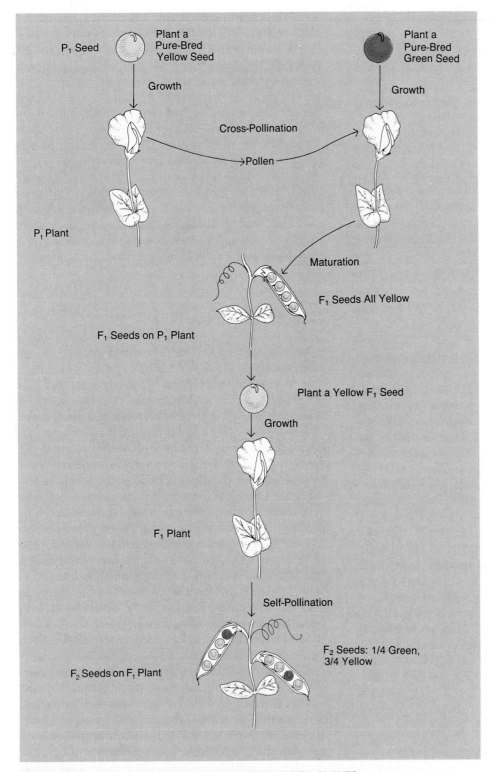

Figure 10-5   HOW SEED COLOR IS INHERITED IN PEA PLANTS.
Mendel crossed plants in the P_1, or parental, generation grown from yellow seeds with plants grown from green seeds. All the progeny from this cross in the F_1 (first filial) generation bore yellow seeds (peas). Mendel then planted the F_1 yellow seeds and allowed the resulting plants to self-pollinate and produce the F_2 (second filial) generation. He counted about three yellow seeds for every green seed in the F_2 generation pea pods. Since the green trait was absent in the F_1, but "returned" in the F_2, Mendel reasoned that it must have been present but hidden in the F_1 generation.

Since only one type of an allele is present in a homozygote, all gametes formed by a homozygous parent throughout its life must be identical with respect to that particular gene.

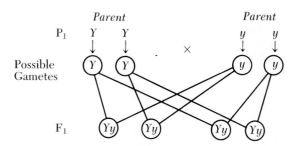

According to Mendel's hypothesis, each of the F₁ progeny receive a Y allele from its yellow-seeded parental strain and a y allele from its green-seeded parental strain. All members of the F₁ progeny then have one Y and one y and are designated Yy. These plants are termed **heterozygous** ("different pairs") for the gene affecting seed color and are called *heterozygotes*. Thus, although all the F₁ progeny produce yellow seeds, those progeny contain a recessive allele for green seed color in every one of their cells. That is, there is a difference between their *genotype* and their *phenotype*. The **genotype** is the genetic make-up of a cell or an organism, whereas the **phenotype** is the cell or organism's appearance—the expression of those genes. In the F₁ yellow-seeded plants, the genotype, Yy, contains alleles for both yellow and green seed color, but the phenotype, yellow seed color, is an expression of only the dominant allele; the recessive allele remains hidden or masked. These recessive alleles could then reappear in the genotypes of three-fourths of the progeny of the F₂ generations.

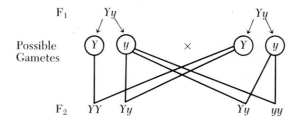

Although three-fourths of these F₂ progeny could have a recessive allele (y) in their genotype (Yy, Yy, yy), only one-fourth would show the recessive phenotype (yy), since the y allele would be masked in Yy individuals by the dominant Y allele. Other examples of phenotypic characters in peas are pod color, plant height, and flower placement. In fact, each of the fourteen traits that Mendel studied is the phenotypic expression of an underlying genotype.

Probability and Punnett Squares

One reason Mendel was able to predict the existence of dominant and recessive traits was his knowledge of probability. He knew, for instance, that if you toss a coin in the air it has an equal chance—probability—of landing heads or tails. If you toss two coins, there are four probabilities: two heads; two tails; one head and one tail; or one tail and one head. Thus the probability of two heads (or two tails) is 1 in 4, while the probability of one head and one tail is 2 in 4. Perhaps Mendel's observation that one-quarter of the F₂ generation displayed green seeds was the key that led him to the conclusion that two independent factors are obeying the rules of probability as traits are passed from one generation to the next—it is as though green and yellow "coins" are being tossed during pea reproduction.

Figure 10-6 shows one means of displaying probability and genetic-trait relationships. This method, called a **Punnett square,** may be used for three, four, or more coin tosses or genetic traits. With four coins tossed in the air, for example, the chance of achieving four heads on the throw is 1 in 16. The various boxes in the Punnett square indicate the probabilities of seeing each of the possible head–tail combinations or, in the case of genetic factors, allele combinations. Thus Punnett squares may be drawn with the alleles from one parent along one side (representing the classes of possible gametes) and those from the other parent along the other side (Figure 10-6). Crossing each pair of alleles allows all the boxes of the Punnett square to be filled and so to display in pictorial form all the possible combinations and to calculate the ratio of results. In the simple case of parents with genotypes YY and yy, all the boxes of the square will show genotype Yy. Try preparing a Punnett square crossing parents with genotypes Yy and yy.

The Law of Segregation

So far, we have followed Mendel's reasoning that each pea plant must have two alleles for each trait but that each parent donates through its gamete only one allele to each individual in the next generation. Apparently, Mendel reasoned, the number of alleles must be reduced during formation of gametes so that the fusion of two gametes produces offspring with two, not four, alleles. We know from Chapter 9 that just this kind of reduction occurs during meiosis and the formation of gametes. Thus every time pollen (or sperm in animals) or eggs are produced, the meiotic reduction divisions ensure haploid gametes with just one allele for each trait.

With this in mind, it is easier to understand the results of the self-pollination cross of a heterozygote ($Yy \times Yy$); such a cross is equivalent to that of two Yy plants with

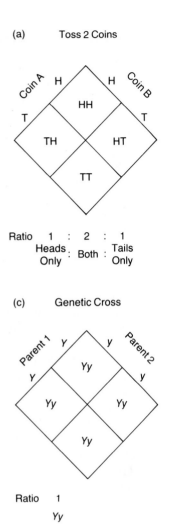

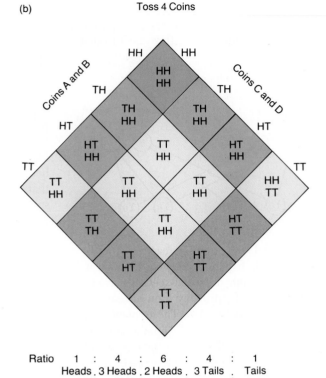

Figure 10-6   THE PUNNETT SQUARE: DETERMINING GENETIC PROBABILITIES.

Using a Punnett square, all the various combinations of genetic factors can be plotted, and the results read off quite easily.

(a) Suppose you toss two coins into the air. Coin A can fall heads up (H) or tails up (T); these probabilities are plotted along one side of the square. Coin B can fall in the same ways, and its probabilities are plotted along the other side of the square. The Punnett square shows that when both coins are tossed together, one-fourth of the throws will result in both coins landing heads up, one-half will result in one head and one tail, and one-fourth of the throws will result in two tails, for a combined 1:2:1 ratio.

(b) If you toss four coins instead of two, the results are more complicated, but just as predictable. In (a) we determined the probable results for two tossed coins; these are plotted on one side of the larger Punnett square. The same probable results for the second two coins are plotted on the other side of the square. The Punnett square then shows all the probable combinations of heads and tails when all four coins are tossed together.

(c) Genetic combinations can be predicted in the same way. If a parent with alleles of YY for a specific genetic trait is mated to another parent with yy alleles for the same trait, the Punnett square shows the gene combinations that might result.

those alleles. Because each F_1 parent cell, being heterozygous, carries two different alleles for each trait, two types of gametes are produced: those carrying the dominant Y allele for yellow seeds and those carrying the recessive y allele for green seeds. If these gametes occur in equal ratios in both pollen (or sperm) and eggs, and if a random combination of gametes with these alleles occurs at fertilization, one-quarter of the F_2 progeny will receive a Y from both parents (YY), two-quarters will receive a Y from one parent and a y from the other (Yy), and one-quarter will receive a y from both parents (yy), as illustrated in the Punnett squares in Figure 10-7. The genotypic ratio will be 1 YY to 2 Yy to 1 yy, but because the phenotype of YY and Yy is the same, the phenotypic ratio of yellow-seed progeny to green-seed progeny will be 3:1. These ratios are statistical averages, based on large numbers of crosses. In a small sample, the ratios might be somewhat different.

The results of Mendel's experiments on dominant and

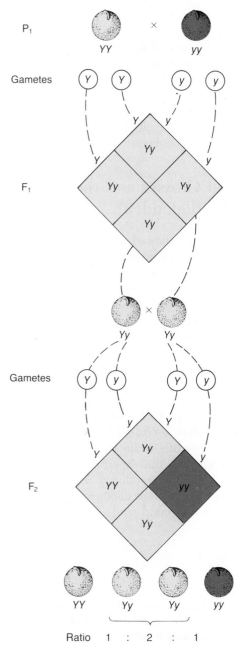

Ratio   1   :   2   :   1

Figure 10-7   MENDEL'S LAW OF SEGREGATION.
When gametes form, the two alleles for a genetic character segregate so that an individual egg or sperm carries only a single allele. For example, in a cross between a homozygous yellow-seeded pea plant (*YY*) and a homozygous green-seeded plant (*yy*), each parent produces only one type of gamete: *Y* or *y*. All the F$_1$ offspring will produce yellow seeds and will be *Yy* for the trait, because they receive one allele from each parent. They can, in turn, produce two kinds of gametes; half of the F$_1$ eggs will carry a *Y* and half will carry a *y*, as will the sperm cell. Because these gametes can combine randomly, the F$_2$ progeny will tend to occur in the genotype ratio of 1 *YY* to 2 *Yy* to 1 *yy* and in the phenotype ratio of 3 yellow-seeded plants to 1 green-seeded plant.

recessive inheritance led to what is now known as Mendel's first law: the **law of segregation.** According to this law, individuals carry two discrete hereditary units (alleles) affecting any given character; one allele for each trait is received from each parent, and together they form the allele pair. During meiosis, these two alleles segregate, or become separated, from each other. One allele for every character is then incorporated into each maturing gamete and is transmitted during fertilization in an unaltered state to the next diploid generation. The new diploid individual has in every cell nucleus two alleles for each character, one from each parent. During development of the new organism, the alleles exert their influence on the phenotype of the new individual.

Science historians believe that Mendel actually may have formulated his main idea, the segregation of discrete units with dominant and recessive tendencies, *well before* he carried out his experiments. His studies of all seven pea-plant characters generated precise numbers in the expected 3:1 ratio, as shown in Table 10-1. Historians wonder whether Mendel—or perhaps his well-meaning, loyal assistants—may have stopped counting pods or seed coats when the expected number was reached. Alternatively, certain counts may have been unconsciously biased—that is, a wrinkled pod may have been identified as smooth. Regardless of the present uncertainty about their precision, Mendel's studies were extremely meticulous for his day, and he indeed proved experimentally the first principles of genetics.

Table 10-1   **MENDEL'S RESULTS IN CROSSES BETWEEN PEA PLANTS DIFFERING IN SINGLE CHARACTERS**

P$_1$ Characters	F$_1$	F$_2$	F$_2$ Ratio
Round or wrinkled seed	All round	5,474 round; 1,850 wrinkled	2.96:1
Yellow or green seed	All yellow	6,022 yellow; 2,001 green	3.01:1
Gray or white seed coat	All gray	705 gray; 224 white	3.15:1
Red or white flowers	All red	705 red; 224 white	3.15:1
Inflated or pinched pod	All inflated	882 inflated; 299 pinched	2.95:1
Green or yellow pod	All green	428 green; 152 yellow	2.82:1
Axial or terminal flower	All axial	651 axial; 207 terminal	3.14:1
Tall or short stem	All tall	787 tall; 277 short	2.84:1

Test Crosses

Support for Mendel's law of segregation came from another test he conducted. If this law is correct, then among a number of yellow pea seeds in the F$_2$ generation, there should be two genotypes present. One-third should by *YY*, while two-thirds should be *Yy*. But if both genotypes produce identical yellow seeds (the same phenotype), how can this hypothesis be tested? Mendel's solution was the **test cross,** mating the plant of known phenotype but unknown genotype to a homozygous recessive plant for the trait in question. A homozygous recessive was used so that the contribution of the unknown parent, be it dominant or recessive, would be obvious in the phenotype of the offspring. Let us see how this works.

If a F$_2$ yellow seed (*Yy* or *YY*) is grown into a plant and is crossed with the plant from a green seed *(yy)*, one of two ratios can be observed in the progeny: the cross can yield all yellow seeds or half yellow seeds and half green seeds, as illustrated in Figure 10-8. That is, if the F$_2$ yellow seed had the homozygous dominant genotype *YY*, each of the test-cross progeny would receive one *Y* (from the yellow-seeded parent) and one *y* (from the *yy* test-cross parent), and all offspring would have the yellow-seeded phenotype and the heterozygous genotype *Yy*. By inference, the genotype of the F$_2$ yellow-seeded plant could be certified as *YY*. But if the F$_2$ yellow seed had the heterozygous genotype *Yy*, in the test cross, half the offspring would receive one *Y* from the yellow-seeded parent and half would receive one *y* from the same parent. Thus half the test-cross progeny would be

yellow *(Yy)*, and the other half would be green *(yy)*, with one recessive allele from each parent. The F$_2$ yellow-seeded parent could then be certified as *Yy*.

Test crosses are one method still used today to determine unknown genotypes. In general, if all of the test-cross offspring are the dominant color (or other trait), then the unknown genotype is homozygous dominant. If only 50 percent of the offspring show the dominant phenotype, then the unknown genotype is heterozygous.

Dihybrid Crosses and the Law of Independent Assortment

So far, we have followed only one trait at a time during a cross between two organisms. However, in a typical organism, each cell contains pairs of alleles for hundreds or thousands of traits. Let us go, as Mendel did, one step further toward the real situation by considering how two traits are inherited relative to each other.

Pea seeds have both a color (yellow or green) *and* a shape (round or wrinkled). Mendel designed experiments to study how two characters—shape and color—might interact during inheritance. After determining that round seeds are dominant to wrinkled seeds, he considered what might happen if a true-breeding strain with round yellow seeds (*RRYY* genotype) was crossed with a true-breeding strain with wrinkled green seeds (*rryy* genotype). In this experiment, he found that the entire F$_1$ generation produced round yellow seeds—totally in keeping with the dominance of the yellow-color allele (*Y*) and the round-seed allele (*R*):

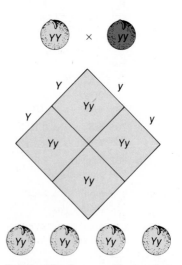

If all the offspring are yellow seeded, the yellow parent must be *YY*

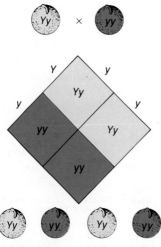

If half the offspring are yellow seeded and half are green seeded, the yellow parent must be *Yy*

Figure 10-8   TEST CROSSES: TRACING A PARENT'S UNKNOWN GENOTYPE.
Mendel invented the test cross to determine unknown genotypes. He cross-bred a yellow-seeded plant of unknown genotype to a homozygous recessive for the trait in question, here *yy* (green). The phenotypic ratios of the test-cross progeny reveal whether the genotype of the yellow parent is *YY* or *Yy*.

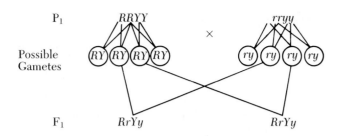

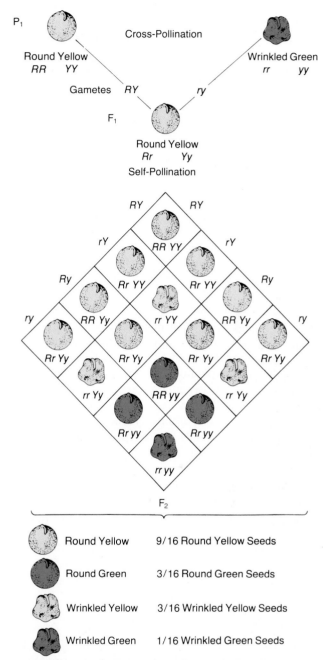

Next Mendel predicted that when the F_1 progeny were allowed to self-pollinate and produce an F_2 generation, one of two things would happen.

1. The traits for shape and color would always segregate together, producing two kinds of gametes coupled in the same way they were in P_1—that is, *RY* and *ry*; or

2. At least some of the gametes would carry a nonparental, or *recombinant*, genotype (*Ry* or *rY*) and would produce F_2 progeny with totally new phenotypes, such as round green seeds or wrinkled yellow seeds.

Mendel designed an experiment to test which hypothesis was true. He performed **dihybrid crosses** (crosses between parents that differ in two characters, in this case, seed color and shape). He invariably observed that four phenotypes appeared in the F_2 generation, including two phenotypic combinations not observed in either parent (Figure 10-9). In order for this to happen, the alleles of the two genes for color and shape must have segregated *independently* of each other during gamete formation in the F_1 generation. That is, four types of eggs and four types of pollen were formed, not just two. Instead of gametes carrying only *RY* and *ry*, the alleles for seed color and seed shape must have reassorted independently so that different gametes carried *RY*, *ry*, *rY*, or *Ry*. Random recombination during fertilization would produce an F_2 generation with $4 \times 4 = 16$ combinations of alleles among the offspring. These sixteen combinations would yield nine different genotypes and four different phenotypes in a 9:3:3:1 ratio: nine plants with round yellow seeds (a parental phenotype), three with wrinkled yellow seeds (a recombined phenotype), three with round green seeds (the other recombined phenotype), and one with wrinkled green seeds (the other parental phenotype), as shown in Figure 10-9. In Mendel's F_2 plants, he counted 315 round yellow seeds, 108 wrinkled yellow seeds, 101 round green seeds, and 32 wrinkled green seeds, a result very close to the 9:3:3:1 ratio expected from random segregation in any dihybrid cross.

It is possible to predict how many of the sixteen possible allele combinations in a dihybrid cross will produce a

Figure 10-9   DIHYBRID CROSSES: A CASE OF INDEPENDENT ASSORTMENT.

When pea plants with purebred round yellow seeds (*RRYY*) are mated to pea plants with purebred wrinkled green seeds (*rryy*), all F_1 progeny have round yellow seeds (*RrYy*). If alleles for the two characteristics always segregated together during meiosis, then only two kinds of gametes (*RY* and *ry*) would result, and all progeny would display only the two parental phenotypes in a 3:1 ratio. As Mendel discovered, however, the genes for the two characters segregate, or "assort independently." They produce four kinds of gametes (*RY*, *Ry*, *rY*, and *ry*), and the F_2 generation displays four phenotypes in a 9:3:3:1 ratio. The two parental combinations of traits are represented, but two new recombinant types (round green seeds and wrinkled yellow seeds) also appear among the F_2 progeny.

given phenotype—say, round green seeds. Since the round allele (R) is dominant, three-fourths of the F_2 progeny will be round, and since green (y) is recessive, only one-fourth will be green. Because the two characters act independently, we can simply multiply their individual probabilities to determine their likelihood of occurring together: $\frac{3}{4} \times \frac{1}{4} = \frac{3}{16}$. If we want to know how many of the progeny will have wrinkled green seeds, we can multiply $\frac{1}{4} \times \frac{1}{4}$ (since both traits are recessive) and predict that only $\frac{1}{16}$ will show the doubly recessive phenotype. This handy formula becomes very useful when three, four, or more traits are considered at the same time, because a Punnett square for three or four traits has 64 or 256 boxes, respectively. Thus, one can predict phenotypes mathematically without laborious diagramming: simply multiply the individual probabilities, no matter how many traits and alleles are being analyzed.

Let us take another example using three pea-plant characters: seed color (Y or y), seed shape (R or r), and plant height (T or t). The probability that a plant would have at least one dominant allele for each of three characters (the genotype would include YRT)—that is, it would be tall and have round yellow peas—is $\frac{3}{4} \times \frac{3}{4} \times \frac{3}{4} = \frac{27}{64}$. The probability that a plant would be homozygous recessive for each of three characteristics (yrt)—be short and have wrinkled green seeds—is $\frac{1}{4} \times \frac{1}{4} \times \frac{1}{4} = \frac{1}{64}$. The probability of its being tall (dominant) with round (dominant) green (recessive) seeds is $\frac{3}{4} \times \frac{3}{4} \times \frac{1}{4} = \frac{9}{64}$, and so on. Try determining the probability of a pea plant having the dominant alleles—Y, R, T, and W (red flower color)—or the four recessive alleles—y, r, t, w (white flower color).

The results of the dihybrid experiments led Mendel to formulate what is now called Mendel's second law, the **law of independent assortment,** which states that different characters are inherited independently of one another. In other words, inheritance of a dominant or a recessive allele of one gene, such as for yellow or green seeds, has nothing to do with the way the alleles for other genes are inherited, such as those for smooth or wrinkled seeds or short or tall stems. (In reality, as we shall see in Chapter 11, there are important exceptions to this rule.)

Incomplete Dominance

The characters and alleles used to establish Mendel's two laws are lucky choices because they are so straightforward. For each of the seven characters that Mendel studied, he found that the dominant allele completely masked the recessive allele in the phenotype of the organism. But nature had surprises in store for Gregor Mendel as well as for many later geneticists. The reason

is that the dominant–recessive relationship is not the only possible interaction between alleles of a gene. When a true-breeding red-flower strain of snapdragon (RR) is crossed with a true-breeding white-flower strain (rr), the flowers produced in the F_1 plants (Rr) are *pink*, not red. Shades of August Weismann . . . this would seem to be a perfect case of blending inheritance, contrary to Mendel's laws. However, after further crosses, it is evident that the law of segregation does indeed hold for snapdragon color. The expected segregation of flower color occurs when the pink F_1 snapdragons are self-pollinated to produce F_2 progeny (Figure 10-10). In this generation, the usual 1:2:1 ratio obtains, but each genotype has a distinctive and recognizable phenotype. This is because the heterozygote (Rr) has its own intermediate phenotype (pink), which does not reflect the complete dominance of the red-color allele. The existence of red and white flowers in the F_2 generation again supports the concept of genetic integrity. The alleles are not altered or merged in any way by their presence in the cells of pink-flowered plants.

The "mixing" responsible for production of pink flowers occurs not at the level of genotype, since the alleles for red and white flower color remain unchanged, but at the level of the phenotype. One possibility is that one R allele may make half as much red pigment as two R alleles. Thus the heterozygote Rr will have only half as much red pigment as the homozygote RR, whereas rr makes no red pigment. The less intense red pigmentation therefore allows the white allele r (the absence of pigmentation) to "show through," and the human eye perceives pink. This phenomenon is called **incomplete dominance** since both alleles exert an effect and jointly produce an intermediate phenotype. Many organisms display characters with incomplete dominance, including the mild-frizzle fowl and the palomino horse. A mild-frizzle is a chicken that has one parent with normal feathers and one with brittle, curly feathers (Figure 10-11). A palomino has one white parent and one that is either light chestnut or sorrel. Incomplete dominance hints at greater genetic complexities than Mendel found with his simple experiments. And, indeed, there are greater complexities, which we will consider in Chapter 11. Now let us resume unraveling the puzzle of inheritance through still more hypotheses, experiment, and clever intuition.

MENDEL'S IDEAS IN LIMBO: A THEORY BEFORE ITS TIME

In 1866, Mendel presented the results of his experiments and his two fundamental laws of heredity in a

(a)

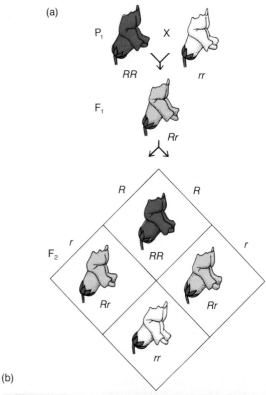

(b)

Figure 10-11   INCOMPLETE DOMINANCE IN THE MILD-FRIZZLE CHICKEN.
Many mutations, such as frizzle, affect the ways that feathers develop; here they protrude and fail to form the usual, smooth contours of normal chickens.

Figure 10-10   FLOWER COLOR IN SNAPDRAGONS: A CASE OF INCOMPLETE DOMINANCE.
(a) When red (*RR*) and white (*rr*) snapdragons are crossed, the F$_1$ generation has all pink flowers (*Rr*). In the F$_2$ generation, one-half of the plants have pink flowers, one-fourth have red flowers, and one-fourth have white flowers. This 1:2:1 phenotypic ratio directly reflects the genotypic ratio (1*RR*:2*Rr*:1*rr*) and reveals that the red gene is incompletely dominant over the white, causing a plant with *Rr* alleles to be pink rather than red. (b) This gorgeous display depends on a number of genes and alleles for snapdragon colors.

clearly written paper published by the Brünn Society for Natural History. He presented volumes of supporting data and showed how his ideas could explain the numerous less complete observations made earlier by other plant breeders. He argued quite cogently that the key to understanding his laws lay in the mechanisms of gamete formation in the parent plant. (This, we now know, includes the process of meiosis, although, of course, Mendel was unaware of chromosomes, much less meiotic processes.) Mendel also presented test-cross data as evidence that the two or four types of gametes were formed in equal numbers, as predicted by the rules of segregation.

Despite the excellence of Mendel's research and the clarity of its presentation, his paper was not understood in his day, and for many years it did not come to the attention of other biologists who could have appreciated its meaning. The mathematical presentation of results and the physical interpretations that Mendel provided were simply not a language understood by nineteenth-century biologists. Even now, when science is a well-organized enterprise and communication among scientists is relatively easy, the introduction of a thoroughly new way of seeing an old problem sometimes requires lengthy argument and one-on-one discussion before the ideas can be effectively transferred.

Mendel's work was finally championed when Hugo de Vries in Holland and Carl Correns in Germany achieved results identical to Mendel's. While preparing their data for publication, they discovered Mendel's paper and gave him the well-deserved credit for priority. Correns's paper, "G. Mendel's Law Concerning the Behavior of Progeny of Varietal Hybrids," was published in 1900, the same year as de Vries's paper and sixteen years later after Mendel's death—fully thirty-five years after Mendel had first taken his results to the scientific community.

The rediscovery of Mendel's laws opened the way to a new and exciting field. Other scientists soon found that these laws apply to birds, mammals, insects, and other organisms as well as to pea plants. The science of genetics was under way.

CHROMOSOMES AND MENDELIAN GENETICS

In 1902, soon after the rediscovery of Mendel's work, Walter Sutton in the United States and Theodor Boveri in Germany independently suggested that Mendel's hereditary units might be located on chromosomes. Both had observed the segregation of chromosomes during meiosis and their apparent reshuffling and independent assortment in the gametes. Sutton, for instance, studied cells in the testes of grasshoppers and noted that the two chromosomes of a pair closely resembled each other in shape. Moreover, members of each pair came together during meiosis but then segregated to different cells that later gave rise to separate gametes—the chromosomes behaved just like Mendel's theoretical factors, or the alleles of later geneticists. An abundance of careful observations led Sutton and Boveri to espouse the **chromosome theory of heredity.**

In our explanation of Mendel's laws, we have seen how the halving of chromosome number to produce haploid gametes is consistent with Mendel's theory of the distribution of hereditary factors during gamete formation. Figure 10-12 shows how this works, using two pairs of pea-plant chromosomes. In the P_1 generation, the female parent has two chromosomes with R alleles and two with Y alleles (the determinants of round yellow seeds); the male parent has two chromosomes with r alleles and two with y alleles (the determinants of wrinkled green seeds). In the F_1 generation, the two chromosomes bearing Y and R are derived from the female parent, while the homologous chromosomes bearing y and r are from the male parent. Now imagine that the F_1 individuals grow, mature sexually, and begin to produce gametes. At the first meiotic division in the F_1 generation, the two pairs of homologous chromosomes may align in either of two ways on the metaphase plate. In the first alignment (on the left in Figure 10-12), the chromosomes bearing Y and R segregate into one cell, while

those bearing y and r go to the other. This type of segregation preserves the association of the dominant alleles (Y and R) and of the recessive alleles (y and r) originally present in the two parents and thus is referred to as a *parental association*. In the other type of alignment on the metaphase plate (on the right in Figure 10-12), the chromosomes bearing Y and r segregate from those bearing y and R at the first meiotic division; in this case, *nonparental*, or *recombinant*, *associations* of alleles are present in the gametes and can lead to recombinant phenotypes. If the two types of alignment of chromosomes on the metaphase plate occur with equal frequency, the result is independent assortment of the two alleles for a particular trait.

The work of Sutton and Boveri showed merely a correlation between chromosome movements and segregation patterns of Mendel's factors. But proof that the hereditary units lie on chromosomes could be obtained only by experiment. Such experiments were conducted by geneticists at Columbia University, especially Thomas Hunt Morgan and his students, in the early 1900s. They worked with *Drosophila melanogaster*, a type of fruit fly. This small, delicate insect is a popular research subject because it is easily maintained in the laboratory, reproduces quickly, and has only four pairs of chromosomes—three pairs of autosomes, or non-sex chromosomes, and one pair of sex chromosomes (XX in females and XY in males), as shown in Figure 10-13. Earlier researchers had found that, as one would expect of homologous chromosomes, the X and Y chromosomes of the male fruit fly segregate from each other at the first meiotic division, producing two types of sperm: (1) those with an X chromosome and three autosomes, and (2) those with a Y chromosome and three autosomes. Because the female has two X chromosomes, all eggs carry one X chromosome and three autosomes. Thus, if an egg (with its X) is fertilized by a sperm with an X chromosome, it develops into a female fly (XX); if it is fertilized by a sperm with a Y chromosome, it becomes a male fly (XY). The elucidation of the fruit fly's sex-chromosome system at the turn of the century made it clear that the presence of a pair of differently shaped chromosomes residing in the cells of most animals correlates with the regulation of sexual phenotype, sexual physiology, and sexual behavior—a spectrum of specific gender-related traits. The implication was powerful: hereditary traits were somehow linked to the chromosomes, the "colored bodies" inside the cell nucleus.

Morgan's group searched, as had Mendel with pea plants, for *Drosophila* strains with true-breeding characteristics that could be crossbred and studied. The first trait they studied was eye color. Normal fruit flies have bright red eyes, but one day they discovered a single white-eyed male in the colony. Morgan guessed that white eyes were a spontaneously occurring change (a

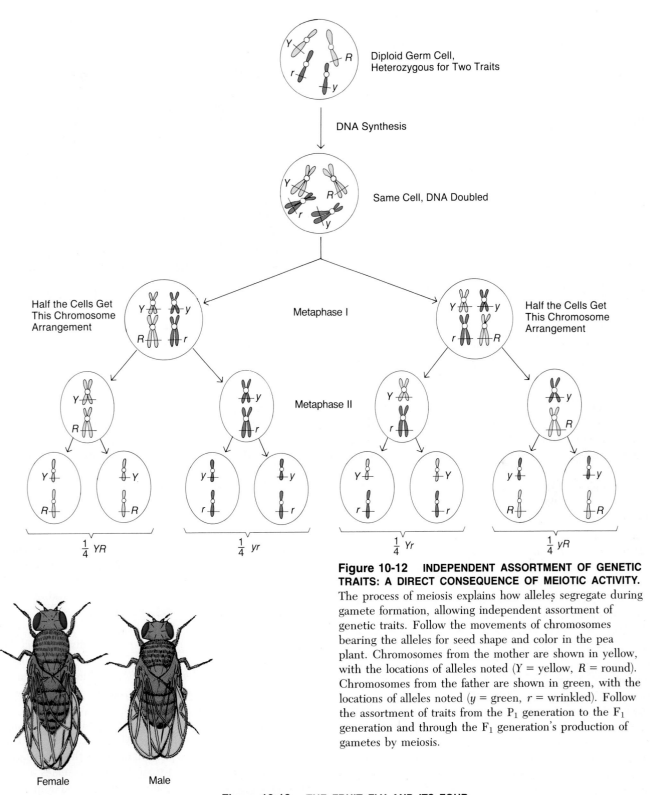

Diploid Germ Cell, Heterozygous for Two Traits

DNA Synthesis

Same Cell, DNA Doubled

Half the Cells Get This Chromosome Arrangement

Metaphase I

Half the Cells Get This Chromosome Arrangement

Metaphase II

$\frac{1}{4}$ YR $\frac{1}{4}$ yr $\frac{1}{4}$ Yr $\frac{1}{4}$ yR

Female   Male

Figure 10-12   INDEPENDENT ASSORTMENT OF GENETIC TRAITS: A DIRECT CONSEQUENCE OF MEIOTIC ACTIVITY.
The process of meiosis explains how alleles segregate during gamete formation, allowing independent assortment of genetic traits. Follow the movements of chromosomes bearing the alleles for seed shape and color in the pea plant. Chromosomes from the mother are shown in yellow, with the locations of alleles noted (Y = yellow, R = round). Chromosomes from the father are shown in green, with the locations of alleles noted (y = green, r = wrinkled). Follow the assortment of traits from the P_1 generation to the F_1 generation and through the F_1 generation's production of gametes by meiosis.

Figure 10-13   THE FRUIT FLY AND ITS FOUR CHROMOSOMES.
One reason *Drosophila melanogaster* is well suited for genetic studies is that it has only four chromosomes. Both males and females have pairs II, III, and IV. Pair I, the sex chromosomes, determines the insect's gender. Females have two X chromosomes, while males have one X and one Y.

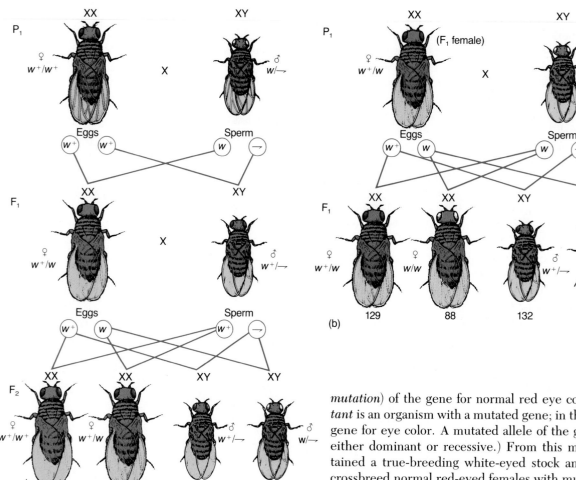

Figure 10-14 *DROSOPHILA* **EYE COLOR: A SEX-LINKED TRAIT.**
Eye color in fruit flies is not inherited according to simple Mendelian law. (a) All the F₁ progeny from a red-eyed female (♀) mated to a white-eyed male (♂) have red eyes. This proves that red is dominant to white. However, when F₁ males and females are crossed, the expected 3:1 phenotypic ratio of red-eyed to white-eyed flies is not produced. Instead, all females have red eyes, while half of the males have red eyes and half have white eyes. This result can be explained if the genetic factor for eye color lies on the X chromosome (inherited by a male offspring only from the mother), and if no corresponding allele occurs at all on the Y chromosome.

In the figure, w^+ represents the allele for red eyes, and w represents the allele for white eyes. Since the allele is carried on the X chromosome, a male will have only one allele. For females, with two alleles, w^+, or red eye, is dominant. If a female inherits the white-eyed allele from both parents (b), she will have white eyes.

mutation) of the gene for normal red eye color. (A *mutant* is an organism with a mutated gene; in this case, the gene for eye color. A mutated allele of the gene can be either dominant or recessive.) From this male, he obtained a true-breeding white-eyed stock and began to crossbreed normal red-eyed females with mutant white-eyed males. Morgan quickly discovered that this mutant trait did not act in the normal dominant–recessive fashion that might be predicted from Mendelian laws.

By crossing red-eyed females with white-eyed males, Morgan obtained an F₁ generation with red eyes; clearly, the white-eyed trait was recessive and the red-eyed trait, dominant. However, when the red-eyed F₁ progeny were bred together, the F₂ generation did not turn out as expected. Rather, there were 3,470 red-eyed offspring and 782 white-eyed. Of these offspring, all the females had red eyes, while half the males had red eyes and half had white. For some reason, white eyes seemed to be solely a male trait, though that later proved not to be so (Figure 10-14).

After a series of crosses, recrosses, and careful analyses of the phenotypic patterns, Morgan figured out that this unusual inheritance pattern parallels the inheritance of the X chromosome. Morgan knew that males always receive their single X chromosome from their mother, but females received an X from each parent. Thus, the male's eye-color phenotype is determined by the eye-color allele carried on the X chromosome. If the X chromosome carries a dominant red allele (symbolized $w+$), the male offspring will have red eyes; if it carries a white allele (w), the male will have white eyes (Figure 10-14).

This explanation would hold only if the Y chromosome carries *no* allele for eye color or *always* carries the white allele. Later studies showed the first assumption to correct: the Y chromosome has no eye-color allele. With no corresponding allele on the Y chromosome to block the recessive allele for white eyes, the recessive allele always shows up in an $X^w Y$ male (the superscript w indicates the allele carried on the X chromosome). Among the female offspring, eye-color alleles act as a normal dominant–recessive pair. A female with a dominant allele on each of her two X chromosomes, $X^{w+} X^{w+}$, displays red eyes. A heterozygous female, $X^{w+} X^w$, also shows red eyes. Only a female with a recessive allele for eye color on each X chromosome, $X^w X^w$, has the white-eye phenotype.

From these experiments, Morgan concluded that the eye-color gene must be carried on the X chromosome. This was the first proof that chromosomes do carry the hereditary units, and this research laid a solid foundation for the genetic research that followed over the next seventy-five years. Once geneticists knew where to look for the hereditary factors that Mendel had followed so carefully, they could—and did—determine how those factors operate at the molecular level. Morgan's studies of eye color in fruit flies also had a more immediate effect; they were the first exploration of **sex-linked traits.** Thousands of such traits are now known in various organisms and include several in our own species (as we shall see in Chapter 14).

Another peculiarity of the inheritance of white eyes in *Drosophila* led to yet another important discovery. In 1916, the geneticist Calvin Bridges observed that 1 in every 1,000 fruit flies displays a phenotype not predicted by the dominant–recessive sex-linked character of red and white eyes. Bridges discovered that during meiosis in white-eyed females, the homologous X-chromosome pair occasionally fails to segregate, or disjoin. He called this **nondisjunction** of the X chromosome (Figure 10-15). Some of the female's eggs therefore receive two X chromosomes, while others receive none at all. If an egg lacking an X chromosome is fertilized by a sperm with an X, the offspring will be a sterile XO male, and its eye color will correspond to the allele carried on the sperm's X. (It will be a male because it takes two X's to produce a female phenotype; hence, both XY and XO yield males.) With this finding, there was no longer any doubt that each chromosome carries a single specific allele for a given gene; of course, large numbers of alleles, one each for many different genes, are on each chromosome. Bridges's exploration of the nondisjunction phenomenon was accepted as the final proof of Sutton's and Boveri's chromosome theory of heredity and was, in addition, an important forerunner of modern techniques, such as amniocentesis, for screening early human embryos (Chapter 15).

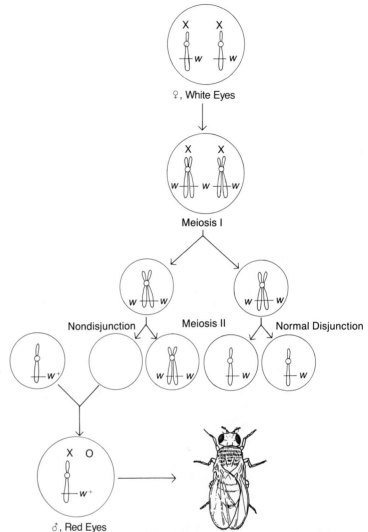

Figure 10-15   NONDISJUNCTION: A FAILURE TO SEGREGATE.
The homologous X chromosome pair in females sometimes fails to "disjoin" or segregate during meiosis, and thus one egg will receive two X chromosomes while another receives none at all. This situation is called nondisjunction.

If an egg lacking an X chromosome is fertilized by a sperm that carries an X chromosome, the offspring will be a sterile XO male, and its eye color will correspond to the allele carried on the sperm's X chromosome, in this case, red eyes. This proves that the gene for eye color is associated with the X chromosome.

A CENTURY OF PROGRESS

In this chapter, we have traced the birth of genetics. This field is little more than a century old, yet it spans an enormous range of ideas. For most of human history, people observed and even manipulated the similarities between parents and offspring, yet they could not explain what they saw. Not until Weismann severed mouse tails and Mendel grew pea plants were old notions like pangenesis and gemmules toppled, and with them, the idea of blended inheritance. Shortly after Mendel's work was accepted, the role of chromosomes in heredity was established by Morgan and Bridges; within a few decades, the structure and function of genes began to be explained with great clarity. It has been like solving a puzzle: from pangenesis to gene manipulation in just over a century. But, as we shall see, just when most of the pieces of the puzzle seemed to be in place, the advent of genetic engineering in the 1980s has revealed wholly unsuspected levels of complexity in the hereditary material of eukaryotic organisms. Even so, the discoveries of the scientific explorers highlighted in this chapter are a solid foundation for understanding the exciting puzzles of inheritance.

SUMMARY

1. One of the early theories of heredity was *pangenesis*, which held that characteristics were passed from one generation to the next in the form of fluids representing the various body parts. It was not until the 1880s that Weismann formulated the *germ plasm theory*, which states that only gametes are involved in transmitting traits to progeny.

2. According to *blending theories* of inheritance, the heritable traits of both parents fuse, thereby losing the distinct characteristics of each.

3. Mendel disproved blending theories when he discovered that pea plants possess units of heredity, now called *genes*, for each of seven characters he studied. For each character (color, height, and so on) Mendel found alternative forms of the gene; these are called *alleles*. One allele for a character can be *dominant* and the other, *recessive*. The terms P_1 (*parental*), F_1 (*first filial*), and F_2 (*second filial*) label the generations of organisms crossed in genetic tests.

4. An individual is said to be *homozygous* for a trait if it has two identical alleles for that character; an individual is *heterozygous* for a trait if it has two dissimilar alleles (commonly, a dominant and a recessive allele are present).

5. *Genotype* is the genetic makeup, or gene content, of a cell or an organism; more precisely, it is the allele content. *Phenotype* is the physical appearance and properties of a cell or an organism and is an expression of the genotype.

6. According to Mendel's first law, the *law of segregation*, one unit of allele for each trait is inherited from each parent; the separation, or segregation, of each allele pair occurs during meiosis. One method Mendel used to prove this law was the *test cross*, in which he mated plants of unknown genotype to homozygous recessive plants for the trait in question. *Punnett squares* can be used to easily summarize the results of crosses.

7. Mendel performed *dihybrid crosses* (crosses between plants that differ in two distinct characters) to arrive at his second law, the *law of independent assortment:* the alleles of genes governing different characters are inherited independently of each other.

8. A seeming exception to Mendel's laws is *incomplete dominance*, is which progeny show a phenotype intermediate between that of the parents, such as pink snapdragons from red and white parents. However, reappearance of the parental types in subsequent generations bears out Mendel's laws.

9. Sutton and Boveri advanced the *chromosome theory of inheritance*, which states that genes are carried by and inherited as parts of chromosomes. Proof of this theory was obtained from work with *Drosophila* sex chromosomes by Morgan and his students. *Sex-linked traits* are attributable to genes located on sex chromosomes.

10. Bridges's work with *nondisjunction* of the X chromosome added the final proof that specific chromosomes carry specific alleles.

KEY TERMS

allele
blending theory of heredity
chromosome theory of heredity
dihybrid cross
dominant
F_1 (first filial) generation
F_2 (second filial) generation

gene
genotype
germ plasm theory of heredity
heterozygous
homozygous
incomplete dominance
law of independent assortment
law of segregation

nondisjunction
pangenesis
P_1 (parental) generation
phenotype
Punnett square
recessive
sex-linked trait
test cross

QUESTIONS AND PROBLEMS

1. Mendel discovered that the units of heredity occur in pairs which segregate at meiosis. What are these units of heredity now called? On what cellular structures are they found?

2. Let's call the dominant allele for red flower color (in peas) *W*, and the recessive allele for white flower color *W*. If a pea plant has the genotype *WW*, what is its phenotype? If the genotype is *ww*, what is the phenotype? What if the geneotype is *Ww*?

3. What are the possible genotypes of a red-flowered pea plant? A white-flowered plant? Which of these genotypes are homozygous? Heterozygous? What gametes will each type produce?

4. When Mendel crossed pea plants that grew from round yellow seeds with pea plants that grew from wrinkled green seeds, what were the phenotypes of the F_1 hybrid seeds? When the F_1 hybrid seeds were grown, and the flowers self-pollinated, what kinds of seeds were produced? Were they like the original parents only, or were there also round green seeds and wrinkled yellow seeds? Explain this result.

5. A couple has two children. What is the probability that the first is a girl? A boy? What is the probability that the second is a girl? What is the probability that both are girls? Both boys? One of each sex? Do your probabilities add up to 1?

6. Most living lobsters are brown, but a few are blue. Crosses have shown that the blue color is due to a recessive allele. Let's call the brown allele *B* and the blue allele *b*. What is the genotype of a blue lobster? What gametes would it produce? What is the genotype of a blue lobster's brown mother? What gametes would she produce? If a blue lobster

has two brown parents, and the parents produce many offspring, what fraction of them would you expect to be blue? If a blue lobster is mated to its brown parent, what fraction of their offspring would have the genotype *bb*? What fraction would have the genotype *BB*? If two blue lobsters mate, what fraction of their offspring will be blue? Brown?

7. A brown-eyed man whose mother was blue-eyed marries a brown-eyed woman whose father was blue-eyed and they produce three blue-eyed children. What are the genotypes of the brown-eyed man and woman? What is the probability that their fourth child will be brown-eyed?

8. Wild red foxes occasionally produce a prized silver-black pup in a litter. Such pups when mated breed true. How would you explain the infrequent occurrence of these exceptions?

9. Consider two independent characters, normal skin color *C* versus albino *c* and normal blood *T* versus thalassemia *(t)* (anemia). Show by a Punnett square the expected ratio of a large number of children produced from parents heterozygous for both traits.

10. In the fruit fly, *Drosophila melanogaster*, vestigial wings and scarlet eyes are caused by recessive genes on different chromosomes. If a vestigial scarlet male is crossed to a wild-type female, what will be the phenotype of the F_1? If the F_1 progeny mate among themselves, what would be the ratios of the F_2 offspring? Show the genotypes and phenotypes expected for each class. Use a Punnett square to analyze the F_2.

11. Brown eyes *(B)* are dominant to blue eyes *(b)*. Show gentoypic and

phenotypic ratios of children from the following parental types:
(a) $BB \times bb$
(b) $Bb \times bb$
(c) $Bb \times Bb$

In a family where the mother is blue-eyed *(bb)* and the father is brown-eyed *(Bb)*, which genotypes would be possible for each offspring with the following phenotypes:
(d) two children both blue-eyed: 1 boy, 1 girl
(e) three brown-eyed children: 2 girls, 1 boy
What is the probability that a sixth child will be blue-eyed?

SUGGESTED READINGS

Ayala, F. J., and J. A. Kiger, Jr. *Modern Genetics*. Menlo Park, Calif.: Benjamin, Cummings, 1980.

An excellent general text for this and the succeeding chapters on genetics.

Harrison, D. *Problems in Genetics with Notes and Examples*. Reading, Mass.: Addison-Wesley, 1970.

Solving problems like those in this workbook is an excellent way to come to understand the basis of heredity.

Mendel, G. *Experiments in Plant-Hybridisation*. Cambridge, Mass.: Harvard University Press, 1965.

Mendel's famous paper plus biographical notes about the father of genetics.

Srb, A. S., K. D. Owen, and R. S. Edgar.: *General Genetics*. 2d ed. San Francisco: Freeman, 1965.

The classic genetics text for most of today's geneticists. It gives an excellent account of Mendel's work, and its problem sets are particularly useful.

11
MENDEL MODIFIED

Had Mendel studied coat color in dogs, eye color in fruit flies, or the genetics of Manx cats, he might never have devised his laws of inheritance; the rules might not have emerged from the exceptions.

John Postlethwaite (1983)

A Manx Cat: Heterozygotes—Special Characteristics; Homozygotes—Dead.

Manx cats are a striking breed of domestic cat developed on the small Isle of Man in the Irish Sea. These animals are peculiar because their hindquarters are higher than their shoulders, they have a tendency to hop, and many of them have no tails. Cat fanciers find Manx cats to be notoriously difficult to breed. If a Manx is mated to a normal, long-tailed cat, half the offspring are Manx and half have normal tails; the Manx is heterozygous for a dominant mutation causing short tail, tall hindquarters, and hopping gait. According to Mendelian genetics, if a Manx is heterozygous dominant, the mating of two Manx cats should yield three Manx kittens for each normal, long-tailed kitten. However, when this cross is made, the cat breeder will see two Manx kittens for each normal, long-tailed one. What accounts for this departure from Mendel's laws? Do Mendel's rules hold for peas and fruit flies but not for cats? Or is there something special about the way Manx traits are inherited that alters the expected 3:1 ratio to 2:1?

This chapter investigates some important inheritance patterns, such as that of the Manx cat, that appear to violate Mendel's laws. Understanding the reasons for these apparent contradictions provides a good illustration of the way science actually advances. The many pieces of scientific puzzles, especially in genetics, are not necessarily assembled in a particular logical order. Availability of the right organism, chance observation of a new mutant trait, a shrewd guess about a way to proceed experimentally, and other similar factors are part of the complex trail leading to major scientific discoveries.

Mendel's work certainly revealed some basic relationships of heredity. But it was left for others, just a few decades later, to piece together more of the puzzle; in fact, wholly unsuspected pieces of this puzzle have been revealed only in the 1980s after invention of recombinant DNA biotechnology (as we will see in Chapter 14). So this chapter is as much about how science works as it is about the complexities of heredity not seen in the elegant experiments that Gregor Mendel performed in his abbey garden.

The analysis of exceptions to Mendel's laws by geneticists in the first half of the twentieth century yielded important insights into the arrangement of genes on chromosomes, the interaction of genes with one another, and the role of mutation as a source of genetic variation. These concepts are now so fundamental to the entire study of biology that it is difficult to imagine a time when they were unknown. However, the elucidation of each fact was a hard-fought victory for the pioneers of modern genetics, as we shall see.

HOW GENES ARE ARRANGED ON CHROMOSOMES

We discussed in Chapter 10 how William Sutton, Theodor Boveri, Calvin Bridges, and Thomas Hunt Morgan proved that chromosomes carry the units of heredity, the genes. Genes, as we saw in Chapter 9, are relatively short stretches of DNA that determine physical traits. They are strung together on the very lengthy coiled DNA molecule that forms the backbone of each chromosome. The establishment of chromosomes as the bearers of genes gave a physical basis to the inheritance "factors" described by Mendel's laws. Because all organisms have many traits and only a few chromosomes, it is obvious that each chromosome carries hundreds or thousands of genes. For example, fruit flies have only 4 pairs of chromosomes but more than 5,000 genes, and humans have 23 pairs of chromosomes but about 50,000 genes. Consequently, even before these numbers were established, geneticists began to accept the idea that many different genes reside on each chromosome.

This inference ultimately helped explain some observations made by Morgan and other early geneticists. They found that whereas most genes obey the law of segregation, many pairs of genes do not follow the patterns predicted by Mendel's law of independent assortment. Some traits appear almost never to assort independently of one another. In a family, for example, traits A and B, or traits a and b, might tend to be inherited together, but traits A and b or a and B might only rarely appear in the family. The reason for this unexpected inheritance pattern is that many genes are arranged on few chromosomes, and some genes that lie on the same chromosome move together during meiosis and do not separate. Inheritance patterns among genes are described in terms of their **linkage,** or the degree to which genes on chromosomes are inherited together. There are three possibilities: (1) a chromosome's genes might assort independently, according to Mendel's second law (nonlinkage); (2) all the genes on one chromosome might be inherited as a single unit (complete linkage); or (3) the genes might be inherited in some intermediate fashion (partial linkage). Figure 11-1 illustrates these three possibilities, the last two of which violate Mendel's laws.

Figure 11-1 shows a cross between two organisms, each of which has a pair of homologous chromosomes bearing alleles (or versions) of two genes, A and B. Either the dominant or the recessive alleles occur at a

P₁

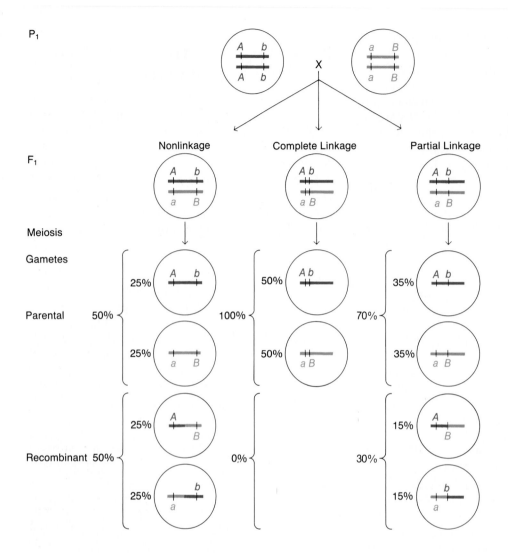

Figure 11-1   LINKAGE AND NEIGHBORING GENES.
Genes on one chromosome can show different patterns of inheritance. With nonlinkage, the first possibility, genes that reside on the same chromosome may be inherited according to the law of independent assortment. In this example, parental and recombinant offspring occur in equal frequencies. Second, the two genes on the chromosome may always be inherited together and appear completely linked, with no recombinants occurring. The third possibility is the intermediate pattern of partial linkage, in which some parental and some recombinant types occur, but in unequal percentages and with mostly parental types.

given location, or **locus,** on each chromosome. Of course, only one allele of the gene occurs at the locus of that gene on each homologous chromosome. Each parent in Figure 11-1 has a homozygous, or matched, pair of alleles for each gene. That is, that parent on the left is *AbAb*—homozygous dominant for the *A* gene and homozygous recessive for the *B* gene. The parent on the right is *aBaB*—homozygous recessive for the *A* gene and homozygous dominant for the *B* gene.

When meiosis occurs in the F₁ individuals, Mendel's second law predicts the first possibility—that the gene pairs will assort independently and that gametes will form with four genotypes in equal proportions: two original, *parental* genotypes, *Ab* and *aB;* and two novel, *recombinant genotypes, Ab* and *ab.* (A **recombinant genotype** or **phenotype** is one that appears in an offspring but was not present in either parent.) If that occurred, the genes, in this case *A* and *B*, would be *unlinked.* If the second possibility were true—that genes on the same chromosome were always inherited together—gametes

would form with only two genotypes: *Ab* and *aB.* Such genes would be completely linked, and both types of gametes would be of the parental genotype; no recombinant gametes would form. Finally, if the third possibility were true, *partial linkage*—an intermediate state between complete linkage and no linkage—might take place, forming gametes with four genotypes (two parental, two recombinant) in unequal proportions. In the example of partial linkage shown in Figure 11-1, each of the parental genotypes is found 35 percent of the time—more often than the 25 percent predicted by nonlinkage (independent assortment), but less often than the 50 percent predicted by complete linkage. Likewise, each recombinant genotype forms 15 percent of the time—less often than predicted by nonlinkage, but more often than predicted by complete linkage.

In answer to the question about what happens to genes on chromosomes during meiosis, it turns out that all three situations—nonlinkage, complete linkage, and partial linkage—occur in nature. Nonlinkage is common. For instance, the seven characters that Mendel studied in peas are not linked. They assort independently and produce F_2 offspring that have both parental and recombinant traits in equal frequencies. Chromosomes do not break up into little pieces during meiosis; rather, genes for the traits Mendel studied that were on different chromosomes assorted independently because the chromosomes moved as independent units during meiosis.

Complete linkage also occurs in nature. Thomas Hunt Morgan and his students at Columbia University in the 1920s explored one instance of complete linkage in *Drosophila* fruit flies in which black body color and purple eye color always occur together. The genes for these traits lie close to each other on the same chromosome and tend to be inherited together as a "package," for reasons we will see shortly.

Partial linkage is often seen in nature in fruit flies, humans, and many other organisms. Early explorations of this inheritance pattern led to several significant developments in the field of genetics, including the mapping of the exact locations of genes on chromosomes.

Crossing Over and Recombination

As has happened so often in genetics, researchers turned to the tiny fruit fly to study the phenomenon of partial linkage. Recall from Chapter 10 that Morgan's group found that a gene for eye color in *Drosophila* lies on the X chromosome and is therefore a sex-linked trait. The research group found that a number of mutant alleles besides that for eye color are sex-linked as well. They noted, for example, that genes for wing length and body color are sex-linked; normal alleles of these two genes produce normal-length wings and a tan body, respectively, and mutant alleles of the same genes produce miniature wings and a yellow body. These early geneticists used two of the characters, eye color and body color, to study the inheritance of partially linked genes that occur together on the X chromosome.

Before proceeding, we should point out that *Drosophila* has its own system of nomenclature to denote normal and mutant traits. The most common unmutated allele of any characteristic is called the **wild-type allele** and is designated by a superscript plus (+). The mutant allele is designated by a letter: a capital letter if the mutant is dominant to the wild-type and a lower-case letter if the mutant is recessive to the wild-type. For example, the recessive mutant white eye color is designated w, and the wild-type brick-red eye color is symbolized w^+.

Figure 11-2 explains the inheritance of the traits for body color and eye color on the X chromosome. Notice that the X chromosome is larger than the Y chromosome, that the alleles for body color (y or y^+) and the alleles for

eye color (w or w^+) appear on the X chromosome, and that the Y chromosome has no allele for either body or eye color. If a female fruit fly with a yellow body and red eyes (X^{yw^+}/X^{yw^+}) is mated to a male with a tan body and white eyes (X^{y^+w}/Y), all the F_1 daughters will have a tan body and red eyes (X^{yw^+}/X^{y^+w}). When these daughters are mated to any male (in Figure 11-2, the example used is a male with X^{yw}/Y genotype), the F_2 sons will inherit one of their mother's X chromosomes and their father's Y chromosome. Since the Y chromosome lacks the genes for body and eye color, the phenotypes of the F_2 sons will be determined by the genotypes of the X chromosomes they inherit from their mother.

Interestingly, 98.5 percent of these F_2 males have the phenotypes of the two original parents in roughly equal frequencies, but the other 1.5 percent are of two phenotypes: either both mutant traits together (yellow body and white eyes) or both wild-type traits together (tan body and red eyes). Furthermore, these two recombinant phenotypes are present in roughly equal numbers. This is clearly an example of partial linkage, since the parental and recombinant phenotypes occur in such different proportions (98.5 percent and 1.5 percent instead of the 50 percent and 50 percent expected with independent assortment).

What happens during meiosis to account for the appearance of the recombinant phenotypes? The answer is *crossing over*, the peculiar "trading" of pieces of chromosomes that is an important part of meiosis (Chapter 9). True independent assortment cannot take place because, as we saw, it requires nonlinkage, and the genes for body color and eye color in *Drosophila* are physically linked on the X chromosome. Nevertheless, recombinant genotypes can arise, albeit at low frequency, as a result of crossing over during prophase of meiosis I, when pieces of chromosomes are exchanged.

Figure 11-3 shows the crossing-over process during meiosis I in *Drosophila* eggs. During prophase I, the chromatids of homologous chromosomes pair closely; breakage of two nonsister chromatids then occurs by chance somewhere along their length, followed by a crossed rejoining of the broken chromatid arms derived from the original nonsister chromatids. If the break and repair happen to occur between the locus of the body-color gene and the locus of the eye-color gene, some distance away on the same chromatid, the crossing over will produce two recombinant chromatids, one carrying the genes for tan body and red eyes, and the other, for yellow body and white eyes. Subsequent normal segregation of these chromatids into gametes yields eggs that carry either the recombinant sets of alleles or the original sets of parental alleles.

As Figure 11-3 shows, one recombination event yields two parental and two recombinant gametes, or 50 percent recombinants. Yet in their study, Morgan's group

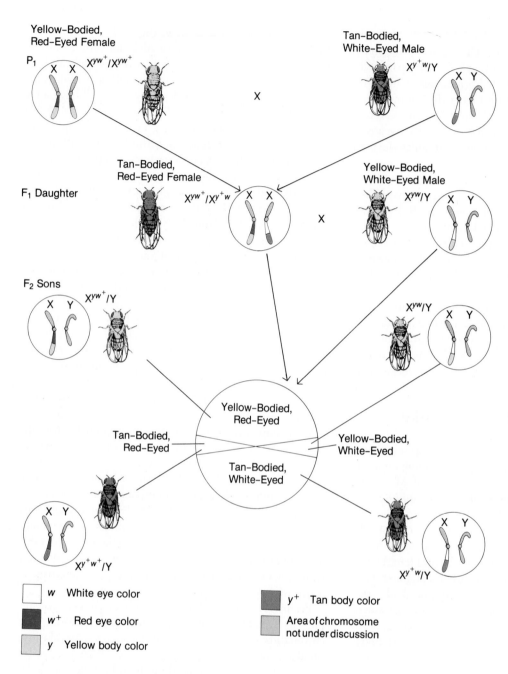

Yellow-Bodied,
Red-Eyed Female

P_1 X^{yw^+}/X^{yw^+}

Tan-Bodied,
White-Eyed Male

X^{y^+w}/Y

X

Tan-Bodied,
Red-Eyed Female

F_1 Daughter X^{yw^+}/X^{y^+w}

Yellow-Bodied,
White-Eyed Male

X^{yw^+}/Y

X

F_2 Sons X^{yw^+}/Y

X^{yw}/Y

Yellow-Bodied,
Red-Eyed

Tan-Bodied,
Red-Eyed

Yellow-Bodied,
White-Eyed

Tan-Bodied,
White-Eyed

$X^{y^+w^+}/Y$

X^{y^+w}/Y

Figure 11-2   FRUIT-FLY COLORS: A STUDY IN PARTIAL LINKAGE.
The alleles for eye color and for body color are on the X chromosome in *Drosophila*, but not on the Y. In a cross between a yellow-bodied, red-eyed female and a tan-bodied, white-eyed male, the F_1 daughters will receive one X chromosome from each parent and thus demonstrate the dominant form of each gene in their phenotype.

The F_2 sons, however, will receive their X chromosomes only from the mother. If the genes for body color and for eye color were linked, one would expect to see only the phenotypes of the maternal grandparents. But this is not the case. About 98.5 percent of the F_2 males demonstrate the phenotypes of the P generation, and 1.5 percent demonstrate recombinant phenotypes. This shows that the genes for eye color and for body color are tightly, but not completely, linked.

☐	*w*	White eye color
⬛	*w⁺*	Red eye color
▨	*y*	Yellow body color

⬛	y^+	Tan body color
▨		Area of chromosome not under discussion

found only 1.5 percent recombinants in the cross shown in Figure 11-2. Why? It is because crossing over between the sites of the *y* and *w* genes does not occur in every cell at prophase I of meiosis. In fact, it is a relatively rare event.

The relative frequency of the two recombination phenotypes among the progeny—in this case, fruit flies with yellow body and white eyes or tan body and red eyes—reflects how often recombinant gametes are formed during meiosis. This, in turn, suggests the frequency of crossing over. **Recombination frequencies,** then, are measures of how often crossovers occur between particular gene loci. They can be used in a technique called **recombination analysis** to find out how close two genes lie to each other. We can reason that just as a short garden hose gets fewer kinks and knots than a long one, the closer two genes lie on a chromosome, the less likely it is that a break and crossover event will chance to occur between them; thus their recombination frequency is low. Conversely, as the distance between genes on a chromosome increases, the likelihood of a crossover between them increases; thus their recombination fre-

Figure 11-3   GENETIC CONSEQUENCES OF ONE CROSSING-OVER EVENT.
In this crossing-over event on the X chromosome of a *Drosophila* fruit fly, the genes are represented by the eye color or body color that they determine. (a) Chromosomes of a heterozygous fly with genotype yw^+/y^+w. (b) During interphase, the chromosomes are duplicated, and they appear paired in prophase I. While they are closely paired, chromatids may exchange parts, (c). In this case, the exchange occurred between the genes for eye color and body color. (d) The result is a set of chromosomes in which alleles have recombined. Normal segregation of these chromosomes into gametes results in eggs or sperm that carry either the original set of parental alleles or the recombinant sets of alleles. (Normal chromosomes carry many more genes than are represented here.)

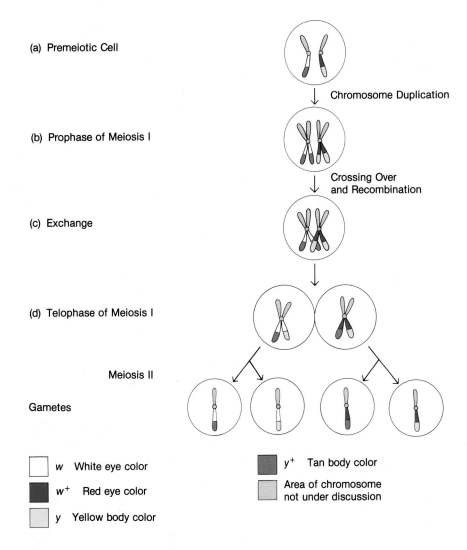

(a) Premeiotic Cell

Chromosome Duplication

(b) Prophase of Meiosis I

Crossing Over
and Recombination

(c) Exchange

(d) Telophase of Meiosis I

Meiosis II

Gametes

□ *w* White eye color	■ y^+ Tan body color
■ w^+ Red eye color	▨ Area of chromosome not under discussion
▨ *y* Yellow body color	

quency increases until independent assortment occurs.

Recombination analysis soon led to an important development in the field of genetics—the creation of gene maps. Let us see how this development came about.

Gene Maps

When Thomas Hunt Morgan's group performed crosses of the type shown in Figure 11-2, they arrived at recombination frequencies for several different pairs of traits carried on the fruit fly's X chromosome. Alfred H. Sturtevant, an undergraduate working with Morgan, realized that these data could be used to construct a **gene map** of the X chromosome that would show the relative distances between various genes and where they lie along the chromosome.

In his mapping experiments, Sturtevant reasoned that the frequency of crossing over between any two genes on a chromosome would be directly proportional to their distance from each other: the closer together, the lower would be their recombination frequency; the farther apart, the higher. Sturtevant assigned 1 map unit, or measure of distance between genes, to be equal to 1 percent of recombinant gametes. Genes that were 5 units apart would therefore have a recombination frequency of 5 percent. The map units are called *centimorgans*, in honor of his mentor, T. H. Morgan.

For his mapping, Sturtevant conducted experiments with several mutants and got the results listed in Table 11-1. With these frequencies, he could assign map distances between various pairs of genes. What he did not know, however, was the order of those genes on the chromosome. So he compared sets of three genes at a time, such as those for vermilion colored eyes (*v*), miniature wing size (*m*), and white eyes (*w*). He could tell from the recombination frequencies that *v* and *m* are 3.0 centimorgans apart on the chromosome and that *v* and *w* are 30.0 centimorgans apart. However, there could be

Table 11-1   **RECOMMBINATION FREQUENCIES FOR SOME SEX-LINKED GENES IN FRUIT FLIES**

Crosses of Mutant Genes		Frequency of Recombinants
Yellow body (y)	and white eyes (w)	1.0
Yellow body (y)	and vermilion eyes (v)	32.2
Yellow body (y)	and miniature wings (m)	35.5
Vermilion eyes (v)	and miniature wings (m)	3.0
White eyes (w)	and vermilion eyes (v)	30.0
White eyes (w)	and miniature wings (m)	32.7
White eyes (w)	and rudimentary wings (r)	45.0
Vermilion eyes (v)	and rudimentary wings (r)	26.9

Source: A. H. Sturtevant, "The Linear Arrangement of Six Sex-Linked Factors in *Drosophila,* as Shown by Their Mode of Association," *Journal of Experimental Zoology* 14 (1913): 43–59.

two possible orders of the three genes based on those distances:

Which order is correct? Does *w* lie to the *right* or to the *left* of *v* and *m*? Sturtevant reasoned that if *w* is located to the left of *v* and *m*, then the recombination frequency between *m* and *w* would be 30.0 units + 3.0 units = 33.0 units. But if *w* is located to the right of *v* and *m*, then the recombination frequency between *m* and *w* would be 30.0 units − 3.0 units = 27.0 units. One look at Table 11-1 shows that the frequency of recombination between *w* and *m* is actually 32.7 units, or almost 33.0. Thus Sturtevant concluded that *w* lies to the left of *v* and *m* and that the order of the three genes is

Using this same technique, he was able to map five genes (*y*, *w*, *v*, *m*, and *r*) occurring on the same *Drosophila* chromosome.

There was great excitement in Morgan's lab the morning that "young boy Sturtevant" presented his genetic map. One of the main implications of his work was that it served as the first proof that genes on chromosomes have a *linear relationship.* That is, they lie in "single-file" order along the chromosome. Sturtevant's linear model provided a framework for all the genetics research that followed, including the discovery that genes and chromosomes are made up of linear DNA molecules.

With Sturtevant's technique, Morgan and his colleagues went on to map genes on each of the four *Droso-*

phila chromosomes. The technique provided a reliable means of placing genes in their proper order on maps, but questions remained: Did the actual physical distances between real genes on real chromosomes correspond directly to distances derived from recombination data? Or were they more like the early explorers' maps of North America, showing plains, mountains, lakes, and rivers in the right positions relative to one another but the wrong distances apart in actual miles?

To answer the questions, researchers turned to a special type of chromosome. In certain insect cells (including some in *Drosophila* fruit flies), DNA is replicated over and over, but neither the separation of daughter strands nor nuclear division takes place. Consequently, as many as 1,000 DNA threads line up side by side, forming giant chromosomes, such as those shown in Figure 11-4. These giant interphase chromosomes are called **polytene chromosomes.** (*Polytene* means "many-threaded.") They are about 100 times as long as normal mitotic chromosomes, and when stained, they are seen to contain bands and interbands—stripes across the chromosome where chromatin is either more or less dense. The sequence and size of the bands are characteristic for each region of each giant chromosome. Because of this banding pattern, researchers can letter and number the regions to produce what is called a **cytogenetic map** of *Drosophila* chromosomes. In fact, geneticists

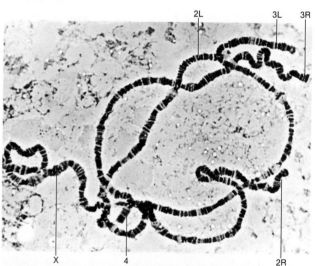

Figure 11-4   GIANT DROSOPHILA CHROMOSOMES: A BOON TO GENE MAPPING.
These giant chromosomes from the salivary gland of *Drosophila* (magnified about 360 times) have characteristic bands of chromatin that are large and distinct enough for researchers to study. Specific genes have been assigned to specific bands on the chromosome (Figure 11-5). All four *Drosophila* chromosomes can be seen here: the X chromosome is on the left; the left and right arms of the second and third chromosomes are labeled on the right; and the small fourth chromosome lies in the lower center.

Figure 11-5   GENE MAPS OF *DROSOPHILA* CHROMOSOMES.

Thomas Hunt Morgan's laboratory showed that genes on chromosomes have a linear relationship and lie in a specific order on a chromosome. Further work on stained *Drosophila* chromosomes showed that the locations of particular genes can be pinpointed and labeled. Genetic and cytogenetic maps of one end of the X chromosome are shown here. A stained X polytene chromosome, to the left, shows the banding pattern. The distances on the map do not represent exact distances on the chromosome. Since this is the X chromosome, the traits indicated are sex-linked traits.

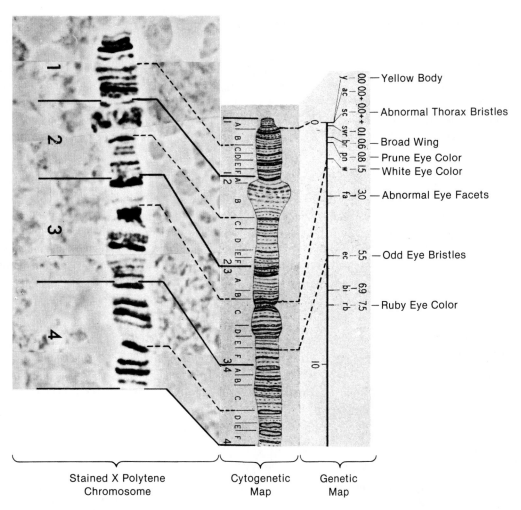

Stained X Polytene Chromosome   Cytogenetic Map   Genetic Map

have been able to assign specific genes to particular bands on the stained polytene chromosomes and thus to locate them on the map. For instance, if a gene is absent due to a deletion and a specific band also is absent, it is likely that the band is the site of the gene. Alternatively, if sections of chromosomes are rearranged so that bands are reordered, and recombination analyses reveal that genes are reordered in the same way, the band–gene correlation is strengthened. This sort of gene assignment is shown in Figure 11-5 for some genes of the X chromosome of *Drosophila*.

Armed with these two maps—Sturtevant's recombination gene map and the cytogenetic map—we can look for a correspondence. We find that the two maps show the same order of genes but not an exact correspondence of the distances between genes. The simplest explanation is that recombination is probably not an exact and simple function of distance along a chromosome, at least not in polytene chromosomes. The point is that both maps show the same genes arranged in the same order, both support the notion of a linear arrangement of genes on chromosomes, and both mapping techniques allowed

geneticists to locate the positions of genes on chromosomes and thus to probe more deeply into the mystery of how genes determine physical traits.

The example of two techniques and scientific approaches converging from different directions on one problem is not a rare one; indeed, major advances often are dependent on lines of research that at first seem unrelated, but to a perceptive scientific eye, fit together in some way that leads to such important conclusions as the linear ordering of genes on chromosomes. So, the elucidation of gene function, made possible in part by these maps, was one of the most significant advances in the history of biological science, as we shall see.

HOW GENES ACT AND INTERACT

So far, we have seen how genes are linked on chromosomes, how linked genes behave during meiosis, and how an analysis of this behavior led to the creation of gene maps. This work, of course, was based on the principles that Mendel derived from his work in his monastery garden. But as we suggested earlier, genes in nature do not always act as predictably as do the seven characteristics that Mendel studied in pea plants. Consider, for example, the traits of eye color and hair color in humans. People's eyes range from pink (albinos) to almost black, with light and dark shades of blue, violet, green, gray, and brown in between. Hair ranges from white to black, with dozens of intermediate shades of yellow, red, and brown. Clearly, eyes and hair are not simply one phenotype or another, as pea plants are either tall or short and peas are either yellow or green. For eye color and hair color in humans, as well as for other traits in other organisms, several genes or alleles of those genes can affect the same phenotype. Let us see how genes act to bring about phenotypes; then we will consider how they interact to bring about multiple phenotypes.

How Genes Act

How do genes actually influence the color of a seed or of a person's eyes? The expression of contrasting traits often is due to the different effects of specific alleles on the organism's embryonic development. In Chapter 13, we will see that genes control functions by causing the production of specific proteins, which are of three main types: some play *structural* roles inside a cell, on a cell surface, or in the immediate environment of a cell; others act as *enzymes*, facilitating the myriad biochemical reactions that make up metabolism; and still others are *regulators* of genes that coordinate the timing and patterning of developmental processes. The interaction of many genes and many proteins during embryonic development culminates in a leaf, a hand, an eye, or a cat's tail.

Genes exist as either dominant or recessive alleles. The frequently encountered normal, or wild-type, allele usually is dominant to the rare mutant form. A recessive allele often has lost part or all of its ability to perform the function of the normal allele. In a heterozygote, one copy of the dominant allele may provide enough of a given gene's normal function to support the development of a normal phenotype. Thus in a pea plant heterozygous for round seed trait (Rr), the single R allele allows enough of a specific enzyme to be synthesized so that sufficient starch is manufactured to give a firm, round appearance to the seed. In a homozygous recessive plant (rr), the protein controlled by the r allele is not enzymatically active, so that not enough starch is produced to make the seed plump, and it appears wrinkled. Note that the gene and its product, the enzyme, do not act directly, but indirectly, on what we see—the plump, round seed or the shriveled seed.

The insight that genes control specific biochemical functions also allows us to understand the phenomenon of incomplete dominance, which we encountered in Chapter 9. When white snapdragons (rr) are crossed with red snapdragons (RR), the heterozygous progeny (Rr) are pink. We can explain the intermediate phenotype if the R allele provides a function (say, an enzyme) necessary for pigment production, and the r allele does not provide that function. With two doses of the enzyme in the RR homozygote, enough pigment would be produced to yield a red flower. But one dose of the enzyme in the Rr heterozygote would allow only half as much pigment to be produced, and thus the flower would appear pink. In cases of incomplete dominance, such as the pink snapdragon, the limiting factor in the phenotype is the amount of active gene product formed by the dominant allele. In the more common cases of complete dominance, one dose of gene product from the dominant allele is sufficient to ensure that the dominant phenotype occurs in a heterozygote, even though the recessive allele makes no contribution to phenotype.

Genes with More Than Two Alleles

Mendel was concerned with only two alternatives for each particular trait in pea plants. But a gene often can have 3, 6, 12, or even 100 alleles, all of which confer a slightly different phenotype for the same trait. Of course, only two of the possible alleles are present in any individual of a diploid species. A group of alleles determining many forms of the same trait is called a **multiple allelic series.** For example, dozens of sex-linked alleles of the white eye-color gene determine the palette of eye colors in *Drosophila*, ranging from deep brown through brilliant red to pure white (Figure 11-6). And in humans, the A, B, AB, and O blood types are determined by three alleles. How do these alleles interact? Let us look at the multiple allelic series that determine eye color in fruit flies and blood type in people.

Drosophila *Eye Color*

In fruit flies, the different shades of red eye color are based on several recessive alleles that resulted from mutations of one sex-linked gene—white, apricot, eosin (shown in Figure 11-6), coffee, blood, cherry, and a host of others. Females homozygous for any of these mutations—say, coffee-colored eyes—can be crossed with males carrying a different mutation from the allelic series—say, apricot-colored eyes. The F_1 heterozygotes usually express a mutant phenotype that is intermediate in color between the two parental alleles; this incomplete dominance results in the production of varying *amounts* of pigment, not in the type of pigment present. Because the phenotype is intermediate, neither allele

blood that a person can safely be given in a blood transfusion depends on his or her blood type: A, B, O, or AB. If blood type is determined incorrectly in the hospital or is ignored, a person can die as surely from a blood transfusion as he or she can from the original wound or disease.

One function of A, B, and O alleles is to govern the presence or absence of molecular markers or nametags on the surfaces of red blood cells. In actuality, such nametags are glycoproteins that are *antigens* (substances that can evoke an immune response). Thus every red blood cell of a type A person has A antigens but not B antigens. A type B person has red blood cells with B antigens only; a type AB person, both A and B anitgens; and a type O person, neither A nor B antigens. These ABO blood types are determined by a multiple allelic series. The ABO blood-group gene has three alleles that code for blood types, as shown in Table 11-2. The allele I^O is reminiscent of the *Drosophila* white-eye gene (w) in that it yields the O phenotype, which adds neither A nor B antigens to red blood cells. Note that I^O is recessive and that both I^A and I^B are dominant to it. Furthermore, I^A and I^B alleles are fully expressed when they occur together. Hence, they are said to be **codominant,** and both phenotypic traits are present (an $I^A I^B$ person has an AB blood type).

Medical difficulties can arise because the body spontaneously makes proteins that will attack foreign blood-group antigens—in other words, blood-group substances that it does not possess. These proteins, called *antibodies,* act by binding to specific foreign molecules (antigens), causing them to clump together (Figure 11-7). For example, in an $I^A I^A$ person, anti-B antibodies are present in the blood fluid, and in an $I^B I^B$ individual, anti-A antibodies are present. In a person with the $I^A I^B$ genotype, neither antibody is present, and with the $I^O I^O$ genotype, both anti-A and unit-B antibodies are present. If an A type individual mistakenly receives a transfusion of B blood, a massive clumping of blood cells results as the anti-B antibody binds to the B antigen on the transfused cells. An error like this can be fatal.

Human *Rh blood groups* represent another set of alleles affecting blood chemistry. About 85 percent of adult humans have a molecular marker called the Rh

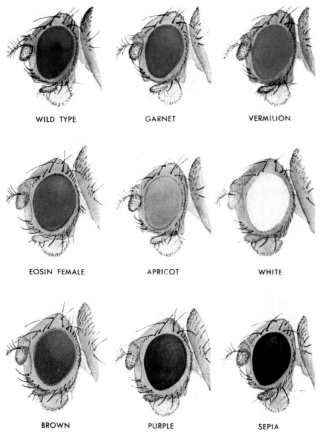

Figure 11-6 A PALETTE OF EYE COLORS IN DROSOPHILA.
The wild-type *Drosophila* eye is red, but a variety of sex-linked mutations can produce other eye colors, only some of which are shown here.

WILD TYPE GARNET VERMILION
EOSIN FEMALE APRICOT WHITE
BROWN PURPLE SEPIA

appears able to perform the normal function provided by the wild-type allele for brick-red eyes. Thus we can deduce that each mutation must affect the *same* genetic function required for normal eye color—that is, they must be different alleles of the same gene. However, each allele affects eye color in a slightly different way. The white-eye allele appears to cause a complete loss of normal gene function and thus an absence of eye pigment, while the other alleles cause a less drastic alteration of the normal gene function, and so the eyes contain different amounts of pigment. Because these mutations are members of a multiple allelic series, they are designated as alleles of the *white* mutated gene by the notations w, w^e, w^a, and so on. In fact, over 100 alleles of the white-eye gene have been discovered.

Human Blood Type

The multiple allelic series for blood types affects the very survival of many people every day. The type of

Table 11-2	MULTIPLE ALLELES AND ABO BLOOD TYPE
Blood Type	**Genotype**
A	$I^A I^A$, $I^A I^O$
B	$I^B I^B$, $I^B I^O$
AB	$I^A I^B$
O	$I^O I^O$

Donor Blood Type ⟶

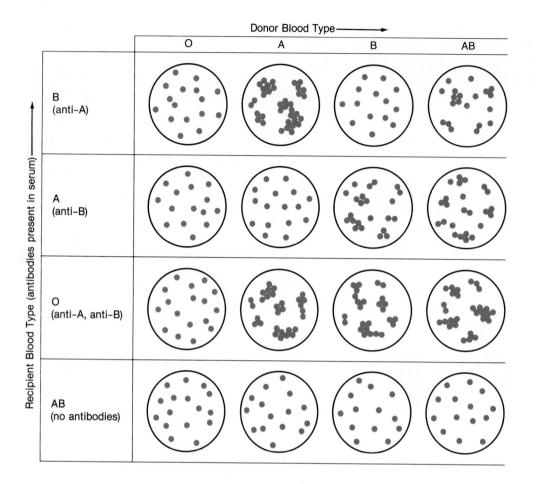

Recipient Blood Type (antibodies present in serum)

Figure 11-7   BLOOD GROUPS AND TRANSFUSIONS.
This diagram shows the consequences of the genetics of blood groups. Blood types of the donor are shown along the top axis, and blood types of recipients are shown on the left axis. The types of antibodies present in the serum of recipients of each blood group are in parentheses. The various consequences of mixing the four types of serum (B, A, O, AB) with blood cells on each of the four types are shown in the petri dishes. For instance, anti-A clumps A and AB cells, both of which carry A nametags (antigens) on their surfaces.

factor on their red blood cells and are classified as Rh positive (Rh$^+$). (The Rh comes from *Rhesus*, the genus of monkey in which this blood factor was first detected.) Rh negative (Rh$^-$) persons lack this factor because they are homozygous recessive for an allele that does not produce the RH$^+$ factor. However, if an Rh$^-$ person is exposed to Rh$^+$ blood, his or her immune system responds over ensuing weeks by making antibodies that will react with the Rh antigen factor on the red blood cells and cause clumping. The production of such antibodies sets the stage for the deadly response of clumping any foreign cells that happen to bear the antigenic marker. This condition is called Rh disease.

Rh disease can result from a blood transfusion, but more commonly, it follows the birth of an Rh$^+$ baby to an Rh$^-$ mother (the father is Rh$^+$, hence the baby is heterozygous for the Rh factor). Problems do not usually occur with a first-born Rh$^+$ child to an Rh$^-$ mother. However, some fetal cells containing Rh factor may cross the placenta during the birth process and enter the mother's blood vessels. Although only a moderate amount of antibodies form with this first pregnancy, the mother's immune system retains a "memory" of the event; the memory corresponds to populations of cells primed specifically to recognize the Rh factor. In later

pregnancies, higher and higher levels of antibody production take place each time the Rh$^-$ mother is exposed to Rh$^+$ blood from a fetus during birth. During a subsequent pregnancy, these antibodies can cross the placenta and cause the fetus's red blood cells to clump, creating a potentially fatal disease called erythroblastosis, which affects the late fetus or newborn.

About 35 percent of potential Rh-factor fatalities can be prevented by almost total replacement of the newborn's Rh$^+$ blood with blood free of antibodies against the Rh$^+$ factor. This prevents the newborn's red blood cells from being clumped by the mother's antibodies. During the following few days, the newborn's bone marrow produces new red blood cells with Rh$^+$ markers, but they are safe because the mother's antibody is no longer present. Another increasingly common means of preventing Rh disease is to inject the Rh$^-$ mother with antibodies at the birth of an Rh$^+$ child; the antibodies clump any fetal red blood cells with Rh$^+$ markers that have entered her bloodstream, thus removing them and preventing her immune system from forming antibodies that could damage future babies.

Now let us consider why the Rh trait is an example of multiple alleles. Careful examination of blood from a variety of Rh$^+$ individuals has revealed between thirty-five and forty distinguishable Rh-factor phenotypes. One explanation for the genetic basis of Rh phenotypic variability suggests that any one of three alleles—*C*, *D*, and *E* (or the recessives, *c*, *d*, and *e*)—may reside at the site of the Rh locus on the chromosome. The critical allele in this set is *D*, because it makes the molecules on the surface of the red blood cells maximally antigenic (that is, recognizable by an Rh$^-$ mother). Thus an Rh$^-$ woman who bears several children fathered by a male with allele *D* (even if his other Rh allele is the recessive *c* or *e*) is most susceptible to Rh disease and development of antibodies against fetal red blood cells. We know about the Rh and ABO multiple allelic series because they make such obvious markers on the red blood cell surface and exert such profound effects on life and health. There are probably many other multiple allelic series with more subtle effects that have not yet been recognized but will turn out to impinge on the normal development and function of the body.

Complementary Genes

We have seen how several alleles of a single gene can affect a fruit fly's eye color or a human's blood type. Sometimes, however, one phenotype is generated by the action of two or more separate **complementary genes.** These are genes whose products must act together to produce a given phenotype; hence, both genes must be present and active. An example of complementary genes occurs in human albinism. Figure 11-8 shows a human couple, both of whom are homozygous for recessive alleles causing the albino phenotype, which is characterized by the absence of pigment in skin, eyes, and hair and sometimes by the presence of crossed eyes. Notice that their son does not have the albino phenotype; he has normal pigmentation in his skin, eyes, and hair, and uncrossed eyes. The explanation for this phenomenon is that two complementary genes affect normal pigmentation. Each gene controls the production of one enzyme, and both enzymes are required in order for pigment to form.

Each parent in this example must be homozygous recessive for only one of the two complementary genes. Thus we will say that the mother is normal for *A* but lacks (is homozygous recessive for) *B*, whereas the father lacks *A* but is normal for *B*. Since each lacks one critical enzyme in the two-step pigment-forming process, pigment cannot be generated, and each shows the albino phenotype. However, their son apparently inherited one normal dominant allele of both genes, *A* from his mother and *B* from his father in our example. Thus both complementary genes could exert their necessary influence on the phenotype, so the boy's pigmentation is normal.

This example of complementary genes helps us under-

Figure 11-8   ALBINISM AND COMPLEMENTARY GENES.
These two parents show albino traits, to varying extents, yet their son is normal. Since normal pigmentation is controlled by two complementary genes (say, genes 1 and 2), each parent must be homozygous recessive for one of the genes. The father could be normal for gene 1, but aberrant for gene 2 (he is slightly pigmented), and the mother mutant gene 1, but normal for gene 2 (she is unpigmented). The son thus inherited one normal allele of each gene from each of his parents and has normal pigmentation.

stand that many aspects of adult phenotype result from biochemical pathways with many steps and from complex series of developmental events, each step or event dependent on products of a separate gene. It is no wonder, then, that mutation in any one of a number of complementary genes can affect an organism's phenotype.

Epistatic Genes

We have discussed two types of gene interactions, multiple allelic series and complementary genes. Let us consider a third type, **epistasis** (meaning "to stand over"), in which the effects of one gene override or mask the effects of other, entirely different genes.

Consider domestic cats, for example. Cats have two gene pairs that determine the appearance of the coat: one for coat color and one for coat pattern. The coat-color gene is *A*. An *AA* or *Aa* cat is *agouti*; each hair is gray with a yellowish band or tip. Agouti is the typical grayish-brown color of a wild rabbit or rodent. An *aa* cat is pure black. The separate coat-pattern gene is *T*. A cat with *TT* or *Tt^b* has a *mackerel-tabby* coat pattern, with vertical curving black stripes. A *t^{b}t^b* cat has a *blotched-tabby* coat pattern, with broad bands in swirls and whorls.

When a homozygous agouti, blotched-tabby female (*AAt^{b}t^b*) is mated to a pure-black male of genotype *aaTT*, all F$_1$ kittens are *AaTt^b*, as shown in Figure 11-9. These

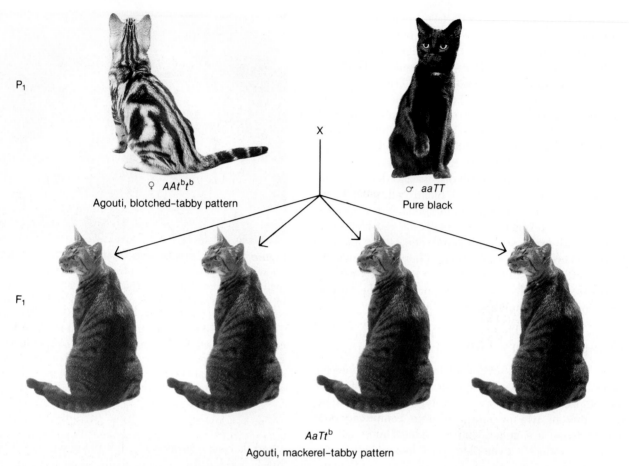

P_1

♀ AAt^bt^b
Agouti, blotched–tabby pattern

X

♂ $aaTT$
Pure black

F_1

$AaTt^b$
Agouti, mackerel–tabby pattern

Figure 11-9   CATS' COATS: A DEMONSTRATION OF EPISTASIS.
Two genes determine coat appearance in cats: A, for coat color; and T, for coat pattern. An AA or Aa cat is agouti; an aa cat is all black; a TT or Tt^b cat has a mackerel-tabby pattern; and finally a t^bt^b cat has a blotched-tabby coat pattern. When an agouti, blotched-tabby female (AAt^bt^b) is mated with an all-black male of genotype $aaTT$, all F_1 kittens are $AaTt^b$. They exhibit the dominant mackerel pattern even though neither parent appears to be mackerel. The reason for this unexpected coat pattern inheritance is epistasis—one gene masking the expression of a *different* gene for an entirely different trait. In this case, the male cat's coat-color genotype (aa) is epistatic to the tabby-pattern genotype (TT). It is not apparent that the black cat carries the genotype for mackerel-tabby coat pattern. However, when the F_1 kittens inherited the dominant agouti color, the dominant mackerel-tabby coat pattern was unmasked.

F_1 kittens represent a peculiar situation: all the offspring show the dominant mackerel-tabby pattern, yet neither parent appears mackerel. The only parent with a coat pattern has the recessive blotched-tabby phenotype. The kittens show this unexpected pattern of inheritance because the aa coat-color genotype in the pure-black tomcat masks a coat-pattern gene from his mother. The coat-color genotype aa is epistatic to the tabby-pattern gene T or t^b. Looking at a black cat, then, we really cannot tell if its genotype is $aaTT$, aat^bt^b, or $aaTt^b$. It could be a mackerel tabby or a blotched tabby disguised by epistasis.

Epistasis should not be confused with dominance. Epistasis is an interaction in which one gene masks the expression of a *different* gene for an entirely different trait. Dominance is an interaction in which one allele masks the expression of another allele of the *same* gene.

LETHAL ALLELES AND PLEIOTROPY

Most of the genes that we have discussed—for eye color, plant height, coat pattern, and so on—affect the way an organism looks but may not be crucial to its survival. Other genes, however, perform such essential functions during an organism's development that certain allelic forms can have a life or death influence. Mutated genes that are capable of causing death are called **lethal alleles.** Depending on the nature of the essential function that is missing, death may occur at any time in life: early in the embryo, later in embryonic growth and development, during infancy, or even at some time during adulthood.

THE COMPLEX GENETICS OF "MAN'S BEST FRIEND"

Consider, for a moment, the sizes and shapes of dogs: an average St. Bernard weighs 100 times as much as a typical Pekingese. An Irish wolfhound on its hind legs can peer over a tall person's head, while a Chihuahua can rest in the palm of the hand. A dachshund trots about on short, stubby legs, and a greyhound gallops on very long, narrow ones. The list goes on and on, yet all breeds of dogs belong to only one species (Figure A). How did all this remarkable variation come about?

"Man's best friend" is also the oldest domesticated animal, whose breeding is controlled by selecting and mating individuals with desired traits. The earliest evidence of animal domestication comes from a 12,000-year-old tomb at a site called Ein Mallaha in northern Israel. In the grave, archeologists found the bones of a wolf/dog puppy buried alongside the bones of an elderly person. Zoologists believe that all modern dog breeds—collectively classified under the species *Canis familiaris*—are descendants of the wolf, *Canis lupus*. The remains at Ein Mallaha cannot be clearly distinguished as either wolf or dog because they display traits of both species. However, since even a wolf cub can be easily "imprinted" to bond with a person, there is little question that the small skeleton found curled inside the tomb is that of a pet.

The great range of characters that we see in modern dogs is the result of at least 12,000 years of selective breeding. This range far exceeds that in any species of wild animal, and, as one might expect, the characters are determined by a variety of genes, many of which have complex interactions.

Coat color in dogs is probably the best example. Fifteen genes are involved throughout the dozens of dog breeds. Some are epistatic over others; some occur in multiple allelic series; some are incompletely dominant over others; and a few operate in normal Mendelian fashion. Figure B shows the phenotypes and genotypes for coat color in one breed, the pointer: suppose the *B* gene determines red, yellow-red, or tan coloration; the *E* gene, shades of brown on the coat, nose, and balls of the feet; the *T* gene, total coat color (the homozygous recessive form [*tt*] yields a mostly white, mottled coat); and the *S* gene, small spots, or "speckles." When *B* occurs with *ee*, the pointer has reddish coat markings. When *E* occurs with *bb*, the dog has brown coloration. When both *B* and *E* are present, the dog has black coat markings. And when *bb* and *ee* occur together, the dog has almost lemon-yellow markings. The size and type of the markings depend on *T* and *S*.

The genotypes have been worked out for many—

Figure A
Despite their differences, all domestic dogs are the same species, *Canis familaris*. Theoretically, at least, this St. Bernard and Chihuahua could breed!

but not all—characters in dogs. To do so usually requires crossing dogs of different breeds; for example, a short-legged bassett hound might be bred with a long-legged German shepherd to study the genetics of leg length. However, most breeders who sell and show dogs are interested in maintaining pure lineages that display narrow, rigidly determined sets of characteristics; they create crossbreeds only occasionally, out of curiosity. For this reason, ironically, the genetics of the beloved dog are less well understood than are those of the fruit fly, *E. coli* bacteria, or bread molds!

Figure B
Four genes determine the complex array of coat colors in pointers. A dash means that the allele at that position could be either dominant or recessive.

(a) bbeett
(b) bbE-tt
(c) B-eett
(d) B-E-tt
(e) B-eeT-
(f) B-E-T-
(g) bbE-ttS-
(h) B-E-ttS-

(a)

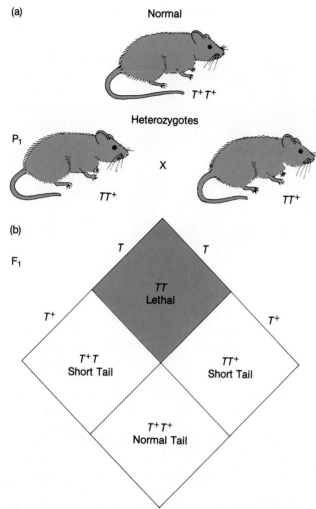

(b)

Figure 11-10   SHORT TAILS AND LETHAL ALLELES IN MICE.

A gene for tail length in mice displays an unusual inheritance pattern. (a) Mice with the recessive genotype (T^+T^+) have normal-length tails, whereas heterozygotes (TT^+) have a shorter tail. (b) When two heterozygotes mate, the F_1 phenotypic ratio is 2:1 rather than the expected Mendelian ratio of 3:1 because mouse embryos with the homozygous dominant genotype (TT) die before birth.

If some lethal alleles are fatal to early embryos, how do we know that they exist? In the late 1920s, a mutation named *brachyury* arose in a laboratory mouse, causing one of its offspring to have a particularly short tail. When this mutant mouse matured, biologists determined that it was heterozygous for the short-tailed trait and that this allele is dominant to the normal, long-tailed allele. A line of progeny was established from this short-tailed mutant and was crossed with various wild-type mice, all with normal tails. No matter what was tried, however, offspring that were homozygous dominant for the short-

tailed trait could not be isolated from these crosses. Put simply, the crosses would be expected to yield a classical 3:1 ratio (which also can be written 1:2:1 when we are talking about $TT:TT^+:T^+T^+$; since the T^+T^+ and TT^+ phenotypes cannot be distinguished, they are lumped as 3:1). But in the breeding experiments, there was always a 2:1 ratio of short tails to normal tails (Figure 11-10). After puzzling over these unexpected results, geneticists realized that the observed 2:1 phenotypic ratio could be generated from the expected 1:2:1 genotype ratio if one-quarter of the progeny were missing from the F_1 generation. Because the 1's in the 1:2:1 ratio are homozygotes, geneticists concluded that mice that are homozygous for the short-tailed trait must die early in development. It was later directly observed that such embryos do die and are resorbed in the mother's uterus.

With this knowledge of lethal genes, we can understand why Manx cats, discussed at the beginning of this chapter, are so difficult to breed. They are, like the mutant mice, heterozygotes for a recessive lethal gene. Presence of one copy of the mutant allele produces the Manx characteristics (the allele is dominant for the visible phenotype). However, the presence of two mutant copies is lethal (they behave as if recessive—with two copies needed—for the lethal trait). Hence no pure breeding line can be developed.

After studying the Manx trait and other cases of lethal alleles, geneticists generalized that lethal genes that persist in natural populations in the wild are always recessive. If they were dominant, then both homozygotes and heterozygotes would die. Of course, many lethal genes that are dominant may arise by mutation, but such mutations disappear immediately due to the death of both heterozygotes and homozygous dominants.

The Manx allele seems to affect two phenotypes: when heterozygous, it causes a shortened tail; when homozygous, it causes death. In fact, the two phenotypes are related, since in the dying embryos, not only does the tail fail to form, but the entire posterior of the embryo develops abnormally. This correlation suggests that an individual gene can affect several traits. Even Mendel recognized this effect, now known as **pleiotropy**, when he observed that a single hereditary unit seems to determine simultaneously (1) whether a pea flower is red or white, (2) whether or not each leaf has red coloration where it joins the stem, and (3) whether the seed coat is gray or white (Figure 11-11).

Later researchers noticed other examples of single genes that affect several traits. For example, as we noted earlier, humans who are homozygous for the albino gene not only have white hair, pink eyes, and pale skin, but also sometimes have crossed eyes; the allele causes defects in the way that nerves from the eye connect with targets in the brain. *Drosophila* males with the mutation for yellow body color have poor mating success with nor-

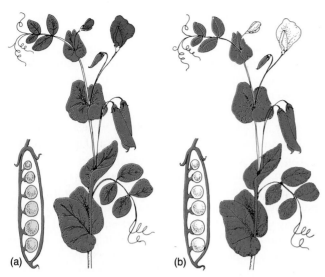

Figure 11-11   PLEIOTROPY: INDIVIDUAL GENE AFFECTING SEVERAL TRAITS.
Mendel observed three traits in the pea plant that seemed to be controlled by just one gene. (a) A plant with red flowers has red coloration where the leaf joins the stem and has gray seed coats. (b) A plant with white flowers has no coloration where the leaf joins the stem and has white seed coats.

mal females and yet can mate quite easily with homozygous yellow females; thus both body color and mating behavior are affected by this one gene.

Pleiotropic genes are interesting to researchers because studying them can reveal fundamental interconnections between processes or traits that on the surface appear unrelated. One well-known example of pleiotropy is the gene that causes sickle cell anemia, a human disease that usually results in death before age twenty. Hemoglobin, the affected protein, is present in red blood cells and carries oxygen to the body tissues. Sickle cell disease is caused by a mutated gene, Hb^S, for the production of hemoglobin (the normal allele is Hb^A). The Hb^S allele affects several traits and has at least five readily observed phenotypic conditions: (1) red blood cells shaped like sickles instead of normal disks (Figure 11-12); (2) severe and ultimately lethal anemia (reduced number of red blood cells); (3) pain in the abdomen and joints; (4) enlarged spleen; and (5) resistance to malaria in heterozygous individuals.

Figure 11-12   RED BLOOD CELLS, NORMAL AND SICKLED.
Normal red blood cells (bottom) are shaped like fat disks with a dent in each side. Sickled red blood cells (top) have elongated, irregular shapes because of abnormal hemoglobin molecules that have aggregated into fibers. These cells are magnified about 6,900 times.

At first, these phenotypic characteristics seem so unrelated that it is difficult to imagine that they are due to a change in the same gene function. However, when the sources of the phenotypic traits are examined carefully, each can be traced back to the same fundamental defect: a *single* substitution of one amino acid for another in hemoglobin. The mutant abnormal hemoglobin molecules tend to join together to form long fibers of hemoglobin; these fibers then distort the shape of the red blood cell, causing it to sickle. Because the job of the spleen is to remove old and damaged red blood cells from the circulating blood, it becomes enlarged and overworked from removing the many sickled cells. This removal also leads to severe anemia, since there are fewer red blood cells left to carry oxygen. Finally, the shape of the sickled cells causes them to block blood flow in tiny capillaries, starving nearby body cells of oxygen and leading to pain in the abdomen and joints.

Individuals who are heterozygous for the sickle cell trait (Hb^A/Hb^S) produce both normal and abnormal hemoglobin molecules. Their red blood cells do not sickle under normal conditions of low oxygen availability, such as when the blood gives up its oxygen to hard-working cells in the arm or leg muscles. But sickling does occur if available oxygen declines when malaria parasites are present in the red blood cells. The spleen then removes and destroys the sickled infected cells while allowing the normal-shaped, uninfected cells to pass through unharmed. As a result, the spleen may become enlarged, the blood may become somewhat anemic, but the individual is resistant to malaria.

The fact that the sickle cell gene confers resistance to malaria is the key to its preservation during human evolution despite its deleterious effects. Consider what happens to a population that includes some sickle cell heterozygotes and that lives in a malaria-prevalent region.

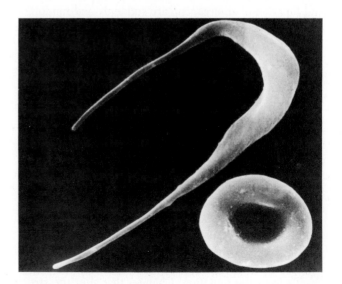

Individuals homozygous for the trait die from the sickle cell disease, and many of those homozygous for the normal gene die from malaria. The sickle cell heterozygotes, however, usually survive both conditions. Then, ironically, the protection from malaria is so marked that it ensures that the sickle cell gene, with all its negative pleiotropic effects, will not be eliminated from the population. This is a good example of the unexpected in science: Who would have guessed that a gene that is so dangerous to homozygous individuals would be so advantageous for heterozygotes? It took biochemists, electron microscopists, x-ray crystallographers, population geneticists, and physicians to unravel this fascinating story.

MUTATION: ONE SOURCE OF GENETIC VARIATION

The rules of inheritance and their exceptions can be studied only if a gene has alternative forms, as in the case of white versus red eyes in fruit flies or white versus red pea flowers. How do these different forms of the same gene arise? They are the direct consequence of **mutation,** a change in the chemical structure of a gene or the physical structure of a chromosome.

Many mutations in the genetic information are deleterious to the organism, leading to a competitive disadvantage or to death. Imagine, for example, the chaos that would result if a few critical words in the instruction manual for operating a complex machine were suddenly changed—if instead of saying "Turn right control knob to the left," it said "Turn right snob control left" or "Turn lortnoc bonk thgir left." The machine might fail to start, might run badly, or might even destroy a delicate circuit when operated with the wrong set of instructions. The same is true when the genetic instructions for the form and function of an organism are suddenly and randomly mutated. Occasionally, however, a genetic alteration lends some survival advantage to the organism. Thus the sickle cell allele increases a heterozygous person's chances of surviving in malaria regions. When such advantages arise as a result of mutation, the organism may survive, pass the mutation to its offspring, and thereby spread the trait in the population. Thus mutation is one of the primary sources of biological diversity, providing the raw material for the evolutionary process.

"Mutation" is a word that encompasses several processes. A **point mutation** alters the properties of a single gene and creates new alleles ("knob" to "snob" is analogous to a point mutation). The molecular nature of point mutations will be discussed in Chapter 13, but we already have seen a point mutation in the sickle cell gene, in which a substitution of one amino acid in hemoglobin results from the mutation. Changes that alter the structure of chromosomes—**chromosomal mutations**—involve rearrangement of blocks of genes in the chromosome, not alteration at a point in one gene. Several types of chromosomal mutations are shown in Figure 11-13. These include *gene deletion; gene duplication; inversion,* or flipping, of genes; and *translocation,* or movement of a group of genes from one place to another.

Point and chromosomal mutations were first defined and studied successfully in *Drosophila* and in corn plants during the early days of genetics. However, detecting them in a systematic way was not easy. To begin with, mutations are rare events. In addition, the detection of point mutations is particularly difficult because most are recessive and are not observable except in the homozygous condition. Mutations that occur on sex chromosomes are detected somewhat more easily because a mutated recessive allele on the X chromosome will be expressed in individuals that are XY.

An important discovery provided a fundamental tool for probing genetic mutations. Herman J. Muller, another member of Thomas Hunt Morgan's research group at Columbia University, found that mutations can be *induced* by physical and chemical factors in the environment acting on the genetic material. Muller devised a system for measuring the **mutation rate** in *Drosophila*—that is, how frequently mutations arise naturally in a given population. Because mutations that alter the phenotype may range from the quite obvious to the almost imperceptible, the detection of a slight mutation often depends on which geneticist does the experiments and observations. In order to avoid such judgments, Muller chose to measure the appearance of only sex-linked lethal mutations—mutations that cause the death of the affected fly before adulthood. He found that new lethal mutations arise spontaneously in about 2 of every 1,000 X chromosomes tested in each generation of flies. But because the X chromosome of *Drosophila* contains perhaps 1,000 essential genes, the mutation rate of each individual gene is much lower—somewhere between 10^{-5} and 10^{-6} mutations per gene per generation.

Muller received the Nobel Prize for a subsequent discovery. When he exposed male fruit flies to x-rays and then tested their offspring for lethal mutations, he found that the frequency of mutations was greatly increased. Furthermore, the more powerful the x-ray dose, the more lethal mutations occurred. Muller's techniques were later used to discover that certain chemicals called **mutagens** are also capable of causing mutations. The first of many chemicals shown to be mutagenic was mustard gas, a chemical-warfare agent widely used in World War I. Since then, hundreds of natural and synthetic mutagens have been identified. Because genetic mutations cause many severe human diseases (as we will see in Chapter 15), a means of identifying potential causes of such mutations is extremely important.

A link also was discovered between mutagens and **car-**

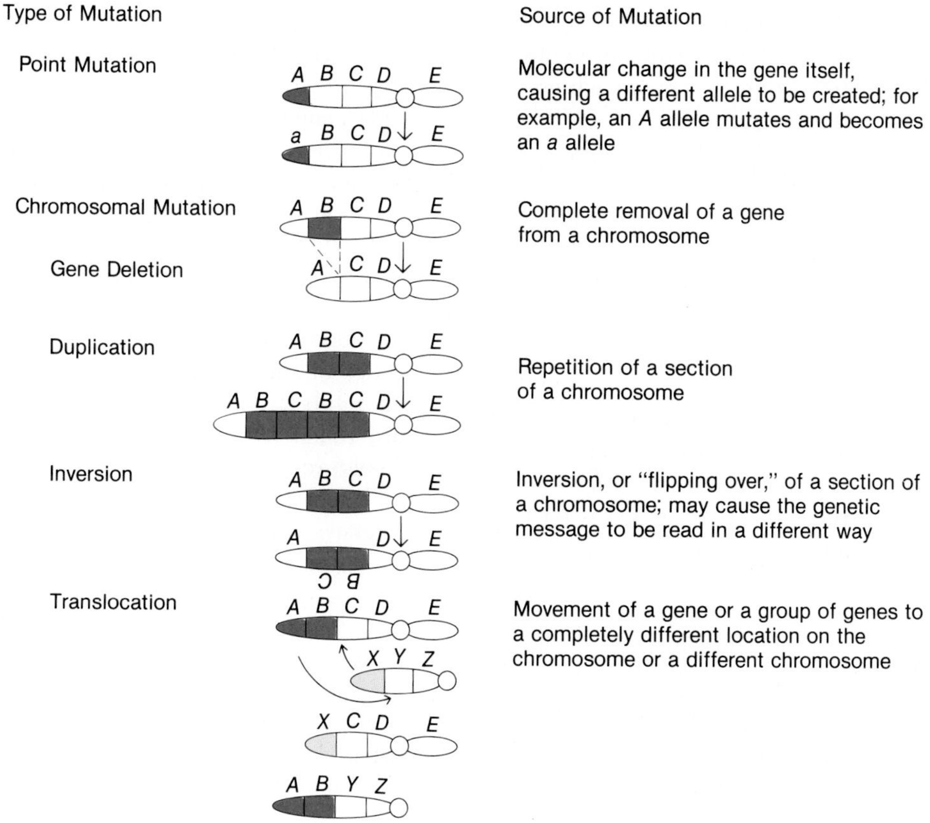

Type of Mutation

Point Mutation

Chromosomal Mutation

Gene Deletion

Duplication

Inversion

Translocation

Source of Mutation

Molecular change in the gene itself, causing a different allele to be created; for example, an *A* allele mutates and becomes an *a* allele

Complete removal of a gene from a chromosome

Repetition of a section of a chromosome

Inversion, or "flipping over," of a section of a chromosome; may cause the genetic message to be read in a different way

Movement of a gene or a group of genes to a completely different location on the chromosome or a different chromosome

Figure 11-13   MUTATIONS AND CHROMOSOME REARRANGEMENTS.
Several processes can alter the arrangement of genes on a chromosome. Of course, individual genes are much shorter relative to the length of the chromosomes than in these diagrams.

cinogens—agents that can cause cancer. A test that employs bacteria of the genus *Salmonella* is now widely used to screen chemicals for possible mutagenic effects. More than 75 percent of the agents shown to be carcinogenic through long, complicated studies on animals also show up as genetic mutagens of bacteria in this fast, simple screening test. Therefore, the *Salmonella* test, or **Ames test** (after its originator, Bruce Ames of the University of California at Berkeley), is being used, along with chromosome tests such as those Muller devised, to monitor the thousands of physical and chemical agents to which people are exposed every day in their homes, in their workplaces, and out of doors.

Ultimately, biologists may have a good understanding of the process of genetic mutation, the role of mutations in cancer and other diseases, and even the rate of evolutionary change based on the frequency of mutations in nature. Modern geneticists engaged in these studies are, so to speak, standing on the shoulders of giants. Based on the profound theories and remarkable discoveries of such pioneers as Mendel, Weismann, Sutton, Morgan, Sturtevant, Bridges, and Muller, modern geneticists have begun to unlock some of life's most complex secrets. One of the most important secrets was the chemical nature of the gene, the subject of Chapter 12.

SUMMARY

1. Genes on the same chromosome are said to be *linked* if they segregate together during meiosis; they are unlinked if they segregate independently. Genes on separate chromosomes are unlinked. Crossing over during prophase of meiosis I can cause pieces of chromatids to be exchanged, producing recombinant chromatids. Recombination analysis involves study of the recombinant genotypes and phenotypes and al-

lows conclusions to be drawin about the placement of alleles on chromosomes.

2. Genes that are close together on a chromosome recombine as a result of crossing over less frequently than do genes that lie farther apart. By analyzing *recombination frequencies* between genes (as the percent of recombinant gametes), Sturtevant was able to construct the first *gene*

map, a representation of the order of and distances between the *loci* of genes on a chromosome.

3. Sturtevant showed that genes are arranged on chromosomes in a linear fashion. Subsequent analysis of the banding patterns of giant *polytene chromosomes* in *Drosophila* has yielded *cytogenetic maps* of the fruit-fly chromosomes that bear out Sturtevant's findings of gene order.

4. Because genes control specific biochemical functions, many dominant alleles are forms of genes that function normally, whereas most recessive alleles are forms of genes that have lost part or all of their ability to function.

5. Some genes exist in many alternative forms, including a *wild-type allele* and variations on it; such groups are called *multiple allelic series*. Genes with many alleles include those that code for eye color in fruit flies and for blood groups in humans. The AB blood type is the result of the *codominance* of two alleles, both of which are fully expressed in the phenotype.

6. *Complementary genes* are sets of genes whose products must act together to produce a given phenotype; thus all genes must be present and active. The pigmentation of skin, eyes, and hair in humans is determined by two complementary genes. If either of the genes is absent or abnormal, the albino phenotype results.

7. Masking of the phenotypic effects of one gene by an entirely different gene is called *epistasis*. In cats, for example, the homozygous recessive alleles for black coat mask the gene for mackerel-tabby or blotched-tabby coat patterns.

8. The presence of certain mutated alleles can lead to an organism's death. Such *lethal alleles* usually are recessive; heterozygotes for a lethal allele may show an unusual phenotype, as does the Manx cat. When present, lethal alleles occur in all the cells of an organism (as do all alleles) and can cause death at characteristic times in the organism's life cycle.

9. One gene sometimes affects several traits; this phenomenon is called *pleiotropy*. The sickle cell gene, for example, affects the shape of red blood cells and can indirectly cause enlargement of the spleen, anemia, and pain in the abdomen and joints. Because the sickle cell gene also confers resistance to malaria, it is perpetuated in populations inhabiting malaria-prevalent regions.

10. *Point mutations* alter the properties of single genes; changes that alter the structure of chromosomes are called *chromosomal mutations*. In nature, *mutations* occur randomly and spontaneously, but in low frequency; in the laboratory, they may be induced by physical or chemical factors (*mutagens*). Because there is a high correlation between mutagens and *carcinogens*, screening devices such as the *Ames test* are used to monitor physical and chemical agents in the environment for their potential danger to humans.

KEY TERMS

Ames test

carcinogen

chromosomal mutation

codominant allele

complementary gene

cytogenetic map

epistasis

gene map

lethal allele

linkage

locus

multiple allelic series

mutagen

mutation

mutation rate

pleiotropy

point mutation

polytene chromosome

recombinant genotype

recombinant phenotype

recombination analysis

recombination frequency

wild-type allele

QUESTIONS AND PROBLEMS

1. Mendel crossed true-breeding peas producing round, yellow seeds (*RRYY*) with peas producing wrinkled, green seeds (*rryy*). All the F_1 seeds were round and yellow. What were the genotypes of the gametes in the F_1 flowers? Were there recombinant genotypes? Did the recombinant types represent 50 percent of the gametes and show independent assortment? Or did they represent less than 50 percent and show linkage of the two genes? If alleles for two genes assort independently (are unlinked), how could you tell if the two genes are on the same or different chromosomes?

2. Are some pea plants male and others female or are both egg and sperm produced on a single plant? Would you expect pea plants to have sex chromosomes like fruit flies or humans? Many of the first mutations isolated by geneticists were sex-linked. Why would this have been true? Most of the mutant human genes that were first assigned to a specific chromosome were on the X chromosome. Why would mutant genes appear on this chromosome rather than on one of the other 22 chromosomes?

3. From which parent did Gregor Mendel inherit his X chromosome? Mendel's mother was a normal female with two X chromosomes, one from each parent. Her parents inherited them from their parents. Which of her four grandparents could *not* have contributed either of her X chromosomes? Which of Gregor's eight great grandparents could *not* have been the source of his X chromosome?

4. Diagram the process of meiosis using two pairs of chromosomes and three pairs of alleles.

$$\begin{array}{ccc} a & b & c \\ \vdash\!\!\!-\!\!\!\!-\!\!\!\!-\!\!\!\!-\!\!\!+ & & \end{array} \quad \begin{array}{ccc} A & B & C \\ \vdash\!\!\!-\!\!\!\!-\!\!\!\!-\!\!\!\!-\!\!\!+ & & \end{array}$$

Show how the chromosomes assort independently. And show how recombination may occur between two linked genes.

5. What are homologous chromosomes? In what sense are the X and Y chromosomes homologous? In what sense are they not homologous?

6. Human ABO blood types are determined by three alleles at one locus: I^A, I^B, and I^O. I^A and I^B are codominant. Explain what this means. I^O is recessive. Is it possible for a normal individual to carry all three alleles? Explain. What gametes will someone with type AB blood produce? What gametes will someone with type O blood produce? Is it possible for a normal gamete to contain more than one allele of a given gene?

7. A fairly common sex-linked trait in humans is green color blindness. The allele for normal vision is dominant. In the United States 6 percent of males are green color-blind, but fewer than 1 percent of females. Explain why there are more color-blind males than females. Does every color-blind male have a color-blind parent? Does every color-blind female?

8. There are alleles of several sex-linked genes in humans that produce lethal diseases. One of these is severe combined immune deficiency (SCID), which leaves infants with no immune system and subject to death by massive infection, usually within their first year of life, and invariably before adulthood. No female infant has ever been seen with this condition; explain why all victims of this disease are male.

9. Humans, mice, cats, and other animals have several dominant alleles that produce skeletal abnormalities (short, absent, or kinked tails, or short fingers and toes) in the heterozygotes. When two affected individuals mate, only 2/3 of their offspring are affected instead of the expected 3/4. In one respect these alleles are recessive. Explain.

10. Short-tailed sheep never breed true; they produce both short-tailed and long-tailed lambs in a 2:1 ratio. Matings of short-tailed sheep with long-tailed sheep produce equal numbers of short-tailed and long-tailed lambs. The short-tail trait is due to a mutant allele at one locus. Which allele is dominant? What genotypes are possible for a long-tailed sheep? Short-tailed? What genotypes and phenotypes would you expect to result from the two types of crosses mentioned above?

11. In *Drosophila*, the normal brick-red eye color is formed by two pigments, bright red and brown. A brown-eyed mutant fails to make the bright-red pigment, and a bright-red-eyed mutant fails to make the brown pigment. A mutation causing brown eyes, *bw*, occurs on the second chromosome. A mutation causing bright red eyes, called scarlet, *st*, occurs on the third chromosome. If brown females are mated to scarlet males, what will be the frequencies of the phenotype and genotypes of: (a) the F_1, and (b) the F_2?

12. If the map distance between *a-b* is 7 and *b-c* is 46, what frequency of recombinant individuals would be expected between *a-b*, *b-c*, and *a-c*? If the only available information is the distance between *a-c*, what could be concluded about linkage between *a* and *c*?

The progeny from a mating of parents *AB/ab* and *ab/ab* produce the following offspring: *AB* 96 *ab* 84 *Ab* 8 *aB* 12. What is the map distance between *a* and *b*?

13. A Leghorn chicken with a large single red comb is homozygous for recessive alleles for two unlinked recessive genes (*aabb*). It is mated with a hen, and the F_1 progeny are mated. The F_2 progeny display four comb types (walnut, rose, pea, and single, in a ratio of 9:3:3:1). Show the probable genotypes of each type of comb type and include the dominant alleles. Use a Punnett square.

14. Consider the following table of sex chromosome constitutions.

	XY	X	XX	XXY
Drosophila	male fertile	male sterile	female fertile	female fertile
People	male fertile	female sterile	female fertile	male sterile

What is the main function of the Y chromosome in *Drosophila*? In humans?

SUGGESTED READINGS

AYALA, F. J., and J. A. KIGER, JR. *Modern Genetics.* Menlo Park, Calif.: Benjamin/Cummings, 1980.

This is the standard of excellence for genetics texts. Easy to follow and covers all the major concepts.

ROBINSON, R. *Genetics for Cat Breeders.* New York: Pergamon Press, 1977.

This is a book that nonbiologists can understand, yet it gives a good picture of the complexities of cat genetics.

STRICKBERGER, M. W. *Genetics.* 2d ed. New York: Macmillan, 1976.

Some geneticists feel that this is the clearest account of Mendelian genetics and its aftermath as described in this chapter.

STURTEVANT, A. H. *A History of Genetics.* New York: Harper & Row, 1965.

The story of genetics from Mendel to molecules, written in a fascinating manner.

12

DISCOVERING THE CHEMICAL NATURE OF THE GENE

The fundamental biological invariant is DNA.
Jacques Monod
Chance and Necessity (1970)

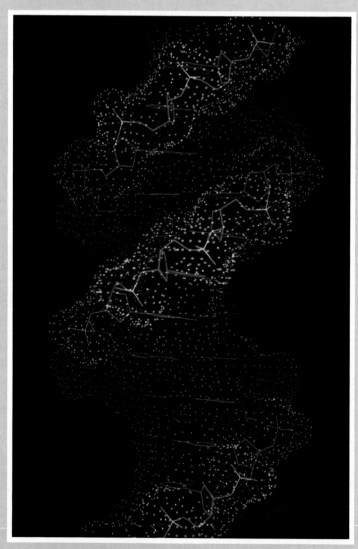

DNA, life's hereditary material, as seen by a computer.

The discovery of the structure and activity of genes was one of the greatest intellectual adventures of modern science. It spanned a century, involved dozens of biologists on two continents, and focused on experimental subjects ranging from blood cells to bacteria and from bread molds to cardboard cutouts. The result of this protracted and far-reaching scientific exploration was no less than an understanding of the universal code of life—the master molecules that carry and pass along complex instructions for the physical characteristics and activities of all living things.

This search involved two avenues of inquiry pursued by two groups of researchers: geneticists and biochemists. For decades, they worked on what first appeared to be entirely unrelated questions and were, for the most part, unaware and uninterested in each other's findings in this area. Not surprisingly, they have been compared to miners digging for the same vein of gold from opposite sides of a mountain.

Geneticists were "tunneling" toward the mystery of gene activity. From the kinds of studies we discussed in Chapters 10 and 11, geneticists knew that genes reside on chromosomes and are copied precisely as part of chromosome duplication during cell division. They knew that these microscopic bits of chromosomal material somehow bring about the phenotypes of organisms—the colors of flowers and eyes, the shapes of pea pods and noses, the biochemical characteristics of cells. Finally, geneticists knew that genes occasionally change, or mutate, in such a way that copies of the altered form are inherited during subsequent cell divisions and new phenotypes arise in the cells or organisms. What geneticists did not know was *how genes act* to bring about these effects and whether the physical structure of genes held the answer. For more than four decades, they delved deeper and deeper toward a solution.

During the same decades, biochemists were studying a class of biological molecules that had been discovered in the nucleus of cells—nucleic acids—and were trying to determine their properties and functions in living things. Gradually, the two sets of "miners" came to realize that they might be heading for the same pay dirt. In 1953, with the work of James Watson and Francis Crick, the two lines of inquiry finally converged: the structure of DNA, which these young scientists unraveled, explained the chemical nature of genes and suggested how genes are replicated and function.

We already have discussed the nature of DNA, chromatin, and chromosomes in some detail in Chapters 3, 6, and 9. In this chapter, we go back to a time well before those details were known and genes were associated with DNA, and retrace the great adventure for ourselves. In the first section, we will explore the genetic research that proved that each gene acts by directing the synthesis of one polypeptide. Then we will see how bio-

chemists identified DNA as the essential material in genes and deduced its structure—the double helix that bears the universal code of life.

GENES CODE FOR PARTICULAR PROTEINS

One fundamental question about the newly forged concept of the gene was: How do genes affect an organism's phenotype? The first scientist to address this question successfully was Sir Archibald Garrod, an English physician whose work revealed a connection between genes and enzymes.

In 1902, Garrod described in detail an inherited human disease called alkaptonuria. Victims of this disease excrete in their urine a compound, *homogentisic acid*, that turns dark when the fluid is exposed to air. This chemical can build up in the victim's eyes and cartilage, creating vision problems and severe arthritis and sometimes turning the ears and nose black. Garrod showed that the disease is inherited as a simple recessive trait according to Mendel's laws. What is more, he hypothesized that homogentisic acid builds up in alkaptonuria victims because the normal enzymatic reaction needed to break down the acid is blocked. Garrod suggested that because the sufferers are homozygous recessives, they lack a normal dominant allele needed for this enzymatic reaction to function.

In his writing, Garrod did not specifically state that a defective gene was responsible for the absence of a particular enzyme, but this concept was implied, and as such, Garrod's was the first suggestion that genes and enzymes are related. Moreover, Garrod was also the first scientist to link genes with specific chemical reactions in the body. Unfortunately, Garrod's thinking was so advanced that the significance of his ideas was not fully understood for thirty years. At the time, Mendel's work was just beginning to be appreciated, and there was insufficient understanding of genetics to explain how a recessive allele could block a normal metabolic pathway. This issue was not, in fact, addressed until the late 1930s, when geneticists began probing the connection between cellular biochemistry and mutated genes in a more direct way.

The One-Gene–One-Enzyme Hypothesis

Three decades after Garrod's insightful work, a team of geneticists began to reexplore the same terrain: the American George W. Beadle and the Russian-born Boris

Ephrussi sought to discover the specific chemical reactions directed by particular genes in *Drosophila*. They showed that wild-type eye color in fruit flies depends on a series of biosynthetic reactions, each of which is controlled by a different gene.

Now one step closer to uncovering the mysterious link between genotype and phenotype, Beadle and his colleague Edward L. Tatum went on to discover, in a series of classic experiments, an association between individual genes and individual enzymes. This was another important piece in the puzzle of how genes determine an organism's physical traits, but not the last one. It did, however, form the foundation for the work that ultimately solved the mystery, as we shall see.

In 1941, when Beadle began working with Tatum at Stanford University, the two decided to reverse the approach that had worked so well in early genetic studies: rather than look for particular chemical reactions directed by particular genes in fruit flies, they decided to switch to an organism whose enzymes and chemical reactions were already understood and try to induce mutations in genes that would block these familiar reactions. They reasoned that if an organism with a newly mutated gene suddenly also lacked an enzyme needed for one step in a biosynthetic pathway, then they could assume that each step has its own enzyme and, in turn, its own gene. Perhaps in this way they could explain how thousands of genes yield the thousands of proteins that govern an organism's phenotype. The idea that each gene codes for a particular enzyme was called the **one-gene–one-enzyme hypothesis.** It sounded quite reasonable at the time, but was it correct?

As their research model, Beadle and Tatum chose the red bread mold *Neurospora crassa*, a fungus that had been used for genetic studies. *Neurospora* has several advantages as a research subject. First, it is easy to maintain in the laboratory. It grows quickly from spores inside a test tube, producing dense reddish fungal tufts. The entire life cycle from spore to spore takes only ten days. Second, *Neurospora* is haploid throughout most of its life cycle; it has seven single chromosomes, not seven pairs, and thus recessive mutated alleles are not masked by dominant alleles on the homologous chromosome. Third, *Neurospora* can reproduce either sexually or asexually. Therefore, genetically identical generations can be produced during vegetative (asexual) reproduction. In this process, the fungus releases *conidia*, tiny spores that are dispersed and give rise to identical fungi. During sexual reproduction, a different type of spore, called an *ascospore*, forms in a fruiting body. This process is diagrammed in Figure 12-1. Ascospores are produced in tiny cases called *asci* (singular, *ascus*); there is no room for spores to shift around, and therefore the order in which the spores occur in the asci reflects the precise distribution of chromatids, bearing different alleles, during the first and second meiotic divisions of the nucleus. Thus the products of meiosis can be observed directly, without reliance on complicated statistical analysis. Finally, and perhaps most important, *Neurospora* can grow on a simple synthetic medium containing no complex biochemicals but just a few inorganic salts, sugar, and the vitamin biotin (a vitamin of the B complex). From this so-called *minimal medium*, the fungus can synthesize all the more complex compounds its cells require.

Beadle and Tatum's plan was to cause mutations in *Neurospora* that would eliminate a mutant individual's ability to synthesize a specific biochemical that it needed in order to grow. Such mutants, called *auxotrophs*, would be able to grow on an enriched medium containing all the amino acids, vitamins, nucleotides, complex sugars, and so on, but they would not be able to grow on the minimal medium because they would lack a necessary enzyme. If each mutated gene could be linked to abnormality in one enzyme, the direct link between genes and enzymes could be established.

To conduct the experiment, the researchers first irradiated conidia with x-rays to induce genetic mutation. Sexual fruiting bodies grown from these conidia then were crossed with cells from unirradiated (wild-type) conidia to produce diploid zygotes, which, in turn, underwent meiosis within the asci and produced haploid ascospores. Beadle and Tatum cultured the ascospores on a rich medium. Those that did not grow contained lethal mutations. Those that grew on the rich medium were retested in the kind of minimal medium that supports normal *Neurospora*. Those that could not grow on the minimal medium clearly had mutated genes that interfered with some metabolic pathway.

The next task was to determine *which* of the many biochemicals in a cell a particular mutant was unable to make. Each mutant was tested in a minimal medium with *one* specific complex molecule added to it—an amino acid, or a sugar, or a vitamin, and so on—until a wide range of substances had been tested (Figure 12-2). Most tests were unsuccessful. But if the spore grew, the investigators, by chance, had supplied the nutrient that was missing due to the mutated allele. In this way, Beadle and Tatum isolated a number of mutant strains, each of which needed a single additional nutrient in order to grow on a minimal medium.

Now the researchers were in a position to ask whether the nutritional deficiencies were due to single mutant alleles and, more specifically, to the action of genes on individual enzymes. To find out, they crossed the mutant strains with normal *Neurospora*, allowed the F_1 progeny to produce ascospores, and cultured all the spores produced within one ascus. If four spores were always normal and four were always mutant, this 1:1 ratio would show that only a single allele had changed.

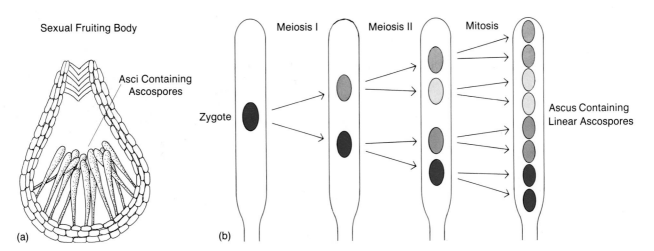

Sexual Fruiting Body

Asci Containing
Ascospores

Zygote

Meiosis I

Meiosis II

Mitosis

Ascus Containing
Linear Ascospores

(a)

(b)

Figure 12-1   UNIQUE GENETIC SUBJECTS: ASCOSPORES OF THE RED BREAD MOLD, *NEUROSPORA CRASSA.*

(a) The sexual fruiting body of *Neurospora* contains spore cases (asci), within which ascospores develop. (b) Ascospores are produced. *Neurospora* is ideally suited for genetic studies because the spores are arranged within the spore case in a linear order reflecting their meiotic origin, an order that is maintained during subsequent mitosis. Because of this unique characteristic, the genetic make-up of each zygote can be deduced directly without statistical analysis. If color genes are contributed by both parental nuclei to the diploid zygote nucleus, those genes might segregate during meiosis as shown by the blue and gray shading. (c) This squashed fruiting body shows the eight spores in each case. Note the light and dark spores, reflecting the meiotic distribution of alleles that led to those colors.

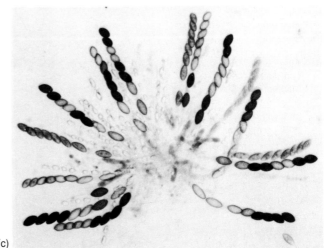

(c)

The results were consistent with Mendel's first law: the mutated allele segregated from the normal one at meiosis, producing a 1:1 ratio of spores within each ascus (Figure 12-3).

Beadle and Tatum's experiment proved that each mutation interferes with the pathway for a specific essential nutrient. Further studies showed that, indeed, each mutated gene affects a single enzyme. The one-gene–one-enzyme hypothesis appeared to be correct—at least for a time. For their important contribution to genetics, Beadle and Tatum were awarded the Nobel Prize in 1958.

In the years since the famous *Neurospora* studies, many such analyses of specific genes and enzymes have been carried out in a range of organisms. This work has helped biologists define the precise steps in dozens of complex metabolic pathways. Some researchers found that one nutrient can correct a defect in more than one mutated gene. This is not a contradiction of the one-gene–one-enzyme hypothesis, but an indication that many cellular substances are formed by a series of reactions, catalyzed by enzymes, called a metabolic pathway (Chapter 4). That is, enzyme 1 catalyzes the reaction $A \rightarrow B$, and enzyme 2 catalyzes the reaction $B \rightarrow C$. Normal, wild-type alleles for both enzymes are needed to produce compound C, whereas mutations in *either* allele will cause the mutant phenotype—that is, "no C." A mutant for enzyme 1 can be "cured" phenotypically by adding either compound B or compound C to the medium, but a mutant for enzyme 2 can be cured only by adding compound C (the end point) to the medium.

Complementation tests have also been used to study pathways. For instance, in *Neurospora*, two nuclei bearing two mutant alleles that affect the same trait can be introduced into the same cytoplasm. If the mutant phenotype persists, the two mutations must affect one gene that codes for an enzyme that catalyzes a step in a biosynthetic pathway; if the wild-type phenotype reappears, researchers can conclude that the mutations affect different genes in the pathway (the normal allele of one

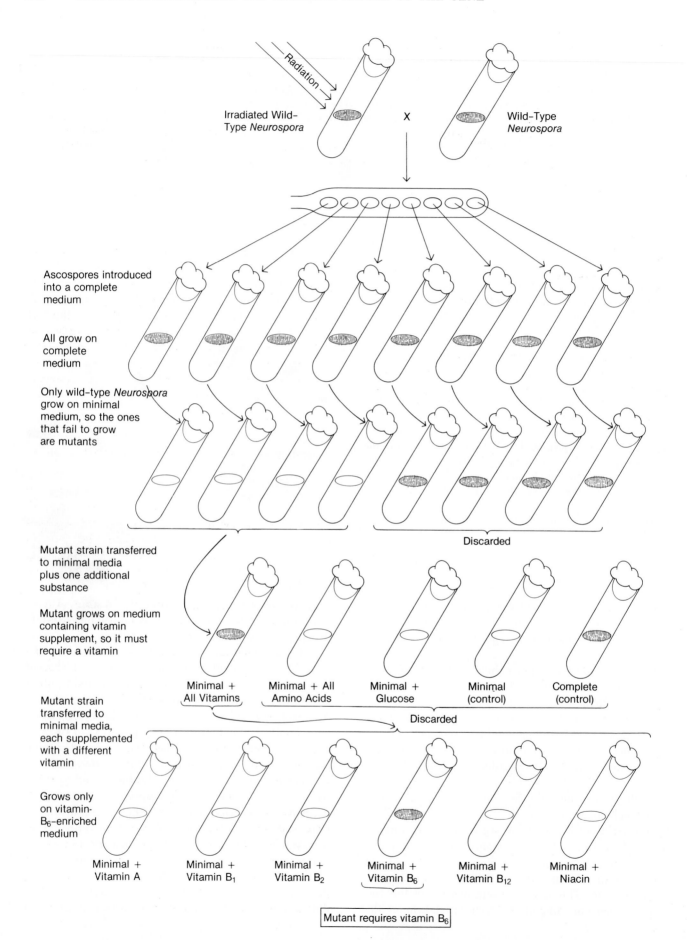

Irradiated Wild-Type *Neurospora*

X

Wild-Type *Neurospora*

Radiation

Ascospores introduced into a complete medium

All grow on complete medium

Only wild-type *Neurospora* grow on minimal medium, so the ones that fail to grow are mutants

Discarded

Mutant strain transferred to minimal media plus one additional substance

Mutant grows on medium containing vitamin supplement, so it must require a vitamin

| Minimal + All Vitamins | Minimal + All Amino Acids | Minimal + Glucose | Minimal (control) | Complete (control) |

Discarded

Mutant strain transferred to minimal media, each supplemented with a different vitamin

Grows only on vitamin-B$_6$-enriched medium

| Minimal + Vitamin A | Minimal + Vitamin B$_1$ | Minimal + Vitamin B$_2$ | Minimal + Vitamin B$_6$ | Minimal + Vitamin B$_{12}$ | Minimal + Niacin |

Mutant requires vitamin B$_6$

Figure 12-2   DETECTING *NEUROSPORA* MUTANTS.

Beadle and Tatum invented an ingenious method for detecting nutritional mutants in the red bread mold. Wild-type *Neurospora* ordinarily grows on a minimal medium, but nutritional mutants lack the ability to synthesize one or more nutrients. Beadle and Tatum irradiated wild-type *Neurospora* to induce mutations, and then mated the irradiated culture with an untreated wild-type culture to produce haploid ascospores. They then introduced ascospores into a complete medium, where all grew. When samples of each colony were transferred to a minimal medium, only the wild-type *Neurospora* grew. Transferred to minimal media that contained varying additional substances, the mutant strain grew only on the medium containing a vitamin supplement. In the final stage of the experiment, Beadle and Tatum transferred the mutant strain to minimal media containing different vitamins. Their results showed the mutant strain could not synthesize vitamin B_6.

gene gives rise to one step in the pathway, and the normal allele of the other gene allows the second step to proceed).

Another discovery was that the mutation of one gene sometimes can alter two nutrient requirements in an organism. Suppose, for example, that

$$A \xrightarrow{\text{Enzyme 1}} B \text{ and } B \xrightarrow{\text{Enzyme 2}} C + D$$

If the mutation renders the cell unable to make A into B, neither C nor D will be made.

These extensions of Beadle and Tatum's work, plus the *Neurospora* research itself, were crucial to understanding gene activity. When they began in 1941 to build on the earlier work of Garrod and of Beadle and

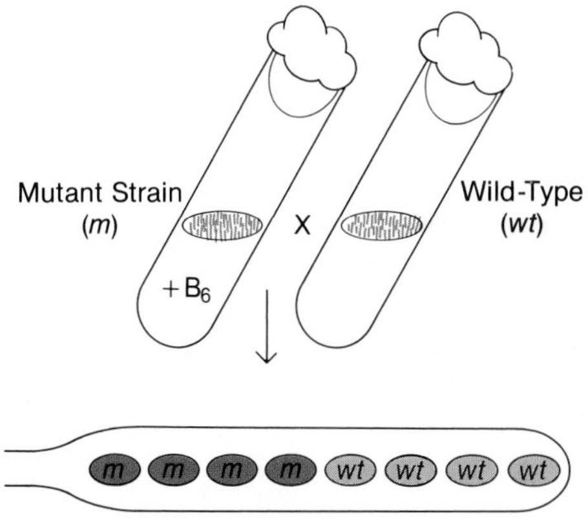

Ephrussi, Beadle and Tatum knew only that genes somehow are involved with biochemical reactions and hence presumably with metabolic pathways. But it was the *Neurospora* work that showed clearly that each gene affects one enzyme in a metabolic pathway; it was the best evidence to date of the way genes affect phenotype as well as an experimental means of studying enzyme-catalyzed metabolic pathways. Not long afterward, however, new experiments expanded the one-gene–one-enzyme concept into a more universally applicable understanding of gene function.

The One-Gene–One-Polypeptide Hypothesis

Beadle and Tatum's hypothesis about genes and enzymes raised a new question: Since enzymes are proteins, is there a link between genes and other types of proteins as well? In 1949, Nobel laureate Linus Pauling, a chemist, provided some insights that ultimately helped refine the one-gene–one-enzyme concept into the *one-gene–one-polypeptide hypothesis*, on which our current, well-documented understanding of the genotype–phenotype link is based.

Pauling studied the role of the blood protein hemoglobin in the disease sickle cell anemia. We saw in Chapter 11 that sickle cell anemia is inherited as a recessive Mendelian trait and that the red blood cells of its sufferers assume a sickle shape under conditions of low oxygen availability. Pauling wondered whether the abnormal allele could affect the hemoglobin molecules within the red blood cells in a way that caused those cells to sickle. He and his colleagues compared hemoglobin from the red blood cells of normal individuals, of homozygous recessive individuals who had a severe form of the disease, and of heterozygous individuals who did not show disease symptoms. The comparison was made by means of *electrophoresis*, a technique that separates molecules on the basis of their electric charge. If sickle cell hemoglobin is structurally different from normal hemoglobin, it might have a slightly different electric charge and hence migrate differently in an electric field.

Figure 12-3   PROVING THE LINK BETWEEN MUTANT GENES AND ENZYME PRODUCTION.

Beadle and Tatum crossed a *Neurospora* nutritional mutant that required vitamin B_6 with a wild-type and found that a single mutated allele was responsible for the nutritional deficiency. After the cross, they cultured the eight spores produced by the F_1 progeny within a single ascus. The 1:1 ratio of mutated to normal spores showed that only one allele had changed.

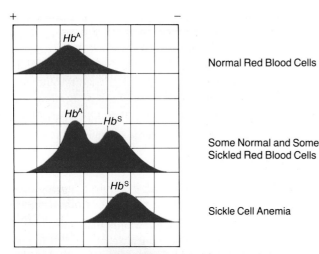

Figure 12-4   HEMOGLOBIN S FROM VICTIMS OF SICKLE CELL ANEMIA COMPARED TO NORMAL HEMOGLOBIN.
Hemoglobin from a normal person (Hb^A/Hb^A), a heterozygous person (Hb^A/Hb^S) with sickle cell trait, and a homozygous person (Hb^S/Hb^S) with sickle cell anemia moves differently in an electric field. During electrophoresis, normal hemoglobin moves closer to the positive pole, whereas the sickle cell hemoglobin moves closer to the negative pole. The reason is that the sickle cell mutation changes a negatively charged amino acid to an uncharged amino acid, causing the hemoglobin S to be less negatively charged and thus to be attracted less to the positive pole.

As shown in Figure 12-4, Pauling's results confirmed that the hemoglobin of recessive homozygotes is different from normal hemoglobin and that heterozygotes have a mixture of normal and mutant hemoglobin. This discovery suggested that the sickle cell gene causes a chemical alteration in the hemoglobin protein molecule that, in turn, alters the entire red blood cell. This finding also automatically extended the Beadle and Tatum hypothesis from "one gene–one enzyme" to "one gene–one protein," since hemoglobin is a protein, but not an enzyme.

In 1956, the British researcher Vernon Ingram, working at the Massachusetts Institute of Technology, modified this hypothesis. In additional hemoglobin studies, Ingram found that the mutant molecule in sickled red blood cells differs from normal hemoglobin by only one amino acid. Normal hemoglobin consists of four polypeptide chains: two identical alpha (α) chains and two identical beta (β) chains. Ingram discovered that in sickle cell hemoglobin, both α chains are normal, but both β chains are mutant. Each abnormal β chain contains the amino acid valine in the place where normal

chains have glutamic acid. It seems incredible that the substitution of 1 amino acid among the approximately 150 in the β chain can lead to such a serious and complex disease.

At first this finding seemed to strengthen the idea that one gene codes for one protein—in this case, hemoglobin. The inference was that a gene somehow specifies the sequence of amino acids in a whole protein. But if one gene codes for one protein, then one gene should code for both the α and β chains of the hemoglobin molecule. Furthermore, research revealed that a form of thalassemia, another hereditary blood disease due to a gene unlinked to the sickle cell anemia gene, is a result of a substituted amino acid in the α chain rather than in the β chain of hemoglobin. This finding suggested that production of the complete hemoglobin molecule must be under the direction of two genes, one for each type of chain. Therefore, each gene must control the synthesis of a polypeptide chain rather than of an entire protein (except, of course, when the entire protein *is* a polypeptide). By the late 1950s, then, the theory of gene activity had been refined to the **one-gene–one-polypeptide hypothesis:** each gene brings about the formation of one polypeptide chain (whether in an enzyme or in a structural protein) and in this way influences an organism's physical traits—blood type, hair texture, flower color, way of breaking down nutrients, and so on. This idea has been confirmed by thousands of studies, and now it is more a law than a hypothesis. We will see precisely *how* genes code for polypeptides in Chapter 13.

Summarizing the quest to understand gene activity, we can say that it began around 1900 and spanned fifty years before reaching its goal. This line of research worked "backward" from an organism's phenotype to the activity of its genes, and as we have just seen, it uncovered

1. The role of genes in general cellular metabolism (Garrod)

2. The role of genes in enzyme function (Beadle and Tatum)

3. The role of genes in determining protein structure (Pauling)

4. The role of genes in the amino-acid sequence of specific polypeptide chains (Ingram)

Even before geneticists had acquired this full understanding of gene activity, however, they had become aware of a second important line of research—the study of nucleic acids—which had its roots in the nineteenth century. The intersection of these separate efforts to study gene activity and nucleic-acid structure ushered in a new era in biological science.

THE SEARCH FOR THE CHEMISTRY AND MOLECULAR STRUCTURE OF NUCLEIC ACIDS

The study of nucleic acids had its start in 1869, when a young Swiss physician named Johann Friedrich Miescher began buying used bandages at a local surgical clinic in Tübingen, Germany. His idea was to study human cells, and, indeed, the discarded bandages provided a large supply of pus from which he obtained white blood cells. When Miescher treated the cells with pepsin, a digestive enzyme, to dissolve proteins, he found that the total amount of material within the nucleus shrank to some degree but that most of the contents remained largely intact during this enzymatic attack. Clearly, the remaining undissolved nuclear material was not protein; he called it "nuclein." By 1871, Miescher and other researchers had located the same substance in egg yolk, salmon sperm, yeast cells, and nuclei from bird and snake tissues. Miescher eventually suggested that nuclein was unlike a protein in behavior or chemical composition. By 1900, his work was well accepted, and the compound he had discovered was being called *nucleic acid*. Miescher had uncovered a fourth class of biological molecules, in addition to carbohydrates, proteins, and lipids.

A few years later, Robert Feulgen, a German chemist, developed a staining technique, which came to be known as *Feulgen staining*, that dyes nucleic acids a deep red. When he used the technique to stain whole cells, he found that the red dye became concentrated in the chromosomes within the nucleus. Thus Feulgen concluded that chromosomes are largely made up of the newly identified nucleic acid. Researchers in other labs utilized Feulgen staining with many types of cells and always found nucleic acid in the cell nucleus. A general feeling arose among some researchers that nucleic acid might carry hereditary information. However, next to nothing was known about the structure of nucleic acid, and many scientists questioned whether the substance could convey the complex and variable information that somehow produced the immense spectrum of cell types, body types, and species in the living world.

To confuse the issue, work in several labs had demonstrated that the nucleus also contains some protein, since the protein breaks down and the nucleus shrinks when it is treated with pepsin. Because much more was known about proteins than about nucleic acids, including the fact that the large protein molecules are built from about twenty types of amino acids and have great structural diversity, many biologists chose to believe that the genes themselves must be made of protein and are merely surrounded by nucleic acid in some configuration. They thought that these "protein genes" must somehow act as patterns for the formation of all the enzymes and structural proteins in an organism. However, a different view began to emerge in the late 1920s; it suggested that genes are made not of protein, but of nucleic acid.

Genes: Nucleic Acid, Not Protein

The first real insight into the structure of genes came from an unexpected result of a well-controlled experiment, as do so many scientific breakthroughs. In 1928, a British bacteriologist named Frederick Griffith was attempting to develop a vaccine against pneumococcus, the bacterium that causes pneumonia (vaccines are discussed in detail in Chapter 32). In the process, Griffith inadvertently discovered a substance that could change one form of bacterium into another, with a different set of traits—a strange but obvious sign of genetic activity. How did he do this?

At that time, before the advent of penicillin and other antibiotics, pneumonia claimed thousands of lives each year. A vaccine seemed like the major hope for preventing such deaths, and Griffith studied two strains of pneumococci to try to develop one. The *virulent*, or disease-causing, strain of pneumococci secretes a polysaccharide coat called a *capsule* around its cell wall. The capsule protects the bacterium from destruction by the host animal's immune system, allowing it to multiply and attack the host's cells. Virulent pneumococci form large, smooth, glistening colonies and are designated "S," for smooth (Figure 12-5). The other strain Griffith studied lacks a capsule and is nonvirulent. These harmless cells

Figure 12-5   COLONIES OF ROUGH AND SMOOTH PNEUMOCOCCI.

The nonvirulent R (rough) strain of pneumococcus, on the left, is compared with the virulent S (smooth) strain, on the right (both magnified about 1,640 times). The S strain secretes a capsule around its cell wall that protects it from a host animal's immune system.

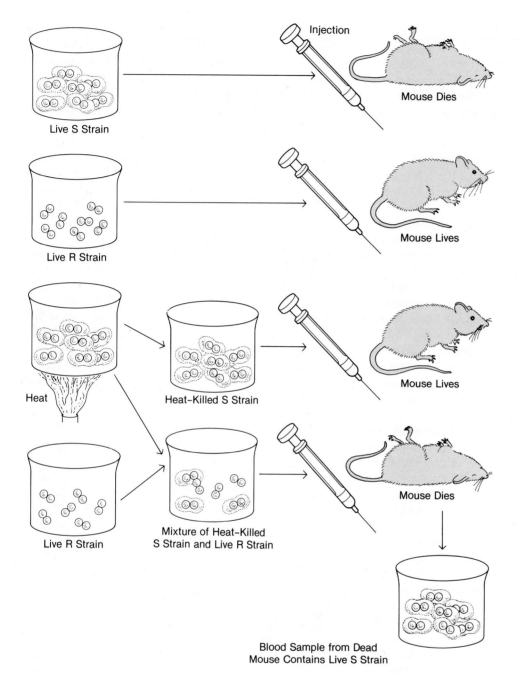

Figure 12-6   GRIFFITH'S EXPERIMENT WITH THE R AND S STRAINS OF PNEUMOCOCCI.
When a researcher injects mice with live S-strain bacteria, the rodents die, whereas mice injected with live R-strain bacteria live. This shows that the S strain is virulent. When the S strain is killed by heat and then is injected, the mouse lives. However, when the heat-killed S strain is mixed with the live R strain and injected, the mouse dies. A blood sample from the dead mouse shows that mixing the heat-killed S strain and the R strain transforms the live R strain into an S strain. Experiments years after Griffith's proved that the DNA of the heat-killed virulent strain was incorporated into the nonvirulent strain, rendering it virulent—and deadly.

form small, rough colonies and are designated "R," for rough.

Griffith infected mice with the two strains and observed three expected patterns, as illustrated in Figure 12-6. The mice injected with live S-type pneumococci died from pneumonia, while those injected with live R-type bacteria survived. Mice also lived if injected with S-strain pneumococci that had been killed—the dead bacteria were unable to cause the disease. Curiously enough, however, Griffith found an unexpected fourth pattern in his well-conceived controls: when injected with a mixture of dead S-strain and live R-strain bacteria, the mice died of pneumonia. The dead mice were found to be teeming with *live*, virulent S-strain pneumococci. How could this be? The live R-strain bacteria were harmless, and the dead S-strain bacteria certainly could not have come back to life! Griffith hypothesized that some material—perhaps hereditary material—from the dead S-strain cells must have transformed the harmless R-strain cells into the virulent S strain. But no one knew what kind of hereditary material could accomplish such a trick or how it might work.

During the next few years, two pieces of evidence answered these questions. When researchers repeated the pneumococcal experiments in vitro (in the test tube; *in vitro* is the Latin term for "in glass") rather than in mice (*in vivo*), they showed that the transformation of nonvirulent into virulent bacteria does not require the presence of the killed S-strain pneumococci themselves; simply an *extract* of the fluid surrounding the dead cells is sufficient to cause a transformation. Obviously, some chemical in the extract is responsible. In 1944, a group of scientists at Rockefeller University set out to identify the substance.

Oswald T. Avery, Colin MacLeod, and Maclyn McCarty took the extract from dead virulent bacteria and gradually removed one chemical compound after another—first the proteins, then the carbohydrates, next the lipids—each time testing the ability of the material to transform nonvirulent into virulent pneumococci. Finally, there was virtually nothing left in the extract but a fine, clear viscous thread that Avery could pick up on a glass stirring rod. When dissolved, this threadlike substance could transform R-strain into S-strain bacteria. Chemical analysis showed the viscous substance to be millions of molecules of a nucleic acid called *deoxyribonucleic acid*, or *DNA*.

Avery's team had isolated the very material—the clear, viscous DNA—that could change an organism with one set of physical traits into an organism with an entirely different set of physical traits. One could assume that such strong evidence would at last convince the biological community that genes are made of DNA. But it did not. A well-known biochemist named P. A. Levene had published results of research on the chemical composition of DNA that clouded Avery's findings, at least for a time. Levene had discovered that DNA contains four nitrogenous bases (which we will consider in more detail shortly) and that each of the four bases is attached to a sugar molecule and a phosphate group in a compound he called a *nucleotide*. Levene was correct about the bases and the structure of nucleotides. However, he made one erroneous assumption: he believed that the bases, which occur in roughly equal proportions, must be arranged in groups of four and that these groups of four must be arranged in a simple, repeating sequence along the length of the molecule.

Despite the careful work of Avery's team in isolating DNA as the transforming factor in pneumococci, biologists still could not accept that a molecule such as Levene described, with a chattering repetition of the same four chemical constituents, could possibly contain complex genetic information. Thus many biologists still thought that the twenty amino acids of proteins could serve as a richer information code for the thousands of proteins in organisms than could DNA.

The dispute over the composition of genes—proteins versus DNA—went on until the early 1950s, when a series of brilliant experiments solved the question permanently. Alfred Hershey and Martha Chase at the Carnegie Institution devised a clever means of testing the chemical nature of genes and established that the carrier of hereditary information is indeed DNA and not protein.

The Carnegie researchers chose for their experiments the common intestinal bacterium *Escherichia coli* and a certain class of bacterial viruses, or *bacteriophage* ("phage" for short), that specifically infects and destroys *E. coli* cells. Four critical pieces of information had been established about bacteriophage viruses, designated the T-even (T2, T4, etc.) viruses: (1) they have a distinctive shape, with a head, a tail, and tail fibers that could be easily identified by the electron microscope, then a new tool (Figure 12-7a); (2) they possess genes; (3) they are composed solely of protein and DNA; and (4) they attach to the outside of the *E. coli* cell, and although most of the bacteriophage body remains outside the cell (the head, tail, and tail fibers), within about twenty-five minutes, the bacterium bursts, releasing hundreds of complete, new bacteriophages. The assumption at the time was that the phages somehow send their genes into the cell to direct the production of new virus particles. Because phages consist only of protein and DNA, whichever was injected into the bacteria had to be the genetic material.

Hershey and Chase set out to determine which substance was inserted by bacteriophages into *E. coli* cells—not an easy task considering the submicroscopic realm in which they were exploring; a bacteriophage next to a bacterium is like a tadpole next to a walrus. But Hershey and Chase had one chemical clue; they knew that DNA contains phosphorus, whereas most proteins do not, and that most proteins contain sulfur, whereas DNA does not. Thus they could exploit radioactive isotopes of the two elements as tracers. (Such isotopes were one of the benefits stemming from the nuclear-science industry of World War II.) Hershey and Chase grew one set of *E. coli* cells on a culture medium containing a radioactive isotope of phosphorus (^{32}P) and another set of cells on radioactive sulfur (^{35}S). Bacteriophages growing inside the bacteria incorporated the ^{32}P into their DNA and the ^{35}S into their proteins. Thus the phages in one culture had ^{32}P-radioactively labeled DNA, whereas those in the other culture had ^{35}S-radioactively labeled proteins, and all were "hot"—emitting high-energy particles.

Once the infected cells had burst, the released radioactive bacteriophages were collected and washed free of unincorporated isotopes. The two groups of hot phages were then used to infect unlabeled *E. coli* cells in isotope-free media. Within a few minutes after mixing the phages and the bacteria, the phages had inserted their

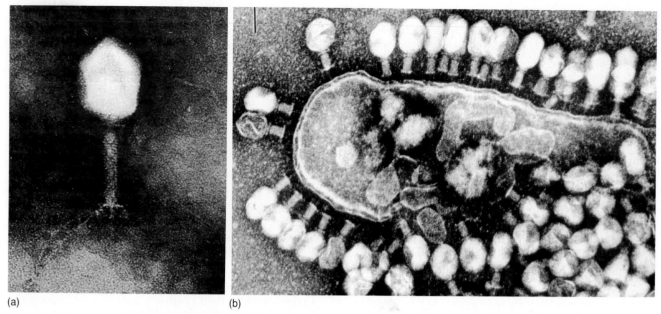

(a)   (b)

Figure 12-7   THE BACTERIOPHAGE VIRUS.
(a) This electron micrograph (magnified about 164,000 times) reveals the structure of a T-even bacteriophage. The head contains DNA and sits above the tail, with its visible tail fibers. DNA passes down the core of the tail to enter a bacterium.
(b) This electron micrograph (magnified about 95,000 times) captures T4 bacteriophages injecting DNA into an *E. coli* cell.

genes, leaving behind empty viral coats that clung to the *E. coli* walls. Hershey and Chase agitated the two cultures in blenders to knock off the empty viral particles and suspend them in the surrounding fluid. When they analyzed the contents of the blender containing phages originally labeled with ^{32}P, they found ^{32}P associated with the *E. coli* cells, not with the fluid containing the empty phage protein coats, showing that DNA had entered the cells (Figure 12-8). In the other blender, they found ^{35}S in the medium but not in the cells. Thus phage protein had not entered the bacteria. Because the researchers knew that phage genes had entered the *E. coli* cells and that DNA, not protein, also had entered, they concluded that genes must be composed of DNA, not protein—a conclusion that the scientific community finally accepted.

Chemical Composition of DNA Revealed

The Hershey–Chase experiment proved that genes are made of DNA, but not how the DNA subunits—or nucleotides—are arranged or how DNA can carry genetic information. Another study helped provide answers.

Recall that P. A. Levene had correctly identified the basic unit of nucleic-acid structure, the nucleotide. As we saw in Chapter 3, each DNA nucleotide contains a five-carbon sugar called *deoxyribose*, which is attached

to one of the four bases that Levene identified: **adenine (A)**, **guanine (G)**, **cytosine (C)**, or **thymine (T)**. Each **base** is a ring structure composed of carbon and nitrogen, as can be seen in Figure 12-9. Adenine and guanine, double-ring structures, are called **purines**; thymine and cytosine, single-ring structures, are called **pyrimidines.** The molecule made up of a base plus a sugar is called a **nucleoside**; *nucleotides* have an additional component, a phosphate group. In each long molecule of DNA, a phosphate group links the five-carbon sugar of one nucleoside to the five-carbon sugar of the next nucleoside in the chain. This phosphate bonding creates a sugar–phosphate backbone, with the nitrogenous bases protruding (Figure 12-9).

Between the time of Avery's work showing that DNA can transform R-strain into S-strain pneumococci and Hershey and Chase's blender experiments proving that genes are made of DNA, the American biochemist Erwin Chargaff performed new chemical studies of DNA that underscored the significance of the Hershey–Chase experiments. Chargaff found that the four bases do not occur in identical proportions in the DNA of different species. The DNA in human cells contains about 30 percent adenine, 30 percent thymine, 20 percent guanine, and 20 percent cytosine. The DNA in sea-urchin cells, however, contains about 32 percent A, 32 percent T, 17.5 percent G, and 17.5 percent C. Chargaff immediately noticed that the percentages of adenine and thymine are always equal, as are the percentages of guanine

Is the genetic material protein? . . .

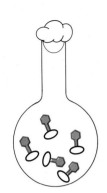
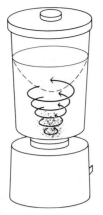

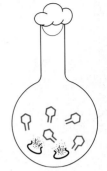

Bacteria infected with bacteriophage that have been labled with ^{35}S, a constituent of proteins

Suspension churned in blender to separate virus particles from bacteria

Suspension centrifuged; ^{35}S does not appear in pellet with bacteria but stays with virus particles

New generation of virus particles produced contains no ^{35}S

. . . or is the genetic material DNA?

The genetic material is DNA

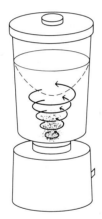

Bacteria infected with bacteriophage that have been labeled with ^{32}P, a constituent of DNA

Suspension churned in blender to separate virus particles from bacteria

Suspension centrifuged; most ^{32}P appears in pellet with bacteria, not in fluid with virus particles

New generation of virus particles produced contains ^{32}P

Figure 12-8   THE HERSHEY–CHASE BLENDER EXPERIMENTS: PROOF THAT THE GENETIC MATERIAL IS DNA, NOT PROTEIN.
Hershey and Chase employed bacteriophages, which consist only of DNA with a protein coat. Sulfur is a constituent of many proteins, but not of DNA, and phosphorus is a constituent of DNA, but not of most proteins. The researchers used radioactively labeled sulfur and phosphorus to discover whether protein or DNA was incorporated into the bacteria, and then into the next generation of virus. The parts of this figure show the steps of the experiment. Since the new generation of bacteriophages contained radioactive phosphorus, the genetic material must be DNA.

and cytosine, even though the proportions vary among organisms. The constant and variable features of DNA became known as **Chargaff's rules**: first, [A] = [T] and [G] = [C]; second, the ratio ([A] + [T])/([G] + [C]) is constant within a species but varies among species.

Chargaff's study superseded Levene's: clearly, DNA does not contain a simple repetition of the four bases but, instead, different percentages of the bases in the

various species of organisms, and perhaps different arrangements of those bases. These differences were highly suggestive, considering the chain of evidence about nucleic acids that had been forged in the decades before Chargaff formulated his rules:

1. Miescher discovered nucleic acids in the nuclei of many cell types.

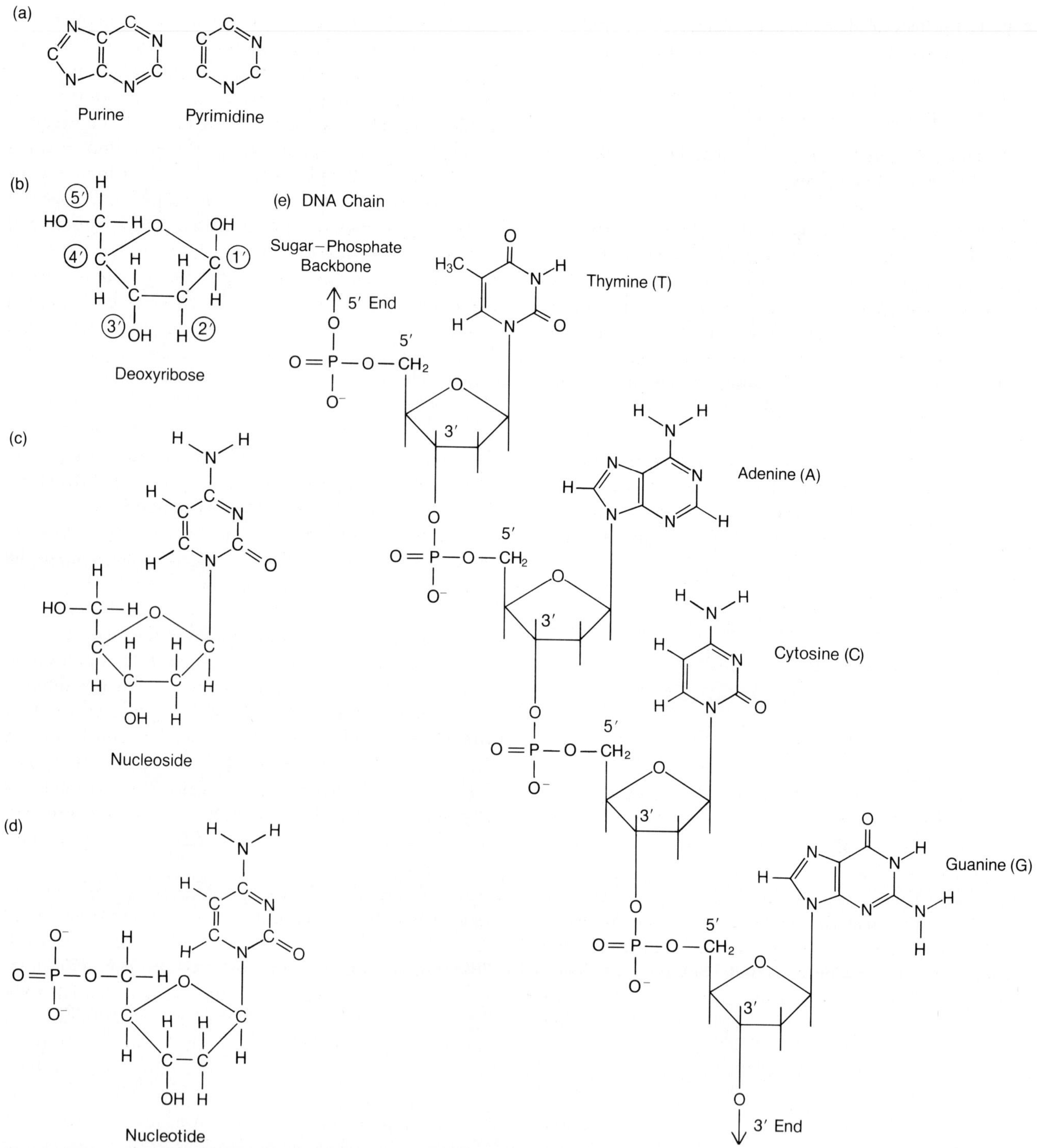

Figure 12-9   THE STRUCTURE OF DNA AND ITS SUBUNITS.
Each of the nitrogenous bases (a), in the shape of purine (adenine and guanine) and pyrimidine (thymine and cytosine) rings, is joined to deoxyribose sugar (b). (The carbons are numbered 1' to 5' by convention.) The resulting structure is called a nucleoside (c). The nucleoside plus a phosphate group make up a nucleotide (d). A DNA molecule is built from chains of nucleotides (e). A phosphate group links the 5' carbon on the sugar of one nucleoside to the 3' carbon on the sugar of the next nucleoside in the chain. This arrangement creates a phosphate–sugar backbone with the nitrogenous bases protruding as side groups.

2. Griffith found that a specific biological material can transform harmless bacteria into virulent ones.

3. Avery, MacLeod, and McCarty ascertained that Griffith's substance is DNA.

4. Hershey and Chase proved that DNA makes up the genes themselves.

With Chargaff's additional evidence that the structure of DNA might be varied enough to account for the genetic variability among species, the stage was set at last to understand the genotype–phenotype connection. One set of biologists had discovered that each gene brings about the formation of one polypeptide chain. And now another set of researchers had revealed that genes are made of DNA. Clearly, DNA somehow acts to create the polypeptide chains of proteins, which, in turn, act structurally or enzymatically to express an organism's specific phenotype. But how? Biologists had only to discover DNA's exact structure and mode of action to truly understand the mechanisms of inheritance at the molecular level.

IN SEARCH OF THE MOLECULAR STRUCTURE OF DNA

A race to discover the structure of DNA began in several research labs around the world once Chargaff's work and Hershey and Chase's experiments revealed the molecule's true significance. The convergence of the two lines of research on the nature of genes finally occurred with the work of two young researchers, the American biochemist James D. Watson and an English physicist turned molecular biologist named Francis Crick. Their discovery of the molecular structure of DNA led directly to an understanding of how genes carry hereditary information.

Watson and Crick met at the Cavendish Laboratory at Cambridge University in England. Both shared a fascination with DNA and a strong motivation to be the first to disclose the structure of nucleic acid. Taking a theoretical route, rather than designing and performing experiments, they carefully analyzed all the existing evidence on DNA structure and built models of cardboard and metal that could account for the observable facts. As we shall see, their approach paid off.

Watson and Crick concentrated on four important pieces of evidence. First, Levene's and Chargaff's work indicated that DNA molecules are long, thin polymers containing four types of nucleotides linked by phosphate bonds. Second, Chargaff's first rule states that [A] = [T]

and [G] = [C]. Third, physical evidence from other laboratories affirmed that purified DNA forms a viscous solution like egg white that, when moderately heated, becomes nonviscous (more watery). It was known that moderate heating is not sufficient to break the covalent bonds in the sugar–phosphate backbone. Since heating clearly changes the physical nature of DNA from viscous to nonviscous, a set of weaker chemical bonds that can be broken by moderate heating also must be present and necessary to maintain the normal structure of DNA. Finally, Linus Pauling had found that polypeptide chains are often held in the shape of an α-helix by hydrogen bonds, which can be broken by moderate heating, and he had suggested that DNA might conform to the same pattern.

Further evidence that DNA has a helical configuration came from photographs of crystallized DNA taken by Maurice Wilkins and by Rosalind Franklin at King's College, London, using a technique called **x-ray diffraction.** With this technique, researchers can determine the spatial arrangement of atoms in molecules—the lengths and angles of chemical bonds, and the distances atoms lie from one another throughout the molecule. X-rays are bounced off the parallel planes of atoms in a crystal, and a photographic image is generated and analyzed. However, when the technique is used on giant, complex molecules like DNA, the photographic evidence is difficult to decipher. Nevertheless, Franklin wondered whether the shadows and markings that can be seen in Figure 12-10a mean that DNA is a helix, with the phosphate backbone on the outside and with a uniform diameter of about 2.0 nm. On the basis of the photos, she also estimated possible distances between adjacent turns of a hypothetical helix.

Armed with the various types of evidence, Watson and Crick began trying to build three-dimensional models of DNA that would incorporate all the information at hand. They appreciated the power of using precise scale models to solve a complex, three-dimensional spatial problem, because that is what Pauling had done to discover the α-helical structure of polypeptide chains. "In place of paper and pencil," they later wrote, "the key tools were a set of molecular models superficially resembling the toys of preschool children." Watson and Crick made models of the individual nucleotides, taking into account the size of atoms, the lengths and angles of bonds, and so on. But the work was tedious and frustrating—there were dozens of possible ways to fit together the bases, sugars, and phosphates, and at first, none seemed to agree with all the observed data and criteria.

Wilkins's and Franklin's measurements of the x-ray diffraction photographs were critical in implying two repeating features of DNA, one occurring every 3.4 nm and the other, every 0.34 nm. Watson and Crick surmised that the shorter measure, 0.34 nm, might repre-

(a)

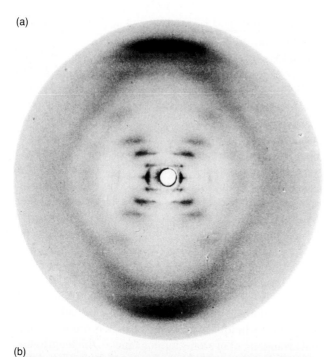

(b)

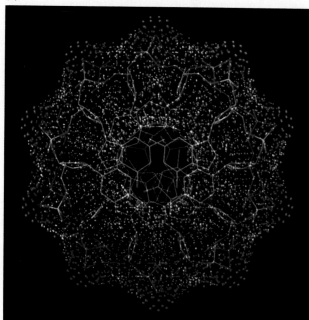

Figure 12-10   X-RAY DIFFRACTION PATTERN FOR A FIBER OF DNA.

(a) Photos of x-ray diffraction patterns such as this one produced by DNA fibers were interpreted to mean that DNA is a helix, with the phosphate backbone on the outside and the nitrogenous bases on the inside. (b) Precise measurements of bond distances in DNA in varying states, plus computer modeling techniques, allow creation of computer-generated views of the genetic material.

Figure 12-11   THE DISCOVERERS OF DNA STRUCTURE. James Watson (left) and Francis Crick in 1953 with one of the first models of DNA.

sent the distances between the stacked bases of the nucleotides, so they tried arranging the molecules in their cardboard working models into the suspected helical shape, with turns of 3.4 nm long and 2.0 nm wide. Imagine the flash of excitement and elation the young James Watson must have felt when one morning he suddenly saw that the cardboard pair of adenine and thymine has exactly the same shape as the pair of guanine and cytosine. That was the key! The two sets of base pairs could extend across the inside of a **double helix**, much like rungs on a ladder. It was obvious, then, that DNA is not a single helix, as are many proteins, but *two* chains entwined about each other in a double helix.

At once, Watson and Crick realized that Franklin had been correct: the phosphates and sugars are arranged like the outer rails on a ladder (Figure 12-11). As the first complete molecular model was pieced together, it became clear that the nitrogenous bases fit perfectly in place between the "rails." Moreover, because of the way the bases fit together, the double-helical "ladder" is twisted to yield a helix with a 3.4 nm repeat.

In the base pairs, if A was linked to T and G was linked to C by weak hydrogen bonds, the viscous–nonviscous behavior observed in heated DNA solutions could be explained. At low temperatures, the hydrogen bonds would cause the double helix to be stiff, and so the solution would be viscous. But at higher temperatures, the hydrogen bonds would be broken, and the two helices would come apart; as a result, the DNA would be less stiff, and the solution, too, would be less viscous.

Watson and Crick's models showed clearly that a pair of double-ring purine bases sitting side by side is too large for the 2.0 nm distance between rails observed in the x-ray diffraction photos. At the same time, the two single-ring pyrimidines are too small to fill the space when placed side by side. The only combination that fits is one purine hydrogen-bonded to one pyrimidine. However, Watson and Crick found that both purine–pyrimidine combinations are not possible: adenine–cytosine and thymine–guanine do not fit together correctly to allow hydrogen-bonding and still yield the 2.0 nm diameter measured by Franklin. The only combinations of paired bases that work are adenine–thymine and guanine–cytosine (Figure 12-12). This regular pairing coincides perfectly with Chargaff's first rule: [A] = [T] and [G] = [C]. With marvelous insight, Watson and Crick had solved the mystery of the primary genetic molecule in all living organisms on Earth.

Another question next arose: If the base pairing is restricted to A–T and G–C, how can DNA carry a wide variety of genetic information? The answer, according to another of Watson and Crick's brilliant deductions, lies in the *sequence* of base pairs along the length of the double helix. The A–T and G–C base pairs can occur in *any* sequence along the molecule; thus the huge number of possible sequences can encode the genetic information for the huge variety of proteins and enzymes responsible for each organism's phenotype. On a gross level, in fact, the overall genetic differences among species can be reflected in the *ratio* of A + T to G + C. The number of paired nucleotide bases is extremely large in a long DNA molecule and is colossal in a species' full set of chromosomes. Humans, for example, have about 10 *billion* base pairs in the DNA of their forty-six chromosomes. For this reason, a seemingly infinite number of base-pair sequences are possible—just one of those orders characterizes each species' DNA. This immense

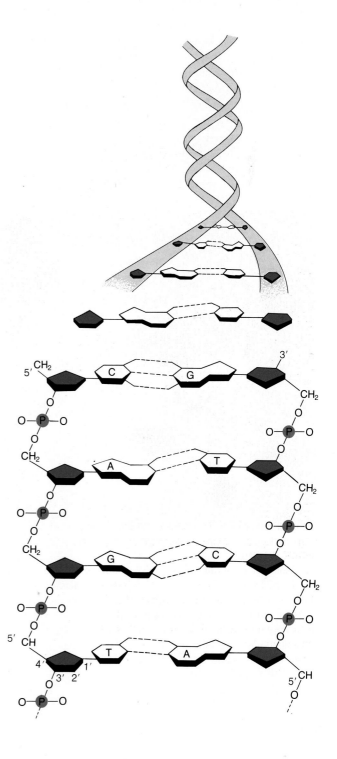

Figure 12-12   DNA DOUBLE HELIX: THE CODE OF LIFE. DNA consists of two sugar–phosphate backbones running in opposite orientations and arranged as a double helix, with the nitrogenous bases on the inside. These four bases are paired in just two possible combinations—adenine–thymine and cytosine–guanine. Here the DNA double helix is depicted as two ribbons winding around each other (top) and enlarged to show the base pairs as the rungs of a winding staircase between the helices (bottom). Notice that the two chains are arranged in opposite directions because the 5' end is "up" on one chain and is "down" on the other chain.

"ZIGZAG DNA": A CLUE TO GENE ACTIVITY

It has been more than thirty years since Watson and Crick pieced together the structure of DNA—a double helix that twists to the right. During that time, other researchers have turned up more than twenty variations that also twist to the right. They differ from the classic *B-form,* or *B-DNA,* which Watson and Crick described, in that the bases are (1) stacked more closely; (2) tilted at different angles; or (3) have a more tightly wound helix. Recently, a researcher discovered a variety of DNA in which the backbone zigzags, and the entire molecule twists to the *left,* not to the right (Figure A). The unusual structure of this new form, called *Z-DNA,* may help explain how genes function.

Alexander Rich and his colleagues at the Massachusetts Institute of Technology believe that Z-DNA occurs in nature and is not an artifact of DNA studies performed in the laboratory. They confirmed this conclusion with evidence of Z-DNA in normal fruit flies. Rich hypothesizes that regions of Z-DNA alternate with regions of B-DNA and that the Z-form can switch configuration to become the B-form when certain chemical changes occur causing Z-DNA to become unstable. He suggests that the genes in the "switched" regions might then be turned on and begin to be translated so that protein can be synthesized.

If this is true, researchers could locate Z-DNA regions in chromosomes and study their conversion to B-DNA to learn more about gene activity. In addition, they might be able to prevent deleterious genes from operating by preventing the chemical changes that precede the switch from the Z-form to the B-form. Although Rich's team and other researchers have a great deal more to learn about Z-DNA, the discovery provides a potential tool for studying genes and DNA in action.

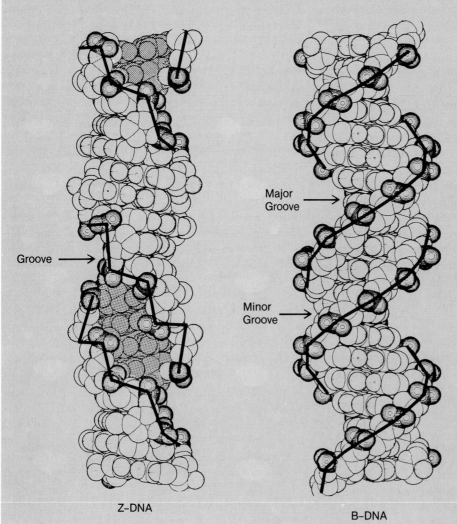

Groove →

Major Groove →

Minor Groove →

Figure A
Left-spiraling Z-DNA is compared to the classic right-spiraling B-form. The Z configuration favors unwinding of the DNA to permit transcription.

Z-DNA

B-DNA

sequence diversity can clearly explain the huge variety of organisms on Earth.

The model proposed by Watson and Crick was tremendously appealing to other biologists, since it accounts both for the observed chemical and physical properties of DNA and for the information-carrying, replicational, and mutational properties of genes (as we shall soon see). In Crick's words, there is an "intrinsic beauty of the DNA double helix"—it is a "molecule which has style." In 1962, nine years after constructing their DNA model, Watson and Crick shared the Nobel Prize for this work with Wilkins, whose x-ray diffraction studies, as well as those of Franklin, had provided the critical measurements. (Franklin died in 1958, at the age of 37, before the prize was awarded.)

At last, the "vein of gold" had been reached from both sides of the "mountain": geneticists and biochemists had a working hypothesis for the structure and function of genes that could be applied to numerous other biological problems, as we shall see. The discovery of DNA structure launched the new discipline of molecular genetics, which has developed into one of the most important and exciting fields in all of science.

HOW DNA REPLICATES

Like all the best theories of science, Watson and Crick's model not only looked back to explain available evidence, but also looked ahead to predict the results of experiments yet to be performed. As we have seen, the double-helical ladder with base-pair rungs accounts for many of DNA's chemical and physical properties, and it explains how nucleic acid can carry complex genetic information. However, genes have two additional properties: (1) they normally replicate quite faithfully, and (2) they sometimes become mutated and then pass along this altered message to subsequent generations. Could the Watson–Crick model account for these two properties of genes as well?

In their theoretical paper, published in 1953, Watson and Crick predicted a "copying mechanism" by which the double helix might faithfully replicate itself. If the hydrogen bonds that connect adenine to thymine and guanine to cytosine were broken, the DNA molecule would "unzip" down the middle. Because base pairs are complementary, these single strands of DNA, each with attached single bases, could serve as *templates*, or patterns, on which new complementary strands could form. If free nucleotides existed in the surrounding medium, they would be linked by means of hydrogen bonds to the complementary bases on the two chains. Free adenine nucleotides would bond to thymines on the single

strand, and free thymines would bond to adenines on the strand. Conversely, free cytosines would link to guanines on the template strands, and vice versa. The sequence of base pairs on a forming chain would be complementary to that on the chain serving as a template: A's opposite T's, G's opposite C's, and so on. And the forming chain, when complete, would have the exact sequence as the original strands that had "unzipped." Thus, the "unzipping" of a double-helical DNA molecule would result in the formation of two daughter molecules, each with a "parent" strand, and each with a new, "daughter" strand. Most important, the two new double-stranded molecules would retain the original sequence of nucleotide base pairs found in the parent molecule, and thereby the encoded information in the intact double helix is preserved from one generation to the next.

Implicit in this explanation is a means of understanding gene mutation. If an error in base pairing occurred during DNA replication, so that an incorrect (noncomplementary) nucleotide became inserted into the forming strand, the new nucleotide sequence would be different from the original. This, in turn, would change the encoded information and could be transmitted to new generations in later division and DNA replication cycles. Therefore, an alteration in the sequence of nucleotides in a DNA molecule induced by x-rays, chemicals, or simple chance mispairing of the nitrogenous bases could bring about a genetic mutation.

Is DNA Replication Semiconservative?

The mode of replication predicted by the Watson–Crick model, in which each strand of the "unzipped" molecule serves as a template for a new strand, can be called **semiconservative replication.** This term is derived from the notion that each newly formed DNA molecule will have one strand that is intact, or conserved from the original parent molecule, and one newly synthesized complementary strand, made on the conserved template (see Figure 12-13). This seemed like a very reasonable assumption to many biologists. However, another copying mechanism might be just as likely. First, DNA replication could be *conservative*—the double-stranded parent molecule might somehow replicate without unraveling and separating. Therefore, each new daughter molecule would have not one, but *two* new strands. Alternatively, replication could be *dispersive*—the double-stranded parent chains might break into pieces, with new segments filling in the gaps. In each case, the proportion of A to T and G to C would remain the same, and the genetic results would be similar. Which of the three models is correct?

Figure 12-13   THE SEMICONSERVATIVE REPLICATION OF DNA.
The "unzipping" of a double-stranded parent molecule results in the formation of two daughter molecules, each consisting of a parent template strand and a newly synthesized strand.

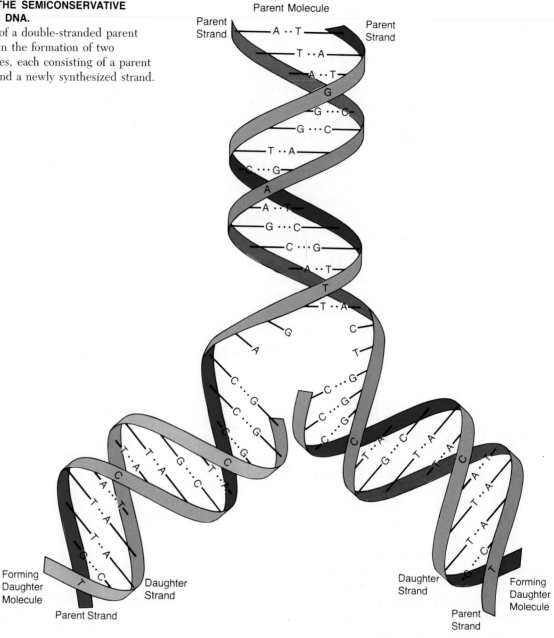

Two young scientists at the California Institute of Technology, Matthew Meselson and Franklin W. Stahl, designed a test to determine which of the three possibilities actually occurs, and they confirmed Watson and Crick's semiconservative theory. Their experiments were quite novel, and are worth considering in detail.

They began by growing twelve consecutive generations of *E. coli* on a growth medium containing a heavy isotope of nitrogen, ^{15}N. The word "heavy" derives from the extra neutron in the nucleus of each ^{15}N atom. By the twelfth generation, virtually all the normal nitrogen molecules (^{14}N) in the cells' proteins and nucleic acids

had been replaced by heavy nitrogen. DNA molecules containing ^{15}N can be distinguished from those containing ^{14}N on the basis of their densities in a system called a *cesium chloride* (CsCl) *density gradient*. CsCl solution in a tube, when spun in a high-speed centrifuge, will disperse so as to be densest near the bottom of the tube and progressively less dense toward the top. When heavy DNA containing ^{15}N is spun with the CsCl, it "bands" at a lower level in the tube than does the lighter, normal DNA with ^{14}N. The two kinds of DNA form visible bands, detectable with an ultraviolet light, that can be marked on the centrifuge tube, as shown in Figure 12-

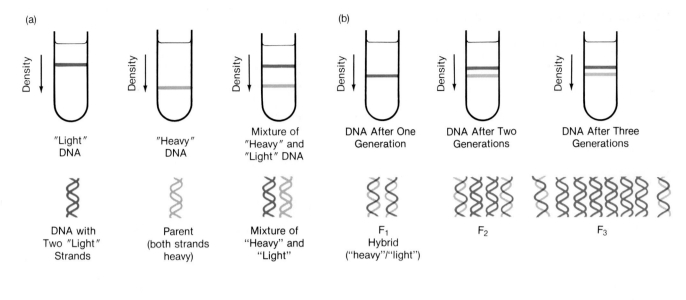

Figure 12-14   THE MESELSON–STAHL DNA-REPLICATION EXPERIMENT: PROOF OF SEMICONSERVATIVE REPLICATION.
Meselson and Stahl used DNA molecules, some "light" (containing normal ^{14}N nitrogen) and some "heavy" (containing heavy ^{15}N nitrogen) to elucidate DNA's mode of replication. (a) The results of centrifugation of DNA containing two light strands, two heavy strands, and a mixture of DNA containing all-heavy and all-light strands. (b) The results of centrifugation of three successive generations (F_1, F_2, and F_3) of DNA produced by the replication of heavy DNA in a ^{14}N growth medium. (c) An overview of three generations of semiconservative DNA replication of original heavy DNA in a light medium.

14a. Meselson and Stahl used a CsCl density gradient to establish a baseline density for heavy DNA from the specially grown *E. coli* cells.

In the second step of the experiment, they took a batch of *E. coli* with 100 percent ^{15}N DNA and suddenly switched it to a growth medium containing only normal nitrogen (^{14}N). After allowing the cells to go through only one cell-division cycle (one period of DNA replication), they removed some cells in order to extract their DNA and measure its density. The remaining cells were allowed to go through a second cell-division cycle with another period of DNA replication, and then a third cycle. After each cell generation, Meselson and Stahl measured the densities of DNA. The results are shown in Figure 12-14b. After the first cell division, all the DNA had a density exactly intermediate between the densities of heavy DNA and normal DNA. This finding was consistent with the hypothesis that each double-stranded molecule contained one ^{15}N strand and one ^{14}N

strand, making their combined density midway between the two original densities. Not only did this step verify the semiconservative copying pattern that Watson and Crick had proposed, but it also eliminated the conservative hypothesis, which predicted that no hybrid DNA should form and that the first generation would include some DNA molecules containing only ^{15}N and others containing only ^{14}N. The second and third generations of the cells also revealed DNA densities consistent with semiconservative replication.

In the third step of the experiment, Meselson and Stahl heated the DNA samples from the first generation to break the hydrogen bonds between the base pairs and thereby separate the DNA double helices into single strands. When analyzed on the CsCl gradients, the single strands contained either *all* heavy or *all* normal nitrogen. This eliminated dispersive replication as the copying mechanism because it would have resulted in both strands of each parent DNA double helix being mixtures of ^{15}N and ^{14}N. The researchers concluded that the copying mechanism for DNA could not possibly be either fully conservative or fully dispersive; it was indeed semiconservative, as Watson and Crick had theorized.

DNA Replication in *E. coli*

Meselson and Stahl went beyond their in vitro studies, and the theory of DNA replication, to see how DNA actually occurs in living things—beginning with the familiar biological subject *E. coli*. Their work and that of others have shown that DNA replication in bacteria requires a series of specific enzymes and proteins. The enzyme **DNA polymerase,** for example, links free nucleotides as they line up on the template formed by the original strand of the parent molecule. Several other enzymes are involved in other aspects of the replication process. For instance, unwinding protein assists in the initial separation of the entwined helical strands; other enzymes repair damaged DNA or eliminate incorrectly paired bases, and still others join ends of broken chains. The biochemical complexity of the replication process probably evolved to ensure fidelity between the parent and the daughter DNA molecules. Consider what would happen if the error frequency were even 1 per 1,000 bases per generation. An average gene is about a 1,000 base pairs long; therefore, at that rate of error, most genes would mutate every time a cell divided. Thus the

system must be much more accurate. Because of the special enzymes, only 1 error occurs in every 1 billion to 10 billion base pairs, even though replication may occur at the rate of 1,000 nucleotides a second.

Another fact revealed by studies of DNA replication in *E. coli* helps explain why DNA is so stable in cells and is not degraded in a reverse reaction by the same enzymes that catalyze its formation. The key to this stability is that the two phosphate groups on the free nucleoside triphosphate building blocks of which DNA is formed are connected with high-energy bonds, just as in ATP. When DNA polymerase helps attach the free nucleotide to the lengthening chain along the template of the parent DNA strand, the two high-energy phosphates are released. Thus the reaction is highly exergonic. It should be no surprise, then, that DNA polymerase cannot easily catalyze the reverse reaction and degrade DNA—much energy would be needed to free each nucleotide and attach two high-energy phosphates. So living cells pay a price in making DNA—the expenditure of energy—but they get a big dividend—much greater stability of their precious genetic material.

It turns out that the *E. coli* chromosome is a circular DNA double helix. Therefore, one might expect such a circle to be replicated in one direction from one starting point, the way a baseball player always runs the bases counterclockwise after hitting a home run. However, studies show that DNA replication in *E. coli* starts at one point on the circle, called the origin of replication, and

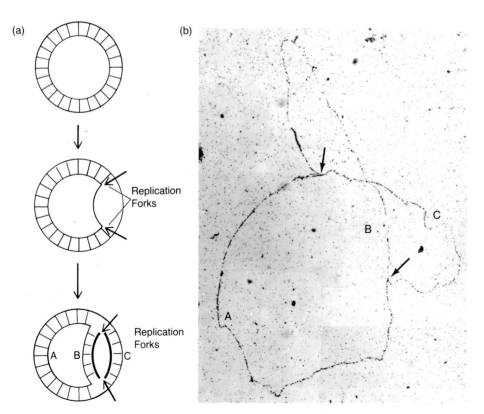

Figure 12-15   DNA REPLICATION IN *E. COLI*. When the DNA in an *E. coli* chromosome replicates, a bubble-shaped region forms and expands until the leading edges meet on the opposite side of the circle. (a) The two strands separate at the origin of replication. The bubble formed by the replicating B and C strands will expand around the circular chromosome until they meet on the other side. (b) An autoradiograph (produced by the exposure of a radioactive substance to photographic film in a dark place) of a replicating chromosome (magnified about 200,000 times). Compare the labeled areas in the autoradiograph with the labeled areas in (a) in order to see the replication bubble.

proceeds around the circle in both directions at the same time. The two so-called **replication forks** are like two base runners leaving home plate in opposite directions and meeting at second base. This two-directional replication creates a bubble-shaped region that grows increasingly larger in both directions until the leading edges meet on the opposite side of the circular chromosome (Figure 12-15).

Just what happens at replication forks? An *unwinding protein* located at each such fork causes the double helix to unwind at that site. As a result, the two original strands of the DNA double helix separate so that they can serve as templates for construction of daughter strands.

What causes replication to occur in this manner? Interestingly, the two strands of a double helix are oriented in opposite directions. In fact, Watson and Crick's original model showed that the phosphate in the sugar–phosphate backbone forms a link between the 5′ carbon of one sugar and the 3′ carbon of the adjacent sugar. Thus, while one strand of the DNA molecule is oriented in a 5′ to 3′ direction, the complementary strand has a 3′ to 5′ direction. DNA polymerase can add a new nucleotide only to one end of each growing chain; specifically, the phosphate group on the 5′ carbon of the incoming nucleotide can be added only to the free hydroxyl group on the 3′ carbon of the last sugar on the growing DNA chain (Figure 12-16). Therefore, *a DNA chain lengthens only in the 5′ to 3′ direction.*

Looking at the replication fork shown in Figure 12-17a, you will see that on one of the strands—the *leading strand*—the 5′ to 3′ direction leads *into* the fork, but on the other strand—the *lagging strand*—it leads *away* from the fork. To accommodate this asymmetry, DNA synthesis occurs differently on the two strands of the separating double helix. On the leading strand, synthesis is *continuous;* nucleotides are added constantly in the 5′ to 3′ direction as the replication fork moves along the strand (Figure 12-17b–e). On the lagging strand, however, DNA synthesis is *discontinuous* because the 5′ to 3′ direction leads away from the replication fork. Small DNA fragments called **Okazaki fragments** are synthesized, in the 5′ to 3′ direction, but are not joined together right away. This process begins a short distance from the replication fork on the lagging strand where synthesis of Okazaki fragments commences with the formation of a ten-nucleotide RNA primer (Figure 12-17a). This primer initiates DNA synthesis in the 5′ to 3′ direction, away from the fork (Figure 12-17c). When an Okazaki fragment has grown to about 150 nucleotides, the RNA primer is removed by an enzyme and replaced with DNA; another enzyme, *polynucleotide ligase*, joins the two Okazaki fragments. By this time, the fork has moved further to the right, exposing more bases. A new Okazaki fragment forms and ultimately is joined to the previously

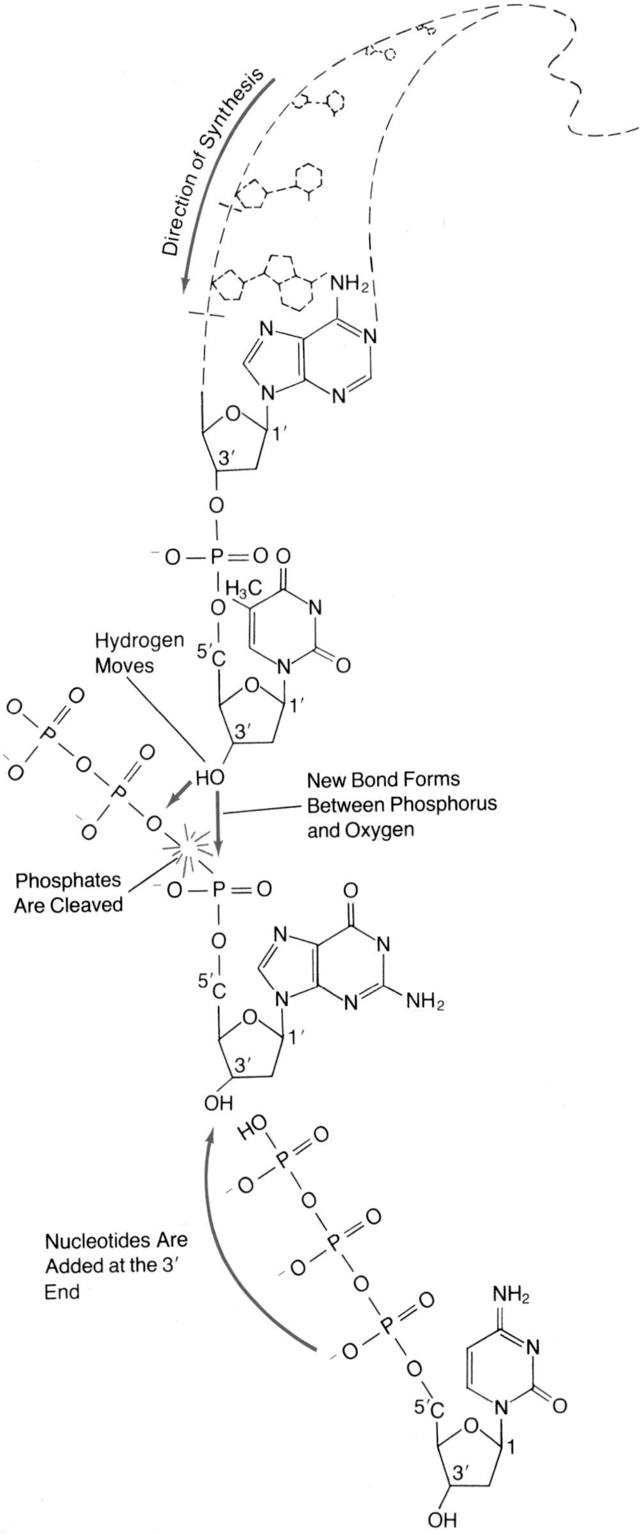

Figure 12-16   GROWTH OF A DNA CHAIN IN THE 5′ TO 3′ DIRECTION.

When DNA is replicated, additional nucleotides are joined by DNA polymerase to the lengthening chain along the template of the parent DNA strand. The enzyme can add a nucleotide's phosphate group only to the free hydroxyl group on the 3′ end of a DNA chain. For this reason, DNA replication can occur only in the 5′ to 3′ direction.

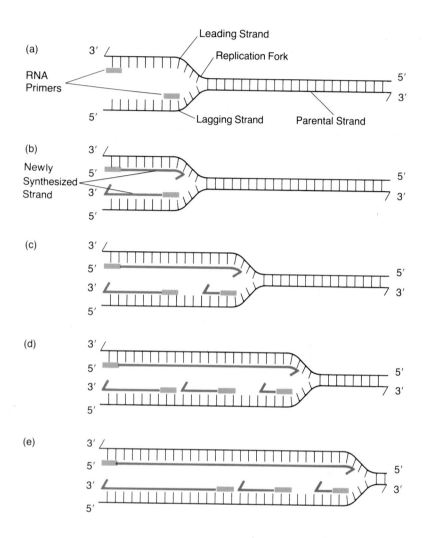

Figure 12-17   THE PROCESS OF DNA REPLICATION.

(a) DNA synthesis commences with the formation of a short RNA primer. (b) As the fork moves to the right, the *leading* strand is synthesized continuously in the 5' to 3' direction, or toward the moving fork. The *lagging* strand also is synthesized in the 5' to 3' direction, but away from the fork. (c) As more bases are exposed by the rightward movement of the fork, a second RNA primer forms on the lagging strand extending in the 5' to 3' direction away from the fork and forming an Okazaki fragment. (The leading strand continues to be synthesized as before.) (d) The Okazaki fragment lengthens until it meets the previously formed fragment. Meanwhile, a new primer appears, and a new Okazaki fragment begins to form. (e) The initial RNA primer is removed and replaced with DNA; polynucleotide ligase then joins the two DNA fragments into a single longer DNA strand. This process is repeated over and over until the entire DNA molecule has been replicated.

formed fragment. This "backstitching" process repeats over and over on the lagging strand as the replication fork progresses until the DNA molecule is completely replicated.

DNA Replication in Eukaryotes

E. coli is a prokaryotic organism, lacking a separate nucleus and the more densely packed, protein-rich chromosomes characteristic of eukaryotic cells (Chapter 9). Nevertheless, studies of DNA replication in eukaryotes suggest that the same general principles operate as just described for *E. coli*. The long DNA molecules of eukaryotes, which are supercoiled into chromosomes, have many points—from several hundred to more than a thousand—from which replication proceeds in two directions simultaneously. This creates many replication bubbles, as can be seen in Figure 12-18, and ensures that the huge length of eukaryotic DNA can be replicated in just a few hours instead of the forty-five or so

days that would be required if one replication site were used. Finally, it is clear that replication of eukaryotic DNA is semiconservative, just as in prokaryotes. The mechanisms of DNA replication appear to be universal, to apply to all DNA-containing organisms on Earth.

LOOKING AHEAD

Our knowledge of the mechanisms of heredity grew tremendously in just fifty years, due to the ingenuity of twentieth-century biologists. From two separate lines of inquiry came an understanding of the activity of the gene and the chemistry of DNA. The link between genes and phenotype was traced to the level of polypeptide coding, and the substance of genes was traced to DNA. These lines of research converged when scientists discovered the elegant structure of the DNA molecule, which, in turn, explained (1) its capacity for carrying genetic information, (2) its mode of replication, and (3) its great sta-

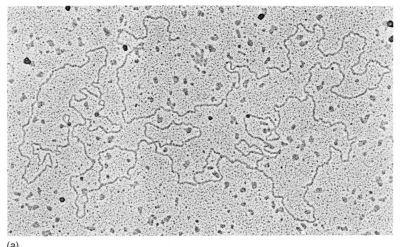

(a)

(b)

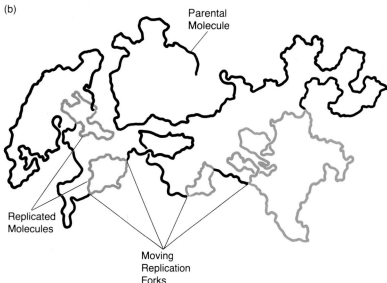

Figure 12-18   DNA REPLICATION IN EUKARYOTIC CHROMOSOMES: INITIATION AT MANY SITES SIMULTANEOUSLY.
In a eukaryotic cell, the replication of DNA takes place concurrently at hundreds of sites on a chromosome. This greatly decreases the time required for a chromosome to replicate its entire length of DNA. (a) In this electron micrograph (magnified about 37,000 times), four regions of DNA replication can be seen. (b) A replication fork lies at each end of each region.

Parental
Molecule

Replicated
Molecules

Moving
Replication
Forks

bility from generation to generation. As we shall see in subsequent chapters, the historic intersection of genetics and molecular biology led to rapid advances in many other fields of biology and gave all students of biology a greater understanding of the unity of all life forms at the most fundamental levels.

SUMMARY

1. Archibald Garrod, describing a disease called alkaptonuria, was the first researcher to suggest that genes are responsible for specific chemical reactions in the body. He called genetic diseases "inborn errors of metabolism."

2. Working with *Neurospora,* Beadle and Tatum confirmed that genes somehow code for specific enzymes and thus formulated the *one-gene-one-enzyme hypothesis.*

3. The idea was later refined by Pauling and Ingram, based on their research on hemoglobins, to the *one-gene-one-polypeptide hypothesis.*

4. In 1869, Miescher was the first to isolate a substance that he called "nuclein" in the nuclei of a wide variety of cells. Feulgen staining revealed that chromosomes are largely made up of this material.

5. Griffith's work on pneumococcal bacteria suggested that a substance in dead virulent strains can transform living nonvirulent strains into virulent ones—that is, change their phenotype. Avery, MacLeod, and McCarty showed this substance to be DNA.

6. Levene found that DNA contains four nitrogenous *bases* and that each base is attached to a sugar mol-

ecule and a phosphate group in a nucleotide.

7. A dispute over whether DNA or proteins carry the genetic material was resolved by the experiments of Hershey and Chase, who used bacteriophages (which consist solely of protein and DNA) to prove that DNA alone is the carrier of genetic information.

8. Each DNA *nucleoside* contains the five-carbon sugar deoxyribose, which is attached to one of four bases: *adenine, guanine, cytosine,* or *thymine.* A and G are double-ring structures called *purines;* C and T are single-ring structures called *pyrimidines.* The addition of a phosphate group makes the unit into a nucleotide.

9. Knowing the components of the DNA molecule was not enough—its structure still had to be elucidated. Evidence leading to the piecing together of the structure of DNA included (a) *Chargaff's rules,* that quantities of [A] = [T] and of [G] = [C] and that each species has a different ratio of [A] + [T] to [G] + [C]; (b) Pauling's suggestion that DNA might have a helical structure held in place by hydrogen bonds; and (c) Wilkins's and Franklin's *x-ray diffraction* studies, suggesting a helical DNA shape with given distances between the coils.

10. Using this information, Watson and Crick were able to construct a three-dimensional model of the DNA molecule, based on the insight that DNA is a *double helix,* with two outer sugar–phosphate chains and rungs formed by nucleotide pairs. Paired nucleotides, which always occur as A–T or G–C, are linked by hydrogen bonds.

11. Watson and Crick hypothesized that the sequence of the bases carries the genetic information and that both accurate gene replication and occasional gene mutation can be explained if each chain of the original molecule separates and acts as a template for production of a new chain.

12. The Meselson–Stahl experiments with *E. coli* confirmed Watson and Crick's hypothesis of *semiconservative replication* and, with it, the double-helical model.

13. Because the enzyme *DNA polymerase* can join nucleotides only to the 3' carbon of a growing DNA chain, DNA replication proceeds in a *5' to 3' direction* simultaneously along the two separating strands. Replication occurs at *replication forks,* which produce a bubble-shaped structure in the circular bacterial chromosome and many bubble shapes in the eukaryotic chromosome. The leading strand of the replication fork is synthesized as a continuous chain and the lagging strand is synthesized discontinuously in short segments called *Okazaki fragments,* which are later joined together by polynucleotide ligase.

KEY TERMS

adenine

base

Chargaff's rules

cytosine

DNA polymerase

double helix

5' to 3' direction

guanine

nucleoside

Okazaki fragment

one-gene–one-enzyme hypothesis

one-gene–one-polypeptide hypothesis

purine

pyrimidine

replication fork

semiconservative replication

thymine

x-ray diffraction

QUESTIONS AND PROBLEMS

1. How did Garrod apply Mendel's laws, derived from experiments with peas, to human disease? How did Beadle and Tatum's work on *Neurospora* substantiate Garrod's ideas?

2. What is the genetic material in pneumococcus? What is the genetic material in T$_2$ bacteriophage?

3. What are Chargaff's rules? How did the Watson–Crick model for DNA explain Chargaff's rules?

4. Explain what is meant by "semiconservative replication." Draw a molecule of double-stranded "heavy" DNA. Now draw two progeny molecules after growth for one generation on "light" medium. Now draw four progeny molecules after two replications on "light" medium. If Meselson and Stahl had been able to separate the two DNA strands without breaking them, what would they have found after one generation? After two generations? What fraction of the DNA would have been in each band?

5. What feature of Watson and Crick's model suggested a method by which DNA could copy itself?

6. What are the subunits of a nucleotide? Which (if any) of these contain sulfur? Phosphorus? Nitrogen? Carbon? What is meant by polarity of DNA chains? Which subunits of nucleotides form the backbone of DNA? Which subunits form hydrogen bonds between the two chains? What are the precursors of DNA synthesis, and how are they different from nucleotides?

7. In most prokaryotes, replication of the circular chromosome begins at just one point, while in most eukaryotes, DNA replication begins at many points. Why do you think eukaryotes need more?

8. Draw a diagram of replicating DNA with a single replication fork. Label the leading strand, lagging strand, Okazaki fragments, RNA primer, and 5′ and 3′ ends of the growing chains. Indicate the name and function of at least three of the enzymes involved.

9. Why is it advantageous to use haploid organisms, rather than diploid, for mutagenesis experiments? In what way is this like using sex-linked mutations described in Chapter 11?

10. What similarities do you see between the mutant selection procedures used in *Neurospora* and natural selection?

11. Four *Neurospora* mutants fail to grow on minimal medium, but they will grow on medium containing the amino acid arginine. Some of these mutants will grow when substances related to arginine are added to the medium, while others do not. What is the suggested metabolic pathway of arginine? (+ = will grow, − = won't grow.)

Mutant	Minimal	Citruline	Ornithine	Arginine
1	−	+	+	+
2	−	+	+	+
3	−	−	+	+
4	−	−	−	+

12. If all dangerous mutagens caused by human beings were removed from the environment, mutations would still continue to arise. Why?

ESSAY QUESTION

Describe the experiment of Avery, MacLeod, and McCarty and the experiment of Hershey and Chase. Did the latter actually demonstrate something new, or did both experiments show convincingly that DNA was the genetic material? Can you think of possible reasons why Hershey and Chase's results were accepted immediately, whereas those of Avery, MacLeod, and McCarty were not?

SUGGESTED READINGS

AYALA, F. J., and J. A. KIGER, JR. *Modern Genetics.* Menlo Park, Calif.: Benjamin/Cummings, 1980.

This fine introductory book is strong on topics covered in this chapter, especially the function of DNA in heredity.

JUDSON, H. F. *The Eighth Day of Creation: The Makers of the Revolution in Biology.* New York: Simon and Schuster, 1979.

A semipopular account of the history of DNA research; it is fascinating reading.

STRYER, L. *Biochemistry.* San Francisco: Freeman, 1975.

The section on the molecular basis of heredity is easy to follow and very comprehensive.

WATSON, J. D. *The Double Helix.* New York: Atheneum, 1968.

Here is Watson's humorous and serious view of the great adventure; Crick does not agree with it all, but judge for yourself.

WATSON, J. D., and F. H. C. CRICK. "Molecular Structure of Nucleic Acids: A Structure for Deoxyribose Nucleic Acid." *Nature* 171 (1953): 737–738.

This may be the greatest two-page paper in the history of science; it is the beginning of molecular biology.

13

TRANSLATING THE CODE OF LIFE: GENES INTO PROTEINS

Proteins are, in the final analysis, the executors of each organism's inheritance.

John Cairns, "The Bacterial Chromosome,"
Scientific American (1966)

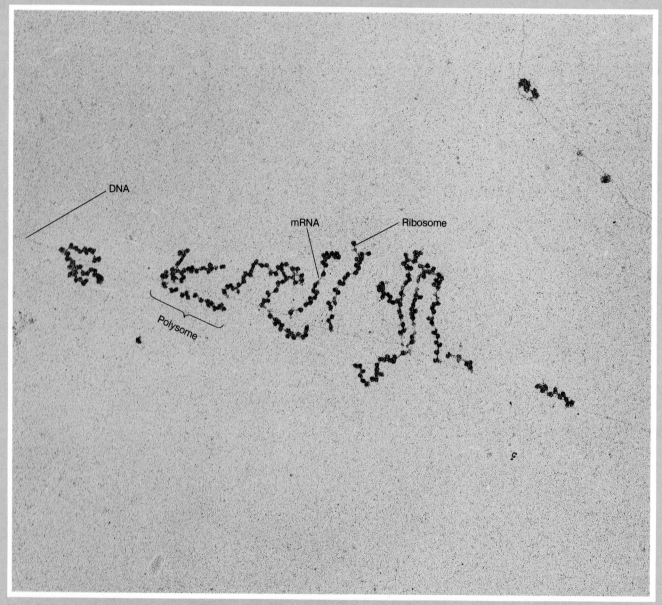

Heredity in action.
DNA, the genetic material, is seen being transcribed into the messenger, mRNA (magnified about 64,000 times). Ribosomes on the mRNA mark sites where proteins encoded by the DNA are manufactured.

When an important scientific puzzle, such as the chemical composition of the gene, is finally solved, the solution does not engender satisfied silence, but a barrage of new questions. Biologists knew by 1953 that genes are made of DNA and that each gene carries the information for building a polypeptide. But what steps lie between the genetic information for building a specific polypeptide chain or protein and the actual construction of the protein itself? Within a decade, the *central theme of molecular biology* had been defined: genetic information is stored in a linear message on nucleic acids and is expressed in a corresponding, linear sequence of amino acids in proteins.

Underlying that succinct description of gene function was the discovery that the information in DNA is stored in a code and that the code is universal; the same one is used in the same way in nearly every living thing on Earth. This revelation was like finding a key for translating the universal language of life. It not only explains how genes act as blueprints for the building of proteins and thus how genotype controls phenotype at the molecular level, but also establishes in an evolutionary sense the close chemical relatedness of all organisms. Moreover, all modern genetic engineering and cancer research is built on the foundation of knowledge that we describe in this chapter: how proteins are synthesized from the information coded in genes and then assume specific roles in an organism's cellular functioning and overall survival.

A brief definition of the gene will simplify our study of the DNA code and its translation into protein. In functional terms, a *gene* is the DNA that codes for a specific RNA. The RNA, in turn, usually codes for a polypeptide. Some RNAs play an active but indirect role in building proteins. Nevertheless, the key relationships are DNA → RNA → protein and one-gene–one-polypeptide. We will see why as we explore the elucidation of the universal genetic code.

GENETIC INFORMATION MUST OCCUR IN CODE

Francis Crick once referred to nucleic acids and proteins as the "two great polymer languages." As we have seen, the "languages" have different alphabets: nucleic acids have a four-letter alphabet—the nucleotide bases A, T, G, and C—whereas the protein alphabet consists of the twenty amino acids. In both languages, the sequence of "letters" conveys the meaning of the message,

just as the words in this book convey meaning. But superficially, at least, the alphabets do not seem to correspond with each other. One of the first questions that biologists asked in the 1950s about the newly characterized DNA molecule was: How does the four-letter genetic alphabet translate into proteins, with their twenty different amino-acid "letters"? They quickly formulated an answer—a *code* based on *sets* of DNA bases.

Simple logic shows why such a code is necessary. If each of the four genetic "letters"—a C, let us say, or a G—coded for a single amino acid, the code could specify only four amino acids. Clearly, the relationship between the two alphabets cannot be a one-to-one correspondence. Instead, the information in DNA must occur in the form of a code in which a set of nucleotide bases corresponds to a single amino acid. If a set contained two consecutive bases, such as AT or TC, there would be 4^2, or 16, possible pairs of bases to serve as DNA code words (for example, AT, TC, AG, GC, and so on) for specific amino acids. This is a few short of the actual twenty amino acids found in proteins. However, if *three* consecutive nucleotide bases acted together to specify one amino acid, there would be 4^3, or 64, letters in the coded alphabet (AAA, AAT, ATA, ATG, and so on). That would be more than enough to accommodate a language of twenty amino acids such as the one that occurs in the protein molecules produced by cells.

The next-larger possibility—a set of four nucleotide bases coding for each amino acid—could specify 4^4, or 256, different amino acids. This number is, of course, more than ten times larger than the actual twenty "letters" in the protein alphabet. Considering nature's normal economy, so generous a margin of coding power seemed very unlikely. Thus soon after Watson and Crick identified the structure of DNA, biologists advanced the hypothesis that three consecutive nucleotides might code for a single amino acid. A name was even coined—**codons**—for the hypothetical sets of nucleotides. As we will see, research confirmed that a three-nucleotide codon *does* determine each amino acid. However, experimental proof of the codon hypothesis was not to come until researchers had answered questions about the "two great polymer languages":

1. Is the code linear?

2. Does the coded message lie in the *order* of the bases in DNA?

3. If the answers to these questions are "yes," how is the linear order "read"?

With the knowledge of how genetic information is coded, scientists could unravel the mystery of how genes actually work to produce proteins.

The Genetic Code Is Colinear

Before any secret message—genetic or otherwise—can be translated from a cryptic language to a known language, a cryptographer must determine whether the letters in the coded message correspond in a *linear fashion* to the letters in the solution. Such an arrangement is said to be *colinear;* for example, in the code FDMDSHBR = GENETICS, each letter of the solution is represented in the code by the letter that precedes it in the alphabet. A code that is not colinear could read backward or skip letters. Early in their attempt to understand the genetic code, biologists, like cryptographers, had to figure out whether the order of nucleotide bases in the DNA molecule corresponds in a colinear or a noncolinear way to the order of amino acids in a protein.

Colinearity, in genetic terms, would mean that codons in a 1, 2, 3, 4, 5, 6 sequence on the DNA molecule corresponded to amino acids a, b, c, d, e, f on the resulting protein, instead of some other amino-acid sequence, such as b, f, c, d, a, e.

To establish whether or not the genetic code is colinear, during the early 1960s Charles Yanofsky and his colleagues at Stanford University performed experiments using a gene and its corresponding polypeptide pair in *E. coli. E. coli* produces an enzyme called *tryptophan synthetase*, which is necessary for the biosynthesis of the amino acid tryptophan (abbreviated trp). This enzyme is made up of two polypeptide chains, A and B, which are coded for by the trp A$^+$ and trp B$^+$ genes. Yanofsky and his co-workers identified a large number of mutant *E. coli* strains in which tryptophan synthetase activity was abnormal. Using the gene-mapping technique of recombination analysis (Chapter 11), they were able to locate the mutated genes on a map of the *E. coli* chromosome (Figure 13-1). Yanofsky also determined the sequence of the 268 amino acids in the enzyme's A polypeptide chain and the amino-acid sequences of several inactive forms of the enzyme synthesized by strains with mutated trp A genes. Each mutant A polypeptide

had an altered amino acid somewhere along the chain. Comparing the position of each altered amino acid and the position of each mutation site in the trp A gene on the chromosome map, Yanofsky found a correspondence. Each mutation produced a change in one of the amino acids on the normal polypeptide chain. This comparison made it clear that the sequence of bases in DNA is colinear with the sequence of amino acids in the polypeptide.

The trp A experiment also supported the hypothesis that more than one nucleotide is needed to code for a single amino acid. Yanofsky observed that mutations at two distinct but close sites in the trp A gene (see arrows in Figure 13-1) could bring about the substitution of either of two different amino acids at the same site along the enzyme's A polypeptide. He reasoned that the two mutations must reside within the *same* DNA codon but affect different nucleotides. Thus the trp A experiments proved two things: *that nucleotides and amino acids are colinear and, in keeping with the codon hypothesis, that more than one nucleotide codes for each amino acid.*

The Nature of the Genetic Code

With this knowledge that genes and polypeptides are colinear, geneticists quickly surmised that the *meaning* of the genetic message might lie in the sequence of nucleotides, which, in turn, might correspond directly to the sequence of amino acids (the primary structure) of polypeptides. Thus the **genetic code**—the molecular "grammar" relating nucleic-acid bases to amino acids—might be colinear, sequential, and based on codons—and therefore, based on *the sequence of codons.*

Several experiments confirmed these suspicions and revealed four fundamental principles about the genetic code in virtually all organisms.

1. The code does not overlap—that is, adjacent codons do not share nucleotides.

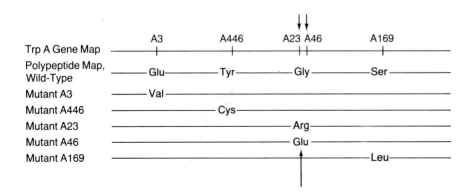

Figure 13-1   MAP OF THE TRP A GENES IN *E. COLI* AND OF THE CORRESPONDING AMINO-ACID SEQUENCE IN THE POLYPEPTIDE CODED FOR THE TRP A GENE. Polypeptides of five mutant trp A genes are shown; they are identical except at the sites indicated. Thus A3 has a single amino-acid change of a val for the wild-type glu. Both the A23 and A46 mutations alter the amino acid at the same position.

2. The code is deciphered by reading frames—that is, adjacent sets of bases called codons.

3. The code is degenerate—that is, different codons can code for the same amino acid.

4. Three nucleotides make up each codon.

Understanding these principles of the genetic code in turn helped biologists to solve the mystery of protein synthesis and, as we shall see, to determine which codons translate into which amino acids.

The Genetic Code Is Nonoverlapping

It was clear from Watson and Crick's work that the "rungs of the ladder" of the DNA molecule form an extremely long series of the same four bases: A, T, G, and C. Groups of three bases in a row on either strand of the double-stranded molecule in a row might function as codons. However, it was conceivable that the sets of three might overlap, thereby functioning as an *overlapping code*. To visualize what that would mean, consider that the nonsense word "hatend" actually contains four real three-letter words that overlap: "hat," "ate," "ten," and "end." If the code overlapped fully and a single nucleotide were altered, say, from C to G, three codons would be affected, as Figure 13-2a shows. The result of this single alteration would be three incorrect amino acids that could be detected in the mutated organism. If the code overlapped partially, a single alteration would lead to two incorrect amino acids (Figure 13-2b). If the code were nonoverlapping, then a single alteration would lead to just one incorrect amino acid (Figure 13-2c).

Biologists observed that mutations affecting just one nucleotide alter only a single amino acid. On the basis of this and other evidence, they determined the first principle of the genetic code: that it is *nonoverlapping*.

Deciphering the Genetic Code

If one creates a long word composed of nonoverlapping short words—"thecatinthehat"—reading and understanding the individual words depends on knowing where to start reading each one. The same is true for a nonoverlapping genetic code, but how is it read correctly within a cell? We have seen that the helical backbone of the DNA molecule is a continuous chain of sugar and phosphate groups, spanned by an uninterrupted series of paired nucleotide bases. No structure is known to mark off the molecule into genes, codons, or other units of information. Yet somehow, inside a living cell,

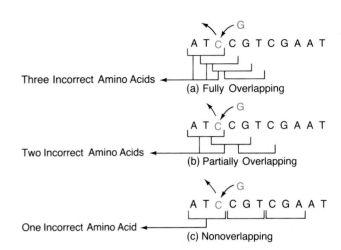

Figure 13-2   THREE WAYS OF READING A NUCLEOTIDE SEQUENCE.

(a) If a nucleotide codon consists of three nucleotides that overlap fully, and a single nucleotide is then altered by a point mutation, three amino acids will be affected by the change of that single nucleotide. (b) If the codons overlap partially, a single point mutation will result in two incorrect amino acids. (c) If the nucleotide codons do not overlap at all, a single nucleotide substitution will affect only the codon of which it is a part. This in fact is what biologists have observed experimentally.

the information is read in appropriate units that code for individual amino acids. Clearly, reading must begin at the first nucleotide of each codon (ATT), not at the second or third, just as a correct reading of "hat" begins at the *h*, not at the *a* or the *t*. And the reading of a long coded message for a specific polypeptide chain must have a starting codon—like the first word in a sentence—and an end codon—like the period.

Francis Crick conceptualized the reading of the genetic message in terms of a **reading frame.** He and his colleagues used a chemical mutagen (Chapter 11) that caused insertion or deletion of a single nucleotide pair in a normal sequence of base pairs in the DNA double helix of a bacteriophage. They guessed that the DNA content might be translated into protein by reading out the nucleotide order in one direction from a fixed starting point and moving either to the right or to the left from the starting point along a single DNA chain, but not in both directions.

To illustrate the reading-frame concept, let us consider an example devised by George Beadle and Muriel Beadle. A sequence of the four letters *A*, *D*, *I*, and *N* might look like this:

DADANDNANAIDANNDANAIDIDA

If the sequence is read in sets of three letters, starting at the left, the message becomes

DAD AND NAN AID ANN DAN AID IDA

A strange command to assist Ann and Ida, but understandable. The reading frame for translating the entire coded message is set by beginning at the first *D* and reading groups of three letters—*triplets*—at a time.

In a similar way, if the nucleotide sequence

ACCCGGGTTTAAACCCCC

is read in three-nucleotide codons beginning at the first *A*, it becomes

ACC CGG GTT TAA ACC CCC

trp   ala   gln   ile   trp   gly*

which we now know codes for the amino acids listed under the codons.

Now, returning to the Beadle example, if a letter is deleted, say, the first *A*, there will be a shift in the reading frame for the entire message:

A

D DANDNANAIDANNDANAIDIDA

so that it now gives the nonsense message

DDA NDN ANA IDA NND ANA IDI DA

Similarly, if a chemical mutagen causes a deletion of the first *C* in the second codon of the nucleotide sequence we examined above, the following would result:

C

1 deletion: ACC GGG TTT AAA CCC CC

trp   pro   lys   phe   gly†

The deletion would cause a shift in the reading frame from CGG to GGG—a process that Crick called a **frameshift mutation.** The first triplet is still normal, but the five subsequent ones are altered codons that specify different amino acids from those specified by the original triplets. Obviously, the resulting polypeptide would be radically changed.

If the deletion of a nucleotide can cause a frameshift, a similar shift should occur following a mutation that leads to an insertion of one nucleotide. When Crick induced this insertion-type mutation in bacteriophage that had suffered a frameshift-deletion mutation, he found that the original reading frame was restored:

1 deletion and   C   (T)
1 insertion: ACC GGG TTT TAA ACC CCC

trp   *pro*   lys   ile   trp   gly

Note that the only different amino acids, pro and lys, are the ones between the original deletion in the second codon and the insertion in the fourth codon. All the other amino acids are correct because the normal reading frame has been restored. In fact, such a protein might function quite normally if the deletion and insertion in the gene are not too far apart and if the substitute amino acids are not in critical positions for the protein's function.

In other instances, however, mutations might so disorder the nucleotide sequence as to disrupt the proper reading frame entirely, either altering the majority of amino acids or halting protein synthesis altogether. Both could lead to the death of the cell.

In the nucleotide sequence we used as an example, two mutations—one deletion and one insertion—restored the proper codon sequence. In others, as we shall see, it takes three mutations to restore the sequence. Nevertheless, they all follow the first two principles of the genetic code: in all organisms, the genetic code is nonoverlapping and is deciphered by reading frames that begin at a specific point and move in just one direction, so that the coded message conveys the appropriate sequence of amino acids in the finished polypeptide chain.

The Genetic Code Is Degenerate

The third principle of the genetic code is that the code is *degenerate;* that is, more than one codon can code for a given amino acid. If the codon hypothesis is correct, and the code is read in triplets, then there are sixty-four possible codons but only twenty amino acids. Do forty-four codons code for nothing at all, or can an amino acid be specified by more than one codon? Biologists borrowed the concepts of "degenerate" and "nondegenerate" codes from the cryptographer's lexicon to characterize these possibilities. In a *degenerate* code, each amino acid would be specified by several codons. In a *nondegenerate* code, each amino acid would be specified by a single unique codon, and the remaining forty-four codons would be noncoding, or *nonsense codons,* or have some other, unknown function.

Francis Crick's frameshift experiments demonstrated that the genetic code indeed is degenerate. If the code were nondegenerate, almost 75 percent of all mutations should halt protein synthesis because synthesis presumably would stop at the first nonsense codon, and 73 percent of the time (or forty-four times out of sixty-four), a mutated triplet would turn out to be a nonsense codon. Crick's team concluded that the most frequent result of inducing a mutation is the insertion of a *different* amino

---

*These abbreviations stand for tryptophan, alanine, glutamamine, isoleucine, glycine.

†The abbreviations pro, lys, and phe stand for proline, lysine, and phenylalanine, respectively.

acid rather than no amino acid at all (which would halt further synthesis of that protein). Therefore, most or all of the sixty-four codons must code for one amino acid or another. As a cryptographer would say, the code must be degenerate, and any of several codons may specify each of the twenty amino acids. By now, the concept of a degenerate code has become well established and accepted, and its biological significance determined.

Since most of the sixty-four codons specify one or another amino acid rather than "nonsense," many types of mutations simply cause substitutions of inappropriate for appropriate amino acids into a protein instead of blocking the synthesis of the protein entirely. This is important, since a halt in synthesis of an essential protein often can lead to the death of a cell or an organism.

Codons Are Really Nucleotide Triplets

The frameshift experiments had one final and important result: they proved beyond doubt that codons are indeed triplets of nucleotides—the fourth principle of the genetic code. We saw earlier that one deletion mutation followed by an insertion mutation can restore the original reading frame. In a sense, one insertion and one deletion "cancel out." Crick also showed that a second deletion or insertion following a first (in other words two fewer or two additional nucleotides in a DNA chain) fails to restore a reading frame. Thus two insertions do not "cancel out" each other, nor do two deletions. However, Crick and others found that deletion or insertion of *three* bases does restore the reading frame:

```
        ACC  CGG GTT TAA ACC CCC
             ┌─┐
             │C│
             └─┘
1 deletion:  ACC  GGG TTT AAA CCC CC
            ┌─┐        ┌─┐
            │C│        │T│
            └─┘        └─┘
2 deletions: ACC  GGG TTA AAC CCC C
           ┌─┐ ┌─┐  ┌─┐
           │C│ │G│  │T│
           └─┘ └─┘  └─┘
3 deletions: ACC  GGT TAA ACC CCC
```

The deletion of three nucleotides in this example results in a protein that is one amino acid too short and has one substitute amino acid (encoded by the second triplet); conversely, addition of three nucleotides results in a protein that is one amino acid too long, so long as the three do not function as a nonsense codon. These experiments also reconfirmed that the direction in which codons are "read" makes a difference; ACC, for example, codes for a different amino acid than does its inverted form, CCA.

The genetic code yielded to biologists' experiments, just as mysterious messages give way to the decoding techniques of cryptographers. The genetic code is nonoverlapping; it is read in appropriate frames; it is degenerate; and it consists of codons made up of three consecutive nucleotide bases along the DNA strand. Nearly all organisms share a universal genetic language coded according to these principles, just as their proteins are made up of an equivalent but separate language of amino acids. The translation of one language into the other is our next topic. In Chapter 19, we will discuss the implication of the universal code for the relatedness of all living things.

DNA, RNA, AND PROTEIN SYNTHESIS

The structure of DNA and the way it encodes information were profound revelations themselves, but their discovery had an even more important ramification: the final solution, at the molecular level, of how genes are translated into biological molecules—how a genotype can produce a phenotype.

We have seen that the sequence of codons along a strand of DNA specifies the sequence of amino acids in polypeptides. But the DNA molecule is a double helix, composed of two strands. It turns out that in the region of a gene being read out, only one of the two strands is transcribed into RNA, as we shall soon see. The other strand of that gene usually is not read. One might assume that the synthesis of proteins takes place right along one of the DNA strands itself, coiled in the nucleus, the way a machine part is stamped directly from a metal template. However, direct translation of DNA to protein does not take place. Biochemists established in the 1960s that protein synthesis does not even occur in the nucleus of eukaryotic cells. *Proteins are made only in the cytoplasm.* In the mammalian red blood cell, for example, the entire nucleus is ejected from the cell as it matures, yet protein synthesis can continue for hours. Because the DNA that codes for most of the cell's proteins is found only in the nucleus, this fact alone establishes quite clearly that protein synthesis does *not* take place directly on the genes of nuclear DNA.

Additional evidence for cytoplasmic protein synthesis comes from experiments in cell-free systems. When prokaryotic and eukaryotic cells are broken up and homogenized into mixtures containing no intact cells (hence cell-free), amino acids continue to be incorporated into polypeptides. What is more, this incorporation takes place in the absence of DNA. If an enzyme that hydrolyzes DNA molecules is added to the mixture, protein synthesis continues unabated. However, if an enzyme that digests *ribonucleic acid* (RNA) is added to the mixture, incorporation of amino acids into polypeptides ceases immediately. Other studies have shown that cells that are especially active in protein synthesis contain unusually large numbers of ribosomes, organelles that are built from unique types of RNA (Chapter 6).

It was this type of evidence that turned scientific attention to the function of RNA in cells. As we saw in Chapter 3, RNA is a class of nucleic acid that is similar to DNA, except that each RNA molecule contains the sugar ribose instead of the sugar deoxyribose. Furthermore, RNA is usually single stranded instead of double stranded, even though it too is constructed from sequences of nucleotides. There is one additional difference: in RNA, the pyrimidine **uracil** (U) replaces thymine, which is found only in DNA. Thus the RNA bases are A, U, G, and C, rather than A, T, G, and C.

The evidence from the cell-free systems suggested that RNA might act as a messenger. If RNA somehow carried a copy of the genetic sequence inscribed on DNA out of the nucleus and into the cytoplasm, it might then be used as a template during protein synthesis to determine the sequence of amino acids in a polypeptide. This, in fact, turned out to be correct. However, RNA was found not to be a single homogeneous substance, but several different types of molecules, each with a specific role in the translation of genetic information into protein structure. Let us take a closer look at the three main types of RNA: messenger RNA, transfer RNA, and ribosomal RNA.

Messenger RNA

The genetic information encoded in DNA is read out during the process of **transcription** into several types of RNA. **Messenger RNA (mRNA)** is a molecular emissary that carries information from the DNA in the nucleus to the sites of protein synthesis in the cytoplasm. Messenger RNAs are single-stranded chains of nucleotides that are synthesized on single strands of the DNA molecule, which uncoils and separates just as it does during DNA replication. In order for transcription to occur as mRNA forms on a DNA template, nucleotides floating free in the nucleus must be linked to the growing RNA strand in a sequence precisely complementary to that of DNA, except that uracil takes the place of thymine and occupies all locations opposite adenine, as shown in Figure 13-3. As a result, a complementary copy of the DNA strand is made; each set of three adjacent bases of a gene is transcribed into a codon along the mRNA. Enzymes called *RNA polymerases* do the linking. They join the high-energy, triphosphate form of the free-floating ribonucleotides, adding them to the growing chain in the 5′ to 3′ direction, just as a new strand of DNA is assembled during replication (Chapter 12).

What determines where transcription will begin? Specific DNA sequences called **promoters** serve as the binding site for RNA polymerase near each gene. Transcrip-

tion proceeds from that site. But there is a difference in the product in eukaryotic versus prokaryotic cells. In eukaryotic cells, once the genetic information has been transcribed into what is called a *primary transcript*, the RNA is altered by processes that we will describe later (Chapter 17). As a result of those alterations, true mRNA is formed and passes to the cytoplasm through a nuclear pore. In prokaryotic cells, mRNA is transcribed directly from DNA and is used directly in **translation**, the process of protein synthesis. Thus, there is no distinctive primary transcript and no mRNA alteration steps in prokaryotes.

Transfer RNA

The second type of RNA is **transfer RNA (tRNA)**. These small polynucleotide chains have been likened to carriages or tugboats that transport individual amino acids to the sites of protein synthesis. Different types of transfer RNA molecules can bond only to specific amino acids; there is at least one unique kind of tRNA for each of the twenty kinds of amino acids. Transfer RNA molecules are the smallest RNA molecules, containing about 80 nucleotides as compared with 1,000 or more in most mRNAs. Like mRNA, tRNA is synthesized in the nucleus and then passes into the cytoplasm; synthesis occurs on genes that code for the various types of tRNA. A key feature of all tRNA molecules is their looped shape (Figure 13-4), which results from the presence of a region where the RNA strand has folded back on itself to form a double strand. As in double-stranded DNA, hydrogen bonds between pairs of bases stabilize the double-stranded portion. Figure 13-4 illustrates the distinctive cloverleaf shape of the tRNA molecules.

There are two especially important sites far apart on the tRNA molecule. At the 3′ end of the chain is the **aminoacyl attachment site,** where a specific amino acid attaches to the tRNA molecule. This site—a kind of molecular "hook"—always consists of the sequence CCA for all twenty kinds of tRNA. An enzyme specific to each tRNA molecule joins the correct amino acid to this site. The tRNA then has a specific amino-acid cargo "in tow," tugboat style, with a covalent bond. Some distance away on the tRNA molecule lies the **anticodon,** three unpaired bases that form the arc of one "leaf" of the cloverleaf. There is a specific anticodon for each amino acid—the anticodon is a kind of three-letter name tag for the cargo on the other end (as UAC for methionine). Each type of anticodon can bind only to a specific, complementary three-base codon on an mRNA molecule (UAC binds only to AUG, for instance). Consequently, a tRNA can transport a particular amino acid to an appropriate site on the mRNA, a site that was copied during transcription from the codon sequence of the DNA (TAC in

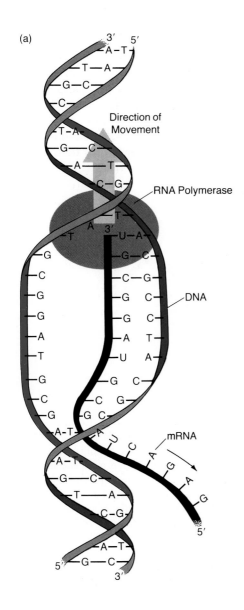

(a)

DNA codes for AUG in mRNA). The tRNAs, with amino acids attached, form hydrogen bonds with the complementary mRNA codon sites. The amino acids they carry are then held in just the right position to be added to the growing polypeptide. Obviously, tRNAs are involved in "reading" the genetic message encoded in mRNA and helping to translate it into a protein.

Ribosomal RNA

Ribosomal RNA (rRNA) molecules, along with a variety of proteins, make up the ribosomes. These tiny globular organelles can be likened to a movable vise that holds mRNA and tRNA molecules in just the right position so that amino acids on the tRNAs can be joined together. The task of the ribosome is to ensure assembly of amino acids in a specified sequence into a protein molecule. That sequence, as we have seen, is provided by

Figure 13-3   TRANSCRIPTION OF GENES INTO RNA.
(a) As the DNA double helix unwinds at the site of RNA polymerase attachment, mRNA is transcribed from one of the chains. The "matching" of complementary bases—U to A, G to C—is the true information-transfer step. In another gene, the other strand of DNA might be transcribed. (b) Transcription is seen in this electron micrograph (magnified about 56,500 times). DNA forms a continuous thread along which RNA polymerase molecules move; the farther they move along the DNA of these genes, the longer the chain of newly synthesized RNA that protrudes to the side. Many RNA chains are being synthesized simultaneously from this stretch of DNA.

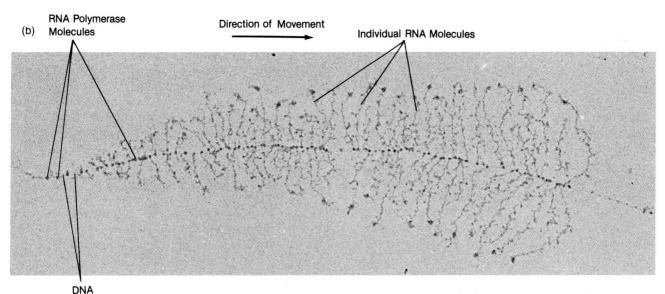

(b) RNA Polymerase Molecules   Direction of Movement   Individual RNA Molecules

DNA

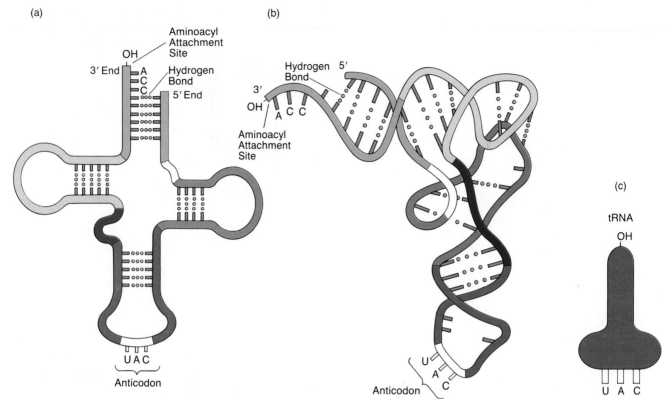

Figure 13-4   TRANSFER RNA MOLECULES: THE CARRIERS OF AMINO ACIDS TO THE SITES OF PROTEIN SYNTHESIS.
There is a different kind of tRNA for each of the twenty amino acids, but all tRNA molecules have a similar shape and structure. (a) Each tRNA has four short sections in which hydrogen bonds bind complementary base pairs and form double-helical regions that stabilize the molecule. Notice the unpaired triplet at the aminoacyl attachment site. This triplet, CCA, is the same for all tRNAs. The loop at the bottom contains the anticodon triplet sequence, which serves as the name tag for the amino acid that the tRNA specifies. For met–tRNA, the anticodon is UAC. (b) The tRNA molecule as it is actually folded. (c) A simplified symbol to be used in subsequent figures, which represents the various forms of tRNA.

the coded information in the mRNA molecule, the emissary from DNA. Ribosomes are made up of three types of RNA; genes for two of these types are located in the nucleolus (Chapter 6), whereas multiple copies of the gene for the third type are found at many sites along various chromosomes. Proteins and rRNA molecules assemble to form the two major subunits of ribosomes in the nucleolus. These large molecular aggregates or subunits then migrate to the cytoplasm, where the two types of subunit attach to mRNA molecules to form functioning ribosomes.

In bacteria, ribosomes consist of one large and one small subunit designated 50S* and 30S; the complete organelle is called a 70S ribosome (Figure 13-5). In eukaryotic ribosomes, the subunits are somewhat larger (60S and 40S) and the complete organelle is an 80S ribo-

some. When not involved in protein synthesis, the two types of ribosomal subunits exist separately in the cytoplasm. During protein synthesis, an mRNA molecule forms a complex with a small ribosomal subunit; then the large subunit joins, and, if other factors are present, polypeptide synthesis can begin. Several ribosomes often move along a single mRNA molecule; the whole complex is called a *polysome*, or polyribosome (Chapter 6). Ribosomal RNA performs a structural role, partially comprising the movable vises or ribosomes, which, in turn, hold the mRNA and tRNAs, as the protein chain elongates during protein synthesis.

During the translation process of protein synthesis, the three types of RNA function *together:* mRNA serves as a copy of the genetic information; tRNA shuttles amino acids to the growing chain of polypeptides; and rRNA is contained in the ribosomal subunits that combine to function as the actual site of protein synthesis. The discovery of these roles was a necessary prelude to understanding the synthetic process itself.

---

*S, the Svedberg unit, is a measure of size based on rates of sedimentation in a high centrifugal field. Sedimentation rates are not additive, hence the complete ribosome is 70S rather than 80S (30 + 50).

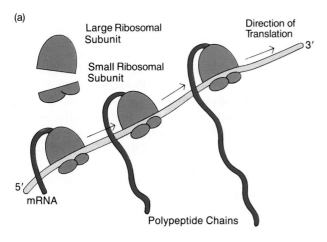

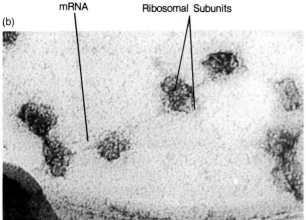

Figure 13-5   RIBOSOMES: MOVABLE SITES OF PROTEIN SYNTHESIS.
(a) Ribosomes are made up of ribosomal RNA (rRNA) and proteins, and they consist of two subunits that are unequal in size. When ribosomes are not engaged in protein synthesis, the subunits reside separately in the cytoplasm, joining together again for the next round of protein synthesis. The ribosomes function as movable vises that hold molecules in the right position so amino acids can be linked together into polypeptides during protein synthesis. (b) An electron micrograph (magnified about 560,000 times) of a polysome from a red blood cell. A polysome is a group of ribosomes on an mRNA.

PROTEIN SYNTHESIS: TRANSLATING THE GENETIC CODE

We come at last to the actual translation of the genetic code—not simply by curious genetic-code cryptogra-

phers, but by nature. Within each living cell, genetic information first is transcribed into mRNA, and translation then occurs as proteins are synthesized. Protein synthesis involves four general steps, which were elucidated by hundreds of experiments in the 1960s:

1. Before synthesis actually begins, **amino-acid activation** takes place; amino acids are activated, or joined to tRNAs, by high-energy bonds.

2. Next, **initiation** of protein synthesis occurs when a ribosome, an mRNA, and two tRNAs bearing specific amino acids bind together.

3. Polypeptide **elongation** takes place as amino acids are joined one by one to the lengthening chain.

4. Protein synthesis **termination** occurs when a "stop signal" within the genetic message is reached on the mRNA molecule.

Before considering each step in detail, it is worth emphasizing a difference between prokaryotic and eukaryotic cells. The absence of a membrane-bound nucleus in the former allows ribosomes and tRNAs to attach to an mRNA even before the synthesis of that molecule is complete. As seen in the micrograph at the beginning of this chapter, the result is simultaneous transcription and translation. This never occurs in eukaryotes, since newly synthesized RNA must be altered, and the mRNA must leave the nucleus prior to initiation of polypeptide synthesis. In eukaryotes, transcription and translation are distinct both in space and in time. With this in mind, let us turn to the steps of translation.

Step 1: Amino-Acid Activation

Translating the nucleotide sequence of the mRNA to the amino-acid sequence of a protein begins with amino-acid activation. This process takes place in the cytoplasm and creates a pool of amino-acid-bearing tRNA molecules ready to take part in polypeptide synthesis. During amino-acid activation, the attachment of specific amino acids to appropriate tRNAs is catalyzed by enzymes called **aminoacyl-tRNA synthetases.** The enzyme creates a covalent bond between the carboxyl end (–COOH) of the amino acid and the 3′ end of the tRNA, forming aminoacyl-tRNAs. This is a high-energy bond resulting from the cleavage of ATP, as shown in Figure 13-6. The process of amino-acid activation prepares amino acids for participation in polypeptide synthesis and helps to provide some of the energy used later in the formation of the peptide bond between an amino acid and the elongating polypeptide chain.

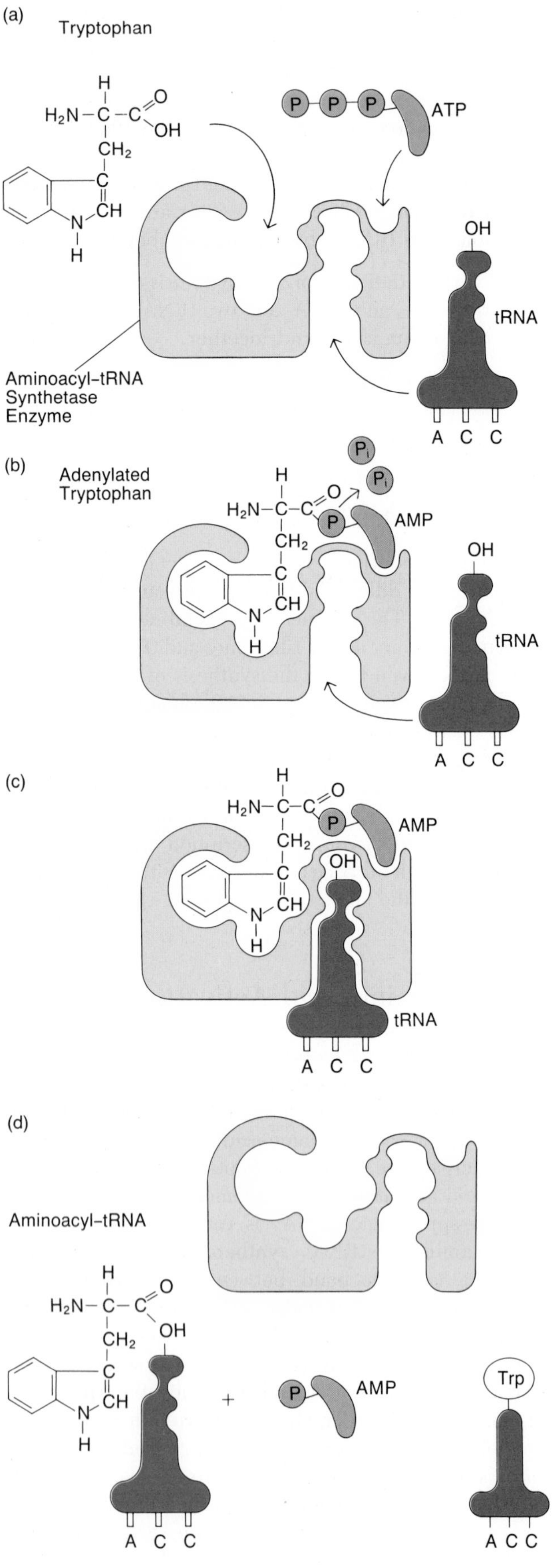

(a)

Tryptophan

Aminoacyl–tRNA
Synthetase
Enzyme

(b)

Adenylated
Tryptophan

(c)

(d)

Aminoacyl–tRNA

Step 2: Initiation

Polypeptide synthesis actually begins with the attachment of a small ribosomal unit to a ribosome binding site on the mRNA molecule. At this binding site is a special codon (AUG), which functions as a "start signal" for protein synthesis, much as a capital letter signals the beginning of a sentence. The small ribosomal subunit already carries a special type of *initiator tRNA*, the anticodon of which pairs with the initial AUG codon in the mRNA molecule (Figure 13-7a). (Note that methionine [met] is always on that tRNA, and is the amino-terminal residue of each newly made protein; later methionine and other amino acids may be removed to leave still other amino acids on the N-terminal end.) Next, a large ribosomal subunit binds to the smaller one, forming a complete ribosome attached to the mRNA. The ribosome has a groovelike site in which the initiator aminoacyl-tRNA nestles. This is called the **"P" site** (for "peptidyl"). The adjacent **"A" site** (for "aminoacyl") on the ribosome is vacant. Another aminoacyl-tRNA can bind to the mRNA codon located at the A site. Various ions, molecules, and tRNAs diffuse in and out of the A site randomly, until by chance an aminoacyl-tRNA approaches that has an anticodon complementary to the mRNA codon at the A site. This aminoacyl-tRNA then binds.

Now the A and P sites are filled by tRNAs, each bearing an amino acid. These amino acids are sitting next to each other, ready to be joined together.

Step 3: Elongation

Peptide bond formation between the adjacent amino acids is the beginning of the elongation step. The reaction is catalyzed by the enzyme **peptidyl transferase,** which is part of the large ribosomal subunit. This enzyme breaks the bond holding the first amino acid (met, at the P site) to its tRNA and attaches that amino acid

Figure 13-6   AMINO-ACID ACTIVATION: PRELUDE TO PROTEIN SYNTHESIS.

Amino-acid activation occurs when enzymes called aminoacyl-tRNA synthetases catalyze the attachment of a specific amino acid to the appropriate tRNA. First, the amino acid, tryptophan in this illustration, and ATP (a) join specific sites on the synthetase and (b) react. ATP is hydrolyzed, inorganic phosphate released, and AMP is joined to the amino acid. (c) This adenylated amino acid then is joined to the specific tRNA that fits in another site on the synthetase. (d) Finally, AMP and the newly created aminoacyl-tRNA molecule are released separately from the enzyme. Energy from the ATP remains in the aminoacyl-tRNA molecule and helps to drive peptide bond formation.

instantly by a peptide bond to the second amino acid, still bound to its tRNA at the A site, as shown in Figure 13-7b. The tRNA at the P site leaves immediately. The formation of the peptide bond creates a peptidyl-tRNA molecule, which then moves from the A site to the newly vacated P site, carrying the mRNA with it. The distance moved is equivalent to three bases—that is, one codon. This process frees the A site on the ribosome so that still another aminoacyl-tRNA can bind with the third codon along the mRNA. A new peptide bond then forms between the second and third amino acids, and the chain becomes one amino acid longer.

The cycle of the action of the enzyme peptidyl transferase, followed by movement of the mRNA along the ribosome, leads to assembly of the entire polypeptide. Each step in the elongation process depends on the splitting of several high-energy phosphate bonds. That is one reason why protein synthesis, which goes on continuously in most living cells, is an energetically expensive process.

Step 4: Termination

Elongation continues until a *termination codon*—like the period at the end of a sentence—is reached on the mRNA molecule. When this stop signal enters the vacant A site on the ribosome, peptidyl transferase, the release factor, hydrolyzes the peptidyl-tRNA to release the completed polypeptide (Figure 13-7c). The ribosomal subunits dissociate from the mRNA and are free to participate in another round of protein synthesis.

When reviewing the translation process, keep in mind that the small ribosomal subunit, where the mRNA codon and tRNA anticodon meet and pair, carries out the task of *ordering* the amino acids according to the mRNA sequence. The large subunit has the task of *linking* amino acids to the elongating polypeptide chain. Polysomes are formed after one ribosome has moved along the mRNA, leaving space for a second ribosome to bind and begin synthesis of a second polypeptide (even though the first may not be complete). The polysome complex may remain free in the cytoplasm or may attach to the membrane surface of endoplasmic reticulum (in eukaryotic cells) or plasma membrane (in prokaryotic cells). Recall from Chapter 6 that the presence or absence of a special "leader" segment of the polypeptide determines which alternative occurs.

Protein synthesis is one of the primary functions of every cell. It occurs with amazing speed. A bacterial cell can generate an average-size protein, containing about 400 amino acids, in close to 20 seconds. And, of course, thousands of such proteins may be synthesized simultaneously on different mRNAs and ribosomes. Such speed helps explain how so many bacterial species can have generation times of just 20 minutes and how young seedlings can grow one inch or more in a day. Figure 13-8 gives an overview of transcription, translation, and related processes in a bacterial cell. Since prokaryotes have no membrane to separate the DNA from the ribosomes before mRNA has been completely transcribed, the ribosomes in the figure have already bound to the end of the mRNA and have begun to translate it. In eukaryotes, the nuclear membrane forms a partition that physically separates the processes of transcription in the nucleus and translation in the cytoplasm. Note that in eukaryotic cells, the initial RNA transcript is altered before leaving the nucleus as mRNA. Only then do ribosomes join it and does protein synthesis begin.

Determining the steps of protein synthesis was a triumph for modern biology. It led to the central understanding of molecular biology, which we mentioned earlier: genetic information is stored in a linear message on nucleic acids and is expressed in a corresponding linear sequence of amino acids in proteins. And much additional research followed from these ideas, opening wide avenues for future research.

CRACKING THE GENETIC CODE

By the early 1960s, biologists were in a position somewhat analogous to that of Egyptologists before 1799. The latter had by then discovered many examples of Egyptian hieroglyphics, had determined that they represented a pictorial language, and had painstakingly interpreted some elements of the language. However, they could not read an entire message in hieroglyphics. In a similar way, biologists in the early 1960s knew that the language of genes is translated during protein synthesis into the language of proteins, and they knew most of the letters of the genetic alphabet. But they could not read the genetic language written in nucleotide triplets—that is, they could not match a given codon (say, ATT or GTC) with a given amino acid. In the summer of 1799, a French soldier made an incredible chance find at a site in Egypt: the Rosetta stone, a large inscribed tablet that ultimately yielded the secrets of hieroglyphics. Biologists, too, found the key to the genetic code through deep insight and hard work.

The first "crack" in the genetic code was entirely accidental—the unexpected result of a carefully controlled experiment. Two researchers at the National Institutes of Health, Marshall Nirenberg and J. Heinrich Matthaei, were studying protein synthesis in what are called cell-free systems. A cell-free system is a test tube solution containing materials extracted from cells. These ri-

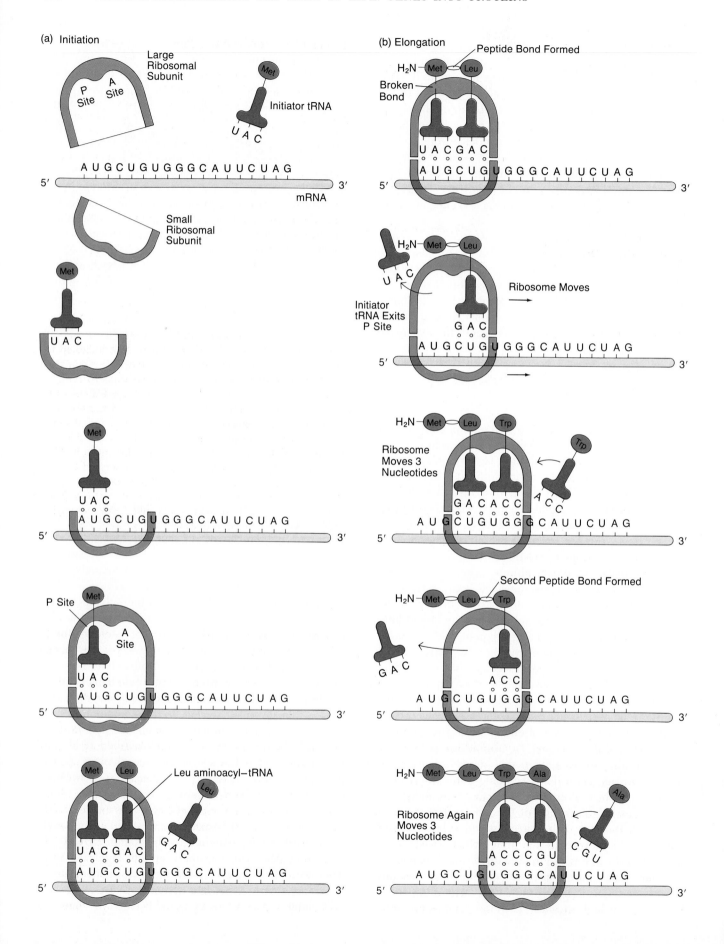

(a) Initiation

(b) Elongation

(c) Termination

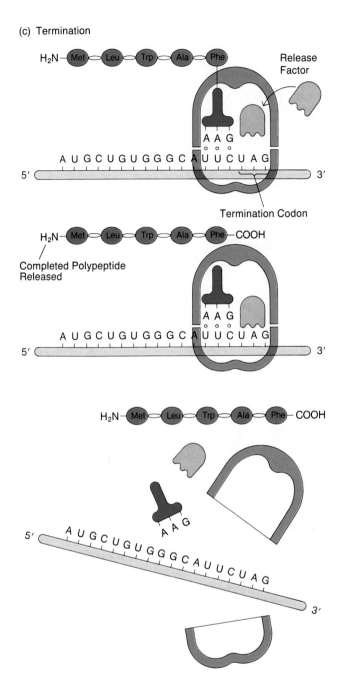

Figure 13-7   THE THREE STEPS OF PROTEIN SYNTHESIS.

After amino-acid activation there are three steps to protein synthesis: (a) initiation, (b) elongation, and (c) termination. At the start, all required components are freely soluble in the cytoplasm of the cell. (a) Initiation. Small ribosomal subunits each bind an aminoacyl-tRNA, including some that bind the initiator tRNA. To start initiation, a small ribosomal subunit carrying an initiator tRNA binds to an mRNA so that its tRNA anticodon binds to an AUG sequence on the mRNA; the AUG is the start codon that codes for met. Next, a large ribosomal subunit with an A and a P site takes position so that the initiator tRNA is in its P site. The adjacent A site is open and can receive another aminoacyl-tRNA, in this case one carrying leucine since the leucine tRNA anticodon is complementary to the leucine codon (CUG) on the mRNA. (b) Elongation. The enzyme peptidyl transferase breaks the bond between the first amino acid, met, and its tRNA; a peptide bond is immediately formed between the first (met) and second (leu) amino acids and the initiator tRNA exits the P site. (At this point the N-terminal end of the peptide is indicated by H_2N-.) The ribosome moves along the mRNA a distance of three nucleotides so the second tRNA molecule (carrying met-leu) is in the P site. Then a new aminoacyl-tRNA with a Trp attached moves into the A site. Peptidyl transferase breaks the bond between the second amino acid (leu) and its tRNA; another peptide bond is formed between that amino acid (leu) and the third one (trp), adjacent to it. The ribosome then moves another three nucleotides and the process is repeated. (c) Termination. The ribosome ultimately reaches a termination codon on the mRNA. The release factor binds to the stop codon and causes peptidyl transferase to release the completed polypeptide. The ribosomal subunits dissociate from the mRNA.

bosomes, amino acids, RNAs, and other substances, in the cell-free system, are capable of incorporating free amino acids into polypeptides despite the absence of living cells. In one experiment, Nirenberg and Matthaei included two ingredients as controls that proved to have a dramatic effect. One was a synthetic RNA called polyuridylic acid (poly-U), which is simply a long chain of RNA containing the base uracil, and the equivalent of dozens of UUU codons: UUUUUUUUUUUUUUUUUUUU. . . . The other ingredient was a high concentration of magnesium ions,

which, we know today, allows polypeptide elongation to proceed in a system despite the absence of the normal mRNA initiation codon, AUG.

The researchers never guessed that the poly-U, their synthetic substitute for natural mRNA included as a control, would direct the synthesis of a polypeptide. To their astonishment, they found that a polypeptide— polyphenylalanine—*was* generated in the special cell-free system with poly-U. Even more remarkable was the fact that although all twenty amino acids were available in the test tube, the peptide made contained *only* phenylalanine. This result indicated that the triplet UUU must be the codon that specifies the amino acid phenylalanine (phe). The genetic message UUU, UUU, UUU, was translated into the polypeptide phe, phe, phe. The researchers soon created other synthetic messengers, including polycytosine (CCCCCCCCCCCC) and polyadenine (AAAAAAAAAAAA). Poly-C was found to

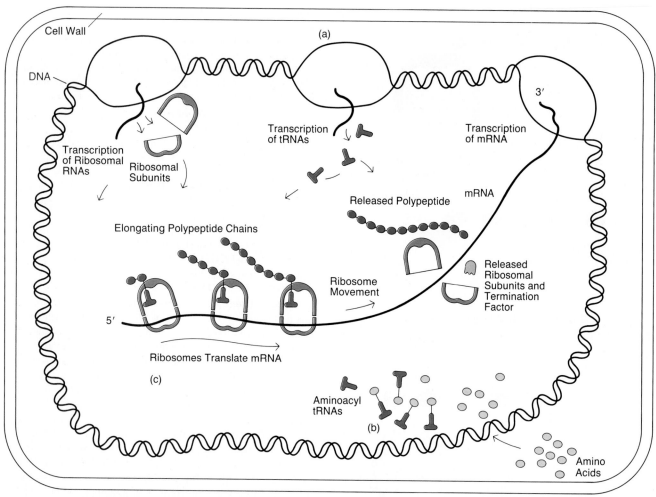

Figure 13-8   AN OVERVIEW OF THE TRANSCRIPTION AND TRANSLATION PROCESSES IN A BACTERIAL CELL.
(a) The genetic message is transcribed from DNA to mRNA. Thus, the ribosomal RNAs and at least thirty-two different tRNA molecules specific for the twenty amino acids are transcribed from the DNA of the cell. Then mRNAs are transcribed. Even as the mRNA is synthesized, (b) activated amino acids on aminoacyl-tRNAs and small and large ribosomal subunits combine with the mRNA (c) to translate the message and synthesize proteins. Transcription can take place simultaneously at many locations on the DNA molecule, and translation can take place simultaneously at multiple sites on the mRNA molecule as ribosomes move along it. (For clarity, the various components are not drawn to scale.)

direct the synthesis of polyproline, and poly-A was found to code for polylysine, suggesting that CCC = proline and AAA = lysine.

Soon after this serendipitous finding, an organic chemist at the University of Wisconsin, Har Gobind Khorana, developed a technique for chemically synthesizing RNA triplets with all sixty-four possible codons, as well as short mRNAs of known nucleotide sequence. By adding these synthetic molecules in cell-free systems, Khorana and others were able to identify the meaning of all sixty-four codons. The genetic code they deciphered is shown in Figure 13-9. This codon dictionary shows the

extreme degeneracy of the genetic code: six codons specify leucine; six code for serine; four, for proline; and so on. Notice that only two amino acids, methionine and tryptophan, are represented by single codons. Also notice that three codons—UAA, UAG, and UGA—function as termination codons.

Since the development of the genetic-code dictionary, researchers have been able to synthesize specific genetic sequences and proteins and to use them to delve far deeper into the processes of transcription and translation. They even have learned to alter the genes in living organisms, as we shall see in Chapter 14.

A SHY GENETICIST AND JUMPING GENES

In 1951, the geneticist Barbara McClintock made a discovery for which she was awarded the Nobel Prize in Physiology or Medicine. However, that well-deserved recognition did not come until 1983. Her discovery—that some genes are mobile rather than permanently fixed in place along the chromosomes—was so startling in its day that it was largely ignored until the rest of the field caught up to McClintock's level of insight decades later. McClintock's unusual research career has certain parallels with that of Gregor Mendel, as well as timely lessons for the field of biology.

McClintock started studying the genetics of maize, or Indian corn, just twenty years after Mendel's work had been rediscovered and not long after Thomas Hunt Morgan, Alfred Sturtevant, Calvin Bridges, and others had done their classic work with *Drosophila*. McClintock originated many of the techniques still used to study maize chromosomes and made some important early discoveries, including the cell's nucleolar-organizer region and the role of crossing over in the exchange of genetic information—a role that proved that plant chromosomes carry hereditary information.

In the mid-1940s, McClintock came to Carnegie Institution's Cold Spring Harbor Laboratory on Long Island, and there, living and working for many years in virtual isolation, she made her most important discovery. McClintock was interested in the genetic basis for color variations in maize kernels and leaves. She noticed that the pigmentation patterns were usually inherited intact but that mutations occasionally occurred, causing abrupt shifts in the pigmentation phenotype of the offspring ears of corn. For six years, she studied maize chromosomes meticulously and arrived at the inescapable conclusion that certain control elements along the chromosome can *move* from one position to another and, in so doing, bring about the phenotypic shifts.

Specifically, McClintock found that a pair of genes along the ninth chromosome, the *Ac* (activator) gene and the *Ds* (dissociation) gene, act in concert to turn on and off the genes that control color in the kernels. A signal from the Ac gene causes the Ds gene to jump to new positions along the chromosome, thus inactivating neighboring genes. This, in turn, causes abrupt changes in kernel pigmentation once the ears of corn develop.

In 1951, McClintock published this work and presented the paper at an important symposium held

Nobel laureate Barbara McClintock with her favorite genetic subject.

each summer at Cold Spring Harbor. According to colleagues, however, the work was neither comprehended nor accepted. As a result, the shy geneticist withdrew and continued to work in silence on the movable, or "transposable," genetic elements of maize, neither publishing nor presenting her findings further. Two decades passed before "jumping genes" were discovered to be ubiquitous in nature by researchers using the modern techniques of molecular genetics. Only then did the scientific community finally come to appreciate McClintock's data and conclusions.

In certain ways, McClintock followed in Mendel's footsteps, sequestered in her lab and quietly tending her corn plants each summer. She backed her theory with hundreds of individual experiments and never wavered in her belief that genes can indeed move about, despite both the dogma that genes stay put except for meiotic crossing over, and the indifference with which her work was greeted. Some molecular geneticists were slow to accept McClintock's ideas because they lacked her experience with and understanding of the whole organism and because some also tended to discount studies that rely on less sophisticated methods than their own. It is fortunate, nevertheless, that the significance of her work was recognized before the end of her sixty-year career in genetics. McClintock's story should inspire research committees, funding agencies, and biologists in general to consider carefully all high-quality scientific work, even if the results seem anomalous at the time.

	AGA							UUA			AGC									
	AGG							UUG			AGU									
GCA	CGA						GGA	CUA				CCA	UCA	ACA			GUA			UAA
GCC	CGC						GGC	CUC	AUA			CCC	UCC	ACC			GUC			UAG
GCG	CGG	GAC	AAC	UGC	GAA	CAA	GGG	CAC	AUC	CUG	AAA		UUC	CCG	UCG	ACG		UAC	GUG	UGA
GCU	CGU	GAU	AAU	UGU	GAG	CAG	GGU	CAU	AUU	CUU	AAG	AUG	UUU	CCU	UCU	ACU	UGG	UAU	GUU	
Ala	Arg	Asp	Asn	Cys	Glu	Gln	Gly	His	Ileu	Leu	Lys	Met	Phe	Pro	Ser	Thr	Trp	Tyr	Val	stop

Figure 13-9   THE GENETIC CODE: ROSETTA STONE OF MOLECULAR GENETICS.
Each of the twenty amino acids is represented by at least one three-base codon. The genetic code is degenerate because most of the amino acids are specified by several codons. The degeneracy of the genetic code serves a protective function; the many codons for a single amino acid often are quite similar; for example, four of the six codons for leucine begin with CU, no matter what the third base. Because of this similarity, a point mutation in the third place will not lead to an incorrect amino acid being placed in a protein. Note that AUG, which codes for methionine, is a start signal for protein synthesis. There are three stop signals: UAA, UAG, and UGA.

THE CONCEPT OF THE GENE REFINED

With their knowledge of the patterns of inheritance, the structure of DNA, the nature of the genetic code, and the translation of DNA into protein, biologists finally could develop a complete working concept for the gene.

As we have seen, the earliest definition of the gene was Mendel's (even though he did not use the word "gene"): a factor that specifies one of two alternatives for a given physical (phenotypic) characteristic. According to this definition, the gene is a unit of hereditary function. When geneticists began to study genes more closely and realized that a gene could be converted to any of several forms (alleles) by mutation, they decided that the gene is also a unit of mutation.

When biologists further realized that genes are arranged on chromosomes, they thought that any two mu-

tations that occurred at different locations and that could recombine must represent different genes. Thus the gene was thought to be a unit of recombination. (We now know that recombination can occur between any two nucleotides and therefore need not involve the entire gene.)

How, then, shall we define "gene"? The most up-to-date definition of the gene, based on the one-gene–one-polypeptide hypothesis, is that the gene is *a unit of DNA that codes for a polypeptide or a structural RNA molecule.* This definition is supported by data on DNA and proteins generated by new techniques of gene manipulation (described in Chapter 14). Notice that our modern definition is functional, like Mendel's, even though it is far more specific. A gene controls a specific function via a polypeptide or RNA, but it also controls a larger, more general function—the expression of an organism's physical characteristics.

In later chapters, we will see that the concept of a gene has been modified still further. This is because, in eukaryotes there often are numerous lengths of DNA that are transcribed into RNA but are *not* translated into polypeptides. Furthermore, some genes can code for more than one polypeptide because of ways that RNA is processed in cell nuclei.

Regardless of the changing definitions of the gene and our broadened understanding of gene structure, one unifying concept stands out: all living organisms, from *E. coli* to moss on damp rocks to soaring eagles, share a genetic code, with similar molecular structure and the same basic form of translation into proteins. This knowledge helps us appreciate the profound unity of all life, and, as we shall see, evolutionary relationships among all species.

SUMMARY

1. Yanofsky and his group demonstrated that the sequence of gene *codons* is colinear with the amino acids in the corresponding polypeptide.

2. Additional research revealed several important aspects of the *genetic code:* (1) it is nonoverlapping; (2) it is

translated in a fixed *reading frame;* (3) it is degenerate, meaning that several codons can specify the same amino acid; and (4) the codons are nucleotide triplets along one of the chains of a DNA double helix. There are sixty-four possible codons to code for the twenty amino acids in

organisms. Among the evidence for these conclusions are *frameshift mutations,* which involve deletion or addition of nucleotides to DNA so that the reading frames of the gene are shifted.

3. The genetic code is nearly universal, reflecting common descent of

all organisms from a single ancestral source.

4. Proteins are made only in the cytoplasm. Thus mechanisms must exist in eukaryotes for the DNA instructions to be carried from the nucleus to the cytoplasm. This function is fulfilled by RNA, which differs from DNA in that it usually is single stranded, contains the sugar ribose, and has the base *uracil* in place of thymine.

5. There are several types of RNA, all of which are made as complementary copies of the DNA nucleotide sequence. The copying process is called *transcription*. *Messenger RNA (mRNA)* serves as a template for protein synthesis. Each of the types of *transfer RNA (tRNA)* picks up a specific amino acid at its *aminoacyl attachment site;* another tRNA site, the *anticodon*, then can bond to the complementary codon of the mRNA. *Ribosomal RNAs (rRNA)*, in combination with a number of proteins, make up the ribosomes. They are sites where proteins are assembled.

6. Protein synthesis, or *translation*, involves four steps: (1) in *amino-acid activation*, a specific amino acid becomes attached to a specific tRNA molecule; (2) at *initiation* of polypeptide synthesis, the small ribosomal subunit and the initiator tRNA bind the mRNA at the initiator codon, where they are joined by the large ribosomal subunit and the second tRNA; (3) *elongation* of the polypeptide occurs as the ribosome moves along the mRNA molecule from codon to codon and as peptide bonds are formed to link the amino acids into a polypeptide (the enzyme *peptidyl transferase* catalyzes the linkage process); (4) *termination* of protein synthesis takes place when a termination codon is reached on the mRNA molecule.

7. The small ribosomal subunit has the primary task of ordering the amino acids according to the mRNA codon sequence, whereas the large ribosomal subunit has the primary task of linking amino acids to the elongating polypeptide chain.

8. Nirenberg and Matthaei cracked the genetic code when they discovered that a long artificially prepared mRNA molecule composed solely of uracil directed the synthesis of a polypeptide that contained only phenylalanine. Later Khorana synthesized RNA triplets of all possible codons, allowing elucidation of the complete genetic code.

9. The gene has been defined as a unit of hereditary function, a unit of mutation, a unit of recombination, and a unit that codes for a polypeptide or a structural RNA.

KEY TERMS

amino-acid activation

aminoacyl attachment site

aminoacyl-tRNA synthetases

anticodon

A site

codon

elongation

frameshift mutation

genetic code

initiation

messenger RNA (mRNA)

peptidyl transferase

promoter

P site

reading frame

ribosomal RNA (rRNA)

termination

transcription

transfer RNA (tRNA)

translation

uracil

QUESTIONS AND PROBLEMS

1. A synthetic polymer of random sequence containing the bases U and G in a 5:1 ratio was made. In this polymer, the triplet UUU should occur five times more often than triplets containing two Us and a G, and 25 times more often than triplets containing one U and two Gs; UUU should occur about 125 times more often than GGG. Nirenberg and Matthaei used this polymer to direct polypeptide synthesis in the cell-free system. Then they measured the proportions of the various amino acids in the polypeptide chains. Now look at the codon table (Figure 13-9) and determine which amino acid(s) was (were) most abundant. Which were present at about ⅕ that amount? Which were the least abundant?

2. Examine the genetic code (Figure 13-9). Do the codons have polarity? Is UUG different from GUU? Which way do we read them: 5′ to 3′, or 3′ to 5′? In which direction is mRNA translated into protein?

3. Khorana and his co-workers devised a method of synthesizing RNA molecules of *known* sequence: for instance, UGUGUGUGUG. . . . What two codons are present in this polymer? When this RNA is used to direct protein synthesis in a cell-free system, what two amino acids would you expect to find in the polypeptides? What would be their sequence? In the repeating RNA molecule AUCAUCAUCAUC . . . , what three codons are present? Which amino acids do they code for? If the code were overlapping, what sequence of amino acids would you expect in the polypeptide? If the code is not overlapping, but instead is translated from a fixed starting point in groups of three, what

amino-acid sequences would you expect to find in the three possible polypeptides that could be formed? On genuine mRNA, how do the ribosomes "know" where to start translation?

4. Bacteriophage T$_4$ lysozyme contains 164 amino acids in a single polypeptide chain. What is the minimum number of base pairs in the lysozyme gene? This is the base sequence of part of the mRNA for T$_4$ lysozyme:

5'-AGGAGGUAUUAUGAAUAUAU
UUGAAAUGUUACGUAUA . . . 3'

Write the base sequence of the DNA strand from which the mRNA was transcribed. Below it write the base sequence of its complementary strand. Does one of these match the mRNA? Now pretend you are a ribosome. Bind to the A- and G-rich ribosome binding site at the 5' end of the mRNA, find the nearest initiation codon, and start translating the message into protein. What are the first five amino acids? What amino acid is encoded by the initiation codon? Can this amino acid occur within a polypeptide as well as at the beginning?

5. Suppose a mutation has occurred in the DNA in Problem 4 so that the underlined U is changed to C. How will that mutation affect the amino-acid sequence? How many amino acids can be changed by a single base change?

6. This is the amino-acid sequence of a section in the middle of the lysozyme molecule; the corresponding codons on the mRNA are written below:

. . . thr-lys-ser-pro-ser-
leu-asn-ala-ala-lys-ser-
glu-leu-asp

. . . ACA AAA AGU CCA UCA
CUU AAU GCU GCU AAA UCU
GAA UUA GAU . . .

A mutant strain #1 contains a frameshift mutation in the DNA corresponding to the underlined region: one of the As is deleted. Write the mutant base sequence and the corresponding amino-acid sequence for this region. How many amino acids can be changed by a single frameshift mutation?

7. During protein synthesis, do the amino acids pair up with their corresponding codons in the mRNA? Or does tRNA pair with the codon? Does each codon actually specify an amino acid, or does it specify the tRNA?

8. Which of the following descriptions applies to DNA, which to RNA, and which to both? Contains a five-carbon sugar; contains phosphate; contains the base adenine; contains the base uracil; contains the base thymine; is synthesized from a DNA template; is synthesized in the 5' to 3' direction; in eukaryotes is made in the nucleus.

9. Many different enzymes and ribosomal proteins are involved in the synthesis of any particular protein. If each protein requires, say, 20 different types of proteins for its synthesis, would each of those proteins require 20 more, and each of those 20 more, until millions of proteins were required to make a single protein? Why or why not? How can a cell make many thousands of different kinds of proteins?

10. Which of the following statements applies to mRNA? To rRNA? To tRNA? To all three? Leaves the nucleus in eukaryotes after it is transcribed; it is transcribed from DNA; it is transcribed in the 5' to 3' direction; it contains an amino-acid attachment site and an anticodon; it carries the blueprint for synthesis of a specific polypeptide; it acts as an adapter between an amino acid and its codon; it combines with a number of proteins to form a structure which

acts as a movable vise that holds the blueprint and the subunits for protein synthesis.

ESSAY QUESTION

1. Define the word "gene." Incorporate all the features of a gene that you have learned about so far.

SUGGESTED READINGS

These three review articles summarize the evidence on the major events of protein synthesis.

CASKEY, C. T. "Peptide Chain Termination." *Trends in Biochemical Science* 5 (1980): 234–237.

CLARK, B. "The Elongation Step of Protein Biosynthesis." *Trends in Biochemical Science* 5 (1980): 207–210.

HUNT, T. "The Initiation of Protein Synthesis." *Trends in Biochemical Science* 5 (1980): 178–188.

These four excellent review articles summarize the genetic code and colinearity.

CRICK, F. H. C. "The Genetic Code." *Scientific American*, October 1962, pp. 66–74.

———. "The Genetic Code III." *Scientific American*, October 1966, pp. 55–62.

NIRENBERG, M. W. "The Genetic Code II." *Scientific American*, March 1963, pp. 80–94.

STRYER, L. *Biochemistry.* 2d ed. San Francisco: Freeman, 1981. A fine, well illustrated summary of the topics discussed in this chapter.

YANOFSKY, C. "Gene Structure and Protein Structure." *Scientific American*, May 1967, pp. 80–90.

14

BACTERIAL GENETICS, GENE CONTROL, AND GENETIC ENGINEERING

As the relentless revolution in genetics continues year after year, it becomes clear that we are learning enough about life's molecular dance to begin to become its choreographers.

Boyce Rensberger, "Tinkering with Life," *Science 81* (November 1981)

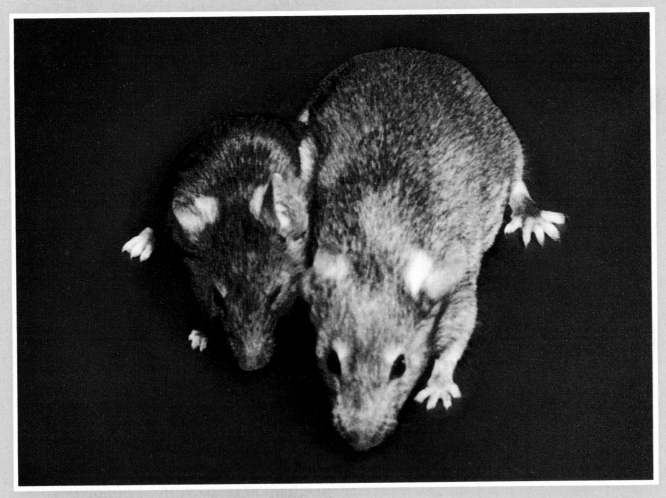

GENETIC ENGINEERING.
One of these two twenty-four-week-old mice is normal; the other contains the human growth-hormone gene in the chromosomes of all its cells.

The world will beat a path to the door of the person who builds a better mousetrap, as the familiar saying goes. But how about the person who builds a better *mouse?* A team of scientists from several universities has produced mice that grow twice as fast as normal because they make greater-than-normal quantities of growth hormone, a protein that stimulates proliferation of muscle, cartilage, and bone cells. The unimaginative might say: Fine, but who needs giant mice? However, it is a conceptually small step from bigger mice to faster growing cattle, sheep, rice, or corn.

The scientists who created the giant mice accomplished their feat using what has been called **genetic engineering,** procedures dependent on **recombinant DNA technology.** This technology involves the manipulation and transfer of genes from one organism to another in the laboratory and is the most exciting advance in genetics since Watson and Crick described the structure of DNA. In a single decade, recombinant DNA techniques have led to results unimaginable even in 1980, such as giant mice and bacteria that can make human insulin. The promises of genetic engineering are even greater: cereal crops that can fix nitrogen from the air into biologically useful compounds; bacteria that can detoxify industrial pollutants; and blue-green algae that can generate massive amounts of milk protein (casein) for the world's hungry, to name but a few.

The techniques for selecting genes and transferring them from one organism to another grew directly from work on the genetics of bacteria. This work, in turn, was a logical outgrowth of the genetic discoveries we considered in Chapters 12 and 13. There we learned that a gene is a portion of DNA that codes for a polypeptide or for structural RNA. We know how genes are transcribed and are translated into polypeptides. In this chapter, we see how they are regulated—turned on and off. This control helps us understand how a prokaryotic cell can use nutrients from its environment efficiently, and it tells us how genes are used in the right eukaryotic cell at the right time. In fact, the key to understanding gene regulation in prokaryotes was the discovery of the various ways that bacteria can exchange genes in nature. Answers of great significance soon emerged. And once geneticists understood the transfer, location, and control of genes in bacteria, they were able to study these phenomena in higher organisms. Eventually, they learned to transfer genes between species in the laboratory.

In this chapter, we describe how genes are transferred between bacteria in nature and how they are controlled. We discuss the more complex structure, control, and expression of eukaryotic genes and the recombinant DNA techniques that grew from such genetic studies. Finally, we explore the promises of, and concerns about, the purposeful manipulation of genes. The path from Watson and Crick to recombinant DNA was traversed in just twenty years; one can scarcely imagine what two more decades of progress in this field will bring. But it is no exaggeration to predict that recombinant DNA technology will have an impact on human society equivalent to that of the steam engine, the electric light, and the use of antibiotics to fight disease. We are truly at the dawn of a new age in medicine and agriculture because of our new-found ability to transplant genes.

BACTERIA EXCHANGE GENES SEXUALLY AND ASEXUALLY

As we have seen, bacteria were key organisms in the elucidation of DNA structure and gene expression during protein synthesis, just as *Drosophila* played a key role in studies on non-Mendelian patterns of inheritance. Quite logically, geneticists wanted to develop a gene map of the circular chromosome of a common bacterium such as *E. coli*, just as they had mapped many genes on the four chromosomes of the fruit fly. Such gene maps of various bacterial species, along with the new understanding of gene behavior at the molecular level, would allow scientists to study the development of bacterial phenotypes in unprecedented detail. However, there were two significant stumbling blocks to mapping genes and their expression in bacteria: (1) In order to study how genes are passed from one generation to the next, one must have two pure-breeding strains that differ in some easily recognized trait. Whereas pea plants and fruit flies have flowers, eyes, and so on that vary in size, shape, and color, bacteria are too small to exhibit easily visible phenotypic traits. (2) The passage of genes from one generation to the next—so-called transmission genetics—in bacteria differs from that in higher organisms. A bacterium has a single circular chromosome; it lacks a nuclear envelope; and the bacterial cell, lacking microtubules, spindles, and multiple chromosomes, does not undergo mitosis before it divides in two. The single, circular chromosome is duplicated, of course, with one copy passing to each daughter cell during the splitting of the cytoplasm in two, so-called *fission.* Finally, bacteria do have a means of exchanging genes sexually, even though there is no meiosis. In bacteria, therefore, Mendel's rules apply in principle but not in major details.

Detection of Bacterial Phenotypes

Geneticists had to devise special ways to study phenotypes and gene transmission in bacteria, but those techniques eventually paid great dividends. Because they could not rely on visible phenotypic traits, geneticists learned to use biochemical characteristics in their study of genetic variation in bacteria. Bacteria exhibit three kinds of biochemical phenotypes that can be detected and compared: (1) resistance to or sensitivity to certain antibiotics, (2) abnormal nutritional requirements, and (3) the inability to use certain compounds for growth.

Although normal *E. coli* cells cannot grow in the presence of antibiotics such as streptomycin, mutants can be isolated that are *resistant* to streptomycin. The procedure for isolating these mutants is summarized in Figure 14-1. A bacterial suspension is first treated with a chemi-

cal or physical agent (such as x-rays) known to cause mutations. Millions of tiny bacterial cells then are spread on a dish that contains several substances: *agar*, a rubbery gel derived from seaweed; a *minimal medium* mixed into the gel—a simple nutrient solution of salts, glucose, and ammonia that will allow normal bacteria to survive (these chemicals provide inorganic ions plus sources of energy, carbon, and nitrogen); and the antibiotic streptomycin. Once the cells are spread on the culture medium and incubated, all the normal cells will die, but any cells with the mutated gene for resistance to the antibiotic will survive and produce a colony. So even though millions of bacteria are not mutated by the x-rays or chemicals and do not survive in the streptomycin-treated cultures, chances are good that one or two mutant cells will divide repeatedly and form visible colonies on the surface of the agar. These mutant bacteria can then be used for subsequent genetic analysis, since they obviously have a gene that is different from the wild-type gene for the trait of sensitivity to streptomycin.

To identify mutant bacteria with nutritional deficiencies, researchers use a different technique. Normal bacterial cells can make all the sugars, amino acids, nucleotides, fatty acids, and enzyme cofactors they need for growth from the few chemicals present in a minimal medium. Some mutants, called **auxotrophs,** are unable to make all the necessary growth factors from minimal medium. Because auxotrophs cannot survive on minimal medium, geneticists use the method shown in Figure 14-2 to isolate them.

Suppose, for example, that a deleterious mutation occurs in a gene that codes for an enzyme needed to make the amino acid tryptophan. Although such a mutated cell could not grow on a minimal medium, it could be "rescued" by the addition of tryptophan to the medium. But then how could the mutant cell be separated from the hundreds of thousands of normal cells also growing in the tryptophan-enriched medium? As Figure 14-2 shows, the separation process involves a procedure called *replica plating*. Bacteria are grown on a petri dish with tryptophan-supplemented medium on which both mutant and normal bacteria will form colonies. A piece of velvet is pushed gently onto the first dish, and some bacteria from each colony stick to it. The velvet is then

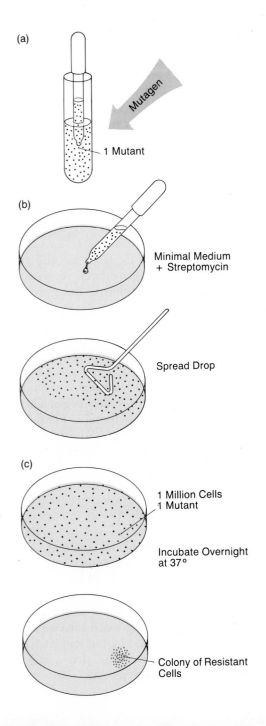

(a)

Mutagen

1 Mutant

(b)

Minimal Medium + Streptomycin

Spread Drop

(c)

1 Million Cells
1 Mutant

Incubate Overnight at 37°

Colony of Resistant Cells

Figure 14-1   ISOLATING ANTIBIOTIC-RESISTANT MUTANTS.

(a) A suspension of normal bacteria is treated with a mutagen to cause mutations. (b) The researcher then spreads the bacteria on a petri dish containing agar with minimal medium, plus the antibiotic streptomycin. (c) Although the normal cells will die, any bacterial cells with the mutated gene for resistance to streptomycin will survive and multiply into colonies.

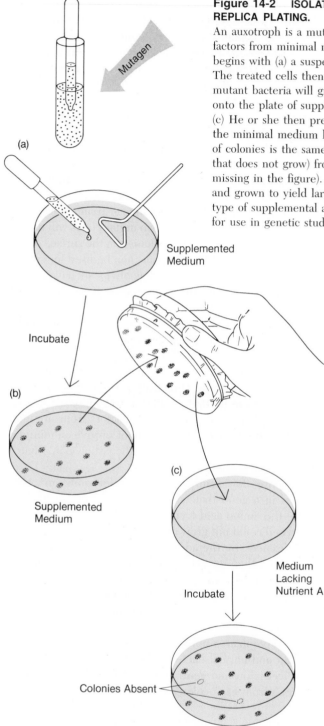

Figure 14-2   ISOLATING NUTRITIONAL MUTANTS BY REPLICA PLATING.

An auxotroph is a mutant bacterial cell that cannot make all its own necessary growth factors from minimal medium. The laboratory procedure for isolating such mutants begins with (a) a suspension of normal bacterial cells that is treated with a mutagen. The treated cells then are allowed to grow on supplemented medium. Both normal and mutant bacteria will grow colonies. (b) The researcher next pushes a piece of velvet onto the plate of supplemented medium; cells from some of the colonies will stick to it. (c) He or she then presses that piece of velvet against the surface of a plate containing the minimal medium lacking a specific nutrient, say, A. Because the spatial distribution of colonies is the same on both plates, one can assume that any colony missing (that is, that does not grow) from the minimal-medium plate is an auxotroph (two colonies are missing in the figure). Such mutant colonies can then be retrieved from the first plate and grown to yield large numbers of the mutated auxotrophic cells. Depending on the type of supplemental and minimal media used, any type of auxotroph can be isolated for use in genetic studies.

tryptophan. Using this procedure, bacterial geneticists have isolated auxotrophs for hundreds of biochemicals required for cell metabolism.

A third type of mutant bacteria used in genetics research cannot use certain nutrients. Normal *E. coli* cells grow just as readily on many other kinds of sugars as they do on glucose in a minimal medium. Bacteria have enzymes that can convert more complex sugars, such as lactose, into simpler, more readily usable ones, such as glucose. But if the gene that codes for one of those enzymes is defective, the mutant bacterium cannot grow on a lactose medium even though it can still grow on a minimal glucose-containing medium. As we shall see, mutants that can grow on glucose but not on lactose played a major role in studies on gene control.

Bacterial Gene Transmission

Once bacterial geneticists had isolated mutants for different biochemical phenotypes, they began to study how different traits are inherited in bacteria. For many years, it seemed to them that bacteria lead a celibate life—it was difficult to detect mating and subsequent exchange of genes between bacterial strains. In 1946, however, Joshua Lederberg, a student, and his professor, Edward Tatum, performed an elegant experiment at Yale University demonstrating that "sex" occurs in bacteria.

Lederberg and Tatum worked with two strains of bacteria, which were auxotrophic for different substances. For strain A to grow, both the amino acid methionine and the vitamin biotin had to be added to the minimal medium. For strain B to grow, the minimal medium had to be supplemented with the amino acids threonine and leucine. Since strain A could make threonine and leucine but not methionine and biotin, its genotype could be represented as met$^-$ bio$^-$ thr$^+$ leu$^+$, and strain B would then have the genotype met$^+$ bio$^+$ thr$^-$ leu$^-$.

pressed against the surface of a second petri dish, this one containing a minimal medium lacking tryptophan on which the mutant cannot grow. Because the spatial distribution of the colonies is the same on both plates, any colony that fails to appear on the minimal medium but does appear on the supplemented plate can be assumed to be a mutant for an enzyme needed in the synthesis of

Neither strain could grow on pure minimal medium, which lacks amino acids and vitamins. When Lederberg and Tatum mixed the two strains, they found that 1 in every 10 million cells could grow on minimal medium. The surviving cells must have had the genotype met$^+$ bio$^+$ thr$^+$ leu$^+$, since they could make all the amino acids and vitamins necessary for growth. Therefore, they must have received some genes from strain A and some from strain B. The meaning of the results was clear: the bacteria were exchanging genes.

Since those pioneering experiments were done, researchers have discovered that bacteria exchange genetic information in several ways. Among them are the processes of *conjugation, transformation,* and *transduction.*

Conjugation

The kind of gene exchange discovered by Lederberg and Tatum is called **conjugation.** In bacterial conjugation, DNA is transferred from one cell to another by direct contact. The donor cell, analogous to a male sexual organism, produces a *sex pilus,* a cytoplasmic bridge between the cells through which DNA is transferred to the recipient cell, analogous to a female organism (Figure 14-3). Male cells contain the ("fertility") **F factor,** a piece of DNA that codes for the sex pilus. Male cells are therefore designated F$^+$, while recipient cells are designated F$^-$. The F factor is a set of genes on a small circle of double-stranded DNA called a **plasmid** that is separate from a bacterium's single chromosome. Plasmids have been called minichromosomes; they carry a handful of genes that usually are not necessary for the bacterium's survival. Importantly, however, plasmids can replicate autonomously in the cell's cytoplasm. As we shall see later, those remarkable little DNA circles were the key to genetic engineering.

During conjugation, one DNA strand of the circular F factor plasmid is nicked, or broken, and then this single strand is transferred through the sex pilus to the recipient F$^-$ cell (Figure 14-4). The one strand of DNA left in the donor is replicated and becomes double stranded once again. In the F$^-$ cell, the complementary DNA strand is synthesized to form a double-stranded-plasmid DNA molecule (an **F factor plasmid**) which once again takes the form of a circle. Since the F factor has been transferred to the recipient cell, that cell is converted to an F$^+$. Maleness is contagious in bacteria!

The usefulness of the F factor plasmid to a bacterial geneticist depends on a few additional features. Occasionally, the F factor plasmid becomes inserted into the bacterial chromosome so that the two form one double-stranded circle of DNA. Thereafter, when that cell and its progeny divide, the full chromosome will be duplicated; hence, any genes that the F factor plasmid carried

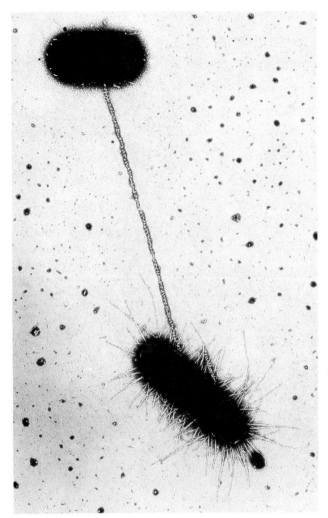

Figure 14-3   BACTERIAL CONJUGATION.

The *E. coli* cells shown in this electron micrograph (magnified about 18,000 times) are joined by a sex pilus during conjugation. DNA is transferred from the male, or F$^+$, cell (bottom) to the female, or F$^-$, cell (top) through the pilus. An F$^+$ cell can conjugate with more than one cell simultaneously.

will be inherited by all the progeny of that cell. The full lineage of dividing bacterial cells will contain both the F factor genes and the chromosomal genes (Figure 14-5). So the F factor plasmid can exist in two forms: free in the cytoplasm (extrachromosomal) or integrated into the bacterial chromosome. Genetic elements that can replicate in a bacterial host independently of the chromosome or that can integrate into the bacterial chromosome and replicate with it are called **episomes.**

The integration of the F factor plasmid into a bacterial chromosome converts the F$^+$ cell into an **Hfr cell,** so called because it exhibits a *high frequency of recombination* when mixed with genetically different F$^-$ cells. An Hfr cell, like an F$^+$ cell, is a donor during conjugation

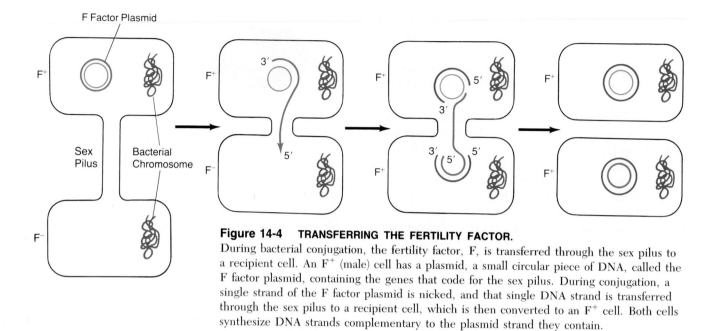

Figure 14-4   TRANSFERRING THE FERTILITY FACTOR.
During bacterial conjugation, the fertility factor, F, is transferred through the sex pilus to a recipient cell. An F$^+$ (male) cell has a plasmid, a small circular piece of DNA, called the F factor plasmid, containing the genes that code for the sex pilus. During conjugation, a single strand of the F factor plasmid is nicked, and that single DNA strand is transferred through the sex pilus to a recipient cell, which is then converted to an F$^+$ cell. Both cells synthesize DNA strands complementary to the plasmid strand they contain.

because it forms a pilus, but with the important difference that when the F factor of the Hfr cell enters the recipient cell, *it can take the entire bacterial chromosome with it.* The recipient cell thus will have two chromosomes, with two copies of most bacterial genes. If the donor and the recipient are genetically different, any recombination between the two **genomes**—the full set of genes on each organism's chromosomes—can be de-

tected, because traits from both original types will be displayed. This is in fact what happened in Lederberg and Tatum's experiments. They had unwittingly chosen an Hfr and an F$^-$ strain for their experiments, and this bit of serendipity led to a means of mapping the bacterial chromosome.

It takes a substantial period of time—about ninety minutes—for an entire strand of a bacterial chromosome

Figure 14-5   INTEGRATING THE F FACTOR PLASMIDS.
When the F factor plasmid is integrated into a bacterial chromosome, the F$^+$ cell is converted into an Hfr cell. The F factor plasmid can exist in two forms, either separate from or integrated into the bacterial chromosome. The integration of the F factor plasmid into a bacterial chromosome of an F$^+$ cell causes the resulting Hfr cell to exhibit a high frequency of recombination when mixed with F$^-$ cells. The reason is that when the F factor plasmid is passed to the F$^-$ cells through sex pili, it takes along much of the bacterial chromosome. The new cell thus will have two copies of most bacterial genes, and recombination can occur between the two sets.

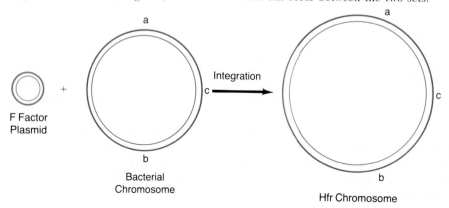

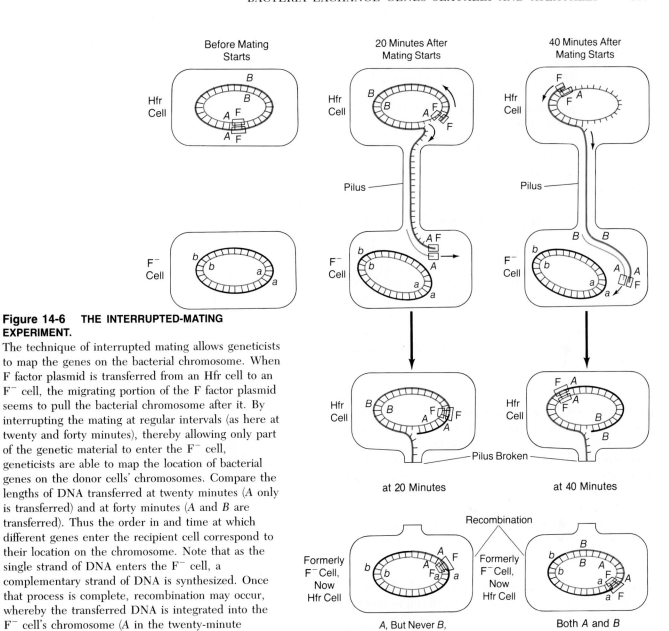

Figure 14-6   THE INTERRUPTED-MATING EXPERIMENT.
The technique of interrupted mating allows geneticists to map the genes on the bacterial chromosome. When F factor plasmid is transferred from an Hfr cell to an F$^-$ cell, the migrating portion of the F factor plasmid seems to pull the bacterial chromosome after it. By interrupting the mating at regular intervals (as here at twenty and forty minutes), thereby allowing only part of the genetic material to enter the F$^-$ cell, geneticists are able to map the location of bacterial genes on the donor cells' chromosomes. Compare the lengths of DNA transferred at twenty minutes (A only is transferred) and at forty minutes (A and B are transferred). Thus the order in and time at which different genes enter the recipient cell correspond to their location on the chromosome. Note that as the single strand of DNA enters the F$^-$ cell, a complementary strand of DNA is synthesized. Once that process is complete, recombination may occur, whereby the transferred DNA is integrated into the F$^-$ cell's chromosome (A in the twenty-minute experiment; A and B in the forty-minute case).

to be transferred from an Hfr to an F$^-$ cell. Since the F factor plasmid seems to force the strand of chromosomal DNA to which it is attached into the F$^-$ cell, the first chromosomal genes to be transferred are those nearest the site where the F factor plasmid is inserted into the cellular chromosome. The distance between a gene and the F factor plasmid is reflected in a corresponding time required for the chromosome to pass through to the point where the gene has entered the F$^-$ cell. Consequently, geneticists can *map* the genes on the bacterial chromosome by mixing Hfr and F$^-$ cells, allowing sex pili to form, and then interrupting the conjugation process at progressively longer time intervals. In such an *interrupted-mating* experiment, the order in which and

time at which different genes enter the recipient cell correspond to their location on the chromosome. Figure 14-6 shows how this is done in the laboratory.

Transformation and Transduction: Indirect Gene Transfer

The transfer of genetic material between bacteria also can take place even if the cells do not come into direct contact, as they do in conjugation. During the process of **transformation,** DNA that has been released from one cell into the surrounding medium is taken up by another cell. The release of DNA is usually the result of cell disruption or cell death. Recall from Chapter 12 the experiments in which nonvirulent rough-type pneumo-cocci were transformed into virulent smooth-type pneumococci by taking up DNA released by smooth cells that had been disrupted in a blender. These transformation experiments provided one early clue that DNA is the hereditary material. We shall see later that the transformation process is now employed by genetic engineers to help transfer genes from one organism to another.

In **transduction**, DNA is carried from one bacterium to another by a virus. Most *bacteriophages* (viruses that infect bacteria) consist of a strand of DNA surrounded by a protein coat (Chapter 12). When new phages are being assembled in the cytoplasm of a host bacterial cell, phage coats encapsulate copies of viral DNA formed in the infected bacterium; mistakes sometimes occur, so that a coat encloses a piece of the bacterial chromosomal DNA along with phage DNA (Figure 14-7). The new phage can inject the bacterial DNA into another bacterium. If the injected DNA recombines into the DNA of the recipient bacterial cell, *phage-mediated transduction* has occurred.

Transposons Move Genes About

Another type of gene exchange in bacteria deserves mention. In 1955, a serious epidemic of bacterial dysentery broke out and presented physicians with an even more distressing situation than in a normal outbreak. A strain of the bacterium *Shigella dysenteriae* became resistant simultaneously to several of the antibiotics being used to treat the patients. The genes, or **R factors**, that provided resistance to these drugs were not on the bacterial chromosome but were carried on ("resistance") **R factor plasmids**, pictured in Figure 14-8. To make matters worse, some of these plasmids also carried genetic elements that enabled the rapid transfer of the R factor plasmids from one bacterial cell to another. Multiple drug resistance thus spread like wildfire through the infecting bacterial population, causing a medical crisis. Fortunately, the dysentery epidemic eventually abated.

The genetic elements responsible for the transfer of these drug-resistant genes between plasmids are called *transposons*. They are DNA segments that owe their mobility to *insertion sequences*, short stretches of DNA that can insert a gene into a variety of chromosomal sites. Transposable genetic elements have been discovered in eukaryotes as well as in prokaryotes, functioning in both to move genes about. Thus in both fruit flies and corn, transposons may move from one site to another among the chromosomes and thereby alter gene activity. We will see later that a high proportion of eukaryotic DNA may be made up of transposable elements and that these elements may contribute to the origin of new genes and even of new species. Although the eukaryotic genome was once believed to be very stable, this discovery tells us that it may not be nearly as constant as once thought.

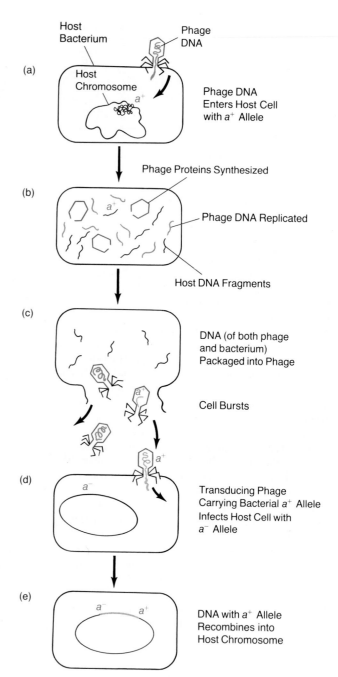

Figure 14-7   PHAGE-MEDIATED TRANSDUCTION.
(a) Phage DNA enters the host bacterial cell. (b) The phage DNA replicates and is transcribed; phage proteins begin to be synthesized. One result is that the host DNA breaks into fragments. (c) DNA becomes packaged inside phage heads; occasionally pieces of bacterial DNA, in addition to phage DNA, are enclosed in phages. (d) When a phage containing bacterial DNA infects a cell, recombination may occur between the DNA of the donor bacterium carried by the phage and the host cell DNA. (e) If the recombined donor DNA contains an allele different from that of the host cell (e.g., a^+ or a^-), transduction can be recognized because the new gene product produces an altered phenotype in the recipient cell and its progeny.

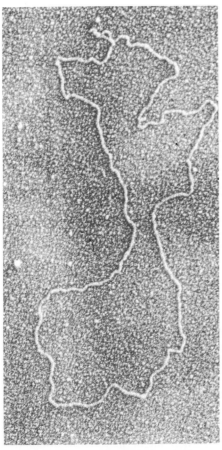

Figure 14-8   AN R FACTOR PLASMID.
This circular DNA molecule (magnified about 82,000 times) contains genes that confer resistance to antibiotics. Antibiotic resistance can be transmitted from bacterium to bacterium by transduction. During an epidemic, this type of infectious resistance can have dire medical consequences.

Figure 14-9   A GENE MAP OF *E. COLI*.
The genes on this map represent only a small fraction of the 1,000 known genes of *E. coli*. Indeed, there is enough DNA in the circular chromosome to code for 5,000 polypeptides. The minutes around the inner circle signify times of entry into a cell during interrupted-mating experiments. Note that genes for enzymes involved in related tasks often occur close together on the genome—for example, the lac site, which is really three genes that code for polypeptides involved in lactose utilization.

USE OF DNA TRANSFER PROCESSES TO MAP PROKARYOTIC GENES

Geneticists have used conjugation, transformation, and transduction to help them map about 2,000 of the estimated 3,000 genes in *E. coli*. Some of these genes are shown in the map in Figure 14-9. Notice that the map represents the circular molecule of double-stranded DNA that comprises the *E. coli* chromosome. One interesting revelation of this gene map is that genes for enzymes involved in related tasks are frequently grouped close together. For example, five genes for tryptophan (trp) synthesis appear at "25 minutes" on the map, and three genes for lactose (lac) metabolism appear at "10 minutes."

Not long after the first bacterial gene maps were produced, careful experiments revealed that the clustering of related genes allows precise control of gene expression, so that a cell makes tryptophan or breaks down lactose at the right time. Thus the study of how bacteria exchange genes in nature led directly to a basic understanding of genetic processes and, later, to methods for the deliberate transfer of genes between organisms.

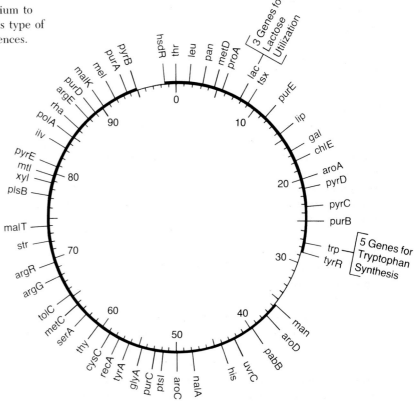

BACTERIAL GENES: SUBJECT TO PRECISE REGULATORY CONTROL

The clustering of related bacterial genes seemed remarkable at first because mapping of fruit-fly chromosomes had revealed that the many genes governing eye color or wing shape are scattered on separate chromosomes. Why, then, would all the bacterial genes related to a given trait occur in a single place? The answer turned out to be that in prokaryotes, clustering of functionally related genes makes it physically possible for certain regulatory molecules to control expression of those batteries of genes, not just single genes. These processes are called **gene regulation.**

Regulation of gene expression is crucial to the efficient utilization of a cell's limited resources, so at any given moment, the cells produce only those materials it needs. An efficiently functioning cell should not produce the enzymes for metabolizing lactose if that sugar is absent from the medium; likewise, it would be wasteful for a cell to make tryptophan-synthesizing enzymes if tryptophan is readily available. Cells that have the ability to produce certain enzymes only when they are needed can divide a little faster than cells that squander their resources making unnecessary proteins. Prokaryotes have evolved several important regulatory mechanisms. The first such mechanism that geneticists studied in detail governs the utilization of lactose by an *E. coli* cell.

The Operon Model of Gene Regulation

We have said that an efficient cell produces a given enzyme only when that protein is needed. But how is the supply of the enzyme turned on or off so the cell's precious energy is not wasted making an unnecessary product? There are several levels of regulation; some act on existing enzyme molecules, and others, on the production of new molecules. As we saw in Chapter 4, an enzyme's activity can be regulated at the level of the enzyme itself, through the presence or absence of cofactors, or by degradation of the enzyme molecules. Another mechanism involves end-product inhibition, in which the product of a biochemical pathway provides negative feedback and acts on the first enzyme in that pathway to inhibit overproduction of the product. In prokaryotes, the amount of enzyme present in a cell is usually controlled at the level of mRNA synthesis. The details of regulation at this level were elucidated in the late 1950s and early 1960s by the French geneticists François Jacob and Jacques Monod. In their work with

E. coli, Jacob and Monod found a cluster of adjacent genes that function together to make the enzymes for breaking down lactose. They called this cluster of genes an *operon.* Let us see how the operon works to regulate lactose metabolism.

Regulation of Lactose Metabolism

Lactose is a disaccharide that is broken down by the enzyme β-galactosidase into its two constituent sugars, galactose and glucose. If no lactose is present in the cell's environment, fewer than ten molecules of this enzyme are found in an average *E. coli* cell. However, if lactose becomes available to that same cell (such as occurs with the bacteria growing in your intestine after you have eaten a bowl of ice cream), each cell will soon contain several thousand molecules of the β-galactosidase needed to break down the lactose. Two other lactose-pathway proteins (β-galactoside permease, which transports lactose, and thiogalactoside transacetylase) increase in quantity at the same time. Clearly, the presence or absence of lactose is what regulates the amount of enzyme produced. Thus lactose is called the *inducer,* and the three proteins are called **inducible enzymes** or proteins.

An important clue about the way an inducer works was that the regulation of all three proteins seemed to be coordinated so that when one of the proteins doubled in amount, so did the other two. However, studies showed that each of the three proteins is encoded by a separate *structural gene*—a gene coding for a polypeptide—since mutations can affect them individually. The three genes occur in a linear cluster along the *E. coli* chromosome and are labeled z (for β-galactosidase), y (β-galactoside permease), and a (β-galactoside transacetylase). (Strictly speaking, the y gene product is a transport protein, not an enzyme; for convenience, however, we will refer to the three products as enzymes, since their activities were measured, not the actual quantity of protein present.) Thus the designation $z^- \, y^+ \, a^+$ represents a mutant *E. coli* cell that does not show β-galactosidase activity but shows activity in the other two enzymes specified by genes y and a. The order of the genes within the cluster is $z \, y \, a$.

Because most mutations affect only one of the three enzymes, it came as a real surprise to Jacob and Monod when they found a mutant in which all three were affected simultaneously. This mutant makes large amounts of all three enzymes even in the absence of lactose and is called a *constitutive mutant.* Thus constitutive mutants do not require the presence of the normal inducer in order to make inducible enzymes. Jacob and Monod concluded that there is a genetic element distinct from the three structural genes that in some way governs the expression of all three simultaneously and that this genetic element was missing in the mutant. They labeled as i the regulatory gene responsible for the inducibility of the enzymes. Wild-type bacteria are thus $i^+ z^+ y^+ a^+$, and constitutive mutants are $i^- z^+ y^+ a^+$.

The important question then became: How does the regulator gene i interact with lactose, the inducer, to control the synthesis of the three enzymes? The simplest hypothesis was that the i gene codes for a lactose **repressor** substance that, in the absence of lactose, goes to the z, y, and a genes and blocks, or *represses*, their expression. Adding lactose might alter or inactivate the repressor so that the z, y, and a proteins can then be made.

Another way to state the hypothesis is to say that the inducer (lactose, in this case) inactivates the repressor, while the repressor inactivates transcription of the genes coding for enzymes. Therefore, if the inducer is absent, the repressor inhibits the production of inducible enzymes. Also, according to this same hypothesis, in the constitutive i^- mutants, the altered repressor molecule is not effective at blocking the expression of the three structural genes, whether or not lactose is present.

Jacob and Monod tested this hypothesis in a clever experiment. They used F^+ cells to create bacteria with two sets of the i and z genes, each on a different DNA molecule (that is, on the bacterial cell's chromosome or on the inserted plasmid), coding for the repressor and for β-galactosidase. One DNA molecule had $i^- z^+$, and the other had $i^+ z^-$. The question then was: Can the i^+ gene on the one DNA molecule regulate the z^+ gene on the other molecule? The researchers found that the heterozygous cells behaved as normal bacteria, making β-galactosidase only when lactose was present. They concluded that a signal from the i^+ gene on one DNA molecule traveled to the other molecule. This was evidence that the i gene codes for some type of repressor molecule that can travel by diffusion to the z gene and somehow physically block the transcription of that gene and production of the enzyme β-galactosidase, just as Jacob and Monod had hypothesized.

But if the repressor is a signal in the cell that can regulate the z gene, there must be a genetic element near the z gene that can receive the signal. Jacob and Monod called this hypothetical regulator gene the **operator**, designated by the symbol o. Mutations in the operator gene were soon discovered. When mapped according to the procedures described above, the operator mutations were found to be located at sites very near the z gene. Furthermore, operator mutants tended to be constitutive; in other words, lactose proteins were produced even in the absence of inducer. Experiments with $o^- z^+ / o^+ z^-$ heterozygotes showed that the o^+ gene does *not* code for a protein, as does the regulator gene i, which codes for the repressor molecule. This is an important difference in the way the regulator (i) and the operator (o) genes work.

These experiments led Jacob and Monod to propose the operon model for gene regulation, illustrated in Figure 14-10. The essential features of the operon model are

1. a *regulator* gene that codes for a *repressor* molecule that can travel to another site and act as an "off" signal;

2. an *operator* gene that receives the "off" signal from the repressor;

3. a nearby set of protein-coding genes, all of which are transcribed onto the same mRNA molecule; and

4. the **promoter site** (p), adjacent to and partially overlapping the operator, where RNA polymerase binds to initiate mRNA synthesis.

An **operon** is defined as the operator plus the protein-coding genes it controls, in this case, $o z y a$.

The lactose, or lac, operator works like this. The regulator gene makes a repressor protein that can move to the site of the operator gene, bind to it, and shut down the transcription of the structural genes in the operon. The repressor can exist in either of two shapes. If there is no lactose (inducer) available, the repressor molecule assumes a form that can bind to the operator. As can be seen in Figure 14-10, if the repressor is on the operator, the promoter (p) is also partially covered, so the RNA polymerase is physically blocked from binding to the promoter; thus no mRNA can be transcribed and the three protein molecules cannot be synthesized by the cell. When lactose is present in the environment, the repressor binds to this inducer and assumes the alternative shape. In this other conformation, the repressor can no longer recognize and bind to the operator. The promoter is thus left uncovered, RNA polymerase binds to the site, and the genes for the enzymes are transcribed.

Because the operon model of gene regulation works at the level of mRNA synthesis, this type of regulation is referred to as *transcriptional level regulation*. We will encounter translational level regulation in eukaryotes; it also occurs in some prokaryotic operons.

Catabolite Repression

The repressor/operator mechanism is simple and elegant, and it explains why a normal *E. coli* cell makes only as much β-galactosidase as is necessary to break down the available lactose. What would happen, however, if an *E. coli* cell were presented with both lactose and glucose? The metabolic quandary for the cell might be akin to that of a child asked to choose between ice cream and fudge. There are good reasons, however, for the cell to break down glucose *before* lactose. Glucose is immediately useful for metabolism, whereas lactose must first be subjected to enzymatic action before its component sugars are in usable form. Thus bacteria that can use all

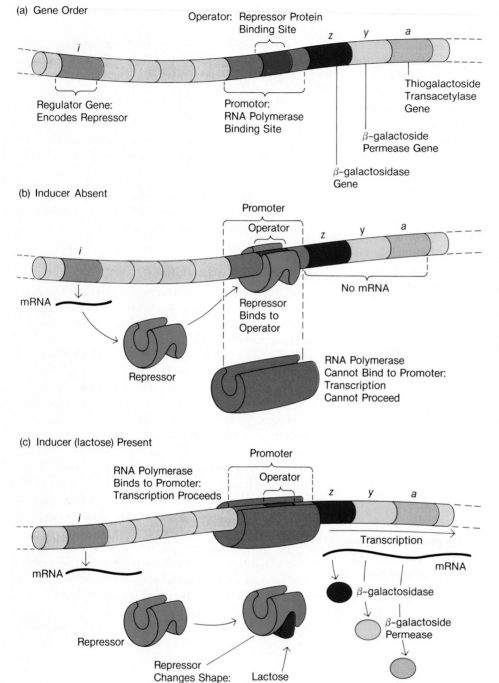

(a) Gene Order

Operator: Repressor Protein Binding Site

Regulator Gene: Encodes Repressor

Promotor: RNA Polymerase Binding Site

Thiogalactoside Transacetylase Gene

β-galactoside Permease Gene

β-galactosidase Gene

(b) Inducer Absent

Promoter

Operator

mRNA

Repressor

No mRNA

Repressor Binds to Operator

RNA Polymerase Cannot Bind to Promoter: Transcription Cannot Proceed

(c) Inducer (lactose) Present

Promoter

Operator

RNA Polymerase Binds to Promoter: Transcription Proceeds

Transcription

mRNA

mRNA

Repressor

Repressor Changes Shape: Cannot Bind to Operator

Lactose Binds to Repressor

Lactose (inducer)

β-galactosidase

β-galactoside Permease

Thiogalactoside Transacetylase

Translation of Three Proteins

Figure 14-10 AN INDUCIBLE OPERON: THE LACTOSE OPERON.
(a) The gene order. Note that the operator and the promoter overlap. (b) Inducer (lactose) absent. Repressor protein, coded by the *i* gene, is synthesized, recognizes the operator, and binds to it. This physically blocks the access of RNA polymerase to the promoter; hence, mRNA is not made, and lactose-metabolizing enzymes are not formed. (c) Inducer (lactose) present. When lactose is present, it functions as an inducer that binds to the repressor and causes it to change shape. This new shape does not allow binding to the operator. Thus, RNA polymerase has free access to the promoter, and transcription occurs, followed by translation to produce the enzymes. The cell can now use lactose as a source of energy and carbon.

the available glucose before starting to make the enzymes to catabolize (break down) lactose grow faster and have a competitive advantage over cells that break down lactose first. Given the selective advantage of using glucose before lactose, it is not surprising that a mechanism exists in *E. coli* that allows cells to use glucose whenever it is present in the medium, even if other energy sources, such as lactose, are present. Glucose represses the appearance of the enzymes that catabolize other sugars. This mechanism, called **catabolite repression,** is shown in Figure 14-11.

Catabolite repression depends on the fact that RNA

Figure 14-11   CATABOLITE REPRESSION.
The formation of lactose-metabolizing enzymes is inhibited in the presence of glucose. (a) A site along the DNA called the CAP (the catabolite gene activator protein) binding site (CBS) is immediately adjacent to the promoter. (b) When glucose is absent, large amounts of cAMP are present in the cell. The cAMP binds to CAP, causing it to shift to a shape that can bind to the CBS. This facilitates binding of RNA polymerase to the promoter, and transcription of the structural genes for lactose-metabolizing enzymes is speeded up. (c) When a large amount of glucose is present, the amount of cAMP is reciprocally small. CAP cannot bind to CBS, and RNA polymerase binds poorly to the promoter. Hence little RNA is made, and few enzyme molecules are formed.

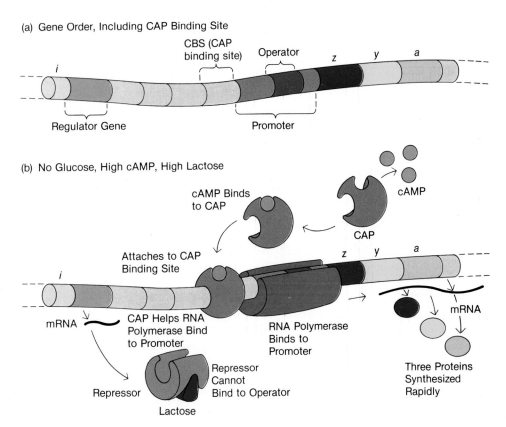

(a) Gene Order, Including CAP Binding Site

(b) No Glucose, High cAMP, High Lactose

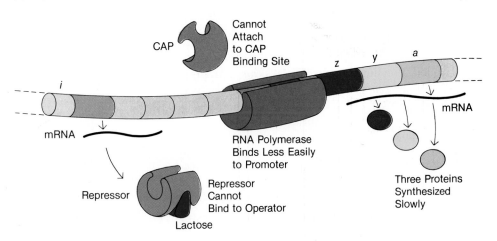

(c) High Glucose, Low cAMP, High Lactose

polymerase binds to a promoter site much better if another protein, called *catabolite gene activator protein* (CAP), is bound to a special DNA site nearby (this occurs when glucose is absent). CAP acts as a guide, getting RNA polymerase started down the pathway of transcribing the z gene faster than if the polymerase had to find the way itself. The presence of glucose indirectly causes CAP to change shape. The altered CAP does *not* bind well to its site; in turn, RNA polymerase does not bind as well to the lactose promoter site. The synthesis of β-

galactosidase is therefore reduced, and the cell cannot catabolize lactose as well—all because glucose is present and alters CAP. Actually, CAP does not bind directly to glucose. Rather, CAP is sensitive to molecules of the chemical messenger *cyclic adenosine monophosphate* (cAMP). Levels of cAMP rise in the cell when glucose is absent; the cAMP binds to CAP and alters that molecule's shape.

We can see now that the β-galactosidase genes can operate at three levels of activity: low, medium, and high. These levels depend on the presence or absence of lactose and glucose, and are governed by three separate proteins: the lac repressor, RNA polymerase, and CAP. The lowest level of expression takes place in the absence of lactose: since no substrate is present, no lactose-catabolizing enzymes are needed, and the lac repressor is bound to the lac operator, blocking the binding of RNA polymerase. The highest level of gene expression occurs when lactose but no glucose is present in the medium: lactose makes the repressor detach from the operator, freeing the promoter so that RNA polymerase can bind. The CAP-cAMP complex will bind near the promoter site, further encouraging RNA polymerase action. Finally, a middle level of gene expression occurs when both lactose and glucose are present. No repressor is bound; no CAP is bound. Thus RNA polymerase is free to transcribe the gene, but at a reduced rate compared with that when CAP is present. So CAP provides a simple slow–fast switch, a molecular toggle switch that is thrown by glucose to alter utilization of lactose.

It is important to realize that this mechanism of regulation via CAP works not only on the lac operon, but also on several other operons coding for enzymes that break down other sugars.

Positive and Negative Control

We have discussed the regulation of the lactose operon in *E. coli* in detail, and at this point, it may seem like a very complex system. However, there are really just two general modes of control at work, one positive and one negative, analogous to the way the speed of a car is controlled. Pressing the brake pedal slows a car; pressing the accelerator speeds the car. In a similar way, when CAP binds to the DNA, it speeds the transcription rate. This is *positive control*: an action is followed by a positive response. On the contrary, when the lactose repressor acts by binding the operator, it slows the transcription rate. This is an example of *negative control*: an action is followed by a negative response. The presence of both kinds of control—like the brakes and accelerator of a car—renders the cell exquisitely sensitive to external governing conditions.

Repressible Enzymes

The highly efficient mechanisms of the repressor/operator system ensure that *E. coli* cells will make enzymes to break down a particular sugar only when that sugar is the best available substrate in the environment. Such enzymes are considered *inducible*, since an inducer substance, such as lactose, is instrumental in controlling their production. It also would be efficient, however, for a cell to make the enzymes that *synthesize* a particular amino acid, vitamin, or nucleotide only if that substance is *not* already available in the medium. Enzymes whose synthesis is repressed when their pathway

end product is present are called **repressible enzymes.** Regulation of repressible enzymes is similar to that of inducible enzymes except for a new wrinkle or two. Let us use the example of the tryptophan (trp) operon to show how this works.

The trp operon is a repressible system consisting of genes for five enzymes that convert a particular starting substance to the amino acid tryptophan (Figure 14-12). Transcription of the trp operon utilizes a repressor encoded by a gene some distance away from the trp operon. Like the lac repressor, the trp repressor can exist in either of two shapes: one shape allows binding to the operator and blocks transcription; the other shape does not allow binding to the operator so that gene expression can occur. If tryptophan is present, it binds to the trp repressor, causing it to change shape. The resulting complex (unlike the lactose-lac repressor complex) can bind to the operator and repress transcription (an example of negative control). In the absence of tryptophan, the repressor cannot bind, and enzymes for producing this necessary amino acid are synthesized.

To summarize the control of inducible and repressible enzymes, both systems rely on the activity of allosteric regulatory proteins—that is, proteins that assume different shapes depending on the binding of other substances (Chapter 4). In an inducible system, when an inducer such as lactose binds to the repressor, the resulting complex cannot bind the operator, and lactose-degrading enzymes are formed. In a repressible system, when an end product such as tryptophan binds to the repressor, the complex does bind to the operator, and tryptophan-synthesizing enzymes are not produced. In both cases, when the repressor is bound to the operator, RNA polymerase has no access, and no transcription occurs.

The two systems together control enzyme production in a way that is truly economical: only when a substrate such as lactose is available does the cell produce enzymes to break down that sugar, and only when a necessary substance such as tryptophan is available to the cell in ready-made form does the cell stop producing the enzymes needed to make its own supply of the amino acid. The elucidation of the operon model explained the control of genes in prokaryotes in a way that both underscored this beautiful economy of nature and served as a guide for studying gene regulation in eukaryotes.

GENE REGULATION IN EUKARYOTES

As soon as these elegant systems of gene induction and repression were explained in bacteria, speculation ran high that gene regulation in eukaryotes might be similar.

Figure 14-12   A REPRESSIBLE OPERON: THE TRYPTOPHAN (TRP) OPERON.
(a) The regulator gene lies far from the operon containing the five structural genes (*E, D, C, B, A*) that it controls. (b) When tryptophan is present, it binds to the repressor, causing the complex to assume a shape that can bind to the operator. This blocks access of RNA polymerase to the promoter; thus transcription cannot occur and tryptophan-synthesizing enzymes are not made. (c) When tryptophan is absent, the repressor cannot bind to the operator, and the promoter is free to bind RNA polymerase. Transcription ensues, tryptophan-synthesizing enzymes are made, and the bacterium begins to supply itself with this amino acid.

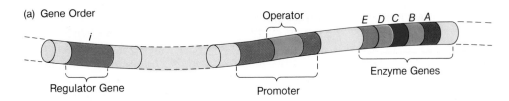

(a) Gene Order

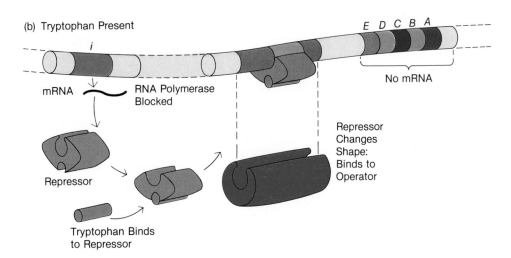

(b) Tryptophan Present

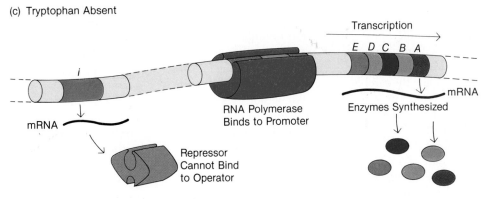

(c) Tryptophan Absent

Although fantastic progress has been made in the late 1970s and early 1980s, the revelations have been of a sort completely unpredicted by the scientists who had studied prokaryotic cells. The new knowledge points to profound differences in the organization of the genomes of prokaryotes and eukaryotes, rather than a common regulatory strategy.

Researchers attempting to probe gene control in eukaryotes were faced with two enigmas: eukaryotes seemed to have too much DNA when compared with prokaryotes, and the majority of newly manufactured RNA seemed to be destroyed before it could leave the nucleus. Let us consider each enigma and the facts that emerged.

Different organisms have very different amounts of DNA in their haploid genomes, as you can see by looking at Table 14-1. In terms of the numbers of nucleotide base pairs, mammals have about 500 times as much DNA as *E. coli*; but they probably make only 50 times as many proteins. The conger eel has almost 100 times as much DNA per haploid genome in each of its cells as do you and I, but surely no more genes are required to specify the proteins in this amphibian than we humans require. So what is all this extra DNA doing in eukaryotes?

Part of the answer to this question has come from DNA *dissociation/reassociation* experiments, in which a solution of DNA is treated so that the double helices are

Table 14-1   **AMOUNT OF DNA IN HAPLOID GENOMES**

Organism	Base Pairs
Plasmid	
pBR322	4.3×10^3
Virus, including bacteriophage	
SV-40	5.2×10^3
Lambda	4.6×10^4
Bacteria	
Diplocccus pneumoniae	1.8×10^6
Escherichia coli	4.1×10^6
Insect	
Drosophila melanogaster	1.8×10^8
Mammals	
Mouse	2.2×10^9
Human	2.8×10^9
Amphibian	
Conger eel	1.5×10^{11}
Plants	
Corn	6.6×10^9
Lily	3.0×10^{11}

separated into two strands; then the strands are allowed to come back together in a process called *hybridization*. When only two copies of a gene are present, it takes a long time for them to chance to collide and reanneal (form a double helix). If there are multiple copies of a gene, random meetings will occur more frequently; if there are dozens of copies of a gene, reannealing may be quite rapid. Researchers were surprised to find that the reassociation curve for a mammal is not as simple as those for prokaryotes, but instead consists of a complex curve with three distinctive portions (Figure 14-13). One portion of the curve results from the presence of multiple copies of a small number of DNA sequences. This *highly repeated DNA* consists of about 10 percent of the genome in mice—about 1 million copies of a repeating sequence some 300 bases long. Highly repeated DNA, sometimes called satellite DNA, is located in chromosomes mainly near the centromere and may play some role in aligning chromosomes during mitosis.

An additional 20 percent of mouse DNA is *moderately repeated DNA;* this includes DNA sequences that are present in 30 to 10,000 copies per haploid genome. At least some of these sequences code for genes with which we are familiar, including those for rRNAs, tRNAs, and the histone proteins. In every body cell of the clawed frog, *Xenopus*, for example, there are about 800 copies of each gene coding for each of the large types of ribosomal RNAs.

The remaining 70 percent of DNA reassociates as though there were only one copy per haploid genome; geneticists call it the *unique DNA*. It includes most of the genes for proteins such as insulin, ovalbumin, and growth hormone. Nevertheless, the presence of so much repetitive DNA in eukaryotes helps explain why eukaryotes seem to have too much DNA—much of it is simply repetitious.

Another feature that distinguishes eukaryotic from prokaryotic DNA is the presence of *multiple gene families* in eukaryotic DNA. Many eukaryotes contain families of genes—genes related in both structure and function. Clusters of genes are found for the five histones, for the several different hemoglobin molecules, and for the many antibody genes. Scientists believe that these multiple gene families arose by gene duplication during evolution. Once there were two diploid copies of the original gene (four alleles), a mutation in one pair of alleles would not harm the organism possessing it, since the other pair could maintain the original function. Thus the base sequences of the two diploid copies were "free" to become different from each other, and new but related gene functions were able to develop. It is important to note that even though the members of multiple gene families may be close together on the chromosome, and though their expression and function may in some way be related, these are *not* operons of the kind found in prokaryotes, since each of the slightly different genes may be regulated independently. These gene families also help to explain the mysterious "excess" DNA in eukaryotes and they underscore the evolutionary versatility of the eukaryotic genome.

The "Pieces" of Eukaryotic Genes

The second enigma—why most RNA is destroyed before leaving the eukaryotic cell nucleus—was explained by the discovery of two facts: (1) most eukaryotic genes occur in "pieces," and therefore (2) the "pieces" of mRNA encoded by them must be joined together or processed in a special way.

Researchers compared the structure of an mRNA and its corresponding gene in hybridization experiments. First, they isolated mRNA. Next, they caused the DNA for the corresponding gene to dissociate into single strands. Then the mRNA was allowed to *hybridize*—bind to—its complementary DNA sequence, and the resultant complexes were examined under the electron microscope. If genes and mRNAs are colinear, as are genes and polypeptides (Chapter 13), then the RNA and DNA should hybridize perfectly over the entire length of the mRNA.

When the gene for hemoglobin was hybridized to hemoglobin mRNA, a remarkable fact emerged. A single mRNA molecule hybridized at three distinct places on the gene, but the two intervening stretches of gene did not hybridize with the mRNA. Instead, they formed loops

Figure 14-13 REASSOCIATION CURVES: DETERMINING THE AMOUNT OF REPETITIVE DNA AN ORGANISM CONTAINS.
(a) For simple situations such as phage lambda and *E. coli*, the curve is simple. (b) Mammals, however, have huge numbers of copies of some DNA sequences (the rapid part of this graph), moderate numbers of other DNA sequences (the intermediate region), and single copies of some DNA sequences (the slow region).

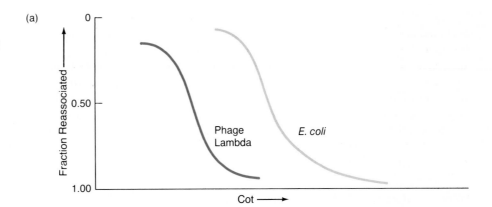

(a)

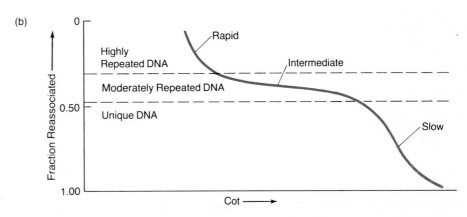

(b)

Cot = Initial DNA Concentration × Time for Reassociation

of DNA (Figure 14-14). The reason is that nucleotides of these two segments of DNA do not pair with mRNA nucleotides. Subsequent work has proved that these nonpairing regions are not represented in a functional mRNA molecule; thus they do not code for protein. The nonpairing regions of the DNA are called **introns,** or *in*tervening sequences. The protein-coding parts of the gene are called **exons,** since they are *ex*pressed.

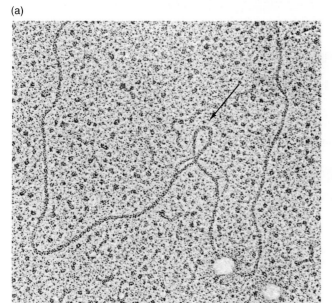

(a)

Figure 14-14 INTRONS AND EXONS.
Exons can be viewed experimentally by isolating mRNA and causing it to bind to DNA. (a) Globin mRNA (the protein part of hemoglobin) has been hybridized to a long sequence of DNA containing the globin gene (magnified about 73,000 times). The vertical loop in the center is an intervening sequence in the globin gene. (b) A simple representation of an intervening sequence loop is shown; the mRNA binds only to the exon portions of the gene.

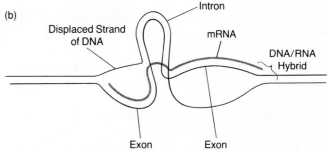

(b)

Some genes have many introns; the collagen gene, for example, has fifty, and the fibronectin gene (a cellular glue) has forty-eight, most of which are much longer than the exons of that gene. Introns are characteristic of nearly all eukaryotic genes but are not usually found in prokaryotic ones (some occur in archaebacteria). At least some of the unexpectedly large amount of DNA in eukaryotes is due to the presence of introns, which are sometimes very long. The origins and functions of introns are still a major mystery.

Eukaryotic mRNA: Special Processing

The discovery of introns and exons also helped explain why a high proportion of the RNA that is transcribed from DNA is destroyed before it leaves the nucleus of a eukaryotic cell.

We saw earlier that in prokaryotes, all the genes that code for a string of enzymes in a metabolic pathway occur in a cluster called the operon on the circular chromosome. All those genes are transcribed into one long

mRNA message that can be translated into several different polypeptides sequentially as ribosomes move along the mRNA "tape." However, in eukaryotes, this is not true—the genes for the enzymes in a related pathway are separate, are transcribed into separate mRNAs, and are translated separately. What form, then, does the mRNA take, considering the presence of intervening sequences (introns) between pieces of the structural gene (exons) itself?

It turns out that the introns are transcribed along with the exons of a gene into a single long RNA molecule called the *primary RNA transcript* (Figure 14-15). Before this molecule leaves the nucleus, the portions of RNA corresponding to the introns are cleaved enzymatically from the primary transcript. Then the exon-encoded RNA sequences are spliced together to form an RNA molecule, which after further processing passes into the cytoplasm. Only when fully processed is it considered to be a mature mRNA molecule. Meanwhile, the excised RNA, which corresponds to the introns, is degraded. The degradation of intron RNA molecules helps

Figure 14-15   INTRONS AND EXONS IN EUKARYOTIC CELLS.
Eukaryotic genes have translated (exon) and untranslated (intron) segments. In this drawing of the ovalbumin gene, notice that the untranslated introns are substantially longer than the translated exons. The full gene is transcribed into a primary RNA transcript. Then, processing includes capping, adding a tail, and splicing; during splicing, intron-coded regions are deleted as exon-coded ones are spliced together. The final mature mRNA is translated into ovalbumin after it passes into the cytoplasm. Note that the original gene region is 7,700 base pairs long, although only 1,800 or so comprise the final mRNA molecule. Both the cap and the poly-A tail remain untranslated when protein is synthesized on the mRNA. The cap and tail sequences are probably protective and may play a role in facilitating translation.

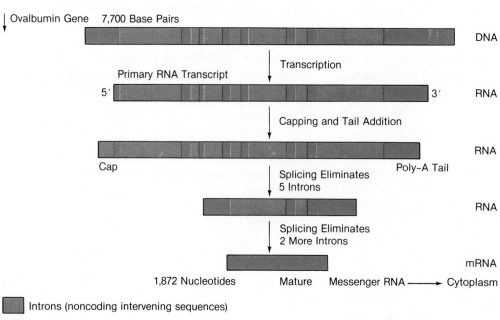

explain why so much newly synthesized RNA is destroyed rapidly and never leaves the eukaryotic nucleus.

The origins and basic functions of introns remain enigmatic. Introns have been found that actually code for viral enzymes. Others code for proteins that help remove introns from primary RNA transcripts. Still others catalyze their own removal from the primary transcript, in the absence of protein enzymes. Some introns form loops or lariats, reminiscent of bacterial plasmids. The elucidation of the origins of introns is likely to tell us something profound about the evolution of DNA function on Earth.

The final processing of RNA, after intron deletion and exon splicing, involves two specific modifications: a "cap" consisting of a special nucleotide and a number of other nucleotides is added to the 5′ end of the RNA, and a *poly-A tail* composed of 100 to 200 adenosine residues is added to the 3′ end. These modifications may make the mature mRNA less likely to be degraded and may facilitate its translation.

The extensive processing of RNA occurs only in eukaryotic cells and is possible, in part, because transcription and translation are separated in time and in space. In prokaryotes, one end of a message loads on ribosomes and tRNAs and begins to be translated into protein even before the other end has been completely transcribed. But as we have just learned, much takes place between the transcription of DNA into a primary transcript and the formation of protein in a eukaryotic cell's cytoplasm. RNA processing probably plays a key role in regulating transport to the cytoplasm and translation of eukaryotic messages. Hence, eukaryotic genes can be controlled at additional levels—processing or transport—that do not appear to exist in prokaryotes.

How are eukaryotic genes turned on and off at appropriate times? The answer to this question is not yet fully known, but some progress has been made by studying gene expression in developing embryos. Since it would be difficult for us to consider the results of such studies without first exploring the basic processes of developing organisms and cells, we will return to the subject of eukaryotic gene control in Chapter 17.

We should mention here, however, one relevant technical advance in molecular genetics. Methods have been invented recently that enable scientists to determine rapidly the sequence of bases in the DNA molecule. These methods have revealed that eukaryotic genes tend to have certain base sequences at specific sites that may act as promoters, determining where along chromosomal DNA transcription of a gene begins; other sequences determine where along primary RNA transcripts processing enzymes will cleave transcripts to excise intron RNAs. The ultimate explanation of gene regulation in eukaryotes requires understanding how these recognition sequences work.

RECOMBINANT DNA TECHNOLOGY

In the course of mapping the bacterial chromosome and studying the transfer and control of genes in nature, molecular biologists amassed far more than just information. They also developed a special set of laboratory techniques that give them unprecedented power to manipulate genes in new ways. These techniques, collectively called *recombinant DNA technology*, helped uncover a good deal of what we know about eukaryotic gene regulation (Chapter 17)—a process of considerable interest to medical researchers. But more than that, recombinant DNA techniques accord molecular biologists the rather awesome ability to move genes from one chromosome to another and to create totally new genomes and altered organisms such as giant mice and bacteria that express human genes. Let us look more closely at these techniques of gene transfer, and then see how they were used to "build a better mouse."

General Steps in a Gene-Transfer Experiment

A biologist generally uses recombinant DNA techniques to find an interesting gene or other piece of DNA, from either a bacterial or a eukaryotic chromosome, and then move it into the genome of a different organism. There the expression of the foreign gene or gene segment, set in a different background, may be studied in detail. In addition, a polypeptide encoded by the inserted foreign gene either may be harvested in quantity or may cause a desired effect in a host organism. An example of the former is the harvesting of insulin from bacteria into which human insulin genes have been transferred. An example of the latter is the giant mouse; the gene for rat growth hormone was transferred to normal mice, causing them to grow rapidly. We shall look at both examples later in the chapter.

Finding an interesting piece of DNA, such as one that codes for a certain medically valuable protein, is harder than it sounds, especially if the scientist wants to remove and produce millions of copies of a single gene from a eukaryotic nucleus containing many chromosomes bearing more than 1 million genes. To find this needle in a haystack, the researcher must do several things: (1) cut all the DNA from the donor cell into small pieces—some containing whole functioning genes; (2) make many copies of all these pieces of DNA by introducing the DNA into special vehicles (usually plasmids) and then inserting these vehicles into prokaryotic "factory" cells (usually bacteria such as *E. coli*) that act as hosts. Once inside the host cells, the foreign genes begin to be expressed as foreign polypeptides. The researcher must (3) identify the host cell that can suddenly make the polypeptide that is encoded by the desired gene (the "needle"), and discard all the bacterial cells that express undesired genes (the "haystack") transferred from the original donor cell; (4) grow huge numbers of the host cells (each containing the desired gene) as a "gene factory." The gene later can be removed from these host cells, still in its plasmid vehicles, and be transplanted by another set of steps into another host—say, a mouse, a frog, or a tomato plant.

The Tools of the Genetic Engineer

What tools do modern researchers have for locating and transferring a piece of DNA? Special "cut-and-paste" enzymes, molecular vehicles, or *vectors*, for making copies of DNA and inserting it into a host, and molecular probes for finding the desired gene in its new host cell are some of the available tools. Let us look at each.

Molecular Scissors and Genetic Glue

In 1973, a class of enzymes called **restriction endonucleases** was first applied to gene research by Stanley Cohen at Stanford University and Herbert Boyer at the University of California at San Francisco. Enzymes of this class recognize specific nucleotide sequences along DNA molecules and cut both complementary DNA strands of the double helix at those specific sites. Table 14-2 shows some of the sequences that these enzymes recognize and cut. These bacterial enzymes cleave DNA double helices at the sites indicated. The resultant ends

---

Table 14-2   **RESTRICTION ENDONUCLEASES AND STICKY ENDS**

Eco R I
5′   –G–A–A–T–T–C–3′
3′   –C–T–T–A–A–G–5′

Eco R II
5′   –N–C–C–N–G–G–N–3′
3′   –N–G–G–N–C–C–N–5′
                              (N = A or T)

Hin d III
5′   –A–A–G–C–T–T–3′
3′   –T–T–C–G–A–A–5′

Bam HI
5′   –G–G–A–T–C–C–3′
3′   –C–C–T–A–G–G–5′

---

have a single stranded tail, the nucleotide sequence of which is shown as TTAA for the left region of the Eco R I case and AATT for the right region.

The interesting thing about restriction endonucleases is the *way* they cut the double-stranded DNA molecule: many of them make a staggered cut in the two strands and leave "sticky" ends (Figure 14-16). These sticky, or *cohesive*, *ends* are stretches of unpaired nucleotides that can form hydrogen bonds with other cohesive ends having the appropriate complementary sequence:

Staggered cut   —G|A A T T C—
                —C T T A A|G—
Cohesive ends —G   A A T T C—
              —C T T A A   G—

Another class of enzymes called **DNA ligases** can act as molecular paste to rejoin the complementary cohesive ends of DNA fragments.

How would a researcher use these cut-and-paste enzymes? The DNA from a rat, let us say, could be cut with restriction endonucleases into hundreds or thousands of fragments, some containing complete, functioning genes. Since restriction endonucleases cleave DNA at specific sites, all the DNA fragments would have cohesive ends with the same specific complementary sequences. Because of this, the fragments could pair up with other DNA fragments cut by the same restriction endonuclease at each place that same nucleotide sequence appears. The paired sticky ends hold the fragments in place for ligase to do its job. DNA ligase enzyme then could be added to re-form the sugar-phosphate links between cohesive ends. The result would be a re-paired double helix. Sometimes researchers simply cut rat DNA into fragments and rejoin the fragments to each other in a different order. Alternatively, DNA ligase enzymes can be used to paste fragments of rat DNA into molecular vectors.

Molecular Vectors

Molecular vectors are pieces of DNA that can carry a foreign gene into a host cell, where such genes are replicated along with the cell's own DNA. Plasmids, which we encountered earlier in the chapter, are typically used as vectors. These naturally occurring circular strands of bacterial DNA carry genes for traits of use to bacteria, such as the F factor genes and antibiotic resistance, but have several features that make them ideal tools. First, they can be cleaved by restriction endonucleases at the same sequences recognized and cleaved in longer strands of DNA. Cleaving a plasmid in this way will "cut the ring" so it opens and has two sticky ends. If DNA ligase is supplied to a mixture of "cut" plasmids and rat DNA fragments, let us say, the cohesive ends of the

Figure 14-16 RECOMBINANT DNA TECHNIQUES.
The creation of complementary sticky ends in the plasmid DNA and in the foreign DNA to be inserted is the key to this technology. Once the recombinant plasmid is created with the ligase enzyme, it is introduced into host cells. Millions of copies can then be made as both the cells divide repeatedly and plasmids are replicated in each of them.

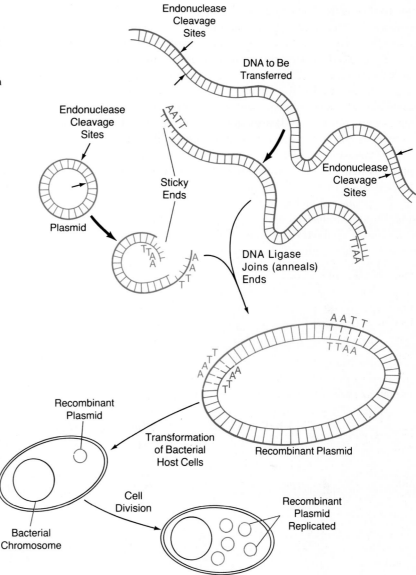

plasmids and the fragments will match complementary sequences and be joined by the ligase. Thus the rat DNA will be "spliced" into the re-formed plasmid rings. This totally new genetic entity, a combination of both bacterial and rat DNA, is called a **recombinant DNA molecule.**

A second feature that makes plasmids an ideal vector is that they can be introduced into host cells via transformation and therefore can carry a foreign gene into a host cell like a molecular Trojan horse. A third critical feature is that plasmids can replicate inside the host cell. An isolated bit of DNA cannot do this. Some plasmids actually replicate tens or hundreds of times inside a host cell—a natural mechanism that can generate extra copies of the complete recombinant DNA molecule (the foreign gene along with the plasmid DNA). Finally, since plasmids often carry genes for antibiotic resistance, they

provide a means of screening for recombinant DNA molecules. A scientist can mix plasmids carrying foreign genes with bacteria lacking plasmids, wait for transformation to take place, and then spread the mixture on agar-filled petri dishes containing a deadly antibiotic. Only transformed bacteria—those that incorporated the plasmids bearing antibiotic-resistance genes and, coincidentally, the foreign DNA fragments—will grow.

At this point in the experiment, it is time for the researcher to locate the "needle in the haystack," the foreign gene of interest in one transformed bacterial cell among hundreds. To do this, a molecular probe is needed.

Molecular Probes

Transformed bacteria allowed to grow on agar-filled petri dishes will divide and redivide into **clones**—

colonies of cells. Every cell in a colony is genetically identical to the original parental cell that yielded that particular clone. After several hours, a researcher will have a set of petri dishes dotted with hundreds of round bacterial colonies (the "haystack"), each a clone of a different transformed cell containing a recombinant DNA molecule. However, only about 1 in 1 million of the clones will contain the desired gene—say, the gene for rat growth hormone. The rest of the clones will contain other, undesired fragments from the original rat genome, which was chewed into pieces by restriction endonucleases. To find the desired gene, the researcher needs a *probe*—a molecule that in some way recognizes specifically the desired gene.

The most characteristic feature of any gene is its sequence of nucleotides. This sequence can be recognized by a stretch of RNA with a complementary sequence during the process of **DNA–RNA hybridization**, a technique we described in our discussion of introns and exons. In this technique, DNA is heated to break the weak hydrogen bonds that hold together the complementary strands. Radioactively labeled RNA is then supplied, and if any of its sequences of bases are complementary to regions on the unzipped DNA strands, the two linear molecules will pair. In the case of the gene for rat growth hormone, radioactively labeled mRNA for that gene would be isolated from cells of a rat's pituitary gland. The many clones of transformed cells prepared earlier would be heated and exposed to the radioactive growth-hormone mRNA. Any hybridization occurring between the radioactive mRNA probe and a DNA sequence must be due to DNA sequences for the rat growth-hormone gene. The clone of bacterial cells containing that particular stretch of DNA would be isolated and grown in mass culture; billions of cells could eventually be harvested. Finally, the plasmids containing the gene for rat growth hormone would be removed from these cells and treated to isolate the growth-hormone gene itself. At last, the biologist would have a pure preparation of billions of copies of the desired gene, uncontaminated by copies of other genes.

With it, he or she could then "engineer" a new organism. Let us see how recombinant DNA technology was applied in experiments to create a giant mouse, and then see how it will be used widely in coming decades for chemical, pharmaceutical, and agricultural research and in industry.

Giant Mice from Gene Transfers

To produce the giant mouse shown at the beginning of this chapter, researchers used the techniques we just outlined to locate and produce billions of copies of a gene from mice that codes for metallothionein, a protein produced mainly by the liver that helps prevent poisoning from heavy metals such as zinc. They then cleaved the metallothionein gene with a restriction endonuclease so as to remove a regulatory region. The regulatory region was joined to the gene for rat growth hormone they had cloned using plasmids and *E. coli* host cells (Figure 14-17). This produced a *fused gene* that was then injected into a fertilized mouse egg and apparently became integrated into the chromosomal DNA. When cultured in a petri dish, the egg developed normally into an early-stage mouse embryo, and a copy of the injected DNA was present in every one of its cells. The embryo was then implanted into the reproductive tract of a foster mother and allowed to develop normally.

The researchers wondered whether the fused gene was regulated as though it were the metallothionein gene, being produced by the liver, or the growth-hormone gene itself, being produced by the pituitary gland. The answer was that when the mouse was fed zinc, its liver made rat growth hormone, and great quantities of it. That is why the mice in the experiment grew so large.

The experiment showed that the signals for controlling the metallothionein gene reside within the portion of the gene included in the fused-gene construction. Investigators are even now seeking to isolate those sequences more precisely so they can learn for eukaryotic cells what they have learned from the lac operon about regulation of genes in prokaryotes. And while the giant-mice experiments had implications for understanding eukaryotic genes, it had even more important implications for human medicine, as we shall see.

GENETIC ENGINEERING: PROMISES AND PROSPECTS

In a sense, people have practiced genetic engineering for thousands of years, allowing genes to recombine naturally in domesticated plants and animals and then choosing those individuals with the best new combination of desirable characteristics as parents for the next generation. Some anthropologists consider this kind of genetic manipulation of domesticated organisms to mark the dawn of civilization.

The early human inhabitants of Central America used a small native grass as a food source because of its edible seeds. Over the course of 5,000 years, they selected varieties—with new chance combinations of genes—that had progressively more seeds. Today this plant is modern corn, one of the mainstays of world agriculture (Figure 14-18). This ancient kind of genetic engineering relies on chance mutation, random genetic recombination, and normal reproduction for the origin and propagation

Figure 14-17 HOW GENES ARE INCORPORATED INTO A MOUSE GENOME.
(a) Injection of DNA, such as the recombined growth hormone (GH) gene. The zygote (magnified about 450 times) is held in place with the pipette, seen at the right, while the injection microneedle on the left penetrates the male pronucleus. (b) This illustration outlines the genetic engineering steps by which the rat growth-hormone gene is placed under control of a mouse gene, and then introduced into a fertilized egg. The growth-hormone gene later acts in excess and a giant mouse results.

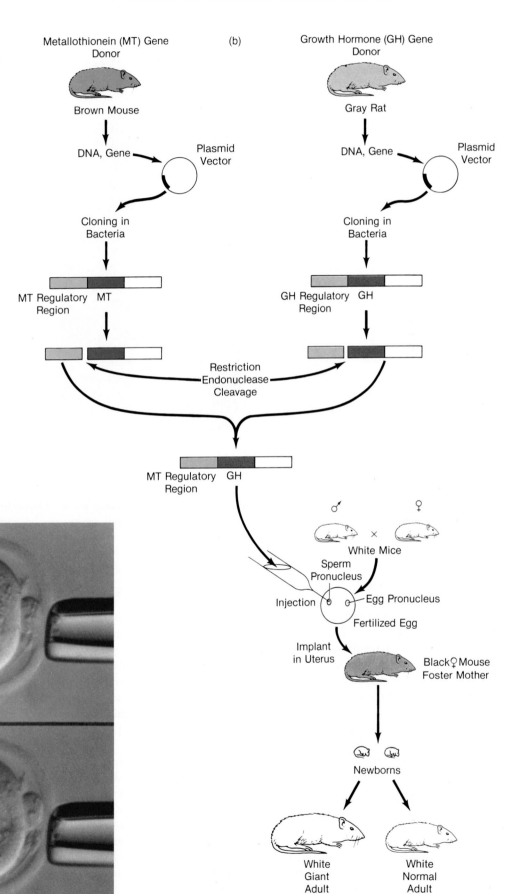

Figure 14-18   MODERN CORN: A RESULT OF GENE SELECTION BY CHANCE AND BY FARMERS. Although human consumers affected the evolution of corn by choosing to plant varieties with the best and most edible seeds, the 5,000-year "genetic engineering" of corn resulted mostly from chance mutation, random genetic combinations, and normal reproduction.

of new varieties. The new technology avoids elements of chance: it alters a gene in a defined and desired way in the test tube and then introduces it into another organism's genome.

There are basically three strategies for using recombinant DNA technology for our own benefit. The first is to put a gene for a desired protein into a microorganism such as a bacterium or yeast. The genetically engineered microbe then may produce the protein in large amounts. The second strategy involves putting engineered genes into domestic species to improve their growth rate, production of desired products such as fibers or food, or protein quality. The third strategy—gene therapy—involves putting engineered genes directly into people to cure genetic disease. Let us look at the status of each application, as well as the ethical problems raised by some of these new approaches.

Microbial Production of Valuable Gene Products

When techniques of gene splicing were developed in the mid-1970s, one of the first applications proposed was bacterial synthesis of human insulin, the peptide hormone produced by the pancreas that helps regulate blood-sugar levels. Diabetics do not secrete enough insulin, but many can be successfully treated with daily insulin injections. Previously, insulin was isolated from cattle, hogs, or other mammals and used in the treatment of diabetes, but it had several disadvantages, including the possibility of provoking an adverse immunological reaction. Scientists at several corporations have used recombinant DNA techniques to produce bacteria that can make human insulin and have improved the production process so that bacterially produced human insulin is now commercially available.

Many other proteins can be produced by means of bacterial synthesis. A partial list includes interferon, an antiviral and perhaps antitumor protein; growth hormone to treat people with genetic defects that retard growth and to treat burn victims; urokinase and tissue plasminogen activator enzymes that help to reduce blood clots; and rennin, an enzyme from calf stomach used in the manufacture of cheese. The list will continue to grow as industrial experience with the new technology increases.

Other organisms besides bacteria may be suitable hosts for genetic engineering plasmids. Like *E. coli*, yeasts may be particularly suitable because they can be engineered to secrete the desired proteins into a liquid culture medium, thus making purification easier. Since

AUTOMATED GENETICS

Like so many other fields in recent years—manufacturing, banking, even baking—genetics research has become automated; two new kinds of machines have been introduced: one for synthesizing genes and the other for sequencing proteins. They each can do in one day what used to require from six to eight months of tedious effort by geneticists and technicians. While these machines will facilitate and accelerate the pace of genetic engineering research, they may, some fear, present certain risks as well.

The first instrument, the DNA/RNA synthesizer (nicknamed the "gene machine"), automatically produces gene segments with a sequence of nucleotides specified by the researcher. He or she simply types the sequence into a keyboard, and a computer chip inside the machine then stores the information. Other microprocessor chips control a set of reservoirs and valves. Each reservoir contains one type of nucleotide in solution—A, T, C, G, or U. When signaled, a valve opens and releases a tiny drop of one nucleotide solution into a container. There, the bases are joined chemically, one by one, in the appropriate sequence, and a gene segment of up to thirty nucleotides can be made. This segment then can be linked to other segments made in the same way, and an entire functioning gene thereby is synthesized.

The gene machine costs about $30,000, which sounds expensive, but in research terms it is quite a bargain. Besides, before the synthesizers were available, a geneticist wishing to avoid the long labor of making a gene segment from scratch could easily pay $10,000 for one microgram of the material from a supply company.

The second instrument is the protein sequencer. To use it, a researcher places a protein sample into the machine and, through a series of chemical reactions, amino acids are automatically "snipped off" one by one and identified by liquid chromatography (Chapter 5). In about a day, instead of a month or more, the entire amino-acid sequence is revealed and recorded.

The gene machine and the protein sequencer can be used in tandem to clone a gene and then produce a desired protein. An unknown protein with a useful property—say, a hormone or an antibody—can be analyzed, and once the sequence is known, a corresponding genetic sequence can be typed into the DNA synthesizer. The researcher then harvests the synthetic gene for the protein, and perhaps introduces it into a bacterial cell or other host organism to produce the protein in quantity.

These "automated genetic engineers" are already being used in various ways and hold real commercial promise. For example, synthesized gene segments are often radioactively labeled and then introduced into cells, where they bind to a corresponding sequence on the chromosome and thus help reveal the position of a specific gene. DNA segments are also being combined to create totally new sequences that do not occur in nature and that code for novel proteins. They may prove useful for medical and agricultural applications. For instance, combination antibodies and combination antibiotics (that inhibit two diseases) have been made successfully. The automated operation of the instruments will make possible the large-scale industrialization of genetic engineering. And the machines have one additional selling point: they require no expertise in genetics or experience with sophisticated lab techniques. Practically anyone can use them. But therein, according to some observers, lies the danger. If someone wanted to create a deleterious gene, a toxic protein, or some other harmful biological material for illicit use, it would be far easier to do so with these machines than without.

The promises of technology—electronic, medical, communications, and so on—are never entirely free of risks and societal costs, and it seems that biotechnology is no exception. Most biologists would probably agree, however, that the potential for misuse of these machines is minimal compared with the potential for a true scientific, medical, and agricultural revolution.

---

yeasts are eukaryotes, they also can make important modifications to the proteins, such as addition of sugars, that bacteria cannot. This trait can be useful in the production of vaccines against serious diseases such as hepatitis, since such vaccines have carbohydrate parts.

Other possible single-celled hosts for eukaryotic genes are photosynthetic cells. By inserting into cyanobacteria the genes for the enzymes that synthesize casein (milk protein) and lactose (milk sugar), we may someday replace milk cows. Such cyanobacteria could be grown in massive illuminated vats; they would use the energy of sunlight to manufacture milk protein and sugar. The result would not be milk, but it could prove to be a marvelous dietary supplement for a hungry world.

Genetic Alteration of Agricultural Species

The giant mice we discussed earlier are prototypes for improving farm animals. Introducing engineered growth-hormone genes into cattle might decrease their rearing time and could conceivably increase the amount of meat produced for the amount of feed consumed by the animals; how these hormones might affect the quality of meat has not yet been determined. Growth hormone also increases milk production. In experiments in which cows were treated with bovine growth hormone obtained from genetically engineered bacteria, milk production increased 12 percent. Application of the mouse techniques directly to farm animals will be difficult, but given enough research, that accomplishment is just a matter of time.

An unexpected result from the "supermouse" experiments was the very large amount of growth hormone present in the blood of the engineered mice. Concentrations of growth hormone were as much as 100 times higher than those found in cultures of bacteria that were genetically engineered to produce growth hormone. Therefore, we might be able to carry out "genetic farming" of such valuable gene products as growth hormone or interferon in large animals instead of in bacteria. This approach might be preferred to bacterial production of such proteins as insulin, since the desired protein product must be modified in a way specific for eukaryotes, such as cleavage by enzymes or addition of sugars.

The genetic engineering of plants will undoubtedly have the greatest impact of all on civilization. Growing plants is much more efficient than raising livestock, since plants make their own food by photosynthesis using the sun as an energy source, whereas animals must obtain their energy by consuming plants. The problem with plants as a human food source is the quality of the proteins. As we will see in Chapter 34, most plants are low in certain amino acids, which limits their food value for people. If genes for animal-quality protein could be introduced and expressed successfully in plant crops, human nutrition in many parts of the world could be improved.

Considerable progress has been made in the genetic engineering of plants. One advance was based on a type of natural engineering that certain bacteria have been conducting for hundreds of millions of years. A soil bacterium called *Agrobacterium tumefaciens* infects certain broad-leafed plants, transfers several pieces of DNA from a plasmid to the host cells, and causes the growth of a tumor called a crown gall. The transferred DNA, called *T-DNA*, is found in the cell nucleus of every crown gall. Scientists at Washington University have modified this plasmid, and its T-DNA regions, so it can carry foreign genes into host plant cells, where those genes are inte-

grated into the host chromosomes. Thus the gene for a yeast enzyme, alcohol dehydrogenase, was transferred by T-DNA to tobacco cells. Full tobacco plants grown from those cells all contained the yeast alcohol dehydrogenase gene, as did the cells of the F_1 generation grown from their seeds. That is a spectacular finding: it shows that a foreign gene, once introduced, may be bred by standard agricultural practices in huge numbers of host plants. The procedures used in this process are not important here, but we can appreciate the elegance and precision with which researchers can now manipulate genes and insert them into plant cells.

How will an advance such as the development of plant-engineering methodology affect agriculture? The first gene inserted may be one that gives plants great resistance to pesticides and to herbicides or allows them to make their own pesticides. A growing field of engineered plants could be treated with high levels of such agents and be kept weed- and pest-free. This would no doubt save labor and increase productivity. However, the environmental consequences of applying such high levels of agricultural chemicals must also be considered.

Gene complexes that control drought or cold resistance, natural pest resistance, nitrogen fixation (a process by which atmospheric nitrogen is built into organic molecules), and other such traits will also surely be sought and perhaps transferred to food-crop plants. Most important, work is already under way to get T-DNA to transfer genes into the plant group that includes cereal crops (wheat, rice, barley), which are so central to the human diet. If animal-type proteins could be produced in large quantities by such engineered cereal plants, then science will have taken a major step toward feeding the world's burgeoning human population.

Human Gene Therapy

Serious genetic diseases, including sickle cell anemia, thalassemia, cystic fibrosis, and hemophilia, affect about 1 percent of all children. These genetic conditions impose a large burden on the affected child, on his or her family, and on society. It is now possible to diagnose many genetic diseases before birth, so that prospective parents have the option to terminate pregnancy or to bear a child with a disabling illness. It will not be too many years before another option will be available: masking the effects of the mutant gene with a normal gene.

There are two possible approaches to gene therapy. The first is to add the normal gene to only a subset of the somatic cells of the body (ones in which the gene will be expressed). This addition would affect the phenotype of the individual but not the genotype of sperm or egg cells. Federal regulations so far prohibit gene changes on the germ line. Perhaps in the not-too-distant future, much potential human suffering will be prevented by transferring genes to bone marrow cells and other somatic cell types.

Preventing the Proliferation of Dangerous Bacteria

As soon as recombinant DNA techniques became feasible, scientists acted to set strict safeguards to prevent accidents with recombined organisms. What would happen if genes responsible for the pathogenicity of the plague bacterium or those of a cancer-causing virus were to be cloned in the laboratory strain of *E. coli*, which is distantly related to strains present in all human intestines? If such altered bacteria infected a person, would they cause the plague or cancer? By the late 1970s, tight regulations were in place, and experiments under carefully controlled conditions were undertaken to deliberately clone cancer-causing genes in animals and to reintroduce them into sensitive animals. The results showed that the engineered bacteria used for recombinant DNA technology were safe to work with, given levels of care routinely followed by competent bacteriologists. In fact, in some cases, it is safer to clone and study individual genes from pathogenic organisms or viruses in *E. coli* than to handle the intact organism or virus itself. Moreover, work on the crown gall T-DNA plasmids and others suggested that gene transfer by plasmids may have been going on commonly in nature for millions of years, so regulations governing experiments were subsequently relaxed. Nevertheless, the self-policing action of responsible scientists has set a standard of ethical behavior that is applicable whenever science generates potentially hazardous situations.

Humanistic Concerns

The ethical implications of the new technology are a second major issue for society to address.

There is no doubt that gene therapy for persons suffering from genetic diseases will soon be possible—simply witness the mouse that can make rat growth hormone. In important respects, adding normal genes to a human embryo or adult is not ethically different from doing kidney or bone-marrow transplants, both long-accepted medical procedures. Faced with the critically ill person with a genetic disease, it probably will become the accepted practice to add normal genes to his or her body. That can be done without introducing them into the germ cells, a process that is much more complex to consider ethically and morally.

Geneticist Howard Schneiderman, a faculty member of the University of California and vice president of one of the largest chemical companies in the United States, has confronted these issues:

> I believe—and many hard-headed scientists agree with me—that with the new biotechnology, almost anything that can be thought of can ultimately be achieved—new organisms, new limbs and organs, new treatments for disease, new ways of controlling pests, crops which produce their own pesticides, disease-free domestic animals, whole new industries that will sell products that even today cannot be imagined, let alone made. But if society is to reap social benefits from biotechnology and if industries are to realize financial rewards, we must understand and deal with not only the scientific and technical questions that confront us, but the social questions as well.*

Schneiderman emphasizes that the social, moral, and ethical issues posed by some applications of biotechnology can be resolved. However, it will require scientists working together with responsible science writers and the media to develop an informed public opinion on such matters. Only then can the almost limitless potentials of recombinant DNA technology be fully realized.

*Speech presented at a conference on Biotechnology: Research to Reality, October 26, 1981, in New York City.

SUMMARY

1. The mechanisms by which genes are passed from one generation to another differ between bacteria and eukaryotes because prokaryotes lack a nuclear envelope and do not undergo true mitosis and meiosis.

2. To study phenotypes in bacteria, geneticists use the traits of antibiotic resistance, requirement of specific nutrients, and the inability to use certain nutrients (as in *auxotrophs*).

3. Bacteria can exchange genetic information in three ways. During *conjugation*, two cells make direct contact; a donor, or male, cell forms a sex pilus; and DNA is transferred. The genes for forming a sex pilus are carried on a *plasmid* called the F fac-

tor plasmid, which can integrate into the chromosome. Geneticists can map genes on bacterial chromosomes by timing the passage of DNA between conjugating cells. During *transformation*, DNA released from one cell (or by a researcher) into the surrounding medium is taken up by another cell. During *transduction*, DNA is carried from one bacterium to another by a bacterial virus, or bacteriophage. *Episomes* are genetic elements that can replicate independently of the cell's chromosome.

4. In prokaryotes, clusters of related genes make it possible for limited numbers of regulatory molecules to control at the same time the expression of sets of genes. This control is the process of *gene regulation*, and it commonly includes function of gene clusters called *operons*.

5. The lactose operon in *E. coli* is a cluster of adjacent genes involved in the cell's use of lactose. Three genes—z, y, and a—code for proteins involved in the use of lactose. Lactose itself acts as an inducer of the transcription of these genes. Two additional regulatory elements, i and o, take part; i codes for a *repressor* substance, while o is the *operator* site on the operon DNA. The *promoter site* is adjacent to and partially overlaps the o gene.

6. The i gene makes a repressor that can bind to the operator if there is no lactose present; when bound to the operator, the repressor partially blocks the promoter site and prevents transcription of the z, y, and a genes into an mRNA molecule. Therefore, when lactose is not available, the cell does not make enzymes for breaking down this sugar.

7. If both lactose and glucose are available, the cell uses glucose first. Glucose indirectly causes a catabolite gene activator protein (CAP) to change shape, and it no longer binds well to the promoter site. This slows the transcription of the z, y, and a genes and the synthesis of enzymes for utilizing lactose. The phenomenon is called *catabolite repression*.

8. Two types of controls regulate a bacterial cell's use of lactose. When lactose is present and glucose is absent, CAP binds well to the promoter, and gene transcription takes place; this is positive control. When lactose is absent, the repressor protein binds the operator gene, and transcription slows; this is negative control.

9. Lactose enzymes are *inducible*, since an inducer substance ultimately governs their production. *Repressible* enzymes are those in which the end product of their metabolic pathway builds up and blocks further synthesis of the enzyme. The five enzymes that convert a precursor substance to tryptophan are repressible enzymes; transcription of the genes coding for these enzymes stops when tryptophan is present.

10. Eukaryotic chromosomes have repetitive DNA sequences and a great deal of DNA that does not code for proteins. Within the genes themselves, there are regions that are expressed, *exons*, and intervening sequences that are not expressed, *introns*. Both exons and introns are transcribed into the primary RNA transcript; then the intron regions are cleaved out, and the exons are spliced together before a functional mRNA is made. RNA processing is probably involved in eukaryotic gene control.

11. *Recombinant DNA technology* is a set of techniques for transferring genes from one organism to another. *Restriction endonucleases* and *DNA ligases* are cut-and-paste enzymes that allow a researcher to cleave DNA in such a way that cohesive ends remain, and to rejoin DNA fragments into a re-paired double helix. Vectors such as the F factor plasmid can carry foreign DNA into a host cell. A plasmid into which a foreign gene has been spliced is a new genetic entity, a *recombinant DNA molecule*. The technique of *DNA–RNA hybridization* is used as a molecular probe to locate the specific, desired foreign gene sequence in *clones* of bacterial host cells.

12. In the giant-mouse experiment, the gene for rat growth hormone was fused to the control region of the mouse gene for metallothionein (a detoxifying liver protein). This fused gene was injected into a fertilized mouse egg, and the resulting mouse produced very abundant amounts of rat growth hormone and grew extremely large.

13. Genetic engineering probably will be applied in three ways: (1) foreign genes are already being introduced into microbes and the gene product, such as human insulin or interferon, harvested; (2) genetically altering livestock or food-crop species so they grow faster, are resistant to disease, cold, chemicals, or drought, and have improved protein quality; (3) replacing mutant human genes to cure genetic diseases such as sickle cell anemia and hemophilia. All three possibilities hold great promise and pose certain ethical dilemmas for society.

KEY TERMS

auxotroph

catabolite repression

clone

conjugation

DNA ligase

DNA–RNA hybridization

episome

exon

F factor

F factor plasmid

gene regulation

genetic engineering

genome

Hfr cell

inducible enzyme

intron

operator gene

operon

plasmid

promoter site

R factor

R factor plasmid

recombinant DNA molecule

recombinant DNA technology

repressible enzyme

repressor

restriction endonuclease

transduction

transformation

QUESTIONS AND PROBLEMS

1. What is an operon? In which organisms would you expect to find operons? Draw a diagram of a specific operon, and describe how the various regulatory molecules control transcription in this operon. Define the following terms: "constitutive," "inducible," "repressor," "promoter," "operator."

2. A mutation sometimes occurs in the lac repressor gene so that the mutant repressor no longer can bind the inducer, but it can still bind to the operator. What phenotype would you expect this mutant strain to have? Suppose a different mutation changes the repressor so it can no longer bind to the operator, whether or not it is bound to inducer. What phenotype would this mutant strain have?

3. A mutation occurs in the operator so it can no longer bind repressor. What phenotype does this mutation produce? If you construct a partial diploid with that mutation, what phenotype will it have?

4. The promoter is the site near the beginning of a gene or operon where RNA polymerase binds. Would you expect promoters for many different genes to have the same base sequence or many different sequences (in other words, does RNA polymerase recognize a specific sequence)? If the lac promoter mutated so it could no longer bind RNA polymerase, what phenotype would the cell have? What phenotype would a partial diploid have? If the promoter mutated so it could still bind RNA polymerase, but not as efficiently as the wild-type sequence, what phenotype would the mutant cell have?

5. What are F factors? What are F^+ and F^- cells? What are Hfr cells?

6. The genome of E. coli contains 4×10^6 base pairs. Assuming that the "average" gene contains 1,200 base pairs, and most of the genome represents genes, how many genes might you expect E. coli to have? The genome of bacteriophage lambda contains 5×10^4 base pairs. How many genes would you expect to find? A human sperm or egg contains 3×10^9 base pairs. If most of the genome represented genes, how many genes would humans have? Eels have 1.5×10^{11} base pairs. Does it seem likely that eels have 50 times as many genes as humans? What kinds of evidence suggest that eukaryotes have lots of "extra" DNA?

7. A strain of E. coli contains plasmids that carry two genes for antibiotic resistance: one for resistance to streptomycin (Str^R) and the other for ampicillin (Amp^R). These genes make enzymes which are secreted by the cell and which destroy the antibiotics. The plasmid DNA has a single restriction site for restriction endonuclease Eco R I, in the Str^R gene, and a single restriction site for the restriction endonuclease Hin d III, in the Amp^R gene. You wish to use this plasmid as a cloning vector for a piece of broccoli DNA. First you isolate lots of plasmid DNA and broccoli DNA; then you cut both kinds of DNA with the enzyme Eco R I. Next you mix the two kinds of DNA together. How will you join the broccoli DNA to the plasmid DNA? After joining the DNAs, you transform an antibiotic-sensitive strain of E. coli with this DNA and plate the cells on petri plates containing ampicillin. What types of cells will grow? What types of cells will not grow? Some of the cells that grow will contain a plasmid that is carrying some broccoli DNA. A prokaryotic gene with the plasmid DNA in it will not make its product protein. Using that fact, how could you distinguish between colonies that contain a plasmid with inserted broccoli DNA from colonies that contain a plasmid with no broccoli DNA?

8. Transduction experiments were done using bacteriophage. The phage were grown on a bacterial strain containing the alleles $x^+ y^+ z^+$, and then used to infect a strain that was $z\ y\ z$. Among the y^+ transduced cells, 90 percent were x^+ and 20 percent were z^+. Among the x^+ transduced cells, 90 percent were y^+ and only 10 percent were z^+. What is the order of the x, y, and z genes?

9. Gene-transfer experiments were done using four different Hfr strains of E. coli (strains in which the F fac-

tor plasmid had integrated in different parts of the chromosome, and possibly in different directions relative to the host cell chromosomes). Conjugation between the Hfr and F⁻ strains was interrupted at various times, and the order of gene transfer was determined. The following orders of genes were found:

Hfr strain 1: A C T E R
Hfr strain 2: T E R I U
Hfr strain 3: A B M U I
Hfr strain 4: M U I R E

What is the order of all these genes on the bacterial chromosome?

10. What are some of the ways in which eukaryotic mRNA differs from that of prokaryotes?

11. Why would a geneticist who is isolating bacterial mutants deficient in the ability to make one amino acid use a medium that lacks only that one amino acid?

12. Whenever the sugar arabinose is present, a bacterium produces an enzyme A that degrades the sugar. Mutations that eliminate the activity of enzyme A are found in two different genes. How could you tell which gene coded for the enzyme and which for a repressor protein?

ESSAY QUESTION

1. Does the discovery of introns change your ideas about what a gene is? What are gene families? Discuss their role in evolution.

SUGGESTED READINGS

ALBERTS, B., et al. *Molecular Biology of the Cell.* New York: Garland, 1983.

The best of the current cell biology books has a good explanation of gene regulation, the types of excess DNA in eukaryotes, and recombinant DNA technology.

AYALA, F. J., and J. A. KIGER, JR. *Modern Genetics.* Menlo Park, Calif.: Benjamin/Cummings, 1980.

This up-to-date book has a good, detailed treatment of gene regulation.

CHAMBON, P. "Split Genes." *Scientific American*, May 1981, pp. 60–71.

This article is a clear treatment of the early work on introns and exons.

FEDOROFF, N. V. "Transposable Elements in Maize." *Scientific American*, June 1984, pp. 84–98.

Here is a discussion of the jumping genes that gained Barbara McClintock a Nobel Prize; the molecular mechanisms in corn and other species are described.

GLOVER, D. M. *Genetic Engineering.* New York: Methuen, 1980.

Nonbiologists will have no trouble understanding this book.

JACOB, F., and J. MONOD. "Genetic Regulatory Mechanisms in the Synthesis of Proteins." *Journal of Molecular Biology* 33 (1961): 318.

This classic paper describes the elegant experiments that led to the concept of the operon and other features of gene function, including messenger RNA.

NOVICK, R. "Plasmids." *Scientific American*, December 1980, pp. 102–27.

This is a clear account of the molecular circles that will change medicine and agriculture.

15

HUMAN GENETICS

*The capacity to blunder slightly is the real marvel of DNA.
Without this special attribute, we would still be anaerobic
bacteria, and there would be no music.*

Lewis Thomas,
The Medusa and the Snail
(1979)

ALBINISM IN SHI-MO-PA-VI NATIVE AMERICANS.

Although we are all members of one species, human beings display remarkable differences. Adults can be less than 3 feet tall or more than 7, and they can weigh under 100 pounds or well over 400. Hair, eyes, and skin have a wide variety of hues. Facial features vary greatly from race to race and from individual to individual. Most families have at least a few obvious genetic characteristics they can trace back through many generations: near-sightedness, red hair, baldness, cleft chin, or perhaps a rare but harmless trait, such as extra fingers and toes, hairy ears, or a white forelock. A few families seem to be cursed with more serious disorders, such as color blindness, albinism, hemophilia, or Huntington's disease, all of which are based on genetic mutations. Such pronounced variations among members of our own species and the underlying patterns of inheritance that explain them are the concerns of human genetics.

Most of the principles that we have discussed in our survey of genetics from Mendel to gene splicing apply equally to human genetics: some human genetic traits are inherited in simple Mendelian fashion, but most human traits, like those of all other complex animals, seem to be determined through more complex gene interactions. And human genes are as subject to mutation and manipulation as those of other species. Although the basic principles of inheritance are the same, the study of human genetics until recently has been unusually difficult because human beings present unique challenges as research subjects and, as a result, require a set of special study methods. As we will see, however, methods of genetic engineering have already ushered in a new area of human genetics. Research in human genetics has important implications for the treatment and prevention of many disorders through genetic analysis, screening, and counseling.

In this chapter, we shall focus on the major methods used in the study of human genetics and on some of the best-understood genetic diseases, applying many facts and principles from earlier chapters. We shall also discuss possibilities for repair of defective genes and some of the legal, ethical, and financial issues surrounding genetic screening and counseling. We begin by considering the special challenges in studying the genetics of our own species.

HUMANS AS GENETIC SUBJECTS

Why do people present a particular challenge as subjects of genetic research? There are three main reasons, which could be labeled "the marriage gamble," "the missing F_2," and "the small-sample syndrome."

One of the geneticist's most important tools, as we

have seen, is the deliberate cross between two genetically interesting adults. In a standard cross, both the phenotype, or physical characteristics, and the genotype, or set of alleles for the trait in question, are known for both adults. In a test cross, the phenotype is known for both parents, but the genotype is certain for only one of them and is revealed by the appearance of the offspring. These deliberate mating techniques are applied to organisms from bacteria to buffaloes but, of course, cannot be used on human beings. We mate for various reasons, but satisfying the curiosity of genetic researchers is not one of them. Human matings thus are not genetically controlled, except in the sense that couples sometimes choose not to reproduce on the basis of genetic counseling.

Second, as we have seen, a true F_2 generation is by definition the result of a cross between members of the F_1 generation. In human terms, this would mean the mating of a brother and a sister, which is an increasingly rare occurrence due to cultural taboos. Because of this "missing F_2," a geneticist virtually never sees results of a cross between people who vary in only one of a few genes of interest, as would be the case with true F_2 genetic subjects. Geneticists studying humans therefore must reconstruct "statistical F_2 generations" through data gathered on many unrelated families that display a certain gene of interest.

Finally, a given human family rarely produces enough children to represent a statistically significant sample. Most genetic subjects produce tens, hundreds, or even thousands of offspring within the time frame of a genetic study. In contrast, people have generation times of twenty-five years or more and only rarely produce as many as ten offspring; the current average in the United States is fewer than three children per family. To counteract the "small-sample syndrome," geneticists again must gather and pool data from many families in which the parents appear to have the same genotype—say, both heterozygous for a given pair of alleles.

The challenge is to recognize genotypes without the standard tools of the deliberate cross and the true F_2 generation. How, then, does a geneticist learn the genotype for a trait of interest and whether that trait is determined by one pair of alleles or by more than one pair? The answers lie in the specific methods of human genetics.

WHAT ARE THE METHODS OF HUMAN GENETICS?

Because of the unique nature of human subjects, geneticists must rely on several indirect research methods to reconstruct inheritance patterns and to diagnose, predict, and sometimes treat genetic diseases and defects.

These indirect methods include pedigree analysis, karyotyping, biochemical analysis, somatic cell genetics, in situ hybridization, and twin studies. Let us consider each method in some detail.

Pedigree Analysis, or Constructing Family Trees

Although it may sound like a document that comes with a puppy, a **pedigree** is a formal representation of a set of traits for all members of a family lineage. Such an analysis traces a particular trait over several generations and identifies, if possible, the first individual who manifested the trait. Once the pedigree is established, the geneticist may be able to deduce whether the trait in question is inherited according to Mendel's laws and whether it is dominant or recessive. As we saw in Chapters 10 and 11, a dominant allele usually expresses itself (*AA, Aa*) and does not skip generations; that is, offspring will not express a trait determined by a dominant allele unless one or both parents have that allele. A recessive allele *can* skip generations. Normally, a trait determined by a recessive allele usually expresses itself only in homozygotes (*aa*); occasionally, however, a dominant allele is masked by some other allele or process, so that its recessive partner is expressed in the phenotype.

A pedigree analysis for the occurrence of a dominant trait called *achondroplastic dwarfism* in a large family is shown in Figure 15-1a. Achondroplastic dwarfs have a developmental abnormality in which the long bones of the limbs do not reach normal size, although the bones of the trunk and head do. This condition does not affect intelligence, physical maturity, or personality development. In medieval times, dwarfs sometimes were employed as court jesters or buffoons (Figure 15-1b). The pedigree traces this trait through four generations and indicates (in blue) the male *propositus*—the person with the trait who first came to the attention of the person constructing the pedigree.

In the first generation, a male dwarf married a normal-sized woman and produced seven children, including

Figure 15-1   ACHONDROPLASTIC DWARFISM.
(a) A family pedigree for the occurrence of achondroplastic dwarfism. Individuals with the trait are indicated in color. Use the symbols below to read the pedigree. (b) A portrait of an achondroplastic dwarf entitled *Giacomo Favorchi* and painted by the Dutch artist Karel van Mander around 1600.

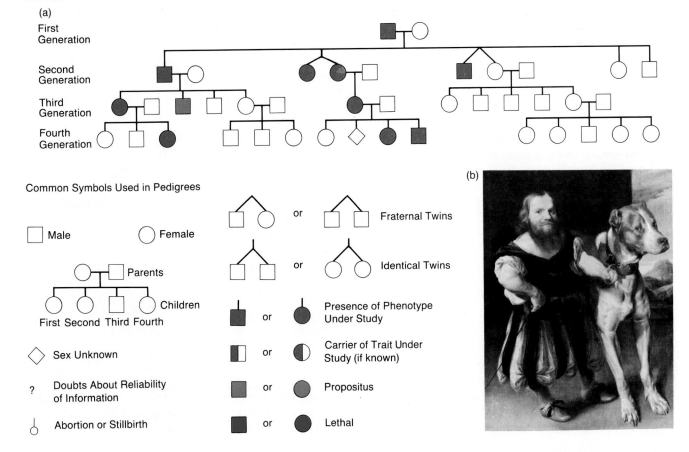

one pair of monozygotic (identical) twins and one pair of dizygotic (fraternal) twins. Four children were dwarfs (green symbols), and three were normal. Because the dwarf husband produced both dwarf sons and dwarf daughters, achondroplastic dwarfism must be carried on an autosome—a chromosome other than a sex chromosome. Can you deduce why neither the X nor the Y chromosome of the dwarf husband could have carried the mutant allele? By noting where dwarfs appear in the second and subsequent generations of such pedigrees, geneticists reasoned that the trait cannot be recessive.

Using a pedigree like this, a geneticist can begin to determine the genotype of the family members, even without deliberate crosses and test crosses. Based on the first and second generations, a geneticist could hypothesize that the trait is an autosomal dominant and that the dwarf parent in the first generation was heterozygous, having one dominant allele causing dwarfism and one normal allele. The progeny of the third and fourth generations turn out to be consistent with this hypothesis. The trait reappears in every generation irrespective of the sex of the dwarf parent, which supports the assumption of dominant inheritance. In addition, the marriage of a normal woman from this lineage to a normal man produced no dwarfs over two more generations, which is also consistent with dominant inheritance, although it does not, by itself, rule out recessivity. By using such pedigrees from a number of unrelated families whose members include achondroplastic dwarfs, geneticists have been able to confirm that the trait is determined by a dominant allele on an autosome. Furthermore, geneticists have discovered that the mutant allele for this trait is lethal when homozygous (Chapter 11); the homozygous offspring of two achondroplastic dwarfs die before birth, so that only heterozygotes survive and become dwarfs.

By amassing and comparing thousands of family pedigrees for various traits, geneticists have been able to determine hereditary patterns for a number of human genetic disorders, some of which we will examine later in the chapter. Such well-established patterns make it possible to counsel prospective parents on the chances of producing a child with a deleterious trait that runs in one or both families.

Karyotyping: Chromosomes Can Reveal Defects

Another method for determining whether a given human trait—usually a deleterious or an abnormal one—might have a genetic cause is to examine the chromosomes themselves. This process, called **karyotyping**, involves staining chromosomes during mitotic metaphase, photographing them, and cutting out the images

and arranging them on the basis of size and shape. Furthermore, certain staining procedures reveal banding patterns like those found in the polytene chromosomes of fruit flies (Chapter 11).

By karyotyping human chromosomes, geneticists have shown that several syndromes involving abnormal embryonic development are linked to specific abnormalities in the number or structure of chromosomes. For example, **Down's syndrome** is a genetic disorder that occurs in 1 of every 600 newborns. This syndrome is characterized by a group of traits that can include mental retardation; heart defects; a short, stocky body; and distinctive

Figure 15-2   DOWN'S SYNDROME: AN EXTRA COPY OF CHROMOSOME 21.

(a) The karyotype of a human with Down's syndrome. The extra copy of chromosome 21, one of the smallest chromosomes, causes all of the deleterious traits characterizing this genetic disorder. (b) A sixteen-year-old girl with Down's syndrome, or trisomy 21.

(a)
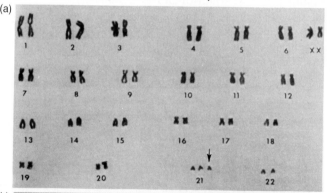

(b)

eyelid folds. Karyotype analyses reveal that individuals with this syndrome have an extra chromosome 21, as shown in Figure 15-2a. Since its victims have three copies of chromosome 21, Down's syndrome is also called *trisomy 21*. The extra chromosome 21 is a result of **nondisjunction,** or the failure of homologous chromosomes to separate properly during meiosis. We shall talk more about nondisjunction later in this chapter.

Biochemical Analyses Reveal Mutations

At present, the most direct tool for determining the genotypes that underlie human phenotypes is the analysis of specific metabolic pathways. When the genetic system being studied is the short, direct pathway between a single gene and a single gene product, such as an enzyme or antigen, a geneticist often can determine both whether the allele of the gene is a dominant or a recessive mutation from the normal, wild-type allele and how the mutant allele is expressed. Ironically, the traits that can be studied most effectively through biochemical analyses are often uninteresting to the nongeneticist. They usually center on the blood factors (ABO, Rh, or MN, all sets of marker molecules on the surfaces of red blood cells) or on relatively rare metabolic defects and diseases. More visible traits—such as eye color, skin color, stature, or weight—often derive from two or more interacting gene pairs influenced by developmental and environmental factors and are more effectively studied with family pedigrees. Biochemical analysis can therefore be regarded as a potentially powerful technique but as yet only for certain traits.

One particularly deadly disorder that can be detected through biochemical analysis is **Tay-Sachs disease.** Children born with this disease are homozygous for a recessive allele that prevents production of the enzyme *hexosaminidase A.* They inherit one recessive allele from each parent—heterozygous *carriers* who do not show the trait themselves. Normally, hexosaminidase A is required for lipid metabolism; in its absence, fatty deposits accumulate and cause the nervous system to degenerate in a process that begins a few months after birth and worsens steadily. At eighteen months, the infant is usually deaf and blind and suffers seizures. Tragically, death occurs before age five.

When blood-serum levels of hexosaminidase A are measured, the results differ among normal people, Tay-Sachs carriers (heterozygotes), and infants with the disease (homozygotes), as shown in Figure 15-3. Such analyses make carriers of the recessive allele fairly easy to identify so that potential parents can learn their chances of producing a diseased child. Tay-Sachs disease, although rare, is most common among Jews from central

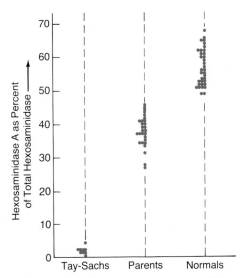

Figure 15-3 BIOCHEMICAL ASSAY OF TAY-SACHS DISEASE.

Tay-Sachs disease is caused by the body's inability to produce the enzyme hexosaminidase A, the result of a homozygous recessive allele. Heterozygous carriers can be detected by the biochemical test the results of which are shown in this diagram. The percentage of hexosaminidase A is very low in homozygous victims of the disease, and the heterozygous carriers of the disease ("parents") show a value between those of normal individuals and of victims.

Europe and their descendants, a group referred to as Ashkenazic Jews; one in twenty-eight Ashkenazic Jews are heterozygous for the trait. Non-Jewish Canadians also suffer from an identical Tay-Sachs disease, but it stems from a completely different mutation, a deletion from the hexosaminidase gene. In recent years, screening programs to detect carriers have helped decrease the occurrence of Tay-Sachs disease in the United States and other countries.

Mapping Human Genes: Somatic Cell Genetics

In 1911, the first sex-linked human trait, color blindness, was traced to a gene located on the X chromosome. By the early 1970s, only one additional gene had been mapped on an autosomal human chromosome. In a standard genetic subject such as *Drosophila,* deliberate crosses are made to produce offspring with desired genotypes, and such crosses facilitate gene mapping. Since controlled matings cannot be used in humans, and since pedigree analysis is not a precise enough tool to locate genes easily, a different method was needed to map human genes. Such a method was developed in 1971,

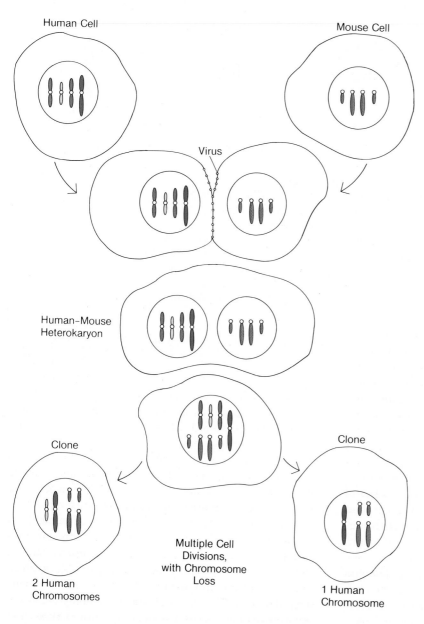

Human Cell

Mouse Cell

Virus

Human–Mouse Heterokaryon

Clone

Clone

2 Human Chromosomes

Multiple Cell Divisions, with Chromosome Loss

1 Human Chromosome

Figure 15-4   MAPPING GENES WITH SOMATIC CELL TECHNIQUES.
Researchers grow a human somatic cell in culture in the presence of a mouse cell. The chromosomes are easily distinguishable, since many human chromosomes have the centromere in the middle, whereas all mouse chromosomes have the centromere at the end. An inactivated virus binds to the surface of the cells and allows them to fuse. The result is a heterokaryon in which the two nuclei may later fuse. As the heterokaryons carry out mitotic cycles, they tend to lose human chromosomes. Ultimately, different subclones may contain different human chromosomes; by identifying which gene products are still made by a heterokaryon containing a specific human chromosome, a gene can be shown to be on that chromosome. For simplicity only one copy of each chromosome is shown in these cells.

and in the decade that followed, 400 human genes were assigned to particular chromosomal locations. Today, gene mapping using these methods continues at a rate of about four genes per month.

The new method of chromosome mapping effectively by-passes human reproduction as a means of producing new genotypes because it relies on somatic cells, such as skin and liver cells, rather than on eggs and sperm. This field of research is called **somatic cell genetics.** In this approach, two genetically distinct, diploid somatic cell types are grown in tissue culture and fused by means of inactivated viruses that bind to the cell surfaces (Figure 15-4). The fused cell, with its two genetically distinct, diploid nuclei, is called a *heterokaryon* (meaning "different nucleus"). These two nuclei may fuse to make a tetraploid *cell hybrid,* which grows and divides normally to make a clone of cells.

Fortunate "accidents" sometimes occur as the somatic cell hybrids divide: individual chromosomes may be lost from the cells. If mouse and human cells are fused to make hybrids, the chromosome loss can be controlled to a certain extent, so that the hybrids eventually have mostly mouse chromosomes plus one or a few human chromosomes of random type. One clone of cells might have human chromosomes 2, 12, and X, while another might have 7 and 9. By identifying which human genes or gene products are still found in the human–mouse cell hybrid, with, say, its single remaining human chromosome, geneticists can be sure that particular genes are on individual human chromosomes.

The technique of somatic cell genetics allows certain defects to be diagnosed before a child is born or, in some cases, to be predicted before the child is conceived. Using the techniques of *amniocentesis* or of *chorionic*

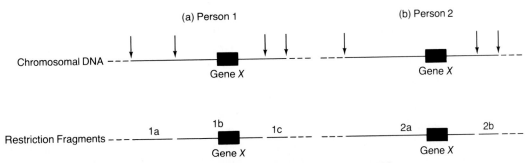

Figure 15-5   RESTRICTION ENDONUCLEASES: MAPPING HUMAN GENES.
(a) Person 1 has four characteristic cleavage sites on the DNA of a specific chromosome that will produce three fragments of different sizes after cleavage (1a, 1b, 1c). (b) Person 2 is a genetic variant in whom one of the enzyme-recognition sites has been eliminated by mutation. Because there are only three cutting sites in this region of the chromosome of person 2, only two fragments of DNA are obtained (2a, 2b). Fragment 2b is the same fragment as 1c, but fragment 2a is equal to 1a + 1b. The sizes of the fragments can be detected by observing their behavior when subjected to electrophoresis. The large amount of variability in sizes of restriction fragments from one person can allow their use in DNA fingerprinting and in the mapping of human genes.

villi sampling (described later in this chapter), cells can be obtained from the developing fetus to be analyzed, or somatic cells can be taken from prospective parents who are suspected carriers of a deleterious recessive gene. The somatic cells are grown in culture and are analyzed for the presence of a gene or DNA sequence called a *biochemical genetic marker* that is known to be closely linked to a mutant gene of interest. Such a biochemical marker must be (1) readily detectable in cultured somatic cells, (2) closely enough linked on a chromosome to the deleterious allele so that recombination between the two is rare, and (3) present on the particular chromosome that bears the mutant allele but absent from normal chromosomes.

Such markers also can be located with *restriction endonucleases*, enzymes that cleave DNA at specific sites (Chapter 14), creating DNA fragments of different lengths. If the base sequence is changed by mutation at one of these sites, the expected cleavage will not occur, and the DNA fragments will not be the predicted lengths (Figure 15-5). If such a site is very close to a deleterious allele, its presence serves to identify the allele, too.

Sickle cell anemia can be diagnosed in cells from a fetus using this technique. After somatic cell DNA is "digested" with a particular restriction endonuclease, the normal allele for the β hemoglobin chains is found on a DNA fragment that is about 7,600 base pairs long. This is similar to fragment 1b in Figure 15-5. In people of African descent who carry the mutant sickle cell allele, the mutant β hemoglobin allele is carried on a 13,000-base-pair fragment, analogous to fragment 2a in Figure 15-5.

The power of restriction endonuclease cleavage has been exploited by Alec Jeffreys of Leicester University in England to develop "DNA fingerprints" that are just as unique as the whorls on the fingertips of each human. It turns out that there are large numbers of short, repeated sequences of bases in human DNA, so-called minisatellites; furthermore, the number of these minisatellites varies substantially from one individual to the next. Consequently, human DNA cleaved by restriction endonucleases yields immense polymorphism (variability) in the lengths of restriction fragments. These **restriction fragment polymorphisms** have been dubbed **RFLPs** (pronounced "ruflups"). Jeffreys found ways to probe this diverse array of RFLPs, and he comes to the conclusion that the probability that two human individuals will have identical sets of restriction fragments is about 5×10^{-19}—essentially zero! In the practical world, this allows police or others to prove without doubt the identity of the human source of blood (even old, dried blood), of sperm in sexual assaults, or of parents of a child. The first such use in Britain yielded the happy proof that a small Ghanaian boy was fathered and mothered by a particular couple and was, in fact, entitled to live in England. DNA fingerprinting will surely revolutionize forensic (law-related) investigations.

Still another powerful new gene-mapping technique employs the tools of the molecular biologist to make copies of mRNA and DNA molecules that can serve as probes. By means of the process, called *in situ hybridization*, a chromosome preparation otherwise used for karyotyping can be treated chemically, so that either the mRNA or the DNA region coding for a protein such as sickle cell hemoglobin will bind specifically to the chromosomal gene for that protein. If the probe is radioactive or is tagged with a fluorescent dye, the site of the gene can be identified. This procedure, which is much quicker than older techniques, is beginning to be used to map human chromosomes; for instance, the genes for three color-receptor proteins (opsins) have been mapped in this way (Chapter 38). The result of widespread use of in situ hybridization will be a precise map for many genes coding for many proteins.

By analyzing somatic cells, family pedigrees, and other data, geneticists can correctly identify couples at risk of conceiving a child with sickle cell disease. If more were known about other mutated genes and surrounding genetic markers, accurate predictions could be made for other genetic defects as well. At this point, however, the powerful tools of somatic cell genetics and in situ hybridization cannot yet be used for prenatal diagnosis of most genetic diseases.

Reverse Genetics

The new technologies of DNA sequencing, gene cloning, and controlled protein manufacture now permit a "bottom up," or **reverse genetics**. Classical genetics, from peas to fruit flies to humans, starts with phenotypic traits and works back to the gene. Molecular biologists now can work in the opposite direction, starting with long stretches of chromosomal DNA that can be shown to be inherited in exactly the same way a given human disease is inherited. Next, a specific gene in that DNA is identified, the protein it encodes is manufactured, and, finally, the role of the protein in generating the disease is sought. The first success with reverse genetics involved a gene on the human X chromosome that causes a disease abbreviated CGD in which victims suffer chronic severe bacterial infections. A long DNA segment from the X chromosome yields a gene that, in turn, produces a protein responsible for CGD. Reverse genetics is now being used to search for the genes and gene products that cause cystic fibrosis, Huntington's disease, and other deadly maladies. Nowhere are the marvels of molecular biology more welcome.

Nature, Nurture, and Twin Studies

The final method of human genetics we shall consider is the *twin study*, which can help determine the influence of environment on gene expression.

Differences among people are not entirely attributable to genes; environment also plays a role in the expression of many traits. The extent to which either genes or the environment affect a certain trait has come to be called the "nature–nurture" question. Obviously, the expression of an individual's genotype can be strongly affected by the environment in which he or she develops as an embryo, a child, and an adolescent. Factors such as nutrition, sanitation, disease, and availability of medical care can influence the eventual phenotypic traits of the adult. For example, even genetically identical monozygotic twins (derived from a single egg) can grow up looking fairly different if one ate poorly as a child or contracted a serious disease, while the other ate normally and stayed healthy. There are undoubtedly many subtle influences as well, such as presence or absence of siblings or parental neglect or affection.

Geneticists clearly cannot deprive children of food or affection to test the effects of these nongenetic, environmental influences on adult phenotypes. Thus they rarely know to what degree the profound variations we can easily see among people are based on nature or nurture. In addition, most traits are **polygenic:** they are controlled by several genes. Height, body weight, and scores on IQ tests are examples of polygenic human traits. Since environmental factors may also influence polygenic traits, human geneticists face a formidable challenge when they study many of the most common and interesting traits.

One way to approach the nature–nurture problem is to examine a given phenotypic trait in identical twins (Figure 15-6). Geneticists assume that any differences between two genetically identical individuals probably arise from environmental influences on gene expression. If this is true, comparing identical twins reared apart in very different environments since infancy with identical twins raised together should reveal the influence of nurture or home environment on a particular trait.

Consider the variation among individuals for height, weight, and IQ shown in Table 15-1. Monozygotic twins who are reared in different environments are much closer in height than are dizygotic twins (derived from two eggs) or nontwin siblings, regardless of whether they are reared together. Height therefore appears to have a strong genetic component little influenced by environment. However, weight is a different story: identical twins reared separately show greater weight variation than do identical twins reared together; and both the former group and fraternal twins approach the variation seen in normal siblings. Therefore, environment, including learned eating habits, seems to affect weight to a much greater extent than it does height.

Table 15-1   **DIFFERENCES BETWEEN TWINS***

Trait	Monozygotic Twins		Dizygotic Twins	Nontwin Siblings
	Reared Together	Reared Apart		
Height	1.7 cm	1.8 cm	4.4 cm	4.5 cm
Weight	4.1 lbs	9.9 lbs	10.0 lbs	10.4 lbs
IQ score (Stanford-Binet)	5.9	8.2	9.9	9.8

*Based on fifty pairs each of monozygotic twins reared together, of dizygotic twins, and of nontwin siblings, and on nineteen pairs of monozygotic twins reared apart.

Source: H. H. Newman, F. N. Freeman, and K. J. Holzinger, *Twins: A Study of Heredity and Environment* (Chicago: University of Chicago Press, 1937).

(a)

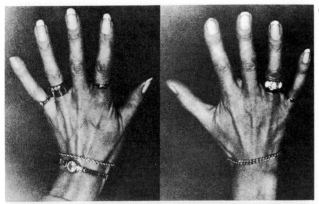

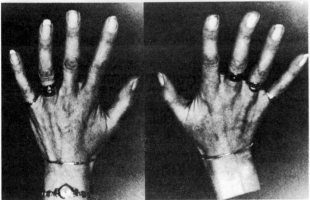

(b)

Figure 15-6   IDENTICAL TWINS.

(a) It is difficult to separate nature from nurture (heredity from environment) in studying identical twins. However, the female identical twins whose hands are shown here were separated at two weeks of age and reunited after thirty-eight years. At the time of their first meeting, both twins were wearing almost identical jewelry—a coincidence with a very low probability! (b) The twins in this picture were not told how to pose for the photographer. Notice that each set of twins has a characteristic posture and placement of hands. Genetics clearly plays a role in these examples of very complex behavior associated with tastes, body posture, and movement.

So far, twin studies have shown that the kinds of traits measured by traditional IQ tests are more strongly influenced by environmental factors, such as parental stimulation and type of schooling, than by genotype. However, even IQ studies of identical twins must be interpreted with caution, since twins raised together, whether monozygotic or dizygotic, probably have much more similar total environments than does a pair of nontwin siblings in the same home who spend less time together. Despite such complexities, some people make exaggerated claims linking IQ to race. Too often, such persons ignore the nurture component of this complex trait. Furthermore, many psychologists have pointed out that IQ tests are a measure not of "intelligence," but of skills that correlate well with performance in school.

One correlation that cannot be denied is the close relationship between low IQ and unusually low birth weight. Whether a result of poor maternal nutrition, maternal cigarette smoking or alcohol consumption, or premature delivery, infants who are small and thin are likely to have lower IQs than heavier, full-term babies. No doubt, the relationship reflects the degree to which higher brain centers develop in the fetus prior to the transition to life outside the womb. Thus environment during the initial development of the brain plays a significant role years later in the IQ trait of such individuals.

Despite the keen interest in the genetic basis of human traits, our understanding is still incomplete. But as today's new methods of analysis and genetic engineering are applied to humans, we can anticipate the elucidation of the genetic basis for many of the thousands of human traits and variations.

CHROMOSOMAL ABNORMALITIES: A MAJOR SOURCE OF GENETIC DISEASE

Geneticists now know that successful embryonic development requires the normal complement of human chromosomes. Until the 1950s, however, the methods of identifying individual chromosomes were so imprecise that geneticists could not even agree on the number of chromosomes in human cells. As karyotyping and other techniques became available, forty-six chromosomes were confirmed. By now, geneticists have carefully examined the chromosomes of normal, malformed, and diseased adults, newborns, and embryos. As a result, they have determined that many specific disorders are directly attributable to chromosome defects. Let us see how variations in chromosome number or shape affect human phenotype.

Abnormal Numbers of Autosomes

Deviation from the normal diploid number of forty-six chromosomes is the most frequent type of genetic defect in humans. The most common deviation, called **aneuploidy,** is the absence of one or more chromosomes, or the presence of one or more extra chromosomes. Aneuploidy often results in the death of the embryo at an early stage. Other forms of aneuploidy involve duplication or deletion of parts of chromosomes, so that chromosome content is changed, even when chromosome number is not.

The entire complement of gene pairs represented on a diploid chromosome set, without additions or deletions, appears to be essential for normal development. It is "normal" for our species that *about 45 percent of all embryos* are spontaneously aborted during early pregnancy (indeed, since many early embryos may not be detectable, the figure may be considerably higher). Nearly half of these aborted embryos are very abnormal; in turn, half of these abnormal embryos exhibit some deviation from the normal diploid chromosome number. Figure 15-7 classifies the types of chromosomal abnormalities. In *monosomics,* only one partner of a given chromosome pair is present. In *trisomics* (such as individuals with trisomy 21), three rather than two chromosomes of a given type are found. *Triploids* are individuals who have three entire sets of chromosomes

Figure 15-7   CHROMOSOME ABNORMALITIES.

Just two pairs of homologous chromosomes comprise the diploid state of this cell line. Note in the abnormalities the differing numbers of either chromosome type. In this rearrangement, a full chromosome arm has been exchanged between a large and a small chromosome. In other cases, only parts of chromosome arms may be exchanged or be rearranged.

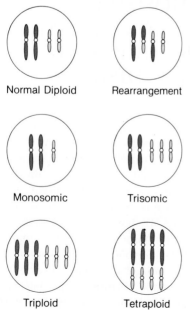

(sixty-nine chromosomes) instead of two sets, and *tetraploids* have four sets (ninety-two chromosomes). Unusual numbers of chromosomes such as these apparently cause imbalances in genetic activity that lead to altered metabolism and abnormal development in the embryo, commonly leading to spontaneous abortion.

Many forms of aneuploidy result from nondisjunction during meiosis in the mother. As the mother's age increases, spontaneously aborted embryos show an increased incidence of aneuploidy (Figure 15-8a), making it clear that as a woman nears or passes age forty, it becomes riskier for her to bear children. One reason for the high rate of nondisjunction in older women is that all

Figure 15-8   FREQUENCY OF DOWN'S SYNDROME AND MOTHER'S AGE.

(a) The danger of bearing a child after age 40 or so is evident from the steep rise in the curve. (b) Aneuploidy in this case results when one chromosome (number 21) is "sticky," so that both copies are delivered to one gamete; the other gets none.

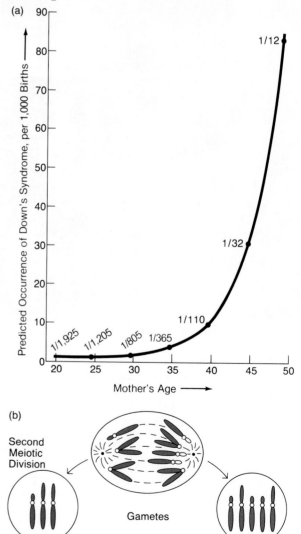

human eggs become arrested in prophase of meiosis I while the female is herself still an embryo. Meiosis is completed in individual eggs only after ovulation and fertilization by a sperm. Apparently, forty or fifty years in the arrested state is simply too long, and nondisjunction of chromosome 21 and others becomes more and more likely. One of every twelve babies born to mothers aged fifty suffers from Down's syndrome. Aneuploidy also occurs in male testes, but it has not been found to be directly correlated with age, as is the case with females.

Although most cases of aneuploidy are lethal, babies can survive with extra chromosomes 21, 13, and 18. All such infants have physical defects, however. As we have seen, trisomy 21 produces Down's syndrome. *Trisomy 13,* which yields individuals with harelip, cleft palate, and various eye, brain, and cardiovascular defects, occurs in about 1 in 10,000 live births and usually causes death within the first three months after birth. *Trisomy 18,* which is observed in about 1 in 5,000 births, causes malformation in virtually every organ system. About 80 percent of babies born with an extra chromosome 18 are females, and virtually all affected infants of both sexes die within a few months.

Abnormal Numbers of Sex Chromosomes

In contrast to embryos with too many or too few autosomes, those with aneuploidy of the sex chromosomes frequently survive to birth and adulthood. The various patterns that can occur are listed in Table 15-2. The presence or absence of the Y chromosome determines the route of sexual development in mammals, including humans. XX and XO embryos develop into females, whereas XY or XXY embryos develop into males.

Table 15-2 **ABNORMAL NUMBERS OF SEX CHROMOSOMES**

Sex Chromosomes	Syndrome	Frequency at Birth
Females		
XO, monosomic	Turner's	1 per 5,000
XXX, trisomic		
XXXX, tetrasomic	Metafemale	1 per 700
XXXXX, pentasomic		
Males		
XYY, trisomic	Normal	1 per 1,000
XXY, trisomic		
XXYY, tetrasomic		
XXXY, tetrasomic	Klinefelter's	1 per 500
XXXXY, pentasomic		
XXXXXY, hexasomic		

Individuals with monosomy of the X chromosome (XO) occur in about 1 in 5,000 live births. Nondisjunction in either the mother or the father can lead to an XO offspring, who displays **Turner's syndrome.** Individuals with this syndrome have external female organs but degenerate ovaries lacking germ cells. Since ovarian hormones are absent, puberty does not occur, and secondary sexual characteristics do not develop.

Nondisjunction of the sex chromosomes during meiosis can also produce XX eggs and either XY or YY sperm. Fusion of a normal X or Y gamete with one of these abnormal gametes can lead to the production of XXY, XXX, or XYY offspring. The XXY pattern, which occurs in about 1 in 500 live births, is called **Klinefelter's syndrome.** Individuals with this syndrome have the external sex characteristics of males but are usually sterile. Klinefelter's syndrome often goes undetected in childhood, since the individual appears normal until the onset of puberty, when partial breast development and other female characteristics appear, and sperm are not produced.

Some near-normal-appearing men have small testes and are sterile. They prove to have XX chromosomes and once were viewed as a possible exception to the rule that a Y chromosome is needed in male mammals. Molecular cloning now reveals, however, that a short segment of Y-specific DNA is located at the end of one of the X chromosomes in XX males. This DNA includes the gene that causes testes to form instead of ovaries.

Men who have an extra Y chromosome—XYY males—do not appear to suffer any negative consequences. At one time, it was thought that an extra Y chromosome predisposed such men to aggression and criminal tendencies, but statistics do not bear out this hypothesis. Nevertheless, uninformed lawyers sometimes have used this theory in their defense of XYY clients.

X Chromosome Inactivation

Interestingly, regardless of the number of X chromosomes in a normal mammalian cell, *only one is genetically active;* the others become inactivated during embryonic development. This seems to equalize the amounts of the products of X-linked genes in males and females. This means that the single X chromosome in every somatic cell of a male mammal but only one of the two X chromosomes in every somatic cell of a female mammal is active. A female's second X and any additional X chromosomes become dense masses of chromatin that no longer function. Under the microscope, they appear as dark-staining bodies in the nuclei of interphase cells (Figure 15-9). These inactivated X chromosomes are called **Barr bodies.** In normal (XX) females, every somatic cell has one Barr body; normal male (XY) so-

(a)

(b)

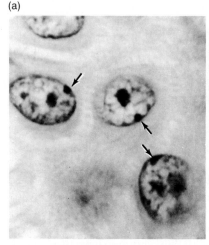
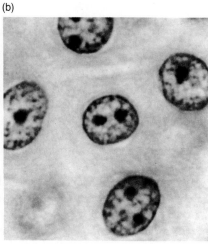

Figure 15-9   BARR BODIES: CONDENSED X CHROMOSOMES.

(a) Stained nuclei from a female mammal (magnified about 3,000 times). A Barr body is a compacted mass of chromatin comprising a nonfunctional X chromosome. Barr bodies are seen located near the nuclear envelope of each of these cells (arrows). Any individual with more than one X chromosome per cell has a Barr body for each inactive chromosome. (b) Nuclei from a male mammal (magnified about 3,000 times). No Barr bodies are evident, as the only X chromosome present is uncompacted. The large, darkly stained masses near the centers of both female and male nuclei are nucleoli.

matic cells have none. Cells of XO females also lack Barr bodies, while Klinefelter males may have one Barr body.

The presence of Barr bodies in every somatic cell of a normal female implies that there must be a regulatory mechanism that can shut down RNA transcriptional activity of a whole, specific chromosome. That mechanism acts randomly so that the X chromosome inherited from the father is inactivated in some cells, and from the mother in others. The result is **mosaicism,** the condition in which some cells of an individual's body express one phenotype, while others express an alternative one. Thus the result of X inactivation is that every organ in a human female's body is really a mosaic of cell clones with respect to her X chromosomes: in some clones, the X inherited from her mother is active, while in others, the X from her father is functional. In some mammalian embryos and in adults related to kangaroos (marsupials; Chapter 25), X chromosome inactivation is not random, and X's inherited from the father are inactivated in cells of certain tissues (this may help avoid recognition and attack by the immune system). The molecular mechanism of X inactivation may involve either chemical modification of DNA or changes in the supercoiling of chromatin due to the presence of proteins. Perhaps X inactivation will reveal the general means by which genes are inactivated during development—a topic we explore in Chapter 17.

Translocations, Deletions, Duplications

We have discussed genetic abnormalities in which entire chromosomes are duplicated or deleted as a result of nondisjunction. Sometimes, however, only *part* of a chromosome is altered, lost, or moved during cell division. These partial changes are the result of chromosomal *translocation, deletion,* and *duplication.*

Translocation

One form of Down's syndrome arises when most of a chromosome 21 is translocated, or moved, to the tip of another large chromosome, usually chromosome 14 (Figure 15-10). The karyotype of an affected person will show the normal complement of forty-six chromosomes, but one of the chromosomes 14 will bear the translocated 21 and the other chromosome 21 will be shorter than normal. People who have the long chromosome 21 and the short chromosome 21 appear normal because they have two functionally normal chromosomes 14 and two chromosomes 21. However, at the time of meiosis in the gonads of such individuals, four types of gametes can be produced, so that some of the offspring will have the syndrome or die as embryos. Geneticists study the cells of a child with Down's syndrome and of its parents to determine whether the nuclei contain forty-seven chromosomes (trisomy 21, which arises sporadically) or forty-six chromosomes (fused chromosomes 14 and 21). Armed with such information, the parents can assess the probability that they will produce additional children with Down's syndrome.

Some translocations can arise in certain somatic cells while the rest of the body cells retain normal chromosomes. One puzzling example occurs in patients who suffer from a form of bone-marrow cancer called chronic myelogenous leukemia. In 80 percent to 90 percent of these cases, the cancerous bone-marrow cells have karyotypes that show a translocation of part of chromosome 22 to the end of the long arm of chromosome 9. The rest of the body cells of such patients have normal karyotypes. Geneticists believe that this specific translocation may arise spontaneously in a normal bone-marrow cell and create a clone of cells that proliferates to form the cancer. Chromosomal abnormalities such as this in clones of somatic cells are properly classified as *somatic mutations,* since they do not occur in germ cells and thus cannot be passed on to the next generation.

Deletions and Duplications

Occasionally, parts of chromosomes are deleted entirely, usually in the germ cells of the parents. Such chromosome abnormalities are rarely observed in live infants, since the loss of a number of genes usually leads to severe defects and death in utero. A loss of material from the short arm of chromosome 5 leads to the *cri du chat* ("cat's cry") *syndrome*, so called because the infant's cry is distinctly catlike. Only about 200 cases of this syndrome have been reported; its frequency is probably less than 1 in 100,000 births. The affected individuals are mentally retarded but may survive through adolescence. This syndrome illustrates the importance of gene balance—having exactly two copies of all autosomal genes—for normal development.

Loss of parts or all of human chromosome 22 may lead to tumors in the nervous system (acoustic neuroma and meningioma). This results in absence of a normal allele and leaves on the remaining chromosome 22 a mutant allele that elicits the cancerous growth.

Chromosome duplications are sometimes a consequence of deletions, at least in *Drosophila* and other organisms that can be studied genetically. Thus the deleted portion of one chromosome may become attached to its normal homologous partner, as occurs during the translation in Down's syndrome. Subsequently, when gametes form, some will have the deletion, while others will carry the deleted chromosomal piece; of course, that piece *duplicates* an existing region of the homologous chromosome, so two copies of that region are present in the haploid sperm or egg nucleus. Duplications have been difficult to study in human chromosomes, and little is known about their effects; but improved staining methods may make them easier to study in the future.

HUMAN GENETIC TRAITS: A SURVEY OF THE "CLASSICS"

Like all other complex organisms, human beings have a host of normal traits determined by wild-type alleles and dozens of abnormal traits based on dominant and recessive mutations. To a nongeneticist, a survey of the most extensively studied human traits—such as color blindness, hemophilia, albinism, and others we are about to discuss—looks like a strange compilation of defects and diseases with little relevance to the average healthy adult. It is true that human geneticists often focus on mutated alleles that produce abnormal traits and diseases rather than on the genetic basis for such traits as hair color, nose shape, or foot width. Why is this?

We have seen that humans are uniquely challenging genetic subjects, since breeding experiments are not ethically acceptable. Recall, too, that Mendel in his research on peas, and Morgan in his experiments with fruit flies, had to work with normal or mutant traits that could be seen or measured in some convenient way. The same considerations apply to the "indirect" genetics research on humans. And so such traits as color blindness and bleeder's disease are studied as sex-linked mutations; abnormally short fingers and Huntington's disease prove to be dominant mutations that are expressed in irregular ways; and so on. Furthermore, although the classic human genetic traits occur rarely, they usually have medical significance and substantial impact on those who inherit them. With all of this in mind, let us examine a number of classic trait models.

Figure 15-10   TRANSLOCATION OF A PORTION OF CHROMOSOME 21 TO CHROMOSOME 14.
Nuclei of the female gametes are shown; when these gametes are fertilized with a sperm carrying normal chromosomes 14 and 21, the offspring are as indicated. The lethal situation arises from the absence of the chromosome arm. In the balanced translocation situation the gamete has a full set of genes since both chromosome arms are present, although one is translocated to chromosome 14.

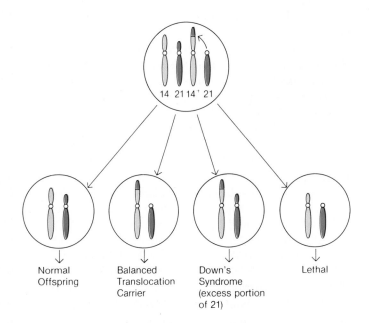

(a)

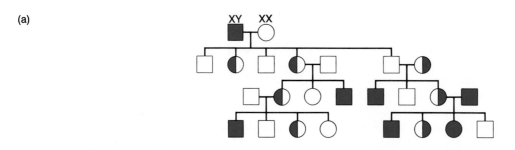

(b)

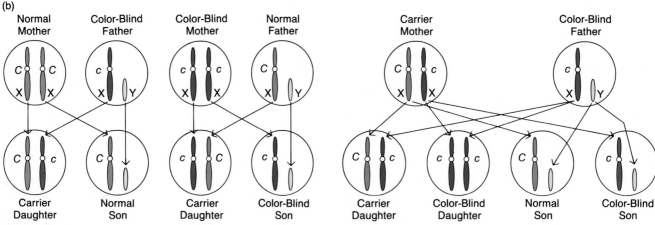

Figure 15-11   INHERITANCE OF COLOR BLINDNESS.
(a) Typical pedigree of a family in which the father was color blind and the mother was normal. Use the key to common pedigree symbols in Figure 15-1 to distinguish which individuals were normal, which were color blind, and which were carriers. (b) Modes of transmission of the recessive *c* (color-blind) allele.

Sex-Linked Traits

Among the first human traits to be carefully studied in the early days of genetics were those that seemed to be transmitted by human sex chromosomes. Sex-linked inheritance patterns have been used both to analyze the X and Y chromosomes themselves and to help locate genes for specific traits.

The fact that males are heterozygous, having one X and one Y chromosome, makes it relatively easy to identify sex-linked traits, since many genes carried by the X are not carried on the much smaller Y (some genes do occur on both the X and Y). The Y chromosome is transmitted directly from father to son in every generation and never to daughters. The only character that geneticists have unequivocally assigned to the Y chromosome is the trait of maleness itself. A father (XY) transmits an X chromosome to his daughter; similarly, a mother (XX) passes an X to any son. Such X chromosomes can carry many traits and some significant mutations.

In the late eighteenth and early nineteenth centuries, physicians realized that red-green color blindness (the inability to perceive red, green, and sometimes yellow) shows a regular pattern of inheritance. This type of color blindness affects about 8 percent of men in most Caucasian populations, but only 1 percent of women. If a color-blind man marries a woman with normal color vision, virtually all their children are normal. But if a color-blind woman marries a normal man, all the sons are color blind and all the daughters are normal. The normal daughters of a color-blind mother can transmit their mutant, recessive allele to their sons (Figure 15-11a). These observations were explained only after the inheritance of white eyes in *Drosophila* was noted to act identically to that of human color blindness and was found to be associated with the X chromosome (Chapter 10). A similar explanation was then proposed for color blindness in humans—that is, color blindness is a recessive mutation of a gene on the X chromosome; hence, it is an X-linked trait (Figure 15-11b). Recombinant DNA technology has now revealed the presence of two genes on the X chromosome and one on an autosome that code for the proteins that absorb light of different colors (see Chapter 38).

Another, much more serious X-linked trait is **hemophilia,** a disease in which a protein that is necessary for normal blood clotting is missing. Hemophiliacs can bleed to death from even the slightest wound. The genetics of hemophilia was partially understood as long ago as the second century A.D. According to the Talmud, the younger brothers of boys who have bled to death following circumcision are exempt from that ritual, and the

(a)

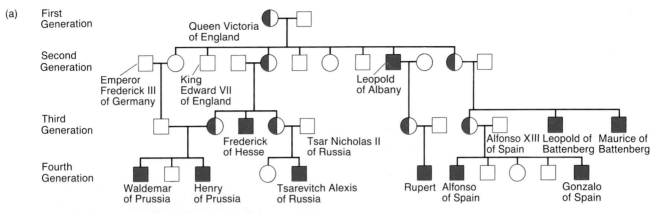

First Generation

Second Generation

Third Generation

Fourth Generation

(b)

Figure 15-12   HEMOPHILIA IN THE ROYAL FAMILIES OF EUROPE.
(a) Pedigree for the hemophilia trait in descendants of Queen Victoria. The original mutation probably arose in the queen herself. (b) Queen Victoria with members of her immediate family in 1894.

male cousins who are sons of the mother's sisters are also recognized to be in danger. The frequency of hemophilia among males is about 1 in 10,000, and until the twentieth century, few afflicted males survived adolescence.

For several generations, hemophilia has plagued the royal families of Europe descended from Queen Victoria of England. Her pedigree (Figure 15-12a) has been thoroughly investigated, and there is no evidence for hemophilia in her mother's line, nor was her father directly affected. Furthermore, her husband, Prince Albert, could not have been the source of the mutated gene, since it would have rendered him a hemophiliac as well. The phenotypes of her descendants clearly indicate that she must have been heterozygous, having on one X chromosome a normal dominant allele for the blood-clotting protein, and on the other X chromosome, a mutant recessive allele for hemophilia. Since the genetic disease did not precede Victoria, a mutation probably arose during the meiotic divisions that produced the egg or sperm by which she was conceived, or in an early mitotic division of her germ line cells.

It is interesting that females who are heterozygous for the hemophilia gene are normal phenotypically, since they have only one active X chromosome, plus an inactive Barr body in each cell. Would not the X with the

normal dominant allele for blood clotting be just as likely to be inactivated as the one with the mutant recessive allele for hemophilia? The answer is that X chromosome inactivation occurs only after the human female embryo has grown for many hours to the size of about 1,000 cells. Since X inactivation is nearly random, the normal X would be inactivated in about half the cells, while the abnormal X would be inactivated in the other half. Because cells with an active normal X make a sufficient amount of the protein necessary for blood clotting, the heterozygous female does not suffer from the disease.

Similarly, mosaicism due to X chromosome inactivation operates in females who are heterozygous for color blindness. Such mosaicism is confirmed by the presence of both normal and abnormal cone cells (the cells in the retina responsible for color vision) in each eye. Because about one-half of the cone cells of such females are normal, resulting from the presence of X chromosomes bearing the active wild-type allele, they usually have normal color vision.

A striking example of female mosaicism in which a recessive mutant phenotype is not entirely masked is found in women who are heterozygous for an X-linked trait called *anhidrotic ectodermal dysplasia*, the genetic cause of the "toothless men of Sind" in West Pakistan. Men with this trait lack sweat glands and teeth and have almost no body hair. Heterozygous females, however, are missing some teeth; have thin, sparse hair; and lack sweat glands in patches of their skin (Figure 15-13). Obviously, each such deficiency represents a mutant patch on the "mosaic body." Normal teeth, hair, and sweat glands occur only where cells have the wild-type allele on their active X chromosomes and where the mutant allele is on the Barr body. The vacant sites occur

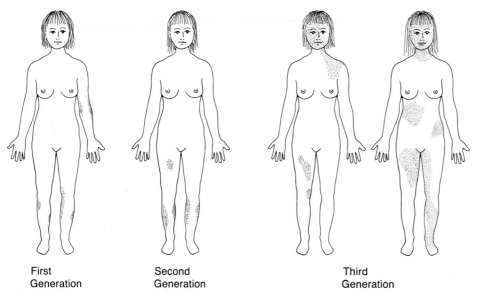

First
Generation

Second
Generation

Third
Generation

**Figure 15-13   MOSAICISM RESULTING FROM X
CHROMOSOME INACTIVATION.**
If a woman's two X chromosomes carry different alleles, one
set of alleles will be expressed in some cells, whereas the
other X chromosome's alleles will be expressed in the rest
of the cells. A family carrying the allele for anhidrotic
ectodermal dysplasia shows mosaicism among the women in
three generations. The colored areas indicate regions in
which the women have no sweat glands due to the presence
there of an active mutant X chromosome. The two women
in the third generation are identical twins. Notice that the
X chromosomes are not expressed in identical patterns.

where cells chance to have the mutant allele on their
active X chromosomes.

Recessive Traits on Autosomal Chromosomes

Most abnormal human traits are specified by recessive
alleles on autosomal chromosomes. Some of the mutant
recessive alleles yield harmless phenotypic traits. These
include tiny pits just above the ears; a cowlick that grows
in a counterclockwise whorl (clockwise is dominant); dry,
brittle ear wax (sticky ear wax is dominant); and even
hairy elbows. Unfortunately, some of the most debilitating human abnormalities are also caused by recessive
alleles of autosomal genes. What makes the situation
even worse is that adults who are heterozygous for a
deleterious recessive allele do not show the phenotypic
trait and hence usually do not suspect that they are carriers of the recessive mutation until they have produced
a child with the defect. A few disorders of recessive autosomal alleles are associated with specific enzyme deficiencies and are thus more amenable to analysis. We will
discuss several of them here.

Phenylketonuria (PKU)

A number of well-known human defects are caused by
abnormal metabolism of the amino acid phenylalanine
(Figure 15-14). For instance, **phenylketonuria (PKU)** is a
serious disease that results from the deficiency of the
enzyme phenylalanine hydroxylase, which converts
phenylalanine to tyrosine. Newborn babies with PKU
have abnormally high levels of phenylalanine in their
blood serum and excrete high levels of phenylpyruvate
(a breakdown product of phenylalanine) in their urine.
Children with PKU become severely mentally retarded,
seldom progressing beyond the age of about two in mental development; they have abnormal muscle tone and
body movement, pale skin, light hair, and blue eyes due
to low levels of tyrosine. In Northern European populations, about 1 child in 10,000 is born with PKU, and
until recently, this disease was responsible for about 1
percent of the severely retarded patients in mental institutions.

Once geneticists and physicians learned that high levels of phenylalanine in blood serum forecast impending
brain damage in an infant, they developed screening
tests and treatment programs for newborns. Treatment
involves eliminating most sources of phenylalanine from
the diet, thereby preventing most of or all the PKU
symptoms. Treated children develop near-normal intelligence and behavior. The dietary treatment of PKU illustrates the effects of both nature and nurture on phenotype. A special diet alters the cellular environment, so
that phenylalanine is excluded and normal phenotype
develops. Without such environmental change, the mutant genotype would be expressed as an abnormal phenotype.

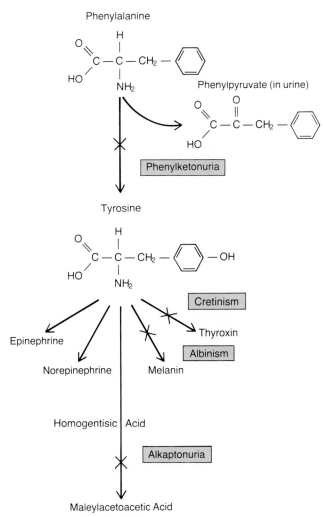

Figure 15-14 THE METABOLIC PATHWAYS INVOLVED IN PHENYLALANINE METABOLISM.
The colored boxes show the diseases or states that result from the errors in the metabolic pathways. The various normal products of this metabolic sequence provide important molecules for nerves, pigment cells, and the thyroid gland. Not all the enzymatic steps between the various substances are shown. Cretinism is characterized by dwarfing and mental problems and is due to a thyroid gland deficiency in hormone secretion. In alkaptonuria, a substance in urine turns black on exposure to air.

Albinism

Another mutant recessive gene involving the same biochemical pathway as PKU causes **albinism,** or the deficiency of pigment in the skin, eyes, and hair. Albinism shows up in 1 person of every 30,000. In certain inbred Native American groups, albinism occurs in as many as 1 in 240 individuals, perhaps because albinos have been regarded with high esteem in these cultures, and thus families with a history of albinism may tend to have more children than normal. Inbreeding among rel-

atives in the small populations may also be a factor. In the most extreme form of albinism, the enzyme tyrosinase is completely absent. This enzyme is required to begin the conversion of tyrosine to the dark pigment melanin, which creates most of the light and dark shades of the hair, skin, and eyes. A person lacking tyrosinase cannot synthesize melanin and thus has no colored pigment granules to produce these shades (the granules are present but are colorless).

The skin of albinos is extremely sensitive to sunlight—exposure to the sun leads to roughness and wrinkling and to a high risk of skin cancer. Their eyes are also sensitive to light and appear pink because of light reflecting from underlying blood vessels. Most albinos, including the rare albino tigers bred by Indian maharajas (Figure 15-15), exhibit *nystagmus,* an involuntary jerking of the eyes, and many suffer from eye defects that involve abnormal wiring of the nerves extending from the eyes to the brain. Almost total blindness can result. No known treatment can overcome the enzymatic defects in albinos.

Anemias

Two types of anemia (low red blood cell count), both caused by recessive genes, are quite prevalent in certain

Figure 15-15 CROSSED EYES IN A PARTIAL ALBINO TIGER.
Not only are the eyes commonly crossed and a pale blue color, as in this female, but also nerve connections between the eyes and the brain are abnormal.

human populations. The recessive allele for sickle cell anemia, as we saw in Chapter 11, occurs in high frequency in malaria-endemic regions because heterozygous carriers of the mutant allele are more resistant to malarial infection than are individuals with two normal alleles.

Another anemia prominent in malaria-ridden areas is **thalassemia,** which is actually a group of similar diseases first described in people who live near the Mediterranean Sea. These anemias are one of the most serious health problems worldwide—well over 100,000 children die of these diseases each year. Thalassemias are characterized by a marked reduction in or absence of synthesis of either the α or the β chain of hemoglobin. Because the synthesis of only one chain is impaired, the other chain builds up within the red blood cells, causing them to burst. The most common form of the disease is β-thalassemia major. If untreated, a homozygous recessive individual usually dies before age ten. If the sufferer receives periodic transfusions of normal red blood cells, he or she can survive to early adulthood, but then the victim usually dies.

The new techniques for analysis of DNA sequences have revealed a completely new, hitherto unexpected type of mutation that accounts for β-thalassemia. In a mutated β-thalassemia gene, a point mutation in which one nucleotide is substituted for another occurs at the junction between one intron and one exon of the β-globin structural gene. This prevents correct processing of primary RNA transcripts, so that no final mRNA for β-globin passes to the cytoplasm for use. It is no wonder that β-globin is lacking! In some forms of β-thalassemia, all or parts of regions near the globin gene are missing because of deletions. The molecular basis of thalassemias has only recently been discovered, as a result of work involving molecular cloning of human genes in bacteria (Chapter 14). As a result, normal human globin genes

have been transferred into the germ line cells of mice suffering from thalassemia; new generations of mice produced from such cells have thalassemia reduced or eliminated altogether. This is a model of what conceivably could be done to correct the β-thalassemia mutation in humans. These powerful techniques will soon yield precise molecular explanations for other human mutations. They may also allow geneticists to design strategies for repairing the defects at the level of molecules, thereby relieving incalculable human suffering.

Dominant Traits and Variations in Gene Expression

As in other organisms, some mutations of human genes are dominant rather than recessive. Many mutant dominant alleles, such as those determining cleft chin and baldness, are harmless. Others, such as the mutant allele causing extra fingers and toes, are bothersome but not life threatening. Still others, however, can be lethal, such as the gene for Huntington's disease, which leads to grave deterioration of the nervous system. Mutant dominant alleles not only have a range of consequences, but also reveal important ways that gene expression may vary. Let us examine a few classic examples of genetic traits based on dominant alleles.

Pattern Baldness

A common dominant trait is pattern baldness, a condition in which heterozygous males (*Bb*) experience premature hair loss on the top and front of the head. The men of the prestigious Adams family, shown in Figure 15-16, are typical examples. Pattern baldness shows a variation in gene expression called **sex-influenced inheritance:** heterozygous males exhibit the trait, whereas heterozygous females may have thinning hair but do not become bald. High levels of testosterone, the male hor-

Figure 15-16   PATTERN BALDNESS IN THE ADAMS FAMILY.

(a) John Adams, the second president of the United States, fathered (b) John Quincy Adams, the sixth president, who, in turn, fathered (c) Charles Francis Adams, a diplomat and the father of (d) Henry Adams, a historian.

(a)   (b)   (c)   (d)

(a)

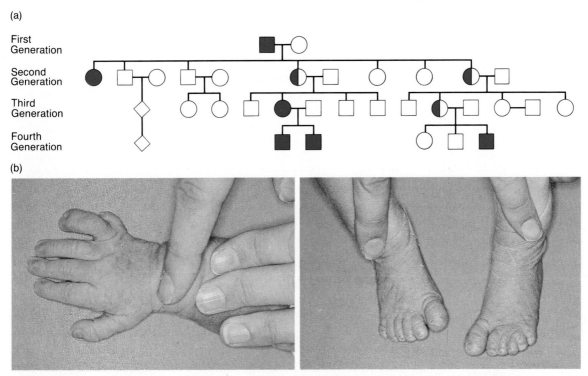

(b)

Figure 15-17 POLYDACTYLY—TOO MANY FINGERS AND TOES: AN EXAMPLE OF INCOMPLETE PENETRANCE.
(a) Polydactyly is caused by a dominant allele that is not always expressed. In this pedigree, three individuals (indicated by half-colored symbols) must have been carriers of the trait because they passed it on to their offspring. They did not exhibit the trait themselves even though it is dominant. (b) Polydactyly of the hands and of the feet.

mone, apparently are required for expression of the *B* gene; accidentally castrated males with the *Bb* genotype retain their hair unless they are treated with testosterone, in which case they become bald. Similarly, heterozygous females become bald if they are treated with testosterone to control certain cancers.

Polydactyly

People with the mutant dominant allele for extra digits, or *polydactyly*, can have from five to seven digits on their hands or feet. The polydactyly gene shows **variable expressivity**, or differences in the way the gene is expressed among individuals with identical genes. For example, one family member may have six fingers on the right hand, five on the left, six toes on one foot, and seven on the other. An older brother, however, might have seven toes on one foot, five on the other foot, and five fingers on both hands.

It is not known why the gene for polydactyly is expressed differently among the four limbs of an embryo. However, as we will see in Chapter 17, polydactyly itself results from abnormal activity of embryonic tissues found at the end of the developing paddle-shaped arm or leg bud. In particular, cells that die in the embryo as part of normal digit development fail to do so; the result

is extra fingers or toes.

Polydactyly is a good example of a dominant mutation that shows **incomplete penetrance**: the dominant allele is present but is not expressed at all in certain individuals. The pedigree in Figure 15-17a depicts a dominant pattern of inheritance with incomplete penetrance: three individuals (shown in green) had the normal five digits per limb even though they passed the mutant allele to their descendants and thus must have had it themselves. The basis for incomplete penetrance of a gene is unknown.

Huntington's Disease

Huntington's disease is a tragic affliction that involves a progressive deterioration of the brain cells, leading to muscle spasms and, after ten to fifteen years, death of the victim. The allele for Huntington's disease is an autosomal dominant. In most sufferers, the first symptoms begin to appear between ages thirty-five and fifty. Because most victims have reproduced before signs of the disease appear, their children must live with the knowledge that they may have inherited the dominant trait. Huntington's disease is a lethal mutation, but does not die out in human populations because its expression is delayed until after reproductive age.

HUNTINGTON'S DISEASE: STALKING ELUSIVE GENES WITH A NEW TECHNIQUE

Researchers using recombinant DNA techniques have learned to detect the genes for the hereditary condition called Huntington's disease. The disease is incurable, the cause is unknown, and until 1983, there was no way to detect victims before their first symptoms occurred. The new techniques will allow physicians to identify probable victims far earlier than ever before. This will not only help alleviate the anguish of some potential Huntington's sufferers, but may eventually eliminate the devastating disease itself.

Until recently, Huntington's victims have first known they were afflicted when their personalities started to change and their movements became jerky and clumsy—symptoms which appear sometime between the ages of 30 and 50. For the next 25 years or so, the victims suffer a progressive, irrevocable deterioration of the brain leading to constant writhing, blurred speech, irrationality, and eventual death. Huntington's disease is caused by an autosomal dominant allele; thus up to half of the children of an affected parent will inherit the mutant allele and develop the disease (Figure A). Since most victims have already had children by the time the disease begins to manifest itself, their offspring live in constant fear of suffering the same fate.

In 1983, Dr. James Gusella of the Massachusetts General Hospital in Boston, decided to compare the DNA of Huntington's disease victims with DNA from family members who were unaffected. He obtained DNA from mutant and normal white blood cells and added restriction endonucleases (Chapter 14) to cleave the DNA into millions of fragments of different lengths—each fragment carrying a different gene. He reasoned that one of those fragments must contain the mutant allele for Huntington's disease, and he hoped that the allele would be on a fragment longer or shorter than the fragment bearing the normal allele so the two alleles could be distinguished. Such a range of fragment size, also called a *restriction fragment length polymorphism (RFLP)*, might then allow Gusella to show that every person with, say, the short

fragment developed the disease. With this genetic marker, he would have a new means of identifying victims early in their lives.

In order to locate which of the millions of DNA fragments contained the Huntington gene, Gusella prepared fragments of 12 different pieces of human DNA and labeled them radioactively to act as *probes*. These probes would hybridize to any complementary piece of DNA and could then be detected because of their radioactivity.

When Gusella used these probes on DNA from Huntington patients and DNA from their unaffected family members, and compared the results, he found that the afflicted people always had a restriction fragment of short length for one of the 12 probes (called G-8), while the normal family members lacked the short fragment. With the G-8 probe, then, Gusella could predict just by looking at a person's DNA whether or not the mutant allele for Huntington's disease was present.

The net result of Gusella's discovery is that the 100,000 Americans at risk for Huntington's disease may learn their fate if they wish to do so. If one partner in a married couple has the Huntington's allele, they may choose to forgo having children. If enough couples do this, the emotional and physical burden of Huntington's disease could be cut drastically within a few generations. Another option for parents at risk is prenatal diagnosis and possible termination of the pregnancy if the Huntington's allele is present. If it is not present, the child that is born can later know that he or she will remain free of the disease, and as an adult can have children without fear of transmitting the mutant allele.

Gusella's method involving RFLPs as genetic markers closely linked to human diseases should have two major consequences: (1) it should allow the cloning of the Huntington's gene itself so that researchers can study the cause of the disease; and (2) it should be possible to adapt the technique to cystic fibrosis, muscular dystrophy, and other genetic diseases. In the

TREATING GENETIC DISEASES

One of the great dreams fostered by advances in genetics and genetic engineering is the hope for successful treatment of people born with genetic defects and even the correction of such defects *before* birth. Theoreti-

cally, physicians could treat genetic diseases by altering the phenotype of the affected person or by changing the actual genotype. So far, only the phenotype has been successfully altered, but attempts have been made to correct genotype as well, and such efforts undoubtedly will achieve successes in the future.

Mutant phenotypes can be altered in a number of ways. Eyeglasses and false teeth are so commonplace that we often fail to appreciate the gravity of genetically

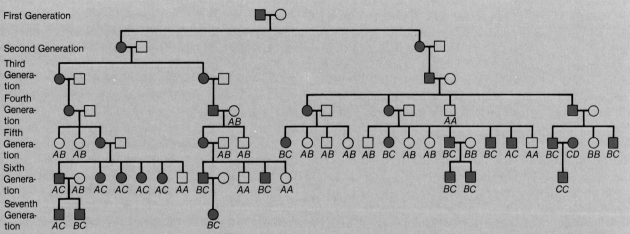

Figure A   PEDIGREE OF A FAMILY WITH HUNTINGTON'S DISEASE.
The complete pedigree begins in the early 1800s and involves 3,000 individuals. Circles indicate females and squares, males; open symbols represent normal individuals and filled symbols, victims. Note that all victims have a parent who was also affected. The genotype of all individuals who were surviving at the time of the analysis appears beneath the individual's symbol. Since people are diploid, each person can have two alleles. Only people with the dominant C allele have the disease.

next decade, thousands of lives will be touched by this new technique.

As hopeful as this sounds, however, genetic screening carries certain ethical complications. Learning that one has inherited the currently untreatable Huntington's allele could be a devastating blow that brings on depression and, some genetic counselors fear, even suicide. Some potential Huntington's victims have already refused the chance to be screened, preferring uncertainty with hope to certitude with despair. Some observers argue that society has a right to know: insurance companies might insist on the information, they point out, and so might spouses of people at risk. But there is a real danger that a young person found to have the allele will be labeled by society and passed over for scholarships, loans, jobs, and relationships.

One goal of genetic research is to correct or replace defective genes like the Huntington's allele in the germ cells—an extension of the work that produced giant mice (Chapter 14). However, certain observers fear that as soon as biologists learn to alter disease-producing genes in human eggs or sperm, some parents will inevitably seek changes in the genes for intelligence, height, obesity, or other nonlethal characters.

Obviously, researchers are rapidly gaining the ability to find and manipulate human genes, but the applications of this remarkable ability must be carefully considered and debated. One hopes, however, that regardless of any future restrictions which may be applied, the benefits of modern genetic research will always be extended to the individuals and families who suffer from tragic diseases like Huntington's.

caused vision problems or lost teeth to our human ancestors. And, as we have seen, babies with PKU can be put on diets that are low in phenylalanine. Extra digits can be surgically removed from people with polydactyly, and insulin injections are used to regulate sugar metabolism in diabetics. In each of these cases, the phenotype becomes more normal, either temporarily or permanently; but the mutant genes are unchanged in normal body cells as well as those that give rise to eggs and sperm,

and thus the mutant alleles can still be passed along to new generations.

Altering the genotype is a bigger challenge, but researchers are making rapid progress toward gene therapy—the insertion of normal genes into mutant cells in order to block genetic diseases. Children born with a defect of the immune system called severe combined immunodeficiency cannot mount an immune response against any invading microorganisms. Unless they are raised in artificial, absolutely germ-free rooms with no human contact and with sterilized food, clothing, and toys, they die at an early age. These children lack the gene that codes for the enzyme adenine deaminase, and the mutation is expressed in bone cells. Attempts to cure these children will involve removing their bone-marrow cells and treating those cells in the laboratory with DNA containing the normal gene for adenine deaminase. The theory is that some of the cells will take up the normal gene and incorporate it into their chromosomes; they will have become *transformed* with the normal gene. Transformed cells will then be reinjected into the patient's bone marrow, where they will grow, produce the normal enzyme, and cure the symptoms of the disease. This series of procedures has already been accomplished successfully in mice, and currently is being attempted in monkeys. Human trials are planned under the rigorous guidelines for this kind of recombinant DNA research set forth by the National Institutes of Health. It is possible that molecular gene therapy will become the cost-effective way to treat severe genetic diseases; thus it will be cheaper to implant cells with a normal allele than to cover the repeated hospitalizations of cystic fibrosis or sickle cell anemia patients.

Although this kind of genotype "cure" might alter affected somatic cells, the patient's eggs or sperm would still retain the mutant allele, and the disease could still be transmitted to offspring. Curing a genetic disease by actually exchanging the mutant allele in the egg or sperm for a normal allele seems at this point to be a procedure for the twenty-first century. When the technique is eventually perfected, it will surely raise ethical questions, since it will represent true "human engineering."

GENETIC COUNSELING

One of the primary applications of human genetics is **genetic counseling,** providing advice to couples who are at risk for producing a genetically defective child or who have had such a child and want to avoid this heartbreaking situation in future births.

Let us consider the case of a couple with a child born with a genetic defect. Before counseling begins, the specific defect or medical syndrome must be identified precisely as karyotyping, blood typing, and various immunological and biochemical tests are carried out. Next, a family pedigree is constructed. This pedigree may confirm the medical diagnosis and may suggest a mode of inheritance as well as identify other family members who may be carriers of the mutant allele. The question of whether to inform these other carriers is a difficult one. There is no simple ethical position on whether it is right or wrong to provide unsolicited information to such individuals.

Once medical and genetic information has been gathered, a counselor can assess the couple's statistical risk of producing another child with the same defect. If the defect is believed to be caused by a single recessive or dominant gene, the risk can be calculated using Mendel's laws. If the defect is polygenic or the result of a chromosomal abnormality, the degree of risk may best be estimated on the basis of accumulated medical data. For most genetic defects, the only options open to couples are to avoid future pregnancies or to accept the risk and take a chance.

Many genetic defects can be detected prenatally by the process of **amniocentesis,** in which about 20 milliliters of the amniotic fluid surrounding the fetus is removed by needle and syringe. Some fetal epithelial cells and even a few fetal blood cells can be recovered from the fluid by centrifugation. A number of genetic defects can be detected by biochemical tests on the amniotic fluid, but most require direct analysis of the fetal cells. Sex of the fetus is indicated by the presence or absence of a Barr body. If the fetus is male and its mother is a carrier of a severe X-linked defect, the parents might choose to abort it. In such a case, however, there is a 50–50 chance that the fetus is normal, which presents a difficult moral problem.

About forty genetic defects affecting metabolic pathways can be detected using amniocentesis; some of these disorders are listed in Table 15-3. If the geneticist discovers chromosomal abnormalities or homozygous recessive alleles for deleterious traits, the parents can consider aborting the fetus. Because fetal genotypes cannot yet be altered, abortion is the only "treatment" when severe genetic defects are detected. However, many people find this solution unacceptable.

A new technique called **chorionic villi sampling** can be performed earlier in pregnancy than amniocentesis, may be safer, and yields diagnostic information in less time. The physician takes a sample of the embryo's placental tissue; then the cells are cultured and assayed, since they are genetically identical to fetal cells. There is one current weakness in the test: all paternal X chromosomes are inactivated in fetal placental cells, which makes it difficult to test for deleterious alleles on an X chromosome donated by the father.

One alternative to the procedures of amniocentesis, chorionic villi sampling, and abortion is screening whole populations for certain genes. Random screening of groups of young adult Jews or blacks has identified heterozygotes for Tay-Sachs disease or sickle cell anemia. Even with this information, however, two heterozygotes can choose to take the risk of conceiving—and sometimes they and the offspring lose. Although some experts regard mass screening as an important solution to the problem of genetic defects, such screening is limited to a few defects and has itself been controversial, since an individual's privacy is invaded, and carriers may be un-

Table 15-3 **SOME INHERITED BIOCHEMICAL DISORDERS THAT CAN BE DIAGNOSED PRENATALLY IN CULTURED FETAL CELLS**

Error in the Metabolism of	Disorder	Metabolic Defect	Brief Description of Phenotype
Amino Acids	Maple sugar urine disease (AR)*	Deficiency of enzymes needed in the breakdown of some amino acids	Poor development, convulsions, and early death; urine has maple-sugar odor; diet therapy seems promising
	Cystinosis (AR)	Accumulation of cystine in cells	In severe form, kidney function is impaired, leading to poor development, rickets, and death in childhood
Sugars	Galactosemia (AR)	Deficiency of an enzyme needed in the metabolism of galactose (derived primarily from milk)	Liver and eye defects, mental retardation, and early death if untreated; restrictive diet can control adverse symptoms
Lipids	Tay-Sachs disease (AR)	Deficiency of an enzyme needed in the breakdown of a complex lipid, allowing it to accumulate in nervous tissue	Progressive physical and mental degeneration, paralysis, blindness and death in infancy
	Adrenogenital syndromes (AR)	Deficiency of enzymes needed in the synthesis of sex hormones and of steroids controlling salt and water balance	In the most common form, dehydration and early death; females often experience masculinization
Purines	Lesch-Nyhan syndrome (XR)	Deficiency of an enzyme involved in the metabolism of purines	Mental retardation, muscular spasms, and compulsive self-mutilation; possible survival into adulthood
Complex Polysaccharides	Hurler syndrome (AR) Hunter syndrome (XR)	Defects of connective tissue	Dwarfism, grotesque facial features, and mental retardation; Hunter syndrome is less severe
Heme	Porphyria (AD)	Deficiency of an enzyme needed in the synthesis of heme, a component of hemoglobin	Attacks of abdominal pain sometimes accompanied by impairment of mental functions

*AR = autosomal recessive, AD = autosomal dominant, XR = X-linked recessive

Source: A. P. Mange and E. J. Mange, *Genetics: Human Aspects* (Philadelphia: Saunders, 1980), p. 581.

justly stigmatized. The field of bioethics grapples with societal concerns such as these. Ultimately, however, individuals must form opinions and make up their own minds on such issues as amniocentesis, abortion, carrier screening, and mutant-gene replacement. Even far-reaching advances in genetic manipulation cannot abrogate personal choice and personal responsibility when the issue involved is human life.

SUMMARY

1. Among the methods used to study human genetics are *pedigree* analysis, *karyotyping* of chromosomes, *biochemical analysis*, *somatic cell genetics*, *in situ hybridization*, and *twin studies*.

2. Most variable human characteristics reflect action of both genotype and environment during development, childhood, adolescence, and adult life. Most human traits are due to *polygenic* effects, ones in which several genes interact to produce the trait.

3. Techniques of molecular genetic engineering permit *reverse genetics*, going from the gene to the gene product that causes a human genetic disease. Such techniques also allow DNA fingerprinting by analyzing *restriction fragment polymorphisms (RFLPs)*. These and related procedures are leading to a mapping of the human genome and to procedures for correcting genetic diseases in humans.

4. Some genetic disorders are attributable to chromosomal defects. *Aneuploidy* of autosomes involves losses or gains in chromosome number. Other abnormalities include trisomy, triploidy, and tetraploidy, all of which cause imbalances in genetic activity and thus alterations in development. Many forms of aneuploidy, including *Down's syndrome* (trisomy 21), result from *nondisjunction* of one or more chromosomes during meiosis of the egg cells. Most types of aneuploidy lead to abnormal development in the embryo and spontaneous abortion.

5. Aneuploidy of sex chromosomes can create individuals with *Turner's syndrome* (XO), *Klinefelter's syndrome* (XXY), or the XYY condition.

6. In mammalian cells, only one X chromosome is active; in normal (XX) females, the inactivated X can be seen in the nuclei of interphase cells as a mass of chromatin called a *Barr body*. Every female mammal is a *mosaic* with some cells having an active paternally derived X and others an active maternal X chromosome.

7. Some genetic disorders arise from translocations, deletions, or duplications within chromosomes.

8. Among sex-linked traits in humans are color blindness, *hemophilia*, and anhidrotic ectodermal dysplasia.

9. Disorders of recessive alleles of autosomal genes that are associated with specific enzyme deficiencies include *phenylketonuria (PKU)* and *albinism*. Other recessive disorders that are prevalent in certain populations are sickle cell anemia and the *thalassemias*.

10. Dominant alleles determine such traits as pattern baldness, polydactyly, and Huntington's disease. Pattern baldness is a good example of *sex-influenced inheritance*. Polydactyly is a good example of *variable expressivity* of genes and of *incomplete penetrance*.

11. Efforts to treat genetic diseases focus on altering phenotype, but it is probable that in the near future, treatment will involve altering genotype.

12. In *genetic counseling*, couples are advised on the probability of producing a genetically defective child. If a child already has been conceived, techniques such as *amniocentesis* or *chorionic villi sampling* can be used to determine the genetic health of the fetus.

KEY TERMS

albinism

amniocentesis

aneuploidy

Barr body

chorionic villi sampling

Down's syndrome

genetic counseling

hemophilia

incomplete penetrance

karotyping

Klinefelter's syndrome

mosaicism

nondisjunction

pedigree

phenylketonuria (PKU)

polygenic trait

restriction fragment
   polymorphisms (RFLPs)

reverse genetics

sex-influenced inheritance

somatic cell genetics

Tay-Sachs disease

thalassemia

Turner's syndrome

variable expressivity

QUESTIONS AND PROBLEMS

1. A man claimed that industrial exposure to pesticides caused a mutation in his germ line, causing his son to be born with hemophilia. There was no history of hemophilia in his or his wife's family. Is it likely that the son inherited the defective allele from his father? Why or why not?

2. Cystic fibrosis is the most common severe genetic disease in white populations, occurring once in every 2,000 live births. Victims are homo-

zygous for this recessive allele. What is the probability that the child of a normal couple will suffer from cystic fibrosis? If a couple has an affected child, what is the probability that their next child will be affected?

3. A normal couple has a son with Duchenne muscular dystrophy, a sex-linked fatal disease affecting about one out of every 3,000 male infants. What is the probability that their next son will be affected? Their next daughter? What is the probability that their daughter will be a carrier? Which individuals have a 1 in 4 or greater chance of being carriers: The mother? The father? The father's mother? The mother's father? The mother's mother? The mother's sister? The mother's sister's daughter (her niece)? Explain.

4. How many chromosomes per cell does a Down's syndrome (trisomy 21) victim have? A victim of Turner's syndrome (monosomy X)? A person with both Down's and Turner's syndrome?

5. Consider a woman who is heterozygous for the X-linked condition called anhidrotic ectodermal dysplasia; some patches of her skin lack sweat glands and other patches are normal. Suppose she is also heterozygous for variants of G6PD (glucose-6-phosphate dehydrogenase), an enzyme encoded by a gene on the X chromosome. Many alleles of the gene produce variants of the enzyme. You examine the enzyme produced in different skin patches. One normal patch contains variant A and one abnormal patch contains variant B. Would you expect all the normal patches to contain variant A? Or would some contain variant B? Explain.

6. Assume that height is determined by three genes (A, B, and C) which are unlinked, and that a person of average height (male: 5′9″; female: 5′4″) has three "tall" and three "short" alleles. Each tall allele adds 3 inches to the person's height. A person of average height could have the genotype AaBbCc, or AAbbCc, or aaBBcc. What would be the height of a woman with genotype aabbcc? A man with genotype AABBCC? The height of a woman whose genotype is aaBBcc? What alleles would be present in the gametes she produces? What is the height of a woman whose genotype is AaBbcc? What alleles would be present in gametes she produces? If two people of the genotype aaBBcc had children, what would be the adult height of their shortest daughter? Their tallest daughter? If a very short person (aabbcc) and a very tall person (AABBCC) had children, what would their heights be? Their genotypes? Could parents of average height produce a son 6′6″ tall? (In answering this question, assume that nutrition does not affect height.)

ESSAY QUESTIONS

1. Each normal female mammal has one active X chromosome per cell (the other is inactivated) and so does each normal male. Can you speculate on why a second sex chromosome is necessary for normal development, even though (in the case of the X) most of that chromosome's genes are turned off early in embryonic life?

2. If people married at random with respect to height, skin color, hair color, blood type, nose length, and other traits, would differences disappear after a few generations?

SUGGESTED READINGS

CAVALLI-SFORZA, L. *Elements of Human Genetics.* 2d ed. San Francisco: Freeman, 1977.

This may be the clearest concise account of the topic for the novice.

EPSTEIN, C., and M. GOLBUS. "Prenatal Diagnosis of Genetic Diseases." *American Scientist* 65 (1977): 703–711.

This is a good, if a bit dated, account of an important part of modern medicine.

McKUSICK, V. A. "The Royal Hemophilia." *Scientific American,* February 1965, pp. 88–98.

Here is a fascinating account of the distribution of a recessive gene through the royal families of Europe.

STERN, C. *Principles of Human Genetics.* 3d ed. San Francisco: Freeman, 1973.

This book, by one of America's great geneticists, contains many classic facts, theories, and techniques.

SUTTON, H. E. *An Introduction to Human Genetics.* 3d ed. Philadelphia: Saunders, 1980.

Here is another easy-to-read book for the initiate.

WALTERS, L. "The Ethics of Human Gene Therapy." *Nature* 320 (1986): 225–27.

Here is a balanced account of the options and the ethical issues we confront.

16
ANIMAL DEVELOPMENT

There is no part of the future foetus actually in the egg, but yet all the parts of it are in it potentially. . . . The perfect animals are made by Epigenesis, or superaddition of parts.
William Harvey, *De Generatione Animalium* (1651)

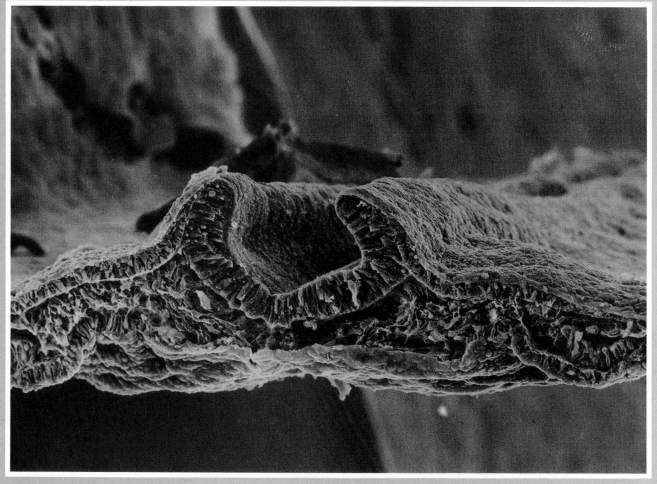

Development in process: A flat sheet of cells folds upward to form the hollow, tubular brain and spinal cord of a chicken embryo (magnified about 425 times).

At conception, you were a tiny sphere, one-tenth of a millimeter in diameter. Newly created by the union of sperm and egg, you were swept down a long, dark tunnel into the warmth and safety of the uterus. There, in the following nine months, you enlarged thousands of times, to a length of 45 to 50 centimeters. You developed eyes, arms, legs, internal organs, and vocal chords capable of piercing cries. How did something so complex and wonderful happen so easily and naturally from a single cell?

All animals begin as a single cell and go through an arduous period of development that includes a proliferation of cells, the differentiation of those cells into functional body parts, and then the growth of the whole body. During these developmental processes, genetic information is interpreted and used, along with other kinds of information, by the developing cells to construct the species-specific phenotype. A fascinating interplay of control processes, involving developmental information

and genetic information goes on during the embryonic stages as cells divide, move about, and assemble into organs we recognize as heart and limb and brain.

In this chapter, we will examine the basic processes of sexual development in detail: (1) how one cell becomes many; (2) how the many new cells organize into functional units, and how these units develop at correct locations in the body; (3) how cells mature and take on different forms to fulfill different functions; and (4) how the developing organism grows many thousandfold both before and after birth. Figure 16-1 provides an overview of this sequence in the salamander, a common relative of frogs and toads. The sequence really begins with egg and sperm production, proceeds with fertilization and the stages of embryonic development, and continues into adulthood and aging. Some terms used in the figure may be unfamiliar to you at this point, but as you learn more about the individual steps, you will find the figure to be a useful reference for putting each process in context. The

Figure 16-1   OVERVIEW OF THE EMBRYONIC DEVELOPMENT AND LIFE CYCLE OF A SALAMANDER.
The new individual is created when haploid sperm and egg produced by the parents come together at fertilization. The result is a diploid cell (zygote). Cleavage divides the single-celled zygote into the blastula, a cluster of many smaller cells, and segregates the embryonic cytoplasm in precise ways. In gastrulation, cell populations are rearranged, and the basic body plan of the species becomes evident for the first time. Part of the ball of embryonic cells sinks inward, forming a slit called the blastopore; the mass elongates, the future spinal column develops, and the head and tail appear. Organogenesis—organ formation—is taking place: the eyes, with their round lenses, begin to form; and the liver, kidney, and other organs are developing internally. The integration of these processes is critical, so that various developing parts and organs have proper sizes and orientations. After hatching, the organism increases its size through growth. Ultimately, the tadpole larva undergoes metamorphosis into a juvenile salamander. Adulthood is reached, at which time the body's aging eventually diminishes the function of some cell populations, leading to death from failure of an organ or a system. The orchestration of incredibly complex life cycles, such as this one, is among the most remarkable of life's properties.

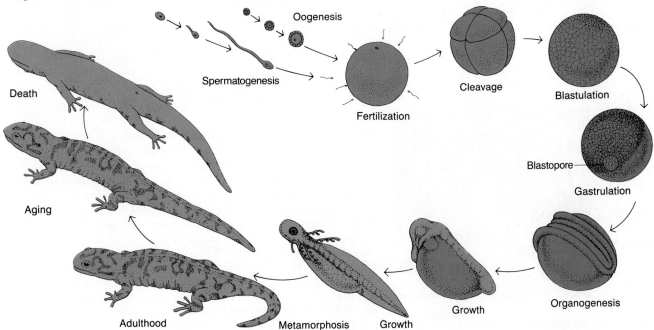

sexual reproduction of the salamander is very different from the asexual reproduction we discussed in Chapter 9—the process based on mitosis (never meiosis) that produces genetically identical offspring.

In the course of this chapter and the next, we will be looking at the developmental stages of many animal species, including sea urchins, frogs, fruit flies, and chickens. Each of these various creatures has embryos with certain characteristics that make it easier for us to study particular aspects of development. Sea-urchin eggs, for example, are exquisitely transparent, allowing us to see cells move during the earliest developmental stages. *Xenopus* (African clawed frog) eggs have biochemical properties that are easily examined. And *Drosophila* eggs have such thoroughly studied genes and chromosomes that their developmental roles can be more easily understood than those of other organisms. The rules and processes discovered by experiments on these creatures also apply to human embryos, which we examine in greater detail in Chapter 18. Other aspects of development are covered in later chapters (particularly Chapters 25 and 26, on invertebrates and vertebrates), so one object of this chapter is to provide a background for the more specialized information included there. The development of plants is examined in Chapter 28.

PRODUCTION OF SPERM AND EGGS

The coming together of sperm and egg triggers the development of a new animal. What are the characteristics of these two unique cells that enable them to give rise to a new individual? How are sperm and eggs produced, and how do they function?

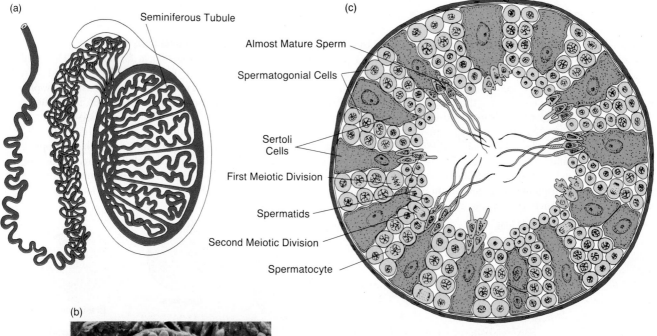

(a)

Seminiferous Tubule

(c)

Almost Mature Sperm

Spermatogonial Cells

Sertoli Cells

First Meiotic Division

Spermatids

Second Meiotic Division

Spermatocyte

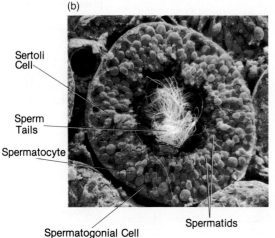

(b)

Sertoli Cell

Sperm Tails

Spermatocyte

Spermatogonial Cell

Spermatids

Figure 16-2   THE TESTIS: A LIVING FACTORY.
Year after year, the testis churns out hundreds of millions of sperm cells, nearly all of which are perfect in form and function. (a) The coiled seminiferous tubules are visible in this cutaway diagram. Spermatogonia lining each tubule give rise to spermatocytes, which eventually mature into spermatids and then sperm with tails. (b) A scanning electron photomicrograph (magnified about 300 times) of a cross section of a seminiferous tubule. The cells visible in the wall are in various stages of meiosis or sperm maturation. The central cavity contains mature sperm that do not swim (the tails only are visible). From *Tissues and Organs: A Text-Atlas of Scanning Electron Microscopy* by Richard G. Kessel and Randy H. Kardon, W. H. Freeman and Company © 1979. (c) A simplified diagram of the seminiferous tubule wall, showing the various cell types and stages of sperm production.

Organisms that reproduce sexually manufacture sex cells called *gametes:* **sperm** in males and **ova** (eggs) in females. The process of sperm production is called **spermatogenesis;** the process of egg maturation is called **oogenesis.**

Spermatogenesis

Sperm are generated in male organs called **testes** (singular, *testis*). As Figure 16-2 shows, testes are composed primarily of hollow *seminiferous tubules*, which are twisted about like spaghetti in a dish. It is in the walls of these tubules that the sperm develop. The sperm originate from the **gonial cells** *(spermatogonia)*, which undergo repetitive cell divisions to yield huge numbers of cells that develop into sperm. More specifically, spermatogonia undergo mitotic divisions to produce *spermatocytes*, which then divide meiotically, reducing the adult male diploid ($2n$) genome to the haploid ($1n$) chromosome number (Figure 16-3). Starting at the tetraploid ($4n$) state, two successive meiotic divisions produce four $1n$ spermatids. The heads of the spermatids are embedded in *Sertoli cells*, a group of helper cells that may contribute vital materials to the maturing spermatids.

A tail begins to grow on each spermatid and protrudes toward the central cavity of the seminiferous tubule. The nucleus becomes more compact, and the Golgi complex gives rise to a storage granule called the *acrosome* (Figure 16-4). (Recall from Chapter 6 that the normal task of the Golgi complex is to receive newly synthesized materials from the rough endoplasmic reticulum, process them, and pass them on in granules for storage and secretion; the formation of the acrosome is just a specialized case of the same phenomenon.) Through this maturation process, the sperm cell takes on a shape and acquires biochemical attributes that will allow it to fertilize an egg, its main function.

If fertilization is to be successful, a sperm cell must be able to swim to the egg, penetrate its surface, and insert the haploid male genome into the egg cell. The motive force for swimming comes from the tail, or *flagellum*, which forms on one end of the spermatid. The flagellum undulates with a whiplike motion to drive the sperm through sea water, fresh water, or the fluids of the female's reproductive tract. The energy that powers the flagellum comes from many mitochondria, which are wrapped tightly around its base, as shown in Figure 16-4. In that position, they are ideally situated to deliver ATP, the biochemical fuel whose energy-rich phosphate bonds are "burned" by the swimming sperm.

Because sperm may have to swim considerable distances to reach the egg, carrying only a compact payload in a small, streamlined head is advantageous. Thus the

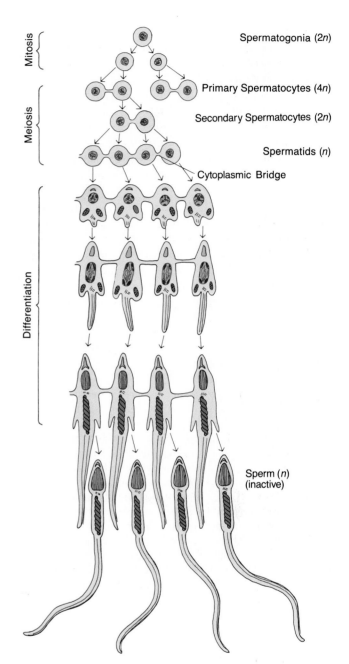

Figure 16-3   SPERMATOGENESIS.

During meiosis, the diploid ($2n$) spermatogonia produce haploid ($1n$) spermatids. As the spermatids differentiate and mature, the tail grows, the nucleus changes, and a storage granule forms at the anterior end. A cohort of four spermatids is shown here, interconnected by bridges of cytoplasm; in reality, cohorts of eight are more common. The sequence shown here occurs in the wall of a seminiferous tubule (Figure 16-2); the spermatogonia are located periferally, at the outer wall of the tubule, and the mature sperm are released into the central cavity of the tubule. Although not pictured here, huge Sertoli cells envelop the spermatocytes, spermatids, and differentiating sperm; as a result, the Sertoli cells can supply nutrients or other substances to aid the maturation process.

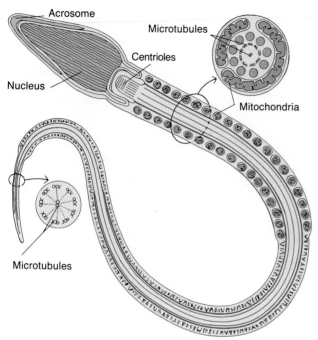

Figure 16-4   THE SPERM.
The sperm is a streamlined, self-propelled torpedo, built to transport its genetic cargo a long distance, penetrate a tough barrier, and deliver its haploid genome safely inside the egg cytoplasm. These longitudinal and cross sections show the structures responsible for accomplishing the tasks in a mammalian sperm. The acrosome is the storage site for the enzymes that will digest the egg's protective surface and allow sperm penetration. The nucleus contains the haploid chromosomes in a highly condensed state—the smaller and more compact the genetic cargo, the less energy is required for the sperm to transport it. The centrioles are involved in building the tail. The spiral of mitochondria is packed tightly around the microtubules, which are the backbone of the tail. Here ATP generated by mitochondrial enzymes will fuel the whiplike tail movements during swimming.

head contains only the acrosome and the nucleus. The acrosome carries enzymes needed to bore through the egg's protective layers and contains proteins that are used to produce a needlelike structure, also for penetrating to the egg. The nucleus becomes compact after meiosis is complete; all gene activity is repressed, and the haploid chromosomes are packed for transport in a crystal-like form.

The mature sperm cell that is shed into the cavity of the seminiferous tubule thus possesses a tail, a nucleus containing inactive haploid chromosomes, and a front end specialized for fertilization. These completed sperm do not swim under their own power—they become motile only after being expelled from the male's body.

Oogenesis

Ova are generated in the female's **ovaries.** The gonial cells (*oogonia*) of many female animals complete their normal mitotic divisions early in life (well before birth in mammals). The cells, called *oocytes*, then enter a state of arrest in early meiosis, a pause that lasts for years in some species (Figure 16-5). In humans, meiotic arrest continues until puberty, when about one egg per month is released and matures—a pattern that continues for the rest of the woman's reproductive life. Thus in many species, oogenesis (shown in detail in Figure 16-6) is more a maturation process of preexisting cells than of production and maturation, as is spermatogenesis. The period of oocyte maturation and growth (the box in Figure 16-5) is followed by a final ripening, ovulation, and the first meiotic division. Then, if fertilization occurs, the second meiotic division takes place, and development proceeds. This figure emphasizes that the final two meiotic divisions are not necessary parts of egg production or maturation.

The structure and function of the egg differ significantly from those of the sperm because the egg provides not only genes, but also a place for them to reside and operate. In addition to containing the haploid set of female chromosomes, most animal eggs (1) contain nutrients to support early development, (2) contain regulatory molecules that can turn genes on and off or affect other aspects of embryonic development, (3) have protective layers, and (4) have mechanisms that enable them to respond appropriately to contact with sperm. Because of the egg's complexity and its contribution to the embryo's development, it is important for us to take a closer look at its structure and contents.

Egg Structure

Whereas sperm of most animal species are about the same length—50 micrometers or so—eggs vary tremendously in diameter and mass. A human egg is only about one-tenth of a millimeter across; a shark egg is 3 centimeters; and an ostrich egg is about 6 centimeters, making it perhaps the largest living single cell on Earth. Eggs of different species vary in size largely because they contain widely divergent quantities of stored nutrients. In addition, some have special protective coats and contain different kinds of information for directing and carrying out embryonic development.

Other types of cells commonly help the egg to meet its many specialized tasks. Virtually all developing animal ova are surrounded by helper cells, either *nurse cells* or *follicle cells.* Nurse cells synthesize proteins and nucleic acids (mRNA and ribosomes composed of ribosomal RNAs and proteins) and transfer them to the cytoplasm of the developing oocyte, where they are stored for use

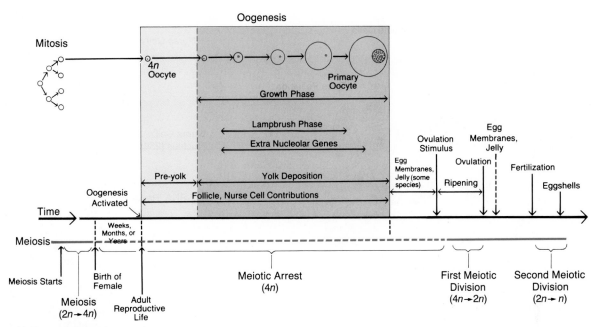

Oogenesis

Figure 16-5   OVERVIEW OF OOGENESIS.

Although this overview is based largely on the frog *Xenopus*, it includes events for birds, mammals, and humans as well. The relative timing of the phenomena described in the text is shown. Meiosis starts in the ovary of the female when she is an embryo. Then, in adult life, single eggs or groups of eggs in the ovary commence the oogenesis sequence. The total egg-maturation period in *Xenopus* is six to eight months; in other species, that time period might be several weeks, a month, or over a year long. Oogenesis, represented in the box, occurs during meiotic arrest and includes special processes. One is the lampbrush phase, during which egg chromosomes are active in mRNA synthesis; numerous extra copies of ribosomal RNA (nucleolar) genes also may be made and used during this period. In some species, egg coatings and membranes are applied after oogenesis is complete and at various times prior to ovulation; in others, they are completed in the oviduct after fertilization. In species with hard, impervious eggshells, fertilization is internal, within the female's reproductive tract, so that the shell must be added later. Note that oogenesis starts only after the meiotic process has started and been arrested—hence the maturing egg has a tetraploid (4n) amount of DNA, whereas the maturing spermatid (Figure 16-3) has a haploid (n) amount. The actual meiotic divisions in the egg occur only at ovulation and fertilization.

Figure 16-6   STAGES OF OOGENESIS.

The oogonia undergo mitosis, often early in life while the female is still an embryo. Next, the oocyte enters meiosis by synthesizing DNA and becoming 4n. Meiosis then arrests, perhaps for decades. Individual eggs then grow, as in Figure 16-5. Only in response to ovulation does the tetraploid genome of the oocyte begin to be reduced to the haploid state, as we saw in Chapter 9. Though not shown here, fertilization is necessary in many species to trigger the second meiotic division. The first polar body usually divides in synchrony with the second meiotic division of the oocyte (or zygote in the case of fertilization). The three polar bodies degenerate.

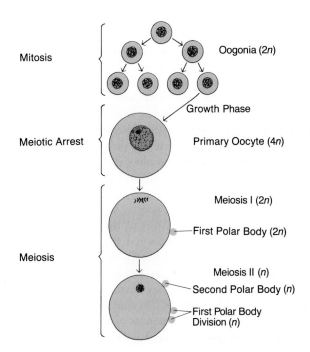

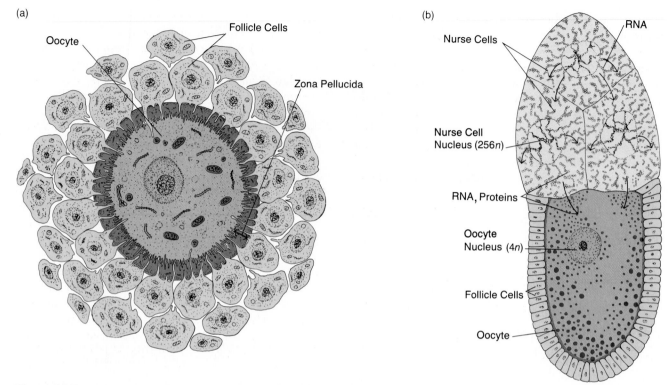

Figure 16-7   HELPER CELLS AND THE TRANSFER OF SUBSTANCES INTO THE OVUM.
(a) An immature mammalian oocyte surrounded by follicle cells. The long, fingerlike projections from the follicle cells contact the surface of the egg and are sites for transferring substances to the egg cytoplasm. The halo of hydrated sugar polymers (zona pellucida) will form a protective covering for the egg after release from the ovary. (b) An insect oocyte surrounded by small follicle cells. The egg's cytoplasm is directly open to the cytoplasm of the giant nurse cells at the top. RNA manufactured in the nurse-cell nuclei, each of which contains many sets of chromosomes, may pass directly to the egg cytoplasm for storage. Or some of the RNA may be used in the nurse-cell cytoplasm to synthesize proteins, which are then passed to the oocyte. The follicle cells also pass substances to the oocyte or add protective layers to its surface.

after fertilization (Figure 16-7). Follicle cells also may transfer materials to oocytes and plaster various protective substances on their surfaces.

Other cells also help the egg produce the nutrients that it stores as the **yolk.** Yolk material, including proteins, carbohydrates, and lipids, usually is manufactured in digestive-gland cells within the mother's body. In hens, for example, yolk protein is produced in liver cells and then is released into the bloodstream, which carries the nutrient molecules throughout the bird's body. In the ovaries, the yolk materials are preferentially absorbed into the developing oocyte, where they accumulate as the egg matures. Most of the yellow yolk in a hen's egg is made and deposited in this manner. The total quantity of yolk varies greatly among species. Bird eggs have a huge amount; as a result, the embryo can carry out a long and complex development even when isolated within the eggshell. In contrast, most mammalian eggs store only tiny amounts of yolk, since the embryo is continuously supplied with nutrients from the mother's blood.

A final contribution to egg structure provided by other cells is a set of protective coatings. These coatings are built up by follicle cells or as the fertilized egg is transported through the female's reproductive tract. In hens, for example, the "white" and the shell of the egg are secreted by cells lining the **oviduct,** the tube through which the egg travels after leaving the ovary. The egg white, or *albumen,* is a special solution of salts and proteins that, among other things, protects the embryo from bacterial infection. Some shells laid down around eggs of different species are tough, pliable, and leathery, whereas others are brittle and hard. All shells, however, limit the loss of water and other vital materials from the embryo and improve its chances of surviving in turbulent sea water, a grassy nest, or a desert burrow.

Manufacture of Molecules

During the maturation process, some oocytes manufacture many of their own supplies, including a large quantity of RNA. As such an oocyte matures, the meiotically arrested, duplicated chromosomes ($4n$) usually go through a period of great activity called the *lampbrush phase* (see Figure 16-5). The chromosomal backbone unravels at hundreds of sites, so that regions composed of certain genes loop outward from the main chromosomal axis. These loops give the chromosome its brushlike appearance. Large amounts of mRNA are being made on each loop. After processing (Chapter 14), this mRNA is passed to the egg cytoplasm, where most of it is stored in inactive form; only after fertilization is the mRNA employed in building proteins for the embryo. Such storage of inactive mRNA never occurs in prokaryotic cells—in those cells, mRNAs are translated into protein immediately.

As eggs mature, a huge number of ribosomes are manufactured in the nucleus (10^{12} ribosomes in a single egg of the African clawed frog, *Xenopus*, for example), and these ribosomes are then passed to the cytoplasm for storage. The ability of a single egg to manufacture such huge numbers of ribosomes truly stretches the imagination. Donald Brown of the Carnegie Institution of Washington calculated that even if all the chromosomal genes for the major classes of ribosomal RNA (rRNA) were functioning continuously at top speed, it would take 465

years for a frog oocyte to manufacture 10^{12} ribosomes. How, then, can this manufacture occur in just five or six months, the length of oogenesis in *Xenopus*? The answer lies in the special process of **gene amplification,** illustrated in Figure 16-8. During this process, about 1 million extra copies of the ribosomal genes are made. These extra copies act as templates during oogenesis, so that the huge number of rRNA molecules and the resultant ribosomes may be generated, thereby permitting oogenesis to occur in a reasonable length of time.

These ribosomes and mRNA from the oocyte nucleus—along with the RNA, yolk, and other substances manufactured by nurse cells or digestive-gland cells—are said to be *maternal* in origin. This means that genes and chromosomes in the mother's somatic (body) cells, as well as in the oocyte, code for the components in and around the egg. The use of the word "maternal" also distinguishes these ribosomes, mRNAs, and proteins from others made after fertilization, when the embryo's own genes, including paternal ones, begin to play a role.

Because the maternal proteins and nucleic acids form the bulk of the egg cell itself, they can exert profound influences on the development of the embryo. In particular, the positions in the egg cell of mRNAs, ribosomes, and proteins can affect the organization of the embryo—its very body plan. The female also adds jellies, egg whites, and shells around the oocyte, the molecular makeup of which is dictated by the mother's genes. The maternal genes and the spatial distribution of their products provide critical developmental information that will govern much of the course of embryonic development. Thus the mother contributes considerably more than a haploid genome to the embryo.

FERTILIZATION: INITIATING DEVELOPMENT

The process that initiates development and unites the nuclei of male and female gametes is **fertilization.** Fertilization may be either external—in sea or fresh water—or internal—in the female's reproductive tract. But how do the gametes meet—how does a sperm "find" an egg?

Mating behavior is the main strategy that helps ensure that egg and sperm will meet. For external fertilization, seasonal synchronization of *spawning* (release of gametes) is essential, so that the relatively short-lived egg and sperm are released into water at the same time. Males and females of a particular species of marine worms, for example, swarm in the surface waters off Cape Cod around the time of the first full moon in July. These worms release their gametes into the ocean water in concentrations great enough that many eggs will be fertilized. Salmon, on the contrary, scoop out a depres-

Figure 16-8   GENE AMPLIFICATION IN NUCLEOLI OF A SALAMANDER EGG.

Each nucleolus (the red spots just inside the nuclear envelope) contains several circular strands of DNA. Each circle, in turn, contains numerous copies of the genes that code for the major types of ribosomal RNA (rRNA). A very large quantity of ribosomal RNA may be manufactured during oogenesis because of the large number of nucleoli and the many circles of DNA they contain, each of which is composed of hundreds or thousands of duplicate copies of ribosomal genes. A huge number of ribosomes thus are created and stored in the egg cytoplasm, ready to serve as the protein-factory sites that will spew out building material for the growing embryo. This "prepackaging" of ribosomes saves time, so that cleavage can proceed very rapidly.

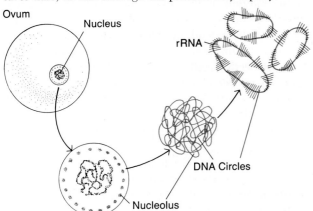

sion in a pebbly streambed; as the female lays thousands of eggs, the male pours forth sperm over them, so the chances of fertilization are increased.

Some aquatic organisms and all terrestrial animals that do not return to water to breed carry out internal fertilization. This often involves the transfer of a *spermatophore*, a packet of sperm that can be stored in the female's body. Male squid insert the spermatophore with a tentacle; male spiders use a leg for the transfer; and a female salamander will pick up a spermatophore left on the ground by a male and insert it herself. Male sharks, insects, and mammals have a copulatory organ, the *penis*, which discharges sperm solution directly into the female's body. This method provides the greatest protection to the sperm. Mating rituals of varying complexity, coordinated by both the nervous and the endocrine systems, accompany the internal-fertilization process in many organisms.

Once a sperm and egg do come together, the fertilization process involves several steps and accomplishes several tasks. First, the sperm penetrates the egg's protective membranes and surface, so that the sperm head is ingested, and the sperm and egg nuclei can fuse. Second, the egg's metabolic machinery is activated, and the completion of meiosis usually takes place. And third, the egg raises barriers to the entry of other sperm, thereby ensuring that only the first sperm to enter provides the paternal genes for the new individual. Let us look at each of these steps in greater detail.

Sperm Penetration

Eggs are universally immobile. Consequently, to meet the egg, sperm must swim through either body fluids if fertilization is internal or water if fertilization is external. Although some eggs may release specific chemicals to attract sperm, most do not. Thus, chance plays a large role in sperm meeting egg and helps explain why millions of sperm are released; the probability is low that any one sperm will encounter an egg.

Even after a sperm does contact an egg's surface, it must overcome several barriers before it can deliver its genetic cargo. First, it must penetrate the protective *jelly coat* that surrounds the egg; then it must pass through the *vitelline layer*, another protective coat; and finally, it must breach the plasma membrane itself.

The first contact of the sperm head with the egg's jelly coat triggers the remarkable and rapid **acrosome reaction.** This reaction consists of two parts: the release of the enzymes stored in the acrosome, and the thrusting forth of a long, rigid filament that stabs through the jelly toward the plasma membrane. The enzymes digest a hole through the transparent jelly coat and the vitelline layer (Figure 16-9). In mice and humans, for instance,

several enzymes are released to digest these and other protective layers. One such enzyme, called *hyaluronidase*, digests the hyaluronic acid that forms the protective matrix for the follicle cells surrounding the mammalian egg. In the second part of the acrosome reaction, a rapid rearrangement of the structural protein actin into a rigid bundle of filaments causes the front end of the sperm to protrude through the covering layers. In this way, the plasma membrane at the tip of the sperm is in a position to bind to the ovum's surface.

The actual fusion of the sperm and the egg plasma membranes immediately triggers a series of processes at the surface of the egg. In one such process, the sperm head is engulfed, as shown in Figure 16-10. The details of this process vary among species, but in each, the haploid male nucleus moves into the egg cytoplasm, and almost immediately, its chromosomes begin to lose their inactive, crystal-like configuration in preparation for joining the female nucleus. The sperm and egg nuclei are moved through the egg cytoplasm toward each other by microtubules.

Completing Meiosis

Strange as it may seem, most fully grown animal oocytes cannot be fertilized. In fact, most have not even completed meiosis. The completion of meiosis must be initiated in some way, and the oocyte surface must undergo final maturation so that the egg can be fertilized. The trigger for these events varies according to species. It may be a chemical signal, such as the hormone *progesterone* acting on the unovulated mature egg in various vertebrates (Figure 16-5, "Ovulation Stimulus"). Or it may be ovulation itself, in response to hormonal cycles or to copulation, that starts the final maturation and meiosis. Frequently, ovulation stimulates only the first meiotic division, whereas the second usually is stimulated by the fusion of sperm and egg. The final reduction division then occurs, and only a haploid set of egg chromosomes remains to join those in the sperm nucleus (the other haploid set of female chromosomes is discarded in the second polar body). When the sperm and egg nuclei come together, the two sets of chromosomes mingle, yielding a diploid set and the genetic coding for a new organism. At this point, the first round of DNA synthesis can begin as the fertilized egg, now called a **zygote,** prepares for mitotic cleavage divisions.

Barriers to Other Sperm

Many sperm may reach the egg simultaneously and adhere to its surface. Once the first sperm has been admitted, however, the egg cannot afford to admit any

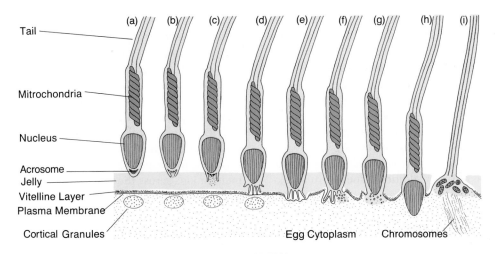

16-9   THE STAGES OF EGG FERTILIZATION BY A SINGLE SPERM.

(a) The sperm approaches. (b) The front of the sperm touches the jelly coat, and the acrosome reaction is triggered. (c, d) The acrosome opens, and digestive enzymes pour forth to begin eating through the jelly coat and vitelline layer. (d, e) Fingerlike projections of the sperm surface (which shoot forward as the result of the presence of internal actin filaments) protrude toward the egg. (f) Fertilization occurs when one of the projections touches and fuses with the egg's plasma membrane. (f, g) The two cell surfaces fuse, and egg cytoplasm flows outward, engulfing the sperm head. (h) The sperm nucleus rapidly responds to this exposure to the egg cytoplasm by showing a decondensation, or uncoiling, of its chromosomes (i) from the crystal-like state to the typical looser configuration. Figure 16-10 shows electronmicrographs of the later stages of the sequence.

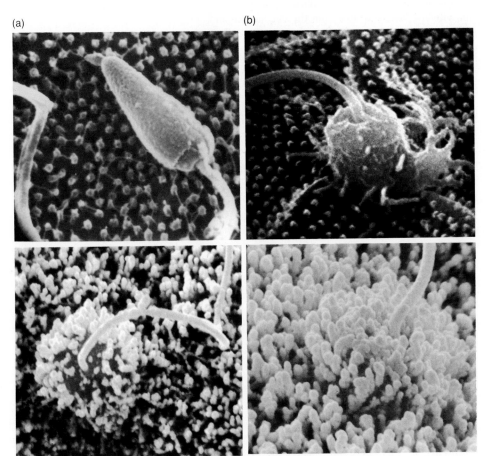

Figure 16-10   SPERM PENETRATING THE EGG SURFACE.

(a) Only a second or so after contact, this sea-urchin sperm adheres to tiny projections on the vitelline layer. (b) The sperm head has penetrated half way through the vitelline layer. (c, d) The vitelline layer has been removed, and the egg's plasma membrane is seen around the site where a sperm just penetrates, the tail still protrudes. These photographs are magnified between about 5,000 and 15,000 times.

more. If several were to gain entrance, the new organism would not be diploid, but would be triploid or polyploid. How, then, is more than one sperm prevented from fertilizing an egg? The answer lies in the egg's **cortical reaction** to the sperm's contact.

Fusion with a sperm's plasma membrane causes rapid changes in the electrical properties of the egg's plasma membrane. As a result, the egg loses its normal internal negative charge relative to the surrounding fluids and becomes positively charged on the inside. This blocks the attachment and entry of additional sperm. In an experimental system that maintains an egg with a constant internal negative charge, sperm after sperm swim up and fuse with an egg. Conversely, if an unfertilized egg is artificially given a positive charge, not even one sperm can fuse with and enter the egg. Loss of normal negative charge is temporary; it lasts only long enough for other

changes to occur in the egg surface that will form a permanent barrier to sperm entry.

Activation is the second step in the cortical reaction of the egg cell. Calcium, sodium, and hydrogen ions begin to move across the plasma membrane, and the intracellular pH rises because hydrogen ions pass out of the cell. As a result of the pH rise, enzymes and proteins change activity: the egg cell undergoes an increase in oxygen consumption, the initiation of protein synthesis, and a variety of other metabolic events. In this way, the movement of ions across the plasma membrane and the elevation in pH activate the dormant egg and get development started.

The final event of the cortical reaction is spectacular to see: the rapid elevation of the *fertilization membrane* from the surface of the egg cell. As Figure 16-11 shows, starting at the point of sperm–egg fusion, a "wave" ap-

Figure 16-11   ELEVATION OF THE FERTILIZATION MEMBRANE.

(a) Numerous sperm surround the jelly-encased egg, but only one has succeeded in reaching and fusing with the egg's plasma membrane. (b) Starting at the point where the sperm and egg plasma membranes fuse, cortical granules fuse with the egg's plasma membrane and discharge their contents. These contents include an enzyme that separates the vitelline layer from the plasma membrane and an enzyme that destroys or modifies sperm receptor-binding sites. (c) As the vitelline layer elevates from the egg surface and loses its sperm-binding sites, other sperm fall away. (d) The final result is an egg cell surrounded by a fertilization membrane to which no additional sperm can adhere. Note that the final egg plasma membrane includes patches of membrane that originally encompassed each cortical granule.

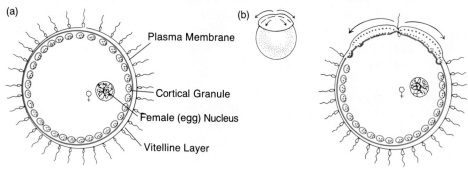

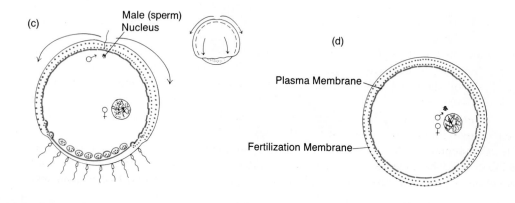

pears to pass over the surface of the spherical egg. At the advancing front of the wave, cortical granules break down, fuse with the egg's plasma membrane, and discharge their contents. Among these contents is an enzyme that separates the vitelline layer from the plasma membrane, thereby permitting the vitelline layer to rise and form the tough fertilization membrane. This often takes place in just ten to twenty seconds and leaves the egg floating free within the protective envelope of the fertilization membrane. It is no coincidence that this membrane is elevated by the time the internal electrical charge of the zygote has returned to its normal negative value; from then on, the fertilization membrane prevents the penetration of additional sperm.

The detailed events of the cortical reaction are being studied in hopes of developing safer, more effective contraceptive methods. So, too, are the events that activate sperm swimming and the details of the acrosome reaction. Swimming, like egg-cell activation, is initiated by the loss of hydrogen ions across the plasma membrane. If a means could be discovered to interfere with any of these processes, researchers might be able to develop important new ways to prevent fertilization.

Parthenogenesis: Sperm Are Not Essential in Some Species

Strangely enough, fertilization is not always required for reproduction. Insects, lizards, certain relatives of lobsters, and even turkeys can reproduce without sperm and normal fertilization. They carry out **parthenogenesis:** a spontaneous activation of the mature egg followed by normal egg divisions and embryonic development— or literally translated, a "virgin birth." In fact, mature eggs of species that do not undergo parthenogenesis in nature can be activated to develop without normal fertilization by pricking them with a needle, by exposing them to elevated calcium-ion levels, or by altering their temperature.

Since parthenogenetic eggs are not fertilized and do not receive male chromosomes, one would expect the resultant organism to have only one haploid set of chromosomes. In some species, however, meiotic division is suppressed, so the haploid number is avoided. In others, meiosis occurs, but an unusual mitosis returns the haploid embryonic cells to the diploid condition, resulting in two sets of *identical* chromosomes, in which both copies of every gene locus are identical. Thus organisms that reproduce parthenogenetically have substantially less genetic variability than do species with chromosome sets from two parents. That may be advantageous if they are well adapted to a relatively stable environment. But in meeting challenges of a variable environment, parthenogenetic species may have less flexibility, which may explain why this form of reproduction is relatively rare.

Fertilization is obviously a momentous event in the life of any individual. It is not, however, as we noted in Chapter 1, the beginning or creation of life. Both sperm and egg cells are alive prior to fertilization and have undergone long and complex maturation processes; thus fertilization is not the creation of life but of a new and unique diploid genome and cell. Thousands of developmental events occur between fertilization and the emergence of the newborn, and further growth and development lie between the newborn and the adult organism. At several points along this developmental continuum, the organism's final form can be altered. Thus although fertilization is an early and critical stage in the life process, it should not be thought of as solely responsible for a new individual.

CLEAVAGE: AN INCREASE IN CELL NUMBER

The major developmental event immediately following fertilization is **cleavage,** a special kind of mitosis, or cell division. Cleavage generates many small cells from the large single-celled zygote, and it distributes, in precise ways, yolk, mRNA, ribosomes, and other materials built into the egg during oogenesis. Cleavage also produces a **blastula,** a sphere of cells that contains a cavity in most species.

During cleavage, the new diploid genome and the chromosomes derived from both male and female parents are duplicated by normal processes of DNA replication and protein synthesis, so that each new cell (**blastomere**) of the blastula receives a full diploid set of chromosomes.

Patterns of Cleavage

Patterns of cleavage depend on several factors. One is the volume of yolk material incorporated into the egg. Figure 16-12 compares the cleavage patterns for three types of organisms. If little yolk is present, as in a mammal egg, the zygote cleaves completely through, forming cells that are more or less the same size (Figure 16-12a). Note that the cleavage pattern goes on relatively independently in different blastomeres, so that instead of going from two cells to four to eight, the pattern may also proceed arithmetically (1, 2, 3, 4, 5, etc.). In mammals, the initial ball of cells produced is called a *morula*. It becomes a blastula when a cavity, the *blastocoele*, forms within the mass.

If somewhat more yolk is present, as in a frog or salamander egg (Figure 16-12b), the cleavage goes on more rapidly in the regions of the embryo that contain less

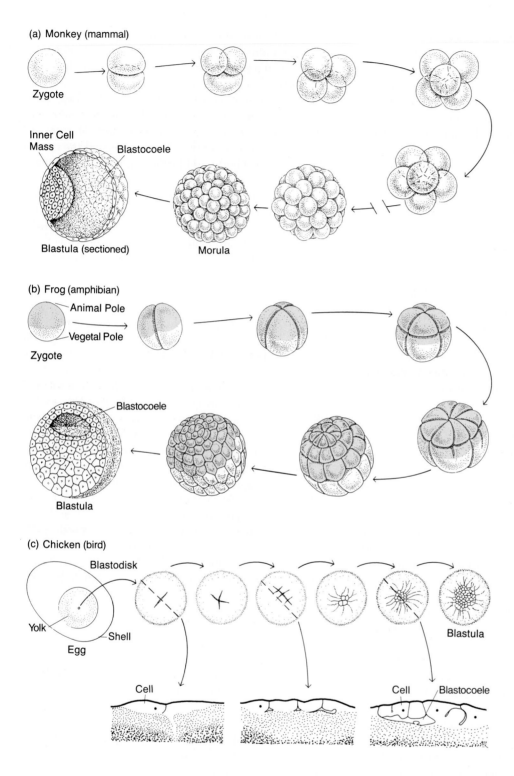

(a) Monkey (mammal)

Zygote

Inner Cell Mass
Blastocoele

Blastula (sectioned)   Morula

(b) Frog (amphibian)

Animal Pole
Vegetal Pole
Zygote

Blastocoele

Blastula

(c) Chicken (bird)

Blastodisk

Yolk
Shell
Egg

Blastula

Cell

Cell   Blastocoele

Figure 16-12   INFLUENCE OF YOLK ON CLEAVAGE PATTERNS IN EGGS.
(a) In a monkey embryo, which has very little yolk, the cleavage process occurs relatively independently in different blastomeres. The morula is a solid ball of cells that becomes a blastula when an inner cavity forms (the blastocoele). The blastula consists of an inner cell mass and the surface layer. (b) Cleavage in an amphibian embryo, which has an intermediate quantity of yolk. Blastomeres at the darkly pigmented end, the animal pole, divide more rapidly than do those at the vegetal pole, where more yolk is located. (c) Cleavage in the flat disk of cytoplasm on a bird zygote as seen from the top and in cross section. The yolk is so large that only the upper island of cytoplasm is shown. Each cleavage plane forms as though the surface were sliced downward with a dull knife. As the small blastomeres arise, a space (blastocoele) forms between them. The dashed lines in the blastodisk drawings show the plane of the cross section below.

yolk. As a result, cells at the less yolky end—the *animal pole*—become smaller during rapid divisions than do those at the yolky end—the *vegetal pole*. The small blastocoele is offset toward the animal pole.

Finally, some zygotes have such a massive amount of yolk that they cannot be cleaved through completely. In

birds, for instance, cleavage divisions are restricted to one tiny island of cytoplasm at the animal pole of the huge egg cell, and the yolky portion remains undivided. The result is a small disk of blastomeres, the *blastodisk* (Figure 16-12c). Only much later in development will cells at the edge of the disk spread over the huge sphere of the yolk, engulfing it within a specialized embryonic membrane, the *yolk sac.*

In addition to the amount of yolk present, cleavage patterns are determined by an organism's evolutionary history. As we shall see in Chapter 25, there are two main lineages of animals. In one lineage (vertebrates, starfish, and related organisms), cleavage is *radial:* the new blastomeres accumulate in even rows, above one another. In the other lineage (worms, clams, snails, and related organisms), cleavage is *spiral:* successive layers of newly arising blastomeres are rotated first counterclockwise and then clockwise (when viewed from the animal pole). We shall discuss these cleavage patterns and their significance in Chapter 25.

Cytoplasmic Distribution During Cleavage

One of the critical outcomes of cleavage in some species is the precise distribution of cytoplasmic components to different blastomeres. This distribution takes place because substances may be arranged in the zygote in such a way that cleavage restricts them to certain blastomeres. In a frog egg, for example, black pigment is concentrated at the animal pole, whereas the vegetal pole is white. This visible distribution is a simple representation of the subtler distribution of organelles and chemical agents (ribosomes, mRNA, enzymes) that may affect the development of different cell types in the embryo.

Experiments with the sea squirt, a urochordate and relative of the vertebrates (Chapter 25), have helped reveal this precise distribution. In sea-squirt zygotes, two crescent-shaped regions appear in the minutes following fertilization, a yellow region and a gray region (Figure 16-13). During subsequent cleavage divisions, the cytoplasm with yellow granules is segregated into certain blastomeres; the gray cytoplasm into others. Hours later, cells that arise from the yellow group give rise to the skeletal muscle of the little tadpole, while the gray group gives rise to the notochord, the main skeletal organ of the tadpole's tail. Biologists wondered what would happen if the zygote's cytoplasm were rearranged. Would the embryo develop normally, or would

moving the gray and yellow segments change developmental patterns? To answer this question, they centrifuged sea-squirt zygotes so that the yellow, gray, and other cytoplasmic substances were redistributed, and then observed the zygotes' development. They found that cleavage still occurs and that muscles and notochord develop. But the arrangement is chaotic—these organs do not appear in normal locations and orientations. Thus when the cytoplasmic determinants are redistributed, the resultant cell types are, too.

This centrifugation experiment reveals that the place where yellow cytoplasm ends up will be the site of muscle development, and the place where gray cytoplasm goes will be the site of notochord formation. Thus these regions of maternally originated cytoplasm contain control agents for specific types of cell maturation. The experiment also indicates that the normal distribution of the control substances is a key to organizing the embryo's body correctly (that is, ensuring that muscles and other organs form at the correct sites). Hence, one form of developmental information built into a maturing oocyte is the spatial arrangement of control substances.

Experiments such as this lead us to liken the surface of the zygote to the surface of the Earth: cleavage divides the zygote into continents (major regions of blastomeres), the continents into countries (smaller subsets of blastomeres), and, finally, the countries into states (groups of blastomeres that have similar developmental fates). Furthermore, these precise subdivisions will affect the "diplomatic" interactions of the cell subsets during later development.

The example of the yellow and gray crescents shows that certain developmental information derived solely from the mother's genes is distributed *unevenly* by the process of cleavage. Another example is the develop-

Figure 16-13   CYTOPLASMIC LOCALIZATIONS AND CELL FATES.

(a) The gray and yellow crescents in this sea squirt zygote are precisely segregated to certain blastomeres during cleavage. (b) At the eight-cell stage, two of the eight cells possess gray-crescent cytoplasm; two, yellow-crescent materials. (c) After gastrulation, a tadpole develops, containing a notochord derived from the gray-cell lineage and muscle cells from the yellow-cell lineage.

(a) Zygote   (b) Eight-Cell Stage   (c) Tadpole Stage

Gray   Yellow
Crescent   Crescent

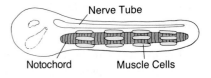

Nerve Tube

Notochord   Muscle Cells

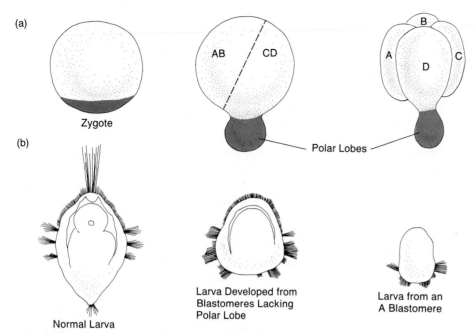

Figure 16-14   DEVELOPMENTAL INFORMATION IN THE CYTOPLASM: THE POLAR-LOBE SYSTEM OF A SNAIL EMBRYO.
(a) Instead of being distributed to all blastomeres, the lobe and its contents are restricted to one line of cells during cleavage; first CD, then D, and, finally, D's descendants. (b) A normal larva, a larva that developed after the polar lobe had been cut away early in cleavage, and a larva that developed from an isolated A blastomere. The maternally derived developmental "information" within the polar lobe was not available to the A blastomere or to the embryo lacking the lobe; as a result, their development is deficient, and they do not survive.

ment of *polar lobes* (not to be confused with polar bodies) in some worm and snail embryos (and their relatives). As Figure 16-14 illustrates, the first cleavage divides the contents of the egg asymmetrically, so that the daughter cells do not have the same developmental potentialities; one daughter cell (CD) receives the polar lobe, and one (AB) does not. If these two cells are experimentally separated from each other, the cell with the polar lobe can go on to form a complete, normal embryo, whereas the cell without the polar lobe forms an aberrant embryo (Figure 16-14b, right two larvae).

Various experiments have shown that the polar lobe contains unique developmental information. For instance, it causes the formation of a special tuft of cilia used in swimming, and it may be responsible for the formation of the embryo's entire posterior portion. This is a case in which maternally derived developmental information contained in the zygote becomes allocated to some blastomeres but not to others. In mammals and birds, developmental information is not partitioned into such obvious lobes or crescents; it actually arises as the cleavage process goes on. Thus the position of a cell late in cleavage determines what that cell will become: those on the inside yield the embryo's body, while those on the outside form protective membranes and ultimately die. Thus developmental information that controls the

genome may result from position (as in mammals) or from preformed molecular determinants that are distributed precisely during cleavage (as in sea squirts and snails).

GASTRULATION: REARRANGEMENT OF CELLS

The process of cleavage makes a large number of small cells from one large egg cell. It results in a sheet of cells, usually one-cell-layer thick, that surrounds either a cavity or a mass of yolk. In oversimplified terms, the adult body of most multicellular animals is a tube on the outside (the skin), a tube on the inside (the gut), and clusters of cells in between (muscles, bones, heart, and so on). Thus, a process is needed to change the blastula into a complex three-dimensional organism with inner, outer, and middle layers. This change occurs during **gastrulation.**

During gastrulation, some of the blastomeres located on the surface of the embryo move to the inside, producing the three-layered **gastrula.** As a result, cell popula-

Figure 16-15
GASTRULATION PATTERNS AND THE AMOUNT OF YOLK.
(a) Gastrulation in an *Amphioxus* egg with very little yolk. The flattened plate of cells invaginates—sinks in—to generate the endoderm and ectoderm. Later the mesoderm, not shown here, will form. (b) Gastrulation in a frog, whose eggs have an intermediate quantity of yolk. This frog embryo is cut in half so that the invaginating layer of cells is visible as it moves gradually across the blastocoele space. The dorsal lip of the blastopore is the site where the sheet of invaginating cells rolls inward and disappears from the surface. Invaginated endoderm and mesoderm separate at a later stage. (c) Gastrulation in the flattened embryo of a bird. The tiny embryo rests on the huge sphere of yolk. Cells on the surface of the embryo move centrally toward the groovelike primitive streak and then migrate inward to give rise to the two inner cell layers: endoderm and mesoderm. These movements are similar to those in *Amphioxus* and the frog, although the flattened character of the embryo makes them appear to be different.

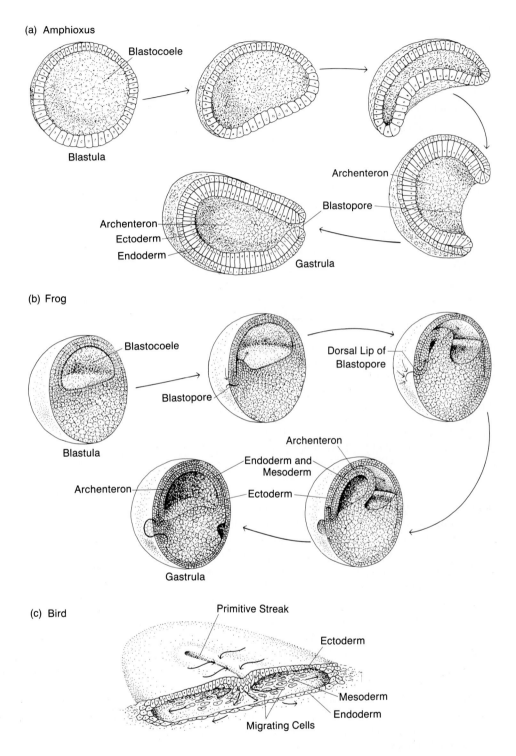

tions originally separated from one another come into contact and interact, leading to normal development of most organs. The three *germ layers* that are produced (so called after "germinal," for "origin") are the **ectoderm** (the outer layer), the **endoderm** (the inner layer), and the **mesoderm** (the layer between the ectoderm and the endoderm). As we shall see shortly, each layer gives rise to specific tissues during the course of embryonic development.

As with cleavage, the movement of cell populations during gastrulation is affected by the quantity of yolk present in an embryo. In organisms with relatively little yolk, gastrulation movements are simple and involve a process of inpocketing called *invagination*. If you were to take a soft rubber ball and push a finger into the side, the "dent" created would be similar to the invaginating zone of a blastula during gastrulation. This opening into the blastula is called the *blastopore*. Figure 16-15a shows how the two-layered condition is achieved in *Amphioxus*, a simple aquatic relative of vertebrates whose eggs have little yolk. The cavity created by invagination is called the *archenteron* ("early gut") because it will become the gut cavity. It is also termed the *gastrocoele*. The formation of mesoderm is more complicated: in *Amphioxus*, individual cells located in the wall of the blastula migrate into the space between the ectoderm and the endoderm. Gradually, populations of these mesodermal cells accumulate and begin to develop such structures as the skeleton. Thus in *Amphioxus*, gastrulation includes two processes: *invagination* of a sheet of cells to form the endodermal gut lining and the *migration* of individual cells to form mesoderm.

With intermediate quantities of yolk, as in frog eggs, a similar invagination process occurs (Figure 16-15b), but it happens near the equator of the embryo, more or less at the junction between the smaller animal-pole cells and the larger vegetal-pole cells, which contain most of the yolk. A slit, the blastopore, marks the site where cells invaginate; the dorsal lip of the blastopore is composed of such moving cells (Figure 16-15b). As gastrulation nears completion, complex movements of the mesodermal and endodermal cell populations finally yield the three-layered gastrula.

When large quantities of yolk are present, as in bird eggs, gastrulation appears to be quite different (Figure 16-15c), but it is actually a variation on the processes and cellular rearrangements found in *Amphioxus* and frog eggs. Cells located on the blastodisk surface move to the inside through a thickened portion of the embryo called the **primitive streak.** The first cells moving downward through the primitive streak become the lower endodermal layer; later ones become mesoderm. Interestingly, mammals—with virtually no yolk in their eggs—retain the primitive-streak method of gastrulation, which originated in reptiles. This is one of the important pieces of supporting evidence for the hypothesis that mammals are descendants of reptiles (Chapter 26). Birds, too, arose from reptiles and their large, yolky eggs.

ORGANOGENESIS: FORMATION OF FUNCTIONAL TISSUES AND ORGANS

Once gastrulation is complete, the next phase of development begins; it is **organogenesis,** the formation of the body's organs and tissues. During organogenesis, the cells brought inside the embryo or left on its surface during gastrulation become specialized to form the ner-

vous system, the limbs, the kidneys, and other tissues. The ectoderm develops into the epidermis and its appendages (hair, skin, glands, and so on), the nervous system, and the sense organs. The mesoderm yields the connective tissues of most organs, the circulatory and immune systems, the kidneys and gonads, and the skeleton and muscles. The endoderm forms the lining of the gut and the epithelial portion of such internal organs as the liver, thyroid, lungs, and pancreas. Table 16-1 summarizes this process.

Even as the three germ layers arise, their cells are already committed to forming the cell types appropriate to each layer. Many complex processes, however, must operate to elicit normal development of the various tissues and organs. These processes and their results comprise two major events: **morphogenesis,** the change in shape of cells and cell populations; and **differentiation,** the maturation of cells so that they may perform separate functions, either in the embryo itself or in later stages of development.

Although we distinguish between morphogenesis and differentiation, the two really go hand in glove. Differentiating cells assume characteristic shapes to suit their functions: muscle cells become long and spindly; red blood cells take on disk shapes; nerve cells branch in complex ways. By the time the cells have acquired their final shapes, they are built into their separate tissues so that their biochemical and cellular functions may be carried out efficiently for the benefit of the entire organism.

Morphogenesis depends on the ability of cells and cell populations to assume specialized shapes and positions in the developing embryo. Take, for example, the process of **neurulation**—the formation of the neural tube, from which the brain and spinal cord later develop. Figure 16-16 shows the basic steps in this morphogenetic process. The cells start out in a flat and somewhat thickened sheet, called the *neural plate.* Through changes in cell shape, the right and left edges begin to fold upward toward each other. When they meet, the tissues fuse, and the resulting tube is pinched off from the adjacent ectoderm. The result is the hollow brain and spinal cord of the vertebrates. Through similar processes, the other cells of the embryo become arranged into tissues and organs that will be functional parts of the whole organism.

The second major aspect of organogenesis is the differentiation of cells into a functionally mature condition. Cell groups become different from one another and from what they were earlier in embryonic development. They manufacture unique sets of molecules that account for their particular function: sets of contractile proteins for heart-muscle cells; chemical transmitters for nerve cells; hemoglobin for red blood cells; hormones for endocrine cells.

A good example of cellular differentiation occurs in the

Table 16-1   **FATE OF THE GERM LAYERS**

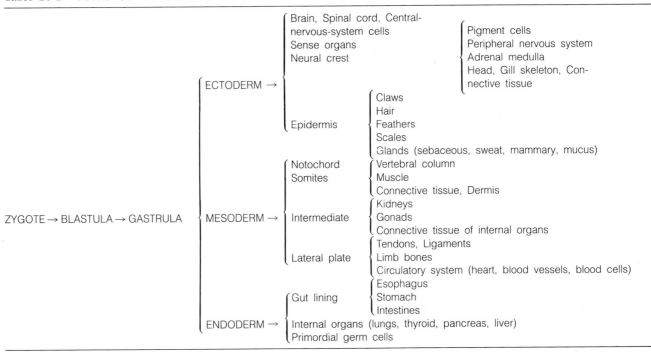

ZYGOTE → BLASTULA → GASTRULA	ECTODERM →	Brain, Spinal cord, Central-nervous-system cells Sense organs Neural crest	Pigment cells Peripheral nervous system Adrenal medulla Head, Gill skeleton, Connective tissue
		Epidermis	Claws Hair Feathers Scales Glands (sebaceous, sweat, mammary, mucus)
	MESODERM →	Notochord Somites	Vertebral column Muscle Connective tissue, Dermis
		Intermediate	Kidneys Gonads Connective tissue of internal organs
		Lateral plate	Tendons, Ligaments Limb bones Circulatory system (heart, blood vessels, blood cells)
	ENDODERM →	Gut lining	Esophagus Stomach Intestines
		Internal organs (lungs, thyroid, pancreas, liver) Primordial germ cells	

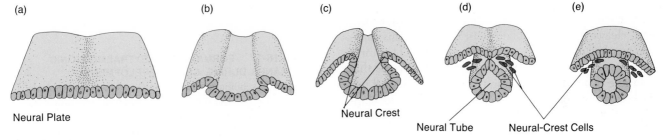

(a)   (b)   (c)   (d)   (e)

Neural Plate

Neural Crest

Neural Tube   Neural-Crest Cells

Figure 16-16   FORMATION OF THE NEURAL TUBE.
The nervous system of primitive organisms was located in the skin. During the evolution of increasingly complex animals, the nervous system sank beneath the surface, perhaps for better protection of its special clusters of cells. Today, in each developing vertebrate embryo is a repetition of this sinking-in process during neurulation. This sequence shows the morphogenesis of a small portion of spinal cord in cross section. (a) Morphogenesis starts with the neural plate, a flat sheet of cells on the surface of the embryo. (b, c) Changes in cell shape cause the plate to curve inward, so that the two sides move toward each other. (d) The cells at the crest of the neural folds (dark color) come together, and the tissues fuse. (e) Finally, the neural tube separates from the overlying ectoderm. (f) A scanning electron micrograph of a stage between (c) and (d). From *Tissues and Organs: A Text-Atlas of Scanning Electron Microscopy* by Richard G. Kessel and Randy H. Kardon, W. H. Freeman and Company © 1979. Organelles within the cells such as microtubules and microfilaments and changing adhesion between cells may be responsible for the shape changes in the neural-plate cells and their movements. Note that the scanning electron micrograph at the beginning of the chapter is at the stage seen in part (c).

(f)
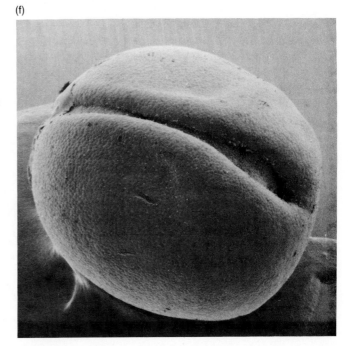

pancreas, where certain digestive enzymes and hormones are made. In mouse embryos, the amount of pancreatic digestive enzymes increases some 10,000-fold during development. Genes that code for these enzymes are turned off completely at first, and then function slowly for a time, so that a small amount of enzymes is synthesized. Later, as the embryonic pancreas grows larger, the rates of mRNA and enzyme syntheses rise dramatically to generate the large quantities present at birth. Most of the increase occurs in differentiating cells that have stopped dividing. Similar increases occur as hemoglobin is produced by red blood cells; antibodies, by certain white blood cells; and other specific products, by various differentiative cells. (We discuss the regulation of gene function in the various cell groups in Chapter 17.)

Another aspect of differentiation is the cells' attainment of *responsiveness*, the ability to be regulated within the organism. A given developing cell type becomes sensitive to hormones, neurons, or other signals and, as a result, can be made to function at the correct time, place, and rate. In other words, it will function as an integral part of the entire organism.

Morphogenesis and differentiation are the culminating processes of the developmental sequence that begins with gamete formation and continues through fertilization, cleavage, and gastrulation. In the rest of this chapter, we shall consider some additional aspects of embryonic development and some developmental events that continue in adult life. Then in Chapter 17, we shall undertake an in-depth analysis of the *control* of development, particularly of organogenesis.

EMBRYONIC COVERINGS AND MEMBRANES

So far, we have seen zygotes and embryos the way scientists in the laboratory see them—isolated from their normal surroundings. But in the real world, the fragile embryo must be protected from heat, drought, predators, the mother's immune system, and other dangers of its own particular environment. It cannot, however, be sealed off from the world completely; it must also have a means of gas exchange, nutrient uptake, and waste disposal.

Most animal embryos are encased in a tough protective membrane, such as the leathery coverings of fish and insect eggs or the shells of snake and bird eggs. Hooks on the surfaces of shells or coverings can hold an egg to a leaf or submerged branch, so that it is not blown or swept away. Egg containers of land-dwelling species are also resistant to drying and prevent water loss from the embryo, but do permit a limited amount of gas exchange. Even embryos that remain within the mother's body need to be encased in protective membranes.

Figure 16-17   FORMATION OF EXTRAEMBRYONIC MEMBRANES DURING CHICK DEVELOPMENT.
(a) A highly schematic drawing of the yolk-sac membrane and the chorion, the outermost membrane, encircling the spherical yolk mass, on which the chick embryo develops. The amnion, or inner membrane, arises as it and the chorion travel over the top of the embryo and fuse. The amnion encloses the embryo in the fluid-filled amniotic cavity. In (a), the allantois begins to expand. This pouchlike membrane, which is formed from the embryo's primitive hindgut, serves as a garbage bag, collecting nitrogenous wastes (uric acid) until the chick hatches. (b) The allantois enlarges greatly and its outer wall fuses with the inner wall of the chorion; the result is the chorioallantois. The chorioallantois acts as a respiratory membrane during the later stages of the chick's development.

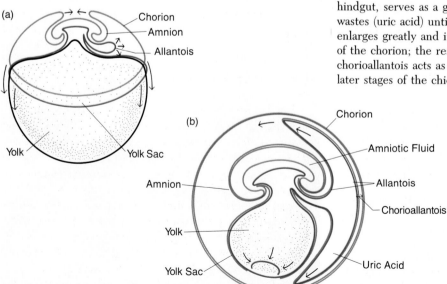

The land vertebrates—reptiles, birds, and mammals—have evolved the most complex set of membranes surrounding their embryos. These *extraembryonic membranes* include the yolk sac, the chorion, the amnion, and the allantois (Figure 16-17). The first of these to form is the **yolk sac,** a membrane that may be thought of as an outgrowth of the embryo's endodermal gut. It completely encloses the sphere of yolk in reptile and bird eggs so food can be absorbed by the developing, enlarging embryo. The yolk-sac wall serves as the first site of differentiation of red blood cells and thus provides the red cells and hemoglobin essential to oxygen binding and transport. The **chorion** and the **amnion** are originally continuous with each other but soon separate. The amnion fills with *amniotic fluid* and forms a cushionlike sac around the embryo. The chorion's functions are best understood in relation to the **allantois,** the fourth membrane, which arises as a pouchlike outgrowth of the embryo's posterior gut. Abundant allantoic blood vessels form and carry nitrogenous wastes from the embryo to the allantois, which serves as a kind of organic garbage bag, storing such wastes until the reptile or bird hatches.

The allantois and the chorion become fused to form the *chorioallantois,* where oxygen diffusing inward through the egg-white fluid is picked up, and where carbon dioxide is given off from embryonic blood vessels. In eggs with hard shells, the chorioallantois is also a site of calcium absorption into the blood, so that calcium ions can be transported into the embryo and used for bone growth and other processes.

We shall encounter these four extraembryonic membranes again in Chapter 18, when we discuss human reproduction. Modifications of the membranes allow the special kind of relationship between mother and fetus that is characteristic of mammals, including humans.

GROWTH: INCREASE IN SIZE

We have seen how the egg cell is divided into an increasing number of smaller and smaller cells during cleavage, how these cells are rearranged during gastrulation, and how the groups of cells differentiate and develop into the tissues and organs of the later embryo. Now let us consider *growth,* the increase in size that is an obvious necessity if a single fertilized egg is to develop into an even larger newborn and if a newborn organism is to develop into an adult.

Growth in Embryos

The rate of growth in an embryo can be astonishing: in just twenty-two days—from fertilization to birth—a rat embryo increases from one cell to *three billion* cells. This developmental explosion is based on the simple, steady mitosis of cells. The fertilized egg divides into two cells; these two, into four, eight, sixteen, thirty-two, sixty-four, and so on. Thirty-one such cell generations precede birth in the rat. The tiny newborn, of course, continues its rapid growth into adolescence and adulthood: within three months, the rat pup's three billion cells become *sixty-seven billion.*

An extreme case of embryonic growth occurs in the blue whale. This marine mammal's egg is less than a millimeter in diameter and weighs a fraction of a gram, but the newborn calf weighs 2,000 kilograms and is 7 meters long. That represents about a 200-millionfold increase in weight.

In the whale as well as in the rat and the human, growth is largely due to the increase in the number of cells, not in the size of individual cells. When the absolute number of cells in the many developing organs increase because of sustained mitotic activity, the organs and the embryo itself grow in size.

In many species, a finite amount of food available to the developing embryo limits its ultimate size before birth. For instance, a chicken embryo inside its shell grows at the expense of the yolk and the "white," which gradually shrink as the number of cells increases. Hatching must occur soon after the yolk has been used up, or the embryo will starve to death. Similarly, in many types of animal embryos, the only food to sustain development is stored in the egg from the start; hence, the late embryo at hatching can be no larger than the egg. Animals that carry out internal fertilization and incubation, however, do not have to depend on a limited food supply. Mammalian development provides the embryo in the womb with seemingly unlimited nutrients from the mother's body rather than from a store of yolk. Study of mammalian evolution suggests that the general timing of mammalian birth is based not on limitations in the food supply, but on such things as the size of the fetal head relative to the birth canal and potential attack by the mother's immune system.

How do embryos that develop free of the maternal body, such as those of frogs and insects, get along with only intermediate or small quantities of yolk? Such animals have a special growth phase in their life cycle. The embryo does not grow, but gives rise to a *larva* that can grow by feeding itself. Larvae typically lead very different lives from adults of the same species: the salamander tadpole is aquatic and feeds on plant material, but the terrestrial adult eats animal material and lives in a burrow on the forest floor or in a streambed; the caterpillar feeds on plant leaves, but the moth flies about in search of flower nectar. A transformation, or **metamorphosis,** intervenes between larval and adult stages. For example, after growing manyfold, the silkworm caterpillar becomes a *pupa* that metamorphoses into the adult moth. In Chapter 37, we shall see how hormones are used to control the metamorphosis of insects and amphibians, as caterpillars become moths and tadpoles become frogs or salamanders.

Growth in Later Stages of Life

Embryonic growth can be impressive, but the most spectacular growth stages of many species take place during the juvenile and adolescent phases of the organism's life cycle. It is then, for instance, that intensive cell division near the ends of the limb bones produces the typical adolescent growth spurt in humans. This rapid growth is controlled by hormones, including the vertebrate's growth hormone. If the timing or quantity of such hormones is abnormal, the result in humans can be a 7-foot giant or a 3-foot dwarf.

Not all animal species stop growing when they reach adulthood. Lobsters continue to grow in spurts year after year by casting off their old external skeletons and growing new, larger ones. Fish such as salmon grow continuously. Sustained adult growth depends on mitosis and increase in cell number, just as does embryonic growth. In fact, there are only rare cases of organ or organism growth by increase in cell size; and they involve polyploidy, multiple copying of the genome in the enlarging cells. The much more common strategy, growth by mitosis, can be either continuous and slow, as in the salmon, or periodic, as when the lobster molts.

Most animal species do stop growing at some point. Growth cessation is also regulated by hormones. An extreme case of the precision with which cell division can be controlled is the tiny fresh-water animal called a rotifer. Adult rotifers apparently have a constant number of cells, down to the number of cells in each organ. For instance, by some counts, the brain has 183 cells; the stomach, 39; and a gastric gland, 6. But the rotifer is an exception; other animal species with a "normal," predictable adult size usually show a good deal of variation in cell number on both sides of a statistical average, as do humans.

Once an animal reaches its typical adult size, cell division slows, so that the number of new cells produced more or less compensates for older cells that die. Thus the organism does not grow in overall size, even though it produces new cells. Consider, for example, the replacement of red blood cells in a normal adult human. Daily production of hundreds of millions of cells in the bone marrow (the core of many bones) occurs at a rate equal to the natural rate of cell death, so that the number of circulating red blood cells remains constant. This cell production is regulated by a protein called *erythropoietin*, secreted by the kidney whenever the blood flowing through the kidney is low in oxygen. After any loss of red blood cells (such as by natural cell death or by bleeding), erythropoietin released by the kidney stimulates stem cells in the bone marrow to increase their already substantial mitotic activity. (**Stem cells** are those that serve as a continuing source of a differentiated cell type, even for the lifetime of an organism.) This increased mitosis generates extra cells that differentiate into erythrocytes (red blood cells). Soon the concentration of red blood cells per milliliter of blood approaches normal values, oxygen carrying capacity is restored and is sensed by the kidney, and erythropoietin secretion falls to a resting level.

As in other systems, including the gonial cells that generate sperm and the basal cells that generate the surface layers of skin, the control factor in erythrocyte production regulates a mitotic stem-cell population, not the final, differentiated cell type itself. Thus if the skin surface is wounded, so that many surface cells are scraped away, there is a local increase in mitosis in the underlying basal layer of stem cells, and the gap is soon filled in.

Biologists do not fully understand the exact mechanisms by which mitosis in adult tissues is regulated to keep the various populations of differentiated, functioning cells constant. The process of controlling growth during embryonic development is even more mysterious. But the fact that all the various embryonic organs—eyes, limbs, heart, and so on—grow in proper proportion and stop enlarging at adulthood suggests that beautifully coordinated controls exist. Biologists and medical researchers are searching for these controls to help them develop new treatments for cell-growth abnormalities as well as cancers, in which growth regulation goes awry.

Regeneration: Development Reactivated in Adults

Adult sea squirts and hydras can be produced from tiny buds; male elk grow new sets of antlers each year; whole new worms may regenerate from just pieces of the body; and some aquarium fish can sprout new fins to replace ones that have been bitten off. These are all cases of **regeneration,** a special type of growth in adults.

Salamanders are the best studied organisms that regenerate lost parts. If, for example, the crystalline lens is removed from the eye of an adult salamander, a new one develops from the nearby iris, the pigmented ring of tissue that surrounds the pupil. Similarly, if the forelimb of a salamander is severed midway between the elbow and the wrist, the stump re-forms the missing forelimb, wrist, and digits within a few months (Figure 16-18). In both cases, cells in the iris or in the stump tissue undergo a process of **dedifferentiation.** That is, they lose their functional phenotype (as pigment, bone, connec-

Days

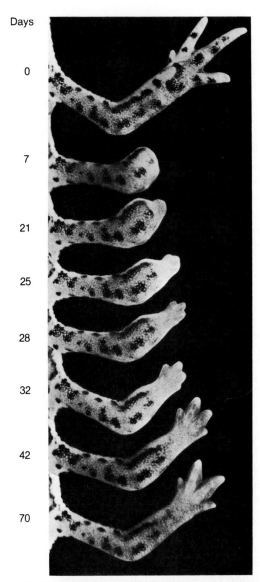

Figure 16-18   LIMB REGENERATION IN A SALAMANDER.
Cells at the end of the stump dedifferentiate and proliferate
intensively. They yield populations that once again
differentiate into bone, muscle, tendon, skin, and so on.
The numbers indicate days since the amputation.

tive tissue, and so on). The dedifferentiated cells divide
rapidly and generate a population of cells that will re-
form the lost parts. The dividing daughter cells then
carry out differentiation and morphogenesis, thereby
forming a new lens or a new forelimb, wrist, and digits.
Note that one differentiated cell never transforms directly into another differentiated cell type; a period of
mitotic divisions always intervenes between the original
differentiated state (the iris) and the new state (the re-
formed lens).

One requirement for the regeneration of salamander
limbs seems to be the presence of a minimal amount of
nervous tissue in the stump near the wound site. Removal of the nerves in a salamander limb stump prevents
regeneration. Conversely, if extra nerve tissue is transplanted into a frog's forelimb (which normally cannot
regenerate), a reasonable amount of regeneration occurs
following amputation. Nerve tissue thus plays a basic
role in setting up the conditions that lead to regeneration. In fact, attempts are under way to stimulate limb
regeneration in mammals, such as rats, by implanting
extra nerves or injecting nerve extracts. One day such
work may allow physicians to stimulate limb regeneration in humans.

A related phenomenon often confused with regeneration is **compensatory hypertrophy,** a kind of temporary
growth response that occurs in such organs as the liver
and kidney when they are damaged. If a surgeon removes up to 70 percent of a diseased liver, the remaining
liver tissue, arrayed in lobes, compensates by undergoing a very rapid rate of mitotic activity (greater than the
fastest growing cancers) until almost all the original liver
mass is restored. Similarly, if one kidney is removed, the
other greatly increases in size to compensate. Note that
the missing kidney (or liver lobe) is not regenerated; it is
the uninjured residual tissue that increases in mass and
cell number.

Clearly, mitosis is regulated in mature liver and kidney cells, as well as in the blood and skin cells already
discussed. Intensive research is under way to discover
what turns mitosis on in kidney or liver tissue following
surgical removal, and then what turns it off again so that
approximately the normal total mass of tissue is obtained. One hypothesis is that differentiated tissue cells
produce inhibitors of mitosis called *chalones.* According
to this theory, when a kidney is removed, the level of
kidney chalone falls. Mitosis then can occur. New kidney
cells are generated that differentiate into mature kidney
tissue, which again gives off chalones that inhibit further
growth. Researchers are exploring the mysteries of regeneration and compensatory hypertrophy in search of
answers that may allow physicians of the future to turn
on these processes in limbs or organs that have been
severed or damaged and thereby restore them to normal
size and functioning.

AGING AND DEATH: BOTH DEVELOPMENTAL PROCESSES

"Death is but an end to dying," wrote the sixteenth-
century French essayist Michel de Montaigne. By

dying, he really meant *aging*, the time-dependent deterioration of many of the body's parts that actually begins early in life. Aging is a characteristic of multicellular organisms, both plants and animals. In contrast, prokaryotes and some single-celled organisms called protists (Chapter 20) may simply divide periodically and thus produce new progeny. In a sense, these cell lineages are immortal. With the increased organismal size and complexity of multicellularity, however, come the mortality of the individual organism and the immortality of the germ line (the gonial cells that produce gametes); the germ cells alone survive, in the gonads of offspring, to contribute to future generations.

The causes of aging—with its graying hair, wrinkled skin, and worn-out organs in humans, for instance—are under intensive investigation, and many theories attempt to explain this process. According to one theory, the fibrous proteins of the connective tissues degenerate. *Collagen*, which gives tensile strength to the connective tissues everywhere in the body, undergoes changes over the years. The collagen fibers become less extensible, more brittle, and more subject to tearing. One result is the sagging skin of the older human. More important is impairment of functions in lungs, kidneys, heart, major blood vessels, and other vital organs as collagen and other connective-tissue proteins deteriorate.

A second theory of aging suggests that cells are capable of only a limited number of divisions. Some highly controversial experiments have led to the idea that an intrinsic limitation in mitotic capacity ultimately leads to a depletion of new, healthy cells in vital tissues; thus, an insufficient number of cells may replace dying cells.

Other theories of aging are based on declines in the immune system. We shall see in Chapter 32 that one of the vital parts of the immune system, the thymus-dependent T-cell system, declines in function after puberty. This decline may contribute to an increased frequency of cancers and debilitating infections. Another phenomenon is increased *autoimmunity*—attack by the immune system on the body's own cells. Rheumatoid arthritis and a variety of other diseases that are prevalent in old age stem from autoimmune responses.

Yet another theory focuses on the accumulation in cells of *lipofuscins*, or aging pigments. For unknown reasons, such pigments are not degraded or discharged from cells, so the volume builds up. For instance, lipofucsin makes up 5 percent of the volume of the heart muscle in an eighty-year-old man. The accumulation is most marked in cells that do not divide during adult life—nerve cells and skeletal-muscle cells. Slowly dividing cells, such as those in the heart, are also affected. Some biologists theorize that these increasingly bulky lipofuscin granules—permanent cellular garbage cans—begin to interfere with the proper functions of cells and so contribute to aging.

These and a variety of other theories have been offered to explain aging. Ultimately, the decreased functional capacity of vital organs leads one of them to fail, and death results. An organism does not die of "old age"; it dies of a bad heart, a deteriorated liver, a ruptured blood vessel in the brain, or another precipitating event. From a biological point of view, natural selection probably has not operated to resist the causes of aging, since reversal of such processes might make no significant contribution to survival of the species and, indeed, might endanger the young by providing competition for food, space, or other resources.

THE CONTINUITY OF LIFE

We have traced the growth and development of an individual organism from zygote through embryonic development, birth, youth and adulthood, aging, and death, as depicted in Figure 16-1. We have seen that the fertilized egg contains genetic and developmental information, but the adult that this information generates is far more complex than the egg. It is clear, however, that the embryonic and adult forms are intrinsically linked; one biologist has even stated that "an adult is just an egg's way of making another egg." This means that over time, it is the germ line—the gonial cells, sperm, and eggs—that passes down through the history of animal organisms. Obviously, adults of each generation are the vehicles—fancy culture chambers, if you will—that carry these germ cells through time. An adult protects its cargo of germ cells by surviving the rigors of natural selection and by reproducing and bearing offspring. Ultimately, of course, every adult dies. The germ line alone survives, resident in each member of a species and destined to be passed on by some individuals to their offspring or to die along with those who fail to reproduce. Thus the gonial cells in the human ovaries and testes are a heritage from one's forebears, a line of cells and genetic information that links each person with all his or her ancient ancestors. Indeed, that linkage extends all the way back to the very first living things on Earth.

SUMMARY

1. *Sperm*, or male gametes, are specialized cells that can swim, penetrate the surface of an egg, and deliver the haploid male genome to the egg nucleus. The process of sperm production, *spermatogenesis*, takes place in the seminiferous tubules of the *testes*, where diploid *gonial cells* (spermatogonia) give rise to spermatocytes, which then divide meiotically to produce haploid spermatids. The mature sperm cell has a head containing an acrosome and a compact nucleus, and it has a flagellum for swimming.

2. *Ova* (eggs), or female gametes, are specialized cells that in some species contain various substances arranged so that a rough developmental pattern of the organism is already present at fertilization. Ova develop in the *ovaries* through the process of *oogenesis*. The gonial cells (oogonia) give rise to oocytes, which remain in a state of meiotic arrest while they enlarge and mature. Other cells—nurse cells, follicle cells, *oviduct* cells, and digestive glands—help the egg to meet its specialized tasks by manufacturing proteins and nucleic acids, producing *yolk*, and adding protective coatings. As the oocyte matures, the process of *gene amplification* generates extra copies of ribosomal genes and allows the oocyte to manufacture and store large numbers of ribosomes. Much mRNA also may be made and stored.

3. The process that unites egg and sperm is called *fertilization*. One step of fertilization is sperm penetration: the sperm head must pass through the several protective coatings of the egg. In the *acrosome reaction*, enzymes are released from the acrosome to digest the egg's protective layers, and a long, rigid filament is formed that stabs through the layers to the egg's plasma membrane. Fusion of the sperm and the egg plasma membranes triggers the *cortical reaction*, which comprises three events: temporary loss of negative electrical potential of the egg, activation of the egg cell, and elevation of the fertilization membrane. By the time the egg and sperm nuclei have come together, the egg's final meiotic reduction divisions will have taken place, so that when the two sets of chromosomes intermingle, a single diploid set will exist. Some species reproduce by *parthenogenesis*, a process in which unfertilized eggs develop normally, even including attainment of the normal chromosome number.

4. *Cleavage* divides the fertilized egg, or *zygote*, into increasing numbers of cells, frequently in a precise manner that distributes developmentally significant control factors to different lines of *blastomeres* (cells) in the *blastula*. Cleavage patterns vary with the amount of yolk present and according to the radial or spiral plan. The examples of the yellow and gray crescents in sea-squirt zygotes and of the polar lobes in snail zygotes show that certain developmental information derived from the mother's genes is distributed in precise but uneven ways during cleavage.

5. *Gastrulation* is the rearrangement of cell populations to generate the three-layered *gastrula*. The three germ layers present after gastrulation are the *ectoderm, endoderm,* and *mesoderm,* which give rise to all the tissues and organs of the body. The movement of cell populations during gastrulation is affected by the quantity of yolk in the embryo. Gastrulation cell movements commonly occur at the blastopore or at its equivalent in birds and mammals, the *primitive streak.*

6. *Organogenesis*, the development of the body's organs and tissues, is achieved through two processes: *morphogenesis*, the attainment of special shapes of cells or cell populations; and cellular *differentiation*, the functional maturation of cells. Morphogenesis depends on cell locomotion and change in shape. *Neurulation* is an example of morphogenesis and involves a flat sheet of cells forming a hollow tube, the brain and spinal cord of vertebrates. Differentiation involves not only acquiring a specialized function, but also developing responsiveness to regulatory signals.

7. As they develop, animal embryos are protected by a leathery or brittle outer covering. Reptile, bird, and mammal embryos have membranes that help provide nutrients, means of gas exchange, waste storage, and protection. These extraembryonic membranes include the *yolk sac*, the *chorion*, the *amnion*, and the *allantois*.

8. Growth results from mitotic activity of cells and accounts for the increase in mass of tissues, organs, and the whole organism. Rates of growth may be very high in the embryonic period and continue to be high until the organism reaches adulthood. Once full growth is achieved, mitosis remains coupled to the rate of cell loss or death. Some species have a larval stage following embryonic development. Larvae undergo *metamorphosis* into the adult body plan.

9. *Regeneration* can restore lost parts in some animal species. Cells usually *dedifferentiate* before they divide to regenerate new tissues or organs. *Compensatory hypertrophy* is growth of certain organs in reaction to loss of tissue of like type.

10. Aging is the time-dependent deterioration of many of the body's organs and tissues. Possible processes involved include degeneration of connective tissues, reduced mitotic ability, decline in the immune system, and accumulation of aging pigments.

KEY TERMS

acrosome reaction
allantois
amnion
blastomere
blastula
chorion
cleavage
compensatory hypertrophy
cortical reaction
dedifferentiation
differentiation
ectoderm
endoderm

fertilization
gastrula
gastrulation
gene amplification
gonial cell
mesoderm
metamorphosis
morphogenesis
neurulation
oogenesis
organogenesis
ovum
ovary

oviduct
parthenogenesis
primitive streak
regeneration
sperm
spermatogenesis
stem cell
testis
yolk
yolk sac
zygote

QUESTIONS

1. What is the function of a sperm cell? Draw one, indicate its size, label its structures, and describe their functions.

2. Do the same for an egg cell.

3. What fraction of the chromosomes in a fertilized egg is contributed by the sperm? by the egg? What fraction of the mitochondria is contributed by the sperm? by the egg? What fraction of the cytoplasm is contributed by the sperm? by the egg?

4. What is gene amplification? Give an example.

5. Describe the important events that take place at fertilization.

6. What is the significance of the yellow and gray crescents that appear in sea-squirt zygotes soon after fertilization?

7. From which cell layer (ectoderm, mesoderm, or endoderm) is each of the following tissues derived: hair, bones, nervous system, skin, muscles, outer layer of abdominal organs, germ cells, gut lining, connective tissue?

8. Explain how erythropoietin secretion in the kidney regulates the production of red blood cells. What triggers secretion of erythropoietin? Would you expect the inhabitants of a village high in the Andes to have more red blood cells than people who live near sea level? Why?

9. Describe the function of each membrane surrounding the embryo of a reptile, bird, and mammal.

10. What are stem cells?

ESSAY QUESTIONS

1. In Gregor Mendel's time, many scientists believed that offspring (of any species) inherited traits only from their fathers. The mother was considered a passive incubator. Mendel showed that both parents contribute equally. What is the origin of the chromosomes in a fertilized egg (zygote)? What is the origin of its mitochondria? Do mitochondria contain genes? What is meant by maternal inheritance? Is there such a thing as paternal inheritance?

2. If you wished to design a perfect birth-control agent, what processes or properties of sperm, eggs, or fertilization would be good targets for your agent?

3. Are germ cells immortal? Are mitochondria?

SUGGESTED READINGS

BALINSKY, B. I. *An Introduction to Embryology*. 6th ed. Philadelphia: Saunders, 1982.

This is the classic text, at least for descriptive material.

BROWDER, L. W. *Developmental Biology*. 2d ed. Philadelphia: Saunders, 1984.

This book is especially strong on the earlier phases of development and emphasizes the molecular basis of development.

EPEL, D. "The Program of Fertilization." *Scientific American* 237 (November 1977): 128–140.

One of the pioneering contributors to our knowledge of fertilization summarizes and illustrates the fascinating world of sperm and eggs.

17

DEVELOPMENTAL MECHANISMS AND DIFFERENTIATION

Among the vertebrates as a whole—fishes, amphibians, reptiles, and mammals—there is some variation in cell type, but the key to the different organization of all these forms does not lie in the cells as such; it lies in how these basic building units are arranged in space during development.

Lewis Wolpert,
"Pattern Formation in Biological Development,"
Scientific American (October 1978)

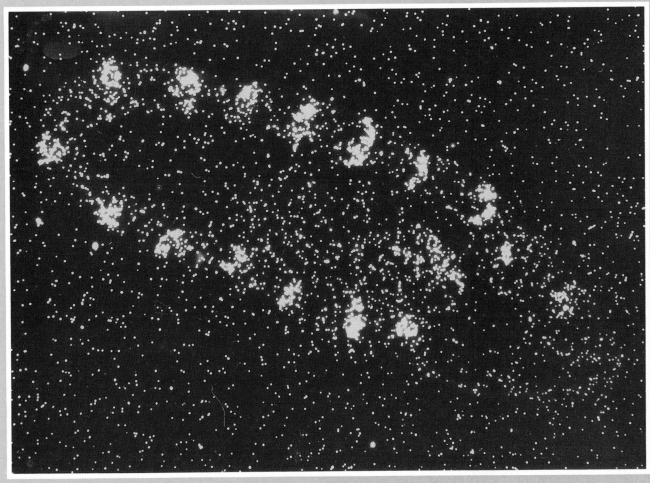

Sites of regulatory gene function in an early fruit fly embryo (magnified about 400 times).

The human hand has been called the most sophisticated tool on earth, and for good reason. It allows the performance of such intricate and highly delicate tasks as embroidery, diamond cutting, piano playing, eye surgery, and soldering of microcircuits. A detail of the statue of *David* by Michelangelo (Figure 17-1) serves as a symbol of the hand's perfection of function and of shape. The shape of the adult hand is no more extraordinary, however, than is the *development* of that appendage or of the limbs, feet, internal organs, and other parts of the body from a single fertilized egg. Part of this remarkable phenomenon can be seen in Figure 17-2, which shows the formation of the hand in the human embryo.

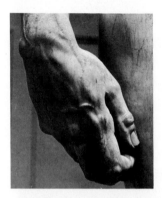

Figure 17-1   DAVID'S HAND, BY MICHELANGELO. Precisely coordinated developmental processes generate the many parts of a hand, all arrayed in correct relative positions and oriented to the body so that specific functions can be carried out.

(a)

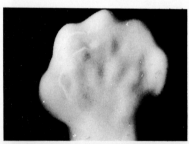

(b)

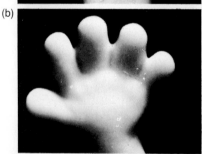

(c)

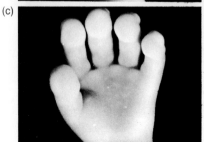

Figure 17-2   THE HUMAN HAND: FROM PADDLE TO SOPHISTICATED TOOL IN A FEW WEEKS. Gestational ages: (a) 5 weeks; (b) 6 weeks; (c) 7 to 8 weeks.

In Chapter 16, we learned about the processes that underlie the development of specialized body parts. We saw that sexual reproduction has an unvarying sequence that begins with one cell, proceeds by means of cleavage to a mass of cells, and develops specialized organs and tissues through cell rearrangement, morphogenesis, and differentiation. The question now is: What controls and coordinates this complex sequence of events? Although they are still far from having all the answers to this question, biologists have learned a great deal about the causal mechanisms of development. In this chapter, we shall look at how individual cells differentiate and begin to function and how cells and cell populations are integrated so that the shape, size, and orientation of each organ are correct. We shall also examine the possible mechanisms of gene regulation during development, since the genotype is the source of biological information passed from one generation to the next. The real business of development is the translation of genotype into phenotype.

DETERMINATION: COMMITMENT TO A TYPE OF DIFFERENTIATION

We saw in Chapter 16 that cells in embryos differentiate and participate in morphogenesis. Before doing so, most cell lines in the embryos of higher animals undergo a remarkable process: cells in such lines become *committed*, or *determined*, for one particular type of differentiation, such as to form bone, muscle, lung, or another specific differentiated cell type. Figure 17-3 illustrates the progressive commitment of several types of cell lines. **Determination** is the final selection in cells of a single developmental pathway from among several alternatives.

The timetable for determination varies from cell type to cell type. In some cases, it takes place soon after fertilization. For example, as soon as the pole plasm of a frog egg is incorporated into some blastomeres, these cells are determined as the germ line. (Pole plasm is composed of nucleic acid and protein and is built into certain sites of the oocyte during oogenesis.) Later, when those primordial germ cells and their progeny take up residence in the ovary or testis, they become the gonial cells, which will produce eggs or sperm. The pole plasm acts as a determinant in this case, just as the yellow- and gray-crescent materials and the polar lobe do in the sea squirt and the snail, as we saw in Chapter 16.

Most other cells undergo a more gradual process by which their options are reduced. The covering layer of

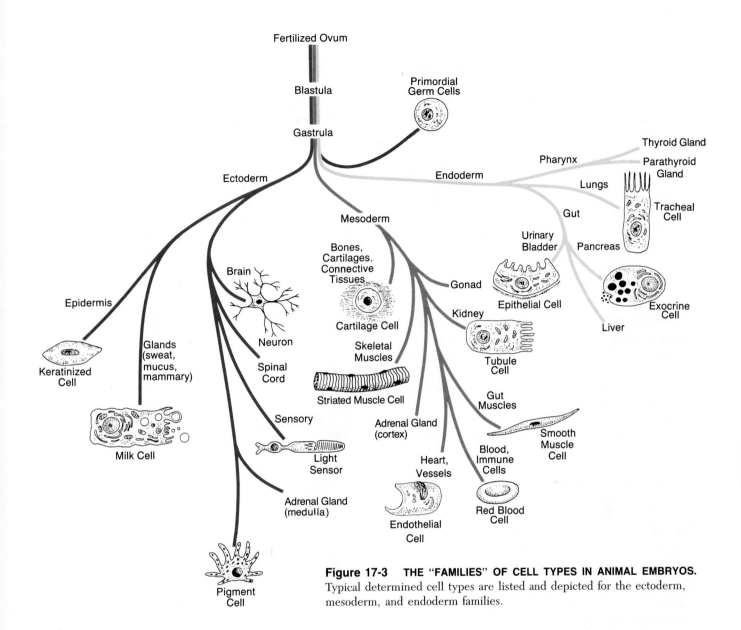

Figure 17-3   THE "FAMILIES" OF CELL TYPES IN ANIMAL EMBRYOS.
Typical determined cell types are listed and depicted for the ectoderm, mesoderm, and endoderm families.

the embryo, the ectoderm, is committed at an early stage to forming the ectodermal lineage, or family, of cell types. Later, different subpopulations of ectoderm become the brain, the lens of the eye, the sweat glands, and many other types of cells (Figure 17-3; review Table 16-1). In these cases, determination comes about through a series of steps: first as ectoderm, then as brain cells, and finally as one or another type of specific nerve cell used in seeing, moving, feeding, or some other brain function.

Once a subpopulation of cells is determined, the state is usually permanent. Take the example of one of the ectodermal subpopulations—the pigment cells from the eye of a mouse or chick embryo a few days before birth or hatching. These cells are already determined and have differentiated, so that they are black with pigment.

As is the case with most fully differentiated cells, they divide within the organism only at rare intervals to compensate for cell death. If the pigment cells are removed from the embryo, separated from one another, and placed in a dish of nutrient medium, they may be induced to divide again and again, month after month, year after year. During this time, they remain determined, but they dedifferentiate; they are colorless, since their rapid division in the nutrient medium leaves no time for producing the black pigment. If the cells are eventually allowed to cease their rapid division and to differentiate once more, what do they do? They once again produce pigment granules. Why? The determined state of pigment cells has remained stable through these hundreds or thousands of divisions, even though the differentiated state, with its black pigment phenotype, has not been expressed. The same holds true for other cell types: the determined state is extraordinarily stable and usually does not change as a result of mitosis. Because this state is passed from one cell to its progeny during mitosis, it is clear that the determined condition is an *inherited* state of gene control.

The stability of the heritable, determined condition is an important cellular property. It is one reason that cells of long-lived organisms remain true to type. Imagine the chaos if your skin suddenly started to form eye lenses or your liver started to produce lung cells. Despite, the normal permanence of determination, however, changes or mistakes in cell type can occur.

Changes in Determination

In Chapter 16, we saw that when the lens is removed from the eye of a salamander, the nearby pigmented iris cells dedifferentiate, divide, and give rise to a population that differentiates anew as lens cells. In this case, the pigmented iris state *changes* to the lens state, which shows that determination can be modified. Recall, however, that a period of mitosis intervenes between the two states: it is the progeny of the iris cells that become lens cells—direct transformation from one determined state to another rarely, if ever, occurs within one cell generation.

Fruit flies provide another good example of change in a population of determined cells. Inside fruit-fly larvae are little sacs of cells called **imaginal disks.** ("Imaginal" comes from the word "imago," which means "an adult

insect"; hence, these are disks that eventually mature into the adult parts.) Each disk is composed of cells determined to become a leg, an eye, a wing, a genital apparatus, or some other part of the adult fly. The late Swiss developmental biologist Ernst Hadorn investigated the determined state of the imaginal disks by causing their cells to divide again and again. To do so, he used the

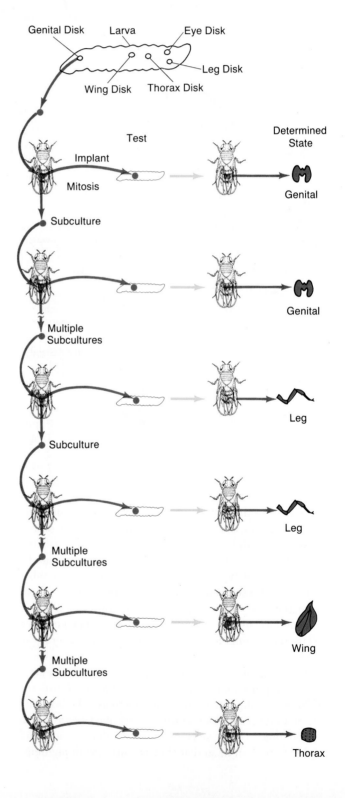

Figure 17-4 TRANSDETERMINATION SEQUENCES.
A tiny fragment of a genital imaginal disk is removed from a fruit-fly larva and is implanted in the body cavity of an adult female. The cavity serves as a kind of culture vessel, and the fragment enlarges greatly. A tiny piece is "tested" by implanting it in another larva, which ultimately gives rise to an adult fly, in whose abdomen is the differentiated tissue derived from the implant. Another piece of the large imaginal-disk fragment from the fruit-fly abdomen is subcultured in another adult female. After the second fragment has grown large, a piece can be subcultured in still a third adult female—and so on, for serial culture. At any time, the subcultures may be "tested" for their determined state by implanting fragments in fruit-fly larvae. If the original genital-disk cells divide enough times during a series of the subcultures, the disk cellular progeny transdetermine into alternative structures. The sequence is not random: the first "mistake" is to form leg or antenna; from that state, the next change is to form wing; and that, in turn, can transdetermine to form thorax. The reason that transdetermination occurs in a nonrandom sequence of cell types is not understood.

abdominal cavities of a series of adult flies as "culture chambers" (Figure 17-4). After serial transplantations of genital disk cells in abdominal cavities of the adult flies, he was surprised to find that once a sufficient number of divisions had taken place among the progeny of the determined cells of the genital disk, these cells suddenly gave rise to *leg* tissue. The disk cells in such a case are said to be **transdetermined;** that is, they shift from one state of determination (genital) to another (leg). The new determined state proved stable for many more months of culture and rounds of mitosis. But once again, if divisions continued long enough, another transdetermination occurred, and in place of leg tissue, *wing* tissue formed. The reasons for transdetermination are still not understood. But related experiments on vertebrate embryos tell us more about determination.

The Stability of Determination Depends on the Cytoplasm

Robert Briggs and Thomas King at Indiana University used a different approach—**nuclear transplantation**—to study determination. They removed the diploid nucleus from a frog zygote and injected a nucleus from one of various types of embryonic cells into the zygote to see what would happen. If the donor nucleus came from a frog blastula cell, the injected zygote usually developed normally, even to the tadpole or adult frog stage. This result indicates that even though the frog blastula cells are partially determined, they still carry in their nuclei a full set of genes capable of directing complete development.

John Gurdon used similar procedures on zygotes of *Xenopus*, the African clawed frog. He found that nuclei from fully differentiated gut cells and even from adult cell types could direct development of a zygote to tadpole or adult stages in which every normal cell type was present. In fact, these nuclei could be used as donors to establish clones of frogs, all derived from the initial nuclear transplant. These results suggest that in a determined and differentiated cell, all the genes required for other determined states and for modes of differentiation are still present.

These nuclear-transplantation experiments have another important message: when the nucleus of a determined cell is removed and exposed to the cytoplasm of another cell type, the original state of determination is lost. Thus although determination is an exceedingly stable condition, it is dependent on the cytoplasm of the determined cell type; exposure to a different cytoplasm (such as that of the zygote) "deprograms" the nucleus and permits it to participate in the entire developmental process.

Biologists still do not know the basis for determina-

tion, transdetermination, or cytoplasmic influences on determination. Nevertheless, research proves beyond doubt that determined cells retain the genes required for other types of differentiation, although under most circumstances, these genes are never again used. Fortunately for adult fruit flies, African clawed frogs, humans, and other organisms, tissues and organs are stable and can function normally through the span of adult life.

DIFFERENTIATION: BUILDING CELL PHENOTYPE

Simply being determined does not make a cell functional in an embryo or in an adult. Determined cells must undergo **differentiation,** or maturation. First we will look at what triggers differentiation; then we will describe the unique characteristics of differentiated cells.

The Causes of Differentiation

In 1921, one of the great experiments of biology identified a major developmental strategy for the control of cell differentiation and organogenesis. This mechanism is **tissue interaction,** which its discoverers called "induction." The German biologists Hans Spemann and Hilde Mangold showed that the prospective mesodermal cells, which move to the inside of the embryo during gastrulation, have remarkable properties. Spemann and Mangold transplanted a mass of cells from the dorsal-lip area of an amphibian blastopore to the surface of another gastrula, as depicted in Figure 17-5. Although the results varied with the details of specific experiments, the researchers found that a second embryonic axis formed at the transplantation site; that is, a duplicate brain and spinal cord developed on the belly of the host embryo. Spemann and Mangold concluded that the implanted dorsal-lip cells *induced*, or directed, the surrounding host ectoderm to form the new brain and spinal-cord axis. In other words, those ectodermal cells became determined to form additional nerve or other cell types that they normally would not have formed. Spemann called the dorsal-lip tissue an *organizer*, since it appeared to organize a set of axial organs (brain, spinal cord, vertebral column) in a host embryo.

Subsequent experiments yielded similar results in birds and mammals. Mesodermal cells that migrate through the primitive streak of a bird or mammal embryo can be transplanted beneath ectoderm that normally gives rise to skin. Within a few hours, a brain or

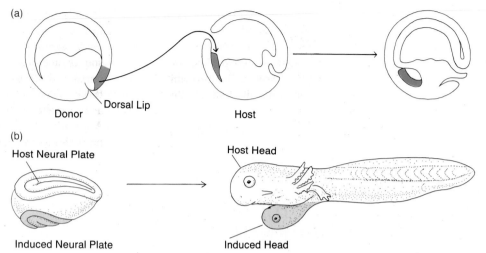

Figure 17-5   THE ORGANIZER-INDUCTION EXPERIMENT.
(a) Tissue from the dorsal lip of the forming blastopore is grafted from a donor embryo into the blastocoele of a host. Gastrulation proceeds, and the implant is pressed closely against the host's belly ectoderm.
(b) The ectoderm responds to the inducing tissue by forming a brain, and, indeed, the entire head and anterior trunk may develop.

spinal cord begins to form from the nearby ectodermal cells in the embryo. Thus Spemann's organizer activity operates in all land vertebrates.

Six decades of research have failed to show unequivocally how the organizer-inducing mesoderm acts on ectoderm. Biologists have implicated nucleic acids, proteins, and mechanical effects, among others, but no consensus has emerged. Nevertheless, it is absolutely clear that tissue interaction must occur in a developing embryo, or the brain and spinal cord will not form. Furthermore, Spemann's Nobel Prize-winning experiment spawned many others demonstrating the essential role of inductive-tissue interactions during the development of almost every complex organ in animal embryos.

A prime example of inductive-tissue interaction occurs in the lenses of the eye, which develop only where hollow outpouchings of the embryonic brain—the *optic vesicles*—contact the head ectoderm, as shown in Figure 17-6. If an optic vesicle is surgically removed from the developing embryo, the lens usually fails to form on the side of the head from which the vesicle was removed. Furthermore, if that optic vesicle is transplanted so that it contacts ectoderm on the top of the embryo's head, the ectoderm at that spot forms a lens. Ectodermal cells become determined as lenses and employ genes that they normally would not use. It is as though the tissue interaction includes a process of instructing ectoderm to form a lens instead of following its normal fate.

Similarly, mesoderm commonly acts on an overlying ectoderm or underlying endoderm to support development of hair, mammary glands, lungs, thyroid, pancreas, teeth, and virtually every other organ of a vertebrate's body. Mesoderm is an inducer in these cases because the organ in question does not form if mesoderm is not present. This is another example wherein genetic information alone is not sufficient to achieve normal devel-

opment. These inductive-tissue interactions, like cytoplasmic determinants (Chapter 16), comprise the developmental information that controls determination, differentiation, and, as we will see, morphogenesis.

Inductive-tissue interactions also ensure proper spatial relationships among developing organs so that a viable body organization is constructed during development. If eye and lens were to develop independently of each other, for example, great precision in organ formation would be required for them to align in order that they could function properly in the adult. However, the fact that a lens forms only where the optic vesicle happens to contact head ectoderm guarantees that the two organs will be aligned correctly. In fact, a broad area of the head ectoderm constitutes the *lens field*, an area in which all cells have the ability to respond to the inductive stimulus of the optic vesicle; it is as though the optic vesicle has a broad target, and a "hit" anywhere in it generates the lens. Thus the embryo (and resultant adult) is assured of precise alignment of lens and eye. Induction as a control process therefore allows for a certain amount of developmental "noise"—a degree of imprecision of the sort one might expect in any living system.

Although biologists are still investigating the chemical nature of most tissue interactions, they have discovered some molecular signals that influence differentiation and morphogenesis. Examples of these substances include the protein hormone *prolactin*, which acts on the mammary gland to cause milk synthesis, and the protein erythropoeitin, which can stimulate bone-marrow stem cells to produce red blood cells (Chapter 16). Both agents can induce differentiation and sometimes morphogenesis in specific tissue cells; one day, we may learn what causes determination earlier in the histories of these same cell lines.

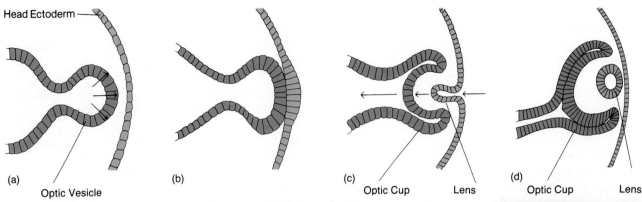

Head Ectoderm

(a)   Optic Vesicle   (b)   (c)   Optic Cup   Lens   (d)   Optic Cup   Lens

(e)

Figure 17-6   INDUCTIVE-TISSUE INTERACTION IN LENS FORMATION.
(a) A lens forms from the embryo's head ectoderm at the site where the optic vesicle and the ectoderm come into close contact as in (b). (c) Then, both the optic cup and the lens sink inward. (d,e) The hollow lens separates from the overlying ectoderm, and a space opens between the lens and the layers of the optic cup, as both the diagram and the scanning electron micrograph (magnified about 140 times) show.

Characteristics of Differentiated Cells

Although the precise signals leading to differentiation are not fully understood, one thing is certain: once a cell begins to differentiate, it takes on specific characteristics that enable it to perform its particular functions. Let us examine these characteristics.

1. *A differentiated cell makes and uses a specific set of proteins that enables it to carry out its functions.* It is this spectrum of proteins that really defines the differentiated condition of a cell type. Some proteins carry out function, whereas others govern shape—both of which are basic features of a differentiated cell. Proteins also serve as enzymes and govern the metabolism of cells. Red blood cells are an example of cells that require specific proteins to function. The red blood cell is much more than a bag of hemoglobin (the protein that carries oxygen; Chapter 28). This disk-shaped cell must have special enzymes to maximize its ability to carry carbon dioxide and to manufacture molecules that control the binding of oxygen to hemoglobin. Furthermore, the red blood cell has a characteristic cytoskeleton that affects the flexibility of the cell and thus the way it behaves when moving through tiny blood vessels. Characteristic recognition molecules, sugar–protein complexes, lie on the red blood cell's surface (and identify the cell's blood group: A, B, O, M, or N). So red cells exemplify the fact that all differentiated cell types possess specific proteins (Table 17-1).

2. *A differentiated cell is metabolically active.* To maintain metabolic activity, cells usually have "house-

Table 17-1   **MAJOR SPECIFIC PROTEINS OF DIFFERENTIATED CELLS**

Cell	Protein
Red blood cell	Hemoglobin (oxygen carrier)
Pancreatic B cell	Insulin (hormone)
Pancreatic exocrine cell	Trypsin (digestive enzyme) Lipase (digestive enzyme) Amylase (digestive enzyme)
Thyroid cell	Thyroglobulin (hormone storage)
Epidermal cell	Keratin (structural protein)
Dermal fibroblast	Collagen (structural protein)
Muscle cell	Actin (contractile protein) Myosin (contractile protein) Troponin (contractile protein)
Pigment cell	Tyrosinase (pigment enzyme)
Immune B cell	Immunoglobulin (antibody)
Oviduct cell	Ovalbumin (egg-white protein)

keeping" proteins in addition to specialized proteins. These essential proteins include enzymes that help maintain general cell activity (such as glycolytic enzymes; Chapter 7), as well as structural proteins like tubulin.

3. *As it matures, a differentiating cell assumes a char-*

acteristic shape that enables it to function effectively in the tissue to which it belongs and in the organism as a whole. As we saw in earlier chapters, proteins help determine cell shape. Obvious examples are tubulin, the building block of microtubules, and actin and myosin, which form some of the filament systems of the cytoplasm (Chapter 6). New evidence suggests that the tubulins, actins, and other cytoskeletal proteins may differ slightly from cell type to cell type. In other words, there is a family of actin genes: one actin gene may be used in a sperm cell to produce the actin of the acrosomal filament; another may be used in a muscle cell; still another, in a nerve cell. Researchers do not yet know whether actual differences in cell shape can be partly explained by differences of only a few amino acids in cytoskeletal proteins. It is clear, however, that the arrangement of these cytoskeletal components, in combination with the ways in which cells are glued to their neighbors, does control cell shape.

Why is the shape of a differentiated cell important? The answer lies in the link between form and function that we encounter so often in biology. Notice in Figure 17-7 how the cluster of secretory cells in a mammalian pancreas is oriented around a hollow cavity, into which the digestive enzymes are released. The shape and arrangement of a secretory cell allow the enzymes to be secreted only at one end of the cell; from there, the released enzymes are channeled toward the intestine, where they participate in digesting food. The important thing to remember is that the changes in shape and

alignment of individual differentiated cells and of cell populations are essential if the mature cells are to play their appropriate role in the activity of the tissue, and so of the whole organism.

4. *A differentiated cell often ceases to undergo further cell division.* Thus differentiation is frequently a terminal condition. Red blood cells, skeletal-muscle cells, and nerve cells are examples of differentiated cells that have completely lost the capacity for mitosis; once matured, they cannot divide into daughter cells. Many other kinds of differentiated cells also remain in a postmitotic state until they die. They are then replaced by newly formed and differentiating cells.

Why do many types of mature cells divide infrequently at most? One answer may be that cells whose function depends on highly ordered cytoplasm or on special shapes may divide only on rare occasions, thereby minimizing the disruption of cellular and tissue architecture caused by mitosis. Imagine the chaos in the brain if the billions of individual cells, each with hundreds or thousands of connections, were to withdraw their nerve-cell processes in order to round up to divide; their progeny would have to reestablish connections, an extremely complicated process.

The great bulk of specific proteins in blood, muscle, and nerve cells—as well as the proteins in epidermis, the immune system, and other cell types—are manufactured *after* the cell has ceased to divide. Thus a general strategy in multicellular organisms is that (1) a determined cell divides, (2) its progeny then differentiate,

Figure 17-7   CELL SHAPE AND ARRANGEMENT IN RELATION TO TISSUE FUNCTION.
Clusters of secretory cells near ducts in the pancreas and other secretory glands facilitate the release of secretory proteins. (a) A section through a cluster of secretory cells and the duct to which they are attached. (b) The large black spots in the electron micrograph (magnified about 13,000 times) are secretory granules packed with digestive enzymes destined for secretion. The enzymes were made on the cell's rough endoplasmic reticulum and packaged in the Golgi apparatus into the granules.

(a)

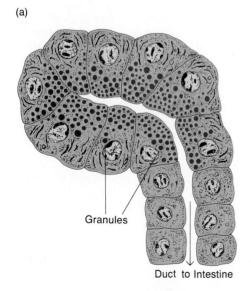

Granules

Duct to Intestine

(b)

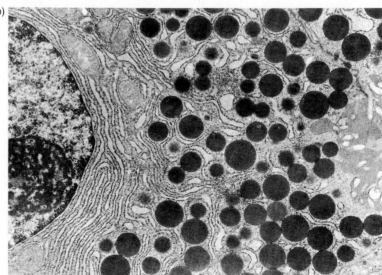

Table 17-2 **AVERAGE LIFE SPANS OF HUMAN DIFFERENTIATED CELL TYPES**

Cell Type	Life Span (days)
Intestinal lining	1.3
Stomach lining	2.9
Tongue surface	3.5
Cervix	5.7
Stomach mucus	6.4
Cornea	7
Epidermis—abdomen	7
Epidermis—cheek	10
Lung alveolus	21
Lung bronchus	167
Kidney	170
Bladder lining	333
Liver	450
Adrenal cortex	750
Brain nerve	27,375$^+$ (75$^+$ years)

forming a large number of protein molecules, and (3) these cells function for days, weeks, months, or years, depending on the kind of cell and the species of organism of which it is a part.

The average survival times of some human cell types are shown in Table 17-2. The range is enormous—from 1.3 days in the intestine to more than a year in the liver and a lifetime in the brain. It is remarkable that mitosis in each of these organs and cell types is closely coupled to the rate of cell loss. Mitosis is most frequent in intestinal and skin cells, rare in liver cells, and nonexistent in nerve cells. This linking of division rate to the life span of cells is indeed a special strategy that is one of the most important means of ensuring the stability of cell types in multicellular animals. Without coordination between mitosis and cell loss, organs and tissues could vary willy-nilly in shape, size, and function.

Nevertheless, there is a price to be paid for complete nondivision in a cell population: gradual depletion of cell populations. A certain amount of cell death is apparently unavoidable. Cell death without replacement in the nervous system helps account for the slower reaction time, lower learning capacity, and general decrease in nervous-system functioning in aging humans, whether they are farmers, physicians, ballet dancers, politicians, or professors.

MORPHOGENESIS: THE ORGANIZATION OF CELLS INTO FUNCTIONAL UNITS

Every functioning organ and differentiated cell has a characteristic shape. We saw earlier, for example, how the shape and arrangement of secretory cells in the pancreas allow them to secrete into a duct system. The exciting advances of cell biology in the 1970s revealed the underlying mechanisms responsible for the shapes and arrangements of embryonic cells. The many diverse types of morphogenesis in embryos are produced by just a few different mechanisms: (1) the movement of single cells, (2) the interaction between moving or stationary cells and extracellular substances, (3) the movement of cell populations, (4) localized relative growth, (5) localized cell death, and (6) the deposition of massive amounts of extracellular materials. Let us look at each phenomenon in more detail.

Single-Cell Movements

Individual cells may migrate from one site to another in embryos and thus set up populations that will form organs. The best studied case of cell migration in embryos involves *neural-crest cells*, which arise near the site where the neural tissue closes into a tube in vertebrate embryos (Chapter 16). Figure 17-8a shows the regular pathways that neural-crest cells follow into the head and face, downward to the gut, outward to the skin, and so on. These cells take up residence in a variety of locations, come under the influence of local controlling environments, and proceed to develop into a variety of tissues, such as the adrenal gland, pigment cells of the skin, most of the peripheral nervous system (sensory receptors and nerves that control the internal organs), and much of the face and head.

The case of the neural-crest cells shows us that an important component of morphogenesis is cell locomotion and guidance in embryos. But how do the cells "know" where to travel and by what route? Beginning to move, guiding cell movement, and stopping in the correct location are functions affected by the *adhesion* of cells to one another and to the extracellular environment. Experiments reveal that cells of a given type tend to adhere better to one another than to other cell types. For example, when pigment and cartilage cells are randomly intermingled, they sort out into separate groups of like cells: the pigment cells stick to other pigment cells; the cartilage, to cartilage. This implies that cells have recognition markers on their surfaces—molecular name tags, so to speak—that may help keep cells of similar type glued together in the body's tissues.

Interaction Between Moving and Stationary Cells and Extracellular Substances

The locomotion of single cells is also affected by huge extracellular polymers composed of proteins and sugars.

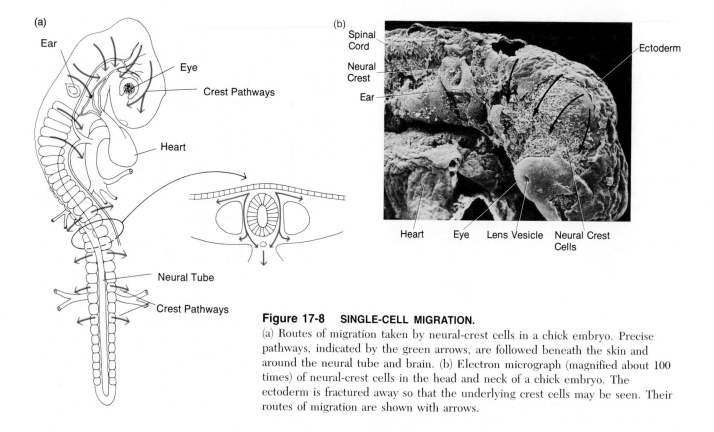

Figure 17-8   SINGLE-CELL MIGRATION.
(a) Routes of migration taken by neural-crest cells in a chick embryo. Precise pathways, indicated by the green arrows, are followed beneath the skin and around the neural tube and brain. (b) Electron micrograph (magnified about 100 times) of neural-crest cells in the head and neck of a chick embryo. The ectoderm is fractured away so that the underlying crest cells may be seen. Their routes of migration are shown with arrows.

Some of these substances—fibronectin, laminin, collagen, and hyaluronic acid—provide a substratum, or ground substance, to which cells can stick and on which they can move as a slug moves on a trail of mucus.

A good example of this type of movement is seen in the developing cornea, the outermost portion of the eyeball. Neural-crest cells migrate to the very edge of the cornea but do not enter it. Then a layer of corneal cells manufactures and secretes hyaluronic acid. Immediately, the neural-crest cells migrate into the hyaluronic-acid-filled matrix of the cornea, as shown in Figure 17-9. Then, surprisingly, the enzyme hyaluronidase is secreted; that enzyme digests the hyaluronic acid, and the neural-crest cells are left stranded in the corneal matrix, where they proceed to differentiate into connective-tissue cells.

The development of the cornea and other organs demonstrates that cell migration into or out of an area is controlled by the presence or absence of essential extracellular molecules. On a grander scale, controlled production, secretion, deposition, and destruction of such molecules can be the key to coordinating the movements of particular cell types throughout the embryo.

The role of extracellular substances in morphogenesis is not just to enable cell movement, however. In the process of interaction between moving cells and these substances, there is another critical outcome: the application of mechanical forces in the embryo. To under-

stand the nature of such forces, one must realize that the extracellular materials in the living embryo are not solid; instead, they comprise a deformable meshwork of fibers (collagens, fibronectins) and gel (hyaluronic acid) plus feltlike sheets of epithelial basement membranes (the sheets to which most epithelial layers attach). When a cell moves, it adheres to this substratum of substances and pushes backward so that it can move forward—just as the wheels of a dirt bike push backward and deform sand as the bike moves ahead. In this way, embryonic cells exert tension or *traction* on their substratum and can actually deform the network of fibers into parallel arrays. In fact, a stationary cell may use its motile machinery solely to deform the meshwork on which it sits. The biologists Albert Harris and David Stopak have shown that such *traction-induced* morphogenesis may account for the development of aligned collagen bundles, such as those in tendons and ligaments. In turn, individual muscles usually develop along tracts of such aligned fibers because the muscle-forming cells position themselves parallel to the oriented fiber bundles.

The traction effect may be even more important than cell locomotion itself; this is because most mesodermal cells do not move very much in intact embryos, but they do exert traction, orient the substratum, and so establish tendons, ligaments, muscles, and oriented "roadways" upon which elongating blood vessels and other motile populations may move. A simple action at the cell level—

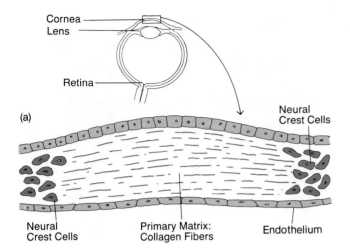

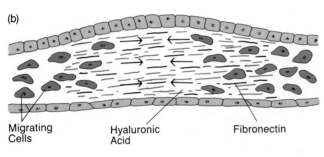

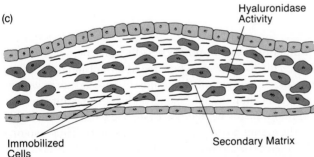

Figure 17-9 THE ENVIRONMENTAL CONTROL OF CELL MIGRATION.
(a) Neural-crest cells fail to enter the primary matrix in the cornea, which is composed mainly of collagen fibers. (b) After hyaluronic acid is secreted by the endothelium, the cells migrate in (fibronectin also appears). (c) The enzyme hyaluronidase is secreted, the hyaluronic acid is degraded, and the neural-crest cells remain trapped in the secondary matrix.

sticking to and pulling on the substratum—thus can bring about important mechanical events and morphogenetic ordering in the embryonic body.

Cell-Population Movements

Entire populations of cells, such as epithelial sheets, may fold inward or outward during the formation of lungs, kidneys, eyes, brain, and other organs. We encountered this type of folding in our descriptions of optic and lens vesicle formation and neural-tube development (Chapter 16). These foldings are caused by the activity of cellular cytoskeletal organelles, microtubules, and microfilaments, as well as by adhesion.

Microtubules may cause a cell to elongate; microfilaments may narrow one end of a cell. Figure 17-10 shows how such action may change the shape of a sheet of cells. For example, if cells are cemented together on their lateral surfaces by junctions (Chapter 6), the narrowing of individual cells causes the entire population to change shape so that the sheet buckles or bulges. Interfering with the function of microtubules or microfilaments by administering certain drugs prevents the normal changes in shape of individual cells; as a consequence, the morphogenesis of the tissue and organ is halted.

One of the major discoveries of modern cell biology is that just a few cytoskeletal organelles are employed in a variety of ways to control and alter cell shape and movement. It seems very likely that microtubules, microfilaments, and cell-to-cell adhesion are instrumental in the morphogenesis of almost all organs in multicellular animals. No doubt, many mutations that alter the shapes of organs or organisms during evolution actually operate at the level of these organelles.

Localized Relative Growth

Relative growth of specific groups of cells is another mechanism that underlies morphogenesis. Theoretically, this growth could be achieved in two ways: by increasing mitosis at a specific site or by decreasing mitosis around the site. Both possibilities do, in fact, occur naturally in embryos. In the earliest stages of vertebrate limb formation, mitosis slows in the surrounding body wall, while higher rates are maintained at the site of arm or leg formation. The alternative—localized sites with high mitotic rates—occurs in the branching mammalian lung. However, very little is known about how the increase or decrease in local mitotic rate is controlled. This is an area that is ripe for investigation, not only because of its role in normal development, but also because the mechanisms that control normal mitosis are likely to be among those that go awry when cancer cells begin to grow at unchecked rates.

Localized Cell Death

Ducks have webbing between their toes; chickens and humans do not. The reason is that massive **cell death** occurs in the mesoderm between the toe bones of the chick and human embryos; the overlying skin ectoderm

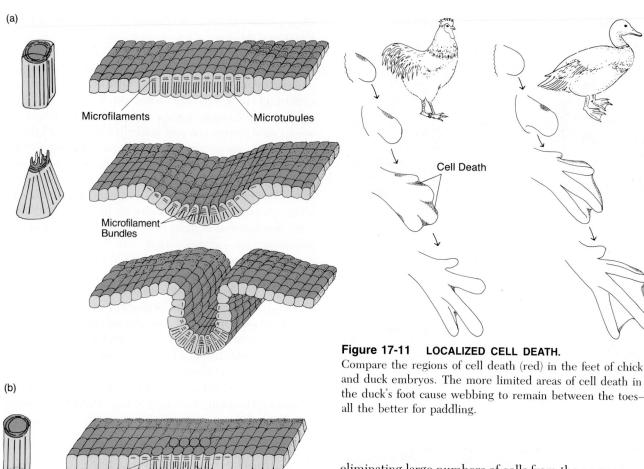

Figure 17-11   LOCALIZED CELL DEATH.
Compare the regions of cell death (red) in the feet of chick and duck embryos. The more limited areas of cell death in the duck's foot cause webbing to remain between the toes—all the better for paddling.

Figure 17-10   A BASIC PROCESS OF MORPHOGENESIS: NARROWING OF INDIVIDUAL CELLS.
(a) Cells may become wedge-shaped and so contribute to a sheet forming a groove; this occurs as the central nervous system forms in vertebrates. (b) Cells may become conical and thus cause a pit to form; this occurs when the lens of the eye pinches off from the head ectoderm.

therefore sinks inward between the toes (or fingers), as shown in Figure 17-11. Because comparable cell death does not occur in ducks, the mesoderm and overlying ectoderm form webs. The mesodermal cells occasionally do not die in human and other vertebrate embryos; such individuals are born with skin stretching between the fingers or between the toes.

Cell death is a mechanism for sculpting digits, shaping the shoulder, separating bones from one another, and

eliminating large numbers of cells from the nervous system. This may seem to be a peculiar method of building organs, since tens of thousands of cells are first generated and then killed. Nevertheless, the strategy is a common means of achieving morphogenesis in such different organisms as insects and mammals. Thus two options—a positive one (mitosis) and a negative one (cell death)—are available as means of shaping organs and regions of the body.

Deposition of Extracellular Matrix

Massive quantities of **extracellular matrix** (Figure 17-12) contribute to the shape of the bones, cartilage, tendons, corneas, cavities of the eyes, and embryonic heart, among others. Tough filamentous molecules such as collagen or huge space-filling sugar polymers—*glycosaminoglycans*—that absorb water are the main components of this matrix. Embryonic cells synthesize and secrete these molecules; by doing so, they build the matrix that surrounds cells in the bones, corneas, and so on.

The six basic processes we have just described are employed, to varying extents, in the morphogenesis, or shaping, of each organ and tissue. Cell movement and traction may be particularly crucial in the development of one organ; cell death, in that of another. Mutations

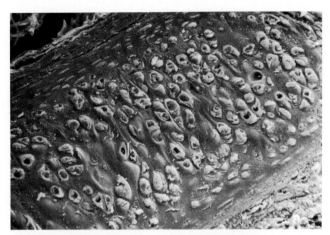

Figure 17-12   EXTRACELLULAR MATERIALS.
This electron micrograph (magnified about 300 times) shows the matrix of cartilage that surrounds and separates these individual cartilage cells from one another. Control of the amount and the position of the extracellular matrix in this embryonic bone helps govern the length, width, and thickness of the bone. From *Tissues and Organs: A Text-Atlas of Scanning Electron Microscopy* by Richard G. Kessel and Randy H. Kardon, W. H. Freeman and Company © 1979.

operating on the basic mechanisms of morphogenesis are the source of new shapes in organs or organisms that are recorded in the evolutionary history of plants and animals on Earth. Thus in some ancestral bird egg, the failure of the morphogenetic mechanism of cell death to fully separate the toes of the embryo could have led to webbing. Webbing, in turn, allowed better paddling by adult aquatic birds and their progeny, enabling them to survive by exploiting a life that included feeding in water. Similar arguments may be made for the role of the other morphogenetic mechanisms in relation to evolution.

PATTERN FORMATION: THE VERTEBRATE LIMB

Having considered the mechanisms of morphogenesis, let us see how they are integrated to yield a complex organ or the basic body plan of an entire organism. **Pattern formation** is the gradual emergence of the body plan: one head; two, four, six, eight, or more legs, depending on the species; dorsal (upper), ventral (lower), anterior (head end), posterior (tail end), right, and left sides, again depending on the species; and placement of all the organs and their positions relative to the whole. Consider the consequences, for instance, if hands and feet developed in random orientation on the ends of limbs, or if one arm protruded from your chest and the

other extended from your back. Fortunately, such morphological disasters almost never happen, and we shall soon see why.

In Chapter 16, we discussed the importance of the *position* of control factors, such as the yellow and gray crescents and the polar-lobe contents in the zygote. Precisely oriented cleavage divisions segregate these factors into certain blastomeres but not others, and cells that receive the control factors differentiate because of the presence of these factors. Those various cell groups are arrayed so that the tissues and organs they subsequently form will be in the correct positions relative to one another in the embryo and adult.

A related process affecting placement may occur in the first minute after the sperm fertilizes the egg: establishment of the body's three axes of symmetry. (An axis of symmetry is a line or plane connecting two regions.) These axes are the *anterior–posterior* (head–tail), *dorsal–ventral* (top–bottom), and *right–left*. To be more specific about how axes affect organ placement and orientation, let us investigate the development of the forelimb, a complex organ with its own axes.

The first forelimb axis to consider is proximal–distal (proximal is near the main body; distal away from it): upper arm, forearm, wrist, and hand, with palm, thumb, and digits. For ease of understanding, we will refer to parts of the human arm, even though most experimental work has been done on chicken-embryo wings and mouse-embryo forelegs. The different segments of the forelimb arise in an embryo in a sequence starting from the shoulder and progressing toward the digits. Lewis Wolpert, a leading British biologist, has proposed that the parts of the limb arise from the **progress zone,** a special set of mesodermal cells located just beneath the tip of the elongating limb bud. Biologists think that the first group of cells left behind the progress zone as it is moved passively outward during limb growth is assigned a "positional value" that may be called "upper arm." The next set to arise is assigned the positional value of forearm; the next is the wrist; and so on. Figure 17-13 shows this progression in a bird wing; analogous processes go on in mammals, including during the development of the human arm. It is in this way that the correct proximal–distal sequence of arm parts is ensured. Once a population of cells has assumed a positional value—say, upper arm—other processes cause the appropriate bones, muscles, tendons, and so on to develop.

What about the anterior–posterior (thumb–little finger) axis? At the posterior junction between the limb bud and the body wall is a region called the **zone of polarizing activity (ZPA).** The most posterior digit (the little finger, or fifth digit) forms near the ZPA. Anterior to this, middle digits develop. Finally, at the most anterior position, the thumb (or first digit in other vertebrates) forms.

Imagine what would happen if an extra ZPA were

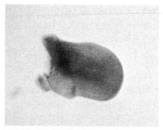

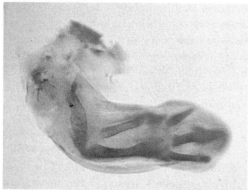

Figure 17-13   LIMB DEVELOPMENT IN A BIRD.
The progress zone is located just beneath the tip of the bud, shown at the top. Already at that stage, the positional values for the upper wing and the forewing have been assigned. Immense growth of the wing occurs as the various bones, muscles, and other components of the wing differentiate.

would develop. Clearly, "posterior" can be defined as *near the ZPA.*

Now that we have seen how the proximal–distal and the anterior–posterior axes are controlled, let us consider control of the dorsal–ventral (palm side–back of the hand side) axis. Does the mesoderm (the tissue that forms the limb's bone, muscle, and blood vessels) control this axis, or does the ectoderm (the source of the epidermis, skin glands, and so on)? One way to find out is to treat an early limb bud with certain digestive enzymes (trypsin) so that the ectodermal covering can be removed from the mesodermal core. The ectoderm then can be rotated 180° and rapidly put back in contact with the mesoderm. The tissues heal together in a few minutes. When this experimental limb grows out, the palm

Figure 17-14   THE ZPA AND DIGIT REVERSAL.
(a) ZPA tissue normally grows at the posterior part of the limb bud, where the bud meets the body wall. If extra ZPA tissue is grafted to sites A, B, and C of a chick embryo, it induces the formation of extra digits (whose numbers are shown to the right). These digits always are in inverted order, since the posterior-most digit is nearest to the ZPA. (b) An adult woman's hand on which the anterior, extra sets of digits formed instead of a thumb. Their order is reversed, just as though a ZPA activity acted as at site A in the chick experiment.

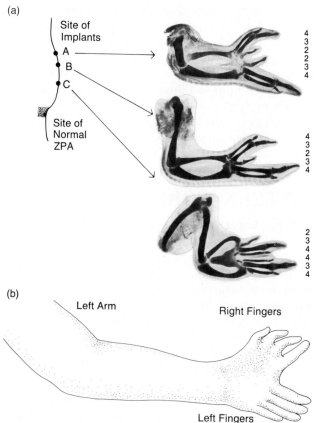

grafted near the front of a young limb bud. As you might guess, extra digits would form, in reverse order, as shown in Figure 17-14. Thus the digits of the whole limb might have the sequence 543345—5 being the little finger and 3, the middle finger. The reason is that the implanted ZPA causes the digits that form near it to be posterior in type (5 and 4), while the host ZPA continues its normal action. So there would be two "posteriors" for this limb, and the duplicated and reversed set of digits faces up, the digits curve up, and the back of the hand faces down. The experimental procedure of rotating the embryonic ectoderm reverses the dorsal–ventral orientation of the limb. Thus biologists assume that in a normal limb bud, the ectoderm imposes the dorsal–ventral axis on all the limb parts.

The progress zone (proximal–distal), the zone of polarizing activity (anterior–posterior), and the ectoderm (dorsal–ventral) act together to coordinate the development of the limb. An error in one of their functions may lead to extra digits, missing parts, or incorrect orientations. But consider the long-term consequences of such errors: they could provide the variations in limb structure that are the very source of evolutionary change. No doubt the evolution of the basic five-digit limb of land vertebrates into bat wings, horse legs, and whale flippers (Chapter 1) was possible because mutations affected the control systems of limb-axis formation. Thus evolutionary alterations in body structure result from heritable changes in the developmental process of individual organisms.

REGULATORY GENES AT WORK DURING PATTERN FORMATION

Pattern formation—of a limb, the central nervous system, or, indeed, the basic body plan (one head, two arms, two legs, and so on)—must ultimately be based on gene activity. New work on the regulatory genes, which control pattern formation, is one of the most exciting areas of biological research, for the results begin to show us how genes participate in building a body. Moreover, as we will see in Chapter 43, these same genes may play a paramount role in evolution of new body plans and types of organisms.

As we have seen, structural genes code for proteins, whereas other genes code for the nonmessenger RNAs (transfer, ribosomal). But comparing the products of structural genes in a bird's forelimb and hindlimb reveals few, if any, differences: the contractile proteins (actin, myosin) are the same, as are the enzymes, collagens, and so on. Individual bones, tendons, cartilages, blood vessels, and skeletal muscles may be shaped differently, but their cells and molecules are virtually identical. The difference between the upper-arm and the upper-leg bones (humerus and femur, respectively), the thumb and the big toe, and the wing and the leg is shape—the way populations of differentiated cells are put together. New work suggests that a novel class of genes, regulatory genes, determines shape by controlling morphogenetic processes. Regulatory genes appar-

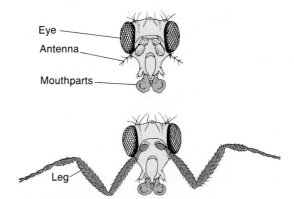

Figure 17-15   REGULATORY GENE MUTATION AT WORK.
This is an extreme form of the antennapedia effect. Two full legs have grown from the sites on the head where antennae normally form.

ently control batteries of structural genes, switching them on and off.

Let us examine the best studied regulatory genes in our favorite genetic workhorse, *Drosophila*. In this insect, biologists have identified a complex of genes called *antennapedia*, each of which can control the pattern of an appendage. For instance, one mutant gene causes a pair of perfect legs to grow out of the head in place of the two antennae (Figure 17-15). For that to occur in the cells that normally form an antenna, a battery of hundreds of leg genes must be switched on so that the cells generate the new pattern, a leg protruding from the head. Yet that whole battery of genes is switched on as a result of one mutation in one regulatory gene.

In the *Drosophila's bithorax* gene complex, one mutation alters the development of a whole body segment so that many changes occur, including the formation of an extra pair of wings. (By "segment," we mean head, thorax, or abdomen area. Only the thorax normally develops wings; in bithorax flies, two thoracic segments develop.) This and much other evidence indicates that genes in the antennapedia and bithorax complexes act as simple switches that turn on and off batteries of genes that, in turn, control morphogenetic processes. In fact, the sites where such regulatory genes are actually "turned on" to transcribe RNA can now be visualized. The photograph at the beginning of the chapter, in fact, shows that a regulatory gene, *engrailed*, is active only in the posterior segments of a fruit fly's embryo.

A fascinating, but still puzzling feature of regulatory genes such as bithorax or engrailed is the presence of a *homeobox*, a 180-base-pair DNA sequence. This same sequence has been found in sea urchins, frogs, mice, and humans. Some work suggests that the protein encoded by the homeobox functions to regulate batteries of genes involved in morphogenesis and differentiation. There is remarkably little difference in the amino-acid sequences of homeobox proteins, implying great conservatism in these developmental regulators through much of the animal kingdom.

Regulatory genes, therefore, may prove to be keys to understanding how an organism is constructed by developmental processes. And they may be keys to explaining some changes in body structure that are the basis of evolution. An example is the duplication of body segments, the hallmark of evolution of organisms whose bodies are built of many like segments (such as earthworms, lobsters, flies, and their kin). Some bithorax mutant genes and other regulatory genes show clearly how single mutations can change segment order or number and, in turn, the appearance of the whole body.

Regulatory genes, acting as they do *early* in development, can have profound effects on pattern and body organization. This is a major new idea. Geneticists originally thought that mutations in hundreds of genes, perhaps occurring over thousands or millions of years, were needed to generate so complex a characteristic as a new pair of legs. In reality, as Thomas Kaufman of the University of Pennsylvania has said, "minor gene change can result in an alteration in development that allows a radical change in adaptation: a small key can open a large door."* It should be no surprise, therefore, that there is 99 percent sequence identity between proteins (structural genes) of chimpanzees and of humans. A relatively small number of regulatory genes, acting early on morphogenetic processes, may well be the key that produces a Michelangelo and not a chimp.

GENE REGULATION IN DEVELOPMENT

The processes of differentiation, morphogenesis, and pattern formation depend on proteins and other macromolecules. In order for these proteins to be manufactured by developing cells, messenger RNAs must be available. Some are maternally manufactured RNAs, built into the egg during oogenesis; others are transcribed from genes in the nuclei of embryonic cells. Beginning at about the time of gastrulation, maternal mRNAs no longer suffice, and gene transcription must be activated, or normal development will halt. Activation is caused and coordinated by such diverse things as polar-lobe or yellow-crescent materials and tissue inter-

actions of the kind seen in the lens of the eye, the lungs, and other organs. Similarly, the action of the progress zone or the zone of polarizing activity in a limb bud establishes conditions leading to the regulation of certain genes in subpopulations of limb cells that cause the proteins of limb cells and tissues to be made.

As we learned from the experiments with nuclear transplantation and transdetermination, each determined body cell usually retains all the original genetic material present in the zygote. Yet once a cell is differentiated, it uses only some of those genes to manufacture the specific proteins it needs to fulfill its particular function. What, then, turns on the correct genes for making that cell's characteristic proteins and keeps the unnecessary genes turned off?

Biologists have amassed a catalog of plausible explanations of the way genes are regulated during development. Hundreds of experiments have narrowed the list of gene-control mechanisms to a finite number of possibilities, all of which act at certain times in certain cell types.

In this section, we shall examine in detail the most likely control mechanisms, from which a more complete explanation of gene regulation during development will emerge someday. We have organized these mechanisms according to the cellular level at which they operate: (1) the level of DNA processing, (2) the level of RNA synthesis and processing, and (3) the level of protein production and modification.

DNA Processing

Mechanisms of gene control that can occur at the level of DNA processing include gene amplification and gene rearrangement. Both mechanisms have been found to occur only in very specific instances.

Gene Amplification

Classical genetics assumes that there is one copy of a given gene in the haploid chromosome set and, therefore, two copies in the usual diploid cell. We learned in Chapter 16, however, that the genes that code for the structural RNA of ribosomes are *amplified* several thousandfold in maturing oocytes of organisms such as *Xenopus*, the African clawed frog. As a result, huge numbers of ribosomes can be manufactured in reasonable amounts of time. Are there also multiple copies of genes that code for proteins?

Biologists have studied several cell types in search of amplification in the genes that code for specific proteins. Their strategy is to analyze DNA in cells that make huge quantities of a single protein, such as red blood cell pre-

---

*R.A. Raff and T.C. Kaufman, *Embryos, Genes, and Evolution* (New York: Macmillan, 1983).

cursors, which manufacture hemoglobin, and silkworm glandular cells, which produce silk. These measurements have typically yielded the same result: no extra copies of the genes were found. However, in 1980, Alan Spradling and Antony Mahowald at Indiana University made an important discovery: extra copies of three structural genes are made in ovarian follicles of female fruit flies just a few hours before the follicle cells must manufacture large quantities of the three proteins that make up the tough, protective layer—the chorion—around the egg cell. Therefore, structural genes that code for proteins may also be amplified. As in oogenesis, quantity and time are the factors responsible for this amplification. Because extra copies of the genes are made, enough DNA templates become available to permit a larger amount of mRNA to be synthesized for the chorion proteins in a very short time. Then, using this massive mRNA reserve, the three proteins can be produced rapidly and assembled around the egg as it moves down the oviduct. Although this case establishes the principle, most differentiated cells do not have to make extra copies of specific genes when they differentiate.

Gene Rearrangement

A unique mechanism in the cells of an embryo's immune system can shift pieces of genes around in differentiating cells to produce a variety of combined genes. The details of the mechanism are described in Chapter 32, where we explore how the body produces *antibodies*, proteins that help destroy foreign substances in the body. It is sufficient to say here that this mechanism, which appears to operate only in the developing immune system, can bring close together pieces of a gene originally located far from one another in the chromosomes to produce a combined gene. This process, called **gene rearrangement,** allows a relatively small number of gene pieces to be rearranged in a great variety of ways to generate millions of types of antibody proteins. Perhaps because immense variability in structure is not a property of most structural and enzymatic proteins, this gene-rearrangement process is not a common developmental mechanism.

RNA Synthesis and Processing

The question one continually encounters in developmental biology is: If all body cells contain all the genes inherited from both parents, how does each type of differentiated cell use only a tiny fraction of the full complement? Certainly gene amplification and gene rearrangement at the level of DNA processing do not account for most types of cell differentiation. The answer lies instead in **differential gene activity**—the function-

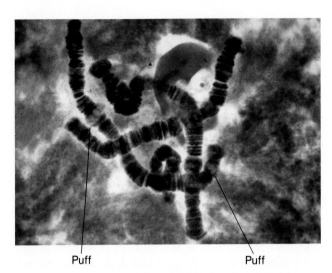

Puff   Puff

Figure 17-16   POLYTENE CHROMOSOMES OF INSECTS.
This photograph made with a light microscope (magnified about 500 times) shows genes at work. The many stained bands are unpuffed inactive genes. The light blue expanded "puffs" are sites of intense RNA synthesis.

ing of some genes and not others—during the synthesis and processing of RNA.

Production of mRNA

Differential gene activity can actually be observed in the huge *polytene chromosomes* of certain insect cells (Chapter 10). The DNA double helix of a normal fruit-fly chromosome is duplicated about 1,000 times in such polytene chromosomes. Furthermore, the duplicated DNA helices are aligned next to one another so that the same genes are situated side by side; this yields the characteristic banding patterns of polytene chromosomes readily seen with a light microscope, as shown in Figure 17-16. When a specific genetic locus on a *Drosophila* polytene chromosome becomes active in synthesizing mRNA, we see what is called a "puff." A puff is a site where the multiple copies of the DNA helices are somewhat unraveled from one another, allowing the RNA polymerase enzyme responsible for RNA synthesis to operate efficiently and to make RNA. The puffs are analogous to the loops of lampbrush chromosomes, which are active genes in maturing oocytes (Chapter 16).

Two interesting puffing patterns appear during development of fruit flies. First, different puffs are active at different times during the development of a single cell type, indicating that genes are being turned on and off. We can observe the second aspect of puffing by looking at different cell types at the same time. For instance, at one specific time, certain sets of chromosome bands in salivary-gland cells are puffed, whereas in developing kidney cells, a different set of bands is puffed. Different genes are being transcribed in these two tissues, just as we would predict, because different proteins characterize the tissues.

Polytene chromosomes and puffing actually let us see evidence of genes at work. Of course, the same basic phenomenon goes on in differentiated cells of all animals and plants that are only diploid and not polytene. Even though we cannot see the "puffing" at single haploid genetic loci with a light microscope, the phenomenon takes place there as well. It is generally true, therefore, that some genes are active while most are inactive in a cell, and, as a result, a specific set of mRNAs is manufactured to support a given type of differentiation.

Intensive studies are under way to establish the basis for differential gene activity in chromosomes. One line of research concentrates on special base sequences in the DNA near genes. Sequences called *promotors* or *enhancers* are sites where RNA polymerase, or so-called transcriptional control factors, bind and thereby facilitate transcription from the gene. Such sites are located "upstream" from a gene. (Since RNA polymerase reads in a "downstream" direction, that is, from the 5' toward the 3' end of a gene, "upstream" means, in the DNA, continuous with the 5' end of the gene.) After a transcription factor binds to the specific DNA-base-sequence region, it somehow interacts with other proteins and RNA polymerase bound at the transcription initiation site of the gene. Some researchers envision a multiprotein assembly that may induce a conformational change in the DNA, such as unwinding, so that RNA polymerase can begin its transcriptional march along the gene.

An example of enhancer function is the three enhancers located near the α-fetoprotein gene in mice. Evidence suggests that one enhancer acts primarily in fetal yolk-sac cells, another in liver cells, and the third in gut cells. As a result, the same structural gene can be turned off or on at different rates in the three cell types—the key is the type of control factor present and which enhancer it binds.

Whereas this work emphasizes the importance of enhancer and promotor sites on DNA, other experiments focus on the proteins present on chromosomes that may regulate gene transcription. We have already discussed the basic proteins of chromosomes, the *histones* (Chapter 9), which are required for the coiling of DNA into nucleosomes. Because few differences have been found among the histones in various types of differentiated cells, these proteins cannot be the source of specific gene activation or repression. In contrast, the other major class of chromosomal protein is quite diverse and includes up to 500 types of molecules that remain behind after all the histone has been extracted from chromosomal material. Called **nonhistone chromosomal (NHC) proteins,** these molecules include a variety of enzymes, such as RNA polymerases and DNA polymerases, the catalysts that assist in the manufacture of mRNA, rRNA, and DNA itself. Other NHC proteins with no known enzymatic function may regulate specific genes during determination and differentiation. There is a good deal of variability among the NHCs in different cell types, which is consistent with the theory that different NHC proteins are involved in gene regulation.

How might these NHCs act? When NHC proteins bind to DNA, the binding of histone is altered, and the beaded nucleosome structure may change, especially at sites where DNA is being used as a template for RNA synthesis (puff sites). The major question is whether and how NHC proteins could govern the specificity of gene transcription. Let us use the ovalbumin gene to suggest an answer.

Ovalbumin is an egg-white protein manufactured in the oviducts of birds. The hormones estrogen and progesterone can act as triggers to turn on ovalbumin synthesis in the oviducts of female birds. If radioactive progesterone is applied to oviduct cells, the marked trigger molecule can be traced after it diffuses through the plasma membrane. New results suggest that the steroid hormone passes through the cytoplasm and into the nucleus; there the hormone quickly binds to a receptor molecule, a protein, and causes an allosteric change in its shape. The two "complexed" molecules (progesterone plus the receptor) become associated with certain sites on the chromosomes, as presented schematically in Figure 17-17. Special NHC proteins on chromosomes apparently bind to the hormone–receptor complex. That binding, in turn, may trigger gene activity at those chromosome sites so that mRNA can be manufactured and then ovalbumin protein made.

Suppose we combined the NHC proteins from ovalbumin-synthesizing cells in the hen oviduct with chromosomal material (DNA plus histones) from a liver cell. In this experiment, the liver DNA is tricked into responding to the reproductive-hormone trigger (progesterone) and even into making the message for egg-white protein—something a liver cell normally would not do. Thus nonhistone chromosomal proteins are likely to be a link in the gene-regulation sequence that is the basis of differentiation and probably are responsible for the transcription of the hemoglobin gene in a red blood cell, the ovalbumin gene in an oviduct cell, and so on.

Processing of RNA

As we saw in Chapter 14, one of the completely unexpected results of using the new recombinant DNA technology has been the finding that individual genes of eukaryotic cells are actually separated into discrete sections (exons) interrupted by sequences of DNA called introns. This arrangement, even though normal and common, appears at first glance to create a problem for the cell. If functional mRNAs are to be derived from the DNA template, it is essential that the intron portions be excluded. Thus if a given primary mRNA transcript is not proc-

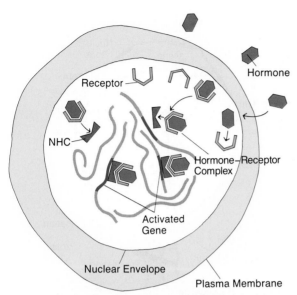

Figure 17-17   GENE ACTIVATION BY NHC, PROTEINS, STEROID HORMONES, AND RECEPTORS.
The steroid hormone enters the cell and binds to a receptor in the nucleus. Then the hormone-receptor complex binds to specific NHCs and the full set of molecules binds to specific sites on the chromosomes and triggers RNA transcription there. In the case shown, three genes will be activated, and three proteins are produced by the one hormone regulator. The sizes of the genes, receptors, and hormones are greatly exaggerated for clarity. Previous work suggested that steroid hormone receptors occur in the cytoplasm; that now seems unlikely.

essed so that introns are removed and special end sequences added, the mRNA never leaves the nucleus and therefore cannot be translated into protein. This processing of mRNAs in a cell's nucleus therefore provides another source of control for gene expression, since it can prevent mRNAs from being exported to the cytoplasm.

The availability of RNA sequences derived from different exons also allows for the possibility of combining, at the level of RNA, the information encoded in different genes. Thus exons coding for pieces of different digestive enzymes could conceivably be spliced together in a pancreatic cell to yield a "new" digestive protein. Some evolutionary biologists believe that various protein families have evolved in this way. We shall return to this idea later in the book when we discuss evolution.

Another exciting aspect of RNA processing modifies the basic one-gene–one-polypeptide rule. In antibody-producing cells, an RNA primary transcript is made with a series of exons. Two of them code for the terminal portion of the antibody molecule. If one of the exons is joined to the main set of exons, so-called *m*-type antibody is synthesized. If, instead, the other terminal exon is joined to the final mRNA, *g*-type antibody is made. So this is a case of *one gene–two polypeptides*, the produc-

tion of *m* or *g* depending on the processing of the RNA transcript in the nucleus.

Storage and Stability of mRNAs

Imagine that a processed mRNA for a specific protein moves into the cytoplasm of a differentiated cell. Is this passage enough to guarantee that the protein encoded by the mRNA will be produced? No, it is not. Although the mRNAs of most differentiating cell types are used soon after they enter the cytoplasm, this is not necessarily the case. Recall from Chapter 16 that eggs can store a large quantity of inactive mRNA for months at a time before using it during the cleavage period. Storage of inactive mRNAs ("masked messengers") also occurs in dried brine-shrimp embryos and in a variety of plant seeds and pollen (Chapter 27). Mechanisms exist to permit activation and use of these masked mRNAs at specific times in certain cells but not in others. How such choices are made is still a mystery, although factors such as the yellow-crescent materials seem to be likely candidates for the role of control agents.

Another significant feature of mRNAs is their stability. Although the mRNAs for housekeeping proteins have relatively short lifetimes (a few hours), the mRNAs for specific proteins in differentiated cells—hemoglobin, ovalbumin, silk, and so on—have lifetimes of about fifty to several hundred hours. The source of this great stability is unknown, but the phenomenon has important consequences. It is a primary reason why most genes for specific proteins do not have to be amplified (have extra copies made). The great stability of the specific protein mRNAs allows these molecules to be used as templates again and again, hour after hour. The result is that huge quantities of hemoglobin, ovalbumin, silk, and similar specific proteins may be manufactured on the long-lived mRNAs, even though only two gene copies are present in each diploid cell.

Protein Synthesis

We have been discussing the ways in which genes are regulated during development so that appropriate proteins are made by differentiating cells, thereby enabling the cells to assume their specific functions in the developing embryo. We know at this point that once RNAs are transcribed, processed, and transported from the cell nucleus into the cytoplasm, protein synthesis can begin. Obviously, amino acids, energy in the form of ATP, and various enzymes must be present in order for protein to be assembled. In addition, some agents in the cytoplasm may affect protein synthesis. A good example is the synthesis of *globin*, the protein part of hemoglobin. The nonprotein part of the molecule, *heme*, binds and carries oxygen and usually is bound to globin. If the red blood cell lacks the precursor of heme (a molecule called *hemin*), a rapid inhibition of globin synthesis takes place. This relationship exemplifies synthetic coordination: the protein globin is made only if its cofactor, heme, is also made. In a broader context, this example demonstrates that cells are not forced to use an mRNA in protein synthesis simply because it is in the cytoplasm. Thus *control of translation* is another point at which gene expression can be regulated.

Yet another mechanism affecting the appearance of specific proteins during differentiation is *modification* of newly manufactured proteins. This process sometimes involves chopping off and discarding pieces of a protein. Or it may entail the addition of sugars to certain of the protein's amino-acid side chains; recall from Chapter 6 that this is a common practice within the cell's endoplasmic reticulum and Golgi complex. Furthermore, charged groups, such as phosphate, may be added to modify the properties of the protein. All these alterations are caused by still other proteins (enzymes): hence these modifier proteins must be present beforehand in order for protein modification to proceed correctly. When the various alterations are complete, the protein is finally ready to be used in cells of the developing embryo.

A final means by which levels of a specific protein are regulated in a cell is **degradation.** As soon as macromolecules, such as proteins, are manufactured within cells, they are subject to chance destruction. In fact, the larger the protein, the greater the possibility that it will be destroyed. Thus the actual amount of a protein or an enzyme in a differentiated cell is the result of a balance: synthesis generates new protein, while degradation removes some. The balance between the two processes establishes the protein level. By speeding up or slowing down protein degradation, therefore, gene expression can be regulated indirectly.

Discoveries in the mid-1980s reveal that the stability and integrity of proteins are under precise genetic control and are not due solely to chance destruction. Short-lived proteins have certain amino acids at their N-terminal (amino) end: (1) arginine, lysine, aspartic acid, leucine, and phenylalanine give protein half-lives (the time it takes for half the protein to disappear) of 2 to 3 minutes; (2) tyrosine and glutamic acid give half-lives of 10 minutes; and (3) isoleucine and glutamine give 30 minutes. Long-lived proteins, ones that persist for 20 hours or more, have methionine, serine, alanine, threonine, valine, or glycine in the N-terminal site. Other experiments reveal the presence of PEST regions (P = proline; E = glutamic acid; S = serine; T = threonine) in proteins with short half-lives. These remarkable findings imply that cells have evolved a mechanism to ensure rapid turnover of some proteins and persistence of others. Clearly, this phenomenon complements transcriptional, processing, and translational controls in governing the quantities of housekeeping and differentiative proteins in cells.

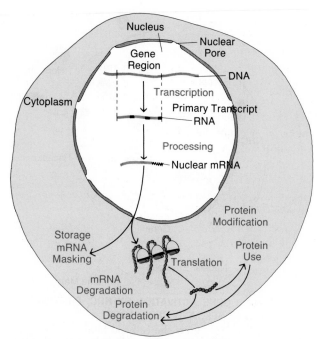

Figure 17-18   CONTROL POINTS FOR GENE EXPRESSION.
Each step or process labeled in green may serve as a potential control point in the overall expression of the genetic information on chromosomes as functional proteins.

To sum up what we know about gene regulation during development, differential gene activity is the basis for cellular differentiation and morphogenesis. Figure 17-18 summarizes the mechanisms—the catalog of possibilities—that may be used to regulate gene expression during development. Included are mechanisms for making additional copies of genes, rearranging genes, and transcribing them; for processing RNA in the nucleus; for storing or using mRNA in the cytoplasm; for controlling the use of mRNAs with more than one factor; for modifying newly made proteins; and for balancing the synthesis and degradation of proteins.

How these mechanisms are actually applied in any given cell, tissue, or organ—how whole sets of genes are turned on and off by regulatory genes to account for the complex development of an eye or a limb axis—is still a major mystery of developmental biology. In contrast, a related long-standing mystery of cell behavior and gene regulation is finally yielding to new techniques—it is cancer and its causes.

CANCER: NORMAL CELLS RUNNING AMOK

Cancer is a scourge of plants and animals—including humans—and has been for millions of years. Paleontologists have discovered signs of cancer in the bones of dinosaurs and of hominids (ancient relatives of humans) from Java and in Egyptian mummies. Since 1930, cancer has become tragically common in our society: one of every four Americans will develop cancer, due to smoking. Since the invention of recombinant DNA technology, scientists have made great strides in understanding the basic mechanisms of cancer and have uncovered fundamental links among certain kinds of viruses, cellular genes, and normal growth factors. Together, they add up to the diversion of normal cellular machinery to bad ends—and to some, if not all, of the 100 or more diseases we collectively call "cancer."

The term for these dread diseases comes from the Greek word *karkinos*, which means "crab," and derives from the large, red clawlike arteries that feed the relentless growth of cancerous cells. In an animal with cancer, cellular growth controls function abnormally, and huge numbers of cancerous cells can accumulate, either as solid masses called **tumors** or as circulating cancer cells. Tumors known as *carcinomas* can arise in the epithelial sheets covering the outer and inner surfaces of the body, while tumors called *sarcomas* arise in connective tissues. Circulating cancers called **leukemias** and **lymphomas** arise in the blood-forming cells of the bone marrow and lymph nodes.

Cancer usually follows a set program: first, one or more healthy cells are altered or *transformed* into cancerous ones; second, these transformed cells divide into clones of cells that make up the tumor or the circulating cancer cells; third, the cancerous cells invade neighboring tissue; and fourth, the cells may **metastasize,** or break away from the tumor and spread to a distant site in the body. A breast tumor, for example, often arises in a milk gland. If it is *benign*, it remains as a lump and does not move into surrounding tissues. If it is *malignant*, however, it invades nearby ligaments, fat, and underlying muscle tissues, and cells may metastasize and lodge in adjacent lymph nodes, setting up new cancerous sites.

Cancerous cells continue to carry out many cellular functions in a normal way, but the transformation process brings about several profound changes. The most striking characteristic is uncontrolled growth. The cell's surface properties and migrating behavior also change, and cancer cells are abnormally long-lived. Some scientists think of cancer as "a wound that overheals itself." When a common injury occurs—a cut, say—the cells near the wound are stimulated to grow rapidly, and when the torn edges meet after two or three days, the rapid mitosis stops. In cancerous cells, however, mitosis goes on and on because the normal regulatory events that would lead to arrested growth do not occur. The net result of uncontrolled mitosis and long-lived cells is an enlarging tumor, leukemia, or other malignant condition that disrupts normal tissues, ultimately to the point of killing the organism.

What Causes Cancer?

Clearly, the transformation of a healthy cell to a cancerous one is the key event in the development of cancer. But what causes it to take place? Scientists have spent many decades looking for *carcinogens*—cancer-causing agents—and have amassed a long list, including unusual irradiation, chemicals, viruses, and genes.

Excessive exposure to ultraviolet light in sunlight, x-rays, or radioactive emissions (such as at Hiroshima and Nagasaki) can cause cancers. Thousands of chemicals are also carcinogens in some species. These include the hydrocarbon tars in cigarette smoke; substances in the blackened parts of charred meat, breads, and other foods; such dietary additives as red dye #2 and the synthetic sweetener saccharin; such industrial solvents as benzene and carbon tetrachloride; and such other industrial materials as vinyl chloride and asbestos fibers. Dozens of viruses are known to cause animal and plant cancers such as leukemias in chickens, mice, cats, and cattle; mammary tumors in mice; and kidney tumors in frogs. Some of these viruses carry cancer-causing genes, or **oncogenes** (which we will discuss shortly), while others do not. Finally, abundant statistical evidence shows that families can pass on genetic tendencies toward cancer. Some families have histories of breast cancer among female members, while others have high rates of melanomas (deadly pigment-cell cancers) or of prostate cancer among males.

An obvious question emerges from this cataloging of environmental and genetic factors: How do genetic tendencies or exposure to substances lead to the malignant transformation of cells? Although thousands of scientists worldwide have sought the answer, only in the 1980s have the first hints emerged.

Proto-Oncogenes in Cells

A major recent discovery shows that certain normal cellular genes may participate in cancer initiation. Every normal embryonic and adult cell has genes that code for the proteins and enzymes involved in mitosis, locomotion, adhesion to other cells, receptors for growth factors or hormones, and so on. Some of these genes, however, also behave as **proto-oncogenes;** that is, they can cause cancers when altered in some way by mutations or chromosome translocations. A type of human bladder cancer, for example, can result from a single nucleotide substitution in a human gene that causes the amino acid glycine to be present in the protein product instead of valine. In cancer patients with Burkitt's lymphoma, a piece of chromosome 8 is attached to chromosome 14; that piece apparently contains an oncogene called *myc* (pronounced "mick"), which is inactive in normal cells when on chromosome 8, but is activated in the new chromosomal site on 14. Cancer, and high probability of death, are the result. In still other cancers, another control region from viral genetic material becomes inserted near a cell's proto-oncogene; the result is transformation of the cell into the cancerous state. This is one way viruses lacking oncogenes themselves can still cause cancer. Finally, cellular proto-oncogenes may be amplified, that is, be present in multiple copies. This occurs, for instance, in the HER-2/neu, the apparent result being much greater likelihood that breast cancer in the human woman will recur and even lead to death.

It is not farfetched to imagine a connection between the long list of carcinogens and the activation of the proto-oncogenes that may already be present inside normal cells. Chemicals, radiation, or viral infections may induce such mutations, chromosome translocations, or DNA insertions in proto-oncogenes, and these events, in turn, may trigger the cancerous transformation of a cell.

Viral Oncogenes Are Derived from Cells

A startling discovery made in the early 1980s revealed that the cancer-causing genes in many viruses actually originate from proto-oncogenes in animal cells, and not in the viruses themselves. Biologists have studied several dozen viruses that carry oncogenes, and have found that when one, such as the Rous sarcoma virus, infects a host cell, the viral DNA is incorporated into the cell's chromosomal DNA (Figure 17-19). There, the viral DNA oncogene is reproduced each time the cell goes through a division cycle. The virus genome itself may also be duplicated, yielding hundreds of new viral particles, each carrying a copy of the oncogene.

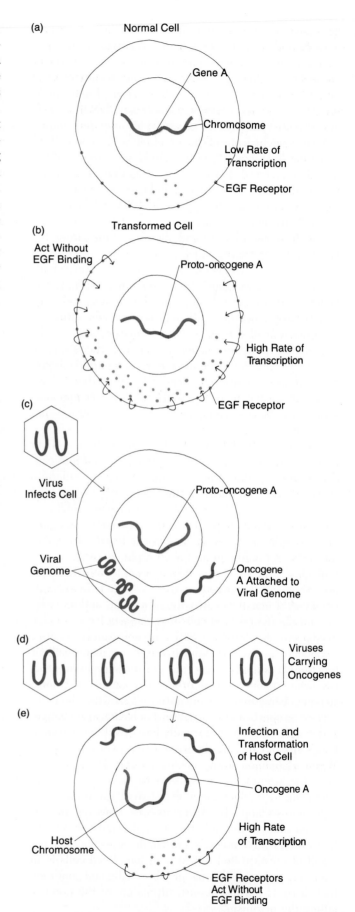

Figure 17-19   ONCOGENES, VIRUSES, AND CANCERS.
(a) A normal cell with gene A produces modest quantities of EGF (epidermal growth factor) receptor and grows slowly when EGF binds. (b) A change has occurred in the regulatory region for A, rendering it a proto-oncogene; much EGF receptor is made and it functions in the absence of EGF. The cell divides rapidly in an uncontrolled fashion. (c) A normal virus infects a cell. (d) By chance, copies of proto-oncogene A become attached to viral genetic material and are incorporated into viruses. Those cellular genes may mutate and become distinctive viral oncogenes. (e) When such viruses infect a normal cell, the oncogene may be incorporated in the host genome, transforming the cell. In this hypothetical case, its product is shown as the EGF receptor that functions in the absence of EGF and causes uncontrolled cell division.

A natural assumption might be that oncogenes originate in the viruses that contain them. However, studies of sixteen cancer-causing viruses show that the nucleotide sequence of the oncogene corresponds quite closely to those of sixteen cellular proto-oncogenes. For example, the oncogene of Rous sarcoma virus (which causes tumors in chickens and is abbreviated *src*, pronounced "sarc") is present in *normal* chicken cells, where it is arranged in exons and introns, as are other normal cellular structural genes. Surprisingly, genes almost identical to *src* are present in the DNA of humans, other mammals, and even fish. Thus the *src* DNA sequence has been conserved during evolution and makes essentially the same protein in all vertebrates. Yet the same gene, altered only slightly and injected by the virus, can cause cancer.

Biologists have concluded from these findings that the *src* gene originated in a cell, not in a virus, and that copies of it may have become inserted into the viral genome. *Src* and other oncogenes are thus no more than passengers in viruses—pieces of cellular DNA going along for the ride, performing no useful function for the virus and its replication cycle but, because of their cellular origin, capable of doing great harm in a virus-infected cell.

Oncogene Products: Proteins Related to Normal Growth Factors

Yet another recent finding helps explain the uncontrolled growth of a transformed cell—why, in other words, the cancerous cell runs amok. Scientists searching for the products of oncogenes found several things about them: (1) these proteins usually become associated with the host's cell surface or cytoskeleton; (2) they are present in large quantity; (3) they usually act as enzymes that catalyze the addition of phosphate groups to tyrosine residues in cellular proteins (hence they are called **tyrosine phosphokinases**). All sorts of activities carried out by proteins are alterable by phosphate addition; hence an oncogene product—in plentiful supply—may literally trigger a cascade of events by phosphorylating a variety of cellular proteins that have available tyrosine groups. It is as though a cellular growth substance normally present in small amounts is suddenly present in

abnormally large quantities, and so the cell's behavior goes awry. No wonder a single functioning oncogene can trigger so complex a disease as cancer.

Perhaps the most intriguing part of the findings about oncogene products is that the protein of the *src* gene is almost identical to a major portion of a normal cell-surface receptor for epidermal growth factor (EGF)—a hormone that controls cell growth. Normally, this hormone must bind to the receptor in the plasma membrane and activate it (by phosphorylating a tyrosine residue) before cell growth is stimulated. It turns out that the protein coded for by the *src* oncogene *acts as a receptor that does not have to be activated by EGF*. When the oncogene is turned on and the protein is produced, it functions like an activated EGF receptor *that cannot be turned off*. This puts the cell in a state of perpetual growth and division. The cell is thus tricked and shows all the features of a transformed malignant cell, including metastasis. Studies show that the protein products of other oncogenes resemble growth factors and that this *src* model probably represents the general case.

The pieces of the great cancer mystery are falling together fairly quickly, and we can see that a predisposition toward a certain type of cancer may be based on the presence of proto-oncogenes or on rearranged chromosome pieces. To paraphrase the oncogene researcher Robert Weinberg, humans carry the seeds of their own cancers in their genes. Radiation, chemical carcinogens, and viruses may act on preexisting cellular genes or proto-oncogenes, or oncogenic viruses may insert the deadly baggage they picked up from other cells. The common result may be the abnormal use of normal cellular molecules, such as in the *src* product.

These findings are exciting, but they raise some worrisome questions about developing cures for cancer. For instance, drugs could be designed to act against the oncogenic tyrosine phosphokinase, but these drugs almost certainly would have side effects on the normal, essential tyrosine phosphokinase present in every cell. Much more research and ingenuity on the part of biomedical researchers will be required to solve such problems. However, we are now in a much better position to solve them than we were a decade ago because of the important information about oncogenes. The underlying causes of cancer are at last being unraveled in a stunning series of experiments that underscore the sophistication of modern science.

SUMMARY

1. *Determination* is the commitment of a cell to a specific type of differentiation. Once a cell population is determined, the state is usually permanent and is passed on to the cells' progeny. However, changes or mistakes in the determination of a cell type can occur, as experiments showing *transdetermination* of *imaginal-disk* cells in fruit flies have shown. *Nuclear-transplantation* experiments also demonstrate that the determined state can change as the result of exposure of the nucleus of a determined cell to the cytoplasm of a different cell type.

2. Determined cells undergo *differentiation*, or maturation. Differentiation may be triggered by inductive-*tissue interactions*, by control factors built into eggs, or by certain regulatory molecules.

3. A differentiated cell (a) makes and uses a specific set of proteins; (b) has "housekeeping proteins" to help keep it metabolically active; (c) assumes a characteristic shape; and (d) may cease to undergo further cell division.

4. The many diverse types of morphogenesis (at the cellular or organ level) are produced by just a few different mechanisms: the movement of single cells; the interaction between moving or stationary cells and the extracellular fibers or matrix to change those substances chemically or mechanically; the movement of cell populations; localized cell growth; localized *cell death*; and the deposition of *extracellular matrix*.

5. The importance of control factors in development and differentiation can be seen in *pattern formation*. One aspect of pattern formation is the development of the body's three primary axes (anterior–posterior, dorsal–ventral, and right–left) and the development of limbs,

each with its own sets of axes. In the arm, for example, the *progress zone* controls proximal–distal (upper arm–digits) development, the *zone of polarizing activity* (ZPA) controls anterior–posterior (thumb–little finger) development, and the ectoderm controls dorsal–ventral (palm–back of the hand) development. These three mechanisms act together to coordinate the development of the entire arm and hand.

6. Regulatory genes control batteries of other genes that are responsible for pattern formation. The body plan, its number of segments or limbs, and the shape and position of parts are controlled by regulatory genes.

7. Biologists have compiled a list of explanations of the ways genes can be regulated during development. These mechanisms may differ from one cell type or embryo to another, but all act at certain times in various cells. Thus gene regulation may occur at the level of DNA processing, RNA synthesis and processing, or protein synthesis and processing.

8. Mechanisms of gene regulation that occur at the level of DNA processing include gene amplification (the making of extra copies of a gene) and *gene rearrangement*. Neither strategy is very common, however.

9. The primary mechanisms of gene regulation in development seem to lie in the *differential gene activity* that occurs in the synthesis and processing of mRNA. Such differential gene activity can be observed in the chromosome puffs of polytene chromosomes. *Nonhistone chromosomal (NHC) protein* varies in differentiated cell types and has been implicated in gene regulation. NHC proteins are thought to govern the specificity of gene transcription of mRNA. Gene regulation may also occur in the processing of RNA tran-

scripts (so that RNA sequences encoded by introns are deleted), and in the storage and activation of mRNAs. Such processing can allow one gene to encode for more than one polypeptide.

10. Mechanisms of gene regulation at the level of protein synthesis and storage include the control of translation through the availability (or unavailability) of necessary precursors, enzymes, energy, or cofactors; the modification of newly manufactured proteins; and the speeding up or slowing down of *protein degradation*.

11. Cancers are diseases in which normal cells are transformed into a state of rapid mitosis, long life, and altered surface properties. Cancerous cells can accumulate, either as *tumors* or as circulating cancers. Some cancers *metastasize*, or set up new sites in the organism.

12. Normal cells of all organisms may have *proto-oncogenes*, prospective cancer-causing genes, which when acted on by environmental factors (carcinogenic chemicals, radiation, or viruses) may cause normal cells to become cancerous.

13. Cancer-causing viruses may carry *oncogenes*, cancer-causing genes originally derived from animal or plant cells.

14. Cellular oncogenes may produce variations on molecules used normally in cell-growth control. For example, the EGF receptor of the *src* gene tricks cells into uncontrolled cancerous growth. It seems safe to conclude that although cancers may have many causes, most of these may act at the levels of DNA and chromosomes.

KEY TERMS

cell death
determination
differential gene activity
differentiation
extracellular matrix
gene rearrangement
imaginal disk
leukemia

lymphoma
metastasis
nonhistone chromosomal (NHC)
protein
nuclear transplantation
oncogene
pattern formation
progress zone

protein degradation
proto-oncogene
tissue interaction
transdetermination
tumor
tyrosine phosphokinase
zone of polarizing activity (ZPA)

QUESTIONS

1. Compare determination and differentiation. In what sense is determination inherited?

2. What proteins within a cell control cell shape?

3. Some differentiated cells divide rapidly, some slowly, and some not at all. What are some examples of rapidly dividing cells? slowly dividing cells? nondividing cells? Is the rate of mitosis related to the survival time of the cells?

4. What mechanisms give rise to morphogenesis? Give a specific example of each mechanism.

5. At what stage in the development of an embryo are the axes of symmetry established? Describe how each axis is set up in the development of a forelimb.

6. What is a structural gene? What is a regulatory gene?

7. Each polytene chromosome contains about 1,000 DNA molecules lined up side by side. How do "puffs" relate to these multiple copies?

8. Give specific examples illustrating at least five methods by which expression of genes can be regulated.

9. Give an example of an inductive-tissue interaction.

10. Proto-oncogenes have been found in yeast, insects, fish, birds, and mammals. They code for factors that are essential to normal growth and development. Explain how agents that damage DNA may cause cancer. Explain how viruses may cause cancer.

ESSAY QUESTIONS

1. Explain how mistakes in development (in nature or induced in the laboratory) can help us understand normal development. Give some specific examples.

2. How can mutations in regulatory genes generate a longer earthworm, an insect with six pairs of wings, or a human with nine fingers on each hand?

SUGGESTED READINGS

BISHOP, J. M. "The Molecular Genetics of Cancer." *Science* 235 (1987): 305–311.

Here is a fine summary of oncogenes, proto-oncogenes, and the role of DNA damage in cancer.

DAVIDSON, E. *Gene Activity in Early Development.* 2d ed. New York: Academic Press, 1976.

This is high-powered molecular biology and high-powered thinking about the subject.

GURDON, J. *The Control of Gene Expression in Animal Development.* Cambridge, Mass.: Harvard University Press, 1974.

The nuclear-transplantation experiments are just one of the fascinating topics explored in this book.

HOPPER, A., and N. HART. *Foundations of Animal Development.* Oxford: Oxford University Press, 1985.

An up-to-date work that includes experimental and descriptive materials.

RAFF, R. A., and T. C. KAUFMAN. *Embryos, Genes, and Evolution.* New York: Macmillan, 1983.

An excellent treatment of regulatory genes and how development and evolution are related.

TRINKAUS, J. P. *Cells into Organs.* Englewood Cliffs, N.J.: Prentice-Hall, 1984.

The leading scholar of cell movements in embryos wrote this provocative and complete treatment of the major issues in developmental biology.

WESSELLS, N. K. *Tissue Interactions in Development.* Menlo Park, Calif.: Benjamin/Cummings, 1977.

An easy-to-understand, well-illustrated treatment of inductive-tissue interactions.

WOLPERT, L. "Pattern Formation in Biological Development." *Scientific American* 239 (April 1978): 154.

The originator of the positional-information concept writes about the chick-wing system and other pattern problems.

18
HUMAN REPRODUCTION AND DEVELOPMENT

The child lives inside its mother for nine months, floating weightlessly in a dark wet world of amniotic fluid. At delivery, it will literally be pressed and pushed out into a very different world. Not even the pearl diver returning to the surface experiences such a dramatic change.

David H. Ingvar, Stig Nordfeldt, and Rune Petterson, *Behold Man* (1974)

The cord of life, the umbilical cord, carries this 4.5 month human fetus's blood to and from the placenta, where oxygen and nutrients are added and carbon dioxide and wastes removed.

Reproduction is a basic attribute of all organisms, whether single cells or multicellular creatures. Chapters 16 and 17 described the general features of animal development and the control processes that allow a species's genotype to be translated into its phenotype. Although development is "center stage" in reproduction, the whole process includes the behavior, physiology, and anatomy of the adult males and females, whether sea urchins, frogs, or humans. Our task in this chapter is to focus on reproduction in *Homo sapiens*, not only because of the subject's intrinsic interest, but also because we know more about the hormones, biochemistry, anatomy, physiology, and behavior involved in human reproduction than we do for any other species.

Nowhere are human biology and behavior more intricately linked than in reproduction and sexual behavior. As we shall see, sex has deep roots that stem from the basic reproductive process of cells. This process was elaborated on in the physiology and anatomy of our ancestors in the animal kingdom and was reinforced in *Homo sapiens* by instinctive and learned behaviors that link the individual to the survival of the species.

In this chapter, we shall examine the full gamut of human reproduction. After briefly comparing reproduction in humans and other vertebrates, we will describe the male and female reproductive systems and their functioning, and then focus on human prenatal development and birth. In the course of our discussion, we will touch on such topics as human sexuality, hormonal control of reproductive functioning, the mechanisms of various methods of birth control, the origins of gender differences, multiple births, and milk production and lactation (secretion).

REPRODUCTION IN THE VERTEBRATES: ANATOMY AND STRATEGY

To fully understand reproduction in humans, we must consider the anatomical and behavioral contexts established first in other vertebrates. Through the evolution of multicellular organisms, from invertebrate ancestors to the first fishes, on to mammals, and finally to primates and humans, the basic sexual use of haploid male and female gametes is preserved, and with it, the two evolutionary advantages of sexual reproduction: first, genes may recombine during the crossing-over event of meiosis; and second, the random meeting of one sperm and one egg generates a unique new diploid set of genes. The variability derived from new combinations of genes is a crucial means of permitting lines of organisms to adapt to changing environments on our planet.

Much of the actual anatomy and physiology that ensures successful human reproduction originated and was shaped in our ancestors, the reptiles and first mammals. It was in the reptiles that a system was perfected for producing sealed, desiccation-resistant eggs. These innovative eggs had the four basic embryonic membranes that still characterize every human embryo, as well as a flat embryo that developed and underwent gastrulation atop a huge yolk mass. This same gastrulation mechanism is still seen in human embryos, even though the yolk has been absent for perhaps 100 million years in our lineage. Early mammals apparently evolved with the ability to retain the developing embryo inside the female's body for long periods. During this gestation time, the embryo was nourished and supplied with oxygen, yet it was guarded from attack by the mother's immune system. And, after birth, the ancient mammals nourished their newborns with milk from mammary glands, just as primate mothers may do today.

The close link between human reproductive biology and sexual behavior also has roots in vertebrate evolution. The drive to reproduce truly dominates the lives of many vertebrates, as illustrated by the salmon's fateful spawning run and the seasonal matings of deer, bear, and whales. Females of most mammalian species come into heat, or **estrus** (the period of sexual receptivity), at the same time each year. Estrus is timed so that either (1) the young will be born when the environment and weather make their survival most likely (for example, elk estrus and mating might take place in October so that the young are born in April) or (2) the young will be born early in the year (when conditions are not optimal for them) so that several months later, when the young have grown and are demanding the most milk, maximal forage will be available to the lactating mother.

Estrus in primates follows several patterns. Individual female apes and monkeys tend to be asynchronous in entering estrus; this means that mating and resultant births take place over much of the year. Such females will mate only when in estrus, and the chances of achieving pregnancy are thereby increased. Human females show a less distinctive estrus phase and can reproduce year round. It is significant that human females can engage in sexual activity independently of reproduction; no longer is sexual behavior closely tied to ovulation. The source of this immensely important innovation may be physiological or a consequence of the human brain's higher centers having gained ascendancy, in a sense, over the sorts of behaviors that in other organisms are controlled primarily by hormones, reflexes, and instincts. Among other things, this separation of sex from reproduction may be an important component of the long-lasting pair bond between one female and one male that has evolved into marriage and that supports the highly dependent human offspring. Thus, in an indirect way "sex for pleasure" contributes to the transmission of culture that is the key to evolution of our own species.

With this background, let us look more closely at the reproductive anatomy and physiology of human males and females.

THE MALE REPRODUCTIVE SYSTEM

In order to participate successfully in reproduction, human males, like their reptilian and mammalian ancestors, must produce sperm and fluids to carry and protect the sperm, must discharge the sperm within the female reproductive tract so that it may be transported toward the egg, and must have appropriate hormonal and behavioral mechanisms to support and control these processes. The male system consists of several internal and external structures needed to fulfill these functions. The main structures of the male reproductive tract are shown in Figure 18-1. Their functions will become clear as we describe their roles in accomplishing the tasks of sperm production and egg fertilization.

Production and Transport of Sperm

As we saw in Chapter 16, the *testes* are the sites of sperm production. Most of the tissue in the testes consists of the *seminiferous tubules*; it has been estimated that, if unraveled, the tubules of a single human testis would extend more than twice the length of a football field. Because the ideal temperature for production of viable human sperm is about 34°C, which is below normal body temperature (37.5°C), the testes hang suspended in the saclike *scrotum* outside the hotter abdominal cavity. Muscles elevate or lower the testes, depending on outside air temperature.

Mature sperm are moved from the seminiferous tubules into the *epididymis*, a tube in which they are stored prior to being released into the vas deferens. The *vas deferens* is a tube that passes from the epididymis into the body cavity and curves around to ultimately join the urethra. It is in the vas deferens that sperm become motile in preparation for their journey up the female's duct system toward the egg. It is also the vas deferens of each testis that is cut and tied off in a *vasectomy*, the permanent male sterilization surgery.

The contractions of smooth-muscle cells wrapped around the walls of the vas deferens help to move the

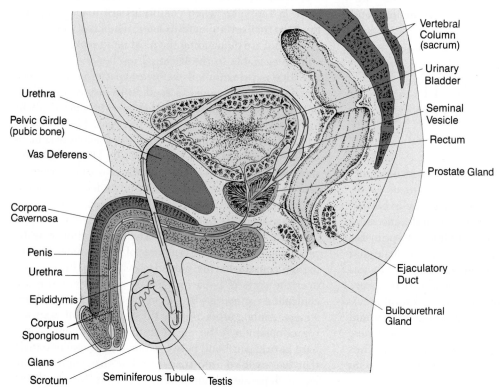

Figure 18-1   THE MALE REPRODUCTIVE SYSTEM.
The route of sperm passage from the testis to the outside world is shown by the colored arrows.

sperm along. As the sperm travel through this tube, several glands add secretions that provide the sperm with nutrients and other materials. The *seminal vesicles* and the *prostate gland* contribute sugars, fatty acids called **prostaglandins,** and other substances that increase the pH so that the sperm suspension becomes alkaline. The resultant fluid is called **semen.** The alkalinity of the semen is important in neutralizing the acidic environment of the female's vagina, which can be acidic enough to inhibit sperm.

Sperm and semen next pass through the *urethra*, the tube that leads through the **penis,** the male copulatory organ, which originated in the reptiles. The bulk of the penis is composed of three spongelike masses of tissue: two *corpora cavernosa* and one *corpus spongiosum*, as shown in Figure 18-2. The end of the penis, called the *glans*, has a slightly larger diameter than the shaft and has an abundance of sensory-nerve endings.

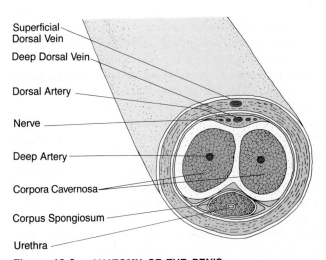

Figure 18-2   ANATOMY OF THE PENIS.
In this cross section, you can see the three corpora within the penis. Each corpus can fill with blood and so generate an erection.

The Male Sexual Response

When the male is stimulated by various sexually related sensory inputs, the spongy tissues of the penis fill with blood, the veins draining those tissues are squeezed shut, and the trapped blood causes *erection*, the swelling of the whole organ. Erection occurs in the first phase—the **excitement phase**—of the four-phase human sexual response identified by the sex researchers William Masters and Virginia Johnson.

During erection, the *bulbourethral glands* secrete into the urethra a slightly alkaline lubricating fluid. This fluid neutralizes any residual acidic urine in the urethra and provides lubrication for movement of the penis in the vagina. Because the first discharges of the bulbourethral fluids may carry some sperm along, pregnancy can result even if the male withdraws his penis from the vagina at this early stage.

During the **plateau phase** of the male sexual response, blood pressure, heart rate, and breathing rate increase, and the testes enlarge from blood engorgement. Sexual excitement reaches its peak at the **orgasmic phase,** when a series of brief, rhythmic muscular contractions causes *ejaculation*—the jetlike expulsion of semen and sperm into the upper vagina. A typical male ejaculum consists of about 3 to 4 milliliters of semen, containing 300 to 500 million sperm. During the final phase, **resolution,** the penis returns to its unaroused size, and bodily processes such as heart rate and respiration gradually return to normal. Following orgasm, many men go through a refractory period, lasting from a few minutes to several hours, during which they are relatively unresponsive to sexual stimulation and may not easily experience another erection.

The Role of Hormones in Male Reproductive Function

Before a male can mature and function sexually, special regulatory chemicals called *hormones* must come into play. Male sex hormones are collectively called **androgens.** The hormones that travel from the brain and pituitary gland to the testes (or ovaries in females) are the *gonadotropins*. The *hypothalamus*, a region in the brain (Chapter 40), passes two releasing hormones (small peptides) to the *anterior pituitary*, an endocrine gland that lies at the base of the brain not far from the hypothalamus. These releasing factors trigger secretion of two gonadotropins, **luteinizing hormone (LH)** and **follicle-stimulating hormone (FSH),** into the blood. LH acts in male embryos and newborns on the *interstitial cells* located between the seminiferous tubules of the testes. The interstitial cells respond to LH by manufacturing and secreting **testosterone,** an androgen, into the blood. The other pituitary hormone crucial to male reproductive function, FSH, acts on the seminiferous tubules to support sperm production.

As puberty approaches, testosterone causes various body changes, including the development of the **secondary sexual characteristics:** the penis, testes, and related glands enlarge and become sexually functional; body hair grows; the voice deepens; and general body shape changes as muscle and bone growth occurs. Although most of us are unaware of it, there are structural and biochemical differences between the kidneys, liver, salivary glands, and muscles of adult men and women. These differences are due to the activity of testosterone or its metabolic products in males. Without testosterone, most of the bodily features commonly considered masculine would not appear, and, like Peter Pan, the male would remain boyish forever.

The maintenance of male secondary sexual characteristics depends on the continuous presence of testosterone. These are called **activational effects** of the hormone. In contrast, certain developmental processes require only a brief exposure of the embryo or newborn to the hormone at a critical time; these are **organizational effects** of the hormone. Male sexual behaviors are due to both kinds of hormonal action: thus testosterone triggers permanent changes in the embryonic and newborn brain and spinal cord; and when present continuously after puberty, this androgen activates those nerve circuits built years before so that masculine behaviors result.

THE FEMALE REPRODUCTIVE SYSTEM

Like the male, the adult human female must perform several tasks in order to participate successfully in reproduction. She must produce eggs, prepare the uterus for the embryo, engage in appropriate sexual behavior, meet the needs of the embryo as it develops, give birth to the child, nourish the newborn child, and develop hormonal and behavioral patterns that support these functions. The main structures of the female reproductive system are shown in Figure 18-3. In this section, we shall describe how these structures enable the female to carry out the tasks necessary to bring a baby into the world.

Production and Transport of the Egg

Human eggs, like those of other animals, are produced in the *ovaries*. All the eggs that a woman will ever release are already present in her ovaries late in her own embryonic development. Of the initial 2 million or so oocytes, only about 400 will ever mature and be released, usually at the rate of one a month, from puberty to menopause. As Figure 18-4 shows, each egg matures surrounded by a layer of helper follicle cells. As the egg reaches maturity, the fluid-filled follicle bulges from the surface of the ovary. At a hormonal signal, **ovulation** occurs: the swollen follicle ruptures, releasing the egg. Soon after the egg has been expelled, the remnants of the follicle collapse in the ovary and form a **corpus luteum** ("yellow body," named for its appearance).

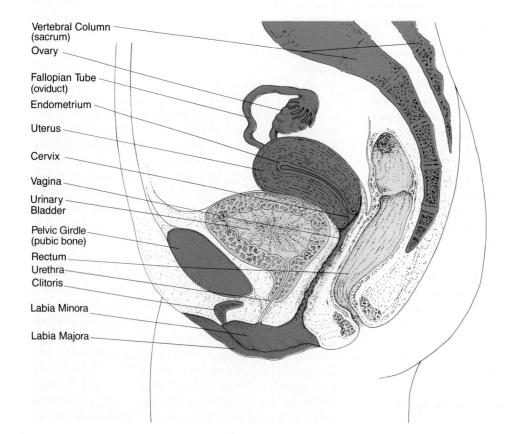

Vertebral Column (sacrum)
Ovary
Fallopian Tube (oviduct)
Endometrium
Uterus
Cervix
Vagina
Urinary Bladder
Pelvic Girdle (pubic bone)
Rectum
Urethra
Clitoris
Labia Minora
Labia Majora

Figure 18-3   THE FEMALE REPRODUCTIVE SYSTEM.
The egg passes from the ovary to the upper Fallopian tube, where it may meet sperm that have been carried from the vagina through the cervix, uterus, and lower portion of the Fallopian tube.

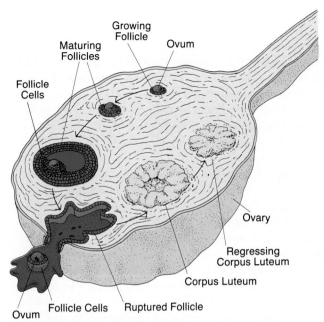

Figure 18-4   GROWTH OF THE FOLLICLE.
This idealized section through a mammalian ovary shows
the stages of follicular maturation, ovulation, and formation
of the corpus luteum.

The ovulated egg is caught by the motile, fingerlike
projections at the opening of the **Fallopian tube,** or *oviduct.* The Fallopian tubes are not connected directly to
the ovaries, which are held in place in the abdominal
cavity by ligaments. Thus the egg must traverse a gap
between the ovary and the fringes of the tube. Once the
egg is safely inside the tube, the beat of hairlike cilia and
the rhythmic waves of muscle contractions move the egg
down the Fallopian tube and into the muscular, pear-
shaped **uterus** (Figure 18-5). About ten days before ovulation, the uterine lining—the **endometrium**—has
begun preparation to receive a fertilized egg by thickening and filling with tiny blood vessels. If the egg is not
fertilized, it passes through the uterus and is discharged,
and the thick uterine lining subsequently is shed in the
process called **menstruation.**

Occasionally, transport of a fertilized egg down the
Fallopian tube fails, and an *ectopic pregnancy* results.
That is, the embryo begins to develop outside the
uterus, usually in the Fallopian tube itself but sometimes in the ovarian follicle or even in the abdominal
cavity.

At the same time that the egg is being transported
downward, various events may be facilitating the
sperm's upward journey. Near the time of ovulation, an
important change takes place in the **cervix,** the base of
the uterus that also serves as the upper end of the **vagina,** the muscular tube that leads from the uterus to the
outside of the body. The levels of various sex hormones
present near the time of ovulation cause the mucus of
the cervix to become much less viscous than it is during
the rest of the monthly cycle. As a result, sperm can
move through more easily, en route to a chance meeting
with the egg far up in a Fallopian tube. In addition, both
the prostaglandins in the semen and **oxytocin,** a small
peptide hormone released by the female's posterior pituitary gland during sexual intercourse, cause uterine
contractions, which help move semen and sperm toward
the Fallopian tubes and the egg. Tracts of cilia in the
lower Fallopian tube may actually beat upward, thereby
helping the swimming sperm to move up toward the
egg. These various transport processes are rapid; it is
estimated that human sperm may reach the upper one-
third of the Fallopian tube in as little as five to thirty
minutes following ejaculation. Even though 300 to 500
million sperm are released, the rigors of the journey, the
dilution of semen by the fluids of the female system, and
the fact that about half the sperm may chance to travel
up the Fallopian tube lacking the egg mean that only a
few hundred to a few thousand sperm may reach the
level of the descending egg.

Figure 18-5   CILIA LINING THE FALLOPIAN TUBES.
The electron micrograph (magnified about 2,000 times)
shows the tiny, beating cilia on cell surfaces of the Fallopian
tubes that propel the egg downward and perhaps propel the
sperm upward.

BIRTH CONTROL AND ABORTION

All human societies have, at one time or another, developed mechanisms of birth control. The impact of too many children on the family, tribe, or population has led to methods as diverse as infanticide (permitting the newborn to die) and the sophisticated pharmacology of the "pill."

Many societies employ a "natural" type of birth control: women nurse their babies for many months or even for years. Prolactin, which is released in response to nursing, inhibits the release of LH (and perhaps FSH), which may lead to the suppression of ovulation and the menstrual cycle. This is an effective—although not guaranteed—means of birth control that can be used to regulate the spacing of children; a new child can be conceived only after an older one ceases to nurse frequently. Recent data on women in Bangladesh, Guatemala, and the United States indicate that women who breast-feed exclusively (without occasional bottle feedings) may not ovulate for from fourteen to twenty months. If bottle supplements are used, the less intensive nursing permits the ovarian cycle to resume.

Table A lists the major methods of birth control, or *contraception*, as practiced in the Western world. Mechanical means of interfering with fertilization may be employed by either sex partner. Condoms are thin, rubber sheaths designed to fit snugly over the erect penis and to retain the sperm. Diaphragms are filled with a sperm-killing jelly and inserted in order to cover the cervical opening. A cervical cap does the same job and is left in place permanently except during menstruation. The intrauterine device, or IUD, is inserted by a physician into the uterus; there, the IUD interferes with the implantation process. Unfortunately, a very low percentage of women unknowingly expel their IUDs and thus may become pregnant if they engage in sexual activity. IUDs do have rare side effects, such as minor pelvic inflammation or uterine bleeding.

Surgical methods of contraception are virtually permanent and, of course, are the most foolproof procedures. Whereas vasectomy of a male is a brief, minor operation, tubal ligation of a female is more complex; but new procedures make the operation easier and relatively risk-free. Both vasectomies and ligations leave the man or woman free to participate normally in sexual intercourse, but without the possibility of conceiving a child.

Any process dependent on hormones, including ovulation, is subject to potential control by drugs. The "pill" is a minute dose of estrogen and progesterone that is taken daily by a woman between the fifth and twenty-sixth days of the menstrual cycle. The two hormones inhibit FSH and LH secretion and so prevent the growth of follicles and ovulation. Unfortunately, the pill may cause some women to suffer side effects that mimic aspects of pregnancy (tender breasts, nausea, and so on). And even though a very, very small percentage suffer more severe problems with their vascular systems, numerous studies show that the pill is *much less dangerous* to a woman's health and vascular system than is pregnancy.

The "rhythm" method of birth control depends on knowledge of the time of ovulation so that the partners may abstain from sexual intercourse for the period from two days before ovulation to one-half day afterward. A woman must take her temperature daily in order to detect the 0.5°F elevation that signals ovulation. Safety argues that the couple should probably avoid sexual congress for about a week around the time of ovulation. Besides that problem, about 15 percent to 20 percent of women have such irregular cycles that the rhythm method is simply not reliable.

Because of the immense difficulties for humankind that unchecked population growth is sure to bring, research is under way to find safer, more effective, and easier-to-use contraceptive procedures. Development of a male "pill," for instance, is being eagerly sought. Can you think of some aspects of sperm maturation or function that might be susceptible to chemical inhibition?

Abortion is not an effective long-term form of birth control but has been practiced for centuries as a way to terminate occasional unwanted pregnancies. Induced abortion was described in Chinese writings from 4,000 years ago and by such philosophers as Aristotle and Plato. Almost three out of every five implanted embryos abort for "natural" reasons; hence, abortion is in that sense a fairly frequent phenomenon in nature.

Table A   **BIRTH-CONTROL METHODS**

Device or Strategy Employed	Effect	Failure Rate*	Advantages	Disadvantages
Methods Used by Female				
Abortion	Prevents completion of development	0	Effective	Expensive; morally objectionable to some
Tubal ligation (cutting and tying of oviducts)	Prevents egg transport	0–1	Effective	Irreversible
Oral contraceptives (the "pill")	Suppresses ovulation	0–3	Effective; reversible	Expensive; requires daily action; possible side effects
Intrauterine device (IUD)	May prevent implantation of embryo in uterus	0–3	Effective; reversible	May cause extensive bleeding; may be expelled and lost
Diaphragm	Inhibits sperm survival and/or transport	10–30	Safe; reversible	Must be inserted and removed regularly
Vaginal foams and jellies	Inhibits sperm survival and/or transport	8–40	No prescription required	Effective only if applied immediately before intercourse; messy
Diaphragm plus spermicidal	Inhibits sperm survival and/or transport	0–6	Safe; effective; reversible	Same as diaphragm
Vaginal douche	Inhibits sperm survival and/or transport	30–50	Inexpensive; can be used "after the fact"	Ineffectual; may actually promote conception in some cases
Methods Used by Male				
Vasectomy (cutting of vas deferens)	Prevents sperm release	0–1	Simple; effective	Irreversible
Condom	Prevents transfer of sperm to vagina	7–15	Simple; reversible; may prevent VD spread	Expensive; interrupts sexual activity; may leak or break
Premature withdrawal	Prevents transfer of sperm to vagina	15–25	No cost	Requires strong will; frustrating; ineffectual
Methods Used by Couple				
Rhythm method (abstinence from intercourse around time of ovulation)	Prevents contact of egg and sperm	15–35	No cost; acceptable to Roman Catholic Church	Requires extensive study and effort to be effective; ineffective if periods irregular

*Figures given are best and worst estimates of the number of undesired pregnancies per year per 100 couples using the methods as their sole form of birth control.

Several types of abortions are performed surgically. Early in pregnancy, the cervix may be dilated, or gradually widened, so that an evacuation, or suction, device may be inserted to remove the embryo (the procedure is called "dilation and evacuation," or "D and E"). Alternatively, a curette, or scraping instrument, may be used. Abortions of fetuses older than sixteen weeks is a slightly more complicated process involving injection of a salt solution that triggers expulsion of the fetus.

Abortion is a complex issue for society, since for various persons, it involves basic biology, a life, freedom of choice, ethics, religion, and the law. The arguments for and against abortion are beyond the scope of this book.

Female Genitals

The human female external genitals, collectively called the **vulva**, consist of the *labia majora*, two major folds of skin; the *labia minora*, two inner folds; and the *clitoris*, a small, highly sensitive structure that, like the penis, becomes enlarged and erect during sexual excitement. The labia protect not only the clitoris, but also the openings of the urethra and of the vagina, the female organ of copulation. The walls of the vaginal chamber are thin and distensible so that they can expand to encompass the erect penis and to permit the passage of a baby at birth. In most young girls, the vaginal opening is partially covered by a thin membrane, the *hymen*, which may be ruptured during normal strenuous activities or may be stretched or broken during sexual activity. Also present near the vaginal opening are the *Bartholin's glands*, whose functions are not understood.

The Female Sexual Response

The phases of the sexual-response cycle in women parallel those in men—excitement, plateau, orgasm, and resolution. As in the male, the female's blood pressure, heart rate, and breathing rate increase during the excitement phase, and her breasts and nipples may also swell, as do the labia minora. During the plateau phase, the clitoris, which has become engorged, retracts upward so that its hypersensitive tip is better protected from further stimulation, and the outer third of the vagina swells and thickens into the *orgasmic platform*, which more effectively grips the penis.

A woman's orgasm is characterized by rhythmic contractions of vaginal muscles, an intensive "suffusion of warmth," and a hyperawareness of the clitoral–pelvic area. Whereas men usually experience a sexually unresponsive refractory period, many women are capable of experiencing several orgasms without intervening rest periods. Following orgasm, the woman enters the resolution phase, during which the heightened physiological processes return to normal, and she may have a sensation of well-being and warmth. The extraordinarily pleasurable sensations of orgasm may be one way of explaining why human females possess the unique capability to engage in sex when reproduction itself is impossible—at times of the month when no egg is mature and the uterus is unprepared for an embryo.

The Role of Hormones in Female Reproductive Function

Reproductive cycles and sexual functioning in female mammals are coordinated by a variety of hormones secreted in various parts of the body. These hormones regulate the **menstrual cycle,** the cyclic preparation of the uterus to receive an embryo, and the **ovarian cycle,** during which eggs mature and ovulation occurs.

The monthly preparation of the uterine lining for the fertilized egg normally begins at puberty with **menarche,** the first menstruation. Day 1 of each cycle begins when the menstrual flow starts, a process in which the old lining of the uterus breaks down and is sloughed off. As Figure 18-6 shows, beginning on about day 5, the endometrium begins to build up again. All these changes depend on several gonadotropins and **gynogens,** or female sex hormones. One, FSH, is secreted in minute quantities by the pituitary gland and stimulates an ovarian follicle to commence intensive growth and the egg inside to mature. It is not known why only one egg and follicle respond each month. At the same time, LH released from the pituitary gland stimulates the follicle to manufacture and secrete **estrogen,** a gynogen. Among other functions, estrogen causes the uterine wall to grow 2 to 6 millimeters thick and to become rich in blood vessels.

Ovulation is triggered about midway through the menstrual cycle by a burst of LH from the pituitary gland. The corpus luteum (Figure 18-4), the mass of cells formed from the collapsed follicle soon after the egg is released, secretes yet another gynogen, **progesterone,** in addition to estrogen. Progesterone acts on the wall of the uterus to cause a final "ripening" in the form of increased development of tiny blood vessels and glands. Progesterone also inhibits the growth of the numerous immature follicles in the ovary, thereby preventing additional eggs from maturing, being released, and perhaps being fertilized following sexual activity. Thus the first egg has an optimal chance of completing its development without competition from younger embryos in the same uterus.

The normal ovulated egg can be fertilized only within twenty-four hours of release. If it is not fertilized, it dies within a few days. About eleven days after ovulation, if the egg has not been fertilized the corpus luteum begins to regress, thereby cutting off the supply of progesterone. This leads to the resorption and sloughing off of the endometrium—menstruation. Despite extensive research, it is still not clear what causes the corpus luteum to degenerate and to lower progesterone production so dramatically. Research on nonhuman female mammals suggests that prostaglandins produced after ovulation decrease corpus-luteal functioning and hence progesterone secretion. These declines are crucial, of course, since they trigger not only menstruation, but also renewed pituitary-gland secretion of FSH and LH. As a result of the availability of LH, a new cycle starts.

Figure 18-6   MONTHLY CYCLES OF HORMONE LEVELS, THE OVARY, AND THE UTERUS IN THE HUMAN FEMALE. Pituitary FSH and LH control ovulation and production by the ovary of estrogen and progesterone. The endometrium grows thicker in response to estrogen and progesterone. If fertilization does not occur, progesterone levels fall, and the endometrium breaks down. If any embryo is present, it secretes human chorionic gonadotropin (HCG), which keeps the progesterone level high, and menstruation does not occur.

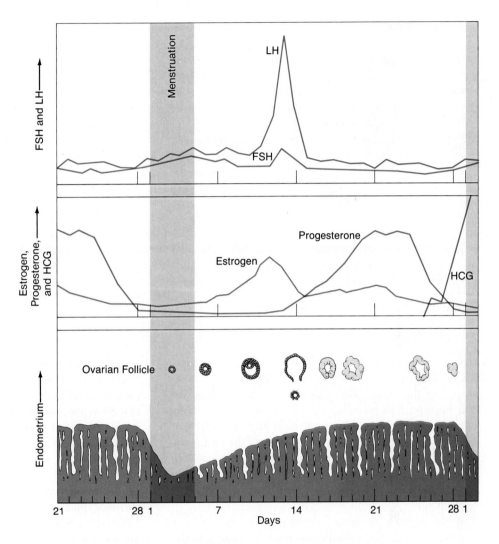

As in males, female secondary sexual characteristics develop at puberty in response to the presence of sex hormones. Estrogen secretion in females causes breast development, changes in body proportions and fat deposition, hair growth, initiation of ovulation and menstruation, and the female sexual response.

Some aspects of female behavior are affected by the presence of sex hormones in the blood. Toward the end of each menstrual cycle, some women experience irritability, depression, or slight nausea. This so-called premenstrual tension may stem from low levels of progesterone and estrogen. Several studies also imply that prostaglandins may contribute to painful cramping during menstruation. More dramatic emotional, behavioral, and physiological changes may accompany the **menopause,** when the ovaries cease to function. When a woman reaches about age forty-five to fifty-five, the ovaries lose their sensitivity to FSH and LH, they stop making normal amounts of progesterone and estrogen, and the monthly menstrual cycles cease.

ORIGINS OF SEX DIFFERENCES: FEMALENESS AND MALENESS

The ultimate control of gender and sexual development resides in the chromosomes. Human females have two X chromosomes; males, an X and a Y. Therefore, genes on the Y chromosome in male embryonic and adult cells are not present in female cells, and this difference has clear developmental consequences. In both male and female embryos, the early development of the gonads, internal duct system, and external genitals is identical. Figure 18-7a shows this *indifferent stage* of development. Let us look at what happens to the various organs from this point onward.

An early step in gender development occurs when the gonial cells, the source of eggs or sperm, migrate to the indifferent gonad, which consists of a *cortex* on the sur-

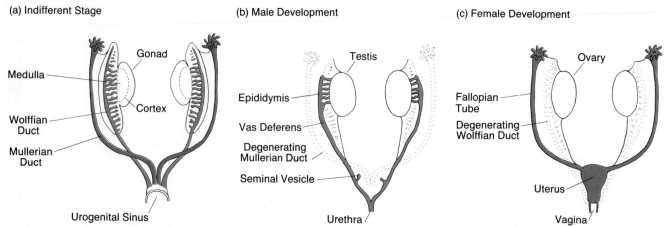

Figure 18-7   DIFFERENTIATION OF MALE AND FEMALE REPRODUCTIVE STRUCTURES IN HUMAN EMBRYOS.
(a) The indifferent stage, at about five weeks' gestation. (b) Male development begins at about the seventh week. Note that the Wolffian ducts develop into the epididymis and vas deferens, while the Mullerian ducts degenerate. (c) Female development begins at about the eleventh week. Here the Mullerian ducts develop into the Fallopian tubes, uterus, and vagina, while the Wolffian ducts degenerate.

face and a *medulla* within. If the developing embryo has XX sex chromosomes, the cortex develops into the main tissues of the ovary. The cells surrounding each gonial cell become follicle cells, and the nearby connective-tissue cells become interstitial cells that manufacture estrogen in response to LH. Conversely, if the embryo is XY, the cortex regresses, and the medulla develops into a testis. Gonial cells take up residence in the seminiferous tubules, where they may begin sperm production later in life.

If the indifferent gonad develops into an ovary, a nearby tube called the **Mullerian duct** develops into the Fallopian tubes, uterus, and upper portion of the vagina. If the indifferent gonad forms a testis instead, embryonic **Wolffian duct** forms the epididymis and the vas deferens for later sperm storage and transport. In females, the Wolffian duct regresses and disappears, whereas in males the Mullerian duct vanishes.

The external genitals also have an early indifferent stage when they are identical in the two sexes. A tiny swelling called the phallus forms the glans of the penis or the clitoris as development proceeds. Swellings surrounding the tiny phallus become either labia or scrotum, depending on the sex of the fetus. The differentiation of these external genitals depends on the direction taken by the developing gonads and the hormonal signals they produce and receive.

Hormonal Control of Sexual Development

If the primitive gonads are removed from a male and a female embryo, both will develop a Mullerian-duct sys-

tem (Fallopian tubes, uterus, and vagina). The two Wolffian ducts, which normally would yield the epididymis and the vas deferens, degenerate, even though normal XY chromosomes are in the cells of those structures in the castrated male embryo. The external genitals of these gonadless embryos also develop as female structures. For this reason, biologists call the female body type the "neutral sex," since it develops in the *absence* of sex hormones and irrespective of whether XX or XY chromosomes are present.

What controls the development of male organs in a normal XY embryo? Three substances play a role. First, if a Y chromosome is present, testosterone is produced in the early gonad. This hormone causes each Wolffian duct to form an epididymis and a vas deferens. Second, some testosterone enters cells in the pelvic region and is modified into *dihydrotestosterone*, a different hormone that brings about masculinization of the external genitals (formation of penis and scrotum rather than vulva). Third, still another hormone, **Mullerian inhibiting substance (MIS)**, does what its name implies—it causes the Mullerian system to regress in males so that Fallopian tubes, uterus, and upper vagina do not develop. In short, the three hormones cause the internal male duct system to develop, the internal female duct system to be repressed, and external genitals to become masculine.

Sex hormones also affect the brain. Specifically, testosterone travels to the brain of a male embryo, where it is taken into and acted on by certain developing nerve cells. In the cells, it is changed by enzymes into estrogen. This estrogen causes clusters of nerve cells in the brain and lower spinal cord to mature in such a way that they will control such male behaviors as sexual responsiveness and the thrusting reflex of sexual intercourse.

(Although we commonly think of estrogen as a female hormone, it is the active agent that functions organizationally to cause male brain and spinal-cord centers to arise in the embryos of mammals and birds.) One action of this testosterone-derived estrogen is to permanently suppress in males the ability of the hypothalamus to control monthly menstrual and ovulatory cycles with LH and FSH. In female embryos, the absence of this estrogen effect permits the hypothalamus to develop in such a way that it can activate these cycles later, at puberty.

To summarize, normal femaleness is due to the absence of testosterone and MIS in the embryo, and normal maleness is due to the action of both testosterone and MIS in the embryo. Thus both positive and negative hormonal control agents influence sexual development in males. As a result of these two kinds of controls, the basic mammalian tendency toward the female type of body and nervous system is altered into maleness in XY individuals.

Intersexes—Hormones Gone Wrong

Any time chemical control factors are manufactured and secreted, opportunities for error exist. Errors in hormonal control of sexual development can produce individuals with both male and female sex characteristics, or **intersexes.** One type of intersex is the true **hermaphrodite,** an individual with both testes and ovaries. Such a person's genitals are usually masculine, but developed breasts may also be present. This type of intersex is quite rare. More commonly, the person is a **pseudohermaphrodite,** having the gonads of one sex and the external genitals of the other. How do intersex individuals arise?

There appear to be several causes of pseudohermaphroditism. In one type, a genetic female embryo (XX) develops ovaries, but during embryonic development, her adrenal glands begin to produce testosterone. Because every female embryo has a full set of androgen receptors on her cells, the testosterone succeeds in producing male external genitals and internal ducts (epididymis and vas deferens). Of course, no source of MIS is present, so the Mullerian system forms the Fallopian tubes, uterus, and upper vagina.

The second type of pseudohermaphrodism, called *testicular feminizing syndrome*, occurs when a genetic male embryo (XY) develops testes, but the nearby tissues are insensitive to the testosterone that the testes produce. As a result, female external genitals form. In most pseudohermaphrodites, the genitals tend to be incompletely formed at birth, so that it is difficult to determine on the basis of appearance whether the child is male or female.

Although these developmental outcomes are relatively uncommon, they reflect the fact that all kinds of subtle gradations in sexual development and behavior can occur in humans; all of us passed through a neutral-sex stage and went beyond it to varying degrees, depending on the intricate hormonal spectrum that shaped our bodies and brains. Prejudice against individuals who may not fit the stereotypes of "male" or "female" has no biological or ethical justification.

PREGNANCY AND PRENATAL DEVELOPMENT

Now that we have considered how human reproductive organs work and how they develop, we shift our focus to the central process of reproduction: the nine-month pregnancy period, during which the woman's body nourishes and protects the embryo as it grows to a full-term baby.

Events of Prenatal Development: From Zygote to Baby

Pregnancy is arbitrarily divided into **trimesters,** periods of three months each. The first trimester, beginning at fertilization, is the time when most of the embryo's organs take form. The other two trimesters are mainly growth periods for the organism, then called a **fetus.**

The First Trimester

The route followed by a fertilized egg is shown in Figure 18-8. After fertilization high in the Fallopian tube, the zygote goes through several cleavages as it is transported down the tube. It becomes a solid ball of cells, the *morula* (Latin for "mulberry," which the cluster resembles), and by the fourth day develops into the 50- to 100-cell blastula stage, called a *blastocyst*. As we saw in Chapter 16, the mammalian blastocyst is characterized by an outer layer of cells—the **trophoblast**—and an *inner cell mass.*

The next stage occurs when the blastocyst enters the uterus and adheres to the uterine wall. It undergoes **implantation,** a process during which the trophoblast cells invade the endometrium, securing the embryo to the wall of the uterus. Implantation usually is completed eleven to twelve days after fertilization; from then on, the woman is considered pregnant.

The implanted embryo sends an important hormonal signal into the blood of the mother. The trophoblast cells secrete **human chorionic gonadotropin (HCG),** which stimulates the corpus luteum to continue secreting progesterone and estrogen. These hormones, in turn, do

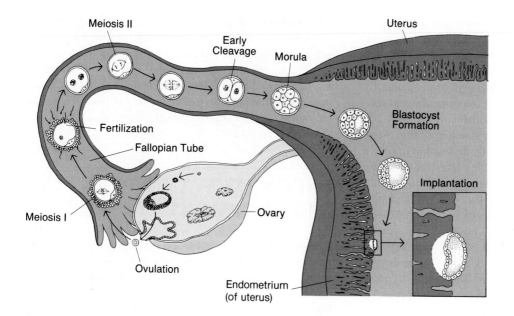

Figure 18-8   THE JOURNEY AND DEVELOPMENT OF THE FERTILIZED EGG. The egg is fertilized in the upper Fallopian tube, undergoes cleavage while traveling down the tube, and finally implants in the endometrium of the uterus.

two things: they suppress release of FSH and LH by the pituitary gland, thereby preventing maturation and ovulation of another egg; and they stabilize the uterine wall to prevent menstruation and loss of the embryo (recall that it is a decline in progesterone that precipitates the sloughing off of the uterine lining at menstruation).

Not all of the HCG that passes into the mother's blood is used in her ovaries. A substantial amount is excreted in her urine. A variety of chemical procedures can be used to detect the HCG and so give an early test for pregnancy.

One of the surprising features of mammalian development is that most of the cells of the early embryo make no contribution to the embryo's body, and instead give rise to membranes. It is only the inner cell mass that gives rise to the whole embryonic body. These cells become arranged in a flat sheet, which subsequently undergoes gastrulation using a primitive streak, precisely as do embryos of reptiles and birds (Chapter 16).

Once gastrulation has taken place, the remainder of the first trimester is devoted to organogenesis. The eyes, the brain and nervous system, the limbs, and most of the other organ systems develop. The embryo's heart begins to beat; like that of hummingbirds and other diminutive animals, it contracts several times per second. Only much later in prenatal development does the heartbeat slow to a rate closer to that of adult humans. Regulatory events and inductive-tissue interactions that take place as the organs form make the embryo particularly susceptible to foreign chemical agents (drugs, alcohol, nicotine, and so on) and to diseases such as rubella (German measles). The latter can cause severe abnormalities of the eyes, ears, brain, and heart of an embryo, while various

drugs may lead to malformations in these or other organs (as thalidomide, which causes limb malformation).

The embryo grows substantially during the first trimester, as shown dramatically in Figure 18-9. For instance, at eight weeks, it weighs 1 gram and is 2.3 centimeters long, but only a month later, it weighs as much as 15 grams and is about 9 centimeters long. By then, its leg muscles can execute the first weak kicks, and its facial muscles produce the first movements of the developing lips, cheeks, and eyebrows. These muscle twitches are not intentional or due to anything resembling thought processes, because the brain is still extremely primitive.

The Second and Third Trimesters

Growth is spectacular during the second and third trimesters. The pregnant woman is usually conscious of fetal movements from the fourth month or so. By then, the fetus's bones have begun to calcify, the brain has added millions of new cells, and the circulatory and respiratory systems are preparing for the remarkable transition from being a fetus immersed in warm fluid to being an air-breathing newborn infant. By about the end of the fifth month, the fetus's circulatory and respiratory systems have developed enough to give it a chance of surviving if born prematurely—although only a moderate chance, even with intensive hospital care. During the eighth and ninth months, the final stages of organ development are completed, so that all the essential organs will be fully functional at birth. During these last two months, the weight of the fetus doubles. Such rapid growth is supported by the feeding efficiency of the **placenta,** the organ that sustains the embryo and fetus

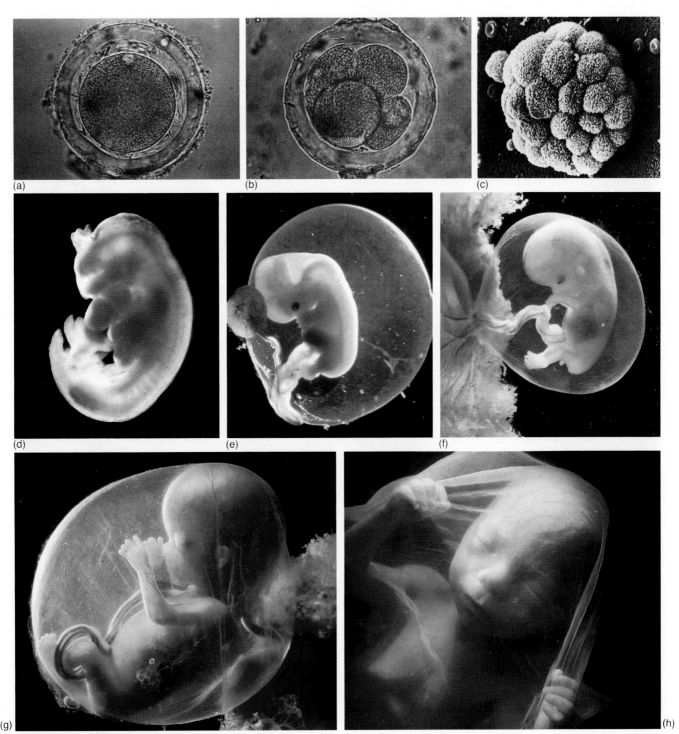

Figure 18-9   HUMAN EMBRYONIC DEVELOPMENT.

These are selected stages of the 9-month developmental odyssey that produces a baby. (a) A fertilized human zygote near the time of the first cleavage division. (b) The 4-cell stage. This embryo would still be traveling down the Fallopian tube. (c) The blastula stage at about the 64-cell stage. This hollow sphere can carry out implantation into the endometrium of the uterus. (d) Between the fourth and fifth weeks of gestation. The major organs are forming; the bright red, blood-filled heart is just below the lower jaw and nearby gill slits (not visible). (e) The embryo at 6 to 7 weeks. The digits are beginning to form on the paddlelike limbs, the eye is pigmented and its clear lens can be seen, and the large brain is evident. The umbilical cord extends from the belly and the fluid-filled amnion surrounds the embryo. (f) At 9 weeks. The fetus is recognizable as a primate, the limbs are elongated, and ears are clearly visible. (g) At about 3 months. The fetus is about 8 centimeters in length and weighs about 28 grams. Its muscles begin moving the limbs and body. (h) At 4 months. The fetus is some 16 centimeters long, weighs 200 grams, and has begun the intensive growth leading to birth.

throughout the pregnancy and through which an exchange of gases, nutrients, and wastes takes place between the maternal and the fetal systems.

The Placenta: Exchange Site and Hormone Producer

The lengthy, intimate maternal–fetal relationship characteristic of mammals is possible because of embryonic membranes that originated in reptiles: the amnion, the yolk sac, the allantois, and the chorion. The latter two give rise to the embryonic parts of the placenta. In mammals, the *amnion* is a fluid-filled sac that surrounds the embryo, cushioning and protecting it. The rupture of the amnion, or "bag of waters," is often the signal that delivery is imminent.

Mammals also develop the *yolk sac*, an evolutionary remnant retained partly because the walls of the yolk sac are the first site where red blood cells develop and begin early in development to transport oxygen between the placenta and the embryo. The yolk sac also gives rise to the primordial germ cells, which migrate to the primitive, indifferent gonad, later to form the gonial cells of the ovary or testes.

The outermost membrane, the *chorion*, arises from the trophoblast, the outer layer of the early embryo involved in implantation, as shown in Figure 18-10. The chorion grows thousands of fingerlike projections called **chorionic villi,** which provide a huge surface area for the purpose of carrying out exchange with the mother's endometrial blood vessels. Each villus eventually houses a tiny blood vessel, part of the embryo's vascular system. This embryonic trophoblastic tissue and maternal endometrial tissue together make up the placenta.

The walls of the maternal vessels near the chorionic villi break open in humans, so that maternal blood bathes the outer surface of the villi (Figure 18-11). Consequently, the substances exchanged between fetal and maternal blood must travel only a short distance by diffusion—about 0.002 millimeter—during late pregnancy.

Many substances are exchanged at the placenta. Oxygen moves from maternal to fetal red blood cells. Carbon dioxide diffuses from fetal to maternal blood and is excreted by way of the mother's lungs. Nutrients pass from the mother's blood plasma into the fetus's. Waste products move from fetal to maternal blood plasma and are filtered away by the mother's kidneys. In addition, certain antibodies and hormones may pass across the placental barrier so that they may act in the fetus or newborn.

The mature placenta is roughly disk shaped, about 20 centimeters across, and about 6 centimeters thick. It has a surface area for exchange that is equivalent to half the

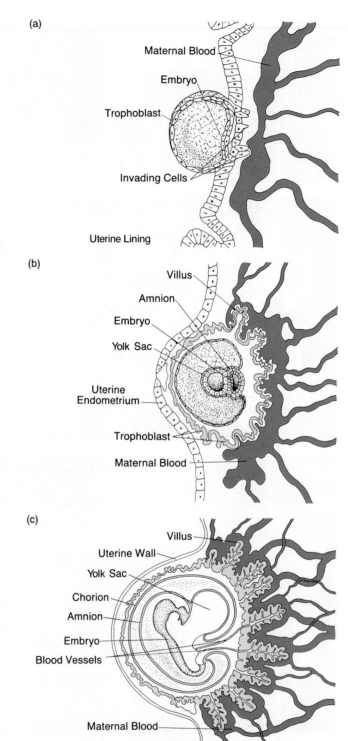

Figure 18-10   THE ORIGIN OF THE PLACENTA.
(a) The blastocyst implants in the endometrium. (b) The trophoblast is surrounded by uterine tissue, and trophoblastic villi of the placenta have started to form. (c) The number of villi and their surface area continue to increase to meet the needs of the enlarging embryo.

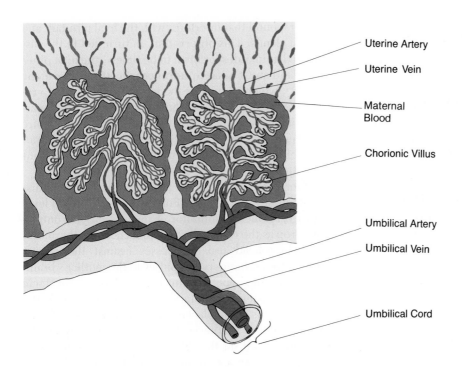

Figure 18-11   MAGNIFIED VIEW OF BLOOD-VESSEL RELATIONSHIPS IN A HUMAN PLACENTA.
The surface of the fetal villi is bathed directly in maternal blood, so that efficient exchange of gases, nutrients, and wastes is assured.

Uterine Artery

Uterine Vein

Maternal Blood

Chorionic Villus

Umbilical Artery

Umbilical Vein

Umbilical Cord

size of a tennis court—16 square meters. About 600 milliliters of the mother's blood flows through the placenta each minute during the later part of pregnancy, supplying the impressive amounts of food and oxygen that permit fetal growth. The placenta remains connected to the abdomen of the fetus by the **umbilical cord,** in which the fetus's umbilical arteries and vein spiral about each other. (These spiraled vessels are the source of the red-and-white–striped design of the barber pole, the medieval advertisement for the surgeon–barbers of the day.) The umbilical blood vessels are all that human embryos retain of the *allantois*, the fourth embryonic membrane of our reptilian and avian relatives, which utilize it as a storage site for wastes. Because mammalian embryonic wastes pass into the mother's blood instead of being stored, the allantois never develops extensively, but its blood vessels form the critically important umbilical arteries and vein.

The placenta has several other functions besides being a site of exchange. For one thing, it serves as a barrier. In humans and all other mammals, there is no true mixing of the maternal and fetal bloods; a barrier of fetal tissue always separates the two types of blood and prevents certain kinds of molecules and cells from passing between mother and fetus. If the surface of the early placenta were not covered with a protective coating of molecules (one of which is thought to be HCG), the embryo might be recognized by the mother's immune system as foreign genetic tissue. This is because half of the embryo's genes (and hence proteins) come from the father and so are foreign to the mother's immune system. If not for the protective function of the placenta,

the embryo might be rejected shortly after implantation, as though it were a foreign skin graft or a heart transplant.

The placenta also functions as a self-insurance system for the embryo—it is almost the equivalent of a pituitary gland because it secretes a variety of hormones. Toward the end of the first trimester, the embryo's production of HCG dwindles. This drop in production might endanger the embryo, since HCG stimulates the corpus luteum to continue secreting progesterone and estrogen (the stabilizers of pregnancy). Just when HCG levels drop, however, the placenta itself begins to manufacture and secrete large quantities of progesterone and estrogen. The embryo truly protects itself by this strategy, since these two hormones help to sustain the endometrial portion of the placenta and also inhibit FSH and LH secretion, which, in turn, prevents the beginning of a new cycle of ovulation and menstruation. The timing of the switchover to placental progesterone and estrogen secretion is critical for maintaining pregnancy, and miscarriage (spontaneous abortion of the embryo) often happens at this time because placental hormones fail to fully take over soon enough.

BIRTH: AN END AND A BEGINNING

About 266 days after fertilization, the human infant is born. When the birth process starts, the fetus's vascular

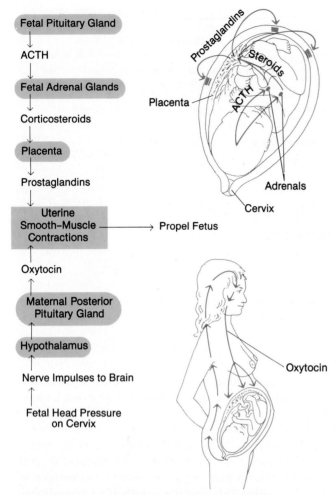

Fetal Pituitary Gland

↓

ACTH

↓

Fetal Adrenal Glands

↓

Corticosteroids

↓

Placenta

↓

Prostaglandins

↓

Uterine Smooth-Muscle Contractions ——→ Propel Fetus

↑

Oxytocin

↑

Maternal Posterior Pituitary Gland

↑

Hypothalamus

↑

Nerve Impulses to Brain

↑

Fetal Head Pressure on Cervix

Placenta

Prostaglandins

Steroids

ACTH

Adrenals

Cervix

Oxytocin

Figure 18-12   A HUMAN BIRTH IS TRIGGERED.
Hormones from the fetus, the placenta, and the maternal brain must act together at the time of birth.

and pulmonary systems—the heart, blood vessels, and lungs—are poised to switch from placenta to air as a source of oxygen. The mother's uterine muscles are keyed to act, and the cervix is set to dilate, or open. **Relaxin,** a hormone from the ovaries and placenta, acts toward the end of pregnancy to loosen the junction of the bones in the front of the mother's pelvis, so that they are better able to spread apart, allowing the baby to pass through without injury.

Changing hormonal levels are believed to initiate the process of birth, and the baby itself plays a key role in starting this chain of events. This conclusion was first suggested by one of nature's experiments. Idaho sheepherders found that pregnant ewes that fed on a weed (*Veratrum californicum*) in high pastures often died because their lamb fetuses remained unborn and grew to two or three times the usual size. Biologists investigating this phenomenon found that the weed contains an alka-

loid compound that interferes with the fetal pituitary and adrenal glands. This observation, in turn, led to the discovery that in order for birth to be timely, the fetal pituitary gland must secrete **adrenocorticotropic hormone (ACTH),** which stimulates the fetal adrenal glands to release steroid compounds. These compounds then apparently signal maternal cells in the placenta to manufacture and secrete two prostaglandins that are powerful stimulators of uterine-muscle contractions; these contractions expel the fetus from the uterus.

Other hormonal changes also precede birth, as diagrammed in Figure 18-12. For example, the blood level of progesterone, an inhibitor of smooth-muscle contraction, drop as the time of birth approaches. Furthermore, the baby's head usually presses against the cervix, thereby triggering nerve impulses to the mother's brain. That causes the hypothalamus to start release of oxytocin from the posterior pituitary. Oxytocin works with the prostaglandins of the placenta to stimulate waves of muscle contractions in the walls of the uterus, forcing the baby downward on the dilating cervix. The rate of the contractions increases from a few each hour in the early stages of labor to about one every two or three minutes. Ultimately, a series of very powerful contractions and a strong "pushing" by the woman push the head and body of the infant through the cervical opening, down the vagina, and out into the world.

After the newborn emerges, uterine contractions continue and expel the placenta and attached membranes—now called the **afterbirth.** The umbilical cord is tied and cut after delivery. (In mammals other than humans, the mother usually bites through the cord.) Muscle contractions in the cord itself and blood clots in the umbilical blood vessels protect the newborn from excessive bleeding. The newborn usually is examined, is given several brief tests of functioning, and receives preventive treatment for any infections it may have been exposed to in the birth process. The mother's body slowly returns to normal, although it takes four to six weeks for her uterus to shrink from a weight of 1,000 grams to its normal 60 grams.

It is sobering to realize how "inefficient" is the entire process of producing a newborn baby. If we start with 1,000 human eggs and expose them to sperm, 840 will be fertilized. Of these, about 690 will implant, but after seven days, only 420 will have survived. Six weeks after fertilization, about 370 will be alive; finally, 310 may actually be born alive. Clearly, much can go wrong with the reproductive process. If spontaneous abortions and resorptions of embryos did not occur, however, the incidence of infants born with severe developmental problems would be much higher. Fortunately, natural processes see to it that the great majority of fetuses that do survive to term are born free of difficulties.

IN VITRO FERTILIZATION: A GOING CONCERN

When Louise Brown, the first "test-tube baby," was born to a British couple in 1978, the happy event made headlines around the world. This reproductive feat was accomplished by removing an egg from the mother's ovary, combining it with the father's sperm in a glass dish, and reinserting the live, two-day-old embryo into the mother's uterus. Since that historic birth, *in vitro fertilization* (*IVF*, literally "fertilization in glass") has received far less media attention, but has grown steadily from a bold experiment to an established medical procedure performed in clinics around the world. Nevertheless, both the medical profession and the public continue to grapple with ethical issues surrounding the manipulation of human eggs and embryos.

In vitro fertilization was developed as a way to help infertile couples reproduce. An estimated 500,000 American women lack Fallopian tubes or have blockages of these vital conduits, even though they do have normal ovaries and uterus. Since fertilization usually occurs in the tubes, these women normally cannot conceive a child unless the tubes can be opened surgically. Failing that alternative, IVF can be used to circumvent the blockage.

By the mid 1980s, more than 100 commercial IVF clinics were in operation or planned, and many were handling 200 to 400 patients a year. This level of demand is impressive, considering the costs and success rates of the procedure. Each attempted fertilization costs a couple more than $5,000 and has only a 15 percent to 20 percent chance of success. Therefore, it may take several attempts before a pregnancy is achieved. Even so, having children is a high priority for many couples, and IVF is the only alternative when standard fertility treatments fail.

In light of the steep costs of IVF, one of the first issues to emerge was who will pay the bills—the couple or their insurance company (and its pool of policyholders)? So far, the practical answer has been both, with the couple paying the major share and insurance paying part of the hospital costs. The American Fertility Society is establishing standards for new IVF clinics to try to protect both paying parties and physicians in this high-risk field of medicine.

By far the biggest ethical issue regarding IVF concerns what types of embryo manipulations should be allowed. At present, IVF technology is sophisticated enough to allow three procedures in addition to the external fertilization of a woman's egg by her husband's sperm: (1) the egg can be donated by a woman other than the wife, or the sperm can be donated by a man other than the husband; (2) an embryo can be produced from a couple's own gametes, but implanted in a second woman, who carries the baby to term; or (3) a couple's embryos can be frozen, stored, and used later, whether by them or by an entirely different infertile couple. While these three procedures are not being attempted in the United States, some Australian and British clinics announced their intention to do so—an announcement that was met by public concern. In both countries, ethics committees were quickly established to consider and control the freezing of human embryos and transfers involving third parties.

The Ethics Advisory Board of the U.S. Department of Health and Human Services also met to consider such issues and made several recommendations, including the following: human embryos are entitled to profound respect, but not necessarily to the full legal and moral rights of a person; embryos can be experimented on, but these laboratory subjects cannot later be implanted; and all implanted embryos must be derived from married couples. So far, all IVF clinics in the United States are privately owned, and while there are no state or federal laws governing their operation, they do appear to be abiding by the ethical guidelines noted above, according to recent reports. Clearly, human reproductive technology is advancing quickly and making possible exchanges of living cells that challenge our traditional notions of conception and parenthood. Society must watch and carefully judge developments in this field, and informed citizens with biological training can play an important leadership role in that process.

MULTIPLE BIRTHS

Twins are born about once in every 90 human births; triplets, once in every 7,500 human births; and quadruplets, only once in every 435,000 births. What biological processes are responsible for these relatively rare events? Human multiple births may arise in two ways: (1) dizygotic (fraternal) twins arise by fertilization of two eggs that happen to be ovulated during a single monthly

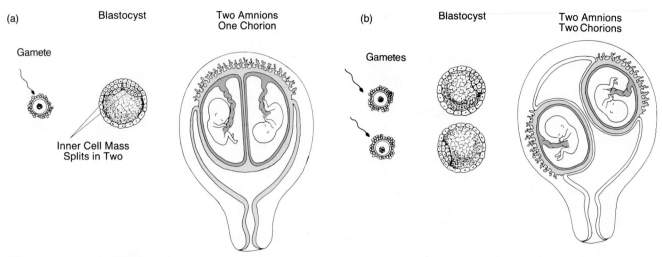

Figure 18-13   TWINNING IN HUMANS: MONOZYGOTIC AND DIZYGOTIC TWINS.
(a) Each inner cell mass is the source of an embryo. There is a single chorion (placenta) for identical twins. (b) The two chorions (placentas) of dizygotic twins normally remain separate. They may fuse; however, and as a result, the blood of the two embryos may mix. This situation can produce abnormalities in sex development, particularly a masculinization of a female embryo if its twin is a male.

cycle, and (2) monozygotic (identical) twins arise by fertilization of one egg followed by separation of the early cleavage-stage embryo into two (or more) developing systems (Figure 18-13).

Multiple births are common in many other animals. An informative example is the nine-banded armadillo of the southern United States and Mexico. The fertilization of one egg by one sperm yields one cleaving embryo in which the inner cell mass develops four primitive streaks, gastrulation sites, each of which produces a normal embryo. In another armadillo species, an average of eight embryos arise in this manner from a single fertilized egg. Thus the fertilization event in armadillos culminates in four or eight genetically identical siblings.

It is extraordinary, considering this process, that these armadillo embryos and young *are not precise physical copies of one another*. Each may arise from slightly different numbers of cells, and since mitosis may yield varying numbers of differentiated cell types, each individual may look or function a little differently from the others. This slight variability is also found among most human monozygotic siblings, emphasizing the fact that *genes are not the sole controlling factors in development*. There are nongenetic influences in every biological system: the cytoplasm, the environment within the egg, embryo, or uterus; and the external environment after birth may all contribute to variability among individuals.

A condition opposite to twinning has been demonstrated experimentally by causing several embryos to fuse and yield a single individual. For example, a cleavage-stage embryo from a purebred strain of white mice may be fused with a similar embryo from a purebred brown strain, as shown in Figure 18-14. The aggregate of cells is implanted in the uterus of the host mother. When born twenty days later, the chimera (combined creature) proves to be a random mixture of cells from the white and brown strains. The pelt may be striped; the heart, liver, brain, and other organs may be composed of patches of each strain cell. (In Chapter 32, we will learn why these cells do not display immune rejection.) A similar mixed organism has been created with three embryos from six parents.

These experiments, as well as natural ones involving multiple births, demonstrate that it is not biologically accurate to regard fertilization as the step that "creates" a new individual. At fertilization, an already living system is given impetus to develop; but that system may form one, two, or more individuals or—experimentally—only portions of individuals. Morphological, physiological, and behavioral individuality appears only over the course of embryonic development and after birth. Fertilization creates only the potential for the individuality that develops later.

MILK PRODUCTION AND LACTATION

Once the baby is born and no longer is nourished by the placenta, the mother can take over feeding the infant by producing and **lactating** (secreting) milk. **Mammary glands,** which are an evolutionary innovation of mammals (Chapter 26), evolved from glands in the skin. They manufacture a highly nutritious mixture of fat, protein, and carbohydrate. After undergoing identical initial de-

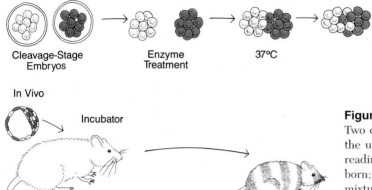

In Vitro

Cleavage-Stage
Embryos

Enzyme
Treatment

37°C

In Vivo

Incubator

Figure 18-14   CHIMERIC MICE: ONE MOUSE FROM TWO.
Two cleavage-stage embryos are fused and are implanted in
the uterus of an incubator female hormonally induced into
readiness for harboring an embryo. A normal embryo is
born; seen here as an adolescent, all its organs have
mixtures of cells from the brown and the white donor
embryos.

velopment in female and male embryos, the mammary
glands mature further in females at puberty. However,
not until a woman becomes pregnant does the final
growth and maturation take place in the milk-secreting
cells and ducts, in response to progesterone and estrogen.

In the fourth or fifth month of pregnancy, the mammary glands begin to synthesize and store small quantities of a remarkable yellowish fluid, **colostrum,** which
will be the first food of the newborn. Colostrum contains
an abundance of maternal antibodies. In newborn horses
and a few other mammals, antibodies in colostrum can
be absorbed across the wall of the newborn's gut without
being digested; foals receive a massive infusion of infec-
tion-fighting antibodies in this way. In humans and many
other mammals, however, there is less need for the massive amount of antibodies in colostrum. This is because,
in the later stages of human pregnancy, the placenta permits certain maternal antibodies to pass into the fetal
blood (the process is called *passive immunization*). The
baby is therefore born with a dose of antibodies against a
variety of dangerous foreign substances. How, then,
does the infant further benefit from the antibody-rich
colostrum? Biologists believe that the antibodies react
with bacteria and viruses in the newborn's gut and cause
infectious agents to clump so that they can be expelled
from the baby's body with the feces. Colostrum also has
a high protein content, which combats diarrhea. Thus,
ingesting it provides several benefits to the newborn.

A few days after birth, the synthesis of true milk is
stimulated in the woman's breasts by the infant's suckling and by a pituitary hormone, **prolactin.** Once the
placenta has been separated from the newborn at birth
and expelled from the mother, progesterone and estrogen from the placenta can no longer inhibit the release of
prolactin. After milk production starts, an intimate mutual relationship between mother and infant begins (Figure 18-15). The baby instinctively sucks on the nipple,

Figure 18-15   NURSING IN HUMANS.
The nursing relationship between a mother and her child is
psychological as well as physiological.

sending sensory-nerve impulses to the mother's brain that cause prolactin and oxytocin to be released from the pituitary gland. Prolactin stimulates more milk production, and oxytocin stimulates milk secretion. Interestingly, the mother's uterine muscles also contract during nursing in response to oxytocin, just as they did during labor and delivery. Both uterine and mammary smooth-muscle cells have oxytocin receptors that trigger contractions.

The human breast will continue to produce milk as long as nursing continues; when nursing stops, the resultant accumulation of milk acts as a negative-feedback signal on prolactin secretion, and the mammary–pituitary-gland system shuts down. Nevertheless, once a woman has nursed an infant, she will always retain the capacity to produce milk again, even after menopause. In fact, some women who are well beyond child-bearing age make an important contribution to their societies by serving as wet nurses to infants in their tribes.

LOOKING AHEAD

It is fitting that this portion of the book should end with the emergence of the new organism. So far, we have discussed how atoms and molecules become organized into cells and how cells yield organisms, and we have examined heredity, genotype, and the processes of development that yield phenotype from genotype. In the next section of the book, we address another question about life: not how the individual arises and develops, but how life itself began and how it then diversified into the myriad microbes, fungi, plants, and animals that populate our planet.

SUMMARY

1. Sperm are produced in the seminiferous tubules of the testes, are stored in the epididymis, and then pass through the vas deferens, prior to being discharged through the urethra. Accessory glands such as the seminal vesicles and the prostate contribute sugars, *prostaglandins*, and other substances to the nutrient-rich, alkaline *semen* in which the sperm are suspended.

2. In the male sexual response, the *excitement phase* is characterized by blood accumulation in the *penis*, producing an erection. In the *plateau phase*, bodily functions such as heart rate and respiration increase, until the *orgasmic phase* is reached. After orgasm and ejaculation of the semen and sperm, the male body returns to the preexcitement condition; this is called the *resolution phase*.

3. *Androgens* (male sex hormones such as *testosterone*) and pituitary-gland hormones (*luteinizing hormone [LH]* and *follicle-stimulating hormone [FSH]*) govern sperm production, sexual maturation, and many aspects of male sexual behavior. Testosterone may exert *activational effects* on male *secondary sexual characteristics* (effects that occur in the brain during puberty and adulthood) or *organizational effects* (effects that result from a brief exposure at a critical time in embryonic life).

4. Eggs released from the ovary during *ovulation* in response to a burst of LH pass down the *Fallopian tube* and, if fertilized, *implant* in the *endometrium* of the *uterus*. If the egg is not fertilized, the endometrium is shed in the process called *menstruation*.

5. The female external genitals, collectively called the *vulva*, consist of the labia majora, the labia minora, the clitoris, and openings to the urethra and the *vagina*. The *cervix* is at the junction of vagina and uterus.

6. The female sexual response parallels the male's, going through the excitement, plateau, orgasmic, and resolution phases. However, women do not usually experience a refractory period after orgasm and thus are capable of multiple orgasms. *Oxytocin* released during orgasm causes uterine contractions that help propel sperm toward the Fallopian tubes and any ovulated eggs.

7. *Gynogens* (female sex hormones) control the monthly *menstrual* and *ovarian cycles*, which recur from the time of *menarche*, the first menstruation, to that of *menopause*, the cessation of ovarian function. Two pituitary-gland hormones are involved: FSH stimulates ovulation, and LH stimulates production of *estrogen* by the ovarian follicle. Estrogen causes the endometrium to thicken and mature. After ovulation, the follicle becomes a *corpus luteum* and begins to produce *progesterone*, which causes a final "ripening" of the endometrium and inhibits growth of additional follicles. In the absence of pregnancy, the corpus luteum regresses, progesterone levels drop, and the endometrium is shed. The decline in progesterone also triggers renewed secretion of FSH and LH, starting the cycles once again.

8. Both male and female embryos progress to the indifferent stage of sexual development, during which the embryo has both *Wolffian* and *Mullerian ducts*. Then, if testosterone and *Mullerian inhibiting substance (MIS)* act (as is the case in normal males), the gonads, sex

ducts, and genitals develop into the male type. If neither of these substances act, these structures develop into the female type. *Intersexes—hermaphrodites* and *pseudohermaphrodites*—are produced when there are errors in hormonal control of this sexual differentiation in the embryo.

9. Shortly after fertilization, the zygote is transported down the Fallopian tube and becomes implanted in the endometrium. The *trophoblast* cells of the implanted embryo secrete *human chorionic gonadotropin (HCG)*, which stimulates the corpus luteum to continue producing progesterone and estrogen.

10. The first *trimester* of prenatal development is characterized by formation of most of the body parts and organs, at least in primitive form. The second and third trimesters are marked by intense growth of the *fetus*, by completion of organ and tissue development, and by preparation for birth and life outside the womb.

11. Embryonic membranes such as the amnion protect the embryo and fetus as it develops. Embryonic trophoblastic tissue and maternal endometrial tissue form the *placenta*, the organ that sustains the embryo throughout pregnancy and to which it is connected by the *umbilical cord*. Exchange of oxygen and carbon dioxide, nutrients, and waste products occurs through diffusion across the surface of the *chorionic villi*. The placenta's other functions include providing a barrier to certain types of molecules and secreting hormones that maintain pregnancy.

12. Birth results from hormonal and neuronal signals between the fetus and the mother and includes expulsion of the placenta, then called the *afterbirth*, from the uterus. The hormone *relaxin* allows bones at the front of the pelvis to spread apart during birth.

13. Multiple births may arise from fertilization of two or more eggs that happen to be ovulated during a single monthly cycle (fraternal siblings) or from fertilization of a single egg followed by separation of the early cleavage-stage embryo into two or more developing systems (identical siblings).

14. In the fourth or fifth month of pregnancy, the mother's *mammary glands* begin to produce *colostrum*, the newborn's first food. A few days after birth, true milk synthesis and *lactation*, milk secretion, are stimulated in the mother's breasts by the infant's suckling and by a pituitary-gland hormone, *prolactin.*

KEY TERMS

activational effect
adrenocorticotropic hormone (ACTH)
afterbirth
androgen
cervix
chorionic villus
colostrum
corpus luteum
endometrium
estrogen
estrus
excitement phase
Fallopian tube
fetus
follicle-stimulating hormone (FSH)
gynogen
hermaphrodite
human chorionic gonadotropin (HCG)
implantation
intersex
lactation
luteinizing hormone (LH)
mammary gland
menarche
menopause
menstrual cycle
menstruation
Mullerian duct
Mullerian inhibiting substance (MIS)
organizational effect
orgasmic phase
ovarian cycle
ovulation
oxytocin
penis
placenta
plateau phase
progesterone
prolactin
prostaglandin
pseudohermaphrodite
relaxin
resolution
secondary sexual characteristic
semen
testosterone
trimester
trophoblast
umbilical cord
uterus
vagina
vulva
Wolffian duct

QUESTIONS

1. What are the purposes of meiosis?

2. Why are the testes of many mammalian species located in a scrotum?

3. Describe the journey of an egg from the ovary to the uterus. Describe some methods of birth control that prevent or interrupt this journey, and explain how they work.

4. Describe the journey of an individual sperm from the testis until it fertilizes an egg. Describe some methods of birth control that prevent or interrupt this journey, and explain how.

5. Explain how gender is determined in the embryo. What are indifferent gonads? Explain what is meant by the neutral sex. In what ways does the Y chromosome contribute to sex determination? the X chromosome?

6. If a zygote does not contain a Y chromosome, can it develop into an embryo? What if it does not contain an X chromosome? Explain.

7. Would you expect a rare individual with two X chromosomes and one Y (XXY) to be male or female, or in between? What about an individual with one X and no Y (XO)?

8. Explain the terms "zygote," "morula," "blastula," "embryo," "fetus."

9. What tissues give rise to the placenta? The placenta functions as an exchange site for dissolved gases, nutrients, and wastes; as a barrier; and as a complex endocrine organ. Explain each of these functions.

10. What is the role of human chorionic gonadotropin (HCG) in pregnancy? Where is this hormone produced? What makes detection of HCG a useful test for pregnancy?

ESSAY QUESTIONS

1. In what sense is a pregnant mammal an incubator? Would you say that the relationship between the pregnant mammal and the embryo or fetus is symbiotic? Or is it parasitic?

2. What is an individual? Are identical twins separate individuals? Is a chimera (formed by the fusion of two or more embryos) a single individual?

3. Can a zygote think, reason, or feel pain? Can a newborn infant? Can a college student? When do you suppose these capacities develop?

SUGGESTED READINGS

AUSTIN, C. R., and R. V. SHORT. *Embryonic and Fetal Development.* Cambridge: Cambridge University Press, 1972.

A marvelous treatment of normal and abnormal human development, the role of genes and mutations in birth defects, and embryonic membranes.

BEACONSFIELD, P., G. BIRDWOOD, and R. BEACONSFIELD. "The Placenta." *Scientific American,* August 1980, pp. 94–102.

This lucid, well-illustrated account is a superb explanation of this complex organ.

GOLANTY, E. *Human Reproduction.* New York: Holt, Rinehart, and Winston, 1975.

The first place to go for the broad picture—from behavior to biochemistry.

KATCHADOURIAN, H. *Human Sexuality: Sense and Nonsense.* New York: Norton, 1979.

This is a broad introduction to educate and to dispel myths.

MASTERS, W. H., and V. E. JOHNSON. *Human Sexual Response.* Boston: Little, Brown, 1966.

A classic, written in clinical terms and describing in detail this important mechanism.

Part
THREE

ORDER IN DIVERSITY

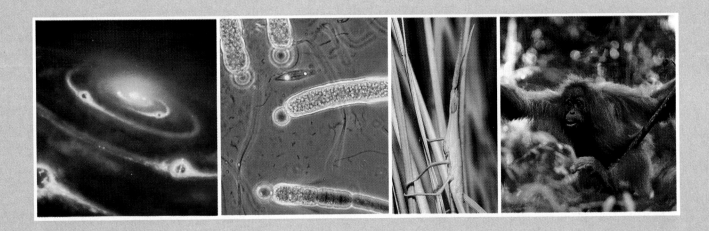

Just as astronomers are fascinated but puzzled by the events that led the universe to form, biologists face a profound mystery when they consider how life may have originated on our planet. They are certain that life arose as a by-product of the physical and chemical conditions and forces that existed early in the Earth's history, and they have theories for how biological molecules, and later single cells, formed and began to reproduce. They are also certain that once present, living things have acted upon and changed the physical world, leading to vast continuing changes of biological and physical features around the world. Many challenging questions nevertheless remain about life's origins and early evolution.

There is little doubt that once the first cells emerged, they gave rise to the splendid diversity of life forms—the millions of species of microbes, fungi, plants, and animals—that have inhabited Earth during its four-billion-year history. Biologists have mapped out the evolutionary relationships between the major groups using fossils and other forms of evidence, and they have cataloged and studied the enormous diversity of life forms in great detail.

Organisms can be categorized as producers, consumers, or decomposers of organic matter. Within each general type and each family lineage, we may see trends toward more and more complex body forms—complexity that brings new ways of living, new adaptations to changing environments. The flowering plants and the warm-blooded animals are sometimes viewed as pinnacles of complexity, structure, activity, and complex interrelations with other organisms and the environment. However, the single-celled alga floating in the sea, the fungus on a rotting log, or the insect chewing a leaf are just as well adapted to their own places in nature as the more complicated plants and animals, and share with them the identical set of fundamental life properties. This sharing reflects the common descent of all organisms from the first pioneers on the early Earth, and helps reveal the evolutionary process by which the millions of later species arose.

19

THE ORIGIN AND DIVERSITY OF LIFE

We are indeed the stuff of which stars are made. Life may be so associated with carbon that at some point, we may be able to make a generalization that life is a property of the carbon atom.

Cyril Ponnamperuma, *The Origins of Life* (1972)

Mounds of photosynthetic cyanobacteria, so-called stromatolites, similar to 2.7-billion-year-old fossils.

How did life begin on Earth? This simple question has spawned an active area of scientific research in recent times as well as mythologies and religious beliefs throughout much of human history. The scientific answers that have emerged provide what is perhaps the ultimate unifying concept in biology. Most modern biologists believe that the splendid diversity of life forms on our planet—both alive and extinct—evolved from simple ancestral cells that lived billions of years ago and that arose, in turn, from nonliving substances by a process of **chemical evolution.** The exact details of that process are still not clear, and there is inherent difficulty in trying to explain the forces at work in the progressive evolution of cells. Nevertheless, biologists believe that chemical evolution involved (1) the increasing organization of naturally occurring small organic molecules into macromolecules; (2) the interaction of these macromolecules in dynamic pathways and assemblies; (3) the coalescence of these interacting molecules into cell-like structures; and, eventually (4), the emergence of true cells, the basic units of life. In this chapter, we will consider many theories and findings that provide the current understanding of this evolution of matter from the nonliving to the living.

An important part of our discussion will center on the geological history of our planet itself, for the chemical evolution that preceded life was a natural and inevitable consequence of the physical and chemical forces that shaped the Earth from its beginning more than 4.6 billion years ago. Thus the story of life's origins really begins with the formation of Earth. After discussing the formation of our planet and the chemical evolution of life, we shall consider the subsequent alterations of the Earth's surface features. Topographic changes such as the drifting of continents, the rising of mountains, and the movement of glaciers have greatly influenced the evolution of diverse life forms by creating a changing variety of habitats.

Finally, we shall discuss the classification into groups of the millions of types of organisms that evolved after cells emerged on the early Earth. The categorization, or *taxonomy,* of the living world helps us understand the physical similarities and evolutionary relationships among individual organisms. That understanding will guide us through Chapters 20 to 26, as we consider the diverse forms of life, one kingdom at a time. Thus we begin in this chapter to examine the single continuum of matter that stretches from cosmic dust to rocks and planets to life in all its exquisite, interrelated forms.

A HOME FOR LIFE: FORMATION OF THE SOLAR SYSTEM AND PLANET EARTH

As inconceivably ancient as our world may be, there was a time when the Earth and the sun did not exist. The sequence that led to life on Earth is hypothesized to have begun with what cosmologists call the **Big Bang**—a monstrous explosion of a ball of extremely hot gases that occurred about 18 billion years ago. Sometime after the blast, all the atoms in the universe formed, 99 percent of them hydrogen and helium and 1 percent all the heavier elements. The cooling and condensing of clouds of this matter during vast periods of time as they hurled (and continue to hurl) outward through space are credited with having led to the formation of the stars and planets.

Astronomers now think that as a star forms, the crushing gravitational force of the compressing gases becomes so great that in a sense the ball ignites. Our sun formed in this way, about 5 billion years ago, from massive clouds of hydrogen and other elements, its central region reaching temperatures of about 20×10^6 °C. The sun's extraordinary heat caused various heavier elements to be made from the hydrogen and helium that predominated prior to star formation, and clouds of this matter were expelled from the sun, with the lightest elements traveling farthest. Local clusters of gases, dust, and larger particles of matter, unevenly distributed in space, tended to set up gravitational fields that attracted other matter expelled from the sun, as well as additional elements produced at the centers of much older, hotter stars, and then distributed in our galaxy in giant clouds of gas and dust (Figure 19-1). This process eventually led to the formation of the planets of our solar system, including the planet Earth.

The best current estimates suggest that Earth and its moon had aggregated as solid bodies by 4.6 billion years ago. As Earth formed, its heaviest elements spiraled inward as part of a gravitational compaction process, whereas lighter elements remained nearer the surface, with the lightest—hydrogen (H_2) and helium (He)—forming the first atmosphere. The weak gravity of the new planet failed to hold this low-density atmosphere in place, however, and nitrogen (N_2) and carbon compounds, such as carbon monoxide (CO), carbon dioxide (CO_2), hydrogen cyanide (HCN), methane (CH_4), and so on, were lost, along with H_2 and He. The compaction

(a)

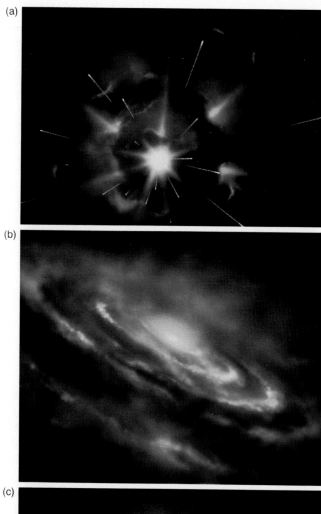

(b)

(c)

Figure 19-1   FORMATION OF THE SOLAR SYSTEM.
(a) The burst of a supernova starts the process that created the sun and planets. The sun is believed to have formed at the center of a huge disk of rotating gas (b), with the planets condensing from gases and rings of matter at varying distances from the sun (b, c).

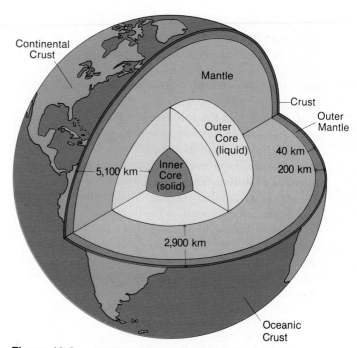

Figure 19-2   EARTH'S INTERNAL STRUCTURE.
The inner core is solid, the outer core fluid. The thick, semisolid *mantle* is surrounded by solid crust that is divided into several layers. The outer crust is thickest in continental regions, thinnest in oceanic regions. In relation to the Earth's diameter (12,800 kilometers), it is a very thin skin everywhere.

process, plus radioactive decay, generated immense heat inside the new planet, producing a molten **core** that consists mainly of iron (Fe) and nickel (Ni), the commonest heavy elements on Earth (Figure 19-2); today the innermost core is solid. Intense upwellings of gases and heavier elements from the core resulted in widespread volcanic and hot-spring activity and the "outgassing" to help form the second atmosphere. In addition, millions of high-velocity impacts of the young Earth with planetesimals (bodies from space of all sizes) caused immense outgassing and, in the view of some scientists, could account for the total H_2O content of the atmosphere and of the oceans. The atmosphere that formed dates from about 4.4 billion years ago and at that time probably consisted of CO, CO_2, H_2, N_2, H_2O vapor, and little or no free oxygen. As atmospheric temperatures fell—and water could exist in liquid form, not just as vapor—hot, torrential rains began falling, and the first oceans appeared.

By about 3.9 billion years ago, some of the lighter nongaseous elements and compounds near the Earth's surface had formed an outer rock "skin," the **crust**, which averages some 26 to 39 kilometers in thickness. Rocks of that early crust are visible today in Greenland

Figure 19-3   A PRESENT-DAY EXAMPLE OF "OUTGASSING" AND OUTFLOW OF MOLTEN INNER EARTH MATERIALS.
Volcanic activity such as that shown here occurs repeatedly from Mauna Loa on the island of Hawaii.

and western Australia. Even that long ago, the basic geological patterns of our planet apparently were established, with huge lava upwellings producing land masses that were surrounded by oceans (Figure 19-3). Finally, cycles of water evaporation from the oceans, subsequent cloud formation, and condensation contributed to the global weather patterns we still observe. Rain falling over the continents resulted in erosion and slow geological alterations. As eroded materials flushed into the oceans, the sea waters became progressively more mineralized and salty.

Given this dynamic setting and a history of formation somewhat similar to those of Mercury, Venus, and Mars—the other three inner, dense planets of the solar system—why did life emerge *here*, and not apparently on the others? One answer may lie in the Earth's temperature, size, composition, and distance from the sun. It is not too cold for complex molecular processes to proceed spontaneously and for much of the water to remain liquid; nor is it so hot that complex organic polymers are degraded (denatured) and water can exist only as vapor, as occurs on the searing surfaces of Mercury and Venus. Smaller planets have a gravitational force too weak to hold much atmosphere, while the huge, gaseous outer planets, such as Jupiter and Saturn, have gravity so strong and atmospheres so dense that penetration of sunlight is poor. Earth, then, has temperatures, liquid water, and other physical parameters of an acceptable range for the type of life that originated here.

Do these "ideal" conditions make it *automatic* that life would arise on Earth? Many scientists feel that life was an inevitable by-product of the physical forces at work on the early Earth: as the Harvard biologist George Wald once put it, given enough time, "the impossible becomes possible, the possible becomes probable, and the probable becomes inevitable."* Nevertheless, an element of chance was surely involved, since so many physical and chemical parameters had to fall within certain ranges in order for life as we know it to begin here.

THE QUESTION OF HOW LIFE BEGAN ON EARTH

With our reconstruction of Earth's history, we see a stage set for the drama of life to begin. Perhaps 600 million years have passed since the planet formed. Sterile continents rise above salty seas warmed by sunlight, by outgassing, and by the molten-rock layers below the Earth's crust. Volcanoes spew out lava and release gases (especially CO_2 and H_2O vapor) into the atmosphere, which, in turn, is held like an invisible cloak by the forces of gravity. Intensive ultraviolet wavelengths, as well as visible light, penetrate the thin atmosphere and the oceans to depths of nearly 20 meters. The CO_2 in the atmosphere helps create the greenhouse effect, in which heat is trapped in the atmosphere and at the Earth's surface.

How could life arise in such a stark environment? Traditionally, there have been three types of answers to this question. The oldest hypotheses for the origin of life are the creation myths, each affirming that life was created by a supreme force or being, although the age of Earth and conditions at the time of creation are different from one myth to the next. A second group of hypotheses suggests that life began elsewhere in the universe and "infected" Earth, so to speak, as microbes that drifted to our planet or arrived in such "capsules" as meteorites or dust from space (even today, some 1,000 tons of meteor material enter the Earth's atmosphere every year). While the first kind of answer may be spiritually satisfying to many people, it cannot be tested or disproved, and thus it lies outside the realm of science. The second "infection" hypothesis is under investigation, but is not judged probable by most scientists.

The major current hypothesis holds that life arose spontaneously on the early Earth by means of chemical evolution from nonliving substances. There is no easy or certain way to verify events that took place billions of years ago, when our planet was very different, but scien-

---

*George Wald, "The Origin of Life," *Scientific American* (August 1954), pp. 45–53.

tists have re-created many of those original conditions in the laboratory and have discovered a great deal of evidence—geological, chemical, biological—to support the hypothesis. In the next sections, we review that evidence and present a step-by-step model for how chemical evolution may have occurred.

THE EMERGENCE OF ORGANIC AND BIOLOGICAL MOLECULES

A simple analysis of traits shared by all present-day forms of life reveals that, regardless of shape and size, each organism is constructed of the same building blocks—a few types of amino acids, sugars, nucleic acids, and lipids. Therefore, these organic molecules themselves are the logical starting point in the search for life's origins.

Until 150 years ago, scientists believed that organic molecules could be produced only by living organisms. Then chemists discovered that, given an energy source, relatively simple carbon-containing compounds, such as urea, can be synthesized from other substances. Now we know that if energy and a combination of light elements, including carbon, are available, many kinds of organic compounds will form spontaneously. Thus, although "seeding" of the ancient Earth by meteorites that bore organic molecules is probable, it seems most likely that the massive quantities of organic material on the early Earth could have formed only on the planet itself. What chain of events could have produced the first organic molecules?

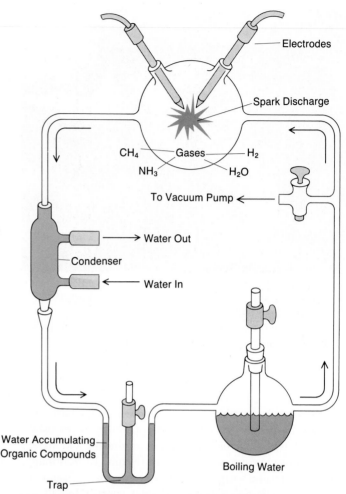

Figure 19-4   MILLER AND UREY'S APPARATUS.
By successfully producing organic compounds under conditions approximating the atmosphere of the primitive Earth, Miller and Urey provided support for the theory that life began by chemical evolution.

The Formation of Monomers

During the 1920s, two scientists made a startling speculation about the formation of organic compounds on the primal Earth. The Russian Aleksandr I. Oparin and the Briton J. B. S. Haldane theorized that energy—in the form of ultraviolet light, heat, radiation, or massive displays of lightning—could have catalyzed the formation of small organic molecules, or *monomers*, from atmospheric gases. Oparin's and Haldane's predictions were probed experimentally in the 1950s, when biochemists at the University of Chicago, Harold Urey and his graduate student Stanley Miller, devised a clever test. Using special laboratory apparatus, they re-created the hypothesized atmospheric conditions on early Earth. Miller connected two round flasks, as shown in Figure 19-4. In the upper flask, he placed an "atmosphere" of methane (CH_4), ammonia (NH_3), water vapor (H_2O),

and hydrogen gas (H_2)—compounds that, at the time of the experiment, were thought to have made up the early atmosphere. He filled the lower flask with a reservoir of water—a miniature ocean. To simulate an energy source such as lightning, electric sparks were discharged into the upper flask for an entire week. At the end of the experiment, the primal "sea" in the bottom flask had collected large quantities of amino acids and simple sugars as well as tarry residues.

Current geological evidence suggests that 4 billion years ago, the Earth probably did not have an atmosphere like the one in Miller's flask, but one closer to the modern atmosphere minus free oxygen. Nevertheless, Urey and Miller's findings are not negated by this new evidence. Experiments using various combinations of CO_2, CO, N_2, and H_2O and such energy sources as ul-

OUTER SPACE: THE SOURCE OF ORGANIC PRECURSORS?

Ever since the famous experiment performed thirty years ago by Harold Urey and Stanley Miller, most biologists have accepted the idea that organic molecules could, and probably did, form in the Earth's atmosphere or oceans and then undergo chemical evolution. More recent studies by astronomers and chemists, however, suggest an extraterrestrial origin for many of those critical compounds.

In the early 1970s, astronomers discovered the presence of enormous clouds of organic molecules in the arms of our galaxy, which spiral outward. Because these clouds are permeated with dust particles that absorb light, they were completely invisible to standard telescopes and were discovered only after radio and infrared telescopes were perfected. Since the clouds' detection, more than fifty types of organic molecules have been identified in them, from simple CS, CN, CH, and CO to methyl and ethyl alcohol, formaldehyde, amino-acid precursors, and straight-chain carbon compounds. Based on these data, astronomers have concluded that organic molecules are ubiquitous in the universe and that they are present in huge quantities.

A logical question arises: How could organic matter from space have ended up on Earth early in its history? The answer, some contend, is that for almost 1 billion years, organic compounds showered down on our planet in the form of *carbonaceous chondrites*— small meteorites that contain a spectrum of organic molecules similar to those detected in the galactic clouds. Like the moon, which still bears scars in the form of ancient impact craters, the Earth was bombarded by asteroids, comets, meteors, and meteorites from 4.2 to about 3.3 billion years ago. One team of astronomers has calculated the total mass of these objects to have been at least ten times the mass of the present-day oceans. Large objects would have vaporized on impact, according to Eugene Shoemaker and Ruth Wolfe of the United States Geological Survey in Flagstaff, Arizona, but smaller objects, such as carbonaceous chondrites, would have survived—and along with them their load of complex organic molecules. Most of Earth's carbon compounds thus were *delivered*, the team contends, not made in the atmosphere or the oceans.

Although there is no firm proof of this hypothesis, there is interesting confirming evidence: recent analysis of a meteorite that fell on Australia in 1969 revealed all five of the nitrogenous bases (adenine, guanine, cytosine, thymine, uracil) that make up DNA and RNA, as well as numerous amino-acid precursors. This finding does not prove that the actual substrates of chemical evolution came in meteorites nor that sufficient quantities were present to engender the earliest forms of life. However, it does confirm that such molecules could have come from any of a number of sources and that such laboratory reconstructions as that of Urey and Miller need not even be invoked in order to show how chemical evolution could have begun.

---

traviolet light, heat, and electric discharges continue to yield more than 100 kinds of organic monomers—indeed, such experiments have produced almost all the building blocks of living cells.

The implication of these experiments is clear: under conditions of heat, humidity, energy, and raw materials similar to those probably present on the Earth billions of years ago, amino acids, sugars, fatty acids, and nucleic-acid bases would have formed readily. And the presence of low levels of NH_3, dissolved in fresh and sea waters, would have stabilized these organic building blocks and prevented them from degrading. Thus the stage could have been set for more complex molecules to appear.

The Next Step: Polymers

At this point in our reconstruction of conditions and processes on the early Earth, the landscape is still life-less, but an impressive array of organic monomers has formed in the atmosphere and has been washed by rain into the soil as well as into warm lakes, rivers, and salty seas. What would be the next step toward life's origin? Probably it was the spontaneous linking of monomers into such polymers as proteins and nucleic acids.

Until recently, biologists believed that the concentrations of organic molecules on the early Earth were high, and they pictured the oceans and seas as a rich "primordial soup." However, this now seems unlikely because it has been shown that ultraviolet light tends to degrade such molecules. Therefore, the ancient oceans and seas probably contained only marginally higher concentrations of organics than they do now, and it probably was only at sites of high concentration that polymerization could have occurred.

Recall from Chapter 3 that polymerization is a kind of "zipper chemistry" and that its individual steps are not necessarily complex, even if the final product is a large, complicated molecule. Where could polymerization have occurred? Current research favors the hypothesis that clay or rock surfaces were a more likely site for polymer formation than were free solutions. Clay crystals, in fact, are rather like genes in that they can reproduce (new crystals), can show variation, and can, in a sense, "evolve" into new crystal structures. Clays also can serve as sites of high local concentrations of organic monomers. In fact, organics such as amino acids can act as primitive catalysts of clay-crystal growth and can alter crystal structure. At the same time, the positive and negative charges within stacked sheets of clay or mica can act as simple catalysts to promote the linking of amino acids into polypeptides (so-called **proteinoids**). Such clays are also sites where polynucleotides (early RNAs) might have formed from bases. That is especially important, as RNA is believed to be the first kind of self-replicating informational macromolecule. The only water that is present as polypeptides form arises during the formation of peptide bonds. Hence, a watery solution is not essential to polymer formation. Even more intriguing is the finding that L-amino acids and D-sugars bind to certain clays, whereas their stereoisomers do not (recall from Chapter 3 that the atoms of certain molecules can be arranged in mirror-image configurations). Because the macromolecules of all living cells on Earth contain only L-amino acids and D-sugars, the binding of these molecules to clay is indeed a fascinating coincidence that may help explain why living things contain just that combination of stereoisomers.

Where would energy come from to drive organic polymerization of polypeptides in clays? Heat from the Earth's core or heat or ultraviolet light from the sun could have been the energy source that drove polymerization reactions. Alternatively, molecules of ATP, which forms in Urey-Miller chambers, could have supplied the chemical energy for polymerization. If ATP is mixed with amino acids and various condensing agents, amino-acid adenylates (the activated amino acids used in protein synthesis) form. Such molecules may then undergo a slow, spontaneous polymerization to form polypeptides. Thus the "first step" in the synthesis of protein could have occurred without ribosomes or tRNAs. But even in the absence of ATP, application of extra heat to a batch of drying amino acids will produce polypeptides with 200 or more amino-acid units. (Interestingly, freezing of amino-acid solutions also can create such polymers.) Finally, as we will see, RNAs can form without any energy input.

From Polymers to Aggregates

If, as many biologists believe, polypeptides and perhaps even short chains of nucleotides collected in clays, ground waters, or small pools, what might the next step toward life have been? Three model aggregates have been generated in the laboratory in order to study whether properties of living cells would become evident.

First, the researcher Sidney Fox found that heating a mixture of dry amino acids and exposing it to water causes the formation of tiny spheres of proteinlike polymers, which he called **proteinoid microspheres** (Figure 19-5). Billions of such spheres can be produced from 1 gram of amino-acid mixture, and some spheres even form chains, very much like certain modern bacteria. Each individual sphere has an outer layer of water and protein molecules and an aqueous interior that may show movement somewhat like cytoplasmic streaming. These spheres can take up and concentrate other molecules from the surrounding solution, can fuse to form larger spheres, can shrink or swell osmotically, and behave as though they have a selective barrier at their surface, even though no lipid is present. Under certain conditions, they produce small buds, form junctions with adjacent spheres, or divide into smaller spheres. This is truly a remarkable finding, since no lipid is required for the membranelike behavior these microspheres display.

Years before Fox published his studies in the early 1970s, Aleksandr I. Oparin had taken another approach and found that solutions of various polymers derived from contemporary cells, such as proteins plus carbohydrates or proteins plus nucleic acids, form polymer-rich droplets—so-called **coacervates** (Figure 19-6). Coacervates not only are reminiscent of tiny cells, but also, under certain circumstances, will divide into smaller spheres. They also will preferentially absorb other large molecules. Oparin actually created self-growing coacervate systems with droplets made of RNA plus protein and other systems that appear to carry out a simple form of electron transport.

The third structure to be generated in the laboratory is the **liposome,** a spherical lipid bilayer that forms easily if phospholipids at the correct concentration range are shaken in an aqueous solution. Proteins may also be embedded in the lipid, just as occurs in the fluid-mosaic plasma membrane of all cells. Such model droplets surround an aqueous internal space, an ideal site to harbor nucleic acids, proteins, building blocks, or energy sources.

Liposomes, coacervates, and proteinoid microspheres are laboratory constructs, of course. Since microspheres form without polymers provided from preexisting cells, they are a more probable analogue to early cell-like structures than are coacervates. However, they lack lipids, an invariant component of the surface of every living cell. Perhaps some aggregate with the combined properties of microspheres and liposomes was an ancient precursor of true cells.

The most important conclusion of these experiments is

(a)

Figure 19-5   PROTEINOID MICROSPHERES.
(a) A group of proteinoid microspheres produced in the laboratory of Sidney Fox and seen magnified about 3,000 times with a scanning electron microscope. (b) A few tiny buds are seen on these microspheres, which are magnified about 14,000 times; when released, such buds may fuse to form larger microspheres.

(b)

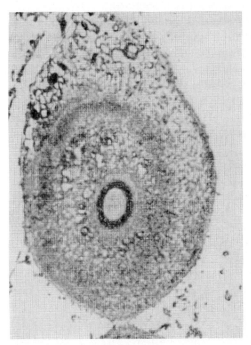

Figure 19-6   COACERVATES.
An electron micrograph (magnified about 10,000 times) of a microsphere in which the inner portion of the proteinoid has been replaced with the lipid α-lecithin. This may also be referred to as a complex coacervate droplet.

that they indicate that lifelike polymers have an intrinsic tendency toward higher-order structures when energy is available; the chemistry of life is truly a self-ordering process.

From Aggregates to Cell-like Structures

Despite the increase in complexity from monomers to polymers to aggregates of polymers, there is a profound difference between even the most complex aggregate of molecules and the living cell. By considering the origins of reproduction and information transfer and then the origins of metabolism, we can gain some hints at least of further steps in the appearance of cells on Earth.

Origins of Reproduction and of Translation Machinery—RNA Leads the Way

The key issue in understanding how cellular reproduction started is to explain how molecular reproduction began. Proteinoids, RNA, and DNA deserve our attention. An intriguing finding is that proteinoids formed under laboratory conditions have a nonrandom composition: certain amino acids tend to be linked to one another, but not to other amino acids. Thus different proteinoids can have acidic, basic, hydrophobic, hydrophilic, and other properties. This nonrandom association results, no doubt, from stereochemistry—the way the electrically charged building blocks can fit together—and not as a result of any direction by RNA or DNA.

Significantly, many proteinoids formed in the laboratory have catalytic properties. Thus their three-dimensional shape, which arises as the amino-acid chain folds up, may produce the equivalent of an active site on an enzyme molecule. This is indeed a remarkable intrinsic feature of folded protein chains, a feature that increases the likelihood that biochemical pathways could have developed among the early polymers.

RNA is now considered to be the first informational macromolecule. In a remarkable discovery, Thomas Cech and his colleagues at the University of Colorado proved that RNA can function as a catalyst. This finding revises the long-held notion that the only biological catalysts are protein enzymes. We now know that various tRNAs, mitochondrial RNAs, and nuclear RNAs all have sequences that can splice (cut) themselves out of a longer RNA molecule. Furthermore, certain of the excised RNA pieces can function enzymatically as **ribozymes** in building new RNA molecules on an RNA template. All this occurs in the absence of proteins, as well as with no heat, ATP, or GTP added to drive the reactions.

Today the enzymatic activity of RNAs is used by cells in splicing out introns from exons and in processing precursor molecules into mature tRNAs, rRNAs, and mRNAs. Long ago, catalytic RNA could have assembled new RNAs from nucleotides in pools or clays. In fact, the mechanisms involved from the very start include self-removing and self-inserting catalytic RNAs that function just the way transposons move genes from one cell to another today. This is the equivalence of sex—transmission of genetic material from one cell to another. One imagines, then, an "RNA world" of increasing complexity with self-replicating molecules, sexlike exchanges of base sequences, and, finally, a kind of natural selection favoring certain of those molecules more than others

Even before the catalytic activity of RNA was discovered, it was shown that RNA can form spontaneously under conditions that were probably present on the ancient Earth. Furthermore, if a spontaneously formed single-stranded RNA (or DNA) molecule is added to a mixture of nucleotides, hydrogen bonding occurs between the purine and pyrimidine bases of the free nucleotides and the complementary bases of the chain (for example, guanine pairs with cytosine). Then addition of a condensing agent will initiate a slow polymerization of the bound nucleotides, so that a complementary chain forms. Separation of these two chains yields two new templates for another round of "replication." Given the vastness of geological time, single- and double-stranded nucleic acids probably arose by just such processes. Some may have had the enzymatic activity of ribozymes and so the processes of RNA replication, and evolution could have started (Figure 19-7). Clearly, sophisticated protein polymerase enzymes were not needed when nucleic acids first replicated on Earth.

Some RNA molecules formed by these processes in a test tube are hairpin-shaped, like modern transfer RNA, and can be extremely stable. Besides this, short RNA sequences, carrying an amino acid at one end, will bind to complementary sites along a large nucleic-acid molecule. The small RNA functions like a tRNA, and the larger one like an mRNA template. Equally surprising is the finding that spontaneously produced proteins rich in

Figure 19-7   RNA: THE FIRST INFORMATIONAL MOLECULE?
The remarkable catalytic and template capacities of RNA suggest that it may have served as the critical link in the origin of life. First, RNA could serve as a template that can be reproduced, including processes leading to RNA variation and evolution. Catalytic RNA, ribozymes, is the key to that process. Some RNAs might also serve as templates for protein manufacture. Finally, DNA copies of the RNA templates could have arisen, thereby yielding the extremely stable DNA genetic material common to most organisms on Earth.

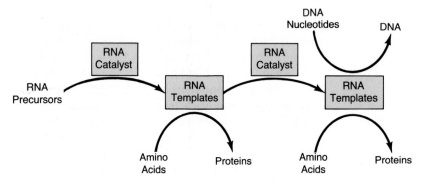

basic amino acids will complex with RNA molecules and form a primitive ribosomelike structure, even one with some ability to stimulate polypeptide polymerization. Perhaps, 4 billion years ago, such processes led to the development of the translation machinery—tRNAs, mRNAs, and ribosomes—to make proteins that are copies of RNA templates.

DNA usually is regarded as the "last step" in the origin of informational macromolecules. RNA can serve as a template on which a complementary DNA strand can be assembled (using an enzyme called reverse transcriptase [Chapter 20]). Perhaps just such a process occurred long ago and the extremely stable storage site for biological information, the DNA double helix, resulted. It is noteworthy that the introns and exons of today's eukaryotic DNA may have been present from the very start; those regions of DNA are merely copies of the intron and exon regions associated with catalytic sites in the more ancient RNA. Though we are learning much more about the three types of biological informational macromolecules, we still do not know when the genetic code itself arose— that is, the association of each amino acid with corresponding triplets of bases in nucleic acid.

An important unresolved issue is whether the first replicating RNAs, or even DNAs, were free in water solution (perhaps in clays) or whether compartmentalization of nucleic acids and other interacting molecules had occurred within protein spheres. Scientists who believe that nucleic acids were free coined the term *naked genes* for free, replicating biological macromolecules. Sooner or later, nucleic acids did take up residence within membrane-bounded domains. The second mystery in this sequence is when a lipid became a prominent component of such aggregates. When a lipid and protein raincoat, grossly similar to that on every living cell, came to surround nucleic acids that had replication and information-transfer abilities, a major step toward life was taken.

Origins of Metabolism

Microspheres or naked genes in the primordial waters or ground waters would have depended on an external supply of purine and pyrimidine bases, amino acids, sugars, and lipids to reproduce and maintain structural integrity. As such naturally formed monomers were "consumed," first by nonliving aggregates and later by more and more cell-like structures, those raw materials would have become increasingly scarce. It is possible that variations (mutations) in catalytic proteinoids allowed the use of a simpler and less scarce raw material, Y, instead of the rare compound X under these conditions, to the benefit of the mutant generations; recall that primitive catalytic active sites arise when proteinoid chains fold

into three-dimensional shapes. Likewise, a new enzyme, y, the chance product of a certain gene alteration, might have allowed an ancient cell-like structure to form the necessary X from the raw material Y:

$$Y \xrightarrow{y} X$$

In time, Y also might have become scarce, but the formation of yet another enzyme, z, and the ability to carry out a second chemical conversion might still have allowed X to be made:

$$Z \xrightarrow{z} Y \xrightarrow{y} X$$

It is believed that the first metabolic pathways, even forerunners of the Krebs cycle, might have evolved in this stepwise fashion. Figure 19-8 shows a simple reaction chain—a metabolic pathway—involving acidic and basic proteinoids and copper ions. If the acidic proteinoid is absent, acetic acid does not form by reaction 2; if copper is absent, neither reactions 3 nor 4 occur.

The compartmentalization of interacting molecular chain reactions within cell-like structures no doubt facilitated the emergence of true metabolism. When metabolism did appear and when a true form of reproduction was present, the transition to life finally may have been made. As we have seen, the "step" from naturally occurring compounds to the complex, integrated set of processes we call life was really not a single "step" at all, but a continuum of increasing complexity from monomers to polymers to polymer aggregates to cell-like structures to living cells. We cannot say precisely where and when

Figure 19-8   A SIMPLE METABOLIC PATHWAY.
Each of the reactions numbered 1 to 4 is catalyzed by a different proteinoid or metal–proteinoid complex. Pyruvic acid formed from the first reaction can be a precursor of either the amino acid alanine or acetic acid, which in turn is a precursor of other molecules.

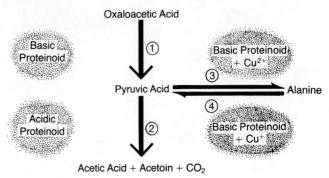

"life" began near the end of that sequence. The important thing is that the chemical evolution of life could well have happened, as demonstrated by laboratory experiments and geologic evidence, in just the sort of sequence we have described, governed by the physical and chemical processes and laws at work throughout the universe, then and now. Life indeed seems to have been a natural and inevitable consequence of those fundamental processes.

THE EARLIEST CELLS

The oldest fossils that most scientists agree are remains of once-living organisms consist of rod-shaped, cell-like structures found in Australian rocks dated at about 3.5 billion years old. Others dated at 3.4 billion have been found in southern Africa (Figure 19-9). In fact, organic molecules in even older rocks (dated at 3.8 billion years) from western Greenland may have had a biological origin. Some of the 3.4-billion-year-old fossils show extensive sets of spherical, photosynthetically active cells that may have lived in colonies. Perhaps the

most surprising feature of all these fossils is their great age—recall that the Earth's crust itself stabilized about 3.9 billion years ago. The stages of chemical evolution we have described must have begun even before that, when the planet was very young.

Heterotrophs First, Autotrophs Later

The first cells were in all probability *anaerobic*—able to survive in the absence of free oxygen—and *heterotrophic*—unable to make organic nutrients from simple inorganic precursors. These early *heterotrophs*, or "other feeders," consumed the amino acids and polypeptides, nucleotides, sugars, and other carbon-containing compounds that had formed spontaneously in the atmosphere and the primitive waters of the Earth. These first heterotrophs had to be anaerobes because ancient mineral deposits and other evidence indicate that the Earth's second atmosphere contained less than 0.1 percent of the O_2 present today (21 percent by volume). A few heterotrophic anaerobes still survive. Among them are bacteria of the genus *Clostridium*, one species of which causes botulism. As mentioned earlier, competition for increasingly scarce organic compounds would

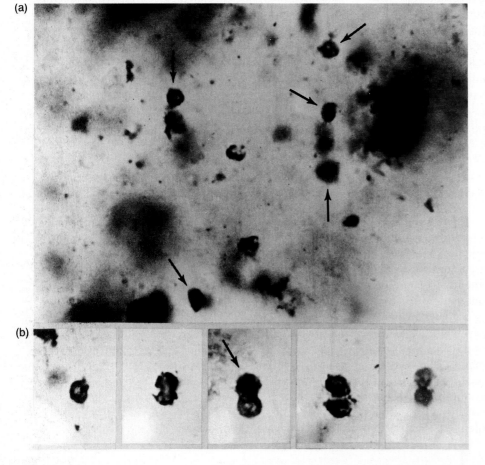

(a)

(b)

Figure 19-9   EARLY CELL-LIKE FOSSILS.
These fossilized microorganisms were found in rocks more than 3.4 billion years old from the Swaziland System in southern Africa. (a) A thin section of the chert in which several fossilized cells are indicated by arrows. (b) A selection of these microfossils in various stages of cell division.

Figure 19-10   TIME PERIODS AND SELECTED EVENTS IN THE EARLY EVOLUTION OF LIFE AND THE BIOSPHERE. Living cells arose early in the history of the Earth. Eukaryotes are relative newcomers.

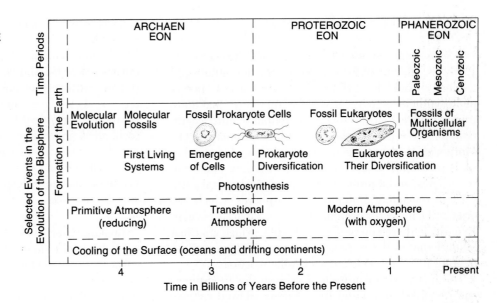

have favored evolution of metabolic pathways. It is also likely that early anaerobic heterotrophs evolved the ability to use ATP as an energy intermediary to couple and break down nutrients for synthesis of cellular molecules. True cells having such capacities soon would have depleted the store of monomers formed by nonbiological processes. Indeed, all life might have died out in a blitz of competition if it were not for one critical set of innovations: those leading to photosynthesis.

Sometime more than 3.4 billion years ago, mutations apparently occurred giving cells the ability to exploit new energy sources. Structural proteins and enzymes inherited from earlier cells became capable of trapping energy from sunlight, probably at first to generate ATP by means of photophosphorylation (Chapter 8) and later to store that energy in carbon compounds, the sugars. These new abilities would have been a distinct advantage. Cells with such attributes no longer depended on limited supplies of spontaneously formed organic compounds for their energy because they could create their own nutrients from inorganic precursors. These mutant cells were the first *autotrophs,* or "self-feeders."

The emergence of autotrophs has been dated to 3.4 billion years ago by the discovery of fossils with $^{12}C:^{13}C$ ratios very similar to those found in photosynthetic cells today; the same ancient rocks contain chlorophyll, the primary light-absorbing molecule of today's photosynthesizers. Photosynthetic cyanobacteria (very similar to species alive today) had evolved by 2.7 billion years ago. Remains of these early cells formed **stromatolites,** minutely layered rock mounds. Stromatolites are still being formed on Earth, for example in warm, salty inlets on the western coast of Australia, as shown at the beginning of the chapter.

The evolution of photosynthesis had immensely significant consequences for the planet Earth and its inhabitants, the chief one being the addition of massive amounts of molecular oxygen to the atmosphere. As we saw in Chapter 8, oxygen is a by-product of photosynthesis. It has been hypothesized that oxygen released by the early autotrophs reacted with iron dissolved in sea water, causing ferric oxide to precipitate to the ocean floor and form immense deposits, which we see today as red beds of rock. However, new results show that light acting on iron compounds may have caused the deposits. One way or the other, in a very real sense, the oceans "rusted."

Autotrophs continued to photosynthesize and release oxygen after most of the dissolved iron had oxidized. Significant quantities of O_2 started to accumulate in the oceans and in the atmosphere by 2 billion years ago. Some of the atmospheric O_2 was converted by sunlight to ozone (O_3), which began to collect in a high-altitude layer, the **ozone layer.** This layer screened out much of the ultraviolet light that had been searing the Earth's surface since the planet formed. Because of the ozone layer's protection from the intense ultraviolet rays of the sun, cells could begin to live closer to the surfaces of oceans and lakes and even on moist land surfaces. Ironically, the slow change to an oxidizing atmosphere with an ozone layer also slowed the atmospheric production of monomers. The organic molecules so essential in prebiological evolution and in supporting heterotrophs could no longer be produced in great quantities without the energy source of ultraviolet light. No doubt, under these conditions, heterotrophs consumed the existing organic compounds more quickly than new compounds accumulated.

The Emergence of Aerobes

Because atmospheric CO_2 is consumed during photosynthesis, the primary carbon source for autotrophs gradually would have diminished as a consequence of photosynthetic activity. It is clear, therefore, that autotrophs inadvertently created a potentially lethal condition for themselves—depletion of atmospheric CO_2. Moreover, the free oxygen released during photosynthesis is poisonous to anaerobic cells, including photosynthetic ones, and it probably disrupted biochemical pathways in remaining proteinoid microspheres, which had originated in an environment having no free oxygen. Eventually there emerged *aerobic* cells, which are capable of living in the presence of oxygen and even of exploiting it in very efficient metabolic reactions.

The increasing quantity of oxygen in the atmosphere and dissolved in sea and fresh waters allowed *cellular respiration* to originate. (As we saw in Chapter 7, cellular respiration is a series of steps, "tacked on to" the initial glycolysis pathway, that burns compounds produced during glycolysis in the presence of oxygen.) One immense advantage of respiration is that a cell harvests eighteen times as much energy as it does when breaking down the same nutrients by means of glycolysis and the additional steps of fermentation alone. That extra stored energy permits much higher rates of growth and reproduction. And, for the first time, respiration allowed one cell to actively prey on other cells, since a cell with the new pathway can derive more energy by digesting another cell than it uses up while hunting it down. Thus cells with the respiratory pathway became the first consumers of other cells and started the food chains that now dominate so much of the Earth's ecology.

But cellular respiration had another major consequence: the beginning of the *carbon cycle*. During cellular respiration, carbon compounds are largely oxidized to CO_2. Thus that gas can return to the atmosphere to help replenish the CO_2 reservoir used for photosynthesis. In this way, atmospheric CO_2 replenished by cellular oxidation ensures a carbon source for photosynthesis and a supply of an essential raw material for autotrophs.

Eukaryotes

Once aerobic cells appeared and the ozone layer screened much ultraviolet light from the sun's penetrating rays, the variety of organisms increased tremendously in the sea, in fresh water, and on land. But these cells were all *prokaryotic*—lacking a membrane-bound nucleus, membrane-bound organelles, a cytoskeleton, and so on (Chapter 5). Not for another billion years or so would *eukaryotic* cells—having a true nucleus and other membrane-bound organelles—appear. The oldest fossil eukaryote was discovered in rocks dated at 1.5 billion years old.

The most widely accepted theory about the origin of eukaryotic organelles is **endosymbiosis,** a hypothetical process during which one type of prokaryotic cell is literally "swallowed" by another and is retained as a self-reproducing resident that functions as an organelle. Mitochondria and chloroplasts are believed to have originated in this way, as will be described in Chapter 20. The host cell and its guest organelles together formed viable organisms with specialized structures to carry out specific functions.

Looking back over the origin-of-life model now accepted by most biologists, it seems likely that two-thirds of the Earth's total history passed before the first eukaryotic cells appeared. During the early part of that time span, a lifeless planet gave rise quite rapidly, through natural chemical processes, to monomers, polymers, polymer aggregates, more and more cell-like structures, and, finally, true, living cells. The early photosynthesizers, in turn, altered our planet, transforming a reducing atmosphere into an oxidizing one with an ozone screen against ultraviolet light and eventually eliminating the conditions in which life arose. These events are summarized in Figure 19-10. Once these massive transformations had taken place, an incredible diversity of multicellular species appeared, as we shall see. But the Earth's own geological evolution had such a profound effect on the organisms that came to inhabit our planet that we shall first discuss these interesting events.

THE CHANGING FACE OF PLANET EARTH

The history of Earth is inextricably entwined with the history of life. Since forming 4.6 billion years ago, the Earth itself has not been static. Changes in land masses, the seas, and climate have altered the habitats of organisms and, in turn, influenced biological evolution. Let us look at that chain of effects in more detail.

Continents Adrift on a Molten Sea: Tectonic Activity and Geological Eras

The Earth we stand on seems monolithic and immutable, but it is actually very dynamic. Our planet's outer molten metal core is surrounded by a hot, semisolid **mantle** and a lighter, solid crust. The crust, which is about 5 kilometers thick over the sea floors and perhaps

33 kilometers thick over the continents, is divided into massive plates. In a process called **continental drift,** these plates move slowly as convection currents in the mantle lead to upwelling and additions of new crustal materials at certain sites. The concept of continental drift was developed by the German meteorologist and geologist Alfred Wegener in 1912; although some evidence supported the radical idea, it was not until new techniques for making measurements of the Earth's magnetic field and of sea-floor rocks became possible in the 1950s that the hypothesis began to be accepted.

The building of crustal plates, so-called **plate tectonics,** began more than 4 billion years ago. Since that time, about fifteen plates (including eight major plates) have formed and drifted about on the semisolid mantle. Upwelling of new materials near mid-ocean ridges produces new crustal areas and pushes the plates apart so that collisions occur. When the edges of two plates collide, one edge may be driven beneath the other, causing mountain ranges to elevate; the Himalayas, Alps, Andes, and Rockies originated in this way.

Using evidence of rock layers in the crustal plates, geologists have divided the past 550 million years into three *geological eras:* Paleozoic, Mesozoic, and Cenozoic. That immense span of time preceding the Paleozoic and during which life began is called the Precambrian period (see the geological time scale inside the front cover). The boundaries of geological eras correspond to discontinuities in rock layers (strata), fossil types, and other features that suggest major geological change. The transition time from one era to the next can be dated fairly precisely from the amounts of radioactive isotopes present in rocks. Subdivisions of geological eras, called *geological periods,* are dated in the same ways.

During the Paleozoic, the present continents of Europe and North America at first lay far apart but were pushed together by plate-tectonic activity (as evidenced by rocks of the Silurian period found in Newfoundland, Scotland, and Greenland). During the Permian period, two previously separate land masses—a northern continent, *Laurasia,* and a southern one, *Gondwana*—had converged to form **Pangaea,** a single giant land mass (Figure 19-11). The equator crossed the Laurasian portion of Pangaea and tropical forests of early land plants grew in abundance there. Their remains form the extensive coal beds of Europe and North America. Shallow seas advanced over vast land areas and retreated many times during the Paleozoic, creating special habitats and then removing them.

At the transition between the Paleozoic and the Mesozoic eras, major changes apparently occurred in mantle convection as plate movements and volcanic activity increased and Pangaea broke apart. That time was also marked by the most massive extinctions of living creatures that ever took place on Earth. The distribution of the continents continued to change, and it was not until 25 million years ago—a brief interval in the Earth's long history—that the general continental configuration that we see today was attained.

Accompanying these momentous plate movements were climatic changes that greatly affected living organisms. Sites on the drifting continents went from wet to dry, from cold to hot, and back again numerous times as their latitudes shifted and as their altitudes rose from sea floor to lofty peaks and were eroded again over time. Cyclic alterations in the Earth's orbit and variations in the sun's output of energy also produced complicated local shifts in weather and periods of glaciation, with dramatic consequences on the habitats of organisms. Even the length of a day varied because of these cyclic changes; in the Silurian period, for example, the day was 21 hours and the year was 421 days!

Such massive environmental changes influenced the evolution of specific groups of organisms, and as we shall see in subsequent chapters, the dominant species during each major geological period were in many ways shaped by the prevailing climates and terrains. But keep in mind that the geological record reveals average climates over millions of years. Just think of the variations in ten, or a hundred, or a thousand years. And consider that even brief, minor fluctuations in climate, sea level, and habitat may have major consequences for individual organisms. In Yellowstone National Park, for instance, fossil remains of tree trunks show that some twenty successive forests of huge trees grew at one site and were covered by volcanic ash, one after the other. Yet the immense time frame in which that succession occurred is but the briefest moment in the scale of geological history. Therefore, to understand life and its evolution on Earth, we must endeavor to visualize both the vast, slow changes in land, sea, and weather that affect the evolution of species and the daily, seasonal, and cyclic changes that affect individual lives. Furthermore, we must remember that the Earth itself changes continuously even today, and as its landscapes and climates alter, so must its living inhabitants.

TAXONOMY: CATEGORIZING THE VARIETY OF LIVING THINGS

Just as the Earth's changes are chronicled in its rocks and recorded in the geological time scale, the evolutionary history of organisms is evident in the fossil record. The diversity of those fossils and also of living organisms has been systematically cataloged. Attempts to catego-

Figure 19-11   CONTINENTAL DRIFT OVER THE PAST 380 MILLION YEARS.
(a) The northern continent of Laurasia and the southern continent of Gondwana were separate land masses through the Devonian period of the Paleozoic era 380 million years ago. (b) They fused to form Pangaea during the Permian period. (c) and (d) Pangaea split apart during the Mesozoic era with Laurasia comprising what is now North America, Europe, and Asia; Gondwana later formed South America, Africa, Australia, and Antarctica. Note the remarkable drifting of the continents during the past 50 million years (d); for instance, see how India has drifted north toward Asia, and how Australia is moving northward toward the equator. Arrows indicate direction of drift.

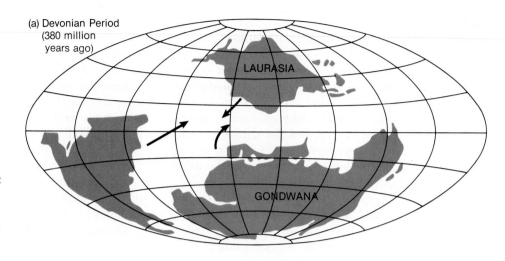

(a) Devonian Period (380 million years ago)

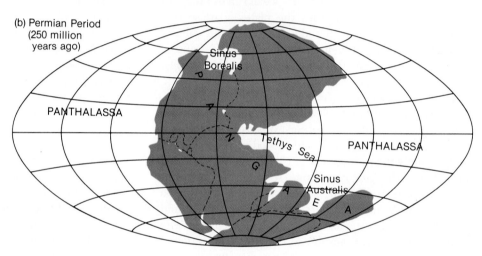

(b) Permian Period (250 million years ago)

rize the grand diversity of life predate Aristotle. But it was a Swede, Carl von Linné, who made the first major contribution to **taxonomy**—the classification of organisms. Let us examine Linné's scheme, for it is the one on which our modern classifications are based.

The Binomial System of Nomenclature

In the mid-eighteenth century, Linné, who used the Latin form of his name, Carolus Linnaeus, wrote *Systema Naturae*, a landmark book because of the way Linnaeus named and organized plants and animals known at that time. First, he employed a system—called the **binomial system of nomenclature**—by which he assigned all organisms a name using two Latin words, as Latin was the one language shared by all educated people of Europe at the time. The first term denotes the **genus**—a grouping of very similar organisms—to which the organism belongs (note, the plural of genus is **genera**). The second term denotes the organism's **species**—a specific

group of closely related organisms within a genus. The full species binomial (two-part) name always includes genus and species; for example, the modern domesticated horse is labeled *Equus caballus*, while its close relative, one of the zebras, is *Equus burchelli*.

Today, groupings do not stop at genus and species, but include broader *taxonomic groups*—arrangements of organisms into hierarchical classifications based on similarities. Similar genera are placed in the same **family;** similar families, in the same **order;** similar orders, in the same **class;** similar classes, in the same **phylum** (or **division** in botany); and, finally, similar phyla or divisions, in the largest, most inclusive category—the **kingdom.**

The binomial system and the use of the higher taxa (each **taxon** is one category in a system of taxonomy) have proved to be such useful tools for organizing similar life forms that biologists have kept the system, with some modification, and continue to assign Latin or Latin-sounding names to the various taxa, even though the classification criteria used today are different from the ones on which Linnaeus relied. (Table 19-1 shows the

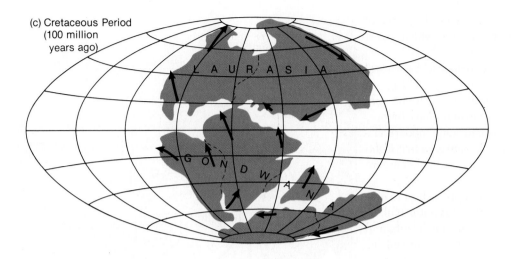

(c) Cretaceous Period (100 million years ago)

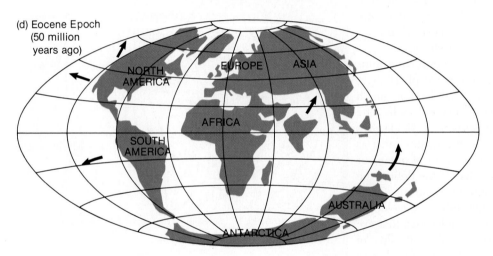

(d) Eocene Epoch (50 million years ago)

taxonomic classifications of humans, honeybees, corn, and the common edible field mushrooms.)

Taxonomy and Darwin's Theory of Evolution

Linnaeus had no alternative in the 1750s but to use a strictly morphological approach in defining his taxa. Indeed, the word "species"—from the Latin word for "appearance" and "sort"—was coined by the Englishman John Ray to classify organisms on the basis of their visible characteristics. Hence, in Linnaeus's system, the more structural traits shared by different organisms—the number and placement of limbs; the shapes of leaves, flowers, or internal organs; and so on—the closer their taxonomic relationship.

Two major factors have altered the basis on which taxonomies are constructed and the uses to which they are put. The first was Charles Darwin's theory of evolution. The second was the advent of new techniques to

Table 19-1 **THE CLASSIFICATION OF ORGANISMS**

Taxonomic Level	Human	Honeybee
Species	*Homo sapiens*	*Apis mellifera*
Genus	*Homo*	*Apis*
Family	Hominidae	Apidae
Order	Primates	Hymenoptera
Class	Mammalia	Insecta
Subphylum	Vertebrata	
Phylum	Chordata	Arthropoda
Kingdom	Animalia	Animalia

Taxonomic Level	Corn	Mushroom
Species	*Zea mays*	*Agaricus campestris*
Genus	*Zea*	*Agaricus*
Family	Poaceae	Agaricaceae
Order	Commelinales	Agaricales
Subclass	Monocotyledoneae	
Class	Angiospermae	Basidiomycetes
Subdivision*	Spermatophytina	
Division*	Tracheophyta	Mycota
Kingdom	Plantae	Fungi

*Botanists use the terms "division" and "subdivision" instead of "phylum" and "subphylum."

delineate physiology, embryology, and biochemistry of living creatures—the hidden things on which overt structure depends.

Darwin's theory of evolution in response to natural selection (Chapters 1 and 42) forced a major change in the definition of "species." The point at which one species diverges into two, according to evolutionary theory, is when they can no longer interbreed. One modern criterion for defining species, then, is *reproductive isolation*—the inability to breed successfully with other organisms, even closely related ones.

This biological definition of species applies best to certain sexually reproducing animals. But it is not useful for many groups of microorganisms, fungi, and plants and even for some animals. This is because such organisms reproduce primarily by asexual means; hence, the concept of reproductive isolation involving sex does not apply. So reproductive isolation, like most criteria for species, is applicable in some, but not all, cases.

The second major elucidation of the relationships among organisms comes from comparative physiology, embryology, and biochemistry, as well as from much more sophisticated morphological analysis itself. The sharing of metabolic pathways or properties, similarities in nerve function, means of processing wastes (such as from nitrogen metabolism), and properties of organs or tissues are among the evidence that helps define the degree of relationship among groups of organisms. Careful study of the stages passed through by developing embryos often reveals unsuspected similarities that are not evident in adults—for instance, the gill slits of human embryos help establish our close relationship to all other vertebrates. At a finer level, molecular taxonomy compares amino-acid sequences of proteins or base sequences of DNA or RNA from different organisms. In fact, the day is not far off when sequencing of DNA bases will be so rapid that scientists will be able to analyze the full haploid genome of organisms in this way. Finally, the use of electron microscopes and other sophisticated probes of tissue, cellular, and subcellular structure has expanded greatly our concept of "morphology." For instance, only the scanning electron microscope can reveal minute layers and grooves in teeth of the ancestors of *Homo sapiens;* the grooves help establish relationships among the fossil types. Application of these various techniques varies from one line of organisms to the next. Nevertheless, they provide important ways to distinguish among creatures for which the reproductive-isolation criterion is not useful.

Taxonomy and Evolution

Evolutionary thought and theory since the time of Darwin have created a new function for taxonomic classi-

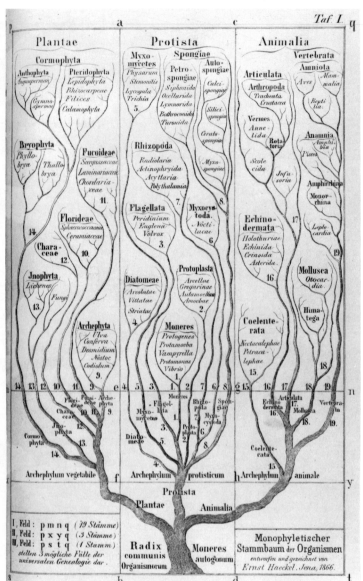

Figure 19-12   GERMAN BIOLOGIST ERNST HAECKEL'S PHYLOGENETIC TREE, FIRST PUBLISHED IN 1866.

fication; taxonomy provides a framework for studying evolutionary relationships. Common sense suggests that pin oaks and blue oaks are more closely related to each other than to cactuses or yellow pines. By applying all the available modern criteria to define the precise taxonomic identity of an organism, insight emerges about the actual degree of relatedness between that organism under study and all its relatives, near and distant.

The overwhelming evidence that all organisms on Earth are related by common descent from some of the first cells has led to methods for tracing the family trees of living and fossil organisms. Some phylogenetic trees, such as the one shown in Figure 19-12, are based on the morphological evidence used to define species. Figure

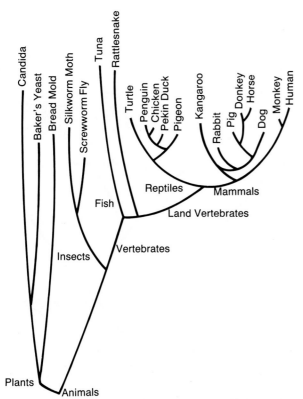

Figure 19-13   A PHYLOGENETIC TREE BASED SOLELY ON THE AMINO-ACID SEQUENCE IN CYTOCHROME c.
The distance between two species is indicated by the distance down the line from one species to a junction and then up the line to another species. This total distance shows the degree of difference in amino-acid sequence between the two species. For example, the distance from turtle to human is much greater than from turtle to pigeon; many more amino-acid differences are found in the first case than the second.

19-13 shows a different sort of phylogenetic tree, in this case based on amino-acid differences in one protein, the respiratory pigment cytochrome *c*. Analogous trees can be drawn for other proteins or for DNA itself—the process has been called *molecular taxonomy.*

Besides allowing construction of various types of phylogenetic trees, modern taxonomy provides a convenient new way to discuss related organisms. Ideally, forms collected into one taxon—whether it be a species, family, phylum, division, or whatever—can be considered to have evolved from one ancestral species. Thus all the robin species in the genus *Turdus* evolved from one ancestral *Turdus* species. They can be called **monophyletic,** meaning that they share one ancestral source. A group of monophyletic organisms is a **clade,** and the study of how closely related groups branched and separated from one another is called **cladistics.** An illustration of this view is provided in the *cladogram* in Figure 19-14.

Intellectually, cladistics and molecular taxonomy seem a long way from Linnaeus's scheme of the mid-1700s. Yet, despite continual refinement of the criteria for classifying organisms, Linnaeus's organizational scheme remains the basis for taxonomy because it is logical and simple and because it serves other uses, such as the study of evolutionary relationships.

THE FIVE KINGDOMS

Linnaeus's lifetime of work resulted in classification of several thousand species. At the highest taxonomic level, all were in either **Plantae** or **Animalia,** the only two kingdoms recognized by the brilliant naturalist. The organisms in these two kingdoms were large enough to be studied with the naked eye. Many were green and immobile and seemed to need only water to live—these plants and their supposed degenerate relatives, the fungi, composed the kingdom Plantae. Other organisms were not green, in general moved, and consumed food—these were Animalia.

But what of the tiny microorganisms discovered since the microscope was invented? And what of single-celled organisms that are both green and capable of swimming (such as euglenoids [Chapter 21])? Almost 200 years of discovery and controversy over the problem cases culminated in what is now viewed as the best compromise classification scheme—the five kingdoms defined by R. H. Whittaker in 1963.

As depicted in Figure 19-15, Whittaker placed each of the 2 million or so defined types of living and fossil organisms into one of five kingdoms. Bacteria and cyanobacteria, the single-celled prokaryotes, are assigned to the kingdom **Monera** (Chapter 20). Single-celled eukaryotes, including protozoans and some types of algae, are grouped in the kingdom **Protista** (Chapter 21). Fungi, such as molds and mushrooms, have their own kingdom, **Fungi** (Chapter 22). Multicellular algae and land plants make up the kingdom Plantae (Chapters 23 and 24), and multicellular animals form the kingdom Animalia (Chapters 25 and 26).

Whittaker's simplifying scheme has truly rendered order to the study of life's vast diversity. But evolution has not operated to fit the neat schemes of human minds. There are, in fact, certain taxonomic groups that are extremely difficult to place, even at the kingdom level. We will discuss in detail, for instance, some algae that might legitimately be placed in either Protista or Plantae. Furthermore, the kingdoms probably are not true clades—that is, taxonomic groups derived from one ancestral type. On the contrary, the Monera include so-called archaebacteria, so different from more common forms that

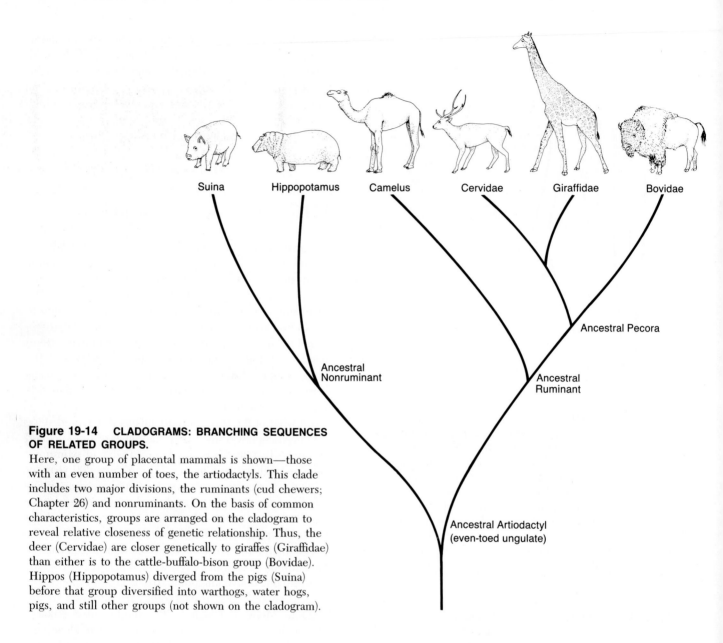

Figure 19-14   CLADOGRAMS: BRANCHING SEQUENCES OF RELATED GROUPS.
Here, one group of placental mammals is shown—those with an even number of toes, the artiodactyls. This clade includes two major divisions, the ruminants (cud chewers; Chapter 26) and nonruminants. On the basis of common characteristics, groups are arranged on the cladogram to reveal relative closeness of genetic relationship. Thus, the deer (Cervidae) are closer genetically to giraffes (Giraffidae) than either is to the cattle-buffalo-bison group (Bovidae). Hippos (Hippopotamus) diverged from the pigs (Suina) before that group diversified into warthogs, water hogs, pigs, and still other groups (not shown on the cladogram).

most biologists now assign them to a separate "sixth" kingdom. Separate protistan groups gave rise to the major types of fungi. Similarly, Animalia almost certainly contains organisms derived from different protists. Thus the set of characteristics that defines so huge and broad a group as a kingdom is in good part an arbitrary convenience, helpful for certain things but not for all that a taxonomy might be called on to do. We have to be conscious of what a taxonomy does and does not tell us. In this, cladistics is a useful addendum to the classification processes begun so well by Linnaeus and extended so successfully by Whittaker. Furthermore, taxonomy is by no means a dead science; even today, new types of organ-

isms are being discovered, such as the black frog and bearded catfish shown in Figure 19-16. Both were discovered on one expedition and are so different from known relatives that each is placed in a new genus.

As we study the members of the five kingdoms in detail in the next seven chapters, we will retrace the evolutionary pathways that produced the many remarkable forms that have inhabited Earth, in all their intricacy and beauty. During our journey through biological history, however, keep in mind the ultimate relatedness of all organisms by their common descent from the first cells and their probable origin by means of chemical evolution.

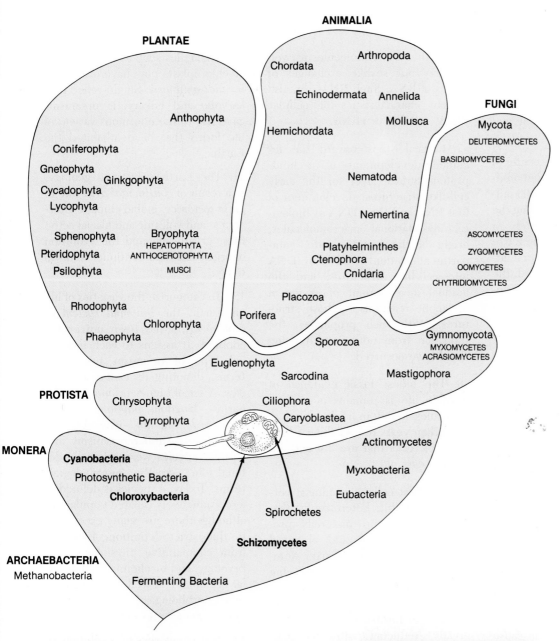

PLANTAE

ANIMALIA

FUNGI

Anthophyta

Coniferophyta
Gnetophyta
Cycadophyta   Ginkgophyta
Lycophyta

Sphenophyta   Bryophyta
Pteridophyta   HEPATOPHYTA
                ANTHOCEROTOPHYTA
Psilophyta   MUSCI

Rhodophyta
        Chlorophyta
Phaeophyta

Chordata   Arthropoda

Echinodermata   Annelida
Hemichordata   Mollusca

            Nematoda

            Nemertina

        Platyhelminthes
        Ctenophora
            Cnidaria

        Placozoa
Porifera

Mycota
DEUTEROMYCETES
BASIDIOMYCETES

ASCOMYCETES
ZYGOMYCETES
OOMYCETES
CHYTRIDIOMYCETES

PROTISTA

Sporozoa

Euglenophyta
        Sarcodina   Mastigophora

Chrysophyta   Ciliophora

Pyrrophyta   Caryoblastea

Gymnomycota
MYXOMYCETES
ACRASIOMYCETES

MONERA

Cyanobacteria

Photosynthetic Bacteria

Chloroxybacteria

Spirochetes

Schizomycetes

Actinomycetes

Myxobacteria

Eubacteria

ARCHAEBACTERIA

Methanobacteria   Fermenting Bacteria

**Figure 19-15
WHITTAKER'S
CLASSIFICATION
SYSTEM.**
This system of
classifying living and
fossil organisms as
members of five
kingdoms was originally
proposed by R. H.
Whittaker in 1963.
With some variations it
remains the basic
taxonomic system in use
today. The principal
change is the addition
of the kingdom
Archaebacteria as a
separate major group of
prokaryotes.

**Figure 19-16   NEWLY
DISCOVERED
ORGANISMS.**
Both this black frog and
bearded catfish were
found on a recent
expedition in tropic
jungles.

SUMMARY

1. Most biologists agree that evidence strongly favors the theory that life arose spontaneously on the Earth between 3 and 4 billion years ago through normal chemical and physical processes.

2. The sun, the star at the center of our solar system, formed about 5 billion years ago. Within about 400 million years, the planets, including the Earth, developed from clouds of matter circling the sun. As the Earth cooled and condensed, its first atmosphere and its layers formed: the solid and liquid *core;* the hot, semisolid *mantle;* and the thin, solid *crust.* Outgassing from the hot interior produced the Earth's second atmosphere, which included CO, CO_2, H_2O vapor, N_2, H_2, and perhaps other gases, but little or no O_2.

3. The Earth's composition, size, temperature, and distance from the sun provided a special set of conditions compatible with the origin and maintenance of life. The process of chemical evolution led to that life.

4. Laboratory experiments demonstrate that under physical and chemical conditions probably similar to those on the early Earth, such organic monomers as amino acids, sugars, fatty acids, and nucleic-acid bases form.

5. The same conditions can cause some of those monomers to form polymers—*proteinoids* and nucleic acids. Polymers in solution can easily form such aggregates as *proteinoid microspheres.* Experimental production of *coacervates* and *liposomes* demonstrates still other small, nonliving systems with properties of precellular forms.

6. When supplied with appropriate nucleotides, certain RNAs act enzymatically as *ribozymes* and can build other RNA molecules as well as carry out sexlike exchanges of pieces of RNA. Such RNAs can also display characteristics of cellular tRNA, mRNA, or rRNA.

7. Critical steps toward life included development of a lipid-protein surface layer for the early cell-like structures; development of first RNA and then DNA as biological informational macromolecules; origin of the genetic code; compartmentalization of RNA or DNA into cell-like structures; and the gradual development of chains of metabolic reactions in those structures. With such properties, the transition from nonliving to living could have occurred.

8. The oldest fossil evidence of possible life is remains of organic molecules found in rocks dated at 3.8 billion years old. Cellular fossils occur in rocks that are 3.4 and 3.5 billion years old.

9. The first cells were almost certainly anaerobic heterotrophs that consumed organic materials produced by physical and chemical processes on the early Earth. Autotrophs emerged later, producing their own nutrients and altering the atmosphere by releasing O_2. The *ozone layer* formed soon thereafter, greatly reducing ultraviolet-light penetration and so permitting life to survive in shallow water and on the land surface. Cellular respiration appeared after free O_2 became available and resulted in the return of CO_2 to the atmosphere. Aerobic cells, having the abundant energy supply derived from cellular respiration, could feed on other cells.

10. Remains of the oldest eukaryotic cells have been found in rocks that are 1.5 billion years old. Eukaryotic cell organelles such as mitochondria or chloroplasts may have originated by *endosymbiosis.* Single-celled prokaryotic and eukaryotic organisms gave rise to the enormous variety of life forms that have inhabited the Earth.

11. The Earth's changing geology, as a result of *continental drift* due to *plate tectonics,* global climatic alterations, and volcanic and glacial activities, greatly affected the evolution of organisms, and continues to do so today.

12. To categorize the varieties of life on Earth, the *binomial system of nomenclature* has been universally adopted. It assigns to each individual type of organism two Latin terms, denoting its *genus* and *species.* A set of higher taxonomic categories—*kingdom, phylum, division, class, order, family*—is used to further classify similar organisms into groups. Species were originally defined in strictly morphological terms. Today, a species is defined as a reproductively isolated population, although there are many exceptions to that strict definition. Evidence from comparative physiology, embryology, and biochemistry (molecular taxonomy), as well as high-resolution morphological studies, helps define species and higher *taxa.*

13. *Taxonomy* reveals evolutionary relationships among organisms. *Clades* are taxonomic units derived from a common ancestor.

14. Phylogenetic trees have been used to depict evolutionary relationships graphically. The system used in this text arranges all organisms into five kingdoms: *Monera, Protista, Fungi, Plantae,* and *Animalia.* Some of these kingdoms are not *monophyletic;* hence, they are not true clades.

KEY TERMS

Animalia

Big Bang

binomial system of nomenclature

chemical evolution

clade

cladistics

class

coacervate

continental drift

core

crust

division

endosymbiosis

family

Fungi

genus (genera)

kingdom

liposome

mantle

Monera

monophyletic

order

ozone layer

Pangaea

phylum (phyla)

Plantae

plate tectonics

proteinoid

proteinoid microsphere

Protista

ribozyme

species

stromatolite

taxon

taxonomy

QUESTIONS

1. The first atmosphere around the newly formed earth consisted of which gases? What happened to this atmosphere? What was the origin of the second atmosphere? How has the atmosphere changed and what was responsible for that change?

2. What was the origin of salts in the oceans?

3. In the laboratory, organic molecules are readily produced by passing energy through a mixture of the gases that probably were present in the atmosphere 4 billion years ago. Name some kinds of molecules that have been produced in these experiments. What energy sources have been used in these experiments? What energy sources may have been available when organic molecules first formed on the Earth?

4. What are proteinoid microspheres, coacervates, and liposomes? In what ways are they analogous to living cells?

5. How does RNA function as an enzyme? What roles did RNA play in the chemical evolution that preceded life on Earth?

6. Is it more likely that the first living organisms on Earth were heterotrophs or autotrophs? aerobes or anaerobes? What was their source of energy? What was their source of carbon?

7. The early autotrophs consumed the carbon dioxide in the atmosphere. How was that carbon dioxide replenished?

8. When oxygen started accumulating in the Earth's atmosphere, some of it was converted to ozone. What energy source is screened out by ozone? How did the presence of ozone affect the evolution of life on Earth?

9. What is meant by a clade? Is each kingdom a clade? Is each phylum a clade? each species? Are all living things on Earth related?

10. Name the five kingdoms of organisms in Whittaker's classification system. To which kingdom do you belong? a mushroom? a pine tree?

ESSAY QUESTIONS

1. Has the Earth attained its ultimate form, or is it still changing? What is your evidence? What about the living creatures on the Earth?

2. Consider the consequences if the ozone layer that formed around the upper atmosphere of the Earth 2 billion years ago had screened out *visible* light. How would this have affected the evolution of life on Earth? How would it have affected our atmosphere? Now imagine that ozone had not formed. What would the major consequence be for life on Earth?

SUGGESTED READINGS

DICKERSON, R. E. "Chemical Evolution and the Origin of Life." *Scientific American*, September 1978, pp. 70–86.

A general easy-to-understand account. Another *Scientific American* article, "Cytochrome *c* and the Evolution of Energy Metabolism," by Dickerson (March 1980, pp. 137–53), discusses the evolution of photosynthesis and respiration in ancient cells.

DOBZHANSKY, T., F. J. AYALA, G. L. STEBBINS, and J. W. VALENTINE. *Evolution.* San Francisco: Freeman, 1977.

The chapters on taxonomy by these eminent scientists are superb.

GILBERT, W. "The RNA World." *Nature* 319 (1986): 618. E. G. NISBET. "RNA and Hot-Spring Waters." *Nature* 322 (1986): 206. F. H.

WESTHEIMER.   "Polyribonucleic Acids as Enzymes." *Nature* 319 (1986): 534–35. R. LEWIN. "RNA Catalysis Gives Fresh Perspective on the Origin of Life." *Nature* 231 (1986): 545–46.

Four summaries of the new work on RNA as enzyme and life precursor.

MAYR, E. "Biological Classification: Toward a Synthesis of Opposing Methodologies." *Science* 214 (1981): 510–16.

A world leader of evolutionary thought demonstrates how all the various approaches to taxonomy may contribute to our understanding of evolution.

WHITTAKER, R. H. "New Concepts of Kingdoms of Organisms." *Science* 163 (1969): 150–60.

Here is the famous hypothesis about kingdoms, including discussion of the problems.

20

MONERA AND VIRUSES: THE INVISIBLE KINGDOM

The most important discoveries of the laws, methods, and progress of Nature have nearly always sprung from the examination of the smallest objects which she contains.
Jean Baptiste de Lamarck, *Philosophie Zoologique* (1809)

Photosynthetic prokaryotic cells live in the hot water of these mineralized terraces in Yellowstone National Park.

If you were to lightly touch your index finger to the surface of a culture dish filled with a semisolid nutrient medium and then keep the dish warm overnight, by morning, there probably would be colonies of bacteria on the surface, growing in the whorled configuration of your fingerprint. This simple exercise reflects the fact that every square centimeter of the human skin is covered with thousands of microscopic bacteria. Other sets of bacterial species live in the human digestive tract, mouth, nose, and respiratory passages. In one sense, then, the human body can be viewed as a set of small, interacting ecosystems inhabited by millions of microscopic residents. And the larger world around us is similarly inhabited: some drops of water or a pinch of soil collected from anywhere on Earth will teem with thousands to millions of individual bacteria.

Bacteria are the most prominent, but not the only, members of the kingdom Monera. *Cyanobacteria*—blue-green photosynthetic cells—and *archaebacteria* are other types we shall survey in this chapter. The 2,000 or more species of monerans play important roles in breaking down organic matter, in recycling soil nutrients, in fermenting foods, and in manufacturing food by photosynthesis. Some cause animal and plant diseases, and many have been harnessed by human industry for food and drug production. In addition, bacteria are used in leather and textile processing, in sewage treatment, and in a variety of other operations that keep our civilization running. Furthermore, different lineages of ancient monerans probably gave rise to the other kingdoms of living things—the protists, fungi, plants, and animals. It is quite literally the case that without monerans, we would not be here: they are our ancestors, and they are our critically important ecological neighbors, whose functioning is essential to keeping the Earth habitable for us.

All monerans are **prokaryotes** and, as such, lack a nuclear envelope, other membrane-bound organelles, and a cytoskeleton (Table 6-1). A hallmark of all monerans is their exceedingly rapid growth and reproduction. In good part, this growth is based on a metabolic diversity that allows monerans to survive in a wide variety of nutrient conditions and physical environments. Monerans can use many kinds of raw materials in living, and they do so while drawing on an unexpected range of energy sources to drive their metabolism. Their metabolic diversity often allows dissimilar types to interact, so that the "wastes" of one serve as "food" for another. The result may be a primitive kind of cooperation that supports survival where it might not otherwise be possible. We shall begin to appreciate the diversity of monerans later in this chapter as we survey the major types.

Monerans, as well as many eukaryotic cell types, are susceptible to invasion by *viruses*. These complex packets of genetic material and protein subvert the host cell's molecules and metabolism to meet their own needs. And ultimately, they often kill host cells. Viruses are not true cells, and are not members of any of the five kingdoms of living organisms. Nevertheless, their simplicity of structure and possible origin from prokaryotic cells make this the appropriate chapter to consider viruses, their reproductive cycles, their deadly effects on hosts, and their possible evolutionary origins.

MONERA: TINY BUT COMPLEX CELLS

Although each major type of moneran has unique characteristics, all share certain features that are most easily exemplified by bacteria. So we will study the general features of all monerans as they occur in bacteria and then look at the special properties in the diverse types.

Perhaps the best way to begin to visualize bacteria is to go back to the colonies growing in your fingerprint whorls on a culture medium and to focus at high magnification on individual moneran cells (Figure 20-1). One of the first things you may notice is the variety of sizes and shapes of different bacterial types. Bacteria range in size from 1 micron to 10 microns long (Figure 20-2). Most eukaryotic cells are about ten times longer than a medium-size bacterium. However, because of the range of shapes, comparing cell *volume* is a more accurate measure of size. Most bacteria have volumes of about 1 to 5 cubic microns. In contrast, the smallest eukaryotic yeast cells have volumes of about 20 to 50 cubic microns, while the smallest algal cells are 5,000 cubic microns.

The size differences between prokaryotic and eukaryotic cells can be explained by cell architecture. Even the smallest eukaryotic cell contains a nucleus, one or more mitochondria, ribosomes, and usually a few other organelles. The lower limit on cell size is dictated by the necessity to contain these organelles—a eukaryotic cell could not be smaller without giving up vital internal structures. For monerans, on the contrary, the lower limit on cell volume is set by the number and sizes of molecular assemblies: ribosomes, clusters of enzymes, and the circular DNA strand that serves as the cell's single chromosome. The smallest monerans (indeed, the smallest known living cells), called *mycoplasmas*, have no outer cell wall, contain half as much DNA as do larger bacteria, and usually are unable to move actively. Yet mycoplasmas, with their bare minimum of biological molecules, manage to survive quite successfully in the mucous membranes of the respiratory and urogenital tracts of mammals and other hosts.

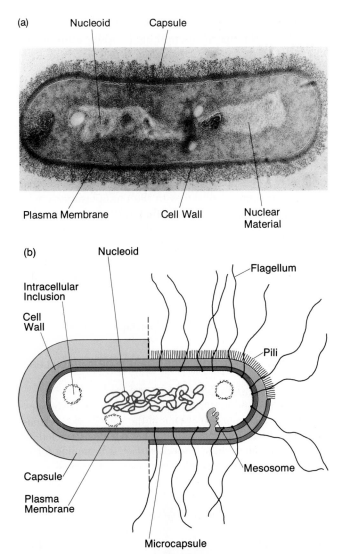

Figure 20-1   LIFE IN THE INVISIBLE REALM: A TYPICAL BACTERIAL CELL.
(a) This electron micrograph of a bacillus, magnified about 29,000 times, shows the plasma membrane surrounded by the cell wall and capsule. The nucleoid region appears transparent, and the dense-staining cytoplasm contains many ribosomes. (b) In this composite drawing of a bacterial cell, a capsulated half cell appears on the left, and a flagellated, noncapsulated half cell is shown on the right. The same types of cytoplasmic components are found in both.

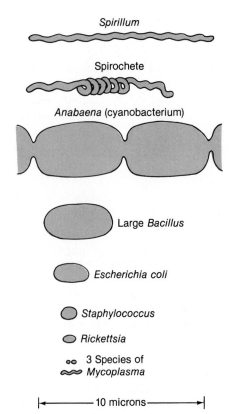

Figure 20-2   MAJOR TYPES OF PROKARYOTIC CELLS.
Several of the prokaryotes we will discuss in this chapter are drawn to scale to show the range of sizes that exists.

Bacterial Cell Structure

One feature that is common to most monerans is a cell wall. Perhaps because their total content of ions and small molecules makes monerans hypertonic to the fluids in which they live, a cell wall is present to resist swelling. This wall also gives shape, support, and protection to the plasma membrane and cytoplasm within. The typical bacterial cell wall is about 5 to 10 nanometers thick and is composed of huge carbohydrate and protein chains called **peptidoglycan polymers** (or *murein*). Within the polymers, short peptides cross-link and stabilize the main chains of sugar molecules.

Some bacteria have walls made up of only a single, thick layer of peptidoglycans outside the plasma membrane. Others, including *Escherichia coli*, have a second, outer layer of lipopolysaccharide that is a true lipid bilayer surrounding a thick peptidoglycan layer. The traditional method of distinguishing among bacteria, *Gram staining*—so-called after Christian Gram, the Danish physician who invented it in 1884—exploits this major difference. The procedure is to stain the cell wall with a dye called crystal violet (an iodine solution) and a stain called safranine. The bacterial cells with a single peptidoglycan wall take up crystal violet and appear purple under the light microscope. These are called **Gram-positive bacteria**. In contrast, **Gram-negative bacteria**, such as *E. coli*, take up safranine and appear red, as the micrograph in Figure 20-3 shows. Because of their cell walls, Gram-negative bacteria are much more resistant than are Gram-positives to attack by antibiotics and by

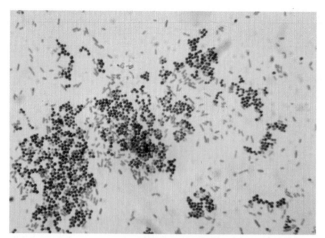

Figure 20-3   A MIXED SMEAR OF BACTERIAL CELLS STAINED BY THE GRAM PROCEDURE.
The Gram-positive *Staphylococcus aureus* cells (magnified about 2,600 times) have retained the crystal-violet dye (and stain purple), whereas the *Escherichia coli* cells, which are Gram-negative, do not retain crystal violet and instead show red dye.

lysozyme (recall from Chapter 4 how lysozyme hydrolyzes carbohydrate chains, thereby disrupting cell walls).

Some bacterial species have an additional protective structure surrounding the cell wall: a thick, mucuslike polysaccharide coating called the *capsule* (Figure 20-1). This coating helps the cell resist attack from a host organism's immune system by preventing white blood cells from attaching to the bacterial cell wall and then immobilizing and engulfing the cell. The presence of this capsule makes some strains of bacteria more infectious than others.

The cell walls of certain intestinal-tract bacteria, such as *E. coli*, may bear yet another outer structure, made of hundreds of projections called *pili* (Latin for "hairs"). These threadlike pili, shown in Figure 20-1b, are 1 to 2 microns long and are composed of a protein called pilin. Pili help the bacterial colony adhere to surfaces inside a host cell and also are used in the sexual process of bacterial conjugation (Chapter 14). Although pili are advantageous for adhesion and mating, they are something of an Achilles heel as well, since the tiny viral parasites we shall study later in this chapter can stick to the pili and thus gain entrance to the bacterial cell.

Beneath the cell wall lies the plasma membrane (Figure 20-1). As in eukaryotic cells, the membrane serves as a selective barrier to the import and export of various substances. However, the bacterial plasma membrane lacks cholesterol and has different fluidity properties from animal and plant plasma membranes. Enclosed within the membrane is the cell's cytoplasm, in which are dissolved various enzymes, RNA molecules, sugars, amino acids, fats, and so on. Bacterial cells lack the cytoskeletal support system found in eukaryotic cells and so

cannot actively move their plasma membrane and cell surface. But the plasma membrane provides a surface to which certain organelles can attach to carry out such processes as secretion or DNA replication.

A circular chromosome is present in a dense area within the cytoplasm called the *nucleoid*. During DNA replication, the chromosome remains attached to the inner surface of the plasma membrane. Ribosomes associated with mRNAs that code for proteins that will be exported from the cell also attach to the plasma membrane. This is analogous to the attachment of active mRNA–ribosome complexes to the endoplasmic reticulum in eukaryotic cells. For other mRNAs—those that code for cytoplasmic proteins—the ribosome complexes remain free in the cytoplasm.

Projecting from the plasma membrane into the cell is a convoluted, whorled membranous structure called the *mesosome* (Figure 20-4). The mesosome may be a source of new membrane during bacterial cell division; it may act as a crude mitochondrion, perhaps containing cyto-

Figure 20-4   ELECTRON MICROGRAPH OF A BACILLUS CELL.
In this bacillus (magnified about 30,000 times), the membranous whorls of the mesosome are clearly continuous with the plasma membrane. One of the pale nuclear regions is associated with the mesosome. (The indentions labeled "septum" are sites where the cell surface is beginning to extend inward across the cell to separate it as the cell begins to divide; see Figure 10-7.)

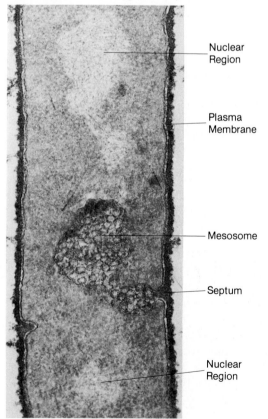

Nuclear Region

Plasma Membrane

Mesosome

Septum

Nuclear Region

chromes involved in cellular respiration; and it may serve as a simple type of "chloroplast" in photosynthetic bacteria. Recall that critical aspects of both respiration and photosynthesis depend on the spatial ordering of enzymes and electron transport molecules in membranes (Chapters 7 and 8). Both the mesosome and the plasma membrane appear to provide a membranous surface on which such molecules are arranged. Hence the functions dependent on ordering in membranes can take place in the absence of traditional eukaryotic membranous organelles.

Bacterial Cell Movement

Some bacteria cannot actively propel themselves and so move only as wind, water, or other organisms displace them. But many others can move toward light and dissolved food and oxygen and away from high concentrations of chemicals, toxins, and waste products. The structure that enables most of these motile cells to move is one of the loveliest pieces of molecular engineering in the world: the bacterial flagellum (Figure 20-5).

Bacterial flagella look and function quite differently from eukaryotic flagella, since they possess neither tubulin proteins nor the 9 + 2 arrangement of microtubules found in eukaryotic flagella (Chapter 6). In fact, the flagellum itself is not much larger than a single microtubule in a eukaryotic cilium or flagellum. Bacterial flagella are not covered by the plasma membrane—they are truly extracellular in the sense of protruding outward beyond the plasma membrane.

Whereas eukaryotic flagella lash back and forth because of the bending of microtubule doublets, bacterial flagella *rotate* at about 100 revolutions a second either clockwise or counterclockwise, like the propeller of a ship. When the rotation is clockwise, the cell moves forward in one direction. When it is counterclockwise, the cell tumbles chaotically. This remarkable rotation can

occur because the bacterial flagellum is attached to the cell by means of a unique socket, depicted in Figure 20-6. The shaft of the flagellum, called the *filament*, is a

Figure 20-6   STRUCTURE OF THE BACTERIAL FLAGELLUM.
(a) Unlike analogous eukaryotic flagella and cilia, the bacterial flagellum is a solid structure that extends through the cell wall and is anchored in the plasma membrane by protein rings (for simplicity, the four protein rings are shown as a disk and a rotating ring). The flagellum moves by means of the propellerlike rotation of the inner ring of proteins, which, in turn, rotates the attached rod, hook, and filament. Movement of hydrogen ions powers the rotation. (b) An electron micrograph of an *E. coli* flagellum, magnified about 422,000 times. The filament and various of the protein rings can be seen.

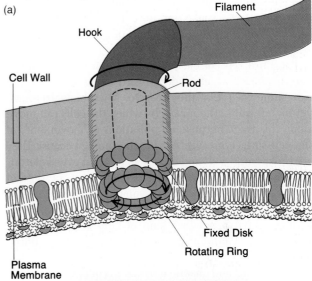

Figure 20-5   BACTERIAL FLAGELLA AND PILI: ORGANELLES OF LOCOMOTION AND EXCHANGE.
The number and arrangement of flagella are characteristic of bacterial species. This dividing *Salmonella anatom* cell (magnified about 10,000 times) has two flagella and hundreds of pili.

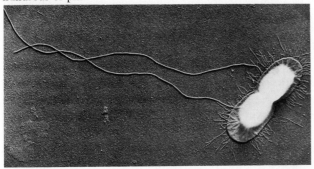

hollow structure composed of subunits of the protein flagellin wound into a long helix. The filament is attached to a midpiece, the *hook*, which, in turn, is connected to a shaft that extends through the cell wall and a protein disk before attaching to a ring of protein molecules. The ring is the site where the force for moving the flagellum is generated. The ring is the "motor" that responds to changing levels of a hydrogen-ion gradient across the plasma membrane and so causes the attached flagellum to rotate clockwise or counterclockwise. Thus the cell moves forward or chaotically.

Two other means of bacterial movement are known. Helically shaped bacteria called *spirochetes* have two unusual protein fibrils called axial filaments, each of which is attached to one end of the cell. The two wrap around the cell, overlapping in the center. As these filaments repeatedly lengthen and shorten, the cell is able to lash, snake, or bore forward. No one yet understands the exact mechanics of axial-filament function.

The third type of bacterial movement is found in the *myxobacteria,* which glide forward along a polysaccharide slime track that they secrete. Microbiologists have yet to determine the means by which the cells can move along this track because no pulsations or movements of the cell surface are visible. Furthermore, because they are prokaryotic cells and so lack an internal cytoskeleton, myxobacteria are thought to be incapable of the type of movement that characterizes amoebae or eukaryotic cells (Chapter 6). Our first hint about mechanisms of gliding is that a peculiar kind of fatty acid containing sulfur (called *sulfonolipid*) is essential for the gliding movement to occur.

The forward and chaotic movements of flagellated bacteria can be controlled so that aerobic bacteria move toward areas of high concentration of oxygen, while anaerobic bacteria move toward areas of minimal oxygen. In addition, bacteria move away from toxic chemicals and high concentrations of their own gaseous waste products. These behaviors are automatic. One photosynthetic bacterium, for example, will remain inside a lighted spot on a dark background. The instant the moving cells pass across the boundary between light and darkness, they tumble and effectively turn back, away from the edge of darkness. The reason is simple biochemistry: ATP production from photophosphorylation (Chapter 8) drops off sharply as the cell enters the dark zone. This rapid decrease in ATP apparently causes changes in the H^+ gradient near the flagellum so that it reverses its direction of rotation. The cell tumbles into the light once more. Other environmental factors, such as levels of different chemicals or nutrients, may also cause such forward or chaotic movements.

Bacterial Reproduction

The speed with which bacteria can reproduce is astonishing. The small size and comparative simplicity of prokaryotic cells allow many types to divide and their daughter cells to undergo new reproductive cycles as often as every twenty minutes. If unlimited space and nutrients were available to a single bacterium and if all its progeny divided at this rate, after just forty-eight hours, there would be some 2.2×10^{43} cells. The mass of bacteria would weigh 24×10^{24} tons, or 4,000 times the weight of the Earth.

In nature, bacterial growth usually is checked by limited nutrients and the build-up of waste products. Nevertheless, these startling figures illustrate how masses of bacteria can spring up rapidly in contaminated food or untreated wounds.

Bacteria reproduce by nonsexual means. Asexual reproduction can involve budding—in which a new individual arises from an outgrowth, or bud, on an existing cell—but it most often occurs by means of cell division. After examining this basic reproductive process, called *binary fission,* we will turn to the special genetic exchange processes that give meaning to the phrase "sex in bacteria" and to spore formation, a strategy for survival.

Binary Fission

Binary fission is the division of prokaryotic cells into two virtually identical daughter cells. Bacterial cell division is analogous to eukaryotic mitosis and cytokinesis, but the mechanisms are quite different because prokaryotes lack microtubules, a spindle, multiple chromosomes, and actin to form a contractile ring. In bacteria, attachment of the chromosome to the plasma membrane ensures segregation of the two chromosomes into the two daughter cells.

Binary fission begins when the cell starts to elongate and the old cell wall starts to break down at a site close to the middle of the cell, as illustrated in Figure 20-7. Just inside this region, the circular bacterial chromosome is attached to the plasma membrane. At about the same time that the cell wall begins to degrade, enzymes nick open one of the two strands of DNA, allowing the DNA strands to unwind, separate, and be replicated. After DNA replication begins, the original attachment point to the plasma membrane becomes two points. New membrane is inserted between these two points, so that they move apart. New cell-wall materials begin to assemble at this location and extend inward from the sides of the cylindrical cell, with the plasma membrane proceeding inward first, to form a *septum.* The process continues until the cell has divided in two, with each daughter cell

(a)

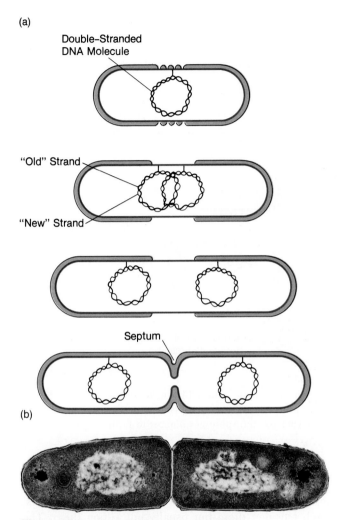

Double-Stranded
DNA Molecule

"Old" Strand

"New" Strand

Septum

(b)

Figure 20-7   BACTERIAL CELL DIVIDING.
(a) This series shows binary fission in a bacterial cell. As DNA replication begins, the cell wall ruptures, and new plasma membrane is added between the two sites where the DNA molecules are attached to the membrane. The plasma membrane then folds inward, and new cell-wall material is added in the septum to complete the division of the cell. (b) A bacterial cell after the septum has formed, magnified about 33,000 times.

containing a circular chromosome attached to the plasma membrane.

Sexual Processes

Although many bacteria apparently reproduce only through asexual binary fission, others carry out exchanges of genetic information reminiscent of sexual processes in eukaryotes. Strictly speaking, there is no

true sex in bacteria of the sort that includes meiosis, gamete fusion, and so on. We described the three bacterial "sexual" processes in Chapter 14, but we shall review them briefly here.

During *transformation,* DNA fragments released from one cell are taken up by and incorporated into the DNA of another cell. During *transduction,* pieces of DNA from a host cell's chromosome are picked up by a virus and carried to the recipient cell, into whose DNA they become incorporated (Figure 14-7). In both these processes, there are no new progeny, but a new genome is created. *Conjugation* is more like the familiar sexual transfer of genetic material in animals and plants. A donor male cell (designated F^+) forms a sex pilus with a recipient female cell (designated F^-). The F factor plasmid (small ring of DNA) in the male bacterium is replicated and then travels over the conjugation bridge to the female bacterium (see Figure 14-4). In the course of this process, the plasmid may take along with it portions of the male's chromosome. Although conjugation involves direct genetic transfer between paired cells, it is still not reproduction in the sense of eukaryotic reproduction because there are no progeny.

The three sexual transfer processes are costly; in each, the donor cell dies. However, the evolutionary advantages of new gene combinations apparently outweigh the costs. The frequency with which such processes occur in nature is not known, although bacterial viruses, the vehicles of transduction, are commonly found in soil, water, and air. In the laboratory, conjugation occurs when a bacterial population is large, the supply of available nutrients is dwindling, and metabolic wastes are accumulating. Under such conditions, the premium is not on producing more cells but on generating new genotypes that may have a better chance for survival under the new stringent environmental conditions.

Spore Formation

Pseudosexual processes are not the only means by which bacteria increase their chance of surviving under unfavorable conditions. Many species can form tough resting cells, or **endospores.** Worsening environmental conditions trigger the formation of a heavily encapsulated spore within another cell and initiate a series of steps that begins with chromosome duplication and ends with spore release (Figure 20-8). The spore contains only those cellular components necessary to start active metabolism once a favorable environment is encountered: genetic material, some ribosomes, enzymes to manufacture ATP and to carry out protein synthesis, amino acids, sugars, trace elements, and a minimum of water.

Spores have enabled the monerans to populate virtu-

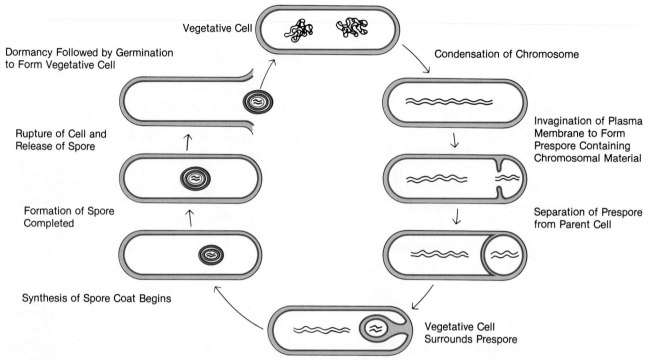

Figure 20-8   THE FORMATION OF AN ENDOSPORE: A SURVIVAL CAPSULE.
The successive stages of spore formation within a bacterial cell. A copy of the genetic material is enclosed in one end of the cell, thus forming the prespore. The vegetative cell expands around the prespore and the spore coat is synthesized. Ultimately the vegetative cell bursts, liberating the spore. Following a period of dormancy, the spore germinates to form a new vegetative cell.

ally every pinch of soil and drop of water on Earth. They can survive for long periods and through incredibly harsh circumstances. When 1,300-year-old anthrax spores are activated, they can still initiate a deadly infection. Other bacterial spores have been found 27,000 meters high above the Earth's surface and in the deepest ocean rifts. Some spores can survive more than an hour of being boiled or years of being locked in polar ice. The spores of one unusual bacterium can survive being bombarded with up to 500,000 rads of ionizing gamma radiation. Still others can survive being baked in hot desert sands, germinating only when it rains. Endospores are truly an important evolutionary survival mechanism—although not the only one that bacteria use. The versatile metabolism of bacteria is the key to their adaptation to wide-ranging environments and external stresses.

Bacterial Metabolism

It has been said that if there is an energy-yielding chemical reaction in nature, there is some species of bacterium to take advantage of it. Bacteria use a number of different metabolic strategies to harvest energy and to fuel their internal processes. *Heterotrophic bacteria* consume waste products and dead organic matter or

exist as **parasites,** organisms that feed on living hosts, often harming them but usually not killing them. The ability of heterotrophic bacteria to break down a wide range of organic substances makes them important to ecological cycles, but they are significant in causing certain diseases as well.

Autotrophic bacteria use energy and inorganic substances from the environment to manufacture the necessary organic building blocks and substrates for ATP production. They derive energy from either photosynthesis or oxidation of inorganic compounds. Thus there are two types of autotrophs: *photoautotrophs* and *chemoautotrophs.*

Photoautotrophs include the green sulfur bacteria and two kinds of purple bacteria, sulfur and nonsulfur. None contains the chlorophylls *a* and *b* of higher plants; modified chlorophylls are present. The purple bacteria look purple because they have yellow and red carotenoids plus a bluish-gray pigment called *bacterio-chlorophyll.* This compound absorbs infrared wavelengths and thus can function in deep waters, into which such wavelengths penetrate. Because these cells are true prokaryotes, no true chloroplasts are present; instead, bacterio-chlorophylls occur on membranous tubes, stacks of flattened sheets, or vesicles.

MICROBES AT SEA: SOME LIKE IT HOT

An exciting recent discovery has shattered a long-standing idea about bacterial life. For decades, thermophilic bacteria, such as those living in the bubbling mud around the geysers in Yellowstone National Park, have exemplified for biologists the extreme boundary conditions for life. These cells exist at temperatures that would quickly kill virtually all other life forms. Within the last few years, however, scientists have found in deep ocean vents extraordinary bacteria that seem to thrive in environments several times hotter than those around thermal springs and actually stop growing when "cooled" to the mere boiling point of water.

Much to everyone's surprise, John Baross and Jack Corliss of Oregon State University and Jody Deming of Johns Hopkins University found chemoautrophic bacteria in water samples obtained from tall sulfide chimneys almost 3,000 meters below the ocean surface near Baja California (Figure A). The mineral-laden water spewing from these "black smokers" is 350°C (about 662°F) and remains liquid rather than vaporizing due to the tremendous pressures at those depths. Remarkably, the bacteria discovered by the team grow and divide at 250°C and cease reproducing when the temperature falls below 100°C. Thus the growth of microbes is clearly not limited by temperatures that were long thought to denature proteins. Given sufficient pressure such temperatures can be tolerated.

So far, the vent microbes have raised more questions than they have answered. What kinds of structural proteins, enzymes, membranes, and other biological components can withstand such intense pressures and temperatures? Do these cells have normal DNA? Are their gene sequences similar to those of other archaebacteria, suggesting a common ancestry? Or are the vent bacteria genetically distinct from other prokaryotes? If so, do these microbes represent a throwback to the earliest cells that formed in severe habitats on the early Earth? And could, as the researchers speculate, new cells be forming from non-

Figure A   **BLACK SMOKER BACTERIA.**
These cylindrical cells (magnified about 3,700 times) thrive at extraordinarily high temperatures and pressures.

living precursors today by means of chemical evolution in deep ocean vents? Finally, will evidence of such biogenesis be found in future voyages to the ocean abyss?

We will not know the answers to these questions until return trips are made and until Baross and his co-workers develop routine systems for cultivating and studying the rare organisms in the lab. But studies of the extreme thermophiles, living in what the essayist Lewis Thomas has called "a hell of a place," will certainly deepen our understanding of basic life processes.

Bacterial photosynthesis is an anaerobic process. Both purple and green sulfur bacteria use molecules of H_2S (instead of H_2O) from mud sediments as electron donors for photosynthesis. The sulfur bacteria live in foul-smelling mud or in deep ocean sediments. In purple nonsulfur bacteria, such organic compounds as alcohols and fatty acids serve as electron donors. The nonsulfur bacte-

ria can grow in the presence of oxygen and thus are not as ecologically restricted as are the sulfur users.

Chemoautotrophs can live in extremely harsh environments and use some surprising substrates to obtain energy. These bacteria have the unique ability to oxidize inorganic compounds needed to build organic molecules. Thus, they trap energy released from such chemi-

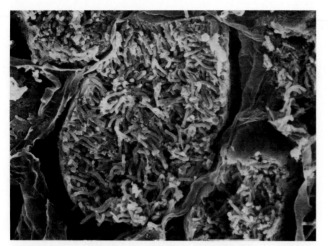

Figure 20-9   NITROGEN-FIXING BACTERIA.
Here a root nodule (magnified about 1,500 times) is shown cracked open to reveal many symbiotic *Rhizobium* bacteria. These bacteria use sugars stored in plant roots as an energy source and in return provide the plant with a continuous supply of fixed nitrogen that can be used to build proteins. From *Living Images* by Gene Shih and Richard G. Kessel, Science Books International, 1982. Reprinted by permission of the present publisher, Jones and Bartlett Publishers.

cals as hydrogen sulfide gas, elemental sulfur, ferrous iron, ammonium ions, and nitrites. Bacteria that oxidize nitrogen and sulfur are critical to the cycling of these elements in the Earth's environment.

The process of *nitrogen fixation*—whereby nitrogen is taken from the air, oxidized, and built into amino acids and nucleotides—is the critical step in the Earth's nitrogen cycle. It involves action by free-living bacteria and cyanobacteria, and by bacterial symbionts that are associated with the roots of various plants (Figure 20-9). ("Symbiosis" means "living together, with mutual benefit.") The *Rhizobium* bacterium, for instance, fixes nitrogen in the roots of peas, clover, alfalfa, and other legumes. It is estimated that up to 200 million tons of nitrogen are fixed each year, mostly by various free-living or symbiotic monerans. We will discuss the nitrogen cycle further in Chapter 44.

Regardless of their energy source or their ability or inability to fix nitrogen, all bacteria have an extremely high rate of metabolism. Because of their microscopic size and high surface-to-volume ratio, bacterial cells can rapidly import metabolic precursors and export wastes. Bacteria that ferment milk sugar (lactose) break down 1,000 to 10,000 times their own weight of this substance each hour. It would take a human thirty to forty years to metabolize his or her own weight in lactose. And even if a person *could* burn food close to the rate of bacterial metabolism, the heat given off would be lethal.

The small size of bacteria coupled with this metabolic speed and the range of available energy substrates have a

great deal to do with the success of the kingdom Monera. This success is reflected not only in the spectrum of energy sources used by bacteria, but also in the range of oxygen they can tolerate. Species that are **obligate anaerobes** grow only in the absence of free oxygen and employ only fermentation (rather than aerobic respiration) to generate ATP. The green and purple sulfur bacteria are obligate anaerobes, as is *Clostridium botulinum*, the agent that causes botulism, a usually fatal disease brought on by eating food containing toxins produced by this bacterium. A different class, the *microaerophylls*, grows best in low levels of oxygen. **Facultative anaerobes** grow with or without oxygen. At the other end of the oxygen spectrum, **obligate aerobes** must have oxygen for metabolic processes, or their growth and survival are impeded. These variations in the oxygen tolerances of bacteria may reflect the evolution of microorganisms during a period on Earth when oxygen levels in the oceans and atmosphere were slowly changing.

MICROORGANISM DIVERSITY: THE VARIETY OF MONERANS

Now that we have examined the major features common to all prokaryotic monerans, it is time to survey the real world of these organisms. The monerans are a widely divergent group—so much so, in fact, that biologists do not agree on which types of organisms should be included in this kingdom. Two groups are always considered monerans: the bacteria, or *schizomycete* subkingdom, and the *cyanobacteria* subkingdom (formerly called blue-green algae). Two other distinctive groups related to monerans are the *chloroxybacteria*, which resemble both the cyanobacteria and the chloroplasts of green plants, and the archaebacteria, which includes methanobacteria. We will consider each of these four types of cells and then discuss the viruses, the large group of disease-producing noncellular agents composed of only a protein coat and a nucleic-acid core.

Subkingdom Schizomycete: The Bacteria

Because bacteria, or **schizomycetes,** do not reproduce sexually, they cannot be classified into species using the same criteria employed for eukaryotes (recall from Chapter 19 that different species cannot interbreed). Thus from the time of Anton van Leeuwenhoek, bacteria were assigned to taxonomic groups solely on the basis of shape. With the development of the Gram-staining and

Table 20-1   **MAJOR GROUPS OF SUBKINGDOM SCHIZOMYCETE**

Phylum (Examples)	Form	Mode of Movement	Mode of Reproduction	Mode of Nutrition	Ecological Role	Other Distinguishing Factors
Eubacteria (*Escherichia coli, Streptococcus, Staphylococcus, Myobacterium tuberculosis, Clostridium botulinum*)	Rod, sphere, spiral	Flagella, wriggling	Binary fission	Chemoautotroph, photoautotroph, heterotroph	Decomposer, symbiont, disease agent	Rigid cell wall, form endospores
Spirochetes (*Spirocheta, Treponema pallidum, Leptospira*)	Extremely long, flexible spiral	Twisting	Binary fission	Heterotroph	Symbiont (in mollusks), decomposer, disease agent (syphilis)	Obligate anaerobes (many forms)
Myxobacteria	Filament, large and short rods	Gliding (mechanism unknown)	Short filaments break off, binary fission	Heterotroph, chemoautotroph	Decomposer (especially of complex polysaccharides)	Flexible cell wall, some form fruiting bodies
Actinomycetes (*Streptomyces, Actinomyces*)	Branching multicellular filaments	Nonmotile	Spores, fragmentation, binary fission	Heterotroph	Decomposer (lipids, waxes), disease agent (tuberculosis, leprosy)	Many are moldlike in appearance
Rickettsiae, Chlamydiae	Small (0.5 by 1.1 μm)	Nonmotile	Binary fission	Heterotroph	Disease agent (typhus, Rocky Mountain spotted fever)	Intracellular parasites, rigid cell wall
Mycoplasmas	Smallest free-living cells	Nonmotile	Binary fission	Heterotroph	Disease agent (pleuropneumonia)	Plasma membranes contain cholesterol, no cell wall, many are intracellular parasites

other techniques, however, species can be differentiated by their chemical, metabolic, and genetic differences. With this consideration in mind, let us look at the major types of bacteria: eubacteria, spirochetes, myxobacteria, actinomycetes, rickettsiae, and mycoplasmas. We will consider these groups to be phyla. Table 20-1 summarizes key characteristics of each group.

Eubacteria

Eubacteria, or "true" bacteria, are the most abundant and diverse group of prokaryotes and were the basis of much of the general description of moneran cells to this point. They traditionally have been classified by shape. A rod-shaped bacterium is called a *bacillus;* a sphere-shaped bacterium is called a *coccus;* and a spiral-shaped bacterium is called a *spirillum.* The names for these shapes often are incorporated into the Latin genus

names, making the names a bit more descriptive and more easily understood. For example, *Bacillus anthracis* is the rod-shaped bacterium that causes anthrax in animals and people. Members of the genus *Staphylococcus,* common inhabitants of the human body, form irregular clusters of spherical cells. The bacterium that causes strep throat, a species of *Streptococcus,* divides repeatedly along the same axis, forming a long chain of cocci. Both cocci and bacilli move by means of flagella. In contrast, many members of the genus *Spirillum* wriggle forward with a corkscrew motion. However, one group of spirilla, the vibrios, is propelled by flagella. This group includes *Vibrio cholerae,* the disease agent of cholera. Representative eubacteria are shown in Figure 20-10.

Spirochetes

The **spirochetes** are bacteria with a spiral or corkscrew shape, as shown in Figure 20-11. Many are free-living

(a)

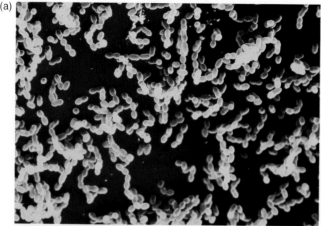

(b)

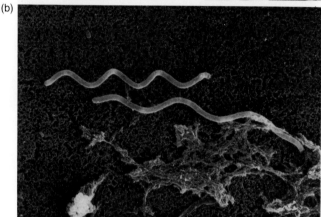

Figure 20-10   COMMON EUBACTERIA.
These organisms include (a) spherical cocci (*streptococcus mutans*) and (b) helical spirilla (both magnified about 1,500 times). (b) From *Living Images* by Gene Shih and Richard G. Kessel, Science Books International, 1982. Reprinted by permission of the present publisher, Jones and Bartlett Publishers.

aquatic cells. In liquid, spirochetes move by snakelike lashing movements, while in denser substances, they bore forward in a screwlike manner. The best known spirochete is *Treponema pallidum*, the causative agent of syphilis.

Myxobacteria

The **myxobacteria** are small, unflagellated, rod-shaped cells that, as we have noted, glide along slime tracks. They occasionally swarm together to form *fruiting bodies* (Figure 20-12), where some myxobacterial cells develop into *myxospores* capable of surviving harsh environmental conditions. Myxobacteria tend to live in soil, where they play an important ecological role in the breakdown and consumption of dead organic matter.

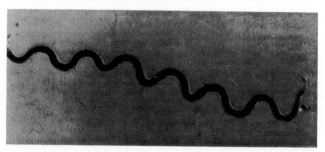

Figure 20-11   THE SPIROCHETE *TREPONEMA PALLIDUM.*
This spiral bacterium is responsible for syphilis (magnified about 10,000 times).

Actinomycetes

A large number of types of diverse filamentous organisms called **actinomycetes** form the fourth schizomycete group. Branching chains of cells are formed because daughter cells arising during binary fission remain attached to one another, forming a multicellular network, or "filament" (Figure 20-13). The actinomycetes include species that can cause human diseases, such as leprosy and tuberculosis. This group also includes many species of the genus *Streptomyces*, the microorganisms from which antibiotics such as streptomycin, erythromycin, and the tetracyclines are derived. Certain actinomycetes form mutually beneficial symbiotic relationships with the roots of nonleguminous plants and perform the important task of nitrogen fixation.

Figure 20-12   FRUITING BODIES FORMED BY MYXOBACTERIAL CELLS.
Myxobacteria (*stigmatella durantiaca*) secrete the extracellular materials that compose the stalks and then lie dormant in the spherical caps. The caps protrude above the soil surface, which facilitates the dispersal by wind of released spores. This electron micrograph is magnified about 1,500 times.

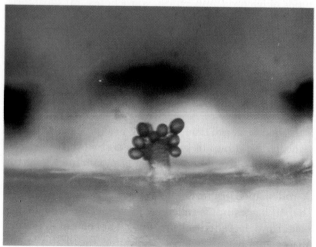

(a)

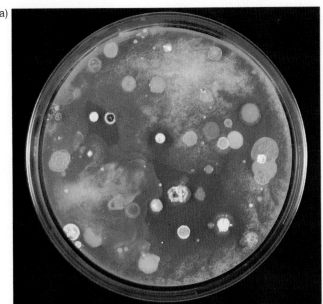

(b)

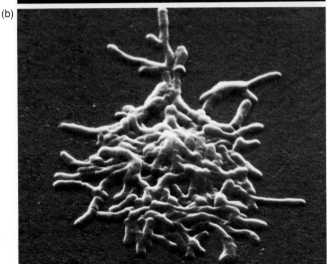

Figure 20-13   SOME ANTIBIOTIC-PRODUCING ACTINOMYCETES.

(a) In the mixed bacterial culture shown here, clear zones free of bacterial growth can be seen around antibiotic-producing bacterial colonies, such as *Streptomyces*. (b) The scanning electron micrograph reveals a small actinomycete microcolony (magnified about 6,750 times).

Rickettsiae

Rickettsiae are among the tiniest prokaryotes and live as parasites in two alternating hosts: arthropods (such as lice, ticks, and fleas) and mammals or birds. Although rickettsiae do not harm their arthropod hosts, they can be transmitted by means of the bites of fleas, ticks, or lice to mammals, where they can cause such serious diseases as typhus and Rocky Mountain spotted fever. Rickettsiae are so small and simple because they have lost

functions carried out by their hosts' cells. This is a general rule for parasites: Why go to the biological expense of maintaining genetic information, enzymes, and so on for processes that are done just as well (or better) by a host? We shall encounter this principle again in Chapters 21 and 25 when we discuss protistan and animal parasites.

Mycoplasmas

Mycoplasmas are the smallest living cells yet found. They lack rigid cell walls, grow to form filaments, globules, rings, and other shapes in colonies, and are usually aerobic. They cause such diseases as pleuropneumonia and are resistant to antibacterial agents such as penicillin.

Subkingdom Cyanobacteria: The Blue-green Photosynthesizers

The second major group of monerans is the subkingdom Cyanophyta, which includes 200 species formerly called blue-green algae and now referred to as **cyanobacteria** (from the Greek word *kyanos*, meaning "blue"). Cyanobacteria occur in a wide range of habitats, from salt and fresh water to soil, where they produce some of the compounds responsible for "earthy" odors.

Some of the oldest fossilized cells of any kind found on Earth appear in sediments 2.7 billion years old and resemble modern cyanobacteria. As mentioned in Chapter 19, masses of these simple organisms living in primeval seas are believed to have released much of the oxygen that allowed the development of cellular respiration, the establishment of the carbon cycle, and the origination of many of the life forms that followed. It is no exaggeration to say that these tiny colorful cells were indeed architects of the new world of rapid metabolism, which was a precondition for the large size and the complex form and function of plants and animals.

Cyanobacteria are single rod-shaped or spherical cells that occur in clusters or in long filamentous chains, as shown in Figure 20-14. The individual cells are somewhat larger than typical bacteria, although not as big as eukaryotic cells. Cyanobacterial cell walls, like those of other bacteria, are made of peptidoglycan polymers. The walls are surrounded by a protective coating of gelatinous material that is often poisonous to herbivores. The cells are truly prokaryotic, but they lack flagella and axial filaments and appear to glide, like myxobacteria, or to spin longitudinally.

Cyanobacteria contain chlorophyll *a*, carotenoids, and red and blue pigments called *phycoerythrins* and *phycocyanins*. Together, these pigments impart to the cells colors that range from yellows and reds to violets and

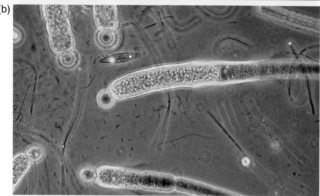

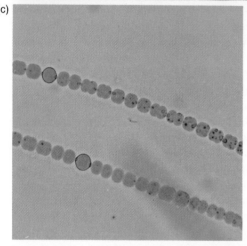

Figure 20-14   PHOTOSYNTHETIC POND SCUM.
(a) Colonies of cyanobacteria (blue-green algae) often form a scum on the surface of polluted ponds. Two common types of filamentous cyanobacteria are (b) *Gleotrichia* (magnified about 100 times) and (c) *Anabaena* (magnified about 75 times).

deep blues. The photosynthetic pigments are associated with layers of membranes called *thylakoids*, which are quite different from and simpler than the chloroplasts in plant cells. As cyanobacteria give off oxygen during photosynthesis, gas bubbles form and buoy the filaments and matlike colonies of aquatic species near the water's surface, where they can absorb light more efficiently.

Many cyanobacterial species are capable of nitrogen fixation. This process depends on an enzyme called *nitrogenase*, which happens to be destroyed by oxygen. To overcome this problem, nitrogen-fixing bacteria form *heterocysts*, cells with thick walls that impede oxygen entry (Figure 20-15). Nitrogenase in heterocysts uses ATP generated by photosystem I (Chapter 8) plus electrons to reduce N_2 to NH_3 (ammonia). The fixed nitrogen (or the amino acids into which it is incorporated) is then passed to adjacent cells through plasmodesmata; in turn, photosynthetic products pass into the heterocysts for their use. The division of labor by two cell types—those for photosynthesis and those for nitrogen fixation—plus the plasmodesmata connections are characteristics that imply a primitive kind of multicellularity in colonies of cyanobacteria.

The ability to carry out photosynthesis as well as nitrogen fixation means that many cyanobacterial species can survive in harsh environments that provide only water, inorganic minerals, air, and light. Some species can live on the sides of damp flowerpots, on bare rock surfaces, on tree trunks, in deserts, and even on polar-bear fur, to which they impart a greenish tinge. A few species have particularly resistant cell walls that enable them to survive extreme environments: thermal hot springs up to 75°C; alkaline water and soil up to pH 11; acidic sulfurous soil with a pH of 0.5; and highly concentrated salt solutions (25–30 percent NaCl). Some cyanobacteria that have adapted to extremes of temperature, salinity, or pH may represent the kinds of cells that lived in similarly harsh circumstances on the early Earth, especially at times when atmospheric oxygen levels were very low.

Chloroxybacteria: Possible Ancestors to Chloroplasts

The third, unique group of monerans, the **chloroxybacteria,** were discovered in 1976 in the bodies of tunicates, small marine animals that live on rocks or pilings (some species are called sea squirts). In 1986, a filamentous, free-living species was discovered to be a dominant kind of plankton in shallow lakes in the Netherlands. A new phylum, Chloroxybacteria (also called Prochlorophyta), was created for these light-green, single-celled organisms. Some biologists consider these cells to be a separate subkingdom of monerans.

(a)

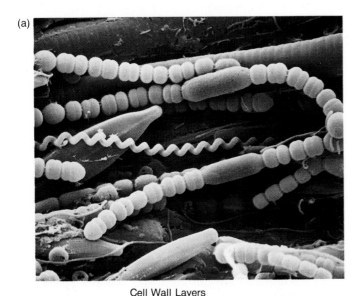

Cell Wall Layers

(b)

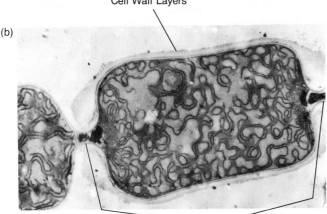

Plasmodesmata

Figure 20-15   HETEROCYSTS IN *ANABAENA*.
(a) In these *Anabaena* filaments (magnified about 1,500 times), elongate, cylindrical heterocysts are interspersed between smaller, normal photosynthetic cells. (b) A heterocyst (magnified about 11,000 times) has extra wall layers to impede oxygen entry and plasmodesmata to connect it to adjacent photosynthetic cells, thus facilitating the passage of fixed nitrogen or amino acids between the cells.

The cells are definitely prokaryotic, yet they contain photosynthetic pigments similar to those in green algae and higher plants: both chlorophylls *a* and *b* and yellow and orange carotenoids. In some respects, chloroxybacteria resemble chloroplasts of plants and green algae, yet in others they are similar to cyanobacteria. Even so, it is hypothesized that prokaryotic green bacteria such as the chloroxybacteria were engulfed by early cells, giving rise to chloroplasts (we will return to this theory later in the chapter).

Archaebacteria: An Independent Kingdom?

The fourth group of monerans are the **archaebacteria.** These organisms are so different from schizomycetes and cyanobacteria that some biologists prefer to classify them as an independent kingdom. Other scientists classify them as a subkingdom or phylum of monerans. Archaebacteria include cells that produce methane, others that are sulfur-dependent, and a third group that lives in conditions of high salt in their environment.

The methane bacteria (or **methanobacteria**) are anaerobes that derive energy from coupled reactions in which H_2 is oxidized to H_2O (yielding energy) and CO_2 is reduced to CH_4 (methane gas). Many methane bacteria live in decaying vegetation in streams, lakes, and bogs, where they generate "swamp gas." A few species live in the intestinal tracts of cattle, which belch the methane gas. Like the cyanobacteria, many methanobacteria species are also adapted to extreme environments. Some live thousands of fathoms deep on the sea floor. One species lives only in hot sulfur springs, while another inhabits only smoldering piles of coal-mine tailings. Perhaps, 3 billion or more years ago, equivalent extreme environments were the habitats of archaebacteria, the ancestors of today's methanobacteria.

Sulfur-dependent archaebacteria (**sulfobales**) may thrive at high temperatures (80°–95°C), and may live either by oxidizing or reducing sulfur. One remarkable species normally uses carbon dioxide as a carbon source and oxidizes sulfur with oxygen to provide an energy source. However, those cells contain a DNA plasmid which upon amplification lets the cells switch to reducing sulfur with hydrogen; the resultant H_2S serves as an energy source.

Archaebacteria living in high salt environments, the **halobacteria,** may also live at very high temperatures in such places as volcanoes (**thermoproteales**). Some species have introns in their tRNA and rRNA genes, a feature also found in some sulfur-dependent archaebacteria. This characteristic of archaebacteria is shared, of course, with eukaryotic cells but not with eubacteria, which lack introns.

Recent studies of archaebacterial protein and nucleic-acid sequences have led the biologist Carl R. Woese to conclude that these organisms had an origin different from that of eubacteria and cyanobacteria. Archaebacteria have several unique enzymes. They lack the cytochrome molecules so essential for electron transport in most other kinds of cells; their transfer RNA and ribosomal RNA have unique sequences of bases; their RNA polymerase is like that of eukaryotes and unlike that of eubacteria; nucleosomelike structures (again like eukaryotes) are present; they have distinctive plasma-membrane lipids; and their cell walls are not made of peptidoglycan polymers. Of special significance is the presence of a ribosomal structural protein that is much more similar to that in eukaryotic ribosomes than to those in other bacteria. All these findings suggest that archaebacteria are as different from other prokaryotes as prokaryotes are from eukaryotes. Based on these molecular differences, Woese concludes that some extremely primitive ancestral cell populations gave rise to three groups: the archaebacteria, the eukaryotes, and the monerans. Many additional studies are needed to test this hypothesis, but the archaebacteria certainly represent an important link in the study of the earliest forms of life on Earth.

BACTERIA AND HUMANS

So far, we have noted the evolutionary success of the various types of prokaryotes, their ubiquity, and their role in soil-nutrient cycles and in the breakdown of dead organic matter. This last process is especially useful in the treatment of human waste. The bacterial decomposers usually need only O_2 and moisture to reduce sewage sludge or a compost heap to a small volume of innocuous organic matter that can be used as a fertilizer or soil conditioner. Bacteria function in other areas of great importance to humans as well: the production of food and the spread of disease.

Bacteria and Food Production

Were it not for bacteria, the human diet might be missing some favorite foods and beverages. Cheese, butter, sour cream, yogurt, buttermilk, and other dairy products are produced by bacterial action on milk. In particular, *Lactobacillus* species and *Streptococcus lacti* degrade lactose and release the lactic acid that gives many dairy products their sharp flavor. Other lactic-acid bacteria are used to ferment vegetables and fruits into pickles, sauerkraut, and ripe olives. Acetic-acid bacteria produce vinegar from wine or fruit juices by a slow oxidative process.

Of course, other bacteria can spoil food products. For this reason, milk must be pasteurized by heating it to 60°C, canned foods must be processed with high heat and pressure, and many foods must be kept frozen or refrigerated to retard bacterial growth.

Bacteria and Disease

Slightly more than a century ago, bacteria had a far greater impact on daily life and health than they have today. A respiratory infection or a small infected wound then could mean death. Having a tooth pulled or a limb amputated was a life-threatening event. In those days, before refrigeration and sterilization, improper salting of meats, boiling of sausages, or storing of grains could easily poison the unsuspecting consumer. It was not until the late nineteenth century that scientists realized that microorganisms cause disease and produce potentially deadly toxins that spoil food. The German physician Robert Koch was the first to establish procedures for determining whether a microorganism (which we know today includes bacteria, viruses, and eukaryotic microbes) is a **pathogen**—an agent that causes a specific disease (Figure 20-16). Koch understood that identifying a pathogen requires showing that the agent is always present at the time of the disease and that a pure culture of the organism causes the same disease to appear in a healthy host.

The work of Koch and later that of Louis Pasteur, who proposed the germ theory of disease, led to the adoption of sanitation measures that have contributed to the lengthened life span of humans and the ability to sustain huge human populations in densely crowded urban environments. Table 20-2 lists a number of human diseases caused by bacteria and the main ways in which these

Figure 20-16   A MILESTONE IN MEDICINE.
The medical treatment of microbial diseases began with Robert Koch's experiments, which showed that pathogenic bacteria cause diseases such as cholera. Koch is seen here, to the left, in his tent near Lake Victoria in Africa during an expedition in 1906. Koch's assistant, Dr. Panse, on the right, died on this search for the cause of African sleeping sickness.

Table 20-2 HUMAN BACTERIAL DISEASES, BY MODE OF TRANSMISSION

Mode of Transmission	Disease	Causative Agent
Fecal contamination	Typhoid fever	*Salmonella typhi*
	Cholera	*Vibrio cholerae*
	Dysentery (shigellosis)	*Shigella*
	Food poisoning	*Clostridium botulinum*
		Salmonella
		Staphylococcus species
	Traveler's diarrhea	*Escherichia coli* strain
Animal bites	Typhus	*Rickettsia typhi*
	Q fever	*Coxiella burnetii*
	Plague	*Pasturella pestis*
	Tularemia	*Francisella tularensis*
	Rocky Mountain spotted fever	*Rickettsia rickettsia*
Exhalation droplets	Diphtheria	*Corynebacterium diphtheriae*
	Tuberculosis	*Mycobacterium tuberculosis*
	Meningitis	*Neisseria meningitidis*
	Scarlet fever, rheumatic fever	*Streptococcus* species
	Pneumonia	*Streptococcus pneumoniae*
	Tonsillitis	*Streptococcus*
	Whooping cough	*Bordetella pertussis*
Direct contact	Gonorrhea	*Neisseria gonorrhoeae*
	Syphilis	*Treponema pallidum*
	Puerperal fever	*Streptococcus* species
Wounds	Tetanus	*Clostridium tetani*
	Gas gangrene	*Clostridium perfringens*

diseases are spread. While some spread slowly, such as those requiring human contact, others can spread with frightening speed. Periodically, outbreaks of dysentery or of *Salmonella* poisoning spread among groups of people living together, as in college dormitories and hospitals. These diseases may even affect people who have eaten in a particular restaurant. If traced to the source, such cases might involve a food handler who chanced to pass the bacterium to lettuce leaves, silverware, and so on; from there, the microbe entered the unknowing people. The dense population of contemporary urban areas makes such person-to-person spread more and more likely.

The actual symptoms of bacterial disease can result from destruction of host cells and tissues, from irritation by bacterial waste products, from exaggerated immune responses to the presence of foreign cells, or from reactions to bacterial toxins. These toxins can be released by living bacteria, in which case they are called **exotoxins,** or they may be liberated when bacterial cells die and burst, in which case they are called **endotoxins.** Most exotoxins are proteins, whereas endotoxins usually are lipid–polysaccharide materials from the burst cell wall. For the bacterium, the toxins it produces often inhibit other cell types and therefore help reduce competition for nutrients and other resources.

Some bacteria—such as the agents responsible for diphtheria, scarlet fever, and botulism—release disease-causing toxins only when they are infected by specific viruses. Botulin, the toxin formed by *C. botulinum*, is extraordinarily lethal: eating a single infected string bean or even momentarily taking a mouthful of contaminated home-canned green peppers and spitting it out has resulted in agonizing asphyxiation and death.

Most bacterial diseases can be treated effectively with antibiotics and other drugs, many of which are produced by other monerans or by fungi. Antibiotics may interfere with DNA, RNA, or protein synthesis; ribosome function; membrane integrity; or a variety of other essential aspects of a sensitive moneran species. Some antibiotics prevent synthesis and assembly of bacterial cell walls; such drugs have virtually no effect on animal or plant cells because the plant cell wall and the animal cell glycocalyx are made of different substances from bacterial cell walls. Some bacterial diseases can be prevented by *vaccination*, the process by which a person's immune system is stimulated to set up a permanent defense response against a given type of disease-causing bacterium or virus. We will discuss this approach in Chapter 32.

One critical problem now facing human medicine is the tendency of many types of bacteria to develop resistance to antibiotics. Chance mutations can arise, conferring drug resistance to a few cells. These unchecked cells can quickly reproduce into large, infectious populations. Drug resistance also can be transferred from cell to cell by plasmids and pieces of DNA. Some particularly worrisome bacteria, such as *Neisseria gonorrhoeae* (the agent of gonorrhea), develop multiple drug resistances and cannot be controlled by any known antibiotics. A great deal of research is currently directed toward developing new antibiotics and finding other ways to deal with drug-resistant strains of bacteria.

VIRUSES: NONCELLULAR MOLECULAR PARASITES

It has been said that viruses lie on the threshold of life, straddling the shadowy line between living and nonliving. **Viruses** are not cells, but particles of genetic material and protein that can invade living cells, commandeer their metabolic machinery, and in this way reproduce. Furthermore, viruses evolve by the same basic mechanisms that operate in living cells, including especially adaptive selections of their genetic variations. That is surely a characteristic of life. Although they are not classified as monerans and are not similar structurally, we consider viruses here because of their microscopic size and capacity for inducing disease.

Biologists in the late nineteenth century suspected that agents much smaller than bacteria can cause disease. The agents were thought to be extremely tiny cells. Not until 1935 was the first virus purified and studied, however. It was then that Wendell Stanley set out to chemically isolate the agent responsible for a serious disease of tobacco leaves. He squeezed the juice from 2,000 pounds of tobacco leaves and from the juice extracted a residue that could be purified to form pure, needlelike crystals. These crystals could be dissolved in a neutral solution or recrystallized in an acidic solution. When the crystalline form was dissolved and applied to tobacco leaves, it could cause disease. The particles were named *tobacco-mosaic virus (TMV)*. As crystals, they were clearly not living cells but some sort of inert entity. Later tests showed TMV to be rod-shaped particles composed of more than 2,000 identical protein molecules forming a coat around a core of RNA, the genetic material of this type of virus. With this discovery, Stanley launched modern *virology*, the study of viruses.

Structure of Viruses

Many viruses have been identified in the last half-century, largely through use of the electron microscope.

(a)

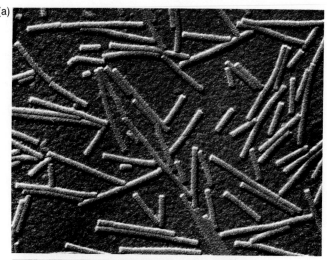

(b)

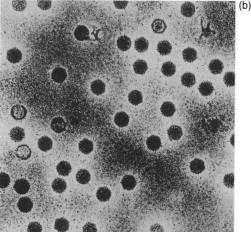

(c)

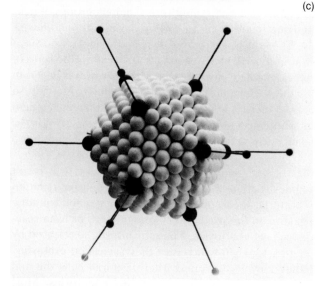

Figure 20-17   VIRUSES: A VARIETY OF PARTICLE SHAPES.

(a) The rod-shaped tobacco-mosaic virus, TMV (magnified about 92,000 times). (b) Negatively stained electron micrograph of adenovirus (magnified about 59,000 times). (c) A model of the adenovirus shows some of its twenty sides, each an equilateral triangle.

Some viruses, such as TMV, are rod-shaped crystals (Figure 20-17a). Others, such as the agents that cause colds and fever blisters, are shaped like geodesic domes (Figure 20-17b and c). Still others, such as those that cause influenza, are somewhat globular, with spiky proteins over their surfaces. The bacterial viruses, or bacteriophages, we examined in Chapter 12 are tadpole-shaped, with a hexagonal head, a tail, and a set of leglike tail fibers.

Regardless of overall configuration, each extracellular virus particle, or **virion,** is composed of an outer protein coat, or *capsid,* and an inner core of nucleic acid (Figure 20-18). This core can be single- or double-stranded DNA or RNA and can contain from five to several hundred genes. Together, the capsid and nucleic-acid core are called the *nucleocapsid.* A number of animal, plant, and bacterial viruses—including the adenovirus pictured in Figure 20-17b—are also surrounded by an *envelope,* a compact membranelike structure composed of a lipid bilayer and tightly packed viral proteins.

Reproduction of Viruses

To reproduce, viral nucleic acid must enter a host cell. Some viruses inject their genetic core into the host, using for energy ATP molecules derived from a former host cell and stored in the tail (Figure 12-7). The protein capsid left outside the cell is degraded as the viral genetic material begins functioning in the host cell. An alternative strategy is used by certain animal viruses such as the Semliki Forest virus (Figure 20-19), which has "spike proteins" on its surface that bind to surface receptors on a host cell. Binding triggers the cell to engulf the intact virion. The vesicle containing the virion is rapidly shunted to one of the cell's lysosomes, where the low pH (about 5) causes the vesicle to fuse with the surrounding lysosomal membrane, thereby ejecting the nucleocapsid into the cell cytoplasm. This occurs so rapidly that the lysosomal enzymes apparently do not have time to degrade the virion's capsid. Next, the viral RNA is released and begins to function.

Many viruses are *nonvirulent* on entering a host cell; that is, they do not kill their host. In this situation, the DNA can become inserted into the host's chromosome. The inserted DNA is called a *provirus.* Following insertion, the viral DNA is replicated along with the host's chromosome each time the host cell passes through a growth and division cycle. This process is called **lysogeny** (Figure 20-20). One of the inserted viral genes causes synthesis of a repressor protein, which then inhibits transcription of mRNAs from the viral genes responsible for virulent behaviors. During lysogeny, then, the virus is said to be dormant, or *temperate.* The repressor substance may also change the host cell so that

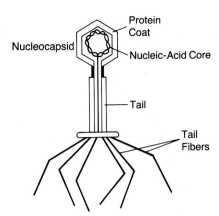

Figure 20-18   A TYPICAL VIRUS PARTICLE.
This bacteriophage has a head region composed of a protein coat and nucleic-acid core, and a tail. Tail fibers are involved in attachment to the bacterium that is to be invaded.

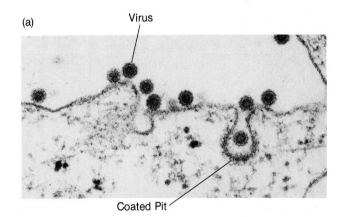

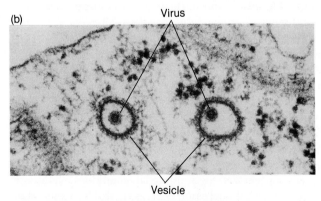

Figure 20-19   SENDING THE VIRAL MESSAGE.
The viral genome (DNA or RNA) can enter the host cell in different ways. The entry of Semliki Forest virus's genetic information is a more complex process than that of bacteriophage injection. (a) The viral particles first attach to receptors on the host's plasma membrane and then are taken into the cell by way of a "coated pit." (b) Within the cell, the vesicle containing the virus particle fuses with a lysosome in a process that leads to rapid release of the viral RNA into the cell. (Both electron micrographs are magnified about 83,000 times.)

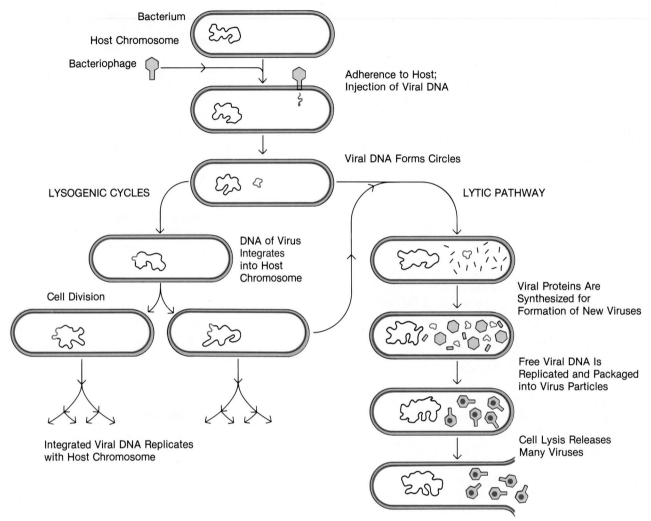

Figure 20-20   VIRAL LIFE CYCLE: REPRODUCTION AT THE HOST'S EXPENSE.
During the lysogenic cycle, the viral DNA is integrated into the host cell DNA and is replicated each time the host cell divides. In this way, the viral genome (now called a provirus) is passed on to successive generations of bacterial cells. Proviruses may become virulent at any time and enter the lytic pathway. In the lytic pathway, viral proteins are synthesized and assembled, along with viral nucleic acid, into numerous viral particles. The host cell is ruptured as the viruses are released.

similar viruses cannot infect it. In a sense, the original virus immunizes its host and ensures itself a private "home."

Under certain environmental conditions, however, the amount of repressor decreases, the lysogenic state breaks down, and the viral DNA is reproduced independently of the host chromosome. The viral genes are transcribed, and viral proteins—including viral-coat protein—are made. Mature viruses assemble, and the host cell bursts, releasing the newly replicated viral particles. This process of *virulent* (cell-killing) behavior is called the **lytic pathway.** Note that in the lytic pathway, there is no hint of cell division, mitosis, or other such processes found in prokaryotic and eukaryotic reproduction.

The Semliki Forest virus provides an intriguing exam-

ple of how viral lysis occurs in animal cells. Messenger RNA for the viral spike protein is translated on the host cell's rough endoplasmic reticulum. From there, the viral spike protein is shunted via the Golgi complex to the cell's plasma membrane. Spike proteins aggregate as part of new envelopes that coalesce about new nucleocapsids. Thus a piece of the cell's plasma membrane, but containing viral protein, is the ultimate coat of the new virus particle. This explains the remarkable puzzle of how a virus—normally only protein and nucleic acid—can obtain such a complex raincoat as a lipid bilayer.

The exact sequence of use of viral genetic material depends on whether the genome consists of DNA or RNA. If the viral nucleic acid is DNA, it acts directly as a template for the synthesis of viral DNA and mRNAs. These mRNAs, in turn, are translated using the host's ribosomes, and viral proteins appear. Some of these proteins are new capsid molecules; others are enzymes that aid in the replication of viral DNA; and still others degrade the host cell's DNA to provide a source of nucleotides. A somewhat different sequence is followed if the viral genetic material is RNA (Figure 20-21). A special viral enzyme called *reverse transcriptase* is synthesized in infected cells and proceeds to make DNA molecules complementary to the viral RNA genome. These DNA molecules can then be used for transcribing viral mRNAs

and, in turn, viral proteins. Ultimately, copies of the full RNA genome are made and packaged into new virus particles.

In summary, we see that viruses are utterly dependent on the living cells that they parasitize to provide an environment in which the viral genome and viral proteins can be manufactured. Although the biochemistry of viral nucleic-acid and protein synthesis is identical to that of host cells, the energy and all other essentials for viral duplication come from cells. Viruses alone, then, do not display the full spectrum of characteristics we associate with life (Chapter 1).

Figure 20-21   ROUS SARCOMA VIRUS.

This RNA virus that induces cancer in chickens has in its genome the genetic information encoding viral-coat proteins as well as the oncogene *src*, which codes for a protein believed to be involved in inducing the cancerous state in the infected cell. In this diagram, you can see that the viral genome is copied into DNA by the enzyme reverse transcriptase and then is integrated into the host cell's DNA. Transcription of the viral DNA yields RNA molecules that can be translated into viral-coat proteins (as spike proteins) and the *src* product. The transcription of the viral DNA also serves to produce many copies of the viral RNA genome. The result of infection by the Rous sarcoma virus is the reproduction of the virus and the induction of the cancerous state in the host cell. After the viral genome and proteins have been assembled, the new viral particles exit from the host cell by outward budding (exocytosis); during this process, a capsule derived from the plasma membrane and including viral spike proteins surrounds the virus.

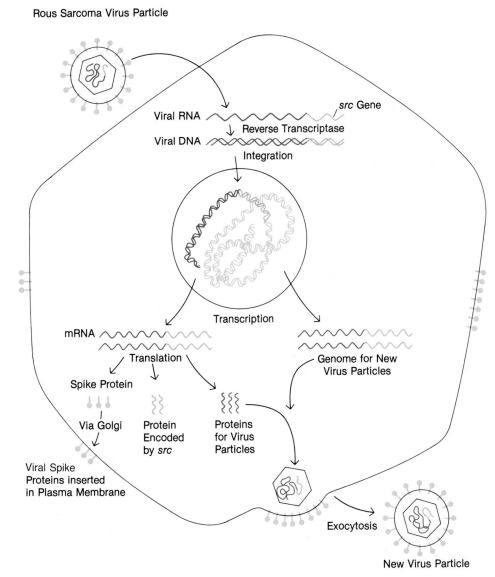

Rous Sarcoma Virus Particle

Viruses and Disease

Viruses cause various diseases in plants and animals by several mechanisms. The most obvious is based on the lytic cycle: once the virus replicates, the host cell is destroyed. The virions released can infect and destroy nearby host cells or can be spread to other individual organisms. Another disease-producing mechanism involves a different form of cell lysis: some viruses can cause the host cell to release lysosomal enzymes into its cytoplasm, resulting literally in digestion from inside out. Still other viruses can cause disease by making toxins that inhibit cell metabolism. Table 20-3 lists several human viral diseases.

An intriguing aspect of viral infection is that it is often tissue specific. That is, only certain tissues of a plant or an animal can be infected by a given virus. We know now that vulnerable cells carry molecules on their surfaces that can act as receptors: they bind virus particles that happen to bump into them. Viruses do not enter cells that lack receptors. This is why, for example, cold viruses localize in the respiratory tract, where receptors are present, but do not infect brain cells or heart cells, which lack receptors. On a higher level, the complementarity of virus and receptor governs whether a given organism will be susceptible at all to infection by a given virus. For example, the flowering plants (angiosperms) are susceptible to viral infection, whereas conifers (pines and their relatives) are not.

Table 20-3   **SOME HUMAN VIRAL DISEASES**

Disease	Virus
Measles	Paramyxovirus
Rubella (German measles)	Rubella virus
Atypical pneumonia	Paramyxoviruses, types 1–3; orthomyxoviruses, types A, B, and C
Common cold	Coryza viruses; rhinoviruses
Influenza	Orthomyxovirus, types A, B, and C
Hepatitis	Hepatitis A virus
Herpes	Herpes simplex, types 1 and 2
Mononucleosis	Epstein-Barr virus
Poliomyelitis	Polio virus
Encephalitis	Semliki Forest virus
Mumps	Paramyxovirus
Smallpox	Smallpox virus
Rabies	Rhabdovirus
Dengue fever	Togavirus (flavivirus)
Yellow fever	Togavirus (flavivirus)

How can we explain the apparently maladaptive phenomenon of cells possessing receptors that can contribute to their own death? In fact, it is likely that the surface molecules perform other functions for the cells that make them. And probably it was the viruses that evolved coat proteins that bind to cell-surface molecules. So, although we call them "receptors," the molecules may not be receptors in terms of their real cellular function.

Some disease-causing viruses may persist in the host for many years, becoming alternately active and inactive. Examples of this persistent pattern are the herpes viruses, carried permanently by most people. One type, herpes simplex type 1, produces cold sores (also called fever blisters) when it becomes active. Herpes simplex type 2 is the causative agent in genital herpes conditions. The virus lies dormant, probably in nerve cells, and then becomes lytic when the person suffers physical or psychological stress. Sunlight, x-rays, cigarette tars, and carcinogenic chemicals may also trigger viral activity.

Fighting viral diseases has long been a concern of medical science and a puzzling problem: How can one "kill" an inert particle? Antibiotics, so potent against bacteria, are usually ineffective against viruses. One reason is that viruses lack the cell walls that antibiotics often act to destroy in bacteria. Furthermore, since viruses depend almost completely on host-cell machinery for reproduction, anything that will destroy viral reproduction is also likely to destroy the host cell. Nevertheless, a handful of new antiviral drugs seems to be effective in preventing viral reproduction in some patients. (See the box.)

The vertebrate body's immune system sometimes may mount an attack on viral infections, but that process usually is slow, so that the infection nevertheless may be severe. Vaccines are effective against polio and smallpox viruses, but common viruses, such as those that cause colds and influenza, mutate so regularly that vaccines that are effective against this year's strain rarely provide protection against next year's mutant.

One promising approach to combating viral diseases is the use of **interferons,** naturally occurring proteins liberated by mammalian cells after they have been infected by viruses with double-stranded RNA as genetic material (or DNA viruses that cause such unusual RNA to be made). Interferons do not rescue the infected cell that makes them; rather, the proteins represent a strategy for survival of the whole organism. Thus when interferons are released by an infected cell, they bind to nearby infected cells, where they stimulate the production of antiviral proteins that prevent viral reproduction. Until recently, interferon cost $178 million an ounce to isolate and purify—much too costly for common use. But new production techniques based on genetic engineering promise a much cheaper and greater supply for both studies and treatments.

BEATING VIRUSES AT THEIR OWN GAME

The challenge faced by researchers attempting to design new drugs to fight viruses is how to kill an enemy that is noncellular. As we have seen, a virus usurps a cell's molecular machinery for its own reproduction. Since viruses cannot be killed by disrupting membranes or interfering with metabolism, as is possible with bacterial pathogens, the only means of attack is to block viral reproduction in some way. Luckily, there are several "Achilles heels" in the viral reproductive process, and researchers are working to exploit them.

Tampering with Viral Enzymes

Some viruses rely entirely on the host cell's enzymes; others, such as herpes virus, cause the cell to make many viral enzymes. The second kind of virus would be susceptible to a drug that inhibits viral enzymes directly or a drug that is converted to active form by viral enzymes.

One of the few viral drugs currently available, *acyclovir*, resembles the nucleic-acid base guanine and acts in both ways. One viral enzyme phosphorylates acyclovir, converting it to active form. This active substance then binds to DNA polymerase, preventing it from catalyzing the addition of nucleotides to DNA. In this way, acyclovir prevents the replication of the virus; a cell not infected by virus will not be affected by the drug. The drug is being used successfully to treat genital herpes, shingles, and cold sores. Other similar drugs are being developed.

Tying up Viral Genes

Another approach is to make short stretches of RNA or DNA having a nucleotide sequence that is complementary to a section of the virus's own genetic material. These stretches can be introduced into infected host cells to bind the viral genes and prevent their being replicated or transcribed into proteins. Researchers have created sequences that strongly inhibit the activity of Rous sarcoma virus, which causes tumors in animals.

Fighting Viruses with Interferons

Part of the body's natural defensive reaction against a virus is the production of *interferons*, a group of small glycoproteins. Made and released by virus-infected cells, interferons act on neighboring cells to render them resistant to a wide range of viruses. Many scientists predict that large quantities of interferons will soon be available through cloning in bacterial hosts. Interferons produced on a small scale in this way already have proved effective experimentally in fighting the common cold.

Synthesizing Antiviral Vaccines

Although vaccines against pathogens such as polio virus and rabies virus have been available for many years, there is a distinct risk involved in administering them: active virus particles sometimes are accidentally included, and patients contract the very diseases they sought to avoid. A new technique involves synthesizing short peptides that match the outermost regions of globular viral proteins—those regions that normally are recognized by the host organism's immune system and that trigger an immune reaction.

By injecting only these synthetic peptides into rabbits, biologists have immunized the animals against later attacks by real virus particles with corresponding protein regions. This technique provides safe vaccines that are effective against a range of pathogenic viruses. Some biologists foresee a day not long from now when such vaccines, as well as drugs that work by blocking viral genes or enzymes, will be standard medical tools that spare us from viral diseases as effectively as antibiotics now fight bacterial infections.

Viruses and Cancer

As we learned in Chapter 17, recombinant DNA technology has provided strong evidence linking viruses and cancer. Malignant cells often contain viral genetic material, either free or integrated into the cell's chromosomes. So-called *proto-oncogenes* are genes found in "normal" animal cells, including human ones, that are virtually identical to genes in the viruses that can cause cancers. Whereas the normal cellular gene product is beneficial to the cell, the presence of the viral oncogene alters normal cell behavior, the result being uncontrolled division and other abnormalities. Some biologists now believe that many organisms carry in their genomes these dangerous oncogene residents; "spontaneous" leukemias, connective-tissue sarcomas, and other life-threatening cancers may be the result.

A main line of study of cancer viruses involves cellular transformation. Rous sarcoma virus transforms chicken cells (Figure 20-22), in the same way that simian virus 40 transforms mammalian cells. The cells change shape and their surface-marker molecules become altered. Perhaps as a result, normal constraints on cell division and locomotion are eliminated. Such cells often divide rapidly

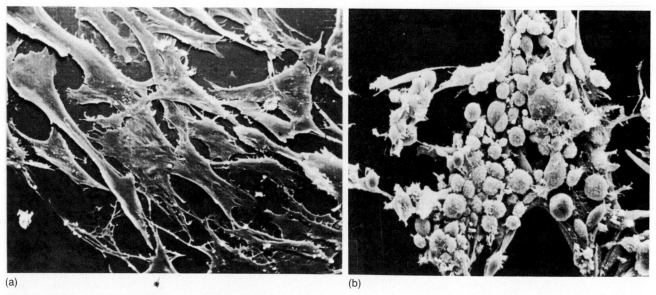

(a)   (b)

Figure 20-22   CHICKEN CELLS TRANSFORMED.
These scanning electron micrographs (magnified about 750 times) show chicken connective-tissue cells, or fibroblasts, grown in culture and transformed by the Rous sarcoma virus. (a) Normal fibroblasts have a flattened appearance and adhere to the culture dish. (b) Infected cells become rounded during the transformation—a shape change believed to be due to (or at least initiated by) the production of a protein encoded by the *src* gene in the viral genome.

and crawl over and under one another; normal cells cannot do this. True cancer cells frequently display similar behavior as they metastasize (spread from one site to others). Conceivably, such surface changes may render the infected cell invisible to the body's immune system, so the malignancy goes unchecked.

Research on transformed cells and oncogenes provides a new window on a great curse of humankind. Treatments and, ultimately, prevention of cancers may emerge as we learn more about the perversion of normal cell processes by oncogenes.

EVOLUTION OF MONERANS, VIRUSES, AND EUKARYOTIC CELLS

In Chapter 19, we examined theories of and evidence for the emergence of simple prokaryotic cells from non-living substances on the early Earth. A natural question now arises: What are the relationships among prokaryotes, viruses, and eukaryotes? Answering this question involves considerable speculation because the fossil record of early cells is fragmentary at best (these cells were so small and so soft that they did not make many easily discovered fossils). Nevertheless, an internal biochemi-

cal record laid down in the structures of metabolic pathways of living species can be used to help reconstruct the evolutionary history.

As outlined in Chapter 19, the first cells on Earth were prokaryotes. Almost certainly, they were simple anaerobic heterotrophs that consumed preformed organic nutrients from the environment. They probably lacked walls, flagella, and spore-forming ability. These specialized prokaryotic structures evolved later, as did photosynthetic capability. Aerobic autotrophs probably evolved from the early anaerobic cells and changed the Earth's atmosphere by liberating free O_2. Later, aerobic heterotrophs appeared. How could these various prokaryotic cells have evolved into eukaryotic cells?

The **endosymbiont theory** may provide an answer (Figure 20-23). According to this theory, one or more simple progenitor cell types led to three early forms: the monerans, the archaebacteria, and a prospective eukaryotic cell type without membrane-bound organelles but with the multiple chromosomes and other common biochemical attributes of eukaryotes. The eukaryotes then evolved by means of a mutually beneficial association, or *symbiosis*, between the prospective eukaryote and one or more types of bacteria. One such symbiont was probably an efficient aerobic bacterium that was engulfed but not digested and came to function in the host cell's metabolism. That new resident was the ancestor of the mitochondrion, and the cell containing it could have been the progenitor of protistan, animal, and plant cells. Evi-

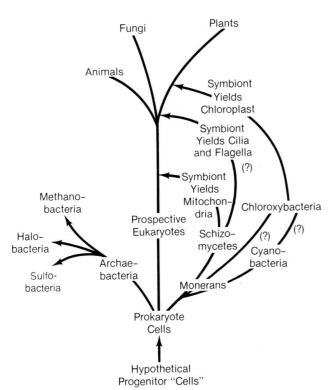

Figure 20-23   RELATIONSHIPS BETWEEN PRESENT-DAY PROKARYOTES, AND THE ORIGIN OF THE EUKARYOTIC CELL.
The separate lineages leading to archaebacteria and monerans are shown. Eukaryotic cells are believed to have acquired mitochondria and chloroplasts by establishing permanent symbiotic relationships. The origin of cilia and flagella remains very uncertain.

dence from rRNA sequences suggests strongly that the symbiont that gave rise to mitochondria was a purple eubacterium related to the rickettsias.

A second proposed step is much more controversial and uncertain; it suggests that microtubule-based cilia and flagella may have evolved after the attachment of spirochetes or bacteria called spiroplasmas to the prospective eukaryotes. The attached undulatory surface could then have evolved into the flagella and cilia of various eukaryotic cell types.

Finally, a third step occurred, this theory hypothesizes, in the lineage leading to plants. A second engulfment, this time of a photosynthetic aerobic prokaryote—similar to a chloroxybacterium—established a symbiont that evolved into the chloroplast.

Evidence in support of the endosymbiont theory in-

cludes the fact that all mitochondria and chloroplasts of eukaryotic cells possess the kind of ribosomes found in bacteria, tRNAs that differ from cytoplasmic ones, and circular strands of DNA similar to bacterial chromosomes. In addition, these organelles reproduce independently during the cell-division cycle. They do not function completely independently at all times, however. Genes in the eukaryotic nucleus code for several proteins that are manufactured in the cytoplasm and then enter mitochondria and chloroplasts, where they are essential to normal functioning. For instance, nuclear genes code for enzymes that produce ten essential amino acids and carotenoid pigments within chloroplasts. Early eukaryotic cells that had acquired protomitochondria but not photosynthetic endosymbionts may have become animal cells, whose descendants remain unable to synthesize those essential amino acids that are made in the chloroplasts of plant cells. The evidence for the endosymbiont theory of the origin of eukaryotes is so persuasive that the biologist F. J. R. Taylor has said, "The eukaryotic 'cell' is a multiple of the prokaryotic 'cell.' "*

Before the accumulation of evidence favoring the endosymbiont theory, biologists wondered whether viruses resemble primitive life forms that evolved into the first prokaryotic cells. While intriguing, this idea is difficult to defend. Since all viruses depend on cells to reproduce, how could they have predated cells? An alternative explanation for the origin of viruses suggests that they may be "escaped genes"—fragments of host-cell chromosomes that survived at first without a protein coat. Perhaps the earliest viruses resembled modern **viroids**—tiny particles of RNA with no protein coat that cause several types of plant and animal diseases. Another hypothesis suggests that viruses are highly evolved parasites that have lost all but the barest essentials for genetic survival.

A definitive explanation for the evolution of prokaryotes, eukaryotes, and viruses will be difficult to attain, but now seems possible because of the great advances in the techniques of molecular biology. Studies in this area will continue to point up relationships among these groups and to enlarge our understanding of the invisible kingdom Monera, the ancestral kingdom of all other life on Earth.

*F. J. R. Taylor, "Implications and Extensions of the Serial Endosymbiosis Theory of the Origin of Eukaryotes," *Taxon* 23 (1974): 5–34.

SUMMARY

1. The 2,000 or so kinds of members in the kingdom Monera are the smallest, oldest, and most numerous organisms on Earth. All monerans are *prokaryotes*, with bacteria being the most abundant and most common examples.

2. Prokaryotic cells are approximately one-tenth to one-hundredth the volume of average eukaryotic cells. Eukaryotic cells must be larger because they contain various organelles and molecular assemblages not found in prokaryotic cells.

3. Bacterial cells have a rigid outer wall made of *peptidoglycan polymers*. Some also have additional outer layers of lipids and polysaccharides. Variations in cell-wall structure, detectable with Gram-staining techniques, account for differential susceptibility to antibiotics and lysozyme between *Gram-positive* and *Gram-negative* bacteria.

4. Although bacteria lack a nucleus, they have a nucleoid where the single chromosome is located. Also unique to bacteria is a convoluted membranous structure called the mesosome, which may serve a variety of functions.

5. Many bacteria move by means of a flagellum anchored in a freely rotating socket. Other motile bacteria are equipped with axial filaments, which repeatedly lengthen and shorten, or the cells glide along a track of self-secreted slime.

6. Given unlimited space and nutrients, many types of bacteria can produce a new generation every twenty minutes. Bacteria reproduce most commonly by budding and by *binary fission*. They also exchange genetic information by several processes: transformation, transduction, and conjugation. Many species form tough, resistant resting cells called *endospores*, which are capable of surviving harsh environmental conditions.

7. Bacteria employ a wide range of metabolic strategies. Heterotrophic bacteria use organic compounds as substrates or exist as *parasites*. *Photoautotrophs* carry out photosynthetic processes using compounds other than O_2 as electron donors. *Chemoautotrophs* oxidize inorganic compounds to obtain energy. Bacteria also vary in their oxygen tolerance. *Obligate anaerobes* grow only in the absence of oxygen; *facultative anaerobes* can grow with or without oxygen; and *obligate aerobes* require oxygen to survive.

8. Among the true monerans are the schizomycetes and the cyanobacteria. The subkingdom *schizomycetes* includes some 1,600 species of bacteria classified into the main phyla: *eubacteria, spirochetes, myxobacteria, actinomycetes, rickettsiae,* and *mycoplasmas.*

9. *Cyanobacteria* are rod-shaped or spherical cells that often occur in filamentous groups. They contain chlorophyll *a*, carotenoids, and red and blue pigments and they carry out photosynthesis. In some species of this subkingdom, nitrogen fixation takes place in special cells called heterocysts. Cyanobacteria can live in harsh environments that provide only water, inorganic minerals, air, and light.

10. The *chloroxybacteria* are prokaryotic cells that photosynthesize by means of chlorophylls *a* and *b*, as do green algae and higher plants.

11. *Archaebacteria* include methane producers (*methanobacteria*), extreme salt dwellers (*halobacteria*), sulfur-dependent species (*sulfobales*), and species that live at high temperatures (*thermoproteales*). Methanobacteria derive energy from coupled reactions that oxidize H_2 to H_2O and reduce CO_2 to CH_4 (methane gas). Because of the many fundamental biochemical differences between archaebacteria and other monerans, this group is now classified separately as the kingdom archaebacteria.

12. Bacteria aid in the production of many foods, including cheese, butter, sour cream, sauerkraut, and vinegar. They are also major *pathogens*, which produce disease through the destruction of tissues, irritation by waste products, and release of toxins either while still alive (*exotoxin*) or at their death (*endotoxin*).

13. *Viruses* are noncellular, metabolically inert particles composed of a core of RNA or DNA, a protein capsid, and sometimes a membranous outer envelope. Viral genetic material may become incorporated into a host cell's DNA where it is replicated with the host chromosome. This is the process of *lysogeny*. Viruses may reproduce independently of the host cell and ultimately kill it; this is the *lytic pathway*. Viruses may cause disease by destroying the host cell or by causing production of toxins. Viruses have been implicated in certain cancers. Oncogenes in such viruses are almost identical to cellular genes; research results imply that cancer may be a normal cell function gone wrong.

14. According to the *endosymbiont theory*, the eukaryotes probably arose as a result of the establishment of at least two symbiotic relationships in different lines: (a) engulfed aerobic purple eubacteria evolved into mitochondria and (b) engulfed photosynthetic aerobic prokaryotes, probably related to chloroxybacteria, evolved into chloroplasts.

KEY TERMS

actinomycete

archaebacteria

binary fission

chemoautotroph

chloroxybacteria

cyanobacteria

endospore

endosymbiont theory

endotoxin

eubacteria

exotoxin

facultative anaerobe

Gram-negative bacteria

Gram-positive bacteria

halobacteria

interferon

lysogeny

lytic pathway

methanobacteria

mycoplasma

myxobacteria

obligate aerobe

obligate anaerobe

parasite

pathogen

peptidoglycan polymer

photoautotroph

prokaryote

rickettsia

schizomycete

spirochete

sulfobales

thermoproteales

virion

viroid

virus

QUESTIONS

1. How could you use light to "guide" a flagellated, photosensitive bacterium through a tiny "maze" on a culture dish? What happens in the cell to account for this behavior?

2. Draw a "family tree" of the monera, methanobacteria, and eukaryotic plant and animal cells. Indicate some of the critical events hypothesized to account for the origins of the various groups.

3. Using two host cells diagram the events during infection by an RNA and a DNA virus; include the lysogenic and lytic cycles.

4. What kind of prokaryote converts carbon dioxide into its own building materials? What energy source do such organisms use?

5. Which prokaryotes break down organic compounds in decaying plant material, convert some of these into building materials for their own use, and derive energy from the breakdown of these organic compounds?

6. What microorganism uses organic material as a carbon source, but derives its energy elsewhere?

7. What features distinguish prokaryotes from eukaryotes?

8. Why is it thought that the first living organisms were anaerobic?

9. Review Chapter 1 and list the attributes of living organisms. Which of these attributes (if any) apply to viruses? Would you say that viruses are living organisms, nonliving particles, or incomplete organisms?

10. Draw diagrams of a bacterial cell dividing. Draw diagrams of mitosis in a eukaryotic cell. Why do you suppose eukaryotic cells go to all the trouble of mitosis? What is the role of the cell membrane in binary fission?

ESSAY QUESTION

1. All eukaryotic cells contain organelles and other structures that are thought to have evolved from various prokaryotes. Summarize the evidence supporting this theory.

SUGGESTED READINGS

FREDRICK, J. F., ed. "Origins and Evolution of Eukaryotic Intracellular Organelles." *Annals of the New York Academy of Science* 361 (1981). See also D. YANG, et al. "Mitochondrial Origins," *Proceedings of the National Academy of Science, U.S.A.* 82 (1985): 4443–47.

An up-to-date series of papers on all aspects of these intriguing hypotheses.

KLUYVER, A. J., and C. B. VAN NIEL. *The Microbe's Contribution to Biology.* Cambridge, Mass.: Harvard University Press, 1956.

This classic of biological literature consists of a series of essays about microbes and makes pleasant, easy reading for the serious student.

LURIA, S. E., J. DARNELL, D. BALTIMORE, and A. CAMPBELL. *General Virology.* New York: Wiley, 1978.

This general text covers most aspects of viral structure, function, genetics, and evolution.

STANIER, R., and M. DOUDOROFF. *The Microbial World.* 4th ed. Englewood Cliffs, N.J.: Prentice-Hall, 1984.

The classic, comprehensive text that has educated more microbiologists than any other.

WOESE, C. "Archaebacteria." *Scientific American*, June 1981, pp. 98–122. See also R. A. GARRETT. "The Uniqueness of Archaebacteria." *Nature* 318 (1985): 233–35.

Here are Woese's ideas on this separate lineage of microbes.

21
PROTISTA: THE KINGDOM OF COMPLEX CELLS

*The minute living Animals exhibited in . . . this Work, will
excite a considerable Mind to admire in how small a Compass
Life can be contained, what various Organs it can actuate,
and by what different Means it can subsist. . . .*

Henry Baker, *Employment for the Microscope* (1753)

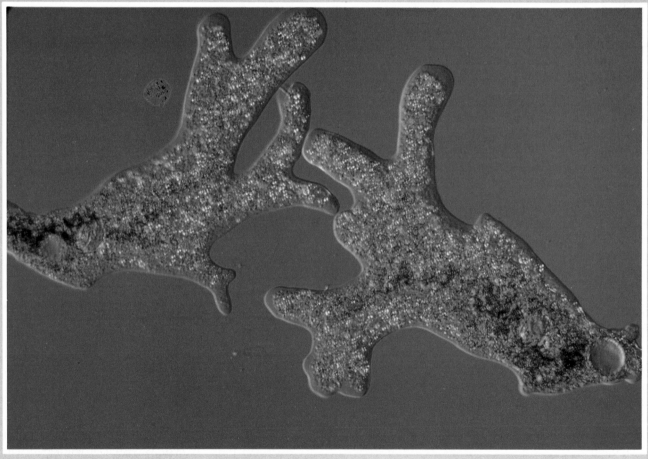

Like minute, many-armed fencers, two single-celled amoebae extend and retract their pseudopods in an encounter only seen by human eyes through a microscope.

The kingdom Protista comprises a marvelously diverse collection of organisms. It contains mostly single-celled eukaryotes that occur in an almost endless variety of shapes and life styles: plantlike cells that can swim; funguslike decomposers that creep toward food sources; animal-like cells that cannot move at all; cells that look like bells, eels, pincushions, fans, stars, shells, or snow-flakes.

Some species of protists form colonies at certain stages in their life cycle. Others are colonial most of the time, and with few exceptions their cells are identical and do not adopt specialized functions, unlike the cells of multi-cellular organisms. Despite these colonial forms, the single-celled state predominates, allowing such organisms to be classified as protists.

Whereas monerans have no nucleus, cytoskeleton, or internal membranous organelles, protists have all these eukaryotic features. The transition from prokaryotic to eukaryotic cells proved to be a momentous one in the life of the Earth, for two reasons. First, single eukaryotic cells can be much more complex in structure, function, behavior, and ecology than can any prokaryotic cells. Second, eukaryotic cells had the potential to form multi-cellular organisms that could evolve to the large size and diverse life styles of present-day fungi, plants, and animals. In fact, as we look at the characteristics of the protistan groups, we will see that some of them may be thought of as "testing grounds" for the life styles of higher organisms. Some of these "tests" were so successful that the protists of the Precambrian era were the ancestors of the multicellular plants, fungi, and animals that have since populated our planet.

Although many eukaryotic protists live in the same environments as monerans, the protistan way of life is substantially more complex. Moreover, protists face the same variety of external circumstances encountered by larger and more complex organisms. They must carry out the basic functions of feeding, digestion, excretion, respiration, coordination, and locomotion, using only minute organelles rather than the differentiated tissues, organs, and organ systems found in multicellular organisms. And given that an organelle—a flagellum, for instance—can never be the equal of a tissue or an organ, such as a leg or wing, protists represent the greatest complexity that has been attained by single-celled organisms.

PROTIST CLASSIFICATION: BY LIFE STYLES

Perhaps the most interesting characteristic of the protists is that they do not fit our commonsense categories

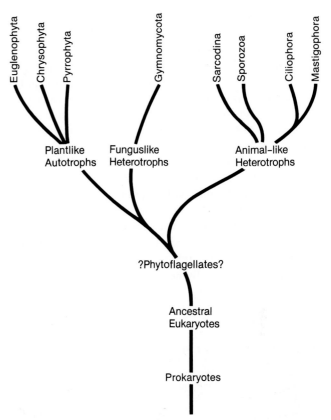

Figure 21-1   PHYLOGENETIC TREE OF THE IMPORTANT PROTISTAN GROUPS.
It is not certain whether a single ancestral eukaryotic type gave rise to all protists. Some biologists favor the idea that a flagellated photosynthetic cell gave rise to the major protistan groups as shown here; others suspect that the lineages leading to funguslike and animal-like protists arose separately from the line leading to plantlike protists.

of "plant" or "animal." They seem to straddle the divisions among plant, fungus, and animal, perhaps because they are descendants of cells that lived before the emergence of the immediate ancestors of those organisms.

What principles can we use to set up a classification scheme for the protists? A convenient method is to categorize them into three arrays, according to life style (Figure 21-1): (1) the *plantlike autotrophs* are "producers" that generate food molecules by means of chloroplasts and photosynthetic pigments; (2) the *funguslike heterotrophs* are "consumers" and "decomposers" that feed on preexisting organic matter; (3) and the *animal-like heterotrophs* are "consumers" that feed on whole bacteria, other protists, or the cells of multicellular organisms.

Within these three broad arrays there are, of course, the phyla that comprise any kingdom. At present there is a taxonomic controversy as to how best to describe the phyla. For example, biologists L. Margulis and K. Schwartz recognize twenty-seven phyla, whereas J. Corliss recognizes forty-five. Furthermore, the distinctions used by these experts utilize characters that are quite technical in terms of molecular constituents of cells and of cellular fine structure. This is especially true among many of the animal-like protists. What will be presented here is a selection of key protistan groupings. They are keys in the sense that they represent one of the three major protistan life styles, they illustrate well the variety of protistan adaptations, and they very often are of significant importance to humans. Because these groupings may or may not coincide with phyla recognized by experts we will simply present them as groupings sharing certain important features.

The phyla listed in Table 21-1 are the main subgroups in the kingdom Protista. Although plantlike, funguslike, and animal-like protists are ecologically diverse, they share the basic eukaryotic cell structure: (1) a membrane-bound nucleus; (2) multiple dissimilar chromosomes that include many proteins as well as DNA; (3) ribosomes and nuclear RNA processing as in eukaryotic cells, not as in most monerans; (4) complex and diverse membrane-bound organelles in the cytoplasm; and (5) a cytoskeleton, which provides the cell with form and structure and allows most protistan cells to move.

As we survey the protists and classify them by life style, keep in mind that the kingdom Protista itself and many of the taxonomic groups in it are artificial constructs. The fact that two protists have the same life style or the same structure is not in itself evidence that they descended from a common ancestor. Rather, single-celled organisms of many different origins would (and

Table 21-1   **PROTISTAN PHYLA**

Group or Phylum	Representative Members	Means of Locomotion	Cell Wall	Habitat	Other Characteristics
Plantlike Protists					
Euglenophyta	*Euglena*	Flagella	None	Mostly fresh water	Chlorophylls *a* and *b* plus carotenoids
Pyrrophyta (dinoflagellates)	*Gymnodinium Protogonyaulax*	Flagella	Rigid cellulose wall	Mostly marine; some parasitic or symbiotic	Chlorophylls *a* and *c* plus carotenoids
Chrysophyta (golden-brown algae, diatoms)	*Ochromonas*	Some with flagella	May have none; may have rigid pectin or pectin and silica wall	Fresh water and marine; some terrestrial	Chlorophylls *a* and *c* plus carotenoids
Funguslike Protists					
Gymnomycota (slime molds)	*Dictyostelium*	Pseudopods	None	Terrestrial	Multinucleate plasmodium (true slime molds) Multicellular slug (cellular slime molds)
Animal-like Protists					
Mastigophora (zooflagellates)	*Trichonympha Trypanosoma*	Flagella; some with pseudopods	None	Mostly symbiotic or parasitic	Parasitic forms have complex life cycle with multiple hosts
Sarcodina (amoebae, foraminiferans, radiolarians)	*Amoeba Actinosphaerium*	Pseudopods; some with flagella	None	Marine and fresh water; terrestrial; some parasitic	Many secrete elaborate shells of silica or calcium compounds
Sporozoa	*Plasmodium*	Adult forms nonmotile	None	Parasitic	Complex life cycle with sporelike stage and multiple hosts
Ciliophora	*Paramecium*	Cilia	None	Marine and fresh water	Macronuclei and micronuclei; most complex single-cell organism
Caryoblastea	*Pelomyxa*	Amoeboid	None	Fresh water	Lacks mitochondria

did) tend to evolve highly similar structures, chemistries, life styles, and so on because they all carried out the same biological functions with essentially the same equipment.

THE PRODUCERS: PLANTLIKE PROTISTS

Many protists are primarily plantlike, in the sense that they contain chlorophyll and the metabolic machinery required to produce their own food by photosynthesis. Indeed, some biologists estimate that one type of photosynthetic protist, the diatoms, contributes more oxygen to the atmosphere than do all land plants combined.

Immense numbers of microscopic autotrophic protists float near the surface of fresh and salt waters around the world. These masses of cells are one type of **phytoplankton,** a group that includes many different photosynthetic, usually single-celled, organisms that generate huge quantities of biomass (the total mass of living substance in one or more organisms) as well as of atmospheric oxygen. Phytoplankton are the first link in aquatic food chains: phytoplankton are eaten by minute animal larvae and other small organisms, which, in turn, are eaten by larger invertebrate and vertebrate predators. There are almost 12,000 species of phytoplankton, many of which are not protists. Almost all of them carry out photosynthesis (like plants) and can move about (like animals) by means of one or more flagella. In this sense, they straddle the division between plant and animal life styles. The three most common phytoplankton phyla of protists are: *Euglenophyta*, or euglenoids; *Pyrrophyta*, or dinoflagellates; and *Chrysophyta*, or golden-brown algae.

Euglenophyta

The euglenoids have long been a taxonomic puzzle. Early zoologists observed their motility and called them plantlike animals. Early botanists, noting their ability to perform photosynthesis, called them animal-like plants. Modern biologists simply group them with other photosynthetic protists.

The phylum **Euglenophyta** is named after its most characteristic genus, *Euglena*—beautiful bright green unicellular organisms that swim about watery habitats propelled by a whiplike flagellum (Figure 21-2). There are about 800 species of euglenoids in 40 genera, many of which live in fresh water that contains a great deal of dissolved organic matter. The majority are photosynthetic autotrophs, but some are heterotrophs that digest or engulf food particles.

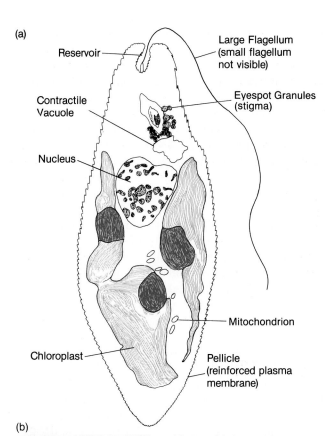

Figure 21-2   EMERALD *EUGLENA*: A TYPICAL PROTIST.

Various species of this protistan genus function in plantlike or animal-like ways. Euglenoids have typical eukaryotic features, including chloroplasts, mitochondria, and a nucleus. (a) This labeled drawing depicts the structures visible in the transmission electron micrograph (b), which is magnified about 2,000 times.

Euglenoids are animal-like because they do not have the cellulose walls of plants. The lack of a rigid and prominent cell wall permits euglenoids great flexibility of shape—a characteristic that is particularly useful for a mud-dwelling species that travels by wriggling rather than by swimming. In the absence of a cell wall, the plasma membrane would be a fragile interface with the world if it were not strengthened by underlying protein bands called the *pellicle*.

The typical euglenoid cell is oval, with a slightly pointed posterior end. An open-ended reservoir is often found at the anterior end. Because there is no cell wall and the cell is hypertonic to its environment, the normal tendency is for water to enter the cell. Euglenoids are able to counteract this tendency by means of a *contractile vacuole*, which takes up excess cytoplasmic water and discharges it into the reservoir. From there, it is expelled into the environment.

The reservoir also serves as an attachment site for one or two flagella. These whiplike organelles are not like the flagella of monerans. Instead, they have basal bodies and the 9 + 2 arrangement of microtubules with dynein side arms characteristic of all eukaryotic flagella (Chapter 6). In *Euglena,* one flagellum is long, with a fringe of tiny hairlike filaments on one side, while the other flagellum is short. The beating of the large flagellum propels the tiny creature through liquid.

A *Euglena* cell also has a tiny, orange, light-sensitive eyespot called the *stigma.* When light stimulates this spot, a chain of processes is set off that triggers the flagella to propel the cell toward the light.

Autotrophic euglenoids need only light, dissolved CO_2, and minerals in order to carry out photosynthesis and manufacture carbohydrates. Instead of manufacturing starch (as plants do) or glycogen (as animals do), euglenoids build their carbohydrates into a unique substance called *paramylum.* A mass of this material is generated in the cytoplasm outside the chloroplasts. The pigments found in euglenoid chloroplasts are chlorophylls *a* and *b* and various carotenoids. As we will see, differences in the type of chlorophylls present in protistan cells reflect differences in the evolutionary origins of the "producer" protists.

Sexual reproduction has never been observed in euglenoids. Instead, the cells reproduce mitotically. A *Euglena* cell, for example, divides longitudinally, producing two cells that are mirror images of each other. Mitotic division is followed by cytokinesis similar to that seen in animal cells: a contractile ring of actin filaments pinches inward from the cell surface and squeezes the cell in two.

Euglenoids sometimes divide so rapidly that the independent growth and division of chloroplasts cannot keep up. This rapid division can result in permanently colorless strains, since once chloroplasts have been lost during cell division, they cannot be regained. These strains then act as heterotrophs, consuming organic food molecules. Successful lineages of such "degenerate" euglenoids have been found in natural habitats, where they swim about and function much like animal cells. Clearly, the distinction between producer and consumer is very slight among the protists of the phylum Euglenophyta.

Pyrrophyta

The second phylum of protistan phytoplankton is **Pyrrophyta,** the fire plants. Because these organisms are responsible for so-called red tides, "fire plants" is an apt name for the group; however, their common name, **dinoflagellate,** is also quite descriptive. This name means "spinning cell with flagella," and, indeed, these unicellular protists have a unique set of flagella that causes them to spin as they swim. One flagellum runs around a circumferential groove like a belt; another lies in a longitudinal groove and projects past the cell like a tail, as shown in Figure 21-3. The first flagellum causes the cell to spin, while the second one propels it forward or backward. The grooves in the cell also separate the rigid cellulose wall into regularly shaped plates, giving it an armorlike appearance. The different arrangement of flagella in euglenoids and dinoflagellates shows that evolution led to the same basic organelle being used in quite different ways to achieve the same end, locomotion.

Most of the 1,000 or so dinoflagellate species have contractile vacuoles and chloroplasts. These chloroplasts

Figure 21-3   ARMORED PROTISTS: THE DINOFLAGELLATES.
Dinoflagellates usually possess two flagella: the first, lying in a circumferential groove, causes the organism to spin, and the second, lying in a longitudinal groove, propels the organism forward or backward. These features are visible in the scanning electron micrograph (magnified about 3,000 times) of a species called *Protogonyaulax catenella.*

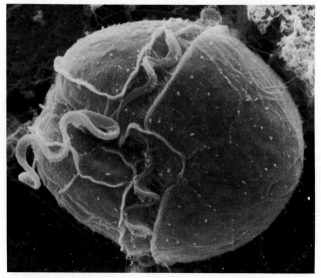

contain chlorophylls *a* and *c*, indicating that the dinoflagellates evolved independently of the euglenoids, which have chlorophylls *a* and *b*. The Pyrrophyta also possess certain carotenoid compounds that cause the cells to appear gold, brown, or red, rather than green. Autotrophic dinoflagellates, the second most common type of phytoplankton (after the Chrysophyta), create tremendous amounts of both biomass (living matter) and oxygen. Some photosynthetic dinoflagellates live symbiotically in the bodies of marine animals. Enough light penetrates the host animal's tissues to allow photosynthesis. As a result, the host corals, sea anemones, flatworms, and even giant clams can obtain many of the nutrients they need from these producer organisms residing in their tissues; in turn, the dinoflagellates benefit by being protected in the animal's body.

Some dinoflagellates lack chloroplasts and survive as free-living heterotrophs that can engulf organic matter and other cells. Some of these nonphotosynthetic consumer species produce *trichocysts*, tiny threadlike poisoned darts that can be released to protect the cell or to kill prey. Other nonphotosynthetic dinoflagellates live within the bodies of various marine organisms. Lacking chlorophyll, these protists cannot manufacture carbohydrates to supply the host. They function, then, as true parasites, deriving benefit and harming the host but giving nothing in return.

Glowing Waves and Red Tides

People who live in coastal areas often observe two phenomena caused by dinoflagellates. At night, the ocean may seem to burn with a greenish glow. Where the water is disturbed by the action of waves or by the splashes of a midnight swimmer, it can shimmer as though it were releasing a barrage of tiny sparks. At other times, offshore waters can turn rusty or bloody red, a phenomenon we call *red tides*. These tides contain toxins that poison fish, shellfish, and occasionally people.

Dinoflagellates can cause glowing waves because some species have an enzyme system that can emit light. When dense populations, or *blooms*, of these cells occur in a given area, the bioluminescent mechanism makes the water glow and the edges of waves sparkle.

Toxic red tides occur after the appearance of dense blooms of certain species, such as *Gymnodinium* and *Protogonyaulax* (formerly called *Gonyaulax*). The toxins these cells produce are usually low molecular weight nonprotein compounds that act as nerve poisons or cause red blood cells to lyse (break open). Fish, barnacles, clams, and other organisms that feed on phytoplankton or pump water and suspended cells through their respiratory gills may be killed by the dinoflagellates themselves or by the build-up of dinoflagellate toxins in their tissues. Furthermore, an oyster, a mussel, or a clam with

such toxin in its tissue can poison a person who eats it. For this reason, collection and consumption of shellfish are often banned in many coastal areas during warm summer months, the time of dinoflagellate blooms. Some types of red tides do not contain toxins but still can be deadly. When blooms occur, they can kill fish and other marine life by physical asphyxiation (clogging up gills, for example) or by so depleting the oxygen content of water that respiration is impeded.

Both toxic and nontoxic dinoflagellate blooms occur worldwide in temperate and tropical waters. The reasons for the periodic blooms of red-tide organisms are not known, but they may include water temperature, upwelling nutrients, trace metals, and the presence of numerous *cysts*—resting stages of dinoflagellates ready to germinate under the appropriate conditions. Cysts tend to accumulate in great abundance in "seed beds," sites where red tides are frequent occurrences.

Like the euglenoids, dinoflagellates usually reproduce by asexual cell division. Sexual reproduction has been observed in some species and may produce the cysts that appear to provide the seed populations for red-tide blooms.

Chrysophyta

The *golden-brown algae* and *diatoms*, the most abundant phytoplankton organisms, are surely among the most beautiful members of the kingdom Protista. Diatoms have glasslike cell walls sculpted in thousands of geometric and faceted shapes. The other chrysophytes are a golden color. Both types are so common that a gallon of sea water may contain millions of them. Of the 6,000 to 10,000 species assigned to the phylum **Chrysophyta** by different taxonomic schemes, some live in fresh water and some in salt water; some are unicellular, and some live in colonies; most reproduce asexually, but some produce flagellated gametes that carry out sexual reproduction.

Golden-Brown Algae

Locomotive structures and cell walls differ greatly among the **golden-brown algae**. Some lack a cell wall, and thus can creep along on solid surfaces in much the same way as amoebae and animal cells do (Chapter 6). Others possess a cell wall made largely of *pectin* (rather than cellulose), a complex carbohydrate compound that stiffens plant cell walls by gluing together cell-wall components. The pectin-walled algae usually have two flagella. Golden-brown algae contain chlorophylls *a* and *c* as well as carotenoids, characteristics shared with the Pyrrophyta. The golden color comes from a carotenoid pigment called *fucoxanthin*. Chrysophytes store food molecules in a unique polysaccharide, unlike that used by euglenoids.

Although most golden-brown algae are photosynthetic and live in aquatic habitats, a few species can grow on damp rocks or tree trunks; a few types can even survive desiccation and dispersal by wind. One unusual species, *Ochromonas malhamensis*, is an ideal example of a cell that straddles the divisions among biological life styles, for it can function as a producer, consumer, and decomposer. Placed in water in the sunlight and given a few salts and sugars, *Ochromonas* will carry out photosynthesis. Denied sunlight but given a broth of organic molecules that includes proteins and nucleic acids, this alga will act as a decomposer, digesting the organic matter in the manner of a fungus (Chapter 22). And, kept in the dark with certain types of bacteria added to the culture, *Ochromonas* cells will behave like animal cells and engulf and digest the prokaryotes. This amazing versatility not only has great adaptive advantages for *Ochromonas*, but also illustrates that real organisms living in nature need not conform to the simplifying classification schemes devised by biologists.

Diatoms

Recall that the immense numbers of **diatoms** in the sea may be the greatest source of atmospheric oxygen. Like the golden-brown algae, the diatoms are colored by fucoxanthin. The cell walls of diatoms contain both pectin and silica, a compound found in sand and glass. This rigid, glassy shell appears intricately sculpted with tiny holes, channels, and patterns (Figure 21-4). These are actually pores through which nutrients, minerals, gases, and wastes can pass back and forth between the cell and the environment. After the death of the cell, its hard silica wall may sink to the bottom of the aquatic habitat. Over time, such cell walls accumulate into layers of a crumbly white sediment called **diatomaceous earth.** The gritty powder made from these deposits makes an excellent abrasive and absorbent that is used in swimming pool filters, toothpastes, detergents, fertilizers, and a range of other products.

Diatom reproduction is as interesting as the cells are beautiful. The cell walls are constructed like tiny boxes

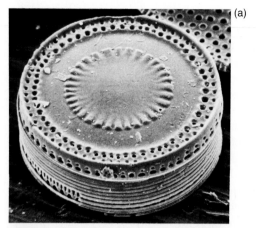
(a)

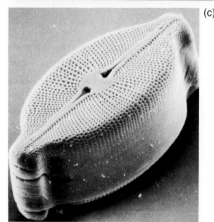
(b)

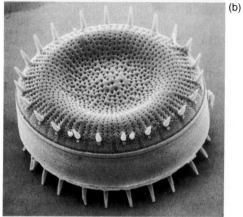
(c)

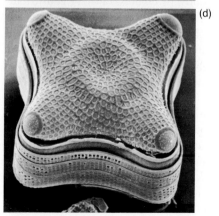
(d)

Figure 21-4   BIOLOGICAL ARCHITECTURE AT THE SINGLE-CELL LEVEL.

Here are representative diatoms, viewed with the scanning electron microscope. The species are (a) *Melosira sulcata* (magnified about 1,000 times); (b) *Stephanodiscus astrea* (magnified about 1,800 times); (c) *Navicula perpusilla* (magnified about 1,600 times); and (d) *Amphitetras antediluviana* (magnified about 800 times).

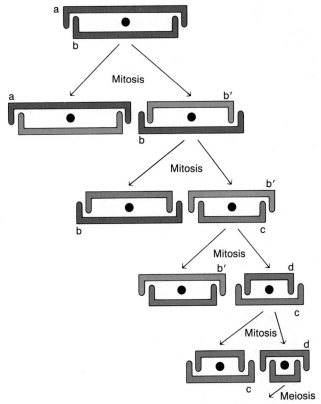

Figure 21-5   DIATOMS: DIMINISHING REPRODUCTION.
Diatoms' cell walls fit together like a box and a tight-fitting lid. When the cells divide, these parts separate from each other, and new cell-wall material is manufactured and assembled. Diatoms can manufacture only boxes, never lids. As a result, both the box and the lid of the parental cell become the lids of the progeny. This unusual division pattern results in a progressive size reduction of the cell lineage; after several divisions, the very small diatom cells can no longer undergo mitosis and, instead, enter a sexual phase in which they carry out meiosis.

with close-fitting lids (Figure 21-5). During cell division, the upper and lower portions of the cell wall separate from each other, and each half, whether "box" or "lid," grows a slightly smaller new "box" inside itself. When each daughter cell divides, both of its progeny make a still smaller box. Each successive mitotic division results in a line of cells with smaller and smaller cell walls, until the cells become so small that sexual reproduction is triggered. The now minute cell undergoes meiosis, producing just one flagellated gamete that swims away. When two gametes encounter each other, they unite, forming a diploid *auxospore*, which secretes a new cell wall of the original large size with the typical box–lid construction. Then the whole sequence starts again. Apparently, no new diatom cell has ever acquired the

ability to make a half cell wall larger than the one it inherits during mitosis. So this bizarre process of building only ever-smaller "boxes" has gone on for hundreds of millions of years.

Biologists sometimes classify another unicellular class, the *yellow-green algae*, or *Xanthophyta*, with the diatoms and golden-brown algae. The Xanthophyta are somewhat like golden-brown algae but lack fucoxanthin and chlorophyll *c*.

THE DECOMPOSERS: FUNGUSLIKE PROTISTS

Several types of heterotrophic protists derive food and energy by breaking down dead organic matter—fallen logs, leaves, or small food particles—in damp soil. This funguslike behavior suggests that simple decomposers like these may have given rise to some of the complex multicellular fungi such as mushrooms and puffballs, which we will examine in Chapter 22. One group of predominantly single-celled fungi, the Protomycota, straddles the line between protists and fungi. We group the protomycotes with the fungi because the structure of these cells can be understood more easily in the context of fungal cells. In this chapter, we shall limit our discussion to an unusual group of funguslike protists, the **Gymnomycota**, or *slime molds*.

The phylum Gymnomycota includes two distinct classes of slime molds. ("Slime" refers to the glistening appearance of these protists, and "mold" comes from their funguslike characteristics.) These classes are the **Myxomycetes,** or true slime molds, and the **Acrasiomycetes,** or cellular slime molds. Although their life cycles are very different, both can exist as single, motile cells, and both can form large, sluglike masses. The myxomycete mass is one huge cell with many nuclei. The acrasiomycete mass, to the contrary, is a true aggregate of individual cells.

Myxomycetes: True Slime Molds

The life cycle of the **true slime molds** has a mature stage characterized by the **plasmodium,** a whitish or brightly colored mass that may be either highly branched and fan-shaped or a solid slimy layer. Plasmodia may be tens of centimeters long and can weigh 40 to 50 grams. Each plasmodium is actually a continuous cytoplasm containing many diploid nuclei that arise by mitotic division of existing nuclei. This creature may be thought of as a giant multinucleate cell or as an "acellular" organism. The slimy myxomycete mass glides for-

(a)

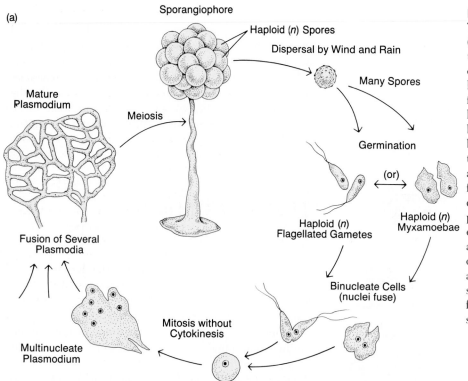

Sporangiophore

Haploid (n) Spores

Dispersal by Wind and Rain

Many Spores

Mature
Plasmodium

Meiosis

Germination

(or)

Haploid (n)
Flagellated Gametes

Haploid (n)
Myxamoebae

Fusion of Several
Plasmodia

Binucleate Cells
(nuclei fuse)

Mitosis without
Cytokinesis

Multinucleate
Plasmodium

Diploid (2n) Zygote

Figure 21-6   LIFE CYCLE OF TRUE SLIME MOLDS.
(a) The haploid (n) spores germinate under favorable environmental conditions, producing flagellated haploid cells. They may fuse to form a diploid zygote; most, however, lose their flagella and become myxameobae, cells that both feed and divide. Subsequently, two myxameobae may fuse to yield a diploid zygote. Mitosis then follows, but in the absence of cytokinesis, a large, multinucleated plasmodium forms. The plasmodium enlarges through mitotic divisions and growth, or through fusion with other plasmodia. It ultimately forms a sporangiophore containing haploid spores. (b) Sporangiophores, or fruiting bodies, of plasmodia of a species of *Diachea.*

(b)

ward, using its cytoskeleton to carry out locomotor movements through rotting leaves, wood, or grass. As it moves, it ingests woody debris, whole bacteria, or organic molecules decomposed from debris by the action of enzymes secreted by the plasmodial cells.

It is the plasmodium's task to carry out a special type of sexual reproduction. When the plasmodium encounters a nutrient-poor area or adverse environmental conditions, it halts and sends up a series of vertical stalks bearing spore capsules (Figure 21-6). Inside these "fruiting bodies," or *sporangiophores,* the many diploid nuclei undergo meiosis. As the sporangiophore capsules bulge with rapidly produced haploid spores, they come to resemble golf balls perched on tees. The haploid nu-

clei within the sporangiophore become surrounded by plasma membranes and cell walls. Eventually, the spores are released, are dispersed by wind or rain, and may remain quiescent for months or years (up to sixty!). Then, when conditions are favorable, germination occurs.

Germination produces a new generation of single flagellated, single haploid cells that swim about and may fuse to form a diploid zygote. Alternatively, the cells may lose their flagella and become **myxamoebae** that glide about feeding on plant debris or bacteria. They can thus be regarded as being on the borderline between fungi and animals. Myxamoebae are true protistan cells that divide and may survive for long periods if food is available. Myxamoebae may also fuse, to form diploid zygotes. A zygote arising by either mechanism develops into a new plasmodium. As the cytoplasm of the giant single-celled plasmodium grows, the nuclei divide mitotically, but cytokinesis does not follow. The nuclei usually are genetically identical; however, some plasmodia arise from the fusion of two or more smaller plasmodia (or even of myxamoebae), allowing some genetic dissimilarity within an individual plasmodium.

Acrasiomycetes: Cellular Slime Molds

The life cycle of the other type of slime mold, the **cellular slime mold,** also has an individual haploid-cell

(a)

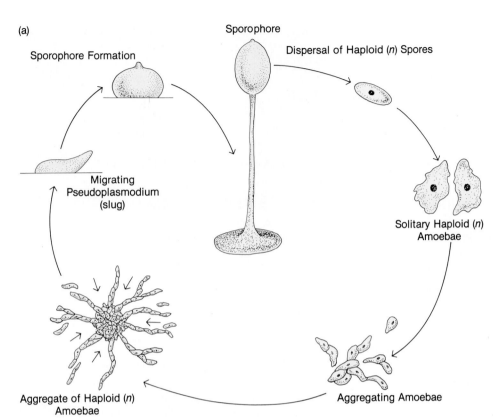

Sporophore Formation

Sporophore

Dispersal of Haploid (*n*) Spores

Migrating Pseudoplasmodium (slug)

Solitary Haploid (*n*) Amoebae

Aggregate of Haploid (*n*) Amoebae

Aggregating Amoebae

Figure 21-7   LIFE CYCLE OF THE CELLULAR SLIME MOLD *DICTYOSTELIUM.*
(a) The life cycle begins when haploid spores are released from the fruiting body, or sporophore, and individual amoebae emerge. Solitary feeding amoebae are motile, but they begin to aggregate when starved. Cyclic AMP production by aggregating cells is the coordinating signal to other cells. The aggregate forms a migrating slug or pseudoplasmodium. The latter develops into a fruiting body. This species does not have a diploid stage. (b) Aggregating amoebae (white dots) of *Dictyostelium*, magnified about 10 times; the aggregation centers are the clear areas.

(b)

Figure 21-7. Each aggregate mass forms a sluglike **pseudoplasmodium.** This aggregation process occurs in response to a biochemical signal in the form of pulses of the chemical cyclic AMP. This signal compound is released from the posterior end of advancing cells and diffuses in the environment. Those behind the leaders move toward the higher concentrations of cyclic AMP, secreting more of the chemical as they go for others to follow. In this way, cells "follow the leaders" toward centers of aggregation.

The pseudoplasmodium moves forward in a coordinated fashion even though its thousands of cells are distinct individuals. Eventually, the pseudoplasmodium settles down and develops a stalked fruiting body, or *sporophore.* Tough-walled haploid spores with diverse genotypes are formed, are dispersed, germinate, and give rise to new individual haploid cells. Diploidy and meiosis are also known to occur in some species of cellular slime molds.

The two types of slime molds are quite different: acellular slime molds carry out the sexual process and produce a huge, multinucleated cell mass. Cellular slime molds are asexual and form a large mass by cell aggregation. Curiously, most cells of the pseudoplasmodium and most nuclei and cytoplasm of the plasmodium perish: only a subset gives rise to the next generation, just as only the germ line of eggs and sperm survives among fungi, plants, and animals. There is a difference, how-

stage. These cells also glide along the ground, engulfing food particles, debris from forest litter, or bacteria. They undergo normal eukaryotic mitosis and cytokinesis, and large populations of single amoeboid cells can often result. When the food in an area is depleted, the amoebae begin to stream together into clumps, as illustrated in

ever: the many cells of a pseudoplasmodium that die are not genetically identical to the subset that becomes spores.

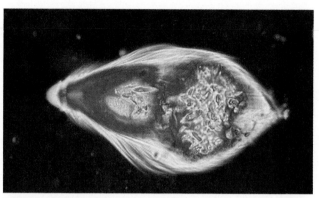

Figure 21-8   *TRICHONYMPHA:* **THE TERMITE'S HELPER.** This zooflagellate (magnified about 380 times) inhabits the termite's hindgut. The ability of *Trichonympha* to break down cellulose enables termites to use wood as a food source.

THE CONSUMERS: ANIMAL-LIKE PROTISTS

Some of the most versatile single-celled organisms on Earth are the **protozoans** ("first animals"), protists that are primarily animal-like in their life style. "Protozoan" is a general descriptive term, like algae, and is not a taxonomic category. Like animals, protozoans are consumers, cells that feed on other living cells or on food particles. Among the protozoans are many special forms and uses of organelles, a variety of locomotive patterns, and life styles that seem to have been testing grounds for those of multicellular animals. Their single cells exist in a spectrum of complex shapes, from bells and horns to eels, intricate shells, and blobs. There are four common-sense groupings of protozoans. The first three (Mastigophores, Sarcodines, and Sporozoans) each include a number of phyla. The fourth (Ciliophores) is a true phylum. A convenient distinction between the groups is their means of locomotion (or lack of movement): the *Mastigophora* possess flagella; the *Sarcodina* display amoeboid movement; the *Sporozoa* are nonmotile in the adult form; and the *Ciliophora* have numerous cilia that beat in a coordinated fashion. These four groups include the most complexly structured single cells of any organisms on Earth.

Despite differences in their modes of locomotion, all protozoans have two features in common: (1) food vacuoles, which contain digestive enzymes that break down ingested bacteria or food particles, much as lysosomes do in animal cells; and (2) contractile vacuoles, which collect and then discharge excess cytoplasmic water to maintain osmotic balance. Both types of organelles demonstrate that protozoans are at once both single cells and fully functioning organisms with mechanisms that are analogous to those in much more complex living things.

Mastigophora

The Greek word *mastigos* means "whip," and, indeed, many members of the phylum **Mastigophora,** or **zooflagellates,** move about by means of whiplike flagella. However, like the versatile plantlike protists, the zooflagellates also use other methods of locomotion. Some species use their cytoskeleton to move by means of **pseudopods** ("false feet") instead of flagella. The pseudopods are extensions of the cell surface that surge forward to pull the organism along or to engulf food particles, and then retract into the body.

Zooflagellates are the simplest and perhaps most primitive of all protozoans. They have no cell walls, and most forms live as parasites or as harmless symbionts inside other organisms. A typical and economically significant zooflagellate is *Trichonympha*, a genus of cellulose-digesting protozoan that resides in the gut of termites (Figure 21-8). Without these digestive-tract organisms, termites could not use wood as food and would be unable to degrade fallen forest trees or tunnel through wooden floors and building foundations.

Perhaps the best known and most dreaded zooflagellate is *Trypanosoma gambiense*, the cause of one form of African sleeping sickness. The discovery of this agent took many years because the organism has such a complex life cycle. In 1857, British physician David Livingstone was working in the Zambezi basin in Africa. After much observation, he concluded that a disease of cattle and sheep that the local residents called "nagana" was transferred by the tsetse fly from wild game animals to domesticated ones. Forty years later, a British army surgeon, David Bruce, proved that wild animals in that region constitute a vast reservoir for *Trypanosoma* protozoans, the cause of nagana.

Since the turn of the century, 125 trypanosome species, including *Trypanosoma gambiense*, have been found to infect almost 400 species of mammals. Many of these zooflagellates are long, wriggling, eel-like cells, each with a flagellum and an undulating membrane on its dorsal surface (yet another variation of protistan locomotion). The cell biology of these trypanosomes is remarkable in other ways. For instance, mitochondria are absent for much of the life cycle. However, near the base of the flagellum an unusual structure is present; it contains 100 "maxicircles" and up to several thousand "mini-circles" of DNA. The maxicircles allow a mitochondrion to appear when the trypanosome is resident in one of its hosts, the tsetse fly. Trypanosomes are also noteworthy because they contain large numbers of chromosomes (108 or so in *T. brucei*). Only asexual reproduction was known until 1985; meiosis and fertilization seem to occur only when the trypanosome is in a tsetse fly. Much remains mysterious about inheritance and sex in trypanosomes, since there is considerable variation in the number and size of minichromosomes, even between successive generations of a single clone of tryp-anosomes.

Although the trypanosome cells may release toxins into a host's bloodstream, wild animals are not killed by it. However, in humans and domesticated animals, the protozoans can invade the brain and spinal cord and ulti-mately kill the host. *Trypanosoma gambiense* invades its hosts in the cycle shown in Figure 21-9. The *secondary host*, also called the transfer organism or the *vector*, is the tsetse fly, within which the trypanosomes develop, multiply, and infest the salivary glands. When the fly bites a mammal, the protozoans are passed through the insect's mouthparts into the *primary host's* body.

Molecular analyses done in the 1980s have revealed utterly unsuspected reasons why *Trypanosoma gam-biense* is so successful at infecting mammals, including humans. It turns out that each trypanosome cell in the bloodstream of a host is coated with about 10 million molecules, every one an identical variable surface glyco-protein (VSG; "variable" because there are many differ-ent kinds; "glyco-" because of complex sugars on the protein backbone). Just as the host's immune system mounts an attack directed against that VSG, thereby kill-ing 99 percent of the trypanosomes, an evasion tactic is put into effect. In a few trypanosomes, a gene for a dif-ferent VSG is activated, the result being the build-up of a huge population of trypanosomes in which all are coated with the new VSG! The battle is waged again and again, with the parasite remaining one step ahead of the host's defensive system. Finally, trypanosomes invade the brain and kill the host.

The progressive switching among 1,000 to 2,000 VSG genes is only one novelty. Imagine the surprise of mo-lecular biologists when they discovered that the 1,000 to 2,000 inactive VSG genes are located at internal sites along chromosomes. To be activated, one VSG gene ap-parently is copied, and the copy is moved to an activa-tion site always found at the end of a chromosome. Only then is mRNA made on that gene copy, the particular VSG synthesized, and the new protective coat assem-bled.

The gene-transfer process is a new discovery for eu-karyotic genetic material. And the sequential use of dif-

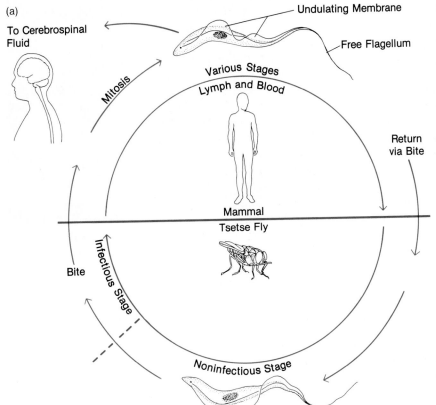

(a)

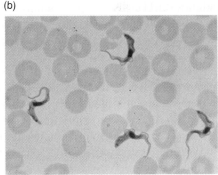

(b)

Figure 21-9   LIFE CYCLE OF *TRYPANOSOMA*.

(a) Tsetse flies transmit *Trypanosoma gambiense* to mammals, including humans. Large numbers of the protozoans arise by mitosis; ultimately, some infect the brain and spinal cord, killing the host. "Stumpy" forms of the trypanosomes can be picked up by another tsetse fly; the organisms reproduce and mature in various parts of the fly's gut (hindgut, proventriculus, salivary glands) and then may be passed to a new mammalian host. (b) Trypanosomes (magnified about 700 times) among red blood cells of an elderly missionary.

ferent VSG genes is a neat strategy to allow more successful parasitism. We know about these processes only because of recombinant DNA technology and new means of recognizing and studying proteins. The results are a step toward designing better medical protection for the 1 million people who each year become victims of African sleeping sickness. But in the meantime, it has been necessary to mount massive programs to eliminate tsetse flies. Nature is difficult to subdue, however: in some areas that are finally free of tsetse flies after years of aerial spraying, new blood-sucking insect vectors are beginning to appear.

Sarcodina

The group **Sarcodina** provides further examples of differences in locomotion between closely related protists. These protozoans range from the amoebae, with their constantly changing shapes, to beautiful hard-shelled organisms that resemble diatoms. The prototypal member of this group is *Amoeba proteus*, pictured in Figure 21-10. This irregularly shaped fresh-water organism uses its pseudopods both for moving and for engulfing bacteria, other protists, and sometimes even small multicellular animals. Another sarcodine, *Naegleria gruberi*, switches between amoeboid locomotion and flagellated movement as environmental conditions vary.

Several common sarcodines form hard, intricate shells of calcium or silicon salts about their soft cell bodies (Figure 12-11). One type, members of the genus *Actinosphaerium*, resemble tiny pincushions. Long, narrow, needlelike cytoplasmic extensions protrude through their shells, each "needle" supported by a complex double spiral of microtubules and containing actin and myosin. Food particles adhere to sticky secretions on these special pseudopods, are ingested and transported to the cell body, and finally are digested by enzymes. Thus in this protozoan, molecules (actin, myosin, and tubulin of the microtubules) and cell organelles that usually are used for locomotion are instead used for feeding.

Two other major types of shelled sarcodines are the **foraminiferans** and the **radiolarians**. Both are marine protozoans and members of the **zooplankton**—nonphotosynthetic protists living at the ocean's surface. They can live as single cells, as colonies, or as multinucleate individuals that can reach several millimeters in diameter (reminiscent of the plasmodium of acellular slime molds). Radiolarians secrete delicately patterned silicon-containing shells, through which they extend pseudopods that capture food and draw it inward for intracellular digestion. Foraminiferans secrete a calcium-containing shell that can look like a minute chambered nautilus, or spiral sea shell.

(a)
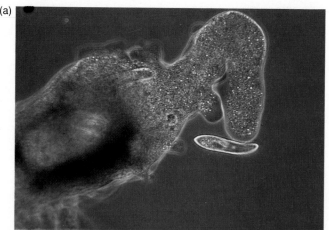

(b)
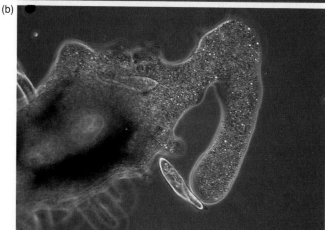

(c)
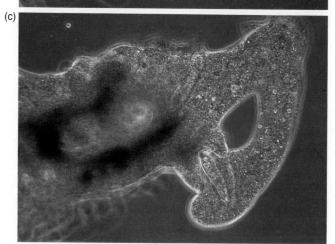

Figure 21-10   AMOEBA: THE PROTEAN PROTIST.
Sarcodines such as *Amoeba* use pseudopods for locomotion and feeding. (a, b, c) An amoeba (*Pelomyxa carolinensis*) capturing prey, which in this case is a ciliate, *Paramecium* (all magnified about 100 times).

Figure 21-11   SARCODINES: TINY EXOTIC SHELLS.
The living cells manufactured these intricate, hard skeletal assemblies (magnified about 800 times).

Both radiolarian and foraminiferan shells fall to the ocean bottom after the cells die. In depths of less than 4,000 meters or so, these shells collect into sedimentary deposits and eventually form rocks. (In deeper water, the high CO_2 content dissolves the shells.) Radiolarians help form chert and other siliceous rocks, while foraminiferans contribute to limestone—the white cliffs of Dover, England, were formed by millions of generations of foraminiferans. Because limestone deposits sometimes occur near phytoplankton sediments that have formed oil, petroleum companies often use foraminiferan fossils as clues to potentially rich drilling sites.

Sporozoa

A third grouping of protozoans is the **Sporozoa,** named for a sporelike stage of their life cycle. All sporozoans are parasites, inhabiting mammals, birds, fishes, insects, other invertebrates, and some plants. The adult forms are nonmotile, although immature cells can move about in many ways, including by flagella or pseudopods. The various sporozoans have such diverse locomotive patterns and life styles that the group probably is not a single evolutionary unit; instead, its members most likely arose from both sarcodines and mastigophores.

The many sporozoan-induced diseases are typified by malaria, caused by *Plasmodium falciparum* and related *Plasmodium* species. Study of the *Plasmodium* life cycle

MODERN METHODS FOR FIGHTING MALARIA

Malaria is considered by many to be the world's number-one disease: each year, more than 1 million African children under the age of five die from it; an additional 150 million people suffer the recurrent chills and fever of malaria; and almost 2 billion people in more than 100 tropical countries are in danger of exposure. Despite decades of effort to control its spread, malaria is making a dramatic resurgence, due to the drug resistance of strains of malarial plasmodia and the pesticide resistance of *Anopheles* mosquitoes. One strategy is to breed strains of mosquitoes that are resistant to *Plasmodium* infection. But that effort is just begun. Meanwhile, plasmodial strains resistant to quinine-based drugs—the first line of defense for humans—have forced physicians to employ drugs with dangerous side effects. To fight these growing threats, researchers are trying to answer a straightforward but difficult question: How do malarial parasites recognize as appropriate targets the red blood cells they infect and destroy?

The emerging answer seems to center around recognition by the merozoites of specific molecules on the surface of red blood cells. Two types of surface glycoproteins—sialoglycoproteins and glycophorins

A, B, and C—seem to be sites at which merozoites can invade. Research also has revealed what is probably the actual receptor site on the large, complex glycophorin molecules: a sugar called N-acetylglucosamine, which is present at the ends of the molecules. Once the tiny merozoite binds to the sugar, it somehow can navigate through the membrane and multiply inside the red blood cell.

Biologists studying this system envision a means of blocking such invasions by preparing monoclonal antibodies (Chapter 32). These, they think, would complete for and bind to the same receptor sites as the merozoites and, in that way, prevent entry by the malarial parasites. Just this strategy has worked for a kind of trypanosome which attaches to fibronectin (Chapter 6) and then invades human cells. Antibodies to the attachment site or peptides that mimic it both prevent invasion by the parasite. Although this has not yet been accomplished for malarial plasmodia, the technology is increasingly available, and identifying the receptor was an important first step. In parallel research, other biologists are trying to prepare standard vaccines against this dangerous and growing disease threat.

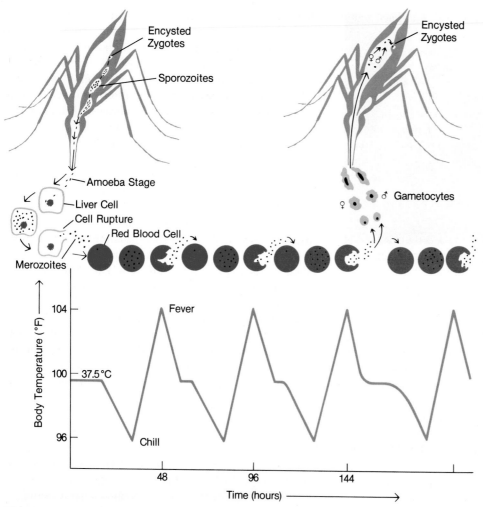

Figure 21-12 LIFE CYCLE OF *PLASMODIUM VIVAX*.
The life cycle of this and other malaria-causing sporozoan parasites of the genus *Plasmodium* begins when sporozoites enter the bloodstream with the saliva of a female *Anopheles* mosquito. The amoeba stage develops in liver cells and merozoites are freed into the blood. They enter red blood cells, reproduce, and cause the cells to burst. The reproduction of *Plasmodium* and the lysis of the red blood cells in this case take place in about 48 hours. Liberated merozoites enter other red blood cells and the cycle begins again. This produces the cyclical patterns of chills and fevers seen in the graph. Note that the temperature peak corresponds to the bursting of thousands of red blood cells. The *Plasmodium* cells eventually produce male and female gametocytes. If a malaria victim is bitten by a female *Anopheles* mosquito, some gametocytes enter the mosquito and fuse to form zygotes, which bore into the mosquito's gut lining where they develop into cysts. These cysts give rise to sporozoites which ultimately are stored in the mosquito's salivary glands. The next blood meal for the mosquito injects the sporozoites into a new mammalian host and the cycle starts again.

illustrates the complexities typical of sporozoan existence. The life cycle of this parasitic sporozoan begins when an infected female *Anopheles* mosquito bites a human or other mammal (Figure 21-12). Filamentous sporozoites harbored in the insect's salivary glands pour through the mouthparts and enter the victim's bloodstream, along with an anticoagulant, which prevents the blood from clotting and allows the insect to drink freely. The sporozoite cells lodge in the liver and reproduce

asexually, generating a large number of tiny round *merozoites*, which pass into the bloodstream and enter the red blood cells. There they reproduce synchronously, and at intervals of forty-eight or seventy-two hours, depending on the *Plasmodium* species, they lyse massive numbers of red blood cells. The lysis releases merozoites and toxins, lowers the oxygen-carrying capacity of the blood, and produces a period of high fever. The merozoites released from the red blood cells invade still other red blood cells, reproduce asexually, and once again cause lysis and new fever cycles.

Ultimately, the *Plasmodium* cells become sexual, and immature male and female gametocytes are produced. However, these cells can mature into sperm and eggs only in the gut of a female *Anopheles*. Thus an insect must bite a malaria victim with gametocytes in his or her blood in order to become infected. Following fertilization, zygotes develop in the wall of the mosquito gut; a new generation of amoebalike sporozoites appears and invades the salivary glands, ready to be transmitted to a new host.

Ciliophora

The most complex single-celled organisms on Earth are members of the phylum **Ciliophora,** commonly referred to as **ciliates.** These largely free-living aquatic protozoans derive their name from the many cilia distributed in rows, in bands, or uniformly across the cell surface (Figure 21-13). The first microorganism ever discovered was a ciliate: in the seventeenth century, Anton van Leeuwenhoek saw a species of *Euplotes* with his hand-held microscope. He considered the clumps of cilia to be legs, and he called the cells "animalcules."

Ciliates have several specialized organelles whose functions are analogous to those of organs in multicellular organisms. The cell surface may have a pellicle of protective plates, but there is no cell wall exterior to the plasma membrane. The cortical region of the cell can have musclelike contractile fibers, called *myonemes*, and structural microtubules. These "muscles" and "bones" interconnect the bases of the abundant individual cilia, somehow allowing clumps or rows of cilia to beat with beautiful coordination so the cell can swim forward or backward. The coordination and direction of beat of the cilia—forward or backward—are controlled by currents of calcium ions and cyclic nucleotides (cyclic GMP). We will learn about those same regulatory substances later in this book when we study hormones, messengers, and nerve function.

A typical ciliate such as *Paramecium* also can contain other specialized organelles. Trichocysts, the toxin-bearing harpoonlike structures also found in certain dinoflagellates, help the cell adhere to substrates, or they can be fired at prey or predators. *Paramecium* also has an oral groove leading to a *cytostome*, as shown in Figure 21-14. The beating of the cilia around the cytostome sweeps or draws food particles into the cell. From the base of this tiny mouth, food vacuoles pinch off and join vesicles containing lysosomal digestive enzymes. An anal pore expels undigested wastes.

A fascinating aspect of ciliate behavior is the ability to change shape in response to substances released by predators. *Euplotes*, for instance, changes from a lemon-shaped cell measuring 22 by 61 micrometers to a flattened disk with "wings and ridges" measuring 41 by 100 micrometers. This oddly shaped cell is more difficult to consume. The shape is maintained, even during cell division, as long as the warning chemical is present.

Figure 21-13   CILIATES: PROTISTS WITH MANY CILIA.
These are the most complexly structured free-living single cells on Earth. These cells belong to the genera (a) *Didinium* (magnified about 375 times); (b) *Vorticella* (magnified about 350 times); and (c) *Bursaria* (magnified about 50 times).

(a)　　　　　　　　　　(b)　　　　　　　　　　(c)

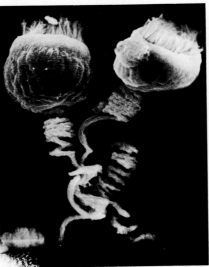

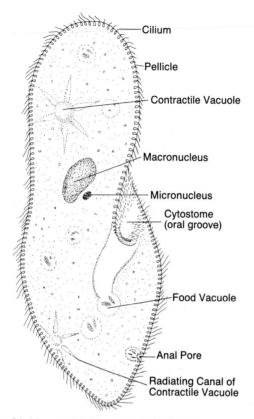

Figure 21-14   ANATOMY OF *PARAMECIUM*.
The major internal organelles are shown. The contractile vacuoles are involved in water discharge from the cell.

The highly complex structures found only in ciliates suggest that ciliates are unique products of evolution. This is nowhere better illustrated than by the two types of nuclei in these creatures. Each cell has at least one *micronucleus* and at least one *macronucleus*. Each micronucleus contains a diploid set of chromosomes. These chromosomes and the micronucleus function only during reproduction. Micronuclei therefore are not sites of RNA synthesis and are not involved in ongoing cellular metabolism. Macronuclei are highly polyploid (they contain many sets of chromosomes), and their chromosomes are the sites of RNA synthesis. The macronuclei thus contribute to cellular metabolism and maintenance by providing the information needed to manufacture enzymes and structural proteins.

When a *Paramecium* cell divides in a kind of asexual reproduction, each micronucleus undergoes normal eukaryotic mitosis, with a spindle apparatus. The macronucleus does not have a spindle; instead, chromosomes pass to daughter cells, apparently at random, along tracks of microtubule bundles. We might expect this randomness to cause problems, but it does not because there are so many copies of each chromosome in the macronuclei that each daughter cell is virtually certain to receive at least one copy of each chromosome.

We saw in Chapters 14 and 20 that bacteria may exchange genetic information through the sexual process of conjugation. A much more elaborate conjugation process evolved independently in ciliates (Figure 21-15). Two ciliates align and join near their oral grooves. The diploid micronuclei of both cells undergo two meiotic divisions; all the haploid nuclei except two per cell disintegrate. Then, in a remarkable exchange process, each cell passes one micronucleus to the other. The transferred micronucleus fuses with the resident one, restoring diploidy. Meanwhile, the macronuclei of both cells disintegrate. The diploid nuclei divide mitotically at least once. (The actual number of divisions depends on what is needed to restore the original number of micro- and macronuclei. It, therefore, differs from species to species. Here the simplest condition of one micronucleus and one macronucleus is under consideration.) The conjugants now separate and in each cell one product of the nuclear division remains diploid and becomes a micronucleus. The other product becomes polyploid and then an RNA-producing macronucleus. This restores to the cell its typical nuclear organization.

Developmental biologists have been fascinated with ciliates because they provide the clearest example of how cytoplasmic inheritance contributes to the formation of cellular structures. For example, the development of the ciliary system depends on cytoplasmic rather than nuclear inheritance. In experiments with *Stentor* species, it was found that a piece of excised cell cortex and cytoplasm can regenerate an entire new cell, provided that a micronucleus is included. However, no cilia form unless at least a portion of a row of cilia, with the underlying basal bodies and cytoskeletal interconnections, is present on that piece of cortex. The micronucleus with the cell's genetic information by itself does not provide the information needed to support development of the complete ciliary system. This is another illustration of the point made in Chapter 17 that the nuclear DNA of a species cannot by itself produce a cell, a ciliate, or a cat.

Caryoblastea

The last phylum of protists, **Caryoblastea,** has only a single species, but it raises fascinating questions. *Pelomyxa palustris* is an amoebalike cell discovered living in muddy pond sediments. It is remarkable in two respects. First, it lacks mitochondria. It is inhabited by

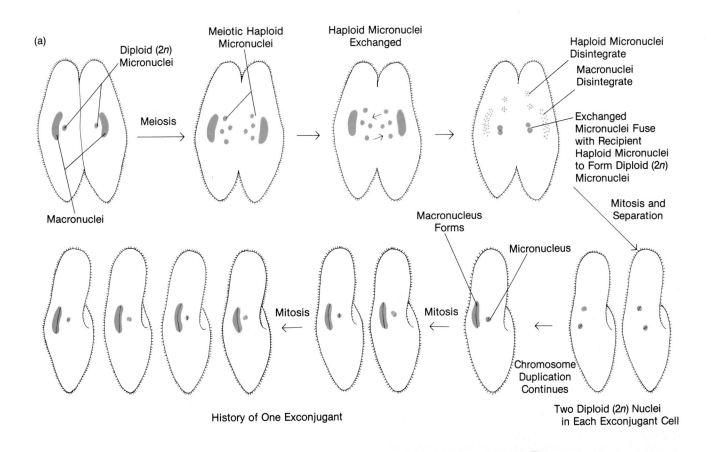

Figure 21-15   CONJUGATION IN CILIATES.

(a) Conjugating individuals pair and undergo an elaborate exchange of micronuclear material. The micronuclei first undergo meiosis; then haploid micronuclei are exchanged. The macronuclei disintegrate, as do micronuclei that fail to fuse. As shown, the exchanged micronucleus fuses with a micronucleus of a recipient cell, yielding a diploid nucleus. New micronuclei then arise and some enlarge to produce a highly polyploid macronucleus. Each exconjugant cell divides twice, producing four progeny cells. Thus the original two cells yield eight progeny, each with chromosomes and genes from the parental cells.
(b) Conjugation may involve side-to-side contact between two cells or only contact near their ends, as shown in this scanning electron micrograph of conjugating *Tetrahymena*, which is magnified about 5,500 times.

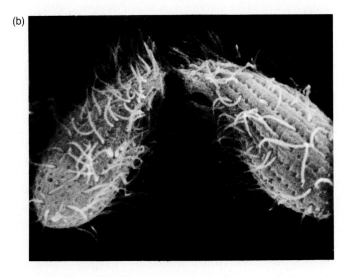

two types of symbiotic bacteria which conceivably play the role of mitochondria. Second, *Pelomyxa* never undergoes mitosis. Instead, after duplicating its chromosomes, *Pelomyxa*'s nucleus pinches in two in a process reminiscent in some ways of events during bacterial binary fission (Chapter 20). Clearly, *Pelomyxa* is an extremely primitive organism at the borderline between prokaryotes and eukaryotes. We can only wonder whether ancient relatives of this protist became colonized by nonsulfur purple bacteria (see Chapter 20) that gave rise to mitochondria; if so, those colonized cells may have been the source of other groups of eukaryotes.

PROTISTAN EVOLUTION

According to some protozoologists, the ancestors of the spectrum of modern protozoan species may have been zooflagellates. The zooflagellates could have given rise to the sarcodines and to amoeboid forms by losing flagella. Ciliates may be highly modified flagellates, with many cilia having replaced a few flagella. The symbiotic and parasitic protozoans appear to have been derived from various free-living species. But where did the stem group of zooflagellates come from? Some biologists suggest that phytoflagellates (resembling euglenoids) evolved first; perhaps one line that lost its chloroplasts resulted in the animal-like protists. Another line, upon loss of chloroplasts, could have yielded the funguslike protists (review Figure 21-1).

If the entire kingdom Protista arose from flagellated phytoplankton, then what was the source of those cells? The best answer so far relies on the endosymbiont theory—the idea that moneran cells were colonized by mitochondrionlike and chloroplastlike prokaryotic symbionts. As we discussed in Chapter 20, there are strong arguments for this theory. It is not clear, however, whether all protists arose from one type of stem cell. Nonphotosynthetic zooflagellates may represent one independent lineage, which was colonized solely by a mitochondrionlike symbiont. Could *Pelomyxa*, with its absence of mitochondria and chloroplasts, represent such a stage? A separate lineage might have received two colonizations, by chloroplastlike and mitochondrionlike symbionts, to yield the photosynthetic protists.

Several evolutionary trends are apparent among the protists. One involves cellular diversity and the exploitation of different ways of life as producer, decomposer, or consumer. Each such life style depends on biochemical and structural specializations, such as specific photosynthetic pigments, unique sets of digestive enzymes, contractile vacuoles, intricate cell walls of silica or calcium compounds, and even the bizarre specializations for parasitism. These features are, of course, the products of evolution, of changes over time that yield the combination of characteristics for each life style.

A second evolutionary feature of protists is diversity in locomotion. Pseudopods, flagella, cilia, or undulating membranous sheets propel these remarkable cells, sometimes in directed ways toward light, chemicals, or food. Some of the protists we have examined move by one mechanism; some, by another; and some, not at all. Nevertheless, there seems to be a trend toward complex, coordinated cell-surface activity among some flagellated and ciliated forms. Because their cell surfaces enable them to move quickly in response to environmental stimuli, flagellates and ciliates can function as successful predators. Thus these locomotor patterns, and the cytoskeletal bases for them, are adaptations to the life styles and to the environments in which particular species live. In the next chapters, we will turn to the multicellular descendants of diverse protists—fungi, plants, and animals—and see where evolution has led.

SUMMARY

1. The kingdom Protista is composed primarily of single-celled eukaryotic organisms that function as producers, decomposers, or consumers of food materials.

2. The plantlike protists include the photosynthetic cells that make up *phytoplankton*, which serves as the first link in aquatic food chains and as a major source of atmospheric oxygen. The three phyla of producer protists are the *Euglenophyta*, the *Pyrrophyta*, and the *Chrysophyta*.

3. The euglenoids comprise photosynthetic autotrophs (and some heterotrophs) that lack a rigid cell wall and are propelled by whiplike flagella. Reproduction in euglenoids is asexual.

4. The pyrrophytes, or *dinoflagellates*, have unique flagella that cause them to spin as they swim, are mostly photosynthetic autotrophs, may bear trichocysts, and are responsible for red tides.

5. The chrysophytes include the *golden-brown algae* and the *diatoms*. Both contain the pigment fucoxanthin and have pectin in their cell walls. The cell walls of diatoms also contain silica, and sediments of the walls, called *diatomaceous earth*, are used commercially as an abrasive and absorbent. Diatoms have an unusual method of reproduction, in which cell-wall "lids" form new "boxes."

6. Funguslike protistan decompos-ers of the phylum *Gymnomycota* include the *Myxomycetes* (*true slime molds*) and the *Acrasiomycetes* (*cellular slime molds*). Both have single-celled amoeboid stages, but the true slime mold forms a multinucleate *plasmodium*, whereas the cellular slime mold forms an aggregate of individual cells, or *pseudoplasmodium*.

7. The animal-like, consumer protists are the *protozoans*, which include the four groups *Mastigophora*, *Sarcodina*, *Sporozoa*, and *Ciliophora*.

8. The mastigophores are *zooflagellates* that move about by means of flagella. Many—such as *Trichonympha*, which inhabits the gut of

termites—live within multicellular hosts. Some cause diseases, such as African sleeping sickness. Wholly novel ways are used to move genes and express variable surface glycoproteins by *Trypanosoma* parasites.

9. The sarcodines include amoebae, *radiolarians*, and *foraminiferans* (the latter two are components of *zooplankton*). All move by means of *pseudopods*, and some have hard, intricate shells with openings through which pseudopods can project to capture food.

10. All sporozoans have a sporelike stage in their life cycle. The adults are nonmotile, but the immature cells can move in diverse ways, including by means of pseudopods or flagella. They are parasitic, such as the *Plasmodium* species that cause malaria, and frequently have two or more hosts.

11. The *ciliates* move by using large numbers of coordinated cilia. They are exceedingly complex structurally and possess both micronuclei and macronuclei. Ciliates can engage in a sexual-conjugation process.

12. The single type of *Caryoblastea*, *Pelomyxa*, lacks mitochondria and does not undergo mitosis. Perhaps it represents the borderline between prokaryotic and eukaryotic amoeboid cells.

13. The many types of protists probably evolved from flagellated cells. Some may have had chloroplasts, while others did not. These cells, in turn, arose from several types of moneran cells, perhaps by endosymbiosis. Various protists appear to be the stem groups for the multicellular fungi, plants, and animals.

KEY TERMS

Acrasiomycete
Caryoblastea
cellular slime mold
Chrysophyta
ciliate
Ciliophora
diatom
diatomaceous earth
dinoflagellate
Euglenophyta

foraminiferan
golden-brown alga
Gymnomycota
Mastigophora
myxamoeba
Myxomycete
Pelomyxa
phytoplankton
plasmodium
protozoan

pseudoplasmodium
pseudopod
Pyrrophyta
radiolarian
Sarcodina
Sporozoa
true slime mold
zooflagellate
zooplankton

QUESTIONS

1. In what ways are some protists like plants? like fungi? like animals?

2. Name some examples of phytoplankton.

3. If a planetary disaster wiped out all the Earth's phytoplankton, what would be the effect on fishes? on sea mammals? on sea birds? on the atmosphere?

4. Diatom cell division would seem to be a dead end because each division gives rise to a smaller daughter cell, enclosed in a silica box. The cells cannot grow larger. How is this problem overcome in diatoms?

5. How does the pseudoplasmodium of a cellular slime mold (Acrasiomycete) differ from the plasmodium of a true slime mold (Myxomycete)?

6. What environmental signals trigger spore formation in the plasmodium?

7. What group of protists is responsible for red tides? chalk deposits? diatomaceous earth? cellulose digestion in termites? African sleeping sickness? malaria?

8. Give examples of ciliate structures that are analogous to legs, anchors, darts, muscles, and the mouth.

9. Describe some protistan methods of locomotion.

10. Diagram the stages of conjugation in ciliates. What are the functions of micronuclei? of macronuclei?

ESSAY QUESTIONS

1. If a photosynthetic protist loses its chloroplasts, would you say that it is defective? Or would you say that it is cured of a parasite? Can the freed chloroplasts live on their own? Can the protist survive the loss of the chloroplasts? Can the protist generate new chloroplasts? Now consider a protist that loses its mitochondria. Answer the same questions.

2. African sleeping sickness and malaria both are evolving and staying "ahead" of human defenses. Explain how.

SUGGESTED READINGS

CORLISS, J. "The Kingdom Protista and Its 45 Phyla." *Biosystems* 17 (1984): 87–126.

Here is a detailed discussion of all the protists.

CURTIS, H. *The Marvelous Animals.* Garden City, N.Y.: Natural History Press, 1968.

This little gem describes the world of protists.

DOBELL, C. *Anton van Leeuwenhoek and His "Little Animals."* New York: Dover, 1962.

Here is a pleasant view of the life, times, and work of the discoverer of the world of protists.

FARMER, J. N. *The Protozoa.* St. Louis: Mosby, 1980.

Here is a superb text on all aspects of the single-celled nonphotosynthesizers.

HANSON, E. D. *Understanding Evolution.* New York: Oxford University Press, 1981.

One of America's preeminent biology teachers writes especially well here on protists and evolutionary relationships.

HARDY, A. *The Open Sea.* Part 1, *The World of Plankton.* Boston: Houghton Mifflin, 1965.

This is a fascinating description of the protists and small invertebrates found near the sea's surface.

22
FUNGI: THE GREAT DECOMPOSERS

*Yeasts, molds, mushrooms, mildews, and the other fungi
pervade our world. They work great good and terrible evil.
Upon them, indeed, hangs the balance of life; for without
their presence in the cycle of decay and regeneration, neither
man nor any other living thing could survive.*

Lucy Kavaler, *Mushrooms, Molds, and Miracles* (1965)

Sturdy, helmeted decomposers of the forest floor: *Boletus luritus* mushrooms rise silently into the damp air.

Now that we have explored the world of single-celled prokaryotes and eukaryotes, we enter the realm of multicellularity, beginning with the fungi, some of the simplest multicellular organisms. Although the 175,000 or more species in the kingdom Fungi are an unheralded and often overlooked group of organisms, they are a critical link in the web of life on Earth. Fungi lack chloroplasts and so cannot carry out photosynthesis. Nor are they motile consumers, like animals. Instead, they derive nutrients by breaking down the waste products and remains of other organisms. Thus they are called decomposers. As fungi grow and metabolize and when they die, essential inorganic and organic materials are returned to the food chains. Without fungi, our planet might be meters deep in the remains of long-dead plants, since fungi are the most abundant decomposers of cellulose and lignin.

Fungi can provide other benefits as well. Some types of fungi, such as mushrooms and morels, are used as foods by humans; yeast is used to produce bread, beer, and wine; and some molds give blue-veined cheeses their pungent odor and flavor. In addition, we manufacture antibiotics from the kinds of molds often found on oranges and melons.

On the other side of the coin, the life style of many fungi makes them the single largest cause of plant diseases: agricultural crops and gardens are attacked by no fewer than 5,000 fungal species, causing millions of dollars in damage each year. Fungi cause further massive damage to stored grains, breads, fruits, vegetables, and even foodstuffs that are produced with the aid of fungi. Some fungi also may live as parasites in humans and other animals, causing a wide range of problems—from ringworm, athlete's foot, and other skin conditions to allergies, yeast infections, and diseases of the blood, the lungs, and other organs. Furthermore, fungi can attack and decompose almost any organic material, including leather, textiles, paper, wood, glue, drugs, paint, and cardboard.

Clearly, fungi are critical for ecology and the turnover of nutrients in ecosystems. They affect all ecosystems and nearly all other life forms. In this chapter, we shall describe the basic characteristics of these "great decomposers" that set them apart in their own kingdom, and then we shall examine some representative members of the major classes of fungi to see how they differ from one another in morphology (shape and structure) and modes of reproduction. As part of tracing the evolution of fungi, we will also learn how they have become part of the ultimate symbionts, the lichens.

As we will see, the modes of reproduction are related to the complex life cycles of fungi. It is useful to recall from Chapter 9 that a life cycle is all the changes between one adult stage and the next. A life cycle may involve asexual, or vegetative, reproduction, or it may involve the sexual events of meiosis, gamete formation, and fertilization. The haploid stage is predominant in most fungi. During sexual reproduction, fusion of the haploid gamete nuclei yields the diploid stage; but meiosis soon follows so that the major free-living organisms are haploid. However, in the most complex fungi (as mushrooms), a dikaryotic stage, equivalent to the diploid states of plants and animals, alternates with the haploid portion of the life cycle. But this will become clearer as we describe specific types of fungi and see the marvelous ways in which these seemingly simple organisms exploit their environments to live so successfully.

CHARACTERISTICS OF FUNGI

Members of the kingdom Fungi share several characteristics that set them apart from other kinds of organisms, including an innovative form of digestion and a unique structure.

All fungi consume molecules manufactured originally by other organisms—they are *heterotrophs.* Fungi obtain nutrients by secreting enzymes into their environment. The enzymes digest leaves, fruit, or other organic substances. The fungi then absorb the nutrient molecules that result from the breakdown of the organic matter. This **extracellular digestion** is carried out by most animals, of course, but they do so within a gut. The particular battery of enzymes a given fungus secretes largely governs what the organism may use as a food source. Once absorbed, nutrients are processed metabolically and stored primarily as glycogen; this is similar to animals but not to plants, which store sugars as starch. It is noteworthy that the two types of slime molds discussed in Chapter 21 do not carry out extracellular digestion; cells of both types engulf food particles or absorb small molecules from the environment. This is one reason that they are not classified as fungi.

The way a fungus obtains nutrients defines its basic life style. Fungi that derive food molecules by decomposing dead matter are called **saprobes.** They include bread molds, mushrooms, morels, and other fungi that grow on organic matter—from fallen trees to dung. Fungi that obtain nutrients from living plants and animals are *parasites* and may be found on or in just about every uni- or multicellular creature. Finally, many fungi are *symbionts* with roots of higher plants; as such, they aid the plants and derive nutrients in return.

As with all other forms of life, the basic structure of fungi is fundamental to their success. Except for yeasts and a few other unicellular forms, fungi are built with a main *thallus* (body), composed of cellular filaments

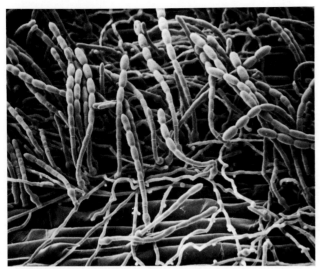

Figure 22-1 HYPHAE: BASIC FUNGAL UNITS.
Hyphae of the fungus *Erysiphe graminis* growing on barley shown magnified about 600 times. Hyphae are cytoplasmic tubes and are seen in this powdery mildew as the small fibers at the bottom. Reproductive structures, conidiophores, grow upward from the hyphae and appear here as the structures with constrictions extending toward the top of the photograph.

called **hyphae** (Figure 22-1). You can see individual hyphal filaments in molds that grow on bread or oranges. The most common fungal form (a mushroom) is, as we will see, a specialized reproductive structure that is actually a mass of hyphae packed tightly together. Each hypha, whether in a mushroom or growing alone, consists of a tubular cell wall that surrounds cytoplasm, which contains many of the usual eukaryotic organelles (mitochondria and so on; however, no basal bodies or centrioles are present in many fungi) and one or more nuclei. There are no separate differentiated cell types in hyphae. Nevertheless, some hyphae in certain species become specialized to form **rhizoids,** which anchor the fungus much like roots anchor a plant. Other hyphae may form **haustoria,** feeding structures that penetrate the living cells of other organisms and absorb nutrients from them (Figure 22-2). Still other hyphae may be specialized for spore production. Thus fungi, as multicellular organisms, do display localized functional specialization.

The cell walls of most hyphae are made of *chitin* (a carbohydrate polymer composed of glucosamine) and other organic molecules. Chitin is also found in the external skeletons of such animals as beetles, lobsters, and black widows, as we shall see in Chapter 25.

Fungal hyphae may or may not be cellular in nature. Thus the hyphae may or may not be **septate**—have cross walls, or *septa* (singular, *septum*), that segregate independent cells, each with at least one nucleus (Figure 22-3a). In the so-called "lower fungi," there are no septa, and the cytoplasm streams freely through the branched hyphal channels (Figure 22-3b). Such fungi have multiple nuclei in one mass of cytoplasm and are said to be **coenocytic.** In the "higher fungi," the electron microscope reveals cross walls, but even these are perforated by pores, which allow cytoplasmic continuity and streaming between separate cellular compartments (Figure 22-3c). Thus although fungi themselves are unable to carry out true locomotion, the cytoplasm and small organelles within the tubular hyphae stream freely. Most fungi can therefore be described as masses of multinucleate cytoplasm arranged as hyphae.

Fungal Growth Patterns

Hyphae grow only by elongating at the tip and by forming branches some distance behind the tip. Although proteins and other essential macromolecules are synthesized throughout the cell, cytoplasmic streaming carries the products to the tips for assembly into new cell wall, plasma membrane, and so on. Such growth patterns generate a filamentous network, or **mycelium,** as shown in Figure 22-4. Continual growth at the tip is highly adaptive for fungal life styles. As useful food sources are digested and nutrients are depleted by absorption, the growing hyphal tip literally moves on. It penetrates into undigested organic material so that a fresh supply of nutrients becomes available. Fungal metabolic wastes also accumulate in regions around older

Figure 22-2 HAUSTORIA: TAKING IN NUTRIENTS.
Haustoria grow from hyphae and penetrate the living cells of other organisms. Nutrients absorbed from the host cells are passed back into the fungal hyphae.

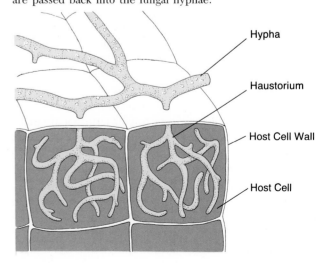

Hypha

Haustorium

Host Cell Wall

Host Cell

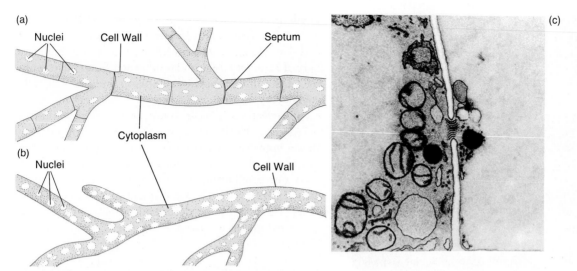

Figure 22-3   **SEPTATE AND NONSEPTATE HYPHAE.**
(a) Septa delimit cell-like domains. (b) In the absence of septa, a hypha is a single cytoplasmic
domain. (c) Septa are perforated with holes, as seen in the electron micrograph (magnified about
23,000 times). Obviously, a variety of substances and organelles can pass through such a large doorway
between cells.

Figure 22-4   **A LIVING POWDER PUFF.**
This fungal mycelium growing on a leaf derives nutrients by
haustoria and uses them to rapidly build these whitish
hyphae that make up the mycelium.

hyphae, but such wastes do not have deleterious effects
because the tip has grown onward.

As the tips of hyphal branches probe forward and as
nutrients are absorbed, more hyphae are generated.
This growth can be so fast that in just five hours or so, a
fungal colony may produce 200 meters of hyphae. The
surface-to-volume ratio of hyphae, even when packed
tightly in the body of a mushroom, is thus very high,

allowing a great deal of opportunity for exchange of nutrients, wastes, and gases between the fungus and the
local environment around the hyphae. (Not surprisingly,
plants have evolved defenses against fungal attack; bean
leaves, for instance, secrete chitinase, an enzyme that
digests fungal cell walls and so inhibits growth of the
fungus.)

Fungal growth depends on mitosis and rapid manufacture of cytoplasm. Fungal mitosis is unique in that it
occurs *within* the nucleus, and it culminates in two nuclei that separate and are moved apart in the hyphal cytoplasm. Some data suggest that the chromosomes of
fungi differ from those of other eukaryotes by having
extraordinarily small quantities of histones, the basic
(pH, that is) proteins involved in DNA's coiling into nucleosomes (Chapter 9). The histones appear to be of the
same types as those in other eukaryotes.

As fungal hyphae grow through a rotting log, soil, or
another substratum, hyphae from genetically distinct
individuals of a species may fuse and thereby give rise to
a single cytoplasm with dissimilar nuclei, a so-called **heterokaryon. Heterokaryosis,** the process by which heterokaryons form, is a special feature of fungi. One of its
great evolutionary advantages is that it brings dissimilar
sets of genes into one mass of cytoplasm. The result may
be a new composite phenotype, combining features of
more than one parental strain. Heterokaryosis thus
serves as an agent for increasing the genetic variability
on which natural selection acts. Not only does this process help to compensate for deleterious mutations in one
parental strain by providing alternative versions of each
gene, but novel combinations of gene products can en-

h'ance survival. For instance, new combinations of digestive enzymes, coded by different parental-strain nuclei, might permit heterokaryons to utilize previously unexploitable food sources.

Spores for Dispersal and Survival

As nonmotile heterotrophs, fungi growing on a rotting log or a moldy piece of bread will, at some point, consume all the available nutrients. How, then, can the species survive and find new sources of nutrients? There must be a more mobile life stage to ensure dispersal to sites where individuals can grow again. This function is filled by *spores,* the reproductive bodies fungi produce. Some spores are borne on specialized **aerial hyphae,** which grow vertically and discharge spores into the air in various ways. Other spores form internally, as in puffballs, and then are discharged explosively (Figure 22-5). Individual spores can be composed of one or more cells, can have one or more nuclei, and can be formed sexually or asexually. Thus, either a haploid or a diploid chromosome number may be present, depending on the species and on the stage in the life cycle when the spores are produced. Figure 22-5b shows spores greatly enlarged. Their real size is a diameter of 5 to 50 micrometers, although some may attain lengths of several hundred micrometers.

Spores are quite variable in form, even within one species, and the numbers an individual fungus produces can be truly amazing. One average-size commercial mushroom can give rise to about 16 billion spores. The giant puffball is even more impressive: at maturity, its brown powdery interior is filled almost entirely with a loose network of as many as 7 trillion spores (Figure 22-6). Spores are so light that they can be carried for hundreds and even thousands of miles. Because of the enormous number produced and the ease of dispersal, fungal spores are everywhere in our environment.

Spores can be divided into two groups: those that help disperse the species and those that are adapted for survival through periods of unfavorable environmental conditions. *Dispersal spores* are usually short-lived. They are produced in large numbers during active growth of the fungal body and can germinate quickly if environmental conditions are favorable. *Survival (resting) spores* are usually produced in lesser numbers and at a particular time in the life cycle, such as when the growth of the fungus is limited by heat, cold, or drought. Such spores usually have a thick, darkly pigmented cell wall, are derived sexually, and require a period of dormancy (rest) before they will germinate. Survival spores can remain viable for long periods, often years. A typical fungal species forms dispersal spores during its period of active growth, and then generates survival spores as nu-

Figure 22-5   SHOWERS OF SPORES: A FUNGAL STRATEGY.
(a) The discharge of spores, as shown in this earth star, *Geastrum triplex,* can be triggered by the slightest touch, even from falling raindrops. (b) A scanning electron micrograph of fungal spores magnified about 2,400 times. From *Living Images* by Gene Shih and Richard G. Kessel, Science Books International, 1982. Reprinted by permission of the present publisher, Jones and Bartlett Publishers.

Figure 22-6   THE GIANT PUFFBALL.
This *Lycoperdon giganteum* contains an astonishing 5×10^{12} spores by the best estimates. The chances of a spore landing at a favorable site, germinating, and giving rise to a new puffball are minute; hence the trillions of spores.

trients are used up or as weather conditions become unfavorable. These spores may then be carried to a more hospitable area with a fresh source of nutrients.

Clearly, fungi can be considered multicellular organisms, even though there are single-celled stages in their life cycles. However, fungi generally demonstrate relatively little differentiation of cell types into tissues and organs. The rhizoids, haustoria, and spore-producing aerial hyphae are the main regions of specialized structure and function, and might be thought of as prototypes of differentiated structures in a multicellular organism. Indeed, as we will see, certain structures of higher fungi do contain various cell types and what can rightfully be called tissues of specific function. However, it is really only in plants and animals that the number and types of differentiated cells increase dramatically. So it should not be surprising that all fungi, living on the edge of multicellularity, have relatively similar ways of life and structural features. The rest of this chapter is devoted to an examination of variations on the basic fungal plan.

CLASSIFICATION OF FUNGI

As we saw with the protists, classification schemes are necessary for scientific purposes, but sometimes they may be only convenient constructs because biologists do not know enough to properly categorize certain groups according to actual evolutionary relationships. This is surely the case with fungi. In general, fungi are classified according to morphology, methods of reproduction, and modes of spore production. The kingdom Fungi consists of one division, **Mycota,** which is divided into six principal classes. (This division name and the suffix "-mycete" derive from the Greek word *mykes*, meaning "mushroom.") Table 22-1 lists the primary characteristics of the six classes. The 50,000 species of lower fungi comprise three classes: Chytridiomycetes, Oomycetes, and Zygomycetes. Their hyphae lack septa, so they commonly are coenocytic. These fungi form their asexual spores by cleavage of the cytoplasm within a special spore case. In addition, the chytrids and oomycetes have some motile cells during certain stages of their life cy-

Table 22-1   **CHARACTERISTICS OF THE SIX CLASSES OF FUNGI**

Class	Usual Vegetative State	Asexual Reproduction (only mitosis involved)	Sexual Reproduction		Representative Member
			Fusion of:	Resulting in:	
Lower Fungi (coenocytic)					
Chytridiomycete	Haploid	Flagellated spores in sporangia	Flagellated gametes	Resting spores	Chytrids
Oomycete	Diploid	Flagellated spores in sporangia	Gametes in gametangia	Oospores	Water molds, *Phytophthora infestans* (potato blight fungus), *Plasmopara viticola* (downy mildew)
Zygomycete	Haploid	Unflagellated spores in sporangiophores	Gametes in gametangia	Zygospores	*Rhizopus* (black bread mold)
Higher Fungi (septate)					
Ascomycete	Haploid	Conidia on conidiophores	Hyphae	Ascospores in an ascus	*Neurospora* (red bread mold), *Penicillium*, yeasts, truffles, morels, cup fungi, powdery mildews
Basidiomycete	Haploid Dikaryotic	None or conidia on conidiophores	Hyphal tip cells	Basidiospores on a basidium	Mushrooms, puffballs, bracket fungi, rusts, smuts
Deuteromycete (Fungi Imperfecti)	Haploid	Conidia on conidiophores	None known		*Aspergillus*, ringworm

cles. These two classes are sometimes called water molds because the motile cells rely on the presence of water for dispersal. This characteristic, in addition to other features, suggests to some biologists that these two classes of fungi probably represent ancient forms that arose before adaptations evolved for life on land. Indeed, these two types of fungi have such distinctive characteristics that they are occasionally classified as protists rather than fungi. Though the zygomycetes are listed traditionally as lower fungi, they share a number of characteristics with higher fungi. Among these are the absence of centrioles, basal bodies, and flagella built of microtubules. This suggests that zygomycetes are closer evolutionarily to higher fungi than to the other types of lower fungi.

The higher fungi (Figure 22-7) consist of almost 100,000 species in two classes: Ascomycete and Basidiomycete. Both groups are septate and both reproduce sexually by the fusion of hyphae of different mating types to form spores. One stage of spore formation is characterized by the presence of two nuclei in special hyphal cells called **dikaryons.** This feature is unique to the ascomycetes and basidiomycetes; it is not found elsewhere in the biological world.

The sixth class, Deuteromycete, or Fungi Imperfecti, is a grab bag of more than 25,000 species in which sexual reproduction either does not occur or has not been discovered. They are called imperfect for this reason, not because they are biologically deficient in any way. The deuteromycetes are incredibly diverse and do not represent a single lineage.

Now that we have outlined the basic characteristics and types of fungi, let us look at each class to see how representative species live and reproduce.

Chytridiomycetes: The Interface Between Protists and Fungi

Some mycologists (biologists who study fungi) believe that members of the class **Chytridiomycete** are the simplest and most ancient fungi. Some chytrids are aquatic, and some are parasites on algae or on other fungi. Most, like the one pictured in Figure 22-8, are unicellular, but a few species form chains of cells and an undifferentiated body made up of hyphae with chitinous cell walls. The characteristic feature of the 750 species of chytrids is the formation of two types of motile cells: (1) spores formed asexually within a spore case, or **sporangium;** and (2) gametes formed in separate male and female **gametangia** (gamete containers). Gametes are propelled through the environment by a posterior, whiplike flagellum of the microtubular type, one of the characteristics that suggests the chytrids are protists and not fungi.

Most chytrid cells are haploid throughout the life cycle, with the exception of the zygotes. These motile cells are diploid and can act as resting cells, becoming dormant during unfavorable periods. When conditions

Figure 22-7 HIGHER FUNGI: A TYPICAL TOADSTOOL, *AMANITA MUSCARIA.*
Individuals of this species can be wider, redder, or somewhat shorter than this specimen, but all are equally poisonous to humans.

Figure 22-8 LOWER FUNGI: A CHYTRID CELL.
In this light micrograph (magnified about 400 times) of a *Spizellomyces acuminatus* cell from a pure culture of the typical chytrid grown in the laboratory, one can see the characteristic spherical vegetative cell body and branching rhizoids.

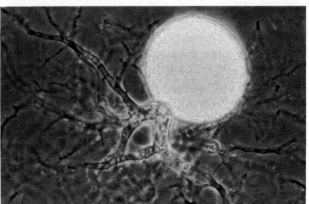

grow favorable, these cells divide meiotically to form motile haploid *zoospores*, which settle onto appropriate substrata and grow into cell chains, somewhat resembling hyphae.

Oomycetes

The oomycetes make up a diverse and unusual class, least like the other fungi and somewhat like algae. Some species in the class **Oomycete** are single-celled organisms, and others form branching hyphae. Some are aquatic, and others live in soil. Some are saprobes, and others are parasites. However, traits that all oomycetes share include few or no cross walls in their hyphae and an appearance resembling colorless forms of certain algae (Chapter 21). Like algae, but unlike other fungi, typical oomycetes have cell walls containing cellulose instead of chitin. The oomycetes produce motile asexual zoospores that contain centrioles and basal bodies and are propelled by a pair of microtubular flagella. During

most of the life cycle, oomycete nuclei are diploid, not haploid. Finally, all oomycetes have basic histone proteins in their chromosomes. Although some biologists have interpreted these characteristics to mean that the oomycetes evolved from algal ancestors, it may represent a case of parallel evolution, in which similar properties arose independently in organisms with similar life styles.

Sexual reproduction in this class of fungi also resembles that in certain algae and gives the group its name. The prefix "oo-" refers to eggs: the oomycetes produce large, immobile egg cells. Both the egg and the male gamete (*protoplast*) develop in gametangia. During fertilization, the *antheridium* (male gametangium), which surrounds or is attached to the *oogonium* (female gametangium), passes a protoplast through a pore or tube into the oogonium (Figure 22-9a). Ultimately, the egg and the protoplast nuclei fuse, and the diploid oospore is formed. Figure 22-9b shows the life cycle of a typical oomycete, *Saprolegnia*, a fish parasite.

Figure 22-9   LOWER FUNGI: OOMYCETES.
(a) Structures for sexual reproduction. An antheridium (top left structure) is seen in contact with an oogonium to the right. This *Sapromyces androgynus* is magnified about 650 times. (b) Simplified life cycle of the oomycete *Saprolegnia*. Asexual reproduction, seen at the bottom, involves the production of motile zoospores by the diploid mycelium. Those zoospores grow into new mycelia. Sexual reproduction involves the male antheridium and the female oogonium, seen here in enlarged view at the top. Haploid protoplasts in the antheridium fuse with haploid egg nuclei to form diploid oospores. These oospores germinate after a period of dormancy to form new mycelia.

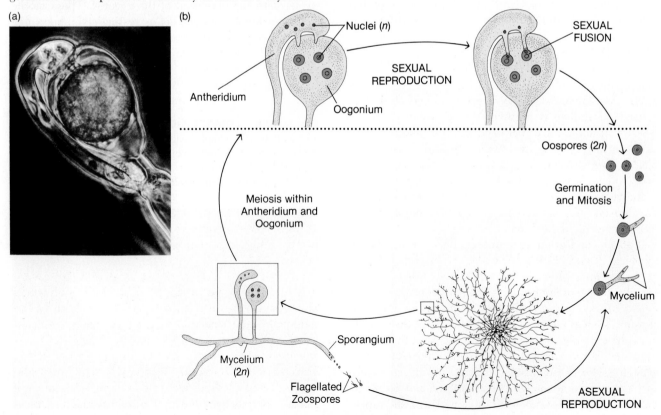

Although the oomycetes are referred to as water molds, the water molds such as *Saprolegnia* make up only part of this class. Several terrestrial plant parasites belonging to this group have had a significant impact on human history. One of these fungi is *Phytophthora infestans*, the agent that causes late blight in potatoes. This fungus hit the potato crop in Ireland particularly hard between 1845 and 1847, when wet, cool weather provided optimal growth conditions. Since potatoes were the staple food of most of the Irish population at that time, the blight caused widespread famine. More than 1 million people died as a result. Many of those who survived—one-third of the population—left Ireland during the fifteen years following 1845, mostly for the United States and Canada.

Another oomycete, *Plasmopara viticola*, which causes downy mildew on grapes, almost destroyed the French wine industry around 1880. As with the potato blight fungus, dispersal spores are responsible for the rapid, devastating spread of this plant pathogen. Fortunately, the world's first chemical antidote to plant disease—the copper-containing Bordeaux mixture—was discovered, and some of the vines of Bordeaux, Burgundy, and other regions were saved.

Zygomycetes

The members of **Zygomycete**, the third class of lower fungi, often are grouped with the ascomycetes and basidiomycetes as true fungi. Even so, the zygomycetes share certain traits with the oomycetes. Like the chytrids and oomycetes, the zygomycetes form asexual spores in a sporangium. However, these spores are nonmotile; they have no flagella and are dispersed by air rather than in water. The zygomycetes are entirely terrestrial: most live as saprobes that digest and absorb dead plant and animal matter in the soil; the few parasitic species feed on small soil organisms, insects, or plants. Their hyphae have few or no septa, so that the many nuclei are in a single mass of cytoplasm. The zygomycetes include the pin molds commonly found growing on fruit and bread. A typical example is the black bread mold *Rhizopus*.

Rhizopus and other zygomycetes have an interesting life cycle (Figure 22-10a). The black walls of the sporangia break open when the spores mature, releasing dozens of spores that are carried by air currents to potential food sources. After a spore germinates on its new substratum, such as a slice of bread, it sends out three types of hyphae: *stolons*, or lateral stemlike hyphae that form a network on the surface of the bread; rhizoids, which grow into the bread and there carry out digestion, absorb nutrients, and anchor the mycelium; and *sporangiophores*, or stalks that bear the sporangia. Sexual reproduction occurs when the hyphae of two different mating strains (designated + and − because they are not usually morphologically distinguishable as male and female) grow in close proximity and contact each other at their hyphal tips. Cross walls form behind the zones of contact, forming gametangia that function as sex organs. One gametangium contains + gametes and the other, − gametes. First the gametangia fuse; then their haploid gametes fuse to produce a cell that develops into a thick-walled *zygosporangium* which functions as a survival spore. The term "zygomycete" is derived from this structure. After a period of dormancy and when conditions are favorable, the haploid nuclei in the zygosporangium fuse to yield a true diploid zygote. Soon thereafter meiosis occurs; the survival spore germinates; and a new sporangium is produced, borne on a single hypha (Figure 22-10b). Haploid spores are released from the sporangium and are blown about. Those that settle on a new food source germinate and begin a new life cycle. As with the chytrids, the zygote is the only diploid cell in the life history of the zygomycetes.

Although unseen on a stroll through any meadow or forest, certain zygomycetes deep in the soil play a critical ecological role as symbionts with the roots of the vast majority of plants. Such associations of roots and the thin filaments of fungi are called **mycorrhizae,** which means "fungus roots." Individuals of many plant species infected with mycorrhizae grow more successfully in poor soils, particularly soils deficient in phosphates, than do plants without mycorrhizae. This is because the branching microscopic filaments of the mycelium fan out beyond the root, adding a huge amount of surface area for water and mineral-nutrient absorption. Mycorrhizae apparently supply the host plant with substantially more nutrients, especially phosphorus, than it can absorb through its own roots alone. The plant more than repays its debt to the fungus by supplying it with photosynthetic products that are the raw materials for fungal cell metabolism. Almost all plant species can form mycorrhizae, and, indeed, it is estimated that 80 percent of all land plants develop mycorrhizae with one fungus or another. All the common garden plants, including onions and strawberries, form mycorrhizae with zygomycete species. And we will see that many trees and other plants form just as important mycorrhizae with certain basidiomycetes.

Paleontologists have made the unexpected discovery that fossils of some of the earliest multicellular land plants have associated mycorrhizae fossils. Perhaps the soils on the early land masses lacked decaying vegetation and were poor in nutrients, so that fungal symbiosis would have been highly adaptive, if not critical, to the initial colonization of land by plants. Without doubt, the numbers and kinds of plants that have inhabited the Earth during the past 400 million years would have been

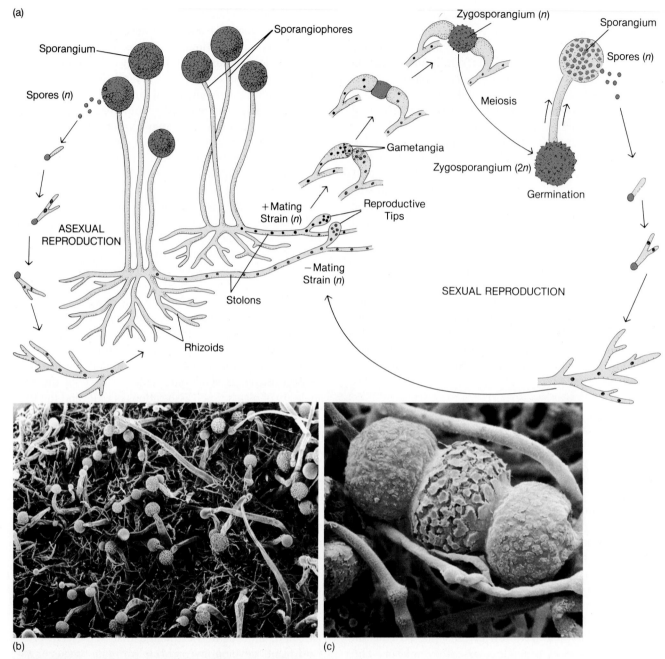

Figure 22-10 STAGES OF THE ZYGOMYCETE LIFE CYCLE.
(a) In sexual reproduction, opposite mating strains (+ and −) fuse at the tips of the hyphae. A zygosporangium forms and remains dormant for a period. Then, after the two haploid nuclei fuse to yield a diploid zygote nucleus, meiosis takes place. Later, the zygosporangium germinates and a new sporangium grows upward. It releases haploid spores which develop into mycelia. Asexual reproduction involves the discharge of haploid spores from the haploid sporangia that form on the hyphae. (b) Typical spherical sporangia of the pin mold *Mucor* (magnified about 85 times). In life such sporangia are colorless at first, but later turn black. (c) A mature zygosporangium between the adjacent reproductive tips, as seen by scanning electron microscopy (magnified about 600 times).

(a)

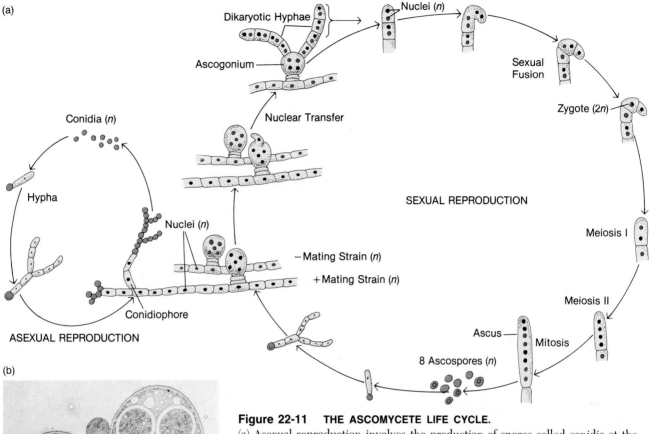

Figure 22-11   THE ASCOMYCETE LIFE CYCLE.
(a) Asexual reproduction involves the production of spores called conidia at the ends of aerial hyphae, or conidiophores (review Figure 22-1). These spores are usually dispersed by wind action and form new mycelia. Mycelia may also be generated from ascospores through a type of sexual reproduction that involves an unusual delayed fertilization. When the hyphae of two mating strains contact each other, the nuclei are brought together within the ascogonium, an ovarylike structure. Binucleate hyphae emerge from the ascogonium, each cell remaining dikaryotic, with separate nuclei originating from both the parental mating strains. At the tips of the dikaryotic hyphae, a special cell forms in which the nuclei fuse and yield a diploid nucleus. This cell and its progeny undergo meiosis and mitosis to produce eight ascospores, which then germinate and form new hyphae. (b) A cross section through two asci (magnified about 2,100 times) shows forming ascospores before their release.

radically different if mycorrhizae had not evolved to aid in nutrient and water uptake.

Ascomycetes

Ascomycete is the largest class of fungi, containing some 30,000 free-living species. An additional 18,000 species grow symbiotically with algae to form lichens (discussed later in this chapter). The name means "sac fungus" and is derived from the *ascus,* a spore sac to be described later. Most ascomycetes are saprobes, but the class also includes many important parasites of living plants, including several types of powdery mildews and the fungi that cause Dutch elm disease, chestnut blight, and peach leaf curl. A few ascomycetes are unicellular (the yeasts), but most are composed of filaments in which the nuclei are haploid. The ascomycetes, like other higher fungi, are septate. The hyphal septa are perforated, however, so the cytoplasm is continuous between neighboring cells.

Asexual reproduction involves the generation of spores called **conidia,** which come in many shapes and sizes and may be multicellular. The word "conidium" is derived from the Greek word *konis,* which means "dust"—an appropriate label because these spores are so tiny and numerous. Conidia develop not within enclosed sporangia, but usually on the tips of specialized aerial hyphae called *conidiophores* (Figure 22-11). Held aloft

ELM EPIDEMIC IN DUTCH?

Plant researchers are applying some creative approaches to eradicating a fungal pest that has decimated the elm populations of North America: Dutch elm disease. This normally fatal tree disease causes leaves to yellow and wither and interferes with the critical passage of water upward through the trunk (Figure A). The fungus that causes Dutch elm disease is *Ceratocystis ulmi*, the infectious spores of which are carried from tree to tree by bark beetles, which tunnel through the tender wood just below the bark (Figure B). The boring of uncontaminated bark beetles can damage elm trees but not kill them—the fungus is the real pathogen, leaving millions of elms first leafless and then lifeless.

Traditionally, biologists and forestry managers have used three means of controlling the spread of such diseases. One is strict quarantine—prohibiting the importation of diseased trees or wood. Quarantines of agricultural products and raw materials are taken quite seriously today, but in the past, they were sometimes lax. Dutch elm disease was first noticed in the Netherlands in 1919, and in 1930, it was reported in forests around Cleveland, Ohio. American biologists warned against the further importation of elm logs from Europe, but their advice was ignored. By 1975, the disease had spread all the way across the North American continent.

A second means of fighting plant epidemics has been to quickly remove and destroy diseased or susceptible organisms. Finally, chemical insecticides and fungicides can be applied to reduce the populations of bark beetles and fungus. However, at a cost of $100 per tree per year, this spraying approach is useful only for individual landscape trees, not for groves or forests. By the mid-1950s, all three of these methods were being utilized, but Dutch elm disease continued to spread.

Two current lines of investigation may provide badly needed weapons for battling the epidemic. One promising tactic has been the manipulation of *pheromones*, chemical signals an organism produces that stimulate a physiological or behavioral response in another member of the same species. Female bark beetles release one such odorant, an aggregation pheromone, when they land on elm wood that is suitable for colonization. The pheromone then attracts large numbers of males, which attack the tree. Once an elm tree is dying from either Dutch elm disease or other causes, it too gives off odorants that attract beetles.

Researchers have capitalized on both these facts in

Figure A **DUTCH ELM DISEASE.** One elm survives, but the others have succumbed to the fungal disease. Normally all the elms in an area of infection will die or must be destroyed.

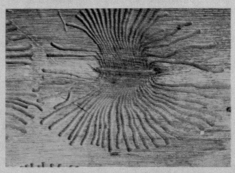

Figure B **BARK BEETLES TUNNEL IN AN ELM.** Borings through the trunk destroy parts of the vascular system, but it is the fungus that kills the tree.

a clever way, creating "trap trees." They inject diseased or unwanted "weed" elms with a chemical called cacodylic acid, which dries out the inner bark and causes the release of odorants that signal "dying tree" to a hungry beetle. In addition, they spray the tree with artificial aggregation pheromones. Together, these scents attract huge numbers of bark beetles, which then become trapped on sheets of sticky paper attached to the tree.

The second new technique employs a strain of the bacterium *Pseudomonas syringae*, which produces a natural *antimycotic*, or fungus-killing chemical analogous to antibiotics made by other prokaryotes. When released on a grove of trees, the *Pseudomonas* can kill fungal spores so effectively that the infection rate is reduced from nearly 100 percent to as low as 2 percent of the trees. Current efforts are aimed at developing a bacterial strain that can survive cold winters, as well as spread to the trees' highest branches, where Dutch elm disease usually starts.

by the conidiophores, the conidia can be dispersed by air movements to new food sources. Such airborne conidia are often a source of allergy in humans.

The sexual stages of most ascomycetes are less conspicuous than the asexual stages. For instance, many fungi that cause plant diseases in temperate regions generate huge numbers of conidia during the host plant's growing season, but the germinated conidia form the microscopic sexual stages only as autumn approaches, when the environment and host's physiology are changing.

The sexually produced *ascospores* are formed in a little sac called an **ascus** (Figure 22-11). Before an ascus and ascospores can develop, the hyphae of different mating strains must fuse. This brings together haploid nuclei so that a diploid stage can form—the only such stage in the ascomycete life cycle. The diploid cell immediately undergoes meiosis to give rise to four haploid cells, which, in turn, divide mitotically to produce eight tough-walled haploid ascospores. The ascus eventually bursts, freeing the ascospores.

The typical sexual cycle can be seen in the well-studied red bread mold *Neurospora crassa* (recall the studies of Beadle and Tatum, described in Chapter 12). Like many other ascomycetes, *Neurospora* forms specialized hyphal branches that function as sex organs. Nuclei from the donor mating strain (+) pass into the hyphae of the

receptive (−) mating strain through the fused hyphal tips. Thus the resultant cells are dikaryons, cells with two nuclei. These haploid nuclei undergo a number of mitotic divisions inside a cell called the *ascogonium*, which will produce one or more asci. In each ascus, fusion of the + and − nuclei occurs, producing the diploid chromosome complement.

The groups of asci derived from each ascogonium are usually surrounded by protective hyphae, forming a structure called the **fruiting body.** Fruiting bodies can be flask-shaped containers with a pore or channel, spherical containers with an opening, or disk- or cup-shaped. In the cup-shaped fruiting bodies, such as the cup fungi (Figure 22-12), the asci are exposed on the upper, or inner, surface. A few related species, such as the morels and false morels, have stalked fruiting bodies crowned by bell-shaped, saddle-shaped, convoluted, or pitted tissue that contains the asci (Figure 22-13). The morels are highly prized as gourmet delicacies, but many of the false morels contain toxic compounds that can be fatal. The truffle is another prized edible ascomycete.

Another group of ascomycetes that people eat are the single-celled yeasts. Yeasts usually reproduce by asexual *budding*—pinching off of small haploid cells from large older haploid ones. Sometimes, however, two cells unite, form a diploid nucleus, and eventually produce an

Figure 22-12   CUP FUNGI.

(a) If magnified, this specimen of *Peziza aurantia* would resemble the diagram (b). The cup fungus is composed of thousands of hyphae, the primary business of which at this stage is sexual reproduction. Fusion of the male and female organs leads to formation of the dikaryotic cells; that, in turn, allows sexual fusion (top), which is followed by meiosis and ascospore formation.

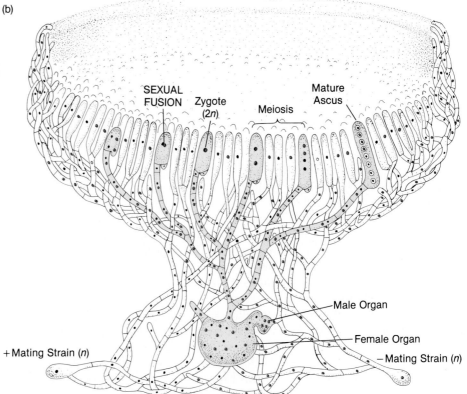

TRUFFLES AND MAMMALIAN SEX ATTRACTANTS

The ascomycete that gourmet cooks most prize is the truffle. The truffle's fruiting body is a dense, spore-containing structure. Fungal species called the *Tuberales,* which live in mycorrhizal symbiosis with the roots of trees (especially oaks and beeches), produce these fruiting bodies underground. Spores can be released only after the truffle slowly decays or a burrowing animal breaks it apart.

Most species of truffle have pungent odors and flavors that attract squirrels, deer, pigs, and other mammals. These odors and flavors range from a cheese–garlic bouquet to the smells of sewer gas and rancid bacon. Needless to say, gourmets prefer the first. Only six or seven of Europe's seventy truffle species are edible, and virtually all of North America's sixty species taste and smell terrible. However, some truffle enthusiasts are trying to grow truffles from imported spores in wooded regions of northern California and in Texas where the soils and climate are similar to those in the truffle regions of southern France and Italy. Others are transplanting small oaks, reputedly infected with truffle spores, to countries other than France.

In Europe, truffle hunters have traditionally used pigs to sniff out and dig up edible truffles. Why will pigs, with furious intensity, dig as deep as three feet to seek out truffles? Researchers have found that truffles contain a highly volatile pig sex attractant. The substance is androstenol, a steroid that is related to testosterone. Androstenol is manufactured in the testes of boars, transported in the blood to the salivary glands, and then released in the saliva during premating behavior. The truffles that humans find so delicious apparently release this steroid into the soil,

thereby attracting the female sow and ensuring that the spore case (ascocarp) will be broken open in timely fashion and the spores scattered about.

But that is not the full story. The human testes also produces androstenol, and sweat glands in the male's armpits secrete it. Studies conducted by psychologists have shown that exposure of either women or men to the musky–nutty odor of the pure compound causes them to score women in photographs as more beautiful than do people not exposed to the compound. Whether such responsiveness is related to the highly prized taste of truffles is still a matter of speculation, but it does seem possible that subconscious stimulation of the brain's sex centers may be part of our taste for, and appreciation of, truffles.

These findings about pigs, humans, and truffles relate to the general phenomenon of coevolution (Chapters 24 and 46), wherein evolutionary adaptations of one species are exploited for other purposes by another species. The natural compounds of plants and fungi may have evolved for defense or for other aspects of survival; yet those compounds may affect the behavior of certain animals to the benefit of those plants or fungi. The black truffles, for instance, are rooted out by deer in response to a still unknown compound. It probably was over millennia that these fungi gradually came to manufacture a mammalian odor signal; those fungi that happened to make the compound attracted the mammal, had their spore cases broken open as a result, and thereby, we reason, may have gained a reproductive advantage over fungi dependent solely on decay of the ascocarp to free the spores. This is yet another example of a unique solution to survival.

ascus and ascospores. Yeasts such as *Saccharomyces cerevisiae* are enormously important in human food production because of their ability to ferment sugars and release ethyl alcohol and carbon dioxide. Such yeasts are used to make wines and beer; similarly, the carbon dioxide that metabolizing yeasts give off causes bread dough to rise.

But ascomycetes can be destructive, too. The parasite *Claviceps purpurea,* which infects rye grain, causes a serious disease in humans and domesticated animals if bread is baked from contaminated grain. The fungal hyphae develop into a hard, tumorlike growth on rye called **ergot.** When ground with the rye, ergot releases a toxic chemical. Ergot poisoning is characterized by hallucinations, convulsions, uterine contractions, and tissue damage in the extremities. Historians believe that some of the "witches" executed during the late seventeenth

century in Salem, Massachusetts, may actually have been suffering from ergot poisoning—and so may have been their accusers. During the Middle Ages, thousands of Europeans died from this disease, which was called "St. Anthony's fire." Interestingly, *Claviceps purpurea* now is grown for medical purposes. In small quantities, ergot alkaloids can be used to treat migraine headaches and to induce childbirth. Ergot is also the source of the hallucinogenic drug lysergic acid diethylamide (LSD).

Ascomycetes include various *Penicillium* species, fungi that yield the antibiotic that has saved millions of human lives (Figure 22-14). In addition to producing penicillin, species of *Penicillium* are used to ripen and flavor Camembert, Roquefort, and various other blue-veined cheeses. (It is said that the Roquefort fungus was

Figure 22-13   ASCOMYCETES: THE LARGEST AND MOST DIVERSE CLASS OF FUNGI.
This common morel, *Morchella esculenta*, is typical of the class.

(a)

(b)

discovered by a boy who nibbled on some cheese that turned moldy after he left it in a cave near Roquefort, France.)

Basidiomycetes

Common mushrooms; coral fungi; puffballs; boletes, or pore mushrooms; and bracket, or shelf, fungi are members of **Basidiomycete,** the second class of higher fungi (Figure 22-15). Most basidiomycetes form visible

Figure 22-14   PENICILLIN: UGLY TO THE EYE, BUT A LIFESAVER TO THE BODY.
Pharmaceutical chemists derive the life-saving drug penicillin from the *Penicillium* mold, often seen growing on the surface of rotting citrus fruits such as this tangerine.

(c)

Figure 22-15   BASIDIOMYCETES: FANTASTIC DIVERSITY IN SHAPE AND COLOR.
All basidiomycetes are constructed of hyphae. (a) The coral fungus (*Clavaria* species), (b) the boletus (*Boletus edulis*), and (c) the shaggy mane fungus (*Coprinus comatus*) all demonstrate the many ways hyphae can be welded into different shapes.

PENICILLIN PRODUCTION FOR HUMANKIND

In 1929, Alexander Fleming noted that the growth of colonies of the bacterium *Staphylococcus* on a petri dish was inhibited in the vicinity of a mold contaminant. The mold, a *Penicillium* species, was producing a diffusible substance that had an inhibitory effect. Such was the discovery of the first antibiotic, penicillin; the same phenomenon of growth inhibition has been used to search for other antibiotics.

Commercial production of antibiotics got its start during World War II. By 1944, penicillin was available for use in saving lives on the Allied battlefronts. However, it was in the decade following the war that the greatest strides were made in improving yields so that commercial production of penicillin became feasible. In the thirty-five years after the commercial development of penicillin, biologists discovered more than 4,000 antibiotics. These antibiotics are produced mainly by fungi and by the prokaryotic actinomycetes (see Chapter 20). Annual world production is now more than 100,000 tons, and annual sales revenue in the United States is greater than $1 billion.

Production of low-cost antibiotics depends on high-yielding microbial strains that can be easily grown in large quantities. Such industrial strains have been obtained by a process of "directed evolution," a selection process the laboratory personnel carry out. They create mutations in high-producing strains isolated from natural populations by exposing the microbes to chemicals such as nitrogen mustard or by irradiating them with x-rays or ultraviolet light. The mutants are then screened for increased antibiotic production or other desired traits, such as rapid growth in liquid culture. The following table illustrates how these methods, used in the 1950s by scientists at the Northern Regional Research Laboratories (NRRL) of the United States Department of Agriculture and at the University of Wisconsin, dramatically improved penicillin yield.

The initial strain, isolated from a moldy melon, made about twice the amount of penicillin that the Fleming isolate produced. The following three strains, representing a ninefold increase in production, were naturally occurring mutants or biologists created them with irradiation. Wis BL 3-D10, although producing less penicillin than Wis Q176, from which it had been derived, did not produce the yellow pigment chrysogenin, which had contaminated the purified antibiotic preparations made from previous strains. Biologists then derived all subsequent strains from Wis BL 3-D10. Many other mutants, each with some, usually slight, improvement in attributes, followed Wis 47-1564 to yield the modern strains the pharmaceutical industry employs. These represent a total improvement in penicillin productivity of about fifty-five-fold over the Fleming strain. Similar procedures, together with genetic recombination by crossing of high-yielding strains, are still used in selecting the best microbial genotypes for producing new antibiotics for the marketplace.

Penicillium chrysogenum Strain Number	Origin	Penicillin (units/ml submerged culture)
NRRL 1951	Moldy cantaloupe	100
NRRL 1951-B25	Natural variant of previous strain	250
NRRL 1951-B25-X1612	X-ray mutant of previous strain	500
Wis Q176	Ultraviolet mutant of previous strain	900
Wis BL 3-D10	Ultraviolet mutant of previous strain	675
Wis 47-1564	Natural variant of previous strain	900
Modern strains	Chemical mutagens	>3,000

fruiting bodies. The familiar mushroom is the conspicuous fruiting body of a huge hyphal mass that penetrates the soil or another substratum. Some basidiomycetes, such as the rusts and smuts that parasitize plants, do not form conspicuous fruiting bodies.

The basidiomycetes differ from the ascomycetes because of their **basidiocarp** (the "mushroom" we see on a forest floor) and because of their more lengthy dikaryotic stage in the life cycle. The basidiocarp of gilled mushrooms is actually a dense mass of dikaryotic hyphae.

(a)

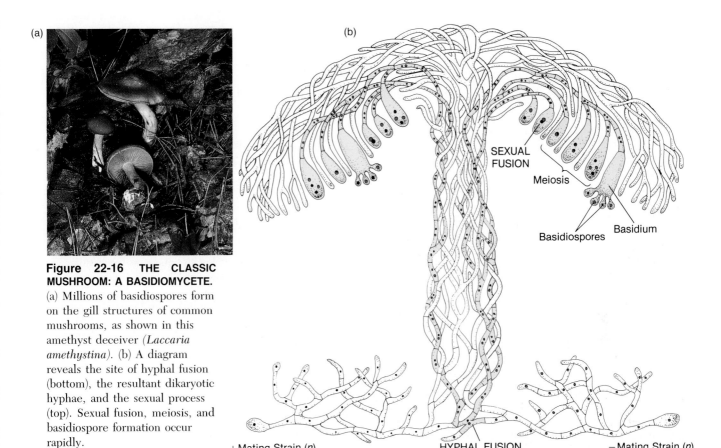

(b)

SEXUAL
FUSION

Meiosis

Basidiospores   Basidium

+ Mating Strain (*n*)   HYPHAL FUSION   − Mating Strain (*n*)

Figure 22-16 THE CLASSIC MUSHROOM: A BASIDIOMYCETE.
(a) Millions of basidiospores form on the gill structures of common mushrooms, as shown in this amethyst deceiver *(Laccaria amethystina)*. (b) A diagram reveals the site of hyphal fusion (bottom), the resultant dikaryotic hyphae, and the sexual process (top). Sexual fusion, meiosis, and basidiospore formation occur rapidly.

Club-shaped **basidia** line the surfaces of the gills on the underside of the mushroom cap (Figure 22-16). In most species, each basidium bears four *basidiospores* on the tips of spikelike processes (Figure 22-17). Each basidiospore is haploid and results from meiotic divisions of a diploid cell at the tip of the basidium.

The life cycle of a typical basidiomycete is considerably more complex than are the life cycles of members of the fungal classes we have already discussed. The stages of this typical life cycle, for a gilled mushroom, are shown in Figure 22-18.

When a basidiospore is blown or carried to a suitable substratum, it germinates to produce a haploid mycelium. This mass cannot undergo sexual reproduction until it contacts another haploid mycelium of the same species but different mating type. When such contact

Figure 22-17   BASIDIOSPORES FROM A COMMON MUSHROOM.
Note the four basidiospores "hanging" from the basidium at the left center of this scanning electron micrograph (magnified about 2,200 times). Above and to the right is a basidium that has already shed its four basidiospores.

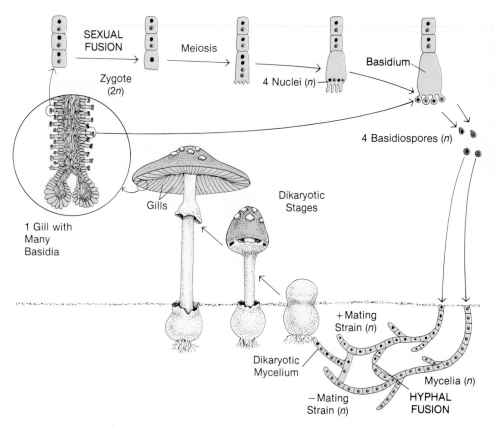

Figure 22-18   LIFE CYCLE OF A MUSHROOM, A TYPICAL BASIDIOMYCETE.
The aboveground portion of a basidiomycete is the dikaryotic reproductive structure; below ground lies a mat of mycelia. Haploid mycelia of different mating strains make contact and fuse, giving rise to a "fusion hypha." This grows rapidly, producing the aboveground structure or basidiocarp, housing basidia. In each cell of the growing fusion hypha, the two parental nuclei remain separated, but the nuclei in the basidia eventually fuse, undergo meiosis, and basidiospores form. These are dispersed by wind or rain; they then germinate and, once again, they produce underground mycelia.

does occur, the hyphae fuse, but nuclear fusion and meiosis do not immediately follow. Instead, the "fusion hypha" grows for a prolonged period and forms a new mycelium. Nuclei from the two mating strains remain separate but in close proximity, and each new cell that forms during the growth of the hyphae is dikaryotic. Only in the basidia do these haploid nuclei fuse to form diploid nuclei. Once fusion has occurred, meiosis follows, and the haploid basidiospores are produced.

Many mushroom species grow unseen for several years as huge dikaryotic mycelia in soil or wood until environmental conditions foster the development of fruiting bodies and basidiospores. Such environmental requirements as specific levels of moisture, hospitable temperature, adequate carbon dioxide, and minimal light need to be met if fruiting bodies are to be induced. Once induced, growth of the mushroom can occur very quickly. In the spring, for example, a lawnful of mushrooms can appear overnight. For some basidiomycetes, however, this growth takes several weeks and requires different environmental conditions from those that brought about induction of the fruiting body stage. For example, many wood- and dung-inhabiting fungi require an absence of light for the induction process, but depend on light for fruiting body maturation.

Only a few basidiomycetes reproduce asexually through dispersal of conidia. Huge numbers of sexually derived basidiospores are usually produced when growing conditions are optimal and germinate quickly when they land on suitable food material. Mycelia also become established quickly in wood, soil, or another substratum before unfavorable conditions set in.

Many of the large gilled mushrooms and the boletuses seen on the floor of forests and in other wooded areas grow in mycorrhizal association with forest trees. One species of puffball that grows well in poor soils is inoculated around the roots of nursery tree stocks that are used to reforest regions that have poor soil.

Deuteromycetes (Fungi Imperfecti)

Mycologists have described almost 25,000 fungal species that appear to lack modes of sexual reproduction. For this reason, they do not fit into any of the classes that we have described, so they are classified together as **Deuteromycetes,** or **Fungi Imperfecti.** It is a noteworthy group on several counts, one of which is that it includes many of the predatory fungi.

Most of the deuteromycetes are known to reproduce asexually by means of conidia. Thus they are related to the ascomycetes and may have lost the ability to reproduce sexually during their evolution. A few species may have a sexual phase under certain environmental conditions never duplicated in the laboratory. When a sexual stage is discovered in the life cycle of one of these species, it is usually reclassified as an ascomycete genus or, more rarely, as a basidiomycete.

Many of the most familiar and commercially important molds are deuteromycetes, including species of *Aspergillus*. *Aspergillus niger* is used to produce citric acid, which is used commercially in enormous quantities to make soft drinks, jams, jellies, salad dressings, and many other foods. *A. tamarii* and *A. oryzae* are used to make soy sauce, sake, and a number of other fermented products. Enzymes produced by cultured fungi are used to make corn syrups and high-fructose sugar syrups and to clarify fruit juices and wines.

One dangerous fungus in this genus is *Aspergillus flavus*, which produces an extremely potent compound, *aflatoxin*. Aflatoxin was discovered in the 1960s in moldy peanut meal that killed thousands of cattle and poultry in England. Stored peanuts are particularly susceptible to *Aspergillus* infection if they are not dried properly, but this fungus also attacks corn, millet, rice, and many other grains and seeds. The United States government sets stringent limits on the amounts of aflatoxin that can be present in human food or livestock feed because it is one of the most potent carcinogens known. A number of other *Aspergillus* species can cause respiratory and other diseases in humans and animals, particularly birds.

Other deuteromycetes are predators in a sense: they grow lassolike hyphae with which they can ensnare and digest tiny worms. This behavior does not depend on any sort of nervous system. Instead, hyphal cells built in rings are stimulated by the touch of the soil worm so that an extremely rapid (0.1 second) osmotic change causes the cells to expand and trap the worm. Other hyphae soon grow into the prey and secrete digestive enzymes. Some fungi use sticky substances to trap a variety of small aquatic or terrestrial animals. Thus not only animals and carnivorous plants evolved predatory strategies to obtain food. Although we do not usually think of fungi as having differentiated cells, the lasso and the glue-secreting hyphae indicate that "special" cell functions are possible in fungi.

LICHENS: THE ULTIMATE SYMBIONTS

The gray, orange, and other colored encrustations we commonly see on rocks and trees are actually alive—they are **lichens,** composite organisms built from one species of fungus and one or two species of algae (Figure 22-19). About 90 percent of a lichen mass is fungal, and 10 percent is algal. Lichens generally grow in three

forms: *crustose* forms are flat and cling tightly to the substratum; *fruticose* lichens grow upright or hang from a branch or rock; and *foliose* types look like leaves, having lobes that can be lifted from the rock, tree bark, or other surface on which they grow. Just below the surface of the lichen thallus, whatever form it assumes, the algal species grows as a thin layer of single cells, entwined by the fungal hyphae. The discovery that lichens are usually among the first successful inhabitants of newly exposed marine rocks suggests that lichens may have been an early invader of the Earth's land areas. Today, an important ecological role of lichens is to break down rocks, thereby starting the process leading to soil production.

There are some 18,000 lichen species, most of which contain their own unique fungal species—usually ascomycetes but sometimes deuteromycetes or basidiomycetes. Only about 30 species of algae are found as part of these thousands of lichens, however. The most common algal components are members of the eukaryotic green-

Figure 22-19   LICHENS.
Lichens are composite organisms formed from the close association between fungal hyphae and algae. (a) Crustose lichens are common sights on rocks or trees in areas free of heavy air pollution. (b) Fruticose lichens can resemble greenish beards hanging from limbs or ledges.

(a)

(b)

algal genera *Trebouxia* and *Trentepohlia* or are prokaryotic cyanobacteria of the genus *Nostoc*. The difficulty in classifying lichens can be appreciated if you consider that a single fungal species will produce radically different lichen morphologies when paired with different algae. In the past, such varied forms were even defined as different genera.

Studies using radioactive tracer molecules such as $^{14}CO_2$ have shown that the algal partner provides the lichen with organic compounds (sugars), the products of photosynthesis. Furthermore, lichens that contain *Nostoc* cyanobacterial species can fix atmospheric nitrogen (N_2) in addition to carbon dioxide. Thus such lichens can grow while taking up almost no nutrients from the substratum. In fact, the majority of lichens derive most of their necessary water and minerals from rain water and air, allowing them to live where few other fungi could survive, including on bare rocks, barren soils, and tree trunks and in regions ranging from Arctic tundra to sunbaked desert. Moreover, lichens can literally "dry out," surviving with only 2 to 10 percent water by weight (compared to the 70 to 90 percent water by weight in most organisms). Such desiccated lichens cease photosynthesis and enter a resting condition, which can be broken in minutes as the lichen rapidly imbibes water after a rainstorm.

Lichen reproduction is not well understood. One hypothesis is that fungal ascospores are produced and dispersed. As mycelia sprout from germinating ascospores, nearby algal or cyanobacterial cells may be captured on contact to create the new composite organism. Asexual reproduction, on the other hand, is common among lichens and involves fragmentation of the thallus: specialized, easily detachable fragments or powdery bits of thallus, containing both the fungal and the algal components, are broken off and are carried by wind or splashing raindrops to a new site of attachment. Once in place, most lichens grow very slowly—from 0.1 to 10 millimeters per year in diameter. The low, relatively constant rates of lichen growth have allowed them to be used for dating rocks and human artifacts. The giant carved statues on Easter Island in the Pacific, for instance, have been estimated to be 400 years old on the basis of lichen size and growth.

Mycologists have long debated whether lichens represent an association beneficial to both fungus and alga or whether the fungus is simply a parasite on the alga. After all, the fungus does not appear to provide the alga with any essential nutrients. However, there are many environments in which the alga could not survive as a free-living organism, such as on rock faces exposed to high and low temperatures and extreme light intensities. Covered by fungal hyphae containing light-absorbing compounds, the algal cells are protected from excessive sunlight and probably also from drying out. Thus it has been argued that the algal component of lichen benefits from the association by being able to inhabit an ecological niche it otherwise could not exploit. But the fact that many algal or cyanobacterial partners in lichens grow better when alone suggests that in some sense, the fungus is a parasite on its resident partners.

Lichens have a remarkable ability to absorb necessary inorganic nutrients. But their efficiency at acquiring elements such as sulfur and phosphorus—an ability that enhances survival on almost inert substrata—can also be devastating to lichens. For example, the centers of industrialized cities are "lichen deserts," particularly when sulfur dioxide is present in the atmosphere. Some foliose lichen species can accumulate more than 1,000 times the concentration of sulfur present in the atmosphere in just a few months' time. The large quantity of sulfur dioxide is dissolved in the lichen's cytoplasm and forms sulfuric acid, which can damage or kill the lichen. Thus the general appearance and condition of sensitive lichen species ("indicator species") growing at various distances from sources of sulfur dioxide emission can serve as living measures of air-pollution levels.

FUNGAL EVOLUTION

Where did the various fungal groups come from? The most likely possibility is that unicellular eukaryotes without chloroplasts gave rise to several lineages now grouped together as Mycota. The oldest fossil filaments are Precambrian. If they are fungi, and not algae, then fungi must be among the earliest of all eukaryotic organisms. The various fungi may in fact have arisen independently from prokaryotes. The early progenitor cells probably became coenocytic, thereby allowing larger cytoplasmic systems—hyphae—to evolve. Finally, in the higher fungi, the nuclei became at least partially walled off as more complex structures and ways of life evolved. The absence of basal bodies, centrioles, and microtubular flagella from the zygo–asco–basidio complex sets these fungi off from the flagellated lower forms. This suggests a common ancestor for the former complex, and, in fact, that ancestor may also have given rise to red algae (Chapter 23). And it suggests that these higher fungi did not arise from the lower ones, but are independent descendants of prokaryotes. Finally, the presence of the dikaryotic state during their life cycles links ascomycetes and basidiomycetes as close relatives. But throughout this lengthy odyssey, the dependence on organic foods produced by plants, animals, or algae remained a dominant feature of the highly successful fungal life styles. In Chapter 23, we turn to the primary producers of food on Earth—the green plants.

SUMMARY

1. The 175,000 species in the kingdom Fungi carry out *extracellular digestion* by secreting enzymes that digest organic matter, then absorbing the resulting nutrients derived from the food source. Fungi are either *saprobes*, which decompose dead organic matter; parasites, which derive nutrients from living hosts; or symbionts with the roots of higher plants or algal cells.

2. Most fungi are composed of filamentous *hyphae*, whose cell walls contain chitin or, in some species, cellulose. Hyphae may be specialized into structures such as *rhizoids*, which anchor the fungus to a substratum; *haustoria*, or feeding structures; and *sporangia*, in which spores are formed.

3. Hyphae may be *coenocytic* (multinucleate) or *septate* (having cross walls between nuclei). A filamentous network or dense packing of hyphae forms the *mycelium*. If hyphae of two fungi of the same species fuse, *heterokaryons* may form in which genetically different nuclei are found in the same cytoplasm.

4. Fungi produce asexual and sexual spores that either disperse the species or enable it to survive unfavorable environmental periods.

5. The fungi are in one division—*Mycota*—which is divided into six classes. The lower fungi consist of the classes *Chytridomycete*, *Oomycete*, and *Zygomycete*; members of the first two classes are often considered to be protists rather than fungi. The higher fungi are in the classes *Ascomycete* and *Basidiomycete*. The sixth class—*Deuteromycete*, or *Fungi Imperfecti*—is made up of fungi that cannot be placed in the other classes because sexual reproduction has not been observed in them.

6. The chytridomycetes are aquatic or parasitic organisms characterized by the formation of two types of motile cells: asexual spores and gametes, both propelled by a single whiplike flagellum. Most chytrids are unicellular and may be the most ancient fungi.

7. The oomycetes range from single-celled aquatic species to multicellular terrestrial plant parasites. In many ways, they are more like algae than like other fungi: their cell walls usually contain cellulose rather than chitin; they produce motile asexual zoospores; their nuclei are diploid through most of the life cycle; and sexual reproduction involves large, immobile egg cells that fuse with male protoplasts within *gametangia*.

8. The zygomycetes are terrestrial saprobes or parasites. As do the chytrids and oomycetes, they form asexual spores in a sporangium, and they are coenocytic. Sexual reproduction involves the formation of gametangia on the tips of the hyphae after two mating strains contact each other. Their haploid gametes fuse to form a zygospore—a diploid survival spore that germinates to produce a sporangium that, in turn, releases haploid spores.

9. *Mycorrhizae* are associations of zygomycetes or basidiomycetes with plant roots. Mycorrhizae greatly expand a plant's surface area for absorbing water and nutrients and enable it to grow faster in poor soils than it could without the fungus.

10. The ascomycetes, which include many serious plant parasites as well as the cup fungi, truffles, and yeasts, reproduce asexually by means of spores called *conidia*, which develop on the tips of *aerial hyphae* rather than in sporangia. Sexually produced spores—ascospores—form in sacs called *asci*. The asci usually are surrounded and protected by a *fruiting body*, such as a cup-shaped container.

11. The basidiomycetes include the familiar gilled mushrooms, puffballs, and bracket fungi. Sexual spores—basidiospores—are borne on club-shaped structures called *basidia*. The fruiting body, or *basidiocarp*, is constructed of *dikaryotic* hyphae (as is the mycelium), in which each cell contains two nuclei, each of which arose from a different parent. Only in the basidia do the haploid nuclei fuse, so that haploid basidiospores may be produced meiotically.

12. The deuteromycetes, which have no known sexual cycle, include many commercially important genera and some predatory fungi.

13. *Lichens* are composite organisms, each of which is built of one species of fungus (usually an ascomycete) and one of some 30 species of algae or cyanobacteria. The alga provides products of photosynthesis, and probably is protected by the fungal hyphae from adverse environmental conditions. Lichens absorb water, minerals, and sometimes nitrogen to provide the needs of both fungus and alga.

14. The various fungal groups may have arisen independently from prokaryotes. The zygomycetes and various higher fungi seem most closely related, but cannot be traced to lower fungal ancestors.

KEY TERMS

aerial hypha	dikaryon	lichen
Ascomycete	ergot	mycelium
ascus	extracellular digestion	mycorrhiza
basidiocarp	fruiting body	Mycota
Basidiomycete	Fungi Imperfecti	Oomycete
basidium	gametangium	rhizoid
Chytridiomycete	haustorium	saprobe
coenocytic	heterokaryon	septate
conidium	heterokaryosis	sporangium
Deuteromycete	hypha	Zygomycete

QUESTIONS

1. Why are the fungi called the great decomposers? What features contribute to their success?

2. Fungi used to be classified as plants. Can you think of any ways in which they resemble plants and not animals?

3. In some ways, the fungi resemble animals and not plants. What are they?

4. The main "body" of a fungus is usually hidden underground or within living or decaying organisms. What is the structure of this body? What is its function?

5. How do heterokaryons arise?

6. Fungi are nonmotile, yet their spores can disperse over thousands of miles. How do they accomplish this remarkable feat?

7. Fungi can produce spores of two types that have two very different functions. What are these functions? Is either of these functions analogous to the function of bacterial endospores discussed in Chapter 20?

8. What features distinguish the lower from the higher fungi?

9. What are the six classes of fungi? Which of the following features are found in each class?
a. Motile gametes
b. Motile zoospores
c. Cellulose cell wall
d. Septate hyphae
e. Sexual reproduction
f. Mycorrhizal association with plant roots.

10. What are some human diseases caused by fungi? some plant diseases? What are some fungi that we use in food production? That we eat directly? What are some human drugs made by fungi? Some human poisons?

ESSAY QUESTIONS

1. Animals often produce sex pheromones to attract a suitable mate of the same species. We have learned about a novel use of a sex pheromone. Describe how such a "use" might have evolved.

2. Do lichens represent a symbiotic relationship or a parasitic one? Give evidence to support your point of view.

3. Discuss some of the features of mycorrhizae and of lichens. How have these features enabled organisms to live in hostile environments?

SUGGESTED READINGS

ALEXOPOULOS, C. J., and C. W. MIMS. *Introductory Mycology.* 3d ed. New York: Wiley, 1979.

This well-illustrated book is an excellent broad treatment of the fungi.

CHRISTENSEN, C. M. *The Molds and Man: An Introduction to the Fungi.* 3d ed. Minneapolis: University of Minnesota Press, 1965.

This is an easy-to-read account of the fungi, which affect human life in so many ways.

RAY, P. M., T. STEEVES, and S. A. FULTZ. *Botany.* Philadelphia: Saunders, 1983.

The discussion of fungi in this excellent introduction to botany is a good supplement to the discussion in this chapter.

23

THE PLANT KINGDOM: ALGAE AND LOWER LAND PLANTS

The migration of plants from the seas to the dry land surfaces of the earth more than 400 million years ago ranks among the most important events in the history of the biological world.

Sara A. Fultz, in *Botany* (1983)

The sea surface is nearly covered by fronds of giant kelp off the central California coast. A sea otter, feeding on a sea urchin, finds protection and food in kelp forests.

While walking along a quiet, shady forest trail, most people will notice the tall, stately trees, the wild flowers by the trail's edge, and perhaps a familiar bush or herb. Far less often, however, will they notice the "lower plants"—the clumps of mosses, leafy liverworts, ground pines, horsetails, and ferns—that grow on damp slopes, moist logs, or stream banks. Limited in size and habitat, these plants seldom catch the eye or inspire a strong following. Yet they and their distant aquatic relatives, the algae, represent an evolutionary spectrum ranging from single-celled green algae to giant tree ferns and reaching back in time to the first large organisms that survived on dry land. But it is also clear that the simple lower plants gave rise to the seed plants, including the conifers and the flowering plants—the "higher plants," which today cover vast stretches of the Earth's surface. Growing in towering forests, lush meadows, and cultivated fields, the higher plants now dominate the very habitats where lower plants were once the conspicuous denizens.

In this chapter, we survey the major characteristics of all plants and the main branches of the plant kingdom. We first examine the characteristics and life cycle of a typical plant, and then focus on the major types of lower plants, one by one, considering their life habits and reproductive modes. As we go along, we shall see how the transition from life in the oceans and fresh waters to life on land presented survival challenges to the early plants and how each group evolved mechanisms to cope with these environmental exigencies. Then in Chapter 24, we examine the seed plants, the descendants of successful invaders of the terrestrial world.

THE PLANT KINGDOM AND ITS MAJOR DIVISIONS

Visitors to a park may spot twenty or thirty kinds of mosses, grasses, seasonal flowers, bushes, and trees—an impressive array of species, all in all, but only a tiny sample of the huge and diverse kingdom Plantae. Because of the diversity of its members, the plant kingdom is far from easy to define. Plants usually are multicellular organisms; most contain chlorophyll and thus produce their own organic nutrients; they are usually stationary; and most have alternating haploid and diploid phases in the life cycle. Despite these general characteristics, some organisms classified as plants are single-celled; some lack chlorophyll; some are motile; and some show no alternation of phases.

Clearly, the diversity of the plant kingdom is formidable. Its members include some of the smallest eukaryotes (unicellular algae) and the largest organisms ever to inhabit our planet (giant sequoias). They grow in habitats as different from one another as the Atlantic Ocean, the

Mojave Desert, the Amazonian rain forest, and the Arctic tundra; and range from extremely simple individuals to highly complex ones, with differentiated tissues and specialized organs.

Botanists have identified about 300,000 living plant species. Of those, more than 80 percent are the familiar flowering plants, which include all the major crop and hardwood-tree species and most ornamental plants; less than 5 percent are lower plants, while about 15 percent are pines and their relatives. Table 23-1 lists the major divisions in the plant kingdom and the approximate number of species in each. (Notice that in the plant kingdom, the term "division" is used instead of "phylum.") The divisions Chlorophyta, Rhodophyta, and Phaeophyta are referred to as **algae,** primarily aquatic plants, and the remaining divisions are considered land plants. The earliest plants on Earth were probably simple green algae, and many biologists believe that their descendants became the first groups of land plants; these, in turn, radiated into the rest of the diverse plant kingdom. Some contemporary aquatic plants are not algae, but are thought to be descendants of land plants that evolved once again the capacity to live in water, so we include them in the "land plant" category.

The green, red, and brown algae are different enough from one another and from other plants that they are usually classified in three divisions: Chlorophyta, Rhodophyta, and Phaeophyta. Some alternative classification schemes include these three divisions in the kingdom Protista—along with the golden-brown algae, dinoflagellates, diatoms, and euglenoids—because some algae are single-celled.

Simple land plants that lack an internal system for conducting water and nutrients—a *vascular system*—are often placed in a single division (called Bryophyta) that includes the liverworts, hornworts, and mosses. These so-called *nonvascular plants* share similar life cycles but are physically quite different from one another, so our classification system places them in three classes: Hepatophyta (liverworts), Anthocerotophyta (hornworts), and Musci (mosses). Each member is a small, low-growing plant that is limited largely to moist environments and that lack roots, leaves, and a true internal transport system.

All the remaining groups are *vascular plants*—plants with a specialized vascular system for transporting water and nutrients. They probably arose from algae independently of the nonvascular plants. Most vascular plants have true roots and leaves. The most familiar and economically important are the ferns, conifers, and flowering plants. Older classification systems generally placed all the vascular plants in one division (Tracheophyta), but since recent research has revealed significant differences among them, our system assigns them to nine divisions. Most of the plants in the first four divisions are rather unfamiliar plants that live in moist environments,

Table 23-1   **MAJOR GROUPS OF THE PLANT KINGDOM***

Number of Living Species	Division/Class				
7,000	Chlorophyta (green algae)				Algae
4,000	Rhodophyta (red algae)				
1,500	Phaeophyta (brown algae)				
22,300	Bryophyta			Nonvascular Plants	Land Plants
8,000	Hepatophyta (liverworts)				
300	Anthocerotophyta (hornworts)				
14,000	Musci (mosses)				
13	Psilophyta (whisk ferns)		Seedless Plants	Vascular Plants	
1,000	Lycophyta (club mosses)				
25	Sphenophyta (horsetails)				
10,000	Pteridophyta (ferns)				
100	Cycadophyta	Gymnosperms	Seed Plants		
1	Ginkgophyta				
3 (genera)	Gnetophyta				
550	Coniferophyta				
250,000	Anthophyta	Angiosperms			

*Certain classification schemes other than the one we follow in this book categorize the divisions differently. The division Bryophyta includes the classes Hepatophyta, Anthocerotophyta, and Musci.

and none of them produces seeds: Psilophyta (whisk ferns), Lycophyta (club mosses and quillworts), Spenophyta (horsetails), and Pteridophyta (ferns). The vascular land plants in the other five divisions that *do* produce seeds are the *gymnosperms* (divisions Cycadophyta, Ginkgophyta, Gnetophyta, and Coniferophyta) and the *angiosperms* (division Anthophyta). As we shall see, seed plants dominate almost all terrestrial environments.

THE BASIC PLANT LIFE CYCLE

Despite their tremendous diversity, plants have just one type of reproductive life cycle with many variations—a life cycle that serves as a hallmark of the plant kingdom, distinguishing it from the other four kingdoms of life.

To understand that distinctive life cycle, think first about the life cycle of a human being, which is typical of many animal life cycles. As we saw in Chapter 18, each individual's unique genetic complement is determined at the moment of conception, when the nucleus from the father's haploid gamete (sperm) merges with the mother's haploid gamete (egg). The resulting diploid zygote

divides mitotically, and a baby is eventually born and grows to adulthood. In humans and many other animals, then, the haploid stage is represented solely by the unicellular gametes. Meiosis produces haploid cells from diploid ones; fertilization restores the diploid state.

In contrast, the plant life cycle is characterized by the *alternation* of a haploid phase and a diploid phase (Figure 23-1). In some plants, both haploid and diploid forms are free-living green organisms of substantial size. During a plant's haploid phase, gametes (eggs, sperm, or analogous structures) are formed; thus the haploid organism is called the **gametophyte,** or gamete-producing plant. During the diploid phase, spores are formed; thus the diploid plant is called the **sporophyte,** or spore-producing plant. In ferns, for example, both the gametophyte and the sporophyte phases are independent green plants. In other plants, one phase dominates and becomes the recognizable plant, while the other is reduced and inconspicuous and depends on the dominant one for moisture, nutrition, and physical support. Interestingly, in some plants, such as mosses, the haploid gametophyte is the dominant phase. In other words, the plant we see has but a single set of chromosomes in each cell. In most land plants, though, the diploid sporophyte is the dominant phase. Since haploid and diploid phases are often referred to as generations, the alternating life cycle of plants is called the **alternation of generations.**

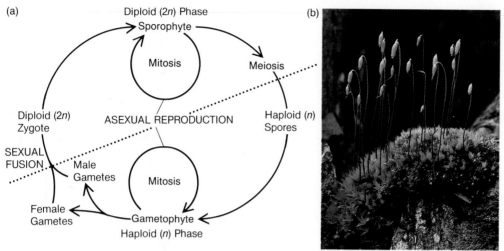

Figure 23-1   GENERALIZED PLANT LIFE CYCLE: ALTERNATION OF GENERATIONS.
(a) Diploid and haploid generations alternate, separated by the events of meiosis and of sexual fusion (the union of gametes). Plant life cycles differ mainly in the length of the two phases; for instance, the sporophyte stage can be very short. Both the sporophyte and gametophyte can reproduce asexually through mitotic divisions. (b) The small, green, leafy structures are moss (*Bryum capillare*) gametophytes. They produced the sperm and eggs that gave rise to the erect yellowish-green sporophytes.

Figure 23-2 shows the life cycle of a moss, beginning with a haploid spore that divides mitotically, to form a filamentous *protonema*, which grows into a leafy moss plant—the haploid gametophyte. Each moss gametophyte produces either eggs or sperm in a specialized structure at the tip of the leaf. Sperm from one plant swim to eggs of another through a film of water, and fertilization takes place, resulting in the diploid zygote. The zygote divides mitotically, developing into the diploid sporophyte. Through meiosis, the mature sporophyte produces a new generation of haploid spores to begin the cycle anew.

Comparing human and moss life cycles, one can see that whereas the adult human is diploid and produces short-lived, haploid gametes that remain as single cells and do not grow into haploid organisms, the diploid plant zygote grows into a diploid sporophyte, which produces haploid spores that *do* develop into haploid organisms—the gametophytes. In a sense, both sporophyte and gametophyte are "adults," one forming asexual reproductive cells (spores) and the other forming sexual reproductive cells (gametes). The plant life cycle differs from that of many animals in other important ways: plants can undergo direct, asexual reproduction by propagation from stems, roots, bulbs, leaves, or other specialized structures. Although a salamander can grow a new limb, a severed limb cannot grow a new salamander. Plants may also enter a resting stage to survive periods of drought, cold, or other conditions unfavorable for growth and reproduction. The basic plant life cycle, with its alternation of haploid and diploid generations, serves

as the main model for understanding the roles of gametes and spores in plant reproduction. We shall encounter variations on this botanical theme in all the plant species (except one, the brown alga *Fucus*), from unicellular algae to towering forest trees. We will discuss more details of plant reproduction and life cycles in Chapters 24 and 28.

THE ALGAE: DIVERSE AQUATIC PRODUCERS

For many people, the word "algae" conjures up an image of slimy green strands in pond water—and little else. These simple plants, however, actually display a stunning variety of shapes, sizes, and colors, and occupy an important place in the plant world as well as in the global economy of interdependent life forms. There are about 12,000 algal species. These diverse plants make up a large proportion of the Earth's biomass because huge numbers of algae can inhabit the vast expanses of ocean on our planet.

Algae are the oldest plant group, and are believed to have been the ancestors of the other aquatic and land plants. As we saw in Chapter 19, the earliest organisms on Earth, 3.5 billion years ago, were most likely anaerobic heterotrophs—single-celled prokaryotic scavengers—living in aquatic habitats. Eventually, these organisms gave rise to single-celled autotrophs that were able to

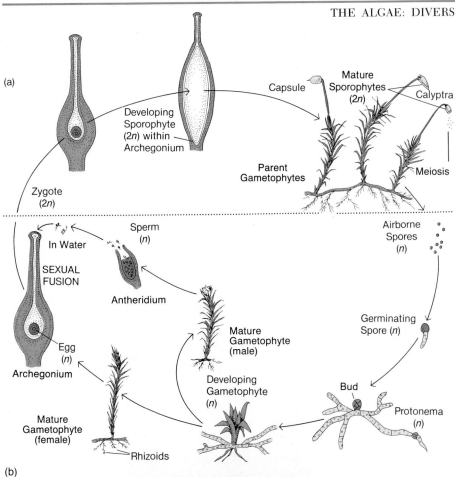

(a)

Developing Sporophyte (2n) within Archegonium

Capsule

Mature Sporophytes (2n)

Calyptra

Parent Gametophytes

Meiosis

Zygote (2n)

In Water

SEXUAL FUSION

Sperm (n)

Antheridium

Airborne Spores (n)

Egg (n)

Archegonium

Mature Gametophyte (male)

Germinating Spore (n)

Mature Gametophyte (female)

Developing Gametophyte (n)

Bud

Protonema (n)

Rhizoids

Figure 23-2   GENERALIZED LIFE CYCLE OF A MOSS.
Leafy gametophytes develop from the haploid protonema, and at the leaf tips produce antheridia (organs where sperm develop) and archegonia (organs where eggs develop). On contact with dew or raindrops, the antheridia burst and release sperm, which, if they chance to swim to the archegonia, can fertilize the eggs. Depending upon the species, the two sexes may be present on the same gametophyte. The diploid zygote remains within the archegonium and divides mitotically. The sporophyte then develops and, following meiosis, produces haploid spores. The calyptra, the protective coating surrounding the capsule, falls off, exposing the capsule, which then ruptures, dispersing the spores. (b) Germinating haploid spores (magnified about 200 times) are the roundish structures seen here; the filamentous protonema grow from the spores.

(b)

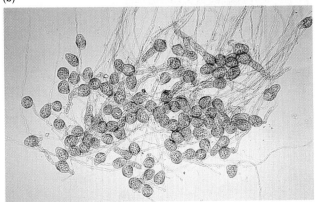

This radiation of algal species (possibly from several different original types) led to both single-celled motile species and large multicellular forms with alternating free-living gametophyte and sporophyte generations. The spectrum of modern species that we will survey here—the green, brown, and red algae that grow as tiny threads, undulating sheets, giant kelps, and tangled seaweeds—reflects this diversity.

produce their own food molecules through photosynthesis. Despite this ability, they remained simple prokaryotes, much like modern chloroxybacteria. Perhaps another billion years crept by before the emergence of eukaryotic autotrophs, with a true nucleus, chloroplasts, and other internal organelles. These photosynthetic aerobes almost certainly gave rise to the first primitive algae, probably in the late Precambrian era, about 600 million years ago.

Over many millions of years, thousands of algal species evolved with the ability to gather sunlight at different water depths, depending on what specific chlorophylls and accessory pigments they contain (Table 23-2).

Table 23-2   **CHARACTERISTICS OF THE ALGAE AND THE LAND PLANTS**

Group	Photosynthetic Pigments	Storage Products
Chlorophyta	Chlorophyll *a*, chlorophyll *b*, carotenoids	Starch
Rhodophyta	Chlorophyll *a*, chlorophyll *d* (some) phycobilins, carotenoids	Floridean starch
Phaeophyta	Chlorophyll *a*, chlorophyll *c*, carotenoids (including fucoxanthin)	Laminarin, lipids, mannitol
Land Plants	Chlorophyll *a*, chlorophyll *b*, carotenoids	Starch

Algae form an integral part of most aquatic ecosystems. (An *ecosystem* is the full set of all types of organisms living and interacting in a given environment.) Unicellular algae are primary producers in nearly all aquatic food chains. They produce nutrient molecules through photosynthesis and serve as food for a great variety of marine and fresh-water protists and animal consumers. Many algal species, especially large-bodied ones, also yield raw materials that people use as food, fuel, glues, and other products, as we shall see.

While most algae never left the water, they should not be regarded as inferior to land plants—just different in several important respects. Their solutions to the basic survival problems common to all plants are determined by the properties of their aquatic habitat. In water, the plant body has very little weight, so internal structural support is less of a problem for aquatic plants than for land plants. Procuring water is also easy: water is obviously available on all sides in an aquatic habitat, and the plant surface is not waterproofed to prevent evaporation, as it is in many land plants. This means that each cell of a multicellular alga can absorb nutrients and exchange gases directly with the water. For these reasons, few algae exhibit any differentiation of specialized tissues, despite their tremendous variation in size and shape.

We shall examine the algae in some detail in the next three sections, especially because they reveal significant evolutionary trends. We can see an interesting progression in their structures, their modes of reproduction, and the relative development of their haploid and diploid generations that helps us understand further specializations in more complex plants. No single lineage of algae shows advances in all aspects of structure and life cycle, so we will look at several lineages—particularly of the green algae. From this broad survey and the continuum of increasingly complex traits among the algae, we can get a sense of the evolutionary trends that led from these simple aquatic plants to the impressive variety of land plants.

Chlorophyta: Likely Ancestors of the Land Plants

Some ancient green algae growing in an ocean inlet or a fresh-water pond were probably the ancestors of the lineage from which all land plants evolved more than 400 million years ago. Today, there are 7,000 species of green algae in the division **Chlorophyta**—most living submerged in fresh water or exposed on moist rocks, tree trunks, or soil. A few, however, inhabit the oceans. Most green algae are small, simple plants shaped like filaments, hollow balls, or flat blades or sheets. Some marine forms, however, may range up to 8 meters in length and 25 centimeters in width. Green algae store food as starch and possess both chlorophylls *a* and *b*—

characteristics shared only with the land plants (Table 23-2) and major evidence for regarding green algae as land plant ancestors.

Siphonous Algae

Individual cells of green algae usually have a single nucleus, but some species have cells containing many nuclei in one large mass of cytoplasm. This group, the **siphonous algae,** includes plants like *Bryopsis*, which grows attached to rocks in clear, shallow water (Figure 23-3). A siphonous alga starts life as a single cell, but as it

(a)

Figure 23-3
BRYPOSIS: **A SIPHONOUS ALGA.**
(a) *Bryopsis pennata* plants growing at 1 meter underwater in Jamaica. (b) This drawing is a detail of the upper tip and shows the gametangia protruding laterally from the main, nucleus-filled portion of the alga.

(b)

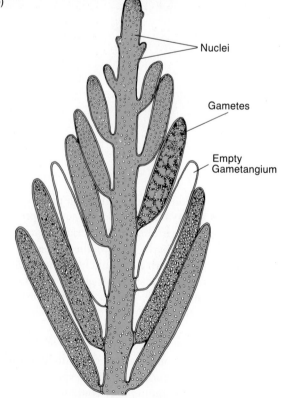

Nuclei

Gametes

Empty
Gametangium

grows, the nucleus divides and redivides, while the cytoplasm does not. The result is a large coenocytic organism. In this respect, siphonous algae resemble many species of fungi (Chapter 22). Some siphonous algae have complex structures, and a few grow to lengths of 8 meters or more. *Acetabularia*, the organism used in the historic experiments that proved that the nucleus is the site of genetic information in a eukaryotic cell (Chapter 9), is a siphonous alga.

Chlamydomonas and Other Volvocine Algae

The best-known green alga, and the best example of sexual reproduction in a single-celled organism, is *Chlamydomanas*, an oval, motile haploid alga propelled by two flagella. This common organism lives in moist soils and fresh-water pools. Typically, each cell is haploid and contains a single large cup-shaped chloroplast; sometimes an eyespot, or *stigma*, to help the cell orient toward light; and one or more spherical organelles called *pyrenoids*, which appear to help produce starch for food storage.

Chlamydomonas can reproduce asexually or sexually (Figure 23-4). During asexual reproduction, the nucleus of the haploid cell divides twice mitotically, creating four daughter cells that develop cell walls and flagella. Enzymes soften the parent's outer wall, and the daughter

cells then escape. These motile cells, or **zoospores,** mature into haploid adults. Notice that nowhere in this process does meiosis or fusion of gametes occur—this is true asexual reproduction.

Under stressful environmental conditions, such as cold, drought, or lack of nutrients, *Chlamydomonas* can reproduce sexually. The haploid parent cells are gametophytes; they undergo mitotic division to produce several **isogametes,** gametes of two mating types of identical size and appearance. Sexual fusion of two isogametes yields a diploid zygote, or **zygospore.** That zygospore develops a thick protective wall and sinks to the bottom of a pond. The mature zygospore is the sporophyte of *Chlamydomonas*, but unlike the diploid generation in many plants, it remains as a single cell. When favorable environmental conditions return, the zygospore divides meiotically to produce four haploid spores that develop into adult, unicellular gametophytes, completing the life cycle. As we shall see, many plants, like *Chlamydomonas*, have evolved resting stages that enable them to survive periods when conditions are unfavorable to growth. Usually, however, it is not the sporophyte itself but the spores or seeds that withstand stressful conditions.

Chlamydomonas is part of a lineage of algae in which individual cells sometimes function together in colonies of from 4 to 50,000 cells. The largest such colonies are *Volvox*, a bright green cluster that spins in the water and

Figure 23-4   LIFE CYCLE OF *CHLAMYDOMONAS.*

Chlamydomonas has a short diploid, sporophyte stage following formation of the zygote. Meiosis in the zygospore gives rise to four haploid spores which can divide to form isogametes; these then participate in sexual reproduction. Alternatively, the haploid spores, the equivalent of gametophytes, may reproduce asexually by repeated mitoses and generate large numbers of progeny. Any of the huge number of asexual haploid cells can enter the sexual cycle.

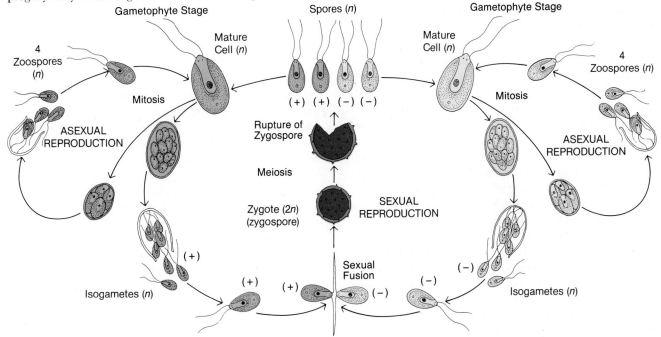

(a)

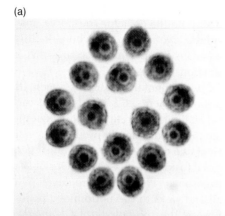

Figure 23-5   COLONIAL ALGAE.

Members of the volvocine line of colonial algae are made up of individual *Chlamydomonas*-like cells connected to each other by cytoplasmic strands. Cell number in each colonial organism varies with species. (a) *Gonium*, the simplest volvocine colony, has between 4 and 32 cells; this one, with 16 cells, is magnified about 3,000 times. (b) *Volvox*, the most complex member of this evolutionary line, has many thousands of cells. Each *Volvox* colony is a sphere with a single layer of cells embedded in a gelatinous matrix. As seen here, new daughter colonies form inside the hollow sphere as a parent cell divides mitotically (magnified about 15 times).

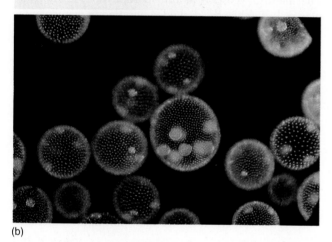

(b)

sparkles like the facets of an emerald. In fact, all colonial green algae are called the **volvocine algae** after this jewellike colony (Figure 23-5). In modern volvocine algae, increasing colony size is accompanied by an ever-greater degree of cell coordination, interdependence, and specialization and by progressively more complex types of reproduction. This trend toward greater specialization of parts and processes occurs throughout plant evolution. In many plant groups, an increase in size is accompanied by a greater division of labor—differentiated tissues with specific roles that act in concert.

Various kinds of volvocine algae demonstrate the relationship between size and division of labor—a feature of virtually all land plants and animals. *Gonium*, the simplest colonial member of the volvocine group, is made up of just four cells (or multiples of four up to thirty-two) surrounded by a sticky envelope constructed of various polysaccharides (Figure 23-5a). As in *Chlamydomonas*, each cell has two flagella that beat in unison, thus allowing the colony to swim. Each *Gonium* cell can reproduce daughter colonies asexually or can form isogametes. A slightly larger and more complex colony is *Pandorina*, a hollow sphere of 8, 16, or 32 cells that has some characteristics not seen in *Gonium* colonies: definite anterior and posterior poles; interdependent cells that cannot

survive separately; and gametes of different sizes—**anisogametes**—produced during sexual reproduction. *Pleodorina* is a still larger colony (32, 64, or 128 cells) featuring distinct cellular specialization; only the large posterior cells take a regular part in reproduction. Some species of this genus exhibit **oogamy**—the development of true **oocytes**, or large, immobile egg cells—in addition to producing swimming male gametes. This level of specialization contrasts with the more primitive **isogamy** of most algal species, in which gametes of the same size are produced and no true egg is present. Oogamy, which emerges in green algae, is also the common type of gamete production in land plants. By far the most complex green alga is *Volvox* (Figure 23-5b). Its large colonies have from 500 to 50,000 cells connected by tiny strands of cytoplasm, and only a few of the cells are exceptionally big and are specialized as oocytes. Like the other colonial algae, it can reproduce asexually, with hollow spherical daughter colonies enlarging mitotically within the central cavity of the parent colony. Such daughter colonies escape through tears in the wall of the parent volvox.

Filamentous Algae

Just as the volvocine line shows a progression in size and reproductive specialization, another line of green algae, the **filamentous algae,** shows specialization in body form and in the alternation of generations. One common type is *Ulothrix*, a genus of mostly threadlike fresh-water algae (Figure 23-6). The main plant body, the haploid gametophyte, is multicellular in *Ulothrix*, and it has a specialized **holdfast** cell, which attaches it to a substratum. The threadlike chains of cells often reproduce asexually; certain cells act as sporangia that produce flagellated zoospores by mitotic division. These haploid spores disperse, settle down, form a holdfast, and divide mitotically to develop into new multicellular haploid gametophytes. Adverse conditions can trigger sexual reproduction, during which one or more cells of the *Ulothrix* filament can act as a gametangium that produces flagellated isogametes by mitotic division. The

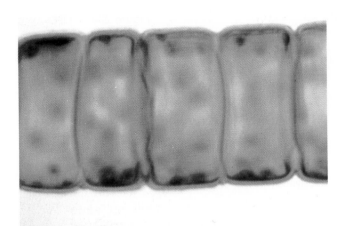

Figure 23-6 **FILAMENTOUS GREEN ALGA *ULOTHRIX*.**
This multicellular alga (magnified about 3,600 times) consists of single nucleated cells joined end to end.

isogametes are released, pair off, fuse (as in *Chlamydomonas*) to form a diploid zygote, and develop into a diploid resting cell, the sporophyte. When conditions improve, the sporophyte undergoes meiosis to produce four haploid zoospores that settle onto a substratum and develop into new multicellular gametophytes, completing the cycle.

A filamentous green alga with an even more complex body type is sea lettuce, or *Ulva*, a delicate ocean-dwelling alga common in tidal pools (Figure 23-7). *Ulva* is the first type of algae we discuss in which both gametophyte and sporophyte are multicellular, as they are in the land plants. Each generation is a conspicuous, free-living organism—a nearly identical flat sheet of cells, a *thallus*, that undulates gracefully in the ocean's swells. Cells of the gametophyte generation can produce zoospores

Figure 23-7 **THE SEA LETTUCE, *ULVA*.**
The mature sporophyte and gametophyte stages of this *Ulva lectuca* are virtually indistinguishable.

asexually (Figure 23-8). These zoospores are released and divide mitotically to form new haploid thalli. Cells of the gametophyte may also produce isogametes with two flagella; the zygote that results from the union of these isogametes divides mitotically to form a sporophyte thallus that sways with the ocean currents and is nearly indistinguishable from the gametophyte. Individual cells of the sporophyte can undergo meiosis to produce four zoospores a piece. These zoospores are released and develop into haploid multicellular gametophytes, starting the cycle over again.

This trait of alternating multicellular generations, as well as the production of chlorophylls *a* and *b* and other characteristics, leads botanists to believe that green algae of unknown types almost certainly gave rise to all the land plants. For algae to have made the transition from life in water to life on dry land, various modifications were necessary for resistance to drying and for internal support. We will see what those modifications were after we discuss the "cousins" of the green algae: the red and brown algae.

Rhodophyta: Photosynthesizers in Shallow and Deep Waters

Most of the 4,000 or so species of red algae in the division **Rhodophyta** are small, delicate marine plants that live in shallow or deep tropical waters. But several genera inhabit fresh-water streams, lakes, and springs, and one genus even thrives in Antarctic waters. A few red algae occur as single cells or as variously shaped colonies, but most are made up of many cells connected end to end in thin filaments (Figure 23-9) or side by side in thin, flat, leaflike sheets.

Red algae have a variety of pigments that absorb well all wavelengths of visible light. These pigments, called **phycobilins,** include the red *phycoerythrin* and the blue *phycocyanin*. They give the plants their characteristic pink to red-black hues and enable them to capture the green and blue wavelengths of light, which filter through deep water. Thus various species can live in waters more than 150 meters deep—five times the depth that the deepest-growing green or brown algae can tolerate. Red algae contain chlorophyll *a* and *carotenoids*, and some also contain chlorophyll *d* rather than chlorophyll *b*, which only the green algae share with the land plants.

Red algae have an additional distinguishing trait: they store carbohydrates in the form of *floridean starch*—layered starch granules built from glucose molecules and located in the cytoplasm, rather than in chloroplasts.

People extract several materials from red algae for commercial use. These include *agar*, which is used as a nutrient medium for many laboratory cultures, and *car-*

(a)

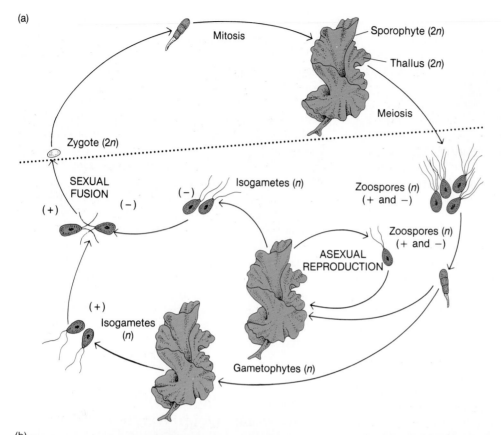

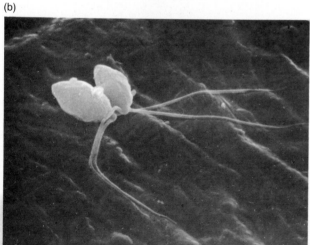

(b)

Figure 23-8   LIFE CYCLE OF THE SEA LETTUCE, *ULVA*.
(a) The sporophyte and gametophytes are morphologically identical, consisting of ultrathin sheets of cells. (b) Sexual fusion of isogametes. The cells (magnified about 3,800 times) have not yet fused but will soon do so to yield the zygote, which will not have the flagella seen here.

rageenan, which is used as a stabilizing agent in paints, cosmetics, ice cream, and other dairy products. Red algae have been grown commercially for hundreds of years in China and Japan for both human and livestock consumption. Various members of the Rhodophyta are significant producers in marine food chains. The **coralline algae,** which can deposit hard calcium carbonate crystals in their cell walls, build coral reefs. The life cycles of most red algae include an alternation of generations and are similar to those of the green algae.

Figure 23-9   THE EXQUISITE CELLULAR ARCHITECTURE OF A RED ALGA.
Most red algae have a filamentous, branched morphology as seen here (magnified about 60 times). Red algae carry on photosynthesis even at great depths.

Phaeophyta: Giant Aquatic Plants

Many brown algae are fairly small, but the division **Phaeophyta** also includes the giants of the algal world—enormous kelps that can reach 100 meters in length. Commonly called seaweed, the brown algae number about 1,500 species, many of which inhabit cool offshore waters. *Sargassum*, however, is found in warm tropical seas and sometimes forms huge floating masses (Figure 23-10). The Sargasso Sea, named for this brown alga, is an area in the Atlantic Ocean northeast of the Caribbean that is infamous for entangling sailing vessels in clogged masses of the plants.

The simplest brown algae have branched, filamentous bodies, but many have more complex forms. Some of their organs are analogous to those in many land plants and represent important examples of parallel evolution: large, flat **fronds** collect sunlight and produce sugars, much as do leaves; the **stipe** provides stemlike vertical support; and the *holdfast* anchors the plant to the bottom, much as do roots (Figure 23-10b). There are localized regions on the fronds where cells can divide rapidly, resulting in extremely rapid growth. These areas are reminiscent of similar growth zones in the tissues of plants with true vascular (conducting) systems. Kelps also have a region that resembles the zone from which new vascular tissue is generated in vascular plants.

Kelps can grow in waters up to about 30 meters deep, although the amount of light available for photosynthesis diminishes rapidly as the water deepens. Kelps overcome the problem of poor light penetration with long stipes and fronds that extend along the ocean surface, where light is much more abundant that it is below. They also have brownish *fucoxanthins*—carotenoid pigments unique to this group—that are quite sensitive to the blue and violet wavelengths, which penetrate deepest in water, and thus enable the plants to survive in a dim environment.

One reason that kelps can grow as long as 100 meters is that their *vessel cells* make up conducting tissue that pipes the products of photosynthesis from plant parts receiving plentiful sunlight to those in deeper water. These cells are similar in some ways to the vascular tissue called *phloem*, which conducts sugars in land plants. However, kelp vessel cells probably did not give rise to phloem: apparently both kelps and vascular plants separately evolved similar solutions to the problem of nutrient transport within the plant body.

Brown algae store food reserves in the form of *laminarin*, a mixture of polysaccharides, lipids, and the sugar mannitol. One substance extracted from the cell walls of brown algae, *alginic acid,* is used commercially as a stabilizer, emulsifier, and coating for paper; as a waterproofer for cloth; and as an additive to improve the

(a)

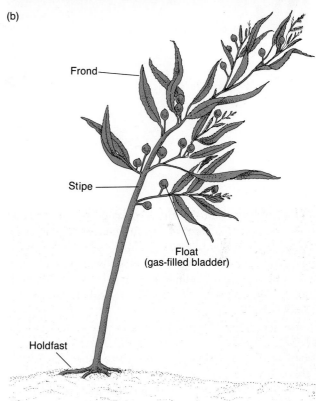

(b)

Frond

Stipe

Float
(gas-filled bladder)

Holdfast

Figure 23-10   *SARGASSUM*, BROWN ALGAE.
(a) This buoyant brown alga is swept into parallel rows at the ocean surface by wind and waves. These species float freely, but many brown algae are anchored by a holdfast. (b) The major structures of an attached specimen of a brown alga are indicated in this typical *Sargassum* plant and include the light-gathering fronds, the stipe, and the holdfast. Such algae that break free of their holdfast join others floating at the surface.

(a)

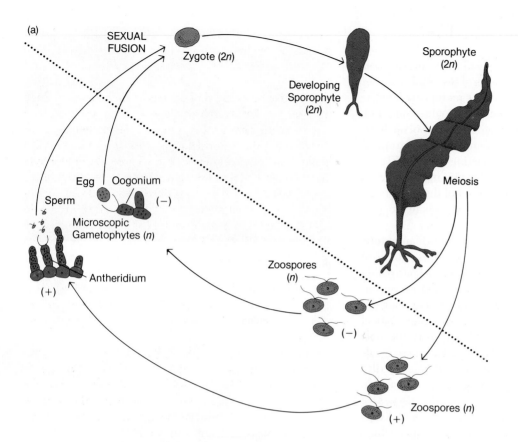

Figure 23-11   LIFE CYCLE OF *LAMINARIA*.
The gametophytes are microscopic filaments that alternate with the large sporophyte generation.

texture of ice cream. In addition, many species concentrate the important nutrients nitrogen, potassium, and iodine from the surrounding water. Because of these properties and their rapid growth rates, large kelps are sometimes farmed commercially and used for livestock fodder and fertilizer and in glue, alcohol, and other products.

The life cycles of most brown algae are similar to that of the green alga *Ulva*, in that both alternating generations are multicellular. For example, the alternating generations of *Laminaria* (Figure 23-11) include male and female haploid gametophytes that are tiny filaments (the male often smaller in diameter than the female). Male gametes develop in an *antheridium*, whereas female gametes arise in an *oogonium*. In contrast to the small gametophytes, the diploid sporophytes—the actual kelp plants—have broad fronds with ruffled edges, a stripe, and a holdfast.

The brown alga *Fucus* and its close relatives, however, have a life cycle that is unique in the plant kingdom and that almost duplicates the life cycle of animals. The diploid plant is a greenish-brown, branching thallus with hollow, gas-filled *bladders* (floats) to support the plant and bumpy-looking chambers called *conceptacles* inside special structures at the tips of the branches. Each conceptacle contains either a sperm-producing organ or an egg-producing organ in which meiotic divisions give rise to gametes. As in animals, no independent haploid gen-

eration is formed; the haploid phase is so reduced that the only representatives are the egg and sperm cells. These gametes are released into the surrounding ocean water, and a sperm and an egg fuse to form a new diploid zygote; the zygote settles down and divides mitotically to produce a new adult plant. Botanists are not sure why *Fucus* lacks the plant hallmark of alternating generations, but the absence underscores algal diversity.

As a group, the algae display many of the specializations we will see in more complex members of the plant kingdom: they have adapted to a variety of environments; all but one show alternation of generations; they carry out both asexual and sexual reproduction; they display cellular division of labor; and some have leaflike, stemlike, and rootlike structures. A similar continuum of characteristics is evident in many groups of land plants, but these specializations are more obvious and dramatic than are those of any kind of alga.

PLANTS THAT COLONIZED LAND

The more than 12,000 species of algae that live today in fresh- and sea-water habitats are relatively simple organisms. When one compares them with the hundreds of thousands of kinds of terrestrial plants, it is clear that the challenges of life on land and the diversity of distinctive habitats there allowed for greatly increased complexity in the descendants of the algae that left the oceans, and thus a profusion of new plant forms. Both the structures and the activities of land plants have changed as part of their adaptation to life out of the water.

Plants apparently colonized the land in the early Devonian period, about 400 million years ago. This was before the breakup of Panagaea, the Earth's single great continent (Chapter 19), when the climate was warm and moist and when episodes of mountain-building activity alternated with erosion and a lowering of average altitudes. This massive shifting of the Earth's surface drained and refilled the vast inland seas again and again. This, in turn, placed a survival premium on the ability of any plant type to cope with changing environments, especially ones in which desiccation (drying out) was common. Some botanists believe that before the early Devonian, multicellular gametophytes arose that could exist for an extended period as free-living plants. Later, it is hypothesized, they evolved a capacity to produce gametes in response to adverse environmental conditions. The union of such gametes yielded a zygote that entered a resting state and was capable of riding out the environ-

mental stress. Some botanists think that the evolution of such drought-tolerant resting zygotes gave rise to the sporophyte phase of the life cycle of the land plants.

The relation of the diploid zygote to the gametophyte may be the key to the origin of the life cycle of land plants. Most green algae simply release their eggs or zygotes into the water. Imagine what would happen if a diploid zygote was retained on the parental gametophyte body and grew there for some time as an embryo. Such an organism would be a diploid, multicellular sporophyte; then, when meiosis occurs, haploid spores could be produced which could mature into the gametophyte phase again. Biologist Linda E. Graham of the University of Wisconsin points out that modern green algae of the genus *Coleochaete* do just this; their zygote is retained and nourished by the gametophyte, and develops as an embryo. Here is a perfect model for the intermediate stage that may have led to the first land plants over 400 million years ago.

Since land plants and green algae share the same photosynthetic pigments and storage products (chlorophylls *a* and *b* and starch granules), botanists believe that the drought-tolerant ancestors of the land plants were closely related to multicellular green algae (Figure 23-12). All land plants share certain additional features: they have a relatively impervious jacket of cells surrounding

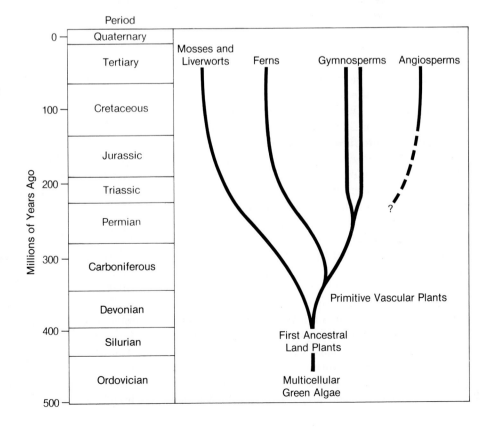

Figure 23-12   EVOLUTION OF THE MAJOR PLANT GROUPS.
The approximate time various plant groups arose is indicated on this geological time scale. The four types of gymnosperms—pines and relatives—are not closely related. The source of angiosperms is not known yet.

the reproductive structures that acts to preserve a sterile environment inside; they have openings through the *epidermis* (outer layer of the plant body) through which exchange of carbon dioxide and water vapor can take place; and both gametophyte and sporophyte generations are multicellular. The ubiquity of these characteristics leads botanists to conclude that the features were also present in the most primitive land plants; thus we can reconstruct a picture of what those ancient pioneers must have been like.

To begin with, the land in the early Devonian period held several advantages over the aquatic environment for the early colonizers. For one thing, the first land plants had few competitors and no predators: dry land was essentially devoid of large life forms. In addition, exchange of gases can occur more readily with air than with water, since the concentrations and rates of diffusion of gases such as carbon dioxide and oxygen are much higher in air than in water.

Despite these advantages, life on land presented stiff challenges not faced by aquatic plants—the result was evolution in the body forms and the physiological activities of land plants, and this evolution contributed to plant diversification. The first and most serious problem that plants faced was desiccation, a problem also faced by the first land animals. To survive, a means of minimizing evaporation while allowing controlled gas exchange was essential. Second, support for the plant body also was essential, since, without the buoyant effect of water, land plants are especially subject to gravity. Third, a means of absorbing moisture and minerals from the soil and transporting them throughout the plant evolved. Finally, a new means of reproducing sexually evolved, since unprotected gamete cells could no longer swim or be transported in the water.

Just what were the specific structures that evolved in the land plants? Waterproofing substances reduced evaporative water loss, and tiny pores, called *stomata*, in the waterproof outer layer arose in most groups; stomata can be opened and closed and thus allow a controlled exchange of gases and water vapor with the surrounding air. In many groups, root systems evolved with large surface areas and the ability to absorb water, ions, and molecules. And in the vascular plants, specialized tissues developed that conduct water and nutrients to every cell of the multicellular plant body and provide strength for vertical support.

Just how good the first terrestrial plants were at absorbing nutrients from soil is not clear. Recent fossil findings show that early vascular plants had as many mycorrhizal associations (Chapter 22) as do modern vascular plants. Land fungi apparently coevolved with land plants. Their symbiotic pairing would have allowed both groups to survive and invade the relatively sterile soils of the early continents—and thus would have been a critical early step in the colonization of land by multicellular organisms.

Our survey of the various divisions of land plants will reveal diverse types that are sometimes beautiful and sometimes a bit bizarre—but are always interesting in themselves. Beyond their inherent interest, however, the various groups of land plants deserve careful discussion because, like the algae, they followed a pattern of progressive specialization of structure that made life on land possible.

Nonvascular Land Plants: Bryophyta

Liverworts, hornworts, and mosses are land plants that lack vascular tissue for conducting water and nutrients. Many botanists think that these plants are more closely related to one another than to any other land plants and thus classify them in a single division, **Bryophyta.** It is true that all **nonvascular land plants** are similar to each other and unique from other groups in having a sporophyte that is partially or completely dependent on the free-living gametophyte for nutrition. Despite this common feature, however, they are very different in most respects, and thus some biologists have classified them in three separate classes: Hepatophyta (liverworts), Anthocerotophyta (hornworts), and Musci (mosses).

All three types of nonvascular land plants are small (the largest species are less than 60 centimeters tall) and usually grow in moist places—in marshes and swamps and on damp rocks or logs, shady forest floors, stream banks, and sometimes the north sides of trees. They are found throughout the world, but a large percentage of them live in rain forests. Their life cycles are quite different from those of vascular land plants, and although the earliest fossils of vascular plants are 50 million years older than those of nonvascular plants, it is unlikely that the latter evolved from the former. These two major groups probably diverged shortly after plants colonized the land and evolved with separate and very different solutions to the challenges of life out of water.

Hepatophyta

Liverworts (class **Hepatophyta**), a name that comes from the Old English words meaning "lobed plant," are small plants that often grow over the soil surface as flat thalli, although a few live in water. Liverworts survive only in moist environments for several reasons. First, the upper surface of the plant has pores resembling the true stomata of vascular plants. However, these pores do not have *guard cells*, the pair of crescent-shaped cells on either side of a stoma that regulate its opening and closing. Thus air containing the carbon dioxide that the plant needs for photosynthesis can always enter, but it can diffuse out just as easily, along with precious water vapor. Furthermore, liverworts lack a true root system, being attached to the soil by hairlike *rhizoids*. Rhizoids generally serve only to anchor the plants, since absorption of water and nutrients commonly occurs directly through the entire lower surface of the gametophyte. Most liverworts also lack water-conducting tissue.

Liverworts bear both a sperm-producing chamber, or antheridium, and an egg-producing chamber or *archegonium*, on the conspicuous gametophyte. An individual gametophyte may bear both male and female structures, or the antheridia and the archegonia may form on different gametophytes. Often the antheridia and archegonia develop directly on the thallus, but they may be elevated on stalks, as in *Marchantia polymorpha*.

During rain or heavy dew, sperm are splashed or washed from a mature antheridium into the vicinity of an archegonium, swim down the neck of the chamber, and fertilize an egg. The resulting diploid sporophyte remains attached to the gametophyte during its entire growth. In this respect, the liverworts and the other nonvascular land plants differ from the green algae, such as *Ulva* or *Ulothrix*, in which the gametophyte and sporophyte generations are nutritionally independent and physically separate from each other. In the mature sporophyte, spores are produced by meiosis in the sporangium. On release, these spores grow into new haploid gametophytes. Asexual reproduction is also fairly frequent in *Marchantia* and other liverworts; small **gemma cups**—analogous to buds—form on the thallus and contain gemmae, which separate from the parent and grow into new plants genetically identical to the parent (Figure 23-13).

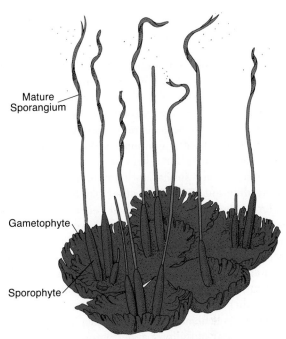

Figure 23-14   *ANTHOCEROS*, A HORNWORT.
The hornwort sporophyte remains embedded in the gametophyte. Drying and twisting of the many sporangia create "horns," which split and release the spores.

Anthocerotophyta

Like liverworts, hornworts (class **Anthocerotophyta**) are inconspicuous residents of moist, shady places, and the gametophyte superficially resembles some liverworts. Gametes are produced in antheridia and archegonia that are sunk into the top surface of the gametophyte. When gametes unite, the zygote develops into a sporophyte that remains attached to the gametophyte (Figure 23-14). The diploid sporophyte produces a long, cylindrical sporangium that splits along two lines of weakness as it dries and twists into "horn" shapes. This drying and twisting expel the spores, which develop into round, flattened haploid gametophytes. Hornworts have internal cavities that are filled with mucilage (a gluelike polysaccharide) rather than with air, as in liverworts. These cavities are inhabited by symbiotic cyanobacteria that fix nitrogen in organic compounds useful to the hornwort and, in return, are provided with a moist, protected environment.

Figure 23-13   LIVERWORT ASEXUAL REPRODUCTION.
Each gemma cup is a site where a new *Marchantia* will arise from the parent plant.

Musci

By far the most numerous and familiar nonvascular land plants, mosses (class **Musci**) form velvety carpets and spongy beds almost anywhere the air is moist and clean. During the moss reproductive life cycle, which we used earlier to demonstrate alternation of generations the sporophyte is nonphotosynthetic and depends on the gametophyte for nutrition (Figure 23-1). In this sense, mosses are similar to liverworts and hornworts and are different from many other plants: in most algae and ferns, the haploid and diploid generations are independent organisms; in conifers and flowering plants, the gametophyte relies on the sporophyte for sustenance.

Moss spores are particularly notable: after being blown about by the wind and landing on a suitable substrate, they produce a tiny threadlike protonema. This very small branched gametophyte bears a striking resemblance to a filamentous green alga and serves as a visual reminder of the probable origin of all land plants. The growing gametophyte forms the familiar moss plant that carpets moist soil and rocks. As the plant reaches maturity within a few weeks, the antheridium or the archegonium forms at the tip of each gametophyte. A sperm reaches the egg inside the archegonium by swimming through a continuous film of water from heavy dew, rain, or spray from a nearby watercourse. This dependence on water is a remnant of an earlier life in some respects and stringently limits the mosses' possible habitats.

Moss gametophytes develop rootlike rhizoids that secure the plant, but as in liverworts, the entire lower portion of the plant takes up water and minerals. The leaflike appendages from the stems are also important for the absorption of water and nutrients. Stems of moss gametophytes often have two systems reminiscent of the major conducting vessels of the vascular plants: elongated cells of two types—hydroids, which aid in the transport of water, and leptoids, which transport nutrients from one part of the plant to another.

With their swimming sperm and relatively inefficient water uptake and transport systems, mosses are rarely able to survive in habitats without regular seasonal moisture. Some, however, can tolerate drying out for considerable periods and so can inhabit areas with dry seasons. Others can live in marginally dry areas because their habit of growing in densely packed mounds creates a high-humidity zone around individual plants. The peat mosses (Sphagnum) can hold great quantities of water; an extensive peat-moss bed can even serve as the source of a small creek. The peat moss harvested from these bogs is used in home gardening to improve water retention, and in some areas of the world, fossilized peat deposits are excavated, dried, and burned as fuel.

The nonvascular land plants we have just described—the liverworts, hornworts, and mosses—are a fairly large and diverse group, but they are limited to small size and moist habitats by their lack of an efficient vascular system, by the absence of stomata that can close to prevent moisture loss, and by their swimming sperm. The next group of land plants, which includes the ferns and horsetails, evolved adaptations that allow larger size and survival in drier habitats. Nevertheless, they reproduce by means of swimming sperm and disperse by means of spores. Dispersion of the new generation through the highly successful evolutionary "invention" of the seed comes only in later groups, as we shall see.

The lower land plants have been compared to amphibians such as frogs and salamanders, since these animals can live on dry land but must lay their eggs in water. Interestingly, the age of amphibians—when these animals were the largest and most successful group on land—coincided with the heyday of the lower plants: the Carboniferous period (Figure 23-12), with its forests of club mosses, horsetails, and ferns—the plants we shall survey next.

Seedless Vascular Plants: Support and Transport, But No Seeds

Colonization of the land was most successful for plants that developed an efficient internal-transport system and vertical support tissue—the **vascular plants.** The main reason for their success was the **tracheid,** a special type of hollow cell with thick, rigid, pitted walls. Occurring end to end in long, rigid pipelike structures, tracheid cells help support plants vertically and provide a system for water transport from the root to other parts of the plant body.

Vascular plants display great physical diversity, but they all share four unique characteristics: (1) two types of specialized conducting tissues: xylem, which conducts water, and phloem, which conducts sugars and other nutrients; (2) a layer of waterproofing material, cutin, on the portion of the plant aboveground to reduce water loss; (3) multicellular embryos that are retained within the archegonium; and (4) a diploid sporophyte that is the dominant stage in the life cycle.

Modern vascular plants are classified in nine divisions. Four consist of vascular plants that do not produce seeds: Psilophyta (whisk ferns), Lycophyta (club mosses), Sphenophyta (horsetails), and Pteridophyta (ferns). These are the subject matter of the rest of this chapter. In Chapter 24, we consider the other five divisions of seed-producing vascular plants. Although some botanists place all nine of the vascular plants in a single division, the group is almost certainly not monophyletic. Thus the fossil record indicates that at least six or seven distinct lines of vascular plants existed not long after the first vascular systems evolved. The various groups probably arose independently of each other.

Psilophyta

The whisk ferns (division **Psilophyta**) are an ancient plant group whose members resemble the earliest vascular land plants. The division includes only two genera of modern plants, *Psilotum* and *Tmesipteris*. The members of the genus *Psilotum* are small, branching plants that resemble whisk brooms, hence the name "whisk ferns" (Figure 23-15). In North America, *Psilotum* is found only in Arizona, Florida, Louisiana, and Texas. *Psilotum nudum* is a greenhouse weed in some places, while *Tmesipteris* hangs from the trunks of tropical tree ferns. The sporophyte of *Psilotum*, which is the conspicuous generation, is composed of only a stem, with no true roots or leaves, although it does have rhizoids and scalelike appendages. *Tmesipteris* has leaflike appendages that may simply be flattened branches rather than true leaves. In both genera, a lateral stem called a *rhizome*, with hairlike rhizoids, serves as an absorptive and anchoring organ. Mycorrhizal fungi are usually present in the rhizoids and serve to increase nutrient availability to the whisk fern. Branched aerial stems sprout from the rhizome and produce three-lobed sporangia.

During reproduction, haploid spores are produced in each sporangium, and when released, the spores grow into gametophytes that are much smaller than the sporophyte. The gametophytes of *Psilotum* lack chlorophyll and are saprophytic, living on decaying soil materials (however, fungi may actually provide nutrients to the gametophytes). The sporophyte is dependent on the gametophyte body for nourishment during the early stages of its development, but it eventually turns green and becomes autotrophic.

Psilotum looks strikingly similar to members of the genus *Rhynia*, the earliest colonizers of land. A few botanists think that the two modern whisk-fern genera are actually small, simple ferns. Whether or not that view is correct, *Psilotum*, with its vascular tissue but no true roots or leaves, exhibits traits intermediate between those of the green algae and the other plants with vascular systems.

Lycophyta

The club mosses and quillworts (division **Lycophyta**) form another ancient plant group that dates back to the Devonian period, almost 400 million years ago. Club mosses and other lycophytes were probably the most numerous plants in the later Carboniferous period, and their decomposed and compressed tissue, collected in

Figure 23-15   *PSILOTUM NUDUM*, A WHISK FERN.
The stemlike and branchlike parts of this whisk fern are clearly visible. Primitive vascular tissue extends throughout such structures.

vast underground coal deposits, provides much of our fossil fuel. Although most species have been extinct for millions of years, three relatively common genera remain: two club mosses (*Lycopodium* and *Selaginella*) and one quillwort (*Isoetes*).

The lycophytes have a conspicuous sporophyte and show the next advance in vascular-plant architecture: *roots*. Unlike the rhizoids of the other land plants we have discussed, roots do more than anchor the plant. They absorb water and minerals from the substratum to which they are attached; then, by means of conducting tissue that interconnects with the vascular system in the stem, they permit these substances to be carried throughout the plant. The leaves of the club moss are similar in appearance to the leaflike appendages of mosses, but these are true leaves, since they contain vascular tissue. The leaves of quillworts are quite different in appearance but also contain vascular tissue.

Sporangia are borne on leaves called **sporophylls** (spore leaves). In the quillworts, any one of the leaves may act as a sporophyll. However, in many species of the

Figure 23-16 *LYCOPODIUM COMPLANATUM*, A CLUB MOSS.
Spore-producing strobili are clearly visible at the top of these lycopods growing in the leaf litter on the shady forest floor.

club-moss genera *Lycopodium* and *Selaginella*, sporophylls are grouped on one axis to form the **strobilus,** the club-shaped, spore-producing organ that gives lycophytes their common name (Figure 23-16). Some species of lycopods produce only one type of spore, and are considered to be **homosporous** (same spores). The identical spores give rise to gametophytes that produce both antheridia and archegonia. Other species produce two different types of spores; they are **heterosporous** (different spores). The smaller type, or **microspore,** differentiates into a male gametophyte that bears only antheridia. The larger type, or **megaspore,** differentiates into a female gametophyte that bears only archegonia. Water is still necessary for reproduction in these plants, since the sperm produced by the antheridia must swim to the archegonium in order to fertilize the egg. Thus like the other "lower plants," lycophytes tend to be found in moist habitats.

Sphenophyta

During the long Carboniferous period, when Earth's land masses were covered by vast, steamy swamps, the horsetails (division **Sphenophyta**) were extremely common plants, growing globally in a wide range of sizes and forms. Fossilized remains of these plants—especially of tree-sized *Calamites*, which reached heights of 10 to 15 meters—formed part of the coal we use today. This once large division of plants is now represented by only one genus with about twenty species: *Equisetum*, the common horsetail or "scouring rush" (Figure 23-17). Scouring rushes were sometimes used for scrubbing clothes and pots before modern soaps and steel wool were invented. The stems are abrasive because they contain silica crystals, which stiffen and strengthen them. Horesetails are common in marshy areas, but patches of horsetails, with their whorls of thin, bright green branches, can be seen along railroad tracks and even in fairly dry areas.

The conspicuous plant is the diploid sporophyte. The shoots bear scalelike leaves in clusters around joints in the rough stem. Many species produce a special reproductive stem called the *fertile shoot*. This tan or reddish stem is nonphotosynthetic and bears a terminal strobilus, which produces hundreds of identical spores in clusters of sporangia. Like the other land plants discussed so far, the horsetails have swimming sperm and depend on water for reproduction.

Pteridophyta

Ferns (division **Pteridophyta**) were also prominent during the Devonian and Carboniferous periods, when many types grew as large treelike plants. In some tropical areas today, tall tree ferns still prosper, but the majority of the 10,000 modern fern species are low, graceful

Figure 23-17 *EQUISETUM*, A HORSETAIL.
This diminutive plant is a holdover from the Carboniferous swamps.

plants less than a meter tall (Figure 23-18). Familiar inhabitants of shady forests, ferns usually have lacy fronds, highly divided leaves that grow from horizontal subterranean rhizomes. Young, coiled fern fronds often are referred to as "fiddleheads" or "croziers" and unroll as they grow. The fiddleheads of some species are edible. A fern frond is a true leaf, with vascular tissue that usually branches and rebranches to supply all parts of the leaf with water and nutrients.

The leafy fern plant is the diploid sporophyte (Figure 23-19). In most species, sporangia are formed on the undersurface of the fronds and are grouped in structures calles **sori** (singular, *sorus*). Some species have two types of fronds that look very different from each other: one is large, green, photosynthetic, and does not produce spores, while the other is usually smaller, nonphotosynthetic, and bears spores. In virtually all fern species,

Figure 23-18   FERNS: A SOFT GREEN GROUND COVER.
This dense layer of ferns in a forest lets little light pass through to smaller plants on the soil surface.

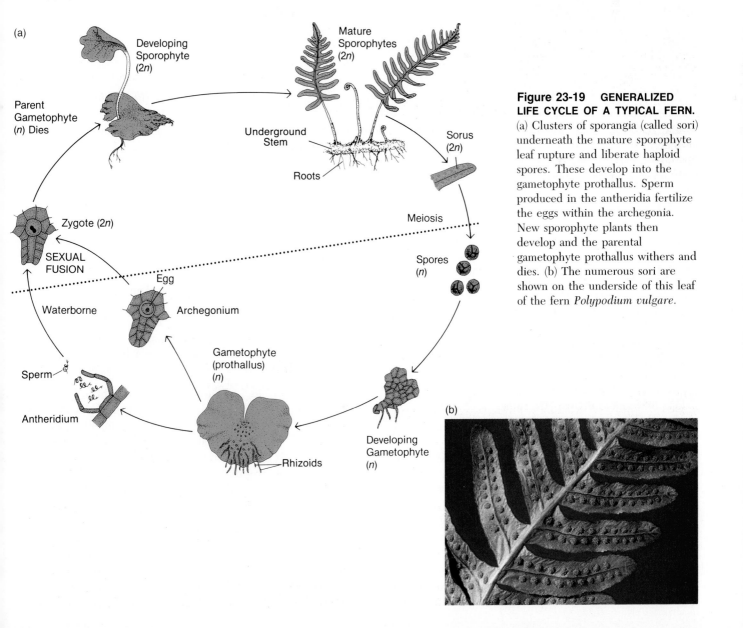

(a)

Developing Sporophyte (2*n*)

Mature Sporophytes (2*n*)

Parent Gametophyte (*n*) Dies

Underground Stem

Roots

Sorus (2*n*)

Zygote (2*n*)

Meiosis

SEXUAL FUSION

Egg

Spores (*n*)

Waterborne

Archegonium

Sperm

Gametophyte (prothallus) (*n*)

Antheridium

Rhizoids

Developing Gametophyte (*n*)

(b)

Figure 23-19   GENERALIZED LIFE CYCLE OF A TYPICAL FERN.
(a) Clusters of sporangia (called sori) underneath the mature sporophyte leaf rupture and liberate haploid spores. These develop into the gametophyte prothallus. Sperm produced in the antheridia fertilize the eggs within the archegonia. New sporophyte plants then develop and the parental gametophyte prothallus withers and dies. (b) The numerous sori are shown on the underside of this leaf of the fern *Polypodium vulgare.*

these spores are similarly sized homospores. The tiny spores are discharged into the air, sometimes forcefully, by a springlike mechanism in the sporangium wall. The haploid spores eventually settle and grow into tiny, thin, heart-shaped gametophytes, each of which is called a *prothallus*. The prothallus generally produces both antheridia and archegonia. It is photosynthetic and has tiny rhizoids that imbibe water. Because the gametophyte is delicate and because the sperm must swim in a continuous film of water to the archegonium to fertilize the egg, ferns are most common in moist environments.

The lower land plants we have considered—liverworts, hornworts, mosses, whisk ferns, club mosses, horsetails, and ferns—are descendants of ancient groups that thrived and diverged in the Devonian and Carboniferous periods, when the global climate was much warmer than it is today and competition from seed plants had not yet developed. While seedless vascular plants possess most of the basic characteristics required for success on land, their dependence on a film of water for swimming sperm has kept them from invading the drier continental areas. Only the next step in plant evolution, the seed, could break that final barrier to colonization of the land.

Today, many of the lower plants are rarely noticed, are overshadowed—literally and figuratively—by the larger, more colorful seed plants, and are little used in agriculture and commerce. Yet these groups represent the transition from water to land and from simple algae to more complex plants with specialized internal structures. The hallmark of the plant kingdom—the alternation of generations—evolved first among them. While the algae and lower land plants may be the more obscure members of the plant kingdom, a hike down a quiet forest path or along an ocean beach in search of these organisms is, in a sense, a walk back through biological history.

SUMMARY

1. Members of the kingdom Plantae are quite diverse, but most plants share four characteristics: (a) multicellularity, (b) chlorophyll, (c) nonmotility, and (d) alternating haploid and diploid generations.

2. The plant kingdom is composed of about 300,000 species, classified in 13 divisions.

3. Plants have a unique life cycle based on the *alternation of generations*. A diploid *sporophyte* divides meiotically to produce haploid spores that give rise to haploid *gametophytes*. Such gametophytes produce haploid gametes that fuse to produce diploid zygotes, which develop into sporophytes. In some plants, the sporophyte and the gametophyte are free-living organisms; in others, one generation—usually the sporophyte—is dominant, and the other depends on it for water, nutrients, and support.

4. *Algae*, a large group of aquatic plants, are the oldest plants and make up a large percentage of the Earth's biomass. Green algae (*Chlo-rophyta*) are mostly small freshwater plants. Like land plants, green algae have chlorophylls *a* and *b* and store food as starch. *Siphonous algae* are large coencytic organisms. *Volvocine algae* display a spectrum of increasing size; cell coordination, specialization, and interdependence; and reproductive complexity, with haploid *zoospores* and diploid *zygospores*. Some larger colonies of volvocine algae exhibit *oogamy*—reproduction by means of immobile eggs and motile sperm—rather than the *isogamy* of most algae, whose gametes are the same size and just fuse with each other. *Filamentous algae* show specialization in body form and in alternation of generations. The reproductive pattern displayed by filamentous algae underlies the life cycle of all land plants.

5. Red algae (*Rhodophyta*) are small, delicate aquatic plants that can live at a variety of depths, including deep waters, because of the *phycobilins* and other pigments they contain. They store carbohydrates as floridean starch, are important to marine food chains, and have some important commercial uses.

6. Brown algae (*Phaeophyta*), including giant kelps, contain pigments that can absorb light filtered through deep water. Many brown algae have body parts analogous to those of land plants: leaflike *fronds*, a stemlike *stipe*, and a rootlike *holdfast*. Male gametes are produced in *antheridia*, female gametes in *oogonia*. Kelps have vessel cells that conduct photosynthetic products through the plant. Brown algae store nutrients as laminarin. Brown algae are harvested for fodder, food, and chemicals. *Fucus* and its close relatives are unique in the plant kingdom in having a life cycle that lacks alternation of generations.

7. Land plants are thought to have evolved from green algae that colonized the shores of ancient seas some 400 million years ago. Plant adaptation to terrestrial environments involved mechanisms for "waterproofing" plant surfaces, allowing exchange of gases through these surfaces, supporting the plant body, absorbing moisture and minerals from the soil, transporting substances through the plant body, and allowing fertilization and dispersion

of spores and zygotes. Retention of the zygote on the gametophyte and evolution of a multicellular sporophyte were key to the origin of the life cycle of land plants.

8. *Nonvascular land plants*—liverworts (*Hepatophyta*), hornworts (*Anthocerotophyta*), and mosses (*Musci*)—are terrestrial plants classified in the division *Bryophyta* that lack specialized vascular tissue for conducting water and nutrients through the plant. Gametophytes are the conspicuous generation of the life cycle and have structures analogous to roots, stems, and leaves. Gametes are produced in either the antheridium (sperm) or the archegonium (egg), and spores are produced in sporangia. Water is necessary for reproduction to occur.

9. *Vascular land plants* are terrestrial plants with two kinds of vascular tissue made up of *tracheid* cells: xylem and phloem. Other basic characteristics include a layer of cutin to prevent water loss and multicellular embryos that are retained within the archegonium on the diploid sporophyte, which is the dominant generation.

10. Whisk ferns (*Psilophyta*) have a stem, but no roots or leaves, only hairlike rhizoids and scalelike appendages. Whisk ferns resemble the earliest land plants. Club mosses and quillworts (*Lycophyta*) have stems, leaves, and roots. The club mosses also have a club-shaped *strobilus*, made up of grouped *sporophylls* (leaves that bear sporangia). Horsetails (*Sphenophyta*) have stems strengthened by silica and bear whorls of needlelike branches, scalelike leaves, and spore-producing strobili. Ferns (*Pteridophyta*) have wide, flat, highly subdivided fronds. They produce *sori*, clusters of sporangia, on the underside of the fronds, and heart-shaped prothalli (gametophytes) develop from the spores. In all four divisions of vascular land plants water must be present for reproduction to take place.

KEY TERMS

alga	homosporous	Rhodophyta
alternation of generations	isogamete	siphonous alga
anisogamete	isogamy	sorus
antheridium	Lycophyta	Sphenophyta
Anthocerotophyta	megaspore	sporophyll
Bryophyta	microspore	sporophyte
Chlorophyta	Musci	stipe
coralline alga	nonvascular land plant	strobilus
filamentous alga	oocyte	tracheid
frond	oogamy	vascular plant
gametophyte	oogonium	volvocine alga
gemma cup	Phaeophyta	zoospore
Hepatophyta	phycobilin	zygospore
heterosporous	Psilophyta	
holdfast	Pteridophyta	

QUESTIONS

1. What feature of the life cycle is unique to plants?

2. What are some important characteristics of the green algae? the red algae? the brown algae?

3. What is meant by parallel evolution? Give some examples of parallel evolution in the plant kingdom.

4. From which group of algae are the land plants thought to have evolved? What is some evidence to support this relationship?

5. What are some of the problems faced by land plants (but not by water plants)? How have these problems been solved or not solved in the course of evolution?

6. What are the major types of nonvascular land plants? Describe each type.

7. What are the major types of seedless vascular plants? Describe each type.

8. Compare the nonvascular and the vascular land plants. What are

some major differences in their basic structures and life cycles?

9. The nonvascular land plants and the seedless vascular plants share one feature that makes them dependent on a moist environment for part of their life cycles. What is that feature?

10. Choose any alga, any nonvascular land plant, and any seedless vascular plant, and diagram their life cycles.

ESSAY QUESTION

1. What are some factors that limit the size of nonvascular water plants? nonvascular land plants? vascular land plants? What are some advantages of large size? What are some disadvantages?

SUGGESTED READINGS

DYER, A. F., ed. *The Experimental Biology of Ferns.* New York: Academic Press, 1979.

All about the ferns in the words of the experts.

FOSTER, A. S., and E. M. GIFFORD. *Comparative Morphology of the Vascular Plants.* San Francisco: Freeman, 1974.

This is the major reference for both the nonvascular and the vascular plants.

GRAHAM, L. E. "The Origin of the Life Cycle of Land Plants." *American Scientist* 73 (1985): 178–186.

A good account of the key transition to plant life on land.

RAY, P. M., T. STEEVES, and S. A. FULTZ. *Botany.* Philadelphia: Saunders, 1983.

This is an excellent, critical treatment of the entire field of botany and an effective discussion of the divisions of the plant kingdom discussed in this chapter.

WATSON, E. V. *Mosses.* Oxford Biology Reader, no. 29. Burlington, N.C.: Carolina Biological Supply, 1972.

A fine brief account of the mosses.

24
THE SEED PLANTS

*Exquisitely beautiful, and unlike anything else that we have,
is the first white lily just expanded in some shallow lagoon
where the water is leaving it,—perfectly fresh and pure,
before the insects have discovered it. How admirable its
purity! how innocently sweet its fragrance! How significant
that the rich, black mud of our dead stream produces the
water-lily,—out of that fertile slime springs this spotless
purity!*

Henry D. Thoreau, *Journal*, v, 283 (June 19, 1853)

" . . . out of that fertile slime springs this spotless purity!" Water lilies.

The majority of terrestrial landscapes—whether the scene is an alpine meadow, a tropical rain forest, rolling farmland, a tall stand of timber, a scorched desert, or an urban park—have a common denominator: their primary biological component is seed plants. The current geological era, the Cenozoic, has been called the age of mammals and angiosperms, or flowering plants; indeed, angiosperms constitute more than 95 percent of all living seed plants and over 80 percent of the plant species on Earth.

Most land animals rely on seed plants either directly or indirectly for cover and food. Human beings are particularly dependent, since virtually all our grains, fruits, vegetables, beverages, and spices come from seed plants. We use fibers and wood from seed plants for clothing, shelter, and fuel. And we extract chemicals from seed plants for use in medicines; for such raw materials as rubber, resins, and turpentine; and for such psychoactive drugs as tobacco, marijuana, opium, cocaine, and alcohol.

Besides the economic significance of seed plants, the tallest, oldest, and most massive organisms on Earth are members of this group, and fossil evidence shows that seed plants have been the most widespread, diverse, and abundant land plants for 250 million years. Clearly, seed plants are among the most successful living things and are of great interest to biologists.

This chapter will discuss the taxonomy and evolution, major types, and life cycles of seed plants.

CLASSIFICATION OF THE SEED PLANTS

As we mentioned in Chapter 23, the five divisions of seed plants are Cycadophyta (cycads), Ginkgophyta (ginkgoes), Gnetophyta (gnetinas), Coniferophyta (conifers), and Anthophyta (angiosperms). Together, the members of the first four divisions are called *gymnosperms*. The fifth division—composed of the *angiosperms*, or flowering plants—is the largest, with six times more members than all the other plant divisions combined. Although structurally very different from one another, all seed plants share certain traits:

1. The *megasporangium*, the chamber in which meiosis takes place and the female gametophyte subsequently develops, is protected by one or two layers of skinlike tissue known as the *integument*.

2. Both the male and the female gametophytes of seed plants are smaller than those of other vascular plants, and the female gametophytes depend on the sporophyte for water, nutrients, and protection.

3. Sperm do not have to swim through a film of water to reach the egg, but reach it by other means.

4. Like ferns, most seed plants produce relatively large leaves with several, sometimes many, veins (or, in the case of pines and their relatives, have huge numbers of narrow leaves, the needles).

In addition, seed plants have the same type of life cycle, which we will discuss as we consider each group, and they evolved several structures that proved to be a winning formula for life on dry land. These include *pollen*, or movable male gametophytes; plant embryos; and **seeds**—complex units that store nutrients and a developing embryo and can withstand severe environmental conditions.

EVOLUTION OF THE SEED PLANTS

Our discussion of plant evolution in Chapter 23 left off during the Carboniferous period—the age of amphibians and coal-forming swamps—almost 350 million years ago (Figure 23-12). Forests of giant tree ferns, club mosses, and horsetails dotted the flat, marshy terrain of the giant continent Pangaea, which included what is now North America. Eventually, plant by plant, those ancient trees fell into the soft earth, becoming the fossil fuels of today. It was during the long Carboniferous that the first seed-producing plants arose.

Paleobotanists (scientists who study the evolutionary history of plants) have found large numbers of fossilized fern species and individuals in layers of Carboniferous rock. They discovered that many of those ferns actually produced seeds. These *seed ferns* were the first seed plants to evolve. They form a now-extinct class of gymnosperms.

The next geological period—the last in the long Paleozoic era—was the Permian, when the early reptiles evolved, and three more groups of gymnosperms appeared: conifers, cycads, and ginkgoes. These plant groups are thought to have arisen and diverged from a simple fernlike ancestor in a landscape such as the one pictured in Figure 24-1. If the earliest land plants were, like the amphibious animals, tied to water for reproduction (as we saw in Chapter 23), then these later groups of gymnosperms were analogous to the evolving reptiles—capable of withstanding drought and cold and of reproducing without abundant standing water. The global climate during the Permian and the next period, the Tri-

Figure 24-1   THE GYMNOSPERMS OF THE PERMIAN PERIOD.
The evolution of the gymnosperms beginning in the Permian period paralleled the early diversification of the reptiles—both were freed from dependence on standing water for reproduction. The conifers, cycads, and ginkgoes replaced the seed ferns as the climate on the Earth's land masses became cooler and drier.

assic—the first period in the Mesozoic era—eventually cooled. The vast inland seas drained off and dried up as the continental crusts buckled upward and mountains formed. The conifers, cycads, and ginkgoes—better suited to the changing environmental conditions—gradually replaced the seed ferns. But while these massive climatic and geological changes were taking place, the seed ferns and true ferns dominated landscapes for 70 million years.

The "golden age of gymnosperms" stretched from 350 million to 100 million years ago and coincided with the rise and fall of the dinosaurs. Some of those beasts were herbivores, plant eaters, and grazed on gymnosperm species similar to some that survive today. Among the most abundant were the cycads, which resembled certain modern palm trees.

Beginning about 120 million years ago, as the predominance of the dinosaurs was coming to an end, flowering seed plants began to appear, as judged by fossil evidence. Paleobotanists believe that they diverged from a low, shrublike gymnosperm and survived in higher, drier areas (Figure 24-2). Giant ferns, club mosses, cycads, ginkgoes, and conifers continued to crowd the wetter lowland habitats. The fossil record reveals that as of 100 million years ago, thousands of angiosperm species had rapidly evolved and spread into the same type of lowland areas and, in fact, had populated much of the recently separated North American continent. Today, angiosperms dominate most ecological zones. A few hundred species of the Mesozoic gymnosperms did survive and prosper, but now they are the most abundant large plants only where the land is dry; the soil, poor; the average temperature, relatively low; or the elevation, high. Most of the cycads and ginkgoes became extinct as the flowering plants, the angiosperms, invaded new territories. Coincident with the grand diversification of angiosperms was the corresponding abundance and diversification of mammals, birds, and flying insects; we shall see how these animals and the flowering plants have coevolved.

Figure 24-2   THE ANGIOSPERMS OF THE CRETACEOUS PERIOD.
The angiosperms are believed to have evolved rapidly, competing favorably for some of the terrains of the gymnosperms.
Common by 100 million years ago, the mid-Cretaceous period, flowering plants coevolved with mammals, birds, and flying
insects.

Let us consider each group of seed plants separately, to see how their characteristics help to explain this dramatic evolutionary history.

GYMNOSPERMS: "NAKED-SEED" PLANTS

The gymnosperms include some of the most familiar, majestic, and economically important members of the plant kingdom: the pines, spruces, redwoods, Douglas firs, and other so-called evergreens. The divisions Cycadophyta, Ginkgophyta, Gnetophyta, and Coniferophyta are called **gymnosperms**—a term that means "naked seed" and refers to the fact that the ovules and seeds of these plants are exposed on the surface of the sporophyte. Despite this common trait, the four types of gymnosperms look very different from one another, and the evidence strongly suggests that they evolved from

separate kinds of ferns and are not closely related phylogenetically. Therefore, some biologists think of the gymnosperms simply as a catchall group containing all the seed plants that are not angiosperms.

Cycadophyta

The nearly 100 species belonging to the division **Cycadophyta,** the **cycads,** are a fascinating group of primitive seed plants that is native only to tropical and subtropical areas (Figure 24-3a). Some species, however, especially the familiar sago palm (*Cycas revoluta*), are cultivated outdoors in warmer climates or grown indoors as house plants. *Zamia pumila* is the only species native to the United States, growing fairly commonly in the sandy-soiled woodlands of Florida. Cycads, with their short, thick stems and crowns of palmlike leaves, do resemble palms superficially. True palms, however, are angiosperms and are only distantly related to cycads.

Figure 24-3   THE CYCADS.
(a) The cycad sago palm (*Cycas media*) is so named because of its palmlike foliage. (b) Cones of a cycad. The tip of one cone is ruptured so that the sizes of individual, red-colored seeds can be seen.

Cycads grow as separate male and female plants, each of which produces a characteristic type of conelike strobilus near its apex (Figure 24-3b). Because each adult cycad produces only one type of gametophyte, the species is said to be **dioecious** (literally, "two houses"). Cycads have a typical gymnosperm life cycle: inside the strobili of female plants, **megaspores** are produced within the **megasporangium,** and these develop into the female gametophytes (**megagametophytes**). Egg cells then form within the many **ovules**—megasporangia surrounded by one or two layers of **integument** tissue— inside the strobili. Within the strobili of male plants, **microspores** are produced in **microsporangia,** and ga-

metophytes (**microgametophytes**) develop. Before the microgametophytes are mature, they are released as **pollen grains.** These encased, immature male gametophytes are blown about by the wind, and some chance to adhere to a strobilus that has produced a female gametophyte. There, the pollen grains mature and fertilize the eggs. After fertilization, the zygote develops into an embryo, the immature sporophyte, and the integuments surrounding the ovule become fleshy and fruitlike, forming seeds. These seeds remain in the conelike strobilus until they are dispersed, often by an animal, which eats the strobilus and then excretes the seeds. Cycad strobili and seeds can require up to ten years to mature; this extremely slow growth rate may help explain the decline of cycads relative to more rapid-growing seed plants.

Ginkgophyta

Some 25 million years ago, a few species of **ginkgoes** (division **Ginkgophyta**) grew abundantly throughout North America, Europe, and Asia. Today, just one species, *Ginkgo biloba*, the "maidenhair tree," survives (Figure 24-4). Chinese and Japanese monks may have inadvertently saved this species from extinction by planting it around their temples. The tree grows 25 meters tall and produces graceful, fan-shaped leaves. It is **deciduous;** that is, like many angiosperm trees, its leaves fall in the autumn, after turning a luminous golden color.

The leaves of the dioecious ginkgo are resistant to smoke and air pollutants; because of its resistance and bright fall foliage, the ginkgo is commonly planted along urban streets. The naked round seeds produced by the *ovulate,* or ovule-bearing, "female" tree have a fleshy outer layer that in late summer begins to overripen and produce butyric acid, a substance with an odor similar to that of rancid butter. For this reason, ovulate trees are seldom planted for ornamental purposes; this role is reserved for the pollen producers, which do not bear fruit.

Gnetophyta

The small but diverse group of gymnosperms called **gnetinas** (division **Gnetophyta**) has only about seventy living species, in three genera: *Gnetum, Ephedra,* and *Welwitschia.* About thirty-five species are in the genus *Gnetum*—mostly tropical vines and trees with large, leathery leaves. Most of the thirty-five or so species of *Ephedra,* in contrast, are densely branched shrubs that bear tiny leaves and grow in arid regions around the world (Figure 24-5a). The alkaloid *ephidrine,* derived from some species of this genus, acts much like adrenaline and is used in certain asthma and nerve-stimulant drugs.

(a)   (b)

Figure 24-4   *GINKGO BILOBA:* THE ONLY SURVIVING SPECIES OF THE DIVISION GINKGOPHYTA.
(a) A lane of the ancient survivors growing in a New Jersey field. (b) A close-up view of the leaves and a ripe fruit of *Ginkgo*.

One of the strangest-looking plants on Earth, the gnetina called *Welwitschia mirabilis*, is found in the Namib Desert of southwestern Africa (Figure 24-5b). This plant has a huge, carrotlike taproot that grows several feet deep in the sandy soil of the Namib and can store gallons of water. (A taproot is a single, major root that tends to grow straight downward into the soil.) The exposed part of the plant is a low, concave, woody crown with strobilus-bearing branches in the center and two large strap-shaped leaves that trail across the ground and

often become frayed as they are blown about on the rough surface by the wind, which gives the plant the appearance of having many leaves.

Coniferophyta

Cone-bearing plants, or **conifers** (division **Coniferophyta**), include many familiar and economically important species that grow in vast forests around the

(a)   (b)

Figure 24-5   THE GNETINA.
(a) The Mormon tea, *Ephedra* species, growing in Canyonlands National Park, Utah. The tiny leaves cannot even be seen on these narrow, cylindrical stems. (b) The gnetina *Welwitschia mirabilis*, which is found in the Namib Desert of southwestern Africa. The large taproot stores gallons of water, and the plant's two leaves become frayed so that they come to look like many leaves.

Figure 24-6   A LANDSCAPE DOMINATED BY CONIFERS.
Lodgepole pine, Engelman spruce, and other evergreen conifer species constitute the forests near the Teton mountains in western Wyoming.

world (Figure 24-6). These plants are often called evergreens because most species remain green year-round—although many other plant groups do so as well, and several conifer species are actually deciduous, including the larches of the genus *Larix*, the bald cypress (*Taxodium distichum*), and the dawn redwood (*Metasequoia glyptostroboides*). In total, there are between 500 and 600 contemporary conifer species.

Conifers are important to human commerce as well as being fascinating because of their size or age. The paper on which this book is printed is made from conifer wood pulp, and the building you live in no doubt contains framing plywood and other building materials derived from conifers. Millions of acres in cold, mountainous regions are dominated by pines, spruces, firs, cedars, or other conifers. Decorative landscaping often includes smaller conifers, such as yews, hemlocks, junipers, and larches. In addition, the tallest, oldest, and most massive plants on Earth are conifers. Coast redwoods (*Sequoia sempervirens*) thrive in moist, cool conditions; one has grown to a height of almost 113 meters (372 feet). One scrubby bristlecone pine (*Pinus longaeva*), found at an elevation of 10,700 feet in western Nevada, has been estimated to be 4,900 years old. And the 2,500- to 3,000-year-old General Sherman Tree, a giant sequoia (*Sequoiadendron giganteum*) growing in the Sierra Nevada of California, has a circumference of 31.3 meters (102.6 feet) at its base.

The leaves of many conifers are similar to those of pine (Figure 24-7): long, narrow, somewhat flattened needles with a thick waterproof *cuticle*, or surface layer, that contains many stomata; all these modifications minimize water loss in dry environments. Because the stomata can be opened or closed, gas exchange and water vapor loss can be regulated. Needles also have internal vascular

(a)

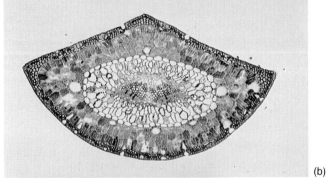

(b)

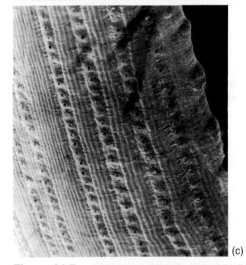

(c)

Figure 24-7   CONIFER LEAVES AND CONES.
Most conifers have needlelike leaves. (a) Needles, two brown- and green-colored ovulate cones, and the small pollen-bearing cones (top) can be seen on this branch of a pine. (b) Cross section through a pine needle (magnified about 25 times). Numerous stomata are visible as white "holes" in the thick, brown-colored, waterproof cuticle. The internal vascular system (whitish tubes) is surrounded by a regularly arranged ring of cells (the endodermis). Can you spot the six resin ducts just outside the endodermis? (c) Close-up of a pine needle. The surface of this needle (magnified about 60 times) from a Monterey pine is waxy and waterproof. The regular rows of stomata are clearly visible, as are the parallel ridges covering the underlying pipelines of the needle's vascular system.

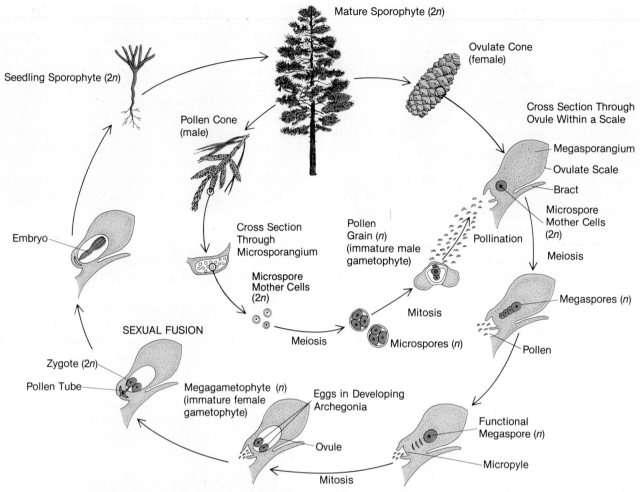

Figure 24-8   LIFE CYCLE OF A PINE, A TYPICAL CONIFER.
The mature sporophyte (tree) produces both ovulate (female) and pollen-bearing (male) cones. The pollen grains, or immature male gametophytes, develop from the microspores within the microsporangia of the male cones. When released, some pollen may chance to adhere to a female megasporangium (upper right). Each scale on a female cone is a megasporangium containing a megaspore mother cell. That cell yields haploid megaspores, which develop into archegonia, each housing one egg cell. As these processes in the developing archegonia go on, a very slow process pulls the pollen grain's pollen tube through the micropyle. The pollen tube enables the sperm nuclei to fertilize the eggs within the archegonia. Although several eggs may be fertilized, usually just one develops. After fertilization, the ovule and its embryo within develop into the seed. The seed eventually germinates, and a seedling (sporophyte) is produced.

tissue to transport water, minerals, and sugars, and the leaves of many conifers have ducts to transport *resins*—fragrant gummy substances that may protect the leaves and plant from insects.

The life cycle of the pine, diagrammed in Figure 24-8, is typical of most conifers. The tall, conical adult tree is the diploid sporophyte and produces both male and female cones. Ordinarily, the small pollen-bearing cones are borne on the lower branches, and the large ovulate cones are borne on the higher branches. Because both pollen-bearing cones and ovulate cones are found on the same individual, pines are considered to be **monoecious** ("one house"), in contrast to the dioecious cycads and ginkgoes.

A pine tree's ovulate cone is composed of tough, woody scales radiating from a central axis; each scale bears a pair of ovules close to the central axis, where they are out of sight and protected from harm. Below these scales are small, stiff structures called *bracts*. Ovulate cones range in size from 0.8 centimeter long (white cedars) to as much as 61 centimeters long (sugar pines) when the seed has matured. Ovulate cones also vary considerably in structure. Pine cones are probably familiar to you, but you may not have realized that the red and blue "berries" of yews and junipers are actually fleshy ovulate cones.

Each ovule is composed of a megasporangium enclosed in one or two integuments with an opening at one end called the **micropyle.** Within the megasporangium, a diploid **megaspore mother cell** divides meiotically to produce four relatively large haploid megaspores. The three megaspores nearest the micropyle degenerate, leaving one functional haploid megaspore, which divides mitotically and develops over a year or so into the multicellular, haploid female gametophyte. Housed within the ovule, this megagametophyte is nonphotosynthetic and depends completely on the sporophyte for nourishment. It develops two or more archegonia, each of which contains one haploid egg.

Pollen-bearing pine cones are composed of soft tissue as opposed to hard woody tissue; in many species such cones are colored red or yellow, and consist of a central axis surrounded by sporophylls (Figure 24-8). Each sporophyll bears two microsporangia, inside of which diploid **microspore mother cells** give rise through meiotic divisions to haploid microspores. Each microspore divides mitotically to produce a four-celled pollen grain, an immature male gametophyte that will later mature and deliver sperm to the eggs. The microspore wall develops two winglike projections that facilitate pollen distribution by wind.

Pollen-bearing cones are generally short-lived, remaining on the tree only long enough to produce huge numbers of pollen grains in the spring. Although each pollen grain is microscopic, so many are produced that the air downwind from a pollen-shedding tree can appear golden. This cloud of conifer pollen is blown about by spring breezes. Individual pollen grains that chance to fall on an ovulate cone and sift down into the open scales become fixed to a sticky fluid secreted by the maturing ovule and exuded through the micropyle. This reception of pollen by the ovulate cone is called **pollination.** The sticky fluid later dries and contracts, thereby pulling the pollen through the micropyle and inside the ovule's integument. At the time of pollination and **germination,** or reactivated growth of the pollen grains, the megasporangium has not yet produced a megaspore. Megasporogenesis does not occur until about a month later.

Following pollination, the developing male gametophyte matures over the next fifteen months, digesting a path through the integument and extending a **pollen tube** toward the developing female gametophyte and one of the archegonia that it will contain at maturity. When the pollen tube reaches the egg cell in the archegonium, much of its cytoplasm and two nonmotile haploid sperm nuclei are transferred into the egg. Fertilization occurs when one of the sperm nuclei fuses with the egg nucleus. The other sperm nucleus then degenerates.

In a typical conifer, well over a year passes between pollination and fertilization. During the second summer, the newly fertilized egg (zygote) develops into an embryo inside the ovulate cone. Although each megagametophyte contains several archegonia and the egg that

each produces is often fertilized, usually just one embryo develops fully. The one successful embryo enlarges by mitotic divisions within the ovule to form a tiny plant, the new sporophyte, with an embryonic root and several embryonic leaves. The ovule's integument hardens and develops into the protective *seed coat.* The female gametophyte tissue remains as a food source for the developing embryo. By late in the second summer, the scales of the new woody ovulate cone open, and the mature seeds fall to the ground. Many of these seeds will be broken open and consumed by hungry squirrels and other forest animals, for the nutrients stored in the seed are just as useful to animals as to the tiny developing plant. Indeed, conifer seeds are sometimes collected, roasted, and eaten by people as "pine nuts." The seeds that survive unscathed lie dormant through the fall and winter; the following spring, a small percentage germinates successfully and grows, completing the life cycle.

In summary, the gymnosperms are a small but diverse group, ranging from "living fossils" (cycads and ginkgoes) to bizarre forms (gnetinas) to common and economically important species (conifers). They include the tallest, oldest, and most massive plants on Earth. Their reproductive cycle can be quite prolonged, with pollination taking place in the spring; fertilization, the next spring; and seed dispersal, the late summer of the second year. They differ from the flowering plants, as we shall see, in subtle but important ways.

ANGIOSPERMS: THE APEX OF PLANT DIVERSITY

While gymnosperms are usually the dominant vegetation of the world's cold and high-elevation forest areas, **angiosperms** (division **Anthophyta**) are the most common and conspicuous species in tropical and temperate regions (Figure 24-9). Even where gymnosperms are the dominant vegetation, however, angiosperms are ecologically important to various animals, fungi, and other organisms. More than 95 percent of all living seed plants and over 80 percent of all the plant species on Earth are angiosperms. They range in size from the giant *Eucalyptus marginata,* which may be more than 100 meters tall, to some duckweeds like *Wolffia microscopica,* which is barely 1 millimeter across. From flowering plants come virtually all our food crops (rice, corn, wheat, beans, tomatoes), herbs and spices (pepper, cinnamon, sage), beverages (tea, coffee, fermented drinks), cloth, animal fodder, and many dyes, medicines, and psychoactive drugs.

The word "angiosperm" means "seed in a container," which refers to the specialized structure that encloses

Figure 24-9   LANDSCAPE DOMINATED BY ANGIOSPERMS.
Sunlight, warm spring rains, and rich soil near Sequin, Texas, produce this luxurious growth of phlox (*Phlox drummond,* foreground) and Texas blue bonnets (*Lupinus subcranasas,* background).

the ovules and the seeds of all members of this division. There are two major classes of angiosperms: the **dicotyledons** (class Dicotyledon) number between 170,000 and 190,000 species, and the **monocotyledons** (class Monocotyledon) total 50,000 to 60,000 species. These names derive from the number of embryonic seed leaves, or **cotyledons,** which serve as storage sites for the nutrients that sustain growth of the newly germinated plants. The embryos of dicotyledons have two seed leaves, while the embryos of monocotyledons have only one seed leaf. Other differences between these two groups are summarized in Table 24-1.

Special Adaptations of Angiosperms

Why have the angiosperms been so successful? First of all, they have the basic adaptations for life on land common to vascular plants: a waterproof covering of the waxy substance cutin to reduce water loss; stomata with guard cells to regulate the exchange of water vapor, carbon dioxide, and oxygen with the atmosphere; a vascular system for the transport of water, minerals, and nutrients throughout the plant; and specialized tissues for support of the plant body. Angiosperms also have the adaptation—shared with the gymnosperms—of reproducing by means of seeds (Figure 24-10).

Angiosperms have some unique characteristics as well. The ovules and developing embryos of angiosperms are protected within a specialized structure that is part of the **flower.** Thus angiosperms are often simply called "flowering plants." Flowers are the keys to these plants' success since their colors, odors, and shapes serve to attract insects, birds, and mammals that inadvertently transfer pollen from male to female flower parts. That free transportation makes the efficiency of

fertilization far greater than fertilization which depends on chance distribution of pollen or male sex cells by wind or water, as in the gymnosperms and lower plants. In fact, some biologists now believe that flowers evolved primarily as structures to disperse pollen so that fertilization would be most efficient. In addition, however, part of the flower's seed-protecting "container" develops into the **fruit,** a mature ovary or group of ovaries that surrounds the seeds and aids in their dispersal and protection. Fruits are sometimes luscious and colorful, sometimes dry and inconspicuous, but virtually all angiosperms produce some type of fruit.

Another crucial angiosperm adaptation is rapid growth, particularly of reproductive structures: in a barley plant, for example, pollination, growth of the pollen tubes, and fertilization all take place in less than one hour, in marked contrast to the year or more these stages require in pines. Growth of the entire plant can be quite fast, too. Iowans claim they can hear the corn growing in the rustling of the leaves on a warm summer day. Some bamboos can grow as much as three feet in one day.

Rapid growth requires a high rate of photosynthesis, which, in turn, requires a relatively large leaf surface and explains why most angiosperms have broad leaves rather than needles. But large leaves increase the demand for water, since a great deal of water can evaporate through the open stomata as carbon dioxide enters for photosynthesis. Nearly all the angiosperms have a specialized cell type in their vascular system—the *vessel cell*—that is not found in most gymnosperms and other

Table 24-1   **MONOCOTS AND DICOTS: DIFFERENCES OBVIOUS AND SUBTLE**

Characteristic	Dicotyledons	Monocotyledons
Usual Arrangement of Flower Parts (sepals, petals, stamens)	Multiples of Fours or Fives	Multiples of Threes
Number of Cotyledons (seed leaves)	Two	One
Usual Pattern of Leaf Venation	Network	Parallel
Usual Arrangement of Vascular Bundles in Young Stem	Circle	Scattered
Usual Presence of Secondary, or Woody, Growth	Present	Absent

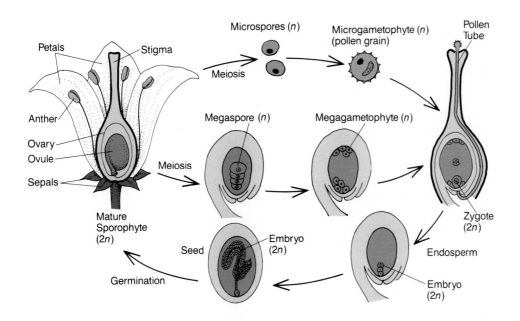

Microspores (n)

Microgametophyte (n)
(pollen grain)

Pollen
Tube

Petals

Stigma

Meiosis

Anther

Ovary
Ovule

Megaspore (n)

Megagametophyte (n)

Sepals

Meiosis

Mature
Sporophyte
(2n)

Zygote
(2n)

Seed

Embryo
(2n)

Endosperm

Germination

Embryo
(2n)

Figure 24-10   THE ANGIOSPERM LIFE CYCLE.
The life cycle of a typical angiosperm is very similar to that of a typical gymnosperm (Figure 24-8). The mature sporophyte plant produces flowers. Meiosis yields male (microspore) and female (megaspore) haploid cells, which develop into the microgametophyte (pollen grain) or the megagametophyte. A process called double fertilization yields a diploid zygote and endosperm, which become enclosed in the seed (Chapter 28). Germination and growth produce the mature sporophyte again.

vascular plants and that allows for much more efficient water conduction. Veins, sites of major conducting vessels, are prominent parts of angiosperm vascular systems. We will consider vessel cells, veins, and vascular systems in some detail in Chapters 27 and 29.

Flowers: Key to the Angiosperm Life Cycle

Of all the adaptations seen in angiosperms, the flower is the most distinctive. Flowers not only have been critical to the success of these seed plants, but also have assumed significance to humans at times of birth, death, and marriage as symbols of beauty, purity, fertility, affection, and renewal.

A flower can be thought of as a central axis with four specialized whorls of distinctive parts: sepals, petals, stamens, and carpels (Figure 24-10). The *petals* are often large and colorful and are very important in attracting insects, birds, and other animals to carry away pollen. Male *anthers* (a part of the *stamens*) are each a group of four chambers in which pollen grains develop. The female part, the *carpel*, has a *stigma*, the sticky top surface to which pollen grains adhere, and an enlarged base, or *ovary*, containing developing ovules. Angiosperm ovules are structurally similar to gymnosperm ovules: each is a megasporangium surrounded by one or two integuments with a micropyle for fertilization. We will discuss these structures in detail in Chapter 28.

An important feature of angiosperm reproductive strategy is the *endosperm*, a nutrient store that sustains the developing plant embryo. The endosperm does not develop within the ovary until fertilization actually occurs and a zygote exists. In this respect, flowering plants are analogous to placental mammals, whose young receive virtually all their nourishment from the body of the

mother *after* fertilization. In many gymnosperms, in contrast, most of the food reserves are built up from female gametophyte tissue *before* fertilization. If the gymnosperm egg is not fertilized, the food stored in the gametophyte is wasted, unless it can be resorbed. In a sense, gymnosperms are analogous to reptiles and birds, whose females invest substantial metabolic energy in manufacturing and storing food in all eggs—energy that is wasted on those eggs that are never fertilized.

Pollination and Pollinators: Key to the Reproductive Success of Angiosperms

While flowers provide protective enclosures for the developing ovules and seeds, they confer another novel and equally important advantage over the gymnosperms' exposed seeds and cones: greater insurance that pollination will occur and thus greater reproductive success. Just what takes place?

After the flower's pollen grains have developed, the anther splits open, and the mature pollen grains are dispersed by wind or carried by **pollinators**—insects, birds, or mammals—to the sticky tip of the carpel on a flower of the same species. The pollinated flower is often not the one that originally produced the pollen, although self-pollination does occur in many species. (Recall from Chapter 10 that Mendel had to manipulate pea flowers to prevent self-pollination.)

Most flowers attract animal pollinators that act as inadvertent couriers for the tiny pollen grains. Indeed, flowers evolved as attractants for pollinators. Most pollinators benefit directly from the flowers they pollinate by consuming sugary nectar or protein-rich pollen. Unwittingly, their bodies become dusted with pollen grains, which they carry from flower to flower and from plant to plant as they move around in search of food. An individ-

THE ORIGIN OF CORN: AN UNSOLVED MYSTERY

Corn is America's largest crop. It is also the most important food crop in the Western Hemisphere, and a close third to wheat and rice as the largest annual harvest worldwide. The 400 million metric tons of corn grown each year support billions of farm animals and nations of people. Yet the ancestry of this valuable staple is shrouded in mystery—a mystery with surprising implications for modern agriculture.

By the time Columbus landed in the New World, native Central American agriculturists had selected and bred more than 200 varieties of corn, including yellow, red, and blue field corn; yellow and white sweet corn; and popcorn. This accomplishment has been called the greatest feat in the history of agriculture because there is no nondomesticated plant in nature—at least not today—that closely resembles corn as we know it. The indigenous people of the Americas must have started with wild grasses of some kind, since corn is in the grass family. But which grasses? And when? Archeologists have found small ears of corn (Figure A) amid Indian remains dating to 7,000 years ago, and many biologists agree that the initial selection and breeding of corn's progenitor must have taken place between 8,000 and 15,000 years ago. However, this still leaves the mystery of which progenitor plant the farmers actually used.

A lively scientific debate has sprung up around this question centering on theories about corn's ancestry: (1) some biologists believe that the progenitor was a grass called *teosinte*, which still grows wild in Mexico, Guatemala, and Honduras; (2) others believe that a true wild-corn species, now extinct, gave rise to both modern corn *and* teosinte; (3) still others hypothesize that the Indians crossbred the extinct wild corn with teosinte and that the resultant hybrid gave rise to corn.

The most prevalent theory is the first, despite the enormous differences between the two plants; teosinte has a small spike of triangular kernels with very hard shells (Figure B), while corn bears a large ear of squarish kernels (Figure C); and teosinte has numerous thin stalks and many spikes (Figure D), while corn has only one stalk and bears only one or two ears.

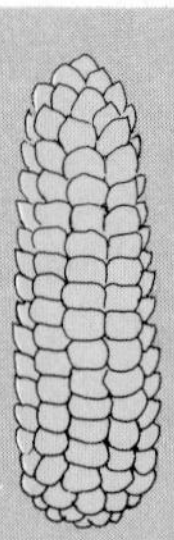

Figure A
An archeological specimen of corn, dated at 7,000 years old.

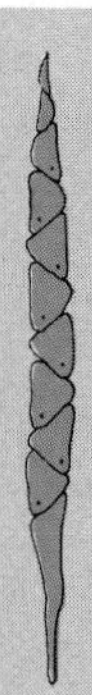

Figure B
The teosinte spike, one row of hard triangular kernels.

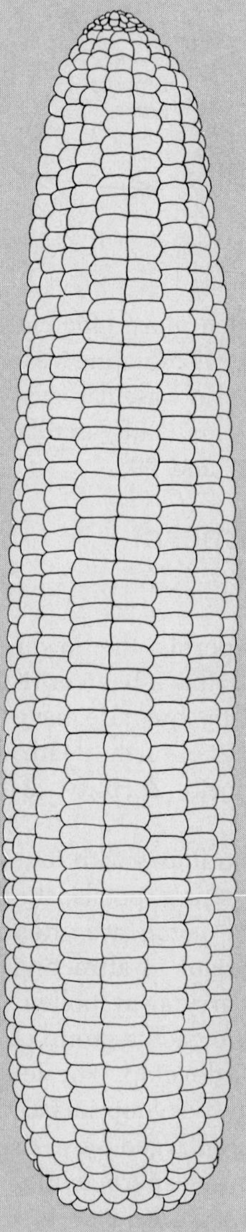

Figure C
An ear of modern domesticated corn.

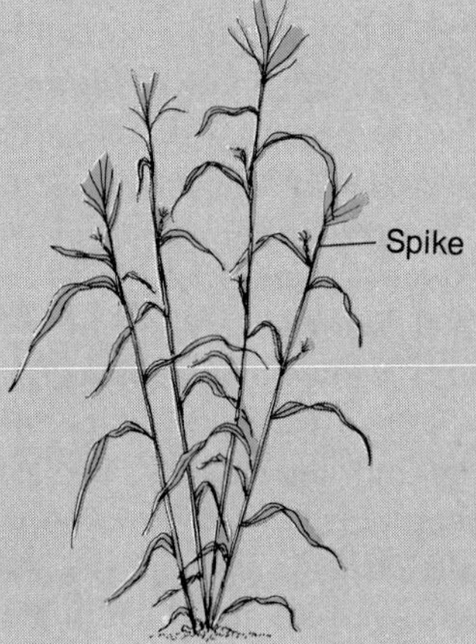

Spike

Figure D
The teosinte plant.

ual plant that is isolated from others of its species may be easily spotted by a flying or walking animal scouting for a favorite food source and thus may still have a good chance of achieving pollination. Some angiosperms, including grasses and oaks, reverted to the wind pollina-tion used by gymnosperms, even though they probably evolved from angiosperm ancestors that were insect-pollinated. This reversion is a curious phenomenon, however, because wind pollination is a hit-or-miss prop-osition that depends on wind speed and direction and on

Much as the ancestry of corn may seem like an academic argument with little relevance, the question bears directly on the long-term survival of modern hybridized corn. Amazingly, more than 70 percent of the entire United States corn crop derives from only six parent lines. Thus literally billions of corn plants are close to identical—a lack of diversity that means an insect, fungal, or viral pest could arise through mutation and devastate a large percentage of the crop in short order. If the ancestor to corn turns out to be teosinte instead of an extinct progenitor, then there would be a living reservoir of genes in the form of a sturdy Latin American weed that still grows wild and undergoes natural selection. Plant breeders hope that with advanced techniques of biotechnology, such genes could be transferred to modern hybrid corn to yield more vigorous plants.

An exciting possibility along these lines was discovered fairly recently by a Mexican undergraduate student of botany. Rafael Guzmán hiked into a hilly region in the state of Jalisco and found a perennial form of teosinte that grows up from the same roots year after year. If this trait for perennial growth could be transferred to corn, farmers could grow the crop without costly annual plowing of soil and sowing of seed. The savings in diesel fuel for farm equipment would be an estimated $300 million a year.

A great deal more work will be needed in the search for corn's historical roots. But in the meantime, many biologists have called for the preservation of wild teosinte in parks, gardens, and preserves. It would be tragic if a priceless reservoir of wild genes disappeared before the mystery of corn's ancestry could be solved and before those genes—if appropriate—could be used to strengthen our vulnerable number one crop.

Figure 24-11   FLOWERS AND THEIR ANIMAL POLLINATORS.

The apparent coevolution of flowers and their animal pollinators—often birds, bees, beetles, and mammals—has involved the development of flower structures, colors, and odors linked to the body shape, color vision, and odor preference of animal pollinators. (a) Pollen from a banana plant dots the head of this fruit-eating bat; (b) a bumblebee and a butterfly share the feast, but carry away the pollen of this thistle; and (c) this oddly striped fly feeds on a daisy before carrying pollen to another flower.

the plants' locations and distance from one another. To compensate for these factors, a wind-pollinated plant must produce 1 million pollen grains for each successfully fertilized ovule, while an insect-pollinated plant produces only about 6,000.

A fascinating coevolution of flowers and their pollinators has produced specialized physical characteristics in both that help ensure the exchange of nutrients for the sake of pollination (Figure 24-11). Various species of bees, butterflies, moths, bats, and birds have mouthparts of the right shape and length to fit the flower parts of particular plant species. Plants, in turn, have specialized characteristics for attracting appropriate pollinators and ensuring that pollen is transmitted from flower to flower within the same species. Biologists believe that these interlocking characteristics, like all other evolutionary adaptations, emerged gradually and conferred an advantage on both plants and their pollinators, so that they were retained and spread through related species over time.

One striking example of coevolution can be seen in the interaction of scarlet gilia, which grows in the mountains of Arizona, with two pollinators, hummingbirds and hawk moths. Scarlet-gilia flowers, which have a long tubular shape, are red early in the summer. Hummingbirds are especially able to sense color at the red end of the spectrum and have long, narrow beaks and tongues that are ideal for delving into tubular flowers. By August, the hummingbirds migrate away. Pollination continues, however, because individual plants and the population as a whole produce pink and even white flowers—perfect for attracting hawk moths in the evening hours. Thus the plants have evolved color-shifting as an adaptation to the two types of pollinators. The result is a longer flowering season and more reproductive success for the plant species.

As another example, flowers pollinated by bees are often blue or yellow and tend to be fragrant; bees can perceive blue and yellow but not red and can detect sweet odors from great distances. Hence bees tend to visit flowers with these characteristics, instead of flowers such as scarlet gilia. Bees can also detect light in the ultraviolet range; it is probably not a coincidence, therefore, that many flowers have complex markings visible to humans only under "black" ultraviolet light. The marsh marigold, for example, has a series of ultraviolet-reflecting or -absorbing flower markings called honey guides that direct bees to the store of nectar hidden inside the flower. Most such markings are invisible to humans, but on a few flowers, like the lemon-yellow toadflax, the spots are visible to both humans and bees.

Flowers have evolved certain mechanical adaptations to ensure that only pollinators with appropriate body structures can accomplish the pollen exchange while feeding. For example, only strong bumblebees (as opposed to smaller insects) can move a plate that blocks the path to the nectar in certain species of sage. When a bee presses against this plate, two levers drop down and apply pollen to the bee's back. The bee, now carrying its thick pollen coat, must pass beneath the curved, sticky tip of the next sage flower's carpel, thus pollinating it before pushing on that plant's plate and getting more nectar. Alfalfa flowers have a set of organic levers called sexual columns. When a bumblebee trips these levers, it is showered with pollen as though from tiny slingshots, and this pollen clings to the insect's stubby bristles as it

flies away to the next alfalfa flower.

The range of flower shapes is much greater today than it was 100 million years ago, as represented in the fossil record. The earliest angiosperms were probably small bushes bearing flowers similar to that of the modern Magnolia family (actually the Chloranthaceae); such flowers were probably radially symmetrical (they are the same in each direction from the flower center and do not have right and left halves), and on the basis of fossilized pollen and anthers had wind-blown or beetle-assisted fertilization. Many of the flowers of species that have evolved during the Cenozoic tend to be small, to be *bilaterally symmetrical* (one axis through the flower identifies right and left halves), to have fused flower parts, to occur in clusters, or to have some combination of these traits. Some botanists compare *Hepatica*, a modern woodland plant, to an orchid in order to show the sorts of evolutionary changes that took place in flowers during the Cenozoic (Figure 24-12).

Figure 24-12   EVOLUTIONARY TRENDS IN FLOWER SHAPES.

(a) Early flowers of the Chloranthaceae may have resembled the modern liverleaf, or *Hepatica*. The flower is large and symmetrically arranged around the reproductive structures, has separate petals, and bears both male and female flower parts. (b) Evolutionary trends are visible in the orchid, which shows bilateral symmetry, has few petals, and bears fused reproductive organs (not visible). Other flower species may be imperfect (either male or female), may have fused petals, and may grow in a cluster called an *inflorescence*.

(a)

(b)

Fruits: An Important Angiosperm Evolutionary Strategy for Seed Dispersal

Just as flowers increase the likelihood of pollination, so the shapes and sizes of fruits help ensure the dispersal of seeds from the parent angiosperm to new locations. A fruit is the product of a ripened ovary and often develops along with associated parts such as the receptacle, the part of the stem to which the flower is attached. After fertilization, the wall of the ovary often undergoes modification to become a fruit wall, or *pericarp*. The pericarp can be hard and dry, as is the fibrous husk of the coconut; thin and papery, as in the tiny fruits of the elm tree; or fleshy and moist, as in apricots and peaches.

It is sometimes advantageous for a parent plant simply to drop its seeds to the ground, directly below its outstretched branches. This often occurs in *annuals*, plants that live for just one year; since the sporophyte dies at the end of the yearly cycle, its growth site becomes vacant, and the dropped seeds have a good chance of attaining the foothold necessary for survival when they germinate. An advantage, of course, is that the spot has already proved to be good for the growth and reproduction of the parent. For some types of annuals, however, and for *perennials*—plants that survive and reproduce over several years—it is more advantageous for the seeds to be dispersed from the parent plant and to germinate elsewhere, in a location where the new plants will not compete with the parent for nutrients, light, and moisture.

Thus evolution has tended toward a minimization of competition—and the result is dozens of kinds of fruit shapes and sizes (Figure 24-13). Some fruits are structurally modified to facilitate seed dispersal by wind. For example, winged maple fruits ("squirts") flutter as they are blown through the air; tiny dry dandelion fruits are suspended beneath plumelike tufts that can be blown long distances; and fruits of squirting cucumber split open and catapult their seeds through the air. Other fruits disperse seeds by water. The coconut, for example, is buoyant and will float for weeks or even months as ocean currents disperse the encased embryo hundreds of miles from the parent palm tree.

A fleshy, palatable fruit wall is another major means by which fruits can ensure seed dispersal. Seeds inside edible fruits usually have hard, acid-resistant coats that allow them to pass undamaged through an animal's digestive tract. Such fruits—cherries and grapes, for example—have colors and flavors that attract birds and mammals. These animals consume the fruit and later deposit the seeds in a distant location along with a bit of organic fertilizer. The same animal that swallows seeds and inadvertently transfers them to a new location may also carry other seeds on its fur or feathers. Many fruits,

Figure 24-13   FRUIT SHAPES AND SIZES.
The variation in shape, size, and type of fruits is important for the dispersal of the seeds contained within. The fruit of the dandelion (*Taraxacum officinale*) (a) is modified for dispersal by wind, while that of the coconut palm (*Cocos nucifera*) (b) is adapted to dispersal by ocean currents. Note the pile of coconuts beneath this tree. Too luscious to resist, these Royalann cherries (*Prunus* species) (c) and blackberries (*Rubus rubrisetus*) (d) have seeds resistant to the acids in the digestive tracts of animals; thus the seeds can be dispersed by the birds and mammals that eat the fruits. The soft, sweet cherry, peach, apple, or other fruit tissue is digested, of course, and is the reason a variety of animals, including people, have evolved a "taste" for such fruits and berries.

such as that of the cocklebur, have hooked spines that allow them to "hitch a ride" to a new spot by clinging to mobile passers-by. Of course, they fall off the animal by chance and so, like those deposited in guano or dung or wafted by the wind, may or may not wind up in a hospitable site for germination, but many times they do.

LOOKING AHEAD

The fruits and flowers of angiosperms represent a high point in plant evolution and are the result of an immensely long progression of increasing specialization that began with the first algae. So, too, the ingenious symbiotic relationships between plants and animals reflect a slow progression of interdependences over vast stretches of time. In the next chapter, we will go back in Earth's history to explore the origins and evolution of the invertebrate animals, some of which came to act as plant pollinators. We should remember, however, that the emergence of the diverse and ubiquitous primary producers—the plants—made possible the evolution of all other multicellular forms of life.

SUMMARY

1. Seed plants are classified as either *gymnosperms,* "naked-seed" plants, or *angiosperms,* flowering plants.

2. All seed plants (a) produce *seeds,* the *ovule* containing the plant embryo and nutrients; (b) produce multicellular female gametophytes (*megagametophytes*) in a *megasporangium* protected by sterile *integument* tissue; (c) have reduced male (*microgametophyte*) and female gametophytes that are entirely dependent on the sporophyte; (d) have sperm that do not require water to reach the egg; and (e) have relatively large leaves and efficient vascular systems for internal transport.

3. The first seed plants, the seed ferns, were abundant during the Carboniferous period and are now extinct. The gymnosperms, the major plants at the time of the dinosaurs, flourished during the Mesozoic and remain the primary vegetation in cooler, higher habitats. During the past 100 million years, the angiosperms have evolved vast diversity and now are the primary plant types in most land habitats.

4. The gymnosperms are classified in four divisions: *Cycadophyta* (the *cycads*), *Ginkgophyta* (the *ginkgoes*), *Gnetophyta* (the *gnetinas*), and *Coniferophyta* (the *conifers*).

5. The conifers, or cone-bearing plants, include the tallest, oldest, and most massive plants on Earth. Although most conifer species are evergreen, a few are *deciduous,* dropping their leaves in the fall. Conifers are the most common forest trees in cold, dry, and mountainous regions.

6. The sporophyte (adult tree) of a typical conifer produces both male and female cones; it is said to be *monoecious,* in contrast to the cycads, for example, which are *dioecious* (grow as separate male or female plants). The ovulate (female) cone bears ovules in which diploid *megaspore mother cells* divide meiotically into haploid *megaspores,* one of which divides mitotically into the multicellular female gametophyte. Each megagametophyte has two or more archegonia, each of which contains an egg. The pollen-bearing (male) cone bears *microsporangia* in which diploid *microspore mother cells* divide meiotically into haploid *microspores,* each of which divides mitotically into a four-celled *pollen grain,* the immature male gametophyte. After being released, pollen grains are dispersed by the wind. Those that land on an ovulate cone enter through the *micropyle* and then extend a *pollen tube* toward one of the archegonia of the female gametophyte. Fertilization takes place after sperm nuclei reach the egg through the tube.

7. The angiosperms, which make up the vast majority of modern plant species, are classified in one division: *Anthophyta.* They have enclosed ovules, in contrast to the gymnosperms. Angiosperms are divided into two classes—the *dicotyledons* and the *monocotyledons*—depending on the number of *cotyledons,* or seed leaves, the embryo has. Angiosperms produce *flowers* and *fruit,* both of which help ensure reproductive success. Flowers have four whorls of parts: sepals, petals, stamens, and carpels. Stamens are pollen-producing structures and carpels include an ovary in which ovules develop.

8. The shapes and structures of flowers help ensure *pollination* by attracting specific insect, bird, or mammal species that will inadvertently carry pollen to flowers of the same species. Animal *pollinators* and the plants they pollinate have coevolved complementary structures and behaviors that aid both organisms.

9. The shape and characteristics of a fruit—the mature ovary surrounding the seeds—help ensure seed dispersal by adaptations to fluttering on wind (maple "squirt"), floating on water (coconut), clinging to fur or feathers (cocklebur), or being eaten and excreted by hungry animals (cherry).

KEY TERMS

angiosperm

Anthophyta

conifer

Coniferophyta

cotyledon

cycad

Cycadophyta

deciduous

dicotyledon

dioecious

flower

fruit

germination

ginkgo

Ginkgophyta

gnetina

Gnetophyta

gymnosperm

integument

megagametophyte

megasporangium

megaspore

megaspore mother cell

microgametophyte

micropyle

microsporangium

microspore

microspore mother cell

monocotyledon

monoecious

ovule

pollen grain

pollen tube

pollination

pollinator

seed

QUESTIONS

1. What are the main characteristics of the seed plants?

2. What is pollen? How is it formed? How does fertilization occur in conifers?

3. Diagram the life cycle of a pine.

4. How much time elapses between pollination and fertilization in a pine? in barley?

5. How did evolution of the seed-plant life cycle free plants from dependence on standing water?

6. The "golden age of gymnosperms" coincided with the dominance of which land animals? Which lived during the "golden age of angiosperms"? Can you draw some parallels between the animal and the plant groups?

7. What characteristics would you look for in a newly discovered specimen to determine whether it was a seed plant or a nonseed plant? a gymnosperm or an angiosperm? a monocot or a dicot?

8. What is the biological role of flowers? How does a plant benefit by attracting insects? How do the insects benefit? What is the biological role of fruit? Describe some methods of seed dispersal. Seedlings of berry plants often are found beneath trees. Explain why.

9. Draw a flower and label the parts. Explain where the female gametophyte forms; the male gametophyte. When do the gametophytes mature?

10. Define the following terms:
   a. endosperm
   b. micropyle
   c. pollen tube
   d. ovule
   e. megasporangium
   f. microspore mother cell
   g. coevolution
   h. monoecious
   i. deciduous

ESSAY QUESTION

1. Many plants produce psychoactive and addictive drugs—such as caffeine, nicotine, opium, and tetrahydrocannabinol—some of which are deadly poisons. Many plants produce bitter or corrosive substances. Some of these have medicinal value for humans. Why do plants manufacture these substances? Do they benefit the plants in some way?

SUGGESTED READINGS

DUDDINGTON, C. L. *Evolution in Plant Design.* London: Faber & Faber, 1969.

This book's chapters on pollination mechanisms may be the best yet written.

EAMES, A. J. *Morphology of the Angiosperms.* New York: McGraw-Hill, 1961.

The primary source book on the evolution, reproduction, and morphology of the flowering plants, written by one of the preeminent plant morphologists of the twentieth century.

MILNE, L., and M. MILNE. *Living Plants of the World.* New York: Random House, 1975.

Marvelous illustrations and discussions of many of the most important plants, including angiosperms and gymnosperms.

MIROV, N. T., and J. HASBROUCK. *The Story of Pines.* Bloomington: Indiana University Press, 1976.

A wonderfully readable, popular treatment of the biology and aesthetics of the gymnosperms.

RAY, P. M., T. A. STEEVES, and S. A. FULTZ. *Botany.* Philadelphia: Saunders, 1983.

This text's sections on the higher plants, plant reproduction, and pollination are especially strong.

25

INVERTEBRATES: EVOLUTIONARY DIVERSITY

*From protozoans to sea stars, from squid to insects,
invertebrates represent the essential diversity of animal life itself.*
Robert H. Barth and Robert E. Broshears,
The Invertebrate World (1982)

Splattered like a Jackson Pollock canvas, the nudibranch mollusk *Chromodoris banksi* crawls over a reef.

A distinct majority of multicellular species on Earth today are animals. The marvelous diversity of the plant kingdom seems modest compared with the wealth of animal types: there are four animal species for every plant species. And of the 2 million species of animals alive today, 97 percent are **invertebrates,** the subjects of this chapter.

"Invertebrate" is a term of convenience, as are "alga" and "nonvascular plant." It refers to a collection of organisms as different from one another as earthworms, butterflies, lobsters, and octopuses—which bear no close genetic, anatomical, or phylogenetic relationship to one another. "Invertebrate" is really a catchall term used traditionally to encompass all animals that lack backbones (the backbone, or vertebral column, is a hallmark of the *vertebrates*, the group to which humans belong).

In this chapter, we continue our survey of life forms—which began with the monerans and went on to the pro-

tists, fungi, and plants—with the kingdom Animalia and, specifically, with its real majority, the invertebrates. We focus here on the main types of invertebrates and on the major features of each group, whether those characteristics are anatomical structures, physiological adaptations, modes of reproduction, or behavior patterns. We do not attempt a complete comparative treatment of every invertebrate phylum, but instead provide an overview aimed at showing what is special about the main groups, how they may be related, and how they may have arisen and diversified during evolution.

One way to make sense of the fantastic scope of animal diversity is to draw up a list, such as Table 25-1, showing the thirty-three animal phyla. Of these, we shall discuss only the major phyla. Nevertheless, animals of the minor phyla are successful survivors of hundreds of millions of years of natural selection and evolution and should not be overlooked. Moreover, members of these minor

Table 25-1   **ANIMAL PHYLA***

Group	Development	Body Cavity	Symmetry
PORIFERA / Mesozoa / PLACOZOA		Ancestral Acoelomate	
CNIDARIA / CTENOPHORA			Radially Symmetrical
PLATYHELMINTHES / Gnathostomulida / NEMERTINA		Acoelomate (no cavity)	Bilaterally Symmetrical
NEMATODA / Nematomorpha / Acanthocephala / Kinorhyncha / Loricifera / Gastrotricha / Rotifera / Entoprocta		Pseudocoelomate	
ANNELIDA / Sipuncula / Echiura / Priapulida / Pogonophora / Pentastomida / Tardigrada / ARTHROPODA / Onychophora / MOLLUSCA	Protostome (mouth develops from blastopore)	Coelomate	
Phoronida / Ectoprocta / Brachiopoda			
ECHINODERMATA / Chaetognatha / HEMICHORDATA / CHORDATA	Deuterostome (anus develops from blastopore)		

*Those phyla in capital letters are the major phyla and are discussed in this and the next chapter.

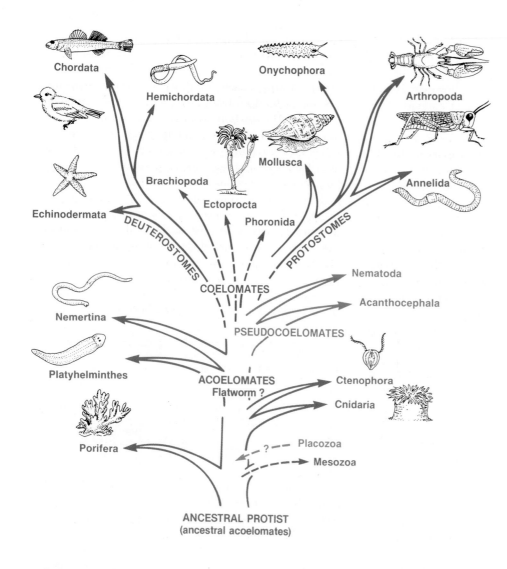

Figure 25-1 A PHYLOGENETIC SCHEME FOR THE EVOLUTION OF ANIMALS. All the major phyla discussed in this book (and shown in capital letters in Table 25-1) appear in this tree; there are many minor phyla, some of which are also included here. The phyla may be grouped on the basis of their body-cavity (coelom) type or they may be grouped as protostomes and deuterostomes, depending on whether their mouth or anus develops from the embryonic blastopore. These distinctions are indicated here by the different colored branches.

phyla often play major roles in the economy of nature—in the ecology of a beach, marsh, reef, or mountain meadow. A second way to understand the diverse animal kingdom is to construct a rough family tree, such as Figure 25-1, which gives a quick overview of the possible relationships among the major phyla and their origins. While there is, unfortunately, no single, agreed-on phylogenetic tree, the scheme used here is widely accepted by biologists and should serve as a useful reference throughout our discussions. Notice, for example, that Porifera (the sponges) is a separate lineage, off the mainstream of animal evolution. Also note that the tree has two main branches, *protostomes* and *deuterostomes*. The protostomes include most worms, snails, clams, crayfish, insects, spiders, and their relatives. The deuterostomes include starfish, sea squirts, and the vertebrates—fishes, amphibians, reptiles, birds, and mammals. We will define these two main branches in detail later.

The evolution of these various multicellular animals has not been orderly and sequential. A number of groups arose independently from different protists, and various groups diverged or converged in body structures or functions, or died out. Our knowledge of these tortuous evolutionary paths is based on scattered fossil evidence and on evolutionary products—animals that are alive today. We shall point out as we go along which interpretations remain tentative and which relationships are more certain.

Perhaps the primary feature shared by all 2 million animal species is multicellularity. This fundamental condition that animals share with plants (Chapters 23 and 24) both imposes restrictions and confers advantages far beyond those of the unicellular state. We will begin our survey of the invertebrates by examining the evolutionary importance of multicellularity, and then consider the major invertebrate phyla one by one.

MULTICELLULARITY: OPPORTUNITIES AND CONSEQUENCES

All animals, invertebrates and vertebrates alike, share a critical feature: multicellularity. This characteristic allows organisms to exploit their environments in ways impossible for bacteria, protists, and the unicellular fungi and algae. Multicellularity gives animals the potential for (1) large size; (2) greater mobility; (3) a stable, controlled internal environment; and (4) relative independence from the harsh and changeable external environment. Once multicellularity evolved, diversification of the invertebrates and, as we have seen, of the plants occurred from a number of single-celled progenitor lines.

Despite the immense diversity of animals, there are constraints on the body organization of any such multicellular organism. First, because of surface-to-volume ratio, described in Chapter 5, cells of most animals tend to be similar in size: thus the difference between a flea and an 18-meter-long giant squid is the number of cells, not the size of cells. Second, movement in large animals requires more powerful force generators than tiny cilia and flagella; muscles have evolved instead. As we survey the invertebrate phyla, we will see the evolution of more and more sophisticated muscle systems and skeletons as an evolutionary theme. Third, cells within multicellular organisms need oxygen, nutrients, means of waste removal, and so on. We shall see in our survey that specialized tissues and organs evolved to meet these needs, including circulatory and respiratory systems, mechanisms for conserving water, and organs for eliminating wastes. Fourth, the vast majority of cells in multicellular organisms are internal and so are not directly exposed to environmental stimuli. Localized sensory systems evolved, providing the entire organism with a capacity to detect light, sound, chemicals, pressures, and magnetism—all rich sources of information that affect survival. Fifth, the complexity of multicellular organisms with specialized systems is often such that coordination and integration of the various systems and of the organism's responses are essential. Nervous and hormonal systems fulfill these functions. Finally, the evolution of "new" features in the various lines of multicellular animals is constrained by molecular, genetic, and developmental "conservatism." Thus there are limitations on the degree to which cells of embryos can form novel structures or organs or can perform wholly new functions. In other words, organisms do not simply start from scratch when a new enzyme, cell type, or organ evolves. Instead, as we learned in Chapter 17, the genes, molecules, cells, and organs that are already present are modified—perhaps step by step, perhaps in major ways—as regulatory genes act; and, as a result, the "new" character develops.

This stepwise modification will be apparent throughout our survey of invertebrates. We will see that once major new adaptations arose in a stem group of organisms, variations on that adaptation appeared in the derivative groups, and great diversification could take place. The first adaptations to appear were usually basic tissues and organs that fulfilled fundamental needs, such as locomotion, feeding, and digestion. Only later did special organs for excretion, respiration, circulation, and coordination emerge. But "later" is a relative term; you will see as we move up the scale of organizational complexity that most of the major animal organs arose in simple and ancient invertebrates.

PORIFERA: THE SIMPLEST INVERTEBRATES

Our survey of the invertebrates starts with sponges, members of the phylum **Porifera**. Numbering some 5,000 living species, these aquatic creatures can be as small as a rice grain or as large as 2 meters across. Sponges are **sessile**—they are permanently attached to rocks, pilings, and other hard surfaces, usually in relatively shallow salt water but sometimes in deeper ocean water or in fresh-water lakes and streams. Sponges are hollow **filter feeders**—organisms that extract microscopic food particles, such as protists, algae, and bacteria, from streams of water pumped through their bodies. Though some are vaselike, and therefore have *radial symmetry*, others have irregular shapes and have no axes of symmetry. Water enters the sponge through hundreds of perforations called *incurrent pores*. This flowing water, or *feeding current*, is strained of food particles, and then the water leaves the body cavity, the *atrium*, through one or more large *excurrent pores*, each of which is called an *osculum* (Figure 25-2). A sponge 1 centimeter in diameter and 10 centimeters high pumps 22.5 liters of water through its body in a day.

Not all cells in a sponge face the same environment: some border the inner cavity or atrium; some face the external world. This difference in position has permitted evolution of differences in structure and function of cell groups. Sponges are the most primitive animals with a true *tissue level of organization* (we shall see that other invertebrates display the organ level of organization). The division of labor among sponge tissues supports a larger body size and more complex organization than are found in any protist.

As Figure 25-2 shows, the body of a sponge has three

(a)

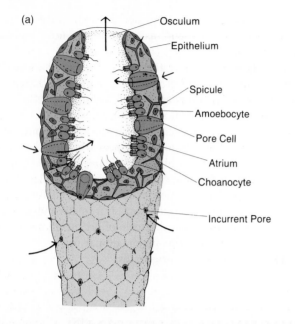

(b)

Figure 25-2   SPONGES: THE SIMPLEST MULTICELLULAR ANIMALS.

(a) The structure of a simple sponge is shown in this diagram of a partially sectioned animal. The flow of the feeding current, or water carrying food particles (shown by the arrows), is through the incurrent pores into the central cavity, the atrium. The beating of flagella on the choanocytes creates a current that drives the water out of the atrium through the osculum. Food is absorbed by cells lining the atrium and passed on to the other cells in the mesenchyme. (b) A cluster of tube sponges growing on a reef in the Caribbean Sea. Each large osculum is located at the top. The small black dots on the inner wall lining the atrium of each sponge are openings of pore cells.

tissues, or cell layers: (1) the outer *epithelium*, a sheet of cells held together by a set of cell-to-cell junctions (Chapter 5); (2) the inner layer that includes *choanocytes*, or collar cells, in an epithelium; each choanocyte has a long flagellum surrounded by a delicate

collar of microvilli; and (3) the *mesenchyme* tissue, between the two epithelial sheets, made up mostly of *amoebocytes*, amoebalike cells. The feeding process is primarily carried out by two differentiated cell types: large, modified epithelial cells called *pore cells*, perforated by the incurrent pores, and the flagellated choanocytes. The pore cells open and close to regulate the flow of feeding current into the sponge, while the current is actually drawn in by the beating of the choanocytes' flagella. Food particles are ingested by the cells lining the atrium. They are digested there or passed to mesenchyme cells for digestion.

Sponges do not have bones or muscles, but some specialized cells and cell products provide analogous functions. Amoebocytes can synthesize a type of collagen that forms a fibrous internal "scaffolding" that is reminiscent of a skeleton. This scaffolding, which remains when the animal dies, is the porous commercial "sponge" used by humans when taking a bath or washing a car. Amoebocytes also synthesize *spicules*, needles made of calcium carbonate or silicon that provide rigidity, support, and protection by making the sponge body a less palatable meal for other animals. Spicule shapes are used to classify sponge species in taxonomic groups. Cells near the incurrent pores and near channel-like branches of the atrium of certain species can behave like the smooth-muscle cells of higher organisms. That is, because of their many internal microfilaments, these specialized cells can contract, thereby regulating the diameter of the pores and channel openings.

In addition to tissues that provide support and allow simple movement, many sponges have cells with long, thin extensions similar to those found on nerve cells. No evidence has yet been found of nerve-cell activity, but some sponge cells may have a primitive capacity to communicate like nerves. Coordinated cell contractions near pore openings may somehow be governed by the nerve-like cells.

Interestingly, sponges can be dispersed into single cells by squeezing them through a fine mesh. The individual cells then can reaggregate and reorganize a new sponge, just like the original. Huge sugar–protein polymers on the sponge cell surfaces are the glue that holds the cells together. These polymers are also involved in the reaggregation process. Study of cell recognition and glue molecules in sponges has helped cancer researchers investigate why human cancer cells do not adhere well to each other.

The ability of sponges to regenerate following experimental dissociation into a suspension of single cells may be related to their capacity to reproduce asexually, since both processes depend on mitosis. Many sponges can form buds, small masses of cells in which the various sponge tissues develop; the buds break away, settle onto rocks or pilings, and form new sponges. But sponges are hermaphrodites that can also reproduce sexually. Although no sex organs are present, amoebalike cells can undergo meiosis and form either sperm or eggs. Since meiosis is present in these most primitive animals, as well as in plants, it is reasonable to conclude that meiosis, with all its wonderful complexity of chromosome movements and crossing over, arose far earlier in still more ancient organisms. Sperm liberated into the sea enter other sponges, are engulfed, and are carried to an egg in the mesenchyme. Fertilization leads to the development of an embryo that forms a ciliated, free-swimming larva or, in some species, a larva that crawls on the sea bottom. Ultimately, the larva attaches to a solid object; forms tissues, pores, and channels; and begins to function as a sponge.

The unique features of sponge development plus their special cell and tissue structure lead some biologists to classify sponges as a separate subkingdom, the *Parazoa*, and the rest of the animals in the subkingdom *Metazoa*. Here, as is more commonly done, we place sponges in the phylum Porifera, where they represent an independent example of evolution from protistan ancestors to the multicellular condition. In a sense, sponges are an evolutionary dead end; they are not believed to have given rise to any other group of multicellular organisms. We will discuss the possible evolutionary origin of the sponges a bit later when we consider the origins of multicellularity, but it is clear that their origin was separate from that of other invertebrates.

RADIAL SYMMETRY IN THE SEA

The next invertebrates we consider are the phyla Cnidaria and Ctenophora (the C's are silent). These phyla are noteworthy for a special kind of symmetrical shape and for their tissue level of organization. These simple aquatic creatures include jellyfishes, corals, comb jellies, and hundreds of other simple and often beautiful species. Their body plan shows **radial symmetry:** they look circular when viewed from above or below, and certain structures radiate in all directions from the center toward the periphery like spokes from the hub of a wheel. Their upper and lower surfaces are usually not alike. Such radially symmetrical animals may be sessile, or they may be **pelagic**—drifting about freely in the water. Both cnidarians and ctenophorans show a higher level of tissue organization than do the sponges, but they lack complex organs, such as kidneys or a heart.

Cnidaria: Alternation of Generations in Animals

Some of the most delicate and colorful animals in the seas and in fresh-water streams are members of the phy-

lum **Cnidaria,** including hydras, jellyfishes, sea anemones, and corals. Cnidarians come closer than any other animal group to displaying the common life strategy of plants because they have a kind of alternation of generations. Many species have a sessile stage with a hollow, elongated body, the **polyp,** which subsequently forms a pelagic (free-floating) **medusa,** a radial jellyfish drifting freely in the sea (Figure 25-3). Polyps have been compared to flexible, open-ended bottles. The medusa is rather like an inverted polyp with a wider opening. Both forms have a simple organization: a three-layered body wall; a single opening, the mouth, surrounded by tentacles; and a central gut, or *coelenteron*. The phylum is sometimes called Coelenterata ("hollow gut") after this cavity.

Differentiation of cells into separate tissues is apparent in the body-wall layers: the external *epidermis* ("outer skin"), the inner *gastrodermis* ("stomach skin"), and the middle layer of jellylike *mesoglea* ("middle glue"), which may or may not contain cells. The epidermis and the whorl of tentacles typical of cnidarians bear highly differentiated stinging cells called *cnidocytes*. Each of these cells houses a remarkable organelle, the *nematocyst*, consisting of a hollow capsule that contains a coiled, harpoonlike filament (Figure 25-4). The capsule may have osmotic pressures equal to 140 atmospheres (over 2,000 lbs. per sq. in.). When triggered to discharge, the harpoon everts, shooting outward, rotating as it goes, and reaching lengths from 5 micrometers to 1 millimeter. The nematocyst filament may be barblike, for hooking fish; ropelike, for lassoing prey; or sticky, for ensnaring the unsuspecting food. Sharp penetrating nematocyst filaments are often coated with toxins that immobilize prey. A single puncture wound might not deter a small fish or other potential meal, but a barrage of "harpoons" from dozens of nematocysts can entrap, paralyze, and kill the prey.

This nematocyst response is apparently specific to certain stimuli. If you touch the tentacle of a cnidarian with the tip of a sterile glass rod, nothing will happen. But if you first rub the rod in animal extracts and then touch the tentacle, the nematocysts will fire. Certain small fishes, such as the clown fish, swim among the tentacles of sea anemones and are stung—but to no avail because they have evolved immunity to the nematocyst toxin. In fact, a given species of fish is attracted to a species of anemone because that cnidarian secretes special compounds that trigger approach by the fish as well as behavior beneficial to the anemone.

While sponges are limited to absorbing tiny food particles and digesting them within individual cells, cnidarians have a more advanced way of obtaining nutrients, **extracellular digestion.** After a cnidarian swallows a food morsel, digestive enzymes are secreted by gastrodermal gland cells, and partial digestion takes place in the coelenteron. The nutrients are then ab-

(a) Polyp Body

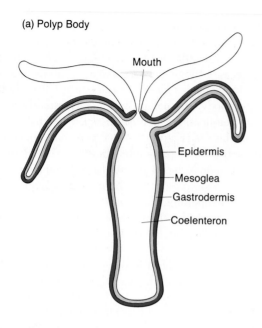

(b) Medusa Body

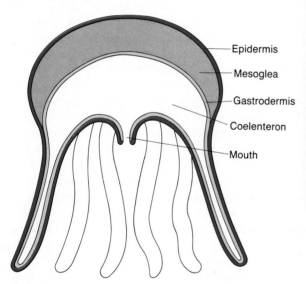

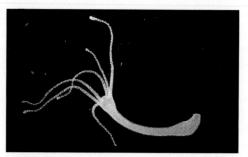

Figure 25-3   THE TWO BASIC BODY PLANS OF CNIDARIA.
(a) The polyp body form includes three tissue layers and tentacles protruding upward. Although the organism looks perfectly stable, in fact, new cells produced in growth zones move upward and outward in the tentacles and downward in the main body region. (b) The free-floating medusa is buoyant and swims by frequent pulsatile contractions; its tentacles hang downward in the water and, like those of the polyp stage, contain numerous stinging cells.

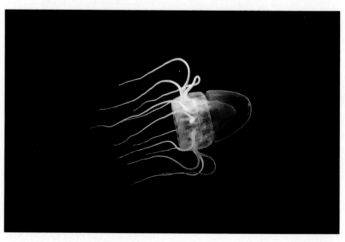

sorbed by gastrodermal nutritive cells. Extracellular digestion occurs in most animals, but cnidarians are the simplest creatures that employ it. Some tiny food pieces and nutrients are also ingested whole by nutritive cells and then broken down by internal lysosomal enzymes in vacuoles. This is the typical intracellular digestion seen in protists and in the individual cells of animals. Some cnidarians also derive nourishment from algal cells that live in their epidermis: light penetrates the transparent body surface of the polyp or medusa, photosynthesis occurs, and some of the resultant carbohydrates pass to the host cells as food.

Cnidarians can capture and swallow prey because they have evolved cells that are analogous to muscles and a supporting structure that is analogous to a skeleton. In-

deed, the cells of the body wall function in these capacities. The outer portion of many epidermal and gastrodermal cells has normal epithelial-cell shape, but the inner end, near the mesoglea, is drawn out into a long "tail" process packed with bundles of contractile microfilaments. Such cells are called *epitheliomuscular cells* (Figure 25-4). The microfilament bundles in the tails of the gastrodermal cells are oriented around the body and around each tentacle. These can contract like tiny circular muscles. The epidermal-cell tails are oriented and contract like longitudinal muscles.

To do work effectively, muscles must pull against a fixed structure, such as bone. Cnidarians lack bone; however, water trapped in the coelenteron when the animal closes its mouth acts as a **hydroskeleton,** a noncompressible mass that can change shape but not volume. Thus as the circular gastrodermal tails contract, the body or tentacle elongates like a squeezed balloon. Conversely,   if   the   longitudinal   muscle   tails   of

(a)

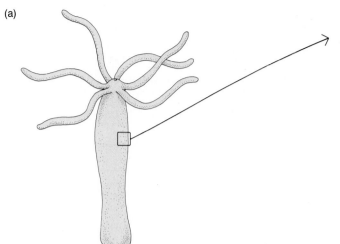

(b)

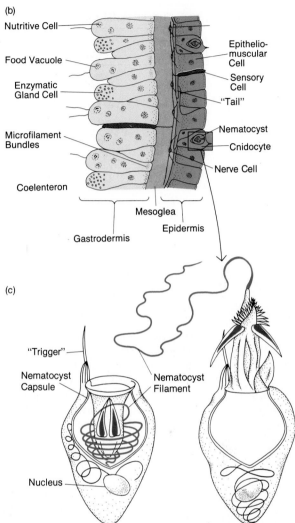

Nutritive Cell

Food Vacuole

Enzymatic
Gland Cell

Microfilament
Bundles

Coelenteron

Epithelio-
muscular
Cell

Sensory
Cell

"Tail"

Nematocyst

Cnidocyte

Nerve Cell

Gastrodermis

Mesoglea

Epidermis

Figure 25-4   THE CNIDARIAN SPECIALIZED EPITHELIAL AND MESOGLEAL TISSUES.

(a) *Hydra* is a typical cnidarian. (b) A detailed section of the body wall. The outer layer, or epidermis, consists of several cell types, including the epitheliomuscular cells, which are involved in the animal's movement, and cnidocytes, which contain the nematocysts for protection and trapping prey. The cells of the gastrodermis are concerned primarily with digestive processes. This layer encloses the coelenteron. The middle mesoglea may lack cells, as shown here, or it may be cellular, especially in medusas. (c) A nematocyst is seen in the "cocked" (left) and the "fired" (right) positions. When discharged, the nematocyst can either entangle the prey or release a toxic substance that will immobilize it.

(c)

"Trigger"

Nematocyst
Capsule

Nematocyst
Filament

Nucleus

epitheliomuscular cells contract, the body or tentacles shorten like a squashed balloon. A 20-millimeter-long hydra, for example, may shorten to 0.5 millimeter when the contraction occurs. Movement or locomotion based on a hydroskeleton against which opposing sets of muscles act is a basic strategy for locomotion among invertebrates and apparently originated in the ancestors of cnidarians.

The diverse activities of cnidarians are possible only because the organisms possess another evolutionary innovation—*nerve cells* arranged in a primitive nervous system, the *nerve net*. Although no true brain or coordination center is present, the nerve net includes at least six types of nerve cells, including sensory ones and ganglion ones. The net is arrayed in such a way that the radially symmetrical animal can swim, move its tentacles, or respond to stimuli from any side of its body. Several complex behavioral and physiological processes depend directly on nerve cells, which originated in the radially symmetrical invertebrates.

One characteristic sets cnidarians off from all other animals: a kind of perpetual rejuvenation, a living fountain of youth. Quite literally, cnidarians such as hydras *never grow old*. Why not? If we mark cells that arise by mitosis in a growth zone just beneath the mouth and ring

of tentacles, we see in ensuing days that the cells move. Some move slowly toward the edge of the mouth (a twenty-day passage); others move up the tentacles (four days), where they drop off from the tips; and still others move toward the base (eight to twenty days). In fact, this process goes on all the time—new cells arise continuously in the growth zone and form the body tissues. Cells in every tissue, including the nerve net, are continuously moving toward sites where the aged cells die and drop away. The hydra we see today will literally be a "new" creature three weeks later. This is a different cell behavior indeed from that found in most animals, with their nondividing differentiated cells, some of which age, lose function, and contribute to senescence, the aging process of the whole organism.

Reproduction in members of the phylum Cnidaria can be asexual or sexual. Polyp stages typically grow buds that mature into new polyps, and whole colonies can form in this asexual manner. But polyps can also develop buds that mature into medusas, the free-swimming sexual stage. These medusas develop true gonads, from which haploid sperm and eggs are released into the sea

Table 25-2   **CLASSES OF THE PHYLUM CNIDARIA**

Class	Number of Species	Representative Animals	Habitat	Other Characteristics
Hydrozoa	~3,700	Hydra	Most marine; a few fresh water	Polyp and medusa; polyp stage often colonial
Scyphozoa	~200	Jellyfish	Marine	Free-swimming; large medusa; polyp reduced
Anthozoa	~6,100	Coral	Marine	No medusa; sessile; often colonial

Figure 25-5   TWO BODY FORMS AND TWO LIFE STYLES IN HYDROZOANS.

(a) The sessile colonial hydrozoans are called hydroids; an example is *Obelia*, which consists of feeding polyps and reproductive polyps. This photograph of *Obelia* shows the translucent tentacles of the feeding polyp and the vaselike reproductive polyps containing developing medusas. (b) Hydroid colonies that develop by asexual budding produce male and female medusas on the reproductive polyps. The medusas reproduce sexually, generating sperm and eggs, which, after fertilization, give rise to a larval planula. In *Obelia* the life cycle consists of a sessile asexually reproducing colony alternating with a free-swimming solitary form that carries out sexual reproduction.

water. After fertilization, embryos develop into **planulae**—small, elongated, solid, cellular larvae propelled by cilia. Eventually, a hollow space appears in the planula's endoderm layer and forms the gut cavity, or coelenteron. The planula attaches to a hard substratum and gradually converts into a little polyp, thus completing the alternation of generations. Note that unlike plants, both asexual polyp and sexual medusa are diploid creatures; only the short-lived gametes are haploid.

The degree to which the polyp or the medusa stage dominates the life cycle varies greatly and is a major means of classifying a cnidarian as a member of the class **Hydrozoa, Scyphozoa,** or **Anthozoa** (Table 25-2). Hydrozoans usually carry out alternation of generations (Figure 25-5). Some members of the class, such as hydras, occur

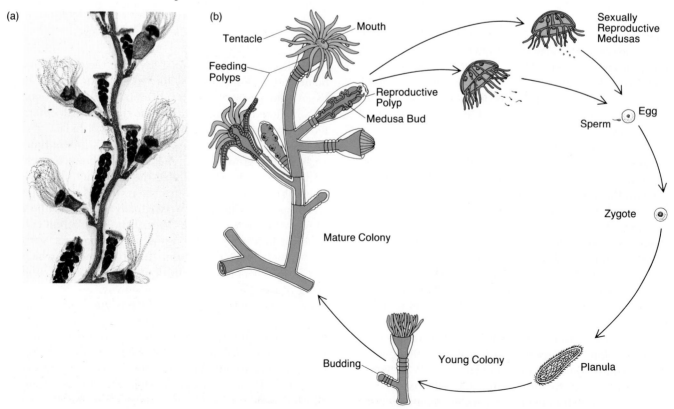

Figure 25-6   THE PORTUGUESE MAN OF WAR: AN UNUSUAL HYDROZOAN.
This *Physalia physalis* is pelagic at all stages of its life cycle. The form seen here is really a giant polyp stage floating free in the water.

limestone encasement around itself. Coral polyps can build up slowly into giant reefs tens or hundreds of kilometers long, but the living cnidarians inhabit only the narrow outer crust. As planulae settle and polyps grow, underlying polyps are entombed by the ever-expanding encasements.

The simple, tissue level of organization of the cnidarians strongly suggests that they are properly regarded as the simplest animals near the main evolutionary line leading toward the flatworms and the higher animals (Figure 25-1).

Figure 25-7   BEAUTY IN THE DEEP.
(a) This sea anemone (*Anthopleura* species), with its whorls of flexible tentacles, typifies the radial life style. (b) The otherworldly shapes of living pillar coral colonies (*Dendrogyra cylindricus*) attract colorful reef fish . . . and human skin divers. This pillar coral is growing off the coast of Belize, Central America.

(a)

mainly in the polyp stage and exist for only short times as medusas and planulae. Others lack the polyp stage altogether; these medusas may represent the most primitive stem group of the cnidarians, with polyps and a sessile life style evolving in later members. Hydrozoans are a confusing group because the polyp stage of some members is not always sessile. The deadly Portuguese man-of-war, with its long stinging tentacles, is a true asexual polyp that produces groups of male and female medusas that look like bundles of deep-blue grapes (Figure 25-6). The female medusas break free from the polyp, and their eggs, once fertilized by the still-attached males, form planulae that develop into new free-floating Portuguese men-of-war.

The second class of cnidarians, scyphozoans such as jellyfish, spend the majority of the life cycle as free-swimming medusas. The asexual polyp stage may be brief, inconspicuous, or absent. The tentacles of jellyfish—so ephemeral and gently billowing, yet so deadly to prey—can reach 40 meters in length. In the waters off the eastern seaboard of Australia, some fifty human deaths have resulted from encounters with jellyfish tentacles. (In contrast, typical jellyfish encountered on Atlantic and Pacific seaboards of the United States are not a danger to humans, as their toxins are not so deadly.)

Anthozoans such as sea anemones and corals skip the medusa stage entirely (Figure 25-7). They reproduce both by asexual budding and by sexual processes that yield planulae. Sea-anemone polyps can creep slowly along rocks, but corals remain fixed in place. Coral polyps live in colonies, and each animal secretes a hard

(b)

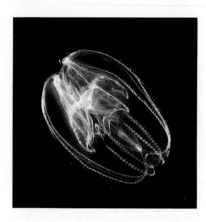

Figure 25-8 TRANSPARENT COMB JELLIES— INHABITANTS OF SURFACE WATERS OF THE SEAS. *Bolinopsis chuni*, a representative ctenophore (comb jelly), swims through the warm waters of Sydney harbor, Australia.

Ctenophora: The Comb Jellies

Although living in the same environment as the pelagic stages of cnidarians, members of the phylum **Ctenophora** meet the demands of free-floating life in the sea in different ways. This suggests that this second group of radially symmetrical animals evolved independently of the cnidarians. The ninety species of medusalike ctenophores, or comb jellies, are unique in that all have eight rows—the combs—of heavily ciliated cells running from pole to pole on the transparent body (Figure 25-8). These cilia provide the organism with a means of locomotion.

Most ctenophores feed using their two tentacles, which are coated with sticky substances that entrap food particles and hold them until the tentacles can be moved slowly to the mouth. The ctenophore gut has discrete parts, reflecting a more complex body design than is found in the cnidarians. Nerve cells arrayed in a nerve net are present, and, like some jellyfish, ctenophores have a sense organ that detects when the animal has tipped over and must right itself. This organ coordinates the beating of the combs.

The exact relation of ctenophores to cnidarians is unknown. On the one hand, general similarities in radial symmetry, in nerve nets, and in other physiological characteristics suggest a common origin. But on the other, the unique combs, the absence of alternation of generations, and a pattern of embryonic development unlike that of cnidarians are consistent with the hypothesis that ctenophores arose independently of the cnidarians.

THE ORIGINS OF MULTICELLULARITY

Now that we have briefly examined the multicellular sponges and the two phyla of simple organisms with ra-

dial symmetry, we can ask how multicellularity arose. Biologists have advanced two hypotheses. According to the *colonial theory*, flagellated protistan cells aggregated to yield either colonies resembling the hollow, ciliated blastula or gastrula stage of animal embryos (Chapter 16) or solid organisms with a ciliated surface resembling the planula of sponges and cnidarians. In contrast, the *syncytial theory* suggests that ciliated, single-celled protists containing many nuclei became "cellularized" when plasma membranes walled off those nuclei into separate cells. Most biologists are not persuaded by evidence to support the syncytial theory and thus look to the aggregation of discrete protistan cells as the most likely explanation.

Both regeneration and asexual reproduction are common among sponges and cnidarians. Those processes, which basically depend on mitosis, may be remnants of an earlier process in which single flagellated cells divided mitotically but remained attached to each other. Colonies could have built up in that way, and such a process would have produced colonial cells that were genetically alike. This theory overcomes one of the problems with most colonial hypotheses—that genetically distinct cells somehow came together to form a colony. Such "mixed genetics" is difficult to understand, though we did encounter it in cellular slime molds (Chapter 21).

What did the original multicellular colonies give rise to? Zoologists hypothesize that they evolved into solid creatures resembling the planulae of cnidarians. Such ancient planulae could, in turn, have been the ancestors of cnidarians. Ctenophores may have arisen independently by essentially similar processes. Many sponges also have solid, planulalike larvae. However, so much is unique about sponges that it has been concluded that they arose independently, probably by a process in which protozoans called *choanoflagellates* generated multicellular creatures (choanoflagellates resemble the choanocytes of sponges).

There is one intriguing puzzle that may give us a perspective on the early, planula stage of multicellularity. In the warm Mediterranean Sea lives what may be the most primitive free-living multicellular animal, the **plakula** (Figure 25-9). This three-layered marine worm was first described in the 1880s. Given the generic name *Trichoplax*, it was classified with the flatworms. A German zoologist working in the late 1960s, K. G. Grell, reinvestigated the little worm and placed it in its own phylum, **Placozoa**. The creature, only 1 to 3 millimeters in diameter, glides over the sea floor, driven by surface flagella. As it goes, it assumes a variety of irregular shapes in an amoeboid fashion. *Trichoplax* shows no hint of a front end or of right and left sides, features seen in planulae and in even the most primitive worms. Electron microscopy reveals upper and lower flagellated epithelial layers of distinct cell types in plakulae, and a middle zone reminiscent of the mesoglea of cnidarians.

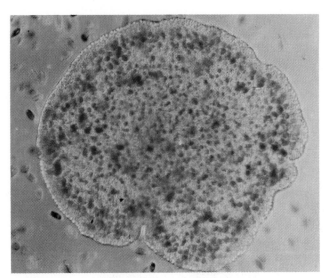

Figure 25-9 *TRICHOPLAX:* **THE SIMPLEST MULTICELLULAR ANIMAL.**
Trichoplax adhaerens (here magnified about 34 times) was first thought to be some form of larva, but was later recognized to be an adult organism. It is now placed in the phylum Placozoa, and may possibly be a primitive type of organism that gave rise to higher animal groups.

Although there is absolutely no true gut, the entire central region of the plakula's body sometimes elevates to form a temporary gutlike cavity. And surprisingly, cells lining this cavity secrete enzymes that digest trapped algae and protists. This is extracellular digestion in a creature more primitive than the sponges, which lack it.

No one is sure where to place the plakula in phylogeny. Some zoologists suggest that it is not far removed from the hypothetical planula that gave rise to most multicellular animals. Others argue that it is very much like sponge planulae and so may be on the Porifera side branch of animal evolution. The plakula is a prime candidate for analysis of DNA sequences to allow comparison with the DNA of sponges, cnidarians, and flatworms. Such analysis should help elucidate relationships among these close relatives of the Earth's first colonial multicellular animals.

ORGANS AND BILATERAL SYMMETRY

Some of the radially symmetrical animals we have discussed can move using ciliary combs or musclelike cells, but none moves predominantly in a single direction. The flatworms are the most primitive animals that carry out unidirectional locomotion. They do so because they possess a front end—a site where the mouth, major sense organs, and primary integrating center of the nervous system are located. The anterior end, with its special organs, is the first part of the animal to encounter new environments, food, and predators. This front-end specialization is called **cephalization,** or head formation.

Imagine an ancient organism swimming forward or crawling or gliding along a solid surface—several factors would operate that could affect the evolution of body organization. For instance, the lower, or ventral, surface of animals that crawl would become specialized for locomotion and perhaps for feeding; the upper, or dorsal, surface, for protection or camouflage. As such a creature moves forward in one predominant direction, it would generally experience the same thing on each side; thus its right and left sides would tend to be alike. Indeed, animals that move forward and are cephalized have evolved the condition of **bilateral symmetry:** their right and left sides are mirror images of each other. Most members of the animal kingdom have these characteristics.

Platyhelminthes

The first group of organisms we encounter with both cephalization and bilateral symmetry are the flatworms, classified in the phylum **Platyhelminthes** (Figure 25-10). Flatworms are also the first group organized at the *organ level* rather than at the tissue level. These creatures range from 1 millimeter to 30 centimeters in length. Like all the animals we will study from here on, including humans, the body and its organs develop from three embryonic layers: the ectoderm, mesoderm, and endoderm (Figure 25-11). The ectoderm becomes the surface epithelium, which is often ciliated and responsible for locomotion in many species. The mesoderm is much more complex than the cnidarian's mesoglea and gives rise to true muscle fibers, excretory organs, and reproductive organs. The endoderm comes to line the digestive cavity. But not all types of organs are present in flatworms, which do not have respiratory and circulatory systems. That is one reason these organisms are so thin and flat—diffusion is the means by which dissolved gases, nutrients, and wastes enter, move through, and leave the body.

The almost 13,000 species of platyhelminths are divided into three classes: the free-living Turbellaria and the parasitic Trematoda and Cestoda (Table 25-3).

Turbellaria: Free-Living Flatworms

The members of the class **Turbellaria** include the most primitive bilaterally symmetrical invertebrates. The simplest of these, the *acoels,* are less than 1 milli-

(a)

(b)

Figure 25-10 MARINE FLATWORMS: BRILLIANT MARKINGS AND BILATERAL SYMMETRY.
(a) This tiger flatworm, a marine turbellarian, inhabits the warm shallow waters around the Hawaiian Islands. (b) This flatworm, *Thysanozoon flavomaculatum*, literally flaps through the hot waters in the Gulf of Aqaba in the Red Sea.

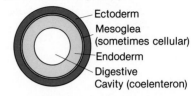

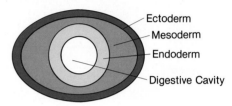

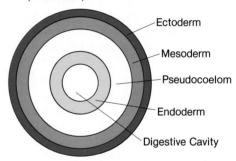

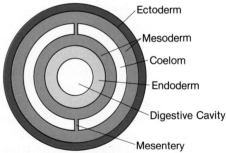

Figure 25-11 EVOLUTION OF THE BASIC ANIMAL BODY PLAN.
The evolution of body organization progressed from (a) a body with two epithelial layers plus mesoglea to (b and c) three distinct tissue layers surrounding a digestive cavity to (d) a coelom (body cavity) surrounded by mesoderm cells.

meter long and lack a true gut and any hint of a body cavity. All turbellarians have an outer epidermis and a middle *parenchyma* (dense connective tissue) with true cells. All but the acoels also have an inner endodermis (gut lining). In fact, the gut of turbellarians varies widely. Whereas in acoels the mouth leads directly to the parenchyma, other turbellarians, such as *Planaria* (Figure 25-12), have a pharynx that leads to complexly lobed sacs called the gastrovascular cavity that holds food during digestion. No anus is present, however, so that wastes must be discharged from the mouth.

The smallest acoels are propelled by cilia on their ventral surface, where mucus-secreting cells literally "grease" the way as they glide along. Larger turbellarians are dependent on true muscle-fiber cells for movement.

Planaria and other turbellarians have a more complex nervous system than do radially symmetrical animals. Near the anterior end of the body lie two *eyespots* for sensing light, while another organ senses gravity. Both

Table 25-3   **CLASSES OF THE PHYLUM PLATYHELMINTHES**

Class	Number of Species	Representative Animals	Habitat	Other Characteristics
Turbellaria	~3,000	Planarian	Most marine; some fresh water; a few terrestrial	Predator and scavenger; ciliated body surface
Trematoda	~6,000	Fluke	External and internal parasite, usually of vertebrates	Suckers attach to host; may require intermediate host
Cestoda	~3,400	Tapeworm	Internal parasite of vertebrates	Scolex attaches to host; no digestive system; proglottids break off after fertilization; one or more intermediate hosts

connect to nearby aggregations of nerve cells, or **ganglia** (singular, *ganglion*), that may be thought of as primitive brainlike organs. From them, nerve trunk lines or cords extend in ladderlike fashion down opposite sides of the body's longitudinal axis.

Each individual turbellarian has both male and female reproductive organs, a condition called hermaphroditism. When two adults mate, they exchange sperm and thus fertilize each other's eggs. This cross-fertilization ensures greater genetic diversity than does self-fertilization.

Most turbellarians have an efficient excretory system for ridding the body of excess water. In particular, cilia on *flame cells* beat to drive a watery filtrate down the ducts, from where it can leak to the exterior through pores in the epidermis. We will describe this system in Chapter 35.

Trematoda and Cestoda: Parasitic Flatworms

Parasites are organisms that live on or within other organisms and derive nutrients from their hosts. Para-

Figure 25-12   TURBELLARIANS: PRIMITIVE FLATWORMS.
Flatworms are unsegmented but have multiple organs. (a) The body organization of a planarian flatworm has various organ systems, including the digestive, nervous, and reproductive systems shown here separately. (b) Cross-sectional view through the region of the pharynx shows the three-layered, acoelomate body plan. The gastrovascular cavity, which is seen at the left in (a), is also seen in cross section in (b).

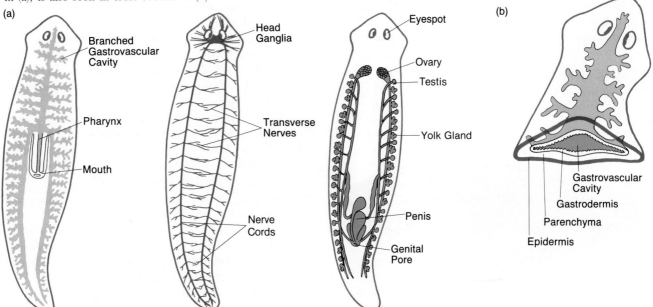

sitic flatworms feed repeatedly on their host's body or food reserves. They grow and reproduce at the expense of the host but contribute nothing to the host as compensation. The host may be debilitated by a parasitic infection but is usually not killed, since the host's death represents a real disaster for the parasite, just as though its "home" suddenly burned down. Finding a new "house"— an organism of a susceptible species—is usually impossible for adult parasites, since the mode of transmission from one host to another is usually a small reproductive stage of the parasite. For these reasons, much of the biology of parasites is based on large-scale reproductive capacity: continuous or periodic production of huge numbers of gametes and offspring, some tiny fraction of which will chance to find a new host and so perpetuate the species.

Flukes, members of the class **Trematoda,** are parasitic flatworms that inhabit a variety of vertebrate tissues (Figure 25-13). They range from less than 1 millimeter to 7 meters in length and usually have suckers for attaching to the host organism and a covering, the *cuticle,* composed of a cytoplasmic *syncytium* (a multinucleated cytoplasm) that is resistant to attack by the host's enzymes. Flukes feed on and can extensively damage liver, muscle, lung, and other tissues. Many flukes have complex life cycles with alternating hosts. Blood flukes of the genus *Schistosoma* alternate between human and freshwater snail hosts and cause a severe human disease called *schistosomiasis.* Because of the ubiquity of infected snails in tropical areas, schistosomiasis kills more humans in the tropics than does any other disease. Chinese liver flukes have humans as primary hosts and snails and fishes as secondary hosts. Mature flukes, about 1 centimeter long, live in ducts of the human liver. Developing fluke eggs pass through the bile ducts into the intestine and then out in the feces. On entering fresh water, the larval flukes infect snails. Following asexual reproduction in snail tissue, free-swimming stages of

some species can penetrate directly through the skin of a wading human; the free-swimming stage also infects fish muscle. If a person eats this infected fish raw, the flukes invade various human tissues, where they grow and ultimately lay eggs, beginning the cycle anew. A key feature in the ability of schistosomes to avoid attack by their snail hosts is that the parasites have evolved cell surface markers identical immunologically to markers on the snail's cells. Hence, they are apparently not recognized as being "foreign" and are not killed by the host's immune system (Chapter 32).

Unlike flukes, tapeworms are members of the class **Cestoda.** These denizens of the vertebrate intestine have, in the course of evolution, lost their own gut and must absorb every nutrient from the host's digestive tract through their own epidermal wall, which is resistant to the host's digestive enzymes. The abundance of nutrients in the gut of a cat or a human permits rapid growth of the tapeworm's body to many meters in length. The cestode's body is made up of a head unit, the *scolex,* followed by a continuously lengthening chain of units called *proglottids* (Figure 25-14), each of which is essentially a "packaged gonad" containing up to 80,000 eggs, as well as sperm. The proglottids mature, break off, and pass from the host organism in its feces. Proglottids can be picked up by an intermediate host, such as a pig. If a person eats inadequately cooked pork from an infected animal, the tapeworm can establish itself in a new human host.

Evolution of Flatworms

It is not certain whether acoels, the simplest bilaterally symmetrical animals, were the ancestors of all flatworms or whether their rudimentary gut and organ systems represent an evolutionary simplification from more complex organs typical of other flatworms. The turbellarians surely have diversified into complex free-living types and also gave rise to the parasitic trematodes and cestodes.

Today, the various intermediate hosts of the parasitic flatworms are invertebrates—mollusks, arthropods, echinoderms. However, these animals were probably the original primary hosts of the parasites, and the relationship was established almost 600 million years ago in the Cambrian period. Later, as vertebrates appeared and diversified, a new kind of "secondary" host became available. Finally, the appearance of higher vertebrates (birds and mammals) provided the flatworm parasites with highly mobile and long-lived hosts with warm and remarkably constant internal environments—an ideal home. It is believed that by about 60 million years ago,

Figure 25-13   PARASITIC FLUKES.

The giant liver fluke, *Fasciola hepatica,* is one of the many parasitic flatworms. It is seen here magnified about 2.5 times.

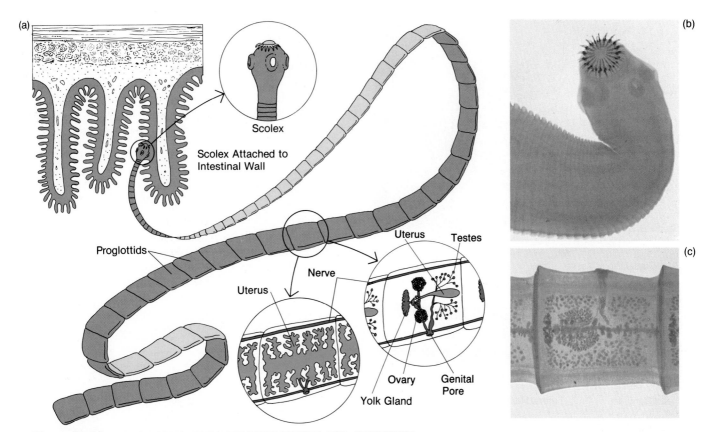

Figure 25-14 TAPEWORMS: EVOLUTIONARY DESIGN FOR ABSORBING NUTRIENTS.

(a) The tapeworm's scolex is shown embedded in a host's intestinal wall. (b) This photo (magnified about 4 times) of a stained tapeworm (*Taenia serrata*) shows the clublike scolex or head unit with its anterior ring of dark hooks and its many round suckers. Both structures help the animal attach to a host's intestinal wall. The worm's body is made up of proglottid segments, each of which contains primarily reproductive organs. The circled enlargements of one proglottid show either male or female organs; both can be discerned in the photograph (magnified about 6 times) in (c).

birds and mammals had been adopted as hosts, but for a different stage in the parasite's life history from the stage that parasitized invertebrates. As still new types of mammalian hosts evolved—monkeys, apes, humans—the flukes and tapeworms parasitized them as well.

Nemertina: Two-Ended Gut and Blood Vessels

We now begin to move from the simplest worms to the first of a number of separate lines. This first lineage includes a major innovation: a one-way gut with openings at both ends. Members of the phylum **Nemertina** (or *Rhynchocoela*), or ribbonworms, live on most marine coastlands, hidden under rocks, shells, and seaweed or burrowed into the sand and mud. Some are so long and thin that they are called "bootlace" worms. The 650 species of these worms have locomotory, nervous, and excretory systems similar to those in flatworms. Their similar physiology suggests that ribbonworms may be derived from flatworms. However, nemertines also have important physical differences that have allowed them to reach greater size than the flatworms. These include a proboscis, both a mouth and an anus, and blood vessels.

The ribbonworm's anterior end consists of a long hollow structure, the *proboscis*, that can be extended and used to lasso small prey. The proboscis may also be used as a tactile organ. More important, ribbonworms have a complete one-way *digestive system*, with mouth and anal openings at opposite ends and with specialized regions—an anterior esophagus and a posterior intestine—where different phases of extracellular digestion and nutrient

(a)

Figure 25-15   RIBBONWORMS: INNOVATIONS IN BODY FORM.

The acoelomate ribbonworms (nemertines) show the evolution of a circulatory system involving vessels and a two-ended gut. (a) Living specimens of the genus *Lineus* living in the Caribbean off Panama. (b) The diagram shows the body organization of this three-layered acoelomate worm.

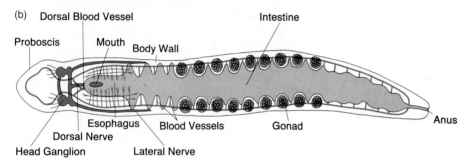

(b)

absorption take place (Figure 25-15). Undigested wastes pass out through the anus. The great advantage of this one-way gut is that partially digested materials are not mixed with undigested wastes. For the first time in animal evolution, the processes of digestion, absorption, and excretion are separated.

Finally, a major innovation for all later animals evolved first in ribbonworms: blood vessels run along the body's longitudinal axis, carrying a clear fluid. While the cells in this clear "blood" do not seem to have respiratory pigments such as hemoglobin to carry oxygen, the fluid nevertheless transports nutrients digested in the gut to cells in the body wall and carries dissolved oxygen, carbon dioxide, and various metabolic wastes to and from all body regions. Since no heart is present, there is no real blood circulation, but movement of the body wall and contractions of the vessels themselves cause a slow to-and-fro movement of the blood fluid. Because of their internal-transport system, nemertines can be thicker and longer than flatworms, which depend on diffusion alone for movement of dissolved gases, nutrients, and wastes. The combination of blood vessels, with their capacity to serve local tissues, and a unidirectional gut, with its greater efficiency at processing nutrients, allows free-living ribbonworms to reach lengths up to 30 centimeters—much longer than the turbellarians.

Nemertines display a spectacular capacity for regeneration. When grasped by a predator or a curious biologist,

the body of a large ribbonworm will fragment into a dozen or more pieces. Each piece will regenerate the missing anterior end, posterior end, or both. Such fragmentation is in fact the common means of asexual reproduction in some ribbonworm species. Sexual reproduction also occurs: parenchymal cells produce gametes that are released into the sea, where fertilization takes place. The free-swimming larvae that develop are unlike those of any other invertebrate group.

Nematoda: Emergence of the First Body Cavity

A second lineage derived from flatworms has not only a one-way gut, but also a primitive body cavity, a space that separates the body wall from the gut, making independent movement of the two possible. This lineage is the phylum **Nematoda,** or roundworms. Roundworms are the most abundant animals on Earth—there may be some 5 billion roundworms in the upper 3 inches of an acre of sandy beach or rich farm soil. Billions of nematodes are also found in every acre of ocean, lake, or river bottom. Biologists estimate that there are anywhere from 10,000 to 500,000 roundworm species, feeding on every conceivable source of organic matter—from rotting substances to the living tissues of other invertebrates, vertebrates, and plants.

Despite their great abundance, nematodes are less complex in some ways than are flatworms and ribbonworms. They lack blood vessels and therefore remain quite narrow in diameter, since diffusion alone must distribute all their essential commodities throughout the body. Locomotion is also quite limited: nematodes lack both cilia and muscles running in a circular direction around the body wall. However, longitudinal muscles attach to the inside of the surface cuticle—a relatively rigid covering made in part of tough fibrous collagen. Because of this arrangement, the animal can move only by means of lengthwise contraction—bending and thrashing back and forth, somewhat reminiscent of a fish on a riverbank.

Why, then, are nematodes so successful? The great abundance of individuals and species in widespread habitats appears to be based on their efficient digestive system. This system has both a mouth and an anus, as in the ribbonworms, and has specialized gut regions for digestion and absorption. The gut lies cushioned in a fluid-filled cavity rather than packed inside a solid layer of cells and connective tissues. This cavity is called the **pseudocoelom** (meaning "false body cavity") and is a carry-over into adulthood of the embryonic blastocoel (Chapter 16). The pseudocoelom is bounded on one side by mesodermal tissue and on the other either by the gut tissue or by the epidermis (Figure 25-16). The pseudocoelom functions basically like the true body cavity of all higher animal phyla, in allowing the gut to be cushioned in fluid and the gut and body wall to move independently of each other. The true *coelom* is bounded by mesoderm but it functions identically to the pseudocoelom.

Parasitic nematodes cause serious medical and agricultural problems. Disease-producing hookworms, pinworms, and filarial worms are members of this phylum. All species of complex plants and animals can be the host for at least one type of nematode parasite; botanists calculate that nematodes consume about 10 percent of all crops grown each year. Of the fifty roundworm species that parasitize humans, one of the most common and dangerous is *Trichinella spiralis*, the agent of *trichinosis*. People usually pick up the *Trichinella* parasite by eating wild game and, secondarily, pork. *Trichinella* roundworms living in the hog intestine mate and produce larvae that ultimately infest the animal's muscles, so that even a small piece of pork can contain tens of thousands of the worms in cysts. If not killed by thorough cooking of the meat, the worms produce larvae that bore into human muscles, especially those in the tongue and around the eyes, ribs, and diaphragm. Heavy infestations can nearly destroy these muscles, with potentially lethal consequences.

A summary of characteristics of nematodes and the other major invertebrate phyla discussed to this point is presented in Table 25-4. It is among these animals that

Figure 25-16   NEMATODES: A "FALSE BODY CAVITY," THE PSEUDOCOELOM.
The nematode body organization in (a) longitudinal section and (b) cross section. The body cavity surrounding the gut first appears as the pseudocoelom of the nematode worms, and it makes possible movement of the two-ended gut independent of the body wall. A nematode's maximum diameter is limited because it has no blood vessels and must depend on diffusion and stirring of the coelomic fluids to move nutrients, wastes, and dissolved gases around.

(a)   (b)

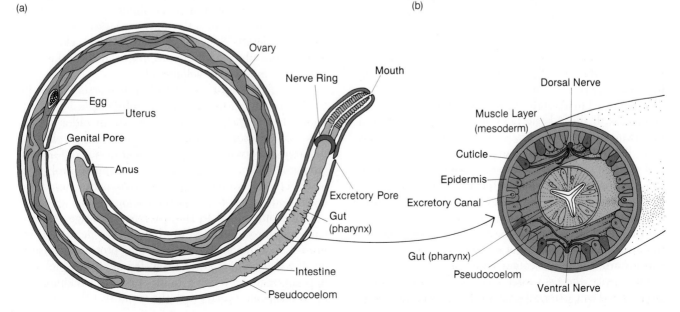

Table 25-4   **SOME PRIMITIVE INVERTEBRATE PHYLA**

Phylum	Number of Species	Representative Animals	Level of Organization	Symmetry	Digestion	Circulation	Gas Exchange	Waste Disposal
Porifera	~ 5,000	Sponges	Tissue, loosely arranged	Radial or asymmetric	Intracellular	Diffusion	Diffusion	Diffusion
Cnidaria	~ 10,000	Jellyfish	Tissue	Radial	Gut, with one opening (mouth); intra- and extra-cellular	Diffusion	Diffusion	Diffusion
Platyhelminthes	~ 13,000	Planarian	Organ	Bilateral	Gut, with one opening	Diffusion	Diffusion	Flame cells and ducts
Nemertina (Rhynchocoela)	~ 650	Ribbonworms	Organ	Bilateral	Complete gut, with two openings (mouth and anus)	Pulsating longi-tudinal blood vessels	Diffusion	Excretory canals with flame cells
Nematoda	~ 10,000 +	Roundworms	Organ	Bilateral	Complete gut, with two openings (mouth and anus)	Diffusion	Diffusion	Excretory canals

we see the transitions from radial to bilateral symmetry, the evolution of a one-way gut, and the beginnings of the coelomic body cavity.

THE TWO MAJOR ANIMAL LINEAGES: PROTOSTOMES AND DEUTEROSTOMES

All animals more complex than roundworms incorporate and build on the one-way digestive tract with an opening at both ends and the true body cavity, or *coelom*, bounded by mesoderm. One might suppose that all animals with these basic body designs derived from a single ancient group of progenitors. However, it seems more likely that different lines at the flatworm level of organization produced the two great assemblages of species, the protostomes and deuterostomes.

As is often the case, the embryology of these two lines tells us a great deal about their evolution. Recall from Chapter 16 that gastrulation in animal embryos involves the invagination of a sheet of endodermal cells. As they sink inward, these cells come to line the gastrula's newly formed cavity, the gastrocoel, which becomes the adult's gut cavity. The gastrocoel has a single opening, the *blas-*

topore. But if there is just one opening, does it become the embryo's mouth or anus? The answer is that in one major animal lineage, the blastopore becomes the mouth, and a separate orifice breaks through to become the anus. Animals with this developmental pattern are called **protostomes** (meaning "first mouth") and include the three groups we shall consider next—the segmented worms, arthropods, and mollusks. In the other animal lineage, the **deuterostomes** (meaning "second mouth"), the blastopore becomes the anus, and a second opening yields the mouth. Starfish and their relatives, along with a few minor phyla and all vertebrates, display the deuterostome mode of development.

Although biologists named these two assemblages by the fate of the blastopore during embryological development, that characteristic is not necessarily older in evolutionary terms nor as important as other major physical differences between the groups. The major differences are variations in (1) cleavage patterns and control processes of the early embryo, (2) the origin of the mesoderm layer, and (3) the origin of the coelom.

Protostomes tend to show a spiral cleavage pattern, while deuterostomes tend to have radial cleavage (Figure 25-17; Chapter 16). In addition, protostomes carry out the *determinate* type of development, in which cell lineages tend to receive quite specific developmental instructions very early, with the result that there is little

Nervous System	Reproduction	Other Characteristics
Irritability of cells	Asexual, by budding; sexual (most hermaphroditic); larvae swim by cilia or crawl	Filter feeders; skeleton of collagen plus spicules
Nerve net	Asexual, by budding; sexual (sexes separate)	Hydroskeleton, cnidocytes
Anterior ganglia; ladder-type system; simple sense organs	Asexual, by fission; sexual (hermaphroditic)	No body cavity
Anterior ganglia; two nerve cords; simple sense organs	Asexual, by fragmentation; sexual (sexes separate)	No body cavity; proboscis for defense and capturing prey
Simple brain; dorsal and ventral nerve cords; simple sense organs	Sexual (sexes separate)	Pseudocoelom

(a) Protostomes: Spiral Cleavage

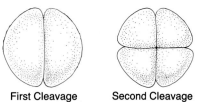

First Cleavage   Second Cleavage   Third Cleavage

(b) Deuterostomes: Radial Cleavage

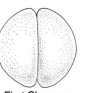

First Cleavage   Second Cleavage   Third Cleavage

Figure 25-17   CLEAVAGE PATTERNS OF ANIMALS.
The protostomes and the deuterostomes have different patterns of cleavage. (a) In the protostomes, cleavage is spiral; in the case shown, the third cleavage is in the clockwise direction, in terms of the position of the four small blastomeres, seen here from the animal pole of the embryo. (b) In the deuterostomes, cleavage is radial, with the four small, upper blastomeres arising directly above the larger, lower blastomeres. This difference in cleavage patterns is apparent from the third cleavage on.

capacity of the embryo to "regulate," to compensate for lost or injured parts. In contrast, deuterostome embryos carry out processes in which developmental instructions arise gradually in the *regulative* type of development. As a result, the embryo can make up for lost or damaged cells. In protostomes, mesoderm is derived from cells that migrate into the blastocoel near the blastopore. In deuterostomes, pouches of mesoderm arise from the endoderm of the archenteron wall. (There are, however, important exceptions to these generalizations.) Finally, the coelomic (body) cavity has different origins in the two groups. In many protostomes, it forms by a splitting of the mesodermal mass, as shown in Figure 25-18; in deuterostomes, the coelom develops directly from the cavity of the early mesodermal pouches.

The developmental differences between protostomes and deuterostomes provide us with two general sets of characteristics that correlate with each other and lead to the hypothesis that there are two major animal lineages. However, increasing numbers of biologists are skeptical about the usefulness of the distinction: there is so much variation among both protostomes and deuterostomes that the differences between the groups are blurred, and the degree of relatedness of different phyla in each group is highly variable. Be that as it may, let us turn to the first of the groups traditionally considered to be protostomes—the segmented worms.

Protostomes

Annelida: The Segmented Worms

The familiar earthworm is among the 8,000 or so species in the phylum **Annelida,** or segmented worms (Table 25-5). These protostomes inhabit soil, salt-water, and fresh-water environments. The word "annelid" means "tiny rings" and, indeed, the annelids' most striking external feature is the many body segments fused together in a linear arrangement (Figure 25-19). In addition, however, they have a true coelom and a closed circulatory system. There are three major classes of segmented worms: **Oligochaeta, Polychaeta,** and **Hirudinea.**

The 3,000 species of oligochaetes include earthworms and other soil and fresh-water worms that range from tiny animals less than 1 millimeter long to giant tropical worms 2 meters long and up to 40 millimeters in diameter. Almost all the 5,000 polychaete species are marine worms, many of which are brightly colored. Some build U-shaped tubular chambers in the sand and may bear a feathery crown (Figure 25-19c). The hirudineans are the 300 species of leeches that live in fresh-water streams and lakes and in the foliage and leaf litter of moist tropical forests. Many kinds of leeches are carnivorous and feed on virtually any source of animal protein, including

(a) Protostome

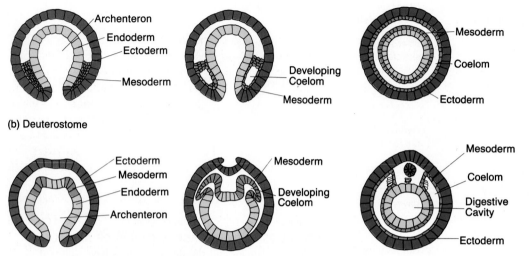

(b) Deuterostome

Figure 25-18   BODY CAVITIES: TWO PATTERNS OF DEVELOPMENT.
(a) In the protostomes, a solid mass of mesodermal cells splits to form the coelom. (b) In the deuterostomes, hollow pouches of mesodermal cells arise, and the cavities of those pouches become the coelom.

Table 25-5   **CLASSES OF THE PHYLUM ANNELIDA**

Class	Number of Species	Representative Animals	Habitat	Body Segments	Reproduction	Other Characteristics
Oligochaeta	~3,000	Earthworm	Terrestrial and fresh water	No parapodia; has setae	Hermaphroditic	Reduced head
Polychaeta	~5,000	Lugworm; sea mouse	Most marine	Usually parapodia; has setae	Separate sexes; trochopore larva	Well-developed head
Hirudinea	~300	Leech	Terrestrial and fresh water	No setae	Hermaphroditic	Posterior and anterior suckers allow bloodsucking

Figure 25-19   ANNELIDS: THE SEGMENTED WORMS.
Segmented annelids have many forms. (a) Some, like the polychaete *Nereis*, consist of up to hundreds of nearly identical segments. (b) In this *Chaetopterus variopedatus*, as in earthworms, certain segments differ structurally and functionally. The ones near the anterior end of this specimen (right) are expanded and specialized for feeding. Note the large tube of debris in which the worm resides. (c) Some polychaetes are tube worms with glorious crowns of feathery tentacles which act as respiratory gills and bear many cilia that help collect microscopic food particles. This Caribbean polychaete, *Spiro branchus*, has unfurled its spiraling crown. If irritated by nearby movement, the worm can quickly retract this showy crown and hide in its tube.

(a)   (b)   (c)

fresh blood. In fact, leeches make such a powerful anti-coagulant of human blood that both that chemical and living leeches are being used by physicians when induced bleeding is beneficial.

Many annelids are abundant and ecologically important—more than 8,000 small oligochaetes have been found per square meter of lake sediment or rich farm-land. Along with about 800 larger earthworms in that same square meter of land, these annelids serve as biological "plows," ingesting and aerating the soil and making it richer and easier for penetration by plant roots.

Segmentation: An Adaptation for Size and Specialization Organization of the body into a series of segments—**segmentation**—has functional and evolutionary importance. The function in annelids relates to key body structures that are repeated in each segment; these include muscles for locomotion, blood vessels, kidneylike excretory organs, and clusters of nerve cells (ganglia) arrayed along a major ventrally situated nerve cord. This type of segmentation appears to be associated with the evolution of increased body size, as simple addition of similar units yields longer and larger organisms. It also relates to moving and generating thrust against the earth, water, or air. In addition, in some annelids and in many annelid descendants, certain individual segments can be highly specialized and bear intricate mouthparts, antennae, legs, or even wings—organs with dramatically different combinations of cell types, forms, and functions. Thus the segmentation that arose in annelids and that was re-tained in other groups gives the potential to increase body size, to move efficiently, and to perform various specialized tasks.

Coelom and Circulatory System Annelids are the first animal phylum with a true **coelom:** a fluid-filled cavity that is completely lined by mesoderm and in which many of the internal organs are suspended. Depending on the species, the coelom can function as a hydroskeleton; a site of gas, nutrient, and waste circulation; a place where gametes are discharged before they exit the body; and a temporary reservoir for excretory products. In earthworms, the full coelom consists of a series of discrete chambers sealed off from one another by the sheets of tissue that separate adjacent body segments (Figure 25-20). The fluid in each coelomic compartment, together with the fluid in the gut cavity, acts as a hydroskeleton: when body-wall muscles contract in one segment of the body, the incompressible fluid transmits the force. Thus when the circular muscles of a segment contract (and the longitudinal ones relax), the segment lengthens; conversely, when the longitudinal muscles contract (and the circular ones relax), the segment shortens and swells outward. Figure 25-21 shows how sequential contractions in the segments produce coordinated waves of elongation and swelling, resulting in net movement forward of the whole worm. As these contractions are coordinated into waves by segmental nerves, bristlelike structures, or *setae*, on each segment are pushed against the soil, and the worm can inch for-

Figure 25-20 ANNELIDS: THE TRUE BODY CAVITY.
The annelids, represented here by the earthworm, *Lumbricus*, are the first true coelomates. The body is segmented, with most body parts, excluding the gut (pharynx, crop, gizzard, intestine), appearing repeatedly in most body segments. (a) This drawing shows these organs, but excludes a few other systems that will be discussed in later chapters. (b) The familiar garden earthworm, with slime coat glistening and reproductive segments bulging.

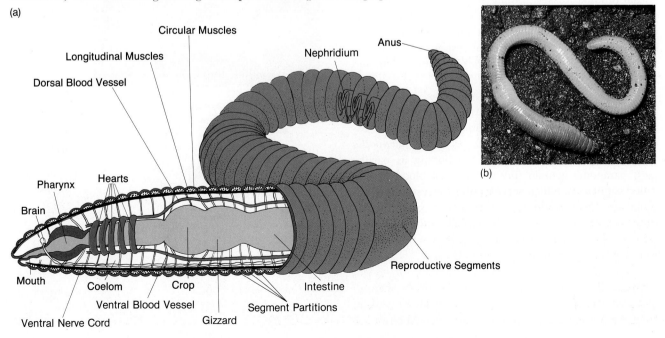

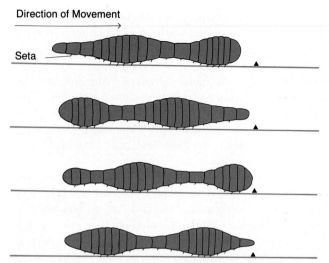

Figure 25-21 LOCOMOTION IN SEGMENTED WORMS.
Contractions of circular muscles cause body segments to elongate and narrow (as shown at this earthworm's front end in the first two drawings). Then, after the ventral surface adheres to the substratum, the longitudinal muscles contract, shorten those same segments, and thereby pull that section of the body forward. These alternating processes of circular and longitudinal contractions may occur simultaneously at several regions along the body. Note the movement from left to right relative to the arrowhead.

ward. A great advantage of the coelom is that these body-wall movements and the animal's locomotion can go on independently of gut movements and food processing, since the gut lies free within the coelom.

Annelids have another major adaptive feature not found in the other worms we have discussed: a *closed circulatory system* in which blood circulates entirely within tubelike vessels, rather than partly in vessels and partly bathing the body tissues themselves. Closed circulatory systems can operate at higher pressures and faster rates of flow than can open systems. This important innovation supports higher rates of metabolism and larger body sizes (Chapter 31).

In earthworms and other oligochaetes, oxygen and carbon dioxide are exchanged through the body surface itself. Polychaetes have an additional site for gas exchange: pairs of flanged appendages called *parapodia* (meaning "part legs"). Besides functioning as oarlike legs, parapodia contain tiny networks of blood vessels called capillaries, which serve as sites for respiratory gas exchange. Overall, the closed circulatory system was a critical evolutionary innovation that allowed many annelid worms to attain a larger size on average than do the roundworms, ribbonworms, and flatworms.

Organs for Waste Removal The large body of annelids that can arise because of the circulatory system creates a problem: the wastes of cellular metabolism are

too abundant to handle by diffusion outward through the body wall alone. In earthworms, pairs of organs called *nephridia* ("small kidneys") collect wastes, ions, and water in most body segments and excrete them. As we will see in Chapter 35, the type of filtration, reabsorption, and excretion that takes place in individual nephridia occurs in similar structures of vertebrate kidneys.

Annelid Reproduction: Asexual and Sexual Reproductive strategies that evolved in simpler worms also characterize annelids. Polychaetes reproduce asexually when a single body segment breaks free and regenerates an entire worm. Alternatively, the entire body can break into separate segments, or local buds can form a set of segments to one side of the body. Both types develop into new adult worms. In some polychaete species, sex-

Figure 25-22 THE TROCHOPHORE LARVA: HALLMARK OF PROTOSTOMES.
(a) This scanning electron micrograph of a trochophore larva of *Serpula vermicularis* (magnified about 375 times) characterizes the annelids. Beating of the many cilia around the larva's waist propels the organism through sea water. (b) Development of a worm from the larva. Repetitive developmental processes occurring at the posterior end of the larva generate the many body segments.

(a)

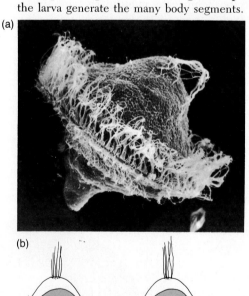

(b)

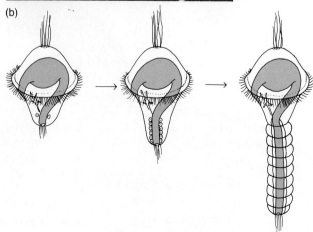

Figure 25-23   DIVERSITY OF ARTHROPODS.

(a) Velvet mite (*Trombidium* species). (b) Leg joints are clearly visible in this fearsome-looking spider, a rusty kneed tarantula (*Brachypelma smithii*) on display at the Cincinnati Zoo. Tarantulas are the largest and longest-lived spiders. Contrary to much lore, the bite of most species is no more harmful than a bee sting. (c) A lichen katydid (*Markia hystrix*) from the Costa Rican Cloud Forest Preserve is camouflaged to escape predation. (d) A scurrying mass of fiddler crabs (*Uca pugrax*); the males have the enlarged right pincers. (e) A snout beetle (*Otiorhynchinae episomus*) has an iridescent sheen as it sits awaiting a meal. (f) Huge wings marked with eyelike markings to frighten predators are sported by this atlas moth (*Coseinocera hercules*).

ual reproduction takes place between male and female, while in other species both mates are hermaphroditic and exchange sperm, as do the flatworms. The ciliated larva of aquatic polychaetes is called a **trochophore** and is just like that found in mollusks and other marine protostomes (Figure 25-22a). Segments form from one end of the larva (Figure 25-22b).

All earthworms are hermaphrodites; when two in breeding condition encounter each other on the surface of the soil, they position themselves head to tail and become "glued" together by mucous secretions from special reproductive segments. They part after mutual fertilization, leaving behind a group of zygotes inside a protective cocoon formed by the sloughed mucus. Twelve weeks later, the young worms emerge. The annelids' diverse reproductive repertoire no doubt has been adaptive for the group's success. However, the success of the segmented worms is overshadowed by that of a phylum of more complex animals, the arthropods, which almost certainly arose from ancient annelid progenitors.

Arthropoda: Exploiting Segmentation

There are almost 1 million species in the phylum **Arthropoda** (meaning "jointed legs")—more than the combined number of species in all the other animal phyla. As might be expected, this huge group of protostomes is highly diverse and includes the crustaceans, insects, spiders, and mites (Figure 25-23). The arthropods have a staggering diversity of life styles and habitats as well. Moreover, one kind of arthropod, millepedelike organisms, provide the first evidence of animals on land in late Ordovician times.

Arthropods share certain basic features with the annelids, including bilateral symmetry, segmentation, and a similar nervous system. However, they also display three major differences: (1) arthropod segments are highly modified for different functions, especially those segments that bear jointed appendages; (2) body segmentation is largely lost internally, and the coelom is replaced with the **hemocoel,** a blood-filled cavity; and (3) the body is armored with a rigid **exoskeleton** that pro-

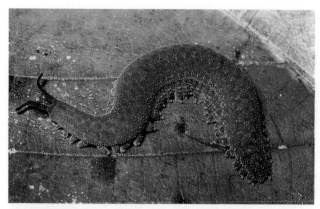

Figure 25-24   *PERIPATUS:* A LIVING FOSSIL.
Peripatus has the soft body of an annelid and the jointed limbs of an arthropod. This velvety species is found in the rain forests of Costa Rica.

vides protection and serves as a firm site to which muscles attach and against which they pull instead of pressing inward against a hydroskeleton. Of special importance are the armored, jointed appendages (hence the phylum name) that serve as movable limbs to propel the animal by pushing against the air, ground, or water.

These basic arthropod characteristics seem far removed from those of the annelids. However, alive today in the tropics is a group of animals that appears to be intermediate between annelids and arthropods. Velvety, caterpillarlike animals of the phylum **Onychophora** display an obvious combination of annelid and arthropod traits. Members of the most common genus, *Peripatus*, are called "walking worms" because of their short, stubby legs and wormlike locomotion (Figure 25-24). They have a soft cuticle and a segmented excretory system, as do the annelids, but they also have a hemocoel and a tubular respiratory system of the sort we will examine in arthropods. Biologists believe that onychophorans are an early offshoot of the line that led from primitive annelids to arthropods.

The Advantages of a Hard Skeleton The arthropod's tough external skeleton, or exoskeleton, is much more than simple armor plating. It is a cuticle that contains mostly chitin but also glycoproteins (sugar–protein complexes), waxes, and lipids that can seal body fluids in and water out, especially in land- and air-dwelling arthropods. In one class of arthropods, the *crustaceans* (lobsters, crabs, shrimp, crayfish), the cuticle is impregnated with calcium carbonate crystals, making it very hard. In all arthropods, the cuticle remains thin and flexible at the joint regions of the legs and of the body segments so the animal can move and bend its body.

The evolution of a hard exoskeleton was accompanied

by modifications in the locomotory muscles and in the growth process. Recall that annelids and other worms have layers of circular and longitudinal muscles. In contrast, arthropods have discrete bundles of muscle tissue attached to the inner surface of the cuticle on each side of the joints. The arrangement provides leverage so that muscle contractions can move the legs, wings, antennae, and body segments either rapidly or forcefully, instead of slowly and sequentially, as in annelids.

Despite its advantages, the very rigidity of the exoskeleton presents a problem: it is literally a nonexpandable "container"—an obvious block to the organism's growth. As such a rigid exoskeleton evolved, arthropods evolved the capacity to molt, or cast off the exoskeleton, at intervals throughout juvenile life and early adulthood (Figure 25-25). Molting involves several steps. Before the existing exoskeleton is shed, the epidermis beneath it forms a new exoskeleton—larger, folded, and soft. The old hard covering then splits, and the animal emerges in its soft cuticle. The animal immediately expands its new soft exoskeleton by means of pressure (water or air), and the new exoskeleton begins to harden by oxidation or by the deposition of calcium carbonate. Molting is a time of danger for arthropods, since they are especially vulnerable to predation during the intervals when they have a soft exterior. Molting is also a complex developmental process that must be coordinated precisely. This coordination is accomplished by a kind of organ we have not yet discussed—a hormone-secreting *endocrine organ*. We shall see in Chapters 36 and 37 that both the nervous and the endocrine systems secrete chemicals that control the activity of other body cells. Especially in the arthropods, these two systems are central to the regulation of the entire complex organism.

Figure 25-25   MOLTING HARD EXOSKELETONS.
A cicada (*Tibecen* species) dries its electric blue wings after shedding its old, constricting exoskeleton, upon which it rests.

High Metabolic Capacity: Vital for Rapid Movement
Evolution of the exoskeleton and muscles for locomotion allowed the disappearance of the hydroskeleton in arthropod ancestors. The arthropod coelom is represented by only small pouches near the gonads. The embryo's blastocoel gives rise to the hemocoel, a cavity filled with the clear-looking *hemolymph* fluid that bathes most of the internal body tissues. This bloodlike fluid circulates through various vessels and cavities mainly as a result of stirring by muscle contractions. Hemolymph circulation in arthropods is described more fully in Chapter 31.

The main task of hemolymph is to transport gases, nutrients, and metabolic wastes so that the needs of actively metabolizing and functioning tissue cells are met. In arthropods such as spiders and lobsters, oxygen is transported by *hemocyanin*, a bluish respiratory pigment dissolved in the hemolymph, whereas in one kind of insect (midges), hemoglobin carries out that task. (We will see that insects do not use hemolymph for gas transport.)

In general, arthropod tissues function at high rates of metabolism, especially when the body is warm. High metabolism means a high ATP requirement, which, of course, means a high oxygen demand at the cellular level. Various arthropods evolved three main types of respiratory organs for exchanging oxygen and carbon dioxide which meet that demand. Insects have a branching network of blind-ended tubes called *tracheae*, which carry air deep into the body, near every cell. Spiders, mites, and scorpions have *book lungs*, which are leaflike plates inside hollow chambers invaginated from the body surface. Aquatic arthropods such as crabs and crayfish have *gills*, which are highly branched, feathery structures over which water flows. Tracheae and gills are described in Chapter 33. The waste products of cellular metabolism are processed and expelled by arthropods in various ways, as described in Chapter 35.

Variations on the Arthropod Body Plan   The many attributes of the basic arthropod body plan have allowed a great evolutionary diversification of forms and the exploitation of many ways of life. This variety is reflected in the four subphyla: Trilobita, Chelicerata, Crustacea, and Uniramia (Table 25-6).

Members of the subphylum **Trilobita,** one of which is pictured in Figure 25-26a, were very abundant during the early Paleozoic era, from about 600 to 500 million years ago. Although all trilobite species are extinct, they left millions of marine fossils of flattened, oval-shaped animals with three body regions: the *head, thorax,* and *abdomen.* Each region was segmented and bore identical appendages. From this early condition, subsequent arthropods show a reduced number of appendages and specialization of those that remain for a variety of uses.

Chelicerata, the second subphylum of arthropods, includes horseshoe crabs, which often wash up on Atlantic coastal beaches (Figure 25-26b); the ubiquitous *arachnids,* or spiders, mites, ticks, and scorpions; and several rare and extinct groups (Figure 25-27). Chelicerates have only two body regions: the anterior *cephalothorax* (a fused head and thorax) and the posterior abdomen. This group's most distinguishing feature is the presence of *chelicerae:* pincerlike or clawlike mouthparts that replace the first pair of walking legs possessed by the trilobites, *Peripatus,* and some other early arthropods. The second pair of legs, *pedipalps,* also are used in feeding. The oversized pincers of scorpions are pedipalps (Figure 25-27a). Legs tend to be missing altogether on the abdominal segments—a trait that the chelicerates share with the insects. Instead, their walking legs are attached to the cephalothorax. Spiders have four pairs of walking legs on that body segment. Spiders also have silk glands for wrapping eggs in a silk cocoon and for building webs, as well as venom glands for stinging and paralyzing prey.

Table 25-6   **SELECTED SUBPHYLA AND MAJOR CLASSES OF THE PHYLUM ARTHROPODA**

						Appendages		
Subphylum	Class	Number of Species	Habitat	Body Divisions	Gas Exchange	Antennae	Mouthparts	Legs
Chelicerata	Arachnida	~57,000	Most terrestrial	Cephalothorax, abdomen	Book lungs or tracheae	None	Chelicerae, pedipalps	Four pairs on cephalothorax
Crustacea	Crustacea	~25,000	Marine or fresh water, but a few terrestrial	Cephalothorax, abdomen	Gills	Two pairs	Mandibles, two pairs of maxillipeds*	One pair per segment or less
Uniramia	Insecta	~900,000	Most terrestrial	Head, thorax, abdomen	Tracheae	One pair	Mandibles, maxillae*	Three pairs on thorax; (+ wings)

* Maxillipeds and maxillae are specialized mouthparts.

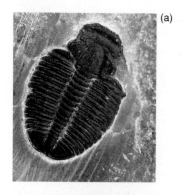

(a)

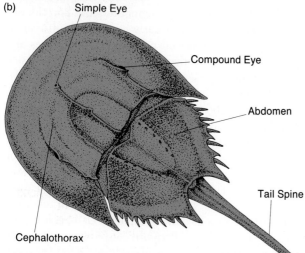

(b)

Simple Eye

Compound Eye

Abdomen

Tail Spine

Cephalothorax

Figure 25-26   A FOSSIL TRILOBITE AND A LIVING CHELICERATE.
(a) The head is at the top of this fossil trilobite. Trilobites flourished for over 300 million years, from Cambrian to Permian times. (b) *Limulus*, the horseshoe crab. The segmental body plan is obvious only in a ventral view of this animal. Because this living creature is almost identical to fossils hundreds of millions of years old, it is often referred to as a living fossil.

(a)

(b)

Figure 25-27   PREDATORY ARACHNIDS: LIFE IN A WORLD OF VIOLENCE.
(a) These scorpions (*Centruroides* species) are in fact sparring during a mating dance. The needlelike tip of the scorpion abdomen is highly poisonous to many types of animals. The pincers are also effective offensive weapons.
(b) A pair of black widow spiders (*Latrodectus mactans*) hang suspended in their glistening web. The larger female has a red hourglass marking on the ventral surface of the abdomen, while the smaller male has a harlequin white-and-red pattern. Both produce a nerve poison that paralyzes their prey and can seriously harm or even kill a human. The female will kill the male after mating.

Crustacea, the third subphylum, includes some 25,000 species of shrimps, crabs, lobsters, barnacles, and sow bugs that are noteworthy for the great variety of their body forms (Figure 25-28). Most crustaceans are aquatic. A distinctive feature of this subphylum (and class) is the presence of specialized appendages. In particular, they have *two* pairs of sensory antennae on the head, which are modified for touch and detection of chemicals (Figure 25-29). The next set of appendages, the *mandibles,* are jawlike mouthparts for biting and chewing. The crustacean thorax is often covered by a hard, protective *carapace* that overhangs the side of the body, even extending completely around it. The carapace and the rest of the cuticle usually are calcified as well as colored in various ways by pigment cells.

Uniramia is a subphylum that includes centipedes, millipedes, and insects. Unlike crustaceans, which have so-called *biramous* appendages (meaning that the end of each appendage is branched and jointed, as in the pincer claws of lobsters), all uniramians have unbranched appendages (hence the subphylum name, which means "single appendage"). Insects, the most abundant uniramian type, have the same body regions as trilobites: head, thorax, and abdomen (Figure 25-30). The thorax always bears three pairs of legs and sometimes one or two pairs of wings. A typical insect, such as the grasshopper, has a tracheal system for gas exchange (Chapter 33). It also has a hemocoel, circulatory vessels, and a pump—the heart—to move the colorless hemolymph. Specialized appendages are used for flying and for hopping hundreds of times its own length and height.

(a)   (b)

Figure 25-28   CRUSTACEANS: A VARIETY OF BODY FORMS.

(a) This tropical crustacean, the regal lobster (*Enoplometopus vanuato*), is brilliantly colored and has jointed appendages for pinching, walking, swimming, and sensing the environment. (b) These chitons (*Chiton fulvus*) are primeval-looking marine crustaceans. Each has a thick, rubbery foot covered by a series of eight overlapping plates that allow the animal to roll up and present its enemies with a hard protective surface.

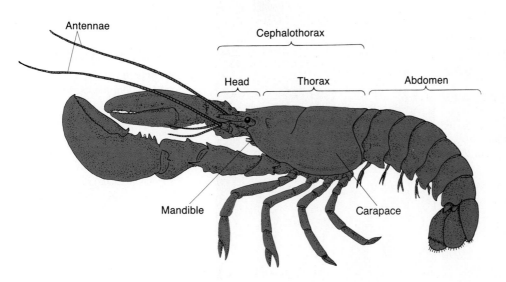

Figure 25-29   SPECIALIZED APPENDAGES: THE HALLMARK OF THE CRUSTACEANS.

Crustaceans such as the American lobster, *Homarus*, have 19 pairs of appendages, including antennae, mouthparts, and legs specialized for feeding, walking, and swimming.

Figure 25-30   INSECT BODY FORM.

Insects have six legs and several distinct body segments. The main insect body segments are the head, thorax (bearing the limbs and wings), and abdomen. The three thoracic segments of this grasshopper are indicated with Roman numerals, and the eleven abdominal segments with Arabic numbers. Some of these insect structures will be discussed in later chapters.

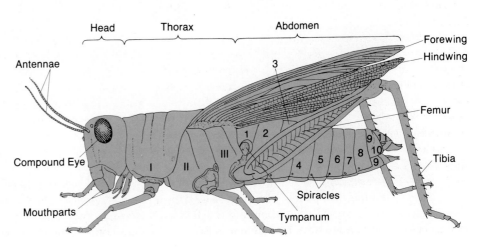

Insecta: Diversity and Success   The 900,000 species in the class **Insecta** are amazingly diverse in appearance, and they thrive in virtually every terrestrial and freshwater habitat (Figure 25-31; Table 25-7). Insects vary in size from tiny gall midges only 80 micrometers long to extinct giant dragonflies with a wingspread of 60 centimeters. Most contemporary insects are of a modest size that allows for exploitation of a vast array of microhabitats, such as the underside of leaves, the inner bark layer of trees, and the spaces between feathers or hairs on birds and mammals. One group can even tolerate salt well enough to live on the ocean surface. What makes insects so successful as a group? There are three answers: (1) adaptations for life on land, on water, and in the air; (2) versatile styles of development that include distinct phases of the life cycle; and (3) complex nervous, endocrine, and sensory systems.

An insect's hard, impervious cuticle prevents water loss in terrestrial habitats and water gain in aquatic habitats. Thus when an insect is on land or in the air, water vapor cannot evaporate through the cuticle. Alternatively, when an insect is immersed, water cannot inadvertently enter the body and affect osmotic balance. The tracheal system is a second important adaptation for all insects; it carries oxygen to all parts of the body, a function carried out in other animals by the circulatory system and oxygen-transporting cells. The insect muscular system is yet another example of special adaptations: the muscles are striated (striped; Chapter 39) and arranged to exert great force or to attain great speed. Such muscles permit insects to run extraordinarily fast, to carry many times the body weight, and to fly on wings that beat up to 1,000 strokes a second. This muscle power and speed give insects a mobility and range on land, in

(a)

(b)

Figure 25-31   BIZARRE AND BEAUTIFUL INSECTS.

(a) A dobsonfly (*Corydalus* species) is a formidable predator. (b) This desert grasshopper from the Sudan is well camouflaged when at rest on these reeds. (c) Like construction cranes, these giraffe beetles (*Apoderus giraffa*) survey the landscape from on high. (d) Like a child's cutout figure, this leaf insect (*Phyllium* species) rests on its food plant.

(c)

(d)

Table 25-7   **MAJOR ORDERS OF THE CLASS INSECTA**

Order	Representative Animals	Mouthparts	Wings	Other Characteristics
Thysanura	Silverfish	Chew	None	Long antennae; three "tails" extend from posterior tip of abdomen; run fast
Collembola	Springtail	Chew	None	Abdominal structure for jumping
Odonata	Dragonfly	Chew	Two pairs: long, narrow, membranous	Predators; large compound eyes; fresh-water nymph
Ephemeroptera	Mayfly	Chew	Two pairs: membranous	Small antennae; vestigial mouth; two or three "tails" extend from tip of abdomen; fresh-water nymph
Orthoptera	Grasshopper	Chew	Two pairs or none: leathery forewings; membranous hindwings	Most herbivorous; praying mantis eats other insects
Isoptera	Termite	Chew	Two pairs or none	Social insects, form large colonies
Dermaptera	Earwig	Chew	Two pairs: short forewings; large, membranous hindwings	Forcepslike appendage on tip of abdomen
Anoplura	Sucking louse	Pierce and suck	None	External parasites of birds and mammals
Hemiptera	True bugs	Pierce and suck	Two pairs: membranous hindwings	Mouthparts form beak
Homoptera	Aphid	Pierce and suck	Usually two pairs: membranous	Mouthparts form sucking beak
Neuroptera	Dobsonfly	Chew	Two pairs: membranous	Predatory larvae
Lepidoptera	Butterfly	Suck	Two pairs; scaly	Larva caterpillars have chewing mouthparts; eat plants, adults suck flower nectar
Diptera	Housefly	Usually pierce and suck	Functional forewings; small hindwings	Maggots larvae; adults often transmit disease
Siphonaptera	Flea	Pierce and suck	None	Legs for jumping; lack compound eyes; external parasites of birds and mammals
Coleoptera	Beetle	Chew	Usually two pairs: forewings modified as heavy, protective coverings	Largest order of insects (300,000 species); herbivorous; some aquatic
Hymenoptera	Bee	Chew and lap or suck	Two pairs or none: membranous	Some social; some sting

the air, and on water unknown in the other animal groups we have discussed.

Versatile developmental strategies allow insect larvae and adults of most species to feed on different types of foods and to thrive under different environmental conditions. Most insect species, including butterflies, beetles, and flies, undergo *complete metamorphosis*. The embryo forms a larva with a radically different body plan than that of the adult; examples are the caterpillar larva of the butterfly and the maggot of the fly. A caterpillar feeds voraciously on plant materials and then in response to a hormonal signal, it *pupates*, or forms a pupa in a cocoon (Figure 25-32). The pupa undergoes metamorphosis; the

butterfly emerges; its wings and cuticle harden; and it flies off. Complete metamorphosis allows the vast majority of insects to pursue radically different life styles as larvae and as adults and thus to adapt to the foods and conditions present during different seasons of the year or stages of the life cycle. Consider, for example, a mosquito whose larva feeds and grows in a temporary pool of water in a rotting tree stump, but whose adult feeds on the blood of deer or humans.

In contrast to the species that undergo complete metamorphosis, a minority of insect species, including grasshoppers and cockroaches, have a developmental pattern called *incomplete metamorphosis*, which involves a se-

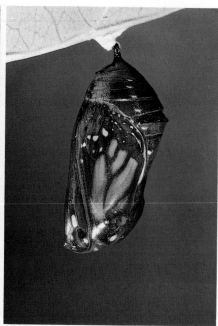

Figure 25-32   METAMORPHOSIS OF THE MONARCH BUTTERFLY (*DANAUS PLEXIPPUS*).

(a) The caterpillar attaches itself to a twig, head down, and then (b) molts its striped skin, revealing the chrysalis already formed below. (c) Developing butterfly wings are visible through the chrysalis wall after one week. (d) Eventually, the fully formed adult monarch emerges and rests as liquid is pumped from the body to expand the wings.

ries of molts. The embryo develops into a *nymph*, a miniature version of the adult, which then feeds, molts, grows, and repeats the sequence again and again. The stages between molts are called *instars;* after about six instars, the adult finally emerges with fully mature organ systems. The species with incomplete metamorphosis do not have radically different life styles between early and late instars.

Finally, the insects' tremendous success depends on complex sensory, endocrine, and nervous systems. Color vision and the capacity to see in the ultraviolet range are among the abilities of *compound eyes,* with their hundreds or thousands of facets. Compound eyes are described more fully in Chapter 38, and, of course, are an element of the insects' coevolution with flowering plants discussed in Chapter 24.

Many insects and other arthropods can hear very well, although primarily in terms of sensitivity and direction and not with the frequency discrimination of our own ears. Male crickets chirping on a warm summer night are using sound to warn other males about encroaching on their territory and to attract females into that territory. In some insect species, hearing depends on vibrations of either drumlike membranes on the abdomen or minute fringes of the antennae, which vibrate like a tuning fork. Antennae and other organs can also be involved in extremely sensitive chemical reception: some male moths can detect chemical signals given off by females from a distance of over 1 kilometer.

Insect sensory organs feed data to a modest-sized brain that is responsible for behavior patterns that are primarily *stereotyped*—automatic and controlled by the activities of specific nerve circuits determined genetically and developmentally. Thus the bee's response to a colored flower, the male moth's homing toward a source of sexual attractants 1,000 meters away, or the ant's methodical cutting and carrying of leaves to the anthill are all programmed behaviors that operate without the same degree of "choice" and "learning" we see in vertebrates. But new evidence reveals real learning in honeybees and behavioral versatility in other insects (Chapters 48 and 49).

Social Insects and Their Complex Societies The small brain and stereotyped behavior pattern of insects have not prevented the evolution of extraordinary behavioral complexity in two groups of insects. In these groups, social dependence has evolved in which individuals rarely if ever live alone—instead, they are members of colonies. These **social insects** include members of the orders *Isoptera* (termites) and *Hymenoptera* (ants, wasps, and bees). Social colonies consist of groups of individuals from several generations and in which some polymorphism exists. That is, there are different body structures associated with *castes*, types of individuals that perform different duties; males, fertile females (queen), and infertile females (workers) are the typical castes. Caste membership results from developmental processes: a queen bee develops from an immature female that is fed royal jelly, a special nutrient produced from pollen by worker bees. Later, the queen's mandibular glands secrete a chemical known as queen substance, which triggers a stereotyped behavior in the workers that feed the queen. They feed female larvae in such a way that the larvae develop into female workers rather than queens. Only when the queen ages does her supply of queen substance lag; soon the workers are no longer inhibited, and they begin to construct royal cells. They feed the eggs in such cells with royal jelly, producing new queens.

We will learn more about the complex behavior of social insects in Chapter 49. However, we should note here that while the full pattern of behavior in a social-insect colony seems amazingly complex, in fact most of its separate component behaviors are stereotyped and are induced by nervous and endocrine activities that are, in turn, determined by genetic and developmental processes. Social-insect behavior is an extraordinary illustration of great complexity being constructed from many simple parts as a result of evolutionary processes.

Arthropod Evolution How are the diverse arthropods related to each other? The answer to that question has changed radically in recent years. Fossils dating from the start of the Cambrian period suggest that at least twenty-four and perhaps as many as seventy distinct types of arthropods had diverged by that time from the segmented-worm level of organization. Of these, the trilobites diversified early and persisted for several hundred million years. Some biologists now conclude, contrary to earlier hypotheses, that the chelicerates, crustaceans, and uniramians arose independently of each other; almost no intermediate fossils have been found between those major groups. However, uniramians such as insects develop in a way that could easily have been derived from animals with a yolky annelid-type egg, and such evidence is consistent with the idea that annelids, *Peripatus*, and insects may be a real evolutionary lineage. In contrast, it seems highly unlikely that the developmental pattern of crustaceans could have arisen from any annelid, and they are probably an independent line. However, in contrast to annelids, all arthropods lack ciliated epithelia, and most use the coelom as a hemocoel; these facts argue that arthropods are indeed a single lineage. For the moment, then, we cannot be sure whether "arthropod" represents a grade of organization or is a true single lineage.

The origins of the uniramia are of particular interest since that group is the first type of invertebrate that arose on land. In fact, the earliest fossil centipedes, found in New York State in 1983, date from the Devonian period, 380 million years ago. Various types of fossilized insects appear in the fossil record only in conjunction with other organisms important for their survival: butterflies appeared only when flowering plants evolved; fleas, only when birds and mammals arose. This implies an evolutionary versatility whereby rapidly reproducing insects coevolve in a sense as other organisms with which they interact diversify. This suggests to many biologists that insects have a much better chance than do vertebrates of surviving the "evolutionary sweepstakes" for terrestrial life on Earth.

Mollusca: Rearrangement of the Protostome Body Plan

Now that we have surveyed the two major segmented protostome phyla, let us turn to a separate and highly successful group of protostomes, one that includes members with the most complex nervous systems and learned behaviors. This is the phylum **Mollusca** (Table 25-8). The 100,000 living species of mollusks include slugs, snails, clams, scallops, octopuses, and squids (Figure 25-33). Approximately 35,000 additional species are extinct. Molluscan life styles are so diverse that the biologist Donald Abbott wrote of them: "Among existing species are those that cling, creep, burrow, walk, leap, float, swim slowly or swiftly, or even occasionally become airborne, jet-propelled for short distances above the water."*

*R. H. Morris, D. P. Abbott, and E. C. Haderlie, *Intertidal Invertebrates of California* (Stanford, Calif.: Stanford University Press, 1980), p. 227.

Table 25-8   **MAJOR CLASSES OF THE PHYLUM MOLLUSCA**

Class	Number of Species	Representative Animals	Habitat	Body Parts	Locomotion
Polyplacophora (Amphineura)	~650	Chiton	Marine	Segmented shell of eight transverse plates; reduced head	Foot
Gastropoda	~55,000	Snail; slug; nudibranch	Marine, fresh water, or terrestrial	Body and shell usually coiled; head with tentacles and eyes	Foot
Bivalvia	~11,000	Clam; oyster; scallop	Marine, fresh water	Two hinged shells compressed laterally; mantle forms siphons; filter feeders	Sessile; shells move
Cephalopoda	~650	Squid; octopus	Marine	Shell internal, external, or missing; foot divided into tentacles and arms usually bearing suckers; complex eyes and brain	Funnel; tentacles

(a)   (b)   (c)

(d)   (e)

Figure 25-33   MOLLUSKS: A LARGE PHYLUM WITH STUNNING DIVERSITY.
Mollusks can be shell-less or shelled, and can live in salt water, fresh water, or on land. (a) In this Pacific pink scallop (*Chlamys hastata*) the fringed mantle can be seen as well as the small blue dots—eyespots. (b) The moon snail, *Lunatia heros*, travels on a broad blue foot and carries a spiraled shell along for protection. (c, d) Nudibranchs [(c) is *Noumea violacea*], or sea slugs, are among the ocean's most colorful animals. Projections from the dorsal surface can serve as gills or house extensions from the digestive system. (e) A giant octopus (*Octopus vulgaris*) photographed from a safe distance in the Red Sea.

Some mollusks are very abundant: samples taken in the Clyde Estuary off the coast of Scotland show that some 30 billion snails live in an area of 6 square kilometers (2.5 sq. mi.). Mollusks not only are varied and abundant, but also are quite different in many ways from the annelids and arthropods. In particular, they lack body segmentation. Because of this dissimilarity, some biologists conclude that mollusks arose from unsegmented flatworm turbellarians independently of the annelid and arthropod origin. However, mollusks do have a trochophore larva like that of annelids, implying some degree of common origin.

If you have ever eaten a fresh clam or a raw oyster, you may have the impression that a mollusk's body is little more than a slippery undifferentiated blob. However, the molluscan body has several distinctive regions, shown in Figure 25-34: (1) a muscular organ called the *foot* lies ventral to the other organs and is used for locomotion and for tightly gripping rocks or other substrata; (2) the *visceral mass* contains the internal organs—heart, digestive system, excretory system; (3) the *head* houses the sensory organs, brain, and mouth; and (4) the *mantle* is a thickened fold of tissue that covers the visceral mass and sometimes overhangs it. The space between mantle and visceral mass is the *mantle cavity;* respiratory gills hang down in this space in aquatic mollusks. Surrounding these organs in most mollusks is a shell that protects the animal, serves as an anchor for muscles, and supports the mantle folds so the cavity around the gills does not collapse. Cells on the outer surface of the mantle secrete shell material that usually includes calcium carbonate crystals. Figure 25-34 makes clear how these various major regions of the body have "shifted" during the evolution of the major molluscan groups.

Mollusks have a true circulatory system with a heart; a few closed vessels that pass through the gills; and an open, blood-filled sinus, or cavity, that bathes internal tissues. They have an oxygen-carrying blood protein called hemocyanin. Muscles in the body wall push against the incompressible blood sinus, using it as a **hemoskeleton.** This causes the foot to protrude from the shell, such as for locomotion. Although the blood-filled spaces surround the gut in spots, they do not constitute a true coelomic cavity. Remnants of the coelom remain near the heart.

Reproduction of primitive mollusks involves fertilization of eggs in the sea water and development of trochopore larvae. Among modern mollusks, all sorts of variations in larval development occur: in snails, for instance, the trochophore stage may occur prior to hatching but it develops into a differently shaped larva that emerges and swims about. In squids and octopuses, fertilization is internal, and the larval stages are suppressed, so that miniature "adults" hatch through the protective membranes.

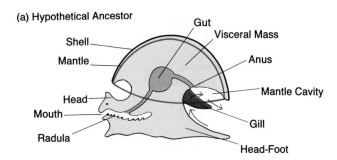

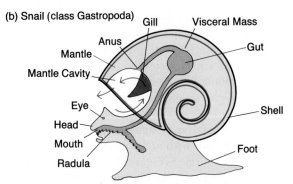

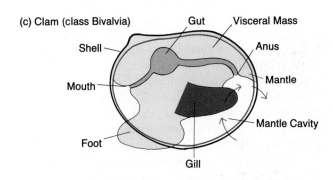

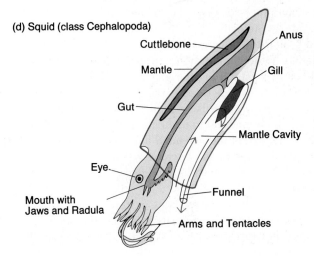

Figure 25-34   MOLLUSCAN BODY PLANS: VARIATIONS ON A THEME.

The position of the mantle cavity changes its relative position in each of these three classes. Water flow (blue) over the gills (purple) is shown.

Variations on the Molluscan Body Plan   There are four major classes of mollusks, and while most follow the generalized body plan, some—like the octopus and squid—are very different.

Members of the class **Monoplacophora** are primitive mollusks thought to have become extinct hundreds of millions of years ago until some living specimens were dredged up from a deep ocean trench near Costa Rica in 1957. These flat, bilaterally symmetrical animals have a single locomotory foot and a single unsegmented shell but six pairs of kidneys, five pairs of gills, eight pairs of foot-retractor muscles, and odd-numbered pairs of other organs. Some zoologists interpret this arrangement of paired organs as a primitive type of segmentation, but others regard monoplacophorans as simple unsegmented mollusks in which varying degrees of organ duplication took place.

Chitons—flattened oval marine mollusks that look thoroughly primeval, with eight armored plates curving across the dorsal surface (Figure 25-28)—make up the class **Polyplacophora** (or *Amphineura*). Chitons have a large foot that is capable of tenacious gripping and the general body features of most mollusks. They also have a *radula*, a tonguelike strap covered with horny, chitinous teeth that scrape off small bits of algae and other marine plants living in shallow intertidal zones.

Gastropoda (meaning "stomach-foot"), the most diverse class of mollusks, includes 55,000 species of snails, periwinkles, whelks, abalone, and related animals inhabiting salt-water, fresh-water, and terrestrial habitats (Figure 25-33b). All gastropods with shells display a characteristic coiling or twisting of the body that begins during the early cleavages of embryonic development. Torsion processes in the larvae and adults cause the visceral mass to coil so that the animal's tail end is rotated 180°, and the anus and gills lie over the head (Figure 25-34). Other modifications prevent solid wastes from the anus from fouling the gills and head. Like chitons, gastropods have a rasping radula to scrape off bits of food. Some species have reduced shells and can swim due to buoyancy and to undulating waves that pass along the thin edge of the extended foot. Others creep slowly along submerged rocks or crawl over land. In land dwellers, the mantle cavity is modified to serve as an air-breathing lung.

The class **Bivalvia** includes oysters, clams, scallops, mussels, and similar animals in which the shell is secreted in two halves called *valves* (Figure 25-33a). The mantle extends downward on either side of the head and foot, enclosing them in a cavelike mantle cavity. Bivalve mollusks lack the radula and tend to be filter feeders that have cilia to drive the feeding current through their sievelike gills.

Some oyster species have regions of the mantle that respond to the presence of irritating foreign materials by forming pearls. The laying down of many concentric layers of inorganic or proteinaceous materials creates black, reddish, or whitish pearls that are highly prized for their lustrous beauty. The most common natural irritants are the larvae of parasitic worms, but well-formed natural pearls are rare. Some oysters encase implanted bits of oyster shell and produce the cultured pearls usually sold today.

Cephalopoda (meaning "head-foot") includes squids, octopuses, chambered nautiluses, and a few other species that externally do not resemble other mollusks (Figure 25-33). The largest invertebrate on Earth is the giant squid. In cephalopods, the foot is modified and terminates in the funnel, and the head is large, central, and gives rise to eight to ten arms or tentacles.

Members of the cephalopod lineage evolved as fast-moving hunters of the sea. Early species, most closely related to today's nautiloids, evolved gas-filled chambers inside their shells that provided buoyancy. The chambers became more elaborate as faster-swimming nautiloids evolved. Nautiloids have shells of spectacular colors and shapes, but in squids, the shell is reduced to an internal *cuttlebone*—a light, flexible structure made of calcified chitin and permeated by tiny chambers filled with fluid that is less dense than sea water. A squid can raise or lower the density (actually, the specific gravity; Chapter 26) of its cuttlebone and hence alter its own buoyancy and depth in water. Octopuses lack shells, external or internal. Most are not buoyant and live on the sea floor or on coral reefs; nevertheless, they can swim well in pursuit of prey for limited distances. Deep-water octopuses (*Cirrothauma*) have full webs between the arms so that, when spread, the buoyant creature can hover like a medusoid jellyfish. Yet, the same remarkable animals have strong fins that generate rapid horizontal swimming.

Squids have a remarkable means of propelling themselves through the water. Besides being torpedo-shaped and thus highly streamlined, squids have a narrow tubelike opening from the mantle cavity (the funnel), from which water can be forcibly expelled in a powerful jet, propelling the body in the opposite direction like a rocket. In fact, the funnel can be aimed at various angles, so that the body can be thrust in a variety of directions. The propulsion backward can be powerful enough to launch the squid right out of the water. As squids "fly," they spread their fins to act as wings for gliding. Thor Heyerdahl, on the *Kon Tiki* expedition, reported seeing the amazing sight of airborne squid, flying "over the raft at a height of four or five feet" and "sailing along for fifty to sixty yards, singly and in two's and three's."*

---

*T. Heyerdahl, *Kon Tiki: Across the Pacific by Raft* (New York: Pocket Books, 1956), pp. 118–119.

The giant squid, the largest living invertebrate, is a denizen of the deep ocean that can reach 18 meters (60 feet) in length and nearly 500 kilograms (1,100 pounds) in weight. It has the largest eyes in the animal kingdom—complex organs that can equal the size of a car's headlight. Only a few dozen of these giants have washed up on beaches and have been studied. Humans have never seen the epic battles between giant squid and the toothed whales that feed on them. However, pieces of great tentacles found in whale intestines and rows of tentacle-sucker marks on whale skin testify to the deadly struggles that must go on thousands of meters below the ocean surface.

Octopuses and squids are capable of the most complex behavior of all invertebrates. One reason is their so-called camera eyes, which have high resolving power, much like our own. But the development and evolutionary origins of these eyes are quite distinct from those of vertebrates. This is one of the best-known examples of *convergent evolution*, the appearance of a functionally equivalent feature in two very separate lineages. Light detection and pattern recognition set demands on living systems that can be met, apparently, in only a few specific ways. Hence, similar visual organs had very different evolutionary origins.

Another reason for the complex behavior is that the brain of a medium-sized octopus is surprisingly large: it may contain nearly 170 million neurons (nerve cells) aggregated into one centralized, integrated mass. The uppermost lobes receive no direct sensory input and instead function, like the human cerebral cortex (Chapter 40), as a learning center. As a result, octopuses can be trained to run mazes, to discriminate between shapes or colors, and to respond to rewards or "penalties" such as a mild electric shock. Thus octopuses show true learned behavior. They are, without doubt, the most intelligent invertebrate animals.

Deuterostomes: A Loosely Allied Assemblage

We turn now from the protostome phyla to the other category of coelomate animals, the deuterostomes (Table 25-9). They include some minor phyla we will not describe, as well as the sea stars and their relatives (Echinodermata), the acorn worms (Hemichordata), and the animals with notochords (Chordata), which are the subjects of Chapter 26.

One might suppose that all deuterostomes are fairly similar in structure because of their common radial cleavage and the other developmental similarities we mentioned earlier. However, the adults of these three phyla are quite different. The groups apparently diverged from one another almost as early as differences arose between the prostostomes and the deuterostomes.

Echinodermata: Radial Symmetry Encountered Once More

Echinodermata (meaning "spiny skinned") is an apt phylum name for the sea stars (starfish), sea urchins, and their relatives. Most of the 6,000 species are relatively sedentary marine creatures that are radially symmetrical as adults but begin life as bilaterally symmetrical larvae. The radial symmetry is no doubt related to the echinoderms' sedentary life style, which brings exposure of all lateral surfaces to similar environments.

As you can see in Figure 25-35, echinoderms have diverse body forms. They include soft, cylindrical sea cucumbers and spiky sea urchins as well as multiarmed sea stars and feathery sea lilies. Echinoderms vary from a few millimeters to 1 meter in diameter and may weigh up to 10 kilograms. Despite their external differences, all possess a unique *water vascular system* that is in-

Table 25-9 **SOME MAJOR DEUTEROSTOME PHYLA**

Phylum	Number of Species	Representative Animals	Habitat	Other Characteristics
Echinodermata	~6,000	Sea star; sea urchin; brittle star; sea cucumber	Marine	Radial, fivefold symmetry; water vascular system; sexual reproduction plus regeneration in some species
Hemichordata	~80	Acorn worms	Marine	Gill slits; larvalike echinoderms; dorsal nerve cord
Chordata	~39,000	Sea squirt; trout; deer	Marine; fresh water; terrestrial	Notochord; gill slits; dorsal, hollow nerve tube

(a)   (b)   (c)

(d)   (e)

Figure 25-35   ECHINODERMS: A WIDE ARRAY OF BODY FORMS SHARING FIVEFOLD SYMMETRY.
All echinoderms have five radiating arms or five markings on the body. (a) An ochre starfish (*Pisaster ochraceus*) feeds on mussels, barnacles or snails. (b) A scarlet brittle star with typical long arms. (c) A sea apple, a rosy-hued sea cucumber with yellow tube feet and bright red branching gills (*Paracucumana tricolor*). Many sea cucumbers are green and bumpy—as the name implies. (d) Sand dollars (*Dendraster excentricus*) have "petal markings" made up of small holes on the test (a type of shell). Tube feet protrude through the holes. (e) These Hawaiian slate-pencil sea urchins (*Heterocentrotus mammillatus*) have large, brilliantly colored spines for protection. Tube feet protruding below pull the animal along slowly.

volved in feeding, locomotion, respiration, and sensory perception.

Echinoderms can move slowly by means of the water vascular system and certain muscles. This system is a modified portion of the coelom that acts as a hydraulic device. The system is filled with slightly modified sea water. In sea stars, special tubular extensions of the system (radial canals) end in rows of water-filled *tube feet*, each tipped with a sucker, on the ventral surface of the arms (Figure 25-36). At the end opposite from the sucker, each tube foot has a small muscular reservoir, the *ampulla*, which can contract, forcing more water into the foot and causing it to extend. When the tip of the foot touches the substratum, the center of the sucker is pulled in, creating suction. Contractions of the ampul-

lae, suckers, and muscles in the tube feet are coordinated so that the tube feet swing forward, grip the surface of a rock, and then swing backward in a kind of stepping motion. As this occurs in numerous tube feet, the sea star moves slowly across the sea floor.

The suckers can hold the animal to a rock despite pounding surf, and they can pull open the valves of a clam. The tube feet of sea stars in the genus *Evasterias* can exert considerable force on a clamshell and maintain that force for several hours. This eventually exhausts the clam's powerful shell muscles and opens the valves. The sea star then everts its stomach and ingests the helpless mollusk, so that enzymatic digestion can occur. Besides enabling movement and grasping, tube feet have the capacity to detect sounds and tastes.

Figure 25-36   THE SEA STAR: AN EXAMPLE OF THE BASIC ECHINODERM BODY STRUCTURE.
(a) The digestive gland has been removed from one arm, revealing the gonad and one branch of the water vascular system. (b) The tube feet are external to the animal, while the other parts of the water vascular system, the ampullae and the ring canal, are internal. The sieve plate is the site where water enters or leaves the water vascular system.

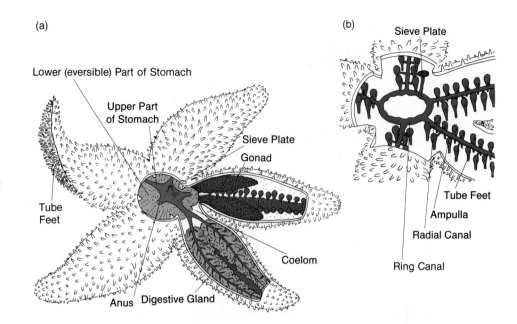

Most echinoderms have a nervous system with a nerve ring near the mouth and main nerves that radiate outward to major portions of the body. Some echinoderms have eyespots at the tip of their arms that can sense light and shadow.

A surprising feature of echinoderms is that, unlike most worms and other protostomes, they lack a kidney system and the ability to regulate in a major way the ion content of their body fluids. Thus they conform osmotically to sea water and are restricted to the marine habitat.

Although sedentary, echinoderms are not defenseless. Small, porous but stiff skeletal plates, each one a separate calcite crystal ($CaCO_3$ and $MgCO_3$), lie in the mesoderm just beneath the outer epidermis; these plates, called *ossicles*, are unique in the animal kingdom and serve as a kind of armor. The plates are sutured together in creatures such as sea urchins, forming a one-piece endoskeleton, the *test*.

Reproduction in echinoderms is by both asexual regeneration and sexual processes. A single starfish arm containing just a small portion of the central disk region will slowly regenerate the rest of the creature, although the process sometimes takes as long as a year. It is no wonder that uninformed clam and oyster fishers have had little success in wiping out predatory starfish by cutting them into pieces and throwing the parts back into the sea.

Sexual reproduction involves typical seasonal breeding in which gametes mature and are released directly into the ocean at one season of the year. Following fertilization and radial cleavage, a larval stage is reached that is quite different from the trochophore of protostomes. These *tornaria* larvae resemble those of the hemichor-

dates, the next deuterostome phylum we will describe. There is substantial variation among deuterostome larvae, just as among protostome ones, so that contrary to older ideas, larval stages are no longer viewed as a simple or reliable means of constructing phylogenies.

Biologists are not sure how the many different-looking echinoderms are related because much of the evolutionary divergence took place before the Cambrian period. Echinoderms that lived after that time left one of the richest fossil records. That record and other evidence suggest that sea stars and sea lilies make up one lineage, sea urchins and sea cucumbers a second, and brittle stars another. A newly discovered class, the Concentricycloidea, or sea-daisies, has a medusalike appearance, retains its embryos until they look like miniature adults, and is most closely related to sea stars. The source of all echinoderms remains a mystery, although a consensus is developing around a hypothetical creeping bilaterally symmetrical animal with five hollow tentacles surrounding its mouth. It is noteworthy that a coelom—the water vascular system—arose in this lineage, just as one did independently in annelids and in mollusks. Such convergent evolution underscores the many advantages of a true body cavity.

Hemichordata: The First Gill Slits

The last group of invertebrates we discuss are the first animals with a body feature that is clearly associated with a common trait of our own phylum, Chordata. It is in members of this invertebrate phylum, **Hemichordata,** that we see the most plausible candidate for creatures that might have yielded both echinoderms and chordates.

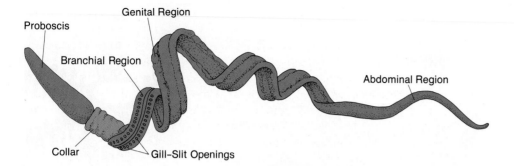

Proboscis
Genital Region
Branchial Region
Collar
Gill–Slit Openings
Abdominal Region

Figure 25-37
HEMICHORDATES: A NOVEL FEEDING MECHANISM.
Acorn worms live in mud burrows on marine beaches. The body structure of this hemichordate is typical of several species. Water, particles of food, and mucus enter the mouth (located near the collar); the water exits the gut through the gill slits.

All hemichordates are marine animals; some are mobile and wormlike (the enteropneusts, or "acorn worms"), while others are sessile and bear tentacles (the pterobranchs). Both types are noteworthy for their paired sets of *gill slits*, openings through the lateral walls of the anterior gut (Figure 25-37). A feeding current driven solely by cilia enters the mouth and exits the body through the gill slits. This feeding mechanism may seem bizarre; however, it gathers enough food to support the needs of acorn worms, which can reach lengths of 150 centimeters. And, more important from the human vantage point, it is the basic feeding mechanism of primitive chordates, our own lineage.

The wormlike enteropneusts have a dorsally situated nerve cord that arises by an infolding, just as occurs in vertebrates (Chapter 16). However, other structures are quite different from those in vertebrates. The name "hemichordate" is actually a misnomer based on a chordlike extrusion of the gut that was once regarded as a *notochord*, the primitive skeletal element of the chordates (Chapter 26). Development in hemichordates yields a ciliated larva like the tornaria of echinoderms and unlike those of protostomes or of the "lower" chordates. Overall, the living groups of hemichordates, particularly the pterobranchs, are thought to be more like the earliest deuterostomes than are the echinoderms. In fact, an ancient pterobranchlike creature may have given rise to separate lineages leading to echinoderms and chordates.

LOOKING AHEAD

In this chapter, we have traced more than 500 million years of evolutionary history and completed our survey of the major types of invertebrates. Although the variety seems immense, we have not even mentioned about one-half of the animal phyla, which include many bizarre, beautiful, and fascinating creatures. Among the major phyla we did discuss, we saw that multicellularity arose in an unknown number of types, probably by colonization, and may have resulted in a planulalike creature. That stage of organization gave rise to the radially symmetrical phyla, perhaps even to the sponges, and to the flatworms, with their bilateral symmetry and cephalization.

These early evolutionary stages show the beginnings of differentiated cell types and of organization of such cells into tissues and organs. Thus among the invertebrates arose the first muscle cells, the pseudocoelom and later coelom, segmentation and its variations in body regions, and various excretory systems. Nervous and sensory systems of diverse complexity evolved, including the highly specialized squid eyes; the octopus brain, which allows for learned behavior; and the social-insect brain, which allows highly organized stereotyped behavior. The invertebrate lines underwent a fascinating evolutionary odyssey, just as did microbes, fungi, and plants: all these lineages of organisms influenced each others' gradual evolutionary change over time and were, in turn, affected by geological events on land, in the water, and in the atmosphere.

We are ready now to consider the final major animal phylum—the one that yielded the vertebrates and humans.

SUMMARY

1. Multicellular animals probably arose either from aggregates of flagellated protistan cells that yielded either blastula- or gastrulalike colonies, or from organisms that resembled the *planula* larva of cnidarians. Multicellularity allows animals the advantages of (a) large size, (b) greater mobility, (c) a stable internal environment, and (d) greater independence from the external environment. With multicellularity came the evolution of muscles; circulatory, respiratory, excretory, nervous, and hormonal systems; sense organs; and behaviors of varying complexity.

2. Of the 2 million species of animals alive today, 97 percent are *invertebrates*, which lack a backbone, and 3 percent are vertebrates.

3. The sponges, the members of the phylum *Porifera*, are hollow, *sessile* (attached to a substratum) aquatic animals organized at the *tissue level* into three layers: the epithelium, the mesenchyme, and the flagellated choanocytes. As a *filter feeder*, the sponge pumps water through its body and extracts microscopic cells or organisms from the flowing water, or feeding current. The sponge's collagen "scaffolding" and hard spicules are reminiscent of a skeleton, and some of its cells may function like nerves. Sponges reproduce both asexually and sexually. They probably arose independently of all the other animal phyla and did not give rise to any other group of animals.

4. Animals that are sessile or *pelagic* (free floating) tend to be *radially symmetrical* (look circular when viewed from above or below). Those that move in one predominant direction tend to be *bilaterally symmetrical* (have left and right sides that are mirror images of each other) and *cephalized* (possess a head).

5. The life cycle of members of the phylum *Cnidaria* (hydras, jellyfishes, and corals) may be spent as a sessile, asexually reproducing *polyp* or as a pelagic, sexually reproducing *medusa*, or as both. Many species show a kind of alternation of generations: the polyp forms buds that yield the medusa, whose fertilized eggs develop into planulae, which, in turn, develop into polyps. Both forms have a three-layered body plan; a mouth surrounded by tentacles; and a gut, the colenteron. Cnidarians are distinguished by their stinging cells, the cnidocytes, each of which houses a harpoonlike filament for trapping prey. Their water-filled colenteron can act as a *hydroskeleton*, against which the muscles act. Other characteristics include *extracellular digestion* and a nerve net, a primitive kind of nervous system. *Hydrozoans* show typical polyp and medusa stages; *scyphozoans* are limited to the medusa form; and *anthozoans* live as the polyp stage.

6. The comb jellies, members of the phylum *Ctenophora*, are freefloating sea creatures that are medusa-shaped and have eight rows of ciliated cells—the combs—which propel them through the water, as well as two feeding tentacles.

7. The *plakula*—a small, flattened marine worm with dorsal and ventral surfaces but no head and no other axis of symmetry—is classified in the phylum *Placozoa*. It is uncertain whether it is related to sponges or is on the main evolutionary line toward all higher animals. They are remarkably like the solid, ciliated planula larva, and may be the source of the flatworms and other higher animals.

8. Flatworms, which make up the phylum *Platyhelminthes*, are organized at the *organ level*, are bilaterally symmetrical, and have a gut with one opening. The anterior end of the free-living species of the class *Turbellaria* (planarians) contains *ganglia*, aggregations of nerve cells, from which nerve cords extend down the sides of the body. Eyespots sense light, and another organ senses gravity. Hermaphrodites, turbellarians reproduce by exchanging sperm. The species of the classes *Trematoda* (flukes) and *Cestoda* (tapeworms) parasitize vertebrates. The pared-down tapeworm has no gut and is composed of little more than a chain of proglittids—segments containing gonads with sperm and eggs.

9. Ribbonworms, members of the phylum *Nemertina*, are characterized by their lassolike proboscis. They are the most primitive animals with a complete, one-way digestive system—with a mouth at one end and an anus at the other—and with blood vessels, which contract to move their clear blood to and fro.

10. *Nematoda*, the phylum of roundworms, are both more and less complex than the ribbonworms. Although nematodes have neither blood vessels nor circular muscles, they do have a *pseudocoelom*, a fluid-filled cavity in which the one-way gut lies. Although most roundworms are free living, the parasitic species cause plant and animal diseases, including trichinosis.

11. The animals with a *coelom*—a fluid-filled cavity surrounded by mesoderm in which many of the internal organs are suspended—are divided into two lineages: the *protostomes* and the *deuterostomes*. In the protostomes—the segmented annelid worms, arthropods, and mollusks—the blastopore becomes the mouth, the embryonic cleavage

pattern is spiral, development is determinate, and the coelom forms by a splitting of the mesoderm. In the deuterostomes—the echinoderms, hemichordates, and chordates—the blastopore becomes the anus, the embryonic cleavage pattern is radial, development is regulative, and the coelom forms from the mesodermal pouches.

12. The worms of the phylum *Annelida* are characterized by the *segmentation* of the body into units, each of which contains muscles, blood vessels, excretory nephridia, and ganglia. Annelid locomotion depends on the coelom, acting as a hydroskeleton, and on alternate coordinated contractions of the segmental circular and longitudinal muscles. Reproduction can be asexual or sexual—the latter involving males and females or two hermaphrodites, which exchange sperm. A *trochophore*, a free-swimming larva, develops in typical marine forms.

13. The phylum *Arthropoda* includes spiders, crustaceans, insects, and mites. It is by far the most diverse animal phylum, with 1 million species adapted for life in the oceans, fresh water, most terrestrial habitats, and the air. Arthropods have specialized body segments, especially noticeable in limb structure and function. Most have a rigid *exoskeleton*, an open circulatory system (blood vessels that drain into an open sinus), complex sense organs, and a nervous system that allows complicated, rather stereotyped behavior, especially among the *social insects*. Most insects undergo complete metamorphosis as the larva pupates and becomes an adult, while others undergo incomplete metamorphosis, developing through a series of nymphal stages and recurrent molts. Arthropods probably are not a single evolutionary lineage.

14. Chitons, snails, clams, octopuses, squids, and their relatives—members of the phylum *Mollusca*—are protostomes that apparently never were segmented. Mollusks have variations on a four-part body plan: (a) the foot, (b) the visceral mass, (c) the head, and (d) the mantle. Portions of the circulatory system may function as a *hemoskeleton*. The class *Bivalvia* (clams, oysters, mussels) and some species in the classes *Gastropoda* (snails and abalone) and *Cephalopoda* (chambered nautilus) have shells. The cephalopods—especially the squids and octopuses—have the largest and most complex brain among invertebrates; octopuses, in particular, are capable of learned behavior.

15. Members of the phylum *Echinodermata*—sea stars, sea urchins, sea cucumbers, and brittle stars—tend to be radially symmetrical as adults. They have a unique water vascular system that serves as a hydroskeleton. The sea stars' water vascular system includes tube feet that allow locomotion, grip substrata, and pull open bivalve shells. Many echinoderms have hard calcareous ossicles beneath the skin for protection and support. Reproduction is by normal sexual processes; regeneration can occur in some types.

16. The phylum *Hemichordata* consists of the mobile, wormlike enteropneusts and the sessile, tentacled pterobranchs. Adults of these marine animals possess gill slits—paired exits for the feeding current through the wall of the gut.

KEY TERMS

Annelida	Echinodermata	medusa	Polyplacophora
Anthozoa	exoskeleton	Mollusca	Porifera
Arthropoda	extracellular digestion	Monoplacophora	protostome
bilateral symmetry	filter feeder	Nematoda	pseudocoelom
Bivalvia	ganglion	Nemertina	radial symmetry
cephalization	Gastropoda	Oligochaeta	Scyphozoa
Cephalopoda	Hemichordata	Onychophora	segmentation
Cestoda	hemocoel	pelagic	sessile
Chelicerata	hemoskeleton	Placozoa	social insect
Cnidaria	Hirudinea	plakula	Trematoda
coelom	hydroskeleton	planula	Trilobita
Crustacea	Hydrozoa	Platyhelminthes	trochophore
Ctenophora	Insecta	Polychaeta	Turbellaria
deuterostome	invertebrate	polyp	Uniramia

QUESTIONS

1. Which of the phyla discussed in this chapter include individuals with radial symmetry? bilateral symmetry?

2. What are the alternating generations found in many species of the phylum Cnidaria? Do these stages differ in genetic make-up, chromosome number, life style, or form? Which form produces gametes? Describe briefly the three classes of cnidarians. What sorts of nervous system, muscular system, and skeletal system are found in the cnidarians?

3. What are the major evolutionary innovations seen in each of the worm phyla?

4. What are some of the differences between the protostomes and the deuterostomes? Which phyla are considered to be protostomes? Which are considered to be deuterostomes?

5. The animals in the phylum Arthropoda are thought to have evolved from those in Annelida. In what ways do arthropods differ from annelids? What are the subphyla in the phylum Arthropoda?

6. Uniramia arose on land. What classes are included in this subphylum? Describe some of their adaptations to land. What other arthropods are land dwellers?

7. Mollusks are not segmented. In this respect, they differ from the annelids and arthropods, yet they are similar to annelids in one important respect. What is it? Describe the major classes in the phylum Mollusca. Which contain individuals with a radula, external shells, no shells, twisted or coiled shells, a large head and brain, large eyes?

8. Describe the water vascular system of echinoderms. What are the functions of this unique system?

9. Which of the invertebrate phyla mentioned in this chapter are multicellular? Which carry out extracellular digestion?

10. Most animals that are bilaterally symmetrical move in a head-first direction. What are some examples of bilaterally symmetrical invertebrates described in this chapter?

ESSAY QUESTIONS

1. In what ways do humans resemble a sponge? an insect? a starfish? To which are you more closely related? Explain.

2. How would you define a "generation"? Compare the alternation of generations found in plants with that found in the cnidarians. Do the "generations" in either type of organism differ genetically?

3. Discuss parasitism as an evolutionary strategy as seen in worms. What are its advantages and disadvantages?

SUGGESTED READINGS

BARNES, R. D. *Invertebrate Zoology.* 5th ed. Philadelphia: Saunders, 1987.

A standard work that is highly respected.

BARRINGTON, E. J. *Invertebrate Structure and Function.* Boston: Houghton Mifflin, 1967.

A leading authority on invertebrates discusses evolution and function in particular.

BUCHSBAUM, R. *Animals Without Backbones.* 2d ed. Chicago: University of Chicago Press, 1976.

Written at an elementary level, this is a widely used book.

DALY, H. V., J. T. DOYEN, and P. R. EHRLICH. *Introduction to Insect Biology and Diversity.* New York: McGraw-Hill, 1978.

A good, broad view of the vast insect world and its evolution.

GOREAU, T. F., N. I. GOREAU, and T. J. GOREAU. "Corals and Coral Reefs." *Scientific American,* August 1979.

A lively account of the building of the great reefs by tiny polyps and their symbiotic algal associates.

ROPER, C. F. E., and K. J. BOSS. "The Giant Squid." *Scientific American,* April 1982, pp. 96–105.

A good picture of the ecology and structure of these huge, brainy creatures.

26
VERTEBRATES AND OTHER CHORDATES

In a comparative study of the vertebrates, it is rather surprising to find how few entirely new structures appear between the fishes and the mammals, for almost invariably, the finished structures of one class of vertebrates can be traced through earlier classes, where they can be seen in simpler form, although they sometimes are performing an entirely different function.

Leverett Allen Adams, *An Introduction to the Vertebrates* (1938)

A baby Galápagos tortoise (*Testudo elephantopus*) breaking from its egg.

It is sobering to realize that the invertebrates—the corals, nematodes, earthworms, beetles, barnacles, and others that we described in Chapter 25—make up more than 99.9 percent of the individual animals alive on Earth today. When most people think of "animals," they do not think of insects or worms; they think of nature's remaining 0.1 percent, the **vertebrates**—the fishes, amphibians, reptiles, birds, and mammals that have internal bony skeletons and usually a backbone. This is natural enough, for we ourselves are vertebrates, and many people keep other so-called higher animals for pets, raise them for food and labor, hunt them, study them in laboratories and zoos, and observe them in their natural habitats for their beauty and complex behavior.

Vertebrates are a diverse lineage whose basic body design and physiology have, in the course of evolution, given rise to the largest animals that have ever lived, in addition to fierce predators, skillful long-distance fliers, fast runners, and creatures with complex behavior. One of the 58,000 vertebrate species has even flown to the moon, traveled to the floor of the deepest ocean abyss, and modified the planet Earth—for good or ill from a human perspective—more dramatically than any kind of organism before it.

The vertebrates evolved late in our planet's history. If that history were compressed into a single year, the first primitive fishes would appear about November 20. Their descendants would first crawl onto land as late as November 30. Reptiles would become the dominant land animals about December 7; mammals would be common by December 15; and our own genus, *Homo*, would arise no earlier than 11:00 P.M. on December 31. In contrast, the first photosynthesizers—single-celled algae—would be found in the seas in May, and the first bacteria would appear in March.

Despite the relative newness of the vertebrates on our ancient planet, their evolution spans nearly 600 million years and is well documented; ancient fishes, dinosaurs, and other vertebrates left behind millions of fossilized bones and teeth in rock strata all over the world. From this fossil record, we can see the emergence of a few major structural innovations allowing swifter movement, larger size, and more efficient reproduction. Such innovations were followed by **adaptive radiation,** the development from an ancestral group of a variety of forms adapted for different habitats and ways of life, such as walking, climbing, burrowing, and flying.

In this chapter, we will examine the major structural and functional innovations of the vertebrates and of their immediate relatives among the chordates. This will allow us to trace the emergence and radiation of the major groups. In the process, we will be following the evolutionary pathway leading to *Homo sapiens* and the derivation of most of the tissues and organ systems that we will describe in Chapters 31 to 40. It should become clear as we discuss the diversity and beauty of vertebrates why some people consider them to be evolution's most magnificent animal experiment.

CHARACTERISTICS OF THE CHORDATES

The phylum **Chordata** includes just three subphyla—Urochordata, Cephalochordata, and Vertebrata—encompassing animals as different as sea squirts and squirrel monkeys. Despite their diversity, all chordates display the same four physical characteristics at some time during their life cycle: (1) *gill slits*, (2) a *notochord*, (3) a *tail and blocks of muscles*, and (4) a hollow *nerve cord* (Figure 26-1). In some, the characteristics remain throughout adult life. In others, one or more of the characteristics may be present only in the embryo.

Gill Slits: Feeding and Respiration

One of the most ancient common chordate features is the presence of **gill slits.** These are paired openings through the lateral walls of the anterior gut in the region

Figure 26-1 MAJOR CHORDATE FEATURES.
Chordates are characterized by gill slits, a notochord (which may be incorporated into the vertebral column), a dorsal hollow nerve cord, and a postanal tail with segmental muscles.

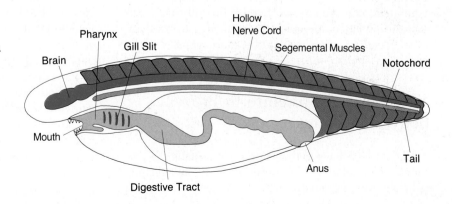

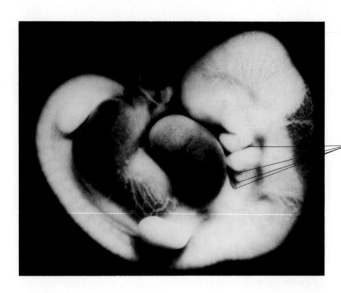

The gill slits in this human embryo (magnified about 50 times), as in other mammals, will largely disappear in the adult. However, parts of the embryonic gill apparatus mature into parts of the ear, of major blood vessels, and other structures.

Gill Slits

immediately posterior to the mouth cavity—the **pharynx.** As in the hemichordates (Chapter 25), the earliest chordate gill slits acted as exits for the feeding current that entered the mouth. Since the feeding current in both hemichordates and nonvertebrate chordates is propelled only by cilia on cells in the gill-slit walls, it is a weak current at best, capable of carrying only small food particles. In living primitive chordates, and probably in ancient ones too, the tiny particles become trapped in mucus produced by a special region of the pharyngeal floor, the *endostyle*; from there, the particles and mucus pass to the rear portion of the gut for digestion.

Although the gill slits originally evolved as exit sites for the feeding current, they became associated with respiratory gas exchange in later chordates—the first vertebrate fishes. Today, some chordates (such as trout, sharks, and most other fishes) retain gill slits into adulthood; others (including reptiles, birds, and mammals) lose them in adult stages. But in every chordate embryo—including human embryos—gill slits and associated structures do develop (Figure 26-2). They are true evolutionary "remembrances of things past," structures that develop because the genetic and developmental information required for them continues to be expressed in generation after generation. Even if the gill slits disappear during late embryonic development, parts of them are retained for other uses in the adult (as parts of the jaws, ears, and so on).

Notochord: A Structure to Prevent Body Shortening

All chordates have as their primary internal longitudinal skeletal element the **notochord,** a stiff but flexible rod that runs the length of the bilaterally symmetrical animal just ventral to the nerve cord. The notochord develops in every chordate embryo and is retained in many adults. There is nothing similar to the notochord in any of the invertebrates discussed in Chapter 25. The notochord is really a kind of *hydroskeleton*: fluid-filled vacuoles in the notochord cells exert pressure against a surrounding sheath of tough collagen fibers, just as a pressurized inner tube keeps a rubber bicycle tire firm. Also, like an inner tube, the notochord can bend up and down and from side to side, but it resists shortening. This resistance is the structural feature that allows the unique swimming locomotion characteristic of aquatic chordates: blocks of muscles contract to bend the body from side to side. Without the notochord, these contractions would shorten the entire body and would not propel it with the same efficiency.

Virtually all modern vertebrates have a set of bones, the *vertebral column*, that develops around the notochord to protect the nearby nerve cord and major blood vessels and to provide additional stiffening and support for the body. In land vertebrates, whose body support is not provided by surrounding liquid, the vertebral column serves several functions that could not be met by the notochord alone.

Muscle Blocks: The Basis for Swimming Locomotion

All the ancient chordates (and many modern chordate species) lived in water and depended on swimming, made possible by the sweeping of a tail that extends beyond the anus—hence the name *postanal tail*. Blocks of muscles, called **myotomes,** make up the bulk of the tail in a swimming chordate and also occur more anteriorly along both sides of the body. Each myotome can be regarded as a body segment, although this is a very different kind of segmentation from that seen in annelids, where many of the internal organs are repeated in each segment. Chordate segmentation is mainly muscle-mass segmentation related to the evolution of this special mode of swimming (note that the vertebral column, the backbone, is also segmented in the sense that there is one vertebra per pair of myotomes). The key feature is that each myotome on each side of the body can contract as a unit. Thus myotomes on one side of the body contract in a head-to-tail sequence and cause a wave of bending to sweep from head to tail on that side. Myotomes on the opposite side then contract sequentially, and a wave of bending occurs from head to tail in the opposite direction. These alternating waves sweep the sides of the body one way and then the other, causing the body to push against a noncompressible material—the water. This propels the chordate in a manner quite unlike the muscular locomotion seen in an earthworm's body, a snail's foot, or a grasshopper's leaping legs. As we will see, forms of muscle-driven locomotion other than swimming and types of tails not involved in locomotion evolved in many land vertebrates.

Nerve Cord: Coordination of Movements

Sequential muscle movements require coordinated control. All chordates thus have a long, hollow **nerve cord,** the *spinal cord,* situated just dorsal to the notochord. Nerves run from the spinal cord to each myotome (and so are segmental in that sense) and to other parts of the body. The spinal cord acts as a switchboard for coordinating the swimming movements of aquatic chordates and the limb-driven locomotion of terrestrial vertebrates. A brain need not be connected to the spinal cord to get such coordinated movements. However, directional swimming led to *cephalization*—the localization of sense organs in the animal's anterior end—and, with it, the enlargement of the anterior spinal cord to form the brain.

Let us now turn to the three subphyla that make up the phylum Chordata to see how the gill slits, notochord, muscle blocks, and nerve cord have varied in structure and function in our ancestors and relatives.

THE NONVERTEBRATE CHORDATES

The simplest chordates are bottom-dwelling marine organisms whose life habits resemble those of many of the invertebrates we examined in Chapter 25. In fact, members of two of the three chordate subphyla—Urochordata and Cephalochordata—are nonvertebrates: although they have the flexible notochord and gill slits, they lack a bony skeleton and a vertebral column. Both urochordates and cephalochordates feed by means of cil-

iated gill slits and swim at some time during their life cycle by means of the characteristic chordate tail.

Urochordata

Members of the subphylum **Urochordata**—the commonest type being the sea squirts—are small marine organisms that usually spend their adult lives permanently attached to rocks or other hard surfaces, such as pilings in harbors. Notice in Figure 26-3b that the urochordates have an outer tough and leathery *tunic* (which is why they are also called *tunicates*). Inside the tunic and body walls lies the pharynx, a large basket-shaped structure that hangs within the body cavity and is perforated by hundreds of ciliated gill slits. The beating of the cilia draws the feeding current into the pharynx through the mouth, or incurrent siphon. Once in the pharyngeal cavity, the water and suspended food particles pass through a moving sieve, a mucous sheet produced by cells in the endostyle, a groove in the pharyngeal wall. Food particles mired in the mucus pass into the gut, where they are digested and the nutrients are absorbed. The feeding current that has passed through the mucous sieve is swept through the gill slits and exits from the excurrent siphon, along with metabolic and digestive wastes. Sea squirts got their name from the periodic expulsion of water through this siphon.

You may have noticed in Figure 26-3b that adult urochordates have gill slits but lack the other chordate traits: hollow nerve cord, notochord, and tail. However, these features are present in the animal's free-swimming larval stage, the *tadpole,* shown in Figure 26-3c (in fact, the term "urochordate" comes from "urochord," another name for the tadpole's notochord). The tadpoles never feed; rather, they swim near the ocean surface for a short time—twenty-four hours or so—and then settle on some object, attach to it, and undergo metamorphosis. Within minutes, the tail is retracted, and its parts are degraded. Concomitantly, the immature pharynx begins to develop, and the gill slits begin to function. So the remarkable urochordates are like primitive fish as larvae but are quite different from any vertebrate as adults.

Cephalochordata

The second subphylum of nonvertebrate chordates, **Cephalochordata,** contains only one type of organism, the lancelet (formerly called amphioxus). It is the adult body of this small marine animal, which lives half-buried in the sand of shallow ocean bays and inlets, that exhibits all four chordate characteristics (Figure 26-4). Lancelets are filter feeders; water taken in through a mouth ringed by sensory tentacles enters a long pharynx perforated by

Figure 26-3   UROCHORDATES: EARLY CHORDATES.
(a) A living urochordate, *Ascidia* species. The incurrent (top) and excurrent (at the side) siphons are easy to see in this tunicate found on a reef off the Philippines. (b) This cut-away view of the adult shows how the feeding current passes through the gill slits and then out the excurrent siphon. (c) The chordate characteristics are apparent in the tadpole. Note in particular the dorsal nerve cord and the notochord. The tail muscles are not truly segmented in these organisms.

(a)

(b)

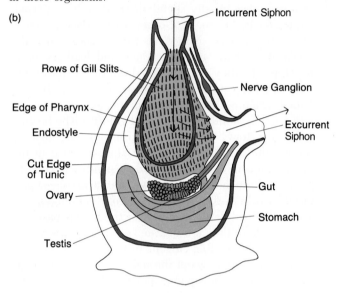

Incurrent Siphon
Rows of Gill Slits
Nerve Ganglion
Edge of Pharynx
Excurrent Siphon
Endostyle
Cut Edge of Tunic
Ovary — Gut
Testis
Stomach

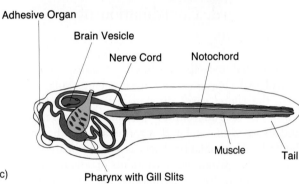

Adhesive Organ
Brain Vesicle
Nerve Cord   Notochord
Muscle
Tail
Pharynx with Gill Slits

(c)

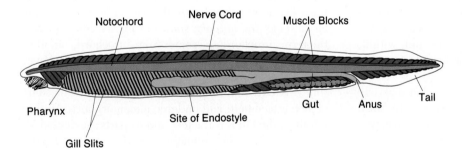

Notochord   Nerve Cord   Muscle Blocks

Pharynx
Gut   Anus   Tail
Site of Endostyle
Gill Slits

Figure 26-4   CEPHALOCHORDATES: A VERTEBRATE RELATIVE.
The typical chordate characters are easy to spot in this adult lancelet (*Branchiastoma lanceolatum*). Note the dorsal, hollow nerve cord (lacking a brain), the notochord, the gill slits, the segmental muscle blocks, and the postanal tail.

more than 100 ciliated gill slits. Just as in the urochordates, food particles that enter the mouth are trapped by mucus from the endostyle and are passed along to the gut. Although lancelets swim only sporadically, the adults have a notochord that goes all the way to the tip of the head (hence the name cephalochordate), a nerve cord (but no brain), myotomes, and a tail. However, electron microscopical evidence shows that the lancelet notochord actually contains contractile, musclelike fibers; thus it is unlike that found in any other chordate, supporting the conclusion that the lancelet lineage was an early offshoot of the line that led to the vertebrates.

JAWLESS FISHES: THE FIRST VERTEBRATES

The earliest members of the subphylum **Vertebrata** were fishes of the class **Agnatha,** which means "jawless." Their fossils appear first in late Cambrian period rocks, 540 million years old. They lived as filter feeders in shallow ocean waters for at least 150 million years, into late Devonian times, and included the *ostracoderms*, or bony-skinned fish. Two major "innovations" distinguish these first vertebrates from their more primitive relatives, the urochordates and the cephalochordates. First, the feeding current in these animals was powered by muscles rather than by cilia; second, their bodies were covered by plates made of bone lying within the skin. Although they are considered vertebrates, only a notochord, not a vertebral column, was present. Fossils provide a good deal of evidence about these ancient jawless fish.

Bony-Skinned Fish and Modern-Day Descendants

The very first known vertebrates, the ostracoderms, had a surprisingly complex body, brain, and set of sensory organs (Figure 26-5). Obviously, this complexity did not arise in a single rapid step, but came about slowly through the evolution of predecessors that lacked bone and therefore did not yield fossils. Thus we have no fossil record of how their complex body parts arose. Fossils of many different kinds of ostracoderms show that each type had a mouth and a series of paired gill slits. Between adjacent gill slits were muscles and skeletal gill-support pieces called *branchial arches*. When the muscles contracted, water was pumped out of the pharynx through the gill slits, and fresh water entered the mouth as a result. Biologists studying this feeding process in the

larval forms of modern jawless fishes have discovered that the muscles pump more water (containing suspended food particles) than can be propelled by cilia on the gills of the more primitive chordates. More food intake meant that larger body size was attainable by the jawless fishes.

The ostracoderms also had a bony brain case, the head shield, surrounding a complex brain with the three parts found in every subsequent vertebrate: the forebrain, midbrain, and hindbrain. So-called *cranial nerves* led from the brain to the gill-slit muscles, to two eyes and a third pineal eye (Chapter 37), and to internal ears that were used to sense body position in the water. In other words, the marvelously complicated eyes, ears, and nervous system that are so vital to all vertebrate life were fully formed in these first known vertebrates.

The ancient armored ostracoderms are survived today by the jawless lampreys (Figure 26-6) and hagfish, both of which are classified in the subclass *Cyclostomata*, which means "round mouthed." Unlike the earlier ostracoderms, cyclostomes lack bone and have skeletons of *cartilage*, a pliable tissue formed of cells embedded in a fibrous, extracellular matrix composed largely of collagen and sugar–protein polymers. Lampreys are parasites that feed on other fish by attaching to an animal's side with the suction-cup action of their mouths, cutting a wound in the host's skin with their rasping tongue, and

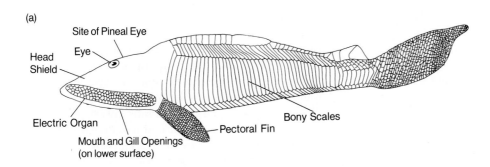

(a)
Site of Pineal Eye
Head Shield
Eye
Electric Organ
Mouth and Gill Openings (on lower surface)
Pectoral Fin
Bony Scales

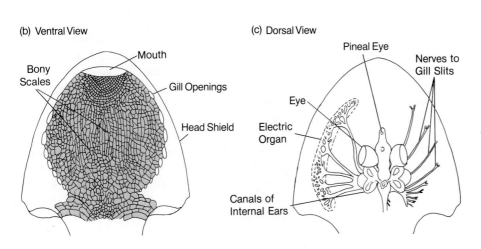

(b) Ventral View
Mouth
Bony Scales
Gill Openings
Head Shield

(c) Dorsal View
Pineal Eye
Nerves to Gill Slits
Eye
Electric Organ
Canals of Internal Ears

Figure 26-5   RECONSTRUCTION OF AN OSTRACODERM.

(a) This agnathan, or jawless fish (*Hemicyclaspis*), is covered by a bony head shield and scales. (b) When viewed from below, the mouth and gill openings are seen, as well as the small bony scales, which indicate that the whole lower surface of the head was flexible and so could be moved during the pumping of the feeding current into the mouth and out the gills. (c) An internal reconstruction of the nervous system. Large nerves to the electric organ extend to the left while nerves to the gills are shown on the right. The eyes, the pineal, and parts of the brain and internal ears are present too.

(a)

(b)

Figure 26-6   A LIVING REPRESENTATIVE OF AGNATHA, THE LAMPREY.

(a) Three lampreys are firmly attached to this unfortunate carp. The lampreys are literally drinking the carp's blood. (b) A close-up view of the lamprey, *Lampetra fluviatilis*, shows the seven gill slit openings.

drinking the victim's blood. The very distantly related hagfish use their circular, jawless mouths to feed on various invertebrates and on dead animal and plant matter that falls to the ocean floor. Closely related to hagfish on the basis of structure are the conodonts. Known only from the fossil record, conodonts were very diverse and abundant in the ancient seas. (It is not certain whether Conodonta is properly classified as Agnatha or as a separate subphylum of the Chordata.) Hagfish, lampreys, and, in fact, all other vertebrates share possession of an important endocrine gland, the *thyroid*. Today, when a filter-feeding lamprey larva undergoes metamorphosis into the adult form, the endostyle that secretes mucus for trapping food transforms into the hormone-producing thyroid gland. Both the thyroid and pituitary originated in jawless fish.

Vertebrate Origins and the History of Bone

The evolutionary relationship among the urochordates, cephalochordates, and jawless vertebrates has long fascinated biologists. Attention has focused on the concept of an ancient free-swimming creature with both the chordate-type tail and gill slits—an animal that resembled the urochordate tadpole but had the ability to feed and reproduce. Zoologists now believe that hemichordates may have developed larvae with the chordate-type tail and that these little creatures yielded three distinct lines: the urochordates, with their sessile adult stage; the cephalochordates, with their unique notochord; and the first jawless vertebrate ostracoderms.

It is only in the last group that bone appeared, providing scientists with the fossil record of these extinct fishes. The developmental processes responsible for bone formation probably arose in the ancestors of ostracoderms, leading to a new kind of differentiated cell—the *osteocyte*, or bone cell—and to embryological events that bring about bone formation. Bone is a remarkable substance, and there are several theories to explain how it evolved. It consists of a matrix of organic fibers composed of collagen molecules, on which grow crystals of an inorganic material called *apatite*, made of calcium and phosphate. Living bone is a highly dynamic tissue that serves as the body's main reservoir of calcium and phosphate. Thus when the blood-calcium level is low, bone apatite becomes soluble to release that vital ion into the blood. When the blood-calcium level is too high, bone apatite is formed, thereby lowering the level of soluble calcium. Because it is also quite hard, bone is useful as a protective encasement, a supportive skeleton, and a site of attachment for muscles.

Biologists long theorized that bone evolved because of its hardness, but a newer theory suggests that the role of bone in storing phosphate may have been the key evolutionary factor. Phosphate stored in bone protects an aquatic organism from the wide fluctuations in phosphate that occur in sea water as a result of seasonal algal blooms. Thus when little organic phosphate is available in the diet, bone phosphate can be released into the blood for use in nucleic-acid synthesis, ATP production, and so on. Despite the uncertainty about how bone evolved, biologists agree that it was critical to the evolution of vertebrates as we know them—imagine, for instance, what vertebrates might be like in the absence of hard skeletal parts for support and protection.

JAWED FISHES: AN EVOLUTIONARY MILESTONE

Even though the earliest vertebrates had a chordate tail, bones, and a "new" mode of feeding (muscular gill slits), they were still limited in one vital sense. The food they ate had to be small—suspended in the water or in the silt of the sea floor. In other words, they could not bite off a chunk of food from a large object. The next notable advance in vertebrate evolution was a new feeding structure that led to a revolution in body size and feeding styles: *jaws.*

The most ancient jawed-fish fossils are members of the class **Acanthodii,** an extinct group that left fossils dating back to the Ordovician period, 435 million years ago (Figure 26-7). Strong evidence suggests that the jaws in acanthodians and their later relatives, the placoderms, arose from the two or three anterior sets of branchial arches, which we first encountered in the ostracoderms. The muscles that move the hinged jaws and the nerves that control these muscles were also once part of the muscular filter-feeding apparatus of the jawless ostracoderms. In a sense, the transition from ciliary gill slits to muscular gill slits was a precondition for the subsequent evolution of jaws. But why are jaws significant?

Jaws literally transformed the world's ecology. Fishes with jaws were no longer limited to filtering small food particles, but instead could eat other fish; graze on large marine algae; or crush worms, clams, and other invertebrates in their mouths. Jawed fishes apparently competed successfully with predatory arthropods and cephalopods. Their new feeding styles allowed the jawed vertebrates to attain a wide range of body sizes, from a few centimeters to several meters in length. One acanthodian species was so large that an adult human can stand erect in its open fossilized jaws. The new feeding styles and body sizes in turn led to complex food chains in which herbivorous fish grazed on plants, carnivorous fish ate the herbivores, and those predators were themselves eaten by still larger carnivorous fish.

In the acanthodians, the gills and gill slits were no doubt involved in gas exchange. That would have been essential, since simple diffusion of respiratory gases across the body surface could not have met the needs of anything but small organisms. There is no more reasonable place for respiratory exchange surfaces to develop than in the gill slits, through which the feeding current of sea water exited the body. Hence, scientists surmise that the gills came to be used for both feeding and respiration in the jawless ostracoderms and that as the jaws took over the feeding function in the acanthodians, the gills remained as respiratory exchange sites.

While the skull and skeleton were bony in these early jawed fishes, the adults had flexible notochords rather than complete vertebral columns. Fossils do show the presence of small, separate bony arches over the spinal cord that eventually expanded and fused with the notochord in later jawed fishes to become the bony spine, or vertebral column.

The ancient acanthodians showed another structural advance: *fins,* or swimming appendages. In the very first acanthodian fossils, the fins were little more than sharp spines, but they must have allowed those early fish to control to some extent their direction in the water. In contrast, many of the early jawless fishes were probably limited to movement along the sea bottom, sucking up mud and food particles like self-propelled vacuum clean-

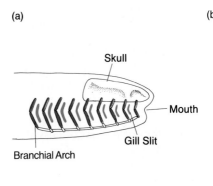

(a)

Skull

Mouth

Gill Slit

Branchial Arch

Jaws

(b)

Figure 26-7   THE EVOLUTION OF JAWS.
Jaws probably arose from the anterior branchial arches of the gill system. (a) The anterior branchial arch skeletal pieces are believed to have become altered to form the upper and lower jaw bones. As that occurred, the number of gill slits decreased. (b) Reconstruction of an early jawed fish, a placoderm (*Coccosteus decipiens*), that possessed true jaws, as seen in (a).

(a)

(b)

Figure 26-8   CHONDRICHTHYES CARTILAGINOUS FISH.
(a) This streamlined black tip shark (*Carcharhinus maculipinnis*) is a voracious predator. Note the gill-slit openings posterior to the eye. (b) A blue spotted stingray (*Taeniura lymma*) rests quietly on this Red Sea reef off Egypt. The greatly expanded pectoral fins are almost like wings that flap through the water.

ers. Fins continued to evolve and change so that in later acanthodians, the fins contained more skeletal elements that could be moved independently of each other. The fin thus became a highly sophisticated control surface, not just a rigid surface like an airplane wing. As a result, later acanthodians and their descendants could maneuver more easily and could swim at a range of depths, speeds, and directions.

It is of great evolutionary significance that two pairs of fins, so-called *pectoral* (anterior) and *pelvic* (posterior) fins, are supported by *girdles*, sets of bones or cartilage that provide a firm base to which muscles can attach and against which the paired fins can move. As we will see, fin girdles were forerunners of the shoulder and hip bones of land vertebrates. So in these bizarre-looking creatures living hundreds of millions of years ago in shallow sea regions, we find the origins of the legs with which we run, the arms with which we write, and the jaws with which we eat and speak.

The early jawed fishes gave rise to all the subsequent jawed vertebrates, including the two major classes of fishes: the cartilaginous fish and the bony fish.

Chondrichthyes: Fast-Swimming Predators

One ancient lineage of jawed, heavily armored fish, the *placoderms* (Figure 25-7), gave rise to **Chondrichthyes.** This class, which contains modern sharks, skates, and rays (Figure 26-8), is distinguished by a lightweight but strong skeleton made only of cartilage.

Cartilaginous fish have huge oil-storing livers. The oil is a source of great buoyancy, since oil floats on water. Thus these creatures can almost be neutrally buoyant— that is, their bodies weigh about the same as an equivalent volume of water. Sharks need expend very little muscular energy to maintain their depth in water, and they are remarkably maneuverable, which contributes to their success as voracious predators.

High speed and agility are not enough, however, for a shark to outmaneuver its prey. Control of direction in the water depends on fins. Sharks have two sets of fins: pectoral fins, anterior to the body's center of gravity, and pelvic fins, posterior to this center. They also have a *caudal* fin, the asymmetrically shaped fin at the end of the tail, with its large dorsal extension. Sharks illustrate how the emergence of fins expanded the repertoire of swimming movements: pectoral fins provide "lift" at the front end that counteracts lift produced by the caudal fin, which thrusts water backward and downward as it sweeps from side to side during swimming. The angle of the pectoral fins can direct the body upward, downward, or sideward as the animal pursues prey or swims in leisurely fashion.

Skates and rays are lovely examples of evolutionary variation on a preexisting theme. The pectoral fins of these creatures are greatly enlarged into "wings" that literally flap through the water, gracefully propelling the animal.

The combination of a light skeleton, a buoyant liver, and paired fins contributed to the successful adaptive radiation of the early chondrichthyans into the modern skates, rays, and sharks, some of the most remarkable predators on Earth. However, these cartilaginous fish have rarely invaded fresh water (for reasons discussed in Chapter 35), and those lines that have, never diversified, as did their bony cousins, the osteichthyes.

Osteichthyes: Adaptability and Diversity

Fish with bony skeletons, members of the class **Osteichthyes,** arose almost 400 million years ago from

Figure 26-9   TELEOSTS, THE MOST DIVERSE FISH.
The Earth's oceans, rivers, and lakes teem with seemingly endless varieties of teleosts. (a) The sailfish (*Istiophorus platypterus*) has a long bladelike upper jaw and a huge dorsal fin that can be raised or lowered out of the way. (b) The sea horse (*Hippocampus reidi*) has a grasping tail, rather like a monkey's, that can cling to undersea algae. A female lays her eggs into a pouch on the male's belly. Below the surface, colorful fish inhabit coral reefs in tropical oceans around the world. Shown here are (c) the flame angelfish (*Centropyge loriculus*) that lives off Mexico's western coast, (d) the masked butterflyfish (*Chaetodon semilarvatus*) of the Red Sea, and (e) the rainbow wrasse (*Laproides phthirophagus*), a common reef fish in both hemispheres. (f) The moray eel (*Gymnothorax* species) with its jagged teeth and well-camouflaged skin lurks between submerged rocks, waiting to strike at its prey.

bony, jawed acanthodians. Two subclasses now predominate: *Actinopterygii*, or spiny-finned fish, and *Sarcopterygii*, or fleshy-finned fish. The actinopterygians include the infraclass *Teleostei*, almost 30,000 species of familiar, modern bony fish ranging from fresh-water trout to sea bass, from brightly colored reef fishes to giant tuna, from eels to sea horses (Figure 26-9).

There is debate over whether teleost fishes arose in fresh or in salt water. However, biologists agree that teleosts owe their great diversity in part to the presence of a *swim bladder*, an internal gas-filled organ that can change volume and allow the animal to be neutrally buoyant at nearly any depth. Since pectoral fins and tail are not essential for lift, body and fin shapes can vary dramatically, and fins can be used as paddles, brakes, or even true gliding wings, as in the flying fish. The physiology of teleosts plays a major role in their ability to populate salt and fresh water and to extract so much dissolved oxygen from water that they can be extremely active. We will discuss various aspects of fish physiology in later chapters.

The second subclass of bony fish, the sarcopterygians, is older than the actinopterygians. Today it is composed of a few surviving species with fascinating adaptations for breathing air and walking; it also includes the extinct species that were the first vertebrates to crawl onto land more than 360 million years ago. The sarcopterygians are distinguished by their thick *lobed* fins, each with a small number of large bones and well-developed muscles, which contrast with the thin, bony fins of teleosts. In addition, two of the three major types of fleshy-finned fish are characterized by internal nostrils, or *nares*, which are openings inside the mouth cavity. The internal nares connect to the nasal cavity and, in turn, to the external nares. The flesh-finned fish have a good sense of smell because a stream of water passing through the external nares carries odors to the nasal cavity, where they can be detected. Nares were not used by these fish to take in air, but that use became commonplace in the descendants of certain fleshy-finned fish—the amphibians and other land vertebrates.

The three types of sarcopterygians—the lungfish, the coelacanths, and the rhipidistian fishes—share some characteristics but are probably not one evolutionary line. *Lungfish* are rare fresh-water fish that live in shallow rivers and lakes in Australia, South America, and Africa (Figure 26-10). They have both gills and lungs.

Figure 26-11   COELACANTH (*LATIMARIA CHALUMNAE*): A FLESHY-FINNED "LIVING-FOSSIL" FISH.
This rare fish closely resembles its ancient predecessors. The fish is about a meter in length and can move slowly forward using its thick-based fins, such as the posterior-dorsal one seen here.

When a lungfish's normal habitat begins to dry up, the fish burrows into the mud, encases itself in a protective mud and mucous "cocoon," and enters a state of low metabolism and oxygen consumption. In this state, the fish can exchange sufficient amounts of oxygen and carbon dioxide through its lungs to survive until the rains return, the waters rise, and the fish can once again emerge from its cocoon. The fleshy lobed fins of lungfish serve as movable stilts that allow the fish, when submerged, to move along the bottom of ponds in a way that resembles walking.

Coelacanths are large (up to 1 meter or so) primeval-looking fish that were thought to have died out 150 million years ago—until one was captured in the deep waters off Madagascar in 1938 (Figure 26-11). Since then, several dozen have been studied, and they are known to possess a fat-filled swim bladder for buoyancy that is analogous to the shark's fat-storing liver.

The final group of fleshy-finned fish, the extinct *rhipidistians*, were probably the ancestors of the land vertebrates. The muscular lobed fins of these fish allowed them to "walk" along the bottom of shallow ocean bays and tidal flats and perhaps even onto the shore. Rhipidistians lived during the Devonian period, about 350 to 400 million years ago, when the land was already occupied by tree ferns, the first large land plants, and by many types of insects and other invertebrates. Those invertebrates may have been the food sources that enticed the rhipidistians—with their lungs and movable lobed fins—to leave the water and move onto land. The oldest fossilized animal tracks of vertebrates yet found were left by rhipidistians in sandstone formations on the Orkney Islands, off the northeastern coast of Scotland. As each animal crawled on its belly in the sand, it left a wide, flat track punctuated on either side by round scrape marks made by the stubby lobed fins. Such

Figure 26-10   A SURVIVOR OF DEVONIAN TIMES: A LIVING AFRICAN LUNGFISH (*PROTOPTERUS ANNECTEUS*).
This fish has two lungs and probably gets over 90 percent of its oxygen from the air. This species can survive drying of its pond by building a cocoon in the mud.

tracks, dated from about 360 to 370 million years ago, are a remarkable record of the first faltering vertebrate steps on land.

Although it is tempting to think of the evolution of the jawless and jawed fishes as linear, that simply was not the case. Diverse lines arose from different jawless ostracoderms; even more diverse lines arose from various lines of acanthodians, or jawed fishes. Thus lineages leading to the fleshy-finned fish, somewhat later to the teleosts, and finally to the chondrichthyans went their separate evolutionary ways (Figure 26-12). During the Devonian period, the age of fishes, the waters literally teemed with a bewildering array of ostracoderms, cyclostomes, acanthodians, early actinopterygians, lungfish, and rhipidistians. As all these vertebrates competed for survival in fresh and marine waters, the first amphibians appeared and lived as their contemporaries.

AMPHIBIANS: VERTEBRATES INVADE THE LAND

Fossilized tracks and bones found in Devonian period rocks show that early fishlike animals of the class **Amphibia** lived as land vertebrates perhaps as long as 350 million years ago. They gave rise to the familiar amphibians—frogs, toads, and salamanders—that can be found today in nearly every fresh-water lake, stream, and pond. The oldest amphibian fossils yet found were *icthyostegids*, animals up to 100 centimeters long that shared many traits with the rhipidistian fishes but that also possessed some unique adaptations for life on land. For example, both icthyostegids and rhipidistians had teeth with peculiar folds in the surface layer; these *labyrinthine* teeth have never been seen in other vertebrates, thus confirming that the early walking fishes and the first true land vertebrates were related. Both groups also had similar skull bones; intact notochords; small bony scales in the skin similar to fish scales; and rows of sensory structures called the *lateral line* (which occurs in ancient and modern fish but not in adult modern amphibians). Thus the ichthyostegids are almost a perfect "missing link."

These early amphibians, like the walking, air-breathing fishes, almost certainly crawled up onto land to exploit the abundant food resources already existing there. But unlike the fishes, they possessed adaptations for life on land: (1) a skeletal and muscular system better adapted to locomotion on land; (2) skin conducive to gas exchange; (3) the ability to return to watery habitats for reproduction; and (4) better sensory equipment for receiving airborne sound waves.

Amphibian Adaptations

Two primary problems confront every land animal—support and locomotion—in the absence of the buoyant effect of water to support body weight. Fossils show that the earliest amphibians already had a reduced number of large bones in their forelimbs and hind limbs—in fact, precisely the same sets found in the limbs of most modern amphibians, reptiles, birds, and mammals. These limbs had powerful muscles to help lift the belly off the ground and to walk, albeit slowly and with the limbs oriented more outward than forward. Such strong bones and powerful muscles have remained part of nearly all the land vertebrates that followed.

As in early fleshy-finned fish, the notochord of the early amphibians was the primary longitudinal skeletal element in the adult. It was surrounded by U-shaped bones—forerunners of parts of vertebrae that evolved in later land vertebrates. The spinal cord of these early land animals was protected by bony arches called *neural arches*. Only in later amphibians did the complete vertebral column arise. In modern amphibians (as in modern fishes, reptiles, birds, and mammals), each individual vertebra is a fused bone made up of a *centrum* (derived from the U-shaped bone) and a neural arch. In adults, the notochord is pinched into pieces that yield *disks*, the shock-absorbing pads between vertebrae.

In a land vertebrate, such as an early amphibian, the full set of vertebrae form a strong "suspension girder," while the attached arches form a hollow canal that surrounds and protects the spinal cord. As the elongate animal stands on all fours, the weight of the trunk is hung beneath the suspension-girder vertebral column. The body weight is then transferred through the pectoral and pelvic girdle to the limbs, and from the limbs to the ground. In all vertebrates, embryonic development of the vertebral column essentially repeats the evolutionary process: the notochord and neural tube form; neural arches and centra develop and fuse to form vertebrae; and the notochord is constricted and made into disks.

While the first challenge for life out of water was bodily support, another challenge involved gas exchange and drying out in the air. Nearly all amphibians possess true lungs but do not have an efficient means of filling and emptying them. Air is pumped into the lungs by swallowing movements of the lower jaw and coordinated opening and closing of the external and internal nares. In fleshy-finned fish those nares aided in the sense of smell, whereas they became the route of air inflow and outflow in amphibians. To compensate for the inefficient air-pumping mechanism, amphibians have a scaleless skin (with embedded blood vessels) that releases most of the body's excess carbon dioxide and absorbs about half the oxygen the animal needs. Since gas exchange must take place by diffusion through a layer of water, most of an

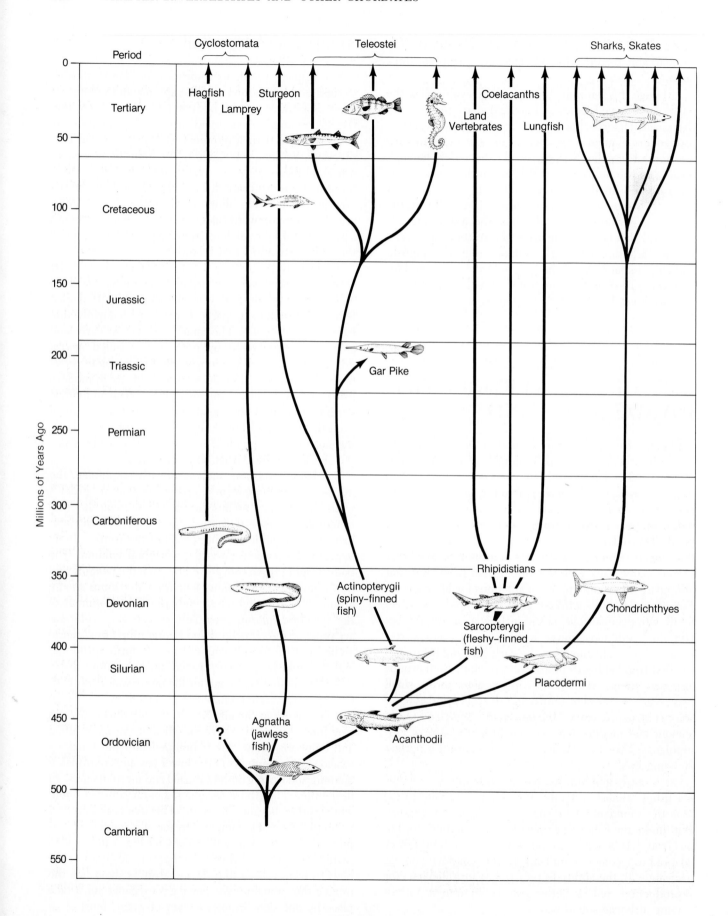

Figure 26-12   THE EVOLUTIONARY RELATIONSHIPS OF THE FISHES.

Though hagfish and lampreys are shown as related cyclostomes, in fact, hagfish may be a separate lineage related to conodonts. The approximate times that the other major lineages arose are shown. It is not certain that acanthodians are the common source of the three major groups of fish types; however, if other forms were involved, their jaws were probably similar to those in acanthodians.

amphibian's skin surface has to remain moist at all times. One undesirable consequence is that a large quantity of water may evaporate from the skin surface, thereby threatening the animal's life. Since evaporation consumes body heat, no amphibian has a high body temperature or a rapid rate of metabolism and the behavioral repertoire that goes with a warm body. It should thus be no surprise that nearly all amphibians are limited to moist habitats and exhibit water-seeking behaviors that minimize evaporation.

Amphibians are tied to moist habitats for another equally compelling reason: their eggs have the same kind of clear, jellylike covering, as do some fish eggs, and can easily desiccate, or dry out (Figure 26-13). Most modern amphibians have to return to water to repro-

Figure 26-13   AMPHIBIAN EGGS: A NEED FOR MOISTURE.

Most amphibians lay eggs in water, but several species lay eggs underground or, as seen here, on a set of leaves in an aerial environment. One major adaptation seen in amphibian eggs is a jellylike coating, which gives protection. In the aerial environment, the eggs are usually surrounded by some type of "nest" such as the foamy mucus seen here secreted by a mass of gray tree frogs (*Chiromantis xerampelina*) that are engaging in group egg laying and fertilization.

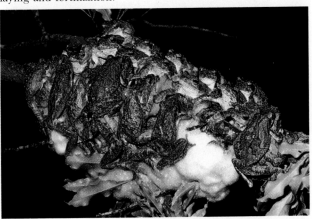

duce, no matter how widely they range on land. The ancient amphibians probably also deposited their eggs in water. In a sense, then, the first land vertebrates were analogous to the first land plants, which, as we saw, had swimming sperm and required at least films of water to reproduce.

Life on land presented yet another challenge to these early vertebrates: sound waves traveling through air are harder to detect than are sound waves in water. Thus without some evolutionary innovations for detecting sounds in air, these animals would have been deprived of critical information about the environment, such as approaching prey, predators, or potential mates. Such adaptations did arise; even the earliest known amphibians had a primitive eardrum and ear bones derived from modified branchial (gill) arches. The trend toward modification of the gill arches and jawbones into ear bones continued in reptiles, birds, and mammals. We can thus trace the upper and lower jawbones of all vertebrates (so essential to feeding) and the bones of the internal ear (so important to sensing environmental information) directly to the gill-support bones of the jawless fishes that swam along the sea floor more than 450 million years ago.

Modern Amphibians

The ancient ichthyostegids may have included several lineages that arose independently from the rhipidistians. Those lineages gave rise to many species that invaded fresh waters in temperate and tropical regions of the Earth. Most of these creatures became extinct millions of years ago, but one lineage generated the first reptiles. Other lineages, fossilized in Permian period rocks about 280 million years old, led directly to the three orders of modern amphibians.

Modern amphibians (Figure 26-14) include the tailless frogs and toads (order *Anura*), derived from one ancestral line; the tailed salamanders and newts (order *Urodela*), of separate derivation; and the legless salamanders (order *Apoda*). Like their ancestors, most modern amphibians live in or near water and have a strictly carnivorous life style; they consume insects, worms, and other animals but usually not plant matter.

Most frogs, toads, and salamanders breed and lay their eggs in water, although some use such damp places as wet earthen burrows or small watery pools in trees. The embryos usually develop in the water, becoming tadpoles that swim about and feed for varying lengths of time, depending on the species. We will see in Chapter 37 how hormones coordinate the metamorphosis from tadpole to frog or salamander. We can see in the life cycle of modern amphibians a representation of one im-

(a)

(b)

Figure 26-14   MODERN AMPHIBIANS.
(a) Golden toads (*Bufo periglenes*) live only along the Continental Divide in the Costa Rican Cloud Forest Reserve. Here, a male (right) sits astride a larger, darker female as a second male (left) looks on. (b) This red spotted salamander (*Notophthalmus viridescens*), like all amphibians, must remain in moist habitats so that its skin won't dry out.

portant stage of vertebrate evolution: the aquatic tadpole's change to a terrestrial adult corresponds to the transition from early fish to early amphibian.

REPTILES: ADAPTATIONS FOR DRY ENVIRONMENTS

About 40 million years after the first amphibians crawled onto the muddy shores, creatures about the size of small dogs were coexisting with the amphibians. These new animals were the *Cotylosauria*, the first reptiles, and are the ancestral group from which the dinosaurs and the modern representatives of the class **Reptilia** stemmed, a class that includes lizards, snakes,

iguanas, crocodilians, turtles, and tortoises. The cotylosaurs were barely distinguishable from the amphibians that also inhabited the dense, steamy forests of the Carboniferous period, about 310 million years ago, and in fact probably evolved from them. During that time of moist, tropical climates, seed plants spread inland from the swamps toward higher, drier land, and large insects appeared. These plants and invertebrates presented a diverse and rich food source for any land vertebrates that could resist desiccation and venture inland in search of them.

Reptilian Adaptations

The first venturesome reptiles, and their contemporary descendants as well, share certain key adaptations: they had a new sort of shelled egg that minimized water loss and hence freed the reptiles from having to migrate to water to lay their eggs. They also had skin that was dry and scaly rather than moist, and they could therefore live in dry environments (Figure 26-15). We will see that dry skin resulted indirectly from evolution of the rib cage and efficient breathing. Finally, reptiles evolved the capacity to greatly restrict water loss during excretion. All three adaptations expanded enormously the range of habitats that the dry-skinned reptiles could colonize.

With their reproductive strategy, the early reptiles were analogous to the first seed plants; both were freed from dependence on free-standing water. The reptilian egg is essentially a pool of water and a supply of food for the developing embryo—all surrounded by membranes and a shell that is either leathery and collagenous or brittle and made of calcium carbonate. It is referred to as a

Figure 26-15   A REPTILE ADAPTATION TO LAND.
A land iguana (*Conolopus pallidus*) defends its territory. This scaly, dry-skinned relative of common lizards, basking in the sun to elevate its body temperature, raises its head in a typical display to signal its territory to intruders. The vegetation nearby on this Galápagos island is representative of that in dry iguana habitats.

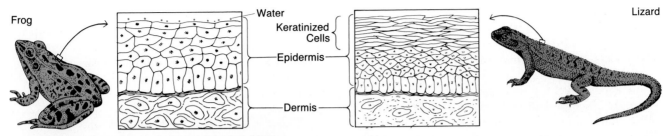

Figure 26-16 THE DEVELOPMENT OF DRY SKIN.
The reptiles' dry, scaly skin was a key to their survival on land. While the frog's skin is thinner and moist, that of the lizard is much thicker and dry. The dead keratinized cells accumulate and lipids in and between them form a barrier to evaporation of body water.

cleidoic, or "sealed off," egg (see the photograph at the beginning of the chapter). The shell is perforated by several thousand tiny pores, which allow oxygen to pass into and carbon dioxide to pass out from the embryo. However, the membranes and shell minimize evaporation as long as the egg lies within a nest or burrow with reasonably high humidity.

Unlike a typical small amphibian egg (1 mm or less), the reptilian egg cell is centimeters in diameter and contains enough yolk to sustain the developing embryo until it can survive on land. Thus when the baby reptile hatches, it is not a larva, but a free-living animal capable of breathing, moving in search of food, and resisting desiccation. Because fertilization must occur before the tough shell is built around the ovum by the female's oviduct (Chapter 16), mating behavior of males and females culminates in internal fertilization, a process carried on by all reptiles and their descendants—the birds and mammals. Despite the marvelous advantages of the cleidoic egg, ninety-three diverse lineages of extinct and living reptiles evolved independently the bearing of live young (viviparity).

Even more important an adaptation than the cleidoic egg is the reptile's dry, scaly skin (Figure 26-16). Its outer layers are composed of dead, dry cells containing keratin, a class of protein, as well as lipids that are a barrier to water loss. The reason that skin can be dry in reptiles is an indirect result of the expandable rib cage, the set of rib bones around the chest region that fills and empties the lungs with great efficiency. As a result, the lungs are the sole site of oxygen and carbon-dioxide exchange. Hence, the skin can be dry and impervious to water loss. Dry, scaly skin, the expandable rib cage, and the restriction of respiratory exchange to the lungs were critical adaptations for inhabiting dry areas. In time, birds and mammals inherited this basic breathing system from their reptilian ancestors, while reptilian scales gave rise to feathers and to hair—both highly efficient insulation materials.

Finally, reptiles evolved with enzymes and kidney functions that allow them to excrete a pasty guano instead of watery urine. As we will see in Chapter 35, this is expensive metabolically, but it is worth the energy expended, since a large quantity of water is retained by the kidneys and digestive tract rather than lost as it is in the amphibians.

The Age of Reptiles

Just as the Devonian period was the heyday of diverse ancient fishes and the Carboniferous period has been called the age of amphibians, the entire Mesozoic era—dominated by flying, swimming, and running dinosaurs—is thought of as the age of reptiles. The cotylosaurs first appeared about 310 million years ago, and by the end of the Paleozoic era, some 85 million years later, they had radiated into a host of reptilian groups that inhabited the shifting continental land masses. One early line of the cotylosaurs gave rise in the late Permian period to the *therapsids*, an equally diverse group that contained the direct ancestors to the mammals. These mammal-like reptiles included species surprisingly like squirrels, hippos, and saber-toothed tigers.

Toward the end of the Permian period, about 230 million years ago, the sea levels fell, continents collided, and temperatures changed, leading to the largest mass extinction in Earth's history: some 96 percent of all animal types—invertebrates and vertebrates—were wiped out. As in later periods of mass extinction, this was a time of shifts in habitat availability. Although many early reptiles and mammal-like reptiles met their demise, one important line, the *thecodonts*, survived and eventually gave rise to the dinosaurs and to the modern crocodilians. The early thecodonts were similar to small lizards with very short forelimbs; they probably ran on their two hind legs and thus were *bipedal* (two-legged). One lineage of early dinosaurs was the bipedal *Ornithischia*. Later members of this group reverted to a four-legged stance, sported head shields, had heavily armored bodies, and were herbivores (Figure 26-17). Another group of dinosaurs, the *Saurischia*, contained mostly bipedal

ARE MASS EXTINCTIONS CYCLICAL?

Several recent theories hold that mass extinctions such as the demise of the dinosaurs occur in regular, predictable cycles rather than as single catastrophic events. Such periodic extinctions might be based on the movements of stars, comets, even the entire solar system as it swings through the Milky Way galaxy. And the implications for evolution are nearly as sweeping: if the theories are true, then the extinction of many species may be due more to bad luck than bad genes.

Two paleontologists from the University of Chicago, David Raup and John Sepkoski, made a startling discovery in 1983. For years they had been compiling and studying the names of extinct marine species on a list that included more than 250,000 organisms. While graphing the patterns of extinction over the last 250 million years, they saw a dozen peaks, each about 26 million years apart. These peaks suggested that some agent, at work over extremely long but regular intervals, could periodically disrupt biological systems on Earth, wiping out large numbers of organisms and opening up a race among the survivors to radiate anew in the depopulated areas.

Astronomers began searching for solar or galactic phenomena that might occur on a 26-million-year cycle; they came up with two interesting possibilities. The first is based on the movement of the solar system through the starry "fingers" of the Milky Way galaxy, a flattened pinwheel of 100 billion stars. The sun com-

Figure A
Arizona's Barringer Crater, nearly 10 kilometers wide, is the remnant of a giant impact some 25,000 years ago.

pletes a rotation around the center of the galaxy once every 250 million years and oscillates up and down through the horizontal plane every 67 million years. Thus every 33 million years or so, our solar system is crossing the galaxy's most crowded central layer of stars. During this passage, some astronomers speculate, massive clouds of dust and gas in this central region might disturb a spherical shell of icy fragments

Figure 26-17   THE MESOZOIC EARTH: A WORLD DOMINATED BY DINOSAURS.

(a) An ostrich dinosaur (*Dromiceiomimus*) grazes in a redwood groove near what is today Trochu, Alberta, Canada. The bipedal stance characterized many dinosaurs; the elongate hind legs, no doubt, permitted rapid running. (b) Two great horned dinosaurs (*Triceratops*) walk beside a lake among bald cypress and gum trees near what today is Saskatchewan, Canada. These creatures evolved from earlier bipedal forms that also lacked the heavy, protective head shield.

(a)

called the Oort cloud that surrounds the solar system. Theoretically, this disturbance could dislodge thousands of ice balls, causing them to plunge toward the sun as comets and causing some of them by chance to strike the Earth in a shower hundreds of thousands of years long. The effect of this shower, according to theory, might be to throw up huge clouds of dust and debris, blocking sunlight, drastically reducing the levels of photosynthesis, and leading to mass extinctions. Luis and Walter Alvarez at the University of California at Berkeley have proposed that this sort of event led to the demise of the dinosaurs. Many large craters attributed to the impact of the ice fragments are still observable on Earth's surface. Interestingly, these depressions, including the Deep Bay Crater in Saskatchewan and the Barringer Crater in Arizona (Figure A), are dated to roughly 28-million-year cycles.

The second astronomical theory suggests that our sun has a dim sister star circling the solar system in an elliptical orbit that brings it near the sun only once every 26 million years. As the hypothetical star nears, its gravitational pull would theoretically disturb the Oort cloud and cause comets to fall like apples, striking our planet (as well as others) and causing mass extinctions. This hypothetical sister star is called Nemesis after the Greek goddess who persecutes the rich, powerful, and proud.

Subsequent analyses by other astronomers and pa-leontologists cast serious doubt on the cyclical nature of the mass extinctions. More important, however, is the concept that unpredictable catastrophes play a major role in shaping the evolutionary history of life on Earth. Organisms have not evolved defenses against such drastic events. One reason is that organisms have life cycles that are minute in comparison with events occurring at intervals of tens or hundreds of millions of years. Life spans are so minute that there is no hypothetical way mutations, natural selection, or any other known mechanism of heredity and evolution could lead to adaptations for such catastrophic events.

The chance bombardment of the Earth by meteorite showers is just an extreme example of the risks inherent in the existence of every living thing on our planet. Those organisms with attributes that allow survival in the cold and reduced light following such an event would be selected to endure in the same way that selection for other traits continually works in favor of some organisms over others. What is most intriguing is that mass extinctions may have opened the way for equally massive radiations of surviving species. In that respect, the concept of large-scale periodic extinctions may be a major new factor in explaining the overall evolution of life on Earth.

(b)

carnivores. The vast diversity of these dinosaurs is represented today only by the crocodilians.

Nearly all dinosaurs were large as adults, weighing at least 25 kilograms and in some cases thousands of kilograms. Some evidence suggests that the larger ones may have been highly active, warm-bodied creatures. Some biologists argue that the heat was self-generated, as in modern warm-blooded birds and mammals. However, most current evidence suggests that these reptiles did not generate their own heat with high ATP consumption (Chapter 35). Instead, these large animals, living in warm Mesozoic forests and swamps, probably absorbed heat from sunlight during the day and then retained a good deal of the heat overnight.

Although the cotylosaurs and their descendants were land animals, many reptilian lineages reinvaded the water and competed successfully with amphibians and fishes. The most fishlike of the ancient reptiles were the *ichthyosaurs*—beautiful, graceful creatures with features that parallel those of modern sharks and porpoises (Figure 26-18). The demands of efficient locomotion in water brought about many such cases of parallel evolution; in general, the aquatic reptiles evolved limbs that functioned as paddles, as in the *plesiosaurs*, or as control planes, like shark fins. Many also evolved flattened fin-like tails for propulsion. And the ichthyosaurs, lacking limbs to walk on land, ceased laying cleidoic eggs and instead retained the developing embryos inside the female's reproductive tract, again like sharks and porpoises.

Just as the seas and lakes were reinvaded by reptiles, so the final habitat on Earth—the air—came to be populated with flying vertebrates, the *pterosaurs*, descendants of the thecodonts (Figure 26-19). A few were the size of pheasants or large bats, but most had large wing spans, some up to 15 meters (48 ft.). The main flight surfaces of pterosaur wings were huge sheets of skin that stretched tautly from a highly elongated fourth digit on the forepaw to the side of the body. This arrangement made for efficient gliding but clumsy walking and running, and also probably led to the evaporation of large amounts of water and heat from the huge flight surface of the wings. The pterosaurs' chest and leg bones were light, and the animals could probably glide well and flap fairly effectively to propel themselves. Overall, however, they were probably somewhat clumsy flying machines that died out and whose niche in the air was later filled by two other types of flying vertebrates that evolved quite independently from the pterosaurs and each other: the birds and the bats.

(a)

(b)

Figure 26-18   REPTILES IN THE SEA.
This ichthyosaur (*Opthalmosaurus* species) has a streamlined shape and forelimbs that are flat fins just as in modern sharks and porpoises. (a) A complete fossil skeleton of a female includes remains of seven unborn embryos, indicating that the cleidoic egg was no longer laid by these sea creatures. (b) An artist's conception of a living ichthyosaur.

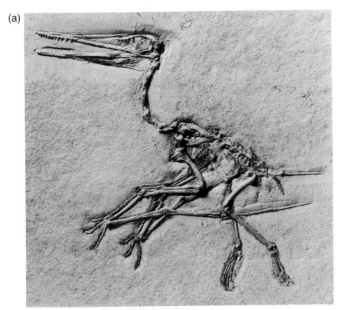

Figure 26-19   FLYING REPTILES.
Pterosaurs were flying vertebrates that descended from the thecondonts. (a) A fossil pterosaur (*Pterodactylus*). Note the greatly elongate fourth digit, which served as the leading edge of the great wings. (b) An artist's conception of this reptile. Pterosaurs died out during the later Mesozoic era and were replaced in the air by the birds and the bats, which evolved along different paths from both the pterosaurs and each other.

During the lengthy Cretaceous period, various dinosaurs became extinct. Then, at the close of the Cretaceous, 65 million years ago, the remaining dinosaurs disappeared. Their demise was not sudden; paleontologists recently found dinosaur bones in Montana that prove that some of the great reptiles were alive 41,000 years

after the purported final extinction. Besides large-sized terrestrial organisms, many types of plankton with calcareous skeletons and tropical reef organisms also became extinct. Many geological theories attempt to account for this pattern of extinctions.

The late Cretaceous period was marked by continental drift, mountain building, and a particularly intense 10,000- to 100,000-year-long episode of volcanic activity. The latter probably depleted atmospheric ozone and led to a prolonged acid rain that could have lowered sea pH and altered oceanic chemistry (hence, plankton dependent on carbonate and alkaline pH were killed). When coupled with a major drop in sea level that resulted in changes in weather patterns and habitat availability (such as may have caused the reef organism extinctions), the intense volcanic activity may be the best explanation of the extinctions.

Alternatively, biologists from Cornell University have suggested that these geological changes resulted in temperature extremes, both daily and seasonal. The onset of widely fluctuating temperatures and the cooling of formerly tropical habitats could have created severe problems for dinosaurs. The little heat they could absorb during the cool days may not have been sufficient to carry them through the cold nights. Thus temperature fluctuations alone might have contributed to their eventual extinction.

Another hypothesis is that an enormous object—perhaps a meteorite—slammed into the Earth in the late Cretaceous period, vaporized on impact, and threw an immense cloud of ash and dust into the atmosphere. This cloud could have lowered temperatures by blocking sunlight. Contradictory evidence suggests, however, that massive emissions of sulfate aerosols from volcanoes (and not dust itself) could have lowered temperatures; hence, the possibility of a meteorite is not needed to explain the events.

Many kinds of animals alive in the late Cretaceous period, at the end of the Mesozoic era, did survive. The survivors included the smaller "cold-blooded" reptiles—turtles, lizards, snakes, and crocodiles—as well as mammals, birds, amphibians, teleost fishes, lungfish, insects, worms, many other invertebrates, plants, fungi, protists, and monerans. As C. B. Officer of Dartmouth College has written, the dinosaurs apparently failed to evolve new groups to occupy the niches vacated when their kin species became extinct, as had occurred so successfully earlier in the Mesozoic era. It was the mammals, with their fur insulation, capacity to generate body heat, and placentas that so rapidly radiated at this time. Perhaps competition from mammals, the intense volcanic activity, sensitivity to fluctuating and cool temperatures, and temperature-dependent sex determination may together explain the dinosaur extinctions.

Modern Reptiles

Three orders of reptiles survived the extinctions at the end of the Mesozoic era (Figures 26-20 and 26-21). The order *Crocodilia* is closest to the dinosaurs, having evolved from the early saurischians. During the Cenozoic, which began about 65 million years ago, the crocodilians radiated once again, becoming the alligators of China and the United States; the caiman of Central and South America; and the crocodiles of Asia, Africa, Australia, and Central America. These rugged, ancient-looking carnivores—vulnerable only to cold, drought, and bullets—spend most of their lives in water but occasionally venture onto land to attack prey as large as small deer.

Tortoises and turtles are members of the order *Chelonia,* which arose from an ancient stem group that survived both the Permian and Cretaceous extinctions. These animals have been successful on land, in fresh water, and in the oceans in good part because of their protective *carapace,* or shell, of bone or leathery skin and their production of cleidoic eggs.

The order *Squamata* includes lizards, snakes, and iguanas. Modern lizards are descendants of a line that originated in the Triassic period and competed successfully with the dinosaurs. Living along the ocean shore and in jungles, temperate regions, and deserts, lizards are among the most versatile of the reptiles. Snakes first appeared late in the Mesozoic era and are essentially legless lizards with modifications for locomotion and for killing and swallowing prey. Snakes slither along by making S-shaped body contractions and by pushing their skin scales against the ground for traction. Many snakes manufacture venom in their salivary glands that can par-alyze a prey animal's nerves or muscles or can lyse (break open) the prey's blood cells to immobilize it. The snake swallows its prey whole with jaws that can open extraordinarily wide because they are not attached to the rest of the skull. It then slowly digests the prey in its expandable stomach.

BIRDS: VERTEBRATES TAKE TO THE AIR

Although some fish can soar on outstretched fins and mammals such as flying squirrels and bats can fly with varying agility, the most familiar flying vertebrates are members of the class **Aves** (Latin for "birds"). These winged animals are believed to have arisen in the Triassic period, about 225 million years ago, just when the dinosaurs were beginning to diversify. The first known bird, *Protoavis*, already had hollow bones and other adaptations for flight. It was very like a small, meat-eating, fast-running, bipedal dinosaur in other respects. Birds arose after the pterosaurs and from a different ancestral lineage (review Figure 26-21).

Another much-studied fossilized bird, *Archaeopteryx*, reveals that a key feature of avian physiology and flight—feathers—was already present in the middle Jurassic (Figure 26-22). Feathers are made of dead cells containing keratin, the same family of protein found in reptilian skin and mammalian hair. Because they are dead, feathers become dehydrated and are exceedingly light for their size. Thus they can be large and form the flight

Figure 26-20   SOME LIVING REPTILES: ANCIENT SURVIVORS IN MODERN HABITATS.
Most reptiles share certain traits such as a dry skin, an expandable rib cage, and cleidoic eggs, but their body forms and life styles are quite diverse. (a) Cobras are predatory snakes with deadly venomous bites. Like all cobras, the Siamese cobra (*Naja* species) shown here assumes an erect position before striking. (b) These salt-water crocodiles (*Crocodylus porosus*) are heavily armed—their teeth are viciously sharp, their jaws are powerful, and their heavy tail can lash sideways to deliver a crushing blow. (c) The aqua-colored chameleon (*Chamaeleo* species) is a small tree-dwelling reptile that feeds on insects, snatching them with a long sticky tongue.

(a)   (b)   (c)

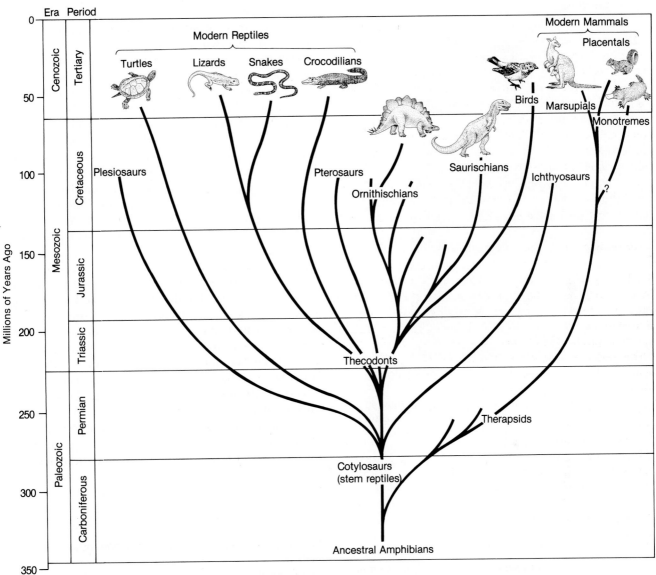

Figure 26-21   THE EVOLUTIONARY RELATIONSHIPS OF THE REPTILES AND THEIR DESCENDANTS, THE BIRDS AND THE MAMMALS.
This complex family tree of the reptiles is really like viewing a bare maple tree in mid-winter—there are large numbers of reptilian types that are omitted, with only the "main branches" shown here. The dinosaurs—ornithischians and saurischians—were especially diverse. The flying pterosaurs and birds are very distinctive lineages. The therapsid-mammal lineage is an ancient one. New evidence suggests that monotremes are closely related to placentals and marsupials.

surface of the wings and tail without weighing much and without being susceptible to water or heat loss. Large birds may have 25,000 feathers, and on each of the flight feathers, up to 1 million tiny hooklike barbules that interlock and create a seamless "fabric" for sailing through the air.

Feathers probably originated as overlapping, fringed extensions from the ends of reptilian scales. These scales may have trapped a layer of stagnant air next to the skin

to keep excess heat *out*. (Some modern lizards living in very hot environments have such extensions on the scales of their backs.) Once present, feathers could have come to serve the opposite purpose in cooler environments: to prevent heat loss from the body. Thus they were probably a precondition for the development of *homeothermy*, or warm-bloodedness, in birds (Chapter 35). Once present for insulation, feathers could have become elongated into flight feathers on the forelimbs

Figure 26-22   ARCHAEOPTERYX: AN ANCIENT FOSSIL BIRD.

The wings, visible at the top, are very much like those of modern birds, in both bone structure and feather structure. This cast, taken from the original fossil, embedded in limestone, is on display in East Berlin's natural history museum. Archaeopteryx was about the size of a crow but it had a lizardlike tail, a feature absent from all modern birds.

be moved rapidly during landing or while grasping prey or can be folded against the body like airplane landing gear so they will not impede flight.

The great skeletal modifications of the forelimbs into wings left birds with no forepaws to manipulate objects in the environment. Not surprisingly, they use the only alternative—their mouths. Consequently, birds have a highly flexible and somewhat elongated neck that allows them to use their beak to build nests, to crack seeds, or to spear tasty insects or minnows. To save weight, no teeth are present on the beak, although embryologists have discovered that some of the genes needed for tooth development are still present on birds' chromosomes and can express themselves in experiments. It is incredible that these reptilian genes are still present and potentially functional after 150 million years of apparent "nonuse" in evolution.

How did such remarkable skeletal modifications as wings arise? Remember that birds evolved from a group of fast-running *bipedal* dinosaurs, whose forelimbs were already free for purposes other than locomotion. Perhaps these dinosaurs, while running to escape from predators, began to hold out their forelimbs and glide into the air in order to reach the safety of a tree branch. Thereby, the ancestral forelimb may have evolved into the wing. We surmise that the forelimbs became organs of flight only after feathers had evolved for insulation and after the feathers on the forelimbs had elongated to provide extra surface area. For this reason, it is not surprising that

and tail. Since then, feathers of specialized shapes, sizes, and colors have evolved, serving birds for flight, camouflage, communication, and courting behavior as well as for insulation.

The prodigious flying ability of most birds is due not only to flight feathers, but also to a skeletal and muscular system that is a marvel of biological engineering. Birds' bones are particularly light and strong; some of the thicker ones are hollow and filled with air. A special design feature is the large pectoral, or breast, muscles that both lower and raise the wings. They are located in the ventral region of the chest, well below the wings (Figure 26-23). A tendon running over the shoulder joint allows such ventrally placed muscles to pull the wings upward. Although both the chest and the thigh muscles of birds are enlarged and powerful, the outer wing muscles and the lower leg and feet muscles are reduced in number and mass. The lower legs and feet are not much more than scaly skin covering bone and tendon. Contractions of the thigh and calf muscles are transmitted to the legs and feet by long tendons, so that the lightweight legs can

Figure 26-23   BIOMECHANICS AT WORK DURING BIRD FLIGHT.

The major flight muscles of birds originate at the keel, a flange of bone projecting ventrally from the sternum. When the pectoralis majors contract, the wings are lowered. When the supracoracoidei contract, the force pulls on the tendons that run over the shoulder joints; as a result, the humeri are pulled upward and the wings are raised. Since the downstroke provides most of the lift and thrust that keep the bird up and propel it, the pectoralis majors tend to be the largest muscles in the flying bird's body.

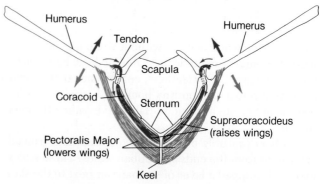

(a)

(b)

Figure 26-24   ADAPTIVE RADIATION OF BIRDS.
The diversity of modern birds is reflected in their bodies and behaviors. (a) An Anden condor (*Vultur gryphus*) stares with its sharp eyes at the mountain meadow as it soars on its giant, spread wings. (b) A spruce grouse (*Dendragapus canadensis*) cocks a wary eye and ear for danger lurking in the thickets. (c) Two huge waved albatrosses (*Diomedea* species) "presenting" during their courtship ritual on a South Sea island. (d) A female horned grebe (*Podiceps auritus*) and two offspring sit on their floating nest and watch for the return of the food-gathering male. (e) Like a skilled weaver at work, a thick-billed weaver finch (*Amblyospiza albifrons*) perches on its roosting platform and weaves its nest in Zululand, South Africa.

(c)

(d)   (e)

birds are the only flying vertebrates with "free" fore-limbs. All other airborne vertebrates—bats, flying foxes, flying squirrels, and the extinct pterosaurs—are believed to have arisen from quadrupeds that initially glided from trees or elevated sites down to the ground. Sheets of skin between the fore- and hind limbs of these creatures became the wing surfaces, and early flight involved a gliding rather than a flapping motion.

The remarkable avian skeleton, with its free forelimbs and light bones, led to adaptive radiation in many directions. Today, Aves is a large, highly diverse class with 8,600 species of water birds, songbirds, and birds of prey, including huge flightless runners, deep divers, and migrators that travel thousands of miles each year (Figure 26-24). Every one of these birds reproduces by laying a cleidoic egg followed by maternal and, often, paternal care. Aves is the only diverse class of vertebrates that has never evolved the capacity to bear live young.

MAMMALS: WARM BLOOD, HAIR, MAMMARY GLANDS, AND A LARGE BRAIN

A true mammal can be distinguished by the production of milk for its young in mammary glands, a high internal body temperature, and the presence of hair. Mammals include the primates, bats, whales, cats, and many other familiar animals. The history of the class **Mammalia**, which includes our own species, *Homo sapiens*, began with the therapsids in the distant Mesozoic era, the age of reptiles. A wide variety of these mammal-like reptiles flourished in the Permian and Triassic periods and fed on plants, insects, fish, amphibians, and small reptiles. Most became extinct, so it is difficult to say exactly where those creatures end and the first mam-

mals begin, but 180-million-year-old fossils are safely classified as those of mammals. Small members of the class lived alongside the dinosaurs for millions of years.

It may seem puzzling, at first, that our mammalian ancestors originated so long ago and yet remained marginal as the dinosaurs flourished. The explanation probably lies in the efficiency of food utilization. Picture two grassy meadows of equal size: in one, today, is a single cow eating grass, while in the other, some 100 million years ago, are eleven cow-sized dinosaurs being supported by exactly the same amount of grass. The difference is metabolism. A full 90 percent of the food energy taken in by a mammal is turned into heat to keep the body warm, leaving only 5 to 8 percent available to make new cells and body tissue. Reptiles, in contrast, turn 60 to 90 percent of food energy into new reptilian biomass. The dinosaurs probably did not have to use food to generate body heat because Mesozoic environmental conditions of high, constant temperature allowed the large-bodied dinosaurs to enjoy high body temperatures and resultant great activity without the metabolic cost of producing the heat themselves. But once global climates cooled in the Cretaceous period, uninsulated dinosaurs could not stay warm, whereas small insulated mammals already had the ability to generate high internal body temperatures, and so were better able to exploit the new environmental conditions.

Most of the early lines of mammals became extinct during the Mesozoic era. However, early in the Cretaceous period, fossils of the primitive and fascinating group of modern mammals, the **monotremes,** are found. Today there are only three species of monotremes: the duck-billed platypus of Australia and two species of spiny anteaters, or echidnas, of Australia and New Guinea (Figure 26-25). All three of these strange species lay small leathery eggs but, like all other mammals, nurse their young with milk after hatching.

Late in the Cretaceous period, the stem groups of other modern mammals appear in the fossil record. They were mouselike creatures resembling modern tree shrews. At that time, the dinosaurs were dying out, and the second radiation of mammals took place. One type, the **marsupials**—pouched animals such as the kangaroo and opossum—evolved into dozens of species on the then-isolated continents of South America and Australia. The other type, the **placentals,** diversified into hundreds of species on the continents of North America, Europe, Asia, and Africa. This group includes horses, dogs, cats, rats, deer, whales, primates, and most of the other familiar mammals, and all have a placenta (Chapter 18). Recently discovered fossils suggest that the monotremes are more closely related to the placentals and marsupials than had previously been thought.

Mammalian Adaptations

Mammals show several important physical innovations. First, as we have mentioned, mammals evolved temperature-control mechanisms that enable them to

Figure 26-25   THE DUCK-BILLED PLATYPUS.
The duck-billed platypus (*Ornithorhynchus anatinus*) is a monotreme, a primitive mammal that both lays eggs and nurses its young with milk produced in mammary glands. This one dives and swims rapidly with its webbed feet.

maintain a high, constant body temperature. The high, constant temperature is possible because of the presence of hair, an excellent form of insulation; stored fat or blubber, a source of both insulation and heat energy; heat-producing metabolic pathways (Chapter 35); and efficient respiratory and circulatory systems, which deliver large quantities of oxygen needed for the generation of heat.

A second significant physical advance in mammals entailed a set of new reproductive structures and processes based on old reptilian designs. In monotremes, the fertilized egg, then the embryo, is maintained in the uterus for about two weeks and grows as it absorbs materials from a nutrient mixture called uterine milk. This process depends on a unique mammalian adaptation—the luteal phase of the ovarian cycle (Chapter 18). During this phase, the corpus luteum of the ovary secretes female steroid hormones that cause the uterus to support the developing embryo. After about two weeks, various membranes and a shell are built around the embryo, and the female platypus or echidna lays a shelled, reptilian-like egg. This cleidoic egg is incubated next to the female's abdomen. A short time later, a primitive newborn chips its way through the eggshell with its tiny egg tooth (a reptilian characteristic), which consists of a sharp beaklike point on the nose that drops off after hatching. The hatchling then begins to suckle milk from its mother's mammary glands. Reproduction in monotremes is thus a combination of reptilian and mammalian processes.

These gestation processes are modified further in marsupials. The embryo develops in the uterus for about three to four weeks, during a longer luteal phase. The young marsupial is only about the size and weight of a small bean when it is born. Because a placenta does not develop in most marsupial species, the embryo cannot grow any larger and still be provided with sufficient oxygen by means of simple diffusion through the uterine fluids. In addition, if the embryo's protective membranes were to erode in order to allow better gas exchange, the mother's immune system would likely attack and kill the embryo. At birth, the blind marsupial—still little more than a fetus—crawls from the birth canal through the mother's abdominal fur and into the *marsupium*, or pouch. There it finds the teat of a mammary gland; attaches to the tip of it; and proceeds to feed, grow, and develop (Figure 26-26). It remains a pouch young for up to five months or more, as it gradually makes excursions to the outside world and finally leaves the pouch for good. Marsupials, then, have eliminated the monotremes' cleidoic egg and retain the embryo inside the mother for a longer time, but most have not evolved a true placenta.

What is it that allows a foal, a human baby, or a newborn whale to be so huge in comparison with a bean-sized marsupial? The secret is, of course, the placenta. Supported hormonally by a lengthy luteal phase and, in fact, by its own hormones (Chapter 18), the placenta meets the growing embryo's needs for gas, nutrient, and waste exchange. The placental embryo is also protected from attack by the mother's immune system and so can

grow thousands of times larger than the marsupial embryo. The placenta is truly a marvelous structure. Consider that in the absence of a placenta, a newborn red kangaroo weighs 0.1 gram, even though its mother may weigh 65 to 90 kilograms; in contrast, a blue whale can weigh 2,700 kilograms (3 tons) at birth—27 million times more than the largest newborn marsupial. Moreover, the placental mammal is much more mature at the time of birth—the foal can stand, and the whale can swim. Placental reproduction has other far-reaching consequences. Among them is the continuing mobility and independence of the pregnant female and, indeed, of the nursing female. She is usually not tied to a nest, as are some reptiles and most birds. The placenta and mammary glands are indeed elegant evolutionary adaptations that lie close to the core of what it means to be a mammal.

Trends in Mammalian Evolution

A fantastic diversification has taken place among mammals during the 65 million years since the Cenozoic era—the age of mammals—began. Simple shrewlike creatures have given rise to today's 26 major orders of mammals that walk, swim, fly, and occupy a diverse range of habitats. The great success of this class during the Cenozoic era can be attributed to several factors.

1. *Increasingly sophisticated temperature regulation.* Moving from the oldest surviving type of mammals, the monotremes, to the marsupials and then to the placentals, there is a trend toward increasingly complex adaptations for regulating body temperature. Because the marsupials and placentals have higher core body temperatures and can better tolerate temperature extremes, they are able to live in a broader range of habitats than are the monotremes, which are limited to warm climates.

2. *Increased body size.* Within nearly every mammalian order, there is a trend toward larger size. This trend is apparently related to heat balance: the larger the body, the lower the ratio of surface area (where heat is lost) to body mass (where heat is produced).

3. *Diversification of tooth shape.* From a basic pattern of four tooth types, mammals have evolved a wide array of tooth shapes and functions that correlate with food sources: wide flat molars and premolars (bicuspids) for grinding grains and grasses or for crushing bones; sharp conical canines for slicing and tearing flesh; and flat, chisel-like inci-

Figure 26-26   AN OPOSSUM IN A POUCH.
The development of young marsupials such as this brush tail opossum (*Trichosurus vulpecula*) is completed in the maternal pouch, where the tiny young remains attached to and is fed by way of a teat. At birth the marsupial is bean-sized and has few well-developed features: its claws and forelimbs, to help it climb a vertical path from the birth canal up the mother's abdomen and into the pouch, and its sense of smell, which may help it locate a teat once inside the pouch. Here, after several weeks in the pouch, the baby is seen with its head poking out of the marsupium.

sors for gnawing on woody plants or grazing on grasses. Some mammals have still further specialized teeth; for example, elephant tusks are modified incisors, while walrus tusks are modified canines. Vertebrate teeth are described in greater detail in Chapter 34.

4. *Elongation and specialization of limbs.* Mammals have developed diversified limb structures that permit many different locomotory styles and behaviors. Antelopes, cheetahs, jack rabbits, and other rapid runners have a lengthened stride based on elongated leg, ankle, and toe bones and on a flexible vertebral column, especially in the hip region. Primates have arms specialized for swinging locomotion, while bats have wings for flying, moles have limbs for digging, and porpoises have streamlined flippers that function like the pectoral fins of sharks.

5. *Increased brain size.* All mammals have an enlarged *neocortex*, the region of the forebrain that controls complex learned behavior. Learning and intelligence reach their highest form in the cetaceans (whales and porpoises) and the primates (monkeys, apes, and humans; Chapter 40).

Modern Mammals

Taxonomists group mammals largely on the basis of their teeth and head bones. The monotremes are placed in the infraclass *Prototheria*, which means "first mammals." The marsupials are in the infraclass *Metatheria* ("middle mammals"), and the placental mammals are in the infraclass *Eutheria* ("true mammals"). The last group is classified into 26 orders, but we will describe only the major ones.

The surviving prototherians (monotremes)—the duck-billed platypus and the spiny anteaters—are more primitive than their mammalian cousins. Besides laying eggs and having less effective temperature regulation than other mammals, they have pelvic girdles and middle-ear bones that are intermediate in form between those of reptiles and of other mammals.

The metatherians (marsupials) have a range of life styles as broad as that of the placental mammals. The radiations occurring in Australia and South America during the Cenozoic era have resulted in forms that graze (kangaroo), climb (koala), burrow (wombat and marsupial mole), eat meat (Tasmanian devil and marsupial wolf), eat grubs and insects (bandicoot and numbat), live in trees (opossum), and even fly (flying phalanger), as illustrated in Figure 26-27. Despite their success in isolation, many South American species became extinct after the Central American land bridge formed and allowed pla-

Figure 26-27   MARSUPIALS: RADIATION INTO MANY TYPES.
Though all marsupials share reproduction using the pouch and mammary glands, they differ widely in life styles. (a) Kangaroos are fast-hopping grazers. Note the joey (or young) of this red kangaroo (*Macropus rufus*) dangling its head and front limbs outside the pouch, while its hindquarters rest safely inside. (b) The koala spends most of its time in trees, usually eucalyptus, feeding on leaves. This young koala (*Phascolarctos cinereus*) clings to its mother though it still may return to the pouch for milk. (c) This common wombat (*Vombatus ursinus*) is a small ground-dweller, with a mixed diet, and strong claws for protection.

cental mammals to invade from the north. Likewise, many Australian marsupials are now threatened by imported placentals, especially rabbits, dogs, and cats.

The eutherians (placental mammals) are the most diverse infraclass of mammals. Here we briefly describe only ten orders (see Figure 26-28 for a representative sampling).

1. *Insectivora:* shrews, moles, hedgehogs. The insectivores, small, nocturnal, insect-eating mammals, are close in body type to the ancestral

(a)

(b)

(c)

(d)

(e)

(f)

Figure 26-28 PLACENTAL MAMMALS: FAMILIAR FUR-BEARERS, WITH A MULTITUDE OF BODY FORMS AND LIFE STYLES.

(a) The long, powerfully muscled arms of this orangutan of Borneo (*Pongo pygmaeus abelii*) allow the intelligent and agile creature to move safely through the trees far above the forest floor. (b) The Masai giraffe (*Giraffa camelopardalis*) inhabits the Serengeti Plain and surrounding East African savannah, grazing on the tender leaves and shoots atop trees. (c) The mountain lion (*Felis concolor*), the largest North American cat, is a stealthy, solitary, nocturnal hunter. (d) Alaskan fur seals (*Callorhinus ursinus*) are streamlined fish eaters, highly maneuverable in the water, but awkward on land. These marine mammals inhabit coastal areas in the far northern and southern latitudes. (e) The American black bear (*Ursus americanus*) is a 200- to 500-pound omnivore that will fish for trout or salmon and browse on berries or roots during the same day. (f) Bats are the most diverse mammals, comprising one-quarter of all mammal species; the greater horseshoe bat (*Rhinolophus ferrum-equinum*) shown here uses echolocation (a type of sonar) to find flying insect prey at night.

group that gave rise to all the modern orders of placental mammals.

2. *Chiroptera:* bats and flying foxes. The bats arose from the insectivore line and have very diverse life styles: blood lappers, fruit eaters, and even fish eaters that skim the surfaces of ponds. Most bats use echolocation—sound production and detection—for navigating at night.

3. *Primatas:* lemurs, tarsiers, monkeys, apes, and humans. The primates share primitive mammalian features with the insectivores and bats. Like insectivores, primates possess limbs, sense organs, and other adaptations for life in the trees.

4. *Rodentia:* mice, rats, beavers, porcupines, squirrels. The rodents make up an incredibly successful and diverse order. They are characterized by chisel-like teeth for gnawing wood, nuts, and grains.

5. *Lagomorpha:* rabbits and hares. The lagomorphs, like their close relatives the rodents, diverged early from insectivores. They make up a very successful order of herbivores adapted for rapid running and prolific reproduction.

6. *Cetacea:* whales, dolphins, and porpoises. The cetaceans include large-brained marine mammals derived as a separate lineage from insectivores. They cannot extract oxygen from water and so must come to the surface to breathe. Their limbs and tails are highly modified for swimming, as is their streamlined body. Toothed whales hunt fish and cephalopods, while baleen whales filter large quantities of crustaceans and plankton from the sea water. The largest animals that have ever lived are the filter-feeding baleen whales.

7. *Carnivora:* dogs, bears, cats, skunks, weasels. The diverse carnivores—meat eaters—are predators. Some live in complex social groups in which the young are protected and taught survival skills. Members of the suborder *Pinnipedia*—seals, sea lions, and walruses—are marine carnivores that eat fish and live primarily in the oceans, but return to land to mate and give birth. Consequently, the pinnipeds have retained forelimb and hind limb bones and girdles that evolved for walking and are not as highly adapted to aquatic life as are the cetaceans.

8. *Proboscidea:* elephants. Asian and African elephants are an order of near ungulates. Like ungulates, they can digest grasses and leaves. They walk on five toes and are distinguished by their trunk. This modification of the upper lip and nostrils allows elephants to lift heavy objects, squirt water, and brush away insects.

9. *Perissodactyla:* horses, zebras, tapirs, rhinoceroses. These ungulates walk on an odd number of toes (one or three). Their special digestion occurs in the intestine.

10. *Artiodactyla:* cattle, sheep, deer, camels. The artiodactyls are ungulates that walk on the tips of their toes, which are usually modified into hooves. The artiodactyls walk on an even number to toes (two or four), and have a special stomach chamber for digestion.

The human species, dating from only about 150,000 years ago (Chapter 50), is a relatively late development in this radiation of eutherians that began with nocturnal, mouselike tree dwellers in the Cretaceous period. The evolution of humans, and all other species of eutherians, continues today with complex, specialized members of eutheria living in aquatic and terrestrial habitats all over the planet. However, despite the complexity of modern humans, other primates, and, indeed, all mammals, our simpler and far more ancient chordate roots can still be easily observed in the human embryo, with its gill slits, notochord, tail and myotomal muscles. And similarly, every chordate, embryo and adult alike, displays in subtle and overt ways a common ancient heritage with other deuterostomes and with even more primitive groups tracing back to the flatworms and the protists. Thus the human species is but one newly sprouted twig on the tree of animal life we surveyed in Chapters 25 and 26, a tree that is representative of but one of the five kingdoms studied in Chapters 19 through 26. Animals, plants, fungi, protists, and monerans have coevolved in innumerable ways since life's origins on Earth some 4 billion years ago. Then, and ever since, the physical and chemical properties of our planet have set limits on how cells and organisms can live. In turn, those cells and organisms have altered the Earth's crust, oceans, and atmosphere during the magnificent history of the living and the nonliving. Now, with this survey of life's variety complete, we can turn in the next sections to a detailed examination of the forms and functions of two major kingdoms, Plantae and Animalia.

SUMMARY

1. The phylum *Chordata* includes three subphyla—*Urochordata*, *Cephalochordata*, and *Vertebrata*. Although they are the most familiar to us, *vertebrates* make up only a small fraction of all animals.

2. At some stage of their life cycle, all chordates possess *gill slits*, a *notochord*, segmental body and tail muscles (*myotomes*), and a dorsal hollow *nerve cord*.

3. Urochordates (tunicates) and cephalochordates (lancelets) are nonvertebrate chordates. They possess the basic chordate characteristics as either larvae or adults, but lack a brain, jaws, a vertebral column, and a bony skeleton.

4. All the primitive chordates have a *pharynx* (the gut cavity posterior to the mouth cavity) with ciliated gill slits. A feeding current is drawn into the pharynx through the mouth, food particles adhere to mucus, and the water exits through the gill slits. The first-known vertebrates, jawless fishes of the class *Agnatha*, used muscle-driven gill slits for filter feeding. These agnathan ostracoderms also possessed bone, a hard, yet dynamic tissue composed of collagen and apatite. Bone serves as the body's main reservoir of phosphate and calcium. The evolution of jaws from the anterior sets of branchial arches in the ancestors of acanthodian fishes had immense evolutionary and ecological consequences.

5. The early jawed fishes, members of the extinct class *Acanthodii* whose fossils date to 435 million years ago, had another important adaptation other than jaws—paired fins. Fins help control movement in the water and allow diverse types of behavior among fish. They also were the precursors of the specialized limbs of land vertebrates.

6. One ancient lineage of acanthodians, the placoderms, gave rise to the cartilaginous fish of the class *Chondrichthyes*, including modern sharks, skates, and rays. Various bony, jawed acanthodians gave rise to members of the class *Osteichthyes*—a wide range of teleosts and other spiny-finned fish, or actinopterygians, and of sarcopterygians, or fleshy-finned fish such as lungfish, coelacanths, and rhipidistians.

7. The extinct marine rhipidistians were probably the ancestors of members of the class *Amphibia*, which arose more than 350 million years ago. Life on land resulted in a strong vertebral column; strong limb muscles, limb bones, and pectoral and pelvic girdles; the ability to respire across the moist skin; reproductive behaviors to allow embryos to develop in water; and modified ear bones for receiving sound waves through air. Modern amphibians include the frogs and toads, salamanders, and legless salamanders.

8. Amphibians gave rise to the early stem reptiles, or cotylosaurs, about 310 million years ago. Members of the class *Reptilia* have dry, relatively impermeable skin; a water-conserving excretory system; and the cleidoic egg, which resists desiccation. The expandable rib cage fills and empties the lungs efficiently and so, indirectly, allows the skin to be dry. That permits reptiles to thrive in dry habitats. The largest land animals ever to live were the dinosaurs, which inhabited the warm lands of the Earth for about 150 million years, but died out at the end of the Cretaceous period, about 65 million years ago. Modern reptiles include crocodiles, turtles, lizards, and snakes.

9. Small bipedal dinosaurs gave rise to the birds, which make up the class *Aves*. Birds have light, compact bodies with a variety of adaptations that permit flight. Their feathers are dead, dry, lightweight structures that serve as insulation and flight surfaces. Bird bones, muscles, and limbs are modified for strong flying, quick movements, light weight, and special functions such as paddling, digging, and so on. Birds maintain a high, constant temperature and generate heat internally.

10. Members of the class *Mammalia* arose from early Mesozoic reptiles, the therapsids, and radiated twice—first during the early Mesozoic era and again during the Cenozoic. Mammals are insulated with hair or fat and can maintain a high, constant body temperature, as can birds. Mammalian groups are distinguished by the ways they reproduce: by leathery cleidoic eggs (*monotremes*), through a period in the pouch (*marsupials*), or by a placenta that supports the embryo's development (*placentals*). All mammals feed their newborn and young with milk from mammary glands.

11. During their evolution, mammals have been characterized by trends toward more sophisticated temperature regulation at a high, constant temperature; larger body size; diversification of tooth shape and function; longer and more specialized limbs; and a larger brain. Of all the mammals living today, the placental mammals are the most diverse.

KEY TERMS

Acanthodii

adaptive radiation

Agnatha

Amphibia

Aves

Cephalochordata

Chondrichthyes

Chordata

gill slits

Mammalia

marsupial

monotreme

myotome

nerve cord

notochord

Osteichthyes

pharynx

placental mammal

Reptilia

Urochordata

Vertebrata

vertebrate

QUESTIONS

1. What are the three subphyla of chordates? What features are common to all?

2. What are the functions of gill slits? How is water driven through the gill slits of nonvertebrates? of vertebrates?

3. How does bone differ from cartilage? What are some advantages of each?

4. The acanthodians were an ancient group of jawed fishes. From what structures did their jaws and jaw muscles probably evolve? How did jaws change the life style of the early fishes? What other important structures did the acanthodians possess, and what present-day vertebrate structures have evolved from them? What was the function of the gills in the acanthodians?

5. The cartilaginous fish (Chondrichthyes) include what present-day fishes? What sort of "flotation device" do they possess? The bony fish (Osteichthyes) include two subclasses: spiny-finned and fleshy-finned fishes. From which group did modern bony fish (teleosts) arise? What sort of "flotation device" do the teleosts possess? What are some advantages of this device? From which group did the amphibians evolve? Members of this group possess internal nostrils. What is their function?

6. Amphibians were the first land-dwelling vertebrates. What are some of the problems they encountered in living on land? What mechanisms do they use for gas exchange? Why are they restricted to moist environments?

7. What are some of the innovations of reptiles that freed them from dependence on a very moist environment? As reptiles evolved, some lineages took to the air, and some returned to the water. Give an example of an ancient flying reptile and of an ancient aquatic reptile. Are there any present-day flying or aquatic reptiles? What are the three orders of modern reptiles?

8. What are some of the functions of feathers? What are some advantages of using dead material rather than living tissue as a flight surface? Birds have several structural innovations that aid in flying, including feathers. What are some others? Birds have scales (as well as feathers), a legacy from their reptilian ancestors. Where would you find the scales on a bird?

9. What attributes are shared by all mammals? Which of these attributes are unique to the mammals?

10. The three lines of present-day mammals are the monotremes, marsupials, and placentals. What are some features that distinguish each group? To which group does each of the following belong?

   a. opossum
   b. bear
   c. kangaroo
   d. duck-billed platypus
   e. human
   f. wolf
   g. lion
   h. mouse
   i. spiny anteater
   j. dolphin
   k. bat
   l. sea lion
   m. zebra
   n. rabbit
   o. squirrel

Can you name a monotreme that is common in the United States? a marsupial?

ESSAY QUESTIONS

1. Mammals have a much more rapid metabolism than do reptiles. What are some advantages of a rapid metabolism? What are some disadvantages?

2. Imagine that fish never evolved pectoral and pelvic fins. What might vertebrate evolution on land have been like as a result? Would human beings be here?

SUGGESTED READINGS

ALEXANDER, R. McN. *The Chordates*. Cambridge, England: Cambridge University Press, 1975.

A superb book with an emphasis on mechanics, physiological function, and evolution.

GANS, C., ed. *Biology of the Reptiles*. 9 vols. New York: Academic Press, 1969–1979.

A compilation of many articles by experts on every aspect of the reptiles.

OFFICER, C. B., A. HALLAM, C. L. DRAKE, and J. D. DEVINE. "Late Cretaceous and Paroxysmal Cretaceous/Tertiary Extinctions." *Nature* 306 (1987): 143–149.

A cogent discussion that may best explain dinosaur extinctions.

ROMER, A. S. *Vertebrate Paleontology*. Chicago: University of Chicago Press, 1966.

The foremost American paleontologist of the twentieth century writes on his lifelong love—the vertebrates.

YOUNG, J. Z. *The Life of Mammals: Their Anatomy and Physiology*. Oxford: Oxford University Press (Clarendon Press), 1975.

The world's preeminent vertebrate zoologist writes on all aspects of mammals.

————. *The Life of Vertebrates*. 3d ed. Oxford: Oxford University Press, 1981.

Surely one of the most influential and best written texts on the vertebrates—a gold mine that includes a marvelous bibliography.

Part
FOUR

PLANT BIOLOGY

Plants are the great producers—many-celled photosynthetic organisms that cover the continents with dense forests, rolling grasslands, and sparse tundra. As plants evolved from single-celled algal precursors and adapted to life on land, numerous specialized structures arose—roots, stems, leaves, fronds, spores, cones, flowers, seeds and their equivalents—and myriad physiological systems and processes developed. We will discuss those structural and physiological adaptations and the survival value they afford plants in a variety of habitats.

We will study the architecture, reproduction and development, physiology, and hormonal systems of angiosperms, or flowering plants—the most complex and diverse of producers. A wide range of adaptations enables angiosperms to survive in mild and harsh cli-

mates around the world and in places with widely differing patterns of soil composition, predators, and light, water, and nutrient availability. Yet the flowering plants share several common solutions to the everyday problems of survival, such as transporting water, minerals, and nutrients; exchanging gases with the environment; producing enough surface area for capturing sunlight; producing gametes; coordinating cyclic events such as flowering and dormancy; protecting against predators and diseases; and providing internal support against gravity.

Two dominant features have shaped the evolution of plant survival systems. First, plants are immobile; virtually all plants are rooted in a given spot for a lifetime, and must be able to survive the full range of environmental insults to which they are subject.

Moreover, their cells remain fixed at their site of origination. This immobility helps explain why physical factors such as diffusion, cohesion, and tensile strength so dominate plant life. Second, plants have zones of embryonic tissue that can be active periodically throughout the organism's life, enabling the plant to generate many new sets of flowers, additional branches, roots, or leaves. This contrasts strongly with the pattern in animals—a specific number of limbs, eyes, and other organs preprogrammed in the genes and formed only once in the embryo. We will consider these differences in detail, as well as the more obvious plant traits—their shapes, colors, rare fragrances, and impressive photosynthetic abilities—that set them apart from the other kingdoms of life.

27

THE ARCHITECTURE OF PLANTS

Of the theory of vegetables, or of the growth, propagation, and nutriment of vegetables, our knowledge is only slight and superficial. A close inspection into the structure of plants affords the best ground for reasoning on this subject, and, indeed, every thing beyond it is little better than mere fancy or conjecture.

George Le Clerc de Buffon, *Natural History* (1821)

Bitternut hickory rings: living historical records of past years' growth.

Seed plants—the gymnosperms and the angiosperms—have an inspiring diversity of sizes, shapes, structures, colors, and life styles. In a single visit to a botanical garden, one might see plants as delicate as violets or as bright as primroses growing in the shadow of furrowed giants like ponderosa pine or Douglas fir. Sandhill sage, with its silvery gray foliage, might be planted just meters from purple hopyard trees or, in a nearby greenhouse, tall philodendrons that produce massive, shiny leaves. And in a far corner of the garden might be a giant saguaro or a barrel cactus, both with no obvious leaves at all—just bulbous stems and fierce-looking needles.

While the ubiquitous and ecologically important seed plants are a marvelously diverse group, they share a number of characteristic physical structures, including leaves, stems, and roots. As with all living things, such shared physical structures are the basis of function. And the study of plant anatomy is a foundation that allows us to more fully understand plant physiology—how plants, grow, reproduce, collect solar energy, transport materials, absorb water, and so on.

The major plant structures aptly demonstrate this merging of form and function. Because of their shapes, their positions on plants, and the types of cells and tissues of which they are composed, *leaves* are well equipped to act as efficient solar collectors—absorbing sunlight and converting light energy to chemical energy. *Stem* tissues are beautifully suited to a different set of tasks—supporting and lifting the leaves toward the light and housing the vascular "pipelines" that transport water and nutrients throughout the plant. Finally, the physical make-up of *roots* allows them to simultaneously anchor the plant, absorb water and minerals, and, often, store carbohydrates.

In this chapter, we will consider the main structures of seed plants and see how, together, they form the smoothly functioning organisms that support virtually all life on land, directly or indirectly. And we will discuss how a host of variations on these basic themes allows these plants—from violets to fir trees—to survive in a nearly endless variety of habitats.

STRUCTURES TO MEET THE CHALLENGES OF LIFE ON LAND

The leaves, stems, and roots (or analogous structures) that function in virtually all vascular land plants arose as evolutionary adaptations to the difficulties posed by life on dry land. The earliest plant cells—the algae—lived in the shallow seas and bodies of fresh water about 600 mil-

lion years ago, and about 400 million years ago, the green algae began to invade the moist fringes of the barren continents. Life on land presented several new challenges to these earliest invaders, including the danger of desiccation; the higher intensity of sunlight than that penetrating watery habitats; rapid shifts in temperature from day to day and often more dramatic shifts from season to season; lack of buoyant support for the plant body; buffeting by rain and wind; and the limited availability of water, needed by swimming sperm in order to reach and fertilize eggs. The leaves, stems, and roots we will discuss in this chapter, as well as the seeds, flowers, and fruits (Chapter 28), evolved over tens of millions of years, and together they enabled plants to adapt to this harsher terrestrial existence and to diverge into more than 300,000 species of land plants.

The earliest plant "invaders" of land lacked roots, stems, and leaves, but botanists have reconstructed a history of how these structures may have evolved by studying fossil evidence as well as the spectrum of modern land plants. The whisk ferns, an ancient plant group dating back almost 400 million years, were the simplest land plants with a vascular transport system. But the equally ancient club mosses and horsetails were the first to show all three structural innovations (leaves, stems, roots) in addition to vascular tissue. Despite these adaptations, all the early land plants (and their modern counterparts) were, in a sense, amphibious, like frogs and salamanders; they required standing water for fertilization and thus were (and still are) relegated to permanently or at least seasonally moist environments.

During the Triassic and Jurassic periods in the Mesozoic era, from about 230 to 120 million years ago, the gymnosperms (conifers, cycads, ginkgoes, and gnetinas) were the most abundant vegetation on land and showed a great evolutionary advantage—the production of seeds. Windborne pollen and seeds eliminated the need for standing water during fertilization, just as internal fertilization and shelled eggs did for the great reptiles. During the Cretaceous period, 120 to 150 million years ago, the angiosperms, or flowering plants, appeared and began to evolve, and in them, seed production became much more complex. While the gymnosperms have naked seeds, the angiosperms have protected seeds that develop inside flowers and fruit. This kind of innovation was so advantageous that the angiosperms quickly spread across the continents. Today, they are the most diverse and abundant plants on Earth—not to mention the major source of food for many animals, including humans. Because they are so prevalent and ecologically important, we will focus mainly on the angiosperms in this chapter and in the next three.

You may recall from Chapter 24 that the angiosperms are divided into two main groups: *monocotyledons* and *dicotyledons*. The structures that differentiate monocots

AGRICULTURE AND HUMAN CULTURE

Few of us realize that much of human culture is a direct outgrowth of agriculture—the deliberate cultivation of plants. Until about 10,000 years ago, all humans lived as roving hunter-gatherers. Then hunter-gatherer tribes in several parts of the world abandoned their nomadic life styles in favor of domesticating useful plants. They cultivated rice in Southeast Asia, wheat in the Middle East, corn and beans in North and Central America, and potatoes in the Andean highlands of South America. After harvest, these crops were easy to store for long periods, and they provided carbohydrates, the major source of calories, as well as many of the vitamins and minerals humans need to stay healthy. These starchy foods became the staff of human life, and their domestication led to many changes in human society.

Perhaps most important, agriculture led to the formation of villages, then towns, and eventually cities. The survival of hunter-gatherer societies required vast territories and, in fact, a few people still exist this way. However, the advent of agriculture led to the establishment of the first permanent settlements and allowed people to remain in one area and produce enough food for their needs. The original settlements almost certainly necessitated the development of architecture and the early forms of sanitary and civil engineering, since people needed permanent dwellings and public buildings, as well as streets, water systems, and bridges.

Another major consequence of settlement was the need for more labor (in the form of larger families) to farm the land. The resulting higher birth rate marked the beginning of the population explosion that now strains the Earth's resources and ecology.

In addition, agriculture probably led to mathematics and written language. The storage of food required accounting methods, and archeological evidence shows that each of the early agricultural peoples developed systems of arithmetic to keep track of their food supplies and population. A large percentage of the hieroglyphics discovered at Egyptian archeological sites is devoted to tabulating food stores and agricultural productivity (Figure A). Many Aztec writings are also thought to be vast ledgers in which the annual harvests of corn and beans were totaled.

Agriculture is closely linked to early astronomy and religion. Calendars were probably created so that farmers could predict the proper times for planting and harvesting. Many ancient peoples had extremely accurate calendars, and astronomers were held in

Figure A   **ANCIENT AGRICULTURE IN EGYPT.**
Hieroglyphics depict the harvesting of cereal grain crops (middle panels), as well as healthy orchards under cultivation (bottom panels).

high esteem. Religions, too, were based on agriculture: many cultures worshiped the sun and soil fertility—a fact that reveals that early farmers knew a great deal about the requirements of plants. Totems of plants and plant gods were common, and some societies appeased their agricultural deities with human sacrifices.

Finally, agriculture required a high level of community cooperation. The terracing of rice fields, the digging of canals and irrigation ditches, and the building of vast food storage cellars depended on coordinated efforts by large groups of people; these and other communal tasks almost certainly led to expanded social skills, and perhaps even to castes and social classes as well. Ironically, agriculture probably led indirectly to the development of taxation and bureaucracy, as means to keep track of the personal and communal wealth represented by stored food.

Those of us who live in modern urban areas tend to forget that our economy is based on agricultural productivity—on the fertile soil and good climate of North America's farm belt, as well as on advanced agricultural science and technology. Plant products are the major exports of the United States to other nations and the ultimate source of much of our country's wealth. Although we no longer worship plants or base our community activities on agriculture, we remain highly dependent on domesticated plants to provide animal feeds; raw materials for textiles, chemicals, and dyes; and, of course, food for ourselves.

(a) Monocot: Rice

(b) Dicot: Potato

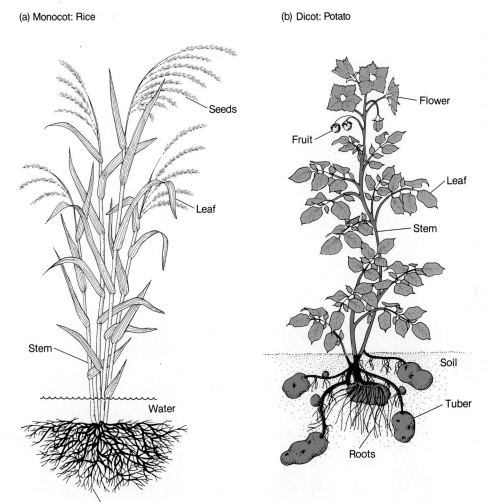

Figure 27-1   TYPICAL MONOCOT AND DICOT PLANTS. (a) Rice is a representative monocot. (b) The potato plant is a representative dicot. Differences in the shapes of leaves and roots, and seeds versus fruits are visible here; however, the many differences in the way cells are arranged can only be seen by looking within the leaves, stems, and roots. The potatoes humans eat are actually storage sites called tubers.

and dicots are shown in the generalized drawings of rice and potato plants in Figure 27-1. The distinctions can be observed directly by comparing palm trees, corn, Bermuda grass, daffodils, and onions (monocots) with maples, sunflowers, ivy, zinnias, and tomatoes (dicots). As we discuss the leaves, stems, roots, and other basic plant structures, we will show how these structures differ—in large and small ways—from monocots to dicots.

LEAVES: LIVING COLLECTORS OF SOLAR ENERGY

Plants trap solar energy in their specialized light-collecting organs, the **leaves.** Human solar engineers would have a herculean task if they set out to match the all-around performance of even the simplest leaf: leaves are biodegradable; they need nothing more than light, water, carbon dioxide, and a few minerals to function; the plants to which they are attached are self-generating through biological reproduction; most kinds of leaves move to track the sun's motion; and they let in light and gases but prevent excessive water loss. Moreover, many leaves are quite beautiful to humans. What mechanical solar-energy collecting device could compare with the graceful shapes and lovely hues of holly leaves, willow leaves, and the silvery appendages of quaking aspen?

The leaf's efficiency at collecting solar energy—the amount of energy harvested compared with the amount of light impinging on a given surface area—depends on the size, shape, and position of the leaf, as well as on the photosynthetic process itself (Chapter 8). Most dicots have leaves with two distinct portions: an enlarged, usually flattened region called the **blade,** and a stemlike portion called the **petiole,** which connects the leaf blade to the stem of the plant. In some plants, such as the

(a)

Petiole

Blade

Figure 27-2 LEAVES: A VARIETY OF SHAPES TO MEET A VARIETY OF NEEDS.

Leaves have two major parts, a blade and a petiole. Leaf blades exist in many complex shapes and are adapted for the conditions under which the plant lives. (a) Larger leaves, such as those of this Costa Rican *Cecropia peltata*, can absorb more radiant energy, but may lose more water than small leaves. (b) Ice plant (*Carpobrotus edulis*) leaves are thick and waxy, and store large quantities of water. (c) The needles of this fir tree (*Abies grandis*) are long and narrow. This shape helps reduce evaporative water loss by diverting rather than catching wind. (d) As leaves form, they may remain tightly clustered and form a dense bunch such as the "head" of a cabbage plant (*Brassica olertacea*). (e) Aquatic leaves may have gas-filled bladders to help them remain near the water's sun-filled surface layers. This rockweed (*Fucus vesiculosus*), gathered from a tidal pool in Washington, has dozens of such floats. (f) Aquatic plant leaves, supported by the water and not subjected to tearing by the wind, can grow to huge proportions. These leaves of water lily (*Victoria amazonica*) are 1 to 2 meters in diameter.

(b)

(c)

(d)

(e)

(f)

marguerite and the Christmas cactus, the petiole is reduced or absent, and the leaf blade connects directly to the stem; such a blade is called **sessile.** Plants are often identified by their leaf shapes, and the range is stunning—from long and narrow to rounded, smooth, lobed, toothed, fluted, finely divided, or needlelike. Figure 27-2 shows just a few examples. Some leaves, including those of the split-leaf philodendron, actually have holes that look as though they were chewed by insects or animals but develop naturally as the plant matures.

Leaves are an interesting—and successful—compromise between the need to collect sunlight and the need to prevent water loss from the plant. In most cases, the broader and flatter the blade, the more light will fall on it and the more radiant energy can be converted to chemical energy by means of photosynthesis.

However, a broad, flat shape has other consequences: much of the impinging sunlight is converted into heat, which causes moisture to evaporate from the leaf's delicate internal tissues; the broader and flatter the leaf, the more surface area there is for such heating and evaporation. A completely sealed, watertight coating on the leaf could prevent such evaporation—however, it could also prevent the exchange of carbon dioxide and oxygen, so critical to photosynthesis.

Leaf Structure

The leaf structure must somehow accommodate the need for a large, light-absorbing surface area, the need for waterproofing to prevent evaporation, and the need for gas exchange through the watertight seal. And, of course, there must be a transport system to carry water and nutrients between the leaves and the rest of the plant. The leaf cells and tissues that evolved to carry out these disparate tasks are as good an example as we shall see of the interplay of form and function in biology. Figure 27-3 illustrates the structures of a typical leaf in cross section. We shall examine each of the structures—starting from the outermost leaf layer, the epidermis, and working inward—to see how the "compromises" have occurred.

Epidermis: Outer Protection

Leaves are protected on both their upper and lower surfaces by the **epidermis,** a layer of cells, usually one cell thick, that cuts down on moisture loss and helps prevent invasion by microorganisms (Figure 27-4). Thus the interior of the leaf may be sterile, just as are the internal body tissues of an animal. In some leaves, the epidermal cells develop as shiny hairs that collectively act as a kind of mirror to reflect light from the leaf surface, thereby reducing heat absorption and subsequent evaporation. Guard hairs may also make the leaf a difficult place for insects to land or to walk (Figure 27-5). The familiar African violet and the deep purple velvet plant have leaves with immense numbers of guard hairs, which make the leaves feel velvety to the human touch. In most leaves, however, the epidermis reduces moisture loss in a different way: epidermal cells secrete a transparent waxy material called **cutin,** which forms an

Figure 27-3   ANATOMY OF A LEAF.

(a) This scanning electron micrograph shows an apple leaf in cross section, magnified about 400 times. (b) The various tissue layers and some of the cell types that make up the leaf. The upper epidermis, with its overlying cuticle, seals and guards the upper leaf surface. Palisade parenchyma cells contain many chloroplasts (not shown) and are the major site of photosynthesis. Air spaces permeate the lower spongy parenchyma. A very small vein is visible in cross section (left side of leaf). The lower epidermis and its cuticle seal off and protect the underside of the leaf. The guard cells on either side of each stoma (not shown in the micrograph) control the movement of gases into and out of the plant.

(a)

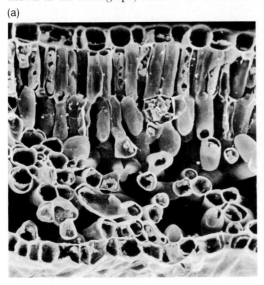

(b)

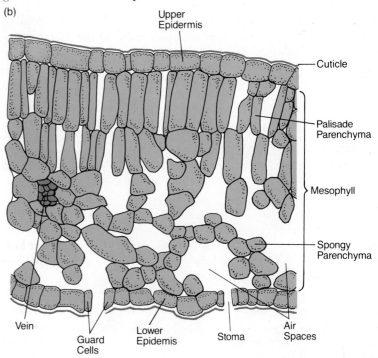

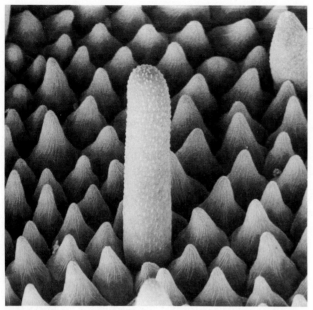

Figure 27-4   THE LEAF SURFACE MAGNIFIED: PRICKLY PROTECTION.

The epidermis of leaves can have complex architecture. This photograph of a lobelia (*Zygadenus venenosus*) leaf (magnified about 400 times) illustrates how the epidermis surface tissue, which serves primarily as a seal against water loss, at the same time can protect the rest of the leaf. In this case, the bumpy leaf surface makes a poor landing strip for juice-sucking insects.

Figure 27-5   GUARD HAIRS.

These incredibly shaped guard hairs are found on the lower surface of a live oak (*Quercus agrifolia*) leaf (magnified about 125 times). Their true diminutive size can be appreciated by noting the sizes of the epidermis cells of the leaf's lower surface.

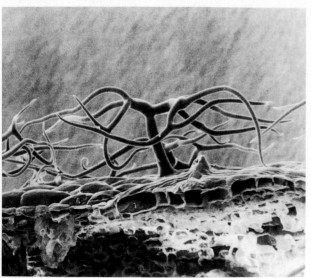

outer layer, or **cuticle,** over the leaf. This waxy layer, and not the epidermal cells themselves, seals moisture in the leaf tissues. In arid regions, the cuticle can be quite thick and can give the entire plant a whitish, grayish, or bluish hue. The blue spruce and several species of juniper, cypress, and eucalyptus have such hues because of their thick cuticles.

Whether a cuticle is thick or thin, it is essentially transparent to incoming light. The epidermal cells below the cuticle are semitransparent, allowing light (including wavelengths appropriate for photosynthesis) to pass through to the inner cellular layers. Most epidermal cells lack chlorophyll; thus they do not carry out photosynthesis, but serve protective and structural functions exclusively.

The epidermis and the cuticle are so effective in reducing water-vapor loss that only about 10 percent of the water that evaporates from a typical plant passes across this outer surface. This effectiveness, however, could have disastrous consequences—preventing the normal exchange of oxygen and carbon dioxide—if the epidermis were not dotted with thousands of tiny pores through which gases, including water vapor, could diffuse passively. These pores are called *stomata*, and each individual pore, or **stoma** (the Greek word for "mouth"), is bounded by a pair of **guard cells,** which regulate the opening and closing of the pore and thus affect the movement of gases into and out of the plant (Figure 27-6). Guard cells usually are kidney-shaped in dicots and are dumbbell-shaped in monocots. They do contain chloroplasts, making them the only photosynthetic epidermal cells. Guard cells change shape by means of a mechanism that we will discuss in Chapter 29; it is their shape change that opens or closes the stomata and thus regulates the passage of air and water vapor in the aboveground parts of the plant.

Stomata are extremely small—only about 100 micrometers long—but they are also extremely numerous. With about 19,000 stomata per square centimeter, a rose leaf might have nearly 250,000 stomata on its lower surface and about 60,000 on its upper surface. These—multiplied by all the stomata on all the leaves of the plant—add up to the major site of evaporation: fully 90 percent of a plant's water loss occurs through the stomata. For every carbon dioxide molecule that enters a leaf through an open stoma, several hundred water molecules exit from that same opening. We will see in Chapter 29 how carbon dioxide gain and water loss are balanced and regulated. But it is clear that the form of the epidermis, with its waterproof cuticle and active pores, is closely tied to the efficient functioning of the leaf.

Mesophyll: Photosynthetic Cell Layers

Moving inward from the epidermis, we encounter the leaf's main photosynthetic tissue, the **mesophyll** (mean-

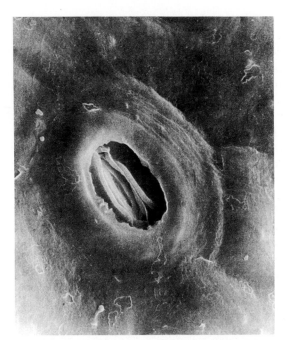

Figure 27-6   REGULATING GAS EXCHANGE.
Stomata regulate the movement of oxygen, carbon dioxide, water vapor, and other gases into and out of the leaf. In this scanning electron micrograph (magnified about 4,100 times), the guard cells of this stoma appear to be closed, thus reducing gas movement in and out of this flowering plant leaf.

ing "middle leaf"; Figure 27-3). Mesophyll is bright green because the cells have many chloroplasts, which contain the green photosynthetic pigments, the chlorophylls. Just as blood looks uniformly red but is really a suspension of red blood cells in yellowish serum, mesophyll cells look uniformly green but actually have clear cytoplasm in which the bright green chloroplasts are suspended. The leaf mesophyll tissue is made up of **parenchyma cells,** which are thin-walled and have large vacuoles. In monocots, the parenchyma cells of the mesophyll are rather similar to one another in structure. In dicots, however, there are two types of parenchyma cells, the **palisade parenchyma** and the **spongy parenchyma,** arranged in two layers (Figure 27-3).

Palisade parenchyma cells are rod-shaped and are arranged vertically in a layer just below the translucent upper epidermis; chloroplasts are abundant in these cells and are in an ideal location to absorb light energy. The palisade layer can be one cell or two cells deep. These cells are tightly packed, with little space around them for the exchange of gases or the circulation of air and water vapor. In contrast, spongy parenchyma cells are irregularly shaped and are loosely arranged in a layer directly below the palisade parenchyma. The ample space around them allows for more contact between their surfaces and gases.

Much of the air entering through the stomata can permeate the spongy parenchyma layer through an air space just above each stoma and through smaller channels that branch into the layer like numerous twisting passageways off a main cave (Figure 27-7). Thus carbon dioxide, oxygen, and other gases can pass quickly into and out of the spongy parenchyma cells by diffusing across the moist cell walls and plasma membranes. The total cell surface area for the exchange of gases is enormous in the loosely packed spongy parenchyma layer; in this sense, it is analogous to the inner surface of a mammal's lung, with its convolutions and minute outpockets that add up to a huge surface area. In the mammalian lung, however, oxygen is taken in and carbon dioxide is given off as a waste product, whereas in leaves, carbon dioxide—the source of carbon for photosynthesis—is taken in and oxygen is given off.

Veins: Pipelines for Support and Transport

If you hold a thin leaf up to a strong light, you can see a pattern of **veins** inside the leaf blade somewhat reminiscent of the ribs in a kite or a fan. The veins in leaves

Figure 27-7   THE KEY TO CARBON DIOXIDE AND OXYGEN DISTRIBUTION WITHIN LEAVES.
The large air spaces of the spongy parenchyma allow gases to diffuse close to the leaf's parenchyma cells. It is across the surfaces of these cells that most of the plant's crucial water, carbon dioxide, and oxygen are exchanged. The loose cellular scaffolding of spongy parenchyma cells seen here (magnified about 435 times) provides a huge surface area for this crucial exchange.

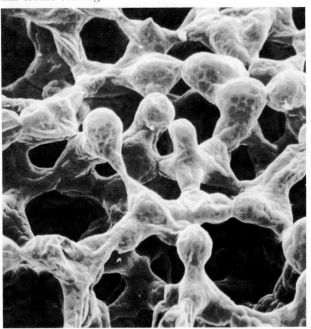

(a)

(b)

Figure 27-8   VEIN PATTERNS IN DICOT AND MONOCOT LEAVES.

(a) A large vein called a midrib usually runs down the center of a dicot leaf, as in the magnolia leaf shown here. Smaller veins branch out in a network from the midrib. These veins are subdivided into smaller and smaller passageways, carrying materials within easy diffusion range of every leaf cell. (b) In a monocot such as this corn leaf, the veins run in characteristic parallel lines from the petiole to the leaf tip.

act not only as a structural framework that helps maintain the shape and stiffness of the blade, but also as a transport system that is continuous with the vascular system in the rest of the plant. The rigid-walled *xylem* cells carry water and minerals and strengthen the veins, while the sievelike *phloem* cells transport sugars and other products of photosynthesis. We will return to these cell types when we discuss plant stems.

Most leaf veins are extremely small in diameter, just as most blood vessels in the human body are very tiny. The veins that can be seen clearly in a leaf are *major veins* formed by the fusion of dozens of small veins into transport pathways like bundles of conduits. In dicots, smaller and smaller veins branch off the major veins in a netlike array that permeates nearly every region of the leaf. The largest vein in a dicot leaf, usually located at the center of the blade, is called the **midrib** (Figure 27-8a). The midrib passes through the petiole, which attaches the leaf to the stem, and joins the vascular systems of leaf and stem. Smaller veins may branch off the midrib or may branch directly outward from the point where the petiole attaches to the stem, without a midrib present. In many monocots, such as iris or corn, veins extend in parallel lines from the petiole to the leaf tip, and midribs are not so apparent (Figure 27-8b). Field guidebooks and botanical keys often include the pattern of leaf venation to help identify plants. Regardless of the pattern, however, all veins function as conduits and help support the leaf blade.

Leaf Adaptations

Consider for a moment the brilliantly colored leaves of a coleus, a rex begonia, or a caladium; the huge leaves of

an elephant's-ear philodendron; and the tiny, fragile leaves of baby tears. It is obvious that leaves come in an artful palette of colors and a range of shapes and sizes. It is less obvious that these myriad forms generally represent adaptations to the demands of a particular environment or reflect the season in which the plant grows. Showy house plants often come from tropical rain forests, where, some botanists believe, their richly colored leaves may help attract pollinators. Extremely large leaves are also more common in jungle regions, where moisture is plentiful and evaporation is not such a problem but where competition among plants for the dim light filtering through the forest canopy favors those with a large leaf surface area.

Where light levels are high and moisture levels are low, leaf adaptations are quite different from those of rain-forest plants. For example, the leaves of many gymnosperms, such as pine needles, are long and thin and have a thick epidermis and cuticle. The needle form not only reduces water loss, but also prevents wind damage; winds can tear a large leaf, but tough, finely divided needle leaves allow the force of the wind to pass between the needles.

Some plants, such as the woodland violet, must contend with two distinctly different environments during the same growing season. When small, fragrant violets first emerge on the forest floor in early spring, only the evergreen conifers forming the forest canopy above them have leaves, and these are slim needles, not broad sun blockers. The majority of the woodland trees are still dormant, and their leaves have not yet emerged. Thus at ground level, where the violets grow, there is intense springtime sunlight, and the first leaves of the violet plant are small. These so-called sun leaves are well adapted to high light intensity and gather light energy efficiently. As the months of spring pass, leaves emerge on oaks, maples, and other broad-leaved trees, shading the ground. Because the violet's sun leaves are not well adapted to capturing the sunlight that filters down through the denser leaf canopy of summer, the plant grows a set of shade leaves; these are larger than sun leaves, with more surface area and a darker green color due to more chlorophyll. The shade leaves can absorb and use more of the dim light that reaches the plant during the summer, and the sun leaves wither and die.

Many plants can modify their leaves to suit changing environmental conditions. While violets do so on a seasonal schedule, some plants produce special kinds of leaves in response to physical stress. For example, if you move a house plant from a shady room to a sunny one, the plant may suddenly drop all its leaves and gradually sprout a whole new set. When newly exposed to the high-intensity light, the plant adjusts in part by producing smaller leaves, presumably better suited to its new environment.

Some leaves are so modified to reduce water loss that they are no longer even recognizable as leaves. For example, the water-storing leaves of the ice plant are so thick that they appear to be stems, and many familiar vegetables—such as cabbage, Brussels sprouts, and onions—consist of thick starch-storing leaves tightly compressed into "heads." In celery, on the other hand, the blades are reduced, but the petiole is greatly enlarged—so much so that we use *it* as food and usually throw away the small leaf blades.

Clearly, the architecture of a typical leaf is well suited to the major functions of that plant organ: absorbing sunlight, minimizing water loss, maximizing gas exchange, and transporting water and nutrients. And the specific shapes, sizes, and colors of leaves help to adapt a plant to its particular environment—whether dry or humid, cold or hot, dim or bright. Let us move on to the second major plant organ—the stem.

THE STEM: SUPPORT AND TRANSPORT

Just as the primitive fishlike amphibians that first crawled onto land had adaptations for internal support and for internal transport, so did many of the early land plants and, in turn, so do their magnificent descendants that grace our planet today. In vascular land plants, the two critical functions of support and transport are carried out by **stems.** And as with leaves, the particular types of stem cells and tissues make these functions possible.

We have seen that leaves are a plant's solar-energy collectors; to absorb sunlight and carry out photosynthesis most successfully, these collectors must be positioned so they can gather light without shading each other. It is the stem that supports the leaves, holding them aloft toward the light. And developmental processes ensure an arrangement of leaves on the stem such that they interfere as little as possible with each other's light-gathering function.

Besides supporting the weight of the leaves, the stem is largely composed of vascular tissue, which ultimately connects the leaves to the absorptive roots anchored in the soil. The stem's dual functions of support and transport are served primarily by the vascular tissue itself, which is not only specialized for moving materials back and forth, but is also reinforced to lend strength to the entire plant. In large plants such as trees, the majority of the stem is the mature, strengthened vascular tissue we call *wood*. We will examine the development and composition of wood in Chapter 28. Here we consider the stems of **herbaceous** dicots and monocots, which normally have short, thin, soft, and nonwoody stems.

Stem Structure

If you cut a cross section of the stem of a dicot, such as a tomato plant, you would see four concentric zones of tissue: epidermis, cortex, vascular tissue, and pith (Figure 27-9a).

Epidermis

Stem epidermis is similar to leaf epidermis; it is a layer of nonphotosynthetic cells, one cell thick; the cells secrete a waxy cuticle to prevent water loss. Stomata, like those in leaf epidermis, also dot the epidermis of herbaceous stems and regulate the diffusion of carbon dioxide and oxygen to and from underlying photosynthetic cells. Stem epidermis is always present in young plants, but older plants often lack this outer layer.

Cortex

Just inside the epidermis lies the second stem zone, the **cortex.** Most of the cortex cells are parenchyma cells—large, thin-walled, regularly shaped structures similar to those in leaf mesophyll. Stem parenchyma cells are usually considered to be unspecialized, but they are quite versatile and can differentiate into a variety of cell types depending on environmental conditions.

Among the cell types derived from parenchyma are **collenchyma** and **sclerenchyma.** The term "collenchyma" comes from the Greek word *kolla*, meaning "glue," referring to the glistening appearance of the cell walls. Unlike undifferentiated parenchyma cells, collenchyma cells have particularly strong primary walls (Chapter 5), thickened at the corners, which lend extra support to the stem. The cells are usually found in sheets

(a)

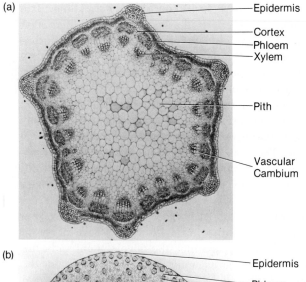

- Epidermis
- Cortex
- Phloem
- Xylem
- Pith
- Vascular Cambium

(b)

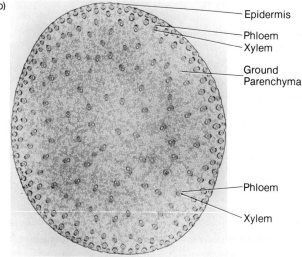

- Epidermis
- Phloem
- Xylem
- Ground Parenchyma
- Phloem
- Xylem

Figure 27-9   VASCULAR BUNDLES IN DICOTS AND MONOCOTS.
(a) The cross section of a dicot (magnified about 20 times) shows the epidermis, the adjacent cortex, and the vascular bundles arrayed around the central pith. Each vascular bundle has a central growth zone (the vascular cambium) with xylem toward the center of the stem and phloem toward the periphery. In the monocot stem (magnified about 6 times) shown in (b), the epidermis surrounds the ground parenchyma. The vascular bundles are scattered throughout this parenchyma, but within each bundle, phloem is external to xylem, as in dicots.

or vertical bands just inside the epidermis. In celery, this cell type makes up the strings of the stalk and contributes to the pleasant crunch you hear when taking a bite. The term "sclerenchyma" comes from the Greek word *skleros*, meaning "hard," and these cells are believed to enable plant organs to withstand various strains, such as gusts of wind, and to strengthen the plant so that as it grows, it can better support the ever-increasing weight of the leaves and lengthening stem.

(a)

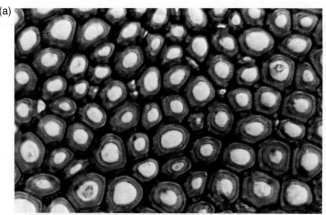

(b)

Figure 27-10   SCLERENCHYMA: A SPECIALIZED CELL TYPE IN THE CORTEX DERIVED FROM PARENCHYMA.
(a) A cross section of fibers from a corn plant (magnified about 375 times). Notice the thick walls of these sclerenchyma cells. Many fibers are actually dead when mature, but their walls continue to give the plant rigid support. (b) A second type of sclerenchyma cell is the sclereid or "stone cell." These sclereids (magnified about 350 times) are from a pear pit; sclereids are also found in nutshells and tree bark.

There are two types of sclerenchyma cells: **fibers** and **sclereids** (Figure 27-10). Both have strong secondary walls composed of cellulose that is deposited as the cell matures. Fibers are extremely long cells and tend to occur in bundles or in cylinders, often toward the periphery of the stem or root. We rely heavily on plant fibers to make baskets, rope, linen fabric, and many other woven goods; the tough cell walls of dead fibers are the reason hemp, sisal, jute, and flax make such good raw materials for weaving. Sclereids—often called "stone cells"—have strong secondary walls, sometimes impregnated with lignin, a noncarbohydrate polymer of ring-structured carbon compounds called phenols. Sclereids occur either in groups throughout the stem cortex or as single cells; individual sclereids give pears their gritty texture. Very hard plant tissues, such as nutshells and peach pits, are made up of thick layers of sclereids.

Vascular Tissue

Moving farther inward, the stem zone lying just interior to the cortex is the **vascular tissue,** made up of two types of tissue: the *xylem* and the *phloem.* Support for the entire plant and conduction throughout it depend on these two tissues. In an alfalfa plant—a herbaceous dicot—xylem and phloem occur together in distinctive **vascular bundles,** which form a single ring just inside the cortex (Figure 27-9a). A layer of cells called the **procambium** separates the xylem from the phloem in each bundle of young stems; in older stems, this layer is called the **vascular cambium.** Notice that within each vascular bundle, the phloem, or nutrient-conducting tissue, is usually exterior to the xylem, or water-conducting tissue.

Pith

The fourth and final zone of stem cells, the **pith,** occurs in the center of a dicot stem (Figure 27-9a). Pith is a storage tissue composed of large, thin-walled parenchyma cells. It provides much of the diameter of a young green stem, but as the stem ages, it plays a progressively less important role.

Monocots have the same four stem structures as in herbaceous dicots, but with some modifications (Figure 27-9b). The epidermal tissue is quite similar in monocots and dicots, but the vascular bundles in most monocot stems are scattered throughout a tissue called the **ground parenchyma,** rather than arranged in a ring. Consequently, there are no discrete zones that can be designated as cortex and pith. Within each vascular bundle, the phloem is exterior to the xylem, as in dicots.

Stem Vascular Tissue

As we have seen, for the stem to support the leaves and conduct water and nutrients between leaves and roots, it must have functioning xylem and phloem. These tissues are obviously critical to the life of every vascular plant, so let us look at them more closely.

Xylem

Xylem is a conducting tissue made up of cells stacked end to end like sections of pipe; these cells, although dead when functional, transport water and minerals, usually from the roots to the stem, leaves, and fruit. The direction of water flow is usually upward, from roots to leaves, but in plants with drooping or trailing branches, such as willows and some types of begonias and fuchsias, the water may actually be pulled downward or sideways.

In angiosperms, there are two types of xylem cells, tracheids and vessel elements (Figure 27-11). **Tracheids** are long, tapering cells that overlap, forming thin tubes within the vascular bundles. The walls of the tapered ends are incomplete, sievelike, and perforated by pits (which typically have membranes). Water is pulled through these pits as it moves through the xylem. (We will see in Chapter 29 how evaporation from the leaves plus cohesion between water molecules "pull" water upward.) **Vessel elements** are much shorter and broader than tracheids and have either blunt end walls with large perforations or no end walls at all. A string of vessel elements, stacked open end to open end, functions like an open pipeline and so allows a freer flow of water than do tracheids. Most angiosperms have both tracheids and vessel elements, while the xylem tissue of other types of vascular land plants (whisk ferns, horsetails, ferns, and gymnosperms) contains only tracheids.

Immature xylem cells look very similar to each other, and like other plant cells, they are completely surrounded and reinforced by a thick cell wall composed of cellulose, lignins, and other polymers. As an angiosperm grows and matures, the cells of the vessels die, and the end walls of each vessel element are digested away completely, creating the pipeline channel for water movement. The mature vessel resembles a narrow pipe or a straw; it is only one cell diameter wide but thousands of cells long. During development of tracheids, the tapered end walls of the dying cells are incompletely digested, leaving porous sieves. When functioning in water transport, both vessels and tracheids are dead; they no longer contain cytoplasm. But the rigid, lignin-containing cell walls remain and continue to conduct water effectively in the vascular system. In Chapter 29, we will describe in more detail the mechanisms of water transport in the xylem.

Phloem

Phloem is a tissue specialized for carbohydrate transport. It has a more complex structure than does xylem (Figure 27-12). In addition, the transport cells of phloem must be alive in order to function. In angiosperms, each column of phloem is composed of **sieve tube elements** stacked vertically end to end to form the **sieve tube.** The end walls, or **sieve plates,** of each sieve tube element are perforated by many pores. At maturity, the sieve tube elements lose their nuclei but retain some living cytoplasm. These strands of cytoplasm line the inner cell surface and pass through the sieve plate connecting the adjoining sieve tube elements. Each mature sieve tube element is associated with a **companion cell,** a nucleated cell derived from the same embryonic precursor cell as the sieve tube element. The companion cell is believed to serve as the supplier of macromolecules—proteins, enzymes, and so on—required for the long-term survival of the cytoplasmic strands.

(a)

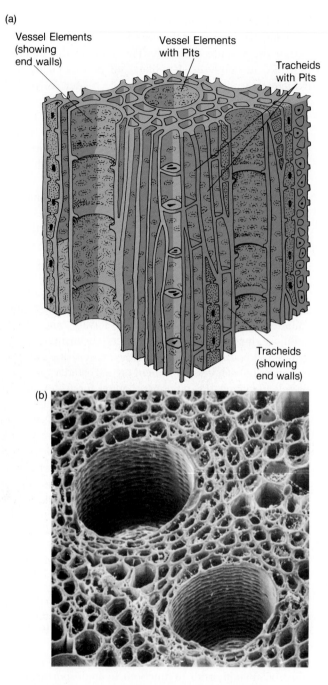

The phloem tubes occur within the same vascular bundles as the xylem conduits, but they transport sugars and other organic materials away from the sites of manufacture in photosynthetic tissue and toward the parts of the plant (roots, woody stems) that do not carry out photosynthesis. In spring, a reverse process can occur in some plants: sugars stored in stems or roots during the winter begin to be transported upward to help fuel the rapid growth of leaves and new stems. The phrase "the sap is rising" refers to this phenomenon, and when it occurs in sugar maples, people sometimes tap the phloem tissue and harvest the rising "sap" to make a rich, sweet syrup. We will return to the mechanism of carbohydrate transport in the phloem in Chapter 29. For our purposes here, the cell types in phloem and in xylem are yet another reminder of the interplay of form and function in plants: the structures of these cells make possible both the conducting of materials through the plant and the strengthening of the stem.

Figure 27-12   THE ORGANIZATION OF VASCULAR TISSUE: PHLOEM.

Phloem is composed primarily of sieve tube elements, with their perforated end walls (sieve plates). The closely associated living companion cells are visible, with their cytoplasm and nuclei.

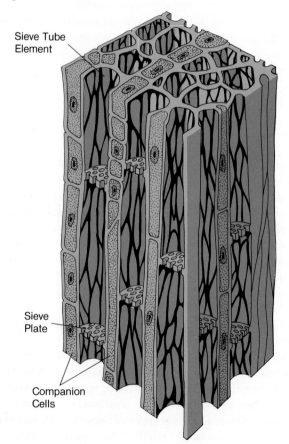

Figure 27-11   THE ORGANIZATION OF VASCULAR TISSUE: XYLEM.

(a) In xylem tissue, there are two major cell types, tracheids and vessel elements. In a woody plant both cell types lose their cytoplasm and remain as dead empty cell walls. Both are arranged in columns end to end, but the vessel elements lose their end walls, while the tracheids retain angled end walls that are thin and perforated. Together the columns of both cell types form the plant's water-conducting xylem tissue. (b) This scanning electron micrograph (magnified about 320 times) shows two vessel elements extending vertically (large holes); perforations can be seen in their side walls and end walls. The smaller cylinders on each side of the vessel elements are tracheids.

Stem Adaptations

While transport and support are the major tasks of stems, these plant structures can also perform a variety of other tasks. In many plants, they serve as storage organs or protect the plant from predators. For example, the stems of sugar cane, rhubarb, and broccoli store carbohydrates produced in the leaves. The potato, which people often think of as a root, is actually an enlarged underground stem modified for starch storage. And bulbs, too, are modified stems: onion and garlic are very short stems with tightly packed, fleshy leaves that store the products of photosynthesis.

Stems that bear thorns (which are actually modified branches) discourage predation by browsing animals. You have probably encountered this common mechanism firsthand while picking roses or blackberries. Many cactus species grow spines (modified leaves) and hairs that help reflect the intense desert sun as well as reduce predation (Figure 27-13). And the thick, photosynthetic, water-storing stems of cactuses and succulents are yet another set of stem adaptations. Interestingly, desert-browsing animals such as camels have coevolved with African and Asian thorny plants and have acquired extremely tough lips and tongues that permit them to graze despite the spines.

THE ROOT: ANCHORAGE AND UPTAKE

In addition to the highly visible leaves and stems, there is another part of the plant—usually hidden from view—that can be many times larger than all the parts seen aboveground: the root system. A tree usually has roots spreading to a radius far larger than its crown, and an alfalfa plant less than 2 meters tall can have roots at least 6 meters deep. **Roots** have two key functions in the overall life of the plant: (1) they absorb water, minerals, and oxygen from the soil; and (2) they anchor the plant firmly in the soil or, if the plant is a vine, to the vertical surface it may be climbing. The vascular system that serves the stem and leaves is continuous in the roots; all the "pipes" are therefore in place for transporting water and mineral nutrients from the soil to the rest of the plant. The strength of the vascular tissue augments the anchoring function, helping the plant to withstand wind, beating rain, and the determined gardener trying to weed the vegetable patch. In addition to absorbing and anchoring, roots may store energy reserves for the plant in the form of starch.

Figure 27-13   CACTUS SPINES: MANY FORMS, TWO ROLES.

(a) These long, sharp spines are sufficiently close together to protect the *Opuntia* species cactus from predation. (b) The incredibly dense, overlapping spines in this *Opuntia echinocarpa* make it nearly impossible to see the underlying tissue; the whitish color of the spines reflects sunlight and helps keep down the plant's internal temperature.

Root Structure

Several root structures make all the activities of the root system possible. Concentric zones of root tissue—epidermis, cortex, endodermis, pericycle, and stele—are similar to those in the stem (Figure 27-14). There are, however, a few important modifications that can help us understand how roots function.

Epidermis

Roots have an outermost zone of cells, the epidermis, that is one cell thick. Unlike the epidermis of leaves and

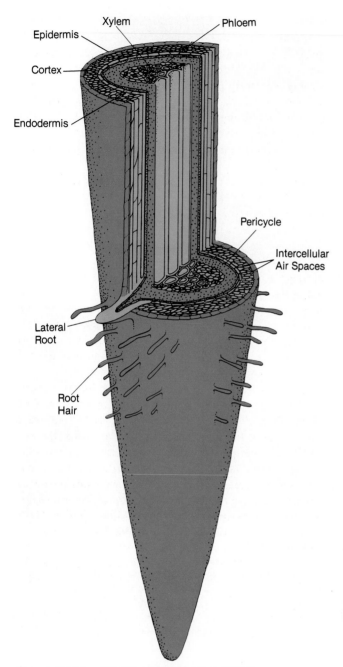

Figure 27-14 MAJOR TISSUES OF A ROOT.
The various cell layers and root hairs are evident in this cross section of a root. Water and ions entering root hairs must ultimately pass through the endodermis cell cytoplasm before entering the pericycle and thus the stele. (The stele includes the pericycle, xylem, and phloem.) The pericycle gives rise to lateral roots, one of which is shown here just starting to form.

stems, the root epidermis does not usually excrete cutin to produce a cuticle (a very thin one may occasionally be detected). The reason is simple: the root is an organ of water absorption—the problem is not preventing water evaporation, but achieving efficient uptake. Most of the

actual absorption takes place through fine, hairlike extensions of individual epidermal cells called **root hairs** (Figure 27-15). Each epidermal cell near the growing root tip produces only one such extension, but each hair may be up to 1 centimeter long, there are so many cells that, collectively, root hairs increase the absorptive surface area of the root tip a hundredfold. One botanist estimated that the number of root hairs on the root system of a four-month-old rye plant exceeded 14 billion and that the surface area equaled 401 square meters—the floor area of a large house. End to end, these root hairs would stretch some 10,628 kilometers! The surface area of the root hairs is quite large compared with the surface area of the same rye plant's leaves (4.6 square meters).

Because the most efficient absorption takes place through root hairs slightly behind the root tip, roots continuously grow into the soil, sending out new root hairs from the newly formed epidermal cells in the probing root tip. Older hairs soon collapse and slough away. The concentrations of sugars and minerals dissolved in the cytoplasm of epidermal cells and root hairs are usually much higher than the solute concentrations in the soil. For this reason, water moves by osmosis across the plasma membrane of root hairs and into the epidermal cells.

Figure 27-15 ABSORPTIVE SURFACE AREA: ROOT HAIRS.
Root hairs are extensions of single root epidermal cells. Their enormous number on this radish root creates a huge surface area for absorbing water and minerals. From *Living Images* by Gene Shih and Richard G. Kessel, Science Books International, 1982. Reprinted by permission of the present publisher, Jones and Bartlett Publishers.

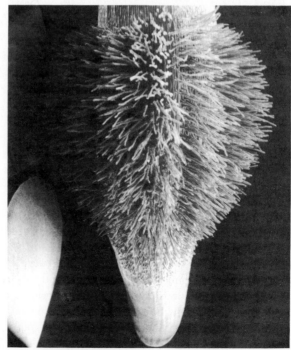

mines how successfully the root system, and hence the plant, will grow. Because root epidermal tissue lacks a cuticle, oxygen from the soil can diffuse easily into the internal tissues of the cortex. Root cortex cells, in turn, are loosely packed, allowing oxygen to move freely toward the stele. If soils are deficient in air and oxygen, the root tissue can become anaerobic and eventually die. This is particularly a problem in lowland or swampy areas with waterlogged soil or in zones with clay or other tightly compacted soils. Farmers and gardeners purposefully cultivate soil to aerate it; earthworms boring through the soil (Chapter 25) do an even better job of natural cultivation.

Types of Roots

Many kinds of animals caught in a storm can seek shelter, but plants must remain rooted to the spot and withstand the buffeting of wind and rain or snow. Thus the anchorage provided by roots is crucial to plant survival, and two patterns of root growth have evolved to facilitate this function (in addition to the task of absorption): most gymnosperms and dicots have a taproot system, while most ferns and monocots have a fibrous root system (Figure 27-18).

Plants with a **taproot system** have one main root extending 1 meter or more underground. Taproots allow the plant to withstand adverse weather as well as to gather water from far below the surface. In addition to the large taproot itself, such plants usually have numerous lateral roots, or smaller branches off the main axis that allow the plant to absorb water and nutrients from a large soil area. The taproot is usually derived from the embryonic root, the **radicle,** which bursts out of the seed. A good example of a plant with a taproot system is the dandelion, which is anchored firmly in the ground by a taproot that is often thicker than the stem itself. As any gardener knows from experience, it does no good to simply cut off a dandelion at ground level; the entire taproot must be exhumed, or the root will quickly sprout a new shoot. Besides anchoring the plant firmly and absorbing water and minerals, taproots store large quantities of starch to support the continuing growth of the shoot during bad weather conditions. Carrots, turnips, and ginseng—a popular Asian import used for tea and herbal medicine—are taproots used by people.

Plants with a **fibrous root system** have many equal-sized roots, arising at the junction of stem and root and fanning out through the soil. The primary root that emerges from the seed dies, and new roots sprout from the base of the stem. These new roots are referred to as **adventitious,** meaning that they arise from an unusual place on the plant—in this case, from organs other than the root. As they penetrate the ground, fibrous roots produce lateral roots and form a dense mat that anchors and supports the plant. Because such roots spread in all directions, plants with a fibrous system are difficult to remove from the soil. Grasses have fibrous root systems and thus make excellent ground-cover plants: they prevent soil erosion by literally tying the soil together with thousands of tiny, tough, threadlike roots.

Figure 27-18 MAJOR PATTERNS OF ROOT GROWTH.

(a) The sand sage (*Artemesia* species) taproot system and (b) the meadow fescue's (*Festuca etatior*) fibrous root system demonstrate that though the former has a very strong central root, both provide great surface area for absorbing water and minerals.

Fibrous root systems can provide a tremendous surface area for absorbing water and minerals. The total length of the fibrous roots of the four-month-old rye grass mentioned earlier was more than 600 kilometers—the distance from Los Angeles to San Francisco—even though the stems averaged only 38 centimeters in height. The surface area of the root system of this plant was 639 square meters—130 times greater than the surface area of the shoot and leaves. Thus the surface area of the root system was about the same as that of a tennis court, while the shoot and leaves provided a surface area less than that of a Ping Pong table. This impressive absorptive area is needed to gather enough moisture to compensate for the steady water loss that occurs through actively photosynthesizing leaves.

Root Adaptations

Plants that live in unusual places often have highly specialized roots. In swampy habitats, for example, trees may develop dozens of **prop roots:** roots that sprout downward from the stem and exist half in the air and half in the waterlogged soil (Figure 27-19). Besides ensuring a sufficient oxygen supply, prop roots provide a particularly firm anchorage despite the soft substratum. Prop roots are evident among the mangrove trees and neighboring species in the Florida Everglades. These trees can withstand hurricane-force winds that would readily knock down trees growing on firm, dry land.

Plants that live in the air also have special root adaptations. "Air plants," or **epiphytes,** usually grow on other

Figure 27-19   PROP ROOTS: ADDING STABILITY AND PROVIDING A BETTER OXYGEN SUPPLY.
This group of red mangrove (*Rhizophora mangle*) trees in shallow water near a Florida key shows the large numbers of prop roots on each plant, and the fact that some prop roots arise from far up on the main trunk.

plants rather than in the soil. An orchid growing on the trunk of a jungle tree can absorb water vapor from the air through its exposed roots. It is noteworthy that roots of some epiphytes grow upward, away from gravity, along the trunks of jungle trees. In so doing they are positioned better to absorb mineral-rich water flowing down the host tree's bark from the rich canopy of leaves far overhead. Vines have also developed specialized roots: ivy stems sprout small roots with "suckers" at each root tip that help the plant cling tenaciously to a wall or a tree trunk.

Finally, we have seen that carbohydrate storage is another specialized root function. Sweet potatoes, beets, carrots, radishes, turnips, and ginseng are roots that fuel the energy needs of these plants and inadvertently provide nutrition for animals, including humans.

LOOKING AHEAD

By surveying the basic elements of plant architecture—the leaves, stems, and roots and the tissues of which they are composed—one can appreciate the close interplay of form and function in the plant world. We have seen that the architecture of leaves strikes a balance among collecting sunlight, exchanging gases, and preventing excessive moisture loss from the plant. The combination of a waterproof cuticle, stomata, and photosynthetic mesophyll tissue, as well as the shapes, sizes, and colors of leaves, achieves this compromise. Stems have zones of tissue, including strengthened parenchyma cells and equally strong transport tissues—the xylem and phloem—that enable this major plant organ to support the weight of leaves, withstand harsh weather, and act as a conduit for the passage of sugars, water, minerals, and other materials. Finally, there are the roots with their various zones of tissue that can absorb and transport water and minerals, store nutrients, and simultaneously keep the plant well anchored to its substratum. Table 27-1 summarizes the major cell types in these plant organs.

It is clear from an evolutionary perspective that such structures have enabled plants to survive the myriad challenges of life on land. And it is equally clear that a plant's day-to-day existence depends on the smoothly integrated functioning of all these parts. In the next three chapters, we shall build on our foundation of plant anatomy to explore how plants develop and grow, how they absorb and transport water and nutrients, and how regulatory molecules coordinate the overall growth, maturation, and survival of plants.

Table 27-1   **MAJOR CELL TYPES OF FLOWERING PLANTS**

| | Cell Types | |
Name	Description	Function
Epidermis	Small living cells; secretes cuticle	Protection of internal cells; prevents water loss; preserves sterility
Parenchyma	Small living cells; usually thin cell walls	Photosynthesis and storage
Collenchyma	Elongated living cells; thick cell walls	Support
Sclerenchyma (fibers and sclereids)	Usually extremely elongated cells; thick, reinforced cell walls; usually dead at maturity	Protection and support
Tracheids and vessel elements	Hollow, thick walls of dead cells	Xylem conducting tubes
Sieve tube elements	Elongated living cells lacking nuclei; thick cell walls; sieve plates with pores connect successive cells to form sieve tube	Phloem conducting tubes
Companion cells	Small, elongated living cells with nuclei; adjacent to sieve tube elements	Involved in transport by sieve tube elements

SUMMARY

1. Vascular seed plants are able to thrive in most of the Earth's land areas because they have evolved with structures for collecting solar energy and converting it into chemical energy (leaves), for lifting and orienting leaves toward air and light (stem), and for anchoring the plant and taking up water and nutrients from the soil (roots). They have also developed a system of vascular conduits for carrying water, minerals, and nutrients throughout the plant.

2. The *leaf* is the plant's primary sunlight collector, and its efficiency depends on the size, shape, and position of the *blade*, and the attachment by the *petiole* to the stem. The outer layer of the leaf, or *epidermis*, protects the plant from moisture loss and invading microorganisms. It secretes a waterproofing material (*cutin*) in a layer, the *cuticle*, and contains specialized pores, or *stomata*, through which gases diffuse. *Guard cells* regulate the opening and closing of stomata.

3. Photosynthesis takes place in the *mesophyll*, a tissue made up of *parenchyma cells* and containing chloroplasts. In monocots, the parenchyma cells are somewhat uniform, but in dicots, they are of two types: *palisade parenchyma* and *spongy parenchyma*.

4. Leaf *veins* transport materials between the leaf and the rest of the plant and act as structural "girders" for the leaf. Vein patterns are branched in dicots (and include a *midrib*) and extend in parallel lines in many monocots.

5. Leaves come in a variety of forms and sizes, the result of adaptations to particular environments. Plants that live in changing environments may have more than one type of leaf.

6. The *stem* provides the plant with support and contains the plant's main vascular tissues. The stem of a *herbaceous*, or nonwoody, plant usually consists of four zones of tissue: the outer epidermis; a photosynthetic zone of *cortex* cells, including *collenchyma* and *sclerenchyma* (*fibers* and *sclereids*) for vertical and radial strength; *vascular tissue* containing xylem and phloem; and, in dicots, a central storage area of *pith*.

7. There are two types of *xylem* transport cells: *tracheids* (found in all vascular plants) and *vessel elements* (found only in angiosperms). When functional in water transport, both vessel and tracheid cells die, and their walls alone remain to carry out the transport function.

8. *Phloem* is composed of *sieve tube elements* stacked end to end to form the *sieve tube*; perforated *sieve plates* occur at the ends of the sieve tube elements. Each mature sieve tube element is associated with a *companion cell*, a living nucleated cell.

9. Xylem and phloem occur together in *vascular bundles*. These can be in rings or scattered throughout the *ground parenchyma*. A layer of cells called the *procambium* separates the xylem from the phloem in young stems; in older stems, this layer becomes the *vascular cambium*.

10. *Roots* anchor the plant, absorb water and minerals, and transport

the absorbed substances to the stem. Like stems, roots consist of several concentric layers: the epidermis, with its *root hairs;* the root cortex, consisting of nonphotosynthetic parenchyma cells and the *endodermis;* and the *stele,* or central cylinder of vascular tissue, which includes the *pericycle,* xylem, phloem, and (in monocots) pith. New *lateral roots* arise from the pericycle.

11. Water and minerals absorbed by root hairs may move through the cortex in adjoining intercellular spaces (the *apoplastic pathway*) or in interconnected cytoplasm (the *symplastic pathway*). Water and ions passing through the cortex must go through the endodermal cell cytoplasm because of the presence of the *suberin*-waterproofed *Casparian strip,* the impermeable lateral walls of endodermal cells. This permits cellular regulation of water and ion movement into the root's vascular system.

12. Root systems may be of two types: a deep *taproot system,* derived from the *radicle* (embryonic root), or a branched *fibrous root system* composed of *adventitious roots.* Roots can store large quantities of carbohydrates in addition to providing tremendous surface area for absorption of water and minerals. Special adaptations such as *prop roots* and the exposed clinging roots of *epiphytes* help plants survive in swampy or nonsoil habitats.

KEY TERMS

adventitious root	fibrous root system	pith	stele
apoplastic pathway	ground parenchyma	procambium	stem
blade	guard cell	prop root	stoma
Casparian strip	herbaceous	radicle	suberin
collenchyma	lateral root	root	symplastic pathway
companion cell	leaf	root hair	taproot system
cortex	mesophyll	sclereid	tracheid
cuticle	midrib	sclerenchyma	vascular bundle
cutin	palisade parenchyma	sessile blade	vascular cambium
endodermis	parenchyma cell	sieve plate	vascular tissue
epidermis	pericycle	sieve tube	vein
epiphyte	petiole	sieve tube element	vessel element
fiber	phloem	spongy parenchyma	xylem

QUESTIONS

1. What three basic vegetative (nonreproductive) structures make up the body of a vascular plant?

2. What is the major function of a leaf?

3. In what ways is a leaf a "compromise"?

4. What is the functional relationship between the leaf epidermis and the leaf cuticle?

5. In what ways may the cuticle of a leaf be modified to protect the leaf against (a) desiccation and (b) predation?

6. Sketch a stoma and guard cells. Does your sketch indicate the presence of chloroplasts? where? What is the function of the guard cells?

7. Sketch and label a cross section of a dicot stem and of a dicot root.

8. What tissue is usually at the center of a dicot root? at the center of a dicot stem? Where is the endodermis in such roots and what is its primary function?

9. Why do roots need a large surface area? What accounts for the enormous surface area of many roots?

10. Describe parenchyma tissue. Where is parenchyma tissue located in leaves? in stems? in roots? What is its function in each of these structures?

ESSAY QUESTIONS

1. Discuss some of the challenges to plants of a terrestrial (as opposed to an aquatic) existence and indicate how vascular plants evolved structures to meet these challenges.

2. Trace the path followed by a molecule of water from the soil through

the root, stem, and leaf of a vascular plant, until it enters the atmosphere. Name the various cells and tissues along the pathway. Can a water molecule pass from root to leaf without passing through cytoplasm?

SUGGESTED READINGS

CRONSHAW, J. *Support and Protection in Plants: Topics in the Study of Life*. New York: Harper & Row, 1971.

This is a straightforward and complete treatment of plant anatomy.

EAMES, A. J. *Morphology of the Angiosperms*. New York: McGraw-Hill, 1961.

A bit dated, but still the best general treatment of plant anatomy.

EPSTEIN, E. "Roots." *Scientific American*, May 1973, pp. 48–55.

A good treatment of mineral and water uptake and transport.

GUNNING, B. E. S., and M. W. STEER. *Ultrastructure and Biology of Plant Cells*. London: Arnold, 1975.

A spectacularly illustrated, complete account of plant cell types.

WILSON, C., W. LOOMIS, and T. STEEVES. *Botany*. New York: Holt, Rinehart and Winston, 1971.

The chapters on plant roots, stems, and leaves are excellent.

WOODING, F. P. B. *Phloem*. Oxford Biology Reader, no. 15. Burlington, N.C.: Carolina Biological Supply, 1977.

This is a well-illustrated, fine introduction to the complexities of phloem function.

28

HOW PLANTS REPRODUCE, DEVELOP, AND GROW

I took an earthenware pot, placed in it 200 lb of earth dried in an oven, soaked this with water, and planted in it a willow shoot weighing 5 lb. After five years had passed, the tree grown therefrom weighed 169 lb and about 3 oz. But the earthenware pot was constantly wet only with rain or (when necessary) distilled water; and it was ample in size and imbedded in the ground; and, to prevent dust flying around from mixing with the earth, the rim of the pot was kept covered with an iron plate coated with tin and pierced with many holes. I did not compute the weight of the deciduous leaves of the four autumns. Finally, I again dried the earth of the pot and it was found to be the same 200 lb minus about 2 oz. Therefore, 164 lb of wood, bark, and root had arisen from water alone.

Jean-Baptiste van Helmont, *Ortus Medicinae* (1648)

A living castaway: after months at sea, a coconut palm sprouts and takes root just above the tide line on a Virgin Islands beach.

For many of us, the very mention of different varieties of flowers and fruits evokes bright colors (delphinium, nasturtium, poinsettia, pansy), pungent fragrances (marigold, rose, gardenia, lily of the valley), and sweet flavors (pineapple, peach, raspberry, plum, honeysuckle). In addition to being beautiful and succulent, these structures are essential for plant reproduction. Many fruits and flowers, however, are simply small, nondescript reproductive organs: they are drab shades of green or brown, and they lack sweet odors or flavors. Some of them even taste bitter or sour or smell like rotting meat. And many kinds of plants reproduce quite successfully, at least part of the time, without employing sexual structures at all.

We know from Chapter 24 that angiosperms have flourished largely because of their flowers, fruits, and seeds. Yet many plants within this group have two distinct modes of reproduction: sexual and asexual. The asexual process, called *vegetative reproduction*, is a kind of cloning that can involve various parts of the plant and result in offspring that are genetically identical to the parent. The study of both modes of angiosperm reproduction and of subsequent growth and development of new individuals helps show in a clear and dramatic way how plants differ from animals.

One aspect of plant development is particularly distinctive: plants have perpetual embryonic centers that produce new organs throughout the life of the individual, whereas most animals form organs only in the embryonic stage of life. In this chapter, we shall see that this unique capacity for renewed growth and development throughout adult life is utilized during flowering and sexual reproduction as well as in modified form during vegetative reproduction. We shall focus on the structures responsible for both types of reproduction, including *flowers*, which can be alluring bait—living advertisements—for animal intermediaries in the sexual process; *fruits*, an angiosperm's insurance for seed protection and dispersal; *seeds*, a kind of living time capsule (for both angiosperms and gymnosperms) containing an embryo and stored food; and *meristems*, the zones of perpetual growth. Finally, we shall follow the sequence of events during the development and growth of a plant embryo into a seedling and then into a mature adult.

Our primary goal is to see how the entire range of plant parts and reproductive processes—from the drab to the glorious and from the asexual to the sexual—helps to characterize the angiosperms (and, in some respects, the gymnosperms) and to make them the highly successful groups they are.

VEGETATIVE REPRODUCTION: MULTIPLICATION THROUGH CLONING

High in the Appalachian Mountains of West Virginia, there is a region where blueberries grow wild in thickets of dense, low bushes. One thicket, in particular, is nearly 1 kilometer in diameter and has a startling feature: it is a single clone—all the bushes, which are connected by underground stems, were derived from the same initial plant and thus are genetically identical. Although the plants flower and produce blueberries containing seeds, the spread of new individuals into immediately adjoining areas is by **vegetative reproduction,** a process in which new plants, genetically identical to each other and to the parent, emerge from the parent's body. Like blueberries, many angiosperms can multiply asexually, or vegetatively; but in addition, all angiosperms are capable of sexual reproduction.

Recall from Chapter 23 that alternation of generations characterizes the plant kingdom and that in angiosperms, the *sporophyte,* the conspicuous, diploid, spore-producing plant—with its leaves, stems, and roots—alternates with the *gametophyte,* the gamete-producing generation. For most kinds of angiosperms in most situations, the sexual phase is retained in the life cycle since it ensures the spread of new, genetically unique individuals. Genetic recombination occurs during meiosis in the gametophyte and during fertilization, when genes of two individuals usually are united; each offspring thus has a unique genetic complement that may give it a slight edge in surviving.

For some types of angiosperms, however, there is a distinct advantage in multiplying asexually—going directly from one sporophyte generation to the next without alternation and relying on mitosis, rather than meiosis, as the basis for reproduction. If an individual plant grows successfully at a particular spot, it is, in a sense, adapted to that place on the Earth's surface. Hence, if the plant can reproduce vegetatively, it can spread rapidly into similar adjoining areas. The plant's single genotype will then be found throughout a large area and may be well suited to the environment at the time the plant spreads. There are some inherent disadvantages to this form of reproduction that we will discuss later. First let us see how plants reproduce vegetatively and how humans make use of vegetative techniques in agriculture.

Vegetative Reproduction in Nature

A wide array of plant parts, including modified stems, leaves, and roots, can give rise to new individuals through vegetative reproduction.

Rhizomes, stolons, runners, bulbs, corms, and tubers are all stem structures that can lead to plant reproduction. **Rhizomes** are subterranean stems that grow laterally from the main shoot. Periodically, a new set of roots and stems emerges from the rhizome, and then leaves develop: the result is a new plant that is a physical exten-

sion of (and thus genetically identical to) the parent. Blueberries (Figure 28-1a) spread by means of rhizomes, and these underground stems are also the reason that cattails can so quickly take over a farm pond or a marsh. One cattail can send out rhizomes that produce more than sixty new shoots in a single summer. Bamboo is an even more spectacular example of reproduction via rhizomes. Entire bamboo forests are often composed of a single individual with hundreds of thousands of offshoot plants. Some bamboo forests in Southeast Asia are reported to be more than 160 kilometers long and contain

(a)

(b)

Figure 28-1   VEGETATIVE REPRODUCTION.
(a) A blueberry (*Vaccinium uliginosum*) bush growing near Ontario, Canada, shows immature whitish berries and the mature, succulent, seed-filled fruit. (b) Lily of the valley (*Convallaria majalis*), growing in a New Jersey marsh, reproduces both vegetatively and sexually by these flowers. (c) Kudzu vines (*Pueraria lobata*) propagate vegetatively by runners and can grow nearly 30 centimeters (about 12 inches) a day in the summer, blanketing trees and fields with millions of leaves. (d) The daffodil (*Narcissus pseudonarcissus*) corms on the left are compressed, nutrient-filled parts of stems. The saffron crocus (*Crocus sativus*) bulbs on the right are modified leaves. (e) The *Kalanchoë daigremontiana* plant (also called mother of thousands) produces genetically identical plantlets in the margins of leaves. Each has tiny roots and can fall to the ground and quickly become established near the base of the parent plant.

(c)

(e)

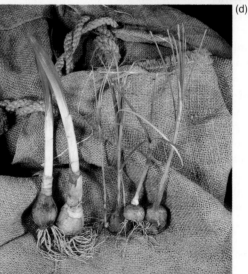
(d)

only the plants of a single clone, all connected by a vast network of rhizomes.

Stolons are branches of aerial stems that grow laterally, touch the ground, and put out roots and stems at those sites. Lily of the valley (Figure 28-1b) grows in dense clumps because of stolons, as does a troublesome weed, the water hyacinth; this species can choke a tropical waterway with its rapid vegetative reproduction based on stolons. Botanists estimate that in a year, 10 water hyacinths could reproduce into more than 600,000 identical plants!

Runners are horizontal stems that grow out of the base of the plant and run along the ground. When certain parts touch down on a bit of unoccupied soil, they sprout a new shoot with leaves and a new root system. Strawberries commonly reproduce through runners, as does a large-leaved, fast-growing vine called kudzu. Kudzu was imported into the United States from its native Japan in the nineteenth century and was planted extensively in the South. Unfortunately, kudzu has few natural predators in this country, and now covers empty fields and the edges of forests with billows of leafy vines (Figure 28-1c). Each plant can send out runners that grow at least 20 meters in a season and start yet more new plants.

Bulbs and **corms** are dense underground stem structures involved in vegetative reproduction as well as carbohydrate storage. Bulbs are compact, conical stems with modified leaves; the "scales" of an onion bulb are such storage leaves. Corms are usually smaller than bulbs and are solid (Figure 28-1d). Iris, tulip, and grape hyacinth grow from bulbs; crocus and gladiolus, from corms. Tiny *bulblets* can form at the base of a bulb, and *cormels* at the base of a corm; they remain attached to the original bulb or corm but can grow separate new plants. Once the bulblet or cormel has grown to full size, it can again sprout new offshoot structures and give rise to new plants.

Tubers are another type of modified stem that can participate in vegetative reproduction. The common white potato is a tuber and is the swollen tip of a stolon. Each tuber bears many "eyes," or buds, which can sprout new stems and roots.

Unlikely as it may seem, leaves too can be reproductive structures. In the *Kalanchoë* plant, dozens of tiny plantlets are produced along the margin of each leaf (Figure 28-1e). These plantlets, complete with leaves and roots, fall to the ground around the parent plant, colonizing any vacant space. Some plants, such as begonia, can be induced to reproduce vegetatively from the leaves, even though they do not usually do so in nature. For example, if a begonia leaf is cut along a vein and is placed on moist soil, small plantlets will form along the cut surface. These can take root and produce a clone of new begonia plants.

Finally, roots can give rise to new plants. Lilac, pop-lar, elderberry, and many types of grasses send out horizontal roots, called **root suckers,** from which new stems and roots can emerge. Whole groves of aspens, turning synchronously to gold in the crisp autumn air, give evidence of being clones linked by root suckers.

Plant Propagation in Agriculture

Humans use seeds to propagate many kinds of plants: sometimes, however, it can take several years for a species to grow from seedling to flower- or fruit-producing adult. More important, the genetic recombination that occurs during sexual reproduction can cause some desirable characteristics to be lost or masked in the offspring. And some hybrids, including pineapple and Marsh grapefruit, do not reproduce sexually at all.

Vegetative propagation is both faster and more certain than seed propagation, since the offspring are clones of the parent. Consequently, farmers, gardeners, and others who raise plants use various techniques to produce plants vegetatively. Bulblets and cormels, for example, are broken off and replanted individually to grow new tulips or crocuses. Commercial banana plants also arise from huge corms, with several "suckers," or new shoots, sprouting from them. Growers cut the corm into heavy chunks, each bearing a sucker, and these chunks produce plants fairly quickly. Tubers, too, can be cut into pieces, with one "eye" per piece, and planted to yield new individuals. In sugar cane and pineapples, plant hormones are used to induce stem cuttings, or *slips*, to sprout roots and hence form new plants.

Another important and widely used technique for propagating woody plants is *grafting*. A horticulturist attaches a stem cutting from a desirable plant—a Peace rose, for example, or a Red Delicious apple—to the root stem of a hardier variety of rose or apple. The small grafted branch contains special growth zones that allow it to grow and develop into a new plant with all the desirable qualities of the original donor, while it remains attached to and permanently dependent on its new root stock. Grafting is such a dependable means of propagation that most commercial fruit grown in the United States comes from trees reproduced in this way.

Advantages and Disadvantages of Vegetative Reproduction

Agriculture clearly depends to a great extent on vegetative reproduction, and as we saw, vegetative reproduction has distinct advantages for plants in nature: a successful plant, well suited to its environment, can spread rapidly into nearby areas, and all the identical offspring are equally well suited to the similar sites. However, the

SYNTHETIC SEEDS: A BRIDGE FROM LAB TO FIELD

Just as the techniques of genetic engineering have been used to produce new sources of drugs, improved livestock, and giant laboratory mice (Chapter 14), they are being applied to plants. This work is proceeding in at least three major areas. Some plant geneticists have been combing the plant kingdom for species with desirable genes—genes for traits such as salt tolerance, cold tolerance, insect resistance, firmer fruit, nitrogen fixation, oil production, and increased protein yield. Others have been working on a type of mustard plant called *Arabidopsis* which may become the fruit fly of molecular botany. Here, and in other species, methods of transferring genes into important crop species are being developed. The methods so far include the transport of gene fragments into cells via bacterial vectors (Chapter 14); the fusion of two different kinds of protoplasts (plant cells whose walls have been removed); the passage of genes into cells through artificial, electrically produced pores in their plasma membranes; and even the removal of DNA from one cell and its injection into another using extremely fine needles. A final set of researchers is working on ways to get engineered plants out of the lab and into the fields using synthetic seeds.

Let us say that a researcher is trying to engineer a better tomato plant by introducing into it genes for herbicide resistance (so weedkillers can be applied) or drought tolerance (so less irrigation water is required). Major soup and tomato-sauce companies, in fact, have been funding just such experiments, and researchers in a few labs have fused protoplasts from tomato plants with protoplasts from tobacco plants in the hope that the disease resistance of the latter will be expressed in the new hybrids. How can researchers make the leap from altered protoplasts to healthy tomato plants growing in rows in the sunshine? Here are some probable steps they will take and some ways synthetic seeds might come into play.

First, the new hybrid cells would be grown in tissue culture on specially prepared growth media so that a lump of undifferentiated tissue (called callus) forms. This callus would be broken up into many small clumps and induced with plant hormones to form *somatic embryos*—embryos derived from cells other than fertilized egg cells. These tiny green embryos could then be grown into large seedlings for transplantation into the field. Ultimately, such plants would produce the thousands of seeds—each with the engineered genotype—needed by farmers to plant huge fields. This actual transfer to the field is where synthetic-seed technology comes in.

Researchers at several universities and biotechnology firms are devising ways of coating somatic embryos with a transparent organic jelly like that on a fish egg and then encapsulating both embryo and gelatinous layer in a thin polymer jacket that will biodegrade once in the soil (Figure A). With this technique, huge numbers of genetically altered tomato or other plants could be produced from a single, original hybrid cell, and then be planted directly in the fields as space-age seeds.

This particular scenario is hypothetical because no one has yet found a good way to induce tomato callus to form somatic embryos. However, such tissue-culture embryos can be induced in lettuce, corn, cotton, soybeans, coffee, and oranges, and plant researchers have already found ways to alter most of these plants by genetic engineering. For instance, corn protoplasts subjected to a high electric field form pores in their plasma membranes through which any desired piece of DNA can pass. Such DNA winds up in the cell's chromosomes—this is the most direct way yet of introducing desired traits into a crop plant genome. Ultimately embryos resulting from such cells may be encapsulated in synthetic seeds.

Synthetic seeds promise to be a valuable tool for plant genetic engineers for three reasons. Most important, they circumvent the need for sexual reproduction between the engineered plant and another adult plant to produce natural seeds. In the course of

presence of this single genotype can be a major disadvantage if the absence of genetic variability renders the clone more susceptible to disease or to changes in the environment. Consider the examples of strawberries and bamboo. Farmers often grow vast fields of genetically identical strawberries. Each plant tends to produce the same size and flavor of fruit—a commercial advantage. However, a single viral infection could destroy the farmer's entire crop. Such disasters also occur in nature

when a new disease organism encounters a susceptible clone.

Another characteristic of plants with a single genotype is that they tend to flower at the same time. After growing and reproducing vegetatively for up to a century, stands of bamboo covering up to thousands of acres may suddenly reproduce sexually by flowering and then die in the same brief period. The loss of such large stands of bamboo greatly upsets the local ecology and affects many

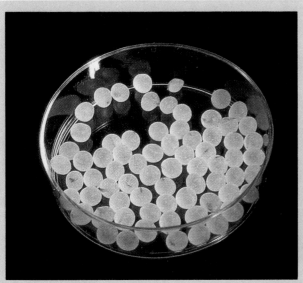

Figure A   **SYNTHETIC SEEDS.**
An organic jelly is used to coat alfalfa embryos generated from callus on undifferentiated plant tissue.

such breeding, the desired and newly inserted trait might not be passed to the F_1 generation. Furthermore, traits from plants so distantly related that they could never successfully interbreed sexually can easily be introduced using bacterial vectors or electric pore formation. The new biotechnology also has the advantage that the gelatinous coating around the somatic embryo can be tailored to suit the needs of the infant plant with nutrients, hormones, pesticides, and fungicides to speed growth and prevent damage. Such made-to-order seeds may indeed be the bridge between lab and field, between the genetic revolution and the agricultural revolution of the 1990s. As such, those seeds may be the greatest hope humankind has yet to meet the food demands of the burgeoning human population.

animal species and human populations that over many decades have grown dependent on the bamboo forest. In some regions of southern China, centuries-old bamboo is now flowering and dying, threatening the already endangered giant pandas, whose main diet consists of young bamboo shoots.

Clearly, there are pluses and minuses to vegetative reproduction, but its prevalence among the angiosperms suggests that in many species, the risks are offset by the

ability to spread quickly and with genetic fidelity. In fact, vegetative reproduction can rapidly spread identical copies of new mutated genes that arise in the plant growth zones which generate rhizomes, stolons, and other vegetative structures. These somatic mutations may be important contributions to the genetic variability and evolutionary success of angiosperms.

SEXUAL REPRODUCTION: MULTIPLICATION AND DIVERSITY

Most angiosperms reproduce sexually, and evidence of this is all around us in spring blossoms, summer produce—and even autumn hay fever—an allergic response to airborne pollen. As in animals, sexual reproduction in plants involves the production and fusion of haploid male and female gametes to create new gene combinations in the zygotes. These processes take place in the flowers of angiosperms and ensure genetic diversity among the seeds and subsequent offspring. Such diversity, in turn, increases the chances that members of a species can be adapted to short-term fluctuations and long-term changes in the environment and thereby increase the chances that the species will persist. No wonder so much of an angiosperm's energy and so many of its nutrients are consumed in producing flowers, fruits, and seeds: these structures overcome the major disadvantages of vegetative reproduction.

Flowers: Ingenious Insurance for Fertilization

Flowers are lovely examples of nature's diversity. Each of the some quarter of a million angiosperm species has a flower uniquely its own in at least some minor detail. Flower appearance is the most obvious; you can probably identify at least a dozen types of common garden flowers—roses, tulips, carnations, pansies, violets, daisies, chrysanthemums, and so on (Figure 28-2). Fragrances differ widely, too, as do modes of pollination—including pollen borne by wind or carried by a particular species of bird, bat, butterfly, beetle, bee, or even ant (Figure 24-16). Despite variations in appearance, fragrance, and pollination, however, all flowers are based on the same organizational plan and have some or all of the same parts, as shown in Figure 28-3. This essential similarity reflects the effectiveness of the basic flower "design" for producing gametes and protecting them throughout fertilization and seed development.

(a)

(b)   (c)

(d)

(e)

Figure 28-2   FLOWERS: SEXUAL STRUCTURES WITH A SEEMINGLY INFINITE VARIETY OF SHAPES, COLORS, AND ODORS.
(a) The male and female parts of this Jamaican rain forest *Lisianthius exsertus* protrude far beyond the petals. (b) Only the tip of the female pistil (a group of fused carpels) is visible in this jack-in-the-pulpit (*Arisaema atrorubens*). (c) Each *Gazania* flower is actually composed of a large number of very small flowers, whose parts cannot be seen at this magnification. The *Gazania* is thus a "composite" head of flowers. Individual flowers within this head mature over several days so different pollinators may fertilize the egg cells. (d) Multiple whorls of colored petals characterize this Christmas cactus (*Schlumbergera* species). (e) When the drably hued petals (top and left) of these prickly pear cactus (*Opuntia* species) flowers open (right), the large central pistil and surrounding anthers are available to flying or crawling insects or birds. (f) Water lilies, such as this lotus (*Nymphaea* species), have carpels fused into a central yellow cylindrical structure with the small protruding stigmas. (g) This strange-looking carrion flower (*Stapelia variegata*) from Africa gives off an odor of rotting animal flesh, thereby attracting insect pollinators.

(g)

(f)

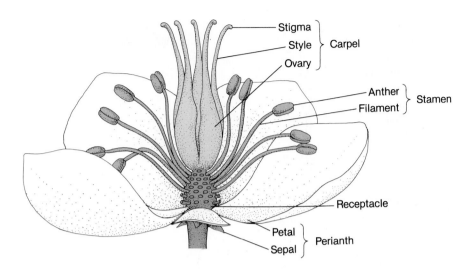

Figure 28-3   ANATOMY OF A TYPICAL DICOT FLOWER.
The stamen consists of the anther and filament. The carpel consists of the ovary, style, and stigma. Several fused carpels constitute a pistil.

Flowers are really a group of modified leaves. The typical flower consists of four whorls of parts: sepals, petals, stamens, and carpels. The *sepals*, or outer ring of floral parts, are often green and photosynthetic. They are attached to the *receptacle*, the reinforced base of the flower that supports the combined weight of the reproductive tissue. Collectively, the sepals are called the **calyx.** The *petals*, another whorl of modified leaves, are often brightly colored—from white to blackish purple and every hue in between—and serve to attract pollinators to the flower. The petals collectively are called the **corolla.** Together, the flower's showy exterior parts are called the **perianth** (calyx and corolla). The interior parts—*stamens* and *carpels*—are the sexual structures involved in producing pollen and eggs.

Pollen Production

The pollen-bearing organs, the stamens, are the plant's male flower parts. They produce the microspores that develop into pollen. Most stamens consist of two parts: the stalklike *filament*, attached to the receptacle; and the *anther*, borne on top of each filament. Typically, each anther contains four *microsporangia*, or *pollen sacs*, chambers in which the pollen grains form. Pollen is eventually released when the anther's peripheral tissue ruptures, and the grains are blown or carried to the female parts of the same flower or other flowers.

Anthers are pollen "factories" that generate enormous quantities of pollen grains, much in the way that a male animal's testes produce vast numbers of sperm. Each of the anther's four pollen sacs is composed of two cell types: *peripheral cells* and *microspore mother cells*. The peripheral, or outer, cells form a sac of tissue around the developing pollen grains (Figure 28-4). The innermost layer of peripheral sac cells, the *tapetum*, supplies food to the developing pollen grains, just as the Sertoli cells in an animal's testes do for developing sperm. In the center of each microsporangium is a large number of microspore mother cells, which are destined to give rise to pollen grains through a multistep process.

Figure 28-4   STRUCTURE OF THE ANTHER.
(a) The anther is the pollen-producing structure of the stamen. (b) A detailed cross section through the anther reveals four pollen sacs (microsporangia). The outer layer of each pollen sac is the tapetum, a set of cells that secretes nutrients into the pollen sac. (c) Microspore mother cells fill each pollen sac.

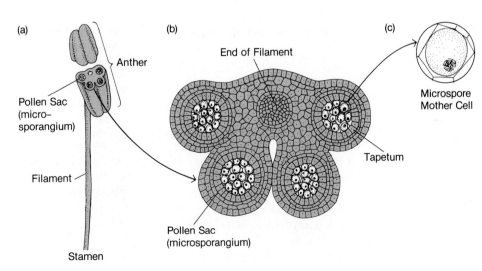

The first step in the development of pollen grains is **microsporogenesis** (Figure 28-5). Each diploid microspore mother cell divides meiotically, producing four haploid microspores, each of which is destined to give rise to a pollen grain. The haploid microspores are unusual in that no plasmodesmata interconnect the cells, and the common cell walls that at first attach the haploid cells to each other are later digested enzymatically, freeing each microspore from even minimal physical connection to other cells. (This is intriguingly different from animal sperm. Recall from Chapter 16 that developing sperm remain connected to each other by cytoplasmic bridges, so that clusters of sperm develop synchronously.)

The second step is **microgametogenesis,** in which the microspores differentiate into functional pollen grains. Maturation of pollen involves mitotic division of each haploid microspore. This occurs in two steps. First, the haploid nucleus and adjacent cytoplasm divide once mitotically to form two cells: the *tube cell* and the *generative cell*, which is enveloped by the tube cell. These two cells, surrounded by a single cell wall, constitute the pollen grain. Second, on germination, a *pollen tube* is formed from the tube cell, after which the generative cell divides to yield the two haploid sperm cells.

During pollen maturation, polysaccharides and proteins are deposited in the pollen cell wall, creating an amazing three-dimensional surface that looks as though it were sculpted by an artist (Figure 28-6). Spikes, knobs, cavities, and craters are created by deposition of *sporopollenin,* an extremely hard carotenoid polymer. Each pollen surface pattern is unique and species-specific and acts as a kind of structural signature. Unfortunately for many humans, the spikes, knobs, and other projections of pollen grains contain proteins that can induce strong allergic reactions. Ragweed, the plant that causes hay fever, has a wonderful scientific name, *Ambrosia,* which suggests the fruit of the gods. Viewed through an electron microscope, however, *Ambrosia* pollen grains look like tiny medieval weapons—a suitable candidate for causing allergic reactions and seasonal human misery.

Egg Production

Just as angiosperm pollen is produced and protected within the flower, so are the eggs. At the center of every flower lies the *pistil*, a vaselike structure consisting of one carpel or several fused carpels (Figure 28-7). The pistil itself is the site of egg-cell production and of fertilization. The pistil consists of three parts: the *stigma*, the pistil's broadened sticky top surface, to which pollen adheres; the *style*, the slender "neck" that connects the stigma to the ovary; and the *ovary*, at the base of the

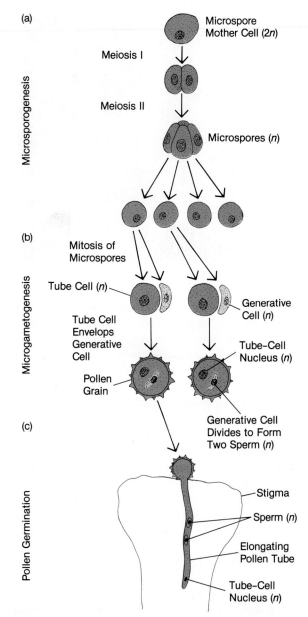

Figure 28-5   POLLEN PRODUCTION AND GERMINATION.
Pollen production has two parts. (a) During microsporogenesis, each microspore mother cell in a microsporangium undergoes meiosis to form four haploid microspores. (b) During microgametogenesis, each haploid microspore undergoes mitosis, forming two differentiated cells: a tube cell and a generative cell. The tube cell then envelopes the generative cell and forms a pollen grain which contains the generative cell and which has a specialized wall. (c) The pollen germinates when it lands on a stigma; the generative cell undergoes mitosis to form two haploid sperm cells that pass down the elongating pollen tube. They will participate in double fertilization. For clarity, mitosis of only two microspores is shown.

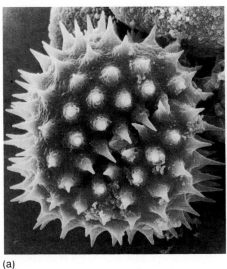
(a)

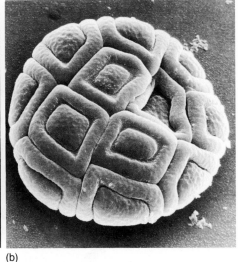
(b)

Figure 28-6   POLLEN GRAINS: UNIQUE FORMS AND SURFACE PROPERTIES. The often elaborate outer protein and sugar coats of pollen grains induce allergic reactions in some people. Different species of plants produce distinctively different pollen. (a) Flowers related to asters (*Aster chilensis*) have "spiky" pollen (magnified about 1,350 times), whereas (b) the stock (*Matthiola incana*) pollen has a smoother, sculptured appearance (magnified about 1,550 times).

pistil, the actual site of egg development. The ovary matures into the fruit. Some plant species, such as the peach, have only one *ovule*, or egg chamber, and thus develop just one seed, which we know as the pit. The fruits of many other plants can have several ovules: string beans and pea pods are good examples, and each "bean" or "pea" within the ripened ovary is a separate ovule.

The production of egg cells is somewhat like that of pollen, but it has unique aspects. As we have seen, thousands of pollen grains result from microsporogenesis followed by microgametogenesis within the male tissues. Within the female tissues, the analogous process, **megasporogenesis** produces only a single egg per ovule (Figure 28-8). Each ovule contains a diploid *megaspore*

mother cell, which undergoes meiotic division to produce four haploid cells. Of these, three degenerate, and the fourth becomes the megaspore, with a haploid nucleus. (Recall from Chapter 16 the similar three ill-fated polar bodies of animal eggs.)

Megasporogenesis is followed by a second analogous process, **megagametogenesis**, during which the female gamete is produced and readied for fertilization. The megaspore undergoes three mitotic divisions of the nucleus alone; cytokinesis does not occur. The result is a single cell, the *megagametophyte*, which usually contains eight haploid nuclei. These eight nuclei separate into two groups of four at opposite ends of the megagametophyte. Then, as if in a dance, one nucleus from each group of four migrates to the center of the cell, leaving

Figure 28-7   OVULE STRUCTURE. (a) Each ovule of a flower contains one megaspore mother cell that will ultimately give rise to the egg cell. (b) The megaspore mother cell is surrounded, protected, and sustained by the nearby ovule cells.

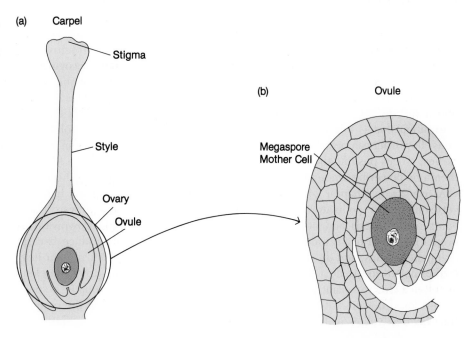

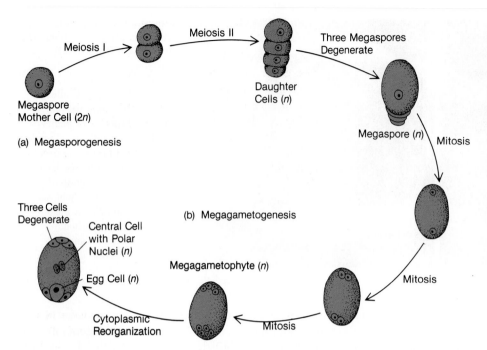

Figure 28-8   EGG CELL PRODUCTION IN A FLOWER'S OVARY.
During megasporogenesis, the megaspore mother cell in an ovule divides meiotically to form four haploid daughter cells. Three of these degenerate, leaving one megaspore with a haploid nucleus. Megagametogenesis follows and involves three mitotic divisions with no cytokinesis. The resultant eight haploid nuclei are then walled off in separate cells. These eight include the egg cell and the central cell (usually binucleate) that will form the endosperm. (Note that in various angiosperms the central cell may contain anywhere from one to fourteen nuclei.)

three nuclei behind at each end. The two centrally located nuclei are called the **polar nuclei.** The other six nuclei become separated from this central region as plasma membranes form and dissect the original megagametophyte's cytoplasm into a central binucleate cell (with the polar nuclei), three small cells at one end, and three small cells at the opposite end. One of the six cells becomes the haploid egg cell, and it remains flanked by two other living cells. The three cells at the opposite end degenerate after fertilization.

Pollination and Fertilization

Spring and early summer are lovely times of the year because once the egg cells inside flower ovaries have matured and are ready for fertilization, the corollas unfurl, and the simultaneous display of bright petals in thousands of flowers can be dazzling. Because each angiosperm species has its own internal "timetable" for egg maturation that is influenced by environmental conditions, flowering can occur throughout the year from the earliest crocus pushing up through the snow to chrysanthemums flowering among drifts of fall leaves.

As we saw in Chapter 24, flowers can function as advertisements to pollinators. The colors, shapes, and odors of the petals and other flower parts entice the proper pollinator to a reward of sugary nectar or protein-rich pollen grains, and the animal inadvertently ferries a load of pollen to the next flower. Flower characteristics are often a clue to a plant's mode of pollination: white or drab-colored flowers with strong, sweet odors often attract beetles; yellow or blue flowers that smell fruity or

spicy draw bees; deep tube-shaped flowers can be pollinated successfully by birds, moths, or butterflies with long beaks or mouthparts; and flowers that smell like dead meat or fish (including skunk cabbage, some milkweeds, and many other species) attract carrion flies. Small, drab flowers with reduced petals and no fragrance are often wind pollinated; the trend in these flowers is toward long, exposed stamens that bear masses of pollen that can easily be blown about or enlarged, sticky stigmas. The flowers of corn plants are a good example.

Whatever its form, once the flower opens, the stigma is exposed, the pollen is released from the anthers, and the pollen grains are blown or carried to an exposed stigma on the same or another flower. The molecules that comprise the surface of the pollen grain interact with proteins and polysaccharides on the stigma, and if they "recognize" each other by a series of reactions, the pollen grain is stimulated to begin growing—to **germinate.** This specificity of molecular matching helps prevent cross-fertilization between unrelated plants. Recall that the pollen grain contains a tube cell and a generative cell. When the pollen grain germinates, a pollen tube, which is a filamentous extension of the tube cell, pokes through the hard wall of the grain (Figure 28-9). This tube then begins to grow down through the stigma and style, toward the ovary. As the pollen tube continues to grow downward, the generative cell divides mitotically to produce two sperm cells, with their haploid nuclei, which remain just behind the growing tip. Ultimately, the tip of the elongating pollen tube passes through the *micropyle*, a specialized opening at the surface of one end of the ovule.

The evolution of pollen tubes, according to some bota-

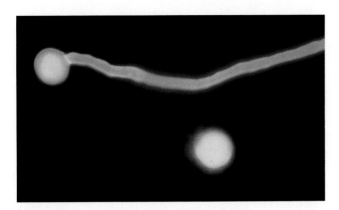

Figure 28-9   POLLEN TUBE ELONGATION.
The elongation of pollen tubes in a plant such as a tomato is an example of spectacular growth by a single cell. Here is a pollen tube (magnified about 95 times) early in its elongation process. The sperm cells in the tube cannot be seen.

nists, solved a big problem for angiosperms and led to additional advantages. They theorize that the ovary evolved to protect the eggs from predation by animal pollinators. In all probability, the evolution of better and better protective ovaries was probably concurrent with the evolution of pollen tubes able to reach and penetrate such ovaries. The growth of some pollen tubes is truly an Olympian performance. Corn pollen, for instance, sticks to one of the silks (the hairlike stigmas and styles at the ends of ears of corn), and in just twenty-four hours, the pollen tube enters the silk, grows down the full length of the strand, and reaches the kernel (ovule)—a distance of up to 30 centimeters in large ears of corn. This amazing growth occurs largely because of rapid elongation of the cell composing the pollen tube. The net result is movement by the tip of the pollen tube some ten times as far in twenty-four hours as fast-moving animal cells can travel in an embryo or a culture dish.

Once the pollen tube has reached the micropyle, the tube's precious cargo, the two sperm nuclei, can be discharged so that fertilization can occur. Angiosperms undergo **double fertilization,** a process unique to this group of organisms and more complex than fertilization in animals. The five participants in double fertilization are the two sperm nuclei from the pollen grain, the nucleus of the egg cell, and the two polar nuclei of the large central cell of the female megagametophyte. When the pollen tube enters the micropyle, the two sperm cells are released through a pore in the wall of the pollen tube. One sperm fuses with the egg, forming a zygote with a diploid nucleus. The second sperm cell penetrates the large central cell containing the two polar nuclei. The three nuclei fuse, forming a triploid ($3n$) nucleus. This triploid *endosperm* cell will develop into the primary source of nutrition for the embryonic plant (Figure 28-10).

Double fertilization is not found even in the angiosperms' nearest relatives, the seed-bearing gymnosperms. In that group, two sperm are released from the pollen tube into the egg, but one degenerates. A diploid zygote results, but a triploid endosperm does not form.

It is remarkable to think that such precise processes can go on inside every flower and, further, that the evolution of these complex and often beautiful structures helps ensure successful sexual reproduction—from the formation and protection of gametes, to the appropriate specificity of pollination, to the elaborate and unique double fertilization.

Figure 28-10   DOUBLE FERTILIZATION.
(a) When the pollen tube enters the micropyle, one sperm cell fuses with the egg cell (b, c), forming the zygote. The other fuses with the central cell containing the polar nuclei (b, c) to form the triploid endosperm nucleus. (d) This endosperm nucleus divides to produce many triploid endosperm nuclei; as that occurs the zygote divides and begins to form the embryo.

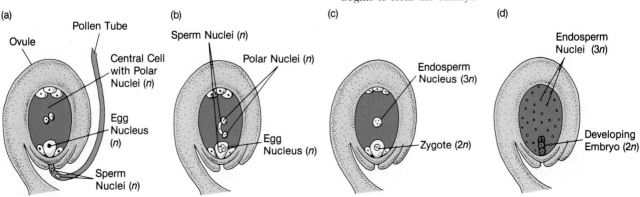

THE DEVELOPMENT OF PLANT EMBRYOS

Once fertilization takes place, many complex events are triggered in the plant ovule, just as they are in the animal egg. As we shall see, the embryonic development of plants and animals is quite similar in many respects but differs in a few significant ways. Four processes occur simultaneously in plants after fertilization: (1) the triploid endosperm cell undergoes many mitotic divisions as nutrient tissue is produced for the developing embryo; (2) the diploid zygote divides and slowly begins to form the embryo; (3) the ovule itself changes to become the seed coat (the "pit"); and (4) the ovary becomes the fruit tissue surrounding the developing seed.

The Endosperm and the Embryo

Within the endosperm, rapid mitotic divisions allow the storage tissue to develop much more quickly than the embryo (Figure 28-10). A process of "laying food by" begins, so that an abundant supply will be available for the embryonic plant. The endosperm absorbs nutrients that are delivered to the ovule by the parent plant's xylem and phloem. Among the many substances found in the endosperm are sugars, starch, amino acids, proteins, and oils.

In grasses and many other monocots, endosperm forms the largest component of the mature seed. The seed itself, as we shall see in more detail, is composed of a tough coat surrounding the stored nutrient tissue and the embryonic plant. Endosperm forms the starchy part of a kernel of corn, for example, which is a member of the grass family. Sweet corn gets its sweetness from the sugars that the endosperm initially receives from the phloem. If corn is picked at the right time, most of the endosperm is sugary; if it is picked too late, however, most of the sugar will have been converted into starch. The endosperm of wheat comprises the stored starch from which we make flour. Grass embryos consume most of the food reserves of the endosperm after they burst forth from their seed coats in the spring. In contrast to such monocots, the embryos of most dicots, such as peas or beans, consume the endosperm while developing and store the nutrient reserves in two embryonic leaves, the *cotyledons.* Thus the first leaves in dicot seedlings are large and plump.

The development of the monocot and dicot embryo itself proceeds more slowly than that of the endosperm. The initial division of the zygote is transverse and establishes the basic polarity of the embryo. The embryo arises from cells at the upper end of the zygote (Figure 28-11). The **suspensor,** a column of cells that connects the embryo to the ovule wall, arises from cells at the opposite end. Mitosis continues in both cell populations; the upper cells produce a ball of cells, the *globular-stage embryo.* The embryo remains attached to the suspensor, which continues to elongate. Suspensor cells feed the embryo by passing nutrients to the embryo's cells; both proteins made by suspensor cells and nutrients from the endosperm are passed to the embryo cells. In microscopic views of young embryos, the suspensor is the prominent structure.

The small globe of cells composing the embryo is not much bigger than the original single embryonic cell. The initial divisions are **cleavage divisions** that divide the embryo cell into smaller and smaller units, though some real growth may occur. Growth of the embryo, as evidenced by a real increase in size, accelerates at the globular embryonic stage, when there are only two types of cells: **inside cells** and **outside cells.** The inside and outside cells are the first examples of differentiation in the embryo.

The small outside cells form the **protoderm,** the embryonic epidermis. The word "protoderm" means literally "first-formed skin." Protoderm also is present in the growth centers of roots and shoots (where it is part of the apical meristems, the growth sites). Epidermis is the analogous tissue in the seedling and adult plant. Each protodermal cell divides so that the new cell wall that forms between each pair of daughter cells is perpendicular to the outside surface of the embryo. The original protoderm is one cell thick, and it remains that way as it expands, increasing the embryo's surface area to allow for internal cell division and growth.

Some of the inside cells form the *procambium,* the layer of cells that gives rise to the embryo's vascular tissue (Chapter 27). The procambium and early vascular tissue form *procambial strands;* in dicots this occurs on two opposite sides of the embryo. This marks the transition from the globular embryo's radial growth pattern (symmetrical in all directions) to the later embryo's bilateral growth pattern (two-sided symmetry).

Thus, by the time the procambial strands appear, we can see the first difference between monocot and dicot embryos. In dicots, two bumps emerge at the top of the embryo above the procambium. These two bumps, which give the embryo a heart-shaped appearance, will form the two **cotyledons,** or seed leaves, characteristic of dicots. In monocots, there is one cylindrical cotyledon, the **scutellum,** which arises directly from the end of the embryo opposite the suspensor (Figure 28-12).

Wedged between the two dicot cotyledons, or at one side of the monocot's scutellum, is the **apical meristem** (Figures 28-11, 28-12, and 28-13). A meristem is an organizing center of undifferentiated, actively dividing cells. Meristems are found only in plants and form zones

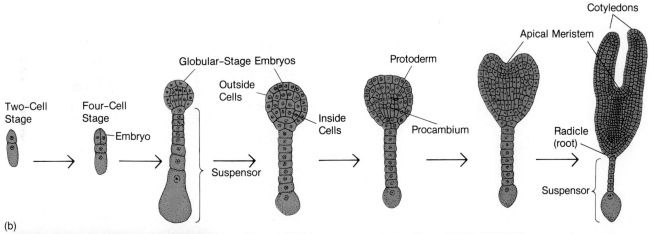

Figure 28-11   DEVELOPMENT OF A DICOT EMBRYO.
(a) Development of both monocot and dicot embryos begins as mitosis of the zygote forms cells that will develop into the embryo and the suspensor. (For simplicity, the surrounding endosperm tissues are not shown.) Cells at the end of the suspensor form the globular embryo that soon has inside cells and outside cells. The outside cells form a protoderm (forerunner of epidermis), and the inside cells form procambium, which gives rise to vascular tissue. At this point, dicot development diverges from monocot development. In dicots the two cotyledons gradually take form, leaving the apical meristem located between these "seedling leaves." (b) A slightly more advanced embryo (magnified about 85 times) of the shepherd's purse (*Capsella bursapastoris*) than that last stage in (a). The two cotyledons are seen at the top and the forming root is at the bottom. The suspensor has been broken off and is not present here.

where cells for new organs can be generated throughout the life of the plant. Meristems have been called "permanent embryos" because of their lifelong activity. They mark a major difference between plants and animals, which normally develop new organs only in the embryo.

The apical meristem lies at the growing tip of the shoot and is responsible for generating—directly or indirectly—all the cells for a plant's leaves, stems, branches, and flowers. Although the apical meristem is the true apex of the plant, it may appear to be lower than a dicot's

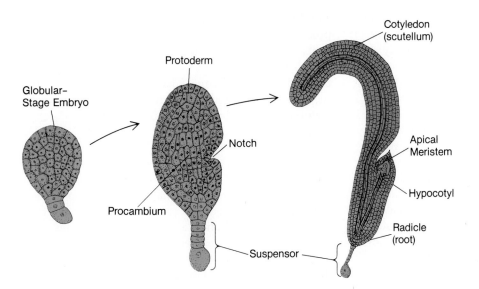

Figure 28-12   LATER DEVELOPMENT OF MONOCOT EMBRYOS.
The early stages of monocot development resemble those in dicots. In later stages, a notch forms on one side of the cylindrical embryo, marking the site of the apical meristem. The single cotyledon, the scutellum, is located above and the root below the apical meristem of monocots.

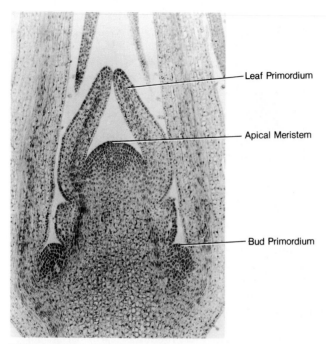

Figure 28-13   APICAL MERISTEMS: ZONES OF ORGAN GENERATION.
This dicot apical meristem (magnified about 16 times) is a growth center giving rise to cells that will form new leaves and stem. The immature leaf primordia of this honeysuckle (*Lonicera* species) are folded over the apical meristem, protecting it. Bud primordia may later become active meristems and sites of stem outgrowth as branches.

— Leaf Primordium

— Apical Meristem

— Bud Primordium

two cotyledons because they elongate very rapidly. In monocots, the apical meristem is found in a notch on the side of the embryo; from that position, it is reoriented in the seedling so that the apex points upward (Figure 28-12).

At the opposite end of the embryo, another important precursor tissue forms: the **root meristem.** Like the apical meristem, this zone of undifferentiated, rapidly dividing cells will continue to generate new cells, but these will grow and mature into functioning root tissue. The root meristem forms at the end of the procambial strands, which traverse the embryonic tissue between the apical and root zones (recall that the procambium is the source of the vascular system). Formation of the shoot and root meristems defines the growth axis of the plant; that is, it establishes the polarity of root and shoot. There is a tendency for the shoot to grow upward and the root, downward. The establishment of the growth axis is critical for normal plant development, since all the adult tissues will be derived from the two meristematic zones. Once the meristems are established, the major events of early plant embryonic development are complete.

SEEDS: PROTECTION AND DISPERSAL OF THE NEW GENERATION

The embryo—a new generation—develops within the seed, and this, in turn, usually develops within the fruit. **Seeds,** the mature ovules, are like time capsules that protect the embryo and the endosperm until the appropriate time for germination. In many species, the establishment of the embryo's meristems is a crucial stage of development; once this occurs, the embryo stops growing, and the *seed coat* forms. This coat can be thin and papery, as in the "skin" on a peanut, or stony, as on a Brazil nut. But it always serves as a tight, protective seal around the embryo, and once the seed coat has formed, the fruit begins to enlarge around the seed.

Most mature **fruit** tissue is derived from the wall of the ovary at the base of the flower, although some fruits, such as the strawberry, are derived from the enlarged receptacle, below the ovary. The ovary wall or the receptacle can become fleshy and succulent, as in apples, pears, tomatoes, and cherries; it can be leathery, as in pea pods and the twisted pods of a locust tree; or it can be papery and dry, as in the "squirts" that flutter down from maple or elm trees (Figure 28-14).

The fruit tissue that surrounds the seeds helps protect them from drying out during early development and, through a number of ingenious adaptations, eventually aids in their dispersal. The fruit can split open, forcibly ejecting the seeds or allowing them to fall to the ground or be blown on a breeze. Milkweed pods are a good example; after the pods dry out and split open, dozens of small, papery seeds bearing gossamer plumes are lifted and carried away by the wind. Succulent fruits such as apples and tomatoes often attract birds or animals that swallow the seeds along with the fruit tissue and later deposit them—unharmed by the digestive process—some distance away. Finally, a number of fruits and seeds have mechanisms that allow them to hitch a ride on an animal's body: some have hooks and barbs that cling to animal hair or feathers, while the seeds of mistletoe and some water plants secrete a sticky material that attaches them to birds' feet.

Seed Maturation

Many kinds of seeds must mature as several processes occur before the seeds are ready to be dispersed; food storage and dehydration are the two main ones. The seeds of most staple food crops have a special period for weight gain in the embryo and accumulation of food in the cotyledons and endosperm. In legumes such as pea-

Figure 28-14   SEEDS AND FRUITS: ADAPTATIONS FOR SURVIVAL.

(a) The wind will soon carry this last milkweed (*Asclepias syriaca*) seed from its pod to some distant site. (b) California buckeye (*Aesculus californica*) trees have hard, nutlike seeds that protect the embryos within. (c) When spring comes and snow melts, the burrs of this burdock (*Arctium minus*) will adhere to the fur, feathers, or clothing of a passing animal and ensure transport of the seeds to new sites. (d) The papery, thin seed cases of the smooth ground cherry (*Prunus* species) will dry and be blown to sites of germination. (e) The sweet fleshy fruits of the twisted stalk rose (*Streptopus roseus*) attract hungry birds, bears, or other animals; the seeds inside the fruit pass through the animals' digestive tracts intact to be deposited elsewhere. (f) These hard *Banksia serratifolia* fruits can float in the sea off Australia and, after drifting ashore, resist opening until fire heats the nut wall.

nuts and beans, the enormous seed that we eat is really two starchy cotyledons, which dwarf the tiny embryo. Such food reserves are crucial, since the seedling must survive on stored food until it becomes photosynthetic and self-sufficient.

Before the cavity within the seed coat is totally filled, weight gain in the cotyledons and endosperm is *wet weight*—that is, mostly water. However, after the coty-ledons or endosperm fill the seed cavity completely, the stage of *dry weight* accumulation occurs, during which the water in the cells is replaced progressively with oils, starch, and proteins. In some legume species, the seeds increase in weight as much as 5 percent each day as they mature. To do so, the entire sugary photosynthetic output of the vegetative plant is diverted to the ovarian tissue and seeds. The adult legume plant may look yellow and pathetic at the time of seed maturation because of this diversion of the photosynthetic products. It is not surprising that mechanisms have evolved to produce seeds with optimal food reserves despite the cost to the adult plant: for many sorts of plants that do not reproduce vegetatively, seeds are the only means of species survival.

During the late stages of maturation of the seed itself, the embryo begins to dehydrate. In many species, the embryo accomplishes an amazing feat: it dehydrates almost completely, yet remains alive. The early embryo is about 90 percent water and 10 percent dry matter (cell walls, membranes, nutrients, and so on). As the seed matures, the synthesis and storage of these materials increase the proportion of dry matter to about 50 percent. After this state is attained, dehydration accelerates so that the proportions of wet and dry weights eventually are reversed, becoming 10 percent wet weight and 90 percent dry. Few other living tissues can withstand such a water loss and still survive.

Seed Dormancy

How does the plant embryo manage to survive in a dehydrated condition? It becomes **dormant.** The cells of the dormant embryo respire at a very low rate and carry out only a small amount of metabolic work. In this resting state, the seed can survive unfavorable environmental conditions, such as a cold winter, a prolonged drought, or both. In some species, the embryo may remain in this resting state for years. There have been several reports of lotus seeds from 200 to more than 400 years old that were still able to germinate. Weed seeds excavated in Denmark germinated after 1,700 years. And lupin seeds frozen deep below the Arctic tundra were still able to germinate after 10,000 years of dormancy!

Most seeds have other dormancy mechanisms besides dehydration. If a seed were only dehydrated, the addition of water and the appropriate temperature for growth would automatically promote germination—the end of dormancy. Under many circumstances, however, this could be disastrous: seeds germinating during a warm spell in January would be killed by the next cold snap. Thus to prevent premature germination, plants living in the Earth's temperate and Arctic zones have developed a requirement that seeds be exposed to periods of cold at or below a certain temperature. For some seeds shed in the fall, this temperature requirement is 4°C, and the time period is several weeks. Only under the correct conditions can the embryo acquire the "message" to germinate the next time the temperature is warm.

But how can a dormant seed "keep track" of time? The answer lies in cellular chemistry. Some seed coats are filled with compounds that inhibit germination and that are gradually inactivated over time. These compounds act like a biochemical hourglass, requiring the embryo to remain dormant until a certain number of months have passed—a number that guarantees germination at a time of year when the chances for survival are statistically greatest. For example, some plants growing in mild climates, such as that of southern California, shed their seeds in the late spring, after the winter rains have ended. Had the seeds been shed earlier, those rains could have leached away plant-growth inhibitors in the seed coat. Instead, the long, hot summer apparently is required to "burn off" the inhibitors. The levels of these inhibitors remain high enough to prevent germination during rare summer rainstorms and to ensure that the seed germinates in the following winter when sustained wetness is likely.

Some seeds have extremely hard (sclerified) seed coats that must be partially abraded away to allow the root to protrude. This abrasion may take place while the seed is blown about over rough ground or after it has been swallowed and is passing through a bird's digestive system. There it may be etched by acidic digestive juices or may be fractured and ground away by small pebbles. Once the seed has undergone this abrasion process and has passed from the bird's body, it is ready to germinate.

GERMINATION AND SEEDLING DEVELOPMENT

Once a seed has undergone the processes that allow germination and is then exposed to water and sufficient warmth, dormancy ends and germination begins. Germination occurs like the eruption of a slumbering volcano. Resumption of growth by the tiny plant resting inside the seed starts when the quiescent embryo imbibes, or takes up, water at a rapid rate, restoring the water content of 80 to 90 percent and causing the embryo's cells and tissues to swell. Imbibing water also establishes conditions of tonicity and solubility that are conducive to enzyme activity; as a result, the embryo's sluggish metabolism soon speeds up so that cell division can resume in the meristematic regions. The pressure caused by the swelling and renewed growth of the embryo cracks and weakens the encapsulating seed coat, which has served to protect the tiny plant. In just a few hours, the cells of the root axis begin to elongate, so that the root region pokes out through the ruptured seed coat. At this point, germination has taken place, and the embryo begins a race to establish itself as a photosynthetic seedling before it runs out of the nutrients stored in the endosperm or cotyledons. The embryo cannot crawl back into its protective shell and wait for better conditions. All systems must be GO, and the sooner the plant can produce its own carbohydrates by photosynthesis, the better.

The first structure to emerge from the rupturing seed coat is the **radicle,** the short length of root that becomes the primary root. The radicle is continuous with the **hypocotyl,** the initial length of stem that emerges from the germinating seed as an arching, white (nonphotosynthetic) structure. The radicle is *positively gravitropic*, which means literally "movement toward gravity"; hence, it grows downward into the soil. As that occurs, the **epicotyl,** or future stem, emerges from the soil and begins to grow upward.

The many thousands of small cells in embryonic root, shoot, and cotyledon undergo a remarkable elongation process as part of germination. Individual cells may lengthen 10 to 100 times, so that a small segment of tissue only 1 millimeter long could elongate to 100 millimeters. This burst of cell elongation pushes the root several centimeters deep into the soil and propels the shoot upward toward air and sunlight.

Early seedling growth differs somewhat between monocots and dicots, as shown in Figure 28-15. In both, the radicle and hypocotyl emerge first. However, monocots have a specialized organ for shoot growth, the **coleoptile.** This protective sheath around the embryonic leaves accomplishes two tasks: it pushes through the soil and it seeks light. Sheathed in and led by the coleoptile, the shoot's growth is oriented upward. Once exposed to light, all the shoot tissues have the capacity to participate in growth toward sources of light (Chapter 30).

Soon after the monocot's coleoptile emerges from the soil, it is split apart by the rapid expansion of the embryonic leaves within. These young leaves push their way into the sunlight, turn green, and begin photosynthesis. Although the monocot's shoot has erupted, the rest of the seed remains underground, and the endosperm and scutellum stay within the seed coat to nourish the growing seedling.

Mobilizing stored food becomes every seedling's top priority. Seeds of barley, a monocot, are a good example. Barley endosperm stores both starch and protein in dead, centrally situated cells. Surrounding that food cache is a layer of living cells called the **aleurone.** When a barley seed germinates, the embryo makes or releases a hormone that stimulates protein synthesis in the aleurone layer. An aleurone-produced enzyme, *amylase*, is secreted into the endosperm at a very high rate, resulting in the rapid conversion of starch to sugars. These sugars are used to manufacture new cell walls during the rapid cell division and elongation following germination. After a few days, the endosperm is totally depleted, and only the cell walls remain visible, much as in a honeycomb from which the honey has been drained.

Dicot seedlings follow a slightly different sequence of

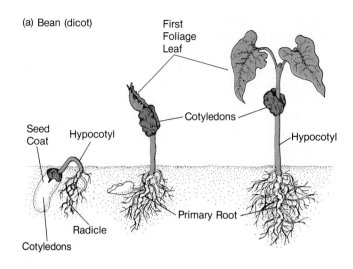

(a) Bean (dicot)

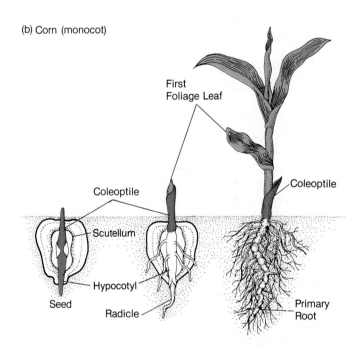

(b) Corn (monocot)

Figure 28-15   THE NEW PLANT: GERMINATION AND SEEDLING GROWTH IN A MONOCOT AND A DICOT.
In both (a) dicots and (b) monocots, seedling growth begins with the emergence of the radicle (root sprout) and the hypocotyl. Whereas the dicot cotyledons of many species will emerge from the soil as the hypocotyl elongates, the monocot's scutellum remains within the seed coat with endosperm to nourish the seedling. In a monocot, the coleoptile leads the shoot through the soil.

events. Recall that most dicots store their food reserves in the cotyledons rather than in the endosperm. In contrast to the central store of endosperm, the cotyledons are living tissue and usually have two important functions during germination: their stored reserves are mobilized and transferred to the rest of the tiny plant, and after expanding, they themselves may carry on photosynthesis. Most cotyledons are ferried up through the soil by the elongating shoot and, once they have broken ground, are stimulated by sunlight to turn green and become the seedling's first photosynthetic organs.

Regardless of whether nutrients are stored in endosperm or in cotyledons, small molecules—sugars, amino acids, fatty acids, and others—arise from the breakdown of starch, proteins, and fats. Meristematic cells use these building blocks to synthesize proteins, nucleic acids, membrane components, and cell-wall constituents. As a result, cell division and elongation and, eventually, photosynthesis proceed, and the root and shoot lengthen. The seedling is on its way to becoming a mature plant.

PRIMARY GROWTH: FROM SEEDLING TO MATURE PLANT

Even after the seedling is well established, development continues. In fact, from this point on, most seed-plant growth is similar, whether angiosperm or gymnosperm. Here is what occurs. The root and shoot meristems generate cells, which grow and differentiate into specialized cell types. Most of the size expansion in stems and roots of adult plants is the result of cell elongation in zones just behind the meristematic regions. Thus cells arising by division in the meristem elongate dramatically to yield real growth in size. This elongation at the tips of the root and shoot is called **primary growth.**

Root Growth

The tips of young roots grow rapidly and continuously. Each tip has a root meristem that produces several tissues. Cell divisions at the front of the meristem yield **root cap cells** (Figure 28-16), which act as a shield to protect the delicate meristem and in most plants are positively gravitropic. In addition, these cells synthesize and secrete slime to ease the root's passage through the soil, among other functions. As the root probes the soil, the root cap cells are continually sloughed off or crushed, so that new ones must be generated continuously.

Root tips are often damaged by even gentle handling. As a result, they cease functioning, and root growth stops. Since the youngest root tissues, near the tips, are sites of root-hair formation and thus of maximal water and mineral absorption, the simultaneous injury of many

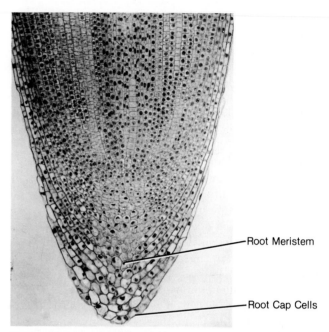

Figure 28-16   THE ROOT TIP: BURROWER THROUGH SAND, LOAM, OR ROCKY SOIL.
The delicate meristem of this onion (*Allium* species) root is protected by root cap cells (magnified about 85 times). These large cells with tough walls are continually abraded and replaced as the root tip pushes through the soil.

root tips can cause severe water stress in a plant. Gardeners must be careful to give plants that have just been transplanted extra water and special care until new root tips can form.

There are three major zones in a rapidly growing root: the meristematic region, the **region of elongation,** and the **region of differentiation** (Figure 28-17). The first two zones allow the root's vertical growth; the third, its development. The meristematic region is a zone of mitotic cell division where growth results from an increase in the number of cells. In the region of elongation, just behind the meristem, are mitotically active precursors of root epidermal cells and a central core of cells that become vascular tissues and cortex. New cells that are left behind the advancing meristematic region of division show growth by an increase in size rather than an increase in number. Once the new cells have increased in length 10- to 100-fold, they begin to differentiate into individual cell types, such as epidermis and xylem. The region of differentiation is usually found 1 centimeter or so behind the meristem. One of the first conspicuous signs of differentiation is the formation of root hairs from the epidermal cells. This occurs in the *root-hair region.*

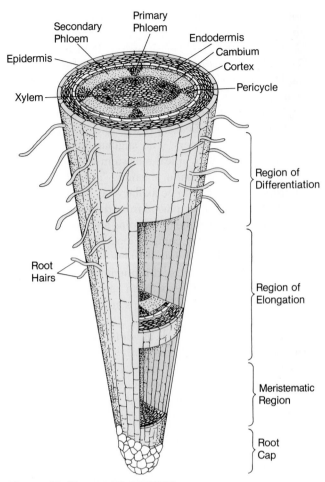

Figure 28-17   ROOT GROWTH.

Three major zones contribute to root growth. New cells arise by mitosis in the meristematic region. They increase in size, and especially in length, in the region of elongation. Finally, the cells differentiate into epidermis and vascular tissue cells in the region of differentiation, where root hairs extend from epidermal cells.

Shoot Growth

Just as in the root system below ground, enlargement of the visible parts of the plant depends on meristematic activity and cell elongation in growth regions. New leaves, stem, and branches are generated by the apical meristem, and the organization of this shoot meristem is slightly different from that of the root meristem. For one thing, the shoot lacks a cap of protective cells because air is not as dense as soil and is not damaging to move through. The apical meristem is considered to sit at the very top of the plant; however, in nature, it is normally covered and protected by many immature leaves folded around it. In preparation for the initiation of a new leaf, a

group of meristematic cells on the apical meristem divides, creating a bulge on the side of the meristem called the **leaf primordium** (Figure 28-18). The overlying epidermal cells divide concurrently, so that new surface cells cover the bulge at all times. As the primordium enlarges, it first resembles a swollen mound and then a flattened pad that begins to develop the leaf features that we later recognize. The midrib of the leaf can be seen as a rounded ridge down the center of the primordium, and

Figure 28-18   APICAL MERISTEM SURROUNDED BY LEAF PRIMORDIA.

(a) This scanning electron micrograph of a succulent, *Graptopetalum paraguageuse* (magnified about 28 times), shows the apical meristem of a shoot as a dome in the center. The meristem cells give rise to the leaf primordia, four of which are seen here in various stages of development (b) Several days later the stem (magnified about 8 times) is elongating and the apical meristem is hidden by the upright, elongating leaf primordia.

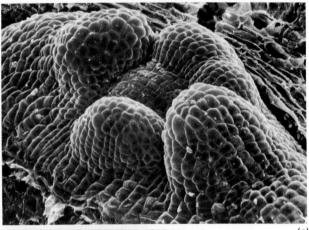

(a)

(b)

bumps at the edges can become lobed or toothed leaf margins. A combination of precisely oriented cell division and expansion gradually sculpts the shape of the leaf primordium. Some leaves, such as those of split-leaf philodendron, undergo the same kind of localized cell death that sculpts the digits or limbs of many animal embryos. However, in most plants, the shapes of leaves, stems, and other organs are strictly the result of differential cell division and oriented elongation.

After a leaf primordium forms on one side of the apical meristem, the next primordium forms a precise distance away, a specific fraction (one-half, one-third, one-quarter) of the way around the cylindrical stem. This process occurs again and again, all new leaves forming at these regular intervals around the stem. The result is a *spiral* or an *opposite* pattern of leaves around the plant stem (Figure 28-19).

Besides giving rise to new leaves, the apical meristem produces the plant's stem tissue—epidermis, cortex, phloem, xylem, and pith. The position at which a leaf arises is called the **node,** and the stem tissue between leaves is the **internode.** There is continuous upward growth of the apical meristem so that successive leaves initially are separated by a small wedge of tissue, all the way around the apex. As the wedge of tissue elongates, it becomes the internodal tissue of the stem. The length and growth rate of the internodal tissue can vary greatly. In a head of cabbage, for example, internodes are extremely short, and the leaves are virtually stacked on top of each other. In many other plants, such as corn, the internodes are quite long, and the leaves are far apart. Compact and distant leaf spacing sometimes can be found on the same plant as a result of variances in the amount of internode elongation, which depends, in turn, on environmental conditions, such as light intensity.

The apical meristem has yet another task besides initiating leaf primordia and producing stem tissue: it is the site of formation of **bud primordia.** These primordia form in the angle between the base of developing leaf primordia and the apical meristem and may grow into branches or form flowers. The angular junction of leaf primordium and shoot apical meristem is called the *axil* (from the Latin word *axilla,* meaning "armpit"). Hence, bud primordia are said to be **axillary.** As the leaf primordium develops and is displaced from the shoot apex, the axillary bud may continue developing and, depending on species, may eventually come to look like the shoot apical meristem and itself initiate new leaf and bud primordia. It is only because of this phenomenon that large, lateral branches of a tree or bush can themselves give rise to numerous smaller branches and they, in turn, to still smaller ones.

Some plant species, such as elms, have well-developed axillary buds, whereas other species, such as palms, rarely do. In many species with a tall, narrow shape, such as Lombardy poplar, even if axillary buds are well developed, they may fail to grow out and are said to be *inhibited.* In other plants, such as the apple tree, axillary buds grow out profusely, and the plant becomes highly branched.

Buds may be **terminal,** or situated at the ends of branches instead of being axillary. The dormant buds at the tips of maple twigs in the spring are an example of terminal buds. Terminal buds are commonly covered

(a)   (b)   (c)

Figure 28-19   LEAF ARRANGEMENTS RESULT FROM PATTERNS OF MERISTEM ACTIVITY.

(a) If both sides of the apical meristem are highly active at the same time, the resultant leaves lie opposite each other (*Gentiana lutea*). (b) If a meristem shows several discrete active regions at one time, multiple leaves at one level of the stem result, in this case a whorl (*Lysimachia quadrifolia*). (c) If first the left, then the right, then the left sides of a meristem are active, a pattern of alternate leaves arises (*Smilacina racemosa*).

with tough scales that help them survive through the winter. Within the scales are the shoot apical meristem and often several sets of well-developed leaves waiting for the proper signal to break through the scales and begin growth. An additional type of bud, also produced by the apical meristem, is perhaps the most familiar: the **flower bud.** As we will see, the apical meristem sometimes ceases production of leaf primordia and itself becomes a floral apex.

To summarize, we can say that a young plant's enormous potential for growth depends on the long-term survival and function of its meristems. These generate growth regions as well as bud primordia that elongate and develop in specific patterns and help shape new organs—stem, leaves, and flowers—and the plant itself.

ANGIOSPERM AND MAMMALIAN DEVELOPMENT COMPARED

In many respects, angiosperms and mammals are the most complex groups of plants and animals. It is remarkable that organisms so distantly related and so complicated should have similar modes of sexual reproduction and embryonic development; yet they do.

Both flowering plants and mammals protect and nourish their embryos with maternal tissue: the mammalian uterus and placenta are analogous to the angiosperm ovary, suspensor, and seed coat. In addition, both groups share certain details of development. The first cleavage of the angiosperm zygote yields a suspensor cell, which gives rise to the suspensor structure that anchors the embryo to the ovarian wall and helps feed it via plasmodesmata junctions. We can draw an analogy between the suspensor and the trophoblast of mammalian embryos. As we saw in Chapter 18, cleavage of the mammalian zygote produces a ball of cells, the outer ones of which yield the outer trophoblast layer, which, in turn, gives rise to the embryonic portion of the placenta, the site of anchorage and feeding. The inside cells of the ball give rise to the inner cell mass. The commitment of maternal angiosperm and mammalian tissues to protecting and nourishing the embryo may mean that relatively few embryos and offspring can be produced. However, these well-protected embryos have a better chance of surviving than they would in the absence of such protection.

The similarities between mammalian and angiosperm development are counterbalanced by major differences. In mammalian embryos, cell movements and rearrangements are critical to the shaping of the organism and its parts. In contrast, plant cells cannot move relative to

each other. Hence, roots, shoots, leaves, and flowers are generated in meristems in different regions of the plant by differential growth and by oriented cell elongations.

The presence of meristems truly distinguishes embryos of higher plants from those of higher animals. In general, embryos of the more complex animals develop the same number of organs as in the adult; these organs are constructed during embryonic development and must last for the animal's lifetime. In plants, on the other hand, the major life strategy involves periodic generation of new organs. As leaves become less efficient with age, these light-collecting organs are replaced by new, more efficient leaves. As shrubs or trees increase so spectacularly in size, the total number of leaves, branches, and limbs increases greatly. And roots grow continuously too, so there is perpetually a fresh zone of absorptive root hairs. A healthy plant, therefore, is always growing and making new organs—leaves, roots, and stems—except when dormant during cold or very hot weather. The plant embryo simply contains the first examples of these organs and, by means of the perpetual growth centers, the meristems, retains the potential to make many more.

Because of this potential, the dividing cell populations of meristems are said to be *totipotent:* each cell is thought of as being able to give rise to a complete new plant or any of its parts. In contrast, most animal cell lines become determined and lose their totipotency in the process of commitment to restricted types of differentiation. This latter strategy in animals is coupled with the setting aside of separate germ cells early in embryonic development, so that at least this single line retains the capacity to give rise to a whole new individual via the gametes. Plants do not need or have germ lines; totipotent cells derived from the meristem are located in the forming flower and give rise to the microspore or megaspore mother cells, which yield pollen or eggs. Their distinctive cells, organelles, and tissues, as well as their capacity for perpetual growth, make plants truly fascinating organisms.

SECONDARY GROWTH: THE DEVELOPMENT OF WOOD AND BARK

Wood is one of the most common and useful materials to humans. We frame buildings and construct smooth, hard floors with it; and we use wood to create furniture, utensils, sculptures, and heat energy. Wood is so common that we sometimes forget it was once live tissue. What is wood, exactly? And how does it fit into the sequence of events we call plant development?

Figure 28-20   RINGS IN WOOD: A RECORD OF XYLEM GROWTH RATES DURING DIFFERENT SEASONS OF THE YEAR.

(a) During the late summer, the water-carrying xylem cells are becoming dormant, and few are produced from the cambium. These are represented by narrow dark rings, the late wood, in this three-year old woody stem. In the spring, the vascular cambium is reactivated and begins to produce many large xylem cells. These account for the lighter colored broad xylem bands, the early wood. The vascular rays are lines of parenchyma cells involved in lateral transport of minerals and water. The phloem rays conduct the products of photosynthesis in the vertical-transport system. (b) A cross-cut of a 39-year-old larch (*Larix* species) tree. Note the darker heartwood in the center, a region composed of older dead xylem. The outer rings of younger growth are called sapwood because they are composed of xylem that still actively conducts water. Note the asymmetric shape of the rings and the trunk itself; perhaps environmental factors led growth to be faster on the left than the right side of this tree.

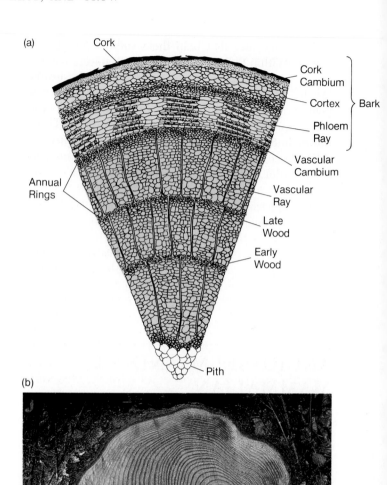

We have seen that all the tissues in a young seedling arise by primary growth from root and apical meristems. As a nonherbaceous plant matures and grows taller, its stem also begins to grow laterally, increasing its diameter—its girth. This thickening of the stem, or **secondary growth,** enables the plant to withstand the added load of branches and leaves, as well as wind, rain, gravity, and other environmental forces. The new, secondary tissues are generated in two growth regions, collectively called **lateral meristems.** The regions are specifically the *vascular cambium* and the *cork cambium.*

Vascular Cambium

We have discussed the critical transportation roles of the vascular tissues, xylem and phloem, as well as their importance in providing physical support (Chapter 27). As a plant matures, individual cells of both xylem and phloem cease to transport materials; phloem elements usually function for only a year or two before dying, and xylem elements often fill up with materials that diminish the vital flow of water and nutrients. The cylinder of generative cells called the vascular cambium overcomes the problem of vascular-tissue maturation by producing new xylem and phloem cells and, in the process, adding girth to the plant. The vascular cambium is a cylindrical zone of actively dividing cells that lies between the xylem and the phloem (where the procambium was located in the early vascular bundles) in both the stem and the root. Mitotic divisions in the vascular cambium produce new, undifferentiated cells. Those on the inner side of the cambium become secondary xylem, or **wood,** while those on the outer side differentiate into secondary phloem, one component of *bark.*

Each spring, the vascular cambium, dormant over the winter, is reactivated and begins producing new xylem and phloem. Spring xylem cells are large in diameter and constitute **early wood,** while summer xylem cells are much smaller and form the **late wood.** The difference in cell size probably results from the availability of water: larger cells form in the moist spring; smaller ones, in the drier summer. The difference in cell size between early and late wood is quite apparent and results in the distinct rings we can see in cross sections of tree trunks (Figure 28–20). A count of tree rings actually reveals the number of times the vascular cambium has been activated to produce early wood, the lighter part of each ring. The rings of younger xylem, nearer the periphery of the trunk or stem, are called the **sapwood;** sapwood xylem still functions in water conduction. A darker region, at the center of the trunk, is composed of dead, nonconducting xylem and is called **heartwood.** This is the older xylem that is often clogged with various substances and no longer transports water and nutrients. Thus the novice trying to collect sap for maple syrup would have little success if he or she bored deep into the trunk—the old, dead heartwood. The material we call wood consists of both the heartwood and the sapwood portions of tree trunks.

Since the diameters of the stems, branches, and roots increase each year as plants grow older, transport *across* the tissue—**radial transport**—becomes as much of a problem as **vertical transport** between roots and leaves. The vascular cambium contains two cell types that function in the formation of cells of the vertical- and radial-transport systems. The **fusiform initial cells** produce the cells of the vertical-transport system—the xylem and phloem cells. Fusiform initial cells are oriented vertically, so that when they divide, the new xylem or phloem cells are oriented vertically. In contrast, **ray initial cells** are squarish cells that produce **vascular rays,** which function as the system of lateral transport (Figure 28-20). These rays are lines of parenchyma cells extending from the center of the stem toward the periphery. Minerals and water in the vertical-transport system can be diverted into the rays, which then distribute the materials laterally to the sapwood and to living cells of the bark. The ray cells also serve as storage depots for excess starch.

Cork Cambium

The other type of lateral meristem that produces secondary growth in plants is the **cork cambium,** a layer of cells just beneath the epidermis (Figure 28-20). As the trunk and roots increase in diameter due to the activity of the vascular cambium, the original epidermis and cortex become too small to encompass the expanding trunk or roots and split open. Cell divisions in the cork cambium then produce a secondary layer, the **cork,** which replaces the epidermis. Cork cells become impregnated with suberin, the same waxy substance that helps prevent water loss from young stems and roots. Such waterproofing, of course, also inhibits the passage of oxygen and carbon dioxide into and out of the plant. Therefore, in older cork-covered stems and roots, only certain spots in the cork, called **lenticels,** allow gas exchange. Lenticels serve the same functions as stomata in the epidermis of young stems and are often visible as dark spots on tree trunks, branches, or roots (Figure 28-21).

The cork cambium develops from the cortex in stems and forms the pericycle in roots. In certain trees, if one layer of cork cambium is destroyed, the phloem parenchyma cells can give rise to a new one. This is why the bark of some old trees is layered, resulting in bark patterns characteristic of particular species. The outer cork layer from cork oak trees can be harvested periodically in sheets or round plugs and is used to make bulletin boards, floor tiles, stoppers for wine bottles, and so on.

The familiar **bark** of the woody stem includes all the tissues outside the vascular cambium: the cork, cork cambium, cortex, and phloem (Figure 28-20). The outer bark (cork tissue) is dead, while the inner bark is alive.

(a)

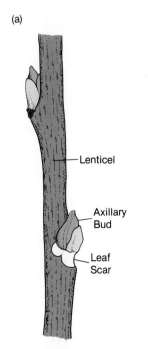

Lenticel

Axillary
Bud

Leaf
Scar

(b)

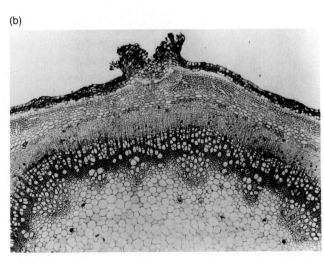

Figure 28-21   LENTICEL.
(a) Lenticels, which allow gas exchange through cork-covered stems and roots, are easily seen on this slender stem from a tree. (b) A transverse section of a lenticel, magnified about 11 times, from the stem of an elderberry (*Sambucus canadiensis*) vine.

Most trees cannot withstand the loss of the bark layer because the phloem is lost, too. Thus the removal of a ring of bark all the way around the tree, called *girdling,* is often fatal. Of course, if the underlying vascular cambium remains undamaged, new phloem can be produced, but a scar may mark the site of stripping. Deer, porcupines, and other animals sometimes girdle young trees by feeding on tender bark.

Just as the growth of new tissue from meristems adds up to a kind of perpetual development in most plants, the activities of the vascular cambium and the cork cambium in woody plants contribute to ever increasing size and the laying down of wood. The giants of the living world—the redwoods, up to 100 meters tall, and the sequoias, nearly 30 meters in circumference—developed in this way. And they retain an organic history of the environment through which they stood—the harsh winters, wet springs, and dry summers—in the growth rings of their wood.

PLANT LIFE SPANS AND LIFE STYLES

Throughout our discussions of plants, we have mentioned herbaceous species, which stay green, and others that grow woody and often much larger. Clearly, plants have different growth habits, and anyone who has purchased flower or vegetable seeds probably has encountered three terms that characterize these differences: "annual," "biennial," and "perennial."

In **annual plants,** the seed germinates in the spring; the plant grows; the flowers and seeds develop; and the plant dies—all within a single growing season (Figure 28-22a). The entire cycle is repeated annually, with only new sets of seeds surviving from one year to the next. Many weeds, grains, flowers, and garden vegetables are annual plants. The life span of the annual allows the plant to spread rapidly and to produce seeds within a relatively short period.

Annual plants may live for just a few weeks or for nearly an entire year. Much of their energy goes toward flowering and producing seeds. Thus the plant body grows only large enough to support the maturation of the seeds; very little secondary growth or wood and bark production occur, since those tissues are unnecessary investments for a small plant that is programmed to die after reproducing. As soon as the seeds appear on an annual, the plant typically begins to age and die.

Biennial plants are much less common than annuals and require two years from the time the original seed germinates to the production of a new generation of

seeds. Parsley and foxglove are typical biennials (Figure 28-22b). The plant body grows and becomes established in the first season; in the second year, the plant flowers, produces seeds, and dies soon thereafter.

One might wonder why annuals and biennials die in just a year or two, when other plants live much longer. The reason is simple: they cease to grow. And this occurs because the apical meristem is converted into a flower. When the apical meristem is making leaves and inter-

Figure 28-22   ANNUALS, BIENNIALS, AND PERENNIALS.
(a) Annual plants such as these poppies (*Papaver* species) live just one growing season; the seeds they produce survive to repeat the cycle. (b) Biennial plants such as foxglove (*Digitalis purpurea*) require two years from seed germination to production of the next generation of seeds. The plants then die at the end of the second season. (c) Perennials such as these blossom-covered plum (*Prunus cerasifera*) trees can survive indefinitely because their flower production does not use up a significant portion of the plant's apical meristem tissue.

nodes, the meristematic tissue is not used up, since it continuously regenerates itself through cell division. The stimulus to produce a flower, however, causes a dramatic increase in the use of the meristem. The leaflike parts of the flower, the sepals and petals, are differentiated from all around the base of the apical meristem rather than from a single side. The stamens and carpels are then derived from the remaining meristematic cells. The flower represents the transformation of the meristem from vegetative to sexually reproductive growth, and the whole meristem is "sacrificed" to make the flower. There are no dividing cells left to make more leaves or stems. As soon as all the shoot meristems (apical and bud) have flowered, the annual or biennial plant is doomed.

Perennial plants, in contrast, live for many years, and some, such as trees, have large bodies with a considerable degree of secondary growth. Other perennials, such as grasses and herbs, may not be particularly woody or large (Figure 28-22c). Perennials often require several years of growth before they can flower and produce fruits and seeds. Flowering in perennials does indeed "consume" apical meristems; however, this does not affect the plant's survival because flowering typically involves only a small subset of the lateral branches, not the entire complement of meristems. The perennial plant can produce leaves and flowers for many years from different subsets of lateral branches arising each year.

It could be said that perennial plants have an indefinite life span—only a disaster can kill them. However, as their stem vascular tissues age, many woody plants become less and less efficient at water and nutrient transport. As roots grow larger and longer, water must be transported farther and farther through the vascular system to the leaves. Size in a sense becomes the plant's enemy, as does the cumulative damage caused over the years by weathering, diseases, and predators. Most woody perennials have a typical life span of ten to fifty years, although many trees can survive for hundreds of years. A few exceptional trees, such as the giant sequoias of California and the bristlecone pines of Nevada and California, are thousands of years old.

Annuals, biennials, and perennials share the potential for having new, chance mutations in meristem cells incorporated into flowers formed from such meristems, and so into pollen and eggs. Thus one branch of a large tree or even one flower on a herbaceous plant could produce gametes with a different genotype than the rest of the plant. This is potentially a major source of genetic variability that is not present in animals with germ lines, for in them somatic mutations do not enter sperm or eggs and so make no direct contribution to evolution.

LOOKING AHEAD

Thus far in our discussion of plant biology, we have surveyed the structure of roots, stems, and leaves. And we have seen how plants grow, develop, and reproduce, sometimes vegetatively through a host of plant parts and sometimes sexually through flowers, fruits, and seeds. We have also seen that plants have a special form of development quite different from that of animals—a perpetual development based on embryonic centers that leads to renewed growth throughout adult life, whether programmed for one season, two, or a thousand. In the next chapter, we will see how plants live day to day—in particular, how they absorb and transport water, minerals, and sugars. And from this, we shall form an increasingly detailed picture of how plants have evolved ways of coping so successfully with Earth's many and varied environments.

SUMMARY

1. During *vegetative reproduction*, plants genetically identical to each other and to the parent arise from the parent's body. Such plants can increase their numbers rapidly in favorable environments, but the population of a species is more vulnerable to being eliminated by disease or by a shift to unfavorable conditions.

2. Vegetative reproduction can involve a number of plant parts, including *rhizomes*, or subterranean stems; *stolons*, or aerial stems; *runners*, or horizontal stems from the plant's base; *bulbs*, conical underground stems with modified leaves; *corms*, solid underground stems; and *tubers*, dense structures with budding "eyes"; leaves; and *root suckers*.

3. Sexual reproduction in angiosperms involves *flowers* that attract the appropriate pollinator. Flowers are modified leaves consisting of sepals (collectively, the *calyx*); petals (collectively, the *corolla*); stamens, the pollen-bearing organs; and carpels, which fuse to form the pistil, site of egg production. Together, the calyx and the corolla are the *perianth*.

4. Most stamens have a stalklike filament holding up the anther. Pollen is produced in four microsporangia on each anther. Pollen production involving *microsporogenesis*, during which the diploid microspore mother cell divides meiotically to produce four haploid microspores, and *microgametogenesis*, during which each microspore divides mi-

totically to form the tube cell and the generative cell, encased together in the pollen grain.

5. The pistil is composed of the stigma, to which pollen sticks; the style, connecting the stigma to the ovary; and the ovary, site of egg production and fertilization. Egg production involves *megasporogenesis*, during which the diploid megaspore mother cell divides meiotically to produce four haploid cells, three of which degenerate. The fourth is the megaspore; during *megagametogenesis*, mitotic divisions of the megaspore produce one cell, the megagametophyte, with eight nuclei. Further steps yield a central cell with two *polar nuclei*, the egg, and five other cells.

6. When a pollen grain adheres to a stigma, the pollen is stimulated to grow, or *germinate*. A pollen tube grows down the stigma and style, enters the ovule, and two sperm are released. One fertilizes the egg, forming the zygote; the other fuses with the cell containing polar nuclei, forming the triploid endosperm cell. This unique process is called *double fertilization*.

7. The zygote divides and one cell gives rise to the *suspensor*, which attaches the embryo to the ovarian wall. The other cell derived from the zygote gives rise to a ball of cells—the globular stage embryo—which remains attached to one end of the suspensor. *Cleavage divisions* in the embryo are followed by differentiation into *inside cells* and *outside cells*. Outside cells form embryonic epidermis *(protoderm)*; some inside cells give rise to embryonic vascular tissue as well as the seed leaves—two *cotyledons* in dicots and one *scutellum* in monocots.

8. The endosperm serves as the source of nutrients for the embryo inside the seed coat. (Dicots generally store nutrients in the cotyledons.)

9. *Seeds*, the mature ovules, develop within the *fruit*, which is usually derived from the wall of the ovary. Fruits—fleshy, leathery, or dry and papery—protect the seeds and aid dispersal. Seeds must undergo food storage and dehydration before dispersal. The embryo enters a *dormancy* period until conditions become favorable for germination.

10. Germination begins when the seed takes up water, swells, and meristematic activity commences. The seed coat cracks, and via cell elongation, the *radicle* (root), *hypocotyl* (nonphotosynthetic stem), and *epicotyl* (future stem) enlarge and emerge. The hypocotyl is protected by covering *root cap cells* as it probes into the soil.

11. The light-seeking *coleoptile* protects the monocot shoot pushing upward through the soil. Endosperm and scutellum nourish monocot seedlings, as do newly photosynthetic leaves. The *aleurone* layer provides enzymes for food mobilization. Such food is used in energy metabolism and cell-wall manufacture after germination. Nutrient reserves in the cotyledons nourish dicot seedlings; the cotyledons become the first photosynthetic leaves.

12. *Primary growth* involves cell elongation at the tips of roots and shoots. A rapidly growing root has a meristematic region (zone of cell division), a *region of elongation* (zone of cell-size increase), and a *region of differentiation* (zone of tissue cell specialization).

13. *Apical meristem* produces new leaves and branches. A leaf arises from a *leaf primordium* or bulge on the side of the meristem. The leaf arises at the *node;* the stem tissue between leaves is the *internode*. Branches arise from bulges called *bud primordia*. Bud primordia may be *axillary, terminal,* or *flower* in type. *Root meristems* function like apical meristems and give rise to cells which form all the cell types in roots, including new epidermal ones with root hairs.

14. *Secondary growth*, or thickening of stem, involves *lateral meristems*, which generate xylem and phloem, and *cork cambium*, which produces a layer of *cork*.

15. *Fusiform initial cells* in the vascular cambium produce the cells of the *vertical-transport* system (xylem and phloem). *Ray initial cells* generate *vascular rays*—parenchyma cells in the *radial-transport* system.

16. *Wood*, or secondary xylem, in a thick stem shows annual growth rings due to cell-size differences in *early wood* and *late wood*. Functional xylem near the stem or trunk periphery is called *sapwood;* the central *heartwood* is dead xylem.

17. Waterproof cork replaces the epidermis in a thick trunk. *Lenticels* allow gas exchange through the cork. *Bark* is made up of cork, cork cambium, cortex, and phloem.

18. *Annual plants* mature rapidly, produce seeds, and die within a single growing season. *Biennial plants* do the same over a two-year period. In both, when apical meristems give rise to flowers the plant stops growing and dies. *Perennial plants* produce new meristems each year and are long-lived.

KEY TERMS

aleurone	apical meristem	bark	bud primordium
annual plant	axillary bud	biennial plant	bulb

calyx
cleavage division
coleoptile
cork
cork cambium
corm
corolla
cotyledon
dormancy
double fertilization
early wood
epicotyl
flower
flower bud
fruit

fusiform initial cell
germination
heartwood
hypocotyl
inside cell
internode
late wood
lateral meristem
leaf primordium
lenticel
megagametogenesis
megasporogenesis
microgametogenesis
microsporogenesis
node

outside cell
perennial plant
perianth
polar nuclei
primary growth
protoderm
radial transport
radicle
ray initial cell
region of differentiation
region of elongation
rhizome
root cap cell
root meristem
root sucker

runner
sapwood
scutellum
secondary growth
seed
stolon
suspensor
terminal bud
tuber
vascular ray
vegetative reproduction
vertical transport
wood

QUESTIONS

1. Give examples to show how (a) roots, (b) stems, and (c) leaves may be used by specific types of plants for vegetative (asexual) reproduction.

2. Give examples to show how vegetative reproduction is employed in modern agriculture.

3. What are the advantages to a plant species of vegetative reproduction? sexual reproduction? Which has the greater evolutionary advantage for the species? Explain.

4. Name the four whorls of modified leaves that make up a typical flower. For each, briefly describe both the structure and the function.

5. Describe the roles of the following in the life cycle of an angiosperm: pollen tube, polar nuclei, double fertilization, seed coat.

6. Briefly describe each of the following parts of an angiosperm embryo:
   a. protoderm
   b. procambium
   c. cotyledon
   d. scutellum
   e. apical meristem

7. Meristematic cells are said to be *totipotent*. Define this term.

8. Where are meristems found in the plant body? What role does each play in growth, development, and/or reproduction of the plant?

9. What is the physical and metabolic condition of a dormant seed? How may seed dormancy be broken? How is the breaking of dormancy regulated in the seed?

10. List, in chronological order, the major steps in seed germination and seedling development.

ESSAY QUESTIONS

1. Compare and contrast (a) primary plant growth from apical meristems, and (b) secondary plant growth from lateral meristems.

2. Explain the steps in the formation of (a) a pollen grain (male gametophyte), and (b) an ovule (female gametophyte) in a typical angiosperm. How many haploid nuclei does each have at maturity? Describe the role of seed dormancy in a typical angiosperm.

SUGGESTED READINGS

BEWLEY, J. D., and M. BLACK. *Physiology and Biochemistry of Seeds.* New York: Springer-Verlag, 1978.

A series of volumes that includes everything you need to know about seeds and their germination.

BRISTOW, A. *The Sex Life of Plants.* New York: Holt, Rinehart and Winston, 1978.

This popular book does an excellent job of explaining the subject.

CLOWES, F. A. L. *Morphogenesis of the Shoot Apex.* Oxford Biology Reader, no. 23. Burlington, N.C.: Carolina Biological Supply, 1972.

A fine account of apical-meristem function.

ESAU, K. *Anatomy of Seed Plants.* 2nd ed. New York: Wiley, 1977.

A complete treatment of secondary growth and wood formation, as well as an extensive coverage of flower structure and function.

29
EXCHANGE AND TRANSPORT IN PLANTS

*From its Office, which is To feed the Trunk . . . the sap must
also, in some Part or other, have a more especial motion of
Ascent.*

Nehemiah Grew, *The Anatomy of Plants: With an
Idea of the Philosophical History of Plants* (1682)

A grisly fate for a fly stuck to sweet-smelling nectar. Action potentials like an animal's traveled down this sundew's *(Drosera rotundifolia)* "tentacles," causing them to bend and entrap the hapless insects so they can be digested and their nitrogenous compounds absorbed by the carnivorous flower.

In 1628, the English physician and scientist William Harvey published a book that describes how an animal's blood circulates because of the heart's pumping action. Naturally enough, botanists in Harvey's time began to search for a pumping mechanism and transport vessels in plants analogous to the heart and blood vessels in animals. Although they were disappointed in this search, by the 1720s, they had learned that water travels from the roots to the leaves. However, the details of internal-transport processes in plants were not worked out until the twentieth century. This chapter describes the principles of transport and exchange in plants and explains how all plant organs obtain the gases, nutrients, minerals, and water essential for life. As we examine these processes, it will become evident that basic physical properties and events—adhesion, cohesion, evaporation, osmosis, and so on—lie at the core of the fluid movements so essential to the continuing life of every plant. After considering the basic ways in which plants meet their physiological needs, we will examine the theories that attempt to explain transport in plants.

PLANT STRATEGIES FOR MEETING BASIC NEEDS

All organisms, plants and animals alike, interact with their environments in a variety of ways: they exchange gases, acquire nutrients and water, and eliminate waste products. Plants generally expend much of the energy they harvest from the sun in building new cells, cell walls, and so on. Since plants are stationary, they do not expend energy on locomotion; but a cost and danger is that plants cannot escape local environmental stresses. A seed may travel passively a long distance; but once it has germinated, the resulting plant is established for its full life in a specific location, where it may have to cope with external conditions ranging from heat and drought to extreme cold, snowstorms, and downpours.

Plants have evolved strategies that are inexpensive energetically for exchanging gases, moving food and water, and eliminating wastes (Figure 29-1). For example, their surface area is often enormous in order to ex-

Figure 29-1 BASIC PLANT LIFE STYLES: STRATEGIES FOR SUCCESS.
Nearly all plants remain rooted in one place, grow continuously during clement seasons, and exchange gases and water passively with the environment. Plants have evolved special structures to meet the basic needs for exchange and transport: stomata, lenticels, and spongy leaf tissue for air (O_2 and CO_2, in particular) exchange and distribution; roots and root hairs for water and mineral absorption from the soil; and xylem and phloem for internal transport of water and nutrients. Plants generate a lower volume of waste products than do animals because of the recycling of nitrogen and other minerals and the tendency to sequester wastes in special ducts or vacuoles. For simplicity, entry and exit of the various substances are shown on separate leaves.

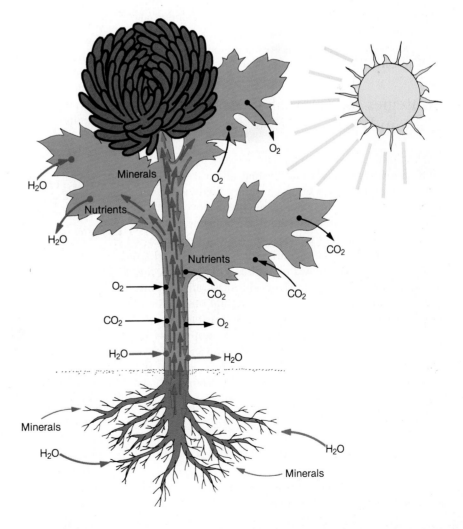

change sufficient quantities of gases, water, and nutrients, all of which are distributed diffusely in the environment. By "energetically inexpensive," we mean processes that do not depend on a large-scale expenditure of ATP or other high-energy compounds that are metabolically costly to form. Thus plants tend to use passive processes—processes that operate by the physical rather than the chemical laws of nature.

Strategies for Gas Exchange

The green parts of a plant exchange gases through thousands of stomata. In the nongreen parts of the shoot system, limited gas exchange occurs through the epidermis or lenticels. The roots, too, must take up oxygen from the soil (Chapter 28) in addition to absorbing water and minerals.

All the normal gases in air diffuse into and out of the stomata. But there is net movement of carbon dioxide inward, since that gas is used up in photosynthesis. And, of course, oxygen liberated by photosynthesis diffuses back out into the atmosphere. Once carbon dioxide or other gases have entered the leaf or stem, they need diffuse only short distances through intercellular air spaces in order to reach all tissue cells. Thus there is no true circulatory system of the sort common in many animals to carry gases long distances within plants.

Strategies for Internal Transport of Liquids

Although gases do not circulate actively in plants, we know from Chapters 27 and 28 that all vascular plants have two sets of vessels in which watery fluid moves: the *xylem*, where water and mineral nutrients move; and the *phloem*, where organic molecules are transported. Biologists consider each of these separate vessel systems to be a pipe line. Thus water continuously enters the xylem in the roots, passes upward, and is lost through the leaves; sugars and other nutrients enter the phloem from their site of manufacture in the leaves and other photosynthetic sites and are removed in the meristems, stems, and roots either for use or for storage in nonphotosynthetic sites. We will see later that water and sugar movements depend on passive physical processes that are exploited by the structure of plant cells. There is little coordinated cell activity to cause water or sugar to move, as is common in animals.

Strategies for Coping with Wastes

All organisms produce waste products, some of which are potentially toxic. In general, plants have fewer ex-

cretory problems than do animals. Excess food is stored temporarily in parenchyma tissue in roots and tubers or as starch in chloroplasts of photosynthetic cells. Some plants also convert carbohydrate to fat, a less bulky form of stored carbon. Nitrogenous wastes are not a major problem in plants. Plants acquire nitrate (NO_3^-) from the soil and manufacture their own amino acids. Nitrogen is usually a limited nutrient in the soil for reasons we will discuss later, so plants are rarely favored with an excess. Moreover, the continuous growth of leaves and roots demands a steady supply of amino acids and other nitrogen-containing organic molecules. Plants are often so short on nitrogen that much of their new growth is supported by recycling amino acids and other nitrogen compounds from older tissue (lytic enzymes in vacuoles may contribute to this process). A yellowing leaf often is undergoing a process of dissolution, during which its existing proteins are broken down to amino acids and are shipped to the growing portions of the plant and incorporated into new proteins.

Plants do make some toxic waste products, such as tannins and phenolics (both small molecular derivatives of phenol, the basic ring compound of lignin), that could injure the cytoplasm of living cells. However, plants have no means of excreting such products through the leaves or stem. Therefore, some individual cells must act as storehouses for noxious products. Resins and tars, for example, may be sequestered, without being broken down, in the central vacuole of a living cell or secreted (as in conifers) into a resin duct, where they remain indefinitely. These long-term storage strategies are reminiscent of the formation of lipofuscin granules, the modified lysosomes of aging animal cells that also store waste products until the cell dies (Chapter 6). However, most wastes in plants are recycled—turned into new cytoplasmic components or stored for future use.

Now that we have outlined the basic plant strategies, let us see how these strategies function on a daily basis.

TRANSPORT OF WATER IN THE XYLEM

Have you ever had the experience of returning home after a few days' absence to find your favorite plant wilted in its pot, only to perk up within minutes after a thorough watering? How can drying and wetting have such dramatic effects on a plant? After all, most animals do not change shape and droop when thirsty. To understand why plants wilt or become erect when supplied with water, we need to review the principles of osmosis and water movement across membranes introduced in Chapter 5.

Principles of Water Movement

Recall from Chapter 5 that water moves across a semipermeable membrane from a region of low solute concentration to one of high solute concentration. Thus if a cell has a high solute concentration relative to the cells or solution surrounding it, water will move into the cell by *osmosis*. The solute concentration either inside or outside the cell is measured in terms of **solute potential.** (In Chapter 5, we referred to solute concentrations; scientists working on plants find solute potential to be a more convenient concept.) As water moves inward, the cell begins to swell. Assuming that the solute potentials of the cell and the external solution remain unequal, one of two things can now happen: if enough water moves in, the cell can burst; or water can be forced out of the cell so that it does not burst. How might this happen? As more water moves into the cell, pressure is exerted against the cell wall, and the cell becomes *turgid*, or stiffened and distended. The *turgor pressure*, or pressure exerted by water in the cell, increases. Imagine that the turgor pressure of water inside the cell becomes greater than the pressure outside. When this happens, water is forced out of the cell and moves down a gradient in pressure. So water continues to move into the cell because of a difference in solute potential, but, at the same time, once the cell has reached its maximum turgid size, water is forced out of the cell down a pressure gradient. When there is no net difference in the amount of water entering or leaving the cell, the cell is said to be at equilibrium with its environment. At equilibrium, the turgor pressure acting outward precisely counterbalances the tendency of water to pass inward due to the cell's solutes. These kinds of experiments show that two values affect the overall water status, or **water potential,** of a cell: the turgor pressure and the solute potential (more precisely, the term **osmotic potential,** calculated as a pressure, is used instead of solute potential). These two values are related to the water potential by the following equation:

$$\text{water potential} = \text{turgor pressure} + \text{osmotic potential}$$

By convention, the water potential of pure water is set at zero. Any solutes dissolved in the water make it less pure, and the value of the water potential becomes a negative number. Normal plant tissues have a negative water potential (say, -5 bars [kilograms per square centimeter]); this results from a small positive turgor pressure (say, $+5$ bars) added algebraically to a larger negative osmotic potential (-10 bars). Consequently, water tends to enter the cell.

If a plant cell with a water potential of zero (water tends to neither enter nor leave) is placed in a hypertonic solution, water will leave the cell and pass into the medium, causing the cell to become *plasmolyzed*. In plasmolyzed cells, the cytoplasm shrinks (because of water loss from the large vacuole) and pulls away from the rigid cell wall (Figure 29-2). If the cell is placed in a *hypotonic* medium (whose solute potential is greater than that in the cell), the cell takes up water from the medium, thereby becoming turgid. Thus returning to the example of the wilted house plant, we see that its water potential is more negative than that of soil because it has lost water. No wonder, then, that the plant takes up water quickly when it becomes available and that turgidity is rapidly restored. We shall soon see that turgor pressure and solute potential are principles critical to our understanding of how plants take up the water and nutrients they must have to survive.

Figure 29-2   SWELLING AND SHRINKING: BASIC RESPONSES OF PLANT CELLS.
Plasmolysis occurs when a cell loses water and the volume occupied by the cytoplasm and its vacuole is reduced. The plant vacuole contains mostly water and may occupy up to 90 percent of a cell's volume. This water can be lost without permanent damage to the cell. The plasma membrane pulls away from the cell wall, and the cell shrinks. If the cell is returned to a solution of higher water potential (hypotonic), it takes up water and returns to its normal shape and volume.

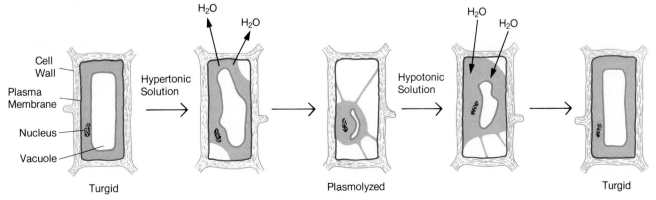

Theories of Upward Movement

How does water get from the roots to the top of a tall tree? Botanists have puzzled over this question for many years. They have long known that normal atmospheric pressure can support a column of water approximately 10 meters high. Conceivably, atmospheric pressure alone could raise water to the top of most low plants, bushes, and short trees. However, it could not account for water reaching the tops of oaks, much less lofty redwoods—thus the botanists' dilemma.

Any explanation of water flow in plants must account not only for the heights that water reaches, but also for the rapid rate of water movement sometimes observed—as much as 1 meter a second. Several possibilities have been suggested for the motive force behind water movement in the xylem. Among them are theories based on root pressure, capillary action, and transpiration.

Root Pressure

According to the **root-pressure theory,** water pressure builds up in the roots and pushes water upward toward the leaves. In plants such as the tomato, considerable root pressure can be demonstrated by removing the stem just above the roots and attaching a tube that contains mercury and that measures atmospheric pressure (rather like a barometer). The mercury in the column rises as it is pushed upward by water from the roots. If the roots are killed by heat or other damage, the water ascent stops.

This theory suggests that living root cells, carrying out normal metabolic processes, are required to maintain root pressure. In fact, root pressure probably is a result of the active uptake of mineral ions by the roots and the transfer of those ions into the root xylem. This energy-dependent ion uptake increases the water potential within the xylem. As a result, water ultimately passes through the endodermis layer, with its Casparian strip (Chapter 28), and enters the root xylem. From there, it is forced upward through the stem xylem by more water entering below. The result can be seen in some plants, as water droplets sometimes form on pores at the edges of leaves. This droplet formation, or **guttation,** results from the root pressure and water flow upward (Figure 29-3).

Although root pressure might be sufficient to explain bulk water movement in a few plants such as the tomato, most large plants have root pressures that are much too low to force water to flow upward to significant heights. Tremendous pressure would be required to force water to rise to the top of a large tree; yet if a branch is sawed off, water does not rush out of the xylem. Botanists conclude, therefore, that the root-pressure theory cannot account for water movement in the stems of most plants.

Figure 29-3   GUTTATION: SEEPAGE OF WATER FROM A LEAF.
When minerals become highly concentrated in the root xylem, so much water enters and passes upward in xylem vessels that water is forced out of the leaf margins, as in these strawberry (*Fragaria ananassa*) and grass leaves.

Capillary Action

A second suggestion for how water might move in plants is **capillary action,** the tendency of water to move upward in a thin tube. Upward movement within a tube occurs because of the *cohesion* of individual water molecules to each other and the *adhesion* of water to the sides of the tube. In plants, the thin tubes are the *tracheids* and *vessel elements,* the cells that, stacked end to end, make up the xylem. The thin strands of cellulose and other polysaccharides that compose the walls of xylem tracheids and vessel elements allow water to adhere tightly and creep upward. However, the physical limitation to such upward capillary flow is only about 0.5 meter. Thus, like root pressure, capillary action could explain water movement in small plants but is clearly insufficient to explain water movement in larger plants or trees.

Transpiration

The best hypothesis for how water makes its extensive upward climb in plants is the **transpiration-pull theory** (also called the **cohesion–adhesion–tension theory**). Several experimental observations suggest that water is pulled up through the xylem from the top of the plant rather than pushed up from the roots. As we have noted, when a xylem vessel in the stem is cut or punctured, water does not flow out, as would be expected if it were pushed up from the roots; rather, air moves into the stem to fill the space vacated by water that has moved upward in the stem above the cut. This implies that the water is pulled through the plant from above. Other experiments reveal that the rate of fluid movement in the xylem is greatest when the plant is in bright sunlight and is losing much water through evaporation; this rate of fluid movement falls at night, when evaporation slows.

Several other observations and experiments have established that **transpiration**—evaporative water loss through the leaves—is crucial to water movement upward in the xylem. Water is constantly lost when stomata on the surface of leaves and stems are open, allowing gas exchange with the atmosphere. The loss can be explained in terms of the water-potential equation. Water in the atmosphere is at a much lower concentration (a more negative water potential) than it is in the interior of the leaf. Thus when the stomata are open, water moves toward the atmosphere. As water vapor departs the leaf by the process of transpiration, more water moves from the root to the leaf to replace the lost water. This process is illustrated in Figure 29-4.

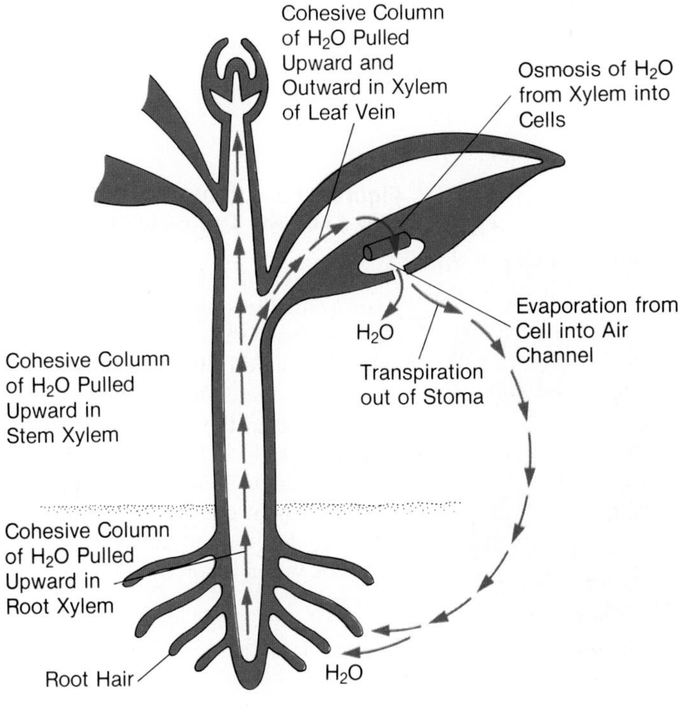

This transpiration-pull theory depends on properties of water molecules. In the xylem, the water column exists as a long liquid "chain" held together by the high cohesion among individual water molecules. The hydrogen bonds that link water molecules have a tensile strength (Chapter 2) of 140 kilograms per square centimeter (about 2,000 lb. per sq. in.)—more than enough to support a thin column of water 100 meters high. When a water molecule in the leaf's air channels changes from the liquid to the gaseous state and transpires through a stoma, the entire column in that xylem tube is drawn upward a tiny amount and another water molecule then moves into the roots. As a result of one water molecule pulling on another, the entire water column in the xylem is stretched, like a piece of elastic. Plant physiologists say that water in the xylem is under *tension*. This process of pulling the water column upward goes on as long as the soil contains enough moisture to maintain the continuous flow.

If upward water movement through the xylem requires an unbroken column of water molecules, what happens if this column is broken? We saw that a cut in the xylem, as by an ax or the breaking of a branch, causes air to be pulled in. Freezing may do the same thing, and is one reason why many plants cannot tolerate cold climates. Air bubbles form near an injury, enlarge, and move up the water column, eventually blocking the cut vessels' entire diameter. The nearby plant tissue usually responds by sealing off the injured section with resins and tars. These substances help prevent infection by fungi or bacteria and stop water from seeping out of the area. The pits on the side walls of the xylem vessels allow water rising from the roots to move laterally and thus by-pass the sealed area and continue its upward movement in nearby uninjured vessels.

Figure 29-4   TRANSPIRATION: KEY TO UPWARD WATER MOVEMENT IN PLANTS.
When the air surrounding a plant is saturated with water vapor, water will not evaporate from the leaves, since their immediate environment has such high water potential. In a dry atmosphere, however, water will evaporate from the leaves. Water is taken up through the roots of the plant by osmosis and moves up the stem to the leaves, where it evaporates into the intercellular spaces of the leaf. Then water vapor leaves the leaf through the stomata. Water moves up because of its cohesive properties; that is, diffusion from cells into the leaf's gas-filled spaces sets up a transpirational pull, which helps pull the column of water upward. Adhesion of the water column to the walls of xylem vessels also plays a role in the process.

Regulation of Water Loss

If the upward water movement in a plant depends primarily on the loss of water from leaves, then that loss must be carefully regulated. Every day, plants regulate the amount of water vapor lost from the leaves by the opening and closing of stomata. The two guard cells bordering a stomatal pore have thick walls on the side next to the stomatal opening and much thinner, more flexible walls on the opposite side. Most biologists believe that stomata open and close in response to the movement of ions across the guard-cell membranes. When ions such as potassium (K^+) and chloride (Cl^-) enter the guard cells, water follows by osmosis. Resultant turgor pressure causes the cells to swell and bulge sideways, pushing the thin, flexible walls outward and pulling the thick, inner walls apart (Figure 29-5). Because this bulging occurs on both sides of the pore, the pore opens wider, permitting gas exchange by diffusion—water vapor leaves, net CO_2 flows inward, and net O_2 flows outward.

Experiments show that there is about a twentyfold increase in K^+ ions in guard cells when they swell. Cells can tolerate such an excess of positively charged ions because Cl^- ions balance about one-third of the K^+ charge, and organic acids (with $-COO^-$ groups) balance the remainder. These acids arise by degradation of starch that occurs slowly as the guard cells swell.

What causes the guard cells to gain or lose K^+ ions and therefore to open or close stomata and, in turn, to regulate transpiration and photosynthesis? The clearest answers come from experiments with protoplasts prepared from isolated guard cells. Enzymes are used to remove the thick, cellulosic cell walls, leaving the naked guard cell in its flexible plasma membrane—this is the *protoplast*. Experimental findings imply that the strong-est environmental signal to cause guard-cell shrinking is low water potential within the leaf. This fits with the so-called midday closure of stomata frequently observed in many plants; such closure occurs when leaf cells lose turgor, a condition that arises when water loss through stomata exceeds water intake through the roots. How is a leaf's water potential communicated to the guard cells? Certain evidence hints strongly that the plant hormone *abscisic acid* (Chapter 30) accumulates in leaves subjected to water stress. Furthermore, abscisic acid applied to guard cells causes them to close their stomata. However, the accumulation of abscisic acid does not fully explain stomatal regulation because the pores also open when the concentration of CO_2 is low inside the leaf. This tends to occur when photosynthesis is most intense—during periods of high light intensity. Finally, guard cells have pigments that absorb blue or red light; when stimulated, these pigments activate electron transport processes that may affect K^+ ion movements and the degree of guard-cell swelling. The intact plant in nature is subjected to varying light, humidity, and other conditions that can affect guard-cell behavior. Hence, all these interrelated environmental factors may help regulate guard cells.

Stomatal opening and closing is not alone in governing the movement of water vapor, CO_2, and O_2 across the leaf surface. Movement of these gases also depends on (1) the concentration gradients of these gases from inside to the outside of the leaf, (2) the diffusion properties of the gases, and (3) the wind speed across the leaf surface (for instance, faster moving air carries water vapor away more rapidly). The relative humidity of air is usually well below 100 percent, so that evaporation and transpiration are favored. There is usually little absolute difference between the low CO_2 levels outside and inside the leaf.

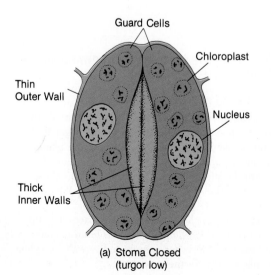

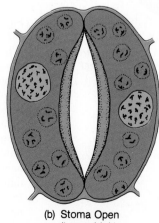

Figure 29-5   STOMATA: OPENING AND CLOSING THE GUARDIAN GATES.
(a) When turgor is low in both guard cells, the stoma is closed. (b) When ions such as K^+ and Cl^- enter the guard cells, water follows by osmosis, increasing the turgor pressure within the cells and causing them to bulge sideways, pulling their thick inner walls apart. This movement opens the pore, and air containing carbon dioxide can enter, or water vapor and oxygen can exit.

Guard Cells

Chloroplast

Thin
Outer Wall

Nucleus

Thick
Inner Walls

(a) Stoma Closed
(turgor low)

(b) Stoma Open
(turgor high)

However, as CO_2 is used up by photosynthesis within the leaf cells, a *sink*, or absorption region, is created within the leaf toward which more CO_2 diffuses.

Despite the elegant regulatory machinery of stomata, water loss from plants through transpiration can be staggering. A corn plant cultivated in full sun can lose 2,500 grams of water a day through leaves that collectively weigh only 500 grams. Thus it is easy to see why thousands of acres of such crops consume huge volumes of irrigation water as the plants transpire and photosynthesize rapidly to grow and reproduce. In the next section, we will see that transpiration is also involved in the movement of sugar in the phloem.

TRANSPORT OF SOLUTIONS IN THE PHLOEM

Transport of solutes in the phloem is called **translocation.** It involves the movement of carbohydrates, primarily disaccharides, and other organic molecules from the photosynthetic leaf cells to the leaf's phloem and from there throughout the plant. Phloem also transports organic molecules from storage areas such as parenchyma in the roots and shoots to other areas of the plant in need of these molecules. Finally, ions move in the phloem as part of the solute distribution process. The phloem is a living tissue composed primarily of sieve tube elements, which have lost their nuclei but retain some cytoplasm, and companion cells, which retain their nuclei.

What Is Translocated in the Phloem?

Individual sieve tube elements are nearly impossible to study in an intact plant. As soon as the phloem is injured, transport stops in that area, and one can no longer sample the materials passing through. Fortunately, nature has provided botanists with the perfect tool for sampling the contents of the phloem: aphids. No manufactured needle is as fine and as accurate a probe as the *stylet*, or feeding tube, of an aphid. These small insects feed by boring into individual sieve tube elements of stems or leaves. Then they suck in nutrient solutions flowing in the phloem. The stylet does not inhibit the sieve tube element's function. Therefore, translocation continues unabated, as does the aphid's "free lunch."

To study the fluid contents of phloem, plant physiologists allow aphids to attach to a plant and penetrate sieve tube elements with their stylets. The aphids are then anesthetized with ether, and their bodies are cut away, leaving only the stylet protruding from the phloem (Figure 29-6). The phloem fluid oozes out of the stylets and is collected and analyzed.

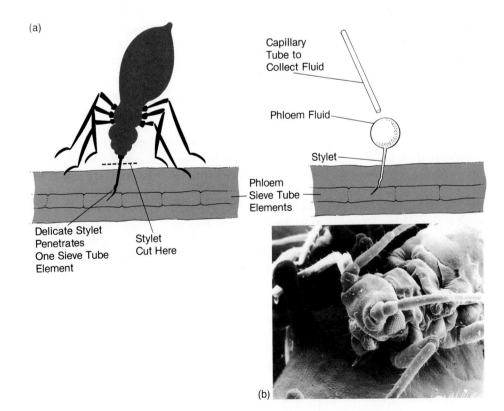

Figure 29-6   APHIDS: BIOLOGICAL TOOLS FOR STUDYING PHLOEM.

If the contents of the sieve tube elements in phloem are to be sampled, the cells must not be damaged. Otherwise, nutrient transport will cease. (a) Aphids have a delicate feeding tube called a stylet that they plunge into a sieve tube element like a hypodermic needle into an animal's vein. Researchers employ the aphid's stylet by severing the body of a feeding aphid and leaving the stylet in the sieve tube. The phloem contents continue to seep out of the stylet and can be collected and analyzed. (b) An aphid (magnified about 50 times) is shown with its feeding tube inserted into a marigold stem.

(a)

Delicate Stylet Penetrates One Sieve Tube Element

Stylet Cut Here

Capillary Tube to Collect Fluid

Phloem Fluid

Stylet

Phloem Sieve Tube Elements

(b)

Studies of this sort have revealed that in most plants, the fluid transported in the phloem is a concentrated solution: although normally lower, up to 30 percent of the volume can be dissolved materials. The major component is sucrose, a disaccharide and indirect product of photosynthesis, but other sugars and organic compounds can be found, along with low concentrations of nitrogen compounds. Nitrate and sulfate, absorbed by the roots as inorganic ions, are transported through the xylem to the leaves, where photosynthetic cells convert them into organic compounds, primarily amino acids. These amino acids are then transported in the phloem to support protein synthesis in other parts of the plant.

Kinetics of Transport in the Phloem

Although aphids have been invaluable for tapping the phloem fluids, they have not been reliable tools for studying the direction in which and the rate at which substances move through phloem—the *kinetics* of phloem transport. In order to use aphids for this purpose, researchers would have to "train" aphids to position themselves at prescribed locations along the stem—a rather difficult trick. Instead, researchers have learned to study the direction and rate of material flow by adding radioactive tracers to the phloem, either by injection or by exposure of the leaves to $^{14}CO_2$. An individual photosynthetically active leaf can be exposed by enclosing it in an airtight bag and then introducing radioactive CO_2. Within a few minutes, the leaf chloroplasts incorporate the radioactive carbon dioxide into sugars, which begin to be exported from the leaf cells. The movement of radioactive sugars can be traced by removing pieces of leaf, stem, and root tissue at various intervals and determining the amount of radioactive carbon that has reached the sample tissue.

The typical experimental observation is that the radioactive compounds exit the leaf, enter the stem, and are translocated *downward*. However, a fraction of the newly synthesized compounds also moves up the stem. Developing fruits—growing bean pods, for example—are sites toward which translocation is directed; thus most of the photosynthetic product from nearby leaves is transported toward the fruit. This observation has led to the concept that leaves are *sources* of carbon compounds and that other parts of the plant are *sinks* for those compounds. Obviously, there are strong and weak sinks. For example, growing regions are strong sinks, even if they are situated at the top of the plant, away from normal downward phloem transport. A final feature of the source–sink scheme is that energy is required to "load" at the source and to "unload" at the sink. ATP supplies that energy.

These studies make it clear that translocation is a much more complex phenomenon than is transpiration. Whereas xylem is a one-way transport system, open at the top, phloem is a two-way transport system in which organic compounds can flow either up or down the plant, depending on the location of strong sinks. Flow in sieve tube elements located side by side can actually be in opposite directions, at least for short periods. Furthermore, different organic molecules may be translocated at different rates.

Radioactive tracers have shown that the first few radioactive sugar molecules that exit a leaf may arrive at a given measuring point very quickly, sometimes moving at a rate of more than 200 centimeters an hour. (Rates 40,000 times that of simple diffusion have been measured in cotton plants.) The majority of the radioactive molecules move more slowly, with transport rates of 50 to 100 centimeters an hour. But this *bulk flow*, or mass movement of phloem fluid, is rather rapid movement for a system that has no apparent pump and is carrying a viscous sugar solution through thin, partially blocked pipe lines composed of sieve tube elements with perforated end walls—the *sieve plates*. What accounts for this rapid transport?

Mass Flow Theory: Transport in the Phloem

Several theories have been proposed to explain the rate and mechanisms of phloem translocation. The generally accepted one, however, is the mass flow theory. The other, less strongly supported ones involve energy-dependent processes.

According to the **mass flow theory**, sugar produced by photosynthesizing cells in leaves is loaded into the sieve tube elements by the companion cells, creating a high solute concentration. The phloem is thus hypertonic (that is, has a low water potential or high solute potential) relative to neighboring xylem cells, and water moves by osmosis from the xylem to the phloem. This increases the phloem cell's water content and creates a high turgor pressure within the phloem. But the phloem cells cannot expand past a certain size because of their rigid cell walls. The only open route to relieving the building pressure is through the perforated sieve plates. Therefore, the mass flow theory suggests that the elevating water pressure in the leaf is relieved by the flow of water and organic substances through the leaf phloem and then through the stem phloem. This model is demonstrated in Figure 29-7.

Experiments show that translocation occurs down a gradient of turgor pressure. The turgor pressure in the phloem is highest in the small vascular bundles of the leaf, where sugars are pumped into the sieve tube elements by the companion cells. As the sugary phloem

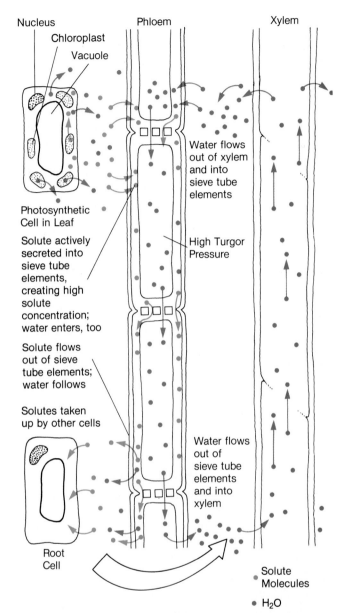

Nucleus
Chloroplast
Vacuole

Photosynthetic Cell in Leaf

Solute actively secreted into sieve tube elements, creating high solute concentration; water enters, too

Solute flows out of sieve tube elements; water follows

Solutes taken up by other cells

Root Cell

Phloem

Xylem

Water flows out of xylem and into sieve tube elements

High Turgor Pressure

Water flows out of sieve tube elements and into xylem

Solute Molecules

H₂O

Figure 29-7   THE MASS FLOW THEORY AND NUTRIENT TRANSPORT.
Sugar, which acts as a solute, is produced by photosynthesizing cells and passes into the sieve tube elements of the phloem. As more solutes enter those tubes, the resultant higher solute concentration causes water to flow into the sieve tube elements from neighboring xylem cells. Since the volume of the sieve tube elements is restricted by their cell walls, the continuously enlarging fluid volume in the sieve elements is reduced by passage of fluid through the sieve plates. As a result, a flow of water and nutrients is set up through the leaf phloem and into the stem phloem, and thus to the root phloem. This translocation of water and nutrients occurs down a water pressure gradient from the leaves to the roots. Solutes are supplied to stem and root cells, as shown. Water that passes down the phloem passes out of the sieve tube elements and then joins the water that enters root hairs in order to return upward in the xylem, thereby completing the cycle.

contents are carried out of the leaves, the turgor pressure in the phloem gradually falls. This is because nonphotosynthetic cells of the stem actively take up disaccharides from the passing phloem fluid to support their own cellular metabolism. This loss of sugar renders the phloem solution progressively more dilute. Turgor pressure falls progressively as the phloem's fluid contents move downward toward the roots or upward toward the meristems, and more and more sugar is removed. The final carbohydrates are removed from the phloem in the roots. The water in the phloem, now at a high water potential, moves into the xylem to be recirculated toward the top of the plant. This is a remarkably efficient strategy: water transferred from xylem to phloem in leaves creates the turgor pressure in the phloem required to move the sugar out of the leaf. Transport both up the xylem and down the phloem is ultimately traceable to the transpiration from the leaves, which carries water upward in the first place (of course, photosynthesis and solute build-up also play roles).

The mass flow theory is very attractive, but there are some minor problems with it. Some plant physiologists believe that there is not enough difference between the turgor pressures in the leaf phloem and in the stem and root phloem to create a flow of sugars. Furthermore, the perforated end plates of the sieve tube elements may offer considerable resistance to the flow of a viscous sugar solution. Hence, some believe that the turgor-pressure differential may account for the *direction*, but not the rapid speed, of sugar flow. Despite these problems, the mass flow theory joins the transpiration pull theory as the best explanation so far of a vital plant process.

HOW ROOTS OBTAIN NUTRIENTS FROM THE SOIL

House plants sometimes produce only small, pale leaves and fail to flower. Such plants, growing in a limited amount of frequently watered soil, are probably starved for the nutrients that are essential to build nucleic acids, enzymes, chlorophyll, and cell walls. A few drops of commercial plant food added to the leached soil soon sets everything to right. But what *is* plant food?

Mineral Requirements and Uptake

Complete plant foods contain **minerals,** inorganic substances such as nitrogen, potassium, and phosphorus, plus traces of iron, zinc, manganese, and magnesium. Thus plant foods do not contain proteins, fats, and carbo-

Table 29-1   **ESSENTIAL MINERALS FOR HIGHER PLANTS**

	Major Functions	Typical Compounds
Macronutrients		
Nitrogen (N)	Major ingredient in many compounds	Amino acids, purines, pyrimidines, porphyrins, hormones
Phosphorus (P)	Energy transfer; structural	Sugar phosphates, ATP, GTP, nucleic acids, coenzymes, many proteins
Potassium (K)	Osmotic relations; protein conformation and stability; stomata; enzyme cofactor	Potassium ion
Magnesium (Mg)	Enzyme activation; pigments and ribosome stability; nucleic acid synthesis	Chlorophyll
Calcium (Ca)	Enzyme activation; cell walls; permeability of membranes	Calcium pectate
Sulfur (S)	Active groups in enzymes and coenzymes	Amino acids, coenzymes, vitamins
Iron (Fe)	Active groups in enzymes and electron carriers	Cytochromes, peroxidases
Micronutrients		
Copper (Cu)	Enzymes, photosynthesis	Enzymes for oxidation reactions; plastocyanin
Manganese (Mn)	Photosynthesis; cellular respiration	Cofactor for many enzymes
Molybdenum (Mo)	Nitrogen fixation; nitrate reduction	Nitrate reductase, nitrogenase
Zinc (Zn)	Enzyme activation	Carbonic anhydrase, auxin synthesis
Sodium (Na)	Enzyme activation; osmotic relations	Sodium ion
Cobalt (Co)	Nitrogen fixation	Vitamin B_{12} in nitrogen-fixing micro-organisms only
Silicon (Si)	Structural	Hydrated silicon dioxide
Chlorine (Cl)	Photosynthesis; ion and electrical balance	Chlorinated alkaloids
Boron (Bo)	Sugar transport	Borate ion

hydrates, as do the foods of heterotrophs. These minerals are essential nutrients that cannot be obtained by the plant through gas exchange or photosynthesis. Table 29-1 lists the minerals that are *macronutrients* (required in large amounts) and *micronutrients* (required in smaller amounts) for plants. If a plant is deficient in any one of these mineral nutrients, it will exhibit specific symptoms. Figure 29-8 provides examples of such deficiency symptoms in the tobacco plant.

The minerals required for plant growth originate from the Earth's rocks. Over eons, rocks weather—they gradually wear away into smaller and smaller pieces or into dissolved compounds or ions. This process contributes to the composition of soil, which also contains organic materials. Minerals are dissolved in soil water, which is taken up in the roots and transported in the xylem. Although the water enters the roots by simple diffusion in response to a water-potential gradient, mineral uptake may occur in two ways: passive or active. In passive uptake, minerals may diffuse across a membrane in response to a difference in solute potential on the two sides

of the membrane. Passive uptake is not a particularly reliable method for obtaining minerals, however, since there must be a gradient for solute uptake to occur. Active uptake of solutes requires the input of energy, most often in the form of ATP, as well as pump enzymes, or carrier molecules, located in the plasma membrane (Chapter 5). By means of active transport, plants can accumulate minerals and ions against a concentration gradient, moving even trace quantities of minerals from the soil into the cytoplasm.

Recall that water enters roots primarily through the tiny root hairs. These, however, are not the sole sites of mineral entry. In fact, minerals apparently diffuse into the fluid within cell walls of a root, the apoplastic pathway (Chapter 27). Any cell of the root cortex, epidermis, or endodermis has the capacity to actively transport mineral ions from the dilute solutions within the apoplastic space of the cell walls into the cell's cytoplasm, the symplastic pathway. Individual pump enzymes may be specific for each kind of ion. Ions absorbed into cells of the cortex and epidermis follow the symplastic pathway;

Figure 29-8   THE EFFECTS OF MINERAL DEFICIENCIES ON TOBACCO PLANTS.
A plant that is deficient in a mineral nutrient shows specific symptoms, usually in the most rapidly growing tissues, such as new leaves. Plant (a) was given a complete nutrient solution. Each of the other plants was deprived of one nutrient: (b) nitrogen; (c) phosphorus; (d) potassium; (e) boron; (f) calcium; and (g) magnesium.

eventually they reach the root's endodermis by traveling through the plasmodesmata—the cytoplasmic connections between plant cells—and are transferred from the endodermis into the xylem (Figure 29-9).

The xylem carries the dissolved minerals passively upward, presumably because of the transpiration-pull effect; in this way, the nutrients are distributed throughout the plant. Leaf and stem cells then remove ions from the xylem solution using active transport pump enzymes similar to those used by the root cells to remove ions from soil water. The ATP used to power these pumps is ultimately derived from photosynthesis. In roots, translocated products of photosynthesis can be used to generate the ATP. As a result, ATP allows the crucial pumps to grind away hour after hour, day after day, transporting the critical nutrients across cell plasma membranes.

Rate of Mineral Uptake

The rate at which plants can take up various minerals depends on the concentration of the minerals in the soil and on the soil's pH. Some soils are rich in inorganic nutrients, whereas others are deficient in one or more. The solubility of each mineral changes over a range of pH values. Most plants grow best in a slightly acidic soil, in which mineral nutrients are most soluble. At more extreme high or low pH values, most of the mineral nutrients in soil are tied up as insoluble salts and are not available to plants.

The composition of a soil can affect the rate of mineral uptake in a number of ways. Sandy soils are composed predominantly of large particles; consequently, they have large air spaces, but retain little water and there-

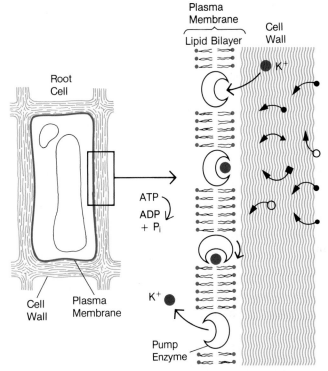

Figure 29-9   PLANT ROOTS: MINERAL UPTAKE BY ACTIVE TRANSPORT.
Carrier proteins that are pump enzymes exist in plasma membranes. Each pump enzyme is specific for a particular ion and requires ATP in order to function. When sufficient mineral ions are present, in this case K^+, in the extracellular space, they diffuse throughout the cell wall. Ions then bind specifically to pump enzyme proteins. If ATP is present, the carrier protein is able to move the ion into the cell, either by creating a channel or by changing its own conformation. In either case, the process requires the expenditure of energy in the form of ATP.

fore provide few dissolved minerals. Clay soils, composed mainly of very fine particles, have little air space and hold much more water and dissolved mineral ions. The amount of organic material—decaying vegetation, or compost, for example—also has an important effect on mineral availability. In soils that are deficient in organic material, most of the dissolved calcium reacts with carbonate ions in the soil to form an insoluble salt, calcium carbonate. In the presence of organic material, calcium ions form complexes with organic acids and remain available to plants. For this reason, gardeners often "compost" their grass clippings and leaves—pile them up to decompose so they can return both organic and inorganic matter to the soil. On a larger scale, farmers usually use mineral-containing fertilizers to recondition and replenish soils depleted year after year by crops.

Nitrogen Uptake and Fixation

Nitrogen (N) needs present a special problem for plants, since unlike other inorganic nutrients, which are abundantly derived from the weathering of rocks, nitrogen must be derived from the atmosphere. Because nitrogen is a component of the subunits used during the synthesis of protein, RNA, and DNA, the amount of available nitrogen is usually the most important single factor limiting plant growth.

Ironically, 80 percent of our atmosphere is N_2, yet green plants cannot assimilate molecular nitrogen directly. Instead, plants must take in nitrate (NO_3^-) from the soil and transport it to the leaves, where an enzyme called *nitrate reductase* converts the oxidized form of the element back to reduced nitrogen, or ammonium ion (NH_4^+). This conversion is another energy-dependent process. But there is an exception to this usual sequence: some herbaceous plants can use NO_3^- itself in root cells during manufacture of amino acids and other nitrogen-containing organic compounds and then transport those substances upward in either phloem or xylem.

How does atmospheric N_2 become oxidized into NO_3^- in soil? Such conversion of N_2, or **fixation,** is accomplished by three naturally occurring mechanisms: (1) abiotic fixation, (2) fixation by free-living soil microorganisms, and (3) fixation by means of certain bacteria that live symbiotically with plant roots.

Abiotic N_2 fixation (that is, fixation by nonliving processes) is the least important of these mechanisms and takes place in the atmosphere. Flashes of lightning provide an immense but localized energy source that can break the strong $N\equiv N$ bonds in N_2, allowing the spontaneous formation of nitrogen oxides (NO_2^-, NO_3^-) and ammonia (NH_3). During rainstorms, these compounds are washed down into the soil, where the nitrates can be taken in directly by plants.

Biological nitrogen fixation involves conversion of N_2 to NH_4^+. It is carried out by free-living soil microorganisms. Species of the bacterial genera *Azotobacter* and *Clostridium*, among others, and certain cyanobacteria can reduce N_2 into NH_3 by means of the enzyme *nitrogenase*. A second group of soil microbes, including species of *Nitrosomonas* and *Nitrobacter*, carry out the process of **nitrification,** by which they oxidize NH_3 to NO_2^- and NO_3^-, the latter being directly available for absorption by plant roots.

The second naturally occurring case of biological nitrogen fixation involves symbioses between bacteria and plant roots and produces by far the most fixed nitrogen each year. *Legumes*, a large group of plants that includes peas and beans, can form symbiotic relationships with particular species of *Rhizobium*, a genus of symbiotic nitrogen-fixing bacteria that forms the NH_3 (Chapter 20). The host plant then uses the NH_3 to form amino acids and other compounds. Although rhizobia can live freely in the soil, most do not fix nitrogen unless they are living in the roots of a legume.

The mutually beneficial association between bacteria and plants actually begins in the root hair (Figure 29-10). The bacteria attach to the outer surface of a root hair, causing that tiny absorptive structure to curl into the shape of a shepherd's crook. The curled root hair then manufactures an *infection thread*, a channel through which the bacteria can enter the interior of the root. This response is very different from the reception most microbes encounter when attempting to penetrate roots. Because most microbes are potential pathogens or parasites, root cells protect against their entry by producing toxic chemicals such as phenolics. Once rhizobia have passed along the infection thread, they enter the cortex of the root, begin to divide rapidly, and build up a large population. Soon many cortical cells are invaded. At the same time, the root cells near the invaded hairs proliferate and create **root nodules,** small, hard lumps on the root surface that encapsulate the nitrogen-fixing factory (Figure 29-11).

Within the root nodule, a remarkable differentiation of root and bacteria takes place. The free-living bacteria change their metabolism and appearance and become **bacteroids,** no longer able to survive outside the root nodules. The bacteroids make large amounts of the enzymes required for nitrogen fixation. One of these enzymes uses the trace metal molybdenum as an essential cofactor—a good example of why a variety of rare metals must be present in soils for full plant nutrition. Rhizobial nitrogen-fixing enzymes require a nearly anaerobic environment in which to reduce nitrogen. This presents a problem, since root cells are normally aerobic and use oxygen for respiration. But the root cells of the nodule secrete a pigment–protein complex called *leghemoglobin* (meaning "hemoglobin of legumes") that

(a)

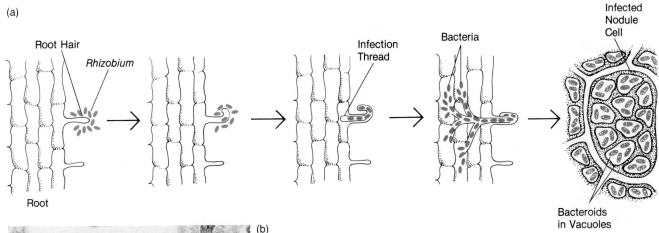

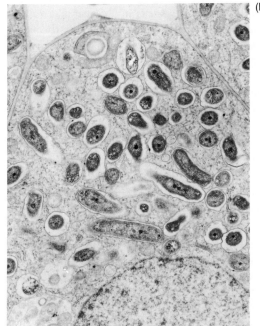

(b)

Figure 29-10   *RHIZOBIUM:* **A NITROGEN-FIXING FACTORY LIVING WITHIN LEGUME ROOTS.**
(a) *Rhizobium* bacteria invade the plant, and a mutually beneficial relationship is established in which *Rhizobium* supplies the legume with nitrogen, while the legume supplies the bacterium with carbohydrates. In this series, the bacteria form an infection threat that enables the microbes to enter root cells. Those cells then divide to form the nodule, and all daughter cells are filled with bacteroids. (b) *Rhizobium* bacteroids within the vacuoles of a soybean root cell (magnified about 7,600 times). The cell's nucleus is at the bottom. The bacteroids are isolated in the membrane-bound clear spaces in the cytoplasm.

can bind oxygen very tightly. The bacteroids are bathed in a leghemoglobin solution that soaks up and binds oxygen diffusing toward them, thereby keeping the free-oxygen concentration low enough to allow nitrogen fixation. Root nodules look pinkish red because of their high concentration of leghemoglobin.

Production of this pigment–protein complex is itself a fascinating case of symbiosis. The protein portion of the pigment—a globin—is encoded in the legume's DNA and is synthesized by the plant cells. The heme group, to which oxygen binds, is apparently made by the rhizobium. The two molecules combine, and both contributors benefit.

Legumes supply the bacteroids with carbohydrates. The bacteroids then metabolize these compounds to reap the large amount of energy needed to reduce N_2 to NH_3. In an inadvertent exchange for their "room and

Figure 29-11   NITROGEN-FIXING ROOT NODULES.
Each of these nodules on this pea plant (*Pisum sativum*) root (magnified about 6 times) is composed of thousands of cells, all of which contain *Rhizobium* bacteroids.

(a)

(b)

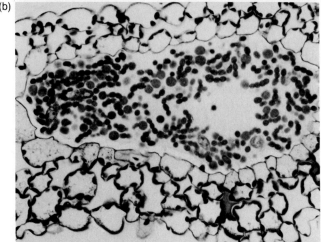

(c)

Figure 29-12   SYMBIOSIS BETWEEN FERNS AND CYANOBACTERIA.

(a) A frond of the water-dwelling fern *Azolla caroliniana* (magnified about 10 times). Inside many of these parts of the frond are found cyanobacteria. (b) A section through an *Azolla* frond (magnified about 200 times) shows the membrane-bound compartment that contains many nitrogen-fixing *Anabena azollae* cyanobacterial cells. (c) Chinese farmers often plant these ferns to help fertilize their rice fields because they release a large amount of nitrogen into the soil when they die. Here we see living rice plants (the erect, spikelike leaves) growing through so dense a mat of *Azolla* fronds that the pond surface cannot even be seen.

board," the bacteroids produce NH_3 as a waste product, which the plant then assimilates and uses as a source of reduced nitrogen. Interestingly, amino acids are often manufactured in root cells themselves, so that nitrogen is transported as part of those organic compounds, not as NH_3, in the xylem.

Certain nonleguminous plants have microbial symbionts that fix nitrogen. Western mountain lilac, alder shrubs, and the Eastern sweet gale are typical examples; they can survive in rocky soils containing little nitrogen. Some types of water-dwelling ferns have cyanobacteria of the genus *Anabena* living in specialized compartments in their leaves (Figure 29-12), making nitrogen fixation possible in those organs. Chinese farmers fertilize rice plants by cultivating water ferns in a dense mat around each rice plant in their flooded fields. When the fields are drained, the ferns die and release large amounts of organic nitrogen into the soil. As with the legumes, the water ferns and other such plants pay a high price to their symbionts—organic carbon compounds—in return for an equally important resource, reduced nitrogen. We shall return to the issue of nitrogen in Chapter 44, where the cycling of the element from the atmosphere to the biological world and back to the atmosphere will be discussed.

Mycorrhizae

The more that biologists study roots in their natural environment, the more they are learning about the association of roots with soil microorganisms. Most plant families have members that can form long-term associations with specific types of fungi (Chapter 22). The fungi invade or cover the roots with threadlike hyphae, the basic structures of all fungi, often completely surrounding them like a sheath. This symbiotic association, called a *mycorrhiza*, is beneficial to the plant, since fungi are more efficient than plant root hairs at absorbing mineral nutrients from the soil. In fact, mycorrhiza-covered roots often lack root hairs altogether and depend on the fungi to handle absorption. As we saw in Chapter 22, mycorrhizal associations are not well understood, but it is believed that the fungus derives carbohydrates and organic substances from the roots while contributing to the uptake of water and perhaps the conversion of soil minerals to a form the plant can more readily use.

CARNIVOROUS AND "SENSITIVE" PLANTS

Carnivorous plants, such as the Venus's-flytrap and the sundew (Figure 29-13), meet some of their nitrogen requirements in an unusual way: they trap and digest insects. When an insect lands on the special leaves of Venus's-flytrap, the leaves spring shut and release enzymes to digest the victim; amino acids (and their nitrogen) are then absorbed. But obtaining part of its nitrogen this way is costly to the plant. Enzymes to digest the insect must be manufactured and secreted. Mechanisms must be present in the leaves to absorb soluble nutrients from the prey, so that they can be distributed to other parts of the plant. Clearly, this is getting nitrogen the hard way, so it is not surprising that only a few rare plants evolved the capacity to do so.

Perhaps the most curious thing about carnivorous plants is how they move rapidly enough to catch their prey. Most plants, if touched, do not respond visibly, since plants do not possess muscles or a nervous system. However, a few types of plants *do* respond to touch by means of *action potentials*—changes in electrical potential that travel from the stimulated area, such as a leaf, toward the base of the leaf and the stem. The action potential "travels" because adjacent cells become sequentially depolarized (Chapter 36). We will see that the same phenomenon characterizes the animal's nerve impulse and that changes in membrane permeability and ion flow are the keys to action potentials. Unlike that in animal nerves, however, the action potential in plants triggers a change in turgor pressure in certain cells rather than muscle contraction or similar responses.

The sundew attracts insects by sweet nectar secreted on the tips of sticky club-shaped tentacles that jut upward from the leaves. When a fly lands on one of the tentacles, it sets off an action potential that travels down the tentacle, stimulating closure of the tentacles and entrapment of the fly. The Venus's-flytrap works on a similar principle, except that the lower leaves are equipped with toothed margins and a hair-trigger apparatus that stimulates the leaves to close if the hairs are disturbed in a certain way. The result is that the two halves of the leaf fold together, the marginal teeth interlock, and the insect is trapped. Both the sundew and the Venus's-flytrap can be fooled into closing by touching them lightly with a fingertip or a stick.

Although it does not feed on insects, *Mimosa pudica* (Figure 29-14) is another plant that responds to touch by means of action potentials. When a leaf is touched, an electrical stimulus travels quickly down the leaf to the leaf base, where cells in the axil of the petiole rapidly lose water so that their turgor pressure falls. Because the normally turgid axil cells hold the leaf upright, this water loss causes the leaf to droop rapidly. The whole process takes only a few seconds. If the touch stimulus is strong enough or is repeated often enough, the action potential travels to other leaves and soon the entire plant looks wilted.

The folding response of the mimosa may be an evolu-

Figure 29-13   DEADLY INSECT TRAPS.
Carnivorous plants harvest needed nitrogen by capturing and digesting insects and by absorbing nitrogen from their bodies. (a) This Venus's-flytrap (*Dioneae muscipula*) produced action potentials in response to the touch of the fly and immediately snapped shut to trap its prey. (b) The Cape sundew (*Orosera capensis*) traps insects such as this black mushroom fly with the blobs of sticky fluid on the ends of glandular hairs (magnified about 150 times). Digestion of insects by both plants is an energetically expensive process. Living in nitrogen-poor swamps and bogs, these plants have evolved the capacity to manufacture and secrete the necessary digestive enzymes.

tionary adaptation to avoid being eaten; if a grazing deer or a hungry insect touches the mimosa bush, the plant folds up and the leaves are more difficult to find. Because mimosa can grow in windy, dry regions, an alternative hypothesis is that its leaves fold on sufficient agitation by wind, thereby greatly reducing evaporation from the stomata.

(a)   (b)   (c)

Figure 29-14   SENSITIVITY WITHOUT NERVES: A SPEEDY DEFENSIVE REACTION.
The *Mimosa pudica* (a) responds to touch (b) with a fast drop in turgor pressure. An action potential is produced on touch; turgor pressure drops as a result, and the leaves droop. If the touch is repeated, the action potential travels to other parts of the plant; eventually, all the leaves on the plant droop (c). This response may have been an evolutionary response to prevent predation. Mimosa are also bitter tasting—another protection from animals.

These kinds of rapid plant movements are radically different from animal movements. Whereas plants move their leaves by the rapid alteration of turgor pressure in critically placed cells, animals move their legs, wings, or jaws by harnessing the contractility of the actin–myosin protein system. Ironically, plant cells also contain actins and myosins. However, these proteins in plants do not move individual cells or organs relative to each other; rather, they are involved in cytoplasmic streaming, as in the companion cells of the phloem. The absence of cell movement is no doubt due to the rigid, all-encompassing cell walls that forever anchor the individual plant cell in one spot. The inability to exploit the actin–myosin system for cell and organ movement is the price that plants pay for the strength, protection, and capillarity of their cell walls.

LOOKING AHEAD

We have seen in this chapter and the two previous ones how plants are constructed, how they reproduce and grow, and how they carry out crucial day-to-day processes. Utilization of physical forces and minimal expenditure of energy characterize plants and their cells. The transpiration-pull theory and the coupling of upward water flow in the xylem and the nutrient-distribution system in the phloem are good examples of this unwitting parsimony. The result is efficient distribution of water, minerals, photosynthetic products, and other nutrients through a plant as tall as a redwood or as specialized as a desert cactus. But in all plants, these various activities must be coordinated. Hormones, the subject of Chapter 30, are the coordinating agents.

SUMMARY

1. Plants have several fundamental characteristics: they photosynthesize, of course, and, in addition, remain in a single location, exchange gases and water passively with the environment, and store or recycle wastes.

2. Gases are exchanged through the stomata in the photosynthetic plant parts and through the epidermis or lenticels in the nonphotosynthetic parts of the shoot. And the roots take up oxygen from the soil. Once in the plant, gases diffuse to all cells.

3. Plants usually cannot excrete their own toxic wastes; thus they store some indefinitely and break others down and recycle them for use in new cells and tissues.

4. Water transport in the xylem relies on several principles of water

movement, including osmosis and *water potential*, which in a cell is equal to turgor pressure plus *solute (osmotic) potential*. When a cell loses water or when the concentration of solutes increases in the cell relative to the surrounding cells or medium, the water potential becomes more negative, and the cell takes up water from the medium.

5. Explanations for upward movement of water in plants include the *root-pressure theory* and *capillary action*. According to the first, water pressure builds up in the roots and pushes water upward. *Guttation*, droplet formation on leaves, may be one result. According to the second, water molecules cohere to each other and adhere to the xylem walls, and water is able to creep upward. Neither theory explains how water can reach the tops of tall trees.

6. The explanation that best accounts for the extensive upward climb of water in many plants is the *transpiration-pull theory*, also called the *cohesion–adhesion–tension theory*. According to this theory, *transpiration*—water loss through leaves—causes water to be pulled up through the xylem as the column of water molecules cling together (cohesion) and adhere to the cell walls (adhesion). As water molecules change to the gaseous state and transpire out stomata, the whole column may be visualized to move upward a minute distance and equivalent numbers of water molecules are taken in by the roots.

7. Regulation of water loss is governed by the opening and closing of the stomata, which is a response to movement of ions across guard-cell membranes. Movement of gases through the stomata also depends on the concentration gradients of these gases, their diffusion properties, and wind speed across the leaf surface.

8. *Translocation* of sugars and other organic substances through the phloem generally proceeds from sites of synthesis in leaves and stems (sources) to sites of use or storage in meristems, stems, roots, and flowers (sinks). Energy is expended during loading at sources and unloading at sinks.

9. Transport of solutions in the phloem is well explained by the *mass flow theory*. It is generally held that translocation depends on the tendency of water to flow from the xylem to the phloem in leaves because of the high solute concentration there due to photosynthesis; this drives the leaf phloem's contents into the stem's phloem, from where the contents move downward or occasionally upward (to sinks).

10. Plants require several *minerals*, inorganic substances that originate in the Earth's rocks. Minerals are dissolved in soil water and must be taken up by the roots and transported in the xylem. Essential nutrients may be passively or actively transported by root cells. Passive transport involves simple diffusion; active transport is accomplished by individual pump enzymes in the plasma membrane and requires expenditure of ATP by both root cells and mineral-requiring cells throughout the plant.

11. The rate at which plants can take up various mineral nutrients depends on the concentration of the nutrients in the soil and on the soil pH. Acid soil promotes mineral uptake. In addition, plants can take up more minerals and grow better in clay soils, which have fine particles and hold water, than in sandy soils.

12. *Nitrogen fixation* can be carried out through nonbiological means (for example, lightning flashes) and by both free-living and symbiotic microorganisms. Plants must depend primarily on nitrogen-fixing microorganisms to convert atmospheric N_2 into NH_3 (and NH_4^+). Some microbes carry out *nitrification*, converting NH_3 to NO_2^- or NO_3^-. In legumes, symbiotic nitrogen-fixing bacteria invade root hairs and become encapsulated in *root nodules*, where they change their metabolism and appearance to become *bacteroids*. Root-nodule cells secrete leghemoglobin to keep free-oxygen concentrations low enough to permit the bacteroids to convert atmospheric nitrogen to ammonia. As part of the symbiotic association, the bacteroids receive carbon compounds. Many other plants are symbiotic with certain fungi, forming mycorrhizae; the fungi are more efficient than root hairs at adsorption, and take up abundant quantities of minerals.

13. *Carnivorous plants* can obtain nitrogen by trapping and digesting insects. Trapping sometimes involves quick movement of leaves in response to touch; such movement is brought about by electrical action potentials, which trigger rapid loss of turgor pressure in cells that control the position of the leaves.

KEY TERMS

bacteroid	guttation	nitrogen fixation	translocation
capillary action	mass flow theory	root nodule	transpiration
carnivorous plant	mineral	root-pressure theory	transpiration-pull theory
cohesion–adhesion–tension theory	nitrification	solute (osmotic) potential	water potential

QUESTIONS

1. Use the process of *osmosis* to explain why spraying fresh water on green vegetables in a supermarket both makes them "crisper" and increases their cost when sold by the pound. What might happen if they were sprayed with salt water?

2. Explain briefly how each of the following physical phenomena may play a role in the upward movement of water in the xylem:
   a. osmosis
   b. capillary action
   c. adhesion
   d. cohesion
   e. transpiration

3. What experimental evidence suggests that transpiration is a primary process contributing to water movement in the xylem?

4. Explain how the movement of potassium (K^+) and chloride (Cl^-) ions into and out of guard cells regulates the opening and closing of stomata and thus the loss of water vapor from leaves.

5. Are the stomata open or closed when their guard cells are *turgid*? Explain.

6. Explain the difference between *transpiration* and *translocation*.

7. Describe how phloem cells may operate to cause flow of the phloem contents.

8. How do plant physiologists use (a) aphids and (b) the radioactive tracer $^{14}CO_2$ to study the activity of the phloem? What have these studies revealed about the phloem contents and its movement?

9. What accounts for the several types of symbiotic relationships between plant roots and bacteria or fungi?

10. How are waste products handled by plants?

ESSAY QUESTIONS

1. Name the seven macronutrients and the nine micronutrients that are essential for optimum plant growth. For each nutrient, indicate its function(s) in the plant.

2. Nitrogen fixation is essential to maintain higher plant and animal life on earth. Why is nitrogen fixation essential to the growth of higher plants? How are various bacteria involved in nitrogen fixation? Describe the very important symbiotic relationship between leguminous plants and bacteria of the genus *Rhizobium*?

SUGGESTED READINGS

BAKER, D. A. *Transport Phenomena in Plants*. New York: Chapman & Hall, 1978.

This book covers the entire subject and is especially good on the various hypotheses for phloem function and on root function.

BEEVERS, L. *Nitrogen Metabolism in Plants*. London: Arnold, 1976.

This brief discussion covers just about all aspects of nitrogen uptake and utilization.

HEWITT, E. J., and T. A. SMITH. *Plant Mineral Nutrition*. New York: Wiley, 1975.

A full account of nutrient requirements and deficiency diseases of plants.

RAY, P. M., T. STEEVES, and S. A. FULTZ. *Botany*. Philadelphia: Saunders, 1983.

This book includes an excellent treatment of transport phenomena as well as a good list of references on the subject.

ZIMMERMAN, M. H., and C. L. BROWN. *Trees: Structure and Function*. New York: Springer-Verlag, 1971.

Although a bit dated, this book remains one of the best overall treatments of water movements in plants.

30
PLANT HORMONES

All observers apparently believe that light acts directly on the part which bends, but we have seen with the above described seedlings that this is not the case. Their lower halves were brightly illuminated for hours, and yet did not bend in the least towards the light, though this is the part which under normal circumstances bends the most. It is a still more striking fact, that the faint illumination of a narrow stripe on one side of the upper part of the cotyledons of Phalaris determined the direction of the curvature of the lower part. . . . These results seem to imply the presence of some matter in the upper part which is acted on by light, and which transmits its effects to the lower part.

C. Darwin and F. Darwin, *The Power of Movement in Plants* (1888)

Plant hormones at work: Cabernet sauvignon grapes ripening in the hot California sun of early autumn.

There is a spectacular stand of virgin redwood trees in Muir Woods National Monument, a park just north of San Francisco's Golden Gate Bridge. A display near the park entrance shows a cross-sectional slab cut from a towering coast redwood at the time the tree was felled. The circular slab is about 10 feet across—not cut from the largest of redwoods, by any means, but bigger than a large restaurant table or a tractor tire. Starting at the middle and then every foot or so across the slab, a few of the hundreds of growth rings are labeled and painted white, making the slab look like a huge target. The center label shows that the tree was a small sapling in A.D. 909. It was a strong young tree in 1066, the year the Battle of Hastings was fought. It was sturdy and tall when the Magna Carta was signed in 1215 and had grown into an impressive pillar standing in an untouched forest by 1492. The tree was a venerable giant by 1776, and it had added another 12 inches of girth, ring by ring, by the time loggers cut it down in 1930 to build California bungalows.

During the millennium when the tree grew on its hillside near the California coast, it withstood innumerable storms, periods of drought, cold snaps, swarms of insects, and airborne fungal parasites and saprophytes. Each day for 1,000 years, it confronted a unique combination of sunlight, water, nutrients, and weather. Without a system for controlling growth and internal functions to meet these environmental challenges, the tree never could have had such a long life. Neither, for that matter, could a zinnia live out its short life span from seed to adult to seed in your summer garden. All plants require coordinated control over growth and internal processes. The substances that provide much of this control are plant hormones. **Hormones,** whether in plants or in animals, are substances manufactured in minute quantities in one part of an organism that produce effects in other parts of the same organism. More specifically, plant hormones must be extractable from plant tissue, must work in very low concentrations, and must have the same biological effect on all members of a plant species.

Hormones are less specific in plants than they are in animals. In animals, specific glands, such as the pituitary or ovary, make particular kinds of hormones (Chapters 18 and 37). Most plant tissues, however, are capable both of making and of responding to each of the various plant hormones. Plants have five major classes of hormones, each of which has many different effects on plant growth and metabolism. This is because the specificity of hormonal action depends on the way that a particular tissue-cell type responds; hence, a given hormone can inhibit growth in one cell type and promote it in another cell type. Much less is known about the hormone receptors of plant cells than about the analogous receptors of animal cells. It is likely that plant cells do have specific hormone receptors, though only a possible one for auxin is identified. For this reason, we will use the term "responsive cell" rather than "hormone receptor." The search for responsive cells and hormone receptors is difficult because the "target" cells in a given experiment can often both manufacture and react to the hormone being studied. Also, individual plant cells are harder to study than are animal cells because their rigid cell walls make isolating and studying individual cells very difficult. Nevertheless, it is possible to measure the effects of plant hormones, and botanists have learned much about their actions.

The five classes of plant hormones are the auxins, the gibberellins, the cytokinins, abscisic acid (and its breakdown products), and ethylene. The next five sections describe the activities of these hormones in detail.

AUXINS: CELL ELONGATION AND PLANT MOVEMENTS

Like animals, plants move in response to environmental signals. Their movement, however, is usually imperceptibly slow. Changing position by the millimeter, leaves bend toward light, roots push downward under the influence of gravity, some flowers trace the path of the sun across the sky each day, and some other types of flowers close tightly at night. These movements, called **tropisms,** are accomplished primarily by changes in tissue-cell size and, sometimes, number.

In about 1880, Charles Darwin and his son Francis performed the first recorded experiments on tropisms, and their results were published in the Darwins' book *The Power of Movement in Plants.* The Darwins wondered why plant stems bend toward the light—a phenomenon called **phototropism**—and, in particular, what part of the house plant in the window of their library directed the stem to bend toward the light. They designed a simple experiment to study this response: they grew some oat seeds on the window sill. When the coleoptiles (the sheaths covering the embryonic shoot) emerged, the Darwins put tiny black caps on the tips of some coleoptiles and tiny transparent glass caps on others. They then observed that the glass-capped coleoptiles began to bend toward the light, whereas the black-capped ones grew straight up, apparently unstimulated by the light coming through the nearby window. Coleoptiles from which tips were removed also failed to bend toward the light. This experiment is illustrated in Figure 30-1. The elder Darwin concluded that only the coleoptile tip is able to respond to light and that it produces a substance or transmits a signal that causes the stem to curve toward the light.

Figure 30-1   THE DARWINS' DISCOVERY OF PHOTOTROPISM.
Charles and Francis Darwin discovered that phototropism occurs only in coleoptiles with tips exposed to light. The Darwins masked parts of coleoptiles of grass and oat seedlings and found that the tips had to be exposed to light if directed growth (phototropism) was to occur. A transparent glass cap did not interfere with the response, but an opaque cap over the tip did inhibit the phototropism. Covering the stem had no effect.

More important, Went further showed that if the agar block was placed asymmetrically on the decapitated coleoptile—off center, to one side or the other—the coleoptile bent as it grew, with more growth on the side with the attached agar block (Figure 30-2); that is, the stem

Figure 30-2   WENT'S AGAR-BLOCK EXPERIMENT.
Went showed that a diffusible growth factor controls seedlings' phototropism. Went cut the tips from a group of dark-reared seedlings and placed the tips on agar blocks, hoping that the "signal" would move into the agar block. (a) It did—when the agar block was placed to cover the end of a decapitated coleoptile, the coleoptile grew straight upward. (b) When Went then placed the agar containing the "growth factor" on one side of the coleoptile or the other, the coleoptile elongated more on the side with the block; in other words, the plant curved away from the side with the block. Went gave the name auxin to the growth factor.

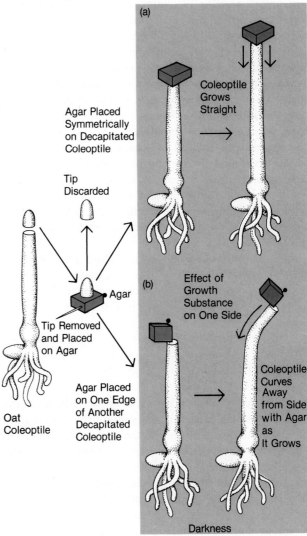

It was not until some forty years after the Darwins published their results that the Dutch botanist F. W. Went carried the tests further. Went reasoned that a signal must move from the tip of a seedling and cause the stem to bend. Using oats, as had the Darwins, he set out to isolate this signal by removing the tips from coleoptiles that had grown in the dark and placing the tips (in the dark) on agar—a gelatinlike material derived from algae that forms a rubbery solid that can retain nutrients. To begin with, of course, all Went's operations had to be done in the dark—otherwise he could not have drawn conclusions as to whether a signal agent or light hitting the little seedlings caused an effect. Went reasoned that if the signal agent moved out of the tip into the agar block, he should be able to substitute the agar for the coleoptile tip on another seedling. When he placed the block symmetrically on a decapitated coleoptile (still in the dark), the coleoptile grew straight up during the next few hours, just as would an intact coleoptile over the same time span (Figure 30-2). Went concluded that the coleoptile tip contains a diffusible substance capable of stimulating growth in basal regions of the coleoptile.

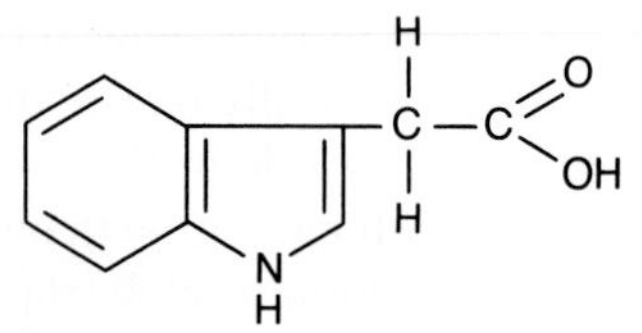

Figure 30-3   AUXIN: HORMONE WITH DIVERSE EFFECTS.
Auxin, indole-3-acetic acid (IAA), was the first plant hormone to be isolated.

bent away from the block. Since this bending in the dark was the same as that observed in intact coleoptiles in the light, Went concluded that light must affect the distribution of the signal agent, so that the side of the stem away from the light receives more signal and therefore grows faster, resulting in a curving of the coleoptile in the opposite direction—toward the light. Went called the substance **auxin,** from the Greek word *aulin,* which means "to increase."

A few years after Went's experiments, the substance from the seedling tips was purified and identified as indole-3-acetic acid (IAA). Its structure, shown in Figure 30-3, shows that it is hydrophobic, the result being easy passage through the plasma membrane of cells. The availability of pure auxin encouraged many plant scientists to study the phenomenon of coleoptile curvature more critically and to investigate other auxin-related effects. Went's original hypothesis was soon confirmed: when a plant is exposed to light, auxins produced near the tip accumulate on the side of the stem away from the light and cause it to grow quickly, thus bending the plant toward the light source. Auxins promote growth by stimulating cell elongation; as the cells on the nonilluminated side lengthen, the stem must curve toward the shorter (illuminated) side.

The group of related hormones called auxins has also been shown to be involved in cell enlargement, axillary-bud development, fruit development, leaf abscission, cambial activity, and root growth. Let us consider these effects.

Effects of Auxins

Auxins seem to act in surprisingly contradictory ways in the same plant: whereas they stimulate cell enlargement in some tissues, they suppress it in others. As we have pointed out, one effect of auxins produced near the apical meristem of the stem's growing tip may be to increase growth on just one side of the stem, causing it to bend. Conversely, the main shoot of many plants produces auxins that inhibit the growth of lower branches, resulting in **apical dominance,** the tendency of the main

shoot of a plant to predominate over all the others. An A-shaped plant—a fast-growing young pine tree, for example—exhibits apical dominance. A short, squatty, box-shaped plant, such as a boxwood shrub with many actively growing lateral branches of about the same length, does not exhibit apical dominance. To promote the growth of side branches in a normally A-shaped plant, gardeners often prune the main growing tip. This releases the axillary buds from the inhibition of apical dominance exerted by auxin produced in the main stem (Figure 30-4). Perhaps you have watched a gardener pinch back the tips of chrysanthemum plants in summer so that in autumn profuse blooms will grow from the induced lateral branches.

Auxins are also instrumental in fruit development. These hormones are often produced by young embryos and speed the maturation of fruit tissue around the seed. The amount of auxin generated by the developing embryos often determines the ultimate size of the fruit. The strawberry is a good example. Each dark spot on a strawberry is a seed, and the conical red strawberry "fruit" we eat results from the fusion of the many tiny fruits surrounding the seeds. Normally, the combined hormone output of the dozens of minute embryos inside the seeds stimulates the development of the familiar plump, sweet-tasting fruit. However, if all the developing ovules except one are removed from a strawberry, the resulting fruit is misshapen, with succulent flesh around only the remaining seed, as shown in Figure 30-5. If all the seeds but one are removed and auxin is applied, an almost normal fruit develops. This experiment provides good evidence that auxins normally produced in embryos promote fruit development.

If you were to orient a normal seedling so that its shoot and root axes were horizontal rather than vertical, you would soon see the elongating shoot tip bend upward and the growing root tip, downward. These movements are **gravitropisms,** or responses to gravity. Experiments reveal that auxins and another class of plant hormones, the *gibberellins,* build up on the lower side of a shoot lying horizontally. Because of this hormone accumulation, greater cell elongation occurs on the lower surface, so the tip bends upward, away from the Earth. When the growing shoot is oriented vertically once again, hormone concentrations equalize across the shoot, and straight vertical growth proceeds. (Unless, of course, light falls on one side and causes the plant to bend in the direction of the light source.)

Gravitropism is thought to be somewhat more complex in roots than in shoots. The root cap must be intact for the downward curvature of a root to occur. If the root cap is cut off, a root may continue to elongate horizontally. Although the exact mechanism is not yet understood, auxins may cause extra cell elongation on the upper side of a horizontally placed root, or they may

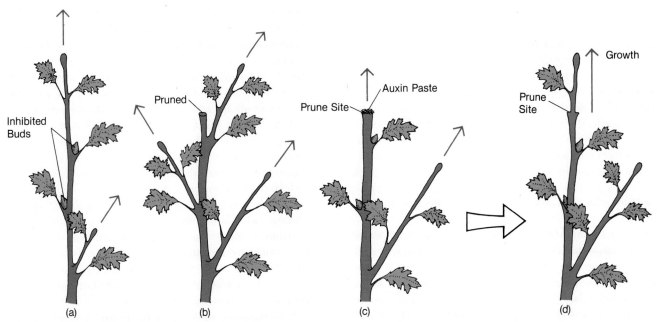

Figure 30-4   APICAL DOMINANCE: THE INFLUENCE OF AUXIN.
(a) Apical dominance is the tendency of the main shoot of a plant to predominate over lateral growth. However, if the apical meristem has grown far enough from the lower axillary buds (as at bottom), the buds may begin to grow. (b) Apical dominance can be altered by pruning (cutting off) the main growing shoot. When this is done, axillary buds produce branches. (c) If auxin paste is applied to the area from which the main shoot was pruned, (d) the tip of the main shoot both will grow and will reassert apical dominance, showing that auxin is produced in the apical meristem and is the controlling factor.

inhibit cell elongation on the lower surface. Either would yield the downward-curving response. No one knows yet whether auxin and root caps are involved when roots of certain aerial plants grow upward away from gravity.

How Auxins Work

The Darwins' experiment on phototropism demonstrated that coleoptiles sheathed in a black cap will not curve toward a light source, whereas those sheathed in a

Figure 30-5   AUXIN: A SUBSTITUTE FOR STRAWBERRY (*FRAGARIA* SPECIES) OVULES IN PROMOTING FRUIT DEVELOPMENT.
(a) A normal strawberry. (b) A strawberry from which all but one of the seeds were removed. (c) A strawberry from which all but one of the seeds were removed and that was then treated with a paste containing auxin; it develops almost normally.

(a)   (b)   (c)

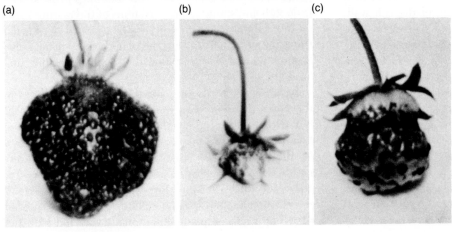

transparent cap will. Botanists have since found indirect evidence that a pigment present in the coleoptile tip specifically receives the light stimulus; when pigment molecules on the illuminated side of the stem receive light, auxins are transported down the opposite, dark side of the stem. There, as we saw, auxins promote cell elongation, causing the seedling to curve toward the lighted side. Despite years of searching for the pigment in the coleoptile tip, it has never been purified, and just one additional fact is known: light at the blue end of the spectrum (wavelengths shorter than 500 nm) is the most effective stimulator of the pigment molecules. Other wavelengths of visible light have no effect on this system.

Botanists would like to know whether auxin is preferentially transported to the dark side of a stem when the pigment receives blue light or whether auxin is transported to *both* sides and is simply destroyed on the illuminated side. The path and rate of auxin movement can be measured by applying a radioactively labeled auxin solution to the cut end of a stem. Most auxins move downward and can travel against a concentration gradient at a velocity of about 1 to 1.5 centimeters per hour. (Some studies suggest that this movement is in association with plasma membranes or cell walls, not through xylem and phloem.) In other words, auxin may move rapidly downward through tissues that already contain more of the hormone than does the tip itself (of course, if auxin moves in the cell walls and if hormone already present is inside the cells, the movement is not truly against a concentration gradient). This bias in auxin movement is called *polar transport*. In stems kept in the dark, the downward polar transport is equivalent on all sides of the stem. In stem sections illuminated on just one side, however, most of the auxin appears to be transported across the stem to the dark side by an unknown mechanism.

Despite the mysteries about auxin transport, plant biologists do know how auxin promotes cell elongation and thus stem curvature. When auxin reaches a responsive cell, it stimulates the cell to *expand*. Auxin's critical action is to indirectly cause loosening of the cell wall; then the cell's internal turgor pressure stretches the wall, allowing expansion. This expansion is oriented by cellular microtubules or other intracellular factors, so that an auxin-treated cell can increase in length more than tenfold. Auxin does not act directly to loosen the cell wall; rather, it promotes the secretion of hydrogen ions from the cell. These protons lower the pH within the cell wall; that, in turn, may increase activity of enzymes that disrupt some of the bonds holding the components of the wall together. When these bonds are broken, the cellulose fibers in the wall can slip past one another so that the cell's turgor pressure pushes the wall into a new shape.

Recent research suggests that acid (H^+) release may also play a crucial role in gravitropisms, just as it does in cell elongation. Specifically, it has been found that tiny amounts of acid are released into the soil from the zone where cells elongate in a vertical root. In a root placed horizontally, acid is secreted from the upper surface, while rapid uptake of acid can occur at the lower surface, as illustrated in Figure 30-6. Because changes in acid movement precede the root's asymmetrical growth and curvature, some botanists have proposed that the effects of auxin are closely related to acid secretion and resultant cell-wall loosening.

Plant physiologists looking for this same relationship between auxin and acid release in a horizontally positioned shoot found that acid is secreted from the lower side, which grows faster and causes the stem to curve upward (Figure 30-6). In a vertical shoot exposed to light on one side, the dark side releases acid after the opposite side has been illuminated.

These various results begin to reveal a consistent pattern of auxin function: the presence of auxin affects the release of hydrogen ions, which, in turn, bring about cell-wall loosening and cell elongation. The resulting asymmetrical growth in shoots and roots causes the basic plant tropisms in response to light and to gravity. More study is needed to explain auxin transport, as well as auxin involvement in fruit development and apical dominance.

GIBBERELLINS: GROWTH PROMOTERS

A disease that affects an important staple crop, rice, led to the discovery of a second class of growth-promoting hormones, the **gibberellins.** In Japan and other rice-growing areas, rice plants begin as small, sturdy seedlings and grow into chest-high mature plants in flooded fields. Sometimes, however, the seedlings grow abnormally tall and spindly, and the stems of the mature plant are so frail that when the tips are heavy with rice seeds, the wind eventually knocks the plant over. Japanese rice farmers named this condition the "foolish-seedling disease."

Japanese scientists studying the "foolish" seedlings in the 1930s discovered not only the cause of the disease, but also an entirely new class of plant hormones. They found that a fungus infects the seedlings and produces a chemical that causes abnormal elongation of the internodes. The researchers named this chemical gibberellin, after the Latin name for the fungus, *Gibberella fujikuori.* The steps in the scientists' research showed first that a normal plant intentionally infected with the

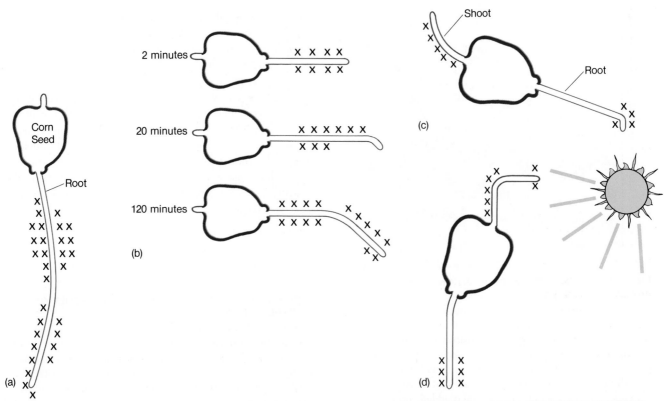

Figure 30-6   THE ROLE OF ACID PRODUCTION OR RELEASE IN RESPONSE TO AUXIN IN ROOT AND SHOOT TROPISMS.
(a) Tiny amounts of acid (represented by x's) are released into the soil from all sides of the region of elongation in a corn root growing downward. (b) When a root is reoriented horizontally, most acid is released from the region of elongation as the shoot turns to grow downward. The same root is shown at three different times after the root reorientation. Shoot growth patterns and patterns of acid release show that acid is released on the side of the shoot that elongates as the shoot curves (c) away from gravity, or (d) toward the light. The acid may play a role in the hormone-induced loosening of cell walls and thereby allow these tropisms.

fungus gets the disease. Second, extracts of the fungus applied to normal rice plants also transmit the disease. Those extracts contain gibberellin; and, in fact, more than seventy related gibberellin compounds have now been identified as occurring naturally in higher plants. The most active appears to be gibberellic acid (GA_3); its structure is shown in Figure 30-7. Today, biologists also know that, unlike auxin, gibberellins increase *both* the size and the number of internodal cells, and as a result, increase stem length.

Genetic dwarf plants, which remain small with short stem internodes despite the application of auxin, can be induced to grow by applying gibberellins. At first, botanists could not tell what was physiologically different about naturally occurring dwarf plants. They considered several hypotheses, including (1) that the dwarfs make too little of their own gibberellins, or (2) that their cells are not receptive or responsive to gibberellins. To investigate, researchers treated dwarf corn plants with gibberellin and found that in some plants, the internodes

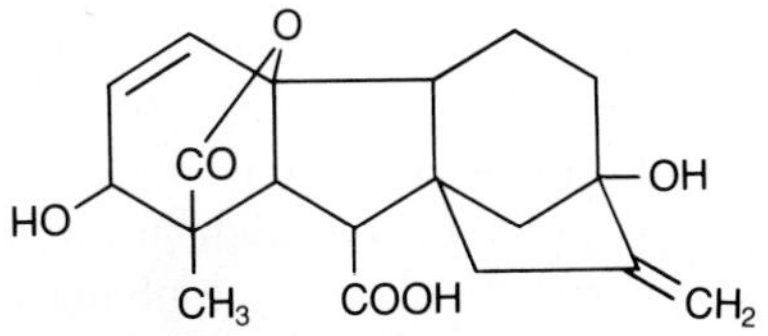

Figure 30-7   GIBBERELLIC ACID: STIMULATOR OF GROWTH.
Gibberellic acid is one of a class of more than seventy related compounds, the gibberellins. They have multiple effects, but are especially active in stimulating plant growth.

did indeed elongate and that the amount of growth depended directly on the amount of gibberellin added, as shown in Figure 30-8. This finding indicates that some genetic corn dwarfs are responsive to gibberellins, and it is now known that many dwarfs have a lower amount of the hormone than do normal-size plants.

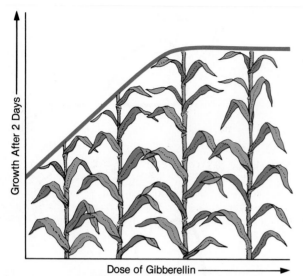

Figure 30-8   GROWTH IN PROPORTION TO THE AMOUNT OF GIBBERELLIN PRESENT.
Biologists discovered a feature of gibberellin action when investigating the cause of dwarfism in corn. Does dwarf corn lack gibberellins? Or is it insensitive to them? When researchers treated dwarf corn with varying doses of the growth hormone, dwarfism was overcome and normal growth occurred. Furthermore, the amount of growth depended on the amount of gibberellin added, up to a maximum concentration.

In addition to growth, various aspects of plant development can be regulated by gibberellins. Many biennials and some annuals grow as a tight cluster of leaves (Chapter 28). Cabbage is a good example. The stem of such a plant displays almost no internodal elongation; thus the leaves are tightly packed in a roundish head, or *rosette*. When the plant is ready to flower, however, it *bolts*—that is, it sends a shoot high in the air. Bolting can be induced in rosette plants by applying a small dose of gibberellic acid. Unusual effects, such as the 10-foot-tall cabbage shown in Figure 30-9, can be created this way. High doses of gibberellins induce both stem elongation and the flowering that takes place after bolting. The hormones are actually present and functioning in the rosette stage, but their concentrations are too low to trigger bolting and flowering.

Gibberellins have some effects similar to those of plant auxins. For example, the growth rate of some fruits depends on the concentration of gibberellic acid. This hormone promotes the enlargement of apples and peaches and can substitute for the presence of a developing embryo, the source of the normal stimulant for fruit growth. Many fruits that are responsive to treatment with gibberellins are not sensitive to auxins, and vice versa.

Fruit growers interested in increasing fruit size by applying hormones must first determine which of the two growth-promoting hormones will aid fruit development in the species they are cultivating.

Although gibberellic acid and auxins have similar effects, auxins and gibberellins are bound by different responsive cells in the plant. Nevertheless, both hormones are required for normal growth. This dependence of plant growth on more than one regulatory chemical is much like that in humans and many other vertebrates, which have growth hormones, thyroxin, sex steroids, and other agents affecting various growth processes in different tissues.

Figure 30-9   TOWERING CABBAGES: A RESULT OF GIBBERELLIN TREATMENT.
Cabbage ordinarily has almost no internodal elongation despite the presence of a low level of gibberellins. Instead, as seen at the left, the leaves are packed in a tight head. Treatment with higher levels of gibberellin induced these plants to elongate, producing spectacular stems instead of tight rosettes.

CYTOKININS: CELL-DIVISION HORMONES

We have seen that auxins and gibberellins induce plant growth by separate mechanisms. We also know, however, that plant growth involves cell division and an increase in absolute cell number. A third class of hormones regulates cell division in plants: the **cytokinins.** This group of chemical compounds is related to adenine, one of the bases common to all nucleic acids (Figure 30-10).

One of the first cytokinins to be discovered was found in coconut milk. Why would anyone be analyzing coconut milk? During the 1940s and 1950s, plant biologists had problems growing isolated plant embryos and bits of tissue in culture. Even though cultured tissues enlarged to a certain extent by cell elongation, they did not undergo many cell divisions. Folke Skoog and his associates at the University of Wisconsin thought of culturing the plant embryos in a dose of coconut milk, which is a type of liquid endosperm. They reasoned that the milk must be rich in all the factors required for the growth of coconut embryos, so why not those of other species as well? Their guess was correct: the growth and cell division of virtually all cultured plant tissues were greatly improved by the addition of coconut milk. Cytokinins were the reason.

Effects of Cytokinins

Cytokinins appear to have little or no effect on plant growth in the absence of auxins. Instead, the ratio of cytokinin to auxin determines the rate of cell division in plants. Cells in culture either grow larger and then divide or elongate but do not divide. The first type remains undifferentiated, while the second type takes on specialized roles. When cytokinins alone are added to a culture of undifferentiated plant cells, nothing happens. When auxin alone is added, the cells elongate. When both hormones are added, the cells divide. If the concentration of cytokinin is higher than that of auxin, the

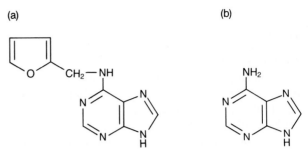

Figure 30-10   CYTOKININS: STIMULATORS OF CELL DIVISION.
Cytokinins cause plant cells to divide, and thus plant regions to elongate or enlarge. Shown is the structure of (a) kinetin, a cytokinin, and (b) adenine, a related compound.

rapidly dividing cells tend to differentiate into shoot or leaf tissue. If the concentration of auxin is higher, roots or disorganized lumps of cells called **callus tissue** tend to form. Clearly, the normal growth and division of cells require the presence of both hormones in a well-regulated balance.

Besides promoting cell division in the presence of auxin, cytokinins delay aging, or *senescence*. When leaves senesce, they lose much of their chlorophyll, their proteins are degraded into amino acids, and many of their organic constituents are exported to the stem. Senescence can be retarded by applying cytokinin to the surface of a leaf, as shown in Figure 30-11. The effect of cytokinin can readily be observed after applying the hormone to only one spot on a senescing leaf: the yellow leaf will sport a single bright green region. Botanists think that cytokinin may spare leaf tissue from aging by making the leaf a more effective sink for nutrients—that is, by promoting nutrient transport to the area. A decrease in cytokinin in older leaves may lead to the senescent state because once used up, vital minerals and organic nitrogen compounds are not replenished.

Figure 30-11   CYTOKININS: INHIBITORS OF SENESCENCE IN A LEAF.
Cytokinins not only promote cell division but also delay aging. When cytokinins are applied to an area of a leaf, that spot remains healthy, while the remainder undergoes normal senescence. Cytokinins may act by promoting transport of nutrients to the treated area, thereby slowing the degradation of treated leaf tissue, a normal aspect of aging.

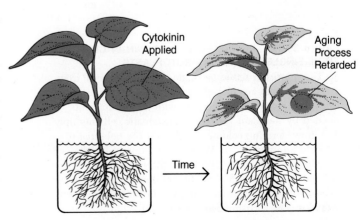

Cytokinins, Auxins, and Plant-Tissue Cultures

The forestry and agricultural industries invest much time and money in the perfection of plant-tissue culture techniques. The aim is vegetative production of desirable stock. Because cytokinins and auxins are responsible for both cell division and tissue growth, they are the two major hormones used in culturing plant cells and tissues. Today, plant-tissue cultures are readily established for most species; these cultures can be maintained for decades, cell generation after cell generation, without losing the ability to grow and divide.

Scientists can culture plant cells and tissues much more successfully than animal cells. A few plant-tissue cultures have been maintained continuously for thirty years, and many survive up to ten years. Most mammalian-cell cultures, on the other hand, survive for less than one year, and those that live longer are almost always altered by viruses. The ability to differentiate and undergo morphogenesis also distinguishes cultured plant cells from cultured animal cells. As we have seen, treatment with various ratios of cytokinins and auxins will persuade plant cells to develop into different structures: roots, shoots, leaves, or even whole plants. Cultures of animal cells, in contrast, rarely retain the ability to rebuild complex organs, such as an eye or a leg, al-though some may re-form simpler tissues, such as cartilage nodules or clusters of secretory cells.

With certain plant species, it is possible to grow an entire plant in culture from one cell, or *protoplast* (Figure 30-12). The ability to regenerate a whole organism from a single cell is a reflection of the cell's *totipotency* (Chapter 28). Totipotency is common in plant cells and means that each cell, while differentiated in the adult, retains all the information required to make an entire plant.

Totipotency has not been demonstrated in cultured normal mammalian cells. Most developing animal cells retain the complete set of genetic information, but as we learned in Chapter 17, this is not enough to guide the construction of an entire organism: epigenetic, or developmental, information is also required. This type of information resides in the cytoplasm of the fertilized animal egg and in properties of the early embryo and acts to regulate the fate of cell behavior during early development (Chapter 17). Recall, for instance, substances in the cytoplasm of the egg responsible for establishing the germ-cell lineage in a frog or a fruit fly. Such substances are not present in typical differentiated animal cells. Hence, an experimenter cannot reconstruct a bat wing, a lizard eye, or a human heart from a clone of cultured cells derived from those organs. The fact that a whole plant can arise from a clone of a single cell, or protoplast,

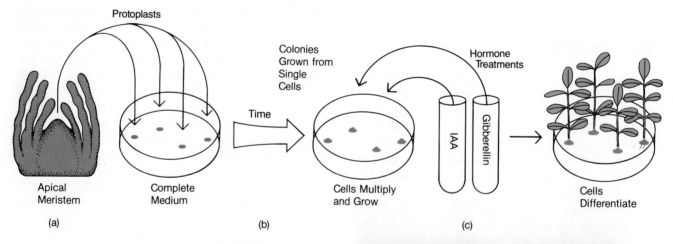

Figure 30-12   ENTIRE PLANTS GROWN IN CULTURE FROM SINGLE CELLS, OR PROTOPLASTS.

(a) Tiny pieces of tissue or single cells (protoplasts) from apical meristem are placed under sterile conditions on an agar plate containing a medium with all the nutrients needed by the plant. (b) The meristematic cells multiply and grow, forming colonies, each with the potential to develop into a plant. (c) With the addition of hormones in specific amounts and proportions, the meristematic tissue can be induced to differentiate and develop into mature plants. (d) Petri dishes showing varying stages of growth of cultured corn and sunflower clones.

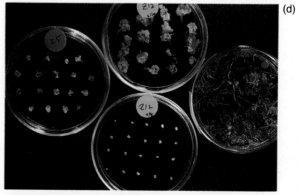

indicates a basic difference between plants and animals; the essential epigenetic information can be generated anew by the proliferating plant-cell clone.

Differences between plant- and animal-cell cultures might be expected solely on the basis of the very different growth mechanisms in adult plants and animals. In animals, all the major organs usually form during embryonic development, and no new ones are normally produced later in life. In plants, however, the growth of new organs never ceases. Meristems proliferate constantly (or at least seasonally), so that stems elongate and branch, new leaves grow, and roots form ever more complex shapes. Furthermore, plants have no distinct germ-line cells, as do many animals; normal meristematic cells give rise to flowers, and cells located by chance at certain sites in the flower yield the ovum or pollen lines. It is no surprise, then, that plant cells in culture organize new meristems and produce either root or shoot, depending on the hormonal environment. If conditions favor extremely fast growth and cell division, callus tissue results. If such a disorganized lump is given appropriate hormones, it can form many organized structures, each growing from a meristem that develops on the surface of the lump.

ABSCISIC ACID: THE GROWTH-SLOWING HORMONE

Some plant hormones promote growth and cell elongation (auxins and gibberellins), and others stimulate cell division and retard senescence (cytokinins). One could almost predict, therefore, the existence of a hormone that suppresses a plant's natural tendency to grow, as might be needed during droughts or cold winters. And, indeed, there is one: **abscisic acid** (Figure 30-13). This name is derived from the word "abscission," which means "cutting off" or "removing." In fact, as we will see, this is not nearly as apt a term for the hormone as is *dormin*, the alternative name.

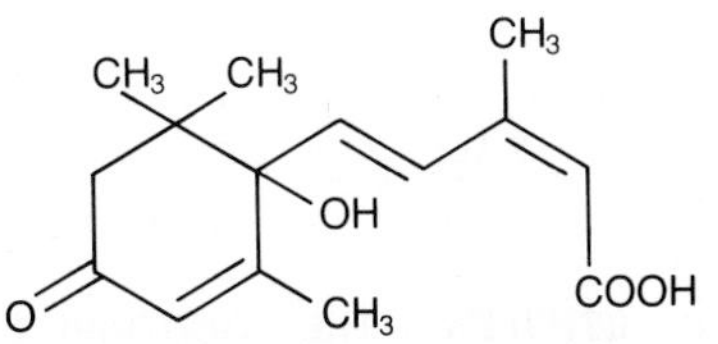

Figure 30-13   ABSCISIC ACID: HORMONE THAT SLOWS GROWTH.
Abscisic acid suppresses a plant's tendency to grow. The hormone is found in dormant bulbs and seeds, and is present in trees at the end of each growing season.

Abscisic acid is found in dormant bulbs and seeds and in some fruits, leaves, and other tissues. Its presence can trigger changes that prepare a plant for leaf drop, for winter dormancy, or simply for general slowed growth. In some plants, abscisic acid helps control flowering. In all plants, it works in a delicate balance with growth-promoting hormones to govern appropriate responses to seasonal and short-term environmental changes.

Abscisic acid was actually discovered twice: it was studied first in the 1940s, when a group of British scientists prepared an extract from birch leaves that could arrest the growth of seedlings of birch trees and other plants. They hypothesized that this new compound caused dormancy in trees, since levels of the hormone increased in trees late in the growing season and appeared to promote bud dormancy during the winter. The second discovery of abscisic acid took place in the 1960s at the University of California at Davis. Scientists isolated sufficient amounts of the hormone from cotton bolls (fruits of cotton plants) in order to determine the compound's chemical structure. This structure proved to be identical to that of the tree-dormancy hormone studied by the British researchers.

Effects of Abscisic Acid

Abscisic acid has a remarkable range of effects. The British scientists noted that it stimulates buds to form a set of outer leaves that become tough, protective bud scales to prepare for winter dormancy (hence, the name dormin). Abscisic acid also acts as a general inhibitor, especially in response to stress. For instance, drought causes an increase in abscisic-acid concentration in leaf tissue and, in turn, the closure of leaf stomata to prevent further water loss (Chapter 29). Abscisic acid is often an antagonist to the positive effects of the other plant hormones on plant growth. Whereas gibberellic acid stimulates the production of hydrolyzing enzymes in endosperm, for example, abscisic acid inhibits this response, thereby acting as a check on the release of the seed's food energy. That, in turn, prevents the increased protein and nucleic-acid synthesis that is required for embryonic development.

Abscisic acid is active in other ways in seed dormancy. For instance, it builds up in maturing seeds and suppresses root and shoot elongation in the embryo. *Stratification* (treatment of seeds at low temperatures) appears to be necessary for the breakdown of abscisic acid in many seeds. Presumably, the same thing occurs in seeds "asleep" under snow or ice as they await the spring thaw to germinate. Some plants exhibit a lethal mutation called **vivipary,** in which root and shoot elongation occurs while the embryo is still in the flower. A viviparous seed is doomed because, dangling high above the ground, its roots cannot become established in the soil. Furthermore, such germination occurs at the wrong time of year. Vivipary can sometimes be suppressed by applying abscisic acid to the developing seeds. This suggests, but does not prove, that the hormone may act in normal, unmutated seeds to check premature growth processes.

Finally, abscisic acid accelerates senescence and in some plants promotes **abscission,** the normal separation of leaf, fruit, or flower from the plant. For instance, in cotton bolls, the largest amounts of this hormone are found in the ovary bases at the time of normal fruit drop. Abscisic acid normally is produced in the leaves, probably in the chloroplasts, and is shipped throughout plants via translocation in the phloem. Older tissues accumulate abscisic acid by an unknown mechanism; as a result, senescence and dormancy are accelerated in those tissues. For example, placing a drop of abscisic-acid solution on a healthy green leaf results in the rapid formation of a yellowed, aged spot—just the opposite effect of a drop of cytokinin on a senescent leaf. Finally, it is noteworthy that despite its name, abscisic acid is not primarily responsible for abscission. Indeed, as we shall see, plant physiologists believe that other hormones may be more commonly involved in that process.

Interaction of Abscisic Acid and Other Hormones

Abscisic acid interacts with the growth-promoting hormones to regulate plant responses. However, almost all of abscisic acid's growth-suppressing effects can be offset by one of the other plant hormones. Perhaps the most confusing feature of plant-hormone biology is that each of the growth-promoting hormones counteracts only *a few* of abscisic acid's inhibiting effects: some can be reversed only by auxins; others, only by gibberellins; and still others, only by cytokinins. Most animal hormones do not inhibit each other's action, and when inhibition does occur, there is a one-to-one correspondence between positive and negative signals; for example, insulin and glucagon are, respectively, positive and negative regulators of sugar metabolism (Chapter 37). At this point, botanists cannot predict which growth-promoting hormone will counteract a particular effect of abscisic acid. Although the specific interactions are complex, the overall meaning of this give-and-take system is clear: the various plant hormones work together to control growth so that a plant even as tall as a towering redwood can meet the numerous physical challenges it faces during any given season, by slowing down or speeding up growth at appropriate times.

ETHYLENE: THE GASEOUS HORMONE

The common expression "One rotten apple spoils the barrel" is most often applied to disobedient children, but it is actually based on an accurate observation: one ripening fruit does produce a volatile compound that accelerates the ripening of nearby fruit. This volatile substance is the hormone **ethylene.** At normal temperatures, ethylene is a gas and has a simple chemical formula: $H_2C{=}CH_2$ (Figure 30-14). The presence of ethylene in air causes increased respiration, which, in turn, leads to the changes in fruit composition that transform a hard, acidic, inedible fruit into a sugary, ripe one.

Ethylene production by a plant can be triggered by localized increases in auxin concentration and perhaps by other physiological factors as well. Ethylene released into the air can then act on the whole fruit or plant; its increase is accompanied by a sharp rise in metabolic activity and a hastening of the condition we call *fruit ripening.*

The ability of ethylene to accelerate and regulate ripening has several commercial applications. Green bananas are shipped to the United States from Central America and ripened later with ethylene treatment. At home, underripe fruits can be ripened by enclosing them in a plastic bag with a small slice of potato or apple—such cut tissues produce ethylene and will promote ripening.

The natural production of ethylene is sometimes a problem, since fruits may ripen too quickly and flowers may spoil in storage. Stanley Burg, who has studied the role of ethylene in fruit ripening and flower fading, developed a storage system for fruits and flowers that slows the ripening process. His controlled-atmospheric storage system consists of a humidified chamber kept at one-tenth normal atmospheric pressure, with air continuously flushing the chamber. Under such storage conditions, ethylene produced by fruits and flowers is quickly removed from the surrounding air; as a result, an apple slice remains juicy, without browning, for days, and

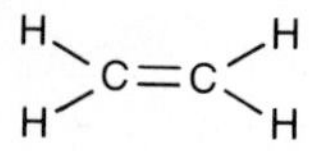

Figure 30-14   ETHYLENE: THE FRUIT-RIPENING HORMONE.
This natural substance is commonly used to ripen bananas and tomatoes that are shipped unripe to minimize damage and waste.

roses and carnations can be kept fresh for six months. Controlled-atmospheric storage also reduces the oxygen concentration to about 2 percent (one-tenth normal), so that aerobically dispersed organisms such as mildews grow poorly, yet there is still sufficient oxygen to inhibit anaerobic microbes that might cause fermentation.

In addition to promoting fruit ripening, ethylene influences plant growth. Ethylene inhibits the transport of auxin across the stem or root—a movement important during the plant's tropic responses to light and gravity. Because ethylene is produced in areas of high auxin concentration yet inhibits its transport, ethylene may serve as part of a negative-feedback loop that prevents excess auxin accumulation. The ratio of ethylene to auxin seems to influence whether cells elongate or expand radially in response to auxin. Because auxin can stimulate ethylene formation, many of the effects originally attributed to auxin may actually be due to ethylene.

INTERACTION OF PLANT HORMONES

All five major types of plant hormones interact in complex ways to control the physiology and structure of the entire plant. The general areas of production are known for all five. Auxins and gibberellins are made in the young growing regions of the shoot, particularly in the leaves. Cytokinins are usually produced in the roots. Abscisic acid is generated by cells experiencing stress, particularly water stress. Ethylene may be produced wherever there is high concentration of auxin. However, biologists still have not proposed a general theory to explain how the hormones produced in the various tissues and locations interact to make a plant responsive to environmental stresses.

Although there is no single model, there is a generally accepted principle of hormonal interaction in plants: two or more positive hormonal signals are required for growth, but one negative signal can prevent further growth. For example, cell elongation requires both auxins and gibberellins, two growth-promoting signals. Also, cytokinin can sometimes stimulate cell division, but the new cells arising from division must be exposed to auxin, too, if they are to elongate and produce normal tissue-level growth (root or shoot elongation). Ethylene and abscisic acid can act as single negative signals to halt growth processes. However, for every example of growth inhibition by abscisic acid, combinations of auxin, gibberellin, and cytokinin can reverse the negative impact. These positive effects of auxins, gibberel-

lins, and cytokinins on growth occur because each hormone stimulates different aspects of growth. The evolution of this complex control system provides plants with a flexible repertoire of responses for meeting the challenges of living from day to day and from season to season.

Let us consider two specific responses brought about by hormonal interactions in plants: falling leaves and germinating seeds.

Leaf Fall

The red, gold, and orange leaves of autumn create one of the most enchanting signs of the changing seasons, as they flutter in the wind and cover the ground. Like so many other striking events in nature, leaf fall has a complex, integrated underlying chemical and physical basis. It might seem logical to suppose that leaf fall, or abscission, is triggered by abscisic acid. This is true in only a few plants, however. In most, the levels of auxins and ethylene appear to bring about leaf fall, as well as the dropping of ripe fruit and unfertilized flowers.

At the base of the leaf petiole, flower stalk, and fruit stem, a special zone of cells called the **abscission layer** forms in response to hormonal signals (Figure 30-15). When the production of auxin declines in the leaf blade, for example, the abscission layer begins to form. An increase in ethylene production also appears to promote such formation. Thus auxin is an inhibitor of abscission, while ethylene is a promoter. The cell walls in the abscission layer are broken down, creating a weak area. The weight of the leaf or a gust of wind can cause the leaf to separate from the stem and fall. The same basic mechanism occurs in flowers and fruits. When auxin production decreases in the flower and ethylene levels increase, an abscission layer forms in the floral stalk, and in time, the flower drops off. In ripe fruit, mature seeds decrease their production of auxin, and ethylene levels (associated with fruit ripening) rise. An abscission layer forms in the fruit stem, and eventually the fruit drops.

Seed Germination

Biologists better understand the relationship between the two hormones needed for seed germination. The control of germination in barley seeds, for example, is mediated by gibberellic acid and abscisic acid. As we saw in Chapter 28, a critical step in barley-seed germination is mobilization of starch stores in the endosperm. Mobilization begins when the embryo secretes gibberellic acid. This hormone, in turn, stimulates the aleurone layer, which surrounds the endosperm, to produce and

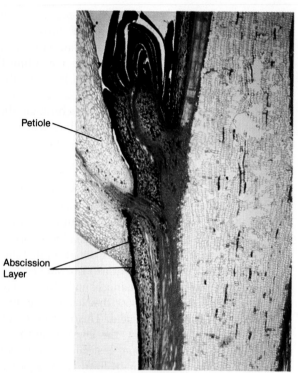

Petiole

Abscission
Layer

**Figure 30-15   FORMATION OF ABSCISSION LAYERS:
A RESPONSE TO HORMONAL SIGNALS.**
The abscission zone is a layer of structurally weak cells in or
near the petiole. Enzymes are synthesized that weaken the
cell walls in this zone to such an extent that the leaf's own
weight or a gust of wind can cause the leaf to break free. A
decrease in the production of auxin and an *increase* in that
of ethylene cause the abscission layer to form. In this view
of a maple (*Acer* species) branch, the vertical black line on
the left is the abscission layer; the leaf's petiole extends
toward the left and an axillary bud is seen at the top, left.

secrete an enzyme (α-amylase) that digests the starchy
endosperm. To do so, gibberellic acid causes an increased synthesis of the messenger RNA for α-amylase
while decreasing synthesis of ribosomal RNA. If the seed
is exposed experimentally to abscisic acid, synthesis of
α-amylase mRNA is inhibited while rRNA synthesis is
stimulated. Both hormones thus appear to act at the
level of gene transcription in the nuclei. Scientists hypothesize that in nature abscisic-acid content rises if the
seed experiences water stress—a soil-moisture content
too low to support seedling growth. Thus the presence of
a growth-inhibiting hormone may prevent the seed from
germinating when conditions are unfavorable for the
seedling to become established.

CONTROL OF FLOWERING

Flowers are major features of the angiosperm reproductive cycle, as are fruits and seeds. We know, simply
from observing the changing seasons each year, that
most temperate-zone plants flower in spring or summer.
However, each species seems to have an internal calen-
dar that tells it when to bloom. Crocuses often break
through the snow, and tulips emerge a few weeks later.
Apricot trees flower before apple trees; Kalanchoë, before black-eyed Susans; strawberries, before radishes;
and all these plants, before chrysanthemums, which unfurl in early autumn (Figure 30-16). The timing of a
plant's reproductive cycle is obviously critical to the pollination of its flowers and hence to the production of
fruits and seeds at appropriate times, so that the plant's
genetic message can be successfully dispersed and so
that the plant has an optimal chance of surviving until
the next growing season. But how does a plant "measure" the seasons and flower at an appropriate time?

The major factors in this internal timekeeping are the
plant's developmental program and environmental cues.
These factors influence hormonal activity, which can
trigger flowering, germination, growth, dormancy, senescence, and so on. In many long-lived species such as
trees, spring flowering is determined by the previous
year's growth of floral buds—unexpanded flowers that
will bloom in the first growth of the new season. Each
spring, the glorious colors of blossoming dogwood,
cherry, plum, and other trees give testimony to the development of floral buds of the previous season. Quite a
different strategy is employed by some annual plants,
such as zinnias, that produce floral buds after a fixed
amount of vegetative growth; their flowering is also
highly programmed. Environmental cues such as temperature and light can hasten or delay programmed flowering; a good example is winter wheat, in which a cold
winter accelerates the time of flowering. But just how
does the environment trigger the activity of plant hormones so that flowering occurs at appropriate times for
the various species? The relative lengths of day and
night, a light-sensitive pigment, and possibly another
unidentified hormone are involved.

Photoperiodism

By the turn of the twentieth century, botanists knew
that some plants require specific environmental cues to
flower. The significance of light was recognized in 1920
by W. W. Garner and H. A. Allard. They showed that
the number of hours of light in a given day—the day
length, or **photoperiod**—is crucial in determining

whether many plants will flower. In the continental United States, the day length in a twenty-four-hour period changes from about nine hours to fifteen hours between the winter and summer solstices. This is equivalent to an average change of two minutes a day. Thus from an ecological point of view, the photoperiod in a twenty-four-hour day provides an accurate measure of the changing seasons.

Garner and Allard classified plants into three groups, depending on the effects of photoperiod on flowering: *short-day* plants; *long-day* plants; and *indeterminate*, or *day-neutral*, plants. Short-day plants require a short photoperiod—fewer than some critical number of hours for each species—in order for flowering to be triggered. Long-day plants require a long photoperiod—more than some critical number of hours for each species. The flowering of day-neutral plants appears to be unaffected by the number of hours of light in a day.

Later research found that this explanation of photoperiodism was not precisely correct, and it was modified in 1938 by Karl Hamner and James Bonner. They performed a very simple experiment: they exposed cocklebur, a short-day plant, to a ten-hour (short) day and a fourteen-hour night and observed that the plants flowered as expected. However, if they interrupted the fourteen-hour dark period with a brief flash of light, the cocklebur did not flower. Consequently, they determined that it is the length of the night period that is critical to a short-day plant (Figure 30-17). Applying hindsight, short-day plants should really have been called long-night plants. Hamner and Bonner also showed that long-day plants actually require a short, uninterrupted night; in other words, they are short-night plants. If, for instance, a long-day plant is kept in just a few hours of light and many hours of darkness, it will not flower. But what happens if that long night is interrupted by a brief period of light? Flowering occurs because the plant is tricked into behaving as if the interrupted long night were actually a short night. In summary, short-day plants require some minimum critical period of uninterrupted darkness in order to flower, whereas long-day plants are induced to flower when the period of light exceeds some critical length and the periods of darkness are relatively short.

Figure 30-16   BLOOMING SEASONS.
As shown in these photographs taken at five to six-week intervals from spring through early fall, each species in this garden, or indeed in a wild meadow, follows its own internal, genetically controlled program that leads to flowering. External cues such as day/night length and temperature are essential for the realization of the internal programs. Various animal species that both transfer pollen for plant fertilization and exploit flowers for food must also be tuned in to the yearly flowering schedules seen here.

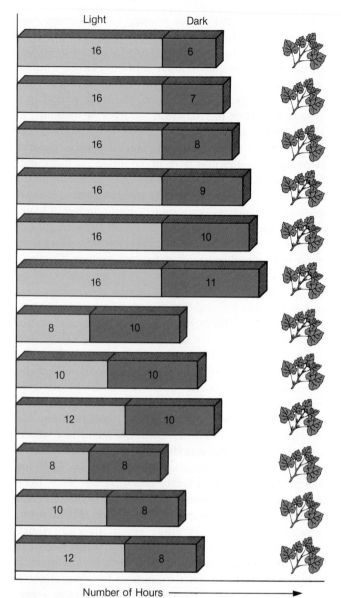

Light	Dark
16	6
16	7
16	8
16	9
16	10
16	11
8	10
10	10
12	10
8	8
10	8
12	8

Number of Hours ⟶

Figure 30-17   FLOWERING: A RESPONSE TO SPECIFIC PHOTOPERIODS.
Experiments with a cocklebur in controlled-light chambers show that plants must experience nights of certain lengths before flowering is induced. For the cocklebur, the minimal night length is nine hours. No matter what time the night begins, or how long the day is, only an *uninterrupted* period of darkness, at least nine hours long, will induce flowering.

Once the day-length concept had been proposed, botanists observed the effects of the photoperiod on flowering in many kinds of plants and found that the number of treatment periods required to induce flowering can vary widely. Some short-day (long-night) plants—rice and cocklebur, for example—require only a single long night to induce flowering. Some long-day (short-night) plants—

such as dill weed, spinach, and white mustard—require only a single short night. Other plants, however, require several days with the appropriate hours of light or darkness reflecting their need for several treatment periods. Plants such as winter wheat, soybean, and orchids also require *cool* nights in addition to the appropriate photoperiod before they will flower. Finally, there are day-neutral plants—such as cucumber, peas, corn, and onion—whose flowering appears unaffected by day length. Significantly, most specific photoperiod requirements of different plants do not depend on temperature (the requirement of some plants for cool temperatures is a separate phenomenon, not related to measuring day or night length). This makes sense, since temperatures vary so much between day and night; direct temperature effects on the photoperiodic response would make it difficult for a plant to keep track of day or night length. In this respect, plants are like animals whose biological "clock" runs at the same rate, irrespective of varying daily temperature.

The Phytochrome-Pigment System

Interrupting the required long night of so-called short-day plants with a flash of light gave plant researchers a simple experimental system for determining the quality and quantity of light that somehow allows plants to measure photoperiods. Thus the brief-flash treatment in the middle of the night might employ different wavelengths or intensities of light. This experimental system also led to the discovery of a pigment called **phytochrome**, which actually has two components: a simple protein and the light-absorbing-pigment molecule.

Phytochrome exists in two conformations, or chemical forms—phytochrome red (P_r) and phytochrome far-red (P_{fr})—that provide the plant with a basis for detecting the dark period. Each form can be converted into the other, but each absorbs a different wavelength of light. P_r absorbs red light (about 660 nm), and the absorbed light energy converts the phytochrome molecules to the P_{fr} form. P_{fr} absorbs far-red light (about 730 nm, beyond the visible spectrum), and the light energy from this absorption converts the pigment–protein complex back to P_r (Figure 30-18). Plants synthesize the pigment and the protein parts of the phytochrome separately. When the units are first joined, they form the P_r molecule. During daylight hours, the P_r absorbs light energy and is converted to P_{fr}. Sunlight is much more energetic in the red part of the spectrum than in the far-red part, so by the end of the day, about 95 percent of the phytochrome molecules have been converted to the P_{fr} form. P_{fr} is unstable, however, and during the night, it is enzymatically converted back to P_r. Once formed, P_{fr} is the active agent that may induce some responses in a minute or

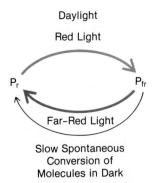

Daylight

Red Light

Far-Red Light

Slow Spontaneous
Conversion of
Molecules in Dark

Figure 30-18   LIGHT ABSORPTION BY PHYTOCHROME PIGMENT: THE PLANT'S MEANS OF DETECTING LIGHT AND DARK.
Phytochrome exists in two forms: phytochrome red (P_r), which absorbs light at about 660 nanometers, and phytochrome far-red (P_{fr}), which absorbs light at the far-red end of the spectrum, about 730 nanometers. During daylight hours, the P_r absorbs light energy and is converted to P_{fr}. P_{fr} is unstable and is converted back to P_r during the night, or more quickly, by far-red light. Somehow, the amounts of each form of the pigment allow plants to measure the number of hours of day and night.

less—opening of stomata is an example. But the main action of P_{fr} is apparent in twenty to thirty minutes, when new mRNAs begin to appear, with the result that more than forty different enzymes increase in activity in various cell types of leaves or stems. Flowering is the result.

The discovery of phytochrome explained how plants detect day length but not how they *measure* it and thus flower at the appropriate time. It would be easy to conclude from the experiments that the time-limiting dark reversion of P_{fr} to P_r is the plant's biological clock. Unfortunately, that is too simple a hypothesis. The enzymatic reversion of P_{fr} back to P_r in the dark takes only about three hours in both short-day and long-day plants. Therefore, phytochrome cannot be the clock that regulates photoperiodism; some other factor must be involved. Plant biologists suspect that this missing factor is an internal clock that somehow regulates daily and seasonal patterns of overall activity within the plant, just as animals have circadian rhythms (that is, rhythms with roughly twenty-four-hour cycles) and seasonal cycles.

Although flowering is partly under the control of phytochrome, this pigment also appears to be involved in a range of other plant responses, including seed germination. For example, lettuce seeds require light to germinate. If planted too deep in the soil, where light cannot penetrate, the seeds fail to germinate. Exposing seeds to red light allows their germination. This exposure—even a brief one—converts all the phytochrome pigment

within the seed to the far-red form. If, however, the seeds treated with red light are then exposed to just a few minutes of far-red light, they will not germinate. The far-red treatment can be reversed by another exposure to red light; the seeds will again germinate. Red–far-red reversal can be continued for many cycles.

To be effective, the light-reversal treatments must be given between a few seconds and two or three hours after the initial red-light treatment. After several hours, however, the seed will have "processed" the red-light signal and become committed to germination, despite subsequent exposure to far-red light. In a plant seed treated for long lengths of time with red light, all the phytochrome is converted to P_{fr}. This form of phytochrome stimulates germination by activating cellular respiration and other processes that cannot be reversed after a few hours of activity.

Florigen: One Flowering Hormone or Many?

A Russian plant physiologist named M. H. Chailakhian, working in the 1930s, tested different parts of plants to determine which are actually involved in measuring the photoperiod. In one series of masking experiments, he placed a light-tight barrier (mask) just below the apex of a chrysanthemum, a short-day plant. He then removed all the leaves above the barrier and exposed the two parts of the plant—the apex and the remaining leaves—to two different photoperiods. He found that only short-day (long-night) treatments of the *leaves* resulted in flowering. This experiment demonstrated that it is the young leaves below the apex, not the apical meristem, that are involved in detecting photoperiod with phytochrome. Without its young leaves, a plant cannot measure day length; but with just one or a few leaves, it can.

Chailakhian also performed a simple experiment showing that a chemical signal can be transmitted from one plant to another and somehow pass along information about day length. He induced one set of plants to flower with the appropriate light conditions and kept another group of plants under noninductive conditions. He then grafted a piece of an induced plant to each noninduced plant. Following the grafting, he found that all the plants flowered. He observed that the flowering stimulus can travel across a graft union, from plant to plant, and can convert a flowerless apex to one with flowers.

Chailakhian suggested that a "flower-making" hormone, which he called **florigen**, was being transmitted from the leaves of the induced plants to the flower-bud meristem tissue of the noninduced plants, where the hormone stimulated the formation and opening of flow-

PLANT HORMONES, MANAGING PLANT GROWTH, AND RELIEF OF GLOBAL HUNGER

In the past few decades, environmentalists have opposed the use of many agricultural chemicals—especially petroleum-based compounds such as DDT, dieldrin, and 2,4-D—that have proved to be long-lasting, highly toxic, and ecologically damaging. One result of this effort has been to promote the belief among some segments of the public that all agricultural chemicals are harmful and unnecessary. The plain truth remains, however, that the human species is struggling desperately to feed its burgeoning populations: observers predict that during the next thirty years, we will have to grow as much food as was produced during *all of previous human history* if we are to stave off massive hunger. And without growth regulators and herbicides, many of which belong to the five classes of plant hormones, this goal will never be reached. Just a few of the agricultural uses of hormones are mentioned here.

Promoting Flowering

One of the first large-scale applications of plant hormones was on Hawaii's pineapple crop. Because pineapples are day-neutral plants, each plant matures and flowers at its own rate, and the flowering of an entire field can take months. This means that fruit ripening is also staggered, and thus harvesting is labor-inten-

sive and expensive. Pineapple growers learned that if ethylene is sprayed before flowering begins, all the plants in an entire field can be induced to bloom simultaneously and, later, to produce mature fruit at the same time. Spraying with ethylene is now the standard technique in the pineapple industry.

Increasing Fruit Size

Grape growers make good use of gibberellins. They found that when grapes grow in tight clusters of small fruits, the bunches are easily infected—and destroyed—by fungi. However, spraying the grape vines with gibberellins just after the flowers open causes some ovules to abort, reduces the number of grapes by about 30 percent, and allows the remaining fruit to grow larger and resist rotting. Plants of the green Thompson seedless grapes sold in grocery stores are now routinely sprayed with gibberellins.

Accelerating Fruit Ripening

Some grape growers produce raisins with green and purple grapes, and rely on ethylene to help accelerate the ripening of the grapes. Because farmers make raisins by drying grapes in the sun, a grape that matures early in the season is more likely to dry thoroughly than one that matures late.

ers. His grafting experiments demonstrated that enough florigen was made in one plant induced by the correct light–dark regimen to promote flowering in many grafts. He also found that the flowering stimulus could travel through a set of plants grafted together in a series (Figure 30-19).

Florigen produced in a long-day plant can induce flowering in a short-day plant, and vice versa. The florigen produced in the leaves of an induced plant, however, takes several hours to be transported in significant quantities to the rest of the plant. Thus if plants are induced and then the leaves are removed too soon, no flowering occurs.

Despite the strong circumstantial evidence for the existence of a flowering hormone, scientists have not yet been able to purify or chemically characterize florigen. The fact that the rate of movement of florigen out of the leaves can be measured by removing leaves suggests that a signal does appear to move from one place to another. The current view of Chailakhian and other scientists is that not one but a number of compounds, including per-

haps gibberellins, may interact to promote flowering. One finding favoring that view is that a plant must be "competent" to respond to florigen, an ability based on several other factors, including developmental status, temperature, and nutrient and water availability. The saga of florigen is an excellent example of well-designed experiments that to date leave an important aspect of plant biology unresolved.

LOOKING AHEAD

In this chapter, we have seen that the reproductive behavior, growth, and internal processes of plants are complex and well coordinated. Much of that coordination is based on hormones, natural internal cycles, and physical factors—such as water, temperature, and light. A range of hormones—triggered and distributed in response to directional light, gravity, day or night length, and other factors—helps allow plants to produce new

Thinning the Number of Fruits

Some commercial fruit trees produce so many individual fruits that each is far smaller than the maximum size it could reach if the competition for resources were less intense. Fruit growers have found that if they spray the trees with auxins a few weeks after flowering, 20 to 50 percent of the fruit drops off. This technique is also helpful for trees such as peach, olive, and plum that tend to overproduce in one year and underproduce in the next. Thinning with auxin allows a uniform, high-quality crop every year.

Removing Leaves, Controlling Growth, and Killing Weeds

Millions of dollars are spent each year for plant hormones used as defoliants (leaf removers) and herbicides (weed-killers). Defoliants—usually auxins, but sometimes synthetic chemicals—were used during the Vietnam War to clear areas of vegetation so American pilots could see roads, bridges, and enemy troops more easily during bombing raids. Defoliants are also used by electric-power companies to remove leaves that entangle power lines. Another substance, Atrinol, blocks auxin production and as a result branches of trees such as sycamore lengthen only about one-quarter as fast as normal—an advantage when sprayed on trees growing beneath power lines. Auxins, as growth regulators, can be used to repress the growth of weeds, and thus they form one important class of herbicides, along with many kinds of unrelated synthetic compounds. Some auxin-based herbicides are sprayed between crop rows, and others, called preemergents, are sprayed in early spring to prevent the germination of weed seeds and their emergence from the soil. Eliminating weeds has been found to increase the yield of a crop by 25 to 50 percent. Another class of herbicides, called postemergents, can be sprayed on highway shoulders and railroad rights of way to completely clear the areas.

It is difficult to imagine how major jobs such as weeding and clearing vast fields could be done without plant-growth regulators. If human labor and farm equipment were used to remove weeds by hoeing and digging, or if workers thinned fruit crops by hand, the cost of raising food would skyrocket.

Chemicals—plant hormones as well as synthetics—will be an increasingly important aspect of the agricultural landscape over the next thirty years, and the major task of chemists and plant biologists will be to make those compounds as effective and as safe as possible.

Figure 30-19   FLORIGEN: THE MYSTERIOUS STIMULUS FOR FLOWERING.
Florigen can move through a series of grafts to induce flowering in plants that have not been exposed to the appropriate photoperiod. Although the exact nature of florigen is not known, this experiment, performed by M. H. Chailakhian, showed that the flowering stimulus can travel from a leaf given the appropriate photoperiod (at left) through a series of grafted plants, to stimulate flowering in all of the plants.

Nonflowering Photoperiod

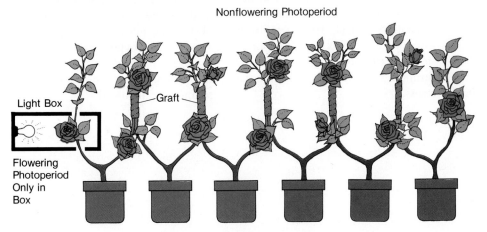

Light Box

Graft

Flowering Photoperiod Only in Box

All Grafted Plants Flower

leaves, shoots, roots, or flowers—sometimes over many years. Indeed, the coordinated control of these hormones enables a plant such as the coast redwood to survive and reproduce despite innumerable environmental variations over 1,000 years or more—truly a remarkable biological system.

SUMMARY

1. Plant *hormones*—substances produced in extremely low concentrations in one part of a plant that bring about effects in another part—are less specific than animal hormones, and there are fewer known types. In plants, the specificity of hormonal action depends on how a particular tissue or cell type responds to a hormone. And, unlike animal tissue, most plant tissue can both make and respond to each of the hormones.

2. *Auxins,* one of the five major classes of plant hormones, are generally produced in the plant's growing apex and regulate plant *tropisms,* or movements, by stimulating cell elongation. Plants exhibit *phototropism,* after auxins accumulate on the nonilluminated side, stimulate faster growth, and cause the plant to bend toward the light. *Gravitropisms,* in which the shoot grows upward and the root, downward, are also due to auxins stimulating cell elongation.

3. Auxins may also inhibit growth, producing such effects as *apical dominance,* the tendency of the main shoot of a plant to predominate over all the others. Auxins also promote fruit development, leaf fall, meristematic activity, and root growth.

4. Auxins stimulate cell elongation by promoting the secretion of hydrogen ions from the cell, lowering the pH in the cell wall, and facilitating enzymatic weakening of the cell walls. The turgor pressure then can push the wall into a new shape. Such release of hydrogen ions is probably also involved in the gravitropic responses controlled by auxins.

5. *Gibberellins* are a class of more than seventy growth-promoting hormones that cause cell and stem elongation, bolting in the rosette stage of long-day plants, and increase in fruit size.

6. *Cytokinins* are a class of hormones that, in combination with auxins, promote cell division. The successful stimulation of mitosis as well as regulation of tissue and organ development depends on a proper ratio of cytokinin to auxin: with higher ratios stimulating shoot and leaf cells and lower ratios stimulating root cells or *callus tissue* (disorganized lumps of cells) to form. Cytokinins also delay cell aging, or senescence, possibly by promoting nutrient transport to cells.

7. *Abscisic acid* is a hormone that suppresses plant growth; helps maintain dormancy of seeds and tissues; and speeds leaf senescence and, in some plants, *abscission*—the separation of leaf, fruit, or flower from the plant. This growth suppression is critical for the plant's survival through periods of environmental stress.

8. Another plant hormone, *ethylene,* is a gas at normal temperatures. It accelerates fruit ripening by stimulating tissue respiration and plays a role in growth patterns by inhibiting auxin transport.

9. The five types of hormones work together in complex ways to control the physiology of the plant. In general, two or more positive hormonal signals are required for growth, while one negative signal can prevent growth.

10. Leaf fall results from the production of an *abscission layer,* which forms when auxin levels decline and ethylene levels increase. Control of germination in seeds involves the interaction of gibberellic acid and abscisic acid.

11. Induction of flowering depends on *photoperiod,* or day length. Plants detect day length with a pigment–protein complex called *phytochrome,* operating in the young leaves. They measure day or night length by means of an unknown internal clock that is compensated for changing temperatures.

12. Flowering is most likely controlled by several chemical—perhaps hormonal—factors. In addition to the hypothetical hormone *florigen,* plant age, water availability, temperature, and season also seem to be involved.

KEY TERMS

abscisic acid	callus tissue	gravitropism	tropism
abscission	cytokinin	hormone	vivipary
abscission layer	ethylene	photoperiod	
apical dominance	florigen	phototropism	
auxin	gibberellin	phytochrome	

QUESTIONS

1. Construct a table that lists the five major types of plant hormones, and include the primary modes of action at the cell level (mitosis, elongation, and so on) and actions on the entire plant.

2. Why can we say correctly—at least for now—that certain of auxins' effects occur outside cells, whereas gibberellins' effects occur within the nuclei of aleurone cells in seeds?

3. Why would abscisic acid be better named "dormin"?

4. Explain the basic mechanisms of photoperiodism in flowering. How is the requirement for a dark period of minimal length established?

5. How is each form of phytochrome converted into the other by light, and what happens when that occurs?

6. Why do researchers suspect that florigen is not gaseous? that ethylene is?

7. Describe how a leaf separates from the stem. What hormones may be involved?

8. Why is the temperature compensation of the plant photoperiod system an important adaptation for most plants on our planet?

9. What is the generally accepted principle of hormone interaction in plants?

SUGGESTED READINGS

EVANS, L. T. *Daylength and Flowering of Plants*. Menlo Park, Calif.: Benjamin-Cummings, 1975.

This is a fine treatment of photoperiodism and of flowering in general.

HILL, T. A. *Endogenous Plant Growth Substances*. Studies in Biology, no. 40. Baltimore: University Park Press, 1973.

A window on the history of the discovery of plant hormones and on the scientific methodology involved.

KENDRICK, R. E., and B. FRANKLAND. *Phytochrome and Plant Growth*. Studies in Biology, no. 68. Baltimore: University Park Press, 1976.

A short, but useful account of the phytochromes.

MOORE, T. C. *Biochemistry and Physiology of Plant Hormones*. New York: Springer-Verlag, 1979.

This is a fine, brief summary of hormonal action in plants.

NICKELL, L. G. *Plant Growth Regulators—Agricultural Uses*. New York: Springer-Verlag, 1982.

This is an excellent summary of the ways in which plant hormones may be used in agriculture.

Part
FIVE

ANIMAL BIOLOGY

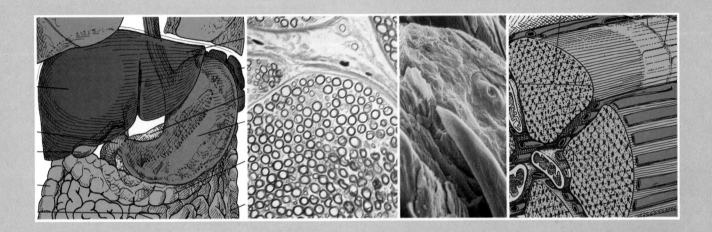

Animals—the great multicellular consumers—must be able to find and ingest food, to digest it, and to distribute the materials to all body cells. They must also detect and respond to features of the environment—oxygen, salinity, temperature levels—as well as find mates, avoid enemies, and defend their own territories. These needs have shaped much of the animal's architecture and physiological functioning.

Animals have several major internal systems. Unlike plants, most animals are mobile and many of their cells are capable of locomotion as well. Skeletal parts and muscles provide support and allow movement of the body regions and appendages, while mobile cells move about during embryonic development to shape new organs, and movable cells play novel roles in the adult. Excretory tubules and organs such as the kidneys expel wastes and help regulate the internal balance of water and salt in body fluids. The lungs and air passages, or equivalent structures, function in gas exchange. The circulatory system conveys dissolved gases, fluids, nutrients, and ions throughout the body. The gut tube digests foods, and in many animals, an immune system stands ready to defend the body against invading microbes, chemicals, or foreign agents.

The more numerous an animal's internal systems, operating simultaneously to ensure survival, the more the organism requires coordinating mechanisms; two important ones have evolved: the nervous system, a point-to-point network of specialized cells capable of rapid communications and control, and the hormonal system, a set of organs that secretes chemical messengers that can govern and regulate the activities of other physiological systems. The nervous and hormonal systems complement each other, and together they orchestrate the events of the life cycle as well as the behaviors that allow an animal to feed, grow, mate, reproduce, play, and defend itself. Fittingly, this part ends with a study of the most intricately organized structure that humans have yet discovered in the universe: the human brain, biological tissue with the capacity for language, politics, music, art—and for the act of discovery itself.

31

THE CIRCULATORY AND TRANSPORT SYSTEMS

*To
The most illustrious and indomitable
Prince
Charles
King of Great Britain, France and Ireland
Defender of the Faith
Most Illustrious Prince!*

The heart of animals is the foundation of their life, the sovereign of everything within them, the sun of their microcosm, that upon which all growth depends, from which all power proceeds.
William Harvey, Dedication of *The Circulation of the Blood* (1628)

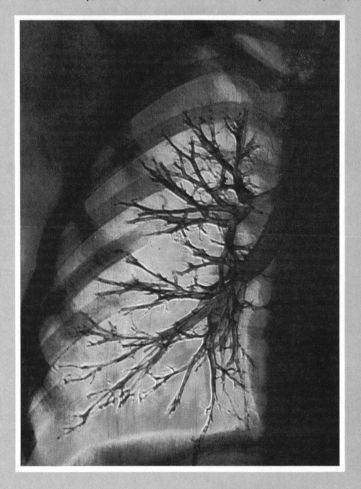

The branching "tree" of arteries in the right human lung as revealed by x-ray photography after injection of a radio-opaque dye.

In our survey of life forms in Chapters 20 to 26, we saw the marvelous diversity of multicellular organisms, from algae, mosses, coelenterates, and flatworms to the largest and most complex plants and animals. Central to both the evolution and the day-to-day existence of any multicellular organism are systems for transporting nutrients, oxygen, and other essential raw materials to and from each of the body cells. We have now studied the transport systems of the vascular plants and have seen how phloem and xylem function. Analogous transport systems in animals are the subject of this chapter.

Transport systems are an appropriate place to begin our series of chapters on animal physiology because they are involved in virtually every physiological process in large multicellular animals. We will encounter transport systems again and again as we discuss how animals carry out certain types of defense against disease, distribute oxygen, digest food, excrete wastes, deliver hormones, move their bodies, and regulate their internal temperature and physiological environment.

In small, simple, relatively sluggish animals, transport systems can be little more than cavities filled with fluid that bathes the tissues and is stirred by the creature's own movements. Larger, more active animals, however, need specialized structures and fluids to supply tissue cells with ions, nutrients, and perhaps gases, and to carry away metabolic wastes at a rate appropriate to the animal's level of activity. These are the elements of **circulatory systems.** In vertebrates, the circulatory system includes (1) an energetic, muscular pump, the *heart*; (2) a "plumbing system" of hollow vessels, the *arteries* and *veins,* which branch and ramify throughout the body; and (3) a "microcircuitry" of extremely fine vessels, the *capillaries,* which link arteries to veins and which are interwoven into virtually every tissue of the body. The capillaries are the actual sites of exchange of materials en route to or from the tissue cells.

In this chapter, we will first consider the general problems of material transport in multicellular animals and then the two general solutions: simple diffusion and a circulatory system. Next, we will explore in detail the structures of the vertebrate's circulatory system and how they function to continuously transport materials in the blood. We will then see how the flow of blood is controlled so that the organism's needs are met appropriately, during activity or rest. Finally, we will discuss the fluid of life—blood—and another important fluid— lymph. The ancients believed that these body fluids, or "humors" as they called them, convey our states of health, disease, and temperament. As you will find in this and subsequent chapters, in a very real sense they do.

STRATEGIES FOR TRANSPORTING MATERIALS IN ANIMALS

The living state is dynamic, and as we have seen, all living cells require raw materials and generate metabolic wastes. The transport of vital materials in single-celled organisms, such as monerans and protists, depends entirely on simple diffusion of materials from areas of high concentration to those of low. Without special delivery systems, nutrients, dissolved gases, and ions diffuse back and forth across the cell wall and plasma membrane that separate the fluid environment in which the cell lives from the cell's interior. Diffusion also accounts for distribution within the cytoplasm of most cells. But diffusion is a relatively slow process: in the absence of stirring, it may take oxygen about three hours to diffuse 1 centimeter through water, and since speed of diffusion varies as the square of distance, moving 2 centimeters requires nine hours. Obviously, diffusion can suffice as a transport mechanism in living organisms only when the distances that materials must be carried are extremely short. How, then, does this fact affect small organisms and how are the problems of transportation solved in larger, more complex multicellular animals?

Relying on Diffusion: Limitations on Body Size and Shape

Multicellular animals that rely on diffusion for transport must be porous; they must be flat; or they must have a very thin body wall. Porifera (sponges), for example, are penetrated by complex sets of channels that carry sea water or fresh water throughout the organism (Figure 25-2). One surface of cells lining these channels is directly bathed with this water, which carries food particles, dissolved gases, and ions. Internal cells not directly bathed by the water are rarely more than 1 millimeter from it, so materials diffuse through a layer at most two to three cells thick. This distance is so short that transport is rapid enough to meet the ionic and metabolic needs of each sponge cell.

The flatworm *Planaria* has more complex structures and functions than does a sponge, but because of the extreme flatness and thinness of its body, every cell is close enough to either the body surface or the fluid-filled gut cavity that cellular needs can be met by simple diffusion (Figure 25-12). Similarly, some cnidarians have such a thin, gossamer body wall that, despite an often large body size, materials can diffuse between every cell and the surrounding sea water.

Circulatory Systems

All multicellular animals that are not porous, flat and thin, or very thin-walled have a separate circulatory system to transport essential materials. In such a system, fluid (such as blood or hemolymph) moves through the organism in a circular flow. This movement of the fluid is called **bulk flow** because masses of fluid move (as opposed to substances diffusing through nonflowing fluids).

Circulatory systems are either open or closed, depending on a set of physical structures. **Open circulatory systems** are found in insects, certain mollusks, and several other kinds of invertebrates. In an open system, a fluid such as hemolymph flows through part of the body in a major vessel or vessels, and then drains through slits in the vessel wall into large open spaces called *sinuses*. The fluid in these sinuses then contacts tissues or organs directly, and materials can move into or out of individual cells via diffusion. In insects, the hemolymph moves anteriorly in the vessel, is discharged into sinuses in the head, and then seeps in a posterior direction, bathing internal organs as it goes (Figure 31-1a). The fluid in the sinuses eventually reenters the vessel through openings in its wall, and the circuit is completed.

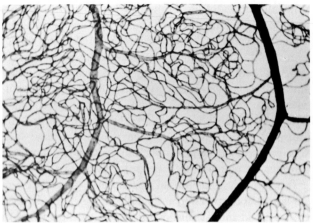

Figure 31-2   BLOOD CAPILLARY NETWORK.
These blood vessels (magnified about 55 times) were isolated intact from the light-sensing tissue of a human eye. A small venule lies to the left; a small arteriole is to the right; and capillaries lie in between.

In a **closed circulatory system,** such as in earthworms, squids, octopuses, and all vertebrates, blood moves through a continuous set of interconnected vessels: the arteries, capillaries, and veins (Figure 31-1b). **Arteries** are relatively large vessels that carry blood *to* the tissues. The smallest arteries are continuous with **capillaries,** tiny, thin-walled vessels interwoven throughout the various body tissues (Figure 31-2). Nutrients, oxygen, and ions carried in the blood diffuse through the walls of the capillaries into the extracellular fluid that bathes the outside of tissue cells, and from there, those substances enter the cells. Metabolic wastes (carbon dioxide, lactic acid, ions, and so on) leave the cells, enter the extracellular fluid, and diffuse from it into the capillaries. Capillaries are continuous with the smallest **veins,** vessels that carry the blood *away from* the tissues. Veins, in turn, feed the blood through a pump, the heart, back into the arteries, completing the circuit.

(a) Open Circulatory System

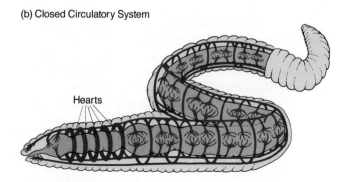

Hearts

(b) Closed Circulatory System

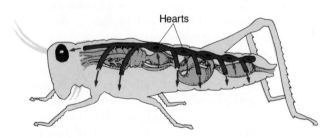

Hearts

Figure 31-1   OPEN AND CLOSED CIRCULATORY SYSTEMS.
(a) In animals with an open circulatory system, such as the grasshopper, fluid is pumped by a heart (or a series of hearts) located along the major dorsal vessel to different parts of the body by way of vessels. The fluid exits the vessels, bathes body cells, and drains back into the vessel. (b) In the annelid's closed circulatory system, which is typical of many animals, the blood is contained within a system of vessels and is pumped through the system by one or more hearts.

Why Fluids Flow in One Direction in Circulatory Systems

Despite their structural differences, all circulatory systems share certain features that explain why the fluid flows, rather than pools up and stagnates, and why it moves in a circuit rather than in random directions. Blood flows because of the contractions of four types of muscles:

First, the walls of some vessels contain muscle tissue that contracts sequentially in *peristaltic waves* to sweep fluid along.

Second, specialized regions of muscle tissue, or **hearts,** act as pumps. Vertebrates have one large heart, whereas insects and other invertebrates can have several small hearts that act like booster pumps for the circulating fluid. The rhythmic contraction of these pumps and the peristaltic waves of contraction in the walls of some vessels push against the volume of circulating fluid and create pressure, the **blood pressure.** Blood pressure varies, depending on whether a circulatory system is closed or open. In a closed system, the pressure is often quite high, and the flow of blood is rapid. Since materials can be delivered quickly, animals with fast-flowing closed systems can often metabolize rapidly and move quickly. In an open system, in contrast, the hemolymph or other fluid is pumped from vessels into a large open sinus, so the blood pressure is low and the flow of fluids is slow, just as the flow of water slows as a stream empties into a lake. The metabolism and movements of animals with open circulatory systems are often sluggish, with one important exception: the insects. Oxygen is supplied to an insect's tissue cells through a unique system of hollow air pipes, the *tracheae*, which branch throughout the body and deliver large amounts of oxygen to actively metabolizing tissues. Thus the insect's open circulatory system is not responsible for gas transport. We will consider tracheae in detail in Chapter 33.

As an animal moves, the contractions of large body muscles massage the blood vessels (and, if present, the sinuses), stirring the fluid and augmenting its circulation through the body.

Finally, the flow of fluid into and out of local regions of the circulatory system is controlled by tiny muscles—*sphincters*—that act as floodgates to shut off or open up vessels, depending on the need for materials in a local tissue. Sphincter muscles encircle vessels and decrease their diameter as they contract.

Circulatory systems also share a feature that explains why their contents flow in a circuit. Blood or other circulating fluid flows in one direction because *valves*, flaps of tissue that extend into the vessels from the inside walls, close when fluid begins to flow in the reverse direction (Figure 31-3). Valves in the heart help ensure that blood can leave only via arteries and enter only from veins.

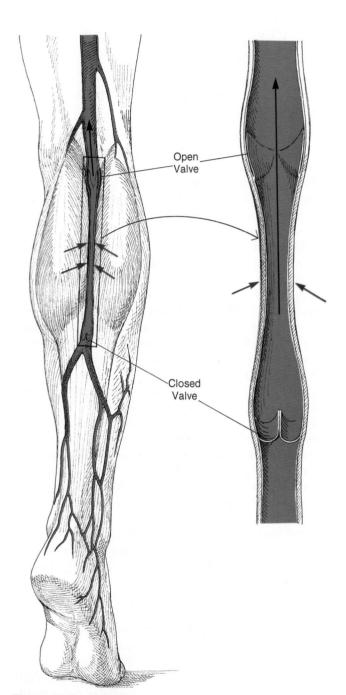

Figure 31-3   VALVES.

Valves in veins ensure that blood flows toward the heart by preventing backflow. The movement of venous blood is caused by the heart's pumping action as well as by compression of veins due to the contractions of nearby skeletal muscles (colored arrows). Here, as the calf muscles contract, the vein is compressed; the upper valve remains open while back pressure closes the lower valve, thereby preventing flow toward the foot.

Thus valves prevent backflow and guarantee the one-way movement of fluids.

Circulating Fluids and the Diffusion of Materials

Even though fluid carrying essential materials is circulated by the bulk-flow process, diffusion still operates at the level of individual cells. Thus circulation may speed the *delivery* of materials to the sites where they will be used, but it does not alter the basic physics of how substances enter and leave cells.

Animals with circulatory systems do differ, however, in how materials are carried to their tissue cells. In an animal with an open circulatory system, the **hemolymph,** like an internal sea, bathes the outside of tissue cells. The hemolymph, in other words, forms the animal's *extracellular fluid*, which fills the nooks and crannies between individual cells and resides in the body cavity (pseudocoelom or coelom; Chapter 25). This extracellular fluid is roughly similar in composition to the fluid within individual cells (the *intracellular fluid*). It is colorless and consists mostly of water with dissolved nutrients, ions, gases, and some macromolecules. Many of these substances move from the extracellular fluid into tissue cells whenever their concentrations are higher outside the cells than inside; this is due to diffusion down concentration gradients (Chapter 5). (Certain substances, however, require energy-dependent transport across the plasma membrane.) In turn, substances move out of the intracellular fluid, across the plasma membrane, and into the extracellular fluid whenever their concentrations are higher inside the cell than outside. For example, if a cell has been metabolizing actively and has only a low level of oxygen remaining, more of that dissolved gas will diffuse from the extracellular fluid into the cell. Conversely, active metabolism may produce a high level of the waste product carbon dioxide within the cell; that substance diffuses out into the extracellular fluid, where its concentration is much lower.

An animal with a closed circulatory system has a third fluid, **blood,** in addition to its intracellular and extracellular fluids. (In Chapter 40, we will encounter yet a fourth fluid found in the nervous system.) Blood flows through the closed network of arteries, capillaries, and veins, and consists of two phases: a watery component, the *plasma*, which contains organic molecules, ions, and dissolved gases, as well as a solid component, the *cells*, among which are the red blood cells, which carry most of the oxygen in the circulatory system. Materials in the blood diffuse first through the capillary walls into the extracellular fluid, and from there, they may diffuse into the interiors of cells. For instance, oxygen carried by red blood cells diffuses through the thin capillary walls into the extracellular fluid only if the level of oxygen in the

extracellular fluid is lower than in the blood. From there, the oxygen diffuses passively into tissue cells if the oxygen concentration within those cells is lower than in the extracellular fluid. All such movements are based on concentration gradients (and, as we shall see, on fluid pressures).

The important thing to remember, then, about the delivery of materials to and from cells in an animal with a circulatory system is that it always involves extracellular fluid, even if the animal has blood and a closed circulatory system. In an animal with an open circulatory system, this diffusion is a one-step process: from the extracellular fluid to the intracellular fluid or back in the opposite direction. In an animal with a closed circulatory system, in contrast, the extracellular fluid acts as a middleman between the blood and the intracellular fluid. We can summarize the difference this way, using the transport of oxygen and carbon dioxide as examples:

Open circulatory system:

$$\text{extracellular fluid} \xrightarrow{O_2} \text{cells} \xrightarrow{CO_2} \text{extracellular fluid}$$

Closed circulatory system:

$$\underset{\text{(in capillary)}}{\text{blood}} \xrightarrow{O_2} \text{extracellular fluid} \xrightarrow{O_2} \text{cells} \xrightarrow{CO_2} \text{extracellular fluid}$$

$$\xrightarrow{CO_2} \underset{\text{(in capillary)}}{\text{blood}}$$

In both open and closed systems, a delicate balance exists between the various body fluids. Diffusion of commodities occurs only if their concentrations differ within adjacent fluids. Moreover, many types of molecules cannot enter or leave the tissue cells unless they are actively pumped across the plasma membrane by means of certain enzymes.

The Discovery of Circulation by William Harvey

Today, we take it for granted that blood flows throughout the human body in a circuit, propelled by the pumping of the heart. For more than a millennium, however, most scholars and scientists believed that blood flowed away from the heart in the veins and ebbed passively back to the heart in the same veins. It was not until the early 1600s that the English physician William Harvey disproved the so-called ebb-and-flow theory. "I began to think within myself," he wrote in 1628, "whether it [the blood] might have a sort of motion, as it were, in a circle." This thought was based on experimental proof that laid the foundation for modern physiology.

Let us consider a prediction of the ebb-and-flow theory. According to this idea, blocking a vein, say, on the

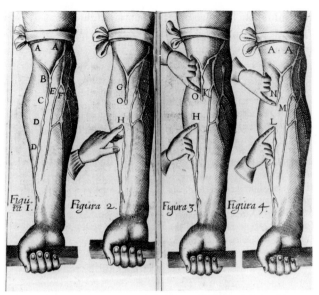

Figure 31-4   WILLIAM HARVEY'S DRAWINGS OF HIS EARLY INVESTIGATIONS OF THE CIRCULATORY SYSTEM. These drawings illustrate the direction of the venous blood return and the existence of valves to prevent backflow of blood. In Figure 1, the veins swell due to the tourniquet. In Figure 2, the finger has stroked the full vein from H upward to O; then, on removing the finger, the vein between H and O remains empty because the blood cannot flow peripherally in the vein past the valve located just there at O.

inner surface of the forearm, should still allow the ebb and flow of blood to continue up to the point of blockage. What Harvey did in about 1610 is so simple that you could do it while sitting at your desk. He put a tourniquet on his upper arm and watched the veins in his lower arm swell with blood. This swelling in itself contradicted the ebb-and-flow theory, since if blood traveled from the heart in the veins, it could not have passed the blockage point to reach Harvey's lower arm. Harvey then placed a finger on a vein in his forearm, and stroked upward toward his shoulder. When he removed his finger, the blood did not flow peripherally back toward his wrist, as the ebb-and-flow theory predicted (Figure 31-4). It could not do so, of course, because the veins are dammed up by valves. These valves, which appear in an arm bound with a tourniquet as slightly swollen spots along the veins, allow blood to flow only toward the heart. Thus the valves in veins prevent blood from ebbing away from the heart, or flowing "backward" in Harvey's simple experiment.

Next, using dogs, snakes, and other animals, Harvey cut out or occluded many large veins and arteries. He observed the same phenomenon again and again: blood flows away from the heart in arteries and toward the heart in veins; it simply does not ebb and flow in the same veins. Harvey went on to prove that blood flows from the right side of a mammal's heart to the lungs, returns from the lungs to the left side of the heart, and then passes to the rest of the body. Not knowing about gases such as oxygen and carbon dioxide or about gas exchange in the lungs, Harvey made the reasonable guess that blood went to the lungs to be cooled.

Harvey not only founded the science of physiology with the first physiological measurements on living animals, but also set a standard of scientific methodology for Renaissance Europe with his deductive reasoning. Harvey measured the maximum volume of blood that can be pumped from the heart with each beat; in humans, that volume is about 2.5 fluid ounces (about 70 ml) per heartbeat. Being conservative, he then calculated that if only 1 fluid ounce was pumped with each beat of the heart, 60 heartbeats in a minute would pump 60 ounces; thus every hour, the heart would pump 3,600 ounces, and every day, it would pump 86,000 ounces—about 27 times the weight of a 200-pound person! Based on his calculations, Harvey deduced that that much blood could neither be produced nor used up every day; therefore, it must circulate out in the arteries and back in the veins, over and over, as long as life continues. He concluded that there must be tiny connections between the arteries and the veins, forming a true circulatory system. However, lacking a microscope, he was unable to see those connections. Not until thirty-three years after Harvey published his conclusions did the Italian scientist Marcello Malphigi use an early microscope to observe blood flowing through capillaries in a frog's lung. The discovery of these minute connecting vessels "hidden from Harvey's eyes but not his reason"* was the final proof that blood does indeed circulate through a closed system.

THE VERTEBRATE CIRCULATORY SYSTEM

In the 360 years since Harvey's work, physiologists have learned a great deal about the structures and functions of the vertebrate circulatory system. The functioning of that system depends on the heart, the pump that drives blood through the vessels. It is appropriate, therefore, to study the heart first and then to consider the circuit—the miles of arteries, capillaries, and veins that pervade the body.

---

*R. B. H. Wyatt, *William Harvey* (London: Leonard Parsons, 1924).

The Heart

Harvey's experiments proved that the heart of a land vertebrate is an energetic pump with thick muscular walls that contract forcefully, expelling blood into the arteries. An adult human's heart, for example, is about the size of a man's closed fist, and its muscular walls contract inward like a fist squeezing shut. Harvey also showed that mammalian and avian hearts are double pumps, divided in half longitudinally (Figure 31-5), and that the two circuits in the closed circulatory systems of birds and mammals both begin and end in the heart. As mentioned earlier, Harvey observed that blood moves from the right side of the heart through the lungs and back to the left side of the heart (one circuit) and then from the left side of the heart to the rest of the body,

returning once more to the heart's right side (second circuit). This may seem like an odd solution to the engineering problem of circulating fluid throughout the body. Why not have two hearts in series or parallel instead of one split in half connecting two circuits? The evolution of the heart provides the answer.

In *Amphioxus* (Chapter 26), the cephalochordate relative of the vertebrates, the heart is merely a specialized portion of a single major anterior blood vessel. In this region, waves of contraction in the muscular vessel wall

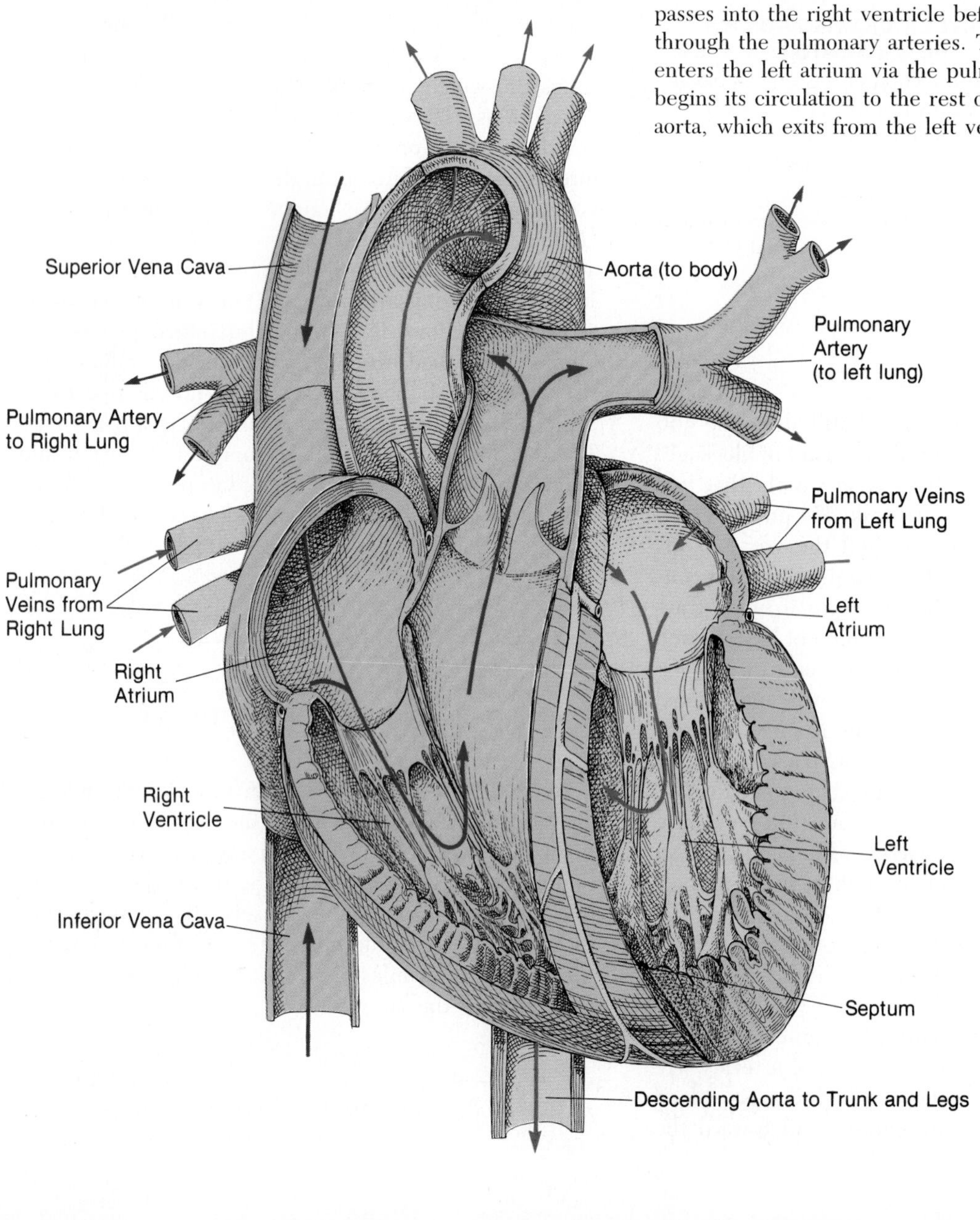

Figure 31-5   THE HUMAN HEART: PULMONARY CIRCULATION AND SYSTEMIC CIRCULATION.
Mammals have a "double" circulatory system. The veins that carry blood from the body tissues to the heart (blue) enter the right atrium. The venous, deoxygenated blood passes into the right ventricle before going to the lungs through the pulmonary arteries. The oxygenated blood then enters the left atrium via the pulmonary veins. The blood begins its circulation to the rest of the body through the aorta, which exits from the left ventricle.

Superior Vena Cava

Aorta (to body)

Pulmonary Artery (to left lung)

Pulmonary Artery to Right Lung

Pulmonary Veins from Left Lung

Pulmonary Veins from Right Lung

Left Atrium

Right Atrium

Right Ventricle

Left Ventricle

Inferior Vena Cava

Septum

Descending Aorta to Trunk and Legs

Figure 31-6 EVOLUTION OF THE HEART AND CIRCULATORY SYSTEM IN VERTEBRATES.
The major events in evolution can be seen in the development of the four-chambered heart and double circulation of birds and mammals from the very different four-chambered heart and single circulation of fish. Oxygenated blood is red; deoxygenated blood, blue; and a mixture of oxygenated blood and deoxygenated blood, purple. Blood is mixed in the single amphibian ventricle because it receives both oxygenated blood from the lungs via the left atrium and deoxygenated blood from the body tissues via the right atrium. In reptiles, the septum between the ventricles is usually incomplete, and the base of the aorta is split into two vessels that lead to the right or left arches of the aorta. In mammals and birds, the ventricle is fully separated into two halves, and only one aorta exits the heart.

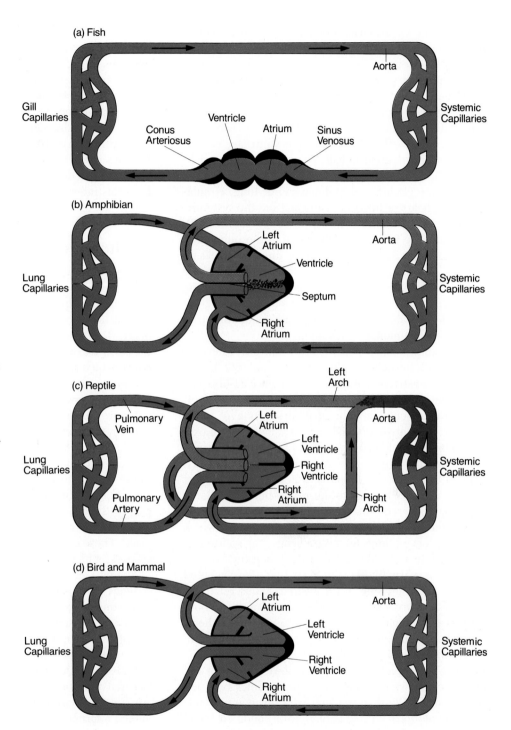

"knead" the blood fluid along anteriorly, similar to the way in which slow peristaltic waves move food or wastes through the intestines (Chapter 34). Many vertebrate fish species have the same sort of contractile vessel, but it is divided into four chambers that come one after the other in posterior-to-anterior sequence: the *sinus venosus*, the *atrium*, the *ventricle*, and the *conus arteriosus* (Figure 31-6a). Evidence indicates that during the course of evolution, this basic four-chambered heart

arose as a means to elevate blood pressure in a stepwise fashion. The thin-walled sinus venosus is only powerful enough to expand the atrium, but the somewhat thicker-walled atrium can contract forcefully enough to expand the ventricle, and so on. The function served by this expanding and stretching of the muscular walls is summarized by **Starling's law:** the more the cardiac muscle is stretched, the more vigorously it responds and the larger the volume of blood that can be pumped per contraction.

Thus stretched heart chambers can supply more blood to an active animal.

In land vertebrates, the four basic chambers seen in the fish heart have been reduced to two by the loss of the sinus venosus and the conus arteriosus. Blood returning to the heart at higher pressure allowed this simplification. Although land vertebrates have only two basic heart chambers, the atrium and the ventricle, each is split to varying degrees in half. This separation into right and left heart chambers permits blood to flow through two separate circuits: the **pulmonary,** or **lung, circulation** (right heart → lungs → left heart) and the **systemic,** or **body, circulation** (left heart → body → right heart). As we shall see in more detail, blood in the pulmonary circulation loses carbon dioxide in the lungs and picks up oxygen. This oxygenated blood is then pumped to the body tissues via the systemic circulation; after delivering oxygen, it returns to the right heart once again high in carbon dioxide and low in oxygen, ready for another pass through the pulmonary circulation.

Adult amphibians, which carry out gas exchange through both the skin and the lungs (Chapter 26), have a heart with an atrium and a ventricle (Figure 31-6b). The amphibian atrium is split in half, but the ventricle is not. This arrangement allows a small proportion of blood from the right atrium (which carries a low level of oxygen) to mix with blood from the left atrium (which carries a high level of oxygen) when the two blood supplies pass through the ventricle. Blood is routed so that the most oxygen-rich portion goes to the brain, but even that is somewhat lowered in oxygen content because of the mixing in the ventricle. In reptiles (Figure 31-6c), which depend solely on lungs for gas exchange, the splitting of the heart's two chambers is carried a step further: the ventricle is partially separated so that deoxygenated blood in the right side of the heart remains relatively unmixed with oxygenated blood in the left side. In both birds and mammals, separation of the ventricular heart chamber is complete: a septum prevents any mixing in the ventricle, as Figure 31-6d shows. Besides preventing mixing, the complete partition of the heart into right and left halves allows the blood pressure to differ between the two sides. We shall see later why this is important.

This evolutionary progression—from modified blood vessel to four-chambered organ—is evident in the ontogeny (embryological development) of every contemporary vertebrate. That is, the early human heart is first a single pulsating vessel, then a line of four chambers, as in a fish. The human embryo's sinus venosus, for instance, is for a while the pacemaker of heartbeat, just as it is in every adult fish. Later, in all birds and mammals, the sinus venosus and the conus arteriosus disappear, and the remaining atrium and ventricle split into right and left halves. This developmental sequence is a famous case of ontogeny recapitulating phylogeny or, more precisely, evolutionary history.

The result of the evolutionary process we have described is a marvelous organ that begins to pump blood early in an organism's development and continues to do so throughout its lifetime. At rest, for example, the human heart, pumps a teacupful (4 fluid ounces) of blood with each three beats, and every fifteen minutes, it pumps enough blood to fill a car's gasoline tank. In a day, the contraction of this fist-sized organ could fill seventy barrels, and over a lifetime, *18 million barrels.* Clearly, our existence depends on the steadfast pump that circulates this internal stream of life day after day, year after year.

Evolution of the Major Vessels Leaving the Heart

As the vertebrate heart evolved, structural alterations also occurred in the major blood vessels that leave the heart. The three main arteries in land vertebrates are the *carotid artery*, which goes to the head and brain; the *aorta*, which supplies blood to most of the body; and the *pulmonary artery*, which connects the heart and lungs. Let us see how these arose.

Figure 31-7 shows that the primitive tubular heart of vertebrate ancestors pumps blood directly into the ventral aorta, the largest vessel, which, in turn, leads into six paired blood vessels, running through gill arches. These are called the **aortic arches** (numbered 1 to 6 in Figure 31-7). In adult fishes, five pairs of aortic arches pass to the gills and break up there into capillary beds, the networks of tiny vessels where gas exchange takes place. These capillaries lead back into larger vessels, the arteries, which enter the dorsal aorta.

Among land vertebrates, which lack gills, significant changes take place in this pattern of arches. During embryonic stages, all six arches develop, but then some degenerate, while others assume critical roles. Arch 3, the carotid arch, develops into the carotid artery; arch 4, the systemic arch, becomes the aorta; and arch 6, the pulmonary arch, becomes the pulmonary artery. The fact that the apparently useless arches 1, 2, and 5 still arise in embryos shows that the genetic and developmental information required to form these vessels is still present and used.

Many biologists believe that simplification of the aortic arches and full separation of the pulmonary and systemic circulations were essential steps in the evolution of the diverse land vertebrates. No bird or mammal could sustain its high rate of metabolism, high oxygen demand, and high level of activity without a high-pressure, fast-flowing internal transport system in which oxygenated and deoxygenated bloods are never mixed. The trend toward this system is evident in lung fishes, and a key factor in its development probably operated in the first amphibians that walked on land. For them, the lack of

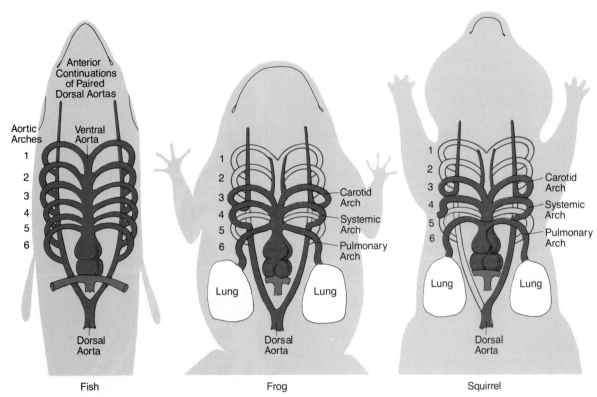

Figure 31-7 EVOLUTION OF THE AORTIC ARCHES.
The aortic arches (1-6) of fish became modified or lost as gills disappeared in land vertebrates. Although the organization of these arches in frogs and mammals differs from that of the fish, the genetic and developmental information required to form the "basic fish arrangement" exists, and in the course of embryonic development, even in mammals, the entire set of six arches arises and then three arches disappear. This is indicated by the outlined vessels shown in frogs and mammals. In adults, of course, the vessels labeled arches here are the corresponding arteries. (The pulmonary arch, for example, becomes the pulmonary artery.) In the squirrel, the fourth systemic arch is on the left (the view in the figure is ventral); the right half leads to the subclavian artery (not shown).

support against gravity that a watery habitat provides meant that the animals had to expend more muscular energy to lift their bodies off the ground and to move about. This energy is nothing more than large quantities of ATP; hence the need for a more efficient delivery system so that more oxygen and nutrients can be supplied to cells for ATP manufacture.

Blood Circuits

Now that we have explored that marvelous pump, the heart, and the associated aortic arches, let us follow the full circulatory pathway that the blood takes through the human body, as shown in Figure 31-8. Blood carried to the body tissues by the arteries delivers oxygen and nutrients to the capillary beds. Then dark red, deoxygenated blood moves into tiny veins, continues on into larger ones, and eventually reaches the two largest veins in the body: the *posterior vena cava*, which carries blood from the legs and most of the body, and the *anterior vena cava*, which carries blood from the head, neck, and

arms. In humans, these are called the *inferior vena cava* and the *superior vena cava*, respectively. The two venae cavae empty into the right atrium of the heart. The muscular walls of this chamber contract, and because valves prevent backward movement of blood into the venae cavae, the blood is forced from the right atrium into the right ventricle. Contraction of this chamber, in turn, sends the oxygen-depleted venous blood through the pulmonary arteries to the lungs (one of the few places where arteries regularly carry deoxygenated blood; note also that valves prevent backflow from ventricle to atrium). In the lungs, arteries branch to finer and finer **arterioles,** and the blood eventually passes through capillary beds, where carbon dioxide is released and oxygen is taken up. The newly oxygenated blood then passes into the pulmonary veins and is carried to the left atrium of the heart. (Note that this is one of the few instances in which veins carry highly oxygenated blood.) The contraction of the left atrium sends blood past valves into the left ventricle, from which it is pumped into the **aorta.** This largest artery in the body feeds into a number of major arteries, including the *coronary artery,*

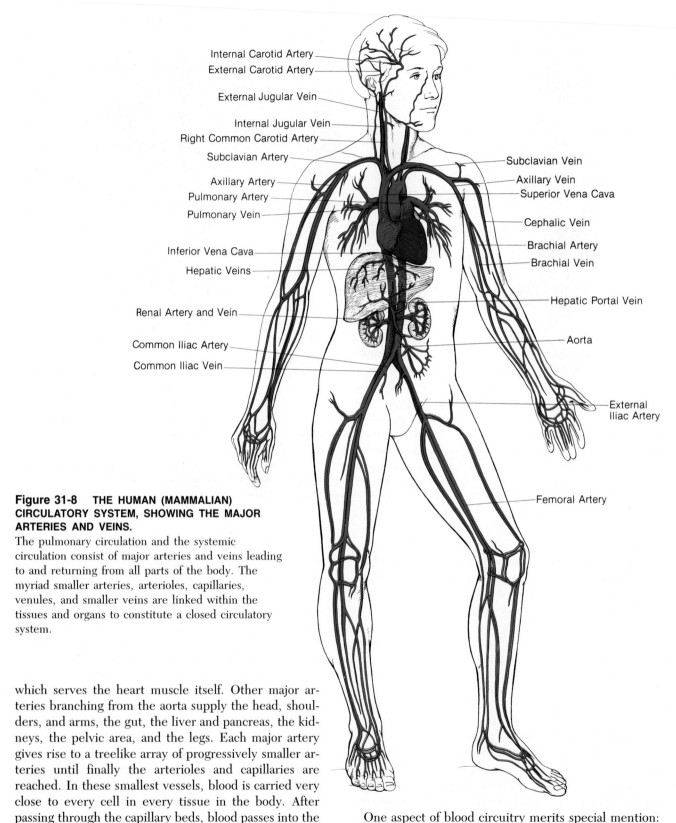

Internal Carotid Artery
External Carotid Artery
External Jugular Vein
Internal Jugular Vein
Right Common Carotid Artery
Subclavian Artery
Axillary Artery
Pulmonary Artery
Pulmonary Vein
Inferior Vena Cava
Hepatic Veins
Renal Artery and Vein
Common Iliac Artery
Common Iliac Vein

Subclavian Vein
Axillary Vein
Superior Vena Cava
Cephalic Vein
Brachial Artery
Brachial Vein
Hepatic Portal Vein
Aorta
External
Iliac Artery
Femoral Artery

Figure 31-8   THE HUMAN (MAMMALIAN) CIRCULATORY SYSTEM, SHOWING THE MAJOR ARTERIES AND VEINS.
The pulmonary circulation and the systemic circulation consist of major arteries and veins leading to and returning from all parts of the body. The myriad smaller arteries, arterioles, capillaries, venules, and smaller veins are linked within the tissues and organs to constitute a closed circulatory system.

which serves the heart muscle itself. Other major arteries branching from the aorta supply the head, shoulders, and arms, the gut, the liver and pancreas, the kidneys, the pelvic area, and the legs. Each major artery gives rise to a treelike array of progressively smaller arteries until finally the arterioles and capillaries are reached. In these smallest vessels, blood is carried very close to every cell in every tissue in the body. After passing through the capillary beds, blood passes into the "mirror image" of the arterial system—the treelike array of **venules** (tiny veins), larger veins, and so on until it arrives once again where it started—the venae cavae. Each day, every drop of the 5,000 milliliters of blood in an adult human's body makes this circuit 1,000 times!

One aspect of blood circuitry merits special mention: *portal vessels* carry blood from one capillary bed to another. For example, a major vessel, the *hepatic portal vein*, connects the capillary beds in the wall of the intestine—the site of nutrient absorption—directly to capillary beds in the liver. This arrangement allows many of the absorbed products of digestion to be processed immediately in the liver without first making an entire circuit of the body. Blood leaves the liver via the *hepatic veins*, which join the posterior (inferior) vena cava en route to the heart.

Structure of Arteries and Veins

The structure of arteries and veins—the "pipes" through which the blood travels—allows them to function as important reservoirs for blood pressure and blood volume. Arteries, the vessels that carry oxygenated blood from the heart, have relatively thick walls, composed of an inner lining called the *endothelium*, a middle layer of smooth-muscle cells, and an outer layer of connective tissue, which contains collagen and springy elastin fibers (Figure 31-9). In general, the closer an artery is to the heart, the more muscular and elastic are its walls.

The largest artery in the body, the aorta, illustrates how muscles and elastin fibers work together to transport blood, and why contraction and elasticity are so important. Each time the heart's muscular left ventricle contracts, a pulse of oxygenated blood is driven at high pressure down the aorta. This pulse of blood causes the aortic walls to expand quickly, and potential energy is therefore stored in the expanded walls. The elastic walls of the aorta then respond as would a stretched rubber band: they contract inward, exerting force on the blood and helping to propel it. Overall, the expansion and contraction of the major arteries take much of the high pressure "shock" out of the strong pulses of blood that leave the heart; the flow is smoothed out or damped, so that great fluctuations in pressure do not damage the delicate capillaries at the ends of the arteries. Equally important, the storage of potential energy by the aorta and other arteries enables them to serve as a *pressure reservoir*, a source of continuous, even pressure to help propel blood through the body.

Like arteries, veins have a three-layered wall of endothelial cells, smooth muscle, and connective tissue (Figure 31-9). However, vein walls are much more flexible than arterial walls. This flexibility allows veins to distend and increase greatly in volume so that a large volume of blood—up to 70 percent of the total blood in the body—may accumulate in them. Because of this distension and storage of blood, the veins are called a *volume reservoir*. In large part, the degree of contraction of the smooth-muscle layer in venous walls governs how much a vein is distended. The pliability of veins also permits the contraction of skeletal muscles that surround veins in the arms and legs to compress the veins and so to push blood along past the one-way valves toward the heart (Figure 31-3). Increased contraction of venous walls (and hence lessened distension) is also important after injuries. If considerable blood loss occurs, such contraction can sub-

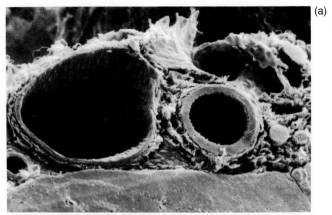

(a)

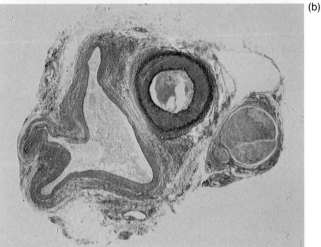

(b)

Figure 31-9   ARTERIES AND VEINS.
Arterial and venous walls consist of three layers: endothelium, smooth muscle, and connective tissue. In general, veins have thinner walls and are much less rigid than arteries. (a) Scanning electron micrograph (magnified about 170 times) of a vein (left) and its companion artery (right) in cross section. The vein has a larger diameter and thinner wall. From *Tissues and Organs: A Text-Atlas of Scanning Electron Microscopy* by Richard G. Kessel and Randy H. Kardon, W. H. Freeman and Company © 1979. (b) Stained section (magnified about 10 times) showing the wall structures of adjacent vein (left) and artery (upper center). Coagulated blood lies inside both. The irregular shape of the vein reflects the flexibility of its walls. A vein like this can be compressed by nearby body muscles (not shown).

stantially reduce the total volume of the venous system, thus keeping the remaining blood near normal pressures and rates of flow in the heart muscle and brain.

Capillary Beds

As arteries carry blood toward the tissues, they narrow into arterioles, which, in turn, branch and narrow fur-

RISK FACTORS IN OUR NATION'S BIGGEST KILLER

America's biggest killer is *atherosclerosis*, or "hardening of the arteries." The odd name of this disease comes from the Greek words meaning "gruel" and "hardening," and it aptly describes the condition: a build-up of yellowish fatty deposits called *plaque* inside the arteries, beginning in late childhood and continuing throughout life. Like corrosion in water pipes, plaque can accumulate until it seriously impedes the flow of blood (Figure A). Reduced blood flow, in turn, can lead to the formation of a blood clot that obstructs an artery completely and can cause a heart attack or stroke. This is particularly true because plaque accumulates preferentially in the arteries of the heart (coronary arteries) and neck (carotid arteries). Atherosclerosis accounts for half of all deaths each year and has inspired a great deal of research into causes and preventive measures. From this research has emerged a fairly complete list of risk factors, which can help an individual assess his or her own chances of developing atherosclerosis and make appropriate life-style changes.

Diet

For years, researchers and physicians suspected that a diet high in cholesterol could increase one's risk for developing atherosclerosis. However, the proof came only recently with a ten-year, $150 million study by the National Heart, Lung, and Blood Institute (NHLBI). By following a group of middle-aged men with elevated levels of cholesterol in the blood, NHLBI researchers found that the incidence of atherosclerosis and of heart attacks and strokes could be lowered by decreasing blood cholesterol through diet

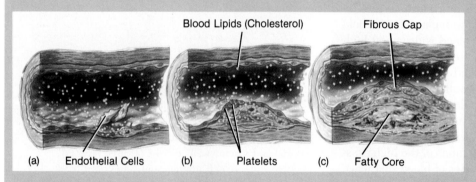

Figure A
(a) Smoking, high blood pressure, diet, and other factors can damage the epithelial lining of arteries and lead to the accumulation of lipids—the start of a plaque deposit. (b) Blood platelets amass, cause a cap of cells to form above the plaque, and isolate the plaque within the arterial wall. This leads to the narrowing of the entire artery. (c) If the cap breaks and lipids from the fatty core combine with blood-clotting factors, a clot can form, block the artery, and lead to a stroke or heart attack.

or drugs or both. For those who already have elevated cholesterol levels, drugs such as cholestyramine may be required. But for those with normal levels who wish to avoid atherosclerosis, it is essential to cut down on foods high in cholesterol and saturated fat. Table 1 lists the main dietary culprits.

Family History

A tendency to develop atherosclerosis can be inherited, and, according to researchers at NHLBI, if your mother, father, brother, or sister died of a heart attack or stroke before age sixty-five, you are at in-

ther into the capillaries (Figure 31-10). Individual capillaries are about 1 millimeter long and about 3 to 10 micrometers or less in diameter—small enough so that red blood cells have to squeeze through one at a time. Large white blood cells often temporarily block their passage.

Each capillary has extremely thin walls composed primarily of a single layer of endothelial cells. Beds of capillaries permeate virtually all tissues of the body, and, as we have noted, it is in these beds that the real action takes place—the transfer of materials carried by the

Table 1 FOODS HIGH IN CHOLESTEROL AND SATURATED FAT

	Cholesterol (in milligrams)	Saturated Fat (in grams)
Pork Brains (3 oz.)	2,169	1.8
Beef Kidney (3 oz.)	683	3.8
Beef Liver (3 oz.)	372	2.5
One Egg	275	1.7
Shrimp (3 oz.)	128	0.2
Roquefort Cheese (3 oz.)	78	16.5
Beef Frankfurter		
(30 percent fat, 3 oz.)	75	9.9
Prime Ribs of Beef (3 oz.)	66.5	5.3
Ice Cream		
(10 percent fat, 1 cup)	59	8.9
Lobster (3 oz.)	46.5	0.075
Doughnut (3 oz.)	36	4.0
Whole Milk (1 cup)	33	5.1
Butter (1 tbsp.)	31	7.1
Creamed Cottage Cheese		
(1 cup)	31	6.0
French Fries (3 oz.)	20	6.0
Milk Chocolate Bar (3 oz.)	18	16.3

All figures courtesy of the United States Department of Agriculture

creased risk. Some families display a rare genetic disorder called *hypercholesterolemia,* in which blood-cholesterol levels are six to eight times higher than normal. Homozygotes usually die of heart disease by their early twenties, while heterozygotes live longer, but still suffer at least 5 percent of all heart attacks in persons under age sixty.

Personal Habits

Both smoking and a sedentary life style are risk factors for developing atherosclerosis. Among other ef-fects, smoking is believed to cause damage to the thin endothelial layer that lines the arteries. It is at such points of damage that plaque begins to form and then accumulates. Research on nonhuman primates reveals that regular exercise helps maintain coronary arteries with larger diameters, even when the test group is fed a high-cholesterol diet and the control group is given a low-cholesterol one.

Medical Conditions

Both high blood pressure and diabetes increase one's chances of developing atherosclerosis. As with smoking, these conditions can damage arteries and lead to sites of plaque formation. In many people, the conditions can be controlled through drugs and diet, including limiting the intake of salt and sugar.

Behavior

Psychologists and psychiatrists have found common behavior patterns in many victims of heart disease. The most often cited is the "Type A personality," a constellation of behaviors and attitudes that includes competitiveness, aggressiveness, and a sense of constant time pressure and of heavy responsibilities. A subset of Type A's also shows what is called "generalized hostility." This includes a suspicious, unfriendly, argumentative, or antagonistic attitude toward others; a tendency to find fault with oneself and others; an intense need to win at games and sports; and a tense demeanor, even when smiling. Research also shows that people who are vengeful in their hostility have a significantly higher chance of suffering heart attack than do other Type A's, and so do people who take a markedly cynical view of the world.

Statistics show that each individual risk factor—smoking, let us say—increases one's chances of developing atherosclerosis by a factor of two or three. In turn, two risk factors, such as smoking and a fatty diet, increase one's chances three- or fourfold. The good news is that lowering or eliminating the risk factors really works, even after atherosclerotic plaques have begun to clog the arteries.

blood to the extracellular fluid, and from there, to the individual body cells. Conversely, substances from cells pass to the extracellular fluid and, in turn, into the capillaries. Thus it could be said that the main function of the circulatory system occurs only in the capillaries and that the heart and other vessels exist solely as "plumbing" to serve these sites of exchange.

The capillaries are so numerous and pervade the human body tissues so completely that their total length, if joined end to end, would exceed 60,000 miles, or

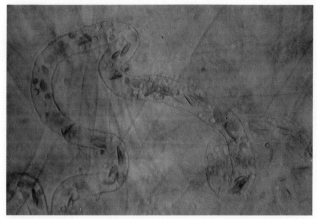

Figure 31-10   CAPILLARIES.
Red and white blood cells (magnified about 550 times) are visible in a capillary in the webbed foot of a living frog.

roughly 1.5 times around the Earth! In most tissues, no cell is more than three or four cell diameters from a capillary—a short enough distance to allow even the slow process of diffusion to provide sufficient amounts of critical substances. Capillary beds have a complex struc-

ture, including main channels that are open most of the time to blood flow and smaller channels in which blood flow is controlled by tiny sphincter muscles that encircle the entrance to local networks of capillaries. These muscles act like miniature floodgates that can contract and close off the flow of blood through particular tissues or can relax and allow blood to pass through. We shall discuss the reasons for such blood-flow control mechanisms a bit later in the chapter.

A capillary's thin walls are the key to its function. Substances can diffuse in either direction between the blood and the extracellular fluid through a capillary's single layer of endothelial cells (Figure 31-11). This movement of substances can take one of four routes: (1) fluid can be ingested by tiny pinocytotic vesicles on one side of the endothelial cells and discharged on the other side; (2) fluid and other substances can diffuse through narrow clefts between endothelial cells at sites lacking tight junctions; (3) the inner and outer plasma membranes of individual endothelial cells may fuse at specific sites to form true holes through the cells, such as occurs in kidney capillaries, which filter blood to generate urine; and (4) lipid-soluble materials can diffuse directly across the plasma membrane, through the cytoplasm, and out the other side. The route taken by a substance depends on its chemical properties.

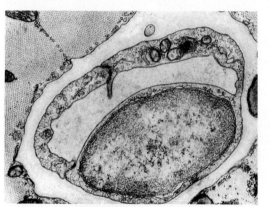

(a)

Figure 31-11   CAPILLARY WALL: KEY TO FUNCTION
The capillary wall consists of a single layer of endothelial cells enclosing the central cavity which may be no wider than one cell diameter. (a) The electron micrograph (magnified about 14,000 times) shows a capillary wall consisting of two cells (the nucleus of only one of these can be seen). This is a continuous type of capillary (b) which has an uninterrupted endothelium in contrast to a type of capillary (c) which has minute pores closed only by a thin piece of basal lamina. The routes of transport—1, 2, 3, and 4—are described in the text discussion.

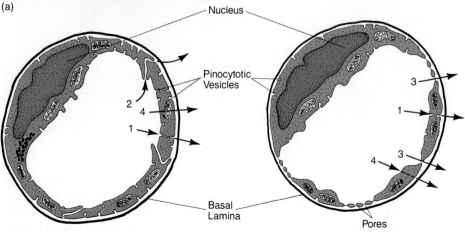

(b)   (c)

Blood Pressure

You have no doubt had your blood pressure taken with an inflatable cuff wrapped tightly around your upper arm. What was being measured was the force that the blood directs outward against a blood vessel's wall. Blood pressure is critical to the functioning of the circulatory system for the simple reason that fluid flows through a tube in response to pressure differences between the two ends. Just as soda moves up a straw toward the low-pressure area created by sipping, blood moves through the circulatory system due to pressure differences between the two sides of the heart as well as within the capillary beds. The blood pressure measured by a nurse or doctor reflects these differences. The measurement is made in millimeters of mercury (mm Hg)—that is, the height to which a column of mercury would rise if pushed on by the blood at that pressure. Human blood is at its highest pressure, about 120 mm Hg, as it leaves the left ventricle during contraction of the heart, or **systole** (pronounced, "sis-toe-lee"). Blood in the heart is at its lowest pressure, about 80 mm Hg, during relaxation of the heart muscle, or **diastole** (pronounced, "die-ass-toe-lee"). Blood pressure of 120/80 is considered normal or average in human males. A reading of 260/180 might be normal in the aorta of a giraffe, where high pressure is needed to send blood up through the long neck to the brain.

Blood-pressure readings taken in one part of the body are not the same as those taken elsewhere in the body. Indeed, when blood reaches a giraffe's brain, it has about the same pressure as blood reaching your brain. As blood flows through the ever-narrowing arteries, friction between the blood and the vessel walls lowers the pressure. The drop in pressure is greatest in the arterioles. By the time blood reaches the capillaries, the pressure is very low, averaging perhaps 10 or 15 mm Hg. If it were much higher, the fragile capillary might burst.

In birds and mammals, blood pressure is substantially lower on the right side of the heart (which supplies the lungs) than on the left; it drops even lower as blood reaches the fantastic system of branching capillaries in the lungs. The pressure there is very low—perhaps 6 mm Hg—and results in relatively less fluid being driven from the blood through the thin capillary walls into the lungs. The lungs are thereby kept relatively "dry" so that carbon dioxide and oxygen do not have to diffuse through thick layers of liquid during gas exchange.

Fluid Balance and the Diffusion of Materials

The movement of critical substances from the blood to the extracellular fluid "middleman" to the cells, and vice versa, is based on two kinds of pressure: blood pressure and osmotic pressure. These forces also play an important role in the delicate balance among the three fluids—the body's *fluid balance*.

The net effect of blood and osmotic pressures is the movement of fluid *out* of the capillaries and into the surrounding extracellular fluid. Let us see why. Blood pressure at the arterial end of capillary beds, about 30 mm Hg, is higher than in the surrounding tissue; thus water is driven out of the blood plasma through the narrow clefts between the endothelial cells of the capillary walls and into the extracellular fluid. At the same time, *osmosis* (movement of water across semipermeable membranes from solutions of low concentration to those of high concentration) leads to water movement in the opposite direction. This is because the blood contains proteins that give it an osmotic pressure of about 25 mm Hg. This is called **colloidal osmotic pressure** because of the presence of the proteins, or colloids (Figure 31-12). Tissue fluids outside the capillaries lack most of these colloids, and so have an osmotic pressure of only about 4 mm Hg. The difference ($25 - 4 = 21$ mm Hg) in osmotic pressure represents a net tendency for water to enter the blood at the high-pressure, or arterial, end of capillaries. This net pressure inward, however, is more than offset by the blood pressure pushing outward against the capillary walls—about 30 mm Hg. The result is a net outward pressure in the arterial-end capillaries equal to $30 - 21$, or 9 mm Hg. It is this net outward pressure that causes the watery contents of the blood to move out through the capillary walls and into the extracellular fluid.

What happens, then, at the low-pressure, or venous, end of the same capillaries? The osmotic pressures here are the same as at the arterial end, giving rise to a net inward pressure of 21 mm Hg. However, friction between moving blood and the huge total surface area of the capillary walls lowers the blood pressure to about 15 mm Hg or lower by the time the blood reaches the venous end of a capillary. Subtracting 15 mm Hg outward pressure from 21 mm Hg inward pressure gives 6 mm Hg net inward pressure. As a result, fluid at the venous end of a capillary moves back into the capillary from the tissues.

Can you see the puzzle here? Net pressure of 9 mm Hg pushes fluid out at the high-pressure end of a capillary bed, whereas a net pressure of only 6 mm Hg pushes fluid back in at the low-pressure end. Since the two pressures are not in balance, there is a continuous net movement, every second of every day, of fluid out of the capillaries. So why doesn't all the fluid gradually leave the blood and accumulate as extracellular fluid? The reason is that fluid is absorbed in a special system of collecting ducts and returned to the blood vessels. We shall describe this *lymphatic system* at the end of the chapter.

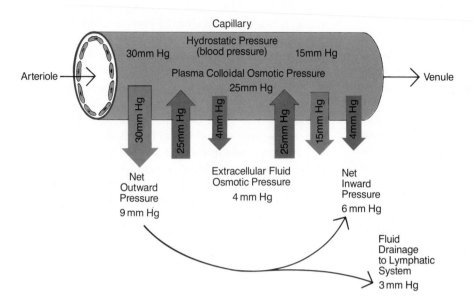

Figure 31-12   FLUID MOVEMENTS NEAR CAPILLARIES.
In many capillaries there is a net flow of fluid from the blood to the tissue fluid. This depends on the hydrostatic pressure of blood and the osmotic pressure differences between blood and extracellular fluid. Because the hydrostatic pressure falls as blood travels through a narrow capillary, water tends to exit at the high-pressure end and enter at the low-pressure end. Any net loss drains to the lymphatic system.

REGULATION OF BLOOD FLOW

The circulatory system must do more than simply pick up materials and deliver them to the tissues. The rate of that material transport must be appropriate to the animal's needs and activities. A strenuously exercising animal, for example, needs quick transport to and from the actively metabolizing muscle cells, while an animal at rest after a meal needs greater blood flow to the stomach and intestines. The rate at which materials are transported is closely tied to the speed of the flowing blood. This speed is based partly on the size of blood vessels, but it is also controlled by the frequency of heartbeats, the amount of blood pumped per beat, and the activity of capillary sphincter muscles. Let us look first at the influence of vessel size on the regulation of blood flow.

Blood Flow and Vessel Size

The speed of blood flow to any spot in the circulatory system depends on the *total* cross-sectional area of the vessels at that spot. Imagine water flowing through a pipe built of sections with different radii or through an irregularly shaped riverbed. The water will slow down on entering an enlarged region and speed up on entering a constricted region. The same principle applies to blood vessels. The aorta and venae cavae have smaller cross-sectional areas in comparison with all the arterioles, capillaries, and venules added together. This means that the aorta and venae cavae are "constricted" regions, whereas the total capillary cross-sectional area is like a wide place

in a pipe or like a wide, shallow region of a river. The rate of blood flow is therefore fastest in the aorta and venae cavae and slowest in the capillaries.

If we turn from all the vessels added together to just a single vessel, we find that the velocity of blood flow varies directly as a function of vessel radius. Specifically, reduction of a vessel's radius by one-half reduces the rate of blood flow through it to one-sixteenth the original speed. Thus in the body's biggest pipe—the aorta—blood flows about 330 millimeters per second, whereas it moves about 1 millimeter per second in an average capillary. This diminished rate of flow in the capillaries is highly desirable, since it permits much more time for exchange of gases, nutrients, and wastes.

As blood leaves the capillaries and passes into the veins, the total cross-sectional area through which it flows is reduced, so the rate of blood flow accelerates even though the blood pressure remains low. As we mentioned earlier, return flow of blood to the heart is aided by the kneading action of body muscles on the veins and by one-way valves scattered along the length of many veins. You have no doubt heard of soldiers fainting after standing at attention too long. This is because inactivity allows large quantities of blood to collect in the leg veins; not enough blood is returned to the heart to be pumped to the brain, and without enough oxygen in the brain, the soldier faints. The same thing happens if a person remains seated upright for a long period, such as on a cross-continental air flight. Sudden standing may result in dizziness because so much blood has accumulated in the venous reservoir of the legs that an insufficient volume is available to supply the brain. As we shall see, however, this situation is quickly rectified by the control system that regulates activity of the heart and blood vessels.

Control of the Heart's Output

The muscle tissue in the heart of a vertebrate or mollusk is truly unique. Whereas most body muscles contract only after they are stimulated by a motor nerve, heart-muscle contractions—heartbeats—are generated by the heart tissue itself. The heart is said to have a *myogenic beat*—one that will continue even if every nerve serving the heart is severed. In a developing embryo, the heart begins to beat even before any nerves grow to it.

The basic source of the myogenic heartbeat lies in every heart-muscle cell, since each can beat spontaneously. However, a special lump of modified heart-muscle cells beats a bit faster and so sets the pace. These cells splay out over the walls of both atrial chambers. This lump of cells, the heart's own pacemaker, is called the **sinoatrial node,** or S-A node. These cells can contract like muscle cells and yet can also conduct electrical impulses of the same type carried by nerve cells (Chapter 36). A special set of conducting fibers carries impulses to another node, which is located at the base of the right atrium near the ventricles; this is the **atrioventricular node,** or A-V node (Figure 31-13). Finally, another set of conducting fibers, the *bundle of His* (pronounced "hiss," as by a snake), extends from the A-V node into the muscular walls of both ventricles, so that a wave of contraction can be triggered synchronously in all the ventricular muscle cells.

The S-A node has an interesting evolutionary history: it is a remnant of the sinus venosus, the most posterior of the four chambers of the early vertebrate heart. In modern fish and amphibians, cells in the walls of this chamber act as the heart's pacemaker. In reptiles, birds, and mammals, only the S-A node remains.

How are heartbeats triggered? At regular intervals, usually less than 1 second apart in humans, the S-A node initiates an electrical impulse by spontaneously *depolarizing;* that is, the electrical charges inside and outside the cells reverse. This impulse spreads over the S-A fibers and causes the atrial muscle cells to contract, thereby propelling blood into the ventricles. As the atria contract, the impulse originating in the S-A node travels toward the A-V node, arriving there about 100 milliseconds later. By then, the ventricles are already filled with blood. The A-V node fires rapidly, sending impulses over the bundles of His to most of the muscle cells of the ventricular walls. The A-V depolarization impulse spreads virtually instantaneously from one heart-muscle cell to the next because the cells are interconnected by numerous gap junctions that readily pass electrical current. Thus the whole of the large ventricular muscle mass contracts at once, just slightly later than the contraction of the atria.

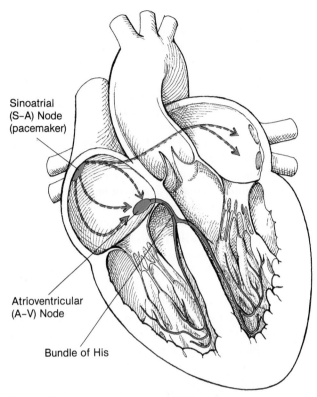

Figure 31-13   MYOGENIC HEARTBEAT.
The coordinated contractions of the heart are regulated by the pacemaker, the sinoatrial (S-A) node. This region of specialized muscle initiates electrical impulses that spread rapidly across the two atria, which then contract almost simultaneously. When the wave of excitation reaches the atrioventricular node (A-V), it stimulates the fibers in the bundle of His; excitation then travels to all parts of the ventricular walls, causing contraction. Contraction of the ventricles occurs after contraction of the atria as a result of relatively slow conduction to the atrioventricular node.

The massive electrical activity associated with each contraction is easily measured using an electrocardiograph. Beating of the heart also makes the characteristic "lub-dub" sound that can be heard with a stethoscope. This sound originates in the valves that open and close between heart chambers during systole (contraction) and diastole (relaxation). The "lub" corresponds to the closing of the two valves between the atria and the ventricles, while the "dub" corresponds to the closing of the two valves between the ventricles and their arteries (the aorta and the pulmonary artery). Heart valves that malfunction let blood seep back in the wrong direction, causing characteristic "murmurs" that can be detected by a stethoscope. Other heart abnormalities cause electrical alterations that can be picked up by an electrocardiograph.

Although the heartbeat is intrinsic to the S-A and A-V nodes, the *rate* of contraction is often slowed or speeded by nerve signals in order to meet changing demands. These extra controls on a muscle with self-generating contractions have been likened to the effects of spurs and reins on a horse, which nevertheless tends to set its own pace. Normally, the human heart beats about sixty-five to seventy times a minute and pumps about 70 milliliters of blood from the left ventricle with each contraction; this amount is referred to as the *stroke volume.* **Cardiac output** is the product of beat rate and stroke volume: 65×70 milliliters, or about 5 liters of blood per minute. If cardiac output is greater than needed to meet the body's immediate demands, the beat rate and stroke volume can be lowered by the controls that serve as "reins." Nerve impulses liberate a chemical neurotransmitter, *acetylcholine,* which (1) inhibits spontaneous discharge of the S-A node, thus slowing the heart rate, and (2) inhibits transmission through the A-V node so that the ventricle beats only once for each two or three times the atrium beats. Conversely, when more cardiac output is needed, different types of nerves (the "spurs") release the neurotransmitter *norepinephrine,* which (1) speeds the rate of S-A discharge and thereby increases the heart rate, (2) increases conduction speed through the A-V node, and (3) acts on heart-muscle cells to increase the force of ventricular contraction.

The workings of these cardiac controls can be demonstrated through simple experiments. If we take a mammal with a resting heart rate of sixty beats per minute and inhibit the nerve that liberates acetylcholine at the heart (the vagus nerve), we will see that heart rate immediately speeds to a new resting rate of seventy-five beats per minute. If we then reactivate the nerve, the resting rate returns to about sixty beats per minute. Significantly, these results tell us that the resting rate of the mammalian heart is not the rate intrinsic to the S-A node pacemaker but is a *slower* rate governed by the nervous system. The slower rate is sufficient to meet the body's resting demands, but as exercise begins, the inhibition can be overcome (the "reins" eased) by shutting off the release of acetylcholine. The heart then speeds to its myogenic rate. If exercise becomes even more vigorous, the "spurs" are used: the second type of nerve releases norepinephrine, and the heart rate speeds even more. Thus mammals have a two-step system for increasing cardiac output, which enables them to meet peak demands for oxygen and nutrients.

Regulation of Blood Flow Through the Vessels

Cardiac output is just one influence on the amount of blood coursing through the blood vessels. Blood flow into a given capillary bed or into the veins is also under the direct control of the nervous system. In the brain and heart, capillaries are virtually always open, so these master organs get a steady supply of oxygen and nutrients. Elsewhere in the body, however, sphincter muscles can tighten and thereby cut down blood flow through a particular capillary bed. At any one instant, most of these muscles are contracted, and only 30 to 50 percent of the body's capillaries are open to blood flow. It would be very dangerous if all opened simultaneously, since the combined length of the capillaries—60,000 miles—can hold 1.4 times the total volume of blood in the body. Only by closing off a high proportion of capillaries at any given time can blood be kept at adequate pressures and volumes in the remainder of the circulatory system.

Contraction of the capillary sphincters and the walls of veins in response to nerve impulses is called **vasoconstriction.** Vasoconstriction creates a dynamic control over capillary beds in a local area. First one bed is closed, then another when the first reopens, and so on so that local cellular needs are met, but the total volume of blood accumulated in the full set of capillaries is never too great. Vasoconstriction also helps stabilize blood flow during a change in body position, such as standing up after lying in bed. As you climb out of bed, blood pressure rises in your legs and falls in your head as gravity pulls the blood downward. This contributes to the increased blood volume in leg veins we mentioned earlier. To counteract this condition and increase the amount of blood available to the heart and brain, vasoconstriction greatly reduces blood volume in leg veins and also shuts down some capillary beds. Likewise, as a giraffe lowers its head to drink or an antelope lifts its head to sniff for predators, coordinated vasoconstriction must go on continuously. This coordination is a feature of terrestrial vertebrates; pooling of blood due to gravity does not occur in aquatic vertebrates immersed in and supported by water.

Capillaries that are closed off reopen as the sphincter muscles relax. This relaxation is called **vasodilation.** Vasodilation allows blood to flow into muscle-tissue capillary beds so that a muscle's oxygen demand is met, into skin capillaries so heat can be released from the body, and so on. The human body employs one special type of vasodilation as a primitive form of communication. When a person is embarrassed or stressed, acetylcholine may be released from nerve endings in the cheeks, causing vasodilation and an infusion of bright red blood that feels warm. In a light-skinned person, this shows up as blushing.

In addition to vasoconstriction and vasodilation, blood volume and rates of flow are also affected by a hormone that is secreted by heart cells. We will discuss these atrionatriuretic factors in Chapter 37, and we will learn how they reduce blood volume within minutes of release.

Brain Regulation of the Circulatory System

We have seen that nerves can regulate the heart and blood vessels, but what controls the nerves? An area of the brain stem called the **vasomotor center** receives input from a variety of sites in the body that monitor blood pressure, oxygen and carbon-dioxide levels, blood pH, body temperature, and other physiological indicators. The vasomotor center then coordinates the heart and blood vessels to keep the levels of those substances or properties within normal ranges. In general, this is accomplished by activating vasoconstriction or vasodilation to control blood flow. During vigorous exercise, for example, the vasomotor center coordinates an increase in breathing rate, heartbeat, blood flow, and sometimes heat loss. A coordinated response is also necessary when a mammal becomes chilled; breathing rate must increase, body heat must be conserved or generated, and the flow of blood must be diverted from the skin to the body core. The vasomotor center interacts with brain centers that regulate breathing and temperature to produce these coordinated responses. When an animal is in danger, the vasomotor center can also trigger release of the hormone *epinephrine*. This hormone mobilizes the "fight-or-flight" response and raises the cardiac output by causing vasodilation in the heart and skeletal muscles and vasoconstriction in many other areas. Thus the animal is prepared to respond quickly to the threat.

The vasomotor center sometimes responds to the anticipation of potential danger, causing the skin to become flushed and the heart to race even if no actual threat exists. In much the same way, the vasomotor center can respond to the idea of exercise even before it begins. For example, immediately before a sprinter starts a race, the vasomotor center "unconsciously" opens the capillary sphincter muscles in the legs. This provides increased oxygen *before* the leg muscles actually do their strenuous work.

The functioning of the circulatory system is beautifully coordinated by the vasomotor center so that the blood pressure and the levels of oxygen, carbon dioxide, and hydrogen ions in the blood remain relatively constant whether exercise is heavy, light, or absent. One might suppose, for example, that the carbon-dioxide levels would rise or the oxygen levels fall *before* the heart rate changes. Instead, the heart rate changes *first* so that despite heavy exercise, levels of hydrogen ions, carbon dioxide, and oxygen in the blood remain normal. How is the need for a change in heart rate signaled? The answer is that muscle and joint movements occurring during exercise are reported to the brain by sensory nerves. The vasomotor center then becomes activated, and it, in turn, speeds heart rate, increases return of the blood through the veins, and coordinates the various responses that maintain steady levels of ions and blood gases *during* the exercise, not lagging behind it.

BLOOD: THE FLUID OF LIFE

We have now considered the pump, the plumbing, and the control systems for circulation. But what of the blood itself—the reason that all the circulatory "hardware" exists? The remarkable life-sustaining blood is a dynamic solution containing ions, nutrients, waste products, hormones, other substances, and several kinds of cells. The solid fraction of blood—blood cells and platelets—makes up about 40 to 50 percent of the blood's volume and is suspended in a yellowish fluid called plasma.

Plasma: The Fluid Portion of Blood

Plasma is water that contains various types of dissolved substances, including oxygen, carbon dioxide, and nitrogen gases. (Most of the oxygen and carbon dioxide, however, is carried inside the red blood cells.) Plasma also carries the ions and nutrient molecules needed by individual tissue cells. Positive and negative ions (cations and anions) make up about 1 percent of plasma by weight. The cations include sodium (Na^+), potassium (K^+), calcium (Ca^{2+}), and magnesium (Mg^{2+}). The anions include chloride (Cl^-), bicarbonate (HCO_3^-), sulfate (SO_4^{2-}), and phosphate (HPO_4^{2-}, $H_2PO_4^-$). Because the compound NaCl is the most common substance in plasma, blood is often compared to dilute sea water. The levels of Na^+ and Cl^- as well as the levels of bicarbonate, potassium, and calcium are particularly important for the health of the organism and its individual cells, as we shall see in later chapters.

Also dissolved in plasma is a variety of small organic molecules, including glucose, amino acids, nucleic-acid bases, certain fats, cholesterol, and vitamins. These molecules can be derived from food digested in the gut, or they can be secreted by cells, particularly in the liver. In addition, plasma carries away the wastes of cellular metabolism, such as lactic acid, urea, uric acid, and ammonia.

Proteins make up about 8 percent of the total volume of blood plasma. Individual types of proteins play special roles: as hormones, as carriers of other hormones or vitamins, as constituents of the immune system, or as part of the blood-clotting system. Many blood proteins are manufactured and secreted by the liver. The full complement of proteins in the blood brings about the colloidal osmotic pressure mentioned earlier in this chapter. Another general effect of blood proteins is to act as a buffer that damps out changes in blood pH—a phenomenon we shall address in Chapter 35.

The Blood's Solid Components

Blood Cells

The most numerous blood cells suspended in plasma are the red blood cells, or **erythrocytes.** A single milliliter of mammalian blood contains about 5 million of them, and at any given time, 25×10^9 red blood cells are circulating through the human body. Each erythrocyte is disk-shaped, with a thickened, relatively rigid periphery and a thinner, extensible central region that can bulge forward like a spinnaker sail as the cell moves through a narrow capillary (Figure 31-14). Throughout life, cellular differentiation occurs in the *bone marrow*, the blood-forming tissue of a mammal. During this process, red blood cells lose their nuclei but retain mitochondria and a few other organelles. The cytoplasm in a mature red blood cell is filled with hemoglobin molecules—300 million per cell. *Hemoglobin* is the agent responsible for

Figure 31-14   BLOOD CELLS.
Erythrocytes (red blood cells) are the most numerous. In mammals, they do not have a nucleus and are seen as biconcave disks, magnified about 2,250 times in the scanning electron micrograph. Leukocytes (white blood cells) consist of several types; the cells with bumpy surfaces visible in the midst of the erythrocytes are leucocytes. From *Tissues and Organs: A Text-Atlas of Scanning Electron Microscopy* by Richard G. Kessel and Randy H. Kardon, W. H. Freeman and Company © 1979.

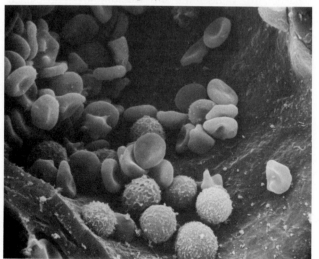

binding and carrying oxygen in the blood, as we will describe in Chapter 33.

Less numerous, but equally important, are the white blood cells, or **leukocytes,** which are involved in the body's defense system and are also suspended in the blood plasma. The average healthy person has about 7,500 leukocytes per milliliter of blood; but that number may rise rapidly in response to infection or disease (Figure 31-14). There are three types of white blood cells: monocytes, granulocytes, and lymphocytes. **Monocytes** mature into *macrophages*, phagocytic cells that, moving in amoeboid fashion, can pass out of the capillaries and through the tissue spaces, where they act as scavengers of debris. **Granulocytes** are full of granules that contain enzymes or other substances involved in inflammatory and allergic reactions. The **lymphocytes,** which are described in more detail in Chapter 32, are immune cells that attack foreign substances, such as bacteria and viruses. Some lymphocytes secrete *antibody* molecules—proteins that bind the foreign substances.

Throughout adult life, blood cells continually die and are replaced. Both white and red blood cells survive about 120 days. Every second of your adult life, about 2 million new blood cells are produced in the central marrow of your bones, and an equivalent number die. The fatty bone-marrow tissue contains many *stem cells* that give rise to either red or white blood cells. One specific type of marrow stem cell is the *megakaryocyte*. Pieces of the peripheral cytoplasm of megakaryocytes break free and form the motile masses of membrane-bound living cytoplasm called **platelets,** which play an important role in blood clotting.

Platelets and Blood Clotting

A modest cut or scrape does not normally endanger life due to bleeding because blood *clots*, forming a semisolid mass that seals off the walls of ruptured blood vessels. When a rupture occurs, blood plasma, blood cells, and platelets leak into the surrounding tissue spaces. Contact between platelets and tissue components such as collagen fibers causes some platelets to break open and leak several agents, including serotonin and thromboplastin. *Serotonin* is a small molecule that stimulates contraction of smooth-muscle cells in the walls of nearby arterioles, thereby causing the vessels to narrow and cut off the blood flowing toward the wound site. *Thromboplastin* is an enzyme that stimulates blood clotting by initiating a remarkable cascade of enzymatic reactions (Figure 31-15a). It acts on one of the globular proteins contained in blood plasma, *prothrombin*. In the presence of sufficient levels of Ca^{2+} ions, thromboplastin catalyzes a massive conversion of prothrombin to *thrombin*, yet another enzyme. Thrombin then acts on another plasma protein, *fibrinogen*, catalyzing its conversion to

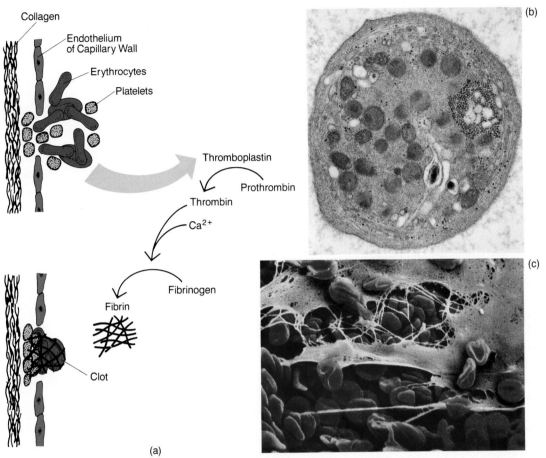

Figure 31-15   PLATELETS AND BLOOD CLOTTING.
The steps in blood clotting. (a) The leakage of blood platelets and erythrocytes out of the capillary leads to their contacting collagen. The platelets adhere, spread on the collagen, and release the contents of their granules, thereby initiating a sequence of reactions leading to the formation of fibrin strands and a clot. (b) Electron micrograph of a platelet (magnified about 22,000 times). (c) A network of fibrin (magnified about 1,250 times) that covers a wound site and traps red blood cells and platelets (not shown).

fibrin. Fibrin makes up the fibrous strands that compose a clot and gives the clot tensile strength. Platelets and some blood cells pouring from the ruptured blood vessel are trapped within the three-dimensional fibrin meshwork as the clot forms (Figure 31-15c). Platelets contain cytoskeletal contractile proteins very much like the actin and myosin of muscle cells; contracting platelets pull the clot into a firmer, tougher configuration.

The sequence of clotting just described is highly simplified, since more than a dozen additional factors are actually involved. Nevertheless, we can see that clotting is a good example of *amplification* in a biological system: a relatively small event triggers a stepwise series of larger and larger events. In the case of clotting, a small wound generates a large amount of thrombin and an immense local supply of fibrin. As a result, most instances of bleeding can be rapidly quelled, and the organism can survive its wound. Fortunately, the body has

its own defenses against the efficiency of the clotting process. The endothelial cells lining every blood vessel have surface molecules that inhibit thrombin or other clotting factors. This helps assure that clots do not form along uninjured vessel surfaces or extend all the way across vessels near wound sites.

The marvels of clotting are not without their price. A number of diseases or genetic defects may affect any step in this complicated series of reactions and lead to poor clotting or the complete inability to clot. Hemophiliacs, for example, lack one critical factor in the clotting process. In other persons, the clotting process itself can lead to injury. In the underexercised legs of convalescing patients and in victims of circulatory system diseases, clots can form and later break free from their sites of origin. When such clots drift through the vascular system and lodge in a capillary bed of the heart, they can trigger a heart attack; when they travel to the brain, they can cause a stroke. In 1987, genetic engineering has produced medically useful quantities of an enzyme that occurs naturally in trace amounts; this plasminogen activator factor digests clots and literally can reverse some heart attacks if it is injected within a few hours.

THE LYMPHATIC SYSTEM AND TISSUE DRAINAGE

Vertebrates have developed an extensive second system of vessels, called the **lymphatic system,** which runs roughly parallel to the venous half of the circulatory system. This system has two main functions: it drains excess extracellular fluid that bathes body cells, and it houses important parts of the immune system. (Lymph vessels in the wall of the intestine serve a third function: they absorb fats from digested foods. It is not known why fats pass into the lymph vessels rather than into blood vessels.) The role of the lymphatic system in immune responses is discussed in Chapter 32. Here we will focus on the system's structure and on its drainage function.

As we saw earlier in this chapter, the high blood pressure present at the arterial end of capillary beds in mammals and birds causes a net flow of water from the blood into the spaces between cells. It is the job of the lymphatic system to drain this fluid and to return it to the venous system to prevent tissue swelling. Interestingly, however, the presence of lymphatic systems in fishes, amphibians, and reptiles—creatures with low blood pressure—indicates that lymphatic ducts did not arise originally as a feature of high-pressure circulatory systems. Other factors must have contributed to the development of lymphatic systems.

The structure of the human lymphatic system is shown in Figure 31-16. Extracellular fluid enters the system through pores in tiny, thin-walled, blind-ended *lymphatic vessels* reminiscent of blood capillaries. Some proteins that have leaked from the blood plasma into the tissue spaces may enter the lymphatic vessels. These vessels merge into larger and larger lymphatic vessels, and the fluid, called *lymph*, flows through them away from the tissues and toward the heart. At various sites in the body, the lymphatic vessels pass through *lymph nodes*, compact meshworks of connective tissue that filter particles from the lymph and produce some of the cells of the immune system. The tonsils, for example, are lymph nodes. Lymphatic vessels continue to merge into larger and larger channels and ultimately join at the *thoracic duct*, which pours lymph into a major vein. The mixture of lymph and blood then enters the superior vena cava and passes to the heart. In humans, about 11

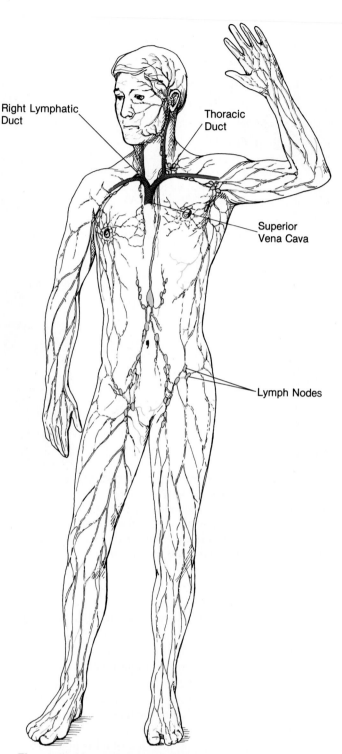

Figure 31-16   THE HUMAN LYMPHATIC SYSTEM.
The lymphatic system is an auxiliary system of vessels that drains via the thoracic duct and the smaller right lymphatic duct into major veins that join the superior vena cava. Lymph is filtered as it percolates through the lymph nodes located at many sites in the body.

milliliters of lymph flow into the blood each hour, although during certain illnesses, the flow may reach the fantastic rate of 900 milliliters per kilogram of body weight per hour.

Lymph is propelled by several mechanisms. In mammals, body muscles compress the lymphatic-vessel walls, kneading the lymph along, while valves like those in veins prevent backflow. In addition, larger lymphatic vessels have smooth muscle along their walls that contracts in waves (rather like peristalsis along the intestine; Chapter 34) and thus propels the watery lymph solution. Most other vertebrates—fish, frogs, birds—have *lymph hearts*, local aggregations of muscle cells that compress lymphatic vessels and pump the fluid through the lymphatic system.

If lymphatic vessels are blocked, water and other substances cannot be properly drained from the tissues, and severe swelling may occur. In the disease *elephantiasis*,

the vessels are blocked by tiny parasitic worms. The affected extremity—an arm, a leg, the testes—can enlarge to monstrous proportions.

LOOKING AHEAD

The lymphatic system is remarkable and critical to the survival of warm-blooded creatures. Still more remarkable, however, is the circulatory system itself—the complex and highly coordinated transport network that provides animal cells with the materials they need for metabolism. As we shall see, that system underlies all other physiological functions, just as roads and railways tie together the commerce of a country. Circulation is truly central to the life processes of most animals.

SUMMARY

1. Diffusion is sufficient to transport commodities into and out of cells over short distances. Long-distance transport, however, requires energy-dependent bulk flow.

2. *Bulk flow circulatory systems* can be open or closed. *Open circulatory systems* are present in many multicellular invertebrates; *closed circulatory systems* occur in some invertebrates and in all vertebrates. *Blood* moves through a closed system in *arteries*, which carry blood to tissues; *capillaries*, tiny, thin-walled vessels that are sites of material exchange; and *veins*, which return blood to the heart.

3. Fluid circulates because (a) some vessel walls contract in waves; (b) the *heart* contracts rhythmically, like a pump; (c) body muscles knead the vessels and augment blood flow; and (d) sphincter muscles control the flow to and from local areas. Valves in the heart and vessels prevent backflow, and ensure movement of blood in a single direction.

4. The evolution of the heart can be studied among various vertebrates. The fish heart has four chambers: the sinus venosus, the atrium,

the ventricle, and the conus arteriosus. Land vertebrates retain the atrium and the ventricle, and in birds and mammals, the two chambers are completely divided into right and left halves. This divided arrangement is associated with separate *pulmonary* (lung) and *systemic* (body) *circulation*.

5. The major arteries of land vertebrates evolved from the ventral aorta and *aortic arches* of fishes. Six arches still develop in vertebrate embryos; in mammals and birds three degenerate, while the carotid arch becomes the carotid artery; the systemic arch becomes the largest artery, the *aorta*; and the pulmonary arch becomes the pulmonary artery.

6. In mammals and birds deoxygenated blood circulates from capillaries in body tissues to *venules* to veins and eventually to the heart's right atrium. It is pumped into the right ventricle and then into the pulmonary arteries and lungs, where gas exchange occurs. Oxygenated blood passes through the pulmonary veins to the heart's left atrium. It is then pumped to the left ventricle, and from there into the aorta, which

leads to smaller major arteries. These branch into smaller and smaller arteries that finally deliver blood to the capillaries in body tissues.

7. Substances can move across capillary walls between blood and extracellular fluid via (a) pinocytotic vesicles; (b) diffusion between endothelial cells; (c) passage through holes in such cells; and (d) diffusion across such cells' plasma membranes.

8. *Blood pressure* is the force of blood against a vessel's wall. Blood pressure is highest during contraction of the heart, or *systole*. It is lowest when the heart is relaxed, or *diastole*.

9. Net fluid loss from capillaries is counterbalanced by absorption of tissue fluid by the lymphatic vessels. The *lymphatic system* transports this fluid back to the circulatory system.

10. The vertebrate heart has a self-generated, myogenic beat, due to intrinsic nodes and electrical impulses. The heartbeat can be slowed down or speeded up when nerve

impulses trigger the release of certain neurotransmitters.

11. The *vasomotor center* in the brain stem controls the rate of heartbeat as well as blood flow through peripheral vessels. The firing of nerves causes *vasoconstriction* and *vasodilation*. Some vasoconstriction of veins and capillaries is essential at all times, since the total volume of the blood vessels far exceeds the volume of the blood plasma.

12. *Blood* contains plasma, red and white blood cells, and platelets. *Plasma* carries ions, maintained in a balanced concentration, as well as small organic molecules, proteins, hormones, vitamins, and wastes. Individual red blood cells, or *erythrocytes*, contain millions of hemoglobin molecules, and carry most of the oxygen delivered to local tissues. White blood cells, or *leukocytes*, include *monocytes*, *granulocytes*, and *lymphocytes*, involved in the body's immune response. *Platelets* are masses of membrane-bound cytoplasm involved in blood clotting.

KEY TERMS

aorta	colloidal osmotic pressure	platelet
aortic arch	diastole	pulmonary (lung) circulation
artery	erythrocyte	sinoatrial (S-A) node
arteriole	granulocyte	Starling's law
atrioventricular (A-V) node	heart	systemic (body) circulation
blood	hemolymph	systole
blood pressure	leukocyte	vasoconstriction
bulk flow	lymphatic system	vasodilation
capillary	lymphocyte	vasomotor center
cardiac output	monocyte	vein
circulatory system	open circulatory system	venule
closed circulatory system	plasma	

QUESTIONS

1. What is the function of the circulatory system? The heart? The large arteries and veins? The capillary beds?

2. What role does diffusion play in single-celled, small, and large animals in the distribution of gases and nutrients to individual cells?

3. What is the main difference between open and closed circulatory systems?

4. Compare the hearts of fish, amphibians, reptiles, birds, and mammals. Describe the embryonic development of the mammalian heart.

5. What is the function of aortic arches in fish? What are portal vessels? How are veins and arteries similar in structure? How are they different?

6. Imagine you are a red blood cell in a capillary of your smallest left toe. Draw a road map and describe your journey through the bloodstream and back to that toe.

7. Where does the lymph originate? Where does it join the blood system? How is it propelled through mammalian lymphatic vessels?

8. Using numbers, explain the source of excess tissue fluid.

9. What controls the heart's intrinsic myogenic beat? What is the evolutionary origin of this node? What other nodes and fibers are involved in regulation of heartbeat, and how do they function? What roles does the nervous system play in regulating the heartbeat at rest? During moderate work? And during heavy exercise?

10. Why is vasoconstriction vital for every mammal? What happens if it fails?

11. Where do mammalian blood cells form? What are stem cells? What are the functions of the red cells? White cells? Platelets? How are the platelets formed?

12. Diagram the interaction of the various hormones and enzymes that lead to formation of a clot at a wound site.

13. When is blood pressure in a human highest? Lowest? Where is rate of flow fast? Slow? Why?

ESSAY QUESTION

1. When severe blood loss occurs due to accident, what other body changes take place? Would water be a good replacement for the lost blood? Salt water (saline)? Plasma? Explain.

SUGGESTED READINGS

Annual Review of Physiology.

Every year, this important series prints papers on cardiovascular control, capillaries, lymph, and other components and processes of circulatory system anatomy and physiology.

BOURNE, G. H., ed. *Hearts and Heart-Like Organs.* Vol. 1. New York: Academic Press, 1980.

A collection of detailed papers with much information.

ECKERT, R., and D. RANDALL. *Animal Physiology.* 2d ed. New York: Freeman, 1983.

A superb book, easy to read and very well illustrated—a first choice for this and the next seven chapters.

32
THE IMMUNE SYSTEM

The remarkable capacity of the immune system to respond to many thousands of different substances with exquisite specificity saves us all from certain death by infection.
Martin C. Raff, "Cell Surface Immunology,"
Scientific American (May 1976)

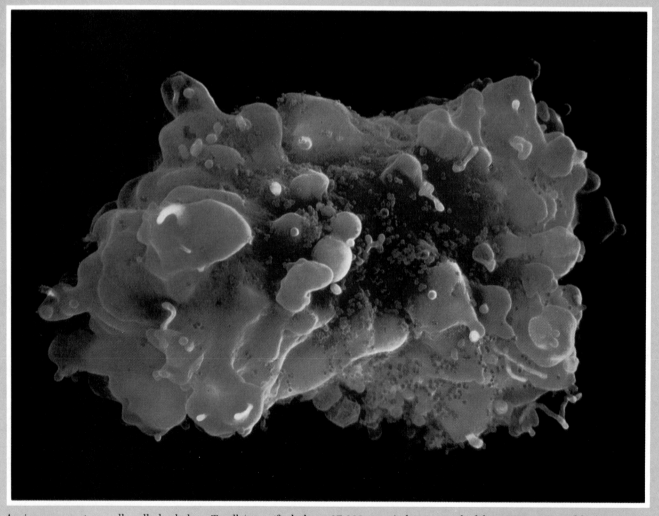

An immune system cell, called a helper T cell (magnified about 27,000 times), being attacked by AIDS viruses (blue).

Every environment on Earth is literally teeming with bacteria, viruses, yeasts, molds, toxins, and other harmful substances, some of which can cause potentially fatal diseases in plants and animals. Yet most humans, as typical animals, live in fairly good health for six or seven decades, suffering diseases for only short periods and recovering fully. The reason is that we are each defended by an **immune system**—a standing army of 1 trillion white blood cells, 100 million trillion special protein molecules called *antibodies*, and a few small organs. Together, the components of the immune system protect the body from foreign substances and from abnormal cells that arise within. The immune system's marvelous protective effects can be seen in a person's spontaneous recovery from a cold or the flu and in the usually lifelong immunity following childhood diseases such as mumps, measles, and chicken pox. However, the immune response also can be harmful, as people who suffer from allergies or arthritis or whose bodies reject desperately needed organ transplants demonstrate so clearly.

The body has various lines of defense, some general and some highly specific in their response to foreign substances. Nonspecific defenses include physical barriers, such as skin and mucus, and the tendency of a wound to become inflamed. The immune system carries out the body's specific defenses and enables it to recognize individual foreign substances, such as ragweed pollen or *Salmonella* bacteria, and then to mount a coordinated attack by white blood cells and antibodies.

We shall see that the mechanism for recognizing foreign and "domestic" body substances and subsequent attack or defense depends on molecular fit—the binding of molecules with reciprocal shapes—just as the critical functioning of enzymes depends on molecular fit. The immense diversity of molecular fit displayed by the immune system depends on a unique process of gene shuffling that occurs in only two types of white blood cells. This gene shuffling allows the production of billions of uniquely targeted, defensive cells and antibodies that remain ready to react against specific foreign substances should they be encountered. Specific immune responses are a truly amazing evolutionary adaptation to life among potentially harmful microbes and molecules. This fact, and the immediate relevance of much immunological research to the treatment of human diseases, make the immune system a fascinating subject indeed.

COMPONENTS OF THE IMMUNE SYSTEM

The immune system is not a separate, clearly delineated set of structures that acts solely for the body's defense. Although this system does include cells and organs that function only to provide immune responses, it also includes some that play several roles in the body. Thus components of the immune system include several classes of blood cells; organs such as the thymus, spleen, and lymph nodes; and soluble circulating proteins, or antibodies. Let us consider each component of the body's "standing army."

White Blood Cells

As we saw in Chapter 31, there are two main kinds of blood cells, red and white. Fully 99 percent of the circulating cells are red blood cells. Nevertheless, the remaining 1 percent includes 1 trillion white blood cells, all of which participate in the body's defense. One class of white blood cells, the *granulocytes*, characterized by granules in their cytoplasm, takes part in the body's inflammatory response to infections, as we shall see. The other two classes of white blood cells are macrophages (mature monocytes) and lymphocytes (Figure 32-1). **Macrophages** engulf and digest most foreign particles and initiate many immune responses by trapping foreign substances and "presenting" them to lymphocytes. Macrophages wander in the extracellular spaces of many different tissues and are found in abundance in the liver,

Figure 32-1   CELLS OF THE IMMUNE SYSTEM.
Scanning electron micrograph magnified about 2,000 times of several lymphocytes (round cells) and macrophages (center). These two cell types interact to initiate many immune responses. Two large granulocytes (mast cells) are also shown.

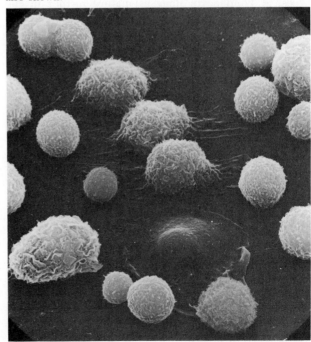

spleen, and lymph nodes. *Lymphocytes* can attack invaders directly or indirectly by secreting attack proteins called antibodies. There are two major kinds of lymphocytes: **T cells** and **B cells.** Whereas T cells (or T lymphocytes) congregate at a site of infection and mount a direct attack on foreign organisms or tissues, B cells (or B lymphocytes) do not localize at the infection site; instead, they synthesize antibody molecules, which perform the attack function. The third type of lymphocyte is the large, granular white blood cell called a **natural killer (NK) cell,** which can directly attack certain viruses and virus-infected tumor cells, may also attack bacteria, and seems to play other basic roles in the immune system.

The Varied Sources of White Blood Cells

In all adult vertebrates, stem cells in bone marrow give rise to red blood cells, as well as to various types of white cells. (A *stem cell* is a determined cell that can divide repeatedly to yield progeny that differentiate in a particular way.) This is why physicians sometimes transplant bone marrow into patients who have received massive doses of radiation, such as occurred at the Soviet Chernobyl nuclear power plant. Such irradiation destroys bone marrow stem cells, and often nothing short of bone marrow replacement can save the victims' lives. After T lymphocyte stem cells arise in healthy bone marrow, they migrate to the thymus where they mature further, especially in response to hormones produced by thymus cells. B lymphocyte stem cells arise in the bone marrow or liver of mammalian embryos. In avian embryos, B lymphocyte lineages mature further in the bursa of Fabricius, an appendixlike organ near the posterior gut. Subsequently, bursa-derived B lymphocytes, and their equivalents in mammals, become localized in **peripheral lymphoid tissues,** the lymph nodes, spleen, and gut-associated appendix, tonsils, adenoids, and intestinal Peyer's patches. T cell lineages can sometimes be found in the spleen, as well. The names T and B lymphocyte refer to *thymus-derived* and *bursa-derived* (but note again that many vertebrates do not have a true bursa; their B cells come from the bone marrow). (See Figure 32-5, which shows the independent origins of the B and T lymphocyte systems.) Abnormalities in the bursa result in the absence of B cells, whereas abnormalities in the thymus leave an organism deficient in T cells. We will discover much more about the defensive role these cells play.

Organs

The discrete organs of the immune system are the lymph nodes, spleen, and thymus (Figure 32-2), each structurally quite different from the next and having its own important functions.

Lymph Nodes

We saw in Chapter 31 that a system of lymphatic vessels helps drain excess tissue fluid. These vessels also carry white blood cells to and from the tissues; large pores in the walls of lymphatic capillaries allow white cells to pass into tissue spaces at the site of an infection. The tiny, nearly translucent lymphatic vessels appear to merge into thickened masses—pinkish-gray beads of tissue called **lymph nodes.** As shown in Figure 32-2, lymph nodes occur at many sites in a mammal's body, with clusters in the neck, armpit, abdomen, and groin. Lymph nodes are dynamic tissue masses, containing millions of lymphocytes that constantly arrive and populate the nodes while older cells die or pass through the lymphatic vessels as they return to the circulatory system (Figure 32-3).

Lymph fluid is filtered as it penetrates a lymph node; particulate materials such as bacteria and debris from dead cells are removed through this filtering. Macrophages take up residence in the lymph nodes and ingest foreign particles as they float by. Such macrophages may also bind foreign substances onto long, fingerlike protrusions of their cell surface. T lymphocytes resident in the node may then "recognize" the foreign substance carried by the macrophage, proliferate, and migrate through the lymphatic vessels to the site of the infection where the foreign substance originates.

Most lymph nodes are only 1 centimeter in diameter. But when an infection occurs in the body, the nearest lymph nodes often swell greatly and become tender as filtering increases and lymphocytes proliferate. Thus during a bout with the flu, you may have noticed tenderness or soreness in some of the lymph nodes of your neck or armpit. The adenoids and tonsils in the throat region are themselves made up of lymph-node tissue and serve as a site for destruction of bacteria and viruses entering the nose and mouth. These glands sometimes become so inflamed from the activity of resident lymphocytes that they must be removed. Lymph nodes near a tumor tend to trap cancer cells that have broken free from the tumor; thus they must often be irradiated or removed along with the tumor itself so that the trapped cancer cells cannot proliferate further.

Spleen

Another organ of the immune system is the **spleen,** an oval, flattened deep-purple organ the size of a small beefsteak (in large mammals), located near the liver (Figure 32-2). The spleen's main functions are to store red blood cells, destroy aged red cells, and recycle iron from hemoglobin molecules. However, the spleen also

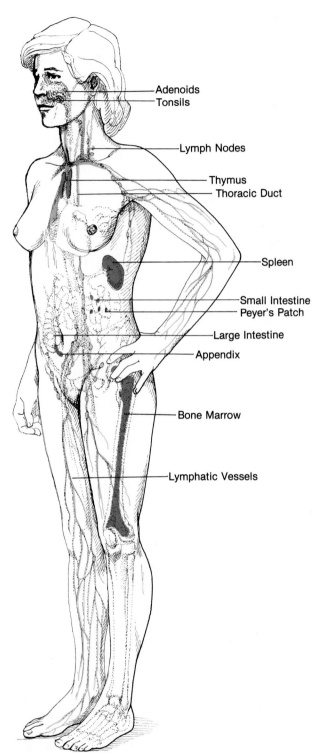

Figure 32-2   THE IMMUNE SYSTEM: THE BODY'S GUARDIAN.

The primary organs of the immune system are the lymph nodes (yellow), spleen, thymus, and bone marrow. Marrow forms inside of most bones. Lymph nodes are associated with lymphatic vessels so that lymph can percolate through the nodes.

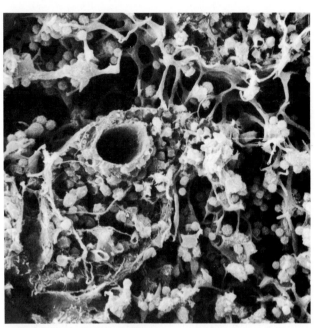

Figure 32-3   LYMPH NODE.
Scanning electron micrograph (magnified about 585 times) of part of a lymph node, showing numerous lymphocytes amid a network of connective-tissue fibers. From *Tissues and Organs: A Text-Atlas of Scanning Electron Microscopy* by Richard G. Kessel and Randy H. Kardon, W. H. Freeman and Company © 1979.

contains many lymphocytes and macrophages, and it filters lymph in much the same way that lymph nodes do. T lymphocytes that proliferate in response to foreign substances on spleen macrophages are also released into the circulatory system to migrate to the site of an infection. The spleen is occasionally removed after injury; that is not too deleterious, since T and B lymphocytes are still found elsewhere in the body.

Thymus

The **thymus** is a small, spongy, spherical organ that sits behind the breastbone in adult mammals (Figure 32-2). In present-day embryos—and, indeed, during the evolution of vertebrates—the thymus arises from the gill slits. As we have mentioned, determined stem cells from the bone marrow migrate to the thymus, where they are acted on by hormones to differentiate further as T cells.

The thymus is essential to development of the immune system. Humans and mice born without a thymus lack the capacity to carry out certain immune responses dependent on T lymphocytes—in particular, fighting viral and bacterial infections and rejecting tissue grafts. The thymus is largest and most active during childhood. It shrinks after puberty, and with age, the activity of T lymphocytes decreases, accompanied by increased susceptibility to infections by agents that T cells would normally combat.

LINES OF DEFENSE: NONSPECIFIC AND SPECIFIC

When a foreign organism or substance enters the body of a mammal, bird, or other vertebrate, two types of defenses can be mounted: nonspecific and specific. *Nonspecific* defense mechanisms function in a similar way regardless of the nature of the foreign substance they serve to eliminate. The skin and the sticky mucous membranes, for example, impede the entrance of bacteria or foreign molecules into the body, and the sweeping action of cilia along the air passages in mammals helps prevent the bacteria from adhering to and penetrating body cells. Similarly, secretions such as tears, mucus, and sputum contain *lysozyme*, an enzyme that digests the cell walls of bacteria. The skin of the toad *Xenopus* has yielded antimicrobial peptides called *magainins*. Magainins, like insect defensive peptides called *cecropins*, seem to lyse bacterial cells, perhaps by inserting in their plasma membranes. If foreign substances do gain entrance to the body, they are ingested and destroyed by macrophages in the blood, liver, and lymph nodes. In a mouse, these cells can completely clear the blood of injected carbon particles within fifteen to twenty minutes. Yet another nonspecific response, inflammation, can be mounted to fight invaders.

The Inflammatory Response

We have all experienced an **inflammatory response** when a minor cut became infected. The site of the wound becomes red, puffy, and tender. What exactly is going on here? If you cut your finger, nearby granulocytes immediately release *histamine*. This molecule increases the permeability of local capillaries and relaxes the sphincter muscles surrounding the capillaries at the wound site. This increases blood flow to the area and causes fluid, serum proteins (serum is the fluid that exudes from clotted blood), antibodies, platelets (Chapter 31), macrophages, and complement proteins (which help burst foreign cells) to leak from damaged blood vessels into the intercellular spaces. Puffiness results. The clotting mechanism (Chapter 31) is activated in order to prevent undue loss of blood plasma or cells through the ruptured capillary walls. Some of the substances released at wound sites act as attractants for macrophages (lymphocytes, too, may be attracted to initiate an immune response). When those cells arrive, they release enzymes that destroy and inactivate many kinds of bacteria introduced into the wound. As the inflammatory response proceeds, a large number of macrophages accumulate and ingest bacteria as well as debris from cells that have been killed around the puncture (Figure 32-4). Eventually *pus* may accumulate; this yellowish sub-

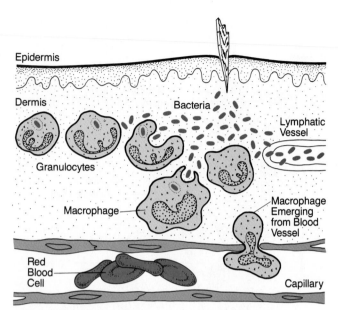

Figure 32-4   INFLAMMATION: THE NONSPECIFIC DEFENSE MECHANISM.
Penetration of the first line of the body's defense, the epidermis of skin, and introduction of a foreign agent (for example, bacteria) elicits inflammation. The inflammatory response involves the migration of granulocytes and macrophages to the site of the wound; they first release histamine, while the latter cells engulf the foreign particles and damaged cellular material. This may be easily accomplished, and healing occurs rapidly; or the result may be a more general bacterial invasion if bacteria at the wound site succeed in invading lymphatic vessels or blood vessels.

stance is an aggregation of macrophages, granulocytes, tissue cells, dead cells, dead or dying bacteria, and extracellular fluid.

In most cases, the foreign agent is destroyed completely, and the inflammation gradually subsides as the materials are degraded and removed via the lymphatic and blood vessels. However, some dangerous foreign cells such as pneumococci cannot be eliminated by the nonspecific defense system, and a specific immune response is called into play. The inflammatory response is the basic defense mechanism in most invertebrates and in chordates. Mollusks and arthropods, for instance, have inflammatory responses and can destroy most of the invading foreign particles. Insects also produce families of antimicrobial, lytic peptides. It is only in the chordates, however, that the highly specific defense system, the immune system, operates.

The Immune Response

Specific defense mechanisms are the exclusive province of the immune system: T and B lymphocytes act in response to specific invading organisms, and NK cells respond to viruses, to transplanted tissues, and perhaps to other foreign agents. Whereas many kinds of bacteria elicit similar nonspecific inflammatory responses, the immune system reacts in a unique way to each species of bacterium or virus. That is why, for example, a child who has the mumps becomes "immune" to subsequent infection by the mumps virus but not to the agents that cause measles or whooping cough. The immunity the child develops is *specific* to the mumps virus. (We will discuss how immunity is acquired later in the chapter.)

The immune response differs from inflammation and other nonspecific defense mechanisms in four ways: *specificity, diversity, memory,* and *tolerance.* We shall describe specificity in the section on antibodies. Here, let us focus on the other three hallmarks of the immune response.

Diversity

The environment surrounding every eukaryote is teeming with bacteria and viruses. These microbes evolve rapidly, with new strains generated frequently. Yet the vertebrate immune system has the ability to defend against longstanding strains as well as those that have never existed before. Despite the enormous diversity of foreign agents that can invade the body, the vertebrate immune system can usually respond to the challenge with a "tailor-made" defense.

Memory

The immune system retains a "memory" of the foreign agents it has reacted to in the past. For example, when a child comes down with a case of chicken pox, it usually takes a week to ten days for the youngster's body to mount a full response to the invading virus. This first reaction is called the *primary response*. Once the child has recovered, no signs of the infection and of the internal immune "battle" remain. However, if the child is exposed, even years later, to this highly contagious disease, the immune response is much stronger and more rapid than it was the first time. This faster, more vigorous reaction to the second exposure is called the *secondary response*. The immune system seems to "remember" the earlier exposure to the same infectious agent. With chicken pox, in fact, this memory response is so strong that a second infection simply cannot get started—the individual has become *immune* to the disease. In contrast, the inflammatory response does not show this more potent secondary response or any hint of memory; each cut or abrasion elicits basically the same sort of nonspecific inflammatory response.

Tolerance

A critical aspect of the immune response depends on tolerance—that is, the immune system's capacity to rec-

ognize the body's own molecules and cells, to distinguish "self" from "nonself." Nonself—foreign cells, tissues, or substances—is reacted against vigorously, whereas self—one's own cells, tissues, and molecules—is generally not reacted against: instead it is *tolerated.* Cells of the immune system may also be able to recognize and attack cancerous cells—"self gone wrong." However, the ability to distinguish self from nonself sometimes goes awry, resulting in attack against the body's own healthy cells (for instance, causing arthritis). The inflammatory response, in contrast, makes no distinction between self and nonself. When granulocytes and macrophages release their destructive enzymes, both invading bacteria and host cells may be damaged.

The immune response to a nonself substance can be one of two types. It may be humoral, involving B cells and antibodies, or it may be cellular, involving T cells. Let us see next how specificity, diversity, memory, and tolerance operate in the humoral immune response.

ANTIBODIES AND HUMORAL IMMUNITY

In 1888, the American researchers E. Roux and D. Yersin observed that laboratory animals with immunity to diphtheria toxin had a substance in their blood that could neutralize the toxin. By transferring this blood substance to a nonimmune animal, they could cause the transfer of immunity itself. Because this substance was found in the blood fluid, immunity conferred by such substances was termed **humoral immunity,** based on the ancient term for body fluids, "humors." The substances that act against foreign bodies such as bacteria are now called **antibodies,** but not until the 1930s were antibodies identified as globular blood proteins.

Biologists now know that antibodies are produced by B lymphocytes in response to the presence of certain foreign entities in the body. Such foreign substances are called **antigens,** and they include bacterial toxins, viral coat proteins, surface proteins on foreign cells, and carbohydrates on the surfaces of bacteria or foreign red blood cells (recall the A, B, and O name tags). An antigen is any substance that elicits an immune response; in humoral immunity, antigens stimulate B cells to produce antibodies (Figure 32-5).

More than fifty years ago, the researcher Karl Landsteiner found that antibodies usually "recognize" small clusters of atoms arranged in a precise shape that is perceived as "foreign" to the body. Recognition occurs because the antibody binds to the cluster of atoms (also called *antigenic determinants*). Such antigenic determinants occur in literally millions of configurations on the

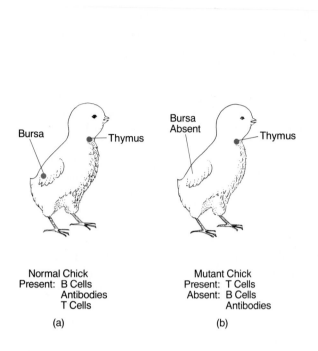

Normal Chick
Present: B Cells
        Antibodies
        T Cells

(a)

Mutant Chick
Present: T Cells
Absent: B Cells
        Antibodies

(b)

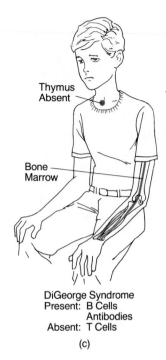

DiGeorge Syndrome
Present: B Cells
        Antibodies
Absent: T Cells

(c)

Figure 32-5   AGENTS OF THE IMMUNE SYSTEM: B LYMPHOCYTES AND THEIR ANTIBODIES AND T LYMPHOCYTES.

(a) A normal chick has a bursa of Fabricius that gives rise to B cells; (b) in the absence of a bursa, the bird produces no B cells, and hence no antibodies are in the blood. The bird's T lymphocytes, however, are normal, as they arise from the intact thymus. (c) An opposite defect in humans, DiGeorge syndrome, involves the absence of a thymus, and the resultant lack of T lymphocytes. The bone marrow (the equivalent of the chick's bursa) is normal, and produces B cells and, in turn, antibodies. These findings illustrate the separate origins of the B and T lymphocyte systems.

surfaces of every virus, bacterium, and protozoan and on the cell surfaces of every plant, fungus, and animal on Earth. Fragments of disrupted cells, cellular secretions, and complex structures such as pollen grains also bear antigenic determinants on their surfaces. Antigens are usually proteins or polysaccharides that include antigenic determinants as part of their structures.

Using synthetic antigens produced in the laboratory, Landsteiner found that animals could make antibodies against such new compounds to which they could never possibly have been exposed. But how could this be? How could antibodies be made against antigens that had probably never existed in nature before that time? Landsteiner discovered another puzzling fact: animals injected with synthetic antigens that differed from each other in minute ways made unique antibodies that bound to (that is, "recognized") these different antigenic determinants. This meant that the different antigens elicited antibodies with slightly different *binding specificities*. In other words, the *binding sites* on the antibody molecules must be highly specific, just as are the active sites on enzyme molecules. But, again, how could this be explained? Because there are millions of possible antigens, a single animal must somehow be capable of making millions of different antibody molecules. What is the nature of the binding site that allows for such great specificity and diversity? And how do the antibodies recognize the antigens? The answer lies in antibody structure.

Antibody Structure: Variable and Constant Regions

There are several classes of antibodies, but the basic structure of an antibody molecule has the shape of the letter Y (Figure 32-6a). Each antibody molecule is composed of four polypeptide chains: two identical **heavy chains** and two identical **light chains.** The four chains are linked by sulfur bridges called *disulfide bonds* (Chapter 3). Each "arm" of the Y consists of one light chain and part of one heavy chain, and the "stem" is formed by the rest of the two heavy chains. The hinge region of the Y allows the two arms to move closer together or farther apart. Each arm region of the Y serves as an antigen binding site.

Suppose we purified two different antibodies (of the IgG type, discussed below) from mouse blood: one directed against tetanus toxin; the other, against diphtheria toxin. If we compared their amino-acid sequences, we would find that one end of the heavy chains is identical in both antibodies, and one end of the light chains is identical in both—these are the **constant regions** of the polypeptide chains and form the "stem" ends of those chains. But the amino-acid sequence at the opposite ends of the light and heavy chains in the arms of the diphtheria antibody would be different from the sequence in the arm end of the tetanus antibody. These

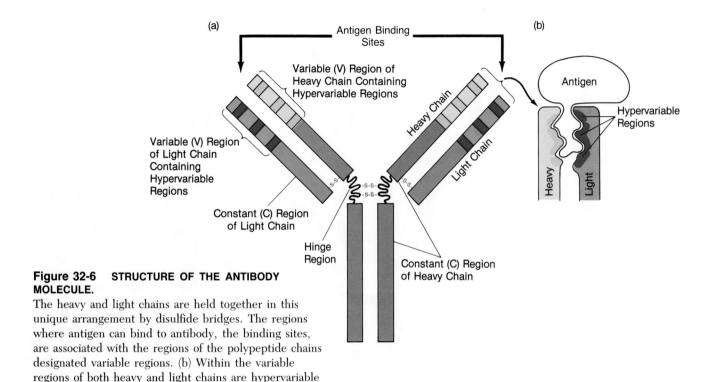

Figure 32-6   STRUCTURE OF THE ANTIBODY MOLECULE.
The heavy and light chains are held together in this unique arrangement by disulfide bridges. The regions where antigen can bind to antibody, the binding sites, are associated with the regions of the polypeptide chains designated variable regions. (b) Within the variable regions of both heavy and light chains are hypervariable regions in which the amino-acid sequences vary from one antibody type to another; as a result, the shape and properties of the binding sites vary, and the specificity for antigens differs accordingly.

are the **variable regions** of the polypeptides. Thus the end of each arm of the Y-shaped antibody molecule is composed of the variable region of a light chain and the variable region of a heavy chain.

The significance of the variable regions is that an antibody binding site for an antigen lies between the variable portions of a heavy and a light chain. A pocket or crevice between the two variable regions has a configuration complimentary to that of the antigenic determinant to which the antibody binds (Figure 32-6b). This antigen–antibody binding is similar to enzyme–substrate binding: there is a "lock-and-key" fit. Different kinds of antibodies have slightly different amino-acid sequences in their variable regions; these sequences account for slight differences in the shape and properties of the binding site, giving different antibodies their specificity for differently shaped antigens. Thus the specificity of an antibody is determined by the amino-acid sequences of its variable regions. Careful analysis reveals, in fact, that there are three **hypervariable regions** situated at intervals in the variable portions of both light and heavy chains. The amino-acid variations in these regions account for the different binding-site specificities of antibodies. Since any one antibody is assembled from two identical light chains and two identical heavy chains, it

follows that both arms of any single antibody molecule are precisely alike and have specificity for the same kind of antigen.

Whereas the variable regions (more precisely, the hypervariable regions) determine an antibody's binding specificity, the constant region determines the *class* of antibody to which the molecule belongs. There are five major classes of antibody proteins, or **immunoglobulins:** IgG, IgM, IgA, IgD, and IgE (Figure 32-7a). Each class of immunoglobulin is capable of carrying out its own specific functions. When B cells participate in the primary response to an antigen, they initially synthesize IgM, each molecule of which looks like a cluster of five Y-shaped antibody molecules joined together by their constant chains and by another protein chain (Figure 32-7b); this leaves the ten antigen binding sites exposed. Because of their large size, IgMs do not pass through undamaged capillary walls and are restricted to the blood plasma. However, the multiple binding sites on IgM permit formation of large aggregations of antigen bound to antibody that can easily be ingested by macrophages. For this reason, only a small amount of IgM need be secreted early in an immune response to serve as an efficient protective agent. Later in the same immune response, B cells synthesize and secrete a large quantity of IgG (each Y-shaped), which has the same binding specificity as the earlier secreted IgM. IgG is much more stable than IgM and can persist for long periods when circulating in the blood plasma.

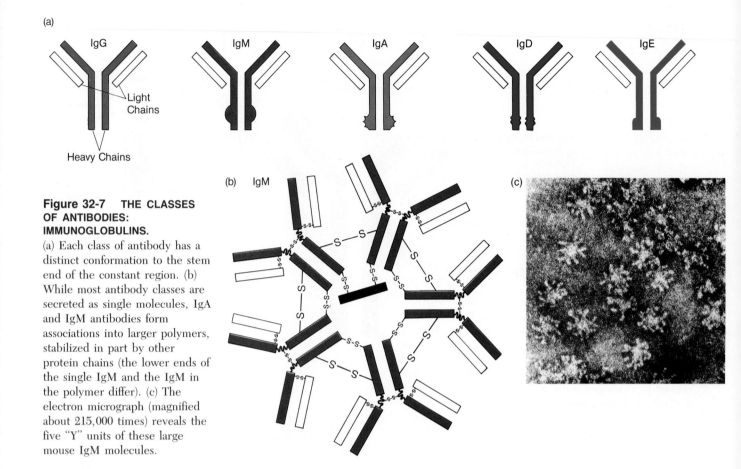

Figure 32-7   THE CLASSES OF ANTIBODIES: IMMUNOGLOBULINS.
(a) Each class of antibody has a distinct conformation to the stem end of the constant region. (b) While most antibody classes are secreted as single molecules, IgA and IgM antibodies form associations into larger polymers, stabilized in part by other protein chains (the lower ends of the single IgM and the IgM in the polymer differ). (c) The electron micrograph (magnified about 215,000 times) reveals the five "Y" units of these large mouse IgM molecules.

IgG and IgM antibodies are the most abundant antibodies in plasma and are directed against the vast unpredictable array of antigens that may be encountered in a lifetime. They help protect animals from invaders by three mechanisms: precipitation, agglutination, and the complement response. Since each antibody molecule can bind to two or more antigen molecules, large aggregates of antigen–antibody complexes may form, like everybody in a room holding hands. *Precipitation* of these clumps of molecules may inactivate the antigen, because it no longer has access to cells. If the antigens are on the surface of invading foreign cells, the invading cells may be joined together by antibody molecules holding one cell with one arm of the Y and another cell with the second arm. This causes the cells to stick together in groups—the process of *agglutination.* Both precipitation and agglutination can stimulate phagocytosis by macrophages to clean up the antigen–antibody complex.

The binding of IgG or IgM antibodies to a cell can also activate the **complement response,** a series of enzymatic reactions that causes the invading cell to burst. Some of the complement proteins bind to the antigen–antibody complex on the surface of the invading cell. A cascade of reactions occurs in which one complement protein enzy-

matically activates another; finally, a protein called *perforin* assembles to form open pores in the invader's plasma membrane. This disrupts the cell's osmotic balance as ions enter and leave freely; as a result the cell usually swells and dies.

Less is known about the other three classes of immunoglobulins. IgA molecules are present in tears, mucous secretions, and saliva, all of which are produced in areas of the body where foreign microbes are likely to enter. IgA reacts with the surface antigens on bacteria and reduces the likelihood of bacterial invasions through the body's natural orifices. IgDs are found on the surfaces of lymphocytes and play a role in their activation. IgEs are bound to mast cells and play an important role in allergic responses (see below).

Studies of immunoglobulins in sharks, bony fishes, and amphibians have yielded some interesting evidence regarding the evolution of the immune system. In these aquatic animals, B cells and IgMs exist, but not the other immunoglobulin classes. In addition, these organisms have a relatively low diversity of IgMs—500,000 different binding sites for antigens compared with perhaps 18 billion in mammals. Clearly, IgM is the oldest immunoglobulin type. Only in the birds and mammals did IgGs, other antibody classes, and the T-cell system evolve.

Development of Antibody Specificity: The Clonal Selection Theory

We have seen that the immune system can make antibodies to combat millions of diverse antigens. In addition, the immune system's memory provides a specific, more vigorous secondary response to a given antigen that occurs even years after the primary response. Furthermore, switches in antibody class occur, as the early predominance of IgM and, later, of IgG during an immune response. What are the cellular and genetic bases for this specificity, memory, diversity, and switching of antigens?

In the 1950s, Sir Macfarlane Burnet, Niels Jerne, and other immunologists proposed the **clonal selection theory** to integrate and explain all these characteristics of the immune response. Perhaps the best way to describe this hypothesis is to follow the life history of a B cell as explained by the theory.

B cells originate from embryonic stem cells that become determined and proliferate in the bursa of birds and in the bone marrow of mammals. During these developmental processes, a stage called the pre-B cell becomes irreversibly committed to making only one heavy-chain variable region. At a later stage, called the **naïve B cell,** only one light-chain variable region can be made, and as a result the particular heavy and light chains can combine to form a specific antibody. This committed naïve B cell and all of its progeny constitute a *clone* of identical cells that can manufacture antibody with only one binding-site specificity (Figure 32-8). The early commitment processes are said to be antigen-independent because the antigen itself is *not* present when the pre-B and naïve B cells become committed; that process is like determination (Chapter 17) in that it involves the choice of genes that are used later—in this case, the genes coding for the particular light- and heavy-chain variable regions.

Figure 32-8   CLONAL SELECTION THEORY.
During the development of the immune system, millions of different committed naïve B cell types are produced. Antigens need not be present for this to occur. Each committed naïve B cell has unique antibody receptors on its surface. In the course of life, the chance appearance of an antigen that binds to a particular naïve B cell causes that cell to divide and differentiate, forming a clone of cells that produce identical antibodies that bind the stimulating antigen. First IgM is made, and later IgG. If a different antigen happened to appear, a different naïve B cell would be "selected," and a different clone would be generated. The "selected" B-cell clone also gives rise to a set of memory cells that can carry out secondary responses.

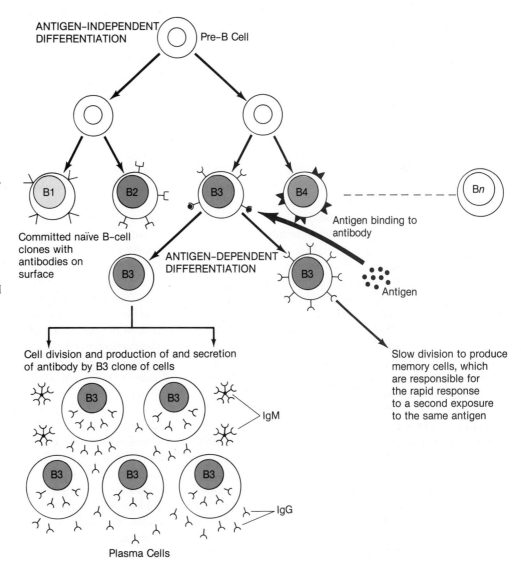

The selection of a set of variable regions by these immature B cells is a random process. As it goes on in the bone marrow or bursa of an embryo, millions of different committed pre-B cells arise, each with a single specificity. Yet all this takes place in the embryo when no exposure to foreign antigens is occurring. The ability of the immune system in the newborn, juvenile, or adult to respond to millions of different antigens is due to the fact that there is a high probability that for a given antigen, there is at least one preexisting clone that produces antibodies with some capacity to bind to it.

The committed pre-B cell divides and gives rise to naïve B cells that have on their surfaces a special form of IgM and IgD, each with precisely the same binding sites. These so-called naïve B cells are commonly found in the lymph nodes of adults and sometimes in the blood as they disperse from bone marrow to the nodes. Since they are not actively secreting antibody they are thought of as being in a resting state. The 100,000 or so surface IgM and IgD molecules on such cells may act as *receptors* for an antigen. Suppose a foreign microbe enters the body and ultimately lodges in a lymph node; its antigenic determinants bind only to naïve B cells whose surface immunoglobulins happen to have specificity for that antigen. This process is called "clonal selection." The particular committed B cell is being "selected" for maturation, but others that do not bind are not. The "selected" naïve B cell divides to generate a clone. Rapidly, a series of divisions occurs in the B cell and its progeny. Then differentiation occurs in the new daughter cells as the cytoplasm increases in volume, much rough endoplasmic reticulum appears, and antibody molecules are synthesized. After about four days, a huge number of large B cells, now called **plasma cells,** have been generated, each actively secreting an immense number of antibody molecules (up to 2,000 per second for the 3- to 4-day life of the cell). Since, as a result of clonal selection, one resting immature B cell gives rise to many plasma cells of precisely the same specificity, this process is termed *clonal expansion.* (Later in this chapter we will discuss another way that naïve B cells can become activated by T cells.)

When the resting B cell is stimulated to divide by binding an antigen, it actually gives rise to two types of cells: a huge number of actively secreting plasma cells and a smaller number of **memory cells.** Memory cells, which look like resting naïve B cells, do not secrete antibody. Rather, their function is to circulate in the blood or reside in the lymph nodes, ready for future exposure to the antigen for which they are specific. The existence of these cells accounts for the speed and effectiveness of the secondary immune response. Whereas the primary response involves only a few naïve B cells giving rise to clones that secrete primarily IgM, the secondary response involves stimulation of existing memory B cells

that quickly divide to become plasma cells, which then rapidly produce a large amount of antibody, primarily IgG molecules.

Antibody Diversity: A Matter of Gene Shuffling

An intriguing paradox surrounds the question of antibody diversity. As we have seen, an individual bird or mammal can have literally millions of different kinds of antibodies, each with a uniquely shaped binding site based on the variable sequences of amino acids in the arm end of the light and heavy chains of the molecule. If each variable region were coded for by its own gene, an organism would need millions of separate genes to make antibodies. Yet the total genetic complement of most mammals is only about 1 million genes, only a fraction of which code for immunoglobulins—clearly not enough to account for the stunning diversity of amino-acid sequences displayed by antibody binding sites. In recent years, geneticists have found the explanation for this paradox: a unique process of shuffling takes place among segments of certain genes in the nuclei of the early stages of B cells. This process accounts for the immense diversity that is the key to the body's defense against a host of invaders.

First, researchers wondered just how diverse antibodies and their genes actually are. Data from studies of mouse and human immunoglobulin genes suggest that there are about 7,500 possible amino-acid sequences for the light-chain variable regions. And there may be about 2.4 million for the heavy chain. Because a given cell expresses one light-chain variable region and one heavy-chain variable region, the total number of different binding sites that can be formed is 7,500 × 2.4 million, or 18 billion possible combinations. No one is sure whether that many actually form or whether certain sequences (and hence binding-site shapes) are favored. Nevertheless, any genetic explanation would have to account for at least the possibility of 18 billion combinations. In fact, the current theories do just that. Let us see how.

Recall from Chapter 14 that the gene sequences that code for most eukaryotic polypeptides consist of exons (expressed regions) separated by introns (intervening sequences) of DNA. The genes coding for immunoglobulin chains also occur in pieces; however, the many sequences that code for the variable regions are separated from the sequences that code for the constant regions by long stretches of DNA. When the pre-B and naïve B cell stages become committed to making a single type of antibody, DNA rearrangements occur: one of the segments coding for the heavy-chain variable region moves next to a sequence coding for the heavy-chain constant region. Likewise, the sequence coding for the light-chain variable region moves next to a sequence coding for the light-chain constant region (Figure 32-9). These two rearrangements of DNA give rise to one complete light-chain gene and one complete heavy-chain gene. Extremely specific nuclear enzymes, endonucleases, actually cleave DNA at precise spots to allow the recombination process to occur.

A key to antibody diversity is that the process of moving any one of the large number of variable sequences next to a single constant sequence is *random*. This occurs for both the light and the heavy chains. Thus a great variety of constant–variable combinations are made, each in a different pre- or naïve B cell. Once a certain variable region has moved next to a constant region for the light chain, and the same process has occurred for the heavy chain, the cell is committed to expressing that pair of constant and variable genes. Thus the cell becomes committed to making one particularly shaped binding site.

It turns out that the gene sequences that code for the variable region also occur in several pieces. The sequences that give rise to the light-chain variable region occur in two pieces: the V (for "variable") segment and the J (for "joining") segment. For the heavy chain, the complete variable region is assembled from three different gene segments: the V, the J, and the D (for "diversity"). There are multiple copies of each of these segments, each coding for a slightly different amino-acid sequence. For example, the human chromosome that

carries the heavy-chain genes has about 80 V segments clustered together, at least 50 D segments clustered together, and 6 J segments in a group. When the variable-region sequences are moved next to a constant-region sequence, 1 V, 1 D, and 1 J segment are randomly selected, combined, and moved as a group. Based on this random association of different segments, a total of $80 \times 50 \times 6$, or 24,000, different heavy-chain variable regions can be formed. In addition, when these three segments are joined, the joining can occur in about 100 slightly different ways, yielding even more total variations ($24,000 \times 100 = 2.4$ million). Similarly, the different kinds of V and J segments in light-chain sequences yield at least 750 light-chain possibilities. The fact that there are about 10 different ways to join the V and J segments greatly increases—to 7,500—the number of possible variable regions that can be formed.

To put all this together, there are at least five factors involved in generating antibody diversity in mammals:

1. Many V, D, and J segments are encoded in the genome and are passed from generation to generation by the germ cells.

2. When a B cell passes through the pre- and naïve stages, it becomes committed, and the various segments for both heavy (V, D, J) and light (V, J) chains are randomly combined.

3. The joining of the segments can occur in slightly different ways to give rise to additional variation.

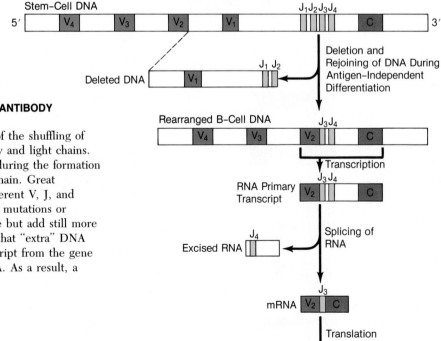

Figure 32-9   GENE SHUFFLING AND ANTIBODY DIVERSITY.
Antibody diversity is largely the result of the shuffling of gene sequences that code for both heavy and light chains. This figure shows the shuffling process during the formation of a gene for an immunoglobulin light chain. Great variability is achieved by combining different V, J, and constant regions; processes such as local mutations or variable joining sites are not shown here but add still more diversity in final gene structures. Note that "extra" DNA remains so that the RNA primary transcript from the gene must be spliced to yield the final mRNA. As a result, a piece of RNA is excised.

IMMUNE SYSTEM VERSATILITY: TWO PROTEINS FROM ONE GENE AND ANTIBODY CLASS SWITCHING

The remarkable way gene segments are rearranged in early B and T lymphocytes is just part of the amazing story of the immune system's dynamic genetic information. Two other processes are equally surprising.

Recall that during an immune response, a "selected" clone of naïve B cells first synthesizes IgM antibodies. Let us look at that process more closely. Naïve B cells have a unitary IgM antibody on their cell surface, but not the huge, normal molecule con-

Figure A   **TWO PROTEINS FROM ONE GENE.**
The single primary transcript for IgM can be spliced in either of the two ways shown here. If the M (membrane) sequence is moved next to C, mRNA for membrane-bound IgM is made. If the S (secretion) sequence remains next to C, then mRNA for secreted IgM is made.

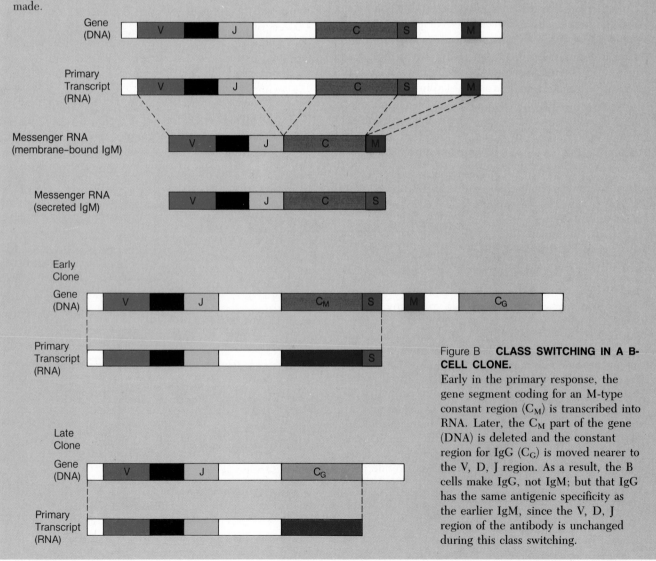

Figure B   **CLASS SWITCHING IN A B-CELL CLONE.**
Early in the primary response, the gene segment coding for an M-type constant region (C_M) is transcribed into RNA. Later, the C_M part of the gene (DNA) is deleted and the constant region for IgG (C_G) is moved nearer to the V, D, J region. As a result, the B cells make IgG, not IgM; but that IgG has the same antigenic specificity as the earlier IgM, since the V, D, J region of the antibody is unchanged during this class switching.

structed of five joined molecules. Instead, the primary transcript of RNA transcribed from the IgM gene is spliced to form an mRNA that codes for a hydrophobic amino-acid sequence (M) on the stem end of the IgM constant chain (Figure A). The resultant hydrophobic tail inserts the small IgM in the cell's plasma membrane, causing it to become membrane-bound.

When that same B-cell lineage is activated by an antigen to begin producing and secreting antibody, a novel thing happens. The identical primary RNA is spliced differently, so that the secreted IgM protein molecules have a different tail(s). They assemble into groups of five to form the large IgM complex. So here is a case of *two proteins from one gene*. Alternative splicing yields the membrane-bound small IgM or the secreted large IgM.

An equally novel event occurs later in the history of the same active B-cell clone. After the primary response is mounted and IgM is synthesized, B cells undergo a rearrangement of gene segments. The segment of DNA coding for the M type constant region is removed, so that the original V, D, J variable region is joined instead to the G type constant region. From then on, the B cells can only make IgG; the B cell has the identical combining site specificity (due to the V, D, J region), but it is now a different class, IgG, that is secreted as the smaller antibody which predominates in blood plasma (Figure B).

Gene splicing, the production of two proteins from one gene, and the rearranging of genes during the history of a clone are additional surprises in the behavior of the immune system. Still more versatility will no doubt be revealed in the body's remarkable guardian.

4. The binding sites are formed by random combinations of different heavy and light chains.

5. Finally, a process of actual somatic mutation appears to go on in the portions of genes that code for the hypervariable regions of heavy and light chains. Thus new DNA base sequences not inherited from either parent suddenly appear in the variable-region DNA during the origin of pre-B cells, as well as later during clone development.

These incredible gene-shuffling processes mean that only a modest proportion of the individual's genome need be devoted to coding for immunoglobulin genes, and yet the organism can be protected by an army of different antibody-producing clones, each clone able to produce its specific antibody aimed at invading antigens. Such gene shuffling has been compared to filing slightly different notches in billions of keys so that each one opens a different lock. In this case, the matching of key and lock allows for the defense of the organism against a diversity of foreign intruders. Not all vertebrates enjoy the same degree of antibody diversity as mammals. In sharks, multiple sets of V, D, J, and C segments are present in the zygote's genome; the main source of antibody diversity comes from slight variations arising when DNA between the V and D, the D and J, and the J and C is spliced out as immune-system cells arise. Even so, the shark is well protected by its restricted arsenal of antibodies.

Monoclonal Antibodies: A Miracle for Medicine

Understanding of B-cell clones and antibody production has led to the recent development of a procedure that promises to be as important to human medicine as the discovery of penicillin: the production of **monoclonal antibodies**—that is, highly specific antibodies produced by a single clone of B cells.

Normally, even when a single antigen enters the body, it is recognizable by a number of naïve B cells, determined to form antibodies with slightly different binding sites. One reason is that most macromolecular antigens may have a number of antigenic determinants on their surface. As a result, not one but a whole set of B-cell clones is selected, and a mixture of antibody molecules is produced. Each of these antibodies binds with slightly different affinity to the antigen molecules (just as enzymes may have different binding affinities for related types of substrates). In 1977, two researchers at Cambridge University, Georges Kohler and Cesar Milstein, discovered a way to obtain large amounts of just a single type of antibody by taking advantage of the unusual characteristics of two types of cells.

ANTIBODIES TO ORDER: MODERN MAGIC BULLETS

In 1901, the German physician Paul Ehrlich began an important search with a remark that has since become famous: "We must learn," he said, "to shoot microbes with magic bullets." He and his co-workers screened 606 different arsenic compounds, looking for one that could kill the spirochete that causes syphilis but, at the same time, leave unharmed the patient who swallows the compound. After eight years, they found such a "magic bullet," a substance they called salvarsan. With it, they saved many lives and, in essence, launched the field of antibiotic research. Today, physicians and researchers have a new type of magic bullet with a far greater range of effects, infinitely more specificity of action, and the potential to study, diagnose, and treat a number of serious diseases. This magic bullet is the monoclonal antibody.

The immunologists Cesar Milstein and Georges Kohler were able to successfully fuse spleen cells and myeloma cells to yield the hybridoma—a totally new cell type with the immortality of a cancer cell and the capacity to produce antibodies of a single type (Figure A). These monoclonal antibodies have been likened to guided missiles because they can hit specific biological targets. It is this trait that gives them their broad applicability in medicine. Milstein and Kohler first reported hybridomas in 1975. By 1984, when Milstein and Kohler won the Nobel Prize along with Niels Jerne, monoclonal antibodies were being used in laboratories and hospitals all over the world.

One obvious application has been medical research. Monoclonal antibodies can be produced, tagged with radioisotopes or fluorescent dyes, and used to identify specific biological molecules and structures, such as hormones and hormone receptor sites on cell membranes. Researchers have used monoclonal antibodies in the basic study of diabetes, cancer, heart disease, rheumatoid arthritis, allergies, and the function of the brain, among other lines of research.

Diagnosis of diseases and conditions has been another significant application of monoclonal antibodies.

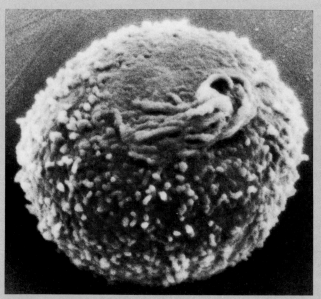

Figure A   **THE HYBRIDOMA CELL THAT RESULTS FROM THE FUSION OF SPLEEN B LYMPHOCYTE AND A CANCEROUS MYELOMA CELL.**
This type of cell (here magnified about 8,000 times) divides and produces a clone of cells, all secreting a single kind of antibody.

For example, several commercial pregnancy tests for use in hospital laboratories and at home are now based on the ability of targeted monoclonals to detect the presence of human chorionic gonadotropin. The tests are from three to twenty times more sensitive than conventional tests, and thus pregnancy can be positively determined as early as ten days after conception.

Monoclonal antibodies have also been prepared against a large number of viruses (including the AIDS-causing virus), bacteria, fungi, and parasites to aid in the diagnosis of infectious diseases. Various types of sexually transmitted diseases, including gonorrhea, herpes, and chlamydia, once required three

First, they injected a mouse with an antigen; next, they removed the mouse's spleen and obtained B cells from it, some of which were directed against the antigen that they had injected. Then they fused individual spleen B cells with a special kind of "immortal" cell, called a *myeloma*—a cancer cell that divides very rap-

idly. Each resultant fused cell, called a *hybridoma*, by Kohler and Milstein, contained both the spleen B-cell and myeloma chromosomes (Figure 32-10). Significantly, a hybridoma expresses the characteristics of both cell types. Thus each hybridoma clone can synthesize a huge amount of a single IgG antibody directed against a

to six days for positive identification, and the conventional microbial techniques sometimes showed false positives or negatives. With monoclonal techniques, laboratory technicians can determine if there is infection with a high degree of accuracy (85 to 99 percent) in only fifteen to twenty minutes. Thus appropriate treatment can begin almost immediately.

The diagnosis of a host of other conditions, including anemia, pituitary insufficiency, infertility, and prostate and liver cancers, is also based on monoclonal antibodies targeted to specific compounds and capable of detecting their levels in the body. A variation on this targeting is the discovery of *abzymes*, antibodies whose binding site behaves enzymatically and catalyzes a specific reaction involving the bound antigen. This may open the door to designing abzymes with any desired specificity.

Finally, monoclonals have been used effectively to treat diseases. Physicians have employed the targeted antibodies to help counter the rejection of transplanted organs, to rectify immune deficiency diseases, and to fight cancers. By targeting antibodies to bind to marauding T cells, physicians can "mark" these T cells for destruction by the immune system, so that they cannot cause a life-saving organ transplant to be rejected. Doctors have treated children with congenital immune deficiency by transplanting bone marrow from a donor and injecting monoclonals that destroy T cells directly. This procedure prevents the T cells in the donor's marrow from attacking the patient and allows time for the marrow to produce new T cells more compatible with the host's tissues. The treatment provides a real, functioning immune system in a child born without one. And doctors have also treated a few cancer patients experimentally by linking monoclonal antibodies to radioactive compounds or poisons such as ricin, a deadly extract from castor beans. The antibody can deliver the lethal materials directly to cancer cells and leave healthy tissues unharmed—true magic bullets, far more effective than Ehrlich could ever have imagined.

specific antigen and can divide again and again to yield a clone of millions of tiny, living factories, all producing this same antibody in huge quantity. Techniques are then used to identify the particular hybridoma clone that makes the desired antibody directed against the injected antigen (the other clones are discarded).

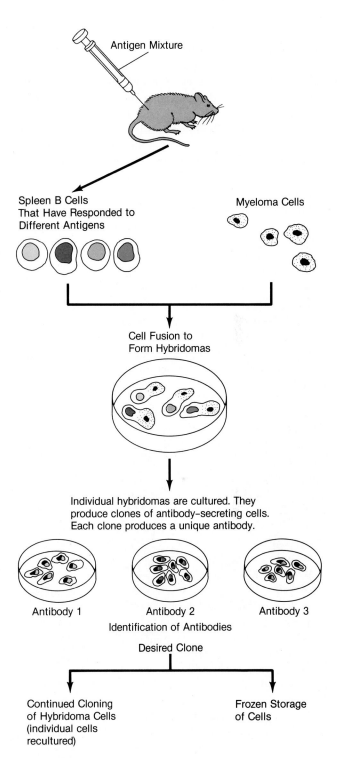

Figure 32-10   THE PRODUCTION OF MONOCLONAL ANTIBODIES.
Antigen-stimulated spleen cells are fused with myeloma cells, yielding "immortal" hybridomas. Each of them secretes a single, "monoclonal" antibody. Once the hybridoma secreting the desired antigen is identified, it is cloned to generate millions of antibody-secreting cells that yield the huge quantity of a single antibody needed in medicine or science. Some hybridoma cells may be stored frozen and later cloned for antibody production.

The product of these hybridoma factories, the monoclonal antibody, is a pure reagent directed against a specific target. Such antibodies have been prepared against rabies and influenza viruses; against the group-B streptococci, which are dangerous to newborn babies; and against antigens on the surfaces of tumor cells, making the monoclonal antibodies exquisitely sensitive tools for diagnosing cancer. Finally, some monoclonal antibodies have been prepared and bound to cell surfaces so that when a scientist adds the complement proteins, the cells are killed.

The medical applications of monoclonal antibodies grow by the month. We can predict with confidence that hitherto unobtainable diagnostic sensitivity and accuracy will result from Kohler and Milstein's procedure. And therapeutic monoclonal antibodies, used as targeted drugs to seek out specific cells or substances, will improve the treatment of many diseases and reduce side effects. There are few instances in which basic research involving a seemingly bizarre and esoteric experiment—in this case, the fusion of dissimilar types of cells—has had such an immediate impact on human medicine and health.

T CELLS AND CELL-MEDIATED IMMUNITY

The other major type of immune response besides humoral immunity is **cell-mediated immunity**—the direct attack of foreign cells or substances by T lymphocytes. T-cell cytoplasm contains free ribosomal aggregates but very little endoplasmic reticulum because, unlike B cells, T cells do not manufacture and secrete a large amount of antibody protein. Instead, they have antibodylike receptors on their surfaces that enable them to recognize and respond to specific antigens.

Immunologists divide T lymphocytes into three main types: *killer T cells,* which attack foreign cells and substances directly; *helper T cells,* which assist certain B cells in producing antibodies and killer T cells in mounting an attack; and *suppressor T cells,* which seem to play a role in control of the immune system and in tolerance (Figure 32-11). Each class of T cell has distinct surface name tags by which it can be identified. All three classes arise from stem cells that have been acted on by the thymus and its hormones.

It is probable that T-cell maturation is roughly like that of B cells. Thus committed immature T cells are present and available for a clonal selection process similar to that of B cells. Once the immature T cell has recognized the foreign antigen to which it can respond, a period of mitosis yields a large clone of T cells. Those cells with killer properties accumulate at the site of an infection, where they attack the foreign cell or material bearing the antigen. Thus T cells attack some bacteria, virus-infected cells, tissue transplants, parasites, and cancer cells. The actual attack by T cells may involve activating the complement response; but T cells also synthesize perforins and insert those pore-forming proteins in the foreign cell's plasma membrane so that it dies.

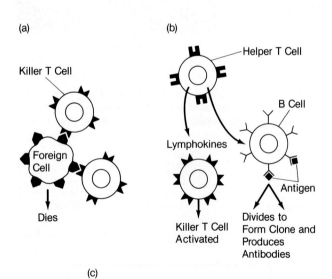

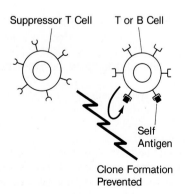

Figure 32-11   TYPES OF T CELLS.
(a) Killer T cells are activated by macrophages and, after forming a clone, directly attack a foreign antigen. (b) Helper T cells augment the function of B and T cells by secreting lymphokines that enable B cells to carry out a full set of responses to antigens and killer T cells to attack foreign antigens. (c) Suppressor T cells apparently slow and halt the production of B or T cells once an immune response successfully eliminates foreign antigens. Suppressor T cells also may help maintain tolerance of self antigens. Thus, as shown here, they may block formation of active T or B cell clones. If this inhibition ceases, autoimmune disease may arise.

Macrophages must participate in the T-cell attack, first by playing a role in bringing the antigen and immature T cell together. Specifically, the foreign antigen is held on the surface of the macrophage by another molecular marker. A helper T cell must recognize both the foreign antigen *and* the macrophage marker molecule (a histocompatibility marker, described below); the T cell is activated as a result and also because the macrophage secretes a **lymphokine,** a lymphocyte-activating protein. As a result, the helper T cell secretes other lymphokines that activate killer T cells and possibly naïve B cells with the identical antigenic specificity. These lymphokines are in a sense amplifying signals that mobilize more and more killer T cells or B cells to respond to the foreign invader. The essential nature of lymphokines is clear from the fact that some naïve B cells may require a lymphokine molecular messenger—in addition to contact with the foreign antigenic substance itself—in order to become activated to divide and to produce a clone of mature B cells, which soon begin to secrete antibody molecules. Thus in order for a given kind of antibody to be made against an antigen, both a helper T cell and a B cell must recognize that antigen as foreign.

Once a foreign substance has been eliminated, there is no longer a need for the immune response. It is thought that suppressor T cells may play a role in turning off immune responses or at least in limiting their magnitude and duration. Another, more widely investigated role of suppressor T cells is the active prevention of immune responses against self antigens. Thus suppressor T cells may be involved in the tolerance process.

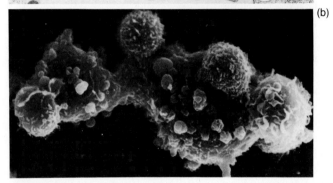

Figure 32-12   NATURAL KILLER (NK) CELLS.
(a) Structure of a human NK cell (magnified about 10,000 times), showing the large nucleus to the left, and the granules of varying size to the top and right. (b) Two large target cells with a number of smaller NK cells (magnified about 2,800 times) bound to their surface (as at top). The small bulbous objects on the surface of the target cells are indicators of their imminent death.

NATURAL KILLERS: THE THIRD CLASS OF LYMPHOCYTES

In the early 1980s, a third class of lymphocytes was discovered: natural killers (NK) (Figure 32-12). These large, granular lymphocytes, which arise in bone marrow, appear to lack binding sites for antigens on their surfaces. However, extensive lysosomal qualities are associated with their granules. NK cells, which make up about 10 percent of the body's lymphocytes, can rapidly attach to and lyse tumor cells while leaving normal cells unharmed. They apparently use perforins to build pores in tumor cell membranes, thereby causing death. NK cells are also found in the epithelial layers of the lungs, cervix, intestine, and male reproductive tract. In these locations, NK cells may participate in surveillance for foreign agents and can directly attack viruses and virus-infected cells.

Experiments with a strain of mice that lacks NK cells have contributed to our knowledge of how these cells function. If tumor cells are injected into such mice, tumors quickly develop. However, if NK cells from normal mice are injected *before* the tumor cells, no tumors appear in most cases. If the NK-deficient mice are exposed to radiation, they frequently develop leukemia; but if NK cells are injected into radiated mice, most of the animals will not develop that cancerous disease. Finally, NK cells seem to be involved in graft rejection in some unknown way.

Immunologists are not sure how NK cells can distinguish potentially cancerous cells or how they relate to T and B cells. Nevertheless, intensive research is showing that NK cells are certainly a line of defense against cancer and probably take part in many other immune responses.

AUTOIMMUNE DISORDERS: A BREAKDOWN IN TOLERANCE

Under certain circumstances, the immune system can become stimulated to attack the body's own cells in a process called an **autoimmune response**; this reaction occurs only when the normal phenomenon of tolerance breaks down. The result can be devastating diseases, such as arthritis and lupus erythematosus.

Tolerance, the recognition of one's own tissues as self rather than as nonself, begins during development (Figure 32-13). If cells from one adult mouse are injected into another adult mouse, the transplanted cells are usually rejected and killed. But if the recipient is a fetal mouse, the transplanted cells are not rejected. Imagine an experiment in which cells from a single donor are injected into a mouse fetus. Then, many months later, the identical kind of foreign cell is transplanted into the adult that grew from the fetus. In this case, the adult recipient does not reject the transplant. The reason is that tolerance to that precise type of foreign cell was

Figure 32-13   TOLERANCE: RECOGNITION OF SELF.
Tolerance appears to be an acquired characteristic of the immune system due to the presence of antigen(s) during the immune system's development. This adult white mouse received an injection of blood cells from a brown mouse at the time of its birth (or during fetal life). Now, as an adult, it accepts a graft of skin from the same genetic strain of brown mouse and fails to attack that skin. The mouse's immune system was rendered tolerant to tissues from the brown strain because the brown mouse antigens were present as its immune system matured.

induced in the embryo as a result of the first experimental injection. Research suggests that during late embryonic development, the immune system takes an inventory of normal antigens already present in the body; all such antigens are thereafter defined as self. If transplanted foreign cells bearing their unique antigens on their surface are present at the time of the tolerance inventory, then they, too, will be considered self forever after.

Tolerance appears to be based on a failure of self antigens to stimulate development of clones of T or B cells. Several mechanisms are hypothesized to be responsible for this phenomenon. One is *clonal deletion*, in which clones of maturing B and T cells with reactivity against the millions of self antigens may be selectively destroyed or inactivated during embryonic development. Another mechanism that may lead to tolerance may simply be the physical masking of certain antigens on the surface of some normal cells so that those antigens are not exposed to the immune system when it is developing. Thus such masked antigens are not apparently included in the inventory of self. For example, tissues inside the eye and cells inside the seminiferous tubules of the testes are apparently not exposed to cells of the developing immune system. If an adult suffers a penetrating wound of the eye or testis, within a week or two, a massive immune attack takes place on internal eye or seminiferous-tubule cells. As a result, the eye or testis itself can be destroyed. It is as though those masked cells were never inventoried during embryonic development and are mistakenly recognized as nonself by immune cells that see them for the first time during adult life.

Suppressor T cells may bring about tolerance in yet another way. If one takes a mouse that is tolerant to foreign antigen A, removes some of the mouse's T cells, and injects them into another mouse, the second mouse immediately becomes tolerant to antigen A. It will not be able to mount an immune attack on antigen A, even though it should be able to. That tolerance is transferred by the injected suppressor T cells is shown by an experiment in which the suppressor T cells are killed prior to injection—other types of T cells cannot transfer tolerance to antigen A or other antigens.

When the elaborate system of tolerance breaks down, a number of grave medical problems can result. Either T or B cells may suddenly begin to attack normal cells or cell products and may produce a life-threatening situation. Victims of the disease myasthenia gravis suffer paralysis and sometimes die because their B cells generate antibodies that attack an enzyme essential to the functioning of the nerve–muscle junction. Another disease, lupus erythematosus, strikes women in their late teens and can cause swelling and inflammation of the joints and connective tissue, similar to that in rheumatoid arthritis. These autoimmune diseases can also involve attack on internal organs—kidneys, lungs, and heart—leading to death.

The most common autoimmune disease is arthritis, a word that means "joint inflammation" and is used to describe more than 100 separate conditions. Arthritis in its many forms is the number-one crippling disease in the United States, affecting more than 30 million people. Rheumatoid arthritis is a severe inflammation of connective tissue that can affect the joints, heart, lungs, blood vessels, and spleen. Numerous studies are currently under way to determine how the immune system causes this and other types of arthritis and to improve treatment of the diseases.

Another autoimmune disorder, rheumatic heart disease, results from an unfortunate coincidence. B cells mistake an antigen normally present on the surface of all heart-valve cells for an antigen on streptococcus bacterial cells. During a bout of strep throat, usually in childhood, B cells may overcome the bacterial infection, but later proceed to attack the heart-valve cells, causing permanent damage.

Although autoimmune diseases are under intense investigation, it is not yet clear in most cases what triggers the attack by T or B cells. Currently, treatment approaches include immunosuppressant drugs and, in severe cases, blood filtering to remove lymphocytes.

IMMUNE DEFICIENCY

The importance of the immune system is demonstrated most dramatically by individuals born without certain components of the system and by people who lose their immune response. Studies of such **immune deficiency**, in fact, helped reveal the roles of various lymphocytes.

An early indication that the T- and B-cell systems are independent parts of the immune system came from studies of mammals and birds that lack immunoglobulins (antibodies) in their blood. It was discovered that these animals do not have B cells. The same condition is found in birds in which the bursa of Fabricius fails to develop (Figure 32-5). Such results linked bursa, B cells, and antibodies together. But these same animals lacking B cells and antibodies can promptly reject a foreign-skin graft. Their T-cell system is normal. This finding helped lead to the discovery and elucidation of the T-cell system. The roles of both T- and B-cell systems were later confirmed in laboratory animals and in humans with precisely the opposite immune condition—circulating antibodies (and presence of B cells) but no ability to reject foreign-tissue grafts or carry out T-cell–based reactions. These individuals had a deficient thymus and so lacked the T-cell system.

A few individuals have even been born with no immune response at all. Most victims die from infections shortly after birth. However, one baby born in Texas in 1971 was quickly diagnosed and confined to a germ-free chamber, in which he survived for more than a decade to become the longest-living person lacking both T- and B-cell systems. He died on leaving the chamber.

In 1981, a startling new immune deficiency disease was detected: *acquired immune deficiency syndrome*, or *AIDS*. Most of the early cases in the United States involved young homosexual men, users of intravenous drugs and their sexual partners, hemophiliacs, and persons from regions of Africa where the disease originated. Today, nearly the whole population is at risk. AIDS is the result of a viral infection. A high proportion of the victims have human T lymphotropic virus in their bodies. The virus apparently destroys helper T cells so that they cannot help killer T cells or B cells to mount attacks against foreign antigens. Although killer and suppressor T cells and B cells are present in AIDS victims, the deficit in helpers prevents their normal function. Some understanding of AIDS comes from experiments on cats infected with feline leukemia virus, similar to human T lymphotropic virus. Such cats are killed by diseases that result from the suppression of T cells more than by the leukemia virus itself. Because these cats and human AIDS victims show vastly reduced B- and T-cell responsiveness, their bodies are exceedingly susceptible to the host of pathogens—bacteria, protozoans, molds, yeasts, viruses, toxins—encountered in everyday life. Human victims sometimes suffer numerous localized and systemic infections and often succumb to pneumonia or to Kaposi's sarcoma, a rare cancer of the blood vessels. The mortality rate has been more than 70 percent of those afflicted with AIDS. An intense research effort is under way to learn more about the cause and treatment of what is the most dangerous disease since the black plague. AIDS, more than any disease, emphasizes the critical role of a normal immune system in day-to-day survival.

IMMUNE PHENOMENA AND HUMAN MEDICINE

The breakdown in natural tolerance that leads to autoimmune disorders is a relatively rare event. The marvelous immune system usually operates to protect rather than attack the body and, in fact, can be stimulated in various ways to provide still broader defense. Let us explore a few natural and induced effects of the immune response on human health.

Active Immunity: Protection Based on Past Exposures

Not many years ago, most children in the United States came down with the so-called childhood diseases, including mumps, measles, chicken pox, and rubella (German measles). Most adults who had those diseases are immune to catching them again. Why is that? The answer is that a reservoir of memory cells specific to the pathogens that cause the particular diseases persists throughout life, able to mount a quick defense (secondary response) should the pathogens return. This type of disease immunity—based on prior exposure to pathogens—is called **active immunity** (Figure 32-14).

Active immunity can be achieved without actually contracting a particular disease. Vaccinations and booster shots for the childhood diseases listed above and for polio, tetanus, diphtheria, whooping cough, and other communicable diseases are now routinely administered in many societies. Vaccines contain dead or attenuated (harmless) forms of the disease-causing virus or bacterium, which, in turn, stimulate the generation of memory cells without actually causing the disease.

The concept of vaccination dates to the late eighteenth century and the English physician Edward Jenner. Jenner noticed that some milkmaids had developed scars on their hands and arms after milking cows infected with cowpox—scars that resembled those of the often deadly human disease smallpox. Following a hunch, Jenner gathered some pus from a cowpox sore on a milkmaid's hand, dipped a needle in it, and made a scratch on a young boy's skin. Several months later, Jenner gave the boy a dose of smallpox microbes, but no disease developed. (Such a direct experiment would never be allowed today!) The boy had acquired immunity as a result of the cowpox vaccination (from the Latin word *vaca*, meaning "cow"). Jenner went on to vaccinate thousands of Londoners, protecting virtually all of them from smallpox. Jenner's procedure worked because cowpox and smallpox are caused by viruses so similar to each other that the human immune system cannot distinguish between them; when smallpox virus invades the body, the cowpox-induced memory cells generate clones of B cells that attack the smallpox virus.

Although vaccination ushered in a new era of preventive medicine and contributed to the lengthening of human life expectancy, vaccines are not without risk. There have been instances in which vaccines were prepared with viruses that were not completely inactivated, and, as a result, recipients have developed polio or other serious diseases. Working with pathogenic viruses can also be dangerous for the people who prepare vaccines. A solution to these problems is currently under development: synthetic vaccines. Using new techniques, scientists construct short peptides that mimic a small region of a virus's protein coat. These short peptides serve as anti-

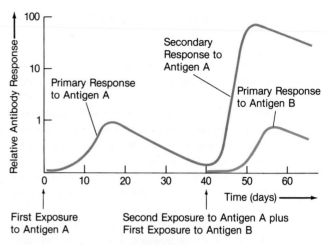

Figure 32-14   ACTIVE IMMUNITY:
The purple curve shows the primary and secondary responses induced by the first and second exposures to antigen A. The very rapid and more extensive secondary response results from the presence of a population of memory cells generated during the first encounter with antigen A. This phenomenon is the principle behind active immunity. Note that exposure to antigen B, or to other antigens, is not affected by the history of exposure to antigen A.

genic determinants and can stimulate antibody production and confer active immunity, thereby eliminating the need to inject whole virus particles. Guinea pigs injected with a synthetic peptide nineteen amino acids long that corresponds to a region of the antigenic determinant on the virus that causes foot-and-mouth disease developed immunity to even huge, lethal doses of the active virus. A combination of three peptides that mimic parts of malarial parasites has elicited complete protection of injected monkeys; this may be the basis for a vaccine against malaria that could save millions of human lives. Finally, synthetic peptides that mimic diphtheria toxin, hepatitis virus, and other human pathogens have been tested as well, and promise eventually to add far greater safety to the active immunization of humans and domesticated animals.

Passive Immunity: Transfer of Antibodies

In some mammals, certain antibodies can cross the placenta from the maternal blood into the fetal blood and so confer immunity to the fetus and newborn. This is an example of **passive immunity,** in which antibodies are not stimulated within an individual but are provided indirectly. Other antibodies are passed to the newborn in colostrum (the first milk) and can protect the nursing infant until its own immune system begins to operate more fully (Chapter 18).

A third form of passive immunity results from injecting a patient with *gammaglobulin,* a mixture of IgG molecules directed against diverse antigens. Some of these IgGs have binding sites specifically directed against potential disease-causing agents and can protect the patient even though they are not produced by his or her own immune system. For example, if a child is bitten by a dog, a physician will inject gammaglobulin into the child's body to produce passive immunity to tetanus bacteria that may have entered the puncture wound. IgGs specific to the tetanus organisms, present in the mixed gammaglobulin, will then immediately attack those dangerous cells. A doctor may also inject killed tetanus bacilli (to produce active immunity) if the child has not had tetanus shots, in order to provide long-term protection. If the dog has not been vaccinated against rabies, the child must also receive treatment to combat the rabies virus.

Passive immunity can also be provided against the bite of certain venomous snakes by injecting *antivenins* (once called antivenoms) prepared in a special way. First, a horse is inoculated with a sublethal dose of snake venom, and the animal generates a massive amount of IgG specifically directed against the particular snake-venom antigens. The IgG is collected from the horse's blood and preserved, ready to inject immediately into the victim of a snake bite. The antibodies in the antivenin can go to work immediately to neutralize the snake toxin; otherwise, several days would elapse before the person's body generated its own immune response—and the toxin could have killed the victim in the meantime.

Allergy: An Immune Overreaction

Sensitivity to dust, cat dander, ragweed and other pollens, strawberries, rhubarb, mussels, and countless other "foreign" agents results from the action of the class E immunoglobulins. IgE molecules are bound to the surface of *mast cells,* which are a type of granulocyte abundant in connective tissue, especially in the skin, lymph nodes, intestines, lungs, and membranes of the eyes, nose, and mouth. Dusting and vacuuming, for example, can disturb minute mites that are carried up with the dust particles and can enter a person's respiratory tract. When the airborne mites contact the IgEs on mast cells in the lung or tracheal tissue, these granulated cells are triggered to release their granules explosively by exocytosis (Figure 32-15). Mast-cell granules contain heparin, which helps prevent blood clotting, and histamine, which increases capillary wall permeability. The outpouring of these substances results in leakage of fluid from capillaries into tissue spaces; as a result, tissues

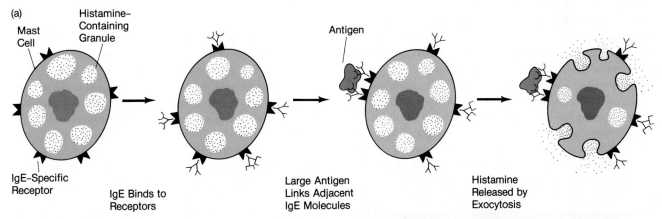

(a) Mast Cell / Histamine-Containing Granule / IgE–Specific Receptor / IgE Binds to Receptors / Antigen / Large Antigen Links Adjacent IgE Molecules / Histamine Released by Exocytosis

Figure 32-15   ALLERGY.
(a) Mast cells release histamine by exocytosis in response to large antigenic agents, such as pollen grains, mites, and cat dander. The mast cells have cell surface receptors for IgE and therefore can bind a range of IgE molecules, specific for different antigens. (b) A mast cell (magnified about 2,500 times) in the process of releasing its granules. A substance such as pollen grains has bound to antibodies on this mast cell's surface, triggering massive release of the cell's granules.

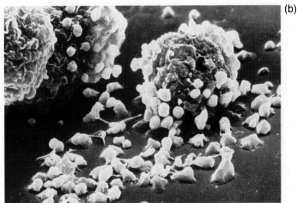

(b)

swell, especially in the respiratory tract. Coughing, sneezing, and mucous secreting also occur—all aimed at expelling the foreign invader. There is also a loss of systemic blood pressure because of capillary leakage. A severe reaction of this type is called *anaphylaxis*.

Such localized symptoms may be irritating, but the result is positive—the invader is removed. Unfortunately, in more than 35 million Americans, this mast-cell system tends to overreact, producing **allergies**—chronic reactions induced by any number of benign environmental stimulants, from pollen grains to milk, eggs, wheat, peanuts, shellfish, and berries. Because mast cells are so widespread in the body, allergic reactions are correspondingly diverse, ranging from the sneezing and coughing of hay fever to the severe wheezing and breathing difficulties of asthma, to such digestive reactions as diarrhea, vomiting, or cramping.

The most serious allergic reactions trigger **anaphylactic shock,** a life-threatening state during which the mast cells in many tissues discharge the contents of their granules simultaneously. When an individual who is severely allergic to bee or spider toxins or to drugs such as penicillin is exposed to the offending antigen, he or she can experience rapid fluid loss from blood vessels all over the body and, as a consequence, reduced blood flow to the brain and heart muscle, and difficulty in breathing. This condition can be fatal unless epinephrine is quickly injected. This hormone counteracts the symptoms of anaphylactic shock by rapidly constricting blood vessels, thereby cutting off the massive leakage of blood fluid into the tissue. Because the IgE reaction is immediate, people who have severe allergic reactions to bee stings, drugs, or certain foods are usually advised to keep epinephrine handy in case of an emergency.

Treatment of mild allergies usually includes administration of antihistamines, drugs that counteract the effects of histamine released from mast-cell granules on the surrounding tissues. Allergy researchers are currently investigating more fundamental ways to disarm the allergic response by removing the "firing pin," IgE. One approach is to chemically "tie up" the IgE molecules on the surface of mast cells so they do not respond to specific antigens, such as ragweed pollen or penicillin. Another approach is to chemically suppress the production of IgE in the first place. In mice, IgE suppression reduces allergic responses. Some researchers predict a time, perhaps a decade from now, when allergies can be very effectively treated or stopped entirely in one of every ten people who now suffer them.

Tissue Compatibility and Organ Transplants

Since about 1960, transplantation of organs such as kidneys has become an increasingly important medical procedure. However, physicians have found that unless the donor and the recipient are closely related, the recipient's immune system is likely to reject a piece of skin, a kidney, or a transplanted heart. This process of rejection and subsequent death of the transplant depends on the fact that all body tissues are labeled antigenically and can be recognized as self or nonself. Researchers have, in fact, discovered antigenic protein labels called *histocompatibility* antigens (*histo-* means "tissue"), as well as a group of genes that code for these labels. These so-called **major histocompatibility complexes** contain at least four closely linked genes in humans and as many as fifty alleles in total—a diversity that explains why two people chosen at random are so unlikely to have the same constellation of alleles and, in turn, compatible tissue types.

Obviously, transplantations do not go on in nature. What, then, is the biological role of histocompatibility genes? Antigen recognition and stimulation of lymphocytes to carry out immune responses may be their major role. Histocompatibility genes code for proteins that appear on most cell surfaces throughout the body. In immune-system cells (T and B lymphocytes and macrophages), these proteins are involved in the ability to recognize foreign substances. For instance, when macrophages accumulate at a wound site and devour debris, they also stimulate the proliferation of T cells. In order for this stimulation to occur, the histocompatibility proteins on the macrophages and T cells must match. In addition, the histocompatibility proteins function as receptors that bind the foreign antigen in a remarkably slow and still not understood process. A T cell then recognizes the complex composed of the foreign antigen bound to the histocompatibility protein. In fact, the whole basis for distinguishing self from nonself has been linked to the fact that histocompatibility proteins may bind self molecules more tightly than foreign molecules.

Some histocompatibility complex genes specifically regulate the level of the immune response; hence, they are called *Ir* genes. Animals with deficient functioning of the *Ir* genes have defective helper and suppressor T cells. Perhaps through the influence of these T lymphocytes on other immune-system cells, such as B cells or killer T cells, the *Ir* genes regulate the entire immune response, and with it the body's day-to-day protection.

Though the proper functioning of the histocompatibility genes is clearly necessary, their earliest evolutionary role is not known. Recent studies reveal that a fundamental role of such genes may not be immune-system function, but sexual acceptance. Mice so genetically similar that they differ in just a few of the dozens of histocompatibility alleles can distinguish each other, probably on the basis of odor. Male mice prefer to mate with females whose histocompatibility genes (and perhaps individual odors) are quite different from their own—a system that favors outbreeding. Researchers have also hypothesized that chimpanzees, fruit flies, bees, and frogs can distinguish kin from unrelated individuals by the reception of odor. Biologists now seek to discover whether odor molecules and highly specific olfactory receptor molecules are like antigens and antibodies of the immune system.

Histocompatibility genes also may affect human reproductive success. Couples that share many cell-surface antigens (that is, histocompatibility alleles) tend to have a high rate of spontaneous abortion; early in pregnancy, the embryo may fail to be recognized by the mother as foreign and hence is not protected by a blocking process that normally shields fetal antigens from the mother's immune system. Consequently, her lymphocytes eventually attack the placenta, and the embryo is rejected.

Perhaps histocompatibility antigens on cell surfaces—discovered through tissue transplantation, which, of course, never occurs in nature—will turn out to have an evolutionary origin unrelated to the immune system. Instead, the far more fundamental phenomena of indi-

vidual identity, sexual behavior, and reproduction may depend on unique combinations of molecular name tags on every body cell.

PERSPECTIVE

The immune system is a sort of microcosm of biology, with numerous cells and organs that can sometimes act independently but most often interact in a coordinated way to bring about a systemic response—the protection of the entire organism. Yet the immune system also has unique properties—in particular, the gene shuffling that leads to the antibody diversity so critical to fighting invaders with specific targeted weapons. Because the immune system has both universal and special features and has important implications for medicine as well, biologists will continue to investigate it intensively and to develop it as a model for how all cells behave within complex eukaryotic animals.

SUMMARY

1. The *immune system* includes white blood cells, a set of organs (lymph nodes, spleen, and thymus), and antibody molecules.

2. White blood cells include granulocytes, *macrophages*, *T* and *B lymphocytes*, and *natural killers (NK)*. In adult vertebrates, white blood cell lineages arise from stem cells in the bone marrow. *T lymphocytes* arise in the thymus and are found in such *peripheral lymphoid tissue* as the spleen, and in the blood. *B lymphocytes* arise in the bone marrow or liver of the embryo and in peripheral lymphoid tissues of adults. In birds, B lymphocyte lineages mature in the bursa of Fabricius and then are found in peripheral lymphoid tissue of adults.

3. *Lymph nodes* are small, thickened masses of tissue at certain places along the system of lymphatic vessels. Millions of lymphocytes and macrophages populate the lymph nodes, which filter lymph. Macrophages can ingest filtered foreign

matter, and stimulate T or B lymphocytes to "recognize" a foreign substance. *Lymphokines*, protein messengers, are involved in macrophage activation of lymphocytes.

4. The *spleen* is a small, oval organ located near the liver that stores macrophages, lymphocytes, and red blood cells; removes aged red cells from the circulation; recycles iron from hemoglobin; and filters lymph.

5. The *thymus* is a spongy, spherical organ behind the breastbone; it produces a hormone and is a site of T-lymphocyte differentiation.

6. The body's two defensive systems are nonspecific or specific in nature. Nonspecific defense mechanisms include the physical barriers provided by skin, mucous membranes, and cilia, the antimicrobial peptides and the *inflammatory response*. Specific immune defense mechanisms involve B lymphocytes and antibodies in *humoral immunity*; T lymphocytes in *cell-mediated*

immunity; and the *natural killer (NK)* cells. The immune response is characterized by diversity, memory, tolerance, and, of course, specificity.

7. *Antibodies* are globular, blood proteins produced by B lymphocytes in response to the presence of an *antigen*. An antigen is any substance that elicits an immune response, such as bacterial toxins, viral coat proteins, surface proteins on foreign cells, and carbohydrates on bacterial or foreign blood-cell surfaces.

8. Type IgG antibody molecules each have two identical *heavy chains* and two identical *light chains*, each with *constant* and *variable regions*. A pocket or crevice, the binding site, is formed between the *hypervariable regions* of light and heavy chains and has a configuration complementary to that of a specific antigen. Thus the specificity of an antibody is determined by the variable regions, each with a unique amino-acid sequence. The constant region determines an antibody's class: IgG,

IgM, IgD, IgA, or IgE. Antibodies, killer T cells, and NK cells may kill a target cell by assembling pores in the cell's plasma membranes. This may involve the *complement response* of plasma proteins.

9. The *clonal selection theory* states that during early development, cells pass through pre-B and naïve stages during which they become committed to making antibodies with only one specifically shaped binding site. All cells derived from such an original determined naïve B cell form a clone. When a foreign substance enters the body, naïve B cells with appropriate binding sites are stimulated to divide rapidly and repeatedly—thus they are "selected" clones. The resulting clone of *plasma cells* actively secretes an immense number of antibody molecules that bind the foreign antigenic substance. Antigen-stimulated immature B cells also give rise to *memory cells* which remain available throughout the individual's life to mount a strong secondary response to the original antigen.

10. In the nuclei of pre- and naïve B cells, rearrangement takes place among segments (V, J, D, constant) of the genes coding for heavy- and light-chain variable and constant regions; these various rearrangements and local mutations contrib-ute to the immense diversity of variable regions and, hence, antibody binding specificities. Even greater diversity results from the fact that any one of thousands of light chains and any one of millions of heavy chains are combined to produce the antibody of a given B cell.

11. *Monoclonal antibodies* are formed by creating a hybridoma from an antigen-stimulated B cell and a myeloma cell. The hybridoma divides into a clone, which then produces a huge number of identical IgG antibody molecules, all specific for a given antigen. Monoclonal antibodies can be used in many ways for medicine and research.

12. Clonal selection also operates in T cells in response to the presence of a foreign substance. There are three kinds of T cells: killer T cells, which attack foreign substances; helper T cells, which produce lymphokines that activate antibody production and killer T cell attack; and suppressor T cells, which control the immune system and turn off the immune response.

13. Natural killers are a third class of lymphocytes. They attack tumor cells, viruses, and virus-infected cells.

14. *Autoimmune response* diseases, such as arthritis, result from a break-down in tolerance. Tolerance depends on the recognition of one's own tissues or molecules as self rather than nonself; tolerance arises as the immune system develops.

15. *Active immunity* is defense based on exposure to pathogens, through either disease or vaccination. *Passive immunity* results from the transfer of antibodies, from mother to fetus and newborn or via deliberate injection.

16. *Allergies* represent a reaction of the immune system's mast cells to a foreign substance such as pollen or a particular food.

17. Unless the donor is closely related to or carefully matched to the recipient of a transplanted organ, immunological rejection usually will occur, based on histocompatibility antigens on the surfaces of all tissue cells. These antigenic name-tag proteins serve as receptors for antigens and are essential as macrophages and helper T cells help activate killer T cells and B cells to respond to foreign antigens. Histocompatibility proteins may have originated as part of individual recognition and sexual selection processes.

KEY TERMS

active immunity

allergy

anaphylactic shock

antibody

antigen

autoimmune response

B cell (B lymphocyte)

cell-mediated immunity

clonal selection theory

complement response

constant region

heavy chain

humoral immunity

hypervariable region

immune deficiency

immune system

immunoglobulin

inflammatory response

light chain

lymph node

lymphokine

macrophage

major histocompatibility complex

memory cell

monoclonal antibody

naïve B cell

natural killer (NK) cell

passive immunity

peripheral lymphoid tissue

plasma cell

spleen

T cell (T lymphocyte)

thymus

variable region

QUESTIONS

1. Describe each type of cell that is part of the immune system and explain its functions. Which possess antigen binding sites?

2. Describe the organs of the immune system and explain their functions.

3. What are antibodies? Which cells synthesize them and how? Does each antibody-producing cell make only one kind of antibody? A few kinds? Or many kinds? Draw a diagram of an IgG antibody and label all of its parts. Why do antibodies have two identical halves?

4. Give examples of nonspecific defense. Describe some features of specific defense. What types of animals exhibit specific and nonspecific defense?

5. What is clonal selection? Give an example, including its relation to a primary and a later secondary response.

6. Discuss the factors that contribute to antibody diversity, from genes through antibody assembly.

7. What are hybridomas? What are monoclonal antibodies? Do they differ from ordinary antibodies? Why not simply clone B cells in tissue culture and use the antibodies they produce?

8. Severe combined immune deficiency (SCID) is an X-linked disease in which the individual lacks B and T cells. What sort of prognosis would you expect for an individual with SCID? What kind of treatment might cure SCID? What is AIDS? What is wrong with the immune system of AIDS victims?

9. What is an antigen? What is meant by active immunity? Passive immunity? Autoimmunity? Tolerance?

10. What is anaphylactic shock? How can it result from a bee sting or eating egg yolk by an allergic person? What hormone do doctors use to treat it, and how does that substance counter the effects of anaphylactic shock?

ESSAY QUESTIONS

1. The study of abnormalities or diseases often leads to an understanding of normal biological processes. Discuss an example from the immune system.

2. The major histocompatibility complex (MHC) consists of a set of closely linked genes; each has many possible alleles, and all of an individual's alleles are expressed as histocompatibility antigens (HLA antigens) on the surfaces of most body cells. What is the maximum number of alleles a normal individual can express for a particular gene? Explain why very few individuals have the same set of HLA antigens you have. If your parents have different alleles for all the HLA antigens, what is the probability that any particular sibling (other than an identical twin) has all the same HLA antigens you have? Remember, the MHC genes are very closely linked.

SUGGESTED READINGS

COOPER, M., and A. LAWTON. "The Development of the Immune System." *Scientific American*, May 1974, pp. 59–72.

This article summarizes the many sites where the cells of the immune system arise and mature.

HOOD, L. E., I. L. WEISSMANN, and W. B. WOOD. *Immunology*. 2d ed. Menlo Park, Calif.: Benjamin-Cummings, 1982.

A very complete treatment of all aspects of immunology.

ROITT, I. *Essential Immunology*. 4th ed. Oxford, England: Blackwell Scientific, 1980.

One of the best general texts in this fascinating field.

TONEGAWA, S. "The Molecules of the Immune System." *Scientific American*, October 1985, pp. 122–131.

The discoverer of gene shuffling summarizes antibody structure and function.

33

RESPIRATION: THE BREATH OF LIFE

Like a fountain, the living organism retains its improbable configuration by borrowing sources of energy from the world around it and by conferring and re-conferring organisation upon the matter which is ceaselessly flowing through it. And, in order to do this, to exploit the energy resources of the substances it borrows from the outside world, the cell must have oxygen.

Jonathan Miller, *The Body in Question* (1979)

Air, the breath of life even for mammals that return to the sea.

Have you ever been held under water just a little too long by a playful friend? Then you know the feeling of panic that comes with not being able to get enough air when you need it. "Breath of life" is more than just a well-turned phrase; we and the vast majority of other multicellular organisms are utterly dependent on a supply of oxygen. This ubiquitous gas is needed to meet the collective demands of our individual cells as they consume oxygen during the transfer of stored energy from nutrients to ATP, the universal energy currency. Deprived of oxygen long enough, our cells and organs are damaged and eventually die.

The process by which organisms exchange gases with the environment is called **respiration.** Respiration includes the intake of oxygen, the transport and delivery of oxygen to cells, and the removal and release of carbon dioxide. A whole organism's respiration is not to be confused with cellular respiration (Chapter 7), during which mitochondria manufacture ATP, as glucose and other organic molecules are metabolized.

We saw in Chapter 31 that in most large animals, oxygen and carbon dioxide are transported to and from every cell in the miles of blood vessels. But how does oxygen enter the circulatory system, and how does carbon dioxide then leave the body quickly? The answer is that most multicellular animals have special sets of respiratory organs, such as gills or lungs, as well as cilia or muscles, to help pump in a steady supply of air or of water containing dissolved oxygen. In this chapter, we shall examine the respiratory organs in detail and see how they function in concert with the circulation of blood to ensure a constant supply of oxygen to all cells, whether the organism is resting or exercising strenuously. (In Chapter 29, we examined analogous organs and processes in plants.) Table 33-1 shows how dramatically the need for oxygen increases during exercise.

We will begin our consideration of respiration with the physics of moving gases and the influences of temperature, diffusion rates, and other factors on an organism's oxygen supply. More than any other physiological system, the respiratory system shows that life processes must conform to the chemistry and physics of matter.

We will then survey the kinds of respiratory systems that have evolved in the animal kingdom. As we move from the simplest organs to the most complex, we will see three recurrent structural themes: the expansion of surface area; the maintenance of a wet surface for gas exchange; and the development of efficient ventilating or pumping mechanisms. Next, we will discuss the marvelous pigments that carry gases in the blood so efficiently. And finally, we will consider the control of respiration and look at special adaptations to life at high altitudes and other environments in which oxygen levels are low.

By looking first at cellular respiration in Chapter 7 and now at respiration in whole organisms, one can begin to understand what the French chemists Antoine Laurent Lavoisier and Pierre Simon Laplace meant when they wrote in 1780, "Life is a combustion." For most organisms, the flame of life will not burn without oxygen.

RESPIRATION AND THE PHYSICS OF GASES

When a commercial airliner is being prepared for takeoff, flight attendants demonstrate the use of oxygen masks that pop out of overhead compartments if the air pressure in the cabin should happen to drop while the plane is aloft. But why are airplane cabins pressurized in the first place? The answer is that air pressure affects the efficiency of every breath we take and, in turn, the intake of oxygen that is so critical to life.

Air pressure is created by the weight of the gases surrounding the Earth in a layer several hundred miles thick. The seven-mile-thick zone closest to the surface contains 75 percent of these gases. Every square inch of the ocean and land surface near sea level is pressed on by a column of air 1 inch × 1 inch × about 200 miles high. This column weighs about 14.7 pounds; thus the pressure of this weight of gases, called **atmospheric pressure,** is 14.7 pounds per square inch at sea level. The standard way to measure atmospheric pressure is to see how high that column of air will push a column of water in a U-shaped tube (10 m at sea level) or a column of mercury, Hg (about 760 mm at sea level) (Figure 33-1).

Since air contains a mixture of different gases, the full measurement of 760 mm Hg is the total pressure of that mixture in air. Of these various gases, each contributes only part of the column's weight and thus only part of the pressure exerted by the column of air. Oxygen occupies approximately 21 percent by volume and thus exerts a pressure of 0.21 × 760, or 160 mm Hg. Nitrogen occupies approximately 78 percent and exerts a pressure of

Table 33-1 **OXYGEN USE UNDER CONDITIONS OF REST AND ACTIVITY**

Organisms	Weight	Oxygen Consumption (ml/kg per hour)
Butterfly	0.3 gram	600 at rest; 100,000 while flying
Mouse	20.0 grams	2,500 at rest; 20,000 while running
Human male	70 kilograms	200 at rest; 4,000 while working maximally

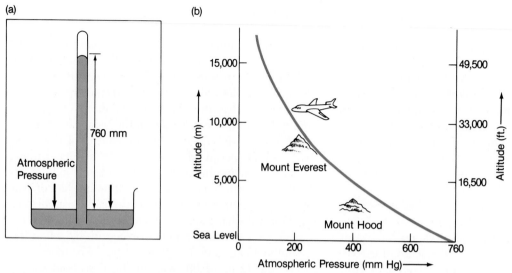

Figure 33-1   MEASURING ATMOSPHERIC PRESSURE.
(a) At sea level, the atmospheric pressure will support a column of mercury 760 millimeters high. (b) The relationship between altitude and atmospheric pressure, as measured by a barometer. At higher altitudes, the atmospheric pressure is lower, and thus the height of the column of mercury that can be supported by the atmospheric pressure becomes shorter.

0.78×760, or 593 mm Hg. Carbon dioxide, at 0.03 percent, has a pressure of 0.2 mm Hg. These figures represent the **partial pressures** exerted by each gas and are abbreviated P_{O_2}, P_{N_2}, P_{CO_2}, and so on. Rarer atmospheric gases, such as hydrogen and helium, exert much smaller partial pressures.

Returning to the commercial airliner, now in flight at 35,000 feet, let us say, what would happen if a cabin wall were punctured? The air in the cabin, maintained artificially at an air pressure similar to that of Denver (about 5,000 feet above sea level), would rush out into the thin low-pressure air at that altitude (about 180 mm Hg). Although the air remaining in the cabin would still contain 21 percent oxygen, the *quantity* of available oxygen and other gases would drop dramatically at such low pressure (Figure 33-2 shows the reduced partial pressure, $0.21 \times 180 = 37.8$ mm Hg). Soon there would be too few oxygen molecules in breaths of cabin air to meet the needs

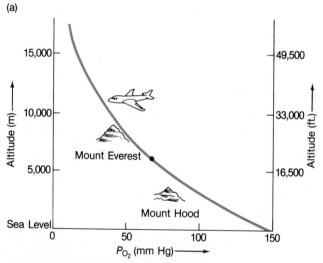

Figure 33-2   PARTIAL PRESSURE OF OXYGEN AND ALTITUDE.
(a) Most unacclimated humans will lose consciousness because of the low oxygen levels at altitudes around 6,000 meters (60 mm Hg or less), although people do live permanently at this altitude. At the summit of Mount Everest, a person acclimated to high altitude can survive for a few hours breathing the air with oxygen at a partial pressure of about 40 mm Hg. (b) Scientists laboring upward, high on Mt. Everest.

THIN AIR AND THE HUMAN BRAIN

Throughout much of human history, people have been obsessed with climbing tall mountains. This grand obsession led Sir Edmund Hilary and Tensing Norgay to conquer Mount Everest in 1953 and twenty-five years later spurred Peter Habeler and Reinhold Messner to achieve the same feat without the aid of supplementary oxygen. One small but important branch of physiology deals with the effect of high altitude on humans, such as these famous adventurers, as well as on other animals. And within that field, one particularly intriguing question is: What happens to the human brain at extreme altitudes?

"Extreme altitude" is defined by many in the field of high-altitude physiology as 5,800 meters, or 18,850 feet. At such height, an organism is besieged by bitter cold, high winds, low humidity, and high levels of solar and ultraviolet radiation. By far the most important physical challenge, however, is *hypoxia*, or low levels of oxygen reaching the body tissues. The percentage of oxygen in air actually remains the same— 21 percent—from sea level to 110,000 meters of altitude. However the barometric pressure drops steadily with increasing height, so that on top of a 3,500-meter mountain, such as Oregon's Mount Hood, the amount of oxygen present is only 65 percent that at sea level, and at the summit of Mount Everest (8,850 meters, or nearly 29,000 feet), it is only 32 percent. At 5,800 meters, where the range of extreme altitude begins, the amount of oxygen is about 50 percent that at sea level.

Physiologists have measured a number of distinct effects of extreme altitude on the human body. Most people experience a sharp decrease in appetite, breathlessness even at rest, muscular fatigue, exhaustion on attempting any sort of exertion, and a falling off of mental capacity. In a card-sorting test carried out at 4,570 meters, subjects were found to have a significant reduction in the speed of their reactions and in their ability to make decisions. In word-association tests and tests of visual patterns, experimenters also found that memory is noticeably impaired.

As a climber presses on to greater heights, brain tissue is further deprived of oxygen, and bizarre mental states often occur. Two climbers who reached the top of the highest peak in the Western Hemisphere, Aconcagua in Argentina (6,920 meters), suffered frank hallucinations in which they saw highway equipment, skiers, trees, and dead mules on the mountain's summit. Three climbers on Mount Everest reported the strong sensation of a phantom companion. And when Hilary removed his oxygen mask for ten minutes while resting at the peak of the Earth's tallest mountain, he reported that his vision began to dim menacingly and brightened again only after he replaced his oxygen mask.

On their daring assault of Everest, Habeler and Messner carried no oxygen and thus had no such reprieve. At the summit, Habeler was overtaken by what he recalls as a powerful sense of euphoria: "that nothing could happen to me. Undoubtedly, many of the people who have disappeared forever in Everest's summit region had also fallen victim to this treacherous euphoria."[*] This feeling passed quickly, and Habeler suddenly feared that he was suffering severe brain damage. He commenced sliding thousands of feet down the mountainside and reached base camp in less than an hour. Unfortunately, Habeler did notice carry-over effects after the ascent, including lapses of memory and nightmares. In fact, physiologists are concerned that permanent brain damage has occurred in a number of climbers, probably due to massive death of nerve cells in the brain.

Considering the effects of hypoxia on the brain, Thomas Hornbein, an anesthesiologist who climbed Everest in the 1960s, has concluded that "the brain . . . rather than the exercising muscle is the organ ultimately limiting function"[†] at an extreme altitude like the slopes of Mount Everest.

[*] Peter Habeler, *Everest Impossible Victory* (London: Arlington Books, 1979), pp. 179–80.
[†] Thomas Hornbein, "Everest without Oxygen," *Hypoxia, Exercise, and Altitude: Proceedings of the Third Banff International Hypoxia Symposium* (New York: Alan R. Liss, 1983), pp. 411–12.

of the human brain, and the passengers would quickly lose consciousness unless they donned oxygen masks and breathed in the gas.

For all organisms—from our endangered airplane passengers to barnacles on the bottom of a boat—it is not the *proportion* of oxygen available in the environment but the *quantity delivered to the tissues* that is so critical and that must be controlled. Several factors affect the amount of a particular gas, such as oxygen or carbon di-

oxide, in the extracellular fluid of an animal. These factors include the partial pressure of the gas in air, the solubility of the gas in water, the temperature, and the diffusion rate of the gas.

The tendency of a gas to enter an adjacent liquid increases as the partial pressure of the gas rises. Partial pressures are rather like concentrations in that sense. In a glass of carbonated soda, for example, the partial pressure of carbon dioxide (P_{CO_2}) in the liquid is higher than

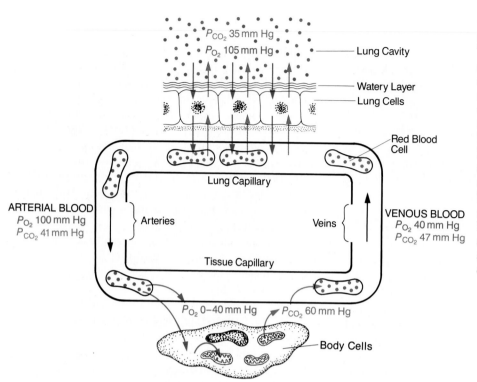

Figure 33-3   GAS EXCHANGE IN THE LUNGS AND TISSUES. The delivery of oxygen to body cells (red) and the elimination of carbon dioxide from lung cells (blue) involve diffusion into and out of the alveolar lung cells, tissue fluids, capillaries, and red blood cells. Oxygen from the lungs diffuses into the blood, and CO_2 diffuses in the opposite direction. When oxygenated blood passes through a tissue capillary, O_2 diffuses out of the blood into the extracellular fluid and then into nearby cells; again, CO_2 moves in the opposite direction.

in the air above it. That is why the CO_2 diffuses out of the liquid and forms gas bubbles that rise to the surface. Likewise, if P_{CO_2} in a cell is higher than in the surrounding extracellular fluid, the dissolved CO_2 molecules will diffuse outward from the cell to that fluid (Figure 33-3). If CO_2 levels in the extracellular fluid build up so that P_{CO_2} is higher than in the blood, CO_2 will diffuse into the blood plasma and be carried away. Conversely, since P_{O_2} in air is high, O_2 will diffuse from air to blood, from blood to tissue fluid, and from tissue fluid to cell, as long as P_{O_2} is successively lower in each of these compartments.

The solubility of gases in liquids is a second important factor governing O_2 and CO_2 availability and movement in organisms and cells. **Solubility** is defined as the amount of a gas that will dissolve in a specified volume of liquid at specified temperatures and a pressure of 1 atmosphere (760 mm Hg). The solubility of gases in liquids is usually quite low. For example, in 100 milliliters of blood plasma, there is only 0.5 milliliter of O_2, whereas in 100 milliliters of air, there is forty-two times more (21 ml). The low solubility of gases in liquids has had several effects on the evolution of large complex organisms, the most significant being the evolution of specialized pigments, such as hemoglobin, that can carry a large amount of O_2 in the blood. We shall describe other effects of low gas solubilities in the next section.

The influence of temperature on gases is familiar to most of us. Take a glass of ice-cold water and let it stand a

few hours. As the water warms to room temperature, it can hold less dissolved gas; therefore, small bubbles of gas form on the inside walls of the glass. For organisms, the temperatures of the surrounding environment, of the respiratory organs, of blood (if they possess it), and of the cells are crucial determinants of gas contents and the ease of gas diffusion into and out of cells. If, for instance, a cold-water fish, such as a trout, moves into warm water (which holds less dissolved oxygen), the fish will be unable to take in as much oxygen as before through its gills even though its rate of metabolism rises in the warm water and its cells require more oxygen. Clearly, it must quickly escape this situation or perish.

The fourth factor that influences the amount of gas in a fluid is the rate of diffusion. In order for O_2 to enter an organism's tissues or cells, it must dissolve in the fluid surrounding the cell and then it must diffuse through that fluid to the cell surface. The diffusion rate of a gas dissolved in liquid is much slower than the diffusion rate of the same gas in air. In an unstirred liquid, it takes an entire week for O_2 to diffuse the same distance that it can travel in only a second by diffusion in air. This slow rate of diffusion severely limits the distance that cells can be from a source of O_2. An organism dependent solely on diffusion of O_2 and CO_2 through its body surface cannot be larger than about 0.5 millimeter in radius, or, as we saw in Chapter 31, it must be very flat or must be porous, like a sponge. Larger, thicker organisms use bulk flow to transport gases and other substances (Chapter

31). Breathing air into lungs and pulling water across gills are methods of bulk flow that bring volumes of gaseous or dissolved O_2 to special respiratory surfaces, where O_2 can be picked up by the blood and carried to all tissue cells. Thus animals with lungs or gills depend on the bulk flow of air or water and the bulk flow of an oxygenated liquid (blood) to supply each cell.

RESPIRATORY ORGANS: STRUCTURES FOR EFFICIENT GAS EXCHANGE

We have seen that the physical limitations on the movement and contents of gases in liquids are quite stringent—so severe, in fact, that they have affected the sizes and shapes of entire organisms and their respiratory mechanisms. A primary respiratory characteristic of multicellular animals is the presence of a gas-exchange organ with a large enough surface area that sufficient intake of oxygen and release of carbon dioxide can occur to meet the needs of all body cells. However, as the body's gas-exchange needs increase, the enlargement of the respiratory organ's surface area has limitations—in terrestrial air breathers, the larger the surface, the more water the body loses by evaporation. In aquatic organisms, the larger the respiratory surface area, the more salts that may be lost or taken in, and thus the greater the problem of maintaining osmotic balance. We will discuss these osmotic problems in Chapter 35. Here we will survey the four main types of respiratory systems that have evolved in animals: the wet body surface of small organisms; the *gills* of aquatic creatures; the *tracheae* of insects and related animals; and the *lungs* of mainly terrestrial animals.

Wet Body Surface as a Site for Gas Exchange

We saw in Chapter 31 that the cell membranes of single-celled and colonial organisms, such as prokaryotes, fungi, sponges, and small flatworms, can provide adequate exchange of oxygen and carbon dioxide by means of diffusion alone. Some relatively large multicellular organisms also exchange gases solely across moist body surfaces, without special respiratory organs. In earthworms, marine worms such as tubifex, lungless salamanders that inhabit ice-cold mountain streams, and frogs that live submerged in deep mountain lakes, oxygen diffuses through the moist body surface and into blood capillaries that lie just below the skin surface (Figure 33-4). The blood then carries the oxygen to deeper-

lying tissues and cells. Carbon dioxide follows the reverse pathway: it moves from body cells to extracellular fluid, to blood vessels, to skin capillaries, and then diffuses out through the moist skin. Thus respiration in this second group of organisms involves both diffusion and bulk-flow transport. The rates of metabolism are low enough in these animals that no specialized, highly efficient respiratory organs are needed. However, there is an ecological price: such organisms can only live where their skins will stay moist constantly, and thus they are excluded from very dry habitats on land. In addition, they may be vulnerable to predators, since their skins cannot be covered with impermeable armor. With the other types of respiratory organs—gills, tracheae, and lungs—a compromise was reached, in a sense: the gas-exchange surface remains moist and permeable, but the animal itself develops an impermeable outer layer.

Gills: Aquatic Respiratory Organs

Many kinds of aquatic animals cannot carry out sufficient gas exchange through their skin and so have specialized gas-exchange organs called **gills** (Figure 33-5). These external respiratory organs provide both a large surface area and a very short distance for oxygen and carbon dioxide to diffuse between the surrounding water and the blood. Thus gills have a rich supply of blood separated from the fresh or sea water by just a few thicknesses of cells of the capillary wall and gill epithelium.

Figure 33-4 **SKIN AS A RESPIRATORY ORGAN.**
The Lake Titicaca frog (*Telmatobius culcus*), which lives submerged for weeks in the depths of Lake Titicaca, has small lungs but relies when submerged exclusively on gas diffusion through the skin for obtaining oxygen and eliminating carbon dioxide. The folds of skin on the body and legs provide increased surface area for gas exchange.

Figure 33-5   GILLS: ORGANS FOR RESPIRATION IN WATER.

The richly vascularized external gills of an albino mudpuppy (*Necturus maculosus*) are easy to see.

Blood in the capillaries picks up O_2 that diffuses the short distance from the water into the gill epithelial cells and then through the capillary wall. Carbon dioxide diffuses in the opposite direction, through the capillary-wall cells, the gill epithelial cells, and then out into the water (CO_2 may actually move as HCO_3^-, bicarbonate, but the path is identical).

Gills of fish take many forms: the gills of some marine worms are flat and paddle-shaped; starfish have small tube-shaped gills; fish and certain mollusks have complex, finely divided gills with enormous surface areas. Large, complex, fast-swimming creatures, such as squid, trout, and sharks, have delicate gills that are hidden beneath the body surface and thus protected from injury.

The vertebrate gill system is derived from the gill slits used as exit ports for the feeding current in hemichordates and lower chordates (Chapters 25 and 26). Respiratory exchange at the gills arose secondarily in the fishes.

Gills of fish have a special design feature that involves the routing of water and blood—these fluids actually flow in opposite directions, and this improves oxygen uptake. A trout, for example, performs repeated pumping movements of its jaws and opercula (an operculum is the flaplike side of a bony fish's head that covers the gill region) that pull water into the mouth and force it past the gills and out the opercular openings. At the rear of the mouth cavity, in the pharynx, water passes sideways over the gills' gas-exchange surfaces from their medial to their lateral portions (Figure 33-6). Simultaneously,

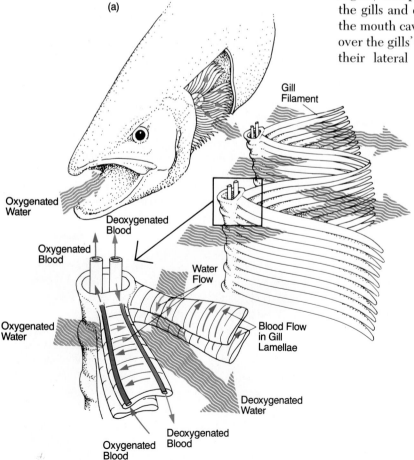

(a)

Gill Filament

Oxygenated Water

Deoxygenated Blood

Oxygenated Blood

Water Flow

Oxygenated Water

Blood Flow in Gill Lamellae

Deoxygenated Water

Oxygenated Blood

Deoxygenated Blood

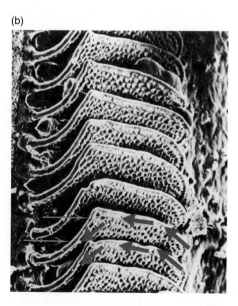

(b)

blood in the gill capillary beds flows in the opposite direction—from the lateral edges of the gills toward the medial. This opposite flow of two fluids separated by a permeable interface is called a **countercurrent exchange system** (Figure 33-7). Countercurrent flow leads to far more complete uptake of O_2 from water to blood than would occur if both flowed in the same direction. Without countercurrent exchange, a fish could theoretically extract only 50 percent of the dissolved O_2 at best; with it, some fish can extract up to 85 percent of the O_2 dissolved in sea water. Countercurrent flow is an engineering principle of such great versatility that vertebrates use it in various organ systems to bring about more efficient exchange of lactic acid, heat, and salts, as well as oxygen (Chapter 35).

Aquatic creatures must expend substantial energy pumping a large amount of water over their respiratory surfaces so they can "harvest" enough O_2. Comparing a trout with an air-breathing lungfish (which we will discuss later in this chapter), the trout must pump its opercula 400 times for each single breath taken by the lungfish. Substantially more muscular work is required for the trout to ventilate its gills—10 to 20 percent of its total body energy—than for the lungfish to ventilate its lungs—1 to 2 percent of its total body energy. This effort required by animals that extract O_2 from water can be explained not only by the low solubility of O_2 in water, but also by the density of water, which is 800 times that of air—more muscular work is required to move dense water than to move low-density air. In some aquatic animals, locomotor movements aid the pumping of water through the gills: fast-moving fishes such as mackerel

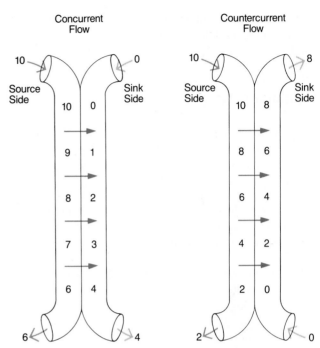

Figure 33-7   COUNTERCURRENT FLOW: AN EFFICIENT TRANSFER METHOD.

Compare the degree to which units of a commodity—such as oxygen, lactic acid, or heat—are transferred when fluids flow in the same (concurrent) direction or in opposite (countercurrent) directions. In this example, only four of ten units are transferred in the concurrent exchange system (40 percent efficiency), whereas eight of ten units are transferred (80 percent efficiency) in the countercurrent exchange system. The source is the site from which a commodity moves and the sink is the site to which it goes. During countercurrent flow, maximal transfer may reach 95 percent or more of the commodity; this is because the source side is always a bit higher than the sink side, even as shown here, when the sink has picked up 6 or 7 "parts" of the commodity being transferred.

Figure 33-6   GAS EXCHANGE IN THE FISH GILL.
(a) Water containing dissolved oxygen enters the mouth cavity and then passes between the gill filaments. The transfer of oxygen from the water to the blood takes place at the gill filaments, where the capillaries in the gill lamellae are located. The flow of water between adjacent gill filaments (turquoise arrows) is in the opposite direction from the flow of blood (red arrows). This type of countercurrent exchange system facilitates maximum extraction of oxygen from water. (b) This scanning electron micrograph (magnified about 56 times) shows a plastic cast of the capillary beds found in a set of gill lamellae lying next to each other on the top of a gill filament. The junctions with the artery that carries oxygenated blood away from the gill are seen at the left. The overlying gill tissue has been digested away to reveal this capillary's huge lacy surface area for O_2 and CO_2 exchange. As seen here, blood flows from right to left (red arrows), and water (turquoise arrows) in a countercurrent direction.

swim with their mouths open in "ram-jet" fashion, causing a large amount of water to move over their gill surfaces. Such fishes will literally suffocate if their nonstop, rapid swimming is prevented.

Gills represent a distinct evolutionary advance over simple diffusion through wet body surfaces. The surface area for exchanging gases remains wet in gills but is greatly expanded, and in many aquatic animals, muscles pump great quantities of water past the gills. Nevertheless, as animals began to colonize land, gills could not be used because their exquisitely thin, pliable exchange surfaces collapse together when not supported by water, preventing efficient exchange of gases. Different respiratory organs have evolved that allow such exchange with air to occur, and this set the stage for the invasion of land.

Tracheae: Respiratory Tubules

Insects and most terrestrial arthropods possess **tracheae** (singular, *trachea*): air-filled hollow tubes that branch and rebranch into a fine network of air passages that penetrate the animal's body and deliver air directly to tissues deep in the interior. In some species, air sacs connected to the internal passages serve as temporary air reservoirs. Like gills, tracheae provide a large surface area for gas exchange and a short diffusion path to the capillaries and tissue fluids; unlike gills, tracheae arise from developmental ingrowths rather than outgrowths.

On examining almost any insect, one can see a set of tiny "portholes" that run in a line down the animal's side. These portholes are closable vents called *spiracles,* which lead into the tracheae. The tiniest tracheae deep in the tissues are called *tracheoles,* and they eventually reach close to every body cell (Figure 33-8). The tip of each tracheole, sometimes called an "air capillary," is usually filled with a watery fluid. When an insect is resting, air travels into the open spiracles, moves throughout the body in the tracheae, and reaches the fluid-filled portion of the tracheoles. Oxygen diffuses into the fluid, then through the walls of the tracheoles, and finally into the hemolymph, which bathes the surfaces of most tissue cells as it moves through the insect's open circulatory system.

When an insect is flying, jumping, or running, and its cells are metabolizing rapidly, the fluid drains from the tracheole tips, and air is able to penetrate farther into the empty tracheoles and deliver O_2 faster to the tissues because the diffusion pathway through the residual liquid is shorter. The drainage results from a simple osmotic trick: the actively metabolizing cells release small organic products of metabolism (amino acids, sugars, urea, lactic acid, and so on) into the hemolymph, raising that fluid's osmotic pressure and thus drawing water out of the tracheole tips and leaving them open for air to enter more deeply. Tracheal tubes also carry CO_2 away from the hemolymph and tissue cells and release it through the spiracles.

In active insects, such as bees and grasshoppers, the efficient exchange of gases via the tracheal system depends on the compression and kneading of the tubes or storage air sacs by the insect's muscles and other organs as it flies, runs, or jumps. Pumping action brought about by muscle contractions ensures that gases in the tracheae do not stagnate when the insect is at rest, and then the greater muscle activity of locomotion further increases O_2 supplies just as the insect becomes more active.

Lungs: Intricate Air Sacs for Gas Exchange

The other evolutionary innovation for gas exchange was the **lungs,** hollow, usually branched internal respiratory organs that are connected through passageways to the outside air and form interfaces with the circulatory system. Lungs arise during embryonic development as outgrowths of the gut wall in terrestrial vertebrates (amphibians, birds, and mammals), in many aquatic vertebrates (certain fresh-water fishes and all aquatic reptiles, birds, and mammals), and in some invertebrates (land snails and certain spiders). Lungs vary in structure from smooth-walled, balloonlike sacs to highly subdivided organs with tortuous inner folds, branches, tubes, and sacs. The anatomy depends on the organism's oxygen needs: smooth inner walls offer a smaller surface area for gas exchange, while subdivided walls offer a greater sur-

Figure 33-8   INSECT TRACHEAE: HOLLOW TUBULES FOR EFFICIENT RESPIRATION.

The respiratory system of insects consists of tracheae which open to the exterior by way of spiracles. (a) The tracheae branch to form smaller, hollow tracheoles, which convey gases to and from all parts of the body. Ventilation is based mostly on compression of the air sacs through muscular action of the abdomen. (b) The terminal portions of tracheoles are filled with fluid through which gases must diffuse. In active tissues, such as muscles, some of this fluid is drawn osmotically out of the tracheole tip; this in turn brings more oxygen-containing air closer to the tissue cells.

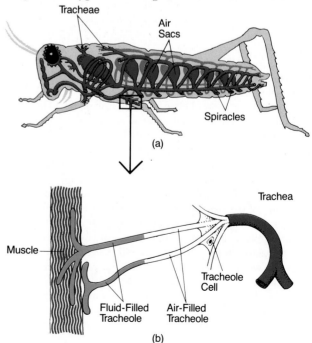

face area capable of meeting a high demand for O_2. Frogs, salamanders, and lungfish have relatively low rates of metabolism and low O_2 demands, and consequently have simple saclike lungs. Reptiles have more complex lungs, with a larger surface area, than do amphibians and lungfish, and birds and mammals have highly convoluted lungs, with a huge surface area up to forty times greater than the total surface area of the skin (Figure 33-9). Irrespective of surface area, the actual gas-exchange site of all lungs is the watery layer covering the innermost layer of lung cells. Since that moisture is subject to evaporation and must be steadfastly maintained, the drying out of the lungs is always an obstacle to life in the air.

Once evolved, air breathing had physiological advantages over gill respiration, since air holds forty times more O_2 than does water, and since an animal with lungs consumes far less energy respiring than does an animal with gills. However, the situation is somewhat different for CO_2 transport. CO_2 is much more soluble in water than is O_2, and it diffuses in water more rapidly. Because of this high solubility and fast diffusion, many aquatic animals that take in O_2 through the lungs still give off CO_2 through the skin (or gills, if present). The CO_2 diffuses from abundant skin capillaries across the skin and directly into the water. Even while an amphibian is sitting on a riverbank, its skin is so moist that CO_2 can diffuse directly out into the air. However, this strategy

Figure 33-9   LUNG EVOLUTION: INCREASING SURFACE AREA.
During the evolution of land vertebrates, the lung surface area for gas exchange became progressively larger and more convoluted. In the diagrams of (a) amphibian, (b) reptilian, and (c) mammalian lungs, the surface area for exchange of gas with the circulatory system is indicated in blue. The black dots and lines in (a) and (b) represent blood vessels. The portion of mammalian lung shown here is but a small fraction of a full lung; the bronchiole is a terminal branch of the treelike trachea.

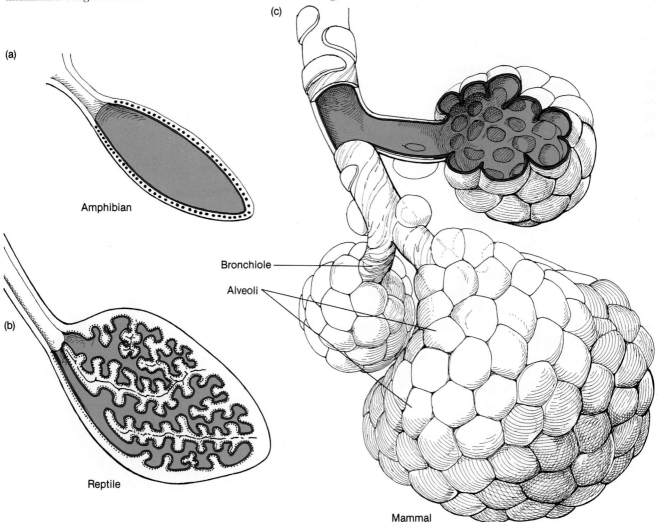

limits such creatures to habitats and life styles in which they can constantly replenish the moisture on their skin. Animals with a body covering that resists desiccation, such as reptiles, birds, and mammals, exchange nearly all O_2 and CO_2 across the respiratory surfaces of the lungs and have ventilatory mechanisms to bring large quantities of air to those surfaces. In the next section, we will consider the structure and function of the human respiratory system; then we will examine the respiratory system of birds.

THE HUMAN RESPIRATORY SYSTEM: THE HOLLOW TREE OF LIFE

As long as life persists, the demand for oxygen never ceases; we breathe in and out, day and night, from our first gasp at birth to our dying breath, relying on a complex respiratory system (Figure 33-10). We shall describe the air pathway in human lungs and the means by which the lungs are ventilated, but keep in mind that the generalizations apply as well to all mammals, birds, and reptiles.

The Air Pathway

As air enters the body, it is filtered, humidified, and warmed by different parts of the *upper respiratory tract*, sometimes called the conducting portion. The *nostrils*, the familiar openings to the nose, are lined with small hairs that act as filters to prevent large airborne objects from penetrating into the respiratory system. The nasal cavities (hollow chambers behind the nostrils) warm and humidify the air and collect airborne particles on a mucous layer. Tiny cilia protruding from epithelial cells lining the nasal cavities sweep the mucus and trapped particles back toward the throat or out the nose.

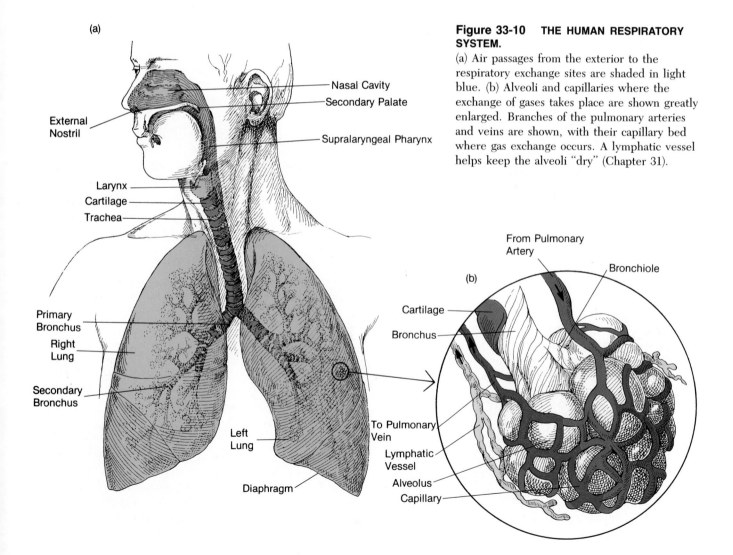

Figure 33-10   THE HUMAN RESPIRATORY SYSTEM.
(a) Air passages from the exterior to the respiratory exchange sites are shaded in light blue. (b) Alveoli and capillaries where the exchange of gases takes place are shown greatly enlarged. Branches of the pulmonary arteries and veins are shown, with their capillary bed where gas exchange occurs. A lymphatic vessel helps keep the alveoli "dry" (Chapter 31).

A set of structures beyond the nasal cavities allows us to continue breathing while chewing and to make vocal sounds. The nasal and mouth cavities are separated from each other by a shelf called the *secondary palate*; food being chewed passes below it toward the esophagus, while air entering the nasal cavities passes above it toward the back of the mouth. Air, liquids, and food pass through a chamber behind the tongue, the **pharynx.** To prevent food or liquid from entering the windpipe, or **trachea,** a flap of tissue, the *epiglottis,* executes well-coordinated movements during swallowing. These movements temporarily seal off the **larynx,** the boxlike entrance to the trachea and lung system, so that food and liquids cannot enter the air passages.

The larynx, or voice box, is the main site of sound production in most mammals. Two ligaments called *vocal cords* can be stretched to varying extents across the opening to the larynx like the sheet stretched on top of a kettledrum. Small muscles vary the position and tension of the two vocal cords, and as air is exhaled from the lungs and passes between them, the resulting vibrations create basic sounds. In humans, these sounds are then modified by the tongue and lips to form consonant sounds and by the *supralaryngeal pharynx,* the chamberlike region with flexible walls posterior to the base of the tongue. Humans alone have this region, and with it, we alone are able to shape cruder vocalizations into vowel sounds (*a, e, i, o, u*), components of every human language.

The trachea is a long tube leading to the lungs. The inner surface of the trachea is lined with ciliated epithelial cells that produce mucus, while the outer wall contains rings of cartilage that keep the tube open at all times, something like the wire helix that keeps a vacuum-cleaner hose open. The trachea branches into two primary **bronchi** (singular, *bronchus*), which, in turn, branch into secondary bronchi, which branch still further into small *bronchioles.* These bronchial tubes form a system of hollow air ducts that resemble an upside-down tree with branches that end in clusters of tiny blind-ended cavities called **alveoli** (singular, *alveolus*) (Figure 33-10b). The branching air passages and tiny cuplike alveoli together make up what we think of as the "lungs." Clearly, our lungs are not hollow sacs, like an amphibian's, but are a mass of moist, delicate tissue enclosing miniature air spaces. Adult human lungs contain about 750 million alveoli, with a total surface area for gas exchange of 80 or more square meters—over forty times the surface area of the skin and larger than a badminton court. Within the walls of each alveolus run numerous capillaries that receive blood from the pulmonary arteries and drain into the pulmonary veins. Lymphatics are also present to keep the lungs "dry" (Chapter 31).

It is in the thin-walled alveoli that O_2 is delivered to the blood and CO_2 is released from it. Inhaled air travels through the upper respiratory tract, bronchi, and bronchioles, and eventually reaches the alveoli. Oxygen from the air then diffuses a distance of just 0.3 to 2.5 micrometers to reach the blood in the capillaries, about the same distance separating sea water and blood in a mackerel's gills. The O_2 must pass through three layers to reach the blood plasma: (1) a thin film of fluid containing a soapy material called *surfactant,* which lowers the fluid's surface tension (cohesion between surface molecules [Chapter 2] and prevents the alveoli from collapsing; (2) the thin layer of alveolar epithelial cells; and (3) the endothelial cells of the capillary walls. Once in the blood plasma, most O_2 passes into red blood cells, is bound to hemoglobin, and is transported throughout the body. The importance of the surfactant in the lung fluid is well illustrated by cases of premature newborns whose immature lung cells do not yet secrete surfactant. Without surfactant, the alveoli collapse, and the infant suffocates. Hormonal treatments are used to speed lung-cell maturation in such newborns.

Most O_2 passes into blood plasma by simple diffusion, but some studies indicate that a carrier molecule may transport about 25 percent of the O_2 via facilitated diffusion (Chapter 5). The carrier appears to be an iron-containing cytochrome called P450 that binds to O_2 reversibly and carries it across the epithelial-cell cytoplasm.

How the Lungs Are Filled

Humans and other mammals fill and empty their lungs by means of a group of bones and muscles that work together like a bellows or suction pump. This mechanism almost certainly evolved from a simpler system, and the evolution of ventilation—the process of filling and emptying internal respiratory organs—will help us understand the marvelous mechanism in humans. Lungfish and amphibians employ a set of valves in the nostrils and the glottis (the opening into the trachea) and coordinated movements of the lower surface of the mouth cavity to push air backward into the lungs (Figure 33-11). This *force pump system* can supply the modest oxygen needs of a cold-blooded frog or salamander, but not the larger demands of terrestrial animals with higher body temperatures. The amphibian's force pump mechanism is so inefficient that, as we noted earlier, the skin must be used as a major auxiliary site of gas exchange.

The biology of land vertebrates is different from that of amphibians. Whereas the force pump of amphibians pushes air into the lungs, a *suction pump action* operates in land vertebrates because of an innovation found in reptiles, the *expandable rib cage.* When a lizard expands its rib cage, the volume of its chest increases, thereby lowering pressure on the outer surface of the lungs.

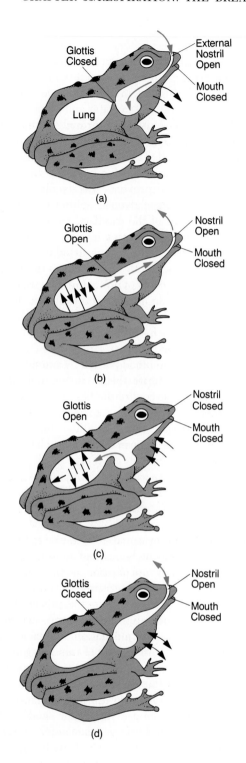

Figure 33-11   AMPHIBIAN VENTILATION: A FORCE PUMP SYSTEM.

(a) Air (blue arrows) is taken into the frog's oral cavity through the open nostrils as the mouth floor is lowered. (b) The glottis then opens, and air from the lungs exits, passing along the dorsal side of the mouth cavity. (c) The external nostrils are then closed, and, as the mouth floor is raised, air is forced from the cavity through the glottis and into the lungs. (d) The glottis is then closed, and air can be drawn into and expelled from the mouth cavity in an oscillatory manner until the cycle begins again.

has been added in mammals. The human lungs lie within the *thoracic cavity,* a space enclosed by the ribs and separated from the abdominal cavity, which lies below, by the diaphragm (Figure 33-12). In an erect human, the ribs curve forward from the backbone and meet at the breastbone, or *sternum,* and a diagonal set of muscles, the **external intercostals,** stretches between each pair of ribs. Both the external intercostals and the diaphragm contract during the inhalation of a breath of air. As the intercostals shorten, the rib cage is raised up and out. Simultaneously, the diaphragm, which is dome-shaped at rest, flattens and pulls downward toward the feet. (In a mammal standing on all four legs, the ribs move down and forward, and the diaphragm flattens toward the tail.)

During inhalation, these muscle actions expand the volume of the thoracic cavity, lowering the air pressure inside the cavity and outside the lung walls below atmospheric pressure. As a result of this natural suction, air flows in through the nostrils, down the trachea, bronchi, and bronchioles, and into the alveoli. Exhalation is largely a passive process brought about by relaxation of the external intercostals and the diaphragm. This relaxation is called *elastic recoil,* and it results in a decrease in chest volume and an expulsion of much of the residual air in the lungs.

During active exercise, however, a second set of muscles between the ribs, the **internal intercostals,** contracts and forcibly expels more residual air from the lungs. This "makes room" for a larger volume of air with each inhalation. (Note that this explanation is a simplified summary; in reality, the roles of the internal intercostals and of the diaphragm are more complex.)

Why is forcible expulsion necessary when oxygen demands are high? A normal inhalation takes in a volume of air equivalent to only 10 percent of the total volume of the trachea and lungs, or about 0.5 liter of a total 5 liters. This 10 percent capacity is called the **tidal volume.** Forcible exhalation and increased inhalation can boost that figure to 80 percent of total lung volume, or about 4 liters. This 80 percent capacity is called the **vital capacity.**

Consequently, air is drawn into its lungs by a kind of suction. Because this process is much more efficient than the force pump at bringing air into the body, all gas exchange can be restricted to the lungs, the skin can be dry, and life styles unattainable by amphibians can be pursued.

Birds and mammals also use the rib-cage suction pump, and an auxiliary sheet of muscle, the **diaphragm,**

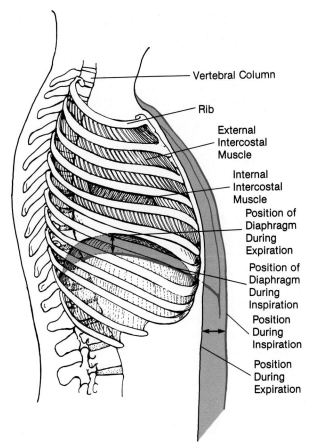

Figure 33-12   VENTILATION IN MAMMALS: A SUCTION PUMP SYSTEM.
The movements of the diaphragm and rib cage bring about changes in the volume of the thoracic cavity, thus sucking air into the lungs or forcing it out again. The downward movement of the diaphragm as it contracts, coupled with the upward and outward movement of the ribs due to contraction of the external intercostal muscles, increases the volume of the thoracic cavity. The reduced pressure within the thoracic cavity, due to the increase in volume, allows higher pressure air outside the body to flow into the lungs. During exhalation (expiration), the relaxation of the diaphragm and external intercostal muscles (which is sometimes aided by contraction of the internal intercostal muscles) returns the diaphragm and ribs to their resting position. The volume of the thoracic cavity decreases, the pressure on the outer surface of the lungs increases, and air is forced out of the lungs. The position of the diaphragm and rib cage during inhalation (inspiration) is shown in red. Only a portion of the intercostal musculature is shown.

Residual air always remains in the mammalian lung system—100 percent can never be expelled since the rib cage cannot collapse completely. Consequently the lungs can never be filled with 100 percent fresh air.

RESPIRATION IN BIRDS: A SPECIAL SYSTEM

Birds have a special system of *air sacs* that makes fresh air almost continuously available to the sites of gas exchange. This provides the animals with the large quantities of oxygen needed for long-distance, high-altitude flights and helps explain an otherwise puzzling phenomenon: if you had the wherewithal to climb to the top of Mount Everest, you might be standing in the freezing wind and inhaling from a tank of compressed oxygen, just as a flock of bar-headed geese fly overhead at 9,200 meters, flapping vigorously and respiring unassisted as they make their yearly migration over the Himalayas. All birds have a set of lungs and auxiliary air sacs that make such feats possible by providing them with highly efficient gas exchange and very light bodies.

The paired lungs of birds are relatively small and have the consistency of dense sponges. However, nine or more hollow air sacs attached to the lungs fill much of the body cavity. These **air sacs** are like balloons that lighten the body and serve as reservoirs for air that will later enter the lungs (Figure 33-13).

When a bird inhales, air in the posterior air sacs flows into the tiny *air capillaries* of the lungs, the only sites where gas exchange can occur. At the same time, fresh air entering the mouth passes through the trachea and, without O_2 exchange, collects mainly in the posterior air sacs. During exhalation, some of this still fresh air passes forward from the posterior sacs into the air capillaries for gas exchange. Interestingly, biologists studied this system by delivering a squirt of pure O_2 into the huge trachea of an ostrich and then tracing the flow of the gas first to the posterior air sacs and then, on the next breath, anterior to the lungs. They found that air flows continuously, during both inhalation and exhalation, in one direction through the lungs. Therefore, O_2 and CO_2 exchange occurs in the lungs' air capillaries during both inhalation and exhalation.

In addition to continuous flow, another special feature of bird lungs is a crosscurrent arrangement of blood capillaries and air capillaries (Figure 33-13b). Whereas countercurrent flow is in opposite directions, 180 degrees apart, the **crosscurrent exchange system** is like a latticework of capillaries at 90 degree angles. The crosscurrents of blood and air permit a high percentage of O_2 to pass from air to blood. This system is also so efficient at removing CO_2 from blood that there is a possibility of the blood's pH rising to dangerous levels. This is counteracted by the addition of CO_2 from the anterior air sacs to inhaled air en route to the posterior air sacs so that blood passing through the lungs is not excessively depleted of this gas.

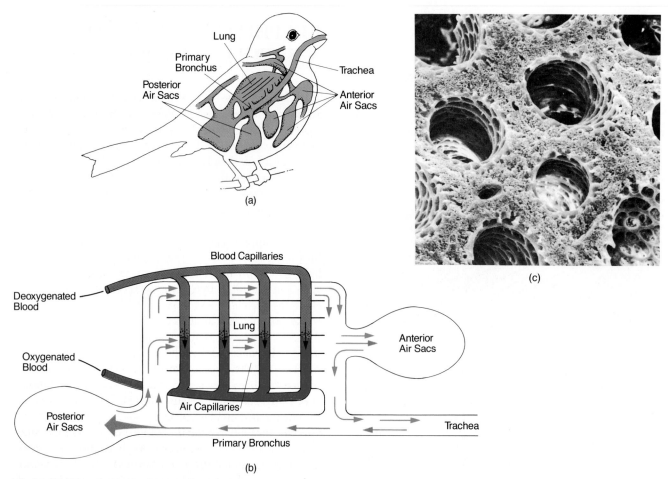

Figure 33-13   AVIAN LUNGS AND AIR SACS.
In birds, oxygenated air passes over the respiratory surfaces in the lungs during both inhalation and exhalation. (a) The placement of the sacs and lungs. (b) Inhaled air passes through the primary bronchus, and most enters the posterior air sacs as well as the lung (orange arrows). During this passage, air that was in the posterior sacs moves through the lung's air capillaries (orange arrows) so that exchange takes place. During exhalation, the air in the posterior air sacs passes through the lung en route to the external environment (turquoise arrows). Again, exchange goes on during exhalation. Blood flow is shown diagrammatically to move in crosscurrent fashion to the direction of air flow. (c) Scanning electron micrograph (magnified about 43 times) of sites in the lung where gas exchange takes place. The numerous, oval-shaped holes seen in the walls of these large bronchi are the entrances and exits to some of the millions of air capillaries.

The combination of continuous unidirectional air flow and crosscurrent exchange, along with other physiological systems, enables birds to do the intense work of flying, even at high altitudes, and still maintain a high body temperature. Rapid breathing also contributes to such performance; a Venezuelan hummingbird, the sparkling violet-ear, breathes 330 times per minute at sea level and 380 times per minute at high altitudes. No mammal can accomplish such a feat.

In addition to augmenting O_2 uptake in the lungs, a bird's many air sacs lower the body weight relative to its size. Branches of the air sacs even extend into the large wing and thigh bones, replacing marrow and rendering the broad, strong bones light in weight. Because much of the body and some large bones are filled with air instead of tissue and tissue fluid, a bird is able to stay aloft while doing much less work and using less energy than is required for other vertebrates to run, climb, or swim. Without such economy of motion, the 2,000-kilometer-plus migrations common in birds would not be possible.

SWIM BLADDERS IN FISH: BUOYANCY ORGANS DERIVED FROM LUNGS

Bony fishes such as trout, tuna, and sea bass have an organ called the **swim bladder,** which is somewhat analogous to a bird's air sacs. The swim bladder is derived from lungs and was probably possessed by early types of bony fishes (teleosts). A swim bladder is a gas-filled sac much like a long, enclosed balloon running below the spine. It is a hydrostatic organ—one that enables the fish to adjust its buoyancy so it will not sink at a given water depth, thus allowing it to expend less energy while hovering effortlessly in one spot. The volume of the swim bladder can be increased or decreased by the *red gland* and the *oval gland.* The former allows O_2 and other gases to pass from the blood into the bladder, whereas the latter removes gas from the bladder (Figure 33-14).

As a fish swims downward into water of increasing pressure, the swim bladder is compressed more and more, so that the fish tends to sink—it is negatively buoyant. In order to reestablish neutral buoyancy at the new depth, the red gland secretes O_2 into the swim bladder, in a sense blowing it up like a balloon. Conversely, if the fish swims upward into water of lower pressure, the bladder expands, carrying the fish toward the surface like a cork. The fish must swim actively to counteract this dangerous positive buoyancy. The oval gland then lets blood carry gas away from the swim bladder, so that neutral buoyancy is achieved at the shallower depth.

The swim bladders of bony fishes and the air sacs of birds are examples of parallel evolution: these organs arose completely independently of each other during evolution, yet both serve as sites where air replaces solid tissue in the body and so makes the body lighter. Evidence from fossils and living primitive bony fishes suggests that the lungs were present in ancient fishes long before swim bladders appeared. Thus, contrary to an old assumption, the swim bladder did not give rise to the lungs, but the lungs of ancient air-breathing fishes probably gave rise to the swim bladder.

RESPIRATORY PIGMENTS AND THE TRANSPORT OF BLOOD GASES

Gills, tracheae, and lungs are amazingly well suited for bringing masses of air or oxygenated water into the body and carrying away waste carbon dioxide. But gases must also be transported simultaneously to and from the billions of cells in a complex organism. That task is carried out automatically and continuously by the circulatory system. We now examine the special pigment molecules that do most of the actual transporting and consider the principles that govern the life-sustaining activities of these pigments.

Pigment Molecules

We know that oxygen has a low solubility in water—only about 0.5 milliliter of O_2 will dissolve in 100 milliliters of water. Blood plasma is a watery solution, yellow in color, that contains dissolved salts, sugars, and proteins; in it are suspended the red and white blood cells. The low solubility of O_2 in water means that blood plasma cannot carry nearly enough O_2 to meet the total needs of body cells if their metabolism is high. The evolutionary solution to this problem is special **respiratory pigments,** molecules, which in some animals circulate freely in the blood but in many animals are contained in the blood cells. These complex pigment molecules bind O_2 reversibly, taking it up when the partial pressure of O_2 is high and releasing it when P_{O_2} is low. In some invertebrates, the primary respiratory pigment is the copper-containing protein *hemocyanin.* In vertebrates, various marine worms, the mud-dwelling larvae of some flies, and even certain beans, the pigment is the iron-containing protein *hemoglobin.*

Figure 33-14   FISH BUOYANCY AND THE SWIM BLADDER.

Buoyancy in many bony fishes (teleosts) is controlled by adding oxygen to or removing oxygen from the swim bladder. This exchange is achieved by two glands: the red gland, which secretes oxygen (green arrows), and the oval gland, which reabsorbs oxygen from the swim bladder (blue arrows). These glands are associated with nets of capillaries.

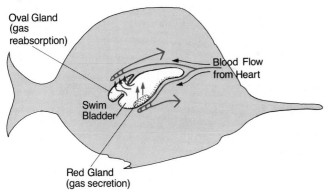

Oval Gland (gas reabsorption)

Blood Flow from Heart

Swim Bladder

Red Gland (gas secretion)

We saw in Chapter 3 that a hemoglobin molecule is composed of four polypeptide chains, each attached to a ringlike, iron-containing *heme* group (Figure 3-25). Oxygen binds specifically between the iron atoms and a particular amino acid in the protein part of the four polypeptide chains. Hemoglobin with four bound O_2 molecules is called *oxyhemoglobin*, HbO_2.

Hemoglobin is manufactured in red blood cells and gives these cells their red color. In vertebrates, hemoglobin molecules are always retained within red blood cells. This contrasts with worms and many other invertebrates, in which hemocyanin molecules circulate in the blood. The retention of hemoglobin in red blood cells, rather than floating freely in the blood plasma, has several advantages:

1. The hemoglobin molecules are always close to enzymes and other factors in the red blood cell's cytoplasm that maintain or vary the pigment's binding properties (as we shall soon see).

2. Bound within cells, hemoglobin does not add to the colloidal osmotic pressure of blood plasma (Chapter 31), which tends to draw fluid from tissues. If hemoglobin circulated freely, an extremely high blood pressure would be needed to balance the colloidal osmotic pressure and to drive fluids back out of the capillaries and into the extracellular fluid.

3. Because a red blood cell is about the same diameter as a capillary, the streaming of these "sacs" of hemoglobin through capillaries stirs up layers of stagnated plasma, located near the capillary walls, that would otherwise diminish gas exchange with the tissues. This turbulence makes exchange of O_2, CO_2, nutrients, and wastes more efficient.

Hemoglobin picks up O_2 in regions where the P_{O_2} is high and releases it in regions where the P_{O_2} is low, facts which neatly explain why the hemoglobin in blood picks up O_2 from the lungs and releases it in the tissue capillary beds. The release of O_2 follows the splitting, or *dissociation*, of HbO_2 into Hb and O_2. The tendency for hemoglobin to bind O_2 under different partial pressures is represented by the S-shaped **oxygen dissociation curve**, such as those shown in Figure 33-15a. In the human lung, where P_{O_2} is about 110 mm Hg, about 98

Figure 33-15   OXYGEN DISSOCIATION CURVE.
Oxygen dissociation curves reflect the amount of oxygen that is bound to hemoglobin at different partial pressures of oxygen. (a) Typical oxygen dissociation curves of mice and humans. When the partial pressure of oxygen is low, usually in the capillaries and tissue fluids, oxygen dissociates from hemoglobin, and, consequently, the percent saturation falls. In the lung's alveolar capillaries, the partial pressure of oxygen is high, and hemoglobin becomes fully saturated—it is loaded with oxygen. In general, larger mammals such as humans have oxygen dissociation curves to the left of those of smaller animals such as the mouse. Curves to the left reflect hemoglobins with greater affinity for oxygen than those for which the curves are to the right. A curve to the right indicates that a lower percent saturation of hemoglobin with oxygen will prevail at any particular partial pressure of oxygen in the tissues; this means that more oxygen is given up or delivered to the tissues. As a result a mouse has much more oxygen delivered to a given amount of tissue than does a human. (b) Fetal hemoglobin has a greater affinity for oxygen than does maternal hemoglobin. Oxygen that is released from the maternal hemoglobin in the placenta will be taken up by the fetal hemoglobin and carried to the fetal tissues.

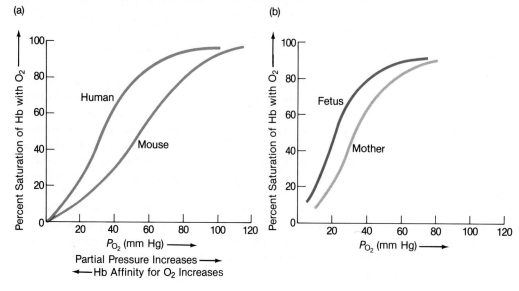

percent of the heme groups in the hemoglobin molecules will bind O_2. When this happens, the hemoglobin is said to be 98 percent *saturated*. Moving left on the graph, we find the typical P_{O_2} in resting tissues of large mammals to be about 30 mm Hg. At this P_{O_2}, the hemoglobin can be only 50 percent saturated. The difference between the two saturation figures—98 − 50, or 48 percent—represents the amount of bound O_2 that is released by the blood when it travels from the lungs to typical capillary beds. Clearly, the venous blood may still carry substantial amounts of O_2 on its return to the heart, since the hemoglobin molecules inside red blood cells normally do not release all their O_2. If exercise or intensified metabolism in a tissue causes the local P_{O_2} to drop from 30 mm Hg to 15 or 10 mm Hg, more HbO_2 will dissociate (98 − 15, or 83 percent), releasing a larger proportion of O_2 to the deprived tissue cells.

Why is the oxygen dissociation curve S-shaped rather than a straight line? The answer lies in the properties of the hemoglobin molecule itself and in the fact that it changes shape as O_2 molecules bind to it. If you had a test tube filled with human hemoglobin in solution, the partial pressure of oxygen in the air above the test tube would have to be raised to 20 to 25 mm Hg before P_{O_2} in the fluid would be high enough for the first heme group on each dissolved hemoglobin molecule to bind an O_2 molecule. This is represented by the lowest, flat part of the curve. The first O_2 binding causes a change in the position of the polypeptide chains—an allosteric change—such that the next heme groups bind easily with only slight increases in the air's P_{O_2}. This is represented by the sharply rising part of the S. This steep slope has important consequences: first, it means that even a modest increase in oxygen partial pressures at the gills or lungs results in rapid loading of O_2 by hemoglobin; second, and conversely, it means that a relatively small drop in P_{O_2} in metabolizing tissues, in the vicinity of the 50 percent loading point on the oxygen dissociation curve, will result in a massive unloading of oxygen from hemoglobin.

In small mammals, oxygen dissociation curves tend to be shifted to the right relative to the curves of large mammals (Figure 33-15a). This means that HbO_2 dissociates to Hb and O_2 at higher tissue P_{O_2}, and therefore more O_2 is delivered to the tissues even though they already contain relatively high levels of O_2. This extra delivery of O_2 allows a mouse, for example, to manufacture the huge amount of ATP required to keep its small body warm. This need is critical in mice, since, compared with elephants, let us say, they have a much larger surface-to-volume ratio and thus are subject to greater heat loss. The curve is shifted to the left in mammalian fetuses (Figure 33-15b). A fetus faces something of an oxygen crisis: it has no direct access to O_2 but, at the same time, has a substantial O_2 demand due to its rapid metabolism and growth. Because fetal hemoglobin has a higher affinity for O_2 than does that of the mother, O_2 tends to leave the mother's blood as it passes through the placenta and enters the fetal red cells, where it binds to the higher affinity fetal hemoglobin. The hemoglobin in fetal red cells then moves from the placenta to fetal tissues, where it gives up the O_2 because of the still lower P_{O_2}.

Control of Hemoglobin Function

The binding of oxygen by hemoglobin is affected by several environmental factors, including blood pH, the presence in red blood cells of organophosphate compounds, and blood temperature. These effects are understood most easily if we speak of hemoglobin's *affinity* for oxygen. Affinity refers to the degree of binding between one molecule and another, such as between O_2 and hemoglobin or a substrate and an enzyme. As shown in Figure 33-15, affinity is higher toward the left side of the oxygen dissociation curve: despite lower P_{O_2}'s there, hemoglobin still binds oxygen. To do so, it must have a high affinity. Conversely, affinity is low to the right, where P_{O_2}'s must be quite high before all four hemes on a hemoglobin molecule bind oxygen. Since the heme groups of varying types of hemoglobin are identical, it is solely different amino acids at certain sites in hemoglobin that determine the degree of affinity, or the response to blood pH, organophosphates, or temperature.

Blood pH

During their normal lives, cells metabolizing aerobically give off carbon dioxide. The level of that waste product tends to rise in the extracellular fluid and the blood. The CO_2 combines with water in the blood plasma or within red blood cells to form carbonic acid, H_2CO_3, which dissociates to H^+ and HCO_3^-:

$$CO_2 + H_2O \rightarrow H_2CO_3 \rightarrow H^+ + HCO_3^-$$

The presence of additional hydrogen ions tends to lower the pH of the blood plasma and red blood cells, and this acidity is said to shift the oxygen dissociation curve to the right. This means that at a given tissue P_{O_2}, the hemoglobin can bind less O_2. This shift to the right, called the **Bohr effect** (after its discoverer), has important implications for the uptake of O_2 in the lungs and its release in the tissues.

Note in Figure 33-16 that at a given P_{O_2} (say, 40 mm Hg) and at a pH of 7.4, hemoglobin will be about 75 percent saturated. Let us say that this represents the situation in a person at rest. When he or she begins to exercise, the CO_2 level rises, and the blood pH drops to

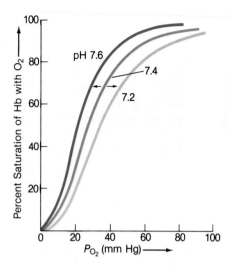

Figure 33-16   THE BOHR EFFECT: BLOOD PH INFLUENCES ON THE BINDING OF OXYGEN TO HEMOGLOBIN.
A lowering of blood pH causes a shift of the oxygen dissociation curve to the right and reflects a reduced affinity of hemoglobin for oxygen. Because of this shift, oxygen will be released from hemoglobin under conditions encountered in the tissue capillaries. This helps ensure the release of oxygen where it is required. Conversely, in the lungs, carbon dioxide is released, H^+ concentration is reduced, pH rises, and the curve shifts to the left. In this high-affinity state, hemoglobin binds oxygen more avidly. The pHs shown are hypothetical.

7.2. As a result, the saturation level of hemoglobin at the same P_{O_2} is only about 60 percent. Just such a drop in pH occurs as blood passes through the capillary beds of actively metabolizing tissues, where in addition the P_{O_2} may be only 20 mm Hg or less. The shift in the oxygen dissociation curve enhances the basic release of O_2 due to low partial pressures in the tissues; thus, O_2 is released where it is needed by the actively metabolizing cells. In the lungs, just the opposite occurs; first, when hemoglobin carrying bound H^+ binds O_2, the H^+ is given up; that H^+, as well as H^+ from plasma proteins, combines with HCO_3^- and as a result, CO_2 is given off into the air:

$$H^+ + HCO_3^- \rightarrow H_2CO_3 \rightarrow CO_2 + H_2O$$

What this means is that with a decrease in H^+, the blood pH rises, the oxygen dissociation curve shifts back to the left, and hemoglobin once again has a higher affinity for O_2. Thus the Bohr effect brings about enhanced uptake of O_2 by hemoglobin as the red blood cells pass through the lungs and an enhanced release of O_2 as they pass through tissues.

During every circuit of the blood through the body, this cycle is repeated: in the tissues, CO_2 builds up, H^+ increases, pH drops, hemoglobin has a lower affinity for O_2, and O_2 is released from the red blood cells and diffuses to the tissue cells; in the lungs, CO_2 is released, H^+ decreases, pH rises, hemoglobin has a higher affinity for O_2, and O_2 is taken up by the red blood cells, which continue their passage through the circulatory system.

Organophosphate Compounds

More important than the Bohr effect is the presence in red blood cells of organophosphate compounds, which also shift the oxygen dissociation curve to the right. In birds, the organophosphate ATP binds to hemoglobin and leads to lower-affinity O_2 binding and thus to greater delivery of O_2 to body tissues. In mammals, a substance called **DPG** (2,3-diphosphoglycerate), which is normally manufactured in red blood cells, causes the curve to be shifted to the right. Without DPG, human hemoglobin has a 50 percent saturation point at just 19 mm Hg, a level that probably would not allow us to maintain a high body temperature and function normally. But when DPG binds to hemoglobin, the 50 percent point shifts to the normal level of 29 mm Hg—a dramatic and relatively huge shift. This means that hemoglobin can release O_2 to the tissues even though P_{O_2} is quite high there already. Thus the amount of O_2 available in all the tissues of a mammal is much greater because of the DPG effect.

Temperature

The temperature of blood is yet another influence on the affinity of hemoglobin for oxygen; as temperature increases, affinity decreases. Within blood flowing through a warm area, such as the heart, the kidneys, or an actively working muscle, HbO_2 releases O_2 more easily than in a cool area. Conversely, as blood flows through the lungs, its temperature drops because evaporation within the small alveoli consumes much heat. This temperature drop in the lungs causes the oxygen dissociation curve to shift to the left, so that hemoglobin has a higher affinity for O_2, a perfect strategy for the lungs. The temperature effect on O_2 binding has been an important aspect of vertebrate evolution. If red blood cells from a cold-blooded frog could be placed in a mammal's circulatory system at 37°C, the oxygen dissociation curve would shift so far to the right that the frog hemoglobin molecules could not bind O_2 and deliver it to the tissues. Conversely, human hemoglobin in a frog's cold body would bind O_2 so avidly, with a radical leftward shift of the curve, that it could not release its load of O_2 under the conditions found in the frog's tissues. Critical to the evolution of reptiles, birds, and mammals were the mutations leading to amino-acid substitutions in hemoglobin that allowed that molecule to function well at high body temperatures. If those mutations had not occurred, birds and mammals could not generate the large amount of heat necessary to maintain body temperatures at 37° to 40°C, and reptiles could not behave in ways that frequently boost their temperature to that level.

Carbon Monoxide

The heme groups of hemoglobin molecules normally bind only oxygen, but they can also bind another substance of similar shape and properties: carbon monoxide (CO). Carbon monoxide actually binds more avidly to hemoglobin than does O_2 and increases the affinity of hemoglobin for both CO and O_2 molecules that are bound. The hemoglobin then cannot release any O_2 as it passes through the tissue capillaries. This explains why people exposed to CO, such as from automobile exhaust, can be asphyxiated even though O_2 is present in the air.

To summarize, the degree of binding between hemoglobin and oxygen depends on P_{O_2} and can be shifted by pH, organophosphate compounds, and temperature. As blood passes through tissues, pH falls, while temperature and DPG rise—all three work together to cause O_2 to be released more completely. Then, in the lungs, pH rises, temperature falls, DPG decreases a bit, and O_2 is bound more effectively.

Myoglobin: A Molecular Oxygen Reservoir

An animal's body will often store a critical material for use under conditions of high demand; for example, fuel for cell metabolism is stored in the all-too-familiar form of adipose tissue—fat. Oxygen, too, can be stored by means of a special respiratory pigment called **myoglobin,** which is abundant in animal muscle (*myo-* from Greek, meaning "muscle"). The structure of myoglobin (Figure 33-17) is roughly similar to that of one of the four globin subunits that make up a hemoglobin molecule. Each myoglobin molecule contains a single heme group and iron atom. Myoglobin has a higher affinity for O_2 than does hemoglobin. Therefore, when blood flows through a muscle capillary bed, O_2 is transferred from hemoglobin to myoglobin. When muscle cells are resting or are moderately active, P_{O_2} remains higher than 20 mm Hg, and myoglobin retains its O_2. But when the animal exercises so hard that the blood alone cannot supply all the O_2 needed, P_{O_2} will fall below 20 mm Hg in muscle cells and O_2 will dissociate from myoglobin, making O_2 available in the cells. Myoglobin also serves as a "ferryboat" for O_2. New studies show that O_2 bound to myoglobin diffuses faster than does free O_2. Thus myoglobin may ferry O_2 to the mitochondria faster than it could diffuse

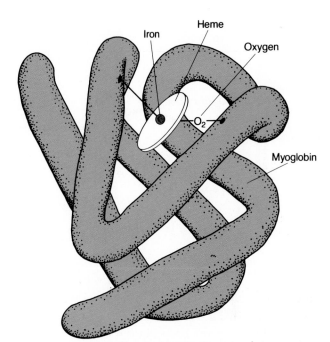

Figure 33-17   MYOGLOBIN: THE STORAGE PIGMENT.
Myoglobin consists of a single polypeptide chain with one heme group for binding oxygen. The storage pigment is present in muscle, where it binds oxygen at low partial pressures and surrenders its oxygen to the muscle cell when oxygen levels are severely depleted due to intense contractions.

on its own. Once there, of course, the O_2 is quickly used in metabolism.

Abundant myoglobin gives muscles a red hue. In sharks and mackerel, for example, a red strip of muscle fibers running down the fish's sides is the primary muscle tissue that contracts when the fish swims at normal speeds. The abundant O_2 from the myoglobin in these muscles helps support and sustain this activity. The rest of the fish's body muscles, which appear white, go along for the ride—they contract only when a burst of speed is needed. Similarly, the red breast muscles of ducks can support sustained flight, whereas the white breast muscles of domestic turkeys, lacking myoglobin, allow for only short bursts of activity. Myoglobin is also abundant in the muscles of diving mammals, such as whales and porpoises, and helps sustain oxidative metabolism even while they are submerged for long periods. Clearly, the whole biology of a vertebrate can be affected by the level of myoglobin in its muscles.

Carbon Dioxide Transport: Ridding the Body of a Metabolic Waste

Carbon dioxide, a by-product of oxidative metabolism, diffuses out of cells and into the capillaries because P_{CO_2}

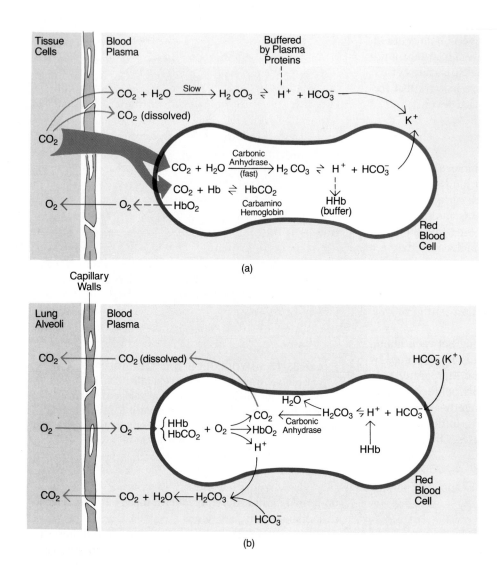

Figure 33-18   TRANSPORT OF CARBON DIOXIDE IN THE BLOOD.
Most carbon dioxide in the blood is transported as HCO_3^-, especially the large amount formed rapidly in red blood cells due to the activity of the carbonic anhydrase enzyme. Some CO_2 is bound to hemoglobin (carbamino hemoglobin), and a small proportion moves as dissolved gas. The reactions occurring in the blood capillaries (a), where CO_2 is taken up, are reversed in the lungs (b) as CO_2 is eliminated into the alveoli. A key feature of this system is that it operates to keep blood H^+ levels low and pH near 7.35. If H^+ builds up in blood plasma or red blood cells, breathing speeds; this eliminates CO_2 faster and shifts the various reactions so more H^+ combines with HCO_3^-. Note that O_2 diffuses into the red blood cells in lung capillaries and binds to hemoglobin (HbO_2) after HHb has surrendered its H^+. In the tissues, the HbO_2 gives up O_2, which diffuses into tissue cells.

is higher in the tissues than in the blood in the venous end of the capillary bed (Figure 33-18a). Some CO_2 dissolves in the blood plasma and is carried to the lungs unchanged. Some dissolved CO_2 reacts slowly with water in the plasma to form carbonic acid and then dissociates into HCO_3^- (bicarbonate) and H^+ ions, as we noted earlier. Plasma proteins act as *buffers* to bind the H^+ (buffers are substances that help prevent wide swings in pH). In terrestrial vertebrates, most of the CO_2 diffuses into the red blood cells, where the enzyme *carbonic anhydrase* greatly speeds up the initial reaction

$$CO_2 + H_2O \rightarrow H_2CO_3$$

Then during the secondary reaction

$$H_2CO_3 \rightarrow H^+ + HCO_3^-$$

H^+ is released. The presence of the H^+ ions lowers the pH, triggering the Bohr effect in the red blood cells. The

resulting acidity could affect enzymes and proteins within the red blood cells and plasma, but it does not because hemoglobin acts as a buffer. In this case, certain amino-acid side chains of the hemoglobin molecule bind to and transport H^+ ions. At the same time, the bicarbonate ion (HCO_3^-) is neutralized by potassium ions; some bicarbonate remains in the red blood cells, but most diffuses out and is transported in the blood plasma. The blood pH remains at about 7.35 because of buffering in the plasma and red blood cells, even though a large quantity of CO_2 is being carried. Finally, about 20 percent of the CO_2 that diffuses into the blood from tissue cells actually binds to amino groups ($-NH_2$) on hemoglobin and is carried in the red blood cell in bound form. Thus hemoglobin acts as a carrier of both H^+ and CO_2.

In the air within the lungs, P_{CO_2} is lower than in the blood. This lower P_{CO_2} shifts the chemical equilibrium of the circulating blood (Figure 33-18b): the hemoglobin side chains release H^+, which can then reassociate with the HCO_3^- that diffuses into the red blood cells from the plasma. Carbamino hemoglobin also dissociates as O_2 is bound. This once again yields H_2CO_3, which in the presence of carbonic anhydrase dissociates into H_2O and CO_2. The CO_2 diffuses out of the red blood cell into the plasma, then through the walls of the capillaries in the lungs, and finally into the alveoli. Then, during exhalation, the CO_2 is expelled through the bronchial passages and out through the nose or mouth. This series of chemical reactions is quite speedy: a human red blood cell enters a lung capillary and completes its CO_2 delivery in only 250 to 500 milliseconds, at the same time picking up O_2 and carrying it off toward the tissues.

CONTROL OF BREATHING

Although we humans can consciously take deep breaths to relax or consciously expel air forcefully to inflate a balloon, our normal rhythmic breathing cycle is automatic. But what regulates it? A region in the brain of all vertebrates, the *medulla*, contains the *respiratory center*, a group of nerve cells that automatically triggers periodic contractions of the muscles which bring about ventilation. In fish, these muscles are associated with the jaw and opercula; in amphibians, they are mainly lower-jaw muscles; in mammals, they are the intercostal muscles and diaphragm. The repeated and synchronous signals from the respiratory center regulate an organism's ventilatory movements and hence its breathing rate so that P_{CO_2} and P_{O_2} in the blood remain constant, regardless of the rates of consumption or production of oxygen and carbon dioxide in the body.

The respiratory center of aquatic vertebrates responds mainly to O_2 level in the body, because O_2 tends to be in short supply underwater and because CO_2 diffuses away so easily. In air, O_2 is abundant, and CO_2 removal is the problem; thus in terrestrial vertebrates, CO_2 level mainly drives the respiratory center. In fact, the center's rate also varies to help keep blood pH constant. If excess H^+ builds up, a land vertebrate breathes faster; this increases the reaction rate of H^+ with HCO_3^- to yield CO_2 and H_2O (CO_2 is breathed out and removed from the body). Conversely, if blood pH becomes too alkaline, the animal breathes slower, and CO_2 builds up, reacts with water, and yields H^+ to lower the pH again.

The respiratory center receives information on P_{O_2} from chemoreceptor nerve endings in the *aortic bodies* found in the wall of the aorta and from nerve endings in the *carotid bodies* located in the carotid arteries (Figure 33-19; endings are also present in the pulmonary artery walls). (*Pressure receptors* are present, too, and report blood pressure and so the work being done by the heart;

thus during exercise, heart rate rises, blood pressure rises, and the organism breathes faster to sustain such activity.) Chemoreceptor nerve endings in the walls of the medulla monitor P_{CO_2} and pH in the extracellular fluid that bathes the nerve endings in that portion of the brain. Changes in P_{O_2}, P_{CO_2}, and pH stimulate the respiratory center to slow or speed the rhythmic discharge of nerve impulses to the ventilatory muscles.

When a person takes fast, deep breaths without exercising (hyperventilates), he or she expels a great deal of CO_2, thereby lowering the CO_2 content of the blood; this raises blood pH, and in turn, raises pH of the extracellular fluid within the tissue near the respiratory center. As a result, respiratory-center nerve firing is slowed and the breathing rate drops, so that P_{CO_2} builds up and pH falls again to normal ranges. If a person exercises vigorously, blood and tissue P_{O_2} will drop and P_{CO_2} will increase (pH drops), triggering more frequent respiratory-center nerve firing and hence faster, deeper breathing. These mechanisms govern breathing rates even if a person tries to overcome their control: breath holding is invariably interrupted by an involuntary gasp for air, while hyperventilation ends in a fainting spell so that a normal, slower breathing rate can quickly be restored.

Surprisingly, new experiments have revealed that factors other than P_{CO_2}, pH, and P_{O_2} may govern the breathing rate during exercise. For example, the rate of breathing can vary widely during exercise, yet blood P_{CO_2} and P_{O_2} remain remarkably constant. If there are no changes in these parameters, what triggers more rapid breathing? Similarly, at the start of vigorous exercise, the breathing rate increases quickly—well before CO_2 builds up. Conversely, when exercise is suddenly stopped, the breathing rate drops before normal P_{O_2} is restored. The explanation for these phenomena is that special receptors monitor the degree of contraction of limb muscles. These stretch receptors in the tendons and joints of the arms and legs actually trigger a change in breathing rate to fit the level of exercise, and not simply to correspond to blood chemistry. This response is called the **mechanoreceptor reflex.** It ensures that blood gases are kept at normal levels during and after exercise, and avoids the undesirable consequences for blood and body chemistry of letting P_{O_2}, P_{CO_2}, and pH levels fluctuate. Instead, the changes are anticipated and are compensated for ahead of time.

Recent studies suggest that the respiratory center may shut off temporarily during sleep, particularly in elderly persons. Under most circumstances, increasing P_{CO_2} reactivates the center, but some older people may die in their sleep if the center is slow to reactivate.

Other research shows that improper brain control of breathing may explain sudden infant death syndrome, which kills 8,000 babies per year and accounts for about

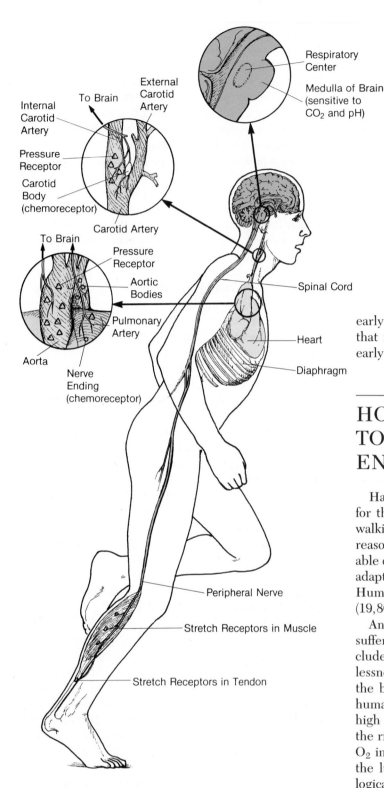

Figure 33-19   RESPIRATORY RATE AND CHEMORECEPTORS AND MECHANORECEPTORS. Chemoreceptors are sensitive to O_2 levels (carotid body, aortic bodies, and pulmonary artery) as well as to CO_2 levels and pH (medulla). Mechanoreceptors measure stretch in certain body muscles and tendons, and pressure receptors in the blood vessels also participate in respiratory control, especially during exercise. The sites of these various sensory receptors are shown. The respiratory center itself is situated in the medulla.

early brain-stem maturation are now being devised so that infants at risk for the syndrome can be identified early.

HOW ANIMALS ADAPT TO OXYGEN-POOR ENVIRONMENTS

Have you ever hiked in high mountains and found that for the first day or two, simply carrying a backpack or walking up a mild grade left you gasping for breath? The reason is that the air at high altitudes has so little available oxygen that organisms cannot survive there without adaptations of the circulatory system and hemoglobin. Humans, in fact, can live and work even at 950 meters (19,800 ft.) high in the Andes; how do they do it?

An out-of-breath hiker in a high-mountain area may suffer altitude sickness: a spectrum of maladies that includes headache, extreme fatigue, nausea, and breathlessness. These symptoms result from a drop in P_{O_2} in the blood. For unknown reasons, the DPG content of human red blood cells increases within a few hours at high altitudes, shifting the oxygen dissociation curve to the right and making it easier for hemoglobin to release O_2 in tissues, but harder for hemoglobin to bind O_2 in the lungs. This seemingly maladaptive human physiological response is quite different from that in birds and other mammals at high altitudes. For instance, high-flying birds and the llamas of the South American Andes have oxygen dissociation curves that are shifted to the left, resulting in better binding of O_2 in the lungs—not worse (Figure 33-20).

A person who lives at sea level and goes to the mountains for several months will gradually become accli-

half of all infant deaths in the United States. Some biologists believe that slow development of the medulla and its respiratory center may impair the arousal response needed to wake a child if breathing stops. And researchers also think that this response may be depressed further by infections of the nose, throat, or bronchi. Tests of

(a)

(b)

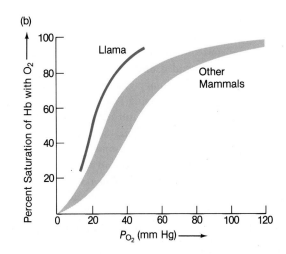

Figure 33-20   LIFE AT HIGH ALTITUDE.
(a) Animals that live at high altitudes, such as these llamas, alpacas, and sheep, generally have hemoglobin with a higher affinity for oxygen than the hemoglobin of similar animals living at lower elevations. (b) This fact is reflected in the oxygen dissociation curve. The llama's curve is located to the left of curves for mammals of comparable size living at lower altitudes.

mated to high elevations through a number of physiological changes. The number of red blood cells increases, and thus there is more hemoglobin available to carry O_2. The rate of breathing and heartbeat also increase to facilitate gas exchange and delivery. None of these changes is permanent: when the traveler returns to sea level, his or her original physiological functioning will gradually be restored.

A person who lives an entire lifetime in the high mountains will develop enlarged ventricles of the heart for driving large volumes of blood. The Quechua Indians of the high Andes, for example, tend to be small in stature, yet they have large chests enclosing significantly larger hearts and lungs than those of South American Indians living at lower elevations. Their enlarged hearts represent a permanent adaptation to life in an oxygen-poor environment.

Adaptations can also be seen in deep-diving marine mammals that may be deprived of the opportunity to breathe for up to an hour. Most have a huge number of very large red blood cells, large quantities of myoglobin in their muscles, and hemoglobin with an oxygen dissociation curve shifted to the right, making it easier to release oxygen to the tissues during a dive.

Marine mammals have two other adaptations that are part of the **diving reflex,** a complex set of responses to swimming downward in the sea. The first is the shunting of blood away from most regions of the body during a dive, and the second is the slowing of the heartbeat. When a whale dives, blood flow is diverted from other regions mainly to the heart muscles, the brain, and, to a lesser degree, the skeletal muscles. The skeletal muscles can incur a large **oxygen debt:** they can continue to con-

tract for locomotion even after all stored O_2 is depleted. The energy for such contractions comes from glycolysis (Chapter 7), so lactic acid is formed and builds up in the muscle tissues. When the whale surfaces, it exhales its moist oxygen-depleted air, which condenses to form the "spout" (Figure 33-21). A great deal of CO_2 is blown off in the spout, and O_2 is rapidly loaded in the hemoglobin passing through the lungs. Contrary to what might be predicted, marine mammals do not surface because they run out of oxygen; instead, they come up when falling blood glucose levels get very low. The second part of the

Figure 33-21   SPERM WHALE (*PHYSETER MACROCEPHALUS*) "SPOUTING" IN THE SEA OF CORTEZ, MEXICO.
The spout is actually water droplets that condense when hot, humid air is exhaled. The whale inhales immediately after exhaling the spout.

diving reflex, the slowing of the heartbeat (also called *bradycardia*), occurs in all vertebrates. Thus the heart of a marine mammal and of a terrestrial vertebrate slows automatically when the head is immersed; conversely, a fish's heart slows when the animal is removed from water and held in air. Perhaps the slowing is an adaptation to ensure less rapid depletion of O_2 until the animal can return to its normal source of oxygen.

You may wonder why marine mammals, birds, and reptiles have not developed gills instead of all these special physiological diving adaptations. One reason is the low solubility of O_2 in water: gills simply could not extract enough O_2 from water to supply the needs of aquatic mammals, birds, and reptiles. In addition, hemoglobin would have to have such a high affinity for O_2 to pick it up from water that it could not deliver much to the tissues. Thus the whole pattern of cellular metabolism would have to be stepped down. In short, the high level of metabolism and the general activity characteristic of reptiles, birds, and mammals are superbly well served by lungs, by their hemoglobins, and by the breathing of oxygen-rich air. In fact, so well matched are these animals and their respiratory mechanisms that one might regard the evolution of lungs as a one-way process: lungs and ventilation mechanisms may evolve to become better adapted to particular ways of life, but wholly new systems to carry out the respiratory functions are unlikely to arise in these vertebrate lineages.

SUMMARY

1. *Respiration* is the process by which organisms exchange O_2 and CO_2 with the environment.

2. Air exerts a weight on the Earth's surface that is measured as *atmospheric pressure*. Individual gases in the air—N_2, O_2, and so on—exert *partial pressures* that are proportional to their concentrations in air. Dissolved gas also has partial pressure in liquid. Gases have different solubilities in liquids; in general, solubility increases as temperature falls, but it decreases as temperature rises. This has important consequences for O_2 and CO_2 movements in cellular and extracellular fluids.

3. Simple diffusion of respiratory gases can meet the needs of only very small or very flat organisms. In large organisms, bulk-flow processes are required to distribute sufficient quantities of O_2 and CO_2 to the body tissues. However, diffusion occurs at both ends of the process—when the gases are taken up from air or water and when they are released in the tissues.

4. Respiratory organs can be external, as are *gills*, or internal, as are *tracheae* and *lungs*. Both types have large surface areas for gas exchange and short diffusion pathways for O_2 and CO_2 to travel between the outer air or water environment and the blood or tissue fluids.

5. Gills take many forms but always function as a site for exchange of dissolved gases with water. *Countercurrent exchange systems* in fish gills and other animal organs greatly increase the efficiency of exchange of O_2, heat, salts, and other substances.

6. Air-breathing organisms expend much less energy obtaining O_2 from air than aquatic organisms do to pick up O_2 from water via gills.

7. Insects have a system of tiny, hollow tubes called *tracheae* that transport air deep into the body. There O_2 dissolves in fluid and passes into the hemolymph, which bathes cell surfaces. The tracheal tubes also carry CO_2 away from cells and out of the insect's body.

8. Lungs are hollow, branched internal respiratory organs that receive air from the outer environment, and can be smooth and saclike or very convoluted. Blood transports O_2 from the lungs to tissues and CO_2 in the opposite direction. The actual site of gas exchange is a moist layer covering lung cells.

9. In mammals, air is filtered, humidified, and warmed as it enters and passes through the nostrils, nasal cavities, mouth cavity, and *pharynx*. The epiglottis seals off the *larynx*, or entrance to the *trachea* (windpipe) and lungs, during swallowing. Vibration of the vocal cords in the larynx produces sounds humans can modify into certain consonants or vowels.

10. In mammals, the respiratory tree includes the trachea, which branches into two primary *bronchi*, then secondary bronchi, and bronchioles, before ending in cuplike *alveoli*, where gas exchange takes place.

11. The muscle-powered expandable rib cage and the *diaphragm* of mammals change the volume of the thoracic cavity and draw in air or push it out through a suction-pump–elastic-recoil system. During exercise, muscles expel more residual air from the lungs, so that 80 percent of lung volume (*vital capacity*) can be filled with fresh air, instead of the usual 10 percent (*tidal volume*).

12. *Air sacs* lighten a bird's body and are part of a reservoir system that allows the lungs to be flushed completely with fresh air. Birds carry out continuous *crosscurrent* gas exchange.

13. *A swim bladder* is a hydrostatic organ that enables a bony fish to ad-

just its buoyancy. The red and oval glands increase or decrease the volume of the swim bladder after the fish descends or ascends.

14. Hemoglobin is an iron-containing *respiratory pigment* that binds O_2 when P_{O_2} is high (as in lungs) and releases it when P_{O_2} is low (as in tissues).

15. Blood pH, organophosphates in the red blood cells, and blood temperature alter the binding characteristics of hemoglobin for O_2. By shifting the *oxygen dissociation curve*, they make more efficient the binding of O_2 in the lungs and the release of O_2 in the tissues. The *Bohr effect* involves a decrease in the affinity of hemoglobin for O_2 as pH falls and increased affinity when pH rises.

16. *Myoglobin* is a respiratory storage pigment similar in form to one hemoglobin polypeptide chain. Myoglobin is found mostly in muscle tissue and maintains a reserve supply of O_2 that can be released under conditions of high O_2 demand and low local P_{O_2} levels.

17. Carbon dioxide is transported in blood plasma and red blood cells, partially as dissolved CO_2, but mostly as the bicarbonate ion, HCO_3^-. Hemoglobin is the blood's primary buffer and carries the H^+ arising during production of HCO_3^-.

18. The brain's respiratory center responds primarily to changes in P_{CO_2} and pH and, to a lesser degree, to changes in P_{O_2} in the blood and body fluids. The center can regulate ventilation rates to maintain P_{O_2} and P_{CO_2} within a set range of values. Rates of breathing during exercise are controlled primarily by *mechanoreceptor reflexes*, and not by altered blood chemistry.

19. Different animals adapt to oxygen-poor environments through a variety of mechanisms, including production of more red blood cells and thus more hemoglobin; faster breathing and heart rates; enlarged heart ventricles and lungs; shunting of blood to vital organs; and the *dive reflex*, a slowing of heartbeat when the face of a terrestrial vertebrate is immersed in water.

KEY TERMS

air sac
alveolus
atmospheric pressure
Bohr effect
bronchus
countercurrent exchange system
crosscurrent exchange system
diaphragm
diving reflex
DPG (2,3-diphosphoglycerate)

external intercostal
gill
internal intercostal
larynx
lung
mechanoreceptor reflex
myoglobin
oxygen debt
oxygen dissociation curve
partial pressure

pharynx
respiration
respiratory pigment
solubility
swim bladder
tidal volume
trachea
tracheae
vital capacity

QUESTIONS

1. The process by which animals exchange gases with the environment is called _____. What does this process entail? How does it differ from cellular respiration?

2. What is atmospheric pressure at sea level? How high (in mm) will this pressure push a column of mercury (Hg)? At sea level, what is the partial pressure of oxygen (P_{O_2}) in mm Hg? Of nitrogen?

3. Name four important physical factors that affect the amount of O_2 available to an organism's cells at a given time.

4. What four major types of respiratory systems have evolved in the animal kingdom? What are two problems that must be overcome as a moist, expanded surface area for gas exchange evolves?

5. What is meant by (a) countercurrent and (b) crosscurrent flow of water or air and blood? Compare the morphology and efficiency of these two exchange systems. Which group(s) of animals display each type?

6. Through what structures do insects take in air? How is air delivered to an insect's body cells?

7. Birds' lungs allow for continuous flow of air. Why is this significant? Contrast the avian respiratory system with the mammalian one.

8. Name three important respiratory pigments found in the animal kingdom. What is the general chemical structure of each? What group(s) of organisms produce these pigments, and in what cells or tissues?

9. What chemical and physical factors control hemoglobin function? Briefly discuss the effect of each.

10. How is CO_2 carried in mammalian blood? What is the role of carbonic anhydrase in CO_2 transport?

ESSAY QUESTIONS

1. Discuss oxygen dissociation curves. On the same graph, draw curves for

   a. normal human hemoglobin (Hb)
   b. human Hb plus DPG
   c. chilled human Hb
   d. fetal human Hb
   e. myoglobin

Explain the significance of each curve.

2. Three structural "themes" are common to all animal respiratory systems:

   a. expansion of surface area
   b. maintenance of a wet surface for gas exchange
   c. a ventilating or pumping mechanism

Describe how each theme is manifested in insects, fish, birds, and mammals.

SUGGESTED READINGS

ECKERT, R., and D. RANDALL. *Animal Physiology.* 2d ed. New York: Freeman, 1983.

The chapters on respiration, circulation, and hemoglobin are superb.

FISHMAN, A. P., section ed. "Respiratory Physiology." *Annual Review of Physiology* 45 (1983): 391–451.

Four articles on the regulatory roles of mechanoreceptors, carbon dioxide, and oxygen during exercise and rest.

RANDALL, D. J. "Gas Exchange in the Fish." In *Fish Physiology*, vol. 4, edited by W. S. Hoar and D. J. Randall. New York: Academic Press, 1970.

This article is a good summary of gill function in respiration.

SCHMIDT-NIELSEN, K. *How Animals Work.* Cambridge, England: Cambridge University Press, 1972.

This good general book is a fine source to amplify the concepts discussed in this chapter.

STEEN, J. B. "The Physiology of the Swim Bladder." *Acta Physiologica Scandinavica* 58 (1963): 124–137.

An excellent treatment of red gland and oval gland functions.

DIGESTION AND NUTRITION

Now good digestion wait on appetite, And health on both!
William Shakespeare, *Macbeth* (III.iv)

Galápagos gourmet: A giant tortoise (*Geochelone elephantopus abingdoni*) grazes on a Pinta Island cactus.

Animals must feed. They must take in prefabricated nutrients for metabolism and growth because, unlike photosynthetic organisms, they cannot manufacture all necessary nutrients from raw materials. The tiny amoeba engulfing a bacterium; the bright green aphid sucking juice from a plant stem; the coyote tearing the soft belly of a jack rabbit; the college student munching popcorn—all are carrying out one of life's most basic strategies: the acquisition of nutrients. All organisms, plants included, ultimately depend on external sources of energy to sustain life processes. But autotrophs can make their own organic nutrients from inorganic precursors, whereas heterotrophs must gather and rely on food in its many forms. We have already discussed energy storage and energy processing by autotrophs. In this chapter, we will focus on the intake and processing of nutrients by animals.

Feeding usually involves both **ingestion**—the taking of food pieces into the body—and **digestion**—the chemical breakdown of foods into compounds small enough to cross cell membranes and to be utilized for energy or by being modified and incorporated into tissues. We often think of ingestion as biting and chewing, since that is how we feed ourselves. But there are a number of other ingestion mechanisms as well. Such mechanisms include rasping, sucking, or crushing food, swallowing it whole, and filtering it from water. Digestion is a multistep process that includes the chemical breakdown of food pieces into large organic molecules, the breakdown of these molecules into smaller compounds, and the absorption of these compounds and their transformation into molecules that can be used in cellular metabolism. From these molecules, the body generates ATP and forms building blocks for biosynthesis, as we have seen in Chapter 7.

We will begin this chapter by examining the modes of ingestion and digestion in the simplest heterotrophs, bacteria and protists and then consider the more complex systems of invertebrates and vertebrates. We will see an evolutionary trend from the intracellular breakdown of food molecules to their extracellular breakdown in special gastric chambers, and we will consider the human digestive system in detail. We will then discuss the chemistry of digestion and absorption of nutrients and the coordination of ingestion and digestion so that an organism has a constant supply of energy even if it feeds periodically. Finally, we will discuss nutrients—carbohydrates, fats, proteins, vitamins, and minerals—that animals must consume and the roles these essential substances play in cellular metabolism and health. Throughout this chapter, we shall see that digestion is as vital as are circulation, immunity, and respiration to an animal's daily survival.

ANIMAL STRATEGIES FOR INGESTION AND DIGESTION

Many animals spend the majority of their time acquiring food. Indeed, one might call feeding the universal pastime (Figure 34-1). The mouth, teeth, intestines, or other major structures of an animal's body usually reflect the way it harvests food, the type of food it eats, and the way it digests that food. Let us look at the fascinating spectrum of mechanisms for ingesting and digesting food among simple and complex organisms.

Intracellular Digestion in Simple Organisms

Many kinds of protozoans as well as internal parasites and certain marine invertebrates take in nutrients or food particles directly through the body walls. Nutrients can be absorbed or engulfed directly by body-wall cells and are then broken down by *intracellular digestion*—that is, within each cell by means of digestive enzymes contained in lysosomes.

Parasites such as tapeworms absorb nutrients that have been partially digested in their host's intestines. These nutrients are actively transported across the plasma membrane of each cell of the tapeworm's body and then are digested intracellularly. Experiments with radioactive tracers show that some aquatic invertebrates, such as certain free-living marine worms, absorb dissolved glucose, amino acids, and other organic molecules directly from sea water and soft bottom sediments. In this way, they can satisfy most, if not all, of their metabolic needs.

Protozoans commonly ingest particulate food and digest it internally. A specialized example is *Paramecium*. This single-celled organism has an ingestion organelle, the gullet, into which tiny food particles are swept by beating cilia (Figures 21-13 and 21-14). When a particle arrives at the base of the gullet, a food vacuole pinches off and carries the particle into the cytoplasm. Smaller vesicles within the cytoplasm fuse with the vacuole and transfer digestive enzymes to it. The enzymes degrade the food particle, hydrolyzing it into sugars, fats, amino acids, nucleic-acid bases, inorganic ions, and so on. These nutrient molecules then move from the food vacuole into the cytoplasm. Finally, the food vacuole travels to the inner side of the cell surface, fuses with a special area called the anal pore, and excretes undigested debris. This sequence is similar to the way a lysosome within an animal's cell digests materials taken in by phagocytosis. Some planktonic relatives of *Paramecium* have discriminatory digestive vacuoles. The chloroplasts of ingested algae remain undigested and continue to function as internal carbohydrate and oxygen factories.

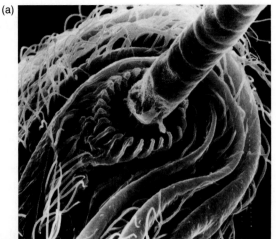

Intracellular and Extracellular Digestion in Simple Animals

Animals that rely on intracellular digestion alone must have bodies that are very small or very flat, and they are confined to habitats that provide a rich supply of food molecules. This is because diffusion is the primary means of distributing nutrients, raw materials, or dissolved gases within the body. Larger size and more complex body organization are possible for animals such as flatworms and cnidarians (jellyfish) because they have an internal gut chamber called the *gastrovascular cavity* ("gastro" because it serves digestion; "vascular" because it also functions like the circulatory system). Within such cavities, ingested water and food particles and dissolved gases can travel passively throughout the body.

In the flatworm *Planaria* and in cnidarians such as hydra, the gastrovascular cavity is a blind-ended chamber with a single opening through which food enters and wastes exit (Figure 25-12). Once flatworms or cnidarians ingest food particles, enzymes are secreted into the gut, and the particles are partially broken down into macromolecules. Because this sort of breakdown process takes place in the *lumen*, the hollow part of the gut chamber, rather than within individual body cells, it is called *extracellular digestion*. In the flatworm and cnidarians, extracellular digestion is incomplete; cells lining the gut phagocytize the macromolecules and within these gut cells, the nutrients are degraded further by enzymes.

Extracellular Digestion in Complex Animals

In most animals, complete extracellular digestion takes place in a two-ended gut called the **alimentary canal,** which extends from mouth to anus. Food makes a long, one-way passage through the lumen of this tube, and the lumen is continuous with the outer world through the two openings. The alimentary canal, also

Figure 34-1   EATING: THE UNIVERSAL PASTIME.
(a) A ciliate protist ingesting a cylindrical cyanobacterial cell (magnified approximately 500 times). (b) A chameleon (*Chameleo jacksoni*) striking at its insect prey by unfurling its long sticky tongue with lightning speed. (c) A nymph stage of a great waterbug (*Lethucerus* species) has killed this minnow and is sucking juices from its body. (d) A manatee, a mammal like ourselves, feeding on sea plants.

called the *digestive tract,* or *gut,* has discrete regions that carry out specialized tasks: breaking up food, storing it temporarily, digesting it chemically, absorbing digested nutrients, reabsorbing water, storing wastes, and, finally, eliminating wastes.

The Earthworm Gut

The digestive tract of the earthworm serves as a simple example of the two-ended gut with specialized regions and functions (Figure 34-2). The earthworm's mouth, like that of all animals, is modified so that food is ingested in small enough pieces that digestion is possible. The worm's mouth operates like a suction tube to draw in dirt and food particles (mostly fragments of dead leaves) no greater than a few millimeters in diameter. After passing through the **pharynx,** ingested materials travel into a tubular passageway, the **esophagus,** where an alkaline secretion is added. The soil and food are stored in the **crop,** a thick-walled chamber. From there, the mixture passes to the **gizzard,** a larger muscular chamber for processing the food into smaller pieces. In the gizzard, hard, toothlike projections from the inner walls function like tiny millstones to pulverize and grind the food. Such pulverizing increases the food's total surface area so that extracellular enzymes can more effectively break it down into sugars, fats, amino acids, and other nutrients. The enzymatic breakdown of food and the absorption of nutrients take place in the earthworm's **intestine,** the long, tubelike remainder of the alimentary canal. Secretory cells in the intestine wall produce and excrete digestive enzymes into the lumen. The inner wall of the intestine has a single, large infolding that increases the surface area for nutrient absorption. In addition, each epithelial cell in the gut lining has tiny hairlike projections, called **microvilli** (singular, *microvillus*), that expand the gut's absorptive surface even further. Blind-ended pouches called **caeca** (singular, *caecum*) extend from the intestine and hold food for extended periods of time, so that enzymes have time to act and absorption can be fairly complete.

While in the gut, food and the molecules derived from it are suspended in a watery fluid. Since most terrestrial animals, including the earthworm, cannot afford to lose much body water, most water is reabsorbed by the terminal portion of the intestine. Finally, fairly dry waste matter is excreted through the anus. Water reabsorption is particularly important for residents of arid habitats. And if reabsorption is disrupted for a long enough period, any terrestrial animal will become dehydrated and die from the water loss.

Insects: Variations on the Basic Plan

The earthworm's basic digestive structures and processes are also characteristic of snails, lobsters, sea cucumbers, spiders, and relatives of each of those animals. A good example of diversity built on that basic plan is provided by the insects, with their range of specialized mechanisms for ingesting and digesting. Many insects feed on plant or animal juices and so have a piercing, needlelike proboscis that can penetrate flowers, stems, leaves, or skin and suck in only fluid (Figure 34-3). Some, such as mosquitoes, have glands that secrete anticoagulants so that a steady flow of fluid will be maintained. Other insects have strong mandibles made of chitin for cutting tiny chunks of wood (as do termites) or leaves (as do leaf-cutting ants) or for biting animal prey. Various insects have caeca or other storage chambers in the gut that store food for digestion over long time periods; this allows irregular and widely spaced feedings to take place, as opposed to continuous feeding. Female mosquitoes, for example, store mammalian blood in thin-walled expandable caeca. The blood from a single feeding can be held for up to a week in these reservoirs and be passed into the alimentary canal as needed. Honeybees that feed on nectar and pollen have special "lips" in the gut called the "honey stopper," which pick out pollen grains and transfer them to the hindgut for protein digestion, while the watery nectar is held in an anterior portion of the gut, where it is converted to honey. As in the earthworm, the terminal portion of the insect intestine reabsorbs water so that large quantities are not lost during excretion of wastes.

Figure 34-2   THE TWO-ENDED GUT: A MAJOR EVOLUTIONARY DEVELOPMENT.
Most bilaterally symmetrical animals have a digestive tract with separate openings at the mouth and the anus. This earthworm gut also has specialized regions that carry out different functions.

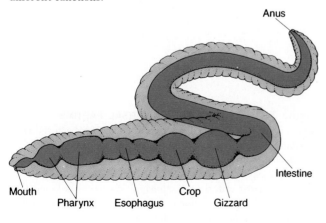

Anus

Intestine

Mouth   Pharynx   Esophagus   Crop   Gizzard

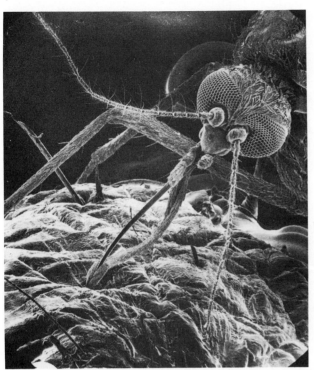

Figure 34-3   INSECT MOUTHPARTS.
Insects have an array of mouthparts that are adapted for biting, sucking, and chewing. This mosquito's syringelike proboscis (magnified about 22 times) is an ideal shape for piercing this human skin and sucking blood.

walled sac that can expand to hold the large, infrequent meals consumed by carnivores. Partial digestion occurs in the stomach, and then the food is passed into the *small intestine*, where further digestion and nutrient absorption occur. Less time is required to digest animal tissues than plant tissues, so a carnivore's intestine tends to be shorter than an herbivore's.

The snake provides an interesting example of the interaction of ingestion and digestion mechanisms in carnivores. Snakes swallow foods whole—birds' eggs, mice, and even (in the case of huge tropical snakes) small pigs and deer. A snake's teeth are back-curving grappling hooks that help move the unchewed food down into the gut and prevent it from moving forward, out of the mouth. The alimentary canal acts slowly and thoroughly, requiring a week or more to crush and digest the prey and absorb its nutrients.

Not all large animals have teeth. Birds lack teeth, probably to reduce body weight for flight. When they eat meat (worms, insects) or plant material (seeds), the food passes into the muscular gizzard, in which pebbles are stored. The food is ground between the pebbles into fine particles. The so-called baleen whales also lack teeth. Baleens include the largest animals on Earth, the blue whales. They and their relatives such as humpback whales survive by filtering krill, small crustaceans, and

Feeding Styles of Meat-Eating Vertebrates

The alimentary canal of meat-eating vertebrates is similar in principle to that of earthworms and insects, except that the mouth is usually outfitted with sharp teeth. Exclusive meat eaters are called **carnivores,** and carnivorous vertebrates use their teeth to tear other organisms into pieces of ingestible size (Figure 34-4). They then often swallow the meat hastily in big chunks. Some carnivores, such as alligators and toothed whales, have only pointed, conical teeth for piercing and tearing. Others, including sharks, have piercing teeth combined with razor-sharp serrations like a saw blade. Mammals such as cats and dogs typically have piercing *canines* in the front of the mouth, to rip apart flesh, and broader *molars* with raised points in the rear, to grind flesh and bones into small pieces. Carnivores need not chew and grind their food extensively before swallowing, since animal flesh is not surrounded by the tough cellulose walls found in plants, which, as we will see, must be ground up.

The tongue and muscular walls of the pharynx and esophagus propel the large chunks of meat through the upper alimentary canal and into the **stomach,** an elastic-

Figure 34-4   CARNIVORE TEETH: FOR TEARING AND CRUSHING.
Teeth, such as a lion's, puncture, tear, and crush the flesh and bones of the animal's prey. But such teeth are of little use in grinding tough plant materials; carnivores thus cannot graze on tough grasses and shrubs.

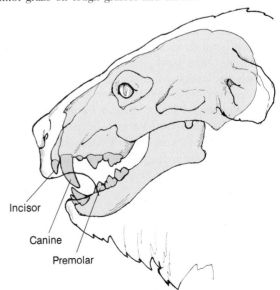

Incisor

Canine

Premolar

Figure 34-5   FILTER FEEDING: A COMMON FOOD-GATHERING TECHNIQUE FROM THE VERY SMALL TO THE VERY BIG.
(a) Mosquito larva (*Culex pipiens*) have combs; (b) flamingos have platelets on their bills; and (c) humpback whales have baleen. These examples reveal the parallel evolution of straining devices to concentrate tiny food particles from water.

plankton from sea water. A ninety-foot blue whale can collect up to 1.5 tons of food a day on fringed sheets of baleen that hang down from the upper gums. Baleen is nothing more than hardened, shredded sheets of the gum epithelial layers. A number of other very different kinds of animals also are filter feeders (Figure 34-5). Flamingos have fringed filters hanging from the upper bill and a deep-sided, curved lower bill, just as do baleen whales. This is a similar evolutionary solution to the same problem—removing small food particles from a dilute medium.

Feeding Styles of Plant-Eating Vertebrates

Most of the Earth's large vertebrates—elephants, giraffes, buffalo—are exclusively plant eaters, or **herbivores.** Most mammalian herbivores are similarly equipped for ingesting plant matter, whether they eat grasses, leaves, or seeds: they have chisel-like *incisors* in the front of the mouth, to gnaw or slice off plant materials, and broad, flat molars in the back, to crush and grind plant cell walls and fibers for long periods of time (Figure 34-6).

To appreciate the remarkable gut structures of large mammalian herbivores, we need to recall a point discussed in Chapter 26. Reptiles, the ancestors of mammals, do not generate their own body heat. As a result, even large reptiles need to eat infrequently: a reptile weighing about as much as an adult human needs the equivalent of only two or three of our meals per week. Humans in developed countries commonly eat from fourteen to twenty-one or more meals per week; but about 90 percent of the food consumed is used to generate body heat. Furthermore, a reptile's intestine is very short compared with that of any mammal, since digestion and absorption in the reptilian gut can go on slowly. The mammalian way of life, with its high body heat and great food consumption, depends partly on a major evo-

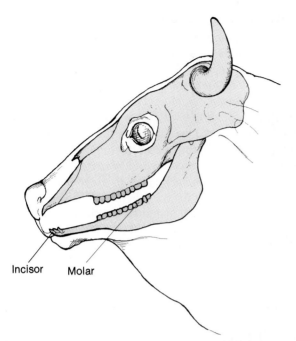

Figure 34-6   HERBIVORE TEETH: FOR GRINDING.
This ox skull provides an example of the chisel-like incisors
that herbivores use to slice off plant materials. The large set
of molars is used in grinding tough plant cell walls.

lutionary event: the emergence of a very long intestine.
And herbivores show an even greater specialization of
the long intestinal tract than do other mammals.

This specialized gut helps explain how herbivores
weighing hundreds of kilograms are able to thrive on an
exclusive diet of grasses and leaves. Herbivores have
symbiotic microbes that break down cellulose even
though the animals themselves lack enzymes to digest
plant matter directly. Thus herbivores reap calories from
plant cellulose—a trick that humans cannot do from the
cellulose in fruits and vegetables. (Other substances in
plants, of course, do yield calories.) When animals lack
their own digestive celluloses, and also lack symbiotic
microbes that make celluloses, any cellulose they eat
simply passes through the alimentary canal as undigested "roughage." In contrast, herbivores harbor in
their guts special monerans or protists that do have such
enzymes. The stomach chambers of large herbivores
such as cows, the large intestine of horses, and the guts
of termites are all places where such symbiotic microbes
thrive. Protists in the termite gut secrete cellulase into
the insect's gut lumen; enzymatic breakdown of cellulose
takes place there, and the insect can then absorb some of
the cellulose subunits. Interestingly, a termite's instinctive behavior guarantees a constant supply of these microbes in the gut: each time the termite molts, and the
old exoskeleton and gut lining are shed, the microbes are

lost. But the termite promptly eats the shed tissues, thus
reinfecting its gut with cellulose digesters. Rabbits also
depend on microbes to digest cellulose; anaerobic bacteria ferment cellulose in a large caecum extending from
the rabbit's intestine. Following digestion, rabbits excrete two kinds of *feces*, or solid wastes: pellets containing undigested waste matter and a viscous fluid containing intestinal bacteria and undigested food from the
caecum. The rabbit consumes the fluid, thus returning
the beneficial bacteria to the digestive tract and giving
the food a second chance to be digested.

Extra stomach chambers containing symbiotic organisms have evolved at least three times in mammals: in
ruminants; in sloths; and in certain marsupials, including
kangaroos and opossums. Let us look at the ruminants to
see how the herbivore's specialized gut was made even
more efficient. Cattle, sheep, deer, and other hooved
herbivores have an elaborate four-chambered stomach in
which plant matter is fermented and digested. Three of
these chambers are considered to be caeca, or storage
sites, and the fourth, the *abomasum*, is the equivalent of
the stomach of most other mammals (Figure 34-7), since
it is the only chamber that secretes digestive enzymes.
Hooved herbivores are called *ruminants* after the largest

**Figure 34-7   THE RUMINANT'S FOUR-CHAMBERED
STOMACH.**
In this cow's digestive system, the green arrows show
the passage of ingested food and the orange arrows show
the passage of rechewed food through the chambers of the
stomach. During the first passage, food enters the reticulum
and the rumen, where it is broken down by microorganisms.
This cud is then regurgitated, rechewed, and swallowed; it
then moves into the reticulum, omasum, and abomasum
(true stomach), where chemical digestion by enzymes takes
place. The digested food then enters the small intestine.
(For clarity, the route of partially chewed food from the
reticulum to the rumen is not shown.)

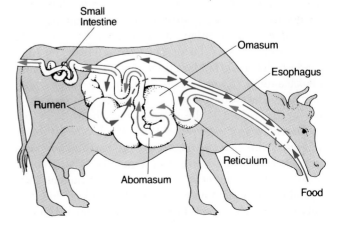

of the three caeca, the *rumen*. When a ruminant, let us say, a cow, eats grass, the partially chewed plant matter passes from the esophagus into the first caecum, the *reticulum*, and then slowly into the rumen, along with a large volume of alkaline saliva (pH 8.5). A cow's daily saliva production exceeds 200 liters, and the alkalinity of the saliva raises the pH in the rumen and reticulum to 6.5—far less acidic than the stomachs of nonruminants. In some species, the rumen is so large that it can hold food and saliva weighing one-seventh of the animal's total body weight. In a large cow, the digestive chambers hold 40 gallons of this semiliquid material. Immense numbers of anaerobic bacteria and ciliate protozoans that live in the rumen and reticulum ferment the grass into its component sugars, fatty acids, and amino acids. These molecules and peptides may be absorbed into the ruminant's bloodstream. As these processes occur, large quantities of methane gas (CH_4) and CO_2 are formed and belched out. Each year, 20 percent of the CH_4 released to Earth's atmosphere comes from domestic ruminants; this amounts to more than 100 trillion kilograms!

Cows can regurgitate from the rumen undigested plant matter, called *cud*, and chew and grind any fibrous material left intact during the first round of digestion. When the cud is swallowed, it is returned to the rumen for renewed bacterial and ciliate digestion, while the smaller and more soluble materials are passed on to the third caecum, the *omasum*, where vigorous churning further breaks up the food and allows more digestion. From the omasum, the food matter moves to the abomasum, where yet more digestion and absorption take place. Bacteria in all four chambers can combine nitrogenous compounds, such as ammonia and urea, with other molecules to form amino acids that the microbes can use for their own protein synthesis and that the cow can absorb in significant quantities along with B vitamins and other nutrients produced by the bacteria. In addition, millions of these microbial cells die each day, are themselves digested, and provide their ruminant hosts with proteins, carbohydrates, and nucleic acids. Ultimately, the largely digested food matter passes along the lengthy small and large intestines, where further digestion and water absorption take place.

In contrast to cows, horses and their relatives exist on cellulose-rich foods but lack the ruminant's special stomach chambers. Symbiotic bacteria are present, but mostly in the posterior part of the large intestine. Although that region of the intestine can absorb some of the nutrients from degraded food, the bacteria themselves are not killed and digested, as in cows, but are lost from the body in the feces. Consequently, a horse's digestive system is not nearly as efficient as a ruminant's at obtaining nutrients from food and from dead symbionts, and therefore a horse must eat more than a cow of similar weight.

The ruminant's digestive pattern has a significant adaptive advantage over the pattern in horses, elephants, rabbits, field mice, and other nonruminant herbivores. These animals feed in meadows or open grasslands most of their waking hours every day, and they must be constantly alert for approaching predators. Ruminants, however, with their four stomach chambers, can consume large amounts of plant matter then retire to safer, more sheltered places to regurgitate and chew their cud. In comparison with herbivores, a large carnivore such as a lion may sleep or remain lethargic for about twenty hours a day. Whether herbivorous or carnivorous, however, all mammals pay the price for generating heat internally; they must feed far more frequently than their reptilian ancestors.

THE HUMAN DIGESTIVE SYSTEM

Humans, many primates, bears, and pigs are **omnivores**: animals that consume both plant and animal matter. Some people eat specialized diets—Eskimos consume mostly meat, blubber, and organs, while Hindus consume only herbaceous foods. However, most humans eat a range of foods and are fully omnivorous. Our digestive system therefore has the mechanical and chemical ability to process many kinds of food. Let us examine the major parts of the human alimentary canal (Figure 34-8) and generally the way they function. The chemical events of food processing are the subject of the following section.

The Oral Cavity

Like many other biological openings, including a plant's stomata and an insect's spiracles, the human mouth is "guarded" by a pair of structures, the upper and lower *lips*. If you have ever tried to give medicine to an unwilling child, you know how tightly the lips can be pressed together to seal the oral cavity against intrusion. The lips are highly vascularized muscular flaps with an abundance of sensory nerves. They help retain food as it is being chewed and are important structures for facial expressions and for human speech.

The oral cavity itself contains the teeth and tongue. As in a number of other omnivores (Figure 34-9), there are four groups of adult teeth: the chisel-shaped *incisors*, used for biting and cutting; the pointed *canines*, used for shredding and tearing; the relatively small *premolars* (or bicuspids), used for grinding; and the large flattened *molars*, also used for grinding. Human teeth represent a mixture of the pointed teeth characteristic of carnivores

Figure 34-8   HUMAN DIGESTIVE SYSTEM: THE PERSONAL FOOD PROCESSOR.
The major organs of the human digestive system are shown here. The liver has been partially cut away to show the stomach and the pancreas. The gall bladder, small and large intestines, and other organs are easily seen. The sublingual and submandibular glands are salivary glands.

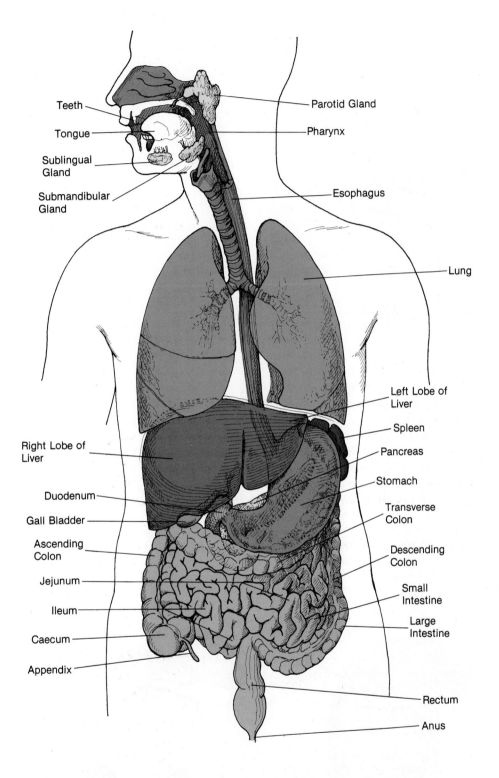

and the chisel-shaped and flat teeth characteristic of herbivores. The mechanical processing of a range of foods is possible because the teeth are covered by enamel, the hardest material in the body, and because amazing force can be exerted by the jaws and teeth—enough, for example, to crack a hard nut.

The *tongue*—an innovation in chordates—is a muscular organ that moves and manipulates food during chewing. It also monitors the texture and taste of foods and helps to alter sounds and form words in humans. Taste buds for salty, sweet, sour, and bitter flavors are distributed in zones on the tongue's surface. The posterior part of the tongue shapes chewed food into a **bolus,** a moistened lump, which is then swallowed.

The oral cavity is continuously bathed by *saliva,* a watery fluid secreted by three pairs of **salivary glands**—the submaxillary, sublingual, and parotid glands—which produce 1 to 1.5 liters of fluid each day in an adult. Saliva

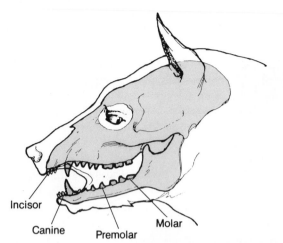

Figure 34-9   OMNIVORE VERSATILITY IN TEETH: AND THUS IN DIET.
Some mammals can be omnivorous because they have teeth for slicing, puncturing, tearing, and grinding. Hence, they are able to consume both animal flesh and plant materials.

humidifies air passing through the mouth toward the lungs; moistens food, thereby contributing to bolus formation; and carries various molecules and ions necessary to digestion. Swallowing is aided by the presence in saliva of *mucus*, a slippery secretion of many tiny glands in the mouth lining. Saliva also contains *amylase*, an enzyme that begins the hydrolysis of starch into sugars. If

you hold a soda cracker in your mouth for a minute or so, a sweet taste will gradually develop as these sugars are released. In addition to mucus and amylase, saliva contains small quantities of thiocyanate ions (SCN^-) and another enzyme that causes these ions to penetrate and kill bacteria entering the mouth. This antimicrobial action helps protect the gut from many infectious organisms.

The Pharynx and Esophagus

In Chapter 33, we noted that both air and swallowed foods and liquids pass from the mouth into the pharynx, the thin-walled chamber at the back of the mouth that leads to both respiratory and digestive tracts. We also saw that the epiglottis must temporarily seal off the opening to the trachea so that swallowed food will enter the esophagus and not the tracheal air passages. Initiation of the swallowing reflex (Figure 34-10) is complex and can be voluntary, but most of the time, it is involuntary. When swallowing begins, sequential, involuntary contractions of smooth muscles in the walls of the pharynx and esophagus propel the bolus or liquid toward the stomach. Coordination of this muscle contraction and the movement of the epiglottis normally prevents aspiration of food into the trachea. However, if a person breathes in suddenly while swallowing, food can become lodged in the windpipe, blocking air from entering the lungs and leaving the victim gasping for breath. The

Figure 34-10   SWALLOWING: A PRELUDE TO DIGESTION.
(a) Prior to swallowing, the chewed and partially digested food begins to form a bolus. (b) and (c) The tongue and epiglottis move in precise coordination during swallowing to press food backward and downward. The epiglottis closes during swallowing and prevents food from entering the respiratory system. The blue arrow represents air; the red arrow the bolus's path.

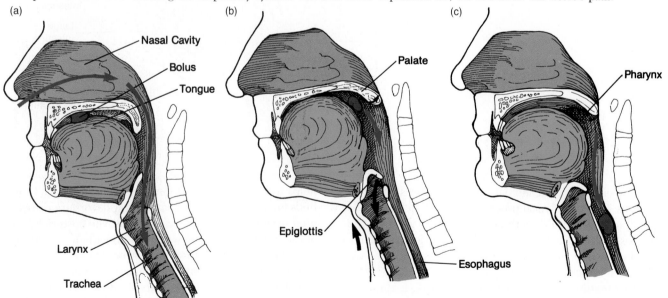

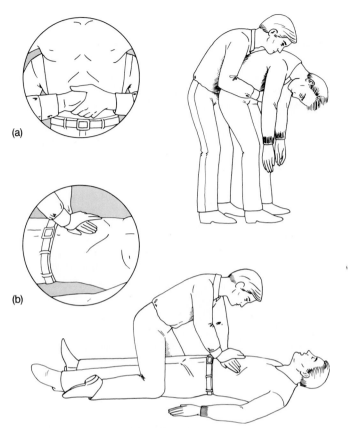

most successful emergency procedure to help a person strangling on food is the *Heimlich maneuver*, pictured and described in Figure 34-11.

As with any organ, the function of the gut is determined by the kinds of differentiated cells it contains. Along most of the alimentary canal, cells are arranged in four layers: the mucosa, the submucosa, the muscularis externa, and the serosa (Figure 34-12). The innermost layer, the *mucosa*, which surrounds the lumen, is composed of an epithelial layer, thin layers of connective tissue, and sometimes smooth-muscle tissue. Surrounding the mucosa is the *submucosa*, a zone of blood and lymphatic vessels, nerves, and fibrous connective tissue. Ducts of mucus-secreting glands in the submucosa lead to the mucosal surface. Next to the submucosa is the *muscularis externa*, a zone of smooth-muscle-fiber bundles. The inner bundles are oriented circularly around the esophagus, while the outer bundles run longitudinally (this arrangement is found in the intestine too). The fourth layer is the *serosa*, made up of more fibrous connective tissue and a moist epithelial sheet called the *peritoneum*. This peritoneal covering lines the entire body cavity and all exposed internal organs. The space it en-

Figure 34-11   THE HEIMLICH MANEUVER.
This technique is an extremely effective means to dislodge an obstruction in a choking person's larynx. Instructions for performing the Heimlich maneuver are now posted in most restaurants, where choking incidents are apt to occur. As illustrated here, the technique can be applied to a choking victim who is in either (a) an upright or (b) a prone position. The basic principle is the same in both cases: applying abrupt and strong pressure to the victim's abdomen in the manner shown here forces a burst of air from the lungs that will usually cause the obstruction to pop out of the victim's windpipe.

Figure 34-12   ANATOMY OF THE GUT.
(a) The mammalian digestive tract has several functional layers. The organization of these layers varies somewhat in different regions of the tract. (b) Glands are commonly situated in the submucosa and pour their secretions into the lumen. This scanning electron micrograph cross section (magnified about 20 times) of the colon shows the mucosa surrounding the lumen, the submucosa, and the longitudinal muscle. From *Tissues and Organs: A Text-Atlas of Scanning Electron Microscopy* by Richard G. Kessel and Randy H. Kardon, W. H. Freeman and Company © 1979.

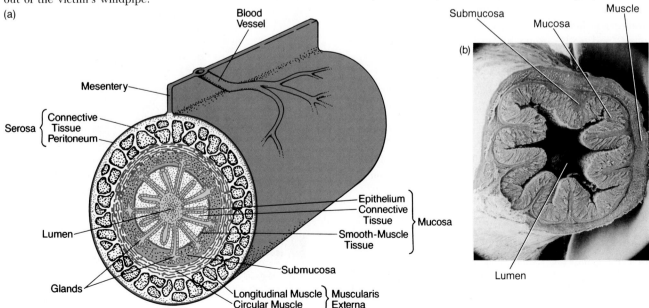

compasses is the coelom (Chapter 25). Double sheets of peritoneum containing blood and lymphatic vessels and connective tissue hang down from the body wall, enveloping and suspending the esophagus, stomach, and other gut parts the way a hammock suspends a person at rest; these sheets are the *mesenteries.*

How do the four layers of the gut wall function to move along swallowed food? A wave of contraction in the smooth-muscle cells of the mucosa and the muscularis externa constricts first one region of the esophagus and then the next. Such waves of contraction, called **peristalsis,** are analogous to squeezing icing from a pastry tube. Peristalsis propels the food bolus through the esophagus toward its junction with the next chamber, the stomach, where a sphincter muscle is normally closed tightly and prevents backflow of stomach contents. A sphincter is a circle of muscle, such as the one surrounding the stomach opening or the anal opening. The sphincter valve at the stomach opens only to allow food to move from the esophagus into the stomach, usually about three times a minute when food is present.

The Stomach

The human stomach is a J-shaped expandable bag in which the initial breakdown of proteins takes place in the presence of a highly acidic solution containing water, mucus, enzymes, and hydrochloric acid (HCl). As can be seen in Figure 34-13, the stomach mucosa contains thousands of *gastric pits,* tiny crevices lined with two types of secretory cells: chief cells and parietal cells. *Chief cells* secrete *pepsinogen,* a precursor of the enzyme **pepsin.** *Parietal cells* secrete a solution containing HCl.

The mucosa and the surfaces of the gastric pits also contain numerous epithelial cells that secrete mucus. This mucus coats the inner lining of the stomach and protects it from the corrosive effects of digestive enzymes and acidic stomach juices. Rapid replacement of

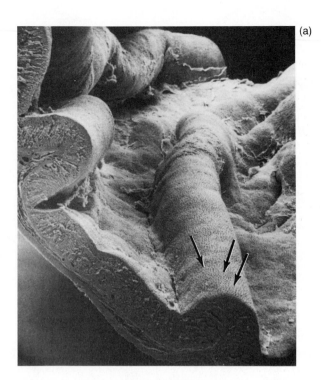

(a)

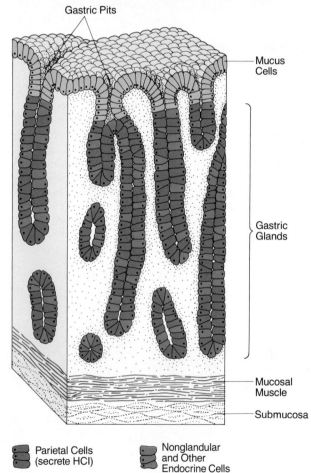

Figure 34-13   THE STOMACH AND ITS INNER SURFACE: LAYERS AND FOLDS.

The structures in the multilayered stomach wall allow that organ to accomplish some preliminary digestion. (a) Openings of thousands of gastric pits are clearly visible as black dots on the mucosa folds in this scanning electron micrograph (magnified about 22 times). Pepsinogen and hydrochloric acid formed in the gastric glands are released through these holes. From *Tissues and Organs: A Text-Atlas of Scanning Electron Microscopy* by Richard G. Kessel and Randy H. Kardon, W. H. Freeman and Company © 1979. (b) These glands are visible in the cross section. The epithelium of the stomach contains numerous mucus-secreting cells. Their mucous secretions protect the stomach lining from digesting itself.

stomach-lining cells also protects against the acid, as does the secretion of acid only when food is present. The surfaces of the upper alimentary tract have a thinner mucous layer than does the stomach; that is why vomiting can cause a burning sensation in those tissues. At a pH of 1.5 to 2.5, gastric juice is the most acidic substance in the body. If the stomach's mucous lining is destroyed or reduced, the pepsin and HCl can begin to digest the stomach lining itself, producing a lesion called a *peptic ulcer.* (Ulcers also form in the small intestine because the surface is not as well protected as that of the stomach.) Although the cause of peptic ulcer is not known, stress is thought to be a major factor, resulting in the production of excess stomach acid when no food is in the stomach. About 10 percent of people in the United States who are autopsied after death are found to have had ulcers, suggesting that the stress effect may be quite common.

What happens in the stomach after a person eats? The act of eating and the presence of food induce the gastric pits to secrete HCl (as H^+ and Cl^-) and pepsinogen. The H^+ ions cause pepsinogen to be cleaved into pepsin, and pepsin, in turn, cleaves additional pepsinogen. As H^+, pepsin, mucus, and fluid mix with and begin to degrade the food, smooth muscles in the stomach wall contract and vigorously churn the mixture. While this is going on, sugars, salts, and alcohol, if present, may be rapidly absorbed across the stomach wall. About three or four hours after a meal, the contents have become a semiliquid mass with the consistency of cream; this substance is called **chyme.** The chyme passes a little at a time through the ringlike valve at the lower end of the stomach, the *pyloric sphincter,* and from there enters the small intestine.

The Small Intestine: Main Site of Digestion

Most digestion of food and absorption of nutrients and water take place in the **small intestine,** a lengthy stretch of gut between the stomach and the large intestine (Figure 34-8). A human's small intestine is only about 4 centimeters (1½ in.) in diameter, but it is 7 to 8 meters (20–25 ft.) long. This is intermediate in length between the small intestines in typical carnivores and in herbivores of similar size, and reflects the human's omnivorous eating habits. Intestinal length has a direct effect on the total surface area available for absorbing nutrients, but still more important are the convolutions and minute projections of the inner gut surface (Figure 34-14). On the ridges and folds of the inner intestinal wall, thousands of tiny, fingerlike villi project from each square centimeter of the mucosa, giving it the appearance of velvet to the unaided eye. Both the folds and the villi are covered by epithelial cells, each bearing numerous mi-

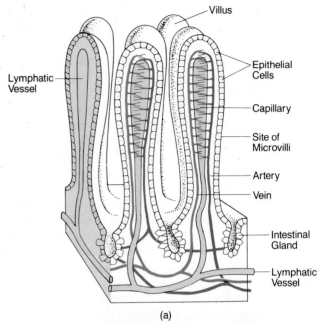

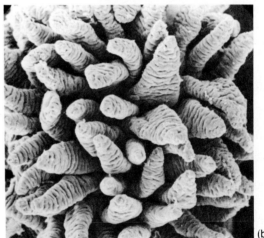

Figure 34-14   VILLI.
The small intestine has a huge absorptive surface area due to fingerlike, cylindrical villi. (a) Longitudinal section of the villi. (b) Scanning electron micrograph of the villi on the inside surface of a monkey's small intestine (magnified about 53 times). Absorbed nutrients pass through the epithelial cells into the blood capillaries (red and blue) and lymphatic vessels (yellow), both of which are in the core of each villus. The villus surface area is further expanded by minute fingerlike projections called microvilli that are found covering the outermost lumenal end of each epithelial cell. The microvilli are too small to be seen in (a) and (b).

crovilli. These minute projections are packed at a density of 200,000 per square millimeter and extend the surface area of the intestinal folds and villi by a factor of 20. The inner wall of the human small intestine thus has a total surface area of some 300 square meters—the size of a tennis court.

Just as the outer skin layers of the body are replaced continuously by new cells, so the inner lining of the intestine is constantly renewed. Recent experiments reveal a high turnover in the epithelial cells covering the villi; new ones are continuously produced near the bases, while old ones drop off the ends. The result is that daily about *17 billion* cells weighing about 250 grams pass into the intestinal lumen. As those cells break up, their lysosomal enzymes are released and may aid in digestion.

The first 30 centimeters of the human small intestine make up the *duodenum*, a region devoted solely to digestion. The next 3-meter segment is the *jejunum*, and the final 4-meter segment, the *ileum*; both carry out absorption of nutrients. The duodenum contains many digestive enzymes. Some of these enzymes are secreted by glands in the duodenal mucosa; others are secreted in the **pancreas** and flow into the duodenum through the pancreatic ducts. The pancreas is a glandular (secretory) organ shaped something like the human tongue that sits just ventral to the stomach (Figure 34-8). The enzymes it secretes break downs fats, proteins, carbohydrates, and nucleic acids. Another duct leads toward the duodenum from the **gall bladder,** a small organ near the liver (Figure 34-8). The gall bladder is a storage sac for the greenish fluid called **bile,** which is produced in the liver. Bile is very alkaline and contains pigments, cholesterol, and bile salts that act rather like detergents to emulsify fats (form them into droplets suspended in water) and aid in fat digestion and absorption.

Both pancreatic juice and bile contain abundant bicarbonate ions (HCO_3^-), which neutralize the acidity of the chyme flowing into the duodenum from the stomach. The pH rises from about 2 to about 7.8—an optimum, slightly alkaline pH for the activity of pancreatic enzymes. Thus while digestion in the stomach is acidic, digestion in the duodenum is alkaline.

As chyme is neutralized and diluted by bile and pancreatic fluids, it becomes a whitish, watery solution called **chyle,** which then moves into the jejunum and ileum. In these regions, absorption of amino acids, sugars, nucleic acid bases, minerals, and water takes place across the surfaces of the epithelial villi (Figure 34-15). Much of this absorption is an active-transport process involving sodium-dependent ATPase pump enzymes that act as carriers (see Chapter 5 for an explanation of how sodium pumping leads to nutrient transport). Large amounts of energy must be expended to maintain the sodium gradient that causes nutrients to move across the intestinal wall. The sometimes fatal disease cholera results when one of the intestinal sodium pumps is inhibited by a microbe. Once absorbed, nutrients pass into blood capillaries inside the villi and drain into the hepatic portal vein, which flows to the liver. A large proportion of the nutrients are taken into the liver cells, where they act as raw materials for the synthesis of pro-

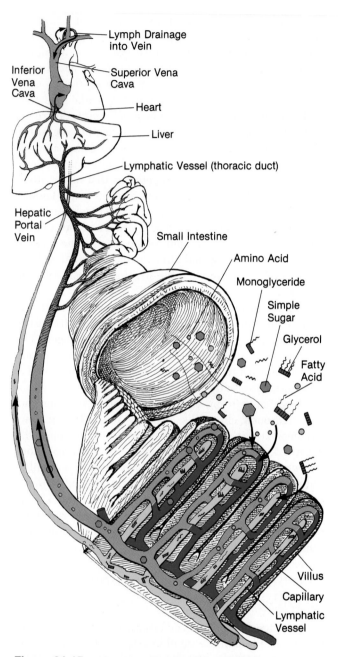

Figure 34-15   MAJOR ABSORPTION OF DIGESTED NUTRIENTS IN THE SMALL INTESTINE.
Sugars and amino acids pass through the intestinal epithelium into the blood capillaries, and fats pass into the lymphatic vessels. The absorbed sugars and amino acids are rapidly transported to the liver, whereas the absorbed fats enter the general circulation through the interconnection of the lymphatic and venous systems.

teins, glycogen, nucleic acids, and other substances.

Some lipids are absorbed undigested through the plasma membranes of the epithelial cells; however, most fats are broken down into fatty acids and monoglycerides (and some glycerol) before being absorbed. Once inside cells of the intestinal wall, triglycerides are resynthesized and coated with phosphoproteins to form minute droplets called *chylomicrons*. These droplets pass out of the cells and into lymphatic vessels called *lacteals*, which are located at the core of each villus. Hydrophilic surface proteins keep the droplets intact as they pass from the lymphatic system into the bloodstream and are transported throughout the body. In muscle, fat, and other tissue cells, the chylomicrons can be converted to other types of molecules, or they can be stored as energy-rich intracellular fat droplets.

A great deal of water is absorbed osmotically through the walls of the small intestine along with the nutrients. The Na^+ gradient used for nutrient absorption sets up the osmotic conditions leading to water uptake. Each day, about 10 liters of fluid pass into the gut. A person drinks about 1.5 liters in the form of water, soft drinks, coffee, and so on, and the other 8.5 liters pass into the gut as saliva, gastric juice, pancreatic secretions, and other fluids. About 9 liters of liquid are absorbed in the small intestine, and most of the rest is absorbed in the large intestine.

The Large Intestine

As the name implies, the human **large intestine** is a sizable segment of the gut, with a diameter of about 6.5 centimeters and a length of 2 meters. Because the inner surface is not convoluted and has no villi or microvilli, the surface area is only one-thirtieth that of the small intestine. The large intestine is joined to the small intestine near a blind-ended sac, the caecum. The human caecum and its fingerlike extension, the *appendix*, are nonfunctional as storage sites and probably represent evolutionary remnants of a larger, functional caecum, such as those found in herbivores. Some evidence suggests that the appendix is a lymphoid tissue. From its junction with the small intestine at the caecum, the large intestine forms a squarish configuration around the folds of the small intestine (Figure 34-8). Also called the *colon*, it ascends the right side of the body cavity, crosses to the left, and then descends (with sections called, respectively, the ascending, transverse, and descending colon). The descending colon terminates in the *anus*, the exit point of the entire alimentary canal.

The role of the large intestine is to absorb water and minerals and to form the feces. As chyle is moved along by peristaltic waves, minerals diffuse or are actively transported from the chyle across the epithelial surface of the large intestine. Water follows osmotically and is returned to the lymph and blood. When water absorption is disrupted, diarrhea results; the chyle passes too quickly and is eliminated as watery feces. When fecal matter moves too slowly and becomes excessively dehydrated, constipation results.

Many bacterial species, including *E. coli*, exist symbiotically in the large intestine. Housed in a warm, moist environment, millions of these cells can consume food materials that resisted digestion previously. As "rent" for their "room and board," they secrete amino acids and vitamin K, which are absorbed by the host along with minerals and water.

The last portion of the descending colon is an expandable storage chamber, the *rectum*, which holds feces until they are eliminated. Feces are composed primarily of great numbers of dead bacteria, undigested food such as plant fibers, debris from sloughed off cells, and other waste products. Strong peristaltic contractions of the large intestine's smooth-muscle layer, combined with the temporary relaxation of the anal sphincter muscle, allows the expulsion of the semisolid feces. The odor of feces stems mainly from bacterial breakdown products and from methane and hydrogen sulfide gases.

The Liver

The largest organ in the human body, the **liver** is truly a versatile workhorse that is an intermediary between digestion and the organism's metabolic needs. It is a reddish-brown, lobed mass that weighs about 3 to 3.5 pounds and sits just under the diaphragm. The liver contains millions of cells called *hepatocytes*, which take up nutrients absorbed in the intestines and which flow into the liver through the hepatic portal vein; in the process, the liver regulates the nutrient content of the blood. Hepatocytes also manufacture several blood proteins, including prothrombin—an enzyme involved in blood clotting (Chapter 31)—and albumin—a plasma protein.

The mammalian liver is a major site for the uptake and conversion of a wide variety of substances, including toxic ones. Excess amino acids are absorbed in the liver and converted to urea, which is then excreted in the urine. Dead red blood cells are collected in the liver, and the hemoglobin is converted to bile pigments. These pigments, *bilirubins* (red) and *biliverdins* (green), color the bile and the feces. Certain enzymes in hepatocytes break down toxins such as alcohol and other drugs. Abusing the body by consuming such chemicals can lead to severe scarring of the liver, a condition called *cirrhosis*. Viral infections, such as hepatitis, can also lead to cirrhosis. Scarring greatly impairs liver function and is the tenth leading cause of death in the United States. The healthy liver absorbs and degrades various types of hormones from the bloodstream. As we will see in Chapter 37, this activity permits sensitive regulation of hormonal levels.

In addition to all these functions, the liver produces *somatomedin*, an agent that controls bone growth; it stores fat-soluble vitamins; it acts as a reservoir for glycogen; and it is involved in maintaining normal blood-sugar levels. Clearly, the condition of the liver is critical to the health of the entire body.

THE CHEMISTRY OF DIGESTION

So far, we have described the organs that chew, grind, churn, and mechanically break down food into smaller and smaller pieces to prepare it for chemical digestion. During the relatively short time span between ingestion of food and excretion of undigested wastes, nutrients must be absorbed for use throughout the body. The chemical part of the digestion process is the breakdown of foods into small absorbable units by enzymes (Table 34-1).

The general strategy of chemical digestion is to break bonds linking monomers of large protein, carbohydrate, and lipid polymers. These reactions most often involve hydrolysis. As we trace the steps of carbohydrate, lipid, and protein digestion, we shall see hydrolysis in action. Another general feature of digestion is that enzymes work in tandem to break down the constituents of foods. That is, they work in complementary fashion to break first one molecular bond and then another.

The breakdown of carbohydrate exemplifies the chemical digestion process. More than half the carbohydrate ingested by humans and other omnivores is starch (amylose and amylopectin) or glycogen. The enzyme amylase in the saliva begins the hydrolysis of amylose, let us say (Figure 34-16). Amylase secreted by the pancreas into the small intestine continues that process, but at an alkaline pH, yielding large amounts of the disaccharide maltose. Much disaccharide passes into epithelial cells on the villi of the small intestine and is digested by a battery of eight disaccharide-splitting enzymes. In addition, the enzyme *maltase*, located in the luminal membrane of cells in the wall of the small intestine, cleaves maltose to produce the monosaccharide glucose. This sugar is then absorbed through the intestinal wall via cotransport with Na^+ ions, as described in Chapter 5. ATP must be cleaved to drive that process.

One digestive enzyme, *lactase*, degrades lactose, the carbohydrate in milk. It is synthesized by infants but not by most adult mammals, since they usually cease drinking milk at the time of weaning. Interestingly, many adult humans can drink unlimited quantities of milk, while others develop cramps, stomach gas, diarrhea, and vomiting if they drink more than a little. These latter individuals are said to be *lactose intolerant*. Anthropologists and geneticists have found that the ability of adults to drink and digest milk evolved within the last 5,000 years or so, and only in populations that traditionally maintained dairy herds, such as cultures in northern Africa, the Near East, and Europe. About 70 to 100 percent of individuals in such long-term dairying regions produce lactase as adults, while 70 to 100 percent of adults from traditionally nondairying regions do not. The retention of the lactase enzyme and tolerance to lactose must have been a selective advantage for children and young adults in parts of the world where animal milk was abundant.

The enzymatic breakdown of fats is similar to that of carbohydrates, but since fats are insoluble in water, they must first be emulsified by bile salts so they remain suspended in the chyle. Then enzymes such as *lipase* (se-

Table 34-1   **SOME MAJOR DIGESTIVE ENZYMES**

Source	Enzyme	Substrate	Product
Salivary glands	Amylase	Starch, glycogen (carbohydrate)	Maltose (disaccharide)
Stomach	Pepsin	Proteins	Peptides
Pancreas	Amylase	Starch, glycogen	Maltose
	Lipase	Triglycerides (lipids)	Fatty acids and glycerol
	Trypsin	Proteins	Peptides
	Chymotrypsin	Proteins	Peptides
	Deoxyribonuclease	DNA	Nucleotides
	Ribonuclease	RNA	Nucleotides
Small intestine	Maltase	Maltose	Glucose (monosaccharide)
	Lactase	Lactose (disaccharide)	Glucose, galactose
	Sucrase	Sucrose (disaccharide)	Glucose, fructose
	Aminopeptidase		
	Carboxypeptidase	Peptides	Amino acids
	Tripeptidase		
	Dipeptidase		
	Nucleases	Nucleotides	Five-carbon sugars and nucleic acid bases

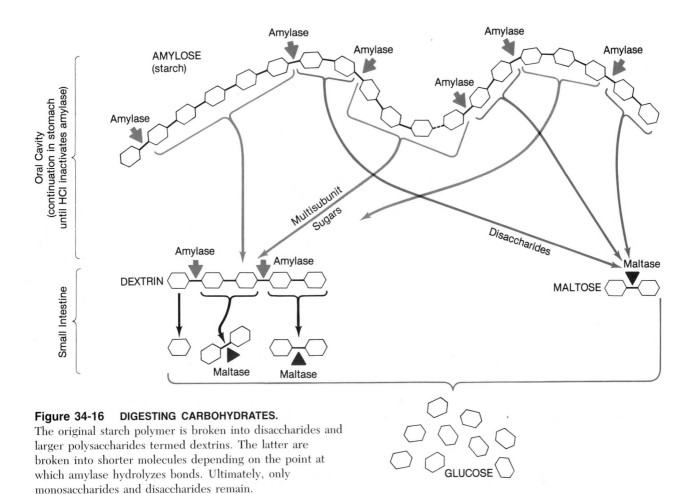

Figure 34-16   DIGESTING CARBOHYDRATES.
The original starch polymer is broken into disaccharides and
larger polysaccharides termed dextrins. The latter are
broken into shorter molecules depending on the point at
which amylase hydrolyzes bonds. Ultimately, only
monosaccharides and disaccharides remain.

creted by the pancreas) hydrolyze the triglycerides into
fatty acids and glycerol (Figure 34-17). Those subunits
are then absorbed across the walls of the small intestine
and are transported to the liver and other organs.

Proteins make up the most formidable class of molecules for chemical digestion because of their great structural variation and their twenty different subunit amino
acids. There are two main types of enzymes that hydrolyze proteins. *Endopeptidases* split bonds somewhere in
the middle of protein molecules, cleaving them into
smaller molecules (Figure 34-18). That step generates
many more amino and carboxyl ends than existed in the
uncleaved protein. Those ends become substrates for
the *exopeptidases*, which remove amino-acid subunits
one or two at a time. Endopeptidases are very specific in
the bonds they cleave; consider, for instance, that the
pancreatic enzyme *trypsin* splits only peptide bonds on
the carboxyl end of two of the twenty amino acids,
whereas *chymotrypsin*, another pancreatic enzyme,
cleaves only on the carboxyl end of four specific amino
acids.

Protein digestion starts in the stomach when pepsin,

an endopeptidase, cleaves food molecules wherever five
particular amino acids occur along protein polymers.
The pancreatic endopeptidases work next, and, finally,
exopeptidases, including dipeptidases, secreted by cells
in the wall of the small intestine, complete the job of
freeing amino acids. Some dipeptides (and even tripeptides and higher polymers), however, are absorbed by
the microvilli and are digested intracellularly.

Additional kinds of food molecules—RNAs, DNAs,
and so on—are broken down by other digestive enzymes
working alone and in series. Together, the enzymes
listed in Table 34-1 ensure that virtually all ingested nutrients are reduced to an absorbable state.

If enzymes can degrade almost any kind of food, why
don't they attack and digest the cells in which they are
formed? There are several good reasons. First, as we
have already mentioned, the gut lining is covered with a
mucous layer that helps resist enzymatic attack. Second,
digestive enzymes are synthesized on mRNAs and ribosomes attached to the rough endoplasmic reticulum.
Hence, they pass through the membranous channels of
the Golgi complex and are packaged into storage gran-

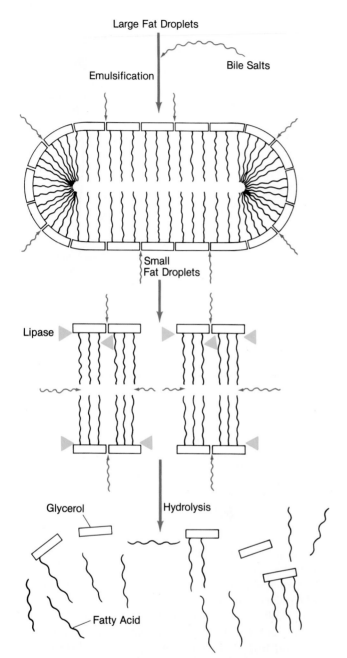

Figure 34-17   DIGESTING AND ABSORBING FATS.
Large fat droplets in the chyle are dispersed by bile salts. Small fat droplets form and are attacked by lipase from the pancreas. The breakdown products (glycerol, fatty acids) cross the plasma membrane of the intestinal epithelial cells that cover the surface of intestinal villi. There the fatty acids and glycerol are reassembled into lipids and released in association with protein. The lipid-protein aggregates enter lacteals or occasionally capillaries in the villus.

ules without ever entering the cell cytoplasm. Third, many digestive enzymes are synthesized as inactive precursors called **zymogens,** which must be cleaved into an active enzyme and a leftover polypeptide. For example, when pepsinogen is secreted from storage granules in the stomach mucosa into the extracellular spaces, H^+ (from HCl) causes cleavage of a forty-two amino-acid region from the end of the zymogen to yield pepsin.

Trypsin is the common activator for all other pancreatic zymogens (as, for example, chymotrypsinogen; Figure 34-19). Trypsin cleaves off a small part of each zymogen to trigger activity. But how does trypsin itself get activated? Cells of the duodenum make a specific peptidase that cleaves a single specific bond in trypsinogen as that zymogen enters the duodenum from the pancreas. During this cleavage, functional trypsin is formed, which then activates more trypsinogen by cleaving off the same peptide. Finally, enough trypsin is made to generate the rest of the digestive enzymes from their zymogens. Despite this safeguard, the protein and lipid digestive enzymes in the pancreas are sometimes activated prematurely. The result is a serious, potentially fatal disease called acute pancreatitis. For some reason, these enzymes become active inside the pancreas and actually begin to digest the organ itself.

COORDINATION OF INGESTION AND DIGESTION

Ingestion and digestion are obviously highly efficient ways to make specific molecules available to all body cells for metabolism, cell growth, and repair. But such a complex system of organs and processes must be controlled for at least two reasons: (1) so that enzymes are produced and secreted, and the gut walls contract and move food along when food is in the digestive tract; and (2) so that an animal senses hunger and is stimulated to hunt, graze, or buy a cheeseburger just at the time its cells need a source of nutrients. Let us see how digestion and feeding are coordinated and controlled.

Control of Enzyme Secretion and Gut Activity

If you are like most people, simply thinking about whatever most delights your palate will bring on a flow of saliva and gastric juices, ready to digest the imaginary delicacy. Ivan Pavlov, a Russian physiologist, discovered this phenomenon in dogs around the turn of the century. He found that dogs will salivate and release gastric juices not only at the sight or smell of food, but also at any

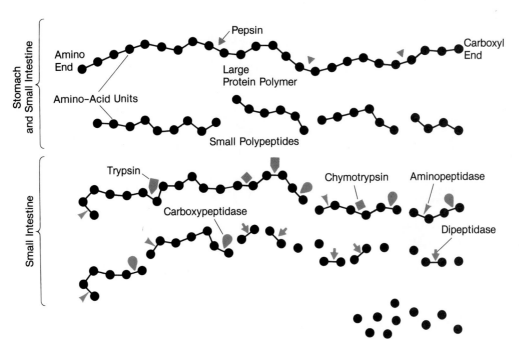

Figure 34-18
DIGESTING PROTEINS.
Pepsin, trypsin, and chymotripsin are endopeptidases (in purple); aminopeptidase, carboxypeptidase, and dipeptidase are exopeptidases (in orange). All of these are enzymes that hydrolyze different peptide bonds. (Water is added during the hydrolysis.) Aminopeptidase, carboxypeptidase, and dipeptidase free individual amino acids. Note in this sequence the sites where the different kinds of enzymes can hydrolyze peptide bonds.

event associated with the delivery of food, such as the sound of the experimenter's footsteps. Psychologists call this *conditioning*. What controls the release of enzymes when feeding is imminent?

The answer is tied to the entire control of digestion. Investigations have revealed that impulses from the brain stimulate secretion of saliva and gastric juices at the sight or smell of food. For instance, a branch of the vagus nerve stimulates cells near the stomach's pyloric sphincter to secrete a digestive hormone called **gastrin** into the bloodstream. It is gastrin that causes the gastric glands in the stomach wall to secrete HCl and pepsino-

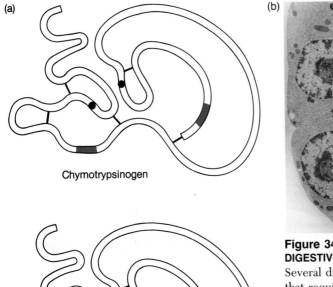

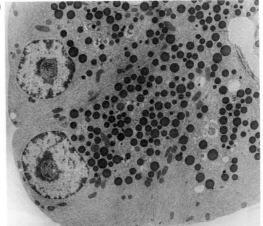

Figure 34-19   THE PANCREAS PRODUCES A BATTERY OF DIGESTIVE ENZYMES.
Several digestive enzymes are synthesized first as inactive zymogens that require activation in the duodenum before they become active digestive enzymes. (a) The single polypeptide of the zymogen chymotrypsinogen. This is converted by the action of trypsin to the active enzyme chymotrypsin; the small peptide regions indicated in red are removed by trypsin. (b) Both precursor zymogens and active enzymes are stored in the dozens of zymogen granules (black circles) in these pancreatic cells before being released into the pancreatic duct (upper right) for delivery to the duodenum (magnified about 2,700 times).

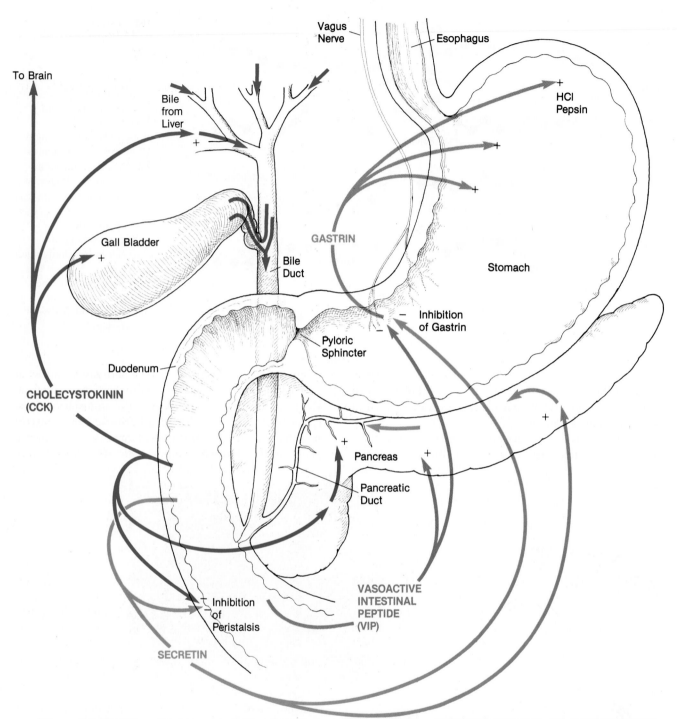

Figure 34-20   HORMONE CONTROL OF DIGESTIVE-JUICE SECRETION BY THE GUT.
Impulses from the brain as well as the presence of food in the gut stimulate the regulatory processes that affect liver, gall bladder, stomach, and pancreatic secretions. Most of the regulatory hormones shown here have both positive (+) and negative (−) effects on different targets. For instance, cholecystokinin (CKK) stimulates bile and pancreatic juice release, but inhibits peristalsis; while vasoactive intestinal peptide (VIP) stimulates pancreatic juice secretion, but inhibits gastrin release.

gen (Figure 34-20). When food is consumed and reaches the stomach, it stimulates further gastric secretion. (This sequence was discovered by investigators who created a special gastric pouch off a dog's stomach from which they could sample gastric juice before, during, and after feeding the animal.)

As we noted earlier, the stomach contents are churned and broken down into chyme and then pass into the duodenum, where the arrival of chyme stimulates secretion of various digestive enzymes by the pancreas. In 1902, physiologists W. M. Bayliss and E. H. Starling wondered what causes pancreatic juices to start flowing. After several fruitless attempts to learn the answer, they made an extract of the duodenal mucosa and injected it into a dog's bloodstream. Immediately, secretions poured from the pancreatic duct. Later, they coined the term "hormone" to refer to a substance produced in one tissue that controls the function of another tissue ("hormone" in Greek means "I arouse"). The Bayliss–Starling substance—the first animal hormone discovered—has since been purified and identified as the protein *secretin*. Biologists now know that during normal digestion, the presence of acid and fats in the chyme causes two digestive hormones to be released by intestinal cells: secretin and *cholecystokinin* (CCK), also called pancreozymin. Both hormones slow contractions of the gut's smooth-muscle layer so that food will be passed along more slowly and energy-rich fats can be fully digested and absorbed (Figure 34-20). Secretin inhibits gastric-juice secretion but speeds pancreatic-juice secretion. CCK stimulates the release of bile from the liver and gall bladder and of digestive enzymes and bicarbonate from the pancreas.

The arrival of chyme in the duodenum has another effect: amino acids freed from protein and sugars freed from carbohydrate raise the concentration of osmotically active molecules in chyme. This elevated concentration plus the stretching of the gut wall when it is full triggers nerve signals that slow the further emptying of the stomach. This inhibition in the release of chyme, called the **enterogastric reflex,** allows the small intestine more time to digest more fully the chyme it already contains. Obviously, the digestive functions of the stomach and intestines require complex coordination and control. Indeed, at least six other hormones are secreted by intestinal cells to control the stomach and pancreas. One of the most interesting is **vasoactive intestinal peptide (VIP),** a hormone secreted by the duodenum when fats are present in its lumen. VIP increases secretion of pancreatic juice and inhibits secretion of gastric juice (and of gastrin), among other actions. VIP, secretin, CCK, and several other intestinal hormones also are synthesized at specific sites in the brain. As we will see in Chapters 37 and 40, they are neuropeptides that alter activity of brain neurons.

Control of Hunger and Feeding

Basic control mechanisms ensure that organisms grow hungry and eat when nutrients are required by the body cells and that they stop eating when sufficient food is ingested. Some biologists believe that an organism senses hunger after cells in the hypothalamus or in the liver respond to lowered levels of glucose, amino acids, or fats in the blood. The animal then starts to hunt, graze, or perhaps open a refrigerator. Eating then distends the stomach and begins to raise the levels of those nutrients in the blood; as a result, the animal stops eating.

Vincent Dethier, working at Princeton University, was interested in knowing whether actual stretching of the stomach by food could inhibit further feeding. Normally, when an insect such as a blowfly drinks a sugar solution, its stomach swells and it stops eating. Dethier wondered what would happen if nerves running from the stomach to the brain of that insect were severed, and the fly was presented with food. A fly treated in this manner continued eating without stopping. Dethier wrote that its abdomen became stretched so fully that all the organs were flattened against the animal's sides. The abdomen, he noted, was so round and transparent that it could almost be used as a miniature magnifying glass. The results are interpreted to mean that inhibitory signals are sent to the brain to stop feeding when cells are stretched as the stomach fills. With the nerves cut, these inhibitory signals could not be sent.

Subsequent experiments with rats, chickens, and various mammals have pinpointed the feeding control mechanism in the hypothalamus of the brain. Some animals in which a surgical lesion is performed on a critical part of the hypothalamus will eat compulsively—rats, for instance, grow to two or three times their normal size. However, not all mammals respond this way. Newer evidence suggests that primary events controlling feeding occur in the liver. From there, it seems, information about levels of various nutrients in liver cells is sent to the brain (as noted earlier, the levels in cells reflect levels in blood flowing to the liver). There it is integrated, and eventually the sensation of hunger appears.

The brain hormone oxytocin (which we encountered in Chapter 18 playing a role during birth) has been found to take a key role in stopping ingestion. Cholecystokinin (CCK) seems to be the trigger that causes the brain's pituitary gland (Chapter 37) to secrete oxytocin. Injecting CCK into a rat that has just begun to feed causes an immediate halt in feeding and a massive release of oxytocin into the blood plasma. Other rats injected with saliva as a control continue to eat and do not show elevated oxytocin levels. Intense aversion to a food, including nausea, may operate in the same way. For instance, sheep ranchers "teach" coyotes to avoid sheep by baiting a sheep carcass with lithium chloride; the coyote that samples the doctored flesh retches violently and usually avoids live sheep thereafter. Administering large doses of CCK to rats causes exactly the same kind of nausea. All these effects of CCK depend on an intact vagus nerve carrying information from the stomach to the brain. Some biologists hypothesize that when an animal has ingested enough to be full or when it consumes nauseating food, the vagus causes release of CCK in the hypothalamus. (As we will see in Chapters 37 and 40, most gut hormones, such as CCK and VIP, are also manufactured and secreted by specific nerve cells in the brain.) That release, in turn, causes oxytocin secretion. By still unknown mechanisms, the animal's behavior then changes—it stops eating.

What an animal chooses to eat is related to hunger, to certain instincts, and to its nutrient needs. When an animal encounters an unknown food, it will usually sample only a small amount and instinctively allow time for any noxious chemicals that may be present to act. If the animal is not sickened, it may then return to eat more of the food. To outfox this "sample-and-wait" strategy, rodent exterminators must first put out wholesome grain for rats, and later put out poisoned grain. If an animal is deprived of a specific nutrient for a long period, it will choose foods rich in this nutrient. This craving, or *selective hunger*, even can transcend a normal repugnance to a food; for instance, even a craving for pure fats or oils might result from a severe deficiency in those vital ingredients of all living cells. The "sweet tooth" some people have may similarly be related to the rich source of calories that foods like honey have provided to humans during their evolution. However, most human cravings probably go well beyond any real shortage in the body and are based on learned behaviors and habit.

Control of Body Fat

During the past few decades, weight-loss diets seem to have become a national fetish in the United States. Dietary restriction clearly is one way that many people try to keep weight down. In addition, recent evidence shows that an organism's weight is strongly governed by factors other than the amount of food it eats. The concept of a *set point* has emerged, a genetically determined level for each individual's body weight and, particularly, for the proportion of fat stored by the body. This set point is analogous to an animal's body temperature, which is automatically maintained within a very narrow range. Tests show that in most people and in various other mammals tested, when the caloric intake is greatly increased, the body does not gain much weight, but burns the excess calories and dissipates them as body heat. When food intake is experimentally lowered, the reverse happens: body weight and fat stores are maintained, and the available calories are used more efficiently. Research also suggests that increased physical activity is the only successful way to change the set point and alter the amount of fat the body stores over an extended period.

The number of fat cells in the human body may play a role relative to the set point. It is likely that fat cells arise early in life and, once present, remain fairly constant in number. Once present, fat cells can vary greatly in size, with those of obese animals being 4 to 5 times larger than normal. Fat cells with so-called α_2 receptors on their surface tend to accumulate fat; for instance, one woman with overactive α_2 receptors on her buttocks and thigh fat cells, through exercise and diet, succeeded in losing 15 percent of her body weight—yet she lost virtually no fat from her buttocks or thighs. Fat cells may signal the brain to control eating by releasing into the blood a protein, adipsin. More adipsin is released when fat is mobilized, as during fasting; and less is present in cases of genetic obesity. We will not know, however, why some mammals store so much food as fat instead of burning it and releasing heat.

Under certain environmental stresses, the set point may shift temporarily. Birds or mammals that migrate or hibernate seem to experience seasonal changes in set point. First, the appetite and the tendency to store fat increase; then, after hibernation or migration begins, the "thermostat for feeding" is lowered, and the body fat is burned slowly. Such temporary alterations are probably due to changes in hormone production. A natural loss of appetite is also part of molting, egg incubation, and defense of territory in many animals. Feeding would compete with such activities, and so it is suppressed, despite abundantly available food and prolonged abstinence. In a female octopus, for example, a secretion of the optic glands suppresses feeding while she broods her eggs; in fact, the female actually starves to death after the young hatch.

An aberration of appetite and feeding in humans, *anorexia nervosa*, cannot easily be related to these natural animal feeding cycles, although the condition is sometimes related to hypothalamic abnormalities. This disease is characterized by minimal food intake over long periods, or by excess eating and induced vomiting, which is unknown in other animals. In most victims of anorexia nervosa, there are deep psychological reasons for the self-induced starvation behavior, and, sadly, the condition can be fatal despite treatment.

NUTRITION

What nutrients must be present in a heterotroph's diet? Heterotrophic bacteria can make all the compounds they need from simple precursors. However, many higher organisms have lost this ability or never had it in the first place. Thus they lack genes for enzymes that catalyze the formation of a number of compounds essential for metabolism. In general, heterotrophs require four types of foods in large amounts: proteins, carbohydrates, fats, and some minerals. These are called **macronutrients.** Heterotrophs also need smaller amounts of vitamins and trace inorganic minerals, the so-called **micronutrients.** Together, these nutrients make up the organism's dietary requirements. Besides these nutrients, animals and all other organisms require water. That vital solvent of life must be ingested or, in a minority of organisms, made metabolically.

Macronutrients: The Basic Foods

In the gut, macronutrients are broken down into subunits that are absorbed, transported in the bloodstream to tissue cells, and used as building blocks, as inputs to various metabolic pathways, or as substrates for storage of chemical energy in ATP. The energy value of food is equivalent to the **calories** produced when the food is oxidized completely. One calorie is the amount of heat required to elevate the temperature of 1 gram of water by 1°C. The kilocalorie (1,000 calories, or 1.0 Calorie) is the unit of measure used most often by nutritionists. When 1 gram of protein or 1 gram of carbohydrate is metabolized, about the same amount of heat is released— around 4 kilocalories per gram. In contrast, fat yields more than 9 kilocalories per gram. Obviously, fats are the most energy-rich macronutrients, and this explains why migrating birds and hibernating mammals lay down stores of fat, not reserves of protein or carbohydrate.

The subunits of macronutrients may be stored within liver, muscle, lung, or other types of cells for future needs. Because the individual subunit molecules would exert an osmotic effect in the cytoplasm, they are converted to large polymers for storage. Carbohydrates are built into glycogen granules (Figure 3-11). Fats can be stored as oil droplets; tissues composed of fat-storing cells are the fat bodies, or blubber. An interesting and universal feature of cells is that protein, unlike carbohydrate and fat, is very rarely stored. Apparently, it takes less energy to store carbon as carbohydrate or fat and then convert those substances to amino acids than it does to manufacture and store large reservoirs of protein directly. One exception is the manufacture of special storage proteins for embryos, such as the yolk proteins in most kinds of bird eggs. When protein is digested, amino acids are acted on by liver enzymes, which remove the amino groups in a process called *deamination.* The carbon atoms are then passed via acetyl coenzyme A to carbohydrates or fats for storage.

Virtually every animal must consume a variety of foods to fulfill its dietary requirements. Each food is composed of a unique mixture of macronutrients, and in turn, each macronutrient has its own unique chemical composition. The fact that macronutrients vary in chemical make-up has a direct effect on what foods an organism must eat to survive. Proteins and fats, in particular, are needed in sufficient amounts to provide amino-acid building blocks

for body proteins and to build cell membranes. Yet many single foods lack certain amino acids or fatty acids, and animals cannot synthesize all of them. Proteins are composed of twenty different amino acids, all essential to life. But most animals are incapable of synthesizing about half of them, and they must be obtained in the diet. The eight **essential amino acids,** those that adult humans cannot synthesize and must obtain through diet, are isoleucine, leucine, lysine, methionine, phenylalanine, threonine, tryptophan, and valine.

The typical, omnivorous human diet in the United States contains poultry, meat, fish, eggs, and milk, each of which usually contains all the essential amino acids. In contrast, the proteins in grains, vegetables, and fruits are usually deficient in one or more of the essential amino acids. Therefore, because the body does not store protein, a vegetarian must be careful to consume foods at *each* meal that contain all the essential amino acids. For example, beans contain lysine, tryptophan, and cysteine, but are low in methionine, whereas corn is a source of methionine but is low in lysine, tryptophan, and cysteine. The Indians of central Mexico have thrived on their traditional diet of corn tortillas and beans because they eat them both at one meal and so have a full set of amino acids available for protein synthesis. Eating corn at one meal and beans at the next does not have the same result: amino acids gained in the first meal cannot be turned into protein and so are diverted to carbohydrate or fat. Then, at the next meal, the same thing happens again. If any essential amino acid is missing from the diet, the body will eventually degrade some of its own proteins in order to make the missing amino acid available. This is costly, because the unneeded amino acids freed by such degradation tend to be deaminated and excreted, and the body can lose more nitrogen than it takes in. Clearly, vegetarian diets must be carefully planned and monitored, or deficiencies can result.

Just as some animals cannot synthesize certain amino acids, some cannot synthesize certain fats. Rats and humans have the ability to convert carbohydrates to various types of fatty acids and lipids for storage or use in cell membranes. However, neither organism is able to synthesize enough of a common fatty acid, *linoleic acid,* which is an essential component of cell membranes. If a sufficient amount of this fat is not consumed in the form of corn, safflower, or other vegetable oils, the animal will suffer from nerve-cell degeneration and severe malfunction of other cell types, and may even die.

Micronutrients

The micronutrients, the vitamins and trace minerals used over and over again in the body's enzymatic reactions, are usually small ions and molecules. They may be needed only in minute amounts, but the body cannot synthesize them rapidly (if at all). Thus they must be taken in the diet.

Vitamins

Vitamins are organic molecules that function as coenzymes or cofactors of enzymes (Chapter 4). The need for vitamins was recognized more than two centuries ago. One discovery began with British sailors on long sea voyages. These men were usually subjected to the same monotonous diet for months on end—salt pork, biscuits, and rum. Frequently, they developed *scurvy*, a condition characterized by poor healing of wounds, excruciating joint pain, anemia, loose teeth, and connective-tissue diseases. It became clear by about 1800 that scurvy was linked to the diet and could be prevented by stocking on board ship lemons, limes (hence British sailors were called "limeys"), and sauerkraut—all of which we now know contain vitamin C, or ascorbic acid. Without this vitamin, cells cannot secrete collagen, the most common protein in the body, and the spectrum of scurvy symptoms results.

Even after people discovered the link between diet and specific diseases, the identification of essential vitamins was difficult for two reasons. First, vitamin needs vary among organisms. Cats and dogs, for example, can synthesize their own vitamin C, whereas humans and monkeys cannot; thus early research results based on studies of various animals were often quite confusing. Second, the quantities of a vitamin needed tend to be so tiny that it is extremely difficult to prove the compound is essential. For example, 1 ounce (28.3 grams) of vitamin B_{12} can supply the daily need of 4,724,921 people! To prove that a vitamin is essential, the researcher must remove it completely from an animal's diet and watch for symptoms of disease. But minute quantities of the vitamin of interest may remain hidden among the macronutrients in food, and thus no deficiency symptoms will show up, even though the vitamin is truly essential. What is more, many animals use their own guts as culture chambers in which bacteria synthesize vitamins such as B_{12} or K; thus there is a built-in vitamin source to confuse experimental results still further. Despite these difficulties, through decades of careful research, nutritionists have developed a list of essential vitamins, their physiological roles in the body, and deficiency symptoms (Table 34-2).

Vitamins are classified as either water soluble or fat soluble. **Water-soluble vitamins,** such as C and the B vitamins, are transported as free compounds in the blood and serve as coenzymes—catalytic factors in the chains of biochemical reactions that are so essential to metabolism. In contrast, **fat-soluble vitamins,** such as A, D, E,

and K, are transported in the blood as complexes linked to lipids or proteins. These vitamins play more specialized roles in particular tissues and physiological activities—vitamin A in the formation of visual pigments, vitamin K in a vertebrate's blood-clotting mechanism, and so on. If excess amounts of water-soluble vitamins are ingested in food or vitamin supplements, they are easily excreted in the urine. However, fat-soluble vitamins cannot pass so easily from the system and can accumulate in toxic levels. For example, Eskimos who eat polar-bear liver can actually consume a lethal dose of vitamin A.

Minerals

Minerals are inorganic molecules that provide ions critical to the functioning of many cellular enzymes or proteins (as iron for hemoglobin). When a zebra, giraffe, or cow licks a salt block or when a person sprinkles food with table salt (NaCl), an essential mineral need is being filled. Organisms require large quantities of sodium (Na) and chlorine (Cl), as well as potassium (K), sulfur (S), calcium (Ca), phosphorus (P), and magnesium (Mg), all of which can be considered to be macronutrients. The micronutrient *trace minerals* include iron (Fe), iodine (I), manganese (Mn), fluorine (F), copper (Cu), zinc (Zn), molybdenum (Mo), selenium (Se), and cobalt (Co). Table 34-3 lists the roles and sources of the major mineral nutrients essential to mammals.

As we saw in Chapter 31, blood plasma and body fluids, both inside and outside cells, contain many types of mineral ions. Since significant amounts of minerals are lost in the sweat, urine, and feces, an animal must replenish its mineral supply regularly. An adult human, for example, must consume about 3 grams of NaCl daily—and even more if jogging, aerobic dancing, or other vigorous exercise leads to heavy sweating. Deficiencies of Na^+, Cl^-, and K^+ seriously disturb the osmotic balance of the blood and body fluids and of all cells themselves. One consequence may be to disrupt propagation of nerve impulses. Calcium is also essential for nerve function, for bone structure, and for cell movements and contraction of cytoplasmic filaments. Phosphate, particularly in the form of ATP, is the cell's vital energy currency. Manganese is needed for normal structure of bones and tendons. Iron is the critical element that binds O_2 in hemoglobin, myoglobin, and cytochromes. Iodine occurs in thyroxine, a hormone that regulates metabolic rate and heat production in mammals. The trace minerals (Table 34-3) appear to act as coenzyme components, but they are present in such minute amounts that, like vitamins, their activities are very hard to study.

SKIN COLOR, DRIFTING CONTINENTS, AND VITAMIN D

Anthropologists believe that our own species, *Homo sapiens*, originated in Africa and that the first populations were dark-skinned. These tenets lead to a logical question: Why did some human populations in the far Northern Hemisphere evolve later with light skins? The answer comes from two rather disparate realms: the movement of continents and the biological production of vitamin D.

During the Mesozoic era, large land masses split apart and drifted to their current positions, forming South America and Africa in the Southern Hemisphere, and Eurasia and North America in the Northern Hemisphere. During that long period, the Atlantic Ocean gradually formed, and with it arose a lucky peculiarity—a water-circulation pattern that carries the warm waters of the tropical Gulf Stream northward, past Newfoundland, Greenland, and Iceland, and toward northern Europe. This sweep of warm water rendered parts of the far northern latitudes green and habitable, and, as a result, human populations were able to migrate from the lower latitudes into these areas. And it was during that migration and early habitation perhaps 75,000 years ago that lightly pigmented skin, eyes, and hair apparently evolved.

Interestingly, the absence of dark pigmentation is directly related to the human body's production of vitamin D. The active form of vitamin D, 3-dehydrocholesterol, is manufactured by human skin cells when they are exposed to certain ultraviolet wavelengths in sunlight. Unless sufficient levels of vitamin D are consumed in the diet or produced in the skin, calcium cannot be absorbed effectively by the intestines, and a person is very likely to develop *rickets*, a disease in which the bones become brittle, misshapen, and easily broken. In India, rickets was once common among upper-class girls because these "delicate" children were kept perpetually indoors. Their poorer counterparts, who often consumed far less nutritious foods, usually escaped the disease because they played outside in the sunshine.

Vitamin D and rickets have a straightforward connection to human skin color. In a tropical region, a light-skinned person can absorb too much ultraviolet light and convert too much precursor to vitamin D. The results can be a condition called *hypervitaminosus*, in which too much calcium is present in the blood plasma, and kidney stones build up, extra pieces of bone form in the tendons, and other problems occur. Dark-skinned people in tropical areas are protected from hypervitaminosus by a simple mechanism: the relatively large amounts of black and brown melanin pigments in their skin and hair absorb the same ultraviolet wavelengths from the sun that convert the precursor to vitamin D (the wavelengths also cause skin cancer). Thus their bodies do not produce a damaging oversupply (nor get skin cancers). Since the Earth's first *Homo sapiens* were dark-skinned peoples in tropical regions of Africa, this physiological mechanism almost certainly operated in them.

But what about their descendants who migrated toward northern Europe and the coastal regions warmed by the Gulf Stream? The farther one travels from the equator, the lower the level of incident sunlight during much of the year. The concentrated melanin pigments in dark-skinned migrants to northern Europe would have absorbed much ultraviolet light and prevented sufficient vitamin D from forming. They surely would have suffered from rickets. Hence many biologists believe that strong selection pressure would have operated for lighter and lighter skins with lesser amounts of melanin. In the extreme cases we see today—the pale blonds of Scandinavia—just the pink cheeks of a baby exposed to sunlight for one hour a day can manufacture enough vitamin D to prevent rickets. The only far-northern peoples with heavily pigmented skins are the Eskimos. And they have traditionally eaten large quantities of fish oils—which are a good source of vitamin D.

It is poignant and tragic that the evolution of skin colors appropriate to latitude and sunlight levels should have led to prejudice, racial discrimination, and cruel persecution over the centuries. Who knows how very different human evolution and history would have been had the continents not drifted to their present configurations and had the Gulf Stream not carried a hospitable climate northward.

Table 34-2   **VITAMINS**

Vitamin	Distribution	Function	Deficiency Symptoms	Primary Sources
Water soluble				
B_1 (thiamine)*	Absorbed from gut; stored in liver, brain, kidney, heart	Formation of coenzyme involved in Krebs cycle	Beriberi, neuritis, heart failure	Organ meats, whole grains
B_2 (riboflavin)*	Absorbed from gut; stored in kidney, liver, heart	Cofactor in oxidative phosphorylation	Photophobia, skin fissures	Milk, eggs, liver, whole grains
B_6 (pyridoxine)*	Absorbed from gut; one-half appears in urine	Coenzyme in amino-acid and fatty-acid metabolism	Dermatitis, nervous disorders	Whole grains
B_{12} (cyanocobalamin)*	Absorbed from gut; stored in liver, kidney, brain	Nucleic-acid synthesis; prevents pernicious anemia	Pernicious anemia, malformed red blood cells	Organ meats, synthesis by intestinal bacteria
Biotin	Absorbed from gut	Protein synthesis; CO_2 fixation; amine metabolism	Scaly dermatitis, muscle pains, weakness	Egg white, synthesis by digestive-tract flora
Folic acid (folacin, pteroylglutamate acid)	Absorbed from gut; utilized as taken in	Nucleic-acid synthesis; red blood cell formation	Failure of red blood cells to mature, anemia	Meats
Niacin	Absorbed from gut; distributed to all tissues	Coenzyme in hydrogen transport (NAD, NADP)	Pellagra, skin lesions, digestive disturbances, dementia	Whole grains
Pantothenic acid	Absorbed from gut; stored in all tissues	Forms part of coenzyme A	Neuromotor and cardiovascular disorders	Most foods
C (ascorbic acid)	Absorbed from gut; little storage	Vital to collagen and ground substance	Scurvy, failure to form connective tissue	Citrus fruits
Para-amino-benzoic acid (PABA)	Absorbed from gut; little storage	Essential nutrient for bacteria; aids in folic-acid synthesis	No symptoms established for humans	
Fat soluble				
A (carotene)	Absorbed from gut; stored in liver	Visual-pigment formation; maintains epithelial structure	Night blindness, skin lesions	Egg yolk, green and yellow vegetables, fruits
D_3 (calciferol)	Absorbed from gut; little storage	Increases calcium absorption from gut; bone and tooth formation	Rickets (defective bone formation)	Fish oils, liver
E (tocopherol)	Absorbed from gut; stored in fat and muscle tissue	Maintains red blood cells in humans	Increased fragility of red blood cells	Green leafy vegetables
K (naphthaquinone)	Absorbed from gut; little storage	Stimulates prothrombin synthesis by liver	Failure of blood-clotting mechanism	Synthesis by intestinal flora, liver

*B-complex vitamins

Dietary Guidelines

A great deal of information and misinformation is published today about the human diet. Some people worry so much about adequate nutrition that they dose themselves with large quantities of vitamins and minerals, whereas others, concerned about overweight, follow patently absurd popular diet plans based on grapefruit, pomegranates, or "junk food." In fact, vitamin and mineral deficiencies are rare in the American population; if they eat a variety of whole foods daily, most people

Table 34-3 **MINERALS**

Mineral	Function	Primary Sources
Essential Minerals		
Calcium	Component of bone and teeth; essential for normal blood clotting; needed for normal muscle, nerve, and cell function	Milk and other dairy products, green leafy vegetables
Chlorine	Principal negative ion in interstitial fluid; important in fluid balance and in acid–base balance	Most foods, table salt
Magnesium	Component of many coenzymes; balance between magnesium and calcium ions needed for normal muscle and nerve function	Many foods
Phosphorus	As calcium phosphate, an important structural component of bone; essential for energy transfer and storage (component of ATP) and for many other metabolic processes; component of DNA, RNA, and many proteins	All foods
Potassium	Principal positive ion within cells; influences muscle contraction and nerve excitability	Many foods
Sodium	Principal positive ion in interstitial fluid; important in fluid balance; essential for conduction of nerve impulses	Most foods, table salt
Sulfur	Component of many proteins; essential for normal metabolic activity	Meat, fish, legumes, nuts
Trace Minerals		
Cobalt	Component of vitamin B_{12}; essential for red blood cell production	Meat, dairy products (strict vegetarians may suffer from cobalt deficiency)
Copper	Component of many enzymes; essential for melanin and for hemoglobin syntheses	Liver, eggs, fish, whole-wheat flour, beans
Fluorine	Component of bone and teeth	Some natural waters, may be added to water supplies
Iodine	Component of hormones that stimulate metabolic rate	Seafood, iodized salt, vegetables grown in iodine-rich soils
Iron	Component of hemoglobin, myoglobin, cytochromes, and other enzymes essential to oxygen transport and cellular respiration	Meat (especially liver), nuts, egg yolk, legumes (mineral most likely to be deficient in diet)
Manganese	Activates many enzymes; as arginase, an enzyme essential for urea formation	Whole-grain cereals, egg yolk, green vegetables
Zinc	Component of at least seventy enzymes, including carbonic anhydrase; component of some peptidases, and thus important in protein digestion; may be important in wound healing and fertilization	Many foods

never need vitamin supplements. Besides malnutrition among the very poor, a major national dietary problem is obesity; it results from too much fat, sugar, and alcohol in the diet and too little exercise, as well as the set-point problems discussed earlier. The United States government guidelines for good nutrition are listed in Table 34-4. In addition, much recent evidence suggests that regular aerobic exercise is important to maintaining ideal weight, reducing stress, and preventing constipation, atherosclerosis (fatty deposits in the arteries), and muscle degeneration.

Table 34-4   **UNITED STATES GOVERNMENT GUIDELINES FOR GOOD NUTRITION**

1. *Eat a variety of foods daily,* including fruits and vegetables; whole grains, enriched breads, and cereals; milk, cheese, and other dairy products; meats, fish, poultry, and eggs; dried peas and beans.

2. *Maintain ideal weight.* Increase physical activity; reduce fatty foods and sweets; avoid too much alcohol; lose weight gradually.

3. *Avoid fats, saturated fats, and cholesterol.* Eat lean meats, fish, poultry, dry peas and beans; use eggs and organ meats sparingly; reduce intake of fats on and in foods; trim fats from meats; broil, bake, or boil—don't fry; read food labels for fat contents.

4. *Consume adequate starch and fiber.* Substitute starches for fats and sugars; eat whole-grain breads and cereals, fruits and vegetables, dried beans and peas, and nuts to increase fiber and starch intake.

5. *Avoid excess sugar.* Limit consumption of sugar, syrup, honey, candy, soft drinks, and cookies. Select fresh fruits or fruits canned in light syrup or their own juices; watch for sucrose, glucose, dextrose, maltose, lactose, fructose, syrups, and honey on food labels.

6. *Avoid excess sodium.* Reduce salt in cooking and at the table. Cut back on potato chips, pretzels, salted nuts, popcorn, condiments, cheese, pickled foods, and cured meats. Watch food labels for sodium or salt contents.

7. *Drink alcohol in moderation if at all.* Limit alcoholic beverage intake (including wine, beer, and liquors) to one or two drinks per day.

*Adapted from *Dietary Guidelines,* United States Department of Agriculture; United States Department of Health and Human Services, 1979.

WE ARE WHAT WE EAT

In this chapter, we examined the digestive organs of many animals, the ways that foods are ingested and digested, and our own nutrient requirements. The similarity of mouths, teeth, guts, and digestive enzymes among many types of animals indicates just how successful these anatomical and physiological means are for breaking up food mechanically and breaking it down chemically. In fact, the diversity of multicellular animals is based in part on the efficiency of the two-ended gut at digesting practically anything an animal can find and swallow, depending on its preferences. We animals not only are "what we eat" on a daily basis, but also have evolved basic body plans and behaviors largely designed around the needs of finding and consuming food.

SUMMARY

1. Feeding usually involves both *ingestion* and *digestion*—chemical breakdown of food.

2. Animals' digestive organs vary in complexity, depending on the size of the organism, its nutrient needs, and whether it is a *herbivore* (exclusive plant eater), a *carnivore* (exclusive meat eater), or an *omnivore* (both plant and meat eater).

3. The *alimentary canal* is the two-ended gut that extends from mouth to anus. Different regions and organs in the digestive tract carry out specialized functions.

4. The teeth and tongue prepare the food for swallowing. The *salivary glands* secrete saliva, which moist-ens food and contains amylase for breaking down complex carbohydrates into sugars.

5. Food moves through the *pharynx* to the *esophagus,* and *peristalsis* in esophageal walls propel it toward the stomach.

6. The *stomach* is commonly the site of further mechanical breakdown of food pieces. Protein digestion begins as *pepsin* attacks the food *bolus.* Pepsin, H^+ (from HCl), mucus, and fluids mix with the bolus to form the *chyme* that continues on to the small intestine.

7. Alkaline enzymatic digestion and absorption of nutrients occur in the *small intestine. Chyle* flows from the duodenum (a digestive region) into the jejunum and the ileum (absorptive regions). Villi and *microvilli* increase the small intestine's absorptive area manyfold. Absorption may involve active (ATP dependent) co-transport with sodium ions (Chapter 5), or other cellular ways to take in molecules of varying composition and size. Nutrient absorption often depends on active transport sodium and potassium pump enzymes.

8. The *pancreas* manufactures digestive enzymes that are secreted into the duodenum. The *liver* is a primary site for use, modification, or storage of absorbed nutrients. It also secretes *bile,* which is stored in the

gall bladder and then secreted into the duodenum. Bile salts emulsify fats and aid in fat digestion.

9. The *large intestine*, or colon, absorbs most of the water and minerals not absorbed by the small intestine. It also forms, stores, and eliminates feces.

10. In vertebrates, the length of the small intestine correlates with whether or not the animal generates heat internally. Symbiotic bacteria (and sometimes protozoans) live in most animals' intestines, and in the specialized stomach *caeca* of ruminant mammals. These bacteria contribute vitamins, sugars, amino acids, and other nutrients to their host organisms.

11. The principle of chemical digestion is that enzymes work in se-

quence to break polymers into smaller units and those molecules, in turn, into monomer subunits.

12. Proteins are digested by endopeptidases and then by exopeptidases into dipeptides or amino acids that can be absorbed. Complex carbohydrates such as starch or glycogen are broken down into smaller and smaller units, and ultimately monosaccharides. Fats are commonly hydrolyzed to fatty acids and glycerol before being absorbed.

13. Nerves and hormones help regulate ingestion and digestion. Eating triggers the initial digestive processes in the mouth (saliva secretion) and stomach (pepsin and HCl secretion).

14. The hormones secretin and CCK control the secretion of pancre-

atic enzymes and juice, as well as the release of bile from the liver and gall bladder. Food movement out of the stomach (*enterogastric reflex*) and through the intestine is coordinated so that there is ample time for digestion and absorption.

15. Proteins, carbohydrates, fats, and some *minerals* are *macronutrients*. All eight *essential amino acids* must be eaten at a given meal if protein synthesis using food-derived amino acids is to occur.

16. *Micronutrients* include *water-soluble* and *fat-soluble vitamins* and trace minerals. Enzyme function, cellular respiration, blood clotting, nerve function, and other vital activities absolutely depend on specific vitamins and minerals.

KEY TERMS

alimentary canal
bile
bolus
caecum
calorie
carnivore
chyle
chyme
crop
digestion
enterogastric reflex
esophagus
essential amino acid

fat-soluble vitamin
gall bladder
gastrin
gizzard
herbivore
ingestion
intestine
large intestine
liver
macronutrient
micronutrient
microvillus
mineral

omnivore
pancreas
pepsin
peristalsis
pharynx
salivary gland
small intestine
stomach
vasoactive intestinal peptide (VIP)
vitamin
water-soluble vitamin
zymogen

QUESTIONS

1. Describe ingestion and digestion, and explain the differences between them.

2. Briefly outline the steps in ingestion and digestion by:
    a. *Paramecium*
    b. *Planaria*
    c. an earthworm

3. Explain extracellular digestion, and give several examples.

4. What are some problems faced by a large herbivore, such as a cow, in getting and processing food? How is a cow's digestive system adapted to deal with these problems.

5. Define the words "herbivore," "carnivore," and "omnivore," and give an example of each. Name one important ingestive or digestive adaptation possessed by each.

6. Name the organs of the human alimentary canal, and describe the role that each plays in digestion.

7. Name two major organs of the human digestive system that are *not* part of the alimentary canal, and describe the role that each plays in digestion.

8. What is the hydrolysis of carbohydrate, fat, and protein, and how is it accomplished during digestion?

9. Give several examples of how the chemical and physical activities of the digestive system are coordinated.

10. List the eight essential amino acids in the human diet, and explain why they are "essential," and why vegetarians must carefully monitor their diets.

ESSAY QUESTIONS

1. All heterotrophs must possess certain general behavioral and chemical adaptations in order to acquire organic molecules. List some of these adaptations.

2. Humans are prey to a number of eating-related disorders, including constipation, lactose intolerance, ulcers, anorexia nervosa, and vitamin deficiencies. Does this mean that we humans are poorly adapted to digest the foods available to us? Discuss.

SUGGESTED READINGS

BARNES, R. D. *Invertebrate Zoology*. 4th ed. Philadelphia: Saunders, 1980.
An excellent survey with good discussions of digestion in invertebrates.

DAVENPORT, H. W. *Physiology of the Digestive Tract*. 4th ed. Chicago: Year Book Medical Publishers, 1978.
A complete treatment of the alimentary canal, mainly of humans and other mammals.

ECKERT, R., and D. RANDALL. *Animal Physiology*. 2d ed. New York: Freeman, 1983.
Includes an excellent overview of the digestive tracts of a variety of animals.

MILNER, A. "Flamingos, Stilts, and Whales." *Nature* 289 (1981): 347.
Filter feeders at work.

MOOG, F. "The Lining of the Small Intestine." *Scientific American*, May 1981, pp. 154–176.
A good review, mainly about the cells and tissues of the small intestine.

35
HOMEOSTASIS: MAINTAINING BIOLOGICAL CONSTANCY

Stability of the internal environment is the condition of free life.

Claude Bernard, *The Way of a Medical Investigator* (1865)

A husky's fur insulation: one means of maintaining internal constancy.

The bodies of animals are more than 50 percent water, and yet kangaroo rats, bluegills, and sharks can live entire lifetimes drinking little or no water. How is this possible?

All complex animals have a salty "internal sea" of body fluids. These fluids stay within a narrow range of values for water and salt concentration appropriate to each species, even though the organism eats salty food, loses water through evaporation, or excretes water and salts in the form of urine. How does this internal sea remain so constant in composition?

Large mammals live in a wide array of habitats, from the polar icecaps to equatorial deserts to temperate forests, and they may encounter dramatic daily and seasonal shifts in temperature. Yet they maintain a high internal temperature that rarely fluctuates. How do they accomplish this feat?

The answer to all three questions is that animals have marvelous mechanisms for **homeostasis**—the maintenance of a relatively constant internal environment despite fluctuations in the external environment. This chapter deals with three separate but interrelated homeostatic systems that enable animals to survive the variations in salinity, water availability, and temperature on our planet.

In previous chapters, we encountered several physiological systems that help maintain homeostasis in an animal's body: the circulatory, lymphatic, respiratory, and digestive systems. In each, one or more body fluids (intracellular, extracellular, blood plasma, lymph) is involved in the system's central function. Those body fluids are also involved in three homeostatic systems we shall discuss in this chapter: the **osmoregulatory system,** which is responsible for governing the levels of water and salt in body fluids; the **excretory system,** which eliminates several kinds of metabolic wastes, usually in the urine; and the **thermoregulatory system,** which maintains an organism's body temperature and/or its responses to shifts in environmental temperature.

In our exploration of these three homeostatic systems, we will first look at mechanisms for regulating internal salt and water balance, including special adaptations of animals that live in particularly dry or salty environments. Next, we will consider how wastes are filtered out of the body fluids and excreted and examine in detail one of the most complex organs ever to evolve—the kidney. Finally, we will describe how animals cope with changes in environmental and body temperatures and are thus able to survive in Earth's wide range of habitats. We shall see how the continuous expenditure of energy is required to maintain a stable internal environment as the three systems function; such expenditures are truly the price of living free in the ocean, in fresh water, or on land.

REGULATION OF BODY FLUIDS

People driving on a freeway, shrimp scudding along in the sea, and mayflies darting above a field in summer are all, in a sense, complex, mobile pools of water. This is not surprising, considering that life originated in salty water and that the earliest organic macromolecules, enzyme systems, nucleic acids, and living cells probably formed in the solvent water. Water is still the primary solvent in all living cells and the chief component of the three body fluids common to multicellular animals with closed circulatory systems (intracellular fluid, extracellular fluid, and blood or hemolymph). But most organisms live in environments where the balance between water and salt is unfavorable to life processes; in fact, only halobacteria, which thrive in highly salty environments, can tolerate very high internal salt concentrations, that is, high *tonicities* (Chapter 5). Therefore, the ability to maintain a stable internal fluid environment by means of **osmoregulation** was one of the most important evolutionary innovations, enabling multicellular animals to carry around their own internal sea.

Osmosis, you will recall from Chapter 5, is the tendency of water to move through semipermeable membranes, depending on the relative concentrations of osmotically active substances, so-called **osmolytes** (ions, small organic molecules, proteins), on the two sides of the membrane. Thus water moves into or out of a cell, depending on the relative concentrations of such substances in the intracellular and extracellular fluids. The majority of invertebrates and vertebrates cannot tolerate very high salt concentrations in their body fluids because of the deleterious effects of salts on protein structure and function. Instead, their bodies display what is called the **compatible osmolyte strategy;** they have, in addition to salt ions, various organic molecules (amino acids or polyalcohols, such as sucrose or glycerol) that raise the total osmotic pressure of the body fluids close to that of sea water. These organic substances do not have deleterious effects on enzymes and structural proteins—hence the term "compatible."

The ratio of different salts in body fluids is also a key to cellular health. This *salt balance* commonly involves higher concentrations of potassium ions in the intracellular than in the extracellular fluid, and higher concentrations of sodium ions in the extracellular than in the intracellular fluid. Maintaining an internal environment that is constant in osmotic pressure and salt balance requires specialized osmoregulatory mechanisms, which vary widely according to the nature of an organism's habitat. Let us look at the problems faced by animals inhabiting fresh water, sea water, and dry land and see how they maintain the proper water and salt balance in body fluids.

Life in Fresh Water

Fresh-water organisms have a curious problem: their body fluids are "saltier" than the surrounding environment. That is, they contain more dissolved salts and organic molecules, and therefore have higher osmotic pressures than the water composing their habitat. Thus the surrounding water is said to be *hypotonic* ("lower tonicity") relative to their body fluids. The result is that pond, lake, or river water tends to move constantly into the animal's body across all permeable surfaces, causing excessive *hydration*, or bloating. At the same time, the organism's own dissolved salts tend to move outward to the pond, lake, or river water (Figure 35-1a); the small organic molecules tend to be retained within the body by cellular plasma membranes. If hydration continued unchecked, the animal's cells could be damaged, and if the outward movement of salt continued, the body fluids would soon be too dilute for life processes to continue.

How do fresh-water animals cope with these problems of salt and water balance? Fresh-water fish have four major adaptations for coping:

1. They almost never drink water.

2. Their bodies are coated with mucus, which helps stem the constant influx of water.

3. They excrete a large volume of water as dilute urine.

4. They can actually take up salt by "active" (energy-requiring) processes from the fresh water that passes through their gills.

Special salt-absorbing cells on the gill surfaces carry out this last task. Thus, the plasma membranes of these special cells bear pump enzymes that transport ions from the area of low concentration (fresh water) to one of much higher concentration (cell cytoplasm). These pump enzymes are ATPases (enzymes that catalyze the hydrolysis of ATP) that use the energy freed by hydrolysis of ATP to carry out their work. Thus the price of continuous salt pumping into the body of the fresh-water organism is expenditure of ATP.

These gill cells represent a lovely case of integrated physiological function. As sodium ions (Na^+) are pumped in, either ammonium ions (NH_4^+) or hydrogen ions (H^+) leave. This is because (as we've seen in several earlier chapters) the pump enzyme requires a counter ion—to transport Na^+ in one direction another positively charged ion must move in the opposite direction. Removal of NH_4^+ is the organism's main means of excreting nitrogen, which we will discuss in more detail later. The H^+ comes from water in the reaction

$$H_2O + CO_2 \rightarrow H_2CO_3 \rightarrow H^+ + HCO_3^-$$

The CO_2, of course, is the product of oxidative metabolism by body cells. The controlled elimination of this H^+ helps regulate body fluid pH. Furthermore, as the

Figure 35-1   WATER AND SOLUTE EXCHANGE IN FRESH-WATER AND IN MARINE TELEOSTS.

(a) The body fluids of fresh-water teleosts are hypertonic to the hypotonic environment, and thus the animals are faced with the problem of losing salts (orange arrows) to the medium and of hydrating as water (blue arrows) tends to move into the animal. To compensate, fresh-water teleosts excrete a large volume of watery urine and actively take up the salts from the water passing through the gills. "Active" uptake or excretion depends on energy, which is usually derived from the hydrolysis of ATP. (b) The body fluids of marine teleosts are more dilute (hypotonic) than the hypertonic sea water and tend to lose water to the environment and to become too "salty." Marine fish obtain water by drinking and minimize water loss by excreting a small volume of highly concentrated urine. Salts that are ingested are eliminated by both active and passive mechanisms (the latter require no energy). Solid arrows indicate active movement of water and solutes; broken arrows indicate passive movements.

(a) Fresh-Water Teleosts

(b) Marine Teleosts

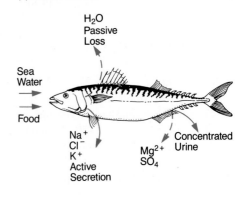

HCO_3^- leaves through the gills, Cl^- enters. Therefore, the transport of Na^+ and Cl^- into the cells is coupled to the excretion of nitrogen (NH_4^+) and HCO_3^- plus H^+. The homeostasis of salts in fresh-water organisms is thus related to the excretion of major waste products— nitrogen and CO_2—as well as to the regulation of pH in body fluids.

Life in Salt Water

The high concentration of salts in sea water presents a radically different challenge to life than does the low salt concentration of fresh water. Many marine invertebrates—such as lobsters, starfish, cnidaria, and annelid worms—have body fluids with osmotic pressure close to that of sea water. Such animals are essentially *isoosmotic* (osmotically the same) relative to the salty ocean; there is little net movement of water into or out of the animal's tissues. Many other marine creatures, however, such as clams and bony fishes, have internal osmotic pressures roughly similar to those of their fresh-water relatives and hence much lower than that of sea water. Sea water is *hypertonic* ("higher tonicity") relative to their body fluids, so water tends to leave their tissues. The result is *dehydration* and the influx of salts, which disturb the normal osmotic and salt balances (Figure 35-1b).

As one might expect, marine bony fishes must compensate for the tendency to lose water by taking some in, and they must rid the body of the excess salt that enters. To do so, they employ adaptations nearly opposite to those of fresh-water fish:

1. They drink sea water frequently to replace water lost by diffusion through permeable body surfaces, such as the gills.

2. They excrete certain salt ions (Mg^{2+}, SO_4^{2+}) in a small volume of concentrated urine.

3. They excrete Na^+ and Cl^- by means of salt-secreting cells on the gills.

The latter cells have a pump enzyme with ATPase activity; other features work with the pump enzyme to move Na^+ ions from inside the cells outward into the salty sea water against a concentration gradient. To do so, ATP must be hydrolyzed to yield the energy that powers the pumping process.

Some fishes encounter both salt and fresh water. An example is the small Atlantic salmon swimming down a fresh-water Newfoundland river and entering the sea for the first time (Figure 35-2). Instead of continuing to pump salt in, as it has done since its embryonic stages, the fish suddenly must rid itself of excess salt. But how? Apparently, new ATPase pump enzyme molecules of the type found in marine fish gills must be synthesized. If synthesis of mRNAs or of new protein is experimentally inhibited in such fish, they cannot adapt to sea water and will die. Years later, of course, these same salmon migrate back from the sea into their home fresh-water streams. As they do, the enzyme pumping direction must revert to that of the young fish—ions must be pumped in from dilute river water.

Sharks and their chondrichthian relatives also live in the salty oceans, but their solution to the dehydration problem involves a different approach from that of other marine fish. Rather than constantly battling against high osmotic pressure by pumping out salt and by drinking water, they have two kinds of organic molecules in their body fluids that raise the osmotic pressure to a level slightly higher than that of sea water: urea, a by-product of nitrogen metabolism, and trimethylamine oxide (TMO). Because its body fluids are rendered slightly hypertonic by these substances, a shark or skate actually *gains* water from the sea across permeable surfaces. Research shows that urea alone in high concentrations denatures proteins and inhibits enzymes; in contrast, TMO stabilizes proteins and activates enzymes. The two together, in a proper ratio, counteract each other and thus raise the osmotic pressure without interfering with proteins. This has been termed the **counteracting osmolyte strategy.** Shark enzymes and proteins have actually evolved special adaptations that allow them to continue functioning in the presence of urea and TMO. Equiva-

Figure 35-2   SALMON AND THE PROBLEM OF CHANGING HABITATS.
These sockeye salmon (*Oncorhynchus nerka*) have salt-balance problems because they migrate when young from fresh-water breeding sites to the ocean. Here they are seen returning to their fresh-water spawning sites from the ocean. In fresh water, the fish must gain salts actively, whereas in sea water, they must excrete salts actively.

lent enzymes or proteins from a bony fish or a mammal are inhibited at the levels of urea and TMO present in sharks. Thus the counteracting osmolyte strategy must have evolved over millions of years in sharks and entailed changes in innumerable molecules and body structures. This evolutionary strategy is so successful, however, that a number of other fishes and invertebrates have evolved the same mechanisms and employ pairs of counteracting osmolytes to raise osmotic pressure of body fluids.

Despite this useful mechanism, the shark must still rid itself of excess salt that enters across the gills and in its food. One means of accomplishing this is a special gland near the rectum, the *rectal gland*, which secretes a highly concentrated NaCl solution. The animal's kidney also removes and excretes additional salts in the urine.

Life on Dry Land

Terrestrial animals have one thing in common with marine creatures—the tendency to dehydrate. Land animals dehydrate not via osmosis, however, but by evaporation of body fluids into the surrounding "ocean of air," a process that can lead to an osmotic imbalance like those of marine animals. Land vertebrates display various adaptations for preventing moisture loss, for taking in water, and for eliminating or conserving salt, depending on the saltiness of the animal's food and water.

A critical stage in the transition from aquatic fish to terrestrial amphibians was the loss of the gills and, with them, that site of gas exchange, nitrogen and H^+ excretion, and salt regulation. As we saw in Chapter 33, frogs' lungs take over part of the gas-exchange function, but their skin is also a site of CO_2 loss, ion exchange, and H^+ excretion. It is only in the dry-skinned reptiles that such homeostatic functions are relegated primarily to the lungs and kidneys. Let us examine these systems in more detail.

Amphibians, the first vertebrates that colonized land, take up water through the skin and across the wall of the bladder, a balloonlike organ for storing urine (Figure 35-3). This uptake counteracts evaporation and prevents osmotic imbalance. Of course, the skin is a site of potential evaporation, but a hormone from the brain can cause water to enter the body through the skin when the animal is on a moist surface or is immersed in water. The bladder of a frog, toad, or salamander is an important water and salt reservoir; the drier the environment, the larger the bladder and the more urine it can store. If an amphibian begins to dehydrate, a hormone causes water to be physiologically "reclaimed" from the bladder and returned to the extracellular fluid. In other land vertebrates—descendants of ancient amphibians—the blad-

(a)

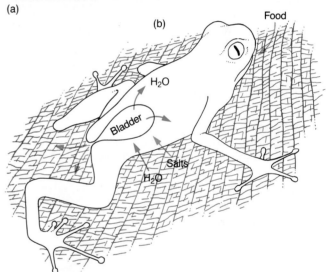

(b)

Figure 35-3   WATER AND SALT UPTAKE IN AN AMPHIBIAN.

(a) A poison arrow frog (*Dendrobates histrionicus*) on a moist leaf. (b) Salts may be absorbed actively, especially through the belly skin in contact with the ground or another substratum, such as a leaf. Water is stored in the bladder, from which it can be recovered; water may enter the body through the skin by passive processes as well.

der is largely a receptacle for urine that will be excreted.

A potential major site of water loss in terrestrial vertebrates is the lungs. In many mammals, the nasal cavities function as a *countercurrent exchange system* to combat such loss (Figure 35-4). During inhalation, air passing through the nasal cavities is warmed by heat from adjacent tissues, but in the process, the temperature of these tissues falls. One result of this is a cold nose—a dog's nose is a good example. The inhaled air is further warmed and humidified in the lungs. Then, as it passes back out during the next exhalation, the warm, moist air flows over the cooler nasal surfaces, and the air gives up some of its heat. As the air cools, much of the water

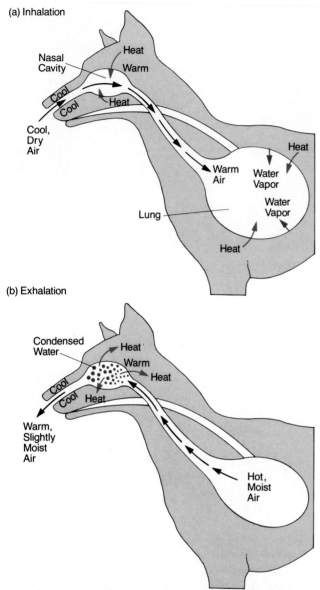

(a) Inhalation

(b) Exhalation

Figure 35-4   WATER RETENTION BY COUNTERCURRENT HEATING AND COOLING.
(a) When a dog inhales, the air passing through its nose is heated and humidified. Simultaneously, its nasal tissues are cooled. (b) The warm, moist air that the dog breathes out gives up heat to the previously cooled nasal tissues; it, therefore, can carry less water vapor, and condensation occurs. This makes the animal's nose wet, and it prevents much water loss in exhaled air.

vapor condenses out on the nasal surfaces and so is not breathed out of the body. In mammals from dry habitats, a cold nose can actually be the key to survival. In the kangaroo rat, camel, and giraffe, for example, more than 50 percent of the water that would be lost via respiratory evaporation is saved by this countercurrent exchange system in the nose. For a giraffe on the East African savanna, this savings can equal 1.5 to 3 liters a day, or as

much as 20 percent of the animal's total need for water.

In most terrestrial vertebrates, the primary means of regulating the osmotic balance of body fluids is to control the amount of water and salt lost in the urine. Thus the kidneys, rather than the gills of fish, are the primary regulatory organs. We shall learn a great deal about the kidneys and control of water loss in the next section. Here, let us look at a special system in reptiles and birds that supplements the activities of the kidneys.

Desert and marine birds and reptiles often build up high salt concentrations in their bodies because they consume salty foods or sea water and lose water through evaporation and in the urine and feces. To rid themselves of this excess salt, these animals have *salt glands*, special secretory organs near the eye or in the tongue that remove excess NaCl from the blood and secrete it as tearlike drops. Perhaps you have seen a seagull shake its beak vigorously to remove salt droplets. Sensors in the wall of the bird's heart apparently monitor the osmotic pressure of the blood and send nerve signals to the brain that trigger activation of the salt glands. Marine birds and reptiles can drink sea water, since ATPase pump enzymes in the salt glands can concentrate a solution that is even saltier than sea water. For each volume of sea water the animal drinks, some of the water and all the salts are excreted. The rest of the water is retained, so the animal achieves a net water gain.

Terrestrial vertebrates without salt glands—including humans—ultimately lose body water by drinking sea water. Our kidney does not have the capacity to produce urine that is saltier than sea water, so for every sip of sea water a human drinks, an even larger volume of urine is excreted to maintain the appropriate salt balance in the body fluids—the source of that extra volume is precious body water. For every liter of sea water a person drinks, 1.75 liters of urine must be produced. Therefore, a shipwrecked person drinking sea water would become rapidly dehydrated and literally urinate to death if he or she could not find a source of fresh water. Unlike the human castaway, marine mammals, descendants of typical terrestrial mammals, evolved remarkable kidneys that can concentrate and excrete urine that is more concentrated than sea water. Baleen whales, for example, can drink sea water and eat salty organisms and still gain, rather than lose, body water.

EXCRETION OF NITROGENOUS WASTES

In previous chapters, we have encountered a number of systems by which animals rid themselves of waste products. Gaseous CO_2 is lost across the tracheal, lung, and skin surfaces, or as HCO_3^- or dissolved CO_2 across the gills. Bacteria, cellulose, and other undigested solid wastes are eliminated through the digestive system. And as we have just seen, excess salt can be excreted by the gills, by special salt glands, and by the kidneys. But what happens to nitrogenous waste products formed in body cells?

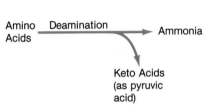

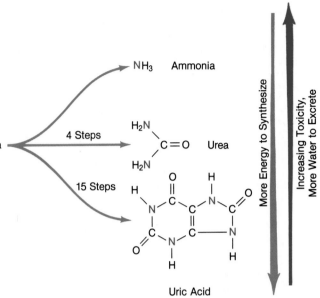

Figure 35-5   THE BREAKDOWN OF AMINO ACIDS INTO NITROGENOUS WASTES.
Nitrogenous wastes form mainly when amino acids are deaminated. The first breakdown product is ammonia, the excretory waste of many animals, particularly aquatic ones. Keto acids, common intermediates in glycolytic pathways and the Krebs cycle, are also formed. In terrestrial animals, ammonia is converted into either urea or uric acid. These processes require energy. This diagram shows the relative toxicity of the three nitrogenous-waste products. The more toxic a compound, the more water is required to dilute and excrete it. Note that only one nitrogen atom is excreted per molecule of ammonia, whereas two are present in urea and four in uric acid.

The breakdown of proteins, nucleic acids, and other nitrogen-containing compounds results in the formation of *nitrogenous*, or nitrogen-bearing, wastes. During the metabolism of proteins, enzymes in the liver remove the amino groups (NH_2) from amino acids in a process called *deamination*. The remainder of the molecules (keto acids) can be converted to sugars and lipids for storage. The released NH_2 groups combine with free H^+ ions to form NH_3 (ammonia) and NH_4^+ (ammonium ions) (Figure 35-5). Ammonia is quite toxic, as anyone who has inadvertently gotten a strong whiff of kitchen ammonia will testify. In fact, no cell can survive in the presence of a high concentration of ammonia. Thus ammonia must be transported or stored in very dilute solution. It, and all the solvent water carrying it, must be excreted in that highly dilute solution; a single gram of nitrogen reduced to NH_3 must be dissolved in 300 to 500 milliliters of water to be flushed from the body without harming cells along the way.

The basic problem for organisms handling nitrogenous wastes is simple: toxicity versus the need to conserve water. When water is abundant—as in typical fresh-water fish—the answer to the problem is equally simple: "dilution is the solution to pollution." But when water is precious—as it is to many marine and most terrestrial animals—the dilution strategy creates problems of severe dehydration and osmotic imbalance. Those animals solve the problem by detoxifying the nitrogenous wastes or by making them insoluble. Both strategies reduce water loss during nitrogen excretion. Let us see the sorts of detoxified and insoluble nitrogenous wastes that are excreted by different animal groups.

Both fresh-water and salt-water fish overcome the problem of NH_3 toxicity by carrying out deamination of amino acids in the gills, rather than in the liver. Therefore, nitrogenous wastes in the form of NH_3 diffuse from the gill cells directly into the respiratory current as it leaves the body. Consequently, body cells never really come in contact with the NH_3, so it need not be diluted and no body water is lost excreting it.

In land animals, enzymes reduce the toxicity of nitrogenous wastes; specifically, liver enzymes manufacture urea or uric acid (Figure 35-5). **Urea** is less toxic than NH_3 because it is not a strong base and can be transported safely in far less water—1 gram in 50 milliliters instead of 300 to 500 milliliters. Urea may be present in both intracellular and extracellular fluids in relatively high levels without being toxic if trimethylamine oxide (TMO) is also present to counteract its effects. In many land animals, urea is removed from the body by the kidneys and is eliminated from the body in fluid urine.

Land animals faced with water shortages often excrete nitrogenous wastes in the form of **uric acid.** One gram of nitrogen in this form can be transported and excreted in only 10 milliliters of water. Uric acid is usually formed in the liver, in a lengthy pathway that requires more than a dozen enzymes and a considerable amount of energy. In

Table 35-1 **MAJOR STRATEGIES FOR MAINTAINING WATER AND SALT BALANCE**

Animal	Environment Concentration Relative to Cells	Urine Concentration Relative to Blood	Major Nitrogenous Waste	Key Adaptation
Fresh-water teleost	Hypotonic	Very hypotonic	NH_3	Absorbs salts through gills
Marine teleost	Hypertonic	Isotonic	NH_3	Secretes salts through gills
Marine chondrichthian	Isotonic	Isotonic	NH_3	Secretes NaCl from rectal gland
Amphibian	Hypotonic	Very hypotonic	NH_3 and urea	Absorbs salts through skin
Marine reptile	Hypertonic	Isotonic	Urea and NH_3	Secretes salts through salt glands
Marine mammal	Hypertonic	Very hypertonic	Urea	Drinks some sea water
Desert mammal	—	Very hypertonic	Urea	Manufactures metabolic water
Marine bird	—	Weakly hypertonic	Uric acid	Drinks sea water; uses salt glands
Terrestrial bird	—	Weakly hypertonic	Uric acid	Drinks fresh water

a lizard, a bird, or an insect, uric acid moves in solution toward the terminal end of the excretory tract; water is reabsorbed across the tract walls, and the compound becomes more and more concentrated along the way, until it crystallizes and precipitates out of solution. Crystallization releases the remaining water serving as solvent, and it, too, can be reabsorbed into the extracellular fluid. Uric-acid crystals do not exert osmotic pressure. Guano, the white semisolid paste excreted by birds and reptiles, is largely uric-acid crystals.

Whether an animal excretes ammonia, urea, or uric acid (or combinations of these nitrogenous compounds) depends to a large extent on water availability in its habitat. For example, desert and other land reptiles excrete uric acid, whereas aquatic reptiles, such as sea snakes, turtles, and crocodiles, excrete urea and even NH_3. Throughout the animal kingdom, the additional energy expense of producing urea and uric acid can be avoided if enough water is available to flush out toxic NH_3 in greatly diluted form. Table 35-1 lists the major nitrogenous wastes of various vertebrates as well as other aspects of their osmoregulation.

In humans and some other animals, uric acid may be formed from the breakdown of nucleic acids (rather than from deamination). If too much uric acid is produced, *gout* can develop. In this condition, uric-acid crystals form in fingers, toes, and other joints, causing severe pain.

In embryos, strategies for excreting wastes also depend on environment. Maturing avian and reptilian embryos, sealed in hard or leathery shells, first produce NH_3, then urea, and, finally, uric acid. Mammalian embryos produce mostly urea, which crosses the placenta into the mother's blood rather than being stored. The mother then excretes these wastes via her own kidneys and urine. Biologists speculate that these strategies are the basis for differences in excretion between adult animals: adult reptiles retain the uric-acid–producing enzymes so essential to reptilian embryos, while adult mammals, like their embryos, lack these enzymes.

Excretory Systems: Organs and "Plumbing" for Waste Removal

We have discussed the common nitrogenous wastes NH_3, urea, and uric acid, and explained briefly that they are excreted by the same organs that regulate osmotic balance in the body fluids. Beginning with the invertebrates and then considering the vertebrates, let us look more closely at the organs and associated structures that rid the body of wastes, ions, and excess water. We will see that the same basic kind of "plumbing"—a tubule with the capability to filter and reabsorb water, ions, and molecules—evolved independently in at least three groups: the annelids, arthropods, and chordates.

Excretory Systems in Invertebrates

The simplest excretory structure found in any animal is the *contractile vacuole* of single-celled protists (Figure 21-2). The vacuole is used primarily to expel excess water, but it also ejects nitrogenous wastes. Small multicellular invertebrates, such as flatworms, have more complex structures for excretion, tubules with *flame cells* (Figure 35-6). At the next level of complexity are the **nephridia,** or excretory organs, of the earthworm (Figure 35-7). Most of the body segments of an earthworm or typical aquatic worm contain a pair of nephridia. Each nephridium is composed of a funnel-like

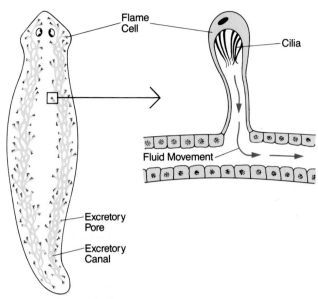

Figure 35-6   EXCRETORY ORGANS IN FLATWORMS.
Planarians have excretory systems with a branching network of nephridial ducts. Excretory fluid is discharged into these ducts by the flame cells and is propelled by cilia. Wastes are then discharged to the exterior through excretory pores.

opening from the body cavity, the *nephrostome;* a coiled *tubule;* an expanded storage portion of the tubule, the *bladder;* and an exit pore through the body wall, the *nephridiopore.* Water, salts, and wastes in the body fluids can enter the nephrostome directly and pass down the tubule. Alternatively, substances in the hemolymph (blood) can move from the blood capillaries that are entwined around the coiled tubule and pass across the tubule wall into the fluid. Water, salt ions, and organic substances may be reabsorbed from the fluid passing through the tubule, depending on the body's needs. For example, if the level of Na^+ is low in the body fluids, it is reabsorbed. The remaining water and wastes collect in the bladder and are excreted through the nephridiopore. As we will see, the functions of nephridia in worms are very like those of the individual functional units of the vertebrate kidney, the *nephrons.*

Grasshoppers and most other insects have several narrow, blind-ended sacs that arise at the junction of the midgut with the hindgut. These sacs, called **Malpighian tubules** (Figure 35-8), are bathed on the outside by hemolymph; the cells of the tubule walls absorb uric acid, K^+, and other substances from the hemolymph and pass them into the tubule by means of active transport. A small amount of water passively accompanies the solutes because of osmosis. The solution drains into the gut and accumulates in the hindgut. There, much of the water

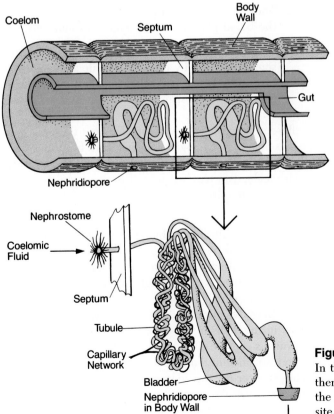

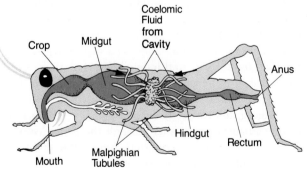

Figure 35-8   EXCRETORY ORGANS IN INSECTS.
In insects, excretory products in the coelomic fluid pass through the walls of the many Malpighian tubules (arrows), and urine is discharged into the hindgut where it is eliminated with the solid fecal waste.

Figure 35-7   EXCRETORY ORGANS IN ANNELID WORMS.
In the earthworm, wastes pass into the coelomic fluid which then enters the excretory organ, the nephridium, through the nephrostome. The bladder forms a reservoir storage site. The excretory fluid (urine) is discharged from the body via the nephridiopore.

and any salts needed by the body are reabsorbed and returned to the hemolymph. Thus the insect's Malpighian tubule, like the worm's nephridium and the nephrons of the land vertebrate's kidney, which we will discuss next, are the sites of nitrogenous-waste removal from body fluids. Each of these structures also helps to maintain the correct water-salt balance in the body.

Excretory Systems in Vertebrates: The Marvelous Kidney

In all reptiles, birds, and mammals, the key to excretion, water and salt balance, and regulation of pH in body fluids is the **kidney,** a blood-filtering and waste-excreting organ. The functional unit of the kidney, the **nephron,** is a tubule with several different parts in which secretion and reabsorption of ions, molecules, and water take place. Some marine bony fishes have the least complex of all nephrons—little more than simple tubular structures. Most fish, however, have more complex nephrons that function appropriately for the animal's environment: in fresh-water fish, the nephrons produce copious urine, but in ocean fish, they produce just a dribble. Amphibian kidneys, called *mesonephric kidneys,* are generally similar in structure to those of fresh-water fish and usually produce a large volume of urine. It is in reptiles that a new kind of kidney develops in the embryo—the *metanephros,* which matures into the adult kidney with many nephrons. Individual mammalian and avian nephrons have a unique section—the loop of Henle, which we will describe shortly. With it come unprecedented abilities to concentrate a hypertonic urine and, in turn, to conserve water for the body. Nowhere in biology is the structure of an organ and its tissues so obviously essential to function as in the kidney. Let us examine the human kidney to learn why.

Due to its immense blood supply, the human kidney is a dark purplish organ. It is about 10 centimeters long and 7 centimeters wide, and has a distinctive shape—rounded on one side and indented on the other (Figure 35-9). Each of the two human kidneys is made of more than 1 million nephrons, and together they perform an astonishing task: every twenty-four hours, all the blood in the human body passes through these two compact organs and their many nephrons some 500 to 600 times! During each pass through the kidneys, the blood's salt and water contents are balanced, and wastes are removed. Of the 700 to 800 liters of blood moving through the human kidneys daily, about 200 liters of fluid leave the blood and enter the nephrons, but only 1.5 liters are excreted as urine. The detailed structure of the nephron is the key to understanding this incredible performance. But to understand why nephrons function the way they do, we need to review kidney structure.

Each kidney has an outer **cortex** region in which the

nephron's main portion is located (Figure 35-9). Internal to the cortex is the **medulla,** where the loops of Henle reside. Ducts passing through the medulla open into the kidney's central cavity, the **pelvis.** Let us examine each part of the nephron as well as these ducts so we can see how the pelvis drains the urine that eventually reaches the bladder.

Each nephron is a long looped unit (unfolded, it would be 30–38 mm long) composed of several parts: a complex, branched, ball-shaped mass of blood capillaries called the *glomerulus,* which is surrounded by a double-walled cup called the *Bowman's capsule;* a slender tube called the *proximal convoluted tubule* extending from the Bowman's capsule; a hairpin-shaped portion called the *loop of Henle;* and a final twisted section called the *distal convoluted tubule,* which joins the *collecting duct.* The many collecting ducts run together like tributaries into a river and then drain into the kidney's *pelvis.* This cavity narrows and leads into the *ureter,* a duct from each kidney that carries urine to the *bladder,* the muscular-walled, balloon-shaped storage organ. A single duct, the *urethra,* exits from the bladder and carries urine out of the body.

In order to remove nitrogenous wastes from the body fluids and to control water content and salt concentration, all vertebrate kidneys operate with the same two-step strategy:

1. They produce a primary filtrate that contains all the blood plasma's small molecules and ions.

2. They reabsorb most of the water and all the substances needed by the body, while allowing urea, toxic substances, and other ions and molecules to pass out in the urine.

Different parts of each nephron carry out these tasks by means of four physiological mechanisms: filtration, reabsorption, secretion, and concentration. Let us consider each mechanism and see how it depends on the structure of the nephron.

Filtration Filtration of blood begins with the delivery of blood by the renal artery, the major vessel to each kidney in mammals. The blood entering the glomerulus of each nephron is under high pressure (about 70 mm Hg), which actually *pushes* a solution of water, salts, urea, traces of protein, and other molecules through holes in the walls of the glomerular capillaries and the Bowman's capsule (Figure 35-10). The filtration mechanism is so efficient that about 25 percent of the water and solutes entering the glomerulus pass across into the Bowman's capsule. This filtration is entirely passive and depends completely on blood pressure; if the blood pressure falls, so may the rate of the glomerular filtra-

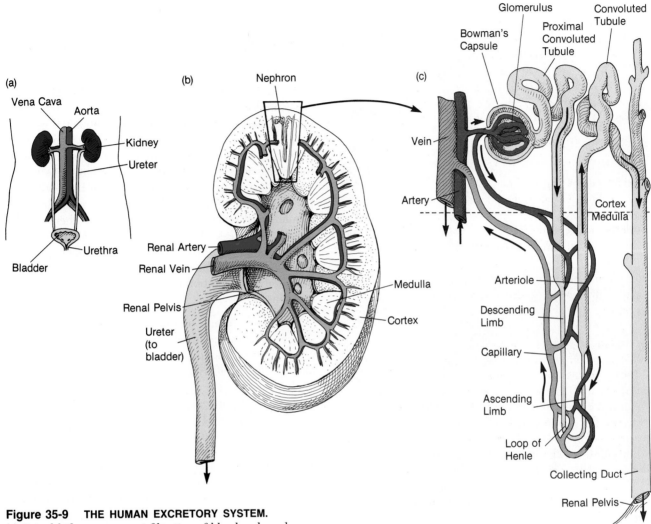

Figure 35-9   THE HUMAN EXCRETORY SYSTEM.
(a) Paired kidneys carry out filtration of blood and produce urine. This fluid passes down the ureters to the urinary bladder, where it is stored before being excreted through the urethra. (b) A longitudinal section through the kidney shows the blood supply to and from the cortex, the actual site of filtration. The glomeruli lie in the outer cortex region, whereas the loops of Henle lie in the medulla region. (c) The nephron, the functional unit of the mammalian kidney, is responsible for the production of urine. The various parts of the nephron are shown, as well as adjacent blood capillaries. The collecting duct is the site of final concentration of the urine.

tion. In moving from the blood to the cavity of the nephron, all substances must pass through a *basement membrane,* a thick layer composed of sugar polymers and proteins and which includes a basal lamina (Chapter 5). The basement membrane acts as a filter and ensures that most of the large molecules in blood plasma do not pass into the cavity of the Bowman's capsule. The solution that does go through is called the **glomerular filtrate.** It

differs from plasma mainly by lacking blood cells and large proteins. It contains not only substances that must be excreted from the body, such as urea, but also useful substances, such as glucose, water, salts, and amino acids, that must not be lost in large quantities and so must be reabsorbed.

Reabsorption   Many useful substances are removed from the filtrate as it passes through the tubular portions of the nephron. Both active (ATP-requiring) and passive processes are involved in the recovery of these substances. Fully 75 percent of the sodium and potassium ions and the amino acids and 100 percent of the glucose and vitamins are actively removed from the filtrate and secreted by the tubule cells into the extracellular fluid surrounding the nephrons. The materials are then taken into the blood by the capillary net that surrounds the tubules. Negatively charged ions, such as Cl^- and HCO_3^-, follow the positively charged Na^+ and K^+ pas-

(a)

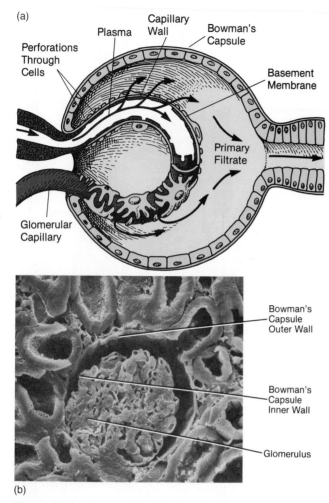

(b)

Figure 35-10 FORMATION OF THE GLOMERULAR FILTRATE.

(a) A highly simplified view of one of the many capillaries in the glomerulus. High blood pressure forces water, ions, and small molecules to filter through the perforations in the walls of the glomerular capillary and Bowman's capsule. The basement membrane is a filter that retains most large molecules in the plasma. (b) A scanning electron micrograph of a glomerulus and Bowman's capsule (magnified about 34 times). From *Tissues and Organs: A Text-Atlas of Scanning Electron Microscopy* by Richard G. Kessel and Randy H. Kardon, W. H. Freeman and Company © 1979.

sively into the extracellular fluid. Most of this reabsorption occurs in the proximal convoluted tubule.

As solutes are removed from the filtrate, it becomes more dilute, and the extracellular fluid becomes more concentrated. To correct this imbalance, water leaves the filtrate passively by osmosis as the filtrate moves through the proximal convoluted tubule and into the descending limb of the loop of Henle. The mechanism—active salt reabsorption with water following passively—is an effective one, and the amount of water saved is dramatic: 75 percent of the original filtrate is reabsorbed before the loop of Henle is reached. The reabsorption

process is not limited to useful substances; some wastes, such as urea, are also reabsorbed. However, enough of the wastes remain in the filtrate so that concentrations in the blood are maintained at lower-than-toxic levels.

Thus the filtration and reabsorption activities of the nephron do not simply remove wastes. They also maintain water and salt balance, and therein lies the important homeostatic function of the kidney. If the concentration of K^+, Cl^-, or some other ion in the blood plasma is too high, the nephron's tubule cells reabsorb fewer of the excess ions from the filtrate, and more are subsequently lost in the urine. Conversely, if the concentration of specific ions in the blood plasma is too low, the tubule cells reabsorb more and return them to the blood, so that less salt is excreted. The homeostatic function of the nephrons and kidneys is obviously crucial and continuous, balancing salt and water despite the animal's ingestion and excretion and despite changes in its environment.

Secretion While reabsorption causes some substances to leave the filtrate and enter the blood, the reverse process also occurs: materials diffuse from the blood in the capillaries into the tubule cells, which actively secrete them into the filtrate. One of the most important substances secreted in this way is hydrogen ions. Secretion of H^+ occurs in both the proximal and the distal convoluted tubules and serves to rid the body of the excess acid that accumulates during normal feeding and metabolism. (Recall that gills do the job of H^+ excretion in many fish.) A second important tubular secretion is of excess K^+. Since that ion is normally reabsorbed in the proximal tubule, the nephron can only rid the body of excess K^+ by secretion from the distal tubules and collecting ducts. Finally, some chemicals that do not naturally occur in the body, such as penicillin and various toxins, are also secreted into the tubules and eventually carried away in the urine.

Concentration A mammal's ability to excrete wastes and salts with minimum water loss depends on the fourth mechanism, concentration. Let us follow fluid through the nephron and see how the concentration process occurs (Figure 35-11).

The process of reabsorption in the proximal convoluted tubule removes both salt and water from the glomerular filtrate and reduces its volume to 25 percent of the original, but it does not significantly change the concentration of salts and urea. As the filtrate flows through the descending limb of the **loop of Henle,** it becomes further reduced in volume and more concentrated. Curiously, cells composing the descending limb cannot carry out active transport of either water or salts. How, then, does the change in concentration occur? The answer is that the descending limb is permeable to water and it (along with the ascending limb) passes through a "brine

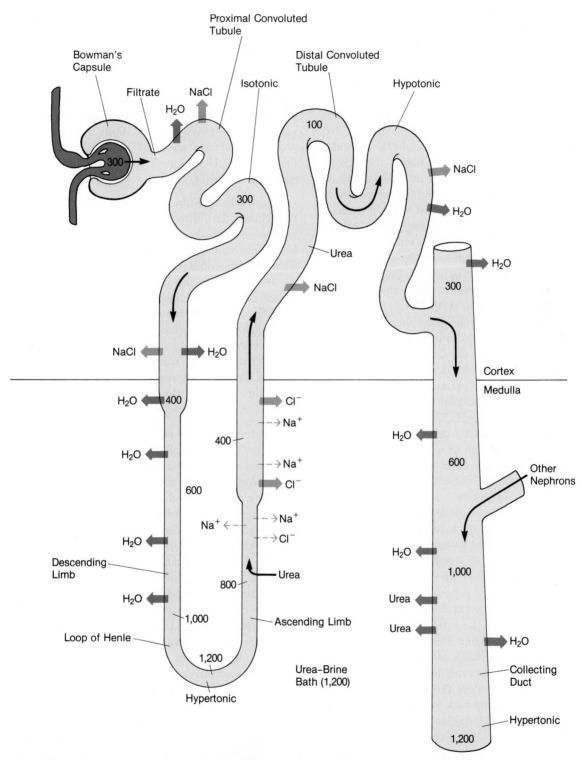

Figure 35-11   URINE PRODUCTION IN THE MAMMALIAN NEPHRON.
This figure follows the movement of filtrate through the nephron. Active transport of ions is shown with solid arrows; passive transport with dashed arrows. The hypothesized active transport of Cl⁻ in the thick-walled portion of the ascending limb of the loop of Henle is the key to NaCl accumulation in the fluid bathing the loop of Henle. Passive movement of urea from the collecting duct builds a very high urea concentration in the fluid surrounding the loop. This, along with the NaCl, gives rise to the extracellular "urea-brine bath." As filtrate passes through the loop bathed with this solution, much water can be recovered and carried away by the blood. Water also may diffuse out of the collecting duct, and a reduced volume of hypertonic urine is thereby produced. The numbers at intervals along the tubules provide a measure of the relative concentration of the filtrate.

bath," a region where extracellular fluid is highly concentrated. Thus water leaves the filtrate in the descending limb by passive diffusion, and the concentration of the filtrate rises three- to tenfold.

Where does this brine bath come from? Several factors account for this zone of high salt and urea concentration around the lower portion of the loop of Henle. The filtrate passes through the descending limb, moves into the hairpin turn of the loop, and then climbs the ascending limb. In the ascending limb, it passes through a thick-walled section where it is currently thought that Cl^- ions are actively transported out of the tubule, and Na^+ follows passively. Here, in contrast to the proximal convoluted tubule, water cannot follow passively because the cells of the ascending limb are impermeable to water. Thus the salt concentration in the extracellular fluid becomes quite high, approaching an osmotic pressure three times that of sea water! This salt is free to reenter the descending limb of the loop of Henle. Therefore, the salt makes a cycle: ascending limb to extracellular fluid to descending limb to ascending limb again (Figure 35-11). Because the flows in the descending and ascending limbs are in opposite directions, this is another case of countercurrent exchange; as a result of its operation, a gradient in extracellular salt concentration is set up, as shown in Figure 35-11. The osmotic pressure of that extracellular brine bath is made still higher because of the abundant addition of urea. That substance moves out of the collecting ducts and concentrations build up in the extracellular fluid so that it is more accurate to refer to it as a "urea-brine bath."

Now let us return to another consequence of salt being removed from the filtrate in the ascending limb: the filtrate becomes *less* concentrated—the opposite of what is appropriate if undue water loss is to be prevented. What is the solution to this dilemma? After the filtrate leaves the distal convoluted tubule, it passes through the **collecting duct.** This duct plunges back through the central region of the kidney where the extracellular urea-brine bath exists; hence, the collecting ducts pass close by the loop of Henle. Because the walls of the collecting duct can be made permeable to water by hormone action, water leaves the filtrate, enters the extracellular fluid urea-brine bath, and diffuses back into the blood. This further concentrates the filtrate remaining in the collecting tubule and reduces its volume to form a highly concentrated, low-volume excretory product—the *urine.* The urine moves down the ureter into the bladder, where it is held until the fullness of this organ stimulates *micturition,* or urination.

The crucial role of the loop of Henle, with its countercurrent exchange mechanism, is made obvious by the fact that only vertebrates with these loops can make urine more concentrated than the blood. The result is that in humans, other mammals, and birds, up to 99

percent of the original glomerular filtrate can be reclaimed. Furthermore, the longer the loop of Henle, the more concentrated the urine that can be made and the more water that is saved. Desert mammals have very long loops of Henle and manufacture the most concentrated mammalian urine. One of the reasons that amphibians are forever tied to aquatic habitats is that their kidneys lack loops of Henle. Only reptiles with their production of uric acid, and mammals and birds with their loops of Henle, were able to colonize the vast expanses of dry land.

Regulation of Kidney Function We have seen how the glomerular filtrate changes from a dilute, high-volume solution in the Bowman's capsule to a dilute, low-volume solution in the distal convoluted tubule, and then to a concentrated and still lower-volume solution as it passes through the collecting tubule. But what if a person has been drinking a large quantity of water, the body fluids become too dilute, and the excretion of dilute, high-volume urine is needed to rid the body of excess water? Or what if a person eats a very salty dinner? What, in other words, regulates kidney function? Several hormones are responsible; some control the concentration of salt in the urine, and others control the amount of water excreted.

Aldosterone, a steroid hormone released by cells in the cortex of the adrenal gland (Chapter 37), governs the rate of salt reabsorption in the distal convoluted tubule. At the same time, it causes salt to be reabsorbed in the salivary glands, sweat glands, and colon. The secretion of aldosterone into the blood is regulated by a series of steps (Figure 35-12). Some conditions, such as reduced salt intake, can cause the osmotic pressure in the blood vessels (the so-called afferent arterioles) leading to the glomeruli to be unusually low or can cause the amount of salt in the glomerular filtrate to be low. When such a situation occurs, cells near the glomeruli release an enzyme called *renin* into the blood plasma (nerves can also cause this release). Renin acts on a blood plasma protein (α-2-globulin) to form a peptide called *angiotensin I,* which another enzyme in turn converts to *angiotensin II.* This circulating peptide hormone triggers constriction of the blood vessels leading into the glomeruli so that less blood enters them and the volume of glomerular filtrate is reduced. Angiotensin II also acts on the adrenal gland to trigger the release of aldosterone, which travels in the blood to the kidneys and acts on the distal convoluted tubules so that more salt is reabsorbed before the filtrate passes into the collecting ducts. Also, more water is reabsorbed osmotically, thereby correcting the low blood pressure that may have triggered renin release in the first place.

A second hormone that regulates kidney function is **antidiuretic hormone (ADH;** also called *vasopressin)*

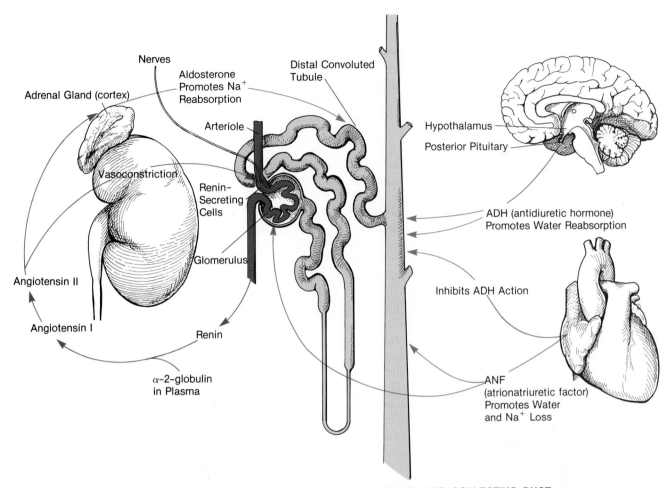

Figure 35-12   HORMONAL CONTROL OF REABSORPTION IN THE NEPHRON AND COLLECTING DUCT.
Decreased blood pressure or low salt in the blood result in the secretion of renin by cells located next to the walls of the arterioles approaching glomeruli. (Nerves leading to such cells may also control this secretion.) Renin initiates a series of reactions in the blood that produces angiotensin II; that compound may cause vasoconstriction of the arteriole to the glomerulus, thereby decreasing glomerular filtration; angiotensin II also causes aldosterone release, which promotes Na^+ reabsorption (and indirectly, water reabsorption). Hypertonic blood traveling to the brain's hypothalamus triggers ADH secretion; as a result water is reabsorbed at the collecting duct, and the blood tonicity falls. Heart atrial cells secrete atrionatriuretic factors (ANFs) in response to high blood pressure; ANFs act directly to increase glomerular filtration, and they inhibit ADH action. As a result, more water and Na^+ are lost. Thus, these various hormones counteract each other and permit precise control of blood volume, tonicity, and pH.

(Figure 35-12). Neurosecretory cells in the hypothalamus fire nerve impulses when the osmotic pressure of the blood plasma is too high. The impulses travel along axons of neurosecretory cells to the posterior pituitary gland, where they cause secretion of ADH into the blood. ADH binds to cells in the walls of the kidney's collecting ducts and increases their permeability to water. The higher the ADH level, the more water passes osmotically out of the filtrate and back into the extracellular fluid, and the less urine is produced. Most of the reclaimed water eventually reenters the blood plasma, diluting it and thereby lowering its osmotic pressure.

This decrease is then detected by cells in the hypothalamus, and soon less ADH is released. Just the opposite occurs if the osmotic pressure of the blood is too low (such as after imbibing too much beer): ADH level falls, less water is reclaimed in the collecting ducts, and copious urine is produced. Here is a classic negative-feedback loop: whichever way the plasma osmotic pressure tends, ADH counteracts, and the condition is corrected to maintain a homeostatic normal condition.

Note the surprising fact that it is *not* the level of nitrogenous wastes in the blood that is monitored or that controls urine formation. However, nitrogenous wastes do contribute to the osmotic pressure of blood plasma and thus indirectly affect the production of ADH.

Changes in blood pressure can also increase or decrease ADH secretion. When you drink water, your blood volume and, consequently, your blood pressure rise. As a result, less ADH is secreted, less water is reabsorbed, more urine passes, and the blood pressure is lower. If you lose water as the result of sweating or bleeding, blood pressure drops, ADH secretion increases, urine production decreases, and the blood pressure once again rises. Blood pressure is monitored by cells in the walls of the heart's atrial chambers. When the atrial walls are stretched excessively due to increased blood volume (pressure), the muscle cells process a precursor hormone and release **atrionatriuretic factor (ANF;** also called atriopeptin) (Figure 35-12). These small peptide hormones cause both water and Na^+ to be excreted in urine. As a result, the blood pressure falls. In part this stems from increased glomerular filtration rates, which result from direct action of ANF on the blood vessels supplying and draining the glomeruli. But also, ANF acts directly on collecting duct cells to inhibit the water retention effects of ADH. Thus ANF and ADH may be viewed as antagonists: the first causes water loss; the second, water retention. Researchers see an intriguing result concerning ANF in humans suffering congestive heart failure. Such individuals have high blood pressure, and abnormally high Na^+ and water retention. One finds in such people just the opposite of what is expected: ANF release is very high, reflecting the heart's "attempt" to protect itself; but, for unknown reasons, the kidneys do not respond properly to the ANF, and Na^+ and water are retained.

ANF also binds to the receptors in the brain and regulates the amount of fluid in the brain and spinal cord cavities (*cerebrospinal fluid*, or *CSF*). Less CSF is secreted when ANF binds, an important adaptation, as the brain lacks lymphatic drainage and is especially sensitive to edema (water accumulation).

Certain foods and drugs suppress the secretion of ADH and thus act as *diuretics*. Coffee, tea, cranberry juice, and watermelon, for example, can have this effect, as can alcohol and certain over-the-counter and prescription drugs. In each case, the result is increased urination. This is usually not harmful unless the diuretic is consumed in large quantities or over an extended time so that the osmotic balance of the body fluids is disturbed. But it is also a reason why drinking beer to quench thirst on a hot day is counterproductive; the alcohol's diuretic action simply increases urinary water loss.

Not surprisingly, the control of kidney function is related to thirst. Drinking behavior can be triggered by the loss of water from the extracellular fluids, by dryness in the mouth and pharynx, by a rise in the blood's osmotic pressure, or by a fall in blood pressure and volume. All such conditions stimulate the *drinking center* in the hypothalamus. Certain physical cues can stimulate drinking and regulate kidney function at the same time; for example, bleeding can lead to intense thirst as well as to the release of ADH and improved water reabsorption in the collecting ducts.

As a homeostatic organ, the kidney also helps to regulate the pH of body fluids in reptiles, birds, and mammals. Recall from Chapter 34 that the rate of breathing is a major factor in controlling the pH of blood and body tissues. The kidney, too, serves to keep the blood plasma at a fairly constant pH of 7.35 to 7.4. To do so, various parts of the proximal and distal convoluted tubules secrete H^+, reabsorb HCO_3^-, and exchange NH_4^+ for K^+ and Na^+—in short, the kidney automatically adjusts the type and number of ions leaving the body so that the basic chemical composition of the body fluids stays within a set range of values. These are the same sorts of exchanges that go on in the gills of a bony fish or through an amphibian's skin. The kidney has had to assume this regulatory function for land animals living in a sea of air.

Proper kidney function is so important for ridding the body of nitrogenous wastes and maintaining pH and salt balance that diseases of these organs are often life-threatening. Inborn defects, infections, tumors, kidney stones, and exposure to toxic chemicals can all diminish kidney function and result in *uremia*—the build-up in the blood of salt, water, Ca^{2+}, K^+, H^+, urea, and other substances normally excreted in urine. Victims of kidney disease or failure often must be placed on an artificial kidney or dialysis machine, which cleanses the blood by pumping it through many meters of cellophane tubing bathed in a saline solution that is osmotically similar to blood plasma. Wastes diffuse across the tubing and are removed before the blood flows back into the patient's vein. Alternatives to this standard treatment are kidney transplantation and new, smaller dialysis instruments that allow the patient more mobility while the blood is being cleansed. The kidneys—no larger than a person's fist—perform so many vital functions and are so difficult to replace that they are clearly a cornerstone of survival and homeostasis.

As one might suspect, the kidney's activities are energy expensive. Active-transport processes involving ATPase pump enzymes are responsible for salt reabsorption in the distal convoluted tubules, for salt pumping out of the ascending limb of the loop of Henle, and for active secretion of uric acid and other molecules. Supporting these homeostatic activities hour after hour, day after day, costs the body a substantial portion of its free oxygen, its food energy, and its supply of ATP. The kidney's maintenance of a stable internal sea at the cost of a continuously high expenditure of energy is one more price that a multicellular organism pays for its survival.

Methods Used for Kidney Research How can physiologists possibly know about the movement of water, ions, and organic molecules into and out of tubules finer than silk thread? The primary way is by *micropuncture*, a process in which a pipette with an exceedingly fine tip is inserted into the Bowman's capsule or one of the tubular portions of the nephron. Samples can then be taken or substances fed into and out of the nephron, and their movements can be traced. Another procedure is to remove a portion of a nephron, thread it on two micropipettes, and pass solutions through it. If the ions in such solutions or the organic solvents are radioactive or bear other types of tracers, precise measurements can be made of the movement of substances across the living cells in the tubular wall. In physiology, as in the rest of biology, researchers can draw conclusions and frame testable theories only when methods such as micropuncture are invented.

HOMEOSTASIS AND TEMPERATURE REGULATION

We have seen that homeostatic mechanisms enable organisms to live in a range of osmotically varied habitats—salt water, fresh water, and dry land. But the Earth's environments vary dramatically in another fundamental way: temperature. In the polar regions, high mountain ranges, and deepest ocean abysses, the temperatures can remain near 0°C year-round. In contrast, daytime temperatures exceeding 40°C are not uncommon in equatorial deserts. And in the Earth's temperate regions, fluctuations in temperature and humidity are frequent. Organisms that have successfully inhabited such places display a set of homeostatic mechanisms that regulate their behavior relative to changes in the external climate or regulate their internal temperatures directly. The nervous, endocrine, circulatory, and other organ systems are involved in temperature regulation. It is no wonder that many biologists regard **thermoregulation** as one of the most complex and highly integrated of the basic physiological processes.

The Impact of Temperature on Living Systems

Every organism's physiological functions are inexorably linked to temperature. Temperature has thus been a strong source of selective pressure on animals. But why is temperature so important? The primary reason is that the rates of chemical reactions vary with temperature: the rate usually doubles for each 10°C rise. Superficially, this might mean that enzymes would catalyze reactions much faster in a mouse, with a normal body temperature of 37°C, than in a snail, at 5°C. In fact, all enzymes have *temperature optima*, temperatures at which they catalyze reactions most efficiently. Enzymes have evolved so that they function optimally in each animal's typical habitat. For example, a digestive enzyme in a trout may function just as rapidly at 10°C as does the enzyme that catalyzes precisely the same reaction in a desert lizard at 34°C. In a sense, enzyme evolution helps overcome the simple relationship between reaction rates and temperature.

Another reason temperature is important is that temperature extremes disrupt basic biological molecules and structures, such as proteins and membranes. If body temperature rises to between 45°C and 50°C, proteins begin to denature; further rises can literally destroy the enzymes that sustain metabolism. When temperatures fall, reactions run more slowly, and weak bonds that hold protein subunits together can rupture as well, causing enzymes to dissociate and cease to function. Furthermore, low temperatures cause the lipids of plasma membranes and organelles such as mitrochondria to change from a fluid to a solid state, particularly in birds and mammals. Such membrane solidification interferes with fundamental cellular processes, such as ion pumping by enzymes, and the functioning of mitochondrial enzymes. These drastic effects of very warm and very cold temperatures set limits on the habitats in which most microbes, plants, and animals can live.

Solutions to Temperature Problems

Animals cope with fluctuations and extremes of temperature in any of three ways:

1. They can occupy a habitat where the temperature remains constant.

2. Their body temperature can fluctuate along with that of their surroundings, but their physiological processes have adapted to that range of temperatures.

3. They can generate heat internally and thus maintain a constant body temperature despite fluctuations in external temperatures.

Animals in the first category live in the deep ocean or deep lakes, where the temperature remains near 4°C all year long. Animals in the second and third categories have a much wider distribution in aquatic and terrestrial habitats.

Animals in these three categories are labeled in several ways, but some of these labels are ambiguous and no longer acceptable to physiologists. "Cold-blooded" and "warm-blooded" are the most common but the least scientific and most confusing terms. These terms stem from the way an animal feels to a person touching it; birds and mammals tend to feel warm, and so were traditionally called "warm-blooded." Most other animals feel cool and were called "cold-blooded." But the body temperatures of many so-called cold-blooded animals at times exceed those of mammals and birds.

Somewhat more precise are the terms "poikilotherm" and "homeotherm." **Poikilotherms** (from the Greek word *poikílo*, meaning "changeable") are animals with a variable body temperature, one that tends to track the temperature of the surrounding environment. Most fish, amphibians, reptiles, insects, and other invertebrates are poikilotherms. **Homeotherms,** in contrast, have relatively constant body temperatures; birds and mammals are examples (Figure 35-13). Even these terms and definitions, however, are inadequate for all cases. A lizard tethered to a stake so it could not crawl away would function as a poikilotherm if we altered environmental temperature by shining a heat lamp on it or by chilling its surroundings. However, in nature, free to move into and out of the sun and to carry out other behaviors, the lizard can maintain a temperature of between 33°C and 35°C all day long; it is a "behavioral homeotherm," meaning that it can behave in ways that keep its body temperature relatively constant. Alternatively, a brittle star or sea cucumber living on the sea floor, where the

water temperature varies little, might have a body temperature of 2°C to 4°C its whole life; that is superb homeothermy—but of a very different sort from that displayed by a wood thrush or mule deer, which generates heat internally and maintains body temperatures between 37°C and 39°C. The same sea cucumber's temperature would vary widely in an environment with fluctuating temperature; there, it would be a true poikilotherm.

Animals can be classified in yet another way, as ectotherms or endotherms. **Ectotherms** derive most of their body heat from the environment rather than from their own metabolism. Reptiles are a good example. **Endotherms**—birds and mammals—produce their own body heat. Of course, chemical reactions in the cells of ectotherms do release a small amount of heat. However, their bodies are not well insulated, and the heat dissipates into the environment. Most endotherms have bodies insulated by fur, feathers, or fat, and so can retain heat, enabling them to maintain high body temperatures (near 40°C) and to keep at least the deep body temperature, the so-called *core temperature*, quite constant. Even then, the endothermic animal's periphery is often much colder and more variable than its core, and the body temperature of some homeothermic endotherms drops radically when they are inactive (a state called torpor) or in the winter (hibernation). Such drops in core temperature are frequently seen in very small birds and mammals.

It is the case that a mammal or bird cannot weigh less than 2 to 3 grams because in a body of lesser mass, heat cannot be generated fast enough to make up for its loss through the animal's relatively large surface area. In fact, all birds and mammals weighing less than 10 grams, such as a hummingbird or a shrew, are *heterothermic*—they can maintain a high body temperature for only brief portions of the day because so much heat is lost.

Obviously, organisms in nature do not always neatly fit the categories defined by biologists. Many common animals can be at once poikilothermic and ectothermic, homeothermic and ectothermic, homeothermic and endothermic, and so on.

Temperature Regulation in Animals

Organisms have a veritable arsenal of responses to the heat and cold of their surrounding environments. These responses can be ecological (where and how the animal lives), metabolic (how its cells and tissues function), behavioral (how the entire organism responds to changes in temperature), structural (anatomical solutions such as insulation), and physiological (specific responses such as shivering or sweating). Let us consider how aquatic organisms, terrestrial organisms other than birds and

Figure 35-13   HOMEOTHERMS AND POIKILOTHERMS: THE RELATIONSHIP BETWEEN BODY TEMPERATURE AND ENVIRONMENTAL TEMPERATURE.
In homeotherms, body temperature remains more or less constant over a range of environmental temperatures. In contrast, the body temperature of a poikilotherm is usually within a degree or two of the environmental temperature.

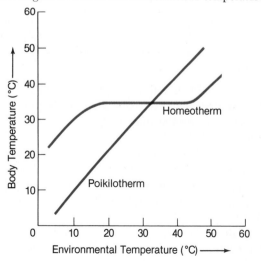

mammals, and birds and mammals carry out thermoregulatory behavior. We separate organisms according to aquatic and terrestrial environments for the simple but compelling reason that temperatures are fairly stable under water but fluctuate dramatically on land, both from day to day and from season to season.

Temperature Regulation in Aquatic Organisms

Water gains and loses heat slowly due to its high specific heat and low conductivity. (Recall that *specific heat* is the amount of heat energy required to raise the temperature of 1 gram of a substance by 1°C. *Conductivity* refers to the degree to which heat energy is conducted through the substance.) In the absence of turbulence, water temperatures at a given depth in oceans, lakes, and rivers tend to fluctuate very little on a daily basis and to remain constant within each depth zone of a deep body of water. Seasonal temperature fluctuations are slow, if they occur at all. Due to the great stability of the aquatic environment, most aquatic invertebrates and vertebrates are ectothermic poikilotherms: they derive heat from the environment, and their body temperatures fluctuate slowly with that of the surrounding water.

Because the environment is stable, the body temperatures of sponges, marine worms, crayfish, squid, shrimp, and most fish are quite unchanging as long as the organism remains at a given depth. In many aquatic organisms, the enzymes that carry out metabolic reactions function optimally and very rapidly at the water temperature where the organism spends most of its time—a tidy outcome of evolutionary processes.

So beautifully are aquatic organisms adapted to the temperatures of their environments that young sockeye salmon, for example, if given a choice of holding tanks filled with water of various temperatures, will cluster in the tank held at 15°C—the optimal temperature for their growth, heart rate, and swimming speed. Some organisms that live in extremely cold water, such as the deep ocean off Antarctica, have "antifreeze" in their blood. Polyalcohols such as sorbitol, glycerol, and trehalose lower the freezing point of blood plasma and other body fluids, and enable fish and other animals to swim about all winter in subfreezing waters in a supercooled state (that is, at a temperature below the normal freezing temperature of a solution).

In an aquatic environment where organisms are fairly limited to specific depths and activity levels, a predator that can move more quickly and range widely through various zones can successfully harvest prey. Diving mammals and birds, such as seals and cormorants, take advantage of this principle. And curiously, certain fish, such as mako sharks and tuna, have evolved mechanisms that permit at least parts of the body to be maintained at a high temperature and activity level. The mako shark and the tuna are both endothermic homeotherms: they derive heat from their own metabolic activities, not from the external environment, and they maintain a fairly constant temperature around the spinal cord and brain. These tissues may be 5°C to 15°C higher than the surrounding water temperature, and sometimes may approach the typical mammalian core temperature of 37°C. Makos and tunas are not insulated, yet they can conserve the heat generated by their major red muscles during swimming, whereas the more common poikilothermic sharks and fish constantly lose heat through their skin and gills. Makos and tunas "trap" heat with a kind of countercurrent heat exchange system built from blood vessels running between the body core and the skin. Specifically, like many animals that live in cold or hot climates, makos and tunas have "miraculous nets" of blood vessels, or **rete mirabile,** in which blood flow in countercurrent vessels can be altered to regulate heat loss or gain (Figure 35-14).

The important difference between tunas and other fish is not the constancy of body temperature, since fish living at a given water depth can also have a constant temperature. Rather, it is the endothermic source of the body heat and the high core temperature that make the "warm-bodied" fish so remarkable. The selective advantage of a high body temperature has been profound for these fish: the power of their muscle contractions can be three times greater at the higher temperatures than those of similar muscles in relatives with cooler bodies. Thus they can swim faster while ranging more widely through various ocean depths in search of prey than can other predatory fish more tied to given water depths or temperatures.

Temperature Regulation in Amphibians, Reptiles, and Insects

Animals with air rather than water as a surrounding medium are subject to marked temperature changes daily as well as seasonally. Land animals have developed various means of coping with these fluctuations. Most terrestrial animals are ectothermic poikilotherms: they derive heat from the environment, and their body temperatures vary with external temperatures. Here we will consider the thermoregulatory adaptations of three such groups: the amphibians, reptiles, and insects.

Amphibians, such as frogs, take in heat from the sunshine and water and absorb warmth radiating from the soil, rocks, and so on. In cold climates, frogs and salamanders have an extremely low rate of metabolism in winter and are quite inactive. Even when more active in the warmer seasons, however, they have a persistent thermoregulatory problem: amphibians must exchange carbon dioxide and oxygen across a moist skin surface,

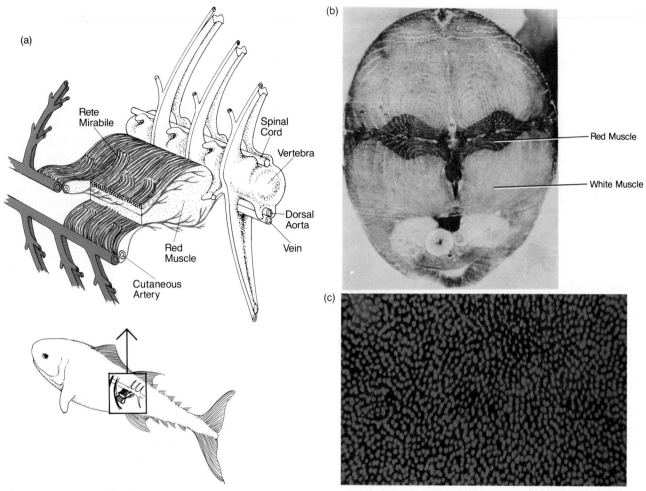

Figure 35-14   WARM-BODIED FISH.
(a) In a few fish, including this tuna, bands of red muscle possess a capillary network known as the rete mirabile. The heat generated by these muscles is not lost because it is transferred in the rete from hot venous blood passing outward to cold arterial blood passing inward from the body surface. This enables the animal's core temperature to remain higher than the environmental temperature. (b) A cross section through the tuna body (reduced about 6 times), in which the red and white muscle regions are obvious. The rete runs in the red muscle on each side of the body. (c) A cross section showing the venous and arterial capillaries of the rete (magnified about 100 times); each dot represents a capillary.

and this moisture layer acts as a natural cooling system. Fully 580 calories of body heat are consumed and lost with each gram of water that evaporates from the skin. This continuous evaporative cooling means that the body temperature of a frog or salamander in air is often lower than the environmental temperature, even on a warm day. Obviously, the problem of water evaporation limits the habitats and activities of amphibians. Most cannot survive in dry areas, in polar regions, or on high mountains. In the warm tropics, however, they are the most abundant vertebrates.

Reptiles, in which an expandable rib cage evolved and allowed efficient ventilation, can have dry rather than wet skin, and so do not lose body heat through respiration-related evaporation from the skin. Terrestrial rep-

tiles seek out sources of heat—direct sunlight, warm rocks, soil—to raise their body temperature. Early in the morning, for example, a lizard will crawl out of its hiding place and turn its body to expose maximum surface area to the sun's rays. As it sits in the morning sun, its skin darkens in response to hormones and thus it better absorbs infrared radiation in sunlight. The lizard may also do "push-ups," pumping its legs up and down to generate heat. Although lizards and other reptiles are classified as poikilotherms and ectotherms, these heat-seeking activities allow them to function as behavioral homeotherms—they can maintain a fairly constant body temperature by means of their own behavior.

When the lizard reaches its preferred body temperature, it will adjust its behavior to maintain that tempera-

Figure 35-15   TEMPERATURE REGULATION IN REPTILES.
By combining a rapid heartbeat with vasodilation, the
Galápagos marine iguana (*Amblyrhynchus oristosis*) gains
heat rapidly as it sunbathes. The skin darkens so that much
radiation can be absorbed. When the iguana dives into the
cold sea water to feed, it reduces its heart rate and
constricts blood vessels in the skin; the combined result is
less heat loss.

ture; it crawls out of the sun and grows pale if it over-
heats, only to seek out a warm spot again when its
temperature drops. The marine iguana shown in Figure
35-15 may dive into the sea to feed or to cool itself.
These behavioral strategies allow reptiles to achieve
homeothermy during much of the day; however, because reptiles are true ectotherms, they do cool off and
become considerably less active after sunset.

Invertebrates, with their vast diversity of life styles
and habitats, are mostly ectotherms and poikilotherms.

In general, they lack the insulation and the behavioral
repertoires of vertebrates. Nevertheless, they do have
metabolic adaptations—enzymes, biochemical functions, and membrane components that remain functional
within the range of temperatures they normally encounter in burrows, in tropical trees, or wherever they live.

A few kinds of invertebrates raise their body temperatures by metabolic means, thereby displaying a form of
endothermy. Certain insects fit this pattern, and their
ability to generate heat is quite remarkable, considering
their small size. How is it that certain species of moths,
flies, bees, dragonflies, and beetles weighing from 100
milligrams to 2 grams can maintain temperatures of 35°C
to 40°C?

Several mechanisms make possible high body temperatures in small insects. First, all such insects have high
temperatures only when active; they are heterothermic.
Second, they may have a kind of insulation on the surface of their thorax. Bumblebees and the sphinx moth
have hairlike bristles on that body segment. Rapidly
contracting wing muscles housed within the thorax generate a large amount of heat during flight, and the insulating bristles help retain this heat within the segment. If
the insect's internal temperature climbs too high, its
heart beats faster, and extra hemolymph flows to the
hairless abdomen so that heat radiates away (Figure 35-
16). Then the heart slows, and abdominal circulation
diminishes as the insect's internal temperature drops.
This shunting of hemolymph to help regulate temperature is possible in part because oxygen is supplied to the
active flight muscles and body tissues via the tracheal
tubes. A third mechanism involves warming a cool insect's body. The sphinx moth undergoes a preflight
"warm-up" and then maintains a remarkably high core
temperature of 41°C, even as air temperatures fluctuate

**Figure 35-16   THE SPHINX MOTH
AND BODY HEAT.**
Temperature changes are recorded at
four points as heat is applied to the
thorax of a tethered moth. As the
moth tries to fly, its thorax
temperature (a) rises rapidly. After
about five minutes of activity, the
temperature is close to 40°C, and the
moth's hemolymph vessels, leading to
the uninsulated abdomen, open up
(b). The temperature then rises in
the moth's abdominal vessels. (c) The
cooling process ensures that the
hemolymph that then flows forward
to the thorax reaches only about
32°C.

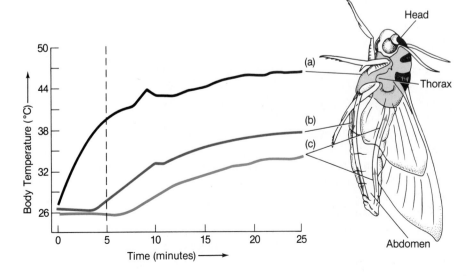

between 3°C and 35°C. In fact, the high temperature in these insects is absolutely essential for flight—moths, bumblebees, and beetles can generate enough lift to fly only when their flight muscles reach 35°C. In order to approach that temperature while sitting on a flower or leaf, a bumblebee shivers, producing frequent minute muscle contractions that generate heat, which is retained in the thorax. Bumblebees are one of the few insects that have an additional capacity to generate heat metabolically. They have a unique metabolic adaptation—the sustained hydrolysis of ATP in thoracic cells. This process generates body heat and allows bumblebees to fly and forage on cool, damp days, when other types of bees are huddled together for warmth in their hive. Despite these various mechanisms for achieving high temperatures, and thus high activity levels, small size has a limiting effect, ensuring that no insect can remain warm around the clock.

Temperature Regulation in Birds and Mammals

The most active and behaviorally complex animals—the birds and mammals—have their large repertoires of activities and can live in habitats all over the world because they are endothermic homeotherms: they maintain constant body temperatures of 35°C to 42°C with internally generated heat.

A vertebrate homeotherm's first prerequisite for maintaining constant temperature is *insulation*. Feathers and fur are superb insulators. They trap a layer of

Figure 35-17.   FEATHER INSULATION.
The thick layer of down feathers keeps these Emperor penguin chicks snug inside their own sleeping bags. Feathers, and hair on huskies and some other mammals, can provide such effective insulation that the animal's metabolism does not have to be turned up unless temperatures go far, far below freezing.

stagnant air next to the body and reduce heat loss from the underlying skin (Figure 35-17). Animals with sparse hair, such as whales and humans, have layers of fat of varying thickness to insulate the body from loss of heat to the water or air. Insulation can also keep heat out: large mammals living in the hottest, driest deserts invariably have very thick, light colored, almost silvery, fur on their backs. Surface temperatures of 70°C and 85°C have been measured on the back fur of camels and Merino sheep, for instance, yet the underlying skin was only 40°C.

The degree of insulation provided by fur or feathers can be varied to suit environmental conditions. When a mammal or bird becomes chilled, its fur or feathers are held erect by tiny muscles so that the layer of air trapped next to the skin becomes thicker. In mammals, this process is called **piloerection.** Piloerection works effectively for a rabbit in the woods, but when it occurs in humans, as an evolutionary holdover, our body hair is so sparse that all we get is goose bumps. In the warmer months, some mammals have thinner pelts or mat down their fur by licking it so body heat can be used up by evaporation.

The problem of overheating—which can be fatal as easily as overchilling—is combatted in many birds and mammals by evaporation of body fluids. Mammals and birds will often pant or take short, shallow breaths to speed evaporation in the lungs. Some mammals also have *sweat glands*, which release a watery salt solution that carries away heat by evaporation. Sweating involves loss of both water and salt and can lead to severe problems if prolonged. A person doing hard physical labor on a hot day can lose up to 4 liters of sweat per hour. In a day's time, the person could lose 10 to 30 grams of salt, or 6 to 18 percent of all the NaCl in the body. The kidneys will compensate by producing a low-volume urine with low salt concentration, but unless fluids and salts are replaced, heat cramps or even heat exhaustion will occur.

Yet another important mechanism for regulating body heat involves vasoconstriction and vasodilation—the opening and closing of capillary beds (Figure 35-18). Vasoconstriction of the skin capillaries prevents most hot blood from passing through them; the temperatures of the extremities and skin surface drop, while those of the central nervous system, heart, and other vital organs remain warm. Vasodilation increases blood flow to the skin, so that heat can be lost more easily, either by *radiation* (transfer of heat waves to the environment) or by *convection* (movement of warmed air away from the skin). As mentioned, many animals in environments with extreme temperatures have rete-mirabile blood vessel systems (Figure 35-14) to further control heat loss or gain. The uninsulated feet and flippers of penguins, walruses, and many other Arctic animals have these countercurrent exchange nets. In these animals, arteries car-

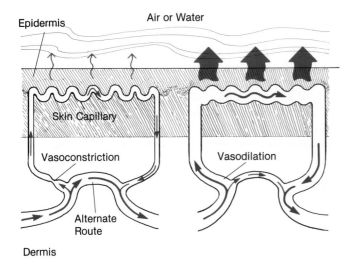

Epidermis
Air or Water

Skin Capillary

Vasoconstriction

Vasodilation

Alternate
Route

Dermis

**Figure 35-18   CONTROLLING HEAT LOSS BY THE
EXTREMITIES.**

Vasoconstriction and vasodilation effectively reduce and
promote, respectively, the loss of heat to the environment
by controlling blood flow to the extremities. Vasoconstriction
reduces the flow of blood to an extremity and hot blood
passes back to the body core. Thus the loss of heat is low.
Vasodilation promotes heat loss by increasing blood flow to
an extremity and allowing heat from the body core to be
carried off by convection.

rying hot blood from the body core toward the
extremities lie side by side with veins carrying cold
blood from the feet or flippers back toward the core (Figure 35-19). Heat is transferred from the arterial blood to
the venous blood, and so it is not carried to the surface,
from where it would be lost. Thus the body core can
remain warm. On overheating, however, the hot arterial
blood is shunted past the countercurrent rete vessels
directly to the feet or flippers, from where heat can be
lost to snow, ice, or cold water.

Another function of the rete mirabile is cooling the
brain when the body is subject to overheating. The
eland, a cowlike animal of dry, hot African regions, has a
rete mirabile in the walls of its nasal passages. Evaporation during inhalation and exhalation cools the venous
blood in the walls of these chambers; that blood flows
counter to the hot arterial blood approaching the brain
and carries away some of the heat. As a result, an eland
can keep a cool head—a cool brain, that is—under a
blistering sun. In fact, recent observations of other
mammals suggest that most of us may have heat exchangers that keep the brain slightly cooler than the core
temperature (Figure 35-20).

Mammals and birds also have behavioral mechanisms
that deal with temperature changes in the external environment. Like reptiles and amphibians, mammals and

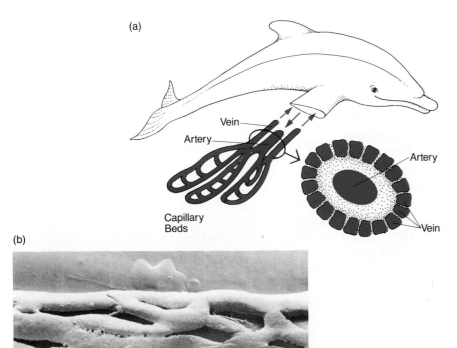

(a)

Vein

Artery

Artery

Capillary
Beds

Vein

(b)

**Figure 35-19   COUNTERCURRENT
MECHANISMS: BIOLOGICAL HEAT
EXCHANGERS.**

Arteries and veins can be arranged
in a way that conserves heat in
appendages in various animals that live
in environments that are less extreme
than the Arctic. (a) In the dolphin
flipper, for instance, the main artery is
surrounded by veins. Heat from the
artery is transferred to the veins,
which carry blood back to the body
core; this blood is warmed. (b) This
scanning electron micrograph
(magnified 12 times) shows a similar
arrangement, with many small veins
covering the central artery in the leg
of a European rook (*Corvus
frigilegues*). Controlling the blood flow
through these vessels allows heat to be
conserved or lost as needed from the
bird's foot, just as occurs in the
penguin standing on ice.

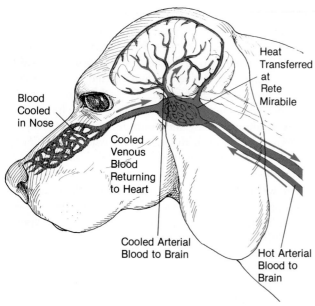

Figure 35-20   HEAT EXCHANGE IN THE NASAL PASSAGES.

The highly vascularized nasal mucous membrane serves another important physiological role in many mammals besides water conservation: brain cooling. As air passes over this surface during inhalation, heat is transferred from the nasal tissues and the blood to the air. The cooled venous blood then flows into a rete-mirabile heat exchanger; arterial blood en route to the brain from the heart gives up heat to the cooler venous blood returning from the nasal tissues.

birds sun themselves or seek shade, as temperature fluctuations dictate. Small homeotherms sometimes share burrows for warmth, and larger range animals huddle together in the wind. Migration to warm climates and hibernation enable many kinds of birds and mammals to survive through the winter. And people alter their clothing, dwellings, and activities to compensate for changes in the weather and climate.

Heat Production in Birds and Mammals   Having considered some general mechanisms by which endotherms respond to changes in and extremes of environmental temperature, we are still left with the question: How do endotherms generate their own internal heat in the first place? There are four main sources of such heat generation, or **thermiogenesis:** (1) muscle contraction, (2) ATPase pump enzymes, (3) brown fat, and (4) metabolic processes. Each time a muscle contracts, millions of ATP molecules are hydrolyzed, releasing heat as a by-product. Voluntary muscular work—such as running, jumping, flying, or stamping the feet—thus generates heat, as does the involuntary muscular process of *shivering.* Enzymatic activity, brown fat, and metabolic processes are sometimes collectively called *nonshivering*

thermiogenesis, since they do not involve muscle contraction, voluntary or involuntary.

Birds and mammals have a unique capacity to generate heat by using enzymes of ancient evolutionary and physiological vintage—the basic ATPase pump enzymes found in the plasma membranes of all cells. Chilling leads to the release of the hormones thyroxine from the thyroid gland and norepinephrine from certain nerve endings. In response to these hormones, certain cells become permeable to Na^+ ions, which then leak in. These ions are then pumped back out by ATPase pump enzymes in the plasma membranes of liver, fat, and muscle cells. In the process, ATP is hydrolyzed, releasing heat. The ions reenter, are pumped out again—and so more heat is released.

The third source of heat production is **brown fat,** a special type of fat present in the bodies of newborn mammals and of adult mammals adapted to cold climates or that hibernate. Deposits of brown-fat cells are found beneath the ribs and shoulder blades and around some major blood vessels. Blood flowing near active brown fat is heated and then carries heat to the brain and the heart muscles, in particular. The brown color comes from immense numbers of mitochondria, with their iron-containing cytochromes (Figure 35-21). Heat is generated when the brown-fat cells oxidize fatty acids. Rather than generating ATP during this oxidation, the process releases as heat the energy that normally would be stored in high-energy phosphate bonds.

The fourth source of thermiogenesis is metabolism. Mammals and birds have a *high basal metabolic rate* that is controlled largely by thyroxine. The intense activities of cell surface pump enzymes, nucleic-acid– and protein-synthesizing machinery, and enzymes of metabolic

Figure 35-21   BROWN FAT: A SOURCE OF METABOLIC HEAT.

This electron micrograph (magnified about 9,800 times) of part of a brown adipose cell shows many large mitochondria surrounding the fat droplets.

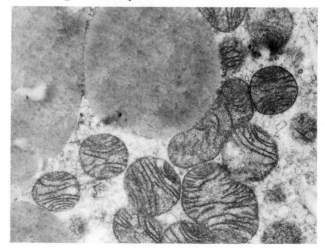

pathways and mitochondria generate heat as an inadvertent but useful by-product.

The Hypothalamus: The Body's Thermostat   We have seen how endotherms generate heat, and we have considered some of their mechanisms for adjusting to changes in outside temperature—shivering, sweating, seeking shade, and so on. But how are changes in the core and surface temperatures of an animal's body measured so that the organism can respond to cold with shivering and to heat with panting, and not vice versa? The answer for mammals and birds is that the portion of the brain called the hypothalamus functions as a thermostat with a specific setting, or **set point,** for each species; the familiar body temperature of 98.6°F, or 37.5°C, is the human set point.

Experiments confirm the role of the hypothalamus in thermoregulation. If this part of the brain is surgically destroyed in a ground squirrel or guinea pig, the animal functions like a poikilotherm—its internal temperature rises and falls along with that of the external environment. Furthermore, tests were done to specifically heat and cool the hypothalamus of a kangaroo rat by implanting in that brain region a *thermode,* an extremely fine probe that can be heated or cooled to precise temperatures. Throughout the experiment, the rat's body temperature remained the same. However, when the implanted thermode lowered the temperature of the hypothalamus from 38°C to 36.5°C, the animal began to shiver, its oxygen consumption rose rapidly (indicating more metabolic heat production), and researchers could detect vasoconstriction and piloerection taking place. When the thermode was warmed, the temperature of the hypothalamus increased, the rat's heat-dissipating arsenal was activated, and it responded by panting, sweating, and showing vasodilation. Physiologists have concluded that this site in the brain is itself capable of

responding to altered temperature and appears to be the **thermoregulatory center.**

One might expect from these results that in nature, an animal's brain (and, of course, its head and body) must warm up or cool down before a response such as sweating or shivering is triggered. But recall that breathing rate increases at the start of exercise in an anticipatory manner. Similarly, temperature responses usually take place very fast; you begin almost immediately to shiver if you walk into a meat locker or to sweat if you sit in a sauna. The explanation is that nerve impulses, especially from cold receptors in the skin and from cold and heat receptors in the walls of major blood vessels and in internal organs, trigger local thermoregulatory processes as well as send data to the brain. Thus if you put your foot in ice water, rapid local temperature defense reflexes coordinated in the spinal column will be initiated long before the blood, the body core, or the hypothalamus change temperature at all. The peripheral control and the thermoregulatory center work together to keep core temperature constant.

A homeotherm's set point can itself be altered at times. At the onset of a fever, for example, the set point rises. This is usually due to circulating compounds called *pyrogens;* some are products of bacteria themselves, while others come from white blood cells that release pyrogens in response to infection by a virus or bacterium or to other systemic conditions. Pyrogens act on the hypothalamus to raise the setting of the natural thermostat. Recent studies show that many pathogens die at elevated body temperatures; fever, therefore, can be regarded as an important defense mechanism that has evolved in mammals for fighting diseases.

During the winter, various homeotherms—including chipmunks, ground squirrels, marmots, and skunks—go into **hibernation,** during which their set points fall to about 5°C (Figure 35-22). They remain in their dens or

Figure 35-22   HIBERNATION AND AROUSAL IN GROUND SQUIRRELS.
The body temperature of a ground squirrel (red) falls close to the environmental temperature after metabolism (blue) is turned down during hibernation. After two weeks, the ground squirrel arouses, and this increased activity allows its blood to be cleansed by the kidneys. After a few hours, metabolism and temperature drop once again.

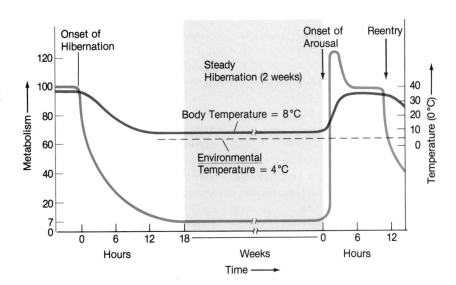

burrows throughout the coldest months in this state of very low metabolism, with slowed heart and breathing rates. Mammals prepare for hibernation by building up fat reserves and growing long winter pelts. Both increased feeding and fur growth are triggered by decreasing day length and by the animals' own biological clocks (Chapters 37 and 48).

During the months of hibernation, the kidneys continue to cleanse the blood, and the animal awakens periodically to excrete wastes. If a hibernator is disturbed or if the air temperature grows so cold that the animal is in danger of freezing, emergency neuronal circuits cause arousal, and the animal warms to normal levels near 37°C.

A variation on hibernation is **torpor,** a temporary lowering of the set point. Bats and hummingbirds, for example, must enter torpor every day when they are inactive. This is because they have small body masses and large surface areas and cannot stay warm without feeding constantly and maintaining high metabolic rates to keep one step ahead of heat loss from the body surface. While resting, they simply cannot stay warm, so they enter torpor, and their normal set points drop to nearly match the temperature of the surrounding air. To arouse from torpor once again before flying off to feed, the animals flex their wings, shiver, and turn up metabolic heat production to raise their body temperature to about 40°C. Clearly, torpor enables small creatures with large surface areas to afford the luxury of homeothermy.

LOOKING AHEAD

We have discussed in this chapter how an animal's internal environment is carefully maintained by homeostatic systems. In particular, we saw how metabolic fires can burn for a lifetime but can be adjusted moment by moment to meet external demands, and how the composition of the salty internal sea is remarkably stable, despite variations in food and water consumption, waste elimination, and moisture or salt loss. In several contexts, we mentioned that nerve cells relay impulses vital to the regulation and integration of homeostatic mechanisms. In the next chapter, we shall focus in detail on the body's electrical network, the nervous system.

SUMMARY

1. Animals have many mechanisms for *homeostasis*—the maintenance of a relatively constant internal environment. Three such mechanisms are the *osmoregulatory system,* which governs water and salt levels in the body; the *excretory system,* which eliminates metabolic wastes; and the *thermoregulatory system,* which maintains an animal's body temperature.

2. Fresh-water animals tend to lose salts and to take in water. To avoid hydration, fresh-water fish rarely drink water, have relatively impermeable body surfaces, excrete much dilute urine, and take up salts through the gills.

3. Marine animals tend to take in salts from ocean water and to lose water. To avoid dehydration, marine fish frequently drink water, excrete a small volume of concentrated urine, and excrete Na$^+$ and Cl$^-$ from the gills.

4. Amphibians can absorb water across the skin and bladder wall. Desert and marine reptiles and birds have salt glands to remove and secrete excess NaCl so they can consume salty foods or water, secrete the salt, and retain the water.

5. Nitrogen, primarily from the breakdown of amino acids and nucleic acids, is excreted as ammonia, *urea,* or *uric acid.* The form excreted depends on the volume of water available for transporting the waste product safely, without causing cell damage.

6. In reptiles, birds, and mammals, the *kidneys* perform *osmoregulation* and secretion of nitrogenous wastes, and they regulate the pH of body fluids (the lungs and breathing also play a role in pH regulation). In such vertebrates, the unit of kidney structure and function is the *nephron,* with its glomerulus, Bowman's capsule, proximal and distal convo-

luted tubules, and the *collecting duct.* Mammalian and avian nephrons also have a *loop of Henle.* The loop of Henle and collecting duct are in the kidney's *medulla,* whereas other nephron parts lie in the kidney's *cortex.* Urine passes into the *pelvis* and from there to the bladder.

7. To make urine, kidneys produce a primary filtrate (the *glomerular filtrate*) and reabsorb most of the water and needed substances, while allowing wastes and unneeded substances to pass from the body. Four physiological mechanisms are involved: filtration of blood through the glomerulus and passage out of Bowman's capsule; reabsorption of useful substances, primarily in the proximal convoluted tubule; secretion of substances, especially H$^+$ and excess K$^+$ into the filtrate, in the proximal and distal convoluted tubules; and concentration of the filtrate in the loop of Henle and the collecting duct; the former generates a high

salt and urea condition in the extracellular fluid around the loop of Henle and through which the collecting duct passes. As a result, water can be reabsorbed passively into the extracellular fluid.

8. *Antidiuretic hormone (ADH)* acts on the collecting ducts to regulate water reabsorption, whereas angiotensin II affects glomerular filtration rate and *aldosterone* production. Aldosterone controls the reabsorption of salt in the nephrons. *Atrionatriuretic factors (ANF)* cause water and salt loss and thereby lower blood pressure.

9. *Thermoregulation* is a complex and important physiological process; it maintains to varying degrees an organism's body temperature despite variations in environmental temperature.

10. *Poikilotherms* generally derive heat from the environment and have body temperatures that track those of the surrounding air, soil, or water. *Homeotherms* have relatively constant core body temperatures (at least part of the time).

11. *Ectotherms* derive most of their body heat from the environment. *Endotherms* generate their own body heat and have high body temperatures.

12. The high, constant body temperature in birds and mammals depends on insulation. It is dependent also on panting, sweating, and vaso-constriction and vasodilation of the peripheral blood vessels, on various specific behaviors, and sometimes on a *rete mirabile*.

13. *Thermiogenesis* involves mainly shivering, enzymatic activity, *brown fat*, and basal metabolic rates of cells.

14. The hypothalamus regulates drinking, feeding, and certain aspects of kidney function. It includes cells of the *thermoregulatory center*, which functions as a thermostat with a *set point*, responds to altered local and peripheral body temperatures, and initiates thermoregulatory physiological reactions. The set point can rise during a fever or fall during *hibernation* or *torpor*.

KEY TERMS

aldosterone
antidiuretic hormone (ADH)
atrionatriuretic factors (ANF)
brown fat
collecting duct
compatible osmolyte strategy
cortex
counteracting osmolyte strategy
ectotherm
endotherm
excretory system
glomerular filtrate

hibernation
homeostasis
homeotherm
kidney
loop of Henle
Malpighian tubule
medulla
nephridium
nephron
osmolyte
osmoregulation
osmoregulatory system

pelvis
piloerection
poikilotherm
rete mirabile
set point
thermiogenesis
thermoregulation
thermoregulatory center
torpor
urea
uric acid

QUESTIONS

1. Identify the substances or body conditions that are regulated by the osmoregulatory, excretory, and thermoregulatory systems. What three body fluids are involved in these regulatory activities?

2. Which of the following characteristics describe fresh-water fish and marine fish?

   a. Lives in a hypotonic environment.

   b. Lives in a hypertonic environment.

   c. Drinks large amounts of water.

   d. Drinks little water.

   e. Excretes a small volume of concentrated urine.

   f. Excretes a large volume of dilute urine.

   g. Takes up salts by active transport across the gills.

   h. Excretes salts by active transport across the gills.

3. Describe briefly the counteracting osmolyte strategy that operates in sharks and related fish to raise the osmotic pressure of body fluids.

4. Identify two groups of animals that excrete nitrogenous wastes as ammonia, two groups that excrete uric acid, and two that excrete urea.

What is the advantage to a flying bird of excreting uric acid rather than ammonia or urea?

5. The _____ is the functional unit of the vertebrate kidney. Sketch this structure and label: glomerulus, Bowman's capsule, proximal convoluted tubule, distal convoluted tubule, collecting duct. What other part of the nephron is found in mammals and birds?

6. Identify the four basic processes the kidney carries out. What force causes blood plasma to pass from the capillaries into Bowman's capsule? What blood components do *not* enter Bowman's capsule?

7. Describe the countercurrent exchange activity in the proximal and distal convoluted tubules of the nephron and its effect on the concentrating ability of the nephron.

8. Identify three hormones that regulate kidney function; discuss the source, mode of action, and effect(s) of each.

9. List and briefly describe five general areas of function in which animals can regulate body temperature.

10. Reptiles are said to be "behavioral homeotherms." Explain.

11. How is heat generated in birds and mammals?

ESSAY QUESTIONS

1. Why must marine vertebrates expend large amounts of ATP energy to keep the "internal sea" and the "external sea" separated?

2. Describe how some insects, reptiles, birds, and mammals maintain a high body temperature. Why do very small birds and mammals often pass the night in a state of torpor?

SUGGESTED READINGS

Annual Review of Physiology.
    Each year, reviews are included on all aspects of kidney, temperature, and other regulatory physiology. The first thing to read to be up to date.

ECKERT, R., and D. RANDALL. *Animal Physiology.* 2d ed. New York: Freeman, 1983.
    Excellent chapters on the regulation of both water–salt balance and temperature.

HAINSWORTH, F. R. *Animal Physiology: Adaptations in Function.* 2d ed. Reading, Mass.: Addison-Wesley, 1984.
    A wealth of interesting examples and a comparative view of physiological adaptations.

HOCHACHKA, P., and G. SOMERO. *Biochemical Adaptations.* Princeton, N.J.: Princeton University Press, 1984.
    This book is destined to be a classic. The chapters on osmolyte–water and temperature adaptations should be read by every biologist.

36
THE NERVOUS SYSTEM

Nervous systems are undoubtedly the most intricately organized structures to have evolved on Earth.

Roger Eckert and David Randall,
Animal Physiology (1983)

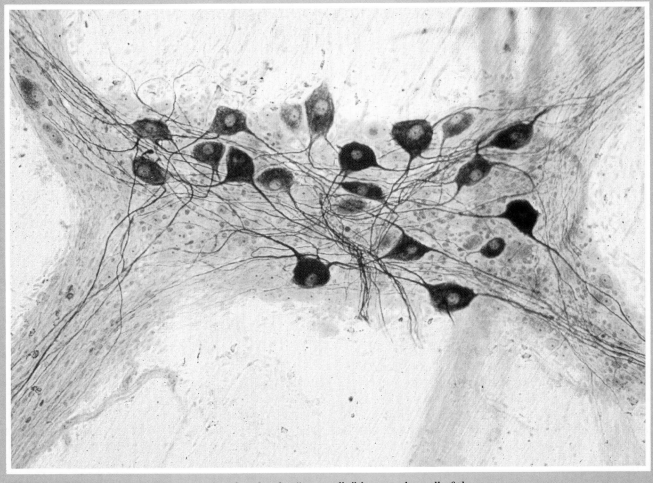

Neurons: a delicate network for electrical and molecular "cross-talk," here in the wall of the gut.

In the summer, if you and your family lived in Eugene, Oregon, you might like to head for the foothills of the Cascade Mountains, about thirty miles away. There, in the shade of the big-leafed maple trees, Fall Creek accumulates in pools about 10 feet deep separated by stretches of wading water. The water is always clear and dark in the shadows, and one can watch the German brown and Dolly Varden trout resting in the shade, facing upstream and gently undulating. Along the bottom creep greenish-brown crayfish, creatures almost the color of the rocks they crawl on. You can try to catch a crayfish by slipping into the water without a ripple, letting the current carry you toward the animal, and then, when it is within reach, making a quick grab for it. But you will find that this little crustacean is much faster than you are; it will suddenly flip its powerful abdomen, dart effortlessly out of your grasp, then swim a few extra strokes to complete its escape.

The behavior of the crayfish as it escapes from danger appears simple. Yet it is complex, indeed; the crayfish must not only detect the danger, but also generate escape reactions that instantly move it in the opposite direction. Even such a "simple" behavior as this one requires the rapid coordination of several body parts. This coordinating role is played by the nervous system.

The many physiological processes discussed in the last several chapters must be coordinated and controlled if an organism is to function in a unified way. Such coordination, or combining of elements into a harmonious whole, is called *integration*. Integration of the activities of various cells, tissues, and organs in a multicellular organism comes about through the functioning of two control systems—the nervous and endocrine systems, which work via the same general principles: they use molecular, ionic, or electrical signals for cell-to-cell communication; and they trigger changes in special *effector* cells and organs, which bring about physiological changes or behaviors in the organism. Traditionally, biologists drew distinctions between the two systems that did not really reflect their shared mechanisms. For example, they considered the endocrine system to act by slow, general distribution of control chemicals, called hormones, to cells and tissues some distance from their point of origin. This traditional view held that the endocrine system works relatively slowly—over minutes, hours, days, or even years—distributing hormones that affect only tissue cells with receptors for those particular hormones. In contrast, the nervous system was characterized by rapid point-to-point communication in which messages consist of volleys of electrical signals (called nerve impulses or action potentials). Individual nerve cells were thought to communicate in fractions of a second only with cells to which they are directly connected.

Modern researchers have blurred the former boundaries between endocrine and nervous activity by identifying literally dozens of "local hormones" that are secreted by nerve cells and that modulate the behavior of other neurons or neuronal circuits. The overlap and similarities between the two systems will become increasingly apparent as we explore nervous-system coordination in this chapter and hormonal coordination in Chapter 37.

We first encountered nervous systems in Chapters 25 and 26. We traced the growing complexity of these systems from the simple *nerve nets* of hydras and jellyfish, to the ventral nerve cord and *ganglia* of annelids and arthropods, to the complicated brains of cephalopods and vertebrates. In each nervous system, whether simple or highly developed, the basic unit is the individual nerve cell, or *neuron*. Therefore, in order to understand the nervous system and the role it plays in an animal's survival, we must understand the structure and function of neurons. And we need to learn about their supportive and insulating cells called glia.

Neurons are electrical switching and conducting devices arranged in a variety of networks that together make up the nervous systems of worms, squids, and monkeys. Imagine thousands, millions, or billions of neurons arranged in chains, nets, or other combinations: that is the staggering reality of the marvelous nervous systems that control the hovering hummingbird, the jet-propelled squid, and the reader of this book. The function of every such network, no matter how complex, depends, in the end, on the properties of neurons. Hence, it is with neuron structure and function that this chapter begins. After discussing the nerve impulse and communication between neurons, we shall turn to nervous systems and review how neurons are organized into chains and networks that coordinate activity in animals as simple as the hydra and as complex as vertebrates. Finally, we shall revisit Fall Creek and the crayfish crawling along the bottom and see how an instinctive reaction—flipping to escape a well-aimed human hand—is controlled and executed.

NEURONS: THE BASIC UNITS OF THE NERVOUS SYSTEM

Essentially, the nerve cell, or **neuron,** is a cellular switching device—an energy *transducer*. ("Transduce" means "to change from one form to another," and we use the word here just as we could say that chloroplasts transduce light energy into chemical energy.) When acted on by chemicals, heat, pressure, or a variety of other energy forms, the neuron transduces that initial input, the stimulus, into an electrical signal—the **nerve**

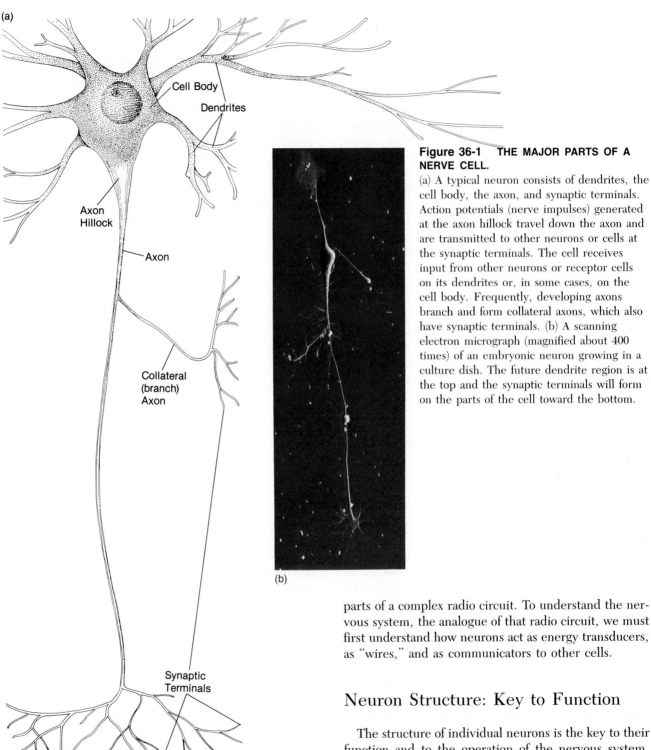

(a)

Cell Body

Dendrites

Axon Hillock

Axon

Collateral (branch) Axon

Synaptic Terminals

(b)

Figure 36-1 THE MAJOR PARTS OF A NERVE CELL.
(a) A typical neuron consists of dendrites, the cell body, the axon, and synaptic terminals. Action potentials (nerve impulses) generated at the axon hillock travel down the axon and are transmitted to other neurons or cells at the synaptic terminals. The cell receives input from other neurons or receptor cells on its dendrites or, in some cases, on the cell body. Frequently, developing axons branch and form collateral axons, which also have synaptic terminals. (b) A scanning electron micrograph (magnified about 400 times) of an embryonic neuron growing in a culture dish. The future dendrite region is at the top and the synaptic terminals will form on the parts of the cell toward the bottom.

parts of a complex radio circuit. To understand the nervous system, the analogue of that radio circuit, we must first understand how neurons act as energy transducers, as "wires," and as communicators to other cells.

Neuron Structure: Key to Function

The structure of individual neurons is the key to their function and to the operation of the nervous system. Neurons have a wide variety of shapes, but nearly all possess certain features critical to function. Four basic structures characterize neurons: each has an "antenna" to receive messages, a "cable" to transmit messages, special endings of the cables to allow communication with the next cell in a circuit, and a maintenance factory to keep all the rest in good repair. In a typical neuron (Figure 36-1), these four structures are the dendrites

impulse, or **action potential.** Neurons not only transduce the energy of a stimulus, but also conduct information in the form of these nerve impulses and communicate with other cells in the nerve network. In a sense, individual neurons are much like individual wires that connect the

("antennas"), the axon ("cable"), the synaptic terminal (cable ending), and the cell body (maintenance site).

Most **dendrites** are relatively short, multibranched, often spinelike extensions of the cell surface. Collectively, they provide a very large surface area for receiving information, which they transmit to the surface of the cell body, from where it may or may not be passed down the axonal cable. Thus the more branched the dendrites, the more separate inputs a neuron can receive. Some kinds of neurons, however, have only a single dendrite, and a few types have none—they receive information messages directly on the surface of the cell body from other neurons.

The *cell body* contains the nucleus and most of the kinds of organelles found in all somatic cells (Chapter 6). Protein synthesis and many metabolic activities are carried out mainly in the cell body, with products passing to the more specialized portions of the neuron for their maintenance.

The **axon** is the neuronal cable that transmits messages in the form of action potentials (nerve impulses) from one point to another in the nervous system. As we will see shortly, the plasma membrane of the neuron is the site of the transmission process. The axon contains a complex cytoskeleton and many mitochondria. It also functions as a "pipeline" that carries molecules manufactured in the cell body to the axonal endings and transports other molecules from the axonal endings back to the cell body. Biologists do not yet know whether this form of molecular transport within the axon plays a role in information transfer, but it does help maintain the health of the neuron and its axonal endings.

The area where the axon joins the cell body is called the *axon hillock*, and it is thought to be the site where action potentials are generated. Individual axons often branch to form *collateral axons*, which can link up to many different nerve or target cells, such as muscle, so that volleys of action potentials may be transmitted to all of them simultaneously.

Axons vary in diameter and length. In mammals, axons may be as little as 1 micrometer in diameter and as large as 20 micrometers. The largest axons in the animal kingdom, found in the giant squid, are up to 1 millimeter in diameter. Since the velocity of transmission of action potentials along axons varies with diameter, differences in size are important to nervous-system function. In length, axons range from only tens of micrometers to 4 or more meters in a giraffe, some of whose axons extend from the spinal cord down to the feet.

As shown in Figure 36-2, axons are not themselves what we commonly refer to as nerves. **Nerves** are actually bundles of many axons; in a single human optic nerve, for example, 1 million axons may be present. The axons in a nerve may run parallel to each other or may be intertwined like the strands of a telephone cable.

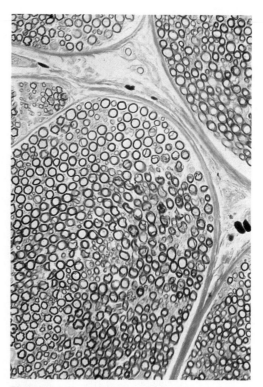

Figure 36-2   NERVES: BUNDLES OF AXONS.
A nerve is like a cable composed of many wires. This cross section of a mammalian peripheral nerve (magnified about 300 times) shows the many axons (white circles) that compose it. Such axons also are called fibers. The dark brown ring around each axon is myelin, the insulating sheath of many axons.

The many axons in a nerve can carry independent messages simultaneously. Thus heat, pain, or other kinds of information received by sense organs in the skin may travel to the spinal cord over some axons, while commands from the brain to the muscles may travel in other axons in the same nerve. It is no wonder, then, that if a person cuts or crushes a major nerve in a finger, an elbow, or a knee, many muscles and sensations may be affected.

The final major neuron structure is the **synaptic terminal,** at the end of the axon. Such a synaptic terminal and the adjacent target cell surface together form the synapse. *Synapses* are sites where one neuron communicates either chemically or electrically with a target cell, which can be another neuron, a muscle cell, or a secretory cell. The synaptic terminals contain packets of the chemical used in cell-to-cell communication, the *neurotransmitter*, as well as secretory machinery and specialized membranes. We will see soon how these components function.

Animal nervous systems also contain **glial cells,** or simply **glia.** Depending on the species of animal and the site in the nervous system, there may be as many as ten times more glia than neurons. Glia provide mechanical support and metabolic aid to neurons; in some species they can sustain an axon for months, even if it is severed from its cell body by injury. The primary direct effect of glia on the neuron's electrical function involves insulation. Glial cells wrap layers of *myelin*, or electrical insulating material, around individual axons. This myelin insulation allows extremely rapid transmission of action potentials (nerve impulses), and thus much faster functioning by myelinated nerves and nervous systems.

Three Primary Types of Neurons

Although most neurons contain the same basic components—the dendrites, cell body, axon, and synaptic terminals—a "typical" neuron is as hard to find as a "typical" vegetable. Morphological differences lie mainly in the various locations and proportions of dendrites and axons. Figure 36-3 shows a few of the many kinds of neurons easily observed in a complex nervous system (all of which will be discussed in this chapter or Chapter 40). The chemistry of these and other neuron types varies too, reflecting, along with the shapes, the determined and differentiated status of the particular neuron type. A survey of the variety does reveal, however, three major classes of neurons that deserve our attention: receptor neurons, effector neurons, and interneurons.

Receptor (sensory) neurons are specialized energy transducers. Each receptor neuron is sensitive to a particular type of stimulus, such as light, pressure, heat, or a specific chemical. The receptor's response to a stimulus is a change in electrical activity. Some sensory organs have evolved that contain vast numbers of receptor cells: perhaps 100 million in an eye, or 20,000 in an ear. Such receptor cells lack axons and pass their information to true sensory neurons, which carry it to interneurons or, occasionally, motor neurons (Figure 36-4).

Effector (motor) neurons transmit messages to muscles—causing them to contract—and to glands—causing them to secrete. Whether in response to specific stimuli or to higher-order commands from the brain (that is, commands leading to complex behavior), everything an animal does—each eye blink or growl—is the direct consequence of coordinated activity in some set of effector neurons. Humans have about 3 million effector neurons.

Interneurons receive information from receptor neurons, sensory neurons, or other interneurons, process it, and send commands on to the effector neurons. Interneurons are arranged in circuits that vary tremendously in complexity and in the number of interneurons they contain, the most complex being tracts of thousands of neurons in a mammalian brain. The interneuron circuits are the sole sites for coordinating sets of motor neurons and so the movement and activity of body parts. Interneuron circuits are also the seat of higher-order processes, such as learning and memory. Thus interneurons are the integration sites of the nervous system. Approximately 98 percent of the 10 billion cells in the human nervous system are interneurons.

The three basic types of neurons are built into a vast variety of circuits. The simplest circuit lacks an interneuron and is a direct connection of a sensory neuron to an effector neuron. Neurons may be arranged in simple chains, or sets of them may *converge, diverge,* or exhibit various kinds of *feedback* (Figure 36-5). Convergence means many neurons feed into just a few or even one neuron. Divergence means the opposite—few neurons feed into many—while feedback loops may involve a single neuron, a few neurons, or whole networks of neurons. We will learn how cell-to-cell communication occurs in each of these circuits after first studying the nerve impulse.

HOW NEURONS SIGNAL

Nerve impulses, or action potentials, are the "language" of the nervous system—a kind of Morse code by which neurons communicate with one another and with other cells. (Note, however, that the Morse code of nerve impulses is all dots and no dashes.) During the nineteenth century, scientists discovered that the signal transmitted by neurons is electrical in character. But it was not until the early part of this century that the German physiologist Julius Bernstein demonstrated that the nerve impulse is an *electrochemical* event involving changes in the electrical polarity of the neuron. He proposed that the basis of the nerve impulse is movement of ions across the nerve cell's plasma membrane, causing the normally negatively charged cell to lose its charge.

Although Bernstein was on the right track, he could not explain what causes the change in membrane permeability that allows ions to flow across the neuronal membrane. Nearly four decades passed while scientists attempted to solve this and related puzzles. Then J. Z. Young discovered in squid two giant axons that proved crucial to future research. Squid axons, which are 1 millimeter in diameter and nearly fifty times bigger than the largest human axons, are large enough to make experimentation relatively easy (Figure 36-6a). In 1939 in England, Alan L. Hodgkin and Andrew F. Huxley undertook a series of experiments in which they inserted electrodes into the squid axon, added and removed ions and other substances from the fluid around the axon, and analyzed the results (Figure 36-6b). They soon realized that the key to the nerve impulse is the axon's plasma membrane and its properties. They were able to demon-

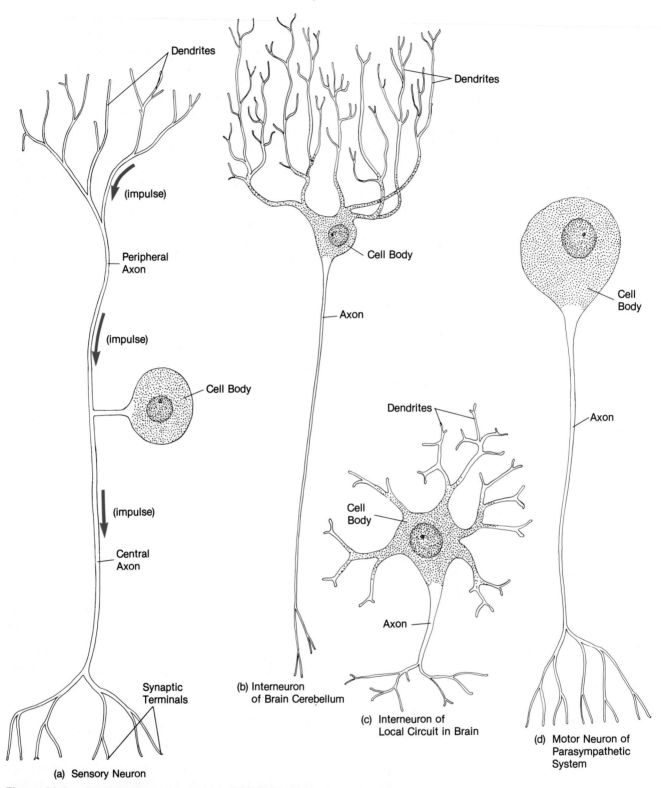

Figure 36-3   MORPHOLOGY: AN AID TO DEFINING TYPES OF NEURONS.
(a) The typical neuron of vertebrate sensory ganglia. (b and c) Brain interneurons show great variation in the number and branching patterns of dendrites, as well as in the length of axons. (d) This motor neuron does not have dendrites and receives information on the cell body; such cells drive certain muscles in the human eye.

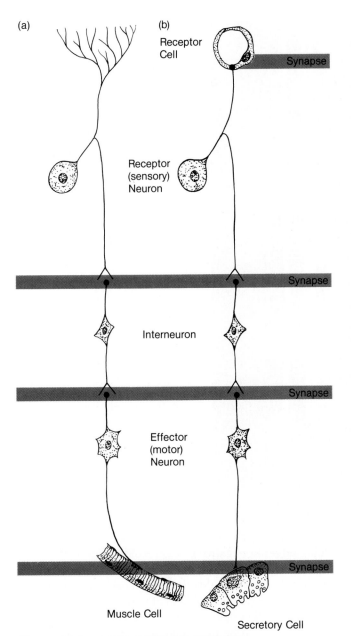

Figure 36-4   BASIC TYPES OF NEURONS.
(a) The simplest type of sensory neuron also functions as a receptor neuron. Sensory neurons form synapses with interneurons or, more rarely, with motor neurons. (b) Some sense organs have special receptor cells that form synapses with sensory neurons. Interneurons both conduct and process information. Effector (motor) neurons form synapses with muscle cells or secretory cells.

Figure 36-5   PRINCIPLES OF NEURONAL NETWORKS.
Networks may have many sites of (a) convergence—where many neurons synapse with a single neuron—and (b) divergence—where one neuron synapses with many others. (c) Circuits may also feed back on themselves, either by (1) chains of neurons or by (2) branches of a cell's own axons providing input to itself.

strate how changes in membrane permeability and subsequent movement of ions produce a nerve impulse and transmit it along the length of the axon.

How the Action Potential (Nerve Impulse) Is Generated

To understand nerve impulses, or action potentials, we have to consider the normal electrical state of the neuron, the so-called resting potential. We also need to see how the resting potential changes to generate the action potential and how the cell returns to the resting state. As we go along, we will see how ion pump enzymes in the neuron's plasma membrane, along with properties of ion channels in that membrane, affect the distribution of Na^+ and K^+ ions inside and outside the nerve cell and thereby "charge" the cell like a battery. As a result, many action potentials can be generated and transmitted. Let us begin by considering the neuron at rest.

Every living cell, whether it is the photosynthetic cell of a plant or the nerve cell of an animal, has a **resting**

(a) Convergence

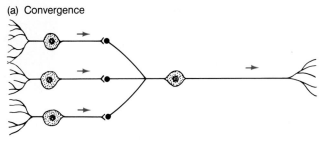

(b) Divergence

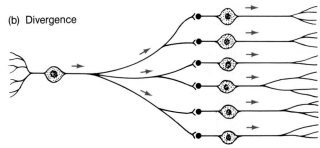

(c) Feedback

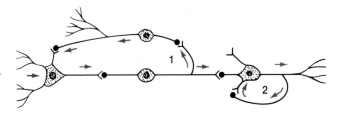

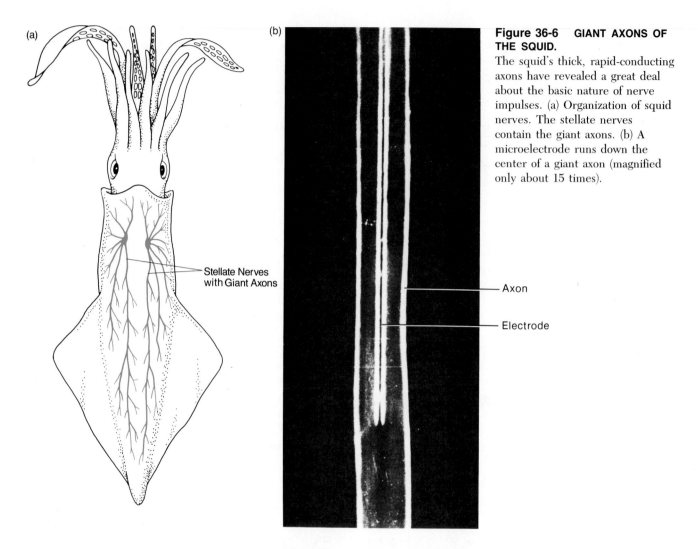

(a)

(b)

Stellate Nerves
with Giant Axons

Axon

Electrode

Figure 36-6   GIANT AXONS OF THE SQUID.
The squid's thick, rapid-conducting axons have revealed a great deal about the basic nature of nerve impulses. (a) Organization of squid nerves. The stellate nerves contain the giant axons. (b) A microelectrode runs down the center of a giant axon (magnified only about 15 times).

potential, a state of electrical charge in which the inside of the cell is electrically negative relative to the outside of the cell. The resting potential is usually due to the relative numbers of Na^+, K^+, and Cl^- ions on the two sides of the membrane and to the relative permeability of the plasma membrane to those ions. What, then, influences ion distribution?

Biologists have found that a neuron's plasma membrane contains a very active $Na \cdot K \cdot ATPase$ pump enzyme that shuttles Na^+ out of and K^+ into the cell when ATP is hydrolyzed (Figure 36-7). This establishes a concentration gradient for Na^+, so that as it accumulates outside the cell, Na^+ tends to leak back in. Na^+ can only leak back slowly, however, due to properties of the protein pores or ion channels through which the Na^+ must pass as it crosses the hydrophobic plasma membrane. The pumping also establishes a concentration gradient for K^+; that ion moves down its gradient and leaks out of the cell more easily than Na^+ moves into the cell because of properties of the ion channels through which it moves. As a result of these differences in Na^+ and K^+ movement, a net positive charge builds up outside the

cell, and a net negative charge, inside (Figure 36-8). At equilibrium, the net outward flow of K^+ equals the net inward flow of Na^+ (this does *not* mean, of course, that there are equivalent numbers of positive ions inside and outside). At that point, the cell is at its resting potential, and it is said to be in a **polarized** state.

A third ion, Cl^-, moves passively across the cell membrane. Because it is attracted to the net positive charge outside the cell and is repulsed by the net negative charge inside the cytoplasm, Cl^-, like Na^+, tends to accumulate outside the cell.

Resting potential is measured in millivolts (mV)—a millivolt is one-thousandth of a volt. For comparison, an AA battery, the type that powers a small cassette player, produces 2,500 mV. In most neurons, however, the resting potential is about -70 mV, reflecting the difference between the net negative charge inside the cell and the net positive charge outside. That may seem like a small charge, but consider that it is close to 0.1 V across a very thin membrane and small space—about 15 nanometers. The same voltage over the same distance in an AA battery would be about 5 million volts!

Figure 36-7   THE NEURON'S ION PUMP.

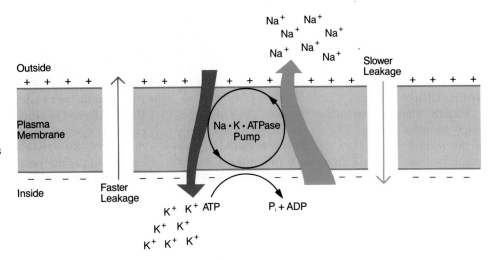

The Na·K·ATPase pump enzyme utilizes energy from ATP to transport Na$^+$ out of the neuron and K$^+$ into it. This mechanism helps establish the resting potential of eukaryotic cells. The pump also restores the normal relative concentrations of Na$^+$ and K$^+$ after many action potentials have passed along an axon. Ion channels (protein pores) through the membrane permit relatively faster K$^+$ leakage outward and slower Na$^+$ leakage inward; these processes also contribute to the resting potential.

Recall that in Chapter 4, we used the word "potential" with respect to stored energy, as in a rock poised at the top of a cliff. The high Na$^+$ concentration outside the neuron is like that rock, poised to move inward; and the high relative K$^+$ concentration inside is like another rock poised to move outward. Thus the polarized, resting potential is a state of stored energy. A word about vocabulary will help us understand an action potential in relation to the resting potential. If the cell's resting potential rises from -70 mV toward 0 mV, the cell becomes *less* polarized. Therefore, the cell is said to be **depolarized** (it has lost some of its polarization). Conversely, if the cell becomes more negative (say, to -90 mV), it is **hyperpolarized** (the prefix "hyper-" means "more" or "excess"). Now let us return to the action potential.

Properties of ion channels in the neuron's plasma membrane are critical to setting up the resting potential and, in fact, to transmitting volleys of electrical signals, or action potentials—the main business of neurons. Ions are hydrophilic and so cannot easily cross the hydrophobic lipid bilayer of a plasma membrane. Ion channels specific to each type of ion may open or close, making the membrane temporarily permeable to the specific ion. Such transient opening of "gates" in the neuron's membrane may be triggered by changes in membrane potential—**voltage-gated channels** open when a certain membrane potential is reached—or may be caused by specific chemicals—**chemical-gated channels** open when a specific chemical is present.

Whereas all cells have resting potentials, neurons and a few other cell types (including muscle cells and certain sensory cells) are unique in that they are *excitable;* that is, the cell membrane responds in an explosive way to an electrical or a chemical signal (the so-called stimulus) by generating an action potential. In such excitable cells, ion channels (gates) that are selective for Na$^+$ open, al-

Figure 36-8   A POLARIZED NEURONAL MEMBRANE.
The action of the Na·K·ATPase pump enzyme and differential permeability of the Na$^+$ and K$^+$ ion channels result in a polarized membrane. It has a net negative charge on the inside and a net positive charge on the outside.

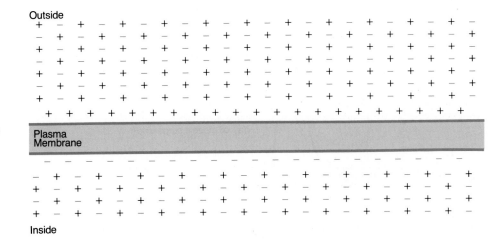

lowing Na$^+$ ions to pass through the membrane at the point of electrical or chemical stimulation. This simple opening of ion channels is the basis of the action potential and, as a result, nerve cells function, whether they are in a feeding hydra or in a Michelangelo painting the Sistine Chapel ceiling.

Exactly what happens when the Na$^+$ gates of a neuronal membrane open? The results are seen most easily if the resting potential of the neuron is raised experimentally from −70 mV to about 0 mV using a microelectrode. Since there are voltage-gated Na$^+$ channels in

the membrane, this change from the resting-potential voltage opens the channels. As a result, some of the excess Na$^+$ ions in the extracellular fluid rush into the cell (because the Na$^+$ concentration is lower inside, and the cell interior has a net negative charge). Enough Na$^+$ ions rush into the cell to depolarize the cytoplasm just inside the membrane to perhaps +40 mV at the site where the channels opened (Figure 36-9a). The Na$^+$ channels remain open for only about 0.5 millisecond before slamming shut, as the pore proteins of which the channel is built enter an inactivated state. The result is the tran-

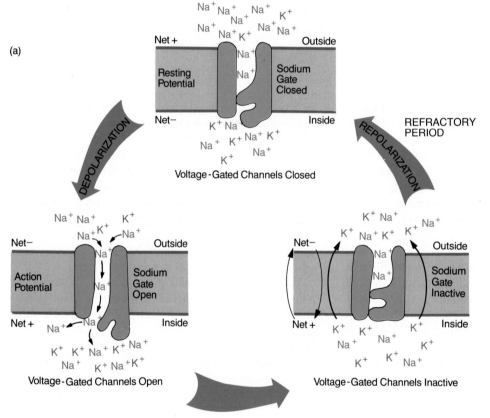

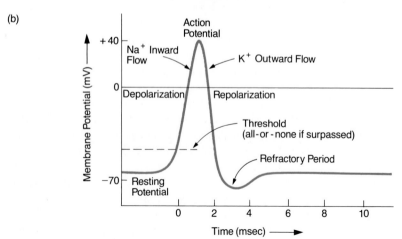

Figure 36-9   THE IONIC AND MOLECULAR BASES FOR AN ACTION POTENTIAL.
(a) Voltage-gated Na$^+$ channels (gates) are closed as the resting potential is maintained. On depolarization above a threshold value, the Na$^+$ channels open, permitting Na$^+$ to flow into the cell. (b) The effects are seen in the oscilloscope recording, as the full action potential is attained. The Na$^+$ channels then close suddenly and enter an inactive state; at the same time, K$^+$ leaks outward through its channels, thereby lowering the potential below the original resting-potential level. The refractory period terminates as the potential returns to the resting level and the Na$^+$ channels revert to the closed, but active state. For simplicity, the K$^+$ channels are not shown.

sient entry of relatively few Na$^+$ ions, but even that small number is sufficient to generate the action potential at that particular site.

During our experiment, the total change between the resting potential (-70 mV) and the peak positive voltage as Na$^+$ rushes in ($+40$ mV) is the action potential of the neuron (in this case, $+110$ mV). When the change from resting potential to action potential is recorded using electrodes, the resulting display on an oscilloscope shows the action potential as a spike (Figure 36-9b). We can appreciate the significance of the Na$^+$ channels slamming shut by imagining what would happen if those channels remained open. Na$^+$ would continue to pour into the neuron and, as a result, would perhaps eliminate the resting potential completely in other parts of the nerve cell.

When the positive charge inside the cell reaches a specific level, voltage-gated K$^+$ channels open, and K$^+$ ions rush out of the cell down their concentration gradient. The exodus of K$^+$ means that less total positive charge remains inside the cell, and thus the potential falls rapidly toward the resting-potential level. Indeed, the charge is briefly reduced below the resting potential of -70 mV. During this **refractory period** (0.5–2 msec), the membrane is unable to react to additional stimulation. Although the refractory period is short, it has great biological significance because it limits the number of action potentials that can be generated per second. During this short time period, the inactivated Na$^+$ channels return to their original configuration, and hence are closed but ready to open again if an appropriate stimulus causes them to do so. The states of Na$^+$ channels during an action potential—closed and active; open; closed and inactivated; and closed and active once more—help us understand the mechanisms of the action potential at the molecular level.

After the action potential has fired, and the membrane has repolarized at a specific site, there is a local imbalance to be corrected: the concentrations of K$^+$ and Na$^+$ have shifted slightly, so that there is now a bit more K$^+$ outside the cell and a bit more Na$^+$ inside the cell than is characteristic of the resting potential. Although the accumulations inside and outside from one action potential are insignificant, they add up after thousands of action potentials pass down an axon. What restores the original ionic balance to the cell? That is the task of the battery charger, the Na·K·ATPase pump enzyme we discussed earlier. This highly active enzyme consumes a large amount of ATP as it pumps Na$^+$ out and K$^+$ in. In turn, neurons need a substantial oxygen supply so that their mitochondria can keep the neurons' ATP reservoir replenished. Given sufficient oxygen and ATP, a neuron's Na·K·ATPase molecules can restore normal ionic balance and thus recharge the battery (the resting potential's charge distribution) that will drive more action po-

tentials. It should not be surprising to learn that when a giant squid axon is poisoned with cyanide, which prevents ATP production, there is *no* immediate effect on action potentials. Thousands of them can be generated before the absence of Na·K·ATPase pump-enzyme activity takes effect. This emphasizes the fact that the action potential, with its changes in state of Na$^+$ and K$^+$ channels, does not consume ATP or oxygen directly and that it is only indirectly dependent on the pump enzyme.

An important feature of all electrically excitable cell membranes is that they initiate action potentials in an *all-or-none* fashion. If a depolarizing stimulus below a certain voltage is applied to the neuron or other cell type, nothing happens. However, if the stimulus reaches a specific **threshold** voltage (0 mV in our example, but different for each cell type), the Na$^+$ channels begin to open, and within a fraction of a millisecond, the full action potential is triggered. What happens if one continues to apply higher and higher voltage stimuli? Once the threshold value is reached, it does not matter how large the stimulus becomes; the same action potential is obtained—say, $+40$ mV and no higher for a given neuron. It is as though the same number of Na$^+$ channels open in response to all stimuli at or above the threshold. We will see that the all-or-none character of action potentials allows their transmission for long distances—from a giraffe's foot to its spinal cord, for example—without fading out.

How the Action Potential Is Propagated

So far, for simplicity, we have considered action potentials at one spot. However, messages in the nervous system commonly consist of volleys of action potentials traveling long distances down axons running within a turkey's leg or a person's arm or a jellyfish's tentacle. For an action potential to travel along an axon, it must be *propagated*, or disseminated, the length of that axon.

Propagation of an action potential (nerve impulse) has been likened to the action of gunpowder igniting adjacent segments of a fuse wire, such as that used to set off dynamite. Unlike a fuse, however, an axon is not permanently "used up" after an impulse has moved along it. Rather, neurons are reusable "fuses." Let us see why.

Impulse propagation depends on the presence of typical voltage-gated Na$^+$ channels at various spots along the axonal membrane (or, as we will see, at special nodes in myelinated axons). At the initial point of depolarization on the membrane, the Na$^+$ channels are caused to open, and an all-or-none action potential is initiated. As Figure 36-10 shows, the local influx of Na$^+$ causes minute **local**

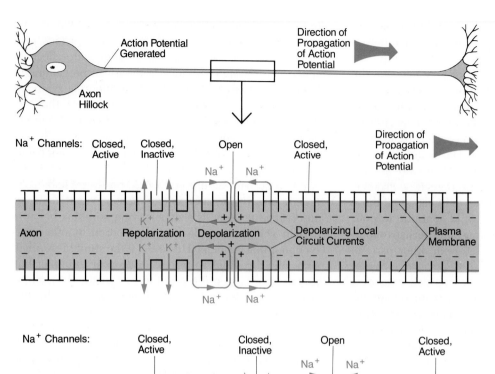

Figure 36-10   PROPAGATION OF AN ACTION POTENTIAL IN AN UNMYELINATED AXON. The action potential is normally generated at the axon hillock. Here we see the action potential at two sites next to each other. Propagation from one site to the next depends on local circuit currents that depolarize above threshold the adjacent membrane and thereby open voltage-gate Na$^+$ channels there. As the action potential sweeps along the axon, the voltage-gated channels successively open, close and become inactive, and become active again at each site. (Voltage-gated K$^+$ channels are not shown.)

circuit currents to flow, thereby depolarizing beyond threshold the adjacent portion of the membrane so that its Na$^+$ channels open. This sequence is the start of propagation. The process continues down the axon, as one after another set of Na$^+$ channels opens, in domino fashion, slams shut into the inactive state, and then returns to the closed condition. The action potential travels at speeds of 1 meter to 100 meters per second and does not fade out or damp in intensity for an obvious reason—at each new site the depolarizing local circuit electric current causes a surplus of Na$^+$ channels to open so that a full all-or-none action potential is produced.

Another feature of a propagating action potential is its unidirectionality; from the site of initiation, usually the axon hillock, the impulse travels down the axon. The brief refractory period during which the Na$^+$ channels are inactive and K$^+$ flows outward accounts for the inability of the action potential to double back on itself toward the cell body.

The speed with which an axon can propagate an action potential varies from only a few centimeters to 120 meters per second. One variable that affects velocity of propagation is axon diameter. Conduction velocity in-

creases only as the square root of diameter; that is, increasing the diameter sixteen times yields only a fourfold increase in speed. As a general rule, the largest axons in an animal's nervous system are located where speed of conduction is most vital. For example, many kinds of fish have a pair of large neurons, called *Mauthner cells,* that have huge axons with diameters up to 200 micrometers extending down each side of the spinal cord. The startle response of fish that allows them to escape predators depends on the rapid rate of conduction within these giant axons. The firing of a Mauthner cell results in a massive contraction of all the muscles on one side of the body, causing the fish to flip violently sideways, away from the lunging beak of an egret or the claws of a crayfish.

Temperature also affects the rate of impulse propagation: as temperature rises, propagation is faster. Thus axons of birds and mammals can be very small in diameter (1 μm or so) and still conduct rapidly because of the animals' high body temperatures. Some large fish, such as the tuna, do not have Mauthner cells because they have heat-retention mechanisms that raise the temperature of their spinal cord to 25°C to 30°C (Chapter 35).

Another factor that allows rapid conduction in the typically small-diameter axons of vertebrates is their insulation by **myelin sheaths.** Myelin is the product of special glial cells (called **Schwann cells** in peripheral nerves and **oligodendrocytes** in the brain and spinal cord). During embryonic development, these glial cells form layer upon layer of specialized plasma membrane that is wrapped around the axons of many vertebrate neurons (Figure 36-11). The multiple lipid layers that make up the myelin sheath widely separate the net positive and negative charges on each side of the axonal membrane. Thus the sheath is like the plastic or rubber insulation around an electrical wire.

At regular intervals along a myelinated axon, the myelin sheath is interrupted, and tiny sections of axonal membrane, called the **nodes of Ranvier,** are exposed. Na^+ channels are abundant at these nodes, but not in the axonal membrane beneath the myelin sheath between nodes. The K^+ channels are found primarily in the axonal membrane at the edges of nodes. When an action potential depolarizes the membrane at one node, the resulting electrical currents pass instantly along the axon to the next node. Thus the impulse moving along a myelinated axon jumps from one node to the next about as fast as it would travel from one site of axonal membrane to the next in an unmyelinated nerve cell. This process of jumping from node to node, illustrated in Figure 36-12, is called **saltatory propagation** (from the Latin word *saltere,* meaning "to leap").

How much difference does the myelin sheath actually make? The largest known unmyelinated axon—that of the squid—is 1 millimeter in diameter and conducts at 20 meters per second. Yet a myelinated axon from a frog, a mere 18 micrometers in diameter, conducts at 40 meters per second. A human optic nerve, about 0.6 centimeter in diameter, contains about 1 million myelinated axons. If the axons were not myelinated, the optic nerve would have to be 15 to 20 centimeters in diameter for the axons to conduct at the speed they do.

To sum up, the action potential, or nerve impulse, is an electrochemical event involving the rapid depolarization and repolarization of the nerve-cell membrane. The process begins at the point where the membrane is depolarized beyond threshold and continues along the axonal membrane. Each depolarization–repolarization cycle initiates a new cycle at the adjacent point on the axonal membrane (or at the next node of Ranvier in a myelinated axon). This basic process is common to nearly all kinds of neurons in all types of animals—mollusks and moths, leeches and lizards, seastars and humans. But as we saw in Chapters 25 and 26, these neurons are, in fact, linked together in nervous systems. What happens when

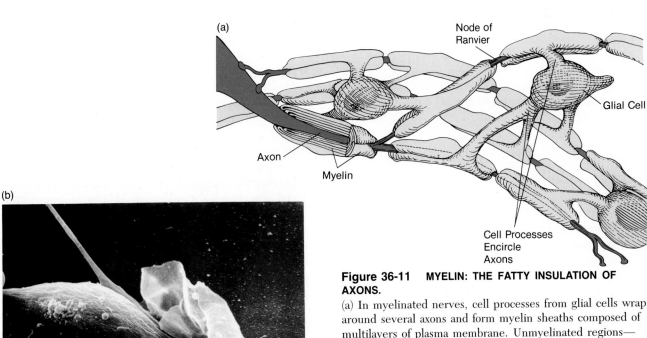

Figure 36-11   MYELIN: THE FATTY INSULATION OF AXONS.

(a) In myelinated nerves, cell processes from glial cells wrap around several axons and form myelin sheaths composed of multilayers of plasma membrane. Unmyelinated regions— the nodes of Ranvier—occur at regular intervals along each axon; the action potential jumps from one node to the next. (b) A living glial cell extends a sheet of membrane upward and encircles the thin, cylindrical axon that runs from the upper left to the lower right in this scanning electron micrograph (magnified about 5,300 times).

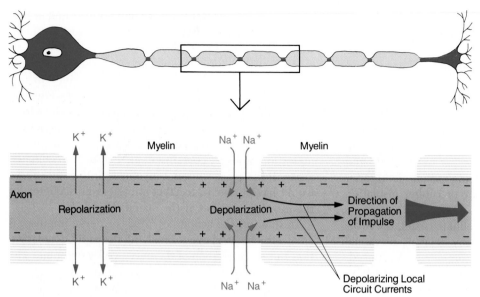

Figure 36-12   PROPAGATION OF AN ACTION POTENTIAL IN A MYELINATED AXON.

The presence of the myelin sheath means that the only electrically excitable axonal membrane is at the nodes of Ranvier. Thus the only sites that can depolarize are the nodes. When a node depolarizes, voltage-gated Na$^+$ channels open, Na$^+$ enters, and the resultant local circuit current "jumps" to the next node of Ranvier (the site of the next depolarization event). This saltatory propagation of the action potential from one node is very rapid. Repolarization occurs in the usual way, with voltage-gated K$^+$ channels at the edges of the nodes opening to allow K$^+$ outflow.

an impulse reaches the synaptic terminals at the ends of axons? How is it conveyed to the adjacent neuron or effector cell?

Transmission of the Action Potential Between Cells

As you read this page, millions of action potentials are being transmitted from one neuron to another in your brain across junctions called **synapses.** Typically, synapses are located at the ends of axons or collateral axons of the so-called *presynaptic* neurons (the neuron whose action potentials move toward the synapse). The dendrite of the *postsynaptic* cell lies on the opposite side of the synaptic junction and receives electrical or chemical stimulation during **synaptic transmission,** the crossing of the synapse by the signal. Typical postsynaptic neurons form synapses with numerous (dozens, hundreds, or thousands) presynaptic neurons, and so can receive independent signals from a variety of sources in any brief time interval.

There are two types of synapses: electrical and chemical. In *electrical synapses*, the plasma membranes of the communicating cells actually touch, so that ion transfer is facilitated (Figure 36-13a). The action potential travels from one neuron to the next via gap junctions, through which ions, small molecules, or electric currents may pass rather freely from cell to cell (Chapter 5). In effect, the electrical synapse consists of a gap junction. Because transmission at a gap junction is extremely fast, electrical synapses are found between neurons and between a few other types of cells where speed of impulse transfer is important. A good example is the muscle cells of the vertebrate heart.

Most electrical synapses are *nonrectifying* (bidirectional). In other words, impulses can move across them in either direction. Nonrectifying synapses are usually part of a circuit that contains at least some *rectifying* (unidirectional) synapses, which allow directional control of information transfer.

In *chemical synapses*, electrical impulses are transduced into chemical signals instead of being transmitted directly (Figure 36-13b). Chemical messengers, called **neurotransmitters,** are released from the presynaptic cell and diffuse across the space between the presynaptic and the postsynaptic cell membranes. The chemical message is then transduced back into an electrical voltage change in the potential of the postsynaptic cell. Chemical synapses are generally slower at transmission than are electrical synapses, but they have a greater capacity to convey information because of possible variations in the type of neurotransmitter, its rate of release, and its rate of breakdown. Because only presynaptic cells release neurotransmitters, chemical synapses transmit information in only one direction—they are rectifying synapses.

(a) Electrical Synapse

(b) Chemical Synapse

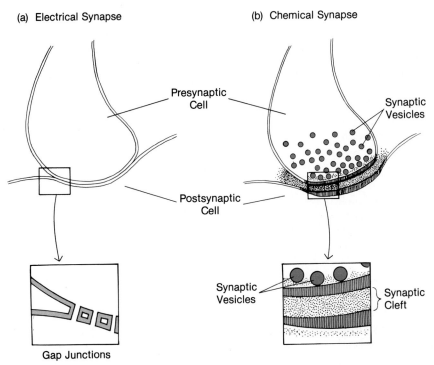

Presynaptic Cell

Postsynaptic Cell

Gap Junctions

Synaptic Vesicles

Synaptic Vesicles

Synaptic Cleft

Figure 36-13   TYPES OF SYNAPSES. (a) An electrical synapse has gap junctions that are sites of cell-to-cell communication. Such junctions pass action potentials in both directions. (b) The chemical synapse has membrane thickenings on the pre- and postsynaptic surfaces, plus a molecular "glue" in the synaptic cleft. Neurotransmitter in synaptic vesicles is released into the cleft and binds to receptors on the postsynaptic surface (not shown). (c) This detail of a synapse between a motor neuron and a skeletal muscle cell shows the important structures. (d) An electron micrograph (magnified about 24,000 times) of the synapse shows how the synaptic vesicles and other structures actually look. A basal lamina, equivalent to the glue in neuronal synapses, is seen between the two plasma membranes.

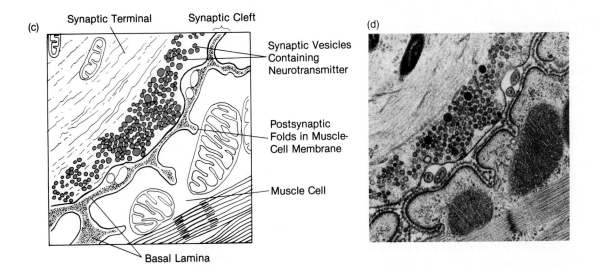

(c)

Synaptic Terminal

Synaptic Cleft

Synaptic Vesicles Containing Neurotransmitter

Postsynaptic Folds in Muscle-Cell Membrane

Muscle Cell

Basal Lamina

(d)

In chemical synapses, the membranes of the two cells are not in direct contact; they are separated by a space—the *synaptic cleft*—so small (15–20 nm across) that it can be seen only with an electron microscope. Sugar–protein polymers form a kind of "glue" holding the thickened presynaptic and postsynaptic cell membranes in place. Just internal to the presynaptic cell membrane are a large number of tiny sacs, called *synaptic vesicles*, which contain neurotransmitter. Dozens—perhaps hundreds—of different chemicals serve as neurotransmitters in various neurons. Among the better understood neurotransmitters are acetylcholine, epinephrine, norepi-

nephrine, dopamine, and serotonin. While different types of neurons release different neurotransmitters, neurons of a single type generally secrete the same neurotransmitter. For example, all motor neurons that drive human skeletal muscles secrete the neurotransmitter acetylcholine. Recently, it has been discovered that some neurons may release more than one neurotransmitter, but they release the same *mixture* of neurotransmitters at all their synaptic terminals.

What is the role of neurotransmitters in synaptic transmission? To find out, let us see what happens when an action potential arrives at a synapse.

The Role of Neurotransmitters

When an action potential reaches a synaptic terminal, it depolarizes the presynaptic cell membrane. This depolarization opens voltage-gated Ca^{2+} channels (which are distinct from the axon's Na^+ channels), allowing a small number of Ca^{2+} ions to flow into the cytoplasm of the terminal from the extracellular fluid surrounding it. Calcium flows in because of a steep concentration gradient. The Ca^{2+} ions, by an unknown mechanism, promote the fusion of synaptic vesicles with the presynaptic cell membrane, and the membrane opens to release neurotransmitter from the vesicles into the synaptic cleft (Figure 36-14). The number of vesicles that fuse and discharge correlates closely with the quantity of Ca^{2+} ions entering the synaptic terminal and, in turn, with the rate at which impulses reach the terminal.

Water-soluble neurotransmitters are like hormones released from an endocrine gland into the bloodstream. But instead of taking minutes to reach a target, neurotransmitter molecules diffuse across the narrow synaptic cleft in a fraction of a millisecond. Some reach and bind to *receptor molecules* embedded in the postsynaptic cell membrane. These receptors are part of the chemically gated channels that are opened only by specific neurotransmitters. The specificity of neurotransmitter–receptor binding ensures that only the "right" signal can trigger the postsynaptic cell and perhaps propagate the action potential.

Receptors are like keys that unlock the ion channels in the postsynaptic cell. When the ion channels associated with the receptors open, ions flow along their concentration gradients into the postsynaptic cell's cytoplasm. Each of the numerous channels that opens in response to neurotransmitter–receptor binding remains open for about 400 microseconds after the neurotransmitter binds; thus only a brief pulse of ion flow can occur. Nevertheless, the influx of ions through the many channels at a synapse can have a positive (excitatory) or a negative (inhibitory) effect on the postsynaptic cell membrane.

At an **excitatory synapse,** the synaptic transmission depolarizes the receiving (postsynaptic) cell. Cation channels (ones that pass positively charged ions) open in

the postsynaptic cell membrane; Na^+, K^+, and some Ca^{2+} flow in; and the electrical potential changes a bit from -70 mV toward 0 mV. This depolarization is called

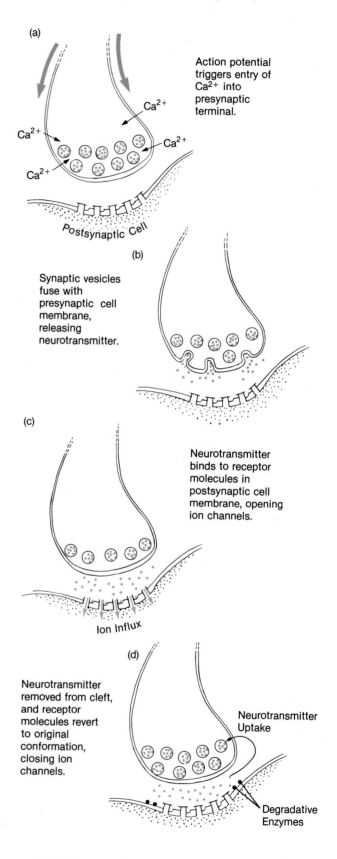

(a) Action potential triggers entry of Ca^{2+} into presynaptic terminal.

(b) Synaptic vesicles fuse with presynaptic cell membrane, releasing neurotransmitter.

(c) Neurotransmitter binds to receptor molecules in postsynaptic cell membrane, opening ion channels.

Ion Influx

(d) Neurotransmitter removed from cleft, and receptor molecules revert to original conformation, closing ion channels.

Neurotransmitter Uptake

Degradative Enzymes

Postsynaptic Cell

Figure 36-14   HOW NEUROTRANSMITTERS WORK.
(a) Arrival of the action potential at the synaptic terminal triggers Ca^{2+} entry into the presynaptic cell cytoplasm.
(b) This, in turn, triggers fusion of presynaptic vesicles with the plasma membrane, thereby releasing neurotransmitter.
(c) Binding of the neurotransmitter—say, acetylcholine—triggers the entry of ions into the postsynaptic cell cytoplasm. In response to this binding, Na^+, K^+, and Ca^{2+} enter at excitatory synapses, and Cl^- enters or K^+ leaves at inhibitory synapses. (d) After a brief time, the neurotransmitter is degraded by an enzyme or is taken back into the presynaptic terminal.

Figure 36-15   PRESYNAPTIC INHIBITION.
One neuron may inhibit the activity of other neurons. For example, neuron 1 cannot transmit across the synapse to neuron 2 if impulses down the axon of neuron 3 act as inhibitors at the synapse.

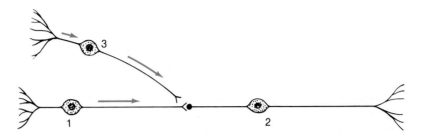

an excitatory **postsynaptic potential (PSP)**. It opens any nearby voltage-gated Na$^+$ channels. We will see momentarily how sets of PSPs and the opening of Na$^+$ channels ultimately may raise the membrane potential above the all-or-none threshold so that an action potential is generated.

An **inhibitory synapse** is one in which the action of synaptic transmission is to *prevent* depolarization of the receiving cell. The neurotransmitter "unlocks" channels selective for ions such as K$^+$ or Cl$^-$. If K$^+$ flows out or Cl$^-$ flows in, the cytoplasm inside the membrane becomes more negative in its potential—it hyperpolarizes so that the potential is even farther from threshold. This is an inhibitory PSP. As a result, the cell is much less likely to generate an action potential—it is "inhibited."

Neuron-to-neuron communication can also be stopped by *presynaptic inhibition*, in which an inhibitory synaptic terminal of one neuron synapses with another neuron's presynaptic terminal (Figure 36-15). When the inhibitory synaptic terminal releases its neurotransmitter, the nearby presynaptic terminal of the other neuron is inhibited, so that it, in turn, cannot release its neurotransmitter. As a result, the postsynaptic cell receives less frequent stimulation.

It is important to understand that any given neurotransmitter is not itself excitatory or inhibitory. For example, acetylcholine excites skeletal muscles by opening Na$^+$ channels, but it inhibits spontaneously active heart-muscle cells by opening K$^+$ channels. The characteristics of the receptor molecules and the transduction sequence, not the neurotransmitter itself, determine the effect on the postsynaptic cell.

We know that action potentials arriving at the synaptic terminal are discrete electrical events. They cause neurotransmitter release and have an excitatory or inhibitory effect on the postsynaptic cell. But what happens to the neurotransmitter itself? It stands to reason that a communication system built around millisecond-long information "bits" would be slowed to the point of malfunction if an essential link in the system took a long time to operate. It is for this reason that the effects of a neurotransmitter on a postsynaptic cell must be very short-lived. Only then will the postsynaptic cell be able to receive new arriving messages. An essential feature of all chemical synapses is the speedy and efficient way in which neurotransmitter molecules are *removed* from receptor sites and the synaptic cleft. Neurotransmitter molecules are removed from the cleft in two ways: they are either rapidly broken down by enzymes on the postsynaptic cell surface or taken back into the presynaptic terminals. Even the transmitter molecules bound to receptor molecules are released and diffuse back into the cleft; they, too, are degraded or taken up. Many nerve gases and insecticides are designed to inactivate the postsynaptic enzymes that break down neurotransmitters. In the absence of functional enzymes, neurotransmitter remains bound to receptor molecules, with a resultant long-lasting excitatory or inhibitory effect. This incapacitates the animal by preventing the continuing transmission of information in circuits of neurons or to effector cells.

Summation and the Grand Postsynaptic Potential (GPSP)

We have seen that transmission of an action potential may result in the release of a neurotransmitter at an excitatory synapse. But is that event and the resultant generation of a PSP sufficient to initiate a true action potential in the postsynaptic cell? The answer is no, and the reason is that there are two few voltage-gated Na$^+$ channels on the dendrites and cell body of most postsynaptic cells to allow the initiation of action potentials. However, most nerve cells have many synapses on their dendrites and cell body, and therein lies the solution. As illustrated in Figure 36-16a, some of these synapses are hyperpolarizing and cause inhibition by means of inhibitory PSPs; a larger number are depolarizing and cause excitation by means of excitatory PSPs. At any one time, many synapses on a neuron's dendrites and cell body may be active. The result is the **grand postsynaptic potential (GPSP)**, which is roughly the summation of all the excitatory and inhibitory PSPs over any short time span. If many excitatory synapses but few inhibitory synapses are active at a given time, it is likely that the combined PSPs will cause sufficient depolarization to surpass the cell's threshold and generate an action potential. The adding together of individual PSPs is called **summation**.

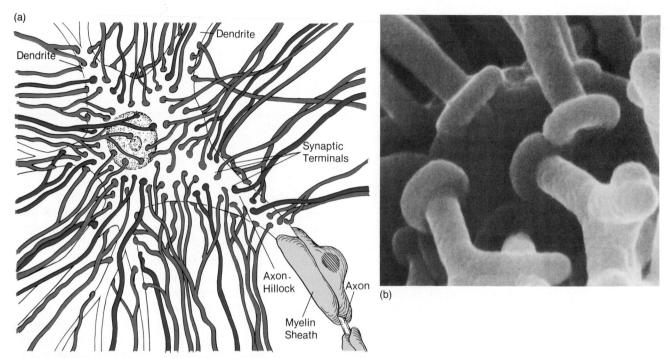

Figure 36-16   THE GRAND POSTSYNAPTIC POTENTIAL AND SUMMATION.
The body of this spinal-cord neuron is virtually covered with excitatory (green) and inhibitory (red) synaptic terminals. (b) A scanning electron micrograph (magnified about 23,000 times) shows such terminals. At any one time, varying numbers of the excitatory and inhibitory synapses fire to generate excitatory or inhibitory postsynaptic potentials (PSPs). They "add" together as the GPSP, and when their total exceeds threshold for this neuron, action potentials are generated at the axon hillock.

Summation occurs over both space and time. *Spatial summation* takes place when a number of excitatory and inhibitory PSPs at different sites are combined. *Temporal summation* is the cumulative effect of a number of PSPs arising at different times. For example, a volley of action potentials coming down the axon of a presynaptic cell may produce excitatory PSPs so rapidly that the increased potential does not have time to fall back to resting-potential levels. The potential might continue to increase until the grand finale is reached—a GPSP sufficient to cause the postsynaptic cell to initiate one or more action potentials.

But how do we relate GPSPs that may have all sorts of values to all-or-none action potentials? That is the role of special sets of ion channels located at the axon hillock. GPSPs on the cell body plasma membrane affect these channels, which translate different levels of the GPSP into a series of action potentials. In other words, they make the firing rate of the postsynaptic cell down its axon *proportional* to the summated stimulus arriving from the presynaptic cell. The higher the GPSP, the faster the impulses are generated, and vice versa.

Let us recount. The Morse code of dot action potentials coming down the axons of many presynaptic cells to a single postsynaptic cell causes neurotransmitter to be released at excitatory and inhibitory synapses. The resultant positive and negative effects summate into a GPSP, and if this is sufficient to surpass the threshold, it triggers a new Morse code of action potentials down the axon of the postsynaptic cell. The process is repeated at the next neuron as it receives input from numerous presynaptic neurons; again, if its GPSP is large enough, action potentials are generated and propagate down its axon. In this way, action potentials pass along a chain of neurons in an animal's nervous system until an effector cell, such as a muscle or secretory cell, is reached. Thousands, millions, and, indeed, tens of millions of such processes go on in nervous systems of varying complexity every second. From this combined activity come behavior, muscle contraction, secretion—in short, the functioning of the animal organism.

It is appropriate to turn now from the single neuron and its nerve impulses to systems of neurons within functioning organisms. Once we see how such systems are organized, we can examine the simple tail-flipping behavior in the crayfish to see quite graphically how neurons account for the survival responses and day-to-day activities we can observe so easily in animals.

THE ORGANIZATION OF NEURONS INTO SYSTEMS

In our survey of the animal kingdom in Chapters 25 and 26, we saw that nervous systems vary in complexity from simple networks in organisms with the least complicated behavior to incredibly intricate living computers—brains—in organisms with the most complicated behavior (Figure 36-17). Even in a small and relatively simple brain, the typical neuron arrangements and connections make a computer seem like a Stone Age tool. Brain tissue, with its millions or billions of interconnecting and interacting neurons, may well be the most complex matter on our planet. Let us take a closer look at how nervous systems are organized, beginning with simple invertebrate nerve nets and progressing to the vertebrate nervous system and brain.

Simple Circuits

All living things are *irritable*, that is, they respond to stimuli (chemical, mechanical, electrical, and so on). The simplest unicellular organism may move away from a noxious chemical or the touch of a glass rod. Such a stimulus usually affects the cell surface directly or indirectly. Specifically, the proportions and distributions of ions may be changed by such stimuli, just as those of Na^+, K^+, Cl^-, and Ca^{2+} are in neurons of multicellular creatures. Nervous systems are really just complex variations on this basic irritability and ionic response.

In multicellular animals, specialized cells evolved to receive and respond to stimuli. Sponges have a few cells that resemble neurons, but they have no true nervous system. The radially symmetrical cnidarians, such as jellyfish and hydras, have **nerve nets.** No centralized control center is present in the net, although there may be a concentration of the network near the mouth (Figure 36-17a). These animals have two circuits, or sets of linked neurons: one, a system of giant fibers, is composed of rapid-conducting neurons that control muscular movements; the other, a set of thin, slow-conducting neurons, controls feeding. Electron microscopy reveals that the chemical synapses in such nerve nets have synaptic vesicles on both sides. In other words, such synapses can conduct in either direction and so are nonrectifying. This makes sense in radially symmetrical organisms, which receive stimuli from all sides and have to send impulses through their nerve net starting from any side.

Figure 36-17   REPRESENTATIVE NERVOUS SYSTEMS.
(a) Cnidarians, such as this hydra, may have a simple nerve net that includes a concentration of network neurons around the mouth. (b) Flatworms may have a simple network or two nerve cords plus a primitive brain formed of aggregations of neurons. (c) Arthropods, such as a grasshopper, have two ventral, solid nerve cords and brain ganglia above the anterior gut. (d) The salamander, a representative vertebrate, has a dorsal, hollow spinal cord and brain.

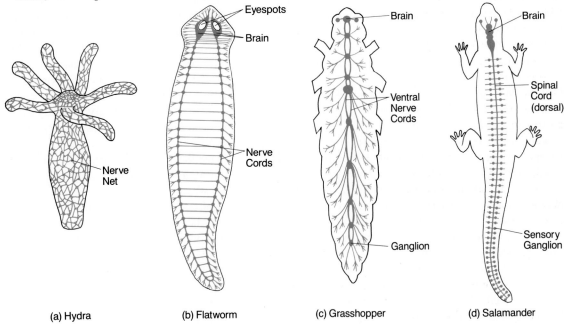

(a) Hydra   (b) Flatworm   (c) Grasshopper   (d) Salamander

Complex Circuits

The next major components of the nervous system to evolve after the nerve net were **reflex arcs** and nervous control centers. Reflex arcs begin with sensory input. This may occur in a sensory receptor cell, in which the actual transduction of energy (light, pressure, and so on) takes place. In that case, a sensory neuron is activated by the receptor cell. Alternatively, the sensory neuron itself may function both as receptor (as for odor, in the human nose) and as neuron to transmit information on the first leg of the reflex arc. The simplest reflex arcs are two-neuron circuits linking a sensory neuron directly to a motor neuron. This type of circuit is quite rare, although some are still present in primates. For instance, such a reflex is responsible for the knee-jerk reflex in humans. Most reflex arcs have one or more interneurons that relay sensory input from sensory neurons to motor neurons and stimulate the motor response. Besides relaying information, interneurons may process it—that is the basis for all complex animal behavior. Thus interneurons may be arranged in circuits of varying complexity that converge, diverge, feedback on themselves, and so on. The more interneurons present in a circuit, the greater the opportunity for increasingly complex information processing and output. In the most complex circuits, multiple excitatory and inhibitory outputs are generated, and behavior can no longer be referred to as a reflex, since processes such as learning may be involved (Chapter 40). Animal brains are, for the most part, massive arrays of interneurons—perhaps millions or billions of highly branched cells wired together to provide a system capable of governing the organism's physiology and overt behavior. In a very real sense, the evolution of complex behavior among animals is the evolution of interneuron properties and circuitry.

Aggregations of the cell bodies of neurons responsible for coordinating functions or behavior in bilaterally symmetrical animals are called **ganglia** (singular, *ganglion*). Ganglia vary greatly in size and complexity; *cerebral ganglia*, or brains, are the most complex ganglia of all. In general, cerebral ganglia are composed primarily of interneurons, whereas ganglia localized in the body segments may contain higher proportions of motor neurons.

The centralization of the nervous system in bilaterally symmetrical animals is also apparent from the formation of one or more nerve cords. Seen even in flatworms and ribbonworms, central nerve cords become prominent features in relatively complex organisms, such as the protostomes (annelids, arthropods, mollusks). Ganglia are located along these animals' ventrally placed nerve cords, often one in each segment of the body (Figure 36-17c). Usually these ganglia control local activity in the body segment, such as leg or wing movements. Axons in the nerve cords carry data in the form of volleys of impulses from sensory organs, such as the taste receptors on the legs of flies (Chapter 38) and sound receptors on the abdomen of moths (Chapter 38), to the cerebral ganglia. As we learned in Chapters 25 and 26, chordates also possess a single longitudinal nerve cord, but this one develops dorsally and does not possess segmental ganglia along its length. Simpler chordates, such as *Amphioxus*, lack a brain, but all vertebrates, of course, have one.

Now that we have reviewed the basic organization of nervous systems, let us examine the most complex one of all—the vertebrate nervous system.

The Vertebrate Nervous System

The complex behaviors characteristic of vertebrates—courting, body language, migration, group hunting, and so on—are possible because their nervous systems, in general, are so highly developed. As Figure 36-18 shows, the vertebrate nervous system is divided into two major parts: the central nervous system and the peripheral nervous system. The brain and spinal cord form the **central nervous system (CNS)**, the integration and control system that ultimately regulates nearly all functions of the body. The **peripheral nervous system (PNS)** innervates all parts of the body and carries information to and from the central nervous system.

The vertebrate's brain is the primary integration and control organ of the body. It receives information, in the form of nerve impulses, from major sense organs in the head as well as from other body regions; the latter information is relayed through the spinal cord. The brain, in turn, communicates with the rest of the body via (1) the cranial nerves, which exit the brain through openings in the skull, (2) again by relaying information through the spinal cord to the PNS, and (3) by controlling various hormone-secreting organs, such as the pituitary gland. The brain's millions or billions of neuronal cell bodies tend to be located near the organ's surface, and, as a result, the exterior of the brain appears gray and is called *gray matter*. Tracts of myelinated axons tend to be found centrally, and because the lipids in myelin appear white, the interior brain region is called *white matter*. We will consider the brain's structure and function in Chapter 40 and its role in behavior, learning, and memory in Chapter 48.

The spinal cord, extending posteriorly from the brain, is hollow and filled with cerebrospinal fluid, which also fills the brain's cavities. In the simplest sense, the spinal cord can be thought of as a relay network between the brain and the motor and sensory nerves running to and from the body posterior to the head. But, in fact, the spinal cord is an integrating switchboard capable of coordinating such complicated activities as walking and running. A cross section of the spinal cord shows the cen-

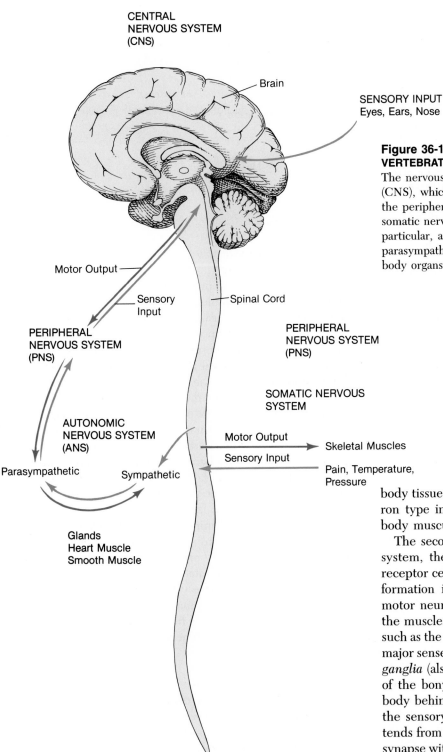

CENTRAL
NERVOUS SYSTEM
(CNS)

Brain

SENSORY INPUT
Eyes, Ears, Nose

**Figure 36-18   SIMPLIFIED SCHEME OF THE
VERTEBRATE NERVOUS SYSTEM.**
The nervous system consists of the central nervous system
(CNS), which is composed of the brain and spinal cord, and
the peripheral nervous system (PNS). The PNS includes the
somatic nervous system, serving the body musculature, in
particular, and the autonomic nervous system, which has
parasympathetic and sympathetic divisions that serve most
body organs in an antagonistic fashion.

Motor Output

Sensory
Input

Spinal Cord

PERIPHERAL
NERVOUS SYSTEM
(PNS)

PERIPHERAL
NERVOUS SYSTEM
(PNS)

SOMATIC NERVOUS
SYSTEM

AUTONOMIC
NERVOUS SYSTEM
(ANS)

Motor Output

Sensory Input

Skeletal Muscles

Parasympathetic

Sympathetic

Pain, Temperature,
Pressure

Glands
Heart Muscle
Smooth Muscle

body tissues. Motor neurons are another important neuron type in the spinal cord; they control much of the
body musculature (Chapter 39).

The second major division of the vertebrate nervous
system, the peripheral nervous system, is made up of
receptor cells; sensory neurons, which carry sensory information inward to the central nervous system; and
motor neurons, which carry control signals outward to
the muscles and organs in various portions of the body,
such as the fingers or the stomach. The PNS includes the
major sense organs (Chapter 38) as well as paired *sensory
ganglia* (also called dorsal root ganglia) arrayed on each
of the bony vertebral column. Sensory neurons of the
body behind the head have their cell bodies located in
the sensory ganglia and have two axons. One axon extends from the cell body to the periphery, where it may
synapse with a sensory receptor cell or, depending on its
particular set of functions, may itself monitor some sensations; thus heat, cold, pressure, pain, and so on are
reported via volleys of impulses moving toward the spinal cord over the sensory axon. The second axon of each
sensory ganglion neuron extends into the spinal cord,
where it synapses with interneurons that feed the information up, down, or across the spinal cord or to effector
cells (as noted above, in rare cases, such sensory axons

trally located gray matter, made up of the cell bodies of
neurons surrounded by white matter composed of myelinated axons (Figure 36-19). Like the brain, the spinal
cord is composed mostly of interneurons; the intricately
connected axons and dendrites of these interneurons
process information. The axons carry information to the
brain or to neurons extending out from the spinal cord to

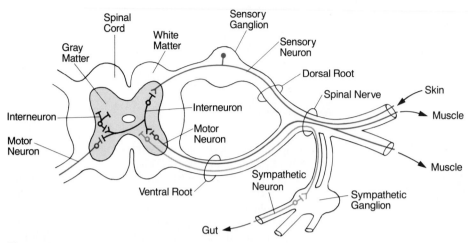

Figure 36-19   THE VERTEBRATE SPINAL NERVE.
This cross section of the vertebrate spinal cord reveals the
relationship between white and gray matter. Sensory
neuronal cell bodies are in the sensory ganglion located on
the dorsal root of the spinal nerve. Motor neurons in the
cord send axons over the ventral root and then out the
spinal nerve to body muscles. Other ventral root axons go
to the sympathetic ganglia, where they drive sympathetic
neurons that innervate a variety of organs. Note the
interneurons in the cord; one synapses with a motor neuron
on the right and also sends a collateral axon across to the
left side where a synapse with a different neuron is formed.

may synapse directly with a motor neuron, thus forming
the simplest reflex arc).

Functionally, the peripheral nervous system is divided into the somatic nervous system and the autonomic nervous system. The **somatic nervous system** includes sensory and motor parts that control the muscles
of the body that we think of as being voluntary, including
those of the arms, legs, neck, throat, lips, tongue, and
eyelids. The **autonomic nervous system** is solely a motor
system and acts generally without a human's awareness
to control physiological functions such as heart rate, digestion, and excretion. The organs affected by the autonomic nervous system are glands (salivary glands and
pancreas, for example), smooth muscles of the blood vessels, and other organs. Certain autonomic functions can
be influenced voluntarily. For instance, people and dogs
can be trained to respond to specific stimuli (sounds,
lights, and so on) and to alter their heart rate or blood-
sugar levels. This process is called *biofeedback*.

The autonomic nervous system, in turn, has two divisions: the sympathetic and the parasympathetic (Figure
36-20). The two divisions often act antagonistically. In
many cases, the **sympathetic nervous system** stimulates
cell or organ function, whereas the **parasympathetic
nervous system** inhibits it. However, there are many
exceptions, especially among the parasympathetic neu-

rons that stimulate such processes as secretion. The two
systems, in general, employ different neurotransmitters.
Sympathetic neurons usually secrete norepinephrine,
which tends to speed up physiological processes; parasympathetic neurons normally secrete acetylcholine,
which tends to slow down the processes. For example,
sympathetic impulses to the heart are excitatory, speeding its beat and increasing the volume of blood pumped.
Stimulation of heart and respiration rates by the sympathetic nervous system occurs during exercise, when oxygen levels fall, or when carbon dioxide levels rise. If a
person's blood pressure becomes too high or carbon dioxide levels in the blood fall quite low, inhibitory parasympathetic impulses act to slow down the heart rate
and lower its output. In fact, the normal resting heart
rate of a person is determined by low-level parasympathetic activity. Thus at any given time, the activities of
the heart and other body organs reflect a delicate balance of inhibitory and excitatory influences.

How would we find these various kinds of peripheral
nerves if we dissected an animal? In vertebrates, pairs of
spinal nerves extend laterally from the spinal cord at the
level of each vertebra. Each nerve contains sensory and

Figure 36-20   INNERVATION OF THE BODY ORGANS.
The autonomic nervous system consists of two sets of motor
systems that generally work antagonistically. Each organ
receives parasympathetic and sympathetic innervation, the
former primarily from cranial nerves that originate in the
brain, and the latter from the spinal cord and the chain of
sympathetic ganglia. In general sympathetic nerves are
stimulatory and parasympathetic ones inhibitory on the
organs. Typical responses of the organs to the two types of
nerve are shown. For convenience, only one sympathetic
chain and set of parasympathetic nerves is shown; in fact
both systems are paired, occurring on both sides of the
body.

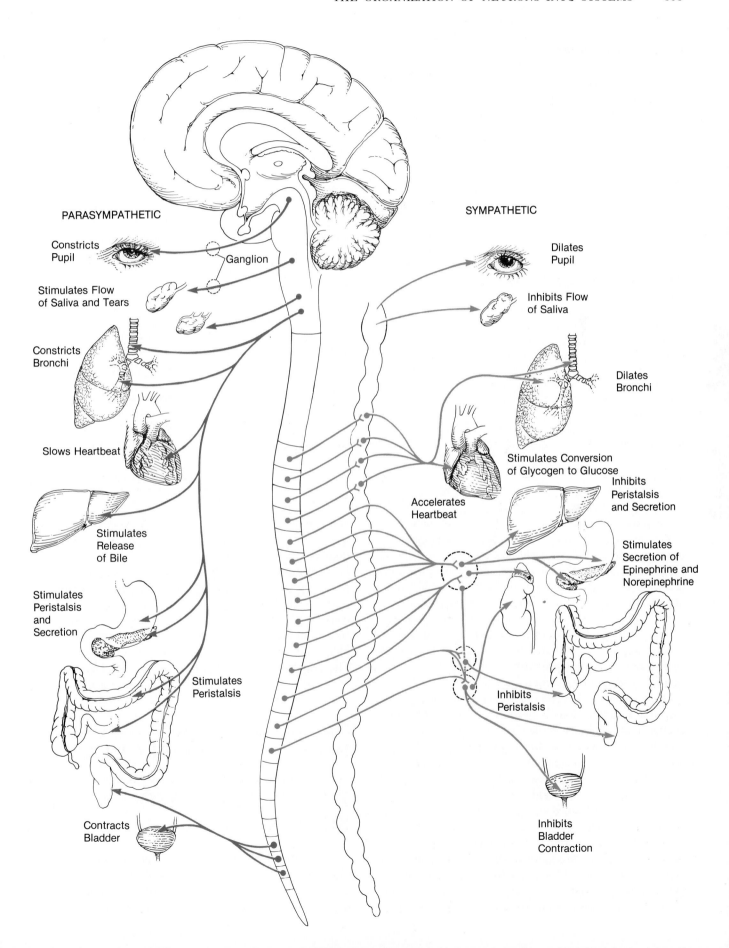

PARASYMPATHETIC

SYMPATHETIC

Constricts
Pupil

Ganglion

Dilates
Pupil

Stimulates Flow
of Saliva and Tears

Inhibits Flow
of Saliva

Constricts
Bronchi

Dilates
Bronchi

Slows Heartbeat

Stimulates Conversion
of Glycogen to Glucose

Inhibits
Peristalsis
and Secretion

Stimulates
Release
of Bile

Accelerates
Heartbeat

Stimulates
Secretion of
Epinephrine and
Norepinephrine

Stimulates
Peristalsis
and
Secretion

Stimulates
Peristalsis

Inhibits
Peristalsis

Contracts
Bladder

Inhibits
Bladder
Contraction

Table 36-1   **MAMMALIAN CRANIAL NERVES**

Number and Name	Sensory Function	Effector Activity
I Olfactory	Smell	None
II Optic	Vision	None
III Oculomotor	Eye position	Eyeball movement, changes in shape of lens, and constriction of pupils
IV Trochlear	Eye position	Eyeball movement
V Trigeminal	Facial skin and teeth	Jaw muscles used in chewing
VI Abducens	Eye position	Eyeball movement
VII Facial	Taste	Facial expression, salivary-gland function
VIII Auditory (vestibu-locochlear)	Hearing, balance, rotation, and movement	None
IX Glossopharyngeal	Taste	Salivary-gland function; pharynx muscles used in swallowing
X Vagus	Sensation in many internal organs: lungs, stomach, aorta, larynx	Parasympathetic fibers to heart, stomach, small intestine, larynx, esophagus, and other organs
XI Spinal accessory	Shoulder position and movement	Shoulder movement
XII Hypoglossal	Tongue position	Tongue movement

motor fibers (another term for axons) of the somatic nervous system, as well as motor fibers of the sympathetic nervous system. The latter run to chains of sympathetic ganglia located near the spinal cord (Figure 36-19). From there, nerves extend more peripherally to the individual organs. Most motor-neuron axons run in the ventral root of the spinal nerve (Figure 36-18), whereas sensory axons run in the dorsal root to the sensory ganglion described above. From there, fibers run into the spinal cord, where they synapse with interneurons (or, occasionally, with motor neurons).

In the head, twelve pairs of cranial nerves carry sensory and/or motor axons between the brain and various organs (Table 36-1). Thus the olfactory organs, eyes, and ears send information to the brain via cranial nerves. Alternatively, parasympathetic axons that run in other cranial nerves serve the heart, lungs, and other organs. Whereas sympathetic ganglia are located far from the organs they serve, parasympathetic ganglia of the cranial nerves are usually found near the organs they control.

We have encountered the function of these various parts of the vertebrate central and peripheral nervous systems many times in the preceding chapters on physiology. And later chapters on the endocrine system, the sensory system, effector organs, the brain, and behavior

will complete the picture. Because the vertebrate brain and spinal cord are so complex and contain so many millions of cells, scientists interested in defining precisely how circuits of nerve cells control behavior have focused primarily on invertebrates, with their relatively simple nervous systems and uncomplicated behaviors. We now consider the Fall Creek crayfish for just that reason; its simple nerve circuitry allows us to see how a specific life-saving behavior—abdomen flipping—is controlled. The crayfish's nervous system and simple reflexive behavior can serve as a model for the immensely more complex circuits in our spinal cords and brains.

HOW NEURONS REGULATE A SIMPLE BEHAVIOR: THE CRAYFISH ESCAPE RESPONSE

Although neurons may be organized into highly intricate systems, the basic principles of organization, like the structure of the neuron itself, apply to all organisms in which neurons have developed. A striking feature of all nervous systems, from jellyfish to humans, is that the sensory-input and motor-output pathways are very nearly the same. What has become increasingly complex is the intervening interneuron circuitry. Certain animal behaviors can be explained in terms of surprisingly simple neuronal circuits and functions. The crayfish escape response is directly traceable to a "hard-wired" circuit of neurons that develops in the embryo. ("Hard-wired" describes a circuit built on genetic and developmental instructions and not dependent on use for its development.) The crayfish's action is called instinctive behavior (Chapter 48) in contrast to learned behavior, which depends on perfection of circuits during a learning process of some kind.

The crayfish evades its predators by flipping its powerful abdomen to propel itself through the water. If attacked from the front, it darts backward. If attacked from the rear, it somersaults forward and away (Figure 36-21). To study this behavior, scientists have broken it down into three stages: (1) rapid flexion of the abdomen, (2) reextension of the abdomen, and (3) swimming away.

We will concentrate on the first two stages and try to answer some basic questions: How does the crayfish "decide" to react? What determines the pattern of the muscular contractions? And how does the nervous system program the order of events?

The first stage begins when a sensory stimulus initiates nerve impulses in either of two pairs of interneurons called escape-command neurons. Even a single impulse in one of these neurons will trigger the full escape response—the abdomen flip. The axons of these interneurons are unmyelinated; since speed is essential, they also have a very large diameter—a feature that is useful to scientists as well as to the animal. The critical stimulus for the crayfish escape response is a rapidly accelerating water current or a brisk tap on its carapace. Normal currents or gentle pinches do not affect the command neurons because the neural network that feeds stimuli to the command neurons is selective.

How does this system work? The hard, protective carapace of the crayfish is covered with thousands of tiny hinged projections, or sensory "hairs." Each hair contains dendrites of sensory receptor neurons that fire impulses when the hair is bent. The axons of these receptor neurons run in nerves to the central nervous system, where they branch and synapse chemically with interneurons. Some of the pathways for pressure and touch converge on the command neurons; others do not. For example, impulses from the sensory receptor neurons in the hairs go to two kinds of sensory interneurons. One kind fires *steadily* in response to a continuous water current, whereas the other fires briefly, only when the water current starts or stops. Only the briefly firing, or phasic, neurons connect with the command neurons.

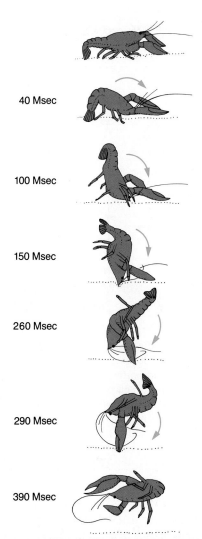

40 Msec

100 Msec

150 Msec

260 Msec

290 Msec

390 Msec

Figure 36-21   CRAYFISH ESCAPE RESPONSE.
Crayfish perform a rapid "somersault" escape response when attacked from behind. Tracings from a high-speed-film sequence demonstrate the crayfish response. The elapsed time after the stimulus was applied is indicated in millisecond. The animal flips over so that its abdomen faces away from the danger, and its pincers confront the attacker. The somersault is a response to impulses from the rear command neurons, which contract abdominal flexors and flair the tail fin.

The nerve impulse produced by any single phasic neuron is too small alone to stimulate a command neuron to fire. But suppose that a predator—a human swimmer, for instance—were to lunge at the crayfish. Such an attack would cause shock waves to travel through the water, and each shock wave would bend many sensory hairs. The result would be the simultaneous firing of many sensory receptor neurons and phasic interneurons. If the sum of these impulses reaches a threshold level, the giant command interneurons will fire.

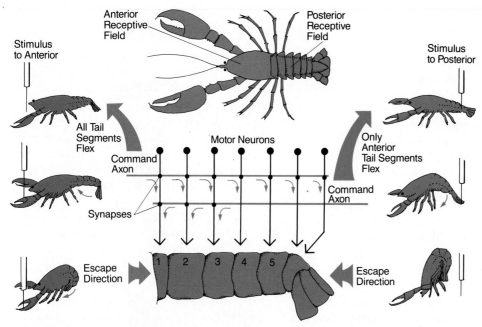

Figure 36-22 SPECIAL NEURONAL CIRCUITS AND THE APPROPRIATE ESCAPE BEHAVIOR BY CRAYFISH.

The anterior (red) and posterior (green) receptive fields connect with the upper (red) or lower (green) command neuron axons shown here. Tapping the head (left) results in the upper command neuron activating the motor neurons leading to all abdominal segments (red arrows). As a result, the tail curls violently and the animal moves backward. Conversely, when the tail is tapped (right), the lower command neuron fires. Only motor neurons leading to segments 1, 2, and 3 are activated (green arrows), and as a result, only those segments contract; this causes the tail to flip, as seen at the right, and the animal moves away to the left as shown.

Now we come to the actual motor response. Why does the crayfish flip backward in response to a frontal attack and forward in response to an attack from the rear? The answer is that only one of the two pairs of command interneurons initiates the first motor response, rapid abdominal flexion. The origin of the stimulus determines which pair of interneurons fires. One pair originates in the brain; it is wired to receive stimuli only from the front half of the animal (Figure 36-22). The other pair originates in the abdomen; it is wired only to receive input from the rear half of the crayfish. By means of these nonoverlapping *receptive fields*, the two pairs of neurons cover the whole animal. Similarly, each pair of command neurons excites a specific set of muscles that constitutes its *motor field*. The responses of these sets of muscles are vastly different. The command center in front excites muscles that cup the tail fin. When a stimulus reaches the frontal field, the abdomen flips into a tight tuck, catapulting the crayfish directly backward. In contrast, the rear command neurons send an impulse to abdominal flexors, muscles that *flare* the tail fin and pitch the crayfish forward. In this way, stereotyped responses brought about by activity of specific nerve cir-

cuits give the crayfish a behavioral repertoire that moves it away from a predator.

Once the initial, rapid flip has been generated, the second stage begins. This stage, reextension of the abdomen, involves a *chain reflex*—a reflex caused by feedback from the flexion.

The rapid movement of the crayfish in the water generates a shock wave of its own. Why don't the command neurons fire again? They do not because the command center also fires inhibitory impulses to every part of the circuit, including the flexor muscles. These inhibitory impulses, which are set off by the initial excitation, arrive a fraction of a second after the excitatory ones.

The final part of the response, the third stage, is controlled by a *central pattern generator* (CPG), which also regulates normal swimming behavior. The crayfish swimming CPG reacts slowly, allowing the first two stages enough time to be completed. When the CPG takes over, the animal is able to swim away. A CPG is a circuit whose output precisely coordinates in sequence and timing a whole series of motor events, such as the dozens of contractions and relaxations that account for each stride of a human's leg and foot.

It is possible that a crayfish receiving a shock wave or a tap will "choose" not to escape. It might scuttle away or try to pinch the attacker. Such variations may appear to be purposeful behavior, but they, too, are brought about by specific neural pathways that scientists are now identifying. For example, the escape response is inhibited if food is nearby; as a result, a crayfish can fight its way through a milling cluster of other crayfish to seize a piece of food, despite the shock waves their movements generate. Once the animal has grasped the food, however, it becomes very excitable. Its abdomen flips vigorously if touched, and it swims rapidly away from the competition. No matter how complex and adaptive these behaviors seem, they are, in fact, due to sets of inhibitory and excitatory activities of hard-wired neuronal circuits. Even so, we will see in Chapters 48 and 49 that animals with such circuits can display adaptive behaviors and learning.

LOOKING AHEAD

In the crayfish escape response we see one model for how a nervous system can generate a reasonably complex behavior. Innumerable such behaviors are added together and ultimately mixed with learning and plasticity (changes in nerve circuits or neurons depending on use) to make up the behavioral repertoire of animals with more complex nervous systems. But nerves alone do not control behavior; in the next chapter, we will see how the endocrine system works with the nervous system to coordinate the full range of physiological functions and behavioral responses. Moreover, we will return to the fact that the endocrine and nervous systems are really variations on a single theme—both utilize chemical signals, receptors, and transduction to generate specific behavioral responses.

SUMMARY

1. The nervous and endocrine systems work together to integrate the functions of multicellular animals. The endocrine system tends to work slowly over long periods of time by secreting hormones; the nervous system works rapidly by transmitting nerve impulses, but neurons also secrete slower-acting hormones that modulate nervous system function. Hence, both systems depend on molecular, ionic, or electrical signals, receptors, and specificity in responding cells to effect changes in target cells.

2. The basic unit of all nervous systems is the *neuron*. Neurons are specialized cells that receive, process, and transmit information. Most neurons have four structures: *dendrites*, which receive information; the cell body, site of metabolism; *axons*, which transmit information; and *synaptic terminals*, which are sites of communication with other cells. Anatomical *nerves* are bundles of many axons.

3. There are three main types of neurons: *receptor (sensory) neurons*, which are special energy transducers (some sensory receptor cells are not true neurons); *effector (motor) neurons*, which transmit messages to muscles and glands; and *interneurons*, which are information processors between sensory and effector neurons. The evolution of interneuron circuits has generated the complex control organs—brains—that make animal behaviors possible.

4. Most neurons transmit signals called *action potentials (nerve impulses)*. The action potential is a "spike" of voltage involving a cycle of *polarization, depolarization*, and repolarization of a site on the neuron's membrane. To evoke an action potential, a stimulus must depolarize the cell beyond a certain *threshold* level above the cell's *resting potential*. Action potentials depend on *voltage-gated channels* in the neuron's membrane that are sites of ion movements. During an action potential, the first event is an opening of channels so that Na^+ flows into the cell, causing the depolarization. The channels then slam shut and are in an inactive state. The action-potential spike then falls as K^+ moves out of the cell at that site. Soon, the channels return to the closed-but-active state so that they can respond to the next depolarization event. After many action potentials have passed down an axon, a $Na \cdot K \cdot ATPase$ pump enzyme reestablishes a normal balance of Na^+ and K^+ inside and outside the cell.

5. The action potential is propagated along an axon without loss of voltage until it reaches the synaptic terminals at the end of the axon. The *refractory period* due to the temporary inactivity of closed Na^+ channels ensures that the nerve impulse does not move back up the axon. The propagation of action potentials involves *local circuit currents* that open new voltage-gated Na^+ channels. The speed with which an axon can propagate a nerve impulse depends on the axon's diameter, on temperature, and on the presence or absence of insulating material. Most axons in vertebrates are wrapped in an insulating *myelin sheath*, multiple layers of plasma membrane that is produced by *glial cells* (specifically *Schwann cells* in peripheral nerves and *oligodendrocytes* in the brain and spinal cord). At intervals along an axon, the myelin sheath is interrupted by *nodes of Ranvier*. Current in these axons jumps from node to node; such *saltatory propagation* is

much faster than that in unmyelinated fibers of the same diameter.

6. The site where messages are transmitted from one neuron to another cell is called a *synapse*. Some synapses are electrical and involve gap junctions; others are chemical. Chemicals called *neurotransmitters* are stored in the synaptic vesicles in the axon's presynaptic terminals; arrival of an action potential at the terminal triggers release of the neurotransmitter into the synaptic cleft between cells. The neurotransmitter binds to a receptor molecule and *chemical-gated ion channels* in the postsynaptic cell membrane respond at *excitatory synapses* by initiating a small depolarization event (the PSP, or postsynaptic potential) or at *inhibitory synapses* by increasing polarization (a negative PSP; such increased negative potential is called *hyperpolarization*).

7. Whether a postsynaptic neuron generates an action potential depends on the sum of numerous excitatory and inhibitory *postsynaptic potentials (PSPs)* it receives spatially and temporally. Spatial and temporal *summation* results in a *grand postsynaptic potential (GPSP)*, which, if above threshold for the neuron, causes it to generate impulses. The rate of impulse generation varies with the size of the GPSP.

8. Neurons are organized into nervous systems ranging from the *nerve nets* of cnidarians to the centralized systems of many invertebrates and vertebrates. The simplest sort of circuit in nervous systems is the *reflex arc*, in which a sensory neuron synapses directly on a motor neuron, or on an interneuron, which in turn synapses on the motor neuron. Aggregations of neuronal cell bodies are called *ganglia*. Complex coordination is under the control of large cerebral ganglia, or brains.

9. The vertebrate nervous system consists of the *central nervous system (CNS)*—the brain and spinal cord—and the *peripheral nervous system (PNS)*, which innervates all parts of the body and transmits information to and from the CNS. The spinal cord is both a relay and an information-processing system because of its many interneurons. The brain has millions or billions of interneurons clustered into areas responsible for various body activities, as well as for complex behavior such as learning and memory. The peripheral nervous system is divided into the *somatic* and *autonomic nervous systems*. The somatic system includes sensory and motor components. The autonomic system includes only effector neurons that control the body's involuntary physiological functioning. It has two subdivisions that work antagonistically: the *sympathetic* and the *parasympathetic*. The cranial nerves carry sensory and motor axons (fibers) and are generally a major part of the parasympathetic system. The spinal nerves carry somatic sensory and motor fibers, and sympathetic fibers.

10. The crayfish escape response serves as an example of nervous-system organization in which seemingly complex behavior is due directly to basic circuits of neurons. These circuits are built during development and do not depend on learning. Hence, the behavior is instinctive or automatic and is triggered by specific stimuli.

KEY TERMS

action potential (nerve impulse)
autonomic nervous system
axon
central nervous system (CNS)
chemical-gated channel
dendrite
depolarization
effector (motor) neuron
excitatory synapse
ganglion
glial cell (glia)
grand postsynaptic potential (GPSP)
hyperpolarization
inhibitory synapse

interneuron
local circuit current
myelin sheath
nerve
nerve net
neuron
neurotransmitter
node of Ranvier
oligodendrocyte
parasympathetic nervous system
peripheral nervous system (PNS)
polarization
postsynaptic potential (PSP)
receptor (sensory) neuron

reflex arc
refractory period
resting potential
saltatory propagation
Schwann cell
somatic nervous system
summation
sympathetic nervous system
synapse
synaptic terminal
synaptic transmission
threshold
voltage-gated channel

QUESTIONS

1. Which two control systems integrate the functions of multicellular animals? Why has the boundary between these two systems recently become blurred?

2. In what way does a nerve cell act as a transducer of energy?

3. Name and sketch a typical nerve cell and label its parts.

4. What are the three basic functional types of nerve cells? Briefly distinguish among them.

5. Define the resting potential of a nerve cell in terms of ions and electrical charges; explain how this potential is maintained.

6. Describe an action potential (nerve impulse) and tell how it is propagated along a nerve cell. How do the states of Na^+ channels help explain the action potential?

7. What variables influence the speed of conduction of the action potential?

8. Identify and describe briefly how the two major types of synapses function. How are excitatory and inhibitory PSPs related to a GPSP, and that, in turn, to initiation of an action potential by a postsynaptic neuron?

9. Identify and describe the major parts of a vertebrate's central nervous system (CNS).

10. Identify and describe briefly the two functional divisions of the peripheral nervous system.

ESSAY QUESTIONS

1. Discuss the role of each of the following in the propagation of action potentials along axons and across synapses: Na^+, K^+, Cl^-, Ca^{2+}, myelin, acetylcholine, ATP, $Na \cdot K \cdot$ ATPase.

2. Discuss the evolution of nervous systems from simple to complex in terms of such structures as nerve net, ganglia, nerve cord, segmental ganglia, cerebral ganglia, CNS, PNS.

SUGGESTED READINGS

ALBERTS, B., et al. *The Molecular Biology of the Cell.* New York: Garland, 1983.

This book includes one of the best simple summaries of the molecular and ionic bases of neuron function.

"The Brain." *Scientific American,* September 1979.

This entire issue is devoted to the brain and includes good, basic articles on neurons and other fundamental components and aspects of the brain.

ECKERT, R., and D. RANDALL. *Animal Physiology.* 2d ed. New York: Freeman, 1983.

A superb treatment of the action potential, synapses, and neurons.

KUFFLER, S., and J. NICHOLS. *From Neuron to Brain.* Sunderland, Mass.: Sinauer, 1976.

A somewhat advanced discussion of the nervous system that gives more detail than does this chapter.

37

HORMONAL CONTROLS

The functioning and the survival of complex organisms would hardly be conceivable, were it not for the existence of these regulatory chemical interactions between cells, tissues, and organs.

Jacques Monod, *Chance and Necessity* (1966)

Sex Hormones at Work. Both the bright red throat skin and the display behavior of this male great frigate bird (*Fregata minor*) on a Galápagos island result from action of androgens, male sex hormones.

An English surgeon named John Hunter suspected in 1771 that glands such as testes and ovaries affected the sex of animals, so he tried an innovative experiment. Slicing an opening into the abdomen of a hen, he inserted a rooster's testis. He then recorded changes in the hen's appearance as it gradually developed male secondary sex characteristics, such as large roosterlike tail feathers. Hunter later wrote about his method "of turning hens into cocks," but he never knew why the transformation happened: that substances called sex hormones were at work.

John Hunter's research was an early demonstration of how the endocrine system functions. The **endocrine system** consists of organs, tissues, and cells that secrete *hormones*, substances that regulate the activities of target cells and thereby help coordinate an organism's physiology. Despite Hunter's pioneering work, more than a century would pass before the British physiologists W. M. Bayliss and E. H. Starling actually defined hormones as the chemicals that are secreted into the blood by specific *glands* (glands are specialized epithelial organs that secrete substances such as hormones, milk, or sweat) and that travel to other parts of the body, where they cause specific cells to change their growth pattern or activity.

The endocrine system is a highly complex communications network that operates along with the nervous system to integrate and control all of an animal's fundamental life processes. Like an individual instrument in an orchestra, each hormone plays a different part in an animal's biology: some hormones trigger and regulate growth; others control sexual maturation or reproductive activity; still others control the secretion of digestive juices after a meal, maintain the chemical balance in blood, or alter an animal's behavior with the changing seasons.

We begin our study of the endocrine system by examining the characteristics of vertebrate hormones and how they exert their effects on target cells. Next, we look briefly at how hormones regulate the life cycles of invertebrates, and then, in more detail, at the major organs and hormones of the vertebrate endocrine system. We then consider some examples of hormonally regulated life processes. Finally, we explore the frontiers of hormone research, including some newly discovered hormones that serve as powerful chemical messengers in the brain. As we proceed, we shall see why the study of hormones and their effects has become one of the most exciting areas of modern biology.

BASIC CHARACTERISTICS OF HORMONES

Every vertebrate possesses two types of glands. One type, *exocrine glands,* secretes substances into ducts that, in turn, empty into body cavities or onto body surfaces (sweat and salivary glands are examples). The second type, *endocrine glands,* have no ducts and instead secrete their chemical products—hormones—directly into the tissue space next to each endocrine cell. Some of these hormones may then diffuse into the bloodstream and be carried throughout the body. Other hormones are secreted in a *paracrine* fashion; they diffuse and affect target cells locally, and do not enter the bloodstream. Finally, *autocrine* secretion involves regulatory substances that act back on the secreting cell itself. The major vertebrate endocrine glands are shown in Figure 37-1.

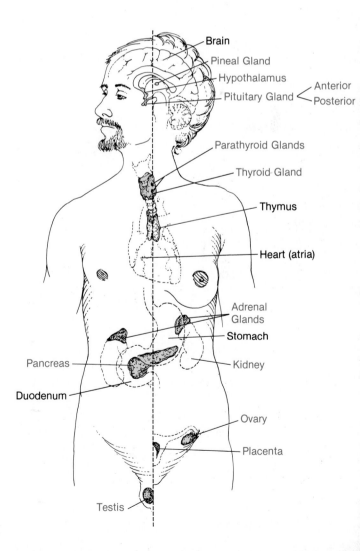

Figure 37-1 THE LOCATION OF HUMAN ENDOCRINE GLANDS.
Those organs labeled in blue are long-recognized endocrine glands, while those shown in black have more recently been found to have endocrine functions.

Dozens of kinds of hormones regulate cell processes, but they all fall into three broad chemical categories (Figure 37-2). Many hormones, such as *insulin* and *growth hormone*, are *proteins*—short-chain polypeptides as well as more complex proteins. Others are *amines*—metabolic derivatives of amino acids. This group includes *epinephrine*, the "adrenalin" that increases pulse rate and affects other physiological changes when a person experiences rage or fear. Still other hormones, including the sex hormones estradiol-17β and testosterone, are *steroids*—complex molecules derived from cholesterol. Interestingly, in a group of vertebrates—monkeys, pigs, frogs, and fish, say—a given hormone may vary slightly in its molecular structure. Biologists use the term "hormone family" to describe a group of chemically related hormones that evolved from a basic hormone molecule. As a result, the members of the hormone family each can carry out somewhat different biological functions in various species (just as can the actin or tubulin families; Chapter 6).

A few principles apply to all hormonal activity, whether it involves proteins, amines, or steroids. The principles are as follows:

1. A given hormone affects only *target cells*. Target cells are responsive to hormones because they possess *receptor molecules* in their membranes, cytoplasm, or nucleus, and *transduction machinery* that is activated by the *hormone–receptor complex* and that thereby carries out the cellular response.

Figure 37-2   CATEGORIES OF HORMONES.

Hormones usually fall into three categories: (a) proteins and small polypeptides, (b) amines, and (c) steroids.

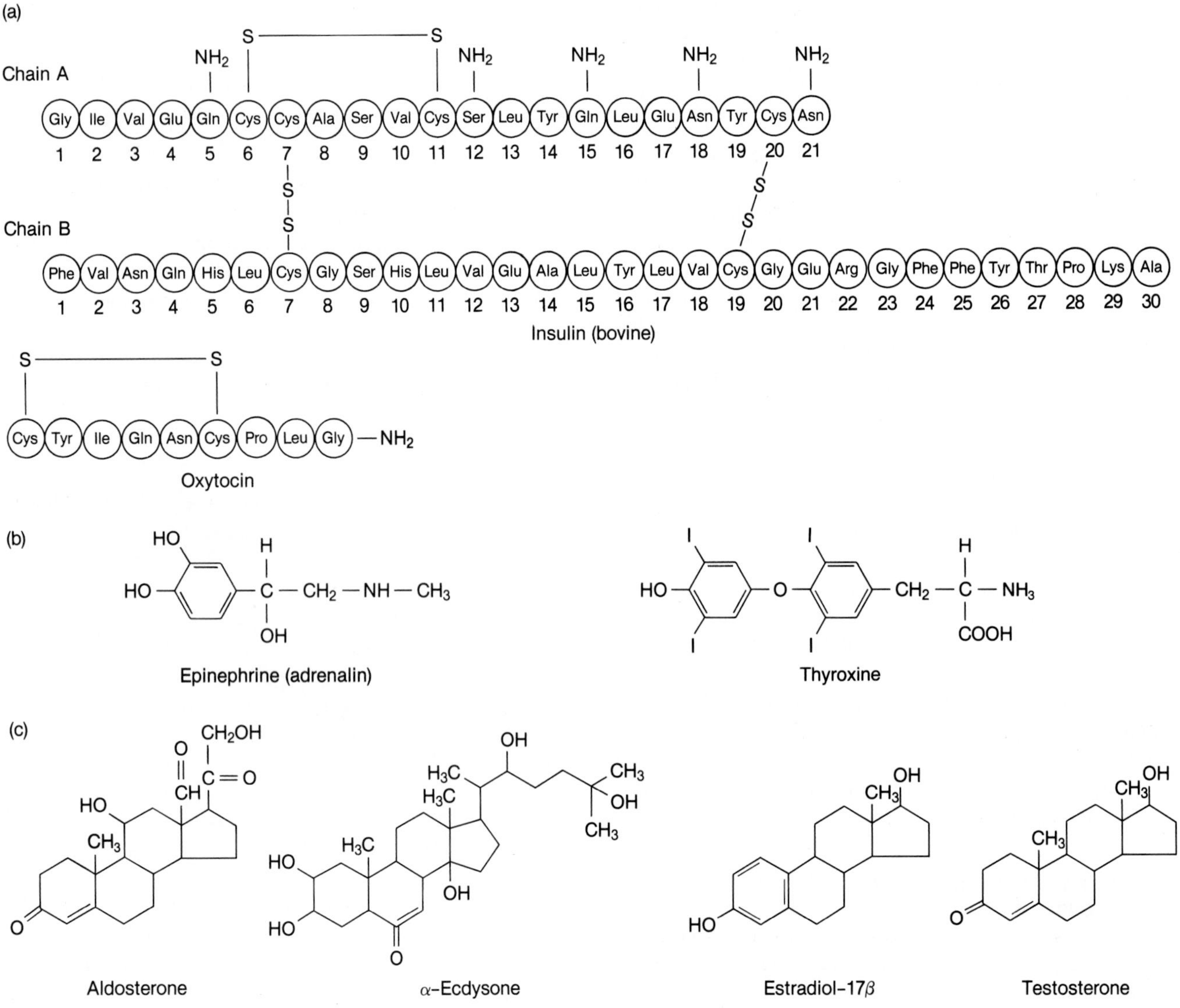

The receptor molecules bind molecules of the circulating hormone as the hormone contacts the cell or as the hormone diffuses into the cell. If cells lack receptors, they do not respond to the hormone. And the way they do respond is solely a function of the transduction machinery that is present in the cell. This means, then, that the specificity of hormonal action is determined by the characteristics of target cells, rather than by the hormone itself. To illustrate, although testosterone from a rooster's testis circulates in the animal's bloodstream and contacts virtually every body cell, only certain sets of cells, such as those in the comb and tail feathers, will respond to the hormone's presence. In a sense, each hormone is like a radio message broadcast on a particular frequency: only those receivers tuned to the frequency will pick up the message.

2. Hormone receptors develop as a cell differentiates. As each cell type of the body matures, it manufactures a particular set of hormone receptor molecules, but not others. As a result, each cell type is regulated by only certain signals—from the pituitary, from the gonads, and so on.

3. Different cells may respond in different ways to the same hormone. Such diversity in target-tissue responses is possible because the transduction machinery of different target cell types "reads" the hormone signal in different ways. Thus each cell type's receptors are linked with different organelles, enzymes, or other cell systems that carry out the cell's response to the hormone stimulus.

4. Some hormones are present much of the time, while others appear only sporadically. Some hormones act steadily or repeatedly over a long period—such as those involved in maintaining homeostasis in body fluids or the growth of long bones during the years of puberty. Compare that long-term action with the brief jolt of epinephrine that stimulates a burst of activity in a gazelle surprised by a hungry lioness.

5. The amount of a circulating hormone is usually governed by negative-feedback control. When the concentration of a given hormone in the bloodstream falls below a certain level, the appropriate endocrine cells become more active and secrete additional hormone molecules into the blood. Conversely, when the concentration rises above a certain level, secretions are inhibited. Some hormones have remarkably complex negative-feedback loops, involving numerous steps and many types of cells.

6. Hormones are usually broken down rapidly. Individual hormone molecules are metabolized rather quickly once they have been taken up by receptor molecules. For instance, when epinephrine binds to receptors on heart-muscle cells, it stimulates stronger muscle contractions, and the hormone–receptor complex is soon degraded by cellular enzymes. The heart cells then return to their former state. Such a finely tuned recovery mechanism is an essential feature of hormone action, for without it, target cells could not be sensitive to changing levels of the hormones that regulate their activities.

These few principles should be kept in mind as we examine individual hormones. Before doing so, however, we need to look more deeply at the transduction process, for that is the key to how cells respond to the regulatory signal.

CELLULAR MECHANISMS OF HORMONAL CONTROL

Once a hormone has been taken up by receptor molecules on the cell surface or within the nucleus of the target cell, the hormone's chemical message is translated into new or different rates of cellular activity. This can happen through one of three mechanisms:

1. A hormone can increase the rate at which other substances enter or leave the target cell.

2. A hormone can stimulate a target cell to synthesize enzymes, proteins, or other substances.

3. A hormone can prompt the target cell's machinery to activate or suppress existing cellular enzymes.

Each of these effects is accomplished by an intricate series of chemical events at the cell surface, within the cytoplasm, and sometimes in the nucleus. Although dozens of hormones interact with an animal's cells, such events follow only two basic sequences: one for steroid and one for nonsteroid hormones.

The Cellular Mechanisms of Steroid Action

Steroid hormones must gain entry into a target cell in order to act. The hydrophobic character of steroids allows them to diffuse through the plasma membrane of target cells. New evidence suggests that once inside, the

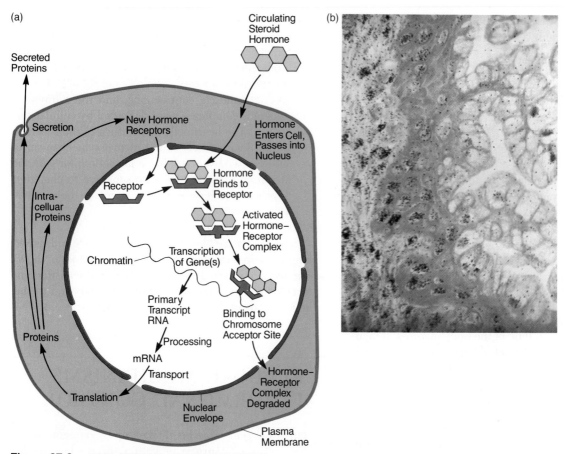

Figure 37-3   HOW STEROID HORMONES WORK.
(a) Steroid hormones usually enter a target cell and become associated with a specific receptor molecule in the nucleus. The hormone–receptor complex then binds to a specific chromosome acceptor site. This results in the activation of a gene or a set of genes. The resultant protein gene products can be secreted into the circulation, can lead to cellular growth or new cellular activities, or can serve as receptor molecules for the steroid. (b) The clusters of black dots show sites where radioactively tagged progesterone, a steroid hormone related to estrogen, has bound to chromosomes in the nuclei of these cells in the cervix of a rat (magnified about 600 times).

steroid binds to specialized receptor molecules located in the nucleus much as a substrate binds to an enzyme or an antigen to an antibody (Figure 37-3). For example, cells of the uterus contain receptors for estradiol-17β, one of the estrogen family of hormones (Figure 37-2c). The bound estradiol forms a hormone–receptor complex, which is then chemically activated; then the complex is thought to attach to acceptor sites on the chromosomes. These sites may have specific nucleotide sequences, and they may be places to which specific proteins bind and, in turn, act as acceptor binding sites for the hormone–receptor complex. The result of the complex's binding to the chromosome is activation of specific genes that synthesize their messenger RNAs; subsequently, these mRNAs are translated into protein. Thus a target cell's specific response to steroid hormones is the production of specific proteins. Soon the hormone–receptor complex is broken down in the nucleus; the gene it activated then returns to its inactive state.

The exquisite specificity of steroid hormone action results from the two binding steps, each highly specific itself; first, the binding of hormone to receptor, and second, the binding of hormone–receptor complex to acceptor sites. Researchers are currently focusing on the proteins that bind to acceptor sites. In addition, investigators have recently discovered short stretches of DNA near hormone-activated genes that may function as highly specific sites where hormone–receptor complexes bind.

In mammals, steroids such as estrogen and progesterone also function in the brain, where they affect both development and function. Besides regulating genes, the steroids may alter the electrical activity of neurons involved in sexual behavior. Such effects in the brain appear to take place at the cell surface, where the steroids affect the neuron's polarization–repolarization behavior. As a result, neuronal circuits function differently, and this alters sexual behavior.

The Cellular Mechanisms of Second Messengers

At least one nonsteroid hormone (*thyroxine*, produced by the thyroid gland; Figure 37-2b) works inside target cells, much as steroids do. Both protein synthesis and cell-surface changes have been observed in different cell types that bind thyroxine. Most nonsteroids, however, have a very different mode of operation than that of steroids and thyroxine. As we see next, their regulatory message is carried into the target cell by chemical go-betweens, so-called second messengers.

Cyclic Nucleotides: The First Set of Second Messengers Discovered

In the early 1960s, Earl W. Sutherland, a scientist at Western Reserve University, won the Nobel Prize for discovering the second-messenger phenomenon. Sutherland knew that the hormone epinephrine (Figure 37-2b) stimulates liver cells to convert glycogen to glucose, which, in turn, fuels cellular metabolism. Sutherland found that instead of crossing the plasma membrane to bind to receptor molecules within the nucleus (as steroids do), epinephrine is able to transmit its regulatory information without entering the target cell.

When epinephrine binds to a receptor molecule in the plasma membrane, the receptor changes shape and binds the so-called G-protein. G-protein then binds the nucleotide GTP (guanosine triphosphate), which allows the G-protein–GTP complex to activate *adenyl cyclase*, an enzyme associated with the membrane. This enzyme, in turn, catalyzes increased production of *cyclic adenosine monophosphate*—or **cyclic AMP** (**cAMP**)—a relative of ATP that is normally found in cells in low concentrations (Figure 37-4a). Only a small number of bound hormone molecules is needed to cause catalytic enzymes to form a large amount of cAMP; in heart tissue, for

Figure 37-4   cAMP: HORMONAL ACTION BY MEANS OF A SECOND MESSENGER.

(a) The action of several hormones involves the following steps: after the hormone binds to the receptor molecule, the G-protein is activated; it, in turn, activates adenyl cyclase, which then converts ATP to cyclic AMP (cAMP). cAMP then influences particular cellular events, notably activation of one enzyme and inactivation of another. (b) The liberation of glucose in response to epinephrine during the fight-or-flight response involves a series of reactions that are initiated by the cAMP system. Note that the activated protein kinase A activates one enzyme (phosphorylase) and inactivates another (glycogen synthetase).

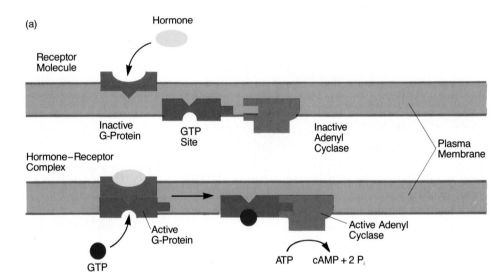

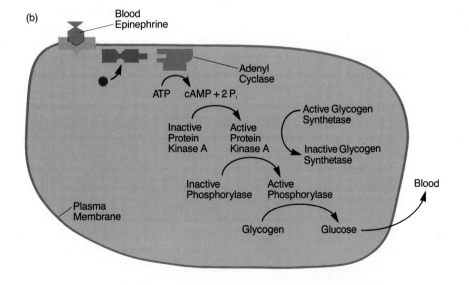

example, cAMP production quadruples within three seconds. Cyclic AMP is an amplified form of the hormone's message; hence, even a tiny amount of hormone can have major effects because of this cAMP amplification. Many hydrophilic hormones—such as proteins, polypeptides, and some amines—act by stimulating cAMP production.

How does cAMP transmit the hormonal signal? Within many types of cells, cAMP acts as an intermediary, or *second messenger*, setting in motion a cascade of chemical interactions (Figure 37-4b). First, cAMP combines with an inactive form of *protein kinase A*, a cytoplasmic enzyme; the cAMP binding activates protein kinase A so that it catalyzes addition of phosphate groups to other cellular proteins or enzymes. This phosphorylation may be the final step in the sequence; as it takes place, the phosphorylated proteins or enzymes are either activated or inactivated, and so carry out the hormone-specific responses. In some cases, however, the phosphorylation activates still another kinase that then activates another enzyme in the chain. This adds one more amplification step. An example of such an amplification cascade occurs after epinephrine binds to skeletal-muscle cells; cyclic AMP is produced and protein kinase A is activated. Protein kinase A phosphorylates glycogen phosphorylase, activating that enzyme to break down glycogen into glucose. Protein kinase A also phosphorylates glycogen synthetase, inactivating that enzyme and thereby preventing it from synthesizing glycogen from glucose. Both processes help to elevate the levels of free glucose in the muscle cells. Calculations of one such sequence indicate that an amplification of up to 10^8 may occur; in other words, the binding of even one molecule of epinephrine (or glucagon in another case) can produce 100,000,000 molecules of glucose!

After the second messenger has done its job, another enzyme, phosphodiesterase, degrades cAMP to AMP, which is used to resynthesize ATP. The cell thus returns to its original state, in which it can once again be sensitive to new hormone molecules that may bind to the cell surface.

Many different cell types can respond to a single non-steroid hormone. Although each type uses the cAMP second-messenger system, the cellular responses can be very different. This was proved by cell biologists who devised an ingenious experiment that demonstrated how different types of cells "interpret" cAMP messages differently. They began by placing heart-muscle cells in a culture dish and adding epinephrine. As expected, cAMP was produced, and the cardiac cells responded by contracting. In another dish, they cultured follicle cells from the ovary and added the pituitary hormone LH (luteinizing hormone; Chapter 18). Again, cAMP was produced, and, predictably, the cells synthesized a specific ovarian enzyme.

Next, heart and follicle cells were cultured together, and the two cell types formed gap junctions with each other. Now came the crucial part of the experiment: when epinephrine was added to the cell culture, the heart cells contracted, and the follicle cells produced the ovarian enzyme! When LH was substituted for epinephrine, the same result was obtained. This meant that when the heart and follicle cells were linked by gap junctions, cAMP manufactured by one cell type could cross the gap junctions to the other cell type and stimulate that cell's internal machinery to produce the appropriate response. With unfailing precision, the cascading sequence of chemical interactions that is unique to each cell type determines how cells respond to the commands issued by nonsteroid hormones and delivered within the cell by cAMP.

In recent years, another second-messenger compound, *cyclic guanosine monophosphate*—or **cyclic GMP (cGMP)**—has been discovered in several cell types, including heart and liver cells. In heart cells, where epinephrine speeds up muscle-cell contraction by way of cAMP, cGMP is a second messenger for the neurotransmitter acetylcholine, and has the opposite effect. Thus when acetylcholine binds to receptor molecules on the cell membrane, cGMP is produced and cell contractions are slowed. Here we see a molecular explanation for the antagonistic actions of parasympathetic and sympathetic nerves on the heart's contractile rate (Chapter 36); different second messengers speed or slow that rate.

Inositol Triphosphate: The New Second Messenger

The early 1980s brought the discovery of a second-messenger system that may be even more important than cAMP, since it helps trigger cell responses to a variety of neurotransmitters, hormones, and growth factors. Numerous substances bind to receptor molecules, with the common effect of releasing Ca^{2+} ions in the cytoplasm. The list includes the neurotransmitters acetylcholine and epinephrine, the hormone vasopressin, substance P (a neuropeptide), epidermal growth factor (Chapter 17), stimulators of mitosis, and many other hormones. Researchers incorrectly thought that calcium was the second messenger in these instances. In fact, the second-messenger role is filled by **inositol triphosphate,** or **IP_3** (inositol is a six-carbon sugar related to glucose).

The plasma membranes of cells contain small amounts of a phospholipid called phosphatidylinositol 4,5-bisphosphate, or PIP_2. When epinephrine, for example, binds to what is called an alpha receptor in the plasma membrane, PIP_2 is immediately cleaved into IP_3 plus **diacylglycerol,** or **DG.** Both substances act as second messengers (Figure 37-5).

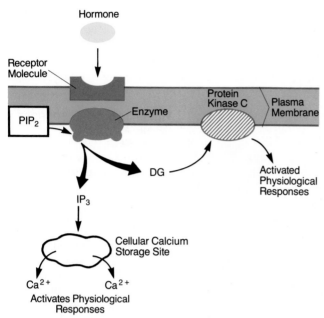

Figure 37-5   IP₃ AND DG: THE NEW SECOND MESSENGERS.

When the hormone or other signaling molecule binds to its receptor molecule, an enzyme cleaves PIP_2 into the second messengers IP_3 and DG. IP_3 liberates Ca^{2+} ions from storage sites, while DG activates protein kinase C. A cascade of cellular events can follow: division, motility, secretion, and so on.

IP_3 instantly triggers the release of Ca^{2+} from cellular storage sites, and that ion acts as a third messenger, causing all sorts of responses, including contraction of the actin and myosin systems, polymerization of macromolecules, and activation of enzymes. Meanwhile, DG activates *protein kinase C*, which phosphorylates a variety of proteins and so activates a different set of responses than those brought about by cAMP and protein kinase A. For instance, insulin secreted by beta cells in the pancreas, aldosterone secreted by the adrenal cortex, and epinephrine secreted by the adrenal medulla are all activated by protein kinase C. (We will encounter each of these hormones later in this chapter.)

The discovery of IP_3 and DG as second messengers helps explain important events in the activation of sperm and eggs (Chapter 16), in the stimulation of cancer cells to divide, and in important aspects of brain function. Furthermore, PIP_2 is cleaved in response to electrical stimulation of cells, as by nerve impulses, as well as in response to hormonal binding to an alpha receptor. This new second-messenger system takes a place with cAMP, and together they enable biologists to understand how most molecular and electrical signals activate specific responses in cells. In the larger sense, these and other discoveries of hormone action revealed the molecular

basis of communication between body cells and regulatory systems. In the next two sections, we explore such regulation and see the ways in which hormones exert control over development, physiology, and behavior. We will begin with what is known about endocrine functions in the invertebrates.

ENDOCRINE FUNCTIONS IN INVERTEBRATES

Hormones control a host of fundamental biological events among the invertebrates, from the periodic molting of exoskeletons to complex reproductive cycles and the astonishing metamorphoses of insects. Invertebrates have both steroid and nonsteroid hormones, the first generally acting at the level of genes and specific protein synthesis, and the second, via cAMP. In fact, it was work on insect steroid hormones, which regulate development, that provided biologists with some of the earliest and best evidence about how steroids act at the level of genes.

Studies of how the silkworm (*Hyalophora cecropia*) metamorphoses into an adult moth provide a particularly clear demonstration of how hormones direct this complex transformation process in an insect. Recall from Chapter 16 that certain insects develop into caterpillars— larvae that feed, grow, and pass through a series of stages, or instars. Finally, they pupate, enter a cocoon, and undergo the radical developmental changes that produce a moth or butterfly. All of these processes are under hormonal control as established by experiments such as that shown in Figure 37-6. Metamorphosis from silkworm pupa to adult moth is controlled by *brain hormone*. This hormone is produced by neurosecretory cells of the brain; such cells in invertebrate and vertebrate brains are electrically excitable and can conduct action potentials; yet the "transmitter" they secrete is a hormone, not a conventional neurotransmitter. When a silkworm secretes brain hormone, the substance sets in motion a series of chemical changes. Brain hormone directly stimulates the *prothoracic gland*, which, in turn, secretes a steroid hormone of the **ecdysone** type called *alpha ecdysone* (Figure 37-2c). Alpha ecdysone is then converted to 20-hydroxyecdysone, which triggers the cellular changes of metamorphosis, even in an isolated abdomen (Figure 37-6). Recall imaginal disks for eyes, wings, legs, genitals, and so on; they are the cell populations triggered to develop by 20-hydroxyecdysone.

It turns out that ecdysones trigger molting (the change from one instar to the next), but that another hormone actually determines what the larva changes into. To understand this process, let us consider the earlier stages of

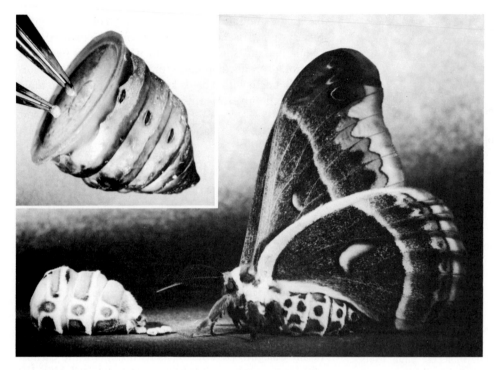

Figure 37-6   THE SOURCES OF METAMORPHIC HORMONES.
(a) Researchers insert pieces of tissue that are sources of brain hormone and ecdysone into a pupal abdomen; the abdomen is then sealed with a plastic plate. (b) The isolated abdomen develops into a perfectly normal posterior section of an adult female. An adult male has been attracted to the abdomen's pheromone. Mating occurred, and the abdomen seen here is laying fertilized eggs.

insect development, as the growing larva passes from one instar to the next. The *corpora allata* are paired endocrine glands located behind the brain that secrete **juvenile hormone (JH),** a terpene derivative, during the caterpillar instar stages. A sufficiently high level of circulating JH prevents metamorphosis from beginning as the larva molts from one instar to the next in response to ecdysones. As long as the levels of both hormones remain high, larva-to-larva molts will occur (Figure 37-7). When other metabolic changes cause the JH level to fall below a certain threshold, but ecdysone levels still remain high, the larva pupates. In the pupa, JH secretion stops altogether; but since ecdysone (and therefore 20-hydroxyecdysone) is still abundant due to the presence of its regulator, brain hormone, the pupa finally develops into an adult. Note that ecdysone (that is, 20-hydroxyecdysone) causes the series of molts to occur, whereas the levels of juvenile hormone determine which pattern of development will take place during those molts—to larva (high JH), to pupa (low JH), or to adult (no JH).

Many insect species follow this hormonally directed developmental pathway; if it is disrupted, they cannot complete their life cycle. In Chapter 47, we shall see that certain plants have evolved defenses against harmful insects that exploit the insects' developmental pattern; the plant cells manufacture and store chemical analogues of insect hormones, which if ingested by a feeding caterpillar prevent that predatory larva from developing into a reproductive adult.

The ecdysone family of hormones is produced by all insects and by many other arthropods; for example, 20-hydroxyecdysone also stimulates molting in shrimp and other crustaceans. That process goes on at periodic intervals of adult life and permits the animal to grow during the brief period between the shedding of the old, hard exoskeleton and the hardening of the new, larger one.

In addition to the hormones identified in insects and arthropods, biologists have discovered a variety of other hormones in a broad spectrum of invertebrates and other organisms. In particular, a number of peptides now known to be secreted by human and other vertebrate brain neurons as neurotransmitter peptides (Chapter 36) are especially widespread. Protozoans, simple invertebrates such as sponges, some plants, and perhaps even some prokaryotes have their life processes influenced in one way or another by such chemicals. This new work shows that the hormones we know best in vertebrates evolved originally as regulatory molecules in much simpler organisms, where their original tasks were, no doubt, quite different.

THE VERTEBRATE ENDOCRINE SYSTEM

Vertebrates have the most complex, but the best understood, systems of hormonal control. For many years, vertebrate endocrine systems, including that of humans,

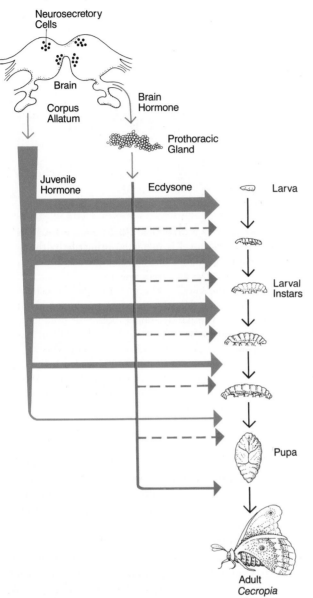

Figure 37-7   HORMONE CONTROL OF INSECT METAMORPHOSIS.

In the final metamorphic event, the pupa is transformed into an adult. This is due to the action of ecdysone, produced by the prothoracic gland in response to brain hormone. The ecdysone is present during much of a larva's life and is responsible for molts—from larva to larva, larva to pupa, and pupa to adult. But it is the level of juvenile hormone that determines what the molting-stage insect develops into. The width of the orange arrows reflects juvenile-hormone levels. Note that the amount of juvenile hormone is less over time; thus the pupa and then the adult develop.

were thought to consist of a series of separate glands (the pancreas and the thyroid, for example) all controlled by a "master" gland (the pituitary). Today biologists recognize, however, that the endocrine system is not so straightforward. As the earliest vertebrates evolved, hormone-producing cells and tissues developed and came to be controlled in several ways. Some, including the thyroid, are directed by sets of nerve cells in the brain, while others function independently of either nerves or the pituitary. In addition, certain hormones once thought to be made exclusively by endocrine organs (insulin is one example) are now known to be manufactured by brain cells as well. Furthermore, brain cells secrete literally dozens of agents that act as neurotransmitters, and other body tissues produce complex substances that engage in hormonelike activity. We will consider some of these substances and their effects on organisms at the end of this chapter.

Two categories of structures are traditionally considered to be part of the endocrine system: (1) organs that combine endocrine functions with other biological activities, and (2) organs whose only function is to secrete hormones. As we survey these organs, we shall look first at the multipurpose organs, such as the pancreas and kidneys, and then at structures that function exclusively in hormonal control, such as the thyroid and pituitary. Finally, we should keep in mind that a portion of the brain, the hypothalamus, both manufactures some classic hormones and indirectly regulates the secretion of others. For this reason, and for its secretion of numerous neuropeptide hormones, the brain may properly be considered part of the endocrine system.

The Pancreas

As we saw in Chapter 34, the *pancreas* is an organ near the stomach that secretes digestive enzymes and bicarbonate ions into the intestine. In fact, it is a dual-purpose organ that has endocrine functions as well as other activities. In the late nineteenth century, a German medical student named Paul Langerhans noted that spherical clusters of cells and adjacent blood vessels were scattered throughout the pancreas. These **islets of Langerhans,** as they came to be called, are composed of endocrine cells that produce **insulin** and **glucagon,** hormones that regulate the glucose levels in the blood, and **somatostatin,** a hormone that inhibits the secretion of insulin and glucagon and glucose absorption by the intestine (Figure 37-8).

Insulin (Figure 37-2a) was first isolated in the 1920s by Frederick Banting and Charles Best at the University of Toronto. Their research showed that dogs with diabetes—a disease characterized by high levels of blood glucose—could be successfully treated with an extract derived

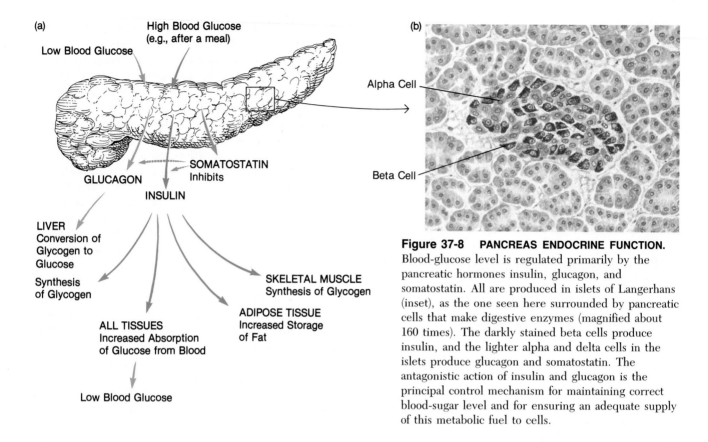

Figure 37-8   PANCREAS ENDOCRINE FUNCTION.
Blood-glucose level is regulated primarily by the pancreatic hormones insulin, glucagon, and somatostatin. All are produced in islets of Langerhans (inset), as the one seen here surrounded by pancreatic cells that make digestive enzymes (magnified about 160 times). The darkly stained beta cells produce insulin, and the lighter alpha and delta cells in the islets produce glucagon and somatostatin. The antagonistic action of insulin and glucagon is the principal control mechanism for maintaining correct blood-sugar level and for ensuring an adequate supply of this metabolic fuel to cells.

from ground-up and filtered islet tissue. Why was this treatment effective? When an animal eats, it digests and absorbs nutrients, and the level of glucose in its blood rises. Normally, *beta cells* in the pancreatic islets of Langerhans respond to an elevated level of blood glucose by secreting insulin, which facilitates the passage of glucose molecules into cells, where they fuel cellular metabolism. Insulin also acts on the liver, where it increases the rate at which glucose is converted into glycogen and at which amino acids are converted into protein. For this to occur, cAMP must be produced in liver cells, followed by phosphorylation of enzymes by protein kinase A so that less glycogen is hydrolyzed to glucose.

Insulin causes the glucose levels to fall. Conversely, when the level of glucose in the blood falls below a certain threshold, pancreatic *alpha cells* begin to secrete glucagon, which accelerates the conversion of liver glycogen to glucose. Once again, cAMP is synthesized in response to this protein hormone, and, as a result, appropriate enzymes are activated. Insulin and glucagon are thus antagonists in that they have opposite effects. The third pancreatic hormone, somatostatin, is secreted by *delta cells*. It inhibits the secretion of both insulin and glucagon and absorption of glucose across the intestinal wall. Somatostatin is released when glucose or amino-acid levels in the blood rise, and by its effects, lowers those levels. If an antibody that binds and thereby inac-

tivates somatostatin is injected into the bloodstream, the level of insulin in the blood rises; this implies that under normal circumstances, somatostatin holds insulin secretion in check. We will see later that somatostatin is also a brain neuropeptide that may act to regulate the levels of glucose in the brain.

Glucose levels in blood are regulated by both positive- and negative-feedback loops that involve the hypothalamus as well as the pancreas. Hypothalamic glucose-sensitive neurons cause signals to be sent along the vagus (parasympathetic) nerves from the brain to the insulin-producing beta cells in the pancreas. Vagal stimulation increases insulin secretion. Other sympathetic pathways go from the hypothalamus via the spinal cord to the islets of Langerhans. Activity of those neurons causes alpha cells to secrete glucagon and beta cells to cease insulin secretion. Finally, both types of nerves may affect the delta cells that secrete somatostatin.

When any of these complex mechanisms goes awry and the pancreas does not secrete sufficient insulin, the level of glucose in the blood and urine rises. The result is *diabetes mellitus*, the condition that Banting and Best observed in dogs. In fact, diabetes can also result if an enzyme destroys insulin too rapidly or if insulin receptors are abnormally few in number. Diabetics tend to lose body water because of excessive urination, to lose weight, and to become exhausted easily; if young, they may face severe problems, such as blindness. Human diabetics usually must have insulin injections or take oral doses of other recently developed drugs that act like insulin to cause glucose to move into body cells so that glucose levels fall in the blood. This alleviates many symptoms of the disease.

The Kidneys

Vertebrate kidneys, like the pancreas, are multipurpose organs. As we discussed in Chapter 35, the *kidneys* not only filter the blood and manufacture urine, but also secrete hormones with strikingly different functions. One of these is **erythropoietin,** a protein hormone that stimulates the production of red blood cells in bone marrow. If a mammal loses more than a small amount of blood, the levels of erythropoietin in its bloodstream rise, and this triggers an increase in the production of red blood cells over the subsequent few days.

The kidneys also contain enzymes that convert vitamin D_3 in the blood into a hormone with the cumbersome name **1,25-dihydroxy-chole-calciferol.** This substance regulates the uptake of calcium in the small intestine; without it, calcium from food present in the chyle cannot be absorbed (see the box in Chapter 34, page 837).

A third hormone is produced indirectly when certain cells in the kidney cortex respond to low levels of salt (hypotonicity) in the blood; the cells secrete the enzyme *renin* into the blood (Figure 35-13). Renin boosts the production of a polypeptide, *angiotensin II*, which, in turn, causes cells in the cortex of the nearby adrenal glands to secrete the steroid hormone **aldosterone** (Figure 37-2c). As we saw in Chapter 35, aldosterone completes this circuit by increasing the amount of sodium (and hence water) recovered from the urine by cells of the distal convoluted tubules of nephrons. Thus the kidney's renin-secreting cells regulate the activity of its tubule cells by indirect means that involve substances in the blood and a hormone made by the adrenals.

The Heart as an Endocrine Organ

The heart has never been considered a source of hormones, but research in recent years has revealed that heart cells produce substances that increase urine production. Muscle cells in the atrial walls secrete *atrionatriuretic factor (ANF)*, short peptides that act on the kidney to cause copious excretion of sodium and water (Chapter 35). To do so, they compete directly with the effects of antidiuretic hormone (ADH) on collecting ducts and also they cause dilation of blood vessels lead-

ing to glomeruli. Both effects lower blood pressure; thus it should be no surprise that ANF hormones are released when blood pressure is too high. The prime signal apparently is the extra stretching of atrial walls due to distention when blood volume (pressure) is too great. Together, ANFs and ADH either raise or lower blood pressure and volume to help maintain fluid homeostasis— a key to the organism's internal physiology. ANFs also act in the brain to help govern the volume of fluid in the cavity of the central nervous system.

The Adrenal Glands

Resting anterior to each kidney is an **adrenal gland** (from the Latin word *ad*, meaning "toward," and "renal," pertaining to the kidney). Each adrenal has an outer *adrenal cortex* and an inner core, the *adrenal medulla*. Both are endocrine tissues, the cortex secreting dozens of steroid hormones and the medulla, two kinds of amine hormones. These hormones are involved in a wide spectrum of regulatory processes. And, as we will see, secretion by the cortex is controlled by chemical signals, whereas the medulla receives direct nerve input.

Hormones of the Adrenal Cortex

The adrenal cortex arises from the same populations of cells that form the kidney; yet the cortex cells produce steroid hormones, known as **corticosteroids** (Figure 37-9). These are divided into three distinct functional groups. One group, the **glucocorticoids,** stimulates the production of glucose from proteins and carbohydrates. In mammals, the most prevalent glucocorticoid is *cortisol*. Glucocorticoids, in conjunction with insulin, are essential to health because they stimulate glucose production in the liver and elsewhere and help to maintain a constant level of glucose in the blood. The glucocorticoids differ from insulin, however, in their long-term action; even when an animal is not taking in food, they ensure that an adequate supply of the sugar will be available as a cellular fuel for ATP production. Not surprisingly, a human who lacks glucocorticoid hormones develops *Addison's disease*, an illness characterized by weakness and weight loss. Glucocorticoids perform other functions as well, including increasing the amount of fat stored in the body and reducing inflammation. *Cortisone* is a well-known anti-inflammatory derivative of cortisol that is used to treat swelling, athletic injuries, arthritis, insect bites, and even poison ivy.

The second group of corticosteroids is the **mineralocorticoids.** We have already seen that the production of one mineralocorticoid, aldosterone, is regulated by a kidney enzyme. In addition to stimulating kidney tissue

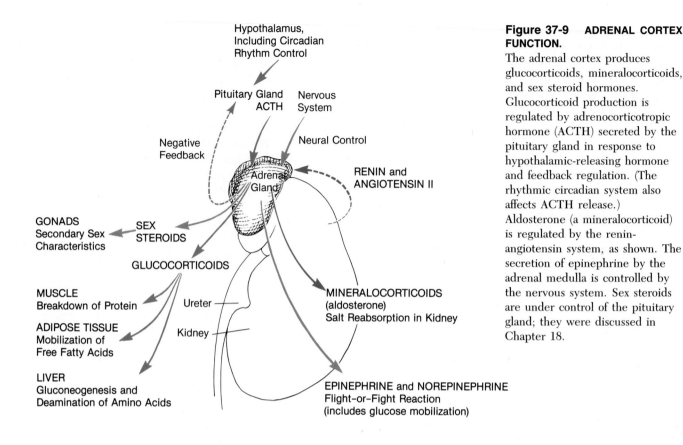

Figure 37-9   ADRENAL CORTEX FUNCTION.

The adrenal cortex produces glucocorticoids, mineralocorticoids, and sex steroid hormones. Glucocorticoid production is regulated by adrenocorticotropic hormone (ACTH) secreted by the pituitary gland in response to hypothalamic-releasing hormone and feedback regulation. (The rhythmic circadian system also affects ACTH release.) Aldosterone (a mineralocorticoid) is regulated by the renin-angiotensin system, as shown. The secretion of epinephrine by the adrenal medulla is controlled by the nervous system. Sex steroids are under control of the pituitary gland; they were discussed in Chapter 18.

to retain sodium ions, aldosterone triggers the excretion of potassium. An excess of aldosterone causes edema (water retention in tissues), high blood pressure, and a severe loss of potassium. If potassium loss continues over a prolonged period, paralysis results as the supply of potassium in muscles is depleted (muscle cells, like neurons, are electrically excitable [Chapter 36]; hence, they need K^+ to maintain a normal state of polarization). Conversely, if aldosterone levels are low, an animal's blood pressure drops precipitously as it eliminates large quantities of urine; the water in the urine comes from the blood so that blood volume falls. Untreated, an aldosterone deficit quickly leads to death.

The last group of corticosteroids secreted by the adrenal cortex is the sex steroids—the *androgens* (male sex hormones) and *estrogens* (female sex hormones). As described in Chapter 18, these steroids affect initial development of sex-related organs and secondary sex characteristics. The hormones also act during and after puberty to regulate maturation and sexual activity. Under normal conditions, the sex steroids are produced mainly by the gonads. On occasion, the adrenal gland overproduces androgens or estrogens. If androgens are produced in a female embryo, the individual often develops a mixed biological gender, with both male and female characteristics. In preadolescent years, the effect of producing too many adrenal sex steroids is precocious puberty and the

development of malelike characteristics. Finally, adult women who suffer from adrenal abnormalities often develop heavy beardlike facial hair.

Hormones of the Adrenal Medulla

Located inside the adrenal cortex, somewhat like the pit of a peach, is the adrenal medulla. This structure has a different developmental origin from the cortex; arising from migratory neural crest cells (Chapter 17), it may really be viewed as an addendum to the sympathetic nervous system (Chapter 36). Its endocrine cells, called **chromaffin cells,** secrete primarily **epinephrine** (adrenalin; Figure 37-2b) and some **norepinephrine** (noradrenalin, which is the common neurotransmitter of the sympathetic nervous system). When released by neurons at synapses, these amine hormones serve as neurotransmitters. When released as hormones by adrenal medulla cells, they produce the so-called fight-or-flight reaction. All humans have experienced the fight-or-flight syndrome in frightening situations: the heart pounds as its rate increases; respiration speeds to supply more oxygen to the muscles, brain, and heart; the pupils dilate to enhance vision; and other more subtle changes occur. Exactly what mechanisms produce these responses? When any vertebrate reacts to an emergency, nerve impulses from centers in the brain stimulate the adrenal medulla to release epinephrine into the blood. The epinephrine then mobilizes the tissues and organs crucial to self-defense or escape. Epinephrine has opposite effects on different tissues, mobilizing some and inactivating others. Whereas blood vessels leading to the brain and skeletal muscles dilate, blood vessels in the skin and kidneys constrict. Consequently, more blood is diverted to the brain, muscles, and rapidly beating heart. These differences are due to different receptors and transduction machinery in the target cell types; as a result, one hormone can act in opposite ways. Although the adaptive advantage of the fight-or-flight response is obvious for animals in nature, the response can be disadvantageous in human society. Studies show that for some people, the tensions of modern life can result in a chronic state of the fight-or-flight response. Because of the sustained secretion of epinephrine, many people develop high blood pressure and other symptoms of unrelenting stress.

The Thyroid and Parathyroid Glands

Thousands of years ago, a Chinese physician treating a patient for a chronically swollen neck—an enlarged thyroid gland, or **goiter**—would prescribe a concoction of wine mixed with charred sea sponge and seaweed, to be drunk at regular intervals. This regimen worked because it corrected a dietary deficiency of iodine, which had caused the patient's thyroid gland to enlarge. The development of a goiter is due to the fact that iodine is an important component of thyroid hormones. The **thyroid gland** is a butterfly-shaped organ located near the esophagus and trachea in air-breathing vertebrates. It is derived during embryonic development from the embryonic gill arch system, and some biologists argue that it is an evolutionary derivative of the endostyle of primitive chordates (Chapter 26).

The thyroid produces the hormones **triiodothyronine** (with three iodine atoms), or simply T_3; **thyroxine** (with four iodine atoms), or T_4 (Figure 37-2b); and **calcitonin**, a protein hormone involved in calcium balance (Figure 37-10). The T_3 and T_4 hormones are derivatives of the amino acid tyrosine, and seem to act similarly on target cells.

Thyroxine has been studied intensively for decades. One of its primary functions is to set an animal's metabolic rate by regulating the rate at which cells consume O_2 in order to synthesize ATP. It also speeds the breakdown of complex carbohydrates and the synthesis of proteins, and slows the rate at which fats are burned. In addition, thyroxine increases heat production in mammals by activating the ATPase pump enzymes on certain cell surfaces; as a result, much ATP is cleaved, and heat is released as a useful by-product.

Both T_3 and T_4 are secreted in response to a third hormone, known as **thyrotropic hormone,** or **TH** (or **TSH,** from its former name, thyroid-stimulating hormone). TH is produced in the anterior pituitary gland, and travels through the blood before binding to receptor molecules on thyroid cells. The regulation of T_3 and T_4 secretion involves a negative-feedback loop that is linked to the levels of thyroid hormones in the blood. For example, when blood levels of thyroxine rise, TH secretion decreases; T_3 and T_4 secretion falls as a result. When blood levels of thyroxine are low, more TH is secreted and thyroid cells are stimulated to increase their output of T_3 and T_4. In addition, the brain can cause TH release and thereby thyroid-hormone secretion; this occurs, for example, when the body needs to produce more heat.

If the control system does not work correctly, the levels of thyroid hormones in the blood can be too high—a *hyperthyroid* condition—or too low—a *hypothyroid* condition. In the first, cellular metabolism accelerates, causing the affected person to lose weight as carbohy-

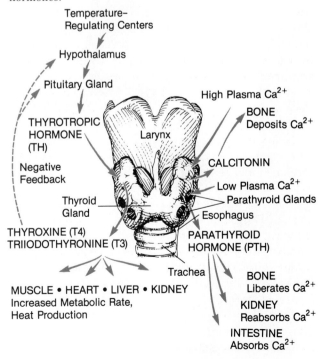

Figure 37-10   THYROID AND PARATHYROID FUNCTION.
Metabolic rate (and heat production in mammals) is regulated by the thyroid hormones which are released into the bloodstream in response to thyrotropic hormone (TH) from the pituitary gland. Plasma calcium levels are regulated through the action of parathyroid hormone (PTH), which is produced by the parathyroid glands, and calcitonin, which is produced by specialized cells that lie between the thyroid follicles. Note that the hypothalamus and neurons are not involved in calcium regulation by those hormones.

drates and fats are consumed. Such people may be hyperactive. Exactly the opposite occurs when there is a thyroid deficit: metabolic and heart rates plummet, and the failure to burn food results in weight gain and sluggishness. Hypothyroidism in a developing fetus, a genetic abnormality that occurs in roughly 1 of every 8,500 births, can result in a condition called *cretinism*. Afflicted children suffer greatly retarded physical and mental growth.

Goiter is another response to a disturbance in the thyroid–pituitary negative-feedback loop, in this case caused by an iodine deficiency. Without iodine, the thyroid gland cannot synthesize T_3 and T_4. In the absence of the negative feedback that results from these hormones circulating in the blood, the anterior pituitary continually secretes a large amount of TH. Such an "overdose" of TH, in turn, causes the thyroid gland to enlarge, as shown in Figure 37-11. The advent of iodized table salt has made this once-common ailment relatively rare.

What about the thyroid's third hormone? In the early 1960s, researchers discovered another function of the thyroid gland. Small clusters of specialized cells in the thyroid secrete calcitonin, a protein hormone that decreases the level of calcium in the blood. Calcitonin acts

Figure 37-11   THYROID GOITER.
The thyroid is enlarged greatly because of continuing TH release from the anterior pituitary, despite the inability of the thyroid to make its hormones.

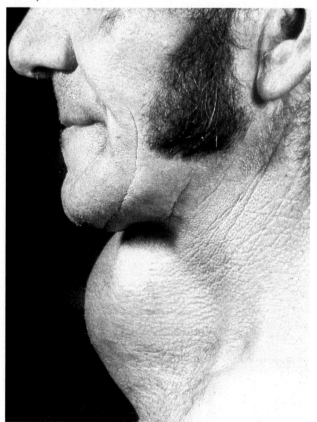

by inhibiting the activity of *osteoclasts*—cells that normally break down bone and release calcium (bone is composed in large part of calcium phosphate salts; Chapter 39). In addition, calcitonin stimulates bone cells (osteoblasts) to bind calcium ions into bone salts (Figure 37-10).

Just the opposite occurs when hormones of the **parathyroid glands** are secreted. These four tiny kidney-shaped glands situated on the surface of the thyroid (Figure 37-10) secrete **parathyroid hormone (PTH)**. PTH, acting when the calcium level in the blood falls, increases both the number and the activity of osteoclasts. They degrade bone salts, freeing Ca^{2+} into the blood plasma. The parathyroids and the calcitonin-secreting cells of the thyroid thus carry out a hormonal tug of war (with no involvement of the nervous system or the pituitary), keeping blood levels of calcium remarkably constant in the process. PTH acts when the levels fall; calcitonin, when they rise.

The Pituitary Gland

No gland in the body has received more research attention than a small ball of tissue on the lower surface of the brain. This **pituitary gland** is nestled in a cup-shaped depression at the base of the skull. Despite the fact that it is no larger than a small marble (about 1.3 cm in diameter) in humans, the pituitary has traditionally been viewed as the vertebrate's "master" endocrine gland because it was thought to act much like a central switching station, regulating the activities of other endocrine glands. More recent research has shown, however, that the pituitary is more like a middleman, receiving regulatory orders directly from the brain and transmitting them to other endocrine organs. It consists of two lobes—the *anterior pituitary* and the *posterior pituitary* (Figure 37-12)—that together produce at least ten hormones. The posterior pituitary arises directly from the embryonic brain wall; it should not be surprising, therefore, that it is a site of neurosecretory cell endings. In contrast, the anterior pituitary arises from the embryo's mouth epithelium; hence, it is not nervous tissue and must receive normal hormonal signals to do its work.

Hormones of the Anterior Pituitary

Hormones are secreted by the anterior pituitary in response to signals received from the hypothalamus. We will discuss that phenomenon later. Here we will concentrate on the anterior-pituitary hormones themselves. These substances either regulate hormone secretion by other organs or act on target tissues directly. They comprise three families of chemically related molecules (Table 37-1). Hormones in each family share amino-acid sequences and reflect the high probability that one an-

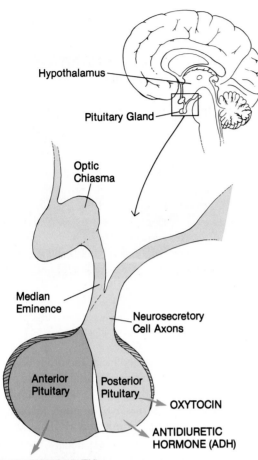

Hypothalamus

Pituitary Gland

Optic Chiasma

Median Eminence

Neurosecretory Cell Axons

Anterior Pituitary

Posterior Pituitary

OXYTOCIN

ANTIDIURETIC HORMONE (ADH)

THYROTROPIC HORMONE (TH)
FOLLICLE-STIMULATING HORMONE (FSH)
LUTEINIZING HORMONE (LH)
PROLACTIN
GROWTH HORMONE (GH)
ADRENOCORTICOTROPIC HORMONE (ACTH)
MELANOPHORE-STIMULATING HORMONE (MSH)
LIPOTROPIN (LPH)

FIGURE 37-12   PITUITARY GLAND FUNCTION.
The pituitary rests in a bony cavity at the base of the skull. Its secretions are controlled by the hypothalamus of the brain. The pituitary has two distinct regions, the anterior pituitary and the posterior pituitary. The posterior pituitary is connected directly to the hypothalamus and is true brain tissue. The anterior pituitary is not brain tissue and is controlled by chemicals carried in the blood from the nearby brain. Both anterior and posterior pituitary hormones are included.

cestral gene for each family evolved to yield the other molecular types. The first family includes the glycoprotein hormones: thyrotropic hormone (TH) and the gonadotropins—*follicle-stimulating hormone (FSH)* and *luteinizing hormone (LH)*. We have already seen how TH regulates thyroid function. Egg production in ovaries and sperm production in testes are triggered directly by FSH and LH. These hormones also regulate the secretion by the gonads of steroid hormones, including estro-

gens and progesterone (high levels in females) and testosterone (high levels in males).

The second family of protein hormones includes prolactin and growth hormone. All vertebrates produce **prolactin,** a substance with various functions in different types of animals. Originally, prolactin may have evolved as a regulator of water and salt balance in fish, a task it still performs. Among birds, it acts during the reproductive season. In some birds, esophageal cells are stimulated by prolactin to produce a cheesy "crop milk," which is fed to hatchlings. In mammals, different sets of cells in the anterior pituitary secrete four prolactin proteins that differ slightly in structure. In females, prolactin stimulates the synthesis and secretion of milk from mammary-gland tissue that has first been "primed" by estrogen and other steroid hormones secreted during pregnancy (Chapter 18). (In some mammals, although probably not in humans, prolactin interacts with LH to help maintain the corpus luteum in the ovary.) Interestingly, prolactin has been found to affect the functioning of human testes as well as ovaries. In some cases, both infertile men and women have been found to have high levels of prolactin in their blood, and both have often responded favorably to a drug (bromocryptine) that inhibits prolactin secretion.

The anterior pituitary also secretes **growth hormone (GH),** which acts both in adults and in maturing animals. GH affects cells in adult, nongrowing tissues in several ways: it augments the incorporation of amino acids into protein, promotes the use of fats instead of carbohydrates for energy, and causes the liver to catabolize glycogen to glucose, thereby raising blood-glucose levels for metabolism. But the hormone's best-known effect—stimulating growth during maturation—occurs indirectly, as it triggers cells in the liver to secrete small peptide hormones called **somatomedins.** One type of somatomedin stimulates the formation of cartilage, especially in the so-called growth plates near the ends of long bones in arms and legs. This is the basis for growth in such bones and occurs during the "growth spurts" that all land vertebrates experience in adolescence. Another type of somatomedin promotes cell division in connective tissues so that they may grow, too. Somatomedins and insulin have quite similar amino-acid sequences, suggesting that the genes for these hormones may have arisen from a single ancestral gene.

Abnormal levels of GH during maturation are responsible for growth abnormalities in humans and other animals. Individuals with a GH deficiency, on the one hand, will be abnormally short, a condition called *pituitary dwarfism.* An excess of GH, on the other hand, causes a pituitary disorder called *pituitary gigantism,* which is characterized by great height but normal body proportions. When surplus growth hormone is secreted after the growth spurt of puberty, the result is *acromegaly,* a condition in which the bones and tissues of the face

SYNTHETIC SOLUTIONS TO GROWING UP SHORT?

A recent advance in genetic engineering—the production of synthetic human growth hormone (HGH)—is bringing new hope to the parents of extremely short children and new inches to those sometimes unhappy kids. At the same time, however, it raises a key pediatric issue. Synthetic human growth hormone works for many children, but is it safe enough? How short must a child be—and by whose standards—to take this compound? And, finally, what about the psychological impacts of this treatment?

Beginning in 1958, natural human growth hormone was painstakingly extracted from human pituitary glands removed from cadavers during autopsies. A child treated with the hormone could expect to grow from 2 to 6 inches taller per year. However, a two-year supply required the extract from 50 to 100 pituitary glands. Thus the hormone was precious and expensive and was given only to "hypopituitary dwarfs"—children with a demonstrated deficiency of their own growth hormone. This left untreated thousands of so-called growth-delayed children—those with no certain hormone deficiency, but with a sluggish growth response that put them in the lowest height percentiles for their age groups (Figure A).

An important change came in the mid-1980s, when a California biotechnology firm succeeded in isolating the gene for human growth hormone, cloning it in bacteria, and producing enough synthetic HGH to market as a pharmaceutical. Clinical trials with the substance showed that hypopituitary-dwarf children grew an average of 4 inches the first year and 3.5 the second, just as they would have if administered natural HGH. Growth-delayed short children grew an average of 1.2 additional inches per year. After two years on synthetic HGH, one third-grade boy had grown 4 inches—but still remained 7 inches shorter than most of the other boys his age, who, of course, kept growing at their normal, faster rates. Thus the treatment is helpful, but only to a point. Many pediatricians suspect that it will accelerate a child's growth but not add inches to his or her predicted maximum adult height, which is based on genes inherited from the parents (Chapter 15). Nevertheless, a very short, slow-growing child may never even approach that predicted height without hormonal treatment and resulting accelerated growth.

Clearly, the availability of synthetic HGH (plus continued supplies of natural HGH) will help many children who might have received no treatment. But what about the possible side effects of such treatment? Gigantism (uncontrolled growth) or acromeg-

Figure A
Growing up: a delicate time psychologically and physically. Correct HGH levels are essential, for body size and proportions will be affected.

aly (abnormally enlarged facial and other bones) is one worry. Another involves potential immune problems. Children given synthetic HGH develop antibodies to it. Conceivably, a massive immune response could block the action of the injected HGH as well as the child's own low natural levels.

And who should receive HGH? Some pediatricians feel that only hypopituitary dwarfs should be treated. Others think that any growth-delayed child below the third percentile for height should also be treated. Still others see this cutoff as arbitrary and wonder why not treat those in the fourth percentile, the tenth, the twenty-fifth—or, indeed, anyone who wants treatment?

There are serious psychological considerations, as well. Many of the children and their parents were happy and satisfied with synthetic-HGH treatment; however, some patients had inflated expectations and reportedly felt like "failures" when treatment only brought them from diminuitive to the short side of normal.

Obviously, a great many ethical issues remain to be pondered about the use of synthetic HGH. However, the application of new medical advances rarely waits for such careful consideration. Observes one professor of pediatrics, "By the time there is enough information to ask, 'Should we be doing it?' we're doing it."*

*Dr. Walter Miller, associate professor of pediatrics, University of California at San Francisco, quoted in *West Magazine (San Jose Mercury News)*, November 11, 1984, p. 35.

Table 37-1   **HORMONES OF THE ANTERIOR PITUITARY**

Hormone	Structure	Target	Action	Regulation*
Thyrotropic hormone (TH)	Glycoprotein	Thyroid gland	Increases release of thyroxine	TRH stimulates release; thyroid hormones inhibit release
Follicle-stimulating hormone (FSH)	Glycoprotein	Seminiferous tubules (male), ovarian follicles (female)	In male, causes production of sperm; in female, stimulates maturation of follicle	Gn-RH stimulates release
Luteinizing hormone (LH)	Glycoprotein	Interstitial cells of ovaries or testes	In female, induces final maturation of follicle, ovulation, and formation of corpus luteum; in male, secretion of androgens	Gn-RH stimulates release
Prolactin	Protein	Mammary glands	Increases synthesis of milk proteins and growth of mammary glands	PRIH inhibits release; increased estrogen and decreased PRIH stimulate release
Growth hormone (GH)	Protein	Bone, fat, other tissues	Increases synthesis of RNA and proteins, transport of glucose and amino acids, lipolysis, and formation of antibodies	GRH stimulates release
Corticotropin (ACTH)	Peptide	Adrenal cortex	Increases secretion of steroids in adrenal cortex	CRH stimulates release; corticosteroids inhibit release
Melanophore-stimulating hormone (MSH)	Peptide	Melanophores and melanocytes	Increases synthesis of melanin, causes darkening of skin	MRIH inhibits release
Lipotropin (LPH)	Peptide	Fat and other cells	Hydrolysis of fats	Not certain

*See Table 37-2 for identification of the abbreviations used here.

and joints become enlarged. Some scholars have postulated that one of the great pharoahs of Egypt, Akhnaton, was agromegalic, based on the bas-relief portrait shown in Figure 37-13.

Members of the third family of anterior-pituitary hormones are polypeptides. It includes **corticotropin (ACTH), melanophore-stimulating hormone (MSH),** and **lipotropin (LPH)**. ACTH regulates corticosteroid production by the adrenal cortex. MSH affects skin pigmentation. Its effects are best understood in lizards, in which it causes the skin to darken by stimulating the dispersion of small granules of brown pigment (melanin) within cells called *melanophores*. As a lizard basks on a rock, the melanophores cause the body surface to absorb more of the sun's infrared heat rays, much as dark clothing will do in sunshine (Chapter 35). If the body be-

comes too warm, MSH ceases to be secreted, and the skin pales again. MSH also influences fur color in mammals such as mice and weasels; if MSH is present when new coat hairs form, they will be dark. Lipotropin has quite a different function. It causes the hydrolysis of fat into free fatty acids and glycerol so that these important precursors are available to cells.

Hormones of the Posterior Pituitary

Whereas the anterior lobe of the pituitary gland manufactures and delivers many hormones into an animal's bloodstream, the posterior lobe secretes only two hormones: **oxytocin** and **antidiuretic hormone,** or **ADH** (also called *vasopressin*) (Figure 37-11). These chemicals are synthesized not by the pituitary itself, but by neuro-

Figure 37-13   PITUITARY MALFUNCTION: AN ANCIENT PROBLEM.
Pituitary disorders, such as that involving the overproduction of growth hormone in an adult, can result in a condition known as acromegaly, in which some parts of the skeleton enlarge under the influence of GH. The Egyptian pharoah Akhnaton is thought to have suffered from this condition.

secretory cells in the nearby hypothalamus. They are then secreted into the bloodstream by way of neurosecretory cell endings in the posterior pituitary as well as elsewhere in the brain.

The octapeptide oxytocin (Figure 37-2a) is best known as a stimulator of contraction in smooth-muscle cells. This occurs, for instance, in a mammal's mammary glands, so that milk is expressed through the nipple. As we saw in Chapter 18, it also stimulates the contraction of uterine smooth muscle during labor or orgasm. It turns out that oxytocin synthesis is not restricted to hypothalamic neurosecretory cells. The hormone is also produced in the thymus, where its role in the immune system is not defined, and in the corpus luteum of the ovary. In reptiles, amphibians, fishes, and birds, a relative of oxytocin, *arginine vasotocin* (AVT), plays an analogous biological role. When a female bluejay is ready to

lay an egg, for example, AVT starts the contractions of her oviduct.

The second posterior pituitary hormone, ADH, has a very different function from oxytocin, even though it is also an octapeptide with a very similar structure. Antidiuretic hormone decreases the production of urine. It exerts its effect by causing the collecting ducts of kidney nephrons to become more permeable to water (Chapter 35). In the absence of ADH, a person's urine flow increases tenfold, resulting in a condition known as *diabetes insipidus*. ADH also has a second function, which is the source of its alternative name, vasopressin. It raises blood pressure by stimulating smooth muscles in the walls of blood vessels to constrict. The release of ADH is triggered whenever blood pressure falls.

Biologists have made an important new discovery about pituitary hormones. The neurosecretory cells that secrete oxytocin and ADH in the posterior pituitary also send axons to sites in other parts of the brain. The researchers do not yet understand what either hormone does when it is released locally in the midst of brain neurons, but effects on local blood flow and oxygen availability seem likely.

The Hypothalamus–Pituitary Connection

The hypothalamus controls nearly all pituitary functions and, through them, many essential processes controlled by hormones. Although the *hypothalamus* makes up a scant 3 percent of human brain tissue by weight, this diminutive structure governs an extraordinary array of physiological and behavioral responses, including hunger, thirst, body temperature, emotional reactions to stress, and sexual stimulation.

We have already seen hints about the way the hypothalamus achieves this complex control. Both ADH and oxytocin are manufactured in grapelike clusters of neurosecretory cells that are embedded in the hypothalamus. Carrier proteins transport the hormones along the nerve cell axons to axonal endings in the posterior pituitary, where they are then stored (Figure 37-14). When the neurosecretory cell bodies are stimulated to initiate action potentials (Chapter 36), those nerve impulses pass down the neurosecretory cell axons and cause the hormones' release (just as neurotransmitters are released at the axonal endings of neurons).

Other hypothalamic neurosecretory cells have axons with a different destination; they extend to an area near the base of the hypothalamus called the *median eminence*, where, on appropriate stimulation, they secrete a variety of small peptide hormones. These substances are carried through a specialized system of capillaries to the neighboring anterior pituitary; there, they regulate the release of anterior-pituitary hormones. Those that stimulate hormone secretion are called **releasing hormones**

(a)
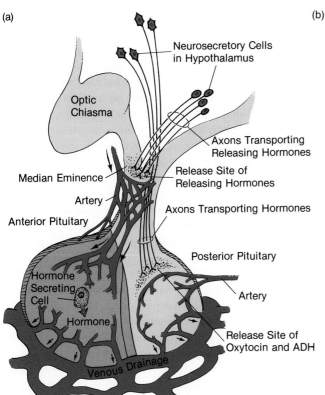

Optic Chiasma

Neurosecretory Cells in Hypothalamus

Axons Transporting Releasing Hormones

Release Site of Releasing Hormones

Median Eminence

Artery

Anterior Pituitary

Axons Transporting Hormones

Posterior Pituitary

Hormone Secreting Cell

Artery

Hormone

Release Site of Oxytocin and ADH

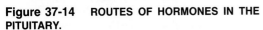
Venous Drainage

Figure 37-14   ROUTES OF HORMONES IN THE PITUITARY.

(a) The releasing hormones produced by the hypothalamus control the production and secretion of anterior-pituitary hormones. The neurosecretory cells discharge their releasing hormones into a capillary system that extends from the median eminence to the anterior pituitary. (b) Terminals of neurosecretory cells in the posterior pituitary gland are seen at the lower left in this electron micrograph (magnified about 18,000 times). The large black spots are secretory vesicles; each contains oxytocin or antidiuretic hormone. The wall of a blood capillary extends down and to the right from the upper left corner; secreted releasing hormones pass through the holes in the capillary wall into the plasma.

(b)

Table 37-2   **RELEASING AND RELEASE-INHIBITING HORMONES OF THE HYPOTHALAMUS**

Hormone	Structure	Action
TH-releasing hormone (TRH)	Peptide	Stimulates TH release
Gonadotropin-releasing hormone (GnRH)	Peptide*	Stimulates FSH and LH release
Prolactin-release-inhibiting hormone (PRIH)	Peptide* (dopamine?)	Inhibits prolactin release
GH-release-inhibiting hormone (GRIH)	Peptide	Inhibits GH release; interferes with TH release
GH-releasing hormone (GRH)	Peptide	Stimulates GH release
Corticotropic-releasing hormone (CRH)	Peptide	Stimulates ACTH release
MSH-release-inhibiting hormone (MRIH)	Peptide	Inhibits MSH release

*One large polypeptide is cleared to yield GnRH plus the so-called GAP peptide that is believed to be PRIH.

(RH), while those that inhibit it are termed **release-inhibiting hormones (RIH).**

Table 37-2 shows that every hormone secreted by the anterior pituitary is controlled by an RH or an RH–RIH combination. For example, prolactin-release-inhibiting hormone (PRIH)—consisting either of a peptide dubbed GAP, or the neurotransmitter dopamine—must cease to be secreted when a female mammal suckles her young in order for prolactin to be released in response to pro-lactin-releasing hormone (PRH). It is as though there are two locks to the prolactin door—the inhibitory signal must cease, and the stimulatory signal must start.

Just as with posterior-pituitary hormones, many of the

releasing hormones are secreted not only in the median eminence but also at other sites in the brain to which hypothalamic neurosecretory cells send their axons. Some of these substances act on brain cells in spectacular ways. For instance, TRH triggers subdued behavior, heat production, neuronal electrical activity, and several other processes that have nothing to do with the pituitary. Such surprising discoveries suggest that many RHs, RIHs, and other classic hormones participate, in ways not yet known, in a range of normal brain functions. Once again, we see that the distinction between the nervous and the endocrine systems is no longer valid: neurons secrete hormones, and hormones affect neurons. An animal's behavior and all its physiology are under the control of this dual system.

HORMONAL CONTROL OF PHYSIOLOGY

As hormones integrate and control an animal's body functions, one of their primary roles is to maintain homeostasis—the optimal steady state of its internal environment (Chapter 35). To illustrate the elaborate interactions of these control processes, we shall look at two phenomena: first, a homeostatic mechanism described briefly earlier—the interplay of insulin and other hormones to keep an animal's blood-sugar levels constant; and, second, the hormonal basis of rhythms in the animal's body.

Control of Blood-Sugar Levels

The hormones epinephrine, glucagon, thyroxine, growth hormone, and glucocorticoids can elevate blood glucose, but only one hormone, insulin, can lower it. To understand how these substances interact, let us look first at what happens when the body receives a sugar "jolt" and then at the hormonal response when the supply of glucose is low.

Imagine for a moment that you have skipped breakfast; you become hungrier and hungrier until about 10:00 A.M. You finally devour a large candy bar. Within minutes, the sucrose in the candy is converted to glucose, which enters your blood plasma. If the glucose level is high enough, beta cells in the islets of Langerhans react by secreting insulin. At the same time, secretion by the adrenal cortex of glucocorticoids decreases. Circulating insulin binds to receptors on skeletal-muscle, fat, and liver cells and stimulates them to take up glucose. As a result, the level of glucose in the blood then falls back toward normal.

Suppose, in contrast, that a mammal cannot find food for several days. During such fasting, blood-glucose levels fall below normal, producing a condition known as *hypoglycemia.* (High blood glucose is *hyperglycemia.*) When such an animal becomes hypoglycemic, secretion of hormones that counteract the condition rises. For instance, glucagon from pancreatic alpha cells stimulates the conversion of liver glycogen into glucose, which is then released into the blood. (The same thing happens when the adrenal gland secretes epinephrine, as during the flight-or-fight emergency response.)

The liver's supply of glycogen is only sufficient to compensate for twelve to twenty-four hours or so of fasting. Thereafter, growth hormone from the pituitary and glucocorticoids from the adrenal cortex stimulate the synthesis of glucose from amino acids in the liver (a process called *gluconeogenesis*). In addition, GH, glucagon, and possibly lipotropin cause fatty acids and glycerol to be converted into glucose. As fatty acids in the liver are oxidized, ketone bodies are produced. These diffuse into the blood and can then be converted to acetyl coenzyme A in muscle cells for synthesis of ATP. In this way, the fasting animal's muscle cells continue to obtain the energy they need to contract. The brain also uses ketone bodies to generate glucose for energy, so that it, too, is able to overcome—at least temporarily—the absence of circulating glucose. Despite these relationships, it is not safe to conclude that prolonged fasting is a safe way for humans to diet.

Rhythms, Hormones, and Biological Clocks

Hormones such as insulin and glucagon operate in response to the chemical changes triggered by food intake or fasting. Other regular hormonal effects on physiology occur as automatic internal cycles. Animals, plants, and even some microbes function on cycles called *circadian* ("about a day") rhythms. Circadian rhythms are daily fluctuations in the way various systems in the organism function. As we shall see in Chapter 48, some complex organisms also have internally controlled *annual rhythms;* the migrations of birds and mammals, for example, depend on hormonal actions and occur on automatic, annual cycles. External cues, such as the daily cycle of light and darkness due to the Earth's rotation, or the changes in season due to the Earth's tilt, may "set" circadian or annual clocks, but the internal rhythm itself is a property of the brain and its hormones. That is, if a fruit fly, ground squirrel, or human is isolated for weeks in constant light or darkness, cycling will persist; sleep cycles, hunger cycles, mitotic cycles in various tissues, and all sorts of other events will continue normally despite the absence of the standard cues of daylight and darkness. Thus the basic rhythms are intrinsic properties of the organism.

An organism's biological clock has a genetic basis. Strains of fruit flies, for instance, can be isolated that have short-day (say, eighteen-hour) clocks or long-day (say thirty-hour) rhythms. The genes responsible for such characteristics presumably do their work in the brain. A diagnostic feature of all such circadian cycles is that their timing is temperature-compensated. This makes sense, since a salamander or fruit fly may be exposed to radically different temperatures during a typical twenty-four-hour day; if their internal clocks ran faster at high temperatures and slower at low ones, time-keeping would be chaotic. But, as shown by the Princeton University biologist C. S. Pittendrigh, the internal clocks are compensated so that despite varying temperatures, they run at a constant rate.

How are circadian rhythms controlled? In complex animals, many hormones are released at preset times during each twenty-four-hour period. Epinephrine levels in human blood and body fluids, for example, are high during the day and low at night. Since epinephrine inhibits cellular mitosis, most mitotic activity in body tissues—including the growth of hair and fingernails—takes place while we sleep.

One source of rhythmic control in many animal species is the **pineal gland,** a pealike nugget of tissue located in the roof of the brain. The pineal gland releases the hormone **melatonin,** a derivative of the amino acid tryptophan, usually when an animal's environment is dark (in humans, it is most abundant in the blood between 11:00 P.M. and 7:00 A.M.). Melatonin, it turns out, may also be made in the eyes, red blood cells, and the hypothalamus of various vertebrates. Once released from the pineal gland in birds, it apparently acts on a variety of neurons and hormone-secreting cells and so acts as a pacemaker signal to synchronize other body activities.

When whole pineal glands or even a few cells from pineal tissues are cultured in a dish in the dark, they release melatonin on a cyclic schedule, just as they do when still embedded in the brain of an animal in an environment with normal changes of light and darkness. Thus cells themselves can have intrinsic rhythmic properties. Other experiments have revealed how the extraordinary built-in rhythmic activity of the brain and pineal gland is related to the animal's "biological clock"—the innate mechanism by which an animal and its tissues carry out basic biological activities on a timetable that is a bit longer than twenty-four hours. Suppose, for example, that a bird is kept in constant light or darkness for an extended period. Despite the absence of external signals, the pineal gland continues to function on a cycle approximating twenty-four hours in length, and most body functions continue to cycle as well (Figure 37-15). In fact, receptors for hormones also may vary cyclically. Female rats, for example, have more estrogen receptors in their brains during the day than at night; we do not know whether this occurs in birds.

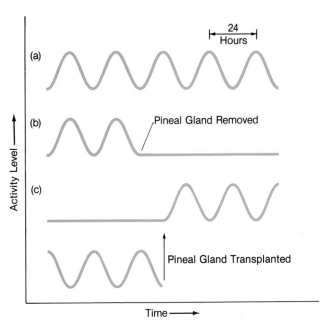

Figure 37-15   CIRCADIAN CYCLES AND THE PINEAL PACEMAKER.

Circadian rhythms are controlled by the pineal gland in birds. (a) Each typical circadian cycle is about twenty-four hours long. They may affect wakefulness and sleep, feeding, hormonal levels, and so on. (b) The cycles cease when the pineal pacemaker is removed. (c) If a bird lacking a pineal gland receives a transplanted pineal gland, it immediately adopts the cycle of the new pineal.

If the pineal gland is removed, however, the bird's internal "schedule" disappears and its biological functioning become arhythmic. Conversely, if a pineal gland is transplanted into an arhythmic bird lacking one, the animal immediately adopts the precise rhythm of the implanted organ.

Although mammals, like birds, have a pineal gland, which secretes melatonin rhythmically, recent studies suggest that more important for the mammalian biological clock are paired regions of the hypothalamus called the **suprachiasmatic nuclei (SCN).** Through exceptionally delicate microsurgery, researchers can separate neurons of this region from the rest of the brain. Like cultured pineal cells, these SCN neurons continue to function on a circadian cycle. However, every other cyclical activity of the mammal's brain, endocrine system, and body ceases (including that of melatonin secretion by the mammal's pineal gland). While the effects of the influence of the SCN are similar to those of the bird's pineal gland, the underlying functioning is not. The pineal gland exerts its influence by secreting melatonin; the suprachiasmatic nucleus sends its regulatory messages via nerve axons to other parts of the brain. But new techniques reveal axons extending from pineal cells in hamsters (a mammal) into the brain near the hypothalamus. This, plus the finding of melatonin in the hypothalamus, suggests that the suprachiasmatic nuclei may, as pacemakers, drive mammalian cycles, but that the evolutionarily ancient pineal cells may still be involved in activating hypothalamic neurons and so influencing cyclical activities. Other techniques show that the mutant *per* gene product in fruit flies can abolish, shorten, or lengthen circadian rhythms. The *per* protein alters gap junction conductivity, leading to the concept that periodic gene function contributes to periodic functional changes in the nervous system. Perhaps similar genes in pineal and SCN cells contribute to circadian pacemaking and the synchronization of a wide array of physiological functions.

HORMONAL CONTROL OF BEHAVIOR

Hormones obviously interact in complex ways to keep many physiological processes coordinated with each other and with the daily rhythms of the Earth's environment. In addition, an animal's overt behaviors often depend on hormonal control. Birds, reptiles, and small mammals, in particular, have reproductive cycles and mating behaviors that are clearly connected with environmental factors such as day length and season. One such animal is the small green anole lizard (*Anolis carolinensis;* Figure 37-16), which undergoes behavioral changes, as well as major alterations in its reproductive organs, with the fluctuating seasons.

During the winter, both male and female anole lizards cluster together under rocks or dead trees. Their gonads (ovaries and testes) are reduced in size, and neither sex exhibits any sexual behavior. In early spring, as the days

Figure 37-16   HORMONAL CONTROL OF BEHAVIOR.
A sexually active anole (*Anolis* species) lizard expanding its dewlap while sitting on a branch in a Costa Rican forest.

lengthen and grow warmer, the males emerge from their winter shelters and begin to fight with each other and establish territories. About a month later, the females emerge, and the anterior pituitaries of both sexes begin to secrete FSH. The gonads begin to grow rapidly and secrete the steroid hormones testosterone and estradiol-17β. The males produce sperm, and increasing testosterone levels cause the male sex structures to enlarge. In the females, estradiol-17β causes the two uteri to grow and stimulates the synthesis and secretion of yolk proteins from the liver (Chapter 16). From mid-April until August, the lizards mate, and the females lay a succession of eggs, one every ten to fourteen days. By the end of August, however, each animal's hypothalamus has stopped secreting gonadotropin-releasing hormone, and the testes and ovaries begin to shrink once again, even if day length is kept artificially long by use of controlled light.

As the lizards engage in their characteristic mating behavior, the males' activities affect the maturation of female gonads. A male attracts a mate by flashing his bright-red throat fan (called a dewlap) as he bobs his head to and fro. This performance acts as a sensory stimulus to females that speeds the growth of the ovaries, presumably by neural messages that are relayed to the hypothalamus, which, in turn, controls the secretion of FSH. For her part, a female will not mate until an ovary contains a large, yolked egg ready to be ovulated. And like many mammals, the female lizard is sexually receptive only when the level of estrogen in her body is high, around the time of ovulation.

Green anole females will not reproduce without the interaction between key seasonal and visual stimuli that trigger high estrogen levels. In fact, mating behaviors of all vertebrates and invertebrates, not to mention dozens of other complex behaviors, involve interaction among diverse sensory inputs, the activity of the hypothalamus and higher brain centers, a variety of hormones, and the physiological and motor responses they produce. But hormones do not stop their work when reproduction occurs; they play major roles in embryonic development, as well.

HORMONES AND DEVELOPMENT: AMPHIBIAN METAMORPHOSIS

As we saw in Chapter 16, hormones influence the maturation and division of specific cell types in embryos. This hormonal control of development persists as different kinds of endocrine cells function through the early stages of an animal's life. The metamorphosis of a tadpole into an adult frog is an excellent example of this developmental control in action.

The tadpole wriggling about in a pond is superbly adapted for the life of an aquatic herbivore, but the environment in which the adult frog lives imposes quite different physiological requirements. Thus when metamorphosis occurs, the tadpole's body undergoes enormous changes. Hind legs begin to develop, and forelimbs soon follow. The tail disappears, literally digested away by lysosomal enzymes. A new visual pigment appears, one suitable for light traveling through air, not water, and the tadpole's hemoglobin is altered to a type that binds less readily with oxygen. At the same time, lungs begin to develop, and the gills are later degraded. The gut shortens, as an adaptation for future carnivorous feeding, and enzymes for digesting protein are synthesized to allow digestion of animal food. Because of a change in liver enzymes, the kidneys start to secrete mostly urea, instead of more toxic ammonia, which must be passed into large volumes of water. The end result of this developmental magic is a frog whose physiology is radically different from that of the tadpole that preceded it.

How exactly do hormones regulate this impressive transformation? The brain and the thyroid gland are two main sources of the hormones that control the tadpole's metamorphosis (Figure 37-17a). The sequence starts when the hypothalamus begins to secrete TRH (TH-releasing hormone). Pituitary cells release TH in response, and that hormone, in turn, triggers the secretion of thyroxine by the thyroid gland. At the same time, the tadpole's pituitary, aided by the secretion of PRIH (prolactin release-inhibiting hormone) by the hypothalamus, slows secretion of prolactin (an antagonist of thyroid hormones).

As thyroxine circulates through the tadpole's bloodstream, it is bound by abundant receptor molecules on cells in tissues affected by metamorphosis—the tail, legs, liver, and so forth. Even a tail treated in a culture dish with thyroxine will regress and its cells will die, just as they would in an intact tadpole (Figure 37-17b). By contrast, cells that remain unchanged as a tadpole undergoes this final stage of metamorphosis may lack thyroxine receptors altogether.

But what causes the hypothalamus to initiate the secretion of the hormones responsible for metamorphosis in the first place? Biologists believe the answer to be changes in the tadpole's environment, such as crowding, falling oxygen content in the water, or limitation of the food supply. As any of these conditions makes survival in the water more difficult, the tadpole's brain initiates metamorphosis.

The control over anatomy and physiology, and thus mode of life, afforded by metamorphosis enables amphibians to exploit their environment in optimal ways. For instance, some salamander species remain at the lar-

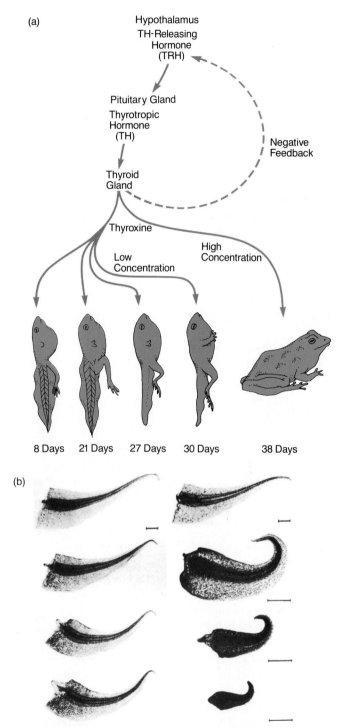

Figure 37-17   AMPHIBIAN METAMORPHOSIS: FROM TADPOLE TO FROG.

During metamorphosis, the tadpole loses some anatomical structures as new ones develop. (a) Biochemical changes, largely the result of thyroid hormone action, are associated with the metamorphosis. (b) Thyroid hormone can have dramatic effects on tail regression. Tails cultured in the absence of T4 (thyrozine) remain unchanged even after 8, 10, and 12 days in culture (left three tails); whereas those cultured in the presence of the hormone undergo rapid regression, as seen after 5, 7, and 9 days (right three tails).

val stage if their pond is not crowded and has adequate oxygen and plentiful food. Their gonads develop, however, and the larvae reproduce. This phenomenon, in which a juvenile form reproduces, is called **neotony.** In a subsequent year, if food in the water becomes very scarce, the hypothalami of the larvae in that generation trigger metamorphosis, and salamanders emerge.

Some salamander species have carried out neotonous reproduction as larvae for thousands of generations and can no longer undergo metamorphosis, even when thyroxine is artificially supplied. Overall, however, the versatility of thyroxine-controlled metamorphosis helps explain why amphibians such as frogs and salamanders are so successful ecologically—in tropical rain forests, for example, they are the most abundant vertebrates of all.

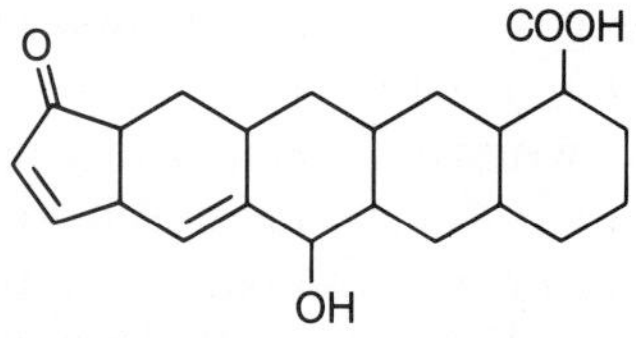

HORMONES: NEW TYPES, NEW SITES, NEW MODES OF SYNTHESIS

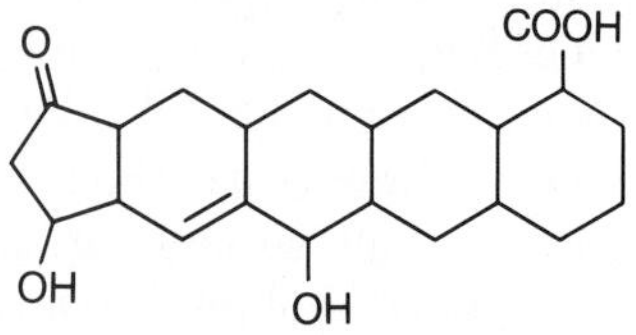

For years, biologists have studied many of the hormones involved in metamorphosis, reproductive cycles, behavior, and physiology. Endocrinology is one of the most exciting areas of biology today because of the many new substances discovered that function as hormones. Among these substances is a fascinating group of chemicals derived from fatty acids: the **prostaglandins** (Figure 37-18). Once thought to be present only in semen, prostaglandinlike substances are, in fact, produced (usually from membrane phospholipids) by virtually all cells.

Unlike most hormones, prostaglandins do not circulate in the blood, but appear in increased quantity in local regions where tissues are disturbed. They are potent stimulators of smooth-muscle contractions and thus may be involved in birth, miscarriage, or blood-vessel constriction in the kidney and other tissues. An excess of prostaglandins acting on human cerebral blood vessels causes them to dilate and press against nerves, producing a headache. Aspirin is an effective reliever for headache pain because it inhibits prostaglandin production.

The major set of new hormones that has revolutionized our view of the relationship between the endocrine and nervous systems is the *brain peptides.* These substances have been identified in the brains of mammals and other vertebrates. We have already learned about oxytocin, which is released from axonal branches of hypothalamic neurons at various specific sites in the brain. Another surprising brain peptide is CCK, a hormone long thought to work only in the gut (Chapter 34). Certain axons of single neurons in the cerebral cortex (the "highest" center of the vertebrate brain; Chapter 40) synapse on tiny blood vessels, where they release

Figure 37-18   PROSTAGLANDINS.

Prostaglandins, a type of hormone, are structurally modified lipids. They are regarded as "local" hormones because their biological effects are usually short range, and the activity is lost over a single circulation cycle. Prostaglandins affect numerous cellular activities, ranging from stimulation of smooth-muscle contraction to blocking hormonally induced cyclic AMP production.

CCK. The hormone then affects local blood flow and the electrical activity of neighboring nerve cells. Conceivably, this relates to the role of CCK in establishing vivid memories (Chapter 34). Still another peptide, VIP (an intestinal polypeptide), is released by other cerebral-cortex neurons.

Some biologists postulate that hormones such as CCK and oxytocin help to maintain normal relationships between nerve-cell activity, blood flow, and cellular metabolism in the brain. Another, very likely function of many brain peptides is to exert relatively long-lasting effects (many seconds, minutes, or longer) on the polarized state of neurons. Thus traditional neurotransmitters do their work in milliseconds, while a brain peptide, perhaps released from the same neuron, has a much more sustained action. This leads to the view that the nervous system is composed of hard-wired circuits that are "tuned" or modulated by brain peptides and certain neurotransmitters. As a result, nerve circuits can be flexible and adaptable, characteristics we associate with learning (Chapters 40, 48). Researchers did not even suspect the possibility of such phenomena until brain peptides were discovered.

It might seem reasonable to assume that substances such as the brain peptides, with their intricate regulatory effects, might be recent evolutionary inventions. This is not the case, however. In fact, most have ancient evolutionary histories; they are found not only in complex mammals, but in invertebrates, unicellular organisms, and plants as well. Of what use is a brain peptide in microorganisms that do not even have nervous systems? Evidence suggests that such substances originated as chemical vehicles for intercellular communication. For instance, yeasts possess a mating pheromone, *alpha factor*, with an amino-acid sequence nearly identical to that of the mammalian gonadotropin-releasing hormone, which stimulates the pituitary to release luteinizing hormone. Some researchers believe that such communication molecules came to be employed as neurotransmitters as nervous systems evolved in cnidaria and wormlike vertebrates. Then, with the evolution of endocrine tissues, many of the peptides present in nerves came to function as hormones. The hormones produced by the endocrine system evolved concurrently with receptor molecules and the transduction processes by which the hormones regulate target-cell activities. In Chapter 40, we will consider the brain peptides and their functions in the mammalian brain in more detail.

While molecular techniques have given us many surprises about hormones, none is more novel than multiple hormone production from single genes. One long mRNA molecule, for example, can be translated into a large polypeptide in pituitary and hypothalamus cells; that polypeptide is then chopped up to yield ACTH, beta-lipoprotein (a hormone that causes fat cells to break down lipids), three forms of MSH, several brain peptides (endorphins, discussed in Chapter 40), and other peptides. In other cells, one polypeptide can be cleaved to yield GnRH and GAP, the prolactin-release-inhibiting peptide. Based on these and other examples, it is clear that the one-gene–one-polypeptide rule worked out so elegantly by geneticists in the 1960s must be modified in hormone-producing cells: one gene can code for several polypeptides, each with its own functions.

These results are compatible with the hypothesis that the profusion of newly discovered hormones, neuropeptides, and tissue growth factors active in organisms are members of relatively few families of protein molecules. The structures and biological functions of family members have diverged over the eons of evolution. It is becoming increasingly clear, therefore, that most vertebrate hormones originated long ago—and for other purposes—in much simpler organisms. From this vantage point, we can view the endocrine systems of more complex animals as exquisite specializations that have come to play intricate roles in the physiology, rhythmic processes, behavior, and development of the organisms that possess them. Moreover, the function of endocrine systems is so intimately linked to that of nervous systems that we can now appreciate that regulation of behavior and physiology is a highly evolved form of intercellular "talk" in molecular, ionic, and electrical languages of differing speed.

LOOKING AHEAD

In the past two chapters we have studied nerves and nervous systems, hormones and endocrine systems. In the next chapters we will first examine input mechanisms—sensory systems—and then output mechanisms—effector motor systems. With this background we will turn to an examination of the brain (Chapter 40) as the control center for both neuronal and hormonal regulation of the body and behavior.

SUMMARY

1. Hormones are chemical messengers that are secreted by cells and glands of the *endocrine system* into the blood and that influence the activities of cells in specific target tissues.

2. Six principles apply to hormonal action: (a) a hormone affects only target cells specific to it; (b) hormone receptors develop as a cell differentiates; (c) different cells develop different transduction mechanisms to hormones and so may respond differently to the same hormone; (d) some hormones act fairly steadily, while others function sporadically; (e) hormonal level is often governed by a negative-feedback loop; (f) hormones are usually broken down rapidly.

3. Chemically, most hormones are proteins (or short-chain polypeptides), amines, or steroids.

4. Steroid hormones bind to specific receptor molecules in the nuclei of target cells. The hormone–receptor complex then binds to acceptor sites on the chromosomes and stimulates specific genes to synthesize messenger RNA, which is then translated into protein.

5. *Cyclic AMP* and *cyclic GMP* are second messengers for many protein, polypeptide, and amine hormones. cAMP combines with inactive protein kinase A, which becomes activated and phosphorylates proteins and enzymes; the proteins and enzymes carry out the hormonal response. In general, cGMP works antagonistically to cAMP. These and other second-messenger systems function as amplification cascades so that binding of a tiny amount of hormone can ultimately trigger a massive biochemical response.

6. Many protein hormones, neurotransmitters, and growth factors and some electrical stimulation act via *inositol triphosphate* (IP_3) and *diacylglycerol* (*DG*) as second messengers. They, in turn, work via

Ca^{2+} ions and protein kinase C to activate major cellular responses.

7. Invertebrates secrete a variety of hormones that affect growth, cell differentiation, and physiology. Two important hormones in arthropods are *juvenile hormone (JH)*, which regulates the stages of development, and *ecdysone* hormones, which are responsible for metamorphosis, molting, and other processes.

8. Beta cells in the pancreatic *islets of Langerhans* secrete *insulin*, which lowers blood-glucose level. Insulin is also produced by some other body cells, notably in the brain. Pancreatic alpha cells secrete *glucagon*, which raises blood-glucose level. *Somatostatin*, a hormone secreted by pancreatic delta cells, inhibits the secretion of insulin and glucagon and the absorption of glucose across the intestinal wall.

9. The kidney secretes *erythropoietin*, which stimulates the production of red blood cells in bone marrow, and *1,25-dihydroxy-chole-calciferol*, which regulates calcium absorption in the small intestine. *Aldosterone*, made by the adrenal glands, regulates sodium, and indirectly water, retention in the kidney.

10. Heart cells secrete atrionatriuretic factor in response to elevated blood pressure. These hormones cause vasodilation and urine production, thereby lowering blood pressure.

11. The cortex of the *adrenal glands* produces *corticosteroids*. One group, the *glucocorticoids*, is primarily involved in maintaining blood-glucose level. The second group, which includes aldosterone, is the *mineralocorticoids*. The third group of corticosteroids is the sex steroids: the androgens are male sex hormones, and estrogens, female sex hormones, although both occur in members of each sex.

12. The adrenal medulla, a tissue analogous to sympathetic nervous tissue, contains endocrine cells, called *chromaffin cells*, that secrete *epinephrine* and *norepinephrine*. These amine hormones mobilize crucial organs and tissues in the fight-or-flight reaction.

13. The *thyroid* gland secretes two amine hormones, *triiodothyronine* (T_3) and *thyroxine* (T_4). They regulate the rate of oxygen consumption and metabolism in cells, and also may play a role in generation of heat. An iodine deficiency can lead to *goiter*, an enlarged thyroid gland, since it disturbs the thyroid–pituitary negative-feedback loop. A third thyroid hormone, *calcitonin*, lowers blood-calcium level, whereas *parathyroid hormone (PTH)*, secreted by the *parathyroid*, elevates blood calcium.

14. The *pituitary gland* consists of two parts: the anterior pituitary and the posterior pituitary. The anterior-pituitary hormones *thyrotropic hormone (TH)*, *corticotropin (ACTH)*, follicle-stimulating hormone (FSH), and luteinizing hormone (LH) regulate endocrine functions in the thyroid, adrenals, and gonads, respectively. The anterior pituitary also is the source of *prolactin*, which stimulates milk synthesis and secretion in mammals, among other functions; *growth hormone (GH)*, which triggers the liver to secrete growth-promoting *somatomedins*; *melanophore-stimulating hormone (MSH)*, which affects skin pigmentation; and *lipotropin (LPH)*, which causes fat hydrolysis.

15. The posterior pituitary secretes *oxytocin* and *antidiuretic hormone (ADH)*, or vasopressin, two hormones synthesized and transported to the posterior pituitary by hypothalamic neurosecretory cells. ADH causes water reabsorption in kidneys, whereas oxytocin stimulates smooth-muscle cell contraction.

16. The hormones of the anterior pituitary either directly influence body tissues or control the functions of other endocrine glands. Anterior-pituitary cells are regulated by the hypothalamus through *releasing hormones* and *release-inhibiting hormones*. These substances pass from the median eminence, near the base of the hypothalamus, to the anterior pituitary to exert their effects.

17. Homeostasis depends on the interaction of many hormones. For example, insulin, epinephrine, glucagon, growth hormone, and glucocorticoids "cooperate" in keeping blood-sugar level constant.

18. In many animal species, certain hormones are released in circadian rhythms, and as a result many physiological processes vary over each twenty-four-hour daily period. The *pineal gland*, which secretes the hormone *melatonin* (whose effects are still unknown), generates such rhythms in birds, and the *suprachiasmatic nucleus (SCN)* of the hypothalamus plays the same role in mammals. Light–dark cycles or other cyclic cues can "reset" the brain's internal pacemaker cells and so the organism's biological clock. That clock is temperature-compensated and "runs" at the same rate even as temperature varies.

19. Hormones also play a critical role in animal development; metamorphosis in amphibians is a dramatic example of hormonal control of a major maturational process.

20. Hormonal chemicals are produced by many cell types. Among these substances are the *prostaglandins*, derivatives of fatty acids that are synthesized when local areas are disturbed. They stimulate smooth-muscle contraction and may be involved in labor and delivery, miscarriage, and headache. Brain peptides are released at specific sites in the brain; chemically, these substances are identical to many of the body's classic hormones, yet they appear to have special, relatively long-lasting effects on neurons. Thus the nervous and endocrine systems are truly variations on one theme—communication among cells by molecules, ions, or electric currents.

KEY TERMS

adrenal gland
aldosterone
antidiuretic hormone (ADH)
calcitonin
chromaffin cell
corticosteroid
corticotropin (ACTH)
cyclic AMP (cAMP)
cyclic GMP (cGMP)
diacylglycerol (DG)
1,25-dihydroxy-chole-calciferol
ecdysone
endocrine system
epinephrine
erythropoietin

glucagon
glucocorticoid
goiter
growth hormone (GH)
inositol triphosphate (IP_3)
insulin
islet of Langerhans
juvenile hormone (JH)
lipotropin (LPH)
melanophore-stimulating hormone (MSH)
melatonin
mineralocorticoid
neotony
norepinephrine
oxytocin

parathyroid gland
parathyroid hormone (PTH)
pineal gland
pituitary gland
prolactin
prostaglandin
release-inhibiting hormone (RIH)
releasing hormone (RH)
somatomedin
somatostatin
suprachiasmatic nucleus (SCN)
thyroid gland
thyrotropic hormone (TH)
thyroxine (T_4)
triiodothyronine (T_3)

QUESTIONS

1. What are the functional and chemical similarities between the endocrine and nervous systems?

2. Substances produced by the endocrine system that exert influence on target cells elsewhere are called _____. Name the three chemical categories into which these substances fall, and give an example of each from vertebrates.

3. List the six principles and the three mechanisms of hormonal action.

4. What are the basic differences between the way in which steroid hormones and nonsteroid hormones act on target cells?

5. Explain the roles of juvenile hormone and ecdysone in the life cycle of a cecropia moth. How would a moth's normal metamorphosis and reproduction be disrupted if juvenile hormone continued to be present in high concentrations in the later larval stages?

6. Describe the role played by pancreatic hormones in glucose metabolism. What other glands and hormones affect the metabolism of glucose?

7. Explain in terms of glands, hormones, and target cells what biologists mean by the fight-or-flight reaction.

8. From what you know of the chemical structure of thyroid hormones, explain how the lack of dietary iodine interferes with basic metabolism.

9. The pituitary gland was traditionally called the "master" gland of the endocrine system, but it is now considered to be a "middleman" in endocrine functions. Explain.

10. Give examples of circadian rhythms. What part of the brain controls circadian rhythms in birds? in mammals?

ESSAY QUESTIONS

1. Define the term "homeostasis" and discuss the role of the endocrine system in the maintenance of homeostasis.

2. Discuss the various ways in which the endocrine system and the nervous system are "variations on one theme."

SUGGESTED READINGS

ALBERTS, B., ET AL. *The Molecular Biology of the Cell.* New York: Garland, 1983.

The treatment of second messengers in this book is excellent.

FROHMAN, L. A. "CNS Peptides and Glucoregulation." *Annual Review of Physiology* 45 (1983): 95–107.

Most people do not realize that nerves affect insulin and glucose secretion; this article tells how.

NISHIZUKA, Y. "Turnover of Inositol Phospholipids and Signal Transduction." *Science* 225 (1985): 1365–1369.

This article reports on the spectacular work on the new second messengers.

SAUNDERS, D. S. "The Biological Clock of Insects." *Scientific American*, February 1976, pp. 114–120.

Major aspects of the clocks that tick in most organisms are summarized.

WHITE, J. D., ET AL. "Biochemistry of Peptide-Secreting Neurons." *Physiological Reviews* 65 (1985): 553–597.

This is a marvelous discussion of most aspects of major brain peptides.

38

THE SENSES

In reality there are as many outside worlds as there are forms of life; bees, dogs, tapeworms, fleas—each contemplates the real world with its own senses and gathers from it whatever is significant for its own existence.
Wolfgang von Buddenbrock, *The Senses* (1958)

Eyes, ears, nose, whiskers, fingers: sensory sharing between two friends.

For many animals, the "world" is nothing more or less than what they can sense—colors, shapes, temperature, and so on. And this perception of the world varies from one type of animal to the next: humans may see the brilliant purple and gold of a desert sunset, feel the polished surface of a pebble on the beach, hear a sonata, taste a summer strawberry, or smell a sprig of sage. The skunk has a much more keenly sensitive nose. It inhabits a veritable universe of odors that signal the presence of a predator, a mate, or a morsel of food. By contrast, a dolphin diving in search of deep-swimming salmon cannot smell at all and has little light by which to see. Instead, this graceful mammal uses sounds and echoes to sense its watery surroundings and its potential prey with amazing accuracy.

In this chapter, we shall describe many of the remarkable structures and processes with which animals perceive their internal and external environments. Often, such sensory systems—specialized for taste, smell, vision, hearing, gravity detection, and so on—are vital to an animal's survival. Through them, environmental cues, such as the availability of food or the presence of an enemy, are communicated to the central nervous system so that the animal can generate an appropriate physical, chemical, or behavioral response. This responsiveness, of course, is not limited to animals; plants, too, respond to light, gravity, and sometimes touch, as we saw in Chapters 29 and 30.

The sensory process starts when sensory receptor cells in the eyes, skin, and other body parts respond to environmental stimuli. For this reason, we shall begin by looking at the types and characteristics of these specialized cells. Each receptor cell responds to a form of energy—chemical, mechanical, or electromagnetic—and we will examine how these various forms of energy are sensed so that the brain may translate such sensations into the taste of an apple, the chirping of a cricket, or the brilliant hue of a morning glory. In so doing, we will encounter a fundamental principle in the operation of sensory systems: when energy acts on a receptor cell, the result is a change in the rate of flow of ions through the cell. It is this alteration in ion flow that produces the sensory "message" sent to the central nervous system.

Finally, as we move from simpler senses—such as taste and smell—to the most complex sense of all—vision—we shall see that the members of each animal species possess sensory equipment attuned to their particular way of life. The eyes of fish in the abyssal sea have evolved ways of detecting the wavelengths of light emanating from special light-producing organs of other fish, while a bee's eyes perceive the ultraviolet wavelengths of light reflected by flowers (Chapter 24), and a hawk's eyes report the rich spectrum of colors in a meadow. The reality of the world—its properties of form, texture, noise, odor, and so on—truly depends on the characteristics of each animal's sensory arsenal.

SENSORY RECEPTORS

All living cells sense and can react to their physical and chemical environment: acids, ions, and hormones affect plant and animal cells. Certain chemicals and light may elicit responses from monerans and protists. In most multicellular animals individual cells or groups of cells have become highly specialized for sensing specific features of an organism's internal and external environments. These receptor cells act as transducers, transforming the energy of the stimulus—the sound of a tinkling bell or the fragrant molecules wafting from freshly baked bread, say—into a form of electrical activity, which is then transmitted to the central nervous system. This transduction process often involves *amplification* of the original energy stimulus; for instance, nerve impulses traveling from the eye to the brain may have 100,000 times the energy of the minute amount of light that stimulates the eye's receptor cell.

Recall from Chapter 36 that most nerve cells have the same basic structure: the axonal end conducts nerve impulses toward a target cell, while the dendritic end is specialized for reception, usually of a chemical (neurotransmitter) or an electrical signal from a presynaptic neuron. Remember, too, that some sensory neurons may themselves function as receptor cells where energy transduction occurs, whereas certain special sense organs have separate receptor cells that synapse with sensory neurons. The dendritic end of both types of sensory receptor cells has unique structures and molecules that respond to light, chemicals, mechanical deformation, or other stimuli. Most such responses alter ion flow through the cell and elicit a particular pattern of electrical activity at the opposite end of the cell. There, an axon may be present, or the cell may communicate directly with dendrites of a postsynaptic cell.

Information about the stimulus encoded in this electrical pattern is then conveyed to the central nervous system—usually the brain—where the information is translated into a particular sensory perception—a sweet taste, a pungent odor, a rough surface, and so on. Note that in this sequence—input of energy to a receptor cell, transduction of energy into altered electrical activity, transmission of information to the central nervous system, and, finally, translation into a perception in the brain—the information sent by a receptor cell is just the beginning. It is the way in which the receiving brain center processes the message that determines the nature of the sensory perception resulting from a stimulus.

Most animals' sensory systems are designed to detect new information. This may occur at the level of receptor cells or centrally, in the brain. For instance, receptor cells have evolved multiple levels of operation for accomplishing this feat: they have a spontaneous level of electrical activity—a *baseline level*—which signals that a particular environmental condition is constant; when the environmental condition varies (becoming "more" or "less"), electrical activity in receptor cells registers this new information by increasing or decreasing. Such a mechanism allows an animal to distinguish between constant conditions that may be of little interest and changing ones that could affect its survival.

"Old" and "new" information may also be distinguished in the brain. For instance, a person's sensory system and brain adapt to the presence of clothing in less than a second; you are fleetingly aware of the feeling of socks when you put them on, but their "touch" soon fades into the mental background, so that you become unaware of their presence. Here, receptors may continue to send in sensory information, but certain brain circuits act as a filter and prevent awareness of that information. However, if a pebble gets into one sock or ants start to crawl between your toes, you would notice immediately. Without this ability to filter information, an animal's nervous system would be inundated with excess sensory input that would have little survival value, while vital information might be lost in surplus sensory "noise."

Sensory receptor cells can occur singly (such as temperature-receptor neurons in skin) or can be clustered in **sense organs**—ears, eyes, and so on. Most of the intricate anatomical parts of sense organs are adaptations that efficiently funnel stimuli—sound waves, light rays, or odorous molecules—to the actual sensory receptor cells. Thus the eardrum, the eye's lens, and the olfactory chamber of the nose have all evolved in response to the physics and chemistry of sensory stimuli in the environment and the form in which such stimuli reach the organism.

Biologists classify sensory receptor cells and sense organs according to the environmental factor they detect. **Chemoreceptors** detect specific molecules and ions and are thus prominent in taste and smell perception. **Mechanoreceptors** monitor pressure, stretching, and bending (shearing) forces; as a result, they sense motion of the body, vibrating air or water (perceived as sound), the pull of gravity, and body acceleration or deceleration. Receptors responsive to different parts of the electromagnetic spectrum are also common in animals. **Thermoreceptors,** for instance, sense infrared heat waves or temperature changes in the skin or the brain. Most animals have also evolved **photoreceptors,** which are sensitive to visible or ultraviolet light. The various types of sensory processes and the senses they register are shown in Table 38-1. Conspicuously missing from this list are the perceptions that animals experience as hunger and thirst—"sensations" that arise as the nervous system monitors the status of an animal's internal environment. Although critical to survival, such phenomena remain poorly understood.

Table 38-1   **SENSES AND SENSORY RECEPTION**

Sensory Process	Sense
Chemoreception	
Ions, chemicals	Gustation (taste)
Chemicals	Olfaction (smell)
Mechanoreception	
Pressure	Hearing, touch
Stretching	Position
Shear forces	Movement
Electromagnetic-radiation reception	
Infrared (heat)	Temperature
Photo (light)	Vision

CHEMORECEPTORS

Virtually every animal on Earth shares the ability to sense chemicals. The two types of chemical perception that biologists understand best are **gustation** (taste) and **olfaction** (smell).

Gustation: The Sense of Taste

Why is a sugar cube perceived to be sweet when a person bites into it? Viewed with a microscope, the human tongue is dotted with 10,000 or so **taste buds,** many of which are located as part of larger swellings called *papillae*. Each taste bud includes clusters of *taste (gustatory) receptor cells*, which provide information about the flavor of a pickle or a peach (Figure 38-1). Such receptor cells lack axons and are not classified as neurons; whether in the mouth, or on an aquatic vertebrate's head skin, they detect relatively high concentrations of dissolved substances in saliva or the surrounding water.

The surface of the human tongue has four types of taste receptors located in zones (Figure 38-2); each zone corresponds roughly to one of four basic taste sensations detected by humans: *sweet, bitter, sour,* and *salty*. However, we can distinguish considerably more than four tastes. For example, nearly all people and many animals can tell the difference among various sugars, although all are sweet. When you eat a slice of roast beef or a tuna sandwich, it is the combined output of various taste receptors, plus information from olfaction, that gives such foods their characteristic flavor.

The outer ends of taste-receptor cells have numerous microvilli and thus a large crinkled surface area, on which the taste-receptor molecules are located. Sensory-neuron endings extend into each taste bud, cupping the receptor cells like fingertips holding an orange. When

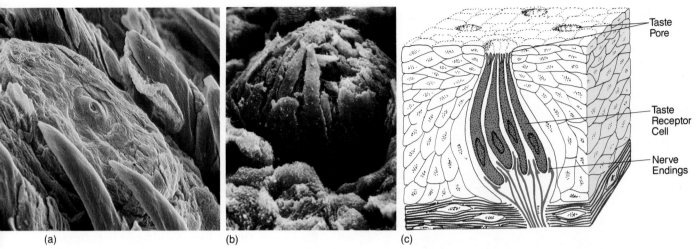

(a)   (b)   (c)

Figure 38-1   TASTE BUDS.
(a) This scanning electron micrograph (magnified about 520 times) shows the surface of a papilla on the tongue from above. The dot in the center is a taste pore that opens onto the taste bud cells below. (b) Surrounding epithelial cells have been removed from the papilla to reveal a taste bud's taste receptor cells (magnified about 1,100 times). (c) The diagram shows the nerve endings that synapse with receptor cells in a taste bud. From *Tissues and Organs: A Text-Atlas of Scanning Electron Microscopy* by Richard G. Kessel and Randy H. Kardon, W. H. Freeman and Company © 1979.

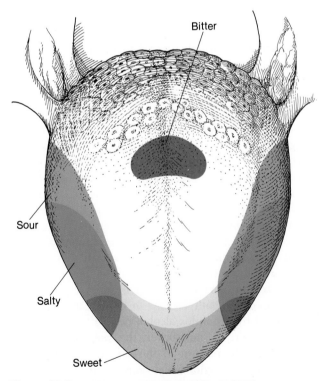

Figure 38-2   TONGUE REGIONS AND TASTES.
Taste buds are distributed on the tongue in zones with sensitivity to sweet, bitter, sour, and salty flavors. These regions overlap each other somewhat.

receptor molecules on the microvilli bind a tasteable substance, such as hydrogen ions (sour), quinine (bitter), or sucrose (sweet), sodium and potassium ions flow into the taste-receptor cell, and neurotransmitter is released at its opposite end. This event, in turn, causes the adjacent postsynaptic sensory neuron to increase the frequency with which it generates and propagates nerve impulses.

Although we do not understand why, many individual taste-receptor cells respond to several types of stimuli (sweet, bitter, and sour, for example), while others bind only with a single class of stimulus molecules. Unlike other sensory cells, taste cells live for only about three days. New ones are continually generated from tongue epithelium, apparently acquiring their taste specificity from the neurons that contact them. Other types of sensory cells, such as those in the eye, are not replaceable and so must last for the organism's lifetime.

In vertebrates, taste may allow an animal to discriminate among foods so that it can accept and eat safe ones and reject potentially harmful ones. The simpler nerve circuits in invertebrates allow biologists to learn how tasting and feeding are linked. An experiment involving the tiny blowfly illustrates such linkage. If a fly is glued to a glass rod and various parts of its body—abdomen, head, and wings—are touched with a fine-tipped pipette filled with a sugar solution, the animal shows no response. Touch the pipette to its legs, however, and taste receptors trigger a reflex that instantly unfurls the fly's proboscis. In nature, a blowfly landing on a food source literally "puts its foot in it," setting the feeding reflex in motion (Figure 38-3).

(a)

(b)

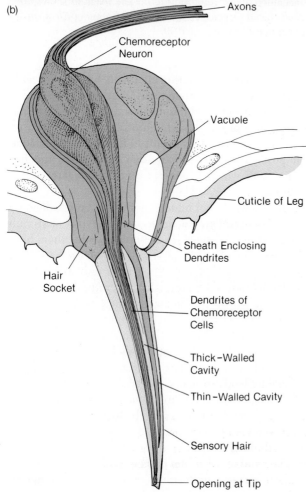

Figure 38-3   A FLY THAT CAN TASTE WITH ITS FEET.
(a) When this green bottlefly lands on a food source, the
feeding reflex is triggered by sensory hairs located along its
legs. (b) A sensory receptor hair is seen to contain several
chemoreceptor neurons with axons leading back to the
central nervous system.

Experiments have shown that each of the hollow *sensory hairs* on a blowfly's legs contains several dozen taste-receptor neurons (in the fly, unlike in vertebrates, taste-receptor cells are true neurons). Each neuron in the blowfly's legs is sensitive to a range of substances (carbohydrates, anions, and cations). Some of these cells serve as early-warning systems, telling the insect to avoid undesirable substances, such as acids and salt. However, a hungry blowfly that steps in a rich sugar solution (1 molar) will feed and search for more such food by flying in a tight "dance" as it explores the surrounding area (it must land periodically, of course, to sample the substrate). If the sugar solution is more dilute—0.5 molar—the dance covers a larger territory, but at 0.125 molar the blowfly simply flies in a straight line; from an evolutionary standpoint, such a meager food source is apparently not worth the effort that would be required to find more of it. The fly's behaviors, all triggered by taste, appear to be adaptations that provide the organism with a maximum return on its food-seeking investment of time and energy.

Olfaction: The Sense of Smell

Whereas the sense of taste primarily assists an animal during feeding and drinking, olfaction serves diverse functions as the animal detects a multitude of airborne or waterborne substances. Often olfactory receptors are sensitive to amazingly low concentrations of odorants, and they convey information that enables an animal to recognize the boundaries of a territory, detect the presence of a predator, and sense imminent changes in the weather. In addition, olfaction can play a critical role in a species' reproductive cycle.

Consider, for instance, the male silk moth, which relies on some 20,000 receptor cells in its fringed antennae (Figure 38-4) to track down the source of molecules of a female sex pheromone, *bombykol*. (Recall that pheromones are substances released by one member of a species that trigger behavior in another member of that species.) The male moth is able to detect as little as 1 molecule of this complex alcohol per 10^{17} molecules of the gases in air. In short order, the male sets out on a zigzagging flight pattern in the direction of a pheromone-releasing female, which may be as far as a mile away. Salmon use a similar sort of highly sensitive olfactory capacity. As they migrate hundreds of miles up mountain river systems, the fish follow olfactory cues that enable them to make choice after choice, selecting the forks in the river that eventually lead to the shallow streams where they spent the early weeks of their lives.

The nasal cavities of vertebrates are lined with a thin sheath of olfactory epithelium in which olfactory neurons are embedded (Figure 38-5). In contrast to taste-recep-

Figure 38-4   THE FEATHERY ANTENNAE OF THE MALE SILK MOTH.
Male silk moth antennae are extremely sensitive receptors for the female pheromone bombykol. This sensitivity enables male silk moths to home in on females over great distances.

tor cells, olfactory receptors of vertebrates are true neurons; the outer end of each receptor is composed of microvilli, which increase surface area and contain the actual receptor molecules. The inner end is an axon, which carries nerve impulses. In fish, stimulus molecules come into contact with these neurons as water flows through the nasal passages; in land vertebrates, air serves as the carrier, but substances must dissolve in the layer of fluid bathing the olfactory receptors in order to be detected. (This is why people who have a cold—and nasal cavities coated with heavy mucus—cannot smell very well.)

Certain vertebrates, such as whales and porpoises, lack olfactory receptors and so have no sense of smell. At the opposite extreme, the olfactory sense of canids, such as wolves and dogs, is astonishingly acute; for example, when a dog leaves the house in the morning and inhales, it exposes approximately 40 million olfactory receptors in each square centimeter of its olfactory epithelium to a world of organic molecules carried on the morning breeze.

Figure 38-5   THE SOURCE OF SMELL: OLFACTORY RECEPTORS IN THE NOSE.
(a) Olfactory receptor cells are located in the epithelium of the nasal cavity; each receptor projects olfactory hairs into the thin layer of mucus that lines the nasal passage. (b) When an odor molecule binds to the receptor, a chain of events involving second messengers ensues; the chain culminates in the relay of patterns of nerve impulses to the olfactory bulb of the brain. From there, other neurons carry the information to higher brain centers.

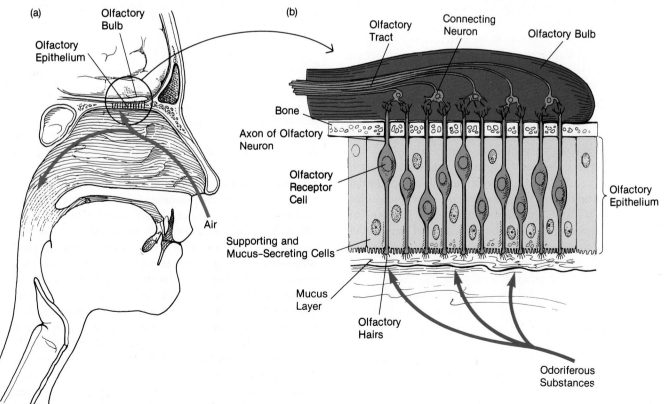

How do olfactory receptors work? One hypothesis is that the microvillar surface of each receptor cell is a patchwork of receptor molecules, which apparently have binding sites with precise configurations. Research suggests that the receptor–substance interaction may be like the lock-and-key interactions of enzyme and substrate or antibody and antigen. That is, the three-dimensional shape of a molecule determines whether it will bind to a given receptor molecule and so to a specific receptor cell. Simple chemical experiments support this view; for instance, adding a methyl (CH_3) group to a certain ring compound changes the compound's odor for humans from spearmint to wintergreen. Other studies show that molecules with different chemical formulas, but approximately the same three-dimensional shapes, tend to smell alike to humans.

Once an odorous substance—the musk in perfume, for example—binds to a glycoprotein receptor molecule, a series of events takes place and changes the baseline rate of nerve-impulse discharge of the olfactory receptor neuron. First, so-called G-protein responds to the receptor that binds the odorous molecule. G-protein activates the enzyme adenyl cyclase, which increases the level of the second messenger, cyclic AMP (Chapter 37). The cyclic AMP (or cGMP) binds to channel proteins, thereby opening the channels and permitting Na^+ to enter the receptor cell. That changes its state of polarization. (Recall that cells normally have a resting potential, say -70 mV, and that either hyperpolarization or depolarization may occur in response to ion entry.) As additional odor molecules bind and the polarized state changes further in response to ion entry, nerve impulses are generated either more rapidly or less rapidly than the baseline rate. Such changed patterns of impulses travel to the *olfactory bulb*, at the anterior end of the brain, where information processing begins. Certain odors—putrid and solvent smells—do not increase cyclic GMP levels, but instead may work via the phosophoinosital second messengers (Chapter 37).

Note that receptor cells respond to stimuli in a graded way—that is, changes in the cell's membrane potential are proportional to the number of odor molecules bound. As a result, the discharge of nerve impulses to the brain is proportional to the intensity and duration of stimuli and to the number of olfactory receptor cells that are affected. This phenomenon of graded response is a common feature of many sensory systems. We will see shortly that other types of sensory receptor cells also employ the G-protein and second-messenger sequence.

MECHANORECEPTORS

Mechanoreceptors may serve simple or incredibly complex functions, but all depend on the same physical

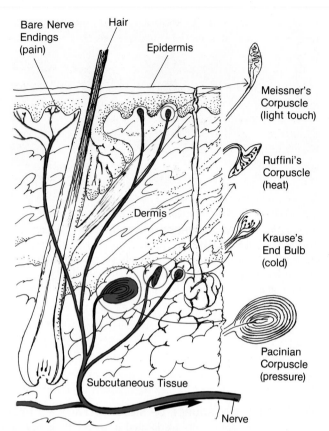

Figure 38-6   SENSORY RECEPTORS IN THE SKIN.
This longitudinal section through a piece of mammalian skin shows the major structures and their sensory capacities: Bare nerve endings in epidermis and dermis report pain. Meissner's corpuscles report light touch (pressure). Pacinian corpuscles report firmer pressure. Ruffini's corpuscles report heat, while Krause's end bulbs report cold. Pacinian corpuscles are classic mechanoreceptors in which onion-shaped layers of cells and fibers surround a mechanoreceptor dendrite; firm pressure downward on the skin, or body or limb movements that stretch or compress the skin, deform the "onion" and thus act on the sensory nerve endings.

principle: distortion of the shape of the mechanoreceptor cell and its plasma membrane. For a simple model, we can look at vertebrate skin (Figure 38-6), which contains two types of mechanoreceptors, *Meissner's* and *Pacinian corpuscles*, both of which are sensitive to tactile pressure. Pacinian corpuscles, for instance, are built of clusters of connective-tissue cells that resemble bunches of grapes; each Pacinian corpuscle contains the dendritic ending of a mechanoreceptor neuron. Firm pressure applied to the skin, whether the steady pressure of gripping a tennis racket handle or the misdirected blow of a hammer, stretches the connective-tissue cells and collagen fibers and pushes the Pacinian corpuscle out of its usual shape. As a result, the neuron's baseline firing pattern is altered, and the brain registers the steady touch or the crushing blow. In contrast, Meissner's corpuscles are much more sensitive and so report information on light pressures such as the gentle touch of a baby's fingers.

In addition to Meissner's and Pacinian corpuscles, skin appears to have relatively simple *pain-receptor* nerve endings (Figure 38-6). They may respond to pressure, temperature, or chemicals produced by tissue damage. As we shall see in Chapter 40, the neural pathways that transmit pain messages to the brain employ a special neurotransmitter known as *substance P*. Finally, skin contains hot (*Ruffini's corpuscles*) and cold (*Krause's end bulbs*) receptors, both of which are located deep in the dermis (Figure 38-6).

Variations on Meissner's and Pacinian corpuscles include mechanoreceptor cells that are wrapped around the base of mammalian body hairs and avian feathers and those within the sensory hairs on the body of an arthropod (such as crayfish, Chapter 36). When wind, flowing water, or other disturbances bend such hairs or feathers one way or another, the receptor cells are distorted, producing altered patterns of sensory impulses. Similarly, the stretch receptors, or **proprioceptors,** of joints, tendons, and muscles function like pressure receptors; how they provide information about the position of limbs, muscle contractions, and related processes will be described in detail in Chapter 39.

The Lateral Line of Fish and Amphibians

Often, sensory receptor cells themselves have specialized structures, such as cilia or microvilli, that, when bent, deform the cell's plasma membrane. Sensory cells of the **lateral line** in fish and amphibians are examples. As Figure 38-7 shows, this line is visible as a series of openings into a canal-like cavity or directly into pitlike depressions. Within each depression of the lateral line are clusters of sensory hair cells, each of which has a single cilium and a cluster of microvilli. The cilium and microvilli are embedded in a flexible, jellylike *cupula* (plural, *cupulae*). As the organism swims or water flows along its sides, the cupulae and the embedded cilium-microvillar complex are bent. Bending in one direction causes the sensory hair cells to release neurotransmitter faster than the baseline rate; bending in the opposite direction slows release below baseline. As a result, adjacent sensory neurons send impulses more frequently or less frequently to the brain. Depending on the species of aquatic vertebrate, the lateral line may detect the presence of nearby objects or organisms, sound vibrations, movements through the water, electric fields, and even minute variations in water temperature or salinity.

In the course of evolution, portions of the lateral line

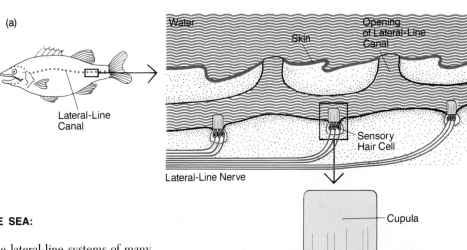

Figure 38-7   SENSATION IN THE SEA: THE LATERAL-LINE ORGAN.

(a) Sensory cells associated with the lateral-line systems of many fish and amphibians detect changes in the velocity of the water currents around the animal, as well as other environmental properties. (b) The hair cells generate nerve impulses in adjacent sensory neurons faster if the cupula is bent in one direction, and slower if it is bent in the opposite direction, thus signaling water flow and speed.

were incorporated into the inner ear of vertebrates, becoming the primary means by which humans and numerous other animals perceive body movements and sound. As we will see next, both of these essential sensory abilities depend on the hair cell.

The Statocyst and Inner Ear: Maintaining Equilibrium and Sensing Movement

When a cat tumbles from a ledge and lands on its feet, how does it "know" up from down? As a tidepool crab is overturned in the surf, how does it sense body position? Most of what is known about how animals monitor their position relative to gravity (that is, "up" or "down" relative to the center of the Earth) and sense motion (acceleration and deceleration) comes from studies of the vertebrate inner ear. Somewhat similar position-detecting systems operate in invertebrates, as well. In both types of organisms, such abilities depend on the deformation of sensory hair cells of the sort present in the lateral line.

The sensory hair cells that enable invertebrates to detect body position and perceive motion lie near structures called **statoliths**—masses of calcium-carbonate crystals, sand grains, or other dense materials. Like a yogi on a bed of nails, each statolith rests on the hairs within a fluid-filled sac called a **statocyst** (Figure 38-8). When the animal moves in position, the relatively heavy

statoliths shift position too; they bend the outer dendritic endings of the underlying sensory hair cells, some in one direction, others in different directions. This, in turn, alters the electrical activity of those cells and the release of neurotransmitter from their axons. Adjacent neurons then generate nerve impulses faster or slower and send a new sensory message to the central nervous system.

For example, the statoliths of a jellyfish adrift in the sea roll from side to side, relieving pressure on some sensory hairs and applying it to others. The changing output from many receptor neurons conveys information about the statocyst's position, and hence the position of the organism's body, relative to the center of the Earth (that is, to gravity). In rapidly swimming invertebrates such as shrimp or squid, the moving statocyst also communicates changes in position and motion as the animal speeds up, slows down, rolls, or turns.

In vertebrates identical functions are carried out within complex structures of the inner ear, fluid-filled chambers known as the *utriculus, sacculus,* and *lagena.* (In mammals, the lagena is called the *cochlea.*) Crystalline structures called **otoliths** (literally, "ear stones") rest on cupula material that overlies patches of sensory hairs on the bottom or side walls of each chamber (Figure 38-9). Like the statoliths of invertebrates, the otoliths shift and bend the sensory hairs to register changes in the position of an animal's head, and thus of its body as well.

In addition to providing information about gravity, the otoliths detect straight-line acceleration and deceleration. Thus as a jet airplane gathers speed for takeoff, passengers sense the craft's linear acceleration because otoliths in their ear chambers move backward while their bodies accelerate forward.

Animals that are free to move in water, on land, or in air must process sensory input about the extraordinary range of movements possible in three dimensions. Whether they scramble up tree trunks or somersault through the air, vertebrates obtain such information largely by way of the inner ear's tubelike **semicircular canals.** The canals measure angular (curving) accelerations and decelerations of the head and body.

To see how the system works, we must look briefly at the anatomy of the semicircular canals. On each side of a vertebrate's head, there are three canals, one oriented horizontally and two oriented vertically. The three are more or less at right angles to each other (Figure 38-10). Each canal is filled with a fluid, called **endolymph,** which moves freely into and out of the ends of each canal, something like water sloshing through a pipe. At one end of each canal is a hollow endolymph-filled chamber, or *ampulla* (the other end enters the utriculus or sacculus described earlier). Extending across each ampulla, much like a hinged door or gate, is a gelatinous cupula;

Figure 38-8   STATOCYSTS: SENSING POSITION AND GRAVITY.
Many invertebrates have statocysts composed of a sphere of sensory hair cells in which statoliths are found. As the body moves in space, the relatively heavy statoliths shift in position, and as a result bend different hairs in different directions. This statocyst is on an antennalike structure of a crab.

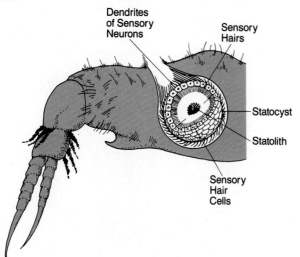

(a)

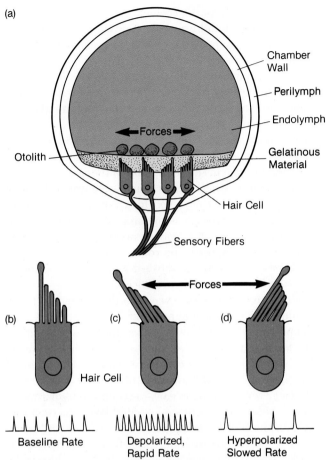

Chamber
Wall

Perilymph

Endolymph

Otolith

Gelatinous
Material

←Forces→

Hair Cell

Sensory Fibers

(b)

Forces

(c)

(d)

Hair Cell

Baseline Rate

Depolarized,
Rapid Rate

Hyperpolarized
Slowed Rate

Figure 38-9   MECHANORECEPTORS OF THE INNER EAR.
(a) Each of the three inner-ear chambers—the utriculus, the
sacculus, and in some species, the lagena—contains sensory
hair cells. The hairs of these cells are embedded in a
gelatinous cupula that is covered with otoliths. Body
position and gravitational force are constantly monitored by
the sensory hair cells as the otoliths shift due to body
movement. (b) In the resting state, the cilia are not bent,
and a baseline nerve-impulse rate is transmitted over the
nerve fibers. (c) When the cilia are bent in the direction of
the largest cilium, the cell depolarizes, and the impulse rate
rises. (d) When the cilia are bent in the opposite direction,
the cell hyperpolarizes, and the impulse rate falls below
baseline.

one end of the cupula is attached to the hairlike extensions of mechanoreceptor cells. How does this system of
pipes, fluid, and sensors operate?

Imagine that a shark coasting straight ahead suddenly
flips its tail so that its head turns to the left toward a
slow-swimming rock bass. When this occurs, endolymph
flows into the horizontal semicircular canal on one side of
the head and out of the horizontal semicircular canal on
the other side. Both flows exert force against the

cupulae, bending them slightly in one direction or the
other and stimulating their attached sensory hairs. In a
series of experiments, Werner Loewenstein, a physiologist at the University of Miami, discovered a mechanism
by which hair cells report which way their sensory hairs

**Figure 38-10   SEMICIRCULAR CANALS: SENSING
ANGULAR ACCELERATIONS AND DECELERATIONS.**
(a) The hollow, fluid-filled canal and chamber system of a
fish's left inner ear. Each of the three semicircular canals
possesses a cupula attached to a set of sensory hair
mechanoreceptor cells. A model semicircular canal is
depicted in (b). As the head moves in a counterclockwise
direction, the walls of the canal move in the same direction
(red arrows). Endolymph fluid in the canal moves in the
opposite, clockwise direction (blue arrows), thereby
displacing the cupula (in this case to the left). That
displacement in turn bends the "hairs" on the sensory hair
cells and changes the rate at which they cause generation of
nerve impulses. If the animal's head and, hence the canal,
stopped moving and rotated clockwise instead, the fluid
would flow counterclockwise, bending the cupula in the
opposite direction and changing accordingly the rate at
which impulses are caused to be generated.

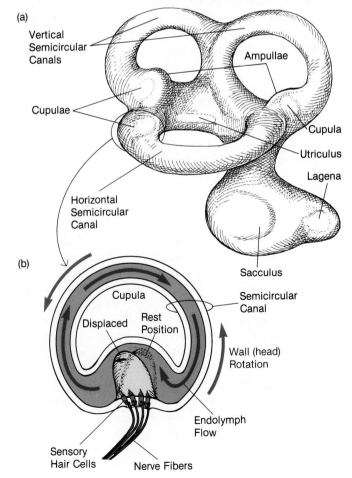

(a)

Vertical
Semicircular
Canals

Ampullae

Cupulae

Cupula

Utriculus

Lagena

Horizontal
Semicircular
Canal

Sacculus

Semicircular
Canal

(b)

Cupula

Displaced

Rest
Position

Wall (head)
Rotation

Endolymph
Flow

Sensory
Hair Cells

Nerve Fibers

are bent. He showed that the hairs' bending in one direction causes the cells to release neurotransmitter more rapidly, whereas their bending in the opposite direction produces slower neurotransmitter release. When the sensory hairs are not bent at all, the cells release neurotransmitter at the baseline rate. As with taste and olfactory receptors, the bending of hair cells produces these variations by altering ion flow through the cell. The rate of neurotransmitter release governs the rate at which postsynaptic neurons generate nerve impulses that are propagated to the brain. That information, and, indeed, all the impulses arriving continuously from all six semicircular canals, is interpreted in the brain along with information from the ear's otolith-chamber systems and the eyes. The result is continuous perception of position and movement.

As information on body movements or changes in position are reported by inner-ear sensory systems, one effect may be to trigger a reflex action in other body parts. For example, if you keep your eyes fixed on a distant point and swing your head back and forth, information from your semicircular canals travels to the brain. After interpretation, motor commands go to the muscles that move your eyes. As a result, the eyes and gaze can remain riveted on the point even though your head continues to move.

Together, the otolith-chamber systems and the semicircular canals provide every vertebrate with a constant stream of information about its position in space and about linear and angular movements. This information permits such marvelous behaviors as the acrobatics of a porpoise, the tumbling of an Olympic gymnast, and the unerring balance of a robin perched on a twig.

The Ear and Hearing

In many vertebrates, sound is detected in the inner ear. An animal's ability to hear, however, depends on the anatomy of the outer and middle ears—the sites where the initial stages of the hearing process take place.

Humans, foxes, bats, and cats have differently shaped external ears, but all are designed to gather high-frequency vibrations (sound waves) traveling through the air and to conduct them inward so that they strike the **tympanic membrane,** or *eardrum,* separating the outer and middle ears (Figure 38-11). This tightly stretched membrane vibrates when sound waves strike it, and it, in turn, sets in motion similar vibrations in a chain of three tiny bones in the middle ear—first the *incus* (sometimes called the hammer), next the *malleus* (anvil), and finally the *stapes* (stirrup).

As the incus, malleus, and stapes vibrate in sequence, they conduct sound waves through the air-filled middle ear to the **oval window,** a thin, taut inner-ear membrane

in the upper wall of the **cochlea.** As this chain of events proceeds, the sound waves are amplified, setting up vibrations that are powerful enough to overcome the inertia of an inner-ear fluid, the *perilymph.*

The mammalian cochlea is the primary structure involved in hearing. As Figure 38-12 shows, it is essentially a set of three tubes, coiled one atop another. In fact, the mammalian cochlea is really just the long lagena of a fish ear twisted into a space-saving spiral. Every time a sound wave sets the oval window vibrating, pressure waves arise in the perilymph in the cochlear chamber behind the oval window. Each wave travels through the upper chamber of the cochlea and, when it reaches the end, passes through an opening into the lower chamber. From there, the waves sweep along the lower chamber, ending up at the **round window** at the chamber's base, which bows outward with each vibration to relieve the pressure (Figure 38-12a).

How are such pressure waves in the perilymph transformed into the sense of hearing? As waves in the perilymph ripple through the cochlea, they travel along the **basilar membrane,** a thin sheet of tissue that lies along the base of the middle cochlear chamber. The **organ of Corti,** composed of thousands of sensory hair cells—about 17,000 in a human ear—rests on the basilar membrane and directly beneath a second layer, the **tectorial membrane** (Figure 38-12b and c). These cells and membranes are bathed in endolymph fluid. The spatial arrangement of these two membranes in the cochlea is the key feature of the ear's sound-sensing architecture (Figure 38-13). Thus, when a given sound enters the ear, its frequency generates pressure waves that displace a particular area of the basilar membrane. Wherever this displacement takes place, microvilli on the overlying hair cells are bent against the tectorial membrane at an angle. This angle of bending, in turn, determines the rate at which the hair cells liberate neurotransmitter at their synapses with auditory neurons. The neurons are thereby stimulated to generate nerve impulses that are propagated down axons that run in the **auditory nerve** to the brain. The result of this simple sequence is an auditory "message" telling the brain that sound of a particular frequency has entered the ear.

Whenever the oval window vibrates inward, the round window flexes outward, thereby equalizing pressure in the inner ear. In addition, the *Eustachian tube,* which connects the middle-ear chamber with the upper pharynx, serves as a pressure-release site, helping to prevent damage to the eardrum, oval window, round window, and other ear parts when an animal is exposed to very loud noises.

The range of frequencies to which an ear is sensitive varies with the length of the cochlea and the properties of its basilar and tectorial membranes. A human can hear frequencies as low as 2,000 hertz and as high as 20,000

COCHLEAR IMPLANTS: THE QUIET REVOLUTION

Sophisticated new devices based on an expanded understanding of the minute and delicate structures of the inner ear are helping thousands of people to hear again. The devices are called *multichannel cochlear implants*, and they work on a simple principle: by-passing a faulty connection between incoming sounds and signals to the brain.

Hearing involves detection of mechanical vibrations in the air—sound waves—that enter the outer ear, are channeled through the tiny bones of the middle ear, and are converted, or *transduced*, to electrochemical signals by the cochlea's sensory hair cells. When vibrations cause the outer ends (the so-called stereocilia) of hair cells to move, extremely low-level electrical changes occur in the cell. As a result, neurotransmitter is released and stimulates the adjacent nerve cell. Altered patterns of nerve impulses are then propagated, travel to the brain, and are interpreted as the reception of sound. It is the simultaneous transduction of sound waves by thousands of hair cells, each "tuned" to different vibrational frequencies, that allows us to hear a broad range of sounds, from high intensity to low, and to distinguish among complex sounds with mixtures of frequencies.

It is easy to see why problems arise if these hair cells are damaged; and, in fact, researchers now know that a number of factors can disrupt the cells irreversibly. One major factor is loud noise: according to one researcher, prolonged exposure to intense sounds "hacks apart the hair cells,"* so that their stereocilia, instead of standing erect, flop over like wet noodles. Another source of damage is high doses of antibiotics, such as gentamycin, neomycin, tobramycin, and streptomycin. These can cause the stereocilia to clump and thus fail to function. Other factors include illnesses—middle-ear infections caused by bacteria and viruses, mumps, measles, meningitis, and, in some cases, oxygen deprivation and high blood pressure.

If hair cells are damaged or die, hearing inevitably declines because hair cells do not divide and so are not regenerated. Sadly, as many as 200,000 people in the United States are deaf because of inner-ear damage. Often, their condition is mistakenly called "nerve deafness" when, in fact, the nerve cells can function but their hair cells can no longer transduce.

Some of these people can now be helped because medical researchers have designed devices called cochlear implants to by-pass the damaged hair cells and to deliver electrical signals directly to the nerves.

*A. J. Hudspeth, quoted in *Science News Magazine*, October 20, 1984, p. 254.

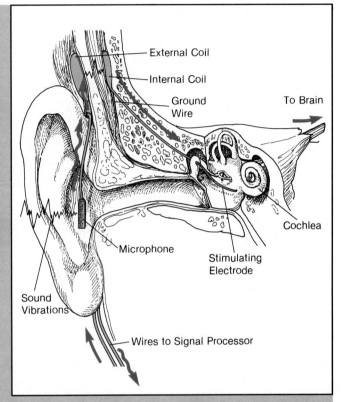

As the figure shows, the patient wears a tiny microphone connected to a signal processor that is the size of a deck of cards and is carried in a pocket. This processor selects the frequencies of sound in the range of human speech and transmits them to an electromagnetic coil on the side of the head. During a surgical procedure, an internal coil is implanted opposite the external coil, and from this, a set of eight or more electrodes runs directly into the cochlea, where hair cells and sensory-nerve cells reside. Sounds picked up by the microphone can then be converted to electrical signals that stimulate inner-ear nerves directly and enable the patient to hear once more.

The quality of that sound is not always high; some patients say that voices sound metallic or as though submerged in water, and the recipient of an implant must often be trained for months before he or she can recognize a ringing phone from a honking horn or a speaking voice. Nevertheless, cochlear implants can release people from social isolation and depression and can enable them to read lips more effectively and to take care of themselves more fully in a dangerous world. Researchers hope that as they learn more about the inner ear and how it transmits and transduces sound, they will be able design much more sensitive implants that will more successfully carry the all-important sound of human voices.

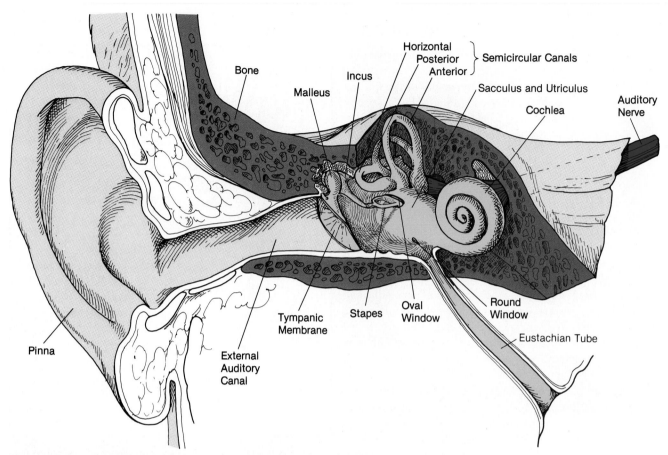

Figure 38-11   THE EAR.
The sense of hearing depends on the anatomy of the ear. The ear has three sections—the external ear, the middle ear, and the inner ear. Sound is channeled through the external ear inward to the tympanic membrane. Sound vibrations pass along the middle-ear bones, vibrate the oval window, and set up similar vibrations in the fluid of the inner ear. Those vibrations pass into the snail-shell-shaped cochlea, where they are detected by sensory hair cells. Pressure is relieved at the round window. The Eustachian tube connects the spaces around the inner-ear structures with the mouth cavity.

hertz; bats and porpoises may detect echoing sounds of 100,000 to 140,000 hertz. Even more amazing is the precision with which ear structures operate: the very loudest noises displace the basilar membrane only about 1 micron, and for the very lowest detectable noises the membrane may move distances 1,000 times smaller than the diameter of a hydrogen atom. Biologists do not yet know how such minute movements can be detected.

Extremely loud noises can permanently or temporarily deafen an animal. For instance, after exposure to an extremely loud sound of a particular frequency—say, one loud note at a rock concert—sensory hair cells at the site along the basilar membrane that is displaced most by that frequency may be damaged or destroyed. Such cells may be depolarized permanently, have disrupted microvilli, or have abnormal coupling of such microvilli to the tectorial membrane. Unfortunately, an increasing number of Americans now in their thirties are experiencing

hearing losses that may be traceable to exposure to very loud music years ago.

Locating Sounds

The mere perception of a sound may be of little use to an animal if it cannot also detect the direction from which the sound comes. Animals use two cues to gain such information: the relative loudness (intensity) of a sound at one ear or the other and the difference in time it takes for a sound to register in one ear and then the other. Small animals, in particular, measure differences in loudness: the ear nearest a sound source registers the sound as louder. For large animals, including humans, the relatively wide space between the ears makes time delay a more accurate gauge. For example, tests with humans have shown that a person can distinguish separate clicks reaching the two ears as little as 0.1 millisec-

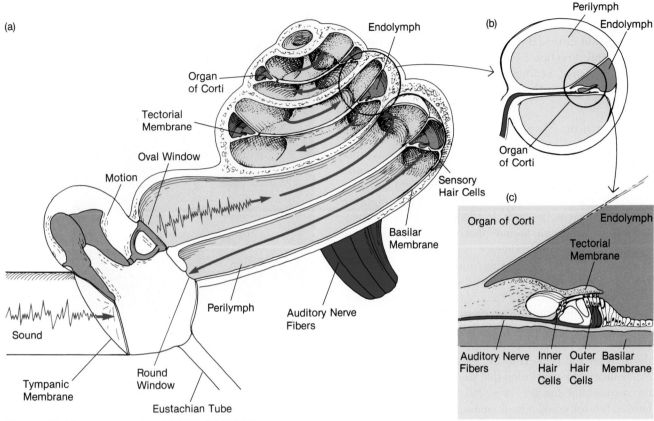

Figure 38-12   STRUCTURE OF THE COCHLEA.
The cochlea of mammals is the space-saving spiral structure in the inner ear that is involved in hearing. (a) This section of a model cochlea shows the tympanic membrane, middle-ear bones, oval window, and round window as sound striking the tympanic membrane sets up compression waves that travel in the perilymph fluid of the cochlear chambers (arrows). (b) and (c) The middle chamber—seen in (a) at the edges of the snail-shell-shaped cochlea—contains the organ of Corti with its sensory hair cells that rest on the basilar membrane. The tectorial membrane rests on the sensory hairs themselves. Compression waves traveling outward from the oval window in the upper chamber, and then back in the lower chamber toward the round window, displace the basilar membrane. That, in turn, moves the sensory hair cells relative to the tectorial membrane, resulting in altered release of neurotransmitter by the hair cells. Patterns of nerve impulses over the auditory nerve (not shown) are changed as a result. The frequencies of sounds entering the ear can be reported because high frequencies displace the basilar membrane maximally in regions near the oval and round windows; progressively lower frequencies displace the basilar membrane progressively farther from those windows and closer to the upper part (apex) of the cochlea (the "tip" of the snail shell).

Figure 38-13   CELLULAR ARCHITECTURE UNDERLYING HEARING.
This scanning electron micrograph shows the organ of Corti with the overlying tectorial membrane removed (magnified about 1,700 times). Each V-shaped set of structures is the group of so-called stereocilia—the sensory hairs—that arise from a single underlying mechanoreceptor hair cell. The single row on the right also arises from sound-detecting hair cells. The four rows of such hair cells run the full length of the basilar membrane and organ of Corti. When sound waves vibrate perilymph in the cochlea, the cells visible here would vibrate up and down (above and below the plane of this page) and thus the stereocilia would be bent as they moved relative to the overlying tectorial membrane. From *Tissues and Organs: A Text-Atlas of Scanning Electron Microscopy* by Richard G. Kessel and Randy H. Kardon, W. H. Freeman and Company © 1979.

ond apart. In general, though, highly visual organisms do not attempt to use hearing to locate a sound source precisely but instead flick their eyes immediately toward the general direction from which a new sound comes. Thus directional hearing is the trigger, but the eyes serve as higher-precision directional devices.

Echolocation: "Seeing" with Sound

Small nocturnal animals, such as shrews and bats, and aquatic mammals, such as porpoises and whales, are often active under conditions of reduced light and low visibility. They have evolved the ability to use sound to sense their way in the dark. In 1793, the Italian scientist Lazzaro Spallanzani found that when he released blinded bats in his laboratory, they could fly freely, avoiding walls and skimming under tables. Furthermore, when blinded bats released out of doors were recaptured, they had bellies full of moths and other flying insects; even though they were unable to see they could easily find and capture food. When Spallanzani plugged the bats' ears with wax, however, the animals were badly disoriented and refused to fly. Despite a series of such experiments, he concluded that some "sixth sense" must be responsible for the bats' behavior; because Spallanzani himself could hear no sounds emanating from the animals as they flew, he reasoned that sound was not likely to be a factor. Biologists using modern instruments can now measure the sounds bats emit from their specialized larynx in the ultrasonic range—well above that detectable by the human ear.

The sound-based system that bats use to navigate and hunt in the dark is known as **echolocation** (Figure 38-14). As the animal flies, it produces a series of sound pulses, at the rate of about 20 per second, each with a frequency ranging from 40,000 to 100,000 hertz—well above the maximum frequency audible to humans. These pulses are unbelievably intense (loud): a few centimeters in front of the bat's mouth, the noise they generate is equivalent to the roaring sound of a subway train we hear when standing just a few feet away. Such pulses of sound bounce off objects in the animal's path—including moths, tree branches, and other obstacles—and produce echoes that the bat is able to hear and analyze.

Once the bat senses an object, the rate of sound-pulse production increases to 50 or 80 per second, so that the animal can fine-tune its hunting or navigational behavior. The result is an amazingly detailed acoustic picture of the bat's world that enables it to fly rapidly through dense forest, distinguish flying insects from tree branches and leaves, or locate its roost site in a cave—all in the blackness of night.

Studying bats has yielded a fascinating example of the evolutionary ripple effect an adaptation like echolocation can have. Echolocating bats often prey on night-flying

Figure 38-14   ECHOLOCATION.
Bats navigate by generating ultrasonic sound pulses that are inaudible to the human ear. Echoes of the pulses provide information about an object's position, direction of movement, and even composition. The huge ear pinnae help this pallid bat (*Antrozous pallidus*) to gather these faint echoes, and may aid in directional location of the echo source.

moths (called noctuids), which, in turn, have evolved tympanic membranes and sensory hair cells in their abdomen that pick up the bats' high-frequency sound pulses. When the moth hears a bat (sometimes as far away as 65 meters), it dives into the grass or makes other evasive maneuvers to avoid being captured. It happens, however, that noctuids can be deafened by infestations of tiny parasitic mites, a condition that could be fatal for both moth and parasite. In fact, these mites have evolved a cooperative behavior in the survival strategy of their host by restricting their infestation to just one ear on a host moth. The moth thus retains its keen hearing sense in the other ear, while its parasitic partners avoid becoming an inadvertent "condiment" in a bat's dinner.

Porpoises, whales, and other marine mammals generate echolocating clicks in their larynx or in the system of air-filled tubes leading to the blowhole. In addition, such animals have an array of adaptations for gathering directional information, including isolation of the inner ear in a foamy cushion so that sound may reach the cochlea only from the eardrum. Perhaps most remarkable, however, are the eerily beautiful moans, whistles, squeaks, and other noises that sea mammals use to communicate and, apparently, coordinate highly complex group behavior.

Interestingly, blind humans are sometimes able to generate interpretable echoes from the sound of a tapping cane or clicks of the tongue. Such sounds can provide excellent directional information, and with practice, some blind persons can learn to distinguish "targets" of different shapes. Overall, however, human attempts to use echolocation are only primitive approximations of the finely honed acoustic abilities that have evolved over millennia in porpoises and bats.

The Electroreceptors of Fish

Thus far we have seen how the basic mechanism by which mechanoreceptors operate—distortion of the shape of a specialized sensory cell—enables animals to detect both sound and motion. A variation of this sensory mechanism in fish is the **electroreceptor cell**—a modified mechanoreceptor of the lateral line that has no cilia and is capable of detecting weak electric currents. Such cells, found in sharks, catfish, and some other species, rest in deep pits along the lateral line; electric current flows into or out of the cell through the plasma membrane that faces the water. As the current flow varies, the rate at which the cell releases neurotransmitter (which, in turn, affects neurons leading to the brain) rises or falls. Some animals that possess these electroreceptor cells can utilize them to detect minute electric currents produced by the functioning muscles and tissues of prey and conducted through the water. The so-called weak electric fish (*Mormyrids*) carry this ability even further: they have an *electric organ* in their tail that generates a continuous stream of weak electric pulses (each lasting a few milliseconds) that travel through the water and are detected by the animal's lateral-line electroreceptors (Figure 38-15). When a rock, plant, crayfish, or other object disrupts the electric field, the fish interprets the disturbance so as to avoid obstacles or find food in murky water.

Magnetism: Detecting the Earth's Magnetic Field

A major feature of mechanoreceptors is that they enable an animal to sense its body orientation with respect to its environment. Few modern biological discoveries have been more surprising than the finding, in the 1970s, that birds, bacteria, and perhaps many other types of organisms orient their bodies to the Earth's magnetic field. Thus as long as nothing interferes with its magnetic sense, a homing pigeon that navigates by the sun on a clear day finds its way just as well in heavy overcast. And, as we will see in Chapter 48, many of the millions of birds that migrate long distances each year

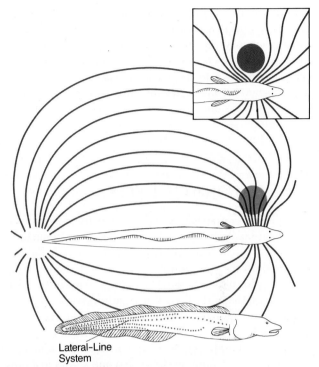

Lateral-Line
System

Figure 38-15   ELECTRIC ORGAN OF THE WEAK ELECTRIC FISH.
Current is produced by the electric organ in the fish's tail and is received by electroreceptor cells located in the lateral-line system (primarily in the head region). The varying conductivity of different objects around the fish leads to changes in the perception of current flow and to a resultant sensing of those objects near the fish. In this diagram, the orange object has a conductivity greater than water so that the current is directed toward the object. The purple object has a conductivity lower than water, and deflects the current away.

use magnetism as a source of compass direction. How can living things perceive a force as elusive as magnetism?

Many birds and at least one mammal—the porpoise—have a small number of tiny crystals of magnetite (a form of iron oxide) in tissues near the brain. Some humans have such crystals in the delicate bones of their sinuses. Likewise, *magnetobacteria* have magnetite crystals in their cytoplasm and swim toward or away from the poles of an applied magnetic field. Precisely how higher organisms use magnetic particles to sense the direction of the lines of magnetic force generated by the Earth's core remains a subject of research. One hypothesis is that when the crystals—in a pigeon's head, say—are aligned parallel to the Earth's lines of magnetic force, perhaps in a north–south direction, mechanical torque (twisting movements) does not act on those crystals. However, if the bird turns to the right or left, weak magnetic forces may tend to twist the crystals back toward north–south, just as happens with a compass needle. This twisting movement may, in turn, stimulate nerve-fiber endings located near the crystals so that the pigeon can sense its direction relative to the planet's magnetic field. Biologists do not yet know whether this sort of mechanoreception process operates or whether some type of electromagnetic sensing underlies a vertebrate's ability to use the Earth's magnetic field as a compass.

THERMORECEPTORS

Whereas a few species can detect electrical fields with lateral-line cells and a few others may react to magnetism, nearly all animals are able to respond to variations in temperature, as we saw in Chapter 35. In various vertebrates, this ability is due to nerve endings in the skin and tongue that alter their firing pattern when the local temperature changes. Warm receptors generate nerve impulses faster as temperature rises; conversely, cold receptors generate impulses faster when the temperature of the nerve endings drops. No one knows just how these cells function, but for clues, biologists have turned to animals with a special capacity to sense temperature—in particular, snakes that hunt their food by detecting temperature changes.

Imagine for a moment a mouse silently scurrying about in search of its dinner on a moonless night. Suddenly, there is a rustling nearby; the mouse freezes and remains perfectly still for minutes. In spite of this precaution, however, a rattler strikes with deadly aim, and the mouse becomes the snake's dinner. Such flawless marksmanship in the dark is dependent on *infrared detectors*, organs located in the snake's *facial pits* that detect the long infrared wavelengths of radiant-heat energy (Figure 38-16).

The thousands of receptor nerve endings in the infrared detectors probably function similarly to the heat-sensitive thermoreceptors on an ant's antennae or on the tip of a human's tongue. But a snake's facial pits are supersensitive in comparison with the warm and cold receptors of an insect or a person; they can detect a temperature increase of only 0.002°C. Consequently, a rattler can sense, within half a second, a mouse whose body temperature is 10°C warmer than its environment if the mouse is within range of about 40 centimeters.

The snake's paired pit organs are aimed toward the front and side, so that by turning its head from side to side, the animal can scan the immediate area and perceive the right and left edges of a warm object, striking at the center with amazing precision. The snake can also sense when an animal is too big to be swallowed whole. Confronted with a coyote, for instance, it will slither away to seek a smaller, safer meal.

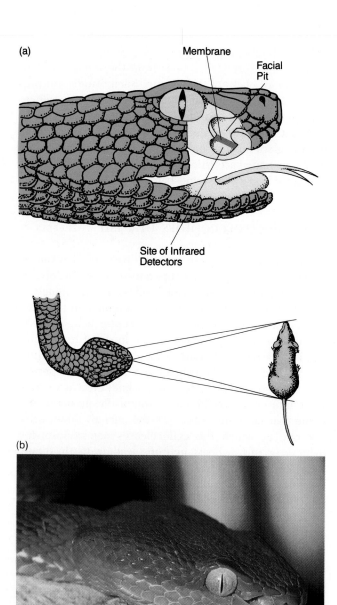

Figure 38-16   THERMORECEPTORS
(a) A pit viper's infrared detectors help it find prey in the dark. The paired infrared detectors are located in structures called facial pits on the pit viper's head. These thermoreceptors can detect minute differences between the infrared level of the background and that given off by a living animal. (b) The facial pit of this bamboo viper, another member of the pit viper family, is visible as a dark line just anterior to the eye.

PHOTORECEPTORS

In Chapter 8, we saw that the photochemical reaction of photosynthesis is driven by certain wavelengths of the electromagnetic spectrum. In fact, only a narrow region of this spectrum can be tapped to provide energy for biological processes; this area includes the visible wavelengths that enable some organisms to see the world around them.

Light-sensitive structures in animals are amazingly diverse. For example, some protists have clusters of light-sensitive molecules in eyespots that allow their owners to move toward or away from a light source. Flagellates, whose light-sensing pigment is masked on one side, will perform a kind of miniature ballet when a light is shined from one side on a culture dish in which the tiny creatures are swimming. As each cell rotates on its long axis as it swims forward, the light receptor is alternately activated and shaded; whenever the receptor enters a shadow, the beat of the cell's flagellum causes the organism to turn toward the light. Soon every cell in the dish has oriented itself so that its eyespot remains facing the light source as it rotates toward the light. Though this cellular dance to light seems unique, the light-absorbing pigments (called opsin) found in organisms as diverse as archaebacteria, unicellular algae, fruit flies, goldfish, and cows share the same amino-acid sequences. The gene probably originated billions of years ago in prokaryotes, reflecting the ancientness of the capacity to absorb light energy.

Among simple multicellular animals such as flatworms, light-sensitive cells are often arranged to provide directional sensitivity to light, as well as to detect on–off changes in light, such as those caused by the moving shadow of a predator swimming overhead. In more complex animals, accessory cells form a lens that focuses light on the underlying light receptors. This arrangement has several advantages:

1. Since the lens gathers and focuses light, it allows *lower intensities* of light to be detected.

2. It allows an animal to sense *direction*, since receptors aimed in one direction may be stimulated while those aimed elsewhere are not.

3. It permits the sensing of *motion*—as when the image of a predator moving across the field of vision is sensed first by one set of receptors, then by another, and so on.

These design features are displayed especially well by the wonderfully complex camera eyes of mollusks and vertebrates and the compound eyes of arthropods.

The Vertebrate Eye

Vertebrate eyes, and those of octopuses and squid, function much like cameras: a single lens focuses light images on a sheet of light receptors (equivalent to the film), and the image is then "developed" by neuron processing in the visual cortex of the brain. A survey of its anatomy will show how the camera eye works.

Typical camera eyes are roughly spherical, like the human eye shown in Figure 38-17. The bulk of the eye is surrounded by a tough, protective sheet of connective tissue, the *sclera*, and by the *choroid*, a layer which contains blood vessels that carry oxygen and nutrients to the eye and remove carbon dioxide and wastes. The **cornea,** the transparent portion of the sphere through which light enters, is specialized skin tissue. It is composed mainly of aligned collagen fibers—somewhat like densely interwoven, see-through threads—that are tough and transparent at the same time. In Figure 38-17, notice also the *aqueous-* and *vitreous-humor* chambers— pressurized fluid-filled spaces that help give the eye its shape.

When light enters the eye, it passes through the cornea and aqueous humor. It may then be halted by the **iris,** or it may pass through the iris's dark central opening—the **pupil**—to the **lens,** a structure built of highly ordered cells (Figure 38-18). In a human eye, the iris is a pigmented ring of tissue (blue, brown, green, gray, or shades in between) that can be opened or closed by radially oriented muscles under reflex control. If you stand in front of a mirror in a darkened room for a few minutes and then turn on the light, you can watch your irises quickly close around the dark pupils to reduce the amount of light entering the eyes. The reverse also occurs: when a vertebrate moves from a brightly lit place to a dark one, the iris rapidly opens to let more light enter the pupil. In addition, the parts of visual receptor cells that absorb light are moved in position as more or less light enters the eye, so that the eye can respond quickly when light levels change. These visual strategies of *light* and *dark adaptation* are vital characteristics of vertebrate eyes, for without them, the eye's light-absorbing molecules, described shortly, would be subject to "bleaching," severely curtailing the capacity to see.

Light rays traveling through the pupil and the lens are focused on the rear surface of a light-sensing epithelial layer, the **neural retina.** An important component of the neural retina in primates, birds, and some reptiles is the **fovea,** a region about 1 millimeter in diameter where maximal *acuity* (high-resolution viewing) is achieved. In all vertebrate eyes, there is an optic disk, or "blind spot," where nerve fibers of the neural retina exit from the eye over the **optic nerve,** which goes to the brain. Finally, surrounding the neural retina is the **pigmented retina,** a layer that absorbs or reflects light that chances

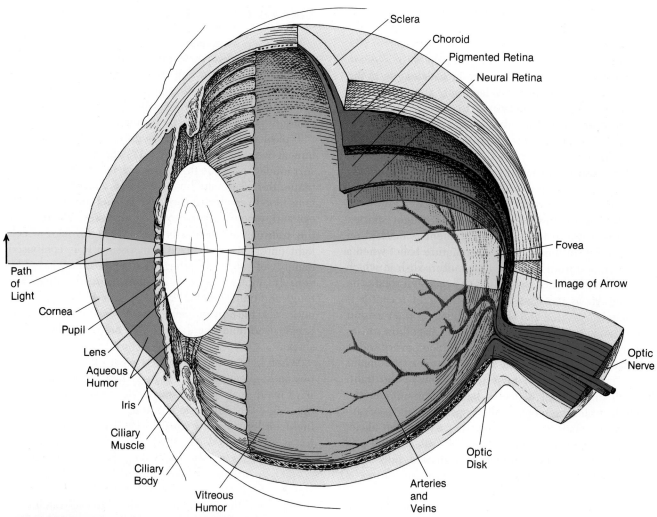

Figure 38-17   ANATOMY OF A MAMMALIAN EYE.
Light—here, the image of an arrow—passes through the cornea and the lens and is focused on the neural retina. The eye's various layers are described in the text. The fovea is the site of highest activity; the optic disk region is a blind spot.

to pass unabsorbed through the neural retina. Before learning how the retina works, let us first examine how focusing takes place.

Forming an Image: Focusing the Eye

When we focus a camera, we actually move the lens toward or away from the film, thereby changing the dis-

Figure 38-18   THE LENS OF THE EYE: A STRUCTURE OF PRECISELY STACKED CELLS.
Geometric stacking of precisely shaped living cells helps give the lens its unique properties of transmitting and bending light. Here a mammalian lens (magnified about 600 times) has been fractured to reveal this internal arrangement; each of the long, thin rectangles is a living cell, stacked together like lumber in a pile. From *Tissues and Organs: A Text-Atlas of Scanning Electron Microscopy* by Richard G. Kessel and Randy H. Kardon, W. H. Freeman and Company © 1979.

tance between lens and film. In nature, fish, amphibian, and snake eyes focus in just this way. In birds and mammals, however, both cornea and lens participate in the focusing process.

First, light rays are bent to varying degrees, depending on the curvature of the cornea. This corneal focusing is usually sufficient to provide a sharp image of distant objects. As it takes place, the lens is held taut in a somewhat flattened position by the elastic ligament of the *ciliary body*, which supports it. When the object being viewed is near, however, the cornea cannot bend the light enough to focus the image on the retina, so the *ciliary muscles* contract, allowing the lens to round up. This increases the angle of bending of the light rays, bringing the image into sharp focus. Birds also bend the cornea to help this process. In land vertebrates, this process—called *accommodation*—goes on constantly as an animal shifts its gaze from one object to another at differing distances.

Some people have abnormalities in the shape or functioning of their cornea or lens that prevent their unaided eyes from focusing images of near or far objects on their retinas. Often, focusing problems develop as a person ages; this is because, over time, the lens loses its elasticity and thus does not round up well when the ciliary muscles contract. As a result, the person cannot focus well on close objects and is said to be farsighted. Conversely, nearsighted persons cannot reduce the curvature of their lens or cornea enough to permit focusing on distant objects.

The Retina

The neural retina, on which an image is focused, is really part of the wall of the brain that is specialized for vision. (Recall from Chapter 17 that the eyes arise from two outpouches of the brain.) The highly complex neural retina contains the visual receptor cells, or photoreceptors, as well as a variety of nerve cells that are the first links in the chain of events by which visual input is processed.

In many vertebrates, including humans, the eye contains two types of photoreceptors, **rods** and **cones,** whose names reflect their characteristic shapes (Figure 38-19). They differ from each other in structure and in the light-absorbing pigments they contain. As rod and cone cells develop in the embryonic eye, both form a highly modified cilium, the *outer segment*, where the visual pigments are assembled in membranes; events during development lead these outer segments to point away from the center of the eye—that is, away from light.

Rods are highly sensitive: they respond even when the amount of light entering the eye is very low, such as at night or in deep water. Cones, on the other hand, react only to higher light intensities and so are most active in daytime light levels. Cones also provide high visual acu-

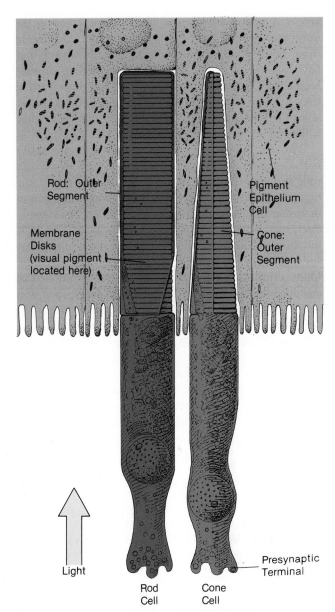

Figure 38-19   VISION AND THE STRUCTURE OF ROD AND CONE CELLS.
The outer segments of both rods and cones are pointed away from the source of light. Rods are extremely sensitive to dim light and are important in night vision (distinguishing black and white). Cones are important for color vision and sharp visual acuity. Neurotransmitter is released spontaneously from the presynaptic terminals when rods and cones are in the dark. The effect of light is to slow that release—a change that is translated into nerve impulses sent to the brain by other cells of the neural retina.

ity (the fovea is made up entirely of cones in some animals) and, as we shall see in the next section, color vision.

The density of cones in an animal's foveas determines how sharp its sight can be. Human foveas have about 160,000 cones per square millimeter; in contrast, the foveas of a hawk contain roughly 1 million cones in the same area, giving the hawk a visual acuity some eight times better than a person's. In fact, birds of prey may have several foveas in each eye, arranged to provide high-acuity vision both straight ahead and downward as they fly. This adaptation for keen visual perception enables the bird to detect a small mouse in the grass while it flies far up in the air.

Whereas humans and other animals that are active both day and night (so-called diurnal animals) have both rods and cones in their retinas, nocturnal creatures and organisms that live in low-light environments, such as deep-sea fish, tend to have predominantly "rod eyes." Such eyes are marvelously adapted to life in the dark, but they usually lack color vision and the ability to resolve fine details, both processes carried out by cones. Many diurnal animals—those of the cat family are an example—also lack color vision. This is one reason why zebras, as potential prey, have what appear to humans to be starkly contrasting black and white markings. To our eyes, with their color vision, zebras stand out against green or tan grasses, but to a lion's eye, a zebra's stripes blend into the background.

Both rods and cones possess millions of molecules of a visual pigment called **rhodopsin.** This pigment is composed of a lipoprotein, **opsin,** and the light-receptor molecule itself, **retinal.** Derived from vitamin A, retinal is a small molecule with a unique property: when it is activated by a photon of light, it undergoes a radical change in shape, forming a structural isomer of the original molecule (Figure 38-20). This is the sole role that light plays in vision; the energy of a photon is converted into molecular motion in retinal. But how does this minute change translate into the image of a rosebud or of letters on a page? The surprising answer is that a process much like hormone binding starts the electrical activity in the nervous system.

Imagine an eye that has not been exposed to light for some hours. The retinal portion of all rhodopsin molecules is bent in an extremely unstable configuration called 11-*cis* retinal. Retinal in this shape sits snugly in a pocket on the surface of the opsin molecule (forming rhodopsin). If a light is flashed at the eye, photons will strike some rhodopsin molecules. If sufficient light energy of the correct wavelength is captured by an 11-*cis* retinal molecule, it isomerizes to a straight configuration called all-*trans* retinal. But the all-*trans* molecule fits poorly in the pocket on opsin, and so the opsin immediately changes in volume and position in the lipid-bilayer membrane in the rod or cone. These changes in both retinal and opsin produce what is called activated rhodopsin. It is at this point that the hormone analogy comes into play.

A single photoactivated rhodopsin molecule instantly triggers activation of about 500 **transducin** molecules. This activated protein immediately binds to an enzyme

Figure 38-20   CHANGES IN THE STRUCTURE OF THE RETINAL MOLECULE.
When light strikes 11-*cis* retinal, the configuration changes to all-*trans* retinal. An enzyme later converts the molecule back to the resting *cis* formation. The effect of light on retinal triggers a chain of events that culminates in hyperpolarization of the rod or cone.

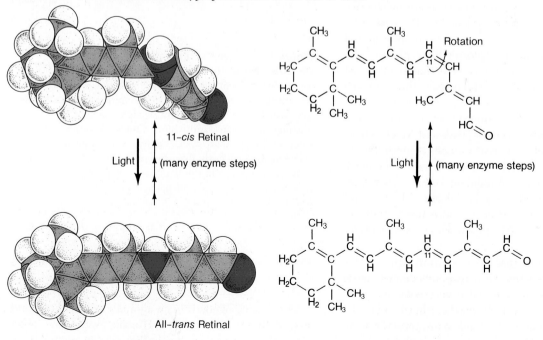

Table 38-2 **SENSORY TRANSDUCTION CASCADES**

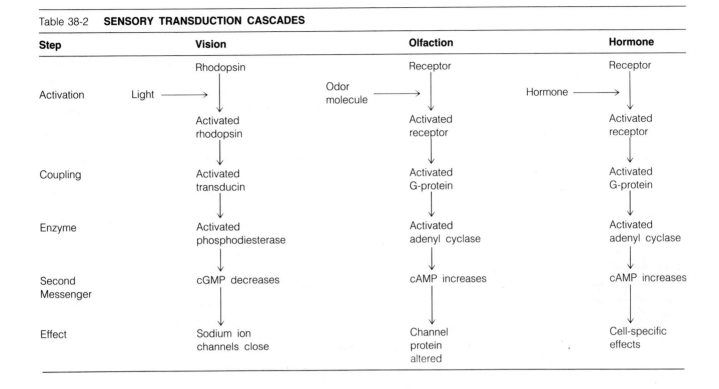

Step	Vision	Olfaction	Hormone
Activation	Light → Rhodopsin → Activated rhodopsin	Odor molecule → Receptor → Activated receptor	Hormone → Receptor → Activated receptor
Coupling	Activated transducin	Activated G-protein	Activated G-protein
Enzyme	Activated phosphodiesterase	Activated adenyl cyclase	Activated adenyl cyclase
Second Messenger	cGMP decreases	cAMP increases	cAMP increases
Effect	Sodium ion channels close	Channel protein altered	Cell-specific effects

called *phosphodiesterase (PDE)*, activating it in turn. PDE hydrolyzes about 400,000 cyclic GMP molecules per second (recall from Chapter 37 that cyclic GMP is a second messenger related to cyclic AMP). The hydrolysis of cGMP removes it from sodium ion channels in the plasma membrane of the outer segment of the rod or cone. Such removal closes the sodium ion channels. All told, the single photon blocks entry of about 1 million sodium ions into the outer segment. The result is hyperpolarization of the visual receptor cell, the initial electrical event in vision.

Compare this sequence with that of olfaction or hormonal action (Table 38-2). The activated rhodopsin or receptor interacts with a signal-coupling protein (transducin or G-protein), all of which are members of a family of proteins with similar amino-acid sequences. Then the signal-coupling protein can act on any of the enzymes that synthesize or degrade the second-messenger cyclic nucleotides. As we saw in Chapter 37, that may control a vast array of cellular events. Note that each step in these sequences is a point of amplification; a minute energy input in the form of a single photon or hormone molecule can cause a cascadelike process leading to altered cell behavior.

Returning to the photoreceptor cell, light has temporarily closed some of its sodium channels on the outer segment so that the cell hyperpolarizes. As a result, the spontaneous release of neurotransmitter (a process that goes on continuously in the dark) is inhibited. The next

cell in line, a true neuron, becomes hyperpolarized too. That, in turn, triggers the third cell to begin generating nerve impulses that carry the message to the brain. We will return to these other cells momentarily.

But what about all-*trans* retinal molecules, which are now bleached and unable to respond to light? Retinal quickly leaves opsin and diffuses to nearby pigmented retina cells where an enzyme restructures it into the 11-*cis* configuration; it then returns to the outer segment and binds to a vacant pocket on opsin, yielding functional rhodopsin once again. During normal vision, millions of these cycles—11-*cis* to all-*trans* to 11-*cis*—go on every second. The huge amount of energy required for this and other aspects of vision leaves the retina critically dependent on a high level of oxygen delivery by the blood so that ATP can be manufactured.

Seeing in Color

The "colors" we perceive in a shimmering rainbow or a peacock's bright plumage are sensations produced by different wavelengths of the electromagnetic spectrum. In addition to a few mammalian species with the remarkable ability to discriminate among wavelengths, many insects, fish, reptiles, and birds have some form of color vision.

The accepted theory of color vision, the **trichromatic theory,** which dates from 1802, states that there are three classes of cones, each with a slightly different type

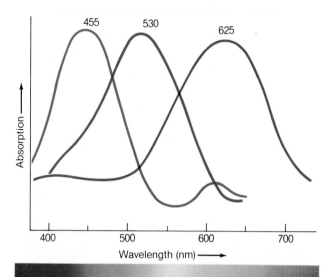

Figure 38-21   CONES WITH DIFFERENT COLOR SENSITIVITIES.
Measurements of cones in a goldfish eye show three different classes of cones. One class absorbs primarily blue wavelengths of light, another absorbs mostly green wavelengths, and the third is most sensitive to red wavelengths of light. These types of measurements serve as the basis for the trichromatic theory of color vision. Each color—say, at 500 nm—is defined by the activity of all three cone types at that wavelength. A different opsin is produced in each of these cone cell types to account for the differing spectral sensitivity.

of opsin molecule in its rhodopsin. Measurements originally made on fish (Figure 38-21), and later on mammals and other vertebrates, show that one cone type is maximally sensitive in the blue wavelengths, another in the green, and the third in the yellow or red; each type of cone responds to lesser extents in other parts of the spectrum. Color of any given hue—say, blue-green—evokes a characteristic level of activity from each of the three cone types; orange light, on the other hand, would yield a different ratio of activities among the cones. It is the complex activity patterns of the three types of cones that the brain interprets as either blue-green or orange.

In 1986, researchers using recombinant DNA techniques proved the correctness of the trichromatic theory. By studying human males, they found several separate genes for the opsins: one for blue-sensitive opsin; one for red-sensitive opsin; and one, two, or three for green-sensitive opsin, depending on the individual. The red and green genes occur only on X chromosomes. Males sometimes lack the red gene, sometimes the green one, and these absences account for the two typical kinds of human color blindness. Women, with their two X chromosomes, are almost always heterozygous for these recessive defects. While not color-blind them-

selves, women who carry the X chromosome lacking the red or green gene are likely to pass it on to half of their sons.

Jeremy Nathans and colleagues at Stanford Medical School also found evidence suggesting that ancient primate ancestors probably had the blue opsin gene and another opsin gene sensitive to long wavelengths (corresponding to the modern red gene). Some time after the New World monkey lineage arose, the African primates apparently had a duplication and gradual divergence of the green gene from the red. The opsin genes may still be evolving in humans: Nathans found a number of men with hybrid red and green genes of the sort that might arise through mistakes in crossing over during meiosis. Perhaps in the future, some selective pressure for improved color perception might expand the numbers of color-sensing opsins in humans or other vertebrates so that they see the world in new colors.

Color vision, which involves only cones in land vertebrates, requires high light intensities. This is why, if you go out of doors at night, only your rods are sensitive enough to function, and the world is composed of black, gray, and white images. (The brighter red taillights of a passing automobile are detected by cones, however.) The condition known as "night blindness" may be due to a deficiency of vitamin A in the diet, resulting in a chronically low supply of retinal in the rods and cones.

Visual Processing in the Retina and the Brain

As images of the world are focused on the retina, rods and cones hyperpolarize to varying degrees, and the result is a stream of neural information that is turned into the sense of vision. To see how this extraordinary transformation takes place, we shall trace a chain of neural-processing steps, beginning with the neurons associated with rods and cones.

Rods and cones form synapses with several types of neurons, most frequently with **bipolar cells.** These bipolar cells, in turn, synapse with **ganglion cells,** neurons whose axons extend along the inner surface of the eye cavity, exit the eye at the optic disk, and travel via the optic nerve to the brain (Figure 38-22).

The axons of ganglion cells are the sole information channels from eye to brain. This means that the wiring of rods, cones, bipolar cells, and other nerve cells in the retina determines the quality of vision. For instance, in the primate fovea (which is the site of highest visual acuity), there may be a simple one-to-one hookup from a cone to a bipolar cell to a ganglion cell. This means that there is point-by-point reporting to the brain of precisely which cones are struck by photons. In most parts of the retina, however, many rods (or cones) synapse with a single bipolar cell, and a number of bipolar cells synapse

(a)

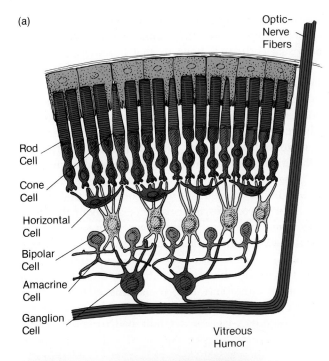

Optic-
Nerve
Fibers

Rod
Cell

Cone
Cell

Horizontal
Cell

Bipolar
Cell

Amacrine
Cell

Ganglion
Cell

Vitreous
Humor

(b)

Figure 38-22   LAYERS OF THE RETINA.
(a) The hierarchy of visual processing by the retina of the
eye. Rods and cones absorb light. They then form synapses
with either horizontal cells or bipolar cells; bipolar cells
continue the processing by synapsing onto amacrine cells
and ganglion cells. Ganglion cells send nerve impulses to
the brain via the optic nerve. Note that a single ganglion
cell receives input from many rods and cones. (b) The
scanning electron micrograph shows the cell layers of the
retina (magnified about 645 times). The outer segments of
rods and cones are visible at the top; ganglion cells are
visible along the bottom.

with each ganglion cell. In a bat, for instance, 1,000 rods
synapse with varying numbers of bipolar cells, all of
which synapse with just one ganglion cell. In the human
eye, some 131 million rods and cones are connected via
bipolar cells to only 1 million ganglion cells. The full set

of rods or cones joined in this way to a single ganglion
cell makes up the **visual field** of that ganglion cell; the
functional consequence is that photons hitting any of the
rods or cones connected to the ganglion cell can trigger it
to generate a nerve impulse.

The larger the visual field of a ganglion cell (that is, the
more photoreceptors that report information to one ganglion cell), the lower its acuity, or resolving power. This
is why night vision, which relies solely on rods, is never
very sharp; images in the nighttime world are constructed of relatively large "patches" (a patch is the large
visual field of a ganglion cell connected to many rods); in
contrast, the patches of the daytime "cone world" are
much smaller, since they arise from only one or a few
photoreceptor cells.

Two other types of neurons that participate in information processing in the retina are *horizontal cells*, which
interconnect rods or cones with bipolar cells, and *amacrine cells*, which interconnect sets of bipolar and ganglion cells. This latticework of cross connections within
the neural retina permits the visual-field pathways of different ganglion cells to influence each other. For example, some ganglion cells generate impulses only when
patterns of light and dark move across their visual fields,
often in specific directions. And lateral "talk" between
pathways can produce heightened contrast or provide
information about an object's contours.

A well-studied instance of this latter phenomenon is
revealed by making recordings from ganglion cells of a
frog retina; some such cells only fire nerve impulses
when a convex edge—corresponding perhaps to the
curve of an insect's wing or body—moves across the
field. If the same curved shape moves in the opposite
direction, it appears to be concave, and the ganglion cell
remains in its resting state. Thousands of such cells in
the retina are wired to respond to all sorts of concave,
convex, straight, and other edges—their combined activities add up to the visual world of the frog with its
insects, water lilies, and other frogs. Furthermore, such
cells allow the frog to identify moving insects as opposed
to still objects, and so to capture food. Still other ganglion cells respond to different shapes and, in turn, contribute to other behaviors. As such examples illustrate,
mechanisms within the eye itself process the basic visual
images striking the photoreceptor cells. Only after this
information has been translated in these ways is the
summary passed on by the ganglion cell axons to the
brain for even more complex processing.

Like all other senses, vision ultimately is the product
of integrating steps in the brain. The large optic nerve
from each eye carries axons of ganglion cells to the *optic
chiasma*, where, in lower vertebrates, all the axons cross
to the opposite side of the brain; in primates and some
other higher vertebrates, up to half of the axons from
each eye do not cross to the opposite side of the brain at

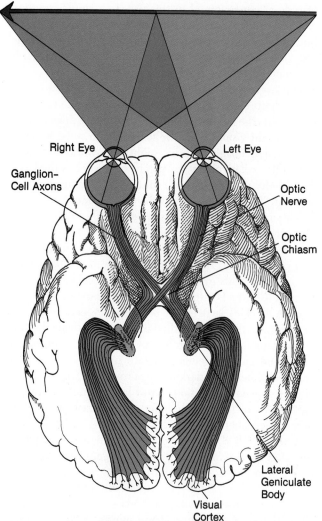

Figure 38-23   THE ROUTES OF VISUAL INFORMATION.
In primates, the left half of both neural retinas is connected by ganglion cell axons to the left side of the visual cortex, while the right halves of both eyes send signals to the right side of the brain. Trace the routing of ganglion-cell axons on this drawing (viewed from below). Information, upon reaching the brain, is processed in the lateral geniculate bodies, and from these is passed to the visual cortex for final processing and interpretation.

the chiasma (Figure 38-23). In mammals, the ganglion cell axons synapse with neurons in regions of the brain called the *lateral geniculate* bodies; from there, nerve axons extend to the right and left **visual cortices** (singular, *cortex*), where the most complex visual processing takes place.

David Hubel and Torsten Wiesel of Harvard Medical School have elucidated some of the truly remarkable properties of the hundreds of millions of cells in the visual cortex, most of which are arrayed in precisely or-

dered columns and layers (see Chapter 40). Working with cats, they found at least four neuronal cell types in the visual cortex: (1) neurons that become active only when bars of light or edges oriented at specific angles—vertical, horizontal, 40 degrees, 70 degrees, and so on—are shined into the eye; (2) neurons that respond to squares or rectangles of light; (3) neurons that are active only when edges move, often in a specific direction; and (4) neurons that respond only when the same spot on both retinas is illuminated. These "binocular" neurons are believed to be involved in depth perception.

More is known about the visual cortex and the manner in which it processes information than about the functions of any other part of the brain. As we shall see in Chapter 40, it is the intricate, highly specific hierarchical arrangement of visual-cortex cells that ultimately enables this remarkable structure to carry out the series of integrative operations that transforms edges, curves, moving patterns, colors, and so on into our sense of seeing Mona Lisa's face or a tennis ball rocketing across the net.

The Cephalopod and Arthropod Eyes

The paired eyes of an octopus and the multiple eyes of a scallop have structures and functions strikingly like those of a vertebrate eye. The similarities are obvious in Figure 38-24. Note, for example, the relative locations of the cornea, pupil, lens, retina, and other parts. Details of focusing and neural processing differ a bit, while another, intriguing difference is anatomical. Because of the way the eye forms in cephalopod embryos, the light receptors point inward toward the center of the eye chamber, whereas just the opposite arrangement is found in vertebrates. In both cases, however, evolution has resulted in a large surface area to array the visual pigments and in a neuron network in the eye for initial processing of visual information.

By contrast, arthropods such as crayfish, spiders, and insects have **compound eyes,** composed of many separate optic units called **ommatidia** (Figure 38-25). Each ommatidium is oriented at a slightly different angle from the others and so is aimed at a different part of the visual world. We know most about ommatidial function in the horseshoe crab. Light passes through a lens into a cluster of sensory cells *(retinular cells)* that have visual-pigment molecules spread on numerous microvilli (called collectively the **rhabdomere**). This is an evolutionary innovation independent from that of vertebrates, in which, recall, modified cilia form the outer segments of rods and cones. When the visual-pigment molecules are stimulated, both cyclic GMP and IP_3 second messengers may act, and the retinular cells depolarize; the changed state of polarization is communicated via gap junctions (elec-

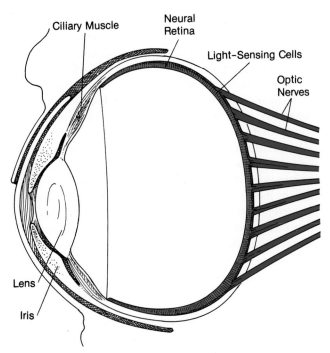

Figure 38-24   PARALLEL EVOLUTION—THE CAMERA EYE OF THE OCTOPUS.
Note how similarly the major structures are located in this octopus eye and in a vertebrate's eye. However, the light-sensing cells are directed inward toward the light and the central chamber—just the opposite of sensory orientation in vertebrates.

trical synapses; Chapter 36) to the dendrite of the single *eccentric cell* of the ommatidium, which, in turn, generates and propagates nerve impulses to the brain.

Each ommatidium samples 2 to 3 degrees of the animal's visual world, compared with 0.02 degree in vertebrates; each retinular cell in an ommatidium thus "sees" a relatively large patch of the world, and high visual acuity is not possible. Instead, arthropods see a grainier,

Figure 38-25   VISION IN A COMPOUND EYE.
(a) Many arthropods have compound eyes which are composed of separate optic units called ommatidia. Each ommatidium in an insect eye samples 2 to 3 degrees of the animal's visual world. Depending on which ommatidia are stimulated by a change in light, an insect can locate and respond to the source of change. (b) Closeup of an ommatidium of the horseshoe crab, showing the single eccentric-cell neuron, which serves all the retinular cells of this optic unit. (c) An electron micrograph of a compound eye from an aphid (magnified about 500 times). Each of the "spheres" in this clump is the outer end of an ommatidium. Obviously, the small number of ommatidia in an aphid's eye means that the world is only seen as a few dozen "patches." In comparison, a fruit fly eye would have over a thousand ommatidia.

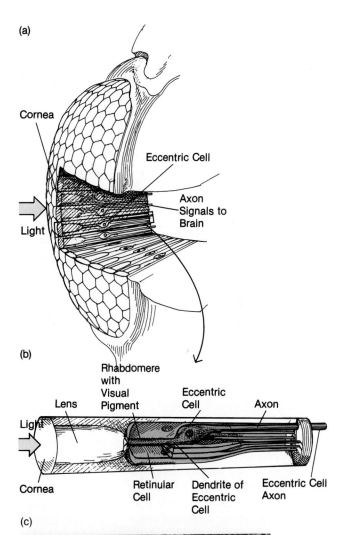

fuzzier version of their surroundings than do species with camera eyes. Interestingly, many arthropods are able to sense ultraviolet light, as well as the plane of polarized light. For this reason, both bees returning to their hive or crabs swimming in shallow water can orient to polarized light patterns in the sky.

LOOKING AHEAD

We have seen in this chapter that animal sense organs have evolved sensitivity to environmental stimuli, such as light, sound waves, and odorous molecules, all of which can alter the ion permeability of receptor cells. As the flow of ions into and out of sensory receptor cells increases or decreases, the resulting depolarization or hyperpolarization begins a sequence of events that transfers information to the central nervous system. Overall, the particular combination of what an animal can see, smell, hear, and feel plays a central role in each species' way of life, generating simple reflexes and complex behaviors that may be critical to survival.

Among the ways animals respond to sensory input are the physical movements necessary to evade a predator, capture prey, dig a burrow, or climb a set of stairs. As we take up the study of muscles in Chapter 39, we shall see in detail how animals are able to carry out such activities, which are based, in large measure, on information gained from sensory processing.

SUMMARY

1. Sensory receptor cells transduce various types of energy into a form of electrical activity. Receptor cells may be true neurons, or they may synapse with sensory neurons. Most sensory receptors have a baseline level of electrical activity that rises or falls, depending on changes caused by the stimulus. Such cells may occur singly or as part of *sense organs.*

2. *Chemoreceptors,* the detectors partially responsible for *gustation* (taste) and *olfaction* (smell), appear to have receptor molecules on their surface. Binding of a specific stimulus molecule or ion leads to changes in a second messenger and produces graded changes in the cell membrane potential that ultimately alter nerve-impulse rates to associated centers in the brain. Taste receptors occur in the *taste buds* on the tongue or on the head skin of various vertebrates and detect dissolved substances in saliva or the surrounding water. Olfactory receptors may be sensitive to extraordinarily low concentrations of odorous molecules.

3. *Mechanoreceptors* respond to deformations of the receptor cell and its membrane by mechanical forces; the state of polarization rises or falls as a result, and in the same or an adjacent sensory neuron, corresponding changes take place in the rate of nerve-impulse generation. For example, Meissner's and Pacinian corpuscles are skin structures that provide a sensory response to pressure. *Proprioceptors* are the stretch receptors in joints, tendons, and muscles. Such receptor cells often have cilia or microvilli that deform the cell's plasma membrane when they are bent. The mechanoreceptors of the *lateral line* in fish and amphibians have this feature; the hairs' bending in one direction causes the cell's depolarization, whereas their bending in the opposite direction produces hyperpolarization.

4. In invertebrates, sensory hair cells that detect body position and motion lie beneath *statoliths,* which are often made of calcium-carbonate crystals within a fluid-filled sac, the *statocyst.* In vertebrates, *otoliths* shift and bend sensory hair cells in chambers of the inner ear and are responsible for sensing body position relative to gravity and linear acceleration and deceleration. Sets of sensory hair cells associated with the *semicircular canals* detect angular acceleration and deceleration; movement of *endolymph* in and out of the canals deflects the sensory hairs.

5. In many vertebrates, sound perception—the sense of hearing—takes place in the inner ear. Sound waves entering the outer ear are conducted inward so that they strike the *tympanic membrane,* or eardrum; resultant vibrations are then transmitted by a chain of tiny bones through the middle ear to the *oval window,* an inner-ear membrane in the upper wall of the *cochlea,* where they set up pressure waves in the inner-ear fluid.

6. The cochlea is a set of tubes, coiled on top of each other. Sound of a specific frequency sets up pressure waves in these chambers that displace maximally a specific site on the *basilar membrane,* thereby bending microvilli on sensory hair cells (composing the *organ of Corti*) against the *tectorial membrane.* This generates activity in associated neurons, whose axons extend via the *auditory nerve* to the brain and "report" the sound frequency that has struck the ear. Pressure waves complete their

circuit of the inner ear as they strike the *round window*.

7. A crucial feature of the hearing sense is the ability to locate the source of a sound; loudness and timing differences of sound registration at one ear or the other are the keys to this capacity. Some animals, such as bats, whales, and porpoises, have evolved highly sensitive mechanisms for *echolocation*, which involves analyzing echoes of ultrasonic sound pulses reflected off objects in the animals' environment.

8. In some animals, mechanoreceptors derived from the lateral line have become specialized to detect weak electric currents. These *electroreceptor cells* are found in sharks, catfish, and weak electric fish.

9. Many types of animals and even some bacteria can "sense" the lines of force of the Earth's magnetic field. Magnetic crystals in such creatures may play a role in the detection system.

10. *Thermoreceptors*, in the facial pits of certain snakes, can detect and locate in space very small differences in infrared (heat) radiation. Although biologists do not fully understand the precise mechanism, the receptor nerve endings in infrared detectors probably function similarly to heat detectors in the skin or tongue of other animals.

11. Nearly all animals have light-sensitive structures or photoreceptors. Mollusks, including cephalopods, and vertebrates have camera eyes. Light enters through the transparent *cornea* and passes through the *iris's pupil* to the *lens*, which focuses light on the photoreceptor cells of the *neural retina*. These specialized cells, known as *rods* and *cones*, have visual pigment molecules *(rhodopsin)* that change shape when they absorb light. *Retinal* changes from the 11-*cis* to the all-*trans* shape and later dissociates from the protein *opsin*. The altered rhodopsin affects *transducin*, which in turn activates an enzyme that reduces cyclic GMP levels in the outer segment of the rod or cone. As a result, sodium ion channels close and the cell hyperpolarizes. Such changes ultimately alter the pattern of nerve impulses generated by *ganglion cells* leading to the brain via the *optic nerve*.

12. Color vision in mammals and some other vertebrates results from sensitivity of three types of cones to different wavelengths of light. The *trichromatic theory* explains how the ratio of activity of the different cones yields sensory input to the brain that can be interpreted as specific colors.

13. Much processing of visual information goes on in the retina prior to reaching the ganglion cells. Each ganglion cell has a *visual field* composed of the photoreceptors to which it is wired. Much more complex processing goes on in the brain, especially in the *visual cortices*. There, different groups of cells respond specifically to information on moving edges, particular shapes, and so on.

14. Arthropod *compound eyes* have separate visual units, called *ommatidia*, which are sensitive to both visible and ultraviolet wavelengths. Since each ommatium sees a relatively large patch of the world, compound eyes have fuzzy vision, unlike camera eyes.

KEY TERMS

auditory nerve
basilar membrane
bipolar cell
chemoreceptor
cochlea
compound eye
cone
cornea
echolocation
electroreceptor cell
endolymph
fovea
ganglion cell
gustation
iris
lateral line

lens
mechanoreceptor
neural retina
olfaction
ommatidium
opsin
optic nerve
organ of Corti
otolith
oval window
photoreceptor
pigmented retina
proprioceptor
pupil
retinal
rhabdomere

rhodopsin
rod
round window
semicircular canal
sense organ
statocyst
statolith
taste bud
tectorial membrane
thermoreceptor
transducin
trichromatic theory of color vision
tympanic membrane
visual cortex
visual field

QUESTIONS

1. What is the basic sequence of nerve-cell activities during the conveyance of a sensory message from a receptor to the brain?

2. What is meant by transduction by a sensory receptor cell? Give an example.

3. What are the three major classes of sensory receptors?

4. Compare gustation in the human and the blowfly.

5. Briefly discuss the structure and function of a dog's olfactory epithelium.

6. Describe Pacinian corpuscles, and outline the role they play in mechanoreception.

7. Compare and contrast the mechanisms that vertebrates and invertebrates use to sense position of the body in space. What is the principle of sensory hair-cell function in sensing position, and in detecting angular or linear accelerations and decelerations?

8. How does the mammalian cochlea detect a specific frequency of sound?

9. Describe the sequence of activities that begin when a photon of light strikes molecules of visual pigment in the outer segment of a rod in the retina.

10. Compare and contrast the camera eyes of vertebrates and cephalopods with the compound eyes of arthropods.

ESSAY QUESTIONS

1. How are modified cilia or hair cells involved in the senses of chemoreception, mechanoreception, and photoreception in animals?

2. What specific adaptations account for
   (a) a dog's olfactory sense?
   (b) a bat's hearing?
   (c) certain fishes' sensitivity to electrical stimuli?
   (d) a pit viper's sensitivity to heat?
   (e) a hawk's visual acuity?

SUGGESTED READINGS

ATTWELL, D. "Phototransduction Changes Focus." *Nature* 317 (1985): 14–15.

A brief summary and list of references on molecular events in the outer segment of rods and cones, including work on cyclic GMP.

BURKHARDT, D., et al. *Signals in the Animal World.* New York: McGraw-Hill, 1967.

A beautifully illustrated description of sensory systems.

ECKERT, R., and D. RANDALL. *Animal Physiology.* 2d ed. New York: Freeman, 1983.

The standard of excellence in physiology books, it is very good on sensory systems.

PARKER, D. "The Vestibular Apparatus." *Scientific American*, November 1980, pp.118–135.

This article gives a fine description of how gravity, motion, and position are sensed in the inner ear.

BOTSTEIN, D. "The Molecular Biology of Color Vision." *Science* 232 (1986): 142–143.

This is an elegant essay on the genes responsible for color vision and color blindness (it includes references to the Nathans work).

SHEPHERD, G. "Welcome Whiff of Biochemistry." *Nature* 316 (1985): 214–215; *Nature* 324 (1986): 17–18; and *Nature* 325 (1987): 389.

Three brief articles on the molecular events during olfaction.

WESSELLS, N. K., ed. *Vertebrates: Physiology.* San Francisco: Freeman, 1978.

A collection of articles from *Scientific American* on sensory systems and other organ systems.

39
SKELETONS AND MUSCLES

If one embarks on a new field one usually does not know where to begin. There is one thing one can always do, and this I did: repeat the work of old masters. I repeated what Kuehne did a hundred years earlier. I extracted myosin with strong KCl and kept my eyes open. . . . we observed that if the extraction was prolonged, a more sticky extract was obtained without extracting much more protein. We soon found that this change was due to the appearance of a new protein "actin." We made threads of the highly viscous new complex of actin and myosin, "actomyosin," and added boiled muscle juice. The threads contracted. To see them contract for the first time, and to have reproduced in vitro one of the oldest signs of life, motion, was perhaps the most thrilling moment of my life.

Albert Szent-Györgyi, "Lost in the Twentieth Century," *Review of Biochemistry* (1963)

Like a pole vaulter using a telephone pole, a red ant carries home a twig.

Tap your flexed knee with a rubber mallet, and your lower leg automatically kicks the air. Lift a load of books, and your biceps bulge. Run a marathon, and muscle fibers in your thighs work for hours to propel your body forward. Muscles and the bones they move accomplish most large-scale forms of work in a vertebrate's daily life. Along with glands and additional organs, they are the body's *effectors*—organs that carry out movements and a variety of other actions so that an animal can breathe, eat, walk, and go about its daily activities. Recall that sensory cells and sense organs send information to the central nervous system. Orders issuing from the CNS, in turn, control the body's effector organs to generate overt behavior or to regulate physiological processes.

The principles and processes by which effectors operate began to evolve early in the history of life on Earth. All eukaryotic cells contain a cytoskeleton with a primitive contractile system that may be used for locomotion, the shuttling of cytoplasmic organelles and fluids, exocytosis, and endocytosis; in animal cells and a few plant cells, contractile filaments carry out cytokinesis, the actual division of cells during the mitotic cycle. Such processes involve the intricate activities of actin and myosin molecules and control by calcium ions (Chapter 6). These same molecules and ions are responsible for the much larger-scale effector movements we see as peristalsis in the intestine, contraction of a biceps muscle, or beating of the heart.

Two principles guide the functions of this universal contractile machinery: (1) coordinated movements of precisely aligned molecules can generate force, and (2) this force results from the shortening of arrays of molecules, not from the diminution of individual molecules. This means that effectors virtually always function by pulling, rather than by pushing.

In general, an animal's muscles pull on its skeleton or on other muscles or tissues. It is appropriate, therefore, to look first at the various types of skeletons that have evolved in animals. Then we will consider the structure of muscles as organs, as well as of muscle cells. With that as background, we turn to the molecular events that underlie muscle contraction and examine how muscles operate as organs controlled by the nervous system. Finally, we will survey the three major classes of muscle and see how they function in vertebrates and invertebrates.

THE ANIMAL SKELETON: A LIVING SCAFFOLD

Animal skeletons have a variety of forms and functions, but all are of two basic types: fluid *hydroskeletons* and *hard skeletons* (Chapters 25 and 26).

Hydroskeletons are volumes of fluid contained by the body cavities of various invertebrates: the gut, pseudocoelom, coelom, vascular system, or water-vascular system. Recall, for instance, that the contraction of circular muscles around the base of a hydra's tentacle compresses the fluid inside and causes the tentacle to elongate. A similar operation enables the earthworm to burrow through soil: waves of contractions in circular and longitudinal muscles compress fluid in the worm's gut and coelom, causing successive portions of the body to elongate, to press against the ground, and to propel the animal forward.

Except for the jet-propelled locomotion of squids and octopods, muscles acting on hydroskeletons do not generally produce effects as rapid or as powerful as those made possible by a hard skeleton. Hard skeletons can be located at the body surface, as exoskeletons, or they can occur internally, as endoskeletons.

Exoskeletons

Exoskeletons include the shells of clams and snails and the hard coverings of lobsters, spiders, and beetles. All serve as protective armor and in many cases act as a permeability barrier between the outside world and the animal's tissues. Exoskeletons tend to be nonliving, acellular deposits of crystallized mineral salts, such as calcium carbonates, or a mixture of organic and inorganic substances, such as chitin. And, depending on their shape and position, they may affect the pattern of an organism's growth. While a mussel shell or clamshell may increase in size as calcium salts are continually deposited at the edges, an arthropod's nonexpandable outer covering prevents much growth, and thus the lobster, insect, or other arthropod must periodically molt in order to expand in size.

All hard skeletons are composed of separate pieces that are *articulated*, or hinged, to permit movement. In some animals, specially shaped sections serve as jaws, pincers, or locomotor appendages. Movement of such structures or parts of the body is possible because joints composed of thinner, flexible exoskeletal materials are strategically located at points of relative movement in an antenna, a claw, a leg, a wing, or a tail.

Looking at a crab or at the imposing scarab beetle shown in Figure 39-1, it is obvious that the animal's muscles must operate from *within* the exoskeleton. Indeed, two sets of muscles are responsible for movement of each skeletal part: one set flexes (bends) and the other set extends (straightens) the parts. Such muscles are called *flexors*, meaning that they bend a limb part, and *extensors*, meaning that they straighten a limb part. A typical, simple arrangement is illustrated in Figure 39-2a, where muscle A is the extensor of the outer leg segment, and muscle B is the flexor. Note the more com-

Figure 39-1   SCARAB BEETLE: ARMORED EXOSKELETON.

The hard, nonliving exoskeleton of the scarab beetle protects soft internal structures. Besides serving as a protective armor, the exoskeleton is impermeable and thereby prevents loss of water to the beetle's environment. The muscles that move parts of such a skeleton are located internally.

plex arrangement in Figure 39-2b, which shows flight muscles located deep within the body that raise or lower the fly's wings.

Endoskeletons

In animals such as rays and sharks, endoskeletons consist of cartilage (Figure 39-3), whereas in most vertebrates, they consist of bone plus cartilage. **Cartilage** is primarily collagen and complex polysaccharides (glycosaminoglycans), while **bone** is predominantly collagen in combination with a large amount of apatite, a calcium and phosphate salt. In contrast to acellular, mineralized exoskeletons, both bone and cartilage contain living, metabolizing cells.

Most vertebrates have two classes of bones in their endoskeletons: those that develop directly as bone within the skin (*dermal membrane* bones) and those that develop first as cartilage and then are transformed into true bone (*endochondral* bones). Outer skull bones are dermal membrane bones, whereas bones in the limbs and the pelvic girdle are endochondral bones. In addition, the skeleton itself is subdivided into the **axial skeleton**—which consists of the skull, the vertebral column, and, in animals that possess them, the ribs—and the **appendicular skeleton**—which is made up of the ante-

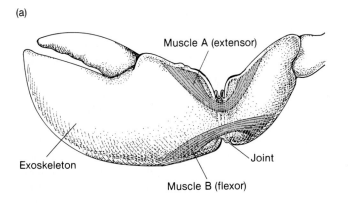

(a)

Figure 39-2   MOVING THE EXOSKELETON: JOINTS AND MUSCLE ATTACHMENTS.

(a) In animals with exoskeletons, muscles function inside the skeletal structure. Muscle A (extensor) extends (straightens) the outer claw segment while muscle B (flexor) flexes (bends) it. (b) Two sets of muscles, the transverse and the longitudinal pairs, work within the fly's thorax to raise and lower the wings. When the transverse muscles contract and the longitudinal muscles relax, the wings are pulled up. Alternatively, the wings are pulled down when the longitudinal muscles contract and the transverse muscles relax. Note that the muscles are *not* attached to the wings, but instead act on the exoskeleton to flex it—because of the fulcrum joint, the wings are forced up and down. To help visualize what occurs, note how the dorsal exoskeleton flexes up and down due to the alternating muscle contractions.

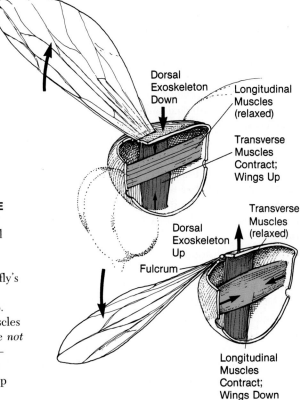

(b)

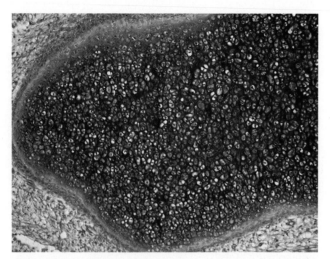

Figure 39-3   CARTILAGE: TOUGH, FLEXIBLE, AND LIGHT SKELETAL MATERIAL.
Cartilage is a matrix secreted by living cartilage cells embedded within the cartilage material. Notice in this light micrograph (magnified about 70 times) that no blood vessels penetrate cartilage, hence dissolved gases, nutrients, and wastes must diffuse through the matrix.

rior *pectoral girdle* and forelimb bones and the posterior *pelvic girdle* and hind limb bones (Figure 39-4).

To understand how parts of an endoskeleton move, we must look first at the structure of **joints,** the regions where individual bones meet. For example, the large bones of the human leg come together at the knee joint, where they are held together by **ligaments**—strong, flexible bands composed primarily of collagen fibers. In addition, bones that move against each other usually have cartilage on their articulating surfaces (Figure 39-5), where the bones rub together. Because cartilage is more pliable and resilient than bone, cartilage in joints acts as a sort of internal pillow to cushion the adjoining bones where they rub against each other. Furthermore, in joints such as the mammalian knee, the whole joint cavity may be filled with a shock-absorbing fluid, called *synovial fluid*, that also acts as a lubricant (the *synovial membranes* hold that fluid in the cavity).

Although joint ligaments are fairly elastic, too great a stress—for example, a blow to the knee during a football tackle—may snap them or pull one end free of the bone. Other joints, although equal in strength to the knee, are considerably more mobile; for instance, the chimpanzee or human shoulder joint (a joint that originally evolved for hanging and climbing in trees) can rotate freely through a complete circle. In contrast, movement of the human knee is highly restricted; it can barely move from side to side or anteriorly beyond a certain angle (hence knee injuries are common among skiers and other athletes), but it has great posterior flexibility.

Some junctions between bones are rigid; as an animal matures, individual bones of the vertebrate skull or of the pelvic girdle grow until their edges meet; then *sutures*—stiff, fibrous, calcified regions—develop to weld the bones together. The fontanel, or "soft spot," on the surface of a baby's head is an area in which the skull bones have yet to fuse.

The Inner Structure of Bone

A sun-bleached antelope skeleton lying on the prairie or a fossilized dinosaur frame on display in a museum can never convey the dynamic nature of living bone. As Figure 39-6 shows, mammalian bones typically have an external layer of extremely hard *compact bone* around a softer center of *spongy bone.* Compact bone is constructed of thousands of cylindrical, densely packed **Haversian systems,** made of multiple layers of bone surrounding a single blood capillary. Within the honeycomblike layers are living bone cells, called *osteocytes;* each osteocyte resides in a separate chamber that is connected by tiny canals to the capillary so that nutrients, wastes, and gases may pass to and from the cell. The outer surfaces of adjacent Haversian systems fuse totally, and together, many such layers give compact bone its hardness.

The less dense spongy bone is often richly endowed with blood vessels in its interior, the *bone-marrow* region. Fat cells (so-called white marrow) or developing blood cells (so-called red marrow, where erythrocytes, lymphocytes, and macrophages arise) fill the spaces between bony struts that extend across the marrow cavity and serve as braces for the surrounding compact bone.

Some bones, such as the long bones of the limbs, must grow substantially as a vertebrate matures—a process that takes place in mammals by way of cartilaginous growth zones present at each end of a bone. Lengthening occurs as new cartilage is produced, and the cartilage, in turn, is replaced by bone. In humans, the growth zones respond to growth hormone and somatomedin (Chapter 37) during the teen-age years, and then regress as a person reaches adulthood.

In living bone, varying levels of the hormones calcitonin and parathyroid hormone act on certain bone cells and cause them either to break down spongy bone—thereby raising calcium and phosphate levels in the blood—or to deposit new bony material. When the latter happens, blood concentrations of calcium and phosphate ions fall. Nerves also may help regulate bone breakdown; axons of sympathetic neurons release vasoactive intestinal peptide (VIP, Chapter 40) in mammalian rib bones and cause calcium release into the blood. In a sense, the bony parts of the skeleton are a reservoir of calcium and phosphate ions that is constantly tapped or replenished as body needs change.

Figure 39-4   THE HUMAN ENDOSKELETON.
The axial skeleton, consisting of the skull, vertebral column, and ribs, is colored yellow in the drawing. The pelvic girdle, the pectoral girdle, and bones peripheral to them compose the appendicular skeleton, and are colored tan. Both sets of bones are made of the same substances, primarily collagen and apatite.

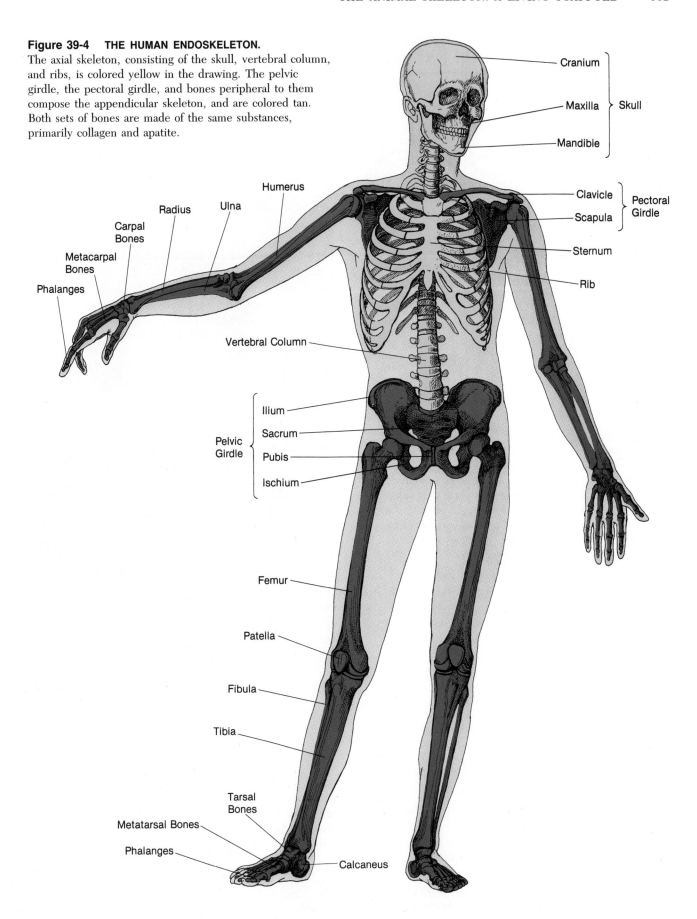

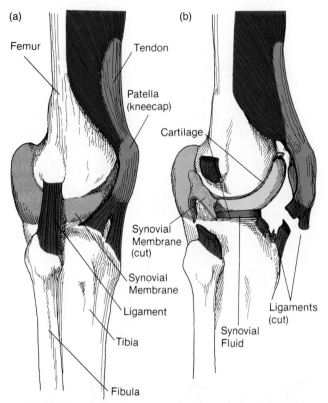

Figure 39-5   THE HUMAN KNEE JOINT: A LIVING HINGE OF CARTILAGE AND LIGAMENTS.
These drawings show a dissection of the human knee joint from the side. (a) The relationship of ligaments holding bones together and tendons attaching muscle to bone. (b) The knee with tendons, ligaments, and synovial membrane cut to reveal the cartilage cushions on the ends of the bones. Synovial fluid bathes the cartilage and also cushions the joint. Here, cartilage is shown as yellow; ligaments, purple; tendons, green; synovial fluid, red; synovial membrane, gray.

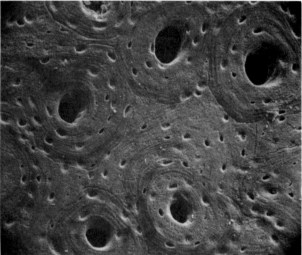

Figure 39-6   THE STRUCTURE OF BONE.
(a) The human femur (thighbone) in longitudinal section, magnified about 1.3 times. The meshwork at the top and the interior of the shaft is spongy bone. The dense bone of the exterior portion of the shaft is compact bone. The main bone-marrow cavity can be seen at the center of the lower bone shaft in this photograph. (b) Electron micrograph showing a cross section of compact bone, magnified about 200 times. The Haversian systems are composed of concentric rings of bone around a blood capillary that occupied the large holes seen here. Individual living bone cells, osteocytes, resided in the small depressions when this bone was alive.

The Skeleton–Muscle Connection

A vertebrate's skeleton may contain many dozens of individual bones, and hundreds of muscles may be involved in moving limbs, fingers, jaws, and other body regions. In turn, the speed and strength of such movements depend on how various sets of muscles interact with each other and the bones they are attached to. We can consider the primate arm as an example of how such muscle-bone connections operate.

When a monkey picks up a banana, the biceps muscle (Figure 39-7) in its arm contracts, flexing the forearm, while the elbow acts as a hinge. The biceps is the *agonist,* or *prime mover*—that is, it is the muscle primarily responsible for the gross movement that raises the fruit

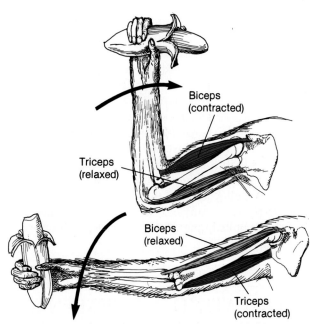

Figure 39-7   MUSCLE ANTAGONISM IN A MONKEY'S ARM.
When a monkey raises a banana to its mouth, the action of bending the arm at the elbow results from contraction of the biceps and relaxation of the triceps. In this motion, the biceps is the agonist or prime mover, while the triceps is the antagonist. Alternatively, when the monkey straightens its arm, the triceps is contracted and thus represents the prime mover, while the biceps is relaxed and becomes the antagonist to this movement. In fact, several other muscles not shown here also participate in such movements.

toward the monkey's mouth. Alternatively, if the triceps muscle—the *antagonist* of the biceps—contracts, it pulls on the ends of the forearm bones, and the forearm extends. The antagonist, then, is a muscle whose action works against that of another muscle, the agonist. In general, simple movements follow this pattern: when a muscle contracts, its antagonist relaxes. Additional muscles serving a joint may contract simultaneously with a prime mover, augmenting or modifying the direction in which a bone moves; these complementary muscles are known as *synergists.*

In more complex muscular operations—the measured finger manipulations of a pianist playing a sonata, for instance—the gradually increasing contraction of prime movers is coupled with balanced relaxation of antagonists. Thus before a finger begins to move, both agonist and antagonist muscles are partially contracted; as one contracts more, the other contracts less. The result is exquisitely controlled high- or low-speed movement of the digit.

In a sense, vertebrate muscles operate as systems of springs or ropes acting on levers—the bones to which

they are attached. It is easier to understand how such systems work if we consider them as an engineer might: one bone moves on another at a *fulcrum* (the support about which a lever turns); the length of bone between the point where a muscle is attached and the fulcrum is the *power arm*, and the length between the fulcrum and the site where work is done (such as a foot or hand) is the *load arm*. Notice in Figure 39-8a that the power arm is relatively long; a result is that the load arm can be moved with substantial force, but not with great speed. If the power arm is short, as in Figure 39-8b, greater speed is possible, but not as much force can be generated.

As animals evolved, such design options were incorporated into legs, wings, flippers, and other appendages: limbs of fast runners, such as cheetahs, have short power arms, and so the limb can be moved very rapidly; powerful diggers or fighters, such as badgers, have relatively long power arms and move their limbs more forcefully

Figure 39-8   THE POWER-ARM/LOAD-ARM CONCEPT.
The power arm is the distance from the fulcrum to the site of muscle attachment; the greater this distance relative to the length of the load arm (from the fulcrum to where the work is done), the greater the power that can be applied to the bone. Conversely, the shorter the power arm relative to the load arm, the faster the bone (the foot in this case) can be moved. (a) The badger's leg is slow but powerful. (b) The cheetah's longer leg sacrifices power to achieve great speed of movement. These legs are drawn to different scales; in reality the cheetah's leg is about three to four times as long as the badger's.

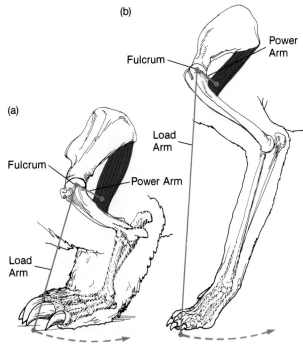

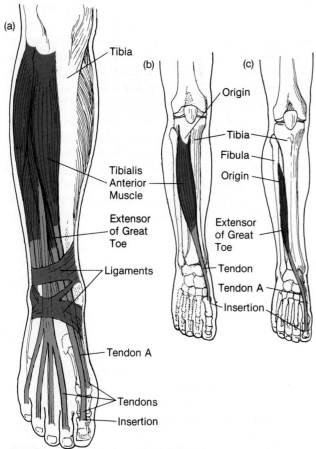

Figure 39-9   MUSCLE ATTACHMENTS AND TENDONS.
(a) In a human leg, the muscles on each side of the tibia end in tendons running under straplike ligaments at the ankle. These then insert on the upper surface of the toes. (b) The tendon at the origin of the tibialis anterior muscle is quite broad; the tendon at the insertion on the upper surface of the base of the great toe is much narrower, concentrating the total force of the muscle's contraction at a small spot. (c) The extensor of the great toe has a long origin on the surface of the fibula and also a narrow insertion on top of the great toe's tip.

but also more slowly. Another example of a long power arm is seen in some humans whose heels are slightly longer than average; people who have this characteristic possess significantly greater jumping ability because their toes can exert more power against the ground.

The sites of muscle attachment to bones are key factors in how bones act as levers. Each muscle has two attachment sites: the *origin*, which acts as an anchor; and the *insertion*, located near the region of movement. The insertion is the site where the force of the muscle's contraction is applied to the bone (Figure 39-9). Muscles may connect directly to the tough connective-tissue layer on a bone's surface or may grade into **tendons**— extremely tough bundles of collagen fibers.

Tendons confer several important advantages on animals. A tendon can concentrate all the contractile force of a large or broad muscle at a small, specific site, usually one that provides greatest leverage for the motion in question. (Thus a muscle with a diameter of 5 centimeters might have its full contractile force applied to a bone via a tendon with a diameter of only a few millimeters.) Tendons can also change the direction of the force generated by a muscle. For instance, a bird's wings are raised by muscles in its lower chest. This arrangement is possible because the tendons from the lower-chest muscles travel over pulley joints of the shoulder bones and then attach to the upper surface of the wing bones. Therefore, the wing is pulled upward when the chest muscles contract (Figure 26-28).

REGULATING MOVEMENT: FEEDBACK CONTROL OF MUSCLE ACTION

The control of muscle activity by the animal nervous system is an awesome task; somehow, the central nervous system detects and keeps track of what hundreds of agonist, antagonist, and synergist muscles are doing each moment, signaling some motor neurons to generate impulses while others remain quiet. Instrumental to this feat of managerial virtuosity is a class of sense organs called *proprioceptors*.

The coordination of body muscles depends first on the perception of position, or proprioception (Chapter 38). Most complex animals have proprioceptors that report on the position of the body or the body's major parts, such as the limbs or head. Vertebrates also have special proprioceptors called **muscle spindles** (or, sometimes, *stretch receptors*) in joints, tendons, and muscles. The muscle spindles are small but remarkable sets of special muscle cells found near the center of each skeletal muscle. Muscle-spindle cells have highly ordered contractile systems at both ends, but not in their central region, where many endings of so-called *1a sensory neurons* reside (Figure 39-10). These sensory neurons play a central role in detecting body position and controlling the muscles.

The 1a sensory neurons that serve muscle spindles form the first part of a reflex arc. The sensory axons enter the spinal cord and form synapses with *α (alpha) motor neurons*. The axons of these cells pass to the very muscle in which the muscle spindle is located. Let us see how the muscle spindle and the reflex arc operate.

Suppose that a chimpanzee's triceps contracts, extending its forearm. The antagonist muscle, the biceps, is stretched as the forearm straightens, and so is the

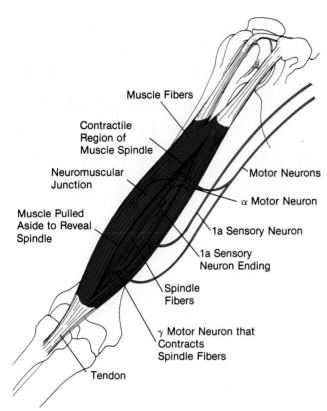

Figure 39-10   MUSCLE SPINDLES.
Muscle-spindle structure. Parts of a skeletal muscle are
pulled aside here to reveal the internally placed muscle
spindle. Each end of the spindle is contractile tissue,
whereas the center is the site of 1a sensory nerve endings.

muscle-spindle within the biceps. The 1a sensory neurons with synapses in the spindle generate impulses at a
slow rate when a muscle is at rest, but as the spindle is
stretched, they are stimulated to send a volley of nerve
impulses to the spinal cord. Such neurons—in this case,
coming from the biceps—usually synapse in the spinal
cord directly with the α motor neurons serving the muscle cells around the muscle spindle—again, in this case,
in the biceps (Figure 39-11). Thus incoming sensory
impulses resulting from a stretched biceps spindle trigger a reflex contraction of the biceps. That contraction
resists the action of the triceps. As the biceps contracts,
its muscle spindle is shortened somewhat, and its 1a sensory output to the spinal cord decreases; as a result, the
motor output causing biceps contraction decreases.

In land vertebrates, this simple, single synapse feedback loop is the key to control of a variety of muscle
functions, including an animal's ability to maintain normal body stance; it is also responsible for the knee-jerk
reflex. But muscle spindles do even more. Innervating
the contractile end regions of the muscle-spindle cells
are small motor neurons, known as γ (*gamma*) *motor
neurons*. These neurons are activated by motor centers

in the brain or by reflex circuits in the spinal cord (as
from pain receptors in the skin). When the γ motor neurons generate impulses, the muscle-spindle cells contract, thereby stretching their central, noncontractile
region (Figure 39-11). This stretch is signaled by the 1a
sensory neuron serving the muscle spindle to the CNS;
that information immediately triggers a reflex, and α
motor neurons cause the muscle itself to contract. Thus
the brain acts indirectly to cause the muscle to contract—
it causes the muscle spindle to contract, and that, in
turn, causes the muscle to contract. It is as though, in
the words of biologist David Kirk, the nervous system
"tricks" the spindle into behaving as if it were being
stretched by the antagonist, a heavy weight, or some
other force.

What good is this elaborate muscle deception to an
organism? One benefit is anticipatory control of contraction. If, for example, a person's eyes perceive that a large

Figure 39-11   MUSCLE-SPINDLE NEURONS.
The 1a sensory neuron monitors the degree of stretch in the
muscle spindle's central portion. It reports to the spinal
cord, and forms synapses on the α motor neuron, as shown.
A reflex arc may thus involve sensory input over the 1a fiber
and motor output over the α fiber. The result is contraction
of the muscle and a shortening of the muscle spindle. Also
notice that the γ motor neurons cause contraction of the
ends of the muscle-spindle fibers; that localized contraction
stretches the central region of the spindle and activates the
1a sensory neurons, thereby triggering the reflex arc and
contraction of the entire muscle.

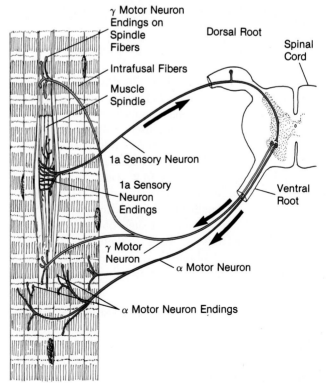

stone is about to be placed in one hand, the brain can activate γ motor neuron pathways so that appropriate finger and arm muscles can be tensed ahead of time. The system also functions like a contraction "thermostat"; when it is "turned up," and γ neuron activity is increased to a higher level, muscles contract more and maintain a higher state of tension. Conversely, a lower level of γ neuron activity means that a muscle will be relatively more relaxed. In these ways, muscle spindles coordinate muscle use and establish the general level of muscle tone that allows the animal to keep its head or body erect, to carry an armful of textbooks, or to perform other tasks that require constant finely honed muscular control. It is important to realize, however, that in addition to the indirect control provided by α motor neurons and muscle spindles, other nerve fibers from the brain synapse directly with α motor neurons to cause muscle contractions that move limbs or body parts forcefully and rapidly.

MUSCLE STRUCTURES AND FUNCTIONS

Muscles that move legs, wiggle toes, or blink eyelids are organs composed of differentiated muscle cells. Muscle cells are also vital components of other organs, such as the gut, lungs, secretory glands, and blood vessels. Though they have different shapes and functions, all a vertebrate's muscle cells can be categorized as smooth, cardiac, or skeletal (Figure 39-12).

Smooth muscle consists of single contractile cells; it is found in the walls of the gut and the uterus, in the linings of certain ducts, and in blood vessels. **Cardiac-muscle** cells are also single and have a much more regular organization of the macromolecules responsible for contraction than do smooth-muscle cells: cardiac-muscle cells are linked together in special ways in the walls of the heart. The third and most abundant muscle type is striated, or **skeletal, muscle.** In it, each "cell" is really a fused set of dozens or hundreds of cells. The resultant muscle cells are usually very long and are known as **muscle fibers;** a large number of such fibers make up the organ we call a "muscle." Because skeletal muscle is the best understood of all muscle types, we will consider its special characteristics in some detail.

Special Features of Skeletal Muscle

Striated-muscle fibers may be classified in several ways; one way is by color—*red* or *white.* Color differences are caused by the muscles' different chemistries and the resulting variations in function (Table 39-1). Red

(a)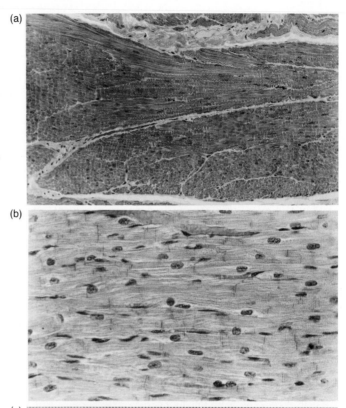

(b)

(c)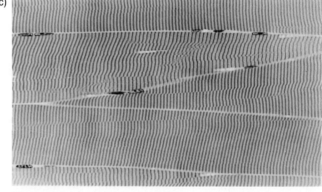

Figure 39-12   MUSCLE TYPES.
(a) Smooth muscle (magnified about 60 times) is composed of individual cells with contractile proteins that are not organized into regular units (sarcomeres). (b) Cardiac muscle (magnified about 100 times) is also composed of individual cells; these are interconnected at special junctions called intercalated discs and have striations due to the ordering of the contractile proteins, actin and myosin. (c) Skeletal, or striated, muscle (magnified about 125 times) is composed of huge multinucleated cells (nuclei are visible between these striated regions); the large quantities of contractile proteins are assembled in sarcomeres aligned so precisely that the cytoplasm takes on this striped, or striated, appearance.

muscle, sometimes called *slow muscle* (or tonic muscle), contains a large quantity of the reddish compound myoglobin, as well as many blood capillaries that run through the muscle tissue. Recall from Chapter 33 that myoglo-

Table 39-1 **PROPERTIES OF RED AND WHITE MUSCLE**

	Speed of Contraction	Myosin-ATPase Action	Blood Supply	Myoglobin	Fuel: Glycogen, Fats	Mitochondria	Functions
Red	Slow (tonic)	Slow	Many capillaries	Abundant	Abundant	Abundant	Sustained contractions
White	Fast (twitch)	Fast	Few capillaries	Little or absent	Little	Few	Brief contractions (fatigues)

bin binds oxygen from circulating blood and acts as a reservoir of stored oxygen molecules. In addition, red-muscle fibers store fat and glycogen, fuels for the manufacture of ATP by the large number of mitochondria that are present. And, as we shall see, the action of myosin as an ATPase mechanoenzyme, which enables a muscle to contract, is quite slow in red muscle. That results in a relatively slow rate of ATP utilization; consequently, the red muscle cell's many mitochondria can keep up the pace by generating ATP by oxidative phosphorylation. All of these properties allow red muscle to contract repeatedly and to resist muscle "fatigue"—a biochemical state in which contraction is inhibited.

White muscle, by contrast, is called *fast muscle* (or twitch muscle). It has little or no myoglobin, is traversed by relatively few capillaries, stores little fat or glycogen, and contains few mitochondria. Its myosin hydrolyzes ATP rapidly—too fast for ATP production even by glycolysis to keep up (1.8 times faster than by oxidative phosphorylation). For these reasons, white muscle soon becomes fatigued as lactic acid builds up from anaerobic metabolism. Therefore, white muscle is most active during brief, intensive flurries of contractions. A domestic turkey's "white-meat" breast muscle is soon exhausted if a fox chases the bird and forces it to take repeated short flights; in contrast, the "dark meat" of a duck's breast muscle allows it to fly over long distances.

The dichotomy of red versus white muscle is really an oversimplification. Individual muscles in birds and mammals may have red fibers in one region and white fibers in another region. Some white fibers have many mitochondria, like red fibers, and so fatigue only slowly. Within limits, sustained use can apparently alter a muscle's characteristics. A marathon runner, for example, will have more of the biochemical properties of red muscle in his or her leg muscles than will a champion weight lifter, whose maximal efforts require the brief but intense activity typical of white-muscle biochemistry.

The number, size, and structure of mitochondria also have important consequences for muscle function; in fact, the muscle cells in a hummingbird's breast muscles or a squid's mantle may contain more than 50 percent of these powerhouse organelles by volume. Insect flight-muscle cells may contain a single, mammoth mitochon-

drion 30 micrometers long and 10 micrometers in diameter—larger than most whole animal cells. But how does the cell's metabolic machinery actually power contractions? Both red and white muscles in vertebrates usually contain **creatine phosphate,** an organic molecule with a high-energy phosphate bond. This storage compound's phosphate group can be transferred enzymatically to ADP, resulting in the rapid formation of ATP. The presence of creatine phosphate thus allows muscle contractions to start and to persist (as ATP is replenished), even when aerobic and anaerobic metabolism is at a low level.

As a red- or white-muscle fiber continues to contract, all the high-energy phosphate stored in creatine may be transferred to ADP. Similarly, the myoglobin oxygen reservoir may be consumed by mitochondria in the highly active cells. When this happens, anaerobic metabolism remains as a source of ATP, a process that, as we saw in Chapter 7, produces a large quantity of lactic acid as a by-product. This accumulated lactate represents an *oxygen debt* that must be "paid off" when the muscle stops working so hard. Various metabolic pathways can use lactate as the cell returns to a normal aerobic state.

Muscle Structure

If you were to cut the sheath of connective tissue that covers the biceps, you would come on bundles containing hundreds or thousands of cylindrical muscle fibers resembling the bundles of straight wires in an underground telephone cable (Figure 39-13). Each of these fibers is a giant single cell, some 5 to 100 micrometers in diameter and several centimeters long. Such enormous cells arise in the embryo after hundreds of precursor cells fuse, each contributing its nucleus, mitochondria, and other organelles. As a result, one muscle fiber may have many hundred functional nuclei.

Packed into the cytoplasm of the muscle-fiber cell are long, cylindrical assemblies of molecules, called **myofibrils.** Myofibrils are built from **sarcomeres,** which are precisely repeating units of muscle proteins and which are the actual sites of contraction. As shown in Figure 39-14, the sarcomeres of neighboring myofibrils tend to

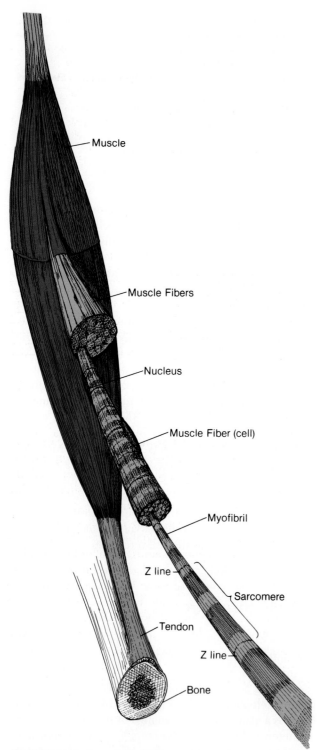

Figure 39-13   STRIATED MUSCLE: FROM ORGAN TO MOLECULES.
A skeletal muscle is constructed of muscle fibers, each of which is a huge, multinucleate cell. Within every muscle fiber are many myofibrils, composed of sarcomeres. Sarcomeres, which are bounded at each end by Z lines, are built of actin, myosin, and several other proteins.

be aligned, so that under a microscope, the muscle fiber appears to be banded, or striated—the origin of the term "striated muscle."

Biologists have labeled the various stripes and bands of sarcomeres for easy reference; note, in Figure 39-14, that the Z *lines* mark the ends of each sarcomere ("Z" stands for the German word *zwischen*, meaning "between"). Extending from each side of the Z lines are *I bands*, which appear lighter. Between the two I bands of a sarcomere, the dark A *band* is located; the A band, in turn, is subdivided by a light, centrally located *H zone*. Finally, the *M line* bisects the H zone, marking the center of the sarcomere. As we will see in the next section, alterations in the spatial relationships of the molecules that make up sarcomere bands generate the force that allows muscles to contract.

THE MOLECULAR BASIS OF MUSCLE CONTRACTION

The banding pattern of a muscle fiber is as useful to a muscle biochemist as a fingerprint is to Sherlock Holmes. A particularly crucial observation is that the widths of certain bands change as a muscle fiber contracts. Specifically, the I bands and central H zones become narrower as each sarcomere shortens—a change that, multiplied by the tens of thousands of sarcomeres in any muscle, produces a contraction. How do such changes in band width take place? To answer this question, we must see how the major muscle proteins are arranged in sarcomeres.

Each sarcomere contains two types of filament: *thin filaments*, which are composed primarily of the protein *actin*, and *thick filaments*, which are made of many *myosin* molecules that function both as a structural protein and as a mechanoenzyme (Figure 39-14; recall from Chapter 6 that mechanoenzymes, such as myosin, dynein, and kinesin, generate mechanical force when ATP is hydrolyzed). Like slender molecular arms, actin thin filaments extend from each Z line into the adjacent sarcomeres, passing through the I bands and well into the A band regions. In contrast, myosin thick filaments are found only in the A band, where they straddle the H zone. The figure shows that both ends of each thick filament are free, while only one end of each thin filament— that nearest the center of the sarcomere—is unattached; the other end is anchored in the Z line. Thus the free ends of the myosin thick filaments and the actin thin filaments overlap—an organizational feature that is vital to the mechanics of muscle movement.

The individual myosin molecules that make up a thick filament function as they do because of their shape and their tendency to bind together. Individual molecules

(a)

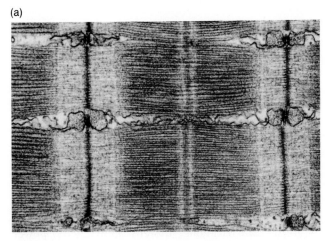

Figure 39-14   SARCOMERES: SITES OF CONTRACTION GENERATION IN STRIATED MUSCLE.
(a) An electron micrograph (magnified about 22,500 times) showing the orderly, repeating sarcomere units within a striated-muscle cell. (b) Notice in the diagram of the structures shown in the electron micrograph that each sarcomere is composed of a series of thick filaments (myosin) and thin filaments (actin). The membranous saclike structures at the end of each Z line are sites where calcium ions are released to trigger contraction.

(b)

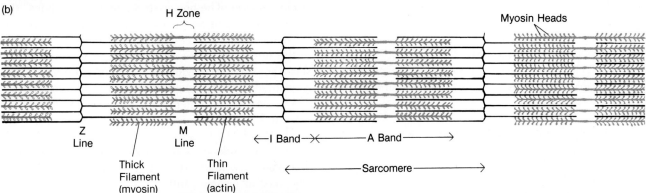

Figure 39-15   HOW ACTIN AND MYOSIN INTERACT.
(a) Myosin molecules have joined to form these two thick filaments (magnified about 128,000 times); the myosin heads project as the tiny white dots. This arrangement allows for the formation of cross bridges with actin molecules. (b) In a muscle cell, actin filaments surround such myosin filaments so that the myosin heads may contact the actin.

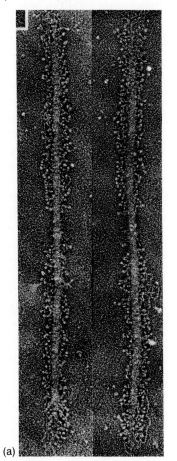

(a)

are composed of two polypeptide chains, each shaped like a golf club; the "shafts," or "handles," of the two polypeptides are twisted around each other, and the "heads" are bent to the sides at hingelike sites. In addition, sets of myosin molecules clump together (that is, their "handles" bind to each other) so that many heads protrude from the resulting double-ended filament, except in the M line region. Because of differences in electrical charges along the "handle" portion of the molecules, the molecules aggregate in a staggered array so that the myosin heads jut out in a spiral pattern at the surface of the thick filament (Figure 39-15a). In a sarcomere, each myosin thick filament is surrounded by actin thin filaments in such a way that the protruding myosin heads and the actin filaments can come in contact with each other when the muscle contracts (Figure 39-15b).

(b)

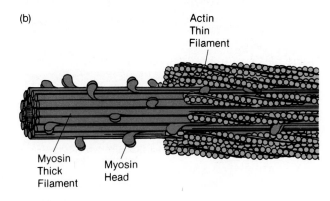

In 1957, the British biologist Hugh E. Huxley, using an early electron microscope, discerned the relationship between actin and myosin filaments, and Sir Andrew F. Huxley (no relation to Hugh) hypothesized how the molecular structure of actin and myosin generates the force for muscle contraction. The two Huxleys then developed the **sliding-filament theory,** which states, in essence, that the myosin heads act as cross bridges between actin and myosin filaments, applying "power strokes," much like an oar pushing on water, that push the actin filaments inward toward the central H zone. As they slide, the actin filaments pull on the Z lines to which they are anchored, and the result is a shortened sarcomere. If the same process goes on in the thousands of sarcomeres along the whole muscle-fiber cell, and in many muscle-fiber cells simultaneously, the muscle itself becomes shorter—that is, it contracts.

How might myosin cross bridges cause muscle fibers to contract? Myosin heads are not just structural protein; they are also active as ATPase enzymes. Figure 39-16

Figure 39-16   THE MYOSIN POWER STROKE.
To begin a contraction event, ATP binds to the myosin head and is hydrolyzed. The head then cocks and binds actin weakly. As ADP and inorganic phosphate (P$_i$) are released, binding of the actin head becomes strong, and the power stroke is applied. That moves the actin filament. Only when another ATP binds is actin released from the head, allowing the cycle to start again.

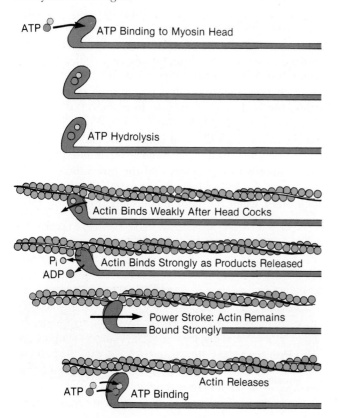

summarizes the events responsible for contraction. First, ATP binds to the active ATPase site on the myosin head. Then ATP is hydrolyzed into ADP plus inorganic phosphate, and both remain bound to the head. The energy released by the hydrolysis "charges" (activates) the myosin head into a "cocked" position, just as a person's thumb might cock the hammer on a pistol. In this cocked position, the head and actin bind weakly together. Then several events occur which help to bring about muscle contraction: the binding of actin to the head causes ADP and inorganic phosphate to be released. As those products exit, the head binds strongly to actin. Virtually simultaneously, the head rocks relative to the myosin backbone, thereby applying the power stroke to actin. The power stroke moves or "slides" the actin filament a distance of 5 to 10 nanometers. The uncocked myosin head remains strongly bound to actin but in the uncharged state. Only a new ATP molecule can start the cycle once again by weakening the myosin-to-actin binding and allowing the head to be released.

An easy way to remember these events is to think of actin and ATP as competitors for binding to the myosin head. When ATP binds to myosin, actin is released; when actin binds strongly to myosin, ADP and inorganic phosphate are released. ATP and ADP obviously play key roles in muscle contraction. First ATP binding releases actin from the head, allowing it to bind to a new position on the sliding actin filament. Second, the hydrolysis of ATP releases energy to cock the head; the head—a mechanoenzyme—uses that energy in turn by applying the power stroke to actin. This cycle occurs rapidly in thousands of heads at each end of a sarcomere as long as ATP is available to weaken head-to-actin binding and to provide energy for the cocking process (and as long as calcium ions are present, as we shall see shortly). Figure 39-17 shows a view of these events with emphasis on the movement of the actin filament by the head's power strokes. Although the basic action of myosin heads is the same in all muscle cell types, there is at least one important variable. Myosin from slow, red muscle tends to have a slow cycle of hydrolyzing ATP. Conversely, myosin from fast, white muscle hydrolyzes ATP rapidly, thus accounting for the fast contractions and also the tendency of the cell to become fatigued.

A major finding in 1987 by James Spudich and colleagues of Stanford Medical School provides strong support for the basic model of contractions (Figure 39-16). They showed that myosin heads, isolated by enzyme digestion of muscle myosin, produced movements in a special test. Therefore, the "hinge" or "handle" parts of the molecule are not needed for force production. Other new work suggests that the power stroke may begin just prior to the release of ADP and inorganic phosphate. These two results support the idea that a simple molecular shape change of the myosin heads is translated at

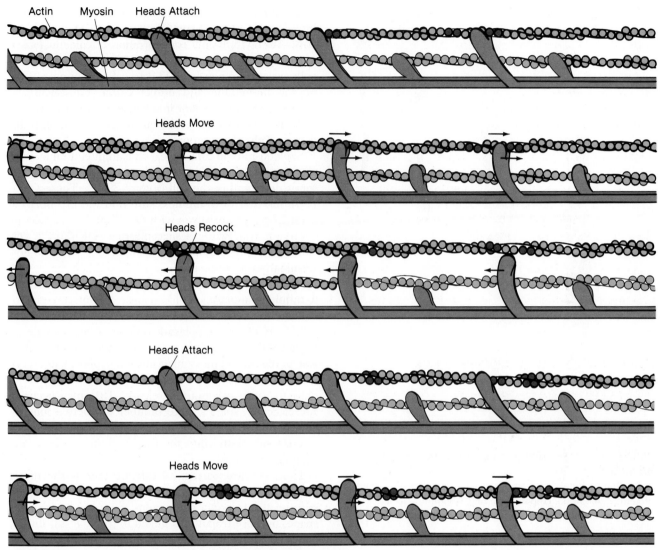

Figure 39-17   THE POWER STROKE AND SLIDING FILAMENTS.
The sequence shown in this diagram illustrates the sliding-filament theory of how myosin head movement causes actin filaments to slide. In the presence of calcium ions and ATP, myosin heads attach to actin filaments, become cocked, and carry out the power strokes that move the actin filaments. Note that actin molecules move progressively to the right in this sequence as the myosin heads make successive power strokes.

higher levels into filaments sliding, muscle cells contracting, a limb moving, and a rabbit dodging a predator's snapping jaws. With muscle contraction, as with so many of life's processes, organization and functioning of molecules allow such hierarchies to exist.

The Chemical Trigger for Muscle Contraction

Muscles are frequently relaxed—their myosin heads are not cycling between cocked and uncocked positions. What is the chemical trigger that initiates a muscle con-

traction? We know from previous chapters that motor neurons or hormones such as epinephrine or oxytocin can cause muscles to contract; how can that occur? To understand the steps between the reception of motor nerve impulses by a skeletal-muscle fiber and the resulting contraction, we need to consider the effects of two additional proteins associated with actin.

As Figure 39-18 shows, an actin thin filament is actually composed of two actin chains twisted into a helix. In the grooves of this helix are molecules of **tropomyosin,** a long, thin protein. When a muscle is at rest, tropomyosin prevents myosin heads from binding actin, probably because it masks the sites at which binding occurs.

Another protein, **troponin,** is situated at regular inter-

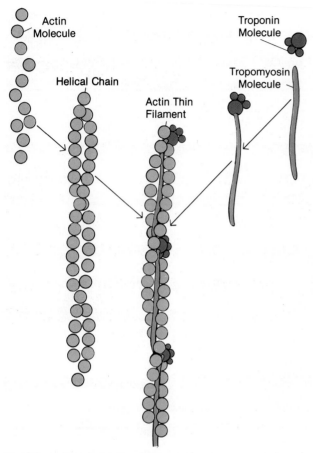

Figure 39-18   ACTIN THIN FILAMENTS: MOLECULAR ASSEMBLIES.
Actin thin filaments are composed mainly of a helical chain of actin monomers. Tropomyosin and troponin associate with the helical actin structure to form actin thin filaments.

vals along the actin thin filament where it binds to tropomyosin molecules and to actin. Troponin is in fact a complex of four proteins, one of which has a critical property: it can also bind calcium ions. Knowing this, we can trace the events that begin when motor nerve impulses reach a muscle.

A special property of muscle cells—which they share with neurons—is that they are electrically excitable; that is, they can initiate and propagate action potentials (impulses). As we saw in Chapter 36, a chemical neurotransmitter, such as acetylcholine, is released when action potentials reach the ending (synaptic terminal) of a motor nerve; when a threshold level of acetylcholine is released at the *neuromuscular junction*, the specialized junction of motor nerve and muscle-cell membranes, an action potential is initiated and propagates over the whole muscle-cell surface. That rapid spread of the potential immediately triggers events *inside* the muscle cell that culminate in sarcomere contraction.

To understand what the action potential does inside a muscle cell, we must consider some facts about calcium ions. The cytoplasmic concentration of calcium ions is normally very low (about 10^{-8} molar) in all eukaryotic cells because of powerful ATPase calcium pumps in the plasma membrane and mitochondria; the calcium is pumped out of the cell or into the mitochondria. Muscle cells have still another calcium reservoir with an interesting mode of operation.

Skeletal-muscle cells have a vital architectural feature: a system of infoldings of the plasma membrane that extends deep within the cell (Figure 39-19). These *transverse tubules*, or **T tubules,** surround the myofibrils at their Z lines. Interwoven with the T tubules is a calcium reservoir, a network of **sarcoplasmic reticulum** (a version of smooth endoplasmic reticulum), whose hollow terminal sacs contain calcium ions. When nerve impulses initiate an action potential in the muscle fiber, the T-tubule membranes act like a maze of miniature electrical conduits, conducting the action potential to the terminal sacs of the sarcoplasmic reticulum. Some evidence suggests that the action potential causes rapid production of the second messenger inositol triphosphate; recall from Chapter 37 that this molecule acts in a variety of cells to cause calcium-ion release. That may occur in muscle cells, since a large number of calcium ions instantly diffuse out of the terminal sacs of the sarcoplasmic retieulum and into the fluid bathing the actin and myosin filaments.

This chemical barrage of calcium ions is the direct trigger of a muscle contraction. As calcium ions bind to sites on the troponin component, the complex changes shape; because troponin is linked to tropmyosin, it, too, shifts position, unmasking the sites on actin that bind myosin heads. Then, like pegs tumbling into slots, the cocked myosin heads bind strongly to actin, and the molecular power strokes begin (Figure 39-20). The more calcium ions released from the sarcoplasmic reticulum, the more troponin binds calcium ions, the greater the sliding of filaments, and the stronger the muscle-fiber contraction.

Turning Muscles Off

If the presence of calcium ions in the cytoplasm of a muscle fiber brings about a contraction, what causes the muscle fiber to relax? When muscle action potentials stop being initiated, the plasma membranes of the muscle cell and the T tubules rapidly return to their normal state of polarization, and the release of calcium ions from the sarcoplasmic reticulum halts. At the same time, ATPase pump enzymes in the membrane of the sarcoplasmic reticulum pump calcium ions back into the terminal sacs. As calcium is withdrawn, troponin reverts to

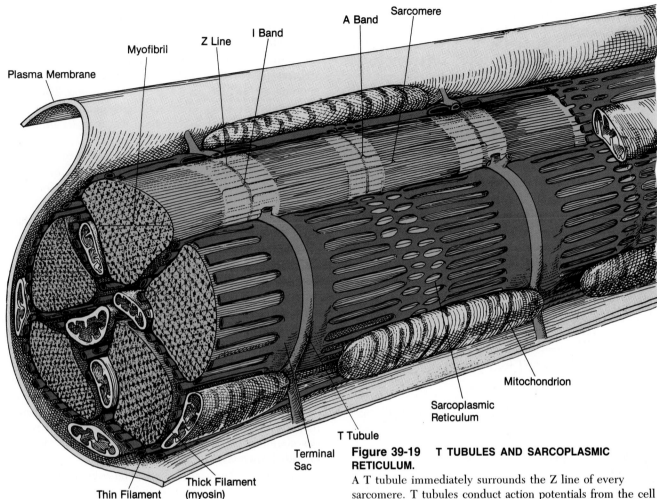

Plasma Membrane

Myofibril

Z Line

I Band

A Band

Sarcomere

Mitochondrion

Sarcoplasmic
Reticulum

T Tubule

Terminal
Sac

Thin Filament
(actin)

Thick Filament
(myosin)

Figure 39-19   T TUBULES AND SARCOPLASMIC RETICULUM.

A T tubule immediately surrounds the Z line of every sarcomere. T tubules conduct action potentials from the cell surface inward and cause calcium ions to be released from the nearby terminal sacs of the sarcoplasmic reticulum. That release stimulates the myosin-actin interaction responsible for contraction and shortening of the muscle cell.

its resting configuration, tropomyosin immediately shifts to cover the myosin-binding sites on actin, and the contraction ends. Finally, actin filaments slide outward again, and the sarcomere returns to its resting length.

Thus as a muscle fiber alternately contracts and relaxes, calcium is conserved—it diffuses out of and is later pumped back into the sarcoplasmic reticulum. ATP, on

Figure 39-20   REGULATION OF MYOSIN BINDING AND MUSCLE CONTRACTION.

(a) The myosin head does not bind to the actin molecule in the absence of calcium. (b) Increased levels of calcium ions in the cytoplasm result in the binding of calcium by one of the troponin component proteins. This binding results in a change in shape of troponin and a shift in the placement of tropomyosin on the actin molecule that uncovers the myosin-binding site. The myosin head then can bind and the muscle can contract. Not all of the subunits of troponin are shown here.

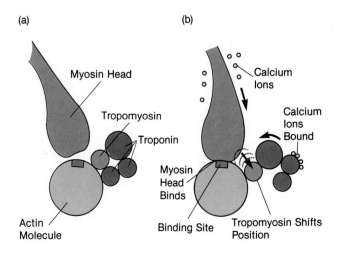

(a)   (b)

Myosin Head

Tropomyosin

Troponin

Calcium
Ions

Calcium
Ions
Bound

Myosin
Head
Binds

Actin
Molecule

Binding Site

Tropomyosin Shifts
Position

the other hand, is consumed during both contraction and relaxation: (1) in altering the strength of binding between actin and the myosin heads and in recharging the heads to make them ready for another power stroke; and (2) in powering the ATPase pump to return calcium ions into the terminal sacs of the sarcoplasmic reticulum.

Muscles that lack a supply of ATP remain contracted. An extreme case develops after a vertebrate dies, when *rigor mortis* sets in. At death, calcium begins to leak from the sarcoplasmic reticulum, causing contractions; soon muscle-cell ATP is fully hydrolyzed to ADP, and actin and myosin filaments remain locked together in their contracted position. Because death also terminates aerobic and anaerobic metabolism, new ATP is not available to release myosin from actin or to drive the pump enzyme that would lower calcium-ion levels again. Only after some hours do the stiff and contracted muscles relax as other degeneration processes dominate over the final contraction of death.

REFINEMENTS OF MUSCLE ACTION: GRADED RESPONSES AND MUSCLE TONE

So far, we have considered the basic chemical events of contraction and have seen how neurons trigger the process. This section focuses on some of the special char-

acteristics of contraction in skeletal-muscle cells and on the way these phenomena vary in muscles that operate as organs.

An important functional feature of most *individual* striated-muscle-fiber cells in vertebrates is that they contract, or "twitch," in an all-or-none fashion. That is, a fiber contracts only when incoming motor nerve impulses exceed a certain threshold that triggers a muscle action potential and liberates calcium ions. Once the threshold has been surpassed, however, further increases in the intensity, rate, and duration of motor nerve impulses produce only a slightly stronger contraction.

In contrast to the individual muscle-fiber cells that make them up, whole vertebrate muscles (which are organs) do not show the all-or-none response; instead, they exhibit *graded responses* to a stimulus. Thus a rabbit's leg muscles might contract fully during a sprint to escape a coyote, and twitch weakly when the animal shivers to generate heat for body warmth. Such variations occur, in part, because the many motor neurons in a typical motor nerve running to a muscle innervate varying numbers of the muscle-fiber cells. If a larger proportion of these neurons conduct impulses to the muscle (so that the all-or-none threshold of more fibers is surpassed), more muscle fibers are called into action, resulting in a stronger overall contraction. Conversely, stimulation of just a few fibers produces a weak contraction of the whole muscle.

Graded responses are also based on *summation*, a condition during which nerve impulses arrive at a muscle before its previous contraction has subsided (Figure 39-

Figure 39-21   SUMMATION AND TETANUS DURING MUSCLE CONTRACTIONS.
Motor neurons electrically stimulate a muscle. When the rate of stimulation is slow, contraction of the muscle occurs by individual twitches. When the rate of electrical stimulation increases, the muscle no longer has time to relax between stimuli. After a period of summation of the individual twitches, a sustained, smooth tetanic contraction results. Ultimately, the muscle fatigues because it no longer has sufficient ATP to maintain contractile processes.

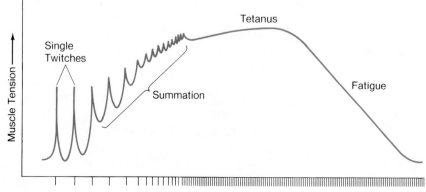

21). Summation results, in part, from the inability of sarcoplasmic reticulum to recover all its secreted calcium ions before new nerve action potentials arrive and cause the muscle cell to initiate action potentials that would release more calcium ions. The strength of summated contractions is always greater than the strength of individual twitches, since twitches are responses to less-frequent stimulation.

Summation often culminates in **tetanus,** a state of sustained maximum contraction. Tetanus is a normal and crucial element of muscle function; in fact, most everyday muscle actions depend on the smooth, strong contractions of tetanus. For instance, when agonist and antagonist muscles in a typist's fingers counteract each other, providing precise muscular control of movement, both muscles are in states of summation. The length of time that high levels of summation and tetanus can be maintained varies with muscle type—briefly in white muscle, much longer in red muscle—because of the differences in the rate at which each type utilizes ATP.

Another important characteristic of muscle function is the ability to achieve **tonus,** or "muscle tone," a condition in which a muscle (or sets of muscles) is kept partially contracted over a long period (thus tonus is not synonymous with the slang meaning of "muscle tone" of a physically fit individual). Tonus is produced when first one set of fibers, then another, and finally yet another is briefly stimulated, so that in the muscle as a whole, some parts are always contracted, though most remain relaxed. The tonus of back, abdominal, neck, and limb muscles enables humans and other terrestrial animals to maintain normal posture in the presence of gravity; the tonus of leg muscles is vital to reducing the total volume of the leg veins and in aiding the blood's return to the heart. This tonus is adjusted continuously as the nervous system reacts to a wealth of sensory information by ordering the appropriate coordinated tonic muscle responses necessary to maintain stance and balance.

Finally, some muscles display *intrinsic rhythmicity*—spontaneous involuntary contractions, such as those of heart muscle that are so vital to life. Other muscle types in animals as diverse as gorillas, lugworms, and locusts also have this feature and so do not need the stimulus of action potentials in order to contract. Periodic involuntary contractions are probably an evolutionary specialization of spontaneous twitch activity—a basic property of all muscle cells.

SMOOTH MUSCLE

After striated muscle, a vertebrate's most abundant muscle type is smooth muscle. Found in the gut wall, the walls of blood vessels (Figure 39-22), the iris of the

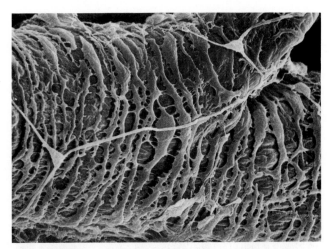

Figure 39-22   SMOOTH-MUSCLE CELLS ENCIRCLING SMALL VEINS.
The contraction of smooth-muscle cells such as those pictured here will narrow the enclosed veins (magnified about 60 times). Axons of autonomic nervous system neurons (Chapter 36) run across these smooth-muscle cells; although not visible here, synapses are common between the axons and smooth-muscle cells.

eye, reproductive organs, and glandular ducts, smooth muscle usually carries out sustained, slow contractions and typically is not under voluntary control.

Smooth-muscle tissue is strikingly simple: its slender, elongate cells have no striations and just a single nucleus (Figure 39-23). Smooth-muscle cells are arranged in sheets in the wall of the large intestine or as straps around blood vessels or ducts. Individual smooth-muscle cells are linked to each other by (1) surface specializations in which peglike protrusions from one cell fit into socketlike depressions in an adjacent cell, (2) abundant collagen fibers, and (3) laterally placed gap junctions that couple groups of cells electrically. This design, in part, allows the contractile force generated in a single smooth-muscle cell to be applied throughout the tissue.

But how do smooth-muscle cells contract if they have no striations? The cytoplasm of these cells is crowded with actin thin filaments, many of which appear to have one end inserted in the inner surface of the plasma membrane (an arrangement that is suggestive of the Z line insertion of actin in sarcomeres). A chemically unique member of the myosin family is also present in smooth muscle, with some molecules organized into thick filaments. The myosin heads in smooth muscle probably function just as they do in the sarcomeres of skeletal muscle: when actin rooted in the plasma membrane is pulled inward, the cell shortens. (A model of how sets of actin are organized into contractile units is seen in Figure 39-23c.) Smooth-muscle cells contract more slowly than do striated-muscle cells and are able to

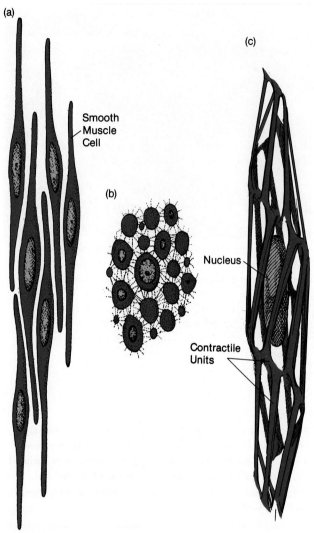

Figure 39-23   SMOOTH-MUSCLE: SLOW, SUSTAINED CONTRACTIONS.

(a) Individual smooth-muscle cells from a frog's bladder wall are visible in this drawing. (b) Here a set of smooth-muscle cells in the wall of a kitten's bladder are cut in cross section; the cells are quite closely packed. (c) This diagram of a smooth-muscle cell reveals a possible arrangement of bundles of actin and myosin—so-called contractile units. Actin is attached to the inner surface of the plasma membrane, but myosin and sarcomeres are not precisely ordered. In reality, the contractile units are probably less regularly arrayed than they are here.

sustain the contraction far longer because metabolism can continuously supply the ATP needed to fuel each cell's contractile activity.

The chemical steps in the excitation of smooth muscle are similar to those for striated muscle, but with several important differences: in the absence of sarcoplasmic reticulum, the large cell surface must act as the site of calcium-ion entry from the tissue fluids bathing the cell.

The cell surface is also the location of the ATPase that pumps calcium ions back out of the cell to terminate contraction. The role of calcium also differs between smooth and striated muscles. In the former, calcium initiates a chain of events leading to the phosphorylation of myosin, which starts the contractile cycles.

Some smooth-muscle cells are innervated by the autonomic nervous system (Chapter 36); in general, sympathetic nerves stimulate contractions, and parasympathetic nerves inhibit them. Smooth-muscle cells also tend to be responsive to hormones: for instance, we saw in Chapter 37 that various hormones can stimulate smooth-muscle contractions directly (the action of oxytocin on uterine muscle during birth in mammals is a prime example). Inositol triphosphate appears to be the second messenger when hormones act on smooth muscle; calcium ions are freed in the cytoplasm as a consequence of this action. Other smooth-muscle cells show spontaneous, frequently rhythmic, contractile activity. Since smooth-muscle cells tend to be interconnected by gap junctions, which pass electrical current from cell to cell via tiny channels, a stimulus that affects one region of cells spreads to others, so that the whole muscle sheet gradually contracts. An example of this mechanism is peristalsis—the slow waves of muscle activity that propel masses of food or fluid through the intestine.

CARDIAC MUSCLE: STRIATED TISSUE WITH SMOOTH-MUSCLE CHARACTERISTICS

The heart is built from the third type of muscle cell—cardiac muscle. Like smooth-muscle cells, each cardiac-muscle cell has a single nucleus, but unlike smooth muscle, cardiac muscle is striated. As befits tissue that is constantly active for the duration of an organism's lifetime, cardiac-muscle cells are liberally supplied with mitochondria.

Cardiac-muscle cells are often branched, with the branches of neighboring cells forming an interlocking system. Depending on the vertebrate species, there is a T-tubule system of varying complexity. There is also an extensive network of sarcoplasmic reticulum close to the plasma membrane that serves as the calcium-ion reservoir for contraction. As Figure 39-24 shows, the ends of cardiac-muscle cells are bound firmly to each other by **intercalated disks**—regions of cell membrane that act both as welds holding cells together and as very leaky junctions (that is, sites where ion or electric currents can

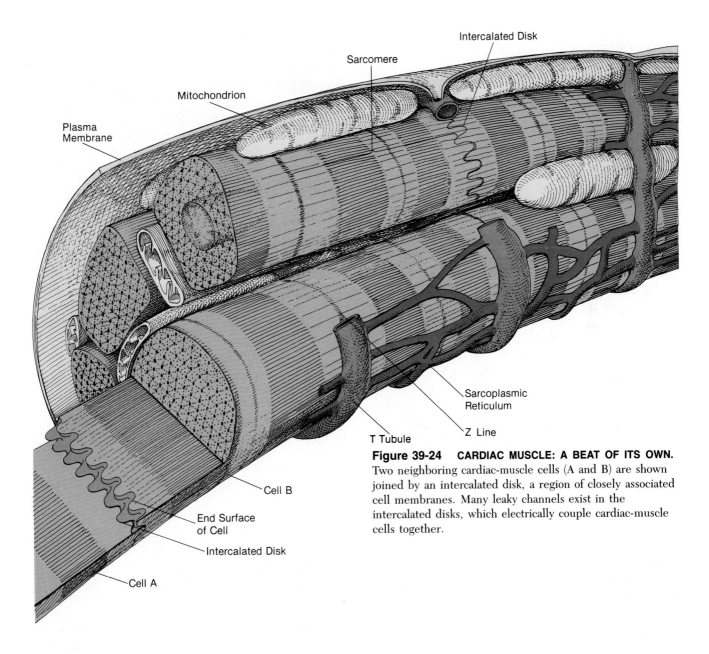

Figure 39-24   CARDIAC MUSCLE: A BEAT OF ITS OWN.
Two neighboring cardiac-muscle cells (A and B) are shown joined by an intercalated disk, a region of closely associated cell membranes. Many leaky channels exist in the intercalated disks, which electrically couple cardiac-muscle cells together.

flow easily). Each disk is made up of layers of folded, reinforced cell membrane where electrical impulses are transmitted, just as across gap junctions of other cell types; it is also the site where actin filaments of the terminal sarcomeres are attached to the cell surface. In one sense the disks are the equivalent of Z lines, since the forces of sarcomere contraction are applied to them.

Placed in a culture dish, isolated single cardiac-muscle cells contract spontaneously at periodic intervals—they truly "beat." Similarly, an intact vertebrate heart free of all nerves has an intrinsic ability to contract rhythmically. This is why a vertebrate's heartbeat is said to have a *myogenic* origin (recall that "myogenic" means "muscle-generated"; Chapter 31); in contrast, the heartbeats of animals such as lobsters, crabs, and spiders are

neurogenic—that is, they are driven by nerves so that if the heart of such a creature is isolated, it will not contract spontaneously.

In Chapter 31, we saw that highly specialized *pacemaker* cells in the heart wall commonly override the intrinsic contraction of individual heart-muscle cells and assume the task of setting the actual heart rate. Furthermore, the rate can be increased by the activity of sympathetic nerves, which produces faster and more forceful contractions (for instance, one reason epinephrine speeds the heart rate is that it causes faster release of calcium ions from troponin and uptake by the sarcoplasmic reticulum). Alternatively, parasympathetic nerves can trigger a slower contraction rate and reduce the strength of contractions.

RUNNING THE MILE: WHAT IS OUR SPEED LIMIT?

During the past two or three decades, sports fans have watched performance records smashed again and again. Experts predicted that no one would ever run a four-minute mile—right up until 1954, when Roger Bannister did it. Since that feat, faster times have been set, only to be superseded a few months or years later. In all, runners have shaved thirteen seconds off Bannister's time and continue to push on to ever faster records (Figure A). But how much faster can a human run the mile and other events? Are there absolute limits to human performance, or will we continue to improve indefinitely, as our nutrition and training methods improve?

Most exercise physiologists believe that, indeed, there *are* limits, based on some fundamental physiological aspects of the human body. For example, the maximum volume of oxygen that a person can consume per kilogram of body weight per minute is an upper limit for aerobic exercise. (Sustained events, such as the marathon, require that the muscles work aerobically, while during short-burst events, such as the 100-yard dash, muscles can work anaerobically and incur an oxygen debt). Measurements show that a world-class marathoner, such as Bill Rodgers, can use about 80 milliliters of oxygen per kilogram of body weight per minute compared with the 45 milliliters of oxygen per kilogram per minute consumed by an amateur jogger. Based on the structure of heart muscle and its maximum pumping speed, some physiologists think that 80 to 90 milliliters per kilogram is the upper limit on aerobic performance. By this standard, marathoners may already be nearing the limits to speed and endurance in their sport.

One exception to this might be an athlete with a huge heart. Swedish researchers found a champion cross-country skier with a massive chest whose heart pumped 36 quarts of blood per minute—twice that of a normal man and 6 quarts per minute more than the current world-class milers. However, this individual's heart was the biggest heart yet studied, and physiologists think it is unlikely that hearts will grow bigger than his—barring some future mutations. Besides, the massive chest needed to house a massive heart is not considered to be the best physique for running.

The strength of bone is another potential limit to human performance—but a distant one. Orthopedic surgeons removed a femur from a cadaver and subjected it to greater and greater compression. It splintered at 1,600 pounds per square inch, which is about twice the force exerted on the leg bones of a 160-

Figure A
What are the limits? Sebastian Coe, straining to the utmost, crosses the finish line in the 1984 Olympics. The biochemistry of such long-distance runners' skeletal-muscle cells sustains prolonged severe use, in contrast to muscles of sprinters.

pound runner and far more than current weight lifters can heft. Even a very large athlete, such as a 7-foot basketball player, does not approach the limits of bone strength when pounding down the court or jumping for a basket.

Yet another physiological limitation on performance is the proportion of fast-twitch to slow-twitch muscle fibers. Marathoners like Rodgers have 80 to 90 percent slow-twitch (a type of red muscle), while sprinters, such as Carl Lewis and Evelyn Ashford, have about 70 percent fast-twitch (a type of white muscle). Record-breaking performances could well be expected from athletes with better ratios—say, 95 percent slow-twitch or 80 to 90 percent fast-twitch. But this is, in part, a question of human genetics, and without some lucky combination, such as the child of two Olympic sprinters or the purposeful genetic manipulation of humans, such a fluke is as unlikely as inheriting a giant heart.

Based on all these considerations, some physiologists are now predicting that the fastest a person can ever run the mile will be about 3:34, or roughly 13 seconds faster than existing records. However, only time will tell whether they are right, or whether some Roger Bannister of the future will smash that record, too.

MUSCLES IN EVOLUTION: THE CONTRACTILE SYSTEMS OF INVERTEBRATES

Studies of skeletal muscle revealed much of what biologists now know about the basic contractile properties of eukaryotic cells; actin and myosin are found nearly everywhere that contractions occur in organisms. Similarly, virtually all contractile systems rely on calcium-ion induction of the myosin power stroke. As animals have evolved, however, differences in the molecular apparatus that underlies contractions have been translated into differences in the muscle function of vertebrates and invertebrates.

The first muscle cells with specialized contractile properties probably evolved from tissue cells whose basic contractile machinery acquired the capacity for stronger contraction. Early forms were likely quite simple—smooth-muscle cells associated with protrusible organs, such as the proboscis and the penis, and later with the gut. Today, such muscles are common in sponges, cnidarians, and many worms and mollusks.

Unlike smooth-muscle cells in vertebrates, invertebrate smooth muscles teem with myosin thick filaments; as a consequence, such cells can produce some dramatic effects. For example, the tentacle muscles in a Portuguese man-of-war can stretch some 21 meters (70 ft) and yet can contract to shorten the tentacle to 14 centimeters (5.5 in.)! A variation is the mollusks' *catch muscle*, in which huge thick filaments and many thin filaments yield cells with a remarkable capability for sustained contraction. The powerful catch muscle of a giant clam, a prime example, can hold the animal's shell firmly closed for up to thirty days at a time (Figure 39-25).

The striated-muscle cell—with its highly organized sarcomeres—apparently arose independently in many invertebrate lines; for instance, it drives the quick, repetitive contractions required to compress the bell-shaped bodies of cnidarians for flotation and movement. Barnacles and octopuses possess striated muscles in only certain parts of their bodies, but in insects, such cells are ubiquitous—they are used everywhere from the gut, with its slow peristaltic movements, to the wings of midges, with their 1,000 beats per second.

The biochemistry of invertebrate muscle action can also vary. Mollusks, for instance, generally lack troponin; myosin itself binds calcium ions and initiates the power stroke that causes actin to slide. Patterns of control differ, too; arthropod muscle cells do not have the all-or-none mechanism but show graded response in which the degree of contraction is proportional to the level of polarization of muscle-cell membranes.

Figure 39-25   MOLLUSKS AND CATCH MUSCLES.
This giant clam (*Tridachna gigantea*), as well as its smaller mollusk relatives, can remain tightly closed because of properties of its very thick myosin filaments and its muscle cells.

Another novel and important feature of arthropod muscle is *stretch activation* of certain insect flight muscles. In insects with slowly beating wings—including dragonflies, locust, and butterflies—the muscles that raise and lower the wings are driven entirely by nerves. Among flies, bees, and beetles, however—insects whose wings beat extremely rapidly—nerve stimulation is only the first step; thereafter, stretch activation takes over. Thus when wing depressor muscles contact, wing elevator muscles are stretched; this stretch stimulates the elevator muscles to contract. As the elevators shorten and raise the wings, the depressors are stretched, which, in turn, triggers their contraction. This cycling allows incredible wing-beat frequencies—in fact, several times higher than the highest rates of action potential propagation in motor nerves. As a result, although nerves in these insects start, stop, or regulate the power of wing beats, they can never function fast enough to account for the beats' extraordinary rapidity.

Biologists still have much to learn about how the muscles of invertebrates and other animals operate; even so, new discoveries are likely to be based on the fundamental elements of contraction—the special organization of actin and myosin molecules and the power stroke of the myosin head. By generating the force to move blood, bone, gastric contents, and entire organisms, such molecular events clearly have played a crucial evolutionary role in shaping animal life styles.

SUMMARY

1. Muscle cells exhibit specializations of structure and function that evolved as variations on the basic contractile ability of the cytoskeleton in all eukaryotic cells.

2. Exoskeletons—nonliving deposits of mineral salts and other materials—provide armor, permeability barriers, and hard surfaces for invertebrate animals. Muscles can move the hinged external skeletons for biting, locomotion, and other activities.

3. The vertebrate's endoskeleton contains *cartilage* (composed of collagen and polysaccharides) and/or *bone* (composed of collagen and apatite). Two major subdivisions of the endoskeleton are the *axial skeleton* (skull, vertebral column, and ribs) and the *appendicular skeleton* (limbs and girdles). *Joints* are the regions where individual bones meet; within joints, bones are held together by *ligaments*—strong, flexible bands composed primarily of collagen.

4. Mammalian compact bones are especially hard because their surface regions are composed of densely packed, fused *Haversian systems.* Spongy bone is an open framework, and may contain central marrow regions. Bones serve as a reservoir of calcium and phosphate ions for the body.

5. Vertebrate muscles operate as systems of springs or ropes acting on "levers"—the bones to which they are attached. Simple movements involve the relative activities of sets of muscles that act in opposition to each other: the agonist, or prime mover, muscle and its antagonist muscle. Muscles may connect directly to the tough connective-tissue layer on a bone's surface or may grade into *tendons*—extremely tough bundles of collagen fibers—which connect to bones.

6. Muscle is divided into three classes: *skeletal* (striated) *muscle,* *smooth muscle,* and *cardiac muscle.* Striated muscle may be red or white (or gradations thereof), reflecting each type's individual chemistry; in vertebrates, both types usually contain *creatine phosphate,* a high-energy organic molecule that provides an ATP fuel reservoir for sustaining muscle contractions.

7. *Muscle spindles,* located near the center of each skeletal muscle, function to report the changing lengths of their parent muscle. Hence, when the muscle is stretched and lengthens, its spindles register this information and start a reflex action that culminates in contraction of the muscle. Conversely, when a muscle contracts, the spindle is stretched less; its sensory output falls, so the muscle itself receives less stimulus to contract.

8. By way of γ (gamma) motor neurons, the nervous system can signal muscle spindles to contract; this stretches the spindles' sensory region and triggers the same reflex arc that causes the parent muscle to contract. This is an alternative means for the brain and spinal cord to elicit muscle contraction. It is important in achieving fine control of muscle activity and in maintaining the *tonus* of muscle contraction required for posture, stance, and balance.

9. Striated muscle is composed of giant, multinucleate cells called *muscle fibers,* which contain long assemblies of molecules known as *myofibrils.* Myofibrils are built from *sarcomeres*—precisely repeating units of the muscle proteins actin and myosin and the sites of muscle contraction.

10. Sarcomeres shorten during contraction because actin thin filaments slide past myosin thick filaments. Myosin heads generate the force that drives this sliding mechanism. This is the central element in the *sliding-filament theory* of muscle contraction.

11. Motor nerve impulses arriving at the neuromuscular junction trigger action potentials in the striated-muscle cell's surface. The action potential spreads to the *T tubules* and *sarcoplasmic reticulum.* As a result, calcium ions are released into the cytoplasm to start the contraction event.

12. When calcium binds to the *troponin* protein complex, another protein, *tropomyosin,* is moved away from actin, thereby uncovering binding sites for myosin heads. As a result, the "power stroke" of contraction can begin.

13. Both calcium ions and ATP must be present if the cycle of myosin-head binding, cocking, and power stroking is to continue. Creatine phosphate, myoglobin, and cellular metabolism play roles in maintaining ATP levels.

14. Vertebrate muscle fibers respond to nerves in an all-or-none fashion. However, graded responses of the whole muscle result as different numbers of fibers are stimulated at any given time. *Tetanus* is the sustained maximum contraction of a muscle and is a vital element in normal muscle functioning.

15. Smooth-muscle cells lack striations, have just one nucleus, and have rather simple, relatively unordered contractile systems. Smooth-muscle cells usually are interconnected by gap junctions. Smooth-muscle contractions are usually slow and may be sustained for a relatively long period without muscle fatigue. Calcium triggers contraction of smooth muscle, in part by causing phosphorylation of myosin.

16. Cardiac-muscle cells have only a single nucleus and typical sarcomeres and are tightly bound to each other by specialized junctions known as *intercalated disks.* Such disks pass ions and electrical cur-

rents, so that the cardiac-muscle cells are coupled electrically.

17. Vertebrate cardiac muscle has an intrinsic rhythmic contractile beat (myogenic beat) that is slowed by parasympathetic nerves or speeded by sympathetic nerves. Cardiac muscle in many invertebrates is driven by nerves, and hence is neurogenic.

18. The first muscle cells were probably of the smooth type, and moved protrusible organs. Later, more highly ordered sarcomeres evolved to allow the greater force and speed of contraction characteristic of cardiac and skeletal muscle.

KEY TERMS

appendicular skeleton
axial skeleton
bone
cardiac muscle
cartilage
creatine phosphate

Haversian system
intercalated disk
joint
ligament
muscle fiber
muscle spindle

myofibril
sarcomere
sarcoplasmic reticulum
skeletal muscle
sliding-filament theory
smooth muscle

T tubule
tendon
tetanus
tonus
tropomyosin
troponin

QUESTIONS

1. Compare exoskeletons and endoskeletons: Which contain living cells? Which grow? Which are jointed? How is each type moved?

2. What are the materials that comprise bone? What materials are present in growth zones? Describe the structures of compact and spongy bone. What are osteocytes and how do they exchange or obtain gases, wastes, and nutrients?

3. Name the major parts of the vertebrate endoskeleton. Explain what ligaments and tendons do, and what materials comprise them. Name the three major types of vertebrate muscles, their locations, and general characteristics. Compare and contrast the three types.

4. What is a muscle spindle and how does it work? What is proprioception?

5. Describe a skeletal muscle fiber. What are myofibrils and sarcomeres? Draw a sarcomere and label its components.

6. What is the function of T tubules in skeletal muscle fibers?

7. What is the role of sarcoplasmic reticulum in skeletal muscle fibers? What structure fulfills this function in smooth-muscle cells? In cardiac muscle?

8. Explain how a muscle fiber contracts; include the roles of ATP, inositol triphosphate, calcium, troponin, actin, myosin heads, and tropomyosin.

9. Describe relaxation of skeletal muscles. Does this process use up calcium or ATP? Why does rigor mortis occur after death?

10. Does the way a single vertebrate skeletal-muscle fiber responds to a stimulus differ from the way a whole muscle responds? Explain. What is tetanus? Tonus?

SUGGESTED READINGS

ADELSTEIN, R. S., and E. EISENberg. "Regulation and Kinetics of the Actin–Myosin–ATP Interaction." *Annual Review of Biochemistry* 49 (1980): 921–956.

This and other volumes of Annual Review are the places to best enter the basic scientific literature on muscle contraction.

ALBERTS, B., et al. *The Molecular Biology of the Cell.* New York: Garland, 1984.

The chapter on contractile systems is excellent.

HUXLEY, A. F. *Reflections on Muscle.* Princeton, N.J.: Princeton University Press, 1980.

One of the pioneers in muscle research recounts the ideas behind our current understanding of these tissues.

KESSEL, R. G., and R. H. KARDON. *Tissues and Organs.* San Francisco: Freeman, 1982.

This book contains superb illustrations of bone, cartilage, and muscle of a variety of types.

40
THE VERTEBRATE BRAIN

The flexibility and appropriateness of animal behavior suggest both that complex processes occur within their brains, and that these events may have much in common with our own conscious mental experiences. To the extent that this proves to be true, many of our ideas and opinions about the relationship between animals and men require modification.

Donald R. Griffin,
The Question of Animal Awareness
(1981)

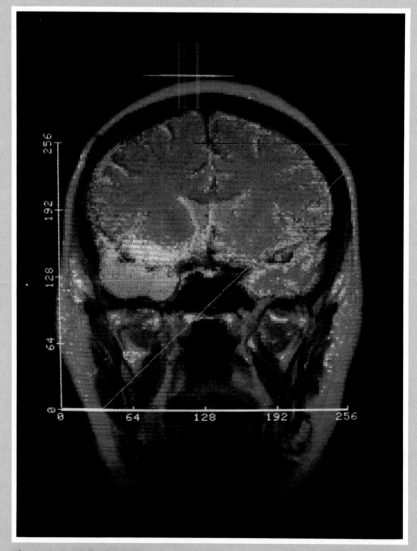

The secrets of the human brain structure are revealed by new technology such as nuclear magnetic resonance imaging. Here is a cross section of a living human brain.

Five thousand years ago, a surgeon peered into the gaping head wound of an Egyptian soldier and saw a gray jellylike mass. Its coarse wrinkles reminded him of the corrugated surface of congealing molten metal. The doctor's account of his observations was later etched in hieroglyphics and became the earliest known written reference to the human brain—the most complex organ ever to evolve.

The human brain contains at least 100 billion neurons and more than ten times that number of supporting glial cells. However, unlike the millions of cells that make up a heart, a muscle, or the lens of an eye, brain cells are not interchangeable replicas of one another. Instead, neurons from different parts of the brain have distinctive shapes, chemistries, and functions. Most are woven into circuits with many axons, dendrites, and synapses; these minisystems or subpopulations of innumerable neurons carry out the brain's varied activities.

Each neuron within a subpopulation interconnects with tens, hundreds, or thousands of other cells. The resulting heterogeneity of neuronal circuits is the source of thinking, learning, and other functions. Moreover, biologists have learned as recently as the mid-1980s that there is substantially more *plasticity*—changeability in the number and function of synapses—than they had ever suspected. Moreover, they have learned that the number of synapses increases in regions of the brain that are under intense use, as during learning. Cellular and circuit heterogeneity and plasticity greatly complicate the efforts of researchers to uncover the mechanisms by which the brain carries out its higher-order tasks of memory and learning. But biologists are now defining "master plans" by which neurons are organized into functional compartments of the brain. Similarly, they are identifying electrical and chemical properties of functioning brain cells, including how neurotransmitters and a newly recognized class of chemicals called *neuroactive peptides* control communication among neurons and the properties of neuronal circuits. These new perspectives provide intriguing hints about two of the brain's most complex tasks: memory and learning. Thus the human brain, the living world's most complex object, is finally beginning to yield its secrets to the powerful new techniques of molecular biology and neurobiology. As we explore outward from Earth into our solar system for the first time, we are successfully exploring inward also, and one day, humans will indeed "know themselves."

STRUCTURE AND FUNCTION OF VERTEBRATE BRAINS

In Chapters 25 and 36, we saw that the nerve nets, ganglia, and nerve cords of various invertebrates are arranged in a variety of patterns. Many invertebrates possess a dominant mass of neurons—a "superganglion," or brain—near the major sense organs, which are situated in the anterior part of the body or head. Unfortunately, biologists know little of how the basic vertebrate brain evolved from the simpler systems of invertebrate ancestors. Indeed, the earliest known vertebrate fossils (fossil agnathans, or jawless fishes) showing remnants of the central nervous system have all the complex structures— brain divisions, cranial nerves, complex sense organs, and other major components—that are found in a modern shark or trout.

In every contemporary vertebrate embryo, the developing brain consists of three large swellings: the forebrain, the midbrain, and the hindbrain; the latter merges into the spinal cord (Figure 40-1). In an adult, these segments give rise to the major brain divisions and

Figure 40-1 THE MAJOR PARTS OF THE HUMAN BRAIN AS SEEN IN AN EMBRYO.
These major parts include the forebrain, midbrain, and hindbrain. Each gives rise to specific subdivisions and regions, as listed in Table 40-1.

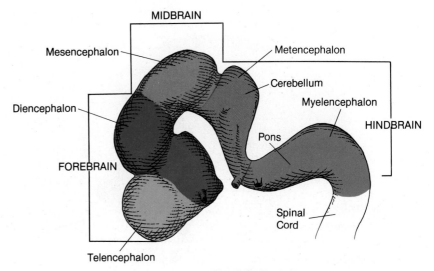

45-Day Stage

parts, which are listed in Table 40-1. The **forebrain** includes an anterior region, the telencephalon, and a posterior region, the diencephalon. The telencephalon includes the olfactory bulbs and the *cerebrum* (Figure 40-2). The diencephalon forms the dorsally situated thalamus, the ventrally placed hypothalamus, and part of the pituitary gland. The **midbrain** forms the mesencephalon and gives rise to the optic tectum in fish and amphibians and to the superior and inferior colliculi in mammals. The large **hindbrain** has two major parts: the metencephalon, which includes the *cerebellum* and the *pons*, and the myelencephalon, which forms the *medulla oblongata*—the site of connection between brain and spinal cord. Each of these brain parts has its own major components, and each component has innumerable clusters of neurons, tracts of axons, and circuits. We will discuss some of the brain parts momentarily.

Table 40-1   **EMBRYONIC DEVELOPMENT OF THE PARTS OF THE BRAIN**

Embryonic Divisions	Subdivisions	Major Adult Regions
Forebrain	Telencephalon	Olfactory bulbs, cerebrum
	Diencephalon	Thalamus, hypothalamus, posterior pituitary gland
Midbrain	Mesencephalon	Optic tectum in fish and amphibians; superior and inferior colliculi in mammals
Hindbrain	Metencephalon	Cerebellum, pons
	Myelencephalon	Medulla oblongata

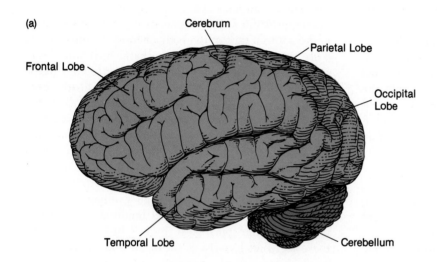

(a)

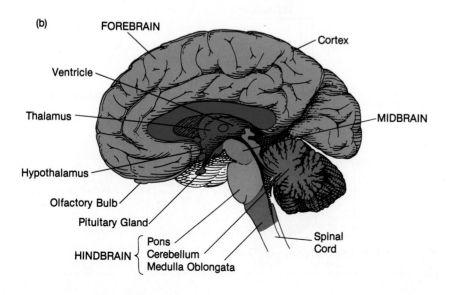

(b)

Figure 40-2   THE HUMAN BRAIN IN SURFACE VIEW AND SECTIONED ON ITS MIDLINE.

The major areas of the cerebrum of the forebrain are called lobes (frontal, temporal, parietal, and occipital). (b) The hindbrain's major regions are shown in green, purple, and pink; the midbrain is orange; and the forebrain is blue.

The entire brain and spinal cord are hollow. Four interior chambers, the *ventricles*, make up the hollow cavities of the brain. The brain and spinal cord are wrapped in protective sheets of connective tissue known as *meninges*, which also line the inside of the skull and the vertebral column. **Cerebrospinal fluid** is one of the body's four distinctive types of fluids. (Recall that the other fluid types are the blood plasma, extracellular fluid, and intracellular fluid.) Cerebrospinal fluid fills the space between the meninges, as well as the ventricular chambers and the intercellular spaces in the walls of the brain and spinal cord. Although there are only a few cupfuls of cerebrospinal fluid, it is vital to normal brain functions; for example, it transports nutrients, oxygen, carbon dioxide, and wastes, and the ions that allow nerve cells to generate and conduct impulses diffuse through it. Components of cerebrospinal fluid are continually reabsorbed into the blood, while new components are transported into it by cells lining the blood capillaries of one of the meninges. The blood capillaries in the brain and spinal cord are unlike those in the remainder of the body. In the mammalian brain, the cells that make up the capillary walls are welded together with continuous tight junctions (Chapter 5), and also lack channels through their cytoplasm. Thus they function as the **blood–brain barrier,** and ensure that while the levels of hormones, amino acids (some of which could act as neurotransmitters), and potassium ions in the blood fluctuate continuously, such changes do not occur on the brain side of the barrier. On the other hand, lipid-soluble substances such as alcohol or nicotine can pass directly through the blood–brain barrier; this explains effects of such substances within seconds of their entry into the blood. Except for lipid-soluble materials, the cerebrospinal fluid can be rigorously regulated to provide the brain's neurons a constant, nonfluctuating extracellular environment. The atrionatriuretic factor (Chapter 37) binds to cells at the blood–brain barrier and may regulate the volume of cerebrospinal fluid.

To understand the brain, we will consider the parts listed in Table 40-2, starting posteriorly where the brain joins the spinal cord. Then we will work forward, both evolutionarily and anatomically, to the cerebrum, with its furrowed cerebral cortex and its tremendous capacity for integrating and storing knowledge.

The Hindbrain: Respiration, Circulation, and Balance

The spinal cord serves as the central nervous system's main "trunk line" for the transmission of sensory and motor information between the body and the brain, as we saw in Chapter 36. The cord and brain merge at the **medulla oblongata,** a thickened stalk at the base of the brain (Figure 40-2). Just anterior to the medulla is the **pons,** and a dorsal swelling, the **cerebellum.** Together with the midbrain, the medulla and pons compose the

Table 40-2 **PARTS OF THE VERTEBRATE BRAIN**

Division	Components	Functions
Medulla oblongata	Lowest portion of the brain stem; continuous with spinal cord. Consists of nerve tracts between spinal cord and brain; its gray matter contains nuclei; its cavity is the site of the fourth ventricle.	Centers that regulate respiration, heartbeat, and blood pressure; reflex centers for swallowing, sneezing, coughing, and vomiting; relays messages between brain and spinal cord.
Pons	Nerve tracts between medulla and other parts of the brain; contains a respiratory center.	Connects and integrates various parts of the brain; helps regulate respiration.
Cerebellum	Two lateral hemispheres above the fourth ventricle.	Controls posture and muscle tone; helps maintain equilibrium and coordinates movements.
Midbrain	Consists of superior and inferior colliculi; red nucleus; cavity is the cerebral aqueduct.	Superior colliculi mediate visual reflexes; inferior colliculi mediate auditory reflexes; muscle tone and posture information integrated by red nucleus.
Diencephalon Thalamus	Contains many important nuclei.	Relay center between spinal cord and cerebrum; messages are interpreted within thalamic nuclei before being relayed to the cerebrum.
Hypothalamus	Nuclei that comprise the ventral floor of third ventricle; hypothalamus connects to pituitary gland.	Centers that control body temperature, appetite, and fluid balance; secretes releasing hormones; produces oxytocin, ADH; helps control emotional and sexual responses.
Cerebrum	Right and left hemispheres, each with a lateral ventricle; frontal, parietal, occipital, temporal lobes.	Site of intelligence, memory, language, and human consciousness. Controls some motor activities.

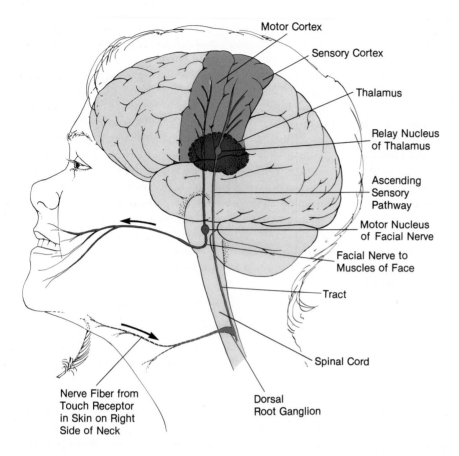

Motor Cortex

Sensory Cortex

Thalamus

Relay Nucleus
of Thalamus

Ascending
Sensory
Pathway

Motor Nucleus
of Facial Nerve

Facial Nerve to
Muscles of Face

Tract

Spinal Cord

Nerve Fiber from
Touch Receptor
in Skin on Right
Side of Neck

Dorsal
Root Ganglion

Figure 40-3   TYPICAL TRACTS AND NUCLEI IN THE CENTRAL NERVOUS SYSTEM.
Tickling the chin with a feather leads to a smile: sensory impulses from the body pass up the spinal cord and are relayed through tracts of axons and a nucleus in the thalamus to a specific site on the cortex called the sensory cortex. Then signals originating in the area of the cortex called the motor cortex pass via tracts and a motor nucleus to facial muscles, producing the grin.

brain stem, forming the base on which the cerebellum and forebrain rest. The continuity of spinal cord and brain stem is apparent in the highly simplified examples of a sensory and a motor "wiring diagram" shown in Figure 40-3. As this figure shows, bundles of nerve axons running within the central nervous system are called *tracts;* clusters of neuronal cell bodies are called *nuclei.*

The medulla oblongata and pons control respiration, circulation, swallowing, and vomiting. They receive sensory information from the spinal cord and from cranial nerves (Chapter 36), as well as instructions from brain centers such as the hypothalamus of the forebrain (the seat of control of temperature regulation, osmoregulation, and so on). Some of the sensory information coming to the medulla is routed via the pons to the nearby cerebellum. The cerebellum regulates a vertebrate's balance and stance and some locomotor movements. Not surprisingly, highly mobile animals—such as birds, bats, porpoises, and monkeys—have greatly enlarged cerebella. The surfaces of their cerebella are pleated into deep folds and fissures to accommodate a much greater surface area and thus space for more neurons. Sensory nerves from the inner ear's gravity and acceleration/deceleration detectors (Chapter 38) are major sources of input to the cerebellum. Similarly, proprioceptors (Chapter 39) in muscle spindles and joints provide a

constant flow of additional information about the positions of the head, trunk, and limbs.

The incredibly complicated tasks of the cerebellum are carried out in good part by an impressive set of cells, the giant, many-branched **Purkinje cells** (named after their discoverer, the Czech physiologist Johannes Purkinje). Arranged in complex networks, each Purkinje cell receives signals across an estimated 200,000 synapses on its highly branched dendrites (Figure 40-4). The output of Purkinje-cell networks is sent via various tracts and nuclei down the spinal cord and causes motor neurons to increase or decrease the varying tensions required in neck, limb, trunk, and leg muscles as a terrestrial vertebrate stands, sits, or moves in complex ways or as an aquatic or airborne vertebrate swims or flies.

The Midbrain: Optic Tecta, Certain Instincts, and Some Reflexes

The midbrain lies just anterior to the pons and cerebellum (Figure 40-2). It may have played an important role in regulating behavior in ancient aquatic vertebrates. Its dorsal region, the *optic tectum* ("tectum" means "roof" or "covering"), is the primary site for processing sensory impulses from the eyes in fish and am-

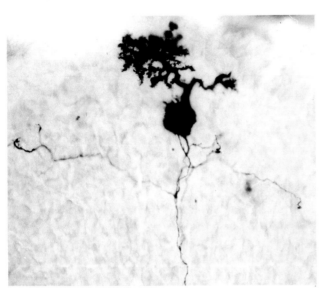

Figure 40-4   PURKINJE CELLS.
Each type of neuron has a unique morphology. This
Purkinje cell (neuron) has thick branching dendrites and a
long, fine axon (bottom) and axon collaterals (magnified
about 420 times).

phibians. The optic tectum may also control some aspects of instinctive behavior in fish and amphibians. As
the brain evolved in reptiles and their descendants,
however, the optic tectum serves to relay information
from the auditory and visual systems to the forebrain. It
also has relinquished most of its control of behavior to
the increasingly complex forebrain. But even in mammals, the midbrain's *superior* and *inferior colliculi* control certain reflexes of the visual and auditory systems.

The Forebrain: Increasing the Complexity of Brain Functions

The forebrain of an amphibian or a fish is a simple,
smooth-surfaced region, traditionally thought to be a
processing center for sensory impulses from the *olfactory bulbs*. However, even in some fish, the forebrain
may control behavior; for instance, the ability of sharks
to perform learned behavior in response to visual cues
depends on the forebrain. The olfactory bulbs form the
anterior ventral portion of most vertebrate brains, reflecting the importance of odor as a sensory modality. (A
few vertebrates, such as whales and porpoises, completely lack olfactory organs and olfactory bulbs and so
they have no sense of smell.)

As reptiles and birds evolved, the size and complexity
of the anterior dorsal part of the forebrain, or **cerebrum**,
expanded, and the activities it regulates came to include
much more complex instinctive behaviors than are found
in fish or amphibians. For example, forebrain neuronal

circuitry enables warblers to migrate using constellations
of stars as a compass and weaverbirds to construct elaborate nests of woven grass. Among mammals, the
forebrain's cerebrum has grown to massive proportions,
with a corresponding increase in learning capabilities.
We will discuss the mammalian and human cerebrum in
the next section.

It is the job of the posterior part of the forebrain, the
diencephalon, to coordinate much of the organism's
physiology. The ventrally placed *hypothalamus* regulates the autonomic nervous system and all the physiological functions it is responsible for (Chapter 36). Thus
panting and sweating in response to overheating, producing and eliminating urine in response to overhydration, and innumerable other physiological processes are
under hypothalamic control. In addition, cell bodies of
neurosecretory cells reside in the hypothalamus, and
some of these cells (located in an area called the *supra-
optic nucleus*) extend their axons to the nearby posterior
pituitary, where the hormones oxytocin and antidiuretic
hormone (ADH) are stored and released (Chapter 37).
Other axons terminate at the median eminence, where
their endings secrete releasing hormones for the many
anterior pituitary hormones. Thus the so-called master
gland, the pituitary, is in a real sense the slave of the
hypothalamus. Another important region of the hypothalamus contains nuclei (neuron clusters) responsible
for gonadotropin release and sexual behavior in mammals. One such nucleus is 2.5 times larger and has 2.2
times the number of neurons in an adult man's brain
than in a woman's. For unknown reasons, still another
portion of the hypothalamus, the suprachiasmatic nucleus (SCN), also has different shapes in males and females. The SCN is the seat of a mammal's "biological
clock" (Chapter 37), and from it, signals originate that
coordinate a wide variety of body activities on roughly a
twenty-four-hour cycle. Among these regular activities
are release of hormones, mitotic activity of body cells,
hunger, and wakefulness.

Dorsal to the hypothalamus is the *thalamus*. In land
vertebrates, this area integrates most kinds of sensory
information and channels it to appropriate parts of the
nearby *cerebral cortex* (described in detail later). The
thalamus also receives and processes signals from the
cerebrum and, in turn, sends signals to the cerebellum,
where balance, posture, and movements are coordinated. Anticipatory messages from a cat's cerebrum, for
example, might indicate that the cat is about to leap on a
bird. The cerebellum must help coordinate that leap,
and the transmission by the thalamus makes that possible. Thalamic signals are also sent to motor nuclei in the
medulla oblongata, so that motor impulses are sent down
the spinal cord to body and limb muscles. Finally, thalamic signals are sent to the heart and other effectors
over cranial nerves, such as the vagus (Chapter 36).

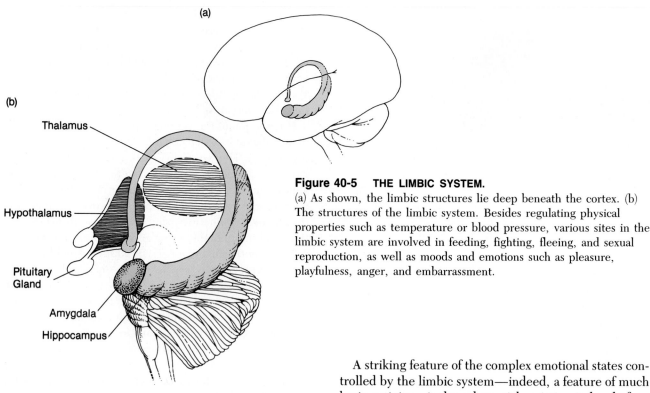

Figure 40-5   THE LIMBIC SYSTEM.
(a) As shown, the limbic structures lie deep beneath the cortex. (b) The structures of the limbic system. Besides regulating physical properties such as temperature or blood pressure, various sites in the limbic system are involved in feeding, fighting, fleeing, and sexual reproduction, as well as moods and emotions such as pleasure, playfulness, anger, and embarrassment.

Emotions are largely controlled by the **limbic system** (Figure 40-5). Part of the thalamus, much of the hypothalamus, and the *amygdala* and *hippocampus* deep within the forebrain make up the limbic system. When electrodes are used to stimulate portions of the limbic hypothalamus, a variety of physical states or emotions are evoked: hunger, thirst, pain, or even complex emotions such as sexual desire, anger, rage, and pleasure. But what is actually involved when a person experiences an emotion such as anger, fear, or embarrassment? The limbic hypothalamus sets in motion a coordinated set of related physiological reactions—blushing, sweating, a more rapid heartbeat, and so on—in response to such experiences. When the limbic system's regulation of mood and desire becomes abnormal, behaviors such as pathological overeating or undereating may be generated. The limbic system contains a reward, or pleasure, center and an avoidance, or punishment, center. If a tiny electrode is implanted in the limbic system's pleasure center, a rat will push a lever to stimulate that electrode and give itself a pleasurable jolt up to 5,000 times an hour. It seems very likely that these centers are important contributors to the "unconscious" drives that control aspects of behavior in all mammals, including humans. Two other limbic-system structures—the forebrain's amygdala and hippocampus—are involved in the highly important process called *short-term memory*, or the temporary storage of memories of recent events that are not yet stored permanently elsewhere in the forebrain.

A striking feature of the complex emotional states controlled by the limbic system—indeed, a feature of much brain activity—is the substantial variation in level of activity. Variations are especially evident when we compare wakefulness and sleep. The reticular activating system enters the picture here.

The Brain's Sentinel: The Reticular Activating System

Embedded in tissues of the medulla and cerebrum are portions of an elaborate but diffuse network of neurons, the **reticular formation** (Figure 40-6). This network receives information from each of the body's sensory systems: its axons ascend from the medulla through the midbrain to the thalamus and descend to the spinal cord, where they inhibit or amplify incoming sensory information or modify motor instructions traveling down the cord.

The **reticular activating system (RAS)** is composed of the reticular formation plus its tracts that lead to the thalamus. In humans, it is sometimes called the "gateway to consciousness." Why? Studies have shown that an electrical stimulus applied to the RAS will cause a sleeping cat to become aroused and alert, while drugs that inactivate the RAS produce sleep or, in humans, loss of consciousness. Imagine for a moment being sound asleep. Would you guess that even when deeply asleep, *all* sensory input (sounds in the night, smells, and so on) reaches the highest levels of your cerebrum responsible for your most complex behavior when awake? That is, in fact, the case, but the barrage of information about

Figure 40-6   THE RETICULAR FORMATION.

The reticular formation is a diffuse network of neurons located in the brain stem. This system monitors information from the sensory systems of the body and, via the reticular activating system (RAS), tracts that lead to the thalamus may activate the higher cortical centers of the brain. The RAS serves to filter sensory input and either dampens or amplifies the signals.

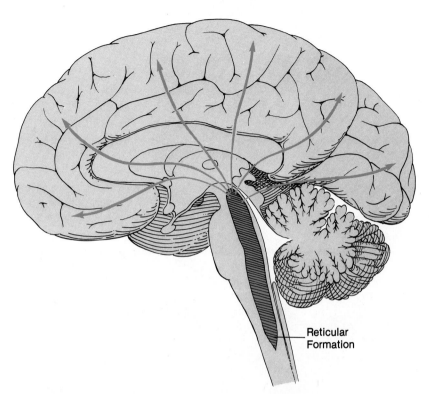

Reticular Formation

odors, sounds, touch, and other stimuli is ignored. That is because the RAS is like a filter that somehow tunes out the higher brain centers. The RAS is also responsible for arousing the higher centers so that the animal awakens and responds or becomes aware of sensory input. During waking hours, the RAS probably acts like a filter in helping the higher centers pay attention to new sensory inputs important for survival while ignoring the flood of other information that is merely "noise" for the system.

The Neocortex of the Cerebrum: The "New Bark"

One hallmark of mammalian evolution was the attainment of a huge brain compared to that of other animals. The tens of millions or billions of neurons added during evolution as the mammalian brain enlarged make up the area of the cerebrum called the **neocortex,** or *neopallium*—literally, "new bark." (Most biologists use the terms "neocortex" and simply "cortex" interchangeably.) The expansion of the mammalian cortex involved overcoming a fundamental problem: the accommodation of billions of new neurons arranged in the typical configuration of the cortex. That configuration consists of a sheet of nerve tissue with an average thickness of 2.5 millimeters in which six precisely ordered *layers* of neurons are present. The evolutionary solution to this problem was folding; the enlarging sheet of cortex is creased and furrowed into hills (*convolutions,* or *gyri*) and valleys (deep *fissures* or shallow *sulci*) of gray matter (Figure 40-2). If all these living hills and valleys in the cortex were pulled

taut into the smooth-surfaced configuration of a shrew's cortex, the human skull would have to be the size of an elephant's to accommodate our complex brain.

The cortex is much like a multilayered sheet of epithelium with an extraordinary and strikingly regular organization. The six layers of neurons, recognizable by neuron shape (Figure 40-7), are composed of **columns** about 20 microns in diameter, each containing up to 100 neurons. The human cortex is estimated to consist of perhaps 400 million columns, with a total of more than 10 billion neurons. To add to this complexity, at least sixty distinct classes of neurons have been described within the various layers and columns. The only place where biologists have even a hint of column function is in the visual cortex. In this brain region, groups of columns lying next to each other all respond to a similar type of stimulus; these similar sets form *slabs.* We will discuss the visual cortex and the function of slabs later.

The cortex has a specialized geography in which three large regions receive sensory input, one from vision, one from hearing, and the third from bodily sensations. A fourth region acts in motor control. Within each cortical region, local groups of neurons are associated with specific parts of the body. For example, each portion of the body's sensory surface—lips, fingers, and so on—is represented on the *sensory cortex* by a set of neurons responsible for processing information from that body area. The tracts of axons carrying sensory information cross the central nervous system so that the left side of the body is reported to the right sensory cortex, and the right side of the body, to the left sensory cortex. Thus if someone steps on your right toe, a small region near the

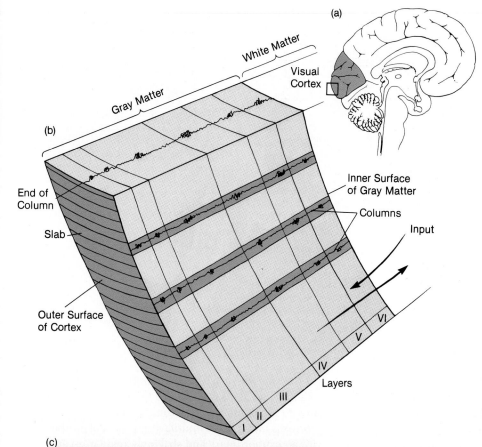

(a)

White Matter

Visual Cortex

Gray Matter

(b)

Inner Surface of Gray Matter

Columns

End of Column

Slab

Input

Outer Surface of Cortex

VI

V

IV

III

II

I

Layers

(c)

Figure 40-7   ORGANIZATION OF THE CORTEX.
The gray matter in the cerebral cortex consists of six layers, each containing neurons with characteristic shapes. Running perpendicularly to the layers are columns of interconnected neurons. (a) The location of a small segment of the visual cortex. (b) This expanded view of a segment of visual cortex has layer I nearest to the brain's external surface, and layer VI adjacent to the white matter, with its axons leading to and from the segment of cortex. Running through the layers are individual columns. Note that in the visual cortex adjacent sets of columns lie within slabs and that all the cells within a set of columns respond to stimuli oriented in the same plane. (c) Neurons have been stained with a dye to make them visible (magnified about 500 times). The neuronal cell bodies seen at the top would be found in a layer, whereas the long axons extend, as part of columns, perpendicular to the surface of the cortex.

center of the left sensory cortex becomes active; if next your right ankle is kicked, a nearby left cortical region activates. If your shin is tapped next, neurons in the next area respond. In effect, there is a maplike representation of the whole body on the sensory cortex, referred to as a *homunculus*, or "little person" (Figure 40-8a). In similar fashion, a map of the visual world—what is seen through the eyes—can be traced out, albeit in a distorted form, on the visual cortex. Just the same sort of auditory map, or homunculus, has been discovered on the auditory cortex of barn owls and mammals. Hence, a topographic organization for hearing has evolved.

Not surprisingly, a motor homunculus can be drawn on the *motor cortex* by noting which muscles twitch or contract when various regions of the motor cortex are stimulated (Figure 40-8b). In general, motor sites on each side of the cortex have the same spatial arrangement as muscles on the opposite side of the body. Distortions in the relative size of body parts in both sensory and motor homunculi reflect the differential density of nerves in various body regions; for instance, the highly innervated, sensitive lips loom grossly large in the homunculus in Figure 40-8a, while the tough skin of the elbow claims only a tiny bit of cortical space.

The four areas of cortex "assigned" to vision, hearing, body senses, and motor functions make up a relatively small proportion of the total cortex in a primate. The vast

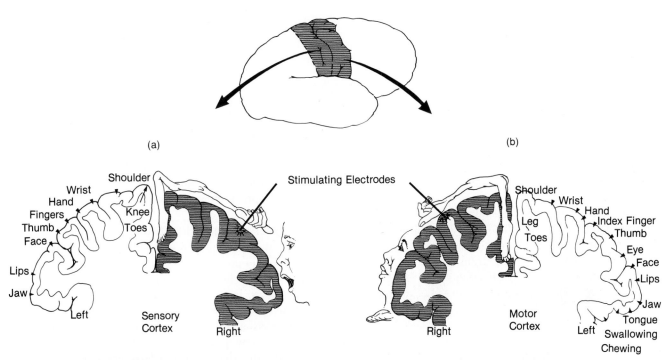

Figure 40-8   THE SENSORY CORTEX AND MOTOR CORTEX.
Cross sections made through a brain in the planes shown here reveal the motor and sensory cortices. (a) If recording electrodes are inserted in the sensory cortex at various sites, electrical activity is detected only when the experimenter touches the area of the body whose sensory input projects to that spot. With the recording electrode at the site indicated in the right sensory cortex, only touching a finger on the left hand would produce an electrical recording. Using this procedure biologists have defined the sensory homunculus. (b) Conversely, stimulating the various parts of the motor cortex with an electrode causes constriction of muscles driven by that region of the cortex. For instance, using the stimulating electrode at the site shown here would cause the left hand to curl up. This process maps the centers of control of body movements and defines the motor homunculus.

remaining cortical area is said to be "unassigned." What is the role of the billions of neurons and columns in this large area? Surprisingly, some dozen representations of the visual cortex map and a half dozen each of the somatic and auditory maps are present. An intriguing result is seen if, for example, the neurons corresponding to the left thumb on the sensory cortex (Figure 40-8a) are destroyed; the corresponding "left thumb" region of the duplicate sensory map disappears even though neurons in that region are not touched. This and other results show that the multiple representation maps are dynamic and change with use. The duplicate maps and the remaining cortex is called the *associational cortex*, this unassigned tissue is the site where the brain's most elusive characteristics and abilities arise—memory, learning, language, and personality. As seen in Figure 40-9, portions of the associational cortex are referred to by anatomical names—the temporal, frontal, parietal, and occipital lobes. In turn, experimental procedures define functional areas of the lobes where memory, speech, and such higher-order phenomena are generated. The saga of brain exploration and discovery rivals the charting of the Earth by early geographers or the heavens by astron-

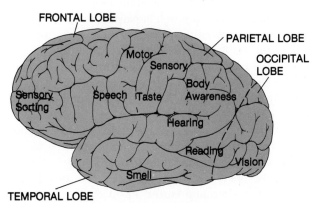

Figure 40-9   THE CORTICAL LOBES.
The portion of the cerebral cortex that is designated the associational cortex is divided into four lobes (labeled in capital letters). Each lobe possesses specific areas of function, some of which are indicated. Together, these areas of function constitute the so-called higher centers of the brain. Mapping of the areas of higher function has been achieved through various means—for example, electrical stimulation.

omers. As with all experimental science, specific techniques have been the keys to unlocking new knowledge.
The next section traces the use of such techniques to
reveal the brain's deepest secrets.

EXPLORING THE LIVING BRAIN

How do we know about the various areas of the cortex
and what they do? Four major research methods have
proved most revealing: (1) creating lesions by destroying
brain tissue, (2) stimulating neurons with electrodes, (3)
recording the brain's electrical activity, and (4) studying
individual neurons and neuronal types. Let us consider
each in turn.

Split-Brain Studies and the Cerebral Hemispheres

Roughly 3,000 years after the Egyptian physician recorded his observations of a soldier's exposed brain, the
Greek physician Galen was the first to witness and document the effects of brain damage, or lesions, while tending the horrible wounds suffered by gladiators in the
Roman circus around 180 A.D. Most severe lesions, then
and now, result in clear behavioral deficiencies. For instance, damage to the left side of the brain in the temporal lobe might lead to speech impairment, while a lesion
in the dorsal cortex may cause loss of motor functions,
such as the ability to move the legs. Not until this century, in the 1920s and 1930s, did neurosurgeons begin to
make systematic discoveries while working with patients
suffering from severe epilepsy—a brain dysfunction in
which normal electrical patterns are temporarily obliter-
ated by the simultaneous discharge of millions of neurons. Without knowing what the exact outcome would
be, the surgeons attempted to cure their subjects with a
pioneering operation that separated the two hemispheres of the cerebral cortex; in effect, they split the
upper brain in two.

A gap, the *median fissure*, normally separates the
brain's two hemispheres; hidden deep within the fissure
is a giant bundle of 200 million axons, the **corpus callosum,** which extends from one hemisphere to the other,
linking the two sides of the brain. The split-brain operations sliced through this giant axonal bundle, with startling results. The epileptic seizures themselves became
much rarer and less severe, but despite the drastic operation, the patients retained almost all their former

mental faculties; only some aspects of memory were
impaired.

It seemed odd that the corpus callosum would play
such a minor role in brain functions and that severing it
would produce so few noticeable effects. The results
remained puzzling until the early 1960s, when Roger
Sperry, who later won a Nobel Prize, and his colleagues
at the California Institute of Technology performed a key
series of experiments on split-brain cats and monkeys.
These researchers found that under controlled laboratory conditions, their animal subjects, as well as split-
brain epileptic humans, acted as if they had two entirely
separate brains.

To understand how this "two-brain" situation was revealed, it will help to recall the special nature of the
visual system possessed by vertebrates with good binocular vision, including humans (Chapter 38). As Figure
40-10 shows, images in the left visual field of both eyes
are conveyed to the right side of the brain, and images in
the right visual field, to the left hemisphere. The experimenters knew from previous studies on the brain's
speech center that an adult's language functions are divided; for most humans, the left brain is specialized for
speech. (This feature is not unique to humans; most
birds sing under control of their left brain and female
mice recognize ultrasonic calls from newborn "pups"
using their left hemisphere.)

Sperry showed split-brain patients two differing
groups of objects simultaneously, including a ring in the
right visual field, "reported" to the left brain, and a key
in the left visual field, "reported" to the right brain.
Next, the written words "ring" and "key" were presented independently to each visual field. The subjects
were then asked to name the objects they had seen. No
matter how hard they tried, they only could say "ring"—
the word for the object "seen" by the left hemisphere.
Their right brain, which saw the key, had no speech
center and had become separated during the split-brain
operation from the left, which allows a normal person to
say "key." Despite their inability to say the word "key,"
when asked to pick the key up with the left hand, subjects grasped the object correctly. Their right brain,
which controls the left side of the body—including its
muscles, arm, and hand—had seen and recognized the
key and could direct a grasp of it, but it could never
cause the person to say that object's name. Sperry eventually wrote of his experiments that each half of a separated brain seemed to have separate and private sensations, with its own perceptions and impulses to carry out
actions.

Much work since Sperry's has revealed a great deal
about how responsibility for the brain's many complex
operations is parceled out. In most people, specific sites
in the left brain are in charge of language comprehension
and speech production; the left brain seems to process

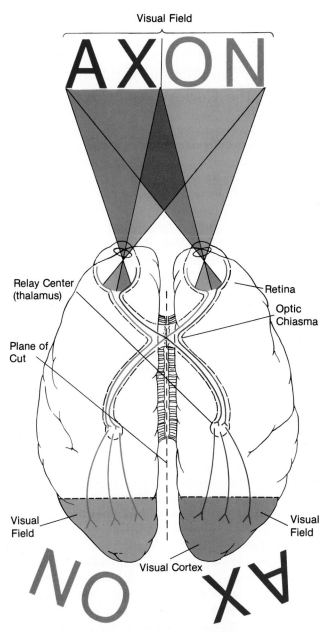

Figure 40-10   SPLIT BRAINS AND VISION.
If a normal person saw the printed word "axon" he or she could say "axon" out loud. However, if a person with a split brain (whose corpus callosum has been severed) sees the same word, the following occurs: since the "ax" is in the left visual field and the "on" is in the right, the patient says "on"; this is all that the left hemisphere, with its speech center, can "see" and report verbally. Though the person cannot utter the word "ax" that is in the left visual field, he or she can point to a picture of the word "ax" with the left hand. The reason for this is that the right brain (which has "seen" and comprehended "ax," but has no way of activating the speech center in the left brain to express it) controls the left hand.

information analytically and in sequences (as in formal logic and mathematics). Conversely, the right brain apparently specializes in spatial relations. For instance, the ability to identify an object by feeling its shape, contours, and texture is a right-brain function. The right brain processes information as a whole—for instance, sets of objects in patterns.

The right brain plays a key role in music recognition. This ability was well demonstrated after the French composer Maurice Ravel had suffered severe damage to his left hemisphere in an automobile accident. After the accident, Ravel could listen to music, criticize it, and express pleasure at hearing it; thus, his music recognition was unimpaired. Yet he could never again compose "the pieces he heard in his head"; the analytical capacity of the left hemisphere is needed for that. Indeed, others with left-hemisphere damage but intact right hemispheres can sing melodies perfectly well. The right hemisphere is crucial in yet another activity: our ability to recognize faces. The physical differences between two faces may be so slight that they are almost impossible to describe verbally, yet most humans can unfailingly pick out the visage of a friend or celebrity, even from a crowd of look-alikes. If, however, the familiar visage is flashed to the right visual field only and so to the left brain, it remains completely unrecognized.

About 95 percent of right-handed people have speech centers located in the left hemisphere (Figure 40-11). About 70 percent of left-handed people do too, but 15 percent have speech centers in the right hemisphere, and 15 percent have bilateral speech centers. Just what are these centers? One is Broca's area, a portion of the

Figure 40-11   SPEECH CENTERS OF THE LEFT CORTEX.
Wernicke's area is the site where the form and meaning of speech are generated and where language comprehension occurs. Broca's area actually controls the adjacent vocalization motor area of the cortex; the latter causes the appropriate facial, tongue, and throat muscles to produce the sounds of speech.

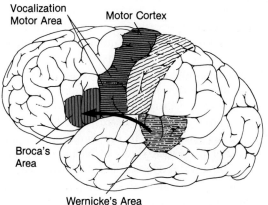

left frontal lobe next to the part of the motor cortex that controls the muscles of the larynx (voice box), tongue, jaw, and lips—all involved in speech production. The nineteenth-century French surgeon and student of the human brain, Paul Broca, described a patient with a lesion in that part of the brain who could not speak, but who could write and read perfectly well; words were there, but not the ability to utter them!

Some people have a lesion in another area of the cortex that allows them to speak, but in a jumbled way, saying things like "I think that there's an awful lot of mung, but I think I've a lot of net and tunged in a little wheat duhvayden."This meaningless sequence is due to a damaged Wernicke's area, a part of the upper, posterior left temporal lobe. Form and meaning of speech seem to originate in Wernicke's area and go from there to Broca's area, where actual speech is generated.

Wernicke's area plays another major role as well: it is not only the site of language generation, but also of language comprehension. Spoken words are reported to the auditory cortex, but then the message goes to Wernicke's area before it can be understood. Damage to the link between these areas leaves the person capable of hearing but quite unable to comprehend spoken words. Amazingly, the words you are reading right now are sent from the visual cortex to a brain site called the angular gyrus, where they are translated into their equivalent as if they had been heard through the ears. That information, in turn, goes to Wernicke's area, and only then can the reading be comprehended. Neuroscientists believe that this indirect processing of the visual (printed) form of words simply reflects the fact that sound communication probably antedated writing by millions of years in human ancestors.

Lest the picture of two speech centers, Broca's and Wernicke's, seem too simple, researchers have amassed abundant evidence showing that damage to or stimulation of a variety of sites on the frontal, temporal,and parietal lobes can all interfere with both speech and comprehension of language. The presence of the speech centers on the left side of most human brains helped establish the term *dominant hemisphere* for the left brain.

Despite the years of experiments emphasizing the left brain's special roles in speech and analytical thinking, it was not until 1968 that biologists noticed and reported what is obvious now to even an untrained eye: the left temporal lobe, especially around Wernicke's area, is substantially larger than the right temporal lobe. This anatomical difference may correspond to millions of extra neurons and arises in the embryo, well before speech develops. Fossil skulls of Neandertals, human ancestors who lived in Europe and the Near East 30,000 to 130,000 years ago, show evidence of a larger left temporal lobe. Perhaps this means those primitive people used primitive speechlike sounds to communicate with each other.

Electrical Stimulation of the Brain: Probing the Cortex

Electrically stimulating specific sites in the brain is a second major method for exploring that complex organ. If you were the subject of such exploration, a neurosurgeon might carefully position an electrode the diameter of a fine hair within your cortex and send a tiny pulse of electric current to the region of neurons around the electrode. Since the brain has no pain receptors, you would not feel any discomfort; instead, the sights, sounds, and smells of a long-lost happy event of childhood might suddenly flash into your consciousness and continue for some time. If the neurosurgeon moved the electrode slightly, some other memory might arise, with all its unique sensations of odor, sound, color, mood, and so forth. Or if the electrode were inserted in still other forebrain sites, you might suddenly feel deep relaxation or uncontrollable rage or overwhelming pleasure. By stimulating various regions of the cortex in studies like this, neurobiologists have been able to identify areas that appear to control some of our most complex behaviors and moods and to store memories as well.

Research on the living human brain is limited to cases where injury or disease has affected a person's functions or has necessitated opening the skull. Therefore, many of the most profound insights in modern neurobiology come from experiments on animals other than humans. For example, when a researcher stimulates certain areas of a cat's limbic system, the animal immediately begins to stalk an imaginary prey. When an object is provided to serve as "prey," the cat's hunting behavior culminates in a typical feline pounce and an energetic bite. Moving the site of electrical stimulation just a few millimeters evokes quite different behavior—like a Halloween cat on a picket fence, the animal reacts to the same prey object with an arched back and a violent hiss, the limbic system's reaction to a serious threat.

In these cases, complex motor and behavioral "programs" unfold in response to the initiating stimulus. A program, here, means an integrated package of outputs that may control sequential use of dozens of muscles, and various organs and relatively long behavioral sequences. Such results imply that the neural bases for all sorts of behaviors are present in the brain, in a sense prepackaged and ready to be called forth by a stimulus in normal daily life.

We referred earlier to the discovery of a pleasure center in the limbic system. Two graduate students working at McGill University in Montreal implanted electrodes in a rat's brain tissue and delivered a small electric shock through the implanted electrode whenever the animal entered a certain corner of the cage. Instead of learning to avoid the corner, however, the rat began to visit it again and again, apparently experiencing the electrical stimulus as pleasurable, not disagreeable. Such stimulation studies have helped to map the limbic system in fairly close detail.

Electrical Recording to Explore the Brain: Sleep and Brain Waves

While lesion experiments and electrical stimulation are useful ways to examine brain function, another successful method involves simply recording the brain's spontaneous electrical activity. In a normal alert brain, the activities of billions of brain cells give rise to a "hum" of electrical impulses that can be detected through the skull and scalp. Such recordings are called *electroencephalograms*, or EEGs (Figure 40-12). A person lying quietly with eyes closed produces *alpha waves*, regular rhythmic electric potentials of about 45 microvolts that occur about ten times a second. They originate in the visual area of the occipital lobes. Mere opening of the eyes causes alpha waves to cease and *beta waves* to commence. Beta waves are irregular rapid waves, dependent in part on visual input, and they are especially active when a mental activity, such as reading a textbook, is carried out. Still other wave patterns appear in abnormal states, such as during an epileptic seizure, or if a tumor or stroke (caused by a ruptured blood vessel in the brain) brings about local brain damage. Other types of brain waves are seen during sleep; EEGs have proved invaluable to the study of sleep and the brain's mysterious secret life during this unconscious state.

Sleep is the most profound change in behavior that most vertebrates regularly experience. If human subjects are confined to a windowless room and kept under constant environmental conditions, they inevitably lapse into the altered brain state we call sleep on some regular schedule. Perhaps every sixteen to eighteen hours, they fall asleep and remain oblivious to the world for about eight or nine hours. Sleep is a component of the physiology of many vertebrates and fluctuates on a circadian rhythm under control of the suprachiasmatic nucleus in mammals.

But what, exactly, is sleep? Researchers have measured certain physiological changes that accompany sleep behavior in various mammals; humans, for example, experience five distinct sleep stages during a typical night's sleep (Figure 40-13). Charted by an EEG, stages 1 to 4 are characterized by increasing amplitude and lower frequency of brain waves. The transition from wakefulness to sleep is sudden, occurring in an instant; once the moment passes, much of the body relaxes: muscle tone, heart rate, blood pressure, and respiration slow down from waking-state levels, while movements in the gastrointestinal tract speed up. This period of *slow-wave sleep* culminates in stage 4, with *delta waves*, about one hour after first falling asleep. During this stage, which lasts about twenty minutes, an individual is usually in such a deep state of sleep that waking him or her may be difficult.

As stage 4 ends, the EEG registers fast, irregular beta waves, a trend that continues until the sleeper appears ready to waken. This is when the fifth and somewhat paradoxical state known as **rapid eye movement (REM) sleep** begins. ("Paradox" here refers to the high rate of brain activity but virtual paralysis of the body's muscles.) During REM sleep, which occurs about every ninety

Figure 40-12 ELECTRICAL RECORDINGS OF THE BRAIN.
Electroencephalograms (EEGs) measure brain waves. (a) A normal adult at rest with electrodes placed at the red dots shows alpha and beta waves. (b) The child never shows alpha waves. (c) An epileptic suffering a small, so-called petit mal seizure has the waves shown here.

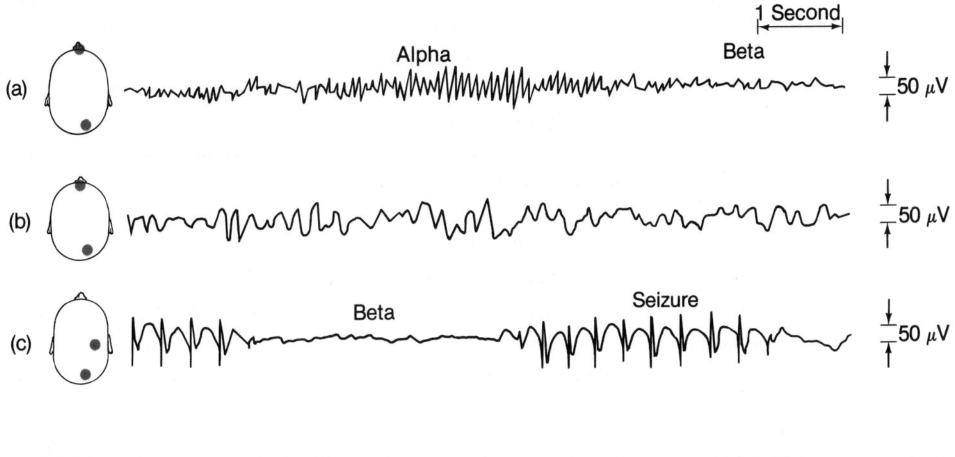

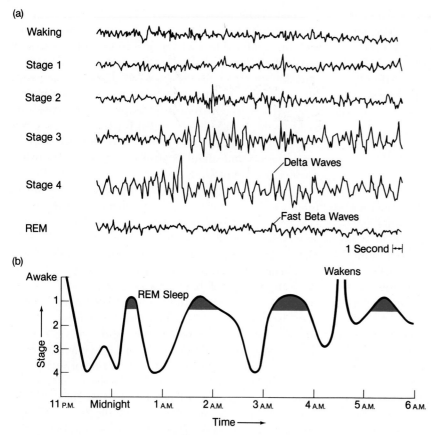

(a)

Waking

Stage 1

Stage 2

Stage 3

Stage 4

Delta Waves

REM

Fast Beta Waves

1 Second

(b)

Figure 40-13   THE STAGES OF SLEEP.
The EEG recordings reveal the remarkable differences between brain waves during the five stages of sleep (stages 1-4 and REM sleep). During stages 1 to 4, the amplitude rises and the frequency falls. The waves during REM sleep are reminiscent of wakefulness, but the eyes show very rapid movements behind closed lids. (b) During a typical night's sleep the subject cycles through the various sleep stages, sometimes reaching REM sleep, sometimes not, and awakening once in the early morning.

minutes, the eyes dart rapidly back and forth beneath closed lids, as if the sleeping person were watching a frantic inner tennis match. Individuals awakened during REM sleep almost always report that they have been dreaming, although dreaming can also occur during stage-4 sleep. Some evidence suggests that norepinephrine release in the reticular activating system (RAS) leads to nerve impulses going to the cortex; perhaps as a result, dreams occur; heart rate, respiration, and blood pressure all rise; and movements in the gastrointestinal tract cease. The likelihood that the sleeper will awaken spontaneously also increases. Oddly, however, throughout REM sleep, all of the body's skeletal muscles, except those of the eyes and ears, are virtually "paralyzed" by powerful inhibition of the spinal motor neurons. It is as if the sleeper were alert but intensely distracted—in effect, the brain appears to be carrying out its own private business, actively suppressing interaction with the external world. That business may, in fact, have to do with memory. Work on rats and other mammals suggests that REM sleep is a time of processing the previous day's information and experiences and of establishing them in the so-called long-term memory (permanent memory). If an animal is awakened and its REM sleep is prevented, tasks learned in the previous period of wakefulness will not be retained.

Just what brain regions start and stop sleep? We saw earlier that the RAS plays a role in arousal. Cells from nuclei in the brain stem sends axons through the RAS to many sites in the forebrain. One set of axons releases serotonin, which inhibits the RAS, and sleep occurs. Other axons release norepinephrine, and the animal is aroused. Presumably, the suprachiasmatic nucleus, the inner "clock," tells these cells when to do their work, and so our sleep–wake cycle remains coupled to darkness and daylight. But that is not the whole story. Both sets of these neurons show slowed activity during the early stages of slow-wave sleep and become virtually silent electrically just before REM sleep is reached (hence, they are called REM-off cells). Still other neurons whose cell bodies are located in the pons always begin firing just before the transition from stage-4 sleep to REM sleep. Subsequently, these REM-on neurons cease generating impulses before the end of REM periods. The inverse activity of the REM-on cells and REM-off cells suggests that these hindbrain neurons may be key players in regulating the still mysterious, but essential, REM sleep.

We do not know the answer to the fundamental biological questions of why animals need to sleep at all and why all mammals have REM sleep. Humans and other animals prevented from sleeping become more and more irritable and are unable to function well. If REM sleep alone is prevented, humans make up for it by carrying out prolonged periods of REM sleep when given the opportunity. It will be intriguing to learn (with the help of electrical recording methods) what kinds of biochemical and neuronal "recovery" processes lie at the heart of sleep and necessitate that daily activity.

Brain Studies at the Level of the Neuron

While the EEG is an excellent indicator of the brain's overall state, the data it provides are statistical averages much too general to reveal the intricate cell-to-cell communications by which the brain processes information and generates commands for actions. We saw in Chapter 36 how biologists used a cell-by-cell analysis to unravel the neuronal networks that enable a crayfish to flip forward or backward in response to a threatening stimulus. A mammal's brain, however, contains 1 million times more neurons than does the nervous system of a crayfish, and thus other methods must be used to explore it.

Several features of the mammalian nervous system simplify studies at the cellular level. The first is *parallel processing* of information. Many parts of the brain are built of enormous numbers of similar circuits; and, in turn, most cells within each circuit type are quite similar to each other. For example, by examining just a few random subsets of the small circuits that connect the neural retina's 125 million rods and 6 million cones to the 1 million ganglion cells, biologists can draw some general conclusions about all those circuits.

The wiring of the neural retina also illustrates a second common property of the nervous system: a *hierarchical organization* based on neuronal networks built on the convergence pattern (Chapter 36). For example, neuron A receives input from many presynaptic neurons. The function of A depends on some pattern or level of activity among its input cells. As we saw in Chapter 38, rods and cones convey their information via bipolar cells and other cell types to the ganglion cells, which, in turn, transmit nerve impulses to the brain over the optic-nerve fibers. Recall that a given ganglion cell is connected in circuits to a number of rods or cones, even to as many as 1,000 or more rods. Only a specific pattern of stimulation by light among those rods or cones will activate a ganglion cell. Ganglion cells are of many types: for instance, an *on-center* ganglion cell generates an impulse only when the center of the set of rods or cones connected to that ganglion cell is illuminated. The impulses from such ganglion cells are sent to a special area of the mammalian thalamus, and from there to layer IV of the visual cortex. The hierarchy phenomenon involving convergence is known best in the visual cortex.

As we have discussed previously, the visual cortex is organized into six layers, each running parallel to the brain's surface and each containing specifically shaped neurons different from those of the other layers. Hundreds of slabs run perpendicular to the brain's surface; and within each slab there are columns of cells also running perpendicular to the brain's surface and extending inward through the gray matter (Figure 40-7). If an experimenter inserts an extremely slender electrode into a neuron in layer IV then shines light in the animal's eye, the neuron will generate impulses revealing that the cell receives input from the eye. Layer IV neurons in one column might prove to receive signals primarily from the right eye, while layer IV neurons in a nearby column receive input from the same spot in the left eye. This alternating arrangement is represented in Figure 40-14 as stripes.

The columns within the visual cortex thus have what is called *ocular dominance:* they are driven by either the right or the left eye. An additional feature of the right- or left-dominated cells within layer IV is that each cell sees only a dot of light—each receptive field in the retina is simply a circular area of cones or rods that transmits information via the various intervening cells to a particular layer IV cell in the brain's visual cortex. Now let us consider the brain's hierarchy for processing visual information using such dots and three types of neurons, the simple, complex, and hypercomplex neurons in the visual cortex layers.

To explore this hierarchy, researchers David Hubel and Torsten Wiesel of Harvard University recorded electrical responses from single cells in the cat's visual cortex as the researchers moved spots of light in the visual field. They found that while layer IV cells see dots of light, these cells in turn stimulate cells in layers V and VI which see oriented lines, edges, and corners. Working together, the cells in the three layers convert the dot patterns that a cat's (or a person's) eyes perceive into a three-dimensional world of moving and stationary lines and shapes.

To understand how the cells interact, imagine looking at the straight edge of a picture frame. The image of that edge falls on the retina and is reported to a set of layer IV cells as a series of dots. Hubel and Wiesel found that those cells in turn activate a layer V neuron (a *simple cell*) which sees a line—the edge of the frame. A given layer V cell may only see the line if it is horizontal; another simple cell may perceive the line only if it is tilted 10 degrees; another if the line is vertical. (At this level, there are **orientation slabs** in the visual cortex, represented in Figure 40-14 by the lines at various angles on the slabs.)

Through this process, the simple cells of the visual cortex can build a line at a particular angle from a row of dots. In turn, *complex cells* in layer V, which sense sets

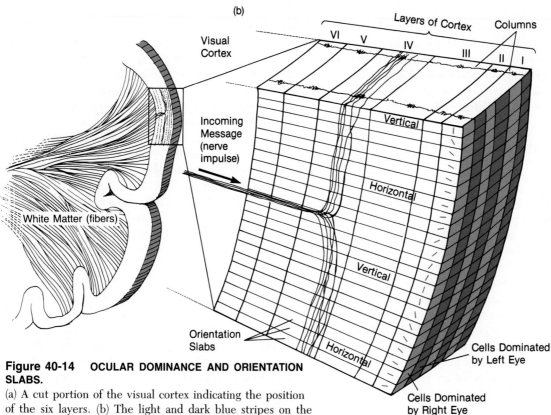

Figure 40-14   OCULAR DOMINANCE AND ORIENTATION SLABS.
(a) A cut portion of the visual cortex indicating the position of the six layers. (b) The light and dark blue stripes on the surface of the cortex to the right of the numbered layers show which of the layer IV and other neurons are driven by either the right or left eye. The expanded view also shows the orientation slabs; the small lines in layer I reveal the angles that a straight line must have in order to be perceived by complex neurons in that slab. Thus the slab marked "horizontal" could see the horizontal upper or lower edge of a picture frame (explained in text), while the slab marked "vertical" could see the frame's vertical sides.

of edges, especially moving edges, may need to receive input in the form of two lines at right angles—a corner of the picture frame, for example, or some other angle. Finally, sets of complex cells activate *hypercomplex* cells in layer VI. They see silhouettes or more complicated sets of lines, especially when moving. Thus layer IV cells perceive dots lined up along the edge of the picture frame, simple or complex cells in layer V convert these dots into a line, and hypercomplex cells in layer VI convert the lines into a corner or the rectangle of the entire frame. This hierarchical functioning means that a hypercomplex cell is active only if its set of complex cells is active; similarly, a particular complex cell is active only if its set of simple cells is active. In a sort of "inverted pyramid" scheme, visual input to a tiny portion of the eye converges from many retinal cells to fewer simple cells, to still fewer complex cells, and finally, to a single hypercomplex cell. Since there are millions of such hypercomplex cells, the brain can assemble in its electrical language of nerve impulses an abstraction of the physical

world the eye sees. The millions of tiny edges and corners—some with "depth" and some without, some moving, some still—are built into the brain's image of a mother's face, a fluttering moth, or a mountain peak.

Visual images are used elsewhere in the brain. For instance, the left temporal lobe of sheep has classes of individual neurons that recognize faces. Different individual cells respond, for instance, to threatening faces (human, sheep dog), sheep faces with horns, or familiar versus unfamiliar sheep faces. How images from the visual cortex are interpreted in these ways is unknown.

Is the precise and highly repetitious order of the visual-cortex column system also found throughout the rest of the cortex? Some research suggests that it is not. For instance, one model of how the cortex functions in learning is based on the analysis of smell. Whereas vision is topographical—involved in locating the position of things—smell and most senses are recognitional—involved in identifying things. Smell, in particular, can be thought of as an associational process: we identify a smell because of the memories it evokes. One pungent odor would immediately mean "orange rind," another fragrance "rose," another "bacon cooking." Work on the olfactory bulb and a portion of the cortex by Gary Lynch of the University of California, Irvine, suggests that unlike the visual cortex, with its hierarchical organization, the olfactory regions may contain what computer scientists call interactive combinatorial networks. That is,

PLASTICITY: DYNAMICS OF BRAIN NEURONS AND CIRCUITS

The adult brain was long regarded as a precisely wired set of permanent circuits, set up in the embryo and newborn. This view was reasonable—but it was wrong. Considerable evidence now suggests that the brain has great plasticity in both its neurons and its circuits. One major source of change is the synapses and dendrites, both places where neurons "talk" to each other.

Synapses are not permanent structures but turn over in a normal, orderly process: older synapses degenerate and are replaced by new ones in both the central and the peripheral nervous systems. A typical synapse in a monkey eye muscle persists for about eighteen days. The constant remodeling of synapses allows for significant alterations in circuits, since more synapses form at sites where neuronal activity is high, and fewer form where neuronal activity is low. Where there is a brain lesion, the normal processes of dendrite degradation occur, but new ones are not formed; hence, a rapid net loss of synapses results. Furthermore, the number of synapses falls in uninjured neurons elsewhere in the brain that are connected to neurons at the lesion site. A return to normal brain function then depends on repairs and synapse information both at the lesion and elsewhere in the circuits.

The number of dendrites and their degree of branching also reflect great plasticity. Dendrites can increase in number from 10 to 23 percent in just those areas where intensive use goes on. For instance, when a "right-handed" rat is trained to reach for food with its left forepaw, there is, within sixteen days, substantially increased dendritic branching in neurons of the motor cortex on the right side of the brain (recall the motor homunculus; Figure 40-8). The corresponding site in the left motor cortex, which controls the right paw, does not increase its dendrites. Researchers have seen similar changes in rats kept in enriched (stimulating) rather than deprived (unstimulating) environments. Animals allowed high levels of social activity and exercise show an increased number of synapses, number and branching of dendrites, number of supportive glial cells, and other parameters of vigorous neurons and neuronal circuits. These sorts of changes with use probably underlie the dynamic nature of the multiple copies of the visual,

sensory, and auditory maps. There is good reason to believe that the same relationships hold for the human brain in cases of enriched or deprived environments, in certain diseases, and during normal aging.

The plasticity of dendrites can also be demonstrated by castrating adult male rats. Very soon, there is a dramatic decrease (56 percent in dendrite length and cell body size of motor neurons in the spinal cord that mediate copulatory actions. Androgen treatment of castrated males soon produces long dendrites and large cell bodies again. It could well be that one way various hormones affect specific behaviors is by such actions on neurons and neuronal networks.

Researchers have observed yet another aspect of brain plasticity during transplantation studies on various mammals in which pieces of embryonic or immature brain from specific sites are inserted into corresponding lesion sites in an adult brain. Pieces of cortex and cerebellum have been transplanted in this way, and scientists have noted that specific types of implant neurons may establish precise, correct connections with the host brain cells. To form such connections, growing axons must extend past apparently available synaptic sites in the host brain, and then form synapses with the correct target cells. Importantly, researchers have succeeded in repairing some major defects in damaged and aged brains by implanting developing nerve tissue. Although the full circuitry is never regained, the implanted neurons do deliver the proper chemicals (neurotransmitters and neuroactive peptides) to the correct areas. Function of those areas thus returns, and motor control may be regained and, in a few cases, complex behavioral functions restored. Implants of neurosecretory cells from adrenal glands and of embryonic brain have been carried out in Sweden and Mexico. Symptoms of Parkinson's disease disappeared in such patients. If implants remain free of attack from the patient's immune system, it seems likely that other brain diseases and minor strokes also will be treated with neuron implants. The medical importance of these transplantation studies for an increasingly aged human population is very great indeed.

each nerve cell's long axon contacts a whole series of cells, one after another. This results in a grid with many feedback loops. The number of connections that any one neuron makes appears to be random. And the properties of synapses of such circuits may change if they are used repeatedly; for instance, a synaptic enzyme may increase

in quantity, and the synapse may become more efficient in operation. The absolute number of synapses and dendrites may also rise with use.

How does this neuronal structure and plasticity of synapses and dendrites relate to olfaction and memory? Lynch recalls that the stem group of the modern mammals included mostly arboreal, shrewlike nocturnal creatures that had large olfactory systems, no doubt used for hunting in the dark. Imagine that they evolved interactive combinatorial networks that associated smells with things. A rabbit might, for instance, smell a musky odor and learn, to its benefit, that the odor comes from a fox. The circuit, perhaps with its increased dendrites, synapses, and feedback loops, generates impulses only when the particular odor enters the nostrils; but when it does, the memory of "fox" is triggered. Finally, Lynch suggests, the use of that circuit without the odor stimulus is really what we mean by a memory, or the idea "fox." Lynch's theories are still new, but they suggest an alternative to assuming that the elegant order of the visual-cortex column system, with its capacity to identify the positions of things, is also the basis for memory and learning.

MEMORY: MULTIPLE CIRCUITS AT WORK?

The hallmark of mammalian behavior is the ability to learn—to learn good places to hunt or graze, dangerous predators to avoid, safe transportation routes, acceptable interspecies behavior, and so on. The basis of such learning is *memory*, the capacity to record and recall past events. To appreciate "past events" in this context, consider what we mean by "the present." Human consciousness of the present, the here and now, involves a constantly changing panorama of sensory input and output. The list is, indeed, a long one—sights, sounds, smells, pain, pressure, gravity, feelings, reasoning—and all these things added together, ever varying in intensity and importance, constitute our conscious awareness. In contrast, then, what are memories? Some involve the recall of the collective sensations and feelings surrounding some past event, such as the events of high-school graduation weekend and walking up front for one's diploma. Other memories seem less direct, for example, the appropriate way to behave in a current situation based on successful behavior in a similar situation in the past. This involves reasoning and is based on learned and recalled behavior from the past. How can such complex processes be explained in cellular, electrical, or molecular terms? A central mystery in the study of memory is whether storage and recall are special, unique processes or whether memory is simply a persistence—albeit a remarkable one—of information processing. In other words, are there circuits and synapses just for memory? Or does the basic circuitry lying between input and output and orchestrating appropriate behavior during any given current moment also happen to store records of past events?

The work of psychologists and surgeons like Wilder Penfield has revealed a great deal about memory. Much information was gained in experiments in which a series of tiny sites on the cortices of epileptic patients were stimulated by electrodes. In one such experiment, when one site was stimulated, a woman stated, "I just heard one of my children speaking . . . and I looked out of my kitchen window and saw Frankie in the yard." Another woman, lying on the operating table in Montreal, said she actually seemed to "hear" the cathedral choir singing on Christmas Eve and "feel" the serenity and beauty of the Amsterdam cathedral—sounds and emotions experienced many years before. Results like these imply that much of our life—sights, smells, moods, and so on—is preserved in our cortex in incredible detail. We are unaware of that stored treasure, however, because we cannot recall it voluntarily. On the other hand we can bring certain kinds of information voluntarily to consciousness—the value of pi, a line from Shakespeare learned for a high-school play, or even some facts from a course. It seems reasonable that most such memories should be difficult to bring to consciousness, lest the brain be bombarded with essentially useless distractions from past life. But what distinguishes one type of memory from another? And how do memories get recorded in the first place?

Evidence from a variety of vertebrates suggests that there are at least three categories of memory: immediate, short-term, and long-term. Immediate memories—the parade of events as one rides to class or work on a bicycle, for example—are stored for just a few seconds. Virtually all these memories fade to obscurity in just seconds unless something unusual happens and calls greater attention to the incident—almost colliding with another bike, for instance. Other events are stored in short-term memory for minutes. If you look up a telephone number and make the call, you may remember the number for a while, but if you try to recall it a few days later, you probably won't be able to. Long-term memory persists for months or years and often involves repeated inputs or particularly vivid single events. If you buy a new lock, for instance, and open it a few times, the combination of numbers will soon easily jump to mind every time you try the lock. No one knows how this kind of learned long-term memory relates to the seeming detailed record of past life revealed by Penfield's electrodes (Figure 40-2 and Figure 40-5).

Much evidence reveals that the hippocampus is involved in learning and memory. If a person's hippocampus is damaged badly or removed surgically, he or she will no longer be able to learn or to recall recent events or new facts and faces. The subject's recall of events that occurred prior to the hippocampal surgery, however, will remain quite normal. If hippocampal neurons are stimulated with electrodes a very large number of times, they commence firing action potentials and may continue to fire for weeks. This activity resembles a phenomenon called long-term potentiation, which occurs when an animal is learning some complex task. Some data suggest that such intense long-term activity by brain neurons may reflect changes in synapses. Thus stimulation of the hippocampus may cause the number of synapses and dendrites to rise and the properties of synapses to change, and both these alterations may represent possible means of storing information for longer periods of time.

Another site of memory formation is the cerebellum, at least with respect to motor responses. A person who suffers damage to the cerebellum may thereafter have to perform consciously each step in a complex motor program that used to be automatic. In raising a glass of water to the lips, for instance, a person with a damaged cerebellum would have to start the motion, direct it, and stop it without smashing the glass against his or her teeth—all in a conscious, step-by-step fashion. This and other evidence shows that learned reflexes involving motor output are stored in the cerebellum. Several studies have shown that if a puff of air is directed at a rabbit's eyeball, the animal will blink. If a sound accompanies each puff of air as it is delivered, the rabbit becomes *conditioned;* that is, the eyelid soon will blink simply in response to the sound. Regions deep within the cerebellum store the memory trace required to elicit this and similar conditioned reflexes. It seems likely, then, that the cerebellum joins the hippocampus as a site for storing certain classes of memories essential to normal daily life. What of more complex memories, however—the lines of a nursery rhyme, the score of a Mozart symphony, or the steps involved in using a word-processing program? Here the cerebral cortex seems to come into play.

In the 1940s, the psychologist Karl Lashley trained animals to do a particular task and then removed one piece after another of the animals' cortices. This removal was to no avail—the obliteration of no single spot alone could destroy the learned behavior. The reason, he and others concluded, is that memories appear to be stored in a redundant and diffuse way in the cortex, with probably no single site for a given memory. Thus they are said to be *distributed,* and not localized.

Biologists do not yet know how immediate and short-term memories (which are probably properties of the hippocampus) are transferred to the cortex. They do know that *consolidation,* or transfer to long-term memory, improves greatly if an event to be remembered has significant physiological consequences. The events or things being experienced may be more likely to enter long-term memory because the levels of norepinephrine from the adrenal gland circulating in the bloodstream go up when an animal is in a stressful, tense, or crucial situation. Perhaps under such circumstances, events are set in motion at the cellular level that gradually affect the number and properties of dendrites and synapses; this, in turn, may be a basis for memory storage.

Biologists are currently teaching various invertebrate animals to perform learned sequences, then are measuring biochemical changes in the neurons of their brains. So far, they have described the appearance of inositol triphosphate, the activation of various protein kinases, calcium-ion entry into neurons, closure of potassium channels, and increased synthesis of proteins such as neurotransmitter receptors. Some work suggests that pre-existing proteins are modified as part of a short-term memory, whereas new genes must be expressed for long-term memory to function. Such changes in neurons and their surfaces, along with changes in dendrites and synapses, may be the underlying basis for the alterations in neuronal circuit activity that is part of the memory process. Other research done in the mid-1980s has provided surprising new hints that related bits of information may be stored together. In one particularly revealing case, a man suffered cerebrovascular damage to his left temporal lobe, with its Wernicke's speech center. After this, he could no longer say the names of a whole class of common objects, namely, fruits and vegetables; but most surprisingly, he could not "access" information in his brain about them. Even though he had that information in his memory, it could not be brought to conscious use. Thus, seeing or holding a peach or tomato elicited no familiarity, and he could neither say their names nor anything about them. However, if another person said the word "peach" or "tomato" or "spinach," the man could easily point to the correct fruit or vegetable. Alternatively, if he read the printed names of fruits or vegetables, he could identify them. Thus only if the key word was supplied verbally or in writing could the man's brain access the categories of information "fruit" and "vegetable" that were intact in his memory. Apparently, the damaged part of his brain had contained circuits that were used to link the sensory picture of the shapes, colors, and flavors of past peaches, tomatoes, and other fruits and vegetables with the memory banks of information about those objects. Another unexpected feature of this man's brain organization is that the bits of information about a related class of objects (fruits, vegetables) appear to be somehow stored and accessed together. One wonders whether other classes of things—automobiles, buildings, chemical formulas— each have their own region and access circuits in the cortex.

Findings like these show that information can be present in the brain but not accessible, and they are reminiscent of Penfield's studies of epileptics and of long-forgotten memories. Such findings are also consistent with the strange phenomenon of prosopagnosia—the inability to recognize the faces of familiar persons, even oneself, in a mirror. Head injuries or strokes in the occipital and temporal lobes, where visual processing occurs, cause this abnormality, and the sight of a person seems to evoke no memory whatsoever. A prosopagnostic patient can often recognize another person or him- or herself by hearing the voice, but still cannot associate the visual image of the face with the voice or with all the memories evoked by the voice. Nevertheless, prosopagnostics do show responses in skin electrical conductance when they see their own or another familiar face. This suggests that the face is being recognized (the memory is there) so that the cortex activates the autonomic nerves responsible for the skin response. Yet there is no consciousness of it by the prosopagnostic person, just as in the case of the man who was unable to gain access to stored information about fruits and vegetables. This fascinating area of research is beginning to reveal new views of how we access memory—a capacity that lies at the very heart of learning in higher animals.

THE CHEMICAL MESSENGERS OF THE BRAIN

While some researchers have been exploring memory and higher brain function with powerful methods, such as tracing electrical patterns and directly stimulating neural tissue, others have found that the brain's intercellular spaces are awash with an unexpected variety of chemicals, including neurotransmitters, hormones, and other substances.

In Chapter 36, we saw that messages between many brain cells are carried by *neurotransmitters*—chemicals that are released from the presynaptic vesicles of one neuron and act on the postsynaptic surface of another neuron. In recent years, investigators have found that individual neurotransmitters predominate in certain local regions of the brain; for instance, norepinephrine is found mainly in the reticular activating system, where it is involved in arousal, maintaining wakefulness, motivation, and other functions. Similarly, dopamine occurs exclusively in certain midbrain cells, where it takes part in neuronal control of complex muscle movements.

Not surprisingly, substances that interact with neurotransmitters also affect body functions. For example, certain antidepressant drugs act by inhibiting enzymes that degrade norepinephrine, so that levels of the neurotransmitter remain constant and a person remains alert and active. Caffeine in coffee and cola drinks inhibits the enzyme that degrades cyclic AMP; a high level of cAMP, in turn, excites various brain systems. Scientists still know little about the effects of the innumerable chemicals that animals ingest (intentionally or unintentionally) and that act on neurons in the brain. However, research

is progressing in the area of **neuroactive peptides** (Table 40-3).

Just a few decades ago, biologists had identified a scant seven neurotransmitters active in brain tissues, including such familiar compounds as norepinephrine and acetylcholine. These substances were thought to function simply by exciting or inhibiting postsynaptic cell surfaces. Today, researchers have a radically different picture of the components of the brain's chemical arsenal; they have detected dozens of additional chemicals—most of which are peptides that act as neurotransmitters or as modulators of neuron activity (Figure 40-15). Perhaps most surprising is the fact that many of these "new" substances were identified long ago elsewhere in the body, where they were shown to act as hormones. One example is angiotensin II, originally isolated in the kidney and believed to regulate blood pressure and now known to act on specific brain neurons to stimulate drinking behavior. Other brain neuroactive peptides include gastrin and cholecystokinin, which we encountered in the gut (Chapter 34), as well as traditional hormones previously thought of as acting elsewhere—for instance, prolactin, insulin, and thyroid-stimulating hormone (Chapter 37).

Where do these powerful peptides come from? Recall from Chapter 3 that a cell's protein-synthetic machinery generates peptides. In a brain neuron, once this synthesis process is complete, a neuroactive peptide is transported down the neuron's axon to a synaptic terminal; it is then stored and secreted when nerve impulses travel down the axon. This pathway is quite different from that of a neurotransmitter such as acetylcholine, which is manufactured directly in the synaptic terminal. Compared with that manufacture, the synthesis and transport of neuroactive peptides is an energetically expensive

---

Table 40-3   **POSSIBLE NEUROACTIVE PEPTIDES PRESENT IN THE BRAIN**

Peptides Common to Brain and Gut Tissues	Hypothalamic-Releasing Hormones
Vasoactive intestinal polypeptide	Thyrotropin-releasing hormone
Cholecystokinin	Luteinizing hormone-releasing hormone
Substance P	Somatostatin
Neurotensin	
Methionine enkephalin	**Others**
Leucine enkephalin	Angiotensin II
Insulin	Bradykinin
Glucagon	Vasopressin
	Oxytocin
Pituitary Peptides	Carnosine
Corticotropin	Bombesin
β-endorphin	Atrionatriuretic factor
α-melanocyte-stimulating hormone	

Figure 40-15 NEUROACTIVE PEPTIDES—MODULATORS OF NERVE ACTIVITY.

These short peptides influence all sorts of complex processes, probably by changing the cell surface properties of neurons so that their responses to neurotransmitters are altered. Note the similarity of amino-acid sequences among some of these peptides. This may reflect their origins from common genes.

undertaking for brain cells. Even so, researchers are finding more and more cases in which individual neurons synthesize and release at synapses both a neuropeptide and a traditional neurotransmitter, such as dopamine or acetylcholine.

Neuroactive peptides and more "usual" neurotransmitters also differ in another important way: whereas neurotransmitters are "hit-and-run" chemicals, affecting a target cell briefly before being quickly degraded by metabolic processes, neuroactive peptides resist destruction and thus act on target cells over prolonged periods of time. The **endorphins** and **enkephalins**—chemicals that mimic the effects of the opium-derived drug morphine and appear to serve as the body's natural painkillers—fall into this category.

Endorphins and enkephalins were discovered when researchers noted an intriguing structural anomaly: they found that certain brain cells have highly specific receptor sites for morphine, even though this substance is not normally present in the body. Reasoning that the brain must itself produce a chemical with an opiatelike molecular structure, J. T. Hughes and H. W. Kasterlitz found the first enkephalins in 1975. As predicted, these enkephalins are peptides five amino acids long and behave chemically like morphine. Within weeks after the enkephalins were discovered, three somewhat longer peptides, endorphins, were shown to have similar properties.

Unexpected in themselves, endorphins, enkephalins, and other neuroactive peptides have continued to surprise biologists by the broad range of roles they appear to play in body functions. Biologists now know that the brain is dotted with sites where specific neurons synthesize specific peptides; moreover, these sites seem to be mutually exclusive—enkephalins are not found where endorphins exist, and other neuropeptides are similarly uniquely distributed. This highly specific arrangement suggests that such peptide substances may prove to regulate much of an animal's biology—from hormone secretion and blood pressure to body temperature and even movement.

Studies in the past few years have revealed yet another difference between classic neurotransmitters and brain peptides; the latter may act by rendering target cells *less able to respond to other signals.* For example, when atrionatriuretic factor is released from nerve endings in the hypothalamus, nearby neurosecretory neurons, which release ADH in the posterior pituitary, are inhibited. Or, when the peptide known as **substance P** reaches a target cell in the cerebral cortex, the action of a neurotransmitter that might otherwise excite the neuron may be blocked. Conversely, neuroactive peptides may prevent an inhibitory neurotransmitter from suppressing neuron activity. Substance P is of great interest medically because it is believed to be the chemical carrier of pain messages in the body's sensory system. When a person pricks a finger deeply with a thumb tack (Figure 40-16), pain receptors in the skin generate nerve im-

pulses over a fast pathway up the spinal cord to the thalamus and then to the sensory and motor cortices. These impulses report the instantaneous, perhaps blinding pain. But in addition, slow-pathway fibers go to the reticular activating system and parts of both the hindbrain and the forebrain. Those pathways cause a prolonged perception of the pain, a "nagging pain" that allows the brain time for a better assessment of the real damage. It is the neuroactive peptide substance P that is released in spinal cord and brain cells, especially of the slow pathways. In the brain, substance P probably affects the emotions associated with pain. And it is there, too, that endorphins act. They modulate pain reception and emotion so that the organism can respond in an adaptive way to eliminate or cope with the causes of pain. For exam-

Figure 40-16   PAIN PATHWAYS.

Pricking one's finger sends pain signals over a sensory cell whose cell body is in the dorsal root ganglion. Substance P is released as the neurotransmitter that acts on spinal cord neurons. These spinal cord neurons carry signals that are distributed in the brain as shown. Brain signals, in turn, pass down the spinal cord ("descending signals") to cause release of enkephalin E, which inhibits discharge of substance P from the sensory neuron presynaptic ending, even though the pain receptor is still stimulated. Less pain is preceived in the brain as a result.

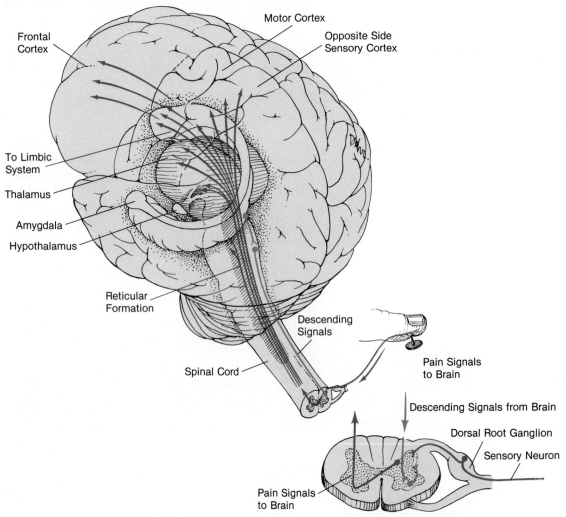

ple, in the case of the thumb tack (Figure 40-16), descending signals from the brain cause spinal cord neurons to release enkephalin E, which inhibits the discharge of substance P by the sensory neuron. The family of endorphins probably acts analogously to modulate other emotions. Researchers propose, for instance, that endorphins may prevent extreme anxieties, irrational fears, and the like. Too little endorphin released at the right place in the brain may allow extreme fears of enclosed spaces (claustrophobia), heights, or snakes to develop.

Although researchers are only beginning to explore the biological consequences of neuroactive peptides, it seems certain that one of the chief functions of such substances is to modulate and refine the continuing flow of nerve activities that govern much of life. Thus the distinction between the nervous and endocrine systems becomes increasingly blurred, so that it is best to consider them not as independent networks, but as parts of a highly organized and integrated continuum. The nerve cell is remarkable for its electrical excitability, its secretory capacities (neurotransmitters, neuroactive peptides), and its intricate axonal and dendritic trees. From such cells, arrayed in networks and circuits ranging from the crayfish's tail-flip circuits to the human's cerebral cortex, come learning, language, and, indeed, the human condition.

Modern neurobiologists are making strides in studying the chemical and electrical operations of brain cells unimaginable only a quarter century ago. The most complex living structure known to exist in the universe—the human brain—still holds deep mysteries. But advances in our understanding of neuroactive peptides and the brain's elaborate architecture suggest that one day we may surmount the greatest scientific challenge of all—the human brain's effort to comprehend itself.

SUMMARY

1. The human brain is an organ composed of at least 100 billion neurons that interact in extraordinarily complex ways. The heterogeneity of its neurons and neuronal circuits is the source of complex brain functions.

2. In its gross anatomy, the brain is divided into three major portions: the *hindbrain*, *midbrain*, and *forebrain*. The *cerebrospinal fluid* fills the cavities of the central nervous system and the tissue spaces in the brain and spinal cord; it is vital to normal brain functions and to transporting gases, nutrients, wastes, and so on.

3. Among the brain's "oldest" parts are the *medulla oblongata*, the *pons*, the *cerebellum*, and the midbrain. The medulla controls basic functions, such as respiration and swallowing, while networks of giant *Purkinje cells* in the cerebellum regulate balance, stance, and some other motor functions. In fish and amphibians, the midbrain processes sensory information from the eyes and may control some instinctive behavior. In reptiles and their evolutionary descendants, it acts as a relay station for sensory information from visual and auditory receptors.

4. More complex brain functions, including many complex reflexes and instinctive behaviors, are carried out by the *cerebrum* and its *neocortex* (cortex). Nearby are parts of the hypothalamus and thalamus, the amygdala, and the hippocampus, all of which compose the *limbic system*. The limbic system controls hunger and thirst as well as anger, fear, and sexual desire. In carrying out its functions, the hypothalamus plays an especially important role via its own hormones and its releasing factors that control the anterior pituitary.

5. The *reticular formation* is an elaborate network of neurons and axons that receives information from all of the body's sensory systems. Part of the reticular formation, the *reticular activating system* (RAS), is the source of arousal and wakefulness and filters diverse sensory inputs so that only vital information reaches the level of consciousness or is acted upon.

6. In evolutionary terms, part of the mammalian forebrain—the neocortex—is a recent development that is responsible for many of the most complex learned behaviors. The cortex is precisely ordered, with six layers of neurons; those neurons also are arranged into columns and, in visual cortex, into orientation slabs. The basic units of operation in the cortex are believed to be the vertically oriented columns of interconnecting neurons. Regions of the cortex are specialized for specific functions and are called the sensory, motor, visual, and auditory cortices. Sensory and motor homunculi are represented on those cortices. Other parts of the cortex are unassigned, associational areas, the sites of higher functions such as memory and learning.

7. The living brain can be studied by destroying brain tissue to create lesions and observing the behavioral consequences, and by recording the brain's electrical activity.

8. Split-brain studies in which the upper cortex of human brains were divided by sectioning the *corpus callosum*—a giant bundle of 200 million axons—reveal that each half of the brain is capable of independent functioning and that each is somewhat specialized for different cognitive tasks. In particular, the left hemisphere of most right-handed

people, and of many left-handed ones too, is the site of the language center and of analytical abilities. The right hemisphere is involved in spatial tasks in particular.

9. Broca's area in the cerebrum controls actual production of speech. Wernicke's area controls what is said. The left temporal lobe is also the place where bits of information in long-term memory are accessed, probably in information categories such as "fruits" or "vegetables."

10. The electroencephalogram and other physiological measures reveal that sleep is a complicated state consisting of at least five stages. In one of these stages—*rapid eye movement* (REM) *sleep*—the brain appears alert even while most of the body's muscles are virtually paralyzed. REM sleep may play a role in laying down long-term memories.

11. Studies of the mammalian visual cortex have revealed the hierarchical relationship of neurons. For instance, each hypercomplex cell is driven by its own set of complex cells; and each of these complex cells is driven, in turn, only when its set of simple cells is active. Each simple cell responds only to a specific pattern of active layer IV cells. The perception of moving or still images and of depth arises from such hierarchies in which a large number of cells at each level drives a smaller number of cells in the next level: the information is distilled, in a sense, in going from level to level.

12. Memories can be immediate, short-term, and long-term. The hippocampus and cerebellum are involved in immediate and short-term memory, whereas the cortex appears to be the site of long-term memory. Memories are associated with alterations in existing proteins, expression of new genes, and changes in neurons and neuronal circuits. Memories are not localized in unique sites in the brain, but are distributed instead.

13. There is substantial plasticity of brain neurons, with dendrites, synapses, and even neurons changing with use and misuse.

14. Central-nervous-system neurons communicate with each other through neurotransmitters and *neuroactive peptides*. Such peptides produce long-lasting effects on synapses and circuits, and add greatly to the subtlety and complexity of information processing and storage in the brain. Some brain peptides are *endorphins* and *enkephalins*—morphinelike substances that appear to be the body's natural painkillers. They and other neuroactive peptides may affect mood and perhaps all other higher-order brain functions.

KEY TERMS

blood–brain barrier	endorphin	midbrain	rapid eye movement (REM) sleep
cerebellum	enkephalin	neocortex	reticular activating system (RAS)
cerebrospinal fluid	forebrain	neuroactive peptide	reticular formation
cerebrum	hindbrain	orientation slab	substance P
column	limbic system	pons	
corpus callosum	medulla oblongata	Purkinje cell	

QUESTIONS

1. The spinal cord and brain are hollow. What are the four hollow cavities of the brain called? What fluid fills them, and what is its function? What are meninges?

2. Draw a diagram of the human brain. Indicate the three major sections, and label the major regions within each section. What are some functions of each region?

3. What is the limbic system and what does it do?

4. Where is the reticular formation, and what is its function? What is the reticular activating system?

5. Describe the structure of the human cerebral cortex.

6. What is a homunculus, which regions of the human cortex have a homunculus, and what do its unusual proportions reflect? What functions are performed by the other, unassigned parts of the cortex?

7. What is the corpus callosum? What structures does it connect? Doctors sometimes treat severe epileptics by severing the corpus callosum. Does this treatment affect their

speech? Reading? Thinking? What is affected?

8. What is the function of Broca's area? Wernicke's area? Does Wernicke's area lie closer to the part of the cortex responsible for hearing? For vision? Did recognition of spoken or of written language arise first in the early evolution of the human brain?

9. Cite some evidence showing that long-term memory is localized in the cortex. Cite some evidence

showing that memory is diffuse, and not localized in a specific spot. Does damage to the hippocampus affect long-term memory? Short-term memory? How does damage to the cerebellum affect memory?

10. What are some differences in production, longevity, and functions between neurotransmitters and neuroactive peptides? What is the significance of such peptides and of the view that the brain is a "neuro-endocrine" organ?

ESSAY QUESTION

With drawings and text, show how information is routed from the neural retina to the visual cortex. Describe the organization of the visual cortex, including ocular dominance and orientation slabs. Explain how a curved series of dots in the retina is translated into a moving line—let's say the edge of a tennis ball rocketing through the air.

SUGGESTED READINGS

ABELSON, P., and E. BUTZ, eds. "Neurosciences." *Science* 225 (1984): 1253–1364.

This is a series of review articles on the nervous system, brain, neuroactive peptides, and molecular approaches to studying the functioning of the nervous system.

ALTMAN, J. "A Quiet Revolution in Thinking." *Nature* 328(1987): 572–573.

New views on the plasticity of maps and circuits in higher brain function.

BLOOM, F. E., A. LAZERSON, and L. HOFSTADTER. *Brain, Mind, and Behavior.* New York: Freeman, 1985.

A superb, easy-to-understand introduction—with marvelous illustrations—to this fascinating topic.

GILINSKY, A. S. *Mind and Brain.* New York: Praeger, 1984.

A thorough treatment of brain function from the psychologist's point of view.

SPRINGER, S. P., and G. DEUTSCH. *Left Brain, Right Brain.* San Francisco: Freeman, 1981.

A clear summary of the two sides of the brain and how they work together.

SQUIRE, L. R. "Mechanisms of Memory." *Science* 232 (1986): 1612–1619.

This, plus an article in *Nature* 322 (1986): 419–422, provide an up-to-date account of the types of memory and their cellular and molecular bases.

Part
SIX

POPULATION BIOLOGY

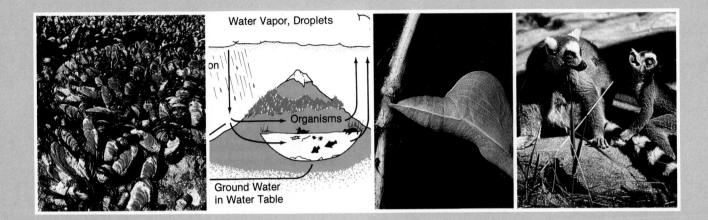

Water Vapor, Droplets

Organisms

Ground Water
in Water Table

The members of a species rarely exist in physical isolation, if for no other reason than their occasional need for genetic exchange. Most, in fact, live in populations—groups that may confer survival value in a number of ways. What's more, species never exist in taxonomic isolation: each is the product of vast natural forces and a lengthy span of geological history, and each is related to many other species—ancestors and descendants—in a lineage that stretches back, ultimately, to Earth's earliest organisms from those creatures alive today. To fully understand biology, one must not stop at the chemistry of life, the genetics of individuals, the diversity of species, and the physiological systems that ensure day-to-day survival. One must also consider the factors that shape how lines of organisms evolve in time: their populations, their ecology—that is, their relationships to the physical and biological worlds around them—and finally, their behavior.

These chapters describe evolution, ecology, and behavior. The modern theory of evolution deals with the mechanisms that allow mutated genes to spread in a population and establish new traits and, eventually, when enough mutations have accumulated, to

produce a new species—a group that cannot interbreed with other, even similar, kinds of organisms. Natural selection, the grim reaper of the living world, is a preeminent factor in evolution, but there are others as well.

To truly know an organism, one must know its ecology—its place in the community of other plants, animals, fungi, and microbes with which it interacts and on which it depends. All individuals, populations, and groups of populations are part of ecosystems. In turn, such systems of interactive biological and physical entities are part of the biosphere, the global system that includes weather, climate, atmosphere, and geography, and which powerfully affects individual organisms and, in turn, may be influenced by their collective activities. Evolution occurs as physical environments gradually change, as sets of organisms coevolve in response to each other, and as the complex ecological webs in which individuals exist change over time.

Animal behavior, an emergent property stemming from the simultaneous activities of all the internal systems—nervous, hormonal, skeletal, circulatory, respiratory, and others—allows exquisitely complex adaptations to the external world. Some animals primarily act on ge-

netically programmed instincts. But most also learn new ways of behaving. Animals can compete, cooperate, or communicate. Sometimes they evolve ways of life unattainable by isolated individuals. The social animals—colonies of insects, flocks of birds, schools of fish, troops of monkeys, human societies—are epitomes of those altered and extended ways of life.

This final part, and the book, end, appropriately enough, with the human species—social organisms for whom learning and cultural evolution go hand in hand with continuing biological evolution. In just fifteen million years, our tree-dwelling primate ancestors gave rise to upright, two-legged creatures with highly dexterous hands, large brains, and the capacity for spoken and written language. The seeds of our species' survival, however, may bear bitter fruit: our populations have exploded, we have wiped out innumerable other species, and we have so altered the physical features of our environment that our own long-term survival is threatened. There is, however, cause for hope, and much of it lies in the future progress of the biological sciences.

41

EVOLUTION AND THE GENETICS OF POPULATIONS

*Unlike physics, every generalization about biology is a slice
in time; and it is evolution which is the real creator of
originality and novelty in the universe.*

Jacob Bronowski, *The Ascent of Man* (1974)

A dawn scene like this of giant tortoises in a pond on Isla Isabela, Galápagos, may have greeted Charles Darwin as he wandered the islands.

In September of 1835, Charles Darwin stood for the first time on the shores of the Galápagos Islands, a cluster of small volcanic upwellings straddling the equator 600 miles west of Ecuador. Despite the HMS *Beagle*'s many stops along both coasts of South America, Darwin had seen nothing to compare with the wind-swept expanses of jet-black lava, the rocky beaches, and the stretches of low scrubby vegetation on these desolate islands. And Darwin became fascinated by the equally unique wildlife he observed: giant tortoises, prehistoric-looking marine iguanas, drab finches of many types, and numerous other animals. This collection of native organisms, so strikingly well suited to their remote islands, helped inspire Darwin years later to formulate a theory that would revolutionize the study of biology and forever alter society's traditional view of humankind's place in nature.

Darwin's theory of **evolution** by *natural selection* became a central unifying concept that allows biologists to probe and understand the structures, functions, and behaviors of modern organisms; to determine how new species may arise; to learn why some species may thrive and diversify, while others die out; and to trace the historical links between groups of organisms over vast stretches of time. We first encountered the ideas of evolution, natural selection, and adaptation of organisms to their environments in Chapter 1, and we have seen the ideas applied again and again in the subsequent discussions of cell biology, genetics, the origin and diversity of life, and plant and animal physiology. In this chapter and the next two, we will focus on the precise mechanisms of evolutionary change. In particular, we will see that while natural selection operates at the level of phenotypes, the gene itself may be viewed as a unit of evolution, and is amenable to detailed study with techniques borrowed from molecular genetics and other branches of modern biology.

In their search for evolutionary mechanisms, biologists have carried out some fascinating detective work. Clues come in an intriguing variety, from fossils that reveal the common ancestry of apparently unrelated groups to minute differences in a spectrum of proteins that can show the speed and distance of evolutionary change. The biologists' major task in studying the formation and extinction of species is to separate chance events from causal factors—a difficult prospect, as we will see.

Contemporary biologists are still asking questions much like those that Darwin pondered as he surveyed the strange life forms of the Galápagos Islands, fossils in Patagonia, and domestic animals in England. We will consider those questions in this chapter and see how the genetics of *populations*, or groups of interbreeding individuals within a species, as well as the study of changes in gene frequencies, can help biologists understand the mechanisms of evolution at work.

THE ORIGINS OF EVOLUTIONARY THOUGHT

As a youth, Charles Darwin was an avid amateur naturalist, and he held to the generally accepted scientific principles of his day. Most nineteenth-century people believed each of the millions of species to be immutable, created in its present form and remaining unchanged over the eons. This ancient belief had been set out in the writings of Plato, Aristotle, and other philosophers and in the biblical book of Genesis. But not all philosophers and not all cultures agreed—in the East, in Africa, and in the Americas, the idea that living things change over time was common. And, as we saw in Chapter 1, natural philosophers such as de Buffon concluded the same thing. It was in the late eighteenth century that evidence consistent with such ideas began to accumulate.

Much of this evidence came from rocky European hillsides and stream beds. There, geologists and amateur fossil hunters were turning up a huge number of ancient bones, shells, and fossilized plant parts that were clearly the remains of bygone forms. Like a parade of the Earth's natural history frozen in layers of stone, plant and animal types unlike any currently alive seemed to appear, diversify, and become extinct. The eighteenth-century French paleontologist Georges Cuvier pioneered the reconstruction of ancient animals from a few fragments of fossilized bone, while the British geologists James Hutton and William Smith were making detailed studies of the Earth's rock layers and the fossils within them. Stimulated by such findings and the obvious implication that organisms change over time, Jean Baptiste de Lamarck, a French zoologist, presented a new evolutionary theory in 1809, the year Darwin was born.

Lamarck and the Inheritance of Acquired Characteristics

To understand Lamarck's ideas and the wide acceptance they gained, we must step back in time. Gregor Mendel's studies on inheritance and Darwin's classic work still lay many decades in the future. As a professor in the Musuem of Natural History in Paris, Lamarck was familiar with the ancient Greek notion that each of an organism's body parts—muscles, heart, toes, eyes, and so on—produced "pangenes." *Pangenes*, or the elements responsible for inheritance, were supposedly pro-

Figure 41-1   THE GIRAFFE'S NECK: LAMARCK'S EXAMPLE FOR THE INHERITANCE OF ACQUIRED CHARACTERISTICS.
The French zoologist Jean Baptiste de Lamarck proposed that in the act of stretching its neck to feed on the leaves of trees, the ancestor of the present-day giraffe acquired a somewhat longer neck. Over many generations, this stretching led to the evolution of the very-long-necked giraffe we know today. Although his theory was incorrect, Lamarck's ideas are a landmark in the development of the theory of evolution.

duced in every organ of the body and then concentrated in the organism's reproductive cells. Thus, they passed the characteristics of the particular body part on to the next generation.

As Chapter 1 explained, Lamarck accepted the idea, put forth by de Buffon and others, that life forms evolve. He proposed that the driving force of evolution is the **inheritance of acquired characteristics.** He believed that organisms change physically as they strive to meet the demands of their environment, and that these changes are then passed to future generations by pangenes. Lamarck also theorized that the inheritance of acquired characteristics is the mechanism by which lower life forms move up the ladder of life to become more complex forms. Lamarck transformed the ladder into an escalator, with individual organisms slowly evolving from primitive to more advanced forms through a series of heritable changes.

A classic example of Lamarckism is the elongation of the giraffe's long neck in response to stretching (Figure 41-1). Lamarck also believed that body parts can be lost because of disuse and cited the "loss" of eyes in moles—animals that spend most of their lives in dark, underground tunnels—as an example. In modern terms, organisms are faced with the imperatives "use it, or lose it" and "use it, and it may get better"!

Today, we know Lamarck's theory to be incorrect because such things as stretching to feed in a tall tree or living in a dark cave have no effect on germ cells, gametes, or heredity. Still, Lamarck's theory does constitute a major landmark in focusing on evolution and, as

Darwin later noted, in declaring that evolution of living things proceeds according to natural laws, not divine intervention. A new generation of scientific thinkers would not only proclaim evolution to be a fundamental characteristic of life, but also describe a mechanism that could explain, for the first time, the overwhelming diversity of life forms. This revolutionary idea was the concept of natural selection.

Darwin and Natural Selection

Charles Darwin was only twenty-two years old when he boarded the HMS *Beagle*, a surveying ship that had been chartered to circumnavigate the globe and map the coasts of South America (Figure 41-2). When the *Beagle* left England in 1831, Darwin, with his scientific training and longstanding interest in natural history, already entertained evolutionary ideas. Five years later, when the ship docked in Falmouth, Darwin had profited immeasurably from his remarkably perceptive eye for the natural world. The theory that would forever change the way both scientists and laymen perceive the living world was taking shape in his mind and notebooks.

Many factors probably contributed to this triumph of reasoning. We know, for example, that before he sailed, Darwin had been avidly reading the first volume of Charles Lyell's *Principles of Geology*. He even arranged to pick up the second volume when the *Beagle* called at Montevideo. Lyell's writings detailed his observations that the Earth's physical landscape underwent long,

Figure 41-2   CHARLES DARWIN AT AGE 31, IN A WATERCOLOR PAINTED BY GEORGE RICHRORD IN 1840.

slow, continuous change. On the basis of the fossil record, Lyell speculated that animal and plant species arose, developed variations, and then became extinct as the ages passed.

Soon after returning to England, Darwin came across an essay published nearly fifty years earlier by the clergyman and economist Thomas Malthus. In his *Essay on the Principles of Population*, Malthus set out the proposition that populations inevitably grow faster than their food supplies. As a result, said Malthus, organisms are forced into a "struggle for existence."

But more than the ideas of Lyell or Malthus, Darwin's exposure to the exotic and varied life forms he encountered during the *Beagle's* 40,000-mile journey influenced his thinking. In Patagonia, Darwin found a naturalist's treasure: fossilized bones of extinct, cow-sized sloths and giant armadillos remarkably similar to the skeletons of smaller forms of these animals living in Central and South America. Darwin saw variation everywhere he looked: plants and animals that were strikingly different from each other in their forms, habitats, and geographical distribution. And Darwin noted the absence of rabbits in the prairie grasslands, for example, even though it would have been an ideal habitat for them.

Nowhere were the plants and animals more striking than in the Galápagos archipelago. Here Darwin discovered a strange species of iguanas—normally terrestrial

reptiles in desertlike environments—that on the Galápagos swam in the sea and ate seaweed in the surf. He also found giant tortoises, whose shells bore inscriptions carved by whalers who had explored the islands a hundred years earlier. In what was probably his most significant observation, Darwin noted that similar animal types show distinctive variations in body form and function from island to island. It took time for Darwin to assimilate all that he had seen in the Galápagos and South America, but in the years following his return to England, two central concepts of his developing theory of evolution crystallized.

First, Darwin realized that the differences among related populations represent adaptations to differing environments. In biology, an **adaptation** is any genetically based feature that results in an individual or species being better suited to some aspect of its environment—jaws for eating particular types of food, insulation to resist freezing temperatures, behavioral instincts to avoid predators such as snakes, and so on. The specializations that Darwin observed in the body sizes, beak shapes, and other features of various finches correlate with the diet and feeding behaviors of each group (Figure 41-3). (Note that although Darwin collected a variety of diverse finches on the Galápagos, he did not describe them in his great book.) What is the significance of adaptation? Darwin reasoned that individuals possessing advantageous adaptations are more likely to outreproduce indi-

Figure 41-3   THE GALÁPAGOS FINCHES: EXEMPLARS OF EVOLUTION BY NATURAL SELECTION.
These skulls of 12 finch species reflect differences in body size and show dramatic variation in beak shape. Consumption of different foods such as berries, hard seeds, or insects is facilitated by the specialized beaks of each species. The tool-using woodpecker finch's skull is shown grasping a "spear."

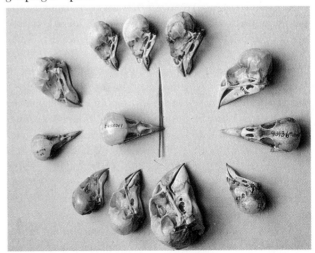

THE FABLE AND FACTS OF DARWIN'S FINCHES

Galápagos finches—those fourteen species of dusky birds with their variously shaped beaks and diets to match—were Darwin's final inspiration, crystallizing his great theory of evolution by natural selection.

Or were they?

Nearly four generations of students have learned this bit of scientific history. But historians are no longer so sure that the finches were Darwin's turning point.

One persuasive piece of evidence comes from Darwin himself and from his magnum opus, *On the Origin of Species*. In it, Darwin describes variations among mockingbirds and other bird species he observed during his long voyage on the *Beagle*. But nowhere does he make the briefest mention of the now-famous finches. Science historians have pondered this odd omission for decades.

One popular explanation holds that Darwin was simply trying to avoid retracing old ground; he had described the finches earlier in his *Journal of Researches* and perhaps felt no need to repeat this evidence when his book was published twenty years later. Some historians remain unconvinced by this explanation, however, because Darwin was a stickler for detail and marshaled huge quantities of data to support all his ideas. Recently, several researchers have proposed an intriguing alternative to this standard explanation. They feel that Darwin's omission of the finches was based on rigorous scientific thinking and revolved around three crucial factors.

First, Darwin mistrusted his own data on the Galá-pagos finches because he collected much of it second-hand, copying most of his notes on the habitats of different finch species from records made by shipmates. Given his penchant for accuracy, Darwin was probably leery of drawing conclusions from this sort of data.

Second, Darwin believed—mistakenly—that eleven of the fourteen finch species ate similar food. Because he believed this, he could not have seen the connection between each species' diet and its beak size and shape, nor could he have concluded that new finch species arose as they became adapted to exploit differing resources in the environment.

Finally, Darwin questioned whether all the Galápagos finches, with their very different characteristics, could have arisen from a single pioneering species. Indeed, it was not until well into the twentieth century that biologists could trace the common ancestry of Darwin's finches with certainty. The British biologist David Lack's book, *Darwin's Finches*, provides overwhelming evidence about the bird's evolution and adaptation.

If legitimate doubts and faulty data caused Darwin to exclude the Galápagos finches from his master work, then an appealing but fanciful legend can finally be laid to rest: the Galápagos finches were not Darwin's ultimate inspiration. In giving up this legend, however, modern biologists have gained a fuller appreciation of Charles Darwin's precise, skeptical, and inquiring mind.

---

viduals lacking the adaptations—*survival to reproduce* is the key.

Second, Darwin recognized that in the Galápagos, the way organisms are geographically isolated onto separate islands provides an opportunity for *reproductive isolation*—the division of a population into groups that do not interbreed (we will discuss this phenomenon further in Chapter 42). Such isolation can, in turn, allow greater and greater differences in form, function, or behavior to accumulate among the reproductively isolated groups. Thus within populations of a species, adaptations may arise because of reproductive isolation. In addition, such isolation and accumulation of greater and greater hereditarily based differences can ultimately lead isolated groups to diverge enough to be classified as different species.

Darwin was not the only scientist to formulate a theory of natural selection, however. Another young British naturalist, Alfred Russel Wallace, had studied plants and animals in Brazil and then in the Malay Archipelago in Southeast Asia (Figure 41-4). Imagine Darwin's shock, surprise, and chagrin when he received a copy of Wallace's paper, which outlined the very theory that had been Darwin's primary focus for decades. Motivated at last to publish his ideas, Darwin did so with great integrity, by having both his own and Wallace's papers read at a meeting of the Linnaean Society in London in 1858. As it turned out, Wallace emphasized competition for resources as the basic factor in natural selection, but Darwin focused on reproductive success—a concept that rapidly became the fundamental idea in evolutionary thought. Furthermore, it was Darwin alone who had amassed thousands of bits of evidence in support of the theory and who had honed it into a comprehensive explanation of evolution in action.

Two decades after the *Beagle* had ended its voyage, Darwin published *On the Origin of Species by Means of Natural Selection, or the Preservation of Favoured*

Figure 41-4   ALFRED RUSSEL WALLACE, CO-DISCOVERER OF EVOLUTION BY NATURAL SELECTION. Working independently of Darwin, Wallace formulated his own theory of evolution by natural selection. Wallace stressed competition for limited resources as a main basis of natural selection, whereas Darwin emphasized competition within populations that tend to expand beyond their food supply.

Races in the Struggle for Life. In this great work, he integrated all the ideas and influences that had shaped his theory. His eloquent summary reads:

> As more individuals are produced than can possibly survive, there must in every case be a struggle for existence, either one individual with another of the same species or with the individuals of a distant species, or with the physical conditions of life. . . . Can it therefore be thought improbable seeing that variations useful in some way to each being in the great and complex battle of life, should sometimes occur in the course of thousands of generations? If such do occur, can we doubt (remembering that many more individuals are born than can possibly survive) that individuals having any advantage, however slight, over others would have the best chance of surviving and of procreating their kind? On the other hand, we may feel sure that any variation in the least degree injurious would be rigidly destroyed. This preservation of favorable variations and the rejection of injurious variations, I call natural selection.*

The kernel of Darwin's argument for **natural selection,** then, is that organisms best adapted to their environ-

---

* Charles Darwin, *On the Origin of Species* . . . , 5th ed. (London: John Murray, 1869).

ment will have an edge in the battle of life, and this edge will tend to increase their chances to survive and reproduce.

On the Origin of Species was an immediate best seller—every copy sold the day the book was published. It outraged the conservative religious leaders of England and some scientists as well. Darwin was personally attacked in the press and from the pulpit, but he greeted those attacks with calm pleas to examine the massive store of evidence he had painstakingly compiled over the course of twenty years. So powerful and reasonable were his ideas that to most scientists of the day, they immediately rang true. The concept of evolution by means of natural selection seemed a revelation that brought all living things into much closer harmony with nature.

The brilliance of Darwin's theory is all the more striking when you consider that he knew nothing of genes or genetics; Mendel did not publish his results until 1866, and his work was not appreciated for many decades. In order to fully understand the mechanisms of evolution, we must step ahead and see how inheritance and evolution intersect at the level of the gene.

VARIATIONS IN GENES: THE RAW MATERIAL OF NATURAL SELECTION

The key to Darwin's concept of a struggle for existence was the idea that some individuals in a species will arise with an "advantage, however slight." Such an advantage might be a longer neck, an enzyme that catalyzes a reaction faster, a leaf that gathers light better, or some other adaptation that increases an organism's chances for survival to reproductive age. And if inherited variations in necks, enzymes, leaves, and other characteristics occur among members of species, then the genes responsible for them must also vary. It seems reasonable, then, that by looking for variations in phenotypes, and the genes that produce them, we can begin our search for the underlying mechanisms of evolution.

We need only look around us to see variation in the morphology of human beings. As Figure 41-5 illustrates, people display an intriguing array of physical features. To the trained eye, members of nearly every other species have this same kind of diversity in shape, size, and features. For evolutionary biologists, the enormous range of morphological variations within a species creates problems: Is a person or a pine tree usually tall because of "tall" genes or because favorable nutrients were available during crucial periods of each organism's growth? The fact is that morphological or other variations in phenotype may or may not reflect variations in an organism's genetic make-up. And even if they do, it is

(a)   (b)   (c)   (d)

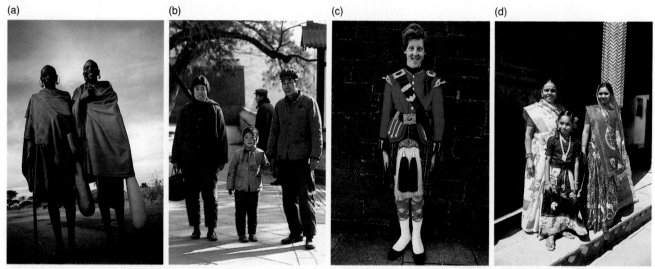

Figure 41-5   THE RACES AND INDIVIDUALS OF *HOMO SAPIENS*: ASTOUNDING VARIATION IN MORPHOLOGY, BUT ALSO IN CULTURE.
(a) Tanzania: pausing for a picture, Masai men. (b) People's Republic of China: going to the park, a Beijing family. (c) Scotland: a smile for the new piper's uniform, an Edinburgh boy. (d) India: three generations of an ancient civilization, Agra women. All one species, all capable of inter-breeding, yet these and many other equivalent differences in morphology have arisen over the millenia in human populations relatively isolated from each other in various places on our planet.

usually difficult or impossible to determine the exact number of gene loci and alleles that are responsible for a particular variation. Recall from Chapter 11 that one gene can affect a number of traits—the phenomenon known as *pleiotropy*. Conversely, several genes can influence a single trait, as in *Drosophila* eye color. These considerations complicate efforts to estimate the degree of genetic variation among individuals of a species.

Given such difficulties, how can one catalog variations in genes that produce differences among organisms, such as the different alleles for eye or flower color? Unfortunately, the methods of Mendel's classic genetics provide information only about genes responsible for visible features—wrinkled, smooth, green, yellow, and so on. What of all the other subtle variations in phenotype that lack an easy genetic explanation? It was not until the 1950s and 1960s, when molecular biologists began unraveling the genetic code and the details of protein structure, that new techniques became available to explore genotype and phenotype at more discriminating levels. This meant, in turn, that a large number of individuals composing populations of a species could be surveyed to ascertain the degree of protein and genetic variation in natural, evolving groups of organisms.

Looking for Genetic Variation: Protein Electrophoresis

Implicit in Darwin's theory is the concept of variation of inherited characteristics. Measurements of such variation can be made by analyzing either the sequences of

amino acids in proteins or the sequences of nitrogenous bases in DNA and other nucleic acids using techniques perfected during the late 1970s and early 1980s. Nevertheless, such techniques are often not practical, since they are expensive and time consuming, and may involve lengthy purification procedures. Thus for researchers interested in surveying a protein and its gene in hundreds or thousands of individuals in a population, a more practical way to measure protein variation is needed. The preferred method has been protein electrophoresis.

Electrophoresis—literally, the "carrying of electricity"—is a technique used to trace the movement of electrically charged particles through a fluid medium. It depends on the fact that the amino-acid subunits of proteins from plant and animal cells may carry a positive or negative charge or no charge at all. Since each kind of protein has a characteristic combination of amino acids, each will tend to move at a different rate in the electrical field—highly charged ones, fast; uncharged ones, slow or not at all. Variations on basic electrophoresis take advantage of the size differences among protein molecules (that is, their numbers of amino acids). As a result, different protein molecules travel along the electrophoretic medium at different rates. Small ones, for example, tend to move quickly, large ones slowly.

Using protein electrophoresis, it is possible to distinguish between enzymes, other proteins, and other charged molecules extracted from tissues. Because a protein's primary structure is determined by the nucleotide sequence of its coding gene, protein variations revealed by electrophoresis serve as a kind of biological window on genetic characteristics of populations.

Electrophoresis involves just a few simple steps. First, protein samples to be assayed are placed in a stripe at one site on a wet sheet or slab of coarse paper, starch, or various gels. Then an electric current is applied, running from a positive electrode to a negative electrode (anode to cathode). Figure 41-6 shows how an electrophoretic gel is used to analyze protein differences in fish.

After electrophoretic data have been obtained, a complex process of interpretation begins. Interpreting results is complicated by the fact that different forms of a protein can be produced in different tissues. These forms are called *isozymes* in the case of enzymes; but structural proteins, such as actins or myosins, also vary from one tissue to another. Electrophoretic patterns can also be affected by processing of proteins—for instance, if sugars are attached to a protein in the cell. These and other kinds of complications mean that electrophoretic analysis cannot easily tell evolutionary biologists the exact amount of protein and genetic variation within a population. It does, however, provide a useful approximation of that variation—the genetic variation that ultimately makes evolution possible.

Using electrophoresis, biologists can get a relatively clear idea of the genotypes of the individuals in a sample—how many are homozygous and how many are heterozygous for each allele that is assayed—indirectly by analyzing its protein product. In order to translate these data into estimates of how much genetic variation actually exists in a population, a researcher usually surveys about twenty or more gene loci by testing their respective proteins. Then the results are summarized in a way that expresses how variable population members can·be.

One method for doing this is to figure the **average heterozygosity** (symbolized by H) of a population. This measure describes the average frequency of individuals that are heterozygous at each gene locus surveyed. Suppose, for example, that you decided to carry out an electrophoretic survey of a population of fruit flies living in a pile of fallen and fermenting apples at a local orchard. The population includes thousands of individuals, but you collect 100 flies at random with a net and take them to the laboratory for analysis. Now, suppose that a survey of 20 different proteins in each fly—representing 20 different gene loci—turns up the following results. At 6 of the loci, 50 of the 100 flies are heterozygous. You know this because two forms of each of the 6 proteins are present on the gel. At 8 of the loci, 10 of the 100 flies are heterozygous, and at the remaining 6 loci, all the flies are homozygous for the allele—that is, only one form of the protein is present for each of the 6.

To determine the average heterozygosity, you would note that 6 loci have a heterozygosity of $^{50}\!/_{100}$, or 0.50; 8 loci have a heterozygosity of $^{10}\!/_{100}$, or 0.10; and the re-

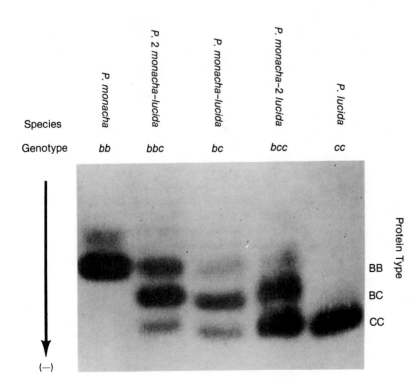

Figure 41-6   ELECTROPHORETIC GEL: REVEALING GENE VARIATIONS.

An electrophoretic gel is produced by placing protein samples on a wet sheet of cellulose, starch, or some similar gel, and then applying an electric current, which induces the proteins to move characteristic distances, determined by both size and electric charge. These patterns are produced from muscle proteins of two different species of *Poeciliopsis* fish and by hybrids of the two species. The protein (alcohol dehydrogenase) is made of two polypeptide chains: in *P. monacha* (left column) it is two B chains, BB, produced by genotype *bb;* in *P. lucida* (right column), it is two C chains, CC, produced by genotype *cc.* Note that the three hybrids of the two parental species show some BB and CC, but also BC protein, proving that both *b* and *c* alleles are present. Patterns such as these are useful in determining the genetic relationship of organisms.

maining 6 loci have a heterogyzosity of $0/100$, or 0. By summing and averaging over all 20 loci

$$H = \frac{(6 \times 0.50) + (8 \times 0.10) + (6 \times 0)}{20} = \frac{3 + 0.8 + 0}{20} = 0.19$$

you would finish up with an average heterozygosity of 19 percent. Another way to view the genetic variation is called **percent polymorphism** of a population. For instance, if 6 + 8 of the 20 loci are heterozygous, the percent polymorphism is 70 percent ($14/20 = 0.7$).

Geneticists have used electrophoresis to estimate the amount of variation in a large number of loci in many species of plants and animals (Figure 41-7). An important conclusion is that the analysis of proteins shows high levels of the genetic variations in populations. This variation makes natural selection possible.

Our own species provides a fascinating example of the extraordinary amount of variation that can exist in a pop-

ulation. Human beings have an average heterozygosity of 6.7 percent, and our DNA includes about 100,000 loci for structural genes (genes that code for proteins). An H value of 6.7 percent means that the average person is heterozygous at about 6,700 structural genes. What does this signify for the genetic diversity of the human species? We know that an individual who is heterozygous at a locus—say, Aa—can produce two different kinds of gametes: half the sperm or eggs carry A, and the other half carry a. The rule is that an individual heterozygous at n gene loci can potentially produce 2^n different gametes. Loci, however, are arranged in linkage groups on chromosomes, which means that they do not assort with complete independence. Hence, n is the haploid chromosome number; for humans, $2^n = 2^{23}$. Although it is substantially smaller than $2^{6,700}$, it is still an unimaginably large number. As a result, the number of different gene combinations is so vast that no two gametes from different people are ever identical. It is no

Figure 41-7   GENETIC VARIATION IN NATURAL POPULATIONS.

A substantial percentage of loci are heterozygous in all these major groups of sexually reproducing organisms. Recombination of those different alleles during sexual reproduction helps account for the variations on which evolution operates.

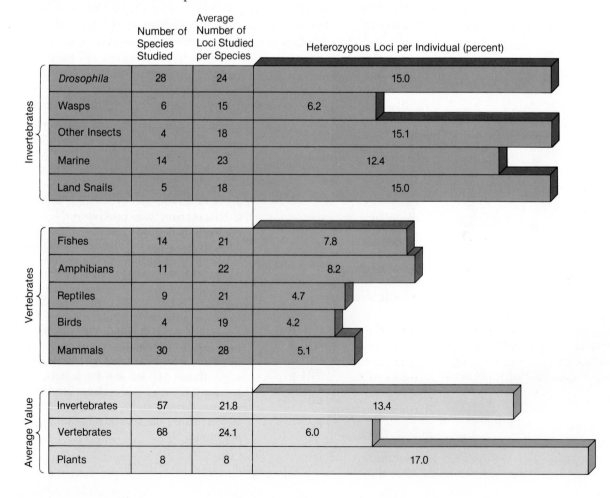

	Number of Species Studied	Average Number of Loci Studied per Species	Heterozygous Loci per Individual (percent)
Invertebrates			
Drosophila	28	24	15.0
Wasps	6	15	6.2
Other Insects	4	18	15.1
Marine	14	23	12.4
Land Snails	5	18	15.0
Vertebrates			
Fishes	14	21	7.8
Amphibians	11	22	8.2
Reptiles	9	21	4.7
Birds	4	19	4.2
Mammals	30	28	5.1
Average Value			
Invertebrates	57	21.8	13.4
Vertebrates	68	24.1	6.0
Plants	8	8	17.0

wonder, then, that with the possible exception of identical twins (derived from a single fertilized egg), the chances that any two human beings will ever be genetically identical are just about zero.

The sheer number of genetically different gametes is only part of the biological insurance that organisms will vary. In diploid organisms—those that reproduce sexually—the crossing over of pieces of chromosomes during meiosis and the random distribution of parental chromosomes to sperm and eggs provide a built-in guarantee that new allele combinations will appear in the genetic heritage of offspring. Further diversity is ensured by the fact that any sperm may chance to meet and fertilize any egg. And, finally, this stockpile of genetic variability is continually replenished by random mutations. Clearly, large natural populations present countless, ever-changing genetic opportunities for natural selection to operate and for evolution to occur.

Figure 41-8   POPULATIONS: GROUPS OF ORGANISMS THAT INTERBREED.
This colony of marmots (*Marmota flaviventris*) in Olympia National Park, Washington, shows typical watchful behavior for hawks or other predators. Three generations of this colony are seen here.

POPULATION GENETICS: THE LINKS BETWEEN GENETICS AND EVOLUTION

Knowing nothing of genes, Charles Darwin was never able to pin down the exact connection between heredity and differences in organisms' characteristics. After the rediscovery of Mendel's work, however, evolutionists were quick to realize that by combining genetics, with its mathematical language, and the concept of natural selection, they could begin to account for phenotypic variations in populations of living things. So, beginning at the turn of the century and continuing today, researchers in the area of biology known as **population genetics** have used mathematical descriptions of genetic phenomena to help them trace evolutionary trends within populations.

The Gene Pool and Gene Frequencies

The basic principle on which population genetics depends is that genes may be viewed as the units of evolution. Put another way, we can say that it is the genetic composition of a population that exists and evolves over time. This genetic composition—all the various alleles of all the genes carried by individuals in a population—is called a **gene pool.**

At this point, you may be wondering what exactly constitutes a "population." For geneticists, a **population** is defined as a group of organisms that interbreed or that have the potential to interbreed. All the rabbits of a particular species living in a mountain meadow qualify as a population, as do the cottonwoods growing in a grove alongside a creek (Figure 41-8). As we shall see in later chapters, populations are dynamic entities that interact with each other and with the environment in many ways. Here we will look at populations in terms of how their gene pools evolve.

One of the most important evolutionary properties of gene pools is their ability to change as a function of space and of time. For example, related populations of trout living in unconnected river systems gradually may develop genetic differences. Or, as one generation gives rise to another, genetic variations may accumulate. These genetic differences will be reflected in the phenotypes of the organisms that carry them. Thus members of a species may look different from each other (just think of variations among different breeds of dogs), function differently, and behave differently because their genes have come to vary in ways that might be detected by protein electrophoresis.

There are two ways to depict the genetic make-up of a population or gene pool. One is to describe the types and relative number of different *genotypes* found in a given population. The other is to list the number and relative frequency of *alleles* in the gene pool. As we noted in Chapter 10, there may be many alleles of a given gene if there have been many mutations of the original gene.

Population geneticists sometimes define evolution as a change in gene frequencies. If so, then to measure evolution—to detect whether it is happening—we have to accurately quantify the frequencies of different alleles.

Table 41-1   **GENOTYPE AND ALLELE FREQUENCIES FOR THE MN GENE LOCUS IN A UNITED STATES CAUCASIAN POPULATION**

Number of Individuals in Each Blood Group				Genotype Frequency			Allele Frequency	
M	MN	N	Total	$L^M L^M$	$L^M L^N$	$L^N L^N$	L^M	L^N
1,787	3,039	1,303	6,129	0.292	0.496	0.213	0.539	0.461

To see how this is done, let us use the example of the MN gene locus in human blood groups. Recall from Chapter 31 that human red blood cells have various marker molecules on their surface. The ABO cell-surface antigens are the ones with which we are most familiar, but the MN surface antigens are present as well. The three possible blood groups, M, N, and MN, are determined by two alleles, L^M and L^N, as shown in Table 41-1.

In a test of 6,129 Caucasians living in the United States, it was found that 1,787 had the M phenotype, and 1,303 had the N phenotype. The remaining 3,039 members of the sample were MN. Therefore, we have all the data needed to calculate the relative frequency of each genotype in the sample population. All we need do is divide the number of individuals with each phenotype by the number of people in the sample. Thus the frequency of the $L^M L^M$ genotype is $1,787/6,129$, or 0.292; the frequency of the $L^N L^N$ genotype is $1,303/6,129$, or 0.213; and the frequency of the $L^M L^N$ genotype is $(6,129 - 1,787 - 1,303)/6,129$, or 0.496. Of course, MM + MN + NN = 1 (although because of rounding, the total will appear to be slightly greater).

In the same way, to determine the allelic frequencies of the MN gene locus, we simply divide the number of alleles found by the total number of alleles in the sample. Now, $L^M + L^N = 1.0$. The total number of L^M alleles in the sample is arrived at by adding the number of alleles in the $L^M L^M$ individuals ($1,787 \times 2 = 3,574$) to the number of L^M alleles in the $L^M L^N$ individuals (3,039), giving a total of 6,613. Dividing this number by the total number of alleles in the sample (the diploid number is $2 \times 6,129 = 12,258$), we obtain $6,613/12,258$, or 0.539. Because $L^M + L^N = 1$, it is then easy to calculate the frequency of the L^N allele: 0.461. Knowing these simple relationships allows us to calculate the proportion of alleles in a population.

It might seem reasonable to assume that in most populations, the relative gene frequencies remain the same. How close does this assumption match reality in a hickory forest or a flock of wild geese? As we shall see in the next section, questions like this led to the formulation of fundamental theoretical principles in population genetics. And these principles, in turn, provided biologists with a yardstick with which to measure real evolutionary changes.

Genetic Equilibrium and the Hardy-Weinberg Law

Many early geneticists thought it reasonable to assume that Mendel's laws for predicting the genotypic frequencies of offspring could also tell us something about the genotype frequencies in populations. It seemed intuitively obvious, for instance, that dominant phenotypes are more abundant than recessives and ought to gradually replace them in a population, But this expectation did not correspond with the abundant evidence of persistent recessive phenotypes (and their underlying genotypes) that researchers could see all around them.

The Engish geneticist R. C. Punnett, whose squares we looked at in Chapter 10, was particularly troubled by the fact that although blue eyes are recessive to brown eyes, there were a great many (perhaps even a majority of) blue-eyed people in England (Figure 41-9). One day, he aired his puzzlement to his cricket partner, the noted mathematician G. H. Hardy, who is reported to have immediately scribbled the explanation on a luncheon napkin. Hardy proposed that while Mendel's laws can predict the outcome of single-pair matings, they indicate nothing about the *proportions* of genotypes in a population. Instead, Hardy speculated, genotypic proportions in a population reflect the frequency of alleles in a population, which is completely independent of the dominance or recessiveness of particular alleles. Furthermore, he suggested, if certain conditions are met, the frequency of alleles in a population will be stable—it will stay in *equilibrium*—from generation to generation. Because this same conclusion was reached independently in the same year (1908) by the German physician W. Weinberg, it is known today as the **Hardy-Weinberg law.**

Hardy and Weinberg set up their "law" so that it describes (in the shorthand of mathematics) what happens to the frequencies of alleles and genotypes in a hypothetical ideal population of sexually reproducing organisms. This ideal population, they said, has to meet the following requirements:

1. *Random mating:* no factors can cause the organisms' choice of mates to be nonrandom.

Figure 41-9   RECESSIVE TRAIT IN A POPULATION.

The proportions of genotypes—say, for hair or eye color—in a population are the result of the frequency of alleles in that population, a frequency that is completely independent of the dominance or recessiveness of particular alleles. The great majority of blue-eyed blondes in Norway, for example, demonstrate that a recessive trait can appear in a majority of a population.

2. *Large population size:* the laws of probability must be able to operate.

3. *No mutations.*

4. *Isolated population:* no exchange of genes with other populations, such as through migration, should be possible.

5. *No natural selection:* no alleles can have reproductive advantage over others.

In any population that meets these conditions, the result is genetic equilibrium: no change in gene frequencies. But when one or more of these conditions is *not* met, changes in gene frequencies may occur. Since by definition, changes in gene pools constitute evolution, then a deviation from the Hardy-Weinberg law tells us that a change in the gene pool has occurred and that evolution has taken place. Thus the law is a means of measuring evolution.

The Punnett square technique can be used to demonstrate how a population might conform to the Hardy-Weinberg law. But whereas we used the Punnett square in Chapter 10 to examine simple crosses between the gametes of one male and one female, we use it here to represent the sperm and eggs of all the males and females in the population. Suppose that our hypothetical population has a dominant allele S, which codes for straight hair, and that this allele makes up 90 percent of the alleles at that gene locus. The other 10 percent of the alleles are the recessive s, which codes for curly hair. Figure 41-10 shows the Punnett square used to analyze the results of random mating; both the sperm population and the egg population have the 90:10 percent S:s ratio.

Assuming that the organisms mate at random, any sperm may have a chance of fertilizing any egg. So to calculate the frequencies of the resultant genotypes (SS, Ss, and ss), we simply do the multiplication steps indicated in the figure. Note that 81 percent of the offspring will be homozygous for the dominant SS genotype, and only 1 percent will be homozygous for the recessive ss genotype. The remaining 18 percent of the organisms of the F_1 generation will be straight-haired heterozygotes— that is, they will show the dominant phenotype but will carry the recessive s allele.

Now, if the F_1 population mates randomly, the genotypic frequencies remain unchanged; they could have arisen only from a set of gametes with the original ratio: 90:10 percent S:s. The recessive alleles are not lost but are "hidden" in F_1 heterozygotes. So the proportion of S to s stays the same from one generation to the next. Furthermore, we can see that even if the ratio is reversed, so that 10 percent of the alleles are S and 90 percent are s, the proportions of the genotypes SS, Ss, and ss in the F_1 generation would be 1, 18, and 81 percent, respectively. Thus 81 percent of the individuals in the population would have the recessive ss genotype and would have curly hair. In this case as well, the 10:90 percent ratio of alleles would be maintained in subsequent generations. We can now see why England could have so many blue-eyed people, even though the trait is recessive.

Hardy and Weinberg showed that the stable gene frequencies in a population could be expressed by an algebraic equation:

$$p^2 + 2pq + q^2 = 1$$

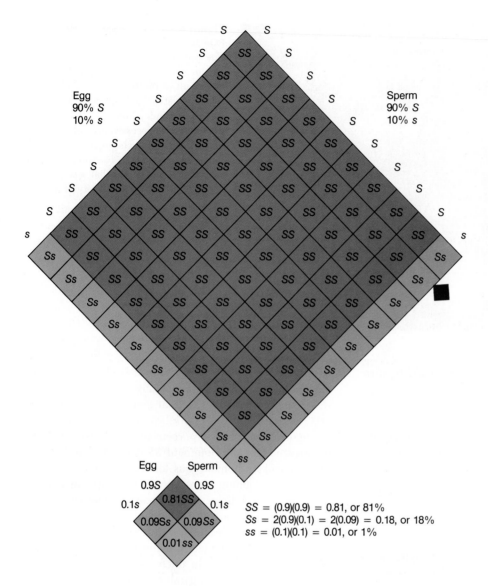

Figure 41-10   THE HARDY-WEINBERG LAW IN TERMS OF THE PUNNETT SQUARE. The Punnett square can be used to indicate how many of each genotype to expect in a population. In this example, 90 percent of the alleles in eggs are dominant *S* and 10 percent are recessive *S*; the alleles in sperm show the same ratio. We can see from the Punnett square that 81 percent of the offspring through random mating will be *SS*, 18 percent will be *Ss*, and 1 percent will be *Ss*. It is not necessary, however, to draw out the 100 crosses of this Punnett square to obtain the information. Follow the multiplication steps in the small Punnett square in the lower left to arrive at the same percentages.

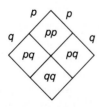

$SS = (0.9)(0.9) = 0.81$, or 81%
$Ss = 2(0.9)(0.1) = 2(0.09) = 0.18$, or 18%
$ss = (0.1)(0.1) = 0.01$, or 1%

The letters p and q stand for the two alleles of a single gene, and a Punnett square explains the origin of the equation:

To see how this equation works, we can consider the frequencies of alleles at the MN gene locus we calculated earlier. Using p to represent the L^M allele frequency (0.539) and q to represent the L^N allele frequency (0.461), we obtain:

$$(0.539)^2 + 2(0.539)(0.461) + (0.461)^2 = 1$$
$$0.29 + 0.50 + 0.21 = 1$$

Notice that the frequencies calculated according to the Hardy-Weinberg equation correspond rather closely to the frequencies of MN blood-group genotypes (0.292, 0.496, and 0.213) shown in Table 41-1. Assuming that American Caucasians were to mate at random with respect to the MN gene locus, we could predict that unless some other factor intervened, these genotype frequencies would be passed on unchanged, generation after generation.

In population genetics, then, equilibrium means that under certain ideal conditions, the relative frequencies of alleles, and therefore of genotypes, stay the same over time. Therefore, we can state the consequences of the Hardy-Weinberg law in three straightforward principles:

1. If individuals in a sexually reproducing population mate at random, genotypic frequencies will be the product of the frequencies of their alleles.

2. If males and females in a population have the same frequencies of alleles, only one generation of random mating will be necessary to achieve equilibrium in genotypic frequencies.

3. In the absence of external factors that affect equilibrium, gene frequencies will remain constant from generation to generation.

MECHANISMS OF EVOLUTION: UPSETTING THE GENE-POOL EQUILIBRIUM

Somehow, unchanging gene frequencies and resultant phenotypes just do not seem to correspond to the changing world of life we see around us. In fact, in real populations of plants and animals, the conditions that Hardy and Weinberg set for maintaining genotypic equilibrium are probably never met. At the very least, gene mutations occur all the time, and many other factors, including migration (which may result in new alleles being brought into a population) and natural selection, also influence the genetic make-up of populations. Thus the Hardy-Weinberg law is actually what logicians call a *null hypothesis*: it is a scenario in which the necessary conditions are never achieved. For biologists, however, the Hardy-Weinberg law has proved to be an invaluable tool, not only as a rigorous standard against which to measure evolutionary changes in populations, but also as a means of demonstrating that evolution is a fundamental and unavoidable characteristic of the living world.

If the null hypothesis is constantly contradicted, implying that evolution occurs, then the factors causing the contradiction are automatically the "causes" of evolution. Included among these factors are nonrandom mating, genetic drift, mutation, migration, and natural selection. We will consider in this chapter and Chapter 42 how each of these factors upsets the Hardy-Weinberg equilibrium.

Nonrandom Mating

One of the most important requirements for gene-pool equilibrium is *random mating*, in which the likelihood that any two individuals in a population will mate is independent of their genetic make-up. For instance, a tabby cat is just as likely to mate with a calico cat as with another tabby cat, and pollen from any one spruce tree is equally likely to blow to any other spruce in a forest, irrespective of their genes. **Nonrandom mating,** on the

other hand, refers to situations where the probability of two organisms mating depends on their phenotypes (and indirectly on their genotypes) (Figure 41-11). Nonrandom mating occurs all the time in nature and can generally be divided into two categories: inbreeding and assortative mating.

Inbreeding refers to mating between relatives at a greater (positive inbreeding) or lesser (negative inbreeding) frequency than would be expected by chance. The most extreme form of inbreeding is self-fertilization, a reproductive strategy found in some species of plants, including Mendel's peas, and in a few animals. Self-fertilizing plants often have flowers that never really open up and in which the pollen-producing anthers grow to physically contact the stigmas with their attached ovaries. Another form of inbreeding is parthenogenetic development, a process in which haploids reproduce haploids, or, as in some turkeys and lizards, normal haploid eggs initiate development but yield diploid females only.

Less extreme forms of inbreeding are found in a variety of other organisms, including humans. Although most human societies have incest taboos that reduce mating between siblings or between parents and siblings, among the Tuaregs, a Berber tribe that inhabits the Sahara, marriage is almost always between first cousins. Such nonrandom unions lead to deviations from the predicted Hardy-Weinberg gene frequencies. Usually,

Figure 41-11   NONRANDOM MATING: A PREFERENCE FOR GENOTYPE.
White and blue lesser snow geese were once believed to be different species. They are now known to be members of a single species (*Anser caerulescens hyperborea*): white is recessive, and both heterozygotes and homozygotes for blue coloration have silvery gray and grayish-brown feathers. The geese mate preferentially with birds of their own color (assortative mating), so there are more homozygotes and fewer heterozygotes than would be expected by the Hardy-Weinberg equation.

animals that inbreed live in small and generally isolated populations.

The major genetic results of positive inbreeding are an increase in the frequency of homozygotes and a decrease in the frequency of heterozygotes within the population. (This is because, in each generation, as heterozygotes mate, they yield a 1:2:1 ratio, the 1's being homozygotes, which because of inbreeding tend to mate with members of the original homozygote population. Thus the heterozygote population, the 2 in 1:2:1, gets smaller with each generation.)

Self-fertilization and significant inbreeding can have either positive or negative consequences for a species. Increased homozygosity means that the diversity of genotypes decreases, but that is not necessarily a problem in an environment that is constant over very long time spans—say, on the deep-sea floor. But in the more common case of environments that vary physically or biologically, inbreeding is generally disadvantageous. For instance, inbred wheat, which is highly homozygous, cannot cope with sustained low soil temperatures. Another result of inbreeding is that deleterious traits may show up because recessive alleles are no longer masked; for instance, inbreeding in wild South African cheetahs due to the population bottleneck has resulted in 10 to 100 times less genetic variability at 200 protein loci than in other mammals. This correlates with poor reproductive rates, high juvenile mortality, up to 71 percent abnormal sperm, and great vulnerability to diseases. These cheetahs face extinction unless more genetic variation can be introduced.

Assortative mating is a special case of nonrandom mating and refers to mating between unrelated individuals of the same species at a greater or lesser frequency than would be expected by chance. For instance, it is well known that tall women marry tall men more often than they marry short men. And the blue and white lesser snow geese shown in Figure 41-11 prefer to choose mates with their same color. Similarly, some species of singing birds, such as white-crowned sparrows, prefer to mate with individuals that sing a local dialect of the species' song. Such preferences clearly contradict the Hardy-Weinberg principle that mating must be strictly random for the genotype of populations to remain stable.

Genetic Drift

One of the major conditions that Hardy and Weinberg set for their law was that it applies only to large populations—at least several thousand individuals. Why is size so important? We have already hinted at the reason: maintaining equilibrium in a gene pool depends on the laws of chance. These laws apply to any situation—whether the flip of a coin or the union of gametes in sexual reproduction—where an event can produce more than one outcome. Imagine, for example, that a child is flipping a penny. The first flip is heads, the second comes out tails, and then there are two tails, one head, and three tails. Out of the first eight flips, six (or 75 percent) have turned up tails. But if the coin is flipped many more times—1,000 or 10,000—the results would come closer and closer to fifty-fifty. Thus chance deviations from random expectations (fifty-fifty) can easily occur, especially if the number of trials (coin flips) is small. When there are only a few cases, a phenomenon known as sampling error may occur; the smaller the sample, the greater the chance that actual frequencies will deviate from expectations.

Sampling error applies to gene frequencies in populations. Suppose that the gene locus for a certain trait in a large population has two alleles and that the frequency of each allele is 0.50 (each allele is present 50 percent of the time). If no external factors intervene, the gene frequencies in the next generation should also be 0.50, according to the Hardy-Weinberg law. But as we saw with our young coin flipper, pure chance may produce deviations from this expected frequency. Such a chance change in gene frequency from generation to generation is called **genetic drift.**

The effect of genetic drift is usually very slight, and in very large populations it may be almost unnoticeable because sampling errors tend to be canceled out when the sampling is large. In a small population, however, genetic drift can result in substantial changes in gene frequencies and even complete loss of an allele from the population in a relatively short period of time. If, for example, an allele—say, z—appears in 10 percent of a population of 10,000 individuals, there would be 1,000 copies of the allele in the total gene pool (scattered among 1,000 individuals). Even if some natural disaster (equivalent to genetic drift) were to wipe out 100 individuals with the allele, there would still be 900 z alleles in the population's gene pool. But if the total population were only 10 individuals, just a single carrier of the z allele would be present in the population. If by chance this individual failed to reproduce, the effects of the genetic drift on the whole gene pool would be dramatic: the allele would be lost from the population. Figure 41-12 traces the way an allele may be lost from a population due to random genetic drift. These examples also show that genetic drift tends to increase the diversity between two isolated populations.

Some of the most dramatic random changes in gene frequencies take place when a few individuals leave a large population and take up residence away from other members of their own species. If these few "pioneers" succeed in starting a new population, its gene pool will reflect only the genotypes that happen to be present in the founders. As a result, the new gene pool may differ

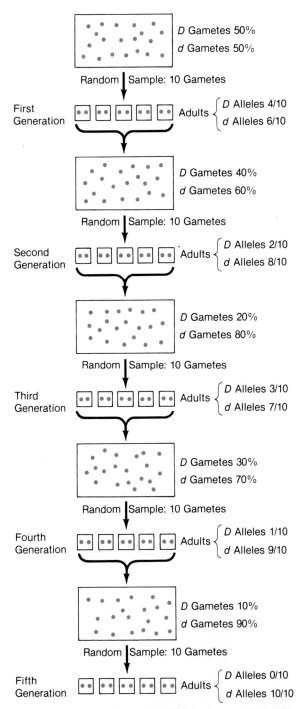

Figure 41-12   LOSS OF AN ALLELE IN A POPULATION DUE TO GENETIC DRIFT.

In this example we start with an equal number of gametes with alleles *D* and *d*, and after five generations all of the *D* alleles are eliminated as a result of genetic drift. For each generation, the gamete pool from the five parental adults is shown; then random selection of gametes from that pool yields the next generation. By this random process, the dominant allele happens to be eliminated, although the recessive allele could as easily have been lost.

substantially from that of the original population; for example, there may be, by chance, a higher frequency of a rare allele or a lower frequency of certain alleles that are common in the parental gene pool. This kind of change in gene frequencies is called the **founder effect** because the characteristics of the new gene pool are determined by the genetic make-up of the founder individuals.

Geologically isolated places such as islands provide an ideal setting for biologists to study the founder effect. For example, the founder effect is probably at work on the small island of Mljet (Meleda) off the coast of Yugoslavia. Colonized centuries ago by a small group of people from the mainland, the island's population remained small (a few thousand people) and, until recently, isolated. The recessive skin disease known as *mal de Meleda* is quite common there, although it is almost entirely unknown elsewhere in the world. One of the early founders must have carried the rare recessive allele for the disease, and the frequency of the allele has increased dramatically in the small population.

A well-documented example of the founder effect can be found among the Amish, a religious sect living in isolated communities in Pennsylvania, Ohio, and Indiana. One group of Amish has a high frequency of an otherwise rare genetic disease, the Ellis–van Creveld syndrome, a type of dwarfism in which the limbs are shortened, extra fingers or toes sometimes develop, and the victim usually dies within a few months of birth (Figure 41-13). The Ellis–van Creveld syndrome is a simple Mendelian recessive, meaning that one out of four children born to two parents who are heterozygous for the trait are likely to be homozygous. In 1964, there were 43 cases of Ellis–van Creveld syndrome among the 8,000 Amish of Lancaster County, Pennsylvania, but none among the Amish of Indiana or Ohio. In fact, more cases have been reported in Lancaster County than in the rest of the world combined! The ancestry of all the people with this genetic syndrome can be traced back to a single couple, Mr. and Mrs. Samuel King, who immigrated to Pennsylvania in 1744. Undoubtedly, either Mr. or Mrs. King was heterozygous for this allele. Since purely by chance at least 1 of the other 200 founders of this Amish population also carried the recessive allele, it became common in the isolated group.

In this example, random genetic drift was also at work. If members of the founding population had mated at random, the incidence of the disease would be expected to be about 1 in 400, yet it is 1 in 14. But it just so happened that in the Pennsylvania Amish community, the Kings and their descendants had larger families than did the other founders (as can be documented by carefully recorded genealogical information). Thus the frequency of the Ellis–van Creveld allele drifted higher as time passed.

A phenomenon related to the founder effect is the bot-

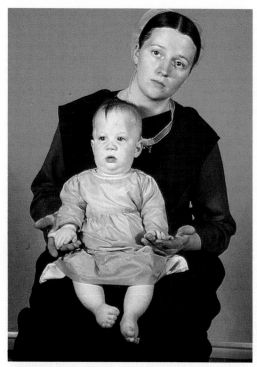

Figure 41-13   AN AMISH CHILD WITH ELLIS–VAN CREVELD SYNDROME.
The child has shortened limbs and six fingers on each hand. All the Amish with this syndrome are descendants of a single couple that helped found the Amish community in Lancaster County, Pennsylvania, in 1744. Because of interbreeding in the isolated community, the recessive trait is now common.

tleneck effect. Among certain flies in New Zealand, for example, the population is very large each summer, but only a small subset of flies survives the harsh cold and dampness of winter. Thus each spring, only a small proportion of the original population (and its gene pool) gives rise to the new generation. Because of this and other population "crashes" (to be discussed in more detail in Chapter 46), a **bottleneck effect** occurs when only a small portion of the original population serves as the sole source of a new population. This apparently occurred in the southern African population of cheetahs (*Acinonyx jubatus jubatus*), with the result that the animals are now so uniform genetically that skin grafts between unrelated animals are not rejected immunologically. In such cases, it is likely that the original and secondary gene pools differ in many gene frequencies, and that there is less genetic variability after the bottleneck. However, Edwin Bryant of the University of Houston ran an experiment in which his bottleneck was 1, 4, or 16 breeding pairs of houseflies out of thousands in the original population. The unexpected result was greatly increased variability in the postbottleneck population. Those houseflies had more variable wing sizes

and shapes, head proportions, and limb dimensions as a result. No accepted explanation of this counterintuitive finding has appeared, but it is clear that population geneticists cannot predict with confidence the consequences of bottlenecks.

Mutation

Mutations—heritable changes in the chemical structure of genes—upset the Hardy-Weinberg equilibrium because they alter alleles. In fact, mutation properly includes certain chromosomal rearrangements of genes and DNA, duplications of genes, followed by divergence of extra copies, moving genes under control of new regulatory elements, and so on. For simplicity, we first will speak of simple point mutations—changes in DNA base sequence, and therefore in the amino-acid sequence of protein. For an individual, mutation in somatic cells (for instance, caused by the ultraviolet of sunlight) may have good or bad consequences. In animals, it will not be passed on to the next generation of the population, since such changes are not heritable. However, when mutations occur in animal germ-line cells so that sperm or eggs carry the altered allele, the population may be affected. In plants, no germ line is present; hence, if a somatic mutation occurs in a cell of the growth zone of a branch, and that mutant later gives rise to a cell lineage that forms a flower, the mutation may be passed on (Figure 41-14). Pink grapefruit arose in just this way on a normal white-grapefruit tree.

Figure 41-14   SOMATIC MUTATIONS.
Since no germ line is present in plants, a somatic mutation, such as one red petal on this chrysanthemum, can be passed on from one generation to the next if cells with the mutation chance to be included in tissue giving rise to pollen or eggs.

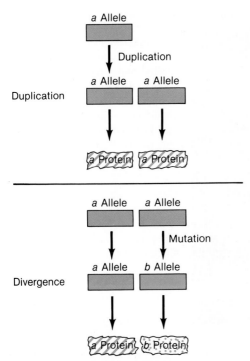

Figure 41-15   DUPLICATION AND DIVERGENCE: GENERATING NEW ALLELES.

The *a* allele is duplicated; each copy produces mRNA that codes for the *a* protein. Since two copies of the *a* allele are present, either one is "free" to evolve without the organism losing the *a* protein. Hence, mutation can generate the *b* allele, which codes for *b* protein. This divergence process can generate families of related genes.

Biologists have discovered an unexpected world of mutations using marvelous new technology for isolating individual genes and determining their nucleotide base sequences. They are seeing that genes may undergo duplication and divergence (Figure 41-15) so that an extra copy of a gene arises by a mutational process. This duplicated copy is then free, in a sense, to evolve: while the original diploid set is still present to produce the phenotype and is still influenced by natural selection, the duplicated allele may have an altered base sequence and thus may yield new proteins with novel properties.

Recall from Chapter 38 that the original gene coding for red-sensitive visual pigment in humans gave rise to the green-sensitive gene and protein, and that adult men can have one, two, or multiple copies of "green genes." Recent research is revealing that in most organisms, many different proteins share amino-acid sequences. This implies that they share pieces of genes and that at one time, the genes may have duplicated or diverged much as the visual pigment genes did in the human eye. The new genetic techniques allow biologists to do the

research equivalent of wiping the dirt from a windowpane and peering into a room never before seen, but filled with new solutions to how hereditary material mutates, duplicates, moves about on chromosomes, yielding the gene-pool variability that is such fertile fodder for natural selection and evolution.

Every time an allele mutates to a different allele, the gene frequencies are changed—that is, evolution occurs. The main significance of mutation for evolution is to provide the raw material on which natural selection or other processes can act to change gene frequencies in a larger way. Mutation creates the new alleles that are a major source of genetic variation.

As it happens, mutations are rare events. In each generation, each gene has on average a 10^4 to 10^9 likelihood of mutating. Since humans have about 10^6 structural genes, each of us probably harbors an average of one new mutation. But most of these mutations are recessive and are not expressed. Still, given enough individuals and enough time, even low natural mutation rates yield ample genetic variability in populations. And despite the extreme rarity of mutation, sooner or later every allele will mutate; hence, even though a genotype may originally be entirely homozygous dominant—*AA*, *BB*, *CC*, and so on—ultimately, *a*, *b*, and *c* will appear, thus increasing genetic diversity. Once again we see that one of the factors that should be unchanging for the Hardy-Weinberg equilibrium to operate is very changeable indeed.

Migration and Gene Flow

Migration into and out of populations occurs all the time. Pollen may be blown for incredibly long distances, jellyfish are carried by ocean currents to new sites before they release their gametes, a new bull elk invades a territory—these are but a few examples of the ways in which new genetic materials can be introduced into populations. When individuals (or other gametes) migrate from one population to another and interbreed with the existing population, **gene flow** takes place (Figure 41-16).

The effect of immigrants on gene frequencies in a population depends on how genetically different the donor population is from the recipient population. As a general rule, gene flow tends to reduce the genetic differences between populations. This effect is just the opposite of genetic drift, which tends to increase diversity between populations (as on an island in comparison with the mainland).

As an example of gene flow, consider the results of the forced immigration of African blacks to the United States. About 300 years ago, the first blacks who came to

Figure 41-16   GENE FLOW AS IT AFFECTS GENETIC DIVERSITY BETWEEN POPULATIONS.
Gene flow between races in the United States has resulted in an increased sharing of alleles. The result is often seen as a combination of phenotypes—as in skin and hair color and in blood types. The individuals in this family illustrate some of the visible traits affected by gene flow.

America from Africa almost certainly had the *R* allele for the rhesus blood type at a frequency of 0.630—the frequency for this allele found among today's blacks living in Africa. The measured frequency among present-day black Americans is 0.446. How did this change in frequency come about? The most likely answer is that gene flow has taken place in the country between blacks and whites, who have a frequency of only 0.028 for the *R* allele. Population geneticists have calculated that the rate of gene flow between American whites and blacks is approximately 3.6 percent per generation over the past ten generations (calculated at 300 years). This rate is significant; given the huge time span and huge number of generations over which plants, animals, and other life forms have existed, gene flow can be seen to have a considerable effect.

LOOKING AHEAD

In this chapter, we have seen that the conditions for the maintenance of a Hardy-Weinberg equilibrium are contradicted again and again. Thus in nature, Hardy-Weinberg stability does not operate, gene frequencies do change, and populations evolve. Evolution is an unavoidable, basic feature of life on Earth. Of the factors that account for changes in gene frequency, we have studied nonrandom mating, genetic drift, mutation, and migration. As we shall discover in Chapter 42, natural selection—first described more than a century ago by Darwin and Wallace—plays a central role, along with these four processes, in shaping the biological characteristics and evolutionary history of every species on Earth.

SUMMARY

1. Many early philosophers and scientists did not believe that organisms change over time. Not until the early nineteenth century did Jean Baptiste de Lamarck present the theory of evolution through the *inheritance of acquired characteristics.*

2. Fifty years after Lamarck's theory, Charles Darwin and Alfred Russel Wallace independently proposed the theory of evolution by means of *natural selection.*

3. An *adaptation* is a genetically based characteristic that better suits an organism to some aspect of the physical or biological environment, and allows the organism a better chance of surviving to reproduce.

4. The more genetic variation in a population, the greater the opportunity for adaptive evolutionary change. Natural selection operates on phenotypic variations that result from underlying genetic variation. Most populations show a high level of phenotypic differences.

5. Protein *electrophoresis* permits biologists to estimate the amount of variation in structural genes of natural populations. That variation is expressed as *average heterozygosity* or *percent polymorphism.*

6. *Population genetics* views genes as the units of evolution. It combines the mechanisms of heredity with natural selection and other evolutionary processes at work in populations in order to describe and explain evolution.

7. A *gene pool* is all the alleles carried by all the individuals in a *population*, or group of interbreeding or potentially interbreeding organisms. Changes in the gene pool over time, reflected in corresponding changes in the phenotype, constitute *evolution*.

8. The genetic make-up of a population can be characterized by frequencies of different alleles or different genotypes. Simple mathematics can be used to calculate the relative frequencies of certain alleles and genotypes in a population.

9. The *Hardy-Weinberg law* describes what happens to the frequencies of alleles and genotypes in a hypothetical ideal population of sexually reproducing organisms. For the gene pool to remain at Hardy Weinberg equilibrium, the following conditions must be met: (a) mating must be random; (b) the population must be large; (c) there must be no

mutations; (d) the population must be isolated; and (e) there must be no natural-selection pressure.

10. Hardy and Weinberg showed that the stable gene frequencies in a population could be expressed in an algebraic equation:

$$p^2 + 2pq + q^2 = 1$$

11. The Hardy-Weinberg law proves to be a null hypothesis; that is, the conditions for equilibrium are rarely or never met. The frequent violations of the conditions are accompanied by measurable changes in gene frequencies, and give no evidence of evolution. The factors that violate equilibrium conditions include nonrandom mating, genetic drift, mutation, migration, and natural selection.

12. *Nonrandom mating* occurs when the probability of mating depends on the phenotypes of the participating individuals. This process can change the frequency of genotypes in a population. Nonrandom mating includes *inbreeding*, mating between relatives at a greater than chance frequency, and *assortative mating*, mating between individuals with certain characteristics (as

height) at a greater than chance frequency.

13. Changes in gene frequency over generations that result from sampling error are called random *genetic drift*. In small populations, genetic drift can result in substantial changes in gene frequency. Extreme cases of genetic drift occur as the result of the *founder effect* and the *bottleneck effect*. Genetic variability either may be higher or lower after such events as a bottleneck.

14. Mutation is a major source of genetic variation and serves as the raw material for evolution. Mutation includes point mutation changes in DNA, as well as processes such as chromosomal rearrangement. Gene duplication and divergence, another category of mutation, can lead to families of related proteins.

15. *Migration* of individuals from one population to another produces *gene flow* between the populations, which tends to reduce the genetic diversity between the populations. Conversely, isolation so that migration is impossible can allow more diversity to accumulate.

KEY TERMS

adaptation	evolution	Hardy-Weinberg law	natural selection
assortative mating	founder effect	inbreeding	nonrandom mating
average heterozygosity	gene flow	inheritance of acquired characteristics	percent polymorphism
bottleneck effect	gene pool		population
electrophoresis	genetic drift	migration	population genetics

QUESTIONS

1. What does adaptation mean? Give some examples.

2. Define the following terms:

   a. gene pool
   b. population

   c. allele frequency
   d. heterozygote
   e. genetic equilibrium

3. Define and give examples of:

   a. genetic drift

   b. founder effect
   c. bottleneck effect

4. How do biologists usually characterize a population's genetic make-up? How do they describe the

amount of genetic variation in a population?

5. Suppose that the S allele in a population of organisms is present at an 80 percent frequency, and the s allele at a 20 percent frequency. After individuals in one generation mate randomly, what will be the frequencies of the genotypes SS, Ss, and ss? If the Hardy-Weinberg conditions apply, what will the genotypic frequencies be after five generations?

6. What conditions must operate to maintain the Hardy-Weinberg equilibrium? Do these conditions occur in nature? That being the case, what can we conclude?

7. If for several generations the members of the population described in question 5 mate only with others of the same genotype rather than randomly, will the genotypic frequencies change? If so, which genotype(s) will become more frequent, and why? Will the frequencies of the S and s alleles change? If so, which allele will increase in frequency, and why?

8. Give some examples of nonrandom mating in nature and among humans.

9. Among the populations that Darwin studied on the Galápagos Islands, what phenomena might have affected gene frequencies when those islands were first colonized? When a species migrated to a neighboring island?

10. Calculate the average heterozygosity in a population of 1,000 sea urchins in which you analyze, using electrophoresis, 30 different proteins from each animal. At 5 of the 30 loci, 300 of the urchins are heterozygous. At 20 loci, 200 are heterozygous. At 4 loci, 400 are heterozygous. And at 1 locus, 100 are homozygous.

ESSAY QUESTION

1. The virtually constant violation of the Hardy-Weinberg law implies that evolution goes on all the time. What conditions would need to prevail to avoid evolution? What is the likelihood that such a situation could exist in natural populations on Earth?

SUGGESTED READINGS

FORD, E. B. *Ecological Genetics*. New York: Wiley, 1965.

This book, written by a leading population geneticist, is a good source from which to learn the basics of this field.

GRANT, V. *The Origins of Adaptation*. New York: Columbia University Press, 1963.

Although dated, this book is a fine discussion of plant and animal adaptations.

KOEHN, R. K., and T. J. HILBISH. "The Adaptive Importance of Genetic Variation." *American Scientist* 75 (1987): 134–141.

An excellent example of electrophoretic analysis that relates genetic variation directly to survival.

LACK, D. *Darwin's Finches*. Cambridge, England: Cambridge University Press, 1947.

This is a famous treatment of the birds that Darwin saw and collected.

STRICKBERGER, M. W. *Genetics*. 3rd ed. New York: Macmillan, 1985.

Many geneticists think that this book is the clearest treatment of population genetics from the student's point of view.

42
NATURAL SELECTION

The essence of Darwin's theories is his contention that
natural selection is the creative force of evolution—not just
the executioner of the unfit.

Stephen Jay Gould,
Ever Since Darwin (1977)

Three hundred fifty million years with remarkably little change: an African lungfish (*Protopterus dolloi*) awaits the drying of its lake prior to burrowing into the mud.

Two dramatically different examples of natural selection are demonstrated by animals living in the Pacific: the strange lungfish of Australia and the diverse fruit flies of Hawaii.

The lungfish is a true "living fossil"—an air-breathing fish that has not changed substantially for 3 million centuries. On a typical January day, midsummer in Australia, you might find one of these odd creatures at the bottom of a pond that has dried up in the scorching heat. Although the caked mud looks thoroughly devoid of life, a lungfish may lie just a few centimeters below the fractured surface in a cocoon of damp earth and mucus. There it rests, breathing very slowly through a tiny air hole and sustained by a drastically slowed metabolism that will quicken only when rains return in autumn and the fish can free itself from its protective containment.

Although the seasonal shifts in the lungfish's environment are pronounced, the yearly pattern has remained essentially the same since the Carboniferous period, and, as a result, natural selection is believed to have had a stabilizing influence on the Australian lungfish. Its anatomical and behavioral adaptations have served the animal so well that an ancestral form encased in the same pond bottom 300 million years ago would resemble its modern successor in most details.

The nearly 500 species of *Drosophila* fruit flies on the Hawaiian Islands represent a marked contrast to the evolutionary stability of the lungfish. Biologists believe that these insects evolved at an astonishing speed that reflects the relatively rapid change in their environment. The islands arose one after the other as the Earth's immense Pacific crustal plate moved across a hot spot in the upwelling lava; the oldest island, Kauai, formed 5 million years ago; the newest, Hawaii, just 700,000 years ago. All 500 *Drosophila* species are considered to be descendants of some intrepid invaders that settled on Kauai after the lava cooled and vegetation took hold. As the newer islands emerged with their slightly differing climates, and as unique subsets of fruit-bearing angiosperms grew and evolved, a radiation occurred among the flies, with each new species adapted for a slightly different way of life, say, in rain forests or on high mountain terrains. Biologists have compared the spectrum of proteins in these Hawaiian flies and have found a remarkable degree of variability—apparently the result of selective pressures that were anything but stable.

Natural selection acts on the organisms' phenotypes in these two examples, as well as on all organisms' phenotypes, at all stages of the life cycle. In so doing, natural selection acts indirectly on genotypes. In Chapter 41, we saw that gene pools change over time due to any combination of five basic processes: nonrandom mating, genetic drift, mutation, migration, and natural selection. Of these, natural selection best accounts for the broad range of adaptations that enable populations of organ-isms such as lungfish and fruit flies to survive in given environments. As Darwin recognized a century and a half ago, natural selection is a key mechanism in the evolution of life.

This chapter focuses on the ways that different selective processes affect the phenotypes of plants, animals, and microbes over time to alter gene frequencies or maintain the genetic status quo. We shall consider three major topics: (1) the concept of *fitness*—the degree to which an organism's characteristics increase its probability of surviving to reproduce; (2) how phenotype affects genotype—that is, how natural selection acting on expressed genetic characteristics affects gene frequencies; and (3) the various types of natural selection. Along the way, we will see how the techniques of molecular biology are used to probe the processes of natural selection and to expand our understanding of evolution.

THE DARWINIAN AND GENETIC MEANINGS OF FITNESS

Charles Darwin's basic argument for evolution by natural selection can be summarized as follows:

1. Some organisms survive to reproduce, perhaps more than once; others do not.

2. Variations exist among individuals and affect the chances that a given individual will survive and reproduce.

3. Offspring resemble their parents and thus tend to inherit those traits that gave their parents a reproductive advantage.

4. Therefore, traits that increase reproductive success tend to increase in frequency from generation to generation.

Darwin made this straightforward case for natural selection before Gregor Mendel published his theory of inheritance and before the concept of the gene became established. In light of what we know about the mechanisms of heredity, we can restate the argument as follows:

1. Variations among organisms affect survival and reproductive success.

2. These variations are heritable (that is, they are genetically determined, at least in part).

3. Genes contributing to new variations and reproductive success tend to increase in frequency over time.

This way of describing evolution emphasizes reproductive success: without it, the fate of individuals and so of their population or species is extinction. Geneticists who study evolving populations use the term **fitness** to define the relative reproductive efficiency of various individuals or genotypes in a population. In essence, the fitness of an individual (or an allele or a genotype) depends on the likelihood that one individual, relative to other individuals in the population, will contribute its genetic information to the next generation. Fitness, then, includes the relative ability of an organism to survive, to mate successfully, and to reproduce—with viable eggs or sperm, zygote, and embryo resulting in a new organism. The familiar phrase "survival of the fittest" relates to this concept, but it has an erroneous connotation. The largest, strongest, and most aggressive organisms do not necessarily have the highest fitness. A much more subtle combination of structure, physiology, biochemistry, and behavior contributes to an individual's fitness and therefore the probability that it will leave offspring.

Let us consider an example of fitness. Suppose that a dominant allele for pure-white coat color gives certain arctic hares a slight advantage over other hares in which an alternative allele—one that confers light-tan coat color—is expressed. Sitting in snow, a pure-white hare is less visible to a fox or an owl and thus white hares survive to reproduce in higher frequency (Figure 42-1). Their pure-white allele thus has a "higher fitness" than does the tan allele (and for this trait, coat color, so do the pure-white hares). Alternatively, in this environment, a dark-tan allele produced by mutation would have a lower fitness than either the white or the light-tan allele. And over time, the selection pressures against such a mutant allele might result in the allele's being eliminated from the population. But the dark-tan allele might confer a higher fitness than the white one on hares living farther south in coniferous forests.

An allele coding for a particular enzyme form in the marine mussel *Mytilus edulis* beautifully illustrates how sensitive an allele can be to natural selection (Figure 42-2). The enzyme in question is an aminopeptidase that cleaves amino acids from peptides, and *Mytilus* species have multiple alleles and corresponding forms of this enzyme. One particular allele, hap^{94}, generates free amino acids so efficiently that all cells in the body can use those amino acids to raise their internal osmotic pressures to levels that match sea water (amino acids are compatible osmolytes; Chapter 35).

Each year, ocean currents sweep little juvenile mussels from the open ocean into Long Island Sound between New York State and Connecticut. There the juveniles settle on rocks and grow to maturity as sessile organisms. An intense selective pressure operates on the little animals, however, as evidenced by the fact that

Figure 42-1   FITNESS: A WHITE ARCTIC HARE IN THE SNOW.

Arctic hares (*Lepus timidus arcticus*) with pure-white fur are less visible against the snow than are hares with tan fur. For this reason, they are more likely to avoid predator attacks and survive to reproduce in higher frequency. Of course, this sort of adaptation is an indirect consequence of predators that use vision for finding prey; if foxes and birds hunted hares by sense of smell alone, the selection of fur color might not occur.

within just 30 kilometers of the ocean source, the frequency of the hap^{94} allele in the young mussels falls from 0.55 to only 0.12. The farther they live from the pure sea water, the more stringently hap^{94} is selected against. The survivors without hap^{94} tend to have alternate alleles that code for variants of the aminopeptidase enzymes that release fewer amino acids. This means that their body fluids and cells have lower osmotic pressures, and are more in balance with the surrounding, more dilute brackish water. Clearly, the hap^{94} allele has high fitness in the sea, but low fitness in more dilute waters.

By having a set of alleles and resultant proteins with different phenotypes and different biological capabilities, the species can survive under a wide range of environmental conditions. Thus multiple alleles are preserved in the full *Mytilus edulis* population, even though

Figure 42-2   MUSSELS AND ALLELE FITNESS.
Mussels (here *Mytilus californianus*) can live in various
environments, such as on this exposed reef in the sea.
Others live in brackish estuaries. Although the individuals
look the same, organisms exposed to such different
environments may have important differences in their
physiology and biochemistry. Alleles such as hap[94] (which
codes for an enzyme that helps balance the mussel cells'
osmotic pressure with that of surrounding water) vary
accordingly in frequency, at least in mussels studied in
Long Island Sound.

the frequency of any given allele will vary widely from
site to site. In other organisms, however, alleles may be
selected against and lost from a species permanently.

In cases where one allele of a pair (say, *A* and *a*) is
eliminated completely from a population, the remaining
allele is said to be **fixed**—that is, it is the only allele for a
given trait in a population because other alleles have
been lost. For instance, in the absence of a tan-coat-color
allele, the white allele would be fixed in a population of
arctic hares. Fixed alleles automatically have a high level
of fitness because they are guaranteed representation in
future generations. But mutations are inevitable; sooner
or later, other alleles will appear, and some may have
higher and others lower fitnesses than the formerly fixed
allele. Hence, the "fixed" state of an allele is not permanent.

At the opposite end of the scale from fixed alleles are
alleles for sterility or for lethal traits; such alleles are
considered to be unfit because the individuals carrying
them would have no chance to reproduce. Between the
two extremes are situations in which a number of alleles
may exist for a given gene locus, as with the multiple
eye-color alleles in *Drosophila* and the numerous hemoglobin alleles in humans. Often, an allele's fitness may
differ from all others, depending on a population's environment. It is also the case, of course, that an individual
offspring inherits a full genotype, the less fit alleles along
with the more fit ones. And the phenotype is generated
by that full set—the whitest coat color is no guarantee by

itself of reproductive success, say, if a digestive enzyme
is defective, nerve circuits are improperly wired, or reproductive behavior is abnormal. Hence, fitness is a concept equally applicable to individual alleles, to the full
genotype, and to individual organisms.

Biologists have recognized some intriguing general
rules involving warm-blooded animals that may reflect
the interaction of geography with fitness of allelic variations in a species. **Bergmann's rule** notes that individuals living in colder parts of their species' range tend to be
larger than individuals living in warmer areas (Figure
42-3); similarly, **Allen's rule** observes that animals inhabiting cold regions tend to have shorter limbs, tails, and
other extremities than do those in warmer regions (Figure 42-4). One plausible explanation for such differences
is the adaptive advantage of thermal efficiency: the ratio
of surface area to volume (Chapter 5) is smaller in a large
animal, so that body heat is retained more effectively;
small ears, short legs, and a stubby tail also help to keep
body surface—and convective heat loss—to a minimum.
While each of these "rules" has many exceptions, such
phenomena underscore the fundamental role that environment can play in determining which of several alleles
is optimum for a population's survival. As we shall see
next, natural selection can be the ultimate arbiter of an
allele's fitness.

NATURAL SELECTION: HOW PHENOTYPE AFFECTS GENOTYPE

Since natural selection tends to eliminate less fit alleles (or genotypes or individuals), selective processes
tend to lead to the increased adaptation of a population
to its environment. Suppose, for instance, that the preferred food of a *Drosophila* population is in the topmost
leaves of trees and on the ground; there is no food in
between. Can natural selection produce fruit flies that
only fly up or down? A simple experiment in which the
researchers acted as the source of selection pressure was
designed to answer this question.

The experiment was set up to select for a genetically
determined behavior known as geotaxis in fruit flies.
Geotaxis is the movement of organisms in response to
the force of gravity—either up (negative geotaxis) or
down (positive geotaxis). (Botanists tend to use the term
"gravitropism" for geotaxis in plants; Chapter 30.) The
experiment's purpose was to see if natural selection
could change existing allele frequencies, and so the phenotype of the populations. Its design was simple. A
funnel-shaped maze like the one shown in Figure 42-5

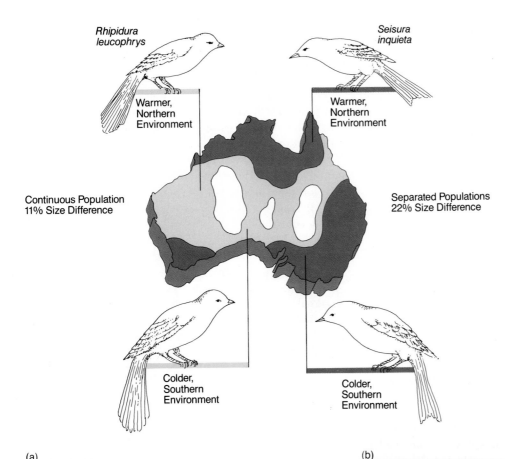

Rhipidura leucophrys

Warmer, Northern Environment

Continuous Population 11% Size Difference

Colder, Southern Environment

Seisura inquieta

Warmer, Northern Environment

Separated Populations 22% Size Difference

Colder, Southern Environment

Figure 42-3   BERGMANN'S RULE: SIZE, TEMPERATURE, AND EVOLUTION.
According to Bergmann's rule, individuals living in a cold climate will have larger bodies than individuals of the same species living in a warm climate. The two Australian bird species shown here illustrate this rule. Note that there is a size difference between northern and southern populations of *Rhipidura leucophrys*, but because the two populations are not separated geographically, there is gene flow between them, reducing the size diversity. In *Seisura inquieta,* the two populations are geographically separated, there is little gene flow, and diversity in size due to climate is more noticeable.

(a)

(b)

Figure 42-4   ALLEN'S RULE: EXTREMITIES, TEMPERATURE, AND EVOLUTION.
According to Allen's rule, animals living in cold climates tend to have shorter extremities (tails, limbs, ears) than animals living in warm climates. This is illustrated by (a) the arctic fox (*Alopex lagopus*) with its short tail, ears, and legs, and (b) the desert fox (*Vulpes chama*) with its longer tail, ears, and legs. These characteristics have obvious adaptive value: body heat is retained more efficiently by short extremities and is dissipated more efficiently by large extremities with their greater surface areas.

was devised, with 105 "decision chambers" connected by one-way valves. Each chamber had two possible exits—one going up and one going down. In order to reach vials of food located at the end of the maze, flies, upon entering, had to make an up or down choice fourteen times. Flies that consistently took the upper passage ended up in the top three or four vials, while those

that always tended toward the lower exits collected in the bottom few vials. The middle vials held flies with no particular tendency for either positive or negative geotaxis.

Starting with a single population presumed to have a variety of alleles for geotactic behavior, the experimenters collected the flies at the top and bottom extremes

Figure 42-5   TESTING POSITIVE AND NEGATIVE GEOTAXIS IN *DROSOPHILA*.
Flies enter the maze at the left and must work their way through a series of chambers in order to reach the food in vials at the right. In each chamber, the fly must make an up or down decision. Twelve consecutive generations of flies that reached the highest vial produced offspring that flew preferentially upward. Likewise, generations of flies that moved downward to reach the lowest vial produced offspring that flew preferentially downward.

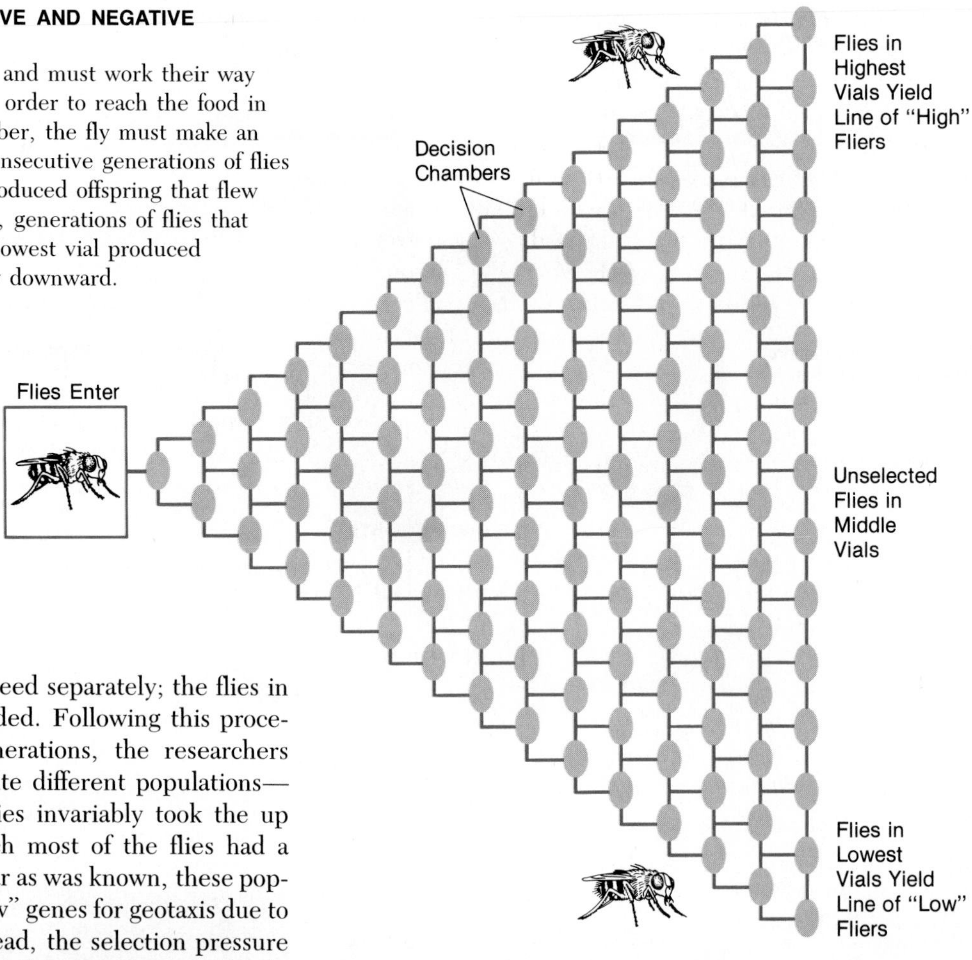

and allowed each group to breed separately; the flies in the middle vials were discarded. Following this procedure for a dozen more generations, the researchers eventually produced two quite different populations—one in which most of the flies invariably took the up exits, and the other in which most of the flies had a strong down preference. As far as was known, these populations did not have any "new" genes for geotaxis due to mutations or gene flow. Instead, the selection pressure (provided by the experimenters) acting on variation already present in the original population resulted in a change in allele frequencies and so in phenotype (behavior).

What would happen if such selection pressure was removed? To answer this question, the experimenters allowed the "high" flies and the "low" flies to mate with each other and reproduce. The original balance of alleles soon was restored: some flies went up, some went down, and the largest proportion took the route in between. This was not the result, however, when Jerry Hirsch at the University of Illinois continued the original selection for high and low fliers for over 600 generations. When selection pressure on those flies was removed, the two populations remained fully stable as high or low fliers. The new, stable phenotype is associated with three major genetic loci and perhaps many more genes in them. The prolonged experimental selection caused (1) an evolutionary change involving stable, inherited differences in behavior (and in other aspects of phenotypes) and (2) changes in groups of genes that underlie those phenotypic characters.

In nature, natural selection influences the genetic composition of a population by working backward or indirectly from the phenotypes produced by those genes. It is the phenotype that survives or perishes, reproduces or leaves no offspring. Thus in the geotaxis experiment, the experimenters chose "high" or "low" phenotypes, not specific genes or alleles. Yet the genetic analyses prove that large-scale genetic changes occurred as a result of prolonged selection of behavioral phenotypes. If evolution is defined as changes in gene frequencies (Chapter 41), then the geotaxis experiments demonstrate directly evolution in response to selective pressure. In nature, where the preferred foods of *Drosophila* are found at many heights, similar selection apparently does not commonly operate. Yet Hirsch's experimental selective pressure results make clear that the capacity for genetic and, hence, evolutionary, change is present. Should the fruit flies' food supply diversify in height in nature, *Drosophila* populations might evolve in response.

TYPES OF SELECTIVE PROCESSES

At this point, we are ready to consider the four major processes that come under the heading of natural selection. Each process is defined by its effect on a population's gene frequencies: some favor the status quo, while others produce change, including the evolution of new species. As we proceed, there are two points to keep in mind. First, natural selection does not always increase the complexity of an organism's structures or behaviors; sometimes, the result is a loss, as with hind limbs during whale evolution and the internal gut during tapeworm evolution. (At the same time, however, evolutionary remodeling has included adaptations that enable the whale to swim in the sea and the tapeworm to absorb nutrients through its body wall.) Second, in these organisms and all others, selective processes operate only on actual phenotypes (and underlying genotypes) available in a population; rather than molding an ideal, "best" organism to fit a given environment, selection "chooses" only from the available options those most likely to survive. For, as we saw in Chapter 1, natural selection is not a generator of new genotype and phenotype, but a mechanism to ruthlessly eliminate the less fit and leave the more fit to compete in an unending quest for reproductive success.

Normalizing Selection: Preserving the Status Quo

Sometimes the effect of natural selection is simple: it preserves what is already present. This **normalizing selection** (also called *stabilizing selection*) is responsible for maintaining the status quo for the lungfish in its cocoon of crusted mud; living in a seasonally shifting but extraordinarily stable long-term environment, lungfish populations have survived the eons not because frequent genetic changes helped them adapt to new environments, but because adaptations which evolved early in their history have remained adequate in their peculiar but stable environment. And, although we have no fossil proof, it seems reasonable to assume that other phenotypes that arose through the normal processes of meiotic recombination, mutation, and random mating were selected against.

Although some biologists claim that it is hard to see selection at work, a famous case of normalizing selection was documented among a population of water snakes (genus *Natix*) living on the islands in Lake Erie. Members of this species normally have plain gray skin, but occasionally some snakes are born with the banded phenotype. These patterned mutants are strongly selected against, however, because they make conspicuous targets for predatory birds. As a result, their chances of surviving to reproductive age and contributing their genes to the next generation are significantly less than are those of their duller compatriots.

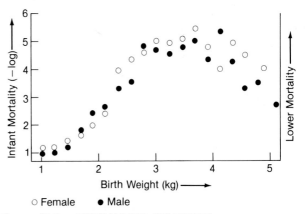

Figure 42-6   NORMALIZING SELECTION.
The graph plots the negative logarithym of human infant mortality as a function of birth weight. Thus the higher the value of the vertical axis, the lower the death rate. Birth weights of about 2.7 to 4 kilograms (6 to 8+ pounds) seem optimal and have the lowest death rates.

Normalizing selection usually operates continuously over many generations. In humans, for example, there is a strong selection pressure for babies to be born at intermediate weights (Figure 42-6). Newborns that are very small (and often premature) have a high death rate. Those that are very large endanger the mother at the time of birth (recall from Chapter 18, the pregnant female sheep that died when their fetuses became too large because they had eaten certain plants that inhibited birth).

An enterprising British scientist, E. B. Ford of Oxford University, showed that normalizing selection does not always follow this pattern. An avid observer of moths and butterflies, Ford kept track of the phenotypes of a population of marsh fritillary butterflies for fifteen years. When he began, the population was small and highly homogeneous in size, body shape, and wing-color patterns. Suddenly, however, a four-year population explosion took place, and during this time, butterflies showed great variation in both color patterns and basic body morphology. Then the population stabilized again, larger than before, with a "new" phenotype that was quite distinct from the original one. Although Ford could not establish the exact cause of the remarkable changes he had recorded, the implication of his observations seemed clear: normalizing selection pressure apparently relaxed for a time—as the four-year burst in numbers and variant phenotypes demonstrated—and then was reestablished. The population went through a period of transient polymorphism and then became less polymorphic as selection once again began to act to hold back variation.

Normalizing selection is especially common in environments that have remained stable through long periods of the Earth's history, such as the open sea and the black, cold, high-pressure regions of the ocean floor. In extremely stable environments like these, it seems likely that the phenotype of organisms whose ancestors have inhabited such places for tens or hundreds of millions of years are already very well adapted for coping with the physical environment. However, if a new predator, competitor, or parasite appears, adaptive changes may well arise in response. Normalizing selection also works in more variable locations—a mountain meadow, a desert sand dune, or other environments that may change over time spans that are short by geological standards but long when measured against the life of an individual.

Directional Selection: Changing Phenotypes

A second category of natural selection, **directional selection,** involves change from one phenotypic property to a new one. The following example illustrates how such directional selection operates. In 1915, oystermen hoisting their barrels in Malpeque Bay in the Gulf of St. Lawrence began to notice that something was awry. Among the plump, healthy mollusks destined to become oysters Rockefeller were others that were frightfully diseased. Small and flabby, these ailing shellfish were afflicted with yellowish pus-filled blisters 0.5 centimeter in diameter. By 1922, the Malpeque disease had all but wiped out the oyster beds; for oysters and oystermen alike, life had become unexpectedly precarious. Then, in 1925, fishermen found that they could again harvest a small catch. By 1940, Malpeque Bay was producing more oysters than ever in its history (Figure 42-7). What is more, the "new" Malpeque oysters could be successfully used to repopulate other areas recently decimated by the disease. Thus in a little more than twenty years, the mollusks had undergone a radical transformation—one that had taken place under the devastating attack of a virulent pathogen. What had happened?

Fortunately for the oysters, some of the 60 million or so offspring produced by each adult in a year carried an allele that conferred resistance to Malpeque disease. The result was directional selection: when environmental conditions favored the survival of individuals carrying the genetic variant, the outcome was an increase in the frequency of that variant in the population.

Note that "directional" does not mean "directed." No force was directing anything; rather, by chance a variant arose or existed, and selective pressure operated to preserve it rather than other alleles. So "directional" refers to a consistent trend over time. In other species, this kind of trend might be toward darker pigmentation, a faster-acting enzyme, larger brain volume, a more extensive root system, or some other trait.

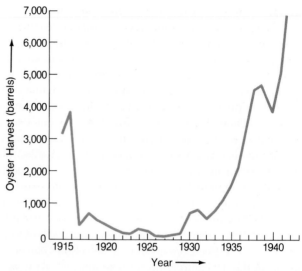

Figure 42-7   DIRECTIONAL SELECTION: OYSTER YIELDS FROM MALPEQUE BAY, 1915–1940.
Malpeque disease reduced the oyster population drastically, but disease-resistant oysters survived and, beginning in 1929, gave rise to the new, large population.

Sometimes directional selection is so dramatic that it can be seen to take place in a population over relatively few generations. The oysters of Malpeque Bay are an example; in their case, the selective pressure was disease, and only the few resistant genetic variants reproduced successfully (bottleneck effect). In fact, such dramatic adaptive change often occurs in response to a sudden and substantial environmental change. One of the classic examples of this kind of directional selection is the case of *Biston betularia*, the peppered moth, and the phenomenon called *industrial melanism* ("peppered" refers to a speckled appearance).

The story of the peppered moth began long before Darwin or his successors puzzled out the workings of natural selection. In eighteenth-century England, butterfly collectors vied for the chance to snare a rare form of this normally light-colored speckled moth—a variant that was black due to melanin pigment. As it happened, trapping a black moth was relatively easy if one had patience. Both forms of the moth spent a good deal of time resting and feeding on the trunks of trees that were often covered with pale, gray-green lichens. For the light moths, the lichens functioned as a background against which their light, peppered wings served as a camouflage, so avian predators and human collectors had difficulty spotting them (Figure 42-8). But for the black moths, their dark wings were no camouflage at all; the moths could easily be noticed against the lighter-colored lichens and captured. As a result of this disadvantage, the frequency of the single dominant allele responsible for the dark form remained low (the selective pressure against the dark color is evident, since both the homozygous dominant and the heterozygous black moths tended to be consumed by predators more than the homozygous recessive light ones).

Then, as the mid-nineteenth-century Industrial Revolution changed Europe, smoke and soot began to darken the vegetation around English factory towns. In heavily polluted regions the dark form of *Biston betularia* began to flourish, while the light form became increasingly rare. Until the 1930s, many collectors and scientists attributed this change to an increase in mutations induced by industrial pollution. Then E. B. Ford proposed that under varying environmental conditions, either the light or dark form would have a selective advantage because one or the other would be more or less visible to birds.

In the 1950s, the entomologist H. B. D. Kettlewell of Oxford University designed several experiments to test Ford's ideas. First, Kettlewell released equal numbers of both light and dark moths in an area where 95 percent of local moths were light. He was able to recapture 12.5 percent of the released light moths, but only 6 percent of the dark. Second, he and his colleagues watched released moths from a blind, especially when both forms rested motionless on tree trunks. In one series of observations, birds ate 26 light peppered moths and 164 dark moths as the two types rested on light, lichen-covered tree trunks. Just the opposite occurred near industrial Birmingham, where Kettlewell recaptured 40 percent of

released dark moths but only 19 percent light moths. And, of 58 moths picked off the blackened trees, 43 were light and 15 were dark. Clearly, predatory birds supplied the selective pressure, and their hunting habits led to one or the other phenotype and its underlying allele being present in higher frequency, depending on the color of the tree trunks. Interestingly, the century-long domination of the black moths may now be ending. As factory emissions are reduced in industrialized regions of England, both lichens on tree trunks and light peppered moths are making a comeback.

In recent years, the techniques of molecular biology have revealed novel, heretofore unsuspected, genetic mechanisms that may underlie some types of directional selection. Many insect populations develop resistance to various insecticides and this represents a kind of directional selection effect. Simple point mutations may modify enzymes such as acetylcholinesterase (active at neuromuscular junctions), so that the insect pest becomes resistant to organophosphate insecticides.

Populations of *Culex* mosquitoes, on the other hand, react quite differently when exposed to deadly organophosphates. These insects have a gene coding for a detoxifying esterase enzyme, and its activity keeps organophosphates at nonpoisonous levels. In fields and swamps treated with the pesticides, the surviving members of *Culex* populations show an *amplification* of the gene so that there are 250 or more copies present in each diploid

Figure 42-8   THE PEPPERED MOTH: A CLASSIC EXAMPLE OF DIRECTIONAL SELECTION.

The peppered moth, *Biston betularia*, occurs naturally in two color variations: the light, speckled moth and the black variant. The light peppered type was much more common in England until the nineteenth century. When on the light-colored tree bark, it was less susceptible to predation than was its darker and more visible relative. As the Industrial Revolution darkened the vegetation of England with pollution, the darker moth came to have more of a selective advantage against predatory birds, and the frequency of the allele for dark pigment increased in the population. Today, as British pollution controls lead to a cleaner countryside, the frequency of the alleles for light peppered moths is again increasing in the population. Both light and dark moths are shown against the normal (a) and the polluted (b) tree trunks.

(a)   (b)

body cell. With this genetic arrangement, the mosquito can make huge quantities of the esterase enzyme, detoxify the organophosphates, and live.

Additional examples of gene amplification have been seen in other arthropods, in mammalian tumor cells, and during normal differentiation of a few cell types (Chapter 17). Biologists do not yet know how so many copies of a single allele are "handled" in evolution: how long they persist, for example, or whether some copies can evolve in independent directions (recall duplication and diversification of alleles; Chapter 41).

In the case of the Malpeque Bay oysters and the *Culex* mosquitoes directional selection took place in rapid response (over just a few decades) to a relatively sudden environmental change. This also occurs in Darwin's finches; the El Niño period of intense rain in 1982–83 resulted in strong selection for small-bodied finches. Dry years lead to selection of large-bodied finches that can consume large, hard seeds. Thus, over just a year or two, selection can act in opposite ways. Even so, the selective process is usually believed to work very slowly over immense periods of time. When this happens, the effect of selection can be detected only by identifying gradual—but clear—directional trends in the fossil rec-

ord. The decreasing number of toes of horses is a classic example (Figure 42-9).

A crucial feature of directional selection is that it can result in shifts in phenotype that allow a population to remain adapted to a gradually changing environment. The gradual cooling of the temperate regions of the Earth during the late Cretaceous period (Chapter 19) or during the periodic ice ages of the last million years provides an example of such slow changes. Survival of plants or animals during such periods might be dependent on directional selection acting on the optimal temperature of their enzymes, the fluidity (melting point) of the lipids in their cellular membranes, and so on.

Diversifying Selection: Producing Variant Phenotypes

A third category of natural selection is really a variation on directional selection, but one in which the outcome is diversity rather than a trend in one direction. Sometimes a population may be faced with new conditions so diverse that no single phenotype can exploit every extreme. If there is sufficient genetic and pheno-

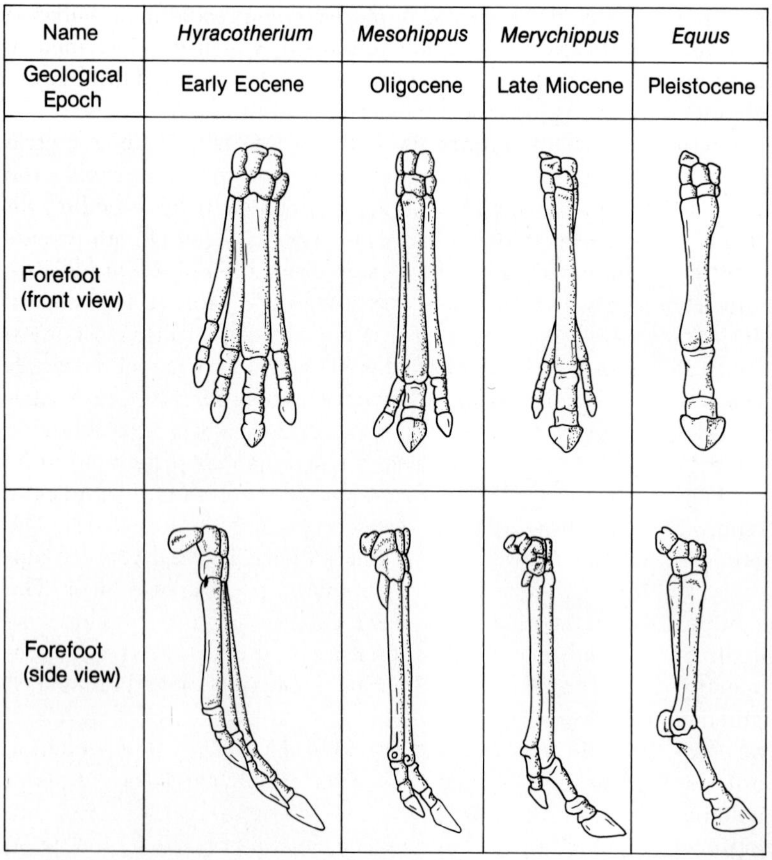

Name	*Hyracotherium*	*Mesohippus*	*Merychippus*	*Equus*
Geological Epoch	Early Eocene	Oligocene	Late Miocene	Pleistocene
Forefoot (front view)				
Forefoot (side view)				

Figure 42-9   DIRECTIONAL SELECTION IN THE FOSSIL RECORD: THE EVOLUTION OF HORSES' FEET.
Horses evolved from four-toed and three-toed browsers to single-toed grazers over about 50 million years (Figure 43-12). Fossil evidence reveals the trend in this directional selection as large horses capable of rapid running evolved from their slower-moving ancestors.

typic variability to allow different selective pressures to operate, **diversifying selection** may occur, resulting in two or more phenotypes, each adapted to some specialized features of a particular portion of the total environment.

The botanist A. D. Bradshaw found diversifying selection operating in a population of bent-grass plants growing in the copper-mining regions of Wales. There, heaps of soil contaminated with copper and other metals dot the landscape. For most plant species in the area, this tainted soil, known as "spoils," is lethal. But while observing "normal" bent-grass plants in clean soils nearby, Bradshaw and his co-workers discovered a variant strain of the plant thriving on adjacent piles of copper-laden dirt. This copper-resistant strain turned out to be highly specialized for its peculiar environment: on uncontaminated soil, it would barely grow at all. Since resistant and nonresistant stands of bent grass are separated from each other by only a few hundred meters, and since bent grass is a cross-pollinating species, the two variants can easily exchange genes. As a result, hybrid seeds containing genes from both types of plant continually sprout in both environments. But natural selection, as always, cannot be avoided; resistant seedlings cannot compete in uncontaminated soil, whereas nonresistant plantlets waste away in the spoils. Neither allele is eliminated in favor of the other, since the two environments provide two permanent "pockets" for the extreme types.

Sometimes an environmental change, such as the appearance of a new predator or a major shift in food supply, disturbs the status quo so much that an extreme form of diversifying selection breaks a population into two or more discrete groups by strongly selecting against the "average" phenotype. Such **disruptive selection** may gradually result in the disappearance of forms that are intermediate between two (or among several) extreme variants that are better adapted to the new conditions. Imagine, for instance, that a breeding pair of dogs, left by visiting sailors on an isolated island in the Galápagos, fed on intermediate-sized marine iguanas. Such selective pressure might result after some years in two surviving iguana populations—one small enough to hide in rock crevices and the other too large for the dogs to handle. The ability of disruptive selection to "split" a species into two or more new species is a possibility we will consider more fully in Chapter 43.

So far, then, we have found that natural selection can shape populations in many ways. Normalizing and directional selection are "conservative": when they operate, variation tends to be reduced, and only a few optimally adaptive phenotypes and underlying genotypes are maintained. Diversifying selection has the opposite effect: variations are maintained or increased simultaneously because more than one phenotype can be optimal

in the nonuniform environment (Figure 42-10). Disruptive selection is "radical": differing variations are favored, and new species may arise. In the real world, these various categories grade into each other; even an experienced geneticist sometimes has difficulty discerning where the effects of one stop and those of another begin.

Balancing Selection: Maintaining Genetic Variation

Sometimes a genetic variation is maintained in a population even though the phenotype it produces is selected against in the overall dynamic of natural selection. Biologists refer to the process that operates to counteract the loss of variant alleles as **balancing selection,** and its result is *balanced polymorphism.* We will look at two forms of balancing selection—heterozygote advantage and frequency-dependent selection.

Heterozygote Advantage

Heterozygote advantage exists when a heterozygote (*Aa*) has a higher fitness than either homozygote (*AA* or *aa*). One of the most famous examples of heterozygote advantage in human populations is the protection against malaria conferred on heterozygous individuals carrying the recessive sickle cell allele, which we described in Chapter 11. Figure 42-11 shows the result of this heterozygote advantage.

Another term for heterozygote advantage is **hybrid vigor.** Plant and animal breeders routinely exploit the genetic benefits of this phenomenon by breeding together two clearly distinct parental lines. If both parents are relatively homozygous at a number of sets of alleles—perhaps as a result of inbreeding to improve a sweet corn crop or a line of prize dogs—then the hybrid corn or the mongrel dog will be heterozygous for many of those sets of alleles. Because any deleterious alleles that have accumulated in the homozygous individuals are suddenly counteracted by other alleles, such offspring tend to be larger, healthier, and—for the corn—more productive than the specialized inbred parents (Figure 42-12). The practice of "outbreeding" in normal populations (Chapter 41) helps ensure a similar state of hybrid vigor. The human incest taboo is a case in point: it is almost certainly a cultural "prescription" that has evolved to protect against the underlying biological danger of too much inbreeding.

New discoveries in molecular biology and immunology are beginning to better explain the basic processes favoring the conservation of heterozygosity. Thus outbreeding and inbreeding may be affected by the genes

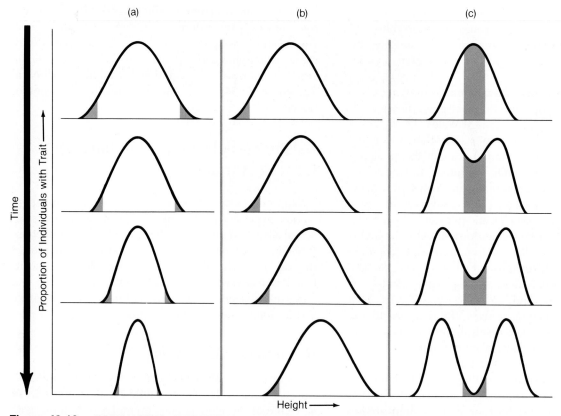

Figure 42-10   NORMALIZING, DIRECTIONAL, AND DIVERSIFYING SELECTION.
A polygenic character such as height of a tree or a giraffe is subjected to different types of selection.
(a) Normalizing, or stabilizing, selection tends to reduce variation so that only the phenotypes best adapted
to some set of circumstances are preserved in a population. (b) Directional selection leads to establishment
of certain phenotypes, especially as a trend over time. Here, the phenotypes at one end of the distribution
are favored, and selection results in the distribution of phenotypes moving to that end. (c) Diversifying and
disruptive selection favor individuals that are most diverse. If the selection is quite strong, two
nonoverlapping populations may result, and new species might arise. The blue areas represent the
phenotypes that are being selected against and are not reproducing as successfully as the others.

Figure 42-11   HETEROZYGOTE ADVANTAGE: MALARIA AND THE SICKLE CELL ALLELE.
Heterozygotes carrying the recessive sickle cell allele of the hemoglobin gene have protection against malignant malaria. (a)
Distribution of malaria caused by the parasite *Plasmodium falciparum* in Africa, the Middle East, and India. (b) Distribution of
the sickle cell allele, which confers protection against malaria. Note the overlap with malaria-infected areas. The recessive
allele is selected for because individuals carrying it are more likely to survive malaria and live to reproduce.

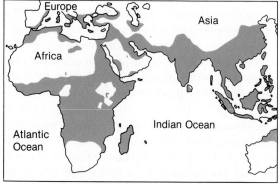

(a) *Plasmodium falciparum*-Caused Malaria

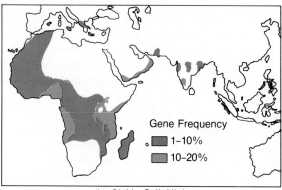

(b) Sickle Cell Allele

(a) (b)

P₁ F₁ P₂   P₁ F₁ P₂

Figure 42-12 HYBRID VIGOR.
(a) Two inbred lines of corn, P₁ and P₂, are crossed and yield an F₁ that is much larger as a plant and produces five times as much corn. (b) The corn produced is also much larger.

coding for cell-surface antigens and for specific odorous molecules involved in sexual reproduction. Fruit flies, bees, frogs, and mammals may all be able to distinguish relatives from nonrelatives on the basis of odor. Most of these animals tend not to mate with close relatives or, if they do, to abort with high frequency embryos of such inbreeding. It is now hypothesized that the histocompatibility gene system responsible for cell-surface antigens (Chapter 32), the involvement of the immune system in embryo abortion, and the use of odor molecules and the olfactory system may have evolved to favor outbreeding and make inbreeding less successful. Thus basic genetic and chemical barriers may underlie complex animal behaviors, including incest taboos, and help to ensure outbreeding and the heterozygosity it produces.

Heterozygote advantage, maintained by natural selection, may have indirect but quite real biological "costs." For instance, the sickle cell allele of the hemoglobin gene contributes to heterozygote advantage by providing resistance to malaria; but that allele is disadvantageous when homozygous, since sickle cell disease results in severe impairment of vital organs and even death. This illustrates **genetic load,** the sum total of those alleles that yield some advantage when they are heterozygous but that are lethal or deleterious when homozygous. The fact that the sickle cell allele exists in even a small percentage of a large population guarantees that some people will be born with the homozygous genotype and thus severe sickle cell disease. Genetic load is a price most kinds of organisms pay for the benefits of those alleles. Each human, for instance, is almost certainly heterozygous for several lethal alleles. In general,

the larger the number of genes in a species, the higher the number of lethal alleles—and the heavier the hidden genetic load.

Frequency-Dependent Selection

The second form of balancing selection is **frequency-dependent selection,** a process that operates when the relative fitnesses of the genotypes in a population vary according to their frequency. In all the examples we have considered so far, we have assumed that natural selection affects a genotype in a consistent way, no matter what its frequency in the population. For instance, individual bent-grass plants resistant to copper will be selected against in normal soils even if their genotype makes up anywhere from 1 percent to 99 percent of the population. But many cases are known in which a phenotype (and its genotype) will be selected against if it is very frequent, whereas an organism with a less common alternative phenotype has a much better chance of surviving to reproduce. One study revealed, for instance, that very common camouflaged snails in Uganda are subjected to much heavier predation by birds than are snails with rare conspicuous phenotypes of the same species. Here, just as with birds feeding on the peppered moth in England, the predator seems to concentrate on a particular phenotype, usually the commonest one. Hence, even though the conspicuous snail or black phenotype of the peppered moth is easier to spot, it is the other more abundant phenotype that is the main food for predators.

These observations show that the fitness of a particular genotype or allele is not independent of other individual organisms in the same species. Those individuals may serve to attract predators, consume nutrients, absorb light (in plants), or affect other members of the species in positive or negative ways. This is why natural selection may be frequency-dependent.

The biologist Lee Ehrman and her colleagues at the State University of New York demonstrated how frequency-dependent selection operates by studying a phenomenon known as **rare-mate advantage** in fruit flies. Ehrman first obtained two collections of a single species of flies, one group gathered in Texas and the other in California. Initially, the flies were put together in equal proportions, and California and Texas flies mated with equal frequencies. Then the workers mixed a few Texans with many Californians and vice versa. In each case, the uncommon males (the "rare mates") were more successful than common males at finding mates. Similar research has shown that after rare mates have become more common due to successful reproduction, they lose their special attraction.

What is the use of a rare-mate advantage? Clearly, it is a kind of balancing selection that provides insurance against the loss of alleles from a population's gene pool. Thus the rarer the males or females of a species become (to a point), the better chance they have as individuals of being reproductively successful. Rare-mate advantage probably operates only in species with complex behaviors. Thus the lone pine on a rocky outcropping or the rare barnacle on the edge of a reef does not benefit from this phenomenon and has a reduced chance of contributing genes to the next generation.

To summarize natural selection to this point, we see that the process can take many forms and produce diverse effects on living organisms. The status quo may be favored in an allele's frequency, in the genotype, or in the phenotype; or trends may occur in one direction or another. Alternatively, it is hypothesized that increasing diversity in genotype and phenotype may mediate the formation of a new species. Considerable research done in the twentieth century supports Darwin's original, perceptive conclusion that natural selection, in all its diverse forms, is a paramount factor in the evolution of all life on Earth.

NEW VIEWS ON EVOLUTIONARY MECHANISMS

Sparked by the powerful techniques of molecular biology and modern insights into genetic processes, today's researchers interested in evolution are asking questions that they could not have imagined just a few decades ago. In this section, we will survey several questions under intense research, including what constitutes an adaptation, whether all variations are the result of natural selection, and whether natural selection works only at the level of individual genes.

The Meaning of Adaptation

Is every aspect of the phenotype—the shape of a leaf on a tree or the nose on a human, the precise color and pattern on a butterfly wing or a flower blossom—adapted for some aspect of survival? One answer to this question is yes: no protein, no appendage, no process will persist or retain its special character in the absence of selective pressure. However, some evolutionary biologists argue against this view, saying that much of what is present in any one individual is not selected "for" or "against"; rather, some characteristics are neutral vis-à-vis natural selection and so cannot correctly be regarded as adaptations. Included in this neutral category might be the prominence of the human chin or nose or the utility of the small toes on a pig's leg which do not touch the ground (Figure 42-13).

Unfortunately, it is extremely difficult to discern which characteristics may be thought of as "neutral" and which may result from selective processes. Nevertheless, biological research has provided some clues. Developmental biology, for example, suggests very strongly that some features present in an adult are there simply because of developmental processes, even though they play no essential role in the adult. Thus the gill-arch system that forms in bird or mammalian embryos yields parts of blood vessels, the inner ears, the lungs, the parathyroids, and so on in an adult bird or mammal. The shape and position of these structures in the adult, as well as aspects of limb and skull shape, may arise solely because of the developmental mechanics of the way organs and tissues are built in embryos. These conclusions suggest a much more conservative view toward adaptations and favor the hypothesis that all features of an organism are not adaptations.

The Neutralist–Selectionist Debate and Molecular Evolution

Concerns about what constitutes an adaptation relate to a scientific argument called the *neutralist–selectionist debate*. This controversy revolved originally around two alternative hypotheses about the evolutionary importance of the protein variations discovered by electrophoresis; but now it applies to other variations too. Recall from Chapter 41 that electrophoretic analyses have shown a high level of genetic variation in the proteins of

Figure 42-13   SOME ADAPTATIONS MAY NOT BE SUBJECT TO NATURAL SELECTION.
Genes and developmental processes may determine the size of a nose; yet as long as this, or any appendage, can carry out its basic function, the specific size and shape may not be subject to natural selection.

many populations. According to biologists in the "selectionist" camp, these variations in protein molecules are preserved because of natural selection. For instance, they may contribute to heterozygote advantage and the resultant vigor of heterozygotes. On the other side are the "neutralists," who believe that the evolution of protein molecules (that is, changes in their amino-acid sequences) is mainly a matter of chance. Neutralists hypothesize that most of the random mutations in the genes coding for proteins are adaptively neutral; that is, they do not affect the function of the protein or the fitness of the organism carrying the mutations. If so, natural selection would function neither to actively remove nor to accumulate such alleles in a population.

Studies designed to explore the merits of these alternative hypotheses are only just beginning. One intriguing set of observations involves cytochrome c (Chapter 7) and other proteins in eukaryotic organisms that have been found to possess a striking property. If we compare cytochrome c amino-acid sequences in tissues from a horse and a snake, or from some other mammal–reptile pair, we find about fifteen differences between the sequences. This corresponds to roughly one amino-acid change—presumably through mutation—per 17 million years since mammals and reptiles diverged from a common ancestor in the Paleozoic era (about 265 million years ago). If we then compare cytochrome c sequences between a horse and an ape, we find about five different amino acids—again, approximately one change per 17 million years since the two orders of mammals diverged, perhaps 90 million years ago. These and many other data imply a very slow and relatively constant rate at which amino-acid changes accumulate in proteins (Figure 42-14a).

Turning from amino-acid changes to nucleotide substitutions in DNA, we can see new views of possible evolutionary mechanisms emerging. Some background information will help us put these new views in perspective. In any given species, a very small proportion—only 5 to 10 percent—of the total DNA codes for protein and is clearly subject to natural selection (Figure 42-14b). The rest of the DNA consists of introns and multiple copies of nucleotide sequences with no known functions.

Since so little of the DNA actually codes for proteins, it is no surprise that even closely related species show numerous differences in the nucleotide sequences in that remaining 90 to 95 percent of the DNA. Chimps and humans, for instance, have about 60 million DNA sequence differences, and the vast majority of these have no effect on phenotype. The biologist Roy Britten from the California Institute of Technology points out that human beings may differ from each other by as many as 5 million nucleotides, and that for every birth, hundreds of new ones probably arise.

Britten and other molecular biologists have carefully studied the nucleotide changes in neutral DNA regions—those that do not code for proteins. One of their most surprising findings is that among different organisms, nucleotide substitution rates can vary fivefold: they found the fastest rates of change in sea urchins, insects, and rodents, and the slowest in primates and some birds. They speculate that rates are fastest where large numbers of DNA replication events take place in the germ line, such as during the many generations of *Drosophila* each year, or the sea urchin's annual huge gamete production. In mammals and birds, DNA repair mechanisms (which correct some mutational events) are probably more efficient, and result in fewer changes

AT THE FRONTIERS: DIRECTIONAL AND NEUTRAL CHANGES IN GENE FREQUENCIES

As an exercise, we could draw a road map of the evolutionary travels of a given tree species or of a particular kind of snail. In it, we could visualize changes in gene frequencies as being associated with adaptations such as tighter cellulose packing and a stronger tree trunk or more adhesive slime so the snail is not swept from a wave-pounded reef. Gene-frequency changes in natural populations reflect the action of mutation, genetic drift, founder effects, natural selection, and other evolutionary processes. As gene frequencies change, phenotypes can change as well, and these visible differences are the basis of adaptations like a stronger trunk or stickier slime that may favor greater reproductive success. Traditionally, biologists have considered gene-frequency changes to be *directional*, since they appear to lead toward adaptations. However, new evidence at a finer level—the level of molecular analysis—suggests that many changes in DNA and genes are not directional. Instead, they lead neither toward nor away from adaptations and have no measurable effect on phenotype. These are called random, or *neutral*, changes.

Such changes can be explained by a simple enzyme example. Many mutations will affect the nucleotide sequence in a gene that codes for a particular enzyme, but have no effect on the organism's phenotype. Let's say that a chance mutation replaces one amino acid in an enzyme with another amino acid of similar properties—say, glycine replaces leucine. If the newly substituted amino acid has no effect on the protein's three-dimensional folding pattern and hence its function, then it probably would not affect the organism's phenotype. Thus the mutation would be neutral. Natural selection cannot "see" the change, and thus it would be absolutely neutral in its effect on reproductive success.

New techniques of DNA sequencing (Chapter 14) have uncovered a large number of changes in single bases of the triplets of the genetic code. (Recall from Chapter 12 that any of several triplets of DNA bases may code for a particular amino acid.) These single base changes have been found in the genes coding for the same protein from various organisms. This significant finding means that an organism's genome may be full of neutral mutations that change the order of nucleotide bases within genes, but do not alter the amino-acid sequence or the gene's activity. These mutations, which occur two to three times more frequently than mutations causing amino-acid substitution, are quite unlike the single amino-acid change that transforms normal adult hemoglobin to the sickle cell type. This is a directional change that does influence natural selection and reproductive success. Very frequent nucleotide substitutions also occur in so-called *pseudogenes*, members of multigene families (as of actin); such pseudogenes may represent duplicated copies of the parental gene type that are in the process of diversifying (Chapter 41).

What do such findings portend for research in the genetics of populations? They indicate that there is much more variability in the genomes and DNA sequences of species than was imagined even a few years ago, and they suggest that much is still to be learned about the importance of neutral mutations in evolution. It may be that much of what we currently believe about some of the genetic mechanisms of evolution will have to be reinterpreted in light of the discoveries of modern genetic research.

accumulating in the neutral (nonprotein-coding) portion of DNA and less random drift. Such reduced mutation rates seem entirely appropriate to the typical reproductive strategy of mammals and birds: producing relatively few young, investing in them great parental attention, and maximizing the survival rates of individual offspring. Studies such as Britten's show clearly that neutral change has gone on continually in all species' DNA and that the rate can vary among species. So far, biologists have not yet reconciled these findings of neutral muta-

tion over evolutionary time and the seemingly constant rate at which amino acids are substituted in proteins, with traditional views of the factors that enable natural selection to operate: variations in climate, predators, competition, local geology, and so on.

Let us return to the neutralists and selectionists. Neutralists use the apparently constant rate of protein evolution as an "evolutionary clock," and argue that random changes lead to alterations in gene frequencies (one per 17 million years or so for cytochrome *c*). For them, se-

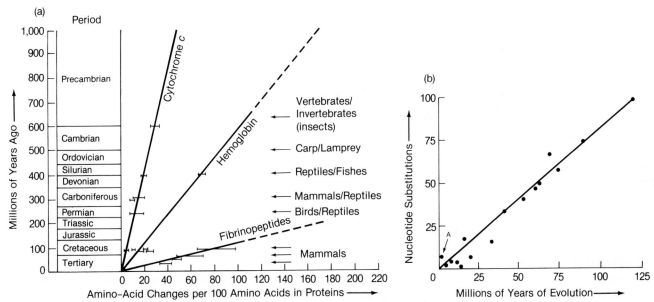

Figure 42-14   THE MOLECULAR EVOLUTIONARY CLOCK: MEASUREMENTS BY AMINO-ACID AND NUCLEOTIDE SUBSTITUTIONS.

(a) The number of amino-acid changes per 100 amino acids in three proteins are plotted against geological time, with the approximate times the different pairs of animal types diverged from common ancestors indicated by arrows. Thus the graph shows that there are about 30 amino-acid differences in each 100 amino acids when cytochromes of vertebrates and of insects are compared; there are about 15 differences per 100 amino acids when cytochromes of mammals and of reptiles are compared. A number of mammalian comparisons yield the points in the Cretaceous and Tertiary periods. Each bar (⊢——⊣) shows the variance in the measurements for that pair of animal types. (b) A curve describing nucleotide substitutions in seven proteins. Each dot represents the time that two mammalian species diverged (as, say, horse and donkey, point A). The line shows a rate of about 0.8 nucleotide substituted per million years. The points falling below the line during the last 25 million years are for primate proteins, suggesting a slower rate of nucleotide substitution in those organisms. This graph shows nucleotide substitutions in DNA coding for proteins; it does not show the much higher rate of nucleotide substitution in the more abundant noncoding DNA discussed in the text.

lective processes play a minor role at best. But selectionists argue that amino-acid-sequence clocks are not fine-tuned; that is, they do not reveal the huge number of mutations in structural genes that occur but do not persist because they are selected out (the argument is that we see only "successes" in the fossil record—a plausible argument, considering the sampling error in the fossil record). In fact, many selectionists interpret exceedingly slow rates of change—such as one per 17 million years—as evidence of continual, intense natural selection: most mutations yield abnormal proteins and so are lethal. Fortunately, the increasingly sophisticated technology of DNA and protein analysis could resolve the debate and tell us the significance of random changes in DNA.

Groups of Genes as Units of Selection

Ever since the rediscovery of Mendel's work, geneticists have assumed that the gene is the "unit" of evolu-

tion. In fact, as we have seen in this chapter, this idea is the intellectual glue that holds together most of the theories of population genetics. The unit of natural selection, in contrast, is commonly said to be the individual—with its full phenotype and underlying genotype. In what sense are genes or groups of genes acted on by selection? Some might argue that the sickle cell allele is a unit of selection: it provides resistance to malaria and thus increases the probability that a heterozygous carrier will survive to reproduce. Other evidence suggests that selection may not always operate on single genes, but rather on groups of genes.

Among the phenomena that have piqued the interest of biologists are cases in which two genes (alleles at separate loci) coding for proteins that interact in biochemical processes occur in the same genomes more often than would be expected by chance. This occurs, for example, with proteins involved in sequential steps in a biochemical pathway or with those coding for structural proteins that function together (actin and myosin, tubulin and dynein are examples). In such cases, selection may have favored the pair of two genes working together in the cellular machinery—after all, their products are both essential parts of the organism's phenotype.

A very different, much more complex, and most puzzling case of multigene selection involves higher-order phenomena, such as animal behavior. Both groups of genes and groups of organisms may serve as the units of selection for such behavior. Bird calls are an illustration. The calls of many birds carry meaning—there is a warning call, a feeding call, a flight call, and so on (Figure 42-15). Unlike bird "songs" used in mating, which may be learned, bird calls are inborn, or instinctive. Even a bird that has been deafened from the time of hatching is able to perfect the calls of its species, although it has never heard them. Thus, calls are inherited in the same way as feather color or beak shape, and it requires many genes to code for the brain circuitry that is involved in the call and that is constructed in the embryonic and early stages of a bird's life.

The ways natural selection might operate as calls (and their nervous-system circuitry) arise have raised questions among biologists. Some bird calls—those that warn of danger—may be interpreted to be unfavorable to the individual that makes them. A bird giving the warning call alerts the flock to the presence of a predator. One view says that the flock may benefit, but the caller may make itself that much more susceptible to predation by attracting the attention of the predator. Many biologists argue that any set of alleles that causes its bearer to carry out such an "altruistic" act would likely be quickly eliminated if genes are the exclusive units of selection. The fact that such "unselfish" traits are common in many species of social animals requires, according to this view, the action of **group selection**—the selective favoring of gene combinations that enhance the probability that a group (a breeding population) will survive. Group-selection theory is another controversial notion that promises to provoke a great deal of research in the future.

Despite this plausible set of arguments, alarm calls by birds are interpreted by other biologists to be favorable to the caller. Thus the alarm caller enlists the aid of others to flee together, thereby reducing the caller's chances of being snared by a predator. Or the alarm

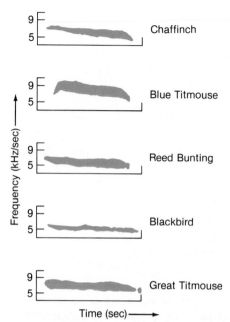

Figure 42-15   BIRD CALLS: INHERITED SIGNALS COMMON AMONG SPECIES.
These sound recordings show the great similarity of the alarm calls of five species of British birds in response to a hawk seen flying nearby. The high-pitched, lengthy calls are believed to be hard to trace to their source. Calls such as these are dependent on complex brain functions, yet they apparently are not learned in most species; they are present because of the way the brain circuits develop, and so are attributable to the activities of many genes.

caller may warn its offspring of danger, thereby improving their chances of surviving to propagate the caller's genes. We will return to these issues in Chapter 49 and attempt to understand individual versus group selection, altruism, and the role of animal behavior in evolution. For now, we can conclude that the concept of natural selection acting on groups of genes or groups of organisms has been hypothesized but not proved.

LOOKING AHEAD

Despite the unanswered questions, what we already know about natural selection, adaptation, and fitness provides satisfactory explanations of how organisms survive and evolve. New concepts in evolutionary theory are built on the foundation laid more than a century ago when Charles Darwin, Alfred Russel Wallace, and other biologists asked a crucial question: What is the origin of species? This is the question that concerns us in Chapter 43, where we shall see how natural selection and other processes may lead to the appearance of new kinds of organisms on Earth.

SUMMARY

1. Natural selection can result in changes in gene frequencies because the individuals carrying variant alleles have differential reproductive success. Natural selection is a major contributor to evolution because it promotes the adaptations that permit differential survival and reproduction.

2. The *fitness* of an individual is a measure of the relative likelihood that the individual will contribute its genetic information to the next generation. If an allele has a high degree of fitness, selection pressures may act against variants of that allele and cause them to be lost from the population. The remaining allele is then said to be *fixed*; but as new variants arise, the fixed state will be lost.

3. *Normalizing selection*, or stabilizing selection, maintains the genotypic status quo. This process involves selection against atypical phenotypes that arise in populations.

4. When adaptation involves a directional shift in phenotype, such as a shift toward disease resistance or darker and darker pigmentation of moths, it is said to result from *directional selection*.

5. *Diversifying selection* results in the coexistence of two or more phenotypes in the same population. *Disruptive selection* is an extreme form of diversifying selection, breaking a population into two or more groups by strongly selecting against the intermediate or average phenotypes.

6. When selection maintains genetic variability in a population, it is called *balancing selection*. One form of balancing selection is *heterozygote advantage*, or *hybrid vigor*, in which organisms with many heterozygous alleles are especially vigorous and reproductively successful. A result of balancing selection is the presence of alleles that are deleterious when homozygous but favorable when heterozygous. The total effect of such deleterious alleles is the *genetic load* of the species. *Frequency-dependent selection*, the other form of balancing selection, occurs when the relative fitnesses of the genotypes in a population vary according to their frequency. For instance, the most abundant phenotype in a species may be most subject to predation; rarer phenotypes (and their genotypes) would be eaten less and have higher fitness. Frequency-dependent selection also occurs when the preferred mates in a population have those genotypes that are low in frequency; this is known as *rare-mate advantage*.

7. It has been suggested that not all phenotypic characteristics are adaptive; some may be neutral and not true adaptations that favor reproductive success.

8. One of the current areas of research in modern biology concerns the neutralist–selectionist debate. According to selectionists, all variations in protein molecules have evolved as the result of natural selection; according to neutralists, many variations may be the result of random mutations that are neutral with respect to survival and reproductive success. Random changes in DNA nucleotides occur at different rates in different species; most are neutral and have no known effect on phenotype.

9. Recent evidence suggests that the gene may not be the exclusive unit of natural selection. Natural selection may operate on many levels of biological organization, from single alleles, to groups of genes, to the individual organism, and, finally, to groups of individuals. *Group selection*, the favoring of genes that enhance a population's survival rate, is hypothesized to explain the existence of so-called altruistic traits that are difficult to account for if selection acts only at the level of the gene.

KEY TERMS

Allen's rule	disruptive selection	frequency-dependent selection	hybrid vigor
balancing selection	diversifying selection	genetic load	normalizing selection
Bergmann's rule	fitness	group selection	rare-mate advantage
directional selection	fixed allele	heterozygote advantage	

QUESTIONS

1. What is fitness?

2. Huntington's disease in humans, which is caused by a dominant allele, usually strikes people during or after their reproductive years and usually leads to death within ten years of onset. Surprisingly, people at risk for the disease (those having an affected parent or grandparent) produce 22 percent *more* children than do people not at risk. What is the effect of the Huntington allele on fitness?

3. List the various types of selection, and indicate whether each is

likely to increase or decrease phenotypic diversity.

4. Contrast heterozygote advantage and genetic load. Why is one so advantageous and the other so disadvantageous?

5. What is the bottleneck effect? Are the individuals that survive a catastrophe (major flood, meteorite impact, fire) the most fit, or does chance play a role in survival and reproduction?

6. Suppose you irradiate a large population of bacteria with ultraviolet light (UV), using a dose of UV that kills 90 percent of the bacteria. Are the 10 percent that survive more resistant to the UV than the 90 percent that were killed? How could you find out? Suppose you irradiate the survivors with the same dose of UV, and you find that 90 percent of those are killed. What determines which individual bacteria will survive and which will die? Repeating this procedure several times, you find that only 10 percent of the bacteria survive each irradiation. But eventually, with only a small remaining fraction of the original population, the same dose of UV kills only 20 percent of the bacteria, instead of 90 percent. Why? If you grow the survivors into another large population and irradiate them, would you expect 10 percent survival, or higher? Explain. What factors can affect survival?

7. Consider a population of mammals in which individuals homozygous for a lethal recessive allele die at birth. Assume that the phenotype of the heterozygotes is identical to that of homozygous normals, and that the mutation frequency is very low. Would it take just one generation to eliminate this allele from

the population? A few generations? Many generations? Why?

8. Consider a population of mammals in which heterozygotes for a lethal dominant allele die at birth. In the absence of mutation, would you expect this allele to be eliminated in one generation? A few? Many? When this sort of allele is present in a real population, what is its source?

9. Consider a population of birds with codominant alleles at one locus, and in which only the heterozygotes can breed. The homozygotes of both kinds are shunned by all, including other homozygotes. What will the frequencies of the two alleles be in the population after one generation? after ten generations? What will the genotype frequencies be for the alleles A_1 and A_2?

10. Biologists usually think of selection as acting on the reproducing (or nonreproducing) individual and its phenotype. What raises the possibility that group selection occurs?

ESSAY QUESTIONS

1. How, as the agent of natural selection, could you obtain a population of brown arctic hares? Drought-resistant maple trees? Flowers pollinated by very small hummingbirds but not large ones? Use other organisms or physical conditions in your selection regime.

2. Biologists have discovered more than 100 insect species that display typical industrial melanism. Explain this phenomenon, and discuss why it is such a widespread response to human polluting of the environment.

SUGGESTED READINGS

BISHOP, J. A., and L. M. COOK. "Moths, Melanisms, and Clean Air." *Scientific American*, January 1975, pp. 90–99.

This article is an up-to-date account of the story of industrial melanism.

BRITTEN, R. J. "Rates of DNA Sequence Evolution Differ Between Taxonomic Groups." *Science* 231 (1986): 1393–1398.

This is a good entry into the literature of molecular evolution.

JOHNSON, C. *Introduction to Natural Selection.* Baltimore: University Park Press, 1976.

This is one of the few books that goes into the topic of natural selection in great depth.

KOEHN, R. K., and T. J. HILBISH. "The Adaptive Importance of Genetic Variation." *American Scientist* 75 (1987): 134–141.

Here is the work on mussels, hap[94], and much more.

MAYR, E. "Evolution." *Scientific American*, September 1978, pp. 47–230.

This issue of *Scientific American* is devoted to evolution. Mayr, a leader in the field, introduces the topic. Included is an excellent discussion of the question of whether all features of an organism are adaptations.

SLATKIN, M. "Gene Flow and the Geographic Structure of Populations." *Science* 236 (1987): 787–792.

An excellent discussion of group selection and gene flow in relation to speciation.

43
THE ORIGIN OF SPECIES

*The ordinary naturalist is not sufficiently aware that, when
dogmatizing on what species are, he is grappling with the
whole question of the organic world and its connection with
a time past and with man.*

Charles Lyell, *Scientific Journals* (1856)

Frozen in time. Fifty million years ago these herringlike fish swam in a shallow sea that covered what is now Wyoming.

In Charles Darwin's day—the Victorian England of the nineteenth century—natural history was a favorite pastime. People watched birds, gathered and pressed plant specimens, and collected all manner of brightly colored feathers, shells, eggs, insects, and minerals. For many, this interest extended to fossil hunting, and both professional and amateur geologists scoured the countryside looking for these peculiar treasures. The rocky outcroppings in southern Wales around Harlech Dome and other sites were a particularly rich source, and hundreds of fossilized bones, teeth, shells, and exoskeletons were recovered from various rock strata in the heavily explored beds.

It became apparent to some observers that thousands of species, long extinct, had lived on the continents and in the ancient seas and that many contemporary species were not represented among the fossil forms. Such evidence implied that the Earth had been a continually changing geological panorama for millions of years and that the living things in that restless tableau were equally changeable rather than static. It is no wonder that such exciting fossil finds sparked revolutionary theories of the Earth's history and inspired Darwin and other biologists to ponder the difficult but fundamental question: How do species originate?

In this chapter, we explore modern answers to this question, drawing on what we have learned about variations in phenotype and genotype, population genetics, and natural selection. We begin with today's definition of *species* and then see how it is linked to the mechanisms by which populations of organisms can diverge into species over time. We will learn how populations can become isolated and how this may lead to divergence of their gene pools so that ultimately, *speciation*—the origin of new species—may take place. We will see how results from molecular and developmental biology help to clarify the interaction of genes in the production of species. Finally, we will enlarge our perspective and consider evolution on a grand scale: the trends and events represented in the Earth's fossil record that reflect the appearance of families, orders, classes, phyla, and kingdoms—the major taxa. By asking if and how the origin of species also generates these larger categories of life forms, we return to the enterprise that stimulated Darwin and other early evolutionists: to discover the biological principles that seek to account for nature's astonishing diversity, past and present.

HOW BIOLOGISTS DEFINE A SPECIES

As we saw in Chapter 19, Carolus Linnaeus and other classical biologists distinguished a "species" as a group of organisms recognizable as a distinct and unique type because of morphological differences from all other life forms. Their criterion for membership in a species was *morphological identity:* close correspondence in physical traits. This notion of species provided a convenient set of conceptual boxes into which biologists could place apparently similar organisms.

Darwin added a powerful new insight about the resemblance among organisms by suggesting that degrees of morphological similarity in general reflect degrees of *common ancestry.* But even with this expanded vision of relationships in nature, taxonomists found that many of the physical traits used to categorize organisms were arbitrary. For example, a naturalist interested in the wing patterns of butterflies might have combined varieties with similar wing markings into one species, while another naturalist might have used a different characteristic, such as body form, as a guideline. No consistent set of biological rules was applied to all types of organisms or at all taxonomic levels.

Although morphological similarity remains an essential criterion for recognizing the members of a fossil species, as well as some contemporary ones, the introduction of population genetics in the early 1900s led to a new way of thinking about species. In particular, the concept of a shared gene pool (Chapter 41) generated a biological definition of species that can be applied to a broad range of life forms. This definition describes **species** as groups of actually or potentially interbreeding populations that are reproductively isolated from other such groups. The "actually" refers to organisms that are members of a population in which breeding is, in fact, taking place. And "potentially" means that individuals could exchange genes if given the opportunity even though they might never actually do so. For example, pollen gathered from fir trees in Washington or Oregon could be used to fertilize a tree of the same species in Norway. Or a Scandinavian person might mate with an Australian aborigine and produce a perfectly normal "hybrid" baby, reflecting the fact that even though contemporary humans are highly variable in their sizes, shapes, and other superficial characteristics, all are members of the same species.

The key to today's biological definition of species is **reproductive isolation:** members of a species can interbreed with each other but they cannot breed with organisms belonging to another species. For instance, a crab and a spider, both arthropods, can never interbreed, nor can a rose and a cherry, both members of the *Rosa* family. Reproductive isolation thus provides a precise standard for determining whether related organisms belong to the same species. Note, however, that this standard applies only to sexually reproducing organisms; for Earth's asexually reproducing creatures—including nearly all prokaryotes, many plants, and some animals—species must still be defined by observable differences in physical traits (morphology, biochemistry, and so on).

In addition to providing a clear definition of species, reproductive isolation can serve as a biological rule of thumb for tracing the relatedness of populations that are in the process of diverging. Consider, for example, the results of an experiment designed to determine whether reproductive isolation can exist between populations that appear similar. Populations of a species of Latin American fruit fly *(Drosophila paulistorum)* were crossed, and the surprising result was identification of at least six subgroups of the tiny flies; all the flies look alike, but members of each group tend strongly to choose mates from within their own group. When members of different subgroups do mate, the result is an evolutionary compromise: the hybrid males are sterile, but the hybrid females are fertile. Although gene flow among the subgroups is limited—and does not happen very often—it is still possible. So the six subgroups are still classified as *D. paulistorum*, although they may be incipient species.

How does one lineage of individuals in a species lose entirely the capacity to breed and exchange genes with other members of that species? The answer hints at how species may arise.

PREVENTING GENE EXCHANGE

Since genes from two individuals are combined at the time of mating and reproduction, we must focus on that critical time in their life histories. We will see that physical aspects of the environment as well as organisms' biological features may serve to restrict gene exchange. We will then turn to remarkable gradients of organisms with slightly differing characteristics—so-called *clines*—that may help reveal the interaction of environment and biology during speciation itself.

The Role of Isolating Mechanisms

Biologists have identified several mechanisms that lead to reproductive isolation, all of which prevent the effective exchange of genes between reproducing individuals. Reproductive isolating mechanisms fall into two major classes: prezygotic and postzygotic. A **prezygotic isolating mechanism** blocks the formation of zygotes; for one reason or another, the male gamete never comes into contact with the female gamete or cannot successfully fertilize it. By contrast, a **postzygotic isolating mechanism** affects a zygote that has already formed. Such a zygote may develop into a nonviable embryo or into a hybrid adult that is sterile. Sterile hybrids are

Table 43-1 **REPRODUCTIVE ISOLATING MECHANISMS**

PREZYGOTIC	**Blockage of formation of viable zygotes**
Ecological Isolation	Life styles using different parts of the environment culminate in speciation
Behavioral Isolation	Behavioral differences result in failure to mate successfully
Mechanical Isolation	Structural or molecular blockage of formation of zygote
Temporal Isolation	Reproduction at different times

POSTZYGOTIC	**Failure of development or of reproduction in individual or descendants**
Hybrid Inviability	Death of hybrids before reproductive capacity is attained
Hybrid Sterility	Inability of hybrids—even though vigorous—to reproduce
Hybrid Breakdown	Reduced viability or infertility in second generation or later generations.

biological "dead ends": they cannot reproduce and hence do not contribute to gene flow (or to much else genetically). Let us take a closer look at how prezygotic and postzygotic isolating mechanisms, summarized in Table 43-1, function in real populations.

Prezygotic Isolating Mechanisms

There are two major types of prezygotic isolation: ecological and behavioral. "Behavior" is used in a broad sense, and includes both mechanical and temporal isolating mechanisms. Thus both the environment and the organisms themselves are involved in these modes of speciation, as the following examples will demonstrate.

Each winter, during the rainy season, the coastal mountains of California are studded with vivid blue and lavender bushes belonging to the genus *Ceanothus*. Because of the geological upheavals that have lifted and twisted the mountains into their current shapes, rocky alkaline soils and rich loams lie side by side. These soil differences are reflected in the genetic constitutions of *Ceanothus* species: *C. jepsonii* grows only on the alkaline soils, while its close relative, *C. vamulosa*, can tolerate a much broader range of soil types. The two species illustrate **ecological isolation;** because each is adapted for a specific soil environment, their genetic differences have grown so great over time that successful cross-fertilization can no longer take place, even though the plants may grow within a few feet of each other.

In Hawaii, two species of fruit flies—so closely related that the females of each group are, to the observer, morphologically indistinguishable—have evolved ecological isolation *par excellence*. Both depend on a scarce re-

SORTING OUT SEXUAL SELECTION

In many animal species, the males and females differ markedly in size, coloration, and body parts, with one sex but not the other displaying antlers, crests, hairy manes, showy tails, shoulder patches, and so on. Naturalists since Darwin's time have considered such differences to be key factors in sexual selection—the subtype of natural selection that bears directly on an individual's ability to attract a mate. Only recently, however, have researchers begun to catalog and understand the many options for sexual selection in the animal kingdom.

Sexual differences can help determine which individual will pass on its genes most successfully, and several field and laboratory studies of birds underscore this conclusion. Red-winged blackbirds, for example, are strongly dimorphic: females are jet black, while males bear beautiful red and yellow shoulder patches visible for many meters (Figure A). Early observers believed that the flashy shoulder patches gave males "sex appeal" in their competition for females. Painstaking field studies have revealed, however, that the male's gaudy shoulders serve mainly as "displays" that help him defend his nesting territory against intruders. Sex enters the picture only indirectly, for females tend to breed most often with males that hold the largest territories (and with them, presumably, the most resources for keeping the young supplied with food).

Many biologists believe that the most striking cases of dimorphisms tended to evolve in species with polygynous breeding (i.e., the males mate with multiple females). More noticeable individuals, they reasoned, would be likely to attract more mates. But not all dimorphisms work this way, and the exceptions disprove the rule. A prime example involves lesser snow

Figure A
(a) A male red-winged blackbird (*Agelaius phoeniceus*) singing and displaying its brightly colored shoulder patches as it advertises its territory. (b) A female of the same species bears little resemblance to her mate.

geese of the Canadian Arctic. These birds show conspicuous variations in plumage; some individuals are pearly white with coal-black wing tips, while others are muted silvery gray and grayish-brown with only

source, the oozing sap of *Myoporum* trees, for food. Thus we might expect that over the years natural selection, perhaps in the form of competition for food, would have eliminated one group of flies in favor of the other. Instead, however, one population of flies has taken the high ground, feeding and breeding on oozing tree trunks, while the other has settled for puddles of sap on the forest floor. The evolutionary result is two "successful" species that have become genetically distinct apparently because of their preferences for the different food locations. We will encounter additional similar examples in Chapter 46, where so-called *competitive exclusion* of one species by another reflects division of an essential environmental resource. In the case of fruit flies and

many other instances of speciation, we do not know the precise physiological basis of reproductive isolation that has arisen during the "division" of resources by the two species.

Populations can also become reproductively isolated because of differences in their behavior. Biologists have discovered such **behavioral isolation** in the complex light patterns displayed by fireflies (Figure 43-1). A female will mate only with a male whose flight paths and twinkling abdomen show the patterns characteristic of her species. Such special features of males result from *sexual selection*—the differential reproductive success that males with certain physical or behavioral characteristics achieve. Thus an awesome collection of other vis-

white heads and necks (Figure 41-11). Unlike the red-winged blackbirds' plumage, however, the two types of lesser-snow-goose plumage are unrelated to gender: both males and females may be of either type. What is more, lesser snow geese do not take several mates—they are strictly monogamous, pairing for life—and they usually select a mate with plumage like their own.

Controlled experiments with lesser snow geese goslings now show that the color of a youngster's parents and siblings is the all-important element in determining mate selection, not some color pattern first viewed in adulthood. Researchers found that dark birds raised in a light family nearly always choose

Figure B
A male bowerbird (*Ptilonorhynchus violaceus*) at the front of his decorated bower in an eastern Australian forest.

light mates, and vice versa. By dyeing a group of "foster parents," one creative researcher inveigled goslings to prefer pink geese! How could this family-oriented color preference confer a selective advantage? One hypothesis holds that birds choosing "like" mates are, in effect, selecting partners genetically close to themselves—a choice which increases the chances that genes like their own will be transmitted to offspring.

Sexual selection need not involve the body itself if behavior is the source of male attractiveness. Adult male "gardener" bowerbirds of New Guinea construct bowers and decorate them lavishly to attract females (Figure B). Both sexes are a drab brown, but the male adorns a mound of sticks and twigs with bits of fungi, colorful flower petals, bright fruits, iridescent butterfly wings, shimmering beetle carapaces, even multi-hued poker chips—all to entice a receptive female. The females appear to rigorously inspect competing bowers before choosing a mate, but once the courtship commences and she is impregnated, she leaves the bower to build a nest and care for the young on her own.

Sexual selection has also led to complex mating displays (Chapter 48) and even to ways in which younger or smaller males can successfully mate despite their subordination to larger males. Thus some male salmon of certain species resemble females and can enter a large male's territory and deposit sperm on newly spawned eggs without being spotted and chased away.

These and other studies are making clear that sexual selection is a multifaceted process and that intricate—sometimes flamboyant—traits can arise from the drive for reproductive success.

ual, chemical, and acoustic signals represents adaptations that make some males more likely to succeed than others in producing offspring. Examples are the peacock's tail, the elk's antlers, insect pheromones, mating croaks of frogs (Chapter 49), cricket "songs," bird songs, and showy courtship displays, such as those of the rather amazing riflebird (Figure 43-2).

One effect of these mating signals is to restrict the exchange of genes to members of the same species or sometimes, even more narrowly, to members of the same population. For example, in areas to the north, south, and east of San Francisco, populations of white-crowned sparrows sing their species' song in different "dialects." A female of the Santa Cruz population recog-

nizes her group's dialect and in experimental situations, preferentially shows mating behavior toward males that sing it. She could breed successfully with a male from the nearby Berkeley or Marin populations, but in nature, this apparently happens infrequently because their songs are slightly different. Conceivably, by cutting off the flow of genes over hundreds or thousands of years, this kind of behavioral isolation might lead to full reproductive isolation and the formation of new species.

Sometimes the differences that isolate populations or species actually prevent mating. In this case, **mechanical isolation** is operative. In flowering plants, variations in the configurations of flowers can discourage cross-pollination by restricting access to pollen-covered anthers or

Figure 43-1   FIREFLY LIGHT PATTERNS: A BEHAVIORAL ISOLATING MECHANISM.
A female firefly will mate only with a male whose flight paths and flash patterns match those characteristic of her species. The patterns of the males of nine species of *Lampyridae* fireflies are illustrated. The dots, ovals, and elongated squiggles and curves represent the flashes; the interconnecting lines trace the flight paths.

to the female reproductive organs. Other blossoms may attract only specific pollinators whose habits ensure that "undesirable" gene exchange will not take place. (Such coevolution of plants and pollinators was described in Chapter 24.) Among the many animals that carry out external fertilization, various echinoderms—sea stars, sea urchins, and sand dollars—provide a good example of mechanical isolation at the molecular level. In these organisms, cross-fertilization may be inhibited because of differences in the fit of the molecules that bind sperm and egg together. The same thing may occur in pollen binding in flowering plants. In animals that carry out internal fertilization, differences in genital size or shape may preclude mating, or molecular incompatibilities to fertilization may occur.

The final type of prezygotic isolating mechanism involves **temporal isolation;** that is, time-related environmental cues that trigger reproductive processes are different for related species. Thus a toad of the species *Bufo americanus* might live in the same woodland as its cousin *B. fowleri*, but the two will probably never interbreed because in each group, the hormones that control the maturation of eggs in ovaries and sperm in testes are produced at different seasons of the year (Figure 43-3). Hence, mating behavior occurs at different times in the two species. Temporal isolation can also prevent gene flow between closely related plant species. On California's wind-swept Monterey peninsula, the pines *Pinus vadiata* and *Pinus muricata* have remained genetically distinct because one species sheds its pollen in early February, while the other does not release its pollen until late April.

All of these prezygotic strategies for deterring the exchange of genes between populations and species work in the same way: they nip gene flow in the bud by preventing the union of gametes. Sometimes, however, matings between members of two species *do* take place; when this happens, postzygotic isolating mechanisms may then prevent a hybrid organism from developing or from becoming a successful reproducer. Remember, of course, that the various prezygotic isolating mechanisms only *may* allow speciation to occur; the latter process does not necessarily occur.

Figure 43-2   SEXUAL SELECTION AND THE MALE RIFLEBIRD.
The dramatic courtship display of the male riflebird attracts females and is a form of behavioral isolation. Female riflebirds (*Ptiloris victoriae*) mate only with males that have the "correct" blue and green colors and that display with wing or tail feathers expanded as shown here.

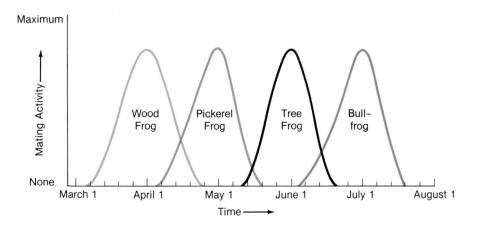

Figure 43-3   TEMPORAL ISOLATION IN BREEDING OF AMPHIBIANS.
Four kinds of frogs are seen to have maximal reproductive behavior at different times; this helps to ensure that interbreeding is reduced or absent.

Postzygotic Isolating Mechanisms

Two major phenomena lead to failure of hybrids to reproduce successfully. The first is **hybrid inviability:** the individual dies before reproducing successfully. The second is **hybrid sterility,** in which the organism develops but is unable to reproduce for some reason. A special case of this involves **hybrid breakdown,** a condition seen when the second and subsequent generations display decreased reproductive success.

A variety of events can result in the death or the sterility of a hybrid plant or animal. For example, sheep and goats can mate, but the hybrid embryo that develops from the fertilized egg dies at an early stage of development. This example of hybrid inviability is usually the result of the failure of the parental genomes to coordinate during the development of the embryo.

Alternatively, a hybrid may develop and mature but be abnormal in some way that keeps it from contributing gametes to subsequent generations. A hybrid seed that germinates but grows into a plant that cannot produce mature flowers is an example. Or a hybrid organism may be a "normal" healthy adult in most respects but produce no functional gametes. Both are examples of hybrid sterility. Although the sex organs of many sterile hybrids—such as the mule, produced by a cross between a horse and a donkey—may appear normal, their production of viable gametes is prevented because homologous chromosomes derived from parents of different species do not pair and cross over correctly during meiosis (Figure 43-4). Alternatively, certain *Drosophila* crosses are sterile because sperm motility is reduced.

In Chapter 42, we saw that when individuals from two genetically distant populations of a single species cross-fertilize, the result is often *heterozygote advantage,* or *hybrid vigor:* offspring of the union may be larger, stronger, and healthier than either of the parents. If these same two populations were to diverge into distinct species, however, the outcome of such a cross might be quite different. Instead of vigorous new offspring,

crosses between the closely related species might end up in hybrid breakdown. Laboratory crosses of fruit flies (*Drosophila pseudoobscura* with *D. persimilis*) have illustrated hybrid breakdown at work: first-generation females are healthy and deposit as many fertilized eggs as do "purebred" flies. However, when second-generation hybrid daughters are mated with males of either parental generation, their offspring are weak and are usually unable to produce viable offspring.

The Role of Clines in Speciation

So far, we have discussed how biologists define a species and how reproductive isolating mechanisms can operate to maintain the differences between gene pools that will ultimately set species apart. Now let us consider some instances in which the chance distribution of populations

Figure 43-4   MULES AND HYBRID STERILITY.
Mules, the hybrid offspring of a female horse and a male donkey, are invariably sterile. While mules are useful to humans for performing heavy hauling work, they cannot breed and become a true species.

throughout certain geographical ranges may give us a glimpse of one way that isolating mechanisms generate new species.

Populations of a species sometimes have clear geographical boundaries—such as the river that runs through a meadow or the shores of an isolated island—but often populations are spread out along a broad geographical range. As we move from one part of this range to another, we often find a gradual change in a characteristic or many characteristics of a species. Such gradual changes are referred to as **clines.** Basically, each local population along a cline seems to evolve adaptations for its particular environment. Counteracting this specialization to local conditions is gene flow from adjacent populations; nevertheless, as we shall see in the following two examples, significant differences between populations of a species can build up along a cline.

Over the years, biologists have traced hundreds of clines in all sorts of species. A typical example of a cline is that of the grass frog, *Rana pipiens*, in which some eleven distinctive geographical races occur in localities between Vermont and Louisiana. Although some interbreeding may occur between members of adjacent frog races, more distant populations are reproductively isolated. The salamanders of the genus *Ensatina*, which inhabit the coastal mountains and Sierra Nevada mountains of California (Figure 43-5), provide another example of a cline. At least seven subspecies of *Ensatina* have been found, each with its own slightly different morphology, skin-color pattern, and adaptations to the specific ecological conditions in its part of the species' range. **Subspecies,** then, are genetically distinct populations of a species with at least some distinctive characteristics. Where any two of the subspecies of *Ensatina* overlap, they sometimes interbreed successfully. But if we were to take two animals from opposite ends of the cline and try to breed them, our neat definition of species would become fuzzy: these salamanders not only look quite different from each other, but also cannot mate at all. By definition, then, they are different species. But at least in theory they can exchange genes, since interbreeding can occur anywhere along the cline. Like the interlocking gears in a gearbox, the "first" and "last" populations are linked by the intermediate groups between them. And in our analogy, although each adjacent pair of gears might mesh perfectly, the gears from each end might

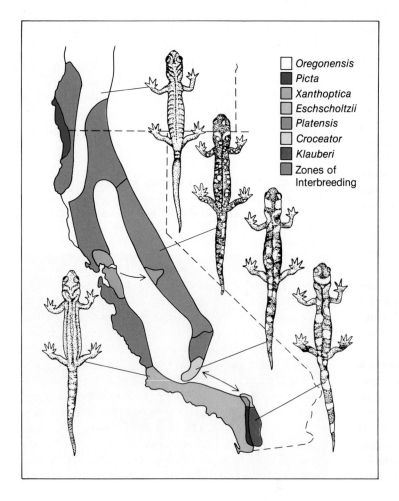

Figure 43-5   THE CLINE OF THE *ENSATINA* SALAMANDERS.

Seven subspecies of this salamander form a cline; subspecies may interbreed with variants living in neighboring or overlapping ranges, but cannot interbreed with individuals from more distant parts of the cline. The cline extends through the coastal and Sierra Nevada mountains of California. The arrows indicate sites where subspecies have "jumped" across geographical regions in which no *Ensatina* subspecies live.

Legend for figure:

☐ Oregonensis
■ Picta
▨ Xanthoptica
▧ Eschscholtzii
▩ Platensis
□ Croceator
■ Klauberi
▨ Zones of Interbreeding

have completely incompatible teeth so that they cannot mesh at all. So, too, the organisms on the ends of clines may not "mesh"—that is, reproduce successfully.

Clines provide dramatic evidence of the beginnings of species formation. Clines apparently arise as an original population spreads from its site of origin and adapts to new and different geographical and ecological settings. Then, if one or more of the intervening groups in a cline were to drop out, geographical isolation would halt any gene flow. Over time, even more independent variation of the separated populations would probably occur. Ultimately, full reproductive isolation would almost certainly result.

BECOMING A SPECIES: HOW GENE POOLS BECOME ISOLATED

During the early part of this century, two hypotheses were suggested for the origin and development of reproductive isolating barriers. One of these, the *by-product hypothesis*, postulated that the isolation of a gene pool happens accidentally as the result of genetic divergence over time. According to this view, populations can become genetically distinct enough to be labeled as separate species in one of two ways: (1) by adapting to different ecological conditions as a result of natural selection (recall disruptive selection; Chapter 42); or (2) as the result of chance events, such as genetic drift, bottlenecks, and so on). The second proposal, the *selection hypothesis*, was based on the view that speciation begins when a single population includes two (or more) groups with different genetic characteristics; hybrids of the two groups might then be less fit than nonhybrid offspring—less able to survive, reproduce, raise young, and so on. In that case, natural selection would favor the development of prezygotic isolating mechanisms that block gene exchanges between the groups.

Recognizing that these two hypotheses are not mutually exclusive, Ernst Mayr at the American Museum of Natural History fused them into what he called the **allopatric** (or geographic) **model of speciation** (Figure 43-6). ("Allopatry" means "living in different places"; "sympatry" means "living in the same place.") According to Mayr's model, species can originate in a two-stage process. In the first stage, populations of an existing species become separated, by chance, by a physical or geographical barrier, such as a mountain range or a newly formed river canyon. Once gene flow has been halted, more and more genetic differences accumulate between the two allopatric populations (an aspect of the by-

product hypothesis). Eventually, these differences become so significant that at least some postzygotic isolation occurs: hybrids of the two populations do not survive or cannot reproduce (the selection hypothesis applies). At this point, the second stage of speciation may occur if, by chance, the physical barrier is removed and the two original populations once again become sympatric. This can happen, for example, if geographical conditions change again (such as a river changing its course) or if members of one or both groups migrate. If hybrids produced by the two parent strains prove to be less fit, they will be selected against; at the same time, characteristics (such as a particular courting behavior) that promote mating may be restricted to members of the organism's own group. In this way, prezygotic isolation can be established.

Since Mayr first proposed the concept of allopatric speciation, biologists have found patterns of variation in populations that correspond with virtually every step of the process he described. Its twin postulates—that physically separated populations diverge genetically and that they then become fully distinct by natural selection—have come to provide convincing explanations for many cases of diversity in the living world.

But differing views of speciation continue to be debated, and some evolutionists have focused on the molecular events—the amount and kind of changes in the genetic material, DNA—that are ultimately responsible for the differences among species. How many gene loci, for instance, must undergo changes before populations become reproductively isolated? Are changes in regulatory genes more or less important than changes in structural genes? These are some of the questions that we will consider next.

THE GENETIC BASES OF SPECIATION

In Chapter 41, we learned that gel electrophoresis of structural proteins and analysis of DNA nucleotide sequences allow us to estimate how much populations can vary in their genetic make-up. This kind of information can also be used to compare the variations in structural genes among different sorts of organisms, between newly formed species, or in populations. As Figure 43-7 shows, as one moves up the taxonomic ladder from population to subspecies to species to genus and so on, related organisms have proportionately fewer of the same structural genes. (The statistic that measures such correspondences is called **genetic identity,** represented in the figure by *I*.) Data like those presented in Figure 43-7 support Mayr's hypothesis that gene pools diverge grad-

(a)

(b)

(c)

(d)

(e)

Figure 43-6   ALLOPATRIC MODEL OF SPECIATION.
Allopatric speciation occurs when a single population of a species (a) becomes separated by physical or geographical barriers, such as the formation of an island, a canyon, or a mountain range. (b) Without gene flow between the populations, genetic differences accumulate as natural selection operates on each group independently. (c) Eventually, the two populations may become unable to interbreed. (d) If the physical barrier between the populations is removed, the two populations may coexist but not reproduce; they have become different species. (e) The Albert squirrel (*Sciurus alberti*) and the Kaibab squirrel (*Sciurus kaibabensis*) are two different species that are believed to have been one population before the formation of the Grand Canyon. The two species now live on opposite sides of this deep chasm, which as seen here, is indeed a forbidding barrier to cross. Allopatric speciation has occurred; the squirrels have developed differently and can no longer interbreed.

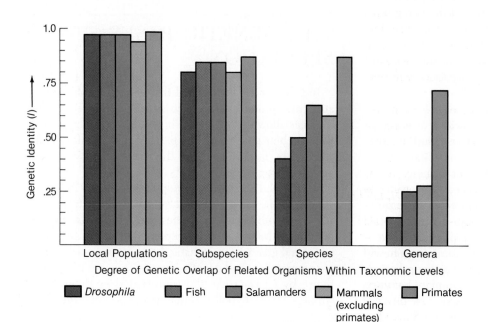

Figure 43-7   GENETIC IDENTITY: A MEASURE OF STRUCTURAL GENES SHARED BY RELATED ORGANISMS.
For related organisms, individuals within local populations have more structural genes in common than do individuals within subspecies. As we move up the taxonomic ladder through species and genera, the related organisms within each group have fewer genes in common at each step. (Data at the genera level are not available for *Drosophila*.)

ually as populations slowly become reproductively isolated; once a group of organisms has attained the rank of species, however, it tends to diverge genetically from related species at a more rapid pace.

A remarkable exception to this generalization is the primate order. Although the human, chimpanzee, gorilla, monkey, and lemur groups diverged from each other between about 8 and 30 million years ago, and the organisms appear morphologically different from each other, the groups remain amazingly alike in their structural genes. (As pointed out in Chapter 42, the 60 million nucleotide sequence differences between chimp and human DNA are neutral mutations, most or all in regions not coding for proteins.) Why, then, are the major differences in the morphology of primate species not mirrored in their structural genes? Part of the answer may lie in the fact that gel electrophoresis measures only the products of structural genes. A growing body of evidence, however, suggests that small changes in special regulatory genes may account for many of the major changes responsible for the origin of species and higher taxonomic groups.

Regulatory Genes and Evolutionary Changes

Recall from Chapter 17 that during an organism's development, regulatory genes act to coordinate blocks of structural genes, so that complex organs or processes arise. If the adult bird is to fly, the embryo's wings must develop with their anterior, posterior, dorsal, and ventral surfaces oriented correctly relative to the body. To function normally, a crayfish or fruit fly must develop the correct number of body segments and appendages. Complex properties such as axes of orientation and body segmentation are due in part to regulatory genes. Furthermore, many biologists believe that changed activity of such regulatory genes may explain important evolutionary alterations in organisms and perhaps speciation itself. Humans and chimpanzees differ primarily in size, relative proportions of body parts, and hairiness. Perhaps alterations in regulatory genes account for such differences; thus chimps and humans continue to share the basic primate body plan and physiology, and their protein sequences are largely identical.

The potential importance of regulatory genes in explaining such facts was suggested as morphologists explored an old and substantial body of embryological evidence gained from observation of human, chimp, and other vertebrate embryos and juveniles. In both chimps and humans, fetuses develop at about the same rate and with amazingly parallel features. After birth, however, human development slows down, while chimps continue to undergo rapid and marked changes in their physical characteristics. The skull "grids" in Figure 43-8 map this

disparity. The form of a human skull alters relatively little as a person matures from infant to adult, but a chimpanzee's skull undergoes a striking transformation. It is as though such transformation has been truncated in humans, so that we retain the embryonic proportions in the adult stage. What causes this developmental difference? According to one hypothesis, the disparity may be due to differences in the timing by which specific regulatory genes switch on (or off) as the young animal develops.

Chance mutations in regulatory genes might also help to explain radical changes in structure that mark the beginning of major new groups of life forms. For instance, although most significant changes in a gene are likely to be detrimental to an organism (a fruit fly with two large legs wiggling where its antennae should be is definitely

Figure 43-8   DEVELOPMENT OF HUMAN AND CHIMPANZEE SKULLS: REGULATORY GENES AND EVOLUTIONARY CHANGE.
Humans and chimpanzee fetuses develop at about the same rate and have very similar skulls. (a) The human skull does not change a great deal in form from infancy to adulthood. (b) But the chimpanzee skull continues to change form as the animal matures, resulting in a markedly different morphology between infant and adult. Since humans and chimps share most of their structural genes and protein sequences, these differences are probably explained by the action of the regulatory genes during development.

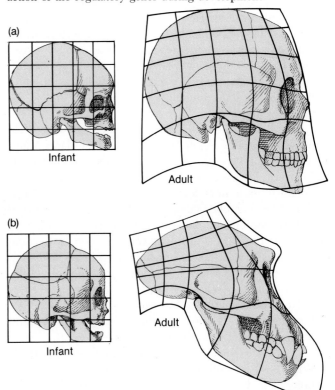

(a)

Infant

Adult

(b)

Infant

Adult

at a disadvantage), some tiny percentage may produce viable "hopeful monsters"—organisms that, although drastically altered, are still adapted to their surroundings and survive to reproduce. More and more data suggest that regulatory genes control major features of the body, such as the number of body segments or pairs of wings in insects. But let us look at other mechanisms that help explain speciation.

Chromosomal Changes

Most biologists have traditionally believed that genetic changes (mutations) that give rise to new species accumulate gradually over many generations and long stretches of time. However, evidence shows that other mechanisms can split populations genetically at a much faster pace. Plants, for instance, may acquire isolating mechanisms in just one generation by way of **polyploidization**—the sudden multiplication of an entire complement of chromosomes. Polyploids come in two types: *autopolyploids*, which result from the multiplication of chromosome sets in a single species ($2n$ going to $4n$ directly, for instance), and *allopolyploids*, which are produced by the combination of chromosome sets from different parental species in a hybrid plant. Often, polyploid individuals are fertile among themselves but reproductively isolated from their diploid parents, since the gametes of a polyploid plant have a different chromosome number from the gametes of the diploid organism (Figure 43-9). Polyploidy also enables plants to sidestep the problems of hybrid sterility: since each chromosome set is doubled, polyploids fulfill nature's requirement for matching pairs of chromosomes during meiosis. And once a polyploid population has been established, it constitutes a new species in which, by definition, the polyploid chromosome number is designated as diploid for that new species.

Polyploids have provided human societies with a tremendous variety of economically important plants, ranging from tobacco and cotton species to species of wheat, sugar cane, and coffee. They are also one of the reasons botanists sometimes find it difficult or impossible to fit the more than 300,000 species of plant life into neatly ordered systems of allopatric speciation.

Chromosome sets in animal cells occasionally become polyploid—for instance, human liver cells do so during puberty—but that never occurs in germ-line cells of the ovary or testis. Whole animals very rarely develop as polyploids. Sometimes, however, animal chromosomes become rearranged in ways that may lead to the rapid development of new species.

In the late 1960s, the Australian biologist M. J. D. White and his colleagues studied neighboring populations of flightless grasshoppers that appeared virtually identical in form but showed clear differences in the con-

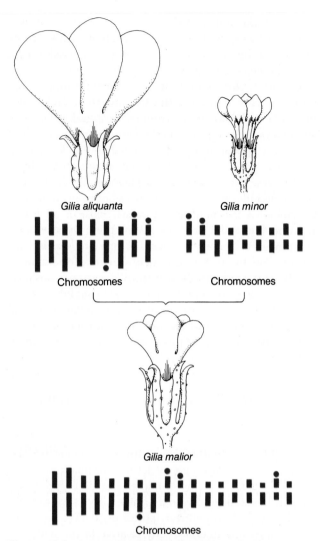

Figure 43-9   INSTANT SPECIATION: POLYPLOIDY IN GILIA.
Plants can form "instant species" through polyploidization, the duplication of an entire set of chromosomes. *Gilia* provides a good example of allopolyploidization, the doubling of chromosomes from two parent plants of different species. The eighteen chromosomes in *G. malior* are no longer exact copies of the nine from each parent species, reflecting changes in the chromosomes subsequent to the "speciation" event. Another type, autopolyploidization, doubles the chromosome sets contributed by a single parent of a given species. Polyploid individuals are reproductively isolated from their parents; since they have a different chromosome number, they cannot interbreed, and so are, by definition, a different species. The small dots at one end of the *Gilia* chromosomes represent end regions separated by constrictions along the chromosomal axis.

figurations of their chromosomes. Techniques were used to stain the chromosomes of dividing cells so that careful assessment could be carried out of the degree of ploidy. The results led White to theorize that in what had once been a single population, a random change in chromosome structure arose that did not result in a lethal zygote or embryo but made those grasshoppers possessing it more fit for a certain portion of the original species' range. The new chromosome arrangement apparently also had the effect of isolating reproductively the two groups of grasshoppers, since hybrids could not develop successfully because of mismatched chromosomes. The result was the splitting of one species into two, even though no geographical barrier prevented gene flow. This is a case of **sympatric speciation,** in which speciation occurs in populations inhabiting the same geographical range. In this kind of situation, therefore, geographical isolation is *not* needed for speciation to occur.

The splitting of the giant panda from the bear lineage involves another case of unusual chromosome rearrangements. The taxonomy of giant pandas has been a puzzle for many decades. While giant pandas resemble bears in many ways (Figure 43-10), they are herbivores that eat primarily bamboo, rather than eating meat and vegetable matter as do bears. Also, unlike bears, the pandas' forelimbs and shoulders are much larger than their hind limbs and hindquarters; they have six digits on their

Figure 43-10   THE GIANT PANDA: A CASE OF FUSED CHROMOSOMES.
Giant pandas *(Ailuropoda melanolenca)* resemble bears, yet have many distinctive features. Most of their chromosomes, for example, seem to be fused pairs of chromosomes that are found singly in brown bears and other bear species.

forepaws instead of five; and they "bleat" rather than roar. Finally, giant pandas have forty-two chromosomes, each with the centromere located in the middle, while bears have seventy-four chromosomes, most with the centromere at one end.

Interestingly, when biologists tested panda and bear chromosomes in special ways, they found that nearly every bear chromosome can be matched with an arm from a giant panda chromosome. They concluded that most giant panda chromosomes represent two bear chromosomes fused at their centromeres. Other data suggest that chromosomes of the earliest bears underwent a massive rearrangement relative to the chromosomes of the primitive carnivores from which they evolved. Using DNA hybridization studies, electrophoretic analysis of over fifty proteins, and immunological comparisons of blood proteins, biologists were able to deduce that chromosome fusion occurred some twenty million years ago as the giant panda line arose.

With this and other work Steven J. O'Brien and his colleagues at the National Cancer Institute helped solve the riddle of the giant panda's phylogeny, and in so doing showed that studies of chromosomes, DNA, and proteins can answer questions about taxonomy and evolution that traditional morphological procedures can never resolve. The results also show that unexpected types of chromosome rearrangements can occur as a major group of animals—in this case, the primitive carnivores—gives rise to new groups such as the omnivorous bears and then the herbivorous giant pandas. Future work must confirm or deny whether the chromosomal fusion step was the key event in the divergence of the giant panda lineage.

Polyploidization, sympatric speciation, and changes in chromosomal structure accentuate a fundamental biological fact: species can originate in a variety of ways, and the alternative pathways leading to reproductive isolation can take thousands of years or only a single generation (the time it takes a diploid to become tetraploid or a tetraploid, octaploid).

In the four billion years since life began, far greater differences than those among species have separated the living world into distinct genera, phyla, and kingdoms. With the processes of speciation as background, we will next consider mechanisms that may account for the origin of the higher-order taxa.

EXPLAINING MACROEVOLUTION: HIGHER-ORDER CHANGES

How do reproductive isolation and speciation, our topics so far, relate to the origins of genera, families, or phyla? There are not even words in our language such as "phylogenization," "genusation," and so forth to describe the processes by which these larger groups are formed. Why not? One reason is that such higher-order taxa are even more abstract categories of the human mind than is species; they are not biological realities in the sense that a species is a tangible group of organisms that can breed with each other. "All echinoderms," "all animals," and "all members of the genus *Felis*" are abstract groups that do not possess a common gene pool or trade genes during reproduction. Another reason is that since Darwin's day, the majority of evolutionary biologists have accepted the hypothesis that **macroevolution**—the major phenotypic changes that occur over evolutionary time—begins with the evolution of new species. (The small-scale changes in gene frequencies, such as mutations in regulatory genes, that are hypothesized to generate species are sometimes termed **microevolution.**)

Macroevolutionary events include alterations in basic body design, such as the acquisition of a closed circulatory system in animals to allow larger body size or of stomata and guard cells in plants to control gas exchange in leaves. Alternatively, macroevolution may involve major changes in physiology, such as the evolution of amphibians from air-breathing fish and of vascular plants from nonvascular ones. Or there may be large-scale changes within a taxonomic group, such as the transition from foreleg to wing in the reptile–bird lineage and the transition from primitive, simple flowers such as magnolias, to later, compound ones such as sunflowers.

All the differences that ultimately distinguish genus from order and family from phylum stem from changes in the gene pools of populations. One way biologists attempt to trace the course of such changes is by reconstructing the evolutionary past.

The Fossil Record

Like a view through a warped windowpane, the fossil record provides biologists with, at best, an incomplete image of the evolution of the Earth's flora and fauna. Most fossils consist of the hard body parts left behind when an organism died—bones, shells, and in plants, tough cell-wall materials. Other fossils consist of burrows, tracks, or other impressions that were left in sediments along ancient lakeshores and creek sides and that were later transformed into stone. Some lines of descent, like those of the modern horse or the transition from reptile to mammal, have been well preserved. But for many others, the evolutionary picture is clouded by the fact that remnants of intermediate forms—say, the presumed "links" between jawed bony fishes and sharks—have so far not been discovered.

There are several reasons for this state of affairs. First, paleontologists estimate that two-thirds of all organisms that have ever lived did not have readily fossilizable parts. Like the invertebrate inhabitants of early seas, they had soft bodies, no rigid skeleton, and no teeth or other hard structures. As a result of the differential preservation of species, we are left with no trace of what must have been a lavish array of life forms. Second, the likelihood that an organism with the required hard parts, such as an ancient reptile or fern of the Paleozoic era, would be preserved in stone depended heavily on where and how it died. Only those very few forms that chanced to die where they would be quickly entombed by mud, silt, resin, or some other stable protective covering and remain undisturbed for millennia appear as fossils in later epochs.

Finally, fossils themselves may be destroyed, altered, or moved about by simple erosion of sedimentary rocks, by the immense pressures of overlying rocks, or by soil dissolution if they happened to be uplifted into the surface soils. Upland, mountainous terrestrial environments are very susceptible to erosion by wind and rain, so that fossils are unlikely to form in such places, and those that are present are likely to be destroyed quickly once exposed to weathering. The best-preserved terrestrial specimens typically arose in regions that were originally lowland basins and plains (Figure 43-11). For the same reason, marine fossils are most abundant in rocks laid down originally in bays, estuaries, and deltas and are less common from sites that were open coasts exposed to the destructive action of waves, tides, and currents. A final problem in interpreting the fossil record is *time-*

Figure 43-11   OPTIMAL SITES OF FOSSILS ARE RARE.
Beartooth Butte in southern Montana is composed of sedimentary rocks that have yielded many early fish fossils, especially from the Devonian period (Chapter 26). Conditions for fossilization were ideal when these rocks were laid down in an ancient inland sea. Now, all of the surrounding sedimentary rocks are eroded away so that the butte stands alone above underlying Paleozoic-era rocks that contain still older fossils. Erosion now exposes the well-preserved fossils to the keen-eyed paleontologist.

averaging—that is, determining the length of time represented in a given sample of fossils. The chance events of death, rapid burial, and movement of remains before or after fossilization make it difficult to precisely define when one organism lived relative to others. The fossils found at a single site could have been from organisms living hundreds, thousands, or even millions of years apart, and thus represent a "time window" with huge dimensions that the researchers must determine. All these facts mean that we gaze at ancient life forms through a warped window that both distorts the view of certain types of ancient creatures and tells us nothing about the much wider variety of organisms that probably have shared our planet.

Estimations of Fossil Age

Biologists trying to reconstruct a possible evolutionary history of the life forms represented in fossil deposits have been able to determine the relative ages of fossils within reasonably precise limits. The best technique for this purpose is *radioisotope dating*, in which isotopes can serve as "radioactive clocks" for measuring the passage of time. In order to apply radioactive-dating techniques to a fossil, it is essential to know three things: (1) the *half-life* of the isotope being measured (the length of time it takes for one-half of a given amount of the isotope to decay); (2) how much of the isotope was originally present in the fossil or the rock containing it (the ratio of naturally occurring mixtures of isotopes of each element is known); and (3) how much of the isotope is left. Under ideal conditions, radioactive dating allows paleontologists to establish the age of a fossil with about 98 percent accuracy.

One of the best known radioisotopes is carbon 14 (^{14}C), which makes up a low but constant fraction of the carbon atoms in living matter (that is, a ratio of ^{14}C to ^{12}C). When an organism dies, no additional carbon (^{14}C or ^{12}C) is added to it. Hence, spontaneous radioactive decay causes the ratio of ^{14}C to ^{12}C atoms to gradually change year after year. Since the half-life of ^{14}C is known to be 5,730 years, scientists can accurately date fossils by measuring how much ^{14}C they contain relative to ^{12}C (that is, if the amount of ^{14}C is one-half that at zero time, the fossil is roughly 5,700 years old; if the ratio is one-quarter of the zero-time level, the fossil is about 11,500 years old, and so on).

There is a fly in this radioactive ointment, however. In terms of the immense spans of geological time, ^{14}C's 5,730-year half-life is quite short. After about 40,000 years, the amount of ^{14}C left in a fossil is so small that it is virtually undetectable. But many major groups of organisms lived hundreds of millions of years ago. Thus for more ancient materials, isotopes with much longer half-lives must be used. Such slower-decaying isotopes, shown in Table 43-2, have been used in combination

Table 43-2 **RADIOISOTOPES COMMONLY EMPLOYED TO DETERMINE THE AGES OF ANCIENT ROCKS**

Parent Element*	Daughter Element*	Half-Life (millions of years)
Uranium 235	Lead 207	713
Potassium 40	Argon 40	1,300
Uranium 238	Lead 206	4,510
Rubidium 87	Strontium 87	47,000

*The parent elements emit radioactive particles and thereby decay into the daughter element. For example, in 713,000,000 years, 1 kilogram of uranium 235 would have decayed to form 0.5 kilogram of uranium 235 plus 0.5 kilogram of lead 207.

with a variety of geological methods to establish the ages of rock formations in many places on Earth. Fossils found in those sites are assumed to be of the same age as the rocks in which they are embedded.

Reconstruction of Evolutionary Lines

Once the ages of groups of fossils have been established, it may be possible to build a **phylogeny**—a description of the lines of descent (or *lineages*) of plants and animals as they lived from one era to the next. A remarkably detailed lineage has been constructed for that noble beast, the horse (genus *Equus*). The unusually complete fossil record for this group of animals has allowed paleontologists to trace forms that resemble each other but also show a sequence of changes through time (Figure 43-12). For instance, the gradual transition in the length of limbs and loss of toes in this lineage is well documented (Figure 42-9). The early "horses," of the genus *Hyracotherium*, were small by modern standards—about the size of a small German shepherd—and had several toes. Then, over more than fifty million years, a series of evolutionary stages led to large animals with one toe with its hoof. Altogether, more than twenty genera of the horse family—of which only *Equus* remains—are represented in a complex evolutionary tree that offers powerful evidence of an ages-long evolutionary continuum.

Unfortunately, fossil collections are not always complete enough to provide the kind of insight into the evolution of a particular type or trait that we have for horses. For example, while fossil representatives exist for all the major categories of echinoderms (sea stars, sea urchins, sea cucumbers, and their relatives), the record has important gaps that make it difficult to map a phylogenetic tree showing unequivocally which organism is descended from which. In situations like this, where the fossil record is spotty but a particular plant or animal type is represented by many living descendants, biologists can infer likely phylogenies in several ways, includ-

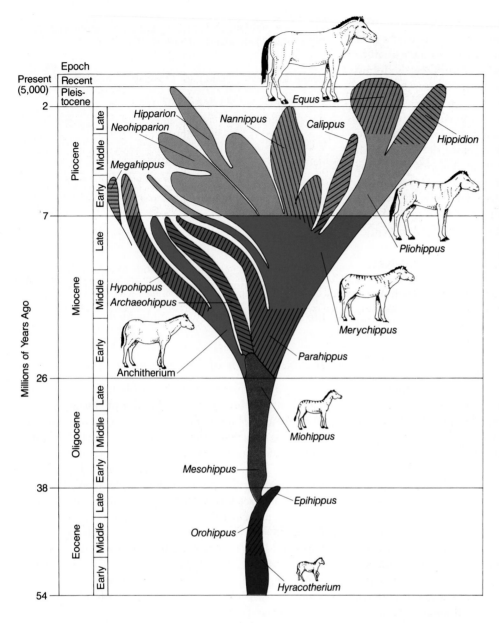

Figure 43-12 THE EVOLUTION OF HORSES. The fossil record for the phylogeny of horses is unusually complete. This figure shows the phylogenetic tree and an artist's re-creations of what some forms of the ancient horses may have looked like. Although the overall trend is toward larger size, genera such as *Nannippus* became progressively smaller again. Note the great diversification of this family tree of horses, most branches of which became extinct.

ing comparison of morphological features and analysis of similarities and differences in chromosomal and biochemical characteristics of living forms.

Parallel, Convergent, and Divergent Evolution

The kernel idea in building a phylogeny is that similarities in body structures, biochemistry, reproductive strategies, and other features of organisms can be used to trace lines of common descent. Doing so can be complicated, however, for extremely different patterns of evolution are found. These patterns are called parallel, convergent, and divergent evolution; failure to distinguish among them has caused much misinterpretation of evolutionary relationships. Let us see why.

Consider porcupines that have evolved independently in Africa and South America (Figure 43-13). More than seventy million years ago, before those two land masses separated and were pushed apart, the ancestral form of today's porcupines resembled a large, furry rat. When the two continents drifted apart, each rodent group evolved independently, but in generally similar environmental surroundings. The intriguing outcome is an example of **parallel evolution:** "New World" and "Old World" porcupines with remarkably similar basic body features, including sharp, hollow spines, even though the two groups have been evolving separately for seventy million years.

Another phenomenon is **convergent evolution,** which occurs when two (or more) distantly related lineages become *more* alike as they evolve similar adaptations (Figure 43-14). For instance, although reptiles, birds,

(a)   (b)

Figure 43-13   PARALLEL EVOLUTION IN PORCUPINES.
(a) The American porcupine, *Coendou prehensilis*, and (b) the Old World porcupine, *Hystrix africaeaustralis*, have a common
ancestor that lived 70 million years ago, before South America and Africa drifted apart. The porcupines have evolved
independently on separate continents to modern forms that are amazingly similar. This is an example of parallel evolution.

Figure 43-14   CONVERGENT EVOLUTION.
Convergent evolution is particularly striking in the body
types evolved for rapid swimming in an aquatic
environment. (a) The porpoise, a mammal; (b) the
Ichthyosaurus, an extinct reptile; (c) the penguin, a bird;
and (d) the shark, a fish evolved independently, yet each
has a shape that creates minimal drag, allows fast
swimming, and employs the anterior pectoral appendage as
a control surface during swimming (see also Chapter 26).

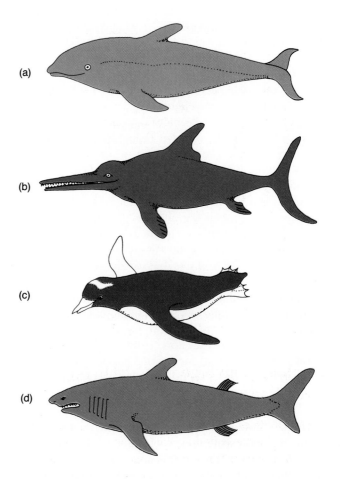

and mammals initially evolved on land, many of their
descendants reinvaded the water. Among many aquatic
mammals are sea lions, seals, porpoises, and whales;
aquatic birds include penguins, auks, and cormorants;
while aquatic reptiles include sea turtles, crocodiles, sea
snakes, and over a dozen types from the Mesozoic era
(age of reptiles). The remarkable likeness of today's porpoises to the extinct *Ichthyosaurus* (Figure 26-18), despite very different ancestry, attests to the strict demands of life in the water placed on fast-swimming
predators: both show sleek, streamlined bodies, dorsal
fins, and substantial tail fins for generating thrust. Even
so, the ichthyosaur's tail moved from side to side to generate propulsion, whereas the porpoise's moves up and
down; here, the same problem, generating thrust, is
solved in different ways during convergent evolution.

Both parallel and convergent evolution point up a per-
plexing problem often encountered in evolutionary biology: deciding whether a particular similarity in form or
function is an example of **homology**—derivation from a
common ancestor, such as a human's arm, a bat's wing,
or a whale's flipper—or of **analogy**—structures that
evolved independently but carry out similar functions,
such as the squid's gliding fin and the wings of birds and
insects. Homologous structures are built in embryos
using the same basic genetic program shared by the related organisms. In contrast, analogous structures serve

SILVERSWORDS: HAWAII'S EXEMPLARS OF ADAPTIVE RADIATION

Adaptive radiation is the height of evolutionary drama: a single ancestral lineage experiences a burst of change over a relatively short time span—sometimes just a few million years—and gives rise to many new species, each suited to its own particular environment. The concept of adaptive radiation helps explain the marvelous diversity among mammals, birds, flowering plants, and other groups, and serves as a major theoretical tool for evolutionary biologists. Until recently, biologists had to rely mostly on fossils to trace instances of adaptive radiation. Now, however, researchers have found adaptive radiation in action in a set of plants that grow in Hawaii's lush rain forests and on the islands' volcanic mountainsides. The subjects of their work are the lovely and varied plants known as silverswords—altogether, twenty-eight species in three genera. If the researchers are right, the silverswords are living proof of adaptive radiation.

The silversword species form what is called the "silversword alliance," and this assembly of related species shows a remarkable diversity of shapes adapted to virtually every type of Hawaiian habitat. For example, the compact, rosettelike silversword *Argyroxiphium macrocephalum* (Figure A) grows in high, dry, barren volcanic soil at altitudes of nearly 10,000 feet on the massive Haleakala Crater on the island of Maui. The plant's narrow, swordlike leaves are covered with a thick carpet of silvery hairs. These probably inhibit water loss and screen the underlying tissue from the intense sunlight at high altitudes. A second member of the alliance, *Dubrutia menziesii* (Figure B), flourishes nearby but while it, too, is adapted to a dry climate, individuals grow not as ground-hugging rosettes but as densely branching shrubs. Yet another silversword species, this one a denizen of moist lowlands, has large, broad leaves, a stem more than one foot in diameter, and stands more than twenty feet tall!

Silversword researchers, exploring evolutionary relationships among such varied species, began with the following premise: the Hawaiian Islands are about 6 million years old, and all the indigenous life forms must have arrived there or evolved there during that span of time. Building on this premise, investigators are beginning to find specific evidence that links the various silversword groups.

Some of the most intriguing findings are genetic. Because 6 million years is just a fleeting moment in geological time, plants descended from a single com-

Figure A
Adaptations for intense sunlight, aridity, and high, cold elevations: the silversword *Argyroxiphium sandwicense*.

Figure B
Silversword diversity: *Dubrutia menziesii* is a bushlike plant with small "swords" and a great ability to withstand high, dry conditions.

mon ancestor might still be able to interbreed. In fact, researchers have found that *all* of the species in the silversword alliance can form hybrids, no matter how unlikely the mating of 20-foot "tree" and 2-foot shrub. What is more, all the hybrids are at least somewhat fertile. This discovery confirms the close genetic relationships between given silversword species, while slight differences in the degrees of fertility help reveal evolutionary distance between them.

Other genetic research focuses on chromosome patterns. Biologists have found that nineteen members of the silversword alliance have twenty-eight chromosomes in each somatic cell, but at least nine species have only twenty-six chromosomes. Researchers Gerald D. Carr, Robert Robichaux, and Donald W. Kyhos identified one twenty-eight-chromosome species as the probable ancestor of the nine twenty-six-chromosome species, all of which, interestingly, grow on the geologically younger islands of the Hawaiian chain.

New information on the evolution of fruit flies on the Hawaiian Islands has relevance for the study of silverswords. The new evidence suggests that the first fly colonization occurred perhaps 40 million years ago on islands that are now 1,000 miles to the northwest of modern Hawaii. The colonization would have taken place when these other islands were located over the volcanic hot spot that is now located under the island of Hawaii. In fact, a series of volcanic cones has probably been forming over that hot spot for 70 million years as much of the Pacific sea floor moves toward the northwest. As each new volcanic island emerged above the Pacific waves, flies from older islands must have been blown to it and formed colonies. Plants, too, could have colonized the first-formed islands, and then their seeds, carried by wind, birds, and ocean currents, might have landed on each new island as it emerged. This sort of serial colonization might have contributed to the silverswords' impressive radiation and suggests that the diversification may have taken much longer than the currently estimated 6 million years.

The tantalizing experiments on silversword radiation continue. We can expect new information on the genetic events that allowed silverswords to thrive in the rain forests, volcanic craters, and other diverse habitats of Hawaii. These studies of adaptive radiation at work should give evolutionary biologists an unparalleled view of one primary pattern of change in the living world.

the same function, but there is no common genetic basis for them; birds, flies, and some kinds of squids can glide through air, but they lack a common winged ancestor and traveled separate evolutionary paths (Figure 43-15). The same conclusion applies to the eyes of insects, octopuses, and vertebrates; while all allow visual perception of the world and employ similar biochemistry to do so, the three types of eyes originated and evolved independently.

The course of evolution is often depicted as a tree, such as that representing the phylogenetic history of the horse in Figure 43-12. In fact, the splitting of one or more "branches" of a single line is one of the most common evolutionary patterns that can be reconstructed from the fossil record. This sort of **divergent evolution,** or **radiation,** has also produced many (perhaps most) modern species and genera. In the Hawaiian Islands for instance, new species have sprouted in astonishing variety from founders blown or washed ashore from distant populations. One example is an unusual family of finchlike birds called honeycreepers that has evolved, probably from a few forebears, into more than twenty species. The honeycreepers have distinctive bills and feeding habits: some feed only on nectar; others, only on fruits, insects, or seeds. And even these specializations are further subdivided: insect eaters are divided into species that pick their victims from the surface of leaves and those that use their sickle-shaped bill to scoop sow bugs from crevices in bark and wood. Different nectar feeders, in contrast, have evolved bills that fit particular types of flowers (Figure 43-16). Radiation virtually always reflects an expanding repertoire of adaptations to new living conditions.

Extinction

Ten thousand years ago, vast herds of horses thundered across the inland prairies of North America—tall, swift-footed remnants of the ancient lineage traced back to *Hyracotherium*. Then, suddenly, the horses vanished, perhaps hunted to extinction. Fortunately other stocks of the genus *Equus* survived in Europe and elsewhere and thousands of years later were reintroduced to both North and South America by Spanish explorers. This local demise of populations, even as vast a one as the original North American horses, probably goes on frequently. And if no relatives survive elsewhere, full extinction occurs. **Extinction** means the permanent loss of species (or higher taxa), some of which may have inhabited the Earth for millennia.

What makes one group of plants or animals become extinct, while others thrive? We know that extinctions have occurred continually over evolutionary time; indeed, extinction has been the eventual fate of nearly all

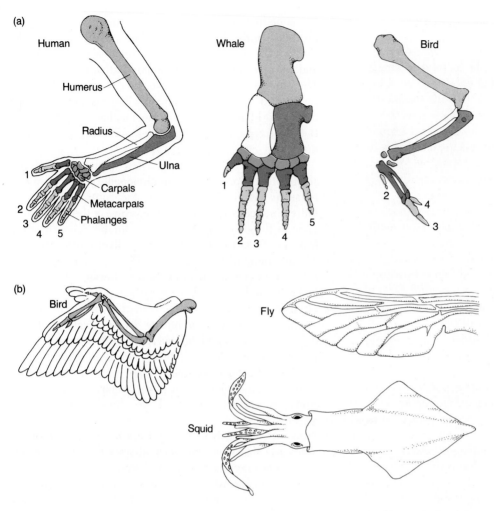

Figure 43-15 HOMOLOGOUS AND ANALOGOUS STRUCTURES.
(a) The forelimbs of a human, whale, and bird are homologous. These limbs derived from that of a common ancestor. The homologous bones—the humerus, radius, ulna, carpals, metacarpals, and phalanges—can be compared in each animal, but the limbs serve vastly different functions in each. (b) In contrast, the wings of a bird and a fly, and the fins of the squid (which are used in a few species for gliding in the air), are all analogous structures. The evolution of each was independent.

species—of the four billion species that have lived on Earth, only about two million are present today. In most cases—the great red elk and the saber-toothed tiger (Figure 43-17) are but two examples out of billions—we can only speculate about the forces or events responsible for the passing of species or of whole lineages of organisms.

By analyzing all sorts of marine and terrestrial fossils, paleontologists have determined that during Earth's protracted history, both a continuous background level of extinctions and a number of *mass extinction* events have taken place. Five mass extinctions are recognized, two of which resulted in enormous species loss. One mass extinction occurred at the end of the Permian period when 96 percent of all then-living marine invertebrate species became extinct, and another occurred at the end of the Cretaceous period, when 60 to 75 percent of all marine species died out (Figure 43-18).

Analysis of the latter event at the boundary between the Mesozoic and Cenozoic eras is crucial to interpreting mass extinction theories. One recent theory, for example, claims that the dinosaurs died out suddenly at this boundary, perhaps as a result of a giant meteorite hitting the Earth. Considerable data, however, show that dinosaur extinctions began in earnest some eight million years earlier, so that by the time the "mass extinction catastrophe" is reputed to have occurred, thirty dinosaur genera had already been reduced to about twelve. Further, fossil remains show that seven to eleven of those twelve genera survived for at least another 40,000 years, and lived among, and perhaps in competition with, the rapidly radiating mammalian ungulates. If we turn to plants, the boundary is marked by disappearance of angiosperm-dominated forests and appearance of fern-dominated, coal-forming "mire" vegetation; that, in turn, was replaced by new angiosperms, but those flowering plants were of very different sorts. This is a good illustration of the consequences of a mass extinction—clearing the environment, in a sense, so that many new species can evolve to fill the empty habitats.

These contradictory findings tell us several things: first, they show how very carefully researchers must interpret data on mass extinctions after tracing each kind of organism through the fossil record in great detail. Sec-

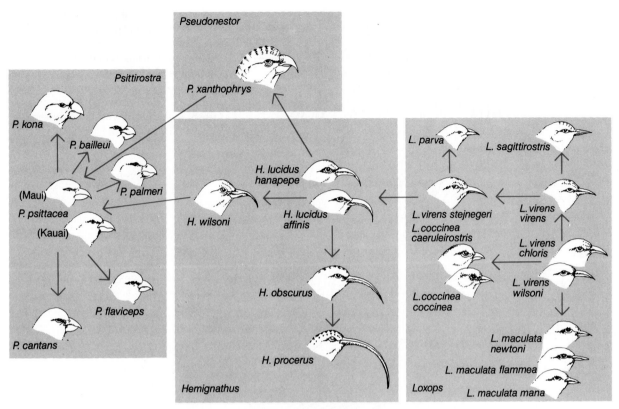

Figure 43-16   ADAPTIVE RADIATION OF HAWAIIAN HONEYCREEPERS.
Divergent evolution of many branches off a single line is called radiation (or adaptive radiation, because the evolution is an adaptation to a variety of living conditions). More than twenty species of Hawaiian honeycreeper have evolved from only a few founders in the Hawaiian Islands. Each species has its own particular bill shape and feeding specialization.

Figure 43-17   FOSSIL SKULL OF AN EXTINCT SABER-TOOTHED TIGER.
The saber-toothed tiger, *Smilodon californicus*, lived in what is now California until about 10,000 years ago. It weighed about 225 kilograms and probably preyed on animals such as mastodons and ground sloths. We can only speculate on the reason this fierce predator became extinct.

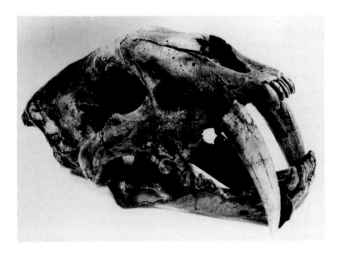

ond, they emphasize that periods of mass extinction were probably not brief catastrophic episodes, but instead may have lasted for huge numbers of years.

Some observations raise the fascinating idea that mass extinctions may occur in cycles many millions of years in length. Though the proposal is now judged incorrect, it is still of interest to ask if mass extinction can influence individual life spans and thus Darwinian natural selection. Biologists have assumed in the past that organisms become extinct because they fail to adapt to some newly arising environmental stress. However, as David Jablonski of the University of Chicago points out, characteristics that arise by the mechanisms of genetic variation and natural selection and which are highly adaptive during normal times may become quite irrelevant when conditions such as a worldwide lowering of sea levels, for example, trigger global changes in weather patterns and subsequent mass extinctions. Then a very different set of characteristics unrelated to normal fitness and survival might hold sway. Consequently, species and their special adaptations may disappear not because of low fitness but because of a new environmental crisis never before experienced.

It is significant that mass extinctions appear to wipe

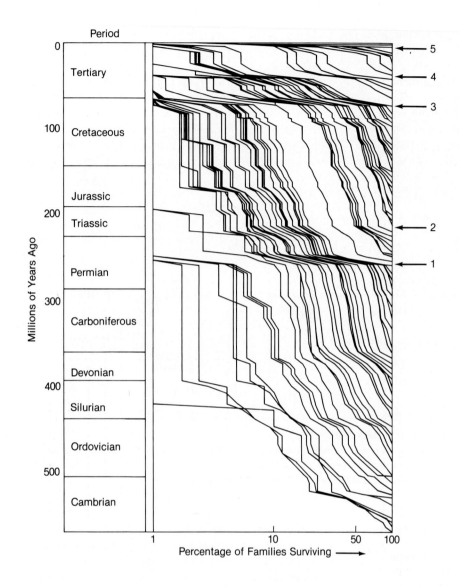

Period

Millions of Years Ago

Tertiary

Cretaceous

Jurassic

Triassic

Permian

Carboniferous

Devonian

Silurian

Ordovician

Cambrian

Percentage of Families Surviving ⟶

Figure 43-18   EXTINCTIONS IN EARTH'S HISTORY.
Here the percentage of 2,316 families of marine organisms surviving at each time period is plotted against geological time. Each line represents a cohort of related families. Times of mass extinctions are seen as sharp drops in the survivorship curves (1, 2, 3, 4, 5).

out many but not all kinds of organisms. Biologists do not yet know the rules of survival—if there are any such rules—for times of environmental crisis. University of Chicago paleontologist David Raup points out that during normal times, widespread geographical distribution and a richness of species within a biological group seem to promote survival. Yet counter to intuition, such properties provide no advantage during times of mass extinction. Hence, mass extinctions may not be "constructive" in the same sense as is Darwinian selection. Randomly spaced, or even cyclical, extinctions tens of millions of years apart would occur at intervals so huge relative to any organism's life history that no individual could experience such events and be selected to survive them. Still, mass extinction events have undoubtedly had major effects on life's major evolutionary outlines. Unpredictable and extensive culling of existing life forms must be considered alongside the ongoing processes of Darwinian variation and selection in constructing an

explanation for how and why life forms have evolved.

Let us now turn to another phenomenon that may have contributed to the panorama of life on our planet.

The Punctuated-Equilibrium Theory

In this chapter and the two previous chapters, we have focused on some of the basic how's of evolution—how gene pools can vary, how natural selection operates, how species arise. But there is another important question we have not yet considered in detail: How fast do *major* evolutionary changes take place? For generations of scientists, the answer has been "very slowly." Immense stretches of geological time are thought to have been marked by an accumulation of gradual changes in species, a kind of steady pulse rate for evolution. But today, there is a hard-fought scientific controversy over this gradualist view, with some biologists suggesting that evolution proceeds by "jumps": radical changes over a short period of time, separated by long periods of stability. This new idea is called **punctuated equilibrium.**

The idea that evolution has two contrasting phases—rapid change (the "punctuations") followed by equilibrium—has been championed by paleontologists attempting to explain apparent gaps in the fossil record as well as instances in which life forms, such as the lungfish and the horseshoe crab, have remained essentially unchanged over hundreds of millions of years. The central

tenet of punctuated equilibrium is that a lineage of organisms arises by some dramatic changes—say, the rapid acquisition of body segmentation in annelids—after which there is a lengthy period with far fewer radical changes taking place. Figure 43-19, for example, shows the fossil history of one type of echinoderm, the sea urchin (Chapter 25). It looks as though this group underwent its first major radiation in the late Cambrian or early Ordovician period: many derivative types evolved with the same basic, common features that allow us to classify them in one phylum. But note that about 225

Figure 43-19   FOSSIL HISTORY OF SEA URCHINS (ECHINODERMS): AN EXAMPLE OF PUNCTUATED EQUILIBRIUM?
The time ranges and classifications of the major sea urchins are shown. The "punctuations" occurred in the Ordovician and Jurassic periods; the "equilibrium" conditions extended through the intervening tens of millions of years. Each vertical line represents a genus or, in some cases, an order of echinoderms that may contain multiple species.

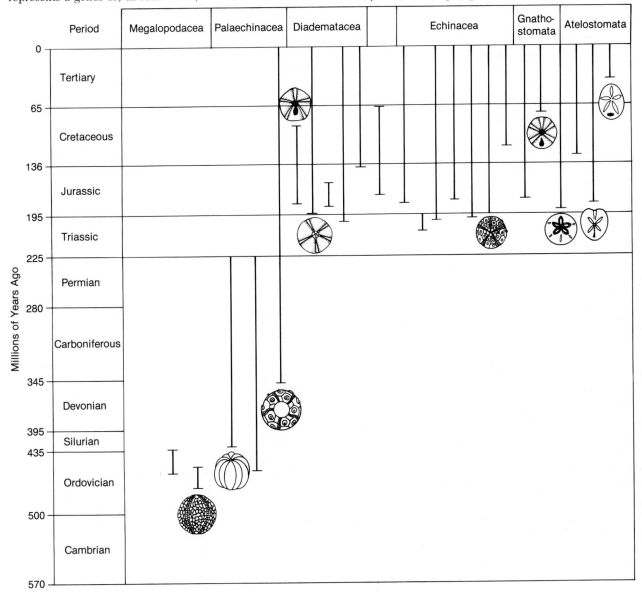

million years ago, many lines became extinct. Soon thereafter, early in the Jurassic period, many new types appear in the fossil record. And still more arose later during the Cretaceous period. One might interpret this record to reflect two "punctuations" in the Ordovician and early Jurassic periods. And the "equilibrium" times would be from the Ordovician through the Triassic and, perhaps, from the Jurassic to today. This record may be consistent with the punctuated-equilibrium hypothesis. To complement such cases of real animals, mathematical analyses show that small to moderate populations undergoing typical Darwinian random, small variations (that is, in equilibrium) do pass through periodic large transitions (punctuations) to new types. This implies that special genetic, developmental, or ecological mechanisms are not needed to yield punctuations.

Many biologists, however, have pointed out problems with the concept of punctuated equilibrium. First, as we saw earlier, the fossil record is frustratingly incomplete. Our modern view of past life on Earth is analogous to what we might see looking through a few scratches and chips on a painted windowpane. Second, our ability to discriminate in time, especially when the evidence consists of fossils hundreds of millions of years old, is quite limited. Even with radioactive-dating techniques, determining fossil ages within five million years might be as accurate as we can get. Such a time period may represent hundreds of thousands or even millions of generations in the life history of a group of organisms.

For these and other reasons, the notion that new groups arise as the result of macroevolutionary "bursts" tends not to be supported when an abundant fossil record is available—one without major time gaps and with an abundance of specimens representing related species. The horse lineage is one example. Another is a study by P. G. Williamson, a paleontologist who examined thousands of fossil clams and snails from the Cenozoic era (the last 65 million years) and found that the thirteen lineages in his study included many species that remained unchanged in morphology over millions of years. Interestingly, Williamson did note that "new" species seem to turn up where geological formations reflected environmental stress—for example, the periodic drying up of lakes in which the mollusks lived.

Even these changes, however, took place over periods of 5,000 to 50,000 years. Today, living snails and clams produce new generations in six months or a year; at this rate, about 20,000 generations would fill the time Williamson allotted for speciation—an ample period for substantial microevolution to occur. In laboratory studies of population genetics, this would be the equivalent of a 1,000-year experiment with fruit flies or a 6,000-year

mouse-breeding test! And think about the case of selecting *Drosophila* that fly up or down in just a few generations (Chapter 42). Clearly, the perspective we take influences the way we interpret facts. On the one hand, the equilibrium of some lineages of organisms may indeed be "punctuated" by environmental changes or other factors; on the other hand, gradualism surely characterizes much of the history of many species.

GETTING TO HERE FROM THERE: THE ROLE OF MICROEVOLUTION IN MACROEVOLUTION

A key question about macroevolution is whether some radical genetic process operates to generate the gross changes in phenotype that distinguish organisms in major taxonomic groups from each other. In normal environmental circumstances, such "hopeful monsters" might be selected against and be lethal. But the chance coincidence of a catastrophic environmental event with such a gross phenotypic change might yield a viable new lineage—a punctuation point in the history of life. However, the known mechanisms leading to genetic variation—nonrandom mating, genetic drift, migration, and the many types of gene and chromosomal change grouped as mutations—seem to offer reasonable explanations for many of the changes we see represented in the Earth's fossil history and in the differences among today's organisms. In general, such differences can be interpreted as microevolutionary adaptations to different ways of life; they can be viewed as variations culled by natural selection that have equipped the organisms that possess them to meet the demands of the particular niches those organisms occupy.

Current explorations of novel genetic and developmental mechanisms, using the new techniques of molecular biology, may broaden the list of potential mechanisms for micro- and macroevolutionary change. Perhaps then we will know whether the sorts of microevolutionary speciation taking place as organisms at two ends of a cline become reproductively isolated can also lead to new genera, families, or even phyla.

In the next four chapters, we move from evolutionary processes to the ecology of organisms and the factors that circumscribe and regulate the environment in which life forms exist. In doing so, we shall see what causes adaptations to persist and account for so many basic biological characteristics of living species.

SUMMARY

1. For most modern biologists, *species* are defined as groups of actually or potentially interbreeding populations that are reproductively isolated from other such groups.

2. Several well-understood mechanisms of *reproductive isolation* can split interbreeding populations into genetically distinct species. *Prezygotic isolating mechanisms* block the formation of zygotes, while *postzygotic isolating mechanisms* result in nonviable or sterile hybrids. In both cases, gene flow is prevented.

3. Prezygotic isolating mechanisms include *ecological isolation*, in which populations become adapted to different environments; *behavioral isolation*, which results from differences in courtship or mating behavior; *mechanical isolation*, in which mating is prevented by incompatible genitals or flower parts or molecules that fail to bind sperm to egg; and *temporal isolation*, in which gene flow is prevented because two species reproduce at different times.

4. Postzygotic reproductive isolation may be the result of *hybrid inviability*, in which the organism dies as an embryo or before reproducing; *hybrid sterility*, in which the interspecific hybrid is viable but sterile; and *hybrid breakdown*, in which hybrid offspring, often in later generations, are weak or sterile.

5. Populations that are spread out over a broad geographical range often are arrayed in *clines*; these are gradual changes in phenotypic characteristics from one end of the range to the other. Local *subspecies* with distinctive characteristics may arise along a cline. When crossed, members of the species at each end of the cline may be unable to produce viable offspring.

6. According to the model of *allopatric speciation*, subgroups of existing species go through an initial period of separation by a physical or geographical barrier; genetic differences between the two groups gradually accumulate, so that if the physical barriers are removed, the subgroups have become reproductively isolated species.

7. The extent of differences between species that are in the process of diverging can be determined through a statistic called *genetic identity*. It has been suggested that small changes in regulatory genes may be responsible for many of the major changes that mark the evolution of species and of higher taxonomic groups.

8. Rapid speciation has been observed in some organisms as a result of chromosomal changes. Rapid speciation in plants is typically due to *polyploidization*—the rapid duplication of chromosome sets. In some animals, changes in chromosome structure may account for new species. Here, *sympatric speciation* occurs even though there is no geographical barrier to prevent gene flow.

9. *Macroevolution* describes the major changes in phenotype that have occurred over evolutionary time. Some macroevolutionary events are identified by careful examination of the fossil record; others are inferred by comparison of related living organisms. Fossil age can be estimated by using radioactive isotopes.

10. *Phylogenies* are reconstructed lines of organisms' descent over evolutionary time. When all the morphological changes undergone by evolving taxa are examined through time, several phylogenetic trends emerge.

11. *Parallel evolution* occurs when two or more lineages evolve along similar lines. *Convergent evolution* is the independent evolution of similar (analogous) adaptations within very distantly related lineages. *Divergent evolution*, or *radiation*, is the common pattern in which lineages split off from a parental group and then diversify into related species. *Homology* refers to phenotypic characters based on the same basic genetic program in related lineages. *Analogy* refers to phenotypic characters that resemble each other in different lineages, but are not based on common descent or genetics.

12. *Extinction* is the complete loss of a species or group of species. At least five mass extinction events have occurred during Earth's history. Species that are highly fit for normal circumstances may be eliminated during mass extinctions since their adaptations may be irrelevant under the new environmental stress. Mass extinctions have in major ways shaped large-scale patterns of life on Earth.

13. According to the *punctuated-equilibrium* theory, evolution proceeds by jumps—radical changes over short periods of time—separated by long periods of stability. Gradualism refers to small-scale evolutionary changes in species, which ultimately might lead to reproductive isolation and speciation.

14. It is not known whether macroevolution proceeds by "punctuations." Nor is it clear whether unique mechanisms are required for either punctuations or macroevolution. Some biologists conclude that normal *microevolutionary* speciation is sufficient to account for macroevolutionary events, including punctuations.

KEY TERMS

allopatric speciation

analogy

behavioral isolation

cline

convergent evolution

divergent evolution (radiation)

ecological isolation

extinction

genetic identity

homology

hybrid breakdown

hybrid inviability

hybrid sterility

macroevolution

mechanical isolation

microevolution

parallel evolution

phylogeny

polyploidization

postzygotic isolating mechanism

prezygotic isolating mechanism

punctuated equilibrium

reproductive isolation

species

subspecies

sympatric speciation

temporal isolation

QUESTIONS

1. Design a test to see whether two newly discovered populations of tropical-rain-forest orchids are distinct species, according to the modern biological definition.

2. Write a brief essay that gives examples of the various types of prezygotic and postzygotic isolating mechanisms.

3. Field biologists in Africa have discovered a cline of zebras with different stripe patterns. How could you test whether the obvious differences in color pattern correlate with differences that might result in speciation?

4. Distinguish between allopatric and sympatric speciation. Cite as examples populations suddenly separated by a Grand Canyon-sized chasm and populations on a remote Pacific island that is large enough to have a variety of local environments.

5. Define analogous structures and homologous structures, and give examples of both.

6. What is convergent evolution? What is parallel evolution? Give examples of both. How do the terms "analogy" and "homology" apply to convergent evolution and parallel evolution?

7. What is macroevolution? Give some examples.

8. Which living things are most

likely to have left fossils? How are fossils dated?

9. What is a radioactive half-life? The half-life of carbon 14 is nearly 6,000 years. In a particular sample, what fraction of the original radioactivity will remain after 12,000 years? after 24,000 years?

10. What kinds of observations does the punctuated-equilibrium theory attempt to explain? Present some arguments for and against the theory. What kinds of data are needed to test the theory?

ESSAY QUESTIONS

1. Explain how gradualism and microevolutionary processes could lead to a group in a new taxonomic category higher than species—say, a different family.

2. Some scientific evidence suggests that a major nuclear war could generate so much smoke and dust that much sunlight and heat would be blocked from reaching the Earth's surface. If such a circumstance resembles the effects of the Earth being hit by a large meteorite, what might consequences be for animal and plant life on our planet?

SUGGESTED READINGS

Bush, L. "Modes of Animal Speciation." *Annual Review of Ecology and Systematics* 6 (1975): 339–364.

In this article are references to the classic books and papers as well as a fine overview of the possibilities.

Gould, S. J. *Ontogeny and Phylogeny.* Cambridge, Mass.: Harvard University Press, 1977.

The originator of the punctuated-equilibrium theory summarizes old ideas in new ways and new ideas in challenging ways.

Mayr, E. *Populations, Species, and Evolution.* Cambridge, Mass.: Harvard University Press, 1970.

One of the great evolutionary biologists of the twentieth century expounds on allopatric speciation and microevolution. He discusses the Panda work in *Nature* 323 (1986): 769–771.

Raup, D. M. "Biological Extinction in Earth History." *Science* 231 (1986): 1528–1533.

Here is a fine discussion of the modern view of extinction, including its possible periodicity and its role in the history of life on Earth.

Stanley, S. M. *Macroevolution.* San Francisco: Freeman, 1979.

This book gives a broad view of major changes in evolutionary history as seen from the punctuationist perspective.

White, M. J. D. *Modes of Speciation.* San Francisco: Freeman, 1977.

Here is a complete view of both plant and animal speciation—an advanced but a first-rate reference.

44

ECOSYSTEMS AND THE BIOSPHERE

. . . Rachel Carson [in Silent Spring] was alerting the world to what has been called the fundamental principle of ecology, namely: we can never do merely one thing. . . . We can never do merely one thing, because the world is a system of fantastic complexity. Nothing stands alone. No intervention in nature can be focused exclusively on but one element of the system.

Garrett Hardin, *Bulletin of the Atomic Scientists* (January 1970)

Tahiti: a multitude of ecosystems, from coral reefs to mountain rain forests, each with its own abundance of species.

The desert pocket mouse seems well adapted to its environment: its tawny-colored fur provides camouflage; the animal can acquire most of the water it needs through the oxidation of molecules from dry seeds and other foods; and to avoid heat and predators, the rodent burrows underground and emerges almost exclusively at night. Evolutionary adaptations such as these arise over generations and millennia, and the individuals are thereby able to exploit their environments and reproduce successfully. But there is also a minute-by-minute interplay of each organism with its own immediate environment. Thus a biologist might find that successful exploitation by a pocket mouse of its small patch of desert depends on food availability this year, the number of predators that chance to be in the area, temperature, wind velocity, soil coarseness, food types and their water content, odor signals left by other mice and by predators, and so on.

Field biologists could draw up an analogous list for many different organisms, whether plant or animal, fungus or microbe. The biological discipline concerned with the interplay of organisms and their home environments is **ecology** (from the Greek word *oikos*, meaning "house"). Although for many, ecology has come to connote the study of environmental pollution and disruption, it has a precise and scholarly scope: study of the interactions that determine the distribution and abundance of organisms.

Such interactions necessarily begin with a constant source of energy—predominantly, the flow and harvesting of life-supporting solar energy through photosynthesis or food gathering. The interactions then extend to chains of dependencies between particular organisms and, ultimately, to complex networks that weave together communities of organisms with the physical and chemical constituents of their world. We saw that life's ancient origin depended on the precise characteristics of the Earth's early environment (Chapter 19). Here we will see the interdependence of life forms and their nonliving surroundings: life changes the world, and the altered planet sets limits on what life can be.

Some branches of ecology *are* concerned with environmental disruption, and in this chapter we will discuss specific instances, such as the rising levels of carbon dioxide and other trace gases and the resultant greenhouse effect, and the impact of phosphate-containing laundry detergents on fresh-water ponds and streams. However, to fully understand the interactions of organisms and their environment, one must take a systematic approach to ecology, as we do in this chapter and in the next two chapters, moving from the most complex level of biological interaction to the smallest and least complex. We begin in this chapter with the *ecosystem*, the broadest view of such interactions. In Chapter 45, we focus on *communities*, all the interacting populations of different species within an ecosystem, and show how changes in community interactions occur over time. Finally, in Chapter 46, we look closely at *populations*, individuals of a single species within the community. Let us begin, then, at the top of the hierarchy with a definition of the ecosystem.

ECOSYSTEMS

The most complex level of biological organization is the ecosystem. An **ecosystem** is a complete life-supporting environment, including the entire community of interacting organisms, their physical and chemical environment, energy fluxes, and the types, amounts, and cycles of nutrients in the various habitats within the system. Ecosystems are both dynamic and incredibly complicated. Nowhere is this better seen than in the largest ecosystem: the Earth itself, or, more precisely, its life-supporting soils, seas, and atmosphere and all the organisms in them—collectively called the *biosphere*. The biosphere is clearly a vast ecosystem, but the term "ecosystem" can also be applied to much smaller environments having natural boundaries that make them more or less self-contained—a meadow within a forest; a beach, with water on one side and grassy dunes on the other; or an island, surrounded by ocean (Figure 44-1). In every one of these cases, however, the boundaries are not sharp, nor are the ecosystems *truly* self-contained: energy must enter; gases, nutrients, and water cycle in and out; organisms come and go in space and in time.

Figure 44-1   GRADING OF ECOSYSTEMS INTO EACH OTHER.
Here, near the tree line high in the mountains, one can find a variety of ecosystems, including these grassy, flower-strewn meadows, open conifer forests, and higher rocky surfaces with little besides lichen surviving.

This blurring of boundaries reflects the fact that every small ecosystem exists within larger ecosystems, and they within still larger ones, until ultimately the largest ecosystem, the biosphere, is reached. Thus a rain forest on the mountain slopes of Tahiti may be viewed as an ecosystem by an ecologist, but it is affected by energy from the sun; moisture evaporated from the oceans and released by clouds as rain; tides generated by the moon; and seeds, microbes, and nutrients carried by winds or ocean currents to the island. John Donne's ". . . no man is an island . . ." may be adapted by the ecologist to ". . . no ecosystem is an island . . . ," for all are part of the marvelous, dynamic biosphere.

Despite their inescapably blurry boundaries, the ecologist must study smaller, circumscribed ecosystems. This is the only way a manageable number of factors and relationships can be analyzed and understood, using rigorous scientific methodology. Indeed, as smaller ecosystems are defined more and more precisely, we are better able to understand the interactions within the larger, more complex ecosystems, and to see how communities, populations, and individual organisms fit into ecosystems.

THE BIOSPHERE

The Earth's surface region—its lands, its waters, and the air above them—forms the **biosphere,** the life-supporting environment for all living creatures, from algae submerged in Arctic tundra ponds to sightless insects scuttling across cave floors. When one includes both the microbes that have been found floating as high as five miles in the Earth's atmosphere and the strange marine animals discovered in deep-ocean trenches, one can quantify the biosphere as the fragile sheath of habitable regions about fourteen miles thick surrounding the huge rocky globe. The biosphere has many parts and subparts—deciduous forest and rain forest, grassland and desert, seacoast and ocean floor—that provide the myriad **habitats,** or actual places where organisms live.

Earth and its biosphere are like an unbelievably large, complex machine with literally billions of ingredients and "moving parts" (carbon, nitrogen, minerals, liquid water, decomposers, photosynthetic organisms, and so on), each setting off untold numbers of chain reactions in the environment every minute of every day. Yet like any machine, absent one crucial factor and it all stops: *energy.* In Chapter 8, we learned how energy from our solar system's blazing central star is trapped by photosynthetic organisms and is used to manufacture the basic organic "foodstuffs" that support virtually all organisms. Without that continual influx of energy from outside the biosphere, nearly the entire complex "machine" at our

planet's surface would grind to a halt. (The exceptions are tiny ecosystems built on heat energy escaping from the Earth's core.) Let us explore the power of sunlight to perpetuate the Earth's life forms by bathing them in a constant flow of usable energy and by establishing climates that determine both the nature and the distribution of all living things.

The Sun as a Source of Energy

The vast majority of energy used by the Earth's life forms comes from the sun in the form of light and heat. The distance between the Earth and the sun determines the amount of solar radiation that our planet intercepts. At the outer edge of our atmosphere, an average of 1.05×10^{10} calories of solar energy falls per square meter (or about 1 million calories per square centimeter) every year. This value is called the **solar constant** (Figure 44-2). Of the sun's energy impinging on the Earth's atmosphere, approximately 9 percent is made up of short, invisible wavelengths in the ultraviolet (UV) part of the spectrum. These wavelengths are so damaging to life that UV radiation probably kept the land and upper regions of the ancient seas free of life on the early Earth (Chapter 19). Today, fortunately, most of the shortest ultraviolet rays are filtered out by a protective screen of ozone (O_3) in our atmosphere. Without this gaseous barrier to UV energy—first generated from the oxygen waste product of early photosynthetic cells—most terrestrial and shallow-water organisms in the biosphere would perish, their organic molecules and genetic material damaged by the energetic UV rays. This is why discovery in 1987 of a large "hole" in the ozone layer over the South Pole is so worrisome.

Of the remaining 91 percent of the sun's energy reaching the Earth's atmosphere, half falls within the range of wavelengths that make up the rainbow spectrum of visible light, while the other half falls within the infrared region—long wavelengths that organisms experience as heat. Because the total amount of energy impinging on the Earth is so great, it might seem reasonable to assume that there is an unlimited supply of energy available for powering life processes. However, more than 95 percent of all the electromagnetic radiation that showers our planet is either absorbed or reflected as it passes through the atmosphere. In addition to the ultraviolet that is screened out by O_3, as much as 35 percent of visible light is simply reflected back into space by clouds, ice, snow, oceans, and other particles in the air and at the planet's surface. Most of the infrared, accounting for nearly half of the solar constant, is absorbed by atmospheric carbon dioxide, water vapor, and the water droplets that form clouds. The net result is a drastic loss of potential energy, so that only about 1 to 4 percent of the

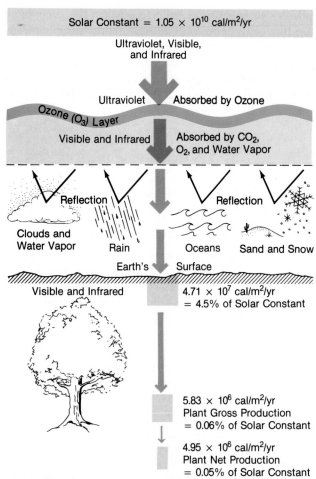

Figure 44-2   PROVIDING ENERGY TO SUPPORT LIFE: AN INEFFICIENT PROCESS.
Less than 5 percent of the solar energy first entering the atmosphere reaches Earth's surface. First, ultraviolet light is largely absorbed by ozone; then other wavelengths are partially absorbed by CO_2, O_2, and water vapor. Finally, reflection from clouds, rain, oceans, sand, and snow takes place. Gross production by all photosynthetic organisms is only about 0.06 percent of the solar constant; respiration then reduces that amount further, yielding the net productivity of plants shown here. Energy is expressed as calories per square meter of the Earth's surface per year.

energy first entering the atmosphere is available to do work at the planet's land and water surfaces.

What happens to the solar energy—heat and light—that finally strikes a terrestrial environment, such as a mountain side or a suburban backyard? To trace that fate, researchers in Michigan measured the photosynthetic activity of a grassland. They found that plants convert to sugars only 1.2 percent of the impinging solar energy (which, recall, is less than 5 percent of that entering the atmosphere). Then 15 percent of this converted

energy is used to fuel the plants' own respiration. In the end, the net conversion of solar energy to plant materials, the so-called net productivity, represents only about 0.05 percent of the solar constant. Other studies have shown that the use of solar energy during photosynthesis by aquatic plants in the Earth's oceans is even less efficient. Clearly, despite the enormous energy source at our solar system's center, usable solar energy is a finite resource for biological systems—limited by the solar constant, absorbed and reflected by living and nonliving parts of the biosphere, and consumed in the life processes of plants. The amount of energy that plants convert and store as the potential energy of organic compounds sets the upper limits of energy available to support all other organisms in the biosphere.

The Sun, Seasons, and Climate

The fact that the biosphere lies at the surface of a rotating, tilted sphere circling the sun has dramatic consequences for weather patterns, local climates, and the ability of different sorts of organisms to live at various places on the Earth's globe. Most of us know that the Earth rotates on its axis as it orbits the sun, thus creating day and night. Furthermore, as Figure 44-3 indicates, the Earth is tilted on its axis, and as the planet moves, this tilt has dramatic effects: twice each year, at the time of the vernal and autumnal equinoxes, the sun's rays fall directly on the equator so that the length of night and day is equal. At the winter solstice, the sun's rays fall directly on the Tropic of Capricorn (23½° south latitude); at the summer solstice, they fall directly on the Tropic of Cancer (23½° north latitude). The most important biological result of this planetary tilt is the uneven heating of the Earth's surface, thereby producing *seasons* and different patterns of *climate* around the globe.

A region's climate is one of its most critical characteristics, largely determining which photosynthetic organisms may survive there; the mix of those producers, in turn, affects the composition of microbial, fungal, and animal life in the area. In this and other ways, climate is a major evolutionary force, imposing constraints and establishing the ground rules for many plant and animal adaptations. Moreover, as the continents have drifted during the millennia relative to the fixed equator and Tropics of Cancer and Capricorn, local climate and weather have inevitably changed. Similarly, as huge mountain chains have pushed upward, only then to erode away, great changes in the continental topography have affected local weather conditions, and so organisms and their survival. This is all part of the long-term interaction between life and the geologically shifting planet on which it arose. Let us see, then, how climate patterns are created by the combined action of four factors: tem-

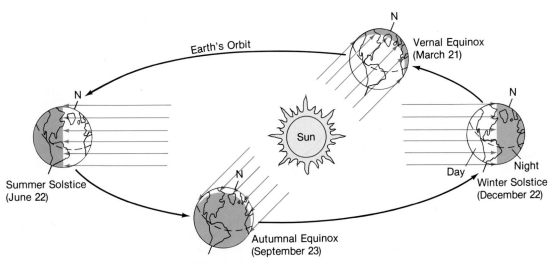

Figure 44-3   GLOBAL TILT AND THE SEASONS.
The Earth is tilted on its axis as it revolves around the sun. The tilt leads to uneven heating of the Earth's surface, and produces seasons and climate patterns around the globe. At the summer solstice (left), the sun's rays hit the earth perpendicularly in the Northern Hemisphere, heating it more and producing summer conditions there. That occurs in the Southern Hemisphere (right) at the winter solstice. The sun's rays fall perpendicularly on the equator at the vernal and autumnal equinoxes. Rotation of the Earth about the axis each 24 hours produces day and night.

perature, wind, rainfall, and topography of the Earth's surface.

Temperature, Wind, and Rainfall

Because the Earth is a sphere, different parts of the globe receive differing amounts of sunlight, as shown in Figure 44-4. In the middle zone, between 23½° north and south of the equator, the Earth's surface receives direct, perpendicular rays during certain times of the year. The highest average temperatures, therefore, are at and near the equator. In contrast, higher latitudes

generally receive the sun's rays at an angle; thus less solar energy reaches each square meter of the Earth's surface in those regions, resulting in lower average temperatures.

This pattern of global temperature would be the same even if the Earth were not tilted relative to the sun. Because it is tilted, the amount of solar energy falling at any one location varies regularly as the Earth circles the sun. Thus we have the splendor of the seasons. As shown in Figure 44-3, the northern latitudes receive the most

Figure 44-4   THE SUN'S RAYS AND THE SPHERICAL EARTH.
Due to the planet's spherical shape, different parts receive different amounts of sunlight. At the equator the sun's rays are shown to be perpendicular to the Earth's surface, and the maximum energy intensity is received. North and south of the equator, the sun's rays hit the Earth obliquely. A given "packet" of sunlight that hits the earth obliquely is spread out over a larger area of the planet's surface than the same packet hitting the earth perpendicularly. The energy intensity at any one such site is thus reduced. In addition, the obliqueness of the rays means that they must travel a greater distance through cloud cover and the atmosphere, further reducing the amount of energy that actually hits the surface of the Earth.

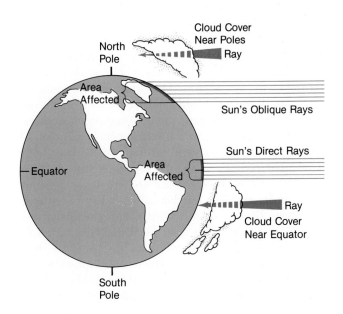

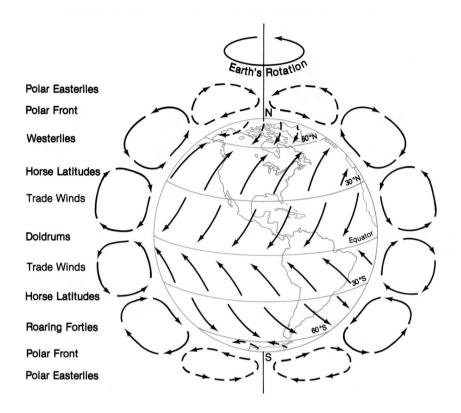

Polar Easterlies
Polar Front
Westerlies
Horse Latitudes
Trade Winds
Doldrums
Trade Winds
Horse Latitudes
Roaring Forties
Polar Front
Polar Easterlies

Figure 44-5   GLOBAL PATTERNS OF AIR CIRCULATION.

Uneven heating and the Earth's rotation cause circulating patterns of air to whirl along the surfaces of land and ocean in the directions shown by the rows of curved arrows drawn on the globe itself. The circular patterns of arrows drawn outside the globe show that air masses are larger near the equator and smaller at higher latitudes (these circles of air are shown in greatly exaggerated size in the drawing). The circles near the equator show that air currents rise in the tropics, spread out north and south, cool, then descend near the horse latitudes. The irregular polar wind patterns are indicated by dotted lines. The common terms for regions of the Earth's surface affected by these air patterns are listed to the left of the globe.

direct sunlight near the time of the summer solstice, the start of the northern summer and the southern winter. At the time of the winter solstice, the southern latitudes receive the most direct sunlight, and they experience summer as the north is in winter.

The uneven heating of the Earth's surface not only causes temperatures to vary, but also sets air masses in motion as *wind*. Consider what happens near the time of the vernal and autumnal equinoxes. Intense solar radiation at the equator heats the Earth's surface. This, in turn, heats the air above, and it rises in massive air currents that ascend from the tropics and lift high into the atmosphere. The heated air spreads out north and south, and because it cools in the process, the air descends to Earth again. As Figure 44-5 illustrates, circulating "cells" of alternately heated and cooled air twirl like atmospheric pinwheels near the land and ocean surfaces.

As this occurs, the Earth is rotating on its axis each 24 hours. In fact, if you stand on the equator your body would be traveling at about 1,000 mph (the Earth is 24,000 miles in circumference and rotates completely every 24 hours). This rotation applies a force on the atmosphere, oceans, and other fluids. This phenomenon—named the **Coriolis effect,** for the French engineer who described it—skews the churning masses of air in a clockwise direction in the Northern Hemisphere and in a counterclockwise direction in the Southern Hemisphere. The outcome of the Coriolis effect is that wind

currents near the Earth's surface moving north and south toward the equator are deflected into new patterns that ultimately come to have a significant impact on ocean currents and climate patterns (Figure 44-6). In latitudes where air masses are rising or falling (moving vertically rather than horizontally), surface air moves very little. Sailors coined both the term "doldrums" to express their dismay at the lack of wind to power their sails in equatorial regions, as well as the expression "horse latitudes" for wind strengths at 30° north and south latitudes, where air masses tend to fall so that there is little horizontal wind; sailors often threw livestock, including horses, overboard to lighten loads and speed progress in those latitudes—hence the name. The roaring forties are latitudes where powerful winds sweep across the Southern Hemisphere near 40° south latitude, uninterrupted by the mountainous land masses that slow similar winds in the Northern Hemisphere. Of course, the topography of land masses, especially when high mountain ranges are present, may influence these wind patterns. Huge lakes, deserts, and other features of continents also may affect wind patterns, so we must include topography as a potentially important influence on climate.

Rainfall is yet another factor helping to determine a region's climate; like temperature and wind, rainfall is determined by the uneven heating of the Earth's surface. The amount of moisture air can hold rises disproportionately with its temperature (Figure 44-7). If a vol-

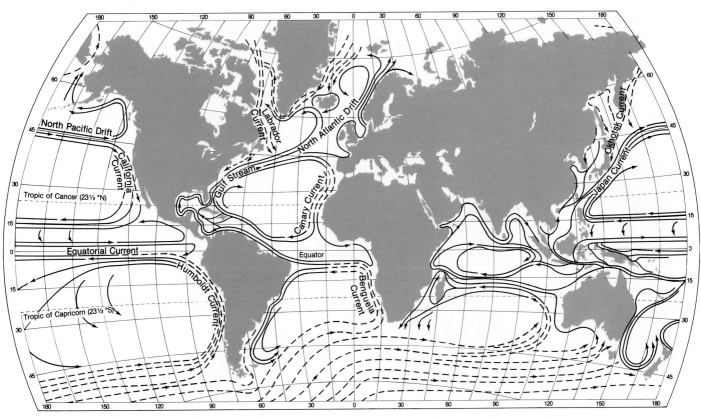

Figure 44-6   OCEAN CURRENTS.
The Coriolis effect, caused by the Earth's rotation, helps generate major ocean currents. Compare the directions of the wind currents in Figure 44-5 with that of the ocean currents shown here; see, for example, the clockwise direction of the California and Canary currents and the counterclockwise direction of the Humboldt and Benguela currents. Ocean currents affect weather of the adjacent land, as, for instance, the warming of England and northern coastal Europe by the Gulf Stream. Names of currents not included here can be found in geography books.

ume of saturated warm air is cooled—like steam from a tea kettle hitting a cool surface—water vapor will condense into liquid droplets. Because air carrying moisture rises and descends at different places on the globe, differing amounts of precipitation (rain or snow) are gener-

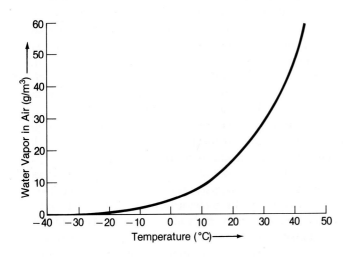

ated in various regions. The tropics, for example, have abundant rainfall because air warmed near the Earth's surface carries much water vapor; on rising, this saturated warm air cools until its water vapor condenses into droplets to form clouds, and, ultimately, rain falls. In contrast, air that has already given up its water content descends around latitudes 30° north and south, where it is both compressed and heated. This is low-humidity air that can hold much more water as it warms near the Earth's surface. Consequently, a great deal of water evaporates from both land and water surfaces; on land,

Figure 44-7   THE RELATIONSHIP BETWEEN AIR TEMPERATURE AND THE AMOUNT OF WATER VAPOR IN AIR.
The graph shows grams of water vapor plotted against temperature for 1 cubic meter of air. It reveals that air holds much more water vapor at higher temperatures. As air cools, the water vapor condenses; the result is dew, mist, rain, or other forms of precipitation.

Figure 44-8   DESERTS OF THE WORLD.
Deserts are always in dry regions, yet their conditions vary substantially depending on local topography, soil, and temperature. Here are three representative desert regions from the western United States: (a) a sandy desert in California's Death Valley characterized by huge, shifting dunes and near absence of vegetation; (b) a sagebrush desert in Nevada has scattered sage and rabbit brush on the stony, arid soil; and (c) a typical Arizona desert where sufficient water is present to support a rich cover of cactuses and drought-resistant angiosperms.

this creates a worldwide band of aridity and even deserts (Figure 44-8).

So far, we have concentrated on latitudes and weather patterns; but what of altitude? Everyone is aware that increasing latitude correlates with cooler and cooler temperatures until the polar regions are reached. But increasing altitude at any latitude can result in the same thing. In fact, altitude is said to "mimic" latitude. Even in the Andes, which straddle the equator, temperature falls as one climbs until polar climatic conditions are reached. Equally important, moist air forced aloft when it blows against mountain ranges tends to drop its water content as rain or snow on the mountains. If winds tend to hit a mountain range from one direction, that side of the mountains may be lush forest and the opposite, downwind side, a desert because it is in a "rain shadow." These parallels between altitude and latitude are reflected in the fact that the range of vegetation and animal types one sees climbing from sea level to 7,000 meters in mountains mirrors that found as one travels from the equator toward the poles.

That organisms living in different parts of the world but under similar climatic conditions are similar in ap-

pearance and physiology reflects an important fact about evolution: essentially similar adaptations must be present if organisms are to survive in the face of similar sets of environmental conditions—say, at high altitude and high latitude. Dozens of plant species found around the eastern Mediterranean have their equivalents in coastal California and northwestern Mexico. A short rainy season, prolonged hot and dry summers, and generally similar topography can be coped with only if plants have such adaptations as desiccation-resistant leaves. And just the same general adaptations are found among analogous plants on certain Chilean, South African, and Australian coasts—all sites where prolonged hot, dry summers place strict demands on plants.

In summary, the dynamics of air and ocean-water movements due to the Earth's rotation, tilt, and orbit around the sun result in global climate and weather patterns. The placement of the continents on the Earth relative to these global patterns, the topography of the continents, and even the characteristics of the sea floor and continental shelves affect climate and thus, indirectly, living organisms. Temperature, precipitation, and wind patterns indirectly determine the predominant vegetation zones of the Earth. The low rainfall and high temperature near latitudes 30° north and south govern the plant types found there (tropical thorn scrubs, savanna grasslands); similarly, at the equator and 60° north, there can be extensive forests (tropical rain forests at the equator and coniferous forests in North America and Eurasia). We would expect the same at 60° south, but due to the pattern of continental drift, no land masses are currently found at that latitude. With all of this information about climate, weather, latitude, and altitude, we are ready to examine the major ecosystems of Earth—so-called biomes—and see how they are characterized by dominant patterns of vegetation.

BIOMES

Whether seen from an orbiting space capsule, a jet airliner, a train, or a Landrover, our world's major visual features are the forests, deserts, grasslands, and waters.

Any one of the Earth's major ecosystems, or **biomes,** may appear quite uniform—hundreds of miles of tropical or coniferous forests, for instance. But within a biome, there may be local peculiarities in geography, climate, and biological history that give rise to isolated areas with unique groupings of small and large plants, animals, fungi, and microbes. At their edges, adjacent biomes usually grade into each other; this is seen on a small scale as one ascends a mountain where areas of tall, lush vegetation merge imperceptibly into zones of lower, drier, more sparse vegetation.

Despite these qualifications, every biome does have characteristic types of organisms. Hence, designating major biomes is a useful device for organizing our planet's vast ecological diversity. Figure 44-9 illustrates the location of major biomes throughout the world.

Figure 44-9 **THE EARTH'S MAJOR BIOMES.**
Each type of biome shown in the map is described in the text. The biomes occur where they do because of global wind, water, and weather patterns in combination with topography and other factors.

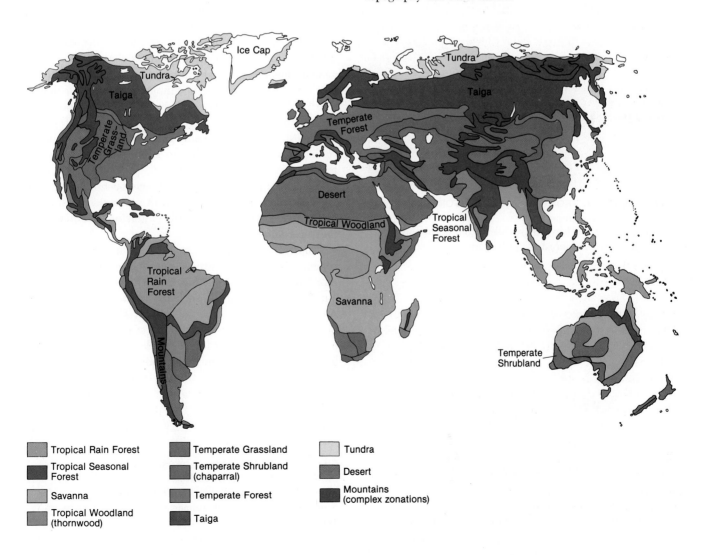

Figure 44-10   A TROPICAL RAIN FOREST IN TRINIDAD.
Tall trees, many vines, and moisture everywhere are typical
of rain forests.

**Figure 44-11   A TROPICAL SEASONAL FOREST IN
AFRICA.**
Photographed when leaves have fallen from deciduous trees,
the tropical seasonal forest is much more open than rain
forests.

**Figure 44-12   SAVANNA GRASSLAND OF THE
SERENGETI PLAINS, TANZANIA.**
Although the numbers of trees can be substantially greater
than shown here, grasses remain the dominant vegetation in
savanna grasslands.

Tropical rain forests occur at low latitudes wherever
rain falls abundantly throughout the year (Figure 44-10).
Warm, wet weather provides perfect growing conditions, so that trees grow to 40 meters or taller. Their
branches and trunks support epiphytes, such as orchids,
and lianas, or vines of various types. The tree foliage, as
well as epiphytes, vines, and many animals, is found in a
"canopy" near the light. The forest floor is deeply
shaded, with just 0.1 percent or less of sunlight penetrating to the floor. Hence, there is often an open, park-
like appearance so that it is easy to walk through, unlike
our usual conception of a jungle. Invertebrate wildlife,
including butterflies and other insects, is exceptionally
diverse, and most species of mammals, such as monkeys,
and of reptiles, such as snakes, live in the trees.

Tropical seasonal forests occur in slightly higher latitudes than rain forests or where topographical conditions
cause locally drier climate (Figure 44-11). Trees in these
forests lose many of their leaves during the dry season
and so are deciduous. In the tropics, the temperature is
fairly constant, and seasonality is expressed in wet–dry
differences rather than cold–warm differences, as in the
temperate zones. Thus, monsoon seasons may have daily
rains measuring several inches, while dry seasons have
much less precipitation, even approaching drought conditions on occasion in places such as central India.

Savannas are tropical grasslands with scattered trees
that occur either where the climate is too dry for dense
forests to grow or where soil conditions and/or periodic
fires maintain a grassland system (Figure 44-12). Usually
the driest savannas have only grasses; with more rainfall,
sufficient water is available to support growth of scattered trees. With more rain still, savannas grade into
tropical woodlands. The most famous savannas in Africa
are grasslands and grasslands with scattered trees; such
places may support huge herds of grazing mammals and
their predators.

Tropical woodlands, or tropical thornwoods, occur in
regions with somewhat more rain than savannas but less
rain than tropical seasonal forests. Thornwoods—bushes
and trees with spines, such as *Acacia* species in the
Americas—dominate the vegetation. The plants survive
on just one or more short rainy seasons each year; a
moist period promotes intensive plant growth, but is followed in the dry season by leaf fall from the plants and
little or no growth.

Deserts occur in subtropical latitudes or where mountains block moist winds. The mountains force moist air upward so that it cools and loses its water vapor as rain or snow on one side of the mountains; the other side, the rain shadow, is left dry. A good example of a topographically caused desert is the thin strip along the west coast of South America; most trade winds from the east are blocked by the long, very high Andes range, so that rains fall on the eastern slopes and support the South American rain forests. Although some deserts are extremely barren, others have seasonal rainfall and contain some vegetation. These deserts usually have diverse kinds of vertebrates such as snakes, kangaroo rats, and perhaps coyotes; invertebrates include scorpions and insects adapted to the dryness. When low temperatures combine with low rainfall, simple deserts result, as in the Great Basin sagebrush-scrub desert of North America; a few major species of shrubs, usually members of the sunflower family, are the major vegetation there. In other deserts, plants such as cactuses store water in their stems and have modified leaves in the form of spines. Some deserts have annuals that germinate, grow, bloom, and produce seeds—all in just the few weeks following a rare rainstorm.

Temperate grasslands are similar to tropical savannas, differing from them only in their lack of trees and cooler, high-latitude climate, which causes differences in soil (Figure 44-13). The climate is usually drier than in forested regions, but not as dry as in deserts. Like savannas, temperate grasslands are subject to frequent fires (often caused by lightning) that repress the growth of trees. Although they appear simple in biological structure, with a single layer of vegetation, temperate grasslands can sometimes support more plant species than

Figure 44-13   A TEMPERATE GRASSLAND ON SOUTH ISLAND, NEW ZEALAND.
This cold, relatively dry, tussock grassland is in the rain shadow of mountains.

Figure 44-14   A TEMPERATE SHRUBLAND: CHAPARRAL IN CALIFORNIA.
Hot, dry climate much of the year dictates the characteristics of the vegetation.

can temperate rain forests. Grasslands are typically poor in bird species but rich in insects and herbivorous mammals that burrow (such as gophers) and run (such as antelope and bison). We know many grasslands by familiar terms: the North American tall-grass *prairies* and short-grass *plains*, Eurasian *steppes*, South African *veld*, and Argentinian *pampas*. Temperate grasslands typically have rich, deep soil—the prime reason they are exploited as the major grain-producing areas of North America and Eurasia.

Temperate shrublands grow in areas with "Mediterranean" climates—that is, hot, dry summers and cool, moist winters (Figure 44-14). Like grasslands, they are subject to frequent fires, and many indigenous plant species such as manzanita bushes are remarkably well adapted to periodic conflagrations. Rabbits, beetles, vultures, and deer mice are typical animal inhabitants. Some temperate shrubs have subsurface root crowns that resprout soon after a fire; others have cones that open and disperse seeds after a fire when the soil is enriched with ash and new seedlings do not have to compete for space. Shrublands are known by such familiar names as California *chaparral*, Mediterranean *maquis*, Australian *heath*, and Chilean *matorral*.

Temperate forests, like grasslands and shrublands, occur to the north and south of subtropical latitudes in the two hemispheres (Figure 44-15). These forests are varied and include deciduous forests, evergreen forests, and even rain forests, depending on prevailing moisture and temperature conditions. But they have far fewer species than do tropical rain forests. Around 60° north latitude, in coastal areas that receive precipitation year-round, forests grow that contain the tallest trees in the world. An example is the coastal redwood forests in northern California, with canopies up to 90 meters high.

Figure 44-15   A TEMPERATE FOREST OF REDWOOD TREES: COASTAL MOUNTAINS OF CALIFORNIA.
The cool, moist climate supports redwoods, ferns, and rapid growth patterns.

Figure 44-16   TAIGA NEAR THE ALASKA RANGE.
Lakes, acres of swamp, and low muskeg plants dominate the flatter areas. Spruce are common conifers.

Temperate forests also occur in seasonal climates. Areas with summer rain and severe winters have forests of deciduous trees, which lose their leaves in the winter to prevent freezing and water loss. Many deciduous forests are species-rich, with varied bird and mammal life. Temperate evergreen forests occur in parts of the southern and western United States where the weather is not too cold or where the climate is routinely dry as in California oak woodlands or Australian eucalyptus forests.

Taiga is sub-Arctic or subalpine needle-leaved forest growing in the coldest zones possible for trees (Figure 44-16). Taiga forms the great conifer forests of the northern United States, Canada, and northern Eurasia. Containing mostly conifers, the forests and woodlands are simple and often dominated by just one or two tree species. In the southern or low-elevation parts of taiga, the trees grow close together in the forests, but in more northerly latitudes or higher elevations, the forest opens, until scattered trees eventually give way to tundra. This is a good illustration of the gradation between biomes. Taiga animals include elk, moose, wolves, jays, grouse, beetles, butterflies, and often an abundance of mosquitoes.

Tundra occurs in the Arctic and on the world's tallest mountains, the Andes and the Himalayas—where high altitudes mimic Arctic latitudes (Figure 44-17). Small shrubs sometimes grow on this treeless plain, but the predominant species are grasses, lichens, sedges, and mosses. Only the upper few inches of soil thaw each summer; the deeper soil remains frozen as *permafrost*. Highly seasonal because of the snows and bitter cold, tundra has both endemic, year-round animal residents (such as the arctic hare and the snowy owl) and summer migrants (such as many songbirds). Tundra is known also for its incredible population of mosquitoes and black flies, which breed seasonally in the numerous moist places in the soil and bogs.

There are also several aquatic biomes. *Fresh-water biomes* include **lentic** (lake and pond) and **lotic** (stream) **communities** occurring worldwide, regardless of climate. Within lentic communities, shore, planktonic, and

Figure 44-17   TUNDRA IN ALASKA, NOT FAR FROM NORTH AMERICA'S HIGHEST PEAK, DENALI.
Low bushes and soft, springy mosses cover the wet soil.

bottom communities tend to blend with one another. In lotic communities, there is less plankton than in lentic communities, but larger invertebrates live on the bottom or attached to rocks in fast-moving streams. In slow-moving rivers, the life forms are more similar to those in lotic communities. Of course, the plant and animal life in aquatic communities varies, depending on latitude, altitude, and other factors that we will discuss at the end of this chapter.

Marine biomes include sandy beaches, intertidal zones, marine mud flats, coral reefs in tropical waters, open-ocean communities, and bottom communities. Sandy beaches are difficult environments for organisms because the sand is always shifting due to wind and waves. Yet beaches are not as barren as they look, and many kinds of crustaceans, worms, and microbes live there. Several community types exist in the intertidal zone, the area between the extremes of high and low tide, depending on the substratum. Rocky intertidal areas are much more stable substrates than beaches. Visitors to tide pools have seen firsthand the great diversity and abundance of sea stars, snails, algae, crabs, and other organisms in such communities. Marine mud flats are home to a range of invertebrate fauna, including crustaceans, worms, and clams. Coral reefs form fringes around tropical islands and are similar to tropical terrestrial communities in supporting a bewildering variety of plants and an even greater variety of animals. **Pelagic** (open-ocean) **communities** are supported primarily by the photosynthesis of phytoplankton, yet innumerable diatoms, invertebrate larvae, fish, and other buoyant creatures are also found there. **Benthic** (bottom) **communities** commonly contain diverse invertebrate types, even on the deep ocean bottom. Some fish survive in waters thousands of meters deep, and microbes, of course, play the role of decomposer in the inky depths.

THE HABITATS OF LIFE: AIR, LAND, AND WATER

We have seen that climatic differences associated with the various biomes largely determine where particular types of organisms may live. At a more fundamental level, however, the three very different states of matter— solid, liquid, and gas—that make up habitats impose constraints on organisms and affect their ways of living.

Air: The Atmosphere

Air temperature can fluctuate rapidly and widely—a basic property that has had important consequences for land-dwelling organisms. They have evolved numerous mechanisms for avoiding, moderating, or tolerating extreme changes in air temperature in their environments (Chapter 35). The molecular composition of air also affects all plants and animals. Recall that "air" is actually a low-density gas composed of 79 percent nitrogen, 21 percent oxygen, and 0.03 percent carbon dioxide. It is also a vast, moving reservoir into which water evaporates and from which organisms' vital needs are met. Since the number of oxygen molecules per liter of air decreases as altitude increases, there are limits on where certain animals can live. For instance, humans can exist reasonably well up to about 6,000 meters, but no permanent human settlements have ever been established at higher elevations. Turning from composition to density, even at sea level, where air is densest, that density is not great enough to provide significant buoyant support against the downward force of gravity; the results have been the evolution of thick cell walls and woody parts in plants, hard exoskeletons in insects, and hard endoskeletons in vertebrates.

The gaseous constituents of air provide important chemical raw materials for life. Carbon dioxide is, of course, the carbon source for photosynthesis. Nitrogen, reduced by nitrogen-fixing organisms, is a necessary component of proteins and nucleic acids. The amount of water vapor in the air (the relative humidity) largely determines rates of evaporation from the skin of a frog, through the stomata of a plant, and so on. Hence, organisms that live in habitats with low relative humidities must have special adaptations for survival. The American pygmy cedar is an extreme example; this denizen of the southwestern American deserts can absorb enough moisture from airborne water vapor alone to sustain life processes. Finally, though we think of rain that forms from water vapor as beneficial for plants, it is true that rain drops may strip the waxy covering layer from leaves, leaving them vulnerable to pollutants, pathogens, and chemicals. Rain also leaches nutrients from leaves and cuts plant productivity as a result. But, at the same time, raindrops may contain dissolved phosphate, potassium, calcium, and other essential plant nutrients. So, with the potential harm comes an even greater good.

Land

Many plants and most animals acquire water and non-gaseous mineral nutrients directly or indirectly from the *soil*. Without living things, the material we know as soil would not exist, for it is a product of the interplay of organisms and the Earth's crust. Living things, such as plant roots, burrowing mammals, earthworms, and fungal hyphae inhabit the soil. They actually are found be-

Figure 44-18   PLANT LIFE AND SOIL pH.
These rhododendrons grow in the wild only when soil pH is
quite acid. The bush shown here is in the Redwood
National Park, California.

tween soil particles, where water and air also occur.
These organisms may alter the soil and add immense
quantities of organic materials to it. In turn, the poros-
ity, texture, and chemical composition of soils affect life
forms.

Soil chemistry and structure determine the amount
and availability to organisms of minerals and water. For
instance, soil pH affects whether Ca^{2+}, PO^{4+}, and other
essential ions are bound tightly to particles or are free to
be absorbed by organisms or leached away in water runoff. Plants tend to be adapted to a specific soil pH; for
example, rhododendrons and azaleas of the northwestern and southeastern conifer forests thrive in acidic soils
that fail to support many other plant species (Figure 44-
18).

Soils contain varying amounts of four main constituents: silt, sand, clay, and humus. **Silt** and **sand** are pulverized rock. **Clay** is a dense material produced by
weathering and is composed of aluminum, silica, and
other minerals. **Humus** is decomposing organic materials. The role of each constituent can best be seen in a
rich soil such as **loam,** which is a mixture of silt, sand,
clay, and humus and which is optimal for plant growth.
Loam has four sizes of particles, from the fine pieces that
make up clay and silt to the coarser particles of humus

and sand. The juxtaposition of these various-sized components creates spaces that contain water and air, with
its precious oxygen required by most roots and soil organisms (Figure 44-19). Water is held between soil particles
by capillarity and by **imbibition,** a process in which
water enters the soil and binds to clay and humus particles. Clay-rich soils imbibe the most water and can become waterlogged, whereas soils with a high proportion
of sand have excellent drainage and much air space.
While the humus in soil imbibes water and subsequently
swells, it shrinks again as water is given up; humus is
therefore an important means of keeping soil from com-
pacting. Humus is also a rich source of nutrients that are
released slowly by decomposition and thus become
available continually for use by soil organisms. A rich
loam has up to 10 percent humus by volume.

Soils, like organisms, undergo a maturation process.
Humification—the process of reducing organic material
to finely divided pieces—occurs in the upper layers.
Below that, **mineralization** takes place; decomposers
turn organic materials into inorganic ones. In rich grasslands, humification is fast and mineralization is slow, so
that abundant organic material accumulates. Exactly the
opposite happens when soils form initially; due to the
paucity of plant material, humification is slow, while
mineralization tends to be rapid. Soils at any one site
also continually change, become mature, and erode
away.

Soil is a restrictive environment for most animals. Relatively little food or oxygen is available, and movement
through densely packed dirt, rocks, sand, and organic
debris is difficult. Burrowing animals adapted to life
underground, however, are more difficult for predators
to find and have one other advantage: although the soil

Figure 44-19   SOIL: SPACES FOR AIR AND WATER.
This scanning electron micrograph (magnified about 250
times) shows the many spaces between soil particles.
Minerals leach into water in those spaces, and can be
absorbed by plant roots.

surface may be subjected to extremes of temperature, creatures living even a few centimeters beneath it are relatively well insulated from both heat and cold. This protective effect enables small mammals, such as the ground squirrel, to survive the cold western winter or arctic hares to live in tundra through the bitter Arctic winters.

Water

In contrast to terrestrial organisms, whose life support comes from both air and soil, aquatic life forms derive food, water, gases, support, protection from temperature extremes, and other basic needs from the surrounding medium. Some waters contain few nutrients, and most may provide 5 percent or less of the amount of oxygen present in air. Water has a density between that of air and soil and buoys up the bodies of plants and animals so effectively that less massive support structures are needed. In addition, a buoyant frond of *Sargassum*, a water flea *(Daphnia)*, or a sea bass can expend less energy per unit of mass to maintain its position than, for instance, a hummingbird hovering in air. Like deep soil, water moderates temperatures because of its large capacity for absorbing and holding heat; temperature extremes occurring in the air are not encountered in oceans and lakes. Very deep waters may be cold—near freezing, in fact—yet their constancy has allowed organisms to evolve enzymes that function very rapidly at such low temperatures. Finally, water itself—a critical requirement for life—is available in unlimited supply.

Although biologists have found animal life even in the inky and ice-cold depths of the great ocean basins, plants can survive only in a top layer of water known as the **photic zone,** where enough light is present for photosynthesis (Figure 44-20). (Bacteria, present at levels of 10^7 to 10^9 per liter of sea water, tend to cluster near the phytoplankton in order to absorb nutrients from them.) Since water both reflects and absorbs light, even the clearest, cleanest ocean water has a photic zone only about 100 meters deep. Most animals living deeper are either carnivores, which eat smaller animals that consume phytoplankton in the photic zones, or scavengers that depend on a steady supply of dead and dying plant and animal bodies drifting down from above.

True miniature ecosystems, however, that do not rely on energy arriving from the photic zone exist in the deep ocean abyss. Communities of giant tube worms, huge clams, crabs, a few fish, and other animals cluster around deep-sea vents, or perforations in the sea floor where superheated mineral-laden water pours out. Biologists now believe that the ultimate source of food that supports these vent communities is the population of chemosynthetic bacteria that oxidize hydrogen sulfide or

Figure 44-20   LIGHT PENETRATION INTO THE PHOTIC ZONE.
Here, light penetrates the sea water around large photosynthetic kelp plants moored to the shallow sea floor.

other compounds spewing from the vents in order to harvest energy to build organic materials. These sulfur bacteria live symbiotically in the tissues of the tube worms and the gills of the clams, while crabs consume the producers directly.

Mineral nutrients and oxygen are precious commodities in aquatic habitats and would be in even shorter supply if bodies of water were static pools of liquid. But lakes and oceans are not static; instead, their waters circulate in steady rhythms of chemical cycling. Constant mixing and wave action help dissolve and distribute atmospheric oxygen throughout the body of water. Horizontal circulation patterns are shown in Figure 44-6, but vertical currents occur as well. In the oceans, such vertical currents are called *upwelling;* in lakes, vertical movement takes place with the fall and spring *overturns.* This vertical movement is especially important because critical nutrients, such as phosphorus, are most soluble in the deeper, colder water; upwelling lifts this nutrient-rich cold water to the photic zone, where it can be exploited by organisms.

ENERGY FLOW IN ECOSYSTEMS

The habitability of every region, zone, and habitat depends on the availability and fluxes of energy. This fact was first appreciated in the 1920s, when Charles Elton, a young biologist at Oxford University, began taking summer trips to Bear Island, off the northern coast of Norway, near Spitzbergen. He wanted to study how animals living in the stark Arctic tundra divide up available food resources. Because only a few kinds of hardy grasses and short, scrubby vegetation grew in the tundra, it was relatively easy for Elton to track the island's larger animal inhabitants.

The most visible animals were Arctic foxes, which caught and ate tundra birds—sandpipers, buntings, and ptarmigan. These birds (which, like Elton himself, spent only summers in the area), in turn, consumed tender leaves and berries of tundra plants or ate insects that had fed on plant parts. Elton described the links between plant, insect, bird, and fox as a **food chain.** He defined the first link of the chain as the trapping of solar energy by a plant via photosynthesis and its conversion to stored chemical energy. The stored energy then passes from the plant to a herbivore and then to a carnivore. Each link in this chain represents a specific **trophic,** or feeding, **level.**

Elton was one of the first biologists to employ levels of energy transfer in describing the structure of biological communities: plants and other photosynthetic organisms are the **producers.** The solar energy they store as carbohydrates and other compounds is transferred to a series of **consumers** at successive trophic levels. An insect consuming a berry is a *primary consumer;* an indigo bunting eating an insect, a *secondary consumer;* and so on. At the end of every food chain, as well as at each of its levels, are the **detritivores** and **decomposers,** mainly bacteria and fungi that break down the remnants of or-

ganic material. In this way, the mineral nutrients remaining in a rotting carcass, a decaying leaf, or an animal's excrement are returned to the soil or water, where they can be taken up anew by producers.

In his Bear Island studies, Elton noticed another characteristic of energy transfer in communities: most food chains are part of more complex systems known as **food webs.** For example, the Arctic foxes did not limit their food intake to one species of tundra bird, creating a single link in a simple chain; they also ate sea gulls, auks, and other marine birds. These birds ate fish, which grazed on barnacles or other invertebrates, which, in turn, may have eaten photosynthetic algae. In winter, the birds migrated to warmer climates, and the foxes ate bits of dead seals left by polar bears, as well as polar-bear dung. Figure 44-21 illustrates this essential feature of food webs: plants, algae, and cyanobacteria on land or in the water become food for many different herbivores, which themselves may become prey for various carnivores in a network of interlocking food chains. The box (on pages 1094–1095) on the Antarctic food web also makes clear that human tampering with a major consumer in a food web can have dramatic consequences in allowing other consumers to expand in numbers.

Ecological Pyramids

Elton's observations about the structure of food chains and food webs revealed a basic principle of ecology: in

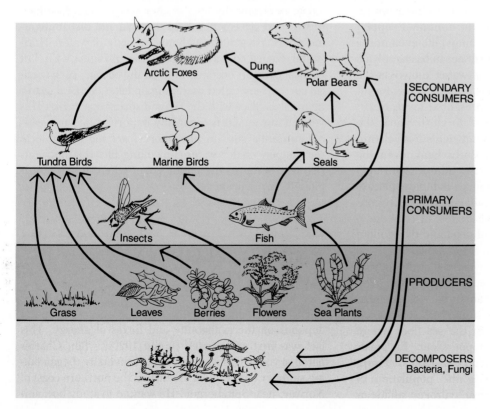

Figure 44-21   FOOD WEB OF ARCTIC BEAR ISLAND.
A food web is a complex system of interlocking food chains. Each organism is positioned according to its trophic level; some, including tundra birds, are both primary consumers (of leaves, berries, and seeds) and secondary consumers (of insects that also fed on berries). Producers, primary consumers, and secondary consumers will ultimately be consumed by the decomposers. Arrows point from each food source to its consumers. This is a highly simplified version of the food web for this tundra and marine biome.

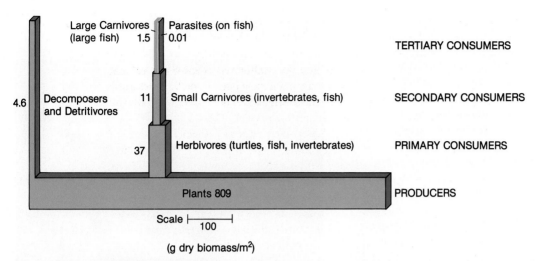

Figure 44-22   PYRAMID OF BIOMASS.
The transfer of energy from one level to the next is never 100 percent efficient, so that as we move from the producers (plants) to the primary consumers (herbivores) to the secondary consumers (small carnivores) to the tertiary consumers (large carnivores), the biomass of each successive group is less. Note the relatively constant biomass of detritus feeders (detritivores) and decomposers at all levels of the pyramid.

virtually every community, there are more plants than herbivores, more herbivores than carnivores, more small carnivores than large ones, and so forth. What accounts for this **pyramid of numbers?** It might seem reasonable to assume that body size is a factor, since in many food webs, body size increases as one moves up the pyramid (that is, little fish are eaten by bigger fish, which are eaten by still bigger fish . . .). If we count all the organisms in an area and convert them to **biomass** (the combined dry or wet weight of all the organisms in a habitat measured in grams) or to energy (the calories or joules [a unit of energy equal to the work done when a current of 1 ampere passes through a resistance of 1 ohm for 1 second] represented by each group of organisms), we usually get a **pyramid of biomass,** like the one shown in Figure 44-22. We can understand this pyramid from what we know of metabolism and the second law of thermodynamics. The transfer of energy from one level to the next can never be even close to 100 percent efficient; in every one of life's millions of chemical reactions there is always loss of heat and some gain in entropy. Hence, with less energy available for each successive trophic level in a food chain, there *must* be less biomass in each succeeding level. Thus a pyramid of decreasing biomass levels can be envisioned.

Not all ecosystems appear to follow the pyramid-of-numbers pattern, but the exceptions give us some important clues about why pyramids of numbers and of biomass exist. In some aquatic systems, for example, ecologists have discovered *inverted pyramids.* If you take a sample of water from the English Channel, weigh the phytoplankton in it, and then weigh the zooplankton

(the tiny animals that eat the phytoplankton), you will find that there are more primary consumers, by weight, than producers. At this point, what is being measured is the **standing crop**—the number of organisms existing at one specific moment. But this standing crop tells us nothing about how fast phytoplankton in the channel have been reproducing or how fast they are being eaten. As it turns out, phytoplankton grow and reproduce quickly but are consumed by the zooplankton as fast as they arise. Zooplankton, on the other hand, survive relatively longer and so accumulate a larger standing crop. Nevertheless, the total mass of phytoplankton consumed by a given mass of zooplankton in their lifetimes would far exceed the consumers' mass—the second law of thermodynamics wins again!

Ecological pyramids are useful as "snapshots" that capture the relative abundance of organisms present at various levels in a given place at one time. In reality, however, an ecosystem is a "moving picture"—the relationships between levels are always changing as energy is transferred from one level to another. We next explore some of the basic characteristics of these energy-transfer interactions.

Generating Biomass: Energy Relationships in the Ecosystem

There are two kinds of production in an ecosystem. The first, **primary production,** is the synthesis and storage of organic molecules during the growth and reproduction of photosynthetic organisms. Significant

ANTARCTIC FOOD WEBS: AN UNCONTROLLED EXPERIMENT

Despite frigid temperatures and interminable winters, the Antarctic region has always supported a rich and complex food web. Living in the icy waters that swirl around Earth's southernmost continent, phytoplankton—the main primary producers—support 750 to 1,350 million tons of krill each year (Figure A). Historically, this titanic krill population supported impressive numbers of filter feeding baleen whales, squid, and fish. And the squid and fish, in turn, sustained the toothed whales, seals, penguins, and other birds (Figure B). During the past 150 years, however, intensive pressure from the whaling industry has reduced the baleen whale population to about 20 to 25 percent of its former level. With their relentless hunting, humans have inadvertently created one of the largest-scale biological experiments in history.

Ecologists measuring population levels of the various species have found that while the consumption of krill by whales has, of course, declined, the krill remain under intense natural harvesting pressure. But from what source? Scientists have several answers. First, they find that seals have increased greatly in number. The crabeater seals of this region, for example, now number about 30 million individuals—66 percent of all the world's seals. Together with other seal species, crabeater seals annually harvest 130 million tons of krill (far more than all remaining whales combined), 10 million tons of squid, and 8 million tons of fish (Table A).

Second, they documented a huge increase in populations of three penguin species. These flightless birds make up 90 percent of the avian biomass (total weight of birds) in the Antarctic, and number some 60 million individuals. These, too, feed on krill and squid by the millions of tons.

Finally, smaller whales have grown more numerous as the larger baleens have been killed off: as blue, humpback, right, and other large whales have disappeared, populations of small minke whales, for example (which are krill feeders), have expanded greatly. Clearly, removing major consumers from the "top" of one portion of the food web has allowed other consumers to exploit newly available food resources. And ecologists have identified many more such impacts on the Antarctic food web due to human interference.

Figure A   ANTARCTIC KRILL.
These primary consumers feed on algae and are themselves the huge "crop" that is grazed by fish, marine mammals, and birds.

Table A   **POSSIBLE CHANGED PATTERNS OF CONSUMPTION OF ANTARCTIC KRILL BY THE MAJOR PREDATORS**

Year	Annual Consumption (tons × 10⁶)	
Predator	1900	1984
Whales	190	40
Seals	50	130
Birds	50	130
Fish	100	70
Cephalopods	80	100
Total	470	470

These recent studies have been only the first phase in the uncontrolled Antarctic "experiment." Now that commercial whaling is being more strictly regulated, scientists will have new opportunities to watch for a return of the food web to its original balance, or to witness the stabilizing of a new order which largely excludes the great baleens from their former position.

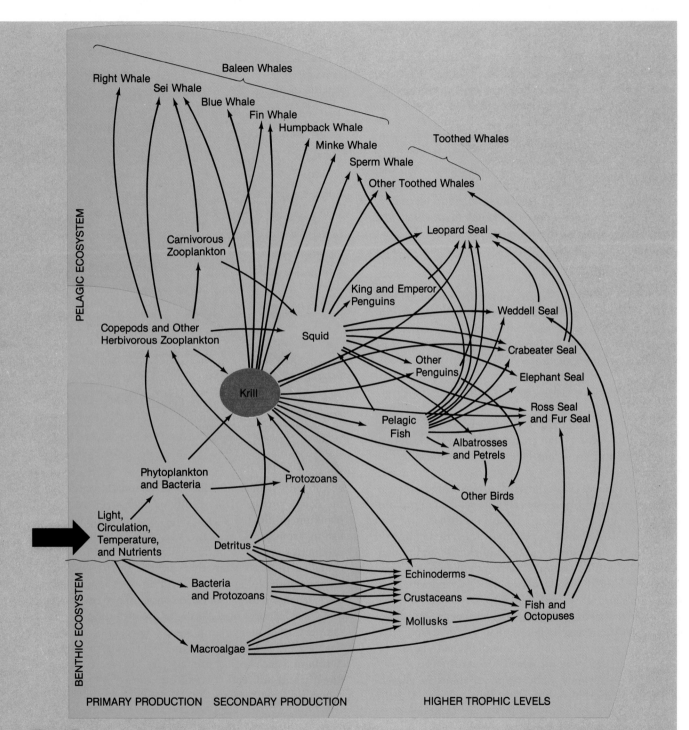

Figure B

The Antarctic food web. Due to the extensive human harvesting of large whales, various seals and birds have become the main consumers of krill and of the squid and fish that feed on krill. If populations of the large krill-eating baleen whales rebound, there will undoubtedly be disturbances in this food web, but we cannot be sure what they will be.

amounts of this energy pass directly to decomposers and detritivores. The rest is passed on to herbivores, which are primary consumers of photosynthesizers. Herbivores use some of that energy for life processes, while the rest is stored in their molecules. The processing and storage of energy by herbivores—which, like photosynthesizers, are living factories—is called **secondary production.** Some of the energy stored in the molecules of herbivores passes, again, to decomposers and detritivores, while the rest is passed on to secondary consumers. (All carnivores are secondary consumers; some animals, including humans, are omnivores, acting as both primary and secondary consumers, depending on whether they eat plant or animal food.) Since all consumers depend ultimately on producers for the energy that keeps them going, the rate at which photosynthetic organisms convert the sun's rays into useful chemical energy determines the amount of energy resources available for all other life forms in the food web.

Of course not all the organic molecules produced by a photosynthesizer (**gross production**) are available to primary consumers. Some are consumed in respiration and metabolism. Therefore, a rate of **net production** can be described by this equation:

net production = gross production − respiration and metabolism

In this sense, a plant can be compared to our national economy; we have a huge gross national product, but we incur many expenses that must be paid in making that product.

Plant ecologists have devised sophisticated ways to measure photosynthetic rates and the rate of net production in a range of ecosystems; typical results are shown in Table 44-1. Notice that production varies, depending on where and under what conditions of climate and rainfall a plant grows. For example, mean production is usually lower on cultivated land than on identical land with trees and other natural vegetation because farmers clear trees, plow under existing vegetation, cull weeds, and occasionally let their land lie fallow. This is true despite irrigation and application of fertilizers to promote large seasonal production of corn, wheat, or other crops.

In aquatic systems, short-term variations in climate and precipitation have less effect than on land, but the availability of nutrients is all-important. Contrary to its image as a fertile cornucopia of life, the open ocean is more like a desert (Table 44-1). Oceans cover 63 percent of the globe, but they account for only one-quarter of the world's net primary production. Why? The main reason is that most oceans lack nutrients—especially nitrogen and phosphorus—that are essential components of living systems. In places like the North Sea (fed by many nutrient-carrying rivers) and offshore from Antarctica and Peru, cold, upwelling water carries a high concentration

Table 44-1   **NET PRIMARY PRODUCTION OF ECOSYSTEMS**

Habitats	Net Primary Production (g/m^2/year)	
Terrestrial	Range	Mean
Tropical rain forest	1,000–3,500	2,200
Temperate evergreen forest	600–2,500	1,300
Temperate deciduous forest	600–2,500	1,200
Taiga	400–2,000	800
Savanna	200–2,000	900
Temperate grassland	200–1,500	600
Tundra and alpine communities	10– 400	140
Desert and semidesert scrub	10– 250	90
Extreme desert	0– 10	3
Cultivated land	100–3,500	650
Aquatic		
Lake and stream	100–1,500	250
Open ocean	2– 400	125
Upwelling zones	400–1,000	500
Continental shelf	200– 600	360
Algal beds and coral reefs	500–4,000	2,500
Estuaries	200–3,500	1,500

From R. H. Whittaker, *Communities and Ecosystems*, 2d ed., Macmillan, 1975.

of nutrients (which are more soluble in cold water) near the ocean surface. This results in high levels of phytoplankton productivity and some of the world's richest fishing grounds. By contrast, one of the biosphere's least productive regions is in the central Atlantic, where horizontal currents are slow, there is little surface-to-bottom mixing, and phytoplankton drift in a thin veil across great reaches of the ocean. Ecologists like to contrast the biological leanness of oceans with the lushness of tropical rain forests, which cover a mere 5 percent of the Earth but yield 28 percent of the world's primary production. A sobering fact is that humans are already consuming directly and indirectly (wood, etc.) nearly 40 percent of the world's net primary production.

The Efficiency of Energy Transfer

We saw that some of the material generated by terrestrial and aquatic primary producers is consumed by herbivores—insects, large animals such as deer and cattle, and grazing zooplankton. The energy accumulated in the growth and reproductive output of these creatures is called *net secondary production*, which can be defined as gross energy intake minus the energy cost of metabolism and other calories defecated as incompletely digested food. The flow of energy from one trophic level to the next is represented in Figure 44-23 for three levels and two energy transfers. This drawing dramatically illustrates two phenomena discussed above: at each transfer of calories from one trophic level to the next, a significant amount of energy is lost to the decomposers (left side) and as heat (right side). Once again, entropy and the second law of thermodynamics are at work.

How much energy is transferred from the cytoplasm of organisms at one trophic level to the cytoplasm of organisms at the next? Ecologists have calculated that from 70 to 95 percent of the calories of net production at one level are lost by the time organisms at the next level have produced their cytoplasm (measured as caloric content) from that food. Or we can say that the efficiency of energy transfer between adjacent trophic levels, called **ecological efficiency,** amounts to between 5 and 30 percent. The efficiency of a particular level-to-level energy transfer depends on how successfully consumers at different trophic levels can find, capture, and eat the available plants or prey; how thoroughly they can digest food and absorb its molecules; and how much energy they expend in building new cytoplasm, maintaining cellular structure and function, and so on.

As energy flows through an ecosystem, the ultimate effect of relatively low ecological efficiencies is to limit the number of trophic levels in a food chain. This limit is usually reached by the fourth or fifth level; beyond that point, harvesting what little total energy remains from lower trophic levels is too inefficient to be biologically worthwhile. This is why a polar bear at the top of the food chain—polar bear → seals → large fish → small fish → shrimp → phytoplankton—has no predators of its own (other than humans, but that is a minor effect). Any animal large enough to capture the bear would have to expend an enormous amount of energy and cover hundreds of square miles of Arctic wasteland for every meal. So the number of trophic levels as well as the pyramids of numbers and biomass in communities reflect the basic energy relationships between trophic levels.

Just as we can follow energy flow, we can trace some individual substances through food chains. An unfortunate example involves toxic industrial wastes and the phenomenon of biological magnification. **Biological magnification** is the process by which materials present in trace amounts in the environment accumulate in organ-

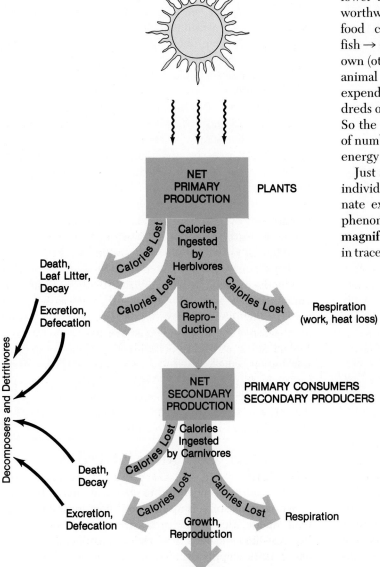

Figure 44-23   ECOLOGICAL ENERGY LOSS. The basic inefficiency of energy transfer leads to much loss of energy between trophic levels. This diagram of the energy flow from plants through secondary consumers shows the many points at which energy is lost due to respiration, decay, and heat dissipation.

ACID RAIN AND DUST: AIRBORNE RECIPE FOR DISASTER?

In recent years, ecologists have issued a series of stern warnings: industry and government must tackle the problem of acid precipitation from the sky before the damage to plants, fish, algae, and entire ecosystems is too great to reverse. At the center of the controversy is sulfur dioxide (SO_2), a gaseous pollutant spewed into the atmosphere by coal-burning power plants, factories, and other industrial sources. Researchers fear that as increasing amounts of this material fall to Earth—either as dry "acid dust" or wet "acid rain"—the damage to living things may reach disastrous proportions.

The effects of acid deposition are sobering and, to many observers, even frightening. The problem begins when airborne SO_2 is picked up by dust particles or is dissolved in water vapor, and forms sulfuric acid. When acid dust or rain comes into contact with plant tissues, metals, cement, and even some plastics, it quickly begins to corrode the surface. Acid rain can be over 1,000 times more acidic than normal rain, and where these levels occur, the impact can be severe. In the heavily industrialized northeastern United States, women have reported acidic raindrops eating holes in their stockings as they walk along city streets.

Acid depositions are becoming a global crisis. In 1982, a West German government task force reported that fully 40 percent of that country's fir trees were either sick or dying. This was not the result of a minor drought, as some German researchers had speculated, but was due to acid rain. In 1981, investigators at Cornell University discovered that more than half of the 400 lakes situated above 2,000 feet in the Adirondack Mountains of New York State were (and continue to be) so acidified that fish can no longer inhabit them. That number has since grown, and researchers estimate that some 50,000 lakes in the United States and Canada are in imminent danger of ceasing to support fish, algae, insect larvae, or aquatic life of any kind. A study of forests in New Hampshire found that the overall rate of tree growth has declined nearly 20 percent from previous levels. One reason for this is that all rain, but especially acid rain, leaches nutrients from leaves of trees and plants and so reduces plant productivity.

Lake and forest destruction are only the beginning. New studies show that acid rain leaches minerals from soil; contaminates reservoirs and leads to chemical reactions that may release lead and other heavy metals from pipes carrying human drinking water; increases respiratory illnesses causing, by some estimates, as many as 120,000 human deaths per year; and reduces agricultural yields by up to 50 percent per year.

Unfortunately, as the cataloging of acid-deposition damage continues, observers note that the United States federal government and industries with the power to curb the problem are resisting change. Such groups have labeled efficient smokestack scrubbers and other new technologies capable of significantly reducing industrial emissions as "unnecessary" or "too expensive." Perhaps one reason that local governments and industries have resisted solutions is that acid sulfates may be carried thousands of miles from their sources by prevailing winds. Acidic residues have been found in pristine wilderness areas of Canada and Hawaii, and even in Arctic ice. Thus one region's pollution is often another distant region's acid precipitation.

An especially disturbing fact is that environmental effects of acid deposition have accumulated in less than thirty years. In many areas, these effects are actually accelerating because the natural buffering capacities of soils and water are nearly exhausted. For many concerned scientists and laypeople alike, the question is no longer whether acid deposition represents a major ecological catastrophe. They are convinced that it does. Their questions now are: Will the catastrophe be stopped? Can it be stopped?

isms. For example, mercury, a poisonous metal, is an industrial by-product often released into rivers and lakes. Some bacteria can add mercury ions to organic groups, forming methyl mercury, which can easily be incorporated by photosynthetic cells. As these cells are eaten by small invertebrate animals, and they, in turn, are consumed by larger animals, the quantity of mercury tends to build up to higher and higher levels in certain animal tissues, such as fat cells; ultimately, what were extremely low concentrations in a lake are high toxic levels in the bodies of large carnivores.

CYCLES OF MATERIALS

Solar energy is converted to chemical energy by photosynthesis and travels up the food chain until it is eventually spent. This would be disastrous were it not for the fact that the sun sends the Earth a continuous supply of energy. The supply of minerals, however, is limited to those already present on or in our planet. Since no more are added, the elements present in living things must cycle and recycle continuously. Living systems have

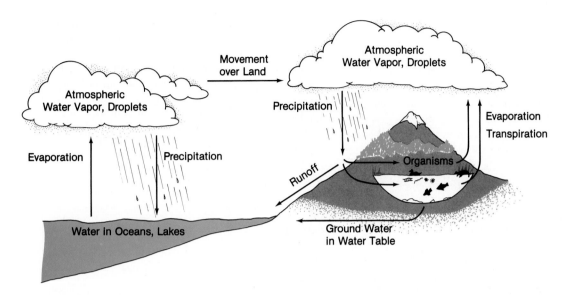

Figure 44-24   THE WATER CYCLE.
Water continuously cycles from oceans and other bodies of
water to the atmosphere and then back again to the Earth.
Water then moves through plants, animals, and other
organisms; from them it returns to the atmosphere. Water
also enters the water table or returns to the sea and lakes as
runoff.

adapted to this limited supply and have evolved complex
processes for taking in and giving off the chemical constituents of life. In this section, we will look at the cycles
of four such substances among the most vital for living
creatures: water, carbon, nitrogen, and phosphorus.

The Water Cycle

Water, water everywhere—three-fourths of our planet's surface is covered by oceans, lakes, rivers, and
ponds of water, the universal solvent of biological reactions. As Figure 44-24 shows, the global water cycle, or
hydrologic cycle, begins when precipitation falls as rain
or snow. Some rain evaporates as it falls, and the rest
enters the Earth's ecosystems in a variety of ways. Water
on the land surface may be taken up directly by plants
and animals, then remain with them for varying periods.
Other water runs off into lakes or streams or percolates
down through layers of soil and rock until it reaches the
water table—the lowermost layer of water in the Earth's
crust, underlain by solid rock. After days, months, or
even hundreds of years, such ground water may be taken
up by the roots of plants, seep into streams, or flow in
underground conduits to the sea. Of course, a portion of
water in lakes, rivers, streams, and oceans enters aquatic
organisms, where it also remains for varying times. Most
fresh water derived from precipitation ultimately flows
back to the seas. The final leg of the global cycle is evap-

oration: moisture evaporates from bodies of water, from
moist soils, or through the stomata of plants or the surfaces of respiratory organs of animals; the water vapor
rises into the atmosphere, eventually to return to Earth
as precipitation.

Water thus cycles through the biosphere on a grand
scale—water that evaporates from the Indian Ocean may
fall to Earth as snow in Iceland. Once a molecule of
water evaporates from a pine needle, from the surface of
a creek, or from your sweaty brow, it is highly unlikely
that the same molecule will ever pass through the same
local ecosystem again.

The Carbon Cycle

The cycling of carbon dioxide to organic compounds
and back to CO_2 affects not only life itself, but also
Earth's weather and geology. In Chapter 8, we described how cyanobacteria, algae, and plants reduce carbon from gaseous CO_2 to generate sugars, from which
other complex molecules can be manufactured. In this
carbon cycle, some carbon molecules are returned to the
biosphere almost immediately as living cells respire and
liberate CO_2; other carbon passes from producers to consumers and then up the food chain; still more is eliminated as organic wastes into the environment.

Of the carbon originally fixed by plants, not all is re-
spired or excreted as waste. As leaves fall to the ground,
and as fungi, animals, or plants die, carbon-rich residues
build up in the soils or water systems. Some of these
residues are temporarily removed from the cycle when
plant material (and some animal material) accumulates at
the bottoms of ocean basins and lakes or in stagnant
swamps. There, if the tissues are not broken down by
aerobic or anaerobic organisms, and if geological conditions are appropriate, oil and coal deposits may form
from the carbon-containing residues.

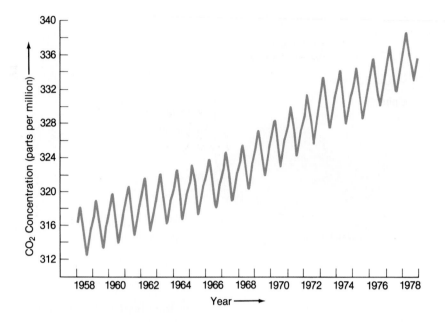

Figure 44-25   ELEVATING CARBON DIOXIDE LEVELS: A GROWING THREAT. This twenty-year record reveals a substantial increase in atmospheric CO_2, as measured on a mountaintop in Hawaii. The burning of fossil fuels worldwide is freeing CO_2 faster than plants can fix it or the sea can dissolve it. The annual cycling is probably due to seasonal differences in photosynthetic fixation.

In recent centuries, huge quantities of carbon stored in this way have gradually been returning to the biosphere as fossil fuels are burned. Much of this CO_2 rises high into the atmosphere in tropical regions and then spreads poleward at high altitudes. The result has been a general rise in the CO_2 content of the world's atmosphere; a typical record of this rise is seen in Figure 44-25. These measurements were made on top of a mountain in Hawaii, surely a long way from cities, industry, and large numbers of people. No one is sure of the consequences of a CO_2 rise for our planet's future climate. Why? Because CO_2 reflects a large amount of long-wavelength infrared radiation, it acts as a sort of heat blanket in the atmosphere; normal atmospheric amounts help to regulate temperatures on the Earth's surface by holding in some of the heat that would otherwise be radiated from the Earth back toward space. Some scientists have speculated that an increase in atmospheric CO_2 as the result of heavy long-term burning of fossil fuels may raise surface temperatures by creating a *greenhouse effect* (a trapping of heat as under the glass of a greenhouse). That could cause worrisome shifts in global weather patterns, including rainfall. Major regions of the globe, including the major food-producing areas, might become hot and even desertlike. Sea levels would rise from the melting of polar ice caps. Other scientists are not so pessimistic—perhaps, they say, increased CO_2 will help produce a photosynthetic boom in global vegetation. It is possible, in fact, that a CO_2 greenhouse effect is the explanation of the lush polar forests that grew down to 85° south latitude during the Cretaceous period and early Cenozoic era. Today's recorded rise in atmospheric CO_2 is huge—about 15 percent in 100 years—relative to the constancy of typical global cycles, but the future is truly an unknown. Prudence suggests, how-

ever, that we should begin to reduce the burning of fossil fuels soon, since once a greenhouse effect is started, there will be little opportunity for humans to reverse its consequences except indirectly, over very long periods of time, by ceasing to burn fossil fuels. The importance of caution is emphasized by an examination of global air temperatures for the past 134 years. Five of the nine warmest years since 1850 occurred after 1978; and the three warmest of all were 1980, 1981, and 1983. Though CO_2 cannot be blamed directly, it is surely a strong suspect.

Carbon is also removed from the global carbon cycle as relatively insoluble calcium carbonate ($CaCO_3$), deposited in the vast numbers of shells and exoskeletons of marine invertebrates or zooplankton. Carbon locked in the rocks (such as limestone and marble) formed of such sediments can be returned to the cycle only when the rocks are uplifted by geological changes in the Earth's crust, eroded by the action of wind and rain, and rendered soluble once again.

Biogeochemical Cycles: Nitrogen and Phosphorus

Some of the most important nutrients required by living systems must be obtained through complex cycles involving chemical interactions between organisms and the Earth. Two of these *biogeochemical cycles* provide organisms with the critical nutrients nitrogen and phosphorus.

The Nitrogen Cycle

Figure 44-26 shows the basic steps in the **nitrogen cycle.** First, atmospheric nitrogen is "fixed"—in-

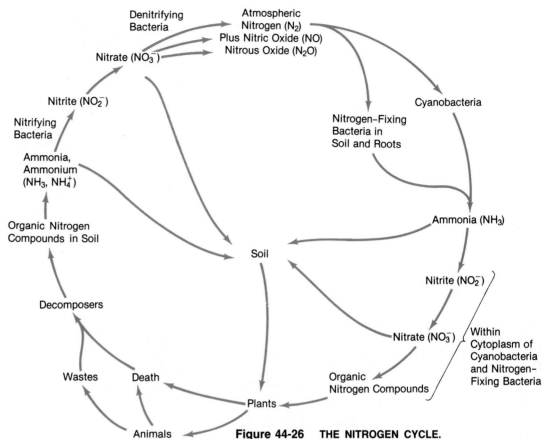

Figure 44-26   THE NITROGEN CYCLE.
Nitrogen, an essential element in the macromolecules within living cells, is cycled through bacteria, plants, and animals, and then back through bacteria to the atmosphere. Cyanobacteria or nitrogen-fixing bacteria form ammonia (NH_3) that is converted to nitrite and nitrate, and ultimately organic nitrogen compounds such as amino acids are synthesized. Plants and animals use the fixed nitrogen, but later produce wastes or die. Decomposers in the soil then consume these wastes and release organic nitrogen. Nitrifying and denitrifying bacteria in the soil generate a series of compounds, including nitrates, which remain in soil, and nitrogen (N_2), which replenishes the atmosphere. Some nitrate is lost to deep-sea sediments (not shown).

corporated into nitrogen compounds—by certain bacteria and various cyanobacteria (and to a limited extent by the action of lightning and through industrial processes; Chapters 20 and 29). It can then be utilized by plants and, through them, by animals, fungi, or microbes. Ultimately, N_2 is returned to the atmosphere by microorganisms.

Notice that at each "end" of the nitrogen cycle, microbes play crucial roles. During **nitrogen fixation,** specialized bacteria and cyanobacteria in soil and aquatic ecosystems reduce N_2 to ammonia (NH_3); to do so, they use the nitrogenase-enzyme complex and consume over twelve ATPs for each two NH_3 molecules produced. Ammonia is then used by those organisms to synthesize nitrates (NO_3) and then organic nitrogen compounds. Leguminous plants such as peas, with their symbiotic bacteria, not only fix nitrogen for direct use, but also produce enough that nitrates escape into the surrounding soil—the reason why farmers often plant legumes in rotation with other crops, since with their rhizobial symbionts, they may fix several hundred pounds of fertilizing nitrogen per year in an acre of farmland.

When a plant or animal dies and begins to decay, or when an animal excretes its wastes, decomposer organisms break down nitrogen compounds, forming ammo-

nia (which forms NH_4^+, or ammonium, ions). This is advantageous biologically, since the charged ion tends to remain bound in soil, in contrast to uncharged NO_2, which readily leaches away in water. Other soil bacteria then complete the cycle by carrying out **nitrification,** the oxidation of NH_4^+ to nitrite and then nitrate, which can be used by plants or microbes. Finally, still other prokaryotes carry out **denitrification,** the reduction of nitrate and nitrite to gaseous N_2, NO (nitric oxide), or N_2O (nitrous oxide or laughing gas), all of which pass into the atmosphere, thereby completing the cycle.

The Phosphorus Cycle

Phosphate is a necessary ingredient of every cell on Earth: it is in ATP, nucleic acids, enzymes, structural proteins, and many membranes. But unlike nitrogen gas and carbon dioxide, molecules of phosphorus compounds are not found floating in the atmosphere; instead, all the phosphorus required by living systems must enter biological pathways as soluble phosphate ions (PO_4^{3-}), dissolved originally by the action of water on rocks in the Earth's crust. Phosphates in soil or in solution in lakes and oceans are incorporated by plants, microorganisms, and phytoplankton, and thus enter the food chain. The **phosphorus cycle** is shown in Figure 44-27.

The rocky origin of phosphate and the fact that much of the Earth's supply ends up in deep-sea sediments (carried there by river systems) make it a relatively scarce commodity—a state of affairs that limits the primary productivity of both land and aquatic environments and explains why phosphates are in great demand as agricultural fertilizers. We saw earlier that in the oceans, the most productive regions are those where cold waters that contain more phosphates upwell and support rich phytoplankton in the photic zone. The increased productivity of phytoplankton, in turn, provides more food for marine herbivores, which eventually support large fish populations.

But how does phosphorus return to the continents again in order to be used by plants and thus perpetuate the cycle? The process goes on very slowly in the sea, where phosphate-containing sediments form rocks over millions of years; then those rocks must be thrust up-

Figure 44-27   THE PHOSPHORUS CYCLE.
Phosphorus cycles in our biosphere as the phosphate ion (PO_4^{3-}). It cycles through water, soil, living systems, bones, some exoskeletons such as shells, and waste materials including guano from birds and reptiles. The green arrows indicate the direct reutilization cycle; the blue arrows show PO_4^{3-} entering shallow marine sediments from which some may be recaptured by the marine food chain (leading to fish, birds); but most is lost to deep marine sediments (the "indirect cycle"). These sediments ultimately become rock that must emerge on the continents again in order to be eroded and once more yield phosphate. Some bones, teeth, and shells are degraded and reenter the direct reutilization cycle.

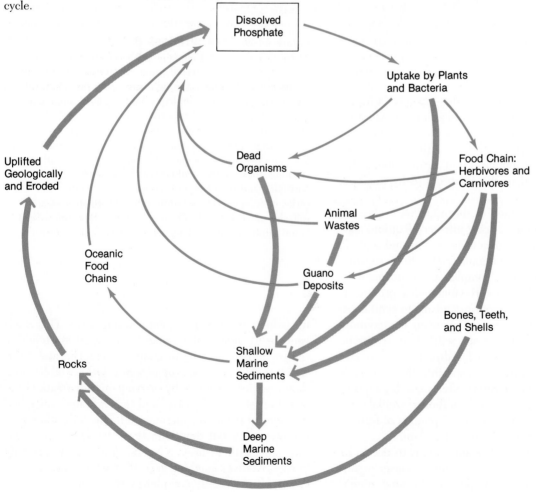

ward as mountains, only to erode away and yield PO_4^{3-} again. Fortunately for life as we know it, there are three shortcuts in the cycle that help return much of the phosphate for reuse by terrestrial organisms, thereby avoiding its multimillion-year residence in rocks. First, PO_4^{3-} derived from animal wastes and dead organisms of all types can be taken up directly by plants and microorganisms; this short-circuits the cycle through rocks and helps sustain a high level of phosphate in the biosphere. Second, humans harvest some marine organisms directly for food and fertilizer and thus eventually return a small fraction of needed phosphate ions from the sea to the soil. Overall, however, this minor bit of recycling cannot begin to meet the needs of modern agriculture, which consumes huge quantities of phosphates to increase the productivity of food plants. The third, and greater, source of phosphates used in agriculture is *guano*—the droppings of marine birds that feed on phosphate-rich ocean plants and animals. At the nesting sites of birds on small islands off the coast of Peru and in other parts of the Pacific, guano accumulates several meters thick. In the last half of the twentieth century, such deposits have come to be regarded as white gold mines of the raw materials for agricultural fertilizers.

Even with the efficiency of PO_4^{3-} reutilization and with the aid of marine birds, the phosphate balance between the land and the oceans is tipped in the oceans' favor. One study estimates that the world's rivers discharge 14 million tons of phosphate each year, whereas sea birds, mining, and fisheries return only 70,000 tons per year back to the land. As the need for agricultural fertilizers continues to climb, the implication seems clear: before long, we will have to mine the seabeds for phosphates, even though the cost will be immense.

CLOSE-UP OF A LAKE ECOSYSTEM

Now that we have discussed the major biomes, energy transfer and trophic levels, and the major mineral cycles, let us see how a relatively self-contained ecosystem, a lake, operates. Lakes have obvious natural boundaries, and fresh-water organisms have little direct interaction with living things on shore. A lake has a surface, a shoreline, and a bottom, in addition to the water mass itself. The physical and chemical properties of water, especially light penetration, govern much of what goes on in a lake. Rock and soil chemistry, both of the bottom and of the drainage flowing into the lake, also affects life there. Photosynthesis by phytoplankton or water plants (such as water lilies) occurs only down to a depth where approximately 5 percent of the sunlight hitting the surface can penetrate. Hence, free oxygen is produced only above that depth. The edges of lakes where light reaches the bottom is called the **littoral zone** (Figure 44-28); plants in the littoral zone are rooted in the lake bottom and grow to the surface, and whirligig beetles, water striders, and many protists and microbes inhabit the water or its surface. The zooplankton in the littoral zone

Figure 44-28   LAKE ZONES.
Lake zones are defined by the amount of light available and of oxygen and biomass produced. (a) The littoral zone lies along the shoreline, where light reaches the bottom and rooted plants grow. The limnetic zone lies in the middle of the lake, beginning at the edge of the littoral zone, and extends to the depth beyond which light is insufficient to support a rate of photosynthesis greater than the rate of respiration. In the profundal zone, there is no effective light penetration. (b) Reflection Lake, Washington, has a typical shallow littoral zone, in the foreground, and the deep-water limnetic zone farther from the shore.

(a)
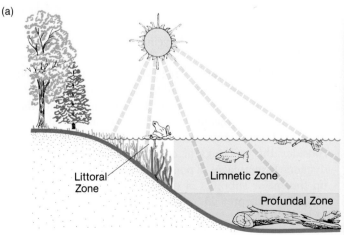

(b)

tend to be less buoyant crustaceans, such as water fleas. The littoral zone is bordered by another rich community in the sand and mud near the water's edge.

The center of a lake, down to the depth where O_2 production by photosynthesis just equals respiratory utilization of O_2 by the collective life forms, is called the **limnetic zone.** The limnetic zone is inhabited by dinoflagellates; algae such as *Volvox* and *Euglena*; nitrogen-fixing cyanobacteria, such as *Anabena*; buoyant crustaceans, called copepods; and rotifers, which graze on the algal organisms. Below the depth of effective light penetration is the **profundal zone,** where decomposition exceeds production of organic materials. The energy input to these deeper lake waters comes in the form of organic materials or dead organisms that sink from the upper water levels. These materials contribute to the lake bottom and are decomposed by bacteria and fungi or are eaten by invertebrates such as blood worms, phantom larvae of dipteran flies, small clams, or certain fish. Such deep waters are usually oxygen-poor or even anaerobic.

Animals that graze on photosynthetic phytoplankton often move vertically in lakes each day. They spend daylight hours in the darker depths and then ascend to feed on phytoplankton each night. The time of full moon, in fact, may be a dangerous one for copepods and rotifers, for if they rise to feed then, small sardinelike fish may feast on the tiny, glistening bodies; this may be so efficient a harvesting process that the population sizes of the crustaceans fluctuate cyclically each month in synchrony with the moon's (lunar) phases.

In winter, the coldest water in a lake is at the surface, where it may freeze into an ice layer. The densest water, at 4°C, is found near the lake bottom; there, trace heating from the underlying Earth may occur. The layering of water at different temperatures is called *stratification.* During a prolonged cold winter stratification tends to disappear as the lake approaches an even temperature. As ice melts in the spring, the surface waters sink as they warm toward 4°C and become denser. Winds also churn the waters, the overall result being movement of deeper waters to the surface, in a process called spring overturn. An autumn overturn also occurs. These phenomena literally stir the soup, since dissolved minerals and nutrients are carried upward and oxygen is carried downward in the process.

Not surprisingly, the physical and chemical features of a lake have direct consequences for its productivity and inhabitants; especially important is the redistribution of nitrates and phosphates, which tend to be the limiting minerals in all fresh waters. The clear, gorgeous blue waters of some deep mountain lakes reflect very low biological productivity because essential mineral nutrients are lacking. Low-productivity lakes are termed *oligo-*

trophic. In contrast, *eutrophic,* or "well-nourished," lakes are good producers of organic materials and organisms. These lakes tend to be shallow; to have high turnover rates of phosphorus, nitrogen, and other nutrients; and to be inhabited by a marvelous variety of microbes, algae, plants, and animals. Without interference from human civilization, many lakes with good drainage may remain in a eutrophic state for very long periods.

A process called **eutrophication** is a natural event in the history of many lakes. In typical eutrophication, a lake becomes richly productive, and more and more organic debris and silt accumulate; the lake becomes shallower, and the spectrum of organisms that live in it changes as it fills in and becomes a bog, and then, perhaps, a meadow. This natural succession is often speeded up by humans; as a result, the word "eutrophication" has come to connote the processes that follow on addition by human activity of nutrients to a lake so that productivity rises faster than normal. This occurs if phosphates from laundry detergents enter a lake; the normal minute amounts of phosphates that limit growth are supplemented massively, and disaster results. Excess algae (especially cyanobacteria) and aquatic plants thrive, choking the open-water areas. The extra vegetation rots, consuming oxygen and making life for fish and many invertebrates impossible. Human sewage, drainage from fertilized farmlands, and water from industries are sources of various pollutants and nutrients—in particular, nitrogen and phosphorus compounds—that greatly speed algal and plant growth.

LOOKING AHEAD

In this chapter, we have seen how the characteristics of our planet and its biosphere—the Earth's tilt, rotation, geography, topography, geochemistry, atmosphere, and climate patterns—contribute to where and how organisms can live. Ecosystems, from the biosphere to the major biomes to smaller ecosystems such as lakes, are incredibly complex tapestries of interacting organisms, energy transfer, and cycling minerals. Truly each part in an ecosystem seems to affect and be affected by nearly every other part. These nonstop chain reactions constitute the inescapable interaction of life and nonlife that characterizes our planet. But to understand the complex interactions in the living world more fully, we must peel back another layer of the ecological onion and consider the workings of *communities* of organisms—dynamic assemblages of populations of species that make up the major subunits of ecosystems. This will be our subject in Chapter 45.

SUMMARY

1. *Ecology* is defined as the study of the interactions that determine the distribution and abundance of organisms.

2. The largest and most complex level of biological organization is the *ecosystem*. The Earth itself is the largest ecosystem, but the word "ecosystem" also refers to *biomes*, the major vegetation regions on our planet, as well as to units such as lakes, meadows, or forests.

3. The Earth's land surface, the oceans and fresh water, and the atmosphere above them form the *biosphere*—the life-supporting environment for all living creatures. Organisms in the biosphere are directly affected by a variety of physical and chemical phenomena; one of the most fundamental of these is the constant flow of energy from the sun that powers nearly all life processes.

4. The amount of solar radiation intercepted by our planet is determined by our distance from the sun and is known as the *solar constant*. Much of the solar energy that strikes Earth is reflected back to space or is absorbed in the atmosphere by oxygen, ozone, water vapor, and carbon dioxide. As a result, only about 1 to 5 percent of the energy first entering the atmosphere ever reaches the planet's land and water surfaces.

5. The sun not only provides energy for the metabolism and growth of organisms, but also drives global climate. Day and night, seasons, different patterns of climate, and uneven heating of the planet's surface, all result from the combination of the Earth's tilt, its rotation each twenty-four hours, and its circling of the sun. The climate of a region is a critical characteristic; it determines what kind of biome will be present—that is, the sorts of plant and animal life that may populate a given area. In this way, climate is a major force in the evolution of life because organisms must be adapted to climatic conditions, or they cannot survive.

6. There are nine terrestrial biomes: *tropical rain forest, tropical seasonal forest, savanna, desert, temperate grassland, temperate shrubland, temperate forest, taiga,* and *tundra*. There are also several aquatic biomes: fresh-water biomes include *lentic* (lake and pond) *communities* and *lotic* (stream) *communities*; marine biomes include sandy beaches, intertidal zones, coral reefs, *pelagic* (open-ocean) *communities*, and *benthic* (bottom) *communities*.

7. The potential *habitats* for life—the places organisms live and grow—are the atmosphere, land, and water. Each imposes constraints on life forms. Organisms that live in contact with air must be able to tolerate or cope with temperature extremes. The low quantity of oxygen at higher altitudes limits the elevation at which most organisms can live. The low density of air has resulted in the evolution of strong structural support against the pull of gravity.

8. On land, many plants and most animals must acquire water and nongaseous mineral nutrients directly or indirectly from the soil. The amounts of these chemicals in soil depend on soil structure and chemistry. The mix of *sand, clay, silt,* and *humus*—decomposing organic matter—known as *loam* is the optimal soil for plant growth. Soils themselves change over time due to the activity of organisms that inhabit them and to climatic conditions.

9. The food, physical support, gases, and water needed by aquatic organisms are provided by the surrounding medium. Water contains relatively low concentrations of nutrients, however, and much less oxygen than does air. The upper layer of water, with sufficient light penetration to support photosynthesis by aquatic plants, is known as the *photic zone*.

10. Energy flows through ecosystems by way of *food chains* that are part of larger interlocking *food webs*. Food chains are organized into *trophic levels*; photosynthetic organisms are the *producers* at the first level, and the solar energy, transformed and stored as carbohydrates and other compounds in plant tissues, is transferred to *consumers* at successive trophic levels. Energy may also be transferred from each level to *decomposers* and *detritivores*. The last link in every food chain are the decomposers—bacteria and fungi that break down the remnants of organic materials and return mineral nutrients to the soil.

11. Ecosystems often show *pyramids of numbers* and *pyramids of biomass* (total mass of living organisms). Such pyramids reflect the basic inefficiency of all energy transfers, including those from one trophic level to the next. In general, the relatively low *ecological efficiency* of energy transfer in ecosystems (5–30 percent) limits the number of trophic levels in a food chain to four or five. In typical food chains, there are decreasing numbers of larger and larger organisms as one goes up the chain. The number of organisms existing at any one time in a trophic level or population is referred to as the *standing crop*.

12. The two classes of production in an ecosystem are primary and secondary production. *Primary production* consists of the energy stored in molecules of photosynthetic organisms; *secondary production* is that stored in the molecules of consumers. All the energy captured by a photosynthetic organism is its *gross*

production; that ultimately available to consumers is its *net production*.

13. *Biological magnification* refers to the accumulation of trace materials in animal tissues, often to toxic levels, as one moves up a food chain.

14. Whereas energy continuously supplied by the sun flows through ecosystems, materials such as minerals, water, and other nutrients must be recycled continuously. As participants in such cycles, organisms may have a significant impact on their environments.

15. The water cycle, or *hydrologic cycle*, is a global phenomenon; water that evaporates in one place is likely to fall to Earth as rain in another part of the globe. Water passes through organisms, sometimes intact, sometimes split to component hydrogen and oxygen, sometimes produced as metabolic water.

16. Carbon taken up from the atmosphere as CO_2 by photosynthesiz-

ing cells is returned to the air by respiration, combustion, or as organic wastes in a massive *carbon cycle*.

17. In the *nitrogen cycle*, atmospheric nitrogen is fixed by bacteria and cyanobacteria. The fixed nitrogen is utilized by plants and animals and then is returned to the atmosphere through a complex series of chemical steps carried out by microorganisms. Nitrogen fixation involves reduction of N_2 to NH_3. Nitrification is oxidation of NH_4^+ to NO_2^- and NO_3^-. Dentrification is reduction of NO_2^- or NO_3^- to N_2, NO, or N_2O.

18. As part of the *phosphorus cycle*, all the phosphorus required by living systems enters biological pathways as phosphate ions dissolved by water eroding rocks on the Earth's crust. Phosphates enter the food chain when they are taken up from soil or water by producer organisms. Much phosphate from dead organisms and animal wastes can be

reutilized by organisms; there is continual loss, however, to deep marine sediments and, over eons, the Earth's rocks. Currently, an important avenue for the return of phosphates to the land for human agricultural use is from guano—the droppings of sea birds.

19. Lakes are good examples of ecosystems. They can be divided into several zones, including the *littoral zone*—shallow regions at the water's edge where light reaches the bottom—and the *limnetic zone*—deep in the lake's center where light, oxygen, and photosynthetic levels are all low. The deep, least productive part of a lake is called the *profundal zone*. *Eutrophication* is the natural process by which lakes become increasingly productive as they fill with silt and organic debris. This natural process can be speeded up dramatically by human activity.

KEY TERMS

benthic community
biological magnification
biomass
biome
biosphere
carbon cycle
clay
consumer
Coriolis effect
decomposer
denitrification
desert
detritivore
ecological efficiency
ecology
ecosystem
eutrophication
food chain
food web
gross production

habitat
humification
humus
hydrologic cycle
imbibition
lentic community
limnetic zone
littoral zone
loam
lotic community
mineralization
net production
nitrification
nitrogen cycle
nitrogen fixation
pelagic community
phosphorus cycle
photic zone
primary production
producer

profundal zone
pyramid of biomass
pyramid of numbers
sand
savanna
secondary production
silt
solar constant
standing crop
taiga
temperate forest
temperate grassland
temperate shrubland
trophic level
tropical rain forest
tropical seasonal forest
tropical woodlands
tundra

QUESTIONS

1. Charles Darwin wrote one paper on fertilization of orchids by insects (1862) and another on formation of leaf mold by earthworms (1881), in addition to his major work on evolution. Was Darwin also an ecologist? Explain.

2. For each of the following, explain briefly why the entity is or is not an ecosystem.

   a. the Earth
   b. a biome
   c. the biosphere
   d. a lake
   e. the moon

3. What percentage of the solar constant actually arrives at the Earth's surface? What percentage of *that* energy is then fixed during photosyntheses? What happens to the rest of the solar energy?

4. The sun's energy strikes the Earth unevenly. Explain aspects of our planet and its motions that combine to produce what we call climate and seasons.

5. Ecologists recognize approximately a dozen major biomes. Which of the terrestrial and aquatic biomes described in the text can be found in the United States? Where?

6. Earth provides three major habitats for living organisms. Outline the challenges of each habitat that adaptations of organisms help to overcome.

7. Explain how energy flows through an ecosystem, using food webs and ecological pyramids in your answer.

8. In the Bear Island ecosystem, which organisms were producers? Primary consumers? Secondary consumers?

9. Which of four natural cycles (water, carbon, nitrogen, and phosphorus)

   a. depends on a number of microorganism species?
   b. depends in the contemporary world on the activity of sea birds?
   c. has been affected by human activity?
   d. may lead to a greenhouse effect?
   e. is biogeochemical?

10. Explain the difference between oligotrophic and eutrophic lakes. Into which category would a deep, high mountain lake most likely fall? Why? What is eutrophication?

ESSAY QUESTIONS

1. List ten foods that you have eaten in the past week, and for each, describe the probable food chain (or food web) through which the sun's energy passed to you, the "top consumer" in the chain.

2. Choose a mammal, a reptile, a bird, or an insect living in some natural or relatively undisturbed habitat near your home. List all the physical and biological factors that may affect that animal. Could your list ever be complete? Explain. How does this exercise help you to understand the Garrett Hardin quotation at the beginning of this chapter?

SUGGESTED READINGS

"The Biosphere." *Scientific American*, September 1970.

This wonderful issue treats the biosphere and all sorts of cycles: water, energy, and nutrients.

MacArthur, R. H. *Geographical Ecology*. New York: Harper & Row, 1972.

One of the most imaginative ecologists in the United States expounded before his death on the ecological basis for distributions of organisms.

Odum, E. P. *Fundamentals of Ecology*. 3d ed. Philadelphia: Saunders, 1971.

This classic text is a bit out of date, but it remains a superb source of facts and introductory material.

Walter, H. *Vegetation of the Earth in Relation to Climate and the Ecophysiological Conditions*. Trans. J. Wieser. London: English Universities Press, 1973.

This may be the clearest treatment of biomes written yet; it is the book from which to learn why vegetation is where it is.

Whittaker, R. H. *Communities and Ecosystems*. 2d ed. New York: Macmillan, 1975.

A leading ecologist's advanced text on these subjects.

45
THE ECOLOGY OF COMMUNITIES

*It is interesting to contemplate a tangled bank, clothed with
many plants of many kinds, with birds singing on the bushes,
with various insects flitting about, and worms crawling
through the damp earth, and to reflect that these elaborately
constructed forms, so different from each other, and
dependent upon each other in so complex a manner, have all
been produced by laws acting around us.*

Charles Darwin, *On the Origin of Species by
Means of Natural Selection* (1859)

A community beneath the Red Sea surface: soft and hard corals, kelp, sea stars, limpets, dozens of fish species, all
interdependent in myriad ways.

In the spring of 1985, a widespread fire broke out on Isabela Island, the largest of the Galápagos Islands. Firefighters battled the blaze for more than a month as it spread through dry grass and scrubby bushes and, below ground, through a layer of humus and tangled roots nearly 2 meters deep. It eventually claimed more than 100,000 acres, or one-quarter of the island's vegetation, destroyed several stands of the unique Galápagos sunflower tree, and removed much of the habitat of a subspecies of giant tortoise and of a rare sea bird, the darkrumped petrel.

While the fire had marked effects on those individual species, research may well show that the greatest impact was on the Isabela Island **community,** the interacting populations of different species in the area. After a region is denuded by fire, there is a period of regrowth and reestablishment of plant and animal species. But which will come back, and which will never recover? In what order will survivors reappear? How long will the process take? Will the area eventually look just like it did before? Or will it be changed forever by the disaster? The answers depend on the complex interactions of all the community members, and we must study the ecology of the community, not just a single affected species, to find the answers.

In this chapter, we consider the characteristics of communities and see how the interactions among species set limits on the way organisms can live. We will begin with the phenomenon of *ecological succession* and see how every land area and many aquatic environments undergo gradual changes in the mix of species that populate their communities. Then we will discuss some important factors of community structure, including the concept of *niche,* and see, for example, why one community can support five or six kinds of hummingbirds, while another can support twenty or thirty kinds.

Like many biological concepts, "community" is an abstraction. We may refer to an entire biome as a community on a very large scale or to the insect larvae in a tiny pool as a small-scale community. We may consider all the species, plants and animals, in a particular area as a community. Or we may restrict the term to only those species using the same resource, such as seeds in a desert sand dune. In general, we define community as an association of two or more potentially interacting species. The concept of community is a useful abstraction that enables ecologists to unravel the fabric of life in the biosphere and study life's tangled relationships in detail. This endeavor is vitally important, for we are confronted by increasingly serious environmental problems due to the enormous human impact on Earth's ecosystems. How well we come to understand the causes and consequences of changes in the living world—from isolated fires to build-up of carbon dioxide in the world's atmosphere—may well determine the course of our collective futures and those of our descendants.

ECOLOGICAL SUCCESSION: A BASIC FEATURE OF LIFE

Anyone who has traveled to the seashore, to the lakeshore, or through the countryside has observed a cardinal rule of ecology at work: different geographical areas contain different types of biotic communities. As one travels inland from the shore on the North Carolina coast, clumps of grasses dotting the sand dunes give way to wild oats and scrubby brush, which give way to pines farther inland, where the habitat is protected from ocean breezes and salt spray. Just as the observant traveler can see how the terrain and vegetation change from one area to another, one can easily note that the animal life changes as well. What is far less obvious is that each of these areas may itself be home to a series of different communities over the years. A meadow with wildflowers may become a grove of shrub species, which may eventually become a deciduous forest. And an area of forest, perhaps due to a fire or another natural disaster, may again become a field of wildflowers. A lake in a high mountain meadow may become a marsh, and the marsh may become a thicket of willows. After many years, the thicket may become a forest of pines.

As we study communities throughout the biosphere, a basic fact about life in the Earth's habitats becomes clear: the types of species present in a given area, the *species composition,* is nearly always changing—sometimes imperceptibly, sometimes quite rapidly. The process by which gradual changes occur in the composition of species that make up a community is called **ecological succession.** One source of ecological succession is that organisms themselves may alter their environment in such a way that it becomes less habitable for them. Lichens erode and destroy their own rocky substrates, thus encouraging displacement by soil-rooting plants. Certain plants may deplete the soil of nutrients, choking off their food supply and paving the way for plants that are better adapted to poor soils. It is important to note that an area's species composition may change even when the physical environment does not change much at all. We will see why later.

Ecologists describe a succession of communities as either primary or secondary, depending on whether the succession begins on bare rock or on soil that has developed over millennia. Areas beginning **primary succession** have no true soil; the first plants must establish themselves on the base material (Figure 45-1). For example, in 1974, Kilauea Crater on the Big Island of Hawaii erupted, spilling new lava across the landscape. After time, this new substrate has weathered and been colonized by lichens and plants. Succession will eventually create a rich *Acacia* and *Metrosideros* forest upon this new land. **Secondary succession,** on the other hand, begins on developed soils where an area has been

(a)   (b)

Figure 45-1   PRIMARY SUCCESSION: A BEGINNING WITH BARE ROCK.
(a) This huge pile was the result of the largest landslide in recorded history. It occurred in
August 1959, and temporarily dammed the large Madison River in Montana. (b) A lava flow
from the Kilauea volcano on the island of Hawaii: about ten years later, plants are taking root.
Both the landslide and the volcano left acres of bare rock exposed, sites where primary
succession occurs.

cleared of preexisting vegetation by a physical disturbance, such as a fire, shifting river courses, and other
natural phenomena, or the denuding of the landscape by
lumbering or some other human activity.

Primary Succession

The first colonizers in a newly available soil-free area,
such as the lava flow from Kilauea Crater, are often the
lichens and mosses that can subsist on bare rock (Figure
45-2). Both trap dead organic materials, creating the first
layer of humus, and biochemical processes carried out
by lichens are thought by some ecologists to contribute
to the slow breakdown and erosion of rock surfaces.

As pockets of soil develop, hardy plants adapted to life
with limited access to soil-bound nutrients and moisture
appear. These "pioneer species" are often grasses and, in
tropical areas, ferns. The seeds of pioneer species are
blown long distances by the wind or are carried in the
coats of animals far from the site where the parent plant
grew. When a seed chances to fall on a suitable harsh
terrain—such as the lava flow on Hawaii—it germinates,
and the plant matures and reproduces quickly before
other colonists arrive, grow, and use up resources. Because such species seem to specialize in precarious exist-
ences—at least from the human perspective—ecologists
have dubbed them **fugitive species** (Figure 45-3).

As the pioneer plant species die and decay, their tissues add organic debris to the decomposing and weathering base. As time passes, the accumulating soil deep-

**Figure 45-2   PRIMARY SUCCESSION: FIRST
COLONIZERS.**
Primary succession often begins with the growth of mosses
or lichens on bare rock.

ens, making more water and nutrients available for plant
roots and allowing a new series of less hardy species—
other kinds of grasses, shrubs, and small trees, perhaps—
to spread into the area. Although a newly opened habitat
may at first have just a few types of pioneer species, once
soil, nutrients, and moisture become available, the habitat may host a population explosion of individuals from
many species, all taking advantage of the increasingly
available resources.

After time passes, the community consists of a mixture
of middle and late species and is at its most diverse state.

Figure 45-3   FUGITIVE SPECIES: SPECIALISTS IN PRECARIOUS HABITATS.

Fugitive species tend to be poor competitors under crowded conditions, but good pioneers in harsh, newly opened environments, as during the process of secondary succession. They tend to produce seeds that can travel a long way by wind or animals, and they grow well in exposed ground. Wild grasses are ubiquitous pioneer species; as shown here, certain types of daisies also can colonize such sites.

Diversity is increased further if the area is subject to small, local disturbances, such as a tree falling and clearing space in a forest. Within such small newly disturbed patches, early-stage species can colonize and grow, thus adding to the diversity of the overall community. Late-successional species depend on support structures (such as tree trunks) to carry photosynthetic leaves above competing plants toward the sunlight and root structures deep into the soil, where moisture—another limited resource for which plants compete—is more abundant. As the community undergoes successive changes, larger and longer-lived plant species tend to become dominant, and the biomass of the area increases. If left undisturbed, the community may eventually become dominated by species that simply reproduce themselves and are not replaced by new arrivals. This kind of ecological constancy is known as a **climax community.** In theory, the climax community will endure as long as the environment remains stable (Figure 45-4).

A form of primary succession called *aquatic succession* occurs in lakes and ponds. The slow transition from a watery habitat to a terrestrial one begins with the accumulation of silt and organic debris (from decaying phytoplankton and other aquatic organisms) on the lake bottom. This process continues until the lake becomes a semisolid bog. Shrubs may then invade, and, finally, as the soil dries even more, a grove of trees can grow. Ultimately, nothing more than a barely detectable depression in a forest, meadow, or prairie may remain of the original lake (Figure 45-5).

Secondary Succession

Instances of primary succession are relatively rare because lava flows, massive rockslides, and other devastating events are themselves unusual. Secondary succession, however, is common because most of the Earth's land surface has been covered with developed soils for millennia. Hence, if existing plants are destroyed, there is abundant opportunity for secondary succession; vacated soils, perhaps rich in humus and nutrients and already harboring dormant seeds, are an ideal "cradle" for the new plant colonizers.

Secondary succession occurs when rivers change their channels, a process that causes substantial lateral erosion

Figure 45-4   STAGES OF SUCCESSION.

A hypothetical time flow goes from left to right in this diagram; typical plants are shown that might be present after the time periods indicated. The specific plants and the ultimate stage reached vary widely from one site to another.

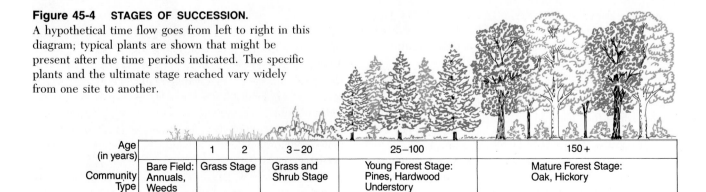

Age (in years)		1	2	3−20	25−100	150+
Community Type	Bare Field: Annuals, Weeds	Grass Stage		Grass and Shrub Stage	Young Forest Stage: Pines, Hardwood Understory	Mature Forest Stage: Oak, Hickory

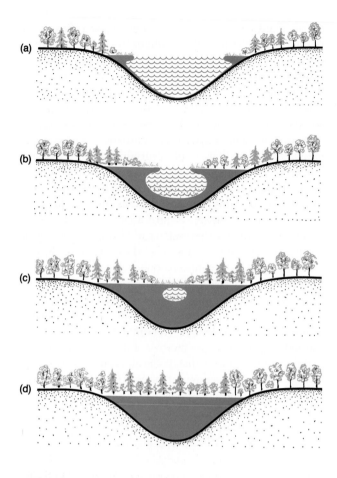

Figure 45-5   AQUATIC SUCCESSION: A DISAPPEARING LAKE.
(a) Organic material encroaches on the edges of a lake and is washed into the lake bottom. (b) A floating mat of organic material eventually closes off the surface, forming a peat bog (c), which then undergoes succession into forest stages (d). (e) Black Pond, in Norwell, Massachusetts, is at the stage of succession illustrated in (b). A mat of sphagnum moss can be found around the pond. Rooted in it are water willows, sedges, cranberry, grasses, orchids, and the carnivorus pitcher plant and sundew.

and deposition of new soils. This process has dominated the vast upper Amazon regions where numerous rivers drain the Andes Mountains. Over the years the rivers have meandered back and forth, with bends eroding forest soils at the rate of twelve meters per year. In time, this wipes out forests, leaves "islands" of surviving trees scattered as mosaics, or so thoroughly denudes areas that secondary succession must start. Traveling from the riverbank inland, the typical stages in that succession are found (as in Figure 45-4 where temperate zone species are shown). As these processes continue, the vast Amazon jungles are broken into a mosaic of small forests of different successional stages on differing soils (which are deposited by the eroding meandering rivers). All of these events contribute to the great biological diversity of the Amazonian forests.

Among the disturbances that can clear vegetation from existing soils are flooding rivers, forest fires, avalanches, and landslides; all can decimate huge areas that have been inhabited by complex communities of plant and animal life. But modern agriculture—including the widespread clearing of land for planting, the clear cutting or heavy harvesting of trees for lumber, and the overgrazing of range lands by livestock—is perhaps the most prevalent of all disturbances.

Sometimes a stretch of farmland or a clear-cut hillside is abandoned and allowed to undergo succession. This type of succession is often termed *old field succession* (Figure 45-6). The many small farms in the southeastern

Figure 45-6   OLD FIELD SUCCESSION: A FOREST RETURNS.
In this photo of old field succession in North Carolina, plow furrows are still visible, even though the field is already in the pine-forest stage of succession. When this "field" was still under yearly cultivation, natural succession was held in check.

Figure 45-7   ADAPTATION AND SECONDARY SUCCESSION.

The jack pine (*Pinus banksiana*) has evolved a seed-dispersal adaptation that gives it an advantage in beginning secondary succession. The jack pine cone opens in response to heat (such as from a forest fire), and the pine is thus one of the first species to sprout after a fire takes place.

United States that failed during the Great Depression in the 1930s provide a vivid example of old field succession: fields that were once planted to corn and other crops were first overgrown with weeds and grasses, and then, within a few years, sprouted pine trees. Oaks and hickory trees—the probable next stage in the succession—are usually not far behind the pines.

Succession can result in "lateral" movement of vegetation types. Since the last ice age 10,000 years ago, the edges of various forest types have "migrated" north at rates of one to eight kilometers per generation. Seeds, of course, are the means of dispersal and of invasion of the tree species.

Some species live in areas where a particular type of natural disturbance—a flood or fire, let's say—occurs frequently; these species may have acquired adaptations that facilitate rapid recovery and give them a head start over other species. In dry shrublands and grasslands, where fires are common, secondary succession begins when certain plants resprout from subsurface root crowns. Others, such as the jack pine shown in Figure 45-7, produce cones that open in response to heat and drop their seeds—often the first harbingers of renewed life—on the scorched earth.

An important feature of fugitive species that can appear early in secondary succession is that they grow quickly and waste little energy in developing "unnecessary" woody support structures. Instead, large portions of their great productivity go into reproductive structures, such as seeds or underground storage tubers—adaptations that also serve as the primary sources of food for billions of people. In fact, many of the food crops that have been domesticated by humans are early-stage fugi-

tive species. Wheat, oats, potatoes, sugar cane, and beets all come from wild stocks that required bare and sunny ground for seed germination. One of the prices humans must pay for continually reaping this harvest is the enormous investment of energy and labor required to keep the community in an early-successional stage by tilling, clearing, and weeding frequently (Figure 45-8).

In contrast to the early-successional species, we have seen that many types of late-successional plant species develop large support structures to give them an adaptive advantage when competition for sunlight and water is intense. Because trunks, large branches, and roots require energy and time for growth and development, late-successional species tend to grow slowly and are less productive over a given time span than are early-successional species. However, once mature, trees such as coffee, walnut, or apple produce large amounts of fruits and seeds, often ones harvested and eaten by humans.

One might be tempted to conclude that succession proceeds in an orderly, preordained sequence, with each successional stage giving way predictably to the next, more stable stage until the area achieves its climax community. This, however, is not necessarily the case. The exact sequence of species that appears during either secondary or primary succession can be highly variable. The search to explain this variability and to understand why succession occurs at all led to an early attempt to establish simple, universal rules by which the obvious changes in communities over time could be both explained and predicted.

Figure 45-8   CULTIVATION: THE PREVENTION OF NATURAL SUCCESSION.

The prevention of natural succession requires an enormous investment of energy and labor, as any farmer can testify.

Looking for the "Rules" of Succession

In 1916, a young botanist from Nebraska, Frederick Clements, published an analysis of plant succession that set the fledgling science of ecology on its ear. Clements had spent his youth watching weeds, grasses, and other species replace each other as vegetation slowly reclaimed the prairie lands worn bare by wagon wheels and migrating buffalo. Indeed, it seemed to Clements that the vast rolling grasslands were like an immense organism—he used the term "superorganism"—which, once disturbed, worked inexorably to repair itself and reestablish its "perfect" state: the climax community of the virgin prairie. Extending his ideas to other types of plant communities, Clements forged a general theory of succession that incorporated three simple assumptions:

1. In every community, the types of plants that occur in various stages of succession are determined solely by climate.

2. All successions lead eventually to a climax in a series of predictable stages.

3. The course of succession is governed by a sort of grand plan, each new group of species paving the way for the next, so that in a particular area, one, and only one, type of climax community is possible.

A core idea of Clements's theory was that climax communities seem to be more stable (that is, they seem to persist longer) than other successional stages, *and thus they must be better adapted*—in fact, best adapted—to the climate of a particular region. His theory also suggested that species in stages preceding a climax must be less well adapted to that environment and perhaps eventually "sacrifice" themselves to encourage the development of the final, "ideal" community.

Despite these conclusions, the more field work researchers carried out, the more variations from predicted courses of succession they observed. Not only did sites with the same climate sometimes show quite different patterns of species replacement, but often the predicted climax community—usually some sort of woodland—failed to appear, even after long periods of time.

In 1970, the British ecologist Donald Walker tested the predictions of Clements's theory by studying sixty-six lakes and ponds scattered throughout Great Britain for which the stages of aquatic succession—the progression from bog to shrubs and trees—were well documented. Walker dug a trench through the mud and earth that extended outward from the edge of each body of water. He reasoned that as the trench cut progressively deeper and farther inland, it would yield the remains of plant life that had existed at various stages in the succession of each lake or pond. He also reasoned that if Clements's theory was correct, the communities arising as each lake or pond gradually filled in would be, in general, the same.

Walker made an important discovery as he painstakingly worked out and compared the sequences in each plant community. As Figure 45-9 shows, there was wide variation in the sequence of communities from one lake to another: there was no regular and predictable order to succession. Furthermore, the supposed climax species, willow and alder trees, were often replaced by peat bogs—an even longer lasting and more stable community.

Clearly, succession involves far more than preordained progress toward an ideally adapted climax community. New research shows that there are three mechanisms of succession: facilitation, inhibition, and tolerance. Clements's original work involved *facilitation*, in which early species make colonization or growth easier for later species. This, however, may be a relatively rare phenomenon. More common is *inhibition* by

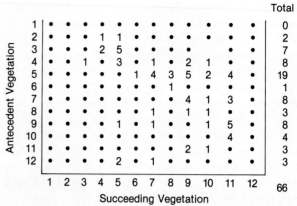

Figure 45-9   WALKER'S STUDY OF LAKE SUCCESSION IN GREAT BRITAIN.
In a study of sixty-six lake communities, Donald Walker used trenches and borings through lakeside deposits to reveal sequences of vegetation. He identified twelve distinctive types of plant communities, ranging from 1—characteristic of open water—to 12—characteristic of the driest areas—and found the succession of these types varied widely, as his tabulated results show. For a given type of antecedent community found by Walker (say, 5 on the left margin), the successor communities he found are listed to the right (in this case, one lake with type 6 vegetation, four with type 7, three with type 8, five with type 9, two with type 10, and four with type 11). Note that whatever the antecedent, there is no single type that always follows (if there were, then vegetation of type 5 would only be succeeded by a single other type, say, type 8).

early species of later ones, so that some disturbance—such as a destructive storm or extensive grazing by herbivores that prefer early-successional species—must occur for succession to proceed. For example, certain shrubs in the chaparral brushlands of the western United States form such dense cover that the seeds of late-successional species either do not germinate or cannot grow in their shade. The later plants can emerge and flourish only when some new disturbance, such as a human's bulldozer, clears away some of the shrubbery. Intermediate between these two are cases when removal of an early species has no effect on the speed with which later species appear. This has been called *tolerance;* the later ones tolerate early ones but are largely unaffected by them. The complexity of these three mechanisms at any one site cannot be overemphasized, since they are affected by soils, local topography, weather, season, herbivores, the chance presence of certain seeds but not others—the list is long, indeed.

Succession involves an immense element of chance. One cannot safely predict how the course of a given succession will run; nor can one consider the so-called climax species to be in some way adaptively "superior" to other species that have preceded them in a succession. Stable associations such as climax communities arise when the environment remains stable and gives the component species an adaptive edge, but they may disappear when the environment changes again. In the end, then, organisms' evolutionary adaptations determine when and how one stage in a succession will give way to the next. For example, in the marine intertidal zone, certain types of algae are adapted to become established quickly on storm-cleared boulders and produce a dense cover. Other algal species, in turn, are adapted to remain dormant beneath this cover and grow only when the cover is reduced by herbivores.

Finally, it is important to remember that communities are *not* entities striving for "ecological balance." Succession reflects the struggle for light, nutrients, space, food, and so on, carried out by the variously adapted individual organisms in their struggle to survive to reproduce.

Animal Succession

If the plants in an area change in successional sequence, it seems reasonable that many of the animals that depend on them might also. In fact, some animals—such as herbivorous insects that breed, pupate, or feed on a particular species of flowering plant—are so closely associated with certain plants that when such species are replaced in a succession, the animals disappear, too. An early study of the correlation between successions in plant and animal communities was carried out in 1911 by Victor Shelford of the University of Chicago. Shelford

found in a series of different-aged ponds near the southern shore of Lake Michigan that the animal communities underwent succession. Not only did he find different species of fish in different ponds, but also different species of insects in the dunes around each pond and varying types of mammals and reptiles inland from each pond (Figure 45-10).

Intrigued by this work, the zoologist L. L. Woodruff decided in 1912 to see if he could find evidence of animal succession in a totally different type of aquatic system: the murky artificial soup created by putting a quantity of pond water and a handful of hay into a laboratory flask. Within a few weeks, Woodruff's novel approach to studying succession paid off. He watched as species after species of protozoans—amoebae, paramecia, tiny flagellates, and other forms—replaced each other in "blooms" at the upper surface of the flask. Woodruff's experiment confirmed that animal succession takes place and always proceeds in the same sequence in a simple laboratory culture. Nevertheless, successions are hard to predict in nature.

THE STRUCTURE OF COMMUNITIES

The hallmarks of the succession phenomenon are series of communities in time. To really understand communities, however, we need a time slice—an in-depth study of one community at one time and, in particular, of its structure. What exactly is meant by "community structure"? When ecologists define a community, they generally consider one or more of the following four categories:

1. The physical appearance of the community—the *physiognomy*, or physical form that the component species take

2. The *relative abundance* of species—the number of rare species relative to common species, or the number of individuals of various species, especially with respect to dominant or rare species

3. The number of species, or *species richness*

4. The *niche* structure of the community—the ecological roles or "occupations" of component species and how they resemble or differ from each other

For the remainder of this chapter, let us consider each of these four factors in some detail.

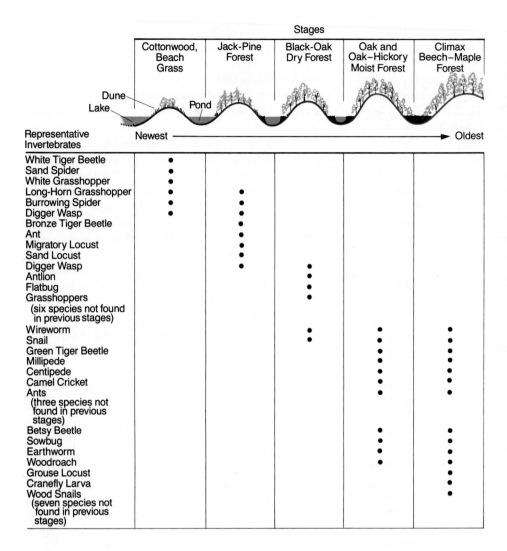

Figure 45-10 ANIMAL SUCCESSION IN A SERIES OF DUNES. As you read the diagram from left to right, the Lake Michigan sand dunes grow older and older—a fact reflected by their plant species. The table lists the patterns of invertebrates and plants found on those dunes. Note that none of the animal species that inhabit the newest dunes, with thin cotton wood and beach grass, inhabit the older dunes with their black oak dry forest. Also note how many new animal types appear as dunes change from black oak dry forest to oak and oak-hickory moist forest.

Representative Invertebrates — Stages (Newest → Oldest):

Representative Invertebrates	Cottonwood, Beach Grass	Jack-Pine Forest	Black-Oak Dry Forest	Oak and Oak–Hickory Moist Forest	Climax Beech–Maple Forest
White Tiger Beetle	•				
Sand Spider	•				
White Grasshopper	•				
Long-Horn Grasshopper	•	•			
Burrowing Spider	•	•			
Digger Wasp	•	•			
Bronze Tiger Beetle		•			
Ant		•			
Migratory Locust		•			
Sand Locust		•			
Digger Wasp		•			
Antlion			•		
Flatbug			•		
Grasshoppers (six species not found in previous stages)			•		
Wireworm			•	•	•
Snail			•	•	•
Green Tiger Beetle				•	•
Millipede				•	•
Centipede				•	•
Camel Cricket				•	•
Ants (three species not found in previous stages)				•	•
Betsy Beetle				•	•
Sowbug				•	•
Earthworm				•	•
Woodroach				•	•
Grouse Locust					•
Cranefly Larva					•
Wood Snails (seven species not found in previous stages)					•

Physiognomy: What Does a Community Look Like?

Most ecologists interested in depicting the physical structure of communities have concentrated on plants. It is easy to understand why: unlike many animals, plants are obvious parts of the landscape, and once they have sunk roots or become attached to the substratum, they stay put. Table 45-1 reviews the major growth forms of plants. We saw in Chapter 44 that the Earth's major terrestrial communities of plants and animals are the biomes and that they are controlled by climate and geography and lack neat boundaries. These major community types instead grade into one another, often along climate gradients. Such gradients—usually of moisture or temperature—are called **ecoclines.**

Table 45-1 **TERRESTRIAL GROWTH FORMS OF PLANTS**

Form	Description	Example
Thallophytes	Small plants without differentiated leaves	Mosses, lichens
Herbs	Plants without woody stems	Grasses, ferns
Shrubs	Small woody-stemmed plants less than 3 meters tall	Yucca, cactuses, chaparral shrubs, broad-leaved shrubs
Trees	Large woody plants more than 3 meters tall	Conifers, oak and other hardwoods, bamboo, tree ferns, palms
Epiphytes	Whole plants that grow on other plants	Bromeliads
Lianas	Woody vines rooted in soil	Passion flower

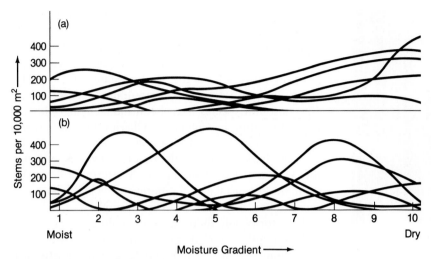

Figure 45-11   GRADIENT ANALYSIS: WHITTAKER'S STUDY OF THE DISTRIBUTION OF PLANT SPECIES.
R. H. Whittaker plotted the distribution of tree species along a moisture gradient from moist ravine locations, which supported temperate forest, to dry southwest-facing slopes, a desert, in (a) the Siskiyou Mountains of Oregon and (b) the Santa Catalina Mountains of Arizona. Abundance is plotted on the vertical axis in terms of the number of tree stems (at least 2 centimeters in diameter) of each species per 10,000 square meters. The data show that the relationships among the species were random in each location; Whittaker found no clear boundaries between biomes.

A major feature of community physiognomy is revealed by viewing the pattern of growth forms in communities along ecoclines. As the environment along a gradient becomes harsher (colder, drier, and so forth), community physiognomy changes from complex to simple. Many plant forms and the complex layering of the various levels of foliage give way to fewer forms, smaller plants, and a single layer of foliage. Just such an ecocline can be seen when hiking up a mountain from a forested valley to the alpine tundra above the tree line.

A prime factor that determines a community's physical form is the particular *combination* of plant species that grow in it. We saw that Frederick Clements considered climax communities to consist of predictable "ideal" combinations of species. Other early ecologists also suspected that these communities would have clear-cut boundaries. To test these ideas, an eminent botanist, R. H. Whittaker, developed a quantitative technique called *gradient analysis*. This technique involves recording the actual distribution of plant species in a region along with data on moisture levels, temperature, and gradients in elevation.

Whittaker carried out his investigations in two mountainous regions in the western United States: the Siskiyou Mountains in Oregon and the Santa Catalina Mountains in Arizona. In both areas, a temperate forest biome gives way to an arid desert landscape along a moisture gradient. In addition to gathering data on soil moisture and temperature, Whittaker meticulously tallied the distributions of various plant and insect species. His findings were unequivocal: both plant and animal species occur at random with respect to most other members of the communities (Figure 45-11), and clear boundaries between communities are nowhere to be found. Instead, communities intergrade.

Relative Abundance of Species

Every community contains a mix of species in which certain individuals are common, others are fairly abundant, and a few are rare. The proportion of individuals of various species in a forest of firs in the Great Smoky Mountains of Tennessee ranges from 70 percent down to 0.001 percent of the total community. By contrast, most of the species in a tropical rain forest in Brazil have about equal population sizes; typical shares of the community range from 0.003 to 1 percent of the total. This feature of *equitability* is typical of tropical communities, whereas communities in other biomes, where one or a few species are often extremely common, are said to show a high degree of *dominance* (Figure 45-12).

Whether dominance or equitability is present may depend on a community's stage of succession. Thus, early-successional stages not only have few species (the pioneers), but usually one or two of the early species

Figure 45-12   DOMINANCE IN A TEMPERATE FOREST.
In contrast to the usual equitability in tropic forests, this lodgepole pine forest in the northern Rockies is composed predominantly of one species; the trees grow at incredible density and extend tall and straight as they seek the light above.

dominate. Later stages tend to accumulate more species, with a more even mix in numbers. Extreme dominance is also found in harsh environments—at high elevations or in hot springs, for example—or in polluted ecosystems. But while these trends are intriguing, they do not reveal the underlying causes for different patterns of species abundance. Answers await future research.

Species Richness: Taking Count of Species

On a global scale, the number of species from different taxonomic groups varies tremendously between one community and the next, as Table 45-2 shows. In the

tropical rain forest biome of Costa Rica, for example, more than 8,000 species of seed plants have been identified, while in the British Isles, with six times the area of Costa Rica, only about 1,600 species are known. Similarly, in the whole of the United States and Canada, there are about 120 species of mosquitoes, yet more than 150 mosquito species have been found living within a ten-mile radius in Colombia. Charles Darwin, Alfred Wallace, and other early naturalists were fascinated by such comparisons, and in the past few decades, community ecologists have been intensely interested in uncovering the reasons for such differences. What determines the number of species in a community?

As it turns out, there are no simple answers to this question, although some provocative insights are emerging from current research. Looking at the Earth on a grand scale, in most major taxonomic groups of land plants and animals, **species richness,** or *species diversity,* can be correlated with latitude. The farther one travels, north or south of the equator, the fewer species are found in various biomes (Table 45-2 and Figure 45-13). Tropical rain forests near the equator are the richest biological areas on Earth. Ecologists have reported up to 200 species of trees in a five-acre patch of rain forest (eight times the number of types that might be found in a vast temperate forest), while entomologists armed with collecting nets have scooped up more than 500 species of tropical insects at a single site. It is therefore extremely alarming that tropical rain forests are

Table 45-2   **EXAMPLES OF SPECIES RICHNESS IN DIFFERENT COMMUNITIES**

Taxonomic Group	Locality (Area)	Number of Species
Seed plants	Costa Rica (18,400 sq. mi.)	8,000
	Coastal California (24,520 sq. mi.)	3,050
	Baja California desert (24,100 sq. mi.)	1,500
	British Isles (120,200 sq. mi.)	1,600
Amphibians	Costa Rica (18,400 sq. mi.)	123
	Florida (54,600 sq. mi.)	42
	California (148,400 sq. mi.)	35
Reptiles	Costa Rica (18,400 sq. mi.)	182
	Florida (54,600 sq. mi.)	30
	California (148,400 sq. mi.)	66

Figure 45-13   SPECIES RICHNESS AND LATITUDE.
This graph shows that maximal numbers of ant species occur near the equator (0°), and that richness falls off sharply as one goes north (left) or south (right) toward the poles.

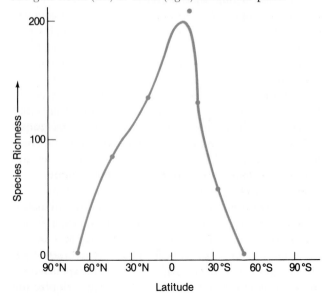

Figure 45-14   SPECIES EQUILIBRIUM MODEL: EXPLAINING SPECIES RICHNESS. The number of species on an island is determined by the balance between the immigration of new species and the extinction of species already present. At the junction of the two lines, the density of species is at equilibrium.

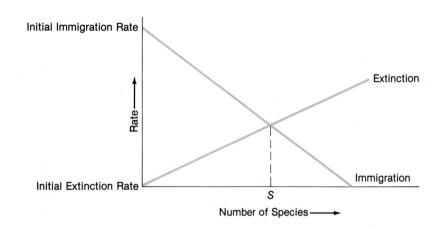

being rapidly destroyed by humans for lumber and agriculture. Each acre of rain forest cleared destroys many plant and animal species, some of which have not even been described yet, and therefore impoverishes the Earth.

The explanation for such large-scale global trends in species richness is likely to be complex. However, if we look at somewhat smaller scales, we may be more successful at understanding why a given number of species occupies a particular area. Let us scale down to the size of an island and consider the work of two theoreticians, Robert MacArthur and Edward O. Wilson, whose ideas, directly or indirectly, stimulated much of today's research in the ecology of communities.

The Species Equilibrium Model

In the 1960s, MacArthur and Wilson began looking at the structures of communities on islands. Ecologists had long observed that large islands tend to have more species of a particular taxonomic group—perhaps birds or ferns—than do small islands. One reason for this might be that larger islands tend to be topographically diverse, with mountains, valleys, and so on, thus providing more habitats for organisms. Topography clearly is not the only reason, however, since large and small islands with similar physical characteristics still show differences in the number of species inhabiting them. (Keep in mind that we are concerned here with the number of *species* rather than with the number of *individuals* in the community.) To explain species richness, MacArthur and Wilson proposed the **species equilibrium model.**

The model's central assumption is that the number of species found on an island is determined by a balance between two factors: (1) the *immigration rate* of "new" species (from other inhabited areas) onto the island, and (2) the *extinction rate* of established species (Figure 45-14). Imagine an island that was formed recently, for example, by a volcanic eruption. As species colonize the

island and accumulate, says the model, some will become extinct—either by chance or because a potent predator or competitor becomes established. The rate of extinction will show a negative correlation with island size: it will be higher on small islands, lower on large islands. (The larger the island, the more individuals of a given species can inhabit it; as we saw in Chapter 41, the larger a population, the less chance it will become extinct.) As a result, large islands will tend to have more species. Eventually, the rates of immigration and extinction will reach equilibrium, and that point will determine the number of species, S (Figure 45-15). At the equilibrium point, in other words, for each species that is newly established, one species will become extinct, *on the average.* The model also predicts that the value of S will be affected by the distance of an island from its source (or sources) of colonists; that is, the farther away the island from the sources, the fewer species will be likely to immigrate, and S will be lower.

The species equilibrium model was tested experimentally on a series of tiny islands of red mangroves—clumps of trees that grow with roots bathed in sea water—off the coast of Florida. Invertebrate species on the islands—mostly insects—were tallied. Then the islands were fumigated, killing the fauna but not the trees, and the process of repopulation was monitored. The results were surprising: within less than a year, each of the islands was repopulated with about the *same number* of species (presumably by immigrants from the mainland) that had earlier inhabited it. In addition, the number of species in each ecological category (herbivore, detritivore, and so on) returned to the previous level. Interestingly, the new communities often contained *different* species from those in the original communities; species identity, in other words, was not predictable, but species richness was. As more time passed, some of the new colonizers disappeared, but they were always replaced by still other arrivals. Finally, smaller islands reached equilibrium with fewer species than larger ones.

(a)

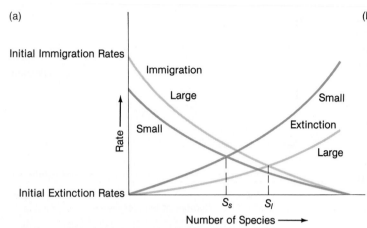

(b)

(c)

Figure 45-15   ISLANDS LARGE AND SMALL: THE EFFECT OF SIZE ON SPECIES EQUILIBRIUM.
(a) Large islands have lower population extinction rates than small islands. For this reason, if all other factors are equal, a large island will have a higher equilibrium species number (S_l) than a small island (S_s). (b) Small islands off the coast of Maine. (c) A large island off the coast of southern California with obvious differences in habitats. These are the sorts of islands that the MacArthur-Wilson hypothesis seeks to explain.

A more recent study of marsh grass in a Florida bay added another crucial piece of data; irrespective of size, islands nearest the mainland were colonized by more species than islands farther away. It is still disputed whether large islands have reached their equilibrium number of species or not.

Other new results show that body size affects colonization. Mammals commonly swim to islands up to a few kilometers offshore. In cold water, body size correlates with ability to reach islands (small bodies lose more heat). Thus shrews only reach islands 700 meters from shore, whereas voles can swim to islands 1 kilometer out in Canada's St. Lawrence River. For long distances, rafting is the means of transport, and then small size is favored. Only rodent-sized placental, flightless mammals have succeeded in reaching the Galápagos and Philippine Islands and Australia.

These findings have established species equilibrium as a fundamental concept in the study of community structure and it may be applicable to communities on mainlands. Ecologists believe that most environments are composed of habitat "islands," like patches in a crazy quilt. Thus an island of woodland in an ocean of grassland will provide suitable habitats for only a certain number of species. The most likely colonists are plants and animals with adaptations that ease *dispersal;* these include fugitive plant species with wind- or waterborne seeds, small animals that may be carried on wind or water currents, and birds or insects capable of flying long distances. Some new arrivals will flourish, and some established species will become extinct. The equilibrium number of species for a particular community will de-

pend partly on the area of that woodland island and partly on its distance from other patches of woodland that serve as sources of colonists. The concept of habitat islands is assuming great practical importance in tropical countries where rain forests are being cut, leaving only scattered islands of trees (Chapter 51).

Species Interactions and Physical Disturbance

If we scale down our view even smaller than the size of islands, we can finally begin to understand the actual processes that may determine species richness and the relative abundance of species. Both of these aspects of community structure—species richness and relative abundance—contribute to the species *diversity* of communities. For example, a forest composed of Douglas fir (98 percent), spruce (1 percent), and pine (1 percent) is technically a three-species community. But it suffers from a high degree of dominance and, to us standing in the forest, it looks very similar to a one-species community, which obviously is very species-poor. In contrast, if each species composed one-third of the forest, the community would be characterized by a high degree of equitability and it would appear much more diverse to us, even though it is still only a three-species community. Tropical rain forests impress us with their diversity not only because they have more species, but also because their relative abundance of species is more equitable than that in temperate forests.

The processes that determine species richness and relative abundances in communities at a local scale are **competition** between species for limited resources, **predation** (including herbivory, disease pathogens, parasites), and **physical disturbance.** Competition is important when there is a limited abundance of resources, such as food, water, or space (e.g., nest sites for birds or space itself for intertidal organisms). When some critical resource is limited, those species that are best at securing it reduce the resource's availability to the other species, which then die, emigrate, or become rare. Eventually the community equilibrates at the number of species that the resources can support. Competition at a local scale, then, tends to reduce species diversity.

Not all communities contain their equilibrium number of species. The disturbances that affect ecological succession are obvious factors in reducing species diversity: a lava flow such as that on Hawaii, or a brush fire, such as that on Isabela Island in the Galápagos, might easily wipe out many individual plants and plant species in a community, and so remove the habitats and food sources of several trophic levels of animals. On the other hand, disturbance, if it is not too frequent or too catastrophic, can increase species diversity by preventing dominant competitors from monopolizing resources. Predation can act similarly to physical disturbance. If predators or disease organisms reduce the populations of competitively superior species, other species that are competitively inferior may be able to persist in the community, thereby increasing species diversity. This is called *predator-mediated coexistence.*

A model called the **intermediate disturbance hypothesis** combines all these ideas (Figure 45-16). At point A, physical disturbance occurs rarely; if it does occur, it is not catastrophic. Similarly, few predators are present, or they are inefficient and have little impact on the prey communities. Under these conditions, the populations of the species in the prey community can grow to high densities and fully exploit their resources, resulting in competition and reduction of the number of species in the prey community. At point C, in stark contrast, either physical disturbance or predation is very intense, reducing prey populations severely and causing the less adapted species to become extinct locally. Again the number of species is low because the community is kept in an early stage of succession by disturbance or because some prey species are eaten to extinction by predators.

When disturbance or predation intensity is intermediate (point B), the community has the most species, thus

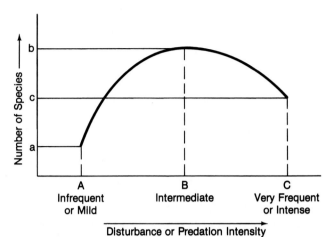

Figure 45-16   THE INTERMEDIATE DISTURBANCE HYPOTHESIS.
Where disturbance or predation is intermediate in intensity (point B), a maximal number of species can exist in a given locality.

the peak in the curve. This is because either physical disturbance or predators reduce the densities of the prey populations, not to extinction, but low enough so that they do not fully exploit their own resources. Since their resources, then, are not limited, competition is not important and many prey species can coexist. Community ecologists now believe that many communities are rarely at competitive equilibrium (represented by point A). For example, it seems probable that predators and parasites on plant-eating insects usually keep these herbivores' densities low enough so that they rarely compete. This may help explain why there are so many herbivorous insects in the world.

The Niche

Every creature has its habitat—in a sense, its "address" in the community. But it also has a **niche**—a role in the community and a way of life more equivalent to a "profession" than to an address. An organism's niche is determined both by physical factors—such as the amount of light, carbon dioxide, and oxygen it needs, and the ranges of temperature and pH it can tolerate—and by biological factors—such as the kinds of food it needs, the diseases it tends to contract, the predators that feed on it, and the competitors that vie for the same limited resources it requires. Studies of the red squirrel, for example, a chattering denizen of North American conifer forests, reveal that the animal transports and eats spruce seeds, fungi, flowers, and other plant materials, and, in the process, spreads them about the forest. All aspects of the squirrel's environment and behavior—from these foods to the trees it climbs, to the effects of its activities on soil and plant life—are elements of its niche.

Yale University biologist G. E. Hutchinson and his student, the pioneering ecologist Robert MacArthur, divided the niche into food- and habitat-related factors. An animal's **food niche,** for instance, includes the types of prey it eats at different stages of its life cycle, the size range of its food items, and the places where it finds the food. Since an animal can only survive near its food resources, the distribution of these resources helps predict where populations of a particular species will be found. Thus the red tree mouse, a species with a specialized food niche—the needles of fir trees—inhabits only conifer forests. An animal with a more generalized food niche, such as a fly-catching swallow, may intercept and consume a variety of insects, almost regardless of species.

An animal's **habitat niche** includes physical factors—temperature, light, salinity, pH, and so on—as well as surrounding vegetation. A plant's habitat niche is mostly determined by physical factors such as light, temperature, soil types, soil moisture and nutrient content, and sizes of soil particles, although competition and herbivory also help define its niche. Organisms are adapted by form, physiology, and behavior to occupy their food and habitat niches. Therefore, these adaptations restrict where populations of a species may or may not survive.

While one can define a species' niche by its food and habitat needs, one can also discuss a given type of niche independent of its occupants. For instance, a "desert seed-eating niche" may be populated by rodents in one desert, ants in another, and birds in a third desert of the world. The "flower nectar-feeding niche" is occupied by hummingbirds in the New World and by unrelated species of sunbirds in Africa, honeycreepers in Ecuador, and honeyeaters in Australia. Interestingly, as Figure 45-17 shows, the single niche occupied by these species has contributed to a striking degree of convergent evolution in the birds' morphology.

Hutchinson and his students at Yale devised yet another way to define the niche concept: by quantifying and graphing it. For instance, the foraging height in trees and the size of insects eaten by a species of bird could provide two axes on a graph. One could create an even more accurate picture of a species' niche by graphing three, four, or many more niche components on separate axes, thereby including information on the range of nest sites, feeding times, types of predators, temperature-tolerance limit, and so forth. The result would be a multidimensional enclosed "space" (called a hypervolume) that describes the niche much more precisely (Figure 45-18).

The niche concept is important to our understanding of communities because various species of organisms may compete with each other for the same kinds of resources in a given area. Their niches are said to overlap, and this competition may limit where and how each species can live, and thus influence the structure of the entire community. (Individuals *within* a species also compete with each other for resources, and we will discuss

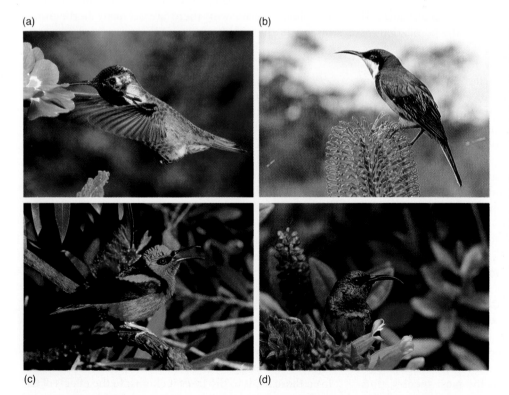

Figure 45-17   THE NECTAR-FEEDING NICHE.
The availability of nectar-laden flowers in four parts of the world resulted in the convergent evolution of four unrelated birds: (a) Costa's hummingbird (*Calypte costae*) in the western United States; (b) the eastern spinebill (*Acanthorychus tenuirostris*), a honeyeater, in eastern Australia; (c) the Ecuadorian honeycreeper (*Cyanerpes cyaneus*), called the Iiwi; (d) the sunbird (*Nectarinia mediocris*) in Africa.

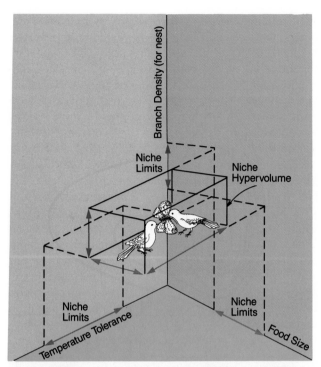

Figure 45-18   HUTCHINSON'S MATHEMATICAL MODEL OF THE NICHE.
The illustration shows three dimensions of a niche: food size, branch density, and temperature tolerance. Try to imagine adding a fourth, a fifth, and more dimensions to this diagram—the more added, the more precisely the niche hypervolume is defined.

the consequences of this type of competition in Chapter 46.)

Biologist Joseph Connell carried out a classic study of competition between two barnacle species that inhabit the shallow waters of a rocky coast. Connell studied one barnacle species, in the genus *Chthamalus*, that tends to live in the shallower waters of the marine intertidal zone, while the other, in the genus *Balanus*, lives in deeper waters of that same zone (Figure 45-19). The two species appear to have clearly separate habitat niches: although the tiny larvae of each species often settle in the other's area, adult *Chthamalus* and *Balanus* are always found at their characteristic depths.

Connell decided to determine the mechanism of their niche separation by clearing barnacles from areas of the intertidal zone and observing repopulation patterns. He discovered that *Balanus* which settled in the upper part of the intertidal zone (areas usually inhabited by *Chthamalus*) dried up and died during low tides. Clearly, the species is not adapted to periodically dry conditions, and the shallower areas cannot serve as part of its habitat niche. Connell also observed that any *Chthamalus* gaining a foothold in the deeper areas were crushed by the larger, faster-growing *Balanus*. If all *Balanus* individuals were removed, however, *Chthamalus* flourished even in the deepest zones. Connell concluded that *Balanus* adults are indeed occupying their full potential niche, whereas *Chthamalus* adults— outcompeted by *Balanus* in areas were they could potentially survive—are not.

Figure 45-19   A CLASSIC CASE OF COMPETITION AFFECTING NICHE STRUCTURE.
Larvae of both *Chthamalus* and *Balanus* barnacle species can settle at any level of these intertidal rocks. *Balanus* fails to survive in the upper range because it cannot resist desiccation and heat. *Chthamalus* ends up restricted to the upper range because it grows more slowly than *Balanus*, which grows over it or displaces it from rocks in the lower zone. The result is a greatly reduced realized niche for *Chthamalus* and near correspondence of the fundamental and realized niches for *Balanus*. (Spring tides are the largest rises and falls of tides and neap tides are the tides that rise and fall least during each lunar month.)

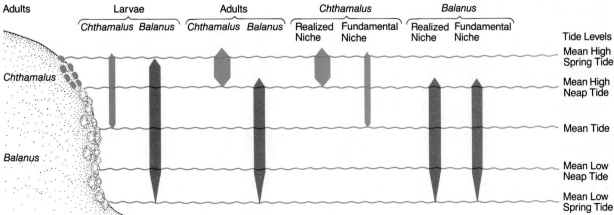

Similar experiments on two species of green algae revealed that one low-growing form can survive anywhere on a given set of intertidal rocks, but tends to be shaded out by a taller species. The taller type, however, cannot cling to the rocks' edges where wave action is most powerful. Therefore, those edges are occupied only by the low-growing form.

Experiments like these on barnacles and algae have led ecologists to formulate two definitions of niche that reflect interspecies competition. A **fundamental niche** is the full environmental range that a species can occupy if there is not direct competition from another species. *Balanus*, for example, occupies its full fundamental niche. *Chthamalus*, however, does not; it occupies a **realized niche,** a niche that is narrowed from the fundamental state by competition. The experiments also illustrate what is known today as the **competitive exclusion principle,** the postulate that no two species can occupy exactly the same niche at the same time in a particular locale if resources are limiting (in the barnacle example, space was limiting). When resources are in short supply so that competition for them is severe, one species will always achieve a competitive edge and eventually exclude the other, at least from part of its fundamental niche.

Competition for resources can have important consequences, not only for species diversity (as discussed above), but also for a species' versatility and for the niche structure of the community. Communities in which competition occurs between species are characterized by *resource partitioning*, that is, every species uses a different aspect of the resource. For example, food is a limited resource to hawks, and hawk communities are composed of species which differ from each other in body size. Since different-sized hawks take different-sized prey, their food niches do not overlap greatly and competition is reduced. In contrast, communities in which resources are rarely limited and competition, therefore, is unimportant often are composed of species which do very similar things. Think of the many different species of plant-eating insects, such as aphids and other insects that suck plant juices—they all seem to occupy very similar niches.

Some species evolve with narrow niches—they can use only one or a few types of food or growth sites—and are termed **specialists.** The blue-gray gnatcatcher is a habitat specialist (Figure 45-20). In contrast, some species have broad niches—they are adapted to life over a wide range of conditions—and are termed **generalists.** The common crow is a generalist that occurs in almost every temperate biome and from wilderness to farm to urban area. Communities made up of competing generalist (broad-niched) species contain few types of organisms. In contrast, specialist (narrow-niched) species may

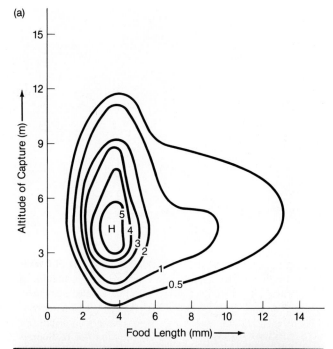

(a)

(b)

Figure 45-20   NICHE OF THE BLUE-GRAY GNATCATCHER: A HABITAT SPECIALIST.

In a study of blue-gray gnatcatchers (*Polioptila caerulea*) inhabiting a narrow niche in a particular community of California oakwood, it was found that they lived primarily within a 12-meter altitude range and ingested prey only in the size range of 1 to 13 millimeters in length. They most frequently capture insects that are 3 to 5 meters above the ground and that measure 3.5 to 4.5 millimeters long. H represents this core part of the niche.

coexist in a community; in effect, they "share the wealth" by dividing the community into more discrete pieces, and the result may be greater species richness than could occur if the species were more generalized.

Niches and the species richness of a community are

Figure 45-21   HUMMINGBIRD DIVERSITY AMID FLOWER DIVERSITY IN TROPICAL FORESTS.
These lovely birds show the range in size, bill structure, and coloration in birds occupying different food niches. This is an example of specialists in feeding niches; each species concentrates on a particular small set of flowers over the course of a year. From left to right: 1. Violet-tailed Sylph (*Aglaiocercus coelestis*) Colombia, Ecuador; 2. Crimson Topaz (*Topaza pella*) Northeastern South America; 3. Wire-crested Thorntail (*Popelairia popelairii*) Colombia, Ecuador, Peru; 4. Tufted Coquette (*Lophornis ornata*) Northeastern South America; 5. Booted Racket-tail (*Ocreatus underwoodii*) Northwestern South America; 6. Horned Sungem (*Heliactin cornuta*) Brazil 7. Black-breasted Plovercrest (*Stephanoxis lalandi*) Brazil, Paraguay, Argentina; 8. Red-tailed Comet (*Sappho sparganura*) Bolivia, Argentina, Peru.

related in still other ways. In every community, characteristics of the physical and biotic environments generate selection pressures that favor either generalist or specialist adaptations. For example, a fluctuating or sparsely available food supply can be most easily exploited by generalists. In the United States, eight broad-niched species of hummingbirds compete for the nectar in seasonal, scattered, temperate-zone flowers. Each species feeds on many types of flowers, and when the nectar supply shrinks in winter, most of the birds migrate south in search of more abundant food. Some of these species have different geographical or habitat distributions, or different breeding seasons or nest sites and feeding times, and so their niches may not overlap in most respects. But the seasonal food supply apparently prevents the species from specializing on one or a few species of flowers.

By contrast, more than 300 hummingbird species are found in the tropical regions of Central and South America, where flowers are available in abundance year-round (Figure 45-21). Although not all 300 species have been studied, their diversity is explained in part by the greater range of microclimates—and therefore habitats—available to these birds. Furthermore, the year-round availability of many types of food has allowed the evolution of highly specialized feeding niches; many species visit only a few species of flowers. Thus the community can include many hummingbird species in a single locale rather than just one or a few.

We can sum up our discussion by stating that the most diverse communities are those in which (1) resources are not limited, so that competition between species is absent or mild; or (2) resources are predictable in time and exploited by specialists. If food resources can always be counted on, specialists as well as generalists will be able to survive in the long term. One difference between temperate and tropical communities is that the tropics seem to have more specialists. Perhaps this is because in the temperate zone, resources are more variable and less predictable, resulting in extinction of many specialists that have evolved there, or in prevention of evolution of specialization in the first place.

LOOKING AHEAD

In this chapter, we have seen that ecological communities have specific physical and biological characteristics. Thus the web of interactions that make up an ecological succession and the niches occupied by species in a community both affect and are affected by organisms' adaptations to their physical and biological environments. Yet in order to fully understand the forces that shape communities, we must also examine the crucial ecological relationships among populations and among individuals within populations. These interactions are the subject of Chapter 46.

SUMMARY

1. A *community* is an association of the interacting members of populations of different species in a particular area.

2. *Ecological succession* is the gradual change in the species composition of communities over time. *Primary succession* begins on rock, lava, or sand, while *secondary succession* begins on developed soil. The first plants in a successional series are usually small, simple forms such as lichens and mosses. Later species are more complex in structure and grow more slowly. Animal species also undergo succession, in keeping with the plants they depend on. *Fugitive species* (weeds, grasses) disperse well and grow fast; they commonly are the early stages of secondary successions.

3. When a community reaches a stable state in which species are no longer replaced by new arrivals, it is said to be a *climax community*.

4. The species involved in each stage of a succession are often not predictable. Local geography and many chance events, including the species composition of surrounding communities and the nature of disturbances, can affect what species succeed in a particular area.

5. The species of plants present in each stage of a succession are well adapted to the physical conditions of their environment. Often, they have adaptations that function to delay the onset of species of the next stage by shading out or otherwise competing with species that will replace them. Alternatively, early-stage plants may facilitate growth of later-appearing species; or the latter may be unaffected by early-species plants.

6. Among factors that play a role in species replacement are the rate at which species are dispersed into a new area, the fact that early-stage species adapted for harsh conditions may moderate the physical environment by their presence, and physical or biotic disturbances.

7. The number of species in a succession tends to be low in the earliest stage, highest in intermediate stages, and lower again in later stages. The physiognomy, or physical structure, of a community tends to increase in diversity and complexity as succession proceeds.

8. Gradations that affect organisms along climatic gradients are known as *ecoclines*.

9. Different communities typically contain different numbers of species; a dramatic example is the higher *species richness* found in tropical rain forests.

10. MacArthur and Wilson, who developed the *species equilibrium model*, explored the complex reasons why species richness varies. They noted that larger islands are richer in species than smaller ones and that both immigration and extinction help account for species richness at a given site. Ecologists are just beginning to understand the factors that determine the relative abundance of species: what proportion of a community is made up of individuals of a particular species. In some communities, one or a few species are dominant, while in others, there is a more equitable mix of species numbers. Dominance seems to be related to the successional stage of the community.

11. Species richness and relative abundance are affected by *competition, predation* (including herbivory), and *physical disturbances*. The *intermediate disturbance hypothesis* seems to explain many cases of community structure, and leads to the conclusion that many communities are rarely at the point of competitive equilibrium.

12. A species' *niche* is its total way of life, including every facet of its ecological, physiological, and behavioral roles in nature. The *fundamental niche* is the full range that a species can occupy, whereas the *realized niche* is the niche that results when competition narrows the fundamental niche. The niche structure of a community is determined by the degree to which its species are *generalists* or *specialists*.

13. If resources are limited and competition is severe, only one species can exist in a given niche at a given time. This is the *principle of competitive exclusion*.

14. In general, relatively few species occur where resources are limited or unpredictable. More species occur where populations are kept below the level at which competition would occur for food, space, and other resources.

KEY TERMS

climax community

community

competition

competitive exclusion principle

ecocline

ecological succession

food niche

fugitive species

fundamental niche

generalist

habitat niche

intermediate disturbance
   hypothesis

niche

physical disturbances
predation
primary succession

realized niche
secondary succession
specialist

species equilibrium model
species richness

QUESTIONS

1. Define "ecological community." What is the relationship between a community and an ecosystem?

2. Which of the following are examples of ecological succession? Explain why.

  a. A lake fills with vegetation, becoming first a bog, then a meadow, and finally a forest
  b. Grasses colonize sand dunes, followed by shrubs and, finally, trees
  c. Lichens begin to grow on bare lava, followed by mosses, grasses, and shrubs
  d. Mosses and lichens colonize rocks scraped bare by a receding glacier; grasses, perennials, and shrubs follow over the years
  e. An abandoned corn field sprouts weeds, then shrubs, and, finally, trees

3. State whether each example of succession in question 2 is (a) primary, (b) secondary, or (c) aquatic, and explain why.

4. A _____ community has apparently reached a "stable state" in which species are no longer replaced by new arrivals.

5. Name the four categories that ecologists often use to define and describe communities, and explain briefly what is included in each category.

6. How is the physiognomy of most communities described? Why?

7. In general, how does species richness differ between tropical and temperate communities? Briefly discuss possible explanations for this phenomenon.

8. Briefly summarize MacArthur and Wilson's island-colonization experiments. What hypothesis(es) were they testing, and what were the results?

9. Give at least two examples of species interactions. How can these interactions affect community structure? How do competition, predation, and physical disturbances affect species richness and relative abundance? In most communities are species at the point of competitive equilibrium? Why?

10. Is it easy to describe the complete niche of an organism? Why? How did Hutchinson and his students attempt to graphically describe a niche?

ESSAY QUESTIONS

1. Discuss the concept of the climax community. How have ecologists' conclusions changed about this concept? In general, what characteristics do modern ecologists expect a climax community to display with regard to physiognomy, species richness, and relative species abundance?

2. Thinking of a city lawn as an ecological community, describe succession, climax, pioneer species, fugitive species, physiognomy, species richness, and competition in relation to that community.

SUGGESTED READINGS

DRURY, W. H., and I. C. T. NISBET. "Succession," *Journal of the Arnold Arboretum* 54 (1973): 331–368. A critical review of this important ecological process.

HUTCHINSON, G. E. *The Ecological Theatre and the Evolutionary Play.* New Haven, Conn.: Yale University Press, 1965.

This book, written by one of the great American scientists, includes a description of the niche hypervolume theory.

MACARTHUR, R., and E. O. WILSON. *The Theory of Island Biogeography.* Princeton, N.J.: Princeton University Press, 1967.

This famous book outlines the species equilibrium model.

RICKLEFS, R. E. *The Economy of Nature.* 2d ed. Newton, Mass.: Chiron Press, 1983.

Some regard this as the best book on ecology. It is strong on succession and community structure.

WHITTAKER, R. H. *Communities and Ecosystems.* 2d ed. New York: Macmillan, 1975. A standard reference since published.

46

THE ECOLOGY OF POPULATIONS

Through the animal and vegetable kingdoms nature has scattered the seeds of life abroad with the most profuse and liberal hand. She has been comparatively sparing in the room and the nourishment necessary to rear them.

Thomas Malthus, *Essays on the Principles of Population* (1798)

As far as the eye can see: a wildebeest population grazing on a Kenyan plain.

For several hundred years, North American fur trappers have sold the pelts of snowshoe hares and lynx to brokers of Canada's Hudson's Bay Company. The company has kept precise records of its purchases since 1800, and these tallies reveal that lynx populations reach a peak—followed by a severe decline or "crash"—every nine to ten years. Populations of the lynx's prey, the snowshoe hare, also show a cycle, with peaks occurring just before those of their predators (Figure 46-1).

There is an obvious explanation for the swings. As the lynx population grows, there are more predators to eat more hares until eventually, the lynx consume their food supply faster than the hares can multiply. Then lynx begin to starve, and the predator population crashes. The swings in the snowshoe-hare populations occur for a different reason. We might predict that if there were no lynx, the hare populations would remain steady. But that proves to be wrong; these animals experience growth–crash cycles even in lynx-free areas. The hares' cycles are based, instead, on underlying peaks and crashes in the amount of the vegetation they consume, and these plant fluctuations, in turn, are due to climatic variations, predatory insects, and so on.

Such cyclic phenomena raise fundamental questions about *populations*, or groups of individuals of the same species. Why does a particular population grow and change in size over time? When it remains steady, what keeps the population at that particular size? Why do populations of lynx, let us say, occur in one place but not another? Just as geneticists depend on Mendel's laws, the Hardy-Weinberg law, and other formal analytical techniques to answer such questions, ecologists use formal mathematical methods to describe populations and to probe for the underlying causes of major events in population ecology.

We start this chapter with the general mathematical curves that describe the growth of populations, and then examine the factors that increase and regulate a population's size. A species' reproductive strategy—whether the females produce thousands of offspring or just a few—plays a central role in population ecology, as we shall see. Then we will focus on competition, predation, and other factors that limit a population's size. Throughout the chapter, we will see how scientists over many decades have scrutinized the details of population ecology and have discovered much about the interplay between an organism, its environment, and the evolution of adaptations that allow it to survive in its surroundings and leave it better able to produce the next generation.

POPULATION GROWTH

In Chapter 41, we looked at populations as a geneticist would: as groups of interbreeding organisms. But ecologists see populations somewhat differently: as groups of organisms that produce and consume resources and that provide or use up nest sites or other habitats. In some ways, a population is like a living thing: over time, it grows from small to large in size; it may persist at a given size for months, decades, or centuries, or it may decrease and increase again; and, just as a snail or a dandelion grows to maturity and finally dies, a population eventually becomes extinct. Like individuals, populations are highly dynamic entities: even as most of an organism's cells die and are replaced, individuals in a population are constantly changing as old ones die and new members arise. And, finally, populations occupy a certain range of habitats and perform specific ecological roles, just as individuals do.

Populations have statistical characteristics, as well. For example, each has a per capita birth rate, known as **natality,** and a per capita death rate, or **mortality.** And environmental conditions, combined with the relative birth and death rates, cause a population to attain a certain *population density*—the number of individuals per unit of area. Each of these elements—birth rate, death

Figure 46-1. PEAKS AND "CRASHES" IN LYNX AND SNOWSHOE HARE POPULATIONS IN CANADA FROM 1845 TO 1935.
As the prey population cycles in numbers, the predators do also, but with a slight lag. These data come from records of pelts bought by the Hudson's Bay Company.

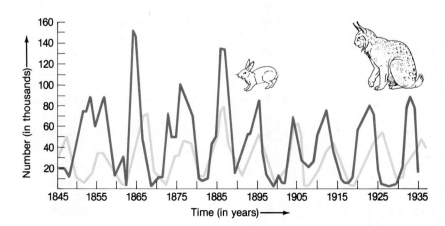

rate, and density—helps determine the total number of individuals in a population at different times in the population's history, and we shall see how as we consider the mathematical curves that represent the two main patterns of population growth.

Exponential Growth

Charles Darwin based much of his theory of evolution on a single concept: a population tends to increase in number, but only the fittest individuals survive. You may recall from Chapter 41 that Darwin drew on the ideas of Thomas Malthus, who had calculated that populations can, in theory, grow in *exponential*, or geometric, leaps (that is, the whole population doubles during some constant period of time). Darwin's famous example involved elephants: using the minimum reproductive rate for a pair of elephants and their offspring (6 young per pair during a 60-year breeding period), he calculated that in 750 years, the descendants of the original pair would number almost 19 million elephants—almost as many elephants as there are people in California today!

This sort of unrestrained exponential rate of growth in a population can be expressed mathematically by the following equation:

$$\frac{\text{change in population size}}{\text{time for that change to occur}}$$
$$= \text{growth rate} \times \text{original population size}$$

or

$$\frac{dN}{dt} = rN$$

where dN stands for change in a population's size, and dt equals the period of time over which the change occurs. Thus dN/dt represents the *rate*—change divided by time—it takes for the change to take place, or the rate at which the population is growing. The growth rate, r, is the per capita growth rate (the birth rate minus the death rate), and N stands for the population size at the beginning of the time period under study. For example, if a population doubles from 2,000 to 4,000 individuals in 4 years,

$$\frac{dN}{dt} = \frac{4,000 - 2,000}{4} = \frac{2,000}{4}$$

Hence,

$$\frac{2,000}{4} = r \times 2,000 = \frac{2,000}{4 \times 2,000}$$
$$r = \frac{1}{4} = 0.25$$

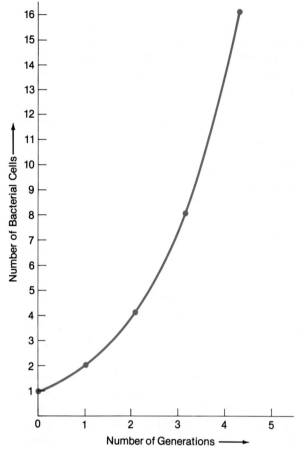

Figure 46-2.   EXPONENTIAL GROWTH OF A BACTERIAL POPULATION.

The overall rate of population growth rises as the number of bacterial cells, each capable of dividing, continues to grow.

Instead of r, the term r_m is sometimes used because it represents the *maximum rate* of population growth—also known as the "intrinsic rate of increase"—per capita in a population when there are no limits on growth, such as competition, predation, and limited resources. An example of an **exponential growth curve** is shown in Figure 46-2. (In reality, we have simplified the equation's solution; dN/dt really stands for the instantaneous rate of change, not the simple rate of change indicated.)

Sometimes a population grows exponentially when it colonizes a new area of habitat suitable for survival and reproduction. This fact was demonstrated in a classic experiment carried out in the 1930s by the Russian population biologist G. F. Gause. Gause cultured the protozoan ciliate *Paramecium caudatum* in a solution to which he added a plentiful, steady supply of food—bacteria that he had cultured on oatmeal gruel. In the beginning, when there were few paramecia, the populations in Gause's test tubes grew rapidly and followed a perfect exponential pattern. Eventually, however, the number of the protozoans leveled off, even though Gause faithfully replenished the daily ration of bacteria.

Gause's experiment demonstrates an important fact: a population cannot grow exponentially for very long. In reality, rapidly growing populations always level off before they become absurdly large, or their size begins to fluctuate up and down. One might guess correctly that the absence of true exponential growth is based on the reason that Malthus proposed: when populations reach high densities, limits on the food supply or some other critical resource or accumulation of wastes inevitably slows population growth. Either the population's reproduction rate slows or the mortality rate increases or both. In some cases, the result of a growth boom will be a "bust," a sharp decline in numbers as individuals die from starvation or disease. (For simplicity, we will ignore the effects of emigration and immigration, although they can be important factors in real populations.)

The United States Coast Guard inadvertently provided ecologists with a dramatic example of a "boom-and-bust" cycle in a reindeer population. In 1944, guardsmen introduced 29 reindeer onto a small island (128 square miles) in the Bering Sea off the coast of Alaska. With abundant food and no predators, the population rose steadily until in 1963, the Coast Guard counted 6,000 reindeer! At that point, however, the island's supply of winter forage for the animals—mostly reindeer moss—gave out. The result was a calamitous population crash. Within three years, only 42 reindeer remained, presumably to start the cycle again.

Carrying Capacity and Logistic Growth

Sooner or later, then, an environment's total amount of resources will determine how many individuals can live in a given area. This concept of an upper limit to population size—which can be reached but never permanently exceeded—is called the **carrying capacity** of a population's environment. Mathematically, it can be expressed by inserting a new term, $(K - N)/K$, into our growth equation:

$$\frac{dN}{dt} = rN\frac{(K - N)}{K}$$

The last term equals:

$$\frac{\text{carrying capacity} - \text{original population size}}{\text{carrying capacity}}$$

The carrying capacity of the environment, K, is simply the maximum number of individuals of a species that

environmental resources will support. When we apply this equation to a population, the result is a growth curve such as the one shown in Figure 46-3. Such a curve is called a **logistic growth curve,** or a *sigmoid* curve, because of its characteristic S shape.

In this equation, we can see that when the population is very small (N is nearly 0, and we are at the left end of the horizontal axis in Figure 46-3), the term $(K - N)/K$ is close to 1. This means that early on, the population grew almost as fast as the exponential rate rN shown in Figure 46-2. However, as N approaches the maximum value of the carrying capacity, K, the term $(K - N)$ gets closer and closer to 0, as does $(K - N)/K$. For example, if $K = 100$, and 98 individuals are present, $(100 - 98)/100 = 2/100$, or 0.02. As a result, the potential population growth (rN) in this almost-filled-to-capacity environment is multiplied by a very small number (or by 0), and population growth levels off: $dN/dt = 0$. It is from this pattern that biologist Paul R. Ehrlich coined the expression "zero population growth."

A logistic growth curve, then, reflects various phases of a population's growth history. At the beginning, when resources are plentiful, growth is rapid and the size of the population increases. When the population reaches equilibrium with available environmental resources (that is, when $K - N$ nears zero, and the carrying capacity is approached), the population may level off at a more or less constant size. The population at this stage is said to be *stable*; birth and death rates (and the effects of migration and immigration) are roughly balanced. However, if a crucial resource is reduced or depleted or if some other

Figure 46-3. LOGISTIC GROWTH CURVE FOR A POPULATION OF *DROSOPHILA* FRUIT FLIES.

An initial lag period leads into a time of constant expansion; then the population size reaches a plateau with little additional growth due to the limitation of certain resources. That plateau approximates the carrying capacity, K, under those conditions.

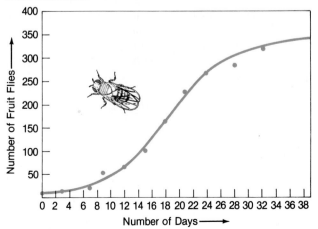

factor, such as a new predator or a change in climate, alters the physical or biological environment, the population may enter a period of decline and may even become extinct. In effect, such a reduction in a resource or an introduction of a predator may yield a new K value—that is, a lower maximum carrying capacity.

Fluctuations in Population Size

Exponential and logistic growth curves seem to reasonably characterize how population size can change after one or a few members colonize a new area. Laboratory cultures of organisms such as bacteria and fruit flies (Figures 46-2 and 46-3) frequently follow these patterns; furthermore, we often see logistic growth in wild-sheep, barnacle, and honeybee populations in nature (Figure 46-4). Like the ecological communities we considered in Chapter 45, however, theoretical population-growth

Figure 46-4. LOGISTIC GROWTH PATTERN OF A NATURAL POPULATION.
This population of an Italian strain of honeybees introduced near Baltimore shows a logistic pattern over a three-month period. The ultimate population size, near K, the carrying capacity, is determined largely by available space within the hives.

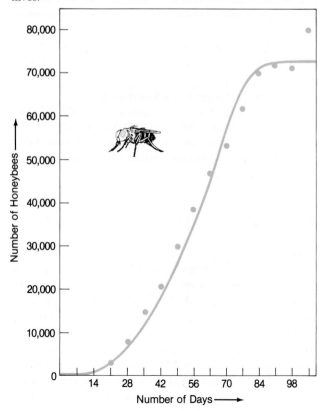

curves are abstractions that only approximate real events. In fact, population sizes often shift, and over time, the number of individuals in any population always changes. For example, even a population that generally shows logistic growth may not level off smoothly when the environment's carrying capacity is reached. Why? When resource limits are approached, time is required for the birth rate to fall and the death rate to rise. This response time is known as a **reproductive time lag.** When growth eventually does decline, the population may begin a sort of ecological roller-coaster ride, alternately exceeding K and falling below it because of lagging reproductive responses. Sometimes these changes are minimal; thus in a stable environment, populations tend to stabilize and fluctuate within a narrow range. In other environmental settings, however, deviations from a hypothetical plateau may be much more extreme. (We explore some of the factors that affect habitat stability later.)

If resources are severely taxed by the initial period of overshooting K, the carrying capacity of the environment may be permanently lowered (Figure 46-5). This is exactly what happened in the 1880s, when thousands of cattle were introduced to arid grazing lands in Arizona and other parts of the Southwest. By 1895, the original animals and their descendants had severely overgrazed the area's vegetation. During the same period, many young trees and shrubs were trampled to death, or the soils became so compacted that seedlings could not sprout. By the turn of the century, thousands of cattle had died of starvation. In addition, the destruction of desert shrubs reduced the environment's carrying capacity for another species—the statuesque saguaro cactus (Figure 46-6). Under normal conditions, seedlings of this long-lived species germinate and grow in the shade of paloverde trees and other plants that shelter the young cactuses from intense sunlight. But as late as 1963, observers reported that in some areas, no saguaros had successfully reproduced in the seventy years since their habitat had been stripped of its protective vegetation.

An environment's carrying capacity and the population it supports may also vary with the time of year. Seasonal fluctuations in the supply of food, water, hiding places, nesting sites, or other crucial environmental resources can be found in many habitats, such as temperate grasslands that are buried in snow in the winter and mountain streams fed by melting snow that run dry by August. As you might expect, many plant and animal populations are adapted to the seasonality of their habitats. They reproduce rapidly during the resource-abundant or climatically favorable season, grow to high densities, and then suffer high, but not complete, mortality during less favorable times (some species migrate, of course, to avoid such consequences). In the Sonoran

(a)

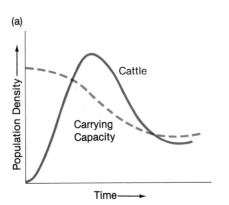

Figure 46-5. OVERSHOOTING CARRYING CAPACITY.
(a) This graph shows the overshoot-crash pattern that develops when herbivores overgraze their range and cause a decline in *K*, the carrying capacity of the environment. *K* may be permanently changed in some cases like this. (b) Cattle grazing on a sage range, Superstition Mountains, Arizona.

(b)

desert of western North America, tiny amphibians known as Couch's spadefoot toads spend most of their lives burrowed in the desert earth to avoid the sun. Not until a July rainstorm wets the soil do the toads emerge from their hiding places to feed and reproduce. Each consumes and stores 50 percent of its body weight in food, and mating begins immediately, with each female laying up to 1,000 eggs in a single night in the newly formed puddles. By morning, the mature toads have disappeared into their burrows once more to avoid desicca-

Figure 46-6. INCIDENTAL VICTIMS OF OVERGRAZING.
Overgrazing of the desert Southwest by cattle led to a reduction in the region's carrying capacity for the statuesque saguaro cactus. Here is a rare stony hillside that has abundant saguaros (*Cereus gigantas*).

tion. Within nine days, the eggs in the slowly drying ponds become tadpoles; soon, the tadpoles metamorphose into minuscule toadlets the size of a human thumbnail. At this point, an ecological sweepstakes begins as the little toadlets leave the water and seek shelter from the heat under twigs and in crevices. After a few days, most will have died from dehydration, but some of the original thousands will have successfully burrowed into the caked desert soil, to await the next rains. Chance obviously plays a major role in this example, as it does in many other instances when environments fluctuate.

We must take a detailed look at the other, less chance-dependent mechanisms that regulate a population's size. First, however, we have to consider two more factors that, like carrying capacity, affect the rate at which a population grows: the population's age structure and reproductive strategy.

The Effects of Age Structure

We have seen that a population's growth curve describes the phases of its growth history, and it can be extremely useful to know something about the current growth phase of a population of plants or animals. Such information about human populations, for example, is vital to government planners concerned with providing Social Security benefits to retired persons or with forecasting the long-range food needs of a region's or nation's population. One way to identify a group's growth phase is to determine its **age structure,** the relative numbers of

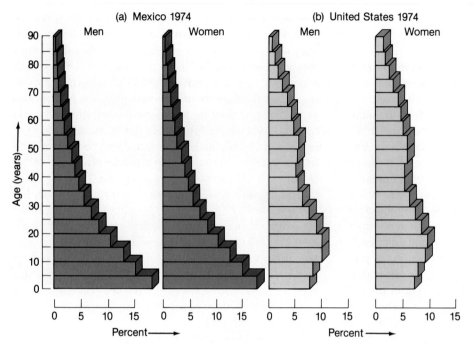

Figure 46-7.   THE AGE STRUCTURES OF POPULATIONS. The age composition is shown for both women and men in (a) a developing country, Mexico, and in (b) the United States. The large number and proportion of young people now approaching child-bearing age suggest that the Mexican population will continue to expand for the foreseeable future.

young, mature, and old individuals in the population. A reproductively active population in a rapid growth phase will have many young individuals (Figure 46-7a), while a declining population will have fewer young members relative to older individuals (Figure 46-7b). For many species that have achieved stable populations, there is a fairly even distribution of individuals among different age groups because the deaths of older organisms are offset by an approximately equal number of births.

Another way to represent the age structure of a population is with a **survivorship curve,** which shows the number of survivors in different age groups. The curves shown in Figure 46-8 are hypothetical representations;

one approximates survivorship of human females and shows that following low rates of mortality in early life, death seldom occurs until the seventh and eighth decades. Contrast that with the curve for typical bony fishes and plants: most organisms die before reproducing. The survivorship curves of many other organisms fall between these two extremes. A more or less diagonal curve, such as the one for birds, implies that there is little age-related mortality. For a bird, for instance, the chance of dying remains about the same at every age once the early period of vulnerability is past (this is in the wild state; the same birds living in the protective confines of an aviary show little death until several decades have passed and senescence is reached). Human survivorship curves and mortality tables are used by insurance companies to estimate life expectancy and to generate the rates charged for policies.

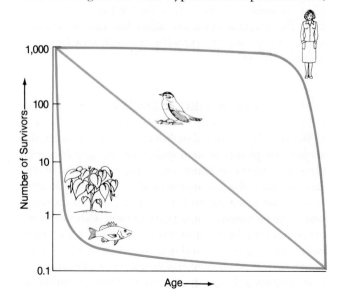

Figure 46-8.   IDEALIZED SURVIVORSHIP CURVES. The first type of survivorship curve (red) is a rough approximation for healthy human females. The second (blue) illustrates survival in birds; it shows no obvious change in rate of mortality with age. The third type (turquoise), for fish and certain plants, shows massive death early in life and then long-term survival for those few organisms that make it through the early crisis years. In nature, the actual curves for human females and birds would show a greater degree of early mortality than do these idealized curves.

Reproductive Strategies:
r and *K* Selection

The reproducing individuals in a population follow a **reproductive strategy,** a complex adaptation that has evolved over millennia for each species. Although the goal is the same for each species—to produce as many offspring as possible that survive to reproduce—there are a variety of reproductive strategies among them. For some species, the strategy is to produce a small number of offspring, each with a high chance of survival; humans, elephants, and whales fall into this category. For other species, such as the oyster, which releases millions of eggs into the sea each season, the tactic is to generate a large number of young each time it reproduces; each offspring, however, has only a small probability of surviving to reproduce. Yet another option, often associated with short-lived species, is to reproduce early in the organism's life span so that many generations arise in a short period of time. Bacteria, microbes, fruit flies, and other small life forms that multiply dramatically in a matter of hours, days, or weeks employ this reproductive strategy.

While they vary from species to species, reproductive strategies generally fit into one of two broad categories: those of *r-selected* species and those of *K-selected* species. Consider an example taken from the open ocean. Females of most bony-fish species (teleosts) lay hundreds, thousands, or millions of eggs; the caviar of sturgeon, lumpfish, or salmon are examples. Only a modest amount of yolk is stored in each egg, and the hatchlings, or *fry,* are tiny—only about 1 centimeter long. Most are snatched up as food by other organisms before they reach adult size, mature, and reproduce. Sharks, on the other hand, usually produce very large, yolky eggs (2–3 cm in diameter). Females of various shark species retain one or a very few embryos in the uterus for gestation periods of six to twenty months. At birth, the small number of young are 10 to 20 centimeters long and are quite able to fend for themselves. As a result, they have a much lower mortality rate than do the minute offspring of bony fishes.

In these examples, oysters and bony fishes follow the reproductive mode of **r-selected species.** That is, they reach reproductive age relatively quickly and produce many offspring, characteristics that give them a high *r* value. The reproductive adaptation exemplified by an *r* strategy is thus one of overwhelming numbers: out of many, a few are likely to survive. But as a result, the offspring are small, and each enters the world equipped with relatively few resources (such as egg-borne food reserves) to help it get an early foothold in the environment. The young are competitively inferior, suffer high mortality, and are likely to be successful only if they are dispersed to unexploited areas. By contrast, sharks are

K-selected species. Their reproductive mode is related to their environment's carrying capacity. As we saw earlier, when a population reaches *K,* its members must compete intensely for limited resources. Under such conditions, the competitive ability of offspring is more important than are sheer numbers of young in making a maximum contribution to the next generation. Thus *K*-selected species tend to grow slowly to large size, to mature gradually, and to produce few young per unit time. However, each offspring represents a considerable investment of parental resources, and because of its larger size, better access to food reserves, and so on, has a greater ability to compete for space, food, and other necessities than do *r*-selected offspring. (However, some *K*-selected shark species are now threatened with extinction because of heavy human fishing and sale of shark steaks in American supermarkets. A fishing industry used to harvesting *r*-selected bony fish cannot operate by the same rules on *K*-selected sharks with their small numbers of offspring.)

In reality, most species do not fall neatly into a distinct *r* or *K* category but have life histories falling somewhere along a gradient between the two. In general, *r*-selected species have adaptations for rapid dispersal, such as seeds borne on the wind or fertilized eggs carried by ocean currents. These adaptations enable the organisms to rapidly colonize unexploited habitats and to reproduce before the habitat is claimed by a competitor or altered somehow to the species' detriment. The fugitive species described in Chapter 45 are *r*-selected populations, and this reproductive characteristic is typical of organisms found in the early stages of ecological succession. *K*-selected species are more common in late-successional stages, when space, light, and other resources are more fully utilized by the community and competition for resources is intense.

LIMITS ON POPULATION SIZE

The size of any population, whether of ants in an ant hill or song sparrows in a hickory wood, is measured in terms of its **population density,** the number of individuals per unit of area. Some populations typically have high densities. The people in a city or the nematodes in rich soil are examples. Groups of hawks, wolves, and other large carnivores tend to have low population densities and to be sprinkled sparsely throughout their habitats.

Whether the populations are dense or sparse, their distribution is usually uneven. In cities of the eastern United States, dozens of starlings congregate on ledges and rooftops in dense flocks. In contrast, it would be

Figure 46-9.   DISPERSION PATTERNS.
Individuals within a habitat display three basic types of spatial dispersion: (a) clumped, as with
elephants, cows, and minnows; (b) uniform, as with hawks and polar bears; and (c) random, as with
dandelions.

unusual to find two hummingbirds on the same limb, since hummingbirds usually keep their distance from one another. The spatial distributions of populations can be classified into three types of *dispersion patterns:* (1) clumped, such as a herd of elephants, a school of minnows, or a stand of pines; (2) uniform or evenly spaced, like bears, coyotes, hawks, or other such territorial animals and some plant types; and (3) random, such as the spacing of dandelions in a lawn, which is entirely independent of the presence or absence of other dandelions (Figure 46-9). Clumped species are most common.

Members of species that clump are usually found as flocks or herds of animals, or groups of plants and animals that have specific resource requirements and hence are aggregated in places where those resources are abundant. Uniformly distributed populations occur when resources are spread thinly and evenly or when

individuals are antagonistic to one another. Such individuals may compete for a limited resource—hummingbirds for nectar or nest sites, or male dragonflies for mating territories around the fringes of a pond. Each individual aggressively defends its territory to ensure access to the resource.

Populations of one species can have different densities in different places. An example is the small bird known as the great tit (*Parus major*), for which the density of breeding pairs is lower in the Netherlands than in England (Figure 46-10). Still other populations fluctuate dramatically in density; rodents such as lemmings or insects such as locusts sometimes disperse across the landscape in a mass migration. For example, from about 1948 to the mid-1960s, much of sub-Saharan Africa was devastated by swarms of grasshoppers and locusts. Then, due largely to droughts, the infestations ceased. But now, in

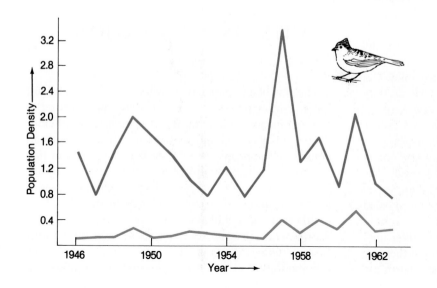

**Figure 46-10.   DIFFERENT DENSITIES IN
DIFFERENT PLACES.**
Breeding pairs of the great tit are found in denser populations in England (red) than in the Netherlands (turquoise).

the late 1980s, rain has not only supported better harvests but also has triggered population explosions of the migratory insects (Figure 46-11). If international efforts at control are unsuccessful, crops will be destroyed and humans will suffer greatly as a result. Both lemming and insect population patterns are examples of clumped distribution. But why do population sizes of different species vary, and why does the density of an individual population sometimes change over time? There must be factors controlling a population's density, but what are they?

The Role of Density

Ecologists generally find that dense populations have lower birth rates, higher death rates, and slower growth rates than less dense populations. Such vital statistics imply that there are negative consequences attached to rising population density. Research has focused on four reasons for this, four so-called **density-dependent factors** that affect population size: predation; parasitism; disease; and, most important, competition.

(a)

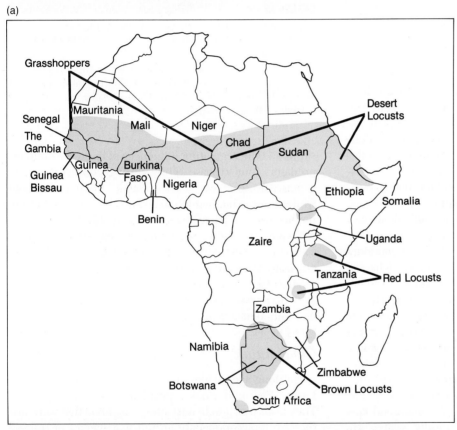

(b)

(c)

Figure 46-11. DRAMATIC FLUCTUATIONS IN POPULATION DENSITY.
Locusts and grasshoppers periodically show immense increases in population size. (a) Following a twenty-year respite, huge populations of grasshoppers and locusts are once again devastating large areas in Africa. (b) These migratory insects can eat their own weight in food each day, and a swarm such as this one in Ethiopia in 1968 can travel up to 200 miles every 24 hours. (c) Here a newly hatched locust dries its wings in the sun while two fifth instar hoppers (above) await their next molt, which in a few hours will yield the migratory, winged stage.

It is not difficult to envision how such mechanisms might operate. As a population of prey species grows denser, for instance, it is a more inviting target for predators; as predators increase in number, they may exert a significant amount of control over the numbers of their prey. Similarly, parasites and pathogens spread more easily among their hosts when the host population is dense. You are more likely to catch the flu in a crowded city than on a farm in a rural area. As its numbers increase, a population will experience greater mortality or reduced reproduction caused by parasites and diseases. And several factors may interact: a tree shaded out by faster growing competitors may become weak and more susceptible to infestation by a fungus. Overall, density-dependent mechanisms may limit a population to a level below its environmental carrying capacity.

Populations in nature often fluctuate much more erratically than one might expect. In locust populations, for instance, there are no predictable relationships among natality, mortality, and density, and thus the population size patterns of change do not seem to be controlled by density-dependent factors. Instead, **density-independent factors** regulate population size independent of predation, competition, parasitism, and disease. Density-independent factors include floods, fire, landslides, unseasonal weather, or other natural catastrophes that may kill many of or all the individuals in a population, regardless of its density. The ultimate density-independent factor may be such things as meteors hitting the Earth or drastic changes in sea level (Chapter 45); mass extinctions may be a result (Chapter 26, box).

This sort of catastrophe occurred in the winter of 1972-1973, when an extremely rare snowfall in the hills above Berkeley, California, wiped out thousands of the trees—some almost a century old—in the area's eucalyptus forest (Figure 46-12). The eucalyptus population seems to be returning to its old growth pattern; ultimately, however, another snow will come, and the population "crash" will be repeated. In fact, plant and animal species living in places periodically visited by disasters may be in a perpetual state of recovery from severe declines. Their evolution is likely to reflect such perturbations if the species continue to survive in the long term. In still other species, combinations of physical factors may trigger an unusual spurt in reproduction and corresponding population growth, as seen in the locusts of Tanzania, whose numbers rise and fall in relation to the area's rainfall.

Density-dependent and density-independent factors probably interact to some extent in all populations. Adverse weather changes and other disturbances in the physical environment occur in almost all habitats and bring about increased mortality. The more severe the disturbance, the greater a population's losses. In contrast, in a stable environment, where physical disturbances are rare, populations grow to densities that are

Figure 46-12.   RESULTS OF AN UNPREDICTABLE NATURAL CATASTROPHE: A FREEZE IN A NORMALLY FROST-FREE AREA.
Thousands of eucalyptus trees died back in the forested hills above Oakland and Berkeley, California, after an unexpected freeze. Now, about 15 years later, trees sprout in the vacant space.

subject to the regulating influences of limited resources, predators, and disease. Only through observation and experiment can we make reliable judgments about which interacting factors are controlling the size of the population at a particular point in its life history.

Ecologists have focused considerable attention in recent decades on two key ingredients in density-dependent population control: competition and predator–prey relationships. The first, competition, is the subject of intense controversy, but let us begin by understanding its basic mechanisms.

Competition

Not long ago, ecologists working in the rain forest of Puerto Rico undertook a peculiar construction project. They built tiny jungle nest sites of bamboo that were just the size to accommodate an unusual species of tropical frog, *Eleutherodactylus coqui* (Figure 46-13). These small amphibians are nocturnally active and retreat to nest sites or hiding places to rest and avoid predators during daylight hours. A census showed that the frog population in the area was quite large, and insects and other preferred food items appeared to be so abundant that they outstripped even the dense frog population. Since food resources seemed not to limit the number of *coqui* frogs in the forest, some other factor must have been at work. The ecologists wondered if it could be space, and this was where the bamboo "houses" came in.

The study area was divided into plots 100 meters square; bamboo shelters were built into some plots, while others were left unchanged as controls. Frog populations in each plot were counted and then retallied at the end of a year. The results of the experiment were

Figure 46-13. *ELEUTHERODACTYLUS COQUI*: COMPETITORS FOR SPACE.
Given extra hiding places, the populations of this tropical frog can grow denser in a tropical forest. In these frogs, only the males hear the "co" of the "coqui" song, while the females hear only the "qui" (Chapter 49).

unequivocal: all the houses in the test plots were occupied, and the original dense frog population had become still denser, while no significant change had occurred in the control plots. Clearly, for this population, K (carrying capacity) was determined by the number of nest sites and hiding places available; the frogs competed for this limited resource.

Here we have a case of competition among members of the same species—**intraspecific competition.** The outcome of such competition may be increased mortality, since in this case, frogs without hiding places are more visible to predators. Other species may evolve adaptations to cope with the stresses resulting from high density and increased competition. One response may be the production of fewer offspring (Figure 46-14). Thus if the food supply is strained, individual female birds in a crowded population may produce smaller clutches, or groups of eggs, at a given time. The female can thus give extra care to a few young, and the offspring are more likely to survive to pass on the parents' genes. Similarly, limitations on light, water, or soil nutrients may cause individual plants in a population to produce less biomass or fewer flowers or seeds. Note that the benefits to the species or populations in all these cases may seem purposeful but are purely incidental, indirect consequences stemming from the functioning of individual organisms in various ways that tend to pass on their genomes.

When several species in a particular area compete for a limited resource, **interspecific competition** occurs.

This kind of competition depends on the densities of all the species involved, and as more individuals divide the resource pie, each competing population generally grows more slowly. Sometimes, one species is able to outcompete its neighbors, thriving where other species become scarce or locally extinct. A simple example is provided by goats freed by sailors on Abingdon Island, one of the Galápagos Islands. The rapidly reproducing mammals have consumed so many plants normally eaten by giant Galápagos tortoises that those aged and ancient inhabitants became extinct on the island in 1960 or so. Another example is that of lizards and spiders, both of which feed on insects. Islands lacking lizards have up to ten times as many spider species. The lizards are both

Figure 46-14.   POPULATION DENSITY EFFECTS ON REPRODUCTION IN BIRDS.
(a) Population density at nesting sites of some bird species can grow quite high, as with these Heermann's gulls (*Larus heermanni*) nesting on an island in the Sea of Cortez, Mexico. (b) The data for growing population density, in this case in great tits, reveal that the fecundity, or production of eggs per pair of birds, drops off dramatically as population density rises.

(a)

(b)

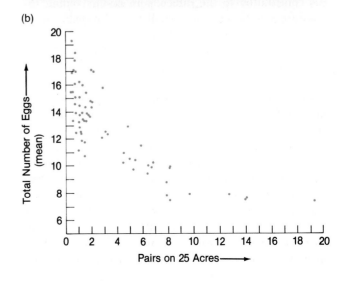

competitors for common prey and predators on their competitors.

Much of our modern theory of population ecology depends on the idea that the stability—or natural equilibrium—of populations in nature is due in large measure to intraspecific and interspecific competition. The essence of the argument is that competition regulates population size and density in ecological communities in ways that help maintain a stable community structure. As we have seen, however, density-independent factors can cause populations to fluctuate dramatically, as when a hurricane rips through the plant and animal communities of a tropical island, or when a population of field mice loses its food supply to a grass fire. For many organisms, especially insects, the combined effects of predators, disease, and disastrous weather keep populations well below the levels where competition would operate.

Some ecologists have proposed that the higher a species' position in the food web (or trophic ladder) of its community, the more likely competition will be an important regulator of its population density. One reason is that predators at the top of the web—such as owls, hawks, and coyotes—tend to consume and therefore compete for the same classes of food. Today, the role of competition in population ecology and community structures is an extremely active and controversial area of research, and many innovative experiments are being carried out to test its significance. An important consideration guiding such research is the recognition of a distinction between the intensity and the importance of competition. Competition may be very intense, let's say, between plants for mineral nutrients; yet that competition may have a minor effect on the long-term ecological or evolutionary success of that plant species; hence, such intense competition is relatively unimportant in the broader context. Conversely, even weak competition may assume great importance in controlling populations evolving through time, if other factors are not limiting. Past confusion of intensity with importance has contributed to the difficulty in understanding the role of competition in control of natural populations in nature.

Predation

Like competition, the interactions of predators and their prey can help regulate population size. Broadly speaking, any organism that feeds on another living creature can be termed a predator. Thus predators include not only large carnivorous hunters, such as lions and eagles, but also grazing herbivores, ranging from tiny planktonic swimmers to bison or giraffes, that consume plant parts but do not kill the plants themselves, and parasites such as wasps that deposit their eggs in the tissues of living caterpillars (Figure 46-15). But while it is relatively easy to identify the predators and prey in natu-

Figure 46-15.   PREDATION UNDER WAY.
Braconid wasps (as *Rogas terminalis*) deposit their eggs on caterpillars, such as this larval sphinx moth (*Manduca quanquemaculatus*). As the eggs mature, cocoons are formed, and wasp larvae eventually emerge and begin to eat the caterpillar alive.

ral communities, it is a bit more difficult to establish general rules for when, how, and to what extent predators regulate prey populations.

In the early 1920s, two mathematicians, Alfred Lotka and Vittora Volterra, independently developed a mathematical model that attempted to describe the relationship between predator and prey populations. The *Lotka-Volterra equation*, as it is called, predicted that such populations would oscillate in a recurring cycle. The reasoning underlying the model is simple: as a prey population grows, its predators have more food available, and their density, after a reproductive time lag, increases, too. Eventually, the predator population becomes so numerous that the prey are eaten faster than they can reproduce; as a result, the prey population declines. After another time lag, the predator population becomes less dense as more and more of its members starve to death. The number of prey then increases, and the cycle begins again.

G. F. Gause, who, you may recall, used populations of protozoans to test Malthus's predictions of population growth, decided to check the Lotka-Volterra equation in a similar way. He set up test-tube cultures of *Paramecium* as prey and *Didinium* (another protozoan) as predator. At first, the paramecia flourished; soon, however, the *Didinium* population expanded rapidly and wiped out its burgeoning food supply. Alas, the predator's fate was to decline and become extinct as well. No cycling of populations took place; both prey and predator were quickly eliminated from the experimental community.

Gause then repeated the experiment with a new twist: he occasionally added some of each protozoan species to the cultures as "immigrants." The populations soon began to oscillate, just as the mathematical model predicted. More recent laboratory studies involving insect populations have produced similar results. In the simple communities inside test tubes and laboratories, researchers can achieve population oscillations that resemble predicted changes, but only as long as immigration or some other environmental feature, such as access to hiding places for prey, is part of the experiment and equation.

Ecologists created some of the early mathematical models of predator–prey interactions after having observed certain populations whose densities alternate in regular cycles of decline and recovery. The classic, but much debated, real-world example of a regular predator–prey oscillation is the cycle involving lynx and hare populations described at the beginning of this chapter.

It can be tricky to study the effects of general predators, which take food from many sources, on natural populations. For instance, we may know that a species of desert snake feeds on kangaroo rats and want to determine the effects of snake predation on rat population density. But rats are prey for owls and other predators as well, and even a careful observer may rarely see a rat actually taken by a predator. And since other regulating factors are involved, too, it is difficult to determine how much of the decline in a rodent population is due to predation and how much is due to interspecific competition for food, to disease, or to emigration to a more favorable habitat. And even if we know what percentage of the population decrease is due to predation in general, we still have to determine how much of it is based on snakes, how much on owls, and so on.

Yet another difficulty in assessing the impact of a generalized predator is that a coyote, for example, may choose its diet from many available prey species. Stomach contents of coyotes trapped in an area of Iowa include traces of grasshoppers, muskrats, deer, chickens, lizards, corn, snails, bark, crayfish, and even other coyotes. Arctic foxes do the same thing. They normally prefer small mammals called lemmings for food, and the two species cycle like the hare and lynx. Each year that the lemmings "crash" the foxes switch to ground-nesting birds such as brent geese—this periodic predation produces a three-year cycle in the goose population.

To complicate matters still more, what appears to be population regulation due to predation may actually be the result of some subtle fluctuation in the environment that ecologists fail to, or cannot, observe. A well-studied instance of predator–prey interaction on Isle Royale, a small island in Lake Superior, illustrates this point. Early studies of the island's moose population—between

Figure 46-16.   FATE OF THE OLD, SICK, OR WEAK.
This moose calf (*Alces alces*) bears the attack wounds of wolves. It escaped once, but no doubt the predators will try again.

600 and 1,000 animals, depending on the season—suggested that the population size was regulated by wolves that hunted moose and other animals. More recent research, however, has cast doubt on this conclusion; it now appears that sodium, an essential micronutrient in the moose diet, may be the important factor limiting the number of moose on the island. Wolves are now believed to cull from the herd mainly those animals that are old, sickly, or too young to defend themselves from attack (Figure 46-16).

It sometimes seems as though the regulatory effects of predation are based on as many variables as there are ecological communities. In general, however, predation may slow or stop the growth of a prey population only when many reproducing members are eliminated. If predators are able to take only aged, weak, or very young victims—as are the wolves on Isle Royale—their effect on the overall density of the prey population may be relatively slight. This is especially true where intended prey can avoid being eaten, like the healthy gazelle that can kill a jackal with a well-timed blow of its hoof, or the agile sparrow that can avoid the lunge of a cat. As we shall see in Chapter 47, both predators and their plant and animal prey have evolved a fascinating variety of adaptations, "designed" as it were by the processes of natural selection to increase each species' chances of survival.

Population Fluctuation and Community Structure

The outbreak of a devastating virus in a monoculture field of corn and the invasion by bark beetles of a forest

BIOLOGICAL PEST CONTROL: DOING WHAT COMES NATURALLY

The negative side effects of pesticides are a growing public controversy. The prices of chemical pesticides have increased along with the public-health costs, and thus the stakes are high in the search for safer means of controlling agricultural pests. As a result, dozens of research teams are actively identifying predators, parasites, and diseases that can be used to advantage in the war against plant pests—a war increasingly waged with biological rather than chemical methods of control.

Ironically, the success of chemical pesticides helped agricultural scientists recognize the importance of biological pest control. Just after World War II, when the chlorinated hydrocarbons DDD and DDT became available at low cost, the chemical industry presented American farmers with a seeming revolution in pest control. Citrus growers, among others, raced to protect their crops with such "miracle" pesticides because their fruit trees were susceptible to damage from mites and scale insects. In addition to dispatching many of these agricultural scourges, however, DDT and similar chemicals killed the natural enemies of the mites and scale insects. Released from such natural controls, populations of cottony-cushion scale, citrus red mites, and other pests exploded in outbreaks that sometimes defoliated trees severely or even killed them altogether (Figure A).

Such unwitting ecological experiments demonstrate the power of natural predation in regulating

Figure A

A lemon infested by red scale mites. Pheromones are now used to survey for incipient red scale population outbreaks so that chemical spray programs can be initiated.

herbivore populations. Biological control may be particularly promising where pests are inadvertently imported to a new area, while their natural enemies are left behind. A case in point is the winter moth, a highly destructive insect introduced into northeastern Canada from Europe in the 1930s. Within twenty years, winter-moth larvae, which feed on leaves and other tree parts, had begun to inflict millions of dollars in damage in Canadian hardwood forests. Furthermore, the spread of the moth to forests and hardwood orchards throughout North America appeared

with only a few tree species are examples of a well-known ecological fact: simple communities are highly vulnerable to the attacks of pests. This correlation inspired a question by ecologists. Can population fluctuations in general also result from simple community structure? The argument supporting this notion goes as follows. In communities with many species and complex food webs, most herbivores feed on many plant species, and each carnivore exploits several herbivore species. As a result of these diversified food sources, if one species suffers a severe population decline or becomes extinct locally, the consumers above it in the community's food web still have an alternative food supply available. By contrast, in a simple community with linear food chains involving a few species, any drastic change in a population at one trophic level may have severe repercussions on the populations above and below it in the trophic hierarchy (Figure 46-17). Such a system is inherently unstable: if a population shrinks rapidly, consumers that depend on it for food may starve; if it grows suddenly, it

may eat itself into oblivion because alternative food sources are unavailable. These observations suggest that communities with high species diversity should be more stable than communities with fewer species; that is, the size of their component populations should fluctuate less.

The idea that stability results from community diversity is disputed by many ecologists, however. Why? One reason is that in nature, high diversity is not always found in combination with complex food webs. In fact, as we shall see in Chapter 47, many plants and animals found in tropical regions have highly specialized adaptations for interacting that result in simple food webs. Nevertheless, communities made up of such plants and animals can be remarkably stable.

For the present, the idea that diversity begets stability seems, at best, oversimplified, and the reverse hypothesis may be closer to the truth: stable environments may beget diversity because they allow rare species to persist. Conversely, in environments where weather or

imminent. Then in 1954, a team of ecologists tracked down two winter-moth predators—a tachinid fly and an ichneumonid wasp—in Europe, and imported them to the affected area. Within six years, the predators effectively held winter-moth populations to acceptable densities. The ecology of the predatory parasites themselves was an important aspect of the program: the fly is most efficient at high moth densities, and the wasp is best when densities fall.

Another success story of biological control involves the diminutive walnut aphid, an insect that originated in Europe and Asia and that invaded the walnut groves of California around 1900. Walnut-aphid infestations mean double trouble: as the aphid sucks leaf juices, it exudes a sugary nectar onto leaf surfaces that attracts damaging fungal growth. Entomologists long believed that aphids resist biological control because they multiply extremely rapidly in the spring and do their damage before natural enemies, such as ladybugs, can catch up with them. One enterprising ecologist, however, reasoned that a species-specific enemy—if one could be found—would probably have evolved its own adaptations to the aphids' life cycle. An extensive search turned up two promising candidates: a highly specialized parasitic wasp that is common in southern France and a different strain of the same wasp species that lives in the hot, dry, central plateau of Iran. Together, the two predators have established a high level of control over walnut-aphid populations—the French variety in the cooler climate

Figure B
A natural predator at work. The aphid (right) is being sucked dry of its body fluids by this green lacewing larva (*Chrysopidae* species).

of northern California and its Middle Eastern cousin in the hotter regions farther south (Figure B).

Biologists continue to study such organisms with potential usefulness in biological-control programs. One of the most important challenges is to expand the field of insect taxonomy, so that likely prospects can be identified. In their search for ways to reduce or replace the use of chemical pesticides, researchers will also be probing some of the fundamental ecological relationships that link organisms in the natural world.

physical disturbances fluctuate, or where populations undergo drastic changes in density for some other reason, perhaps only the best adapted, most densely distributed, and therefore most extinction-resistant species

can survive. Others are winnowed out, and their passing leaves behind a simpler, less diverse ecological community.

(a) Food Web   (b) Food Chain

Secondary Consumers: 13 14 15 | 7 8 9

Primary Consumers (herbivores): 8 9 10 11 12 | 4 5 6

Producers: 1 2 3 4 5 6 7 | 1 2 3

Figure 46-17. FOOD WEBS, FOOD CHAINS, AND COMMUNITY STABILITY.
Food webs are more often stable than simple food chains. (a) If the herbivore in the middle of this food web (species 10) were to decline or become extinct, its predators would have three or more alternative prey species. The food plants eaten by the herbivore (species 3 and 4) are also used by three other consumer species, and this predation might help regulate the densities of 3 and 4 if herbivore 10 declined. (b) If the herbivore amid this set of food chains (species 5) were to decline, consumer species 8 would automatically decline in numbers, and producer species 2 might increase without control. The overpopulation of species 2 might then cause competitive declines in species 1 and 3, which, in turn, would diminish the consumers above them.

HOW POPULATIONS ARE DISTRIBUTED

Just as competition, predation, weather patterns, niche structure, and other elements interact to determine the size of a population within a community, population distribution is the result of many interrelated factors. Some of the factors that affect distribution have to do with a species' adaptations for surviving and reproducing in different physical environments: redwoods and desert yucca have climate-related adaptations that limit their populations to specific localities, while North American coyote populations thrive from Mexico to Canada in a variety of environments. But even within its potential range, no species ever occupies all the area available to it; recall that the fundamental niche can be reduced to a realized niche by a variety of causes, including interspecific competition.

We saw in Chapter 45 how the niches of various species fit together to help define community structure. Here we seek to understand how the species niche helps define population size and distribution. Competition among species for food, breeding sites, or other resources surely is one of the major means of defining limits of the realized niche, and so population size and distribution.

Interspecific competition is most intense when two or more species require many of the same resources—in other words, when they are adapted to similar or identical habitat niches. Among plants, one of the most effective forms of interspecific competition is **allelopathy,** in which chemical substances produced by one species inhibit the germination or growth of seedlings of another species. These chemicals may be inorganic acids or bases that alter soil chemistry, or they may be compounds, such as the phenols produced by chaparral shrubs that wash into the soil surrounding each plant. At least one allelopathic grass species (*Aristida oligantha*) exudes a phenolic acid that acts as an antibiotic and kills nitrogen-fixing bacteria in soil. The grass itself can tolerate a low level of nitrogen, while competing species cannot.

A test of interspecific competition was conducted by the brilliant G. F. Gause, who, as usual, employed paramecia as his experimental organisms. Gause cultured together two species, *Paramecium aurelia* and *P. caudatum*, that eat the same type of bacteria as food. *P. aurelia* multiply at a faster rate than *P. caudatum*; this competitive edge soon resulted in *P. caudatum*'s being eliminated from the "niche," as the population of *P. aurelia* monopolized the common food supply.

Gause then repeated his experiment, this time combining *P. caudatum* with yet another species, *P. bursaria*. Again, the food supply consisted of bacteria consumed by both species, but on this occasion, an unpredicted wrinkle appeared. *P. caudatum* fed on bacteria suspended in the culture solution, while *P. bursaria* consumed only those bacteria that were at the bottom of the test tube. By occupying different food niches, the two species were able to share a limited resource and coexist successfully.

This kind of **resource partitioning** is a common adaptive solution utilized by species that share similar or identical habitat niches. Figure 46-18 shows Robert MacArthur's famous example of resource partitioning: the food niches of three species of warblers in a conifer forest. Except in early summer, when their common food supply is superabundant, the three species feed in different parts of the tree and choose insects and seeds of different sizes and types. These two examples indicate how competition affects the distribution in space of populations—the space resource is divided among them.

Often, closely related species evolve slightly different adaptations for exploiting a limited resource. They show **character displacement**—physical differences in body structures used in the utilization of that resource. For instance, two related species of earth-burrowing lizards

Myrtle Warbler   Bay-Breasted Warbler   Blackburnian Warbler

Figure 46-18. RESOURCE PARTITIONING IN FEEDING NICHES OF WARBLERS.
These three warbler species live in the same habitat but do not compete for food because they feed at different heights in the conifers of northeastern forests. Thus, their food niche has a height component, and there is little direct competition among the three species, at least for food.

in Africa eat a particular type of termite. Where the lizard species occupy the same small area, one species has evolved a larger head and body than the other and feeds on large termites. In places where the lizards' habitats are separate, each species has a small head and body and eats small termites exclusively.

Perhaps the most famous case of character displacement is among the finches of the Galápagos Islands. Although Darwin studied the Galápagos finches, the significance of morphological differences among the fourteen finch species was not fully appreciated until the 1940s, when the British biologist David Lack described in detail how the birds' varying bill shapes correlate with their diet and ways of foraging. From the original seed-eating ground-finch immigrant have evolved five species that feed on different sorts of seeds on the ground; two that feed on seeds in cactuses; six that feed in trees; one a vegetarian; and five insect eaters (one of which behaves like a woodpecker and even excavates insects by using a stick held in its bill). The other birds are warblerlike and generally insectivorous. Isolation on the different Galápagos Islands and competition for food are believed to have been the primary factors leading to this remarkable evolutionary diversification (Figure 46-19). Competition may be with other finches (as witnessed by the rapid increase in small-billed seed eaters in response to the 1983 El Niño rains); or it can be with other kinds of organisms (for example, nectar-eating finches are large-bodied on islands with bees, and small-bodied where no bees are present).

The Hawaiian Islands have witnessed an even more spectacular radiation of honeycreepers, the original founding species (*Loxops virens wilsoni*) having given rise to twenty-two species (Figure 43-16), many with spectacularly different bill and body morphology, as well as feeding habits. Clearly, the combination of relative isolation on islands, competition for food and habitat, and "unoccupied" niches leads to character displacement, speciation, partitioning of the habitat, and, in keeping with the current subject of concern, the size and distribution of populations.

But it is not necessary to be isolated on an island or to show character displacement in order for potentially competitive species to live in a given area, at least temporarily. Predation or weather disturbances also can keep populations of interacting species below carrying capacity, so that resources are plentiful. Along the east coast of the United States, for example, many rocky seashores have storm-tossed boulders on which populations of encrusting marine animals, such as barnacles and periwinkles (snails), coexist with fugitive species of algae and other intertidal organisms. Left alone, such areas might become carpeted with one or two populations of especially effective competitors, such as the barnacle *Balanus* (Chapter 45). Instead, winter storms regularly

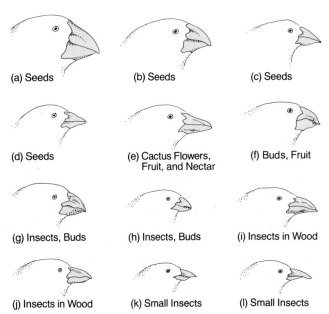

Figure 46-19.   BILL TYPES IN GALÁPAGOS FINCHES.
The labels indicate the types of food each finch eats predominantly. Notice how the bill size and shape correlate with the food type.

(a) Seeds
(b) Seeds
(c) Seeds
(d) Seeds
(e) Cactus Flowers, Fruit, and Nectar
(f) Buds, Fruit
(g) Insects, Buds
(h) Insects, Buds
(i) Insects in Wood
(j) Insects in Wood
(k) Small Insects
(l) Small Insects

Figure 46-20.   COMPETITION, PREDATION, OR DISEASE: LIMITATIONS ON SPECIES' LOCALE.
This graph shows the distribution of (a) sea-urchin herbivores and (b) their algal food along a range of water depths. Notice that the urchins do not occur higher or lower than the depths indicated in part (a), and thus these areas are refuges for the vulnerable algae.

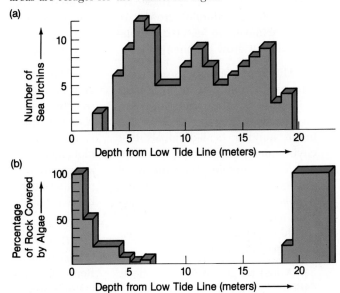

scrape clean patches of substrate so that new space—the usual limiting resource in intertidal communities—is continually made available to colonists.

As is intuitively obvious, predation or disease can act to limit population sizes and distribution. For example, as Figure 46-20 shows, predatory sea urchins are very effective at restricting populations of algae to areas where the urchins are absent. Plants, like animals, are also subject to infection by microorganisms or fungi. Vast numbers of American elm trees found in the eastern and midwestern states by early settlers have been virtually eliminated by a fungus that grows within the conducting cells of the vascular tissue (xylem and phloem). Thus Dutch elm disease limits elm populations, thereby leaving vacant land for other trees. Thus both population and community structures are affected, just as with predators, chance storms, or competition among species.

LOOKING AHEAD

In this chapter, we have continued our analysis of ecology, which began with the biosphere as an ecosystem. We explored biomes and communities and here have seen how growing, changing, and interacting populations of species have a profound impact on community structure. Relationships among species are often molded by features of their respective niches, as well as by predators, disease, and competition for resources in a population's habitat. In Chapter 47, we shall explore some of the adaptations that species have evolved in response to such interactions.

SUMMARY

1. Populations—groups of individuals belonging to the same species—share a set of characteristics: *natality* (per capita birth rate), *mortality* (per capita death rate), and *population density* (number of individuals per unit of area).

2. In theory, a population can grow exponentially, or geometrically, if there are no limits on resources (food, space, water) and if there is no competition or predation, no disease, and no other potentially limiting factors. Such population increase is plotted on an *exponential growth curve*. Because these requirements for unlimited growth are never achieved in nature, populations do not grow exponentially.

3. Different populations are characterized by different intrinsic rates of increase *(r)*, the maximum possible per capita growth rate. The maximum number of individuals that environmental resources can support is the *carrying capacity (K)* of the environment for a particular population. Theoretically, populations growing to carrying capacity should follow a sigmoid curve, or *logistic growth curve*.

4. A logistic growth curve reflects various phases in a population's growth history; when the population reaches equilibrium with available resources, it may level off at a relatively constant size (become stable). The population may decline or become extinct if there is a major change in its environment or available resources.

5. When resource limits are reached, the time it takes for birth rates to fall and death rates to rise is known as the *reproductive time lag*. If a population outstrips its available resources severely, the carrying capacity of the environment may be lowered permanently. In other instances, carrying capacity may fluctuate up and down due to such factors as changes in season.

6. The rate at which a population grows can also be affected by its *age structure*—the relative numbers of young, mature, and old individuals in the group. A *survivorship curve* plots mortality as a function of age in a population, and in the process, it provides information about a population's probable growth rate.

7. All the individuals in a population that reproduce do so according to a particular *reproductive strategy*. Reproductive strategies form a spectrum between two broad categories: *r-selected species* grow to maturity quickly and produce many offspring, but the young suffer high mortality; *K-selected species* mature more slowly and produce few young, but each new individual is relatively well prepared to compete for food, space, and other resources in its environment, and thus mortality is low.

8. The factors that keep populations at their characteristic sizes are both density-dependent and density-independent. *Density-dependent factors* include *intraspecific competition, interspecific competition*, predation, parasitism, and disease. These factors may cause the birth rate to decline or the death rate to rise when the population density is high. *Density-independent factors* include natural catastrophes, seasonal changes, bad weather, and accidents, all of which cause populations to fluctuate erratically.

9. Some populations exhibit regular cycles of decline in density and then recovery; some ecologists think that the interactions between predator and prey populations, combined with reproductive time lag, cause such cycles, although other hypotheses have been suggested.

10. Although some ecologists have

proposed that diverse ecological communities with complex food webs are inherently more stable than simpler communities, this idea is not correct for some populations in nature. A counterproposal is that stable environments beget diversity because they allow rare species to survive.

11. Distribution of a population depends on the food, habitat, physiological requirements of its compo-

nent individuals, interspecific competition, and chance events such as severe storms or disease.

12. Interspecific competition for resources can result in one species' inhibiting the germination or growth of another *(allelopathy)*, or in *resource partitioning*, in which common food sources or habitats are utilized in different ways by competing species.

13. Ecologically similar species that inhabit the same area may evolve morphological differences in the body structures that allow exploitation of a given limited resource. This phenomenon, known as *character displacement*, reflects the evolution of adaptations for slightly different aspects of the niche among otherwise competing species. In fact, it may lead to speciation, as among bird populations on islands.

KEY TERMS

age structure
allelopathy
carrying capacity
character displacement
density-dependent factor

density-independent factor
exponential growth curve
interspecific competition
intraspecific competition
K-selected species

logistic growth curve
mortality
natality
population density
r-selected species

reproductive strategy
reproductive time lag
resource partitioning
survivorship curve

QUESTIONS

1. Theoretically, populations of organisms can increase in size exponentially. What does this mean? Why is this not seen in nature?

2. The equation

$$dN/dt = rN$$

is the curve of exponential increase in an unlimited environment. How does the modification

$$dN/dt = rN\ (K - N/K)$$

make the equation more closely reflect real-life conditions? What do ecologists call the *K* term in the second equation?

3. Give an example to show how the activities of certain members of a community may permanently alter the carrying capacity of a habitat. How might physical factors alter the carrying capacity of a habitat?

4. Different species of organisms exhibit different reproductive strategies. Give examples to show what is meant by *r* selection and *K* selec-

tion. According to some biologists, *r* and *K* selection represent extremes on a continuum. Give some additional examples to show what they mean.

5. Name the three dispersion patterns used to describe the spatial distribution of species, and give a species example for each pattern.

6. Two major categories of factors that limit population size are density-dependent factors and density-independent factors. What do these terms mean? Give examples of limiting factors in each category.

7. Explain why it is complicated and difficult to study the effects of predator activity on the size and structure of prey populations.

8. Discuss the hypotheses that ecologists have proposed to explain the apparent relationship between species diversity and community stability. Which hypothesis is presently in favor, and what evidence supports it?

9. Explain how some plant species employ allelopathy to compete with other plant species.

10. Use examples to show what is meant by the adaptive strategies of resource partitioning and/or character displacement.

ESSAY QUESTIONS

1. What factors influence the carrying capacity of an environment and prevent a population from increasing in size exponentially?

2. Discuss the following generalization: pioneer species that colonize disturbed habitats tend to be *r*-selected species, while the inhabitants of more mature and stable ecosystems tend to be *K*-selected species.

3. Two "visitors" from space hit the Earth. One is a large meteor that generates a huge dust cloud that circles the globe in the atmosphere for

two years. The other, a spaceship, discharges a type of single-celled protozoanlike creature that can outfeed and outreproduce any earthly cell, but only in the hot, dark, wetness of mammalian bloodstreams. They are completely unable to survive in birds. Assess the possible consequences of these two independent events on population sizes of earthly species, using in your essay the terminology and concepts employed by ecologists to explain population levels.

SUGGESTED READINGS

CONNELL, J. H. "Diversity in Tropical Rainforests and Coral Reefs." *Science* 199 (1978): 1302.

In this article are references to Connell's diverse work, as well as good arguments about density-independent factors being critical to population size.

LACK, D. *Darwin's Finches.* Cambridge, England: Cambridge University Press, 1947.

The classic book that explains what Charles Darwin apparently missed but what is one of the primary cases cited to support adaptation to slightly different niches.

———. *Ecological Isolation in Birds.* London: Methuen, 1971.

A more recent summary of this key issue in ecology and evolution.

McLAREN, I., ed. *Natural Regulation of Animal Populations.* New York: Atherton Press, 1971.

One of Hutchinson's students has gathered a group of excellent essays on population dynamics.

WELDEN, C. W., and W. L. SLANSON. "The Intensity of Competition Versus Its Importance." *Quarterly Review of Biology* 61 (1986): 23–44.

Here is an excellent review of a key topic in population biology.

WILSON, E. O., and W. H. BOSSERT. *A Primer of Population Biology.* Sunderland, Mass.: Sinauer, 1971.

This book is an excellent introduction to basic principles and is especially strong on formal aspects of ecological analysis.

47

ADAPTATION: ORGANISMS EVOLVING TOGETHER

*The possibilities of existence run so deeply into the
extravagant that there is scarcely any conception too
extraordinary for Nature to realize.*
Louis Agassiz,
The Structure of Animal Life (1866)

Plant and Animal Adaptations: The sharp spines give protection from most animals, yet this antelope ground squirrel has
evolved a light step and cautious feeding behaviors that allow it to use the fruit of the Buckhorn cholla cactus (*Opuntia
acanthocarpa*) as a source of food and water.

From a distance, the oval streak of white on a birch leaf looks like a splash of bird droppings or a spittle bug's frothy covering (Figure 47-1). Birds hunting for lunch in the area take no notice, since they do not eat droppings or spittle bugs. A closer look, however, reveals long, fine filaments of pearly floss in constant motion. The "floss" is the caterpillar larva of the dagger moth, whittling insatiably at the edges of the leaf. Eventually, when only a small flap of leaf tissue remains, the insect neatly clips it off, removing any potential evidence that a caterpillar is present. The developing larva will soon turn a drab brown and drop to the forest floor to pupate. But during its active eating phase, two adaptations—the flossy disguise and the instinctive leaf-removal behavior—help ensure its safety from keen-eyed avian predators.

Adaptations are genetically based characteristics that result in an individual organism being better suited to some aspect of the environment—and therefore, being more likely to survive to reproduce. In this chapter we examine many adaptations that evolved as trees, caterpillars, birds, and other life forms interacted with each other over long periods of time. Such adaptations to other living things are just as vital to a species' survival as adaptations to temperature range, light, water, nutrient availability, and other physical conditions. Some of the most dramatic adaptations help protect organisms against predators, and they will be our first topic. These adaptations are often reciprocal, with prey and predator acting as two sides in an endless evolutionary tug of war—the predator species gradually evolving a new mode of attack for each defensive adaptation that arises in the prey species.

Biologists Paul R. Ehrlich and Peter H. Raven named this reciprocal phenomenon **coevolution.** A simple illustration might be the evolution in a plant of a noxious-tasting chemical, accompanied by the development in an aphid or another herbivore of an enzyme that inactivates the chemical. Other examples range from the evolution of poisons, venoms, toxins, and thorns to that of bold warning colorations and patterns, with the concomitant evolution in predatory species of structures, behaviors, and enzymes that circumvent the defenses. While coevolution can resemble a tit-for-tat battle of these adaptations, it may also lead to biological joint ventures in which each species benefits to varying degrees. Toward the end of the chapter, we shall consider a spectrum of symbiotic relationships extending from *parasitism* to *commensalism* to *mutualism*—associations in which one species may benefit to another's detriment or in which both organisms may benefit.

It is important to keep in mind that no organism or species can ever deliberately evolve adaptations to help it cope with other life forms. Adaptational structures and strategies come about over many generations as the processes of natural selection alter gene frequencies and

Figure 47-1   CAMOUFLAGE.
Dagger-moth larvae mimic bird droppings or the foam of the spittle bug to deter would-be predators. To avoid leaving signs of leaf damage, which might attract birds, a larva backs down the stem and snips off the remnants of the leaf it has eaten. This may also reduce the toxins the plant musters to defend itself.

resultant phenotypes in populations. Moreover, once evolved, no strategy for defense or escape is ever 100 percent successful; some individuals inevitably fall prey to others or perish as the result of cold, drought, or other physical factors. Adaptations offer no guarantees to individuals, but they do improve the chances that individuals will survive to reproduce their kind. And as we shall see, coevolution and resulting adaptations have produced some of nature's most extraordinary phenomena.

ADAPTATIONS FOR DEFENSE

Nearly all species of plants and animals have adaptations that serve either as defenses against predators or as means of escape. Defense adaptations fall into two main categories: *mechanical defenses* are incorporated into the physical structure of the organism, and *chemical defenses* discourage predation by producing stinging sensations, poisoning, paralyzing, or simply tasting bad.

Mechanical Defenses of Animals and Plants

Most eukaryotes have some form of defense against attack, and the most obvious protective traits are physical structures, such as the large sharp tusks of the elephant, the powerful pincers of the lobster or crayfish, and the stabbing thorns of the cactus or thistle. A few hefty animals, such as polar bears and some large cats, have such effective mechanical defenses that they have few predators or none other than humans. Other animals, such as whales and hippos, escape predation pri-

marily because of their size alone, although their tough skin also helps. Many animals too slow, small, or weak to fight with a predator, including turtles, clams, and pill bugs, have a different type of defense: they retreat from attack behind a bastion of bone, shell, or chitin.

As Figure 47-2 illustrates, plants may possess an extraordinary variety of mechanical defenses against herbivores, including spines, thorns, hairs, and tough leathery seed coats. On the African savannas, the tissues of heavily grazed grasses contain deposits of silica that wear away the teeth of grazing herbivores; however, such animals have become counteradapted to this defense mechanism through the evolution of extra large, hard molars that resist the mineral's abrading action.

Cactuses possess a prime example of mechanical defense in plants. Their spines are modified leaves that protrude from thick pads. The pads, in turn, are modified stems containing the chloroplasts necessary for photosynthesis. Although the prickly defenses of cactus species doubtless deter many potential consumers, coevolution has produced several herbivores that not only eat cactus tissues, but also turn the plants' defensive structures to their own advantage. For example, the pack rat feeds on the pads without difficulty and builds its nest from chunks of the plant with spines pointing outward—quite an effective deterrent to its own predators.

In general, development and maintenance of mechanical and other defensive structures and mechanisms exact a real biological price from the species that possess them. The price is the large number of cells that must be generated by mitosis to build defensive structures (and

(a)   (b)

Figure 47-2   PLANT DEFENSES.
(a) The sharp curved spines of the mammillaria cactus (*Mammillaria microcarpa*) of the southwestern United States, known as the "fishhook cactus," can catch flesh as tenaciously as fishhooks. (b) Thorns and barbs are widely used plant defenses; one pointed example is the prickly stem and leaves of a thistle (*Cirsium* species).

the protein, nucleic acid, and other syntheses to make such cells). Thus energy, nutrients, and time are required for constructing and maintaining defensive structures. Conceivably, such resources would otherwise be available for reproduction or other activities vital to living things. Some biologists argue that a rosebush expends energy to make thorns instead of diverting such energy to produce gametes; similarly, a field mouse trades feeding time for time spent excavating a protective burrow. These sorts of speculations are nearly impossible to test experimentally, even if they make sense anthropomorphically. Let us turn, then, from gross defenses to more subtle chemical ones in animals and plants.

Chemical Defenses of Animals

Nearly 5,000 years ago, an Egyptian priest inscribing the walls of a pharoah's tomb drew the hieroglyph of a pufferfish, an animal whose flesh seemed to the Egyptians to have magical properties. Carefully prepared, a morsel of the fish produced mild euphoria. However, as this ancient calligrapher undoubtedly knew, even the slightest culinary miscalculation could turn such a pufferfish meal into a disaster, causing hallucinations, paralysis, and death. Like hundreds of other marine animals, the pufferfish harbors powerful nerve poisons, or *neurotoxins*, in its tissues. The puffer's chemical weapon is tetrodotoxin, which interrupts the transmission of nerve impulses by binding to neurotransmitter binding sites at synapses. This is just one example of the many defensive chemicals manufactured in the body cells of animals or derived from foods and stored in animals' tissues for later use.

These defensive chemicals are used in a wide variety of strategies for deterring predators. Some animals, such as the bombardier beetle shown in Figure 47-3, produce foul-smelling or irritating sprays to repel potential predators. In the bombardier beetle, oxygen released from hydrogen peroxide (H_2O_2) mixes with a hot (100°C) solution of quinones in an abdominal chamber, creating an explosive combination; the result is a squirt of boiling hot, irritating fluid shot from an aimable "cannon." Other chemical defenses—including the poisons and venoms secreted by some snakes, toads, and stinging insects such as bees and wasps—may inflict pain, cause illness, or even kill the predator. Many species of amphibians, for example, produce toxins in special glands near the skin surface; some of these substances, such as the frog-venom curare, used by people in Central America and Africa to coat the tips of poisoned arrows, can paralyze a mammal within seconds because it blocks transmission across motor end plates, the synapses of nerves on muscles. Of course, chemical defenses do not

Figure 47-3   ONE ANIMAL'S CHEMICAL DEFENSE.
The bombardier beetle aiming its "cannon" forward and spraying its hot defensive secretion. The 100°C heat comes from an exothermic reaction that takes place as phenols are oxidized to quinones. The pressure for the spray comes from oxygen released suddenly from the splitting of hydrogen peroxide. All this occurs in a specialized abdominal chamber that contains the required enzymes and raw material.

provide 100 percent protection: some individuals in each generation are inevitably attacked, damaged, or consumed by predators that discover their prey's noxious qualities too late. In fact, this gives a hint about one way such potent defenses may have evolved in the first place: the sampled individual is not killed outright by the predator, but is dropped or let go after a quick taste. In such cases defensive poisons may have been selected for during evolution *not* because they occasionally kill a predator, but because they taste bad and repel most predators quickly. On the other hand, in cases such as the pufferfish, selection may operate on predators so that they recognize and avoid a deadly meal.

Some animals acquire or supplement their antipredator strategies by borrowing the chemical defenses of other species. For example, hedgehogs (relatives of shrews and moles) not only have a coat of tough, spinelike hairs that provide several inches of sharply pointed mechanical protection, but they may also rub themselves with the bodies of poisonous toads—tipping their "arrows," if you will. Some species incorporate the protective chemicals of others into their own tissues. A well-known example is the beautiful and graceful monarch butterfly. As larvae, monarchs feed on milkweed plants, which contain *cardiac glycosides*, compounds that are poisonous to vertebrates and many insects because of their action on the heart. After pupation, the tissues of the adult butterfly are saturated with the chemicals, and birds that eat a monarch vomit violently.

**Figure 47-4
A FLYCATCHER
(*MYIARCHUS
CRINITUS*)
AT WORK.**
Various birds are
adapted to take
insects on the wing
or off foliage or the
ground. If this bird
happened to catch a
bee, it would
remove the stinger
prior to eating the
bee.

As a result, experienced predators generally avoid eating
monarchs (and other butterfly species that look like
them).

Some predators have evolved behaviors that circumvent a prey's chemical barricade. Some flycatcher birds
have learned to hold bees with their feet and remove the
stingers with their beaks before dining (Figure 47-4).
And a small, ferocious predator, the grasshopper mouse
of the desert Southwest, has a behavioral adaptation that
allows it to dine on the tenebrionid beetle ("stink beetle"), which normally defends itself by ejecting a foul-smelling chemical from the nozzlelike tip of its abdomen. The mouse uses its front paws as pile drivers,
wedging the beetle's backside into the sand, and then
eats its prey head first.

Chemical Defenses of Plants

Plants are potential prey to the world's many herbivores; yet plants obviously cannot flee their predators,
and most plant species are also rather conspicuous, especially those with fruits and flowers designed to attract
pollinating animals. Why, then, do we see thousands of
plant species surviving for the most part *uneaten* everywhere we look? One defense against herbivores is the
basic growth and regeneration strategy of plants: the
ability to produce new limbs, leaves, roots, or reproductive parts throughout the life of the adult plant. This
replaces parts consumed or damaged by herbivores. But
the main type of defense found in plants—and a highly
effective one—is the use of a marvelous assortment of
chemicals.

One passive chemical defense is called **nutrient exclusion**: some plants are simply not worth eating because
their tissues contain only a tiny amount of nutrients,

such as iron and sodium, that are needed in small quantity for plant metabolism but are required in greater
abundance for animal functions. Animals requiring such
substances in comparatively large amounts tend not to
feed on plants that are able to grow under poor nutrient
conditions.

A related plant "defense" results from the fact that
tissues such as stems, trunks, bark, and mature leaves
are low in protein and high in indigestible cellulose.
Because a herbivore that eats protein-deficient plants is
likely to be less healthy and to produce fewer offspring
than one with a protein-rich diet, the vast majority of
herbivores are adapted to consume nonwoody foods.
(Termites, with their symbiotic associations with cellulose-digesting microorganisms, are notable exceptions.)
In addition, the edible tissues of some plants contain
protein-binding chemicals that make existing proteins
unavailable to predators. For example, oak leaves store
phenolic compounds, known as *tannins*, in vacuoles near
the leaf surface. When a caterpillar munches on a leaf,
the vacuoles break open and release their tannins, which
then bind leaf proteins into indigestible compounds. Although the caterpillar may continue to feed on the oak
leaf, its growth will be slowed by protein (that is, amino
acid) deficiency, decreasing the likelihood that it will live
to develop into a reproductive adult. Not surprisingly,
some herbivorous insects have developed adaptations
that outflank this chemical defense. For example, the
larvae of leaf-mining beetles burrow into the interior of
an oak leaf and eat the nutritious inner layers, avoiding
the tannin vacuoles altogether (Figure 47-5).

**Figure 47-5   A DEFENSIVE STRATEGY AGAINST A
PLANT'S OWN CHEMICAL DEFENSE.**
Leaf-mining Amazon beetle larvae (*Galerucine* species)
avoid tannin-filled vacuoles located near the surface of an
oak leaf by burrowing inside the leaf and eating the inner
layers.

TREES AND INSECTS: THE SILENT BATTLE

Trees are the venerable elders of the plant world. They stand rooted in one spot, sometimes for centuries, and endure wave after wave of attack by hungry insects. Botanists have noticed that most leaves on deciduous tree species will have at least one small hole chewed in them by the end of the summer. Yet researchers have also noticed that during an average summer, insects consume less than 7 percent of the total leaf surface in the forest. What keeps insects from simply swarming about the "defenseless" trees and stripping them bare? The answer is that trees are far from defenseless. They produce an arsenal of toxic weapons to stave off insect predators and may even inadvertently emit chemicals that trigger defensive responses of other trees in the silent battle against rampaging bugs.

Several factors—weather, parasites, diseases, and predators such as birds—help control insect populations that feed on forests. Yet somewhat surprisingly, the trees themselves are involved in this control, by producing a wide range of noxious chemicals that repel insects. These include nerve toxins such as pyrethrin and rotenone, which humans extract from plants and use widely as pesticides, as well as other compounds that taste bad, interfere with development, destroy DNA, inhibit digestion, and alter reproduction.

A very common example is the class of compounds called tannins, which are astringent and thus bad-tasting and also inhibit certain enzymes, thus interfering with an insect's digestion of leaf tissue. Sugar-maple leaves are up to 30 percent tannin. But the tannin content differs dramatically from leaf to leaf, and this in itself deters bugs. A caterpillar that eats sugar-maple leaves will have to migrate from one leaf to another, tasting each and searching for a palatable meal—which will contain low tannin levels. Observers have noted that members of one species of caterpillar do just this and reject more than two-thirds of the leaves they sample. During this roving taste-test, such caterpillars are more conspicuous to birds and other predators than they would be if they fed on single leaves. Such caterpillars also expend more energy, grow more slowly, and develop into smaller adults with less potential for reproduction. Apparently, all this results from the simple tree strategy of having variable tannin content in leaves.

According to some recent studies, trees have an even more impressive means of self-defense as well: they step up production of defensive chemicals in response to attack, and some of the chemicals may affect other trees. Jack C. Schultz and Ian Balwin at Dartmouth College tested three tree species for production of tannins or phenols in response to insect attack. When gypsy-moth larvae eat leaves on red oaks, the remaining undamaged leaves produce 200 to 300 percent more tannin. In sugar maples, the undamaged leaves produce 50 percent more tannin. And in poplars, the undamaged leaves produce 123 percent more phenol.

The team traced this stepped-up production of toxins by using radioactively labeled CO_2, which was incorporated into sugars. After a few of the tree's leaves were injured, they could trace the conversion of those radioactive sugars into tannins and other defensive substances. Clearly, a good deal of the plant's photosynthetic product—its store of chemical energy—can go toward self-defense.

Schultz and Balwin found a remarkable and unexpected phenomenon: chemical signals. They raised tree seedlings in pots and isolated them in two clear plastic growth chambers. In one chamber, all the seedlings were left undisturbed. In the other chamber, some seedlings were left alone, while researchers injured the rest by plucking a few leaves every two or three hours. In the plucked trees, as expected, the remaining, undamaged leaves began producing higher levels of defensive chemicals. But Schultz and Balwin made a surprise finding: the untouched seedlings sharing the air space with the plucked trees *also* exhibited higher levels of the defensive compounds within two to three days. The damaged trees obviously gave off some airborne cue and inadvertently triggered their untouched neighbors to increase production of the chemicals.

These results are controversial, but if proven true, should stimulate search for other cases of "communication" between trees. In the meantime, though, the work on defensive chemicals has had an important and lasting effect on our perception of trees: they can never again be regarded as defenseless giants, but rather as well-adapted defenders in a silent battle with insect pests.

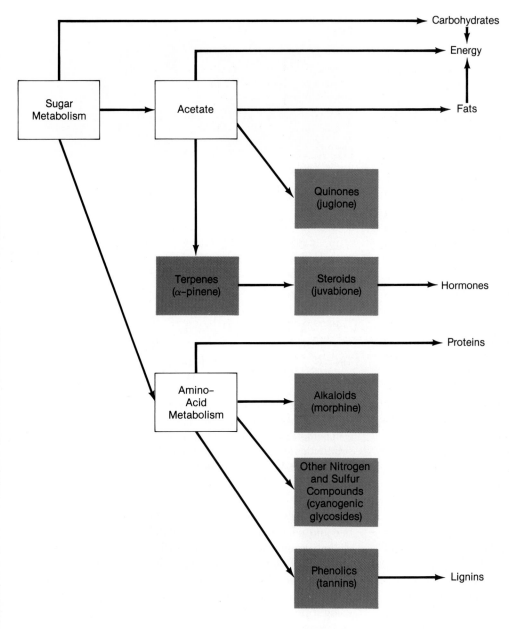

Figure 47-6   SECONDARY COMPOUNDS: METABOLIC BY-PRODUCTS FOR DEFENSE.
Six important classes of secondary compounds are noted in the purple boxes. Derived primarily from carbohydrates and amino-acid metabolism, many thousands of individual compounds are in each of these six classes. *Juglone* is a quinone of walnut trees that inhibits the growth of most other plants. *α-pinene* is a terpene derivative that acts as a fungicide after being produced in response to fungal infection. *Juvabione* is a steroid, produced by balsam firs to guard against one family of insects. *Morphine*, and related atropine and codeine, is made by several plant types and alters animal nerve function. *Cyanogenic glycosides* include, for example, acids produced by cycads, which are carcinogenic or produce nerve degeneration in humans. *Tannins* are a kind of phenolic and include a substance made by nutmeg trees that causes hallucinations in humans and nerve disorders in herbivores.

Plants may also inadvertently communicate with each other to trigger synthesis of defensive compounds. One logical candidate for the message carrier between trees is ethylene gas, a substance produced by damaged plant tissues that also has important effects as a plant hormone (Chapter 30). Of course, it is not correct to conclude that damaged trees "help" or "warn" the others; a passive process dependent on a response to the volatile substance is all that is involved. Nevertheless, that response increases the plant's fitness.

Defensive chemicals such as tannins (a kind of phenolic) are special products of plant metabolism and are known as *secondary compounds* (as opposed to such primary compounds as proteins, nucleic acids, fats, and car-

bohydrates). Secondary compounds are of six major types and most often are produced by short metabolic pathways starting with acetate or amino acids (Figure 47-6). Only a few "new" enzymes need arise in evolution to generate such pathways. Secondary compounds are commonly stored so as not to harm the plant cell that manufactures them; they are released or activated when a herbivore damages that cell within a leaf, a flower, a fruit, or a woody stem. Besides the phenolics, secondary compounds include terpenes in conifers (the source of terpentine), modified terpenes called cardiac glycosides in milkweed, toxic alkaloids in some legumes, quinones, steroids, and a miscellaneous collection of nitrogen and sulfur compounds. Some of these chemicals work by interfering with the growth of insect larvae; for instance, terpenes act like massive doses of juvenile hormone (Chapter 37), keeping a larva in a perpetual state of physiological immaturity and therefore unable to mature to adulthood and reproduce.

Several important examples of secondary defensive compounds can be seen in a large plant family, the legumes. Legumes produce poisonous alkaloids, including nicotine, strychnine, caffeine, colchicine (sometimes used as a treatment for gout), opium, and peyote. These alkaloids act on nerve, muscle, heart, or other vital cell types to do their damage. Legumes also produce several chemical defenses in addition to the alkaloids, perhaps because their seeds tend to be rich in protein and thus provide a tempting food source. One strategy involves a variation on nutrient exclusion: some species produce seeds so small that invading beetle larvae—usually one per seed—run out of food before their development is complete. Other legumes contain chemical inhibitors that block animal digestive processes, similar to the protein-binding compounds of oak trees. A most interesting example of chemical defense is found in certain species of tropical legumes, whose tissues contain *canavanine*, an amino acid not found in proteins that mimics the common amino acid arginine. When most insects consume canavanine, the amino acid is incorporated into their proteins, resulting in severe developmental abnormalities (Figure 47-7) and a lower rate of survival in the insect population. As little as 0.02 percent of canavanine in an insect's diet is enough to adversely affect its development, yet in one legume species, up to 13 percent of the seed's dry weight consists of this one amino acid! However, for one insect predator, the bruchid beetle, the presence of canavanine is an ecological jackpot. The beetle's cells contain an enzyme that distinguishes canavanine from arginine, so that only arginine is utilized to construct proteins (the enzyme joins arginine but not canavanine to arginyl tRNA). The canavanine molecules are broken down enzymatically to yield large quantities of nitrogen for the beetle's metabolism.

Just as some predators have evolved means to overcome the chemical defenses of an animal prey, so have many specialized herbivores evolved adaptations to metabolize, by-pass, or otherwise make use of plant toxins. For example, some members of the cucumber family produce cucurbitacin, a toxin that is poisonous to most insect generalists, which feed on many types of plants. To specialized cucumber beetles, however, cucurbitacin merely serves as a chemical beacon that allows them to locate and feed on the plants amid other types of vegetation. In a different type of counterstrategy, kangaroo rats use their chisel-like teeth to strip the top, salty layer from desert salt-bush plants (*Atriplex*), exposing the edible leaf layers underneath. Insect and vertebrate herbivores without this toothy adaptation are generally unable to exploit *Atriplex* as a food source.

(a)

(b)
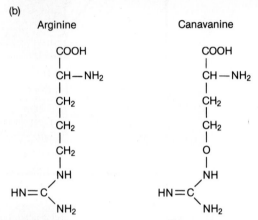

Figure 47-7   DEFENSE BY A TROPICAL LEGUME.
(a) This abnormal larva ate leguminous plant tissue containing canavanine. (b) Only an oxygen-for-carbon substitution distinguishes canavanine from the normal amino acid arginine it replaces in proteins of the larva.

Chemicals originally synthesized by a plant may also become part of a herbivore's own chemical arsenal. As we have seen, the monarch butterfly is impervious to the ill effects of milkweed cardiac glycosides. Similarly, sawflies are able to store the resinous, usually repellent terpenes contained in pine needles or eucalyptus leaves—using a protuberance of their gut as a "holding tank"—and regurgitate a blob of the sticky mass as a defense against predatory birds (Figure 47-8).

The interactions between plants with chemical defenses and their specialized herbivorous predators provide another example of how selection pressures shape the coevolution of prey and predators. It makes sense, for example, that plants living in diverse communities with intense herbivory often have more concentrated toxins—and a greater variety of chemical weapons—than do plants in areas where herbivory is less intense. This is particularly true of tropical rain forests, where researchers have found more types of plant alkaloids than in any other biome in the world.

Insect herbivores may be able to evolve an adaptation for by-passing or detoxifying one defensive chemical but not many independent adaptations for avoiding different chemical defenses. As an example, experiments on *Drosophila* showed that separate populations were able,

BETTER LIVING THROUGH ECOLOGICAL CHEMISTRY

Species in natural communities often interact chemically, and many have evolved chemical defenses against predators. Recent advances in isolating, identifying, and synthesizing such chemicals are starting to pay off in developing life-saving human drugs and effective insecticides. The defenses of certain trees against the leaf-cutting ant are a case in point.

Each day, thousands of leaf-cutting ants leave their nests in search of suitable trees from which to excise bits of leaf tissue. When they find an acceptable plant, worker ants use their sawlike mandibles to cut leaves into pieces and carry them back to the nest. On occasion, these minuscule demolition experts defoliate an entire tree in a Central American forest bit by bit.

As it happens, the ants themselves cannot digest cellulose and other elements of plant tissues. Instead, they harbor a mutualistic partner in their underground nests—a species of fungus that can break down plant tissue into simple nutrients that it uses for its own growth. The ants then harvest a proportion of fungus from their "fungus gardens," thereby getting the nutrients they need. In return, the fungus receives the steady supply of the raw leaf material needed for its own nourishment. Researchers have found that leaf-cutting ants are highly selective in the tree species they choose as food. And chemical sleuthing has revealed the reason.

The leaves of trees avoided by leaf-cutting ants contain a potent fungicide—a defensive adaptation that works against herbivorous ants indirectly, through an antibiotic effect on the ants' ecological partner, the fungus. Pharmaceutical chemists are now testing that fungicide as a possible new antibiotic treatment for fungal infections in humans.

The application of plant chemicals is not new; as many as one-half of the prescription drugs sold today contain plant-derived substances as active agents. Nevertheless, interest in their use is still increasing. For instance, the alkaloid compounds known as pyrethroids, synthesized by chrysanthemums and related flowers as defenses against insects, have become the primary ingredients in an array of commercially produced insecticides. Other plant alkaloids have been found effective against hypertension and heart ailments. A major active ingredient in birth-control pills is extracted from a species of yam, and researchers believe that more than 200 other species of South American plants may contain chemicals with potential functions as contraceptives.

As research continues into the chemical ecology of organisms, investigators will no doubt identify many substances with significant benefits for humankind. Unfortunately, most of the promising plant species inhabit tropical rain forests, ecosystems that are being destroyed so fast by humans that very few will remain by the year 2010. In the future, it will require more than serendipity to find new beneficial plant substances; it will take our collective human willingness to preserve the habitats where potentially useful species reside.

Figure 47-8   THE SAWFLY'S STICKY DEFENSE.
Sawfly larvae gathered in this circular mass have regurgitated around the periphery yellow droplets that contain a repellent substance. The repellent comes from a potentially harmful substance in their own food, eucalyptus leaves.

over many generations, to evolve adaptations to six different plant toxic substances when the toxins were presented individually; however, when all six were offered simultaneously, none of the fly populations evolved the combination of defenses that would be required to allow them to feed on a plant that produced all six toxins. In nature, diverse plant communities duplicate this experiment by presenting herbivores with a battery of chemical defenses to which no single herbivore is likely to become adapted. Thus most herbivores in such communities specialize on one or a few plant species whose chemical defenses are similar. And since there may be only a few members of a particular plant species in a diverse community, food supplies for these herbivores are likely to be hard to find and limited, so each herbivore population is necessarily small.

Thus far, we have focused on chemical defense strategies of animals and plants and have seen how such adaptations are beneficial. And we have seen that the coevolutionary tit for tat has resulted in chemical adaptations of predators and herbivores that counteract chemical defenses. Let us turn now to a very different kind of defense, which depends on sense organs and, in some cases, learning.

Warning Coloration and Mimicry

Many predators and herbivores use vision to find their live food. Furthermore, some may learn to distinguish palatable meals from noxious prey on the basis of visual cues. One result of such phenomena is a set of adaptations in prey categorized as **warning coloration,** or **aposematic coloration.** These terms refer to bright colors or striking patterns in an organism that also possesses chemical defenses. For instance, vivid shades of red, orange, and yellow set off with black markings signal potential predators that bees, wasps, and monarch butterflies are undesirable as food. Similarly, a skunk's black and white stripes serve to warn dogs, foxes, and other predators away from a rude olfactory experience. One of nature's most fanciful aposematic creations is the striped, flag-waving lionfish, shown in Figure 47-9; although its dorsal spines resemble banners waving in the current, they contain a powerful neurotoxin and a hemolytic compound that lyses red blood cells.

Evidently, the bright, contrasting colors and patterns typical of aposematic coloration are so memorable to hunting animals that rely on visual cues that only a few negative experiences are required before the predator learns to eliminate such prey from its diet. In fact, the recognition and avoidance can even be inherent in an

Figure 47-10   MÜLLERIAN MIMICS: BAD-TASTING BUTTERFLY SPECIES.
Members of these four *Heliconius* butterfly species (clockwise from top left: *H. hewitsoni, H. sara theudela, H. doris f. viridis,* and *H. pachinus*) are quite distasteful. Because they resemble each other, a predator that consumes any one of them will learn to stay away from members of all four species.

animal's nervous system and behavior. Some predatory birds, for instance, that were raised from the time of hatching by humans innately recognized the skin patterns of the poisonous coral snake and avoided items similarly marked. Obviously, no prior learning, sampling, or attack by a coral snake was involved. It is reasonable to ask why aposematic coloration and poisonous chemicals occur in the same organisms. One speculation is that genetic variations producing phenotypic "advertisements" of danger to an attacker would be favored and selected.

Interestingly, identical or similar warning colors and patterns may serve *different* aposematic species living in the same area. Figure 47-10 shows a group of bad-tasting butterfly species, all possessing similar black-and-yellow striping patterns. This phenomenon is called **Müllerian mimicry,** after the nineteenth-century zoologist Fritz Müller, who first noticed and described it. Ecologists have speculated that all the species resembling each other benefit because the learning process is simplified for predators. Birds that prey on various butterfly species thus learn one warning coloration pattern and subsequently tend to avoid all the noxious species. This, in turn, may lead to the selection for groups, or *complexes,* of Müllerian mimics: ancestral populations may have shown more variation in color and pattern, but if the most common appearance had been associated with noxious chemicals, it would have provided predators with the most negative experiences. Gradually, the variants from this common aposematic coloration would have become conspicuous and would have fallen to predators in disproportionate numbers, and the Müllerian mimics would have tended to remain as the standard. While

Figure 47-9   LIONFISH: A FLOATING ADVERTISEMENT FOR DANGER.
The warning coloration of the flamboyant lionfish (*Pterois lunulata*) is accompanied by a powerful neurotoxin in its dorsal spines. The striped pattern warns predators away from a bad experience.

such speculations are quite reasonable, it is nevertheless extremely difficult to *prove* why an area's Müllerian mimics resemble each other.

A related adaptation—and testimony to nature's ingenuity—is the phenomenon of **Batesian mimicry.** Among groups of boldly patterned aposematic species that taste bad, there are sometimes palatable species that lack noxious chemicals but have adopted the same warning coloration as the noxious species as a means of self-defense. A contemporary of Müller, the British naturalist Henry Bates, made this startling discovery while observing and recording the rich insect fauna of South America. Later researchers designed studies to test whether the clever adaptation of resembling a noxious model species indeed provides protection from predation—and found that it does.

Jane Van Zandt Brower, while a graduate student with the ecologist G. E. Hutchinson (of niche fame), set out to discover the effectiveness of Batesian mimicry, using the monarch butterfly and its Batesian mimic, the viceroy (Figure 47-11). She first asked the question: Do naïve blue jays that have not fed on either monarchs or viceroys distinguish between the two potential meals? The answer was clear: the jays readily consumed viceroys before their first meal of monarchs. But when a jay ate its first monarch, with its cardiac glycosides, the bird became violently ill. The birds quickly learned not to touch the monarchs. Then came the crucial question: Would experienced birds continue to eat viceroys? The answer was a definite no: blue jays that vomited monarchs avoided the perfectly palatable viceroys; thus Batesian mimicry successfully protects them from predation.

Figure 47-11   BATESIAN MIMICS: THE MONARCH AND THE VICEROY.

The smaller viceroy butterfly (right; *Limenitis archippus*) mimics the larger monarch butterfly (left; *Danaus plexippus*), which is characterized by a strong odor and unpleasant taste. A predator that consumes a monarch, and subsequently becomes ill, stays away from the viceroy as well.

Subsequent studies of other Batesian mimics and their models have also shown that predators rapidly learn to avoid the palatable mimics whenever they are presented in combination with the distasteful models. It seems reasonable that there should not be too many mimics in an area relative to the number of the models; if mimics become too abundant, predators are more likely to learn that they are indeed palatable. Moreover, the models are more likely to be attacked, too, since there are fewer around to reinforce the bad-taste experience. Even so, in both experimental and natural situations, cases have been found in which mimics outnumber models, but predation remains low.

While Batesian mimicry has been most thoroughly studied among butterflies and moths, it is also found in other species. Notable examples are species of milk snakes and the scarlet kingsnake, all of which are nonpoisonous and mimic the appearance of the venomous coral snake (Figure 47-12). Some nonpoisonous snakes even mimic the aposematic *sound* of rattlesnakes by rapidly beating their tails against dry grass or leaves. In fact, the range of mimicry can extend beyond defensive behaviors to include aggressive ones. A good example is

Figure 47-12   REPTILIAN BATESIAN MIMICS.

(a) The nonpoisonous scarlet kingsnake (*Lampropeltis doliata*) is a Batesian mimic of (b) the poisonous coral snake (*Micrurus fulvius*).

(a)

(b)

the firefly, in which the flashing pattern of the female attracts the males of the same species. Females of the genus *Photuris* mimic females of the genus *Photinus* by flashing their luminescent abdomens in the pattern of those fireflies. Any *Photinus* males that respond to this false advertising are seized and consumed by the *Photuris* females.

Mimicry, whether Müllerian or Batesian, is prominent in the web of ecological relationships in natural populations. Let us consider a different strategy with a different set of interesting adaptations.

ADAPTATIONS FOR ESCAPE

In the grand cat-and-mouse game of survival, organisms use another very effective means of self-defense—escape. Some adaptations, such as fast flying or rapid swimming, enable potential prey to flee, while other adaptations allow effective hiding or combinations of the two. Rabbits and foxes, for example, are fast, agile runners, and their dens or burrows provide hiding places. Finally, there are special adaptations that separate predator and prey in space and time, as we will see.

Specializations for rapid escape can be impressive. African ungulates living in grasslands, such as the fleet-footed wildebeest and gazelle, have leg structures and muscles that permit high-speed running. Their limbs are specialized for speed instead of power; for instance, their toes are reduced in number, and those remaining are greatly lengthened. In the coevolutionary game, predators, of course, have responded by evolving an equal capacity for speed. No terrestrial animal is swifter than the cheetah over short distances. Its vertebral column, pelvic girdle, and limbs have become modified to permit exceptionally long, leaping strides and high speed (Figure 47-13). In fact, the cheetah's head and jaw are so reduced in size to save weight for speed that the animal sometimes has trouble killing its captured prey.

African ungulates may also protect themselves in their exposed habitat by *herding*. This is a safety-in-numbers strategy: if animals in a population stay together in a group, the chance that a solitary predator will single out one particular individual is small. In fact, lions, hyenas, and other stalking predators generally obtain their prey from among a herd's stragglers, animals that are aged, ill, or too weak to keep up with the group. The survival value of living in a group is also seen in schools of fish and in flocks of birds. Studies of pigeons, for instance, have shown that the distance at which a group of birds detects the presence of a foraging hawk is directly related to flock size: the more eyes watching, the more efficiently predators are discovered. In some species of territorial birds, even individuals that normally keep

Figure 47-13   ADAPTED FOR HIGH SPEED.
The cheetah's (*Acinonyx jubatus*) body—from its musculature to its skeletal system—is designed for speed, allowing it to sprint up to 60 miles per hour, and thereby to catch fast-moving gazelles and antelopes.

their distance from one another flock together when a predator approaches. An important interpretation of herds, schools, and flocks came from W. D. Hamilton, who coined the term *selfish herd*; what looks like a group of cooperating animals, he theorized, is really a group composed of individuals that come together for their own individual benefit.

Escape in Space and Time

A species can escape from its predators by avoiding them in space or in time. Palatable insects, such as the periodic cicada, may live a major portion of their lives underground, remaining relatively inactive for many years. Then large numbers emerge simultaneously, breed, and produce a new generation of offspring. Immature individuals tunnel underground to repeat the cycle. Although the birds and other animals that feed on cicadas enjoy a temporarily unlimited feast, the total number of insects is so overwhelming that predators have little effect on population density. And, of course, the predator population has not had time to breed to a larger size in order to exploit the temporary cicada supply. The two species of periodic cicadas follow similar patterns—one swarming every thirteen years and the other, every seventeen years (the "seventeen-year locusts," infamous for decimating crops). Because their appearances above ground are so rare, potential predators have not evolved adaptations that enable them to specialize on the insects; however, a fungus that infects adult cicadas is able to remain dormant for long periods and is adapted to become active synchronously with its host.

Plant life cycles may also allow avoidance of predators in space and time. One strategy is to escape specialized

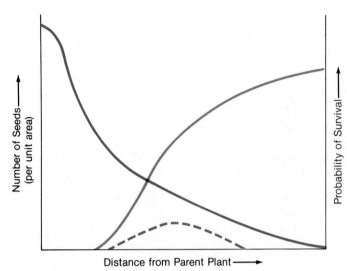

Figure 47-14   ESCAPE IN SPACE.
Distance from the parent plant can affect the survival of certain tree species. The red curve represents dispersal of seeds from the parent plant; the blue curve shows the probability of seed survival. The green line, which represents survivorship, peaks at the optimum distance from the parent plant.

Y-axis (left): Number of Seeds (per unit area)
Y-axis (right): Probability of Survival
X-axis: Distance from Parent Plant

seed-eating predators in time by flowering and seeding only at lengthy intervals, an adaptive tactic similar to that of the periodic cicadas. An extreme case is species of bamboo that flower once every 120 years, producing huge quantities of seeds each time. Since all potential seed predators must eat more often than once every century, they are adapted to utilize other plants as food sources. This escape mechanism, however, necessarily constrains the plant in other ways; it reproduces sexually only once per century, and so does not inadvertently test mutations and other sources of genetic variation nearly as frequently as do more short-lived species. In addition, the bamboo flower must be wind pollinated (no animal pollinators adapt to a plant that flowers so rarely).

Many plant adaptations for seed dispersal (Chapter 24) enable plant species to escape predation in space: the windborne seeds of dandelions, grasses, and maple trees may travel for miles on a summer breeze, far from predators in the local area of parent plants (Figure 47-14). Escape in space is also an important adaptive characteristic of the fugitive species that make up early secondary successional communities (Chapter 45).

Camouflage

The dagger-moth caterpillar, with its spittle-bug disguise, and the lavender-hued flower mantid shown in Figure 47-15 share one of nature's most fascinating es-

cape mechanisms: **camouflage** (also called *crypsis*). Camouflage involves colors or patterns that allow organisms to blend with their background or to appear to be something they are not. These adaptations are believed to be responses to the vision-oriented selection pressure brought by predators that hunt by sight. Similar arguments are made for the evolution of camouflaged predators: prey that do not notice a predator concealed by color or pattern fall easy victim. To be effective, such adaptive fakery must often hide several types of physical features: the organism's bulk, outline or silhouette, and obvious attributes that identify it as a living thing, such as eyes, legs, and antennae.

A common strategy for camouflaging bulk and body outlines is **cryptic coloration**—having the same color or pattern as the background. A gecko lizard, with its mottled skin, is difficult to detect against the bark on which it rests (Figure 47-16a). Similarly, the bodies of tree frogs, leafhoppers, and preying mantids may render them almost indistinguishable from the foliage around them (Figure 47-16b, c). In the deep ocean, numerous fish species are shades of pink and red, colors that are bright in sunlight but are perceived as gray or black in the watery depths, where there is little light. And, as we saw earlier, even the conspicuous vertical black and white stripes of a zebra are a form of camouflage because lions, its major predator, are color-blind. Hence, lions may have trouble discerning zebras against the vertical blacks, whites, and grays of the grassy savanna habitat.

Variations on cryptic coloration include **disruptive col-**

Figure 47-15   CAMOUFLAGE IN ACTION.
The flower mantid (*Pseudocreobota ocellata*) resembles orchids so closely that the mantid's insect prey are caught unawares—and are consumed.

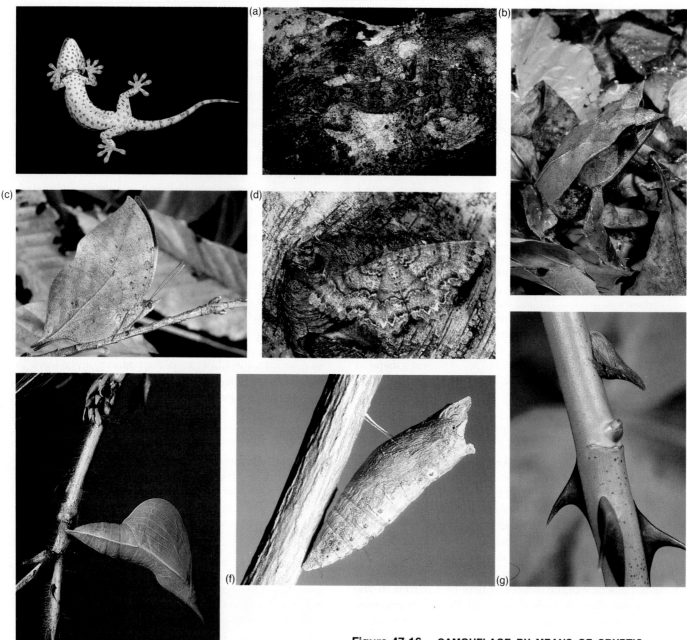

Figure 47-16   CAMOUFLAGE BY MEANS OF CRYPTIC COLORATION.

(a) The gecko (*Gekko* species) shows up easily on this glass pane, but its relative (*Phyllurus cormutus*) is barely discernible when perched on the bark of a tree. (b) This tree frog (*Megophrys monticola*) closely resembles the leaves of the tree on which it sits. (c) This leaf butterfly (*Kallima inachius*) is like the frog in resembling a leaf. (d) An African hawk moth rests on the bark of a tree; the black lines in its wing pattern are an example of disruptive coloration that breaks up the silhouette and makes the animal blend in with the background. (e) A chrysalis of a cloudless sulfur butterfly (*Phoebis sennae*) resembles a leaf hanging from this branch in a Florida swamp. (f) This white tiger swallowtail butterfly (*Papilio eurymedon*) chrysalis resembles a dead twig. (g) The large thorn at the top is actually a tree hopper that mimics thorns.

oration, in which body parts of contrasting colors break up an organism's silhouette (as with the African hawk moth shown in Figure 47-16d), and **countershading.** Countershaded species, including many fish, are typically dark-colored on top and light on the bottom. The effect of this two-tone adaptation is to reduce sharp dark–light contrasts that may provide visual cues to predators. For example, many ocean fish are dark on top, have a silvery stripe on the sides, and are white below. A predator swimming above cannot easily see the back against the darker depths; and one swimming below is unlikely to see the belly because it blends with light penetrating

Figure 47-17   CAMOUFLAGE BY SHAPE.
The African stoneplant (*Lithops villetti*) closely resembles nearby stones.

Figure 47-18   CAMOUFLAGE BY BEHAVIOR.
This European walking stick insect not only resembles a stick, but also behaves like one, spending almost all the time in branches and frozen into the position of a twig. Can you see the head just to the right of center?

the water from above (the silver stripe covers dense red muscle, heart, and liver that would otherwise be visible).

Camouflage may also extend to an organism's shape as well as its color. Examples include insects that are look-alikes for thorns, twigs, and dead, wrinkled leaves (Figure 47-16e–g), and the nubby *Lithops* plant, whose exposed structures may easily be taken for smooth, gray-green pebbles (Figure 47-17).

While camouflaged animals often seem to bear an amazing resemblance to the objects and features they copy, merely looking like a leaf or a piece of wood is not enough. Organisms that possess such adaptations must also *behave* like the things they imitate. For example, the insects commonly called walking sticks are apt to end up rather quickly in a predator's gullet if they rest or feed conspicuously on a bright green leaf. Instead, they spend much of their time on branches, frozen in the angular posture of a twig (Figure 47-18). This behavioral component of camouflage can be as striking as the visual component; insects cryptically colored to resemble leaves or flowers may cling to a green stem and flutter lightly in the breeze or may line up in groups along a stalk in the shape of a flower. Many camouflaged species are nocturnal feeders, remaining motionless during daylight hours, when predators are most active. For others, their disguises make them better predators: the flower mantid pictured in Figure 47-15 not only tends to escape the notice of its own avian predators, but also so closely resembles its floral perch that the smaller nectar-seeking insects it preys on often mistake it for a food source, only to become a tasty meal themselves. Finally, some species that normally may be quite inconspicuous may also possess warning displays, such as the false eyespots of some moths that are believed to frighten predators (Figure 47-19).

Like other protective adaptations, camouflage does not result from any conscious effort on the part of protected species; resemblances, colors, and many of the behaviors are determined by genes and genetically regulated patterns of development. Natural selection can easily account for the spread of such genes, after they arose by chance mutations, in natural populations.

So far, we have been exploring the adaptations of plants and animals evolving together in various types of adversary relationships: carnivore and animal prey, herbivore and plant. But organisms can also live and evolve together in ways that are not so overtly damaging. Like the strategies of defense and escape, such interactions have given rise to remarkable adaptations of form and function. In the rest of the chapter, we will look at a range of these other interactions, all of which are variations of symbiosis.

SYMBIOSIS

Organisms of more than one species that live together in intimate association are said to be symbionts. There are varying levels of **symbiosis,** reflecting the degree of benefit and harm stemming from the association of the

(a)

(b)

Figure 47-19   A FRIGHTENING SURPRISE: WARNING "EYES."

The *Automeris io* moth startles its predators with its prominent eyespots when disturbed: (a) normal resting position; (b) when touched the front wings move forward instantly, revealing the "eyes."

two species. *Parasitism* is one extreme, with one species benefiting greatly and the other being harmed. *Commensalism* also involves benefit to one species but without identified harm to the other. Finally, *mutualism* describes a relationship in which both species benefit. We will see that there are gradations between symbiotic relationships, which are frequent and important associations in natural environments.

Parasitism

Ecologists have dubbed some organisms "live-in" predators: these *parasites* make their living by taking up residence on or within the bodies of their prey. Nearly all plants and animals are prey to parasitic animals or microbes at some time during their life cycle. Such organisms, including protozoans, roundworms, and

tapeworms, are often highly specific in their choice of prey: the host provides nutrients, a protective environment, or other needs that a parasite requires in order to reproduce effectively before its adult stage dies. As we saw in Chapter 25, adaptation to life within a host's gut, muscle, or other organ may involve loss of organs by the parasite. The tapeworm is an extreme case: it has lost its gut and has acquired the ability to absorb nutrients from the host directly through its surface. As this sort of evolutionary adaptation of parasite to host occurs, natural selection acting on the host because of the parasitic infection may actually make individuals in the host population better adapted than if the parasites had never invaded the host; a more versatile immune system is an example.

Among vertebrates, a major defense against parasites is attack by antibodies or T lymphocytes (Chapter 32). Some parasitic species have, in turn, evolved adaptations that avoid such immune-system defenses: they may release chemicals that suppress the host's immune system, or they may have a protective protein coat that mimics that of host cells so that they are not recognized as foreign bodies. Trypanosomes, the protozoans that cause African sleeping sickness (Chapter 21), build up populations in a host that change their surface antigens periodically so that the host's immune system cannot respond fast enough to kill all the invader cells (Figure 47-20). Similarly, some invasive organisms are adapted to "borrow" a surface coat of host proteins immediately on entering their prey.

The most successful parasites do not kill their hosts, since that would destroy their living "hotel," perhaps before the parasite has reproduced successfully. This means that natural selection acts against the most deadly parasites and pathogens. Over time, the more successful invaders are likely to be those that do less harm to host organisms, and the more resistant hosts are the most likely to survive. Hence, there is an evolutionary tendency toward less negative interactions.

A famous instance of selection pressure acting on a pathogen occurred when the myxoma virus was introduced into Australia in 1950 by ranchers who hoped that it would control a population explosion of European rabbits. The rabbits, brought to the continent a century earlier, had multiplied so efficiently that they were stripping vegetation from large areas of range land. The virus was lethal at first and killed most of the rabbits it infected. Within a few years, however, rabbits with genetic resistance to the virus began to appear, while less damaging strains of the pathogen also turned up. In this case, natural selection had worked two ways. First, the few resistant rabbits were selected for, and resistance gradually became established in that population's gene pool. Second, viruses that quickly killed their hosts before blood from the host could infect another rabbit

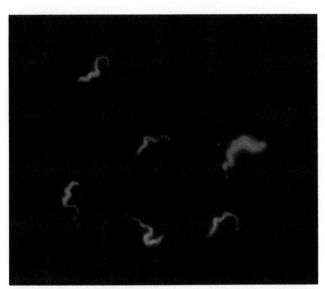

Figure 47-20 PARASITISM.
Trypanosomes are moving targets for the immune system. The green-stained trypanosomes in this human blood sample are descended from cells that entered the bloodstream when this host was bitten by a tsetse fly. Just as the immune system mounts an attack against the "green" lineage, variant trypanosomes arise with different surface proteins (VSGs). The new "red" lineage, as stained by another antibody, persists; but it, too, will ultimately be killed off. Only if the "red" lineage gives rise to still another surface protein will the trypanosomes live on in this particular host.

(myxoma virus is transmitted by mosquitoes) failed to be transmitted. More benign strains, which infected their hosts but were not immediately fatal, persisted. Today, both European rabbits and myxoma virus can be found in Australia, with each adapted to the special requirements of interaction with the other.

Host–parasite relations that may be very stable for long periods of time can be altered radically if the ecological situation changes. The human species has been subject to sudden attacks of virulent diseases—such as bubonic plague, or Black Death—when the environment changed. In the mid-fourteenth century, the high density of population in European cities, unsanitary conditions, and increased commerce and transportation among cities provided an inadvertent but ideal set of environmental circumstances for bacteria of the genus *Pasturella*. The bacteria became instant pathogens, killing one-quarter of the population of Europe within two years and striking four times more before the fourteenth century ended. Even today, human populations have not been exposed sufficiently to *Pasturella* to have evolved adequate defensive reactions to it.

Commensalism and Mutualism

Plunging gracefully in the open sea, the barnacle-encrusted whale in Figure 47-21 is a participant in the interspecific association known as **commensalism.** In this arrangement, one organism lives with or on another, gaining shelter or some other prerequisite for survival and superficially, at least, neither harming nor benefiting its host. Thus hitchhiking barnacles obtain substrate and transportation, while they filter microscopic food particles from passing ocean currents. In this and most cases, it is difficult to be sure that a relationship is truly

Figure 47-21 COMMENSALISM: HITCHHIKING BARNACLES AND BREECHING WHALES.
(a) A breeching humpback whale (*Megaptera novaeangliae*).
(b) The acorn barnacles (*Cyamus scammoni*) and lice on the skin of the gray whale (*Eschrichtius robustus*) take part in commensalism; the barnacles are carried through the water, filtering food along the way, with no apparent harm or benefit to the whale.

(a)

(b)

commensal; for instance, the hard, sharp barnacles very likely create significant turbulence as the whale swims, thereby raising the energy cost of swimming. If so, the relationship merges into parasitism.

Commensal species typically utilize bits of food not consumed by their hosts or obtain food independently of the host. This habit distinguishes them from true parasites, which derive nourishment at their host's expense. Many species of crabs are commensal with other marine organisms; for instance, some species of crab move in with marine worms, whose tubes they use for shelter from predators. Another example of commensalism is the relationship between bromeliads (plants related to cultivated pineapples) and the tropical trees on which they grow (Figure 47-22). Although bromeliads live on the trunks and branches of their hosts, they obtain water and nutrients from the air or bark surface without penetrating host tissues. Even here, however, their presence may impose a cost; the weight of numerous bromeliads filled with rain water after a tropical downpour may break many of the host tree's branches.

The potentially negative effect of bromeliads may, in fact, have contributed to a relationship between ants and trees that protects the trees against bromeliads. Ants of the genus *Azteca* commonly live in *Cecropia* trees, taking advantage of the hollow stems for their nests and feeding on glycogen-rich nodules, Müllerian bodies, found at the base of leaf petioles. Plants, of course, normally store carbohydrate as starch; the production of glycogen, the typical storage carbohydrate of animals, is a strategy to lure the ants. The *Azteca*, in return for room and board, literally scavenge the *Cecropia* trees for new

Figure 47-22   COMMENSALISM IN PLANTS.
Bromeliads (*Neoregelia* species) live on the trunk and branches of their host trees without doing any apparent harm.

small bromeliads; even small mats of bromeliads tied experimentally to *Cecropia* limbs are soon dissected by the ants and dropped to the rain-forest floor (Figure 47-23). Note, in this case, that the ants are not using the bromeliads for food as do the ants with fungus gardens; instead, a behavior has evolved that has indirect benefits for the ants—undamaged trees that produce glycogen.

The *Azteca–Cecropia* relationship clearly benefits both partners: it is an example of **mutualism.** We have already encountered symbiotic relationships between

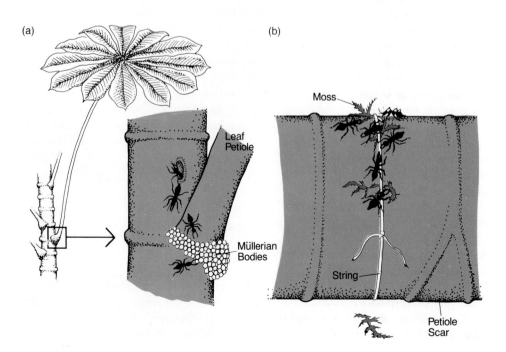

Figure 47-23   MUTUALISM: *AZTECA* ANTS AND *CECROPIA* TREES.
(a) Müllerian bodies provide glycogen to the ants that colonize the hollow stems of the trees. (b) *Azteca*, in turn, keep the trees cleared of bromeliads and other epiphytes, pruning away these mosses tied to the limb or ones that happen to sprout on a limb in natural conditions.

legumes and nitrogen-fixing bacteria (Chapter 44), between fungi and algae in lichens (Chapter 22), and between cellulose-digesting microorganisms and the ruminant mammals (sheep and cattle) or termites whose guts they inhabit (Chapter 34).

Some mutualistic associations may have evolved from age-old relationships between plants and herbivores or between parasitic animal species and their hosts. A celebrated example of intricate plant–herbivore mutualism is the relationship between ants and *Acacia*. House ants live in the hollow thorns of certain species of tropical acacia trees and feed on sugar secretions produced by modified leaves. The ants also harvest beadlike protein-rich bodies (known as *Beltian bodies*) produced at the tips of acacia leaflets at considerable metabolic cost to the plant. In return, the ants perform two indispensable services for their host: they protect the plant from other herbivores by attacking intruders in a massive, stinging frenzy; and they clear the surrounding area of competing vegetation. As a result, the trees thrive in environments in which they would otherwise face intense challenge by herbivores and plant competitors. The relationship between these species of ants and the *Acacia* species is *obligatory* for the trees. For example, when ants are experimentally removed from the acacias, fewer plants survive (72 percent survive with ants and 43 percent without ants), and those plants that do survive grow more slowly than when the ants are present (growth in height over a two-month period was 104 cm with ants, but only 16 cm without ants).

Some associated species, including other species of ants and acacias, have *facultative* relationships: they derive mutual benefits from their association but can survive quite well independently. For example, beetles of the genus *Oncideres* sometimes engage in an intriguing facultative association with mimosa trees. The larvae of this diminutive insect can survive only in dead wood; thus a female first excavates a groove in the bark of a living branch, lays her eggs, and then proceeds to kill the limb by digging a deep, circular trench near where the branch joins the trunk. (This trench is fatal to the branch because it slices through the cambium, cutting off the flow of nutrients through the phloem.) Eventually, the dead limb falls to the ground. But how do lopped-off branches benefit the mimosa? Trees trimmed by *Oncideres* may live as long as 100 years, compared with their twenty- to thirty-year life span when left undisturbed. This may be because such trees grow more slowly, but biologists do not yet know whether the slow-growing, long-lived trees produce more offspring than the short-lived, beetle-free trees. Those data are needed to define the *Oncideres*–mimosa relationship precisely.

In mutualistic associations such as that of ants and acacias, the host species trades food for reduced losses to some other, more damaging predator. Other forms of mutualism enhance the host's ability to reproduce or disperse progeny; prime examples are plant–pollinator and plant–disperser relationships, as we saw in Chapters 23 and 24. Conceivably, these associations also began as predatory–prey interactions, which only much later became transformed, by natural selection and adaptation, into ecological alliances.

LOOKING AHEAD

Throughout this chapter, we have seen that when species share a long coevolutionary history, they may evolve marvelously specialized adaptations for mutually beneficial interactions, for defending against predators, or for circumventing the defenses of their prey. Indeed, we find such highly evolved strategies, along with less intricately linked interactions, in every community of organisms. Coevolutionary relationships are being revealed and investigated more and more frequently, and it is fitting that we conclude these first seven chapters on evolution and ecology with examples of the adaptations that great evolutionary thinkers such as Charles Darwin and Alfred Russel Wallace sought to explain. As we learn more about the precise genetic steps in coevolution, we shall better understand how the process that they described—evolution by natural selection—helps shape the structures, functions, and behaviors of organisms in the natural world. In the next chapters, we turn to animal behavior, in relation to both the kinds of adaptations we have been discussing and the increasingly complex range of behavior that is dependent on evolution of more and more complex nervous systems.

SUMMARY

1. Many adaptations of organisms are directed at aspects of their biotic environment. The adaptive strategies of interacting species are often reciprocal; this phenomenon has been termed *coevolution*. No organism or species can deliberately evolve beneficial adaptations; rather, these adaptations evolve over generations as natural selection acts on variation in phenotypes and indirectly on their underlying genes.

2. Both animals and plants may employ mechanical defenses against predators. Such defenses include horns, claws, and large size or strength in animals, and a variety of spines, thorns, and tough structures in plants.

3. Toxins, venoms, and bad-smelling sprays are examples of animal chemical defenses. Some animals synthesize their own chemical weapons, while others have become adapted to borrow the substances produced by other species.

4. Plants have evolved a variety of chemical defenses against predation, including noxious and poisonous substances stored in plant tissue. In one type of chemical defense, *nutrient exclusion*, plant tissues may contain only minute amounts of nutrients crucial for animal metabolism. Another defense found in potentially nutritious plants involves protein-binding chemicals that make the proteins in leaf cells unavailable to predators. The most important and diverse defensive chemicals in plant tissues are secondary compounds—that is, by-products of the organism's metabolism.

5. Coevolution has also resulted in predators that are adapted to circumvent prey defenses. Carnivores may have behavioral adaptations—such as bird species that remove a bee's stinger—that enable them to consume protected species, while specialized herbivores may metabolize or by-pass plant toxins or use them to locate palatable plants among other types of vegetation.

6. The fact that predators can sometimes learn to recognize potentially noxious prey has led to the evolution of warning characteristics such as *aposematic coloration* (or *warning coloration*). Aposematic species living in the same area may have similar warning colors and patterns—a phenomenon known as *Müllerian mimicry*. Alternatively, *Batesian mimics* are adapted to resemble noxious "model" species, even though they are tasty, edible prey.

7. Escape strategies include physical adaptations, such as long legs that enable an animal to flee, or adaptations that make it possible for an organism to hide from predators. Schooling, herding, or flocking provides safety in numbers. Both plant and animal species may escape predators by avoiding them in space or time. Periodic cicadas and some plants escape in time by being an unpredictable food source; plant adaptations for seed dispersal away from parents—and local predators—allow escape in space.

8. *Camouflage* is an escape mechanism found among animal species whose predators hunt for food by sight. Common adaptations for concealing bulk and body outlines include *cryptic coloration*, *disruptive coloration*, and *countershading*. Camouflaged species must also have behavioral adaptations that complement their physical disguises.

9. *Symbiosis* is an intimate relationship between members of different species. There is a range of symbiotic relationships, including parasitism, commensalism and mutualism.

10. Parasites are predators that have evolved adaptations for living in or on a host. Natural selection acts against the most virulent of these predators; over evolutionary time, the most successful parasites may be those that harm host organisms in minimal ways.

11. In *commensalism*, one organism benefits, while the other appears to be unharmed. *Mutualism* is an association in which both species benefit; an example is the partnership between ants and acacias. Mutualism may be either obligatory, in which neither species can survive without the contribution of the other, or facultative, in which the associated species are able to live independently, although perhaps not as successfully.

KEY TERMS

aposematic coloration
Batesian mimicry
camouflage
coevolution
commensalism

countershading
cryptic coloration
disruptive coloration
Müllerian mimicry
mutualism

nutrient exclusion
symbiosis
warning coloration

QUESTIONS

1. Name selected structures and physiological processes in an oak tree and in an antelope that are probably *not* the result of coevolution. Next, name some features in each that could well result from co-evolution.

2. Could a Batesian mimic be, at the same time, a member of a Müllerian mimic complex? Why or why not? Could a Batesian model be part of such a Müllerian complex? Why or why not?

3. What are plant secondary compounds? Where do they come from? How do typical ones act? And how have animal species responded to secondary compounds in their food species?

4. In a broad sense, camouflage includes a variety of adaptations. Outline the sorts of elements that might be found in an optimally camouflaged prey and equally well camouflaged predator.

5. Imagine a world in which all small animals were highly photosynthetic. What might you predict about plants in such a world? About large–small animal interactions and evolution?

6. Is every structural feature of an organism's body necessarily an adaptation? Explain.

7. Define the three major types of symbiosis, pointing out advantages or disadvantages to the participants.

ESSAY QUESTION

1. "Organisms evolving together" is indeed common on Earth. Can you conceive an organism evolving with absolutely no aspects of its form, function, or behavior being affected by other organisms? Explain.

SUGGESTED READINGS

BROWER, L. "Ecological Chemistry." *Scientific American*, February 1969, pp. 22–30.

This article summarizes the blue-jay–monarch-butterfly case as well as the whole issue of insects that use plant compounds for protection.

EHRLICH, P., and P. H. RAVEN. "Butterflies and Plants." *Scientific American*, June 1967, pp. 104–112.

This is the seminal paper on coevolution—a true landmark.

KLOPFER, P. J., and P. J. HAILMAN. *Behavioral Aspects of Ecology.* Englewood Cliffs, N.J.: Prentice-Hall, 1973.

A simple summary of the complex ways in which behavior operates in the evolutionary context.

LEWONTIN, R. C. "Adaptation." *Scientific American*, September 1978, pp. 213–230.

A thoughtful essay that includes aspects of organisms that are not adaptations.

ROSENTHAL, G. A. "The Chemical Defenses of Higher Plants." *Scientific American*, January 1986, pp. 94–99.

An excellent summary of the strategies of plants and counterstrategies of animals to survive in the ecological war.

48

BEHAVIORAL ADAPTATIONS TO THE ENVIRONMENT

*Beneath the varying behavior which animals learn lie
unvarying motor patterns which they inherit. These behavior
traits are as much a characteristic of a species as bodily
structure and form.*

Konrad Lorenz,
The Evolution and Modification of Behavior (1965)

Animal migration: Behavior for survival. In a "march to motherhood," these pregnant caribou (*Rangifer tarandus*) may travel 300 miles in the Alaskan Brooks Range.

In the late nineteenth century, the French naturalist Jean Fabre recorded the habits of several related species of insects known as digger wasps. Among other details, he noted that bee-hunting wasps always paralyze their prey with a sting to a particular region of the body, while those that prey on crickets or flies invariably attack different vulnerable points. The wasps then stuff their victims inside burrows excavated in the soil, where the prey can serve as living but immobilized food supplies for the wasps' developing offspring.

Fabre also observed that the larvae of each wasp species seem to know just how to consume a paralyzed victim without causing its premature death, employing a precise set of table manners geared to the specific physical characteristics of their prey. This sort of observation suggests to some researchers learning, and it exemplifies the problems that scientists who study behavior must solve. For, as we shall see, it is difficult, indeed, to establish the degree to which complex behaviors include learning, or even what humans call intelligence.

Fabre's careful observations and those of other early naturalists helped set the stage for modern studies of *behavior:* the things that organisms, especially animals, do as they grow, reproduce, seek food, and otherwise interact with their environment. The heritable component of an animal's behavior is subject to natural selection, just as are other aspects of its biology. However, much of behavior is not inherited. Behaviors, both inherited and not, are like many morphological traits and physiological processes—they can be adaptive. That should come as no surprise, since behavior is a direct product of the organism's nervous and endocrine systems employing, as it were, the body's arsenal of limbs, sensory systems, effectors, and other features to act and react.

Biologists have cataloged thousands of behaviors among hundreds of species. In this chapter, we will consider a broad behavioral class, the inborn, or **innate**, behaviors that are sometimes referred to as *instincts*. An instinct is, in a sense, a built-in feature of the nervous system; the digger-wasp larva does not pick up its predation techniques by observing other larvae or adults feeding on paralyzed crickets. A program that brings about effective feeding is a part of its nervous system. We shall begin our study of innate behaviors by examining a subset, so-called *reflexes*, or automatic behaviors brought about by direct links between the nervous system and the body's effectors. Finally, at the opposite end of the spectrum is another class of behaviors—*learning*, or behavior modified on the basis of experience. A young monkey chances to observe a small mammal being attacked by a snake and learns thereby to avoid these slithering predators. In between, there are gradations in instinctive and learned behaviors. And most of the complex behaviors we see in birds and mammals involve a combination of reflexes, instincts, and learning intertwined in complicated patterns. In this chapter, we illustrate two such combinations: navigation and migration. Such complex behaviors are both fascinating and difficult to study, for the behaviorist must endeavor to sort out the influences of reflex, instinct, and learning. We will see how researchers go about this task. With all this as background, we will turn our attention in Chapter 49 to behavior in animal societies. In that chapter, too, the theme of behavior as adaptation will emerge and show that the reproductive success that Darwin recognized as the key to evolution can depend on the behavior of individuals in groups as well as of individuals acting alone.

The subdivision of biology known as **ethology** encompasses the study of animals in nature; it stresses the characteristics of actions that bring food, attract mates, or defend against predators. While ethologists generally do field studies, other behaviorists take a range of approaches from neurophysiological to psychological. Irrespective of the methodology, all behaviorists must contend with a real problem, the human tendency toward *anthropomorphism*—the attribution of human qualities to other species. Thus the wagging tail of a pet dog, the hiss and raised fur of a cat, and many other specific animal behaviors may be seen as signs of "pleasure," "anger," and so on—all of which are subjective and observable human feelings in what we imagine are similar circumstances (Figure 48-1). But there is no evidence that nonhuman animals can "be anxious," "want something," "worry," "love," and so on, or if they can, that such feelings are·significant determinants of their behavior. This assumption has been a great stumbling block for the young science of animal behavior as well as for the study of psychology dealing with humans. We

Figure 48-1 "ANGER" AND "FEAR" VERSUS INSTINCTIVE BEHAVIORS.
Attribution of human emotions or motivations to animals such as this cat showing a threat display does nothing but confuse the science of behavior.

must try to avoid thinking anthropomorphically as we go beyond the natural-history approach of ethology and seek to better understand the evolutionary origins of behaviors and their adaptive roles in the lives of animals.

REFLEX BEHAVIOR

At its most fundamental level, behavior is indistinguishable from the physiological processes that maintain life itself, including homeostatic mechanisms (Chapter 35) that trigger faster cellular respiration when carbon dioxide levels rise or regulate the flow of critical ions into the blood. A single cell such as *E. coli* moves backward or forward (simple behavior) as it detects various compounds by means of thousands of proteinaceous receptors in or on the cell surface. Studies show that such cellular organisms have receptor "systems" sensitive to dozens of chemicals and thus can respond by moving up or down concentration gradients for all of them. Such receptors on single cells and the transduction mechanisms that link the receptors to specific cellular responses (flagellar rotation in *E. coli*, or the hormone–receptor–cAMP–protein-kinase sequence in mammalian cells; Chapter 37) are models for the sensory, nervous, and motor systems of multicellular animals. It is useful, therefore, to examine the simplest type of innate behavior—the reflex.

Reflexes are automatic, involuntary responses to external stimuli. Typically, a reflex behavior takes place when sensory receptor cells communicate a stimulus to the nervous system, which, in turn, sends a set of instructions to effector organs such as muscles. The so-called startle reflex common to many animals illustrates this point: if a garden snail is prodded with a stick, giant nerve fibers convey this touch stimulus simultaneously to nearly all muscle groups, producing an almost instant retreat of the snail into its shell (Figure 48-2). We learned in Chapter 36 that even in humans, such a reflex may be so simple as to be monosynaptic—that is, with direct synapse of the sensory cell on the motor cell. But such an arrangement is atypical; most reflexes, including that in the snail, involve more complex neuronal circuits. Even so, most reflexes can be understood relatively easily as adaptations related to some aspect of survival; hence, their evolutionary significance can be appreciated. One thing is clear: reflexes are fundamental elements from which other simple behaviors can be built. Let us see how by discussing tropisms, taxes, and kineses.

Tropisms, Taxes, and Kineses

One of the first scientists to systematically explore simple reflexes was the late-nineteenth-century physiol-

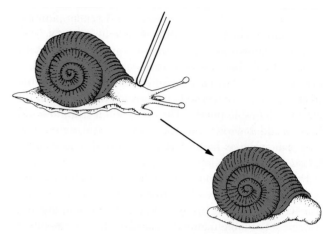

Figure 48-2   THE STARTLE REFLEX.
Touching a garden snail with a stick stimulates its nervous system to initiate muscle contraction automatically. The result is the animal's retreat into its shell.

ogist Jacques Loeb. Loeb was intrigued by the experiments of the botanist Julius von Sachs, which showed how plants orient themselves to light, gravity, moisture, and other stimuli. Loeb wondered if these orientations, or *tropisms* (from the Greek word *trópos*, meaning "to turn"), might apply to animals as well. His research soon revealed that the free-floating larvae of marine barnacles are phototropic—that is, attracted to light—and that phototropic caterpillars presented with food in an unilluminated spot will starve rather than turn their heads into the shadow. Other animals that Loeb tested showed evidence of being either attracted to or repelled by gravity, electric currents, chemicals, and other phenomena. Such tropisms in animals are instinctive reflexes and do not involve learning, although that does not mean that all animal responses to physical features of the environment do not have a learning component. Today, animal tropisms are called **taxes** (singular, *taxis*), to distinguish behavior produced by electrochemical activity of the animal nervous system from the analogous tropisms produced by growth-regulating hormones in plants (Chapter 30).

The maggot of the common housefly, for example, exhibits phototaxis—movement in response to light. The maggot is equipped with primitive photoreceptors on its head, and when the larva is ready to pupate, it begins to crawl while moving its head from side to side. This motion results in the greatest light intensity hitting first one side of the head and then the other side. Each time, the maggot responds by turning its head (and body) more to the darker side. This process is repeated until the maggot eventually faces completely away from the source of illumination. As a result, it crawls into a dark place. There it is less vulnerable to its predators; it got there not by intent or "liking" darkness, but by a program of responses to simple taxes. A similar phenomenon explains how cockroaches scurry out of sight when a light is turned on in a dark room. The neuronal circuits responsible for taxes may turn on or off, depending on an animal's maturity, its physiological state, or the complexity of its sensory and nervous systems. For instance, although the fly larva is negatively phototactic and avoids light, an adult fly may have the opposite response, moving toward light because it is more likely to find food in a lighted area. Although these taxes are apparently reflex-based behaviors, it is hard to use such cases in attempting to understand whether a bat returning to a dark cave as sunrise approaches, or a human miner emerging from a mineshaft late in the day, includes any elements of reflex tropisms along with learning in carrying out such complex behaviors.

The flatworm *Planaria alpina* exhibits yet another example of a taxis: as Figure 48-3 shows, the flatworm orients its body so that it meets a current of water head-on. Presumably, this strategy has evolved because food particles are carried downstream, and such an orientation permits the planarian to intercept them more readily.

Kineses (singular, *kinesis*) are also simple reflexive behaviors, but ones in which the intensity of the animal's response is proportional to the intensity of the stimulus. The single-celled paramecium shows this type of response in the presence of carbon dioxide; at a certain distance from a CO_2 bubble, the microscopic protozoan slows its forward movement in response to the mild acidity surrounding the bubble. If the paramecium moves too close to the bubble, the higher acidity causes its cilia to beat in a way that rotates the animal through an angle of about 30 degrees; having changed course, the single-

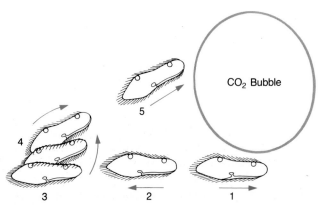

Figure 48-4   KINESIS IN A PROTOZOAN.
When this *Paramecium* nears a noxious object such as a CO_2 bubble, the cilia on its surface reverse their beat, and this causes the animal to back away (1, 2, 3). The organism then turns about 30° to the left and advances (4, 5). On encountering the object again (as at 5), it repeats the reflexive backing and turning.

celled creature continues swimming away from the bubble. This process is repeated until the organism is in the zone of optimum pH once more (Figure 48-4). By adjusting its swimming direction to various pH levels, the paramecium tends to spend most of its time in areas where the decomposer bacteria, which produce acid and serve as its food source, are found.

Uncomplicated behaviors such as taxes and kineses, while automatic, are vital adaptations that enable simple animals to better cope with features of their environment. By so coping, the organism is more likely to survive to reproduce. Even so, behaviorists are discovering more and more complex behavior patterns with experience-based components in so-called simple animals. When we turn to multicellular animals, many behaviors, while seemingly automatic, are clearly much more complex than is a straightforward reflex; such behaviors are often referred to as instincts, and they, too, may serve as important adaptations.

INSTINCTS: INHERITED BEHAVIORAL PROGRAMS

An **instinct** is usually defined as a stereotyped, inherited pattern of behavior that takes place in response to a specific environmental stimulus. When a fawn senses a disturbance in the environment caused by a predator, for example, it instinctively freezes; this instinct makes the fawn less likely to be noticed and attacked. Similarly, the digger wasps that so fascinated Fabre sting their prey

Figure 48-3   TAXIS TO FLOWING WATER.
The flatworm *Planaria* reflexively positions its body so that its head meets the oncoming water, which flows from right to left in this tank. It can thereby better maintain position and also may be better positioned to harvest food particles floating toward it in the current.

in the correct places by instinct: a bee-wolf wasp, for instance, which places its single paralyzing stroke in a tiny soft patch under its victim's "chin," is thoroughly disoriented by a headless bee and so does not lay eggs in the perfectly suitable body. The wasp's nervous system develops in such a way that neuronal circuits are present which cause the wasp to respond to even the first exposure to a prey by carrying out the full stinging and egg-laying repertoire. The instinct does not depend on experience or learning, and the wasp cannot adapt or learn to cope with a headless bee.

For much of the last century, behaviorists have disputed the precise role that instincts play in guiding an animal's behavior. Observations of instinctive behaviors in animals date back to ancient times. During the first half of the twentieth century, however, adherents of behavioral psychology, led by the psychologist B. F. Skinner, stirred up a hornet's nest of controversy by suggesting that nearly all of an animal's activities can be explained as learned—not instinctive—responses to environmental stimuli. In more recent decades, students of ethology have resoundingly rejected this view by providing a wealth of experimental evidence that genetically programmed activities can be important components of animal behavior. Recall, for instance, the crayfish tail-flip escape behavior (Chapter 36). The relatively simple nerve networks of the crayfish account for behavior that at first glance seems to be purposeful, but in reality is a straightforward result of wiring patterns in the nervous system. Even humans exhibit certain kinds of preprogrammed behaviors, as we will see.

The brilliant Austrian naturalist and ethologist Konrad Lorenz carried out several of the classic studies that elucidated the nature of instinct. Some of Lorenz's most famous observations centered on behaviors of the greylag goose, shown in Figure 48-5. In particular, Lorenz noted what are now called **fixed motor patterns,** or *fixed action patterns*—that is, unvarying series of precise physical movements apparently "wired" into the nervous system and hence determined by the animal's genes and the developmental program that builds the brain.

Lorenz found that whenever he displaced a greylag's egg a short distance from its nest, the goose invariably attempted to retrieve it with a characteristic egg-rolling motion of the neck and bill. However, if the egg rolled out of reach entirely or if Lorenz picked it up after the goose had begun its retrieval, the bird nonetheless completed the entire fixed motor pattern of retrieving by moving its bill along the ground and tucking the nonexistent egg between its legs. Clearly, once the goose had noticed the wayward egg and had extended its neck, the fixed pattern was initiated and carried to conclusion. Lorenz theorized that for the greylag, the coordinated muscle sequence associated with egg retrieving is a behavioral unit that is a complex type of instinct.

Figure 48-5   FIXED MOTOR PATTERNS AND IMPRINTING: THE GREYLAG GOOSE.
Konrad Lorenz studied this kind of greylag goose (*Anser anser*) for many years, and from it gained great insight into the concept of fixed motor patterns—seemingly innate behaviors initiated by the nervous system in response to specific stimuli.

Even before Lorenz's work, other researchers had discovered that such stereotyped behavior is a common feature of courtship among birds: each species has its own characteristic pattern of genetically determined behavior that is followed faithfully from generation to generation. Notice, for example, the detailed series of courtship poses assumed by the goldeneye duck, depicted in Figure 48-6.

Today, the genetic basis of such rituals is so well accepted that mating behavior is often used as a characteristic to distinguish among closely related (and morphologically quite similar) species.

Open and Closed Programs

Although Lorenz originally conceived of fixed motor patterns as being entirely innate and somewhat like reflexes, further studies have shown that the picture is a good deal more complicated. Many of the actions of short-lived organisms, such as insects, are **closed programs**—that is, they are entirely "prewired" at birth. In contrast, larger, long-lived animals with more complex brains may exhibit **open programs**—innate motor patterns in which certain elements can be modified by learning. This reflects the general rule of thumb that the lower an animal is on the phylogenetic scale, the greater the fraction of its behavior that depends directly on genes.

A cat, for instance, inherits a basic "pounce-and-disembowel" behavior that can be modified in response to particular situations. Kittens learn to adapt the basic pattern to fit the specific characteristics of available prey

Figure 48-6   FIXED MOTOR PATTERNS IN THE GOLDENEYE DUCK: COURTSHIP COMMUNICATION.

The goldeneye duck (*Bucephala clangula*) exhibits a variety of gestures with a wealth of meaning for other goldeneyes. In these photos, the animal is going through a variety of courtship gestures, trying to stimulate a potential mate to respond. Though at first viewing there seems to be a great variety in these gestures, in fact the sequences are quite precise and have a clear genetic basis.

reasonable conclusion is that such behaviors are inherited as fixed motor patterns that are used in response to mood changes, to the situations that people encounter, and so on. And, obviously, the complex programs that produce such signals can be modified by life experience.

What are the adaptive advantages of open programs over closed programs? Open programs, which are based on more complex central nervous systems, allow an individual to cope better with the unpredictable opportunities and dangers that inevitably arise during its lifetime. The prewired, closed program that brings about the crayfish's highly stereotyped tail flip simply cannot cope with the unforeseen or the rare—with the persistent human skin-diver, for example, intent on capturing a tasty meal. With its open program, the kitten or young cat can benefit from experiences with prey of different sizes or with litter mates during play. Such experiences help expand its repertoire of motor patterns and this, in turn, increases the chance of success during hunting or in avoiding danger. We see, therefore, that open programs can be important adaptations that permit more complex life styles and which may provide greater likelihood of surviving to reproduce—the hallmark of evolutionary success.

Triggering Behavior: Sign Stimuli and Innate Releasing Mechanisms

Most instinctive behaviors are triggered by some environmental event—a visual, auditory, olfactory, or tac-

Figure 48-7   OPEN PROGRAM: A CARNIVORE'S HUNTING BEHAVIOR.

An open program is an innate motor pattern that can be modified by learning and experience. The tendency of this coyote to pounce on prey is a good example of a fixed pattern that must be modified to meet the specific demands of a given hunting experience.

or to conform to the behavior of their parents and litter mates or of the other cats they encounter when young. Thus a kitten, or any hunting mammal, learns by experience that pounce and disembowel requires different techniques for a tiny mouse, a strong ground squirrel, or a vigorous bird such as a large jay (Figure 48-7). By the same token, most human facial expressions are strikingly similar in form and meaning throughout our species, even though people often develop their own individual pattern of using such communication signals. In fact, humans who are born blind still smile, frown, and grimace, without ever having seen another human face. A

tile stimulus that prompts an animal to action. Thus a wood tick (*Ixodes*) begins to bite when it finds itself on a surface with a temperature of 37°C that also exudes the odor of butyric acid—a combination found primarily on the skin of a mammalian host. An event that sparks a behavioral response is called a **sign stimulus,** which, in turn, sets in motion an **innate releasing mechanism** that results in a particular behavior. Once again, a classic series of experiments elucidated how this process works.

Using herring gulls as subjects, the ethologist Niko Tinbergen focused his research on the sign stimuli involved in the feeding behavior of gull chicks. Such a sign stimulus, when produced by another member of an animal's species, is known as a **releaser.** In nature, chicks peep in a stereotyped way that initiates the parents' search for food and tells the parent gulls where to deliver the food. When a parent returns to the nest with food, it holds its bill pointing downward and waves it back and forth in front of the chicks; chicks then peck at the tip of the parent's bill, which is marked with a species-specific red spot (Figure 48-8), and the parent releases the food.

To determine what the exact releaser for the chicks' pecking behavior might be, Tinbergen devised various cardboard and wood models of parent gulls and then observed the sorts of behaviors that the models elicited. After numerous trials, he found that chicks peck whenever the model has both a particular vertical orientation

Figure 48-8   RELEASER: THE GULL'S RED SPOT.
When gulls, and their relatives such as this black albatross (*Diomedea irrorata*), are hatched, they instinctively peck when hungry at moving red spots and vertical bars, features that form a caricature of their parents' bills. Adults, on the other hand, are driven to feed whatever pecks at the tips of their bills. This strategy of sign and countersign not only results in the chicks' getting fed, but also begins a process, similar to imprinting, wherein the parents and their chicks come to recognize each other as individuals.

and a red spot like that on an adult gull's bill. In a similar experiment, the American ethologist Jack Hailman showed that, in fact, the chicks respond specifically to the dot as it moves from side to side (and, in fact, response was better to three dots even on a pencil; Figure 48-9). Both scientists also found that gull chicks peck even more readily when the model is a vertical bar—much like a parent's bill—moving horizontally.

Together, the two stimuli elicit a maximum pecking response. It is interesting to note that the shapes of these releasers—circles and elongate bars—are of the type most readily perceived by receptor cells in the visual systems of higher animals (Chapters 38 and 40).

The colors, forms, odors, courtship displays, and other characteristics of many animals have been shown to release innate behaviors in other individuals. In fact, the reason that such colors, odors, and so on are present at all may be their releaser function. Thus evolution includes not only adaptive responses, but also the development of specific releasers of instinctive behaviors. One special feature of behavioral triggers is that even a crude model, if it has the *essential* characteristics of a releaser, is as potent as the natural sign stimulus in setting a behavior in motion. A famous example, also described by Tinbergen, involves a tiny minnowlike fish, the three-spined stickleback.

When mating season approaches, males of this species begin to develop a long red band on the belly (Figure 48-10), as well as changes in reproductive behavior. Both these changes are physiological in a sense and involve alterations in pigment cells and in the nervous system. They are triggered by environmental factors, such as changing day length or water temperature, that are monitored by the stickleback's nervous system which acts via hormones or nerves to cause the changes in coloration or behavior. As the male fish change color, they also become highly aggressive and stake out small territories; this is followed by the construction of tunnel-like nests. We know that such specific behaviors are dependent on sites in the brain that are activated by the hormones.

Tinbergen and his co-workers at Oxford University found that male sticklebacks recognize other territorial males on the basis of a single sign stimulus: the red belly. In fact, almost any red object, no matter how unfishlike (provided its length is greater than its height), provokes a reaction: territorial males in laboratory aquaria have been known to "attack" a passing red truck seen through the laboratory window. Conversely, very lifelike stickleback models without the red belly elicit little more than passing notice. So these findings show that a releaser, whether the red band on a stickleback's belly or the red spot on a gull's bill, is often just a small part of an organism's total exterior appearance; in many cases the responding organism focuses on just that one specific fea-

Figure 48-9   AN EXPERIMENT IN ANIMAL BEHAVIOR.

The bar graphs at the top show the rate at which herring gull chicks pecked at a model of an adult gull's head (left) and at other substituted variables (a head with no bill; a bill with no head; spots on a moving pencil; a head and bill with no spot; a complete head lacking movement; or a horizontal head). Arrows show the directions of movement. What conclusions would you draw about the necessary releasers for gull chick feeding behavior? (From P. Klopher and J. Hailman, *Control of Developmental Behavior* © 1972. Adapted by permission of Benjamin/ Cummings Publishing Company.)

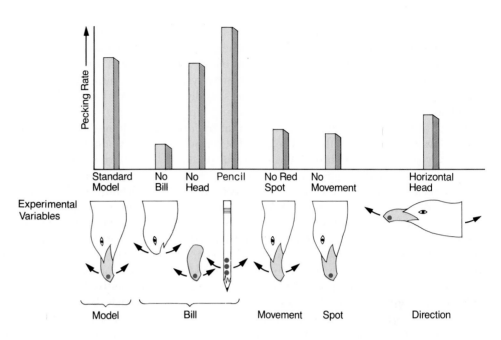

ture, with its color and shape (or, in the case of behavioral releasers, with its precise movements, sounds, and so on).

In laboratory experiments, researchers have discovered an interesting wrinkle on sign stimuli and releasers: animals preferentially respond to a **supernormal stimulus**—one that is in some way more intense or exaggerated than the natural stimulus (such as the red truck for sticklebacks). For example, a greylag goose will react to the largest "egglike" object offered—including beer cans and volleyballs—when it is primed for egg-rolling behavior, ignoring its own egg altogether and rolling and

even incubating the giant egglike objects. Why this is so remains a matter of speculation. It may be that a goose selects larger-than-normal "eggs" simply because the bigger objects exaggerate key features of the stimulus and somehow elicit a stronger response than the real eggs. Since overly large eggs do not occur in nature, perhaps no mechanism evolved to prompt the goose to reject them. Some ethologists theorize that the prominent sexual dimorphisms found in males of many animal species—including the peacock's brilliant tail feathers and the elk's enormous antlers—may be supernormal stimuli, the most extreme forms of which can only be carried by the very best males, that help to increase the chances that a male will successfully attract a mate.

Motivation and Appetitive Behavior: Physiological Readiness for Behavior

We have seen that certain sexual releasers develop in response to changing hormonal levels in the blood; an example is the red belly on sticklebacks. But what hap-

Figure 48-10   RELEASER: THE MALE STICKLEBACK'S RED BELLY.

As the mating season nears for the three-spined stickleback (*Gasterosteus aculeatus*), environmental stimuli such as elevating water temperature or lengthening days cause physiological and behavioral changes. In the male stickleback, this includes the development of a long red band on the belly as well as nest-building activity. The female, seen here at the top, does not become red, but does respond to the male's colorful belly.

pens if a releaser is presented to an "unprepared" organism? Probably the behavior in question will not occur. Indeed, in order for a sign stimulus to exert its effect, an animal must be motivated—that is, it must be in a physiological state that renders it receptive to the stimulus. The term "motivation" as used here is not the anthropomorphism it may seem; it does not suggest that the organism "wants" to do something. The lion that has just fed usually pays little attention to the zebra, for predatory behavior normally requires a state of hunger. Similarly, animals such as the *Anolis* lizards discussed in Chapter 37 participate in reproductive courtship only when they reach sexual maturity and both their nervous systems and their gonads are physiologically ready for reproductive activity. Likewise, the seasonal growth of the elk's antlers or the red marking on the male stickleback act as releasers for others of the species, but only if hormonal changes similar to those that led to the releaser's development have created the necessary state of readiness in the responding animal's brain. If these changes have not taken place, the releaser message will not trigger a response.

Biologists probed the important relationship between internal physiological states and behavior in a common laboratory subject, the ring dove, which is related to the pigeon. Doves typically begin courtship behavior in the spring; mating follows and is, in turn, followed by nest building, egg laying, and incubation as the product of one behavior becomes a releaser for the next. Here is the normal sequence: sixteen days after the eggs are laid, the female dove begins to produce crop "milk," a cheesy substance, in response to prolactin secretion (Chapter 37). After the eggs hatch, the sight of the chicks triggers crop-milk feeding behavior in the adult female dove, and the hatchlings are fed. Finally, after a rigidly programmed waiting period of three days, the adult begins to feed the chicks solid food.

Experimenters probed this sequence in various ways to clarify the actual triggers involved at each step. They confirmed the length of time between egg laying and crop-milk production by placing eggs in a nest before a breeding pair of doves laid their own. This placement appears to have set an internal clock, and sixteen days later, the female produced crop milk. Conversely, a female never produced crop milk if her own eggs were removed from the nest as they were laid. Researchers discovered that the sight of chicks triggered the parents' crop-milk feeding behavior; they did this by placing partially incubated eggs under a newly nesting dove pair that had not seen eggs or chicks. These eggs hatched in fewer than sixteen days, and on seeing the chicks, the parents attempted to feed the young—but with no prolactin-triggered crop milk to offer. Finally, experimenters confirmed the rigid three-day requirement between crop-milk and solid-food feeding. If birds that lack crop

milk were placed with starving chicks, the adults failed to respond to the sight and sound of the starving youngsters. They did not begin feeding the chicks with solid food until three days had passed after they first saw the chicks. Crop-milk feeding in ring doves reveals that reasonably complex sequences of behavior can depend on hormone–brain interactions; in fact, such interactions are common in vertebrate reproductive behavior and may extend to other realms of behavior.

When an animal is physiologically primed to execute a behavior but does not encounter the appropriate stimulus, it may exhibit **appetitive behavior**—activity that increases its likelihood of being exposed to the necessary stimulus. For instance, a warm, hungry western fence lizard will wander along rocks and downed wood; when it sees a small, moving object, this sight stimulates its innate predatory behavior, and the lizard attacks. Some animals become increasingly sensitized to a particular stimulus when they are deprived of it, responding to cruder and cruder approximations of the proper stimulus as time passes. House cats are an example of this *sensitization* phenomenon; descended from wild cats that fed on small birds and rodents, domesticated cats are innately sensitive to the stimulus "small, moving object." In the absence of real prey, a cat will stalk and attack blowing leaves, tennis balls, and even nonmoving objects, such as pebbles. As cat owners know, the animal's threshold can get so low that the cat will chase, capture, and disembowel the air itself.

Habituation

The opposite of sensitization is a phenomenon known as **habituation,** in which repeated exposure to a sign stimulus lessens an animal's responsiveness to it. A sort of "behavioral boredom," habituation is common in higher animals and provides a safety valve that prevents them from repeating the same behavior over and over.

Habituation apparently lowers the sensitivity of the nervous system to sign stimuli. This occurs in the brain and its complex circuits; the sense organs themselves usually retain their normal sensitivity and report information that gradually is ignored due to habituation. The result is that the animal is numbed to the usual, while remaining alert to the special and novel. Mallard ducks display a good example of habituation. These young birds ordinarily react to the stimulus of a hawk's silhouette appearing overhead by fleeing. Ducklings stop this behavior, however, if the silhouette passes overhead many times, but they may continue to eye it suspiciously.

This is also a prime example of the gradation between instinct and learning. There is no doubt that the basic fright response of a young duckling to a hawk silhouette

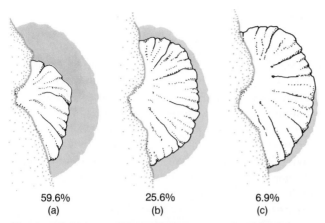

59.6%   25.6%   6.9%
(a)   (b)   (c)

Figure 48-11   HABITUATION: A FORM OF SIMPLE LEARNING.
(a) When prodded, the marine invertebrate *Aplysia* initially contracts its gill almost 60 percent of the way into its mantle. (b, c) As learning takes place, the animal begins to ignore the stimulus and gill contraction becomes less and less complete. The animal learns to suppress the gill-contraction response.

is instinctive. Yet habituation has elements of what we define as learning. Learning, too, has many gradations, as we will see, the simpler ones being difficult to distinguish from habituation in response to repetitive stimuli.

The work of Eric Kandel at Columbia University on the neurophysiological basis of habituation in *Aplysia*, the sluglike marine invertebrate also called the "sea hare," helps provide perspective on the neuronal basis of learning. If some object or movement disturbs an *Aplysia*'s siphon, an instinctive reflex causes the animal to withdraw its muscular gill into its mantle cavity. Normally, *Aplysia* withdraws its gill immediately when the siphon is stimulated. If an investigator repeatedly prods the siphon, however, after only a few such "training sessions," the slug ceases to respond to the stimulus entirely for long periods of time (Figure 48-11). Some biologists view this sort of habituation as a prototype of simple learning because an animal's nervous system must remember not to respond to a particular stimulus. But how does this learning take place? Recent experiments on simple animals—such as *Aplysia*—that are capable of elementary learned behavior have been quite revealing. The portion of *Aplysia*'s nervous system that regulates the gill-withdrawal reflex includes twenty-four sensory neurons that synapse on motor neurons in the gill; axons of the motor neurons, in turn, synapse on nearby muscle cells. In a series of experiments, Kandel showed that as habituation takes place, the amount of neurotransmitter released by the sensory neurons falls steadily, and as a result, the neurotransmitter has a pro-

gressively weaker effect on the motor neurons that stimulate muscle contraction.

The chemical basis for an *Aplysia* nerve cell's "memory" of a stimulus is a long-term change in the ion permeability of the cell membrane, which restricts the flow of substances—especially calcium ions—necessary to trigger the cell's secretion of neurotransmitter (Chapter 36). Although simple, the habituation process in *Aplysia*, together with the crayfish tail-flip circuitry described in Chapter 36, provide tantalizing models for more complex neuronal systems. But we can also include brain circuit plasticity, such as increased numbers of dendrites and synapses in heavily used circuits (Chapter 40), and the modulating effects of neuroactive peptides in beginning to understand the cellular bases for instinctive and learned behaviors.

LEARNING

When a rat masters a maze, a young bird perfects a song, or a child ties his or her shoelaces for the first time, each is exhibiting the results of **learning:** behavior that can have reflexive and instinctive components but that is modified on the basis of experience. Most animals have nervous systems sufficiently complex to allow at least some degree of learning. Unraveling behavior that appears to involve learning and being sure it is not simply a complex series of reflexive and instinctive acts has challenged behaviorists. Consider some difficulties.

By definition, a learned behavior involves changes in response on the basis of prior experiences. But simple physical maturation, for example, can be easily confused with learning. As an animal develops to puberty, altered hormonal levels trigger certain new behaviors, such as sexual play or aggression of young males against older ones. Or consider that some animals can learn in one context but not in another; a rat will learn a maze rapidly if food is received as a reward, but will learn it slowly, if at all, if food is not offered. The experimenter must design tests that are adequate to identify differences in motivation or other variables that might lead to an animal's apparent failure to learn. And, as we will see, learning in some animals may be possible only during **sensitive periods**—times when permanent behavioral repertoires can be built into the nervous system. For instance, young birds of certain species can learn the precise songs of local adult males only during a brief "window" of brain development—between about ten and thirty days after hatching. Finally, there are cases in which an animal really does learn something, yet fails to display it; the learning may emerge only in another context at some other time. The behaviorist studying learning faces a real challenge in coping with such problems;

THE FLEXIBILITY OF ANIMAL BEHAVIOR

A philosopher once said that a person never casts a stone twice in the same stream; and, indeed, in any stream the volume of water, speed of its flow, and direction of its movement change from moment to moment. In at least some respects, all environments are changeable, and this fact presents an interesting puzzle to students of animal behavior. How can an animal survive in a world with limited food and other resources, but a seemingly unlimited capacity to change and threaten? The scientific discipline of behavioral ecology is concerned with just this issue, and recent research is revealing that to cope with the environment, even the simplest animals employ learning to a degree never before suspected.

Classical theory suggests that a great deal of animal behavior is instinctive, particularly in invertebrates and "lower" vertebrates. This, however, leaves the issue of survival in changing environments unresolved. How can such animals adjust to and survive new situations? In 1974, Ernst Mayr of Harvard University proposed a seminal theory of animal behavior, with profound implications for behavioral ecology. He theorized that animals generate behavior not just as many separate actions but as whole coordinated streams of actions that he called *behavior programs.* In Mayr's theory, *open behavioral programs* are highly flexible, sensitive to conditions, and variable, depending on the situation at hand. Conversely, *closed behavioral programs* are inflexible. Whether instinctive or learned, they generate only a small range of behaviors, regardless of conditions.

Taken to extremes, Mayr's theory might seem untenable. A *completely* closed program would be too tight and inflexible to be useful in the real world. For example, even digger wasp stinging behavior must be flexible enough to allow the animals to handle prey of slightly different sizes and in slightly different positions. Likewise, a *completely* open program would be too loose to provide the advantages of advance preparation. A wasp that searched for food without at least some restrictions on where to hunt, what to look for, and how to harvest and eat might well starve. Mayr's theory, however, predicts such flexibility in behavioral programs, as well as outlining a system for which types of behavior should be relatively open and which relatively closed, and which animals, in general, should display each pattern.

Mayr argued that social or *communicative* behaviors are more likely to be relatively closed, to help ensure clear and unambiguous communication. He believed that nonsocial or *resource-directed* behaviors, on the other hand, are more likely to remain relatively open, because environments are so continually variable. Mayr also argued that long-lived animals should have behavioral programs that remain more open than those of short-lived animals, because the former have more opportunities to perform and develop variations on given behavior patterns than the latter.

Terrence Laverty, a doctoral student at the University of Toronto, recently planned an experiment to test some surprising behaviors of bumblebees, and even so, real progress is being made in the study of learning.

In this section, we will look at various types of learning, beginning with some of the simplest—programmed learning and imprinting. Then we will turn to the more complex cases of trial-and-error learning, associative learning, and reasoning.

Programmed Learning

Learning is a mechanism that allows an animal to cope with highly complex phenomena or unpredictable features of its environment. For example, baby herring gulls learn to distinguish their parents from other gulls perhaps in order to be fed. Since mature gulls can look quite different from each other, it is not possible for the young to recognize their parents by instinct alone. Instead, baby gulls come to know their parents in a brief, early sensitive period of **programmed learning** that is initiated by instinct, but that is truly a learning experience.

Honeybees provide a prime example of the complexity of programmed learning as they learn to recognize the species of flower they are harvesting at a particular time, as well as other crucial bits of information. This learned information can be communicated to other bees, but it can be "forgotten," in time scales of hours, to be replaced by new learned information about other flowers as the day or the blooming season continues.

The German zoologist Karl von Frisch and several generations of his students have shown that a forager bee can learn and relate to those back in the hive not only the color, odor, and shape of a flower, but its location, characteristics of nearby landmarks, and the time of day associated with the bee's visit. In learning these flower pahis results lend powerful support to Mayr's theory. Like honeybees, bumblebees fly out from their nests to forage for flower nectar; but, unlike honeybees, they do not communicate information about foraging by dancing when they return to the nest (Chapter 49). During its lifetime, an individual bumblebee forages from many species of flowers that vary greatly in shape and structure, or morphology. One may wonder, therefore, how the bumblebee finds these widely differing flower types without information from other bumblebees. The bumblebee's "flower hopping" behavior is clearly a resource-directed behavior by a short-lived animal. So how much of it is instinctive and how much is learned? And to what extent is its foraging program open or closed?

Laverty reasoned that if the behavior is either closed or instinctive, the bee will not get any better with practice at recognizing flowers of a given morphology (that is, it won't learn to find a new type). To see if this held true, bees raised in the laboratory were allowed to feed only by flying down long tubes to specific flowers. Because the individual bees were marked, Laverty could keep a complete record of their entire lifetime foraging experience.

Laverty found that on the first foraging flight they made, bumblebees took much longer to find the nectar on complex flowers than on simple flowers. It was as if the bees "knew" only to land and search "flower-like" objects. Once landed, they would walk around on a flower, and probe it in various places; they would usually find the nectar eventually, but would some-

times leave without feeding. With subsequent flights, however, the more experience a bee had with a complex flower, the less time it took to find the nectar, until eventually the animal could fly directly to the food source with no exploration.

Most significantly, experience with flowers of one species improved performance on flowers of other species, as long as the flowers had the same general morphology. For example, monkshood and lousewort plants both have complex, bilaterally symmetrical flowers and bees must perform a similar narrow range of movements to reach and harvest the nectar. Experience with monkshood, therefore, aided the bee in later feeding on lousewort.

Laverty's results provide strong evidence that bumblebees learn their foraging behavior not simply by memorizing details of specific flowers, but by learning something deeper and more abstract about how these details are organized.

The bumblebee experiment is just one of a growing number of similar studies that have revealed the learning of complex, open behavior patterns in a wide variety of animals. As Mayr predicted, most of these examples involve resource-directed behaviors, and include activities of mammals, birds, fish, reptiles, amphibians, and a variety of invertebrates. Taken as a whole, these examples demonstrate animals to be amazingly flexible, to have a well-developed ability to learn, and to be more sensitive to their environments than anyone suspected only a few years ago.

Figure 48-12   PROGRAMMED LEARNING: HONEYBEES AND FLOWERS.

As a honeybee *(Apis mellifera)* forager lands on a flower, she learns its color. This pollen collector (note the full pollen "basket" on her right rear leg) memorizes the odor of the sunflower only after landing, and nearby landmarks only as she circles the flower before returning home. She also learns the shape of the flower, calculates its distance and direction from the hive, and registers the time of day when sunflowers are open.

rameters, the foraging bee follows an inflexible, genetically coded routine. Only during the last three seconds before she lands does she learn the color of a flower (Figure 48-12); irrelevant are colors she has seen before that final approach, during her feeding, or when she circles the area before heading back to the hive.

Odor is learned while the bee is on the flower, and landmarks are memorized during the departing circling flight. Moreover, the bee is incapable of learning these classes of information except in the precise contexts permitted by the associated motor programs of landing, feeding, or circling. More recent experiments have shown that newly hatched, naïve honeybees have no predilection for specific flower species. Nevertheless, the rate at which they learn a new flower species depends first on the geometric complexity of the flower and, second, on whether that individual bee has experienced that geometry (not that species) before. Thus, a kind of abstraction—geometric shape—is involved in learning even in the relatively simple brain of a honeybee.

For bees, then, learning is more than a simple, rigidly controlled behavioral operation. Although much is preprogrammed, the final ingredients—the distance, direction, color, and odor of a newly blooming set of flowers—can never be. In addition, the way in which cues are logged into the forager's brain is likewise tightly programmed. Color, odor, geometry, location, landmarks, and time of day are learned separately and at different rates: odor is learned quickly perhaps because it is a reliable, unvarying cue to a flower's identity; color, a bit more slowly; and time of day, slower and last of all. Once learned, these six elements form a coherent set of stored information in the honeybee's brain. The relation *among* the detailed information of that set appears to be a key feature of learning. This is indicated by observations showing that a honeybee can transfer what it has learned about one flower species to another species it has never seen, even though many details of structure are different. If the relationship among the new details is the same as that in the first, "learned" species, then the bee shows transfer of learning. Again, this implies a remarkable level of abstraction for what we used to assume was a relatively simple brain.

Imprinting: A Form of Programmed Learning

Another form of learning is **imprinting**. Konrad Lorenz coined this term to describe a phenomenon long noted by naturalists: on hatching out of an egg, goslings, swans, and similar birds follow the first object they see that satisfies the crudest sign-stimulus requirements of size and pattern of movement. As with programmed learning, imprinting can occur only during a discrete sensitive period after hatching; an appropriate stimulus must be present then if imprinting is to occur. Once imprinted, the young birds faithfully follow the imprinted object until they are mature enough to set out on their own. They ignore even their natural parents as they respond to the imprinted object.

In his early experiments, Lorenz offered himself as a model parent to newly hatched birds; soon, flocks of geese, swans, and jackdaws (crowlike birds) were following him relentlessly about the grounds of his Max Planck Institute.

There are limits to this kind of imprinting in terms of qualities of the model object. For example, certain bird species exhibit preferences in the parental models a hatchling will follow. Thus mallard ducklings respond best to yellow-green objects, while baby chicks imprint best on orange or blue ones. Among wood ducks, hearing—not vision—is the vehicle for imprinting, so that ducklings respond to their mother's quacking.

Imprinting can have lasting effects on behavior quite independent of simple following. For instance, Lorenz discovered that goslings, imprinted on him, on reaching maturity courted him rather than members of their own species. Many birds will rear young of a different species whose plumage is a different color from their own; when the adopted offspring mature, they are more receptive to potential mates of their foster parents' appearance than of their own.

Quite a different kind of imprinting is apparent in young salmon. As the little fish mature into the smolt stage, they imprint on odors in their home stream during a sensitive period that may be only a few hours long. Perhaps four or five years later, the sexually mature adult fish migrate from the sea up long river systems, steering into correct tributaries as they wend their way upstream. To do so, they use the odor on which they imprinted years before—and which they have not smelled in the intervening time—as the guide to their spawning site (Figure 48-13).

Why do animals imprint in the first place? From an evolutionary perspective, imprinting provides a ready mechanism for recognizing one's own species or one's own parents or offspring. This can be a vitally important adaptation. Such a strategy protects parents from wasting their energies on fostering offspring other than their own and helps wandering young identify their parents. And imprinting may help ensure that mating takes place between individuals that can produce viable young.

Latent Learning

A learning mechanism rather more complex than the programmed learning of a honeybee or the imprinting of a bird is **latent learning**. It involves a delay between when a behavior is learned and when it is used. The best examples of this process involve the elaborate songs of perching birds, such as finches, warblers, and sparrows. To begin, consider that primitive bird songs, such as those of the mourning dove or the roadrunner, are completely encoded as inherited fixed motor patterns; thus those songs may be sung perfectly even by an individual

Figure 48-13   IMPRINTING AND THE LONG RETURN HOME.
These red sockeye salmon (*Oncorhynchus nerka*) have been triggered by hormones to turn red, change their behavior, and return to the stream where they began life. It is at a specific time that young fingerling salmon apparently become imprinted on odors present in the stream. They retain the memory for years while growing to large size at sea. This photo shows adult salmon near the final stages of their return, as they almost fill a small stream in Alaska from bank to bank.

that has never heard the song or even by a bird deafened since hatching. In contrast, the complex songs of most perching species are not wholly innate—important parts must be learned. This is seen best in the case of the different dialects sung by white-crowned sparrows, as

studied by Peter Marler and his colleagues at the University of California, Berkeley. (Recall that in Chapter 42, we saw how these dialects enable members of different populations to identify the group to which a potential mate belongs.) Local populations of white-crowned sparrows have evolved their own dialects—each a song built on the basic white-crowned sparrow set of notes, but each also possessing unique sound frequencies, precise intervals between notes, and so on (Figure 48-14). Just how are such dialects used and passed from one generation to the next?

Young nestlings seem to listen to and learn their local dialect when they are between ten and thirty days old; acoustic input either before or after that critical time cannot be learned. During their three-week sensitive period for song learning, both males and females "memorize" the group's song, even though females never normally sing as adults. In the listening and memorizing process, the young birds neither sing nor practice what they hear and apparently never actually try out the fixed motor pattern they will use as adults for the notes of the song. Next, during the late autumn and first winter of life, a young male white-crowned sparrow neither hears nor sings the dialect. Then, in his first springtime, as male sex hormone levels rise in his blood, he begins to sing. Over the course of the next few weeks, he refines his performance, apparently "comparing" his developing song with the pattern "recorded" the previous summer until he produces a perfect copy (Figure 48-15). He then sings the song thousands of times during the late spring and summer. Once learned this way, the dialect be-

Figure 48-14   DIALECTS IN BIRD SONG.
Populations of white-crowned sparrows have developed their own variations on the basic song of the species. These local dialects in Marin, Berkeley, and Sunset Beach are learned early in life. Here sound spectrograph recordings of the songs are plotted with frequency on the vertical axis and time on the horizontal axis.

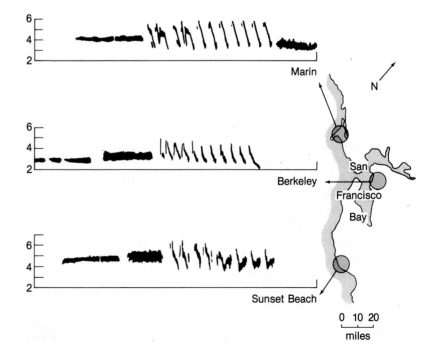

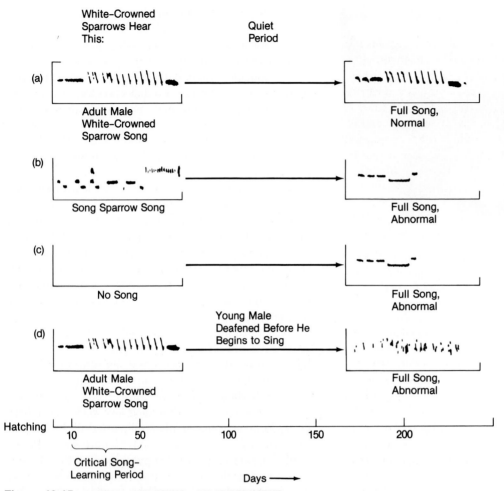

Figure 48-15   ANIMAL LEARNING: AN EXPERIMENT.
White-crowned sparrow young learn specific dialects of their species' song between 10 and 50 days of age. Experimenters exposed groups of young to normal dialect singing (a, d), to the sounds of the song sparrow (b), or to no songs (c). The birds in group (a) developed the dialect they heard as young. Birds in groups (b) and (c) both developed a song unlike that of their parents, revealing that the white-crowned sparrow dialect must be learned. Finally, the birds in group (d) had abnormal songs as adults. This proves that hearing the dialect as a young bird is not enough; the individual must be able to hear itself practice in order to develop the dialect.

comes a fixed motor pattern, and the bird can sing it afterward even if deafened.

The neuronal circuits involved in producing complex bird songs are now being mapped. Undoubtedly, these circuits are set up early in the animal's life, to be activated each spring as sex hormone levels rise to prepare the bird for migration, courtship, nest building, mating, and, indeed, singing. In fact, a female bird that heard a dialect when only a nestling can still call forth that memory; a series of testosterone injections will soon have her doing something that a female white-crowned sparrow never usually does—sing, and not just the general white-crowned sparrow song, but the dialect that she

heard and memorized months or years earlier! Other experiments on canaries show that dendritic branching, numbers of synapses, and even numbers of neurons in a region of the bird's brain responsible for singing, all increase each spring in response to sex hormones. How this plasticity relates to seeming permanence of the song's memory is still a fascinating mystery.

Singing by white-crowned sparrows, mourning doves, and other birds reveals an intricate web of instinctive and learned behaviors. Now we move from these behaviors, which can be linked definitely to hormonal and developmental processes, to examples of even more complex behavior.

Complex Learning Patterns

In general, the more complex an animal's nervous system, the greater the proportion of its behaviors that involve learning. But even organisms such as insects have behaviors that include both closed programs and capacities to be behaviorally adaptive. In comparison, more of a primate's behaviors are influenced by learning and include greater degrees of abstraction. No other organisms are so adaptable, nor so able to apply environmental information to their own needs as the prosimians, monkeys, and apes. Even so, humans and other primates still show a surprisingly large amount of instinctive behavior, such as maternal behavior following childbirth and the series of unfolding behaviors in infants and young children (for instance, the crawling, standing, walking sequence). We humans are not quite as unique as we sometimes like to flatter ourselves. Animals belonging to many other species are capable of relatively complex learning behavior.

Many species exhibit **trial-and-error learning,** also called **feedback learning.** Thus if a behavior elicits a favorable outcome, it is **reinforced,** and the animal will tend to repeat it; alternatively, if the action produces a neutral or negative outcome, it tends to be extinguished. Hence, after one taste of a noxious monarch butterfly, most birds never try to eat another one (Chapter 47). A cat that discovers by accident that brushing against a screen door causes the door to open soon learns to carry out this behavior quite deliberately.

Learning by feedback is the basis for **operant conditioning**—the training method developed mainly by the behavioral psychologist B. F. Skinner—in which an accidental behavior that approximates a desired action is rewarded. A person experienced in operant-training techniques can sometimes train animals to carry out extremely complex strings of behaviors; such training is behind the familiar circus acts in which lions leap through flaming hoops and poodles pirouette in ballerina garb. Most species subjected to such training show frequent behavioral breakdowns into natural behaviors. Much recent work shows that animals are not the kind of blank sheets that Skinner imagined; operant conditioning has many limitations.

The Russian physiologist Ivan Pavlov identified a different kind of learning, known today as **classical** or **Pavlovian conditioning,** or sometimes as **associative learning.** By ringing a bell immediately before each time he fed a dog, Pavlov found that he was eventually able to get the dog to salivate by simply ringing the bell, since the animal had come to associate the bell with being fed. Likewise, ducks in a city pond may come quacking furiously at the sight of a human at the water's edge because they have come to associate this event with bits of bread tossed into the water.

Insight learning or **reasoning** is the ability to use abstractions based on past learning in novel ways, combinations, or situations. This type of learned behavior is apparently available only to animals such as humans and certain other primates that possess highly complex nervous systems. (Cetaceans may also carry out such learning, but evidence is insufficient so far.) A chimpanzee placed in a room with bananas hanging from the ceiling and boxes on the floor demonstrates the classic example of insight learning. The chimp in this situation quickly figures out that it can reach the fruit by stacking the boxes one atop the other. An individual capable of using insight is able to generalize from previous experience, rather than relying on trial and error to cope successfully with a new situation. It is usually very hard for an experimenter to prove that reasoning has occurred, because the animal may have learned from an earlier, analogous experiment. Altering just a few details can sometimes stump the animal and leave it unable to complete the task by "reasoning."

Learning and Cultural Transmission

On occasion, a behavior learned by one member of a population or species is transmitted to the entire group. While this phenomenon is the fundamental aspect of human culture, it occasionally shows up in other species as well (Figure 48-16). One case of cultural spread involves a group of Japanese macaques in which a young, low-ranking female macaque discovered by accident that washing sweet potatoes removes sand from them. As will be described in detail in Chapter 49, this novel behavior, washing sweet potatoes, has become a permanent part of the troop's mealtime behavior that is passed from generation to generation as a kind of cultural inheritance.

Cultural inheritance even occurs in some bird species. Researchers designed an experiment to monitor this process, using a bird species in which parents typically teach their young to recognize the presence of predators by issuing strident "mobbing calls" (Figure 48-17). (In many other bird species, response to predators is instinctive and not learned.) Once such calls are learned, they serve as a releaser for members of a flock to join together and harass the intruder. In the experiment, birds are placed in cages; the birds in one cage are shown a stuffed owl (a natural predator) and the birds in another, a milk bottle. Birds that see the owl become enraged, sound the mobbing call, and attempt to attack the owl through the cage bars. This commotion sets off the birds in the second cage, and they direct their activity to the only "predator" they can see—the milk bottle. In fact, the second group will continue to mob milk bottles, teaching successive generations of offspring the same

Figure 48-16   MILK BOTTLES AND THE TRANSMISSION OF CULTURAL INFORMATION.
In the early 1920s, a blue tit in England discovered how to puncture foil caps on milk bottles (delivered to doorsteps in the predawn hours) and skim off the cream. The practice spread throughout England, invaded the Continent, and was ultimately picked up by a dozen other species of birds, including this great tit *(Parus major)*. The rapid transmission of this piece of cultural information probably results from the fact that young birds closely watch older birds feeding and display a readiness to imitate them. Indeed, chicks that observed a hen trained to peck only, say, green grains out of a mixture of colors will focus on the same color themselves without the slightest encouragement.

Figure 48-17   MOB ATTACKS ON PREDATORS.
Nesting adult birds regularly mob or attack potential predators to drive them away. Here a common barn owl *(Tyto alba)* is harassed. As they attack, the smaller birds' mobbing calls alert other birds of the danger and serve to teach young birds to recognize enemies.

aversion by their example. One researcher has followed this pattern of cultural transmission through six generations in the laboratory, with no indication that the milk-bottle phobia has relaxed, although as far as is known, no milk bottle has ever attacked a bird!

There are no genes in the macaques for washing sweet potatoes, nor in the birds for mobbing milk bottles. Yet through cultural inheritance, specific learned behaviors are passed on generation after generation. A vast amount of human behavior, of course, falls in the same category.

COMPLEX BEHAVIOR: NAVIGATION AND MIGRATION

Some of the most impressive examples of complex behavior are the result of a carefully choreographed interaction between programmed learning and innate behavior. Courtship, nest building, rearing of young, group hunting, and many other behaviors are built on an underlying base of reflex and instinct but with overlays of various types of learning. Nervous systems capable of carrying out complicated behaviors also allow a versatility of response that has important evolutionary and ecological consequences. With a rich repertoire of complex behavior, the animal can cope better with the unexpected, as well as exploit entirely new relationships that may themselves contribute to greater success in surviving to reproduce. Complex behaviors, therefore, can be complex adaptations—not easily predicted from genes and neurons, but just as crucial, perhaps, in perpetuating the species in nature. This adaptive significance can be seen quite clearly in cases of navigation and migration, processes by which certain animals orient themselves in time and space and travel long distances with remarkable accuracy. The mixtures of innate and learned behaviors at work in these processes permit repeated journeys of hundreds or thousands of kilometers to sites where survival is likely in the different seasons.

Let us consider for a moment what is involved in traveling from point A to point B on the Earth—say, from Maine to Florida. If a bird has an innate program for determining direction and distance, true navigation is not necessary; the bird simply has to fly the compass direction south-southwest for ten days at 125 miles per day, and it will arrive in the Everglades. But what if you blindfolded and displaced a Maine-to-Florida migrant 950 miles west, to St. Louis; now, it would clearly be impractical for the bird to reach Florida by trial and error—that is, by flying in all random directions until it finds Miami. If the bird is to reach Florida, it must somehow compensate for its displacement to St. Louis. In practice, this means that it must sense that (1) it is in the wrong spot; (2) its new location is west of the right spot; and (3) its destination can be reached by flying to the east and south. These are the components of true navigation, which requires both an internal *compass* and a *map*, and certain animals are capable of doing it. Let us look at some specific cases.

Navigation

Every summer, individual salmon leave the open ocean, swim into a particular estuary, and travel up river branches until many reach the precise streamlet where they hatched from eggs. As we saw in Chapter 38, olfactory cues guide the fish once it reaches fresh water, but how does a salmon find its home river in the first place? A honeybee leaves the hive on a foraging trip, following a long, circuitous route; when it is ready to return, it makes a direct flight home—the proverbial beeline. Each year, newspapers report the story of a dog, cat, or horse that has traveled hundreds or thousands of miles over unfamiliar territory to return to its home. How do salmon, bees, pets, and other animals find their way? As we shall see, animals have evolved a startling variety of ways of carrying out fixed direction and distance migrations, as well as true **navigation**—movement over novel routes to precise spots. Numerous sensory processes are involved in both, and include olfaction, vision, and perception of the Earth's magnetism.

Once again, Niko Tinbergen performed a classic series of experiments, this time on the navigation abilities of digger wasps on the beaches of Holland. As we saw earlier, these insects lay solitary eggs in burrows and then bring paralyzed prey to feed the larvae. Tinbergen shifted around small objects in the vicinity of the burrows and found that these small changes invariably caused a wasp returning with prey to lose her way. However, if the wasp was in her burrow when Tinbergen moved the objects, she was able to learn the new landscape as she departed and to return successfully. Here we have a case in which the nervous system is triggered to learn every time the wasp leaves the burrow; it "records" anew on each exit the sets of objects and their positions around the burrow. No hint of permanent learning or reasoning is evident, however.

Honeybees also memorize landmarks close to their hives, but when they forage farther away, they navigate through a remarkable series of "readings" involving the position of the sun. As a bee flies about in a variety of directions in search of food, it keeps track of the direction back to the hive. To do so, it uses the sun for orientation (rather like north on a compass). But the bee must compensate for the sun's changing position in the sky;

apparently, both the rate and the direction of the sun's movement are recorded. When the sun disappears from view (as on a cloudy day), the bee automatically switches to a backup system: the patterns of polarized light in the sky. These patterns bear a regular relationship to the sun's position. Finally, when both sun and sky are hidden, bees fall back on yet a third compass system, which enables them to orient to the Earth's magnetic field.

These cases of natural and experimental navigation involve *homing*, return of an individual to its nest site, feeding ground, and so on. Another example, homing pigeons, can be displaced in any direction from their roost; yet most will promptly return. Black bears and grizzly bears, after being captured, sedated, and moved in a vehicle over long distances, nevertheless may return home after release in a matter of days or weeks. It is still a mystery how the bears select a correct course over unknown country. Of course, the creatures must have some compass sense to tell direction—the Earth's magnetic field or the sun are good candidates, as has been established clearly in homing pigeons (Figure 48-18). Even so, having a compass sense that says which way is north is not enough; some map sense also is required so that random wandering is avoided and the animal can follow a reasonably direct route home. Clearly, homing ability is a valuable adaptation so that accidental displacement need not be permanent.

Migration

Nowhere is the ability of animals to navigate seen more dramatically than in the periodic journeys, or **migrations,** made by many species of insects and vertebrates. Animals as different as caribou and monarch butterflies cross huge stretches of terrain, shifting residence from one region to another in response to environmental cues, such as seasonal changes in temperature, food supplies, and rainfall.

The amazingly precise seasonal migrations of birds are perhaps the most extraordinary of these events. Tiny arctic terns migrate from Greenland and Alaska to Antarctica and back again, flying as much as 9,000 miles each way in a quest for perpetual summer. Another spectacular migrant is the greater shearwater, which nests on the tiny island of Tristan da Cunha in the south Atlantic, and then flies in March to Newfoundland, Iceland, and Greenland (Figure 48-19). Finding such precise spots on the surface of the Earth is quite a feat. In one study, of a related species, a female Manx shearwater nesting on the Atlantic coast of England was flown to Boston, tagged, and set free; less than two weeks later, she was found once again in her original nest in the British countryside, despite the fact that no Manx shearwater ever normally flies east and west across the Atlan-

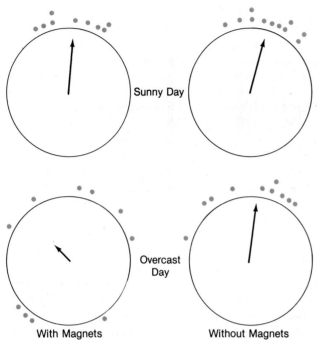

Figure 48-18   MAGNETISM AND SUNSHINE IN PIGEON HOMING.

The dots outside these circles show the direction in which individual homing pigeons flew after being released at the centers of the circles. Home is directly upward on this page. Researchers fastened tiny bar magnets to the backs of some pigeons' heads to disrupt homing based on detecting the Earth's magnetic field. On sunny days (top circles), the magnets had little effect; clearly, the birds can use the sun to orient themselves. On cloudy days, however, pigeons with magnets (lower left circle) had great difficulty in finding their way home, while those without magnets (lower right circle) could still head for home despite not having the sun as a source of compass direction; instead, they apparently sensed the Earth's magnetic field to define compass direction. The arrows represent a statistical average of the chosen directions.

tic. True navigation must have occurred in this spectacular feat of homing.

How do birds navigate with such precision? Experimenters have captured, banded, and released individuals of many species and have found that birds may use solar or stellar cues, landmarks on the Earth's surface, their magnetic sense (Chapter 38), and possibly even low-frequency sound vibrations ("infrasound"), such as those generated by wind upwelling over mountains or hills.

In one famous experiment, researchers captured starlings flying south from northern Germany to their wintering grounds in southern England and France. They trapped the birds in Holland, and then banded and re-

Figure 48-19   GREATER SHEARWATER: LONG-DISTANCE MIGRANT.

The seasonal migrations of birds, such as by this greater shearwater *(Puffinus gravis)*, are incredible examples of the relationship of programmed learning and innate behavior.

leased them in Switzerland, about 460 miles southeast of their capture point. First-year migrants in the group were later recaptured; they had flown southwest parallel to their original course, although doing so had taken them into Spain—far south of their original destination (Figure 48-20). Clearly, they had been flying a fixed, "innate" compass direction and had not been navigating. Interestingly, however, when adults in the group (with experience of the route from previous years) were released in Switzerland, they did not blindly fly southwest toward Spain; instead, they corrected their courses and navigated successfully to the proper wintering grounds, flying over a new route as they went. They could navigate. Apparently, their navigational abilities had matured so that they could return to the wintering grounds learned on their first migration (probably magnetic properties of their path or goal sites, England and France). Thus when released in the wrong place, Switzerland, they adjusted their flight paths to reach the correct goal.

Experiments with other birds have yielded similar results, implying a general pattern: during a bird's first migration, innate factors lead it to fly in a given direction (and for a given distance, as we see shortly). But as it does so, and in the process spends time at both summer and winter sites, learning takes place. Then if a storm (or an experimenter) places the bird off its course, it is able to navigate to the correct destination. Evidence suggests that a bird acquires the information it needs for navigating only during a sensitive period in early life. As with the imprinting phenomenon, once this period is past, new learning is not possible.

Use of the sun to establish compass direction requires an internal clock to compensate for the changing position of the sun as the Earth rotates. In Chapter 37, we learned about the pineal gland and suprachiasmatic nucleus of the brain, the probable sites of the biological clock in birds and mammals. But another clock operates in migrating animals. For instance, willow warblers

Figure 48-20   MIGRATORY BIRDS: LEARNING AND NAVIGATION.

Young starlings *(Sturnus vulgaris)* captured and displaced to Switzerland flew in the same fixed compass direction (southwest) on their first flight; but that carried them to Spain, not the French coast. Adult starlings captured and displaced to Switzerland returned to their normal wintering areas. Somehow past learning of the wintering area or the flight paths was used to permit this true navigation; as adults they are no longer behavioral slaves to fixed compass directions.

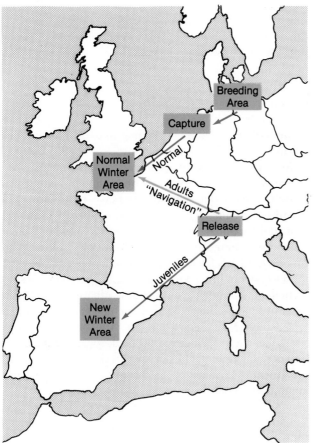

raised in cages under a constant light–dark cycle for seven years or more show intense migratory activity every spring and autumn, even though the birds have never experienced natural seasons as reflected in day lengths or temperatures (although seasonal changes in magnetism conceivably could set their clock). These birds have a *circannual clock* that turns migratory behavior on and off twice each year.

A fascinating part of this work implies that an innate behavior can have extraordinary complexity. The birds not only are active in their cages in the direction of the appropriate migratory destination, but also face in the correct direction for the period of time it would take them to fly the normal route in the wild. First, at the time of autumn migration in Germany, they face southwest for about a month, long enough to fly from Germany to Spain. Then, although still locked securely in their cages, they face southeast for three months, long enough to fly from Spain to South Africa (Figure 48-21). Here, time equals distance: thirty days of flight, at so many miles per day, would take them to their first destination; another ninety days of flight, at a certain number of miles per day, would complete the migration from Spain to southern Africa. Clearly, their innate internal program specifies both direction and distance! How this fantastic capacity is built into genes and a developing nervous system remains a marvelous mystery. This case, by the way, illustrates clearly the dangers of an anthropomorphic approach, for no one suspected that such complex migratory routes, with changes in direction, could be entirely innate and even set off by an internal circannual clock. This is why behaviorists must probe with care exactly which behaviors are innate and which are learned, even in humans.

Perhaps the most important new finding about navigation is that a number of animals can sense the lines of the Earth's magnetic field. Birds, porpoises, honeybees, and even some humans possess tiny crystals of magnetite, a type of iron oxide, near their brains (Chapter 38). It is hypothesized that as the head moves relative to the magnetic lines of force, the nervous system can detect the change so that the animal can define a compass direction (just as the eyes, seeing the constellations around Polaris, the North Star, allow the brain to discern which way is north). Moreover, different sites on the Earth's surface apparently have special magnetic properties; thus it is not too far-fetched to hypothesize that home for many higher vertebrates may be partly determined by cues involving magnetism and gravity. How such information about place is combined with learned aspects of spatial locations, as shown by foraging honeybees and other animals, is yet to be discovered. Even so, for the first time we have hints about map sense—a capacity that surely exists but has never been explained.

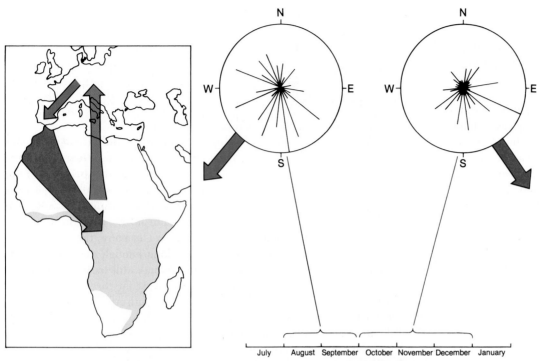

Figure 48-21   DIRECTION AND TIMING OF MIGRATIONS: INNATE MECHANISMS.
Willow warblers normally migrate from Germany to Spain and then to southern Africa. Even birds caged their entire lives will display migratory behavior each autumn, orienting their bodies first to the southwest (during August and September) and then to the southeast (during October through December). Thus, uncaged birds do not need true navigation in order to reach their destinations; they can use innate directions and times of flight to reach them. The shorter, more direct route that willow warblers follow back to Germany in the spring (blue arrow) is also innate. Interestingly, even experienced birds that have flown that shorter route in the spring still fly the longer route via Spain each autumn.

BEHAVIOR IN PERSPECTIVE

We have seen in this chapter that innate and learned behaviors occur at varying levels of complexity; these correlate roughly with the organization of an animal's nervous system. Increasingly complex programs of behavior can be built out of relatively simple and stereotyped stimulus–response events. Experienced-based components of behavior are being found in a wider and wider range of animals and conceivably reflect cellular and molecular plasticity in nervous systems. The behaviors we see in nature and the nervous system that animals possess have been powerfully influenced by natural selection as individuals with certain behaviors have proved better able to meet the demands of the environment than others. Behaviors, then, may be adaptations of paramount importance, helping to explain evolution, just as can physiology or anatomy.

Even in octopuses, cetaceans, apes, and humans—the animals with the most complex brains and learning abilities—it is likely that natural selection, "hard-wiring" of neuronal circuits, chains of stereotyped, hormone-dependent events, neuroactive peptide modulation of brain functioning, and neuronal plasticity, all help to provide the basis for many elaborate behaviors. We next need to turn to other kinds of complex behaviors, especially those involving groups of individuals interacting in social situations. But as we shall see in Chapter 49, social behavior is not limited to animals with complex nervous systems or high intelligence; instead, social behavior is a fundamental aspect of life for species throughout the animal kingdom and can be, in its own way, a major contributor to evolutionary success.

SUMMARY

1. *Ethology* is the study of animals' behavioral interactions with their environment.

2. Animal behavior can be divided into several broad categories: *innate* (inborn) behaviors include *reflexes* and *instincts* both built by genetic and developmental processes, yet occasionally modifiable by experience; and *learned* behaviors that are built on prior experiences. It is often extremely difficult to distinguish innate from learned behaviors. In general, as animal nervous and endocrine systems become more complex, so do the behavioral repertoires they regulate; nevertheless, animals with relatively simple nervous systems may show surprisingly flexible behavioral capacities.

3. The simplest behaviors seen in animals involve reflexes—automatic, involuntary responses to external stimuli. Reflexes may be building blocks for other simple behaviors and include tropisms, *taxes*, and *kineses*.

4. Instincts are stereotyped, inherited patterns of behavior that take place in response to specific environmental stimuli. The Austrian ethologist Konrad Lorenz described instinctive actions as *fixed motor patterns:* unvarying series of precise physical movements genetically and developmentally "wired" into an animal's nervous system.

5. The behavior of animals such as insects that have relatively simple nervous systems includes *closed programs,* which are innate. But some insects show varying degrees of experience-based behavior and so they, like animals with more complex brains, exhibit *open programs:* behavioral patterns that can be modified by learning.

6. An event that sparks a behavioral response is called a *sign stimulus.* Sign stimuli can set in motion *innate releasing mechanisms* that result in a particular behavior. When another member of an animal's species produces a sign stimulus, the phenomenon is known as a *releaser.* Some animals preferentially respond to a *supernormal stimulus*—a sign stimulus that displays characteristics exaggerated beyond the natural stimulus.

7. In order for a stimulus to exert its effect, an animal must be motivated—that is, physiologically receptive to the stimulus's message. If an animal is physiologically ready to execute a behavior but no stimulus is present, it may exhibit *appetitive behavior*—activity that increases the likelihood that the animal will become exposed to the stimulus. Repeated exposure to a stimulus can result in an animal's decreased responsiveness to it, a phenomenon known as *habituation.* Studies of the sea hare *Aplysia* have revealed that simple learning in invertebrates may result from habituation in which a key step is a steady decrease in the amount of neurotransmitter released by sensory neurons. Observations on a variety of animals demonstrate plasticity of neurons and neuronal circuits, and action of diverse neuroactive peptides, that may play roles in behavior.

8. *Programmed learning,* seen in bees, young birds, and other animals, follows a rigid pattern, much as innate behaviors do; thus it is closely tied to the genetic and developmental programs that build the nervous system. Yet such programmed learning is often subtly interwoven with other kinds of learning to yield complex behaviors.

9. Another form of learning is *imprinting,* which Lorenz documented in his research with baby geese. Species subject to imprinting pass through a *sensitive period*—a brief period after birth when the appropriate stimulus may be recorded in the animal's nervous system and attachments to it formed.

10. *Latent learning* is slightly more complex and involves learning at one time, but behaving appropriately at later times. An example is the process by which perching birds memorize the elaborate songs of their species and learn, perhaps months later, to reproduce the complicated song patterns through practice.

11. Complex learning mechanisms include *trial-and-error learning,* *classical* or *Pavlovian conditioning (associative learning),* and *insight learning.* Trial-and-error learning is often called *feedback learning* because it relies on *reinforcement* of accidental behavior. Learning by feedback is the basis for *operant conditioning*—the method of training in which accidental behavior that approximates an action desired by the experimenter is rewarded. Insight learning, or *reasoning,* seen primarily in animals such as higher primates with highly complex nervous systems, is the ability to use abstractions in novel ways, combinations, or situations.

12. Two examples of complex learning are *navigation* and *migration.* Modes of navigation in which learning is involved include recognition of visual landmarks and use of a solar, stellar, or magnetic compass to travel novel routes to targets. Researchers have discovered extremely complex innate migratory behavior in birds, and it can account for yearly migrations over great distances. Nevertheless, true navigation can occur and must involve a map sense, the basis of which is still being defined.

KEY TERMS

appetitive behavior	imprinting	migration	releaser
associative learning	innate	navigation	sensitive period
classical (Pavlovian) conditioning	innate releasing mechanism	open program	sign stimulus
closed program	insight learning	operant conditioning	supernormal stimulus
ethology	instinct	programmed learning	taxis
feedback learning	kinesis	reasoning	trial-and-error learning
fixed motor pattern	latent learning	reflex	
habituation	learning	reinforcement	

QUESTIONS

1. What is a reflex? Taxis? Kinesis? Give a specific example of each.

2. What is meant by instinct? Can an instinct ever be modified? What is the difference between an open and a closed program? Do insects have open (partially open?) programs? Do humans exhibit any instinctive behaviors? Give some examples. What is some evidence that human facial expressions are instinctive, rather than learned?

3. Define the following terms, and give at least one example of each.

   a. sign stimulus
   b. releaser
   c. innate releasing mechanism
   d. supernormal stimulus
   e. appetitive behavior
   f. habituation

4. In ring doves, what external stimulus triggers production of crop milk? What experiments demonstrated this? What stimulus releases the mother's crop-milk feeding behavior? How was this discovered?

5. How is habituation like learning?

6. Give at least one example of culturally transmitted behavior among nonhuman animals. What human behaviors are culturally transmitted?

7. What is imprinting? What is meant by sensitive period? Give some examples.

8. What is latent learning? Is it possible to learn to perform a task without practicing it? Male white-crowned sparrows start singing only after they produce the hormone testosterone. How do we know their song is learned, rather than innate? Do the females learn the song? How do we know?

9. Discuss each of the following types of complex learning.

   a. trial-and-error, or feedback, learning
   b. associative learning
   c. insight learning, or reasoning

Have you ever experienced any of these types of learning? Give examples.

10. Many animals migrate hundreds or thousands of kilometers twice each year. How do they know where to go? Is migration completely instinctive, or are navigational skills required? Give some examples. What does it mean to say "time of flight equals distance" in migratory birds? Does homing require navigation? How does the digger wasp find her burrow?

ESSAY QUESTIONS

1. What role may elaborate courtship displays take in speciation?

2. In the springtime, male robins often attack their own reflection in a window. Is this bizarre action related to a "natural" behavior? Design an experiment to determine what feature of the reflection provokes the attack.

SUGGESTED READINGS

ALCOCK, J. *Animal Behavior, an Evolutionary Approach.* 2d ed. Sunderland, Mass.: Sinauer, 1979.

An excellent general treatment of animal behavior.

GWINNER, E. "Internal Rhythms in Bird Migration." *Scientific American,* April 1986, pp. 84–92.

These are the brilliant observations on circannual clocks and the innate nature of both direction and distance.

LORENZ, K. *King Solomon's Ring.* New York: Crowell, 1952.

A delightful introduction to ethology.

MARLER, P., and W. J. HAMILTON. *Mechanisms of Animal Behavior.* New York: Wiley, 1966.

An excellent book by the dean of American behaviorists. Included is a fine discussion of bird songs.

TINBERGEN, N. *The Animal and Its World: Explorations of an Ethologist.* Cambridge, Mass.: Harvard University Press, 1973.

One of the founders of the discipline of ethology writes about imprinting, aggression, and much more.

49
SOCIAL BEHAVIOR

The cohesiveness and co-ordination of animal societies are often their most striking feature. It is easy to lose sight of the individual in such groups, thus failing to recognize that the forces which shape its behavior are the same as those for more solitary species. Natural selection operates upon individuals and the responses of a social animal to the other members of the group will evolve to its own best advantage.

Aubrey Manning,
An Introduction to Animal Behavior (1979)

High-density social interactions: Plovers (*Pluviales* species) at rest between feedings.

From birth to death, nearly everything a person does requires interaction with other members of the human species. If this were not so, we would each have to produce our own food, clothing, and shelter; face emergencies with no help; and perhaps defend ourselves single-handedly against human attackers or other predators. Clearly, long-term survival would be a chancy proposition if each person had to go it alone.

In reality, virtually no human faces such a bleak scenario. Like a number of other animal species, we humans organize ourselves into **societies**: groups of animals of the same species in which the members communicate and interact in cooperative ways. This cooperation makes social behavior adaptive; being social tends to promote the survival of individuals, and with survival, the likelihood that they will reproduce successfully. Just as a small fish is less likely to be eaten if it swims in a large school, a prairie dog in a colony, a monkey in a troop, or a person in a community derives survival benefit by its social behavior. The individual animal's actions in social situations are built from the combinations of innate and learned behaviors we considered in Chapter 48. Hence, social behavior and its underlying innate and learned components are subject to natural selection, just as surely as are morphological and physiological characteristics.

The study of the biological basis of social behavior is a subfield of biology called **sociobiology.** This term was coined by the eminent Harvard evolutionary biologist Edward O. Wilson, who studies the extraordinary cooperative societies of certain insects and seeks to explain how their complex behavior may have evolved. In fact, sociobiology is, in another sense, the study of behavior from an evolutionary viewpoint. The question is whether and how an individual's social behavior affects the propagation of its genes and those of its closest relatives. To get at this question, we examine the main kinds and characteristics of social behavior, beginning with a survey of the key ecological factors that shape a species' social structure. Next we consider communication, an obvious cornerstone on which social interactions and systems of differing complexity are built. Then a discussion of mating behavior and social adaptations of reproducing individuals will lead us to *altruism*—a phenomenon of special interest to sociobiologists because it seems at first glance to be contradictory to our understanding of natural selection. By altruism, biologists mean self-sacrificing behavior in which individuals appear to do things that benefit other members of their species more than it benefits themselves. We will see that altruism relates to the evolutionary concept of **inclusive fitness.** Thus an individual's total fitness includes not only its likelihood of passing genes directly to its own offspring, but also its likelihood of contributing to the successful passage of identical genes in related animals. In fact, inclusive fit-

ness is really a novel extension of original Darwinian ideas; it offers an interpretive framework that allows us to understand altruism and otherwise puzzling social behaviors in evolutionary terms. This will be especially evident at the end of this chapter, where we explore the characteristics of sociality in insects, mammals, and other vertebrates.

BEHAVIOR AND ECOLOGY

Behavior, like most other aspects of an animal's biology, can be shaped by natural selection. In order for a social system to evolve in a species, the social group must give its individual members competitive advantage so that they are more likely to reproduce successfully. Table 49-1 lists some possible major advantages and disadvantages of sociality. But social behavior is just one aspect of an individual's life—ecological factors such as niche and reproductive strategy also affect behavior in major ways.

Recall from Chapter 46 the concept of *r*-selected and *K*-selected reproductive strategies. Members of *r*-selected species tend to produce many offspring as they exploit an unsaturated but risky habitat. Most young of *r*-selected species die before they reproduce. Many bony fishes fit this category. Although male and female salmon, for example, have intricate migratory abilities, which allow them to home to the stream where they hatched, as well as strong instinctive mating and nest-building behaviors, they still do not protect their thousands of eggs or embryos; thus most of the young are eaten by other animals or die for other reasons, most before they grow to reproductive age. The sorts of social

Table 49-1   **INDIVIDUALS IN SOCIETIES: POSSIBLE PROS AND CONS**

Pros

Better detection of, repulsion of, or escape from predators

Better defense of limited resources

More efficient foraging

Better care of offspring through communal feeding and protection

Cons

More competition with members of the group for space, food, water, and other resources

Higher risk that group members will exploit parental care

Higher risk that group members will kill an individual's progeny

Higher risk that diseases or parasites will spread in the group

Figure 49-1   MOTHER GORILLA WITH HER YOUNG.
Gorillas (*Gorilla gorilla beringei*), like all *K*-selected
animals, invest much energy in the raising, care, and
protection of their offspring. The young mature slowly and
are few in number.

behavior built around parent–offspring interaction tend
not to be seen in *r*-selected species.

In contrast, organisms belonging to *K*-selected species
tend to produce a small number of offspring that mature
slowly and have a good likelihood of surviving to reproduce. Among animals in this category, specific parental
behaviors have evolved so that adults interact with and
care for their offspring; for instance, the mother gorilla,
as seen in Figure 49-1, shows the kind of protective behavior typical of long-lived, social animals. Not surprisingly, social behavior is often characteristic of such *K*-
selected species.

Sociality can be an advantage when group membership offers the strength of numbers. As we saw in Chapter 48, fish in schools, birds in flocks, and grazing animals in herds are usually less vulnerable to predators
than the same individuals would be if isolated. Thus, the
extra eyes, ears, and noses of fellow pronghorn antelope
make the individual less likely to be caught unawares by
a mountain lion or wolves. What works for prey works
for predators, too. Wolves and hyenas hunt in groups,
thereby increasing their chances of obtaining food, even
though they must share the food collectively caught.
Coyotes in the Canadian Rocky Mountains hunt in packs
when their prey is large and cannot be killed by individuals. Farther south, coyotes hunt individually when

their prey is small and provides little meat per kill.
When coyotes hunt large prey at one time of the year
and small prey at other times, they still follow the same
pattern—they hunt large prey in packs and small prey
as individuals. This example challenges the notion that
social behavior is a simple expression of inherited tendencies.

Group social behavior related to surviving to reproduce can be seen in *r*-selected species (including many
fishes), even though they may not exhibit parental care.
K-selected species may show group social behaviors both
with respect to survival and parental care. Thus social
behaviors of varying complexity have evolved, and these
behaviors affect primarily the individual and its survival
to reproductive age or its parent–offspring relationships.
Both survival and parental care relate to inclusive fitness, since both can contribute toward higher likelihood
that an individual's genes will be passed on to subsequent generations.

Although many animals live in groups, social groups
do not automatically mean social interaction. Individuals
may gather in one location because of environmental factors, such as concentrated food sources or aspects of the
species' niche. Caterpillars, from a hatch, all feeding on
a single leaf do not actively cooperate; their close placement is simply the consequence of where a female laid
her eggs. Clearly, social behavior depends on interaction, and to interact, animals must communicate by
some means, whether by chemical signals, visual cues,
or even spoken language. Let us consider communication in more detail, since it is a cornerstone of sociality.

COMMUNICATION

Communication—the transmission of information
from one organism to another—is itself a kind of behavior, one that alters the behavior of another individual or
individuals. Hence, communication must be mediated
by the recipients' sensory organs as well as by the communicators' signaling devices. At its most fundamental
level, communication consists of a sign stimulus that can
release a behavior in another individual (Chapter 48).
The mating odyssey of a male silk moth, for example, is
triggered by a few molecules of female sex pheromone.
The basic building blocks of innate behaviors plus more
intricate learned behaviors have permitted the evolution
of a vast repertoire of socially useful acoustic, visual,
chemical, and tactile signals.

Consider just a few of the astonishingly large number
of ways by which animals communicate with members of
their own or other species. Bird alarm calls warn of the
approaching hawk. In fact, based on convergent evolution, birds of many different species emit alarm calls that

AT THE FRONTIERS: THE BIOLOGY OF HUMAN LANGUAGE

Many of the stunning accomplishments of human culture depend on language—our ability to transfer information about things and events distant in both space and time. Although we tend to think of language as an achievement of our superior human intellect—one that sets us apart from all other animals—our linguistic capability actually depends in good part on innately programmed behavior.

Biologists are virtually certain that our ability to acquire language is innate. For example, all the thousands of human languages draw from the same set of forty consonant sounds (standard English uses about two dozen), and every normal human baby is able to distinguish all forty regardless of whether he or she has heard them before. As a person matures and gains experience in a particular language, the ability to distinguish other consonants fades quickly and may be gone by about two years of age. This points up a primary feature of language acquisition. As with the elaborate songs of perching birds (Chapter 48), learning language involves at least one period in which an individual internalizes the basic elements of his or her native tongue. Thus people who have not heard certain consonant sounds as infants can only rarely distinguish or produce them as adults. It is this phenomenon that leads to the commonplace observation that native Japanese speakers commonly substitute *l* for the unfamiliar English *r*, while speakers of German and French typically replace the English *th* with sounds resembling the English *d* or *z*. But most Americans never master the throaty French *r* and are perplexed even more by the "strange" consonant sounds of Asian and African languages.

The internal guidance of language may go deeper still. The eminent linguist Noam Chomsky maintains that all languages share a "deep structure" that is part of every person's innate language-acquisition program. Consider, for example, how effortlessly even the dullest, most inattentive child acquires a structured vocabulary of thousands of words, while many of the brightest college students, as adults, may have enormous difficulty learning even a limited version of a foreign language.

Finally, many researchers believe that the innate roots of language can be seen in the behavior known as "babbling," which all human babies display, even if they are totally deaf. During babbling, infants who can hear (again, much like young songbirds) learn to control their vocal apparatus to make the sounds of their native language and then to talk. Even if such a child later becomes deaf, he or she is usually able to reproduce sounds learned previously with amazing fidelity—apparently, the motor programs have become "hard-wired," stored permanently in the brain's circuitry.

have very similar sounds (Figure 42-16). Such calls are difficult to locate directionally, and since the calls are so similar to each other, birds of a number of species can benefit from a single bird's alarm communication. Not all birds emit alarm calls; yet ones that do not may still be able to respond to alarms from other creatures, even very different ones. For instance, hummingbirds in the mountains of California make no alarm calls of their own; yet they respond immediately when a chipmunk or ground squirrel chatters an alarm. Birds and small mammals are not the only animals that call or sing. Songs emitted by whales or cicadas, like those of birds, often play a role in mating behavior. The lion's roar, the snake's hiss, the rattler's and peacock's tail rattle, and the human's language deliver other diverse messages.

Each such sound bears information content for friend or foe. That information may be very straightforward and trigger a programmed innate response, a simple learned behavior, or a much more complex pattern of learned behaviors. Alternatively, the information content may be increasingly complex and reach the nuance and subtlety of human language—say, in a pun or a Shakespearean sonnet. Sometimes the information content can be conveyed visually; for instance, the elaborate movements of male and female ducks that are antecedents to reproductive activity. Visual displays such as baring the fangs or threatening attack are important adaptations for defense in a wide range of animal species. Another communication channel is chemical; many animal species produce pheromones or other odor signals in specialized glands or excrete them in waste products, such as feces, urine, or perspiration. Thus black bears, coyotes, and rabbits repeatedly mark the limits of their territories with urine containing pheromones, and many female insects are renowned for producing sexual attractants that can bring males running or flying from great distances. Finally, direct physical contact may serve as a tactile stimulus.

Communication stimuli often have very specific meanings. For example, small birds sound their so-called hawk alarm call only when the caller sees or hears a hawk; the call triggers specific hawk-avoidance behavior. A mother black bear sends her cubs up a tree with a particular open-mouthed tooth click. Small male tropical tree frogs of the species *Eleutherodactylus coqui* repeatedly emit the cry "coqui!" "coqui!" Careful analysis of the acoustic sense of the frogs reveals that the auditory system of females responds primarily to the "-qui" frequencies, while the auditory system of males has almost no sensitivity to "-qui," but hears the "co-" quite well. The "co-" is a territorial warning call between males, while the "-qui" sound attracts females to males. Since the females hear "-qui" but not "co-," their brains do not need to interpret and then ignore useless information (the "co-" warning sound). Similarly, male brains only need respond to the "co-" warning signals. In a sense, the auditory systems of males and females are filters that only report significant information to the brain. This means that the brain can be simple since it does not have to sort out significant from irrelevant information. (Some peripheral filtering also occurs in a mammal's retina and inner ear. Even so, a fuller range of sensory information is reported so that complex brain circuits must decide what matters and what does not.) Research on *Eleutherodactylus* indicates that specific properties of sensory systems and brain circuits are critical components in the specificity of communication. These results cause behaviorists to wonder what other specific communication links between land vertebrates are actually "hard-wired" in the brain. We learned about such hard-wiring of circuits controlling behavior in the crayfish es-

cape response (Chapter 36); now scientists need to search for such circuits in vertebrates.

While certain behaviors may have evolved strictly as communication devices, many others had preexisting biological functions but were modified during evolution to serve a new, communicative purpose. The modification of a functional behavior to include a communicative role is called **ritualization.** For example, many of the courtship patterns of birds, fish, and mammals contain elements of feeding, cleaning, or nurturing activities. In general, mating rituals that incorporate such behaviors are largely instinctive and serve as very effective isolating mechanisms (Chapter 42) because the sequence of behaviors—performed first by the female, then by the male, and so on—is so very precise. There is little chance that an individual from another species could successfully complete the full sequence, so the reproductive isolation results. A classic example, described more than seventy years ago by the biologist Julian Huxley, is the dancelike pair-bonding ceremony in the courtship of a male and female great crested grebe (Figure 49-2). Standing chest-deep in shallow water, the

Figure 49-2   GREAT CRESTED GREBES: PREPARING TO MATE.
This series of elaborate movements is the courtship behavior of male and female great crested grebes. The numbers refer to the order in which the actions of this lengthy sequence occur.

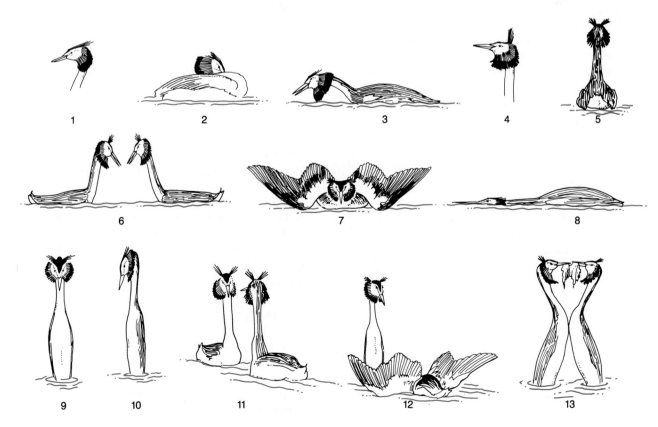

grebes engage in a complex series of movements. One is a ritual "head shaking"—back-and-forth movements of the head, with each bird's bill pointed away from its partner. Like many ritualized actions, this one seems designed to disarm normal aggressive responses by displaying distinctly nonthreatening behavior. Once each bird's instinctive hostility is appeased, the male and the female ritually offer one another beakfuls of waterweeds normally used in nest building. Only when the full series of movements is complete does actual mating take place.

Insect Communication: The "Waggle Dance" of the Honeybee

Like vertebrates, which use visual displays, sounds, and odors to convey messages, many insects communicate through sounds or through pheromones and other chemical substances released as trail markers, attractants, and so on. Honeybees employ numerous chemical signs, but they also communicate extraordinarily complex information using tactile stimuli in an exceptional dance language. Their "waggle dance" is used to describe food sources located between about 80 and 600 meters from the hive.

Recall from Chapter 48 that the studies of Karl von Frisch revealed that a foraging bee is able to memorize details about a food source, including its precise location, geometry, and odor. For this ability to be useful in the social context, the bee must be able to communicate the details to others in the hive. Von Frisch and his coworkers showed that a returning forager communicates the location of particular flowers, for example, by waggling her abdomen in a specific, patterned series of dancelike movements, all done in the dark of the hive.

Positioning herself on the vertical surface of a honeycomb inside the hive, the bee executes the dance to indicate the distance, direction, and richness of the food source. First, distance from the hive to the food source is specified by the length of a straight-line run, known as the "waggle run," during which the bee waves her abdomen from side to side. After executing a semicircular turn, she makes another waggle run, turns in the opposite direction, and so on (Figure 49-3).

Second, the direction of the food source from the hive is indicated by the angle of the waggle run. This astonished von Frisch when he first observed it: the insect translates her visual observation of the direction of the sun into a gravity-guided orientation of her dance on the comb—that is, the straight vertical direction up the wall of the comb represents the direction of the sun from the hive at that time. If the food source is toward the sun, she dances straight up; if it is away from the sun, she dances straight down; if it lies at an angle 30 degrees to the left of the sun, the straight-line run of the waggle dance points 30 degrees to the left of the vertical; and so on.

Recruits that attend the bee's dance feel her movements and learn not only the location of the food but also its odor (which has been absorbed by the waxy hairs on the dancer's body or is present in regurgitated food) and its quality. In addition, the frequency of dancing and the amplitude and rate of abdominal waggling communicate whether a rich or a weak food source has been found. Sometimes the other bees emit a 0.25 second, 300 Hz sound and the dancer stops and offers samples. Figure 49-4 shows variations in the waggle dance typical of several bee species.

One bee's waggle dance imparts substantial information to cohorts in the hive. The other bees use this information but do not depend on it for feeding. Once a bee has visited a feeding site—perhaps originally as a result

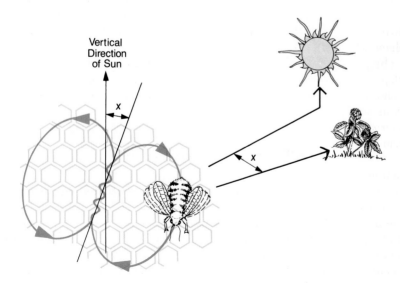

Vertical Direction of Sun

Figure 49-3   THE WAGGLE DANCE OF THE HONEYBEE.
Voiceless organisms often develop elaborate methods of communication: the honeybee's waggle dance is a prime example. As the insect performs the straight-line run, her abdomen waggles; the number of waggles and the orientation of the straight-line run contain information about the distance and direction to a food source from the hive. After completing one waggle run, the bee makes a semicircular turn and does more waggle runs, to ensure that the feeding information is successfully conveyed to and received by the other forager bees in her hive. On a vertically situated honeycomb, straight up is toward the sun; the angle x is the angle between the sun and the food source.

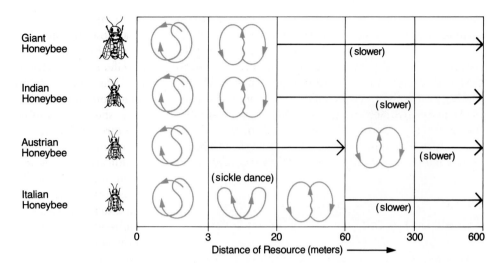

Figure 49-4   DIFFERENT DANCES, DIFFERENT BEES. Honeybee species and subspecies perform waggle dances with different "dialects," moving in varied patterns for different lengths of time. For sources up to 3 meters away, all species use the same dance. For longer distances, the waggle or the novel "sickle" is used.

Giant Honeybee

Indian Honeybee

Austrian Honeybee

Italian Honeybee

Distance of Resource (meters)

of a waggle-dance instruction—she learns a locale map of the area. If captured and released at some other site in the vicinity, the bee can fly directly to the food site without returning to the hive and starting from that point. James Gould of Princeton University has shown that despite its tiny brain, a honeybee can navigate along novel routes, using landmark cues of the locale map to reach its goal—a rich source of sugar. This example alone refutes the old notion that insect brains function solely on innate programs; learning is surely included in their repertoire.

The efficiency and well-being of the entire bee colony depend on such highly developed communication systems. Variations on the dance are used to communicate the location of pollen sources, of water (needed to cool the hive in hot weather), and even of possible new nest sites at the time of swarming.

Primate Communication

Biologists refer to the waggle dance of the honeybee as a language because it has the basic language elements identified by linguists: it is symbolic and refers to objects or events distant in both time and space, and it depends on arbitrary conventions—such as using the sun and the vertical direction as reference points—to convey meaning. Even so, a waggle dance does not begin to approach the amazing complexity and flexibility of human language.

The large brain and complex behavior of nonhuman primates have led a few researchers to attempt to teach various primates to use human language. In recent years, chimpanzees, gorillas, and orangutans have been the subjects of intensive language training, including a major test of linguistic ability—the capacity to use words or symbols in new combinations. In particular, a chimp

named Washoe became a famous user of American Sign Language (Ameslan) hand gestures in place of spoken words. Washoe eventually learned 130 signs and used some word pairs and triplets in meaningful ways; for instance, she invented the sign pairs "water–bird" to describe a swan and "candy–drink" for watermelon. In spite of these feats of creativity, Washoe never achieved anything like the level of language use that a human child with a similar vocabulary displays.

Although the experiments with Washoe and with other apes were initially greeted with enthusiasm because they seemed to show true language capabilities in our primate relatives, they have provoked great controversy. A primary objection is that many of the experiments show only the effects of operant conditioning (Chapter 48), not unlike teaching a pigeon to peck a particular series of typewriter keys in order to obtain food. A second criticism has centered on the researchers themselves, specifically that unconscious prompting of the "correct" responses and overly generous interpretations of word series have contaminated analysis of the results in some cases. At the moment, there is little reason to attribute true language capabilities to animals other than humans. Nevertheless, sound communication is very important in intraspecies primate social behavior, as it is in the social behavior of many other vertebrates.

MATING BEHAVIOR

A major component of inclusive fitness for individuals and species is mating behavior. While the male and female may spend very little actual time together for transfer of sperm and fertilization of eggs, some behavioral steps are likely to precede the mating itself. At one end of the spectrum are cases in which males and females

(a)

(b)

(c)

(d)

inhabit the same area—say, sea stars or sea urchins on a reef—and the only social communication is release by spawning females of a substance that triggers the males to shed sperm. Other cases of external fertilization—by salmon or frogs, for example—include behaviors that bring the individuals close together so that there is greater likelihood of sperm meeting egg. Species with internal fertilization—whether earthworms, dragonflies, salamanders, or warm-blooded vertebrates—usually display more complex mating rituals and means of transferring sperm or sperm packets to the female. We saw in Chapter 37 the roles of the nervous and endocrine systems in lizard reproduction; a chain of largely innate behaviors culminates in internal fertilization. No real social groups are involved in this reproductive behavior, although such factors as territoriality and aggression among competing males may be components of a given species' mating behavior.

Mating behavior tends to be simpler in *monogamous* species than in *polygamous* ones. Among monogamous animals, single males and females may remain together during a full breeding season or longer, while polygamy involves multiple matings with different individuals. Monogamous species tend to have roughly equal numbers of males and females, and there may be little direct competition among males. The two sexes tend to look alike, and males rarely evolve special markings—antlers, showy tail feathers, and so on.

Things are quite different among polygamous animals, with their multiple mates. Mating can involve one male and many females (**polygyny**) or one female and many males (**polyandry**). With so many mates on the scene, competition becomes commonplace, and members of the two sexes tend to look quite distinctive—that is, to display **sexual dimorphism** (Figure 49-5). Dimorphic males may have large canine teeth, huge antlers, luxuriant manes, or other recognition devices, while females are often smaller, drabber, and thus better camouflaged. There tends to be less genetic variability in polygamous groups than in monogamous ones, since during any one

Figure 49-5   SEXUAL DIMORPHISM.
(a) An early autumn snow falls on this bull elk and three females (*Cervus elaphus canadensis*). This male's antlers are of modest size in comparison with ones on older, larger bulls. A few months later in the year the antlers drop off; a new set is grown the following spring. (b) A male fiddler crab (*Uca pugnax*) threatens the photographer with its greatly enlarged pincer claw. (c) A stag beetle (*Lucanidae* species) in a New Guinea rain forest displays its large pincers that open laterally. (d) A rhinoceros beetle (*Oryctes* species) uses this dorsal protrudence for display, not as a weapon.

breeding season, in the case of polygyny, one male may father all the young in the herd. But some advantages must underlie the female's willingness to join a harem, and these include better protection, social benefits of rearing young (for example, other females nurse a lioness's young), and a greater chance of gaining food (by group hunting) and other resources.

For both monogamous and polygamous species, territoriality and aggression are related to reproductive success. Let us consider these subjects in more detail.

Territoriality and Aggression

Territoriality is the defense of a particular feeding or breeding site by a male, generally against other males of the same species. Animals ranging from frogs to lizards to monkeys to some insects establish territories of various types. Perhaps the most familiar examples of territorial animals are songbird species in which males claim areas for mating or nest building and advertise their boundaries by repetitive singing of the species-specific song (Chapter 48). A male bird may emit the song thousands of times, whether or not other males are present to hear or to elicit the warnings. Furthermore, what functions as a warning to other males may be attractive to females, so that they venture into the territory.

The size of an animal's territory depends on many fac-

tors, such as the species' mobility, the identity of the competitors within the territory (of the same or other species), and the function of the territory—nesting or foraging. The territory of male songbirds is usually large enough to support one female and her nestlings. A single male red-winged blackbird, however, can have a territory sufficient for a harem of ten breeding females, and a male vicuña—a relative of the camel that lives in the high Andes—can guard a rich area of grassland that supports up to eighteen grazing females. Conversely, an animal may have only a symbolic territory—a *display court:* white-bearded manakins, for example, have display courts consisting of nothing more than small, bare areas cleared of leaves and debris around saplings. As Figure 49-6 shows, large numbers of such individual territories can occur in close proximity to each other in the jungle. Each male emits explosive sounds as he jumps from perch to perch in the sapling or to the display court and back to a perch. The resulting din from numerous birds displaying simultaneously carries 30 meters or more through the tangled vines and serves to attract females far out of visual range. A female ultimately mates with one of the males in his display court, but then leaves to nest elsewhere. For unknown reasons, certain display courts attract a disproportionate percentage of females (for instance, 75 percent of 438 observed matings in one study occurred in eight adjacent display courts). When researchers removed the most actively

Figure 49-6   MATING TERRITORIES OF WHITE-BEARDED MANAKINS.
The cleared patches beneath the trees are the display courts, or mating territories, of the male manakins. Two females watch as these males jump repeatedly from their perches to the ground and back, all the while emitting loud explosive sounds. Eventually, each female will select a male and enter his territory to mate.

mating male, a lower ranked one replaced him in the favored display court and was soon inseminating numerous females; thus for some reason, the court rather than the individual male that occupies it seems to attract female manakins.

Advertising a territory with scent, sound, or some other signal can have multiple effects. It may warn competitors away, and it may attract females of the species. Dividing the environment into territories can thus affect reproductive processes and influence their success, as well as improve the individual's chances of surviving. Densely spaced territories can overtax a food supply, resulting, for example, in more foxes in a forest–meadow community than birds and mice for them to hunt. The food shortage may affect only adults or both adults and young offspring that must be fed, but eventually the fox density decreases again. Dense territorial spacing is not always a problem, however: penguins and gulls often nest in huge colonies and aggressively defend each relatively small nest-site territory. Those same combative neighbors, however, fly or swim together to gather fish without revealing a hint of the aggression that they display at the nest sites.

What are the consequences for individuals that do not successfully establish a territory? In some species, males that fail to establish a territory do not mature sexually. In others, males mature but may never succeed in reproducing successfully unless a territory becomes available. These may be adaptations that actually increase inclusive fitness, since the individual's lifetime fitness may be greater if it only expends energy in aggression, territoriality, and other aspects of reproduction when adequate resources are available. Thus only in years when suitable territories become vacant do such males mate; then they are more likely to pass on their genes successfully to surviving offspring.

Some species have not evolved the strategy of territoriality. The individuals of such species may interact aggressively as part of the mate-selection process. Still others display aggression in day-to-day adult life as part of the ordering process of social groups. When behaviorists study aggression they are not concerned with the "attack" of a herbivorous deer on an oak sapling or of a coyote on a rabbit. Rather, **aggression** includes overt fighting, displaying, posturing, and so forth—sometimes between animals of different species but more often between members of the same species.

The behaviors that animals use to mark off territories— singing, growling, marking with odors, and the like— may constitute one aspect of aggression. Another frequently involves levels of fighting ranging from **threat displays**—including the baring of fangs or hair rising on the back of the neck (Figure 49-7)—to the full-blown battle of bull-moose antler clashing. Fortunately for the animals involved, most such fighting is carefully stylized,

Figure 49-7   THREAT DISPLAYS: SUBSTITUTES FOR BATTLE.
Aggressive gestures deter intruders. A cat's hiss, a dog's growl, and the bared canine teeth of this olive baboon (*Papio anubis*) are common threat displays. These behaviors are performed as warnings and often are not followed by actual combat.

probably by instinctive programming, so that they do not inflict lethal damage on each other. Threat displays can occur at various times: mainly during breeding season, as among bull moose; or throughout much of the year, as when a zebra stallion continually guards his harem of zebra mares against repeated intrusions of single, young males. And threat displays can play important roles in the establishment and maintenance of social hierarchies, as used by the dominant male chimpanzee in a large troop. Threat displays that do not culminate in violent physical contact confer a great benefit: the warning is made clear without the dangers of real combat. Ultimately, competition is the apparent basis of all such threatening interactions; thus aggressive behaviors can be interpreted as adaptations that render the aggressive individual more likely to reproduce successfully. Once again, this increases inclusive fitness. Regardless of the type of aggression, most animals (humans are a notable exception) do not kill or injure others for sport; to do so would be a dangerous game in which the attacker would be likely to suffer wounds in return, and thus decrease his or her inclusive fitness.

ALTRUISTIC BEHAVIOR

An animal's full repertoire of reproductive behavior includes aggression and territoriality, as we have seen, as well as the species' social structure. Frequently, a number of genetically related individuals remain together in a social group. Sister lionesses, for example, often stay with their mother and perhaps aunts and cousins in a pride. A pair or a few related male lions (brothers, half-brothers, or cousins) usually live with such female groups; periodically, these males are displaced by a new set of younger, stronger males. Zebra colts remain in the herd with their mothers and fathers until reaching sexual maturity. Then, perhaps to decrease the likelihood of being inseminated by the father (which would lead to inbreeding), the young female is separated and taken from the herd by an aggressive young stallion that is establishing his own herd. Within such prides and herds, all sorts of behaviors of the sorts described in this and Chapter 48 may take place between members. In birds, a male and female may pair for a single set of copulations, for one breeding season, or for life. Then both bird parents usually feed the young and participate in their early maturation. But generalizations are difficult because the amount of parental care of offspring varies greatly among all vertebrate classes, including birds.

These examples make it clear that there are endless variations in the make-up of societies among the thousands of animal species that live in social groups. In a surprising proportion of these diverse societies, individuals display so-called altruistic acts toward each other, and these have an interesting impact on inclusive fitness.

Altruism is a special type of behavior that may influence reproductive success. It is defined as self-sacrifice by one member of an animal species that brings benefit to others of the species. A prime example is the sparrow that emits a shrill cry that serves as a signal to the rest of its flock that a hawk is swooping down; that cry is likely to focus the hawk's attention on the sparrow itself. How can such potentially self-destructive behavior be explained? At least some altruistic acts are reputed to stem from so-called selfish genes. Parents that work themselves ragged to feed insatiable offspring or go without food as long as a predator is near are likely carrying out genetically programmed behavior that increases the probability that the parental genes in those offspring will not be destroyed and will be passed on to yet another generation. These animals are not "aware," in any sense, of their genes or of the consequences of their actions. Innate, instinctive responses to predators may seem very "purposive" to the human observer, but in fact they can be explained as behavioral programs triggered by sights, sounds, odors, and so on. Altruistic behaviors toward relatives contribute to inclusive fitness and increase the statistical chances of having genes similar to one's own being passed on to subsequent generations.

Another phenomenon that probably contributes to the evolution of altruistic behavior is *reciprocal altruism,* so named by a leading behaviorist, R. L. Trivers. His theory is that while altruistic acts have immediate reproductive costs for the altruist, by definition, the recipients may inadvertently "repay" him or her at a later date. Thus the alarm call that a bird emits one day may be a temporary source of danger for the caller, but the innumerable alarm warnings that other flock members give when other dangers approach repay the original act of altruism with multiple dividends. Reciprocal altruism thereby raises the inclusive fitness of each participating member of the social group.

Kin Selection

Biologists have found it more difficult to account for some cases of altruism; among these are behaviors lumped under the heading of **kin selection.** The theoretical underpinning of kin selection is a notion that we have discussed briefly—inclusive fitness. Now let us consider the concept in more detail. The scorecard of evolutionary success registers only how many copies of a gene penetrate into succeeding generations, regardless of which individual transports them there. As we saw in Chapter 41, natural selection operates on individuals, but over time, it increases the frequencies of some genes over others. While this has traditionally been interpreted to mean that individuals with the most progeny are favored (most "fit"), in the 1960s the geneticist W. D. Hamilton proposed that the concept of fitness can be expanded to include the total representation of one's genes in future generations, whether they are contributed by children or by *other kin.*

Behaviorists have noted the clearest-cut cases of kin selection in certain social species of hymenopteran insects—ants, wasps, and bees. These insect societies contain a single fertile queen and her sterile daughters. The queen often becomes huge and serves as an immobile, living egg factory that female workers tend. Males leave the colony shortly after they emerge from pupation. Unlike males, which develop parthenogenetically from unfertilized eggs and are thus haploid, females develop from fertilized eggs and are diploid. Each time a queen produces an egg, meiosis distributes 50 percent of her genes into that egg (the remainder are lost in polar bodies). When an egg is fertilized and becomes a diploid female, that individual possesses 50 percent of the queen's genes. But the daughters themselves share 75 percent of their genes with one another (because the father was haploid, each daughter receives 100 percent of his genes, as well as 50 percent of her mother's genes;

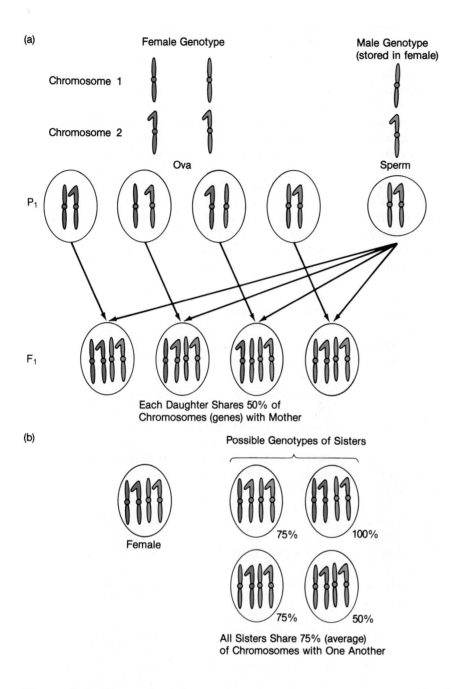

(a)

Female Genotype

Male Genotype
(stored in female)

Chromosome 1

Chromosome 2

Ova

Sperm

P₁

Each Daughter Shares 50% of
Chromosomes (genes) with Mother

(b)

Possible Genotypes of Sisters

Female

75%   100%

75%   50%

All Sisters Share 75% (average)
of Chromosomes with One Another

Figure 49-8   KIN SELECTION AND SHARED GENES IN ANTS, WASPS, AND BEES.
(a) Genetic relatedness: mother to progeny. Meiosis yields four types of ova in a female. Females arising from those eggs share only half their chromosomes with the mother since the other half comes from the father's sperm. (b) Genetic relatedness: sister to sister. The sisters share identical chromosomes from the father plus varying numbers from the mother; on average, the sisters share 75 percent of their genes with one another.

Figure 49-8). The result of this peculiar arrangement is that the genes in a female worker are more likely to be passed on if she helps her mother by raising more of her sisters than if she starts a new colony of her own offspring. In fact, the colonies of some bee, ant, and wasp species are highly cooperative societies in which altruistic sister workers spend their lives tending to the needs of the egg-factory queen.

Not all hymenopterans show the same degree of cooperation. In some species both males and females are quite nonsocial, while in others, most females become reproductive instead of that activity being restricted to a queen. Even so, the sterile worker strategy has evolved independently in eleven types of hymenopterans and in

termites. Clearly, the unusual degree of relatedness (75 percent) among sterile sister workers is a successful strategy for propagating alleles. Although geneticists have yet to identify specific alleles that underlie kin selection, one can reason that the particular alleles that favor sterility and rearing of sisters are passed on more effectively than are alternative alleles which lead to female maturation for mating; hence "worker alleles" are favored, and the mechanism of kin selection spreads.

Returning to the concept of inclusive fitness, it is clear that the female worker's inclusive fitness is substantially higher if her behavior helps genes similar to her own to be passed on to new generations (through her reproductive sisters).

Despite these plausible arguments, it is unlikely that kin selection and sterile sisters were the main factors in the evolution of complex insect societies. For one thing, termites have complex societies, yet both males and females are diploid and reproduce. For another, the queens of most insect societies mate with several males and store sperm from them all. A honeybee queen, for example, copulates between seven and ten times on her single mating flight. This means that sisters produced by such a queen may have any of several fathers, and thus their genetic relatedness averages out to be considerably less than the 75 percent calculated by Hamilton. Perhaps the queen's many daughters are really working to increase their mother's inclusive fitness, so that her genes are passed on successfully to her reproductive progeny.

We have now surveyed several aspects of social systems and seen that social behavior may enhance inclusive fitness and so be favored by natural selection. With this conceptual backdrop, we can now look more deeply at complex social organizations in two groups of animals: insects and mammals.

INSECT SOCIETIES

Biologists have studied insect societies in detail, particularly, as we have seen, those of bees, ants, and other hymenopterans. While the social organization of these animals may be quite complex—and involve frequent contact, division of labor, and interdependence among members—the behavioral repertoires of all insect societies are highly mechanical. Thus great behavioral complexity can be built from simple elements in insects just as can occur in other social species, including humans.

The Social Organization of the Honeybee

As with colonies of other social insects, a honeybee colony consists of 1 queen, 10,000 to 50,000 sterile female workers, and, in spring and summer, 1,000 to 5,000 male drones (Figure 49-9). The labor required to support so many thousands of individuals is rigidly divided; depending on their age, workers clean the hive, tend larvae and the queen, build honeycomb cells and serve as guards, or forage for nectar and pollen. In contrast with this life of constant drudgery, drones have a single function—to mate. Periodically, a virgin queen leaves a hive, mates with drones—thereby receiving and storing a lifetime supply of sperm—and then forms a new hive. Only 1 drone in about 500 ever mates, and with the arrival of cool weather in autumn, these other-

Figure 49-9   COORDINATED BEHAVIOR IN HUGE INSECT COLONIES.

This is a swarm of honeybees *(Apis mellifera)* with thousands of workers, drones, and guards, all of which have specific functions in the colony. Here the buzzing swarm waits in a plum tree while scouts search for a new site for a hive.

wise useless individuals are evicted from the hive and perish.

Unlike other bee and wasp species, in which entire populations may die in winter (except for a mated queen), honeybees are protected from the elements by the positioning of their nest sites in the cavities of trees and, occasionally, in the wall spaces of buildings. They actively regulate the internal temperature and humidity of their nests in various ways. As a result, they begin the spring with an exploding population of workers. When the hive becomes critically overcrowded, it splits in two: the old queen and half the workers leave for a new nest site. Workers again build a comb, in which the queen deposits eggs, and the cycle continues.

Like the information transfer accomplished by a forager's waggle dance, the social coordination necessary for this buzzing corporation to function is achieved through communication. A queen is recognized as such by her pheromones (sometimes called "queen substance"); those molecules not only attract nurse bees (and drones outside the hive), but also are fed to developing females. Queen substance represses development of the ovaries in those females. It also makes them develop into workers. Thus, two mutually exclusive instinctive life styles are possible for any female: that of queen or that of worker, the choice depending primarily on presence or absence of queen substance acting on the genome. As long as a queen is alive and producing queen substance, workers do not build cells for potential rival queens. The pheromone is transferred from worker to worker as they constantly regurgitate and share the contents of their stomachs.

When the queen dies or the colony becomes too large for all members of the group to receive a sufficient quantity of queen substance, the colony automatically begins preparations for swarming. A half-dozen queen cells are built, and the larvae within them are fed a protein-rich diet that causes these otherwise ordinary diploid individuals to develop as queens.

As time for swarming approaches, the old queen begins to produce a pulsating sound signal known as "quacking," which, in turn, elicits a "tooting" sound from any developing queens nearing maturity. This exchange apparently tells the colony that swarming is practicable (that is, that a second queen is available); in addition, the old queen's quacking warns new queens to remain in the safety of their cells. Then a silence descends over the hive until a worker returns and does the dance that informs the colony about the distance and direction to a site for the new hive. The old queen and a swarm leave the hive, whereupon a new queen emerges and stings the remaining queen candidates to death. A day or two later, she flies out to mate, returns to the hive, and begins her career of laying eggs.

The Adaptiveness of Insect Sociality

How do bees and other insects benefit from their social arrangements? Beyond the possible rewards of kin selection, the division of labor seen in insect societies creates several subpopulations of efficient specialists; for example, colonies of army ants, which may contain as many as 20 million individuals, include "soldier" workers with huge hooked mandibles whose exclusive role is defense of the colony (Figure 49-10). Another social species is the dairying ants, in which workers milk "herds" of aphids for their honeydew. The ants protect the aphids from predators, such as ladybird beetles, but exact a price—periodically they tap the feeding aphid's abdomen to stimulate secretion of a drop of sugar-rich honeydew. Another kind of insect agriculturalist is the "weaver" ants of Africa which construct nests of leaves folded over or fastened with silk produced by larvae. Adult workers actually hold the larvae in their mandibles and pass them back and forth, like living tubes of glue, over the seam or fastening point as the larvae exude silk threads from special glands.

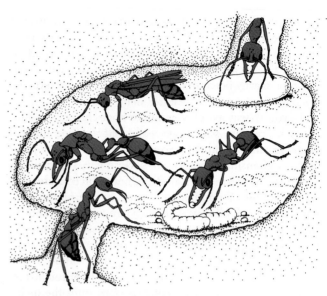

Figure 49-10   INSECT SPECIALISTS.
The bodies and behaviors of individual insects are often specialized in ways that allow individuals to perform unique tasks for the colony. Their huge mandibles, or jaws, allow "soldier" ants to threaten intruders and effectively defend the colony, as well as to manipulate objects such as eggs or larvae.

Among bees, group living makes possible relatively precise maintenance of the nest at optimum temperature and humidity and enables a large population of workers to survive the winter. No single bee or even small group could elevate the temperature of its living space in the way that the tens of thousands can in a hive; as a result, the enzymes and metabolism of bees in large social groups are adapted to the higher temperatures. And, as anyone who has ever disturbed an active beehive has discovered, sociality makes possible effective group defense against even the largest predators.

Despite the many advantages of insect sociality, there are real costs: specialized predators, such as honey buzzards, bee lice, and anteaters, have evolved to exploit the concentrated food supplies that such organizations make available. And diseases or parasites are more likely to spread successfully through dense social groups of suitable hosts.

SOCIAL SYSTEMS OF VERTEBRATES

The intricate social organizations of ant and bee colonies may have arisen partly as a result of the special genetic relationships that link the queen to her horde of sterile workers and the workers to each other. By contrast, although kin selection can be very important in vertebrate societies, such social systems have come about primarily because of the advantages to the individual conferred by group living.

A typical feature of vertebrate social groups is the **social hierarchy** (sometimes called **dominance hierarchy**), in which members possess ranks with varying privileges and responsibilities. An example is the **pecking order** in chickens (Figure 49-11), in which there is a clear-cut line of dominance from the hen at the top, which can peck all others in the coop and is pecked by none, to the hapless hen at the bottom, which can peck none and is pecked by all. Although the details vary among species, in general each animal's position in the social hierarchy is established and repeatedly reinforced by aggressive behaviors, with those higher in the order overcoming those lower down.

Often, the dominant individual in a social hierarchy is a male whose privileges include preferential access to food and mates. In return, this individual may lead and coordinate group activities, such as the course and timing of travel, the length of rest and activity periods, and attacks on prey. Dominance usually also confers the responsibility to act as the primary defender of the group or its territory, and that can be a real cost to the dominant individual. For animals at the bottom of the dominance heap, the story is quite different: such individuals may reproduce infrequently or not at all, starve if food is in short supply, or be excluded from the group altogether. Thus females may reject low-ranking males, or dominant males may actively prevent them from mating.

Figure 49-11 PECKING ORDER: TOUGHEST ON TOP.
Animals display social hierarchies, such as the pecking order of chickens. Those at the top of the hierarchy can peck (and hence dominate) all the others, while those at the bottom can peck no one.

Indeed, in groups of large herbivores and of carnivores, there may be only a single male—such as the zebra stallion and his harem—or a small group of related males—such as lion brothers, half-brothers, and cousins. Frequent aggressive interactions may repeatedly test the individual at the top of a dominance hierarchy, as though a real "survival of the fittest" is being applied. But generalizations are difficult; while the dominant male lion in a pride may mate some three and a half times as often as the most subordinate male, the latter is better off to remain with the pride than to go it alone or to attempt to join another pride. This is because he shares many genes with relatives that rank above him in the hierarchy. Therefore, even though he may not sire many cubs, he contributes to the welfare of the pride and increases his own inclusive fitness through cooperative hunting and other social activities.

Habitats can have strong, indirect influences on how social systems have evolved in vertebrates. In a tropical rain forest, for example, the continuously warm climate has permitted many primates to evolve breeding patterns in which females experience estrus (hormonal and behavioral preparation for breeding; Chapter 18) in any month of the year and quite independently of each other. This pattern helps explain why such societies have year-round dominance hierarchies: the males must compete again and again as individual females enter estrus. This pattern contrasts sharply with the one displayed by many animals in temperate regions: female elk, for example, mate only at times that ensure that the young will be born when their survival is most likely. In the elk society, males often do not stay with the females for most of the year, and they compete aggressively with other males mainly at the time of mating.

Social structure can also reflect an animal's niche. In colonies of ground squirrels living on the plains in the western United States, one can often observe a number of individuals standing high on their haunches and peering about, literally "on watch" for hawks or ground-based predators. (This is a good example of kin selection; among Belding ground squirrels, the females, not the males, are on guard and emit alarm calls. This endangers them, but their offspring and genetic relatives receive benefit from the warning; Figure 49-12.) Observers have also recorded analogous "predator scouting" among large herding herbivores. Similar behavior, however, is not seen in social groups of predators, such as lions. Put simply, an organism's place in the food chain is reflected in both its individual and its social behavior. It does not, of course, "know" its place in the food chain, but its behavior, whether social or not, helps determine that place.

In the next section, we shall look in detail at a representative array of the complex social behaviors that have evolved among mammals, including those of our own order—the primates.

Figure 49-12   BELDING GROUND SQUIRREL "ON GUARD."
Female belding ground squirrels (*Spermophilus beldingi*) are likely to emit alarm calls when they spot hawks or coyotes. This behavior endangers the caller, but the warning benefits both their own young and their genetic relatives. It is an example of kin selection.

SOCIETIES OF MAMMALS

Mammals display a vast array of social organizations. Most large ungulates, for example, live in herds that may number just a few or in the hundreds or thousands. Within this larger classification, the several dozen species of African antelope live in varying social schemes, including (1) monogamous pairs in exclusive territories; (2) single males in a matrix of territories, across which females roam at will to graze; and (3) wandering groups with a strict male dominance hierarchy.

Each of these social structures is ecologically adaptive. Thus animals in the first group, small antelope known as duikers, live in forests, where the stability of the habitat favors permanent territories (Figure 49-13a). Migration does not occur, and pair bonding is associated with permanent division of territorial resources. Wildebeest are typical of the second group. Larger than duikers and adapted to life in the open, grassy savanna, wildebeest roam in herds as a defense against predation. Even so, males persist in defending exclusive territories during the mating season, contesting for the lushest patches of grass and facing off along their mutual borders with display rituals (Figure 49-13b). The resident male has undisputed mating rights with any female in estrus coming onto his turf; since there are more males than land for territories, the excluded males form a bachelor herd with a dominance hierarchy.

The buffalo, the largest of the African ruminants, eats grass too tough and poor to sustain wildebeest (Figure 49-13c). Their large rumen enables buffalo to eat a great

deal and to ferment it longer, thereby obtaining sufficient nutrients from their food. But they must be constantly on the move to find sufficient forage, a fact of life that makes holding territories impossible. Within each herd, males are organized in a strict dominance hierar-

Figure 49-13   VARIATION IN ANTELOPE SOCIETIES.
(a) Small duikers (such as *Cephalophus zebra*) live in pairs in forest territories. (b) Male wildebeests (*Connochates taurino albojubates*) defend territories with this sort of "challenge ritual" in which they kneel and lock horns. Wildebeest herds may number in the thousands. (c) African buffalo (*Synceros caffer*) travel in large herds with a dominant male and a typical dominance hierarchy.

(a)

(b)

(c)

chy, with the top, or "alpha," male having the exclusive right to mate with any receptive female in the group. Since being the alpha male implies greater strength (or more of other qualities required to acquire the alpha ranking), the mating behavior suggests that the "fittest" is most likely to contribute genes to the next generation.

The migrations and social structures of some prey species influence the social groups of predators, such as lions. When zebra and wildebeest have migrated from an area, the local lions, which do not migrate, must feed on small gazelles, each yielding only about 11 kilograms of meat. Since an adult lion needs about 6 kilograms of meat per day to survive, the lion prides disperse, and individuals or pairs must hunt down their small prey. When wildebeest and zebra return, however, prides reassemble, and groups of four or five lions can be seen feeding on single large carcasses once again. Whereas hunting individuals or pairs capture only about 15 percent of the gazelles they pursue, larger prides kill about 40 percent of the animals they stalk collectively. Finally, the larger prides are more successful at fending off groups of scavenging hyenas. So there are a number of advantages to being social and to hunting in larger groups, but the existing food supply determines the maximum size of the pride in a given season.

The Social Organization of Wolves

One of the most intensely studied of all social mammals is *Canus lupus*, the wolf. Once the dominant carnivores of north-temperate North America and Eurasia, wolves often feed on herbivores much larger than themselves: deer, elk, caribou, and moose. However, in order for a predator to hunt large mammals successfully, it either must weigh about as much as its prey—the strategy of tigers and polar bears—or must hunt in groups. The second option is reflected in the basic organizing principles of wolf society, the pack.

Wolf-pack size varies greatly, from a pair to several dozen individuals. Pack size is influenced in part by *economic limits*, as is the lion pride: on the one hand, by the smallest number of animals required to find and kill prey efficiently, and, on the other hand, by the largest number that could be fed by a kill. There are also *behavioral* limits: there appears to be a limit to the number of social bonds that individual animals can form and to the degree of competition that each pack member can tolerate.

The social world of a wolf pack is a microcosm of fascinating and often subtle interactions (Figure 49-14). Males and females are organized in separate dominance orders, with the original mating pair of the pack accorded the alpha position in each ranking; an established pack typically consists entirely of successive generations of their offspring. Within the group, dominance is ac-

Figure 49-14   THE WOLF PACK: DOMINANCE HIERARCHIES IN ACTION.
Note the lowered ears, cowering posture, and gentle pawing of the wolf (*Canis lupus*) on the left as it interacts with a male higher in the pecking order.

knowledged in two ways. One is the behavior of a dominant animal: raising tail and ears, standing broadside to an inferior pack mate, and urinating on territorial markers and even food. The other is the behavior of a subordinate animal which may show a range of submissive behaviors. For instance, acknowledgment of another's more favored status may be expressed by only a slight shift in the position of the submissive animal's ears or by a conspicuous display of whining as the subordinate animal rolls about on its back. Wolves also carry out a striking group ceremony of greeting directed toward the alpha male. In this energetic ritual, pack members crowd together and nuzzle the leader, and then often take part in a group howling session.

The behavior of wolves as social animals is intimately entwined with their ecology as pursuers of large game. Scientists have found that packs grow, divide, and dissolve because the status of pack leadership changes as a result of deaths, pair bonding, and other events, and because the population size and availability of prey fluctuate. In a study of wolves in the isolated environment of Isle Royale (Chapter 45), food supply was a crucial factor: packs grew in good times, and shrank and even dissolved as the supply of game declined. In addition, surviving wolves were forced to change their usual group-hunting style, concentrating instead on small prey, such as beaver, hares, and other animals that are susceptible to capture by a solitary hunter. This illustrates well that wolves, like the coyotes described earlier, probably inherit one life style, but it is flexible. They may live and hunt in packs or as individuals, and it is ongoing "evaluation" of current conditions that triggers the shift from one mode to another.

Primate Societies

Gorillas, chimpanzees, orangutans, and other primates are our closest living relatives in the animal kingdom. Their societies are complex and hint at the origins of our own. We must remember that primate societies have no doubt been evolving over the past 10 to 20 million years, just as our own lineage has evolved over the past 4 to 5 million years.

With the notable exception of *Homo sapiens* and a few other species, nearly all primates are tree-dwelling herbivores and insectivores (Figure 49-15). Among semiarboreal and ground-living species—in particular, gibbons, baboons, macaques, and the highly social chimpanzees and gorillas—several types of stable, complex social systems have been described.

Asian gibbons are among the least social of the higher primates. Living nearly full-time in dense tropical forest, mated pairs and their offspring form the primary social unit. The food resources for these animals, fruits and parts of green plants, are abundant and relatively predictable, and a gibbon family typically defends a territory of about 250 acres. Males and females are nearly identical in size and markings, and both parents take part in territorial defense and care of the young. Thus gibbons are reminiscent of the duiker antelope pairs. Stability of environment and food supplies correlates with the small, stable social group and the filling of available space with breeding pairs.

In extreme contrast, both ecologically and socially, is the hamadryas baboon *(Papio hamadryas)* of the desert grasslands and savannas of northeastern Africa. Baboons are large, dog-faced monkeys with extreme sexual dimorphism. Hamadryas males may weigh more than 50 kilograms, twice the weight of the average female, and they have a characteristic bright-red face wreathed with a lionlike mane. These physical traits are almost certainly related to the hamadryas's peculiar social organization.

In general, baboons are highly social, living in bands of up to several hundred individuals. The hamadryas baboon has elaborated on this theme: its basic social unit is the **harem,** which consists of one male and up to ten adult females and their offspring. The male "overlord" is an autocrat that, because of his superior size and strength, is able to exert virtually total control over the harem's activities. Harems combine to form **bands,** the basic unit for food gathering and defense. Finally, bands unite in a third level of organization, the **troop,** to spend the night sleeping on cliffs or in trees. Troops may contain more than 700 individuals.

Why have the hamadryas evolved this elaborate social structure? Although many of the features of the harem remain unexplained, some ethologists speculate that it is an adaptation to uncertain food supplies, a typical feature of the savanna biome, in which plant productivity varies significantly with climate fluctuations. Thus in lean years, fewer females are impregnated per harem, and the population remains stable relative to available food. Conversely, when food is abundant, relatively more offspring can be produced. The size of hamadryas bands—which varies considerably from year to year—may also be adapted to the richness of food resources at any given time.

Figure 49-15   PRIMATES: SOCIAL AND OFTEN ARBOREAL.
Nonhuman primates, such as these common langurs *(Presbytis entellus),* are usually tree-dwellers that feed on arboreal plant and insect life. Their societies are complex, with a strict social hierarchy and division of labor.

Animal Behavior and *Homo sapiens*

How many and which of the lessons of animal sociality should we humans apply to our own species? As we will see in Chapter 50, perhaps 5 million years or so ago, ancestors of modern humans were ground-dwelling hunters who probably still spent some time in the trees; they fed on both animals and plants and lived in small groups in the bush and on the plains of Africa. After a period of rapid expansion of the brain over several million years, our ancestor *Homo erectus* began using weapons and tools. Then, only 10,000 to 20,000 years ago, the domestication of animals and plants hurled human beings into civilization, with its cities, complex division of labor, and geometrically expanding population. No doubt, cultural evolution has become a dominant feature of human behavior, but there is no reason to think either that the original biological features of human sociality are erased or that our biological evolution has ceased.

Although we may think of culture as being primarily an attribute of our species, its roots are much more primitive. For example, observers of a captive group of Japanese macaque monkeys knew that individual animals would pick up sand-covered sweet potatoes from a beach on their island and eat them, grit and all. One day, a one-and-a-half-year-old female, dubbed Imo, was seen holding a potato in one hand, cleaning off the sand in the water of a brook with the other hand, and then eating the potato. She washed her other potatoes, and soon two infant members of the troop were doing the same. Over several years, washing sweet potatoes spread among closely related families and groups of young playmates, and finally the habit was passed on regularly from mothers to children as they trained them (Figure 49-16). Although the monkeys initially washed their potatoes in only fresh water, they later switched to sea water, perhaps because of its salty taste. Now, the apparent salt lovers dip their potatoes into the sea between each bite.

Imo invented another novel cultural trait when she was four years old. Wheat was usually scattered onto the shoreline for food, along with the sweet potatoes. Over the years, monkeys picked up individual wheat kernels and ate them. Imo learned to gather handfuls of wheat and sand and drop them into the sea, where the heavier sand settles out rapidly and the buoyant wheat can be gathered together in handfuls for easy eating. This troop of macaques now washes its sweet potatoes and separates its wheat from the sand, while other troops do not. There is no gene for washing sweet potatoes or wheat; instead, cultural transmission passes on learned behaviors from one generation to the next. Is not this just what happened with the arrowhead, the spear, the wheel, and, indeed, all other human inventions?

A good place to see the residuum of biological evolution in humans is in the development of human infants,

Figure 49-16   CULTURAL TRANSMISSION: MACAQUES AND SWEET POTATOES.

Here a Japanese macaque (*Macaca* species) washes its potatoes. The behavior, invented by a young female, Imo, is now passed on from one generation to the next. This and another novel feeding behavior, also invented by Imo, are not transmitted genetically, but must be learned by each young macaque.

which suggests strongly that many aspects of early behavior are programmed by genetic and developmental processes. Support for this view comes from the store of innate behaviors and reactions with which a human newborn enters the world. Infants only a few days old cling tenaciously; "walk" if supported; paddle and hold their breath if submerged; search for a nipple and suck rhythmically; and, of course, perform that well-coordinated, communicative motor behavior—crying (Figure 49-17). At four weeks, a baby begins to focus on faces and to smile; blind children direct a winning grin at sound sources, not because they are imitating the parent's smile but because an innate behavior pattern is being triggered. Studies of six-week-old infants show that they innately recognize the consonant sounds of human speech. This list of accomplishments goes on well into childhood, structuring both cognitive and motor development. As a result, human children the world over pass through a strikingly stereotyped sequence of developmental stages.

Considered in the light of our increasing understanding of neuronal plasticity and of how peptides and hormones operate in subtle ways in the brain (Chapter 40), it seems plausible that many of the behaviors that hu-

Figure 49-17   INSTINCTIVE—AND EFFECTIVE—HUMAN COMMUNICATION.
Fixed motor patterns are found in people as well as in other vertebrates. Many aspects of infant behavior are thought to be programmed. The baby's loud cry of distress—whether from hunger or a wet diaper—is a good example.

mans blithely attribute to free will may not be so free after all. We are, in the end, an extraordinary example of primate evolution, one capable of remarkable feats of learning, thought, creative expression, and foible. Nevertheless, much of what we do may ultimately be controlled by the circuits that arise in our brains, by chemicals that act in profound and powerful ways, and by the genes that are responsible for our development. But cultural evolution—from the invention of agriculture, the wheel, and machines to augment our muscles to that of the computer to augment our brains—is uniquely important in the process that has led to contemporary *Homo sapiens*.

The application of sociobiology is an important new biological tool for understanding human culture and its evolution. The theories of the Harvard evolutionary biologist Edward O. Wilson are based on immense sets of animal data, are framed in strict evolutionary terms, and are related to such concepts as inclusive fitness. Wilson and his theories have been attacked by people who fear that those theories could be misused in racist or fascist claims or who believe that human behavior is controlled solely by our conscious will. But sociobiology attempts to explain animal and human behavioral traits without attempting to justify them in philosophical or political ways. The biological justification is simple and sufficient: elevating the individual's inclusive fitness.

To say that evolution and genes play major roles in animal and human behavior is not to say that every specific behavior is rigidly determined by genes. The human brain has an obvious capacity for learning and for generating versatile actions in the face of novel circumstances, just as the simpler brains of many other animals can do too. Yet to deny that genes, brain circuits, and hormones have a significant place in human behavior is essentially to deny evolution. As the behaviorist John Alcock states, enjoying sweet foods, falling in love, wanting to be liked, or learning a language relate to underlying physiological processes. When our bodies need a sugar jolt, we reach for candy or soda pop—not because we are aware that our blood sugar is low, but because the underlying neuronal circuitry triggers sugar-seeking behavior.

Consider another case of underlying biology influencing human behavior. Many South, Central, and North American Indians cook corn meal in alkaline solutions, wash it, and only then use it to make tortillas. This process removes some vitamins and takes extra time. But it also vastly improves the nutritive quality of the corn meal because it increases the amount of the amino acid lysine and favorably changes the ratio of leucine to isoleucine (Table 49-2). Recall from Chapter 34 that many plant foods, including corn kernels, contain proteins that lack amino acids essential to animals. Although we do not know when alkali cooking was invented, *every* Central and South American culture that subsists on corn uses the procedure. This may be a case of convergent cultural evolution. Regardless, humans thousands of years ago, as today, were unaware of the molecules in their foods; those populations that happened to stumble on alkali treatment must simply have been healthier, lived longer, or reproduced better. Even today, individ-

Table 49-2   **AVAILABILITY OF CERTAIN ESSENTIAL AMINO ACIDS AFTER ALKALI COOKING**

Amino Acid	Milligrams of Amino Acid per Gram of Nitrogen	
	Uncooked Corn	Alkali-Treated Corn
Histidine	0.012	0.028
Isoleucine	0.074	0.158
Leucine	0.309	0.358
Lysine	0.045	0.126
Threonine	0.145	0.381
Leucine: isoleucine (ratio)	4.2:1	2.3:1

uals in cultures that have never invented the alkali cooking procedure tend to suffer from the amino-acid deficiency disease pellagara. We do not have to envision a gene for alkali cooking any more than we do for Imo's sweet-potato washing; the evolution of such cultural elements can instead be viewed in the context of sociobiological principles and basic evolutionary theory. With this reasoning in mind, we turn in Chapter 50 to the fossil record to trace, where we can, the many lines of human biological and cultural heritage.

SUMMARY

1. A *society* is a group of animals of the same species in which the members communicate and interact in cooperative ways. Social behavior can be adaptive—that is, it may evolve in situations in which being social tends to promote the survival of individuals so that they are more likely to reproduce successfully than if they live alone. *Sociobiology*—literally the biology of social behavior—is concerned with the evolutionary basis for behavior.

2. All social behavior involves *communication* that alters the behavior of others and is mediated through senses and signals. Animals have evolved an astonishing variety of acoustic, visual, chemical, and tactile stimuli by which they communicate with members of their own or other species.

3. *Ritualization* is the modification of a functional behavior to a communicative role; the courtship rituals of many birds are prime examples—for example, the pair-bonding ceremony of great crested grebes.

4. The waggle dance of honeybees, described by Karl von Frisch, is an extraordinary example of complex insect communication that relies on tactile stimuli. Recruits that attend a forager bee's dance feel her movements and learn not only the distance and direction of a food source from the hive, but also its odor and quality. Once another bee visits the feeding site, learning of landmarks augments the information gained from reading the waggle dance.

5. Language is a symbolic communication system that involves reference to objects or events distant in both time and space. Thus far, no animal other than humans has been shown to have true language capabilities, even though other primates may communicate in complex ways. However, there is a great deal of evidence that many of the basic features of human language ability are genetically programmed.

6. Mating behavior is a key component of behavior in social animals. Polygamy (either *polyandry* or *polygyny*) has profound effects on social structure for a species. It also affects whether males and females are *sexually dimorphic*.

7. *Territoriality* is the defense of a particular feeding or breeding site, generally against other individuals of the same species.

8. *Aggression* refers to behaviors such as fighting and *threat displays* between animals of the same or different species. Most aggressive behavior within a species is highly stylized and thus avoids significant real damage to either party.

9. *Altruism*, a major type of social behavior, is self-sacrifice or restraint exhibited by some members of a social group that benefits other members who are not direct genetic relatives. Altruism carries some cost to the individuals that carry it out. Both altruism and reciprocal altruism are hypothesized to increase *inclusive fitness* and thereby to make more likely successful passage of genes identical to one's own to the next generation.

10. Some altruistic acts are hypothesized to result indirectly from action of hypothetical "selfish" genes. Other altruistic acts may result from *kin selection*—behavior that promotes the survival of genes similar to one's own, whether they are carried by children or by other kin.

11. The social organization of insects is extremely complex, involving frequent physical contact, division of labor, and interdependence among members. The behavioral repertoire of insect societies includes much innate, closed-program behavior. But it also includes learned, open-program behaviors such as in learning the maps of foraging areas. Remarkable openness and plasticity of behavior may develop out of closed parts, not only in insects and other invertebrates, but also in vertebrates and humans. The most complex insect behaviors are seen in bee and ant colonies where altruistic behaviors are common.

12. A typical feature of vertebrate social systems is the *social hierarchy*, or *dominance hierarchy*, in which members possess ranks with varying privileges and responsibilities. One example is the *pecking order* of chickens. Often, the dominant individual in a social hierarchy is a male whose privileges include preferential access to food and mates, and whose responsibilities include a major role in defense of the group.

13. Mammals display a vast array of ecologically adaptive social organizations. For example, African ungulates generally form herds, but depending on available food sources, mating season, and predators, they may live in permanent territories, in

temporary territories, and as wandering groups with strict dominance hierarchies. Wolves can form highly social packs—an organization that enables them to capture prey much larger than themselves. Under other circumstances of food availability, wolves and coyotes may hunt in small groups or alone. This reflects flexibility in using behaviors appropriate to changing circumstances.

14. Primate societies range from small family groups to huge assemblies of individuals organized in mul-tilevel social structures. For example, hamadryas baboons live in basic social units called *harems*, which combine to form *bands*—the units for food gathering and defense. Bands unite to form the largest social unit, the *troop*. Cultural transmission of information and behaviors is especially prominent in monkeys, apes, and humans, though surely not limited to those kinds of animals.

15. As social animals, humans are part of the evolutionary continuum. It is likely that at least some aspects of human behavior are programmed by the nervous and endocrine systems and so have some genetic basis. Support for this view comes, in part, from the store of innate behaviors seen in newborns, including the sucking reflex, smiling and crying, and the recognition of speech sounds. But human behavior, like that of so many other animals, can use closed and open behaviors plus memories of past events so that versatile actions can be taken in novel circumstances.

KEY TERMS

aggression	harem	polygyny	sociobiology
altruism	inclusive fitness	ritualization	territoriality
band	kin selection	sexual dimorphism	threat display
communication	pecking order	social hierarchy	troop
dominance hierarchy	polyandry	society	

QUESTIONS

1. In order to improve your inclusive fitness, what sorts of behavior should you show toward whom?

2. What are some advantages and disadvantages of sociality?

3. How does a worker bee communicate in the darkness of the hive? What information can she pass to other bees? What role does learning play once a worker has left the hive en route to a food source?

4. What are the various ways in which mammals can communicate? Which of those ways do people use?

5. What is sexual dimorphism, and under what conditions is it seen?

6. What roles do social behaviors such as territoriality and aggression play for various species? Does either increase inclusive fitness? If so, how?

7. What is reciprocal altruism, and how can it benefit an individual?

8. Describe common forms of kin selection in social insects. What is a possible explanation for this complex social structure?

9. How are changing ecological circumstances coped with behaviorally by coyotes, lions, or wolves? What does this imply about the degree of closed versus open programs in these animals' behavioral repertoire?

10. What are the types of social organization in primates? What roles do communication, aggression, and cultural evolution play in primate social groups?

SUGGESTED READINGS

ALCOCK, J. *Animal Behavior*. 2d ed. Sunderland, Mass.: Sinauer, 1979.

An outstanding book with emphasis on evolutionary approaches to the study of behavior.

ALEXANDER, R. D. "The Evolution of Social Behavior." *Annual Review of Ecology and Systematics* 5 (1974): 325–383.

A leading behaviorist writes on a key subject.

HAMILTON, W. D. "The Genetical Evolution of Social Behavior." *Journal of Theoretical Biology* 7 (1964): 1–52.

A classic paper that has reshaped thought on evolution.

WILSON, E. O. *Sociobiology: The New Synthesis*. Cambridge, Mass.: Harvard University Press, 1975.

A true classic in the literature of modern biology.

50

HUMAN ORIGINS

We are the products of editing, rather than of authorship.
George Wald

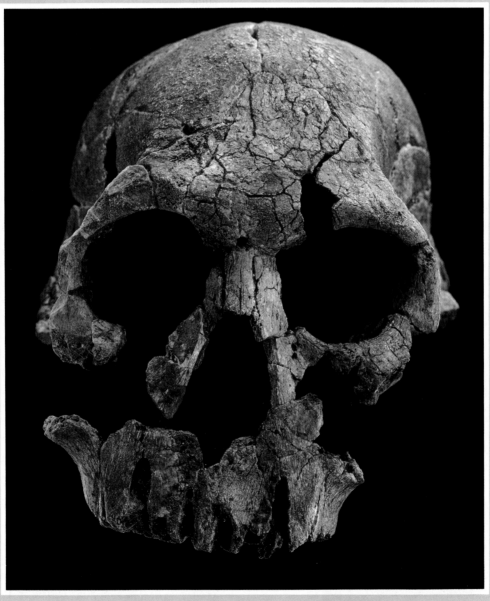

The toolmaker: *Homo habilis*. Known to us only as "1470," this ancestor of modern humans was probably a brown-eyed, hairy, social individual living in the savanna and forests of East Africa from some 2 million years ago.

Paleontologist Mary Leakey watched with anticipation as her co-workers carefully chipped and brushed the last layers of solidified volcanic ash away from three sets of footprints—footprints that were to reveal a visible trail, set in stone, of our early human ancestors (Figure 50-1). From this exciting find in the Laetoli region of Tanzania in 1976, and from other fossils gathered in the same region, Leakey and her colleagues were able to reconstruct an image of the creatures that had tracked through the warm ash that covered an ancient riverbank. About 3.6 million years ago, three figures walked at that site. The largest, on the left, may have been about 4 feet tall; the other two were shorter and smaller—perhaps a female and offspring. Their skin or fur was probably dark brown; their hair was long and shaggy; and heavy brow ridges may have jutted forward above their eyes. The smallest footprints sometimes appear within the outlines of the largest. Was a youngster deliberately walking in a parent's footprints? Where was the group going? How did they live? What did they eat? Did organisms like this really evolve toward modern human beings? These are just some of the kinds of questions that paleontologists try to answer as they search Africa, Asia, and Europe for fossil remains of our early ancestors and for links between them and the apes, monkeys, and other mammals.

Carolus Linnaeus included in the tenth edition of his *System Naturae* (Chapter 19) a regular taxonomic place for humans. Whereas his contemporaries would have listed humans as unique, Linnaeus labeled them *Homo sapiens* and included them within the mammalian order Primates, next to the apes and monkeys. In this chapter, we will briefly survey these and other types of primates and then turn to the fascinating history that led to our species. As we shall see, all primates have bodies that display adaptations for life in the trees. A few species returned to life on the ground, but only one lineage, our own, evolved erect bipedalism, the upright balanced stance on two limbs that allows a mode of locomotion unique among all vertebrates. Erect bipedalism, the remarkable dextrous hand, and an interplay of biological and cultural evolution are believed to have led to the remarkably rapid expansion in size and complexity of the brain in the line leading to modern *Homo sapiens*. These factors in our evolution will be amplified as we trace human origins, and we shall see how the three dim figures from the past, striding across the ash, are representatives of the critical time when bipedalism had been acquired but the brain was still modest in size. We will also see that fossils in the human lineage are relatively rare and are not easy to interpret. This leads to controversy about the evolutionary tree of *Homo sapiens* and about the factors that contributed to our evolution. Nevertheless, great advances are being made as new techniques are successfully used to identify sites likely to yield fossils, to ascertain the age of fossils, and to interpret the fossils in ecological, social, and functional terms.

Figure 50-1   SEQUENCE OF FOOTPRINTS OF THREE INDIVIDUALS OF *AUSTRALOPITHECUS AFARENSIS*.
Mary Leakey and her associates discovered these footprints, along with those of many other types of animals, in 1976 at Laetoli in Tanzania. The footprints were filled with black sand to make them stand out. They were made about 3.6 million years ago when early humans walked across a fresh ash fall from a nearby volcano; a brief rainfall shortly afterward solidified the ash layer, preserving the footprints. The footprints were made by humanlike feet applying weight to the ground as humans do today; and their regular progression indicates a bipedal striding gait basically similar to our own.

THE PRIMATES

Modern **primates** (Table 50-1) exhibit most of the body forms also found in the fossil record (Figure 50-2). **Prosimians** (meaning "before apes") include lemurs, lorises, and tarsiers. Lemurs, and lorises, with their fox-

Table 50-1   **CLASSIFICATION OF SOME LIVING MEMBERS OF THE ORDER PRIMATES**

Suborder	Family	Common Name(s)	Location
Prosimii		Lower primates	
	Lemuridae	Lemur	Madagascar
	Indriidae	Indri	Madagascar
		Sifaka	Madagascar
	Daubentoniidae	Aye-aye	Madagascar
	Lorisidae	Loris	Asia
		Galago (bush baby)	Africa
		Potto	Africa
	Tarsiidae	Tarsier	Asia
Anthropoidea		Higher primates	
Platyrrhini		New World higher primates	
	Cebidae	New World monkeys	Central and
		Howler monkey	South America
		Spider monkey	
		Capuchin monkey	
	Callithricidae	Marmoset	Central and South America
Catarrhini		Old World higher primates	
	Cercopithecidae	Old World monkeys	
	Cercopithecinae	Baboon	Africa
		Vervet	Africa
		Macaque	Asia, Africa
	Colobinae	Colobus monkeys	Africa
		Langur	Asia
Hominoidea		Lesser apes	
	Hylobatidae	Gibbon	Asia
	Pongidae	Great apes	
		Chimpanzee	Africa
		Gorilla	Africa
		Orangutan	Asia
	Hominidae	Humans	Worldwide

like snout, are usually tree dwellers and are closest in structure to the most ancient placental mammals, including insectivores, such as shrews. Tarsiers are vertical climbers and leapers, active nocturnally, and usually about the size of a small rat. The head is bent on the vertebral column, so that the face and eyes point forward when the tarsier sits upright in the trees. Humans have this same head orientation. Typical quadrupeds, such as dogs, do not; their heads extend forward from the vertebral column. The tarsier's snout is shortened, and the visual fields of its enormous eyes overlap for binocular depth perception. Finally, the hand has taken over the tactile functions carried out by the snout in lemurs and more primitive primates.

The **anthropoids** include the monkey, ape, and human lineage. All possess larger brains than do prosimians, and the cerebral hemispheres are particularly large. The **New World monkeys** are *arboreal*, or tree dwelling, with long limbs. Many possess a prehensile tail capable of gripping branches, and some lack a thumb that can be fully opposed against the other digits for gripping. Spider and howler monkeys and other New World monkeys are basically quadrupedal, running about the branches, leaping wildly from tree to tree, and occasionally swinging by the arms. The **Old World monkeys** live in Africa and Asia, tend to be larger than their New World relatives, have tails that are not prehensile, usually have a fully opposable thumb, and possess sexual cycles closely resembling those of apes and humans. Some Old World monkeys, such as vervets and baboons, have become quadrupedal for life on the ground, and the snout has elongated once again. The **apes,** all tailless, include gibbons—a so-called lesser ape—the fantastic acrobats that swing through the trees suspended by one forelimb at a time—a mode of locomotion called **brachiation.** The great apes include chimpanzees, gorillas, and orangutans, the first two of which, like humans, have adaptations for life on the ground. But reflecting their former

(a)   (b)

(c)   (d)

Figure 50-2   TYPICAL LIVING PRIMATES.
(a) A ring-tailed lemur (*Lemur catta*) mother and young reveal the snout, eyes partially directed to the front allowing some stereoscopic vision, longer hind limbs than forelimbs, and somewhat elongated digits. (b) This small tarsier (*Tarsius syrichta*) tends to sit upright in trees; its head is bent at a right angle to the vertebral column as it sits; the large eyes, with the flattened snout, allow good binocular vision; and the elongated fingers and toes allow grasping of branches. (c) Red howler monkeys (*Alouatta seniculus*) grip branches with fingers, toes, and tail. (d) Long, powerful arms, long fingers to wrap around branches, and superb agility mark this high-speed brachiator, a white-handed gibbon (*Hylobates lar*).

life in the trees, they have enlarged, elongate forelimbs that prop up the front of the body when the animals are on the ground.

Although each living primate species is a unique group of animals, specially adapted to its own niche, the study of primates provides us with the perspective we need to identify our distant ancestors and reconstruct our evolutionary history. Our prehistoric ancestors, living millions of years ago, were never identical to today's chimpanzees, baboons, lemurs, or any other modern primate. However, our evolutionary lineage did pass through stages roughly comparable with the prosimian, monkey, and ape body plans. We can therefore use these living primates as a mirror for our past. It is from studies of our primate relatives, combined with analyses of the fossils preserved in geological strata, that we are

able to piece together our own evolutionary history and formulate a picture of our origins in nature.

Early Evolution of the Primates

The primates are one of the oldest extant orders of placental mammals, extending back in time more than 65 million years to the Cretaceous period, near the end of the Mesozoic era (Figure 50-3). The earliest primates, such as the little creature from the late Cretaceous known as *Purgatorius* (named after Purgatory Hill, Montana, where the first fossil of the genus was found), were small, shrewlike creatures. *Purgatorius* was contemporary with at least six species of dinosaurs. The insectivorelike appearance of these fossils has led to con-

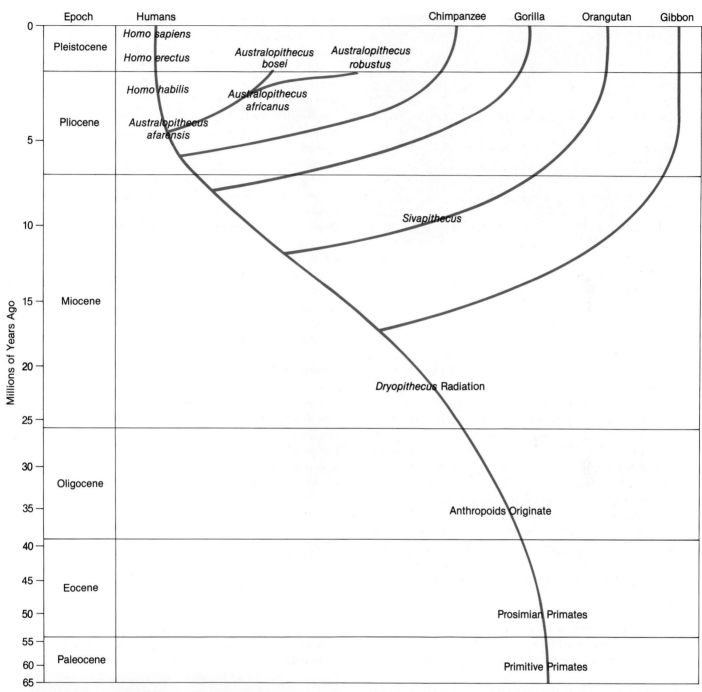

Figure 50-3   THE PRIMATE PHYLOGENETIC TREE.
The major types of primates and times at which different lines diverged are shown here. This tree depicts one interpretation of events of the past 4 million years; alternative interpretations are shown in Figure 50-13.

siderable disagreement among paleontologists as to whether they represent early primates or early insectivores. Anatomically, *Purgatorius* specimens resembled today's living insectivores more than living primates (Figure 50-4); nevertheless, a prudent guess is that they were indeed primates, ones that lived soon after the primate and insectivore lineages had split.

Sixty-five million years ago, at the beginning of the Cenozoic era, during the Paleocene epoch, there was an evolutionary radiation of primitive primates. To the untrained eye, most of these animals would be indistinguishable from small rodents and insectivores; nonetheless, telltale features of the skeletons leave little doubt that these were diversifying primates. At the beginning

Figure 50-4   THE MODERN TREE SHREW.
This Southeast Asian resident (*Tupaia tana*) is generally representative of the types of insectivores from which the earliest primates evolved.

of the Eocene epoch, about 54 million years ago, the first primates with bodily characteristics of modern prosimians appeared. Some of these Eocene fossils so closely resemble living lemurs and tarsiers that they indicate how little evolutionary change has taken place in those lineages since the Eocene. Whereas modern tarsiers, lorises, and lemurs are largely relict groups (isolated survivors of previously widespread groups), inhabiting only islands like Madagascar or being nocturnal on continental mainlands, the Eocene prosimians were extremely numerous and lived in most tropical areas of the Earth. Prosimian fossils, in fact, are among the most common found in some Eocene deposits of Europe and North America—land masses which were still connected at that time. By the end of the Eocene, however, most of these prosimian groups had become extinct. This was probably due to the gradual cooling of the Earth, which eventually resulted in the ice ages of the Pleistocene epoch; to the movement of the continents into different latitudes; and to the evolution of other, well-adapted competitors, such as the rodents. Yet from among these Eocene prosimians evolved the various modern prosimian groups, as well as the earliest members of the Anthropoidea.

The earliest anthropoids appeared in Africa during the Oligocene epoch, about 30 million years ago. Subsequently, they spread across the Old World into South America, which was still considerably closer to Africa than it is today. This spread and geographical separation established the ancestral groups of the New World monkeys, on the one hand, and of the Old World monkeys, apes, and humans, on the other. One of the earliest apelike creatures is known as *Aegyptopithecus* (meaning

"Egyptian ape"); it was about the size of a large house cat and resembled a small New World monkey, such as a squirrel monkey (Figure 50-5).

Much as the Eocene epoch was the heyday of the prosimians, the Miocene epoch saw the flourishing of the apes and Old World monkeys, from 26 to 5 million years ago. During the first half of the Miocene, apes or apelike animals, identified as such primarily on the basis of tooth crowns that resemble those of later apes, occupied much of the tropical and subtropical Old World. They varied considerably in size and appear to have filled many of the niches now occupied by the Old World monkeys. These fossils are generally classified as dryopithecines, after **Dryopithecus** (meaning "forest ape"), the first genus of this group to be named. Dryopithecine skulls have much smaller brow ridges than modern ape skulls, but their teeth have apelike features. A quadruped with apparent ability to run and leap on the ground or in the trees, as well as to stand on its hind limbs, *Dryopithecus* did not have the long, large forelimbs so apparent in modern apes (Figure 50-6).

In the second half of the Miocene, monkeys largely indistinguishable from modern Old World monkeys replaced the dryopithecines across most of Africa and southern Eurasia. These monkeys had evolved teeth and digestive tracts that were better adapted than those of the dryopithecines for eating the abundant grasses and leaves, whereas the apes' diet continued to be restricted to the fleshy parts of plants, such as fruits, berries, and tender shoots. By the beginning of the Pliocene epoch,

Figure 50-5   *AEGYPTOPITHECUS:* AN OLIGOCENE APELIKE PRIMATE.
Side view of the most complete skull of *Aegyptopithecus zeuxis*. Numerous specimens of *Aegyptopithecus* have been discovered in Oligocene deposits in the Fayum Basin, near Cairo in Egypt. These fossils show *Aegyptopithecus* to have been a primitive anthropoid primate, similar to later apes in its dentition, but to New World monkeys in aspects of its brain and limbs.

Figure 50-6 *DRYOPITHECUS:* **A MIOCENE APE.**
Dryopithecus was the first genus of the group of apes that
occupied much of the tropical and subtropical Old World
from about 25 million to 15 million years ago. They filled
niches similar to those of modern Old World monkeys.
(From Eli C. Minkoff, *Evolutionary Biology* © 1984.
Adapted by permission of Benjamin/Cummings Publishing
Company.)

arboreal nature. Virtually all primates, except for the few
that have secondarily adapted to living on the ground,
spend most of their time in trees—eating, sleeping, reproducing, and interacting socially. Trees provide food,
in the form of leaves, buds, fruit, and insects, as well as
protection from most predators. It is not surprising,
therefore, that the first major radiation of primates, in
the late Cretaceous period and the Paleocene epoch, followed closely on the radiation of angiosperms—
flowering plants and trees—in the Cretaceous.

Both locomotion and sensing the environment in the
trees have left legacies on primate bodies. Almost all
primates have highly mobile arms, shoulders, and legs.
Their thumbs and big toes are separate from their other
digits and are capable of grasping objects such as cylindrical branches (Figure 50-7). Elongate claws found in
many reptiles and mammals have been reduced to flattened nails, while the fingertips have become primarily
tactile sites, used instead of the snout to catch insects,
grasp fruit, and so on. The small, light body has narrow
shoulders and hips, so the center of gravity can more
easily be balanced by the primate standing on top of
narrow branches. As Figure 50-7 also shows, the shape
of the tooth arcade and the shapes and sizes of teeth
varied during primate evolution. The tooth differences
correlate with dietary and behavioral features of the organisms in which they appeared. And, of course, the
brain's cerebral cortex increased dramatically in size and
complexity.

Anthropoids living in the trees leap from branch to
branch, reach for fruit or an insect to consume, and carry
out other behaviors dependent on vision. Anthropoids
have evolved a flattened snout, with the eyes facing forward, not to the sides. This provides greater overlap of
the visual fields for better depth perception. Such binocular, stereoscopic vision facilitates catching small, mobile prey such as insects because it permits the accurate
assessment of short distances for the final predatory
strike. It is for this reason that most predators, such as
cats and hawks, have stereoscopic vision, whereas prey
animals, such as mice and deer, have eyes that look more
toward each side than straight ahead (and so can survey a
wide sector of the world around them).

After the dryopithecines had appeared, lineages leading to the large-bodied apes evolved. Early apes apparently carried out a kind of slow vertical climbing, with
the full body weight suspended from the upward-reach-
ing forelimbs and with the feet of the hind limbs gripping the trunk or branch (Figure 50-8a shows a contemporary tarsier doing the same thing). Apes also
suspended their full weight beneath branches (Figure
50-8b) and some developed brachiation, swinging beneath the branches with the full weight suspended from
one arm at a time. The forelimbs of apes became longer
and stronger to support the weight, and also became

about 5 million years ago, the apes descended from
dryopithecines had become greatly limited in their distribution, as are modern apes, and monkeys had taken
over most of the apes' previous niches. Actually, we have
very few fossils of apes more recent than from about 5
million years ago, even though fossils of other Old World
anthropoids continue to be relatively common. This paucity of fossils probably reflects the increasing restriction
of apes to tropical rain forests, where bones seldom fos-
silize in the damp, acidic soil. Also, between 7 and 5
million years ago, there evolved new forms of primates,
the earliest members of the family **Hominidae,** to which
humans belong.

Characteristics of the Primates

The 60-million-year history of our primate ancestors
provided *Homo sapiens* with a major legacy. Since the
processes of evolution, including natural selection, primarily modifies anatomical structures and biochemical
pathways that already exist in a population and do not
start from scratch, the anatomy of our primate ancestors
determined to a large extent our anatomy and physiology
and the resultant possible ranges of behavior.

Much of the body structure of the primates that distinguishes them from other mammals is related to their

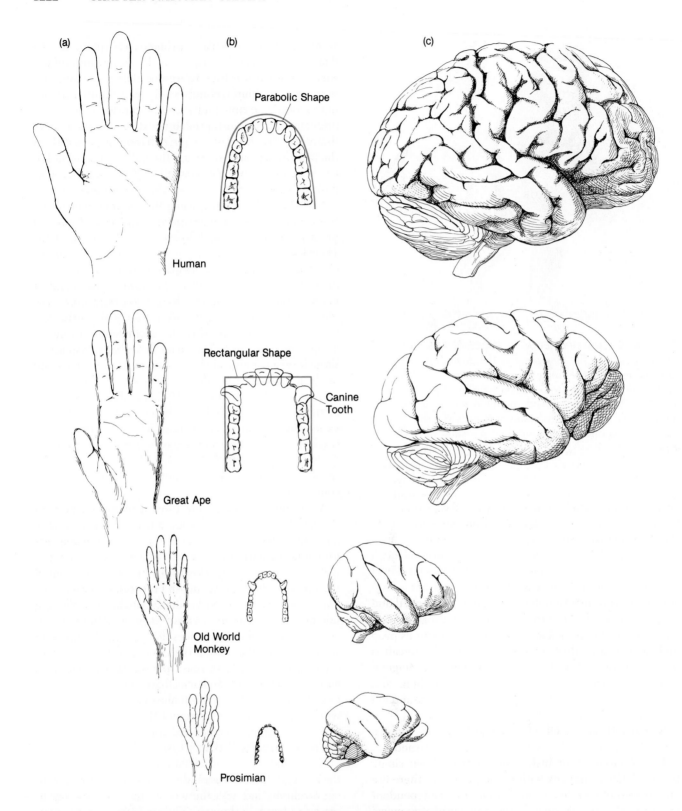

Figure 50-7   HANDS, TEETH, AND BRAINS OF HUMANS AND OTHER PRIMATES.
(a) The opposable thumb (and large toe) arose early to allow gripping of branches from above. The long fingers of monkeys and apes can be wrapped around branches as the body hangs or swings beneath. (b) Note how the shape of the upper jaw has changed among these primates, as have the shape and size of different types of teeth. The large canines of monkeys and apes have been lost in the human lineage. (c) The great increase in surface area of the brain's cerebral cortex is reflected by the folding and fissures in the surface of the ape and human brains. (The brains are not drawn to the same scale; the human brain is much larger in real size than the monkey and prosimian brain.)

(a)

(b)

Figure 50-8   ACCENTUATION OF THE FORELIMBS.
Primates living in the trees frequently hold their forelimbs
in raised positions and use them in tension-bearing ways.
Imagine the stresses and forces you would need to exert
with your legs and arms if in the position of the tarsier
(a), or if you had to carry a little friend the way this
chimpanzee is doing (b).

longer than the hind limbs. The shoulders widened and
became highly mobile too. When modern apes walk
quadrupedally on the ground, their long forelimbs prop
up the front of the body; the animals carry out so-called
knuckle walking, with the greatly elongate fingers,
evolved for gripping branches during climbing and
brachiation, curled inward so that the weight is applied
to the knuckles of the fist.

All primates, but especially the anthropoids, evolved
complex social systems based primarily on learned behavior (Chapter 49). Anthropoids usually have single
births, their newborn are largely helpless, and there is a
long period of growth and maturation that is dependent
on parental care. This means that the primates invest
much time and energy in each offspring, and offspring
have considerable time to learn the complex behaviors
necessary to get along within a social group. This is an
example of *K*-selection, a reproductive strategy centered
on optimizing survival of individual offspring (Chapter
42). In a sense, the evolution of birth at a very immature

stage with prolonged parental care set the stage for the
cultural evolution that is a hallmark of the human lineage. Many primatologists believe that the conditions
within social groups played an important role in the selection for large brains in the apes and humans.

The general primate behavioral patterns we have
listed and the anatomy and physiology that permit them
form the background for human evolution. We, or rather
our ancestors, inherited a complex of genetically determined characteristics from our prehuman primate ancestors, and it is through the modification of these features
that cultural and biological evolution produced modern
humans. It should be emphasized that there never was,
and never will be, anything inevitable about the course
of human evolution. The evolution of all organisms on
Earth is opportunistic and nondirectional. The greatest
portion of human evolution involved hominids that were
adapted to niches considerably different from the one
that *Homo sapiens* fills today. Nevertheless, the complex
of traits adapted for those niches in the trees and grasslands allowed our own lineage to diverge and evolve into
modern humans.

THE TRANSITION FROM APE TO HOMINID

A century and a half after Linnaeus placed humans
among the primates, Charles Darwin published *The
Descent of Man* (1871). In this book, he predicted that
the earliest human fossils would eventually be found in
Africa. He based his prediction on the close anatomical
similarities between humans and the African great apes,
particularly chimpanzees. A century of fossil hunting
worldwide has supported Darwin's prediction of the African origin. Today, we have much besides anatomy to
study; genetics, biochemistry, and behavior of chimpanzees, other apes, and humans overwhelmingly establish
the incredibly close biological relationship among all
anthropoids.

Molecular Relatedness

Humans share a number of anatomical traits with
chimpanzees, including details of the brain, the teeth,
the shoulder and arm, and the thorax, and some aspects
of the legs and feet. These similarities have been supported by the remarkable biochemical similarities of
chimpanzee and human proteins. Table 50-2 shows that
there are very few amino-acid differences among typical
human proteins and those of chimps and gorillas. These
infrequent differences are the sort seen between the

Table 50-2   **DIFFERENCES IN AMINO-ACID COMPOSITION BETWEEN HUMAN PROTEINS AND THOSE OF OTHER MAMMALS***

Protein	Chimpanzee	Gorilla	Orangutan	Gibbon	Macaque	Sheep
α-globin	0	1			4	21
β-globin	0	1		3	8	26
Myoglobin	1	1		1	7	25
Carbonic anhydrase I	0		4		5	

*The numbers mean that there are 0 amino-acid differences between human and chimp α-globin; 1 amino-acid difference between human and gorilla α-globin; 4 amino-acid differences between human and macaque (Old World monkey) α-globin; and so on. Blank spaces indicate that data are not available for some comparisons.

proteins of sibling species, such as dogs and wolves. Data on DNA and blood groups demonstrate corresponding similarities. Figure 50-9 shows the banding patterns of two human chromosomes and corresponding ones in the great apes. In fact, the X chromosomes and four of the autosomes are virtually identical among *Homo sapiens* and the three great apes. Comparison of some 1,000 bands, genetic loci like the bands on *Drosophila* polytene chromosomes (Chapter 10), has revealed a variety of inversions and duplications (note gorilla chromosome 7 in Figure 50-9), but also that humans and chimps clearly have the least differences in banding. These sorts of data have forced a revision of the old idea that the lineage leading to humans split off the lineage leading to the three great apes. To the contrary, evidence now suggests that the line leading to the orangutan split off at least 12 million years ago; the gorilla line separated about 9 million years ago; and, about 6 million years ago, the lines leading to the chimpanzee and to humans diverged (Figure 50-3).

The best reconstruction that paleontologists can piece together at the present suggests that the common ancestor of chimpanzees and humans probably resembled chimpanzees in overall appearance, especially in the skull. But it was undoubtedly smaller than a modern chimpanzee, standing only 3 or 4 feet high and weighing perhaps 50 pounds (23 kilograms), whereas an adult chimpanzee weighs about 100 pounds (45 kilograms). Some paleontologists conclude that these creatures probably lived in open woodland savannas and subsisted on a variety of fruits, nuts, and whatever insects and small game they could obtain, running either on all fours or upright, much as contemporary chimpanzees do. Other scientists point out, however, that humans lack sun-reflecting fur and have subcutaneous fat for insulation, two characteristics never seen in savanna-dwelling animals. We also sweat a great deal and lose much water and sodium to keep cool and also cannot greatly concentrate our urine. In these and other respects we need much more water than do savanna-adapted mammals and other primates. This leads to the idea that our ancestors lived in water-rich environments in African forests.

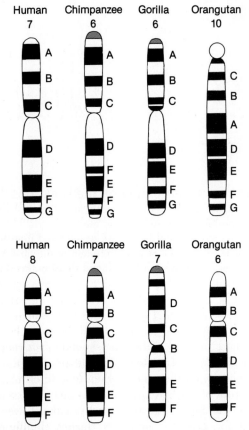

Figure 50-9   BANDING PATTERNS OF HUMAN AND APE CHROMOSOMES.
This figure shows drawings of human chromosome numbers 7 and 8 and the homologous chromosomes in three apes. Note the generally similar distributions of the bands. The A, B, C region of orangutan chromosome 10 is inverted in comparison with humans and other great apes. Gorilla chromosome 7 also shows an inverted region. (These darkened regions are so-called G-bands; they do not represent individual genes.)

Bipedalism

Just what diagnostic feature should paleontologists use to identify the earliest hominid? Anthropologists' best estimate is evidence of an erect, bipedal stance. That is

Figure 50-10 ATTAINMENT OF ERECT BIPEDALISM.
The chimpanzee knuckle walks (a), but can assume a bipedal stance with flexed legs and bent back (b). (c) If its legs are straightened, its body would tilt forward in an unbalanced state never seen in nature. The erect human's legs (d) are considerably longer than the chimp's, and the backbone bends backward in the pelvic region to bring the shoulders and head directly above the feet. (The vertebral columns are shown as solid bars for simplicity.)

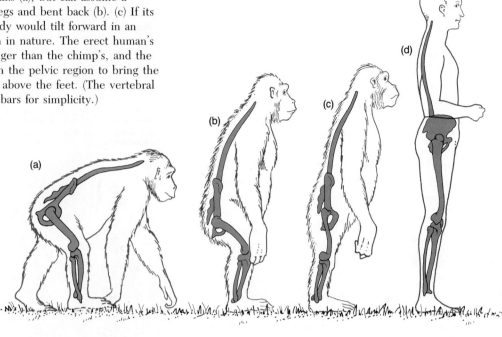

because this stance and the striding bipedal gait evolved well before the large brain, the hand as we know it, and other anatomical features we regard as human. The hominid bipedal stance depends on a special anatomy of the legs and feet, of the pelvic girdle, and of the vertebral column. Only when such features had evolved could the early hominid have moved with the body's center of gravity balanced between the two feet when the legs extend straight down. The result is a very efficient mode of locomotion in which relatively little muscular work must be done to keep the body balanced in an erect state when standing or striding (Figure 50-10). But the nonhuman primates, as well as a variety of other mammals and land vertebrates, that can and frequently do stand or walk on two legs do not have pelvic girdles, lower back structure, or legs that permit them to walk fully erect, with legs directed straight down under the body; instead, as can be seen in Figure 50-10, great muscle work must be done to lift the normally quadrupedal ape or monkey into an upright posture, in which the legs remain bent.

The most probable explanation for the evolution of habitual bipedalism among early hominids is that it is energy-efficient and permits long distances to be covered. Once perfected, it permits an organism to carry food, infants, and perhaps other objects. Later in hominid evolution, bipedalism allowed the hands to be used to carry weapons or simple tools. Those hands, of course, already possessed prehensile fingers and opposable thumbs, for those features had evolved for gripping and locomotion in the trees. Many scientists believe that the

hand and its many uses contributed in major ways to the remarkably rapid expansion of the cerebral cortex in the human lineage. The variety of other behavioral complexities of the human lineage—such as toolmaking, hunting, abstract thought, and elaborate cultural systems—appeared later in human evolution.

Reproductive Behavior and Human Evolution

Bones, muscles, and stance were not the only contributors to human evolution. Many anthropologists have noted that several sexual characteristics and mating patterns appear to be unique to humans: (1) females have permanently enlarged mammary glands (breasts), whereas males have a prominent penis; (2) copulation is usually performed face to face; (3) male and female body and facial hair and fat distributions are quite distinctive; (4) females can copulate independently of ovulation; thus there is a separation of copulation and reproduction (Chapter 18). This behavior by ancient human females is thought to have contributed to the male–female pair bonding that is a key feature of family structure and to the availability of both parents as sources of enculturation for offspring. Unfortunately, such ideas can never really be tested or disproved, since such things as endocrine glands, hormones, and sexual behavior do not make fossils. Even so, it seems reasonable that the unique aspects of human sexual activity did play a role in the origins of family and social groups at some period during early human evolution.

THE EVOLUTION OF THE HOMINIDS

Study of the human fossil lineage has yielded two significant facts: first, that bipedalism preceded brain enlargement; and second, that even after the balanced, bipedal skeleton had evolved, our ancestors retained characteristics appropriate for retreating and climbing into the trees. Let us review the fascinating story of hominid evolution.

The oldest individuals of definite early humans found in the fossil record probably lived not too long after the human lineage had diverged from the chimpanzee line, between 5 and 10 million years ago. The past 4 million years of human evolution can be divided into two overlapping periods, which contain quite different types of humans. The earlier period, which extended from before 4 million years ago to about 1.3 million years ago, was occupied by members of the genus *Australopithecus*. These are the earliest known humans, bipedal in stance and gait, but small in brain. The later period of human evolution began at least 2 million years ago, involved primarily expansion in brain size, and no doubt continues today. It includes individuals whom we place in our own genus, *Homo*.

Australopithecus

Australopithecus (meaning "southern ape") spanned almost 3 million years, and during this time, significant evolutionary modifications took place within its lineage. There is considerable controversy over the precise species designations in the genus; we will adopt the position that there were four species: *Australopithecus afarensis* (which lived from about 4 to about 3 million years ago), *A. africanus* (about 3 to about 2.5 million years ago), *A. boisei* (about 2.5 to about 1.3 million years ago), and *A. robustus* (about 2 to about 1.3 million years ago). These early humans are known from only eastern and southern Africa, where they appear to have been restricted to wooded grassland areas.

Australopithecus afarensis

The fossil remains of the earliest of these creatures, *Australopithecus afarensis*, have been found only within the past decade. The discoveries were made first at Hadar in Ethiopia by Donald Johanson, Yves Coppens, and their co-workers. One of these specimens, "Lucy," named after the song "Lucy in the Sky with Diamonds" (Figure 50-11), is the most complete early fossil discovery, aside from a *Homo erectus* skeleton of much younger age. In 1978, Johanson, Coppens, and Tim

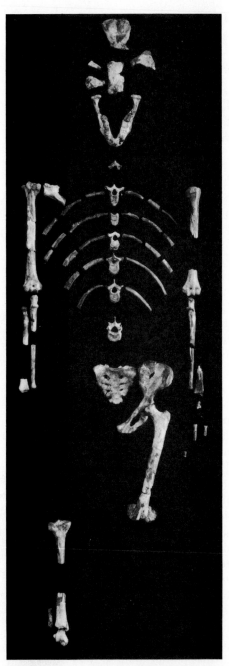

Figure 50-11 *AUSTRALOPITHECUS AFARENSIS:* **THE FIRST HUMAN.**
This remarkably complete 3.5-million-year-old skeleton is of a young female of the species *Australopithecus afarensis*. This skeleton is popularly known as Lucy. It is from fossils such as Lucy that paleontologists are able to reconstruct the overall body size and proportions of *A. afarensis*. The pelvis of this female has the typical human shape; the thigh bone is relatively shorter than our own. Finger and foot remains from other individuals indicate similarities, but not identical configurations, with those of modern humans.

White decided that these fossils represent a new hominid species, *A. afarensis*, named after the Afar region of Ethiopia, where Hadar is located. Other fossils, found at Laetoli in Tanzania by Mary Leakey, are also believed to belong to *A. afarensis*. Enough of the fossils have been found to provide a reasonably complete view of the species's anatomy and general evolutionary relationships. Most paleontologists now regard *A. afarensis* as the stem stock leading to later *Australopithecus* and *Homo* species.

Fossil individuals of *A. afarensis* exhibit a mixture of human and ape characteristics that have been studied in minute detail by several groups of scientists working on precise replicas of the original fossil bones. The skulls appear very apelike, with large projecting faces and small brains—about 450 cubic centimeters (about 1 pint) in volume, which is only slightly larger than a chimpanzee's brain (about 400 cubic centimeters), but about one-third the size of a modern human's brain (average about 1,350 cubic centimeters). The teeth and jaws are also quite apelike, with the molars and premolars arranged in parallel rows, similar to those of chimpanzees but contrasting with the smoothly curved (parabolic) tooth rows of recent humans (Figure 50-12). The front teeth (incisors and canines) are large and slightly projecting, with spaces between the upper canines and incisors to accom-modate the projecting canines from the lower jaw. In contrast to these apelike characteristics, the *foramen magnum*, the hole at the base of the skull through which the spinal cord exits, points downward. It thus indicates that the head sat on top of the vertebral column rather than in front of it, as in all quadrupedal apes. This means that these creatures could stand upright.

The limb skeleton of *A. afarensis* shows a similar mixture of features. The arms are similar to those of both apes and humans; both the shoulder and the elbow have features suggesting that the arms could be held overhead and used to support weight in the trees. The hands are human in overall proportions; the fingers are not quite as long as those of apes, yet they have a degree of curvature like that of apes and could be used for gripping branches. Indeed, the detailed joint structures of the digits are very similar to those of chimpanzees, suggesting that *A. afarensis* lacked the extensive manipulative abilities seen in later humans. The pelvic girdle, in contrast, is very different from that of the apes and is clearly adapted for a balanced bipedal gait—although, again, details of the hip joint suggest mobility of the sort seen in chimpanzees as they grasp branches and move in the trees.

In living humans, the anatomy of the legs, particularly at the hips, knees, and feet, has become highly specialized for walking on only two legs. Actually, we walk pri-

Figure 50-12   TEETH OF APE AND *A. AFARENSIS*.

(a) Chimpanzee's tooth arcades, with a female above and male below. (b) The equivalent tooth arcades of female (above) and male (below) *Australopithecus afarensis* individuals. The two sets are basically similar, and, although it is not obvious from these photos, the canines of both were large. The sexual dimorphism is also apparent. The narrowing of the front of the arcade in the female *A. afarensis* is a harbinger of the condition in both sexes of *Homo sapiens*.

(a)   (b)

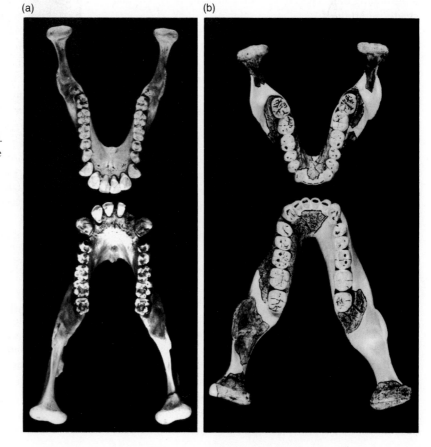

marily on one leg, since 90 percent of the time during walking our weight is on only one foot. Our feet have therefore become semirigid, arched structures with the big toe in line with the other toes. Each foot is a triangular platform over which we balance our bodies as we stride forward first on one foot, then on the other. Our thighs angle in, so that our knees are directly below the center of gravity during walking. And our pelvis has widened to enable the hip muscles to keep the trunk upright above the single support leg. This special configuration of the leg and pelvis is unique not only among primates, but also among mammals.

The fossil remains of A. afarensis indicate that this group, too, possessed legs adapted for bipedalism. The pelvis is short and widened, indicating the same hip-stabilizing mechanism that we have today; the knees angle in toward the midline, and the feet lack the divergent, grasping big toe used by other primates to grasp branches. Nevertheless, the four outside toes, which are more elongate and curved than our own, show similarities to those of apes and, like the curved fingers, suggest that Lucy and her relatives still had adaptations for grasping branches. Furthermore, the thighbone and lower-leg bones are substantially shorter than those of later hominids, reflecting the closeness of A. afarensis to the apes, with their short legs and long arms. One major trend in hominid evolution after A. afarensis was the lengthening of the leg; only then could the fully modern stride be attained.

These anatomical interpretations are supported by Mary Leakey's discovery of the three sets of footprints at Laetoli. Careful analysis of these footprints indicates that body weight and forces of propulsion were transferred to the ground just as in modern humans—whoever made the footprints was truly bipedal, although the unusually short legs meant that the strides must have been short, as well. Since fossils of A. afarensis are known from the same geological deposits at Laetoli, it is assumed that the footprints were made by this type of early human.

Although A. afarensis was definitely humanlike, in the broad sense of the term, we cannot be sure what its behavior was like. Individuals stood only 3.5 to perhaps 4.5 feet high and weighed only 40 to 50 pounds (18–23 kilograms). They probably spent significant portions of time in the trees, perhaps when feeding, at night, or when fleeing danger. The brain shows little of the enlargement in relation to body size that distinguishes later humans from other primates; the pattern of folds (sulci) on the brain surface is like that in apes, not humans. This suggests little likelihood of a significant increase in cognitive abilities over those of a chimpanzee. The small brain size correlates with a small birth canal in the pelvic bones. The hands, with their curved fingers, do not appear to have evolved the fine manipulative abilities associated with tool use, even though we would surmise that

A. afarensis made and used simple tools at least as often as do chimpanzees. As yet, no definite stone or bone tools (other materials would not be preserved) have been found associated with A. afarensis fossils. And there is no evidence that they hunted or systematically scavenged any of the animals whose remains paleontologists find in abundance along with those of A. afarensis— none of these animal bones show definite marks of human butchering. A. afarensis therefore had a way of life that possessed few of the features commonly associated with human cultural behavior.

Australopithecus africanus

The next hominids in the fossil record belong to the species Australopithecus africanus. In 1924, Raymond Dart identified the first australopithecine, the Tuang child, in South Africa. There was great excitement worldwide when Dart published pictures of this first "missing link" between humans and apes. In years since, large numbers of A. africanus specimens have been found. Most share many anatomical features with A. afarensis. They were relatively small bipedal humans with a protruding face and with a brain only slightly larger (450–500 cubic centimeters) than that of A. afarensis. In fact, the anatomy from the neck down changed little. New analyses of hand and foot anatomies are consistent with the hypothesis that these hominids, too, spent time in the trees—perhaps while sleeping or feeding, perhaps for safety.

One important evolutionary change within the Australopithecus lineage was the chewing apparatus. The front teeth lost the slight apelike projection seen in A. afarensis and became entirely human in size and shape. The tooth row took on more of a curved shape. But primarily the cheek teeth—the flat grinding and crushing molars and premolars—became exaggerated in A. africanus. To hold these large teeth and to chew effectively with them, the jaws were heavily built, and the chewing muscles were substantially larger than those of modern individuals. The precise adaptive reason for this heavy dentition is unclear, but these hominids were certainly using their teeth to process coarse, abrasive food, such as tubers. Large, heavy teeth not only are effective for such food, but also wear down slowly and thus last longer.

Australopithecus boisei *and* Australopithecus robustus

Fossil A. africanus are known from about 3 to 2.5 million years ago. Between 2.5 and 2.0 million years ago, two later geographical variants of Australopithecus appeared. A. boisei in east Africa (named for a donor of funds to the Leakey expedition) and A. robustus also in east Africa emerged from the earlier geographical variants of *Australopithecus* (*A. afarensis* and *A. africanus*, respectively). In 1986, a 2.5 million-year-old skull was discovered at Olduvai Gorge; dubbed the "black skull" because of mineral discoloration, the specimen seems to be a precursor of *A. boisei*. The *boisei* and *robustus* species of *Australopithecus* continued the trend toward increasingly large cheek teeth and associated chewing musculature, especially in the *boisei* line. However, the limb skeletons, and hence the implied locomotor and manipulative patterns of these later species of *Australopithecus*, remained the same as those of their predecessors. Their brains did increase slightly, to about 500 to 550 cubic centimeters in volume, but that increase was just a reflection of a slight increase in body size. Figure 50-13 shows three of the possible relationships between the various australopithecines, as well as between the types of *Homo* we will consider in the next section.

There is a tendency to view the various australopithecines as primitive groups of early humans. Their small brains and the lack of evidence for toolmaking or hunting imply that their behavioral system was simpler than that of the genus *Homo*. However, in evolutionary terms they were highly successful. They occupied much of sub-Saharan Africa for at least 3 million years, far longer than the genus *Homo* has been around. And despite their anatomical and adaptive differences from later humans, australopithecines laid the groundwork, in an evolutionary sense, for the emergence of *Homo sapiens*.

The Origin of the Genus *Homo*

By about 2 million years ago, at the beginning of the Pleistocene epoch, a new form of human had appeared in the prehistoric record of Africa. This arrival and its descendants were to change not only the fine ecological balance of their native habitat, but also the patterns of life in much of the world. These were the first members of our own biological group, the genus *Homo*, which in-

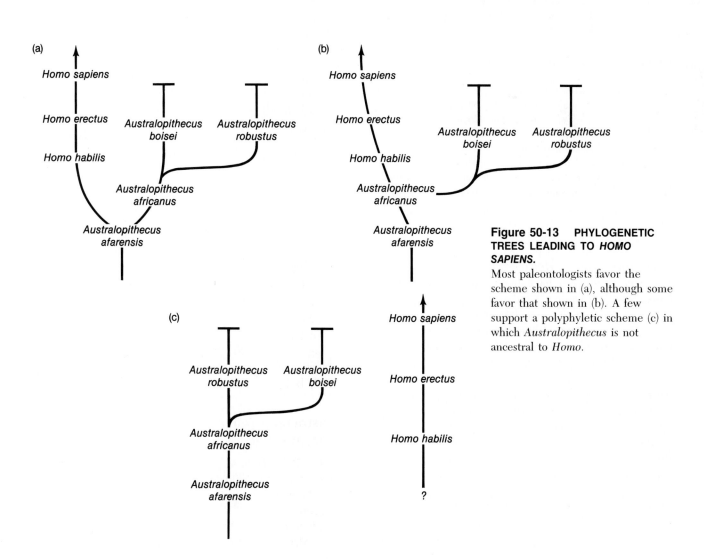

Figure 50-13 PHYLOGENETIC TREES LEADING TO *HOMO SAPIENS*.
Most paleontologists favor the scheme shown in (a), although some favor that shown in (b). A few support a polyphyletic scheme (c) in which *Australopithecus* is not ancestral to *Homo*.

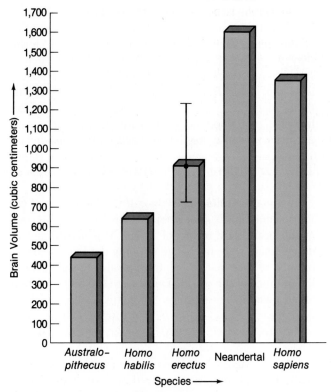

Figure 50-14   RELATIVE BRAIN VOLUMES OF THE HUMAN LINEAGE.
The dramatic trend toward larger brain size is represented by this bar graph, which displays the average brain volumes of a number of fossil specimens. The *Homo erectus* range is particularly great and is indicated with the lines ranging from 725 to 1,225 cubic centimeters; this may reflect the fact that the rate of brain expansion in that species was very high. The Neandertal brain was somewhat larger, on the average, than that of modern *Homo sapiens*.

cludes *Homo habilis*, *Homo erectus*, and *Homo sapiens*. These early humans have been placed in the species **Homo habilis** (meaning "handy man"), named by the late Louis Leakey and his associates on the basis of fossils from Olduvai Gorge. Other specimens of *Homo habilis* have been found, especially at East Turkana in Kenya, by Richard Leakey and his co-workers.

Homo habilis superficially resembled its forebear, *Australopithecus*, but was different in several important aspects. Individuals were larger, standing a full 5 feet tall. The brain was larger, about 750 cubic centimeters on the average, and thus somewhat more than one-half the size of the modern human brain instead of the australopithecine's one-third. But more importantly, while body size increased about 25 percent relative to *Australopithecus*, brain size increased about 50 percent (Figure 50-14). A simple increase in brain size does not necessarily imply an increase in mental abilities, since a larger

body requires more neurons to monitor and control its activities. However, an increase in brain size relative to body mass indicates the presence of additional neurons, over and above those needed to operate basic body functions. The marked increase in brain size in *Homo habilis* relative to *Australopithecus* thus implies an augmentation of mental abilities.

The actual internal organization of a brain is not revealed in fossils. Nevertheless, the earliest brain that shows the human pattern of folds (sulci) is a 2-million-year-old *Homo habilis* specimen, with a volume of about 750 cubic centimeters. Humanlike folding was apparently absent from the enlarged brains of the australopithecine lineage. Biologists would like to know, of course, how the larger brain volume and surface folding relate to such things as the speech centers and the dominant hemisphere (Chapter 40).

The evolutionary origins of *Homo habilis* are currently a subject of considerable debate. The species appears in the fossil record around 2 million years ago, and the preceding 500,000 years yield no diagnostic fossils. Some researchers extend the *Homo* lineage back to 3 million years ago and derive *Homo habilis* directly from *A. afarensis*. Others regard the origin of *Homo habilis* as more recent and identify a population of *A. africanus* as its progenitor. It appears unlikely that *Homo habilis* emerged in its known form much before 2 million years ago, although early crude stone tools, possibly made by *Homo habilis*, have been found in 2.4-million-year-old deposits (forty-eight chipped stone tools, in fact, were found at a single site). The word "possibly" is used intentionally here, since we cannot be sure whether individuals we would identify as australopithecines might not have made these crude stone tools. Only a more complete fossil record will resolve uncertainties about the origin and early history of *Homo habilis*.

Other than the pronounced increase in brain size over and above the modest increase in body size, only a few anatomical features distinguish *Homo habilis* from the earlier *A. africanus*. The face remained large, with relatively large cheek teeth. The front teeth, however, were also large. The trunks and limbs of *Homo habilis* were fully adapted for upright, bipedal locomotion. The hand skeleton of *H. habilis*, however, represents a curious mosaic; it is close to that of recent humans in its thumb and fingertips, suggesting improved manipulative skills; yet the bones of its digits are heavy and somewhat curved, permitting the powerful grasping seen in chimpanzees and gorillas. These characteristics have been interpreted to mean that *Homo habilis* lived on the ground during the day, but retreated to the trees at night to sleep.

Homo habilis apparently performed two characteristic human activities: toolmaking and butchering of large animals. The earliest stone tools are rather crude frac-

(a)

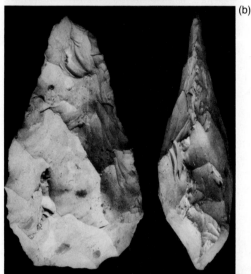

(b)

(c)

Figure 50-15   PRIMITIVE TOOL "INDUSTRIES" OF HUMAN ANCESTORS.
(a) An Oldowan tool excavated in Olduvai Gorge, Tanzania. This stone is a coarse cutting and chopping tool. Its owner made it from a lava pebble by knocking off flakes on a small block of stone. (b) An Acheulian hand ax from France. Tools of this general shape and detail of flaking are found across Africa, Europe, and much of Asia dating from the Pleistocene. The oldest known hand axes are from deposits about 1.5 million years old in East Africa, and they continue to appear in stone-tool industries 100,000 years old. They represent the major category of standardized tool during the Acheulian and were probably a general-purpose chopping and cutting tool. (c) A selection of Mousterian tools, mostly from archeological sites in Western Europe. The tools shown are, from left to right, a backed knife, a piercer, a scraper, and a hand ax. The quality of the workmanship is considerably above that of the Acheulian and Oldowan tools, and there are more specialized, limited-purpose tools.

their human manufacture. The industry that they make up has been called the **Oldowan industry,** after Olduvai Gorge.

It is uncertain whether members of *Homo habilis* actually hunted, or just how far they had changed from being herbivores like australopithecines. They probably did hunt small animals but scavenged from fresh kills of large animals made by other predators, such as lions and hyenas. Their reduced canine and incisor teeth probably made it essential that they use sharp stones to cut skin and meat from their stolen booty. *Homo habilis* definitely butchered both large and small game, as cut marks on the bones indicate. A major addition of meat to their diet would have set them apart from other primates and probably from most australopithecines. It is possible that the beginnings of dependence on tool use and on the exploitation of animal carcasses started a shift in the human adaptive pattern that was to dominate subsequent human evolution.

Homo erectus

From the origins of the Hominidae, around 4–5 million years ago, to the end of the period occupied by *Homo habilis*, about 1.5 million years ago, humans were restricted to sub-Saharan Africa. They lived primarily within the savanna and savanna–woodland belt that extended from Ethiopia to South Africa, a zone that offered abundant plant and animal resources and varied little in temperature during the year. Even today, despite the pressures of agriculture and expanding human populations, this area of the world boasts an extremely rich and varied plant and animal life. It appears that *Homo habilis* lacked the ability to live outside this prehistoric lush environment.

tured rocks that do not look particularly like deliberately fashioned tools (Figure 50-15a). Yet they have been discovered in concentrations associated with broken and cut animal bones at sites in Olduvai Gorge in Tanzania, East Turkana in Kenya, and Omo and Hadar in Ethiopia. Furthermore, their shapes are not randomly fractured, but form patterns of breakage, leaving little doubt as to

Some time around 1.5 million years ago, prehumans passed an adaptive threshold. Within a short period of time, these people extended their geographical range to include all the tropical and subtropical areas of the Old World, excluding the inaccessible islands. Evidence of their presence, in the form of either fossils or primitive tools, appears in strata dated to that time in mountainous areas of Africa, across the North African coast, and across southern Asia into China and Indonesia. In general, the Chinese and Indonesian fossils date from 1.2 to 1 million years ago and thus are younger than the African stock, which is at least 1.6 million years old.

This major geographical expansion, an important indicator of ecological and evolutionary success, was accompanied by anatomical and behavioral changes in the human lineage. A new form of *Homo*, called **Homo erectus,** appears in the fossil record around 1.6 million years ago, and a new stone industry, called the **Acheulian,** appears in the archeological record (Figure 50-15b). Thus these early humans had an ability greater than *Homo habilis* to deal effectively with varied food resources, increased seasonal changes, and greater climatic extremes.

Early *Homo erectus*, at least the males, may have been substantially taller than *Homo habilis*. The most complete human fossil yet found is that of a twelve-year-old boy (as judged by the teeth and the bone growth areas) who stood about 5.5 feet (1.6 meters) tall at death and would have been perhaps 6 feet (1.8 meters) tall if he had lived to maturity (Figure 50-16). The skull is quite massive, considerably more so than many female *Homo erectus* skulls, which implies considerable sexual dimorphism in the species. As Figure 50-16 shows, the arms and legs were quite like those of modern humans, although they were considerably more robust, indicating greater strength and endurance than in modern humans. An intriguing aspect of the boy's skeleton is the extreme narrowness of the pelvic girdle and especially the hole that, in females, serves as the birth canal. This may mean that the newborn of *Homo erectus* were extremely small. We can only speculate about whether that meant a very prolonged period of infant dependence and a restriction in the mother's mobility while the tiny newborn matured. Such parent–offspring relationships could have enhanced the opportunity to pass on learning and culture from one generation to the next. The brain of *Homo erectus* was much larger than that of *Australopitheus*, and it seems reasonable to conclude that it underwent some internal reorganization that permitted more complex tool manufacture and behavior (Figure 50-14). The Acheulian industry, named after the town on St. Acheul, in France, where it was first identified in the nineteenth century, shows a greater variety of tools and finer workmanship than the Oldowan industry.

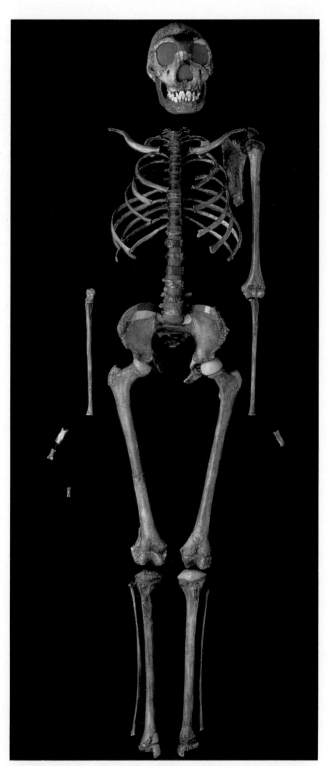

Figure 50-16 *HOMO ERECTUS:* **THE MOST COMPLETE FOSSIL HUMAN SKELETON EVER FOUND.**
This twelve-year-old boy lived 1.6 million years ago. Note the narrow hips, yet very long heads on the two femur bones. The knees could be held in, near each other, when this boy stood. In contrast, an ape's legs are directed straight down or slightly lateral when it stands.

This is particularly evident in the tool that has become a marker for the Acheulian, the hand ax (Figure 50-15b).

The archeological debris of *Homo erectus* indicates a gradual increase in the ability to deal with harsher and more varied environments. It is argued hotly whether *Homo erectus* was primarily a scavenger or a hunter of meat (plants continued to be in the diet). Evidence suggests that hunting may have occurred. An increased reliance on hunting would have enabled them to successfully exploit nontropical regions, where plant foods are seasonably scarce. One of the major technological advances in this process was the control of fire for warmth and cooking. A good campfire provides heat during long winter nights and keeps away potential predators. Furthermore, *Homo erectus* may have discovered that cooked food is easier to eat and is more digestible than raw food. The earliest evidence of campfires dates from about 500,000 years ago, simultaneous with the first habitation of temperate zones in Europe and Asia.

During the 1.3 million years of *Homo erectus*'s existence (from 1.6 to 0.3 million years ago), the species underwent evolutionary change. The face and teeth decreased in size and massiveness (Figure 50-17), while the brain increased in size spectacularly, from about 800 cubic centimeters around 1.5 million years ago to about 1,200 cubic centimeters 500,000 years ago (Figure 50-14). The reasons for this greatly accelerated expansion are not certain, although cultural evolution and even the beginnings of language may have played roles. The gradual alteration in form of the head and face led imperceptibly into the anatomical configuration that we identify as that of archaic forms of *Homo sapiens*, our own species. There is no distinct boundary between *Homo erectus* and *Homo sapiens*. Yet early *Homo erectus* and late *Homo sapiens* are very different creatures so as to warrant separate species designations. A necessarily somewhat arbitrary boundary has been placed at about 300,000 years ago.

Homo sapiens

The last 300,000 years of human evolution have been occupied by two groups of humans: an early group

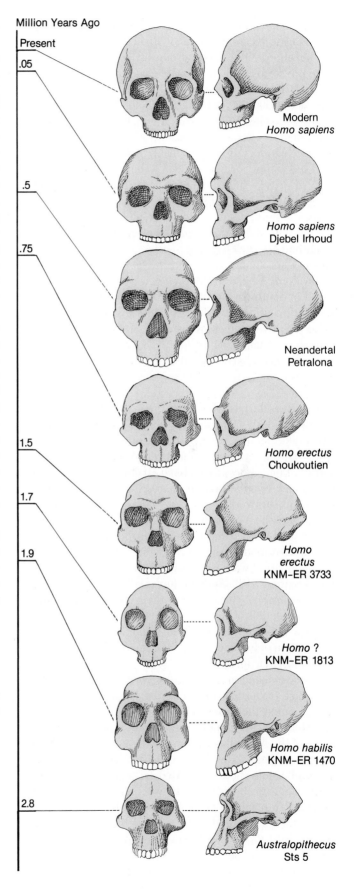

Million Years Ago

Present

.05

Modern
Homo sapiens

.5

Homo sapiens
Djebel Irhoud

.75

Neandertal
Petralona

1.5

Homo erectus
Choukoutien

1.7

*Homo
erectus*
KNM-ER 3733

1.9

Homo ?
KNM-ER 1813

Homo habilis
KNM-ER 1470

2.8

Australopithecus
Sts 5

Figure 50-17   THE TRANSITION TO MODERN HUMANS. The sloping anterior forehead was replaced by the more vertical one, as the brain expanded so greatly. The *Homo habilis* skull has a cranial capacity of 750 cubic centimeters and is the earliest fossil to show the pattern of folds (sulci) on the brain surface similar to those of humans.

known as archaic *Homo sapiens*, which includes the well-known Neandertals (also spelled "Neanderthal"), among others; and a late group of *Homo sapiens* who were physically indistinguishable from modern humans. Archaic *Homo sapiens* occupied most of the Old World from about 300,000 years ago to about 75,000 years ago. In the following 40,000 years a transition took place to yield modern *Homo sapiens*.

The **Neandertals** have probably been more maligned than any other form of human. They owe this fate to their priority in human paleontological studies. The first recognized Neandertal specimen was unearthed in Germany's Neander Valley (the German word *tal* means "valley") in 1856, three years before Charles Darwin published *On the Origin of Species*. Until the 1920s and 1930s, when numerous *Homo erectus* fossils were found in Java and China and *Australopithecus* remains were found in South Africa, little was known of other forms of prehistoric humanity. The Neandertals, therefore, became the archetype of prehistoric—and, by implication, brutish and ignorant—people. Even though their lives may have been harsh, dangerous, and simple in comparison with our own, they do not deserve such a reputation.

The Neandertals lived in Europe and the Near East from about 130,000 to about 35,000 years ago, but Neandertal-like peoples preceded them in the same regions and occupied other areas of the Old World during earlier and the same times. Like *Homo erectus*, all these Neandertal and Neandertal-like people had large faces with projecting brow ridges (Figure 50-17), long low brain cases, and heavily built trunks and limbs. Yet their brains were similar in size and probably in organization to those of modern humans (Figure 50-14). The face, although large, was less massive than that of *Homo erectus* (Figure 50-18). And the limb skeletons were somewhat less robust than those of their hominid ancestors. All these details of anatomy suggest an increasing ability to deal with the environment effectively through an elaboration of learned, cultural behaviors, rather than by pure muscular force.

The associated archeological record supports this observation. The stone tools that the Neandertals left behind, named the **Mousterian industry,** after the site of Le Moustier in France, are more sophisticated than those of the Acheulian industry, with a greater variety of special-purpose tools (Figure 50-15c). The Neandertals routinely built shelters, hunted large game (a wooden spear has even been found between the ribs of a fossil elephant in northern Germany), cared for the aged and infirm for long periods of time, and buried their dead (Figure 50-19). Thus for the first time in the human fossil record, we find among the Neandertals a significant number of elderly individuals, several of whom had suffered broken bones that had healed with deformities

Figure 50-18   AN ARCHAIC *HOMO SAPIENS*.
This Neandertal, shown here without any hair to reveal his facial features and cranium, had a broad face and large projecting brow ridges. The body was quite heavily built. Although large, the face was smaller than that of *Homo erectus*. When compared with earlier *Homo* fossils, one can see a continued progression toward the facial features of modern *Homo sapiens*.

years, if not decades, before the individuals died (the original 1856 Neandertal skeleton, an elderly male, has a broken elbow that prevented him from fully straightening his arm). Yet even the oldest Neandertals lived only to about 50, and few lived more than about 35 years. The Neandertals not only buried their dead, which is one of the reasons we know so much about them (skeletons preserve much better if bodies are buried immediately), but they also made a few simple body ornaments. Together these behaviors indicate an increasing awareness of each other as individuals within a more and more complex social network.

Despite these humanlike behaviors, paleoanthropologists now wonder whether any of the archaic *Homo sapiens* should properly be included in our species. The physical differences are, after all, quite substantial. Clearly, the biological test of species identity, the ability to reproduce, cannot be used as a means of distinguishing between forms known only from fossil remains. For this reason, paleoanthropologists do not know whether various australopithecine "species" and *Homo habilis* individuals living as contemporaries, or indeed even

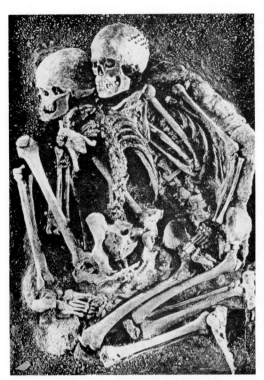

Figure 50-19   NEANDERTAL BURIAL.
The skeletons of an old woman and an adolescent male as they were discovered in the Grotte des Enfants in France, near Monaco on the Mediterranean coast. These two individuals were buried around 60,000 years ago, at the beginning of the Upper Paleolithic. This is one of many burials, some with elaborate grave offerings, from this time in human evolution.

Neandertals and the first modern *Homo sapiens*, could have successfully exchanged genes.

There has been some debate as to whether the transition from archaic, Neandertal-like creatures into modern *Homo sapiens* occurred at one site or widely across the Old World. Both the fossil record and genetic evidence suggest that Africa once again was the source of anatomically modern humans. Analyses of mitochondrial DNA from modern populations across the world imply that all contemporary human mitochondria are derived from women who lived in Africa about 200,000 years ago (mitochondria are passed on solely through eggs, never human sperm). Early forms of modern *Homo sapiens*, sometimes called *Cro-Magnon man*, are found in Europe and the Near East and may have lived for some time as contemporaries of surviving Neandertals. Most likely, early modern humans, who emerged in sub-Saharan Africa around 75,000 years ago, spread gradually northward through Africa, through the Near East, and then across Eurasia. As they spread geographically, they absorbed existing populations of Neandertals and Neandertal-like humans, to produce the geographically variable yet generally very similar populations of early modern humans.

THE ORIGIN AND DIVERSIFICATION OF RECENT HUMANS

The people who evolved from the early modern human populations of the Old World represented a new phase in human evolution. More so than any of their predecessors, they had the ability to adapt to varied and often demanding environments. This can be seen in both their biology and their archeological remains.

The earliest humans of modern appearance were different from Neandertals in several anatomical features. The brain case was higher and rounder, and the brow ridge had all but disappeared (Figure 50-17). The whole body, but particularly the limbs, was markedly less massive, even though these people—men, women, and children—would appear quite athletic if we were to encounter them today. We do not know whether the selection for lighter skin colors in populations living farthest from the equator (see Chapter 34) took place during the spread of early modern humans or earlier in human history. Physical changes continued to their current state as people gradually became less nomadic, adopted agriculture, and became urbanized during the past 10,000 to 15,000 years.

With the arrival of modern-appearing humans, several important behavioral changes took place. There was an elaboration of technology, reflected primarily in tools: many more types of tools were made of varied materials and by more sophisticated techniques than in earlier cultures. Art—in the form of sculpture, painting, and engraving—appeared, suggesting a major elaboration of symbolic forms of communication (Figure 50-20). The use of art, a form of abstraction, was probably accompanied by an increase in verbal-communication ability. The living sites and associated structures became larger and more complex, and human burials became more common and often included elaborate offerings.

These cultural manifestations were accompanied by a general population explosion. This period saw the first major increase in territorial occupation since *Homo erectus* had spread outward from Africa 1.5 million years ago. For the first time, people inhabited the extremely harsh Arctic regions of Eurasia, following the herds of mammoth, woolly rhinoceros, reindeer, and other large game. They expanded across the Bering land bridge into North America, and they crossed open water to occupy New Guinea and Australia. By 12,000 years ago, humans

Figure 50-20   ART: AN ANCIENT FORM OF ABSTRACTION.
Form, perspective, and shadowing all appear in this sophisticated work created by a Stone Age artist some 12,000–15,000 years ago in the Lascaux Caves in France.

Figure 50-21   HUMAN MIGRATION OVER 4 MILLION YEARS IN THE OLD WORLD.
The successive periods of human geographical expansion are based on current fossil evidence. From about 4 million to 1.5 million years ago, humans were restricted to eastern and southern Africa (blue). From 1.5 million to 35,000 or 40,000 years ago, humans spread rapidly across the tropical and subtropical Old World and then slowly moved into more temperate zones (purple). About 35,000 to 40,000 years ago, humans began migrating into Arctic regions, the Americas, New Guinea and Australia, and outlying islands (orange).

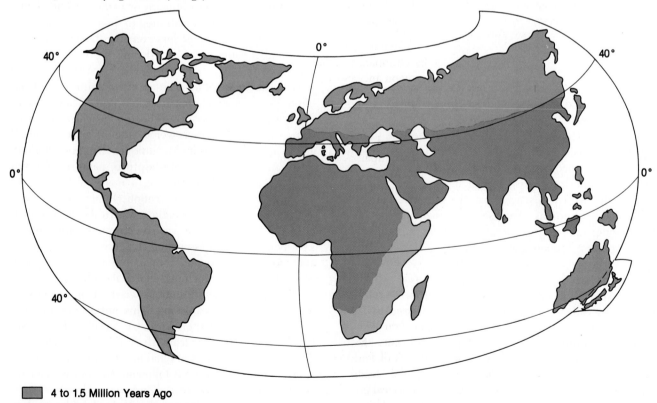

4 to 1.5 Million Years Ago

1.5 Million to 40,000 Years Ago

40,000 Years Ago to Present

had occupied virtually all the inhabitable regions of the Earth (Figure 50-21).

In a species as widely dispersed geographically as *Homo sapiens*, it is only natural to expect genetic differences to have arisen among populations (Chapter 43). The sheer distances between groups, as well as geographical barriers to population movement, would have decreased the rate of gene flow. And populations in different areas of the world must have adapted—to a certain extent biologically, and even more culturally—to different environmental pressures. The product of this adaptation, as with all other dispersed species, was the formation of subspecies, known popularly as *races*.

Long ago, Linnaeus recognized the basic unity of living humanity, despite its currently wide geographical distribution and seemingly varied nature, when he placed all people in one species—*Homo sapiens*. We now know that this biological unity is due to our common evolutionary origins that lead back through geological time through various archaic forms of humans to our nonhuman primate ancestors. The major stages in that

marvelous biological odyssey are listed in Table 50-3. It is intriguing to contemplate that those three australopithecines, striding across the volcanic ash, were unknowingly walking toward us in a very real sense.

Table 50-3   **MAJOR STAGES OF HUMAN EVOLUTION**

Late Miocene 9–5 million years: Divergence of human and chimpanzee lineages, establishment of significant bipedalism.

Pliocene 5–2 million years: Evolution of diverse bipedal, small-brained, open-country (or forest?) australopithecines.

Late Pliocene 2.5–2 million years: Origin of *Homo:* large brained, stone-flaking, more omnivorous, primitive behavior patterns.

Late Pleistocene from 130,000 years: Archaic and later modern *Homo sapiens;* anatomically and probably behaviorally similar to living humans by about 35,000 years ago.

Latest Pleistocene from 10,000 years: Plant and animal domestication; populations increase in density and become sedentary, eventually leading to urbanization.

SUMMARY

1. The *primates* include several groups: the *prosimians* (such as lemurs and tarsiers) and the *anthropoids: New World monkeys* (such as spider and howler monkeys), *Old World monkeys* (such as baboons and macaques), *apes* (gibbons, orangutans, gorillas, and chimpanzees), and the human lineage.

2. The earliest primates—small, shrewlike creatures—appeared more than 65 million years ago. About 54 million years ago, the first primates with bodily characteristics of modern prosimian arose. The earliest anthropoids appeared around 30 million years ago, giving rise to the New World monkeys, the Old World monkeys, the apes, early forms of which are called *Dryopithecus*, and humans.

3. Primates are characterized by adaptations to an arboreal way of life. These adaptations include arms and legs adapted to locomotion through the trees, hands and feet adapted for grasping branches, and stereoscopic vision. This arboreal way of life also may have contributed to the development of complex social systems.

4. Humans and the African apes, the chimpanzee and gorilla, share a common ancestor that lived after the time of the evolutionary split with the ancestors of the Asian orangutan. Subsequently, the gorilla lineage arose and diverged. Thus, humans and chimpanzees share a common ancestor after divergence from all the other ape lineages. It is in the line leading to humans that erect bipedalism evolved as an efficient mode of locomotion.

5. The earliest members of the family *Hominidae*—humans and their direct ancestors—were of the genus *Australopithecus;* they lived between about 4 million and 1.3 million years ago. They combined ape-like skull features with full upright bipedal locomotion, the primary anatomical adaptation of human evolution. This genus is divided into four species—*A. afarensis, A. africanus, A. boisei,* and *A. robustus*—primarily on the basis of differences in skull and in tooth size and shape.

6. *Homo habilis*, the first species, classified in our own genus, appeared about 2 million years ago and was characterized by a significant increase in brain size and the adoption of two human behavior patterns: habitual tool use and butchering of animal carcasses. The stone tools of *Homo habilis* make up the *Oldowan industry.*

7. The first spread of humans outside the African source land began about 1.5 million years ago with *Homo erectus*. Anatomically, *Homo erectus* was little changed from *Homo habilis*, but significant behavioral reorganization seems to have occurred, as is reflected in the advances in toolmaking (the *Acheulian industry*), geographical expansion of the species, and later the use of fire. During the course of its existence, *Homo erectus* experienced a gradual decrease in face and tooth size and a spectacular increase in brain size.

8. The past 300,000 years of hominid evolution have been occupied by two groups of humans: archaic *Homo sapiens* (including *Neandertals*) and modern humans. The large-brained Neandertals, who lived between 130,000 and 35,000 years ago, were much more like modern humans than was previously thought. They used more sophisticated stone tools (the *Mousterian industry*) than did earlier cultures, built shelters, cared for the aged and infirm, and buried their dead.

9. The humans who evolved from archaic *Homo sapiens*, Neandertal and Neandertal-like populations, demonstrated a new ability to adapt to varied and often demanding environments. These humans, anatomically much like ourselves, had appeared by about 75,000 years ago in Africa and they rapidly spread across the inhabitable Earth.

10. The wide geographical spread of recent humans has led to the evolution of human racial differences. These racial variations are superficial and are probably minor adaptations to the environments in which the groups have lived prior to the times of easy long-distance travel.

KEY TERMS

Acheulian industry
anthropoid
ape
Australopithecus
brachiation
Dryopithecus

Hominidae
Homo erectus
Homo habilis
knuckle walking
Mousterian industry
Neandertal

New World monkey
Old World monkey
Oldowan industry
primate
prosimian

QUESTIONS

1. No field of modern biology is more beset by controversy than primate paleontology. Suggest some reasons why.

2. How could life in the trees have contributed to the anatomy of monkeys, apes, and humans?

3. Discuss the origins of human-type erect bipedalism. What evidence of bipedalism can be found in the fossil lineage leading to modern *Homo sapiens*?

4. What features of the hands, feet, or limbs of australopithecines are apelike in character? What does this imply about behavior?

5. Trace the dramatic expansion of the brain in the human ancestral lineage. What types of behavior can we associate with the various brain volumes?

6. What is the significance of the different tool "industries" of the human lineage? What do they imply about the hand? The brain? Cultural evolution?

7. What are the possible relationships of Neandertals to modern *Homo sapiens*?

8. What does the earliest art imply about the early humans?

9. Does evidence of molecular relatedness between primates support deductions based on the fossil record? Explain.

10. What are human races?

SUGGESTED READINGS

CARTMILL, M., D. PILBEAM, and G. ISAAC. "One Hundred Years of Paleoanthropology." *American Scientist* 74 (1986): 410–420.
A first place to start for a balanced account of this controversial field.

DAY, M. H. *Guide to Fossil Man.* 4th ed. Chicago: University of Chicago Press, 1986.
A standby book that has all the facts about our relatives.

GOODMAN, M., R. E. TASHIAN, and J. H. TASHIAN. *Molecular Anthropology.* New York: Plenum Press, 1976.
A consideration of genes and proteins in human evolution.

RICHARD, A. E. *Primates in Nature.* San Francisco: Freeman, 1985.
Here is a fine treatment of how our relatives behave in their social groupings.

WEISS, M. L., and A. E. MANN. *Human Biology and Behavior.* 4th ed. Boston: Little, Brown, 1985.
We are, after all, animals; this book shows how our behavior is based on much that is inherited from our ancestors.

WYMER, J. *The Paleolithic Age.* New York: St. Martin's, 1982.
The last few hundred thousand years mark the origin of *Homo sapiens*; here is what occurred.

51
HUMANKIND AND THE FUTURE OF THE BIOSPHERE: AN EPILOGUE

The eighteenth century was marked by enlightenment, and the nineteenth by industry. The twentieth may go down in history as the ecological century. . . . What the twenty-first century may be labelled—progress or chaos—will depend on how well mankind learns to ensure a proper balance between himself and his environment.

E. Barton Worthington, *The Ecological Century* (1983)

Food for a hungry world: This timeless scene in Nepal might be a thousand or ten thousand years old—human hands and muscles still plant and harvest in much of the world.

The emergence of modern *Homo sapiens* from primate ancestors was an odyssey of biological and cultural evolution. Cultural development, in particular, has changed the human past and will, no doubt, affect its future. Our bodies, brains, and biochemistry remain subject to the same processes of variation and natural selection as those that influenced our ancestors; yet our capacity to intentionally alter some aspects of the environment in which we live—delivering water, heat, and electricity to a home, for example—permits a way of life unknown to other organisms and relaxes certain constraints of natural selection. Our unique relationship to the environment presents us with certain problems and certain opportunities, and these are our final subjects.

FROM HUNTER-GATHERER TO FARMER

The early hominids—*Australopithecus, Homo habilis,* and *Homo erectus*—led, as we saw in Chapter 50, to essentially modern *Homo sapiens* by around 35,000 years ago. Our ancestors migrated from the Old World to lands across the Earth and, by 10,000 years ago, humans had become the most widely distributed land mammal, settled on every continent except Antarctica. Perhaps 5 million people lived at that time, only about 400 generations ago, and population sizes remained roughly constant or increased very slowly. People existed then in small nomadic bands of hunter-gatherers who followed game and collected diverse, seasonally available plants. The modern Lapps of northern Scandinavia probably still live much as those hunter-gatherers did, following giant herds of reindeer on the animals' yearly migrations; using reindeer flesh, hide, and bones for their food, clothing, shelter, tools, and means of transportation; and gathering tundra plants for food, medication, and other purposes. Hunter-gatherers studied in the twentieth century practice birth spacing (birth control that depends on nursing previous young; Chapter 18), abortion, and infanticide. Researchers believe that their populations remain below the carrying capacity of their environments and food supply, and that ancient hunter-gatherer societies probably employed the same methods.

One of the great events of human cultural evolution marked the transition from hunter-gatherer to agriculturist. About 10,000 years ago, the so-called agricultural revolution took place; that is, people began planting seeds, harvesting crops, and domesticating animals. New evidence suggests that agriculture arose independently at many sites (Figure 51-1) and that the earliest farmers planted wheat, barley, and alfalfa in the Fertile Crescent, which extended from the Nile to the Tigris and Euphrates Rivers in what is now eastern Iraq. Simultaneously, agriculture arose in the New World—with the growing of corn, potatoes, and tomatoes—and in the tropical and subtropical Far East—with the cultivation of rice, millet, soybeans, and other crops. Eurasian peoples began to subject camels, dogs, horses, sheep, cattle, fowl, and other animals to the breeding regimens that ultimately led to the animal varieties we use today for food, transportation, work, and companionship.

The agricultural revolution with its widespread environmental modifications had five far-reaching consequences: (1) stable communities at specific sites, (2) career specialization, (3) writing, (4) written numbers, and (5) population expansion. Since planting must be done at specific times of the year, the crop must be tended at regular intervals and harvested, and the seed must be saved for next year's planting, the early farmers were truly tied to the land and no longer needed to, nor could, pursue the migratory hunter-gatherer mode of existence. Land—particularly fertile and desirable tracts—was claimed by tribes, families, and individuals. Continuously inhabited communities arose, with permanent buildings in which to live or to store, distribute, and sell or trade grain. This marked the beginning of architecture of permanent buildings and, later, the early forms of sanitary and civil engineering.

Agriculture usually produces more food than the farm family itself needs. Since some community members were thus freed from food-producing activities, more career specialization became possible than among hunter-gatherers. Some people worked as traders, merchants, predictors of correct planting times, framers of rules and laws, or protectors of the community. This division of labor produced a more efficiently functioning village. In addition, the villagers needed to keep accurate records of the best times to plant particular crops and of the amount of grain planted and bought or sold; this led directly and rapidly to the invention of writing, including the written use of numbers.

By 5,000 years ago, human history was being recorded in a form that subsequent generations have been able to read and understand. Thanks to these records and to our species's intelligence, members of each succeeding generation have benefited from the recorded thoughts and experiences of their ancestors. This great step in human cultural evolution sets *Homo sapiens* apart from all other animals. Thus the seeds of modern society, including the scientific and technological advances that have allowed us to continue modifying our physical environment, were literally sown by the early agriculturists.

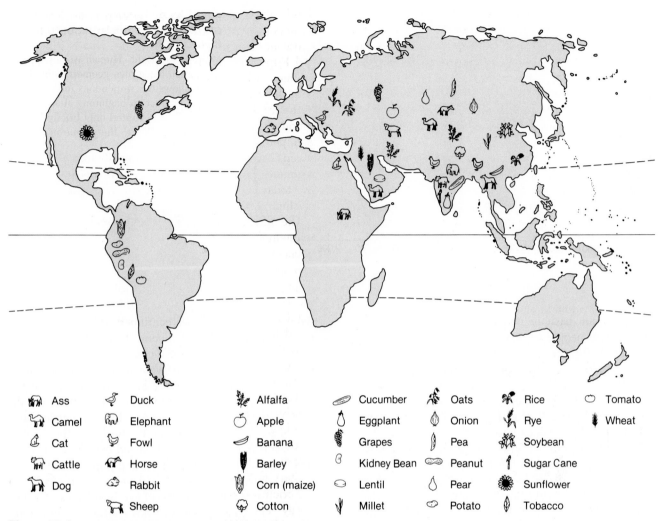

Figure 51-1   AGRICULTURE: A WORLDWIDE INVENTION.
This map shows sites where ancient peoples may have used various plants and animals for the first time agriculturally. Some crops and animals, no doubt, were domesticated independently and simultaneously in several areas. Others, however, were limited to specific continents. Potatoes, tomatoes, and corn, for example, were originally staple crops only in the New World.

THE POPULATION EXPLOSION

The last and, perhaps most serious, consequence of the agricultural revolution was a stimulating effect on the size of the human population. Although increased agricultural efficiency allowed farmers to feed more people, it also required more hours per week than hunting and gathering, and it thus encouraged larger family size to provide extra labor for the tasks of tending fields and herds. Human populations began to grow, increasing 25-fold between about 10,000 and 2,000 years ago. At the onset of the agricultural revolution there were about 133 million people in the world (Figure 51-2); by 1650, there

were about four times as many—roughly 500 million people. Human population doubled during the next 200 years, and then doubled again in only 80 years—reaching 2 billion people in 1930. The next doubling in world-population size took only 45 years. In 1987, the world population reached 5 billion—and it is still growing (Table 51-1).

This enormous population has fanned out across the globe, occupying most of the arable (farmable) land. During the past century, discoveries and advances in science and technology have contributed to further population growth and global occupation. Industrial progress has, among other things, allowed food to be widely distributed via new forms of transportation. Medical discoveries such as the role of bacteria in diseases and drugs

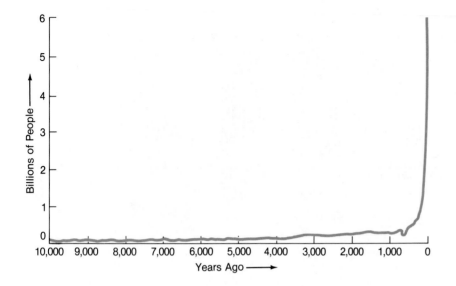

Figure 51-2   WORLD POPULATION GROWTH SINCE THE AGRICULTURAL REVOLUTION.
Malthus was right. Human world population has grown geometrically. This graph shows the stupendous increase in human population beginning about 10,000 years ago, interrupted only briefly by the many deaths due to the plague in the fourteenth century.

to fight them have helped diminish mortality and allowed more people to survive to reproductive age and beyond. At the end of the nineteenth century, for example, the life expectancy for males and females in the United States was below 50 years; today it is 71 and 78, respectively. Improvements in public sanitation made possible by the Industrial Revolution further contributed to falling mortality rates.

In the twentieth century, population pressures have started to ease in some industrialized nations; the birth rate has fallen as families have left farms and women have entered the workplace in unprecedented numbers. Between 1500 and 1700, English women did not nurse their babies, had no natural means of birth spacing as a result, and commonly bore between 12 and 20 babies (30 was not unusual). At the time of the first United States census in 1790, the birth rate was eight births per woman. By 1900, the birth rate had decreased to less than half that, and it has continued to decline steadily (except for the brief "baby boom" during the 1950s when women had an average of 3.8 children each) to the present rate of 1.8 births per woman. In West Germany, Scandinavia, Canada, Switzerland, and other industrialized countries, the birth rate is even lower and is substantially below the replacement rate for a married couple. Were it not for net immigration into the United States and most other industrialized countries, this low birth rate would result in declining population size.

In stark contrast, many less-developed nations in Africa, South America, and Asia have birth rates of 4.0 or higher. The consequences of this continued population growth will be great for the world as a whole: demographers predict that the human population will grow by some 2 billion people between 1975 and 2000—a number equivalent to the entire world population in 1930 (Table 51-1). The primary evidence for this prediction is that the young people in developing countries, where 40 to 45 percent of the population was younger than 15 years old in 1980, will reach reproductive age during the last decades of this century (Figure 51-3). Moreover, if these patterns for the less-developed countries persist, even with substantial changes and improvements in world-population policies, there will be nearly 10 billion people in 2030—when many of the readers of this book will be young grandparents—and 30 billion before the end of the twenty-first century.

No scientist believes that the Earth's ecosystems can support 30 billion people. The United States National Academy of Sciences has estimated that the Earth's carrying capacity for human beings is not much more than 10 billion. This estimate is based on the assumption of an intensively managed world that still preserves some degree of individual freedom. However, assuming that freedom of choice is completely relinquished and that

Table 51-1   **WORLD POPULATION GROWTH OF HOMO SAPIENS**

	Time taken to Reach	Year Attained
1 billion	35,000 years	About 1850 A.D.
2 billion	About 80 years	1930
3 billion	30 years	1960
4 billion	15 years	1975
5 billion	12 years	1987
Projection:		
6 billion	11 years	1998

Source: Population Reference Bureau, Inc., *Population Bulletin* 34 (no.5), p. 4.

(a)

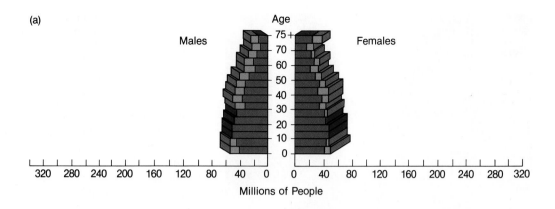

(b)

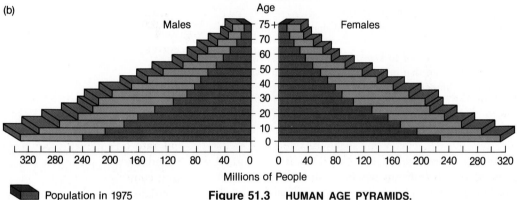

■ Population in 1975

▨ Increase 1975 to 2000

Figure 51.3   HUMAN AGE PYRAMIDS.
These graphs show the numbers of people in various age groups in (a) the developed world and (b) the less-developed world. Males are shown on the left, females on the right. Purple areas show the populations in 1975, and blue areas the change between 1975 and 2000. In 1975, for example, there were 240 million boys between the ages of 0 and 5 years and 233 million girls in the same age range in the less-developed world; by 2000 there will be 330 million boys between 0 and 5 years and 315 million girls. Note the much greater total number of young people of both sexes in the less-developed nations than in the developed world. Reproductive activity by those people will continue to drive upward the populations in less-developed countries.

the vast majority of the world's people live near starvation, the Academy's study predicts an upper limit close to 30 billion, well beyond a stable carrying capacity. These figures illustrate with dramatic clarity what an overwhelming biological factor humans have become during their short tenure on Earth and what serious problems must be confronted in the very near future if we are to maintain an acceptable quality of life.

The age structure of human populations in less-developed countries has significant political as well as biological ramifications. As Table 51-2 illustrates vividly, during the lifetime of a person who was born in 1950 and dies in 2020, the *proportion* of people living in the developed countries will fall from about one-third to about one-sixth of the world population, while the proportion of people living in the mainly tropical, less-developed countries will grow from less than one-half to about two-thirds of the world population (China's proportion remains nearly constant).

The majority of people alive today inhabit the tropics; the percentage of those people in the world population is rapidly growing larger. It is no wonder, for this reason alone, that news of El Salvador, Nicaragua, the Philippines, Africa, and other tropical regions appears more

Table 51-2   **POPULATION ESTIMATES (IN MILLIONS OF PEOPLE) AND PERCENTAGES OF WORLD POPULATION**

	1950	1984	2020
Developed world	832 (33%)	1,166 (24.5%)	1,350 (16.7%)
China	557 (22%)	1,034 (21.7%)	1,545 (19.1%)
Less-developed world (excluding China)	1,136 (45%)	2,561 (53.8%)	5,191 (64.2%)
Total	2,525	4,761	8,086

Source: Population Reference Bureau, Inc., Washington, D.C.

Figure 51-4   AN URBAN FUTURE?
This crowded shopping street in Yokohama is duplicated in a thousand cities worldwide. Faces everywhere, noise, smog, electrical power to keep the city going—such a very different life from that on a rural lane or African forest trail where most humans lived so short a time ago.

and more frequently in newspapers and on television. Citizens, scientists, and government officials in the developed and less-developed countries alike are gradually but steadily becoming more concerned about the problems in tropical regions and are coming to recognize that the burgeoning human population, particularly in Third World areas is crucial to the future of our species (Figure 51-4). Only when the ecological, economic, and political consequences of that area of population increase become more widely appreciated will humanity begin to focus its huge capacity for invention and innovation upon this great challenge.

FOOD AND AGRICULTURE: FEEDING THE BURGEONING MASSES

Increasing human populations will require bountiful food supplies, but there is no agreement on the prospects for feeding the world's people in the future. The Global 2000 study, commissioned by the United States government to predict conditions at the end of the century, calculated that global food production can keep pace with population growth through the year 2000, provided that favorable agricultural conditions can be maintained for the rest of this century. But a more recent study by the Food and Agriculture Organization of the United Nations estimates that there must be a 60 per-

cent increase in world food production by the year 2000 if the world's population is to be fed. Though that figure is judged unattainable by some, data in 1987 showed that world grain harvests doubled between 1960 and 1986. Food production in less-developed countries is increasing at a rate of 4.4 percent annually. And countries such as Bangladesh and India, sites of perennial malnutrition, are either self-sufficient or even exporting food. The reason is the new kinds of crops available to farmers.

In the long run, exactly what will determine the amount of food that can be produced in developed countries? Variables that are especially important include the climate, the amount and intensity of use of arable land, the cost of energy, the distribution of food, the kinds of food people eat, and the quality of the food plants themselves.

Climate

The predicted balance of food supply and population will depend on climate, and most studies assume no significant deterioration. The evidence is not in yet on the potential problem of the greenhouse effect resulting from human production of carbon dioxide. However, data gathered in the mid-1980s suggest that the greenhouse effect is accelerating faster than had been predicted. Many atmospheric scientists are trying to predict the consequences of elevated mean temperatures, but predictions are still uncertain. While lands in the north may become farmable for certain crops, shifts in wind, rain, and temperature patterns are extremely difficult to predict.

Land Use

The Global 2000 study, among others, assumes that people will use the world's arable land much more intensively by the end of this century. Whereas 1 hectare (2.47 acres) produced enough food to support 2.6 people in the 1970s, it will have to produce enough food to support 4 people by the year 2000. But no one is sure whether this can be accomplished, especially in light of the deterioration of the world's present agricultural lands due to erosion, loss of nutrients and soil compaction, rise in salt content, pollution, and desertification—the transformation of farmland into desert. About 6 million hectares are being lost each year to desertification of the Earth's 1.5 billion arable hectares. In the United States, we have to find ways to cut in half the present rate of topsoil loss in order to sustain the present level of productivity. That will require new farming practices, since increasing crop production often accelerates the loss of topsoil.

About 15 percent of the arable land worldwide is irrigated; this is the most productive land. It may be possible to increase the amount of irrigated cropland by the end of the century, but it will be very difficult to maintain such land in continual productivity. Salinity, alkalinity, waterlogging, and other problems must be reduced to preserve such cropland.

Urban encroachment also reduces the amount of good cropland. Cities and industries are often located in valley bottoms on the richest farmland. For instance, the famous Silicon Valley of California used to be one of America's richest regions for growing apricots and other fruits. In both developed and less-developed countries, people and their roads, parking lots, tract houses, cities, and factories are covering arable land at an increasing rate.

Energy

Maintaining high agricultural productivity will require yet another advance: the employment of more energy-intensive technology. Some agricultural economists predict that we will need to use as much as four times more fertilizer, pesticide, and herbicide than we do now. Note in Figure 51-5 the incredibly high energy costs of growing rice, corn, and wheat in the United States compared to growing the same crops in countries like the Philippines, Mexico, and India, where agriculture is more labor-intensive. Beef and milk are included, too, to demonstrate the immense cost of producing human food that is higher on the food chain. The cost of hydrocarbon fuels, both directly for driving farm machinery and indirectly for fertilizer and transporting food, largely determines the costs of energy-intensive agriculture. Rising energy prices in the future could slow food production accordingly.

The Global 2000 study predicts a continuing shortage of energy resources globally, based on a projected rise in energy consumption rates of 58 percent between 1975 and 1990. Per capita consumption rates for oil and coal will rise most dramatically in industrialized countries other than the United States, making these fuels increasingly unavailable to poorer people in developing countries, just as their firewood supplies decline due to deforestation in those areas (Figure 51-6). The result will be that dung and other natural fuels will be burned in cooking fires rather than used in the fields as fertilizers.

So-called alternative energy sources—nuclear power (both fission and fusion), solar energy, wind energy, and others—could help alleviate these shortages if tapped,

Figure 51-6 FIREWOOD AND DUNG: ENERGY FOR COOKING AND WARMTH.
These women return to a village in India bearing heavy loads of branches for the family hearth. Much of the wood and bushes in some areas of the world has been gathered and burned; and much animal dung is collected, dried, and burned, thereby depriving the land of needed fertilizer.

Figure 51-5 ENERGY AND FOOD.
Manual labor in India, Mexico, and the Philippines yields relatively small crops at low energy input. Mechanized agriculture includes fuels, fertilizers, and electrical power; yields may be immense, but the energy input is as well. The basic inefficiency of food chains is seen in the data on beef and milk; fodder corn and other foods are produced with high energy input; yet only a small fraction is converted by cattle into edible beef or milk.

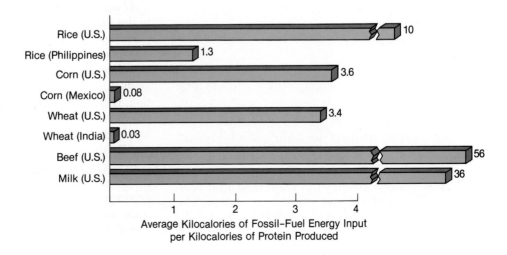

Rice (U.S.) 10
Rice (Philippines) 1.3
Corn (U.S.) 3.6
Corn (Mexico) 0.08
Wheat (U.S.) 3.4
Wheat (India) 0.03
Beef (U.S.) 56
Milk (U.S.) 36

Average Kilocalories of Fossil-Fuel Energy Input per Kilocalories of Protein Produced

(a)

(b)

Figure 51-7   ELECTRICITY FROM SUN AND WIND.
(a) "Solar One," one of the first commercial solar power plants, is an array of solar collectors that transduce the "free" energy of the sun into the electrical power for civilization. (b) Huge blades await the winds of the California desert to power generators and send electricity to western cities and agriculture. Both solar power plants and wind generators are built where there is sun and wind most days of the year.

and according to the report, are likely to be much more widely used, at least in the developed world (Figure 51-7).

Food Distribution

Food distribution is another hurdle to cross on the way to increased food production. Today only about 8 percent of the food eaten in tropical countries is imported. Some such countries have greatly increased production due to new strains of rice, corn, and other crops. We do not know yet how long such high productivity can be maintained as human populations continue to expand. As developed countries increase their food production, means must be found to finance the distribution of that food to less-developed countries that need it and, sometimes, to remote populations within those coun-

tries. As human populations escalate, nations will face difficult ethical issues. For example, how intensively should they farm to produce food for people who are starving when intensive agricultural practices can ruin the land because crops consume minerals and trace elements, and irrigation often washes away some topsoil. Government agencies will have to grapple with such questions, as scientists and technologists try to develop new soil additives and replacements to preserve the productive capacity of the great farmlands of the United States, Canada, and the Soviet Union.

Human Diet

As the world population continues to grow, people's diets will have to change. People in the more densely populated countries will need to reduce their consumption of animal protein, since cattle, sheep, hogs, and poultry will become too costly ecologically to be feasible food sources in many countries.

People already eat more fish each year than beef and poultry combined. It is disquieting that world fish harvests peaked in the 1970s and, due to overfishing and energy costs, have begun to fall despite the use of more and more sophisticated fishing equipment. World fisheries will require good management to correct overfishing and improved technology to maintain their production at present levels. Fish farming will be needed to expand further the availability of fish as a protein source in the human diet.

The extinction of many plant species and the loss of genetic material worldwide could have negative effects on agriculture. Most people eat the food products of no more than a several dozen crops. When new, "improved" crops are introduced in developing countries, the "unimproved" crop plants grown there are often lost. Genetic races of some of the most important crops have been preserved in refrigerated gene banks, but preservation is the exception. The diminished genetic diversity of food crops limits our capacity to develop crops that are resistant to evolving pests. In 1970, a new strain of the southern corn-leaf blight (*Helminthosporium maydis*) destroyed about 15 percent of the corn crop in the United States. Fortunately, plant scientists were able to breed new strains of resistant corn from a "bank" of varieties. But that option is not available for many important crops.

A type of wild corn graphically illustrates the consequence of losing valuable genetic material as plant species continue to become extinct from human-population expansion and lose their habitats in less-developed countries. *Zea diploperennis*, a wild perennial relative of domestic corn, is immune to seven of the most prevalent corn viruses. This species grows only in three small colo-

Figure 51-8   ONE INDIVIDUAL'S CONTRIBUTION.
Rafael Guzmán in a dense stand of *Zea diploperennis*, high in the Sierra de Manantlán mountains of Mexico, just a year after he made the discovery of this wild, disease resistant relation of domestic corn.

nies in the Sierra de Manantlán in the Mexican state of Jalisco and was discovered in 1977 by Rafael Guzmán, a student at the University of Guadalajara (Figure 51-8). Because it grows only on about ten acres, it could easily have become extinct in the wild without ever having been discovered and cultivated. It has already been used to breed a perennial variety of corn resistant to tropical corn viruses. No one knows how many other unidentified species of tropical plants possess genes that could be of immense value to human agriculture; but if wild plants become extinct because of human expansion, we will never know.

The Green Revolutions

Biologists have strong hopes that they can solve the complex problem of feeding the world's people through improvements in plant genetics. By the mid-1960s, a green revolution was already underway with a rapid expansion in the number of acres planted with new high-yield crops. Using traditional plant-breeding techniques, agronomists have bred varieties of wheat, for example, that produce two or more times the traditional harvest—as long as the farmer applies the proper mix of pesticides, fertilizers, and irrigation (Figure 51-9). By using such techniques on large farms in Mexico, farmers were able to quadruple wheat production per acre between 1950 and 1970. These additives, however, are expensive and precious commodities, especially in developing countries. Unfortunately, subsistence farmers cannot afford the high costs of such agricultural aids and they still produce wheat and other crops at the traditional lower yields.

Figure 51-9   TRANSFERRING NEW CROPS TO THE FIELD: THE HARDEST PART OF THE NEW AGRICULTURE.
Here scientists and a farmer in Bangladesh confer about a new rice strain. Molecular agriculture will only realize its potential if this last step is successful.

Today we are on the doorstep of a second green revolution based on genetic engineering techniques. Some results are already seen: Argentinian farmers plant rapid-maturing, cold-resistant corn much farther south toward the Pole; and the North American corn belt is now 250 miles farther north than a decade ago. New strains of rice have raised China's production by 50 percent since 1978 and have given Indonesia, the world's largest importer of rice, three consecutive surpluses. Yields are up dramatically: corn now yields 120 bushels per acre in the American Midwest versus 25 bushels per acre in the 1930s (up to 300 bushels per acre are obtained sometimes!). It is important that the new genetic strains of key crops are in successful, quite widespread use in many less-developed countries.

These successes are encouraging. But there is more to come in the new green revolution. Using methods such as those described in Chapter 14, plant molecular biologists are beginning to develop:

1. Crops that can withstand temperature extremes and thrive in an expanded range of habitats.

2. Crops that are drought resistant, and can grow in relatively arid—and presently unproductive—regions, thus vastly increasing the worldwide acreage available for growing food.

3. Crops that are resistant to insects, fungi, and nematodes, allowing yields per acre to increase as the need for applying pesticides decreases.

4. Crops that fix nitrogen so they can thrive on a lesser amount of fertilizer. This is the most ambitious of the attempts, and involves identifying genes and proteins involved in nitrogen fixation and successfully transferring them into dicot and monocot crops.

5. Crops that make all of the amino acids essential to the human diet, so that corn, wheat, or rice, for example, will have heightened nutritive value and we can decrease our reliance on fish, beef, poultry, and other animal foods. As an initial step toward that goal, scientists transferred the gene for the firefly enzyme luciferase into cells of the tobacco plant (Figure 51-10). Using similar processes, they hope someday to transfer other genes for animal proteins to crop plants.

6. Crops that are living factories capable of harboring the genes for, and producing, insulin, interferon, or other animal proteins of critical importance in medicine.

The second green revolution has enormous potential to augment food supplies substantially for future human populations. It is not easy or inexpensive to develop these new kinds of plants. And, once available, their growth still depends on proper soils, sufficient water, beneficial weather patterns, and other benign environmental conditions, as well as on the availability of energy.

Figure 51-10   GENETIC ENGINEERING OF CROP PLANTS: A GLOWING FUTURE.
Genetic engineers transferred a gene for the enzyme luciferase from fireflies to tobacco cells, using methods described in Chapter 14, and then grew plants from the hybrid cells. The adult tobacco plants glow in the dark when watered with a solution containing the enzyme's substrate, luciferin, proving that an animal protein can be both manufactured and made to function in plant cells.

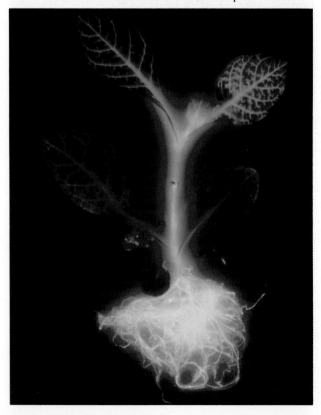

SPECIAL PROBLEMS OF THE TROPICS

In the coming decades, the fate of tropical regions is likely to affect weather patterns and human standards of living around the world. Overpopulation and poverty are driving people to rapidly destroy the zone's rich natural resources—to clear rain forests for agricultural land and for the sale of hardwood timber. By 1980, about 44 percent of the world's tropical forests had been severely disturbed or destroyed. The total area of tropical forests remaining on Earth is approximately equal to all the land west of the Mississippi River in the continental United States. However, people are clearing about 35 acres each minute or an area about equal to half of Iowa each year. At such a rate, *all* tropical rain forests will be gone in 90 years, and most will have suffered great alteration before their disappearance. (Disturbingly, some recent estimates suggest that the rate of destruction may actually be twice as great.)

The consequences of this overwhelming habitat destruction are enormous. The tropics have twice the species richness as temperate regions and, conservatively, support some 3 million kinds of plants, animals, and microorganisms. So far, scientists have cataloged no more than 500,000 tropical organisms, or about 1 species in 6, and some of those only recently. For instance, in the 1970s, scientists discovered and classified the Chacoan peccary *(Catagonus wagneri)*, a piglike mammal of South American forests that can weigh 80 pounds. We know nothing of the remainder—neither their scientifically interesting properties nor their potential utility or harmfulness to humans.

Because most kinds of tropical organisms require the specialized conditions of their natural habitats in order to reproduce, many will become extinct during the decades to come. Some scientists have calculated that more than *1 million* species may become extinct between 1980 and 2010, amounting to the loss of 25 percent of the diversity of life on Earth. In terms of speed, such a catastrophe will be without parallel in biological and geological history. (While more than 90 percent of the Earth's marine species became extinct at the end of the Permian period, that biological cataclysm spanned many millions of years, not just thirty.)

Perhaps the most tragic fact about tropical deforestation is that the cleared lands will be cultivable for just a few years, a mere fraction of a second in evolutionary and geological history. Many tropical soils are very infertile, but they can support lush forests because most of the nutrients in the ecosystem are locked up in the trees and other vegetation. Tree roots tend to spread laterally through the upper 5 centimeters of soil, quickly and efficiently recovering the nutrients from the leaves that fall to the ground. When people cut tropical forests, they usually burn the trees or allow them simply to decay; this releases large amounts of nutrients into the soil, enabling farmers to grow crops for a few years (Figure 51-11). Such slash-and-burn agriculture, also known as shifting agriculture or forest farming, soon depletes the soil, however, and the farmer must cut down other forests and start the cycle again. Unfortunately, no agricultural technology available today allows crops to be grown again once the rich nutrients have been used up in a few seasons. Slash-and-burn agriculture creates such a severe nutrient deficit that it takes about eighty years for the soil to regain its fertility and support a new forest. Reforestation after slash-and-burn agriculture is usually successful only if the cleared patches are relatively small and only if the nearby human population is small enough to allow the eighty-year cycle to take its course. That is now a virtual impossibility because the very countries where slash-and-burn agriculture is practiced have the greatest population densities and rates of increase.

The studies on extinction by Robert MacArthur and Edward O. Wilson help explain why tropical reforestation is so difficult and why slash-and-burn agriculture will, no doubt, lead to unprecedented extinction rates. They pointed out that islands suffer an apparently random loss of species as a result of small island populations

(a)

(b)

Figure 51-11 DESTRUCTION OF TROPICAL FORESTS.
Forest farming or slash-and-burn agriculture accounts for about two-thirds of the destruction of tropical forests, as large numbers of people seek to grow food for themselves. (a) Here Panare Indians in the Guiana highlands of Venezuela clear the forest so crops can be planted. (b) The consequences of slash-and-burn: erosion cracks and carries away soil from a former pine forest area in the Dominican Republic.

and other factors (Chapter 45). Barro Colorado Island, for example, an island of about 6 square miles formed between 1911 and 1914 by the damming of the Rio Chagres, part of the engineering that created the Panama Canal (Figure 51-12), was home at that time to 208 species of breeding birds. In the years since then, at least 51 of these species, or 25 percent, have become extinct, even though the maximum habitat has been carefully

Figure 51-12   BARRO COLORADO ISLAND: A WARNING ON EXTINCTION. Despite the obvious closeness to adjacent jungle, over 50 bird species that formerly lived on this patch of forest are now extinct. Organisms isolated on similar "islands" of vegetation as a result of slash-and-burn argiculture face the same fate.

protected during the entire time and the island is only about 650 feet from the mainland.

By fragmenting mainland habitats into small islandlike patches separated by agricultural land, the extinction of species that could survive temporarily within uncut patches is accelerated. In these small populations, inbreeding becomes an important factor, just as it is in zoos, and incursions by humans inevitably have greater impact as the size of the remaining patches of forest decreases. The lesson is clear: just as species become extinct rapidly on islands where their populations are small, they are likely to become extinct rapidly in isolated patches of forest. As people destroy more and more areas with remaining populations of rare and endangered species, the process is accelerated still further. The consequences are obvious. Few of the original vines, trees, insects, primates, and other organisms that made up the virgin rain forest will be available for reforestation after the soil has regained fertility.

While this loss of diversity is tragic for tropical regions, it may have even more wide-ranging consequences for global climates. More than half of the rainfall in tropical forests is forest-generated; that is, it is due to evaporation from the forests, altered air patterns over the forests, and so on. Thus the precipitation as well as the temperature regime is disrupted severely when the forest is cut. In addition, extensive clear-cutting or destruction of tropical forests will almost certainly alter the Earth's atmosphere and, consequently, the Earth's climate. About twice as much carbon is apparently present in the plant materials of tropical forests as in the Earth's atmosphere, and scientists are puzzled and concerned about how much of this carbon may reach the atmosphere in the form of carbon dioxide, and how quickly. Deforested regions fix less CO_2 into plant tissues than they once did, and tree burning releases more of the gas. Scientists worry that an additional increase in atmospheric CO_2 will lead to the significant warming of global climates, possible melting of the polar icecaps, and shifting of global wind and rain patterns—with resultant effects on agriculture in temperate latitudes. Aside from the loss of species diversity, the potential for climate disruption is a powerful reason for every person and every government to be concerned about the management of tropical forests. This is particularly true for students of biology: as potential scientists and physicians, they may play a profound role in meeting the challenges of overpopulation, food shortages, and environmental degradation.

BIOLOGICAL SCIENCE, SOCIAL RESPONSIBILITY, AND OUR FUTURE

We argued in Chapter 1 that biology is one of the liberal arts, worth studying for its own sake. In addition, the problems facing humankind demand that many bright young people pursue careers in biology. Aside from the academic community, our political leaders, attorneys, businesspeople, and voters in general owe it to themselves and to society to learn at least the rudiments of this fascinating area of human knowledge and its recent advances. Without question, the new biology will assert itself and insert itself farther and farther into the fabric of commerce and society.

This chapter has surveyed a number of the major problems that are likely to confront world society during our readers' lifetimes. The technological solutions that may come from the second green revolution, the widespread use of solar energy, and more effective forms of birth control (Chapter 18) are not sufficient in themselves to meet the challenges. Science and technology can provide possibilities, but sociopolitical processes within and among nations will dictate what actually occurs. Preservation and use of land for food production,

Figure 51-13   OUR ISOLATED WORLD.
Someday humans may communicate with intelligent life elsewhere in the universe. Until that time, however, our species has nothing to rely upon and no place to go but the planet Earth, with its fragile ecological balance. Africa, the nursery of our species, is seen at the upper left.

protection of water quality and availability, and limitation of gaseous effluents are concerns that transcend traditional national boundaries. It is indeed good fortune

that the major new steps in human cultural evolution—the development of the computer and ability to manipulate and control genes—have come at just the time when our species's reproductive capacity is straining the planet's ecosystems so severely. But while scientists and technologists can produce such great advances, informed world leaders and citizens must *exploit* the advances to help the Earth, with its more than 5 billion human inhabitants, to continue on as the marvelous spaceship it has been for the past 4.5 billion years (Figure 51-13).

Where, then, does today's college student fit in? Some of those entering biology in the 1990s will derive satisfaction from inventing and using still newer technologies that help us answer age-old questions. The elucidation of DNA homologies—what the biologist Howard Schneiderman has described as living fossil records—will yield powerful new insights into the evolutionary relationships among organisms. Speculations will be replaced by facts that will allow biologists to reconstruct the history of life with a precision heretofore unimaginable. The mysteries of the human brain will also be unraveled and understood gradually. Ecology, finally yielding to synthesis and general theory, will come closer to center stage as human populations mushroom. The great problems confronting humankind will be overcome, however, only if educated people make the personal commitment to invest lives and careers in the areas of greatest global need. The biological and cultural evolution of *Homo sapiens* has reached new heights with the powerful tools of the computer and genetic engineering. Now it is time to apply these and other tools in an effort to forge a new and different way of living that will sustain the huge human populations of the twenty-first century free from abject hunger, poverty, and disease.

SUGGESTED READINGS

BATIE, S. S., and R. G. HEALY. "The Future of American Agriculture." *Scientific American,* February 1983, pp. 45–53.

   This article presents a good survey of agriculture in relation to economics and ecology.

EDMONDS, J., and J. M. REILLY. *Global Energy: Assessing the Future.* New York: Oxford University Press, 1985.

   This book gives the most comprehensive analysis of energy reserves in the future.

EHRLICH, P., and A. EHRLICH. *Extinction: The Causes and Consequences of the Disappearance of Species.* New York: Random House, 1981.

   An easy-to-understand treatment of what we are doing to nature.

EHRLICH, P. R., A. H. EHRLICH, and J. P. HOLDREN. *Ecoscience: Population, Resources, and Environment.* San Francisco: Freeman, 1977.

   The most complete set of diverse statistics and facts on these topics.

MYERS, N. *The Primary Source.* New York: Norton, 1984.

   A good book from which to learn about tropical forests and their future.

WESTOFF, C. F. "Fertility in the United States." *Science* 234: 554–559, 1986.

   This article provides data on population growth in the industrialized countries and indicates what the consequences will be of changing rates of birth and immigration.

APPENDIX A
The Classification of Organisms

Organisms are classified by taxonomic rules originated in part by Linnaeus and modified each year at international meetings of taxonomists. The following list follows, in general, R. H. Whittaker's five-kingdom scheme (*Science 163* [1969]: 150–160), but with the addition of the kingdom Archaebacteria. The text discussion explains our reasons for placing groups in specific kingdoms in cases where there is lack of agreement on such assignments (as, for example, placement of the slime molds). Groups not described in the text are marked with an asterisk.

KINGDOM MONERA

Prokaryotic, mainly single-celled bacteria and cyanobacteria. Reproduce by asexual fission, but also have genetic recombination processes.

Subkingdom Schizomycete: Bacteria; unicellular or chains; most are heterotrophs; some with a solid molecular flagellum that rotates. Include mycoplasmae, the smallest living cells.

Subkingdom Cyanobacteria: Photosynthetic bacteria; chlorophyll on membranes; some also fix nitrogen. Single cells and colonies. Ancient oxygen liberators.

Subkingdom Chloroxybacteria (Prochlorophyta): Photosynthetic marine bacteria; may be source of chloroplasts.

KINGDOM ARCHAEBACTERIA

Prokaryotes with unique nucleic-acid and biochemical properties.

Phylum Methanobacteria: Methane-producing archaebacteria; obligatory anaerobes.

Phylum Sulfolobales: Sulfur-reducing archaebacteria. Some also tolerate very high temperatures (called thermoproteales).

KINGDOM PROTISTA

Eukaryotic, single-celled organisms (some colonies); heterotrophs and autotrophs; reproduce sexually; protozoans and some algae included. Flagella, cilia of 9+2 microtubule type.

Plantlike Autotrophs

Phylum Euglenophyta: Euglenoid, usually photosynthetic organisms with chloroplasts; most unicellular, but colonies, too.

Phylum Pyrrophyta: Dinoflagellates; photosynthetic phytoplankton, usually single cells with flagella.

Phylum Chrysophyta: Golden-brown and yellow-green algae; photosynthetic single cells and colonies; include diatoms that contribute to sedimentary rocks.

Funguslike Heterotrophs

Phylum Gymnomycota: Slime molds. Phagocytic; sexual and asexual reproduction. Stages with single cells, cell aggregates, and spore-forming structures. (Sometimes classified as fungi.)

CLASS MYXOMYCETE: Acellular slime molds; plasmodium multinucleate. Also single ameoboid cells.

CLASS ACRASIOMYCETE: Cellular slime molds; plasmodium composed of separate cells. Also single ameoboid cells.

Animallike Heterotrophs

Group Mastigophora: Flagellated protozoa; usually reproduce asexually; some parasitic (as, *Trypanosoma*, the cause of African sleeping sickness).

Group Sarcodina: Amoebas; single cells; creeping movement using pseudopods; many marine. Foraminiferans and radiolarians contribute to sedimentary rocks.

Group Sporozoa: Single-celled, nonmotile cellular organisms that parasitize animals (as, *Plasmodium*, cause of malaria). Sporelike stage in life cycle.

Phylum Ciliphora: Unicellular organisms with many cilia; carry out conjugation.

Phylum Caryoblastea: Amoeboid cellular organism that lacks mitochondria. May represent an early stage in eukaryote evolution.

KINGDOM FUNGI

Filamentous heterotrophic organisms with many nuclei in mycelium; nuclei sometimes separated by septae. (The slime molds, Gymnomycota, are sometimes classified as fungi.)

Division Mycota (or Eumycota):

CLASS CHYTRIDIOMYCETE: Chytrids; single celled or small mycelia; gametes and spores with flagella. Parasites or saprobes. Thought to be most ancient mycota.

CLASS OOMYCETE: Water molds; uni- or multicellular organisms that may be terrestrial or aquatic. Asexual spores with flagella; produce large oocytes. Parasites or saprobes.

CLASS ZYGOMYCETE: Bread molds; consume decaying organic materials. Chitin in cell walls. Form mycorrhizae with plants.

CLASS ASCOMYCETE: Cup or sac fungi, molds, yeasts, and relatives. Form asci during sexual reproduction. Most common fungi in lichens.

CLASS BASIDIOMYCETE: Common mushrooms, shelf fungi, rusts, and relatives. Conidial spores; carry out sexual reproduction. Form mycorrhizae with plants.

CLASS DEUTEROMYCETE: Fungi imperfecti; sexual reproductive structures not known. Many are probably ascomycetes.

KINGDOM PLANTAE/ALGAE AND LOWER SEEDLESS PLANTS

Photosynthetic, eukaryotic, and usually multicellular organisms. Some vascular.

Division Chlorophyta: Green algae; both multi- and unicellular photosynthetic and usually aquatic organisms. Some with flagella.

Division Rhodophyta: Red algae; mostly multicellular and marine organisms that lack flagella. Photosynthetic; unique kind of storage starch. Coralline algae have calcareous deposits.

Division Phaeophyta: Brown algae; multicellular photosynthesizers that are mostly marine; some very large and may have vascular tissues. Food stored as laminarin.

Division Bryophyta: Liverworts, hornworts, and mosses; largely nonvascular; alternation of sexual (gametophyte) and asexual (sporophyte) generations. Sporophytes derive food from gametophytes.

CLASS HEPATOPHYTA (OR HEPATICOPSIDA): Liverworts; terrestrial with photosynthetic haploid gametophyte as dominant stage.

CLASS ANTHOCEROTOPHYTA (OR ANTHOCERATOPSIDA): Hornworts; terrestrial with photosynthetic haploid gametophyte that is leafy.

CLASS MUSCI (OR MUSCOPSIDA): Mosses; contain some vascular tissue; with photosynthetic haploid gametophyte as dominant stage of terrestrial and semiaquatic plants.

Division Psilophyta: Whisk ferns; vascular phloem and xylem tissus. Motile sperm must swim in water.

Division Lycophyta: Club mosses, quill worts; vascular plants with leaves, stems, roots. Motile sperm move in water.

Division Sphenophyta: Horsetails; vascular plants with jointed stems and nonphotosynthetic leaves. Tiny gametophytes with sperm that travel in water.

Division Pteridophyta: Ferns; vascular sporophyte dominant stage with feathery fronds; small gametophyte with sperm that travel in water.

KINGDOM PLANTAE/HIGHER SEED PLANTS

Photosynthetic, vascular, nonmotile, with cellulose cell walls. Alternation of generations between haploid and diploid stages of life cycle.

Gymnosperms: Seed plants without flowers; naked seeds (no fruits).

Division Cycadophyta: Cycads; palmlike; individuals with either seed or pollen-producing cones (dioecious); flagellated sperm move through pollen tube.

Division Ginkgophyta: Ginkgoes; large pollen and seed-producing trees (dioecious); sperm swim in pollen tube.

Division Gnetophyta: Gnetinas; Mormon tea; vines, shrubs; nonmotile sperm.

Division Coniferophyta: Conifers; leaves reduced to "needles". Usually moneocious, with both pollen and seed cones on single trees; gametophyte stage greatly reduced to cells in cone. No fruits.

Angiosperms: Flowering plants with seeds enclosed in fruit. Nonmotile sperm move through pollen tube; double fertilization. Unique endosperm nourishes embryo.

Division Anthophyta: Flowering plants; monocotyledon or dicotyledon seed leaves; annuals, biannuals, perennials; herbaceous or woody plants. Gametophyte stage only a few cells.

KINGDOM ANIMALIA

Animals; eukaryotic multicellular heterotrophs. Sexual reproduction with meiosis; some reproduce asexually. Many motile, but some stationary.

Phylum Porifera (sometimes subkingdom Parazoa): Sponges; colonial or solitary, nonmotile, aquatic. Filter feeders; phagocytosis.

*Phylum Mesozoa: Tiny parasites; marine; organs and body composed of a few dozen cells.

Phylum Placozoa (sometimes in subkingdom Parazoa): *Trichoplax;* multicellular but lacks tissues; ciliated; changes shape like amoeba. May be early stage of multicellularity.

*Subkingdom Eumetazoa: Animals with radial or bilateral symmetry and true tissues.

Phylum Cnidaria: Hydroids, jelly fish (medusae), corals, anemones; marine, radially symmetrical with two tissue layers; cnidoblast stinging cells; extracellular digestion.

CLASS HYDROZOA: Hydras, Portuguese men-of-war; alternation of generations in the life cycle; may lack polyp stage or may have only short medusa stage.

CLASS SCYPHOZOA: Jellyfish; mostly medusa stage in life cycle.

CLASS ANTHOZOA: Sea anemones, corals; no medusa stage in life cycle.

Phylum Ctenophora: Comb jellies; marine, radially symmetrical with eight comb rows.

Phylum Platyhelminthes: Flatworms; terrestrial, marine, and fresh water; flat dorsoventrally; some three tissue layers; one opening for digestive system; acoelomate. Asexual and sexual reproduction.

CLASS TURBELLARIA: Planarians; bilaterally symmetrical; organ systems. Asexual fission of body, plus hermaphroditic sexual reproduction.

CLASS TREMATODA: Flukes; parasites, with two or more animal hosts; has a digestive tract.

CLASS CESTODA: Tapeworms; parasites that absorb nutrients and lack a digestive tract. Complex life cycles.

*Phylum Gnathostomulida: Gnathostomulids; minute, wormlike; marine; with a single layer of ciliated epithelial cells. Anterior end has bristlelike sensory cilia.

Phylum Nemertina (Rhynchocoela): Ribbon worms; one-way digestive tract with two openings; marine; lack coelom; protrusible proboscis; circulatory system with blood.

Phylum Nematoda: Roundworms; unsegmented, pseudocoelomate worms with complete gut and longitudinal muscles. Ubiquitous soil organisms that parasitize many plants and animals.

*Phylum Nematomorpha: Hairworms; long threadlike bodies. Free-living in damp soil; mostly fresh water, a few marine.

*Phylum Acanthocephala: Acanthocephalans; small wormlike endoparasites of arthropods.

*Phylum Kinorhyncha: Kinorhynchs; elongated body with cuticle segmented. Less than 1 mm. long; marine.

*Phylum Loricifera: Loriciferans; tiny marine pseudocoelomates; live between sand grains; unique tube comprises mouth.

*Phylum Gastrotricha: Gastrotrichs; elongated body with ciliated ventral surface. Marine and fresh water.

*Phylum Rotifera: Rotifers; anterior end with ciliated crown; posterior end a foot. Microscopic; fresh water, some marine, some inhabitants of mosses.

*Phylum Entoprocta: Entroprocts; body on a stalk; mouth and anus surrounded by a tentacular crown; pseudocoelom; marine.

Phylum Annelida: Segmented worms; protostomes with coelom, complete gut, closed circulatory system, and septae between body segments.

CLASS OLIGOCHAETA: Earthworms; terrestrial and fresh-water aquatic worms; lacking distinct head or external sense organs.

CLASS POLYCHAETA: Marine worms; free swimming; parapodia used in swimming and respiration; well-developed head with sense organs and mandibles.

CLASS HIRUDINEA: Leeches; flattened with reduced segmentation; external parasites or scavengers with suckers.

*Phylum Sipunculida: Sipunculids; cylindrical marine worms with tentacles around mouth.

*Phylum Echiura: Echiurids; cyclindrical marine worms, with large nonretractile proboscis.

*Phylum Priapulida: Priapulids; cucumber-shaped; with a large anterior proboscis; coelom reduced; marine.

*Phylum Pogonophora: Pogonophorans; deepwater marine animals; long body within a chitinous tube. Digestive tract absent.

*Phylum Pentastomida: Wormlike endoparasites of vertebrates.

*Phylum Tardigrada: Water bears; microscopic segmented animals with four pairs of stubby legs. Fresh water and terrestrial in lichens and mosses; a few marine.

Phylum Arthropoda: Segmented animals with exoskeleton of chitin and paired, jointed appendages. Protostomes; with coelom, complete gut, open circulatory system, and dorsal brain. Phylum with largest number of species.

SUBPHYLUM TRILOBITA: Trilobites; extinct, highly diverse group of segmented marine arthropods with well-developed compound eyes.

SUBPHYLUM CHELICERATA: Spiders, ticks, horseshoe crabs, sea spiders; six pairs of appendages, including chelicerae (fang-like biting appendages); two main body regions.

SUBPHYLUM CRUSTACEA: Crabs, shrimps, barnacles, water fleas; aquatic; with two-branched (biramous) appendages; often three pairs of legs; legs on both thorax and abdomen; mandibles.

SUBPHYLUM UNIRAMIA: Centipedes, millipedes, and insects; terrestrial and aquatic with single-branched appendages; up to 200 body segments.

CLASS INSECTA: Largest number of species of any group of organisms; includes bees, ants, termites, dragonflies; 900,000 species described so far. Three body segments; three pairs of walking legs; many have wings; varied mouth parts; compound eyes.

Phylum Onychophora: Peripatus; wormlike; segmented; appendages with claws; trachaea for respiration. Characters of both annelids and uniramia.

Phylum Mollusca: Mollusks; protostomes, usually with an open circulatory system; muscular foot and visceral mass common. Often with external shell. Both coelom and segmentation greatly reduced.

CLASS MONOPLACOPHORA: Neopilina; coelomate, possibly segmented; characters of both annelids and mollusks.

CLASS POLYPLACOPHORA (AMPHINEURA): Chitons; eight skeletal plates embedded in dorsal body; long foot; many gills. Simple body plan. Marine.

CLASS GASTROPODA: Snails, abalone, sea slugs, land slugs; large muscular foot; radula for feeding; often with coiled shell. Marine, fresh water, and terrestrial.

CLASS BIVALVIA: Oysters, clams, scallops; two hinged shells; filter feeders (no radula); reduced head.

CLASS CEPHALOPODA: Octopuses, squids; marine predators; enlarged head that includes tentacles; large, complex brains and eyes; complex behavior. Closed circulatory system and skeleton that may be outside, inside, or not present.

*Phylum Phoronida: Phoronids; marine, wormlike animals with body in a chitinous tube.

*Phylum Ectoprocta: Moss animals; marine; colonial; polyplike bodies with coelom, lophophore, and U-shaped gut.

*Phylum Brachiopoda: Lamp shells; body on a stalk and within two calcareous shells. Marine.

Phylum Echinodermata: Echinoderms; coelomate deuterostomes; five-fold radial symmetry; water vascular system and tube feet; calcareous plates and sometimes spines. Marine.

*CLASS ECHINOIDEA: Sea urchins and sand dollars; rigid skeleton; spines; tube feet.

*CLASS HOLOTHUROIDEA: Sea cucumbers; long cylindrical body; lack tube feet and spines.

*CLASS CRINOIDEA: Sea lilies; cup-shaped body with feathery arms used in filter feeding; mouth and anus on upper disk.

*CLASS ASTEROIDEA: Sea stars; five or more arms; tube feet; mouth and arms on ventral surface.

*CLASS CONCENTRICYCLOIDEA: Sea daisies; deep marine-water inhabitants in rotting wood; tiny, disk-shaped; spines at edge like daisy petals; unique two

rings of tube feet; no gut; embryos develop within gonads. Discovered 1986.

*CLASS OPHIUROIDEA: Brittle stars; long, spiny arms that are very flexible horizontally.

*Phylum Chaetognatha: Arrow worms; marine planktonic animals; dart-shaped bodies bearing fins.

Phylum Hemichordata: Acorn worms and pterobranchs; coelomate deuterostomes; gill slits and dorsal nerve cord; soft bodied; larvae like those of echinoderms. Marine.

Phylum Chordata: Deuterostome vertebrates and relatives; coelomate; gill slits; hollow dorsal nerve cord; notochord endoskeleton; postanal tail; segmental muscles for locomotion.

SUBPHYLUM UROCHORDATA: Sea squirts, tunicates; ciliary filter feeders; gill slits; open circulatory system. Tadpole larva with tail that has notochord and dorsal nerve cord. Adults usually sessile. Marine.

SUBPHYLUM CEPHALOCHORDATA: Lancelet (amphioxus); fishlike body; ciliary filter feeder; segmental body muscles. Marine.

SUBPHYLUM VERTEBRATA: Adults with notochord (ancient fishes, amphibians) or vertebral column; prominent head with major sense organs; closed circulatory system; many with two pairs of appendages.

CLASS AGNATHA: Ostracoderms, lampreys, hagfish; jawless; ancient forms with bone; contemporary forms with cartilage endoskeletons; are parasites or scavengers. Mostly marine.

CLASS ACANTHODII: First jawed vertebrates; paired fins; marine and later fresh water; extinct.

CLASS CHONDRICHTHYES: Sharks, skates; jawed, cartilagenous fish; urea osmoregulators; oils for buoyancy; cartilage endoskeleton; pectoral fins for maneuvering or locomotion. Mostly marine. Internal fertilization and lengthy development.

CLASS OSTEICHTHYES: Jawed bony fish; bony endoskeleton.

SUBCLASS ACTINOPTERYGII: Ray finned teleost fish such as trout and others; swim bladder for buoyancy. Many eggs, embryos.

SUBCLASS SARCOPTERYGII: Lungfish (Dipnoi); coelacanths; and rhipidistians, the source of amphibians. Most with lungs; few eggs, embryos.

CLASS AMPHIBIA: Frogs (Anura), salamanders (Urodela), and wormlike caecilians (Apoda); jawed tetrapods; bony endoskeleton; lungs plus moist skin for respiration. External fertilization with aquatic embryos and larvae; metamorphosis to adult terrestrial form. Terrestrial and aquatic.

CLASS REPTILIA: Turtles, snakes, lizards, and crocodiles; many extinct aquatic, aerial, and terrestrial types, including dinosaurs; jawed tetrapods with dry skin; bony endoskeleton; lungs for respiration. Cleidoic amniotic egg. Uric-acid osmoregulator. Terrestrial, but some marine or fresh water.

CLASS AVES: Birds; jawed tetrapods; feathers; light bony endoskeleton; wings for flight. Cleidoic eggs. Endothermic homeotherms; uric-acid osmoregulators; high pressure circulatory system.

CLASS MAMMALIA: Mammals, including monotremes (Prototheria), marsupials (Metatheria), and placentals (Eutheria); latter group includes 26 orders, one of which is the primates. Jawed tetrapods; bony endoskeleton; hair; endothermic homeotherms; mammary glands to feed newborn; eggs laid by monotremes; others nourish embryo in uterus, usually with placenta; large brains and complex behavior.

APPENDIX B
Some Useful Chemical Measurements

All precise science rests on quantitation and numbers, for mathematics is a language common to all science. The chemist analyzing hereditary material, the biologist measuring water and salt loss in a sweating desert rodent, or the physician interpreting the action of a drug on a patient with kidney disease, all must be concerned with the amounts of substances and the concentrations of solutions. This appendix summarizes the key quantitative measures for such scientists.

DALTON

The dalton is a special unit of measurement for the weights of atoms and atomic particles. These structures are so tiny that expressing their weights in minute fractions of grams is impractical. The atomic weight of the most common isotope of carbon is arbitrarily set at exactly 12 atomic mass units, or 12 daltons. One *dalton* is, therefore, one-twelfth the mass of an atom of ^{12}C.

ATOMIC WEIGHT

Each element occurs not in one form and at one weight, but in the form of isotopes of different weights. Therefore, the standard weight established for atoms of an element is an average figure, called the *atomic weight*. For example, a typical sample of chlorine is 75.4 percent $^{35}_{17}$Cl and 24.6 percent $^{37}_{17}$Cl. Based on the weights of these isotopes and their percentages in a typical sample, the atomic weight of Cl is calculated to be 35.45.

MOLE

A *mole* is Avogadro's number (6.023×10^{23}) of particles of a substance, or 1 gram molecular weight of the substance. A gram molecular weight is the weight in grams equal to the atomic weight of each atom in a molecule multiplied by the number of those atoms. Thus the molecular weight of methane, CH_4, is

$$\underset{\text{atomic weight of H}}{4(1.01)} \quad + \quad \underset{\text{atomic weight of C}}{1(12.01)} \quad = 16.05 \text{ g}$$

Each mole of methane, 16.05 g, contains 6.023×10^{23} molecules.

MOLARITY

Biologists often are interested in the amount of a substance per unit of a solution, such as how much salt is dissolved in the blood. A 1 molar, or $1M$, solution is 1 mole of solute in 1 liter of solution. A $1M$ solution of NaCl is, therefore, 1 gram molecular weight of NaCl in 1 liter of solution:

$$1M = \underset{\substack{\text{atomic weight} \\ \text{of Na}}}{23.0} \quad + \quad \underset{\substack{\text{atomic weight} \\ \text{of Cl}}}{35.5} \quad = 58.5 \text{ g NaCl/liter solution}$$

A $2M$ solution of NaCl is 2(58.5 g) = 117.0 g NaCl/liter solution.

A $0.5M$ solution of NaCl is 0.5(58.5 g) = 29.25 g NaCl/liter solution.

Glossary

abscisic acid: A plant hormone that suppresses growth (Ch. 30).

abscission: The normal separation of a leaf, fruit, or flower from a plant (Ch. 30).

absorption spectrum: The region of the spectrum of electromagnetic energy (usually visible light) that is absorbed by a particular molecule or atom (Ch. 8).

acid: Any substance that gives up or donates H^+ ions in solution, thereby increasing the H^+ content of the solution (Ch. 2).

acrosome reaction: A two-part event that marks the first stage of fertilization of an egg by a sperm, and which permits the sperm to penetrate the egg's outer protective layers (Ch. 16).

actin: A structural protein that with myosin carries out contraction; also called microfilaments (Ch. 6).

actinomycete (ak-ti-no-my-seet): Any of a diverse group of filamentlike bacteria (Ch. 20).

action potential (nerve impulse): A temporary depolarization of the potential across the membrane of a nerve or skeletal muscle cell (Ch. 36).

action spectrum: The range of light wavelengths that triggers a chemical process, as of photosynthesis (Ch. 8).

activation energy: The amount of energy needed to break the bonds between a molecule's constituent atoms (Ch. 4).

active immunity: Immunity to disease based on prior exposure to an antigenic pathogen (Ch. 32).

active site: The small area (or areas) on an enzyme molecule, often a groove or pocket, into which the enzyme's substrate fits in a "lock-and-key" arrangement (Ch. 4).

active transport: The carrier-mediated, energy-dependent movement of materials into or out of cells, especially against a concentration gradient. Active transport differs from diffusion in that it requires expenditure of energy, usually from ATP (Ch. 5).

adaptation: (1) The accumulation of inherited characteristics that make an organism specifically suited to its environment and way of life. (2) A particular genetically based feature, characteristic, or behavior that results in an individual or species being better suited to some aspect of its environment (Ch. 41).

adaptive radiation: The development from an ancestral group of a variety of forms adapted for different habitats and ways of life (Ch. 26).

adenine: One of four nucleotide bases that are fundamental structural components of DNA molecules; classed as a purine and complementary in size to thymine in the DNA double helix (Ch. 12).

adenosine triphosphate (ATP): The primary energy-storage molecule in cells. Large amounts of energy are stored in bonds of the three phosphate groups that make up an ATP molecule's tail (Ch. 7).

adhesion: The tendency of unlike molecules to cling together (Ch. 2).

adrenal gland: Either of two endocrine glands near the vertebrate kidney that secretes a variety of steroid hormones (Ch. 37).

adrenocorticotropic hormone (ACTH): A hormone secreted by the pituitary gland that regulates the adrenals (Ch. 18).

adventitious root: Any new root that arises from an organ other than an existing root (Ch. 27).

afterbirth: The placenta and attached membranes which are expelled from the uterus after the birth of a newborn (Ch. 18).

aggression: Overt fighting, displaying, posturing, etc., sometimes between animals of different species, but more often between members of the same species (Ch. 49).

air sac: In birds, one of numerous balloonlike sacs that are filled with air and lighten the body (Ch. 33).

albinism: A genetically caused deficiency of pigment in the hair, eyes, and skin (Ch. 15).

aldosterone: A steroid hormone released by cells in the cortex of the adrenal gland that regulates salt reabsorption in the kidney (Ch. 35).

aleurone: A layer of living cells that surrounds the food stores of a monocot seedling (Ch. 28).

alga (plural: **algae**): Any of a broad group of simple, mostly aquatic plants; seaweeds are prime examples (Ch. 23).

alimentary canal: A two-ended gut that extends from mouth to anus (Ch. 34).

allantois: A membrane of embryos of land vertebrates. In reptiles and birds, the role of the allantois is to store nitrogenous wastes generated by the embryo's metabolism (Ch. 16).

allele: An alternate form of a particular gene (Ch. 10).

allelopathy: Among plants, a form of interspecific competition in which chemical substances produced by one species inhibit the germination or growth of seedlings of another species (Ch. 46).

Allen's rule: The observation that animals inhabiting cold regions tend to have shorter limbs and other extremities than do those in warmer regions, as an adaptation to minimize convective heat loss (Ch. 42).

allergy: A chronic immune-system response induced by a usually benign environmental stimulant (Ch. 32).

allopatric speciation: The evolution of two or more new species from a single ancestral species as the result of chance geographical separation (Ch. 43).

allosteric enzyme: An enzyme that undergoes reversible changes in shape and in catalytic activity when "control" substances bind (Ch. 4).

α (alpha) helix: A secondary folding pattern of polypeptide chains, in which the protein's amino-acid residues are wrapped in a helical "spiral staircase" shape (Ch. 3).

alternation of generations: The alternating life cycle of plants, which includes both gametophyte (haploid) and sporophyte (diploid) phases (Ch. 23).

altruism: Self-sacrifice by one member of an animal species for the benefit of others of the species (Ch. 49).

alveolus (plural: **alveoli**): Any of the tiny, blind-ended cavities in lungs in which gas exchange takes place (Ch. 33).

amino (uh-**mee**-no) **acid**: An organic molecule consisting of a central carbon atom, an amino group (–NH$_2$), a carboxyl group (–COOH), and a distinctive side chain (R). Amino acids are the monomers from which proteins are built (Ch. 3).

amino-acid activation: The process prior to protein synthesis in a cell in which amino acids are joined to tRNA by high energy bonds (Ch. 13).

aminoacyl attachment site: The position on a tRNA molecule where a specific amino acid attaches (Ch. 13).

amniocentesis (am-nee-oh-sen-**tee**-sis): The removal of amniotic fluid from a pregnant woman's womb, and analysis of fetal cells in the fluid for chromosomal defects (Ch. 15).

amnion: A protective membrane which is filled with fluid and forms a cushionlike sac around a developing embryo (Ch. 16).

anabolism: The chemical reactions involved in the synthesis of essential large and small biological molecules in cells (Ch. 4).

analogy: The independent evolution in unrelated life forms of similar structures that carry out similar functions (Ch. 43).

anaphase: The third stage of mitosis, in which the two sets of chromatids separate and move to opposite ends of the dividing cell (Ch. 9).

anaphylactic shock: A life-threatening allergic reaction during which certain cells of the immune system discharge their contents simultaneously, causing rapid fluid loss from blood vessels throughout the body and reduced blood flow to the brain and heart (Ch. 32).

androgen: The collective term for certain male sex hormones (Ch. 18).

aneuploidy: A genetic condition marked by the absence of one or more chromosomes or the presence of extra chromosomes (Ch. 15).

angiosperm: A flowering plant (Ch. 24).

anisogametes: Gametes of unequal size that are produced by a single species, such as human egg and sperm (Ch. 23).

annual plant: A plant whose life cycle—germination from seed, growth, reproduction, and death—is completed within a single growing season (Ch. 28).

antibody: Globular blood proteins that are produced by B lymphocytes and that bind specifically to foreign antigenic materials in the body (Ch. 32).

anticodon: A set of three unpaired bases on a tRNA molecule that binds to a complementary codon on mRNA (Ch. 13).

antidiuretic hormone (ADH): A posterior pituitary gland hormone that regulates the amount of water allowed to pass from the kidney as urine (Ch. 35).

antigen: Any substance that elicits an immune response (Ch. 32).

aorta: The largest artery in the vertebrate body; carries oxygenated blood away from the heart (Ch. 31).

aortic arch: Any of the paired blood vessels running through the gill arches of vertebrate embryos and adult fish (Ch. 31).

apical dominance: The tendency of the main shoot of a plant to predominate over all others (Ch. 30).

apical meristem: The undifferentiated, actively dividing cells at the growing tip of a plant shoot; such tissue is the source of a plant's leaves, stems, branches, and flowers (Ch. 28).

apoplastic pathway: In a plant root, the "compartment" made up of all extracellular spaces, along with the spaces within cell walls that water can traverse without crossing any plasma membranes (Ch. 27).

aposematic (warning) coloration: Bright colors or striking patterns used as a warning by organisms that also possess chemical or other defenses (Ch. 47).

appendicular skeleton: The portion of the vertebrate skeleton made up of the pectoral girdle and forelimb bones, and the pelvic girdle and hindlimb bones (Ch. 39).

arteriole: A small blood vessel branching off an artery (Ch. 31).

artery: A vessel that carries blood to tissues (Ch. 31).

ascus: In some fungi, a small sac in which sexually produced spores develop (Ch. 22).

A site: One of two groovelike sites on a ribosome where aminoacyl tRNA, the carrier of amino acids during protein synthesis, can bind to mRNA. When both sites are occupied (see also *P site*), the two attached amino acids can be joined together (Ch. 13).

atom: The smallest unit of matter that still displays the characteristic properties of an element (Ch. 2).

atomic number: The particular number of protons in the nucleus of an atom of an element (Ch. 2).

atomic orbital: A cloudlike three-dimensional zone around the nucleus of an atom, in which electrons are confined (Ch. 2).

atomic weight: The sum of an atom's neutrons and protons (Ch. 2).

atrionatriuretic factor (ANF): Any of several small peptide hormones produced by heart atrial cells that increase water and sodium loss in the urine (Ch. 35).

atrioventricular (A-V) node: A bundle of cells that receives electrical impulses from the heart's "pacemaker" (the sinoatrial node) and that in turn triggers waves of contraction in the heart ventricles (Ch. 31).

auto-: From Greek, meaning "self."

autoimmune response: An abnormal process in which the immune system attacks the body's own cells or substances (Ch. 32).

autonomic nervous system: The portion of the peripheral nervous system that controls physiological functions such as heart rate, respiration, and digestion (Ch. 36).

autosome: A nonsex chromosome (Ch. 9).

autotroph: Any organism that uses light energy or chemical energy to manufacture the sugars, fats, and proteins required for cellular metabolism (Ch. 5).

auxin: Any of a group of hormones that stimulate growth of various plant parts (Ch. 30).

auxotroph: A mutant bacterium that, unlike its normal relatives, is unable to manufacture all the materials necessary for its growth on a simple nutrient solution (Ch. 14).

average heterozygosity: A measure of the average frequency of individuals that are heterozygous at each gene locus in a particular genome (Ch. 41).

axial skeleton: That portion of the vertebrate skeleton made up of the skull, the vertebral column, and, in animals having them, the ribs (Ch. 39).

axillary bud: A bud that develops at the junction (the axil) of a leaf primordium and the shoot apical meristem (Ch. 28).

axon: The portion of a neuron that transmits action potentials (nerve impulses) (Ch. 36).

bacteroid: A nitrogen-fixing bacterium that is able to survive only within a plant root nodule (Ch. 29).

balancing selection: The process that operates to counteract the loss of variant alleles in a population (Ch. 42).

Barr body: Any of one or more inactive X chromosomes present in the cells of a normal female mammal (Ch. 15).

basal body: One of several types of microtubule-organizing centers in animal and other kinds of flagellated cells; similar to a centriole (Ch. 6).

basal lamina: A feltlike layer attached to epithelial tissues which consists of highly ordered sugar–protein complexes, along with a specialized form of collagen (Ch. 5).

base: (1) A substance that upon dissociation in water, forms hydroxyl ions (OH^-); such ions commonly combine with hydrogen ions (H^+), thereby raising the pH of the solution. The opposite of an acid, a base has a pH of more than 7 (Ch. 2). (2) A ring structure composed of carbon and nitrogen that serves as one of the chemical building blocks of nucleotides. The nucleotide bases are adenine, guanine, cytosine, uridine, and thymine (Ch. 12).

basidiocarp: The dense mass of hyphae that forms the main body of a mushroom (Ch. 22).

basidium (plural: basidia): One of many club-shaped structures that lines the surfaces of gills on the underside of a mushroom (Ch. 22).

basilar membrane: A thin sheet of tissue that lies over the inner cochlear canal in the vertebrate ear, and upon which rest numerous sensory hair cells (Ch. 38).

Batesian mimicry: A defensive adaptation in which a species lacking chemical defenses evolves the same warning coloration or patterns as a noxious species (Ch. 47).

B cell (B lymphocyte): One of two major types of white blood cells of the immune system; B cells synthesize antibody molecules (Ch. 32).

benthic community: An ocean-bottom marine biome (Ch. 44).

Bergmann's rule: Individuals living in colder parts of their species' range tend to be larger than individuals living in warmer areas (Ch. 42).

β (beta)-pleated sheet: A configuration for the folding of polypeptide chains in which two or more polypeptides lying side by side become crosslinked by hydrogen bonds and form an accordianlike sheet of connected molecules (Ch. 3).

biennial plant: A plant whose life cycle from seed germination to the production of a new generation of seeds requires two growing seasons (Ch. 28).

bilateral symmetry: The animal body plan in which an organism's right and left sides are mirror images (Ch. 25).

bilayer (lipid): The thin, two-layered arrangement of lipid molecules that make up membranes within and at the surface of cells (Ch. 5).

binary fission: The division of prokaryotic cells into two virtually identical daughter cells (Ch. 20).

binomial system of nomenclature: The assignment of names to organisms using two Latin words, the first denoting the genus and the pair denoting the species, e.g., *Homo sapiens* (Ch. 19).

biomass: The combined dry or wet weight of all the organisms in a habitat (Ch. 44).

biome: Any of the Earth's major ecosystems (Ch. 44).

biosphere: The land and waters and the air above them which make up the life-supporting region of the Earth's surface (Ch. 44).

bipolar cell: Any of the specialized neurons in the eye which synapse with rods and cones, and also with ganglion cells which transmit information along the optic nerve to the brain (Ch. 38).

blade: The broad, usually flattened portion of a leaf (Ch. 27).

blastomere: Early embryonic cells of animals (Ch. 16).

blastula: The usually spherelike arrangement of cells around a central cavity or mass of yolk that results when a zygote undergoes cleavage during embryonic development (Ch. 16).

blood: A dynamic, life-sustaining solution in animals with closed circulatory systems containing ions, nutrients, waste products, hormones, other substances, and cells. The cells—blood cells and platelets—are suspended in plasma (Ch. 31).

blood–brain barrier: In the central nervous system, the system of tightly joined, highly impermeable capillary walls that acts to prevent most blood-borne substances from passing easily into the cerebrospinal fluid that bathes CNS neurons. Also inhibits movement in the opposite direction. Lipid-soluble substances can pass the barrier (Ch. 40).

blood pressure: The hydrostatic pressure exerted by the blood in an animal's circulatory system as a result of the rhythmic contractions of the heart and peristaltic waves of contraction in some blood vessels (Ch. 31).

Bohr effect: A decrease in the affinity of hemoglobin for O_2 as pH falls, and an increase in affinity as pH rises (Ch. 33).

bond energy: The amount of energy needed to break a chemical bond (Ch. 2).

bone: A hard, living tissue consisting primarily of collagen and apatite, a calcium and phosphate salt (Ch. 39).

bottleneck effect: A form of genetic drift that occurs when only a small portion of an original population provides the gene pool for a new population (Ch. 41).

bronchus (plural: bronchi): One of the branched passageways through which air enters the lungs (Ch. 33).

bud primordium: A group of meristematic cells that has begun to develop into a bud and will, in turn, grow into a branch or form flowers (Ch. 28).

buffer: A substance that binds hydrogen ions when concentrations of H^+ are high and releases hydrogen ions when concentrations of H^+ are low (Ch. 2).

bulb: A compact, subterranean conical stem having modified leaves in which carbohydrate is stored; participates in vegetative reproduction (Ch. 28).

bulk flow: The movement of a mass of fluid (Ch. 31).

caecum (plural: caeca): a blind-ended pouch that extends from the intestine and holds food for an extended period of time to enhance digestion and absorption of nutrients (Ch. 34).

calcitonin: A peptide hormone that lowers blood calcium levels; produced in the thyroid gland (Ch. 37).

callus tissue: A disorganized lump of plant tissue (Ch. 30).

calorie: The amount of heat required to raise the temperature of 1 gram of water by 1°C. (Ch. 34).

Calvin-Benson cycle: The dark reactions of photosynthesis (Ch. 8).

calyx: The ring of sepals, the outer, usually green parts of a flower (Ch. 28).

capillary: Any of the tiny, thin-walled blood vessels interwoven throughout body tissues (Ch. 31).

capillary action: The tendency of water in a thin tube to move upward (Chs. 2, 29).

carbohydrate: One of a group of carbon compounds including sugars, starches, and cellulose and consisting of a carbon backbone with various functional groups attached. Carbohydrates are the most abundant organic compounds found in living organisms (Ch. 3).

carbon cycle: The natural cycle established by the activities of photosynthesis and respiration in various life forms. Basic steps of the cycle are the fixing of atmospheric CO_2 into carbohydrates via photosynthesis, the use of carbohydrates as fuel by organisms, and the release of waste CO_2 back to the atmosphere (Ch. 8).

cardiac muscle: The specialized striated muscle tissue of the heart (Ch. 39).

cardiac output: A measure of the amount of blood pumped by the heart per unit of time (Ch. 31).

carnivore: An animal that eats only meat (Ch. 34).

carotenoid pigment: A pigment related to vitamin A that appears red, orange, or yellow to the human eye (Ch. 8).

carrier-facilitated diffusion: A process in which specialized proteins act as carriers that transport substances across a cell's plasma membrane (Ch. 5).

carrying capacity: The maximum size population that can be supported by the resources in the population's environment (Ch. 46).

cartilage: A fibrous connective tissue consisting primarily of collagen and complex polysaccharides (Ch. 39).

Casparian strip: Any of the waterproof, suberin-coated walls of endodermal cells in a plant (Ch. 27).

catabolism: The energy-yielding processes in cells, in which molecules are broken down to obtain structural elements, to release energy, or to digest waste products (Ch. 4).

catalyst: Any molecule that increases the rate of a chemical reaction without being used up during that reaction. Biological catalysts are primarily protein molecules known as enzymes (Ch. 4).

cell: The basic structural unit of living organisms; see also *prokaryotic cell* and *eukaryotic cell* (Ch. 5).

cell cycle: The regular sequence during which a cell grows, prepares for division, and divides to form two daughter cells (Ch. 9).

cell-mediated immunity: The direct attack on foreign cells or substances by T lymphocytes (Ch. 32).

cell plate: The layer of membranous sacs that arises to separate the two new daughter cells when a plant cell divides (Ch. 9).

cellular respiration: The oxygen-dependent metabolic process by which cells derive energy (ATP) from glucose and other fuel molecules (Ch. 7).

cellular slime mold: A type of slime mold in which separate amoeboid cells aggregate to form a multicellular slug (Ch. 21).

cellulose: The fibrous structural material of plant cells and wood; a high-molecular-weight polysaccharide composed of long chains of glucose units (Ch. 3).

cell wall: The rigid outer structure that surrounds the plasma membranes of plant cells, bacteria, and some fungi (Ch. 5).

central nervous system (CNS): The brain and spinal cord of vertebrates (Ch. 36).

centriole: A cellular organelle composed of nine microtubular triplets. Centrioles serve as assembly sites for spindle microtubules used in cell division (similar to *basal bodies*) (Ch. 6).

centromere: The constricted area at which two chromatids are attached to one another and to which spindle microtubules are attached; site of satellite DNA (Ch. 9).

cephalization: The formation of a front end or "head" where organs specialized for sensing and feeding are located (Ch. 25).

cerebellum: The region of the vertebrate hindbrain that regulates balance, stance, and some locomotor movements (Ch. 40).

cerebrospinal fluid: The fluid filling the hollow ventricular chambers of the vertebrate central nervous system, as well as the intercellular spaces in the walls of the brain and spinal cord and the spaces between the meninges (Ch. 40).

cerebrum: The anterior dorsal part of the vertebrate forebrain; the site where most sensory processing and complex behaviors are coordinated (Ch. 40).

cervix: The base of the uterus, which also serves as the upper end of the vagina (Ch. 18).

character displacement: In closely related species, the evolution of slight physical differences in structures used to exploit a limited resource (Ch. 46).

chemical energy: Potential energy stored in atoms and molecules and in their bonds (Ch. 4).

chemical-gated channel: An ion-transport channel in the membrane of a neuron that opens when a specific chemical is present (Ch. 36).

chemical reaction: The interaction of atoms, ions, or molecules to form new substances (Ch. 2).

chemiosmotic coupling hypothesis: A model for cellular respiration in which the movement of electrons through the electron transport chain in mitochondria is accompanied by a proton-pumping mechanism, which in turn sets up an "energy gradient." Energy released from this gradient is conserved in the form of ATP (Ch. 7).

chemoautotroph: A bacterium capable of oxidizing inorganic compounds to gain energy for its life processes (Ch. 20).

chemoreceptor: A sensory receptor cell that detects specific molecules and ions; active in taste and smell perception (Ch. 38).

chemotaxis: Movement toward or away from the source of a diffusing chemical (Ch. 6).

chitin: A substance built from nitrogen-containing polysaccharides that is the main ingredient of the hard exoskeletons of lobsters, spiders, houseflies, and their relatives (Ch. 5).

chlorophyll: The principal green pigment active in photosynthesis (Ch. 8).

chloroplast: The chlorophyll-containing plastid organelle in plant cells that is the site of photosynthesis (Ch. 6).

chorion: One of the four extraembryonic membranes surrounding the embryos of land vertebrates. The chorion and the allantois fuse to become the chorioallantois. (Ch. 16).

chrom-, -chrome: From Greek, meaning "color" or "pigmented."

chromaffin cell: Any of the endocrine cells of the adrenal medulla that secretes epinephrine and norepinephrine (Ch. 37).

chromatid: One of the two copies of a chromosome in a cell undergoing division (Ch. 9).

chromatin: The chromosomal substance within the nucleus of a nondividing cell which consists of DNA, histones, and non-histone proteins (Ch. 9).

chromosomal mutation: A change in the physical structure of a chromosome (Ch. 11).

chromosome: A long strand of coiled DNA (deoxyribonucleic acid) that is the site of genes, the genetic information for most organisms. Eukaryotic chromosomes also include many proteins (Ch. 6).

chyle (kile): A whitish, watery solution of partially digested food material produced by the neutralization (by bile and pancreatic juices) of chyme (Ch. 34).

chyme (kyme): The semifluid mass of food material produced by the action of digestive juices in the stomach; the material that passes from the stomach to the small intestine (Ch. 34).

ciliate: A single-celled organism characterized by the many cilia on its surface (Ch. 21).

cilium (plural: **cilia**): A short, hairlike organelle that has a microtubular skeleton and a capacity to beat, thereby propelling a cell through fluid or a fluid past a cell (Ch. 6).

circulatory system: A transport system in animals, consisting of specialized structures such as vessels and a heart and blood, which delivers nutrients and other essential materials to cells and carries away metabolic wastes (Ch. 31).

clade: A group of organisms that shares sets of specific characteristics that reflect common descent; hence, organisms belonging to a single taxonomic unit (Ch. 19).

class: A taxonomic grouping of organisms belonging to related orders (Ch. 19).

cleavage: A specialized form of mitosis (cell division) in embryos in which no cell growth occurs between divisions (Ch. 16).

climax community: A stable ecological community dominated by species that tend not to be replaced by new species (Ch. 45).

cline: A gradual change in one or more characteristics among members of a species over a broad geographical range (Ch. 43).

clonal selection theory: The theory stating that many precommitted types of T and B lymphocytes are present in an organism, and that a newly invading antigen binds to one, thereby "selecting" it for activation (Ch. 32).

clone: An offspring that arises mitotically or asexually and is genetically identical to its parent (Ch. 9).

closed circulatory system: A circulatory system in which blood moves through a continuous set of interconnected vessels (Ch. 31).

closed (behavioral) program: An innate motor pattern that cannot be altered easily by learning (Ch. 48).

co-: From Greek, meaning "a shared condition."

coacervate: A polymer-rich droplet having certain cell-like properties (Ch. 19).

cochlea: A set of spiraled tubes in the vertebrate inner ear; in mammals, the primary structure involved in hearing (Ch. 38).

codominant gene: A gene with several alleles, two of which are equally dominant. Both dominant alleles are fully expressed when they appear together, and both phenotypic traits are present (Ch. 11).

codon: A set of three consecutive nucleotides that code for a single amino acid (Ch. 13).

coelom: A fluid-filled body cavity that is lined by mesoderm and in which a variety of body organs are suspended (Ch. 25).

coenocytic (see-no-**sit**-ik): Having multiple nuclei in one mass of cytoplasm (Ch. 22).

coevolution: A process in which different life forms undergo interrelated, often complementary, evolutionary change (Chs. 1, 47).

cofactor: A special enzyme substrate that binds temporarily to a site on an enzyme and participates in chemical reactions and the formation of products (Ch. 4).

cohesion: The tendency of like molecules to cling to one another (Ch. 2).

coleoptile: A specialized sheath that protects the growing shoot tissue of a monocot embryo (Ch. 28).

collecting duct: The duct associated with a nephron in the vertebrate kidney in which water is removed from the filtrate and returned to the blood (Ch. 35).

collenchyma: A plant cell type having especially strong primary walls that lend extra support to stem tissue (Ch. 27).

colloidal osmotic pressure: Osmotic pressure created by the presence of proteins (colloids) in blood or body fluids (Ch. 31).

colostrum: An antibody- and protein-rich fluid synthesized and stored prior to birth in the mammary glands of a pregnant mammal; when fed to the newborn, it gives immediate resistance to a variety of potentially dangerous microbes (Ch. 18).

commensalism: A form of symbiosis in which members of two species live in intimate association, with one partner benefiting while the other is unharmed (Ch. 47).

community: All of the interacting populations of different species in a particular area (Ch. 45).

companion cell: In a flowering plant (angiosperm), a specialized cell associated with seive tube elements, the cells that make up the transport tissue phloem (Ch. 27).

compatible osmolyte strategy: The presence of nonharmful organic molecules in body fluids of organisms to raise total osmotic pressure without inhibiting proteins (Ch. 35).

compensatory hypertrophy: A growth response sometimes seen in damaged organs, in which cells in remaining healthy tissue divide extremely rapidly so that the affected organ increases in mass and cell number (Ch. 16).

competitive exclusion principle: The postulate that no two species can occupy the same niche at the same time in a particular locale (Ch. 45).

competitive inhibition: A system of enzyme control in which the active site is occupied by a compound other than the normal substrate, thereby preventing the binding of the substrate (Ch. 4).

complementary gene: A gene whose protein product must act with the product of another gene to produce a given phenotype. Human albinism is produced by complementary genes (Ch. 11).

compound: An aggregate of atoms of more than one element (Ch. 2).

compound eye: An eye composed of many separate optic units (ommatidia) (Ch. 38).

concentration gradient: A variation in the concentration of a substance, as from higher to lower (Ch. 5).

condensation reaction: A chemical process which takes place when two monomers join; typically, as a covalent bond forms between them, one monomer loses an –OH group and the other loses a hydrogen atom (Ch. 3).

cone: One of two types of photoreceptors in the eyes of many vertebrates; responsive to higher light intensities, cones provide high visual acuity and color vision (Ch. 38).

conifer: A cone-bearing plant (Ch. 24).

conidium: A spore generated during asexual reproduction in certain fungi (Ch. 22).

conjugation: An exchange of genes in which DNA is transferred from one (usually) bacterial cell to another by way of a cytoplasmic bridge (Ch. 14).

connective tissue: The mesodermally derived, collagen-fiber-containing tissue that surrounds most animal organs and literally connects them together (Ch. 5).

constant region: That portion of an antibody protein that has the same amino-acid sequence as other, different antibodies (Ch. 32).

consumer: An animal that derives energy from eating all or parts of other animals and plants (Ch. 44).

convergent evolution: An evolutionary pattern in which distantly related organisms become more alike as they evolve similar adaptations (Ch. 43).

coralline alga: A red alga that can deposit hard calcium-carbonate crystals in its cell walls, creating a coral reef (Ch. 23).

core (of the Earth): The extremely dense mixture of materials, primarily iron and nickel, that makes up the innermost region of the Earth (Ch. 19).

cork cambium: A layer of cells just beneath the epidermis of a woody plant which produces cork, the outer, nonliving component of bark (Ch. 28).

corm: A solid, subterranean stem structure that serves as a site for carbohydrate storage and vegetative reproduction (Ch. 28).

cornea: In the eye, the transparent portion of the sphere through which light enters (Ch. 38).

corolla: Collectively, the petals of a flower (Ch. 28).

corpus callosum: A giant bundle of nerve axons that links the two hemispheres of the vertebrate brain (Ch. 40).

corpus luteum: Literally, "yellow body"; a cell mass that forms from an ovarian follicle after the egg has been released, and which secretes progesterone (Ch. 18).

cortex: (1) In a plant, the region of the stem underlying the epidermis and composed mainly of parenchyma cells (Ch. 27). (2) The outer region of some animal organs, as the kidney, adrenal gland, or brain; see *neocortex* (Ch. 35).

corticosteroid: Any of the steroid hormones produced in the adrenals; divided into glucocorticoids, mineralocorticoids, and sex steroids (Ch. 37).

cotyledon: An embryonic seed leaf that stores nutrients to sustain the growth of a newly germinated plant (Ch. 24).

co-transport: A form of active transport in which sodium ions and a sugar or amino-acid molecule bind to a carrier that transports them together and discharges them inside the cell (Ch. 5).

counteracting osmolyte strategy: The maintenance of osmotic pressure in body fluids through the presence of pairs of inhibitory and stimulatory substances such as urea and trimethylamine oxide (Ch. 35).

countercurrent exchange system: The flow of two fluids, which are separated by a permeable interface, in opposite directions; the result is much more efficient exchange of commodities that can pass through the interface (heat, O_2, etc.) (Ch. 33).

coupling factor: Large, complex molecules that are sites where the flow of protons through usually impermeable membranes drives the production of high energy compounds (as ATP) (Ch. 7).

covalent (coe-vay-lent) **bond:** A chemical bond between atoms characterized by the sharing of electrons (Ch. 2).

crop: In some animals, a thick-walled chamber that receives and temporarily stores food materials (Ch. 34).

crossing over: The exchange of corresponding pieces of genetic material between a maternal chromatid and a paternal one (Ch. 9).

crust (of the Earth): The rocky outermost layer of the Earth (Ch. 19).

cryptic coloration: Having the same color or pattern as the background; a form of defensive camouflage (Ch. 47).

cultural evolution: The transfer of information from generation to generation of animals by nongenetic means (Ch. 1).

cuticle: The waxy outer layer of cutin on leaves and stems that inhibits the loss of moisture from within (Ch. 27).

cycad: A primitive seed plant; a member of the division Cycadophyta (Ch. 24).

cyclic AMP (cAMP): Cyclic adenosine monophosphate; a compound related to ATP which acts as an intermediate or second messenger in the transmission of signals within target cells (Ch. 37).

cyclic GMP (cGMP): Cyclic guanosine monophosphate; a second-messenger compound (Ch. 37).

cyclic photophosphorylation: A recurring ATP-generating process in photosynthesis that is driven solely by light energy (Ch. 8).

-cyte, cyto-: From Greek, meaning "hollow vessel"; refers to cells.

cytochrome: Any of a class of iron-containing proteins that acts as carriers in the electron transport chain of cellular respiration (Ch. 7).

cytokinesis: The separation of the cytoplasm of a dividing cell into two daughter cells (Ch. 9).

cytokinin: Any of a class of plant hormones that regulates cell division (Ch. 30).

cytoplasm (site-oh-plazm): The semifluid substance that makes up the nonnuclear part of a eukaryotic cell, or the nonnucleoid part of a prokaryotic cell (Ch. 6).

cytoplasmic streaming: The rapid movement of cell cytoplasm to circulate nutrients, proteins, pigments, or other cellular materials or organelles (Ch. 6).

cytosine: A single-ring pyrimidine base that forms a DNA nucleotide; complementary to guanine in the DNA double helix (Ch. 12).

cytoskeleton: The three-dimensional weblike structure that fills the cytoplasm of a cell, and within which organelles are suspended (Ch. 6).

dark reactions: The second-stage reactions of photosynthesis, which do not require light energy to proceed and in which CO_2 is reduced to carbohydrate (Ch. 8).

deciduous: In plants, the property of shedding leaves at the end of the growing season (Ch. 24).

decomposer: An organism that derives energy for its life processes by breaking down remnants of organic materials (Ch. 44).

dendrite: Any of the short, multibranched extensions of the neuron surface that receives input to the cell (Ch. 36).

denitrification: The reduction of nitrate and nitrite to gaseous N_2, NO, or N_2O (Ch. 44).

density-dependent factor: Any factor, such as disease or competition, whose effect on a population's numbers is related to population density and size (Ch. 46).

density-independent factor: Any factor, such as a natural calamity, that regulates population size independent of population density and size (Ch. 46).

depolarization: A change in a cell's electrical state toward a nonpolarized condition (Ch. 36).

desert: A very dry, often rather barren biome found in subtropical latitudes or where mountains block moisture-bearing winds (Ch. 44).

desmosome: An intercellular junction thought to "glue together" the plasma membranes of adjacent cells (Ch. 5).

determination: The final commitment of cells to a single developmental pathway, from among several alternatives (Ch. 17).

detritivore: An organism that obtains food energy by consuming disintegrated organic matter (Ch. 44).

deuterostome: Any member of the lineage of animals in which the blastopore of the developing embryo becomes the anus, while a second opening becomes the mouth (Ch. 25).

di-: From Greek, meaning "two."

diacylglycerol (DG): A second-messenger compound; see *cyclic AMP* (Ch. 37).

diaphragm: A muscular sheet that separates the thoracic cavity from the abdominal cavity in mammals; active in the inhalation-exhalation mechanism of breathing (Ch. 33).

diastole (die-ast-uh-lee): The phase of relaxation of the heart muscle (Ch. 31).

diatom: A phytoplankton organism having sculptured, glasslike cell walls containing silica (Ch. 21).

dicotyledon: An angiosperm in which embryos have two seed leaves (Ch. 24).

differentiation: During development, the phenotypic maturation of cells in both structure and function (Ch. 16).

diffusion: The process by which a dissolved substance moves passively through a fluid (Ch. 5).

digestion: The chemical breakdown of foods into compounds that can be used for cellular metabolism (Ch. 34).

dikaryon: A specialized cell type in fungal hyphae which contains two nuclei and plays a role in spore formation (Ch. 22).

dinoflagellate: A single-celled phytoplankton, often photosynthetic, having a set of flagella that causes the organism to spin as it swims (Ch. 21).

dioecious (dy-ee-shus): Plant species having individuals that are either male or female; hence individuals produce only one of the two types of gametophytes (Ch. 24).

diploid: Literally, "paired"; having two sets of homologous chromosomes in each cell (Ch. 9).

directional selection: Natural selection in which the result is a change from one phenotype to another (Ch. 42).

disruptive coloration: Having body parts of contrasting colors which break up an organism's silhouette (Ch. 47).

disruptive selection: Natural selection in which a population may be separated into discrete groups because the "average" phenotype for a particular trait is strongly selected against (Ch. 42).

divergent evolution (radiation): The evolutionary pattern in which one or more phylogenetic lines branch and rebranch as organisms acquire an expanding repertoire of adaptations to new environmental circumstances (Ch. 43).

diversifying selection: A form of natural selection that helps to generate distinctive phenotypes adapted to specialized features of a portion of the total environment (Ch. 42).

division: In botany, a taxonomic grouping of organisms belonging to similar classes; the equivalent of a phylum (Ch. 19).

DNA ligase: Any of a class of enzymes which can rejoin complementary cohesive ends of DNA fragments (Ch. 14).

DNA polymerase: An enzyme that catalyzes the synthesis of a DNA strand complementary to an original strand of the parent molecule (Ch. 12).

dominant (allele): An allele whose action masks that of another (recessive) allele of the same gene and thereby determines the phenotype of a heterozygous offspring (Ch. 10).

dormancy: A resting state, as of seeds, in which an organism or cell respires at a very low rate and carries out only a small amount of metabolic work (Ch. 28).

double fertilization: In angiosperms, the process in which one sperm from a pollen grain fuses with the small egg cell of the megagametophyte, while the second sperm penetrates the adjoining large endosperm cell containing the two polar nuclei (Ch. 28).

double helix: The "twisted ladder" structure of linked double strands of nucleotides that comprises a DNA molecule (Ch. 12).

DPG (2,3-diphosphoglycerate): A substance in mammalian red blood cells that decreases the binding affinity of O_2 to hemoglobin (Ch. 33).

dynein: A mechanoenzyme, associated with microtubules, that cleaves ATP. Dynein arms may generate the forces that cause cilia and flagella to beat (Ch. 6).

early wood: Xylem cells formed in the early part of a tree's growing season (Ch. 28).

ecdysone: Any of a family of steroid hormones that regulates molting in insects and other arthropods; alpha ecdysone is the precursor of the substance that initiates insect metamorphosis (Ch. 37).

echolocation: A sensory system in which high-frequency sound pulses are used to produce echoes that convey information on the location of objects in an animal's environment (Ch. 38).

ecocline: Variation in the composition of plant or animal communities along a climate gradient—usually of moisture or temperature (Ch. 45).

ecological efficiency: The efficiency of energy transfer between adjacent trophic (feeding) levels in a food chain (Ch. 44).

ecological isolation: The presence of genetic differences between populations that prevent interbreeding and that have arisen as adaptations to particular features of the environment (Ch. 43).

ecological succession: Gradual changes in the composition of the species that make up a community (Ch. 45).

ecology: The study of the interplay of organisms and their environments (Ch. 44).

ecosystem: Any community of interacting organisms, including their physical and chemical environment, energy fluxes, and the types, amounts, and cycles of nutrients in the various habitats within the system (Ch. 44).

ecto-: From Greek, meaning "outer."

ectoderm: The outer layer of cells that arises during cleavage, and later develops, in vertebrates, into the epidermis, the nervous system, and the sense organs (Ch. 16).

ectotherm: An animal that derives most of its body heat from the external environment (Ch. 35).

effector (motor) neuron: Any of the class of neurons that transmits messages to muscles and glands (Ch. 36).

electron: A subatomic particle that bears a negative electrical charge and moves continuously about the nucleus (Ch. 2).

electron transport chain: The final phase of cellular respiration, in which the compounds NADH and $FADH_2$ are oxidized and their electrons passed along a chain of oxidation-reduction steps (Ch. 7).

electrophoresis: A technique used to trace the movement of electrically charged molecules through a fluid medium (Ch. 41).

element: A substance that cannot be decomposed by chemical processes into simpler substances (Ch. 2).

elongation: (1) In all cells, the stage of protein synthesis in which amino acids are added sequentially to the lengthening polypeptide chain (Ch. 13). (2) In plant cells, the increase in the length of cells to cause stem or root growth (Ch. 28).

end-, endo-: From Greek, meaning "inside."

endergonic reaction: Any chemical reaction in which the products have more total energy and more free energy than did the reactants. Endergonic reactions thus require the input of energy from another source before they can take place (Ch. 4).

endocrine system: The animal organs, tissues, and cells which secrete hormones (Ch. 37).

endocytosis: A process through which materials are taken into a cell (Ch. 6).

endoderm: The innermost layer of cells arising during cleavage and gastrulation, and ultimately forming the lining of the gut and other internal epithelial tissues (Ch. 16).

endodermis: In a plant, the innermost layer of the cortex tissue (Ch. 27).

endolymph: The fluid that fills the semicircular canals of the vertebrate ear (Ch. 38).

endometrium: The lining of the uterus (Ch. 18).

endoplasmic reticulum (ER): An array of membranous sacs, tubules, and vesicles within the cell cytoplasm. ER may be rough or smooth; the rough form is studded with ribosomes and is the site of synthesis of proteins that will be secreted or built into membranes (Ch. 6).

endorphin: A neuroactive peptide in the brain that acts as a natural painkiller; see also *enkephalin* (Ch. 40).

endospore: A heavily encapsulated resting cell formed within many types of bacterial cells during times of environmental stress (Ch. 20).

endotherm: An animal that produces its own body heat (Ch. 35).

enkephalin: A five-amino-acid neuroactive peptide synthesized in the brain; along with endorphins, enkephalins are natural opiates (Ch. 40).

enterogastric reflex: A reflex that inhibits the release of *chyme* into the small intestine, thereby allowing the small intestine more time to digest the material it already contains (Ch. 34).

entropy: The amount of disorder in a system (Ch. 4).

enzyme: A member of the class of proteins (and certain RNAs) that catalyze chemical reactions (Ch. 3).

enzyme saturation: A condition in which all of the active sites on one or more available enzyme molecules are occupied by substrate molecules most of the time (Ch. 4).

enzyme-substrate complex (ES): A complex consisting of an enzyme and its reactant (substrate) which is held together by weak bonds. The formation of an ES is the crucial first step in enzyme catalysis (Ch. 4).

epicotyl: The structure in a plant embryo that is the future stem of the seedling (Ch. 28).

epidermis: In plants, a surface layer of cells—usually one cell thick—that reduces moisture loss (from leaves and stems), or is the site of water uptake (in roots) (Ch. 27). In animals, the outermost tissue.

epinephrine: A substance produced by the adrenal medulla which can function both as a neurotransmitter and as a hormone; also known as adrenalin (Ch. 37).

epiphyte: Any of the so-called air plants that usually grow on other plants rather than being rooted in soil (Ch. 27).

episome: A genetic element that can replicate in a bacterial host independently of the chromosome, or that can integrate into the chromosome and replicate with it (Ch. 14).

epistasis (eh-**pee**-stay-sis): A type of gene interaction in which the effects of one gene override or mask the effects of other, entirely different genes (Ch. 11).

epithelium (plural: **epithelia**): A population of animal cells arranged in a sheet. Epithelial tissue lines or covers a variety of internal and external body surfaces (Ch. 5).

equilibrium (chemical): The point in a chemical reaction at which no further net conversion of reactants takes place. When equilibrium is reached, the combined free energy of reactants equals the combined free energy of reaction products (Ch. 4).

erythrocyte: A red blood cell (Ch. 31).

erythropoetin: A peptide hormone secreted by the vertebrate kidney that stimulates the production of red blood cells in bone marrow (Ch. 37).

esophagus: The tubular passageway through which food travels from the pharynx to the stomach (Ch. 34).

essential amino acid: Any of the eight amino acids that adult humans cannot synthesize and must obtain from food (Ch. 34).

ester bonds: Chemical bonds which form between any alcohol and any carboxylic acid (Ch. 3).

estrogen: One of the gynogens, or female sex hormones (Ch. 18).

estrus: The period during which a female mammal is receptive to sexual activity (Ch. 18).

ethylene: The plant hormone responsible for ripening in fruit (Ch. 30).

eubacteria: Literally, "true bacteria"; by far the most abundant group of prokaryotes (Ch. 20).

eukaryotic cell: A cell that possesses a membrane-enclosed nucleus, chromosomes built of DNA and protein, a cytoskeleton, and a variety of membrane-bound organelles (Ch. 5).

eutrophication: The gradual accumulation of organic debris and silt in a lake as the result of high productivity of the organisms that inhabit it (Ch. 44).

evolution: Change in a gene pool (and corresponding phenotype) over time (Ch. 40).

ex-, exo-: From Greek, meaning "outside" or "external to."

excitatory synapse: A synapse in which the synaptic transmission between neurons depolarizes the receiving neuron (Ch. 36).

exergonic reaction: An energy-liberating chemical reaction in which the products contain less total energy and less usable energy than existed in the original reactants (Ch. 4).

exocytosis: A process, often involving a vacuole or vesicle, through which materials are expelled from a cell (Ch. 6).

exon: Any of the protein-coding segments of a gene (Ch. 14).

exoskeleton: The tough, rigid, outer covering (cuticle) that encloses and protects the body of an arthropod (Ch. 25).

exponential growth curve: The steeply climbing curve describing unchecked population growth, in which population size repeatedly doubles (Ch. 46).

extinction: The permanent loss of a species (Ch. 43).

extracellular digestion: Digestion carried out by enzymes secreted outside of cells, as into an organism's gut cavity (Ch. 22).

extracellular matrix: Material made up primarily of space-filling sugar polymers and collagen fibers that contributes to the bulk of bones, cartilage, the eye cavities, and other body parts (Ch. 17).

extrinsic protein: A protein molecule that is attached to the outer surface of the cell's plasma membrane (Ch. 5).

F_1 (first filial) generation: In a genetic cross, the first generation of offspring from a set of parents (Ch. 10).

F_2 (second filial) generation: The offspring from a genetic cross between members of the same F_1 generation (Ch. 10).

facultative anaerobe: A bacterium that can grow with or without the presence of free oxygen (Ch. 20).

FAD (flavin adenine dinucleotide): Along with NAD^+, a coenzyme that carries electrons and hydrogen in a variety of metabolic oxidations and reductions, such as those of the Krebs cycle (Ch. 7).

Fallopian tube (oviduct): One of the two hollow canals that extends from the ovaries to the uterus (in mammals), and that conduct ovulated eggs toward the uterus (Ch. 18).

family: A taxonomic grouping of similar genera (Ch. 19).

fat-soluble vitamin: A vitamin, such as A, D, E, or K, that is transported in the blood as a complex linked to lipids or proteins (Ch. 34).

fatty acid: A molecule consisting of a long chain of carbon atoms attached to an acidic carboxyl group (–COOH). Fatty acids are basic units of fats and oils (Ch. 3).

fermentation: A set of anaerobic reactions in which pyruvate generated by glycolysis is modified to ethanol, lactate, or some other organic end product (Ch. 7).

fertilization: The process that unites the nuclei of male and female gametes and initiates development (Ch. 16).

fibrous root system: A root system consisting of many equal-sized roots (Ch. 27).

filamentous alga: Any of a group of green algae with a specialized, threadlike body form (Ch. 23).

first law of thermodynamics: The physical law stating that energy can change from one form to another form, but can never be created or destroyed. Also termed the law of conservation of energy (Ch. 4).

fitness: The relative reproductive efficiency of various individuals or genotypes in a population (Ch. 42).

fixed allele: An allele that is the sole allele for a given gene in a population, because other alleles have been lost (Ch. 42).

fixed motor pattern: An unvarying series of precise physical movements, thought to have a genetic basis (Ch. 48).

flagellum (plural: flagella): A fine, whiplike, microtubule-containing organelle that undulates to move eukaryotic cells (Ch. 6).

florigen: A plant hormone that stimulates the formation and opening of flowers (Ch. 30).

flower bud: A bud produced by an apical meristem that will develop into a flower (Ch. 28).

follicle-stimulating hormone (FSH): A gonadotropin hormone secreted by the pituitary gland that supports sperm production in males and stimulates egg maturation in females (Ch. 18).

food chain: The chain of events in which plants in a community convert solar energy to a stored chemical form, and that energy (and material) is sequentially transferred to herbivores, carnivores, and decomposers (Ch. 44).

food web: A complex system of numerous food chains (Ch. 44).

foraminiferan: A marine protozoan that secretes a calcium-containing shell (Ch. 21).

forebrain: The anterior region of the vertebrate brain, encompassing the cerebrum, pituitary gland, thalamus, hypothalamus, and olfactory bulb (Ch. 40).

founder effect: An increased or decreased frequency of certain alleles in a new population because the founding members chanced to differ in those alleles from the original population (Ch. 41).

fovea: In primates and some other vertebrates, a region in the neural retina where maximal resolution of images is achieved (Ch. 38).

frameshift mutation: A genetic mutation that arises from a shift in the normal reading frame of a nucleotide sequence (Ch. 13).

free energy: The energy available to do work as a result of a chemical reaction (Ch. 4).

frequency-dependent selection: A process that operates when the relative fitnesses of the genotypes in a population vary according to their frequency (Ch. 42).

frond: The leaf of a fern or large alga (Ch. 23).

fruit: A mature ovary or group of ovaries that surrounds and protects a plant seed, and aids in its dispersal (Ch. 24).

fruiting body: In fungi, the structure that carries sexually produced spores (Ch. 22).

fugitive species: A plant species that typically is one of the first to colonize harsh terrain, and is capable of maturing and reproducing quickly (Ch. 45).

functional group: A cluster of atoms that imparts a similar chemical behavior to all molecules to which it is attached (Ch. 3).

fundamental niche: The full environmental range that a species can occupy if the proper physical and biological conditions are met and if there is not direct competition from another species (Ch. 45).

fusiform initial cell: In a vascular plant, the source of xylem and phloem cells (Ch. 28).

G_1 phase: The period of normal metabolism (gap 1) that is the first phase of the cell cycle (Ch. 9).

G_2 phase: The brief period of cell metabolism and growth (gap 2) that follows the S (DNA synthesis) phase in the cell cycle and is a prelude to cell division (Ch. 9).

gall bladder: A small organ near the liver in which bile is stored (Ch. 34).

gametangium (plural: gametangia): In certain fungi and other organisms, a structure that contains gametes (Ch. 22).

gametophyte: A plant in the haploid, gamete-producing stage of its life cycle (Ch. 23).

ganglion (plural: ganglia): An aggregation of nerve cells (Ch. 25).

ganglion cell: Any of the neurons associated with rods and cones whose axons extend via the optic nerve to the vertebrate brain (Ch. 38).

gap junction: A perforated channel which serves as the primary electrical, ionic, and molecular communication junction between animal cells (Ch. 5).

gastrin: A digestive hormone (also a neuropeptide) secreted in the stomach that causes the secretion of other digestive juices (Ch. 34).

gastrula: The three-layered early embryo that carries out gastrulation (Ch. 16).

gastrulation: The infolding process in the gastrula-stage embryo that creates a complex, three-dimensional organism from the simpler blastula (Ch. 16).

gene: The basic unit of heredity. Residing on chromosomes, genes consist of linked sequences of nucleotides that provide the blueprints for the construction of all cellular proteins and RNAs (Ch. 6).

gene amplification: A process in which extra copies of a gene are manufactured, such as those needed to generate large numbers of ribosomes during oogenesis (Ch. 16).

gene flow: A change in gene frequencies as the result of interbreeding between members of two populations (Ch. 41).

gene pool: All the alleles of all the genes carried by individuals in a population (Ch. 41).

generalist: A species that is adapted to a wide range of living conditions (Ch. 45).

genetic code: The molecular "grammar" in genes that relates nucleic-acid bases in nucleotides to amino acids. The functional units of the code, codons, consist of sets of three nucleotides (Ch. 13).

genetic drift: A change in gene frequency from generation to generation as the result of chance events (Ch. 41).

genetic load: The sum total of those alleles in a genome that yield some advantage when they are heterozygous, but are lethal or deleterious when homozygous (Ch. 42).

genome (genotype): The full set of genes on an organism's chromosomes (Chs. 10, 14).

genus (plural: genera): A taxonomic grouping of very similar organisms considered to be closely related species (Ch. 19).

germination: The reactivation and subsequent growth of a pollen grain or of a plant seed (Ch. 24).

gibberellin: Any of a class of growth-promoting plant hormones that acts to increase stem length (Ch. 30).

gill: Any of the specialized gas-exchange organs of many aquatic animals (Ch. 33).

ginkgo: The maidenhair tree, last remaining member of the division Ginkgophyta (Ch. 24).

gizzard: In birds and some invertebrates, a large muscular chamber in which food particles are pulverized to aid digestion before passing to the stomach or intestines (Ch. 34).

glial cell (glia): A cell in the animal nervous system that supports neurons metabolically and serves as electrical insulating material around axons (Ch. 36).

glomerular filtrate: The solution of wastes and other substances that passes from the blood into the cavity of a nephron (Ch. 35).

glucagon: A hormone that, along with insulin, regulates glucose levels in blood (Ch. 37).

glucocorticoid: Any of a group of steroid hormones produced in the adrenal gland that stimulates the production of glucose from proteins and carbohydrates (Ch. 37).

glycerol: A water soluble organic substance that is one of the basic components of fats and oils (Ch. 3).

glycocalyx: Literally, "sugar coat"; an agglomeration of complexes of sugar polymers, proteins, and sometimes lipids that are found on the surface of animal cells (Ch. 5).

glycogen: A polysaccharide sugar used by animals to store glucose (Ch. 3).

glycolysis: The first phase of energy metabolism in cells. By way of the multistep glycolysis pathway, a single six-carbon glucose molecule is broken down to yield two molecules of the three-carbon compound pyruvate, two molecules of NADH, and two molecules of ATP (Ch. 7).

gnetina (net-eye-na): Any member of a small, diverse group of gymnosperms, including certain tropical vines and arid region shrubs (Ch. 24).

golden-brown alga: Any of a group of phytoplankton having a golden color created by the presence of the carotenoid pigment fucoxanthin (Ch. 21).

Golgi complex: A membranous organelle in the eukaryotic cell cytoplasm that is specialized for the modification, transport, storage, or secretion of proteins (Ch. 6).

gonial cell: Any of the specialized cells in the gonads that gives rise to sperm or to eggs (Ch. 16).

Gram negative bacteria: Bacteria, such as *E. coli*, whose cell walls are surrounded by a lipid bilayer and hence do not take up iodine dye (crystal violet) during staining (Ch. 20).

Gram positive bacteria: Bacteria having a peptidoglycan cell wall, which takes up crystal violet dye and hence appears to stain purple under the light microscope (Ch. 20).

grand postsynaptic potential (GPSP): The summation of all the excitatory and inhibitory PSPs active on a neuron over any short time span (Ch. 36).

granulocyte: A type of white blood cell that holds granules containing enzymes or other substances involved in inflammatory or allergic reactions (Ch. 31).

granum (plural: **grana**): A stack of thylakoids arranged like a pile of coins within a chloroplast (Ch. 6).

ground parenchyma: In a monocot plant, an epidermal tissue in which bundles of vascular tissue are scattered (Ch. 27).

growth hormone (GH): A hormone secreted by the anterior pituitary lobe that acts to stimulate the growth of long bones during maturation (Ch. 37).

guanine: A double-ring purine base that is a fundamental structural element of some DNA nucleotides (Ch. 12).

guard cell: One of a pair of cells which borders each stoma (pore) on a leaf and regulates the opening and closing of the pore (Ch. 27).

gustation: The sense of taste (Ch. 38).

guttation: The formation of water droplets on pores at the edge of a leaf (Ch. 29).

gymnosperm: Any of the broad group of nonflowering seed plants, such as pines and spruces, in which both ovules and seeds are borne on the surface of the sporophyte (Ch. 24).

gynogen: Any of the female sex hormones (Ch. 18).

habitat: The actual place where an organism lives (Ch. 44).

habituation: A phenomenon in which repeated exposure to an environmental stimulus lessens an animal's responsiveness to it (Ch. 48).

haploid: The condition of a cell that contains only one set of parental chromosomes, or of an entire organism whose cells have only one set (Ch. 9).

Hardy-Weinberg law: A series of mathematical statements that describe the conditions necessary for the frequencies of alleles (genes) in a population to remain stable over time (Ch. 41).

haustorium (plural: **haustoria**): In fungi, a feeding structure that penetrates the living cells of other organisms and absorbs nutrients (Ch. 22).

heart: A region or organ made up of specialized muscle tissue that acts as a pump to propel blood or hemolymph (Ch. 31).

heat of vaporization: The amount of heat needed to turn a given amount of liquid water into gas (water vapor) (Ch. 2).

hemocoel: A blood-filled cavity (Ch. 25).

hemolymph: The extracellular fluid in the body cavity that bathes the tissue cells in animals with open circulatory systems. It is grossly similar in composition to intracellular fluid, containing mostly water with dissolved nutrients, ions, gases, and some macromolecules (Ch. 31).

hemoskeleton: A noncompressible, blood-filled space (sinus) against which the muscles of a mollusk push to generate movement (Ch. 25).

herbaceous: In a plant, the property of having a relatively thin, soft, nonwoody stem (Ch. 27).

herbivore: An animal that eats only plant material (Ch. 34).

hermaphrodite: An individual born possessing both testes and ovaries (Ch. 18).

hetero-: From Greek, meaning "different."

heterokaryon: A single cytoplasm with dissimilar nuclei (Ch. 22).

heterosporous: The property in certain plants of having two types of spores, which yield male and female gametophytes respectively (Ch. 23).

heterotroph: An organism that obtains the energy for cellular processes by taking in food consisting of whole autotrophs or other heterotrophs, their parts, or their waste products (Ch. 5).

heterozygote advantage: A condition said to exist when a heterozygote has a higher reproductive fitness than homozygotes (Ch. 42).

heterozygous (hetero-**zye**-gus): The quality of having two different alleles for a particular genetic trait (Ch. 10).

hibernation: A temporary physiological state in which an animal's set point (normal body temperature) falls and metabolism slows dramatically (Ch. 35).

hindbrain: The posterior region of the vertebrate brain that encompasses the cerebellum, pons, and lower portion of the brain stem (medulla oblongata) (Ch. 40).

histone: A protein that bears a net positive electrical charge. The DNA molecules in eukaryotic chromosomes coil around clusters of histones (Ch. 9).

holdfast: An algal cell or tissue specialized to anchor the plant to a substrate (Ch. 23).

homeostasis: The maintenance of a relatively constant internal environment despite fluctuations in the external world (Ch. 35).

homeotherm: An animal that has a relatively constant body temperature (Ch. 35).

homo-: From Greek, meaning "same or similar."

homology: The appearance in related life forms of similar structures or functions, based on the inheritance of the same basic genetic program (Ch. 43).

homosporous: The property in certain plant species of producing a single type of spore, which gives rise to gametophytes having both male and female structures (Ch. 23).

homozygous (homo-**zye**-gus): The quality of having two identical alleles for a particular genetic trait (Ch. 10).

hormone: A substance that is manufactured in minute quantities in one part of an organism and that produces effects in other parts of the same organism (Ch. 30).

humoral immunity: The immunity conferred by antibodies carried in the blood (Ch. 32).

hybrid vigor: A condition in which heterozygotes for one or several genetic traits exhibit a more desirable phenotype than do homozygotes (Ch. 42).

hydr-, hydro-: From Greek, meaning "fluid" or "water."

hydrogen bond: A weak bond formed as a result of the attraction between the oxygen atom of one molecule and a hydrogen atom of another (Ch. 2).

hydrolysis (high-**drol**-i-sis): A "splitting with water" process in which one larger molecule is split into two monomers by the addition of parts of water molecules (Ch. 3).

hydrophilic: The property of being "water-loving." Hydrophilic compounds tend to form hydrogen bonds with water molecules, and thus readily dissolve in water (Ch. 2).

hydrophobic: The property of being "water-hating." Hydrophobic compounds have nonpolar covalent bonds which prevent them from forming bonds with hydrogen and from being electrically attracted to water molecules. Thus such compounds tend to be insoluble in water (Ch. 2).

hydroskeleton: A volume of water trapped within an animal's tissues that is noncompressible and serves as a firm mass against which opposing sets of muscles can act (Ch. 25).

hyper-: From Greek, meaning "over" or "more."

hyperpolarization: An increase in the polarization of a cell (as from −50 to −80 mV) (Ch. 36).

hypertonic: A condition of a solution reflecting the presence of a solute concentration that is higher than that of some other solution (Ch. 5).

hypervariable region: Any of several areas in the variable portions of antibody protein chains that account for the different binding site (antigen) specificities of antibodies (Ch. 32).

hypha (plural: hyphae): The cellular filaments that are the basic structural units of a fungus (Ch. 22).

hypo-: From Greek, meaning "under" or "lower."

hypocotyl: The initial length of stem that emerges from a germinating seed (Ch. 28).

hypothesis: A precisely constructed proposition to explain a scientific question. The systematic testing of hypotheses is the fundamental process that sets science apart from other disciplines and enables scientists to produce accurate, enduring explanations of natural phenomena (Ch. 1).

hypotonic solution: A solution in which the salt concentration is lower than that of another solution (Ch. 5).

imbibition: A process in which water enters soil and binds to clay and humus particles (Ch. 44).

immunoglobulin: A protein antibody molecule (Ch. 32).

implantation: The process during which a newly formed mammalian embryo becomes embedded in the inner wall of the uterus (Ch. 18).

imprinting: A type of learning during a brief, early sensitive period of an animal's life in which a behavioral response is learned to a stimulus with specific characteristics (Ch. 48).

inbreeding: Mating between relatives at a greater or lesser (negative inbreeding) frequency than would be predicted by chance (Ch. 41).

inclusive fitness: A concept in evolutionary theory in which an individual's total fitness includes both its likelihood of passing on its own genes and the likelihood that it will contribute to the successful passing on of the genes of its relatives (Ch. 49).

incomplete dominance: In genetics, a situation in which both alleles of a heterozygous pair exert an effect, jointly producing a phenotype intermediate between the two (Ch. 10).

incomplete penetrance: In genetics, a situation in which a dominant allele is present but is not expressed at all in certain individuals (Ch. 15).

induced fit: An adjustment in a enzyme's shape which improves the fit between the enzyme molecule's active site and its substrate (Ch. 4).

inducible enzyme: A protein synthesized in a cell in response to the presence of a particular inducer substance (Ch. 14).

inflammatory response: The initial defensive response to a wound; characterized by the release of histamine, increased blood flow to the wound site, the arrival of macrophages, and other related events (Ch. 32).

inhibitory synapse: A synapse in which synaptic transmission acts to prevent function of the receiving cell, often by inhibiting depolarization (Ch. 36).

initiation: The beginning of protein synthesis from a transcribed gene. Initiation takes place when a ribosome, an mRNA, and two tRNAs bearing specific amino acids bind together (Ch. 13).

inositol triphosphate (IP$_3$): A second-messenger compound (see *cyclic AMP*) that helps trigger cell responses to a range of neurotransmitters, hormones, and growth factors (Ch. 37).

insight learning (reasoning): The ability to respond correctly to a novel situation by applying information garnered from past experience (Ch. 48).

instinct: A stereotyped, inherited pattern of behavior that occurs in response to a specific environmental stimulus (Ch. 48).

insulin: A hormone produced in the vertebrate pancreas that acts to lower the concentration of glucose in the blood; also found in the brain (Chs. 37, 40).

integument: A tissue that serves as a covering layer, such as the outer covering of a seed (Ch. 24).

intercalated disk: A region of cell membrane in a cardiac-muscle cell that serves to hold adjoining cells together; it also serves as a site through which ion or electric currents can flow (Ch. 39).

interferon: A naturally occurring protein liberated by a mammalian cell after it has been infected by certain viruses. Interferons bind to nearby infected cells, stimulating the production of antiviral proteins (Ch. 20).

intermediate filament: A threadlike structure made of protein which is thought to lend tensile strength to animal cells (Ch. 6).

interneuron: Any of a class of neurons that receive and process input from receptor, sensory, or other interneurons and send commands to other interneurons or to effector neurons (Ch. 36).

internode: The stem tissue between leaves of a plant (Ch. 28).

interspecific competition: Competition between members of several species for a limited resource (Ch. 46).

intestine: The long, tubelike section of the digestive tract where most food digestion and absorption take place (Ch. 34).

intraspecific competition: Competition for one or more resources among members of the same species (Ch. 46).

intrinsic protein: A protein molecule in which part or all of its peptide chain is embedded in the lipid bilayer of a cell's plasma membrane (Ch. 5).

intron: In a eukaryotic gene, a segment of DNA which does not code for protein (Ch. 14).

invertebrate: An animal that lacks a backbone (Ch. 25).

ion (eye-on): An atom that bears a net electrical charge (Ch. 2).

ionic bond: A chemical bond formed when one atom gives up a valence electron and another atom adds the free electron to its outermost orbital. An ionic bond holds its atoms together in an energetically stable unit (Ch. 2).

iris: A pigmented ring of tissue in the vertebrate eye that opens or closes in response to changing light levels (Ch. 38).

islets of Langerhans: Clusters of endocrine cells in the pancreas in which the hormones insulin and glucagon are produced (Ch. 37).

isogamete: In certain plants, a gamete of one of two mating types, identical in size and appearance to gametes of the other mating type (Ch. 23).

isomer: One of two chemical compounds which have identical formulas, but different spatial arrangements of atoms. Isomers often have very different chemical properties as well (Ch. 3).

isotonic solution: A solution which has the same salt concentration as that of a comparison solution (Ch. 5).

isotope: An atom of an element which contains the normal number of protons for atoms of that element, but a different number of neutrons. Thus isotopes of an element have different atomic weights (Ch. 2).

juvenile hormone (JH): A hormone whose presence at high levels in insects prevents metamorphosis (Ch. 37).

karyotype: A display of the number, sizes, and shapes of an organism's chromosomes (Ch. 9).

kidney: A blood-filtering and waste-excreting organ; found in all vertebrates and some invertebrates (Ch. 35).

kilocalorie: The amount of heat energy required to raise the temperature of one kilogram of water by 1°C (Ch. 2).

kinesis (cih-**nee**-sis): A simple reflexive behavior in which the intensity of the animal's response is proportional to the intensity of the stimulus (Ch. 48).

kinetic energy: The energy of motion (Ch. 4).

kinetochore: The portion of a chromosome to which centromere fibers attach during mitosis and meiosis (Ch. 9).

kingdom: The most inclusive taxonomic grouping, such as the classification of all plants into the kingdom Plantae (Ch. 19).

kin selection: Behavior that promotes the reproductive success of closely related individuals (Ch. 49).

Krebs cycle: The fundamental metabolic pathway in cellular respiration. The cycle consists of a series of chemical reactions in which pyruvate (the end product of glycolysis) is oxidized to carbon dioxide, and ATP is generated. Also known as the citric acid cycle (Ch. 7).

K-selected species: A species whose members reach reproductive age relatively slowly, and produce small numbers of young that are well-prepared to compete for food and other resources (Ch. 46).

large intestine: The sizable, nonconvoluted portion of the mammalian intestine that extends from the small intestine to the anus (Ch. 34).

larynx: The boxlike entrance to the mammalian trachea and lung system (Ch. 33).

lateral line: A series of small sensory organs on the sides of fish and amphibians that contain specialized sensory cells that respond to movements, sound vibrations, electric fields, and other stimuli in the water (Ch. 38).

lateral meristem: Either of two tissues, the cork cambium and the vascular cambium, that generates the stem-thickening secondary growth in a nonherbaceous plant (Ch. 28).

late wood: Secondary xylem cells that arise in the latter part of a tree's growing season (Ch. 28).

law of independent assortment: Mendel's second law, which states that different characters are inherited independently of one another (Ch. 10).

law of segregation: Mendel's first law, which states that individuals carry two discrete hereditary units (alleles) for a given character, receiving one from each parent. When sperm or ova are produced, each allele pair is separated and

the alleles are distributed on chromosomes to different gametes (Ch. 10).

leaf: A plant organ specialized for collecting light for photosynthesis (Ch. 27).

leaf primordium: A flattened mound on the side of a plant meristem that will eventually develop into a leaf (Ch. 28).

learning: The modification of behavior on the basis of experience (Ch. 48).

lens: A structure composed of highly ordered cells in the eyes of vertebrates, octopuses and some other animals that focuses light images (Ch. 38).

lentic community: The fresh-water biome represented by a lake or pond (Ch. 44).

lenticel: One of numerous porelike sites in the cork layer of bark at which gas exchange can take place (Ch. 28).

leukocyte: A white blood cell (Ch. 31).

lichen: A composite organism consisting of one fungus species and one or more species of algae (Ch. 22).

ligament: A strong, flexible band of collagen fibers that connects bones (Ch. 39).

light reactions: The first of the two distinctive sets of reactions in photosynthesis in which light energy is required to oxidize water and O_2 is released (Ch. 8).

limbic system: The seat of the emotions and short-term memory in the vertebrate forebrain (Ch. 40).

limnetic zone: The center of a lake, extending to the depth where O_2 production by photosynthesis equals the uptake of oxygen in the respiration of organisms in the lake (Ch. 44).

linkage: A measure of the degree to which genes on chromosomes are inherited together (Ch. 11).

lipid: A member of the class of organic compounds that includes oils, fats, waxes, and other fatlike substances. The main categories of lipids are fats, oils, phospholipids (integral components of cell membranes), and steroids (Ch. 3).

littoral zone: The relatively shallow edges of a lake, where light reaches the bottom (Ch. 44).

liver: A large, lobed organ in vertebrates that regulates the nutrient content of the blood, absorbs and degrades hormones, serves as a reservoir for glycogen, and carries out a variety of other functions vital to health (Ch. 34).

local circuit current: In a neuron, any of the minute, internal electrical currents that flow from the initial point of depolarization and which are involved in propagation of impulses (Ch. 36).

locus: The specific site at which a gene or allele appears on a chromosome (Ch. 11).

logistic growth curve: An S-shaped curve that represents the phases of a population's growth history (rapid initial growth followed by an equilibrium phase which may in turn be followed by a plateau or a decline) (Ch. 46).

loop of Henle: The long, looped portion of a mammalian kidney tubule that helps form a concentrated (hypertonic) extracellular brine bath (Ch. 35).

lotic community: The fresh-water biome represented by a stream (Ch. 44).

lung: One of a pair of hollow, usually branched, internal respiratory organs; connected through passageways to the outside air and also interfacing with the circulatory system (Ch. 33).

luteinizing hormone (LH): A gonadotropin hormone produced in the pituitary which stimulates the production of testosterone in male infants and embryos, and triggers ovulation in mature females (Ch. 18).

lymphatic system: A system of vessels that drains excess extracellular fluid from the spaces around cells and houses important parts of the immune system (Ch. 31).

lymph node: A mass of tissue containing lymphocytes through which lymph is filtered (Ch. 32).

lymphocyte: A cell of the immune system which responds to foreign substances; some lymphocytes secrete antibodies (Ch. 31).

lymphokine: A lymphocyte-activating protein secreted by macrophages and helper T cells (Ch. 32).

lysogeny: The process in which viral DNA is replicated along with a host cell chromosome each time the host cell passes through a growth and division cycle (Ch. 20).

lysosome: A spherical, membrane-bound sac which contains digestive enzymes that can digest most known biological macromolecules (Ch. 6).

lytic pathway: The serial events in which viral genes within a host cell begin to replicate independently, mature virus particles assemble, and the host cell bursts, releasing the particles, which may then infect other host cells (Ch. 20).

M phase: The period of mitosis in the cell cycle, in which the chromosomes become equally apportioned in the two new nuclei (Ch. 9).

macro-: From Greek, meaning "large."

macroevolution: Major phenotypic changes occurring over evolutionary time (Ch. 43).

macronutrient: A nutrient required by a heterotroph in large amounts, such as proteins, fats, carbohydrates, and some minerals (Ch. 34).

macrophage: Any of a class of white blood cells that engulfs and digests foreign particles (Ch. 32).

major histocompatibility complex: A group of genes coding for protein markers that enables body tissues to be recognized by the immune system as self or nonself (Ch. 32).

Malpighian tubule: The blind-ended sac that is the functional unit of an insect's excretory system (Ch. 35).

mammary glands: Glands evolved in mammals that are specialized in mature females for the production and secretion of milk (Ch. 18).

mantle: The hot, semisolid region of the Earth that surrounds the dense core and underlies the lighter, solid crust (Ch. 19).

marsupial: A mammal, such as the kangaroo, that possesses an external pouch in which young are nurtured (Ch. 26).

mass flow theory: A model for the large-scale movement of phloem fluid in plants (Ch. 29).

mechanoreceptor: A sensory receptor cell sensitive to pressure, stretching, and bending forces (Ch. 38).

medulla: The central portion of organs, as of the kidney, which contains the loops of Henle, or of the adrenal gland, where epinephrine is made (Chs. 35, 37).

medulla oblongata (medulla): The area of the vertebrate hindbrain where the brain and the spinal cord merge (Ch. 40).

medusa: The radial, free-floating form of a jellyfish (Ch. 25).

mega-: From Greek, meaning "large."

megagametogenesis: The process in angiosperms during which the female gamete is produced and readied for fertilization (Ch. 28).

megagametophyte: A female gametophyte (Ch. 24).

megasporangium: The structure that contains megaspores (Ch. 24).

megaspore: A large spore that differentiates into a female gametophyte (Ch. 23).

megasporogenesis: The meiotic division of a megaspore mother cell that generates the first stage of a female gamete, the megaspore (Ch. 28).

meiosis: The specialized form of nuclear division in which, following chromosome duplication in a reproductive cell, the diploid parent nucleus divides twice and four haploid daughter cells are formed (Ch. 9).

melatonin: A hormone secreted by the pineal gland; in some animals, affects the internal biological rhythms associated with light and dark ("day" and "night") (Ch. 37).

menstrual cycle: The cyclic preparation of the mammalian uterus to receive an embryo (Ch. 18).

meristem: In plants, an organizing center of undifferentiated, actively dividing cells forming zones where new organs can be generated throughout the life of the plant (Ch. 28).

meso-: From Greek, meaning "middle."

mesoderm: The midlayer (between ectoderm and endoderm) that arises during gastrulation in an embryo. Mesodermal cells give rise to the skeleton, muscles, and circulatory and immune systems, among other structures (Ch. 16).

mesophyll: The major photosynthetic tissue in a leaf (Ch. 27).

messenger RNA (mRNA): The type of RNA (ribonucleic acid) that encodes information from DNA and is translated into corresponding protein structures (amino-acid sequences) (Ch. 13).

meta-: From Greek, meaning "change" or "posterior."

metabolic pathway: A linked series of chemical reactions by which various enzymes catalyze the specific steps needed to construct or break down biological compounds (Ch. 4).

metabolism: The combination of simultaneous, interrelated chemical reactions taking place in a cell at any given time (Ch. 4).

metamorphosis: Among insects, amphibians, and other animals, the developmental transformation from the larval to the adult body plan (Ch. 16).

metaphase: The second stage of mitosis, in which the fully condensed chromosomes become associated with the spindle (Ch. 9).

metastasize (muh-tas-tuh-size): To break away from a tumor and spread to a distant site in the body (Ch. 17).

micro-: From Greek, meaning "small."

microbody: A small, membranous vesicle containing enzymes that break down the waste products of eukaryotic cells (Ch. 6).

microevolution: Small-scale changes in gene frequencies hypothesized to lead to the evolution of distinct species (Ch. 43).

microfilament: Any of the many threadlike actin fibers that make up much of a cell's cytoskeleton (Ch. 6).

microgametogenesis: The second stage in pollen formation, in which microspores differentiate into functional pollen grains (Ch. 28).

microgametophyte: A male gametophyte (Ch. 24).

micronutrient: A nutrient required by a heterotroph in small amounts, such as vitamins and trace inorganic minerals (Ch. 34).

microsporangium: The structure that holds microspores (Ch. 24).

microspore: A small spore that differentiates into a male gametophyte (Ch. 23).

microsporogenesis: The initial step in the development of pollen grains, in which a diploid microspore mother cell divides meiotically (Ch. 28).

microtubule: A long cylindrical tube that serves as bonelike scaffolding to help stabilize the shape of a cell; also, sites along which transport may occur (Ch. 6).

microvillus (plural: **microvilli**): A tiny hairlike projection of epithelial cells that has an actin skeleton (Ch. 34).

midbrain: A centrally located region in the vertebrate brain; in fish and amphibians, the primary site for processing visual input (Ch. 40).

midrib: The largest, centrally located vein in a dicot leaf (Ch. 27).

mineral: An inorganic element or compound (Ch. 29).

mineralocorticoid: Any of a group of steroid hormones produced in the adrenal glands that regulates the retention and excretion of minerals, such as sodium and potassium (Ch. 37).

mitochondrion (my-toe-kon-dree-on; plural: **mitochondria**): A cellular organelle specialized to harvest energy from food molecules and store that energy in ATP (Ch. 6).

mitosis (my-toe-sis): The process of nuclear division in which genetic information is distributed equally to two identical daughter cells (Ch. 9).

molecular formula: A shorthand system used in chemistry to show how many atoms of each type are present in a particular molecule, and whether any of the atoms occur in certain common groups. The molecular formula of water is H_2O (Ch. 2).

molecule: Two or more atoms that have been bound together by chemical bonds (Ch. 2).

mon-, mono-: From Greek, meaning "single."

monocotyledon: An angiosperm in which the embryo has only one seed leaf (Ch. 24).

monoecious (mon-ee-shus): In plants, having both male and female sexual parts on each individual of a species (Ch. 24).

monomer: A simple molecule that can be linked with other monomers to form a more complex polymer (Ch. 3).

monophyletic: In taxonomy, the property ascribed to taxons which share a single ancestral source (Ch. 19).

monotreme: One of a primitive group of modern mammals, including the duckbilled platypus, which lays leathery eggs rather than giving birth directly to live young (Ch. 26).

morpho-, -morph: From Greek, meaning "structure" or "form."

morphogenesis: The developmental process that generates changes in the shapes of cells and cell populations (Ch. 16).

mortality: The per capita death rate of a population (Ch. 46).

mosaicism: A genetic condition in which some cells of an organism express the phenotype of one chromosome, while other cells express the phenotype of the complementary homologous chromosome (Ch. 15).

Mullerian duct: The tube in a female mammalian embryo which develops into the Fallopian tubes, uterus, and upper vagina (Ch. 18).

Mullerian inhibiting substance (MIS): A hormone that inhibits female reproductive structures from forming in a developing male embryo (Ch. 18).

Mullerian mimicry: A phenomenon in which different aposematic species have identical or similar warning colors or patterns; see *aposematic coloration* (Ch. 47).

muscle fiber: A fused set of dozens or hundreds of muscle cells; the basic unit of muscle tissue (Ch. 39).

muscle spindle: Any of the specialized stretch receptors (proprioreceptors) in the joints, tendons, and muscles of a vertebrate (Ch. 39).

mutagen: Any agent capable of causing a mutation (Ch. 11).

mutation: A change in the chemical structure of a gene; see also *chromosomal mutation* (Ch. 11).

mutation rate: The frequency with which mutations arise naturally in a given population (Ch. 11).

mutualism: An arrangement in which two species live in intimate association, to the benefit of both partners; a form of symbiosis (Ch. 47).

mycelium: The dense network of filaments (hyphae) that form a fungus (Ch. 22).

mycoplasma: A minute prokaryote that is the smallest known living cell (Ch. 20).

mycorrhiza: An association of a plant root and fungal filaments which aid the plant in obtaining nutrients (Ch. 22).

myelin sheath: The insulating covering of nerve cell axons, produced by glial cells (Ch. 36).

myofibril: A cylindrical assembly of contractile muscle proteins arranged as sarcomeres in which actin and myosin overlap, found within the cytoplasm of muscle-fiber cells (Ch. 39).

myoglobin: A respiratory pigment abundant in muscle and some other cells (Ch. 33).

myosin: A mechanoenzyme protein that, in the form of thick filaments, interacts with actin to bring about the contraction of muscle cells (Ch. 7).

myx-, myxo-: From Greek, meaning "mucus" or "slime."

myxamoeba: A single, motile haploid cell of the true slime molds (Ch. 21).

myxobacteria: Small, unflagellated, rod-shaped bacterium that moves by gliding along slime tracks (Ch. 20).

NAD⁺ (nicotinamide adenine dinucleotide): One of two important coenzymes (the other is FAD) that serves as an electron and hydrogen carrier in the metabolic oxidations and reductions of the Krebs cycle and other cell processes (Ch. 7).

natural killer (NK) cell: Any of a class of large, granular, white blood cells that can directly attack certain viruses and virus-infected tumor cells, as well as some bacteria (Ch. 32).

natural selection: The differential survival to reproduce of some genotypes; heritable adaptations account for the difference in success (Ch. 41).

negative feedback: A regulatory system in which an increase in the concentration of a substance inhibits the continued synthesis of that substance, and vice versa (Ch. 4).

neocortex (cortex): The outer, sheetlike wall of the mammalian cerebrum, which consists of highly ordered arrangements of neurons that carry out many higher brain functions (Ch. 40).

nephridium (plural: **nephridia**): The excretory organ of an earthworm (Ch. 35).

nephron: A tubule that is the functional unit of the vertebrate kidney (Ch. 35).

nerve: A bundle of many neuronal axons (Ch. 36).

nerve impulse: See *action potential.*

neural retina: The epithelial layer in the eye that contains photoreceptors and on which an image is focused (Ch. 38).

neuroactive peptide: A protein that acts as a neurotransmitter or that modulates neuron activity (Ch. 40).

neuron: A nerve cell (Ch. 36).

neurotransmitter: A chemical messenger between nerve cells (Ch. 36).

neutron: A subatomic particle that carries no electrical charge (Ch. 2).

niche: The total way of life of a species; includes every facet of its ecological, physiological, and behavioral roles in nature (Ch. 45).

nitrification: The oxidation of ammonia (NH_3) to nitrogen oxides (Ch. 29).

nitrogen cycle: The cycling of nitrogen between organisms and the Earth; crucial steps are the fixation of atmospheric nitrogen by bacteria and later the release of N_2 back to the atmosphere through the action of microbes (Ch. 44).

nitrogen fixation: The conversion in plants of atmospheric N_2 to a usable form, NH_4^+ (ammonium ion) (Ch. 29).

node: The position at which a leaf arises on a stem (Ch. 28).

node of Ranvier: Any of the tiny, nonmyelinated sections of a nerve cell axon where Na^+ channels are abundant (Ch. 36).

noncompetitive inhibition: A general system for activating or inhibiting enzyme function in which a substance binds to an enzyme in a way that alters the shape of the active site. It is "noncompetitive" because the control substance does not compete with the substrate to occupy the enzyme's active site (Ch. 4).

noncyclic photophosphorylation: A system of electron flow driven by light energy during photosynthesis in which the electrons pass in a one-way sequence through a series of pigments, proteins, and energy carriers (Ch. 8).

nondisjunction: During meiosis, the failure of the members of a homologous chromosome pair to separate (Ch. 10).

nonrandom mating: A situation in which the probability of two organisms mating depends on their phenotypes (Ch. 41).

nonvascular plant: A plant lacking vascular tissue for conducting water and nutrients (Ch. 23).

norepinephrine: A common neurotransmitter of the sympathetic nervous system; also called noradrenalin (Ch. 37).

normalizing selection: Natural selection whose effect is to preserve the status quo in the genetic makeup of a population (Ch. 42).

notochord: The stiff but flexible rod that runs the length of a chordate, just ventral to the nerve cord (Ch. 26).

nuclear envelope: A flattened, double-layered sac that separates the nucleus from the cytoplasm in a eukaryotic cell (Ch. 6).

nucleic acid: A polymer chain made up of nucleotide subunits that are arranged in a specific linear sequence. The two types of nucleic acids are deoxyribonucleic acid (DNA) and ribonucleic acid (RNA) (Ch. 3).

nucleoid: A dense, unbounded area within a prokaryotic cell that encompasses the cell's single chromosome and serves much like a nucleus (Ch. 6).

nucleolus (plural: **nucleoli**): An organelle associated with a specific chromosome and in which is found the genes coding for the major ribosomal RNAs. It is the site of ribosome manufacture (Ch. 6).

nucleoside: The molecule that is the central component of DNA nucleotides, consisting of a nitrogenous base and a five-carbon sugar (Ch. 12).

nucleosome: A beadlike complex consisting of a portion of a DNA molecule coiled around a cluster of eight histone protein molecules (Ch. 9).

nucleotide: The building block of nucleic acid, made up of a nitrogen-containing base, a five-carbon sugar, and a phosphate group (Ch. 3).

nucleus: Present only in eukaryotic cells, it encloses the chromosomes, is bounded by a membranous envelope, and serves as the source of genetic information for most cellular proteins (Ch. 6).

obligate aerobe: An organism, generally a bacterium, which must have oxygen for its metabolic processes (Ch. 20).

obligate anaerobe: A bacterium which can grow only in the absence of free oxygen, usually carrying out fermentation to generate ATP (Ch. 20).

Okazaki fragment: In replicating DNA, a small, independent fragment of DNA that is synthesized on one of the two separated strands (the lagging strand). Ultimately each Okazaki fragment is joined to its predecessor as the new DNA strand lengthens (Ch. 12).

oligodendrocyte: Any of the myelin-producing glial cells in the brain and spinal cord (Ch. 36).

ommatidium (plural: **ommatidia**): Any of the individual optic units that make up a compound eye (Ch. 38).

omnivore: An animal that consumes both plant and animal matter (Ch. 34).

oncogene: A cancer-causing gene (Ch. 17).

oocyte: A large, immobile egg cell (Ch. 23).

oogamy: The development of true oocytes (Ch. 23).

oogenesis: The process of egg production (Ch. 16).

open circulatory system: A circulatory system in which the circulating fluid is not entirely enclosed within a continuous set of interconnected vessels (Ch. 31).

open program: An innate motor pattern in which certain elements can be modified by learning (Ch. 48).

operator: A regulatory segment of DNA whose role is to receive signals in the form of a repressor substance. Such signals in turn regulate expression of a nearby target gene (Ch. 14).

operon: A unit which regulates prokaryotic gene activity, consisting of an operator plus the protein-coding genes it controls (Ch. 14).

opsin: A lipoprotein that is one component of the visual pigment rhodopsin (Ch. 38).

order: A taxonomic grouping or organisms belonging to similar families (Ch. 19).

organ: A body part composed of several tissues that operate in concert to perform specific functions within the organism (Ch. 5).

organ of Corti: A structure composed of thousands of sensory hair cells situated on the basilar membrane in the vertebrate inner ear; displacement of the cells by sound pressure waves generates the impulses carried by the auditory nerve (Ch. 38).

organelle: A structure within a cell that carries out specific functions; for example, a mitochondrion, ribosome, or microtubule (Ch. 5).

organic compound: Any chemical compound that contains one or more carbon atoms (Ch. 3).

organogenesis: The developmental stage during which discrete organs and tissues form (Ch. 16).

orientation slab: Any of the functional units of the visual cortex sensitive to lines oriented at different angles (Ch. 40).

osmolyte: Any substance that causes osmotic pressure to rise (Ch. 35).

osmoregulation: The process of maintaining a stable internal fluid environment by regulating osmolyte concentrations in body fluids (Ch. 35).

osmosis: The diffusion of water through a semipermeable membrane in response to distribution of osmolytes (Ch. 5).

osmotic pressure: The pressure exerted by a solution separated by a semipermeable membrane from pure water; practically measured as the pressure that must be applied to such a solution to prevent it from gaining additional water through the membrane (Ch. 5).

otolith: Literally "ear stone"; any of numerous crystalline structures in the inner ear of vertebrates that impinge on sensory hairs to register changes in the position of an animal's head (Ch. 38).

oval window: A thin, taut, inner-ear membrane (Ch. 38).

ovarian cycle: The cycle during which eggs mature and ovulation occurs (Ch. 18).

ovary: The female sex organ in which eggs (ova) are generated (Ch. 16).

oviduct: The tube through which a mature animal egg travels after leaving the ovary (Ch. 16).

ovulation: The release of an egg from an ovary (Ch. 18).

ovule: In a seed plant, the structure within which an egg cell forms (Ch. 24).

oxidation: The removal of electrons from an atom or compound (Ch. 7).

oxidation-reduction reaction: A chemical reaction in which one molecule loses electrons (oxidation) while another molecule simultaneously gains electrons (reduction) (Ch. 7).

oxidative phosphorylation: The process by which energy released during oxidation reactions is stored in high-energy phosphate bonds (Ch. 7).

oxygen debt: The condition in which reduced metabolic products (such as lactic acid) comprising the "debt" accumulate due to the inability of oxidative metabolism to function rapidly enough. The debt is paid off when the metabolism that produces reduced products slows (Ch. 33).

oxygen dissociation curve: A graphic measure of the binding of O_2 to hemoglobin under different oxygen partial pressures (Ch. 33).

oxytocin: A peptide hormone released by the female posterior pituitary lobe that causes uterine contractions during sexual intercourse and childbirth; also a neuroactive peptide (Ch. 18).

ozone layer: A high-altitude layer of O_3 which shields the earth from much of the sun's ultraviolet radiation (Ch. 19).

P_1 (parental) generation: In a genetic cross, the two parents which produce the first generation of offspring (Ch. 10).

palisade parenchyma: A tightly packed layer of rod-shaped, chloroplast-filled cells just below the upper epidermis of a leaf (Ch. 27).

pancreas: A long, slender organ near the stomach that secretes digestive enzymes (Ch. 34).

parallel evolution: The evolution in separate but related groups of similar characteristics apparently as a result of exposure to similar environmental or selective conditions (Ch. 43).

parasite: An organism that feeds on its living host, often rendering harm but usually not causing the host's death (Ch. 20).

parasympathetic nervous system: The portion of the autonomic nervous system that, in general, acts to inhibit cell or organ function (Ch. 36).

parathyroid gland: Small endocrine gland located on the surface of the thyroid; the source of parathyroid hormone (PTH).

parathyroid hormone (PTH): A hormone that raises calcium levels in the blood (Ch. 37).

parenchyma cell: Any of the thin-walled plant cells with large vacuoles that make up leaf mesophyll tissue (Ch. 27).

parthenogenesis: A form of reproduction in which no fertilization of an egg takes place. Instead, the mature egg is spontaneously activated and subsequently undergoes normal development (Ch. 16).

partial pressure: The pressure exerted by a gas in air; corresponds to the percentage volume of the gas in a given volume of air (Ch. 33).

passive immunity: Immunity to disease provided indirectly, as in the transfer of antibodies from mother to fetus across the placenta (Ch. 32).

pathogen: An agent that causes a specific disease (Ch. 20).

pedigree: A formal representation of a set of traits for all members of a family lineage (Ch. 15).

pelagic: The quality of drifting about freely in water (Ch. 25).

pelvis: (1) The central cavity of the kidney (Ch. 35); (2) the hip bone of land vertebrates (Ch. 39).

penis: The male copulatory organ (Ch. 18).

pepsin: A digestive enzyme (Ch. 34).

peptide bond: A chemical bond which results from a condensation reaction between the carboxyl group of one amino acid and the amino group of another (Ch. 3).

peptidyl transferase: An enzyme that catalyzes the formation of peptide bonds between amino acids during the synthesis of a polypeptide chain (Ch. 13).

percent polymorphism: The fraction of individuals in a population heterozygous at given gene loci; a measure of the genetic variation in a population (Ch. 41).

perennial plant: A plant whose life cycle typically lasts for a number of years (Ch. 28).

perianth: The showy, exterior parts of a flower (Ch. 28).

pericycle: A circular zone of cells that surrounds the xylem and phloem tissue of a plant root (Ch. 27).

peripheral nervous system (PNS): The entire portion of the vertebrate nervous system other than the brain and spinal cord (Ch. 36).

peristalsis: Wavelike muscular contractions, such as a mechanism that propels food within the digestive tract (Ch. 34).

petiole: The stemlike portion of a leaf which connects the leaf blade to the stem of the plant (Ch. 27).

pH: An expression of the concentration of hydrogen ions in a solution. The pH scale runs from 0 to 14, with acidic solutions having a pH of less than 7 (neutral), and basic (alkaline) solutions having a pH of more than 7 (Ch. 2).

phagocytosis: Literally, "cell eating"; the engulfing of particulate matter by cells (Ch. 6).

pharynx: The region of the vertebrate gut immediately posterior to the mouth cavity (Ch. 26).

phenotype: The observable expression of the genetic makeup (genotype) of a cell or organism (Ch. 10).

phloem: A plant vascular tissue specialized for carbohydrate transport. Unlike the other plant vascular material, xylem, phloem cells must be alive in order to function (Ch. 27).

phospholipid: A lipid molecule in which the glycerol is linked to at least one other molecule containing a phosphate group. Phospholipids are fundamental components of cell membranes (Ch. 3).

phosphorus cycle: The cycling of phosphate between sources in soil, rocks, or water and living organisms (Ch. 44).

phosphorylation: The transfer of a phosphate group (Ch. 7).

photic zone: The top layer of a body of water, where enough light is present for photosynthesis (Ch. 44).

photo-: From Greek, meaning "light."

photoautotroph: A bacterium that contains a modified form of chlorophyll and derives its food energy from photosynthesis (Ch. 20).

photon: A particle of visible light (Ch. 8).

photoperiod: The number of hours of light in a given day (Ch. 30).

photophosphorylation: The production of ATP through the transport of electrons excited by light energy down an electron transport chain (Ch. 8).

photoreceptor: A sensory receptor cell sensitive to visible or ultraviolet light (Ch. 38).

photorespiration: An inefficient form of the dark reactions of photosynthesis in which O_2 accumulates, CO_2 is depleted, and no carbohydrates are generated (Ch. 8).

photosynthesis: A metabolic process in which light energy is converted to chemical energy stored in chemical compounds. Photosynthesis takes place in green plants, algae, and certain protists and bacteria (Ch. 8).

photosystems I and II: The two basic molecular systems for converting light to chemical energy during photosynthesis. Photosystem I tends to absorb light with a wavelength near 700 nm. Photosystem II most strongly absorbs light with a wavelength near 680 nm (Ch. 8).

phototropism: Movement toward light (Ch. 30).

phylogeny: A description of the line of descent of a particular organism (Ch. 43).

phylum: A broad taxonomic grouping of organisms belonging to similar classes; in botany a phylum is termed a division (Ch. 19).

phytochrome: A light-absorbing plant pigment (Ch. 30).

phytoplankton: Microscopic, usually photosynthetic organisms that float near the surface of fresh and salt waters (Ch. 21).

piloerection: The muscular erection of fur or feathers as a mechanism for conserving body heat (Ch. 35).

pineal gland: A pea-sized endocrine gland situated in the roof of the brain; the source of melatonin (Ch. 37).

pinocytosis (pie-no-sy-**toe**-sis): A process of "cell drinking" in which a fluid is taken up in a cell vesicle or is discharged from a cell vesicle to the outside (Ch. 6).

pith: The innermost zone of cells in a plant stem; a storage tissue, especially in a young plant (Ch. 26).

pituitary gland: A small mass of endocrine tissue attached to the lower surface of the brain; made up of the anterior and the posterior pituitary lobes, which together produce at least ten hormones having a variety of effects (Ch. 37).

placenta: The organ in sharks and mammals that connects a developing embryo to surrounding maternal tissue, and through which the fetus may obtain nutrients, give off wastes, and exchange O_2 and CO_2 (Ch. 18).

planula (plural: **planulae**): The ciliated larva produced by the medusa (sexual) stage of various invertebrates; such as by a cnidarian, for instance a hydra (Ch. 25).

plasma: The fluid portion of the blood, consisting of water that contains a variety of dissolved substances including gases, ions, proteins, and antibodies (Ch. 31).

plasma cell: A mature B cell that is actively secreting an antibody in response to a particular antigen (Ch. 32).

plasma membrane: The thin bilayer of lipid and protein molecules that surrounds the cytoplasm of cells (Ch. 5).

plasmid: A small circle of bacterial DNA that is separate from the organism's single chromosome and can replicate independently (Ch. 14).

plasmodesma (plural: **plasmodesmata**): A bridge of membrane-enclosed cytoplasm connecting adjacent plant cells (Ch. 5).

plasmodium: A coenocytic mass of cytoplasm, either branched or solid, that forms the multinucleate body of a true slime mold (Ch. 21).

plastid: An organelle in plant cells that is the site of photosynthesis and of storage of sugars in the form of starch. There are two main categories of plastid; leucoplasts and pigment-containing chromoplasts (Ch. 6).

platelet: A living, motile mass of membrane-bound cytoplasm; platelets circulate in the blood and play an important role in blood clotting (Ch. 31).

pleiotropy (ply-o-trow-pee): A situation in which an individual gene affects several traits (Ch. 11).

poikilotherm: Any animal having a variable body temperature that tends to track the temperature of the surrounding environment (Ch. 35).

point mutation: A mutation that alters a single site in a gene and creates a new allele (Ch. 11).

polar bond: A chemical bond in which atoms share electrons only partially. The electrical charge from the cloud of moving electrons is asymmetrical, and tends to be found near the nucleus of the negative atom in the pair (Ch. 2).

polarization: In a cell, the creation through ion flow of an internal environment having a net negative electrical charge, contrasted with a net positive charge outside the cell (Ch. 36).

polar nuclei: In a megagametophyte of a flowering plant, the two nuclei that migrate to the center of the multinucleate cell prior to the processes that yield a haploid egg cell ready for fertilization (Ch. 28).

pollen grain: A mature male gametophyte (Ch. 24).

pollen tube: After pollination, a pipelike structure extended by the developing male gametophyte toward the egg cell. When the tube reaches the egg, some of its cytoplasm and the sperm are transferred into the egg (Ch. 24).

pollination: The reception of a pollen grain by a female ovule (Ch. 24).

poly-: From Greek, meaning "many."

polyandry: Mating between one female and many males (Ch. 49).

polygenic: Controlled by several genes (Ch. 15).

polygyny: Mating between one male and many females (Ch. 49).

polymer: A large chainlike molecule made up of many simpler units (monomers) which are linked in a specific sequence by covalent bonds (Ch. 3).

polyp: The sessile stage of hydras and some other cnidarians, characterized by a hollow, elongated body (Ch. 25).

polypeptide: A long chain of amino acids linked by peptide bonds (Ch. 3).

polyploidization: The sudden multiplication of an entire complement of chromosomes (Ch. 43).

polysaccharide: A long-chain carbohydrate made up of large numbers of monosaccharides linked by glycosidic bonds (Ch. 3).

polysome: A complex consisting of a single molecule of messenger RNA, plus several ribosomes (Ch. 6).

polytene chromosome: A giant chromosome, consisting of as many as 1,000 DNA double helix molecules, that arises during a prolonged S phase in certain insect cells (Ch. 11).

pons: A region of the vertebrate hindbrain that forms part of the brain stem; along with the medulla, helps control respiration, circulation, swallowing, and vomiting (Ch. 40).

population: A group of individuals of a species that can or do interbreed (Ch. 41).

population density: The number of individuals in a population per unit of area (Ch. 46).

postsynaptic potential (PSP): A transient depolarization of a postsynaptic nerve cell upon receipt of a synaptic transmission (Ch. 36).

primary growth: In a plant seedling, the initial elongation of cells at the tips of the root and shoot (Ch. 28).

primary production: The synthesis and storage of organic molecules during the growth and reproduction of photosynthetic organisms (Ch. 44).

primary succession: The initial establishment of plant life in an area, generally on bare rock (Ch. 45).

primitive streak: A thickened region in reptile, bird, and mammal embryos that serves as the site of gastrulating cell movements (Ch. 16).

procambium: In a young plant, a layer of cells that separates the xylem from the phloem in the vascular bundles of stems (Ch. 27).

producer: In ecology, a plant or other photosynthetic organism that converts solar energy to a stored chemical form usable as food by other creatures (Ch. 44).

profundal zone: In a lake, the deep region where little light penetrates, and hence decomposition of organic material exceeds production (Ch. 44).

progesterone: A female sex hormone (Ch. 18).

prokaryotic cell: A cell that lacks a membrane-bound nucleus and a cytoskeleton. All prokaryotes, including bacteria and cyanobacteria, are single-celled organisms (Ch. 5).

prolactin: A pituitary hormone that stimulates the production of milk in female mammals; it has other functions in such females and in other vertebrates (Chs. 18, 37).

promoter: A specific DNA sequence that serves as a binding site for RNA polymerase near each gene. Transcription proceeds from these sites (Ch. 13).

promoter site: In gene regulation, the site adjacent to an operator where RNA polymerase binds to initiate mRNA synthesis (Ch. 14).

prophase: The first phase of mitosis, in which the chromatin begins to condense and individual chromosomes become visible (Ch. 9).

proprioreceptor: A type of stretch receptor found in joints, tendons, and muscles (Ch. 38).

prostaglandin: Any of a class of fatty-acid hormones present in semen and also produced by virtually all body cells. Prostaglandins stimulate smooth-muscle contractions (Chs. 18, 37).

prot-, proto-: From Greek, meaning "first."

protein: Any member of a class of diverse polymer macromolecules constructed of linked amino acids (Ch. 3).

protoderm: The embryonic epidermis covering a plant embryo; also present in the growth centers of roots and shoots (Ch. 28).

proton: A subatomic particle that bears a positive electrical charge (Ch. 2).

proto-oncogene: A normal gene that can be changed in state to that of an oncogene, as by a mutation or chromosome translocation (Ch. 17).

protostome: Any member of a lineage of animals in which the blastopore of the developing embryo becomes the mouth (Ch. 25).

protozoan: Literally, a "first animal"; any of the single-celled organisms that is primarily animallike in its method of obtaining food and in other characteristics (Ch. 21).

pseudo-: From Greek, meaning "false."

pseudocoelom: The "false," fluid-filled body cavity that is a characteristic of nematodes (Ch. 25).

pseudoplasmodium: A sluglike mass of independent cells that makes up the "body" of a cellular slime mold (Ch. 21).

pseudopod: A retractable extension of the cell surface used for locomotion (Ch. 21).

P site: The grooved site on a ribosome in which an amino-acid–bearing initiator tRNA can bind. Once the P site and its adjacent A site are occupied, protein synthesis on the ribosome can begin (Ch. 13).

punctuated equilibrium: A model of evolutionary change which proposes that evolution proceeds in radical bursts over short periods of time, separated by long periods of stability (Ch. 43).

Punnett square: A box diagram for displaying the possible outcomes of a genetic cross; all of the possible combinations of the parental alleles are shown (Ch. 10).

purine: A nitrogen-containing base with a double-ring structure, such as adenine or guanine. Purines are structural elements of nucleic acids (Ch. 12).

pyramid of biomass: The relationship between the total masses of various groups of organisms in a food chain, in which there is usually less mass—and hence less stored energy—present at each successive trophic level (Ch. 44).

pyramid of numbers: The numerical relationship between groups of organisms in a food chain, in which plants typically outnumber herbivores, herbivores outnumber carnivores, and so on (Ch. 44).

pyrimidine (pie-**rim**-ih-deen): A single-ring nitrogen-containing base that is a structural component of nucleic acids. The bases cytosine, thymine, and uracil are pyrimidines (Ch. 12).

radial symmetry: A body plan that looks circular when viewed from above or below, and in which certain structures radiate outward in all directions from the center (Ch. 25).

radicle: An embryonic root (Ch. 27).

radiolarian: A type of marine protozoan which secretes a delicately patterned, silicone-containing shell (Ch. 21).

rapid eye movement (REM) sleep: A sleep state characterized by a high rate of brain activity, in which the eyes dart rapidly back and forth beneath closed lids (Ch. 40).

ray initial cell: Any of the cells in a woody plant that gives rise to vascular rays, spokelike conduits through which nutrients travel from more central parts of the stem toward the periphery (Ch. 28).

reaction center chlorophyll: A specialized form of the pigment chlorophyll to which light energy must be transferred during photosynthesis (Ch. 8).

reading frame: The specific unit of three nucleotides that is deciphered when a gene is expressed; each reading frame codes for one amino acid in a newly synthesized protein (Ch. 13).

realized niche: The actual portion of a species' potential niche that can be occupied (Ch. 45).

receptor (sensory) neuron: Any of the class of neurons sensitive to a particular type of stimulus, such as light, pressure, heat, or specific chemicals (Ch. 36).

recessive (allele): An allele whose expression is blocked by another, dominant allele of the same gene in a heterozygous individual (Ch. 10).

recombinant genotype (or phenotype): A genotype (or phenotype) that appears in an offspring but was not present in either parent (Ch. 11).

recombination frequency: A measure of how often crossing over occurs between specific gene loci (Ch. 11).

reduction: The addition of electrons to an atom or compound (Ch. 7).

reflex: An automatic, involuntary response to a stimulus (Ch. 48).

reflex arc: A neuronal circuit in which input from sensory neurons is relayed to motor neurons, generally via interneurons (Ch. 36).

refractory period: The time period following passage of an action potential during which the nerve cell membrane is unable to react to additional stimulation (Ch. 36).

regeneration: The regrowth of lost tissue in an adult organism (Ch. 16).

reinforcement: The strengthening of a learned behavior because it produces a favorable outcome (Ch. 48).

release-inhibiting hormone (RIH): Any of the small peptide hormones secreted in the hypothalamus that acts in turn to inhibit the secretion of anterior pituitary lobe hormones (Ch. 37).

releaser: A sign stimulus produced by a member of an animal's own species (Ch. 48).

releasing hormone (RH): Any of the hypothalamic hormones that stimulates secretion of hormones of the anterior pituitary lobe (Ch. 37).

replication fork: A site on replicating DNA at which the two component strands of the helix separate, each becoming a template for the construction of a daughter strand (Ch. 12).

repressible enzyme: An enzyme whose synthesis is repressed when the end product of its pathway is present (Ch. 14).

repressor: A substance which blocks the expression of a gene or genes (Ch. 14).

reproductive isolation: The inability of members of a species to breed with organisms belonging to other species (Ch. 43).

respiration: The process by which organisms exchange gases with the environment (Ch. 33); see also, *cellular respiration.*

respiratory pigment: A pigment molecule that circulates in the blood or resides within red blood cells and is specialized to bind O_2 reversibly (Ch. 33).

resting potential: A state of electrical charge in which the inside of a cell is electrically negative relative to the outside of the cell (Ch. 36).

restriction endonuclease: Any of a class of enzymes that recognizes specific nucleotide sequences along DNA molecules and cuts both complementary DNA strands at those specific sites (Ch. 14).

rete mirabile: A network of blood vessels in which fluids flow in a countercurrent arrangement so that the efficiency of commodity (O_2, heat, etc.) exchange is facilitated (Ch. 35).

reticular activating system (RAS): A network of neurons and nerve tracts in the brain that governs awareness of sensory stimuli and wakefulness (Ch. 40).

reticular formation: The diffuse network of neurons in the mammalian medulla and cerebrum that receives information from sensory systems (Ch. 40).

retina: See *neural retina.*

rhabdomere: In the compound eyes of arthropods, a set of cellular microvilli on which visual pigment molecules are arrayed (Ch. 38).

rhizoid: A specialized filament that serves to anchor fungi and other nonvascular plants to a substrate (Ch. 22).

rhizome: An underground plant stem that grows laterally from the main shoot (Ch. 28).

rhodopsin: The visual pigment present in the rods and cones of vertebrate eyes that is activated by light energy, beginning a series of events that result in vision (Ch. 38).

ribosomal RNA (rRNA): Molecules that, along with a variety of proteins, make up ribosomes (Ch. 13).

ribosome: Any cytoplasmic organelle that serves as a site for the synthesis of amino acids into proteins (Ch. 6).

ribulose bisphosphate carboxylase: A key enzyme that catalyzes the first reaction in the metabolic pathway leading to the reduction of CO_2 in the dark reactions of photosynthesis; probably the most abundant protein found in nature (Ch. 8).

rod: One of two types of photoreceptors in the vertebrate eye; unlike cones, rods are responsive to low light intensities (Ch. 38).

root cap cell: Any of a group of cells at the tip of a root meristem that acts as a protective shield and is positively gravitropic (Ch. 28).

root hair: A hairlike extension of an epidermal cell on a plant root (Ch. 27).

root meristem: In a plant embryo, the precursor tissue that will grow and mature into functioning root tissue (Ch. 28).

root nodule: A small hard lump on the root surface that encapsulates nitrogen-fixing microorganisms (Ch. 29).

root sucker: A horizontal root from which new stems and roots can emerge (Ch. 28).

round window: A thin membrane at the base of the lower chamber of the cochlea that flexes to relieve pressure created by sound waves (Ch. 38).

r-selected species: A species in which individuals reach reproductive age quickly and produce many offspring, relatively few of which survive to reproduce (Ch. 46).

runner: A stem capable of vegetative reproduction that grows horizontally out from the base of a plant and runs along the ground (Ch. 28).

saltatory propagation: The spread of an action potential along a myelinated axon by "jumping" from one node of Ranvier to another (Ch. 36).

saprobe: An organism, such as a fungus, that derives food molecules by decomposing dead organic matter (Ch. 22).

sapwood: The younger xylem that lies near the periphery of a tree trunk or woody stem (Ch. 28).

sarcomere: A precisely repeating unit of actin, myosin, and other proteins; the functional contractile unit of skeletal and cardiac muscle myofibrils (Ch. 39).

sarcoplasmic reticulum: A type of smooth membraneous network in muscle, in which hollow terminal sacs serve as reservoirs for calcium ions that trigger muscle contractions (Ch. 39).

savanna: A tropical grassland biome (Ch. 44).

Schwann cell: Any of the myelin-producing glial cells in peripheral nerves (Ch. 36).

scientific method: The "organized common sense" approach to the study of natural phenomena. The first step in the scientific method, observation, is followed by the development of a hypothesis, systematic experimentation to test the hypothesis, and eventually the formulation of an explanation based on experimental results (Ch. 1).

sclereid: A type of unusually hard plant sclerenchyma cell having a strong secondary wall that is sometimes impregnated with the noncarbohydrate polymer lignin (Ch. 27).

sclerenchyma: A type of plant cell having strong secondary walls of cellulose that enhances the ability of plant organs to withstand physical stresses (Ch. 27).

scutellum: The single, cylindrical cotelydon in a monocot (Ch. 28).

secondary growth: The thickening of the stem in a maturing nonherbaceous plant (Ch. 28).

secondary production: The processing and storage of energy by herbivores; see *primary production* (Ch. 44).

secondary sex characteristic: A sex-related feature arising as a result of hormonal changes around puberty, such as growth of body hair and breast and penis enlargement (Ch. 18).

secondary succession: The initial establishment of plant life on developed soil, as in an area cleared of preexisting vegetation by a physical disturbance (Ch. 45).

second law of thermodynamics: The physical law stating that the energy in any system always decreases as energy conversions take place. The "lost" energy is dissipated as heat, which is a by-product of every energy conversion (Ch. 4).

seed: The complex unit that includes a plant embryo and stored nutrients (Ch. 24).

semen: The sperm-containing fluid produced in the male reproductive system (Ch. 18).

semicircular canal: A fluid-filled, tubelike channel within the inner ear of a vertebrate that measures angular accelerations and decelerations of the head and body (Ch. 38).

semiconservative replication: DNA replication in which each newly formed double-stranded DNA molecule contains one strand conserved intact from the parent molecule (Ch. 12).

sense organ: A cluster of sensory receptor cells, as in eyes, ears, and so on (Ch. 38).

sensitive period: A discrete time during which an animal must learn something to be incorporated into its later behavioral repertoire; similar learning cannot occur after the sensitive period (Ch. 48).

sessile: In animals, the quality of being permanently attached to a fixed surface (Ch. 25).

sessile blade: A leaf blade that is joined directly to the stem; seen in plants where the petiole is reduced or absent (Ch. 27).

set point: A point on a physiological scale about which regulation occurs, as body temperature is maintained near its set point by a bird (Ch. 35).

sex chromosome: An X, Y (or equivalent) chromosome that carries genes involved in sex determination (Ch. 9).

sexual dimorphism: A situation in which members of the two sexes in a species have highly distinctive appearances (Ch. 49).

sieve plate: One of the two end walls of a sieve tube element that are perforated by pores (Ch. 27).

sieve tube: A pipelike arrangement of sieve tube elements in the phloem vascular tissue of an angiosperm (Ch. 27).

sign stimulus: An event that triggers a specific behavioral response in an animal (Ch. 48).

sinoatrial (S-A) node: A lump of modified heart muscle cells that is spontaneously electrically excitable; the "pacemaker" that governs the basic rate of heart contractions in vertebrates (Ch. 31).

siphonous alga: A type of green alga having cells that contain many nuclei within one large mass of cytoplasm (Ch. 23).

skeletal (striated) muscle: Muscle tissue composed of greatly elongated muscle fibers (each of which is built from many fused cells) (Ch. 39).

sliding-filament theory: The model to explain contraction stating that myosin molecules exert force on adjacent actin filaments and cause thin filaments to slide past thick filaments (Ch. 39).

small intestine: The lengthy, convoluted portion of mammalian gut between the stomach and the large intestine; site of most digestion of food and absorption of nutrients and water (Ch. 34).

smooth muscle: Muscle tissue consisting of single contractile cells and having actin and myosin but no sarcomeres (Ch. 39).

social hierarchy: A feature of animal social groups in which members possess ranks with varying privileges and responsibilities. Sometimes called a dominance hierarchy (Ch. 49).

social insect: A member of any of the insect groups, such as ants, bees, and termites, that lives in colonies of related individuals and exhibits complex social behaviors (Ch. 25).

society: A group of animals of the same species in which the members communicate and interact in cooperative ways (Ch. 49).

solar constant: The amount of solar energy (measured in calories) that falls on a square meter of the Earth's atmosphere in one year (Ch. 44).

solubility: A measure of the amount of a solute that will dissolve in a specified volume of solvent at specified temperatures and a pressure of one atmosphere (Ch. 33).

solute: A substance that will dissolve in a solvent; sodium chloride is a solute in water (Ch. 2).

solute potential: A measure of the concentration of solutes either inside or outside a cell; used in regard to plant cells (Ch. 29).

solvent: A substance capable of forming a homogeneous mixture with molecules of another substance (Ch. 2).

somatic nervous system: The division of the peripheral nervous system that controls muscle movements thought of as voluntary (Ch. 36).

somatomedin: Any of several small peptide hormones secreted in the liver; one somatomedin stimulates rapid bone growth during vertebrate adolescence (Ch. 37).

somatostatin: A hormone produced in the pancreas that inhibits the secretion of insulin and glucagon and the absorption of glucose (Ch. 37).

sorus (plural: sori): A cluster of sporangia on the underside of a fern frond (Ch. 23).

species: A group of actually or potentially interbreeding individuals that are reproductively isolated from other such individuals (Chs. 19, 43).

species richness (species diversity): A measure of the total numbers of different species within a community (Ch. 45).

specific heat: The amount of heat needed to raise the temperature of 1 gram of water by 1°C (Ch. 2).

spermatogenesis: The process of sperm production (Ch. 16).

S phase: The stage in the cell cycle during which DNA is replicated and histones are synthesized as a prelude to cell division (Ch. 9).

spindle: A set of microtubules that stretches across a cell from opposite poles during mitosis and that is involved in the separation and movement of the chromosomes (Ch. 9).

spirochete (spy-row-keet): Any of the bacteria that has a spiral or corkscrew shape (Ch. 20).

spleen: An organ whose functions include the storage of red blood cells; as part of the immune system, the spleen also contains lymphocytes and macrophages (Ch. 32).

spongy parenchyma: The layer of loosely arranged cells below the palisade parenchyma in leaf tissue, in which much gas exchange is carried out (Ch. 27).

sporangium: A spore case (Ch. 22).

sporophyll: A leaf bearing spore cases (Ch. 23).

sporophyte: A plant in the spore-producing, diploid stage of its life cycle (Ch. 23).

standing crop: The number of (photosynthetic) organisms existing in a defined place at a particular moment (Ch. 44).

starch: A mixture of polysaccharides (amylose and amylopectin) that serves as the primary nutrient reserve of plants (Ch. 3).

Starling's law: A rule stating that the more the cardiac muscle is stretched, the more vigorously it responds and the larger the volume of blood that can be pumped per contraction (Ch. 31).

statocyst: In invertebrates, a fluid-filled sac containing sensory hairs that register changes in the organism's body position (Ch. 38).

statolith: Any of the movable masses of calcium carbonate crystals or other dense materials in an invertebrate that rests on the sensory hairs of statocysts (organs of balance) (Ch. 38).

stele: The central cylinder of vascular tissue in a root (Ch. 27).

stem: The central, often elongated part of a plant that is composed mainly of vascular tissue and serves both as a structural support for leaves and as a conduit for the transport of water and nutrients (Ch. 27).

stem cell: A cell that serves as a continuing source of a differentiated cell type, such as a bone marrow stem cell that generates red blood cells (Ch. 16).

stereoisomer: A compound in which the constituent atoms may be arranged in one of two ways, each a mirror image of the other. All stereoisomers of a compound share the same chemical properties (Ch. 3).

steroid: Any member of a class of lipid compounds that is composed of four interconnected rings of carbon atoms linked with various functional groups. Some steroids act as vitamins, others as hormones (Ch. 3).

stipe: The stemlike structure that provides vertical support to an alga (Ch. 23).

stolon: A lateral branch of an aerial plant stem that is capable of putting out roots and new stems where it touches the ground (Ch. 28).

stoma (plural: stomata): A pore in the epidermis of a leaf, through which gases, including water vapor, diffuse (Ch. 27).

stomach: An expandable, elastic-walled sac of the gut that receives food from the esophagus (Ch. 34).

strobilus: A spore-producing organ common to club mosses and quillworts (Ch. 23).

stroma: The matrix that surrounds the grana within a chloroplast. Among other constituents, the stroma contains enzymes used in photosynthesis (Ch. 6).

stromatolite: A closely layered rock mound that consists mainly of the remains of ancient photosynthetic cyanobacteria (Ch. 19).

strong bond: A stable chemical bond that can only be broken by the input of a large amount of energy. All covalent and ionic bonds are strong bonds (Ch. 2).

structural formula: A diagrammatic representation of the approximate spatial arrangement of atoms in a molecule that also shows the number of bonds between atoms (Ch. 2).

structural isomer: A compound that has the same molecular formula as another compound, but has different structural properties (Ch. 3).

suberin: A substance impermeable to water that lines the walls of plant root endodermis cells (the innermost layer of the cortex), where two such cells adjoin (Ch. 27).

subspecies: Any subpopulation of a species set apart from the original population by at least some distinctive, genetically derived characteristics (Ch. 43).

summation: The cumulative effect of individual postsynaptic potentials in a neuron (Ch. 36).

suprachiasmatic nucleus (SCN): Either of the paired regions of the hypothalamus thought to govern the innate rhythm of biological activities (the "biological clock") of mammals (Ch. 37).

surface tension: The tendency of a liquid to minimize its surface area (Ch. 2).

suspensor: A column of cells that attaches a plant embryo to the ovule wall (Ch. 28).

symbiosis: An arrangement in which organisms of more than one species live together in intimate association (Ch. 47).

sympathetic nervous system: The division of the autonomic nervous system that, in general, stimulates cell or organ function (Ch. 36).

symplastic pathway: In a plant root, a "compartment" consisting of the collective cytoplasms of all the cells, and the channels (plasmodesmata) between them; one of two pathways for the movement of water and ions (Ch. 27).

synapse: The site where a neuron communicates electrically or chemically with a target cell (Ch. 36).

synapsis: The pairing of homologous chromosomes during the prophase I stage of meiosis (Ch. 9).

synaptinemal complex: A bridgelike structure of proteins and RNA that aligns corresponding regions of homologous chromosomes during the prophase I stage of meiosis (Ch. 9).

systemic (body) circulation: The portion of the circulatory system in which oxygenated blood is pumped by the heart to body tissues and returns through veins to the heart (Ch. 31).

systole (sis-toe-lee): The phase when the heart muscle is contracted (Ch. 31).

taiga: The sub-Arctic or subalpine biome consisting of needle-leaved forests; the coldest zone in which trees may live (Ch. 44).

taproot: In some types of plants, a root that includes one main axis extending underground (Ch. 27).

taste bud: A cluster of taste (gustatory) receptor cells capable of detecting relatively high concentrations of dissolved substances (Ch. 38).

taxis: A reflex movement that orients an animal's body with respect to an external stimulus, such as light (Ch. 48).

taxon: One category in a system of taxonomy (Ch. 19).

taxonomy: The classification of organisms (Ch. 19).

Tay-Sachs disease: A lethal human genetic disorder in which afflicted infants are born homozygous for a recessive allele that prevents the production of an enzyme, hexoaminidase A, required for normal lipid metabolism (Ch. 15).

T cell (T lymphocyte): One of the two major types of white blood cells in the immune system; T cells congregate at sites of infection and directly attack foreign substances, organisms, or tissues (Ch. 32).

tectorial membrane: In the vertebrate inner ear, the membrane which overlies the sensory hair cells of the organ of Corti (Ch. 38).

telophase: The final stage of mitosis, in which the two nuclei form and cytokinesis takes place (Ch. 9).

temperate forest: Any of the forest biomes that occurs north or south of subtropical latitudes; characterized by varied populations of evergreen and deciduous trees (Ch. 44).

temperate grassland: A grassland biome characterized by the absence of trees, a relatively cool, dry climate, and deep, rich soil (Ch. 44).

temperate shrubland: Any of the shrub-dominated biomes typical of areas having hot, dry summers and moist winters (Ch. 44).

tendon: An extremely tough bundle of collagen fibers (Ch. 39).

tensile strength: A measure of the resistance of molecules of a substance to being pulled apart (Ch. 2).

terminal bud: A bud that is produced by the apical meristem and is situated at the end of a branch (Ch. 28).

termination: The completion of protein synthesis that began with transcription of a gene. Termination occurs when a specific codon is reached on an mMRA molecule (Ch. 13).

territoriality: The defense of a particular feeding or breeding site by an animal, usually against other members of the same species (Ch. 49).

testis (plural: testes): The male sex organ in which sperm are generated (Ch. 16).

testosterone: One of the androgens, or male sex hormones (Ch. 18).

tetanus: A state of sustained maximum muscle contraction (Ch. 39).

thalassemia: A group of genetically caused anemias characterized by low or absent synthesis of either alpha or beta hemoglobin (Ch. 15).

theory: A general statement that is usually based on a number of substantiated hypotheses and is designed to explain a range of observations. One feature of a theory is that it can be used to account for future observations as well as for existing findings (Ch. 1).

therapsid: One of a diverse group of ancient reptiles that included the direct ancestors of mammals (Ch. 26).

thermiogenesis: The generation of body heat (Ch. 35).

thermoreceptor: A sensory receptor cell sensitive to infrared heat waves or temperature changes in the skin or brain (Ch. 38).

thermoregulatory center: The brain site (the hypothalamus) that carries out thermoregulation by governing physiological responses to changes in body temperature (Ch. 35).

threat display: A generally instinctive behavioral response, such as the baring of fangs, that serves as a warning to perceived competitors or predators (Ch. 49).

threshold: A level of electrical voltage (polarization) that must be reached before an action potential is initiated in a neuron (Ch. 36).

thylakoid: Any of the flattened sacs within grana, which in turn are the most prominent internal structures in chloroplasts (Ch. 8).

thymine: A single-ring pyrimidine base that is a component of a nucleotide in DNA (Ch. 12).

thymus: An organ located behind the breastbone in mammals; a component of the immune system in which T lymphocytes differentiate (Ch. 32).

thyroid gland: The thyroxine-producing endocrine organ located near the esophagus in air-breathing vertebrates (Ch. 37).

thyrotropic hormone (TH): A hormone produced in the anterior pituitary lobe that regulates the secretion of thyroid hormones (Ch. 37).

thyroxine (T_4; also T_3): A thyroid hormone characterized by the presence of atoms of iodine; regulates metabolic rate (Ch. 37).

tight junction: A seal that encompasses the lateral surfaces of cells in epithelia; they act as barriers to fluid leakage (Ch. 5).

tissue: A grouping of cells that function collectively (Ch. 5).

tonus: A condition in which a muscle is kept partially contracted over a long period (Ch. 39).

torpor: A state in which the body's metabolic rate and activity is lowered temporarily (Ch. 35).

trachea: In animals, the windpipe (Ch. 33); in insects and most terrestrial arthropods, it is the air-filled, hollow, branching tube that carries air to and from the animal's tissues (Ch. 33).

tracheid (tray-kee-id): An elongated hollow cell having thick, rigid, pitted walls; a basic unit of vascular tissue in plants (Ch. 23).

trans-: From Latin, meaning "across" or "through."

transcription: The process in which the genetic information encoded in DNA is deciphered to yield a variety of types of RNA (Ch. 13).

transducin: A protein that interacts with the light-sensitive pigment rhodopsin in a series of light-triggered events leading to visual perception (Ch. 38).

transduction: A mode of indirect gene transfer in the laboratory, in which DNA is carried from one bacterium to another by a virus (Ch. 14).

transfer RNA (tRNA): The form of RNA that is responsible for transporting individual amino acids to sites of protein elongation on a ribosome–mRNA complex (Ch. 13).

transformation: A laboratory process of indirect gene transfer, in which DNA that has been released from one bacterium into the surrounding medium is taken up by another bacterial cell (Ch. 14).

translation: The process in which proteins are synthesized based on information transcribed from DNA (Ch. 13).

translocation: The transport of solutes in the phloem of a plant (Ch. 29).

transpiration: Evaporative water loss through leaves (Ch. 29).

triglyceride: A lipid compound in which three fatty acids are joined to a single molecule of glycerol (Ch. 3).

trochophore: The ciliated larva of an aquatic polychaete worm (Ch. 25).

trophic level: A feeding level in a food chain (Ch. 44).

trophoblast: The outer layer of cells of a mammalian blastocyst-stage embryo (Ch. 18).

tropical rain forest: A biome typified by a warm, wet climate, abundant vegetation, and exceptionally diverse animal life (Ch. 44).

tropical seasonal forest: A forest biome characterized by wet–dry seasonality and in which trees are deciduous during the dry season (Ch. 44).

tropism: Any movement in response to an environmental signal, such as light or gravity (Ch. 30).

tropomyosin: A protein thought to mask the actin binding sites of myosin in a noncontracting muscle (Ch. 39).

troponin: A protein complex that affects tropomyosin function and so helps regulate actin–myosin interaction in muscle cells (Ch. 39).

T tubule: In a skeletal-muscle cell, any of numerous deep infoldings of the plasma membrane that serve as conduits for action potentials that can trigger muscle contractions (Ch. 39).

tuber: A modified plant stem that can reproduce vegetatively, sprouting new stems and roots from budlike "eyes" (Ch. 28).

tubulin: The globular protein that is the structural unit of microtubules (Ch. 6).

tumor: A solid mass of cancer cells (Ch. 17).

tundra: In Arctic and high-mountain regions, a treeless plain in which the predominant plant types are grasses, lichens, sedges, and occasional small shrubs. Deep tundra soil remains frozen year-round as permafrost (Ch. 44).

tympanic membrane: A tightly stretched membrane separating the outer and middle ear which vibrates when struck by sound waves; the eardrum (Ch. 38).

umbilical cord: The thick, ropelike structure that connects the abdomen of a developing fetus to the placenta, and in which a fetus's umbilical arteries and vein spiral about each other (Ch. 18).

uracil: A pyrimidine base which takes the place of thymine in RNA molecules (Ch. 13).

urea: A nitrogenous waste that can be formed during the breakdown of proteins and other nitrogen-containing substances (Ch. 35).

uric acid: A nitrogenous waste product; excreted typically by birds, land reptiles, and insects (Ch. 35).

uterus: The muscular reproductive organ in female mammals in which an embryo may implant and the developing embryo be maintained during pregnancy (Ch. 18).

vacuole: A saclike membrane-bound organelle found within both animal and plant cells. Vacuoles are typically filled with fluids and soluble molecules (Ch. 6).

vagina: The muscular tube that leads from the uterus to the outside of the female mammal's body (Ch. 18).

variable region: Either of the two ends of the arms of a Y-shaped antibody protein. The characteristic amino-acid sequences in the variable regions of an antibody determine the antibody's specificity to a particular antigen (Ch. 32).

vascular bundle: The distinctive grouping of plant vascular tissues, xylem and phloem, that appears as a ring just inside the cortex layer of the stem (Ch. 27).

vascular cambium: In a mature plant stem, the layer of cells that separates xylem and phloem tissues in each vascular bundle in a stem (Ch. 27).

vascular plant: A plant having tissues (xylem, phloem) specialized to conduct water and other materials and to provide vertical support (Ch. 23).

vascular ray: A spokelike line of parenchyma cells that serves as a channel for the transport of nutrients from the center of a woody plant stem outward toward the periphery (Ch. 28).

vasoconstriction: Contraction of the capillary sphincters and the walls of veins in response to action potentials or certain hormones (Ch. 31).

vasodilation: The relaxation of capillary sphincters and walls of veins in response to action potentials, certain hormones, or, in muscles, local small molecules (Ch. 31).

vasomotor center: An area of the hypothalamus that operates to keep blood pressure, blood gases, pH, and other physiological indicators within normal ranges, generally by regulating blood flow through the heart and vessels (Ch. 31).

vegetative reproduction: A process in which new plants genetically identical to each other and to the parent emerge from the parent's body (Ch. 28).

vein: (1) A blood vessel that carries blood toward the heart (Ch. 31). (2) In a leaf, any of the threadlike channels that provides structural support to leaf tissue and serves as transport pathways for the movement of water and nutrients (Ch. 27).

venule: A tiny vein (Ch. 31).

vertebrate: An animal that has an internal bony skeleton and usually a backbone (Ch. 26).

vessel element: One of two types of cells in the xylem tissue of an angiosperm, with structural features that permit the relatively free flow of water (Ch. 27).

virion: A virus particle outside a living cell; composed of an inner core of nucleic acid surrounded by a protein coat (Ch. 20).

viroid: A minute particle of RNA that lacks a protein coat and is capable of causing disease in both plants and animals (Ch. 20).

virus: A particle of genetic material that can invade a living cell and utilize its metabolic machinery to reproduce (Ch. 20).

visual field: The full set of rods or cones linked to each ganglion cell in the eye (Ch. 38).

vitamin: An organic molecule that functions as a coenzyme or cofactor of enzymes (Ch. 34).

voltage-gated channel: A channel through which ions may pass into a neuron that opens in response to changes in membrane potential (Ch. 36).

volvocine alga: Any of the colonial green algae (Ch. 23).

vulva: The external genitals of a female human (Ch. 18).

warning coloration: See *aposematic coloration* (Ch. 47).

water potential: A measure of the tendency of a plant cell to gain or lose water (Ch. 29).

water-soluble vitamin: A vitamin, such as vitamin C and the B vitamins, that is transported as a free compound in the blood (Ch. 34).

weak bond: A chemical bond that is easily broken by the input of a small amount of energy (Ch. 2).

wild-type allele: In *Drosophila*, the most common unmutated allele for any genetic characteristic (Ch. 11).

Wolffian duct: The embryonic structure that develops into the epididymis and the vas deferens if the embryo is male (XY). (Ch. 18).

wood: The hard, cellulose-containing secondary xylem of a nonherbaceous plant (Ch. 28).

xylem: In plants, a pipelike transport tissue made up of dead, cellulose- and lignin-reinforced cells stacked end to end (Ch. 27).

yolk: Material made up of proteins, carbohydrates, nucleic acids, and lipids that is stored in an egg to serve as a nutrient supply for a developing embryo (Ch. 16).

yolk sac: One of the membranes that supports embryos of land vertebrates. In reptile and bird eggs the yolk sac completely encases the large sphere of yolk that serves as the embryo's food supply (Ch. 16).

zonulae adherens: Sites of firm physical contact between cells that are literally "zones of adhesion." (Ch. 5).

zooflagellate: An animallike protozoan that can move about by means of its whiplike flagellum (Ch. 21).

zooplankton: Nonphotosynthetic marine protozoans that live at or near the ocean surface (Ch. 21).

zoospore: The motile reproductive cell that matures into a haploid adult during asexual reproduction in certain algae (Ch. 23).

Zygomycete: A class of lower fungi (Ch. 22).

zygote: A fertilized egg (Ch. 16).

zygospore: The diploid zygote produced by the sexual fusion of two isogametes in certain algae (Ch. 23).

zymogen: Any of numerous inactive precursors of digestive enzymes (Ch. 34).

Credits and Acknowledgments

Illustrators   Dolores Bego, Carol Donner, Marsha Dohrmann, J & R Technical Services, David Lindroth, Judy Skorpil, Vantage Art, Inc.

Photo Editors   Leonora Morgan, Jacquelyn Wong, Christine Carey, Mira Schachne.

Cover photos   Broad-billed Hummingbird (*Cynanthus latirostria*) by Bob & Clara Calhoun/Bruce Coleman; computer generated model of the DNA double helix by Dan McCoy/Rainbow.

Chapter 1   Page 2: W. H. Hodge/Peter Arnold. Figure 1-1: (a) Walter E. Harvey/National Audubon Society—Photo Researchers; (b) James H. Carmichael/Bruce Coleman. Figure 1-2: Manfred Kage/Peter Arnold. Figure 1-3: Richard H. Smith/Photo Researchers. Figure 1-4: (a) Runk/Schoenberger/Grant Heilman; (b) John Colwell/Grant Heilman. Figure 1-5: (a–c) Jerome Wexler/Photo Researchers. Figure 1-6: John Colwell/Grant Heilman. Figure 1-7: (a) Michael C. T. Smith/National Audubon Society—Photo Researchers; (b) Steven C. Kaufman/Peter Arnold. Figure 1-8: Fred Bavendam/Peter Arnold. Figure 1-10: The Bettmann Archive. Figure 1-11: Warner Lambert/Park, Davis & Company. Figure 1-12: Jeff Foott/Bruce Coleman. Figure 1-13: Jonathan D. Eisenback/Phototake. Figure 1-14: David Leah/Science Photo Library—Photo Researchers. Figure 1-16: The Bettmann Archive. Figure 1-17: (a) Joe Munroe/Photo Researchers; (b) Leonard Lee Rue III/National Audubon Society—Photo Researchers. Figure 1-18: Pat and Tom Leeson/Photo Researchers. Figure 1-19: Claude Monet, "Water Lilies," 1920-1922, The Saint Louis Art Museum, gift of the Steinberg Charitable Fund. Figure 1-21: The Bettmann Archive. Figure 1-22: AP/Wide World Photos. Figure 1-23: Grant Heilman. Figure 1-24: Verna R. Johnston/Photo Researchers. Figure 1-25: J. R. Simon/Photo Researchers.

Part I   Page 22: R. B. Taylor/Science Photo Library—Photo Researchers.

Chapter 2   Page 24: Bruce Matheson, Aberdeen, Washington. Figure 2-1: Courtesy of Michael Isaacson, Cornell University and Mitsuo Ohtsuki. Figure 2-3: Don Fawcett, J. Gouranton and R. Folliot/Photo Researchers. Page 28, Figure A: Historical Pictures Service, Chicago. Page 29, Figure B: (a,b,c) From George V. Kelvin, © *Discover* Magazine 4/82, Time Inc.; (c) photo of NMR Scan, Dan McCoy/Rainbow; (d) From Ian L. Pykett, "NMR Imaging in Medicine." Copyright © 1982 by Scientific American, Inc. All rights reserved. Figure 2-10: (a and c) Yoav/Phototake. Figure 2-14: Spencer Swanger/Tom Stack & Associates. Figure 2-15: William Amos/Bruce Coleman. Figure 2-21: Winslow Homer, "The Fishing Net," 1885, courtesy of the Art Institute of Chicago.

Chapter 3   Page 46: Tripos Associates/Peter Arnold. Figure 3-4: (b) Yoav/Phototake. Figure 3-10: (c) Biophoto Associates/Photo Researchers. Figure 3-11: (b) Don Fawcett/Photo Researchers. Figure 3-20: (e) Tony Brain/Science Photo Library—Photo Researchers. Figure 3-21: (b) Robert P. Apkarian, Yerkes Regional Primate Center, Emory University, Atlanta. Figure 3-22: From B. W. Matthews, "Structure of Bacteriophage T4 Lysozyme." Courtesy of Dr. B. W. Matthews, University of Oregon. Figure 3-24: K. R. Porter/Photo Researchers. Figure 3-25: Adapted from R. E. Dickerson and L. Geis, *The Structure and Action of Proteins*, Benjamin-Cummings Publishing Company.

Chapter 4   Page 74: Carol Hughes/Bruce Coleman. Figure 4-2: (a) Diane Rawson/Photo Researchers; (b) Gary Milburn/Tom Stack & Associates. Figure 4-3: (both) Alan Carey/The Image Works. Figure 4-9: "Beaumont and St. Martin" by Dean Cornwall, courtesy of William Beaumont Army Medical Center, El Paso. Figure 4-12: (b) Courtesy of William S. Bennett, Yale University and T. A. Steitz, California Institute of Technology, Pasadena, from W. S. Bennett and T. A. Steitz, *Journal of Molecular Biology* 140 (1980): 211–230. Figure 4-14: Reprinted by permission from *Nature*, vol. 266, p. 332, Copyright © 1977 Macmillan Magazines Limited. Figure 4-15: Norman K. Wessells. Figure 4-16: (both) Reprinted by permission from R. E. Dickerson and I. Geis from *The Structure and Action of Proteins* (Menlo Park, Ca.: Benjamin/Cummings, 1969); illustration copyrighted by Dickerson and Geis, 1969. Page 90, Figure A: A. Kerstitch/Tom Stack & Associates.

Chapter 5   Page 96: Biological Photo Service. Figure 5-1: Norman K. Wessells. Figure 5-2: (a–c) Norman K. Wessells. Figure 5-3: From *Botany* by P. Ray, T. Steeves, and S. Fultz. Copyright © 1983 by CBS College Publishing. Reprinted by permission of Holt, Rinehart & Winston, Inc. Figure 5-6: From Carl P. Swanson, *The Cell*, 3/e, © 1969, p. 76. Adapted by permission of Prentice-Hall, Inc., Englewood Cliffs, New Jersey. Figure 5-8: (a) W. Rosenberg/Biological Photo Service. Figure 5-14: Courtesy of Michael P. Sheetz, Department of Physiology, University of Connecticut Health Center, Farmington, Connecticut. Figure 5-15: (a) Norman K. Wessells; (b) Biophoto Associates/Photo Researchers. Figure 5-16: (a) Biological Photo Service. Figure 5-17: Courtesy of Audrey M. Glauert and Geoffrey M. W. Cook, Strangeways Research Lab, Cambridge, England. Figure 5-18: Barry King, University of California, Davis/Biological Photo Service. Figure 5-19: Adapted from D. E. Kelly, in D. E. Kelly, R. L. Wood, and A. C. Enders, *Bailey's Textbook of Microscopic Anatomy*, 18th Edition, 1984. Copyright © 1984 the Williams & Wilkins Co., Baltimore. Figure 5-20: Courtesy of Norton B. Gilula, Rockefeller University. Figure 5-21: Micrograph by W. P. Wergin, courtesy of E. H. Newcomb, University of Wisconsin/Biological Photo Service.

Chapter 6   Page 124: Norman K. Wessells. Figure 6-1: (b) G. F. Leedale/Biophoto Associates—Photo Researchers; (c) Science Photo Library International/Taurus. Figure 6-2: (b) Don Fawcett and B. Gilula—Photo Researchers. Figure 6-3: (b) Norman K. Wessells. Figure 6-4: (b) Norman K. Wessells. Figure 6-6: Norman K. Wessells. Figure 6-7: (b) Norman K. Wessells. Figure 6-9: J. Burgess, Department of Cell Biology, John Innes Institute, Norwich, England. Page 134, Figure A: M. M. Perry and A. B. Gilbert, *J. Cell Sci.* 39 (1979): 357–372. Figure 6-10: Norman K. Wessells. Figure 6-11: (b) D. W. Fawcett/Photo Researchers. Figure 6-13: (b) Courtesy Daniel S. Friend, University of California School of Medicine, San Francisco. Page 138: From Lewis Thomas, "Organelles as Organisms," from *The Lives of a Cell* by Lewis Thomas. Copyright © 1972 by the Massachusetts Medical Society. Originally published in *The New England Journal of Medicine*. Reprinted by permission of Viking Penguin Inc. Figure 6-15: (a) Norman K. Wessells. Figure 6-16: C. K. Pyne, *Atlas of Cell Biology* (Little, Brown and Co., 1974). Figure 6-18: (a) Norman K. Wessells; (b) courtesy of Lloyd A. Culp, Department of Molecular Biology and Microbiology, School of Medicine, Case Western Reserve University, Cleveland. Figure 6-19: Courtesy of Sidney Tamm. Figure 6-21: (b) Diane Woodrum, North Hills, West Virginia, and Richard Linck, University of Minnesota, Minneapolis.

Chapter 7   Page 150: Sarah King/Robert Harding Picture Library. Figure 7-1: Charles Marden Fitch/Taurus. Figure 7-2: Herve Chaumeton/Bruce Coleman. Figure 7-19: (b) Efraim Racker, Biochemistry and Molecular Biology Department, Cornell University.

Chapter 8   Page 174: Art Wolfe/The Image Bank. Figure 8-2: Courtesy of Lewis K. Shumway, Washington State University. Figure 8-12: (a and b) K. Bendo.

Part II   Page 194: Norman Myers/Bruce Coleman.

Chapter 9   Page 196: (a–d) Andrew Bajer, Department of Biology, University of Oregon, Eugene. Figure 9-2: (a) Carolina Biological Supply Company; (b) W. E. Engler, Courtesy of G. F. Bahr, Washington, D. C. Figure 9-3: Courtesy of O. L. Miller, Jr. from S. L. McKnight and O. L. Miller, Jr., *Cell* 8 (1977): 305–319. Figure 9-4: (a) W. E. Engler, Courtesy of G. F. Bahr, Washington, D. C. Figure 9-5: Courtesy of M. L. Alonso, Laboratory of Cytogenetics, The New York Hospital. Figure 9-7: K. R. Porter/Science Source—Photo Researchers. Figure 9-8: (a–f) Carolina Biological Supply Company. Figure 9-10: T. E. Schroeder, University of Washington/Biological Photo Service. Figure 9-12: (a–f) John D. Cunningham/Visuals Unlimited; (g) Carolina Biological Supply Company. Figure 9-15: Alan Detrick/Photo Researchers. Figure 9-16: (a) Biophoto Associate/Photo Researchers. Figure 9-17: Courtesy of J. B. Gurdon, from "Transplanted Nuclei and Cell Differentiation," *Scientific American* 219 (6) (Dec. 1968).

Chapter 10   Page 218: Darwin Dale/Photo Researchers. Figure 10-2: The Bettmann Archive. Figure 10-3: (b) Steven J. Krasemann/ Peter Arnold. Figure 10-10: (b) Courtesy of W. Atlee Burpee Company. Figure 10-11: From F. B. Hutt, "The Genetics of the Fowl," *Journal of Genetics* 22 (1930): figure 2.

Chapter 11   Page 238: Anne Cumbers/Spectrum Colour Library. Table 11-1: From A. H. Sturtevant, *Journal of Experimental Zoology* 14 (1913): 43–59. Alan R. Liss, Inc., Publisher and copyright holder. Figure 11-4: Paul Roberts, Department of Zoology, Oregon State University, Corvallis. Figure 11-5: Courtesy George Lefevre, Jr., from *The Genetics and Biology of Drosophila*, vol. la. M. Ashburne and E. Novitski, eds., (Academic Press, 1976), Ch. 2. Figure 11-6: From Adrian M. Srb and Ray D. Owen, *General Genetics* (San Francisco, Ca.: W. A. Freeman amd Company, 1957); originally credited to E. M. Wallace, pinx., from Sturtevant and Beadle, *An Introduction to Genetics* (Philadelphia: W. B. Saunders and Co., 1939). Figure 11-8: Victor A. McKusick, Johns Hopkins University; from Mange/Mange, *Human Genetics: Human Aspects* (W. B. Saunders and Co., 1980). Figure 11-9: (Top left) Anne Cumbers/Spectrum Colour Library; (top right and bottom) Walter Chandoha. Page 251, Figure A: Courtesy of Gaines Dog Care Center; Figure B: from Ojuind Winge, *Inheritance in Dogs* (Comstock Publishers, 1950). Figure 11–12: Courtesy of Bruce F. Cameron and Robert Zucker, Miami Comprehensive Sickle Cell Center, Donald R. Harkness, Director.

Chapter 12   Page 258: © R. Langridge, U. C. S. F. Computer Laboratory/ Dan McCoy/Rainbow. Figure 12-1: (c) Courtesy Ronald L. Phillips, University of Minnesota, St. Paul. Figure 12-4: After Linus Pauling, et al. "Sickle Cell Anemia, a Molecular Disease," *Science*, vol. 110, November 25, 1949, pp. 543–548. Figure 12-5: Maclyn McCarty, M. D., The Rockefeller University. Figure 12-7: (a and b) Lee D. Simon/Photo Researchers. Figure 12-10: (a) Courtesy of the Biophysics Department, King's College London; (b) © R. Langridge, U.C.S.F. Computer Graphics Laboratory/ Dan McCoy/Rainbow. Figure 12-11: The Bettmann Archive. Page 274, Figure A: Alexander Rich, Department of Biology, Massachusetts Institute of Technology. Figure 12-15: (b) From John Cairns, Imperial Cancer Research Cancer Fund, London, England; permission from *Cold Spring Harbor Symposium on Quantitative Biology* 28 (1963): 43. Figure 12-18: (a) Courtesy of David R. Wolstenholm from *Chromosoma* 43 (1973):1.

Chapter 13   Page 284: O. L. Miller, B. A. Hamkalo and C. A. Thomas, Jr., "Visualization of Bacterial Genes in Action," *Science* 169 (1970): 392–395. Figure 13-3: (b) O. L. Miller and Barbara R. Beatty, "Portrait of a Gene," *Journal of Cell Physiology* 74 (Suppl. 1):225–232. Figure 13–5: (b) Courtesy Alexander Rich, Department of Biology, Massachusetts Institute of Technology. Page 299, Figure A: Nik Kleinberg/Picture Group.

Chapter 14   Page 303: Courtesy of R. L. Brinster and R. E. Hammer, School of Veterinary Medicine, University of Pennsylvania, Philadelphia. Figure 14–3: Courtesy of Charles C. Brinton, Jr. and Judith Carnahan, Department of Biological Sciences, University of Pittsburgh. Figure 14–8: Courtesy of Stanley N. Cohen, School of Medicine, Department of Genetics, Stanford University. Figure 14–14: (a) Courtesy of P. Leder, Genetics Department, Harvard Medical School. Figure 14–17: (a) Courtesy of R. L. Brinster and Myrna Trumbauer, University of Pennsylvania, School of Veterinary Medicine, Philadelphia. Figure 14–18: Robert S. Peabody Foundation for Archeology, photo by R. S. MacNeish and Paul Mangelsdorf.

Chapter 15   Page 333: Smithsonian Institution National Anthropological Archives. Figure 15–1: (b) Statens Museum for Kunst, Copenhagen. Figure 15–2: (a) Courtesy of Theodore T. Puck, E. Roosevelt Institute for Cancer Research, University of Colorado Medical Center; (b) Terry McKay/The Picture Cube. Figure 15–3: Reprinted by permission of *The New England Journal of Medicine* 283 (1970): 15–20. Table 15–1: From Newman et al., *Twins: A Study of Heredity and Environment*, published by the University of Chicago Press. © 1937 by the University of Chicago. All rights reserved. Figure 15–6: (a) Courtesy of Thomas J. Bouchard, Jr., Minnesota Center for Twin and Adoption Research, University of Minnesota, Minneapolis; (b) Karl Fredga, Department of Genetics, University of Uppsala, Uppsala, Sweden. Figure 15–8: From Arthur P. Mange and Elaine Johnson Mange, *Genetics: Human Aspects*. © 1980 by Saunders College/Holt, Rinehart & Winston. Reprinted by permission of Holt, Rinehart & Winston, Inc. Figure 15–9: (a and b) Courtesy of Murray L. Barr, Health Science Center, University of Western Ontario, London, Ontario. Figure 15–12: (b) Gernsheim Collection, Harry Ransom Humanities Research Center, University of Texas at Austin. Figure 15–13: Permission granted by the C. V. Mosby Co., from Cline, Sidbury, and Richter, *Journal of Pediatrics* 55 (1959): 355–366. Figure 15–15: Courtesy of the National Zoological Park, Smithsonian Institution, Washington, D. C. Figure 15–16: (a and b) The New-York Historical Society; (c) Library of Congress; (d) Massachusetts Historical Society. Figure 15–17: (b) Both, Armed Forces Institute of Pathology. Table 15-3: From Arthur P. Mange and Elaine Johnson Mange, *Genetics: Human Aspects*. © 1980 by Saunders College/Holt, Rinehart & Winston. Reprinted by permission of Holt Rinehart & Winston, Inc.

Chapter 16   Page 358: Norman K. Wessells. Figure 16–10: (a, c and d) Courtesy of Mia Tegner, Scripps Institution of Oceanography; (b) Courtesy of Gerald Schatten, Department of Biological Sciences, Florida State University, Tallahasse. Figure 16–16: (f) R. G. Kessel, H. W. Beams, and C. Y. Shih, "Surface structures of the frog embryo as revealed by scanning electron microscopy," *Anat. Rec.* 175, 489 (1973). Figure 16–18: Courtesy of R. J. Goss, from *Principles of Regeneration*, (Academic Press).

Chapter 17   Page 383: Walter J. Gehring, from *Nature* (Jan. 1985), MacMillian Journals Ltd., London. Figure 17–1: Art Resource. Figure 17–2: (a–c) Courtesy of Carnegie Institution of Washington. Figure 17–6: (e) Norman K. Wessells. Figure 17–7: (b) Don Fawcett/Susumo Ito, and Arthur Kike/Photo Researchers. Figure 17–8: (b) Courtesy of Kathryn Tosney, Biological Sciences, The University of Michigan, Ann Arbor. Figure 17–13: Courtesy of Lewis Wolpert from "Pattern Formation in Biological Development," *Scientific American* (Oct. 1978).

Chapter 29   Page 692: John Shaw/Bruce Coleman. Figure 29-3: William J. Weber/Visuals Unlimited. Figure 29-6: (b) Norman K. Wessells. Figure 29-8: U.S. Department of Agriculture. Figure 29-10: (b) L. Evans Roth/Biological Photo Service. Figure 29-11: Hugh Spencer/ Photo Researchers. Figure 29-12: (a and b) Harry E. Calvert/Battelle-Kettering Research Laboratory; (c) Thomas A. Lumpkin, Agronomy, Washington State University, Pullman. Figure 29-13: (a) D. Lyons/ Bruce Coleman; (b) Jeremy Burgess/Science Photo Library—Photo Researchers. Figure 29-14: (a–c) Jack Dermid.

Chapter 30   Page 711: Alan Pitcairn/Grant Heilman. Figure 30-5: J. P. Nitsch, *American Journal of Botany* 37 (1950):3. Figure 30-9: Courtesy of Sylvan H. Wittwer, Agricultural Experiment Station, Michigan State University, East Lansing. Figure 30-12: (d) Lowell Georgia/Science Source—Photo Researchers. Figure 30-15: Biophoto Associates/Science Source—Photo Researchers. Figure 30-16: (All) Hazel Hankin, Brooklyn.

Part V   Page 732: George Holton/Photo Researchers.

Chapter 31   Page 734: CNRI/Science Photo Library/Photo Researchers. Figure 31-2: Courtesy of Toichiro Kuwabura, Laboratory of Opthalmic Pathology, National Eye Institute, NIH. Figure 31-4: William Harvey, Exercitatio Anatomica de Motu Cordis and Sanguinis, 1648. Figure 31-9: (a) Thomas Eisner, Cornell University. Page 746, Figure A: George Schwenk *Discover* Magazine, 1984, Time, Inc. Figure 31-10: Thomas Eisner, Cornell University. Figure 31-11: (a) D. W. Fawcett/Photo Researchers. Figure 31-15: (b) Courtesy of James G. White, University of Minnesota Medical School; (c) Lennart Nilsson, © Boehringer Ingelheim International GmbH/Bonnier Fakta, Stockholm, Sweden.

Chapter 32   Page 760: Lennart Nilsson, © Boehringer Ingelheim International GmbH/Bonnier Fakta, Stockholm, Sweden. Figure 32-1: Emma Shelton, Laboratory of Biochemistry, National Cancer Institute, Bethesda. Figure 32-3: Richard Kessel, Biology Department, University of Iowa. Figure 32-4: From Ian R. Tizard, *Immunology: An Introduction.* © 1984 by CBS College Publishing. Reprinted by permission of Holt, Rinehart & Winston, Inc. Figure 32-7: (c) Courtesy of R. R. Dourmashkin, from R. R. Parkhouse et al., *Immunology* 18 (1970): 575. Page 774, Figure A: David Scharf/Peter Arnold. Figure 32-12: (a and b) From O. Carper, I. Virtaner, and E. Saksela, *Journal of Immunology* 128 (1982): 2691. Figure 32-13: Leslie Brent, St. Mary's Hospital Medical School, London. Figure 32-14: From Alberts et al., *Molecular Biology of the Cell.* Reprinted by permission of Garland Publishers, Inc. Figure 32-15: (a) From Alberts et al., *Molecular Biology of the Cell.* Reprinted by permission of Garland Publishers, Inc.; (b) Lennart Nilsson, © Boehringer Ingelheim International GmbH/Bonnier Fakta, Stockholm, Sweden.

Chapter 33   Page 786: Steve Lissau/Rainbow. Figure 33-2: (b) John Roskelley/Photo Researchers. Figure 33-4: Victor H. Hutchison, Department of Zoology, University of Oklahoma, Norman. Figure 33-5: E. R. Degginger. Figure 33-6: (b) Warren Burggren, Department of Zoology, University of Massachusetts, Amherst. Figure 33-13: (c) Courtesy of H. R. Duncker, Institute for Anatomy and Cytobiology, Justus-Liebig-University, Giessen, West Germany. Figure 33-20: (a) William E. Townsend Jr./Photo Researchers. Figure 33-21: C. Allan Morgan/Peter Arnold.

Chapter 34   Page 813: Mark J. Jones/Bruce Coleman. Figure 34-1: (a) Robert R. Peck, Institute of Animal Biology, University of Geneva, Switzerland; (b) Cosmos Blank/Photo Researchers; (c) Dwight R. Kuhn; (d) M. Timothy O'Keefe/Bruce Coleman. Figure 34-3: Lennart Nilsson from *Close to Nature* (New York: Pantheon Books). Figure 34-5: (a, left) Thomas Eisner, Cornell University; (a, right) Dwight

Kuhn/DRK Photo; (b, left) photo by Penelope Jenkin from *Life on Earth,* Edmund O. Wilson, Sinauer Associates, 1975; (b, right) Frans Lanting; (c, left) Thomas Eisner, Cornell University; (c, right) Douglas T. Cheeseman, Jr./Peter Arnold. Figure 34-14: (b) R. D. Specian, Department of Anatomy, Louisiana State University, Medical Center, Shreveport. Figure 34-19: (b) Rutter, Pictet, Chirgwin et al., from *Molecular Control of Proliferation and Differentiation* (Academic Press, 1977), p. 205. Figure 34-20: From Hall et al., "Comparative Physiology in High Altitudes," *Journal of Cellular and Comparative Physiology* 8 (1936): 301–313. Alan R. Liss, Inc., Publisher and Copyright Holder.

Chapter 35   Page 843: Spencer Swanger/Tom Stack & Associates. Figure 35-2: Francois Gohier/Ardea, London. Figure 35-3: (a) Heather Angel. Figure 35-13: From Roger Eckert and David Randall, *Animal Physiology,* 2/e. Copyright © 1978, 1983 W. H. Freeman and Company. Reprinted with permission. Figure 35-14: (b) E. D. Stevens, University of Guelph, Ontario; (c) F. G. Carey, Woods Hole Oceanographic Institution. Figure 35-15: Cristopher Crowley/Tom Stack & Associates. Figure 35-16: From Bernd Heinrich, "Thermoregulation in Endothermic Insects," *Science* 185 (August 30, 1974): 747–756. Figure 35-17: Robert Harding Picture Library. Figure 35-19: (b and c) Ulfe Midtgard, The Zoological Institutes, University of Copenhagen, Denmark. Figure 35-21: Courtesy of David Hull, Department of Child Health, University Hospital, Queens Medical Center, Nottingham, Great Britain, from David Hull, *British Medical Bulletin* 22 (1966b). Figure 35-22: From Roger Eckert and David Randall, *Animal Physiology,* 2/e. Copyright © 1978, 1983 W. H. Freeman and Company. Reprinted with permission.

Chapter 36   Page 871: Biophoto Associates/Photo Researchers. Page 36-1: (b) Norman K. Wessells. Figure 36-2: Biophoto Associates/ Photo Researchers. Figure 36-6: (b) A. C. Hodgkin and R. D. Keynes from *Journal of Physiology* 131 (1956). Figure 36-11: Norman K. Wessells. Figure 36-13: (d) Micrograph produced by John E. Heuser of Washington University School of Medicine, St. Louis. Figure 36-16: From E. R. Lewis et al., "Studying Neural Organization in Aplysia with the Scanning Electron Microscope," *Science* 165 (September 12, 1969): 1140–1143. Figure 36-21: From J. Wine. Reprinted with permission. Figure 36-22: From J. Wine. Reprinted with permission.

Chapter 37   Page 900: Heather Angel. Figure 37-3: (b) Walter E. Stumpf, Department of Anatomy, University of North Carolina, Chapel Hill. Figure 37-6: (a and b) Roman Vishniac. Figure 37-8: 1965 CIBA Pharmaceutical Company, Division of CIBA-GEIGY Corporation, reproduced with permission from CIBA Collection of Medical Illustrations by Frank H. Netter, M.D., all rights reserved. Figure 37-11: Biophoto Associates/Photo Researchers. Page 916, Figure A: Bruce Robert/Photo Researchers. Figure 37-13: Photograph by the Egyptian Expedition, The Metropolitan Museum of Art, New York. Figure 37-14: (b) Courtesy of W. W. Douglas, F. R. S., Yale University, from Douglas, Nagasawa, and Shulz, "Sub Cellular Organization and Function in Endocrine Tissues," *Memoirs of the Society of Endocrinology* 19 (Cambridge University Press). Figure 37-16: Carol Hughes/Bruce Coleman. Figure 37-17: (bottom) From R. Weber, *Helvetica Physiologica et Pharmacolgia Acta* 21 (1963).

Chapter 38   Page 928: Barbara Kirk/The Stock Market. Figure 38-1: (a) SIU/Photo Researchers. Figure 38-3: (a) Ken Brate/Photo Researchers. Figure 38-4: J. H. Carmichael/Photo Researchers. Page 939: From Nick Fasciano, *Discover* Magazine © 1987. Figure 38-14: Merlin D. Tuttle/Photo Researchers. Figure 38-15: From H. W. Lissman, "Electric Location by Fishes." Copyright © 1963 by Scientific American, Inc. All rights reserved. Figure 38-16: (a) From R. Igor Gamow and John F. Harris, "The Infrared Receptors of Snakes." Copyright © 1973 by Scientific American, Inc. All rights reserved; (b) Brian Parker/ Tom Stack & Associates. Figure 38-20: R. Hubbard, "Molecular Isomers and Vision," *Scientific American* (June 1967): 64. Figure 38-21: Adapted with permission from Roger Eckert and David Randall, *Animal Physiology*, 2/e. Copyright © 1978, 1983 W. H. Freeman and Company. Figure 38-22: (b) Gene Shih and Richard Kessel, *Living Images*, Science Books International, 1982. Reprinted by permission of the present publisher, Jones and Bartlett Publishers. Figure 38-25: (c) Science Photo Library/Photo Researchers.

Chapter 39   Page 957: Hans Pfletschinger/Peter Arnold. Figure 39-1: Kjell B. Sandved/Photo Researchers. Figure 39-3: Phil Degginger/Bruce Coleman. Figure 39-6: (a) From *Living Images* by Gene Shih and Richard G. Kessel, Science Books International, 1982. Reprinted by permission of the present publisher, Jones and Bartlett Publishers; (b) From *Tissues and Organs: A Text-Atlas of Scanning Electron Microscopy* by Richard G. Kessel and Randy H. Kardon, W. H. Freeman and Company © 1979. Figure 39-8: From Milton Hildebrand, "How Animals Run." Copyright © 1960 by Scientific American, Inc. All rights reserved. Figure 39-12: (a–c) Carolina Biological Supply Company. Figure 39-14: (a) C. Franzini-Armstrong, *Journal of Cell Biology* 47 (1970): 488. Figure 39-15: (a) J. Trinick and A. Elliott from *Journal of Molecular Biology* 131 (1979): 135. Figure 39-22: Fawcett/Vehara/Photo Researchers. Page 978, Figure A: Steven E. Sutton/Duomo. Figure 39-25: Jeff Rotman.

Chapter 40   Page 982: Dan McCoy/Rainbow. Figure 40-4: F. Crepel, N. Delhaye-Bouchaud, J. L. Dupont and C. Sotelo, *Neuroscience* 5 (1980): 333–347. Figure 40-7: (c) Fritz Goro. Figure 40-12: From Gary Vander Ark and Ludwig Kempe, *A Primer of Electroencephalography*, 1970. Reprinted by permission of Hoffman-La Roche, Inc. Figure 40-13: From Floyd E. Bloom, Arlyne Lazerson, and Laura Hofstadter, *Brain, Mind, and Behavior*. Copyright © 1985 Educational Broadcasting Corp. Reprinted with the permission of W. H. Freeman and Company.

Part VI   Page 1008: Adam Woolfitt/Woodfin Camp & Associates.

Chapter 41   Page 1010: Frans Lanting. Figure 41-2: Down House and the Royal College of Surgeons of England. Figure 41-3: Frans Lanting. Figure 41-4: Courtesy of the Department Library Services, American Museum of Natural History, negative number 326696. Figure 41-5: (a, Masai) Luis Villota/The Stock Market; (b, Chinese) Joan Lebold Cohen/Photo Researchers; (c, Scottish) Susan McCartney/Photo Researchers; (d, Indian) Luis Villota/The Stock Market. Figure 41-6: R. C. Vrijenhoek, Rutgers University, from R. C. Vrijenhoek, *Gene Dosage in Diploid and Triploid Unisexual Fishes*; from Clement L. Market, ed., *Isozymes*, vol. IV, *Genetics and Evolution* (Academic Press, 1975). Figure 41-7: Data from R. K. Selander, "Genic Variation in Natural Populations," in *Molecular Evolution* edited by F. Ayala, 1976, pp. 21–46. Figure 41-8: Charles Krebs, Issaquah, Washington. Figure 41-9: J. Messerschmidt/Bruce Coleman. Figure 41-11: C. Allan Morgan. Figure 41-13: Victor McKusick, Johns Hopkins University. Figure 41-14: Tau-San Chou, Plant Geneticist, George J. Ball, Inc. West Chicago. Figure 41-16: Joel Gordon.

Chapter 42   Page 1031: Zig Leszczynski/Animals Animals. Figure 42-1: Richard P. Smith/Tom Stack & Associates. Figure 42-2: Thomas Kitchin/Tom Stack & Associates. Figure 42-4: (a) Breck P. Kent. Figure 42-4: (b) Anthony Bannister/Animals Animals. Figure 42-5: From J. Hirsch. Reprinted with permission. Figure 42-8: (a and b) J. L. Mason/Ardea, London. Figure 42-11: (a and b) From Monroe W. Strickberger, *Genetics*, 3/e. Reprinted with permission of Macmillan Publishing Company from *Genetics*, 3/e by Monroe W. Strickberger. Copyright © 1985 by Monroe W. Strickberger. Figure 42-12: (a and b) Courtesy of J. M. Poehlman, from J. M. Poehlman, *Breeding Field Crops*, Third Edition (Van Nostrand-Reinhold, 1987), Figure 12-4a and b. Figure 42-13: (Peruvian) Owen Franken/Stock Boston; (African)

Ivan Massar/Photo Nats; (Italian) Luis Villota/The Stock Market. Figure 42-14: (a and b) From W. M. Fitch and C. H. Langley, *Protein Evolution and the Molecular Clock*, Fed. Proc. 35, 2092–2097 (1976). Figure 42-15: From Peter Marler. Reprinted with permission.

Chapter 43   Page 1051: Jeff Foott/Tom Stack & Associates. Page 1054, Figure A: (a) Budd Titlow, Naturegraphs/Tom Stack & Associates; (b) Ardea, London. Page 1055, Figure B: C. B. Frith/Bruce Coleman. Figure 43-1: From J. E. Lloyd, *Miscellaneous Publication*, no. 103, 1966, pp. 1-95. Reprinted with permission. Figure 43-2: Hans and Juoy Beste/Ardea, London. Figure 43-3: From Bruce Wallace and Adrian M. Srb, *Adaptation*, 2/e, © 1964, p. 82. Adapted by permission of Prentice-Hall, Inc., Englewood Cliffs, New Jersey. Figure 43-4: Kenneth W. Fink/Bruce Coleman. Figure 43-5: From William D. Stansfield, *The Science of Evolution*. Reprinted with permission of Macmillan Publishing Company from *The Science of Evolution* by William D. Stansfield. Copyright © 1977 by William D. Stansfield. Figure 43-6: (e) (Aberti) Larry Brock/Tom Stack and Associates; (Kaibab) Ian Beames/Ardea, London; (Grand Canyon) Tom Bean/DRK Photo. Figure 43-8: From Richard C. Leontin, "Adaptation." Copyright © 1978 by Scientific American, Inc., All rights reserved. Figure 43-9: Reprinted by permission of Alva Day. Figure 43-10: Art Wolfe/Wildlife Photobank. Figure 43-11: P. G. Wessells. Figure 43-12: From E. Mayr, E. G. Linsley, and R. L. Usinger, *Methods and Principles of Systematic Zoology*. © 1953, McGraw-Hill Book Company. Figure 43-13: (a) E. R. Degginger; (b) Anthony Bannister/Animals Animals. Page 1068, Figures A and B: Gerald D. Carr, Department of Botany, University of Hawaii, Honolulu. Figure 43-16: From G. Ledyard Stebbins, *Darwin to DNA, Molecules to Humanity*. Copyright © 1982 W. H. Freeman and Company. Reprinted with permission. Figure 43-17: Reprinted with permission of Macmillan Publishing Company from *The Science of Evolution*, by William D. Stansfield, © 1977. Figure 43-18: From M. J. Benton, "Mass Extinctions among Nonmarine Tetrapods." Reprinted by permission from *Nature*, vol. 316, p. 811. Copyright © 1986 Macmillan Journals Ltd. Figure 43-19: From E. N. Clarkson, *Invertebrate Paleontology and Evolution*, 1979. Reprinted by permission of Allen & Unwin.

Chapter 44   Page 1077: Nicholas deVore III/Bruce Coleman. Figure 44-1: C. A. Morgan. Figure 44-8: (a) Tom Bean/Tom Stack & Associates; (b) Brian Parker/Tom Stack & Associates; (c) E. R. Degginger. Figure 44-10: Christopher Crowley/Tom Stack & Associates. Figure 44-11: Pierre C. Fisher. Figure 44-12: E. E. Kingsley/Photo Researchers. Figure 44-13: Brian Enting/Photo Researchers. Figure 44-14: Pierre C. Fischer. Figure 44-15: Charles A. Mauzy/Aperture. Figure 44-16: Manuel Rodriguez. Figure 44-17: Dale Johnson/Tom Stack & Associates. Figure 44-18: C. Mauzy/Aperture. Figure 44-19: Entheos. Figure 44-20: Tom Bean/Aperture. Page 1094, Figure A: I. Everson, British Antarctic Survey. Page 1094, Table A: From Richard M. Laws, *American Scientist*, vol. 73, 1985, pp. 26–40. Reprinted with permission of *American Scientist*, Journal of Sigma Xi, The Scientific Research Society. Page 1095, Figure B: From Richard M. Laws, *American Scientist*, vol. 73, 1985, pp. 26–40. Reprinted with permission of *American Scientist*, Journal of Sigma Xi, The Scientific Research Society. Table 44-1: From R. H. Whittaker, *Communities and Ecosystems* 2/e. Reprinted with permission of Macmillan Publishing Company from *Communities and Ecosystems*, 2/e by Robert H. Whittaker. Copyright © 1975 by Robert H. Whittaker. Figure 44-28: (b) Jim Corwin/Aperture.

Chapter 45   Page 1108: Stephen Frink/The WaterHouse. Figure 45-1: (a) J. R. Stacey/USGS; (b) Tom Bean/DRK Photo. Figure 45-2: Brian Milne/Earth Scenes/Animals Animals. Figure 45-3: Tom Stack/Tom Stack & Associates. Figure 45-5: (e) B. Lund, The Nature Conservancy, Massachusetts/Rhode Island Field Office, Boston. Figure 45-6: Jack Dermid. Figure 45-7: James Mason/Ardea, London. Figure 45-8:

Cleveland Museum of Natural History. Figure 50-12: (a and b) Tim D. White. Figure 50-15: (a) Tim D. White; (b) Collection Musee de l'Homme; (c) American Museum of Natural History. Figure 50-16: David L. Brill, courtesy of the National Geographic Society. Figure 50-17: From J. E. Cronin et al. Reprinted by permission from *Nature*, vol. 292, p. 113. Copyright © 1981 Macmillan Magazines Limited. Figure 50-18: British Museum (Natural History), London. Figure 50-19: M. Boule and H. V. Vallois from *Fossil Men* (New York: Dryden Press, 1957), p. 286, courtesy of University of California Library, Berkeley. Figure 50-20: Shelly Grossman/Woodfin Camp & Associates. Figure 50-22: From Erwin Raisz, *General Cartography*, McGraw-Hill, 1938.

Chapter 51   Page 1239: Malcolm S. Kirk/Peter Arnold. Figure 51-1: Data based on *Times Atlas of World History*, edited by Geoffrey Baraclough, Maplewood, N.J.: Hammond, 1978; and *The Random House Encyclopedia*, new revised edition, edited by James Mitchell

and Jess Stein, New York: Random House, Inc. 1983. Figure 51-2: From "Population Age Pyramids: Developed and Developing Regions 1975 and 2000," vol. 34, no. 5. December 1979, p. 6, Population Reference Bureau. Figure 51-3: From "World Population Growth Through History," vol. 34, no. 5, 1979, Population Reference Bureau. Figure 51-4: Bruno J. Zehnder/Peter Arnold. Figure 51-5: After Casper, *Energy Saving Techniques for the Food Industry*, Noyes Data Corporation, Park Ridge, N.J., 1977, as it appears in *Food, Energy and Society* by D. Pimentel and M. Pimentel. Reprinted by permission of Edward Arnold, Ltd., London. Figure 51-6: Robert Harding Picture Library/Peter Arnold. Figure 51-7: (a) Jeff Hunter/The Image Bank; (b) David Madison/Bruce Coleman. Figure 51-8: Courtesy of Hugh H. Iltis, Botany Department, University of Wisconsin, Madison. Figure 51-9: John Paul Kay/Peter Arnold. Figure 51-10: Courtesy of Donald R. Helinski, University of California, La Jolla, photo by Keith V. Wood. Figure 51-11: (a) Jacques Jangoux/Peter Arnold; (b) Sullivan and Rogers/Bruce Coleman. Figure 51-12: Smithsonian Institution, photo no. 86-10915. Figure 51-13: NASA.

Index

Pages on which definitions or main discussions of topics appear are indicated by boldface. Pages containing significant illustrations are indicated by italics.

Metric-English and English-Metric Conversions

Prefixes Used with Units (metric)

Prefix	Symbol	Value	Example
Tera-	T	1,000,000,000,000, or 10^{12}	1 terameter (Tm) = 1×10^{12} m
Giga-	G	1,000,000,000, or 10^9	1 gigameter (Gm) = 1×10^9 m
Mega-	M	1,000,000, or 10^6	1 megameter (Mm) = 1×10^6 m
Kilo-	k	1,000, or 10^3	1 kilometer (km) = 1×10^3 m
Deci-	d	1/10, or 10^{-1}	1 decimeter (dm) = 0.1 m
Centi-	c	1/100, or 10^{-2}	1 centimeter (cm) = 0.01 m
Milli-	m	1/1,000, or 10^{-3}	1 millimeter (mm) = 0.001 m
Micro-	μ	1/1,000,000, or 10^{-6}	1 micrometer (μm) = 1×10^{-6} m
Nano-	n	1/1,000,000,000, or 10^{-9}	1 nanometer (nm) = 1×10^{-9} m
Pico-	p	1/1,000,000,000,000, or 10^{-12}	1 picometer (pm) = 1×10^{-12} m

Length

Metric	=	English (USA)
millimeter (0.001 m)	=	0.039 in
centimeter (0.01 m)	=	0.39 in
meter	=	3.28 ft, 39.37 in
kilometer (1×10^3 m)	=	0.62 mi, 1,091 yd, 3,273 ft

English (USA)	=	Metric
inch	=	2.54 cm
foot	=	0.30 m, 30.48 cm
yard	=	0.91 m, 91.4 cm
mile (statute) (5,280 ft)	=	1.61 km, 1,609 m

Practice Examples:

A foot-long hotdog is 30.5 cm long.
A 32-inch softball bat is 0.8 m long.
A 100-yard long football field is 91 m long.
When driving at 55 mph, you will travel
88.6 km in an hour.

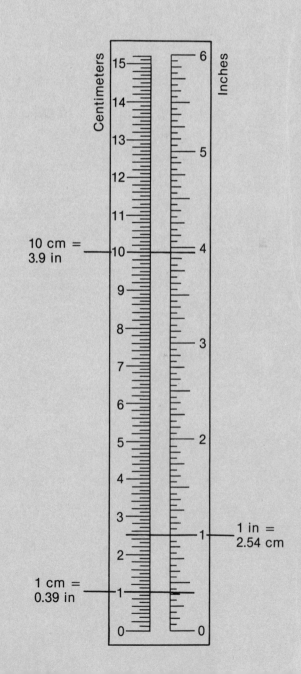